$115.95 per copy, (In United States).
Price subject to change without prior notice.

0015

RSMeans

Building Construction Cost Data

63rd Annual Edition

2005

RS**Means**
Construction Publishers & Consultants
63 Smiths Lane
Kingston, MA 02364-0800
(781) 585-7880

Copyright©2003 by Reed Construction Data, Inc.
All rights reserved.

Printed in the United States of America
ISSN 0068-3531
ISBN 0-87629-750-5

Senior Editor
Phillip R. Waier, PE

Contributing Editors
Barbara Balboni
Robert A. Bastoni
John H. Chiang, PE
Robert J. Kuchta
Robert C. McNichols
Robert W. Mewis, CCC
Melville J. Mossman, PE
John J. Moylan
Jeannene D. Murphy
Stephen C. Plotner
Michael J. Regan
Eugene R. Spencer
Marshall J. Stetson

Senior Engineering
Operations Manager
John H. Ferguson, PE

Editorial Advisory Board
James E. Armstrong, CPE, CEM
Program Manager,
Energy Conservation
NSTAR

William R. Barry, CCC
Cost Consultant

Robert F. Cox, PhD
Assistant Professor
ME Rinker Sr. School of Bldg. Constr.
University of Florida

Roy F. Gilley, AIA
Principal
Gilley-Hinkel Architects

Kenneth K. Humphreys, PhD,
  PE, CCE
Secretary-Treasurer
International Cost
Engineering Council

Patricia L. J
President
Jackson A&

Martin F. Jo
Executive V
Bond Broth

Senior Vice President
& General Manager
John Ware

Product Management
Jane Crudden

Vice President
of Sales
John M. Shea

Production Manager
Michael Kokernak

Production Coordinator
Marion E. Schofield

Technical Support
Thomas J. Dion
Jonathan Forgit
Mary Lou Geary
Gary L. Hoitt
Alice McSharry
Paula Reale-Camelio
Robin Richardson
Kathryn S. Rodriguez
Sheryl A. Rose

Book & Cover Design
Norman R. Forgit

 This book is recyclable.

 The this book is printed on recycled
stock.

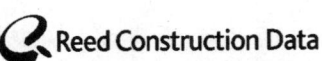 Reed Construction Data

First Printing

Foreword

RSMeans is a product line of Reed Construction Data, Inc., a leading provider of construction information, products, and services in North America and globally. Reed Construction Data's project information products include more than 100 regional editions, national construction data, sales leads, and local plan rooms in major business centers. Reed Construction Data's PlansDirect provides surveys, plans and specifications. The First Source suite of products consists of *First Source for Products*, SPEC-DATA™, MANU-SPEC™, CADBlocks, Manufacturer Catalogs and First Source Exchange (www.firstsourceexchange.com) for the selection of nationally available building products. Reed Construction Data also publishes ProFile, a database of more than 20,000 U.S. architectural firms. RSMeans provides construction cost data, training, and consulting services in print, CD-ROM and online. Reed Construction Data, headquartered in Atlanta, is owned by Reed Business Information (www.reedconstructiondata.com), a leading provider of critical information and marketing solutions to business professionals in the media, manufacturing, electronics, construction and retail industries. Its market-leading properties include more than 135 business-to-business publications, over 125 Webzines and Web portals, as well as online services, custom publishing, directories, research and direct-marketing lists. Reed Business Information is a member of the Reed Elsevier plc group (NYSE: RUK and ENL)—a world-leading publisher and information provider operating in the science and medical, legal, education and business-to-business industry sectors.

Our Mission

Since 1942, RSMeans has been actively engaged in construction cost publishing and consulting throughout North America.

Today, over 60 years after RSMeans began, our primary objective remains the same: to provide you, the construction and facilities professional, with the most current and comprehensive construction cost data possible.

Whether you are a contractor, an owner, an architect, an engineer, a facilities manager, or anyone else who needs a fast and reliable construction cost estimate, you'll find this publication to be a highly useful and necessary tool.

Today, with the constant flow of new construction methods and materials, it's difficult to find the time to look at and evaluate all the different construction cost possibilities. In addition, because labor and material costs keep changing, last year's cost information is not a reliable basis for today's estimate or budget.

That's why so many construction professionals turn to RSMeans. We keep track of the costs for you, along with a wide range of other key information, from city cost indexes . . . to productivity rates . . . to crew composition . . . to contractor's overhead and profit rates.

RSMeans performs these functions by collecting data from all facets of the industry, and organizing it in a format that is instantly accessible to you. From the preliminary budget to the detailed unit price estimate, you'll find the data in this book useful for all phases of construction cost determination.

The Staff, the Organization, and Our Services

When you purchase one of RSMeans' publications, you are in effect hiring the services of a full-time staff of construction and engineering professionals.

Our thoroughly experienced and highly qualified staff works daily at collecting, analyzing, and disseminating comprehensive cost information for your needs. These staff members have years of practical construction experience and engineering training prior to joining the firm. As a result, you can count on them not only for the cost figures, but also for additional background reference information that will help you create a realistic estimate.

The RSMeans organization is always prepared to help you solve construction problems through its five major divisions: Construction and Cost Data Publishing, Electronic Products and Services, Consulting Services, Insurance Services, and Educational Services.

Besides a full array of construction cost estimating books, Means also publishes a number of other reference works for the construction industry. Subjects include construction estimating and project and business management; special topics such as HVAC, roofing, plumbing, and hazardous waste remediation; and a library of facility management references.

In addition, you can access all of our construction cost data through your computer with RSMeans *CostWorks 2005* CD-ROM, an electronic tool that offers over 50,000 lines of Means detailed construction cost data, along with assembly and whole building cost data. You can also access Means cost information from our Web site at www.rsmeans.com

What's more, you can increase your knowledge and improve your construction estimating and management performance with a Means Construction Seminar or In-House Training Program. These two-day seminar programs offer unparalleled opportunities for everyone in your organization to get updated on a wide variety of construction-related issues.

RSMeans also is a worldwide provider of construction cost management and analysis services for commercial and government owners and of claims and valuation services for insurers.

In short, RSMeans can provide you with the tools and expertise for constructing accurate and dependable construction estimates and budgets in a variety of ways.

Robert Snow Means Established a Tradition of Quality That Continues Today

Robert Snow Means spent years building RSMeans, making certain he always delivered a quality product.

Today, at RSMeans, we do more than talk about the quality of our data and the usefulness of our books. We stand behind all of our data, from historical cost indexes to construction materials and techniques to current costs.

If you have any questions about our products or services, please call us toll-free at 1-800-334-3509. Our customer service representatives will be happy to assist you or visit our Web site at www.rsmeans.com

Table of Contents

UNIT PRICES

GENERAL REQUIREMENTS	1
SITE CONSTRUCTION	2
CONCRETE	3
MASONRY	4
METALS	5
WOOD & PLASTICS	6
THERMAL & MOISTURE PROTECTION	7
DOORS & WINDOWS	8
FINISHES	9
SPECIALTIES	10
EQUIPMENT	11
FURNISHINGS	12
SPECIAL CONSTRUCTION	13
CONVEYING SYSTEMS	14
MECHANICAL	15
ELECTRICAL	16
SQUARE FOOT	17

REFERENCE INFORMATION

REFERENCE NUMBERS

CREWS

COST INDEXES

INDEX

Related Means Products and Services

Phillip R. Waier, P.E., Senior Editor of this cost data book, suggests the following Means products and services as **companion** information resources to *Building Construction Cost Data*:

Construction Cost Data Books

Assemblies Cost Data 2005
Square Foot Costs 2005

Reference Books

Unit Price Estimating Methods, 3rd Edition
Managing Construction Purchasing
Successful Estimating Methods: From Concept to Bid
Estimating Handbook, 2nd Edition
Project Scheduling and Management for Construction
Cyberplaces 2nd Ed.: The Internet Guide for Architects, Engineers, Contractors & Facility Managers
The ADA in Practice
ADA Compliance Pricing Guide, 2nd Edition
Cost Planning & Estimating for Facilities Maintenance

Seminars and In-House Training

Unit Price Estimating
Scheduling & Project Management
Advanced Project Management
Facilities Maintenance & Repair Estimating
Plan Reading & Material Takeoff

Means on the Internet

Visit Means at **http://www.rsmeans.com.** The site contains useful *interactive* cost and reference material. Request or download *FREE* estimating software demos. Visit our bookstore for convenient ordering, plus learn more about new publications and companion products.

Means Data on CD-ROM

Get the information found in Means traditional cost books on Means new, easy to use CD-ROM, Means *CostWorks 2005*. Means *CostWorks* users can now enhance their Costlist with the Means *CostWorks Estimator*. For more information see the special full color brochure inserted in this book.

Means Consulting Services

Means Consulting Group provides a number of construction cost-related services including: Cost Estimates, Customized Databases, Benchmark Studies, Estimating Audits, Feasibility Studies, Litigation Support, Staff Assessments and Customized Training.

Means Data for Computer Estimating

Construction costs for software applications.
Over 25 unit price and assemblies cost databases are available through a number of leading estimating and facilities management software providers (listed below). For more information see the "yellow" pages at the back of this publication.

- 3D International
- 4Clicks-Solutions, LLC
- Aepco, Inc.
- Applied Flow Technology
- ArenaSoft Estimating
- ARES Corporation
- Barchan Associates
- BSD – Building Systems Design, Inc.
- CMS – Computerized Micro Solutions
- Corecon Technologies, Inc.
- CorVet Systems
- Discover Software
- Estimating Systems, Inc.
- Entech Engineering
- FAMIS Software, Inc.
- ISES Corporation
- Magellan K-12
- Maximus Asset Solutions
- MC^2 – Management Computer Controls
- Neddam Software Technologies
- Prime Time
- Quest Solutions, Inc.
- RIB Software (Americas), Inc.
- Shaw Beneco Enterprises, Inc.
- Schwab Technology Solutions, Inc.
- Timberline Software Corp.
- TMA Systems, LLC
- US Cost, Inc.
- Vanderweil Facility Advisors
- Vertigraph, Inc.

How the Book Is Built: An Overview

A Powerful Construction Tool

You have in your hands one of the most powerful construction tools available today. A successful project is built on the foundation of an accurate and dependable estimate. This book will enable you to construct just such an estimate.

For the casual user the book is designed to be:

- quickly and easily understood so you can get right to your estimate
- filled with valuable information so you can understand the necessary factors that go into the cost estimate

For the regular user, the book is designed to be:

- a handy desk reference that can be quickly referred to for key costs
- a comprehensive, fully reliable source of current construction costs and productivity rates, so you'll be prepared to estimate any project
- a source book for preliminary project cost, product selections, and alternate materials and methods

To meet all of these requirements we have organized the book into the following clearly defined sections.

How To Use the Book: The Details

This section contains an in-depth explanation of how the book is arranged . . . and how you can use it to determine a reliable construction cost estimate. It includes information about how we develop our cost figures and how to completely prepare your estimate.

Unit Price Section

All cost data has been divided into the 16 divisions according to the MasterFormat system of classification and numbering as developed by the Construction Specifications Institute (CSI) and Construction Specifications Canada (CSC). For a listing of these divisions and an outline of their subdivisions, see the Unit Price Section Table of Contents.

Estimating tips are included at the beginning of each division.

Division 17: Quick Project Estimates:

In addition to the 16 Unit Price Divisions there is also Division 17, Square Foot and Cubic Foot Costs. It contains costs for 58 different building types that allow you to make a rough estimate for the overall cost of a project or its major components.

Reference Section

This section includes information on Reference Numbers, Change Orders, Crew Listings, Historical Cost Indexes, City Cost Indexes, Location Factors and a listing of Abbreviations. It is visually identified by a vertical gray bar on the edge of pages.

Reference Numbers: At the beginning of selected major classifications in the Unit Price Section are "reference numbers" shown in bold squares. These numbers refer you to related information in the Reference Section.

In this section, you'll find reference tables, explanations, and estimating information that support how we develop the unit price data. Also included are alternate pricing methods, technical data, and estimating procedures, along with information on design and economy in construction. You'll also find helpful tips on what to expect and what to avoid when estimating and constructing your project.

It is recommended that you refer to the Reference Section if a "reference number" appears within the section you are estimating.

Change Orders: This section includes information on the factors that influence the pricing of change orders.

Crew Listings: This section lists all the crews referenced in the book. For the purposes of this book, a crew is composed of more than one trade classification and/or the addition of power equipment to any trade classification. Power equipment is included in the cost of the crew. Costs are shown both with bare labor rates and with the installing contractor's overhead and profit added. For each, the total crew cost per eight-hour day and the composite cost per labor-hour are listed.

Historical Cost Indexes: These indexes provide you with data to adjust construction costs over time. If you know costs for a project completed in the past, you can use these indexes to calculate a rough estimate of what it would cost to construct the same project today.

City Cost Indexes: All costs in this book are U.S. national averages. Costs vary depending on the regional economy. You can adjust the national average costs to over 930 locations throughout the U.S. and Canada by using the data in this section. How to use information is included.

Location Factors, to quickly adjust the data to over 930 locations, are included.

Abbreviations: A listing of abbreviations used throughout this book, along with the terms they represent, is included.

Index

A comprehensive listing of all terms and subjects in this book to help you find what you need quickly when you are not sure where it falls in MasterFormat.

The Scope of This Book

This book is designed to be as comprehensive and as easy to use as possible. To that end we have made certain assumptions and limited its scope in three key ways:

1. We have established material prices based on a "national average."
2. We have computed labor costs based on a 30-city "national average" of union wage rates.
3. We have targeted the data for projects of a certain size range.

For a more detailed explanation of how the cost data is developed, see "How To Use the Book: The Details."

Project Size

This book is aimed primarily at commercial and industrial projects costing $1,000,000 and up, or large multi-family housing projects. Costs are primarily for new construction or major renovation of buildings rather than repairs or minor alterations.

With reasonable exercise of judgment the figures can be used for any building work. *For civil engineering structures such as bridges, dams, highways, or the like, please refer to **Means Heavy Construction Cost Data**.*

How to Use the Book: The Details

What's Behind the Numbers? The Development of Cost Data

The staff at RSMeans continuously monitors developments in the construction industry in order to ensure reliable, thorough and up-to-date cost information.

While *overall* construction costs may vary relative to general economic conditions, price fluctuations within the industry are dependent upon many factors. Individual price variations may, in fact, be opposite to overall economic trends. Therefore, costs are continually monitored and complete updates are published yearly. Also, new items are frequently added in response to changes in materials and methods.

Costs—$ (U.S.)

All costs represent U.S. national averages and are given in U.S. dollars. The Means City Cost Indexes can be used to adjust costs to a particular location. The City Cost Indexes for Canada can be used to adjust U.S. national averages to local costs in Canadian dollars.

Material Costs

The RSMeans staff contacts manufacturers, dealers, distributors, and contractors all across the U.S. and Canada to determine national average material costs. If you have access to current material costs for your specific location, you may wish to make adjustments to reflect differences from the national average. Included within material costs are fasteners for a normal installation. RSMeans engineers use manufacturers' recommendations, written specifications and/or standard construction practice for size and spacing of fasteners. Adjustments to material costs may be required for your specific application or location. Material costs do not include sales tax.

Labor Costs

Labor costs are based on the average of wage rates from 30 major U.S. cities. Rates are determined from labor union agreements or prevailing wages for construction trades for the current year. Rates along with overhead and profit markups are listed on the inside back cover of this book.

- If wage rates in your area vary from those used in this book, or if rate increases are expected within a given year, labor costs should be adjusted accordingly.

Labor costs reflect productivity based on actual working conditions. These figures include time spent during a normal workday on tasks other than actual installation, such as material receiving and handling, mobilization at site, site movement, breaks, and cleanup.

Productivity data is developed over an extended period so as not to be influenced by abnormal variations and reflects a typical average.

Equipment Costs

Equipment costs include not only rental, but also operating costs for equipment under normal use. The operating costs include parts and labor for routine servicing such as repair and replacement of pumps, filters and worn lines. Normal operating expendables such as fuel, lubricants, tires and electricity (where applicable) are also included. Extraordinary operating expendables with highly variable wear patterns such as diamond bits and blades are excluded. These costs are included under materials. Equipment rental rates are obtained from industry sources throughout North America—contractors, suppliers, dealers, manufacturers, and distributors.

Crew Equipment Cost/Day—The power equipment required for each crew is included in the crew cost. The daily cost for crew equipment is based on dividing the weekly bare rental rate by 5 (number of working days per week), and then adding the hourly operating cost times 8 (hours per day). This "Crew Equipment Cost/Day" is listed in Subdivision 01590.

Mobilization/Demobilization—The cost to move construction equipment from an equipment yard or rental company to the job site and back again is not included in equipment costs. Mobilization (to the site) and demobilization (from the site) costs can be found in Section 02305-250. If a piece of equipment is already at the job site, it is not appropriate to utilize mob/demob costs again in an estimate.

Overhead and Profit

Total Cost including O&P for the *Installing Contractor* is shown in the last column on the Unit Price pages of this book. This figure is the sum of the bare material cost plus 10% for overhead and profit, the bare labor cost plus total overhead and profit, and the bare equipment cost plus 10% for overhead profit. Details for the calculation of Overhead and Profit on labor are shown on the inside back cover and in the Reference Section of this book. (See the "How To Use the Unit Price Pages" for an example of this calculation.)

General Conditions

General conditions, or general requirements of the contract, should also be added to the Total Cost including O&P when applicable. Costs for General Conditions are listed in Division 1 and the Reference Section of this book. General Conditions for the *Installing Contractor* may range from 0% to 10% of the Total Cost including O&P. For the *General* or *Prime Contractor*, costs for General Conditions may range from 5% to 15% of the Total Cost including O&P, with a figure of 10% as the most typical allowance.

Factors Affecting Costs

Costs can vary depending upon a number of variables. Here's how we have handled the main factors affecting costs.

Quality—The prices for materials and the workmanship upon which productivity is based represent sound construction work. They are also in line with U.S. government specifications.

Overtime—We have made no allowance for overtime. If you anticipate premium time or work beyond normal working hours, be sure to make an appropriate adjustment to your labor costs.

Productivity—The productivity, daily output, and labor-hour figures for each line item are based on working an eight-hour day in daylight hours in moderate temperatures. For work that extends beyond normal work hours or is performed under adverse conditions, productivity may decrease. (See the section in "How To Use the Unit Price Pages" for more on productivity.)

Size of Project—The size, scope of work, and type of construction project will have a significant impact on cost. Economies of scale can reduce costs for large projects. Unit costs can often run higher for small projects. Costs in this book are intended for the size and type of project as previously described in "How the Book Is Built: An Overview." Costs for projects of a significantly different size or type should be adjusted accordingly.

Location—Material prices in this book are for metropolitan areas. However, in dense urban areas, traffic and site storage limitations may increase costs. Beyond a 20-mile radius of large cities, extra trucking or transportation charges may also increase the material costs slightly. On the other hand, lower wage rates may be in effect. Be sure to consider both these factors when preparing an estimate, particularly if the job site is located in a central city or remote rural location.

In addition, highly specialized subcontract items may require travel and per diem expenses for mechanics.

Other factors—
- season of year
- contractor management
- weather conditions
- local union restrictions
- building code requirements
- availability of:
  - adequate energy
  - skilled labor
  - building materials
- owner's special requirements/restrictions
- safety requirements
- environmental considerations

Unpredictable Factors—General business conditions influence "in-place" costs of all items. Substitute materials and construction methods may have to be employed. These may affect the installed cost and/or life cycle costs. Such factors may be difficult to evaluate and cannot necessarily be predicted on the basis of the job's location in a particular section of the country. Thus, where these factors apply, you may find significant, but unavoidable cost variations for which you will have to apply a measure of judgment to your estimate.

Rounding of Costs

In general, all unit prices in excess of $5.00 have been rounded to make them easier to use and still maintain adequate precision of the results. The rounding rules we have chosen are in the following table.

Prices from . . .	Rounded to the nearest . . .
$.01 to $5.00	$.01
$5.01 to $20.00	$.05
$20.01 to $100.00	$.50
$100.01 to $300.00	$1.00
$300.01 to $1,000.00	$5.00
$1,000.01 to $10,000.00	$25.00
$10,000.01 to $50,000.00	$100.00
$50,000.01 and above	$500.00

Final Checklist

Estimating can be a straightforward process provided you remember the basics. Here's a checklist of some of the items you should remember to do before completing your estimate.

Did you remember to . . .
- factor in the City Cost Index for your locale
- take into consideration which items have been marked up and by how much
- mark up the entire estimate sufficiently for your purposes
- read the background information on techniques and technical matters that could impact your project time span and cost
- include all components of your project in the final estimate
- double check your figures to be sure of your accuracy
- call RSMeans if you have any questions about your estimate or the data you've found in our publications

Remember, RSMeans stands behind its publications. If you have any questions about your estimate . . . about the costs you've used from our books . . . or even about the technical aspects of the job that may affect your estimate, feel free to call the RSMeans editors at 1-800-334-3509.

Unit Price Section

Table of Contents

How to Use the Unit Price Pages

The following is a detailed explanation of a sample entry in the Unit Price Section. Next to each bold number below is the item being described with appropriate component of the sample entry following in parenthesis. Some prices are listed as bare costs, others as costs that include overhead and profit of the installing contractor. In most cases, if the work is to be subcontracted, the general contractor will need to add an additional markup (RSMeans suggests using 10%) to the figures in the column "Total Incl. O&P."

1 Division Number/Title
(03300/Cast-In-Place Concrete)

Use the Unit Price Section Table of Contents to locate specific items. The sections are classified according to the CSI MasterFormat (1995 Edition).

2 Line Numbers
(03310 240 3900)

Each unit price line item has been assigned a unique 12-digit code based on the CSI MasterFormat classification.

```
                              ┌── Level One - CSI-MasterFormat Division
                        ┌───── Level Two - CSI

        03300
        03310-240-3900
                              └── Means 12-digit
                                  Line Number
                        └──────── Level Four - Means
                      └────────── Level Three - CSI
```

3 Description
(Concrete-In-Place, etc.)

Each line item is described in detail. Sub-items and additional sizes are indented beneath the appropriate line items. The first line or two after the main item (in boldface) may contain descriptive information that pertains to all line items beneath this boldface listing.

Items which include the symbol **CN** are updated in the Key Material Price Section of *The Change Notice* quarterly publication. To obtain a copy of *The Change Notice* contact RSMeans customer service at 1-800-334-3509.

4 Reference Number
Information

You'll see reference numbers shown in bold rectangles at the beginning of some sections. These refer to related items in the Reference Section, visually identified by a vertical gray bar on the edge of pages.

R03310
-010

The relation may be: (1) an estimating procedure that should be read before estimating, (2) an alternate pricing method, or (3) technical information.

The "R" designates the Reference Section. The numbers refer to the MasterFormat classification system.

It is strongly recommended that you review all reference numbers that appear within the section in which you are working.

Note: Not all reference numbers appear in all Means publications.

03300 | Cast-In-Place Concrete

		03310	Structural Concrete		CREW	DAILY OUTPUT	LABOR-HOURS	UNIT	2005 BARE COSTS MAT.	LABOR	EQUIP.	TOTAL	TOTAL INCL O&P	
240	0010		CONCRETE IN PLACE Including forms (4 uses), reinforcing	R03310 -010										240
	0050		steel and finishing unless otherwise indicated											
	0300		Beams, 5 kip per L.F., 10' span	R03310 -100	C-14A	15.62	12.804	C.Y.	284	440	46.50	770.50	1,050	
	0350		25' span		"	18.55	10.782		214	370	39.50	623.50	860	
	3850		Over 5 C.Y.			81.04	1.382		226	45	.29	271.29	320	
	3900		Footings, strip, 18" x 9", unreinforced			40	2.800		101	91.50	.59	193.09	255	
	3920		18" x 9", reinforced			35	3.200		120	105	.67	225.67	297	
	3925		20" x 10", unreinforced			45	2.489		98	81.50	.52	180.02	236	
	3930		20" x 10", reinforced			40	2.800		114	91.50	.59	206.09	269	
	3935		24" x 12", unreinforced			55	2.036		97	66.50	.43	163.93	211	
	3940		24" x 12", reinforced			48	2.333		113	76	.49	189.49	244	
	3945		36" x 12", unreinforced			70	1.600		93	52.50	.34	145.84	184	
	3950		36" x 12", reinforced			60	1.867		108	61	.39	169.39	214	
	4000		Foundation mat, under 10 C.Y.			38.67	2.896		164	94.50	.61	259.11	330	

Crew (C-14C)

The "Crew" column designates the typical trade or crew used to install the item. If an installation can be accomplished by one trade and requires no power equipment, that trade and the number of workers are listed (for example, "2 Carpenters"). If an installation requires a composite crew, a crew code designation is listed (for example, "C-14C"). You'll find full details on all composite crews in the Crew Listings.

- For a complete list of all trades utilized in this book and their abbreviations, see the inside back cover.

Crews

Crew No.	Bare Costs		Incl. Subs O & P		Cost Per Labor-Hour	
Crew C-14C	Hr.	Daily	Hr.	Daily	Bare Costs	Incl. O&P
1 Carpenter Foreman (out)	$36.25	$290.00	$56.45	$451.60	$32.66	$51.14
6 Carpenters	34.25	1644.00	53.35	2560.80		
2 Rodmen (reinf.)	37.95	607.20	62.60	1001.60		
4 Laborers	26.70	854.40	41.55	1329.60		
1 Cement Finisher	32.85	262.80	48.35	386.80		
1 Gas Engine Vibrator		24.00		26.40	.21	.24
112 L.H., Daily Totals		$3682.40		$5756.80	$32.87	$51.38

Productivity: Daily Output (40.0)/Labor-Hours (2.80)

The "Daily Output" represents the typical number of units the designated crew will install in a normal 8-hour day. To find out the number of days the given crew would require to complete the installation, divide your quantity by the daily output. For example:

Quantity	÷	Daily Output	=	Duration
100 C.Y.	÷	40.0/ Crew Day	=	2.50 Crew Days

The "Labor-Hours" figure represents the number of labor-hours required to install one unit of work. To find out the number of labor-hours required for your particular task, multiply the quantity of the item times the number of labor-hours shown. For example:

Quantity	x	Productivity Rate	=	Duration
100 C.Y.	x	2.80 Labor-Hours/ C.Y.	=	280 Labor-Hours

Unit (C.Y.)

The abbreviated designation indicates the unit of measure upon which the price, production, and crew are based (C.Y. = Cubic Yard). For a complete listing of abbreviations refer to the Abbreviations Listing in the Reference Section of this book.

Bare Costs:

Mat. (Bare Material Cost) ($101)

The unit material cost is the "bare" material cost with no overhead and profit included. *Costs shown reflect national average material prices for January of the current year and include delivery to the job site. No sales taxes are included.*

Labor ($91.50)

The unit labor cost is derived by multiplying bare labor-hour costs for Crew C-14C by labor-hour units. The bare labor-hour cost is found in the Crew Section under C-14C. (If a trade is listed, the hourly labor cost—the wage rate—is found on the inside back cover.)

Labor-Hour Cost Crew C-14C	x	Labor-Hour Units	=	Labor
$32.66	x	2.80	=	$91.00

Equip. (Equipment) (.59)

Equipment costs for each crew are listed in the description of each crew. Tools or equipment whose value justifies purchase or ownership by a contractor are considered overhead as shown on the inside back cover. The unit equipment cost is derived by multiplying the bare equipment hourly cost by the labor-hour units.

Equipment Cost Crew C-14C	x	Labor-Hour Units	=	Equip.
.21	x	2.80	=	$.59

Total (193.09)

The total of the bare costs is the arithmetic total of the three previous columns: mat., labor, and equip.

Material	+	Labor	+	Equip.	=	Total
$101	+	$91.50	+	$.59	=	$193.09

Total Costs Including O&P

This figure is the sum of the bare material cost plus 10% for profit; the bare labor cost plus total overhead and profit (per the inside back cover or, if a crew is listed, from the crew listings); and the bare equipment cost plus 10% for profit.

Material is Bare Material Cost + 10% = 101 + 10.10	=	$111.10
Labor for Crew C-14C = Labor-Hour Cost (51.14) x Labor-Hour Units (2.80)	=	$143.19
Equip. is Bare Equip. Cost + 10% = .59 + .06	=	$ .65
Total (Rounded)	=	$255

Division 1
General Requirements

Estimating Tips

The General Requirements of any contract are very important to both the bidder and the owner. These lay the ground rules under which the contract will be executed and have a significant influence on the cost of operations. Therefore, it is extremely important to thoroughly read and understand the General Requirements both before preparing an estimate and when the estimate is complete, to ascertain that nothing in the contract is overlooked. Caution should be exercised when applying items listed in Division 1 to an estimate. Many of the items are included in the unit prices listed in the other divisions such as mark-ups on labor and company overhead.

01200 Price & Payment Procedures

- When estimating historic preservation projects (depending on the condition of the existing structure and the owner's requirements), a 15-20% contingency or allowance is recommended, regardless of the stage of the drawings.

01300 Administrative Requirements

- Before determining a final cost estimate, it is a good practice to review all the items listed in subdivision 01300 to make final adjustments for items that may need customizing to specific job conditions.
- Historic preservation projects may require specialty labor and methods, as well as extra time to protect existing materials that must be preserved and/or restored. Some additional expenses may be incurred in architectural fees for facility surveys and other special inspections and analyses.
- Requirements for initial and periodic submittals can represent a significant cost to the General Requirements of a job. Thoroughly check the submittal specifications when estimating a project to determine any costs that should be included.

01400 Quality Requirements

- All projects will require some degree of Quality Control. This cost is not included in the unit cost of construction listed in each division. Depending upon the terms of the contract, the various costs of inspection and testing can be the responsibility of either the owner or the contractor. Be sure to include the required costs in your estimate.

01500 Temporary Facilities & Controls

- Barricades, access roads, safety nets, scaffolding, security and many more requirements for the execution of a safe project are elements of direct cost. These costs can easily be overlooked when preparing an estimate. When looking through the major classifications of this subdivision, determine which items apply to each division in your estimate.

01590 Equipment Rental

- This subdivision contains transportation, handling, storage, protection and product options and substitutions. Listed in this cost manual are average equipment rental rates for all types of equipment. This is useful information when estimating the time and materials requirement of any particular operation in order to establish a unit or total cost.
- A good rule of thumb is that weekly rental is 3 times daily rental and that monthly rental is 3 times weekly rental.
- The figures in the column for Crew Equipment Cost represent the rental rate used in determining the daily cost of equipment in a crew. It is calculated by dividing the weekly rate by 5 days and adding the hourly operating cost times 8 hours.

01740 Execution Requirements

- When preparing an estimate, read the specifications to determine the requirements for Contract Closeout thoroughly. Final cleaning, record documentation, operation and maintenance data, warranties and bonds, and spare parts and maintenance materials can all be elements of cost for the completion of a contract. Do not overlook these in your estimate.

01830 Operations & Maintenance

- If maintenance and repair are included in your contract, they require special attention. To estimate the cost to remove and replace any unit usually requires a site visit to determine the accessibility and the specific difficulty at that location. Obstructions, dust control, safety, and often overtime hours must be considered when preparing your estimate.

Reference Numbers

Reference numbers are shown in bold squares at the beginning of some major classifications. These numbers refer to related items in the Reference Section. The reference information may be an estimating procedure, an alternate pricing method or technical information.

Note: Not all subdivisions listed here necessarily appear in this publication.

01103 | Models & Renderings

			CREW	DAILY OUTPUT	LABOR-HOURS	UNIT	2005 BARE COSTS				TOTAL INCL O&P	
							MAT.	LABOR	EQUIP.	TOTAL		
200	0010	**MODELS** Cardboard & paper, 1 building, minimum				Ea.	625			625	685	**200**
	0050	Maximum					1,425			1,425	1,550	
	0100	2 buildings, minimum					835			835	915	
	0150	Maximum				↓	1,875			1,875	2,075	
	0200	Plexiglass and metal, basic layout				SF Flr.	.06			.06	.07	
	0210	Including equipment and personnel				"	.28			.28	.31	
	0300	Site plan layout, minimum				Ea.	1,200			1,200	1,300	
	0350	Maximum				"	2,000			2,000	2,200	
500	0010	**RENDERINGS** Color, matted, 20" x 30", eye level,										**500**
	0020	1 building, minimum				Ea.	1,800			1,800	1,975	
	0050	Average					2,575			2,575	2,825	
	0100	Maximum					4,125			4,125	4,525	
	1000	5 buildings, minimum					3,600			3,600	3,975	
	1100	Maximum					7,200			7,200	7,925	
	2000	Aerial perspective, color, 1 building, minimum					2,575			2,575	2,825	
	2100	Maximum					7,200			7,200	7,925	
	3000	5 buildings, minimum					5,150			5,150	5,675	
	3100	Maximum				↓	10,300			10,300	11,300	

01107 | Professional Consultant

			CREW	DAILY OUTPUT	LABOR-HOURS	UNIT	MAT.	LABOR	EQUIP.	TOTAL	TOTAL INCL O&P	
100	0011	**ARCHITECTURAL FEES**	R01107 -010									**100**
	0020	For new construction										
	0060	Minimum				Project					4.90%	
	0090	Maximum									16%	
	0100	For alteration work, to $500,000, add to fee									50%	
	0150	Over $500,000, add to fee				↓					25%	
200	0010	**CONSTRUCTION MANAGEMENT FEES** $1,000,000 job, minimum				Project					4.50%	**200**
	0050	Maximum									7.50%	
	0300	$5,000,000 job, minimum									2.50%	
	0350	Maximum				↓					4%	
300	0010	**ENGINEERING FEES**	R01107 -030									**300**
	0020	Educational planning consultant, minimum				Project					.50%	
	0100	Maximum				"					2.50%	
	0200	Electrical, minimum				Contrct					4.10%	
	0300	Maximum									10.10%	
	0400	Elevator & conveying systems, minimum									2.50%	
	0500	Maximum									5%	
	0600	Food service & kitchen equipment, minimum									8%	
	0700	Maximum									12%	
	0800	Landscaping & site development, minimum									2.50%	
	0900	Maximum									6%	
	1000	Mechanical (plumbing & HVAC), minimum									4.10%	
	1100	Maximum				↓					10.10%	
	1200	Structural, minimum				Project					1%	
	1300	Maximum				"					2.50%	
700	0010	**SURVEYING** Conventional, topographical, minimum	A-7	3.30	7.273	Acre	16.30	269	18.40	303.70	455	**700**
	0100	Maximum	A-8	.60	53.333		49	1,925	101	2,075	3,125	
	0300	Lot location and lines, minimum, for large quantities	A-7	2	12		25.50	445	30.50	501	745	
	0320	Average	"	1.25	19.200		46	710	48.50	804.50	1,200	
	0400	Maximum, for small quantities	A-8	1	32	↓	73.50	1,150	61	1,284.50	1,925	
	0600	Monuments, 3' long	A-7	10	2.400	Ea.	20	88.50	6.05	114.55	166	
	0800	Property lines, perimeter, cleared land	"	1,000	.024	L.F.	.03	.89	.06	.98	1.47	
	0900	Wooded land	A-8	875	.037	"	.05	1.31	.07	1.43	2.16	
	1100	Crew for layout of building, trenching or pipe laying, 2 person crew	A-6	1	16	Day		545	60.50	605.50	900	
	1200	3 person crew	A-7	1	24	↓		885	60.50	945.50	1,450	

Important: See the Reference Section for critical supporting data - Reference Nos., Crews, & City Cost Indexes

01100 | Summary

01107 | Professional Consultant

			CREW	DAILY OUTPUT	LABOR-HOURS	UNIT	2005 BARE COSTS MAT.	LABOR	EQUIP.	TOTAL	TOTAL INCL O&P	
700	1400	Crew for roadway layout, 4 person crew	A-8	1	32	Day		1,150	61	1,211	1,850	**700**
	1500	Aerial surveying, including ground control, minimum fee, 10 acres				Total					5,700	
	1510	100 acres									9,500	
	1550	From existing photography, deduct				↓					1,370	
	1600	2' contours, 10 acres				Acre					460	
	1650	20 acres									315	
	1800	50 acres									95	
	1850	100 acres									85	
	2000	1000 acres									17.85	
	2050	10,000 acres				↓					11.50	
	2150	For 1' contours and										
	2160	dense urban areas, add to above				Acre					40%	
	3000	Inertial guidance system for										
	3010	locating coordinates, rent per day				Ea.					4,000	

01200 | Price & Payment Procedures

01250 | Contract Modification Procedures

			CREW	DAILY OUTPUT	LABOR-HOURS	UNIT	2005 BARE COSTS MAT.	LABOR	EQUIP.	TOTAL	TOTAL INCL O&P	
200	0010	**CONTINGENCIES** for estimate at conceptual stage				Project					20%	**200**
	0050	Schematic stage									15%	
	0100	Preliminary working drawing stage (Design Dev.)									10%	
	0150	Final working drawing stage				↓					3%	
300	0010	**CREWS** For building construction, see How To Use This Book										**300**
500	0010	**JOB CONDITIONS** Modifications to total										**500**
	0020	project cost summaries										
	0100	Economic conditions, favorable, deduct				Project					2%	
	0200	Unfavorable, add									5%	
	0300	Hoisting conditions, favorable, deduct									2%	
	0400	Unfavorable, add									5%	
	0500	General Contractor management, experienced, deduct									2%	
	0600	Inexperienced, add									10%	
	0700	Labor availability, surplus, deduct									1%	
	0800	Shortage, add									10%	
	0900	Material storage area, available, deduct									1%	
	1000	Not available, add									2%	
	1100	Subcontractor availability, surplus, deduct									5%	
	1200	Shortage, add									12%	
	1300	Work space, available, deduct									2%	
	1400	Not available, add				↓					5%	
600	0010	**OVERTIME** For early completion of projects or where	R01100 -110									**600**
	0020	labor shortages exist, add to usual labor, up to				Costs		100%				

01255 | Cost Indexes

			CREW	DAILY OUTPUT	LABOR-HOURS	UNIT	2005 BARE COSTS MAT.	LABOR	EQUIP.	TOTAL	TOTAL INCL O&P	
200	0010	**CONSTRUCTION COST INDEX** (Reference) over 930 zip code locations in										**200**
	0020	The U.S. and Canada, total bldg cost, min. (Clarksdale, MS)				%					66.80%	
	0050	Average									100%	
	0100	Maximum (New York, NY)				↓					131.90%	
400	0010	**HISTORICAL COST INDEXES** (Reference) Back to 1955										**400**

01200 | Price & Payment Procedures

01255 | Cost Indexes

			CREW	DAILY OUTPUT	LABOR-HOURS	UNIT	2005 BARE COSTS				TOTAL INCL O&P	
							MAT.	LABOR	EQUIP.	TOTAL		
500	0010	**LABOR INDEX** (Reference) For over 930 zip code locations in										500
	0020	the U.S. and Canada, minimum (Clarksdale, MS)				%		31.70%				
	0050	Average						100%				
	0100	Maximum (New York, NY)				↓		162.30%				
600	0010	**MATERIAL INDEX** (Reference) For over 930 zip code locations in										600
	0020	the U.S. and Canada, minimum (Elizabethtown, KY)				%	91.10%					
	0040	Average					100%					
	0060	Maximum (Ketchikan, AK)				↓	143.10%					

01290 | Payment Procedures

			CREW	DAILY OUTPUT	LABOR-HOURS	UNIT	2005 BARE COSTS				TOTAL INCL O&P	
							MAT.	LABOR	EQUIP.	TOTAL		
800	0010	**TAXES** Sales tax, State, average	R01100 -090			%	4.85%					800
	0050	Maximum					7.25%					
	0200	Social Security, on first $87,900 of wages	R01100 -100					7.65%				
	0300	Unemployment, MA, combined Federal and State, minimum						1.92%				
	0350	Average				↓		6.20%				
	0400	Maximum						11.76%				

01300 | Administrative Requirements

01310 | Project Management/Coordination

			CREW	DAILY OUTPUT	LABOR-HOURS	UNIT	2005 BARE COSTS				TOTAL INCL O&P	
							MAT.	LABOR	EQUIP.	TOTAL		
150	0010	**PERMITS** Rule of thumb, most cities, minimum				Job					.50%	150
	0100	Maximum				"					2%	
200	0010	**PERFORMANCE BOND** For buildings, minimum	R01100 -080			Job					.60%	200
	0100	Maximum				"					2.50%	
300	0010	**CONSTRUCTION TIME** Requirements	R01100 -020									300
350	0010	**INSURANCE** Builders risk, standard, minimum	R01100 -040			Job					.24%	350
	0050	Maximum									.64%	
	0200	All-risk type, minimum	R01100 -050								.25%	
	0250	Maximum				↓					.62%	
	0400	Contractor's equipment floater, minimum	R01100 -060			Value					.50%	
	0450	Maximum				"					1.50%	
	0600	Public liability, average				Job					2.02%	
	0800	Workers' compensation & employer's liability, average										
	0850	by trade, carpentry, general				Payroll		18.42%				
	0900	Clerical						.63%				
	0950	Concrete						16.07%				
	1000	Electrical						6.46%				
	1050	Excavation						10.43%				
	1100	Glazing						13.91%				
	1150	Insulation						15.53%				
	1200	Lathing						10.87%				
	1250	Masonry						15.06%				
	1300	Painting & decorating						13.12%				
	1350	Pile driving						22.45%				
	1400	Plastering						14.88%				
	1450	Plumbing						7.96%				
	1500	Roofing						33.14%				
	1550	Sheet metal work (HVAC)						11.46%				
	1600	Steel erection, structural				↓		38.60%				

Important: See the Reference Section for critical supporting data - Reference Nos., Crews, & City Cost Indexes

			CREW	DAILY OUTPUT	LABOR-HOURS	UNIT	2005 BARE COSTS				TOTAL INCL O&P	
		01310 | Project Management/Coordination					MAT.	LABOR	EQUIP.	TOTAL		
350	1650	Tile work, interior ceramic	R01100 -040				Payroll	9.66%				350
	1700	Waterproofing, brush or hand caulking						7.12%				
	1800	Wrecking	R01100 -050					39.48%				
	2000	Range of 35 trades in 50 states, excl. wrecking, min.						2.40%				
	2100	Average	R01100 -060					16.20%				
	2200	Maximum					↓	137.60%				
400	0010	**MAIN OFFICE EXPENSE** Average for General Contractors	R01100 -050									400
	0020	As a percentage of their annual volume										
	0125	Annual volume under 1 million dollars					% Vol.				13.60%	
	0145	Up to 2.5 million dollars									8%	
	0150	Up to 4.0 million dollars									6.80%	
	0200	Up to 7.0 million dollars									5.60%	
	0250	Up to 10 million dollars									5.10%	
	0300	Over 10 million dollars	↓				↓				3.90%	
500	0010	**MARK-UP** For General Contractors for change	R01100 -070									500
	0100	of scope of job as bid										
	0200	Extra work, by subcontractors, add					%				10%	
	0250	By General Contractor, add									15%	
	0400	Omitted work, by subcontractors, deduct all but									5%	
	0450	By General Contractor, deduct all but									7.50%	
	0600	Overtime work, by subcontractors, add									15%	
	0650	By General Contractor, add									10%	
	1000	Installing contractors, on his own labor, minimum	↓						46.90%			
	1100	Maximum					↓		87.30%			
600	0010	**OVERHEAD** As percent of direct costs, minimum	R01100 -050				%				5%	600
	0050	Average									13%	
	0100	Maximum	R01100 -070				↓				30%	
620	0010	**OVERHEAD & PROFIT** Allowance to add to items in this										620
	0020	book that do not include Subs O&P, average					%				25%	
	0100	Allowance to add to items in this book that										
	0110	do include Subs O&P, minimum					%				5%	
	0150	Average									10%	
	0200	Maximum									15%	
	0300	Typical, by size of project, under $100,000									30%	
	0350	$500,000 project									25%	
	0400	$2,000,000 project									20%	
	0450	Over $10,000,000 project					↓				15%	
700	0010	**FIELD PERSONNEL** Clerk average					Week	320	320			500	700
	0100	Field engineer, minimum						765	765			1,200	
	0120	Average						995	995			1,550	
	0140	Maximum						1,150	1,150			1,800	
	0160	General purpose laborer, average						1,050	1,050			1,625	
	0180	Project manager, minimum						1,425	1,425			2,225	
	0200	Average						1,625	1,625			2,525	
	0220	Maximum						1,850	1,850			2,875	
	0240	Superintendent, minimum						1,375	1,375			2,150	
	0260	Average						1,500	1,500			2,325	
	0280	Maximum						1,725	1,725			2,675	
	0290	Timekeeper, average					↓	890	890			1,375	

01320 | Const. Progress Documentation

			CREW	DAILY OUTPUT	LABOR-HOURS	UNIT	MAT.	LABOR	EQUIP.	TOTAL	TOTAL INCL O&P	
200	0010	**SCHEDULING** Critical path, as % of architectural fee, minimum					%					.50%	200
	0100	Maximum					"					1%	
	0300	Computer-update, micro, no plots, minimum					Ea.					450	
	0400	Including plots, maximum					"					1,550	

01300 | Administrative Requirements

01320		Const. Progress Documentation	CREW	DAILY OUTPUT	LABOR-HOURS	UNIT	2005 BARE COSTS				TOTAL INCL O&P	
							MAT.	LABOR	EQUIP.	TOTAL		
200	0600	Rule of thumb, CPM scheduling, small job ($10 Million)				Job					.05%	200
	0650	Large job ($50 Million +)									.03%	
	0700	Including cost control, small job									.08%	
	0750	Large job				↓					.04%	

01321		Construction Photos										
500	0010	**PHOTOGRAPHS** 8″ x 10″, 4 shots, 2 prints ea., std. mounting				Set	292			292	320	500
	0100	Hinged linen mounts					315			315	350	
	0200	8″ x 10″, 4 shots, 2 prints each, in color					340			340	370	
	0300	For I.D. slugs, add to all above					4.09			4.09	4.50	
	0500	Aerial photos, initial fly-over, 6 shots, 1 print ea., 8″ x 10″					690			690	760	
	0550	11″ x 14″ prints					770			770	845	
	0600	16″ x 20″ prints					990			990	1,100	
	0700	For full color prints, add					40%				40%	
	0750	Add for traffic control area				↓	285			285	315	
	0900	For over 30 miles from airport, add per				Mile	5.15			5.15	5.70	
	1000	Vertical photography, 4 to 6 shots with										
	1010	different scales, 1 print each				Set	1,050			1,050	1,175	
	1500	Time lapse equipment, camera and projector, buy					3,675			3,675	4,050	
	1550	Rent per month				↓	545			545	600	
	1700	Cameraman and film, including processing, B.&W.				Day	1,350			1,350	1,475	
	1720	Color				″	1,350			1,350	1,475	

01400 | Quality Requirements

01450		Quality Control	CREW	DAILY OUTPUT	LABOR-HOURS	UNIT	2005 BARE COSTS				TOTAL INCL O&P	
							MAT.	LABOR	EQUIP.	TOTAL		
500	0010	**TESTING** and Inspectional Services										500
	0015	For concrete building costing $1,000,000, minimum				Project					5,200	
	0020	Maximum									41,800	
	0050	Steel building, minimum									5,200	
	0070	Maximum									16,300	
	0100	For building costing, $10,000,000, minimum									33,100	
	0150	Maximum				↓					53,000	
	0200	Asphalt testing, compressive strength Marshall stability, set of 3				Ea.					165	
	0220	Density, set of 3									95	
	0250	Extraction, individual tests on sample									150	
	0300	Penetration									45	
	0350	Mix design, 5 specimens									200	
	0360	Additional specimen									40	
	0400	Specific gravity									45	
	0420	Swell test									70	
	0450	Water effect and cohesion, set of 6									200	
	0470	Water effect and plastic flow									70	
	0600	Concrete testing, aggregates, abrasion, ASTM C 131									150	
	0650	Absorption, ASTM C 127									46	
	0800	Petrographic analysis, ASTM C 295									850	
	0900	Specific gravity, ASTM C 127									55	
	1000	Sieve analysis, washed, ASTM C 136									65	
	1050	Unwashed									65	
	1200	Sulfate soundness				↓					125	

Important: See the Reference Section for critical supporting data - Reference Nos., Crews, & City Cost Indexes

01400 | Quality Requirements

01450 | Quality Control

500			CREW	DAILY OUTPUT	LABOR-HOURS	UNIT	2005 BARE COSTS				TOTAL INCL O&P	500
							MAT.	LABOR	EQUIP.	TOTAL		
1300	Weight per cubic foot					Ea.					40	
1500	Cement, physical tests, ASTM C 150										350	
1600	Chemical tests, ASTM C 150										270	
1800	Compressive test, cylinder, delivered to lab, ASTM C 39										13	
1900	Picked up by lab, minimum										15	
1950	Average										20	
2000	Maximum										30	
2200	Compressive strength, cores (not incl. drilling), ASTM C 42										40	
2250	Core drilling, 4" diameter (plus technician)					Inch					25	
2260	Technician for core drilling					Hr.					50	
2300	Patching core holes					Ea.					24	
2400	Drying shrinkage at 28 days										260	
2500	Flexural test beams, ASTM C 78										65	
2600	Mix design, one batch mix										285	
2650	Added trial batches										132	
2800	Modulus of elasticity, ASTM C 469										180	
2900	Tensile test, cylinders, ASTM C 496										50	
3000	Water-Cement ratio curve, 3 batches										155	
3100	4 batches										205	
3300	Masonry testing, absorption, per 5 brick, ASTM C 67										50	
3350	Chemical resistance, per 2 brick										55	
3400	Compressive strength, per 5 brick, ASTM C 67										75	
3420	Efflorescence, per 5 brick, ASTM C 67										75	
3440	Imperviousness, per 5 brick										96	
3470	Modulus of rupture, per 5 brick										95	
3500	Moisture, block only										35	
3550	Mortar, compressive strength, set of 3										25	
4100	Reinforcing steel, bend test										61	
4200	Tensile test, up to #8 bar										40	
4220	#9 to #11 bar										45	
4240	#14 bar and larger										70	
4400	Soil testing, Atterberg limits, liquid and plastic limits										65	
4510	Hydrometer analysis										120	
4530	Specific gravity, ASTM D 354										48	
4600	Sieve analysis, washed, ASTM D 422										60	
4700	Unwashed, ASTM D 422										65	
4710	Consolidation test (ASTM D2435), minimum										275	
4715	Maximum										475	
4720	Density and classification of undisturbed sample										80	
4735	Soil density, nuclear method, ASTM D2922										38.67	
4740	Sand cone method ASTM D1556										30.17	
4750	Moisture content, ASTM D 2216										10	
4780	Permeability test, double ring infiltrometer										550	
4800	Permeability, var. or constant head, undist., ASTM D 2434										250	
4850	Recompacted										275	
4900	Proctor compaction, 4" standard mold, ASTM D 698										135	
4950	6" modified mold										75	
5100	Shear tests, triaxial, minimum										450	
5150	Maximum										600	
5300	Direct shear, minimum, ASTM D 3080										350	
5350	Maximum										450	
5550	Technician for inspection, per day, earthwork										220	
5650	Bolting										280	
5750	Roofing										255	
5790	Welding										270	
5820	Non-destructive testing, dye penetrant					Day					330	

01450 | Quality Control

			DAILY OUTPUT	LABOR-HOURS	UNIT	\multicolumn{4}{c}{2005 BARE COSTS}	TOTAL INCL O&P	
			CREW			MAT.	LABOR	EQUIP.	TOTAL		
500	5840	Magnetic particle			Day					330	500
	5860	Radiography								495	
	5880	Ultrasonic			↓					340	
	6000	Welding certification, minimum			Ea.					100	
	6100	Maximum			"					275	
	7000	Underground storage tank									
	7500	Volumetric tightness test ,<=12,000 gal			Ea.					400	
	7510	<=30,000 gal			"					660	
	7600	Vadose zone (soil gas) sampling, 10-40 samples, min.			Day					1,500	
	7610	Maximum			"					2,500	
	7700	Ground water monitoring incl. drilling 3 wells, min.			Total					5,000	
	7710	Maximum			"					7,000	
	8000	X-ray concrete slabs			Ea.					200	

01510 | Temporary Utilities

			DAILY OUTPUT	LABOR-HOURS	UNIT	\multicolumn{4}{c}{2005 BARE COSTS}	TOTAL INCL O&P	
			CREW			MAT.	LABOR	EQUIP.	TOTAL		
800	0010	**TEMPORARY UTILITIES**									800
	0100	Heat, incl. fuel and operation, per week, 12 hrs. per day	1 Skwk	100	.080	CSF Flr	6.25	2.79		9.04	11.25	
	0200	24 hrs. per day	"	60	.133		9.40	4.65		14.05	17.60	
	0350	Lighting, incl. service lamps, wiring & outlets, minimum	1 Elec	34	.235		2.32	9.60		11.92	16.80	
	0360	Maximum	"	17	.471		5.05	19.20		24.25	34	
	0400	Power for temp lighting only, per month, min/month 6.6 KWH								.75	1.18	
	0450	Maximum/month 23.6 KWH								2.85	3.14	
	0600	Power for job duration incl. elevator, etc., minimum								47	51.70	
	0650	Maximum				↓				110	121	
	1000	Toilet, portable, see division 01590-400										

01520 | Construction Facilities

			DAILY OUTPUT	LABOR-HOURS	UNIT	\multicolumn{4}{c}{2005 BARE COSTS}	TOTAL INCL O&P	
			CREW			MAT.	LABOR	EQUIP.	TOTAL		
500	0010	**OFFICE** Trailer, furnished, no hookups, 20' x 8', buy	2 Skwk	1	16	Ea.	6,200	560		6,760	7,700	500
	0250	Rent per month					154			154	169	
	0300	32' x 8', buy	2 Skwk	.70	22.857		9,250	795		10,045	11,500	
	0350	Rent per month					163			163	180	
	0400	50' x 10', buy	2 Skwk	.60	26.667		17,000	930		17,930	20,200	
	0450	Rent per month					254			254	280	
	0500	50' x 12', buy	2 Skwk	.50	32		20,300	1,125		21,425	24,000	
	0550	Rent per month					286			286	315	
	0700	For air conditioning, rent per month, add				↓	39.50			39.50	43.50	
	0800	For delivery, add per mile				Mile	1.53			1.53	1.68	
	1000	Portable buildings, prefab, on skids, economy, 8' x 8'	2 Carp	265	.060	S.F.	81.50	2.07		83.57	93	
	1100	Deluxe, 8' x 12'	"	150	.107	"	88.50	3.65		92.15	103	
	1200	Storage boxes, 20' x 8', buy	2 Skwk	1.80	8.889	Ea.	3,175	310		3,485	3,975	
	1250	Rent per month					74.50			74.50	82	
	1300	40' x 8', buy	2 Skwk	1.40	11.429		4,250	400		4,650	5,300	
	1350	Rent per month				↓	105			105	116	
	5000	Air supported structures, see division 13011-200										
550	0010	**FIELD OFFICE EXPENSE**										550
	0100	Field office expense, office equipment rental average				Month	143			143	157	

Important: See the Reference Section for critical supporting data - Reference Nos., Crews, & City Cost Indexes

01520 | Construction Facilities

			DAILY OUTPUT	LABOR-HOURS	UNIT	2005 BARE COSTS				TOTAL INCL O&P	
			CREW			MAT.	LABOR	EQUIP.	TOTAL		
550	0120	Office supplies, average			Month	85			85	93.50	550
	0125	Office trailer rental, see division 01520-500									
	0140	Telephone bill; avg. bill/month incl. long dist.			Month	204			204	224	
	0160	Field office lights & HVAC			"	98			98	108	
900	0010	**WEATHER STATION** Remote recording, minimum			Ea.	5,450			5,450	6,000	900
	0100	Maximum			"	25,500			25,500	28,100	

01530 | Temporary Construction

			CREW	DAILY OUTPUT	LABOR-HOURS	UNIT	MAT.	LABOR	EQUIP.	TOTAL	TOTAL INCL O&P	
700	0010	**PROTECTION** Stair tread, 2" x 12" planks, 1 use	1 Carp	75	.107	Tread	4.80	3.65		8.45	11	700
	0100	Exterior plywood, 1/2" thick, 1 use		65	.123		1.86	4.22		6.08	8.60	
	0200	3/4" thick, 1 use	↓	60	.133	↓	2.82	4.57		7.39	10.20	
900	0010	**WINTER PROTECTION** Reinforced plastic on wood	2 Clab	750	.021	S.F.	.38	.57		.95	1.31	900
	0100	framing to close openings										
	0200	Tarpaulins hung over scaffolding, 8 uses, not incl. scaffolding		1,500	.011		.18	.28		.46	.64	
	0250	Tarpaulin polyester reinf. w/ integral fastening system 11 mils thick		1,600	.010		.78	.27		1.05	1.28	
	0300	Prefab fiberglass panels, steel frame, 8 uses	↓	1,200	.013	↓	.74	.36		1.10	1.36	

01540 | Construction Aids

			CREW	DAILY OUTPUT	LABOR-HOURS	UNIT	MAT.	LABOR	EQUIP.	TOTAL	TOTAL INCL O&P	
500	0010	**PERSONNEL PROTECTIVE EQUIPMENT**										500
	0015	Hazardous waste protection										
	0020	Respirator mask only, full face, silicone				Ea.	194			194	213	
	0030	Half face, silicone					26.50			26.50	29	
	0040	Respirator cartridges, 2 req'd/mask, dust or asbestos					4.65			4.65	5.10	
	0050	Chemical vapor					4.48			4.48	4.92	
	0060	Combination vapor and dust					9.80			9.80	10.75	
	0100	Emergency escape breathing apparatus, 5 min					430			430	475	
	0110	10 min					490			490	535	
	0150	Self contained breathing apparatus with full face piece, 30 min					1,725			1,725	1,900	
	0160	60 min					3,125			3,125	3,450	
	0200	Encapsulating suits, limited use, level A					890			890	980	
	0210	Level B				↓	227			227	250	
	0300	Over boots, latex				Pr.	4.20			4.20	4.62	
	0310	PVC					12.50			12.50	13.75	
	0320	Neoprene					51.50			51.50	57	
	0400	Gloves, nitrile/PVC					5.40			5.40	5.95	
	0410	Neoprene coated				↓	24			24	26.50	
550	0010	**PUMP STAGING**, Aluminum	R01540 -200									550
	0200	24' long pole section, buy				Ea.	360			360	395	
	0300	18' long pole section, buy					276			276	305	
	0400	12' long pole section, buy					187			187	206	
	0500	6' long pole section, buy					100			100	110	
	0600	6' long splice joint section, buy					73.50			73.50	81	
	0700	Pump jack					119			119	131	
	0900	Foldable brace					51.50			51.50	56.50	
	1000	Workbench/back safety rail support					63			63	69	
	1100	Scaffolding planks/workbench, 14" wide x 24' long					580			580	640	
	1200	Plank end safety rail					196			196	215	
	1250	Safety net, 22' long				↓	285			285	315	
	1300	System in place, 50' working height, per use based on 50 uses	2 Carp	84.80	.189	C.S.F.	5.30	6.45		11.75	15.85	
	1400	100 uses		84.80	.189		2.65	6.45		9.10	12.95	
	1500	150 uses	↓	84.80	.189	↓	1.77	6.45		8.22	12	
700	0010	**SAFETY NETS** No supports, stock sizes, nylon, 4" mesh				S.F.	1.03			1.03	1.13	700
	0100	Polypropylene, 6" mesh				↓	1.59			1.59	1.75	

1 GENERAL REQUIREMENTS

01540 | Construction Aids

			CREW	DAILY OUTPUT	LABOR-HOURS	UNIT	MAT.	LABOR	EQUIP.	TOTAL	TOTAL INCL O&P	
700	0200	Small mesh debris nets, 1/4" & 3/4" mesh, stock sizes				S.F.	.74			.74	.81	700
	0220	Combined 4" mesh and 1/4" mesh, stock sizes					2.05			2.05	2.26	
	0300	Monthly rental, 4" mesh, stock sizes, 1st month					.40			.40	.44	
	0320	2nd month rental					.20			.20	.22	
	0340	Maximum rental/year					.90			.90	.99	
750	0010	**SCAFFOLDING**										750
	0015	Steel tubular, reg, rent/mo, no plank, incl erect or dismantle R01540-100										
	0090	Building exterior, wall face, 1 to 5 stories, 6'-4" x 5' frames	3 Carp	24	1	C.S.F.	24.50	34.50		59	80	
	0200	6 to 12 stories	4 Carp	21.20	1.509		24.50	51.50		76	107	
	0310	13 to 20 stories	5 Carp	20	2		24.50	68.50		93	134	
	0460	Building interior, wall face area, up to 16' high	3 Carp	25	.960		24.50	33		57.50	77.50	
	0560	16' to 40' high		23	1.043		24.50	35.50		60	82	
	0800	Building interior floor area, up to 30' high		312	.077	C.C.F.	2.57	2.63		5.20	6.95	
	0900	Over 30' high	4 Carp	275	.116	"	2.57	3.99		6.56	9.05	
	0910	Steel tubular, heavy duty shoring, buy										
	0920	Frames 5' high 2' wide				Ea.	82.50			82.50	91	
	0925	5' high 4' wide					93.50			93.50	103	
	0930	6' high 2' wide					94.50			94.50	104	
	0935	6' high 4' wide					111			111	122	
	0940	Accessories										
	0945	Cross braces				Ea.	17.60			17.60	19.35	
	0950	U-head, 8" x 8"					19.25			19.25	21	
	0955	J-head, 4" x 8"					14.10			14.10	15.50	
	0960	Base plate, 8" x 8"					15.60			15.60	17.20	
	0965	Leveling jack					33.50			33.50	37	
	1000	Steel tubular, regular, buy										
	1100	Frames 3' high 5' wide				Ea.	64			64	70	
	1150	5' high 5' wide					73.50			73.50	81	
	1200	6'-4" high 5' wide					92.50			92.50	102	
	1350	7'-6" high 6' wide					160			160	175	
	1500	Accessories cross braces					16.50			16.50	18.15	
	1550	Guardrail post					16.50			16.50	18.15	
	1600	Guardrail 7' section					8			8	8.80	
	1650	Screw jacks & plates					26.50			26.50	29	
	1700	Sidearm brackets					31			31	34	
	1750	8" casters					36.50			36.50	40	
	1800	Plank 2" x 10" x 16'-0"					47			47	51.50	
	1900	Stairway section					270			270	296	
	1910	Stairway starter bar					32			32	35	
	1920	Stairway inside handrail					58.50			58.50	64	
	1930	Stairway outside handrail					80.50			80.50	88.50	
	1940	Walk-thru frame guardrail					40.50			40.50	44.50	
	2000	Steel tubular, regular, rent/mo.										
	2100	Frames 3' high 5' wide				Ea.	3.75			3.75	4.13	
	2150	5' high 5' wide					3.75			3.75	4.13	
	2200	6'-4" high 5' wide					3.75			3.75	4.13	
	2250	7'-6" high 6' wide					7			7	7.70	
	2500	Accessories, cross braces					.60			.60	.66	
	2550	Guardrail post					1			1	1.10	
	2600	Guardrail 7' section					.75			.75	.83	
	2650	Screw jacks & plates					1.50			1.50	1.65	
	2700	Sidearm brackets					1.50			1.50	1.65	
	2750	8" casters					6			6	6.60	
	2800	Outrigger for rolling tower					3			3	3.30	
	2850	Plank 2" x 10" x 16'-0"					5			5	5.50	

Important: See the Reference Section for critical supporting data - Reference Nos., Crews, & City Cost Indexes

01540	Construction Aids		CREW	DAILY OUTPUT	LABOR-HOURS	UNIT	2005 BARE COSTS				TOTAL INCL O&P	
							MAT.	LABOR	EQUIP.	TOTAL		
750 2900	Stairway section					Ea.	10			10	11	**750**
2910	Stairway starter bar	R01540 -100					.10			.10	.11	
2920	Stairway inside handrail						5			5	5.50	
2930	Stairway outside handrail						5			5	5.50	
2940	Walk-thru frame guardrail						2			2	2.20	
3000	Steel tubular, heavy duty shoring, rent/mo.											
3250	5' high 2' & 4' wide					Ea.	5			5	5.50	
3300	6' high 2' & 4' wide						5			5	5.50	
3500	Accessories, cross braces						1			1	1.10	
3600	U - head, 8" x 8"						1			1	1.10	
3650	J - head, 4" x 8"						1			1	1.10	
3700	Base plate, 8" x 8"						1			1	1.10	
3750	Leveling jack						2			2	2.20	
5700	Planks, 2x10x16'-0", labor only, erect or remove to 50' H		3 Carp	144	.167			5.70		5.70	8.90	
5800	Over 50' high		4 Carp	160	.200			6.85		6.85	10.65	
6000	Heavy duty shoring for elevated slab forms to 8'-2" high, floor area											
6010	incl. erection or dismantle labor											
6100	1 use/month		4 Carp	36	.889	C.S.F.	29.50	30.50		60	80	
6150	2 uses/month		"	36	.889	"	14.80	30.50		45.30	64	
6500	To 14'-8" high											
6600	1 use/month		4 Carp	18	1.778	C.S.F.	43	61		104	143	
6650	2 uses/month		"	18	1.778	"	21.50	61		82.50	119	
755 0010	**SCAFFOLDING SPECIALTIES**											**755**
1200	Sidewalk bridge, heavy duty steel posts & beams, including											
1210	parapet protection & waterproofing											
1220	8' to 10' wide, 2 posts		3 Carp	15	1.600	L.F.	34.50	55		89.50	124	
1230	3 posts		"	10	2.400	"	53.50	82		135.50	187	
1500	Sidewalk bridge using tubular steel											
1510	scaffold frames, including planking		3 Carp	45	.533	L.F.	4.72	18.25		22.97	33.50	
1600	For 2 uses per month, deduct from all above						50%					
1700	For 1 use every 2 months, add to all above						100%					
1900	Catwalks, 20" wide, no guardrails, 7' span, buy					Ea.	138			138	151	
2000	10' span, buy					"	176			176	194	
2800	Hand winch-operated masons scaffolding, no plank											
2810	plank moving not required											
2900	98' long, 10'-6" high, buy					Ea.	28,900			28,900	31,700	
3000	Rent per month						1,150			1,150	1,275	
3100	28'-6" high, buy						35,400			35,400	38,900	
3200	Rent per month						1,425			1,425	1,550	
3400	196' long, 28'-6" high, buy						68,500			68,500	75,000	
3500	Rent per month						2,725			2,725	3,000	
3600	64'-6" high, buy						93,500			93,500	103,000	
3700	Rent per month						3,750			3,750	4,125	
3720	Putlog, standard, 8' span, with hangers, buy						67			67	74	
3730	Rent per month						10			10	11	
3750	12' span, buy						101			101	111	
3755	Rent per month						15			15	16.50	
3760	Trussed type, 16' span, buy						231			231	254	
3770	Rent per month						20			20	22	
3790	22' span, buy						277			277	305	
3795	Rent per month						30			30	33	
3800	Rolling ladders with handrails, 30" wide, buy, 2 step						173			173	191	
4000	7 step						585			585	645	
4050	10 step						765			765	845	
4100	Rolling towers, buy, 5' wide, 7' long, 10' high						1,250			1,250	1,375	
4200	For 5' high added sections, to buy, add						209			209	230	

01540 | Construction Aids

		CREW	DAILY OUTPUT	LABOR-HOURS	UNIT	MAT.	LABOR	EQUIP.	TOTAL	TOTAL INCL O&P
755						**2005 BARE COSTS**				
4300	Complete incl. wheels, railings, outriggers,									
4350	21' high, to buy				Ea.	2,125			2,125	2,325
4400	Rent/month				"	159			159	175
760	**0010 STAGING AIDS** and fall protection equipment									
0100	Sidewall staging bracket, tubular, buy				Ea.	30.50			30.50	33.50
0110	Cost each per day, based on 250 days use				Day	.12			.12	.13
0200	Guard post, buy				Ea.	15			15	16.50
0210	Cost each per day, based on 250 days use				Day	.06			.06	.07
0300	End guard chains, buy per pair				Pair	25			25	27.50
0310	Cost per set per day, based on 250 days use				Day	.12			.12	.13
1000	Roof shingling bracket, steel, buy				Ea.	6.35			6.35	7
1010	Cost each per day, based on 250 days use				Day	.03			.03	.03
1100	Wood bracket, buy				Ea.	13.50			13.50	14.85
1110	Cost each per day, based on 250 days use				Day	.05			.05	.06
2000	Ladder jack, aluminum, buy per pair				Pair	84.50			84.50	93
2010	Cost per pair per day, based on 250 days use				Day	.34			.34	.37
2100	Steel siderail jack, buy per pair				Pair	63			63	69.50
2110	Cost per pair per day, based on 250 days use				Day	.25			.25	.28
3000	Laminated wood plank, 2x10x16', buy				Ea.	42.50			42.50	47
3010	Cost each per day, based on 250 days use				Day	.17			.17	.19
3100	Aluminum scaffolding plank, 20" wide x 24' long, buy				Ea.	690			690	760
3110	Cost each per day, based on 250 days use				Day	2.76			2.76	3.04
4000	Nylon full body harness, lanyard and rope grab				Ea.	207			207	228
4010	Cost each per day, based on 250 days use				Day	.83			.83	.91
4100	Rope for safety line, 5/8" x 100' nylon, buy				Ea.	41			41	45
4110	Cost each per day, based on 250 days use				Day	.16			.16	.18
4200	Permanent U-Bolt roof anchor, buy				Ea.	31			31	34
4300	Temporary (one use) roof ridge anchor, buy				"	24			24	26.50
5000	Installation (setup and removal) of staging aids									
5010	Sidewall staging bracket	2 Carp	64	.250	Ea.		8.55		8.55	13.35
5020	Guard post with 2 wood rails	"	64	.250			8.55		8.55	13.35
5030	End guard chains, set	1 Carp	64	.125			4.28		4.28	6.65
5100	Roof shingling bracket		96	.083			2.85		2.85	4.45
5200	Ladder jack	↓	64	.125			4.28		4.28	6.65
5300	Wood plank, 2x10x16'	2 Carp	80	.200			6.85		6.85	10.65
5310	Aluminum scaffold plank, 20" x 24'	"	40	.400			13.70		13.70	21.50
5410	Safety rope	1 Carp	40	.200			6.85		6.85	10.65
5420	Permanent U-Bolt roof anchor (install only)	2 Carp	40	.400			13.70		13.70	21.50
5430	Temporary roof ridge anchor (install only)	1 Carp	64	.125	↓		4.28		4.28	6.65
780	**0010 SWING STAGING**, 500 lb cap., 2' wide to 24' long, hand operated hoist									
0020	steel cable type, with 60' cables, buy				Ea.	4,450			4,450	4,900
0030	Rent per month				"	445			445	490
0600	Lightweight (not for masons) 24' long for 150' height.									
0610	manual type, buy				Ea.	4,725			4,725	5,175
0620	Rent per month					470			470	515
0700	Powered, electric or air, to 150' high, buy					16,700			16,700	18,400
0710	Rent per month					1,175			1,175	1,275
0780	To 300' high, buy					19,600			19,600	21,600
0800	Rent per month					1,375			1,375	1,500
1000	Bosun's chair or work basket 3' x 3.5', to 300' high, electric, buy					7,350			7,350	8,075
1010	Rent per month				↓	515			515	565
2200	Move swing staging (setup and remove)	E-4	2	16	Move		620	40.50	660.50	1,150
790	**0010 SURVEYOR STAKES** Hardwood, 1" x 1" x 48" long				C	46.50			46.50	51.50
0100	2" x 2" x 18" long				↓	61.50			61.50	68

Important: See the Reference Section for critical supporting data - Reference Nos., Crews, & City Cost Indexes

GENERAL REQUIREMENTS

1

01540	Construction Aids		CREW	DAILY OUTPUT	LABOR-HOURS	UNIT	2005 BARE COSTS				TOTAL INCL O&P	
							MAT.	LABOR	EQUIP.	TOTAL		
790	0150	2" x 2" x 24" long				C	73.50			73.50	81	**790**
	0200	2" x 2" x 30" long				↓	72.50			72.50	79.50	
800	0010	**TARPAULINS** Cotton duck, 10 oz. to 13.13 oz. per S.Y., minimum				S.F.	.49			.49	.54	**800**
	0050	Maximum					.58			.58	.64	
	0100	Polyvinyl coated nylon, 14 oz. to 18 oz., minimum					.48			.48	.53	
	0150	Maximum					.68			.68	.75	
	0200	Reinforced polyethylene 3 mils thick, white					.11			.11	.12	
	0300	4 mils thick, white, clear or black					.14			.14	.15	
	0400	5.5 mils thick, clear					.19			.19	.21	
	0500	White, fire retardant					.18			.18	.20	
	0600	7.5 mils, oil resistant, fire retardant					.19			.19	.21	
	0700	8.5 mils, black					.24			.24	.26	
	0710	Woven polyethylene, 6 mils thick					.48			.48	.53	
	0730	Polyester reinforced w/ integral fastening system 11 mils thick					1.07			1.07	1.18	
	0740	Mylar polyester, non-reinforced, 7 mils thick				↓	1.17			1.17	1.29	
820	0010	**SMALL TOOLS** As % of contractor's work, minimum	R01100 -050			Total					.50%	**820**
	0100	Maximum				"					2%	

01550	Vehicular Access & Parking											
												700
700	0010	**ROADS AND SIDEWALKS** Temporary										
	0050	Roads, gravel fill, no surfacing, 4" gravel depth	B-14	715	.067	S.Y.	3.23	1.89	.30	5.42	6.80	
	0100	8" gravel depth	"	615	.078	"	6.45	2.19	.35	8.99	10.90	
	1000	Ramp, 3/4" plywood on 2" x 6" joists, 16" O.C.	2 Carp	300	.053	S.F.	1.48	1.83		3.31	4.47	
	1100	On 2" x 10" joists, 16" O.C.	"	275	.058		2.11	1.99		4.10	5.40	
	2200	Sidewalks, 2" x 12" planks, 2 uses	1 Carp	350	.023		.80	.78		1.58	2.10	
	2300	Exterior plywood, 2 uses, 1/2" thick		750	.011		.31	.37		.68	.91	
	2400	5/8" thick		650	.012		.39	.42		.81	1.09	
	2500	3/4" thick	↓	600	.013	↓	.47	.46		.93	1.23	

01560	Barriers & Enclosures											
100	0010	**BARRICADES** 5' high, 3 rail @ 2" x 8", fixed	2 Carp	30	.533	L.F.	11	18.25		29.25	40.50	**100**
	0150	Movable	"	20	.800	"	11	27.50		38.50	54.50	
	0300	Stock units, 6' high, 8' wide, plain, buy				Ea.	435			435	480	
	0350	With reflective tape, buy				"	525			525	580	
	0400	Break-a-way 3" PVC pipe barricade										
	0410	with 3 ea. 1' x 4' reflectorized panels, buy				Ea.	305			305	335	
	0500	Plywood with steel legs, 32" wide					72			72	79	
	0600	Telescoping Christmas tree, 9' high, 5 flags, buy					122			122	134	
	0800	Traffic cones, PVC, 18" high					6.05			6.05	6.65	
	0850	28" high				↓	18.45			18.45	20.50	
	1000	Guardrail, wooden, 3' high, 1" x 6", on 2" x 4" posts	2 Carp	200	.080	L.F.	1.02	2.74		3.76	5.40	
	1100	2" x 6", on 4" x 4" posts	"	165	.097		1.85	3.32		5.17	7.20	
	1200	Portable metal with base pads, buy					15.50			15.50	17.05	
	1250	Typical installation, assume 10 reuses	2 Carp	600	.027	↓	1.60	.91		2.51	3.18	
	1300	Barricade tape, polyethelyne, 7 mil, 3" wide x 500' long roll				Ea.	25			25	27.50	
	5000	Barricades, see also division 01590-400										
250	0010	**TEMPORARY FENCING** Chain link, 11 ga, 5' high	2 Clab	400	.040	L.F.	4.65	1.07		5.72	6.75	**250**
	0100	6' high		300	.053		4.48	1.42		5.90	7.15	
	0200	Rented chain link, 6' high, to 1000' (up to 12 mo.)		400	.040		2.94	1.07		4.01	4.89	
	0250	Over 1000' (up to 12 mo.)	↓	300	.053		2.13	1.42		3.55	4.56	
	0350	Plywood, painted, 2" x 4" frame, 4' high	A-4	135	.178		5.30	5.85		11.15	14.85	
	0400	4" x 4" frame, 8' high	"	110	.218		9.90	7.20		17.10	22	
	0500	Wire mesh on 4" x 4" posts, 4' high	2 Carp	100	.160		7.90	5.50		13.40	17.25	
	0550	8' high	"	80	.200	↓	12.10	6.85		18.95	24	

GENERAL REQUIREMENTS 1

01500 | Temporary Facilities & Controls

		01560	Barriers & Enclosures	CREW	DAILY OUTPUT	LABOR-HOURS	UNIT	2005 BARE COSTS				TOTAL INCL O&P	
								MAT.	LABOR	EQUIP.	TOTAL		
400	0010		TEMPORARY CONSTRUCTION See also division 01530										400
800	0010		WATCHMAN Service, monthly basis, uniformed person, minimum				Hr.					9	800
	0100		Maximum									16	
	0200		Person and command dog, minimum									12	
	0300		Maximum				↓					17.70	
	0500		Sentry dog, leased, with job patrol (yard dog), 1 dog				Week					230	
	0600		2 dogs				"					300	
	0800		Purchase, trained sentry dog, minimum				Ea.					1,000	
	0900		Maximum				"					2,500	
		01580	Project Signs										
700	0010		SIGNS Hi-intensity reflectorized, no posts, buy				S.F.	16.40			16.40	18	700

01590 | Equipment Rental

	UNIT	HOURLY OPER. COST	RENT PER DAY	RENT PER WEEK	RENT PER MONTH	CREW EQUIPMENT COST/DAY	
100							**100**
0010 CONCRETE EQUIPMENT RENTAL — R01590-100							
0100 without operators							
0150 For batch plant, see div. 01590-500 — R03310-090							
0200 Bucket, concrete lightweight, 1/2 C.Y.	Ea.	.55	15.35	46	138	13.60	
0300 1 C.Y.		.60	18.65	56	168	16	
0400 1-1/2 C.Y.		.75	25.50	76	228	21.20	
0500 2 C.Y.		.80	30.50	91	273	24.60	
0580 8 C.Y.		4.40	200	600	1,800	155.20	
0600 Cart, concrete, self propelled, operator walking, 10 C.F.		2.10	55	165	495	49.80	
0700 Operator riding, 18 C.F.		3.35	83.50	250	750	76.80	
0800 Conveyer for concrete, portable, gas, 16" wide, 26' long		7.35	117	350	1,050	128.80	
0900 46' long		7.70	142	425	1,275	146.60	
1000 56' long		7.85	150	450	1,350	152.80	
1100 Core drill, electric, 2-1/2 H.P., 1" to 8" bit diameter		1.61	62.50	187	560	50.30	
1150 11 HP, 8" to 18" cores		6.61	83.50	250.80	750	103.05	
1200 Finisher, concrete floor, gas, riding trowel, 48" diameter		4.95	86.50	260	780	91.60	
1300 Gas, manual, 3 blade, 36" trowel		1.05	16.65	50	150	18.40	
1400 4 blade, 48" trowel		1.50	21	63	189	24.60	
1500 Float, hand-operated (Bull float) 48" wide		.08	13.35	40	120	8.65	
1570 Curb builder, 14 H.P., gas, single screw		9.50	195	585	1,750	193	
1590 Double screw		9.90	228	685	2,050	216.20	
1600 Grinder, concrete and terrazzo, electric, floor		2.10	88.50	266	800	70	
1700 Wall grinder		1.05	44.50	133	400	35	
1800 Mixer, powered, mortar and concrete, gas, 6 C.F., 18 H.P.		5.25	102	305	915	103	
1900 10 C.F., 25 H.P.		6.40	120	360	1,075	123.20	
2000 16 C.F.		6.70	143	430	1,300	139.60	
2100 Concrete, stationary, tilt drum, 2 C.Y.		5.25	200	600	1,800	162	
2120 Pump, concrete, truck mounted 4" line 80' boom		21.65	885	2,655	7,975	704.20	
2140 5" line, 110' boom		28.05	1,175	3,510	10,500	926.40	
2160 Mud jack, 50 C.F. per hr.		4.82	115	346	1,050	107.75	
2180 225 C.F. per hr.		7.56	154	461.60	1,375	152.80	
2190 Shotcrete pump rig, 12 CY/hr		11.10	225	675	2,025	223.80	
2600 Saw, concrete, manual, gas, 18 H.P.		3.60	35	105	315	49.80	
2650 Self-propelled, gas, 30 H.P.		7.05	96.50	290	870	114.40	
2700 Vibrators, concrete, electric, 60 cycle, 2 H.P.		.37	10.65	32	96	9.35	
2800 3 H.P.		.56	15.65	47	141	13.90	
2900 Gas engine, 5 H.P.		.95	18.35	55	165	18.60	
3000 8 H.P.		1.35	22	66	198	24	
3050 Vibrating screed, gas engine, 8 H.P.		1.68	49.50	148	445	43.05	
3100 Concrete transit mixer, hydraulic drive							
3120 6 x 4, 250 H.P., 8 C.Y., rear discharge		34.95	550	1,650	4,950	609.60	
3200 Front discharge		40.65	670	2,015	6,050	728.20	
3300 6 x 6, 285 H.P., 12 C.Y., rear discharge		40.60	670	2,010	6,025	726.80	
3400 Front discharge		41.55	680	2,040	6,125	740.40	
200							**200**
0010 EARTHWORK EQUIPMENT RENTAL Without operators — R01590-100	Ea.	1.80	34.50	103	310	35	
0040 Aggregate spreader, push type 8' to 12' wide	"	1.75	31.50	95	285	33	
0045 Tailgate type, 8' wide — R02315-300							
0050 Augers for vertical drilling							
0055 Earth auger, truck-mounted, for fence & sign posts — R02315-400	Ea.	8.30	485	1,455	4,375	357.40	
0060 For borings and monitoring wells		29.20	620	1,855	5,575	604.60	
0070 Earth auger, portable, trailer mounted — R02315-450		1.55	22.50	67	201	25.80	
0075 Earth auger, truck-mounted, for caissons, water wells, utility poles		153.10	3,250	9,770	29,300	3,179	
0080 Auger, horizontal boring machine, 12" to 36" diameter, 45 H.P. — R02455-900		15.85	182	545	1,625	235.80	
0090 12" to 48" diameter, 65 H.P.		22.30	325	970	2,900	372.40	
0095 Auger, for fence posts, gas engine, hand held		.30	4.33	13	39	5	
0100 Excavator, diesel hydraulic, crawler mounted, 1/2 C.Y. cap.		15.80	345	1,030	3,100	332.40	
0120 5/8 C.Y. capacity		19.30	465	1,390	4,175	432.40	
0140 3/4 C.Y. capacity		22.50	490	1,470	4,400	474	

01590 | Equipment Rental

GENERAL REQUIREMENTS — 1

			UNIT	HOURLY OPER. COST	RENT PER DAY	RENT PER WEEK	RENT PER MONTH	CREW EQUIPMENT COST/DAY
200	0150	1 C.Y. capacity	Ea.	27.10	570	1,705	5,125	557.80
	0200	1-1/2 C.Y. capacity		33.55	755	2,260	6,775	720.40
	0300	2 C.Y. capacity	R01590-100	42.35	950	2,855	8,575	909.80
	0320	2-1/2 C.Y. capacity		55.75	1,275	3,845	11,500	1,215
	0340	3-1/2 C.Y. capacity	R02315-300	94.45	2,100	6,320	19,000	2,020
	0341	Attachments						
	0342	Bucket thumbs	R02315-400	2.50	208	625	1,875	145
	0345	Grapples		1	200	600.40	1,800	128.10
	0350	Gradall type, truck mounted, 3 ton @ 15' radius, 5/8 C.Y.	R02315-450	38.35	885	2,655	7,975	837.80
	0370	1 C.Y. capacity	R02455-900	44.25	1,025	3,085	9,250	971
	0400	Backhoe-loader, 40 to 45 H.P., 5/8 C.Y. capacity		8.40	177	530	1,600	173.20
	0450	45 H.P. to 60 H.P., 3/4 C.Y. capacity		10.60	218	655	1,975	215.80
	0460	80 H.P., 1-1/4 C.Y. capacity		13.30	250	750	2,250	256.40
	0470	112 H.P., 1-1/2 C.Y. capacity		18.10	390	1,165	3,500	377.80
	0480	Attachments						
	0482	Compactor, 20,000 lb		4.25	117	350	1,050	104
	0485	Hydraulic hammer, 750 ft-lbs		1.95	68.50	205	615	56.60
	0486	Hydraulic hammer, 1200 ft-lbs		4	133	400	1,200	112
	0500	Brush chipper, gas engine, 6" cutter head, 35 H.P.		5.90	93.50	280	840	103.20
	0550	12" cutter head, 130 H.P.		9.35	150	450	1,350	164.80
	0600	15" cutter head, 165 H.P.		13.40	158	475	1,425	202.20
	0750	Bucket, clamshell, general purpose, 3/8 C.Y.		1	35	105	315	29
	0800	1/2 C.Y.		1.10	41.50	125	375	33.80
	0850	3/4 C.Y.		1.25	51.50	155	465	41
	0900	1 C.Y.		1.30	55	165	495	43.40
	0950	1-1/2 C.Y.		2.05	75	225	675	61.40
	1000	2 C.Y.		2.15	85	255	765	68.20
	1010	Bucket, dragline, medium duty, 1/2 C.Y.		.60	22.50	67	201	18.20
	1020	3/4 C.Y.		.60	23.50	70	210	18.80
	1030	1 C.Y.		.65	25.50	76	228	20.40
	1040	1-1/2 C.Y.		.95	38.50	115	345	30.60
	1050	2 C.Y.		1.05	43.50	130	390	34.40
	1070	3 C.Y.		1.55	58.50	175	525	47.40
	1200	Compactor, manually guided 2-drum vibratory smooth roller, 7.5 H.P.		4.60	143	430	1,300	122.80
	1250	Rammer compactor, gas, 1000 lb. blow		1.60	36.50	110	330	34.80
	1300	Vibratory plate, gas, 18" plate, 3000 lb. blow		1.50	22	66	198	25.20
	1350	21" plate, 5000 lb. blow		1.80	37.50	112	335	36.80
	1370	Curb builder/extruder, 14 H.P., gas, single screw		9.50	197	590	1,775	194
	1390	Double screw		9.90	228	685	2,050	216.20
	1500	Disc harrow attachment, for tractor		.36	60.50	181	545	39.10
	1750	Extractor, piling, see lines 2500 to 2750						
	1810	Feller buncher, shearing & accumulating trees, 100 H.P.	Ea.	19.80	455	1,370	4,100	432.40
	1860	Grader, self-propelled, 25,000 lb.		18.20	400	1,205	3,625	386.60
	1910	30,000 lb.		20.70	485	1,455	4,375	456.60
	1920	40,000 lb.		30.50	730	2,190	6,575	682
	1930	55,000 lb.		40.25	1,025	3,080	9,250	938
	1950	Hammer, pavement demo., hyd., gas, self-prop., 1000 to 1250 lb.		20.15	390	1,170	3,500	395.20
	2000	Diesel 1300 to 1500 lb.		27.80	590	1,765	5,300	575.40
	2050	Pile driving hammer, steam or air, 4150 ft.-lb. @ 225 BPM		6.35	275	825	2,475	215.80
	2100	8750 ft.-lb. @ 145 BPM		8.25	450	1,350	4,050	336
	2150	15,000 ft.-lb. @ 60 BPM		8.60	485	1,455	4,375	359.80
	2200	24,450 ft.-lb. @ 111 BPM		11.35	535	1,605	4,825	411.80
	2250	Leads, 15,000 ft.-lb. hammers	L.F.	.03	1.69	5.06	15.20	1.25
	2300	24,450 ft.-lb. hammers and heavier	"	.05	2.67	8	24	2
	2350	Diesel type hammer, 22,400 ft.-lb.	Ea.	24	620	1,860	5,575	564
	2400	41,300 ft.-lb.		32.20	675	2,025	6,075	662.60
	2450	141,000 ft.-lb.		60.90	1,475	4,430	13,300	1,373
	2500	Vib. elec. hammer/extractor, 200 KW diesel generator, 34 H.P.		26.25	655	1,965	5,900	603

Important: See the Reference Section for critical supporting data - Reference Nos., Crews, & City Cost Indexes

01590 | Equipment Rental

			UNIT	HOURLY OPER. COST	RENT PER DAY	RENT PER WEEK	RENT PER MONTH	CREW EQUIPMENT COST/DAY	
200	2550	80 H.P.		Ea.	44.80	960	2,885	8,650	935.40	**200**
	2600	150 H.P.	R01590 -100		64	1,475	4,455	13,400	1,403	
	2700	Extractor, steam or air, 700 ft.-lb.			14.50	410	1,235	3,700	363	
	2750	1000 ft.-lb.	R02315 -300		16.60	515	1,545	4,625	441.80	
	2800	Log chipper, up to 22" diam, 600 H.P.			38.53	1,250	3,770	11,300	1,062	
	2850	Logger, for skidding & stacking logs, 150 H.P.	R02315 -400		33.80	790	2,375	7,125	745.40	
	2900	Rake, spring tooth, with tractor			7.99	213	638	1,925	191.50	
	3000	Roller, vibratory, tandem, smooth drum, 20 H.P.	R02315 -450		5.05	112	335	1,000	107.40	
	3050	35 H.P.			7.15	203	610	1,825	179.20	
	3100	Towed type vibratory compactor, smooth drum, 50 H.P.	R02455 -900		31.95	560	1,680	5,050	591.60	
	3150	Sheepsfoot, 50 H.P.			32.65	580	1,745	5,225	610.20	
	3170	Landfill compactor, 220 HP			46.75	1,150	3,470	10,400	1,068	
	3200	Pneumatic tire roller, 80 H.P.			8.35	292	875	2,625	241.80	
	3250	120 H.P.			13.60	505	1,515	4,550	411.80	
	3300	Sheepsfoot vibratory roller, 200 H.P.			35.35	855	2,560	7,675	794.80	
	3320	340 H.P.			48.90	1,225	3,670	11,000	1,125	
	3350	Smooth drum vibratory roller, 75 H.P.			13.60	395	1,180	3,550	344.80	
	3400	125 H.P.			17.15	495	1,480	4,450	433.20	
	3410	Rotary mower, brush, 60", with tractor			11.35	230	690	2,075	228.80	
	3450	Scrapers, towed type, 9 to 12 C.Y. capacity			4.12	196	589	1,775	150.75	
	3500	12 to 17 C.Y. capacity			1.51	262	785.60	2,350	169.20	
	3550	Scrapers, self-propelled, 4 x 4 drive, 2 engine, 14 C.Y. capacity			82.40	1,500	4,465	13,400	1,552	
	3600	2 engine, 24 C.Y. capacity			119.45	2,275	6,805	20,400	2,317	
	3640	32 - 44 C.Y. capacity			142.65	2,650	7,935	23,800	2,728	
	3650	Self-loading, 11 C.Y. capacity			40.55	825	2,475	7,425	819.40	
	3700	22 C.Y. capacity			76.40	1,725	5,185	15,600	1,648	
	3710	Screening plant 110 H.P. w/ 5' x 10' screen			21.75	370	1,105	3,325	395	
	3720	5' x 16' screen			23.80	465	1,390	4,175	468.40	
	3850	Shovels, see Cranes division 01590-600								
	3860	Shovel/backhoe bucket, 1/2 C.Y.		Ea.	1.80	55	165	495	47.40	
	3870	3/4 C.Y.			1.85	61.50	185	555	51.80	
	3880	1 C.Y.			1.95	71.50	215	645	58.60	
	3890	1-1/2 C.Y.			2.10	85	255	765	67.80	
	3910	3 C.Y.			2.40	118	355	1,075	90.20	
	3950	Stump chipper, 18" deep, 30 H.P.			4.56	57	171	515	70.70	
	4110	Tractor, crawler, with bulldozer, torque converter, diesel 80 H.P.			16.10	310	930	2,800	314.80	
	4150	105 H.P.			22.10	460	1,385	4,150	453.80	
	4200	140 H.P.			27.25	590	1,770	5,300	572	
	4260	200 H.P.			40.95	985	2,960	8,875	919.60	
	4310	300 H.P.			53.10	1,275	3,850	11,600	1,195	
	4360	410 H.P.			71.95	1,575	4,740	14,200	1,524	
	4370	500 H.P.			95.25	2,125	6,365	19,100	2,035	
	4380	700 H.P.			142.20	3,325	10,005	30,000	3,139	
	4400	Loader, crawler, torque conv., diesel, 1-1/2 C.Y., 80 H.P.			14.95	320	965	2,900	312.60	
	4450	1-1/2 to 1-3/4 C.Y., 95 H.P.			17.35	385	1,150	3,450	368.80	
	4510	1-3/4 to 2-1/4 C.Y., 130 H.P.			23.60	605	1,815	5,450	551.80	
	4530	2-1/2 to 3-1/4 C.Y., 190 H.P.			35.50	840	2,520	7,550	788	
	4560	3-1/2 to 5 C.Y., 275 H.P.			46.95	1,200	3,590	10,800	1,094	
	4610	Tractor loader, wheel, torque conv., 4 x 4, 1 to 1-1/4 C.Y., 65 H.P.			9.85	192	575	1,725	193.80	
	4620	1-1/2 to 1-3/4 C.Y., 80 H.P.			12.50	235	705	2,125	241	
	4650	1-3/4 to 2 C.Y., 100 H.P.			14.20	278	835	2,500	280.60	
	4710	2-1/2 to 3-1/2 C.Y., 130 H.P.			15.25	305	915	2,750	305	
	4730	3 to 4-1/2 C.Y., 170 H.P.			20.95	465	1,390	4,175	445.60	
	4760	5-1/4 to 5-3/4 C.Y., 270 H.P.			34.55	705	2,120	6,350	700.40	
	4810	7 to 8 C.Y., 375 H.P.			58.95	1,250	3,755	11,300	1,223	
	4870	12-1/2 C.Y., 690 H.P.			82.30	1,950	5,860	17,600	1,830	
	4880	Wheeled, skid steer, 10 C.F., 30 H.P. gas			8.80	138	415	1,250	153.40	
	4890	1 C.Y., 78 H.P., diesel			10.35	192	575	1,725	197.80	

GENERAL REQUIREMENTS 1

01590 | Equipment Rental

		UNIT	HOURLY OPER. COST	RENT PER DAY	RENT PER WEEK	RENT PER MONTH	CREW EQUIPMENT COST/DAY	
200	4891	Attachments for all skid steer loaders	R01590 -100						200
	4892	Auger		Ea.	.40	66.50	200	600	43.20
	4893	Backhoe	R02315 -300		.65	109	326	980	70.40
	4894	Broom			.63	105	314	940	67.85
	4895	Forks	R02315 -400		.22	37.50	112	335	24.15
	4896	Grapple			.50	83	249	745	53.80
	4897	Concrete hammer	R02315 -450		.98	163	489	1,475	105.65
	4898	Tree spade			1	167	500	1,500	108
	4899	Trencher	R02455 -900		.68	114	342	1,025	73.85
	4900	Trencher, chain, boom type, gas, operator walking, 12 H.P.			2.75	43.50	130	390	48
	4910	Operator riding, 40 H.P.			8.55	242	725	2,175	213.40
	5000	Wheel type, diesel, 4' deep, 12" wide			46.90	735	2,205	6,625	816.20
	5100	Diesel, 6' deep, 20" wide			63.75	1,575	4,750	14,300	1,460
	5150	Ladder type, diesel, 5' deep, 8" wide			23.10	705	2,115	6,350	607.80
	5200	Diesel, 8' deep, 16" wide			56.80	1,600	4,830	14,500	1,420
	5210	Tree spade, self-propelled			9.60	267	800	2,400	236.80
	5250	Truck, dump, tandem, 12 ton payload			20.45	272	815	2,450	326.60
	5300	Three axle dump, 16 ton payload			27.95	420	1,265	3,800	476.60
	5350	Dump trailer only, rear dump, 16-1/2 C.Y.			4.25	115	345	1,025	103
	5400	20 C.Y.			4.60	132	395	1,175	115.80
	5450	Flatbed, single axle, 1-1/2 ton rating			11.85	56.50	170	510	128.80
	5500	3 ton rating			15.15	91.50	275	825	176.20
	5550	Off highway rear dump, 25 ton capacity			40.40	980	2,945	8,825	912.20
	5600	35 ton capacity			41.30	1,000	3,015	9,050	933.40
	5610	50 ton capacity			53.15	1,250	3,780	11,300	1,181
	5620	65 ton capacity			56.85	1,375	4,105	12,300	1,276
	5630	100 ton capacity			73	1,775	5,325	16,000	1,649
	6000	Vibratory plow, 25 H.P., walking			3.90	58.50	175	525	66.20
400	0010	**GENERAL EQUIPMENT RENTAL** Without operators	R02250 -450						400
	0150	Aerial lift, scissor type, to 15' high, 1000 lb. cap., electric		Ea.	2.30	41.50	125	375	43.40
	0160	To 25' high, 2000 lb. capacity	R02250 -400		2.70	60	180	540	57.60
	0170	Telescoping boom to 40' high, 500 lb. capacity, gas			11.45	268	805	2,425	252.60
	0180	To 45' high, 500 lb. capacity	R02315 -300		12.30	310	925	2,775	283.40
	0190	To 60' high, 600 lb. capacity			14.25	410	1,230	3,700	360
	0195	Air compressor, portable, 6.5 CFM, electric			.42	14.65	44	132	12.15
	0196	Gasoline			.51	22	66	198	17.30
	0200	Air compressor, portable, gas engine, 60 C.F.M.			4.50	32.50	97	291	55.40
	0300	160 C.F.M.			7.80	41.50	125	375	87.40
	0400	Diesel engine, rotary screw, 250 C.F.M.			7.80	83.50	250	750	112.40
	0500	365 C.F.M.			10.35	107	320	960	146.80
	0550	450 C.F.M.			12.55	127	380	1,150	176.40
	0600	600 C.F.M.			21.35	212	635	1,900	297.80
	0700	750 C.F.M.			23.05	222	665	2,000	317.40
	0800	For silenced models, small sizes, add			3%	5%	5%	5%	
	0900	Large sizes, add			5%	7%	7%	7%	
	0920	Air tools and accessories							
	0930	Breaker, pavement, 60 lb.		Ea.	.40	11	33	99	9.80
	0940	80 lb.			.40	12.65	38	114	10.80
	0950	Drills, hand (jackhammer) 65 lb.			.45	15	45	135	12.60
	0960	Track or wagon, swing boom, 4" drifter			35.05	640	1,920	5,750	664.40
	0970	5" drifter			47.70	735	2,205	6,625	822.60
	0975	Track mounted quarry drill, 6" diameter drill			50.45	795	2,380	7,150	879.60
	0980	Dust control per drill			.78	12	36	108	13.45
	0990	Hammer, chipping, 12 lb.			.40	21.50	64	192	16
	1000	Hose, air with couplings, 50' long, 3/4" diameter			.03	5.65	17	51	3.65
	1100	1" diameter			.03	5.65	17	51	3.65
	1200	1-1/2" diameter			.04	7.35	22	66	4.70
	1300	2" diameter			.10	17.35	52	156	11.20

Important: See the Reference Section for critical supporting data - Reference Nos., Crews, & City Cost Indexes

01590 | Equipment Rental

400				UNIT	HOURLY OPER. COST	RENT PER DAY	RENT PER WEEK	RENT PER MONTH	CREW EQUIPMENT COST/DAY	400
	1400	2-1/2" diameter	R02250-450	Ea.	.12	19.65	59	177	12.75	
	1410	3" diameter			.16	27.50	82	246	17.70	
	1450	Drill, steel, 7/8" x 2'	R02250-400		.05	8	24	72	5.20	
	1460	7/8" x 6'			.06	9.35	28	84	6.10	
	1520	Moil points	R02315-300		.03	4.33	13	39	2.85	
	1525	Pneumatic nailer w/accessories			.41	27	81	243	19.50	
	1530	Sheeting driver for 60 lb. breaker			.10	6.95	20.80	62.50	4.95	
	1540	For 90 lb. breaker			.15	10	30	90	7.20	
	1550	Spade, 25 lb.			.35	6.35	19	57	6.60	
	1560	Tamper, single, 35 lb.			.58	38.50	115	345	27.65	
	1570	Triple, 140 lb.			.87	57.50	173	520	41.55	
	1580	Wrenches, impact, air powered, up to 3/4" bolt			.25	7.35	22	66	6.40	
	1590	Up to 1-1/4" bolt			.35	15.35	46	138	12	
	1600	Barricades, barrels, reflectorized, 1 to 50 barrels			.02	2.60	7.80	23.50	1.70	
	1610	100 to 200 barrels			.01	1.93	5.80	17.40	1.25	
	1620	Barrels with flashers, 1 to 50 barrels			.02	3.27	9.80	29.50	2.10	
	1630	100 to 200 barrels			.02	2.60	7.80	23.50	1.70	
	1640	Barrels with steady burn type C lights			.03	4.33	13	39	2.85	
	1650	Illuminated board, trailer mounted, with generator			.65	117	350	1,050	75.20	
	1670	Portable barricade, stock, with flashers, 1 to 6 units			.02	3.27	9.80	29.50	2.10	
	1680	25 to 50 units			.02	3.03	9.10	27.50	2	
	1690	Butt fusion machine, electric			22.35	425	1,270	3,800	432.80	
	1695	Electro fusion machine			8.85	168	505	1,525	171.80	
	1700	Carts, brick, hand powered, 1000 lb. capacity			.30	49.50	148	445	32	
	1800	Gas engine, 1500 lb., 7-1/2' lift			2.77	94	282	845	78.55	
	1822	Dehumidifier, medium, 6 lb/hr, 150 CFM			.68	41.50	124	370	30.25	
	1824	Large, 18 lb/hr, 600 CFM			1.36	82.50	248	745	60.50	
	1830	Distributor, asphalt, trailer mtd, 2000 gal., 38 H.P. diesel			7.35	243	730	2,200	204.80	
	1840	3000 gal., 38 H.P. diesel			8.35	278	835	2,500	233.80	
	1850	Drill, rotary hammer, electric, 1-1/2" diameter			.40	24.50	73	219	17.80	
	1860	Carbide bit for above			.03	5.35	16	48	3.45	
	1865	Rotary, crawler, 250 H.P.			87.35	1,600	4,835	14,500	1,666	
	1870	Emulsion sprayer, 65 gal., 5 H.P. gas engine			1.80	72	216	650	57.60	
	1880	200 gal., 5 H.P. engine			4.85	120	360	1,075	110.80	
	1920	Floodlight, mercury vapor, or quartz, on tripod		Ea.						
	1930	1000 watt			.30	12	36	108	9.60	
	1940	2000 watt			.52	22	66	198	17.35	
	1950	Floodlights, trailer mounted with generator, 1 - 300 watt light			2.45	65	195	585	58.60	
	1960	2 - 1000 watt lights			3.30	108	325	975	91.40	
	2000	4 - 300 watt lights			2.75	76.50	230	690	68	
	2020	Forklift, wheeled, for brick, 18', 3000 lb., 2 wheel drive, gas			13.70	177	530	1,600	215.60	
	2040	28', 4000 lb., 4 wheel drive, diesel			11.75	232	695	2,075	233	
	2050	For rough terrain, 8000 lb., 16' lift, 68 H.P.			15.70	360	1,075	3,225	340.60	
	2060	For plant, 4 T. capacity, 80 H.P., 2 wheel drive, gas			7.80	83.50	250	750	112.40	
	2080	10 T. capacity, 120 H.P., 2 wheel drive, diesel			11.95	157	470	1,400	189.60	
	2100	Generator, electric, gas engine, 1.5 KW to 3 KW			1.75	15	45	135	23	
	2200	5 KW			2.40	21.50	65	195	32.20	
	2300	10 KW			3.95	46.50	140	420	59.60	
	2400	25 KW			7.65	71.50	215	645	104.20	
	2500	Diesel engine, 20 KW			5.95	61.50	185	555	84.60	
	2600	50 KW			11.05	68.50	205	615	129.40	
	2700	100 KW			16.20	78.50	235	705	176.60	
	2800	250 KW			45.65	142	425	1,275	450.20	
	2850	Hammer, hydraulic, for mounting on boom, to 500 ft.-lb.			1.75	63.50	190	570	52	
	2860	1000 ft.-lb.			3.10	102	305	915	85.80	
	2900	Heaters, space, oil or electric, 50 MBH			.92	9.65	29	87	13.15	
	3000	100 MBH			1.66	13.65	41	123	21.50	
	3100	300 MBH			5.33	35	105	315	63.65	

01590 | Equipment Rental

			UNIT	HOURLY OPER. COST	RENT PER DAY	RENT PER WEEK	RENT PER MONTH	CREW EQUIPMENT COST/DAY	
400	3150	500 MBH		Ea.	10.70	50	150	450	115.60	400
	3200	Hose, water, suction with coupling, 20' long, 2" diameter	R02250-450		.02	4	12	36	2.55	
	3210	3" diameter			.03	6.65	20	60	4.25	
	3220	4" diameter	R02250-400		.03	8	24	72	5.05	
	3230	6" diameter			.10	20	60	180	12.80	
	3240	8" diameter	R02315-300		.25	41.50	125	375	27	
	3250	Discharge hose with coupling, 50' long, 2" diameter			.01	3	9	27	1.90	
	3260	3" diameter			.01	4	12	36	2.50	
	3270	4" diameter			.02	5.65	17	51	3.55	
	3280	6" diameter			.06	14	42	126	8.90	
	3290	8" diameter			.31	52	156	470	33.70	
	3295	Insulation blower			.11	7	21	63	5.10	
	3300	Ladders, extension type, 16' to 36' long			.17	28.50	86	258	18.55	
	3400	40' to 60' long			.26	36.50	110	330	24.10	
	3405	Lance for cutting concrete			2.78	105	315	945	85.25	
	3407	Lawn mower, rotary, 22", 5HP			.97	27	81	243	23.95	
	3408	48" self propelled			3.06	100	300	900	84.50	
	3410	Level, laser type, for pipe and sewer leveling			1.27	84.50	253	760	60.75	
	3430	Electronic			.76	50.50	151	455	36.30	
	3440	Laser type, rotating beam for grade control			1.04	69	207	620	49.70	
	3460	Builders level with tripod and rod			.07	12.35	37	111	7.95	
	3500	Light towers, towable, with diesel generator, 2000 watt			2.75	76.50	230	690	68	
	3600	4000 watt			3.30	108	325	975	91.40	
	3700	Mixer, powered, plaster and mortar, 6 C.F., 7 H.P.			1.30	26.50	80	240	26.40	
	3800	10 C.F., 9 H.P.			1.55	41.50	125	375	37.40	
	3850	Nailer, pneumatic			.41	27	81	243	19.50	
	3900	Paint sprayers complete, 8 CFM			.69	45.50	137	410	32.90	
	4000	17 CFM			1.10	73.50	220	660	52.80	
	4020	Pavers, bituminous, rubber tires, 8' wide, 50 H.P., diesel			26.80	775	2,320	6,950	678.40	
	4030	10' wide, 150 H.P.			59.45	1,300	3,895	11,700	1,255	
	4050	Crawler, 8' wide, 100 H.P., diesel			55.60	1,450	4,335	13,000	1,312	
	4060	10' wide, 150 H.P.			66.95	1,875	5,590	16,800	1,654	
	4070	Concrete paver, 12' to 24' wide, 250 H.P.			57.90	1,325	3,995	12,000	1,262	
	4080	Placer-spreader-trimmer, 24' wide, 300 H.P.			87.60	2,025	6,045	18,100	1,910	
	4100	Pump, centrifugal gas pump, 1-1/2", 4 MGPH			2.60	36.50	110	330	42.80	
	4200	2", 8 MGPH			3.25	41.50	125	375	51	
	4300	3", 15 MGPH			3.45	43.50	130	390	53.60	
	4400	6", 90 MGPH			15.75	152	455	1,375	217	
	4500	Submersible electric pump, 1-1/4", 55 GPM			.35	17	51	153	13	
	4600	1-1/2", 83 GPM			.41	19.65	59	177	15.10	
	4700	2", 120 GPM			.43	24.50	73	219	18.05	
	4800	3", 300 GPM			.90	33.50	100	300	27.20	
	4900	4", 560 GPM			6.16	135	405	1,225	130.30	
	5000	6", 1590 GPM			9.27	200	600	1,800	194.15	
	5100	Diaphragm pump, gas, single, 1-1/2" diameter			.78	38.50	116	350	29.45	
	5200	2" diameter			2.70	48.50	145	435	50.60	
	5300	3" diameter			2.75	48.50	145	435	51	
	5400	Double, 4" diameter			3.55	76.50	230	690	74.40	
	5500	Trash pump, self-priming, gas, 2" diameter			2.55	26.50	80	240	36.40	
	5600	Diesel, 4" diameter			6.05	53.50	160	480	80.40	
	5650	Diesel, 6" diameter			18.35	112	335	1,000	213.80	
	5655	Grout Pump			8.30	58.50	175	525	101.40	
	5660	Rollers, see division 01590-200								
	5700	Salamanders, L.P. gas fired, 100,000 BTU		Ea.	1.66	10.65	32	96	19.70	
	5705	50,000 BTU			1.25	7.65	23	69	14.60	
	5720	Sandblaster, portable, open top, 3 C.F. capacity			.40	20	60	180	15.20	
	5730	6 C.F. capacity			.65	29	87	261	22.60	
	5740	Accessories for above			.11	18.35	55	165	11.90	

Important: See the Reference Section for critical supporting data - Reference Nos., Crews, & City Cost Indexes

01590 | Equipment Rental

		UNIT	HOURLY OPER. COST	RENT PER DAY	RENT PER WEEK	RENT PER MONTH	CREW EQUIPMENT COST/DAY
5750	Sander, floor	Ea.	.72	17	51	153	15.95
5760	Edger		.71	25	75	225	20.70
5800	Saw, chain, gas engine, 18" long (R02250-400)		1.20	16	48	144	19.20
5900	36" long		.55	48.50	145	435	33.40
5950	60" long (R02315-300)		.55	50	150	450	34.40
6000	Masonry, table mounted, 14" diameter, 5 H.P.		1.26	54.50	163	490	42.70
6050	Portable cut-off, 8 H.P.		1.30	28.50	85	255	27.40
6100	Circular, hand held, electric, 7-1/4" diameter		.20	7.35	22	66	6
6200	12" diameter		.27	11	33	99	8.75
6250	Wall saw, w/hydraulic power, 10 H.P		2.04	95	284.80	855	73.30
6275	Shot blaster, walk behind, 20" wide		1.42	545	1,635	4,900	338.35
6300	Steam cleaner, 100 gallons per hour		2.25	61.50	185	555	55
6310	200 gallons per hour		3	75	225	675	69
6340	Tar Kettle/Pot, 400 gallon		2.69	51.50	155	465	52.50
6350	Torch, cutting, acetylene-oxygen, 150' hose		1.50	10	30	90	18
6360	Hourly operating cost includes tips and gas		8.10				64.80
6410	Toilet, portable chemical		.11	17.65	53	159	11.50
6420	Recycle flush type		.13	21.50	65	195	14.05
6430	Toilet, fresh water flush, garden hose,		.15	24.50	73	219	15.80
6440	Hoisted, non-flush, for high rise		.13	21.50	64	192	13.85
6450	Toilet, trailers, minimum		.22	36.50	110	330	23.75
6460	Maximum		.66	110	330	990	71.30
6465	Tractor, farm with attachment		10.25	222	665	2,000	215
6470	Trailer, office, see division 01520-500						
6500	Trailers, platform, flush deck, 2 axle, 25 ton capacity	Ea.	4.25	90	270	810	88
6600	40 ton capacity		5.55	125	375	1,125	119.40
6700	3 axle, 50 ton capacity		6	138	415	1,250	131
6800	75 ton capacity		7.45	182	545	1,625	168.60
6810	Trailer mounted cable reel for H.V. line work		4.47	213	639	1,925	163.55
6820	Trailer mounted cable tensioning rig		8.75	415	1,250	3,750	320
6830	Cable pulling rig		55.44	2,375	7,120	21,400	1,868
6850	Trailer, storage, see division 01520-500						
6900	Water tank, engine driven discharge, 5000 gallons	Ea.	5.60	120	360	1,075	116.80
6925	10,000 gallons		7.70	168	505	1,525	162.60
6950	Water truck, off highway, 6000 gallons		49.25	700	2,095	6,275	813
7010	Tram car for H.V. line work, powered, 2 conductor		5.73	116	347	1,050	115.25
7020	Transit (builder's level) with tripod		.07	12.35	37	111	7.95
7030	Trench box, 3000 lbs. 6'x8'		.67	90.50	272	815	59.75
7040	7200 lbs. 6'x20'		.82	136	409	1,225	88.35
7050	8000 lbs., 8' x 16'		.88	147	442	1,325	95.45
7060	9500 lbs., 8'x20'		1.19	199	597	1,800	128.90
7065	11,000 lbs., 8'x24'		1.34	224	671	2,025	144.90
7070	12,000 lbs., 10' x 20'		1.77	294	883	2,650	190.75
7100	Truck, pickup, 3/4 ton, 2 wheel drive		5.75	53.50	160	480	78
7200	4 wheel drive		5.90	63.50	190	570	85.20
7250	Crew carrier, 9 passenger		5.44	92	276.40	830	98.80
7290	Tool van, 24,000 G.V.W.		8.40	91.50	274.80	825	122.15
7300	Tractor, 4 x 2, 30 ton capacity, 195 H.P.		13.65	165	495	1,475	208.20
7410	250 H.P.		18.20	243	730	2,200	291.60
7500	6 x 2, 40 ton capacity, 240 H.P.		17.45	270	810	2,425	301.60
7600	6 x 4, 45 ton capacity, 240 H.P.		22.20	293	880	2,650	353.60
7620	Vacuum truck, hazardous material, 2500 gallon		6.51	300	902	2,700	232.50
7625	5,000 gallon		11.37	400	1,202.80	3,600	331.50
7640	Tractor, with A frame, boom and winch, 225 H.P.		15.75	223	670	2,000	260
7650	Vacuum, H.E.P.A., 16 gal., wet/dry		.27	24	72	216	16.55
7655	55 gal, wet/dry		.60	36	108	325	26.40
7660	Water tank, portable		1	9.35	28	84	13.60
7690	Large production vacuum loader, 3150 CFM		15	600	1,800	5,400	480

Note: R02250-450

400

01590 | Equipment Rental

				UNIT	HOURLY OPER. COST	RENT PER DAY	RENT PER WEEK	RENT PER MONTH	CREW EQUIPMENT COST/DAY	
400	7700	Welder, electric, 200 amp		Ea.	3.79	60.50	181	545	66.50	**400**
	7800	300 amp	R02250 -400		5.28	64	192	575	80.65	
	7900	Gas engine, 200 amp			5.70	28	84	252	62.40	
	8000	300 amp	R02250 -450		7.70	32.50	98	294	81.20	
	8100	Wheelbarrow, any size			.06	10.65	32	96	6.90	
	8200	Wrecking ball, 4000 lb.	R02315 -300		1.85	68.50	205	615	55.80	
500	0010	**HIGHWAY EQUIPMENT RENTAL**								**500**
	0050	Asphalt batch plant, portable drum mixer, 100 ton/hr.		Ea.	55.75	1,275	3,850	11,600	1,216	
	0060	200 ton/hr.			61.85	1,350	4,045	12,100	1,304	
	0070	300 ton/hr.			72.05	1,600	4,785	14,400	1,533	
	0100	Backhoe attachment, long stick, up to 185 HP, 10.5' long			.30	20	60	180	14.40	
	0140	Up to 250 HP, 12' long			.33	21.50	65	195	15.65	
	0180	Over 250 HP, 15' long			.43	28.50	85	255	20.45	
	0200	Special dipper arm, up to 100 HP, 32' long			.88	58.50	175	525	42.05	
	0240	Over 100 HP, 33' long			1.10	73.50	220	660	52.80	
	0300	Concrete batch plant, portable, electric, 200 CY/Hr			10.40	570	1,715	5,150	426.20	
	0500	Grader attachment, ripper/scarifier, rear mounted								
	0520	Up to 135 HP		Ea.	2.75	58.50	175	525	57	
	0540	Up to 180 HP			3.30	75	225	675	71.40	
	0580	Up to 250 HP			3.65	85	255	765	80.20	
	0700	Pvmt. removal bucket, for hyd. excavator, up to 90 HP			1.40	43.50	130	390	37.20	
	0740	Up to 200 HP			1.60	65	195	585	51.80	
	0780	Over 200 HP			1.70	76.50	230	690	59.60	
	0900	Aggregate spreader, self-propelled, 187 HP			37.15	790	2,370	7,100	771.20	
	1000	Chemical spreader, 3 C.Y.			2.20	61.50	185	555	54.60	
	1900	Hammermill, traveling, 250 HP			39.98	1,700	5,120	15,400	1,344	
	2000	Horizontal borer, 3" diam, 13 HP gas driven			3.95	51.50	155	465	62.60	
	2200	Hydromulchers, gas power, 3000 gal., for truck mounting			10.50	187	560	1,675	196	
	2400	Joint & crack cleaner, walk behind, 25 HP			2.05	46.50	140	420	44.40	
	2500	Filler, trailer mounted, 400 gal., 20 HP			6	182	545	1,625	157	
	3000	Paint striper, self propelled, double line, 30 HP			5.15	155	465	1,400	134.20	
	3200	Post drivers, 6" I-Beam frame, for truck mounting			8.15	415	1,250	3,750	315.20	
	3400	Road sweeper, self propelled, 8' wide, 90 HP			23.50	430	1,285	3,850	445	
	4000	Road mixer, self-propelled, 130 HP			29.05	570	1,715	5,150	575.40	
	4100	310 HP			54	1,850	5,570	16,700	1,546	
	4200	Cold mix paver, incl pug mill and bitumen tank,								
	4220	165 HP		Ea.	66.95	1,875	5,600	16,800	1,656	
	4250	Paver, asphalt, wheel or crawler, 130 H.P., diesel			65.65	1,775	5,325	16,000	1,590	
	4300	Paver, road widener, gas 1' to 6', 67 HP			29.65	630	1,890	5,675	615.20	
	4400	Diesel, 2' to 14', 88 HP			39.95	955	2,870	8,600	893.60	
	4600	Slipform pavers, curb and gutter, 2 track, 75 HP			24.85	640	1,920	5,750	582.80	
	4700	4 track, 165 HP			34.75	745	2,235	6,700	725	
	4800	Median barrier, 215 HP			35.35	775	2,325	6,975	747.80	
	4901	Trailer, low bed, 75 ton capacity			7.90	178	535	1,600	170.20	
	5000	Road planer, walk behind, 10" cutting width, 10 HP			2	26	78	234	31.60	
	5100	Self propelled, 12" cutting width, 64 HP			5.30	195	585	1,750	159.40	
	5200	Pavement profiler, 4' to 6' wide, 450 HP			146.05	2,675	8,040	24,100	2,776	
	5300	8' to 10' wide, 750 HP			233.70	4,000	11,980	35,900	4,266	
	5400	Roadway plate, steel, 1"x8'x20'			.06	10	30	90	6.50	
	5600	Stabilizer, self-propelled, 150 HP			27.40	540	1,625	4,875	544.20	
	5700	310 HP			44.55	1,150	3,445	10,300	1,045	
	5800	Striper, thermal, truck mounted 120 gal. paint, 150 H.P.			32.40	485	1,450	4,350	549.20	
	6000	Tar kettle, 330 gal., trailer mounted			2.37	36.50	110	330	40.95	
	7000	Tunnel locomotive, diesel, 8 to 12 ton			20.95	540	1,615	4,850	490.60	
	7005	Electric, 10 ton			20.65	610	1,830	5,500	531.20	
	7010	Muck cars, 1/2 C.Y. capacity			1.50	20	60	180	24	
	7020	1 C.Y. capacity			1.70	28	84	252	30.40	
	7030	2 C.Y. capacity			1.85	32.50	98	294	34.40	

Important: See the Reference Section for critical supporting data - Reference Nos., Crews, & City Cost Indexes

01590 | Equipment Rental

		UNIT	HOURLY OPER. COST	RENT PER DAY	RENT PER WEEK	RENT PER MONTH	CREW EQUIPMENT COST/DAY	
500	7040	Side dump, 2 C.Y. capacity	Ea.	2	40	120	360	40	**500**
	7050	3 C.Y. capacity		2.70	45	135	405	48.60	
	7060	5 C.Y. capacity		3.85	58.50	175	525	65.80	
	7100	Ventilating blower for tunnel, 7-1/2 H.P.		1.22	44	132	395	36.15	
	7110	10 H.P.		1.41	45.50	137	410	38.70	
	7120	20 H.P.		2.30	56.50	170	510	52.40	
	7140	40 H.P.		4.02	83.50	250	750	82.15	
	7160	60 H.P.		6.09	127	380	1,150	124.70	
	7175	75 H.P.		7.79	168	505	1,525	163.30	
	7180	200 H.P.		17.45	245	735	2,200	286.60	
	7800	Windrow loader, elevating		33.40	875	2,620	7,850	791.20	
600	0010	**LIFTING AND HOISTING EQUIPMENT RENTAL** R01590 -100							**600**
	0100	without operators							
	0120	Aerial lift truck, 2 person, to 80' R01590 -150	Ea.	18.60	580	1,740	5,225	496.80	
	0140	Boom work platform, 40' snorkel		9.80	203	610	1,825	200.40	
	0150	Crane, flatbed mntd, 3 ton cap. R02315 -450		12.45	183	550	1,650	209.60	
	0200	Crane, climbing, 106' jib, 6000 lb. capacity, 410 FPM		43.35	1,350	4,070	12,200	1,161	
	0300	101' jib, 10,250 lb. capacity, 270 FPM		48.75	1,725	5,150	15,500	1,420	
	0400	Tower, static, 130' high, 106' jib,							
	0500	6200 lb. capacity at 400 FPM	Ea.	46.50	1,575	4,700	14,100	1,312	
	0600	Crawler mounted, lattice boom, 1/2 C.Y., 15 tons at 12' radius		20.76	570	1,710	5,125	508.10	
	0700	3/4 C.Y., 20 tons at 12' radius		27.68	680	2,045	6,125	630.45	
	0800	1 C.Y., 25 tons at 12' radius		36.90	895	2,690	8,075	833.20	
	0900	1-1/2 C.Y., 40 tons at 12' radius		41.15	930	2,795	8,375	888.20	
	1000	2 C.Y., 50 tons at 12' radius		51.35	1,325	3,975	11,900	1,206	
	1100	3 C.Y., 75 tons at 12' radius		49.10	1,300	3,885	11,700	1,170	
	1200	100 ton capacity, 60' boom		62.80	1,700	5,085	15,300	1,519	
	1300	165 ton capacity, 60' boom		92.20	2,175	6,560	19,700	2,050	
	1400	200 ton capacity, 70' boom		98.10	2,350	7,015	21,000	2,188	
	1500	350 ton capacity, 80' boom		141.90	3,325	9,945	29,800	3,124	
	1600	Truck mounted, lattice boom, 6 x 4, 20 tons at 10' radius		27.76	960	2,875	8,625	797.10	
	1700	25 tons at 10' radius		29.66	1,025	3,070	9,200	851.30	
	1800	8 x 4, 30 tons at 10' radius		36.66	1,075	3,260	9,775	945.30	
	1900	40 tons at 12' radius		32.74	1,150	3,450	10,400	951.90	
	2000	8 x 4, 60 tons at 15' radius		39.81	1,250	3,770	11,300	1,072	
	2050	82 tons at 15' radius		46.50	1,500	4,470	13,400	1,266	
	2100	90 tons at 15' radius		44.08	1,625	4,860	14,600	1,325	
	2200	115 tons at 15' radius		50.55	1,800	5,400	16,200	1,484	
	2300	150 tons at 18' radius		43.10	1,825	5,500	16,500	1,445	
	2350	165 tons at 18' radius		77.25	2,150	6,470	19,400	1,912	
	2400	Truck mounted, hydraulic, 12 ton capacity		32.45	570	1,715	5,150	602.60	
	2500	25 ton capacity		32.60	595	1,780	5,350	616.80	
	2550	33 ton capacity		33.45	625	1,870	5,600	641.60	
	2560	40 ton capacity		31.85	595	1,790	5,375	612.80	
	2600	55 ton capacity		47.05	865	2,595	7,775	895.40	
	2700	80 ton capacity		56	910	2,735	8,200	995	
	2720	100 ton capacity		80.75	2,325	6,945	20,800	2,035	
	2740	120 ton capacity		85.55	2,525	7,550	22,700	2,194	
	2760	150 ton capacity		104.65	3,175	9,535	28,600	2,744	
	2800	Self-propelled, 4 x 4, with telescoping boom, 5 ton		15.50	320	965	2,900	317	
	2900	12-1/2 ton capacity		25.10	510	1,525	4,575	505.80	
	3000	15 ton capacity		25.70	585	1,755	5,275	556.60	
	3050	20 ton capacity		26.80	610	1,835	5,500	581.40	
	3100	25 ton capacity		27.25	595	1,785	5,350	575	
	3150	40 ton capacity		45.20	845	2,540	7,625	869.60	
	3200	Derricks, guy, 20 ton capacity, 60' boom, 75' mast		12.72	330	996	3,000	300.95	
	3300	100' boom, 115' mast		20.60	570	1,710	5,125	506.80	

01590 | Equipment Rental

			UNIT	HOURLY OPER. COST	RENT PER DAY	RENT PER WEEK	RENT PER MONTH	CREW EQUIPMENT COST/DAY	
600	3400	Stiffleg, 20 ton capacity, 70' boom, 37' mast	R01590 -150	Ea.	14.71	425	1,280	3,850	373.70	**600**
	3500	100' boom, 47' mast			23.19	695	2,080	6,250	601.50	
	3550	Helicopter, small, lift to 1250 lbs. maximum, w/pilot	R01590 -100		67.85	2,675	8,050	24,200	2,153	
	3600	Hoists, chain type, overhead, manual, 3/4 ton			.10	2	6	18	2	
	3900	10 ton	R02315 -450		.60	13	39	117	12.60	
	4000	Hoist and tower, 5000 lb. cap., portable electric, 40' high			4.19	191	574	1,725	148.30	
	4100	For each added 10' section, add			.09	15	45	135	9.70	
	4200	Hoist and single tubular tower, 5000 lb. electric, 100' high			5.66	267	801	2,400	205.50	
	4300	For each added 6'-6" section, add			.15	25	75	225	16.20	
	4400	Hoist and double tubular tower, 5000 lb., 100' high			6.06	294	882	2,650	224.90	
	4500	For each added 6'-6" section, add			.17	28.50	85	255	18.35	
	4550	Hoist and tower, mast type, 6000 lb., 100' high			6.56	305	915	2,750	235.50	
	4570	For each added 10' section, add			.11	18.35	55	165	11.90	
	4600	Hoist and tower, personnel, electric, 2000 lb., 100' @ 125 FPM			13.52	815	2,440	7,325	596.15	
	4700	3000 lb., 100' @ 200 FPM			15.40	915	2,750	8,250	673.20	
	4800	3000 lb., 150' @ 300 FPM			17.10	1,025	3,090	9,275	754.80	
	4900	4000 lb., 100' @ 300 FPM			17.73	1,050	3,150	9,450	771.85	
	5000	6000 lb., 100' @ 275 FPM			19.14	1,100	3,300	9,900	813.10	
	5100	For added heights up to 500', add		L.F.	.01	1.67	5	15	1.10	
	5200	Jacks, hydraulic, 20 ton		Ea.	.05	6.35	19	57	4.20	
	5500	100 ton		"	.30	18.65	56	168	13.60	
	6000	Jacks, hydraulic, climbing with 50' jackrods								
	6010	and control consoles, minimum 3 mo. rental								
	6100	30 ton capacity		Ea.	1.65	110	329	985	79	
	6150	For each added 10' jackrod section, add			.05	3.33	10	30	2.40	
	6300	50 ton capacity			2.65	177	530	1,600	127.20	
	6350	For each added 10' jackrod section, add			.06	4	12	36	2.90	
	6500	125 ton capacity			6.95	465	1,390	4,175	333.60	
	6550	For each added 10' jackrod section, add			.48	31.50	95	285	22.85	
	6600	Cable jack, 10 ton capacity with 200' cable			1.38	91.50	275	825	66.05	
	6650	For each added 50' of cable, add			.15	10	30	90	7.20	
700	0010	**WELLPOINT EQUIPMENT RENTAL** See also division 02240	R02240 -900							**700**
	0020	Based on 2 months rental								
	0100	Combination jetting & wellpoint pump, 60 H.P. diesel		Ea.	9.17	272	817	2,450	236.75	
	0200	High pressure gas jet pump, 200 H.P., 300 psi		"	16.39	233	698	2,100	270.70	
	0300	Discharge pipe, 8" diameter		L.F.	.01	.44	1.32	3.96	.35	
	0350	12" diameter			.01	.65	1.96	5.90	.45	
	0400	Header pipe, flows up to 150 G.P.M., 4" diameter			.01	.40	1.20	3.60	.30	
	0500	400 G.P.M., 6" diameter			.01	.47	1.42	4.26	.35	
	0600	800 G.P.M., 8" diameter			.01	.65	1.96	5.90	.45	
	0700	1500 G.P.M., 10" diameter			.01	.69	2.06	6.20	.50	
	0800	2500 G.P.M., 12" diameter			.02	1.30	3.89	11.65	.95	
	0900	4500 G.P.M., 16" diameter			.02	1.66	4.98	14.95	1.15	
	0950	For quick coupling aluminum and plastic pipe, add			.03	1.72	5.15	15.45	1.25	
	1100	Wellpoint, 25' long, with fittings & riser pipe, 1-1/2" or 2" diameter		Ea.	.05	3.43	10.28	31	2.45	
	1200	Wellpoint pump, diesel powered, 4" diameter, 20 H.P.			4.45	157	471	1,425	129.80	
	1300	6" diameter, 30 H.P.			5.81	195	584	1,750	163.30	
	1400	8" suction, 40 H.P.			7.91	267	801	2,400	223.50	
	1500	10" suction, 75 H.P.			10.86	310	936	2,800	274.10	
	1600	12" suction, 100 H.P.			16.25	500	1,500	4,500	430	
	1700	12" suction, 175 H.P.			21.61	550	1,650	4,950	502.90	
800	0010	**MARINE EQUIPMENT RENTAL**								**800**
	0200	Barge, 400 Ton, 30' wide x 90' long		Ea.	16.15	240	720	2,150	273.20	
	0240	800 Ton, 45' wide x 90' long			26.65	345	1,030	3,100	419.20	
	2000	Tugboat, diesel, 100 HP			16.95	168	505	1,525	236.60	
	2040	250 HP			31.45	315	945	2,825	440.60	
	2080	380 HP			73.10	925	2,780	8,350	1,141	

Important: See the Reference Section for critical supporting data - Reference Nos., Crews, & City Cost Indexes

01700 | Execution Requirements

		01740	Cleaning	CREW	DAILY OUTPUT	LABOR-HOURS	UNIT	2005 BARE COSTS				TOTAL INCL O&P	
								MAT.	LABOR	EQUIP.	TOTAL		
500	0010		CLEANING UP After job completion, allow, minimum				Job					.30%	500
	0040		Maximum				"					1%	
	0050		Cleanup of floor area, continuous, per day, during const.	A-5	24	.750	M.S.F.	1.70	20	1.34	23.04	34.50	
	0100		Final by GC at end of job	"	11.50	1.565	"	2.65	42	2.80	47.45	71	
	0200		Rubbish removal, see division 02225-730										

01800 | Facility Operation

		01810	Commissioning	CREW	DAILY OUTPUT	LABOR-HOURS	UNIT	2005 BARE COSTS				TOTAL INCL O&P	
								MAT.	LABOR	EQUIP.	TOTAL		
100	0010		COMMISSIONING Including documentation of design intent										100
	0100		performance verification, O&M, training, min				Project					.50%	
	0150		Maximum				"					.75%	

For information about Means Estimating Seminars, see yellow pages 12 and 13 in back of book

Division Notes

	CREW	DAILY OUTPUT	LABOR-HOURS	UNIT	2005 BARE COSTS				TOTAL INCL O&P
					MAT.	LABOR	EQUIP.	TOTAL	

Division 2
Site Construction

Estimating Tips

02200 Site Preparation
- If possible visit the site and take an inventory of the type, quantity and size of the trees. Certain trees may have a landscape resale value or firewood value. Stump disposal can be very expensive, particularly if they cannot be buried at the site. Consider using a bulldozer in lieu of hand cutting trees.
- Estimators should visit the site to determine the need for haul road, access, storage of materials, and security considerations. When estimating for access roads on unstable soil, consider using a geotextile stabilization fabric. It can greatly reduce the quantity of crushed stone or gravel. Sites of limited size and access can cause cost overruns due to lost productivity. Theft and damage is another consideration if the location is isolated. A temporary fence or security guards may be required. Investigate the site thoroughly.

02210 Subsurface Investigation
In preparing estimates on structures involving earthwork or foundations, all information concerning soil characteristics should be obtained. Look particularly for hazardous waste, evidence of prior dumping of debris, and previous stream beds.

02220 Selective Demolition
The costs shown for selective demolition do not include rubbish handling or disposal. These items should be estimated separately using Means data or other sources.
- Historic preservation often requires that the contractor remove materials from the existing structure, rehab them and replace them. The estimator must be aware of any related measures and precautions that must be taken when doing selective demolition, and cutting and patching. Requirements may include special handling and storage, as well as security.
- In addition to Section 02220, you can find selective demolition items in each division. Example: Roofing demolition is in division 7.

02300 Earthwork
- Estimating the actual cost of performing earthwork requires careful consideration of the variables involved. This includes items such as type of soil, whether or not water will be encountered, dewatering, whether or not banks need bracing, disposal of excavated earth, length of haul to fill or spoil sites, etc. If the project has large quantities of cut or fill, consider raising or lowering the site to reduce costs while paying close attention to the effect on site drainage and utilities if doing this.
- If the project has large quantities of fill, creating a borrow pit on the site can significantly lower the costs.
- It is very important to consider what time of year the project is scheduled for completion. Bad weather can create large cost overruns from dewatering, site repair and lost productivity from cold weather.

02500 Utility Services
02600 Drainage & Containment
- Never assume that the water, sewer and drainage lines will go in at the early stages of the project. Consider the site access needs before dividing the site in half with open trenches, loose pipe, and machinery obstructions. Always inspect the site to establish that the site drawings are complete. Check off all existing utilities on your drawing as you locate them. If you find any discrepancies, mark up the site plan for further research. Differing site conditions can be very costly if discovered later in the project.
- See also Section 02955 for restoration of pipe where removal/replacement may be undesirable. Use of new types of piping materials can reduce the overall project cost. Owners/design engineers should consider the installing construction as a valuable source of current information on piping products that could lead to significant utility cost savings.

02700 Bases, Ballasts, Pavements/Appurtenances
- When estimating paving, keep in mind the project schedule. If an asphaltic paving project is in a colder climate and runs through to the spring, consider placing the base course in the autumn, then topping it in the spring just prior to completion. This could save considerable costs in spring repair. Keep in mind that prices for asphalt and concrete are generally higher in the cold seasons.
- See also Sections 02960/02965.

02900 Planting
- The timing of planting and guarantee specifications often dictate the costs for establishing tree and shrub growth and a stand of grass or ground cover. Establish the work performance schedule to coincide with the local planting season. Maintenance and growth guarantees can add from 20% to 100% to the total landscaping cost. The cost to replace trees and shrubs can be as high as 5% of the total cost depending on the planting zone, soil conditions and time of year.

02960 & 02965 Flexible Pavement Surfacing Recovery
- Recycling of asphalt pavement is becoming very popular and is an alternative to removal and replacement of asphalt pavement. It can be a good value engineering proposal if removed pavement can be recycled either at the site or another site that is reasonably close to the project site.

Reference Numbers
Reference numbers are shown in bold squares at the beginning of some major classifications. These numbers refer to related items in the Reference Section. The reference information may be an estimating procedure, an alternate pricing method or technical information.

Note: Not all subdivisions listed here necessarily appear in this publication.

2 SITE CONSTRUCTION

			CREW	DAILY OUTPUT	LABOR-HOURS	UNIT	MAT.	LABOR	EQUIP.	TOTAL	TOTAL INCL O&P
		**02055	Soils**					2005 BARE COSTS				
150	0010	**BORROW**										150
	0020	Spread, 200 H.P. dozer, no compaction, 2 mi. RT haul	R02315 -400								
	0200	Common borrow	B-15	600	.047	C.Y.	6	1.37	3.12	10.49	12.10
	0700	Screened loam CN		600	.047		21	1.37	3.12	25.49	29
	0800	Topsoil, weed free	↓	600	.047		21	1.37	3.12	25.49	29
	0900	For 5 mile haul, add	B-34B	200	.040	↓		1.10	2.38	3.48	4.30
		**02060	Aggregate**									
150	0010	**BORROW**										150
	0020	Spread, with 200 H.P. dozer, no compaction, 2 mi RT haul	R02315 -400								
	0100	Bank run gravel CN	B-15	600	.047	C.Y.	17.80	1.37	3.12	22.29	25
	0300	Crushed stone, (1.40 tons per CY), 1-1/2"		600	.047		27.50	1.37	3.12	31.99	35.50
	0320	3/4" CN		600	.047		27.50	1.37	3.12	31.99	35.50
	0340	1/2"		600	.047		26.50	1.37	3.12	30.99	35
	0360	3/8"		600	.047		26.50	1.37	3.12	30.99	34.50
	0400	Sand, washed, concrete		600	.047		26.50	1.37	3.12	30.99	34.50
	0500	Dead or bank sand		600	.047		4.13	1.37	3.12	8.62	10.05
	0600	Select structural fill	↓	600	.047		7.90	1.37	3.12	12.39	14.15
	0900	For 5 mile haul, add	B-34B	200	.040	↓		1.10	2.38	3.48	4.30
		**02065	Cement & Concrete**									
300	0010	**PLANT MIXED BITUMINOUS CONCRETE**										300
	0020	Asphaltic concrete plant mix (145 LB per c.f.) CN				Ton	34.50			34.50	38
	0040	Asphaltic concrete less than 300 tons add trucking costs	R02065 -300								
	0050	See 02315-490-0010 for hauling costs									
	0200	All weather patching mix, hot					33.50			33.50	37
	0250	Cold patch					40			40	44
	0300	Berm mix	↓			↓	33.50			33.50	37
		**02080	Utility Materials**									
100	0010	**FIRE HYDRANTS**										100
	0020	Mechanical joints unless otherwise noted									
	1000	Fire hydrants, two way; excavation and backfill not incl.									
	1100	4-1/2" valve size, depth 2'-0"	B-21	10	2.800	Ea.	1,100	86.50	15.85	1,202.35	1,350
	1120	2'-6"		10	2.800		1,100	86.50	15.85	1,202.35	1,350
	1140	3'-0"		10	2.800		1,100	86.50	15.85	1,202.35	1,350
	1300	7'-0"	↓	6	4.667		1,275	144	26.50	1,445.50	1,675
	2400	Lower barrel extensions with stems, 1'-0"	B-20	14	1.714		345	51.50		396.50	460
	2480	3'-0"	"	12	2	↓	495	60		555	640
400	0010	**UTILITY BOXES** Precast concrete, 6" thick										400
	0050	5' x 10' x 6' high, I.D.	B-13	2	28	Ea.	1,525	805	310	2,640	3,275
	0100	6' x 10' x 6' high, I.D.		2	28		1,575	805	310	2,690	3,325
	0150	5' x 12' x 6' high, I.D.		2	28		1,675	805	310	2,790	3,450
	0200	6' x 12' x 6' high, I.D.		1.80	31.111		1,875	895	345	3,115	3,825
	0250	6' x 13' x 6' high, I.D.		1.50	37.333		2,475	1,075	410	3,960	4,825
	0300	8' x 14' x 7' high, I.D.	↓	1	56	↓	2,675	1,600	615	4,890	6,075
	0350	Hand hole, precast concrete, 1-1/2" thick									
	0400	1'-0" x 2'-0" x 1'-9", I.D., light duty	B-1	4	6	Ea.	260	164		424	540
	0450	4'-6" x 3'-2" x 2'-0", O.D., heavy duty	B-6	3	8	"	800	231	72	1,103	1,325
600	0010	**UTILITY ACCESSORIES**										600
	0400	Underground tape, detectable, reinforced, alum. foil core, 2"	1 Clab	150	.053	C.L.F.	1.43	1.42		2.85	3.79
	0500	6"	"	140	.057	"	3.58	1.53		5.11	6.30

Important: See the Reference Section for critical supporting data - Reference Nos., Crews, & City Cost Indexes

02110 | Hazmat Removal & Handling

		CREW	DAILY OUTPUT	LABOR-HOURS	UNIT	2005 BARE COSTS MAT.	LABOR	EQUIP.	TOTAL	TOTAL INCL O&P
300	0010	**HAZARDOUS WASTE CLEANUP/PICKUP/DISPOSAL**									**300**
	0100	For contractor equipment, i.e. dozer,									
	0110	Front end loader, dump truck, etc., see div. 01590-200									
	1000	Solid pickup									
	1100	55 gal. drums				Ea.					220
	1120	Bulk material, minimum				Ton					165
	1130	Maximum				"					550
	1200	Transportation to disposal site									
	1220	Truckload = 80 drums or 25 C.Y. or 18 tons									
	1260	Minimum				Mile					2.50
	1270	Maximum				"					4.40
	3000	Liquid pickup, vacuum truck, stainless steel tank									
	3100	Minimum charge, 4 hours									
	3110	1 compartment, 2200 gallon				Hr.					110
	3120	2 compartment, 5000 gallon				"					110
	3400	Transportation in 6900 gallon bulk truck				Mile					4.75
	3410	In teflon lined truck				"					5.50
	5000	Heavy sludge or dry vacuumable material				Hr.					110
	6000	Dumpsite disposal charge, minimum				Ton					110
	6020	Maximum				"					440

02115 | Underground Storage Tank Removal

		CREW	DAILY OUTPUT	LABOR-HOURS	UNIT	2005 BARE COSTS MAT.	LABOR	EQUIP.	TOTAL	TOTAL INCL O&P
200	0010	**REMOVAL OF UNDERGROUND STORAGE TANKS** R02115-200									**200**
	0011	Petroleum storage tanks, non-leaking									
	0100	Excavate & load onto trailer									
	0110	3000 gal. to 5000 gal. tank	B-14	4	12	Ea.		335	54	389	580
	0120	6000 gal to 8000 gal tank	B-3A	3	13.333			375	240	615	845
	0130	9000 gal to 12000 gal tank	"	2	20	↓		565	360	925	1,275
	0190	Known leaking tank add				%				100%	100%
	0200	Remove sludge, water and remaining product from tank bottom									
	0201	of tank with vacuum truck									
	0300	3000 gal to 5000 gal tank	A-13	5	1.600	Ea.		53	96	149	186
	0310	6000 gal to 8000 gal tank		4	2			66	120	186	232
	0320	9000 gal to 12000 gal tank	↓	3	2.667	↓		88	160	248	310
	0390	Dispose of sludge off-site, average				Gal.					4.40
	0400	Insert inert solid CO2 "dry ice" into tank									
	0401	For cleaning/transporting tanks (1.5 lbs./100 gal. cap)	1 Clab	500	.016	Lb.	1.34	.43		1.77	2.13
	1020	Haul tank to certified salvage dump, 100 miles round trip									
	1023	3000 gal. to 5000 gal. tank				Ea.				550	690
	1026	6000 gal. to 8000 gal. tank								650	825
	1029	9,000 gal. to 12,000 gal. tank				↓				875	1,100
	1100	Disposal of contaminated soil to landfill									
	1110	Minimum				C.Y.					110
	1111	Maximum				"					330
	1120	Disposal of contaminated soil to									
	1121	bituminous concrete batch plant									
	1130	Minimum				C.Y.					55
	1131	Maximum				"					110
	2010	Decontamination of soil on site incl poly tarp on top/bottom									
	2011	Soil containment berm, and chemical treatment									
	2020	Minimum	B-11C	100	.160	C.Y.	5.60	4.91	2.16	12.67	16
	2021	Maximum	"	100	.160	↓	7.25	4.91	2.16	14.32	17.85
	2050	Disposal of decontaminated soil, minimum									66
	2055	Maximum	↓			↓					135

		02210	Subsurface Investigation	CREW	DAILY OUTPUT	LABOR-HOURS	UNIT	2005 BARE COSTS				TOTAL INCL O&P	
								MAT.	LABOR	EQUIP.	TOTAL		
120	0010	**BORING AND EXPLORATORY DRILLING**											**120**
	0020	Borings, initial field stake out & determination of elevations	A-6	1	16	Day		545	60.50	605.50	900	
	0100	Drawings showing boring details				Total		185		185	270	
	0200	Report and recommendations from P.E.						415		415	595	
	0300	Mobilization and demobilization, minimum	B-55	4	6	↓		160	195	355	465	
	0350	For over 100 miles, per added mile		450	.053	Mile		1.43	1.73	3.16	4.11	
	0600	Auger holes in earth, no samples, 2-1/2" diameter		78.60	.305	L.F.		8.15	9.95	18.10	23.50	
	0650	4" diameter		67.50	.356			9.50	11.55	21.05	27.50	
	0800	Cased borings in earth, with samples, 2-1/2" diameter		55.50	.432		14.20	11.55	14.05	39.80	49	
	0850	4" diameter	↓	32.60	.736		22.50	19.70	24	66.20	81.50	
	1000	Drilling in rock, "BX" core, no sampling	B-56	34.90	.458			13.70	28	41.70	52	
	1050	With casing & sampling		31.70	.505		14.20	15.10	31	60.30	72.50	
	1200	"NX" core, no sampling		25.92	.617	↓		18.45	38	56.45	69.50	
	1250	With casing and sampling		25	.640		17.35	19.10	39	75.45	91	
	1400	Drill rig and crew with truck mounted auger	B-55	1	24	Day		640	780	1,420	1,850	
	1450	With crawler type drill	B-56	1	16	"		480	980	1,460	1,800	
	1500	For inner city borings add, minimum									10%	
	1510	Maximum									20%	
200	0010	**CORE DRILLING** Reinforced concrete slab, up to 6" thick slab											**200**
	0020	Including bit, layout and set up										
	0100	1" diameter core	B-89A	28	.571	Ea.	2.40	17.60	3.68	23.68	34	
	0150	Each added inch thick, add		300	.053		.43	1.64	.34	2.41	3.40	
	0300	3" diameter core		23	.696		5.35	21.50	4.48	31.33	44.50	
	0350	Each added inch thick, add		186	.086		.96	2.65	.55	4.16	5.80	
	0500	4" diameter core		19	.842		5.35	26	5.40	36.75	52.50	
	0550	Each added inch thick, add		170	.094		1.21	2.90	.61	4.72	6.50	
	0700	6" diameter core		14	1.143		8.80	35	7.35	51.15	72.50	
	0750	Each added inch thick, add		140	.114		1.49	3.52	.74	5.75	7.90	
	0900	8" diameter core		11	1.455		12	45	9.35	66.35	93	
	0950	Each added inch thick, add		95	.168		2.02	5.20	1.08	8.30	11.45	
	1100	10" diameter core		10	1.600		16	49.50	10.30	75.80	105	
	1150	Each added inch thick, add		80	.200		2.65	6.15	1.29	10.09	13.95	
	1300	12" diameter core		9	1.778		19.20	54.50	11.45	85.15	119	
	1350	Each added inch thick, add		68	.235		3.18	7.25	1.52	11.95	16.40	
	1500	14" diameter core		7	2.286		23.50	70.50	14.70	108.70	151	
	1550	Each added inch thick, add		55	.291		4.04	8.95	1.87	14.86	20.50	
	1700	18" diameter core		4	4		30	123	26	179	254	
	1750	Each added inch thick, add	↓	28	.571		5.30	17.60	3.68	26.58	37.50	
	1760	For horizontal holes, add to above				↓				30%	30%	
	1770	Prestressed hollow core plank, 6" thick										
	1780	1" diameter core	B-89A	52	.308	Ea.	1.59	9.45	1.98	13.02	18.70	
	1790	Each added inch thick, add		350	.046		.28	1.41	.29	1.98	2.82	
	1800	3" diameter core		50	.320		3.51	9.85	2.06	15.42	21.50	
	1810	Each added inch thick, add		240	.067		.58	2.05	.43	3.06	4.30	
	1820	4" diameter core		48	.333		4.67	10.25	2.15	17.07	23.50	
	1830	Each added inch thick, add		216	.074		.81	2.28	.48	3.57	4.96	
	1840	6" diameter core		44	.364		5.80	11.20	2.34	19.34	26.50	
	1850	Each added inch thick, add		175	.091		.96	2.81	.59	4.36	6.10	
	1860	8" diameter core		32	.500		7.75	15.40	3.22	26.37	36	
	1870	Each added inch thick, add		118	.136		1.34	4.17	.87	6.38	8.95	
	1880	10" diameter core		28	.571		10.45	17.60	3.68	31.73	43	
	1890	Each added inch thick, add		99	.162		1.44	4.97	1.04	7.45	10.45	
	1900	12" diameter core		22	.727		12.75	22.50	4.68	39.93	54	
	1910	Each added inch thick, add	↓	85	.188	↓	2.12	5.80	1.21	9.13	12.65	
	1950	Minimum charge for above, 3" diameter core		7	2.286	Total		70.50	14.70	85.20	125	
	2000	4" diameter core	↓	6.80	2.353	↓		72.50	15.15	87.65	130	

2

SITE CONSTRUCTION

		02210	Subsurface Investigation	CREW	DAILY OUTPUT	LABOR-HOURS	UNIT	2005 BARE COSTS MAT.	LABOR	EQUIP.	TOTAL	TOTAL INCL O&P	
200	2050		6" diameter core	B-89A	6	2.667	Total		82	17.15	99.15	147	200
	2100		8" diameter core		5.50	2.909			89.50	18.75	108.25	160	
	2150		10" diameter core		4.75	3.368			104	21.50	125.50	185	
	2200		12" diameter core		3.90	4.103			126	26.50	152.50	225	
	2250		14" diameter core		3.38	4.734			146	30.50	176.50	261	
	2300		18" diameter core		3.15	5.079			156	32.50	188.50	279	
	3010		Bits for core drill, diamond, premium, 1" diameter				Ea.	107			107	118	
	3020		3" diameter					264			264	291	
	3040		4" diameter					294			294	325	
	3050		6" diameter					470			470	520	
	3080		8" diameter					645			645	710	
	3120		12" diameter					1,025			1,025	1,125	
	3180		18" diameter					2,100			2,100	2,300	
	3240		24" diameter					2,800			2,800	3,075	
700	0010	**TEST PITS**											700
	0020		Hand digging, light soil	1 Clab	4.50	1.778	C.Y.		47.50		47.50	74	
	0100		Heavy soil	"	2.50	3.200			85.50		85.50	133	
	0120		Loader-backhoe, light soil	B-11M	28	.571			17.55	9.15	26.70	37	
	0130		Heavy soil	"	20	.800			24.50	12.80	37.30	51.50	
	1000		Subsurface exploration, mobilization				Mile				5.50	6.30	
	1010		Difficult access for rig, add				Hr.				110	126	
	1020		Auger borings, drill rig, incl. samples				L.F.				12.65	14.60	
	1030		Hand auger								18.70	21.50	
	1050		Drill and sample every 5', split spoon								16.50	19	
	1060		Extra samples				Ea.				22	25.50	

		02220	Site Demolition										
110	0010	**BUILDING DEMOLITION** Large urban projects, incl. 20 mi. haul	R02220 -510										110
	0011	No foundation or dump fees, C.F. is vol. of building standing											
	0012		Steel	B-8	21,500	.003	C.F.		.09	.11	.20	.25	
	0050		Concrete		15,300	.004			.12	.15	.27	.36	
	0080		Masonry		20,100	.003			.09	.12	.21	.27	
	0100		Mixture of types, average		20,100	.003			.09	.12	.21	.27	
	0500	Small bldgs, or single bldgs, no salvage included, steel		B-3	14,800	.003			.09	.12	.21	.27	
	0600		Concrete		11,300	.004			.12	.15	.27	.36	
	0650		Masonry		14,800	.003			.09	.12	.21	.27	
	0700		Wood		14,800	.003			.09	.12	.21	.27	
	1000	Single family, one story house, wood, minimum					Ea.				2,525	2,975	
	1020		Maximum								4,400	5,275	
	1200	Two family, two story house, wood, minimum									3,300	3,950	
	1220		Maximum								6,375	7,700	
	1300	Three family, three story house, wood, minimum									4,400	5,275	
	1320		Maximum								7,700	9,250	
	5000	For buildings with no interior walls, deduct									50%		
120	0010	**EXPLOSIVE/IMPLOSIVE DEMOLITION** Large projects,	R02220 -510										120
	0020	no disposal fee based on building volume, steel building		B-5B	16,900	.003	C.F.		.09	.11	.20	.25	
	0100		Concrete building		16,900	.003			.09	.11	.20	.25	
	0200		Masonry building		16,900	.003			.09	.11	.20	.25	
	0400	Disposal of material, minimum		B-3	445	.108	C.Y.		3.09	3.91	7	9.05	
	0500		Maximum		"	365	.132	"		3.77	4.77	8.54	11.05	
130	0010	**BLDG. FOOTINGS AND FOUNDATIONS DEMOLITION**	R02220 -510										130
	0200	Floors, concrete slab on grade,											
	0240		4" thick, plain concrete	B-9C	500	.080	S.F.		2.17	.28	2.45	3.68	
	0280		Reinforced, wire mesh		470	.085			2.31	.30	2.61	3.92	

02220 | Site Demolition

		CREW	DAILY OUTPUT	LABOR-HOURS	UNIT	MAT.	LABOR	EQUIP.	TOTAL	TOTAL INCL O&P	
130	0300	Rods	B-9C	400	.100	S.F.		2.71	.35	3.06	4.61	**130**
	0400	6" thick, plain concrete		375	.107			2.89	.38	3.27	4.91	
	0420	Reinforced, wire mesh		340	.118			3.19	.42	3.61	5.40	
	0440	Rods		300	.133			3.61	.47	4.08	6.10	
	1000	Footings, concrete, 1' thick, 2' wide	B-5	300	.187	L.F.		5.45	3.10	8.55	11.80	
	1080	1'-6" thick, 2' wide		250	.224			6.55	3.72	10.27	14.20	
	1120	3' wide		200	.280			8.20	4.65	12.85	17.70	
	1140	2' thick, 3' wide		175	.320			9.35	5.30	14.65	20.50	
	1200	Average reinforcing, add								10%	10%	
	1220	Heavy reinforcing, add								20%	20%	
	2000	Walls, block, 4" thick	1 Clab	180	.044	S.F.		1.19		1.19	1.85	
	2040	6" thick		170	.047			1.26		1.26	1.96	
	2080	8" thick		150	.053			1.42		1.42	2.22	
	2100	12" thick		150	.053			1.42		1.42	2.22	
	2200	For horizontal reinforcing, add								10%	10%	
	2220	For vertical reinforcing, add								20%	20%	
	2400	Concrete, plain concrete, 6" thick	B-9	160	.250			6.80	.89	7.69	11.50	
	2420	8" thick		140	.286			7.75	1.01	8.76	13.15	
	2440	10" thick		120	.333			9.05	1.18	10.23	15.35	
	2500	12" thick		100	.400			10.85	1.42	12.27	18.40	
	2600	For average reinforcing, add								10%	10%	
	2620	For heavy reinforcing, add								20%	20%	
	4000	For congested sites or small quantities, add up to								200%	200%	
	4200	Add for disposal, on site	B-11A	232	.069	C.Y.		2.12	3.96	6.08	7.60	
	4250	To five miles	B-30	220	.109	"		3.26	7.60	10.86	13.30	
220	0010	**FENCING DEMOLITION**										**220**
	1600	Fencing, barbed wire, 3 strand	2 Clab	430	.037	L.F.		.99		.99	1.55	
	1650	5 strand	"	280	.057			1.53		1.53	2.37	
	1700	Chain link, posts & fabric, remove only, 8' to 10' high	B-6	445	.054			1.55	.48	2.03	2.92	
	1750	Remove and reset	"	70	.343			9.90	3.08	12.98	18.60	
230	0010	**HYDRODEMOLITION**										**230**
	0015	Hydrodemolition, concrete pavement, 4000 PSI, 2" depth	B-5	500	.112	S.F.		3.28	1.86	5.14	7.10	
	0120	4" depth		450	.124			3.64	2.07	5.71	7.85	
	0130	6" depth		400	.140			4.10	2.32	6.42	8.85	
	0410	6000 PSI, 2" depth		410	.137			4	2.27	6.27	8.65	
	0420	4" depth		350	.160			4.68	2.66	7.34	10.10	
	0430	6" depth		300	.187			5.45	3.10	8.55	11.80	
	0510	8000 PSI, 2" depth		330	.170			4.97	2.82	7.79	10.75	
	0520	4" depth		280	.200			5.85	3.32	9.17	12.65	
	0530	6" depth		240	.233			6.85	3.87	10.72	14.75	
240	0010	**MINOR SITE DEMOLITION**										**240**
	0015	No hauling, abandon catch basin or manhole	B-6	7	3.429	Ea.		99	31	130	186	
	0020	Remove existing catch basin or manhole, masonry		4	6			173	54	227	325	
	0030	Catch basin or manhole frames and covers, stored		13	1.846			53	16.60	69.60	100	
	0040	Remove and reset		7	3.429			99	31	130	186	
	0100	Roadside delineators, remove only	B-80	175	.183			5.25	3.05	8.30	11.45	
	0110	Remove and reset	"	100	.320			9.20	5.35	14.55	20	
	0800	Guiderail, corrugated steel, remove only	B-80A	100	.240	L.F.		6.40	1.76	8.16	11.90	
	0850	Remove and reset	"	40	.600	"		16	4.40	20.40	30	
	0860	Guide posts, remove only	B-80B	120	.267	Ea.		7.55	1.75	9.30	13.55	
	0870	Remove and reset	B-55	50	.480			12.85	15.60	28.45	37	
	0900	Hydrants, fire, remove only	B-21A	5	8			264	101	365	510	
	0950	Remove and reset	"	2	20			660	253	913	1,275	
	1000	Masonry walls, block or tile, solid, remove	B-5	1,800	.031	C.F.		.91	.52	1.43	1.97	

Important: See the Reference Section for critical supporting data - Reference Nos., Crews, & City Cost Indexes

SITE CONSTRUCTION 2

			CREW	DAILY OUTPUT	LABOR-HOURS	UNIT	2005 BARE COSTS				TOTAL INCL O&P	
02220		**Site Demolition**					MAT.	LABOR	EQUIP.	TOTAL		
240	1100	Cavity wall	B-5	2,200	.025	C.F.		.74	.42	1.16	1.61	**240**
	1200	Brick, solid		900	.062			1.82	1.03	2.85	3.94	
	1300	With block back-up		1,130	.050			1.45	.82	2.27	3.14	
	1400	Stone, with mortar		900	.062			1.82	1.03	2.85	3.94	
	1500	Dry set		1,500	.037			1.09	.62	1.71	2.36	
	1600	Median barrier, precast concrete, remove and store	B-3	430	.112	L.F.		3.20	4.05	7.25	9.35	
	1610	Remove and reset	"	390	.123			3.53	4.47	8	10.30	
	2900	Pipe removal, sewer/water, no excavation, 12" diameter	B-6	175	.137			3.95	1.23	5.18	7.45	
	2930	15"-18" diameter		150	.160			4.61	1.44	6.05	8.70	
	2960	21"-24" diameter		120	.200			5.75	1.80	7.55	10.85	
	3000	27"-36" diameter		90	.267			7.70	2.40	10.10	14.45	
	3200	Steel, welded connections, 4" diameter		160	.150			4.32	1.35	5.67	8.15	
	3300	10" diameter		80	.300			8.65	2.70	11.35	16.25	
	3500	Railroad track removal, ties and track	B-13	330	.170			4.89	1.87	6.76	9.55	
	3600	Ballast	B-14	500	.096	C.Y.		2.70	.43	3.13	4.65	
	3700	Remove and re-install, ties & track using new bolts & spikes		50	.960	L.F.		27	4.32	31.32	46.50	
	3800	Turnouts using new bolts and spikes		1	48	Ea.		1,350	216	1,566	2,325	
	4000	Sidewalk removal, bituminous, 2-1/2" thick	B-6	325	.074	S.Y.		2.13	.66	2.79	4	
	4050	Brick, set in mortar		185	.130			3.74	1.17	4.91	7.05	
	4100	Concrete, plain, 4"		160	.150			4.32	1.35	5.67	8.15	
	4200	Mesh reinforced		150	.160			4.61	1.44	6.05	8.70	
	5000	Slab on grade removal, plain	B-5	45	1.244	C.Y.		36.50	20.50	57	78.50	
	5100	Mesh reinforced		33	1.697			49.50	28	77.50	108	
	5200	Rod reinforced		25	2.240			65.50	37	102.50	142	
	5500	For congested sites or small quantities, add up to								200%	200%	
	5550	For disposal on site, add	B-11A	232	.069			2.12	3.96	6.08	7.60	
	5600	To 5 miles, add	B-34D	76	.105			2.90	5.50	8.40	10.45	
250	0010	**DEMOLISH, REMOVE PAVEMENT AND CURB**										**250**
	5010	Pavement removal, bituminous roads, 3" thick	B-38	690	.058	S.Y.		1.74	1.20	2.94	3.98	
	5050	4" to 6" thick		420	.095			2.85	1.96	4.81	6.55	
	5100	Bituminous driveways		640	.063			1.87	1.29	3.16	4.29	
	5200	Concrete to 6" thick, hydraulic hammer, mesh reinforced		255	.157			4.70	3.24	7.94	10.75	
	5300	Rod reinforced		200	.200			6	4.13	10.13	13.75	
	5400	Concrete, 7" to 24" thick, plain		33	1.212	C.Y.		36.50	25	61.50	83	
	5500	Reinforced		24	1.667	"		50	34.50	84.50	115	
	5600	With hand held air equipment, bituminous, to 6" thick	B-39	1,900	.025	S.F.		.71	.07	.78	1.18	
	5700	Concrete to 6" thick, no reinforcing		1,600	.030			.84	.09	.93	1.40	
	5800	Mesh reinforced		1,400	.034			.96	.10	1.06	1.60	
	5900	Rod reinforced		765	.063			1.76	.19	1.95	2.93	
	6000	Curbs, concrete, plain	B-6	360	.067	L.F.		1.92	.60	2.52	3.61	
	6100	Reinforced		275	.087			2.52	.78	3.30	4.73	
	6200	Granite		360	.067			1.92	.60	2.52	3.61	
	6300	Bituminous		528	.045			1.31	.41	1.72	2.46	
310	0010	**SELECTIVE DEMOLITION, CUTOUT**										**310**
	0020	Concrete, elev. slab, light reinforcement, under 6 CF	B-9C	65	.615	C.F.		16.70	2.18	18.88	28.50	
	0050	Light reinforcing, over 6 C.F.	"	75	.533	"		14.45	1.89	16.34	24.50	
	0200	Slab on grade to 6" thick, not reinforced, under 8 S.F.	B-9	85	.471	S.F.		12.75	1.67	14.42	21.50	
	0250	8 - 16 S.F.	"	175	.229	"		6.20	.81	7.01	10.55	
	0255	For over 16 SF see 02220-130-0400										
	0600	Walls, not reinforced, under 6 C.F.	B-9	60	.667	C.F.		18.05	2.36	20.41	30.50	
	0650	6 - 12 C.F.	"	80	.500	"		13.55	1.77	15.32	23	
	0655	For over 12 CF see 02220-130-2500										
	1000	Concrete, elevated slab, bar reinforced, under 6 C.F.	B-9C	45	.889	C.F.		24	3.15	27.15	41	
	1050	Bar reinforced, over 6 C.F.	"	50	.800	"		21.50	2.83	24.33	36.50	
	1200	Slab on grade to 6" thick, bar reinforced, under 8 S.F.	B-9	75	.533	S.F.		14.45	1.89	16.34	24.50	

R02220-510 (reference boxes shown at rows 1100, 5010, 0020)

For expanded coverage of these items see *Means Heavy Construction Cost Data 2005*

2

SITE CONSTRUCTION

		02220	Site Demolition	CREW	DAILY OUTPUT	LABOR-HOURS	UNIT	2005 BARE COSTS MAT.	LABOR	EQUIP.	TOTAL	TOTAL INCL O&P	
310	1250		8 - 16 S.F.	B-9	150	.267	S.F.		7.25	.94	8.19	12.30	310
	1255		For over 16 SF see 02220-130-0440 R02220-510										
	1400		Walls, bar reinforced, under 6 C.F.	B-9C	50	.800	C.F.		21.50	2.83	24.33	36.50	
	1450		6 - 12 CF	"	70	.571	"		15.50	2.02	17.52	26	
	1455		For over 12 CF see 02220-130-2500 & 2600										
	2000		Brick, to 4 S.F. opening, not including toothing										
	2040		4" thick	B-9C	30	1.333	Ea.		36	4.72	40.72	61	
	2060		8" thick		18	2.222			60	7.85	67.85	102	
	2080		12" thick		10	4			108	14.15	122.15	185	
	2400		Concrete block, to 4 S.F. opening, 2" thick		35	1.143			31	4.05	35.05	52.50	
	2420		4" thick		30	1.333			36	4.72	40.72	61	
	2440		8" thick		27	1.481			40	5.25	45.25	68.50	
	2460		12" thick		24	1.667			45	5.90	50.90	77	
	2600		Gypsum block, to 4 S.F. opening, 2" thick	B-9	80	.500			13.55	1.77	15.32	23	
	2620		4" thick		70	.571			15.50	2.02	17.52	26	
	2640		8" thick		55	.727			19.70	2.57	22.27	33.50	
	2800		Terra cotta, to 4 S.F. opening, 4" thick		70	.571			15.50	2.02	17.52	26	
	2840		8" thick		65	.615			16.70	2.18	18.88	28.50	
	2880		12" thick		50	.800			21.50	2.83	24.33	36.50	
	3000		Toothing masonry cutouts, brick, soft old mortar	1 Brhe	40	.200	V.L.F.		5.40		5.40	8.20	
	3100		Hard mortar		30	.267			7.15		7.15	10.95	
	3200		Block, soft old mortar		70	.114			3.07		3.07	4.69	
	3400		Hard mortar		50	.160			4.30		4.30	6.55	
	6000		Walls, interior, not including re-framing,										
	6010		openings to 5 S.F.										
	6100		Drywall to 5/8" thick	1 Clab	24	.333	Ea.		8.90		8.90	13.85	
	6200		Paneling to 3/4" thick		20	.400			10.70		10.70	16.60	
	6300		Plaster, on gypsum lath		20	.400			10.70		10.70	16.60	
	6340		On wire lath		14	.571			15.25		15.25	23.50	
	7000		Wood frame, not including re-framing, openings to 5 S.F.										
	7200		Floors, sheathing and flooring to 2" thick	1 Clab	5	1.600	Ea.		42.50		42.50	66.50	
	7310		Roofs, sheathing to 1" thick, not including roofing		6	1.333			35.50		35.50	55.50	
	7410		Walls, sheathing to 1" thick, not including siding		7	1.143			30.50		30.50	47.50	
320	0010	**SELECTIVE DEMOLITION, DISPOSAL ONLY** R02220-510											320
	0015	Urban bldg w/salvage value allowed											
	0020	Including loading and 5 mile haul to dump											
	0200	Steel frame	B-3	430	.112	C.Y.		3.20	4.05	7.25	9.35	
	0300	Concrete frame		365	.132			3.77	4.77	8.54	11.05	
	0400	Masonry construction		445	.108			3.09	3.91	7	9.05	
	0500	Wood frame		247	.194			5.55	7.05	12.60	16.30	
330	0010	**SELECTIVE DEMOLITION, DUMP CHARGES** R02220-510											330
	0020	Dump charges, typical urban city, tipping fees only											
	0100	Building construction materials				Ton					70	
	0200	Trees, brush, lumber									50	
	0300	Rubbish only									60	
	0500	Reclamation station, usual charge									85	
340	0010	**SELECTIVE DEMOLITION, GUTTING** R02220-510											340
	0020	Building interior, including disposal, dumpster fees not included											
	0500	Residential building											
	0560	Minimum	B-16	400	.080	SF Flr.		2.19	1.19	3.38	4.71	
	0580	Maximum	"	360	.089	"		2.44	1.32	3.76	5.25	
	0900	Commercial building											
	1000	Minimum	B-16	350	.091	SF Flr.		2.51	1.36	3.87	5.40	
	1020	Maximum	"	250	.128	"		3.51	1.91	5.42	7.55	

Important: See the Reference Section for critical supporting data - Reference Nos., Crews, & City Cost Indexes

SITE CONSTRUCTION 2

02220 | Site Demolition

		CREW	DAILY OUTPUT	LABOR-HOURS	UNIT	2005 BARE COSTS				TOTAL INCL O&P
						MAT.	LABOR	EQUIP.	TOTAL	
350	**0010**	**SELECTIVE DEMOLITION, RUBBISH HANDLING** R02220 -510								**350**
	0020	The following are to be added to the demolition prices								
	0400	Chute, circular, prefabricated steel, 18" diameter	B-1	40	.600	L.F.	28	16.40		44.40	56.50
	0440	30" diameter	"	30	.800	"	37.50	22		59.50	75.50
	0725	Dumpster, weekly rental, 1 dump/week, 20 C.Y. capacity (8 Tons)				Week					440
	0800	30 C.Y. capacity (10 Tons)									665
	0840	40 C.Y. capacity (13 Tons)				↓					805
	1000	Dust partition, 6 mil polyethylene, 1" x 3" frame	2 Carp	2,000	.008	S.F.	.17	.27		.44	.62
	1080	2" x 4" frame	"	2,000	.008	"	.28	.27		.55	.74
	2000	Load, haul, and dump, 50' haul	2 Clab	24	.667	C.Y.		17.80		17.80	27.50
	2040	100' haul		16.50	.970			26		26	40.50
	2080	Over 100' haul, add per 100 L.F.		35.50	.451			12.05		12.05	18.75
	2120	In elevators, per 10 floors, add	↓	140	.114			3.05		3.05	4.75
	3000	Loading & trucking, including 2 mile haul, chute loaded	B-16	45	.711			19.50	10.60	30.10	41.50
	3040	Hand loading truck, 50' haul	"	48	.667			18.25	9.95	28.20	39.50
	3080	Machine loading truck	B-17	120	.267			7.60	4.52	12.12	16.60
	5000	Haul, per mile, up to 8 C.Y. truck	B-34B	1,165	.007			.19	.41	.60	.74
	5100	Over 8 C.Y. truck	"	1,550	.005	↓		.14	.31	.45	.56
360	**0010**	**SELECTIVE DEMOLITION, SAW CUTTING** R02220 -510								**360**
	0015	Asphalt, up to 3" deep	B-89	1,050	.015	L.F.	.26	.46	.29	1.01	1.30
	0020	Each additional inch of depth		1,800	.009		.06	.27	.17	.50	.66
	0400	Concrete slabs, mesh reinforcing, up to 3" deep		980	.016		.35	.49	.31	1.15	1.47
	0420	Each additional inch of depth	↓	1,600	.010		.12	.30	.19	.61	.79
	0800	Concrete walls, hydraulic saw, plain, per inch of depth	B-89B	250	.064		.32	1.92	1.76	4	5.20
	0820	Rod reinforcing, per inch of depth		150	.107		.44	3.19	2.93	6.56	8.55
	1200	Masonry walls, hydraulic saw, brick, per inch of depth		300	.053		.32	1.60	1.47	3.39	4.38
	1220	Block walls, solid, per inch of depth	↓	250	.064		.33	1.92	1.76	4.01	5.20
	2000	Brick or masonry w/hand held saw, per inch of depth	A-1	125	.064		.26	1.71	.40	2.37	3.39
	5000	Wood sheathing to 1" thick, on walls	1 Carp	200	.040			1.37		1.37	2.13
	5020	On roof	"	250	.032	↓		1.10		1.10	1.71
	9950	See also div. 02210-200 core drilling									
370	**0010**	**SELECTIVE DEMOLITION, TORCH CUTTING** R02220 -510								**370**
	0020	Steel, 1" thick plate	1 Clab	360	.022	L.F.	.18	.59		.77	1.12
	0040	1" diameter bar	"	210	.038	Ea.		1.02		1.02	1.58
	1000	Oxygen lance cutting, reinforced concrete walls									
	1040	12" to 16" thick walls	1 Clab	10	.800	L.F.		21.50		21.50	33
	1080	24" thick walls	"	6	1.333	"		35.50		35.50	55.50

02230 | Site Clearing

		CREW	DAILY OUTPUT	LABOR-HOURS	UNIT	MAT.	LABOR	EQUIP.	TOTAL	TOTAL INCL O&P
100	**0010**	**CLEAR AND GRUB**								**100**
	0020	Cut & chip light trees to 6" diam.	B-7	1	48	Acre		1,350	1,025	2,375	3,225
	0150	Grub stumps and remove	B-30	2	12			360	835	1,195	1,475
	0200	Cut & chip medium, trees to 12" diam.	B-7	.70	68.571			1,950	1,450	3,400	4,600
	0250	Grub stumps and remove	B-30	1	24			720	1,675	2,395	2,950
	0300	Cut & chip heavy, trees to 24" diam.	B-7	.30	160			4,550	3,400	7,950	10,800
	0350	Grub stumps and remove	B-30	.50	48			1,425	3,350	4,775	5,850
	0400	If burning is allowed, reduce cut & chip				↓					40%
	3000	Chipping stumps, to 18" deep, 12" diam.	B-86	20	.400	Ea.		13.85	3.54	17.39	25
	3040	18" diameter		16	.500			17.35	4.42	21.77	31
	3080	24" diameter		14	.571			19.80	5.05	24.85	35.50
	3100	30" diameter		12	.667			23	5.90	28.90	41.50
	3120	36" diameter		10	.800			27.50	7.05	34.55	50
	3160	48" diameter	↓	8	1	↓		34.50	8.85	43.35	61.50
	5000	Tree thinning, feller buncher, conifer									
	5080	Up to 8" diameter	B-93	240	.033	Ea.		1.15	1.80	2.95	3.72

2

SITE CONSTRUCTION

		02230	Site Clearing	CREW	DAILY OUTPUT	LABOR-HOURS	UNIT	2005 BARE COSTS MAT.	LABOR	EQUIP.	TOTAL	TOTAL INCL O&P	
100	5120		12" diameter	B-93	160	.050	Ea.		1.73	2.70	4.43	5.60	**100**
	5240		Hardwood, up to 4" diameter		240	.033			1.15	1.80	2.95	3.72	
	5280		8" diameter		180	.044			1.54	2.40	3.94	4.96	
	5320		12" diameter	▼	120	.067	▼		2.31	3.60	5.91	7.45	
	7000		Tree removal, congested area, aerial lift truck										
	7040		8" diameter	B-85	7	5.714	Ea.		163	96	259	355	
	7080		12" diameter		6	6.667			190	112	302	415	
	7120		18" diameter		5	8			228	134	362	495	
	7160		24" diameter		4	10			285	168	453	625	
	7240		36" diameter		3	13.333			380	223	603	830	
	7280		48" diameter	▼	2	20	▼		570	335	905	1,250	
200	0010	**SELECTIVE CLEARING**											**200**
	0020		Clearing brush with brush saw	A-1C	.25	32	Acre	855	77		932	1,400	
	0100		By hand	1 Clab	.12	66.667		1,775			1,775	2,775	
	0300		With dozer, ball and chain, light clearing	B-11A	2	8			245	460	705	880	
	0400		Medium clearing		1.50	10.667			325	615	940	1,175	
	0500		With dozer and brush rake, light		10	1.600			49	92	141	176	
	0550		Medium brush to 4" diameter		8	2			61.50	115	176.50	220	
	0600		Heavy brush to 4" diameter	▼	6.40	2.500	▼		76.50	144	220.50	275	
	1000		Brush mowing, tractor w/rotary mower, no removal										
	1020		Light density	B-84	2	4	Acre		139	114	253	335	
	1040		Medium density		1.50	5.333			185	153	338	445	
	1080		Heavy density	▼	1	8	▼		277	229	506	670	
300	0010	**SELECTIVE TREE REMOVAL** With tractor, large tract, firm											**300**
	0020		level terrain, no boulders, less than 12" diam. trees										
	0300		300 HP dozer, up to 400 trees/acre, 0 to 25% hardwoods	B-10M	.75	16	Acre		510	1,600	2,110	2,525	
	0340		25% to 50% hardwoods		.60	20			640	2,000	2,640	3,175	
	0370		75% to 100% hardwoods		.45	26.667			855	2,650	3,505	4,225	
	0400		500 trees/acre, 0% to 25% hardwoods		.60	20			640	2,000	2,640	3,175	
	0440		25% to 50% hardwoods		.48	25			800	2,500	3,300	3,975	
	0470		75% to 100% hardwoods		.36	33.333			1,075	3,325	4,400	5,275	
	0500		More than 600 trees/acre, 0 to 25% hardwoods		.52	23.077			740	2,300	3,040	3,650	
	0540		25% to 50% hardwoods		.42	28.571			915	2,850	3,765	4,525	
	0570		75% to 100% hardwoods	▼	.31	38.710	▼		1,250	3,850	5,100	6,125	
	0900		Large tract clearing per tree										
	1500		300 HP dozer, to 12" diameter, softwood	B-10M	320	.037	Ea.		1.20	3.73	4.93	5.95	
	1550		Hardwood		100	.120			3.84	11.95	15.79	19	
	1600		12" to 24" diameter, softwood		200	.060			1.92	5.95	7.87	9.45	
	1650		Hardwood		80	.150			4.80	14.95	19.75	24	
	1700		24" to 36" diameter, softwood		100	.120			3.84	11.95	15.79	19	
	1750		Hardwood		50	.240			7.70	24	31.70	38	
	1800		36" to 48" diameter, softwood		70	.171			5.50	17.05	22.55	27	
	1850		Hardwood	▼	35	.343	▼		10.95	34	44.95	54	
	2000		Stump removal on site by hydraulic backhoe, 1-1/2 C.Y.										
	2040		4" to 6" diameter	B-17	60	.533	Ea.		15.20	9.05	24.25	33.50	
	2050		8" to 12" diameter	B-30	33	.727			22	50.50	72.50	89	
	2100		14" to 24" diameter		25	.960			28.50	67	95.50	117	
	2150		26" to 36" diameter	▼	16	1.500	▼		45	105	150	183	
	3000		Remove selective trees, on site using chain saws and chipper,										
	3050		not incl. stumps, up to 6" diameter	B-7	18	2.667	Ea.		75.50	56.50	132	180	
	3100		8" to 12" diameter		12	4			113	85	198	269	
	3150		14" to 24" diameter		10	4.800			136	102	238	320	
	3200		26" to 36" diameter	▼	8	6			170	127	297	405	
	3300		Machine load, 2 mile haul to dump, 12" diam. tree, add				▼				160	240	

Important: See the Reference Section for critical supporting data - Reference Nos., Crews, & City Cost Indexes

02230	Site Clearing	CREW	DAILY OUTPUT	LABOR-HOURS	UNIT	2005 BARE COSTS				TOTAL INCL O&P
						MAT.	LABOR	EQUIP.	TOTAL	
500 0010	**STRIPPING & STOCKPILING OF SOIL**									**500**
0020	200 H.P. dozer, ideal conditions	B-10B	2,300	.005	C.Y.		.17	.40	.57	.69
0100	Adverse conditions	"	1,150	.010			.33	.80	1.13	1.39
0200	300 HP dozer, ideal conditions	B-10M	3,000	.004			.13	.40	.53	.63
0300	Adverse conditions	"	1,650	.007			.23	.72	.95	1.15
0400	400 HP dozer, ideal conditions	B-10X	3,900	.003			.10	.39	.49	.58
0500	Adverse conditions	"	2,000	.006			.19	.76	.95	1.13
0600	Clay, dry and soft, 200 HP dozer, ideal conditions	B-10B	1,600	.007			.24	.57	.81	.99
0601	Strip topsoil, clay, dry & soft, 200 HP dozer, ideal conditions		1,600	.007			.24	.57	.81	.99
0700	Adverse conditions		800	.015			.48	1.15	1.63	1.99
1000	Medium hard, 300 HP dozer, ideal conditions	B-10M	2,000	.006			.19	.60	.79	.95
1100	Adverse conditions	"	1,100	.011			.35	1.09	1.44	1.73
1200	Very hard, 400 HP dozer, ideal conditions	B-10X	2,600	.005			.15	.59	.74	.87
1300	Adverse conditions	"	1,340	.009			.29	1.14	1.43	1.69
1400	Loam or topsoil, remove and stockpile on site									
1420	6" deep, 200' haul	B-10B	865	.014	C.Y.		.44	1.06	1.50	1.84
1430	300' haul		520	.023			.74	1.77	2.51	3.07
1440	500' haul		225	.053			1.71	4.09	5.80	7.10
1450	Alternate method: 6" deep, 200' haul		5,090	.002	S.Y.		.08	.18	.26	.31
1460	500' haul		1,325	.009	"		.29	.69	.98	1.20

02240	Dewatering	CREW	DAILY OUTPUT	LABOR-HOURS	UNIT	MAT.	LABOR	EQUIP.	TOTAL	TOTAL INCL O&P
500 0010	**DEWATERING**									**500**
0020	Excavate drainage trench, 2' wide, 2' deep	B-11C	90	.178	C.Y.		5.45	2.40	7.85	11
0100	2' wide, 3' deep, with backhoe loader	"	135	.119			3.64	1.60	5.24	7.30
0200	Excavate sump pits by hand, light soil	1 Clab	7.10	1.127			30		30	47
0300	Heavy soil	"	3.50	2.286			61		61	95
0500	Pumping 8 hr., attended 2 hrs. per day, including 20 L.F.									
0550	of suction hose & 100 L.F. discharge hose									
0600	2" diaphragm pump used for 8 hours	B-10H	4	3	Day		96	14.25	110.25	162
0650	4" diaphragm pump used for 8 hours	B-10I	4	3			96	21.50	117.50	170
0800	8 hrs. attended, 2" diaphragm pump	B-10H	1	12			385	57	442	650
0900	3" centrifugal pump	B-10J	1	12			385	63	448	655
1000	4" diaphragm pump	B-10I	1	12			385	86.50	471.50	680
1100	6" centrifugal pump	B-10K	1	12			385	248	633	855
1300	CMP, incl. excavation 3' deep, 12" diameter	B-6	115	.209	L.F.	9.65	6	1.88	17.53	22
1400	18" diameter		100	.240	"	12	6.90	2.16	21.06	26
1600	Sump hole construction, incl. excavation and gravel, pit		1,250	.019	C.F.	.70	.55	.17	1.42	1.81
1700	With 12" gravel collar, 12" pipe, corrugated, 16 ga.		70	.343	L.F.	15.10	9.90	3.08	28.08	35
1800	15" pipe, corrugated, 16 ga.		55	.436		19.30	12.60	3.92	35.82	44.50
1900	18" pipe, corrugated, 16 ga.		50	.480		22.50	13.85	4.32	40.67	51.50
2000	24" pipe, corrugated, 14 ga.		40	.600		27	17.30	5.40	49.70	62.50
2200	Wood lining, up to 4' x 4', add		300	.080	SFCA	12.80	2.31	.72	15.83	18.40
9950	See div. 02240-900 for wellpoints									
9960	See div. 02240-700 for deep well systems									
700 0010	**WELLS** For dewatering 10' to 20' deep, 2' diameter									**700**
0020	with steel casing, minimum	B-6	165	.145	V.L.F.	2.35	4.19	1.31	7.85	10.50
0050	Average		98	.245		4.68	7.05	2.20	13.93	18.40
0100	Maximum		49	.490		12.50	14.10	4.40	31	40
0300	For pumps for dewatering, see division 01590-400									
0500	For domestic water wells, see division 02520-510									
900 0010	**WELLPOINTS** For wellpoint equipment rental, see div. 01590-700 [R02240-900]									**900**
0100	Installation and removal of single stage system									
0110	Labor only, .75 labor-hours per L.F., minimum	1 Clab	10.70	.748	LF Hdr		19.95		19.95	31
0200	2.0 labor-hours per L.F., maximum	"	4	2	"		53.50		53.50	83

SITE CONSTRUCTION 2

SITE CONSTRUCTION — 2

02240 | Dewatering

		CREW	DAILY OUTPUT	LABOR-HOURS	UNIT	MAT.	LABOR	EQUIP.	TOTAL	TOTAL INCL O&P	
900	0400	Pump operation, 4 @ 6 hr. shifts R02240-900									900
	0410	Per 24 hour day	4 Eqlt	1.27	25.197	Day		835		835	1,250	
	0500	Per 168 hour week, 160 hr. straight, 8 hr. double time		.18	177	Week		5,875		5,875	8,850	
	0550	Per 4.3 week month		.04	800	Month		26,400		26,400	39,800	
	0600	Complete installation, operation, equipment rental, fuel &										
	0610	removal of system with 2" wellpoints 5' O.C.										
	0700	100' long header, 6" diameter, first month	4 Eqlt	3.23	9.907	LF Hdr	125	325		450	630	
	0800	Thereafter, per month		4.13	7.748		99.50	256		355.50	495	
	1000	200' long header, 8" diameter, first month		6	5.333		110	176		286	385	
	1100	Thereafter, per month		8.39	3.814		56	126		182	252	
	1300	500' long header, 8" diameter, first month		10.63	3.010		43.50	99.50		143	198	
	1400	Thereafter, per month		20.91	1.530		31	50.50		81.50	111	
	1600	1,000' long header, 10" diameter, first month		11.62	2.754		37.50	91		128.50	178	
	1700	Thereafter, per month		41.81	.765		18.70	25.50		44.20	58.50	
	1900	Note: above figures include pumping 168 hrs. per week										
	1910	and include the pump operator and one stand-by pump.										

02250 | Shoring & Underpinning

		CREW	DAILY OUTPUT	LABOR-HOURS	UNIT	MAT.	LABOR	EQUIP.	TOTAL	TOTAL INCL O&P	
100	0010	**GROUTING, PRESSURE**										100
	0020	Grouting, pressure, cement & sand, 1:1 mix, minimum	B-61	124	.323	Bag	9.10	9.15	2.01	20.26	26.50	
	0100	Maximum		51	.784	"	9.10	22.50	4.89	36.49	50	
	0200	Cement and sand, 1:1 mix, minimum		250	.160	C.F.	18.20	4.54	1	23.74	28	
	0300	Maximum		100	.400		27.50	11.35	2.50	41.35	50.50	
	0400	Epoxy cement grout, minimum		137	.292		126	8.30	1.82	136.12	154	
	0500	Maximum		57	.702		126	19.90	4.38	150.28	175	
	0600	Structural epoxy grout				Gal.	49.50			49.50	54.50	
	0700	Alternate pricing method: (Add for materials)										
	0710	5 person crew and equipment	B-61	1	40	Day		1,125	250	1,375	2,025	
400	0010	**SHEET PILING**										400
	0020	Sheet piling steel, not incl. wales, 22 psf, 15' excav., left in place	B-40	10.81	5.920	Ton	865	200	212	1,277	1,500	
	0100	Drive, extract & salvage R02250-400		6	10.667		385	360	380	1,125	1,425	
	0300	20' deep excavation, 27 psf, left in place		12.95	4.942		865	167	177	1,209	1,400	
	0400	Drive, extract & salvage R02250-450		6.55	9.771		385	330	350	1,065	1,325	
	0600	25' deep excavation, 38 psf, left in place		19	3.368		865	114	121	1,100	1,275	
	0700	Drive, extract & salvage		10.50	6.095		385	206	218	809	990	
	0900	40' deep excavation, 38 psf, left in place CN		21.20	3.019		865	102	108	1,075	1,225	
	1000	Drive, extract & salvage		12.25	5.224		385	177	187	749	910	
	1200	15' deep excavation, 22 psf, left in place		983	.065	S.F.	10.05	2.20	2.33	14.58	17.10	
	1300	Drive, extract & salvage		545	.117		4.31	3.97	4.20	12.48	15.70	
	1500	20' deep excavation, 27 psf, left in place		960	.067		12.65	2.25	2.39	17.29	20	
	1600	Drive, extract & salvage		485	.132		5.60	4.46	4.72	14.78	18.45	
	1800	25' deep excavation, 38 psf, left in place		1,000	.064		18.60	2.16	2.29	23.05	26.50	
	1900	Drive, extract & salvage		553	.116		7.70	3.91	4.14	15.75	19.25	
	2100	Rent steel sheet piling and wales, first month				Ton	206			206	226	
	2200	Per added month					20.50			20.50	22.50	
	2300	Rental piling left in place, add to rental					685			685	755	
	2500	Wales, connections & struts, 2/3 salvage					210			210	231	
	2700	High strength piling, 50,000 psi, add					46.50			46.50	51	
	2800	55,000 psi, add					49.50			49.50	54.50	
	3000	Tie rod, not upset, 1-1/2" to 4" diameter with turnbuckle					1,525			1,525	1,675	
	3100	No turnbuckle					1,175			1,175	1,300	
	3300	Upset, 1-3/4" to 4" diameter with turnbuckle					1,725			1,725	1,900	
	3400	No turnbuckle					1,500			1,500	1,650	
	3600	Lightweight, 18" to 28" wide, 7 ga., 9.22 psf, and										
	3610	9 ga., 8.6 psf, minimum				Lb.	.63			.63	.69	
	3700	Average					.68			.68	.75	

Important: See the Reference Section for critical supporting data - Reference Nos., Crews, & City Cost Indexes

02200 | Site Preparation

SITE CONSTRUCTION **2**

02250		Shoring & Underpinning	CREW	DAILY OUTPUT	LABOR-HOURS	UNIT	MAT.	LABOR	EQUIP.	TOTAL	TOTAL INCL O&P	
400	3750	Maximum				Lb.	.79			.79	.87	400
	3900	Wood, solid sheeting, incl. wales, braces and spacers,										
	3910	drive, extract & salvage, 8' deep excavation	B-31	330	.121	S.F.	1.72	3.47	.39	5.58	7.70	
	4000	10' deep, 50 S.F./hr. in & 150 S.F./hr. out		300	.133		1.77	3.81	.43	6.01	8.35	
	4100	12' deep, 45 S.F./hr. in & 135 S.F./hr. out		270	.148		1.82	4.24	.48	6.54	9.15	
	4200	14' deep, 42 S.F./hr. in & 126 S.F./hr. out		250	.160		1.88	4.58	.52	6.98	9.80	
	4300	16' deep, 40 S.F./hr. in & 120 S.F./hr. out		240	.167		1.94	4.77	.54	7.25	10.10	
	4400	18' deep, 38 S.F./hr. in & 114 S.F./hr. out		230	.174		2	4.98	.56	7.54	10.55	
	4500	20' deep, 35 S.F./hr. in & 105 S.F./hr. out		210	.190		2.07	5.45	.62	8.14	11.45	
	4520	Left in place, 8' deep, 55 S.F./hr.		440	.091		3.10	2.60	.29	5.99	7.80	
	4540	10' deep, 50 S.F./hr.		400	.100		3.26	2.86	.32	6.44	8.40	
	4560	12' deep, 45 S.F./hr.		360	.111		3.44	3.18	.36	6.98	9.15	
	4565	14' deep, 42 S.F./hr.		335	.119		3.65	3.42	.39	7.46	9.75	
	4570	16' deep, 40 S.F./hr.		320	.125		3.88	3.58	.40	7.86	10.25	
	4580	18' deep, 38 S.F./hr.		305	.131		4.13	3.75	.42	8.30	10.85	
	4590	20' deep, 35 S.F./hr.		280	.143		4.43	4.09	.46	8.98	11.75	
	4700	Alternate pricing, left in place, 8' deep		1.76	22.727	M.B.F.	695	650	73.50	1,418.50	1,850	
	4800	Drive, extract and salvage, 8' deep		1.32	30.303	"	620	865	98	1,583	2,150	
	5000	For treated lumber add cost of treatment to lumber										
	5010	See division 06070-400										
500	0010	**SHORING**										500
	0020	Shoring, existing building, with timber, no salvage allowance	B-51	2.20	21.818	M.B.F.	800	590	58.50	1,448.50	1,850	
	1000	On cribbing with 35 ton screw jacks, per box and jack	"	3.60	13.333	Jack	44.50	360	35.50	440	650	
800	0010	**UNDERPINNING FOUNDATIONS** Including excavation,										800
	0020	forming, reinforcing, concrete and equipment										
	0100	5' to 16' below grade, 100 to 500 C.Y.	B-52	2.30	24.348	C.Y.	225	765	171	1,161	1,600	
	0200	Over 500 C.Y.		2.50	22.400		203	705	158	1,066	1,500	
	0400	16' to 25' below grade, 100 to 500 C.Y.		2	28		248	880	197	1,325	1,850	
	0500	Over 500 C.Y.		2.10	26.667		234	835	188	1,257	1,775	
	0700	26' to 40' below grade, 100 to 500 C.Y.		1.60	35		270	1,100	246	1,616	2,275	
	0800	Over 500 C.Y.		1.80	31.111		248	975	219	1,442	2,025	
	0900	For under 50 C.Y., add					10%	40%				
	1000	For 50 C.Y. to 100 C.Y., add					5%	20%				
900	0010	**VIBROFLOTATION**										900
	0900	Vibroflotation compacted sand cylinder, minimum	B-60	750	.075	V.L.F.		2.28	1.55	3.83	5.20	
	0950	Maximum		325	.172			5.25	3.58	8.83	12	
	1100	Vibro replacement compacted stone cylinder, minimum		500	.112			3.43	2.33	5.76	7.80	
	1150	Maximum		250	.224			6.85	4.65	11.50	15.55	
	1300	Mobilization and demobilization, minimum		.47	119	Total		3,650	2,475	6,125	8,300	
	1400	Maximum		.14	400	"		12,200	8,300	20,500	27,800	

02260		Excavation Support/Protection										
200	0010	**COFFERDAMS**, incl. mobilization and temporary sheeting										200
	0080	Soldier beams & lagging H piles with 3" wood sheeting										
	0090	horizontal between piles, including removal of wales & braces										
	0100	No hydrostatic head, 15' deep, 1 line of braces, minimum	B-50	545	.205	S.F.	8	6.65	3.11	17.76	23	
	0200	Maximum		495	.226		8.90	7.30	3.42	19.62	25.50	
	0400	15' to 22' deep with 2 lines of braces, 10" H, minimum		360	.311		9.40	10.05	4.70	24.15	31.50	
	0500	Maximum		330	.339		10.65	10.95	5.15	26.75	35	
	0700	23' to 35' deep with 3 lines of braces, 12" H, minimum		325	.345		12.30	11.15	5.20	28.65	37	
	0800	Maximum		295	.380		13.35	12.25	5.75	31.35	40.50	
	1000	36' to 45' deep with 4 lines of braces, 14" H, minimum		290	.386		13.80	12.50	5.85	32.15	41.50	
	1100	Maximum		265	.423		14.55	13.65	6.40	34.60	45	
	1300	No hydrostatic head, left in place, 15' dp., 1 line of braces, min.		635	.176		10.65	5.70	2.67	19.02	24	

For expanded coverage of these items see *Means Heavy Construction Cost Data 2005*

43

2 SITE CONSTRUCTION

02260	Excavation Support/Protection	CREW	DAILY OUTPUT	LABOR-HOURS	UNIT	2005 BARE COSTS MAT.	LABOR	EQUIP.	TOTAL	TOTAL INCL O&P	
200 1400	Maximum	B-50	575	.195	S.F.	11.45	6.30	2.94	20.69	26	200
1600	15' to 22' deep with 2 lines of braces, minimum		455	.246		16	7.95	3.72	27.67	34.50	
1700	Maximum		415	.270		17.80	8.70	4.08	30.58	38	
1900	23' to 35' deep with 3 lines of braces, minimum		420	.267		19.05	8.60	4.03	31.68	39	
2000	Maximum		380	.295		21	9.50	4.45	34.95	43	
2200	36' to 45' deep with 4 lines of braces, minimum		385	.291		23	9.40	4.40	36.80	45	
2300	Maximum		350	.320		26.50	10.35	4.84	41.69	51.50	
2350	Lagging only, 3" thick wood between piles 8' O.C., minimum	B-46	400	.120		1.78	3.64	.08	5.50	7.90	
2370	Maximum		250	.192		2.67	5.80	.13	8.60	12.45	
2400	Open sheeting no bracing, for trenches to 10' deep, min.		1,736	.028		.80	.84	.02	1.66	2.25	
2450	Maximum		1,510	.032		.89	.96	.02	1.87	2.55	
2500	Tie-back method, add to open sheeting, add, minimum								20%	20%	
2550	Maximum								60%	60%	
2700	Tie-backs only, based on tie-backs total length, minimum	B-46	86.80	.553	L.F.	10.70	16.75	.39	27.84	39	
2750	Maximum		38.50	1.247	"	18.80	38	.87	57.67	82.50	
3500	Tie-backs only, typical average, 25' long		2	24	Ea.	470	730	16.80	1,216.80	1,700	
3600	35' long		1.58	30.380	"	625	920	21.50	1,566.50	2,200	
720 0010	**ROCK BOLTING**										720
2020	Hollow core, prestressable anchor, 1" diameter, 5' long	2 Skwk	32	.500	Ea.	75	17.45		92.45	110	
2025	10' long		24	.667		150	23		173	201	
2060	2" diameter, 5' long		32	.500		280	17.45		297.45	335	
2065	10' long		24	.667		560	23		583	650	
2100	Super high-tensile, 3/4" diameter, 5' long		32	.500		17	17.45		34.45	45.50	
2105	10' long		24	.667		34	23		57	73.50	
2160	2" diameter, 5' long		32	.500		152	17.45		169.45	194	
2165	10' long		24	.667		305	23		328	370	
4400	Drill hole for rock bolt, 1-3/4" diam., 5' long (for 3/4" bolt)	B-56	17	.941			28	57.50	85.50	107	
4405	10' long		9	1.778			53	109	162	201	
4420	2" diameter, 5' long (for 1" bolt)		13	1.231			37	75.50	112.50	139	
4425	10' long		7	2.286			68.50	140	208.50	258	
4460	3-1/2" diameter, 5' long (for 2" bolt)		10	1.600			48	98	146	181	
4465	10' long		5	3.200			95.50	196	291.50	360	

02300 | Earthwork

02305	Equipment	CREW	DAILY OUTPUT	LABOR-HOURS	UNIT	2005 BARE COSTS MAT.	LABOR	EQUIP.	TOTAL	TOTAL INCL O&P	
250 0010	**MOBILIZATION OR DEMOB.** (One or the other, unless noted) R01590 -100										250
0015	Up to 25 mi haul dist (50 mi round trip for mob/demob crew)										
0020	Dozer, loader, backhoe, excav., grader, paver, roller, 70 to 150 H.P.	B-34N	4	2	Ea.		55	112	167	207	
0100	Above 150 HP	B-34K	3	2.667			73.50	175	248.50	305	
0300	Scraper, towed type (incl. tractor), 6 C.Y. capacity		3	2.667			73.50	175	248.50	305	
0400	10 C.Y.		2.50	3.200			88	210	298	365	
0600	Self-propelled scraper, 15 C.Y.		2.50	3.200			88	210	298	365	
0700	24 C.Y.		2	4			110	262	372	455	
0900	Shovel or dragline, 3/4 C.Y.		3.60	2.222			61	146	207	254	
1000	1-1/2 C.Y.		3	2.667			73.50	175	248.50	305	
1100	Small equipment, placed in rear of, or towed by pickup truck	A-3A	8	1			27	10.65	37.65	52.50	
1150	Equip up to 70 HP, on flatbed trailer behind pickup truck	A-3D	4	2			53.50	43.50	97	129	

02305 | Equipment

		CREW	DAILY OUTPUT	LABOR-HOURS	UNIT	2005 BARE COSTS				TOTAL INCL O&P	
						MAT.	LABOR	EQUIP.	TOTAL		
250	2000	Crane, truck-mounted, up to 75 ton (incl both mob & demob) [R01590-100]	1 EQHV	3.60	2.222	Ea.		80		80	120	**250**
	2100	Crane, truck-mounted, over 75 ton	A-3E	2.50	6.400			203	34	237	350	
	2200	Crawler-mounted, up to 75 ton	A-3F	2	8			254	278	532	690	
	2300	Over 75 ton	A-3G	1.50	10.667			340	405	745	960	
	2500	For each additional 5 miles haul distance, add						10%	10%			
	3000	For large pieces of equipment, allow for assembly/knockdown										
	3001	For mob/demob of vibrofloatation equip, see section 02250-900										
	3100	For mob/demob of micro-tunneling equip, see section 02441-400										
	3200	For mob/demob of pile driving equip, see section 02455-650										
	3300	For mob/demob of caisson drilling equip, see 02465-950										

02310 | Grading

		CREW	DAILY OUTPUT	LABOR-HOURS	UNIT	MAT.	LABOR	EQUIP.	TOTAL	TOTAL INCL O&P	
100	0010	**FINISH GRADING**									**100**
	0012	Finish grading area to be paved with grader, small area	B-11L	400	.040	S.Y.		1.23	1.14	2.37	3.14	
	0100	Large area		2,000	.008			.25	.23	.48	.63	
	1100	Fine grade for slab on grade, machine		1,040	.015			.47	.44	.91	1.20	
	1150	Hand grading	B-18	700	.034			.94	.05	.99	1.52	

02315 | Excavation and Fill

		CREW	DAILY OUTPUT	LABOR-HOURS	UNIT	MAT.	LABOR	EQUIP.	TOTAL	TOTAL INCL O&P	
110	0010	**BACKFILL, GENERAL** [R02315-300]										**110**
	0015	By hand, no compaction, light soil	1 Clab	14	.571	L.C.Y.		15.25		15.25	23.50	
	0100	Heavy soil		11	.727	"		19.40		19.40	30	
	0300	Compaction in 6" layers, hand tamp, add to above		20.60	.388	E.C.Y.		10.35		10.35	16.15	
	0400	Roller compaction operator walking, add	B-10A	100	.120			3.84	1.23	5.07	7.20	
	0500	Air tamp, add	B-9D	190	.211			5.70	.93	6.63	9.95	
	0600	Vibrating plate, add	A-1D	60	.133			3.56	.42	3.98	6	
	0800	Compaction in 12" layers, hand tamp, add to above	1 Clab	34	.235			6.30		6.30	9.80	
	0900	Roller compaction operator walking, add	B-10A	150	.080			2.56	.82	3.38	4.79	
	1000	Air tamp, add	B-9	285	.140			3.80	.50	4.30	6.45	
	1100	Vibrating plate, add	A-1E	90	.089			2.37	.41	2.78	4.14	
	1300	Dozer backfilling, bulk, up to 300' haul, no compaction	B-10B	1,200	.010	L.C.Y.		.32	.77	1.09	1.33	
	1400	Air tamped, add	B-11B	80	.200	E.C.Y.		6	2.30	8.30	11.70	
	1600	Compacting backfill, 6" to 12" lifts, vibrating roller	B-10C	800	.015			.48	1.89	2.37	2.81	
	1700	Sheepsfoot roller	B-10D	750	.016			.51	2.04	2.55	3.02	
	1900	Dozer backfilling, trench, up to 300' haul, no compaction	B-10B	900	.013	L.C.Y.		.43	1.02	1.45	1.77	
	2000	Air tamped, add	B-11B	80	.200	E.C.Y.		6	2.30	8.30	11.70	
	2200	Compacting backfill, 6" to 12" lifts, vibrating roller	B-10C	700	.017			.55	2.16	2.71	3.20	
	2300	Sheepsfoot roller	B-10D	650	.018			.59	2.35	2.94	3.49	
	2350	Spreading in 8" layers, small dozer	B-10B	1,060	.011	L.C.Y.		.36	.87	1.23	1.50	
120	0010	**BACKFILL, STRUCTURAL** Dozer or F.E. loader										**120**
	0020	From existing stockpile, no compaction										
	2000	75 H.P., 50' haul, sand & gravel	B-10L	1,100	.011	L.C.Y.		.35	.29	.64	.84	
	2020	Common earth		975	.012			.39	.32	.71	.96	
	2040	Clay		850	.014			.45	.37	.82	1.10	
	2400	300' haul, sand & gravel		370	.032			1.04	.85	1.89	2.52	
	2420	Common earth		330	.036			1.16	.95	2.11	2.82	
	2440	Clay		290	.041			1.32	1.09	2.41	3.20	
	3000	105 H.P., 50' haul, sand & gravel	B-10W	1,350	.009			.28	.34	.62	.80	
	3020	Common earth		1,225	.010			.31	.37	.68	.89	
	3040	Clay		1,100	.011			.35	.41	.76	.98	
	3300	300' haul, sand & gravel		465	.026			.83	.98	1.81	2.33	
	3320	Common earth		415	.029			.93	1.09	2.02	2.61	
	3340	Clay		370	.032			1.04	1.23	2.27	2.93	
	4000	200 H.P., 50' haul, sand & gravel	B-10B	2,500	.005			.15	.37	.52	.63	
	4020	Common earth		2,200	.005			.17	.42	.59	.73	

SITE CONSTRUCTION 2

02315 | Excavation and Fill

		Description	CREW	DAILY OUTPUT	LABOR-HOURS	UNIT	2005 BARE COSTS MAT.	LABOR	EQUIP.	TOTAL	TOTAL INCL O&P	
120	4040	Clay	B-10B	1,950	.006	L.C.Y.		.20	.47	.67	.82	**120**
	4400	300' haul, sand & gravel		805	.015			.48	1.14	1.62	1.99	
	4420	Common earth		735	.016			.52	1.25	1.77	2.17	
	4440	Clay		660	.018			.58	1.39	1.97	2.41	
	5000	300 H.P., 50' haul, sand & gravel	B-10M	3,170	.004			.12	.38	.50	.60	
	5020	Common earth		2,900	.004			.13	.41	.54	.65	
	5040	Clay		2,700	.004			.14	.44	.58	.71	
	5400	300' haul, sand & gravel		1,500	.008			.26	.80	1.06	1.27	
	5420	Common earth		1,350	.009			.28	.89	1.17	1.40	
	5440	Clay		1,225	.010			.31	.98	1.29	1.55	
	6010	For trench backfill, see div. 02315-610 & 02315-620										
	6100	For compaction, see div. 02315-320										
320	0010	**COMPACTION, STRUCTURAL**										**320**
	0020	Steel wheel tandem roller, 5 tons	B-10E	8	1.500	Hr.		48	13.45	61.45	88	
	0100	10 tons	B-10F	8	1.500	"		48	22.50	70.50	97.50	
	0300	Sheepsfoot or wobbly wheel roller, 8" lifts, common fill	B-10G	1,300	.009	E.C.Y.		.30	.61	.91	1.12	
	0400	Select fill	"	1,500	.008			.26	.53	.79	.97	
	0600	Vibratory plate, 8" lifts, common fill	A-1D	200	.040			1.07	.13	1.20	1.80	
	0700	Select fill	"	216	.037			.99	.12	1.11	1.67	
412	0010	**DRILLING ONLY ROCK**										**412**
	0015	Drilling only rock, 2" hole for rock bolts	B-47	316	.076	L.F.		2.24	3.16	5.40	6.90	
	0800	2-1/2" hole for pre-splitting		600	.040			1.18	1.66	2.84	3.64	
	4600	Quarry operations, 2-1/2" to 3-1/2" diameter		715	.034			.99	1.40	2.39	3.05	
416	0010	**DRILLING AND BLASTING ROCK**										**416**
	0020	Rock, open face, under 1500 CY	B-47	225	.107	B.C.Y.	1.89	3.14	4.43	9.46	11.80	
	0100	Over 1500 C.Y.		300	.080		1.89	2.36	3.33	7.58	9.35	
	0200	Areas where blasting mats are required, under 1500 C.Y.		175	.137		1.89	4.04	5.70	11.63	14.55	
	0250	Over 1500 C.Y.		250	.096		1.89	2.83	3.99	8.71	10.80	
	0300	Bulk drilling and blasting, can vary greatly, average				B.C.Y.					5.50	
	0500	Pits, average				"					22	
	1300	Deep hole method, up to 1500 C.Y.	B-47	50	.480	B.C.Y.	1.89	14.15	19.95	35.99	46	
	1400	Over 1500 C.Y.		66	.364		1.89	10.70	15.10	27.69	35	
	1900	Restricted areas, up to 1500 C.Y.		13	1.846		1.89	54.50	76.50	132.89	170	
	2000	Over 1500 C.Y.		20	1.200		1.89	35.50	50	87.39	112	
	2200	Trenches, up to 1500 C.Y.		22	1.091		5.50	32	45.50	83	106	
	2300	Over 1500 C.Y.		26	.923		5.50	27	38.50	71	90	
	2500	Pier holes, up to 1500 C.Y.		22	1.091		1.89	32	45.50	79.39	102	
	2600	Over 1500 C.Y.		31	.774		1.89	23	32	56.89	72.50	
	2800	Boulders under 1/2 C.Y., loaded on truck, no hauling	B-100	80	.150			4.80	6.90	11.70	14.90	
	2900	Boulders, drilled, blasted	B-47	100	.240		1.89	7.10	10	18.99	24	
	3100	Jackhammer operators with foreman compressor, air tools	B-9	1	40	Day		1,075	142	1,217	1,825	
	3300	Track drill, compressor, operator and foreman	B-47	1	24	"		710	1,000	1,710	2,200	
	3500	Blasting caps				Ea.	3.42			3.42	3.76	
	3700	Explosives					.29			.29	.32	
	3900	Blasting mats, rent, for first day					94			94	104	
	4000	Per added day					31.50			31.50	34.50	
	4200	Preblast survey for 6 room house, individual lot, minimum	A-6	2.40	6.667			226	25.50	251.50	375	
	4300	Maximum	"	1.35	11.852			400	45	445	665	
	4500	City block within zone of influence, minimum	A-8	25,200	.001	S.F.		.05		.05	.07	
	4600	Maximum	"	15,100	.002	"		.08		.08	.12	
424	0010	**EXCAVATING, BULK BANK MEASURE** Common earth piled										**424**
	0020	For loading onto trucks, add								15%	15%	
	0050	For mobilization and demobilization, see division 02305-250										
	0100	For hauling, see division 02315-490										

Important: See the Reference Section for critical supporting data - Reference Nos., Crews, & City Cost Indexes

2 SITE CONSTRUCTION

02315 | Excavation and Fill

			CREW	DAILY OUTPUT	LABOR-HOURS	UNIT	MAT.	LABOR	EQUIP.	TOTAL	TOTAL INCL O&P	
424	0200	Backhoe, hydraulic, crawler mtd., 1 C.Y. cap. = 75 C.Y./hr. [R02315-400]	B-12A	600	.027	B.C.Y.		.83	.93	1.76	2.30	**424**
	0250	1-1/2 C.Y. cap. = 100 C.Y./hr.	B-12B	800	.020			.63	.90	1.53	1.95	
	0260	2 C.Y. cap. = 130 C.Y./hr. [R02315-450]	B-12C	1,040	.015			.48	.87	1.35	1.70	
	0300	3 C.Y. cap. = 160 C.Y./hr.	B-12D	1,280	.013			.39	1.58	1.97	2.34	
	0310	Wheel mounted, 1/2 C.Y. cap. = 30 C.Y./hr.	B-12E	240	.067			2.09	1.39	3.48	4.71	
	0360	3/4 C.Y. cap. = 45 C.Y./hr.	B-12F	360	.044			1.39	1.32	2.71	3.58	
	0500	Clamshell, 1/2 C.Y. cap. = 20 C.Y./hr.	B-12G	160	.100			3.13	3.39	6.52	8.50	
	0550	1 C.Y. cap. = 35 C.Y./hr.	B-12H	280	.057			1.79	3.13	4.92	6.15	
	0950	Dragline, 1/2 C.Y. cap. = 30 C.Y./hr.	B-12I	240	.067			2.09	2.71	4.80	6.15	
	1000	3/4 C.Y. cap. = 35 C.Y./hr.	"	280	.057			1.79	2.32	4.11	5.30	
	1050	1-1/2 C.Y. cap. = 65 C.Y./hr.	B-12P	520	.031			.96	1.77	2.73	3.41	
	1200	Front end loader, track mtd., 1-1/2 C.Y. cap. = 70 C.Y./hr.	B-10N	560	.021			.69	.56	1.25	1.65	
	1250	2-1/2 C.Y. cap. = 95 C.Y./hr.	B-10O	760	.016			.51	.73	1.24	1.57	
	1300	3 C.Y. cap. = 130 C.Y./hr.	B-10P	1,040	.012			.37	.76	1.13	1.39	
	1350	5 C.Y. cap. = 160 C.Y./hr.	B-10Q	1,280	.009			.30	.86	1.16	1.40	
	1500	Wheel mounted, 3/4 C.Y. cap. = 45 C.Y./hr.	B-10R	360	.033			1.07	.54	1.61	2.21	
	1550	1-1/2 C.Y. cap. = 80 C.Y./hr.	B-10S	640	.019			.60	.38	.98	1.32	
	1600	2-1/4 C.Y. cap. = 100 C.Y./hr.	B-10T	800	.015			.48	.38	.86	1.15	
	1650	5 C.Y. cap. = 185 C.Y./hr.	B-10U	1,480	.008			.26	.47	.73	.91	
	1800	Hydraulic excavator, truck mtd, 1/2 C.Y. = 30 C.Y./hr.	B-12J	240	.067			2.09	3.49	5.58	7.05	
	1850	48 inch bucket, 1 C.Y. = 45 C.Y./hr.	B-12K	360	.044			1.39	2.70	4.09	5.10	
	3700	Shovel, 1/2 C.Y. capacity = 55 C.Y./hr.	B-12L	440	.036			1.14	1.26	2.40	3.13	
	3750	3/4 C.Y. capacity = 85 C.Y./hr.	B-12M	680	.024			.74	1	1.74	2.23	
	3800	1 C.Y. capacity = 120 C.Y./hr.	B-12N	960	.017			.52	.93	1.45	1.82	
	3850	1-1/2 C.Y. capacity = 160 C.Y./hr.	B-12O	1,280	.013			.39	.75	1.14	1.42	
	3900	3 C.Y. cap. = 250 C.Y./hr.	B-12T	2,000	.008			.25	.63	.88	1.07	
	4000	For soft soil or sand, deduct								15%	15%	
	4100	For heavy soil or stiff clay, add								60%	60%	
	4200	For wet excavation with clamshell or dragline, add								100%	100%	
	4250	All other equipment, add								50%	50%	
	4400	Clamshell in sheeting or cofferdam, minimum	B-12H	160	.100			3.13	5.50	8.63	10.85	
	4450	Maximum	"	60	.267			8.35	14.60	22.95	29	
	8000	For hauling excavated material, see div. 02315-490										
432	0010	**EXCAVATING, BULK, DOZER** Open site										**432**
	2000	80 H.P., 50' haul, sand & gravel	B-10L	460	.026	B.C.Y.		.83	.68	1.51	2.02	
	2020	Common earth		400	.030			.96	.79	1.75	2.33	
	2040	Clay		250	.048			1.54	1.26	2.80	3.73	
	2200	150' haul, sand & gravel		230	.052			1.67	1.37	3.04	4.05	
	2220	Common earth		200	.060			1.92	1.57	3.49	4.65	
	2240	Clay		125	.096			3.07	2.52	5.59	7.45	
	2400	300' haul, sand & gravel		120	.100			3.20	2.62	5.82	7.75	
	2420	Common earth		100	.120			3.84	3.15	6.99	9.30	
	2440	Clay		65	.185			5.90	4.84	10.74	14.35	
	3000	105 H.P., 50' haul, sand & gravel	B-10W	700	.017			.55	.65	1.20	1.54	
	3020	Common earth		610	.020			.63	.74	1.37	1.78	
	3040	Clay		385	.031			1	1.18	2.18	2.82	
	3200	150' haul, sand & gravel		310	.039			1.24	1.46	2.70	3.49	
	3220	Common earth		270	.044			1.42	1.68	3.10	4.01	
	3240	Clay		170	.071			2.26	2.67	4.93	6.35	
	3300	300' haul, sand & gravel		140	.086			2.74	3.24	5.98	7.75	
	3320	Common earth		120	.100			3.20	3.78	6.98	9.05	
	3340	Clay		100	.120			3.84	4.54	8.38	10.85	
	4000	200 H.P., 50' haul, sand & gravel	B-10B	1,400	.009			.27	.66	.93	1.14	
	4020	Common earth		1,230	.010			.31	.75	1.06	1.29	
	4040	Clay		770	.016			.50	1.19	1.69	2.07	

SITE CONSTRUCTION | **2**

02315 | Excavation and Fill

		CREW	DAILY OUTPUT	LABOR-HOURS	UNIT	2005 BARE COSTS MAT.	LABOR	EQUIP.	TOTAL	TOTAL INCL O&P	
432											**432**
4200	150' haul, sand & gravel	B-10B	595	.020	B.C.Y.		.65	1.55	2.20	2.68	
4220	Common earth		516	.023			.74	1.78	2.52	3.09	
4240	Clay		325	.037			1.18	2.83	4.01	4.91	
4400	300' haul, sand & gravel		310	.039			1.24	2.97	4.21	5.15	
4420	Common earth		270	.044			1.42	3.41	4.83	5.90	
4440	Clay		170	.071			2.26	5.40	7.66	9.40	
5000	300 H.P., 50' haul, sand & gravel	B-10M	1,900	.006			.20	.63	.83	1	
5020	Common earth		1,650	.007			.23	.72	.95	1.15	
5040	Clay		1,025	.012			.37	1.17	1.54	1.85	
5200	150' haul, sand & gravel		920	.013			.42	1.30	1.72	2.06	
5220	Common earth		800	.015			.48	1.49	1.97	2.37	
5240	Clay		500	.024			.77	2.39	3.16	3.80	
5400	300' haul, sand & gravel		470	.026			.82	2.54	3.36	4.04	
5420	Common earth		410	.029			.94	2.91	3.85	4.63	
5440	Clay		250	.048			1.54	4.78	6.32	7.60	
5500	460 H.P., 50' haul, sand & gravel	B-10X	1,930	.006			.20	.79	.99	1.17	
5510	Common earth		1,680	.007			.23	.91	1.14	1.35	
5520	Clay		1,050	.011			.37	1.45	1.82	2.16	
5530	150' haul, sand & gravel		1,290	.009			.30	1.18	1.48	1.75	
5540	Common earth		1,120	.011			.34	1.36	1.70	2.02	
5550	Clay		700	.017			.55	2.18	2.73	3.22	
5560	300' haul, sand & gravel		660	.018			.58	2.31	2.89	3.42	
5570	Common earth		575	.021			.67	2.65	3.32	3.94	
5580	Clay		350	.034			1.10	4.35	5.45	6.45	
6000	700 H.P., 50' haul, sand & gravel	B-10V	3,500	.003			.11	.90	1.01	1.16	
6010	Common earth		3,035	.004			.13	1.03	1.16	1.33	
6020	Clay		1,925	.006			.20	1.63	1.83	2.09	
6030	150' haul, sand & gravel		2,025	.006			.19	1.55	1.74	2	
6040	Common earth		1,750	.007			.22	1.79	2.01	2.30	
6050	Clay		1,100	.011			.35	2.85	3.20	3.67	
6060	300' haul, sand & gravel		1,030	.012			.37	3.05	3.42	3.92	
6070	Common earth		900	.013			.43	3.49	3.92	4.49	
6080	Clay		550	.022			.70	5.70	6.40	7.35	
452	**EXCAVATION, BULK, SCRAPERS**										**452**
0010		R02315 -400									
0100	Elev. scraper 11 C.Y., sand & gravel 1500' haul, 1/4 dozer	B-33F	690	.020	B.C.Y.		.66	1.62	2.28	2.78	
0150	3000' haul		610	.023			.74	1.83	2.57	3.15	
0200	5000' haul		505	.028			.90	2.21	3.11	3.80	
0300	Common earth, 1500' haul		600	.023			.76	1.86	2.62	3.20	
0350	3000' haul		530	.026			.86	2.11	2.97	3.62	
0400	5000' haul		440	.032			1.03	2.54	3.57	4.36	
0500	Clay, 1500' haul		375	.037			1.21	2.98	4.19	5.10	
0550	3000' haul		330	.042			1.37	3.39	4.76	5.80	
0600	5000' haul		275	.051			1.65	4.07	5.72	6.95	
1000	Self propelled scraper, 14 C.Y. 1/4 push dozer, sand										
1050	Sand and gravel, 1500' haul	B-33D	920	.015	B.C.Y.		.49	2.01	2.50	2.96	
1100	3000' haul		805	.017			.56	2.30	2.86	3.38	
1200	5000' haul		645	.022			.70	2.87	3.57	4.23	
1300	Common earth, 1500' haul		800	.017			.57	2.31	2.88	3.40	
1350	3000' haul		700	.020			.65	2.64	3.29	3.89	
1400	5000' haul		560	.025			.81	3.31	4.12	4.87	
1500	Clay, 1500' haul		500	.028			.91	3.70	4.61	5.45	
1550	3000' haul		440	.032			1.03	4.21	5.24	6.20	
1600	5000' haul		350	.040			1.30	5.30	6.60	7.75	
2000	21 C.Y., 1/4 push dozer, sand & gravel, 1500' haul	B-33E	1,180	.012			.38	2.22	2.60	3.02	
2100	3000' haul		910	.015			.50	2.87	3.37	3.92	

Important: See the Reference Section for critical supporting data - Reference Nos., Crews, & City Cost Indexes

02315 | Excavation and Fill

			DAILY OUTPUT	LABOR-HOURS	UNIT	2005 BARE COSTS				TOTAL INCL O&P	
						MAT.	LABOR	EQUIP.	TOTAL		
452	2200	5000' haul	B-33E	750	.019	B.C.Y.		.60	3.49	4.09	4.76	**452**
	2300	Common earth, 1500' haul	R02315 -400	1,030	.014			.44	2.54	2.98	3.46	
	2350	3000' haul		790	.018			.57	3.31	3.88	4.51	
	2400	5000' haul		650	.022			.70	4.02	4.72	5.50	
	2500	Clay, 1500' haul		645	.022			.70	4.06	4.76	5.55	
	2550	3000' haul		495	.028			.92	5.30	6.22	7.20	
	2600	5000' haul		405	.035			1.12	6.45	7.57	8.80	
	2700	Towed, 10 C.Y., 1/4 push dozer, sand & gravel, 1500' haul	B-33B	560	.025			.81	2.97	3.78	4.50	
	2720	3000' haul		450	.031			1.01	3.70	4.71	5.60	
	2730	5000' haul		365	.038			1.24	4.56	5.80	6.90	
	2750	Common earth, 1500' haul		420	.033			1.08	3.96	5.04	6	
	2770	3000' haul		400	.035			1.13	4.16	5.29	6.30	
	2780	5000' haul		310	.045			1.46	5.35	6.81	8.10	
	2800	Clay, 1500' haul		315	.044			1.44	5.30	6.74	8	
	2820	3000' haul		300	.047			1.51	5.55	7.06	8.40	
	2840	5000' haul		225	.062			2.01	7.40	9.41	11.20	
	2900	15 C.Y., 1/4 push dozer, sand & gravel, 1500' haul	B-33C	800	.017			.57	2.08	2.65	3.15	
	2920	3000' haul		640	.022			.71	2.60	3.31	3.94	
	2940	5000' haul		520	.027			.87	3.20	4.07	4.84	
	2960	Common earth, 1500' haul		600	.023			.76	2.77	3.53	4.20	
	2980	3000' haul		560	.025			.81	2.97	3.78	4.50	
	3000	5000' haul		440	.032			1.03	3.78	4.81	5.70	
	3020	Clay, 1500' haul		450	.031			1.01	3.70	4.71	5.60	
	3040	3000' haul		420	.033			1.08	3.96	5.04	6	
	3060	5000' haul		320	.044			1.42	5.20	6.62	7.85	
462	0010	**EXCAVATION, STRUCTURAL** R02315 -400										**462**
	0015	Hand, pits to 6' deep, sandy soil	1 Clab	8	1	B.C.Y.		26.50		26.50	41.50	
	0100	Heavy soil or clay		4	2			53.50		53.50	83	
	0300	Pits 6' to 12' deep, sandy soil		5	1.600			42.50		42.50	66.50	
	0500	Heavy soil or clay		3	2.667			71		71	111	
	0700	Pits 12' to 18' deep, sandy soil		4	2			53.50		53.50	83	
	0900	Heavy soil or clay		2	4			107		107	166	
	1100	Hand loading trucks from stock pile, sandy soil		12	.667			17.80		17.80	27.50	
	1300	Heavy soil or clay		8	1			26.50		26.50	41.50	
	1500	For wet or muck hand excavation, add to above				%				50%	50%	
	6000	Machine excavation, for spread and mat footings, elevator pits,										
	6001	and small building foundations										
	6030	Common earth, hydraulic backhoe, 1/2 C.Y. bucket	B-12E	55	.291	B.C.Y.		9.10	6.05	15.15	20.50	
	6035	3/4 C.Y. bucket	B-12F	90	.178			5.55	5.25	10.80	14.30	
	6040	1 C.Y. bucket	B-12A	108	.148			4.64	5.15	9.79	12.80	
	6050	1-1/2 C.Y. bucket	B-12B	144	.111			3.48	5	8.48	10.80	
	6060	2 C.Y. bucket	B-12C	200	.080			2.50	4.55	7.05	8.85	
	6070	Sand and gravel, 3/4 C.Y. bucket	B-12F	100	.160			5	4.74	9.74	12.85	
	6080	1 C.Y. bucket	B-12A	120	.133			4.17	4.65	8.82	11.50	
	6090	1-1/2 C.Y. bucket	B-12B	160	.100			3.13	4.50	7.63	9.75	
	6100	2 C.Y. bucket	B-12C	220	.073			2.28	4.14	6.42	8.05	
	6110	Clay, till, or blasted rock, 3/4 C.Y. bucket	B-12F	80	.200			6.25	5.95	12.20	16.05	
	6120	1 C.Y. bucket	B-12A	95	.168			5.25	5.85	11.10	14.50	
	6130	1-1/2 C.Y. bucket	B-12B	130	.123			3.85	5.55	9.40	12	
	6140	2 C.Y. bucket	B-12C	175	.091			2.86	5.20	8.06	10.05	
	9010	For mobilization or demobilization, see div. 02305-250										
	9020	For dewatering, see div. 02240-500										
	9022	For larger structures, see Bulk Excavation, div. 02315-424										
	9024	For loading onto trucks, add									15%	
	9026	For hauling, see div. 02315-490										

2 SITE CONSTRUCTION

		02315	Excavation and Fill	CREW	DAILY OUTPUT	LABOR-HOURS	UNIT	MAT.	2005 BARE COSTS LABOR	EQUIP.	TOTAL	TOTAL INCL O&P	
462	9030		For sheeting or soldier beams/lagging, see 02250 & 02260 R02315 -400										462
	9040		For trench excavation of strip footings, see div. 02315-610										
490	0010		**HAULING,** excavated or borrow, loose cubic yards R02315 -400										490
	0012		no loading included, highway haulers										
	0020		6 C.Y. dump truck, 1/4 mile round trip, 5.0 loads/hr.	B-34A	195	.041	L.C.Y.		1.13	1.68	2.81	3.56	
	0030		1/2 mile round trip, 4.1 loads/hr.		160	.050			1.38	2.04	3.42	4.35	
	0040		1 mile round trip, 3.3 loads/hr.		130	.062			1.70	2.51	4.21	5.35	
	0100		2 mile round trip, 2.6 loads/hr.		100	.080			2.20	3.27	5.47	6.95	
	0150		3 mile round trip, 2.1 loads/hr.		80	.100			2.76	4.08	6.84	8.70	
	0200		4 mile round trip, 1.8 loads/hr.	↓	70	.114			3.15	4.67	7.82	9.95	
	0310		12 C.Y. dump truck, 1/4 mile round trip 3.7 loads/hr.	B-34B	288	.028			.77	1.66	2.43	2.99	
	0320		1/2 mile round trip, 3.2 loads/hr.		250	.032			.88	1.91	2.79	3.44	
	0330		1 mile round trip 2.7, loads/hr.		210	.038			1.05	2.27	3.32	4.10	
	0400		2 mile round trip, 2.2 loads/hr.		180	.044			1.22	2.65	3.87	4.78	
	0450		3 mile round trip, 1.9 loads/hr.		170	.047			1.30	2.80	4.10	5.05	
	0500		4 mile round trip, 1.6 loads/hr.		125	.064			1.76	3.81	5.57	6.90	
	0540		5 mile round trip, 1 load/hr.		78	.103			2.83	6.10	8.93	11	
	0550		10 mile round trip, 0.60 load/hr.		58	.138			3.80	8.20	12	14.85	
	0560		20 mile round trip, 0.4 load/hr.	↓	39	.205			5.65	12.20	17.85	22	
	0600		16.5 C.Y. dump trailer, 1 mile round trip, 2.6 loads/hr.	B-34C	280	.029			.79	1.45	2.24	2.79	
	0700		2 mile round trip, 2.1 loads/hr.		225	.036			.98	1.80	2.78	3.47	
	1000		3 mile round trip, 1.8 loads/hr.		193	.041			1.14	2.10	3.24	4.05	
	1100		4 mile round trip, 1.6 loads/hr.		172	.047			1.28	2.35	3.63	4.54	
	1110		5 mile round trip, 1 load/hr.		108	.074			2.04	3.75	5.79	7.25	
	1120		10 mile round trip, .60 load/hr.		80	.100			2.76	5.05	7.81	9.75	
	1130		20 mile round trip, .4 load/hr.	↓	54	.148			4.08	7.50	11.58	14.45	
	1150		20 C.Y. dump trailer, 1 mile round trip, 2.5 loads/hr.	B-34D	325	.025			.68	1.28	1.96	2.44	
	1200		2 mile round trip, 2 loads/hr.		260	.031			.85	1.61	2.46	3.06	
	1220		3 mile round trip, 1.7 loads/hr.		221	.036			1	1.89	2.89	3.60	
	1240		4 mile round trip, 1.5 loads/hr.		195	.041			1.13	2.14	3.27	4.07	
	1245		5 mile round trip, 1.1 loads/hr.		143	.056			1.54	2.92	4.46	5.55	
	1250		10 mile round trip, .75 load/hr.		110	.073			2	3.80	5.80	7.20	
	1255		20 mile round trip, .5 load/hr.	↓	78	.103	↓		2.83	5.35	8.18	10.20	
	1300		Hauling in medium traffic, add				C.Y.				20%	20%	
	1400		Heavy traffic, add				"				30%	30%	
	1600		Grading at dump, or embankment if required, by dozer	B-10B	1,000	.012	L.C.Y.		.38	.92	1.30	1.59	
	1800		Spotter at fill or cut, if required	1 Clab	8	1	Hr.		26.50		26.50	41.50	
510	0010		**FILL BY BORROW,** load, 1 mile haul, spread with dozer										510
	0020		for embankments	B-15	1,200	.023	L.C.Y.	6	.69	1.56	8.25	9.35	
	0100		Select fill for shoulders & embankments	"	1,200	.023	"	7.90	.69	1.56	10.15	11.40	
	0201		For hauling over 1 mile, add to above per C.Y., see div 02315-490				Mile				.69	.92	
520	0010		**FILL,** spread dumped material, no compaction										520
	0020		By dozer, no compaction	B-10B	1,000	.012	L.C.Y.		.38	.92	1.30	1.59	
	0100		By hand	1 Clab	12	.667	"		17.80		17.80	27.50	
	0500		Gravel fill, compacted, under floor slabs, 4" deep	B-37	10,000	.005	S.F.	.16	.13	.01	.30	.39	
	0600		6" deep		8,600	.006		.24	.16	.01	.41	.51	
	0700		9" deep		7,200	.007		.39	.19	.01	.59	.74	
	0800		12" deep		6,000	.008	↓	.55	.22	.02	.79	.98	
	1000		Alternate pricing method, 4" deep		120	.400	E.C.Y.	11.80	11.25	.90	23.95	31.50	
	1100		6" deep		160	.300		11.80	8.45	.67	20.92	27	
	1200		9" deep		200	.240		11.80	6.75	.54	19.09	24	
	1300		12" deep	↓	220	.218	↓	11.80	6.15	.49	18.44	23	
	1500		For fill under exterior paving, see division 02720-200										
610	0010		**EXCAVATING, TRENCH** or continuous footing, common earth										610
	0020		No sheeting or dewatering included										

Important: See the Reference Section for critical supporting data - Reference Nos., Crews, & City Cost Indexes

02315	Excavation and Fill	CREW	DAILY OUTPUT	LABOR-HOURS	UNIT	2005 BARE COSTS				TOTAL INCL O&P	
						MAT.	LABOR	EQUIP.	TOTAL		
610 0050	1' to 4' deep, 3/8 C.Y. tractor loader/backhoe	B-11C	150	.107	B.C.Y.		3.27	1.44	4.71	6.60	**610**
0060	1/2 C.Y. tractor loader/backhoe	B-11M	200	.080			2.45	1.28	3.73	5.15	
0090	4' to 6' deep, 1/2 C.Y. tractor loader/backhoe	"	200	.080			2.45	1.28	3.73	5.15	
0100	5/8 C.Y. hydraulic backhoe	B-12Q	250	.064			2	1.73	3.73	4.96	
0110	3/4 C.Y. hydraulic backhoe	B-12F	300	.053			1.67	1.58	3.25	4.29	
0300	1/2 C.Y. hydraulic excavator, truck mounted	B-12J	200	.080			2.50	4.19	6.69	8.45	
0500	6' to 10' deep, 3/4 C.Y. hydraulic backhoe, 6' to 10' deep	B-12F	225	.071			2.23	2.11	4.34	5.70	
0510	1 C.Y. hydraulic backhoe	B-12A	400	.040			1.25	1.39	2.64	3.44	
0600	1 C.Y. hydraulic excavator, truck mounted	B-12K	400	.040			1.25	2.43	3.68	4.58	
0610	1-1/2 C.Y. hydraulic backhoe	B-12B	600	.027			.83	1.20	2.03	2.60	
0900	10' to 14' deep, 3/4 C.Y. hydraulic backhoe	B-12F	200	.080			2.50	2.37	4.87	6.45	
0910	1 C.Y. hydraulic backhoe	B-12A	360	.044			1.39	1.55	2.94	3.83	
1000	1-1/2 C.Y. hydraulic backhoe	B-12B	540	.030			.93	1.33	2.26	2.89	
1300	14' to 20' deep, 1 C.Y. hydraulic backhoe	B-12A	320	.050			1.57	1.74	3.31	4.31	
1310	1-1/2 C.Y. hydraulic backhoe	B-12B	480	.033			1.04	1.50	2.54	3.24	
1320	2-1/2 C.Y. hydraulic backhoe	B-12S	850	.019			.59	1.43	2.02	2.47	
1400	By hand with pick and shovel 2' to 6' deep, light soil	1 Clab	8	1			26.50		26.50	41.50	
1500	Heavy soil	"	4	2	▼		53.50		53.50	83	
1700	For tamping backfilled trenches, air tamp, add	A-1G	100	.080	E.C.Y.		2.14	.35	2.49	3.70	
1900	Vibrating plate, add	B-18	180	.133	"		3.65	.20	3.85	5.95	
2100	Trim sides and bottom for concrete pours, common earth		1,500	.016	S.F.		.44	.02	.46	.71	
2300	Hardpan	▼	600	.040	"		1.09	.06	1.15	1.77	
2400	Pier and spread footing excavation, add to above				B.C.Y.				30%	30%	
3000	Backfill trench, F.E. loader, wheel mtd., 1 C.Y. bucket										
3020	Minimal haul	B-10R	400	.030	L.C.Y.		.96	.48	1.44	1.99	
3040	100' haul	"	200	.060			1.92	.97	2.89	3.99	
3080	2-1/4 C.Y. bucket, minimum haul	B-10T	600	.020			.64	.51	1.15	1.53	
3090	100' haul	"	300	.040	▼		1.28	1.02	2.30	3.07	
620 0010	**EXCAVATING, UTILITY TRENCH** Common earth										**620**
0050	Trenching with chain trencher, 12 H.P., operator walking										
0100	4" wide trench, 12" deep	B-53	800	.010	L.F.		.33	.06	.39	.57	
0150	18" deep		750	.011			.35	.06	.41	.60	
0200	24" deep		700	.011			.38	.07	.45	.65	
0300	6" wide trench, 12" deep		650	.012			.41	.07	.48	.69	
0350	18" deep		600	.013			.44	.08	.52	.75	
0400	24" deep		550	.015			.48	.09	.57	.82	
0450	36" deep		450	.018			.59	.11	.70	1.01	
0600	8" wide trench, 12" deep		475	.017			.56	.10	.66	.95	
0650	18" deep		400	.020			.66	.12	.78	1.13	
0700	24" deep		350	.023			.76	.14	.90	1.29	
0750	36" deep	▼	300	.027	▼		.88	.16	1.04	1.51	
1000	Backfill by hand including compaction, add										
1050	4" wide trench, 12" deep	A-1G	800	.010	L.F.		.27	.04	.31	.47	
1100	18" deep		530	.015			.40	.07	.47	.70	
1150	24" deep		400	.020			.53	.09	.62	.93	
1300	6" wide trench, 12" deep		540	.015			.40	.06	.46	.69	
1350	18" deep		405	.020			.53	.09	.62	.91	
1400	24" deep		270	.030			.79	.13	.92	1.37	
1450	36" deep		180	.044			1.19	.19	1.38	2.06	
1600	8" wide trench, 12" deep		400	.020			.53	.09	.62	.93	
1650	18" deep		265	.030			.81	.13	.94	1.39	
1700	24" deep		200	.040			1.07	.17	1.24	1.85	
1750	36" deep	▼	135	.059	▼		1.58	.26	1.84	2.74	
2000	Chain trencher, 40 H.P. operator riding										
2050	6" wide trench and backfill, 12" deep	B-54	1,200	.007	L.F.		.22	.18	.40	.53	
2100	18" deep	▼	1,000	.008			.26	.21	.47	.63	

SITE CONSTRUCTION **2**

2 SITE CONSTRUCTION

02315 | Excavation and Fill

			CREW	DAILY OUTPUT	LABOR-HOURS	UNIT	MAT.	LABOR	EQUIP.	TOTAL	TOTAL INCL O&P	
620	2150	24" deep	B-54	975	.008	L.F.		.27	.22	.49	.65	**620**
	2200	36" deep		900	.009			.29	.24	.53	.70	
	2250	48" deep		750	.011			.35	.28	.63	.84	
	2300	60" deep		650	.012			.41	.33	.74	.97	
	2400	8" wide trench and backfill, 12" deep		1,000	.008			.26	.21	.47	.63	
	2450	18" deep		950	.008			.28	.22	.50	.67	
	2500	24" deep		900	.009			.29	.24	.53	.70	
	2550	36" deep		800	.010			.33	.27	.60	.79	
	2600	48" deep		650	.012			.41	.33	.74	.97	
	2700	12" wide trench and backfill, 12" deep		975	.008			.27	.22	.49	.65	
	2750	18" deep		860	.009			.31	.25	.56	.73	
	2800	24" deep		800	.010			.33	.27	.60	.79	
	2850	36" deep		725	.011			.36	.29	.65	.87	
	3000	16" wide trench and backfill, 12" deep		835	.010			.32	.26	.58	.76	
	3050	18" deep		750	.011			.35	.28	.63	.84	
	3100	24" deep	↓	700	.011	↓		.38	.31	.69	.91	
	3200	Compaction with vibratory plate, add								50%	50%	
	5100	Hand excavate and trim for pipe bells after trench excavation										
	5200	8" pipe	1 Clab	155	.052	L.F.		1.38		1.38	2.14	
	5300	18" pipe	"	130	.062	"		1.64		1.64	2.56	
640	0010	**UTILITY BEDDING** For pipe and conduit, not incl. compaction										**640**
	0050	Crushed or screened bank run gravel	B-6	150	.160	L.C.Y.	20	4.61	1.44	26.05	30.50	
	0100	Crushed stone 3/4" to 1/2"		150	.160		27.50	4.61	1.44	33.55	38.50	
	0200	Sand, dead or bank	↓	150	.160	↓	4.13	4.61	1.44	10.18	13.20	
	0500	Compacting bedding in trench	A-1D	90	.089	E.C.Y.		2.37	.28	2.65	4	
	0600	If material source exceeds 2 miles, add for extra mileage.				L.C.Y.						
	0610	See 02315-490-0010 for hauling mileage add.				"						

02325 | Dredging

			CREW	DAILY OUTPUT	LABOR-HOURS	UNIT	MAT.	LABOR	EQUIP.	TOTAL	TOTAL INCL O&P	
250	0010	**DREDGING**										**250**
	0020	Dredging mobilization and demobilization., add to below, minimum	B-8	.53	120	Total		3,575	4,450	8,025	10,400	
	0100	Maximum	"	.10	640	"		18,900	23,600	42,500	55,000	
	0300	Barge mounted clamshell excavation into scows,										
	0310	Dumped 20 miles at sea, minimum	B-57	310	.155	B.C.Y.		4.67	4.14	8.81	11.70	
	0400	Maximum	"	213	.225	"		6.80	6.05	12.85	17.05	
	0500	Barge mounted dragline or clamshell, hopper dumped,										
	0510	pumped 1000' to shore dump, minimum	B-57	340	.141	B.C.Y.		4.26	3.78	8.04	10.65	
	0525	All pumping uses 2000 gallons of water per cubic yard										
	0600	Maximum	B-57	243	.198	B.C.Y.		5.95	5.30	11.25	14.90	
	1000	Hydraulic method, pumped 1000' to shore dump, minimum		460	.104			3.15	2.79	5.94	7.90	
	1100	Maximum		310	.155			4.67	4.14	8.81	11.70	
	1400	Into scows dumped 20 miles, minimum		425	.113			3.41	3.02	6.43	8.50	
	1500	Maximum	↓	243	.198	↓		5.95	5.30	11.25	14.90	
	1600	For inland rivers and canals in South, deduct								30%	30%	

02340 | Soil Stabilization

			CREW	DAILY OUTPUT	LABOR-HOURS	UNIT	MAT.	LABOR	EQUIP.	TOTAL	TOTAL INCL O&P	
700	0010	**CALCIUM CHLORIDE**										**700**
	0020	Calcium chloride delivered, 100 LB bags, truckload lots				Ton	460			460	505	
	0030	Solution, 4 lb. flake per gallon, tank truck delivery				Gal.	.98			.98	1.08	

02360 | Soil Treatment

			CREW	DAILY OUTPUT	LABOR-HOURS	UNIT	MAT.	LABOR	EQUIP.	TOTAL	TOTAL INCL O&P	
200	0010	**TERMITE PRETREATMENT**										**200**
	0020	Slab and walls, residential	1 Skwk	1,200	.007	SF Flr.	.28	.23		.51	.67	

Important: See the Reference Section for critical supporting data - Reference Nos., Crews, & City Cost Indexes

02300 | Earthwork

02360 | Soil Treatment

			CREW	DAILY OUTPUT	LABOR-HOURS	UNIT	2005 BARE COSTS				TOTAL INCL O&P	
							MAT.	LABOR	EQUIP.	TOTAL		
200	0100	Commercial, minimum	1 Skwk	2,496	.003	SF Flr.	.30	.11		.41	.50	200
	0200	Maximum		1,645	.005	↓	.45	.17		.62	.76	
	0400	Insecticides for termite control, minimum		14.20	.563	Gal.	11.60	19.65		31.25	43.50	
	0500	Maximum	↓	11	.727	"	19.85	25.50		45.35	61.50	

02370 | Erosion & Sedimentation Control

			CREW	DAILY OUTPUT	LABOR-HOURS	UNIT	MAT.	LABOR	EQUIP.	TOTAL	TOTAL INCL O&P	
450	0010	**RIP-RAP & ROCK LINING**, Random, broken stone										450
	0100	Machine placed for slope protection	B-12G	62	.258	L.C.Y.	25	8.10	8.75	41.85	49.50	
	0110	3/8 to 1/4 C.Y. pieces, grouted	B-13	80	.700	S.Y.	36.50	20	7.70	64.20	79.50	
	0200	18" minimum thickness, not grouted	"	53	1.057	"	15.55	30.50	11.65	57.70	77	
	0300	Dumped, 50 lb. average	B-11A	800	.020	Ton	17.90	.61	1.15	19.66	22	
	0350	100 lb. average		700	.023		25.50	.70	1.31	27.51	30.50	
	0370	300 lb. average	↓	600	.027	↓	30	.82	1.53	32.35	36	
	0400	Gabions, galvanized steel mesh mats or boxes, stone filled, 6" deep	B-13	200	.280	S.Y.	16.90	8.05	3.08	28.03	34.50	
	0500	9" deep		163	.344		26	9.90	3.78	39.68	48	
	0600	12" deep		153	.366		27.50	10.55	4.03	42.08	50.50	
	0700	18" deep		102	.549		35	15.80	6.05	56.85	69.50	
	0800	36" deep	↓	60	.933	↓	59.50	27	10.30	96.80	118	
700	0010	**SYNTHETIC EROSION CONTROL**										700
	0020	Jute mesh, 100 SY per roll, 4' wide, stapled	B-80A	2,400	.010	S.Y.	.69	.27	.07	1.03	1.26	
	0100	Plastic netting, stapled, 2" x 1" mesh, 20 mil	B-1	2,500	.010		.63	.26		.89	1.10	
	0200	Polypropylene mesh, stapled, 6.5 oz./S.Y.		2,500	.010		1.30	.26		1.56	1.84	
	0300	Tobacco netting, or jute mesh #2, stapled	↓	2,500	.010	↓	.07	.26		.33	.49	
	1000	Silt fence, polypropylene, 3' high, ideal conditions	2 Clab	1,600	.010	L.F.	.32	.27		.59	.77	
	1100	Adverse conditions	"	950	.017	"	.32	.45		.77	1.05	
	1200	Place and remove hay bales	A-2	3	8	Ton	52	214	43	309	435	
	1250	Hay bales, staked	"	2,500	.010	L.F.	2.08	.26	.05	2.39	2.75	

02390 | Shore Protect/Mooring Structures

			CREW	DAILY OUTPUT	LABOR-HOURS	UNIT	MAT.	LABOR	EQUIP.	TOTAL	TOTAL INCL O&P	
340	0010	**JETTIES, DOCKS** Floating, recreational, prefabricated galvanized steel with										340
	0020	polyethylene encased polystyrene, no pilings included	F-3	330	.121	S.F.	27	4.19	1.83	33.02	38	
	0200	Pile supported, shore constructed, bare, 3" decking		130	.308		18.45	10.65	4.64	33.74	42	
	0250	4" decking		120	.333		17.80	11.55	5	34.35	43	
	0400	Floating, small boat, prefab, no shore facilities, minimum		250	.160		9.10	5.55	2.41	17.06	21	
	0500	Maximum		150	.267	↓	31	9.20	4.02	44.22	52.50	
	0700	Per slip, minimum (180 S.F. each)		1.59	25.157	Ea.	1,950	870	380	3,200	3,925	
	0800	Maximum	↓	1.40	28.571	"	5,800	990	430	7,220	8,375	

02400 | Tunneling, Boring & Jacking

02441 | Microtunneling

			CREW	DAILY OUTPUT	LABOR-HOURS	UNIT	MAT.	LABOR	EQUIP.	TOTAL	TOTAL INCL O&P	
400	0010	**MICROTUNNELING** Not including excavation, backfill, shoring,										400
	0020	or dewatering, average 50'/day, slurry method										
	0100	24" to 48" outside diameter, minimum				L.F.					640	
	0110	Adverse conditions, add				%					50%	
	1000	Rent microtunneling machine, average monthly lease				Month					85,500	
	1010	Operating technician				Day					640	
	1100	Mobilization and demobilization, minimum				Job					42,800	
	1110	Maximum				"					430,000	

02445 | Boring or Jacking Conduits

		CREW	DAILY OUTPUT	LABOR-HOURS	UNIT	2005 BARE COSTS				TOTAL INCL O&P	
						MAT.	LABOR	EQUIP.	TOTAL		
300	0010	**HORIZONTAL BORING** Casing only, 100' minimum,									300
	0020	not incl. jacking pits or dewatering									
	0100	Roadwork, 1/2" thick wall, 24" diameter casing	B-42	20	3.200	L.F.	67	96	53.50	216.50	284
	0200	36" diameter		16	4		107	120	67	294	380
	0300	48" diameter		15	4.267		157	128	71.50	356.50	455
	0500	Railroad work, 24" diameter		15	4.267		67	128	71.50	266.50	355
	0600	36" diameter		14	4.571		107	137	76.50	320.50	415
	0700	48" diameter		12	5.333		157	160	89	406	520
	0900	For ledge, add								155	190

02450 | Foundation & Load Bearing Elements

02455 | Driven Piles

		CREW	DAILY OUTPUT	LABOR-HOURS	UNIT	2005 BARE COSTS				TOTAL INCL O&P	
						MAT.	LABOR	EQUIP.	TOTAL		
100	0010	**CAST IN PLACE CONCRETE PILES**, 200 piles, 60' long									100
	0020	unless specified otherwise, not incl. pile caps or mobilization									
	0050	Cast in place augered piles, no casing or reinforcing									
	0060	8" diameter	B-43	540	.089	V.L.F.	3.02	2.59	5.90	11.51	13.80
	0065	10" diameter		480	.100		4.80	2.91	6.60	14.31	17.10
	0070	12" diameter		420	.114		6.75	3.33	7.55	17.63	21
	0075	14" diameter		360	.133		9.10	3.88	8.85	21.83	25.50
	0080	16" diameter		300	.160		12.30	4.66	10.60	27.56	32.50
	0085	18" diameter		240	.200		15.20	5.85	13.25	34.30	40.50
	0100	Cast in place, thin wall shell pile, straight sided,									
	0110	not incl. reinforcing, 8" diam., 16 ga., 5.8 lb./L.F.	B-19	700	.091	V.L.F.	5.65	3.09	2.19	10.93	13.60
	0200	10" diameter, 16 ga. corrugated, 7.3 lb./L.F.		650	.098		7.40	3.33	2.35	13.08	16.05
	0300	12" diameter, 16 ga. corrugated, 8.7 lb./L.F.		600	.107		9.60	3.61	2.55	15.76	19.15
	0400	14" diameter, 16 ga. corrugated, 10.0 lb./L.F.		550	.116		11.30	3.93	2.78	18.01	22
	0500	16" diameter, 16 ga. corrugated, 11.6 lb./L.F.		500	.128		13.85	4.33	3.06	21.24	25.50
	0800	Cast in place friction pile, 50' long, fluted,									
	0810	tapered steel, 4000 psi concrete, no reinforcing									
	0900	12" diameter, 7 ga.	B-19	600	.107	V.L.F.	17	3.61	2.55	23.16	27.50
	1000	14" diameter, 7 ga.		560	.114		18.50	3.86	2.73	25.09	29.50
	1100	16" diameter, 7 ga.		520	.123		21.50	4.16	2.94	28.60	34
	1200	18" diameter, 7 ga.		480	.133		25.50	4.51	3.19	33.20	38.50
	1300	End bearing, fluted, constant diameter,									
	1320	4000 psi concrete, no reinforcing									
	1340	12" diameter, 7 ga.	B-19	600	.107	V.L.F.	17.75	3.61	2.55	23.91	28
	1360	14" diameter, 7 ga.		560	.114		22	3.86	2.73	28.59	33.50
	1380	16" diameter, 7 ga.		520	.123		25.50	4.16	2.94	32.60	38.50
	1400	18" diameter, 7 ga.		480	.133		28.50	4.51	3.19	36.20	41.50
	1500	For reinforcing steel, add				Lb.	.80			.80	.88
	1700	For ball or pedestal end, add	B-19	11	5.818	C.Y.	91	197	139	427	570
	1900	For lengths above 60', concrete, add	"	11	5.818	"	95	197	139	431	570
	2000	For steel thin shell, pipe only				Lb.	.69			.69	.76
450	0010	**PRESTRESSED CONCRETE PILES**, 200 piles									450
	0020	unless specified otherwise, not incl. pile caps or mobilization									
	2200	Precast, prestressed, 50' long, 12" diam., 2-3/8" wall	B-19	720	.089	V.L.F.	11.95	3	2.13	17.08	20.50
	2300	14" diameter, 2-1/2" wall		680	.094		15.75	3.18	2.25	21.18	25

Important: See the Reference Section for critical supporting data - Reference Nos., Crews, & City Cost Indexes

			CREW	DAILY OUTPUT	LABOR-HOURS	UNIT	2005 BARE COSTS				TOTAL INCL O&P
		02455 | Driven Piles					MAT.	LABOR	EQUIP.	TOTAL	
450	2500	16" diameter, 3" wall	B-19	640	.100	V.L.F.	22	3.38	2.39	27.77	32	**450**
	2600	18" diameter, 3" wall	B-19A	600	.107		27.50	3.61	3.25	34.36	39.50
	2800	20" diameter, 3-1/2" wall		560	.114		31.50	3.86	3.49	38.85	45
	2900	24" diameter, 3-1/2" wall	↓	520	.123		38.50	4.16	3.76	46.42	53.50
	3100	Precast, prestressed, 40' long, 10" thick, square	B-19	700	.091		8.65	3.09	2.19	13.93	16.90
	3200	12" thick, square		680	.094		10.90	3.18	2.25	16.33	19.55
	3400	14" thick, square		600	.107		12.90	3.61	2.55	19.06	22.50
	3500	Octagonal		640	.100		17.85	3.38	2.39	23.62	27.50
	3700	16" thick, square		560	.114		20.50	3.86	2.73	27.09	31.50
	3800	Octagonal	↓	600	.107		21.50	3.61	2.55	27.66	32
	4000	18" thick, square	B-19A	520	.123		24.50	4.16	3.76	32.42	38
	4100	Octagonal	B-19	560	.114		25.50	3.86	2.73	32.09	37
	4300	20" thick, square	B-19A	480	.133		30	4.51	4.07	38.58	44.50
	4400	Octagonal	B-19	520	.123		28	4.16	2.94	35.10	41
	4600	24" thick, square	B-19A	440	.145		43	4.92	4.44	52.36	60
	4700	Octagonal	B-19	480	.133		40.50	4.51	3.19	48.20	55
	4730	Precast, prestressed, 60' long, 10" thick, square		700	.091		9.10	3.09	2.19	14.38	17.35
	4740	12" thick, square (60' long)		680	.094		11.35	3.18	2.25	16.78	20
	4750	Mobilization for 10,000 L.F. pile job, add		3,300	.019			.66	.46	1.12	1.56
	4800	25,000 L.F. pile job, add	↓	8,500	.008	↓		.25	.18	.43	.61
600	0010	**STEEL PILES**, Not including mobilization or demobilization										**600**
	0100	Step tapered, round, concrete filled									
	0110	8" tip, 60 ton capacity, 30' depth	B-19	760	.084	V.L.F.	6.90	2.85	2.01	11.76	14.35
	0120	60' depth		740	.086		7.80	2.92	2.07	12.79	15.50
	0130	80' depth		700	.091		8.05	3.09	2.19	13.33	16.20
	0150	10" tip, 90 ton capacity, 30' depth		700	.091		8.50	3.09	2.19	13.78	16.70
	0160	60' depth		690	.093		8.75	3.14	2.22	14.11	17.05
	0170	80' depth		670	.096		9.40	3.23	2.28	14.91	18
	0190	12" tip, 120 ton capacity, 30' depth		660	.097		11.65	3.28	2.32	17.25	20.50
	0200	60' depth, 12" diameter		630	.102		11.70	3.43	2.43	17.56	21
	0210	80' depth		590	.108		10.35	3.67	2.59	16.61	20
	0250	"H" Sections, 50' long, HP8 x 36		640	.100		11.85	3.38	2.39	17.62	21
	0400	HP10 X 42		610	.105		13.85	3.55	2.51	19.91	23.50
	0500	HP10 X 57		610	.105		18.75	3.55	2.51	24.81	29
	0700	HP12 X 53	↓	590	.108		17.70	3.67	2.59	23.96	28
	0800	HP12 X 74	B-19A	590	.108		24.50	3.67	3.31	31.48	36.50
	1000	HP14 X 73		540	.119		24.50	4.01	3.62	32.13	37.50
	1100	HP14 X 89		540	.119		30	4.01	3.62	37.63	43.50
	1300	HP14 X 102		510	.125		34	4.24	3.83	42.07	48.50
	1400	HP14 X 117	↓	510	.125	↓	39	4.24	3.83	47.07	54
	1600	Splice on standard points, not in leads, 8" or 10"	1 Sswl	5	1.600	Ea.	77.50	61		138.50	195
	1700	12" or 14"		4	2		113	76.50		189.50	261
	1900	Heavy duty points, not in leads, 10" wide		4	2		120	76.50		196.50	269
	2100	14" wide	↓	3.50	2.286	↓	155	87		242	325
	2600	Pipe piles, 50' lg. 8" diam., 29 lb. per L.F., no concrete	B-19	500	.128	V.L.F.	12.80	4.33	3.06	20.19	24.50
	2700	Concrete filled		460	.139		13.60	4.70	3.33	21.63	26
	2900	10" diameter, 34 lb. per L.F., no concrete		500	.128		15.95	4.33	3.06	23.34	28
	3000	Concrete filled		450	.142		17.60	4.81	3.40	25.81	30.50
	3200	12" diameter, 44 lb. per L.F., no concrete		475	.135		19.75	4.55	3.22	27.52	32.50
	3300	Concrete filled		415	.154		20.50	5.20	3.69	29.39	35
	3500	14" diameter, 46 lb. per L.F., no concrete		430	.149		21	5.05	3.56	29.61	35.50
	3600	Concrete filled		355	.180		22.50	6.10	4.31	32.91	39.50
	3800	16" diameter, 52 lb. per L.F., no concrete		385	.166		23.50	5.60	3.97	33.07	39.50
	3900	Concrete filled		335	.191		26	6.45	4.57	37.02	44
	4100	18" diameter, 59 lb. per L.F., no concrete		355	.180		31	6.10	4.31	41.41	48.50
	4200	Concrete filled	↓	310	.206	↓	30.50	7	4.94	42.44	50.50

SITE CONSTRUCTION **2**

2 SITE CONSTRUCTION

02455 | Driven Piles

			CREW	DAILY OUTPUT	LABOR-HOURS	UNIT	2005 BARE COSTS				TOTAL INCL O&P
							MAT.	LABOR	EQUIP.	TOTAL	
600	4400	Splices for pipe piles, not in leads, 8" diameter	1 Sswl	4.67	1.713	Ea.	57.50	65.50		123	180
	4500	14" diameter		3.79	2.111		75.50	80.50		156	227
	4600	16" diameter		3.03	2.640		93	101		194	282
	4800	Points, standard, 8" diameter		4.61	1.735		64.50	66		130.50	189
	4900	14" diameter		4.05	1.975		89.50	75.50		165	234
	5000	16" diameter		3.37	2.374		109	90.50		199.50	282
	5200	Points, heavy duty, 10" diameter		2.89	2.768		44.50	106		150.50	238
	5300	14" or 16" diameter		2.02	3.960		71	151		222	350
	5500	For reinforcing steel, add		1,150	.007	Lb.	.53	.27		.80	1.06
	5700	For thick wall sections, add				"	.60			.60	.66
650	0010	**TIMBER PILES,** Friction or end bearing, not including									
	0050	mobilization or demobilization									
	0100	Untreated piles, up to 30' long, 12" butts, 8" points	B-19	625	.102	V.L.F.	6.30	3.46	2.45	12.21	15.15
	0200	30' to 39' long, 12" butts, 8" points		700	.091		6.30	3.09	2.19	11.58	14.30
	0300	40' to 49' long, 12" butts, 7" points		720	.089		6.30	3	2.13	11.43	14.10
	0400	50' to 59' long, 13"butts, 7" points		800	.080		6.40	2.70	1.91	11.01	13.45
	0500	60' to 69' long, 13" butts, 7" points		840	.076		7.20	2.58	1.82	11.60	14
	0600	70' to 80' long, 13" butts, 6" points		840	.076		8	2.58	1.82	12.40	14.90
	0800	Treated piles, 12 lb. per C.F.,									
	0810	friction or end bearing, ASTM class B									
	1000	Up to 30' long, 12" butts, 8" points	B-19	625	.102	V.L.F.	9.30	3.46	2.45	15.21	18.45
	1100	30' to 39' long, 12" butts, 8" points		700	.091		9.10	3.09	2.19	14.38	17.35
	1200	40' to 49' long, 12" butts, 7" points		720	.089		9.75	3	2.13	14.88	17.85
	1300	50' to 59' long, 13" butts, 7" points		800	.080		10.75	2.70	1.91	15.36	18.25
	1400	60' to 69' long, 13" butts, 6" points	B-19A	840	.076		14.70	2.58	2.32	19.60	23
	1500	70' to 80' long, 13" butts, 6" points	"	840	.076		18.45	2.58	2.32	23.35	27
	1600	Treated piles, C.C.A., 2.5# per C.F.									
	1610	8" butts, 10' long	B-19	400	.160	V.L.F.	6.80	5.40	3.83	16.03	20.50
	1620	11' to 16' long		500	.128		6.80	4.33	3.06	14.19	17.70
	1630	17' to 20' long		575	.111		6.80	3.76	2.66	13.22	16.40
	1640	10" butts, 10' to 16' long		500	.128		7.75	4.33	3.06	15.14	18.75
	1650	17' to 20' long		575	.111		7.75	3.76	2.66	14.17	17.45
	1660	21' to 40' long		700	.091		7.75	3.09	2.19	13.03	15.85
	1670	12" butts, 10' to 20' long		575	.111		8.35	3.76	2.66	14.77	18.15
	1680	21' to 35' long		650	.098		8.35	3.33	2.35	14.03	17.10
	1690	36' to 40' long		700	.091		8.35	3.09	2.19	13.63	16.55
	1695	14" butts. to 40' long		700	.091		11.95	3.09	2.19	17.23	20.50
	1700	Boot for pile tip, minimum	1 Pile	27	.296	Ea.	19.10	9.85		28.95	37.50
	1800	Maximum		21	.381		57	12.70		69.70	84
	2000	Point for pile tip, minimum		20	.400		19.05	13.30		32.35	43
	2100	Maximum		15	.533		68.50	17.75		86.25	105
	2300	Splice for piles over 50' long, minimum	B-46	35	1.371		47.50	41.50	.96	89.96	121
	2400	Maximum		20	2.400		57	73	1.68	131.68	182
	2600	Concrete encasement with wire mesh and tube		331	.145	V.L.F.	8.95	4.40	.10	13.45	17
	2700	Mobilization for 10,000 L.F. pile job, add	B-19	3,300	.019			.66	.46	1.12	1.56
	2800	25,000 L.F. pile job, add	"	8,500	.008			.25	.18	.43	.61
800	0010	**PILING SPECIAL COSTS**									
	0011	Pile caps, see Division 03310-240									
	0500	Cutoffs, concrete piles, plain	1 Pile	5.50	1.455	Ea.		48.50		48.50	80
	0600	With steel thin shell, add		38	.211			7		7	11.55
	0700	Steel pile or "H" piles		19	.421			14		14	23
	0800	Wood piles		38	.211			7		7	11.55
	0900	Pre-augering up to 30' deep, average soil, 24" diameter	B-43	180	.267	L.F.		7.75	17.65	25.40	31.50
	0920	36" diameter		115	.417			12.15	27.50	39.65	49
	0960	48" diameter		70	.686			19.95	45.50	65.45	80.50
	0980	60" diameter		50	.960			28	63.50	91.50	113

R02455 -900

SITE CONSTRUCTION 2

02455 | Driven Piles

		CREW	DAILY OUTPUT	LABOR-HOURS	UNIT	MAT.	LABOR	EQUIP.	TOTAL	TOTAL INCL O&P	
800											**800**
1000	Testing, any type piles, test load is twice the design load										
1050	50 ton design load, 100 ton test				Ea.				15,000	16,050	
1100	100 ton design load, 200 ton test								19,250	20,300	
1150	150 ton design load, 300 ton test								24,100	25,700	
1200	200 ton design load, 400 ton test								26,200	28,800	
1250	400 ton design load, 800 ton test								30,300	33,700	
1500	Wet conditions, soft damp ground										
1600	Requiring mats for crane, add								40%	40%	
1700	Barge mounted driving rig, add								30%	30%	

		CREW	DAILY OUTPUT	LABOR-HOURS	UNIT	MAT.	LABOR	EQUIP.	TOTAL	TOTAL INCL O&P	
900	0010	**MOBILIZATION**									**900**
0020	Set up & remove, air compressor, 600 C.F.M.	A-5	3.30	5.455	Ea.	146	9.75		155.75	237	
0100	1200 C.F.M.	"	2.20	8.182		219	14.65		233.65	355	
0200	Crane, with pile leads and pile hammer, 75 ton	B-19	.60	106		3,600	2,550		6,150	8,550	
0300	150 ton	"	.36	177		6,000	4,250		10,250	14,300	
0500	Drill rig, for caissons, to 36", minimum	B-43	2	24		700	1,600		2,300	2,825	
0600	Up to 84"	"	1	48		1,400	3,175		4,575	5,650	
0800	Auxiliary boiler, for steam small	A-5	1.66	10.843		290	19.40		309.40	470	
0900	Large	"	.83	21.687		580	39		619	945	
1100	Rule of thumb: complete pile driving set up, small	B-19	.45	142		4,800	3,400		8,200	11,400	
1200	Large	"	.27	237		8,000	5,675		13,675	19,000	

02465 | Bored Piles

		CREW	DAILY OUTPUT	LABOR-HOURS	UNIT	MAT.	LABOR	EQUIP.	TOTAL	TOTAL INCL O&P	
800	0010	**DRILLED CAISSONS** Incl. excav., concrete, 50 lbs. reinf.									**800**
0020	per C.Y., not incl. mobilization, boulder removal, disposal _(R02465 -600)_										
0100	Open style, machine drilled, to 50' deep, in stable ground, no										
0110	casings or ground water, 18" diam., 0.065 C.Y./L.F.	B-43	200	.240	V.L.F.	6.55	7	15.90	29.45	35.50	
0200	24" diameter, 0.116 C.Y./L.F.		190	.253		11.70	7.35	16.75	35.80	42.50	
0300	30" diameter, 0.182 C.Y./L.F.		150	.320		18.35	9.30	21	48.65	58	
0400	36" diameter, 0.262 C.Y./L.F.		125	.384		26.50	11.20	25.50	63.20	74	
0500	48" diameter, 0.465 C.Y./L.F.		100	.480		47	14	32	93	108	
0600	60" diameter, 0.727 C.Y./L.F.		90	.533		73.50	15.55	35.50	124.55	144	
0700	72" diameter, 1.05 C.Y./L.F.		80	.600		106	17.50	39.50	163	188	
0800	84" diameter, 1.43 C.Y./L.F.		75	.640		144	18.65	42.50	205.15	234	
1000	For bell excavation and concrete, add										
1020	4' bell diameter, 24" shaft, 0.444 C.Y.	B-43	20	2.400	Ea.	36	70	159	265	325	
1040	6' bell diameter, 30" shaft, 1.57 C.Y.		5.70	8.421		127	245	560	932	1,125	
1060	8' bell diameter, 36" shaft, 3.72 C.Y.		2.40	20		300	585	1,325	2,210	2,675	
1080	9' bell diameter, 48" shaft, 4.48 C.Y.		2	24		365	700	1,600	2,665	3,225	
1100	10' bell diameter, 60" shaft, 5.24 C.Y.		1.70	28.235		425	820	1,875	3,120	3,800	
1120	12' bell diameter, 72" shaft, 8.74 C.Y.		1	48		710	1,400	3,175	5,285	6,425	
1140	14' bell diameter, 84" shaft, 13.6 C.Y.		.70	68.571		1,100	2,000	4,550	7,650	9,275	
1200	Open style, machine drilled, to 50' deep, in wet ground, pulled										
1300	casing and pumping, 18" diameter, 0.065 C.Y./L.F.	B-48	160	.350	V.L.F.	6.55	10.40	21.50	38.45	46.50	
1400	24" diameter, 0.116 C.Y./L.F.		125	.448		11.70	13.30	27.50	52.50	63.50	
1500	30" diameter, 0.182 C.Y./L.F.		85	.659		18.35	19.55	40	77.90	94	
1600	36" diameter, 0.262 C.Y./L.F.		60	.933		26.50	27.50	57	111	134	
1700	48" diameter, 0.465 C.Y./L.F.	B-49	55	1.600		47	49.50	73.50	170	209	
1800	60" diameter, 0.727 C.Y./L.F.		35	2.514		73.50	78	115	266.50	330	
1900	72" diameter, 1.05 C.Y./L.F.		30	2.933		106	91	134	331	405	
2000	84" diameter, 1.43 C.Y./L.F.		25	3.520		144	109	161	414	505	
2100	For bell excavation and concrete, add										
2120	4' bell diameter, 24" shaft, 0.444 C.Y.	B-48	19.80	2.828	Ea.	36	84	173	293	360	
2140	6' bell diameter, 30" shaft, 1.57 C.Y.		5.70	9.825		127	292	600	1,019	1,250	
2160	8' bell diameter, 36" shaft, 3.72 C.Y.		2.40	23.333		300	695	1,425	2,420	2,950	

02465	Bored Piles	CREW	DAILY OUTPUT	LABOR-HOURS	UNIT	2005 BARE COSTS				TOTAL INCL O&P
						MAT.	LABOR	EQUIP.	TOTAL	
800											800
2180	9' bell diameter, 48" shaft, 4.48 C.Y.	R02465-600	B-49	3.30	26.667	Ea.	365	825	1,225	2,415	3,025
2200	10' bell diameter, 60" shaft, 5.24 C.Y.			2.80	31.429		425	975	1,450	2,850	3,550
2220	12' bell diameter, 72" shaft, 8.74 C.Y.			1.60	55		710	1,700	2,525	4,935	6,200
2240	14' bell diameter, 84" shaft, 13.6 C.Y.			1	88		1,100	2,725	4,025	7,850	9,875
2300	Open style, machine drilled, to 50' deep, in soft rocks and										
2400	medium hard shales, 18" diameter, 0.065 C.Y./L.F.		B-49	50	1.760	V.L.F.	6.55	54.50	80.50	141.55	181
2500	24" diameter, 0.116 C.Y./L.F.			30	2.933		11.70	91	134	236.70	300
2600	30" diameter, 0.182 C.Y./L.F.			20	4.400		18.35	136	202	356.35	455
2700	36" diameter, 0.262 C.Y./L.F.			15	5.867		26.50	182	269	477.50	605
2800	48" diameter, 0.465 C.Y./L.F.			10	8.800		47	272	405	724	915
2900	60" diameter, 0.727 C.Y./L.F.			7	12.571		73.50	390	575	1,038.50	1,325
3000	72" diameter, 1.05 C.Y./L.F.			6	14.667		106	455	670	1,231	1,550
3100	84" diameter, 1.43 C.Y./L.F.			5	17.600		144	545	805	1,494	1,900
3200	For bell excavation and concrete, add										
3220	4' bell diameter, 24" shaft, 0.444 C.Y.		B-49	10.90	8.073	Ea.	36	250	370	656	830
3240	6' bell diameter, 30" shaft, 1.57 C.Y.			3.10	28.387		127	880	1,300	2,307	2,925
3260	8' bell diameter, 36" shaft, 3.72 C.Y.			1.30	67.692		300	2,100	3,100	5,500	7,000
3280	9' bell diameter, 48" shaft, 4.48 C.Y.			1.10	80		365	2,475	3,675	6,515	8,275
3300	10' bell diameter, 60" shaft, 5.24 C.Y.			.90	97.778		425	3,025	4,475	7,925	10,100
3320	12' bell diameter, 72" shaft, 8.74 C.Y.			.60	146		710	4,550	6,725	11,985	15,200
3340	14' bell diameter, 84" shaft, 13.6 C.Y.			.40	220		1,100	6,800	10,100	18,000	22,900
3600	For rock excavation, sockets, add, minimum			120	.733	C.F.		22.50	33.50	56	72
3650	Average			95	.926			28.50	42.50	71	91
3700	Maximum			48	1.833			56.50	84	140.50	181
3900	For 50' to 100' deep, add					V.L.F.				7%	7%
4000	For 100' to 150' deep, add									25%	25%
4100	For 150' to 200' deep, add									30%	30%
4200	For casings left in place, add					Lb.	.67			.67	.74
4300	For other than 50 lb. reinf. per C.Y., add or deduct					"	.71			.71	.78
4400	For steel "I" beam cores, add		B-49	8.30	10.602	Ton	1,400	330	485	2,215	2,600
4500	Load and haul excess excavation, 2 miles		B-34B	178	.045	L.C.Y.		1.24	2.68	3.92	4.83
4600	For mobilization, 50 mile radius, rig to 36"		B-43	2	24	Ea.		700	1,600	2,300	2,825
4650	Rig to 84"		B-48	1.75	32			950	1,950	2,900	3,600
4700	For low headroom, add									50%	
5000	Bottom inspection		1 Skwk	1.20	6.667			232		232	360
950											950
0010	**DRILLED CONCRETE PIERS AND SHAFTS** or Displacement	R02465-800									
0100	Caissons, incl. mobilization and demobilization, up to 50 miles										
0200	Uncased shafts, 30 to 80 tons cap., 17" diam., 10' depth		B-44	88	.727	V.L.F.	16.20	24.50	10.75	51.45	68.50
0300	25' depth			165	.388		11.55	12.95	5.75	30.25	39.50
0400	80-150 ton capacity, 22" diameter, 10' depth			80	.800		20.50	26.50	11.85	58.85	78.50
0500	20' depth			130	.492		16.20	16.45	7.30	39.95	52.50
0700	Cased shafts, 10 to 30 ton capacity, 10-5/8" diam., 20' depth			175	.366		11.55	12.20	5.40	29.15	38.50
0800	30' depth			240	.267		10.80	8.90	3.94	23.64	30.50
0850	30 to 60 ton capacity, 12" diameter, 20' depth			160	.400		16.20	13.35	5.90	35.45	46
0900	40' depth			230	.278		12.45	9.30	4.12	25.87	33
1000	80 to 100 ton capacity, 16" diameter, 20' depth			160	.400		23	13.35	5.90	42.25	53.50
1100	40' depth			230	.278		21.50	9.30	4.12	34.92	43.50
1200	110 to 140 ton capacity, 17-5/8" diameter, 20' depth			160	.400		25	13.35	5.90	44.25	55.50
1300	40' depth			230	.278		23	9.30	4.12	36.42	45
1400	140 to 175 ton capacity, 19" diameter, 20' depth			130	.492		27	16.45	7.30	50.75	64
1500	40' depth			210	.305		25	10.15	4.51	39.66	49
1700	Over 30' long, L.F. cost tends to be lower										
1900	Maximum depth is about 90'										

2 SITE CONSTRUCTION

02510 | Water Distribution

							2005 BARE COSTS			
		CREW	DAILY OUTPUT	LABOR-HOURS	UNIT	MAT.	LABOR	EQUIP.	TOTAL	TOTAL INCL O&P
730 0010	**WATER SUPPLY, DUCTILE IRON PIPE** cement lined R02510-800									**730**
0020	Not including excavation or backfill									
2000	Pipe, class 50 water piping, 18' lengths									
2020	Mechanical joint, 4" diameter	B-21A	200	.200	L.F.	12.50	6.60	2.53	21.63	26.50
2040	6" diameter		160	.250		14.50	8.25	3.16	25.91	32
2060	8" diameter		133.33	.300		16.05	9.90	3.80	29.75	37
2080	10" diameter		114.29	.350		21.50	11.55	4.43	37.48	46.50
2100	12" diameter		105.26	.380		26.50	12.55	4.81	43.86	54
2120	14" diameter		100	.400		34	13.20	5.05	52.25	63
2140	16" diameter		72.73	.550		37	18.15	6.95	62.10	76
2160	18" diameter		68.97	.580		46.50	19.10	7.35	72.95	88.50
2170	20" diameter		57.14	.700		54.50	23	8.85	86.35	105
2180	24" diameter		47.06	.850		70	28	10.75	108.75	131
3000	Tyton, Push-on joint, 4" diameter		400	.100		7.35	3.30	1.27	11.92	14.50
3020	6" diameter		333.33	.120		8.40	3.96	1.52	13.88	16.85
3040	8" diameter		200	.200		11.50	6.60	2.53	20.63	25.50
3060	10" diameter		181.82	.220		18.15	7.25	2.78	28.18	34
3080	12" diameter		160	.250		19.15	8.25	3.16	30.56	37
3100	14" diameter		133.33	.300		21	9.90	3.80	34.70	42
3120	16" diameter		114.29	.350		29.50	11.55	4.43	45.48	55
3140	18" diameter		100	.400		32.50	13.20	5.05	50.75	61.50
3160	20" diameter		88.89	.450		36	14.85	5.70	56.55	68.50
3180	24" diameter	▼	76.92	.520	▼	47	17.15	6.60	70.75	85
8000	Fittings, mechanical joint									
8006	90° bend, 4" diameter	B-20A	16	2	Ea.	126	64.50		190.50	237
8020	6" diameter		12.80	2.500		169	80.50		249.50	310
8040	8" diameter	▼	10.67	2.999		245	96.50		341.50	420
8060	10" diameter	B-21A	11.43	3.500		480	115	44.50	639.50	755
8080	12" diameter		10.53	3.799		495	125	48	668	790
8100	14" diameter		10	4		685	132	50.50	867.50	1,000
8120	16" diameter		7.27	5.502		735	181	69.50	985.50	1,175
8140	18" diameter		6.90	5.797		1,000	191	73.50	1,264.50	1,475
8160	20" diameter		5.71	7.005		2,550	231	88.50	2,869.50	3,250
8180	24" diameter	▼	4.70	8.511		3,150	281	108	3,539	4,000
8200	Wye or tee, 4" diameter	B-20A	10.67	2.999		153	96.50		249.50	315
8220	6" diameter		8.53	3.751		192	121		313	395
8240	8" diameter	▼	7.11	4.501		284	145		429	535
8260	10" diameter	B-21A	7.62	5.249		490	173	66.50	729.50	870
8280	12" diameter		7.02	5.698		610	188	72	870	1,050
8300	14" diameter		6.67	5.997		1,250	198	76	1,524	1,750
8320	16" diameter		4.85	8.247		1,300	272	104	1,676	1,950
8340	18" diameter		4.60	8.696		2,650	287	110	3,047	3,450
8360	20" diameter		3.81	10.499		3,375	345	133	3,853	4,375
8380	24" diameter	▼	3.14	12.739		3,850	420	161	4,431	5,075
8450	Decreaser, 6" x 4" diameter	B-20A	14.22	2.250		104	72.50		176.50	226
8460	8" x 6" diameter	"	11.64	2.749		213	88.50		301.50	370
8470	10" x 6 " diameter	B-21A	13.33	3.001		217	99	38	354	430
8480	12" x 6" diameter		12.70	3.150		245	104	40	389	470
8490	16" x 6" diameter		10	4		885	132	50.50	1,067.50	1,225
8500	20" x 6" diameter	▼	8.42	4.751	▼	1,400	157	60	1,617	1,850
8550	Butterfly valves with boxes, cast iron									
8560	4" diameter	B-20	6	4	Ea.	760	120		880	1,025
8570	6" diameter	"	5	4.800		460	144		604	730
8580	8" diameter	B-21	4	7		760	216	39.50	1,015.50	1,225
8590	10" diameter		3.50	8		1,050	247	45.50	1,342.50	1,575
8600	12" diameter	▼	3	9.333	▼	1,350	288	53	1,691	1,975

SITE CONSTRUCTION 2

For expanded coverage of these items see *Means Heavy Construction Cost Data 2005*

02510 | Water Distribution

			CREW	DAILY OUTPUT	LABOR-HOURS	UNIT	MAT.	LABOR	EQUIP.	TOTAL	TOTAL INCL O&P	
							2005 BARE COSTS					
730	8610	14" diameter	B-21	2	14	Ea.	1,875	435	79	2,389	2,800	**730**
	8620	16" diameter	↓	2	14	↓	2,400	435	79	2,914	3,400	
750	0010	**WATER SUPPLY, POLYVINYL CHLORIDE PIPE**										**750**
	0020	Not including excavation or backfill, unless specified										
	2100	AWWA Class 160, S.D.R. 26, 1-1/2" diameter	B-20	750	.032	L.F.	.45	.96		1.41	2	
	2120	2" diameter		686	.035		1.09	1.05		2.14	2.84	
	2140	2-1/2" diameter		500	.048		1.62	1.44		3.06	4.03	
	2160	3" diameter		430	.056		2.32	1.68		4	5.15	
	2180	4" diameter		375	.064		3.78	1.93		5.71	7.15	
	2200	6" diameter		316	.076		8.15	2.28		10.43	12.50	
	2210	8" diameter	↓	260	.092	↓	13.80	2.78		16.58	19.50	
	3010	AWWA C905, PR 100, DR 41										
	3030	14" diameter	B-20A	213	.150	L.F.	12.60	4.84		17.44	21.50	
	3040	16" diameter		200	.160		16.70	5.15		21.85	26	
	3050	18" diameter		160	.200		21	6.45		27.45	33	
	3060	20" diameter		133	.241		26	7.75		33.75	41	
	3070	24" diameter		107	.299		36.50	9.65		46.15	55	
	3080	30" diameter		80	.400		56.50	12.90		69.40	82	
	3090	36" diameter		80	.400		80	12.90		92.90	108	
	3100	42" diameter		60	.533		109	17.20		126.20	146	
	3200	48" diameter		60	.533		143	17.20		160.20	183	
	4520	Class 150, SDR 18, AWWA C900, 4"		380	.084		2.38	2.71		5.09	6.75	
	4540	8"		264	.121		8.15	3.91		12.06	14.90	
	4550	10"		220	.145		12.30	4.69		16.99	20.50	
	4560	12"	↓	186	.172	↓	17.40	5.55		22.95	27.50	
	8000	Fittings with rubber gasket										
	8003	Class 150, D.R. 18										
	8006	90° Bend , 4" diameter	B-20	100	.240	Ea.	39.50	7.20		46.70	55	
	8020	6" diameter		90	.267		70	8		78	89.50	
	8040	8" diameter		80	.300		136	9		145	163	
	8060	10" diameter		50	.480		235	14.45		249.45	282	
	8080	12" diameter		30	.800		300	24		324	370	
	8100	Tee, 4" diameter		90	.267		54.50	8		62.50	72	
	8120	6" diameter		80	.300		121	9		130	147	
	8140	8" diameter		70	.343		173	10.30		183.30	206	
	8160	10" diameter		40	.600		196	18.05		214.05	243	
	8180	12" diameter		20	1.200		420	36		456	515	
	8200	45° Bend, 4" diameter		100	.240		38.50	7.20		45.70	54	
	8220	6" diameter		90	.267		68.50	8		76.50	87.50	
	8240	8" diameter		50	.480		130	14.45		144.45	166	
	8260	10" diameter		50	.480		199	14.45		213.45	241	
	8280	12" diameter		30	.800		258	24		282	320	
	8300	Reducing tee 6"x4"		100	.240		118	7.20		125.20	141	
	8320	8" x 6"		90	.267		210	8		218	244	
	8330	10" x 6"		90	.267		270	8		278	310	
	8340	10"x8"		90	.267		281	8		289	325	
	8350	12" x 6"		90	.267		320	8		328	365	
	8360	12" x 8"		90	.267		335	8		343	385	
	8400	Tapped service tee (threaded type) 6" x 6" x 3/4		100	.240		42	7.20		49.20	57.50	
	8420	6" x 6" x 3/4"		90	.267		42	8		50	58.50	
	8430	6" x 6" x 1"		90	.267		42	8		50	58.50	
	8440	6" x 6" x 1 1/2"		90	.267		42	8		50	58.50	
	8450	6" x 6" x 2"		90	.267		42	8		50	58.50	
	8460	8" x 8" x 3/4"		90	.267		177	8		185	208	
	8470	8" x 8" x 1"		90	.267		177	8		185	208	
	8480	8" x 8" x 1 1/2"	↓	90	.267		177	8		185	208	

R02510 -800

SITE CONSTRUCTION 2

SITE CONSTRUCTION 2

02510 | Water Distribution

		CREW	DAILY OUTPUT	LABOR-HOURS	UNIT	2005 BARE COSTS MAT.	LABOR	EQUIP.	TOTAL	TOTAL INCL O&P	
750	8490	8" x 8" x 2"	B-20	90	.267	Ea.	177	8		185	208	750
	8500	Repair coupling 4"		100	.240		25	7.20		32.20	39	
	8520	6" diameter		90	.267		38.50	8		46.50	54.50	
	8540	8" diameter		50	.480		84.50	14.45		98.95	116	
	8560	10" diameter		50	.480		143	14.45		157.45	180	
	8580	12" diameter		50	.480		209	14.45		223.45	253	
	8600	Plug end 4"		100	.240		21	7.20		28.20	35	
	8620	6" diameter		90	.267		38	8		46	54.50	
	8640	8" diameter		50	.480		65	14.45		79.45	94	
	8660	10" diameter		50	.480		68.50	14.45		82.95	97.50	
	8680	12" diameter	▼	50	.480	▼	84.50	14.45		98.95	116	

02520 | Wells

		CREW	DAILY OUTPUT	LABOR-HOURS	UNIT	MAT.	LABOR	EQUIP.	TOTAL	TOTAL INCL O&P	
510	0010	**WELLS & ACCESSORIES**, domestic									510
	0100	Drilled, 4" to 6" diameter	B-23	120	.333	L.F.		9.05	28	37.05	45	
	0200	8" diameter	"	95.20	.420	"		11.40	35	46.40	56.50	
	0400	Gravel pack well, 40' deep, incl. gravel & casing, complete										
	0500	24" diameter casing x 18" diameter screen	B-23	.13	307	Total	22,900	8,350	25,800	57,050	66,500	
	0600	36" diameter casing x 18" diameter screen		.12	333	"	24,700	9,025	28,000	61,725	72,000	
	0800	Observation wells, 1-1/4" riser pipe	▼	163	.245	V.L.F.	12.60	6.65	20.50	39.75	47	
	0900	For flush Buffalo roadway box, add	1 Skwk	16.60	.482	Ea.	34.50	16.80		51.30	64	
	1200	Test well, 2-1/2" diameter, up to 50' deep (15 to 50 GPM)	B-23	1.51	26.490	"	515	720	2,225	3,460	4,150	
	1300	Over 50' deep, add	"	121.80	.328	L.F.	13.75	8.90	27.50	50.15	59.50	
	1500	Pumps, installed in wells to 100' deep, 4" submersible										
	1510	1/2 H.P.	Q-1	3.22	4.969	Ea.	330	183		513	640	
	1520	3/4 H.P.		2.66	6.015		390	221		611	765	
	1600	1 H.P.	▼	2.29	6.987		415	257		672	840	
	1700	1-1/2 H.P.	Q-22	1.60	10		1,075	370	375	1,820	2,150	
	1800	2 H.P.		1.33	12.030		1,250	440	455	2,145	2,550	
	1900	3 H.P.		1.14	14.035		1,425	515	530	2,470	2,925	
	2000	5 H.P.		1.14	14.035		1,950	515	530	2,995	3,500	
	3000	Pump, 6" submersible, 25' to 150' deep, 25 H.P., 249 to 297 GPM		.89	17.978	▼	4,250	660	675	5,585	6,425	
	3100	25' to 500' deep, 30 H.P., 100 to 300 GPM		.73	21.918		4,850	805	825	6,480	7,425	
	8000	Steel well casing	B-23A	3,020	.008	Lb.	.59	.24	1.08	1.91	2.21	
	9950	See div. 02240-900 for wellpoints										
	9960	See div. 02240-700 for drainage wells										
520	0010	**WATER SUPPLY WELLS, PUMPS** with pressure control										520
	1000	Deep well, jet, 42 gal. galvanized tank										
	1040	3/4 HP	1 Plum	.80	10	Ea.	585	410		995	1,250	
	3000	Shallow well, jet, 30 gal. galvanized tank										
	3040	1/2 HP	1 Plum	2	4	Ea.	370	163		533	655	

02530 | Sanitary Sewerage

		CREW	DAILY OUTPUT	LABOR-HOURS	UNIT	MAT.	LABOR	EQUIP.	TOTAL	TOTAL INCL O&P	
730	0010	**SEWAGE COLLECTION, CONCRETE PIPE**										730
	0020	See 02630-530 for sewage/drainage collection, concrete pipe										
780	0010	**SEWAGE COLLECTION, POLYVINYL CHLORIDE PIPE**										780
	0020	Not including excavation or backfill										
	2000	10' lengths, S.D.R. 35, B&S, 4" diameter	B-20	375	.064	L.F.	1.92	1.93		3.85	5.10	
	2040	6" diameter		350	.069		3.44	2.06		5.50	7	
	2080	8" diameter	▼	335	.072		5.80	2.15		7.95	9.75	
	2120	10" diameter	B-21	330	.085		8.75	2.62	.48	11.85	14.25	
	2160	12" diameter		320	.087		9.75	2.70	.50	12.95	15.50	
	2200	15" diameter	▼	190	.147	▼	14.75	4.56	.83	20.14	24	

SITE CONSTRUCTION · 2

02540 | Septic Tank Systems

		CREW	DAILY OUTPUT	LABOR-HOURS	UNIT	2005 BARE COSTS				TOTAL INCL O&P
						MAT.	LABOR	EQUIP.	TOTAL	
400	0010 **SEPTIC TANKS**									**400**
	0015 Septic tanks, not incl exc or piping, precast, 1,000 gal	B-21	8	3.500	Ea.	560	108	19.80	687.80	805
	0100 2,000 gallon	"	5	5.600		1,100	173	31.50	1,304.50	1,525
	0200 5,000 gallon	B-13	3.50	16		5,625	460	176	6,261	7,100
	0300 15,000 gallon, 4 piece	B-13B	1.70	32.941		15,600	950	525	17,075	19,100
	0400 25,000 gallon, 4 piece		1.10	50.909		33,000	1,475	815	35,290	39,400
	0500 40,000 gallon, 4 piece	↓	.80	70		38,800	2,025	1,125	41,950	47,000
	0520 50,000 gallon, 5 piece	B-13C	.60	93.333		44,600	2,675	2,525	49,800	56,000
	0540 75,000 gallon, cast in place	C-14C	.25	448		54,500	14,600	94	69,194	83,000
	0560 100,000 gallon	"	.15	746		67,000	24,400	157	91,557	112,500
	0600 High density polyethylene, 1,000 gallon	B-21	6	4.667		865	144	26.50	1,035.50	1,200
	0700 1,500 gallon	"	4	7		1,125	216	39.50	1,380.50	1,600
	1000 Distribution boxes, concrete, 7 outlets	2 Clab	16	1		104	26.50		130.50	156
	1100 9 outlets	"	8	2		281	53.50		334.50	395
	1150 Leaching field chambers, 13' x 3'-7" x 1'-4", standard	B-13	16	3.500		565	101	38.50	704.50	820
	1200 Heavy duty, 8' x 4' x 1'-6"		14	4		390	115	44	549	655
	1300 13' x 3'-9" x 1'-6"		12	4.667		1,375	134	51.50	1,560.50	1,775
	1350 20' x 4' x 1'-6"	↓	5	11.200		910	320	123	1,353	1,625
	1400 Leaching pit, precast concrete, 3' diameter, 3' deep	B-21	8	3.500		370	108	19.80	497.80	595
	1500 6' diameter, 3' section		4.70	5.957		645	184	33.50	862.50	1,025
	2000 Velocity reducing pit, precast conc., 6' diameter, 3' deep	↓	4.70	5.957		310	184	33.50	527.50	665
	2200 Excavation for septic tank, 3/4 C.Y. backhoe	B-12F	145	.110	C.Y.		3.45	3.27	6.72	8.90
	2400 4' trench for disposal field, 3/4 C.Y. backhoe	"	335	.048	L.F.		1.49	1.42	2.91	3.84
	2600 Gravel fill, run of bank	B-6	150	.160	C.Y.	16.15	4.61	1.44	22.20	26.50
	2800 Crushed stone, 3/4"	"	150	.160	"	24	4.61	1.44	30.05	34.50

02550 | Piped Energy Distribution

		CREW	DAILY OUTPUT	LABOR-HOURS	UNIT	MAT.	LABOR	EQUIP.	TOTAL	TOTAL INCL O&P
464	0010 **PIPING, GAS SERVICE & DISTRIBUTION, POLYETHYLENE**									**464**
	0020 not including excavation or backfill									
	1000 60 psi coils, comp cplg @ 100', 1/2" diameter, SDR 9.3	B-20A	608	.053	L.F.	.47	1.70		2.17	3.11
	1040 1-1/4" diameter, SDR 11		544	.059		.87	1.90		2.77	3.85
	1100 2" diameter, SDR 11		488	.066		1.08	2.11		3.19	4.42
	1160 3" diameter, SDR 11		408	.078		2.25	2.53		4.78	6.35
	1500 60 PSI 40' joints with coupling, 3" diameter, SDR 11	B-21A	408	.098		2.25	3.23	1.24	6.72	8.75
	1540 4" diameter, SDR 11		352	.114		5.15	3.75	1.44	10.34	13
	1600 6" diameter, SDR 11		328	.122		16.10	4.02	1.54	21.66	25.50
	1640 8" diameter, SDR 11	↓	272	.147	↓	22	4.85	1.86	28.71	33.50
550	0010 **GAS STATION PRODUCT LINE**									**550**
	0020 Primary containment pipe, fiberglass-reinforced									
	0030 Plastic pipe 15' & 30' lengths									
	0040 2" diameter	Q-6	425	.056	L.F.	3.23	2.16		5.39	6.80
	0050 3" diameter		400	.060		4.23	2.29		6.52	8.10
	0060 4" diameter		375	.064		5.45	2.45		7.90	9.70
	0100 Fittings									
	0110 Elbows, 90° & 45°, bell-ends, 2"	Q-6	24	1	Ea.	32.50	38		70.50	93.50
	0120 3" diameter		22	1.091		34.50	41.50		76	100
	0130 4" diameter		20	1.200		45.50	46		91.50	119
	0200 Tees, bell ends, 2"		21	1.143		39.50	43.50		83	109
	0210 3" diameter		18	1.333		40	51		91	121
	0220 4" diameter		15	1.600		55	61		116	153
	0230 Flanges bell ends, 2"		24	1		13.10	38		51.10	72
	0240 3" diameter		22	1.091		16.50	41.50		58	80.50
	0250 4" diameter		20	1.200		22.50	46		68.50	93.50
	0260 Sleeve couplings, 2"		21	1.143		8.35	43.50		51.85	74.50
	0270 3" diameter	↓	18	1.333		11.85	51		62.85	89.50

Important: See the Reference Section for critical supporting data - Reference Nos., Crews, & City Cost Indexes

02550 | Piped Energy Distribution

		CREW	DAILY OUTPUT	LABOR-HOURS	UNIT	2005 BARE COSTS MAT.	LABOR	EQUIP.	TOTAL	TOTAL INCL O&P	
550	0280	4" diameter	Q-6	15	1.600	Ea.	16.35	61		77.35	110	**550**
	0290	Threaded adapters 2"		21	1.143		10.95	43.50		54.45	77.50	
	0300	3" diameter		18	1.333		19.25	51		70.25	97.50	
	0310	4" diameter		15	1.600		26	61		87	121	
	0320	Reducers, 2"		27	.889		14.60	34		48.60	67	
	0330	3" diameter		22	1.091		16.90	41.50		58.40	81	
	0340	4" diameter		20	1.200		22	46		68	93	
	1010	Gas station product line for secondary containment (double wall)										
	1100	Fiberglass reinforced plastic pipe 25' lengths										
	1120	Pipe, plain end, 3"	Q-6	375	.064	L.F.	5.55	2.45		8	9.80	
	1130	4" diameter		350	.069		9.10	2.62		11.72	13.95	
	1140	5" diameter		325	.074		11.85	2.82		14.67	17.25	
	1150	6" diameter		300	.080		12.20	3.06		15.26	18	
	1200	Fittings										
	1230	Elbows, 90° & 45°, 3"	Q-6	18	1.333	Ea.	40	51		91	121	
	1240	4" diameter		16	1.500		69.50	57.50		127	163	
	1250	5" diameter		14	1.714		162	65.50		227.50	277	
	1260	6" diameter		12	2		164	76.50		240.50	295	
	1270	Tees, 3"		15	1.600		59	61		120	157	
	1280	4" diameter		12	2		87	76.50		163.50	211	
	1290	5" diameter		9	2.667		176	102		278	345	
	1300	6" diameter		6	4		184	153		337	430	
	1310	Couplings, 3"		18	1.333		28	51		79	108	
	1320	4" diameter		16	1.500		72.50	57.50		130	166	
	1330	5" diameter		14	1.714		150	65.50		215.50	265	
	1340	6" diameter		12	2		156	76.50		232.50	287	
	1350	Cross-over nipples, 3"		18	1.333		6.40	51		57.40	83.50	
	1360	4" diameter		16	1.500		7.50	57.50		65	94.50	
	1370	5" diameter		14	1.714		11.15	65.50		76.65	111	
	1380	6" diameter		12	2		11.70	76.50		88.20	128	
	1400	Telescoping, reducers, concentric 4" x 3"		18	1.333		21.50	51		72.50	100	
	1410	5" x 4"		17	1.412		55.50	54		109.50	143	
	1420	6" x 5"		16	1.500		134	57.50		191.50	233	

02580 | Elec/Communication Structures

		CREW	DAILY OUTPUT	LABOR-HOURS	UNIT	MAT.	LABOR	EQUIP.	TOTAL	TOTAL INCL O&P	
100	0010	**RADIO TOWERS**										**100**
	0020	Guyed, 50'h, 40 lb. sec., 70 MPH basic wind spd.	2 Sswk	1	16	Ea.	2,225	610		2,835	3,550	
	0100	Wind load 90 MPH basic wind speed	"	1	16		2,225	610		2,835	3,550	
	0300	190' high, 40 lb. section, wind load 70 MPH basic wind speed	K-2	.33	72.727		6,000	2,550	535	9,085	11,600	
	0400	200' high, 70 lb. section, wind load 90 MPH basic wind speed		.33	72.727		12,100	2,550	535	15,185	18,300	
	0600	300' high, 70 lb. section, wind load 70 MPH basic wind speed		.20	120		17,200	4,200	880	22,280	27,100	
	0700	270' high, 90 lb. section, wind load 90 MPH basic wind speed		.20	120		19,800	4,200	880	24,880	30,000	
	0800	400' high, 100 lb. section, wind load 70 MPH basic wind speed		.14	171		28,900	6,000	1,250	36,150	43,400	
	0900	Self-supporting, 60' high, wind load 70 MPH basic wind speed		.80	30		4,650	1,050	220	5,920	7,150	
	0910	60' high, wind load 90 MPH basic wind speed		.45	53.333		8,350	1,875	390	10,615	12,800	
	1000	120' high, wind load 70 MPH basic wind speed		.40	60		11,400	2,100	440	13,940	16,600	
	1200	190' high, wind load 90 MPH basic wind speed		.20	120		27,600	4,200	880	32,680	38,600	
	2000	For states west of Rocky Mountains, add for shipping					10%					
420	0010	**ELECTRIC & TELEPHONE UNDERGROUND**, Not including excavation										**420**
	0200	backfill and cast in place concrete										
	0400	Hand holes, precast concrete, with concrete cover										
	0600	2' x 2' x 3' deep	R-3	2.40	8.333	Ea.	263	335	66	664	855	
	0800	3' x 3' x 3' deep		1.90	10.526		340	420	83.50	843.50	1,100	
	1000	4' x 4' x 4' deep		1.40	14.286		680	570	113	1,363	1,725	

SITE CONSTRUCTION 2

02580 | Elec/Communication Structures

			CREW	DAILY OUTPUT	LABOR-HOURS	UNIT	2005 BARE COSTS				TOTAL INCL O&P
							MAT.	LABOR	EQUIP.	TOTAL	
420	1200	Manholes, precast with iron racks & pulling irons, C.I. frame									420
	1400	and cover, 4' x 6' x 7' deep	B-13	2	28	Ea.	1,300	805	310	2,415	3,050
	1600	6' x 8' x 7' deep		1.90	29.474		1,600	850	325	2,775	3,425
	1800	6' x 10' x 7' deep	↓	1.80	31.111	↓	1,800	895	345	3,040	3,725
	4200	Underground duct, banks ready for concrete fill, min. of 7.5"									
	4400	between conduits, ctr. to ctr.(for wire & cable see div. 16120)									
	4580	PVC, type EB, 1 @ 2" diameter	2 Elec	480	.033	L.F.	1.29	1.36		2.65	3.44
	4600	2 @ 2" diameter		240	.067		2.58	2.72		5.30	6.90
	4800	4 @ 2" diameter		120	.133		5.15	5.45		10.60	13.80
	5000	2 @ 3" diameter		200	.080		3.64	3.26		6.90	8.85
	5200	4 @ 3" diameter		100	.160		7.25	6.50		13.75	17.70
	5400	2 @ 4" diameter		160	.100		5.75	4.08		9.83	12.35
	5600	4 @ 4" diameter		80	.200		11.45	8.15		19.60	25
	5800	6 @ 4" diameter		54	.296		17.20	12.05		29.25	37
	6200	Rigid galvanized steel, 2 @ 2" diameter		180	.089		16.10	3.62		19.72	23
	6400	4 @ 2" diameter		90	.178		32	7.25		39.25	46.50
	6800	2 @ 3" diameter		100	.160		37.50	6.50		44	50.50
	7000	4 @ 3" diameter		50	.320		75	13.05		88.05	102
	7200	2 @ 4" diameter		70	.229		51.50	9.30		60.80	70.50
	7400	4 @ 4" diameter		34	.471		103	19.20		122.20	142
	7600	6 @ 4" diameter	↓	22	.727	↓	155	29.50		184.50	214
500	0010	**UTILITY POLES**									500
	6200	Poles, wood, preservative treatment, see also div. 16520, 20' high	R-3	3.10	6.452	Ea.	241	258	51	550	705
	6400	25' high		2.90	6.897		255	276	54.50	585.50	750
	6600	30' high		2.60	7.692		281	310	61	652	835
	6800	35' high		2.40	8.333		365	335	66	766	970
	7000	40' high		2.30	8.696		450	350	69	869	1,100
	7200	45' high	↓	1.70	11.765	↓	550	470	93.50	1,113.50	1,400
	7400	Cross arms with hardware & insulators									
	7600	4' long	1 Elec	2.50	3.200	Ea.	114	130		244	320
	7800	5' long		2.40	3.333		133	136		269	350
	8000	6' long	↓	2.20	3.636	↓	153	148		301	390

02600 | Drainage & Containment

02620 | Subdrainage

			CREW	DAILY OUTPUT	LABOR-HOURS	UNIT	2005 BARE COSTS				TOTAL INCL O&P
							MAT.	LABOR	EQUIP.	TOTAL	
610	0010	**PIPING, SUBDRAINAGE, CONCRETE** R02510-810									610
	0021	Not including excavation and backfill									
	3000	Porous wall concrete underdrain, std. strength, 4" diameter	B-20	335	.072	L.F.	1.95	2.15		4.10	5.50
	3020	6" diameter	"	315	.076		2.54	2.29		4.83	6.35
	3040	8" diameter	B-21	310	.090		3.13	2.79	.51	6.43	8.30
	3060	12" diameter		285	.098		6.65	3.04	.56	10.25	12.60
	3080	15" diameter		230	.122		7.60	3.76	.69	12.05	14.95
	3100	18" diameter	↓	165	.170		10.05	5.25	.96	16.26	20
	4000	Extra strength, 6" diameter	B-20	315	.076		2.57	2.29		4.86	6.40
	4020	8" diameter	B-21	310	.090		3.86	2.79	.51	7.16	9.10
	4040	10" diameter		285	.098		7.70	3.04	.56	11.30	13.80
	4060	12" diameter	↓	230	.122	↓	8.35	3.76	.69	12.80	15.80

Important: See the Reference Section for critical supporting data - Reference Nos., Crews, & City Cost Indexes

				DAILY	LABOR-			2005 BARE COSTS				TOTAL	
	02620	**Subdrainage**		CREW	OUTPUT	HOURS	UNIT	MAT.	LABOR	EQUIP.	TOTAL	INCL O&P	
610	4080	15" diameter	R02510 -810	B-21	200	.140	L.F.	9.25	4.33	.79	14.37	17.75	**610**
	4100	18" diameter		↓	165	.170	↓	13.50	5.25	.96	19.71	24	
620	0010	**PIPING, SUBDRAINAGE, CORRUGATED METAL**											**620**
	0021	Not including excavation and backfill											
	2010	Aluminum, perforated											
	2020	6" diameter, 18 ga.		B-14	380	.126	L.F.	2.76	3.55	.57	6.88	9.15	
	2200	8" diameter, 16 ga.			370	.130		3.94	3.64	.58	8.16	10.60	
	2220	10" diameter, 16 ga.			360	.133		4.93	3.75	.60	9.28	11.85	
	2240	12" diameter, 16 ga.			285	.168		5.50	4.73	.76	10.99	14.20	
	2260	18" diameter, 16 ga.		↓	205	.234	↓	8.25	6.60	1.05	15.90	20.50	
	3000	Uncoated galvanized, perforated											
	3020	6" diameter, 18 ga.		B-20	380	.063	L.F.	4.37	1.90		6.27	7.75	
	3200	8" diameter, 16 ga.		"	370	.065		6	1.95		7.95	9.65	
	3220	10" diameter, 16 ga.		B-21	360	.078		9	2.40	.44	11.84	14.10	
	3240	12" diameter, 16 ga.			285	.098		9.45	3.04	.56	13.05	15.70	
	3260	18" diameter, 16 ga.		↓	205	.137	↓	14.40	4.22	.77	19.39	23.50	
	4000	Steel, perforated, asphalt coated											
	4020	6" diameter 18 ga.		B-20	380	.063	L.F.	3.49	1.90		5.39	6.80	
	4030	8" diameter 18 ga.		"	370	.065		5.45	1.95		7.40	9.05	
	4040	10" diameter 16 ga		B-21	360	.078		6.30	2.40	.44	9.14	11.10	
	4050	12" diameter 16 ga			285	.098		7.20	3.04	.56	10.80	13.25	
	4060	18" diameter 16 ga		↓	205	.137	↓	9.85	4.22	.77	14.84	18.20	
670	0010	**PIPING, SUBDRAINAGE, POLYVINYL CHLORIDE**											**670**
	0020	Perforated, price as solid pipe, division 02530-780											
	02630	**Storm Drainage**											
110	0010	**CATCH BASIN GRATES AND FRAMES** not including footing, excavation											**110**
	1600	Frames & covers, C.I., 24" square, 500 lb.		B-6	7.80	3.077	Ea.	239	88.50	27.50	355	430	
	1700	26" D shape, 600 lb.			7	3.429		380	99	31	510	605	
	1800	Light traffic, 18" diameter, 100 lb.			10	2.400		128	69	21.50	218.50	271	
	1900	24" diameter, 300 lb.			8.70	2.759		196	79.50	25	300.50	365	
	2000	36" diameter, 900 lb.			5.80	4.138		425	119	37	581	690	
	2100	Heavy traffic, 24" diameter, 400 lb.			7.80	3.077		193	88.50	27.50	309	380	
	2200	36" diameter, 1150 lb.			3	8		610	231	72	913	1,100	
	2300	Mass. State standard, 26" diameter, 475 lb.			7	3.429		465	99	31	595	695	
	2400	30" diameter, 620 lb.			7	3.429		325	99	31	455	540	
	2500	Watertight, 24" diameter, 350 lb.			7.80	3.077		325	88.50	27.50	441	520	
	2600	26" diameter, 500 lb.			7	3.429		310	99	31	440	525	
	2700	32" diameter, 575 lb.		↓	6	4	↓	665	115	36	816	945	
	2800	3 piece cover & frame, 10" deep,											
	2900	1200 lbs., for heavy equipment		B-6	3	8	Ea.	990	231	72	1,293	1,525	
	3000	Raised for paving 1-1/4" to 2" high,											
	3100	4 piece expansion ring											
	3200	20" to 26" diameter		1 Clab	3	2.667	Ea.	114	71		185	237	
	3300	30" to 36" diameter		"	3	2.667	"	159	71		230	286	
	3320	Frames and covers, existing, raised for paving, 2", including											
	3340	row of brick, concrete collar, up to 12" wide frame		B-6	18	1.333	Ea.	35.50	38.50	12	86	112	
	3360	20" to 26" wide frame			11	2.182		56.50	63	19.60	139.10	180	
	3380	30" to 36" wide frame		↓	9	2.667		70	77	24	171	222	
	3400	Inverts, single channel brick		D-1	3	5.333		78.50	166		244.50	340	
	3500	Concrete			5	3.200		61	99.50		160.50	220	
	3600	Triple channel, brick			2	8		119	249		368	510	
	3700	Concrete		↓	3	5.333	↓	105	166		271	370	
400	0010	**STORM DRAINAGE MANHOLES, FRAMES & COVERS** not including											**400**
	0020	footing, excavation, backfill (See line items for frame & cover)											

SITE CONSTRUCTION **2**

02630 | Storm Drainage

		CREW	DAILY OUTPUT	LABOR-HOURS	UNIT	2005 BARE COSTS MAT.	LABOR	EQUIP.	TOTAL	TOTAL INCL O&P	
400	**0050**	Brick, 4' inside diameter, 4' deep	D-1	1	16	Ea.	320	495		815	1,100	**400**
	0100	6' deep		.70	22.857		445	710		1,155	1,575	
	0150	8' deep		.50	32		565	995		1,560	2,150	
	0200	For depths over 8', add		4	4	V.L.F.	174	124		298	380	
	0400	Concrete blocks (radial), 4' I.D., 4' deep		1.50	10.667	Ea.	255	330		585	785	
	0500	6' deep		1	16		335	495		830	1,125	
	0600	8' deep		.70	22.857		415	710		1,125	1,525	
	0700	For depths over 8', add		5.50	2.909	V.L.F.	41.50	90.50		132	184	
	0800	Concrete, cast in place, 4' x 4', 8" thick, 4' deep	C-14H	2	24	Ea.	445	810	12	1,267	1,750	
	0900	6' deep		1.50	32		650	1,075	16	1,741	2,400	
	1000	8' deep		1	48		920	1,625	24	2,569	3,575	
	1100	For depths over 8', add		8	6	V.L.F.	107	202	3	312	435	
	1110	Precast, 4' I.D., 4' deep	B-22	4.10	7.317	Ea.	675	229	58	962	1,175	
	1120	6' deep		3	10		870	315	79.50	1,264.50	1,525	
	1130	8' deep		2	15		1,025	470	119	1,614	1,975	
	1140	For depths over 8', add		16	1.875	V.L.F.	143	58.50	14.85	216.35	264	
	1150	5' I.D., 4' deep	B-6	3	8	Ea.	720	231	72	1,023	1,225	
	1160	6' deep		2	12		970	345	108	1,423	1,725	
	1170	8' deep		1.50	16		1,225	460	144	1,829	2,225	
	1180	For depths over 8', add		12	2	V.L.F.	159	57.50	18	234.50	283	
	1190	6' I.D., 4' deep		2	12	Ea.	1,175	345	108	1,628	1,950	
	1200	6' deep		1.50	16		1,525	460	144	2,129	2,550	
	1210	8' deep		1	24		1,875	690	216	2,781	3,375	
	1220	For depths over 8', add		8	3	V.L.F.	246	86.50	27	359.50	435	
	1250	Slab tops, precast, 8" thick										
	1300	4' diameter manhole	B-6	8	3	Ea.	165	86.50	27	278.50	345	
	1400	5' diameter manhole		7.50	3.200		325	92	29	446	535	
	1500	6' diameter manhole		7	3.429		405	99	31	535	630	
	3800	Steps, heavyweight cast iron, 7" x 9"	1 Bric	40	.200		12.20	7.05		19.25	24	
	3900	8" x 9"		40	.200		18.35	7.05		25.40	31	
	3928	12" x 10-1/2"		40	.200		16.40	7.05		23.45	29	
	4000	Standard sizes, galvanized steel		40	.200		15.30	7.05		22.35	27.50	
	4100	Aluminum		40	.200		18.90	7.05		25.95	32	
510	**0010**	**PIPING, STORM DRAINAGE, CORRUGATED METAL**										**510**
	0020	Not including excavation or backfill										
	2000	Corrugated metal pipe, galvanized and coated										
	2020	Bituminous coated with paved invert, 20' lengths										
	2040	8" diameter, 16 ga.	B-14	330	.145	L.F.	8.95	4.09	.65	13.69	16.85	
	2060	10" diameter, 16 ga.		260	.185		10.75	5.20	.83	16.78	20.50	
	2080	12" diameter, 16 ga.		210	.229		12.85	6.40	1.03	20.28	25	
	2100	15" diameter, 16 ga.		200	.240		15.65	6.75	1.08	23.48	29	
	2120	18" diameter, 16 ga.		190	.253		20	7.10	1.14	28.24	34.50	
	2140	24" diameter, 14 ga.		160	.300		24.50	8.45	1.35	34.30	41.50	
	2160	30" diameter, 14 ga.	B-13	120	.467		32.50	13.45	5.15	51.10	61.50	
	2180	36" diameter, 12 ga.		120	.467		47.50	13.45	5.15	66.10	78.50	
	2200	48" diameter, 12 ga.		100	.560		72.50	16.10	6.15	94.75	111	
	2220	60" diameter, 10 ga.	B-13B	75	.747		93.50	21.50	11.95	126.95	149	
	2240	72" diameter, 8 ga.	"	45	1.244		140	36	19.90	195.90	231	
	2500	Galvanized, uncoated, 20' lengths										
	2520	8" diameter, 16 ga.	B-14	355	.135	L.F.	6.90	3.80	.61	11.31	14.10	
	2540	10" diameter, 16 ga.		280	.171		7.60	4.82	.77	13.19	16.65	
	2560	12" diameter, 16 ga.		220	.218		8.65	6.15	.98	15.78	20	
	2580	15" diameter, 16 ga.		220	.218		11	6.15	.98	18.13	22.50	
	2600	18" diameter, 16 ga.		205	.234		14.55	6.60	1.05	22.20	27.50	
	2620	24" diameter, 14 ga.		175	.274		21	7.70	1.23	29.93	36.50	

CN

Important: See the Reference Section for critical supporting data - Reference Nos., Crews, & City Cost Indexes

02630 | Storm Drainage

		CREW	DAILY OUTPUT	LABOR-HOURS	UNIT	MAT.	LABOR	EQUIP.	TOTAL	TOTAL INCL O&P
						2005 BARE COSTS				
510	2640	30" diameter, 14 ga.	B-13	130	.431	L.F.	27	12.40	4.74	44.14	54
	2660	36" diameter, 12 ga.		130	.431		44	12.40	4.74	61.14	73
	2680	48" diameter, 12 ga.		110	.509		59	14.65	5.60	79.25	93.50
	2690	60" diameter, 10 ga.	B-13B	78	.718		93	20.50	11.50	125	147
	2780	End sections, 8" diameter	B-14	24	2	Ea.	71.50	56	9	136.50	175
	2785	10" diameter		22	2.182		73	61.50	9.80	144.30	186
	2790	12" diameter		35	1.371		82.50	38.50	6.15	127.15	157
	2800	18" diameter		30	1.600		98	45	7.20	150.20	185
	2810	24" diameter	B-13	25	2.240		143	64.50	24.50	232	284
	2820	30" diameter		25	2.240		272	64.50	24.50	361	425
	2825	36" diameter		20	2.800		360	80.50	31	471.50	555
	2830	48" diameter		10	5.600		745	161	61.50	967.50	1,125
	2835	60" diameter	B-13B	5	11.200		970	320	179	1,469	1,775
	2840	72" diameter	"	4	14		1,750	405	224	2,379	2,800
530	0010	**SEWAGE/DRAINAGE COLLECTION, CONCRETE PIPE** R02510-810									
	0020	Not including excavation or backfill									
	1000	Non-reinforced pipe, extra strength, B&S or T&G joints									
	1010	6" diameter	B-14	265.04	.181	L.F.	4.13	5.10	.82	10.05	13.30
	1020	8" diameter		224	.214		4.54	6	.96	11.50	15.35
	1030	10" diameter		216	.222		5.05	6.25	1	12.30	16.30
	1040	12" diameter		200	.240		6.20	6.75	1.08	14.03	18.45
	1050	15" diameter		180	.267		7.25	7.50	1.20	15.95	21
	1060	18" diameter		144	.333		8.90	9.35	1.50	19.75	26
	1070	21" diameter		112	.429		10.95	12.05	1.93	24.93	33
	1080	24" diameter		100	.480		13.40	13.50	2.16	29.06	38
	2000	Reinforced culvert, class 3, no gaskets									
	2010	12" diameter	B-14	150	.320	L.F.	11.20	9	1.44	21.64	28
	2020	15" diameter		150	.320		14.35	9	1.44	24.79	31.50
	2030	18" diameter		132	.364		15.10	10.20	1.64	26.94	34
	2035	21" diameter		120	.400		19.60	11.25	1.80	32.65	41
	2040	24" diameter CN		100	.480		22.50	13.50	2.16	38.16	48
	2045	27" diameter	B-13	92	.609		28	17.50	6.70	52.20	65.50
	2050	30" diameter		88	.636		30.50	18.30	7	55.80	69
	2060	36" diameter		72	.778		43.50	22.50	8.55	74.55	92
	2070	42" diameter	B-13B	72	.778		59.50	22.50	12.45	94.45	114
	2080	48" diameter		64	.875		73.50	25	14	112.50	135
	2090	60" diameter		48	1.167		118	33.50	18.65	170.15	201
	2100	72" diameter		40	1.400		166	40.50	22.50	229	270
	2120	84" diameter		32	1.750		281	50.50	28	359.50	420
	2140	96" diameter		24	2.333		335	67	37.50	439.50	515
	2200	With gaskets, class 3, 12" diameter	B-21	168	.167		12.25	5.15	.94	18.34	22.50
	2220	15" diameter		160	.175		14.70	5.40	.99	21.09	25.50
	2230	18" diameter		152	.184		18.40	5.70	1.04	25.14	30.50
	2240	24" diameter		136	.206		27.50	6.35	1.17	35.02	41.50
	2260	30" diameter	B-13	88	.636		37	18.30	7	62.30	76
	2270	36" diameter	"	72	.778		55	22.50	8.55	86.05	104
	2290	48" diameter	B-13B	64	.875		90	25	14	129	153
	2310	72" diameter	"	40	1.400		233	40.50	22.50	296	345
	2330	Flared ends, 6'-1" long, 12" diameter	B-21	190	.147		34	4.56	.83	39.39	45
	2340	15" diameter		155	.181		38.50	5.60	1.02	45.12	52.50
	2400	6'-2" long, 18" diameter		122	.230		40	7.10	1.30	48.40	56.50
	2420	24" diameter		88	.318		46.50	9.85	1.80	58.15	68
	2440	36" diameter	B-13	60	.933		84.50	27	10.30	121.80	145
	3080	Radius pipe, add to pipe prices, 12" to 60" diameter					50%				
	3090	Over 60" diameter, add					20%				
	3500	Reinforced elliptical, 8' lengths, C507 class 3									

02630 | Storm Drainage

		CREW	DAILY OUTPUT	LABOR-HOURS	UNIT	2005 BARE COSTS				TOTAL INCL O&P
						MAT.	LABOR	EQUIP.	TOTAL	
530	3520	14" x 23" inside, round equivalent 18" diameter	B-21	82	.341	L.F.	26.50	10.55	1.93	38.98	47.50	530
	3530	24" x 38" inside, round equivalent 30" diameter	B-13	58	.966		47	28	10.65	85.65	106
	3540	29" x 45" inside, round equivalent 36" diameter		52	1.077		60.50	31	11.85	103.35	128
	3550	38" x 60" inside, round equivalent 48" diameter		38	1.474		79	42.50	16.25	137.75	170
	3560	48" x 76" inside, round equivalent 60" diameter		26	2.154		120	62	23.50	205.50	254
	3570	58" x 91" inside, round equivalent 72" diameter	▼	22	2.545	▼	171	73.50	28	272.50	335
	3780	Concrete slotted pipe, class 4 mortar joint									
	3800	12" diameter	B-21	168	.167	L.F.	13.15	5.15	.94	19.24	23.50
	3840	18" diameter	"	152	.184	"	20.50	5.70	1.04	27.24	32.50
	3900	Class 4 O-ring									
	3940	12" diameter	B-21	168	.167	L.F.	13.75	5.15	.94	19.84	24
	3960	18" diameter	"	152	.184	"	18.45	5.70	1.04	25.19	30.50

02640 | Culverts/Manuf Const

		CREW	DAILY OUTPUT	LABOR-HOURS	UNIT	MAT.	LABOR	EQUIP.	TOTAL	TOTAL INCL O&P
810	0010	**OVAL ARCH CULVERTS**										810
	3000	Corrugated galvanized or aluminum, coated & paved									
	3020	17" x 13", 16 ga., 15" equivalent	B-14	200	.240	L.F.	26.50	6.75	1.08	34.33	40.50
	3040	21" x 15", 16 ga., 18" equivalent		150	.320		34	9	1.44	44.44	53
	3060	28" x 20", 14 ga., 24" equivalent		125	.384		49	10.80	1.73	61.53	72
	3080	35" x 24", 14 ga., 30" equivalent	▼	100	.480		59.50	13.50	2.16	75.16	89
	3100	42" x 29", 12 ga., 36" equivalent	B-13	100	.560		88.50	16.10	6.15	110.75	129
	3120	49" x 33", 12 ga., 42" equivalent		90	.622		107	17.90	6.85	131.75	152
	3140	57" x 38", 12 ga., 48" equivalent	▼	75	.747	▼	119	21.50	8.20	148.70	173
	3160	Steel, plain oval arch culverts, plain									
	3180	17" x 13", 16 ga., 15" equivalent	B-14	225	.213	L.F.	14.30	6	.96	21.26	26
	3200	21" x 15", 16 ga., 18" equivalent		175	.274		16.95	7.70	1.23	25.88	32
	3220	28" x 20", 14 ga., 24" equivalent	▼	150	.320		27	9	1.44	37.44	45.50
	3240	35" x 24", 14 ga., 30" equivalent	B-13	108	.519		34	14.95	5.70	54.65	67
	3260	42" x 29", 12 ga., 36" equivalent		108	.519		56.50	14.95	5.70	77.15	91.50
	3280	49" x 33", 12 ga., 42" equivalent		92	.609		66.50	17.50	6.70	90.70	108
	3300	57" x 38", 12 ga., 48" equivalent	▼	75	.747	▼	58.50	21.50	8.20	88.20	107
	3320	End sections, 17" x 13"		22	2.545	Ea.	78.50	73.50	28	180	230
	3340	42" x 29"	▼	17	3.294	"	320	95	36.50	451.50	540
	3360	Multi-plate arch, steel	B-20	1,690	.014	Lb.	.90	.43		1.33	1.65

02660 | Ponds & Reservoirs

		CREW	DAILY OUTPUT	LABOR-HOURS	UNIT	MAT.	LABOR	EQUIP.	TOTAL	TOTAL INCL O&P
600	0010	**SLURRY TRENCH** Excavated slurry trench in wet soils										600
	0020	backfilled with 3000 PSI concrete, no reinforcing steel									
	0050	Minimum	C-7	333	.216	C.F.	6	6.25	2.88	15.13	19.30
	0100	Maximum		200	.360	"	10.05	10.40	4.80	25.25	32.50
	0200	Alternate pricing method, minimum		150	.480	S.F.	12	13.85	6.40	32.25	41.50
	0300	Maximum	▼	120	.600		18	17.30	8	43.30	55
	0500	Reinforced slurry trench, minimum	B-48	177	.316		9	9.40	19.30	37.70	45.50
	0600	Maximum	"	69	.812	▼	30	24	49.50	103.50	125
	0800	Haul for disposal, 2 mile haul, excavated material, add	B-34B	99	.081	C.Y.		2.23	4.81	7.04	8.70
	0900	Haul bentonite castings for disposal, add	"	40	.200	"		5.50	11.90	17.40	21.50

Important: See the Reference Section for critical supporting data - Reference Nos., Crews, & City Cost Indexes

SITE CONSTRUCTION 2

02710 | Bound Base Courses

		CREW	DAILY OUTPUT	LABOR-HOURS	UNIT	2005 BARE COSTS				TOTAL INCL O&P
						MAT.	LABOR	EQUIP.	TOTAL	
200	0010	**ASPHALT-TREATED PERMEABLE BASE COURSE** for roadways									**200**
	0020	and large paved areas								
	0700	Liquid application to gravel base, asphalt emulsion	B-45	6,000	.003	Gal.	2.92	.08	.09	3.09	3.44
	0800	Prime and seal, cut back asphalt		6,000	.003	"	3.45	.08	.09	3.62	4.02
	1000	Macadam penetration crushed stone, 2 gal. per S.Y., 4" thick		6,000	.003	S.Y.	5.85	.08	.09	6.02	6.65
	1100	6" thick, 3 gal. per S.Y.		4,000	.004		8.75	.12	.13	9	10
	1200	8" thick, 4 gal. per S.Y.	▼	3,000	.005	▼	11.70	.17	.18	12.05	13.30
	8900	For small and irregular areas, add						50%	50%		

02720 | Unbound Base Courses & Ballasts

		CREW	DAILY OUTPUT	LABOR-HOURS	UNIT	MAT.	LABOR	EQUIP.	TOTAL	TOTAL INCL O&P
200	0010	**AGGREGATE BASE COURSE** For roadways and large areas									**200**
	0050	Crushed 3/4" stone base, compacted, 3" deep	B-36C	5,200	.008	S.Y.	2.65	.25	.49	3.39	3.83
	0100	6" deep		5,000	.008		5.30	.26	.51	6.07	6.80
	0200	9" deep		4,600	.009		7.95	.28	.56	8.79	9.80
	0300	12" deep	▼	4,200	.010		10.65	.31	.61	11.57	12.85
	0301	Crushed 1-1/2" stone base, compacted to 4" deep	B-36B	6,000	.011		4.17	.33	.49	4.99	5.65
	0302	6" deep		5,400	.012		6.25	.37	.54	7.16	8.05
	0303	8" deep		4,500	.014		8.35	.44	.65	9.44	10.60
	0304	12" deep	▼	3,800	.017	▼	12.50	.52	.77	13.79	15.40
	0350	Bank run gravel, spread and compacted									
	0370	6" deep	B-32	6,000	.005	S.Y.	3.47	.17	.26	3.90	4.36
	0390	9" deep		4,900	.007		5.20	.21	.32	5.73	6.35
	0400	12" deep	▼	4,200	.008	▼	6.95	.25	.37	7.57	8.40
	6000	Stabilization fabric, polypropylene, 6 oz./S.Y.	B-6	10,000	.002		.91	.07	.02	1	1.13
	6900	For small and irregular areas, add						50%	50%		
	7000	Prepare and roll sub-base, small areas to 2500 S.Y.	B-32A	1,500	.016	S.Y.		.51	.59	1.10	1.43
	8000	Large areas over 2500 S.Y.	B-32	3,700	.009	"		.28	.42	.70	.89

02740 | Flexible Pavement

		CREW	DAILY OUTPUT	LABOR-HOURS	UNIT	MAT.	LABOR	EQUIP.	TOTAL	TOTAL INCL O&P
310	0010	**ASPHALTIC CONCRETE PAVEMENT, HIGHWAYS** [R02065-300]									**310**
	0020	and large paved areas								
	0080	Binder course, 1-1/2" thick	B-25	7,725	.011	S.Y.	2.63	.33	.26	3.22	3.69
	0120	2" thick		6,345	.014		3.50	.40	.32	4.22	4.82
	0160	3" thick		4,905	.018		5.20	.52	.41	6.13	6.95
	0200	4" thick	▼	4,140	.021		6.95	.62	.49	8.06	9.15
	0300	Wearing course, 1" thick	B-25B	10,575	.009		1.86	.27	.21	2.34	2.68
	0340	1-1/2" thick		7,725	.012		2.83	.37	.28	3.48	3.98
	0380	2" thick		6,345	.015		3.80	.45	.35	4.60	5.25
	0420	2-1/2" thick		5,480	.018		4.69	.52	.40	5.61	6.40
	0460	3" thick	▼	4,900	.020		5.60	.58	.45	6.63	7.55
	0500	Open graded friction course	B-25C	5,000	.010	▼	1.66	.28	.35	2.29	2.66
	0800	Alternate method of figuring paving costs									
	0810	Binder course, 1-1/2" thick	B-25	630	.140	Ton	34.50	4.06	3.19	41.75	48
	0811	2" thick		690	.128		34.50	3.71	2.91	41.12	47
	0812	3" thick		800	.110		34.50	3.20	2.51	40.21	45.50
	0813	4" thick	▼	850	.104		34.50	3.01	2.37	39.88	45
	0850	Wearing course, 1" thick	B-25B	575	.167		34.50	4.93	3.81	43.24	49.50
	0851	1-1/2" thick		630	.152		34.50	4.50	3.48	42.48	48.50
	0852	2" thick		690	.139		34.50	4.11	3.17	41.78	48
	0853	2-1/2" thick		745	.129		34.50	3.80	2.94	41.24	47
	0854	3" thick	▼	800	.120	▼	34.50	3.54	2.74	40.78	46.50
	1000	Pavement replacement over trench, 2" thick	B-37	90	.533	S.Y.	3.87	15	1.19	20.06	28.50
	1050	4" thick		70	.686		7.65	19.25	1.54	28.44	40
	1080	6" thick	▼	55	.873	▼	12.20	24.50	1.95	38.65	53.50
	1200	For paving projects 300 tons or less add for trucking									

02740 | Flexible Pavement

			CREW	DAILY OUTPUT	LABOR-HOURS	UNIT	2005 BARE COSTS				TOTAL INCL O&P	
							MAT.	LABOR	EQUIP.	TOTAL		
310	1300	See 02315-490-0010 for hauling costs [R02065-300]										310
315	0011	**ASPHALTIC CONCRETE PAVEMENT, LOTS & DRIVEWAYS**										315
	0020	6" stone base, 2" binder course, 1" topping	B-25C	9,000	.005	S.F.	1.38	.16	.20	1.74	1.98	
	0300	Binder course, 1-1/2" thick		35,000	.001		.31	.04	.05	.40	.46	
	0400	2" thick		25,000	.002		.41	.06	.07	.54	.62	
	0500	3" thick		15,000	.003		.63	.10	.12	.85	.97	
	0600	4" thick		10,800	.004		.82	.13	.16	1.11	1.28	
	0800	Sand finish course, 3/4" thick		41,000	.001		.19	.03	.04	.26	.31	
	0900	1" thick		34,000	.001		.23	.04	.05	.32	.38	
	1000	Fill pot holes, hot mix, 2" thick	B-16	4,200	.008		.42	.21	.11	.74	.90	
	1100	4" thick		3,500	.009		.61	.25	.14	1	1.21	
	1120	6" thick		3,100	.010		.82	.28	.15	1.25	1.51	
	1140	Cold patch, 2" thick	B-51	3,000	.016		.50	.43	.04	.97	1.27	
	1160	4" thick		2,700	.018		.95	.48	.05	1.48	1.85	
	1180	6" thick		1,900	.025		1.48	.68	.07	2.23	2.76	

02750 | Rigid Pavement

			CREW	DAILY OUTPUT	LABOR-HOURS	UNIT	MAT.	LABOR	EQUIP.	TOTAL	TOTAL INCL O&P	
300	0010	**PLAIN CEMENT CONCRETE PAVEMENT**										300
	0015	Including joints, finishing and curing										
	0020	Fixed form, 12' pass, unreinforced, 6" thick	B-26	3,000	.029	S.Y.	19.80	.88	.79	21.47	24	
	0100	8" thick		2,750	.032		27.50	.96	.86	29.32	33	
	0110	8" thick, small area		1,375	.064		27.50	1.91	1.72	31.13	35.50	
	0200	9" thick		2,500	.035		32	1.05	.95	34	37.50	
	0300	10" thick		2,100	.042		34.50	1.25	1.13	36.88	41	
	0310	10" thick, small area		1,050	.084		34.50	2.51	2.25	39.26	44.50	
	0400	12" thick		1,800	.049		37	1.46	1.31	39.77	44	
	0500	15" thick		1,500	.059		41.50	1.75	1.58	44.83	50	
	0510	For small irregular areas, add					100%					
	0700	Finishing, broom finish small areas	2 Cefi	120	.133	S.Y.		4.38		4.38	6.45	
	1000	Curing, with sprayed membrane by hand	2 Clab	1,500	.011	"	.47	.28		.75	.96	
	1650	For integral coloring, see div. 03310-220										

02760 | Paving Specialties

			CREW	DAILY OUTPUT	LABOR-HOURS	UNIT	MAT.	LABOR	EQUIP.	TOTAL	TOTAL INCL O&P	
300	0010	**PAINTED TRAFFIC LINES AND MARKINGS**										300
	0020	Acrylic waterborne, white or yellow, 4" wide	B-78	20,000	.002	L.F.	.15	.06	.02	.23	.29	
	0200	6" wide		11,000	.004		.14	.12	.04	.30	.37	
	0500	8" wide		10,000	.005		.20	.13	.04	.37	.45	
	0600	12" wide		4,000	.012		.36	.32	.10	.78	1.01	
	0620	Arrows or gore lines		2,300	.021	S.F.	.59	.56	.17	1.32	1.72	
	0640	Temporary paint, white or yellow		15,000	.003	L.F.	.18	.09	.03	.30	.36	
	0660	Removal	1 Clab	300	.027			.71		.71	1.11	
	0680	Temporary tape	2 Clab	1,500	.011		1.78	.28		2.06	2.40	
	0710	Thermoplastic, white or yellow, 4" wide	B-79	15,000	.003		.67	.07	.06	.80	.91	
	0730	6" wide		14,000	.003		.97	.08	.06	1.11	1.26	
	0740	8" wide		12,000	.003		1.31	.09	.07	1.47	1.66	
	0750	12" wide		6,000	.007		1.94	.18	.15	2.27	2.58	
	0760	Arrows		660	.061	S.F.	1.80	1.64	1.34	4.78	6	
	0770	Gore lines		2,500	.016		1.19	.43	.35	1.97	2.37	
	0780	Letters		660	.061		1.49	1.64	1.34	4.47	5.65	
500	0010	**PAVEMENT MARKINGS**										500
	0790	Layout of pavement marking	A-2	25,000	.001	L.F.		.03	.01	.04	.05	
	0800	Parking stall, paint, white	B-78	440	.109	Stall	3.18	2.95	.88	7.01	9.05	
	1000	Street letters and numbers	"	1,600	.030	S.F.	.61	.81	.24	1.66	2.20	

2 SITE CONSTRUCTION

Important: See the Reference Section for critical supporting data - Reference Nos., Crews, & City Cost Indexes

SITE CONSTRUCTION 2

		02770	Curbs and Gutters	CREW	DAILY OUTPUT	LABOR-HOURS	UNIT	2005 BARE COSTS				TOTAL INCL O&P	
								MAT.	LABOR	EQUIP.	TOTAL		
100	0010		**BITUMINOUS CONCRETE CURBS**										100
	0012		Curbs, asphaltic, machine formed, 8" wide, 6" high, 40 L.F./ton	B-27	1,000	.032	L.F.	.62	.87	.22	1.71	2.27	
	0100		8" wide, 8" high, 30 L.F. per ton		900	.036		.71	.97	.24	1.92	2.55	
	0150		Asphaltic berm, 12" W, 3"-6" H, 35 L.F./ton, before pavement		700	.046		.95	1.24	.31	2.50	3.33	
	0200		12" W, 1-1/2" to 4" H, 60 L.F. per ton, laid with pavement	B-2	1,050	.038		.58	1.03		1.61	2.25	
300	0010		**CEMENT CONCRETE CURBS**										300
	0300		Concrete, wood forms, 6" x 18", straight	C-2A	500	.096	L.F.	2.66	3.18		5.84	7.85	
	0400		6" x 18", radius		200	.240		2.77	7.95		10.72	15.30	
	0410		Steel forms, 6" x 18", straight		700	.069		3.55	2.27		5.82	7.40	
	0411		6" x 18", radius		400	.120		4.13	3.97		8.10	10.65	
	0415		Machine formed, 6" x 18", straight	B-69A	2,000	.024		3.62	.71	.29	4.62	5.40	
	0416		6" x 18", radius	"	900	.053		3.77	1.57	.65	5.99	7.25	
	0421		Curb and gutter, straight										
	0422		with 6" high curb and 6" thick gutter, wood forms										
	0430		24" wide, .055 C.Y. per L.F.	C-2A	375	.128	L.F.	13.50	4.24		17.74	21.50	
	0435		30" wide, .066 C.Y. per L.F.		340	.141		14.70	4.67		19.37	23.50	
	0440		Steel forms, 24" wide, straight		700	.069		5.90	2.27		8.17	10	
	0441		Radius		300	.160		5.90	5.30		11.20	14.65	
	0442		30" wide, straight		700	.069		7.05	2.27		9.32	11.30	
	0443		Radius		300	.160		7.05	5.30		12.35	15.95	
	0445		Machine formed, 24" wide, straight	B-69A	2,000	.024		5.90	.71	.29	6.90	7.90	
	0446		Radius		900	.053		5.90	1.57	.65	8.12	9.60	
	0447		30" wide, straight		2,000	.024		7.05	.71	.29	8.05	9.20	
	0448		Radius		900	.053		7.05	1.57	.65	9.27	10.90	
	0550		Precast, 6" x 18", straight	B-29	700	.080		7.35	2.30	1.20	10.85	12.95	
	0600		6" x 18", radius	"	325	.172		8.40	4.96	2.58	15.94	19.75	
500	0010		**STONE CURBS**										500
	1000		Granite, split face, straight, 5" x 16"	D-13	500	.096	L.F.	9.45	3.14	1.01	13.60	16.30	
	1100		6" x 18"	"	450	.107		12.40	3.49	1.12	17.01	20	
	1300		Radius curbing, 6" x 18", over 10' radius	B-29	260	.215		15.20	6.20	3.22	24.62	30	
	1400		Corners, 2' radius		80	.700	Ea.	51	20	10.45	81.45	98.50	
	1600		Edging, 4-1/2" x 12", straight		300	.187	L.F.	4.72	5.35	2.79	12.86	16.55	
	1800		Curb inlets, (guttermouth) straight		41	1.366	Ea.	113	39.50	20.50	173	208	
	2000		Indian granite (belgian block)										
	2100		Jumbo, 10-1/2" x 7-1/2" x 4", grey	D-1	150	.107	L.F.	1.72	3.32		5.04	6.95	
	2150		Pink		150	.107		2.24	3.32		5.56	7.50	
	2200		Regular, 9" x 4-1/2" x 4-1/2", grey		160	.100		1.56	3.11		4.67	6.45	
	2250		Pink		160	.100		2.14	3.11		5.25	7.10	
	2300		Cubes, 4" x 4" x 4", grey		175	.091		1.49	2.84		4.33	5.95	
	2350		Pink		175	.091		1.56	2.84		4.40	6.05	
	2400		6" x 6" x 6", pink		155	.103		3.86	3.21		7.07	9.15	
	2500		Alternate pricing method for indian granite										
	2550		Jumbo, 10-1/2" x 7-1/2" x 4" (30lb), grey				Ton	98			98	108	
	2600		Pink					130			130	143	
	2650		Regular, 9" x 4-1/2" x 4-1/2" (20lb), grey					110			110	121	
	2700		Pink					150			150	165	
	2750		Cubes, 4" x 4" x 4" (5lb), grey					180			180	198	
	2800		Pink					200			200	220	
	2850		6" x 6" x 6" (25lb), pink					150			150	165	
	2900		For pallets, add					16.50			16.50	18.15	

		02775	Sidewalks										
275	0010		**SIDEWALKS, DRIVEWAYS, & PATIOS** No base	R02065 -300									275
	0020		Asphaltic concrete, 2" thick		B-37	720	.067	S.Y.	3.65	1.87	.15	5.67	7.05

02775 | Sidewalks

		CREW	DAILY OUTPUT	LABOR-HOURS	UNIT	2005 BARE COSTS				TOTAL INCL O&P	
						MAT.	LABOR	EQUIP.	TOTAL		
275	0100	2-1/2" thick	B-37	660	.073	S.Y.	4.63	2.04	.16	6.83	8.45	275
	0300	Concrete, 3000 psi, CIP, 6 x 6 - W1.4 x W1.4 mesh, (R02065-300)										
	0310	broomed finish, no base, 4" thick	B-24	600	.040	S.F.	1.35	1.25		2.60	3.40
	0350	5" thick		545	.044		1.80	1.38		3.18	4.08
	0400	6" thick		510	.047		2.10	1.47		3.57	4.55
	0450	For bank run gravel base, 4" thick, add	B-18	2,500	.010		.41	.26	.01	.68	.89
	0520	8" thick, add	"	1,600	.015		.84	.41	.02	1.27	1.59
	0550	Exposed aggregate finish, add to above, minimum	B-24	1,875	.013		.10	.40		.50	.72
	0600	Maximum	"	455	.053		.35	1.65		2	2.89
	1000	Crushed stone, 1" thick, white marble	2 Clab	1,700	.009		.19	.25		.44	.60
	1050	Bluestone	"	1,700	.009		.21	.25		.46	.62
	1700	Redwood, prefabricated, 4' x 4' sections	2 Carp	316	.051		8.10	1.73		9.83	11.60
	1750	Redwood planks, 1" thick, on sleepers	"	240	.067		5.65	2.28		7.93	9.80
	2250	Stone dust, 4" thick	B-62	900	.027	S.Y.	2.88	.77	.17	3.82	4.54

02780 | Unit Pavers

		CREW	DAILY OUTPUT	LABOR-HOURS	UNIT	MAT.	LABOR	EQUIP.	TOTAL	INCL O&P	
100	0010	**ASPHALT BLOCKS**									100
	0020	Rectangular, 6" x 12" x 1-1/4", w/bed & neopr. adhesive	D-1	135	.119	S.F.	4.22	3.68		7.90	10.25
	0100	3" thick		130	.123		5.90	3.83		9.73	12.35
	0300	Hexagonal tile, 8" wide, 1-1/4" thick		135	.119		4.22	3.68		7.90	10.25
	0400	2" thick		130	.123		5.90	3.83		9.73	12.35
	0500	Square, 8" x 8", 1-1/4" thick		135	.119		4.22	3.68		7.90	10.25
	0600	2" thick		130	.123		5.90	3.83		9.73	12.35
	0900	For exposed aggregate (ground finish) add					.29			.29	.32
	0910	For colors, add					.22			.22	.24
200	0010	**BRICK PAVING** 4" x 8" x 1-1/2", without joints (4.5 brick/S.F.)	D-1	110	.145	S.F.	2.43	4.52		6.95	9.60	200
	0100	Grouted, 3/8" joint (3.9 brick/S.F.)		90	.178		2.78	5.55		8.33	11.45
	0200	4" x 8" x 2-1/4", without joints (4.5 bricks/S.F.)		110	.145		3.08	4.52		7.60	10.30
	0300	Grouted, 3/8" joint (3.9 brick/S.F.)		90	.178		2.84	5.55		8.39	11.50
	0500	Bedding, asphalt, 3/4" thick	B-25	5,130	.017		.34	.50	.39	1.23	1.58
	0540	Course washed sand bed, 1" thick	B-18	5,000	.005		.18	.13	.01	.32	.41
	0580	Mortar, 1" thick	D-1	300	.053		.29	1.66		1.95	2.85
	0620	2" thick		200	.080		.58	2.49		3.07	4.43
	1500	Brick on 1" thick sand bed laid flat, 4.5 per S.F.		100	.160		2.49	4.97		7.46	10.35
	2000	Brick pavers, laid on edge, 7.2 per S.F.		70	.229		2.36	7.10		9.46	13.40
	2500	For 4" thick concrete bed and joints, add		595	.027		.89	.84		1.73	2.25
	2800	For steam cleaning, add	A-1H	950	.008		.05	.22	.06	.33	.47
400	0010	**PRECAST CONCRETE PAVING SLABS**										400
	0710	Precast concrete patio blocks, 2-3/8" thick, colors, 8" x 16"	D-1	265	.060	S.F.	1.18	1.88		3.06	4.16
	0750	Exposed local aggregate, natural	2 Bric	250	.064		5.05	2.26		7.31	9
	0800	Colors		250	.064		5.50	2.26		7.76	9.50
	0850	Exposed granite or limestone aggregate		250	.064		6.10	2.26		8.36	10.20
	0900	Exposed white tumblestone aggregate		250	.064		3.62	2.26		5.88	7.40
410	0010	**PLANTER BLOCKS** Precast concrete, interlocking										410
	0020	"V" blocks for retaining soil	D-1	205	.078	S.F.	4.39	2.43		6.82	8.55
600	0010	**STONE PAVERS**										600
	1100	Flagging, bluestone, irregular, 1" thick,	D-1	81	.198	S.F.	4.63	6.15		10.78	14.45
	1150	Snapped random rectangular, 1" thick		92	.174		7	5.40		12.40	15.95
	1200	1-1/2" thick		85	.188		8.45	5.85		14.30	18.15
	1250	2" thick		83	.193		9.80	6		15.80	19.95
	1300	Slate, natural cleft, irregular, 3/4" thick		92	.174		5.15	5.40		10.55	13.95
	1350	Random rectangular, gauged, 1/2" thick		105	.152		11.15	4.74		15.89	19.50
	1400	Random rectangular, butt joint, gauged, 1/4" thick		150	.107		12	3.32		15.32	18.30
	1500	For interior setting, add								25%	25%
	1550	Granite blocks, 3-1/2" x 3-1/2" x 3-1/2"	D-1	92	.174	S.F.	6.25	5.40		11.65	15.15

Important: See the Reference Section for critical supporting data - Reference Nos., Crews, & City Cost Indexes

SITE CONSTRUCTION — 2

02780	Unit Pavers	CREW	DAILY OUTPUT	LABOR-HOURS	UNIT	2005 BARE COSTS				TOTAL INCL O&P	
						MAT.	LABOR	EQUIP.	TOTAL		
600	1600	4" to 12" long, 3" to 5" wide, 3" to 5" thick	D-1	98	.163	S.F.	5.20	5.05		10.25	13.50	**600**
	1650	6" to 15" long, 3" to 6" wide, 3" to 5" thick	↓	105	.152	↓	2.78	4.74		7.52	10.25	
700	0010	**STEPS** Incl. excav., borrow & concrete base, where applicable										**700**
	0100	Brick steps	B-24	35	.686	LF Riser	9.10	21.50		30.60	42.50	
	0200	Railroad ties	2 Clab	25	.640	↓	2.89	17.10		19.99	29.50	
	0300	Bluestone treads, 12" x 2" or 12" x 1-1/2"	B-24	30	.800	↓	22	25		47	62	
	0500	Concrete, cast in place, see division 03310-240										
	0600	Precast concrete, see division 03480-800										
	4025	Steel edge strips, incl. stakes, 1/4" x 5"	B-1	390	.062	L.F.	3.35	1.68		5.03	6.30	
	4050	Edging, landscape timber or railroad ties, 6" x 8"	2 Carp	170	.094	"	2.56	3.22		5.78	7.80	

02785	Flexible Pavement Coating	CREW	DAILY OUTPUT	LABOR-HOURS	UNIT	MAT.	LABOR	EQUIP.	TOTAL	TOTAL INCL O&P	
250	0010	**FOG SEAL**										**250**
	0012	Sealcoating, 2 coat coal tar pitch emulsion over 10,000 SY	B-45	5,000	.003	S.Y.	.46	.10	.11	.67	.78	
	0030	1000 to 10,000 S.Y.	"	3,000	.005		.46	.17	.18	.81	.95	
	0100	Under 1000 S.Y.	B-1	1,050	.023		.46	.63		1.09	1.48	
	0300	Petroleum resistant, over 10,000 S.Y.	B-45	5,000	.003		.57	.10	.11	.78	.90	
	0320	1000 to 10,000 S.Y.	"	3,000	.005		.57	.17	.18	.92	1.07	
	0400	Under 1000 S.Y.	B-1	1,050	.023		.57	.63		1.20	1.60	
	0600	Non-skid pavement renewal, over 10,000 S.Y.	B-45	5,000	.003		.67	.10	.11	.88	1.01	
	0620	1000 to 10,000 S.Y.	"	3,000	.005		.67	.17	.18	1.02	1.18	
	0700	Under 1000 S.Y.	B-1	1,050	.023		.67	.63		1.30	1.71	
	0800	Prepare and clean surface for above	A-2	8,545	.003	↓		.08	.02	.10	.14	
	1000	Hand seal asphalt curbing	B-1	4,420	.005	L.F.	.34	.15		.49	.60	
	1900	Asphalt surface treatment, single course, small area										
	1901	0.30 gal/S.Y. asphalt material, 20#/S.Y. aggregate	B-91	5,000	.013	S.Y.	.79	.40	.29	1.48	1.79	
	1910	Roadway or large area		10,000	.006		.73	.20	.14	1.07	1.26	
	1950	Asphalt surface treatment, dbl. course for small area		3,000	.021		1.46	.66	.48	2.60	3.14	
	1960	Roadway or large area		6,000	.011		1.31	.33	.24	1.88	2.22	
	1980	Asphalt surface treatment, single course, for shoulders	↓	7,500	.009	↓	.84	.26	.19	1.29	1.53	
300	0010	**LATEX MODIFIED EMULSION**										**300**
	3780	Rubberized asphalt (latex) seal	B-45	5,000	.003	S.Y.	1.05	.10	.11	1.26	1.43	
400	0010	**SAND SEAL**										**400**
	2080	Sand sealing, sharp sand, asphalt emulsion, small area	B-91	10,000	.006	S.Y.	.57	.20	.14	.91	1.09	
	2120	Roadway or large area	"	18,000	.004	"	.49	.11	.08	.68	.80	

02790	Athletic/Recreational Surfaces	CREW	DAILY OUTPUT	LABOR-HOURS	UNIT	MAT.	LABOR	EQUIP.	TOTAL	TOTAL INCL O&P	
400	0010	**SYNTHETIC GRASS SURFACING**										**400**
	0015	Not including asphalt base or drainage,										
	0020	but including cushion pad, over 50,000 S.F.										
	0200	1/2" pile and 5/16" cushion pad, standard	C-17	3,200	.025	S.F.	6.50	.88		7.38	8.50	
	0300	Deluxe		2,560	.031		7.70	1.10		8.80	10.15	
	0500	1/2" pile and 5/8" cushion pad, standard		2,844	.028		9.45	.99		10.44	11.95	
	0600	Deluxe	↓	2,327	.034	↓	10.35	1.21		11.56	13.25	
	0800	For asphaltic concrete base, 2-1/2" thick,										
	0900	with 6" crushed stone sub-base, add	B-25	12,000	.007	S.F.	1.27	.21	.17	1.65	1.91	
500	0010	**SYNTHETIC RUNNING TRACK SURFACING**										**500**
	0020	Running track, asphalt, incl base, 3" thick	B-37	300	.160	S.Y.	12.35	4.49	.36	17.20	21	
	0100	Surface, latex rubber system, 3/8" thick, black	B-20	125	.192		5.55	5.80		11.35	15.10	
	0150	Colors		125	.192		9.85	5.80		15.65	19.85	
	0300	Urethane rubber system, 3/8" thick, black		120	.200		14.80	6		20.80	25.50	
	0400	Color coating	↓	115	.209	↓	17.55	6.30		23.85	29	
600	0010	**TENNIS COURT SURFACING**										**600**
	0020	Tennis court, asphalt, incl. base, 2-1/2" thick, one court	B-37	450	.107	S.Y.	14.45	3	.24	17.69	20.50	

SITE CONSTRUCTION (margin)

2 (margin)

02790	Athletic/Recreational Surfaces	CREW	DAILY OUTPUT	LABOR-HOURS	UNIT	2005 BARE COSTS				TOTAL INCL O&P	
						MAT.	LABOR	EQUIP.	TOTAL		
600	0200	Two courts	B-37	675	.071	S.Y.	7.75	2	.16	9.91	11.75	600
	0300	Clay courts		360	.133		30	3.75	.30	34.05	39
	0400	Pulverized natural greenstone with 4" base, fast dry		250	.192		28	5.40	.43	33.83	40
	0800	Rubber-acrylic base resilient pavement		600	.080		28.50	2.25	.18	30.93	34.50
	1000	Colored sealer, acrylic emulsion, 3 coats	2 Clab	800	.020		3.72	.53		4.25	4.92
	1100	3 coat, 2 colors	"	900	.018		5.50	.47		5.97	6.80
	1200	For preparing old courts, add	1 Clab	825	.010	↓		.26		.26	.40
	1400	Posts for nets, 3-1/2" diameter with eye bolts	B-1	3.40	7.059	Pr.	168	193		361	485
	1500	With pulley & reel		3.40	7.059	"	239	193		432	565
	1700	Net, 42' long, nylon thread with binder		50	.480	Ea.	222	13.15		235.15	265
	1800	All metal	↓	6.50	3.692	"	385	101		486	575
	2000	Paint markings on asphalt, 2 coats	1 Pord	1.78	4.494	Court	64.50	138		202.50	278
	2200	Complete court with fence, etc., asphaltic conc., minimum	B-37	.20	240		13,200	6,750	540	20,490	25,500
	2300	Maximum		.16	300		16,400	8,425	670	25,495	31,700
	2800	Clay courts, minimum		.20	240		15,500	6,750	540	22,790	28,000
	2900	Maximum	↓	.16	300	↓	18,600	8,425	670	27,695	34,100

02800 | Site Improvements and Amenities

02810	Irrigation System	CREW	DAILY OUTPUT	LABOR-HOURS	UNIT	2005 BARE COSTS				TOTAL INCL O&P	
						MAT.	LABOR	EQUIP.	TOTAL		
300	0010	**SPRINKLER IRRIGATION SYSTEM** For lawns										300
	0100	Golf course with fully automatic system	C-17	.05	1,600	9 holes	82,500	56,500		139,000	178,500
	0200	24' diam. head at 15' O.C incl. piping, auto oper., minimum	B-20	70	.343	Head	18.20	10.30		28.50	36
	0300	Maximum		40	.600		42	18.05		60.05	74
	0500	60' diam. head at 40' O.C. incl. piping, auto oper., minimum		28	.857	↓	55	26		81	101
	0600	Maximum		23	1.043		154	31.50		185.50	219
	0800	Residential system, custom, 1" supply		2,000	.012	S.F.	.27	.36		.63	.86
	0900	1-1/2" supply	↓	1,800	.013	"	.32	.40		.72	.97
	1020	Pop up spray head w/risers, hi-pop, full circle pattern, 4"	2 Skwk	76	.211	Ea.	3.36	7.35		10.71	15.10
	1030	1/2 circle pattern		76	.211		3.43	7.35		10.78	15.15
	1040	6", full circle pattern		76	.211		8.60	7.35		15.95	21
	1050	1/2 circle pattern		76	.211		8.60	7.35		15.95	21
	1060	12", full circle pattern		76	.211		12.25	7.35		19.60	25
	1070	1/2 circle pattern		76	.211		12.30	7.35		19.65	25
	1080	Pop up bubbler head w/risers, hi-pop bubbler head, 4"		76	.211		3.50	7.35		10.85	15.25
	1090	6"		76	.211		8.60	7.35		15.95	21
	1100	12"		76	.211		12.25	7.35		19.60	25
	1110	Impact full/part circle sprinklers, 28'-54' 25-60 PSI		37	.432		11.45	15.05		26.50	36
	1120	Spaced 37'-49' @ 25-50 PSI		37	.432		25	15.05		40.05	51
	1130	Spaced 43'-61' @ 30-60 PSI		37	.432		45.50	15.05		60.55	73.50
	1140	Spaced 54'-78' @ 40-80 PSI	↓	37	.432	↓	105	15.05		120.05	139
	1145	Impact rotor pop-up full/part commercial circle sprinklers									
	1150	Spaced 42'-65' 35-80 PSI	2 Skwk	25	.640	Ea.	23	22.50		45.50	60
	1160	Spaced 48'-76' 45-85 PSI	"	25	.640	"	8.85	22.50		31.35	44
	1165	Impact rotor pop-up part. circle comm., 53'-75', 55-100 PSI, w/ acc									
	1170	Plastic case, metal cover	2 Skwk	25	.640	Ea.	148	22.50		170.50	198
	1180	Rubber cover		25	.640		112	22.50		134.50	158
	1190	Iron case, metal cover		22	.727		136	25.50		161.50	189
	1200	Rubber cover		22	.727		147	25.50		172.50	201
	1250	Plastic case, 2 nozzle, metal cover	↓	25	.640		108	22.50		130.50	154

Important: See the Reference Section for critical supporting data - Reference Nos., Crews, & City Cost Indexes

SITE CONSTRUCTION 2

02810	Irrigation System	CREW	DAILY OUTPUT	LABOR-HOURS	UNIT	2005 BARE COSTS				TOTAL INCL O&P	
						MAT.	LABOR	EQUIP.	TOTAL		
300											**300**
1260	Rubber cover	2 Skwk	25	.640	Ea.	115	22.50		137.50	161	
1270	Iron case, 2 nozzle, metal cover		22	.727		157	25.50		182.50	213	
1280	Rubber cover	↓	22	.727	↓	164	25.50		189.50	220	
1282	Impact rotor pop-up full circle comm., 39'-99', 30-100 PSI										
1284	Plastic case, metal cover	2 Skwk	25	.640	Ea.	113	22.50		135.50	159	
1286	Rubber cover		25	.640		125	22.50		147.50	173	
1288	Iron case, metal cover		22	.727		162	25.50		187.50	219	
1290	Rubber cover		22	.727		171	25.50		196.50	228	
1292	Plastic case, 2 nozzle, metal cover		22	.727		122	25.50		147.50	174	
1294	Rubber cover		22	.727		126	25.50		151.50	179	
1296	Iron case, 2 nozzle, metal cover		20	.800		157	28		185	216	
1298	Rubber cover		20	.800		167	28		195	228	
1305	Electric remote control valve, plastic, 3/4"		18	.889		17.05	31		48.05	67	
1310	1"		18	.889		30	31		61	81	
1320	1-1/2"	↓	18	.889	↓	58	31		89	112	
1335	Quick coupling valves, brass, locking cover										
1340	Inlet coupling valve, 3/4"	2 Skwk	18.75	.853	Ea.	42	29.50		71.50	92.50	
1350	1"		18.75	.853		51	29.50		80.50	103	
1360	Controller valve boxes, 6" round boxes		18.75	.853		3.73	29.50		33.23	50.50	
1370	10" round boxes		14.25	1.123		8.15	39		47.15	70	
1380	12" square box	↓	9.75	1.641	↓	16.40	57		73.40	107	
1388	Electromech. control, 14 day 3-60 min, auto start to 23/day										
1390	4 station	2 Skwk	1.04	15.385	Ea.	177	535		712	1,025	
1400	7 station		.64	25		229	870		1,099	1,600	
1410	12 station		.40	40		284	1,400		1,684	2,475	
1420	Dual programs, 18 station		.24	66.667		1,400	2,325		3,725	5,175	
1430	23 station	↓	.16	100	↓	1,825	3,475		5,300	7,450	
1435	Backflow preventer, bronze, 0-175 PSI, w/valves, test cocks										
1440	3/4"	2 Skwk	2	8	Ea.	115	279		394	560	
1450	1"		2	8		120	279		399	565	
1460	1-1/2"		2	8		239	279		518	695	
1470	2"	↓	2	8		249	279		528	710	
1475	Pressure vacuum breaker, brass, 15-150 PSI										
1480	3/4"	2 Skwk	2	8	Ea.	69.50	279		348.50	510	
1490	1"		2	8		91	279		370	535	
1500	1-1/2"		2	8		180	279		459	635	
1510	2"	↓	2	8	↓	167	279		446	620	

02820	Fences & Gates	CREW	DAILY OUTPUT	LABOR-HOURS	UNIT	MAT.	LABOR	EQUIP.	TOTAL	TOTAL INCL O&P	
130											**130**
0010	**FENCE, CHAIN LINK INDUSTRIAL**, schedule 40										
0020	3 strands barb wire, 2" post @ 10' O.C., set in concrete, 6' H										
0200	9 ga. wire, galv. steel	B-80C	240	.100	L.F.	11.80	2.67	.56	15.03	17.75	
0300	Aluminized steel		240	.100		15.15	2.67	.56	18.38	21.50	
0500	6 ga. wire, galv. steel		240	.100		18.50	2.67	.56	21.73	25	
0600	Aluminized steel		240	.100		21	2.67	.56	24.23	27.50	
0800	6 ga. wire, 6' high but omit barbed wire, galv. steel		250	.096		17.95	2.57	.54	21.06	24.50	
0900	Aluminized steel		250	.096		24.50	2.57	.54	27.61	31.50	
0920	8' H, 6 ga. wire, 2-1/2" line post, galv. steel		180	.133		28.50	3.56	.74	32.80	38	
0940	Aluminized steel		180	.133	↓	35	3.56	.74	39.30	45	
1100	Add for corner posts, 3" diam., galv. steel		40	.600	Ea.	82	16.05	3.35	101.40	119	
1200	Aluminized steel		40	.600		98.50	16.05	3.35	117.90	137	
1300	Add for braces, galv. steel		80	.300		22.50	8	1.67	32.17	38.50	
1350	Aluminized steel		80	.300		30	8	1.67	39.67	47	
1400	Gate for 6' high fence, 1-5/8" frame, 3' wide, galv. steel		10	2.400		131	64	13.40	208.40	258	
1500	Aluminized steel	↓	10	2.400	↓	161	64	13.40	238.40	291	

2 SITE CONSTRUCTION

02820	Fences & Gates	CREW	DAILY OUTPUT	LABOR-HOURS	UNIT	MAT.	LABOR	EQUIP.	TOTAL	TOTAL INCL O&P	
130	2000	5'-0" high fence, 9 ga., no barbed wire, 2" line post,										**130**
	2010	10' O.C., 1-5/8" top rail										
	2100	Galvanized steel	B-80C	300	.080	L.F.	10.25	2.14	.45	12.84	15.05	
	2200	Aluminized steel		300	.080	"	11.85	2.14	.45	14.44	16.85	
	2400	Gate, 4' wide, 5' high, 2" frame, galv. steel		10	2.400	Ea.	149	64	13.40	226.40	278	
	2500	Aluminized steel		10	2.400	"	164	64	13.40	241.40	295	
	3100	Overhead slide gate, chain link, 6' high, to 18' wide		38	.632	L.F.	131	16.90	3.52	151.42	175	
	3110	Cantilever type	B-80	48	.667		56.50	19.20	11.10	86.80	104	
	3120	8' high		24	1.333		82	38.50	22	142.50	174	
	3130	10' high		18	1.778		97	51	29.50	177.50	218	
	5000	Double swing gates, incl. posts & hardware										
	5010	5' high, 12' opening	B-80C	3.40	7.059	Opng.	400	189	39.50	628.50	775	
	5020	20' opening		2.80	8.571		545	229	48	822	1,000	
	5060	6' high, 12' opening		3.20	7.500		675	200	42	917	1,100	
	5070	20' opening		2.60	9.231		930	247	51.50	1,228.50	1,450	
	5080	8' high, 12' opening	B-80	2.13	15.002		1,050	430	250	1,730	2,100	
	5090	20' opening		1.45	22.069		1,375	635	370	2,380	2,900	
	5100	10' high, 12' opening		1.31	24.427		1,200	705	405	2,310	2,850	
	5110	20' opening		1.03	31.068		1,800	895	520	3,215	3,925	
	5120	12' high, 12' opening		1.05	30.476		1,750	880	510	3,140	3,825	
	5130	20' opening		.85	37.647		2,250	1,075	630	3,955	4,825	
	5190	For aluminized steel add					20%					
	7001	Snow fence on steel posts 10' O.C., 4' high	B-1	500	.048	L.F.	2.20	1.31		3.51	4.46	
140	0010	**FENCE, CHAIN LINK RESIDENTIAL**, sch. 20, 11 ga. wire, 1-5/8" post										**140**
	0020	10' O.C., 1-3/8" top rail, 2" corner post, galv. stl. 3' high	B-80C	500	.048	L.F.	3.58	1.28	.27	5.13	6.20	
	0050	4' high		400	.060		4.06	1.60	.33	5.99	7.30	
	0100	6' high		200	.120		4.92	3.21	.67	8.80	11.10	
	0150	Add for gate 3' wide, 1-3/8" frame, 3' high		12	2	Ea.	53.50	53.50	11.15	118.15	154	
	0170	4' high		10	2.400		61.50	64	13.40	138.90	181	
	0190	6' high		10	2.400		84.50	64	13.40	161.90	206	
	0200	Add for gate 4' wide, 1-3/8" frame, 3' high		9	2.667		60	71.50	14.90	146.40	192	
	0220	4' high		9	2.667		68	71.50	14.90	154.40	201	
	0240	6' high		8	3		80	80	16.75	176.75	230	
	0350	Aluminized steel, 11 ga. wire, 3' high		500	.048	L.F.	4.64	1.28	.27	6.19	7.35	
	0380	4' high		400	.060		5.95	1.60	.33	7.88	9.40	
	0400	6' high		200	.120		8.40	3.21	.67	12.28	14.90	
	0450	Add for gate 3' wide, 1-3/8" frame, 3' high		12	2	Ea.	64	53.50	11.15	128.65	165	
	0470	4' high		10	2.400		105	64	13.40	182.40	230	
	0490	6' high		10	2.400		132	64	13.40	209.40	260	
	0500	Add for gate 4' wide, 1-3/8" frame, 3' high		10	2.400		72.50	64	13.40	149.90	194	
	0520	4' high		9	2.667		80.50	71.50	14.90	166.90	215	
	0540	6' high		8	3		93	80	16.75	189.75	244	
	0620	Vinyl covered, 9 ga. wire, 3' high		500	.048	L.F.	3.86	1.28	.27	5.41	6.50	
	0640	4' high		400	.060		4.51	1.60	.33	6.44	7.80	
	0660	6' high		200	.120		5.70	3.21	.67	9.58	12	
	0720	Add for gate 3' wide, 1-3/8" frame, 3' high		12	2	Ea.	74.50	53.50	11.15	139.15	177	
	0740	4' high		10	2.400		84	64	13.40	161.40	206	
	0760	6' high		10	2.400		131	64	13.40	208.40	258	
	0780	Add for gate 4' wide, 1-3/8" frame, 3' high		10	2.400		85.50	64	13.40	162.90	208	
	0800	4' high		9	2.667		93	71.50	14.90	179.40	228	
	0820	6' high		8	3		105	80	16.75	201.75	258	
410	0010	**FENCES, MISC. METAL**										**410**
	0012	Chicken wire, posts @ 4', 1" mesh, 4' high	B-80C	410	.059	L.F.	1.63	1.56	.33	3.52	4.57	
	0100	2" mesh, 6' high		350	.069		1.47	1.83	.38	3.68	4.87	
	0200	Galv. steel, 12 ga., 2" x 4" mesh, posts 5' O.C., 3' high		300	.080		2.21	2.14	.45	4.80	6.25	

Important: See the Reference Section for critical supporting data - Reference Nos., Crews, & City Cost Indexes

			CREW	DAILY OUTPUT	LABOR-HOURS	UNIT	MAT.	LABOR	EQUIP.	TOTAL	TOTAL INCL O&P	
02820		**Fences & Gates**										
410	0300	5' high	B-80C	300	.080	L.F.	2.95	2.14	.45	5.54	7.05	**410**
	0400	14 ga., 1" x 2" mesh, 3' high		300	.080		2.35	2.14	.45	4.94	6.40	
	0500	5' high		300	.080		3.25	2.14	.45	5.84	7.40	
	1000	Kennel fencing, 1-1/2" mesh, 6' long, 3'-6" wide, 6'-2" high	2 Clab	4	4	Ea.	370	107		477	570	
	1050	12' long		4	4		445	107		552	650	
	1200	Top covers, 1-1/2" mesh, 6' long		15	1.067		75	28.50		103.50	127	
	1250	12' long		12	1.333		120	35.50		155.50	188	
	1300	For kennel doors, see division 08310-400										
	4500	Security fence, prison grade, set in concrete, 12' high	B-80	25	1.280	L.F.	30.50	37	21.50	89	114	
	4600	16' high	"	20	1.600		36.50	46	26.50	109	141	
	5300	Tubular picket, steel, 6' sections, 1-9/16" posts, 4' high	B-80C	300	.080		23	2.14	.45	25.59	29	
	5400	2" posts, 5' high		240	.100		32	2.67	.56	35.23	39.50	
	5600	2" posts, 6' high		200	.120		36	3.21	.67	39.88	45	
	5700	Staggered picket 1-9/16" posts, 4' high		300	.080		20.50	2.14	.45	23.09	26.50	
	5800	2" posts, 5' high		240	.100		34	2.67	.56	37.23	42	
	5900	2" posts, 6' high		200	.120		35	3.21	.67	38.88	44	
	6200	Gates, 4' high, 3' wide	B-1	10	2.400	Ea.	199	65.50		264.50	320	
	6300	5' high, 3' wide		10	2.400		259	65.50		324.50	385	
	6400	6' high, 3' wide		10	2.400		266	65.50		331.50	395	
	6500	4' wide		10	2.400		310	65.50		375.50	440	
420	0010	**WIRE FENCING, GENERAL**										**420**
	0015	Barbed wire, galvanized, domestic steel, hi-tensile 15-1/2 ga.				M.L.F.	27			27	29.50	
	0020	Standard, 12-3/4 ga.					36			36	39.50	
	0210	Barbless wire, 2-strand galvanized, 12-1/2 ga.					36			36	39.50	
	0500	Helical razor ribbon, stainless steel, 18" dia x 18" spacing				C.L.F.	105			105	115	
	0600	Hardware cloth galv., 1/4" mesh, 23 ga., 2' wide				C.S.F.	47			47	51.50	
	0700	3' wide					45.50			45.50	50	
	0900	1/2" mesh, 19 ga., 2' wide					41.50			41.50	45.50	
	1000	4' wide					40.50			40.50	44.50	
	1200	Chain link fabric, steel, 2" mesh, 6 ga, galvanized					155			155	170	
	1300	9 ga, galvanized					76.50			76.50	84	
	1350	Vinyl coated					64			64	70.50	
	1360	Aluminized					99			99	109	
	1400	2-1/4" mesh, 11.5 ga, galvanized					51.50			51.50	57	
	1600	1-3/4" mesh (tennis courts), 11.5 ga (core), vinyl coated					73			73	80.50	
	1700	9 ga, galvanized					66			66	72.50	
	2100	Welded wire fabric, galvanized, 1" x 2", 14 ga.					44			44	48	
	2200	2" x 4", 12-1/2 ga.					14.95			14.95	16.45	
520	0010	**FENCE, WOOD RAIL** Picket, No. 2 cedar, Gothic, 2 rail, 3' high	B-1	160	.150	L.F.	5.15	4.11		9.26	12.05	**520**
	0050	Gate, 3'-6" wide	B-80C	9	2.667	Ea.	44.50	71.50	14.90	130.90	175	
	0400	3 rail, 4' high		150	.160	L.F.	5.90	4.28	.89	11.07	14.10	
	0500	Gate, 3'-6" wide		9	2.667	Ea.	53.50	71.50	14.90	139.90	185	
	0600	Open rail, rustic, No. 1 cedar, 2 rail, 3' high		160	.150	L.F.	4.61	4.01	.84	9.46	12.15	
	0650	Gate, 3' wide		9	2.667	Ea.	51.50	71.50	14.90	137.90	183	
	0700	3 rail, 4' high		150	.160	L.F.	5.30	4.28	.89	10.47	13.40	
	0900	Gate, 3' wide		9	2.667	Ea.	66	71.50	14.90	152.40	199	
	1200	Stockade, No. 2 cedar, treated wood rails, 6' high		160	.150	L.F.	6.45	4.01	.84	11.30	14.15	
	1250	Gate, 3' wide		9	2.667	Ea.	52.50	71.50	14.90	138.90	184	
	1300	No. 1 cedar, 3-1/4" cedar rails, 6' high		160	.150	L.F.	16.15	4.01	.84	21	25	
	1500	Gate, 3' wide		9	2.667	Ea.	134	71.50	14.90	220.40	274	
	1520	Open rail, split, No. 1 cedar, 2 rail, 3' high		160	.150	L.F.	4.71	4.01	.84	9.56	12.30	
	1540	3 rail, 4'-0" high		150	.160		6.10	4.28	.89	11.27	14.30	
	3300	Board, shadow box, 1" x 6", treated pine, 6' high		160	.150		9.50	4.01	.84	14.35	17.55	
	3400	No. 1 cedar, 6' high		150	.160		18.75	4.28	.89	23.92	28	
	3900	Basket weave, No. 1 cedar, 6' high		160	.150		18.55	4.01	.84	23.40	27.50	
	3950	Gate, 3'-6" wide	B-1	8	3	Ea.	127	82		209	267	

SITE CONSTRUCTION 2

SITE CONSTRUCTION **2**

		02820	Fences & Gates	CREW	DAILY OUTPUT	LABOR-HOURS	UNIT	2005 BARE COSTS				TOTAL INCL O&P	
								MAT.	LABOR	EQUIP.	TOTAL		
520	4000		Treated pine, 6' high	B-1	150	.160	L.F.	11	4.38		15.38	18.90	520
	4200		Gate, 3'-6" wide		9	2.667	Ea.	59	73		132	179	
	5000		Fence rail, redwood, 2" x 4", merch grade 8'		2,400	.010	L.F.	1.01	.27		1.28	1.54	
	5050		Select grade, 8'		2,400	.010	"	3.12	.27		3.39	3.86	
	6000		Fence post, select redwood, earthpacked & treated, 4" x 4" x 6'		96	.250	Ea.	10	6.85		16.85	21.50	
	6010		4" x 4" x 8'		96	.250		11.85	6.85		18.70	23.50	
	6020		Set in concrete, 4" x 4" x 6'		50	.480		12.85	13.15		26	34.50	
	6030		4" x 4" x 8'		50	.480		15.35	13.15		28.50	37.50	
	6040		Wood post, 4' high, set in concrete, incl. concrete		50	.480		7.50	13.15		20.65	29	
	6050		Earth packed		96	.250		5.10	6.85		11.95	16.30	
	6060		6' high, set in concrete, incl. concrete		50	.480		9.60	13.15		22.75	31	
	6070		Earth packed		96	.250		7	6.85		13.85	18.35	

		02830	Retaining Walls										
100	0010		**CAST IN PLACE RETAINING WALLS**										100
	1800		Concrete gravity wall with vertical face including excavation & backfill										
	1850		No reinforcing										
	1900		6' high, level embankment	C-17C	36	2.306	L.F.	59.50	81.50	10.40	151.40	203	
	2000		33° slope embankment		32	2.594		54	91.50	11.65	157.15	214	
	2200		8' high, no surcharge		27	3.074		73.50	108	13.85	195.35	264	
	2300		33° slope embankment		24	3.458		89	122	15.55	226.55	305	
	2500		10' high, level embankment		19	4.368		105	154	19.65	278.65	375	
	2600		33° slope embankment		18	4.611		146	163	21	330	435	
	2800		Reinforced concrete cantilever, incl. excavation, backfill & reinf.										
	2900		6' high, 33° slope embankment	C-17C	35	2.371	L.F.	54	83.50	10.65	148.15	201	
	3000		8' high, 33° slope embankment		29	2.862		62.50	101	12.90	176.40	240	
	3100		10' high, 33° slope embankment		20	4.150		81	146	18.70	245.70	335	
	3200		20' high, 500 lb. per L.F. surcharge		7.50	11.067		243	390	50	683	925	
	3500		Concrete cribbing, incl. excavation and backfill										
	3700		12' high, open face	B-13	210	.267	S.F.	27	7.70	2.94	37.64	44.50	
	3900		Closed face	"	210	.267	"	25.50	7.70	2.94	36.14	43	
	4100		Concrete filled slurry trench, see Div. 02660-600										
200	0010		**INTERLOCKING SEGMENTAL RETAINING WALLS**										200
	7100		Segmental Retaining Wall system, incl pins, and void fill										
	7120		base not included										
	7140		Large unit, 8" high x 18" wide x 20" deep, 3 plane split	B-62	300	.080	S.F.	8.55	2.31	.51	11.37	13.55	
	7150		straight split		300	.080		8.55	2.31	.51	11.37	13.55	
	7160		Medium, ltwt, 8" high x 18" wide x 12" deep, 3 plane split		400	.060		7.55	1.73	.38	9.66	11.40	
	7170		straight split		400	.060		7.55	1.73	.38	9.66	11.40	
	7180		Small unit, 4" x 18" x 10" deep, 3 plane split		400	.060		9.45	1.73	.38	11.56	13.50	
	7190		straight split		400	.060		9.45	1.73	.38	11.56	13.50	
	7200		Cap unit, 3 plane split		300	.080		10.60	2.31	.51	13.42	15.80	
	7210		Cap unit, st split		300	.080		10.60	2.31	.51	13.42	15.80	
	7260		For reinforcing, add								2.20	2.75	
	8000		For higher walls, add components as necessary										
600	0010		**METAL BIN RETAINING WALLS,** Aluminized steel bin, excavation										600
	0020		and backfill not included, 10' wide										
	0100		4' high, 5.5' deep	B-13	650	.086	S.F.	16.30	2.48	.95	19.73	23	
	0200		8' high, 5.5' deep		615	.091		18.75	2.62	1	22.37	25.50	
	0300		10' high, 7.7' deep		580	.097		19.75	2.78	1.06	23.59	27	
	0400		12' high, 7.7' deep		530	.106		21.50	3.04	1.16	25.70	29.50	
	0500		16' high, 7.7' deep		515	.109		22.50	3.13	1.20	26.83	31	
	0600		16' high, 9.9' deep		500	.112		23.50	3.22	1.23	27.95	32.50	
	0700		20' high, 9.9' deep		470	.119		26.50	3.43	1.31	31.24	36	
	0800		20' high, 12.1' deep		460	.122		28.50	3.50	1.34	33.34	38.50	

Important: See the Reference Section for critical supporting data - Reference Nos., Crews, & City Cost Indexes

SITE CONSTRUCTION 2

02830 | Retaining Walls

		CREW	DAILY OUTPUT	LABOR-HOURS	UNIT	MAT.	LABOR	EQUIP.	TOTAL	TOTAL INCL O&P	
						2005 BARE COSTS					
600	0900	24' high, 12.1' deep	B-13	455	.123	S.F.	30.50	3.54	1.36	35.40	40.50	600
	1000	24' high, 14.3' deep		450	.124		31.50	3.58	1.37	36.45	42	
	1100	28' high, 14.3' deep	↓	440	.127	↓	33	3.66	1.40	38.06	43	
	1300	For plain galvanized bin type walls, deduct					10%					
700	0010	**STONE GABION RETAINING WALLS**										700
	4300	Stone filled gabions, not incl. excavation,										
	4310	Stone, delivered, 3' wide										
	4350	Galvanized, 6' high, 33° slope embankment	B-13	49	1.143	L.F.	18.55	33	12.60	64.15	85	
	4500	Highway surcharge		27	2.074		37	59.50	23	119.50	158	
	4600	9' high, up to 33° slope embankment		24	2.333		42	67	25.50	134.50	178	
	4700	Highway surcharge		16	3.500		65	101	38.50	204.50	269	
	4900	12' high, up to 33° slope embankment		14	4		65	115	44	224	297	
	5000	Highway surcharge	↓	11	5.091	↓	93	147	56	296	390	
	5950	For PVC coating, add					20%					
800	0010	**STONE RETAINING WALLS**										800
	0015	Including excavation, concrete footing and										
	0020	stone 3' below grade. Price is exposed face area.										
	0200	Decorative random stone, to 6' high, 1'-6" thick, dry set	D-1	35	.457	S.F.	35	14.20		49.20	60	
	0300	Mortar set		40	.400		35	12.45		47.45	57.50	
	0500	Cut stone, to 6' high, 1'-6" thick, dry set		35	.457		35	14.20		49.20	60	
	0600	Mortar set		40	.400		35	12.45		47.45	57.50	
	0800	Random stone, 6' to 10' high, 2' thick, dry set		45	.356		35	11.05		46.05	55.50	
	0900	Mortar set		50	.320		35	9.95		44.95	53.50	
	1100	Cut stone, 6' to 10' high, 2' thick, dry set		45	.356		35	11.05		46.05	55.50	
	1200	Mortar set	↓	50	.320	↓	35	9.95		44.95	53.50	

02840 | Walk/Road/Parking Appurtenances

		CREW	DAILY OUTPUT	LABOR-HOURS	UNIT	MAT.	LABOR	EQUIP.	TOTAL	TOTAL INCL O&P	
200	0010	**GUIDE/GUARD RAIL** Corrugated stl, galv. stl posts, 6'-3" O.C.	B-80	850	.038	L.F.	15.20	1.08	.63	16.91	19.10	200
	0200	End sections, galvanized, flared		50	.640	Ea.	63	18.45	10.70	92.15	110	
	0300	Wrap around end		50	.640	"	94.50	18.45	10.70	123.65	144	
	0400	Timber guide rail, 4" x 8" with 6" x 8" wood posts, treated		960	.033	L.F.	20.50	.96	.56	22.02	24.50	
	0600	Cable guide rail, 3 at 3/4" cables, steel posts, single face		900	.036		7.25	1.02	.59	8.86	10.20	
	0700	Wood posts		950	.034		9.50	.97	.56	11.03	12.55	
	0900	Guide rail, steel box beam, 6" x 6"		120	.267		23	7.70	4.45	35.15	42	
	1100	Median barrier, steel box beam, 6" x 8"	↓	215	.149		28.50	4.29	2.48	35.27	41	
	1400	Resilient guide fence and light shield, 6' high	B-2	130	.308	↓	22	8.35		30.35	37.50	
	1500	Concrete posts, individual, 6'-5", triangular	B-80	110	.291	Ea.	48.50	8.40	4.85	61.75	71.50	
	1550	Square	"	110	.291	"	52	8.40	4.85	65.25	75	
	2000	Median, precast concrete, 3'-6" high, 2' wide, single face	B-29	380	.147	L.F.	38	4.24	2.20	44.44	50.50	
	2200	Double face	"	340	.165	"	43	4.74	2.46	50.20	57.50	
	2400	Speed bumps, thermoplastic, 10-1/2" x 2-1/4" x 48" long	B-2	120	.333	Ea.	89.50	9.05		98.55	113	
	3030	Impact barrier, UTMCD, barrel type	B-16	30	1.067	"	305	29	15.90	349.90	405	
400	0010	**FENDERS**										400
	0015	Bumper rails for garages, 12 Ga. rail, 6" wide, with steel										
	0020	posts 12'-6" O.C., minimum	E-4	190	.168	L.F.	12.15	6.50	.43	19.08	25.50	
	0030	Average		165	.194		15.20	7.50	.49	23.19	30.50	
	0100	Maximum		140	.229		18.25	8.85	.58	27.68	36.50	
	0300	12" channel rail, minimum		160	.200		15.20	7.75	.51	23.46	31	
	0400	Maximum	↓	120	.267	↓	23	10.30	.68	33.98	44	
800	0010	**PARKING BUMPERS**										800
	0020	Parking barriers, timber w/saddles, treated type										
	0100	4" x 4" for cars	B-2	520	.077	L.F.	2.47	2.08		4.55	5.95	
	0200	6" x 6" for trucks	↓	520	.077	"	5.15	2.08		7.23	8.95	

For expanded coverage of these items see *Means Site Work and Landscape Cost Data 2005*

2 SITE CONSTRUCTION

			DAILY	LABOR-		2005 BARE COSTS				TOTAL	
02840	**Walk/Road/Parking Appurtenances**	CREW	OUTPUT	HOURS	UNIT	MAT.	LABOR	EQUIP.	TOTAL	INCL O&P	
800	0600	Flexible fixed stanchion, 2' high, 3" diameter	B-2	100	.400	Ea.	23.50	10.85		34.35	43	**800**
	1000	Wheel stops, precast concrete incl. dowels, 6" x 10" x 6'-0"		120	.333		38	9.05		47.05	56	
	1100	8" x 13" x 6'-0"		120	.333		44	9.05		53.05	62.50	
	1200	Thermoplastic, 6" x 10" x 6'-0"		120	.333		69	9.05		78.05	90	
	1300	Pipe bollards, conc filled/paint, 8' L x 4' D hole, 6" diam.	B-6	20	1.200		279	34.50	10.80	324.30	370	
	1400	8" diam.		15	1.600		420	46	14.40	480.40	550	
	1500	12" diam.		12	2		550	57.50	18	625.50	715	
	2030	Folding with individual padlocks	B-2	50	.800		675	21.50		696.50	780	
	9000	Parking lot control, see Div. 11156-600										
910	0010	**TRAFFIC BARRIERS, TRAFFIC CONTROL DEVICES**										**910**
	0020	Crash barriers										
	0100	Traffic channelizing pavement markers, layout only	A-7	2,000	.012	Ea.		.44	.03	.47	.71	
	0110	13" x 7-1/2" x 2-1/2" high, non-plowable install	2 Clab	96	.167		18.40	4.45		22.85	27	
	0200	8" x 8"x 3-1/4" high, non-plowable, install		96	.167		16.70	4.45		21.15	25.50	
	0230	4" x 4" x 3/4" high, non-plowable, install		120	.133		2.23	3.56		5.79	8	
	0240	9-1/4" x 5-7/8" x 1/4" high, plowable, concrete pav't	A-2A	70	.343		13.35	9.15	3.47	25.97	32.50	
	0250	9-1/4" x 5-7/8" x 1/4" high, plowable, asphalt pav't	"	120	.200		2.56	5.35	2.03	9.94	13.30	
	0300	Barrier and curb delineators,reflectorized, 2" x 4"	2 Clab	150	.107		1.45	2.85		4.30	6.05	
	0310	3" x 5"	"	150	.107		2.95	2.85		5.80	7.70	
	0500	Rumble strip, polycarbonate										
	0510	24" x 3-1/2" x 1/2" high	2 Clab	50	.320	Ea.	5.55	8.55		14.10	19.45	
920	0010	**TRAFFIC BARRIERS, HIGHWAY SOUND BARRIERS**										**920**
	0020	Highway sound barriers, not including footing										
	0100	Precast concrete, concrete columns @ 30' OC, 8" T, 8' H	C-12	400	.120	L.F.	84.50	4.03	1.51	90.04	101	
	0110	12' H		265	.181		127	6.10	2.27	135.37	152	
	0120	16' H		200	.240		169	8.05	3.01	180.06	202	
	0130	20' H		160	.300		211	10.10	3.77	224.87	253	
	0400	Lt. Wt. composite panel, cementitious face, St. posts @ 12' OC, 8' H	B-80B	190	.168		95	4.76	1.10	100.86	114	
	0410	12' H		125	.256		143	7.25	1.68	151.93	170	
	0420	16' H		95	.337		190	9.55	2.21	201.76	226	
	0430	20' H		75	.427		238	12.05	2.79	252.84	283	

			DAILY	LABOR-		2005 BARE COSTS				TOTAL	
02850	**Prefabricated Bridges**										
210	0010	**BRIDGES, PEDESTRIAN** spans over streams, roadways, etc.										**210**
	0020	including erection, not including foundations										
	0050	Precast concrete, complete in place, 8' wide, 60' span	E-2	215	.260	S.F.	49.50	9.65	6.15	65.30	78	
	0100	100' span		185	.303		54	11.20	7.15	72.35	86.50	
	0150	120' span		160	.350		58.50	12.95	8.30	79.75	95.50	
	0200	150' span		145	.386		61	14.25	9.15	84.40	102	
	0300	Steel, trussed or arch spans, compl. in place, 8' wide, 40' span		320	.175		62.50	6.45	4.14	73.09	84	
	0400	50' span		395	.142		56	5.25	3.35	64.60	74	
	0500	60' span		465	.120		56	4.45	2.85	63.30	72.50	
	0600	80' span		570	.098		66.50	3.63	2.32	72.45	82.50	
	0700	100' span		465	.120		93.50	4.45	2.85	100.80	114	
	0800	120' span		365	.153		118	5.65	3.63	127.28	144	
	0900	150' span		310	.181		126	6.70	4.27	136.97	154	
	1000	160' span		255	.220		126	8.10	5.20	139.30	158	
	1100	10' wide, 80' span		640	.087		69.50	3.23	2.07	74.80	84	
	1200	120' span		415	.135		90	4.99	3.19	98.18	111	
	1300	150' span		445	.126		101	4.65	2.98	108.63	122	
	1400	200' span		205	.273		108	10.10	6.45	124.55	142	
	1600	Wood, laminated type, complete in place, 80' span	C-12	203	.236		42.50	7.95	2.97	53.42	62	
	1700	130' span	"	153	.314		44	10.55	3.94	58.49	69	

Important: See the Reference Section for critical supporting data - Reference Nos., Crews, & City Cost Indexes

		02870	Site Furnishings	CREW	DAILY OUTPUT	LABOR-HOURS	UNIT	2005 BARE COSTS				TOTAL INCL O&P	
								MAT.	LABOR	EQUIP.	TOTAL		
310	0010		**BENCHES**										**310**
	0012		Seating, benches, park, precast conc, w/backs, wood rails, 4' long	2 Clab	5	3.200	Ea.	320	85.50		405.50	485	
	0100		8' long		4	4		670	107		777	900	
	0300		Fiberglass, without back, one piece, 4' long		10	1.600		405	42.50		447.50	515	
	0400		8' long		7	2.286		840	61		901	1,025	
	0500		Steel barstock pedestals w/backs, 2" x 3" wood rails, 4' long		10	1.600		770	42.50		812.50	915	
	0510		8' long		7	2.286		910	61		971	1,100	
	0520		3" x 8" wood plank, 4' long		10	1.600		775	42.50		817.50	920	
	0530		8' long		7	2.286		810	61		871	985	
	0540		Backless, 4" x 4" wood plank, 4' square		10	1.600		755	42.50		797.50	895	
	0550		8' long		7	2.286		715	61		776	885	
	0600		Aluminum pedestals, with backs, aluminum slats, 8' long		8	2		370	53.50		423.50	490	
	0610		15' long		5	3.200		360	85.50		445.50	535	
	0620		Portable, aluminum slats, 8' long		8	2		320	53.50		373.50	440	
	0630		15' long		5	3.200		470	85.50		555.50	655	
	0800		Cast iron pedestals, back & arms, wood slats, 4' long		8	2		290	53.50		343.50	405	
	0820		8' long		5	3.200		845	85.50		930.50	1,075	
	0840		Backless, wood slats, 4' long		8	2		505	53.50		558.50	640	
	0860		8' long		5	3.200		500	85.50		585.50	685	
	1700		Steel frame, fir seat, 10' long	▼	10	1.600	▼	176	42.50		218.50	261	
510	0010		**TRASH RECEPTACLE**										**510**
	0020		Trash receptacle, fiberglass, 2' square, 18" high	2 Clab	30	.533	Ea.	246	14.25		260.25	293	
	0100		2' square, 2'-6" high		30	.533		345	14.25		359.25	400	
	0300		Circular , 2' diameter, 18" high		30	.533		222	14.25		236.25	266	
	0400		2' diameter, 2'-6" high	▼	30	.533	▼	296	14.25		310.25	345	
520	0010		**TRASH CLOSURE** Steel with pullover cover										**520**
	0020		2'-3" wide, 4'-7" high, 6'-2" long	2 Clab	5	3.200	Ea.	840	85.50		925.50	1,050	
	0100		10'-1" long		4	4		1,100	107		1,207	1,400	
	0300		Wood, 10' wide, 6' high, 10' long	▼	1.20	13.333		875	355		1,230	1,525	

02880 | Playfield Equipment

				CREW	DAILY OUTPUT	LABOR-HOURS	UNIT	MAT.	LABOR	EQUIP.	TOTAL	TOTAL INCL O&P	
100	0010		**ATHLETIC OR RECREATIONAL SCREENING**										**100**
	0015		Backstops, baseball, prefabricated, 30' wide, 12' high & 1 overhang	B-1	1	24	Ea.	2,475	655		3,130	3,750	
	0100		40' wide, 12' high & 2 overhangs	"	.75	32		3,100	875		3,975	4,800	
	0300		Basketball, steel, single goal	B-13	3.04	18.421		975	530	203	1,708	2,125	
	0400		Double goal	"	1.92	29.167	▼	590	840	320	1,750	2,300	
	0600		Tennis, wire mesh with pair of ends	B-1	2.48	9.677	Set	1,525	265		1,790	2,075	
	0700		Enclosed court	"	1.30	18.462	Ea.	4,325	505		4,830	5,550	
	0900		Handball or squash court, outdoor, wood	2 Carp	.50	32		3,575	1,100		4,675	5,625	
	1000		Masonry handball/squash court	D-1	.30	53.333	▼	21,700	1,650		23,350	26,400	
110	0010		**TENNIS COURT FENCES**										**110**
	0860		Tennis courts, 11 ga. wire, 2-1/2" post set										
	0870		in concrete, 10' O.C., 1-5/8" top rail										
	0900		10' high	B-80	190	.168	L.F.	11.65	4.85	2.81	19.31	23.50	
	0920		12' high		170	.188	"	13.65	5.40	3.14	22.19	27	
	1000		Add for gate 4' wide, 1-5/8" frame 7' high		10	3.200	Ea.	109	92	53.50	254.50	320	
	1040		Aluminized steel, 11 ga. wire 10' high		190	.168	L.F.	13.85	4.85	2.81	21.51	26	
	1100		12' high		170	.188	"	17.05	5.40	3.14	25.59	30.50	
	1140		Add for gate 4' wide, 1-5/8" frame, 7' high		10	3.200	Ea.	126	92	53.50	271.50	340	
	1250		Vinyl covered, 9 ga. wire, 10' high		190	.168	L.F.	13.65	4.85	2.81	21.31	25.50	
	1300		12' high	▼	170	.188	"	22.50	5.40	3.14	31.04	37	
	1310		Fence, CL, tennis ct, transom gate, single, galv, 4' x 7' x 3'	B-80A	8.72	2.752	Ea.	237	73.50	20	330.50	395	
	1400		Add for gate 4' wide, 1-5/8" frame, 7' high	B-80	10	3.200	"	127	92	53.50	272.50	340	

SITE CONSTRUCTION 2

02880 | Playfield Equipment

			CREW	DAILY OUTPUT	LABOR-HOURS	UNIT	2005 BARE COSTS MAT.	LABOR	EQUIP.	TOTAL	TOTAL INCL O&P	
210	0010	**PLAYGROUND EQUIPMENT** See also individual items										210
	0200	Bike rack, 10' long, permanent	B-1	12	2	Ea.	470	54.50		524.50	600	
	0400	Horizontal monkey ladder, 14' long, 6' high		4	6		610	164		774	930	
	0590	Parallel bars, 10' long		4	6		263	164		427	545	
	0600	Posts, tether ball set, 2-3/8" O.D.		12	2	↓	174	54.50		228.50	277	
	0800	Poles, multiple purpose, 10'-6" long		12	2	Pr.	130	54.50		184.50	228	
	1000	Ground socket for movable posts, 2-3/8" post		10	2.400		92	65.50		157.50	203	
	1100	3-1/2" post		10	2.400	↓	129	65.50		194.50	244	
	1300	See-saw, spring, steel, 2 units		6	4	Ea.	910	109		1,019	1,175	
	1400	4 units		4	6		1,150	164		1,314	1,500	
	1500	6 units		3	8		2,325	219		2,544	2,900	
	1700	Shelter, fiberglass golf tee, 3 person		4.60	5.217		2,550	143		2,693	3,025	
	1900	Slides, stainless steel bed, 12' long, 6' high		3	8		2,275	219		2,494	2,850	
	2000	20' long, 10' high		2	12		3,325	330		3,655	4,150	
	2200	Swings, plain seats, 8' high, 4 seats		2	12		1,025	330		1,355	1,625	
	2300	8 seats		1.30	18.462		1,875	505		2,380	2,850	
	2500	12' high, 4 seats		2	12		1,375	330		1,705	2,025	
	2600	8 seats		1.30	18.462		1,800	505		2,305	2,775	
	2800	Whirlers, 8' diameter		3	8		2,500	219		2,719	3,100	
	2900	10' diameter	↓	3	8	↓	3,025	219		3,244	3,700	
220	0010	**MODULAR PLAYGROUND** Basic components										220
	0100	Deck, square, steel, 48" x 48"	B-1	1	24	Ea.	715	655		1,370	1,800	
	0110	Recycled polyurethane		1	24		645	655		1,300	1,725	
	0120	Triangular, steel, 48" side		1	24	↓	520	655		1,175	1,600	
	0130	Post, steel, 5" square		18	1.333	L.F.	23.50	36.50		60	83	
	0140	Aluminum, 2-3/8" square		20	1.200		21.50	33		54.50	74.50	
	0150	5" square		18	1.333	↓	28.50	36.50		65	88.50	
	0160	Roof, square poly, 54" side		18	1.333	Ea.	895	36.50		931.50	1,025	
	0170	Wheelchair transfer module, for 3' high deck		3	8	"	2,325	219		2,544	2,925	
	0180	Guardrail, pipe, 36" high		60	.400	L.F.	267	10.95		277.95	310	
	0190	Steps, deck-to-deck, 3 - 8" steps		8	3	Ea.	1,275	82		1,357	1,525	
	0200	Activity panel, crawl through panel		2	12		365	330		695	910	
	0210	Alphabet/spelling panel		2	12		690	330		1,020	1,275	
	0360	With guardrails		3	8		1,950	219		2,169	2,500	
	0370	Crawl tunnel, straight, 56" long		4	6		1,150	164		1,314	1,525	
	0380	90°, 4' long		4	6		1,400	164		1,564	1,800	
	1200	Slide, tunnel, for 56" high deck		8	3		1,600	82		1,682	1,900	
	1210	Straight, poly		8	3		227	82		309	380	
	1220	Stainless steel, 54" high deck		6	4		234	109		343	430	
	1230	Curved, poly, 40" high deck		6	4		1,175	109		1,284	1,475	
	1240	Spiral slide, 56" - 72" high		5	4.800		4,125	131		4,256	4,725	
	1300	Ladder, vertical, for 24" - 72" high deck		5	4.800		525	131		656	780	
	1310	Horizontal, 8' long		5	4.800		725	131		856	1,000	
	1320	Corkscrew climber, 6' high		3	8		560	219		779	955	
	1330	Fire pole for 72" high deck		6	4		266	109		375	465	
	1340	Bridge, ring climber, 8' long		4	6	↓	1,050	164		1,214	1,425	
	1350	Suspension	↓	4	6	L.F.	300	164		464	585	
330	0010	**GOAL POSTS**										330
	0020	Goal posts, steel, football, double post	B-1	1.50	16	Pr.	1,525	440		1,965	2,350	
	0100	Deluxe, single post		1.50	16		2,475	440		2,915	3,400	
	0300	Football, convertible to soccer		1.50	16		2,450	440		2,890	3,375	
	0500	Soccer, regulation	↓	2	12	↓	1,950	330		2,280	2,625	
350	0010	**PLATFORM/PADDLE TENNIS COURT** Complete with lighting, etc.										350
	0100	Aluminum slat deck with aluminum frame	B-1	.08	300	Court	39,600	8,200		47,800	56,500	

2

SITE CONSTRUCTION

02880 | Playfield Equipment

		CREW	DAILY OUTPUT	LABOR-HOURS	UNIT	2005 BARE COSTS				TOTAL INCL O&P	
						MAT.	LABOR	EQUIP.	TOTAL		
350	0500	Aluminum slat deck and wood frame	C-1	.12	266	Court	39,600	8,625		48,225	57,000	350
	0800	Aluminum deck heater, add	B-1	1.18	20.339		3,850	555		4,405	5,125	
	0900	Douglas fir planking and wood frame 2" x 6" x 30'	C-1	.12	266		39,600	8,625	.	48,225	57,000	
	1000	Plywood deck with steel frame		.12	266		39,600	8,625		48,225	57,000	
	1100	Steel slat deck with wood frame	↓	.12	266	↓	27,800	8,625		36,425	44,000	

02890 | Traffic Signs & Signals

		CREW	DAILY OUTPUT	LABOR-HOURS	UNIT	2005 BARE COSTS				TOTAL INCL O&P	
						MAT.	LABOR	EQUIP.	TOTAL		
100	0010	**SIGNS** Stock, 24" x 24", no posts, .080" alum. reflectorized	B-80	70	.457	Ea.	55	13.15	7.65	75.80	89	100
	0100	High intensity		70	.457		55	13.15	7.65	75.80	89	
	0300	30" x 30", reflectorized		70	.457		112	13.15	7.65	132.80	151	
	0400	High intensity		70	.457		112	13.15	7.65	132.80	151	
	0600	Guide and directional signs, 12" x 18", reflectorized		70	.457		35.50	13.15	7.65	56.30	68	
	0700	High intensity		70	.457		34	13.15	7.65	54.80	66	
	0900	18" x 24", stock signs, reflectorized		70	.457		40	13.15	7.65	60.80	72.50	
	1000	High intensity		70	.457		40	13.15	7.65	60.80	72.50	
	1200	24" x 24", stock signs, reflectorized		70	.457		49.50	13.15	7.65	70.30	83	
	1300	High intensity		70	.457		49.50	13.15	7.65	70.30	83	
	1500	Add to above for steel posts, galvanized, 10'-0" upright, bolted		200	.160		17.90	4.61	2.67	25.18	29.50	
	1600	12'-0" upright, bolted		140	.229	↓	24	6.60	3.81	34.41	40.50	
	1800	Highway road signs, aluminum, over 20 S. F., reflectorized		350	.091	S.F.	23	2.63	1.53	27.16	30.50	
	2000	High intensity		350	.091		23	2.63	1.53	27.16	30.50	
	2200	Highway, suspended over road, 80 S.F. min., reflectorized		165	.194		23	5.60	3.23	31.83	37	
	2300	High intensity	↓	165	.194	↓	23	5.60	3.23	31.83	37	
300	0010	**TRAFFIC SIGNALS** Mid block pedestrian crosswalk,										300
	0020	with pushbutton and mast arms	R-2	.30	186	Total	54,000	6,675	1,050	61,725	71,000	
	0100	Intersection, 8 signals w/three sect. (2 each direction), programmed	"	.15	373		121,000	13,400	2,125	136,525	156,000	
	0120	For each additional traffic phase controller, add	L-9	1.20	30		6,525	930		7,455	8,675	
	0200	Semi-actuated, detectors in side street only, add		.81	44.444		9,775	1,375		11,150	13,000	
	0300	Fully-actuated, detectors in all streets, add		.49	73.469		33,200	2,275		35,475	40,200	
	0400	For pedestrian pushbutton, add		.70	51.429		18,900	1,600		20,500	23,400	
	0500	Optically programmed signal only, add per head		1.64	21.951	↓	10,900	680		11,580	13,100	
	0600	School flashing system, programmed	↓	.41	87.805	Signal	25,500	2,725		28,225	32,500	
	1000	Intersection traffic signals, LED, mast, programmable, no lane control										
	1010	Includes all labor, material and equip. for complete installation				Ea.	150,000			150,000	165,000	
	1100	Intersection traffic signals, LED, mast, programmable, R/L lane control										
	1110	Includes all labor, material and equip. for complete installation				Ea.	200,000			200,000	220,000	
	1200	Add protective/permissive left turns to existing traffic light										
	1210	Includes all labor, material and equip. for complete installation				Ea.	60,000			60,000	66,000	
	1300	Replace existing light heads with LED Heads wire hung										
	1310	Includes all labor, material and equip. for complete installation				Ea.	40,000			40,000	44,000	
	1400	Replace existing light heads with LED Heads mast arm hung										
	1410	Includes all labor, material and equip. for complete installation				Ea.	80,000			80,000	88,000	

02900 | Planting

02905 | Plants, Planting, Transplanting

		CREW	DAILY OUTPUT	LABOR-HOURS	UNIT	2005 BARE COSTS				TOTAL INCL O&P	
						MAT.	LABOR	EQUIP.	TOTAL		
725	0010	**PLANTING**	R02930 -900									725
	0012	Moving shrubs on site, 12" ball	B-62	28	.857	Ea.		24.50	5.50	30	44	

For expanded coverage of these items see _Means Site Work and Landscape Cost Data 2005_

			CREW	DAILY OUTPUT	LABOR-HOURS	UNIT	2005 BARE COSTS				TOTAL INCL O&P	
							MAT.	LABOR	EQUIP.	TOTAL		
725		**02905 | Plants, Planting, Transplanting**										**725**
	0100	24" ball	B-62	22	1.091	Ea.		31.50	6.95	38.45	56	
	0300	Moving trees on site, 36" ball R02930 -900	B-6	3.75	6.400			184	57.50	241.50	350	
	0400	60" ball	"	1	24			690	216	906	1,300	
		02910 | Plant Preparation										
500	0010	**MULCHING**										**500**
	0100	Aged barks, 3" deep, hand spread	1 Clab	100	.080	S.Y.	2.03	2.14		4.17	5.55	
	0150	Skid steer loader	B-63	13.50	2.963	M.S.F.	226	83	11.40	320.40	390	
	0200	Hay, 1" deep, hand spread	1 Clab	475	.017	S.Y.	.56	.45		1.01	1.32	
	0250	Power mulcher, small	B-64	180	.089	M.S.F.	62	2.38	1.29	65.67	73.50	
	0350	Large	B-65	530	.030	"	62	.81	.59	63.40	70.50	
	0400	Humus peat, 1" deep, hand spread	1 Clab	700	.011	S.Y.	2.13	.31		2.44	2.81	
	0450	Push spreader	"	2,500	.003	"	2.13	.09		2.22	2.47	
	0550	Tractor spreader	B-66	700	.011	M.S.F.	237	.38	.25	237.63	261	
	0600	Oat straw, 1" deep, hand spread	1 Clab	475	.017	S.Y.	.32	.45		.77	1.05	
	0650	Power mulcher, small	B-64	180	.089	M.S.F.	35.50	2.38	1.29	39.17	44	
	0700	Large	B-65	530	.030	"	35.50	.81	.59	36.90	41	
	0750	Add for asphaltic emulsion	B-45	1,770	.009	Gal.	1.74	.28	.30	2.32	2.67	
	0800	Peat moss, 1" deep, hand spread	1 Clab	900	.009	S.Y.	1.69	.24		1.93	2.23	
	0850	Push spreader	"	2,500	.003	"	1.69	.09		1.78	1.99	
	0950	Tractor spreader	B-66	700	.011	M.S.F.	188	.38	.25	188.63	208	
	1000	Polyethylene film, 6 mil.	2 Clab	2,000	.008	S.Y.	.16	.21		.37	.51	
	1100	Redwood nuggets, 3" deep, hand spread	1 Clab	150	.053	"	4.17	1.42		5.59	6.80	
	1150	Skid steer loader	B-63	13.50	2.963	M.S.F.	465	83	11.40	559.40	650	
	1200	Stone mulch, hand spread, ceramic chips, economy	1 Clab	125	.064	S.Y.	6.10	1.71		7.81	9.40	
	1250	Deluxe	"	95	.084	"	9.40	2.25		11.65	13.85	
	1300	Granite chips	B-1	10	2.400	C.Y.	30.50	65.50		96	136	
	1400	Marble chips		10	2.400		115	65.50		180.50	228	
	1600	Pea gravel		28	.857		56	23.50		79.50	98	
	1700	Quartz		10	2.400		148	65.50		213.50	264	
	1800	Tar paper, 15 Lb. felt	1 Clab	800	.010	S.Y.	.31	.27		.58	.76	
	1900	Wood chips, 2" deep, hand spread	"	220	.036	"	1.74	.97		2.71	3.42	
	1950	Skid steer loader	B-63	20.30	1.970	M.S.F.	193	55	7.55	255.55	305	
720	0010	**PLANT BED PREPARATION, SHRUB & TREE**										**720**
	0100	Backfill planting pit, by hand, on site topsoil	2 Clab	18	.889	C.Y.		23.50		23.50	37	
	0200	Prepared planting mix	"	24	.667			17.80		17.80	27.50	
	0300	Skid steer loader, on site topsoil	B-62	340	.071			2.03	.45	2.48	3.63	
	0400	Prepared planting mix	"	410	.059			1.69	.37	2.06	3	
	1000	Excavate planting pit, by hand, sandy soil	2 Clab	16	1			26.50		26.50	41.50	
	1100	Heavy soil or clay	"	8	2			53.50		53.50	83	
	1200	1/2 C.Y. backhoe, sandy soil	B-11C	150	.107			3.27	1.44	4.71	6.60	
	1300	Heavy soil or clay	"	115	.139			4.27	1.88	6.15	8.55	
	2000	Mix planting soil, incl. loam, manure, peat, by hand	2 Clab	60	.267		37	7.10		44.10	52	
	2100	Skid steer loader	B-62	150	.160		37	4.61	1.02	42.63	49	
	3000	Pile sod, skid steer loader	"	2,800	.009	S.Y.		.25	.05	.30	.44	
	3100	By hand	2 Clab	400	.040			1.07		1.07	1.66	
	4000	Remove sod, F.E. loader	B-10S	2,000	.006			.19	.12	.31	.42	
	4100	Sod cutter	B-12K	3,200	.005			.16	.30	.46	.57	
	4200	By hand	2 Clab	240	.067			1.78		1.78	2.77	
810	0010	**LOAM & TOPSOIL**										**810**
	0400	Spread from pile to rough finish grade, F.E. loader, 1.5 CY	B-10S	200	.060	E.C.Y.		1.92	1.20	3.12	4.25	
	0500	Up to 200' radius, by hand	1 Clab	14	.571			15.25		15.25	23.50	
	0600	Top dress by hand, 1 C.Y. for 600 S.F.	"	11.50	.696		21	18.55		39.55	52.50	
	0700	Furnish and place, truck dumped, screened, 4" deep	B-10S	1,300	.009	S.Y.	2.65	.30	.19	3.14	3.57	
	0800	6" deep	"	820	.015	"	3.39	.47	.29	4.15	4.76	

Important: See the Reference Section for critical supporting data - Reference Nos., Crews, & City Cost Indexes

02915 | Shrub and Tree Transplanting

		CREW	DAILY OUTPUT	LABOR-HOURS	UNIT	2005 BARE COSTS				TOTAL INCL O&P	
						MAT.	LABOR	EQUIP.	TOTAL		
200	0010	**GROUND COVER** Plants, pachysandra, in prepared beds	B-1	15	1.600	C	25	44		69	95.50	**200**
	0200	Vinca minor, 1 yr, bare root		12	2	"	26	54.50		80.50	114	
	0600	Stone chips, in 50 lb. bags, Georgia marble		520	.046	Bag	2.35	1.26		3.61	4.56	
	0700	Onyx gemstone		260	.092		17	2.53		19.53	22.50	
	0800	Quartz		260	.092		6.35	2.53		8.88	10.95	
	0900	Pea gravel, truckload lots		28	.857	Ton	24.50	23.50		48	63.50	

02920 | Lawns & Grasses

		CREW	DAILY OUTPUT	LABOR-HOURS	UNIT	MAT.	LABOR	EQUIP.	TOTAL	TOTAL INCL O&P	
310	0010	**SEEDING, GENERAL** R02920-500									**310**
	0020	Mechanical seeding, 215 lb./acre	B-66	1.50	5.333	Acre	510	176	115	801	955	
	0100	44 lb./M.S.Y.	"	2,500	.003	S.Y.	.15	.11	.07	.33	.41	
	0300	Fine grading and seeding incl. lime, fertilizer & seed,										
	0310	with equipment	B-14	1,000	.048	S.Y.	.16	1.35	.22	1.73	2.51	
	0600	Limestone hand push spreader, 50 lbs. per M.S.F.	1 Clab	180	.044	M.S.F.	3.38	1.19		4.57	5.55	
	0800	Grass seed hand push spreader, 4.5 lbs. per M.S.F.	"	180	.044	"	16.65	1.19		17.84	20	
	1000	Hydro or air seeding for large areas, incl. seed and fertilizer	B-81	8,900	.003	S.Y.	.15	.08	.05	.28	.34	
	1100	With wood fiber mulch added	"	8,900	.003	"	.27	.08	.05	.40	.47	
	1300	Seed only, over 100 lbs., field seed, minimum				Lb.	1.10			1.10	1.21	
	1400	Maximum					3.68			3.68	4.05	
	1500	Lawn seed, minimum					1.71			1.71	1.88	
	1600	Maximum					4.47			4.47	4.92	
	1800	Aerial operations, seeding only, field seed	B-58	50	.480	Acre	360	13.85	47.50	421.35	470	
	1900	Lawn seed		50	.480		555	13.85	47.50	616.35	685	
	2100	Seed and liquid fertilizer, field seed		50	.480		430	13.85	47.50	491.35	545	
	2200	Lawn seed		50	.480		625	13.85	47.50	686.35	765	
400	0010	**SODDING**										**400**
	0020	Sodding, 1" deep, bluegrass sod, on level ground, over 8 MSF	B-63	22	1.818	M.S.F.	217	51	7	275	325	
	0200	4 M.S.F.		17	2.353		243	66	9.05	318.05	380	
	0300	1000 S.F.		13.50	2.963		265	83	11.40	359.40	430	
	0500	Sloped ground, over 8 M.S.F.		6	6.667		217	186	25.50	428.50	555	
	0600	4 M.S.F.		5	8		243	224	30.50	497.50	645	
	0700	1000 S.F.		4	10		265	280	38.50	583.50	765	
	1000	Bent grass sod, on level ground, over 6 M.S.F.		20	2		485	56	7.70	548.70	630	
	1100	3 M.S.F.		18	2.222		540	62	8.55	610.55	700	
	1200	Sodding 1000 S.F. or less		14	2.857		615	80	10.95	705.95	815	
	1500	Sloped ground, over 6 M.S.F.		15	2.667		485	74.50	10.25	569.75	660	
	1600	3 M.S.F.		13.50	2.963		540	83	11.40	634.40	735	
	1700	1000 S.F.		12	3.333		615	93	12.80	720.80	840	

02930 | Exterior Plants

		CREW	DAILY OUTPUT	LABOR-HOURS	UNIT	MAT.	LABOR	EQUIP.	TOTAL	TOTAL INCL O&P	
050	0010	**TRAVEL** To all nursery items, for 10 to 20 miles, add				All					5%	**050**
	0100	30 to 50 miles, add				"					10%	
310	0010	**SHRUBS AND TREES** Evergreen, in prepared beds, B & B										**310**
	0100	Arborvitae pyramidal, 4'-5'	B-17	30	1.067	Ea.	42.50	30.50	18.10	91.10	113	
	0150	Globe, 12"-15"	B-1	96	.250		10.85	6.85		17.70	22.50	
	0300	Cedar, blue, 8'-10'	B-17	18	1.778		179	50.50	30	259.50	310	
	0500	Hemlock, canadian, 2-1/2'-3'	B-1	36	.667		22.50	18.25		40.75	53.50	
	0550	Holly, Savannah, 8' - 10' H		9.68	2.479		515	68		583	675	
	0600	Juniper, andorra, 18"-24"		80	.300		15.60	8.20		23.80	30	
	0620	Wiltoni, 15"-18"		80	.300		14.80	8.20		23	29	
	0640	Skyrocket, 4-1/2'-5'	B-17	55	.582		47	16.60	9.85	73.45	88	
	0660	Blue pfitzer, 2'-2-1/2'	B-1	44	.545		24.50	14.95		39.45	50	
	0680	Ketleerie, 2-1/2'-3'		50	.480		31	13.15		44.15	55	
	0700	Pine, black, 2-1/2'-3'		50	.480		38.50	13.15		51.65	62.50	

2 SITE CONSTRUCTION

02930	Exterior Plants	CREW	DAILY OUTPUT	LABOR-HOURS	UNIT	MAT.	LABOR	EQUIP.	TOTAL	TOTAL INCL O&P	
310						**2005 BARE COSTS**					
0720	Mugo, 18"-24"	B-1	60	.400	Ea.	33.50	10.95		44.45	54	**310**
0740	White, 4'-5'	B-17	75	.427		49.50	12.15	7.25	68.90	81	
0800	Spruce, blue, 18"-24"	B-1	60	.400		37.50	10.95		48.45	58	
0840	Norway, 4'-5'	B-17	75	.427		84	12.15	7.25	103.40	119	
0900	Yew, denisforma, 12"-15"	B-1	60	.400		22.50	10.95		33.45	41.50	
1000	Capitata, 18"-24"		30	.800		19.25	22		41.25	55	
1100	Hicksi, 2'-2-1/2'		30	.800		28	22		50	65	
320	0010	**SHRUBS** Broadleaf evergreen, planted in prepared beds									**320**
0100	Andromeda, 15"-18", container	B-1	96	.250	Ea.	22	6.85		28.85	34.50	
0200	Azalea, 15" - 18", container		96	.250		24.50	6.85		31.35	37	
0300	Barberry, 9"-12", container		130	.185		9.75	5.05		14.80	18.60	
0400	Boxwood, 15"-18", B & B		96	.250		26	6.85		32.85	39	
0500	Euonymus, emerald gaiety, 12" to 15", container		115	.209		15.75	5.70		21.45	26.50	
0600	Holly, 15"-18", B & B		96	.250		15.35	6.85		22.20	27.50	
0900	Mount laurel, 18" - 24", B & B		80	.300		48.50	8.20		56.70	66.50	
1000	Paxistema, 9 - 12" high		130	.185		15.75	5.05		20.80	25	
1100	Rhododendron, 18"-24", container		48	.500		27	13.70		40.70	51	
1200	Rosemary, 1 gal container		600	.040		57.50	1.09		58.59	65	
2000	Deciduous, amelanchier, 2'-3', B & B		57	.421		78	11.50		89.50	104	
2100	Azalea, 15"-18", B & B		96	.250		21	6.85		27.85	33.50	
2300	Bayberry, 2'-3', B & B		57	.421		24	11.50		35.50	44.50	
2600	Cotoneaster, 15"-18", B & B		80	.300		14.25	8.20		22.45	28.50	
2800	Dogwood, 3'-4', B & B	B-17	40	.800		22.50	23	13.55	59.05	75	
2900	Euonymus, alatus compacta, 15" to 18", container	B-1	80	.300		18.75	8.20		26.95	33.50	
3200	Forsythia, 2'-3', container	"	60	.400		16.50	10.95		27.45	35	
3300	Hibiscus, 3'-4', B & B	B-17	75	.427		12.90	12.15	7.25	32.30	41	
3400	Honeysuckle, 3'-4', B & B	B-1	60	.400		18.40	10.95		29.35	37	
3500	Hydrangea, 2'-3', B & B	"	57	.421		21.50	11.50		33	41.50	
3600	Lilac, 3'-4', B & B	B-17	40	.800		22	23	13.55	58.55	74.50	
3900	Privet, bare root, 18"-24"	B-1	80	.300		11.40	8.20		19.60	25.50	
4100	Quince, 2'-3', B & B	"	57	.421		18.95	11.50		30.45	39	
4200	Russian olive, 3'-4', B & B	B-17	75	.427		20.50	12.15	7.25	39.90	49	
4400	Spirea, 3'-4', B & B	B-1	70	.343		23.50	9.40		32.90	40.50	
4500	Viburnum, 3'-4', B & B	B-17	40	.800		24	23	13.55	60.55	76.50	
410	0010	**TREES** Deciduous, in prep. beds, balled & burlapped (B&B) — R02930-900									**410**
0100	Ash, 2" caliper	B-17	8	4	Ea.	110	114	68	292	370	
0200	Beech, 5'-6'		50	.640		215	18.25	10.85	244.10	277	
0300	Birch, 6'-8', 3 stems		20	1.600		119	45.50	27	191.50	230	
0500	Crabapple, 6'-8'		20	1.600		155	45.50	27	227.50	270	
0600	Dogwood, 4'-5'		40	.800		66	23	13.55	102.55	122	
0700	Eastern redbud 4'-5'		40	.800		129	23	13.55	165.55	192	
0800	Elm, 8'-10'		20	1.600		109	45.50	27	181.50	220	
0900	Ginkgo, 6'-7'		24	1.333		159	38	22.50	219.50	259	
1000	Hawthorn, 8'-10', 1" caliper		20	1.600		121	45.50	27	193.50	234	
1100	Honeylocust, 10'-12', 1-1/2" caliper		10	3.200		145	91	54	290	360	
1300	Larch, 8'		32	1		93.50	28.50	16.95	138.95	165	
1400	Linden, 8'-10', 1" caliper		20	1.600		101	45.50	27	173.50	211	
1500	Magnolia, 4'-5'		20	1.600		69.50	45.50	27	142	177	
1600	Maple, red, 8'-10', 1-1/2" caliper		10	3.200		154	91	54	299	370	
1700	Mountain ash, 8'-10', 1" caliper		16	2		159	57	34	250	300	
1800	Oak, 2-1/2"-3" caliper		6	5.333		240	152	90.50	482.50	595	
2100	Planetree, 9'-11', 1-1/4" caliper		10	3.200		99.50	91	54	244.50	310	
2200	Plum, 6'-8', 1" caliper		20	1.600		88	45.50	27	160.50	197	
2300	Poplar, 9'-11', 1-1/4" caliper		10	3.200		46	91	54	191	250	

Important: See the Reference Section for critical supporting data - Reference Nos., Crews, & City Cost Indexes

02900 | Planting

02930 | Exterior Plants

			CREW	DAILY OUTPUT	LABOR-HOURS	UNIT	2005 BARE COSTS				TOTAL INCL O&P	
							MAT.	LABOR	EQUIP.	TOTAL		
410	2500	Sumac, 2'-3'	B-17	75	.427	Ea.	22	12.15	7.25	41.40	50.50	410
	2700	Tulip, 5'-6'		40	.800		48	23	13.55	84.55	102	
	2800	Willow, 6'-8', 1" caliper		20	1.600		59	45.50	27	131.50	165	

R02930 -900

02945 | Planting Accessories

			CREW	DAILY OUTPUT	LABOR-HOURS	UNIT	MAT.	LABOR	EQUIP.	TOTAL	TOTAL INCL O&P	
120	0010	**EDGING**										120
	0050	Aluminum alloy, including stakes, 1/8" x 4", mill finish	B-1	390	.062	L.F.	2.21	1.68		3.89	5.05	
	0051	Black paint		390	.062		2.56	1.68		4.24	5.45	
	0052	Black anodized		390	.062		2.96	1.68		4.64	5.90	
	0100	Brick, set horizontally, 1-1/2 bricks per L.F.	D-1	370	.043		.93	1.34		2.27	3.07	
	0150	Set vertically, 3 bricks per L.F.	"	135	.119		2.70	3.68		6.38	8.55	
	0200	Corrugated aluminum, roll, 4" wide	1 Carp	650	.012		.37	.42		.79	1.07	
	0250	6" wide	"	550	.015		.46	.50		.96	1.29	
	0600	Railroad ties, 6" x 8"	2 Carp	170	.094		2.56	3.22		5.78	7.80	
	0650	7" x 9"		136	.118		2.84	4.03		6.87	9.45	
	0750	2" x 4"		330	.048		2.20	1.66		3.86	5	
	0800	Steel edge strips, incl. stakes, 1/4" x 5"	B-1	390	.062		3.35	1.68		5.03	6.30	
	0850	3/16" x 4"	"	390	.062		2.65	1.68		4.33	5.55	
300	0010	**PLANTERS** Concrete, sandblasted, precast, 48" diameter, 24" high	2 Clab	15	1.067	Ea.	555	28.50		583.50	655	300
	0100	Fluted, precast, 7' diameter, 36" high		10	1.600		910	42.50		952.50	1,075	
	0300	Fiberglass, circular, 36" diameter, 24" high		15	1.067		385	28.50		413.50	470	
	0400	60" diameter, 24" high		10	1.600		650	42.50		692.50	780	
	0600	Square, 24" side, 36" high		15	1.067		375	28.50		403.50	455	
	0700	48" side, 36" high		15	1.067		835	28.50		863.50	965	
	0900	Planter/bench, 72" square, 36" high		5	3.200		1,050	85.50		1,135.50	1,275	
	1000	96" square, 27" high		5	3.200		1,650	85.50		1,735.50	1,925	
	1200	Wood, square, 48" side, 24" high		15	1.067		860	28.50		888.50	995	
	1300	Circular, 48" diameter, 30" high		10	1.600		710	42.50		752.50	845	
	1500	72" diameter, 30" high		10	1.600		1,250	42.50		1,292.50	1,450	
	1600	Planter/bench, 72"		5	3.200		2,775	85.50		2,860.50	3,175	
510	0010	**TREE GUYING**										510
	0015	Tree guying Including stakes, guy wire and wrap										
	0100	Less than 3" caliper, 2 stakes	2 Clab	35	.457	Ea.	15.10	12.20		27.30	35.50	
	0200	3" to 4" caliper, 3 stakes	"	21	.762	"	17.85	20.50		38.35	51	
	1000	Including arrowhead anchor, cable, turnbuckles and wrap										
	1100	Less than 3" caliper, 3 anchors	2 Clab	20	.800	Ea.	47.50	21.50		69	85	
	1200	3" to 6" caliper, 4 anchors		15	1.067		68	28.50		96.50	120	
	1300	6" caliper, 6 anchors		12	1.333		84	35.50		119.50	148	
	1400	8" caliper, 8 anchors		9	1.778		96.50	47.50		144	180	

02950 | Site Restoration & Rehabilitation

02955 | Restoration of Underground Piping

			CREW	DAILY OUTPUT	LABOR-HOURS	UNIT	2005 BARE COSTS				TOTAL INCL O&P	
							MAT.	LABOR	EQUIP.	TOTAL		
230	0010	**LINING PIPE** with cement, incl. bypass and cleaning										230
	0020	Less than 10,000 L.F., urban, 6" to 10"	C-17E	130	.615	L.F.	6.90	21.50	.61	29.01	41.50	
	0200	24" to 36"		90	.889		10.95	31.50	.88	43.33	61.50	
	0300	48" to 72"		80	1		17.45	35.50	.99	53.94	75.50	

For expanded coverage of these items see *Means Site Work and Landscape Cost Data 2005*

			DAILY	LABOR-			2005 BARE COSTS			TOTAL
	02955	**Restoration of Underground Piping**	CREW	OUTPUT	HOURS	UNIT	MAT.	LABOR	EQUIP.	TOTAL	INCL O&P
310	0010	**CORROSION RESISTANCE** Wrap & coat, add to pipe, 4" dia.				L.F.	1.46			1.46	1.61	**310**
	0040	6" diameter					1.53			1.53	1.68
	0060	8" diameter					2.37			2.37	2.61
	0100	12" diameter					3.54			3.54	3.89
	0200	24" diameter					6.70			6.70	7.35
	0500	Coating, bituminous, per diameter inch, 1 coat, add					.27			.27	.30
	0540	3 coat					.42			.42	.46
	0560	Coal tar epoxy, per diameter inch, 1 coat, add					.16			.16	.18
	0600	3 coat				▼	.34			.34	.37

	02990	**Structure Moving**									
300	0010	**MOVING BUILDINGS** One day move, up to 24' wide										**300**
	0020	Reset on new foundation, patch & hook-up, average move				Total					9,300
	0040	Wood or steel frame bldg., based on ground floor area	B-4	185	.259	S.F.		7.05	1.77	8.82	12.90
	0060	Masonry bldg., based on ground floor area	"	137	.350			9.50	2.39	11.89	17.40
	0200	For 24' to 42' wide, add				▼					15%
	0220	For each additional day on road, add	B-4	1	48	Day		1,300	330	1,630	2,375
	0240	Construct new basement, move building, 1 day									
	0300	move, patch & hook-up, based on ground floor area	B-3	155	.310	S.F.	6	8.85	11.25	26.10	32.50

For information about Means Estimating Seminars, see yellow pages 12 and 13 in back of book

Important: See the Reference Section for critical supporting data - Reference Nos., Crews, & City Cost Indexes

Division 3
Concrete

Estimating Tips

General
- Carefully check all the plans and specifications. Concrete often appears on drawings other than structural drawings, including mechanical and electrical drawings for equipment pads. The cost of cutting and patching is often difficult to estimate. See Subdivisions 02220 and 03055 for demolition costs.
- Always obtain concrete prices from suppliers near the job site. A volume discount can often be negotiated depending upon competition in the area. Remember to add for waste, particularly for slabs and footings on grade.

03100 Concrete Forms & Accessories
- A primary cost for concrete construction is forming. Most jobs today are constructed with prefabricated forms. The selection of the forms best suited for the job and the total square feet of forms required for efficient concrete forming and placing are key elements in estimating concrete construction. Enough forms must be available for erection to make efficient use of the concrete placing equipment and crew.
- Concrete accessories for forming and placing depend upon the systems used. Study the plans and specifications to assure that all special accessory requirements have been included in the cost estimate such as anchor bolts, inserts and hangers.

03200 Concrete Reinforcement
- Ascertain that the reinforcing steel supplier has included all accessories, cutting, bending and an allowance for lapping, splicing and waste. A good rule of thumb is 10% for lapping, splicing and waste. Also, 10% waste should be allowed for welded wire fabric.

03300 Cast-in-Place Concrete
- When estimating structural concrete, pay particular attention to requirements for concrete additives, curing methods and surface treatments. Special consideration for climate, hot or cold, must be included in your estimate. Be sure to include requirements for concrete placing equipment and concrete finishing.

03400 Precast Concrete
03500 Cementitious Decks & Toppings
- The cost of hauling precast concrete structural members is often an important factor. For this reason, it is important to get a quote from the nearest supplier. It may become economically feasible to set up precasting beds on the site if the hauling costs are prohibitive.

Reference Numbers
Reference numbers are shown in bold squares at the beginning of some major classifications. These numbers refer to related items in the Reference Section. The reference information may be an estimating procedure, an alternate pricing method or technical information.

Note: Not all subdivisions listed here necessarily appear in this publication.

03055 | Selective Demolition

		CREW	DAILY OUTPUT	LABOR-HOURS	UNIT	MAT.	LABOR	EQUIP.	TOTAL	TOTAL INCL O&P	
110	0010	**SELECTIVE CONC. DEMO,** excl saw/torch cutting, load, haul	R02220 -510									**110**
	0050	Break up into small pieces, minimum reinforcing	B-9	24	1.667	C.Y.		45	5.90	50.90	77	
	0060	Average reinforcing		16	2.500			68	8.85	76.85	115	
	0070	Maximum reinforcing	↓	8	5	↓		136	17.70	153.70	230	
	0150	Remove whole pieces, up to 2 tons per piece	E-18	36	1.111	Ea.		42	22	64	97.50	
	0160	2 - 5 tons per piece		30	1.333			50.50	26.50	77	117	
	0170	5 - 10 tons per piece		24	1.667			63	33	96	147	
	0180	10 - 15 tons per piece	↓	18	2.222			84	44.50	128.50	195	
	0250	Precast unit embedded in masonry, up to 1 CF	D-1	16	1			31		31	47.50	
	0260	1 - 2 CF		12	1.333			41.50		41.50	63	
	0270	2 - 5 CF		10	1.600			49.50		49.50	76	
	0280	5 - 10 CF	↓	8	2	↓		62		62	94.50	

03060 | Basic Concrete Materials

		CREW	DAILY OUTPUT	LABOR-HOURS	UNIT	MAT.	LABOR	EQUIP.	TOTAL	TOTAL INCL O&P	
100	0010	**CONCRETE ADMIXTURES & SURFACE TREATMENTS**										**100**
	0040	Abrasives, aluminum oxide, over 20 tons			Lb.	1.02			1.02	1.12	
	0070	Under 1 ton				1.11			1.11	1.22	
	0100	Silicon carbide, black, over 20 tons				1.43			1.43	1.57	
	0120	Under 1 ton			↓	1.54			1.54	1.69	
	0200	Air entraining agent, .7 to 1.5 oz. per bag, 55 gallon lots			Gal.	9.25			9.25	10.20	
	0220	5 gallon lots				9.65			9.65	10.60	
	0300	Bonding agent, acrylic latex, 250 S.F. per gallon				21			21	23	
	0320	Epoxy resin, 80 S.F. per gallon			↓	46			46	51	
	0400	Calcium chloride, 50 lb. bags			Ton	360			360	395	
	0420	Less than truckload lots			Bag	27.50			27.50	30	
	0500	Carbon black, liquid, 2 to 8 lbs. per bag of cement			Lb.	3.71			3.71	4.08	
	0600	Colors, integral, 2 to 10 lb. per bag of cement, minimum				2.32			2.32	2.55	
	0610	Average				2.63			2.63	2.89	
	0620	Maximum			↓	3.02			3.02	3.32	
	0700	Curing compound, solvent based, 400 S.F./gal, 55 gal. lots			Gal.	8.65			8.65	9.55	
	0720	5 gallon lots				10.25			10.25	11.25	
	0800	Water based, 250 S.F./gal, 55 gallon lots				4.20			4.20	4.62	
	0820	5 gallon lots				5.25			5.25	5.80	
	0900	Dustproofing compound, (200-600 S.F./gal.), 55 gallon lots				3.99			3.99	4.39	
	0920	5 gallon lots			↓	4.62			4.62	5.10	
	1000	Epoxy dustproof coating, colors, (300-400 S.F. per coat),									
	1010	or transparent, (400-600 S.F. per coat)			Gal.	25			25	27.50	
	1100	Hardeners, metallic, 55 lb. bags, natural (grey)			Lb.	.60			.60	.66	
	1200	Colors, average				1.05			1.05	1.16	
	1300	Non-metallic, 55 lb. bags, natural (grey), minimum				.37			.37	.41	
	1310	Maximum				.42			.42	.46	
	1320	Non-metallic, colors, mininum				.48			.48	.53	
	1340	Maximum				.77			.77	.85	
	1400	Non-metallic, non-slip, 100 lb. bags, minimum				.48			.48	.53	
	1420	Maximum			↓	.96			.96	1.06	
	1500	Solution type, (300 to 400 S.F. per gallon)			Gal.	6.30			6.30	6.95	
	1550	Release agent, for tilt slabs				7.10			7.10	7.80	
	1570	For forms, average				4.42			4.42	4.86	
	1600	Sealer, hardener and dustproofer, epoxy, 150 S.F., minimum				7.10			7.10	7.85	
	1620	Maximum				15.75			15.75	17.35	
	1630	Solvent based, 200 S.F., minimum				8.70			8.70	9.60	
	1640	Maximum				16.20			16.20	17.80	
	1650	Water based, 250 S.F., minimum				10.25			10.25	11.25	
	1660	Maximum				11.80			11.80	13	
	1700	Colors (300-400 S.F. per gallon)				42.50			42.50	46.50	
	1800	Set accelerator for below freezing, 1 to 1-1/2 gal. per C.Y.			↓	5.65			5.65	6.25	

			DAILY	LABOR-		2005 BARE COSTS				TOTAL	
03060	**Basic Concrete Materials**	CREW	OUTPUT	HOURS	UNIT	MAT.	LABOR	EQUIP.	TOTAL	INCL O&P	
100 1900	Set retarder, 2 to 4 fl. oz. per bag of cement				Gal.	11.55			11.55	12.75	**100**
2000	Waterproofing, integral 1 lb. per bag of cement				Lb.	.85			.85	.94	
2100	Powdered metallic, 40 lbs. per 100 S.F., minimum				↓	1.06			1.06	1.17	
2120	Maximum					2.12			2.12	2.33	
110 0010	**AGGREGATE** Expanded shale, C.L. lots, 52 lb. per C.F., minimum R03310-020				Ton	36			36	39.50	**110**
0050	Maximum				"	48			48	53	
0100	Lightweight vermiculite or perlite, 4 C.F. bag, C.L. lots R03310-030				Bag	6.70			6.70	7.35	
0150	L.C.L. lots				"	7.35			7.35	8.10	
0250	Sand & stone, loaded at pit, crushed bank gravel R03310-050				Ton	18.60			18.60	20.50	
0350	Sand, washed, for concrete					25			25	27.50	
0400	For plaster or brick					13.15			13.15	14.45	
0450	Stone, 3/4" to 1-1/2"					18.15			18.15	20	
0500	3/8" roofing stone & 1/2" pea stone					15.10			15.10	16.60	
0550	For trucking 10 miles, add to the above					5.45			5.45	6	
0600	30 miles, add to the above				↓	13.60			13.60	14.95	
0850	Sand & stone, loaded at pit, crushed bank gravel				C.Y.	22			22	24.50	
0950	Sand, washed, for concrete					13.30			13.30	14.65	
1000	For plaster or brick					36.50			36.50	40	
1050	Stone, 3/4" to 1-1/2"					24			24	26.50	
1100	3/8" roofing stone & 1/2" pea stone					32			32	35	
1150	For trucking 10 miles, add to the above				↓	6			6	6.60	
1200	30 miles, add to the above					14.95			14.95	16.45	
1310	Quartz chips, 50 lb. bags				Cwt.	16.65			16.65	18.35	
1330	Silica chips, 50 lb. bags					9.15			9.15	10.05	
1410	White marble, 3/8" to 1/2", 50 lb. bags				↓	13.75			13.75	15.15	
1430	3/4"				Ton	69.50			69.50	76.50	
200 0010	**CEMENT** Material only R03310-020										**200**
0240	Portland, type I, plain/air entrained, TL lots, 94 lb bags				Bag	8.25			8.25	9.05	
0300	Trucked in bulk, per cwt **CN**				Cwt.	4.25			4.25	4.68	
0400	Type III, high early strength, TL lots, 94 lb bags **CN**				Bag	9.10			9.10	10	
0420	L.T.L. or L.C.L. lots					10			10	11	
0500	White, type III, high early strength, T.L. or C.L. lots, bags					19.85			19.85	22	
0520	L.T.L. or L.C.L. lots					22.50			22.50	25	
0600	White, type I, T.L. or C.L. lots, bags					19.45			19.45	21.50	
0620	L.T.L. or L.C.L. lots				↓	18.40			18.40	20.50	
210 0010	**CRIBBING** See under Retaining Walls, division 02830-100										**210**
220 0010	**CUTTING** Concrete see division 02220-360										**220**
250 0010	**DAMPPROOFING** See division 07100										**250**
300 0010	**EQUIPMENT** For placing conc. see div. 01590-100 R01590-100										**300**
400 0010	**LIFT SLAB** See division 03310-240 R03310-120										**400**
700 0010	**SAWING CONCRETE** See division 02220-360										**700**
850 0010	**WATERPROOFING AND DAMPPROOFING** See division 07100										**850**
0050	Integral waterproofing, add to cost of regular concrete				C.Y.	5.10			5.10	5.60	
870 0010	**WINTER PROTECTION** For heated ready mix, add, minimum				↓	4.50			4.50	4.95	**870**
0050	Maximum					5.65			5.65	6.20	
0100	Temporary heat to protect concrete, 24 hours, minimum	2 Clab	50	.320	M.S.F.	94	8.55		102.55	116	
0150	Maximum	"	25	.640	"	122	17.10		139.10	161	
0200	Temporary shelter for slab on grade, wood frame/polyethylene sheeting										
0201	Build or remove, minimum	2 Carp	10	1.600	M.S.F.	257	55		312	370	

03060	Basic Concrete Materials	CREW	DAILY OUTPUT	LABOR-HOURS	UNIT	2005 BARE COSTS				TOTAL INCL O&P	
						MAT.	LABOR	EQUIP.	TOTAL		
870 0210	Maximum	2 Carp	3	5.333	M.S.F.	310	183		493	625	**870**
0300	See also Division 03390-200										

03100 | Concrete Forms & Accessories

03110	Structural C.I.P. Forms	CREW	DAILY OUTPUT	LABOR-HOURS	UNIT	2005 BARE COSTS				TOTAL INCL O&P	
						MAT.	LABOR	EQUIP.	TOTAL		
300 0010	**EXPANSION JOINT** See division 03150-250										**300**
405 0010	**FORMS IN PLACE, BEAMS AND GIRDERS** R03110-040										**405**
0020	See also Elevated Slabs, division 03310-240										
0500	Exterior spandrel, job-built plywood, 12" wide, 1 use R03110-050	C-2	225	.213	SFCA	3.45	7.10		10.55	14.85	
0550	2 use		275	.175		1.83	5.80		7.63	11.05	
0600	3 use R03110-060		295	.163		1.38	5.40		6.78	9.95	
0650	4 use		310	.155		1.12	5.15		6.27	9.30	
1000	18" wide, 1 use		250	.192		3	6.40		9.40	13.25	
1050	2 use		275	.175		1.65	5.80		7.45	10.85	
1100	3 use		305	.157		1.20	5.25		6.45	9.45	
1150	4 use		315	.152		.98	5.10		6.08	9	
1500	24" wide, 1 use		265	.181		2.76	6.05		8.81	12.45	
1550	2 use		290	.166		1.56	5.50		7.06	10.30	
1600	3 use		315	.152		1.10	5.10		6.20	9.10	
1650	4 use		325	.148		.90	4.92		5.82	8.65	
2000	Interior beam, job-built plywood, 12" wide, 1 use		300	.160		3.82	5.35		9.17	12.50	
2050	2 use		340	.141		1.87	4.71		6.58	9.40	
2100	3 use		364	.132		1.53	4.40		5.93	8.55	
2150	4 use		377	.127		1.24	4.24		5.48	7.95	
2500	24" wide, 1 use		320	.150		2.81	5		7.81	10.90	
2550	2 use		365	.132		1.58	4.38		5.96	8.60	
2600	3 use		385	.125		1.12	4.16		5.28	7.70	
2650	4 use		395	.122		.91	4.05		4.96	7.30	
3000	Encasing steel beam, hung, job-built plywood, 1 use		325	.148		2.73	4.92		7.65	10.65	
3050	2 use		390	.123		1.50	4.10		5.60	8.05	
3100	3 use		415	.116		1.09	3.85		4.94	7.20	
3150	4 use		430	.112		.89	3.72		4.61	6.75	
3500	Bottoms only, to 30" wide, job-built plywood, 1 use		230	.209		4.15	6.95		11.10	15.40	
3550	2 use		265	.181		2.33	6.05		8.38	11.95	
3600	3 use		280	.171		1.66	5.70		7.36	10.75	
3650	4 use		290	.166		1.35	5.50		6.85	10.10	
4000	Sides only, vertical, 36" high, job-built plywood, 1 use		335	.143		4.28	4.78		9.06	12.15	
4050	2 use		405	.119		2.35	3.95		6.30	8.75	
4100	3 use		430	.112		1.71	3.72		5.43	7.70	
4150	4 use		445	.108		1.39	3.60		4.99	7.15	
4500	Sloped sides, 36" high, 1 use		305	.157		4.22	5.25		9.47	12.80	
4550	2 use		370	.130		2.36	4.32		6.68	9.35	
4600	3 use		405	.119		1.69	3.95		5.64	8	
4650	4 use		425	.113		1.37	3.76		5.13	7.35	
5000	Upstanding beams, 36" high, 1 use		225	.213		5.05	7.10		12.15	16.60	
5050	2 use		255	.188		2.83	6.25		9.08	12.85	
5100	3 use		275	.175		2.04	5.80		7.84	11.30	
5150	4 use		280	.171		1.66	5.70		7.36	10.70	

Important: See the Reference Section for critical supporting data - Reference Nos., Crews, & City Cost Indexes

03110 | Structural C.I.P. Forms

		CREW	DAILY OUTPUT	LABOR-HOURS	UNIT	MAT.	LABOR	EQUIP.	TOTAL	TOTAL INCL O&P	
410	**0010**	**FORMS IN PLACE, COLUMNS**										**410**
	0500	Round fiberglass, 4 use per mo., rent, 12" diameter	C-1	160	.200	L.F.	6.40	6.45		12.85	17.10	
	0550	16" diameter		150	.213		7.65	6.90		14.55	19.15	
	0600	18" diameter		140	.229		8.50	7.40		15.90	21	
	0650	24" diameter		135	.237		10.60	7.65		18.25	23.50	
	0700	28" diameter		130	.246		11.80	7.95		19.75	25.50	
	0800	30" diameter		125	.256		12.35	8.30		20.65	26.50	
	0850	36" diameter		120	.267		16.45	8.65		25.10	31.50	
	1500	Round fiber tube, 1 use, 8" diameter		155	.206		1.51	6.70		8.21	12.05	
	1550	10" diameter		155	.206		1.88	6.70		8.58	12.45	
	1600	12" diameter		150	.213		2.17	6.90		9.07	13.15	
	1650	14" diameter		145	.221		3.04	7.15		10.19	14.45	
	1700	16" diameter		140	.229		3.58	7.40		10.98	15.45	
	1750	20" diameter		135	.237		5.50	7.65		13.15	18	
	1800	24" diameter		130	.246		7	7.95		14.95	20	
	1850	30" diameter		125	.256		10.20	8.30		18.50	24	
	1900	36" diameter		115	.278		12.80	9		21.80	28	
	1950	42" diameter		100	.320		30.50	10.35		40.85	50	
	2000	48" diameter		85	.376		39	12.20		51.20	62	
	2200	For seamless type, add					15%					
	3000	Round, steel, 4 use per mo., rent, regular duty, 12" diameter	C-1	145	.221	L.F.	10.45	7.15		17.60	22.50	
	3050	16" diameter		125	.256		11.75	8.30		20.05	26	
	3100	Heavy duty, 20" diameter		105	.305		12.90	9.85		22.75	29.50	
	3150	24" diameter		85	.376		14.15	12.20		26.35	34.50	
	3200	30" diameter		70	.457		16.25	14.80		31.05	41	
	3250	36" diameter		60	.533		17.40	17.25		34.65	46	
	3300	48" diameter		50	.640		24	20.50		44.50	59	
	3350	60" diameter		45	.711		32	23		55	71	
	4000	Column capitals, steel, 4 uses/mo., 24" col, 4' cap diameter		12	2.667	Ea.	191	86.50		277.50	345	
	4050	5' cap diameter		11	2.909		217	94		311	385	
	4100	6' cap diameter		10	3.200		296	104		400	485	
	4150	7' cap diameter		9	3.556		360	115		475	580	
	4500	For second and succeeding months, deduct					50%					
	5000	Job-built plywood, 8" x 8" columns, 1 use	C-1	165	.194	SFCA	2.11	6.30		8.41	12.05	
	5050	2 use		195	.164		1.21	5.30		6.51	9.60	
	5100	3 use		210	.152		.84	4.93		5.77	8.65	
	5150	4 use		215	.149		.69	4.82		5.51	8.25	
	5500	12" x 12" columns, 1 use		180	.178		2.12	5.75		7.87	11.30	
	5550	2 use		210	.152		1.16	4.93		6.09	9	
	5600	3 use		220	.145		.85	4.71		5.56	8.30	
	5650	4 use		225	.142		.69	4.60		5.29	7.90	
	6000	16" x 16" columns, 1 use		185	.173		2.16	5.60		7.76	11.05	
	6050	2 use		215	.149		1.13	4.82		5.95	8.75	
	6100	3 use		230	.139		.86	4.50		5.36	7.95	
	6150	4 use		235	.136		.70	4.41		5.11	7.60	
	6500	24" x 24" columns, 1 use		190	.168		2.45	5.45		7.90	11.20	
	6550	2 use		216	.148		1.35	4.79		6.14	8.95	
	6600	3 use		230	.139		.98	4.50		5.48	8.10	
	6650	4 use		238	.134		.80	4.35		5.15	7.70	
	7000	36" x 36" columns, 1 use		200	.160		2.20	5.20		7.40	10.45	
	7050	2 use		230	.139		1.23	4.50		5.73	8.35	
	7100	3 use		245	.131		.88	4.23		5.11	7.55	
	7150	4 use		250	.128		.72	4.14		4.86	7.25	
	7500	Steel framed plywood, 4 use per mo., rent, 8" x 8"		340	.094		3.01	3.05		6.06	8.05	
	7550	10" x 10"		350	.091		2.40	2.96		5.36	7.25	
	7600	12" x 12"		370	.086		2.81	2.80		5.61	7.45	

R03110 -040
R03110 -050
R03110 -060

CONCRETE 3

3 CONCRETE

03110 | Structural C.I.P. Forms

			CREW	DAILY OUTPUT	LABOR-HOURS	UNIT	2005 BARE COSTS MAT.	LABOR	EQUIP.	TOTAL	TOTAL INCL O&P	
410	7650	16" x 16" R03110-040	C-1	400	.080	SFCA	3.02	2.59		5.61	7.35	**410**
	7700	20" x 20"		420	.076		1.59	2.47		4.06	5.60	
	7750	24" x 24" R03110-050		440	.073		1.49	2.35		3.84	5.30	
	7755	30" x 30"		440	.073		1.35	2.35		3.70	5.15	
415	0010	**FORMS IN PLACE, CULVERT** 5' to 8' square or rectangular, 1 use	C-1	170	.188	SFCA	4.80	6.10		10.90	14.80	**415**
	0050	2 use		180	.178		1.71	5.75		7.46	10.85	
	0100	3 use R03110-050		190	.168		1.34	5.45		6.79	9.95	
	0150	4 use		200	.160		1.15	5.20		6.35	9.30	
420	0010	**FORMS IN PLACE, ELEVATED SLABS** R03110-020										**420**
	0050	See also corrugated form deck, division 05310-300										
	1000	Flat plate, job-built plywood, to 15' high, 1 use R03110-050	C-2	470	.102	S.F.	4	3.40		7.40	9.70	
	1050	2 use		520	.092		2.20	3.08		5.28	7.20	
	1100	3 use		545	.088		1.60	2.94		4.54	6.35	
	1150	4 use R03110-060		560	.086		1.30	2.86		4.16	5.90	
	1500	15' to 20' high ceilings, 4 use		495	.097		1.53	3.23		4.76	6.75	
	1600	21' to 35' high ceilings, 4 use		450	.107		2.12	3.56		5.68	7.90	
	2000	Flat slab, drop panels, job-built plywood, to 15' high, 1 use		449	.107		4.44	3.56		8	10.45	
	2050	2 use		509	.094		2.44	3.14		5.58	7.55	
	2100	3 use		532	.090		1.77	3.01		4.78	6.65	
	2150	4 use		544	.088		1.44	2.94		4.38	6.15	
	2250	15' to 20' high ceilings, 4 use		480	.100		3.73	3.33		7.06	9.30	
	2350	20' to 35' high ceilings, 4 use		435	.110		4.31	3.68		7.99	10.50	
	3000	Floor slab hung from steel beams, 1 use		485	.099		2.17	3.30		5.47	7.55	
	3050	2 use		535	.090		1.52	2.99		4.51	6.35	
	3100	3 use		550	.087		1.30	2.91		4.21	5.95	
	3150	4 use		565	.085		1.19	2.83		4.02	5.70	
	3500	Floor slab, with 20" metal pans, 1 use		415	.116		5.70	3.85		9.55	12.30	
	3550	2 use		445	.108		3.71	3.60		7.31	9.70	
	3600	3 use		475	.101		3.04	3.37		6.41	8.60	
	3650	4 use		500	.096		2.71	3.20		5.91	7.95	
	3700	Floor slab with 30" pans, 1 use		418	.115		5.65	3.83		9.48	12.20	
	3720	2 use		455	.105		3.67	3.52		7.19	9.50	
	3740	3 use		470	.102		3	3.40		6.40	8.60	
	3760	4 use		480	.100		2.67	3.33		6	8.15	
	4000	Floor slab with 19" metal domes, 1 use		405	.119		6	3.95		9.95	12.75	
	4050	2 use		435	.110		3.98	3.68		7.66	10.10	
	4100	3 use		465	.103		3.31	3.44		6.75	9	
	4150	4 use		495	.097		2.98	3.23		6.21	8.30	
	4500	With 30" fiberglass domes, 1 use		405	.119		5.85	3.95		9.80	12.55	
	4520	2 use		450	.107		3.82	3.56		7.38	9.75	
	4530	3 use		460	.104		3.16	3.48		6.64	8.85	
	4550	4 use		470	.102		2.82	3.40		6.22	8.40	
	5000	Box out for slab openings, over 16" deep, 1 use		190	.253	SFCA	3.49	8.40		11.89	16.95	
	5050	2 use		240	.200	"	1.92	6.65		8.57	12.50	
	5500	Shallow slab box outs, to 10 S.F.		42	1.143	Ea.	9.85	38		47.85	70.50	
	5550	Over 10 S.F. (use perimeter)		600	.080	L.F.	1.31	2.67		3.98	5.60	
	6000	Bulkhead forms for slab, with keyway, 1 use, 2 piece		500	.096		1.59	3.20		4.79	6.75	
	6100	3 piece (see also edge forms)		460	.104		1.73	3.48		5.21	7.30	
	6200	Bulkhead form for slab, 4-1/2" high, exp metal, incl keyway & stakes	C-1	1,200	.027		.70	.86		1.56	2.11	
	6210	5-1/2" high		1,100	.029		.78	.94		1.72	2.33	
	6215	7-1/2" high		960	.033		.94	1.08		2.02	2.71	
	6220	9-1/2" high		840	.038		1.62	1.23		2.85	3.70	
	6500	Curb forms, wood, 6" to 12" high, on elevated slabs, 1 use		180	.178	SFCA	1.48	5.75		7.23	10.55	
	6550	2 use		205	.156		.81	5.05		5.86	8.75	
	6600	3 use		220	.145		.59	4.71		5.30	8	
	6650	4 use		225	.142		.48	4.60		5.08	7.70	

Important: See the Reference Section for critical supporting data - Reference Nos., Crews, & City Cost Indexes

03110 | Structural C.I.P. Forms

		Description	CREW	DAILY OUTPUT	LABOR-HOURS	UNIT	2005 BARE COSTS				TOTAL INCL O&P	
							MAT.	LABOR	EQUIP.	TOTAL		
420	7000	Edge forms to 6" high, on elevated slab, 4 use	C-1	500	.064	L.F.	.17	2.07		2.24	3.42	**420**
	7500	Depressed area forms to 12" high, 4 use		300	.107		.78	3.45		4.23	6.25	
	7550	12" to 24" high, 4 use		175	.183		1.06	5.90		6.96	10.35	
	8000	Perimeter deck and rail for elevated slabs, straight		90	.356		13.60	11.50		25.10	33	
	8050	Curved		65	.492		18.70	15.95		34.65	45.50	
	8500	Void forms, round fiber, 3" diameter		450	.071		.62	2.30		2.92	4.26	
	8550	4" diameter		425	.075		.72	2.44		3.16	4.58	
	8600	6" diameter		400	.080		1.09	2.59		3.68	5.25	
	8650	8" diameter		375	.085		1.76	2.76		4.52	6.25	
	8700	10" diameter		350	.091		2.28	2.96		5.24	7.10	
	8750	12" diameter		300	.107		2.95	3.45		6.40	8.65	
425	0010	**FORMS IN PLACE, EQUIPMENT FOUNDATIONS** job built										**425**
	0020	1 use	C-2	160	.300	SFCA	2.67	10		12.67	18.50	
	0050	2 use		190	.253		1.47	8.40		9.87	14.70	
	0100	3 use		200	.240		1.07	8		9.07	13.60	
	0150	4 use		205	.234		.87	7.80		8.67	13.10	
430	0010	**FORMS IN PLACE, FOOTINGS** Continuous wall, plywood, 1 use	C-1	375	.085	SFCA	2.31	2.76		5.07	6.85	**430**
	0050	2 use		440	.073		1.27	2.35		3.62	5.05	
	0100	3 use		470	.068		.92	2.20		3.12	4.45	
	0150	4 use		485	.066		.75	2.14		2.89	4.16	
	0500	Dowel supports for footings or beams, 1 use		500	.064	L.F.	.76	2.07		2.83	4.07	
	1000	Integral starter wall, to 4" high, 1 use		400	.080		.81	2.59		3.40	4.92	
	1500	Keyway, 4 use, tapered wood, 2" x 4"	1 Carp	530	.015		.18	.52		.70	1.01	
	1550	2" x 6"		500	.016		.27	.55		.82	1.15	
	2000	Tapered plastic, 2" x 3"		530	.015		.56	.52		1.08	1.43	
	2050	2" x 4"		500	.016		.71	.55		1.26	1.63	
	2250	For keyway hung from supports, add		150	.053		.76	1.83		2.59	3.69	
	3000	Pile cap, square or rectangular, job-built plywood, 1 use	C-1	290	.110	SFCA	2.36	3.57		5.93	8.15	
	3050	2 use		346	.092		1.30	2.99		4.29	6.10	
	3100	3 use		371	.086		.94	2.79		3.73	5.40	
	3150	4 use		383	.084		.77	2.70		3.47	5.05	
	4000	Triangular or hexagonal, 1 use		225	.142		2.78	4.60		7.38	10.20	
	4050	2 use		280	.114		1.53	3.70		5.23	7.45	
	4100	3 use		305	.105		1.11	3.40		4.51	6.50	
	4150	4 use		315	.102		.90	3.29		4.19	6.10	
	5000	Spread footings, job-built lumber, 1 use		305	.105		1.66	3.40		5.06	7.15	
	5050	2 use		371	.086		.92	2.79		3.71	5.35	
	5100	3 use		401	.080		.67	2.58		3.25	4.75	
	5150	4 use		414	.077		.54	2.50		3.04	4.49	
	6000	Supports for dowels, plinths or templates, 2' x 2' footing		25	1.280	Ea.	5.45	41.50		46.95	70.50	
	6050	4' x 4' footing		22	1.455		10.90	47		57.90	85.50	
	6100	8' x 8' footing		20	1.600		22	52		74	105	
	6150	12' x 12' footing		17	1.882		35	61		96	134	
	7000	Plinths, job-built plywood, 1 use		250	.128	SFCA	3.63	4.14		7.77	10.45	
	7100	4 use		270	.119	"	1.19	3.84		5.03	7.25	
435	0010	**FORMS IN PLACE, GRADE BEAM** Job-built plywood, 1 use	C-2	530	.091	SFCA	1.56	3.02		4.58	6.40	**435**
	0050	2 use		580	.083		.86	2.76		3.62	5.25	
	0100	3 use		600	.080		.63	2.67		3.30	4.84	
	0150	4 use		605	.079		.51	2.64		3.15	4.68	
440	0010	**FORMS IN PLACE, MAT FOUNDATION** Job-built plywood, 1 use	C-2	290	.166	SFCA	2.44	5.50		7.94	11.30	**440**
	0050	2 use		310	.155		.92	5.15		6.07	9.05	
	0100	3 use		330	.145		.65	4.85		5.50	8.25	
	0120	4 use		350	.137		.54	4.57		5.11	7.70	

CONCRETE 3

03110 | Structural C.I.P. Forms

			CREW	DAILY OUTPUT	LABOR-HOURS	UNIT	2005 BARE COSTS				TOTAL INCL O&P
							MAT.	LABOR	EQUIP.	TOTAL	
445	0010	**FORMS IN PLACE, SLAB ON GRADE**									**445**
	1000	Bulkhead forms w/keyway, wood, 6" high, 1 use	C-1	510	.063	L.F.	.80	2.03		2.83	4.04
	1050	2 uses		400	.080		.44	2.59		3.03	4.51
	1100	4 uses		350	.091		.26	2.96		3.22	4.90
	1400	Bulkhead form for slab, 4-1/2" high, exp metal, incl keyway & stakes		1,200	.027		.70	.86		1.56	2.11
	1410	5-1/2" high		1,100	.029		.78	.94		1.72	2.33
	1420	7-1/2" high		960	.033		.94	1.08		2.02	2.71
	2000	Curb forms, wood, 6" to 12" high, on grade, 1 use		215	.149	SFCA	2.19	4.82		7.01	9.90
	2050	2 use		250	.128		1.21	4.14		5.35	7.80
	2100	3 use		265	.121		.87	3.91		4.78	7.05
	2150	4 use		275	.116		.71	3.77		4.48	6.65
	3000	Edge forms, wood, 4 use, on grade, to 6" high		600	.053	L.F.	.27	1.73		2	2.98
	3050	7" to 12" high		435	.074	SFCA	.74	2.38		3.12	4.52
	3500	For depressed slabs, 4 use, to 12" high		300	.107	L.F.	.66	3.45		4.11	6.10
	3550	To 24" high		175	.183		.88	5.90		6.78	10.15
	4000	For slab blockouts, to 12" high, 1 use		200	.160		.73	5.20		5.93	8.85
	4050	To 24" high, 1 use		120	.267		.93	8.65		9.58	14.45
	4100	Plastic (extruded), to 6" high, multiple use, on grade		800	.040		.37	1.29		1.66	2.43
	5000	Screed, 24 ga. metal key joint, see Div 03150-250									
	5020	Wood, incl. wood stakes, 1" x 3"	C-1	900	.036	L.F.	.57	1.15		1.72	2.41
	5050	2" x 4"		900	.036	"	.62	1.15		1.77	2.47
	6000	Trench forms in floor, wood, 1 use		160	.200	SFCA	2.20	6.45		8.65	12.50
	6050	2 use		175	.183		1.22	5.90		7.12	10.55
	6100	3 use		180	.178		.89	5.75		6.64	9.95
	6150	4 use		185	.173		.72	5.60		6.32	9.50
	8760	Void form, corrugated fiberboard, 6" x 12", 10' long		240	.133	S.F.	9.20	4.31		13.51	16.80
450	0010	**FORMS IN PLACE, STAIRS** (Slant length x width), 1 use R03110-050	C-2	165	.291	S.F.	3.66	9.70		13.36	19.15	**450**
	0050	2 use		170	.282		2.01	9.40		11.41	16.85
	0100	3 use R03110-060		180	.267		1.46	8.90		10.36	15.45
	0150	4 use		190	.253		1.19	8.40		9.59	14.40
	1000	Alternate pricing method (0.7 L.F./S.F.), 1 use		100	.480	LF Rsr	5.25	16		21.25	31
	1050	2 use		105	.457		2.89	15.25		18.14	26.50
	1100	3 use		110	.436		2.10	14.55		16.65	25
	1150	4 use		115	.417		1.73	13.90		15.63	23.50
	2000	Stairs, cast on sloping ground (length x width), 1 use		220	.218	S.F.	1.36	7.25		8.61	12.80
	2100	4 use		240	.200	"	.44	6.65		7.09	10.90
455	0010	**FORMS IN PLACE, WALLS** R03110-010									**455**
	0100	Box out for wall openings, to 16" thick, to 10 S.F.	C-2	24	2	Ea.	24.50	66.50		91	131
	0150	Over 10 S.F. (use perimeter) R03110-050	"	280	.171	L.F.	2.07	5.70		7.77	11.20
	0250	Brick shelf, 4" w, add to wall forms, use wall area abv shelf									
	0260	1 use R03110-060	C-2	240	.200	SFCA	2.18	6.65		8.83	12.80
	0300	2 use		275	.175		1.20	5.80		7	10.35
	0350	4 use		300	.160		.87	5.35		6.22	9.25
	0500	Bulkhead, with keyway, 1 use, 2 piece		265	.181	L.F.	2.76	6.05		8.81	12.45
	0550	3 piece		175	.274		3.48	9.15		12.63	18.10
	0600	Bulkhead forms with keyway, 1 piece expanded metal, 8" wall	C-1	1,000	.032		.94	1.04		1.98	2.64
	0610	10" wall		800	.040		1.62	1.29		2.91	3.80
	0620	12" wall		525	.061		2.05	1.97		4.02	5.35
	0700	Buttress, to 8' high, 1 use	C-2	350	.137	SFCA	3.68	4.57		8.25	11.15
	0750	2 use		430	.112		2.02	3.72		5.74	8
	0800	3 use		460	.104		1.47	3.48		4.95	7
	0850	4 use		480	.100		1.21	3.33		4.54	6.55
	1000	Corbel or haunch, to 12" wide, add to wall forms, 1 use		150	.320	L.F.	2.06	10.65		12.71	18.85
	1050	2 use		170	.282		1.13	9.40		10.53	15.90
	1100	3 use		175	.274		.83	9.15		9.98	15.15
	1150	4 use		180	.267		.67	8.90		9.57	14.60

Important: See the Reference Section for critical supporting data - Reference Nos., Crews, & City Cost Indexes

		03110	Structural C.I.P. Forms		CREW	DAILY OUTPUT	LABOR-HOURS	UNIT	MAT.	LABOR	EQUIP.	TOTAL	TOTAL INCL O&P	
455	2000		Wall, below grade, job-built plywood, to 8' high, 1 use	R03110 -010	C-2	300	.160	SFCA	2.37	5.35		7.72	10.90	455
	2050		2 use			365	.132		1.50	4.38		5.88	8.50	
	2100		3 use	R03110 -050		425	.113		1.09	3.76		4.85	7.05	
	2150		4 use			435	.110		.88	3.68		4.56	6.70	
	2400		Over 8' to 16' high, 1 use	R03110 -060		280	.171		4.51	5.70		10.21	13.85	
	2420		2 use			345	.139		1.89	4.64		6.53	9.30	
	2430		3 use			375	.128		1.57	4.27		5.84	8.40	
	2440		4 use			395	.122		1.40	4.05		5.45	7.85	
	2445		Exterior wall, 8' to 16' high, 1 use			280	.171		2.18	5.70		7.88	11.30	
	2450		2 use			345	.139		1.20	4.64		5.84	8.50	
	2500		3 use			375	.128		.85	4.27		5.12	7.60	
	2550		4 use			395	.122		.70	4.05		4.75	7.05	
	2700		Over 16' high, 1 use			235	.204		2.40	6.80		9.20	13.25	
	2750		2 use			290	.166		1.32	5.50		6.82	10.05	
	2800		3 use			315	.152		.96	5.10		6.06	8.95	
	2850		4 use			330	.145		.78	4.85		5.63	8.40	
	3000		For architectural finish, add			1,820	.026		5.40	.88		6.28	7.25	
	4000		Radial, smooth curved, job-built plywood, 1 use			245	.196		2.48	6.55		9.03	12.90	
	4050		2 use			300	.160		1.36	5.35		6.71	9.80	
	4100		3 use			325	.148		.99	4.92		5.91	8.75	
	4150		4 use			335	.143		.81	4.78		5.59	8.35	
	4200		Below grade, job-built plywood, 1 use			225	.213		2.99	7.10		10.09	14.35	
	4210		2 use			225	.213		1.65	7.10		8.75	12.85	
	4220		3 use			225	.213		1.33	7.10		8.43	12.50	
	4230		4 use			225	.213		.97	7.10		8.07	12.10	
	4300		Curved, 2' chords, job-built plywood, 1 use			290	.166		1.95	5.50		7.45	10.75	
	4350		2 use			355	.135		1.07	4.51		5.58	8.20	
	4400		3 use			385	.125		.78	4.16		4.94	7.30	
	4450		4 use			400	.120		.63	4		4.63	6.95	
	4500		Over 8' high, 1 use			290	.166		.94	5.50		6.44	9.65	
	4525		2 use			355	.135		.52	4.51		5.03	7.55	
	4550		3 use			385	.125		.38	4.16		4.54	6.85	
	4575		4 use			400	.120		.31	4		4.31	6.60	
	4600		Retaining wall, battered, job-built plyw'd, to 8' high, 1 use			300	.160		1.85	5.35		7.20	10.35	
	4650		2 use			355	.135		1.02	4.51		5.53	8.10	
	4700		3 use			375	.128		.74	4.27		5.01	7.45	
	4750		4 use			390	.123		.56	4.10		4.66	7	
	4900		Over 8' to 16' high, 1 use			240	.200		2	6.65		8.65	12.60	
	4950		2 use			295	.163		1.10	5.40		6.50	9.65	
	5000		3 use			305	.157		.80	5.25		6.05	9.05	
	5050		4 use		▼	320	.150		.65	5		5.65	8.50	
	5500		For gang wall forming, 192 S.F. sections, deduct						10%	10%				
	5550		384 S.F. sections, deduct					▼	20%	20%				
	5750		Liners for forms (add to wall forms), A.B.S. plastic											
	5800		Aged wood, 4" wide, 1 use		1 Carp	250	.032	SFCA	5.85	1.10		6.95	8.15	
	5820		2 use			400	.020		3.22	.69		3.91	4.61	
	5840		4 use			750	.011		1.90	.37		2.27	2.66	
	5900		Fractured rope rib, 1 use			250	.032		3.69	1.10		4.79	5.75	
	6000		4 use			750	.011		1.20	.37		1.57	1.89	
	6100		Ribbed look, 1/2" & 3/4" deep, 1 use			300	.027		5.40	.91		6.31	7.30	
	6200		4 use			800	.010		1.75	.34		2.09	2.45	
	6300		Rustic brick pattern, 1 use			250	.032		3.69	1.10		4.79	5.75	
	6400		4 use			750	.011		1.20	.37		1.57	1.89	
	6500		Striated, random, 3/8" x 3/8" deep, 1 use			300	.027		3.69	.91		4.60	5.50	
	6600		4 use		▼	800	.010	▼	1.20	.34		1.54	1.85	
	6800		Rustication strips, A.B.S. plastic, 2 piece snap-on											

3

CONCRETE

3 CONCRETE

		03110	Structural C.I.P. Forms		CREW	DAILY OUTPUT	LABOR-HOURS	UNIT	2005 BARE COSTS				TOTAL INCL O&P	
									MAT.	LABOR	EQUIP.	TOTAL		
455	6850		1" deep x 1-3/8" wide, 1 use	R03110 -010	C-2	400	.120	L.F.	4.18	4		8.18	10.85	455
	6900		2 use			600	.080		2.30	2.67		4.97	6.70	
	6950		4 use	R03110 -050		800	.060		1.36	2		3.36	4.60	
	7050		Wood, beveled edge, 3/4" deep, 1 use			600	.080		.10	2.67		2.77	4.26	
	7100		1" deep, 1 use	R03110 -060		450	.107		.14	3.56		3.70	5.70	
	7200		For solid board finish, uniform, 1 use, add to wall forms			300	.160	SFCA	.79	5.35		6.14	9.15	
	7300		Non-uniform finish			250	.192		.74	6.40		7.14	10.75	
	7500		Lintel or sill forms, 1 use		1 Carp	30	.267		2.80	9.15		11.95	17.35	
	7520		2 use			34	.235		1.54	8.05		9.59	14.25	
	7540		3 use			36	.222		1.12	7.60		8.72	13.10	
	7560		4 use			37	.216		.91	7.40		8.31	12.55	
	7800		Modular prefabricated plywood, to 8' high, 1 use		C-2	1,180	.041		1.68	1.36		3.04	3.96	
	7820		2 use			1,200	.040		.92	1.33		2.25	3.10	
	7840		3 use			1,240	.039		.67	1.29		1.96	2.75	
	7860		4 use			1,260	.038		.55	1.27		1.82	2.59	
	8000		To 16' high, 1 use			715	.067		2.19	2.24		4.43	5.90	
	8020		2 use			740	.065		1.20	2.16		3.36	4.69	
	8040		3 use			770	.062		.88	2.08		2.96	4.20	
	8060		4 use			790	.061		.72	2.03		2.75	3.94	
	8100		Over 16' high, 1 use			715	.067		2.63	2.24		4.87	6.35	
	8120		2 use			740	.065		1.45	2.16		3.61	4.96	
	8140		3 use			770	.062		1.05	2.08		3.13	4.40	
	8160		4 use			790	.061		.88	2.03		2.91	4.11	
	8600		Pilasters, 1 use			270	.178		2.73	5.95		8.68	12.25	
	8620		2 use			330	.145		1.50	4.85		6.35	9.20	
	8640		3 use			370	.130		1.09	4.32		5.41	7.95	
	8660		4 use			385	.125		.89	4.16		5.05	7.45	
	9010		Steel framed plywood, based on 100 uses of purchased											
	9020		forms, and 4 uses of bracing lumber											
	9060		To 8' high		C-2	600	.080	SFCA	.37	2.67		3.04	4.56	
	9260		Over 8' to 16' high			450	.107		.37	3.56		3.93	5.95	
	9460		Over 16' to 20' high			400	.120		.37	4		4.37	6.65	
	9475		For elevated walls, add							10%				
	9480		For battered walls, 1 side battered, add						10%	10%				
	9485		For battered walls, 2 sides battered, add						15%	15%				
460	0010	**FORMS IN PLACE, INSULATING CONCRETE**												460
	0020	Forms left in place, S.F. is for both sides												
	1000		Panel system, flat cavity, minimum		2 Carp	960	.017	S.F.	1.84	.57		2.41	2.91	
	1010		Maximum			960	.017		2.76	.57		3.33	3.93	
	1020		Grid cavity, minimum			960	.017		2	.57		2.57	3.09	
	1030		Maximum			960	.017		3	.57		3.57	4.19	
	1040		Post and beam cavity, minimum			960	.017		2.20	.57		2.77	3.31	
	1050		Maximum			960	.017		3.30	.57		3.87	4.52	
	1060		Plank system, flat cavity, minimum			1,920	.008		2.10	.29		2.39	2.75	
	1070		Maximum			1,920	.008		3.14	.29		3.43	3.90	
	1120		Block system, flat cavity, minimum			480	.033		1.96	1.14		3.10	3.94	
	1130		Maximum			480	.033		2.94	1.14		4.08	5	
	1140		Grid cavity, minimum			480	.033		1.96	1.14		3.10	3.94	
	1150		Maximum			480	.033		2.94	1.14		4.08	5	
	1160		Post and beam cavity, minimum			480	.033		1.96	1.14		3.10	3.94	
	1170		Maximum			480	.033		2.94	1.14		4.08	5	
500	0010	**GAS STATION FORMS** Curb fascia, with template,												500
	0050		12 ga. steel, left in place, 9" high		1 Carp	50	.160	L.F.	9.15	5.50		14.65	18.65	
	1000		Sign or light bases, 18" diameter, 9" high			9	.889	Ea.	58	30.50		88.50	111	
	1050		30" diameter, 13" high			8	1		92	34.50		126.50	155	

Important: See the Reference Section for critical supporting data - Reference Nos., Crews, & City Cost Indexes

CONCRETE 3

03110		Structural C.I.P. Forms		CREW	DAILY OUTPUT	LABOR-HOURS	UNIT	2005 BARE COSTS				TOTAL INCL O&P	
								MAT.	LABOR	EQUIP.	TOTAL		
500	2000	Island forms, 10' long, 9" high, 3'- 6" wide		C-1	10	3.200	Ea.	256	104		360	445	500
	2050	4' wide			9	3.556		265	115		380	470	
	2500	20' long, 9" high, 4' wide			6	5.333		425	173		598	740	
	2550	5' wide			5	6.400		445	207		652	810	
750	0010	**REGLET** See division 07710-750											750
800	0010	**SCAFFOLDING** See division 01540-750											800
820	0010	**SLIPFORMS** Silos, minimum	R03110 -030	C-17E	3,885	.021	SFCA	3	.73	.02	3.75	4.45	820
	0050	Maximum			1,095	.073		4.30	2.58	.07	6.95	8.80	
	1000	Buildings, minimum			3,660	.022		2	.77	.02	2.79	3.42	
	1050	Maximum			875	.091		3.70	3.22	.09	7.01	9.15	

03150 | Concrete Accessories

				CREW	DAILY OUTPUT	LABOR-HOURS	UNIT	MAT.	LABOR	EQUIP.	TOTAL	INCL O&P	
080	0010	**ACCESSORIES, ANCHOR BOLTS** J-type, incl. nut and washer											080
	0020	1/2" diameter, 6" long		1 Carp	90	.089	Ea.	.94	3.04		3.98	5.75	
	0050	10" long			85	.094		1.06	3.22		4.28	6.15	
	0100	12" long			85	.094		1.17	3.22		4.39	6.30	
	0200	5/8" diameter, 12" long			80	.100		1.18	3.43		4.61	6.65	
	0250	18" long			70	.114		1.39	3.91		5.30	7.65	
	0300	24" long			60	.133		1.60	4.57		6.17	8.85	
	0350	3/4" diameter, 8" long			80	.100		1.39	3.43		4.82	6.90	
	0400	12" long			70	.114		1.74	3.91		5.65	8	
	0450	18" long			60	.133		2.26	4.57		6.83	9.60	
	0500	24" long			50	.160		2.96	5.50		8.46	11.80	
	0600	7/8" diameter, 12" long			60	.133		2.29	4.57		6.86	9.60	
	0650	18" long			50	.160		3.07	5.50		8.57	11.95	
	0700	24" long			40	.200		3.17	6.85		10.02	14.15	
	0800	1" diameter, 12" long			55	.145		3.32	4.98		8.30	11.40	
	0850	18" long			45	.178		4	6.10		10.10	13.90	
	0900	24" long			35	.229		4.88	7.85		12.73	17.55	
	0950	36" long			25	.320		6.70	10.95		17.65	24.50	
	1200	1-1/2" diameter, 18" long			22	.364		11.85	12.45		24.30	32.50	
	1250	24" long			18	.444		14.10	15.20		29.30	39	
	1300	36" long			12	.667		17.60	23		40.60	55	
	1350	For larger sizes see Division 05090-080											
	8000	Sleeves, see Division 03150-620											
085	0013	**ANCHOR BOLTS** See divisions 04080-070 and 05090-080											085
160	0010	**ACCESSORIES, CHAMFER STRIPS**											160
	2000	Polyvinyl chloride, 1/2" wide with leg		1 Carp	535	.015	L.F.	.22	.51		.73	1.04	
	2200	3/4" wide with leg			525	.015		.30	.52		.82	1.14	
	2400	1" radius with leg			515	.016		.60	.53		1.13	1.49	
	2800	1-1/2" radius with leg			500	.016		2.01	.55		2.56	3.06	
	5000	Wood, 1/2" wide			535	.015		.09	.51		.60	.90	
	5200	3/4" wide			525	.015		.10	.52		.62	.92	
	5400	1" wide			515	.016		.14	.53		.67	.98	
170	0010	**ACCESSORIES, COLUMN FORM**											170
	1000	Column clamps, adjustable to 24" x 24", buy					Set	83.50			83.50	91.50	
	1100	Rent per month						6.45			6.45	7.10	
	1300	For sizes to 30" x 30", buy						116			116	128	
	1400	Rent per month						8.55			8.55	9.45	
	1600	For sizes to 36" x 36", buy						115			115	127	
	1700	Rent per month						11.75			11.75	12.95	
	2000	Bull winch (band iron) 36" x 36", buy						48.50			48.50	53.50	

3 CONCRETE

03150	Concrete Accessories	CREW	DAILY OUTPUT	LABOR-HOURS	UNIT	2005 BARE COSTS				TOTAL INCL O&P	
						MAT.	LABOR	EQUIP.	TOTAL		
170	2100	Rent per month				Set	4.44			4.44	4.88	**170**
	2300	48" x 48", buy					50			50	55	
	2400	Rent per month					5.60			5.60	6.15	
	3000	Chain & wedge type 36" x 36", buy					68.50			68.50	75	
	3100	Rent per month					7.65			7.65	8.40	
	3300	60" x 60", buy					92			92	101	
	3400	Rent per month					10.70			10.70	11.80	
	4000	Friction collars 2'-6" dia., buy					695			695	765	
	4100	Rent per month					68.50			68.50	75	
	4300	4'-0" dia., buy					790			790	870	
	4400	Rent per month					86.50			86.50	95.50	
200	0010	**ACCESSORIES, DOVETAIL ANCHOR SYSTEM**										**200**
	0500	Anchor slot, galv., filled, 24 ga.	1 Carp	425	.019	L.F.	.77	.64		1.41	1.85	
	0600	20 ga.		400	.020		.94	.69		1.63	2.10	
	0800	16 oz. copper, foam filled		375	.021		1.97	.73		2.70	3.31	
	0900	26 ga. stainless steel, foam filled		375	.021		1.23	.73		1.96	2.50	
	1200	Brick anchor, corr., galv., 3-1/2" long, 16 ga.	1 Bric	10.50	.762	C	22.50	27		49.50	66	
	1300	12 ga.		10.50	.762		31.50	27		58.50	75.50	
	1500	Flat, galv., 3-1/2" long, 16 ga.		10.50	.762		24.50	27		51.50	68	
	1600	12 ga.		10.50	.762		46	27		73	91.50	
	2000	Cavity wall, corr., galv., 5" long, 16 ga.		10.50	.762		34	27		61	78	
	2100	12 ga.		10.50	.762		48.50	27		75.50	94.50	
	3000	Furring anchors, corr., galv., 1-1/2" long, 16 ga.		10.50	.762		12.85	27		39.85	55	
	3100	12 ga.		10.50	.762		21.50	27		48.50	65	
	6000	Stone anchors, 3-1/2" long, galv., 1/8" x 1" wide		10.50	.762		96.50	27		123.50	147	
	6100	1/4" x 1" wide		10.50	.762		149	27		176	205	
250	0010	**EXPANSION JOINT** Keyed, cold, 24 ga, incl. stakes, 3-1/2" high	1 Carp	200	.040	L.F.	.58	1.37		1.95	2.77	**250**
	0050	4-1/2" high		200	.040		.70	1.37		2.07	2.90	
	0100	5-1/2" high		195	.041		.78	1.41		2.19	3.05	
	0150	7-1/2" high		190	.042		.94	1.44		2.38	3.28	
	0300	Poured asphalt, plain, 1/2" x 1"	1 Clab	450	.018		.36	.47		.83	1.14	
	0350	1" x 2"		400	.020		1.26	.53		1.79	2.22	
	0500	Neoprene, liquid, cold applied, 1/2" x 1"		450	.018		1.63	.47		2.10	2.53	
	0550	1" x 2"		400	.020		6.25	.53		6.78	7.75	
	0700	Polyurethane, poured, 2 part, 1/2" x 1"		400	.020		2	.53		2.53	3.03	
	0750	1" x 2"		350	.023		7.40	.61		8.01	9.10	
	0900	Rubberized asphalt, hot or cold applied, 1/2" x 1"		450	.018		.44	.47		.91	1.22	
	0950	1" x 2"		400	.020		1.23	.53		1.76	2.18	
	1100	Hot applied, fuel resistant, 1/2" x 1"		450	.018		.76	.47		1.23	1.58	
	1150	1" x 2"		400	.020		1.55	.53		2.08	2.54	
	2000	Premolded, bituminous fiber, 1/2" x 6"	1 Carp	375	.021		.39	.73		1.12	1.57	
	2050	1" x 12"		300	.027		1.37	.91		2.28	2.93	
	2250	Cork with resin binder, 1/2" x 6"		375	.021		1.52	.73		2.25	2.81	
	2300	1" x 12"		300	.027		5.45	.91		6.36	7.40	
	2500	Neoprene sponge, closed cell, 1/2" x 6"		375	.021		1.32	.73		2.05	2.59	
	2550	1" x 12"		300	.027		6.05	.91		6.96	8.10	
	2750	Polyethylene foam, 1/2" x 6"		375	.021		.49	.73		1.22	1.68	
	2800	1" x 12"		300	.027		1.39	.91		2.30	2.95	
	3000	Polyethylene backer rod, 3/8" diameter		460	.017		.04	.60		.64	.97	
	3050	3/4" diameter		460	.017		.06	.60		.66	1	
	3100	1" diameter		460	.017		.07	.60		.67	1.01	
	3500	Polyurethane foam, with polybutylene, 1/2" x 1/2"		475	.017		.60	.58		1.18	1.56	
	3550	1" x 1"		450	.018		1.67	.61		2.28	2.79	
	3750	Polyurethane foam, regular, closed cell, 1/2" x 6"		375	.021		1.19	.73		1.92	2.45	

Important: See the Reference Section for critical supporting data - Reference Nos., Crews, & City Cost Indexes

03150	Concrete Accessories	CREW	DAILY OUTPUT	LABOR-HOURS	UNIT	2005 BARE COSTS				TOTAL INCL O&P	
						MAT.	LABOR	EQUIP.	TOTAL		
250											**250**
3800	1" x 12"	1 Carp	300	.027	L.F.	1.39	.91		2.30	2.95	
4000	Polyvinyl chloride foam, closed cell, 1/2" x 6"		375	.021		1.97	.73		2.70	3.31	
4050	1" x 12"		300	.027		5.50	.91		6.41	7.45	
4250	Rubber, gray sponge, 1/2" x 6"		375	.021		2.46	.73		3.19	3.85	
4300	1" x 12"		300	.027		9.65	.91		10.56	12	
4500	Lead wool for joints, 1 ton lots				Lb.	1.90			1.90	2.09	
5000	For installation in walls, add						75%				
5250	For installation in boxouts, add						25%				
350	0010	**ACCESSORIES, HANGERS**									**350**
0020	Slab and beam form										
0500	Banding iron, 3/4" x 22 ga, 14 L.F. per lb or										
0550	1/2" x 14 ga, 7 L.F. per lb.				Lb.	1.03			1.03	1.13	
1000	Fascia ties, coil type add to frame ties below				C	71			71	78	
1500	Frame ties to 8-1/8"					178			178	195	
1550	8-1/8" to 10-1/8"					190			190	208	
5000	Snap tie hanger, to 30" overall length, 4000 #					420			420	465	
5050	30" to 36" overall length					455			455	505	
5100	42" to 48" overall length					525			525	575	
5500	Steel beam hanger				C						
5600	Flange to 8-1/8"					345			345	380	
5650	8-1/8" to 10-1/8"					288			288	315	
6000	Tie hangers to 24" overall length, 6000 #					390			390	430	
6100	30" to 36" overall length					460			460	510	
6500	Tie back hanger, up to 12-1/8" flange					365			365	405	
8500	Wire, black annealed, 9 ga				Cwt.	105			105	116	
8600	16 ga				"	110			110	121	
400	0010	**ACCESSORIES, INSERTS**									**400**
1000	All size nut insert, 5/8" & 3/4", incl. nut	1 Carp	84	.095	Ea.	3.39	3.26		6.65	8.85	
2000	Continuous slotted, 1-5/8" x 1-3/8"										
2100	12 ga., 3" long	1 Carp	65	.123	Ea.	3.08	4.22		7.30	9.95	
2150	6" long		65	.123		3.98	4.22		8.20	10.95	
2200	8 ga., 12" long		65	.123		9.55	4.22		13.77	17.05	
2250	24" long		65	.123		15.65	4.22		19.87	24	
2300	36" long		60	.133		22	4.57		26.57	31	
2350	60" long		55	.145		33.50	4.98		38.48	45	
7000	Threaded cast										
7100	1/4" diameter bolt	1 Carp	84	.095	Ea.	5.60	3.26		8.86	11.30	
7350	7/8" diameter bolt	"	84	.095	"	8.25	3.26		11.51	14.15	
9000	Wedge										
9050	For 5/8" diameter bolt	1 Carp	60	.133	Ea.	3.71	4.57		8.28	11.20	
9100	For 3/4" diameter bolt	"	60	.133	"	7.95	4.57		12.52	15.85	
9800	Cut washers, "Black"										
9850	5/8" bolt				Ea.	.23			.23	.25	
9950	For galvanized inserts, add					30%					
600	0010	**SHORES** Erect and strip, by hand, horizontal members									**600**
0500	Aluminum joists and stringers	2 Carp	60	.267	Ea.		9.15		9.15	14.25	
0600	Steel, adjustable beams		45	.356			12.20		12.20	18.95	
0700	Wood joists		50	.320			10.95		10.95	17.05	
0800	Wood stringers		30	.533			18.25		18.25	28.50	
1000	Vertical members to 10' high		55	.291			9.95		9.95	15.50	
1050	To 13' high		50	.320			10.95		10.95	17.05	
1100	To 16' high		45	.356			12.20		12.20	18.95	
1500	Reshoring		1,400	.011	S.F.	.38	.39		.77	1.02	
1600	Flying truss system	C-17D	9,600	.009	SFCA		.31	.05	.36	.54	
1760	Horizontal, aluminum joists, 6-1/4" high x 5' to 21' span, buy				L.F.	29.50			29.50	32	
1770	Beams, 7-1/4" high x 4' to 30' span				"	40.50			40.50	44.50	

03150 | Concrete Accessories

		CREW	DAILY OUTPUT	LABOR-HOURS	UNIT	MAT.	LABOR	EQUIP.	TOTAL	TOTAL INCL O&P
600						**2005 BARE COSTS**					
1810	Horizontal, steel beam, adjustable, 4' to 7' span				Ea.	465			465	515
1830	6' to 10' span					600			600	660
1920	9' to 15' span					760			760	840
1940	12' to 20' span					775			775	855
1970	Steel stringer, W8x10, 4' to 16' span, buy				L.F.	16.75			16.75	18.40
3000	Rent for job duration, aluminum joist @ 2' O.C., per mo				SF Flr.	.73			.73	.80
3050	Steel W8x10					.42			.42	.46
3060	Steel adjustable					1.50			1.50	1.65
3500	#1 post shore, steel, 5'-7" to 9'-6" high, 10000# cap., buy				Ea.	310			310	340
3550	#2 post shore, 7'-3" to 12'-10" high, 7800# capacity					330			330	365
3600	#3 post shore, 8'-10" to 16'-1" high, 3800# capacity					360			360	395
5010	Frame shoring systems, steel, 12000#/leg, buy									
5040	Frame, 2' wide x 6' high				Ea.	287			287	315
5250	X-brace					45.50			45.50	50
5550	Base plate					36			36	39.50
5600	Screw jack					115			115	126
5650	U-head, 8" x 8"					18.15			18.15	20
620	**ACCESSORIES, SLEEVES AND CHASES**									
0100	Plastic, 1 use, 9" long, 2" diameter	1 Carp	100	.080	Ea.	.53	2.74		3.27	4.85
0150	4" diameter		90	.089		1.56	3.04		4.60	6.45
0200	6" diameter		75	.107		2.75	3.65		6.40	8.75
0250	12" diameter		60	.133		18.05	4.57		22.62	27
640	**ACCESSORIES, SNAP TIES, FLAT WASHER**, 4-3/4" L&W									
0100	3000 lb., to 8"				C	96			96	106
0200	11" & 12"					115			115	127
0250	16"					122			122	135
0300	18"					121			121	133
0500	With plastic cone, to 8"					89			89	97.50
0600	11" & 12"					105			105	116
0650	16"					111			111	122
0700	18"					116			116	128
1000	5000 lb., to 8"					121			121	133
1150	11" & 12"					141			141	156
1200	16"					155			155	171
1250	18"					152			152	167
1500	With plastic cone, to 8"					152			152	167
1600	11" & 12"					181			181	199
1650	16"					200			200	220
1700	18"					207			207	228
660	**STAIR TREAD INSERTS** Cast iron, abrasive, 3" wide	1 Carp	90	.089	L.F.	7.65	3.04		10.69	13.15
0020	4" wide		80	.100		10.20	3.43		13.63	16.55
0040	6" wide		75	.107		15.30	3.65		18.95	22.50
0050	9" wide		70	.114		23	3.91		26.91	31.50
0100	12" wide		65	.123		30.50	4.22		34.72	40
0300	Cast aluminum, compared to cast iron, deduct					10%				
0500	Extruded aluminum safety tread, 3" wide	1 Carp	75	.107		8.70	3.65		12.35	15.30
0550	4" wide		75	.107		11.65	3.65		15.30	18.50
0600	6" wide		75	.107		17.45	3.65		21.10	25
0650	9" wide to resurface stairs		70	.114		26	3.91		29.91	35
1700	Cement fill for pan-type metal treads, plain	1 Cefi	115	.070	S.F.	1.50	2.29		3.79	5
1750	Non-slip	"	100	.080	"	1.65	2.63		4.28	5.70
850	**ACCESSORIES, WALL AND FOUNDATION**									
2000	Footings, form braces, solid steel, adjustable				Ea.	13			13	14.30

Important: See the Reference Section for critical supporting data - Reference Nos., Crews, & City Cost Indexes

03150 | Concrete Accessories

				DAILY	LABOR-		2005 BARE COSTS				TOTAL	
			CREW	OUTPUT	HOURS	UNIT	MAT.	LABOR	EQUIP.	TOTAL	INCL O&P	
850	2050	Spreaders for footer, adjustable				Ea.	4.45			4.45	4.90	850
	3000	Form oil, coverage varies greatly, minimum				Gal.	4.35			4.35	4.79	
	3050	Maximum				"	6.60			6.60	7.25	
	3500	Form patches, 1-3/4" diameter				C	65.50			65.50	72	
	3550	2-3/4" diameter				"	83			83	91.50	
	4000	Nail stakes, 3/4" diameter, 18" long				Ea.	2.23			2.23	2.45	
	4050	24" long					3.46			3.46	3.81	
	4200	30" long					4.20			4.20	4.62	
	4250	36" long				▼	4.78			4.78	5.25	
860	0010	**WATERSTOP** PVC, ribbed 3/16" thick, 4" wide	1 Carp	155	.052	L.F.	.78	1.77		2.55	3.61	860
	0050	6" wide		145	.055		1.34	1.89		3.23	4.41	
	0500	Ribbed, PVC, with center bulb, 9" wide, 3/16" thick		135	.059		2	2.03		4.03	5.35	
	0550	3/8" thick		130	.062		3.03	2.11		5.14	6.60	
	0800	Dumbbell type, PVC, 6" wide, 3/16" thick		150	.053		1.35	1.83		3.18	4.34	
	0850	3/8" thick		145	.055		2.23	1.89		4.12	5.40	
	1000	9" wide, 3/8" thick, PVC, plain		130	.062		2.60	2.11		4.71	6.15	
	1050	Center bulb		130	.062		5.95	2.11		8.06	9.85	
	1250	Split PVC, 3/8" thick, 6" wide		145	.055		4.03	1.89		5.92	7.35	
	1300	9" wide		130	.062		6	2.11		8.11	9.90	
	2000	Rubber, flat dumbbell, 3/8" thick, 6" wide		145	.055		5.65	1.89		7.54	9.15	
	2050	9" wide		135	.059		8.45	2.03		10.48	12.45	
	2500	Flat dumbbell split, 3/8" thick, 6" wide		145	.055		4.03	1.89		5.92	7.35	
	2550	9" wide		135	.059		6	2.03		8.03	9.75	
	3000	Center bulb, 1/4" thick, 6" wide		145	.055		4.90	1.89		6.79	8.35	
	3050	9" wide		135	.059		9.65	2.03		11.68	13.80	
	3500	Center bulb split, 3/8" thick, 6" wide		145	.055		6	1.89		7.89	9.55	
	3550	9" wide	▼	135	.059	▼	10.35	2.03		12.38	14.55	
	5000	Waterstop fittings, rubber, flat										
	5010	Dumbbell or center bulb, 3/8" thick,										
	5200	Field union, 6" wide	1 Carp	50	.160	Ea.	19.55	5.50		25.05	30	
	5250	9" wide		50	.160		23	5.50		28.50	34	
	5500	Flat cross, 6" wide		30	.267		39.50	9.15		48.65	58	
	5550	9" wide		30	.267		56	9.15		65.15	76.50	
	6000	Flat tee, 6" wide		30	.267		36.50	9.15		45.65	54.50	
	6050	9" wide		30	.267		48	9.15		57.15	67	
	6500	Flat ell, 6" wide		40	.200		33	6.85		39.85	47	
	6550	9" wide		40	.200		44	6.85		50.85	59	
	7000	Vertical tee, 6" wide		25	.320		32	10.95		42.95	52	
	7050	9" wide		25	.320		40.50	10.95		51.45	61.50	
	7500	Vertical ell, 6" wide		35	.229		28.50	7.85		36.35	43	
	7550	9" wide	▼	35	.229	▼	38.50	7.85		46.35	54.50	

03200 | Concrete Reinforcement

03210 | Reinforcing Steel

				DAILY	LABOR-		2005 BARE COSTS				TOTAL	
			CREW	OUTPUT	HOURS	UNIT	MAT.	LABOR	EQUIP.	TOTAL	INCL O&P	
100	0010	**ACCESSORIES** Materials only										100
	0020	See also Form Accessories, division 03150										
	0100	Beam bolsters, (BB) standard, lower, up to 1-1/2" high, plain				C.L.F.	52			52	57.50	
	0102	Galvanized				▼	80.50			80.50	88.50	

03210	Reinforcing Steel	CREW	DAILY OUTPUT	LABOR-HOURS	UNIT	2005 BARE COSTS				TOTAL INCL O&P
						MAT.	LABOR	EQUIP.	TOTAL	
0104	Stainless				C.L.F.	151			151	166
0106	Plastic					91			91	100
0108	Epoxy					125			125	137
0110	2-1/2" to 3" high, plain					68.50			68.50	75
0120	Galvanized					91			91	100
0140	Stainless					208			208	228
0160	Plastic					95			95	105
0162	Epoxy					133			133	146
0200	Upper, standard (BBU) to 1-1/2" high, plain					150			150	165
0210	2-1/2" to 3" high					272			272	299
0300	Beam bolster with plate (BBP) to 1-1/2" high, plain					173			173	190
0310	2-1/2" to 3" high					335			335	370
0500	Slab bolsters, continuous, plain (SB) 3/4" to 1" high, plain					43			43	47
0502	Galvanized					49.50			49.50	54.50
0504	Stainless					69.50			69.50	76.50
0506	Plastic					59			59	65
0510	1" to 2" high, plain					53.50			53.50	59
0515	Galvanized					63			63	69
0520	Stainless					119			119	131
0525	Plastic					71			71	78
0530	For bolsters with wire runners (SBR), add					60.50			60.50	66.50
0540	For bolsters with plates (SBP), add					146			146	161
0700	Clip or bar ties, 16 ga., plain, 3" long				C	11.90			11.90	13.10
0710	4" long					12.60			12.60	13.85
0720	6" long					14.75			14.75	16.20
0730	8" long					15.40			15.40	16.95
0900	Flange clips, expandable flanges, 10 ga., 12" O.C., continuous,									
0910	galvanized, over 500 L.F., 4" to 8"				C.L.F.	44			44	48.50
0920	9" to 12"					60.50			60.50	66.50
0930	17" to 24"					61.50			61.50	68
1200	High chairs, individual, no plates (1 HC), to 3" high, plain				C	64.50			64.50	71
1202	Galvanized					90			90	99
1204	Stainless					184			184	202
1206	Plastic					85.50			85.50	94.50
1210	5" high, plain					99			99	109
1212	Galvanized					129			129	142
1214	Stainless					281			281	310
1216	Plastic					120			120	132
1220	8" high, plain					214			214	236
1222	Galvanized					286			286	315
1224	Stainless					470			470	520
1226	Plastic					269			269	296
1230	12" high, plain					445			445	490
1232	Galvanized					570			570	625
1234	Stainless					855			855	940
1236	Plastic					500			500	550
1240	15" high, plain					830			830	915
1242	Galvanized					970			970	1,075
1244	Stainless					1,425			1,425	1,575
1246	Plastic					900			900	990
1250	For each added inch up to 24" high, plain, add					62			62	68
1252	Galvanized, add					71.50			71.50	79
1254	Stainless, add					80			80	88
1256	Plastic, add					71.50			71.50	79
1400	Individual high chairs, with plates, (HCP), to 5" high, add					284			284	310
1410	Over 5" high, add					325			325	360

100

Important: See the Reference Section for critical supporting data - Reference Nos., Crews, & City Cost Indexes

03210	Reinforcing Steel	CREW	DAILY OUTPUT	LABOR-HOURS	UNIT	2005 BARE COSTS				TOTAL INCL O&P
						MAT.	LABOR	EQUIP.	TOTAL	
1500	Bar chair (BC) for up to 1-3/4" high, plain				C	36.50			36.50	40
1520	Galvanized					42			42	46.50
1530	Stainless					112			112	124
1540	Plastic					74.50			74.50	82
1550	Joist chair (JC), joists up to 6", plain					44			44	48.50
1580	Galvanized					56			56	62
1600	Stainless					80			80	88
1620	Plastic					76.50			76.50	84
1630	Epoxy					185			185	204
1700	Continuous high chairs, legs 8" O.C. (CHC) to 4" high, plain				C.L.F.	85.50			85.50	94.50
1705	Galvanized					105			105	116
1710	Stainless					225			225	247
1715	Plastic					106			106	116
1718	Epoxy					160			160	176
1720	6" high, plain					123			123	136
1725	Galvanized					177			177	195
1730	Stainless					261			261	287
1735	Plastic					169			169	186
1738	Epoxy					215			215	236
1740	8" high, plain					179			179	197
1745	Galvanized					249			249	273
1750	Stainless					310			310	340
1755	Plastic					234			234	258
1758	Epoxy					325			325	355
1760	12" high, plain					440			440	485
1765	Galvanized					530			530	580
1770	Stainless					740			740	815
1775	Plastic					500			500	550
1778	Epoxy					770			770	845
1780	15" high, plain					495			495	545
1785	Galvanized					595			595	655
1790	Stainless					590			590	650
1795	Plastic					565			565	620
1798	Epoxy					1,000			1,000	1,125
1800	For each added 1" up to 24" high, plain, add					42			42	46.50
1820	Galvanized, add					52			52	57
1840	Stainless, add					46.50			46.50	51
1860	Plastic, add					46.50			46.50	51
1900	For continuous bottom plate, (CHCP), add					202			202	222
1940	For upper continuous high chairs, (CHCU), add					202			202	222
1960	For galvanized wire runners, add					190			190	208
2100	Paper tubing, 4' lengths, for #2 & #3 bar					76			76	83.50
2120	For #6 bar					94.50			94.50	104
2200	Screed base, 1/2" diameter, 2-1/2" high, plain				C	196			196	216
2210	Galvanized					203			203	224
2220	5-1/2" high, plain					236			236	260
2250	Galvanized					251			251	276
2300	3/4" diameter, 2-1/2" high, plain					243			243	268
2310	Galvanized					258			258	284
2320	5-1/2" high, plain					300			300	330
2350	Galvanized					320			320	355
2400	Screed holder, 1/2" diam. for 1" I.D. pipe, plain, 6" long					203			203	224
2420	12" long					335			335	370
2500	3/4" diameter, for 1-1/2" I.D. pipe, 6" long					365			365	400
2520	12" long					600			600	660
2700	Screw anchor for bolts, plain, 1/2" diameter					133			133	147

100

CONCRETE 3

	03210	Reinforcing Steel		CREW	DAILY OUTPUT	LABOR-HOURS	UNIT	2005 BARE COSTS				TOTAL INCL O&P	
								MAT.	LABOR	EQUIP.	TOTAL		
100	2720	1" diameter					C	400			400	440	100
	2740	1-1/2" diameter						660			660	725	
	2800	Screw eye bolts, 1/2" x 5" long						152			152	167	
	2820	1" x 9" long						600			600	660	
	2840	1-1/2" x 14" long						1,450			1,450	1,600	
	2900	Screw anchor bolts, 1/2" x up to 7" long						610			610	670	
	2920	1" x up to 12" long						1,975			1,975	2,175	
	3000	Slab lifting inserts, single, 3/4" dia., galv., 4" high						410			410	450	
	3010	6" high						500			500	550	
	3030	7" high						575			575	630	
	3100	1" diameter, 5" high						645			645	710	
	3120	7" high						680			680	750	
	3200	Double lifting inserts, 1" diameter, 5" high						1,275			1,275	1,400	
	3220	7" high						1,350			1,350	1,475	
	3330	1-1/4" diameter, 5" high						1,400			1,400	1,525	
	3500	Sleeper clips for wood sleepers, 20 ga., galv., 2" wide					M	485			485	535	
	3520	4" wide						600			600	660	
	3600	Spacers, plastic for 1" bar clearance, average						74.50			74.50	82	
	3620	For 2" bar clearance, average						90			90	99	
	3800	Subgrade chairs, 1/2" diameter, 3-1/2" high					C	400			400	440	
	3850	12" high						1,125			1,125	1,225	
	3900	3/4" diameter, 3-1/2" high						515			515	565	
	3950	12" high						1,225			1,225	1,350	
	4200	Subgrade stakes, 3/4" diameter, 12" long						415			415	455	
	4250	24" long						560			560	615	
	4300	1" diameter, 12" long						625			625	685	
	4350	24" long						915			915	1,000	
	4500	Tie wire, 16 ga. annealed steel, under 500 lbs.					Cwt.	120			120	132	
	4520	2,000 to 4,000 lbs.					"	113			113	124	
	4550	Tie wire holder, plastic case					Ea.	47.50			47.50	52	
	4600	Aluminum case					"	57.50			57.50	63	
200	0010	**COATED REINFORCING** Add to fabricated & delivered price											200
	0100	Epoxy coated, A775					Ton	360			360	395	
	0150	Galvanized, #3						700			700	770	
	0200	#4						700			700	770	
	0250	#5						685			685	755	
	0300	#6 or over						685			685	755	
	1000	For over 20 tons, #6 or larger, minimum						635			635	700	
	1500	Maximum						765			765	840	
600	0010	**REINFORCING IN PLACE** A615 Grade 60, incl. access. labor											600
	0100	Beams & Girders, #3 to #7 **CN**	R03210 -010	4 Rodm	1.60	20	Ton	800	760		1,560	2,125	
	0150	#8 to #18			2.70	11.852		800	450		1,250	1,625	
	0200	Columns, #3 to #7	R03210 -020		1.50	21.333		800	810		1,610	2,200	
	0250	#8 to #18			2.30	13.913		800	530		1,330	1,750	
	0300	Spirals, hot rolled, 8" to 15" diameter	R03210 -040		2.20	14.545		1,125	550		1,675	2,125	
	0320	15" to 24" diameter			2.20	14.545		1,075	550		1,625	2,075	
	0330	24" to 36" diameter	R03210 -050		2.30	13.913		1,025	530		1,555	2,000	
	0340	36" to 48" diameter			2.40	13.333		970	505		1,475	1,900	
	0360	48" to 64" diameter	R03210 -070		2.50	12.800		1,075	485		1,560	1,975	
	0380	64" to 84" diameter			2.60	12.308		1,125	465		1,590	2,000	
	0390	84" to 96" diameter	R03210 -080		2.70	11.852		1,175	450		1,625	2,050	
	0400	Elevated slabs, #4 to #7			2.90	11.034		850	420		1,270	1,625	
	0500	Footings, #4 to #7			2.10	15.238		760	580		1,340	1,800	
	0550	#8 to #18			3.60	8.889		720	335		1,055	1,350	
	0600	Slab on grade, #3 to #7			2.30	13.913		760	530		1,290	1,700	

Important: See the Reference Section for critical supporting data - Reference Nos., Crews, & City Cost Indexes

03210 | Reinforcing Steel

			CREW	DAILY OUTPUT	LABOR-HOURS	UNIT	2005 BARE COSTS				TOTAL INCL O&P
							MAT.	LABOR	EQUIP.	TOTAL	
600	0700	Walls, #3 to #7	R03210 -010	4 Rodm	3	10.667	Ton	760	405		1,165	1,500	**600**
	0750	#8 to #18		↓	4	8	↓	760	305		1,065	1,325
	0900	Use the following for a rough estimate guide	R03210 -020									
	1000	Typical in place, average, under 10 ton job, #3 to #7		4 Rodm	1.80	17.778	Ton	825	675		1,500	2,025
	1010	#8 to #18			2.70	11.852		845	450		1,295	1,675
	1050	10 - 50 ton job, #3 to #7	R03210 -040		2.10	15.238		810	580		1,390	1,850
	1060	#8 to #18			3	10.667		830	405		1,235	1,575
	1100	50 - 100 ton job, #3 - #7	R03210 -050		2.20	14.545		790	550		1,340	1,775
	1110	#8 to #18			3.10	10.323		810	390		1,200	1,525
	1150	Over 100 ton job, #3 - #7	R03210 -070		2.30	13.913		785	530		1,315	1,725
	1160	#8 - #18		↓	3.20	10		800	380		1,180	1,500
	1200	High strength steel, Grade 75, #14 bars only, add	R03210 -080					60			60	66
	2000	Unloading & sorting, add to above		C-5	100	.560			20.50	6.15	26.65	40.50
	2200	Crane cost for handling, add to above, minimum			135	.415			15.25	4.57	19.82	29.50
	2210	Average			92	.609			22.50	6.70	29.20	43.50
	2220	Maximum		↓	35	1.600	↓		59	17.60	76.60	114
	2400	Dowels, 2 feet long, deformed, #3		2 Rodm	520	.031	Ea.	.33	1.17		1.50	2.29
	2410	#4			480	.033		.59	1.26		1.85	2.74
	2420	#5			435	.037		.92	1.40		2.32	3.31
	2430	#6			360	.044	↓	1.32	1.69		3.01	4.23
	2450	Longer and heavier dowels, add			725	.022	Lb.	.44	.84		1.28	1.86
	2500	Smooth dowels, 12" long, 1/4" or 3/8" diameter			140	.114	Ea.	.69	4.34		5.03	7.90
	2520	5/8" diameter			125	.128		1.21	4.86		6.07	9.35
	2530	3/4" diameter		↓	110	.145	↓	1.50	5.50		7	10.75
	2600	Dowel sleeves for CIP concrete, 2-part system										
	2610	Sleeve base, plastic, for #5 bar, fasten to edge form		1 Rodm	200	.040	Ea.	.33	1.52		1.85	2.86
	2615	Sleeve, plastic, for #5 bar x 9" long, snap onto base			400	.020		.96	.76		1.72	2.31
	2620	Sleeve base, for #6 bar			175	.046		.33	1.73		2.06	3.22
	2625	Sleeve, for #6 bar			350	.023		1.02	.87		1.89	2.55
	2630	Sleeve base, for #8 bar			150	.053		.40	2.02		2.42	3.78
	2635	Sleeve, for #8 bar		↓	300	.027		1.11	1.01		2.12	2.89
	2700	Dowel caps, visual warning only, plastic, #3 to #8		2 Rodm	800	.020		.24	.76		1	1.51
	2720	#7 to #14			750	.021		.56	.81		1.37	1.96
	2750	Impalement protective, plastic, #3 to #7			800	.020		1.74	.76		2.50	3.16
	2760	#7 to #11			775	.021		2.12	.78		2.90	3.62
	2770	#11 to #16		↓	750	.021	↓	2.35	.81		3.16	3.93
	3000	For epoxy dowel anchoring, see Div 05090-300										
700	0010	**SPLICING REINFORCING BARS** Incl. holding bars in	R03210 -070										**700**
	0020	place while splicing										
	0100	Butt weld columns #4 bars		C-5	190	.295	Ea.	3.56	10.85	3.25	17.66	25
	0110	#6 bars			150	.373		5	13.75	4.11	22.86	32
	0130	#10 bars			95	.589		12.30	21.50	6.50	40.30	55.50
	0150	#14 bars		↓	65	.862	↓	19.50	31.50	9.50	60.50	83
	0280	Column splices, bar to bar end bearing										
	0300	#7 or #8 bars		C-5	190	.295	Ea.	32	10.85	3.25	46.10	56.50
	0310	#9 or #10 bars			170	.329		35	12.15	3.63	50.78	62
	0320	#11 bars			160	.350		42.50	12.90	3.85	59.25	71.50
	0330	#14 bars			150	.373		53.50	13.75	4.11	71.36	85
	0340	#18 bars		↓	140	.400		77	14.75	4.40	96.15	114
	0500	Transition bar to bar end bearing, #14 to #18 bar						76			76	83.50
	0520	#14 to #11 bar						58.50			58.50	64
	0550	#10 to #9 bar						36.50			36.50	40.50
	0560	#9 to #8 bar						34			34	37.50
	0580	#8 to #7 bar						33			33	36.50
	0600	For bolted speed sleeve type, deduct							15%			

03210 | Reinforcing Steel

			CREW	DAILY OUTPUT	LABOR-HOURS	UNIT	2005 BARE COSTS				TOTAL INCL O&P
							MAT.	LABOR	EQUIP.	TOTAL	
700	0800	Mechanical butt splice sleeve type w/filler metal, compression									700
	0810	only, all grades, columns only #11 bars R03210-070	C-5	68	.824	Ea.	37.50	30.50	9.05	77.05	100
	0900	#14 bars		62	.903		61	33.50	9.95	104.45	131
	0920	#18 bars		62	.903		111	33.50	9.95	154.45	186
	1000	125% yield point, grade 60, columns only, #6 bars		68	.824		30	30.50	9.05	69.55	92
	1020	#7 or #8 bars		68	.824		32	30.50	9.05	71.55	94.50
	1030	#9 bars		68	.824		33	30.50	9.05	72.55	95.50
	1040	#10 bars		68	.824		35	30.50	9.05	74.55	97.50
	1050	#11 bars		68	.824		42.50	30.50	9.05	82.05	106
	1060	#14 bars		62	.903		53.50	33.50	9.95	96.95	123
	1070	#18 bars		62	.903		77	33.50	9.95	120.45	149
	1200	Full tension, grade 60 steel, columns,									
	1220	slabs or beams, #6, #7, #8 bars	C-5	68	.824	Ea.	10.10	30.50	9.05	49.65	70
	1230	#9 bars		68	.824		11.05	30.50	9.05	50.60	71
	1240	#10 bars		68	.824		12.30	30.50	9.05	51.85	72.50
	1250	#11 bars		68	.824		13.05	30.50	9.05	52.60	73.50
	1260	#14 bars		62	.903		19.50	33.50	9.95	62.95	86
	1270	#18 bars		62	.903		29.50	33.50	9.95	72.95	97
	1400	If equipment handling not required, deduct						50%			
	1600	Mechanical threaded type, bar threading not included,									
	1700	Straight bars, #10 & #11	C-5	140	.400	Ea.	13.05	14.75	4.40	32.20	43
	1750	#14 bars		130	.431		19.50	15.85	4.74	40.09	52
	1800	#18 bars		75	.747		29.50	27.50	8.20	65.20	86
	2100	#11 to #18 & #14 to #18 transition		75	.747		32	27.50	8.20	67.70	88.50
	2400	Bent bars, #10 & #11		105	.533		13.05	19.65	5.85	38.55	52.50
	2500	#14		90	.622		22	23	6.85	51.85	69
	2600	#18		70	.800		36.50	29.50	8.80	74.80	97.50
	2800	#11 to #14 transition		75	.747		21	27.50	8.20	56.70	76.50
	2900	#11 to #18 & #14 to #18 transition		70	.800		35	29.50	8.80	73.30	95.50

03220 | Welded Wire Fabric

			CREW	DAILY OUTPUT	LABOR-HOURS	UNIT	MAT.	LABOR	EQUIP.	TOTAL	TOTAL INCL O&P
200	0010	**WELDED WIRE FABRIC** ASTM A185									200
	0050	Sheets R03220-030									
	0100	6 x 6 - W1.4 x W1.4 (10 x 10) 21 lb. per C.S.F.	2 Rodm	35	.457	C.S.F.	19.35	17.35		36.70	50
	0200	6 x 6 - W2.1 x W2.1 (8 x 8) 30 lb. per C.S.F.		31	.516		25.50	19.60		45.10	60.50
	0300	6 x 6 - W2.9 x W2.9 (6 x 6) 42 lb. per C.S.F.		29	.552		33	21		54	70.50
	0400	6 x 6 - W4 x W4 (4 x 4) 58 lb. per C.S.F.		27	.593		44	22.50		66.50	85
	0500	4 x 4 - W1.4 x W1.4 (10 x 10) 31 lb. per C.S.F.		31	.516		17.70	19.60		37.30	52
	0600	4 x 4 - W2.1 x W2.1 (8 x 8) 44 lb. per C.S.F.		29	.552		21	21		42	57.50
	0650	4 x 4 - W2.9 x W2.9 (6 x 6) 61 lb. per C.S.F.		27	.593		30	22.50		52.50	70
	0700	4 x 4 - W4 x W4 (4 x 4) 85 lb. per C.S.F.		25	.640		54	24.50		78.50	99
	0750	Rolls									
	0800	2 x 2 - #14 galv., 21 lb/C.S.F., beam & column wrap	2 Rodm	6.50	2.462	C.S.F.	28	93.50		121.50	185
	0900	2 x 2 - #12 galv. for gunite reinforcing	"	6.50	2.462	"	31.50	93.50		125	189

03230 | Stressing Tendons

			CREW	DAILY OUTPUT	LABOR-HOURS	UNIT	MAT.	LABOR	EQUIP.	TOTAL	TOTAL INCL O&P
600	0010	**PRESTRESSING STEEL** Post-tensioned in field									600
	0100	Grouted strand, 50' span, 100 kip R03410-090	C-3	1,200	.053	Lb.	1.13	1.85	.10	3.08	4.34
	0150	300 kip		2,700	.024		.63	.82	.04	1.49	2.07
	0300	100' span, 100 kip		1,700	.038		1.13	1.31	.07	2.51	3.43
	0350	300 kip		3,200	.020		1.01	.70	.04	1.75	2.27
	0500	200' span, 100 kip		2,700	.024		1.13	.82	.04	1.99	2.62
	0550	300 kip		3,500	.018		1.01	.64	.03	1.68	2.18
	0800	Grouted bars, 50' span, 42 kip		2,600	.025		.59	.86	.05	1.50	2.08

Important: See the Reference Section for critical supporting data - Reference Nos., Crews, & City Cost Indexes

03200 | Concrete Reinforcement

03230 | Stressing Tendons

			CREW	DAILY OUTPUT	LABOR-HOURS	UNIT	MAT.	LABOR	EQUIP.	TOTAL	TOTAL INCL O&P	
600	0850	143 kip	C-3	3,200	.020	Lb.	.57	.70	.04	1.31	1.79	600
	1000	75' span, 42 kip	↓	3,200	.020		.59	.70	.04	1.33	1.81	
	1050	143 kip	↓	4,200	.015		.53	.53	.03	1.09	1.48	
	1200	Ungrouted strand, 50' span, 100 kip	C-4	1,275	.025		.46	.97	.03	1.46	2.14	
	1250	300 kip		1,475	.022		.46	.83	.03	1.32	1.92	
	1400	100' span, 100 kip		1,500	.021		.46	.82	.03	1.31	1.89	
	1450	300 kip		1,650	.019		.46	.75	.02	1.23	1.77	
	1600	200' span, 100 kip		1,500	.021		.46	.82	.03	1.31	1.89	
	1650	300 kip		1,700	.019		.46	.72	.02	1.20	1.73	
	1800	Ungrouted bars, 50' span, 42 kip		1,400	.023		.40	.88	.03	1.31	1.92	
	1850	143 kip		1,700	.019		.40	.72	.02	1.14	1.66	
	2000	75' span, 42 kip		1,800	.018		.40	.68	.02	1.10	1.59	
	2050	143 kip		2,200	.015		.40	.56	.02	.98	1.38	
	2220	Ungrouted single strand, 100' slab, 25 kip		1,200	.027		.46	1.03	.03	1.52	2.24	
	2250	35 kip	↓	↓	1,475	.022	↓	.46	.83	.03	1.32	1.92	

R03410-090

03240 | Fibrous Reinforcing

			CREW	DAILY OUTPUT	LABOR-HOURS	UNIT	MAT.	LABOR	EQUIP.	TOTAL	TOTAL INCL O&P	
300	0010	**FIBROUS REINFORCING**										300
	0100	Synthetic fibers, add to concrete				Lb.	3.87			3.87	4.26	
	0110	1-1/2 lb. per C.Y.				C.Y.	6			6	6.60	
	0150	Steel fibers, add to concrete				Lb.	.44			.44	.48	
	0155	25 lb. per C.Y.				C.Y.	11			11	12.10	
	0160	50 lb. per C.Y.					22			22	24	
	0170	75 lb. per C.Y.					34			34	37.50	
	0180	100 lb. per C.Y.				↓	44			44	48.50	

03300 | Cast-In-Place Concrete

03310 | Structural Concrete

			CREW	DAILY OUTPUT	LABOR-HOURS	UNIT	MAT.	LABOR	EQUIP.	TOTAL	TOTAL INCL O&P	
200	0010	**CONCRETE, FIELD MIX** FOB forms 2250 psi				C.Y.	72.50			72.50	79.50	200
	0020	3000 psi				"	75.50			75.50	83	
220	0010	**CONCRETE, READY MIX** Normal weight										220
	0020	2000 psi				C.Y.	77.50			77.50	85.50	
	0100	2500 psi					79.50			79.50	87.50	
	0150	3000 psi **CN**					81			81	89	
	0200	3500 psi					82			82	90	
	0300	4000 psi					84			84	92.50	
	0350	4500 psi					86			86	94.50	
	0400	5000 psi **CN**					90			90	99	
	0411	6000 psi					103			103	113	
	0412	8000 psi					167			167	184	
	0413	10,000 psi					238			238	261	
	0414	12,000 psi					287			287	315	
	1000	For high early strength cement, add					10%					
	1010	For structural lightweight with regular sand, add					25%					
	2000	For all lightweight aggregate, add				↓	45%					
	3000	For integral colors, 2500 psi, 5 bag mix										
	3100	Red, yellow or brown, 1.8 lb. per bag, add				C.Y.	21			21	23	
	3200	9.4 lb. per bag, add				↓	109			109	120	

R03310-080
R03310-040
R03310-050

03310 | Structural Concrete

		CREW	DAILY OUTPUT	LABOR-HOURS	UNIT	2005 BARE COSTS MAT.	LABOR	EQUIP.	TOTAL	TOTAL INCL O&P	
220	3400	Black, 1.8 lb. per bag, add				C.Y.	23.50			23.50	26	220
	3500	7.5 lb. per bag, add					98.50			98.50	108	
	3700	Green, 1.8 lb. per bag, add					27			27	30	
	3800	7.5 lb. per bag, add					113			113	125	
240	0010	**CONCRETE IN PLACE** Including forms (4 uses), reinforcing R03310-010										240
	0050	steel and finishing unless otherwise indicated										
	0300	Beams, 5 kip per L.F., 10' span R03310-100	C-14A	15.62	12.804	C.Y.	284	440	46.50	770.50	1,050	
	0350	25' span	"	18.55	10.782		214	370	39.50	623.50	860	
	0500	Chimney foundations, industrial, minimum	C-14C	32.22	3.476		112	114	.73	226.73	300	
	0510	Maximum	"	23.71	4.724		119	154	.99	273.99	375	
	0700	Columns, square, 12" x 12", minimum reinforcing	C-14A	11.96	16.722		287	575	61	923	1,275	
	0720	Average reinforcing		10.13	19.743		455	675	72	1,202	1,625	
	0740	Maximum reinforcing		9.03	22.148		680	760	80.50	1,520.50	2,050	
	0800	16" x 16", minimum reinforcing		16.22	12.330		228	425	45	698	965	
	0820	Average reinforcing		12.57	15.911		385	545	58	988	1,350	
	0840	Maximum reinforcing		10.25	19.512		600	670	71	1,341	1,800	
	0900	24" x 24", minimum reinforcing		23.66	8.453		193	290	31	514	700	
	0920	Average reinforcing		17.71	11.293		345	385	41	771	1,025	
	0940	Maximum reinforcing		14.15	14.134		550	485	51.50	1,086.50	1,425	
	1000	36" x 36", minimum reinforcing		33.69	5.936		170	204	21.50	395.50	530	
	1020	Average reinforcing		23.32	8.576		300	294	31	625	825	
	1040	Maximum reinforcing		17.82	11.223		510	385	41	936	1,200	
	1100	Columns, round, tied, 12" diameter, minimum reinforcing		20.97	9.537		251	325	34.50	610.50	830	
	1120	Average reinforcing		15.27	13.098		415	450	47.50	912.50	1,225	
	1140	Maximum reinforcing		12.11	16.515		635	565	60	1,260	1,650	
	1200	16" diameter, minimum reinforcing		31.49	6.351		225	218	23	466	615	
	1220	Average reinforcing		19.12	10.460		390	360	38	788	1,025	
	1240	Maximum reinforcing		13.77	14.524		590	500	53	1,143	1,500	
	1300	20" diameter, minimum reinforcing		41.04	4.873		224	167	17.75	408.75	530	
	1320	Average reinforcing		24.05	8.316		375	285	30.50	690.50	895	
	1340	Maximum reinforcing		17.01	11.758		590	405	43	1,038	1,325	
	1400	24" diameter, minimum reinforcing		51.85	3.857		209	132	14.05	355.05	450	
	1420	Average reinforcing		27.06	7.391		375	253	27	655	840	
	1440	Maximum reinforcing		18.29	10.935		580	375	40	995	1,275	
	1500	36" diameter, minimum reinforcing		75.04	2.665		210	91.50	9.70	311.20	385	
	1520	Average reinforcing		37.49	5.335		360	183	19.40	562.40	705	
	1540	Maximum reinforcing		22.84	8.757		565	300	32	897	1,125	
	1900	Elevated slabs, flat slab with drops, 125 psf Sup. Load, 20' span	C-14B	38.45	5.410		224	185	18.95	427.95	560	
	1950	30' span		50.99	4.079		232	140	14.30	386.30	490	
	2100	Flat plate, 125 psf Sup. Load, 15' span		30.24	6.878		204	235	24	463	620	
	2150	25' span		49.60	4.194		209	144	14.70	367.70	470	
	2300	Waffle const., 30" domes, 125 psf Sup. Load, 20' span		37.07	5.611		315	192	19.65	526.65	670	
	2350	30' span		44.07	4.720		281	162	16.50	459.50	580	
	2500	One way joists, 30" pans, 125 psf Sup. Load, 15' span		27.38	7.597		395	260	26.50	681.50	870	
	2550	25' span		31.15	6.677		360	229	23.50	612.50	780	
	2700	One way beam & slab, 125 psf Sup. Load, 15' span		20.59	10.102		227	345	35.50	607.50	830	
	2750	25' span		28.36	7.334		209	251	25.50	485.50	650	
	2900	Two way beam & slab, 125 psf Sup. Load, 15' span		24.04	8.652		215	296	30.50	541.50	735	
	2950	25' span		35.87	5.799		181	198	20.50	399.50	530	
	3100	Elevated slabs including finish, not										
	3110	including forms or reinforcing										
	3150	Regular concrete, 4" slab	C-8	2,613	.021	S.F.	1.09	.64	.27	2	2.47	
	3200	6" slab		2,585	.022		1.61	.65	.27	2.53	3.05	
	3250	2-1/2" thick floor fill		2,685	.021		.70	.62	.26	1.58	2.01	
	3300	Lightweight, 110# per C.F., 2-1/2" thick floor fill		2,585	.022		.90	.65	.27	1.82	2.27	
	3400	Cellular concrete, 1-5/8" fill, under 5000 S.F.		2,000	.028		.60	.84	.35	1.79	2.32	

Important: See the Reference Section for critical supporting data - Reference Nos., Crews, & City Cost Indexes

03310 | Structural Concrete

		CREW	DAILY OUTPUT	LABOR-HOURS	UNIT	2005 BARE COSTS				TOTAL INCL O&P
						MAT.	LABOR	EQUIP.	TOTAL	
240 3450	Over 10,000 S.F.	C-8	2,200	.025	S.F.	.57	.76	.32	1.65	2.13 **240**
3500	Add per floor for 3 to 6 stories high [R03310-010]		31,800	.002			.05	.02	.07	.10
3520	For 7 to 20 stories high [R03310-100]		21,200	.003			.08	.03	.11	.16
3800	Footings, spread under 1 C.Y.	C-14C	38.07	2.942	C.Y.	162	96	.62	258.62	330
3850	Over 5 C.Y.		81.04	1.382		226	45	.29	271.29	320
3900	Footings, strip, 18" x 9", unreinforced		40	2.800		101	91.50	.59	193.09	255
3920	18" x 9", reinforced		35	3.200		120	105	.67	225.67	297
3925	20" x 10", unreinforced		45	2.489		98	81.50	.52	180.02	236
3930	20" x 10", reinforced		40	2.800		114	91.50	.59	206.09	269
3935	24" x 12", unreinforced		55	2.036		97	66.50	.43	163.93	211
3940	24" x 12", reinforced		48	2.333		113	76	.49	189.49	244
3945	36" x 12", unreinforced		70	1.600		93	52.50	.34	145.84	184
3950	36" x 12", reinforced		60	1.867		108	61	.39	169.39	214
4000	Foundation mat, under 10 C.Y.		38.67	2.896		164	94.50	.61	259.11	330
4050	Over 20 C.Y.		56.40	1.986		144	65	.42	209.42	260
4200	Grade walls, 8" thick, 8' high	C-14D	45.83	4.364		143	148	15.90	306.90	405
4250	14' high		27.26	7.337		186	249	26.50	461.50	625
4260	12" thick, 8' high		64.32	3.109		129	106	11.30	246.30	320
4270	14' high		40.01	4.999		146	170	18.20	334.20	445
4300	15" thick, 8' high		80.02	2.499		121	85	9.10	215.10	275
4350	12' high		51.26	3.902		128	133	14.20	275.20	365
4500	18' high		48.85	4.094		143	139	14.90	296.90	390
4520	Handicap access ramp, railing both sides, 3' wide	C-14H	14.58	3.292	L.F.	167	111	1.65	279.65	360
4525	5' wide		12.22	3.928		175	132	1.96	308.96	400
4530	With 6" curb and rails both sides, 3' wide		8.55	5.614		174	189	2.81	365.81	490
4535	5' wide		7.31	6.566		178	221	3.28	402.28	545
4650	Slab on grade, not including finish, 4" thick	C-14E	60.75	1.449	C.Y.	104	48.50	.39	152.89	192
4700	6" thick	"	92	.957	"	98	32	.26	130.26	159
4751	Slab on grade, incl. troweled finish, not incl. forms									
4760	or reinforcing, over 10,000 S.F., 4" thick	C-14F	3,425	.021	S.F.	1.07	.65	.01	1.73	2.15
4820	6" thick		3,350	.021		1.56	.67	.01	2.24	2.71
4840	8" thick		3,184	.023		2.13	.70	.01	2.84	3.39
4900	12" thick		2,734	.026		3.20	.82	.01	4.03	4.74
4950	15" thick		2,505	.029		4.01	.89	.01	4.91	5.75
5000	Slab on grade, incl. textured finish, not incl. forms									
5001	or reinforcing, 4" thick	C-14G	2,873	.019	S.F.	1.05	.59	.01	1.65	2.06
5010	6" thick		2,590	.022		1.65	.66	.01	2.32	2.81
5020	8" thick		2,320	.024		2.15	.74	.01	2.90	3.47
5200	Lift slab in place above the foundation, incl. forms,									
5210	reinforcing, concrete and columns, minimum	C-14B	2,113	.098	S.F.	4.92	3.37	.34	8.63	11.05
5250	Average		1,650	.126		5.35	4.31	.44	10.10	13.10
5300	Maximum		1,500	.139		5.75	4.75	.49	10.99	14.25
5500	Lightweight, ready mix, including screed finish only,									
5510	not including forms or reinforcing									
5550	1:4 for structural roof decks	C-14B	260	.800	C.Y.	110	27.50	2.80	140.30	167
5600	1:6 for ground slab with radiant heat	C-14F	92	.783		111	24.50	.26	135.76	158
5650	1:3:2 with sand aggregate, roof deck	C-14B	260	.800		105	27.50	2.80	135.30	162
5700	Ground slab	C-14F	107	.673		105	21	.22	126.22	147
5900	Pile caps, incl. forms and reinf., sq. or rect., under 5 C.Y.	C-14C	54.14	2.069		104	67.50	.43	171.93	222
5950	Over 10 C.Y.		75	1.493		108	49	.31	157.31	196
6000	Triangular or hexagonal, under 5 C.Y.		53	2.113		99.50	69	.44	168.94	219
6050	Over 10 C.Y.		85	1.318		112	43	.28	155.28	192
6200	Retaining walls, gravity, 4' high see division 02830-100	C-14D	66.20	3.021		103	103	11	217	285
6250	10' high		125	1.600		93	54.50	5.80	153.30	194
6300	Cantilever, level backfill loading, 8' high		70	2.857		125	97	10.40	232.40	300
6350	16' high		91	2.198		120	74.50	8	202.50	258

CONCRETE 3

03310 | Structural Concrete

		Description		CREW	DAILY OUTPUT	LABOR-HOURS	UNIT	2005 BARE COSTS				TOTAL INCL O&P	
								MAT.	LABOR	EQUIP.	TOTAL		
240	6800	Stairs, not including safety treads, free standing, 3'-6" wide	R03310-010	C-14H	83	.578	LF Nose	14.80	19.50	.29	34.59	47	240
	6850	Cast on ground			125	.384	"	8.70	12.95	.19	21.84	30	
	7000	Stair landings, free standing	R03310-100		200	.240	S.F.	3.13	8.10	.12	11.35	16.15	
	7050	Cast on ground			475	.101	"	1.84	3.41	.05	5.30	7.40	
450	0010	**INSULATING CONCRETE** See division 03520-250											450
700	0010	**PLACING CONCRETE** and vibrating, including labor & equipment	R03310-090										700
	0050	Beams, elevated, small beams, pumped		C-20	60	1.067	C.Y.		30.50	12.55	43.05	61	
	0100	With crane and bucket		C-7	45	1.600			46	21.50	67.50	94.50	
	0200	Large beams, pumped		C-20	90	.711			20.50	8.35	28.85	40.50	
	0250	With crane and bucket		C-7	65	1.108			32	14.75	46.75	65.50	
	0400	Columns, square or round, 12" thick, pumped		C-20	60	1.067			30.50	12.55	43.05	61	
	0450	With crane and bucket		C-7	40	1.800			52	24	76	106	
	0600	18" thick, pumped		C-20	90	.711			20.50	8.35	28.85	40.50	
	0650	With crane and bucket		C-7	55	1.309			38	17.45	55.45	77	
	0800	24" thick, pumped		C-20	92	.696			19.95	8.15	28.10	39.50	
	0850	With crane and bucket		C-7	70	1.029			29.50	13.70	43.20	60.50	
	1000	36" thick, pumped		C-20	140	.457			13.10	5.35	18.45	26	
	1050	With crane and bucket		C-7	100	.720			21	9.60	30.60	42.50	
	1400	Elevated slabs, less than 6" thick, pumped		C-20	140	.457			13.10	5.35	18.45	26	
	1450	With crane and bucket		C-7	95	.758			22	10.10	32.10	44.50	
	1500	6" to 10" thick, pumped		C-20	160	.400			11.50	4.70	16.20	23	
	1550	With crane and bucket		C-7	110	.655			18.90	8.75	27.65	38.50	
	1600	Slabs over 10" thick, pumped		C-20	180	.356			10.20	4.18	14.38	20.50	
	1650	With crane and bucket		C-7	130	.554			16	7.40	23.40	32.50	
	1900	Footings, continuous, shallow, direct chute		C-6	120	.400			11.20	.40	11.60	17.70	
	1950	Pumped		C-20	150	.427			12.25	5	17.25	24.50	
	2000	With crane and bucket		C-7	90	.800			23	10.65	33.65	47.50	
	2100	Footings, continuous, deep, direct chute		C-6	140	.343			9.60	.34	9.94	15.20	
	2150	Pumped		C-20	160	.400			11.50	4.70	16.20	23	
	2200	With crane and bucket		C-7	110	.655			18.90	8.75	27.65	38.50	
	2400	Footings, spread, under 1 C.Y., direct chute		C-6	55	.873			24.50	.87	25.37	38.50	
	2450	Pumped		C-20	65	.985			28.50	11.55	40.05	56.50	
	2500	With crane and bucket		C-7	45	1.600			46	21.50	67.50	94.50	
	2600	Over 5 C.Y., direct chute		C-6	120	.400			11.20	.40	11.60	17.70	
	2650	Pumped		C-20	150	.427			12.25	5	17.25	24.50	
	2700	With crane and bucket		C-7	100	.720			21	9.60	30.60	42.50	
	2900	Foundation mats, over 20 C.Y., direct chute		C-6	350	.137			3.85	.14	3.99	6.05	
	2950	Pumped		C-20	400	.160			4.59	1.88	6.47	9.10	
	3000	With crane and bucket		C-7	300	.240			6.95	3.20	10.15	14.10	
	3200	Grade beams, direct chute		C-6	150	.320			9	.32	9.32	14.15	
	3250	Pumped		C-20	180	.356			10.20	4.18	14.38	20.50	
	3300	With crane and bucket		C-7	120	.600			17.30	8	25.30	35.50	
	3500	High rise, for more than 5 stories, pumped, add per story		C-20	2,100	.030			.88	.36	1.24	1.73	
	3510	With crane and bucket, add per story		C-7	2,100	.034			.99	.46	1.45	2.02	
	3700	Pile caps, under 5 C.Y., direct chute		C-6	90	.533			14.95	.53	15.48	23.50	
	3750	Pumped		C-20	110	.582			16.70	6.85	23.55	33	
	3800	With crane and bucket		C-7	80	.900			26	12	38	53	
	3850	Pile cap, 5 C.Y. to 10 C.Y., direct chute		C-6	175	.274			7.70	.27	7.97	12.15	
	3900	Pumped		C-20	200	.320			9.20	3.76	12.96	18.25	
	3950	With crane and bucket		C-7	150	.480			13.85	6.40	20.25	28	
	4000	Over 10 C.Y., direct chute		C-6	215	.223			6.25	.22	6.47	9.90	
	4050	Pumped		C-20	240	.267			7.65	3.13	10.78	15.20	
	4100	With crane and bucket		C-7	185	.389			11.25	5.20	16.45	23	
	4300	Slab on grade, 4" thick, direct chute		C-6	110	.436			12.25	.44	12.69	19.35	
	4350	Pumped		C-20	130	.492			14.15	5.80	19.95	28	

Important: See the Reference Section for critical supporting data - Reference Nos., Crews, & City Cost Indexes

03310 | Structural Concrete

		CREW	DAILY OUTPUT	LABOR-HOURS	UNIT	MAT.	LABOR	EQUIP.	TOTAL	TOTAL INCL O&P	
700	**4400**	With crane and bucket	C-7	110	.655	C.Y.		18.90	8.75	27.65	38.50	**700**
	4600	Over 6" thick, direct chute	C-6	165	.291			8.15	.29	8.44	12.85	
	4650	Pumped	C-20	185	.346			9.95	4.06	14.01	19.70	
	4700	With crane and bucket	C-7	145	.497			14.35	6.60	20.95	29.50	
	4900	Walls, 8" thick, direct chute	C-6	90	.533			14.95	.53	15.48	23.50	
	4950	Pumped	C-20	100	.640			18.35	7.50	25.85	36.50	
	5000	With crane and bucket	C-7	80	.900			26	12	38	53	
	5050	12" thick, direct chute	C-6	100	.480			13.45	.48	13.93	21	
	5100	Pumped	C-20	110	.582			16.70	6.85	23.55	33	
	5200	With crane and bucket	C-7	90	.800			23	10.65	33.65	47.50	
	5300	15" thick, direct chute	C-6	105	.457			12.85	.46	13.31	20.50	
	5350	Pumped	C-20	120	.533			15.30	6.25	21.55	30.50	
	5400	With crane and bucket	C-7	95	.758	▼		22	10.10	32.10	44.50	
	5600	Wheeled concrete dumping, add to placing costs above										
	5610	Walking cart, 50' haul, add	C-18	32	.281	C.Y.		7.55	1.56	9.11	13.50	
	5620	150' haul, add		24	.375			10.10	2.07	12.17	18	
	5700	250' haul, add	▼	18	.500			13.45	2.77	16.22	24	
	5800	Riding cart, 50' haul, add	C-19	80	.112			3.03	.96	3.99	5.75	
	5810	150' haul, add		60	.150			4.04	1.28	5.32	7.70	
	5900	250' haul, add	▼	45	.200	▼		5.40	1.71	7.11	10.30	

03350 | Concrete Finishing

		CREW	DAILY OUTPUT	LABOR-HOURS	UNIT	MAT.	LABOR	EQUIP.	TOTAL	TOTAL INCL O&P	
300	**0010**	**FINISHING FLOORS** Monolithic, screed finish	1 Cefi	900	.009	S.F.		.29		.29	.43	**300**
	0100	Screed and bull float (darby) finish		725	.011			.36		.36	.53	
	0150	Screed, float, and broom finish		630	.013			.42		.42	.61	
	0200	Screed, float, and hand trowel		600	.013			.44		.44	.64	
	0250	Machine trowel	▼	550	.015			.48		.48	.70	
	0400	Integral topping and finish, using 1:1:2 mix, 3/16" thick	C-10B	1,000	.040		.06	1.17	.17	1.40	2.03	
	0450	1/2" thick		950	.042		.17	1.23	.18	1.58	2.24	
	0500	3/4" thick		850	.047		.26	1.37	.20	1.83	2.58	
	0600	1" thick		750	.053		.33	1.56	.23	2.12	2.97	
	0800	Granolithic topping, laid after, 1:1:1-1/2 mix, 1/2" thick		590	.068		.18	1.98	.29	2.45	3.51	
	0820	3/4" thick		580	.069		.27	2.01	.30	2.58	3.67	
	0850	1" thick		575	.070		.36	2.03	.30	2.69	3.80	
	0950	2" thick		500	.080		.72	2.33	.34	3.39	4.70	
	1200	Heavy duty, 1:1:2, 3/4" thick, preshrunk, gray, 20 MSF		320	.125		2.38	3.65	.54	6.57	8.70	
	1300	100 MSF		380	.105		2.38	3.07	.45	5.90	7.75	
	1350	For colors, .50 psf, add, minimum		1,650	.024		.34	.71	.10	1.15	1.55	
	1400	Maximum	▼	1,500	.027		.68	.78	.11	1.57	2.06	
	1600	Exposed local aggregate finish, minimum	1 Cefi	625	.013		4.08	.42		4.50	5.10	
	1650	Maximum		465	.017		12.90	.57		13.47	15.05	
	1800	Floor abrasives, .25 psf, aluminum oxide		850	.009		.26	.31		.57	.74	
	1850	Silicon carbide		850	.009		.36	.31		.67	.85	
	2000	Floor hardeners, metallic, light service, .50 psf, add		850	.009		.30	.31		.61	.79	
	2050	Medium service, .75 psf		750	.011		.45	.35		.80	1.02	
	2100	Heavy service, 1.0 psf		650	.012		.60	.40		1	1.26	
	2150	Extra heavy, 1.5 psf		575	.014		.90	.46		1.36	1.66	
	2300	Non-metallic, light service, .50 psf		850	.009		.19	.31		.50	.66	
	2350	Medium service, .75 psf		750	.011		.28	.35		.63	.83	
	2400	Heavy service, 1.00 psf		650	.012		.37	.40		.77	1.01	
	2450	Extra heavy, 1.50 psf	▼	575	.014	▼	.56	.46		1.02	1.28	
	2600	Add for colored hardeners, metallic					50%					
	2650	Non-metallic					25%					
	2800	Trap rock wearing surface for monolithic floors										
	2810	2.0 psf	C-10B	1,250	.032	S.F.	.85	.93	.14	1.92	2.50	
	3000	Floor coloring, dusted on, 0.5 psf, add to above, minimum	1 Cefi	1,300	.006	▼	.63	.20		.83	.99	

03350 | Concrete Finishing

		CREW	DAILY OUTPUT	LABOR-HOURS	UNIT	2005 BARE COSTS				TOTAL INCL O&P
						MAT.	LABOR	EQUIP.	TOTAL	
3050	Maximum	1 Cefi	625	.013	S.F.	2.21	.42		2.63	3.05
3100	Colors only, minimum				Lb.	1.05			1.05	1.16
3120	Maximum				"	3.68			3.68	4.04
3200	Integral colors, see division 03310-220									
3600	1/2" topping using 5 lb. per bag, regular colors	C-10B	590	.068	S.F.	.21	1.98	.29	2.48	3.54
3650	Blue or green	"	590	.068		.21	1.98	.29	2.48	3.54
3800	Dustproofing, silicate liquids, 1 coat	1 Cefi	1,900	.004		.08	.14		.22	.29
3850	2 coats		1,300	.006		.14	.20		.34	.46
4000	Epoxy coating, 1 coat, clear		1,500	.005		.42	.18		.60	.72
4050	Colors		1,500	.005		.42	.18		.60	.72
4400	Stair finish, float		275	.029			.96		.96	1.41
4500	Steel trowel finish		200	.040			1.31		1.31	1.93
4600	Silicon carbide finish, .25 psf		150	.053		.57	1.75		2.32	3.21
325 0010	**CONTROL JOINT**, concrete floor slab									**325**
0100	Sawcut in green concrete									
0120	1" depth	C-27	2,000	.008	L.F.		.26	.06	.32	.44
0140	1-1/2" depth		1,800	.009			.29	.06	.35	.50
0160	2" depth		1,600	.010			.33	.07	.40	.56
0200	Clean out control joint of debris	C-28	6,000	.001			.04		.04	.06
0300	Joint sealant									
0320	Backer rod, polyethylene, 1/4" diameter	1 Cefi	460	.017	L.F.	.02	.57		.59	.86
0340	Sealant, polyurethane									
0360	1/4" x 1/4" (308 LF/Gal)	1 Cefi	270	.030	L.F.	.15	.97		1.12	1.60
0380	1/4" x 1/2" (154 LF/Gal)	"	255	.031	"	.31	1.03		1.34	1.86
350 0010	**FINISHING WALLS** Break ties and patch voids	1 Cefi	540	.015	S.F.	.03	.49		.52	.75 **350**
0050	Burlap rub with grout		450	.018		.03	.58		.61	.89
0100	Carborundum rub, dry		270	.030		.03	.97		1	1.46
0150	Wet rub		175	.046		.03	1.50		1.53	2.24
0300	Bush hammer, green concrete	B-39	1,000	.048		.03	1.35	.14	1.52	2.28
0350	Cured concrete	"	650	.074		.03	2.07	.22	2.32	3.48
0600	Float finish, 1/16" thick	1 Cefi	300	.027		.01	.88		.89	1.31
0700	Sandblast, light penetration	C-10	1,100	.022		.23	.67		.90	1.25
0750	Heavy penetration	"	375	.064		.46	1.97		2.43	3.44
0800	Board finish, see division 03110-455									
0850	Rustication strips, see division 03110-455									
600 0010	**SLAB TEXTURE STAMPING,** buy									**600**
0020	Approx. 3 S.F.- 5 S.F. each, minimum				Ea.	45			45	49.50
0030	Average				"	49.50			49.50	54.50
0120	Per S.F. of tool, average				S.F.	54			54	59.50
0200	Commonly used chemicals for texture systems									
0210	Hardeners w/colors average				S.F.	.45			.45	.50
0220	Release agents w/colors, average					.18			.18	.20
0225	Clear, average					.14			.14	.15
0230	Sealers, clear, average					.12			.12	.13
0240	Colors, average					.15			.15	.17

03370 | Specially Placed Concrete

		CREW	DAILY OUTPUT	LABOR-HOURS	UNIT	MAT.	LABOR	EQUIP.	TOTAL	TOTAL INCL O&P
300 0010	**GUNITE,** dry mix									**300**
0020	Applied in 1" layers, no mesh included	C-8	2,000	.028	S.F.	.25	.84	.35	1.44	1.94
0100	Mesh for gunite 2 x 2, #12, to 3" thick	2 Rodm	800	.020		.32	.76		1.08	1.60
0150	Over 3" thick	"	500	.032		.32	1.21		1.53	2.35
0300	Typical in place, including mesh, 2" thick, minimum	C-16	1,000	.072		.82	2.28	.70	3.80	5.20
0350	Maximum		500	.144		.82	4.56	1.41	6.79	9.55

Important: See the Reference Section for critical supporting data - Reference Nos., Crews, & City Cost Indexes

03300 | Cast-In-Place Concrete

03370 | Specially Placed Concrete

			CREW	DAILY OUTPUT	LABOR-HOURS	UNIT	2005 BARE COSTS				TOTAL INCL O&P	
							MAT.	LABOR	EQUIP.	TOTAL		
300	0500	4" thick, minimum	C-16	750	.096	S.F.	1.32	3.04	.94	5.30	7.20	300
	0550	Maximum	↓	350	.206		1.32	6.50	2.01	9.83	13.75	
	0900	Prepare old walls, no scaffolding, minimum	C-10	1,000	.024			.74		.74	1.10	
	0950	Maximum	"	275	.087			2.69		2.69	4	
	1100	For high finish requirement or close tolerance, add, minimum						50%				
	1150	Maximum				↓		110%				

03390 | Concrete Curing

			CREW	DAILY OUTPUT	LABOR-HOURS	UNIT	2005 BARE COSTS				TOTAL INCL O&P	
							MAT.	LABOR	EQUIP.	TOTAL		
200	0010	CURING Burlap, 4 uses assumed, 7.5 oz.	2 Clab	55	.291	C.S.F.	7	7.75		14.75	19.80	200
	0100	10 oz.		55	.291		11.30	7.75		19.05	24.50	
	0200	Waterproof curing paper, 2 ply, reinforced		70	.229		6	6.10		12.10	16.10	
	0300	Sprayed membrane curing compound	↓	95	.168	↓	5.25	4.50		9.75	12.80	
	0400	Curing blankets, 1" to 2" thick, buy, minimum				S.F.	.35			.35	.39	
	0450	Maximum					.53			.53	.58	
	0500	Electrically heated pads, 110 volts, 15 watts per S.F., buy					4.68			4.68	5.15	
	0600	20 watts per S.F., buy					6.20			6.20	6.85	
	0710	Electrically, heated pads, 15 watts/S.F., 20 uses, minimum					.17			.17	.18	
	0800	Maximum				↓	.28			.28	.31	

03400 | Precast Concrete

03410 | Plant-Precast Structural Concrete

				CREW	DAILY OUTPUT	LABOR-HOURS	UNIT	2005 BARE COSTS				TOTAL INCL O&P	
								MAT.	LABOR	EQUIP.	TOTAL		
100	0011	BEAMS, "L" shaped, 20' span, 12" x 20"	R03410 -030	C-11	32	2.250	Ea.	985	84	45	1,114	1,275	100
	1000	Inverted tee beams, add to above, small beams						15%					
	1050	Large beams						20%					
	1200	Rectangular, 20' span, 12" x 20"		C-11	32	2.250		730	84	45	859	995	
	1250	18" x 36"			24	3		825	112	60	997	1,175	
	1300	24" x 44"			22	3.273		1,025	122	65.50	1,212.50	1,400	
	1400	30' span, 12" x 36"			24	3		1,100	112	60	1,272	1,475	
	1450	18" x 44"			20	3.600		1,400	134	72.50	1,606.50	1,825	
	1500	24" x 52"			16	4.500		1,800	168	90.50	2,058.50	2,375	
	1600	40' span, 12" x 52"			20	3.600		2,050	134	72.50	2,256.50	2,575	
	1650	18" x 52"			16	4.500		2,450	168	90.50	2,708.50	3,075	
	1700	24" x 52"			12	6		2,650	223	120	2,993	3,450	
	2000	"T" shaped, 20' span, 12" x 20"			32	2.250		1,250	84	45	1,379	1,575	
	2050	18" x 36"			24	3		1,400	112	60	1,572	1,800	
	2100	24" x 44"			22	3.273		1,750	122	65.50	1,937.50	2,200	
	2200	30' span, 12" x 36"			24	3		1,875	112	60	2,047	2,325	
	2250	18" x 44"			20	3.600		2,375	134	72.50	2,581.50	2,900	
	2300	24" x 52"			16	4.500		3,050	168	90.50	3,308.50	3,750	
	2500	40' span, 12" x 52"			20	3.600		3,500	134	72.50	3,706.50	4,150	
	2550	18" x 52"			16	4.500		4,150	168	90.50	4,408.50	4,975	
	2600	24" x 52"		↓	12	6	↓	4,500	223	120	4,843	5,475	
210	0010	COLUMNS Rectangular to 12' high, small columns		C-11	120	.600	L.F.	41	22.50	12.05	75.55	97.50	210
	0050	Large columns			96	.750		72	28	15.05	115.05	144	
	0300	24' high, small columns			192	.375		41	13.95	7.55	62.50	78	
	0350	Large columns		↓	144	.500	↓	72	18.60	10.05	100.65	123	
400	0010	JOISTS 40 psf L.L., 6" deep for 12' spans		C-12	600	.080	L.F.	6.90	2.69	1	10.59	12.85	400
	0050	8" deep for 16' spans		↓	575	.083	↓	11.45	2.80	1.05	15.30	18.10	

3 CONCRETE

		03410	Plant-Precast Structural Concrete	CREW	DAILY OUTPUT	LABOR-HOURS	UNIT	MAT.	LABOR	EQUIP.	TOTAL	TOTAL INCL O&P	
400	0100		10" deep for 20' spans	C-12	550	.087	L.F.	20	2.93	1.10	24.03	28	400
	0150		12" deep for 24' spans		525	.091		27.50	3.07	1.15	31.72	36.50	
620	0010		**SLABS** Prestressed roof/floor members, grouted, solid, 4" thick [R03410-030]	C-11	2,400	.030	S.F.	4.02	1.12	.60	5.74	7	620
	0050		6" thick		2,800	.026		4.96	.96	.52	6.44	7.70	
	0100		Hollow, 8" thick **CN**		3,200	.023		5.45	.84	.45	6.74	7.95	
	0150		10" thick		3,600	.020		5.40	.74	.40	6.54	7.65	
	0200		12" thick		4,000	.018		6.15	.67	.36	7.18	8.30	
650	0010		**PRESTRESSED CONCRETE** pretensioned, see division 03400 [R03410-030]										650
	0020		See also division 03230-600										
	0100		Post-tensioned in place, small job [R03410-090]	C-17B	8.50	9.647	C.Y.	560	340	30	930	1,175	
	0200		Large job	"	10	8.200	"	420	289	25.50	734.50	940	
660	0010		**PRESTRESSED** Roof and floor members, see division 03400 [R03410-030]										660
750	0010		**TEES** Prestressed [R03410-030]										750
	0020		Quad tee, short spans, roof	C-11	7,200	.010	S.F.	4.95	.37	.20	5.52	6.30	
	0050		Floor		7,200	.010		4.95	.37	.20	5.52	6.30	
	0200		Double tee, floor members, 60' span		8,400	.009		6.80	.32	.17	7.29	8.25	
	0250		80' span		8,000	.009		8	.34	.18	8.52	9.60	
	0300		Roof members, 30' span		4,800	.015		5.80	.56	.30	6.66	7.65	
	0350		50' span		6,400	.011		6.60	.42	.23	7.25	8.25	
	0400		Wall members, up to 55' high		3,600	.020		8.25	.74	.40	9.39	10.85	
	0500		Single tee roof members, 40' span		3,200	.023		6.50	.84	.45	7.79	9.10	
	0550		80' span		5,120	.014		8.05	.52	.28	8.85	10.05	
	0600		100' span		6,000	.012		12.10	.45	.24	12.79	14.35	
	0650		120' span		6,000	.012		13.20	.45	.24	13.89	15.55	
	1000		Double tees, floor members										
	1100		Lightweight, 20" x 8' wide, 45' span	C-11	20	3.600	Ea.	2,125	134	72.50	2,331.50	2,625	
	1150		24" x 8' wide, 50' span		18	4		2,300	149	80.50	2,529.50	2,875	
	1200		32" x 10' wide, 60' span		16	4.500		3,875	168	90.50	4,133.50	4,675	
	1250		Standard weight, 12" x 8' wide, 20' span		22	3.273		790	122	65.50	977.50	1,150	
	1300		16" x 8' wide, 25' span		20	3.600		1,050	134	72.50	1,256.50	1,450	
	1350		18" x 8' wide, 30' span **CN**		20	3.600		1,350	134	72.50	1,556.50	1,775	
	1400		20" x 8' wide, 45' span		18	4		1,500	149	80.50	1,729.50	2,000	
	1450		24" x 8' wide, 50' span		16	4.500		1,950	168	90.50	2,208.50	2,550	
	1500		32" x 10' wide, 60' span		14	5.143		3,575	191	103	3,869	4,375	
	2000		Roof members										
	2050		Lightweight, 20" x 8' wide, 40' span	C-11	20	3.600	Ea.	1,775	134	72.50	1,981.50	2,250	
	2100		24" x 8' wide, 50' span		18	4		2,375	149	80.50	2,604.50	2,975	
	2150		32" x 10' wide, 60' span		16	4.500		3,925	168	90.50	4,183.50	4,700	
	2200		Standard weight, 12" x 8' wide, 30' span		22	3.273		1,200	122	65.50	1,387.50	1,575	
	2250		16" x 8' wide, 30' span		20	3.600		1,250	134	72.50	1,456.50	1,675	
	2300		18" x 8' wide, 30' span		20	3.600		1,400	134	72.50	1,606.50	1,825	
	2350		20" x 8' wide, 40' span		18	4		1,425	149	80.50	1,654.50	1,925	
	2400		24" x 8' wide, 50' span		16	4.500		1,900	168	90.50	2,158.50	2,500	
	2450		32" x 10' wide, 60' span		14	5.143		3,325	191	103	3,619	4,100	

03450 | Plant-Precast Architectural Concrete

				CREW	DAILY OUTPUT	LABOR-HOURS	UNIT	MAT.	LABOR	EQUIP.	TOTAL	TOTAL INCL O&P	
850	0011		**WALL PANELS** [R03450-010]										850
	0050		Uninsulated 4" thick, smooth gray										
	0150		Low rise, 4' x 8'x 4" thick	C-11	320	.225	S.F.	10.45	8.40	4.52	23.37	31	
	0200		8'x 8' x 4" thick		576	.125		10.35	4.65	2.51	17.51	22	
	0250		8'x 16'x 4" thick		1,024	.070		10.25	2.62	1.41	14.28	17.35	
	0400		8'x 8', 4" thick, smooth gray		576	.125		10.35	4.65	2.51	17.51	22	
	0500		Exposed aggregate		576	.125		13.20	4.65	2.51	20.36	25.50	
	0600		High rise, 4' x 8' x 4" thick		288	.250		10.45	9.30	5	24.75	33	

Important: See the Reference Section for critical supporting data - Reference Nos., Crews, & City Cost Indexes

CONCRETE 3

03450 | Plant-Precast Architectural Concrete

				CREW	DAILY OUTPUT	LABOR-HOURS	UNIT	2005 BARE COSTS MAT.	LABOR	EQUIP.	TOTAL	TOTAL INCL O&P	
850	0650	8' x 8' x 4" thick	R03450 -010	C-11	512	.141	S.F.	10.35	5.25	2.82	18.42	23.50	850
	0700	8' x 16' x 4" thick			768	.094		10.25	3.49	1.88	15.62	19.35	
	0800	Insulated panel, 2" polystyrene, add						.75			.75	.83	
	0850	2" urethane, add						.46			.46	.51	
	1000	20' x 10', 6" thick, smooth gray **CN**		C-11	1,800	.040		15.70	1.49	.80	17.99	20.50	
	1100	Exposed aggregate		"	1,800	.040		19.80	1.49	.80	22.09	25.50	
	1200	Finishes, white, add						1.88			1.88	2.07	
	1250	Exposed aggregate, add						.88			.88	.97	
	1300	Granite faced, domestic, add						22.50			22.50	24.50	
	2200	Fiberglass reinforced cement with urethane core											
	2210	R20, 8' x 8', minimum		E-2	750	.075	S.F.	11.95	2.76	1.77	16.48	19.85	
	2220	Maximum		"	600	.093	"	17.15	3.45	2.21	22.81	27	
855	0010	**PRECAST WINDOW SILLS**											855
	0600	Precast concrete, 4" tapers to 3", 9" wide		D-1	70	.229	L.F.	9.45	7.10		16.55	21	
	0650	11" wide			60	.267		12.50	8.30		20.80	26.50	
	0700	13" wide, 3 1/2" tapers to 2 1/2", 12" wall			50	.320		13.50	9.95		23.45	30	

03470 | Tilt-Up, Precast Concrete

				CREW	DAILY OUTPUT	LABOR-HOURS	UNIT	MAT.	LABOR	EQUIP.	TOTAL	TOTAL INCL O&P	
600	0010	**TILT-UP** Wall panel construction, walls only, 5-1/2" thick	R03470 -020	C-14	1,600	.090	S.F.	3.57	2.99	.62	7.18	9.30	600
	0100	7-1/2" thick			1,550	.093		4	3.09	.64	7.73	9.95	
	0500	Walls and columns, 5-1/2" thick walls, 12" x 12" columns			1,565	.092		4.35	3.06	.64	8.05	10.25	
	0550	7-1/2" thick wall, 12" x 12" columns			1,370	.105		5.80	3.49	.73	10.02	12.65	
	0800	Columns only, site precast, minimum			200	.720	L.F.	26	24	4.98	54.98	71.50	
	0850	Maximum			105	1.371	"	47.50	45.50	9.50	102.50	134	

03480 | Precast Concrete Specialties

				CREW	DAILY OUTPUT	LABOR-HOURS	UNIT	MAT.	LABOR	EQUIP.	TOTAL	TOTAL INCL O&P	
200	0010	**CURBS** Roadway type, see division 02770-300											200
400	0010	**LINTELS**											400
	0800	Precast concrete, 4" wide, 8" high, to 5' long		D-10	28	1.429	Ea.	24	46.50	18.05	88.55	117	
	0850	5'-12' long			24	1.667		68	54	21	143	180	
	1000	6" wide, 8" high, to 5' long			26	1.538		29.50	50	19.45	98.95	130	
	1050	5'-12' long			22	1.818		84	59	23	166	208	
	1200	8" wide, 8" high, to 5' long			24	1.667		35.50	54	21	110.50	144	
	1250	5'-12' long			20	2		92	65	25.50	182.50	228	
	1400	10" wide, 8" high U-Shape, to 14' long			18	2.222		156	72	28	256	315	
	1450	12" wide, 8" high U-Shape, to 19' long			16	2.500		204	81	31.50	316.50	385	
800	0010	**STAIRS**, Precast concrete treads on steel stringers, 3' wide		C-12	75	.640	Riser	64.50	21.50	8.05	94.05	113	800
	0300	Front entrance, 5' wide with 48" platform, 2 risers			16	3	Flight	340	101	37.50	478.50	575	
	0350	5 risers			12	4		530	134	50	714	850	
	0500	6' wide, 2 risers			15	3.200		360	108	40	508	605	
	0550	5 risers			11	4.364		560	147	55	762	905	
	0700	7' wide, 2 risers			14	3.429		375	115	43	533	640	
	0750	5 risers			10	4.800		590	161	60	811	965	
	1200	Basement entrance stairs, steel bulkhead doors, minimum		B-51	22	2.182		700	59	5.85	764.85	870	
	1250	Maximum		"	11	4.364		1,000	118	11.70	1,129.70	1,300	

03510 | Cementitious Roof Deck

			CREW	DAILY OUTPUT	LABOR-HOURS	UNIT	2005 BARE COSTS				TOTAL INCL O&P	
							MAT.	LABOR	EQUIP.	TOTAL		
200	0010	**WOOD FIBER** Lightweight cement system										200
	0050	Plank, beveled, 1" thick	2 Carp	1,000	.016	S.F.	1.82	.55		2.37	2.85	
	0100	Plank, T & G, 1-1/2" thick		975	.016		1.95	.56		2.51	3.03	
	0150	2" thick		950	.017		2.59	.58		3.17	3.75	
	0200	2-1/2" thick		925	.017		2.83	.59		3.42	4.03	
	0250	3" thick		900	.018		3.24	.61		3.85	4.51	
	0300	3-1/2" thick		875	.018		4.76	.63		5.39	6.25	
	0350	4" thick		850	.019		5.35	.64		5.99	6.85	
	1000	Bulb tee, sub-purlin and grout, 6' span, add	E-1	5,000	.005		1.53	.18	.02	1.73	2.01	
	1100	8' span	"	4,200	.006		1.74	.21	.02	1.97	2.29	
250	0010	**CONCRETE CHANNEL SLABS** 2-3/4" or 3-1/2" thick, straight	C-12	1,575	.030	S.F.	4.48	1.02	.38	5.88	6.95	250
	0050	Chopped up		785	.061		4.90	2.05	.77	7.72	9.40	
	0200	6" thick, span to 20'		1,300	.037		5.15	1.24	.46	6.85	8.15	
	0300	8" thick, span to 24'		1,100	.044		5.95	1.47	.55	7.97	9.40	
270	0010	**CONCRETE PLANK** Lightweight, nailable, T&G, 2" thick	C-12	1,800	.027	S.F.	3.14	.90	.33	4.37	5.20	270
	0050	2-3/4" thick		1,575	.030		3.20	1.02	.38	4.60	5.55	
	0100	3-3/4" thick		1,375	.035		4.42	1.17	.44	6.03	7.15	
	0150	For premium ceiling finish, add					10%					
	0200	For sloping roofs, slope over 4 in 12, add						25%				
	0250	Slope over 6 in 12, add						150%				
350	0010	**FORMBOARD** Including sub-purlins										350
	0050	Non-asbestos fiber cement, 1/8" thick	C-13	2,950	.008	S.F.	2.03	.30	.03	2.36	2.78	
	0070	1/4" thick		2,950	.008		2.28	.30	.03	2.61	3.06	
	0100	Fiberglass, 1" thick, economy		2,700	.009		2.53	.33	.03	2.89	3.37	
	1000	Poured gypsum, 2" thick, add to formboard above	C-8	6,000	.009		.80	.28	.12	1.20	1.43	
	1100	3" thick	"	4,800	.012		1.20	.35	.15	1.70	2.01	
900	0010	**WOOD PLANK** Roof decks, see division 06150-600 & 06170-550										900

R05120-250

03520 | Lightweight Concrete Roof Insulation

			CREW	DAILY OUTPUT	LABOR-HOURS	UNIT	2005 BARE COSTS				TOTAL INCL O&P	
							MAT.	LABOR	EQUIP.	TOTAL		
250	0010	**INSULATING ROOF FILL**, lightweight cellular concrete										250
	0020	Portland cement and foaming agent	C-8	50	1.120	C.Y.	80	33.50	14.10	127.60	155	
	0100	Poured vermiculite or perlite, field mix,										
	0110	1:6 field mix	C-8	50	1.120	C.Y.	91.50	33.50	14.10	139.10	167	
	0200	Ready mix, 1:6 mix, roof fill, 2" thick		10,000	.006	S.F.	.51	.17	.07	.75	.89	
	0250	3" thick		7,700	.007		.76	.22	.09	1.07	1.27	
	0400	Expanded volcanic glass rock, with binder, minimum	2 Carp	1,500	.011		.28	.37		.65	.88	
	0450	Maximum	"	1,200	.013		.83	.46		1.29	1.63	

R03520-010

03600 | Grouts

03610 | Construction Grout

			CREW	DAILY OUTPUT	LABOR-HOURS	UNIT	2005 BARE COSTS				TOTAL INCL O&P	
							MAT.	LABOR	EQUIP.	TOTAL		
400	0010	**GROUT** Column & machine bases, non-shrink, metallic, 1" deep	1 Cefi	35	.229	S.F.	10.50	7.50		18	22.50	400
	0050	2" deep		25	.320		21	10.50		31.50	38.50	
	0300	Non-shrink, non-metallic, 1" deep		35	.229		9.35	7.50		16.85	21.50	
	0350	2" deep		25	.320		18.75	10.50		29.25	36	

Important: See the Reference Section for critical supporting data - Reference Nos., Crews, & City Cost Indexes

			DAILY	LABOR-			2005 BARE COSTS				TOTAL	
03920	**Concrete Resurfacing**	CREW	OUTPUT	HOURS	UNIT	MAT.	LABOR	EQUIP.	TOTAL		INCL O&P	
600	0010	**PATCHING CONCRETE**											**600**
	0100	Floors, 1/4″ thick, small areas, regular grout	1 Cefi	170	.047	S.F.	.07	1.55		1.62		2.36	
	0150	Epoxy grout	″	100	.080	″	4.11	2.63		6.74		8.40	
	0300	Slab on Grade, cut outs, up to 50 C.F.	2 Cefi	50	.320	C.F.	6	10.50		16.50		22	
	2000	Walls, including chipping, cleaning and epoxy grout											
	2100	Minimum	1 Cefi	65	.123	S.F.	3.31	4.04		7.35		9.60	
	2150	Average		50	.160		6.60	5.25		11.85		15.05	
	2200	Maximum	↓	40	.200		13.25	6.55		19.80		24	
	2510	Underlayment, P.C based self-leveling, 4100 psi, pumped, 1/4″	C-8	20,000	.003		1.42	.08	.04	1.54		1.73	
	2520	1/2″		19,000	.003		2.56	.09	.04	2.69		2.98	
	2530	3/4″		18,000	.003		3.98	.09	.04	4.11		4.55	
	2540	1″		17,000	.003		5.40	.10	.04	5.54		6.15	
	2550	1-1/2″	↓	15,000	.004		8.25	.11	.05	8.41		9.25	
	2560	Hand mix, 1/2″	C-18	4,000	.002		2.56	.06	.01	2.63		2.91	
	2610	Topping, P.C. based self-level/dry 6100 psi, pumped, 1/4″	C-8	20,000	.003		2	.08	.04	2.12		2.37	
	2620	1/2″		19,000	.003		3.60	.09	.04	3.73		4.13	
	2630	3/4″		18,000	.003		5.60	.09	.04	5.73		6.35	
	2660	1″		17,000	.003		7.60	.10	.04	7.74		8.55	
	2670	1-1/2″	↓	15,000	.004		11.60	.11	.05	11.76		12.95	
	2680	Hand mix, 1/2″	C-18	4,000	.002	↓	3.60	.06	.01	3.67		4.06	

For information about Means Estimating Seminars, see yellow pages 12 and 13 in back of book

CONCRETE **3**

	CREW	DAILY OUTPUT	LABOR-HOURS	UNIT	2005 BARE COSTS				TOTAL INCL O&P
					MAT.	LABOR	EQUIP.	TOTAL	

Division 4
Masonry

Estimating Tips

04050 Basic Masonry Materials & Methods

- The terms *mortar* and *grout* are often used interchangeably, and incorrectly. Mortar is used to bed masonry units, seal the entry of air and moisture, provide architectural appearance, and allow for size variations in the units. Grout is used primarily in reinforced masonry construction and is used to bond the masonry to the reinforcing steel. Common mortar types are M(2500 psi), S(1800 psi), N(750 psi), and O(350 psi), and conform to ASTM C270. Grout is either fine or coarse, conforms to ASTM C476, and in-place strengths generally exceed 2500 psi. Mortar and grout are different components of masonry construction and are placed by entirely different methods. An estimator should be aware of their unique uses and costs.

- Waste, specifically the loss/droppings of mortar and the breakage of brick and block, is included in all masonry assemblies in this division. A factor of 25% is added for mortar and 3% for brick and concrete masonry units.

- Scaffolding or staging is not included in any of the Division 4 costs. Refer to section 01540 for scaffolding and staging costs.

04800 Masonry Assemblies

- The most common types of unit masonry are brick and concrete masonry. The major classifications of brick are building brick (ASTM C62), facing brick (ASTM C216) and glazed brick, fire brick and pavers. Many varieties of texture and appearance can exist within these classifications, and the estimator would be wise to check local custom and availability within the project area. On repair and remodeling jobs, matching the existing brick may be the most important criteria.

- Brick and concrete block are priced by the piece and then converted into a price per square foot of wall. Openings less than two square feet are generally ignored by the estimator because any savings in units used is offset by the cutting and trimming required.

- It is often difficult and expensive to find and purchase small lots of historic brick. Costs can vary widely. Many design issues affect costs, selection of mortar mix, and repairs or replacement of masonry materials. Cleaning techniques must be reflected in the estimate.

- All masonry walls, whether interior or exterior, require bracing. The cost of bracing walls during construction should be included by the estimator and this bracing must remain in place until permanent bracing is complete. Permanent bracing of masonry walls is accomplished by masonry itself, in the form of pilasters or abutting wall corners, or by anchoring the walls to the structural frame. Accessories in the form of anchors, anchor slots and ties are used, but their supply and installation can be by different trades. For instance, anchor slots on spandrel beams and columns are supplied and welded in place by the steel fabricator, but the ties from the slots into the masonry are installed by the bricklayer. Regardless of the installation method the estimator must be certain that these accessories are accounted for in pricing.

Reference Numbers

Reference numbers are shown in bold squares at the beginning of some major classifications. These numbers refer to related items in the Reference Section. The reference information may be an estimating procedure, an alternate pricing method or technical information.

Note: Not all subdivisions listed here necessarily appear in this publication.

04055	Selective Demolition		CREW	DAILY OUTPUT	LABOR-HOURS	UNIT	2005 BARE COSTS				TOTAL INCL O&P
							MAT.	LABOR	EQUIP.	TOTAL	
110	0010	**SELECTIVE DEMOLITION, MASONRY**	R02220									110
	0300	Concrete block walls, unreinforced, 2" thick	-510	2 Clab	1,200	.013	S.F.		.36		.36	.55
	0310	4" thick			1,150	.014			.37		.37	.58
	0320	6" thick			1,100	.015			.39		.39	.60
	0330	8" thick			1,050	.015			.41		.41	.63
	0340	10" thick			1,000	.016			.43		.43	.66
	0360	12" thick			950	.017			.45		.45	.70
	0380	Reinforced alternate courses, 2" thick			1,130	.014			.38		.38	.59
	0390	4" thick			1,080	.015			.40		.40	.62
	0400	6" thick			1,035	.015			.41		.41	.64
	0410	8" thick			990	.016			.43		.43	.67
	0420	10" thick			940	.017			.45		.45	.71
	0430	12" thick			890	.018			.48		.48	.75
	0440	Reinforced alternate courses & vertically 48" OC, 4" thick			900	.018			.47		.47	.74
	0450	6" thick			850	.019			.50		.50	.78
	0460	8" thick			800	.020			.53		.53	.83
	0480	10" thick			750	.021			.57		.57	.89
	0490	12" thick		▼	700	.023	▼		.61		.61	.95
	1000	Chimney, 16" x 16", soft old mortar		1 Clab	55	.145	C.F.		3.88		3.88	6.05
	1020	Hard mortar			40	.200			5.35		5.35	8.30
	1030	16" x 20", soft old mortar			55	.145			3.88		3.88	6.05
	1040	Hard mortar			40	.200			5.35		5.35	8.30
	1050	16" x 24", soft old mortar			55	.145			3.88		3.88	6.05
	1060	Hard mortar			40	.200			5.35		5.35	8.30
	1080	20" x 20", soft old mortar			55	.145			3.88		3.88	6.05
	1100	Hard mortar			40	.200			5.35		5.35	8.30
	1110	20" x 24", soft old mortar			55	.145			3.88		3.88	6.05
	1120	Hard mortar			40	.200			5.35		5.35	8.30
	1140	20" x 32", soft old mortar			55	.145			3.88		3.88	6.05
	1160	Hard mortar			40	.200			5.35		5.35	8.30
	1200	48" x 48", soft old mortar			55	.145			3.88		3.88	6.05
	1220	Hard mortar		▼	40	.200	▼		5.35		5.35	8.30
	1250	Metal, high temp steel jacket, 24" diameter		E-2	130	.431	V.L.F.		15.90	10.20	26.10	38.50
	1260	60" diameter		"	60	.933			34.50	22	56.50	83.50
	2000	Columns, 8" x 8", soft old mortar		1 Clab	48	.167			4.45		4.45	6.95
	2020	Hard mortar			40	.200			5.35		5.35	8.30
	2060	16" x 16", soft old mortar			16	.500			13.35		13.35	21
	2100	Hard mortar			14	.571			15.25		15.25	23.50
	2140	24" x 24", soft old mortar			8	1			26.50		26.50	41.50
	2160	Hard mortar			6	1.333			35.50		35.50	55.50
	2200	36" x 36", soft old mortar			4	2			53.50		53.50	83
	2220	Hard mortar			3	2.667			71		71	111
	2230	Alternate pricing method, soft old mortar			30	.267	C.F.		7.10		7.10	11.10
	2240	Hard mortar		▼	23	.348	"		9.30		9.30	14.45
	3000	Copings, precast or masonry, to 8" wide										
	3020	Soft old mortar		1 Clab	180	.044	L.F.		1.19		1.19	1.85
	3040	Hard mortar		"	160	.050	"		1.34		1.34	2.08
	3100	To 12" wide										
	3120	Soft old mortar		1 Clab	160	.050	L.F.		1.34		1.34	2.08
	3140	Hard mortar		"	140	.057	"		1.53		1.53	2.37
	4000	Fireplace, brick, 30" x 24" opening										
	4020	Soft old mortar		1 Clab	2	4	Ea.		107		107	166
	4040	Hard mortar			1.25	6.400			171		171	266
	4100	Stone, soft old mortar			1.50	5.333			142		142	222
	4120	Hard mortar			1	8	▼		214		214	330
	5000	Veneers, brick, soft old mortar		▼	140	.057	S.F.		1.53		1.53	2.37

Important: See the Reference Section for critical supporting data - Reference Nos., Crews, & City Cost Indexes

04055	Selective Demolition		CREW	DAILY OUTPUT	LABOR-HOURS	UNIT	MAT.	LABOR	EQUIP.	TOTAL	TOTAL INCL O&P	
							2005 BARE COSTS					
110	5020	Hard mortar	R02220 -510	1 Clab	125	.064	S.F.		1.71		1.71	2.66	110
	5100	Granite and marble, 2" thick			180	.044			1.19		1.19	1.85	
	5120	4" thick			170	.047			1.26		1.26	1.96	
	5140	Stone, 4" thick			180	.044			1.19		1.19	1.85	
	5160	8" thick			175	.046			1.22		1.22	1.90	
	5400	Alternate pricing method, stone, 4" thick			60	.133	C.F.		3.56		3.56	5.55	
	5420	8" thick			85	.094	"		2.51		2.51	3.91	

04060	Masonry Mortar		CREW	DAILY OUTPUT	LABOR-HOURS	UNIT	MAT.	LABOR	EQUIP.	TOTAL	TOTAL INCL O&P	
200	0010	CEMENT Gypsum 80 lb. bag, T.L. lots					Bag	23			23	25.50	200
	0050	L.T.L. lots						23			23	25.50	
	0100	Masonry, 70 lb. bag, T.L. lots	CN					6.10			6.10	6.70	
	0150	L.T.L. lots						6.45			6.45	7.10	
	0200	White, 70 lb. bag, T.L. lots						16.35			16.35	18	
	0250	L.T.L. lots						17.45			17.45	19.20	
400	0010	LIME Masons, hydrated, 50 lb. bag, T.L. lots	CN				Bag	5.80			5.80	6.35	400
	0050	L.T.L. lots						6			6	6.60	
	0200	Finish, double hydrated, 50 lb. bag, T.L. lots						7			7	7.70	
	0250	L.T.L. lots						7.75			7.75	8.55	
500	0010	MORTAR	R04060 -100										500
	0020	With masonry cement											
	0100	Type M, 1:1:6 mix	R04060 -200	1 Brhe	143	.056	C.F.	3.56	1.50		5.06	6.20	
	0200	Type N, 1:3 mix		"	143	.056	"	3.19	1.50		4.69	5.80	
	2000	With portland cement and lime											
	2100	Type M, 1:1/4:3 mix		1 Brhe	143	.056	C.F.	4.16	1.50		5.66	6.85	
	2200	Type N, 1:1:6 mix, 750 psi			143	.056		3.41	1.50		4.91	6.05	
	2300	Type O, 1:2:9 mix (Pointing Mortar)			143	.056		3.31	1.50		4.81	5.95	
	2650	Pre-mixed, type S or N						4.07			4.07	4.48	
	2700	Mortar for glass block		1 Brhe	143	.056		7.20	1.50		8.70	10.25	
	2800	Gypsum cement mortar						10.80			10.80	11.90	
	2900	Mortar for Fire Brick, 80 lb. bag, T.L. Lots					Bag	13.50			13.50	14.85	
520	0010	MORTAR FOR MASONRY RESTORATION, See also Division 04060-500					Lb.	1			1	1.10	520
	0050	White					"	1.15			1.15	1.27	
540	0010	MORTAR PIGMENTS, 50 lb. bags (2 bags per M bricks),	R04060 -100										540
	0020	range 2 to 10 lb. per bag of cement, minimum					Lb.	5.80			5.80	6.35	
	0050	Average						7.60			7.60	8.40	
	0100	Maximum						15.10			15.10	16.60	
750	0010	SAND For mortar, screened and washed, at the pit	CN				Ton	12.80			12.80	14.05	750
	0050	With 10 mile haul						13.85			13.85	15.25	
	0100	With 30 mile haul						15.10			15.10	16.65	
	0200	Screened and washed, at the pit					C.Y.	17.75			17.75	19.55	
	0250	With 10 mile haul						19.25			19.25	21	
	0300	With 30 mile haul						21			21	23	
770	0010	SURFACE BONDING CMU walls with fiberglass mortar,											770
	0020	gray or white colors, not incl. block work		1 Bric	540	.015	S.F.	.10	.52		.62	.91	
900	0010	WATERPROOFING Admixture, 1 qt. to 2 bags of masonry cement					Qt.	7			7	7.70	900

04070	Masonry Grout		CREW	DAILY OUTPUT	LABOR-HOURS	UNIT	MAT.	LABOR	EQUIP.	TOTAL	TOTAL INCL O&P	
420	0010	GROUTING Bond bms. & lintels, 8" dp., pumped, not incl. block	R04060 -100										420
	0020	8" thick, 0.2 C.F. per L.F.		D-4	1,400	.023	L.F.	.73	.70	.10	1.53	1.98	
	0050	10" thick, 0.25 C.F. per L.F.			1,200	.027		.83	.81	.11	1.75	2.27	
	0060	12" thick, 0.3 C.F. per L.F.			1,040	.031		.99	.94	.13	2.06	2.66	
	0200	Concrete block cores, solid, 4" thk., by hand, 0.067 C.F./S.F. of wall		D-8	1,100	.036	S.F.	.22	1.16		1.38	2.01	
	0210	6" thick, pumped, 0.175 C.F. per S.F.		D-4	720	.044		.58	1.36	.19	2.13	2.91	

MASONRY 4

4

MASONRY

04070 | Masonry Grout

			CREW	DAILY OUTPUT	LABOR-HOURS	UNIT	2005 BARE COSTS				TOTAL INCL O&P	
							MAT.	LABOR	EQUIP.	TOTAL		
420	0250	8" thick, pumped, 0.258 C.F. per S.F.	D-4	680	.047	S.F.	.85	1.44	.20	2.49	3.34	420
	0300	10" thick, pumped, 0.340 C.F. per S.F.		660	.048		1.12	1.48	.20	2.80	3.71	
	0350	12" thick, pumped, 0.422 C.F. per S.F.		640	.050		1.39	1.53	.21	3.13	4.08	
	0500	Cavity walls, 2" space, pumped, 0.167 C.F./S.F. of wall		1,700	.019		.55	.57	.08	1.20	1.57	
	0550	3" space, 0.250 C.F./S.F.		1,200	.027		.83	.81	.11	1.75	2.27	
	0600	4" space, 0.333 C.F. per S.F.		1,150	.028		1.10	.85	.12	2.07	2.63	
	0700	6" space, 0.500 C.F. per S.F.		800	.040		1.65	1.22	.17	3.04	3.87	
	0800	Door frames, 3' x 7' opening, 2.5 C.F. per opening		60	.533	Opng.	8.25	16.30	2.25	26.80	36	
	0850	6' x 7' opening, 3.5 C.F. per opening		45	.711	"	11.55	21.50	3	36.05	49	
	2000	Grout, C476, for bond beams, lintels and CMU cores		350	.091	C.F.	3.31	2.79	.39	6.49	8.30	

R04060 -100

04080 | Anchorage & Reinforcement

			CREW	DAILY OUTPUT	LABOR-HOURS	UNIT	MAT.	LABOR	EQUIP.	TOTAL	TOTAL INCL O&P	
070	0010	**ANCHOR BOLTS** Hooked type with nut and washer, 1/2" diam., 8" long	1 Bric	200	.040	Ea.	.55	1.41		1.96	2.76	070
	0030	12" long		190	.042		1.17	1.48		2.65	3.55	
	0040	5/8" diameter, 8" long		180	.044		.82	1.57		2.39	3.29	
	0050	12" long		170	.047		.90	1.66		2.56	3.52	
	0060	3/4" diameter, 8" long		160	.050		1.39	1.76		3.15	4.22	
	0070	12" long		150	.053		1.74	1.88		3.62	4.77	
200	0010	**REINFORCING** Steel bars A615, placed horiz., #3 & #4 bars	1 Bric	450	.018	Lb.	.40	.63		1.03	1.39	200
	0020	#5 & #6 bars		800	.010		.40	.35		.75	.98	
	0050	Placed vertical, #3 & #4 bars		350	.023		.40	.81		1.21	1.67	
	0060	#5 & #6 bars		650	.012		.40	.43		.83	1.10	
	0200	Joint reinforcing, regular truss, to 6" wide, mill std galvanized		30	.267	C.L.F.	11.90	9.40		21.30	27.50	
	0250	12" wide		20	.400		14.40	14.10		28.50	37.50	
	0400	Cavity truss with drip section, to 6" wide		30	.267		13.60	9.40		23	29.50	
	0450	12" wide		20	.400		15.45	14.10		29.55	38.50	
650	0010	**WALL TIES** To brick veneer, galv., corrugated, 7/8" x 7", 22 Ga.	1 Bric	10.50	.762	C	5.75	27		32.75	47.50	650
	0100	24 Ga.		10.50	.762		5.25	27		32.25	47	
	0150	16 Ga.		10.50	.762		17.65	27		44.65	60.50	
	0200	Buck anchors, galv., corrugated, 16 gauge, 2" bend. 8" x 2"		10.50	.762		115	27		142	168	
	0250	8" x 3"		10.50	.762		107	27		134	159	
	0670	3/16" diameter		10.50	.762		14.60	27		41.60	57	
	0680	Cavity wall, Z type, galvanized, 6" long, 1/4" diameter		10.50	.762		28.50	27		55.50	72.50	
	0850	3/16" diameter		10.50	.762		16.80	27		43.80	59.50	
	0855	8" long, 1/4" diameter		10.50	.762		31	27		58	75	
	1000	Rectangular type, galvanized, 1/4" diameter, 2" x 6"		10.50	.762		30.50	27		57.50	75	
	1050	4" x 6"		10.50	.762		34.50	27		61.50	79	
	1100	3/16" diameter, 2" x 6"		10.50	.762		21.50	27		48.50	64.50	
	1150	4" x 6"		10.50	.762		23.50	27		50.50	67	
	1500	Rigid partition anchors, plain, 8" long, 1" x 1/8"		10.50	.762		67	27		94	115	
	1550	1" x 1/4"		10.50	.762		132	27		159	186	
	1580	1-1/2" x 1/8"		10.50	.762		93	27		120	143	
	1600	1-1/2" x 1/4"		10.50	.762		221	27		248	284	
	1650	2" x 1/8"		10.50	.762		116	27		143	169	
	1700	2" x 1/4"		10.50	.762		315	27		342	385	

R04080 -500

04090 | Masonry Accessories

			CREW	DAILY OUTPUT	LABOR-HOURS	UNIT	MAT.	LABOR	EQUIP.	TOTAL	TOTAL INCL O&P	
170	0010	**CONTROL JOINT** Rubber, 4" and wider wall	1 Bric	400	.020	L.F.	1.70	.71		2.41	2.94	170
	0050	PVC, 4" wall		400	.020		1.03	.71		1.74	2.20	
	0100	Rubber, 6" wall		320	.025		2.17	.88		3.05	3.73	
	0120	PVC, 6" wall		320	.025		1.28	.88		2.16	2.75	
	0140	Rubber, 8" and wider wall		280	.029		2.79	1.01		3.80	4.60	
	0160	PVC, 8" wall		280	.029		1.50	1.01		2.51	3.18	
	0180	12" wall		240	.033		3.17	1.17		4.34	5.30	

Important: See the Reference Section for critical supporting data - Reference Nos., Crews, & City Cost Indexes

04050 | Basic Masonry Materials and Methods

04090 | Masonry Accessories

			CREW	DAILY OUTPUT	LABOR-HOURS	UNIT	2005 BARE COSTS MAT.	LABOR	EQUIP.	TOTAL	TOTAL INCL O&P	
860	0010	**VENT BOX** Extruded aluminum, 4" deep, 2-3/8" x 8-1/8"	1 Bric	30	.267	Ea.	20	9.40		29.40	36.50	860
	0050	5" x 8-1/8"		25	.320		26.50	11.30		37.80	46	
	0100	2-1/4" x 25"		25	.320		54.50	11.30		65.80	77	
	0150	5" x 16-1/2"		22	.364		47.50	12.80		60.30	72	
	0200	6" x 16-1/2"		22	.364		64	12.80		76.80	90	
	0250	7-3/4" x 16-1/2"		20	.400		50	14.10		64.10	76.50	
	0400	For baked enamel finish, add					35%					
	0500	For cast aluminum, painted, add					60%					
	1000	Stainless steel ventilators, 6" x 6"	1 Bric	25	.320		92	11.30		103.30	118	
	1050	8" x 8"		24	.333		97	11.75		108.75	125	
	1100	12" x 12"		23	.348		112	12.25		124.25	142	
	1150	12" x 6"		24	.333		98	11.75		109.75	126	
	1200	Foundation block vent, galv., 1-1/4" thk, 8" high, 16" long, no damper		30	.267		18	9.40		27.40	34	
	1250	For damper, add					6			6	6.60	
900	0010	**WALL PLUGS** For nailing to brickwork, 26 ga., galvanized, plain	1 Bric	10.50	.762	C	25	27		52	68.50	900
	0050	Wood filled	"	10.50	.762	"	84	27		111	134	

04200 | Masonry Units

04210 | Clay Masonry Units

			CREW	DAILY OUTPUT	LABOR-HOURS	UNIT	2005 BARE COSTS MAT.	LABOR	EQUIP.	TOTAL	TOTAL INCL O&P	
100	0010	**COMMON BUILDING BRICK** C62, TL lots, material only	R04210 -120									100
	0020	Standard, minimum				M	298			298	330	
	0050	Average (select)				"	355			355	390	
300	0010	**FACE BRICK** C216, TL lots, material only	R04210 -120									300
	0300	Standard modular, 4" x 2-2/3" x 8", minimum				M	375			375	410	
	0350	Maximum	CN				465			465	510	
	0450	Economy, 4" x 4" x 8", minimum					720			720	790	
	0500	Maximum					550			550	605	
	0510	Economy, 4" x 4" x 12", minimum					720			720	790	
	0520	Maximum					550			550	605	
	0550	Jumbo, 6" x 4" x 12", minimum					1,125			1,125	1,225	
	0600	Maximum					1,425			1,425	1,575	
	0610	Jumbo, 8" x 4" x 12", minimum					1,125			1,125	1,225	
	0620	Maximum					1,425			1,425	1,575	
	0650	Norwegian, 4" x 3-1/5" x 12", minimum					720			720	790	
	0700	Maximum					715			715	785	
	0710	Norwegian, 6" x 3-1/5" x 12", minimum					720			720	790	
	0720	Maximum					715			715	785	
	0850	Standard glazed, plain colors, 4" x 2-2/3" x 8", minimum					1,150			1,150	1,275	
	0900	Maximum					1,125			1,125	1,225	
	1000	Deep trim shades, 4" x 2-2/3" x 8", minimum					1,150			1,150	1,250	
	1050	Maximum					1,275			1,275	1,400	
	1080	Jumbo utility, 4" x 4" x 12"					1,050			1,050	1,150	
	1120	4" x 8" x 8"					1,075			1,075	1,175	
	1140	4" x 8" x 16"					2,250			2,250	2,475	
	1260	Engineer, 4" x 3-1/5" x 8", minimum					470			470	515	
	1270	Maximum					485			485	535	
	1350	King, 4" x 2-3/4" x 10", minimum					430			430	470	
	1360	Maximum					460			460	505	

MASONRY 4

04210	Clay Masonry Units		CREW	DAILY OUTPUT	LABOR-HOURS	UNIT	2005 BARE COSTS				TOTAL INCL O&P	
							MAT.	LABOR	EQUIP.	TOTAL		
300	1400	Norman, 4″ x 2-3/4″ x 12″	R04210 -120			M	430			430	470	**300**
	1450	Roman, 4″ x 2″ x 12″					430			430	470	
	1500	SCR, 6″ x 2-2/3″ x 12″					430			430	470	
	1550	Double, 4″ x 5-1/3″ x 8″					430			430	470	
	1600	Triple, 4″ x 5-1/3″ x 12″					430			430	470	
	1770	Standard modular, double glazed, 4″ x 2-2/3″ x 8″					970			970	1,075	
	1850	Jumbo, colored glazed ceramic, 6″ x 4″ x 12″					1,425			1,425	1,575	
	2050	Jumbo utility, glazed, 4″ x 4″ x 12″					1,075			1,075	1,175	
	2100	4″ x 8″ x 8″					1,475			1,475	1,625	
	2150	4″ x 16″ x 8″					2,550			2,550	2,800	
	2170	For less than truck load lots, add					10			10	11	
	2180	For buff or gray brick, add					15			15	16.50	

04550	Flue Liners		CREW	DAILY OUTPUT	LABOR-HOURS	UNIT	2005 BARE COSTS				TOTAL INCL O&P	
							MAT.	LABOR	EQUIP.	TOTAL		
250	0010	**FLUE LINING** Including mortar joints, 8″ x 8″	D-1	125	.128	V.L.F.	3.87	3.98		7.85	10.30	**250**
	0100	8″ x 12″		103	.155		5.45	4.83		10.28	13.35	
	0200	12″ x 12″		93	.172		6.60	5.35		11.95	15.40	
	0300	12″ x 18″		84	.190		11.45	5.90		17.35	21.50	
	0400	18″ x 18″		75	.213		16.15	6.65		22.80	28	
	0500	20″ x 20″		66	.242		34	7.55		41.55	49	
	0600	24″ x 24″		56	.286		43.50	8.90		52.40	61	
	1000	Round, 18″ diameter		66	.242		25	7.55		32.55	39	
	1100	24″ diameter		47	.340		39	10.60		49.60	59	

04580	Refractory Brick		CREW	DAILY OUTPUT	LABOR-HOURS	UNIT	2005 BARE COSTS				TOTAL INCL O&P	
							MAT.	LABOR	EQUIP.	TOTAL		
250	0010	**FIRE BRICK** 9″ x 2-1/2″ x 4-1/2″, low duty, 2000° F	D-1	.60	26.667	M	820	830		1,650	2,175	**250**
	0050	High duty, 3000° F	″	.60	26.667	″	1,500	830		2,330	2,925	
260	0010	**FIRE CLAY** Gray, high duty, 100 lb. bag				Bag	42			42	46	**260**
	0050	100 lb. drum, premixed (400 brick per drum)				Drum	52			52	57	

04710	Simulated Brick		CREW	DAILY OUTPUT	LABOR-HOURS	UNIT	2005 BARE COSTS				TOTAL INCL O&P	
							MAT.	LABOR	EQUIP.	TOTAL		
600	0010	**SIMULATED BRICK** Aluminum, baked on colors	1 Carp	200	.040	S.F.	2.50	1.37		3.87	4.88	**600**
	0050	Fiberglass panels		200	.040		2.75	1.37		4.12	5.15	
	0100	Urethane pieces cemented in mastic		150	.053		4.75	1.83		6.58	8.10	
	0150	Vinyl siding panels		200	.040		1.90	1.37		3.27	4.22	
	0160	Cement base, brick, incl. mastic	D-1	100	.160		3	4.97		7.97	10.90	
	0170	Corner		50	.320	V.L.F.	7.50	9.95		17.45	23.50	
	0180	Stone face, incl. mastic		100	.160	S.F.	7.25	4.97		12.22	15.60	
	0190	Corner		50	.320	V.L.F.	8	9.95		17.95	24	

Important: See the Reference Section for critical supporting data - Reference Nos., Crews, & City Cost Indexes

04700 | Simulated Masonry

04730 | Simulated Stone

		CREW	DAILY OUTPUT	LABOR-HOURS	UNIT	2005 BARE COSTS				TOTAL INCL O&P	
						MAT.	LABOR	EQUIP.	TOTAL		
600	0010 **SIMULATED STONE**										600
	0100 Insulated fiberglass panels, 5/8" ply backer	L-4	200	.120	S.F.	9	3.81		12.81	15.85	

04800 | Masonry Assemblies

04810 | Unit Masonry Assemblies

		CREW	DAILY OUTPUT	LABOR-HOURS	UNIT	2005 BARE COSTS				TOTAL INCL O&P	
						MAT.	LABOR	EQUIP.	TOTAL		
040	0010 **ADOBE BRICK** Semi-stabilized, with cement mortar										040
	0060 Brick, 10" x 4" x 14", 2.6/S.F.	D-8	560	.071	S.F.	2.13	2.28		4.41	5.80	
	0080 12" x 4" x 16", 2.3/S.F.		580	.069		3.49	2.20		5.69	7.20	
	0100 10" x 4" x 16", 2.3/S.F.		590	.068		3.17	2.16		5.33	6.80	
	0120 8" x 4" x 16", 2.3/S.F.		560	.071		2.43	2.28		4.71	6.15	
	0140 4" x 4" x 16", 2.3/S.F.		540	.074		2.55	2.36		4.91	6.40	
	0160 6" x 4" x 16", 2.3/S.F.		540	.074		1.94	2.36		4.30	5.75	
	0180 4" x 4" x 12", 3.0/S.F.		520	.077		2.42	2.45		4.87	6.40	
	0200 8" x 4" x 12", 3.0/S.F.	▼	520	.077	▼	2.63	2.45		5.08	6.65	
050	0010 **AUTOCLAVED AERATED CONCRETE BLOCK** Scaffolding not incl										050
	0050 Solid, 4" x 12" x 24", incl mortar	D-8	600	.067	S.F.	1.84	2.13		3.97	5.25	
	0060 6" x 12" x 24"		600	.067		2.47	2.13		4.60	5.95	
	0070 8" x 8" x 24"		575	.070		3.10	2.22		5.32	6.80	
	0080 10" x 12" x 24"		575	.070		4.04	2.22		6.26	7.85	
	0090 12" x 12" x 24"	▼	550	.073	▼	4.71	2.32		7.03	8.75	
100	0010 **BRICK VENEER** Scaffolding not included, truck load lots R04210 -120										100
	0015 Material costs incl. 3% brick and 25% mortar waste										
	0020 Standard, select common, 4" x 2-2/3" x 8" (6.75/S.F.) R04210 -180	D-8	1.50	26.667	M	410	850		1,260	1,750	
	0050 Red, 4" x 2-2/3" x 8", running bond		1.50	26.667		425	850		1,275	1,775	
	0100 Full header every 6th course (7.88/S.F.) R04210 -500		1.45	27.586		425	880		1,305	1,825	
	0150 English, full header every 2nd course (10.13/S.F.)		1.40	28.571		425	910		1,335	1,875	
	0200 Flemish, alternate header every course (9.00/S.F.)		1.40	28.571		425	910		1,335	1,875	
	0250 Flemish, alt. header every 6th course (7.13/S.F.)		1.45	27.586		425	880		1,305	1,825	
	0300 Full headers throughout (13.50/S.F.)		1.40	28.571		425	910		1,335	1,875	
	0350 Rowlock course (13.50/S.F.)		1.35	29.630		425	945		1,370	1,925	
	0400 Rowlock stretcher (4.50/S.F.)		1.40	28.571		430	910		1,340	1,875	
	0450 Soldier course (6.75/S.F.)		1.40	28.571		425	910		1,335	1,875	
	0500 Sailor course (4.50/S.F.)		1.30	30.769		430	980		1,410	1,975	
	0601 Buff or gray face, running bond, (6.75/S.F.)		1.50	26.667		425	850		1,275	1,775	
	0700 Glazed face, 4" x 2-2/3" x 8", running bond		1.40	28.571		1,300	910		2,210	2,825	
	0750 Full header every 6th course (7.88/S.F.)		1.35	29.630		1,225	945		2,170	2,800	
	1000 Jumbo, 6" x 4" x 12",(3.00/S.F.)		1.30	30.769		1,300	980		2,280	2,925	
	1051 Norman, 4" x 2-2/3" x 12" (4.50/S.F.)		1.45	27.586		875	880		1,755	2,325	
	1100 Norwegian, 4" x 3-1/5" x 12" (3.75/S.F.)		1.40	28.571		805	910		1,715	2,275	
	1150 Economy, 4" x 4" x 8" (4.50 per S.F.)		1.40	28.571		790	910		1,700	2,275	
	1201 Engineer, 4" x 3-1/5" x 8", (5.63/S.F.)		1.45	27.586		530	880		1,410	1,925	
	1251 Roman, 4" x 2" x 12", (6.00/S.F.)		1.50	26.667		785	850		1,635	2,150	
	1300 S.C.R. 6" x 2-2/3" x 12" (4.50/S.F.)		1.40	28.571		925	910		1,835	2,425	
	1350 Utility, 4" x 4" x 12" (3.00/S.F.)	▼	1.35	29.630		1,150	945		2,095	2,700	
	1360 For less than truck load lots, add				▼	10			10	11	
	1400 For battered walls, add						30%				
	1450 For corbels, add						75%				
	1500 For curved walls, add	▼					30%				

For expanded coverage of these items see *Means Concrete & Masonry Cost Data 2005*

04810 | Unit Masonry Assemblies

		CREW	DAILY OUTPUT	LABOR-HOURS	UNIT	2005 BARE COSTS				TOTAL INCL O&P	
						MAT.	LABOR	EQUIP.	TOTAL		
100	1550	For pits and trenches, deduct R04210-120						20%				**100**
	1999	Alternate method of figuring by square foot									
	2000	Standard, sel. common, 4" x 2-2/3" x 8", (6.75/S.F.) R04210-180	D-8	230	.174	S.F.	2.88	5.55		8.43	11.60
	2020	Standard, red, 4" x 2-2/3" x 8", running bond (6.75/SF)		220	.182		2.88	5.80		8.68	12
	2050	Full header every 6th course (7.88/S.F.) R04210-500		185	.216		3.35	6.90		10.25	14.20
	2100	English, full header every 2nd course (10.13/S.F.)		140	.286		4.30	9.10		13.40	18.65
	2150	Flemish, alternate header every course (9.00/S.F.)		150	.267		3.82	8.50		12.32	17.15
	2200	Flemish, alt. header every 6th course (7.13/S.F.)		205	.195		3.04	6.25		9.29	12.85
	2250	Full headers throughout (13.50/S.F.)		105	.381		5.70	12.15		17.85	25
	2300	Rowlock course (13.50/S.F.)		100	.400		5.70	12.75		18.45	26
	2350	Rowlock stretcher (4.50/S.F.)		310	.129		1.93	4.12		6.05	8.40
	2400	Soldier course (6.75/S.F.)		200	.200		2.88	6.40		9.28	12.85
	2450	Sailor course (4.50/S.F.)		290	.138		1.93	4.40		6.33	8.85
	2600	Buff or gray face, running bond, (6.75/S.F.)		220	.182		3.06	5.80		8.86	12.20
	2700	Glazed face brick, running bond		210	.190		8.30	6.10		14.40	18.40
	2750	Full header every 6th course (7.88/S.F.)		170	.235		9.70	7.50		17.20	22
	3000	Jumbo, 6" x 4" x 12" running bond (3.00/S.F.)		435	.092		3.62	2.93		6.55	8.45
	3050	Norman, 4" x 2-2/3" x 12" running bond, (4.5/S.F.)		320	.125		3.93	3.99		7.92	10.40
	3100	Norwegian, 4" x 3-1/5" x 12" (3.75/S.F.)		375	.107		2.95	3.40		6.35	8.45
	3150	Economy, 4" x 4" x 8" (4.50/S.F.)		310	.129		3.54	4.12		7.66	10.15
	3200	Engineer, 4" x 3-1/5" x 8" (5.63/S.F.)		260	.154		2.97	4.91		7.88	10.75
	3250	Roman, 4" x 2" x 12" (6.00/S.F.)		250	.160		4.63	5.10		9.73	12.90
	3300	SCR, 6" x 2-2/3" x 12" (4.50/S.F.)		310	.129		4.24	4.12		8.36	10.90
	3350	Utility, 4" x 4" x 12" (3.00/S.F.)		450	.089		3.37	2.84		6.21	8
	3400	For cavity wall construction, add						15%			
	3450	For stacked bond, add						10%			
	3500	For interior veneer construction, add						15%			
	3550	For curved walls, add						30%			
110	0010	**OVERSIZED BRICK**, scaffolding not included										**110**
	0100	Oversized brick, 4" x 4" x 12"	D-8	118	.339	S.F.	.29	10.80		11.09	16.80
	0110	4" x 4" x 16"		155	.258		.29	8.25		8.54	12.85
	0120	4" x 8" x 16"		155	.258		.29	8.25		8.54	12.85
160	0010	**CHIMNEY** See Div. 03310-240 for foundation, add to prices below										**160**
	0100	Brick, 16" x 16", 8" flue, scaff. not incl.	D-1	18.20	.879	V.L.F.	15.90	27.50		43.40	59
	0150	16" x 20" with one 8" x 12" flue		16	1		25	31		56	75
	0200	16" x 24" with two 8" x 8" flues		14	1.143		36.50	35.50		72	94.50
	0250	20" x 20" with one 12" x 12" flue		13.70	1.168		28	36.50		64.50	86
	0300	20" x 24" with two 8" x 12" flues		12	1.333		41	41.50		82.50	108
	0350	20" x 32" with two 12" x 12" flues		10	1.600		48	49.50		97.50	129
	1800	Metal, high temp. steel jacket, factory lining, 24" diam.	E-2	65	.862		186	32	20.50	238.50	281
	1900	60" diameter	"	30	1.867		675	69	44	788	910
	2100	Poured concrete, brick lining, 200' high x 10' diam.					5,600			5,600	6,150
	2800	500' x 20' diameter					9,800			9,800	10,800
165	0010	**CHIMNEY BLOCK** Scaffolding not included										**165**
	0220	1 piece, with 8" x 8" flue, 16" x 16"	D-1	28	.571	V.L.F.	10.85	17.75		28.60	39
	0230	2 piece, 16" x 16"		26	.615		12.50	19.15		31.65	43
	0240	2 piece, with 8" x 12" flue, 16" x 20"		24	.667		22.50	20.50		43	56
170	0010	**COLUMNS** Face brick, includes mortar, scaffolding not included R04210-100										**170**
	0050	8" x 8", 9 brick per course	D-1	56	.286	V.L.F.	3.74	8.90		12.64	17.65
	0100	12" x 8", 13.5 brick		37	.432		5.60	13.45		19.05	26.50
	0200	12" x 12", 20 brick		25	.640		8.30	19.90		28.20	39.50
	0300	16" x 12", 27 brick		19	.842		11.25	26		37.25	52.50
	0400	16" x 16", 36 brick		14	1.143		14.95	35.50		50.45	70.50
	0500	20" x 16", 45 brick		11	1.455		18.70	45		63.70	89.50
	0600	20" x 20", 56 brick		9	1.778		23.50	55.50		79	110

Important: See the Reference Section for critical supporting data - Reference Nos., Crews, & City Cost Indexes

04810	Unit Masonry Assemblies		CREW	DAILY OUTPUT	LABOR-HOURS	UNIT	2005 BARE COSTS				TOTAL INCL O&P	
							MAT.	LABOR	EQUIP.	TOTAL		
170	0700	24" x 20", 68 brick	D-1	7	2.286	V.L.F.	28.50	71		99.50	139	**170**
	0800	24" x 24", 81 brick	R04210 -100	6	2.667		33.50	83		116.50	163	
	1000	36" x 36", 182 brick		3	5.333		75.50	166		241.50	335	
172	0010	**CONCRETE BLOCK, BACK-UP,** C90, 2000 psi	R04220 -200									**172**
	0020	Normal weight, 8" x 16" units, tooled joint 1 side										
	0050	Not-reinforced, 2000 psi, 2" thick	D-8	475	.084	S.F.	.83	2.69		3.52	5	
	0200	4" thick		460	.087		.98	2.77		3.75	5.30	
	0300	6" thick		440	.091		1.44	2.90		4.34	6	
	0350	8" thick		400	.100		1.56	3.19		4.75	6.60	
	0400	10" thick		330	.121		2.19	3.87		6.06	8.30	
	0450	12" thick	D-9	310	.155		2.26	4.81		7.07	9.85	
	1000	Reinforced, alternate courses, 4" thick	D-8	450	.089		1.08	2.84		3.92	5.50	
	1100	6" thick		430	.093		1.55	2.97		4.52	6.20	
	1150	8" thick		395	.101		1.67	3.23		4.90	6.75	
	1200	10" thick		320	.125		2.30	3.99		6.29	8.65	
	1250	12" thick	D-9	300	.160		2.38	4.97		7.35	10.20	
175	0010	**CONCRETE BLOCK BOND BEAM** C90, 2000 psi										**175**
	0020	Not including grout or reinforcing										
	0130	8" high, 8" thick	D-8	565	.071	L.F.	1.64	2.26		3.90	5.25	
	0150	12" thick	D-9	510	.094		2.28	2.93		5.21	6.95	
	0530	8" high, 8" thick	D-8	575	.070		1.53	2.22		3.75	5.05	
	0550	12" thick	D-9	520	.092		2.57	2.87		5.44	7.20	
	2000	Including grout and 2 #5 bars										
	2100	Regular block, 8" high, 8" thick	D-8	300	.133	L.F.	3.27	4.25		7.52	10.10	
	2150	12" thick	D-9	250	.192		4.33	5.95		10.28	13.85	
	2500	Lightweight, 8" high, 8" thick	D-8	305	.131		3.51	4.19		7.70	10.25	
	2550	12" thick	D-9	255	.188		4.62	5.85		10.47	14	
180	0010	**CONCRETE BLOCK COLUMN** or pilaster, scaffolding not included										**180**
	0050	Including vertical reinforcing (4-#4 bars) and grout										
	0160	1 piece unit, 16" x 16"	D-1	26	.615	V.L.F.	14.45	19.15		33.60	45	
	0170	2 piece units, 16" x 20"		24	.667		19.30	20.50		39.80	52.50	
	0180	20" x 20"		22	.727		28.50	22.50		51	65.50	
	0190	22" x 24"		18	.889		39.50	27.50		67	85.50	
	0200	20" x 32"		14	1.143		43	35.50		78.50	101	
182	0010	**CONCRETE BLOCK, DECORATIVE** C90, 2000 psi										**182**
	0020	Embossed, simulated brick face										
	0100	8" x 16" units, 4" thick	D-8	400	.100	S.F.	2.30	3.19		5.49	7.40	
	0200	8" thick		340	.118		3.15	3.75		6.90	9.15	
	0250	12" thick		300	.133		4.14	4.25		8.39	11.05	
	0400	Embossed both sides										
	0500	8" thick	D-8	300	.133	S.F.	3.55	4.25		7.80	10.40	
	0550	12" thick	"	275	.145	"	4.47	4.64		9.11	11.95	
	1000	Fluted high strength										
	1100	8" x 16" x 4" thick, flutes 1 side,	D-8	345	.116	S.F.	2.75	3.70		6.45	8.65	
	1150	Flutes 2 sides		335	.119		3.36	3.81		7.17	9.50	
	1200	8" thick		300	.133		4.34	4.25		8.59	11.25	
	1250	For special colors, add					.28			.28	.31	
	1400	Deep grooved, smooth face										
	1450	8" x 16" x 4" thick	D-8	345	.116	S.F.	1.76	3.70		5.46	7.60	
	1500	8" thick	"	300	.133	"	2.98	4.25		7.23	9.80	
	2000	Formblock, incl. inserts & reinforcing										
	2100	8" x 16" x 8" thick	D-8	345	.116	S.F.	3.10	3.70		6.80	9.05	

MASONRY 4

04810	Unit Masonry Assemblies	CREW	DAILY OUTPUT	LABOR-HOURS	UNIT	2005 BARE COSTS				TOTAL INCL O&P	
						MAT.	LABOR	EQUIP.	TOTAL		
182 2150	12" thick	D-8	310	.129	S.F.	3.84	4.12		7.96	10.50	**182**
2500	Ground face										
2600	8" x 16" x 4" thick	D-8	345	.116	S.F.	3.21	3.70		6.91	9.20	
2650	6" thick		325	.123		3.87	3.93		7.80	10.25	
2700	8" thick		300	.133		4.49	4.25		8.74	11.45	
2750	12" thick	D-9	265	.181		5.30	5.65		10.95	14.45	
2900	For special colors, add, minimum					15%					
2950	For special colors, add, maximum					45%					
4000	Slump block										
4100	4" face height x 16" x 4" thick	D-1	165	.097	S.F.	2.98	3.01		5.99	7.85	
4150	6" thick		160	.100		4.18	3.11		7.29	9.35	
4200	8" thick		155	.103		4.60	3.21		7.81	9.95	
4250	10" thick		140	.114		8.10	3.55		11.65	14.30	
4300	12" thick		130	.123		8.40	3.83		12.23	15.10	
4400	6" face height x 16" x 6" thick		155	.103		3.82	3.21		7.03	9.10	
4450	8" thick		150	.107		5	3.32		8.32	10.55	
4500	10" thick		130	.123		7.55	3.83		11.38	14.15	
4550	12" thick		120	.133		7.95	4.14		12.09	15.05	
5000	Split rib profile units, 1" deep ribs, 8 ribs										
5100	8" x 16" x 4" thick	D-8	345	.116	S.F.	2.14	3.70		5.84	8	
5150	6" thick		325	.123		2.45	3.93		6.38	8.70	
5200	8" thick		300	.133		2.83	4.25		7.08	9.60	
5250	12" thick	D-9	275	.175		3.34	5.45		8.79	11.95	
5400	For special deeper colors, 4" thick, add					.80			.80	.88	
5450	12" thick, add					.69			.69	.76	
5600	For white, 4" thick, add					.80			.80	.88	
5650	6" thick, add					.80			.80	.88	
5700	8" thick, add					.75			.75	.82	
5750	12" thick, add					.69			.69	.76	
6000	Split face										
6100	8" x 16" x 4" thick	D-8	350	.114	S.F.	2.03	3.65		5.68	7.80	
6150	6" thick		325	.123		2.36	3.93		6.29	8.60	
6200	8" thick		300	.133		2.73	4.25		6.98	9.50	
6250	12" thick	D-9	270	.178		3.24	5.55		8.79	11.95	
6300	For scored, add					.20			.20	.22	
6400	For special deeper colors, 4" thick, add					.46			.46	.51	
6450	6" thick, add					.46			.46	.51	
6500	8" thick, add					.41			.41	.45	
6550	12" thick, add					.35			.35	.39	
6650	For white, 4" thick, add					.80			.80	.88	
6700	6" thick, add					.80			.80	.88	
6750	8" thick, add					.75			.75	.82	
6800	12" thick, add					.69			.69	.76	
7000	Scored ground face, 2 to 5 scores										
7100	8" x 16" x 4" thick	D-8	340	.118	S.F.	3.62	3.75		7.37	9.70	
7150	6" thick		310	.129		4.19	4.12		8.31	10.85	
7200	8" thick		290	.138		4.83	4.40		9.23	12	
7250	12" thick	D-9	265	.181		5.70	5.65		11.35	14.90	
8000	Hexagonal face profile units, 8" x 16" units										
8100	4" thick, hollow	D-8	340	.118	S.F.	2.24	3.75		5.99	8.15	
8200	Solid		340	.118		3	3.75		6.75	9	
8300	6" thick, hollow		310	.129		2.49	4.12		6.61	9	
8350	8" thick, hollow		290	.138		3.40	4.40		7.80	10.45	
8500	For stacked bond, add						26%				
8550	For high rise construction, add per story	D-8	67.80	.590	M.S.F.		18.85		18.85	28.50	
8600	For scored block, add					10%					

Important: See the Reference Section for critical supporting data - Reference Nos., Crews, & City Cost Indexes

		04810	Unit Masonry Assemblies	CREW	DAILY OUTPUT	LABOR-HOURS	UNIT	2005 BARE COSTS				TOTAL INCL O&P	
								MAT.	LABOR	EQUIP.	TOTAL		
182	8650		For honed or ground face, per face, add				Ea.	.28			.28	.31	182
	8700		For honed or ground end, per end, add				"	2.26			2.26	2.49	
	8750		For bullnose block, add					10%					
	8800		For special color, add					13%					
184	0010		**CONCRETE BLOCK, EXTERIOR** C90, 2000 psi										184
	0020		Reinforced alt courses, tooled joints 2 sides										
	0100		Normal weight, 8" x 16" x 6" thick	D-8	395	.101	S.F.	1.67	3.23		4.90	6.75	
	0200		8" thick		360	.111		2.48	3.55		6.03	8.15	
	0250		10" thick	↓	290	.138		3	4.40		7.40	10	
	0300		12" thick	D-9	250	.192		3.01	5.95		8.96	12.40	
	0500		Lightweight, 8" x 16" x 6" thick	D-8	450	.089		1.96	2.84		4.80	6.45	
	0600		8" thick		430	.093		2.65	2.97		5.62	7.45	
	0650		10" thick	↓	395	.101		2.99	3.23		6.22	8.20	
	0700		12" thick	D-9	350	.137	↓	4.31	4.26		8.57	11.25	
186	0010		**CONCRETE BLOCK FOUNDATION WALL** C90/C145										186
	0050		Normal-weight, cut joints, horiz joint reinf, no vert reinf										
	0200		Hollow, 8" x 16" x 6" thick	D-8	455	.088	S.F.	1.67	2.81		4.48	6.10	
	0250		8" thick		425	.094		1.80	3		4.80	6.55	
	0300		10" thick	↓	350	.114		2.43	3.65		6.08	8.25	
	0350		12" thick	D-9	300	.160		2.51	4.97		7.48	10.35	
	0500		Solid, 8" x 16" block, 6" thick	D-8	440	.091		1.81	2.90		4.71	6.40	
	0550		8" thick	"	415	.096		2.59	3.08		5.67	7.55	
	0600		12" thick	D-9	350	.137	↓	3.76	4.26		8.02	10.65	
187	0010		**CONCRETE BLOCK, HIGH STRENGTH** Normal weight										187
	0050		Hollow, reinforced alternate courses, 8" x 16" units										
	0200		3500 psi, 4" thick	D-8	440	.091	S.F.	1.24	2.90		4.14	5.80	
	0250		6" thick		395	.101		1.64	3.23		4.87	6.70	
	0300		8" thick	↓	360	.111		2.44	3.55		5.99	8.10	
	0350		12" thick	D-9	250	.192		2.96	5.95		8.91	12.35	
	0500		5000 psi, 4" thick	D-8	440	.091		1.40	2.90		4.30	5.95	
	0550		6" thick		395	.101		2.05	3.23		5.28	7.15	
	0600		8" thick	↓	360	.111		2.77	3.55		6.32	8.45	
	0650		12" thick	D-9	300	.160	↓	4.09	4.97		9.06	12.10	
	1000		For 75% solid block, add					30%					
	1050		For 100% solid block, add					50%					
188	0010		**CONCRETE BLOCK INSULATION INSERTS**										188
	0100		Inserts, styrofoam, plant installed, add to block prices										
	0200		8" x 16" units, 6" thick				S.F.	.85			.85	.94	
	0250		8" thick					.85			.85	.94	
	0300		10" thick					1			1	1.10	
	0350		12" thick					1.05			1.05	1.16	
	0500		8" x 8" units, 8" thick					.70			.70	.77	
	0550		12" thick				↓	.85			.85	.94	
189	0010		**CONCRETE BLOCK, INTERLOCKING**										189
	0100		Not including grout or reinforcing										
	0200		8" x 16" units, 2,000 psi, 8" thick	D-1	245	.065	S.F.	1.77	2.03		3.80	5.05	
	0300		12" thick		220	.073		2.63	2.26		4.89	6.35	
	0350		16" thick	↓	185	.086		3.99	2.69		6.68	8.50	
	0400		Including grout & reinforcing, 8" thick	D-4	245	.131		5.40	3.99	.55	9.94	12.60	
	0450		12" thick		220	.145		6.45	4.44	.61	11.50	14.50	
	0500		16" thick	↓	185	.173	↓	7.95	5.30	.73	13.98	17.55	
190	0010		**CONCRETE BLOCK, LINTELS** C90, normal weight										190
	0100		Including grout and horizontal reinforcing										
	0200		8" x 8" x 8", 1 #4 bar	D-4	300	.107	L.F.	3.63	3.26	.45	7.34	9.45	
	0250		2 #4 bars	↓	295	.108	↓	3.81	3.31	.46	7.58	9.75	

MASONRY **4**

			DAILY	LABOR-		2005 BARE COSTS				TOTAL	
04810	**Unit Masonry Assemblies**	CREW	OUTPUT	HOURS	UNIT	MAT.	LABOR	EQUIP.	TOTAL	INCL O&P	
190	0400	8" x 16" x 8", 1 #4 bar	D-4	275	.116	L.F.	6.20	3.55	.49	10.24	12.80	190
	0450	2 #4 bars		270	.119		6.40	3.62	.50	10.52	13.10	
	1000	12" x 8" x 8", 1 #4 bar		275	.116		5.10	3.55	.49	9.14	11.55	
	1100	2 #4 bars		270	.119		5.25	3.62	.50	9.37	11.85	
	1150	2 #5 bars		270	.119		5.45	3.62	.50	9.57	12.05	
	1200	2 #6 bars		265	.121		5.70	3.69	.51	9.90	12.45	
	1500	12" x 16" x 8", 1 #4 bar		250	.128		8.05	3.91	.54	12.50	15.40	
	1600	2 #3 bars		245	.131		8.10	3.99	.55	12.64	15.55	
	1650	2 #4 bars		245	.131		8.25	3.99	.55	12.79	15.70	
	1700	2 #5 bars	▼	240	.133	▼	8.45	4.07	.56	13.08	16.10	
210	0010	**CONCRETE BLOCK, PARTITIONS**, scaffolding not included R04220-200										210
	0100	Acoustical slotted block										
	0200	4" thick, type A-1	D-8	315	.127	S.F.	2.47	4.05		6.52	8.85	
	0210	8" thick		275	.145		3.69	4.64		8.33	11.10	
	0250	8" thick, type Q		275	.145		5.05	4.64		9.69	12.60	
	0260	4" thick, type RSC		315	.127		3.65	4.05		7.70	10.15	
	0270	6" thick		295	.136		3.65	4.33		7.98	10.60	
	0280	8" thick		275	.145		3.65	4.64		8.29	11.05	
	0290	12" thick		250	.160		3.65	5.10		8.75	11.80	
	0300	8" thick, type RSR		275	.145		3.65	4.64		8.29	11.05	
	0400	8" thick, type RSC/RF		275	.145		4.51	4.64		9.15	12	
	0410	10" thick		260	.154		4.91	4.91		9.82	12.90	
	0420	12" thick		250	.160		5.35	5.10		10.45	13.70	
	0430	12" thick, type RSC/RF-4		250	.160		6.35	5.10		11.45	14.75	
	0500	NRC .60 type R, 8" thick		265	.151		3.90	4.82		8.72	11.65	
	0600	NRC .65 type RR, 8" thick		265	.151		11.15	4.82		15.97	19.60	
	0700	NRC .65 type 4R-RF, 8" thick		265	.151		5.30	4.82		10.12	13.15	
	0710	NRC .70 type R, 12" thick	▼	245	.163	▼	5.80	5.20		11	14.35	
	1000	Lightweight block, tooled joints, 2 sides, hollow										
	1100	Not reinforced, 8" x 16" x 4" thick	D-8	440	.091	S.F.	1.12	2.90		4.02	5.65	
	1150	6" thick		410	.098		1.54	3.11		4.65	6.45	
	1200	8" thick		385	.104		1.89	3.32		5.21	7.15	
	1250	10" thick	▼	370	.108		2.49	3.45		5.94	8	
	1300	12" thick	D-9	350	.137		2.92	4.26		7.18	9.70	
	2000	Not reinforced, 8" x 24" x 4" thick, hollow		460	.104		.80	3.24		4.04	5.80	
	2100	6" thick		440	.109		1.08	3.39		4.47	6.35	
	2150	8" thick		415	.116		1.35	3.59		4.94	7	
	2200	10" thick		385	.125		1.76	3.88		5.64	7.85	
	2250	12" thick	▼	365	.132		2.06	4.09		6.15	8.50	
	2800	Solid, not reinforced, 8" x 16" x 2" thick	D-8	440	.091		.92	2.90		3.82	5.45	
	2900	4" thick		420	.095		1.22	3.04		4.26	6	
	2950	6" thick		390	.103		1.88	3.27		5.15	7.05	
	3000	8" thick		365	.110		2.57	3.50		6.07	8.20	
	3050	10" thick	▼	350	.114		2.91	3.65		6.56	8.75	
	3100	12" thick	D-9	330	.145	▼	4.22	4.52		8.74	11.55	
	4000	Regular block, tooled joints, 2 sides, hollow										
	4100	Not reinforced, 8" x 16" x 4" thick	D-8	430	.093	S.F.	.94	2.97		3.91	5.55	
	4150	6" thick		400	.100		1.40	3.19		4.59	6.40	
	4200	8" thick **CN**		375	.107		1.52	3.40		4.92	6.85	
	4250	10" thick	▼	360	.111		2.15	3.55		5.70	7.75	
	4300	12" thick	D-9	340	.141		2.22	4.39		6.61	9.15	
	4500	Reinforced alternate courses, 8" x 16" x 4" thick	D-8	425	.094		1.04	3		4.04	5.75	
	4550	6" thick		395	.101		1.51	3.23		4.74	6.60	
	4600	8" thick		370	.108		1.63	3.45		5.08	7.05	
	4650	10" thick	▼	355	.113		2.28	3.60		5.88	8	
	4700	12" thick	D-9	335	.143	▼	2.34	4.45		6.79	9.35	

4

MASONRY

04810 | Unit Masonry Assemblies

MASONRY 4

			CREW	DAILY OUTPUT	LABOR-HOURS	UNIT	MAT.	LABOR	EQUIP.	TOTAL	TOTAL INCL O&P	
210	4900	Solid, not reinforced, 2" thick R04220 -200	D-8	435	.092	S.F.	.80	2.93		3.73	5.35	210
	5000	3" thick		430	.093		.92	2.97		3.89	5.55	
	5050	4" thick		415	.096		1.30	3.08		4.38	6.10	
	5100	6" thick		385	.104		1.54	3.32		4.86	6.75	
	5150	8" thick	▼	360	.111		2.31	3.55		5.86	7.95	
	5200	12" thick	D-9	325	.148		3.47	4.59		8.06	10.80	
	5500	Solid, reinforced alternate courses, 4" thick	D-8	420	.095		1.37	3.04		4.41	6.15	
	5550	6" thick		380	.105		1.62	3.36		4.98	6.90	
	5600	8" thick	▼	355	.113		2.39	3.60		5.99	8.15	
	5650	12" thick	D-9	320	.150	▼	3.07	4.66		7.73	10.50	
225	0010	**CONCRETE BRICK** C55, grade N, type I										225
	0100	Regular, 4 x 2-1/4 x 8	D-8	220	.182	Ea.	.34	5.80		6.14	9.20	
	0125	Rusticated, 4 x 2-1/4 x 8		220	.182		.38	5.80		6.18	9.25	
	0150	Frog, 4 x 2-1/4 x 8		220	.182		.37	5.80		6.17	9.25	
	0200	Double, 4 x 4-7/8 x 8	▼	180	.222	▼	.60	7.10		7.70	11.45	
240	0010	**CONCRETE SCREEN BLOCK**										240
	0200	8" x 16", 4" thick	D-8	180	.222	S.F.	1.71	7.10		8.81	12.70	
	0300	8" thick		270	.148		2.53	4.73		7.26	10	
	0350	12" x 12", 4" thick		300	.133		2.05	4.25		6.30	8.75	
	0500	8" thick	▼	330	.121	▼	2.58	3.87		6.45	8.75	
250	0010	**COPING** Stock units										250
	0050	Precast concrete, 10" wide, 4" tapers to 3-1/2", 8" wall	D-1	75	.213	L.F.	15	6.65		21.65	26.50	
	0100	12" wide, 3-1/2" tapers to 3", 10" wall		70	.229		10.50	7.10		17.60	22.50	
	0110	14" wide, 4" tapers to 3-1/2", 12" wall		65	.246		19.85	7.65		27.50	33.50	
	0150	16" wide, 4" tapers to 3-1/2", 14" wall	▼	60	.267	▼	14.50	8.30		22.80	28.50	
	0250	Precast concrete corners		40	.400	Ea.	24	12.45		36.45	45.50	
	0300	Limestone for 12" wall, 4" thick		90	.178	L.F.	13.50	5.55		19.05	23.50	
	0350	6" thick		80	.200		15.75	6.20		21.95	27	
	0500	Marble, to 4" thick, no wash, 9" wide		90	.178		19.25	5.55		24.80	29.50	
	0550	12" wide		80	.200		29	6.20		35.20	41.50	
	0700	Terra cotta, 9" wide		90	.178		4.75	5.55		10.30	13.65	
	0750	12" wide		80	.200		7.80	6.20		14	18.05	
	0800	Aluminum, for 12" wall	▼	80	.200	▼	11.25	6.20		17.45	22	
300	0010	**FACING PANELS** Stone aggregate mounted on plywood,										300
	0020	See division 07440-200										
325	0010	**GLASS BLOCK**										325
	0100	Plain, 4" thick, under 1,000 S.F., 6" x 6"	D-8	115	.348	S.F.	15.25	11.10		26.35	33.50	
	0150	8" x 8"		160	.250		10.15	8		18.15	23.50	
	0160	end block		160	.250		27.50	8		35.50	42	
	0170	90 deg corner		160	.250		28	8		36	42.50	
	0180	45 deg corner		160	.250		13.85	8		21.85	27.50	
	0200	12" x 12"		175	.229		11.90	7.30		19.20	24	
	0210	4" x 8"		160	.250		7.75	8		15.75	20.50	
	0220	6" x 8"		160	.250		9.40	8		17.40	22.50	
	0300	1,000 to 5,000 S.F., 6" x 6"		135	.296		14.95	9.45		24.40	31	
	0350	8" x 8"		190	.211		9.95	6.70		16.65	21	
	0400	12" x 12"		215	.186		11.65	5.95		17.60	22	
	0410	4" x 8"		215	.186		3.66	5.95		9.61	13.10	
	0420	6" x 8"		215	.186		4.10	5.95		10.05	13.55	
	0500	Over 5,000 S.F., 6" x 6"		145	.276		14.50	8.80		23.30	29.50	
	0550	8" x 8"		215	.186		9.65	5.95		15.60	19.70	
	0600	12" x 12"		240	.167		11.30	5.30		16.60	20.50	
	0610	4" x 8"	▼	240	.167	▼	3.60	5.30		8.90	12.05	

4

MASONRY

04810 | Unit Masonry Assemblies

		CREW	DAILY OUTPUT	LABOR-HOURS	UNIT	MAT.	LABOR	EQUIP.	TOTAL	TOTAL INCL O&P	
325	0620	6" x 8"	D-8	240	.167	S.F.	3.77	5.30		9.07	12.25	325
	0700	For solar reflective blocks, add					100%					
	1000	Thinline, plain, 3-1/8" thick, under 1,000 S.F., 6" x 6"	D-8	115	.348	S.F.	13.05	11.10		24.15	31.50	
	1050	8" x 8"		160	.250		7.70	8		15.70	20.50	
	1200	Over 5,000 S.F., 6" x 6"		145	.276		11.65	8.80		20.45	26	
	1250	8" x 8"		215	.186		6.85	5.95		12.80	16.60	
	1400	For cleaning block after installation (both sides), add		1,000	.040		.10	1.28		1.38	2.05	
	4000	Accessories										
	4100	Anchors, 20 ga. galv., 1-3/4" wide x 24" long				Ea.	2.25			2.25	2.48	
	4200	Emulsion asphalt				Gal.	7.30			7.30	8.05	
	4300	Expansion joint, fiberglass				L.F.	.55			.55	.61	
	4400	Steel mesh, double galvanized				"	.25			.25	.28	
350	0010	**GLAZED CONCRETE BLOCK** C744										350
	0100	Single face, 8" x 16" units, 2" thick	D-8	360	.111	S.F.	5.90	3.55		9.45	11.90	
	0200	4" thick		345	.116		6	3.70		9.70	12.25	
	0250	6" thick		330	.121		6.35	3.87		10.22	12.90	
	0300	8" thick		310	.129		7.25	4.12		11.37	14.20	
	0350	10" thick		295	.136		8	4.33		12.33	15.40	
	0400	12" thick	D-9	280	.171		8.60	5.35		13.95	17.55	
	0700	Double face, 8" x 16" units, 4" thick	D-8	340	.118		9.70	3.75		13.45	16.35	
	0750	6" thick		320	.125		10.35	3.99		14.34	17.45	
	0800	8" thick		300	.133		10.70	4.25		14.95	18.30	
	1000	Jambs, bullnose or square, single face, 8" x 16", 2" thick		315	.127	Ea.	10.45	4.05		14.50	17.65	
	1050	4" thick		285	.140	"	11.65	4.48		16.13	19.65	
	1200	Caps, bullnose or square, 8" x 16", 2" thick		420	.095	L.F.	10.10	3.04		13.14	15.80	
	1250	4" thick		380	.105		13.10	3.36		16.46	19.50	
	1500	Cove base, 8" x 16", 2" thick		315	.127		5.95	4.05		10	12.70	
	1550	4" thick		285	.140		5.95	4.48		10.43	13.35	
	1600	6" thick		265	.151		6.35	4.82		11.17	14.35	
	1650	8" thick		245	.163		6.95	5.20		12.15	15.60	
500	0010	**TERRA COTTA** Coping, split type, not glazed, 9" wide	D-1	90	.178	L.F.	6.15	5.55		11.70	15.20	500
	0100	13" wide		80	.200		9.80	6.20		16	20.50	
	0200	Split type, glazed, 9" wide		90	.178		11.85	5.55		17.40	21.50	
	0250	13" wide		80	.200		15.80	6.20		22	27	
	0500	Partition or back-up blocks, scored, in C.L. lots										
	0700	Non-load bearing 12" x 12", 3" thick, special order	D-8	550	.073	S.F.	16.45	2.32		18.77	21.50	
	0750	4" thick, standard		500	.080		4.54	2.55		7.09	8.90	
	0800	6" thick		450	.089		5.40	2.84		8.24	10.20	
	0850	8" thick		400	.100		6.60	3.19		9.79	12.10	
	1000	Load bearing, 12" x 12", 4" thick, in walls		500	.080		4.87	2.55		7.42	9.25	
	1050	In floors		750	.053		4.87	1.70		6.57	7.95	
	1200	6" thick, in walls		450	.089		5.40	2.84		8.24	10.20	
	1250	In floors		675	.059		5.40	1.89		7.29	8.80	
	1400	8" thick, in walls		400	.100		6.60	3.19		9.79	12.10	
	1450	In floors		575	.070		6.60	2.22		8.82	10.65	
	1600	10" thick, in walls, special order		350	.114		15.05	3.65		18.70	22	
	1650	In floors, special order		500	.080		15.05	2.55		17.60	20.50	
	1800	12" thick, in walls, special order		300	.133		15.85	4.25		20.10	24	
	1850	In floors, special order		450	.089		15.85	2.84		18.69	22	
	2000	For reinforcing with steel rods, add to above					15%	5%				
	2100	For smooth tile instead of scored, add					2.20			2.20	2.42	
	2200	For L.C.L. quantities, add					10%	10%				
510	0010	**TERRA COTTA TILE** On walls, dry set, 1/2" thick										510
	0100	Square, hexagonal or lattice shapes, unglazed	1 Tilf	135	.059	S.F.	4.35	1.94		6.29	7.65	
	0300	Glazed, plain colors		130	.062		6	2.01		8.01	9.55	
	0400	Intense colors		125	.064		7	2.09		9.09	10.80	

Important: See the Reference Section for critical supporting data - Reference Nos., Crews, & City Cost Indexes

04810	Unit Masonry Assemblies	CREW	DAILY OUTPUT	LABOR-HOURS	UNIT	MAT.	LABOR	EQUIP.	TOTAL	TOTAL INCL O&P	
540	**0010** WALLS Cavity, brick and CMU incl joint reinf and z-ties										**540**
0200	4" face brick, 4" block	D-8	165	.242	S.F.	3.95	7.75		11.70	16.15	
0400	6" block		145	.276		4.37	8.80		13.17	18.20	
0600	8" block	▼	125	.320	▼	4.44	10.20		14.64	20.50	
650	**0010** WALLS Building brick, including mortar										**650**
0140	4" thick, facing, 4" x 2-2/3" x 8" R04210 -120	D-8	1.45	27.586	M	345	880		1,225	1,725	
0150	4" thick, as back-up, 6.75 bricks per S.F.		1.60	25		345	800		1,145	1,600	
0204	8" thick, 13.50 bricks per S.F. R04210 -180		1.80	22.222		350	710		1,060	1,450	
0250	12" thick, 20.25 bricks per S.F.		1.90	21.053		350	670		1,020	1,400	
0304	16" thick, 27.00 bricks per S.F.		2	20		355	640		995	1,350	
0500	Reinforced, 4" wall, 4" x 2-2/3" x 8"		1.40	28.571		345	910		1,255	1,775	
0550	8" thick, 13.50 bricks per S.F.		1.75	22.857		410	730		1,140	1,550	
0600	12" thick, 20.25 bricks per S.F.		1.85	21.622		410	690		1,100	1,500	
0650	16" thick, 27.00 bricks per S.F.		1.95	20.513		415	655		1,070	1,450	
0660	4" thick, select common, face, 4" x 2-2/3" x 8"	▼	1.45	27.586	▼	400	880		1,280	1,800	
0790	Alternate method of figuring by square foot										
0800	4" wall, face, 4" x 2-2/3" x 8"	D-8	215	.186	S.F.	2.83	5.95		8.78	12.15	
0850	4" thick, as back up, 6.75 bricks per S.F.		240	.167		2.31	5.30		7.61	10.65	
0900	8" thick wall, 13.50 brick per S.F.		135	.296		4.73	9.45		14.18	19.60	
1000	12" thick wall, 20.25 bricks per S.F.		95	.421		7.10	13.45		20.55	28.50	
1050	16" thick wall, 27.00 bricks per S.F.		75	.533		9.60	17		26.60	36.50	
1200	Reinforced, 4" x 2-2/3" x 8", 4" wall		205	.195		2.31	6.25		8.56	12.05	
1250	8" thick wall, 13.50 brick per S.F.		130	.308		4.73	9.80		14.53	20	
1300	12" thick wall, 20.25 brick per S.F.		90	.444		7.10	14.20		21.30	29.50	
1350	16" thick wall, 27.00 bricks per S.F.	▼	70	.571	▼	9.60	18.25		27.85	38.50	
670	**0010** STEPS With select common brick	D-1	.30	53.333	M	355	1,650		2,005	2,925	**670**
700	**0010** STRUCTURAL BRICK C652, Grade SW, incl mortar, scaffolding not incl										**700**
0100	Standard unit, 4-5/8" x 2-3/4" x 9-5/8"	D-8	245	.163	S.F.	4.19	5.20		9.39	12.55	
0120	Bond beam		225	.178		8.05	5.65		13.70	17.50	
0140	V cut bond beam		225	.178		8.30	5.65		13.95	17.75	
0160	Stretcher quoin, 5-5/8" x 2-3/4" x 9-5/8"		245	.163		8	5.20		13.20	16.75	
0180	Corner quoin		245	.163		8.95	5.20		14.15	17.80	
0200	Corner, 45 deg, 4-5/8" x 2-3/4" x 10-7/16"	▼	235	.170	▼	9.30	5.45		14.75	18.55	
750	**0010** STRUCTURAL FACING TILE Scaffolding not incl, standard colors										**750**
0020	6T series, 5-1/3" x 12", 2.3 pieces per S.F., glazed 1 side, 2" thick	D-8	225	.178	S.F.	6	5.65		11.65	15.25	
0100	4" thick		220	.182		7.05	5.80		12.85	16.60	
0150	Glazed 2 sides		195	.205		11.85	6.55		18.40	23	
0250	6" thick		210	.190		10.90	6.10		17	21	
0300	Glazed 2 sides		185	.216		13.60	6.90		20.50	25.50	
0400	8" thick		180	.222	▼	13.65	7.10		20.75	26	
0500	Special shapes, group 1		400	.100	Ea.	5.55	3.19		8.74	11	
0550	Group 2		375	.107		8.45	3.40		11.85	14.50	
0600	Group 3		350	.114		10.55	3.65		14.20	17.15	
0650	Group 4		325	.123		18	3.93		21.93	26	
0700	Group 5		300	.133		26	4.25		30.25	35	
0750	Group 6		275	.145	▼	36	4.64		40.64	46.50	
1000	Fire rated, 4" thick, 1 hr. rating		210	.190	S.F.	11.95	6.10		18.05	22.50	
1300	Acoustic, 4" thick	▼	210	.190	"	13	6.10		19.10	23.50	
1400	For designer colors, add					25%					
2000	8W series, 8" x 16", 1.125 pieces per S.F.										
2050	2" thick, glazed 1 side	D-8	360	.111	S.F.	6.65	3.55		10.20	12.70	
2100	4" thick, glazed 1 side		345	.116		7.60	3.70		11.30	14	
2150	Glazed 2 sides	▼	325	.123	▼	9.65	3.93		13.58	16.60	

MASONRY 4

4

MASONRY

04810	Unit Masonry Assemblies	CREW	DAILY OUTPUT	LABOR-HOURS	UNIT	MAT.	LABOR	EQUIP.	TOTAL	TOTAL INCL O&P	
750 2200	6" thick, glazed 1 side	D-8	330	.121	S.F.	10.70	3.87		14.57	17.65	**750**
2250	8" thick, glazed 1 side	↓	310	.129	↓	12.70	4.12		16.82	20	
2500	Special shapes, group 1		300	.133	Ea.	9.15	4.25		13.40	16.60	
2550	Group 2		280	.143		11.95	4.56		16.51	20	
2600	Group 3		260	.154		12.55	4.91		17.46	21.50	
2650	Group 4		250	.160		27.50	5.10		32.60	38	
2700	Group 5		240	.167		56	5.30		61.30	69.50	
2750	Group 6		230	.174	↓	77.50	5.55		83.05	93.50	
3000	4" thick, glazed 1 side	↓	345	.116	S.F.	8.30	3.70		12	14.75	
3100	Acoustic, 4" thick		345	.116	"	10.75	3.70		14.45	17.50	
3120	4W series, 8" x 8", 2.25 pieces per S.F.										
3125	2" thick, glazed 1 side	D-8	360	.111	S.F.	7.45	3.55		11	13.60	
3130	4" thick, glazed 1 side	↓	345	.116		7.45	3.70		11.15	13.85	
3135	Glazed 2 sides		325	.123		10.70	3.93		14.63	17.80	
3140	6" thick, glazed 1 side		330	.121		10	3.87		13.87	16.90	
3150	8" thick, glazed 1 side		310	.129	↓	14.45	4.12		18.57	22	
3155	Special shapes, group I		300	.133	Ea.	5.30	4.25		9.55	12.30	
3160	Group II	↓	280	.143	"	6.40	4.56		10.96	14	
3200	For designer colors, add					25%					
3300	For epoxy mortar joints, add				S.F.	1.25			1.25	1.38	

04840	Prefabricated Masonry Panels	CREW	DAILY OUTPUT	LABOR-HOURS	UNIT	MAT.	LABOR	EQUIP.	TOTAL	TOTAL INCL O&P	
900 0010	**WALL PANELS** Prefabricated, 4" thick, minimum	C-11	775	.093	S.F.	7	3.46	1.86	12.32	15.75	**900**
0100	Maximum	"	500	.144		9	5.35	2.89	17.24	22.50	
0200	4" brick & 2" concrete back-up, add					50%					
0300	4" brick & 1" urethane & 3" concrete back-up, add				↓	70%					

04850	Stone Assemblies	CREW	DAILY OUTPUT	LABOR-HOURS	UNIT	MAT.	LABOR	EQUIP.	TOTAL	TOTAL INCL O&P	
050 0011	**ASHLAR VENEER** 4" + or - thk, random or random rectangular										**050**
0150	Sawn face, split joints, low priced stone	D-8	140	.286	S.F.	8.40	9.10		17.50	23	
0200	Medium priced stone		130	.308		10.90	9.80		20.70	27	
0300	High priced stone		120	.333		13.90	10.65		24.55	31.50	
0600	Seam face, split joints, medium price stone		125	.320		10.45	10.20		20.65	27	
0700	High price stone		120	.333		12.75	10.65		23.40	30.50	
1000	Split or rock face, split joints, medium price stone		125	.320		10.50	10.20		20.70	27	
1100	High price stone	↓	120	.333	↓	12.85	10.65		23.50	30.50	
100 0010	**BLUESTONE** Cut to size										**100**
0500	Sills, natural cleft, 10" wide to 6' long, 1-1/2" thick	D-11	70	.343	L.F.	11.20	11.35		22.55	29.50	
0550	2" thick	"	63	.381		12.75	12.60		25.35	33.50	
1000	Stair treads, natural cleft, 12" wide, 6' long, 1-1/2" thick	D-10	115	.348		16.55	11.30	4.40	32.25	40	
1050	2" thick		105	.381		17.40	12.35	4.82	34.57	43	
1100	Smooth finish, 1-1/2" thick		115	.348		20	11.30	4.40	35.70	44	
1300	Thermal finish		115	.348		25	11.30	4.40	40.70	49.50	
1350	2" thick	↓	105	.381	↓	28	12.35	4.82	45.17	55	
300 0010	**GRANITE** Cut to size										**300**
0050	Veneer, polished face, 3/4" to 1-1/2" thick										
0150	Low price, gray, light gray, etc. **CN**	D-10	130	.308	S.F.	20.50	10	3.89	34.39	42	
0220	High price, red, black, etc. **CN**	"	130	.308	"	34	10	3.89	47.89	57	
0300	1-1/2" to 2-1/2" thick, veneer										
0350	Low price, gray, light gray, etc.	D-10	130	.308	S.F.	22.50	10	3.89	36.39	44	
0550	High price, red, black, etc.	"	130	.308	"	39	10	3.89	52.89	62	
0700	2-1/2" to 4" thick, veneer										
0750	Low price, gray, light gray, etc.	D-10	110	.364	S.F.	28	11.80	4.60	44.40	54	
0950	High price, red, black, etc.	"	110	.364		45	11.80	4.60	61.40	72.50	
1000	For bush hammered finish, deduct					5%					
1050	Coarse rubbed finish, deduct	↓			↓	10%					

Important: See the Reference Section for critical supporting data - Reference Nos., Crews, & City Cost Indexes

Use of our products can help you obtain credits toward LEED™ Green Building Certification. The LEED™ system was created to evaluate the effects of building design and construction on the environment.

The LEED™ rating system has 6 credit categories from which points can be earned toward certification. Clayton concrete and block products relate specifically to the following LEED™ credits:

Concrete

Produced throughout the state of New Jers[ey] Clayton concrete mixes contain flyash, a po[st] industrial, recycled material contributing [to] MRc4. Since the use of concrete is regional, proje[ct] teams using this product will likely be locat[ed] within 500 miles of where the concrete is man[u]factured and extracted, thus contributing [to] MRc5. Using concrete can also help earn SSc7[.1] because concrete is a high-albedo material.

Take The Leed.™

- **SSc7.1: Landscape & Exterior Design to Reduce Heat Islands (Sustainable Sites Credits)**

- **EAc1: Optimize Energy Performance (Energy & Atmosphere Credits)**

- **MRc4: Recycled Content (Materials & Resources Credits)**

- **MRc5: Local/Regional Materials (Materials & Resources Credits)**

- **EQc4.2: Low-Emitting Materials (Indoor Environmental Quality Credits)**

- **Dc1: Innovation in Design (Innovation & Design Process Credits)**

Architectural Block: Polished, Glazed

Both of these products are manufactured at the Clayton plant in Yardville, NJ and the materials are extracted in Northfield and Millington, NJ and Lehigh Valley, PA.

These products can help earn both points in MRc5 if the project is located within 500 miles of the manufacturing and extraction locations, as well as EQc4.2 by making use of paints and coatings for block unnecessary.

The reflective properties of these products can contribute towards improving energy performance as part of EAc1. Used in conjunction with other design elements, lighting design could be altered to lower the use of light fixtures, thereby lowering the heat generat[ed] by lighting, enabling HVAC systems to be downsized, improving energy performance.

Architectural Block: Split Face, Smooth Colored, Ground

Contributing towards earning MRc5, this block is manufactu[red] at: North Arlington, Edison and Yardville, NJ. The materi[als] are extracted in Jackson and Millington, NJ and Lehi[gh] Valley, PA.

This block product can help earn both points in MRc5 if [the] project is located within 500 miles of the manufacturing a[nd] extraction locations.

Architectural Block: GlasStone™

One of Clayton's newest products, GlasStone™ is unique because of its recycled glass content. This product can help earn MRc4 due to its post-consumer recycled content and post-industrial recycled content. GlasStone™ can also contribute towards earning both points in MRc5 if the project is located within 500 miles of the manufacturing and extraction locations, and EQc4.2 by making use of paints and coatings for block unnecessary. GlasStone's™ reflective qualities could help improve energy performance as part of EAc1 By using this block in conjunction with other design elements, lighting design could be altered to reduce the use of light fixtures and the heat generated from them, enabling HVAC systems to be downsized and improving energy performance.

GlasStone™ is manufactured at the Clayton Plant in Yardville, NJ.

Grey Block

A high post-consumer recycled content, consisting of recovered block and concrete, makes Clayton's Grey Block a unique product, contributing to MRc4.

In New Jersey, Clayton's Grey Block is manufactured at these plants: Saddle Brook, North Arlington, Bayonne, Edison, Belmar/Wall, Lakewood and Barnegat. The materials are extracted in Edison, Belmar, Lakewood and Jackson, NJ, as well as in Lehigh Valley, PA.

Using Grey Block can help earn both points in MRc5 if the project is located wit[hin] 500 miles of the manufacturing and extraction facilities.

		04850	Stone Assemblies	CREW	DAILY OUTPUT	LABOR-HOURS	UNIT	2005 BARE COSTS				TOTAL INCL O&P	
								MAT.	LABOR	EQUIP.	TOTAL		
300		1100	Honed finish, deduct				S.F.	5%					**300**
		1150	Thermal finish, deduct				↓	18%					
		2450	For radius under 5', add				L.F.	100%					
		2500	Steps, copings, etc., finished on more than one surface										
		2550	Minimum	D-10	50	.800	C.F.	76.50	26	10.10	112.60	135	
		2600	Maximum	"	50	.800	"	122	26	10.10	158.10	186	
		2800	Pavers, 4" x 4" x 4" blocks, split face and joints										
		2850	Minimum	D-11	80	.300	S.F.	11.20	9.95		21.15	27.50	
		2900	Maximum	"	80	.300	"	22.50	9.95		32.45	39.50	
		3500	Curbing, city street type, See Division 02770-300										
		4000	Soffits, 2" thick, minimum	D-13	35	1.371	S.F.	31.50	45	14.45	90.95	119	
		4100	Maximum		35	1.371		66.50	45	14.45	125.95	157	
		4200	4" thick, minimum		35	1.371		46	45	14.45	105.45	135	
		4300	Maximum	↓	35	1.371	↓	85.50	45	14.45	144.95	179	
350		0011	**LIGHTWEIGHT NATURAL STONE** Lava type										**350**
		0100	Veneer, rubble face, sawed back, irregular shapes	D-10	130	.308	S.F.	5.50	10	3.89	19.39	25.50	
		0200	Sawed face and back, irregular shapes	"	130	.308	"	5.50	10	3.89	19.39	25.50	
400		0010	**LIMESTONE,** Cut to size										**400**
		0020	Veneer facing panels										
		0500	Texture finish, light stick, 4-1/2" thick, 5'x 12'	D-4	300	.107	S.F.	31	3.26	.45	34.71	39.50	
		0750	5" thick, 5' x 14' panels	D-10	275	.145		33.50	4.72	1.84	40.06	46	
		1000	Sugarcube finish, 2" Thick, 3' x 5' panels		275	.145		18.35	4.72	1.84	24.91	29	
		1050	3" Thick, 4' x 9' panels		275	.145		22.50	4.72	1.84	29.06	34	
		1200	4" Thick, 5' x 11' panels		275	.145		26.50	4.72	1.84	33.06	38	
		1400	Sugarcube, textured finish, 4-1/2" thick, 5' x 12'		275	.145		29	4.72	1.84	35.56	41	
		1450	5" thick, 5' x 14' panels		275	.145	↓	34	4.72	1.84	40.56	46.50	
		2000	Coping, sugarcube finish, top & 2 sides		30	1.333	C.F.	37	43.50	16.85	97.35	125	
		2100	Sills, lintels, jambs, trim, stops, sugarcube finish, average		20	2		55	65	25.50	145.50	187	
		2150	Detailed		20	2	↓	55	65	25.50	145.50	187	
		2300	Steps, extra hard, 14" wide, 6" rise	↓	50	.800	L.F.	19.50	26	10.10	55.60	72	
		3000	Quoins, plain finish, 6"x12"x12"	D-12	25	1.280	Ea.	50	40.50		90.50	117	
		3050	6"x16"x24"	"	25	1.280	"	66.50	40.50		107	135	
500		0011	**MARBLE,** ashlar, split face, 4" + or - thick, random										**500**
		0040	lengths 1' to 4' & heights 2" to 7-1/2", average	D-8	175	.229	S.F.	14.55	7.30		21.85	27	
		0100	Base, polished, 3/4" or 7/8" thick, polished, 6" high	D-10	65	.615	L.F.	13.25	19.95	7.80	41	53.50	
		0300	Carvings or bas relief, from templates, average		80	.500	S.F.	118	16.20	6.35	140.55	161	
		0350	Maximum	↓	80	.500	"	276	16.20	6.35	298.55	335	
		0600	Columns, cornices, mouldings, etc.										
		0650	Hand or special machine cut, average	D-10	35	1.143	C.F.	118	37	14.45	169.45	202	
		0700	Maximum	"	35	1.143	"	253	37	14.45	304.45	350	
		1000	Facing, polished finish, cut to size, 3/4" to 7/8" thick										
		1050	Average	D-10	130	.308	S.F.	19.45	10	3.89	33.34	41	
		1100	Maximum		130	.308		45	10	3.89	58.89	69	
		1300	1-1/4" thick, average		125	.320		28	10.40	4.05	42.45	51.50	
		1350	Maximum		125	.320		56.50	10.40	4.05	70.95	82.50	
		1500	2" thick, average		120	.333		32.50	10.80	4.22	47.52	57	
		1550	Maximum	↓	120	.333	↓	56.50	10.80	4.22	71.52	83	
		2200	Window sills, 6" x 3/4" thick	D-1	85	.188	L.F.	7.30	5.85		13.15	16.95	
		2500	Flooring, polished tiles, 12" x 12" x 3/8" thick										
		2510	Thin set, average	D-11	90	.267	S.F.	9	8.85		17.85	23.50	
		2600	Maximum		90	.267		32.50	8.85		41.35	49.50	
		2700	Mortar bed, average		65	.369		9.15	12.25		21.40	29	
		2740	Maximum	↓	65	.369		30	12.25		42.25	51.50	
		2780	Travertine, 3/8" thick, average	D-10	130	.308	↓	12.40	10	3.89	26.29	33	

MASONRY 4

4

MASONRY

			CREW	DAILY OUTPUT	LABOR-HOURS	UNIT	2005 BARE COSTS				TOTAL INCL O&P	
	04850	**Stone Assemblies**					MAT.	LABOR	EQUIP.	TOTAL		
500	2790	Maximum	D-10	130	.308	S.F.	30.50	10	3.89	44.39	53	**500**
	2800	Patio tile, non-slip, 1/2" thick, flame finish	D-11	75	.320	↓	12.65	10.60		23.25	30	
	2900	Shower or toilet partitions, 7/8" thick partitions				"						
	3050	3/4" or 1-1/4" thick stiles, polished 2 sides, average	D-11	75	.320	S.F.	37.50	10.60		48.10	57.50	
	3201	Soffits, add to above prices				"	20%	100%				
	3210	Stairs, risers, 7/8" thick x 6" high	D-10	115	.348	L.F.	12.75	11.30	4.40	28.45	36	
	3360	Treads, 12" wide x 1-1/4" thick	"	115	.348	"	18.55	11.30	4.40	34.25	42.50	
	3500	Thresholds, 3' long, 7/8" thick, 4" to 5" wide, plain	D-12	24	1.333	Ea.	14.05	42		56.05	79.50	
	3550	Beveled		24	1.333	"	16.30	42		58.30	82	
	3700	Window stools, polished, 7/8" thick, 5" wide	↓	85	.376	L.F.	13.10	11.90		25	32.50	
600	0011	**ROUGH STONE WALL**, Dry										**600**
	0100	Random fieldstone, under 18" thick	D-12	60	.533	C.F.	8.05	16.85		24.90	34.50	
	0150	Over 18" thick	"	63	.508	"	9.70	16.05		25.75	35	
700	0011	**SANDSTONE OR BROWNSTONE**										**700**
	0100	Sawed face veneer, 2-1/2" thick, to 2' x 4' panels	D-10	130	.308	S.F.	17.75	10	3.89	31.64	39	
	0150	4' thick, to 3'-6" x 8' panels		100	.400		17.75	13	5.05	35.80	45	
	0300	Split face, random sizes	↓	100	.400	↓	10.65	13	5.05	28.70	37	
	0350	Cut stone trim (limestone)										
	0360	Ribbon stone, 4" thick, 5' pieces	D-8	120	.333	Ea.	131	10.65		141.65	160	
	0370	Cove stone, 4" thick, 5' pieces		105	.381		131	12.15		143.15	163	
	0380	Cornice stone, 10" to 12" wide		90	.444		162	14.20		176.20	200	
	0390	Band stone, 4" thick, 5' pieces		145	.276		83.50	8.80		92.30	105	
	0410	Window and door trim, 3" to 4" wide		160	.250		71	8		79	90	
	0420	Key stone, 18" long	↓	60	.667	↓	74.50	21.50		96	115	
800	0010	**SLATE** Pennsylvania, blue gray to gray black; Vermont,										**800**
	0050	Unfading green, mottled green & purple, gray & purple										
	0100	Virginia, blue black										
	0200	Exterior paving, natural cleft, 1" thick										
	0500	24" x 24", Pennsylvania	D-12	120	.267	S.F.	9.70	8.40		18.10	23.50	
	0550	Vermont		120	.267		12.60	8.40		21	26.50	
	0600	Virginia	↓	120	.267	↓	14.20	8.40		22.60	28.50	
	1000	Interior flooring, natural cleft, 1/2" thick										
	1300	24" x 24" Pennsylvania	D-12	120	.267	S.F.	6.85	8.40		15.25	20.50	
	1350	Vermont		120	.267		10	8.40		18.40	24	
	1400	Virginia	↓	120	.267	↓	10.25	8.40		18.65	24	
	2000	Facing panels, 1-1/4" thick, to 4' x 4' panels										
	2100	Natural cleft finish, Pennsylvania	D-10	180	.222	S.F.	24.50	7.20	2.81	34.51	41	
	2110	Vermont		180	.222		21	7.20	2.81	31.01	37	
	2120	Virginia	↓	180	.222		23	7.20	2.81	33.01	39.50	
	2150	Sand rubbed finish, surface, add					2.15			2.15	2.37	
	2200	Honed finish, add					4.20			4.20	4.62	
	2500	Ribbon, natural cleft finish, 1" thick, to 9 S.F.	D-10	80	.500		9.10	16.20	6.35	31.65	41.50	
	2700	1-1/2" thick		78	.513		11.80	16.65	6.50	34.95	45.50	
	2850	2" thick	↓	76	.526	↓	14.20	17.05	6.65	37.90	49	
	3000	Roofing, see division 07310-800										
	3500	Stair treads, sand finish, 1" thick x 12" wide										
	3600	3 L.F. to 6 L.F.	D-10	120	.333	L.F.	16.30	10.80	4.22	31.32	39	
	3700	Ribbon, sand finish, 1" thick x 12" wide										
	3750	To 6 L.F.	D-10	120	.333	L.F.	10.75	10.80	4.22	25.77	33	
	4000	Stools or sills, sand finish, 1" thick, 6" wide	D-12	160	.200		7.90	6.30		14.20	18.25	
	4200	10" wide		90	.356		11.80	11.25		23.05	30	
	4400	2" thick, 6" wide		140	.229		12.60	7.20		19.80	25	
	4600	10" wide	↓	90	.356		18.90	11.25		30.15	38	
	4800	For lengths over 3', add				↓	25%					

Important: See the Reference Section for critical supporting data - Reference Nos., Crews, & City Cost Indexes

04800 | Masonry Assemblies

04850 | Stone Assemblies

			CREW	DAILY OUTPUT	LABOR-HOURS	UNIT	MAT.	LABOR	EQUIP.	TOTAL	TOTAL INCL O&P	
900	0010	**WINDOW SILL** Bluestone, thermal top, 10" wide, 1-1/2" thick	D-1	85	.188	S.F.	13.50	5.85		19.35	24	900
	0050	2" thick		75	.213	"	15.75	6.65		22.40	27.50	
	0100	Cut stone, 5" x 8" plain		48	.333	L.F.	10.20	10.35		20.55	27	
	0200	Face brick on edge, brick, 8" wide		80	.200		2.15	6.20		8.35	11.80	
	0400	Marble, 9" wide, 1" thick		85	.188		7.50	5.85		13.35	17.15	
	0900	Slate, colored, unfading, honed, 12" wide, 1" thick		85	.188		15.25	5.85		21.10	25.50	
	0950	2" thick	▼	70	.229	▼	21.50	7.10		28.60	34.50	

04880 | Masonry Fireplaces

			CREW	DAILY OUTPUT	LABOR-HOURS	UNIT	MAT.	LABOR	EQUIP.	TOTAL	TOTAL INCL O&P	
600	0010	**FIREPLACE** For prefabricated fireplace, see div. 10305-100										600
	0100	Brick fireplace, not incl. foundations or chimneys										
	0110	30" x 29" opening, incl. chamber, plain brickwork	D-1	.40	40	Ea.	380	1,250		1,630	2,325	
	0200	Fireplace box only (110 brick)	"	2	8	"	125	249		374	520	
	0300	For elaborate brickwork and details, add					35%	35%				
	0400	For hearth, brick & stone, add	D-1	2	8	Ea.	140	249		389	535	
	0410	For steel angle, damper, cleanouts, add		4	4		98	124		222	297	
	0600	Plain brickwork, incl. metal circulator		.50	32	▼	725	995		1,720	2,325	
	0800	Face brick only, standard size, 8" x 2-2/3" x 4"	▼	.30	53.333	M	380	1,650		2,030	2,950	

04900 | Masonry Restoration and Cleaning

04910 | Unit Masonry Restoration

			CREW	DAILY OUTPUT	LABOR-HOURS	UNIT	MAT.	LABOR	EQUIP.	TOTAL	TOTAL INCL O&P	
600	0010	**NEEDLE BEAM MASONRY** Incl. wood shoring 10' x 10' opening										600
	0400	Block, concrete, 8" thick	B-9	7.10	5.634	Ea.	37	153	19.95	209.95	300	
	0420	12" thick		6.70	5.970		43.50	162	21	226.50	325	
	0800	Brick, 4" thick with 8" backup block		5.70	7.018		43.50	190	25	258.50	370	
	1000	Brick, solid, 8" thick		6.20	6.452		37	175	23	235	340	
	1040	12" thick		4.90	8.163		43.50	221	29	293.50	425	
	1080	16" thick	▼	4.50	8.889		57.50	241	31.50	330	475	
	2000	Add for additional floors of shoring	B-1	6	4	▼	37	109		146	211	
720	0010	**POINTING MASONRY**										720
	0300	Cut and repoint brick, hard mortar, running bond	1 Bric	80	.100	S.F.	.27	3.53		3.80	5.65	
	0320	Common bond		77	.104		.27	3.66		3.93	5.90	
	0360	Flemish bond		70	.114		.28	4.03		4.31	6.45	
	0400	English bond		65	.123		.28	4.34		4.62	6.90	
	0600	Soft old mortar, running bond		100	.080		.27	2.82		3.09	4.60	
	0620	Common bond		96	.083		.27	2.94		3.21	4.76	
	0640	Flemish bond		90	.089		.28	3.13		3.41	5.10	
	0680	English bond		82	.098	▼	.28	3.44		3.72	5.55	
	0700	Stonework, hard mortar		140	.057	L.F.	.35	2.01		2.36	3.46	
	0720	Soft old mortar		160	.050	"	.35	1.76		2.11	3.08	
	1000	Repoint, mask and grout method, running bond		95	.084	S.F.	.35	2.97		3.32	4.91	
	1020	Common bond		90	.089		.35	3.13		3.48	5.15	
	1040	Flemish bond		86	.093		.35	3.28		3.63	5.40	
	1060	English bond		77	.104		.35	3.66		4.01	6	
	2000	Scrub coat, sand grout on walls, minimum		120	.067		2.82	2.35		5.17	6.70	
	2020	Maximum	▼	98	.082	▼	2.03	2.88		4.91	6.60	

MASONRY 4

		04910	**Unit Masonry Restoration**	CREW	DAILY OUTPUT	LABOR-HOURS	UNIT	2005 BARE COSTS				TOTAL INCL O&P	
								MAT.	LABOR	EQUIP.	TOTAL		
750	0010		**SAWING**										750
	0050		Brick or block by hand, per inch depth	D-5	300	.027	L.F.		.94		.94	1.43	
		04930	**Unit Masonry Cleaning**										
220	0010		**MASONRY CLEANING**										220
	0200		Chemical cleaning, new construction, brush and wash, minimum	D-1	1,000	.016	S.F.	.04	.50		.54	.80	
	0220		Average		800	.020		.05	.62		.67	1.01	
	0240		Maximum		600	.027		.07	.83		.90	1.34	
	0260		Light restoration, minimum		800	.020		.07	.62		.69	1.02	
	0270		Average		400	.040		.10	1.24		1.34	2	
	0280		Maximum		330	.048		.13	1.51		1.64	2.45	
	0300		Heavy restoration, minimum		600	.027		.07	.83		.90	1.34	
	0310		Average		400	.040		.11	1.24		1.35	2.01	
	0320		Maximum		250	.064		.15	1.99		2.14	3.19	
	0400		High pressure water only, minimum	B-9	2,000	.020			.54	.07	.61	.92	
	0420		Average		1,500	.027			.72	.09	.81	1.22	
	0440		Maximum		1,000	.040			1.08	.14	1.22	1.85	
	2000		Steam cleaning, minimum		3,000	.013			.36	.05	.41	.61	
	2020		Average		2,500	.016			.43	.06	.49	.73	
	2040		Maximum		1,500	.027			.72	.09	.81	1.22	
	4000		Add for masking doors and windows									.80	
	4200		Add for pedestrian protection				Job					10%	
750	0010		**BUILDING CLEANING** Steam, minimum	B-9	3,000	.013	S.F.		.36	.05	.41	.61	750
	0100		Maximum		1,500	.027			.72	.09	.81	1.22	
	0300		Common face brick		1,750	.023			.62	.08	.70	1.05	
	0400		Wire cut face brick		1,250	.032			.87	.11	.98	1.47	
900	0010		**BRICK WASHING** Acid, smooth brick — R04930-100	1 Bric	560	.014	S.F.	.02	.50		.52	.80	900
	0050		Rough brick		400	.020		.03	.71		.74	1.11	
	0060		Stone, acid wash		600	.013		.04	.47		.51	.76	
	1000		Muriatic acid, price per gallon in 5 gallon lots				Gal.	4.90			4.90	5.40	

For information about Means Estimating Seminars, see yellow pages 12 and 13 in back of book

Important: See the Reference Section for critical supporting data - Reference Nos., Crews, & City Cost Indexes

Division 5 Metals

Estimating Tips

05050 Basic Metal Materials & Methods

- Nuts, bolts, washers, connection angles and plates can add a significant amount to both the tonnage of a structural steel job as well as the estimated cost. As a rule of thumb add 10% to the total weight to account for these accessories.

- Type 2 steel construction, commonly referred to as "simple construction," consists generally of field bolted connections with lateral bracing supplied by other elements of the building, such as masonry walls or x-bracing. The estimator should be aware, however, that shop connections may be accomplished by welding or bolting. The method may be particular to the fabrication shop and may have an impact on the estimated cost.

05200 Metal Joists

- In any given project the total weight of open web steel joists is determined by the loads to be supported and the design. However, economies can be realized in minimizing the amount of labor used to place the joists. This is done by maximizing the joist spacing and therefore minimizing the number of joists required to be installed on the job. Certain spacings and locations may be required by the design, but in other cases maximizing the spacing and keeping it as uniform as possible will keep the costs down.

05300 Metal Deck

- The takeoff and estimating of metal deck involves more than simply the area of the floor or roof and the type of deck specified or shown on the drawings. Many different sizes and types of openings may exist. Small openings for individual pipes or conduits may be drilled after the floor/roof is installed, but larger openings may require special deck lengths as well as reinforcing or structural support. The estimator should determine who will be supplying this reinforcing. Additionally, some deck terminations are part of the deck package, such as screed angles and pour stops, and others will be part of the steel contract, such as angles attached to structural members and cast-in-place angles and plates. The estimator must ensure that all pieces are accounted for in the complete estimate.

05500 Metal Fabrications

- The most economical steel stairs are those that use common materials, standard details and most importantly, a uniform and relatively simple method of field assembly. Commonly available A36 channels and plates are very good choices for the main stringers of the stairs, as are angles and tees for the carrier members. Risers and treads are usually made by specialty shops, and it is most economical to use a typical detail in as many places as possible. The stairs should be pre-assembled and shipped directly to the site. The field connections should be simple and straightforward to be accomplished efficiently and with a minimum of equipment and labor.

Reference Numbers

Reference numbers are shown in bold squares at the beginning of some major classifications. These numbers refer to related items in the Reference Section. The reference information may be an estimating procedure, an alternate pricing method or technical information.

Note: Not all subdivisions listed here necessarily appear in this publication.

05060 | Selective Demolition

		CREW	DAILY OUTPUT	LABOR-HOURS	UNIT	MAT.	LABOR	EQUIP.	TOTAL	TOTAL INCL O&P	
110	0010 **SELECTIVE METALS DEMOLITION** R02220-510										110
	0015 Excludes shores, bracing, cutting, loading, hauling, dumping										
	0020 Remove nuts only up to 3/4" diameter	1 Sswk	480	.017	Ea.		.64		.64	1.14	
	0030 7/8" to 1-1/4" diameter		240	.033			1.27		1.27	2.27	
	0040 1-3/8" to 2" diameter		160	.050			1.91		1.91	3.41	
	0060 Unbolt and remove structural bolts up to 3/4" diameter		240	.033			1.27		1.27	2.27	
	0070 7/8" to 2" diameter		160	.050			1.91		1.91	3.41	
	0140 Light weight framing members, remove whole or cut up, up to 20 lb		240	.033			1.27		1.27	2.27	
	0150 21 - 40 lb	2 Sswk	210	.076			2.91		2.91	5.20	
	0160 41 - 80 lb	3 Sswk	180	.133			5.10		5.10	9.10	
	0170 81 - 120 lb	4 Sswk	150	.213			8.15		8.15	14.55	
	0230 Structural members, remove whole or cut up, up to 500 lb	E-19	48	.500			18.55	16.70	35.25	50	
	0240 1/4 - 2 tons	E-18	36	1.111			42	22	64	97.50	
	0250 2 - 5 tons	E-24	30	1.067			40	20.50	60.50	91	
	0260 5 - 10 tons	E-20	24	2.667			99	40	139	215	
	0270 10 - 15 tons	E-2	18	3.111			115	73.50	188.50	278	
	0340 Fabricated item, remove whole or cut up, up to 20 lb	1 Sswk	96	.083			3.18		3.18	5.70	
	0350 21 - 40 lb	2 Sswk	84	.190			7.25		7.25	13	
	0360 41 - 80 lb	3 Sswk	72	.333			12.70		12.70	23	
	0370 81 - 120 lb	4 Sswk	60	.533			20.50		20.50	36.50	
	0380 121 - 500 lb	E-19	48	.500			18.55	16.70	35.25	50	
	2000 Steel framing, beams, 4" x 6"	B-13	500	.112	L.F.		3.22	1.23	4.45	6.35	

05090 | Metal Fastenings

		CREW	DAILY OUTPUT	LABOR-HOURS	UNIT	MAT.	LABOR	EQUIP.	TOTAL	TOTAL INCL O&P	
080	0010 **ANCHOR BOLTS**										080
	0020 See also divisions 03150-080 and 04080-070										
	0100 J-type, incl. hex nut & washer, 1/2" diameter x 6" long	2 Carp	70	.229	Ea.	.94	7.85		8.79	13.25	
	0110 12" long		65	.246		1.17	8.45		9.62	14.45	
	0120 18" long		60	.267		1.52	9.15		10.67	15.90	
	0130 3/4" diameter x 8" long		50	.320		1.39	10.95		12.34	18.60	
	0140 12" long		45	.356		1.74	12.20		13.94	21	
	0150 18" long		40	.400		2.26	13.70		15.96	24	
	0160 1" diameter x 12" long		35	.457		3.32	15.65		18.97	28	
	0170 18" long		30	.533		4	18.25		22.25	33	
	0180 24" long		25	.640		4.88	22		26.88	39.50	
	0190 36" long		20	.800		6.70	27.50		34.20	50	
	0200 1-1/2" diameter x 18" long		22	.727		11.85	25		36.85	52	
	0210 24" long		16	1		14.10	34.50		48.60	69	
	0300 L-type, incl. hex nut & washer, 3/4" diameter x 12" long		45	.356		1.31	12.20		13.51	20.50	
	0310 18" long		40	.400		1.67	13.70		15.37	23.50	
	0320 24" long		35	.457		2.04	15.65		17.69	26.50	
	0330 30" long		30	.533		2.59	18.25		20.84	31.50	
	0340 36" long		25	.640		2.96	22		24.96	37.50	
	0350 1" diameter x 12" long		35	.457		2.28	15.65		17.93	27	
	0360 18" long		30	.533		2.84	18.25		21.09	31.50	
	0370 24" long		25	.640		3.50	22		25.50	38	
	0380 30" long		23	.696		4.13	24		28.13	41.50	
	0390 36" long		20	.800		4.73	27.50		32.23	47.50	
	0400 42" long		18	.889		5.75	30.50		36.25	54	
	0410 48" long		15	1.067		6.45	36.50		42.95	64	
	0420 1-1/4" diameter x 18" long		25	.640		5.10	22		27.10	39.50	
	0430 24" long		22	.727		6.05	25		31.05	45.50	
	0440 30" long		20	.800		7	27.50		34.50	50	
	0450 36" long		18	.889		8	30.50		38.50	56.50	
	0460 42" long		16	1		9	34.50		43.50	63.50	
	0470 48" long		14	1.143		10.30	39		49.30	72.50	

Important: See the Reference Section for critical supporting data - Reference Nos., Crews, & City Cost Indexes

05090	Metal Fastenings	CREW	DAILY OUTPUT	LABOR-HOURS	UNIT	2005 BARE COSTS MAT.	LABOR	EQUIP.	TOTAL	TOTAL INCL O&P	
080											**080**
0480	54" long	2 Carp	12	1.333	Ea.	12.15	45.50		57.65	84.50	
0490	60" long		10	1.600		13.35	55		68.35	100	
0500	1-1/2" diameter x 18" long		22	.727		7.85	25		32.85	47.50	
0510	24" long		19	.842		9.15	29		38.15	55	
0520	30" long		17	.941		10.35	32		42.35	61.50	
0530	36" long		16	1		11.90	34.50		46.40	66.50	
0540	42" long		15	1.067		13.55	36.50		50.05	72	
0550	48" long		13	1.231		15.25	42		57.25	82.50	
0560	54" long		11	1.455		18.55	50		68.55	98	
0570	60" long		9	1.778		20.50	61		81.50	118	
0580	1-3/4" diameter x 18" long		20	.800		11.85	27.50		39.35	55.50	
0590	24" long		18	.889		13.90	30.50		44.40	63	
0600	30" long		17	.941		16.20	32		48.20	68	
0610	36" long		16	1		18.45	34.50		52.95	74	
0620	42" long		14	1.143		20.50	39		59.50	84	
0630	48" long		12	1.333		23	45.50		68.50	96	
0640	54" long		10	1.600		28.50	55		83.50	117	
0650	60" long		8	2		30.50	68.50		99	141	
0660	2" diameter x 24" long		17	.941		17.70	32		49.70	69.50	
0670	30" long		15	1.067		19.95	36.50		56.45	79	
0680	36" long		13	1.231		22	42		64	89.50	
0690	42" long		11	1.455		24.50	50		74.50	105	
0700	48" long		10	1.600		28	55		83	117	
0710	54" long		9	1.778		33.50	61		94.50	132	
0720	60" long		8	2		36	68.50		104.50	147	
0730	66" long		7	2.286		38.50	78.50		117	165	
0740	72" long	▼	6	2.667	▼	42	91.50		133.50	189	
0990	For galvanized, add					75%					
150	0010 **BOLTS & HEX NUTS** Steel, A307										**150**
0100	1/4" diameter, 1/2" long	1 Sswk	140	.057	Ea.	.06	2.18		2.24	3.96	
0200	1" long		140	.057		.07	2.18		2.25	3.97	
0300	2" long		130	.062		.09	2.35		2.44	4.30	
0400	3" long		130	.062		.13	2.35		2.48	4.35	
0500	4" long		120	.067		.15	2.54		2.69	4.71	
0600	3/8" diameter, 1" long		130	.062		.11	2.35		2.46	4.32	
0700	2" long		130	.062		.14	2.35		2.49	4.35	
0800	3" long		120	.067		.18	2.54		2.72	4.75	
0900	4" long		120	.067		.23	2.54		2.77	4.80	
1000	5" long		115	.070		.28	2.65		2.93	5.05	
1100	1/2" diameter, 1-1/2" long		120	.067		.21	2.54		2.75	4.78	
1200	2" long		120	.067		.24	2.54		2.78	4.81	
1300	4" long		115	.070		.36	2.65		3.01	5.15	
1400	6" long		110	.073		.49	2.77		3.26	5.50	
1500	8" long		105	.076		.64	2.91		3.55	5.90	
1600	5/8" diameter, 1-1/2" long		120	.067		.40	2.54		2.94	4.99	
1700	2" long		120	.067		.44	2.54		2.98	5.05	
1800	4" long		115	.070		.61	2.65		3.26	5.40	
1900	6" long		110	.073		.77	2.77		3.54	5.80	
2000	8" long		105	.076		1.12	2.91		4.03	6.45	
2100	10" long		100	.080		1.39	3.05		4.44	7	
2200	3/4" diameter, 2" long		120	.067		.63	2.54		3.17	5.25	
2300	4" long		110	.073		.88	2.77		3.65	5.95	
2400	6" long		105	.076		1.12	2.91		4.03	6.45	
2500	8" long		95	.084		1.66	3.21		4.87	7.60	
2600	10" long		85	.094		2.16	3.59		5.75	8.80	
2700	12" long	▼	80	.100	▼	2.52	3.82		6.34	9.60	

05090 | Metal Fastenings

		CREW	DAILY OUTPUT	LABOR-HOURS	UNIT	MAT.	LABOR	EQUIP.	TOTAL	TOTAL INCL O&P	
150											**150**
2800	1" diameter, 3" long	1 Sswk	105	.076	Ea.	1.65	2.91		4.56	7	
2900	6" long		90	.089		2.54	3.39		5.93	8.85	
3000	12" long	▼	75	.107		4.78	4.07		8.85	12.55	
3100	For galvanized, add					75%					
3200	For stainless, add				▼	350%					
300	0010 **CHEMICAL ANCHORS**, Includes layout & drilling										**300**
1430	Chemical anchor, w/rod & epoxy cartridge, 3/4" diam. x 9-1/2" long	B-89A	27	.593	Ea.	11.65	18.25	3.82	33.72	45.50	
1435	1" diameter x 11-3/4" long		24	.667		22.50	20.50	4.29	47.29	61.50	
1440	1-1/4" diameter x 14" long		21	.762		43	23.50	4.91	71.41	89	
1445	1-3/4" diameter x 15" long		20	.800		81	24.50	5.15	110.65	134	
1450	18" long		17	.941		97.50	29	6.05	132.55	159	
1455	2" diameter x 18" long		16	1		124	31	6.45	161.45	192	
1460	24" long	▼	15	1.067	▼	162	33	6.85	201.85	238	
1500	Chemical anchoring, epoxy cartridge, excludes layout, drilling, fastener										
1530	For fastener 3/4" dia x 6" embedment	B-89A	27	.593	Ea.	4.58	18.25	3.82	26.65	38	
1535	1" dia x 8" embedment		24	.667		6.85	20.50	4.29	31.64	44.50	
1540	1-1/4" dia x 10" embedment		21	.762		13.75	23.50	4.91	42.16	57	
1545	1-3/4" dia x 12" embedment		20	.800		23	24.50	5.15	52.65	69	
1550	14" embedment		17	.941		27.50	29	6.05	62.55	81.50	
1555	2" dia x 12" embedment		16	1		36.50	31	6.45	73.95	95.50	
1560	18" embedment	▼	15	1.067	▼	46	33	6.85	85.85	109	
340	0010 **DRILLING** For anchors, up to 4" deep, incl. bit and layout										**340**
0050	in concrete or brick walls and floors, no anchor										
0100	Holes, 1/4" diameter	1 Carp	75	.107	Ea.	.08	3.65		3.73	5.80	
0150	For each additional inch of depth, add		430	.019		.02	.64		.66	1.01	
0200	3/8" diameter		63	.127		.08	4.35		4.43	6.85	
0250	For each additional inch of depth, add		340	.024		.02	.81		.83	1.28	
0300	1/2" diameter		50	.160		.08	5.50		5.58	8.65	
0350	For each additional inch of depth, add		250	.032		.02	1.10		1.12	1.73	
0400	5/8" diameter		48	.167		.15	5.70		5.85	9.05	
0450	For each additional inch of depth, add		240	.033		.04	1.14		1.18	1.82	
0500	3/4" diameter		45	.178		.18	6.10		6.28	9.70	
0550	For each additional inch of depth, add		220	.036		.04	1.25		1.29	1.99	
0600	7/8" diameter		43	.186		.21	6.35		6.56	10.20	
0650	For each additional inch of depth, add		210	.038		.05	1.30		1.35	2.09	
0700	1" diameter		40	.200		.24	6.85		7.09	10.90	
0750	For each additional inch of depth, add		190	.042		.06	1.44		1.50	2.32	
0800	1-1/4" diameter		38	.211		.35	7.20		7.55	11.65	
0850	For each additional inch of depth, add		180	.044		.09	1.52		1.61	2.47	
0900	1-1/2" diameter		35	.229		.53	7.85		8.38	12.80	
0950	For each additional inch of depth, add	▼	165	.048	▼	.13	1.66		1.79	2.73	
1000	For ceiling installations, add						40%				
1100	Drilling & layout for drywall/plaster walls, up to 1" deep, no anchor										
1200	Holes, 1/4" diameter	1 Carp	150	.053	Ea.	.01	1.83		1.84	2.86	
1300	3/8" diameter		140	.057		.01	1.96		1.97	3.06	
1400	1/2" diameter		130	.062		.01	2.11		2.12	3.29	
1500	3/4" diameter		120	.067		.02	2.28		2.30	3.58	
1600	1" diameter		110	.073		.03	2.49		2.52	3.91	
1700	1-1/4" diameter		100	.080		.04	2.74		2.78	4.32	
1800	1-1/2" diameter	▼	90	.089		.07	3.04		3.11	4.81	
1900	For ceiling installations, add				▼		40%				
1910	Drilling & layout for steel, up to 1/4" deep, no anchor										
1920	Holes, 1/4" diameter	1 Sswk	112	.071	Ea.	.11	2.73		2.84	5	
1925	For each additional 1/4" depth, add		336	.024		.11	.91		1.02	1.75	
1930	3/8" diameter	▼	104	.077		.13	2.93		3.06	5.40	

Important: See the Reference Section for critical supporting data - Reference Nos., Crews, & City Cost Indexes

			DAILY	LABOR-		2005 BARE COSTS				TOTAL
05090		**Metal Fastenings**				MAT.	LABOR	EQUIP.	TOTAL	INCL O&P
			CREW	OUTPUT	HOURS	UNIT				
340	1935	For each additional 1/4" depth, add	1 Sswk	312	.026	Ea.	.13	.98		1.11	1.89	**340**
	1940	1/2" diameter		96	.083		.14	3.18		3.32	5.85
	1945	For each additional 1/4" depth, add		288	.028		.14	1.06		1.20	2.06
	1950	5/8" diameter		88	.091		.24	3.47		3.71	6.45
	1955	For each additional 1/4" depth, add		264	.030		.24	1.16		1.40	2.33
	1960	3/4" diameter		80	.100		.26	3.82		4.08	7.15
	1965	For each additional 1/4" depth, add		240	.033		.26	1.27		1.53	2.56
	1970	7/8" diameter		72	.111		.31	4.24		4.55	7.95
	1975	For each additional 1/4" depth, add		216	.037		.31	1.41		1.72	2.87
	1980	1" diameter		64	.125		.35	4.77		5.12	8.95
	1985	For each additional 1/4" depth, add		192	.042		.35	1.59		1.94	3.23
	1990	For drilling up, add						40%			
380	0010	**EXPANSION ANCHORS** & shields										**380**
	0100	Bolt anchors for concrete, brick or stone, no layout and drilling									
	0200	Expansion shields, zinc, 1/4" diameter, 1-5/16" long, single	1 Carp	90	.089	Ea.	1.01	3.04		4.05	5.85
	0300	1-3/8" long, double		85	.094		1.11	3.22		4.33	6.20
	0400	3/8" diameter, 1-1/2" long, single		85	.094		1.66	3.22		4.88	6.85
	0500	2" long, double		80	.100		2.05	3.43		5.48	7.60
	0600	1/2" diameter, 2-1/16" long, single		80	.100		2.75	3.43		6.18	8.40
	0700	2-1/2" long, double		75	.107		2.65	3.65		6.30	8.60
	0800	5/8" diameter, 2-5/8" long, single		75	.107		3.93	3.65		7.58	10
	0900	2-3/4" long, double		70	.114		3.93	3.91		7.84	10.40
	1000	3/4" diameter, 2-3/4" long, single		70	.114		5.85	3.91		9.76	12.50
	1100	3-15/16" long, double		65	.123		7.80	4.22		12.02	15.10
	1500	Self drilling anchor, snap-off, for 1/4" diameter bolt		26	.308		.83	10.55		11.38	17.30
	1600	3/8" diameter bolt		23	.348		1.20	11.90		13.10	19.85
	1700	1/2" diameter bolt		20	.400		1.84	13.70		15.54	23.50
	1800	5/8" diameter bolt		18	.444		3.08	15.20		18.28	27
	1900	3/4" diameter bolt		16	.500		5.20	17.15		22.35	32
	2100	Hollow wall anchors for gypsum wall board, plaster or tile									
	2300	1/8" diameter, short	1 Carp	160	.050	Ea.	.26	1.71		1.97	2.96
	2400	Long		150	.053		.31	1.83		2.14	3.19
	2500	3/16" diameter, short		150	.053		.55	1.83		2.38	3.46
	2600	Long		140	.057		.59	1.96		2.55	3.70
	2700	1/4" diameter, short		140	.057		.68	1.96		2.64	3.80
	2800	Long		130	.062		.76	2.11		2.87	4.12
	3000	Toggle bolts, bright steel, 1/8" diameter, 2" long		85	.094		.25	3.22		3.47	5.30
	3100	4" long		80	.100		.37	3.43		3.80	5.75
	3200	3/16" diameter, 3" long		80	.100		.42	3.43		3.85	5.80
	3300	6" long		75	.107		.58	3.65		4.23	6.35
	3400	1/4" diameter, 3" long		75	.107		.47	3.65		4.12	6.20
	3500	6" long		70	.114		.66	3.91		4.57	6.85
	3600	3/8" diameter, 3" long		70	.114		.90	3.91		4.81	7.10
	3700	6" long		60	.133		1.58	4.57		6.15	8.85
	3800	1/2" diameter, 4" long		60	.133		2.33	4.57		6.90	9.65
	3900	6" long		50	.160		3.85	5.50		9.35	12.80
	4000	Nailing anchors									
	4100	Nylon nailing anchor, 1/4" diameter, 1" long	1 Carp	3.20	2.500	C	18	85.50		103.50	153
	4200	1-1/2" long		2.80	2.857		23	98		121	178
	4300	2" long		2.40	3.333		38.50	114		152.50	221
	4400	Metal nailing anchor, 1/4" diameter, 1" long		3.20	2.500		27.50	85.50		113	164
	4500	1-1/2" long		2.80	2.857		37.50	98		135.50	194
	4600	2" long		2.40	3.333		48	114		162	231
	5000	Screw anchors for concrete, masonry,									
	5100	stone & tile, no layout or drilling included									
	5200	Jute fiber, #6, #8, & #10, 1" long	1 Carp	240	.033	Ea.	.24	1.14		1.38	2.04

METALS 5

145

		05090 Metal Fastenings	CREW	DAILY OUTPUT	LABOR-HOURS	UNIT	2005 BARE COSTS				TOTAL INCL O&P	
							MAT.	LABOR	EQUIP.	TOTAL		
380	5300	#12, 1-1/2" long	1 Carp	200	.040	Ea.	.35	1.37		1.72	2.52	380
	5400	#14, 2" long		160	.050		.55	1.71		2.26	3.28	
	5500	#16, 2" long		150	.053		.57	1.83		2.40	3.48	
	5600	#20, 2" long		140	.057		.91	1.96		2.87	4.05	
	5700	Lag screw shields, 1/4" diameter, short		90	.089		.44	3.04		3.48	5.20	
	5800	Long		85	.094		.51	3.22		3.73	5.55	
	5900	3/8" diameter, short		85	.094		.80	3.22		4.02	5.90	
	6000	Long		80	.100		.94	3.43		4.37	6.40	
	6100	1/2" diameter, short		80	.100		1.11	3.43		4.54	6.55	
	6200	Long		75	.107		1.39	3.65		5.04	7.25	
	6300	3/4" diameter, short		70	.114		3.12	3.91		7.03	9.55	
	6400	Long		65	.123		3.78	4.22		8	10.70	
	6600	Lead, #6 & #8, 3/4" long		260	.031		.16	1.05		1.21	1.82	
	6700	#10 - #14, 1-1/2" long		200	.040		.23	1.37		1.60	2.38	
	6800	#16 & #18, 1-1/2" long		160	.050		.31	1.71		2.02	3.01	
	6900	Plastic, #6 & #8, 3/4" long		260	.031		.10	1.05		1.15	1.75	
	7000	#8 & #10, 7/8" long		240	.033		.04	1.14		1.18	1.82	
	7100	#10 & #12, 1" long		220	.036		.13	1.25		1.38	2.08	
	7200	#14 & #16, 1-1/2" long		160	.050		.07	1.71		1.78	2.75	
	8000	Wedge anchors, not including layout or drilling										
	8050	Carbon steel, 1/4" diameter, 1-3/4" long	1 Carp	150	.053	Ea.	.42	1.83		2.25	3.31	
	8100	3 1/4" long		140	.057		.55	1.96		2.51	3.66	
	8150	3/8" diameter, 2-1/4" long		145	.055		.63	1.89		2.52	3.63	
	8200	5" long		140	.057		1.10	1.96		3.06	4.26	
	8250	1/2" diameter, 2-3/4" long		140	.057		.96	1.96		2.92	4.11	
	8300	7" long		125	.064		1.65	2.19		3.84	5.20	
	8350	5/8" diameter, 3-1/2" long		130	.062		1.90	2.11		4.01	5.35	
	8400	8-1/2" long		115	.070		4.05	2.38		6.43	8.15	
	8450	3/4" diameter, 4-1/4" long		115	.070		2.29	2.38		4.67	6.25	
	8500	10" long		95	.084		5.20	2.88		8.08	10.25	
	8550	1" diameter, 6" long		100	.080		7.70	2.74		10.44	12.70	
	8575	9" long		85	.094		10	3.22		13.22	16	
	8600	12" long		75	.107		10.80	3.65		14.45	17.55	
	8650	1-1/4" diameter, 9" long		70	.114		14	3.91		17.91	21.50	
	8700	12" long		60	.133		17.90	4.57		22.47	27	
	8750	For type 303 stainless steel, add						350%					
	8800	For type 316 stainless steel, add						450%					
	8950	Self-drilling concrete screw, hex washer head, 3/16" dia x 1-3/4" long	1 Carp	300	.027	Ea.	.32	.91		1.23	1.77	
	8960	2-1/4" long		250	.032		.49	1.10		1.59	2.25	
	8970	Phillips flat head, 3/16" dia x 1-3/4" long		300	.027		.33	.91		1.24	1.78	
	8980	2-1/4" long		250	.032		.48	1.10		1.58	2.24	
420	0010	**HIGH STRENGTH BOLTS**										420
	0020	A325 Type 1, structural steel, bolt-nut-washer set	R05090 -510									
	0100	1/2" diameter x 1-1/2" long	1 Sswk	130	.062	Ea.	.32	2.35		2.67	4.55	
	0120	2" long		125	.064		.34	2.44		2.78	4.74	
	0150	3" long		120	.067		.44	2.54		2.98	5.05	
	0170	5/8" diameter x 1-1/2" long		125	.064		.50	2.44		2.94	4.92	
	0180	2" long		120	.067		.53	2.54		3.07	5.15	
	0190	3" long		115	.070		.62	2.65		3.27	5.45	
	0200	3/4" diameter x 2" long		120	.067		.74	2.54		3.28	5.35	
	0220	3" long		115	.070		.85	2.65		3.50	5.70	
	0250	4" long		110	.073		1	2.77		3.77	6.05	
	0300	6" long		105	.076		1.24	2.91		4.15	6.55	
	0350	8" long		95	.084		2.27	3.21		5.48	8.25	
	0360	7/8" diameter x 2" long		115	.070		1.10	2.65		3.75	5.95	

Important: See the Reference Section for critical supporting data - Reference Nos., Crews, & City Cost Indexes

				DAILY	LABOR-		2005 BARE COSTS				TOTAL	
05090		**Metal Fastenings**	CREW	OUTPUT	HOURS	UNIT	MAT.	LABOR	EQUIP.	TOTAL	INCL O&P	
420	0365	3" long	1 Sswk	110	.073	Ea.	1.25	2.77		4.02	6.35	**420**
	0370	4" long		105	.076		1.46	2.91		4.37	6.80	
	0380	6" long		100	.080		1.79	3.05		4.84	7.40	
	0390	8" long		90	.089		2.70	3.39		6.09	9	
	0400	1" diameter x 2" long		105	.076		1.52	2.91		4.43	6.90	
	0420	3" long		100	.080		1.67	3.05		4.72	7.30	
	0450	4" long		95	.084		1.84	3.21		5.05	7.75	
	0500	6" long		90	.089		2.32	3.39		5.71	8.60	
	0550	8" long		85	.094		3.73	3.59		7.32	10.50	
	0600	1-1/4" diameter x 3" long		85	.094		3.46	3.59		7.05	10.20	
	0650	4" long		80	.100		1.57	3.82		5.39	8.60	
	0700	6" long		75	.107		4.51	4.07		8.58	12.25	
	0750	8" long		70	.114		5.50	4.36		9.86	13.85	
	1020	A490, bolt-nut-washer set										
	1170	5/8" diameter x 1-1/2" long	1 Sswk	125	.064	Ea.	.71	2.44		3.15	5.15	
	1180	2" long		120	.067		.81	2.54		3.35	5.45	
	1190	3" long		115	.070		.95	2.65		3.60	5.80	
	1200	3/4" diameter x 2" long		120	.067		.96	2.54		3.50	5.60	
	1220	3" long		115	.070		1.09	2.65		3.74	5.95	
	1250	4" long		110	.073		1.23	2.77		4	6.30	
	1300	6" long		105	.076		1.70	2.91		4.61	7.05	
	1350	8" long		95	.084		2.72	3.21		5.93	8.75	
	1360	7/8" diameter x 2" long		115	.070		1.30	2.65		3.95	6.20	
	1365	3" long		110	.073		1.47	2.77		4.24	6.55	
	1370	4" long		105	.076		1.74	2.91		4.65	7.10	
	1380	6" long		100	.080		2.31	3.05		5.36	8	
	1390	8" long		90	.089		3.21	3.39		6.60	9.60	
	1400	1" diameter x 2" long		105	.076		1.75	2.91		4.66	7.15	
	1420	3" long		100	.080		2.01	3.05		5.06	7.65	
	1450	4" long		95	.084		2.23	3.21		5.44	8.20	
	1500	6" long		90	.089		2.81	3.39		6.20	9.15	
	1550	8" long		85	.094		4.13	3.59		7.72	10.95	
	1600	1-1/4" diameter x 3" long		85	.094		4.05	3.59		7.64	10.85	
	1650	4" long		80	.100		4.49	3.82		8.31	11.80	
	1700	6" long		75	.107		5.75	4.07		9.82	13.65	
	1750	8" long		70	.114		7.15	4.36		11.51	15.70	
460	0010	**LAG SCREWS**										**460**
	0020	Steel, 1/4" diameter, 2" long	1 Carp	200	.040	Ea.	.08	1.37		1.45	2.22	
	0100	3/8" diameter, 3" long		150	.053		.23	1.83		2.06	3.10	
	0200	1/2" diameter, 3" long		130	.062		.38	2.11		2.49	3.70	
	0300	5/8" diameter, 3" long		120	.067		.75	2.28		3.03	4.39	
500	0010	**MACHINE SCREWS**										**500**
	0020	Steel, round head, #8 x 1" long				C	2.28			2.28	2.51	
	0110	#8 x 2" long					4.96			4.96	5.45	
	0200	#10 x 1" long					3.25			3.25	3.58	
	0300	#10 x 2" long					6.05			6.05	6.70	
540	0010	**MACHINERY ANCHORS**, heavy duty, incl. sleeve, floating base nut,										**540**
	0020	lower stud & coupling nut, fiber plug, connecting stud, washer & nut.										
	0030	For flush mounted embedment in poured concrete heavy equip. pads.										
	0200	Material only, 1/2" diameter stud & bolt				Ea.	52.50			52.50	58	
	0300	5/8" diameter					58.50			58.50	64	
	0500	3/4" diameter					67.50			67.50	74	
	0600	7/8" diameter					73.50			73.50	81	
	0800	1" diameter					77.50			77.50	85	

R05090
-510

METALS 5

147

		05090	Metal Fastenings	CREW	DAILY OUTPUT	LABOR-HOURS	UNIT	2005 BARE COSTS				TOTAL INCL O&P	
								MAT.	LABOR	EQUIP.	TOTAL		
540	0900		1-1/4" diameter				Ea.	103			103	113	540
580	0010	**POWDER ACTUATED** Tools & fasteners											580
	0020		Stud driver, .22 caliber, buy, minimum				Ea.	310			310	340	
	0100		Maximum				"	495			495	545	
	0300		Powder charges for above, low velocity				C	16.60			16.60	18.25	
	0400		Standard velocity					23.50			23.50	26	
	0600		Drive pins & studs, 1/4" & 3/8" diam., to 3" long, minimum					11.85			11.85	13.05	
	0700		Maximum				▼	46.50			46.50	51	
	0800		Pneumatic stud driver for 1/8" diameter studs				Ea.	2,150			2,150	2,350	
	0900		Drive pins for above, 1/2" to 3/4" long				M	490			490	540	
600	0010	**RIVETS**											600
	0100		Aluminum rivet & mandrel, 1/2" grip length x 1/8" diameter				C	5			5	5.50	
	0200		3/16" diameter					7.70			7.70	8.45	
	0300		Aluminum rivet, steel mandrel, 1/8" diameter					7.30			7.30	8.05	
	0400		3/16" diameter					6.70			6.70	7.35	
	0500		Copper rivet, steel mandrel, 1/8" diameter					6.65			6.65	7.30	
	0600		Monel rivet, steel mandrel, 1/8" diameter					23.50			23.50	26	
	0700		3/16" diameter					68			68	75	
	0800		Stainless rivet & mandrel, 1/8" diameter					11.90			11.90	13.05	
	0900		3/16" diameter					22			22	24.50	
	1000		Stainless rivet, steel mandrel, 1/8" diameter					9.25			9.25	10.20	
	1100		3/16" diameter					16.75			16.75	18.45	
	1200		Steel rivet and mandrel, 1/8" diameter					5.75			5.75	6.30	
	1300		3/16" diameter				▼	8.60			8.60	9.50	
	1400		Hand riveting tool, minimum				Ea.	111			111	122	
	1500		Maximum					212			212	233	
	1600		Power riveting tool, minimum					800			800	880	
	1700		Maximum				▼	2,025			2,025	2,225	
820	0010	**VIBRATION PADS**											820
	0300		Laminated synthetic rubber impregnated cotton duck, 1/2" thick	2 Sswk	24	.667	S.F.	53.50	25.50		79	105	
	0400		1" thick		20	.800		108	30.50		138.50	174	
	0600		Neoprene bearing pads, 1/2" thick		24	.667		20	25.50		45.50	68	
	0700		1" thick		20	.800		41.50	30.50		72	100	
	0900		Fabric reinforced neoprene, 5000 psi, 1/2" thick		24	.667		9.10	25.50		34.60	55.50	
	1000		1" thick		20	.800		18.20	30.50		48.70	74.50	
	1200		Felt surfaced vinyl pads, cork and sisal, 5/8" thick		24	.667		23.50	25.50		49	71.50	
	1300		1" thick		20	.800		42.50	30.50		73	102	
	1500		Teflon bonded to 10 ga. carbon steel, 1/32" layer		24	.667		42	25.50		67.50	92	
	1600		3/32" layer		24	.667		63.50	25.50		89	115	
	1800		Bonded to 10 ga. stainless steel, 1/32" layer		24	.667		75	25.50		100.50	128	
	1900		3/32" layer	▼	24	.667	▼	97.50	25.50		123	153	
	2100		Circular machine leveling pad & stud				Kip	6.15			6.15	6.75	
840	0010	**WELD SHEAR CONNECTORS**											840
	0020		3/4" diameter, 3-3/16" long	E-10	960	.017	Ea.	.37	.65	.27	1.29	1.88	
	0030		3-3/8" long		950	.017		.39	.66	.27	1.32	1.91	
	0200		3-7/8" long		945	.017		.42	.66	.27	1.35	1.95	
	0300		4-3/16" long		935	.017		.44	.67	.28	1.39	1.98	
	0500		4-7/8" long		930	.017		.49	.67	.28	1.44	2.04	
	0600		5-3/16" long		920	.017		.51	.68	.28	1.47	2.09	
	0800		5-3/8" long		910	.018		.51	.69	.28	1.48	2.11	
	0900		6-3/16" long		905	.018		.56	.69	.28	1.53	2.17	
	1000		7-3/16" long		895	.018		.70	.70	.29	1.69	2.34	
	1100		8-3/16" long		890	.018		.77	.70	.29	1.76	2.42	
	1500		7/8" diameter, 3-11/16" long	▼	920	.017	▼	.61	.68	.28	1.57	2.20	

5 METALS

Important: See the Reference Section for critical supporting data - Reference Nos., Crews, & City Cost Indexes

		05090	Metal Fastenings	CREW	DAILY OUTPUT	LABOR-HOURS	UNIT	MAT.	2005 BARE COSTS LABOR	EQUIP.	TOTAL	TOTAL INCL O&P	
840	1600		4-3/16" long	E-10	910	.018	Ea.	.65	.69	.28	1.62	2.26	**840**
	1700		5-3/16" long		905	.018		.74	.69	.28	1.71	2.36	
	1800		6-3/16" long		895	.018		.83	.70	.29	1.82	2.48	
	1900		7-3/16" long		890	.018		.92	.70	.29	1.91	2.59	
	2000		8-3/16" long	↓	880	.018	↓	1	.71	.29	2	2.69	
860	0010	**WELD STUDS**											**860**
	0020	1/4" diameter, 2-11/16" long		E-10	1,120	.014	Ea.	.24	.56	.23	1.03	1.52	
	0100	4-1/8" long			1,080	.015		.23	.58	.24	1.05	1.55	
	0200	3/8" diameter, 4-1/8" long			1,080	.015		.26	.58	.24	1.08	1.59	
	0300	6-1/8" long			1,040	.015		.34	.60	.25	1.19	1.72	
	0400	1/2" diameter, 2-1/8" long			1,040	.015		.25	.60	.25	1.10	1.63	
	0500	3-1/8" long			1,025	.016		.31	.61	.25	1.17	1.71	
	0600	4-1/8" long			1,010	.016		.36	.62	.25	1.23	1.78	
	0700	5-5/16" long			990	.016		.44	.63	.26	1.33	1.90	
	0800	6-1/8" long			975	.016		.48	.64	.26	1.38	1.96	
	0900	8-1/8" long			960	.017		.67	.65	.27	1.59	2.21	
	1000	5/8" diameter, 2-11/16" long			1,000	.016		.43	.63	.26	1.32	1.87	
	1010	4-3/16" long			990	.016		.53	.63	.26	1.42	2	
	1100	6-9/16" long			975	.016		.69	.64	.26	1.59	2.20	
	1200	8-3/16" long		↓	960	.017	↓	.92	.65	.27	1.84	2.49	
880	0010	**WELD ROD**											**880**
	0020	Steel, type 6011, 1/8" dia, less than 500#					Lb.	1.72			1.72	1.89	
	0100	500# to 2,000#						1.55			1.55	1.71	
	0200	2,000# to 5,000#						1.46			1.46	1.60	
	0300	5/32" diameter, less than 500#						1.65			1.65	1.82	
	0310	500# to 2,000#						1.49			1.49	1.64	
	0320	2,000# to 5,000#						1.40			1.40	1.54	
	0400	3/16" dia, less than 500#						1.68			1.68	1.84	
	0500	500# to 2,000#						1.51			1.51	1.66	
	0600	2,000# to 5,000#						1.42			1.42	1.56	
	0620	Steel, type 6010, 1/8" dia, less than 500#						1.90			1.90	2.09	
	0630	500# to 2,000#						1.71			1.71	1.88	
	0640	2,000# to 5,000#						1.61			1.61	1.77	
	0650	Steel, type 7018 Low Hydrogen, 1/8" dia, less than 500#						1.62			1.62	1.78	
	0660	500# to 2,000# CN						1.46			1.46	1.61	
	0670	2,000# to 5,000#						1.37			1.37	1.51	
	0700	Steel, type 7024 Jet Weld, 1/8" dia, less than 500#						1.72			1.72	1.89	
	0710	500# to 2,000#						1.55			1.55	1.71	
	0720	2,000# to 5,000#						1.46			1.46	1.60	
	1550	Aluminum, type 4043 TIG, 1/8" dia, less than 10#						5.95			5.95	6.55	
	1560	10# to 60#						5.35			5.35	5.90	
	1570	Over 60#						5.05			5.05	5.55	
	1600	Aluminum, type 5356 TIG, 1/8" dia, less than 10#						6.60			6.60	7.25	
	1610	10# to 60#						5.95			5.95	6.55	
	1620	Over 60#						5.60			5.60	6.15	
	1900	Cast iron, type 8 Nickel, 1/8" dia, less than 500#						20.50			20.50	22.50	
	1910	500# to 1,000#						18.35			18.35	20	
	1920	Over 1,000#						17.25			17.25	18.95	
	2000	Stainless steel, type 316/316L, 1/8" dia, less than 500#						9.40			9.40	10.30	
	2100	500# to 1000#						8.45			8.45	9.30	
	2220	Over 1000#					↓	7.95			7.95	8.75	
900	0010	**WELDING STRUCTURAL**	R05090 -520										**900**
	0020	Field welding, 1/8" E6011, cost per welder, no oper. engr		E-14	8	1	Hr.	3.44	40	10.15	53.59	87	

METALS 5

05090 | Metal Fastenings

			CREW	DAILY OUTPUT	LABOR-HOURS	UNIT	MAT.	LABOR	EQUIP.	TOTAL	TOTAL INCL O&P	
900	0200	With 1/2 operating engineer	E-13	8	1.500	Hr.	3.44	56.50	10.15	70.09	112	**900**
	0300	With 1 operating engineer	E-12	8	2	↓	3.44	73	10.15	86.59	137	
	0500	With no operating engineer, 2# weld rod per ton	E-14	8	1	Ton	3.44	40	10.15	53.59	87	
	0600	8# E6011 per ton	"	2	4		13.75	161	40.50	215.25	345	
	0800	With one operating engineer per welder, 2# E6011 per ton	E-12	8	2	↓	3.44	73	10.15	86.59	137	
	0900	8# E6011 per ton	"	2	8		13.75	293	40.50	347.25	545	
	1200	Continuous fillet, stick welding, incl. equipment										
	1300	Single pass, 1/8" thick, 0.1#/L.F.	E-14	150	.053	L.F.	.17	2.14	.54	2.85	4.62	
	1400	3/16" thick, 0.2#/L.F.		75	.107		.34	4.28	1.08	5.70	9.20	
	1500	1/4" thick, 0.3#/L.F.		50	.160		.52	6.40	1.62	8.54	13.85	
	1610	5/16" thick, 0.4#/L.F.		38	.211		.69	8.45	2.14	11.28	18.25	
	1800	3 passes, 3/8" thick, 0.5#/L.F.		30	.267		.86	10.70	2.71	14.27	23	
	2010	4 passes, 1/2" thick, 0.7#/L.F.		22	.364		1.20	14.60	3.69	19.49	31.50	
	2200	5 to 6 passes, 3/4" thick, 1.3#/L.F.		12	.667		2.24	27	6.75	35.99	58	
	2400	8 to 11 passes, 1" thick, 2.4#/L.F.	↓	6	1.333		4.13	53.50	13.55	71.18	115	
	2600	For all position welding, add, minimum						20%				
	2700	Maximum						300%				
	2900	For semi-automatic welding, deduct, minimum						5%				
	3000	Maximum				↓		15%				
	4000	Cleaning and welding plates, bars, or rods										
	4010	to existing beams, columns, or trusses	E-14	12	.667	L.F.	.86	27	6.75	34.61	56.50	
920	0010	**STEEL CUTTING**										**920**
	0020	Hand burning, incl. preparation, torch cutting & grinding, no staging										
	0100	Steel to 1/2" thick	E-25	320	.025	L.F.		1	.26	1.26	2.08	
	0150	3/4" thick		260	.031			1.24	.32	1.56	2.56	
	0200	1" thick	↓	200	.040	↓		1.61	.41	2.02	3.33	

Reference: R05090-520

05120 | Structural Steel

			CREW	DAILY OUTPUT	LABOR-HOURS	UNIT	MAT.	LABOR	EQUIP.	TOTAL	TOTAL INCL O&P	
140	0010	**SUBPURLINS**										**140**
	0020	Bulb tees, painted, 32-5/8" O.C., 40 psf L.L.										
	0100	Type 178, max 8'-9" span, 2.15 plf, 2" high x 1-5/8" wide	E-1	4,200	.006	S.F.	1.28	.21	.02	1.51	1.79	
	0200	Type 218, max 10'-2" span, 3.19 plf, 2-1/8" high x 2-1/8" wide	"	3,100	.008		1.49	.29	.03	1.81	2.15	
	1420	For 24-5/8" spacing, add					33%	33%				
	1430	For 48-5/8" spacing, deduct	↓			↓	50%	50%				
180	0010	**CANOPY FRAMING**										**180**
	0020	6" and 8" members	E-4	3,000	.011	Lb.	1.05	.41	.03	1.49	1.93	
220	0010	**CEILING SUPPORTS**										**220**
	1000	Entrance door/folding partition supports	E-4	60	.533	L.F.	17.50	20.50	1.35	39.35	57.50	
	1100	Linear accelerator door supports		14	2.286		79.50	88.50	5.80	173.80	252	
	1200	Lintels or shelf angles, hung, exterior hot dipped galv.		267	.120		11.95	4.63	.30	16.88	22	
	1250	Two coats primer paint instead of galv.		267	.120	↓	10.35	4.63	.30	15.28	20	
	1400	Monitor support, ceiling hung, expansion bolted		4	8	Ea.	277	310	20.50	607.50	885	
	1450	Hung from pre-set inserts		6	5.333		298	206	13.55	517.55	715	
	1600	Motor supports for overhead doors		4	8	↓	141	310	20.50	471.50	735	
	1700	Partition support for heavy folding partitions, without pocket		24	1.333	L.F.	40	51.50	3.39	94.89	140	
	1750	Supports at pocket only	↓	12	2.667	↓	79.50	103	6.75	189.25	279	

Reference: R05120-250

Important: See the Reference Section for critical supporting data - Reference Nos., Crews, & City Cost Indexes

05100 | Structural Metal Framing

05120 | Structural Steel

			CREW	DAILY OUTPUT	LABOR-HOURS	UNIT	MAT.	LABOR	EQUIP.	TOTAL	TOTAL INCL O&P	
220	2000	Rolling grilles & fire door supports	E-4	34	.941	L.F.	34	36.50	2.39	72.89	105	220
	2100	Spider-leg light supports, expansion bolted to ceiling slab		8	4	Ea.	114	155	10.15	279.15	415	
	2150	Hung from pre-set inserts		12	2.667	"	123	103	6.75	232.75	325	
	2400	Toilet partition support		36	.889	L.F.	40	34.50	2.26	76.76	108	
	2500	X-ray travel gantry support		12	2.667	"	137	103	6.75	246.75	340	
250	0010	**COLUMNS, LIGHTWEIGHT** R05120-210										250
	1000	Lightweight units (lally), 3-1/2" diameter	E-2	780	.072	L.F.	2.72	2.65	1.70	7.07	9.40	
	1050	4" diameter	"	900	.062	"	3.99	2.30	1.47	7.76	9.95	
	5800	Adjustable jack post, 8' maximum height, 2-3/4" diameter				Ea.	29			29	32	
	5850	4" diameter				"	46.50			46.50	51	
260	0010	**COLUMNS, STRUCTURAL** R05120-210										260
	0020	Shop fab'd for 100-ton, 1-2 story project, bolted conn's.										
	0800	Steel, concrete filled, extra strong pipe, 3-1/2" diameter	E-2	660	.085	L.F.	29	3.14	2.01	34.15	39.50	
	0930	6" diameter	"	1,200	.047	"	51	1.72	1.10	53.82	60	
	1100	For galvanizing, add				Lb.	.21			.21	.24	
	1300	For web ties, angles, etc., add per added lb.	1 Sswk	945	.008		.88	.32		1.20	1.54	
	1500	Steel pipe, extra strong, no concrete, 3" to 5" diameter	E-2	16,000	.003		.88	.13	.08	1.09	1.27	
	1600	6" to 12" diameter		14,000	.004		.88	.15	.09	1.12	1.31	
	1700	Steel pipe, extra strong, no concrete, 3" diameter x 12'-0"		60	.933	Ea.	108	34.50	22	164.50	202	
	1750	4" diameter x 12'-0"		58	.966		157	35.50	23	215.50	260	
	1800	6" diameter x 12'-0"		54	1.037		300	38.50	24.50	363	425	
	1850	8" diameter x 14'-0"		50	1.120		530	41.50	26.50	598	685	
	1900	10" diameter x 16'-0"		48	1.167		765	43	27.50	835.50	950	
	1950	12" diameter x 18'-0"		45	1.244		1,025	46	29.50	1,100.50	1,225	
	3300	Structural tubing, square, A500GrB, 4" to 6" square, light section		11,270	.005	Lb.	.88	.18	.12	1.18	1.41	
	3600	Heavy section		32,000	.002	"	.88	.06	.04	.98	1.12	
	4000	Concrete filled, add				L.F.	3.14			3.14	3.46	
	4500	Structural tubing, sq, 4" x 4" x 1/4" x 12'-0"	E-2	58	.966	Ea.	144	35.50	23	202.50	246	
	4550	6" x 6" x 1/4" x 12'-0"		54	1.037		236	38.50	24.50	299	355	
	4600	8" x 8" x 3/8" x 14'-0"		50	1.120		510	41.50	26.50	578	665	
	4650	10" x 10" x 1/2" x 16'-0"		48	1.167		950	43	27.50	1,020.50	1,150	
	5100	Structural tubing, rect, 5" to 6" wide, light section		8,000	.007	Lb.	.88	.26	.17	1.31	1.58	
	5200	Heavy section		12,000	.005		.88	.17	.11	1.16	1.38	
	5300	7" to 10" wide, light section		15,000	.004		.88	.14	.09	1.11	1.30	
	5400	Heavy section		18,000	.003		.88	.11	.07	1.06	1.24	
	5500	Structural tubing, rect, 5" x 3" x 1/4" x 12'-0"		58	.966	Ea.	140	35.50	23	198.50	241	
	5550	6" x 4" x 5/16" x 12'-0"		54	1.037		219	38.50	24.50	282	335	
	5600	8" x 4" x 3/8" x 12'-0"		54	1.037		320	38.50	24.50	383	445	
	5650	10" x 6" x 3/8" x 14'-0"		50	1.120		510	41.50	26.50	578	665	
	5700	12" x 8" x 1/2" x 16'-0"		48	1.167		945	43	27.50	1,015.50	1,150	
	6800	W Shape, A992 steel, 2 tier, W8 x 24		1,080	.052	L.F.	23	1.92	1.23	26.15	30	
	6850	W8 x 31		1,080	.052		30	1.92	1.23	33.15	37.50	
	6900	W8 x 48		1,032	.054		46	2.01	1.28	49.29	56	
	6950	W8 x 67		984	.057		64.50	2.10	1.35	67.95	76	
	7000	W10 x 45		1,032	.054		43.50	2.01	1.28	46.79	52.50	
	7050	W10 x 68		984	.057		65.50	2.10	1.35	68.95	77	
	7100	W10 x 112		960	.058		108	2.16	1.38	111.54	124	
	7150	W12 x 50		1,032	.054		48	2.01	1.28	51.29	58	
	7200	W12 x 87		984	.057		83.50	2.10	1.35	86.95	97	
	7250	W12 x 120		960	.058		116	2.16	1.38	119.54	132	
	7300	W12 x 190		912	.061		183	2.27	1.45	186.72	207	
	7350	W14 x 74		984	.057		71	2.10	1.35	74.45	83.50	
	7400	W14 x 120		960	.058		116	2.16	1.38	119.54	132	
	7450	W14 x 176		912	.061		169	2.27	1.45	172.72	192	

05120 | Structural Steel

		CREW	DAILY OUTPUT	LABOR-HOURS	UNIT	2005 BARE COSTS				TOTAL INCL O&P	
						MAT.	LABOR	EQUIP.	TOTAL		
260	8090	For projects 75 to 99 tons, add	R05120-210			All	10%					260
	8092	50 to 74 tons, add					20%					
	8094	25 to 49 tons, add					30%	10%				
	8096	10 to 24 tons, add					50%	25%				
	8098	2 to 9 tons, add					75%	50%				
	8099	Less than 2 tons, add					100%	100%				
300	0010	**CURB EDGING**										300
	0020	Steel angle w/anchors, on forms, 1" x 1", 0.8#/L.F.		E-4	350	.091	L.F.	1.48	3.53	.23	5.24	8.20
	0100	2" x 2" angles, 3.92#/L.F.			330	.097		4.76	3.75	.25	8.76	12.20
	0200	3" x 3" angles, 6.1#/L.F.			300	.107		7.50	4.12	.27	11.89	15.95
	0300	4" x 4" angles, 8.2#/L.F.			275	.116		9.70	4.50	.30	14.50	19.05
	1000	6" x 4" angles, 12.3#/L.F.			250	.128		14	4.95	.33	19.28	24.50
	1050	Steel channels with anchors, on forms, 3" channel, 5#/L.F.			290	.110		5.90	4.26	.28	10.44	14.45
	1100	4" channel, 5.4#/L.F.			270	.119		6.30	4.58	.30	11.18	15.50
	1200	6" channel, 8.2#/L.F.			255	.125		9.70	4.85	.32	14.87	19.75
	1300	8" channel, 11.5#/L.F.			225	.142		13.20	5.50	.36	19.06	25
	1400	10" channel, 15.3#/L.F.			180	.178		17.15	6.85	.45	24.45	31.50
	1500	12" channel, 20.7#/L.F.			140	.229		23	8.85	.58	32.43	41.50
	2000	For curved edging, add						35%	10%			
440	0010	**LIGHTWEIGHT FRAMING**	R05120-235									440
	0200	For load-bearing steel studs see division 05410-400										
	0400	Angle framing, field fabricated, 4" and larger	R05120-245	E-3	440	.055	Lb.	.51	2.12	.18	2.81	4.55
	0450	Less than 4" angles			265	.091	"	.53	3.52	.31	4.36	7.20
	0460	1/2" x 1/2" x 1/8"			200	.120	L.F.	.11	4.66	.41	5.18	8.90
	0462	3/4" x 3/4" x 1/8"			160	.150		.29	5.80	.51	6.60	11.30
	0464	1" x 1" x 1/8"			135	.178		.42	6.90	.60	7.92	13.45
	0466	1-1/4" x 1-1/4" x 3/16"			115	.209		.78	8.10	.71	9.59	16.15
	0468	1-1/2" x 1-1/2" x 3/16"			100	.240		.95	9.30	.81	11.06	18.60
	0470	2" x 2" x 1/4"			90	.267		1.67	10.35	.90	12.92	21.50
	0472	2-1/2" x 2-1/2" x 1/4"			72	.333		2.15	12.95	1.13	16.23	26.50
	0474	3" x 2" x 3/8"			65	.369		3.10	14.35	1.25	18.70	30.50
	0476	3" x 3" x 3/8"			57	.421		3.78	16.35	1.42	21.55	34.50
	0600	Channel framing, field fabricated, 8" and larger			500	.048	Lb.	.53	1.86	.16	2.55	4.09
	0650	Less than 8" channels			335	.072	"	.53	2.78	.24	3.55	5.85
	0660	C2 x 1.78			115	.209	L.F.	.93	8.10	.71	9.74	16.30
	0662	C3 x 4.1			80	.300		2.15	11.65	1.01	14.81	24.50
	0664	C4 x 5.4			66	.364		2.84	14.10	1.23	18.17	30
	0666	C5 x 6.7			57	.421		3.52	16.35	1.42	21.29	34.50
	0668	C6 x 8.2			55	.436		4.16	16.95	1.47	22.58	36.50
	0670	C7 x 9.8			40	.600		5.15	23.50	2.03	30.68	49.50
	0672	C8 x 11.5			36	.667		6.05	26	2.25	34.30	55.50
	0710	Structural bar tee, field fabricated, 3/4" x 3/4" x 1/8"			160	.150		.29	5.80	.51	6.60	11.30
	0712	1" x 1" x 1/8"			135	.178		.42	6.90	.60	7.92	13.45
	0714	1-1/2" x 1-1/2" x 1/4"			114	.211		1.23	8.15	.71	10.09	16.75
	0716	2" x 2" x 1/4"			89	.270		1.67	10.45	.91	13.03	21.50
	0718	2-1/2" x 2-1/2" x 3/8"			72	.333		3.10	12.95	1.13	17.18	27.50
	0720	3" x 3" x 3/8"			57	.421		3.78	16.35	1.42	21.55	34.50
	0730	Structural zee, field fabricated, 1-1/4" x 1-3/4" x 1-3/4"			114	.211		.40	8.15	.71	9.26	15.80
	0732	2-11/16" x 3" x 2-11/16"			114	.211		.93	8.15	.71	9.79	16.40
	0734	3-1/16" x 4" x 3-1/16"			133	.180		1.41	7	.61	9.02	14.75
	0736	3-1/4" x 5" x 3-1/4"			133	.180		1.93	7	.61	9.54	15.35
	0738	3-1/2" x 6" x 3-1/2"			160	.150		2.90	5.80	.51	9.21	14.15
	0740	Junior beam, field fabricated, 3"			80	.300		2.99	11.65	1.01	15.65	25.50
	0742	4"			72	.333		4.04	12.95	1.13	18.12	28.50
	0744	5"			67	.358		5.25	13.90	1.21	20.36	32

5 METALS

		05120	Structural Steel		CREW	DAILY OUTPUT	LABOR-HOURS	UNIT	MAT.	LABOR	EQUIP.	TOTAL	TOTAL INCL O&P	
									\multicolumn{4}{c}{2005 BARE COSTS}				
440	0746		6"	R05120 -235	E-3	62	.387	L.F.	6.55	15.05	1.31	22.91	35.50	**440**
	0748		7"			57	.421		8.05	16.35	1.42	25.82	39.50	
	0750		8"	R05120 -245		53	.453		9.65	17.60	1.53	28.78	44	
	1000	Continuous slotted channel framing system, shop fab, min			2 Sswk	2,400	.007	Lb.	2.71	.25		2.96	3.44	
	1200	Maximum			"	1,600	.010		3.06	.38		3.44	4.05	
	1300	Cross bracing, rods, shop fabricated, 3/4" diameter			E-3	700	.034		1.05	1.33	.12	2.50	3.67	
	1310	7/8" diameter				850	.028		1.05	1.10	.10	2.25	3.23	
	1320	1" diameter				1,000	.024		1.05	.93	.08	2.06	2.92	
	1330	Angle, 5" x 5" x 3/8"				2,800	.009		1.05	.33	.03	1.41	1.79	
	1350	Hanging lintels, shop fabricated, average				850	.028		1.05	1.10	.10	2.25	3.23	
	1380	Roof frames, shop fabricated, 3'-0" square, 5' span			E-2	4,200	.013		1.05	.49	.32	1.86	2.36	
	1400	Tie rod, not upset, 1-1/2" to 4" diameter, with turnbuckle			2 Sswk	800	.020		1.14	.76		1.90	2.62	
	1420	No turnbuckle				700	.023		1.09	.87		1.96	2.76	
	1500	Upset, 1-3/4" to 4" diameter, with turnbuckle				800	.020		1.14	.76		1.90	2.62	
	1520	No turnbuckle				700	.023		1.09	.87		1.96	2.76	
480	0010	**LINTELS**												**480**
	0020	Plain steel angles, under 500 lb.			1 Bric	550	.015	Lb.	.67	.51		1.18	1.52	
	0100	500 to 1000 lb.	CN			640	.013		.66	.44		1.10	1.39	
	0200	1,000 to 2,000 lb.				640	.013		.64	.44		1.08	1.37	
	0300	2,000 to 4,000 lb.				640	.013		.62	.44		1.06	1.35	
	0500	For built-up angles and plates, add to above							.22			.22	.24	
	0700	For engineering, add to above							.09			.09	.10	
	0900	For galvanizing, add to above, under 500 lb.							.28			.28	.31	
	0950	500 to 2,000 lb.							.26			.26	.28	
	1000	Over 2,000 lb.							.21			.21	.24	
	2000	Steel angles, 3-1/2" x 3", 1/4" thick, 2'-6" long			1 Bric	47	.170	Ea.	9.45	6		15.45	19.55	
	2100	4'-6" long				26	.308		17	10.85		27.85	35	
	2600	4" x 3-1/2", 1/4" thick, 5'-0" long				21	.381		21.50	13.45		34.95	44.50	
	2700	9'-0" long				12	.667		39	23.50		62.50	79	
	3500	For precast concrete lintels, see div. 03480-400												
520	0010	**PIPE SUPPORT FRAMING**												**520**
	0020	Under 10#/L.F.			E-4	3,900	.008	Lb.	1.17	.32	.02	1.51	1.88	
	0200	10.1 to 15#/L.F.				4,300	.007		1.16	.29	.02	1.47	1.80	
	0400	15.1 to 20#/L.F.				4,800	.007		1.14	.26	.02	1.42	1.73	
	0600	Over 20#/L.F.				5,400	.006		1.12	.23	.02	1.37	1.66	
560	0010	**PLATES** Structural steel												**560**
	2010	48"-60" wide, 10'-60' long, 5-50 tons, mill prices												
	2100	1/4" thick						Cwt.	43			43	47.50	
	2150	5/16" thick							43			43	47.50	
	2200	3/8" thick							41			41	45	
	2250	1/2" thick							41			41	45	
	2300	3/4" thick							41			41	45	
	2350	1" thick							41			41	45	
	2400	2" thick							42			42	46.50	
	2450	4" thick							44.50			44.50	49	
	2500	6" thick							46.50			46.50	51.50	
	2550	8" thick							46.50			46.50	51.50	
600	0010	**STRESSED SKIN** Roof & ceiling system												**600**
	0020	Double panel flat roof, spans to 100'			E-2	1,150	.049	S.F.	7	1.80	1.15	9.95	12.05	
	0100	Double panel convex roof, spans to 200'				960	.058		11.40	2.16	1.38	14.94	17.70	
	0200	Double panel arched roof, spans to 300'				760	.074		17.50	2.72	1.74	21.96	26	
640	0010	**STRUCTURAL STEEL MEMBERS**												**640**
	0020	Shop fab'd for 100-ton, 1-2 story project, bolted conn's.												

METALS **5**

153

05120 | Structural Steel

			CREW	DAILY OUTPUT	LABOR-HOURS	UNIT	MAT.	LABOR	EQUIP.	TOTAL	TOTAL INCL O&P	
640	0100	W 6 x 9	E-2	600	.093	L.F.	8.65	3.45	2.21	14.31	17.90	640
	0120	x 16	R05120-210	600	.093		15.40	3.45	2.21	21.06	25.50	
	0140	x 20		600	.093		19.25	3.45	2.21	24.91	29.50	
	0300	W 8 x 10		600	.093		9.65	3.45	2.21	15.31	18.95	
	0320	x 15		600	.093		14.45	3.45	2.21	20.11	24	
	0350	x 21		600	.093		20	3.45	2.21	25.66	30.50	
	0360	x 24		550	.102		23	3.76	2.41	29.17	34.50	
	0370	x 28		550	.102		27	3.76	2.41	33.17	38.50	
	0500	x 31		550	.102		30	3.76	2.41	36.17	42	
	0520	x 35		550	.102		33.50	3.76	2.41	39.67	46	
	0540	x 48		550	.102		46	3.76	2.41	52.17	60	
	0600	W 10 x 12		600	.093		11.55	3.45	2.21	17.21	21	
	0620	x 15		600	.093		14.45	3.45	2.21	20.11	24	
	0700	x 22		600	.093		21	3.45	2.21	26.66	32	
	0720	x 26		600	.093		25	3.45	2.21	30.66	36	
	0740	x 33		550	.102		32	3.76	2.41	38.17	44	
	0900	x 49		550	.102		47	3.76	2.41	53.17	61	
	1100	W 12 x 14		880	.064		13.50	2.35	1.51	17.36	20.50	
	1300	x 22		880	.064		21	2.35	1.51	24.86	29	
	1500	x 26		880	.064		25	2.35	1.51	28.86	33	
	1520	x 35		810	.069		33.50	2.56	1.64	37.70	43	
	1560	x 50		750	.075		48	2.76	1.77	52.53	59.50	
	1580	x 58		750	.075		56	2.76	1.77	60.53	68	
	1700	x 72		640	.087		69.50	3.23	2.07	74.80	84	
	1740	x 87		640	.087		83.50	3.23	2.07	88.80	100	
	1900	W 14 x 26		990	.057		25	2.09	1.34	28.43	32.50	
	2100	x 30		900	.062		29	2.30	1.47	32.77	37.50	
	2300	x 34		810	.069		32.50	2.56	1.64	36.70	42	
	2320	x 43		810	.069		41.50	2.56	1.64	45.70	51.50	
	2340	x 53		800	.070		51	2.59	1.66	55.25	62.50	
	2360	x 74		760	.074		71	2.72	1.74	75.46	85	
	2380	x 90		740	.076		86.50	2.80	1.79	91.09	102	
	2500	x 120		720	.078		116	2.87	1.84	120.71	134	
	2700	W 16 x 26		1,000	.056		25	2.07	1.33	28.40	32.50	
	2900	x 31		900	.062		30	2.30	1.47	33.77	38.50	
	3100	x 40		800	.070		38.50	2.59	1.66	42.75	49	
	3120	x 50		800	.070		48	2.59	1.66	52.25	59.50	
	3140	x 67		760	.074		64.50	2.72	1.74	68.96	77.50	
	3300	W 18 x 35	E-5	960	.083		33.50	3.13	1.46	38.09	44	
	3500	x 40		960	.083		38.50	3.13	1.46	43.09	49.50	
	3520	x 46		960	.083		44.50	3.13	1.46	49.09	55.50	
	3700	x 50		912	.088		48	3.29	1.54	52.83	60.50	
	3900	x 55		912	.088		53	3.29	1.54	57.83	65.50	
	3920	x 65		900	.089		62.50	3.34	1.56	67.40	76.50	
	3940	x 76		900	.089		73	3.34	1.56	77.90	88	
	3960	x 86		900	.089		83	3.34	1.56	87.90	98.50	
	3980	x 106		900	.089		102	3.34	1.56	106.90	120	
	4100	W 21 x 44		1,064	.075		42.50	2.82	1.32	46.64	53	
	4300	x 50		1,064	.075		48	2.82	1.32	52.14	59.50	
	4500	x 62		1,036	.077		59.50	2.90	1.36	63.76	72	
	4700	x 68		1,036	.077		65.50	2.90	1.36	69.76	78.50	
	4720	x 83		1,000	.080		80	3	1.41	84.41	95	
	4740	x 93		1,000	.080		89.50	3	1.41	93.91	105	
	4760	x 101		1,000	.080		97	3	1.41	101.41	114	
	4780	x 122		1,000	.080		117	3	1.41	121.41	136	
	4900	W 24 x 55		1,110	.072		53	2.70	1.27	56.97	64	

5 METALS

Important: See the Reference Section for critical supporting data - Reference Nos., Crews, & City Cost Indexes

05120 | Structural Steel

			CREW	DAILY OUTPUT	LABOR-HOURS	UNIT	2005 BARE COSTS MAT.	LABOR	EQUIP.	TOTAL	TOTAL INCL O&P	
640	5100	x 62	R05120-210	E-5	1,110	.072	L.F.	59.50	2.70	1.27	63.47	71.50	**640**
	5300	x 68			1,110	.072		65.50	2.70	1.27	69.47	78	
	5500	x 76			1,110	.072		73	2.70	1.27	76.97	86.50	
	5700	x 84			1,080	.074		81	2.78	1.30	85.08	95.50	
	5720	x 94			1,080	.074		90.50	2.78	1.30	94.58	106	
	5740	x 104			1,050	.076		100	2.86	1.34	104.20	116	
	5760	x 117			1,050	.076		113	2.86	1.34	117.20	130	
	5780	x 146			1,050	.076		141	2.86	1.34	145.20	161	
	5800	W 27 x 84			1,190	.067		81	2.52	1.18	84.70	94.50	
	5900	x 94			1,190	.067		90.50	2.52	1.18	94.20	105	
	5920	x 114			1,150	.070		110	2.61	1.22	113.83	127	
	5940	x 146			1,150	.070		141	2.61	1.22	144.83	161	
	5960	x 161			1,150	.070		155	2.61	1.22	158.83	176	
	6100	W 30 x 99			1,200	.067		95.50	2.50	1.17	99.17	111	
	6300	x 108			1,200	.067		104	2.50	1.17	107.67	120	
	6500	x 116			1,160	.069		112	2.59	1.21	115.80	129	
	6520	x 132			1,160	.069		127	2.59	1.21	130.80	146	
	6540	x 148			1,160	.069		142	2.59	1.21	145.80	163	
	6560	x 173			1,120	.071		167	2.68	1.26	170.94	189	
	6580	x 191			1,120	.071		184	2.68	1.26	187.94	208	
	6700	W 33 x 118			1,176	.068		114	2.55	1.20	117.75	131	
	6900	x 130			1,134	.071		125	2.65	1.24	128.89	144	
	7100	x 141			1,134	.071		136	2.65	1.24	139.89	155	
	7120	x 169			1,100	.073		163	2.73	1.28	167.01	185	
	7140	x 201			1,100	.073		193	2.73	1.28	197.01	219	
	7300	W 36 x 135			1,170	.068		130	2.57	1.20	133.77	149	
	7500	x 150			1,170	.068		144	2.57	1.20	147.77	165	
	7600	x 170			1,150	.070		164	2.61	1.22	167.83	186	
	7700	x 194			1,125	.071		187	2.67	1.25	190.92	211	
	7900	x 230			1,125	.071		221	2.67	1.25	224.92	250	
	7920	x 260			1,035	.077		250	2.90	1.36	254.26	282	
	8100	x 300			1,035	.077		289	2.90	1.36	293.26	325	
	8490	For projects 75 to 99 tons, add						10%					
	8492	50 to 74 tons, add						20%					
	8494	25 to 49 tons, add						30%	10%				
	8496	10 to 24 tons, add						50%	25%				
	8498	2 to 9 tons, add						75%	50%				
	8499	Less than 2 tons, add						100%	100%				
680	0010	**STRUCTURAL STEEL PROJECTS**	R05080-310										**680**
	0020	Shop fab'd for 100-ton, 1-2 story project, bolted conn's.											
	0200	Apartments, nursing homes, etc., 1 to 2 stories	R05090-510	E-5	10.30	7.767	Ton	1,750	291	137	2,178	2,575	
	0300	3 to 6 stories		"	10.10	7.921		1,775	297	139	2,211	2,650	
	0400	7 to 15 stories	R05120-210	E-6	14.20	9.014		1,825	340	107	2,272	2,700	
	0500	Over 15 stories		"	13.90	9.209		1,900	345	109	2,354	2,800	
	0700	Offices, hospitals, etc., steel bearing, 1 to 2 stories **CN**	R05120-220	E-5	10.30	7.767		1,750	291	137	2,178	2,575	
	0800	3 to 6 stories		E-6	14.40	8.889		1,775	335	105	2,215	2,675	
	0900	7 to 15 stories	R05120-225		14.20	9.014		1,825	340	107	2,272	2,700	
	1000	Over 15 stories			13.90	9.209		1,900	345	109	2,354	2,800	
	1100	For multi-story masonry wall bearing construction, add	R05120-230						30%				
	1300	Industrial bldgs., 1 story, beams & girders, steel bearing		E-5	12.90	6.202		1,750	233	109	2,092	2,450	
	1400	Masonry bearing		"	10	8		1,750	300	141	2,191	2,600	
	1500	Industrial bldgs., 1 story, under 10 tons,											
	1510	steel from warehouse, trucked		E-2	7.50	7.467	Ton	2,100	276	177	2,553	2,975	
	1600	1 story with roof trusses, steel bearing		E-5	10.60	7.547		2,075	283	133	2,491	2,925	
	1700	Masonry bearing		"	8.30	9.639		2,075	360	169	2,604	3,100	
	1900	Monumental structures, banks, stores, etc., minimum		E-6	13	9.846		1,750	370	117	2,237	2,700	

05120 | Structural Steel

Line	Description	Ref	Crew	Daily Output	Labor-Hours	Unit	Mat.	Labor	Equip.	Total	Total Incl O&P
2000	Maximum	R05080-310	E-6	9	14.222	Ton	2,900	535	169	3,604	4,325
2200	Churches, minimum		E-5	11.60	6.897		1,625	259	121	2,005	2,375
2300	Maximum	R05090-510	"	5.20	15.385		2,175	575	270	3,020	3,675
2800	Power stations, fossil fuels, minimum		E-6	11	11.636		1,750	435	138	2,323	2,825
2900	Maximum	R05120-210		5.70	22.456		2,625	845	266	3,736	4,675
2950	Nuclear fuels, non-safety steel, minimum			7	18.286		1,750	685	217	2,652	3,375
3000	Maximum	R05120-220		5.50	23.273		2,625	875	276	3,776	4,725
3040	Safety steel, minimum			2.50	51.200		2,550	1,925	605	5,080	6,825
3070	Maximum			1.50	85.333		3,350	3,200	1,000	7,550	10,400
3100	Roof trusses, minimum	R05120-225	E-5	13	6.154		2,450	231	108	2,789	3,225
3200	Maximum			8.30	9.639		2,975	360	169	3,504	4,100
3210	Schools, minimum	R05120-230		14.50	5.517		1,750	207	97	2,054	2,400
3220	Maximum			8.30	9.639		2,550	360	169	3,079	3,625
3400	Welded construction, simple commercial bldgs., 1 to 2 stories		E-7	7.60	10.526		1,775	395	196	2,366	2,875
3500	7 to 15 stories		E-9	8.30	15.422		2,075	580	209	2,864	3,500
3700	Welded rigid frame, 1 story, minimum		E-7	15.80	5.063		1,825	190	94	2,109	2,425
3800	Maximum		"	5.50	14.545		2,375	545	270	3,190	3,850
3900	High strength steel mill spec extras: A242, A441,										
3950	A529, A572 (42 ksi) and A992: same as A36 steel										
4000	Add to A992 price for A572 (50, 60, 65 ksi)					Ton	25			25	27.50
4100	A588 Weathering					"	60			60	66
4200	Mill size extras for W-Shapes: 0 to 30 plf: no extra charge										
4210	Member sizes 31 to 65 plf, add					Ton	5			5	5.50
4220	Member sizes 66 to 100 plf, add						20			20	22
4230	Member sizes 101 to 387 plf, add						40			40	44
4300	Column base plates, light, up to 150 lb		2 Sswk	2,000	.008	Lb.	.96	.31		1.27	1.61
4400	Heavy, over 150 lb		E-2	7,500	.007	"	1.01	.28	.18	1.47	1.77
4600	Castellated beams, light sections, to 50#/L.F., minimum			10.70	5.234	Ton	1,850	193	124	2,167	2,500
4700	Maximum			7	8		2,025	296	189	2,510	2,950
4900	Heavy sections, over 50# per L.F., minimum			11.70	4.786		1,925	177	113	2,215	2,550
5000	Maximum			7.80	7.179		2,100	265	170	2,535	2,950
5390	For projects 75 to 99 tons, add						10%				
5392	50 to 74 tons, add						20%				
5394	25 to 49 tons, add						30%	10%			
5396	10 to 24 tons, add						50%	25%			
5398	2 to 9 tons, add						75%	50%			
5399	Less than 2 tons, add						100%	100%			
5500	Steel domes - see R13128-310										
5700	Steel estimating weights per S.F. - see R05120-220										

05140 | Structural Aluminum

Line	Description	Crew	Daily Output	Labor-Hours	Unit	Mat.	Labor	Equip.	Total	Total Incl O&P
0010	**ALUMINUM**									
0020	Structural shapes, 1" to 10" members, under 1 ton	E-2	1,050	.053	Lb.	2.19	1.97	1.26	5.42	7.20
0050	1 to 5 tons		1,330	.042		2.09	1.56	1	4.65	6.05
0100	Over 5 tons CN		1,330	.042		2.03	1.56	1	4.59	6
0300	Extrusions, over 5 tons, stock shapes		1,330	.042		2.19	1.56	1	4.75	6.20
0400	Custom shapes		1,330	.042		2.23	1.56	1	4.79	6.20
0600	For formed aluminum columns, see div. 05580-200									

05150 | Wire Rope Assemblies

Line	Description	Unit	Mat.	Total	Total Incl O&P
0010	**STEEL WIRE ROPE**				
0020	6 x 19, bright, fiber core, 5000' rolls, 1/2" diameter	L.F.	.80	.80	.88
0050	Steel core		1.06	1.06	1.16
0100	Fiber core, 1" diameter		2.71	2.71	2.98

5 METALS

Important: See the Reference Section for critical supporting data - Reference Nos., Crews, & City Cost Indexes

05100 | Structural Metal Framing

05150 | Wire Rope Assemblies

		CREW	DAILY OUTPUT	LABOR-HOURS	UNIT	2005 BARE COSTS				TOTAL INCL O&P
						MAT.	LABOR	EQUIP.	TOTAL	
0150	Steel core				L.F.	3.09			3.09	3.40
0300	6 x 19, galvanized, fiber core, 1/2" diameter					1.18			1.18	1.30
0350	Steel core					1.35			1.35	1.49
0400	Fiber core, 1" diameter					3.47			3.47	3.82
0450	Steel core					3.64			3.64	4
0500	6 x 7, bright, IPS, fiber core, <500 L.F. w/acc., 1/4" diameter	E-17	6,400	.002		.60	.10		.70	.84
0510	1/2" diameter		2,100	.008		1.46	.30		1.76	2.14
0520	3/4" diameter		960	.017		2.64	.65	.01	3.30	4.09
0550	6 x 19, bright, IPS, IWRC, <500 L.F. w/acc., 1/4" diameter		5,760	.003		.89	.11		1	1.17
0560	1/2" diameter		1,730	.009		1.44	.36		1.80	2.23
0570	3/4" diameter		770	.021		2.50	.81	.01	3.32	4.22
0580	1" diameter		420	.038		4.23	1.49	.01	5.73	7.35
0590	1-1/4" diameter		290	.055		7	2.16	.02	9.18	11.60
0600	1-1/2" diameter	▼	192	.083		8.65	3.26	.03	11.94	15.40
0610	1-3/4" diameter	E-18	240	.167		13.75	6.30	3.32	23.37	29.50
0620	2" diameter		160	.250		17.65	9.45	4.98	32.08	41.50
0630	2-1/4" diameter	▼	160	.250		23.50	9.45	4.98	37.93	48
0650	6 x 37, bright, IPS, IWRC, <500 L.F. w/acc., 1/4" diameter	E-17	6,400	.002		1.12	.10		1.22	1.42
0660	1/2" diameter		1,730	.009		1.91	.36		2.27	2.75
0670	3/4" diameter		770	.021		3.08	.81	.01	3.90	4.86
0680	1" diameter		430	.037		4.89	1.46	.01	6.36	8
0690	1-1/4" diameter		290	.055		7.40	2.16	.02	9.58	12
0700	1-1/2" diameter	▼	190	.084		10.55	3.30	.03	13.88	17.55
0710	1-3/4" diameter	E-18	260	.154		16.75	5.80	3.07	25.62	32
0720	2" diameter		200	.200		22	7.55	3.99	33.54	41.50
0730	2-1/4" diameter	▼	160	.250		29	9.45	4.98	43.43	53.50
0800	6 x 19 & 6 x 37, swaged, 1/2" diameter	E-17	1,220	.013		3.43	.51		3.94	4.69
0810	9/16" diameter		1,120	.014		3.98	.56		4.54	5.40
0820	5/8" diameter		930	.017		4.72	.67	.01	5.40	6.40
0830	3/4" diameter		640	.025		6	.98	.01	6.99	8.35
0840	7/8" diameter		480	.033		7.60	1.30	.01	8.91	10.70
0850	1" diameter		350	.046		9.25	1.79	.02	11.06	13.40
0860	1-1/8" diameter		288	.056		11.40	2.18	.02	13.60	16.45
0870	1-1/4" diameter		230	.070		13.80	2.72	.02	16.54	20
0880	1-3/8" diameter	▼	192	.083		15.95	3.26	.03	19.24	23.50
0890	1-1/2" diameter	E-18	300	.133	▼	19.35	5.05	2.66	27.06	33

05200 | Metal Joists

05210 | Steel Joists

		CREW	DAILY OUTPUT	LABOR-HOURS	UNIT	2005 BARE COSTS				TOTAL INCL O&P
						MAT.	LABOR	EQUIP.	TOTAL	
0010	**OPEN WEB JOISTS**, 40-ton job lots									
0020	K series, horizontal bridging, spans up to 30', minimum	E-7	15	5.333	Ton	1,075	200	99	1,374	1,650
0050	Average		12	6.667		1,225	250	124	1,599	1,925
0080	Maximum		9	8.889	▼	1,450	335	165	1,950	2,350
0130	8K1, 5.1 Lb/LF		1,200	.067	L.F.	3.11	2.50	1.24	6.85	9.15
0140	10K1, 5.0 Lb/LF		1,200	.067		3.05	2.50	1.24	6.79	9.05
0160	12K3, 5.7 Lb/LF		1,500	.053		3.47	2	.99	6.46	8.40
0180	14K3, 6.0 Lb/LF		1,500	.053		3.65	2	.99	6.64	8.60
0200	16K3, 6.3 Lb/LF		1,800	.044		3.84	1.67	.83	6.34	8.05
0220	16K6, 8.1 Lb/LF	▼	1,800	.044	▼	4.93	1.67	.83	7.43	9.25

	05210	Steel Joists	CREW	DAILY OUTPUT	LABOR-HOURS	UNIT	2005 BARE COSTS				TOTAL INCL O&P	
							MAT.	LABOR	EQUIP.	TOTAL		
600	0240	18K5, 7.7 Lb/LF	E-7	2,000	.040	L.F.	4.69	1.50	.74	6.93	8.60	600
	0260	18K9, 10.2 Lb/LF		2,000	.040		6.20	1.50	.74	8.44	10.30	
	0410	Span 30' to 50', minimum		17	4.706	Ton	1,075	177	87.50	1,339.50	1,575	
	0440	Average		17	4.706		1,200	177	87.50	1,464.50	1,725	
	0460	Maximum		10	8		1,275	300	149	1,724	2,075	
	0500	20K5, 8.2 Lb/LF		2,000	.040	L.F.	4.92	1.50	.74	7.16	8.85	
	0520	20K9, 10.8 Lb/LF		2,000	.040		6.50	1.50	.74	8.74	10.60	
	0540	22K5, 8.8 Lb/LF		2,000	.040		5.30	1.50	.74	7.54	9.25	
	0560	22K9, 11.3 Lb/LF		2,000	.040		6.80	1.50	.74	9.04	10.90	
	0580	24K6, 9.7 Lb/LF		2,200	.036		5.80	1.36	.68	7.84	9.50	
	0600	24K10, 13.1 Lb/LF		2,200	.036		7.85	1.36	.68	9.89	11.75	
	0620	26K6, 10.6 Lb/LF		2,200	.036		6.35	1.36	.68	8.39	10.10	
	0640	26K10, 13.8 Lb/LF		2,200	.036		8.30	1.36	.68	10.34	12.20	
	0660	28K8, 12.7 Lb/LF		2,400	.033		7.60	1.25	.62	9.47	11.25	
	0680	28K12, 17.1 Lb/LF		2,400	.033		10.25	1.25	.62	12.12	14.15	
	0700	30K8, 13.2 Lb/LF		2,400	.033		7.90	1.25	.62	9.77	11.55	
	0720	30K12, 17.6 Lb/LF		2,400	.033		10.55	1.25	.62	12.42	14.45	
	1010	CS series, horizontal bridging										
	1020	Spans to 30', minimum	E-7	15	5.333	Ton	1,125	200	99	1,424	1,700	
	1040	Average		12	6.667		1,250	250	124	1,624	1,950	
	1060	Maximum		9	8.889		1,475	335	165	1,975	2,375	
	1100	10CS2, 7.5 Lb/LF		1,200	.067	L.F.	4.69	2.50	1.24	8.43	10.85	
	1120	12CS2, 8.0 Lb/LF		1,500	.053		5	2	.99	7.99	10.05	
	1140	14CS2, 8.0 Lb/LF		1,500	.053		5	2	.99	7.99	10.05	
	1160	16CS2, 8.5 Lb/LF		1,800	.044		5.30	1.67	.83	7.80	9.65	
	1180	16CS4, 14.5 Lb/LF		1,800	.044		9.05	1.67	.83	11.55	13.75	
	1200	18CS2, 9.0 Lb/LF		2,000	.040		5.65	1.50	.74	7.89	9.65	
	1220	18CS4, 15.0 Lb/LF		2,000	.040		9.40	1.50	.74	11.64	13.75	
	1240	20CS2, 9.5 Lb/LF		2,000	.040		5.95	1.50	.74	8.19	10	
	1260	20CS4, 16.5 Lb/LF		2,000	.040		10.30	1.50	.74	12.54	14.80	
	1280	22CS2, 10.0 Lb/LF		2,000	.040		6.25	1.50	.74	8.49	10.35	
	1300	22CS4, 16.5 Lb/LF		2,000	.040		10.30	1.50	.74	12.54	14.80	
	1320	24CS2, 10.0 Lb/LF		2,200	.036		6.25	1.36	.68	8.29	10	
	1340	24CS4, 16.5 Lb/LF		2,200	.036		10.30	1.36	.68	12.34	14.45	
	1360	26CS2, 10.0 Lb/LF		2,200	.036		6.25	1.36	.68	8.29	10	
	1380	26CS4, 16.5 Lb/LF		2,200	.036		10.30	1.36	.68	12.34	14.45	
	1400	28CS2, 10.5 Lb/LF		2,400	.033		6.55	1.25	.62	8.42	10.05	
	1420	28CS4, 16.5 Lb/LF		2,400	.033		10.30	1.25	.62	12.17	14.20	
	1440	30CS2, 11.0 Lb/LF		2,400	.033		6.90	1.25	.62	8.77	10.40	
	1460	30CS4, 16.5 Lb/LF		2,400	.033		10.30	1.25	.62	12.17	14.20	
	2000	LH series, bolted cross bridging										
	2020	Spans to 96', minimum	E-7	16	5	Ton	1,225	188	93	1,506	1,775	
	2040	Average		13	6.154		1,350	231	114	1,695	2,000	
	2080	Maximum		11	7.273		1,600	273	135	2,008	2,375	
	2200	18LH04, 12 Lb/LF		1,400	.057	L.F.	8.10	2.14	1.06	11.30	13.80	
	2220	18LH08, 19 Lb/LF		1,400	.057		12.80	2.14	1.06	16	19	
	2240	20LH04, 12 Lb/LF		1,400	.057		8.10	2.14	1.06	11.30	13.80	
	2260	20LH08, 19 Lb/LF		1,400	.057		12.80	2.14	1.06	16	19	
	2280	24LH05, 13 Lb/LF		1,400	.057		8.75	2.14	1.06	11.95	14.55	
	2300	24LH10, 23 Lb/LF		1,400	.057		15.50	2.14	1.06	18.70	22	
	2320	28LH06, 16 Lb/LF		1,800	.044		10.80	1.67	.83	13.30	15.65	
	2340	28LH11, 25 Lb/LF		1,800	.044		16.85	1.67	.83	19.35	22.50	
	2360	32LH08, 17 Lb/LF		1,800	.044		11.45	1.67	.83	13.95	16.40	
	2380	32LH13, 30 Lb/LF		1,800	.044		20	1.67	.83	22.50	26.50	
	2400	36LH09, 21 Lb/LF		1,800	.044		14.15	1.67	.83	16.65	19.40	
	2420	36LH14, 36 Lb/LF		1,800	.044		24.50	1.67	.83	27	30.50	

CN

5

METALS

Important: See the Reference Section for critical supporting data - Reference Nos., Crews, & City Cost Indexes

05210	Steel Joists	CREW	DAILY OUTPUT	LABOR-HOURS	UNIT	2005 BARE COSTS				TOTAL INCL O&P
						MAT.	LABOR	EQUIP.	TOTAL	
2440	40LH10, 21 Lb/LF	E-7	2,200	.036	L.F.	14.15	1.36	.68	16.19	18.70
2460	40LH15, 36 Lb/LF		2,200	.036		24.50	1.36	.68	26.54	29.50
2480	44LH11, 22 Lb/LF		2,200	.036		14.85	1.36	.68	16.89	19.40
2500	44LH16, 42 Lb/LF		2,200	.036		28.50	1.36	.68	30.54	34
2520	48LH11, 22 Lb/LF		2,200	.036		14.85	1.36	.68	16.89	19.40
2540	48LH16, 42 Lb/LF		2,200	.036		28.50	1.36	.68	30.54	34
3010	DLH series, bolted cross bridging									
3020	Spans to 144' (shipped in 2 pieces), minimum	E-7	16	5	Ton	1,325	188	93	1,606	1,900
3040	Average		13	6.154		1,425	231	114	1,770	2,100
3100	Maximum		11	7.273		1,725	273	135	2,133	2,525
3200	52DLH11, 26 Lb/LF		2,000	.040	L.F.	17.95	1.50	.74	20.19	23
3220	52DLH16, 45 Lb/LF		2,000	.040		32.50	1.50	.74	34.74	39
3240	56DLH11, 26 Lb/LF		2,000	.040		18.65	1.50	.74	20.89	24
3260	56DLH16, 46 Lb/LF		2,000	.040		33	1.50	.74	35.24	40
3280	60DLH12, 29 Lb/LF		2,000	.040		21	1.50	.74	23.24	26.50
3300	60DLH17, 52 Lb/LF		2,000	.040		37.50	1.50	.74	39.74	44.50
3320	64DLH12, 31 Lb/LF		2,200	.036		22	1.36	.68	24.04	27.50
3340	64DLH17, 52 Lb/LF		2,200	.036		37.50	1.36	.68	39.54	44
3360	68DLH13, 37 Lb/LF		2,200	.036		26.50	1.36	.68	28.54	32
3380	68DLH18, 61 Lb/LF		2,200	.036		44	1.36	.68	46.04	51
3400	72DLH14, 41 Lb/LF		2,200	.036		29.50	1.36	.68	31.54	35.50
3420	72DLH19, 70 Lb/LF		2,200	.036		50	1.36	.68	52.04	58
4010	SLH series, bolted cross bridging									
4020	Spans to 200' (shipped in 3 pieces), minimum	E-7	16	5	Ton	1,300	188	93	1,581	1,875
4040	Average		13	6.154		1,475	231	114	1,820	2,150
4060	Maximum		11	7.273		1,750	273	135	2,158	2,550
4200	80SLH15, 40 Lb/LF		1,500	.053	L.F.	29.50	2	.99	32.49	37
4220	80SLH20, 75 Lb/LF		1,500	.053		55.50	2	.99	58.49	65.50
4240	88SLH16, 46 Lb/LF		1,500	.053		34	2	.99	36.99	42
4260	88SLH21, 89 Lb/LF		1,500	.053		65.50	2	.99	68.49	76.50
4280	96SLH17, 52 Lb/LF		1,500	.053		38.50	2	.99	41.49	46.50
4300	96SLH22, 102 Lb/LF		1,500	.053		75	2	.99	77.99	87
4320	104SLH18, 59 Lb/LF		1,800	.044		43.50	1.67	.83	46	52
4340	104SLH23, 109 Lb/LF		1,800	.044		80.50	1.67	.83	83	92.50
4360	112SLH19, 67 Lb/LF		1,800	.044		49.50	1.67	.83	52	58.50
4380	112SLH24, 131 Lb/LF		1,800	.044		96.50	1.67	.83	99	110
4400	120SLH20, 77 Lb/LF		1,800	.044		57	1.67	.83	59.50	66.50
4420	120SLH25, 152 Lb/LF		1,800	.044		112	1.67	.83	114.50	127
6000	For welded cross bridging, add						30%			
6100	For less than 40-ton job lots									
6102	For 30 to 39 tons, add					10%				
6104	20 to 29 tons, add					20%				
6106	10 to 19 tons, add					30%				
6107	5 to 9 tons, add					50%	25%			
6108	1 to 4 tons, add					75%	50%			
6109	Less than 1 ton, add					100%	100%			
6200	For shop prime paint other than mfrs. standard, add					20%				
6300	For bottom chord extensions, add per chord				Ea.	26.50			26.50	29
7000	Joist girders, minimum	E-5	15	5.333	Ton	1,100	200	94	1,394	1,675
7020	Average		13	6.154		1,225	231	108	1,564	1,875
7040	Maximum		11	7.273		1,275	273	128	1,676	2,025
8000	Trusses, factory fabricated WT chords, average		11	7.273		4,000	273	128	4,401	5,000

600

METALS 5

			CREW	DAILY OUTPUT	LABOR-HOURS	UNIT	MAT.	LABOR	EQUIP.	TOTAL	TOTAL INCL O&P
05310		**Steel Deck**									
300	0010	**METAL DECKING** Steel decking R05310-100									300
	0200	Cellular units, galv, 2" deep, 20-20 gauge, over 15 squares	E-4	1,460	.022	S.F.	5.85	.85	.06	6.76	8.05
	0250	18-20 gauge		1,420	.023		6.65	.87	.06	7.58	8.90
	0300	18-18 gauge		1,390	.023		6.85	.89	.06	7.80	9.15
	0320	16-18 gauge		1,360	.024		8.15	.91	.06	9.12	10.65
	0340	16-16 gauge		1,330	.024		9.05	.93	.06	10.04	11.70
	0400	3" deep, galvanized, 20-20 gauge		1,375	.023		6.45	.90	.06	7.41	8.75
	0500	18-20 gauge		1,350	.024		7.80	.92	.06	8.78	10.25
	0600	18-18 gauge		1,290	.025		7.75	.96	.06	8.77	10.35
	0700	16-18 gauge		1,230	.026		8.75	1.01	.07	9.83	11.45
	0800	16-16 gauge		1,150	.028		9.55	1.08	.07	10.70	12.50
	1000	4-1/2" deep, galvanized, 20-18 gauge		1,100	.029		9	1.12	.07	10.19	12
	1100	18-18 gauge		1,040	.031		8.95	1.19	.08	10.22	12.05
	1200	16-18 gauge		980	.033		10.05	1.26	.08	11.39	13.40
	1300	16-16 gauge		935	.034		10.95	1.32	.09	12.36	14.50
	1500	For acoustical deck, add					15%				
	1700	For cells used for ventilation, add					15%				
	1900	For multi-story or congested site, add						50%			
	2100	Open type, galv., 1-1/2" deep wide rib, 22 gauge, under 50 squares	E-4	4,500	.007	S.F.	1.47	.27	.02	1.76	2.12
	2200	50-500 squares CN		4,900	.007		1.14	.25	.02	1.41	1.72
	2400	Over 500 squares		5,100	.006		1.06	.24	.02	1.32	1.61
	2600	20 gauge, under 50 squares		3,865	.008		1.73	.32	.02	2.07	2.49
	2650	50-500 squares		4,170	.008		1.38	.30	.02	1.70	2.07
	2700	Over 500 squares		4,300	.007		1.24	.29	.02	1.55	1.90
	2900	18 gauge, under 50 squares		3,800	.008		2.24	.33	.02	2.59	3.06
	2950	50-500 squares		4,100	.008		1.79	.30	.02	2.11	2.53
	3000	Over 500 squares		4,300	.007		1.61	.29	.02	1.92	2.30
	3050	16 gauge, under 50 squares		3,700	.009		3.01	.33	.02	3.36	3.93
	3060	50-500 squares		4,000	.008		2.41	.31	.02	2.74	3.22
	3100	Over 500 squares		4,200	.008		2.17	.29	.02	2.48	2.94
	3150	For intermediate rib instead of wide rib, deduct					.03			.03	.03
	3160	For narrow rib instead of wide rib, add					.45			.45	.50
	3200	3" deep, 22 gauge, under 50 squares	E-4	3,600	.009		2.02	.34	.02	2.38	2.85
	3250	50-500 squares		3,800	.008		1.62	.33	.02	1.97	2.38
	3260	over 500 squares		4,000	.008		1.46	.31	.02	1.79	2.17
	3300	20 gauge, under 50 squares		3,400	.009		2.36	.36	.02	2.74	3.27
	3350	50-500 squares		3,600	.009		1.88	.34	.02	2.24	2.70
	3360	over 500 squares		3,800	.008		1.70	.33	.02	2.05	2.47
	3400	18 gauge, under 50 squares		3,200	.010		3.05	.39	.03	3.47	4.08
	3450	50-500 squares		3,400	.009		2.44	.36	.02	2.82	3.37
	3460	over 500 squares		3,600	.009		2.20	.34	.02	2.56	3.05
	3500	16 gauge, under 50 squares		3,000	.011		4.02	.41	.03	4.46	5.20
	3550	50-500 squares		3,200	.010		3.21	.39	.03	3.63	4.25
	3560	over 500 squares		3,400	.009		2.89	.36	.02	3.27	3.86
	3700	4-1/2" deep, long span roof, over 50 squares, 20 gauge		2,700	.012		3.78	.46	.03	4.27	5
	3800	18 gauge		2,460	.013		4.88	.50	.03	5.41	6.30
	3900	16 gauge		2,350	.014		3.64	.53	.03	4.20	4.98
	4100	6" deep, long span, 18 gauge		2,000	.016		7	.62	.04	7.66	8.85
	4200	16 gauge		1,930	.017		5.20	.64	.04	5.88	6.90
	4300	14 gauge		1,860	.017		6.70	.66	.04	7.40	8.60
	4500	7-1/2" deep, long span, 18 gauge		1,690	.019		7.65	.73	.05	8.43	9.80
	4600	16 gauge		1,590	.020		5.70	.78	.05	6.53	7.75
	4700	14 gauge		1,490	.021		7.35	.83	.05	8.23	9.65
	4800	For painted instead of galvanized, deduct					2%				
	5000	For acoustical perforated, with fiberglass, add				S.F.	1			1	1.10
	5200	Non-cellular composite deck, galv., 2" deep, 22 gauge	E-4	3,860	.008		1.40	.32	.02	1.74	2.13

Important: See the Reference Section for critical supporting data - Reference Nos., Crews, & City Cost Indexes

5 METALS

05300 | Metal Deck

05310 | Steel Deck

			CREW	DAILY OUTPUT	LABOR-HOURS	UNIT	2005 BARE COSTS MAT.	LABOR	EQUIP.	TOTAL	TOTAL INCL O&P	
300	5300	20 gauge R05310-100	E-4	3,600	.009	S.F.	1.55	.34	.02	1.91	2.34	300
	5400	18 gauge		3,380	.009		1.97	.37	.02	2.36	2.85	
	5500	16 gauge		3,200	.010		2.46	.39	.03	2.88	3.43	
	5700	3" deep, galv., 22 gauge		3,200	.010		1.53	.39	.03	1.95	2.40	
	5800	20 gauge		3,000	.011		1.71	.41	.03	2.15	2.65	
	5900	18 gauge CN		2,850	.011		2.10	.43	.03	2.56	3.12	
	6000	16 gauge		2,700	.012		2.80	.46	.03	3.29	3.93	
	6100	Slab form, steel, 28 gauge, 9/16" deep, uncoated		4,000	.008		.98	.31	.02	1.31	1.65	
	6200	Galvanized		4,000	.008		.87	.31	.02	1.20	1.53	
	6220	24 gauge, 1" deep, uncoated		3,900	.008		1.07	.32	.02	1.41	1.77	
	6240	Galvanized		3,900	.008		1.26	.32	.02	1.60	1.98	
	6300	24 gauge, 1-5/16" deep, uncoated		3,800	.008		1.14	.33	.02	1.49	1.85	
	6400	Galvanized		3,800	.008		1.34	.33	.02	1.69	2.07	
	6500	22 gauge, 1-5/16" deep, uncoated		3,700	.009		1.43	.33	.02	1.78	2.19	
	6600	Galvanized		3,700	.009		1.46	.33	.02	1.81	2.23	
	6700	22 gauge, 2" deep uncoated		3,600	.009		1.89	.34	.02	2.25	2.71	
	6800	Galvanized		3,600	.009		1.85	.34	.02	2.21	2.67	
	7000	Sheet metal edge closure form, 12" wide with 2 bends, galv	E-14	360	.022	L.F.	3.04	.89	.23	4.16	5.20	
	7100	18 gauge	"	360	.022	"	4.12	.89	.23	5.24	6.40	
	7200	16 gauge										
	8000	Metal deck and trench, 2" thick, 20 gauge, combination	R-4	1,100	.036	S.F.	11.70	1.42	.07	13.19	15.40	
	8010	60% cellular, 40% non-cellular, inserts and trench										

05400 | Cold-Formed Metal Framing

05410 | Load-Bearing Metal Studs

			CREW	DAILY OUTPUT	LABOR-HOURS	UNIT	2005 BARE COSTS MAT.	LABOR	EQUIP.	TOTAL	TOTAL INCL O&P	
100	0010	**BRACING**, shear wall X-bracing, per 10' x 10' bay, one face										100
	0120	Metal strap, 20 ga x 4" wide	2 Carp	18	.889	Ea.	22.50	30.50		53	72.50	
	0130	6" wide		18	.889		35	30.50		65.50	86	
	0160	18 ga x 4" wide		16	1		32.50	34.50		67	89	
	0170	6" wide		16	1		47.50	34.50		82	106	
	0410	Continuous strap bracing, per horizontal row on both faces										
	0420	Metal strap, 20 ga x 2" wide, studs 12" O.C.	1 Carp	7	1.143	C.L.F.	57	39		96	124	
	0430	16" O.C.		8	1		57	34.50		91.50	117	
	0440	24" O.C.		10	.800		57	27.50		84.50	106	
	0450	18 ga x 2" wide, studs 12" O.C.		6	1.333		79	45.50		124.50	158	
	0460	16" O.C.		7	1.143		79	39		118	148	
	0470	24" O.C.		8	1		79	34.50		113.50	141	
120	0010	**BRIDGING**, solid between studs w/ 1-1/4" leg track, per stud bay										120
	0200	Studs 12" O.C., 18 ga x 2-1/2" wide	1 Carp	125	.064	Ea.	.97	2.19		3.16	4.47	
	0210	3-5/8" wide		120	.067		1.17	2.28		3.45	4.85	
	0220	4" wide		120	.067		1.24	2.28		3.52	4.92	
	0230	6" wide		115	.070		1.61	2.38		3.99	5.50	
	0240	8" wide		110	.073		2.05	2.49		4.54	6.15	
	0300	16 ga x 2-1/2" wide		115	.070		1.21	2.38		3.59	5.05	
	0310	3-5/8" wide		110	.073		1.47	2.49		3.96	5.50	
	0320	4" wide		110	.073		1.57	2.49		4.06	5.60	
	0330	6" wide		105	.076		2.02	2.61		4.63	6.30	
	0340	8" wide		100	.080		2.57	2.74		5.31	7.10	
	1200	Studs 16" O.C., 18 ga x 2-1/2" wide		125	.064		1.24	2.19		3.43	4.77	

METALS 5

05410 | Load-Bearing Metal Studs

5 METALS

		CREW	DAILY OUTPUT	LABOR-HOURS	UNIT	MAT.	LABOR	EQUIP.	TOTAL	TOTAL INCL O&P
120 1210	3-5/8" wide	1 Carp	120	.067	Ea.	1.50	2.28		3.78	5.20 **120**
1220	4" wide		120	.067		1.58	2.28		3.86	5.30
1230	6" wide		115	.070		2.06	2.38		4.44	6
1240	8" wide		110	.073		2.62	2.49		5.11	6.75
1300	16 ga x 2-1/2" wide		115	.070		1.55	2.38		3.93	5.40
1310	3-5/8" wide		110	.073		1.88	2.49		4.37	5.95
1320	4" wide		110	.073		2.01	2.49		4.50	6.10
1330	6" wide		105	.076		2.59	2.61		5.20	6.90
1340	8" wide		100	.080		3.30	2.74		6.04	7.90
2200	Studs 24" O.C., 18 ga x 2-1/2" wide		125	.064		1.79	2.19		3.98	5.40
2210	3-5/8" wide		120	.067		2.17	2.28		4.45	5.95
2220	4" wide		120	.067		2.29	2.28		4.57	6.10
2230	6" wide		115	.070		2.98	2.38		5.36	7
2240	8" wide		110	.073		3.80	2.49		6.29	8.05
2300	16 ga x 2-1/2" wide		115	.070		2.24	2.38		4.62	6.20
2310	3-5/8" wide		110	.073		2.72	2.49		5.21	6.85
2320	4" wide		110	.073		2.91	2.49		5.40	7.10
2330	6" wide		105	.076		3.75	2.61		6.36	8.20
2340	8" wide		100	.080		4.77	2.74		7.51	9.50
3000	Continuous bridging, per row									
3100	16 ga x 1-1/2" channel thru studs 12" O.C.	1 Carp	6	1.333	C.L.F.	54	45.50		99.50	131
3110	16" O.C.		7	1.143		54	39		93	121
3120	24" O.C.		8.80	.909		54	31		85	108
4100	2" x 2" angle x 18 ga, studs 12" O.C.		7	1.143		79	39		118	148
4110	16" O.C.		9	.889		79	30.50		109.50	135
4120	24" O.C.		12	.667		79	23		102	123
4200	16 ga, studs 12" O.C.		5	1.600		100	55		155	196
4210	16" O.C.		7	1.143		100	39		139	171
4220	24" O.C.		10	.800		100	27.50		127.50	153
300 0010	**FRAMING,** boxed headers/beams									**300**
0200	Double, 18 ga x 6" deep	2 Carp	220	.073	L.F.	5.60	2.49		8.09	10.05
0210	8" deep		210	.076		6.20	2.61		8.81	10.90
0220	10" deep		200	.080		7.55	2.74		10.29	12.55
0230	12 " deep		190	.084		8.30	2.88		11.18	13.60
0300	16 ga x 8" deep		180	.089		7.15	3.04		10.19	12.60
0310	10" deep		170	.094		8.60	3.22		11.82	14.45
0320	12 " deep		160	.100		9.35	3.43		12.78	15.65
0400	14 ga x 10" deep		140	.114		10	3.91		13.91	17.10
0410	12 " deep		130	.123		10.95	4.22		15.17	18.60
1210	Triple, 18 ga x 8" deep		170	.094		9	3.22		12.22	14.90
1220	10" deep		165	.097		10.85	3.32		14.17	17.05
1230	12 " deep		160	.100		11.95	3.43		15.38	18.50
1300	16 ga x 8" deep		145	.110		10.40	3.78		14.18	17.30
1310	10" deep		140	.114		12.40	3.91		16.31	19.75
1320	12 " deep		135	.119		13.55	4.06		17.61	21
1400	14 ga x 10" deep		115	.139		13.65	4.77		18.42	22.50
1410	12 " deep		110	.145		15.10	4.98		20.08	24.50
400 0010	**FRAMING, STUD WALLS** w/ top & bottom track, no openings,									**400**
0020	headers, beams, bridging or bracing									
4100	8' high walls, 18 ga x 2-1/2" wide, studs 12" O.C.	2 Carp	54	.296	L.F.	9.45	10.15		19.60	26
4110	16" O.C.		77	.208		7.55	7.10		14.65	19.40
4120	24" O.C.		107	.150		5.65	5.10		10.75	14.20
4130	3-5/8" wide, studs 12" O.C.		53	.302		11.20	10.35		21.55	28.50
4140	16" O.C.		76	.211		8.95	7.20		16.15	21
4150	24" O.C.		105	.152		6.70	5.20		11.90	15.55

05410 | Load-Bearing Metal Studs

		CREW	DAILY OUTPUT	LABOR-HOURS	UNIT	2005 BARE COSTS MAT.	LABOR	EQUIP.	TOTAL	TOTAL INCL O&P
400										**400**
4160	4" wide, studs 12" O.C.	2 Carp	52	.308	L.F.	11.70	10.55		22.25	29.50
4170	16" O.C.		74	.216		9.35	7.40		16.75	22
4180	24" O.C.		103	.155		7.05	5.30		12.35	16.05
4190	6" wide, studs 12" O.C.		51	.314		14.90	10.75		25.65	33
4200	16" O.C.		73	.219		11.95	7.50		19.45	25
4210	24" O.C.		101	.158		9	5.45		14.45	18.35
4220	8" wide, studs 12" O.C.		50	.320		18.20	10.95		29.15	37
4230	16" O.C.		72	.222		14.65	7.60		22.25	28
4240	24" O.C.		100	.160		11.05	5.50		16.55	20.50
4300	16 ga x 2-1/2" wide, studs 12" O.C.		47	.340		11.05	11.65		22.70	30.50
4310	16" O.C.		68	.235		8.75	8.05		16.80	22
4320	24" O.C.		94	.170		6.45	5.85		12.30	16.20
4330	3-5/8" wide, studs 12" O.C.		46	.348		13.25	11.90		25.15	33
4340	16" O.C.		66	.242		10.50	8.30		18.80	24.50
4350	24" O.C.		92	.174		7.75	5.95		13.70	17.80
4360	4" wide, studs 12" O.C.		45	.356		13.95	12.20		26.15	34.50
4370	16" O.C.		65	.246		11.05	8.45		19.50	25.50
4380	24" O.C.		90	.178		8.15	6.10		14.25	18.45
4390	6" wide, studs 12" O.C.		44	.364		17.55	12.45		30	38.50
4400	16" O.C.		64	.250		13.90	8.55		22.45	28.50
4410	24" O.C.		88	.182		10.30	6.25		16.55	21
4420	8" wide, studs 12" O.C.		43	.372		21.50	12.75		34.25	43.50
4430	16" O.C.		63	.254		17.15	8.70		25.85	32.50
4440	24" O.C.		86	.186		12.75	6.35		19.10	24
5100	10' high walls, 18 ga x 2-1/2" wide, studs 12" O.C.		54	.296		11.35	10.15		21.50	28.50
5110	16" O.C.		77	.208		8.95	7.10		16.05	21
5120	24" O.C.		107	.150		6.60	5.10		11.70	15.25
5130	3-5/8" wide, studs 12" O.C.		53	.302		13.45	10.35		23.80	31
5140	16" O.C.		76	.211		10.65	7.20		17.85	23
5150	24" O.C.		105	.152		7.85	5.20		13.05	16.75
5160	4" wide, studs 12" O.C.		52	.308		14.05	10.55		24.60	32
5170	16" O.C.		74	.216		11.15	7.40		18.55	24
5180	24" O.C.		103	.155		8.20	5.30		13.50	17.30
5190	6" wide, studs 12" O.C.		51	.314		17.85	10.75		28.60	36.50
5200	16" O.C.		73	.219		14.15	7.50		21.65	27.50
5210	24" O.C.		101	.158		10.45	5.45		15.90	19.95
5220	8" wide, studs 12" O.C.		50	.320		22	10.95		32.95	41
5230	16" O.C.		72	.222		17.30	7.60		24.90	31
5240	24" O.C.		100	.160		12.85	5.50		18.35	22.50
5300	16 ga x 2-1/2" wide, studs 12" O.C.		47	.340		13.35	11.65		25	33
5310	16" O.C.		68	.235		10.45	8.05		18.50	24
5320	24" O.C.		94	.170		7.60	5.85		13.45	17.45
5330	3-5/8" wide, studs 12" O.C.		46	.348		16.05	11.90		27.95	36
5340	16" O.C.		66	.242		12.60	8.30		20.90	27
5350	24" O.C.		92	.174		9.15	5.95		15.10	19.35
5360	4" wide, studs 12" O.C.		45	.356		16.85	12.20		29.05	37.50
5370	16" O.C.		65	.246		13.25	8.45		21.70	27.50
5380	24" O.C.		90	.178		9.60	6.10		15.70	20
5390	6" wide, studs 12" O.C.		44	.364		21	12.45		33.45	43
5400	16" O.C.		64	.250		16.65	8.55		25.20	31.50
5410	24" O.C.		88	.182		12.10	6.25		18.35	23
5420	8" wide, studs 12" O.C.		43	.372		26	12.75		38.75	48.50
5430	16" O.C.		63	.254		20.50	8.70		29.20	36
5440	24" O.C.		86	.186		14.95	6.35		21.30	26.50
6190	12' high walls, 18 ga x 6" wide, studs 12" O.C.		41	.390		21	13.35		34.35	44
6200	16" O.C.		58	.276		16.40	9.45		25.85	32.50

METALS 5

05410 | Load-Bearing Metal Studs

		CREW	DAILY OUTPUT	LABOR-HOURS	UNIT	2005 BARE COSTS				TOTAL INCL O&P
						MAT.	LABOR	EQUIP.	TOTAL	
400 6210	24" O.C.	2 Carp	81	.198	L.F.	11.95	6.75		18.70	23.50 **400**
6220	8" wide, studs 12" O.C.		40	.400		25.50	13.70		39.20	49.50
6230	16" O.C.		57	.281		20	9.60		29.60	37
6240	24" O.C.		80	.200		14.65	6.85		21.50	27
6390	16 ga x 6" wide, studs 12" O.C.		35	.457		25	15.65		40.65	52
6400	16" O.C.		51	.314		19.35	10.75		30.10	38.50
6410	24" O.C.		70	.229		13.90	7.85		21.75	27.50
6420	8" wide, studs 12" O.C.		34	.471		30.50	16.10		46.60	58.50
6430	16" O.C.		50	.320		24	10.95		34.95	43
6440	24" O.C.		69	.232		17.15	7.95		25.10	31
6530	14 ga x 3-5/8" wide, studs 12" O.C.		34	.471		23.50	16.10		39.60	51
6540	16" O.C.		48	.333		18.35	11.40		29.75	38
6550	24" O.C.		65	.246		13.15	8.45		21.60	27.50
6560	4" wide, studs 12" O.C.		33	.485		25	16.60		41.60	53.50
6570	16" O.C.		47	.340		19.35	11.65		31	39.50
6580	24" O.C.		64	.250		13.90	8.55		22.45	28.50
6730	12 ga x 3-5/8" wide, studs 12" O.C.		31	.516		32.50	17.70		50.20	63.50
6740	16" O.C.		43	.372		25	12.75		37.75	47.50
6750	24" O.C.		59	.271		17.65	9.30		26.95	34
6760	4" wide, studs 12" O.C.		30	.533		35	18.25		53.25	67
6770	16" O.C.		42	.381		27	13.05		40.05	50
6780	24" O.C.		58	.276		18.95	9.45		28.40	35.50
7390	16' high walls, 16 ga x 6" wide, studs 12" O.C.		33	.485		32	16.60		48.60	61
7400	16" O.C.		48	.333		25	11.40		36.40	45.50
7410	24" O.C.		67	.239		17.55	8.20		25.75	32
7420	8" wide, studs 12" O.C.		32	.500		39.50	17.15		56.65	69.50
7430	16" O.C.		47	.340		30.50	11.65		42.15	51.50
7440	24" O.C.		66	.242		21.50	8.30		29.80	36.50
7560	14 ga x 4" wide, studs 12" O.C.		31	.516		32	17.70		49.70	63
7570	16" O.C.		45	.356		25	12.20		37.20	46.50
7580	24" O.C.		61	.262		17.55	9		26.55	33.50
7590	6" wide, studs 12" O.C.		30	.533		40.50	18.25		58.75	73
7600	16" O.C.		44	.364		31.50	12.45		43.95	54
7610	24" O.C.		60	.267		22	9.15		31.15	39
7760	12 ga x 4" wide, studs 12" O.C.		29	.552		45.50	18.90		64.40	79.50
7770	16" O.C.		40	.400		35	13.70		48.70	60
7780	24" O.C.		55	.291		24.50	9.95		34.45	42
7790	6" wide, studs 12" O.C.		28	.571		57.50	19.55		77.05	93.50
7800	16" O.C.		39	.410		44	14.05		58.05	70.50
7810	24" O.C.		54	.296		30.50	10.15		40.65	49.50
8590	20' high walls, 14 ga x 6" wide, studs 12" O.C.		29	.552		49.50	18.90		68.40	84
8600	16" O.C.		42	.381		38	13.05		51.05	62.50
8610	24" O.C.		57	.281		27	9.60		36.60	44.50
8620	8" wide, studs 12" O.C.		28	.571		60.50	19.55		80.05	97.50
8630	16" O.C.		41	.390		47	13.35		60.35	72.50
8640	24" O.C.		56	.286		33	9.80		42.80	51.50
8790	12 ga x 6" wide, studs 12" O.C.		27	.593		71	20.50		91.50	110
8800	16" O.C.		37	.432		54	14.80		68.80	82.50
8810	24" O.C.		51	.314		37.50	10.75		48.25	58
8820	8" wide, studs 12" O.C.		26	.615		86.50	21		107.50	128
8830	16" O.C.		36	.444		66	15.20		81.20	96
8840	24" O.C.		50	.320		45.50	10.95		56.45	67

05420 | Cold-Formed Metal Joists

		CREW	DAILY OUTPUT	LABOR-HOURS	UNIT	MAT.	LABOR	EQUIP.	TOTAL	TOTAL INCL O&P
100 0010	**BRACING**, continuous, per row, top & bottom									**100**
0120	Flat strap, 20 ga x 2" wide, joists at 12" O.C.	1 Carp	4.67	1.713	C.L.F.	60	58.50		118.50	158

Important: See the Reference Section for critical supporting data - Reference Nos., Crews, & City Cost Indexes

05400 | Cold-Formed Metal Framing

05420 | Cold-Formed Metal Joists

		CREW	DAILY OUTPUT	LABOR-HOURS	UNIT	MAT.	LABOR	EQUIP.	TOTAL	TOTAL INCL O&P
100											**100**
0130	16" O.C.	1 Carp	5.33	1.501	C.L.F.	57.50	51.50		109	144
0140	24" O.C.		6.66	1.201		55.50	41		96.50	125
0150	18 ga x 2" wide, joists at 12" O.C.		4	2		78.50	68.50		147	194
0160	16" O.C.		4.67	1.713		77	58.50		135.50	176
0170	24" O.C.		5.33	1.501		75.50	51.50		127	163
120 0010	**BRIDGING**, solid between joists w/ 1-1/4" leg track, per joist bay										**120**
0230	Joists 12" O.C., 18 ga track x 6" wide	1 Carp	80	.100	Ea.	1.61	3.43		5.04	7.10
0240	8" wide		75	.107		2.05	3.65		5.70	7.95
0250	10" wide		70	.114		2.52	3.91		6.43	8.85
0260	12" wide		65	.123		2.92	4.22		7.14	9.75
0330	16 ga track x 6" wide		70	.114		2.02	3.91		5.93	8.30
0340	8" wide		65	.123		2.57	4.22		6.79	9.40
0350	10" wide		60	.133		3.17	4.57		7.74	10.60
0360	12" wide		55	.145		3.64	4.98		8.62	11.75
0440	14 ga track x 8" wide		60	.133		3.24	4.57		7.81	10.65
0450	10" wide		55	.145		3.98	4.98		8.96	12.10
0460	12" wide		50	.160		4.59	5.50		10.09	13.60
0550	12 ga track x 10" wide		45	.178		5.85	6.10		11.95	15.95
0560	12" wide		40	.200		6.60	6.85		13.45	17.95
1230	16" O.C., 18 ga track x 6" wide		80	.100		2.06	3.43		5.49	7.60
1240	8" wide		75	.107		2.62	3.65		6.27	8.60
1250	10" wide		70	.114		3.23	3.91		7.14	9.65
1260	12" wide		65	.123		3.75	4.22		7.97	10.65
1330	16 ga track x 6" wide		70	.114		2.59	3.91		6.50	8.95
1340	8" wide		65	.123		3.30	4.22		7.52	10.20
1350	10" wide		60	.133		4.06	4.57		8.63	11.55
1360	12" wide		55	.145		4.67	4.98		9.65	12.90
1440	14 ga track x 8" wide		60	.133		4.16	4.57		8.73	11.65
1450	10" wide		55	.145		5.10	4.98		10.08	13.35
1460	12" wide		50	.160		5.90	5.50		11.40	15.05
1550	12 ga track x 10" wide		45	.178		7.50	6.10		13.60	17.75
1560	12" wide		40	.200		8.50	6.85		15.35	20
2230	24" O.C., 18 ga track x 6" wide		80	.100		2.98	3.43		6.41	8.65
2240	8" wide		75	.107		3.80	3.65		7.45	9.85
2250	10" wide		70	.114		4.68	3.91		8.59	11.25
2260	12" wide		65	.123		5.40	4.22		9.62	12.50
2330	16 ga track x 6" wide		70	.114		3.75	3.91		7.66	10.20
2340	8" wide		65	.123		4.77	4.22		8.99	11.80
2350	10" wide		60	.133		5.85	4.57		10.42	13.55
2360	12" wide		55	.145		6.75	4.98		11.73	15.20
2440	14 ga track x 8" wide		60	.133		6	4.57		10.57	13.70
2450	10" wide		55	.145		7.40	4.98		12.38	15.85
2460	12" wide		50	.160		8.50	5.50		14	17.90
2550	12 ga track x 10" wide		45	.178		10.85	6.10		16.95	21.50
2560	12" wide		40	.200		12.25	6.85		19.10	24
200 0010	**FRAMING, BAND JOIST** (track) fastened to bearing wall										**200**
0220	18 ga track x 6" deep	2 Carp	1,000	.016	L.F.	1.31	.55		1.86	2.29
0230	8" deep		920	.017		1.67	.60		2.27	2.77
0240	10" deep		860	.019		2.06	.64		2.70	3.25
0320	16 ga track x 6" deep		900	.018		1.65	.61		2.26	2.76
0330	8" deep		840	.019		2.10	.65		2.75	3.33
0340	10" deep		780	.021		2.58	.70		3.28	3.93
0350	12" deep		740	.022		2.97	.74		3.71	4.42
0430	14 ga track x 8" deep		750	.021		2.65	.73		3.38	4.05
0440	10" deep		720	.022		3.24	.76		4	4.76

METALS 5

165

	05420	Cold-Formed Metal Joists	CREW	DAILY OUTPUT	LABOR-HOURS	UNIT	MAT.	LABOR	EQUIP.	TOTAL	TOTAL INCL O&P	
200	0450	12" deep	2 Carp	700	.023	L.F.	3.75	.78		4.53	5.35	**200**
	0540	12 ga track x 10" deep		670	.024		4.77	.82		5.59	6.50	
	0550	12" deep	▼	650	.025	▼	5.40	.84		6.24	7.25	
300	0010	**FRAMING, BOXED HEADERS/BEAMS**										**300**
	0200	Double, 18 ga x 6" deep	2 Carp	220	.073	L.F.	5.60	2.49		8.09	10.05	
	0210	8" deep		210	.076		6.20	2.61		8.81	10.90	
	0220	10" deep		200	.080		7.55	2.74		10.29	12.55	
	0230	12" deep		190	.084		8.30	2.88		11.18	13.60	
	0300	16 ga x 8" deep		180	.089		7.15	3.04		10.19	12.60	
	0310	10" deep		170	.094		8.60	3.22		11.82	14.45	
	0320	12" deep		160	.100		9.35	3.43		12.78	15.65	
	0400	14 ga x 10" deep		140	.114		10	3.91		13.91	17.10	
	0410	12" deep		130	.123		10.95	4.22		15.17	18.60	
	0500	12 ga x 10" deep		110	.145		13.20	4.98		18.18	22.50	
	0510	12" deep		100	.160		14.65	5.50		20.15	24.50	
	1210	Triple, 18 ga x 8" deep		170	.094		9	3.22		12.22	14.90	
	1220	10" deep		165	.097		10.85	3.32		14.17	17.05	
	1230	12" deep		160	.100		11.95	3.43		15.38	18.50	
	1300	16 ga x 8" deep		145	.110		10.40	3.78		14.18	17.30	
	1310	10" deep		140	.114		12.40	3.91		16.31	19.75	
	1320	12" deep		135	.119		13.55	4.06		17.61	21	
	1400	14 ga x 10" deep		115	.139		14.50	4.77		19.27	23.50	
	1410	12" deep		110	.145		15.95	4.98		20.93	25.50	
	1500	12 ga x 10" deep		90	.178		19.35	6.10		25.45	31	
	1510	12" deep	▼	85	.188	▼	21.50	6.45		27.95	33.50	
410	0010	**FRAMING, JOISTS**, no band joists (track), web stiffeners, headers,										**410**
	0020	beams, bridging or bracing										
	0030	Joists (2" flange) and fasteners, materials only										
	0220	18 ga x 6" deep				L.F.	1.73			1.73	1.91	
	0230	8" deep					2.06			2.06	2.26	
	0240	10" deep					2.42			2.42	2.66	
	0320	16 ga x 6" deep					2.12			2.12	2.33	
	0330	8" deep					2.54			2.54	2.80	
	0340	10" deep					2.97			2.97	3.27	
	0350	12" deep					3.37			3.37	3.71	
	0430	14 ga x 8" deep					3.21			3.21	3.53	
	0440	10" deep					3.71			3.71	4.08	
	0450	12" deep					4.21			4.21	4.63	
	0540	12 ga x 10" deep					5.40			5.40	5.95	
	0550	12" deep				▼	6.15			6.15	6.75	
	1010	Installation of joists to band joists, beams & headers, labor only										
	1220	18 ga x 6" deep	2 Carp	110	.145	Ea.		4.98		4.98	7.75	
	1230	8" deep		90	.178			6.10		6.10	9.50	
	1240	10" deep		80	.200			6.85		6.85	10.65	
	1320	16 ga x 6" deep		95	.168			5.75		5.75	9	
	1330	8" deep		70	.229			7.85		7.85	12.20	
	1340	10" deep		60	.267			9.15		9.15	14.25	
	1350	12" deep		55	.291			9.95		9.95	15.50	
	1430	14 ga x 8" deep		65	.246			8.45		8.45	13.15	
	1440	10" deep		45	.356			12.20		12.20	18.95	
	1450	12" deep		35	.457			15.65		15.65	24.50	
	1540	12 ga x 10" deep		40	.400			13.70		13.70	21.50	
	1550	12" deep	▼	30	.533	▼		18.25		18.25	28.50	
500	0010	**FRAMING, WEB STIFFENERS** at joist bearing, fabricated from										**500**
	0020	stud piece (1-5/8" flange) to stiffen joist (2" flange)										

Important: See the Reference Section for critical supporting data - Reference Nos., Crews, & City Cost Indexes

5 METALS

05420	Cold-Formed Metal Joists	CREW	DAILY OUTPUT	LABOR-HOURS	UNIT	2005 BARE COSTS				TOTAL INCL O&P
						MAT.	LABOR	EQUIP.	TOTAL	
500 2120	For 6" deep joist, with 18 ga x 2-1/2" stud	1 Carp	120	.067	Ea.	2.09	2.28		4.37	5.85 **500**
2130	3-5/8" stud		110	.073		2.30	2.49		4.79	6.40
2140	4" stud		105	.076		2.22	2.61		4.83	6.50
2150	6" stud		100	.080		2.44	2.74		5.18	6.95
2160	8" stud		95	.084		2.51	2.88		5.39	7.25
2220	8" deep joist, with 2-1/2" stud		120	.067		2.29	2.28		4.57	6.10
2230	3-5/8" stud		110	.073		2.48	2.49		4.97	6.60
2240	4" stud		105	.076		2.43	2.61		5.04	6.75
2250	6" stud		100	.080		2.68	2.74		5.42	7.20
2260	8" stud		95	.084		2.88	2.88		5.76	7.65
2320	10" deep joist, with 2-1/2" stud		110	.073		3.23	2.49		5.72	7.45
2330	3-5/8" stud		100	.080		3.53	2.74		6.27	8.15
2340	4" stud		95	.084		3.50	2.88		6.38	8.35
2350	6" stud		90	.089		3.81	3.04		6.85	8.95
2360	8" stud		85	.094		3.86	3.22		7.08	9.25
2420	12" deep joist, with 2-1/2" stud		110	.073		3.42	2.49		5.91	7.65
2430	3-5/8" stud		100	.080		3.70	2.74		6.44	8.35
2440	4" stud		95	.084		3.63	2.88		6.51	8.50
2450	6" stud		90	.089		4	3.04		7.04	9.15
2460	8" stud		85	.094		4.30	3.22		7.52	9.75
3130	For 6" deep joist, with 16 ga x 3-5/8" stud		100	.080		2.42	2.74		5.16	6.95
3140	4" stud		95	.084		2.39	2.88		5.27	7.10
3150	6" stud		90	.089		2.62	3.04		5.66	7.65
3160	8" stud		85	.094		2.76	3.22		5.98	8.05
3230	8" deep joist, with 3-5/8" stud		100	.080		2.68	2.74		5.42	7.20
3240	4" stud		95	.084		2.62	2.88		5.50	7.40
3250	6" stud		90	.089		2.91	3.04		5.95	7.95
3260	8" stud		85	.094		3.11	3.22		6.33	8.40
3330	10" deep joist, with 3-5/8" stud		85	.094		3.67	3.22		6.89	9.05
3340	4" stud		80	.100		3.73	3.43		7.16	9.45
3350	6" stud		75	.107		4.06	3.65		7.71	10.15
3360	8" stud		70	.114		4.22	3.91		8.13	10.75
3430	12" deep joist, with 3-5/8" stud		85	.094		4	3.22		7.22	9.40
3440	4" stud		80	.100		3.92	3.43		7.35	9.65
3450	6" stud		75	.107		4.34	3.65		7.99	10.50
3460	8" stud		70	.114		4.64	3.91		8.55	11.20
4230	For 8" deep joist, with 14 ga x 3-5/8" stud		90	.089		3.48	3.04		6.52	8.55
4240	4" stud		85	.094		3.54	3.22		6.76	8.90
4250	6" stud		80	.100		3.84	3.43		7.27	9.55
4260	8" stud		75	.107		4.11	3.65		7.76	10.20
4330	10" deep joist, with 3-5/8" stud		75	.107		4.88	3.65		8.53	11.05
4340	4" stud		70	.114		4.83	3.91		8.74	11.40
4350	6" stud		65	.123		5.30	4.22		9.52	12.40
4360	8" stud		60	.133		5.55	4.57		10.12	13.20
4430	12" deep joist, with 3-5/8" stud		75	.107		5.20	3.65		8.85	11.40
4440	4" stud		70	.114		5.30	3.91		9.21	11.90
4450	6" stud		65	.123		5.75	4.22		9.97	12.85
4460	8" stud		60	.133		6.15	4.57		10.72	13.85
5330	For 10" deep joist, with 12 ga x 3-5/8" stud		65	.123		5.15	4.22		9.37	12.20
5340	4" stud		60	.133		5.30	4.57		9.87	12.95
5350	6" stud		55	.145		5.85	4.98		10.83	14.15
5360	8" stud		50	.160		6.40	5.50		11.90	15.60
5430	12" deep joist, with 3-5/8" stud		65	.123		5.70	4.22		9.92	12.80
5440	4" stud		60	.133		5.60	4.57		10.17	13.25
5450	6" stud		55	.145		6.35	4.98		11.33	14.75
5460	8" stud		50	.160		7.35	5.50		12.85	16.60

METALS 5

05425	Cold-Formed Roof Framing	CREW	DAILY OUTPUT	LABOR-HOURS	UNIT	2005 BARE COSTS				TOTAL INCL O&P
						MAT.	LABOR	EQUIP.	TOTAL	
100	**0010 FRAMING, BRACING**									
0020	Continuous bracing, per row									
0100	16 ga x 1-1/2" channel thru rafters/trusses @ 16" O.C.	1 Carp	4.50	1.778	C.L.F.	54	61		115	155
0120	24" O.C.		6	1.333		54	45.50		99.50	131
0300	2" x 2" angle x 18 ga, rafters/trusses @ 16" O.C.		6	1.333		79	45.50		124.50	158
0320	24" O.C.		8	1		79	34.50		113.50	141
0400	16 ga, rafters/trusses @ 16" O.C.		4.50	1.778		100	61		161	205
0420	24" O.C.		6.50	1.231		100	42		142	176
200	**0010 FRAMING, BRIDGING**									
0020	Solid, between rafters w/ 1-1/4" leg track, per rafter bay									
1200	Rafters 16" O.C., 18 ga x 4" deep	1 Carp	60	.133	Ea.	1.58	4.57		6.15	8.85
1210	6" deep		57	.140		2.06	4.81		6.87	9.75
1220	8" deep		55	.145		2.62	4.98		7.60	10.65
1230	10" deep		52	.154		3.23	5.25		8.48	11.75
1240	12" deep		50	.160		3.75	5.50		9.25	12.65
2200	24" O.C., 18 ga x 4" deep		60	.133		2.29	4.57		6.86	9.60
2210	6" deep		57	.140		2.98	4.81		7.79	10.80
2220	8" deep		55	.145		3.80	4.98		8.78	11.90
2230	10" deep		52	.154		4.68	5.25		9.93	13.35
2240	12" deep		50	.160		5.40	5.50		10.90	14.50
500	**0010 FRAMING, PARAPETS**									
0100	3' high installed on 1st story, 18 ga x 4" wide studs, 12" O.C.	2 Carp	100	.160	L.F.	5.85	5.50		11.35	15
0110	16" O.C.		150	.107		4.98	3.65		8.63	11.20
0120	24" O.C.		200	.080		4.11	2.74		6.85	8.80
0200	6" wide studs, 12" O.C.		100	.160		7.50	5.50		13	16.80
0210	16" O.C.		150	.107		6.40	3.65		10.05	12.75
0220	24" O.C.		200	.080		5.30	2.74		8.04	10.05
1100	Installed on 2nd story, 18 ga x 4" wide studs, 12" O.C.		95	.168		5.85	5.75		11.60	15.45
1110	16" O.C.		145	.110		4.98	3.78		8.76	11.40
1120	24" O.C.		190	.084		4.11	2.88		6.99	9
1200	6" wide studs, 12" O.C.		95	.168		7.50	5.75		13.25	17.25
1210	16" O.C.		145	.110		6.40	3.78		10.18	12.95
1220	24" O.C.		190	.084		5.30	2.88		8.18	10.30
2100	Installed on gable, 18 ga x 4" wide studs, 12" O.C.		85	.188		5.85	6.45		12.30	16.50
2110	16" O.C.		130	.123		4.98	4.22		9.20	12.05
2120	24" O.C.		170	.094		4.11	3.22		7.33	9.50
2200	6" wide studs, 12" O.C.		85	.188		7.50	6.45		13.95	18.30
2210	16" O.C.		130	.123		6.40	4.22		10.62	13.60
2220	24" O.C.		170	.094		5.30	3.22		8.52	10.80
550	**0010 FRAMING, ROOF RAFTERS**									
0100	Boxed ridge beam, double, 18 ga x 6" deep	2 Carp	160	.100	L.F.	5.60	3.43		9.03	11.50
0110	8" deep		150	.107		6.20	3.65		9.85	12.55
0120	10" deep		140	.114		7.55	3.91		11.46	14.40
0130	12" deep		130	.123		8.30	4.22		12.52	15.65
0200	16 ga x 6" deep		150	.107		6.35	3.65		10	12.70
0210	8" deep		140	.114		7.15	3.91		11.06	13.95
0220	10" deep		130	.123		8.60	4.22		12.82	16
0230	12" deep		120	.133		9.35	4.57		13.92	17.40
1100	Rafters, 2" flange, material only, 18 ga x 6" deep					1.73			1.73	1.91
1110	8" deep					2.06			2.06	2.26
1120	10" deep					2.42			2.42	2.66
1130	12" deep					2.81			2.81	3.10
1200	16 ga x 6" deep					2.12			2.12	2.33
1210	8" deep					2.54			2.54	2.80
1220	10" deep					2.97			2.97	3.27

Important: See the Reference Section for critical supporting data - Reference Nos., Crews, & City Cost Indexes

05425	Cold-Formed Roof Framing	CREW	DAILY OUTPUT	LABOR-HOURS	UNIT	2005 BARE COSTS				TOTAL INCL O&P	
						MAT.	LABOR	EQUIP.	TOTAL		
550	1230	12" deep				L.F.	3.37			3.37	3.71	**550**
	2100	Installation only, ordinary rafter to 4:12 pitch, 18 ga x 6" deep	2 Carp	35	.457	Ea.		15.65		15.65	24.50
	2110	8" deep		30	.533			18.25		18.25	28.50
	2120	10" deep		25	.640			22		22	34
	2130	12" deep		20	.800			27.50		27.50	42.50
	2200	16 ga x 6" deep		30	.533			18.25		18.25	28.50
	2210	8" deep		25	.640			22		22	34
	2220	10" deep		20	.800			27.50		27.50	42.50
	2230	12" deep	▼	15	1.067	▼		36.50		36.50	57
	8100	Add to labor, ordinary rafters on steep roofs						25%			
	8110	Dormers & complex roofs						50%			
	8200	Hip & valley rafters to 4:12 pitch						25%			
	8210	Steep roofs						50%			
	8220	Dormers & complex roofs						75%			
	8300	Hip & valley jack rafters to 4:12 pitch						50%			
	8310	Steep roofs						75%			
	8320	Dormers & complex roofs						100%			
600	0010	**FRAMING, ROOF TRUSSES**										**600**
	0020	Fabrication of trusses on ground, Fink (W) or King Post, to 4:12 pitch									
	0120	18 ga x 4" chords, 16' span	2 Carp	12	1.333	Ea.	65.50	45.50		111	143
	0130	20' span		11	1.455		82	50		132	168
	0140	24' span		11	1.455		98.50	50		148.50	186
	0150	28' span		10	1.600		115	55		170	212
	0160	32' span		10	1.600		131	55		186	230
	0250	6" chords, 28' span		9	1.778		145	61		206	255
	0260	32' span		9	1.778		166	61		227	277
	0270	36' span		8	2		186	68.50		254.50	310
	0280	40' span		8	2		207	68.50		275.50	335
	1120	5:12 to 8:12 pitch, 18 ga x 4" chords, 16' span		10	1.600		75	55		130	168
	1130	20' span		9	1.778		93.50	61		154.50	198
	1140	24' span		9	1.778		112	61		173	219
	1150	28' span		8	2		131	68.50		199.50	251
	1160	32' span		8	2		150	68.50		218.50	272
	1250	6" chords, 28' span		7	2.286		166	78.50		244.50	305
	1260	32' span		7	2.286		189	78.50		267.50	330
	1270	36' span		6	2.667		213	91.50		304.50	375
	1280	40' span		6	2.667		237	91.50		328.50	400
	2120	9:12 to 12:12 pitch, 18 ga x 4" chords, 16' span		8	2		93.50	68.50		162	210
	2130	20' span		7	2.286		117	78.50		195.50	251
	2140	24' span		7	2.286		140	78.50		218.50	276
	2150	28' span		6	2.667		164	91.50		255.50	320
	2160	32' span		6	2.667		187	91.50		278.50	350
	2250	6" chords, 28' span		5	3.200		207	110		317	400
	2260	32' span		5	3.200		237	110		347	430
	2270	36' span		4	4		266	137		403	505
	2280	40' span	▼	4	4		296	137		433	540
	5120	Erection only of roof trusses, to 4:12 pitch, 16' span	F-6	48	.833			26.50	12.55	39.05	54.50
	5130	20' span		46	.870			27.50	13.10	40.60	57
	5140	24' span		44	.909			28.50	13.70	42.20	59.50
	5150	28' span		42	.952			30	14.35	44.35	62.50
	5160	32' span		40	1			31.50	15.05	46.55	65.50
	5170	36' span		38	1.053			33	15.85	48.85	69
	5180	40' span		36	1.111			35	16.75	51.75	72.50
	5220	5:12 to 8:12 pitch, 16' span		42	.952			30	14.35	44.35	62.50
	5230	20' span	▼	40	1			31.50	15.05	46.55	65.50

METALS **5**

05425 | Cold-Formed Roof Framing

		CREW	DAILY OUTPUT	LABOR-HOURS	UNIT	2005 BARE COSTS				TOTAL INCL O&P
						MAT.	LABOR	EQUIP.	TOTAL	
5240	24' span	F-6	38	1.053	Ea.		33	15.85	48.85	69
5250	28' span		36	1.111			35	16.75	51.75	72.50
5260	32' span		34	1.176			37	17.75	54.75	77
5270	36' span		32	1.250			39.50	18.85	58.35	81.50
5280	40' span		30	1.333			42	20	62	87
5320	9:12 to 12:12 pitch, 16' span		36	1.111			35	16.75	51.75	72.50
5330	20' span		34	1.176			37	17.75	54.75	77
5340	24' span		32	1.250			39.50	18.85	58.35	81.50
5350	28' span		30	1.333			42	20	62	87
5360	32' span		28	1.429			45	21.50	66.50	93
5370	36' span		26	1.538			48.50	23	71.50	101
5380	40' span		24	1.667			52.50	25	77.50	109
0010	**FRAMING, SOFFITS & CANOPIES**									
0130	Continuous ledger track @ wall, studs @ 16" O.C., 18 ga x 4" wide	2 Carp	535	.030	L.F.	1.06	1.02		2.08	2.76
0140	6" wide		500	.032		1.38	1.10		2.48	3.22
0150	8" wide		465	.034		1.75	1.18		2.93	3.76
0160	10" wide		430	.037		2.16	1.27		3.43	4.36
0230	Studs @ 24" O.C., 18 ga x 4" wide		800	.020		1.01	.69		1.70	2.18
0240	6" wide		750	.021		1.31	.73		2.04	2.58
0250	8" wide		700	.023		1.67	.78		2.45	3.06
0260	10" wide		650	.025		2.06	.84		2.90	3.57
1000	Horizontal soffit and canopy members, material only									
1030	1-5/8" flange studs, 18 ga x 4" deep				L.F.	1.40			1.40	1.54
1040	6" deep					1.78			1.78	1.95
1050	8" deep					2.15			2.15	2.36
1140	2" flange joists, 18 ga x 6" deep					1.98			1.98	2.18
1150	8" deep					2.35			2.35	2.59
1160	10" deep					2.76			2.76	3.04
4030	Installation only, 18 ga, 1-5/8" flange x 4" deep	2 Carp	130	.123	Ea.		4.22		4.22	6.55
4040	6" deep		110	.145			4.98		4.98	7.75
4050	8" deep		90	.178			6.10		6.10	9.50
4140	2" flange, 18 ga x 6" deep		110	.145			4.98		4.98	7.75
4150	8" deep		90	.178			6.10		6.10	9.50
4160	10" deep		80	.200			6.85		6.85	10.65
6010	Clips to attach facia to rafter tails, 2" x 2" x 18 ga angle	1 Carp	120	.067		.94	2.28		3.22	4.59
6020	16 ga angle	"	100	.080		1.18	2.74		3.92	5.55

05500 | Metal Fabrications

05514 | Ladders

		CREW	DAILY OUTPUT	LABOR-HOURS	UNIT	2005 BARE COSTS				TOTAL INCL O&P
						MAT.	LABOR	EQUIP.	TOTAL	
0010	**LADDER**, shop fabricated									
0020	Steel, 20" wide, bolted to concrete, with cage	E-4	50	.640	V.L.F.	64.50	24.50	1.63	90.63	117
0100	Without cage		85	.376		30	14.55	.96	45.51	60
0300	Aluminum, bolted to concrete, with cage		50	.640		87.50	24.50	1.63	113.63	142
0400	Without cage		85	.376		50.50	14.55	.96	66.01	82.50
1350	Alternating tread stair, 56/68°, steel, standard paint color	2 Sswk	50	.320		146	12.20		158.20	183
1360	Non-standard paint color		50	.320		166	12.20		178.20	204
1370	Galvanized steel		50	.320		165	12.20		177.20	204
1380	Stainless steel		50	.320		246	12.20		258.20	293
1390	68°, aluminum		50	.320		180	12.20		192.20	220

Important: See the Reference Section for critical supporting data - Reference Nos., Crews, & City Cost Indexes

05517 | Metal Stairs

			CREW	DAILY OUTPUT	LABOR-HOURS	UNIT	2005 BARE COSTS				TOTAL INCL O&P
							MAT.	LABOR	EQUIP.	TOTAL	
300	0010	**FIRE ESCAPE**, shop fabricated										**300**
	0200	2' wide balcony, 1" x 1/4" bars 1-1/2" O.C.	1 Sswk	5	1.600	L.F.	43.50	61		104.50	157
	0400	1st story cantilevered stair, standard	↓	.09	88.889	Ea.	1,825	3,400		5,225	8,075
	0700	Platform & fixed stair, 36" x 40"		.17	47.059	Flight	805	1,800		2,605	4,075
	0900	For 3'-6" wide escapes, add to above					100%	150%			
350	0010	**FIRE ESCAPE STAIRS**										**350**
	0020	One story, disappearing, stainless steel	2 Sswk	20	.800	V.L.F.	175	30.50		205.50	247
	0100	Portable ladder				Ea.	53			53	58.50
700	0010	**STAIR**, shop fabricated, steel stringers, safety nosing on treads										**700**
	0020	Grating tread and pipe railing, 3'-6" wide	E-4	35	.914	Riser	128	35.50	2.32	165.82	207
	0100	4'-0" wide		30	1.067		167	41	2.71	210.71	260
	0200	Cement fill metal pan, picket rail, 3'-6" wide		35	.914		192	35.50	2.32	229.82	277
	0300	4'-0" wide		30	1.067		218	41	2.71	261.71	315
	0350	Wall rail, both sides, 3'-6" wide CN		53	.604		147	23.50	1.53	172.03	205
	0400	Cast iron tread and pipe rail, 3'-6" wide		35	.914		205	35.50	2.32	242.82	292
	0500	Checkered plate tread, industrial, 3'-6" wide		28	1.143		128	44	2.90	174.90	223
	0550	Circular, for tanks, 3'-0" wide	↓	33	.970		141	37.50	2.46	180.96	225
	0600	For isolated stairs, add						100%			
	0800	Custom steel stairs, 3'-6" wide, minimum	E-4	35	.914		192	35.50	2.32	229.82	277
	0810	Average		30	1.067		256	41	2.71	299.71	360
	0900	Maximum	↓	20	1.600		320	62	4.06	386.06	465
	1100	For 4' wide stairs, add					5%	5%			
	1300	For 5' wide stairs, add				↓	10%	10%			
	1500	Landing, steel pan, conventional	E-4	160	.200	S.F.	25.50	7.75	.51	33.76	42.50
	1600	Pre-erected	"	255	.125	"	45	4.85	.32	50.17	58.50
	1700	Pre-erected, steel pan tread, 3'-6" wide, 2 line pipe rail	E-2	87	.644	Riser	211	24	15.25	250.25	291
	1810	Spiral aluminum, 5'-0" diameter, stock units	E-4	45	.711		232	27.50	1.81	261.31	305
	1820	Custom units		45	.711		440	27.50	1.81	469.31	535
	1900	Spiral, cast iron, 4'-0" diameter, ornamental, minimum		45	.711		207	27.50	1.81	236.31	279
	1920	Maximum		25	1.280		283	49.50	3.25	335.75	400
	2000	Spiral, steel, industrial checkered plate, 4' diameter		45	.711		207	27.50	1.81	236.31	279
	2200	Stock units, 6'-0" diameter	↓	40	.800	↓	251	31	2.03	284.03	335
	3110	Spiral steel, stock units, primed, flat metal tread, 3'-6" dia	2 Carp	1.60	10	Flight	1,000	345		1,345	1,625
	3120	4'-0" dia		1.45	11.034		1,150	380		1,530	1,875
	3130	4'-6" dia		1.35	11.852		1,275	405		1,680	2,025
	3140	5'-0" dia		1.25	12.800		1,400	440		1,840	2,200
	3210	Galvanized, 3'-6" dia		1.60	10		1,775	345		2,120	2,475
	3220	4'-0" dia		1.45	11.034		1,975	380		2,355	2,775
	3230	4'-6" dia		1.35	11.852		2,175	405		2,580	3,000
	3240	5'-0" dia		1.25	12.800		2,325	440		2,765	3,250
	3310	Checkered plate tread, 3'-6" dia		1.45	11.034		1,250	380		1,630	1,975
	3320	4'-0" dia		1.35	11.852		1,400	405		1,805	2,175
	3330	4'-6" dia		1.25	12.800		1,550	440		1,990	2,375
	3340	5'-0" dia		1.15	13.913		1,650	475		2,125	2,575
	3410	Galvanized, 3'-6" dia		1.45	11.034		2,025	380		2,405	2,825
	3420	4'-0" dia		1.35	11.852		2,275	405		2,680	3,125
	3430	4'-6" dia		1.25	12.800		2,475	440		2,915	3,375
	3440	5'-0" dia		1.15	13.913		2,650	475		3,125	3,675
	3510	Red oak tread on flat metal, 3'-6" dia		1.35	11.852		1,675	405		2,080	2,475
	3520	4'-0" dia		1.25	12.800		1,875	440		2,315	2,725
	3530	4'-6" dia		1.15	13.913		2,025	475		2,500	2,975
	3540	5'-0" dia	↓	1.05	15.238	↓	2,200	520		2,720	3,225
	3900	Industrial ships ladder, 3' W, grating treads, 2 line pipe rail	E-4	30	1.067	Riser	83.50	41	2.71	127.21	168
	4000	Aluminum	"	30	1.067	"	128	41	2.71	171.71	218

METALS 5

05520 | Handrails & Railings

		CREW	DAILY OUTPUT	LABOR-HOURS	UNIT	2005 BARE COSTS				TOTAL INCL O&P
						MAT.	LABOR	EQUIP.	TOTAL	
700	0010	**RAILING, PIPE**, shop fabricated									700
	0020	Aluminum, 2 rail, satin finish, 1-1/4" diameter	E-4	160	.200	L.F.	14.60	7.75	.51	22.86	30.50
	0030	Clear anodized		160	.200		18.10	7.75	.51	26.36	34.50
	0040	Dark anodized		160	.200		20.50	7.75	.51	28.76	37
	0080	1-1/2" diameter, satin finish		160	.200		17.50	7.75	.51	25.76	33.50
	0090	Clear anodized		160	.200		19.50	7.75	.51	27.76	36
	0100	Dark anodized		160	.200		21.50	7.75	.51	29.76	38.50
	0140	Aluminum, 3 rail, 1-1/4" diam., satin finish		137	.234		22.50	9.05	.59	32.14	41.50
	0150	Clear anodized		137	.234		28	9.05	.59	37.64	48
	0160	Dark anodized		137	.234		31	9.05	.59	40.64	51
	0200	1-1/2" diameter, satin finish		137	.234		27	9.05	.59	36.64	46.50
	0210	Clear anodized		137	.234		30.50	9.05	.59	40.14	50.50
	0220	Dark anodized		137	.234		33.50	9.05	.59	43.14	54
	0500	Steel, 2 rail, on stairs, primed, 1-1/4" diameter		160	.200		12	7.75	.51	20.26	27.50
	0520	1-1/2" diameter		160	.200		13.20	7.75	.51	21.46	29
	0540	Galvanized, 1-1/4" diameter		160	.200		16.65	7.75	.51	24.91	32.50
	0560	1-1/2" diameter		160	.200		18.65	7.75	.51	26.91	35
	0580	Steel, 3 rail, primed, 1-1/4" diameter		137	.234		17.90	9.05	.59	27.54	36.50
	0600	1-1/2" diameter		137	.234		19	9.05	.59	28.64	38
	0620	Galvanized, 1-1/4" diameter		137	.234		25	9.05	.59	34.64	44.50
	0640	1-1/2" diameter **CN**		137	.234		29.50	9.05	.59	39.14	49
	0700	Stainless steel, 2 rail, 1-1/4" diam. #4 finish		137	.234		43.50	9.05	.59	53.14	64.50
	0720	High polish		137	.234		69.50	9.05	.59	79.14	93.50
	0740	Mirror polish		137	.234		87.50	9.05	.59	97.14	113
	0760	Stainless steel, 3 rail, 1-1/2" diam., #4 finish		120	.267		65.50	10.30	.68	76.48	91
	0770	High polish		120	.267		108	10.30	.68	118.98	138
	0780	Mirror finish		120	.267		131	10.30	.68	141.98	163
	0900	Wall rail, alum. pipe, 1-1/4" diam., satin finish		213	.150		8.35	5.80	.38	14.53	20
	0905	Clear anodized		213	.150		10.20	5.80	.38	16.38	22
	0910	Dark anodized		213	.150		12.35	5.80	.38	18.53	24.50
	0915	1-1/2" diameter, satin finish		213	.150		9.25	5.80	.38	15.43	21
	0920	Clear anodized		213	.150		11.65	5.80	.38	17.83	23.50
	0925	Dark anodized		213	.150		14.35	5.80	.38	20.53	26.50
	0930	Steel pipe, 1-1/4" diameter, primed		213	.150		7.30	5.80	.38	13.48	18.80
	0935	Galvanized		213	.150		10.55	5.80	.38	16.73	22.50
	0940	1-1/2" diameter		176	.182		7.50	7.05	.46	15.01	21.50
	0945	Galvanized		213	.150		10.60	5.80	.38	16.78	22.50
	0955	Stainless steel pipe, 1-1/2" diam., #4 finish		107	.299		34.50	11.55	.76	46.81	59.50
	0960	High polish		107	.299		70.50	11.55	.76	82.81	99
	0965	Mirror polish		107	.299		83	11.55	.76	95.31	112
780	0010	**RAILINGS, INDUSTRIAL** Welded, shop fabricated									780
	0020	2 rail, 3'-6" high, 1-1/2" pipe	E-4	255	.125	L.F.	17.95	4.85	.32	23.12	29
	0100	2" angle rail	"	255	.125		16.35	4.85	.32	21.52	27
	0200	For 4" high kick plate, 10 gauge, add					3.73			3.73	4.10
	0300	1/4" thick, add					4.80			4.80	5.30
	0500	For curved rails, add					30%	30%			

05530 | Gratings

		CREW	DAILY OUTPUT	LABOR-HOURS	UNIT	MAT.	LABOR	EQUIP.	TOTAL	TOTAL INCL O&P
300	0010	**FLOOR GRATING, ALUMINUM**, field fabricated from panels									300
	0110	Bearing bars @ 1-3/16" O.C., cross bars @ 4" O.C.,									
	0111	Up to 300 S.F., 1" x 1/8" bar	E-4	900	.036	S.F.	9.25	1.37	.09	10.71	12.75
	0112	Over 300 S.F.		850	.038		8.40	1.46	.10	9.96	11.95
	0113	1-1/4" x 1/8" bar, up to 300 S.F.		800	.040		10.50	1.55	.10	12.15	14.45
	0114	Over 300 S.F.		1,000	.032		9.55	1.24	.08	10.87	12.80
	0122	1-1/4" x 3/16" bar, up to 300 S.F.		750	.043		13.50	1.65	.11	15.26	17.90
	0124	Over 300 S.F.		1,000	.032		12.25	1.24	.08	13.57	15.80

Important: See the Reference Section for critical supporting data - Reference Nos., Crews, & City Cost Indexes

5 METALS

05530	Gratings	CREW	DAILY OUTPUT	LABOR-HOURS	UNIT	2005 BARE COSTS				TOTAL INCL O&P	
						MAT.	LABOR	EQUIP.	TOTAL		
300 0132	1-1/2" x 1/8" bar, up to 300 S.F.	E-4	700	.046	S.F.	8.45	1.77	.12	10.34	12.55	**300**
0134	Over 300 S.F.		1,000	.032		7.65	1.24	.08	8.97	10.75	
0136	1-3/4" x 3/16" bar, up to 300 S.F.		500	.064		18.10	2.47	.16	20.73	24.50	
0138	Over 300 S.F.		1,000	.032		16.45	1.24	.08	17.77	20.50	
0146	2-1/4" x 3/16" bar, up to 300 S.F.		600	.053		21.50	2.06	.14	23.70	27.50	
0148	Over 300 S.F.		1,000	.032		19.35	1.24	.08	20.67	24	
0162	Cross bars @ 2" O.C., 1" x 1/8", up to 300 S.F.		600	.053		11.65	2.06	.14	13.85	16.70	
0164	Over 300 S.F.		1,000	.032		10.60	1.24	.08	11.92	13.95	
0172	1-1/4" x 3/16" bar, up to 300 S.F.		600	.053		16.15	2.06	.14	18.35	21.50	
0174	Over 300 S.F.		1,000	.032		14.70	1.24	.08	16.02	18.45	
0182	1-1/2" x 1/8" bar, up to 300 S.F.		600	.053		11.60	2.06	.14	13.80	16.60	
0184	Over 300 S.F.		1,000	.032		10.55	1.24	.08	11.87	13.90	
0186	1-3/4" x 3/16" bar, up to 300 S.F.		600	.053		16.80	2.06	.14	19	22.50	
0188	Over 300 S.F.		1,000	.032		15.30	1.24	.08	16.62	19.10	
0200	For straight cuts, add				L.F.	2.38			2.38	2.62	
0300	For curved cuts, add					3.24			3.24	3.56	
0400	For straight banding, add					2.86			2.86	3.15	
0500	For curved banding, add					3.97			3.97	4.37	
0600	For aluminum checkered plate nosings, add					3.84			3.84	4.22	
0700	For straight toe plate, add					6.95			6.95	7.65	
0800	For curved toe plate, add					8.70			8.70	9.60	
1000	For cast aluminum abrasive nosings, add					4.67			4.67	5.15	
1200	Expanded aluminum, .65# per S.F.	E-4	1,050	.030	S.F.	4.07	1.18	.08	5.33	6.70	
1400	Extruded I bars are 10% less than 3/16" bars										
1600	Heavy duty, all extruded plank, 3/4" deep, 1.8 # per S.F.	E-4	1,100	.029	S.F.	11.40	1.12	.07	12.59	14.65	
1700	1-1/4" deep, 2.9# per S.F.		1,000	.032		12.15	1.24	.08	13.47	15.65	
1800	1-3/4" deep, 4.2# per S.F.		925	.035		14.50	1.34	.09	15.93	18.45	
1900	2-1/4" deep, 5.0# per S.F.		875	.037		16.60	1.41	.09	18.10	21	
2100	For safety serrated surface, add					15%					
320 0010	**FLOOR GRATING PLANKS**, field fabricated from planks										**320**
0020	Aluminum, 9-1/2" wide, 14 ga., 2" rib	E-4	950	.034	L.F.	13.40	1.30	.09	14.79	17.15	
0200	Galvanized steel, 9-1/2" wide, 14 ga., 2-1/2" rib		950	.034		8.65	1.30	.09	10.04	11.95	
0300	4" rib		950	.034		10.40	1.30	.09	11.79	13.85	
0500	12 gauge, 2-1/2" rib		950	.034		11.50	1.30	.09	12.89	15.05	
0600	3" rib		950	.034		13.50	1.30	.09	14.89	17.25	
0800	Stainless steel, type 304, 16 ga., 2" rib		950	.034		23	1.30	.09	24.39	28	
0900	Type 316		950	.034		28	1.30	.09	29.39	33.50	
340 0010	**FLOOR GRATING, STEEL**, field fabricated from panels										**340**
0050	Labor for installing, from ground/floor	E-4	845	.038	S.F.		1.46	.10	1.56	2.73	
0100	Elevated		460	.070	"		2.69	.18	2.87	5	
0300	Platforms, to 12' high, rectangular		3,150	.010	Lb.	1.32	.39	.03	1.74	2.18	
0400	Circular		2,300	.014	"	1.46	.54	.04	2.04	2.61	
0410	Painted bearing bars @ 1-3/16"										
0412	Cross bars @ 4" O.C., 3/4" x 1/8" bar, up to 300 S.F.	E-2	500	.112	S.F.	5.60	4.14	2.65	12.39	16.15	
0414	Over 300 S.F.		750	.075		5.05	2.76	1.77	9.58	12.30	
0422	1-1/4" x 3/16", up to 300 S.F.		400	.140		7.50	5.15	3.31	15.96	21	
0424	Over 300 S.F.		600	.093		6.85	3.45	2.21	12.51	15.85	
0432	1-1/2" x 1/8", up to 300 S.F.		400	.140		6.75	5.15	3.31	15.21	19.95	
0434	Over 300 S.F.		600	.093		6.10	3.45	2.21	11.76	15.10	
0436	1-3/4" x 3/16", up to 300 S.F.		400	.140		9.80	5.15	3.31	18.26	23.50	
0438	Over 300 S.F.		600	.093		8.90	3.45	2.21	14.56	18.15	
0452	2-1/4" x 3/16", up to 300 S.F.		300	.187		11.60	6.90	4.42	22.92	29.50	
0454	Over 300 S.F.		450	.124		10.55	4.60	2.94	18.09	22.50	
0462	Cross bars @ 2" O.C., 3/4" x 1/8", up to 300 S.F.		500	.112		10.75	4.14	2.65	17.54	22	
0464	Over 300 S.F.		750	.075		8.95	2.76	1.77	13.48	16.55	

METALS 5

05530 | Gratings

		CREW	DAILY OUTPUT	LABOR-HOURS	UNIT	MAT.	LABOR	EQUIP.	TOTAL	TOTAL INCL O&P	
340											**340**
0472	1-1/4" x 3/16", up to 300 S.F.	E-2	400	.140	S.F.	16.10	5.15	3.31	24.56	30	
0474	Over 300 S.F.		600	.093		13.40	3.45	2.21	19.06	23	
0482	1-1/2" x 1/8", up to 300 S.F.		400	.140		12.45	5.15	3.31	20.91	26	
0484	Over 300 S.F.		600	.093		10.35	3.45	2.21	16.01	19.75	
0486	1-3/4" x 3/16", up to 300 S.F.		400	.140		16.75	5.15	3.31	25.21	31	
0488	Over 300 S.F.		600	.093		13.95	3.45	2.21	19.61	23.50	
0502	2-1/4" x 3/16", up to 300 S.F.		300	.187		20	6.90	4.42	31.32	38.50	
0504	Over 300 S.F.		450	.124		16.70	4.60	2.94	24.24	29.50	
0690	For galvanized grating, add					25%					
0800	For straight cuts, add				L.F.	2.53			2.53	2.78	
0900	For curved cuts, add					3.48			3.48	3.83	
1000	For straight banding, add					2.90			2.90	3.19	
1100	For curved banding, add					4.48			4.48	4.93	
1200	For checkered plate nosings, add					5			5	5.50	
1300	For straight toe or kick plate, add					7.05			7.05	7.75	
1400	For curved toe or kick plate, add					8.60			8.60	9.45	
1500	For abrasive nosings, add					5.20			5.20	5.70	
1510	For stair treads, see division 05550-700										
1600	For safety serrated surface, minimum, add					15%					
1700	Maximum, add					25%					
2000	Stainless steel gratings, close spaced, 1" x 1/8" bars, up to 300 S.F.	E-4	450	.071	S.F.	31.50	2.75	.18	34.43	40	
2100	Standard spacing, 3/4" x 1/8" bars		500	.064		25	2.47	.16	27.63	32	
2200	1-1/4" x 3/16" bars		400	.080		39.50	3.09	.20	42.79	49	
2400	Expanded steel grating, at ground, 3.0# per S.F.		900	.036		3.11	1.37	.09	4.57	6	
2500	3.14# per S.F.		900	.036		3.29	1.37	.09	4.75	6.20	
2600	4.0# per S.F.		850	.038		4.06	1.46	.10	5.62	7.20	
2650	4.27# per S.F.		850	.038		4.40	1.46	.10	5.96	7.55	
2700	5.0# per S.F.		800	.040		5.20	1.55	.10	6.85	8.65	
2800	6.25# per S.F.		750	.043		6.60	1.65	.11	8.36	10.30	
2900	7.0# per S.F.		700	.046		7.45	1.77	.12	9.34	11.45	
3100	For flattened expanded steel grating, add					8%					
3300	For elevated installation above 15', add						15%				
360	0010	**GRATING FRAME**, field fabricated									**360**
	0020	Aluminum, for gratings 1" to 1-1/2" deep	1 Sswk	70	.114	L.F.	7.05	4.36		11.41	15.60	
	0100	For each corner, add				Ea.	5.55			5.55	6.15	

05540 | Floor Plates

		CREW	DAILY OUTPUT	LABOR-HOURS	UNIT	MAT.	LABOR	EQUIP.	TOTAL	TOTAL INCL O&P	
200	0010	**CHECKERED PLATE**, field fabricated									**200**
	0020	1/4" & 3/8", 2000 to 5000 S.F., bolted	E-4	2,900	.011	Lb.	.76	.43	.03	1.22	1.63	
	0100	Welded		4,400	.007	"	.71	.28	.02	1.01	1.30	
	0300	Pit or trench cover and frame, 1/4" plate, 2' to 3' wide		100	.320	S.F.	20	12.35	.81	33.16	45	
	0400	For galvanizing, add				Lb.	.41			.41	.45	
	0500	Platforms, 1/4" plate, no handrails included, rectangular	E-4	4,200	.008		1.21	.29	.02	1.52	1.88	
	0600	Circular	"	2,500	.013		1.69	.49	.03	2.21	2.79	
700	0010	**TRENCH COVER**, field fabricated									**700**
	0020	Cast iron grating with bar stops and angle frame, to 18" wide	1 Sswk	20	.400	L.F.	53	15.25		68.25	86	
	0100	Frame only (both sides of trench), 1" grating		45	.178		11.65	6.80		18.45	25	
	0150	2" grating		35	.229		18.10	8.70		26.80	35.50	
	0200	Aluminum, stock units, including frames and										
	0210	3/8" plain cover plate, 4" opening	E-4	205	.156	L.F.	33.50	6.05	.40	39.95	48	
	0300	6" opening		185	.173		41.50	6.70	.44	48.64	58	
	0400	10" opening		170	.188		57	7.30	.48	64.78	76	
	0500	16" opening		155	.206		78	8	.52	86.52	101	
	0700	Add per inch for additional widths to 24"					3.11			3.11	3.42	

Important: See the Reference Section for critical supporting data - Reference Nos., Crews, & City Cost Indexes

05540 | Floor Plates

		CREW	DAILY OUTPUT	LABOR-HOURS	UNIT	2005 BARE COSTS				TOTAL INCL O&P	
						MAT.	LABOR	EQUIP.	TOTAL		
700	0900	For custom fabrication, add				L.F.	50%					**700**
	1100	For 1/4" plain cover plate, deduct					12%					
	1500	For cover recessed for tile, 1/4" thick, deduct					12%					
	1600	3/8" thick, add					5%					
	1800	For checkered plate cover, 1/4" thick, deduct					12%					
	1900	3/8" thick, add					2%					
	2100	For slotted or round holes in cover, 1/4" thick, add					3%					
	2200	3/8" thick, add					4%					
	2300	For abrasive cover, add					12%					

05550 | Stair Treads & Nosings

		CREW	DAILY OUTPUT	LABOR-HOURS	UNIT	MAT.	LABOR	EQUIP.	TOTAL	TOTAL INCL O&P	
700	0010	**STAIR TREADS**										**700**
	0020	Aluminum grating, 3' long, 1-1/2" x 3/16" rect. bars, 6" wide	1 Sswk	24	.333	Ea.	47	12.70		59.70	75	
	0100	12" wide		22	.364		69.50	13.85		83.35	102	
	0200	1-1/2" x 3/16" I-bars, 6" wide		24	.333		39.50	12.70		52.20	66.50	
	0300	12" wide	▼	22	.364		64	13.85		77.85	95	
	0400	For abrasive nosings, add					12.35			12.35	13.55	
	0500	For narrow mesh, add				▼	60%					
	0600	Stair treads, not incl. stringers. See also div. 03150-660										
	0700	Cast aluminum, abrasive, 3' long x 12" wide, 5/16" thick	1 Sswk	15	.533	Ea.	105	20.50		125.50	152	
	0800	3/8" thick		15	.533		111	20.50		131.50	159	
	0900	1/2" thick		15	.533		122	20.50		142.50	171	
	1000	Cast bronze, abrasive, 3/8" thick		8	1		455	38		493	570	
	1100	1/2" thick		8	1		595	38		633	725	
	1200	Cast iron, abrasive, 3' long x 12" wide, 3/8" thick		15	.533		92	20.50		112.50	138	
	1300	1/2" thick	▼	15	.533	▼	106	20.50		126.50	154	
	1400	Fiberglass reinforced plastic with safety nosing,										
	1500	1-1/2" thick, 12" wide, 24" long	1 Sswk	22	.364	Ea.	99.50	13.85		113.35	134	
	1600	30" long		22	.364		104	13.85		117.85	139	
	1700	36" long		22	.364		113	13.85		126.85	150	
	2000	Steel grating, painted, 3' long, 1-1/4" x 3/16" bars, 6" wide		20	.400		40	15.25		55.25	71.50	
	2010	12" wide	▼	18	.444		46.50	16.95		63.45	81.50	
	2100	Add for abrasive nosing, 3' long, painted					12.50			12.50	13.75	
	2200	Galvanized					16.75			16.75	18.40	
	2300	Painting 3' treads, in shop, nonstandard paint					2.20			2.20	2.42	
	2400	Added coats, standard paint					.94			.94	1.03	
	2500	Expanded steel, 2-1/2" deep, 9" x 3' long, 18 gauge	1 Sswk	20	.400		22	15.25		37.25	52	
	2600	14 gauge	"	20	.400	▼	27	15.25		42.25	57.50	

05560 | Metal Castings

		CREW	DAILY OUTPUT	LABOR-HOURS	UNIT	MAT.	LABOR	EQUIP.	TOTAL	TOTAL INCL O&P	
200	0010	**CONSTRUCTION CASTINGS**										**200**
	0020	Manhole covers and frames see Division 02630-110										
	0100	Column bases, cast iron, 16" x 16", approx. 65 lbs.	E-4	46	.696	Ea.	102	27	1.77	130.77	163	
	0200	32" x 32", approx. 256 lbs.	"	23	1.391		380	54	3.53	437.53	520	
	0400	Cast aluminum for wood columns, 8" x 8"	1 Carp	32	.250		33	8.55		41.55	49.50	
	0500	12" x 12"	"	32	.250	▼	71	8.55		79.55	91.50	
	0600	Miscellaneous C.I. castings, light sections, less than 150 lbs	E-4	3,200	.010	Lb.	1.58	.39	.03	2	2.46	
	1100	Heavy sections, more than 150 lb		4,200	.008		.77	.29	.02	1.08	1.40	
	1300	Special low volume items	▼	3,200	.010		2.75	.39	.03	3.17	3.75	
	1500	For ductile iron, add				▼	100%					

05580 | Formed Metal Fabrications

		CREW	DAILY OUTPUT	LABOR-HOURS	UNIT	MAT.	LABOR	EQUIP.	TOTAL	TOTAL INCL O&P	
150	0010	**ALLOY STEEL CHAIN**, Grade 80										**150**
	0015	Self-colored, cut lengths, w/accessories, 1/4"	E-17	4	4	C.L.F.	445	157	1.32	603.32	770	

5 METALS

		05580	Formed Metal Fabrications	CREW	DAILY OUTPUT	LABOR-HOURS	UNIT	2005 BARE COSTS				TOTAL INCL O&P	
								MAT.	LABOR	EQUIP.	TOTAL		
150	0020		3/8"	E-17	2	8	C.L.F.	575	315	2.64	892.64	1,200	150
	0030		1/2"		1.20	13.333		905	520	4.40	1,429.40	1,925	
	0040		5/8"		.72	22.222		1,425	870	7.35	2,302.35	3,125	
	0050		3/4"		.48	33.333		2,075	1,300	11	3,386	4,625	
	0060		7/8"		.40	40		3,075	1,575	13.20	4,663.20	6,200	
	0070		1"		.35	45.714		5,050	1,800	15.10	6,865.10	8,775	
	0080		1-1/4"	▼	.24	66.667	▼	9,425	2,600	22	12,047	15,100	
	0110		Clevis slip hook, 1/4"				Ea.	12.95			12.95	14.25	
	0120		3/8"					19.40			19.40	21.50	
	0130		1/2"					32.50			32.50	36	
	0140		5/8"					49.50			49.50	54	
	0150		3/4"					103			103	113	
	0160		Eye/sling hook w/ hammerlock coupling, 7/8"					289			289	320	
	0170		1"					405			405	450	
	0180		1-1/4"				▼	615			615	675	
200	0010	**ALUMINUM COLUMNS**											200
	0020		Aluminum, extruded, stock units, no cap or base, 6" diameter	E-4	240	.133	L.F.	8.25	5.15	.34	13.74	18.65	
	0100		8" diameter		170	.188		11.15	7.30	.48	18.93	26	
	0200		10" diameter		150	.213		14.70	8.25	.54	23.49	31.50	
	0300		12" diameter		140	.229		25.50	8.85	.58	34.93	44.50	
	0400		15" diameter	▼	120	.267	▼	31	10.30	.68	41.98	53	
	0410		Caps and bases, plain, 6" diameter				Set	18.05			18.05	19.85	
	0420		8" diameter					22			22	24.50	
	0430		10" diameter					29.50			29.50	32.50	
	0440		12" diameter					54.50			54.50	60	
	0450		15" diameter					82			82	90.50	
	0460		Caps, ornamental, minimum					207			207	228	
	0470		Maximum				▼	905			905	995	
	0500		For square columns, add to column prices above				L.F.	50%					
	0700		Residential, flat, 8' high, plain	E-4	20	1.600	Ea.	62	62	4.06	128.06	184	
	0720		Fancy		20	1.600		125	62	4.06	191.06	253	
	0740		Corner type, plain		20	1.600		110	62	4.06	176.06	236	
	0760		Fancy	▼	20	1.600	▼	221	62	4.06	287.06	360	
600	0010	**LAMP POSTS**											600
	0020		Aluminum, 7' high, stock units, post only	1 Carp	16	.500	Ea.	33.50	17.15		50.65	63.50	
	0100		Mild steel, plain	"	16	.500	"	32	17.15		49.15	62	
900	0010	**WINDOW GUARDS**, shop fabricated											900
	0015		Expanded metal, steel angle frame, permanent	E-4	350	.091	S.F.	18.75	3.53	.23	22.51	27	
	0025		Steel bars, 1/2" x 1/2", spaced 5" O.C.	"	290	.110	"	12.95	4.26	.28	17.49	22	
	0030		Hinge mounted, add				Opng.	37.50			37.50	41	
	0040		Removable type, add				"	24			24	26	
	0050		For galvanized guards, add				S.F.	35%					
	0070		For pivoted or projected type, add					105%	40%				
	0100		Mild steel, stock units, economy	E-4	405	.079	▼	5.10	3.05	.20	8.35	11.25	
	0200		Deluxe		405	.079	▼	10.45	3.05	.20	13.70	17.15	
	0400		Woven wire, stock units, 3/8" channel frame, 3' x 5' opening		40	.800	Opng.	137	31	2.03	170.03	209	
	0500		4' x 6' opening	▼	38	.842		220	32.50	2.14	254.64	300	
	0800		Basket guards for above, add					188			188	207	
	1000		Swinging guards for above, add				▼	64.50			64.50	71	

05650 | Railroad Track & Accessories

05655 | Railroad Trackwork

			CREW	DAILY OUTPUT	LABOR-HOURS	UNIT	MAT.	LABOR	EQUIP.	TOTAL	TOTAL INCL O&P	
							\multicolumn{4}{c}{2005 BARE COSTS}			
700	0010	**RAILROAD SIDINGS**										**700**
	0020	Car bumpers, standard	R05650-100 B-14	2	24	Ea.	2,525	675	108	3,308	3,950	
	0100	Heavy duty		2	24		4,775	675	108	5,558	6,425	
	0200	Derails hand throw (sliding)	R05650-200	10	4.800		795	135	21.50	951.50	1,100	
	0300	Hand throw with standard timbers, open stand & target		8	6	▼	865	169	27	1,061	1,250	
	0400	Resurface and realign existing track		200	.240	L.F.		6.75	1.08	7.83	11.65	
	0600	For crushed stone ballast, add	▼	500	.096	"	12.55	2.70	.43	15.68	18.45	
	0800	Siding, yard spur, level grade										
	0810	100 lb. rail, new material on wood ties	B-14	57	.842	L.F.	88.50	23.50	3.79	115.79	138	
	1000	Steel ties in concrete w/100 lb new rail, fasteners & plates		22	2.182	"	122	61.50	9.80	193.30	240	
	1200	Switch timber, for a #8 switch, pressure treated		3.70	12.973	M.B.F.	2,125	365	58.50	2,548.50	2,975	
	1300	Complete set of timbers, 3.7 M.B.F. for #8 switch		1	48	Total	8,225	1,350	216	9,791	11,400	
	1400	Ties, concrete, 8'-6" long, 30" O.C.		80	.600	Ea.	85	16.85	2.70	104.55	122	
	1600	Wood, pressure treated, 6" x 8" x 8'-6", C.L. lots		90	.533		35.50	15	2.40	52.90	64.50	
	1700	L.C.L. lots		90	.533		37.50	15	2.40	54.90	66.50	
	1900	Heavy duty, 7" x 9" x 8'-6", C.L. lots		70	.686		39	19.25	3.09	61.34	76.50	
	2000	L.C.L. lots	▼	70	.686	▼	39	19.25	3.09	61.34	76.50	
	2200	Turnouts, #8, incl. 100 lb. rails, plates, bars, frog, switch pt.										
	2300	Timbers and ballast 6" below bottom of tie	B-14	.50	96	Ea.	9,600	2,700	430	12,730	15,300	
	2400	Wheel stops, fixed		18	2.667	Pr.	565	75	12	652	750	
	2450	Hinged	▼	14	3.429	"	765	96.50	15.45	876.95	1,000	
750	0010	**RAILROAD TRACK**	R05650-100									**750**
	0020	Track bolts				Ea.	1.98			1.98	2.18	
	0100	Joint bars				Pr.	60			60	66	
	0200	Spikes				Ea.	.46			.46	.51	
	0300	Tie plates				"	10			10	11	
	1000	Rail, 100 lb. prime grade				L.F.	22.50			22.50	25	
	1500	Relay rail	▼			"	11.25			11.25	12.40	

05700 | Ornamental Metal

05720 | Ornamental Handrails & Railings

			CREW	DAILY OUTPUT	LABOR-HOURS	UNIT	MAT.	LABOR	EQUIP.	TOTAL	TOTAL INCL O&P	
							\multicolumn{4}{c}{2005 BARE COSTS}			
700	0010	**RAILINGS, ORNAMENTAL**, shop fabricated										**700**
	0020	Aluminum, bronze or stainless, minimum	1 Sswk	24	.333	L.F.	26	12.70		38.70	52	
	0100	Maximum		9	.889		249	34		283	335	
	0200	Aluminum ornamental rail, minimum		15	.533		26	20.50		46.50	65.50	
	0300	Maximum		8	1		78.50	38		116.50	155	
	0400	Hand-forged wrought iron, minimum		12	.667		82	25.50		107.50	136	
	0500	Maximum		8	1		245	38		283	340	
	0550	Steel, minimum		12	.667		25.50	25.50		51	73.50	
	0560	Maximum		8	1		76	38		114	152	
	0600	Composite metal/wood/glass, minimum		6	1.333		157	51		208	264	
	0700	Maximum	▼	5	1.600	▼	315	61		376	455	

05730 | Ornamental Formed Metal

			CREW	DAILY OUTPUT	LABOR-HOURS	UNIT	MAT.	LABOR	EQUIP.	TOTAL	TOTAL INCL O&P	
200	0010	**COLUMNS, ORNAMENTAL**, shop fabricated	R05120-210									**200**
	6400	Mild steel, flat, 9" wide, stock units, painted, plain	E-4	160	.200	L.F.	5.80	7.75	.51	14.06	21	

METALS 5

177

		05730	Ornamental Formed Metal	CREW	DAILY OUTPUT	LABOR-HOURS	UNIT	2005 BARE COSTS				TOTAL INCL O&P	
								MAT.	LABOR	EQUIP.	TOTAL		
200	6450		Fancy	E-4	160	.200	L.F.	11.75	7.75	.51	20.01	27.50	200
	6500		Corner columns, painted, plain	R05120 -210	160	.200		10.30	7.75	.51	18.56	26	
	6550		Fancy	▼	160	.200	▼	21	7.75	.51	29.26	37.50	

		05810	Exp. Joint Cover Assemblies	CREW	DAILY OUTPUT	LABOR-HOURS	UNIT	2005 BARE COSTS				TOTAL INCL O&P	
								MAT.	LABOR	EQUIP.	TOTAL		
350	0010	**EXPANSION JOINT ASSEMBLIES** Custom units											350
	0200		Floor cover assemblies, 1" space, aluminum	1 Sswk	38	.211	L.F.	15.45	8.05		23.50	31.50	
	0300		Bronze		38	.211		31.50	8.05		39.55	49	
	0500		2" space, aluminum		38	.211		18.65	8.05		26.70	35	
	0600		Bronze		38	.211		33.50	8.05		41.55	51.50	
	0800		Wall and ceiling assemblies, 1" space, aluminum		38	.211		9.20	8.05		17.25	24.50	
	0900		Bronze		38	.211		29.50	8.05		37.55	47	
	1100		2" space, aluminum		38	.211		15.65	8.05		23.70	31.50	
	1200		Bronze		38	.211		36	8.05		44.05	54	
	1400		Floor to wall assemblies, 1" space, aluminum		38	.211		14.05	8.05		22.10	30	
	1500		Bronze or stainless		38	.211		35.50	8.05		43.55	53.50	
	1700		Gym floor angle covers, aluminum, 3" x 3" angle		46	.174		12.15	6.65		18.80	25.50	
	1800		3" x 4" angle		46	.174		14.35	6.65		21	27.50	
	2000		Roof closures, aluminum, flat roof, low profile, 1" space		57	.140		28	5.35		33.35	40.50	
	2100		High profile		57	.140		34.50	5.35		39.85	47.50	
	2300		Roof to wall, low profile, 1" space		57	.140		15.55	5.35		20.90	26.50	
	2400		High profile	▼	57	.140	▼	20	5.35		25.35	31.50	

		05910	Metal Cleaning	CREW	DAILY OUTPUT	LABOR-HOURS	UNIT	2005 BARE COSTS				TOTAL INCL O&P	
								MAT.	LABOR	EQUIP.	TOTAL		
500	0010	**METAL CLEANING**											500
	6125		Steel surface treatments										
	6170		Wire brush, hand (SSPC-SP2)	1 Psst	240	.033	S.F.	.02	1.03		1.05	1.93	
	6180		Power tool (SSPC-SP3)	"	600	.013		.05	.41		.46	.81	
	6215		Pressure washing, 2800-6000 S.F./day	1 Pord	3,500	.002			.07		.07	.11	
	6220		Steam cleaning, 2800-4000 S.F./day		2,400	.003			.10		.10	.15	
	6225		Water blasting	▼	3,200	.002			.08		.08	.12	
	6230		Brush-off blast (SSPC-SP7)	E-11	6,000	.005		.08	.16	.02	.26	.38	
	6235		Com'l blast (SSPC-SP6), loose scale, fine pwder rust, 2.0 #/S.F. sand		4,000	.008		.15	.24	.03	.42	.62	
	6240		Tight mill scale, little/no rust, 3.0 #/S.F. sand		3,200	.010		.23	.30	.04	.57	.81	
	6245		Exist coat blistered/pitted, 4.0 #/S.F. sand		2,100	.015		.31	.46	.07	.84	1.19	
	6250		Exist coat badly pitted/nodules, 6.7 #/S.F. sand		1,300	.025		.52	.75	.11	1.38	1.95	
	6255		Near white blast (SSPC-SP10), loose scale, fine rust, 5.6 #/S.F. sand		1,700	.019	▼	.43	.57	.08	1.08	1.53	
	6260		Tight mill scale, little/no rust, 6.9 #/S.F. sand	▼	1,400	.023		.53	.70	.10	1.33	1.86	

05910	Metal Cleaning	CREW	DAILY OUTPUT	LABOR-HOURS	UNIT	2005 BARE COSTS				TOTAL INCL O&P	
						MAT.	LABOR	EQUIP.	TOTAL		
500 6265	Exist coat blistered/pitted, 9.0 #/S.F. sand	E-11	1,100	.029	S.F.	.69	.89	.13	1.71	2.39	**500**
6270	Exist coat badly pitted/nodules, 11.3 #/S.F. sand	↓	800	.040	↓	.87	1.22	.17	2.26	3.20	

05950	Paints & Protective Coatings										
650 0010	**PAINTS & PROTECTIVE COATINGS** R05080 -310										**650**
5900	Galvanizing structural steel in shop, under 1 ton				Ton	515			515	565	
5950	1 ton to 20 tons					430			430	470	
6000	Over 20 tons				↓	375			375	415	
6100	Cold galvanizing, brush in field	1 Psst	1,100	.007	S.F.	.07	.23		.30	.48	
6510	Paints & protective coatings, sprayed in field										
6520	Alkyds, primer	2 Psst	3,600	.004	S.F.	.06	.14		.20	.31	
6540	Gloss topcoats		3,200	.005		.05	.16		.21	.34	
6560	Silicone alkyd		3,200	.005		.10	.16		.26	.40	
6610	Epoxy, primer		3,000	.005		.17	.17		.34	.49	
6630	Intermediate or topcoat		2,800	.006		.14	.18		.32	.48	
6650	Enamel coat		2,800	.006		.17	.18		.35	.52	
6700	Epoxy ester, primer		2,800	.006		.34	.18		.52	.71	
6720	Topcoats		2,800	.006		.11	.18		.29	.46	
6810	Latex primer		3,600	.004		.05	.14		.19	.31	
6830	Topcoats		3,200	.005		.06	.16		.22	.35	
6910	Universal primers, one part, phenolic, modified alkyd		2,000	.008		.08	.25		.33	.55	
6940	Two part, epoxy spray		2,000	.008		.20	.25		.45	.68	
7000	Zinc rich primers, self cure, spray, inorganic		1,800	.009		.51	.28		.79	1.07	
7010	Epoxy, spray, organic	↓	1,800	.009	↓	.14	.28		.42	.66	
7020	Above one story, spray painting simple structures, add	↓					25%				
7030	Intricate structures, add						50%				

For information about Means Estimating Seminars, see yellow pages 12 and 13 in back of book

METALS **5**

179

		CREW	DAILY OUTPUT	LABOR-HOURS	UNIT	2005 BARE COSTS				TOTAL INCL O&P
						MAT.	LABOR	EQUIP.	TOTAL	

Division 6
Wood & Plastics

Estimating Tips

06050 Basic Wood & Plastic Materials & Methods

- Common to any wood framed structure are the accessory connector items such as screws, nails, adhesives, hangers, connector plates, straps, angles and holdowns. For typical wood framed buildings, such as residential projects, the aggregate total for these items can be significant, especially in areas where seismic loading is a concern. For floor and wall framing, the material cost is based on 10 to 25 lbs. per MBF. Holdowns, hangers and other connectors should be taken off by the piece.

06100 Rough Carpentry

- Lumber is a traded commodity and therefore sensitive to supply and demand in the marketplace. Even in "budgetary" estimating of wood framed projects, it is advisable to call local suppliers for the latest market pricing.
- Common quantity units for wood framed projects are "thousand board feet" (MBF). A board foot is a volume of wood, 1″ x 1′ x 1′, or 144 cubic inches. Board foot quantities are generally calculated using nominal material dimensions—dressed sizes are ignored. Board foot per lineal foot of any stick of lumber can be calculated by dividing the nominal cross sectional area by 12. As an example, 2,000 lineal feet of 2 x 12 equates to 4 MBF by dividing the nominal area, 2 x 12, by 12, which equals 2, and multiplying by 2,000 to give 4,000 board feet. This simple rule applies to all nominal dimensioned lumber.
- Waste is an issue of concern at the quantity takeoff for any area of construction. Framing lumber is sold in even foot lengths, i.e., 10′, 12′, 14′, 16′, and depending on spans, wall heights and the grade of lumber, waste is inevitable. A rule of thumb for lumber waste is 5% to 10% depending on material quality and the complexity of the framing.
- Wood in various forms and shapes is used in many projects, even where the main structural framing is steel, concrete or masonry. Plywood as a back-up partition material and 2x boards used as blocking and cant strips around roof edges are two common examples. The estimator should ensure that the costs of all wood materials are included in the final estimate.

06200 Finish Carpentry

- It is necessary to consider the grade of workmanship when estimating labor costs for erecting millwork and interior finish. In practice, there are three grades: premium, custom and economy. The Means daily output for base and case moldings is in the range of 200 to 250 L.F. per carpenter per day. This is appropriate for most average custom grade projects. For premium projects an adjustment to productivity of 25% to 50% should be made depending on the complexity of the job.

Reference Numbers

Reference numbers are shown in bold squares at the beginning of some major classifications. These numbers refer to related items in the Reference Section. The reference information may be an estimating procedure, an alternate pricing method or technical information.

Note: Not all subdivisions listed here necessarily appear in this publication.

			CREW	DAILY OUTPUT	LABOR-HOURS	UNIT	2005 BARE COSTS				TOTAL INCL O&P
		06052 | Selective Demolition					MAT.	LABOR	EQUIP.	TOTAL	
110	0010	**SELECTIVE DEMOLITION, WOOD FRAMING**									110
	0100	Timber connector, nailed, small	1 Clab	96	.083	Ea.		2.22		2.22	3.46
	0110	Medium		60	.133			3.56		3.56	5.55
	0120	Large		48	.167			4.45		4.45	6.95
	0130	Bolted, small		48	.167			4.45		4.45	6.95
	0140	Medium		32	.250			6.70		6.70	10.40
	0150	Large		24	.333			8.90		8.90	13.85
	2958	Beams, 2" x 6"	2 Clab	1,100	.015	L.F.		.39		.39	.60
	2960	2" x 8"		825	.019			.52		.52	.81
	2965	2" x 10"		665	.024			.64		.64	1
	2970	2" x 12"		550	.029			.78		.78	1.21
	2972	2" x 14"		470	.034			.91		.91	1.41
	2975	4" x 8"	B-1	413	.058			1.59		1.59	2.48
	2980	4" x 10"		330	.073			1.99		1.99	3.10
	2985	4" x 12"		275	.087			2.39		2.39	3.72
	3000	6" x 8"		275	.087			2.39		2.39	3.72
	3040	6" x 10"		220	.109			2.99		2.99	4.65
	3080	6" x 12"		185	.130			3.55		3.55	5.55
	3120	8" x 12"		140	.171			4.69		4.69	7.30
	3160	10" x 12"		110	.218			5.95		5.95	9.30
	3162	Alternate pricing method		1.10	21.818	M.B.F.		595		595	930
	3170	Blocking, in 16" OC wall framing, 2" x 4"	1 Clab	600	.013	L.F.		.36		.36	.55
	3172	2" x 6"		400	.020			.53		.53	.83
	3174	In 24" OC wall framing, 2" x 4"		600	.013			.36		.36	.55
	3176	2" x 6"		400	.020			.53		.53	.83
	3178	Alt method, wood blocking removal from wood framimg		.40	20	M.B.F.		535		535	830
	3179	Wood blocking removal from steel framimg		.36	22.222	"		595		595	925
	3180	Bracing, let in, 1" x 3", studs 16" OC		1,050	.008	L.F.		.20		.20	.32
	3181	Studs 24" OC		1,080	.007			.20		.20	.31
	3182	1" x 4", studs 16" OC		1,050	.008			.20		.20	.32
	3183	Studs 24" OC		1,080	.007			.20		.20	.31
	3184	1" x 6", studs 16" OC		1,050	.008			.20		.20	.32
	3185	Studs 24" OC		1,080	.007			.20		.20	.31
	3186	2" x 3", studs 16" OC		800	.010			.27		.27	.42
	3187	Studs 24" OC		830	.010			.26		.26	.40
	3188	2" x 4", studs 16" OC		800	.010			.27		.27	.42
	3189	Studs 24" OC		830	.010			.26		.26	.40
	3190	2" x 6", studs 16" OC		800	.010			.27		.27	.42
	3191	Studs 24" OC		830	.010			.26		.26	.40
	3192	2" x 8", studs 16" OC		800	.010			.27		.27	.42
	3193	Studs 24" OC		830	.010			.26		.26	.40
	3194	"T" shaped metal bracing, studs at 16" OC		1,060	.008			.20		.20	.31
	3195	Studs at 24" OC		1,200	.007			.18		.18	.28
	3196	Metal straps, studs at 16" OC		1,200	.007			.18		.18	.28
	3197	Studs at 24" OC		1,240	.006			.17		.17	.27
	3200	Columns, round, 8' to 14' tall		40	.200	Ea.		5.35		5.35	8.30
	3202	Dimensional lumber sizes	2 Clab	1.10	14.545	M.B.F.		390		390	605
	3250	Blocking, between joists	1 Clab	320	.025	Ea.		.67		.67	1.04
	3252	Bridging, metal strap, between joists		320	.025	Pr.		.67		.67	1.04
	3254	Wood, between joists		320	.025	"		.67		.67	1.04
	3260	Door buck, studs, header & access, 8' high 2" x 4" wall, 3' wide		32	.250	Ea.		6.70		6.70	10.40
	3261	4' wide		32	.250			6.70		6.70	10.40
	3262	5' wide		32	.250			6.70		6.70	10.40
	3263	6' wide		32	.250			6.70		6.70	10.40
	3264	8' wide		30	.267			7.10		7.10	11.10
	3265	10' wide		30	.267			7.10		7.10	11.10

R02220
-510

Important: See the Reference Section for critical supporting data - Reference Nos., Crews, & City Cost Indexes

6
WOOD & PLASTICS

			CREW	DAILY OUTPUT	LABOR-HOURS	UNIT	2005 BARE COSTS				TOTAL INCL O&P	
06052 | Selective Demolition							MAT.	LABOR	EQUIP.	TOTAL		
110	3266	12' wide	R02220 -510	1 Clab	30	.267	Ea.		7.10		7.10	11.10	110
	3267	2" x 6" wall, 3' wide			32	.250			6.70		6.70	10.40	
	3268	4' wide			32	.250			6.70		6.70	10.40	
	3269	5' wide			32	.250			6.70		6.70	10.40	
	3270	6' wide			32	.250			6.70		6.70	10.40	
	3271	8' wide			30	.267			7.10		7.10	11.10	
	3272	10' wide			30	.267			7.10		7.10	11.10	
	3273	12' wide			30	.267			7.10		7.10	11.10	
	3274	Window buck, studs, header & access, 8' high 2" x 4" wall, 2' wide			24	.333			8.90		8.90	13.85	
	3275	3' wide			24	.333			8.90		8.90	13.85	
	3276	4' wide			24	.333			8.90		8.90	13.85	
	3277	5' wide			24	.333			8.90		8.90	13.85	
	3278	6' wide			24	.333			8.90		8.90	13.85	
	3279	7' wide			24	.333			8.90		8.90	13.85	
	3280	8' wide			22	.364			9.70		9.70	15.10	
	3281	10' wide			22	.364			9.70		9.70	15.10	
	3282	12' wide			22	.364			9.70		9.70	15.10	
	3283	2" x 6" wall, 2' wide			24	.333			8.90		8.90	13.85	
	3284	3' wide			24	.333			8.90		8.90	13.85	
	3285	4' wide			24	.333			8.90		8.90	13.85	
	3286	5' wide			24	.333			8.90		8.90	13.85	
	3287	6' wide			24	.333			8.90		8.90	13.85	
	3288	7' wide			24	.333			8.90		8.90	13.85	
	3289	8' wide			22	.364			9.70		9.70	15.10	
	3290	10' wide			22	.364			9.70		9.70	15.10	
	3291	12' wide			22	.364			9.70		9.70	15.10	
	3400	Fascia boards, 1" x 6"			500	.016	L.F.		.43		.43	.66	
	3440	1" x 8"			450	.018			.47		.47	.74	
	3480	1" x 10"			400	.020			.53		.53	.83	
	3490	2" x 6"			450	.018			.47		.47	.74	
	3500	2" x 8"			400	.020			.53		.53	.83	
	3510	2" x 10"			350	.023			.61		.61	.95	
	3610	Furring, on wood walls or ceiling			4,000	.002	S.F.		.05		.05	.08	
	3620	On masonry or concrete walls or ceiling			1,200	.007	"		.18		.18	.28	
	3800	Headers over openings, 2 @ 2" x 6"			110	.073	L.F.		1.94		1.94	3.02	
	3840	2 @ 2" x 8"			100	.080			2.14		2.14	3.32	
	3880	2 @ 2" x 10"			90	.089			2.37		2.37	3.69	
	3885	Alternate pricing method			.26	30.651	M.B.F.		820		820	1,275	
	3920	Joists, 1" x 4"			1,250	.006	L.F.		.17		.17	.27	
	3930	1" x 6"			1,135	.007			.19		.19	.29	
	3940	1" x 8"			1,000	.008			.21		.21	.33	
	3950	1" x 10"			895	.009			.24		.24	.37	
	3960	1" x 12"			765	.010			.28		.28	.43	
	4200	2" x 4"		2 Clab	1,000	.016			.43		.43	.66	
	4230	2" x 6"			970	.016			.44		.44	.69	
	4240	2" x 8"			940	.017			.45		.45	.71	
	4250	2" x 10"			910	.018			.47		.47	.73	
	4280	2" x 12"			880	.018			.49		.49	.76	
	4281	2" x 14"			850	.019			.50		.50	.78	
	4282	Composite joists, 9-1/2"			960	.017			.45		.45	.69	
	4283	11-7/8"			930	.017			.46		.46	.71	
	4284	14"			897	.018			.48		.48	.74	
	4285	16"			865	.018			.49		.49	.77	
	4290	Wood joists, alternate pricing method			1.50	10.667	M.B.F.		285		285	445	
	4500	Open web joist, 12" deep			500	.032	L.F.		.85		.85	1.33	
	4505	14" deep			475	.034			.90		.90	1.40	

WOOD & PLASTICS | **6**

			CREW	DAILY OUTPUT	LABOR-HOURS	UNIT	2005 BARE COSTS				TOTAL INCL O&P	
	06052	**Selective Demolition**					MAT.	LABOR	EQUIP.	TOTAL		
110	4510	16" deep	2 Clab	450	.036	L.F.		.95		.95	1.48	110
	4520	18" deep		425	.038			1.01		1.01	1.56	
	4530	24" deep		400	.040			1.07		1.07	1.66	
	4550	Ledger strips, 1" x 2"	1 Clab	1,200	.007			.18		.18	.28	
	4560	1" x 3"		1,200	.007			.18		.18	.28	
	4570	1" x 4"		1,200	.007			.18		.18	.28	
	4580	2" x 2"		1,100	.007			.19		.19	.30	
	4590	2" x 4"		1,000	.008			.21		.21	.33	
	4600	2" x 6"		1,000	.008			.21		.21	.33	
	4601	2" x 8" or 2" x 10"		800	.010			.27		.27	.42	
	4602	4" x 6"		600	.013			.36		.36	.55	
	4604	4" x 8"		450	.018			.47		.47	.74	
	5400	Posts, 4" x 4"	2 Clab	800	.020			.53		.53	.83	
	5405	4" x 6"		550	.029			.78		.78	1.21	
	5410	4" x 8"		440	.036			.97		.97	1.51	
	5420	4" x 6"	1 Clab	270	.030			.79		.79	1.23	
	5425	4" x 10"	2 Clab	390	.041			1.10		1.10	1.70	
	5430	4" x 12"		350	.046			1.22		1.22	1.90	
	5440	6" x 6"		400	.040			1.07		1.07	1.66	
	5445	6" x 8"		350	.046			1.22		1.22	1.90	
	5450	6" x 10"		320	.050			1.34		1.34	2.08	
	5455	6" x 12"		290	.055			1.47		1.47	2.29	
	5480	8" x 8"		300	.053			1.42		1.42	2.22	
	5500	10" x 10"		240	.067			1.78		1.78	2.77	
	5660	Tongue and groove floor planks		2	8	M.B.F.		214		214	330	
	5750	Rafters, ordinary, 16" OC, 2" x 4"		880	.018	S.F.		.49		.49	.76	
	5755	2" x 6"		840	.019			.51		.51	.79	
	5760	2" x 8"		820	.020			.52		.52	.81	
	5770	2" x 10"		820	.020			.52		.52	.81	
	5780	2" x 12"		810	.020			.53		.53	.82	
	5785	24" OC, 2" x 4"		1,170	.014			.37		.37	.57	
	5786	2" x 6"		1,117	.014			.38		.38	.60	
	5787	2" x 8"		1,091	.015			.39		.39	.61	
	5788	2" x 10"		1,091	.015			.39		.39	.61	
	5789	2" x 12"		1,077	.015			.40		.40	.62	
	5795	Rafters, ordinary, 2" x 4" (alternate method)		862	.019	L.F.		.50		.50	.77	
	5800	2" x 6" (alternate method)		850	.019			.50		.50	.78	
	5840	2" x 8" (alternate method)		837	.019			.51		.51	.79	
	5855	2" x 10" (alternate method)		825	.019			.52		.52	.81	
	5865	2" x 12" (alternate method)		812	.020			.53		.53	.82	
	5870	Sill plate, 2" x 4"	1 Clab	1,170	.007			.18		.18	.28	
	5871	2" x 6"		780	.010			.27		.27	.43	
	5872	2" x 8"		586	.014			.36		.36	.57	
	5873	Alternate pricing method		.78	10.256	M.B.F.		274		274	425	
	5885	Ridge board, 1" x 4"	2 Clab	900	.018	L.F.		.47		.47	.74	
	5886	1" x 6"		875	.018			.49		.49	.76	
	5887	1" x 8"		850	.019			.50		.50	.78	
	5888	1" x 10"		825	.019			.52		.52	.81	
	5889	1" x 12"		800	.020			.53		.53	.83	
	5890	2" x 4"		900	.018			.47		.47	.74	
	5892	2" x 6"		875	.018			.49		.49	.76	
	5894	2" x 8"		850	.019			.50		.50	.78	
	5896	2" x 10"		825	.019			.52		.52	.81	
	5898	2" x 12"		800	.020			.53		.53	.83	
	6050	Rafter tie, 1" x 4"		1,250	.013			.34		.34	.53	
	6052	1" x 6"		1,135	.014			.38		.38	.59	

Reference note near 4510: R02220 -510

Important: See the Reference Section for critical supporting data - Reference Nos., Crews, & City Cost Indexes

		06052	Selective Demolition	CREW	DAILY OUTPUT	LABOR-HOURS	UNIT	MAT.	2005 BARE COSTS LABOR	EQUIP.	TOTAL	TOTAL INCL O&P	
110	6054		2" x 4" R02220-510	2 Clab	1,000	.016	L.F.		.43		.43	.66	110
	6056		2" x 6"		970	.016			.44		.44	.69	
	6070		Sleepers, on concrete, 1" x 2"	1 Clab	4,700	.002			.05		.05	.07	
	6075		1" x 3"		4,000	.002			.05		.05	.08	
	6080		2" x 4"		3,000	.003			.07		.07	.11	
	6085		2" x 6"		2,600	.003			.08		.08	.13	
	6086		Sheathing from roof, 5/16"	2 Clab	1,600	.010	S.F.		.27		.27	.42	
	6088		3/8"		1,525	.010			.28		.28	.44	
	6090		1/2"		1,400	.011			.31		.31	.47	
	6092		5/8"		1,300	.012			.33		.33	.51	
	6094		3/4"		1,200	.013			.36		.36	.55	
	6096		Board sheathing from roof		1,400	.011			.31		.31	.47	
	6100		Sheathing, from walls, 1/4"		1,200	.013			.36		.36	.55	
	6110		5/16"		1,175	.014			.36		.36	.57	
	6120		3/8"		1,150	.014			.37		.37	.58	
	6130		1/2"		1,125	.014			.38		.38	.59	
	6140		5/8"		1,100	.015			.39		.39	.60	
	6150		3/4"		1,075	.015			.40		.40	.62	
	6152		Board sheathing from walls		1,500	.011			.28		.28	.44	
	6158		Subfloor, with boards		1,050	.015			.41		.41	.63	
	6160		Plywood, 1/2" thick		768	.021			.56		.56	.87	
	6162		5/8" thick		760	.021			.56		.56	.87	
	6164		3/4" thick		750	.021			.57		.57	.89	
	6165		1-1/8" thick		720	.022			.59		.59	.92	
	6166		Underlayment, particle board, 3/8" thick	1 Clab	780	.010			.27		.27	.43	
	6168		1/2" thick		768	.010			.28		.28	.43	
	6170		5/8" thick		760	.011			.28		.28	.44	
	6172		3/4" thick		750	.011			.28		.28	.44	
	6200		Stairs and stringers, minimum	2 Clab	40	.400	Riser		10.70		10.70	16.60	
	6240		Maximum	"	26	.615	"		16.45		16.45	25.50	
	6300		Components, tread	1 Clab	110	.073	Ea.		1.94		1.94	3.02	
	6320		Riser		80	.100	"		2.67		2.67	4.16	
	6390		Stringer, 2" x 10"		260	.031	L.F.		.82		.82	1.28	
	6400		2" x 12"		260	.031			.82		.82	1.28	
	6410		3" x 10"		250	.032			.85		.85	1.33	
	6420		3" x 12"		250	.032			.85		.85	1.33	
	6590		Wood studs, 2" x 3"	2 Clab	3,076	.005			.14		.14	.22	
	6600		2" x 4"		2,000	.008			.21		.21	.33	
	6640		2" x 6"		1,600	.010			.27		.27	.42	
	6720		Wall framing, including studs plates and blocking, 2" x 4"	1 Clab	600	.013	S.F.		.36		.36	.55	
	6740		2" x 6"		480	.017	"		.45		.45	.69	
	6750		Headers, 2" x 4"		1,125	.007	L.F.		.19		.19	.30	
	6755		2" x 6"		1,125	.007			.19		.19	.30	
	6760		2" x 8"		1,050	.008			.20		.20	.32	
	6765		2" x 10"		1,050	.008			.20		.20	.32	
	6770		2" x 12"		1,000	.008			.21		.21	.33	
	6780		4" x 10"		525	.015			.41		.41	.63	
	6785		4" x 12"		500	.016			.43		.43	.66	
	6790		6" x 8"		560	.014			.38		.38	.59	
	6795		6" x 10"		525	.015			.41		.41	.63	
	6797		6" x 12"		500	.016			.43		.43	.66	
	8000		Soffit, T & G wood		520	.015	S.F.		.41		.41	.64	
	8010		Hardboard, vinyl or aluminum		640	.013			.33		.33	.52	
	8030		Plywood	2 Carp	315	.051			1.74		1.74	2.71	
	9500		See Div. 02220-350 for rubbish handling										

WOOD & PLASTICS | **6**

06052 | Selective Demolition

		CREW	DAILY OUTPUT	LABOR-HOURS	UNIT	2005 BARE COSTS				TOTAL INCL O&P
						MAT.	LABOR	EQUIP.	TOTAL	
120	0010	**SELECTIVE DEMOLITION, MILLWORK AND TRIM** R02220-510									120
	1000	Cabinets, wood, base cabinets, per L.F.	2 Clab	80	.200	L.F.		5.35		5.35	8.30
	1020	Wall cabinets, per L.F.	"	80	.200	"		5.35		5.35	8.30
	1060	Remove and reset, base cabinets	2 Carp	18	.889	Ea.		30.50		30.50	47.50
	1070	Wall cabinets	"	20	.800	"		27.50		27.50	42.50
	1100	Steel, painted, base cabinets	2 Clab	60	.267	L.F.		7.10		7.10	11.10
	1120	Wall cabinets		60	.267	"		7.10		7.10	11.10
	1200	Casework, large area		320	.050	S.F.		1.34		1.34	2.08
	1220	Selective		200	.080	"		2.14		2.14	3.32
	1500	Counter top, minimum		200	.080	L.F.		2.14		2.14	3.32
	1510	Maximum	▼	120	.133			3.56		3.56	5.55
	1550	Remove and reset, minimum	2 Carp	50	.320			10.95		10.95	17.05
	1560	Maximum	"	40	.400	▼		13.70		13.70	21.50
	2000	Paneling, 4' x 8' sheets	2 Clab	2,000	.008	S.F.		.21		.21	.33
	2100	Boards, 1" x 4"		700	.023			.61		.61	.95
	2120	1" x 6"		750	.021			.57		.57	.89
	2140	1" x 8"		800	.020	▼		.53		.53	.83
	3000	Trim, baseboard, to 6" wide		1,200	.013	L.F.		.36		.36	.55
	3040	Greater than 6" and up to 12" wide	▼	1,000	.016			.43		.43	.66
	3080	Remove and reset, minimum	2 Carp	400	.040			1.37		1.37	2.13
	3090	Maximum	"	300	.053			1.83		1.83	2.85
	3100	Ceiling trim	2 Clab	1,000	.016			.43		.43	.66
	3120	Chair rail		1,200	.013			.36		.36	.55
	3140	Railings with balusters		240	.067	▼		1.78		1.78	2.77
	3160	Wainscoting	▼	700	.023	S.F.		.61		.61	.95

06070 | Lumber Treatment

		CREW	DAILY OUTPUT	LABOR-HOURS	UNIT	MAT.	LABOR	EQUIP.	TOTAL	TOTAL INCL O&P
400	0011	**LUMBER TREATMENT**									400
	0400	Fire retardant, wet				M.B.F.	291			291	320
	0500	KDAT					277			277	305
	0700	Salt treated, water borne, .40 lb. retention					133			133	146
	0800	Oil borne, 8 lb. retention					156			156	171
	1000	Kiln dried lumber, 1" & 2" thick, softwoods					88.50			88.50	97.50
	1100	Hardwoods					94.50			94.50	104
	1500	For small size 1" stock, add				▼	11.95			11.95	13.15
	1700	For full size rough lumber, add					20%				

						MAT.	LABOR	EQUIP.	TOTAL	TOTAL INCL O&P
600	0010	**PLYWOOD TREATMENT** Fire retardant, 1/4" thick				M.S.F.	222			222	244	600
	0030	3/8" thick					244			244	268
	0050	1/2" thick					261			261	287
	0070	5/8" thick					277			277	305
	0100	3/4" thick					305			305	335
	0200	For KDAT, add					66.50			66.50	73
	0500	Salt treated water borne, .25 lb., wet, 1/4" thick					122			122	134
	0530	3/8" thick					127			127	140
	0550	1/2" thick					133			133	146
	0570	5/8" thick					145			145	159
	0600	3/4" thick					150			150	165
	0800	For KDAT add					66.50			66.50	73
	0900	For .40 lb., per C.F. retention, add					55.50			55.50	61
	1000	For certification stamp, add				▼	32.50			32.50	36

06090 | Wood & Plastic Fastenings

					UNIT	MAT.	LABOR	EQUIP.	TOTAL	TOTAL INCL O&P
600	0010	**NAILS** Prices of material only, based on 50# box purchase, copper, plain				Lb.	4.99			4.99	5.50	600
	0400	Stainless steel, plain				▼	5.75			5.75	6.30

06090	Wood & Plastic Fastenings		CREW	DAILY OUTPUT	LABOR-HOURS	UNIT	2005 BARE COSTS				TOTAL INCL O&P	
							MAT.	LABOR	EQUIP.	TOTAL		
600	0500	Box, 3d to 20d, bright				Lb.	1.32			1.32	1.45	**600**
	0520	Galvanized					1.44			1.44	1.58	
	0600	Common, 3d to 60d, plain					1.02			1.02	1.12	
	0700	Galvanized					1.39			1.39	1.53	
	0800	Aluminum					3.38			3.38	3.72	
	1000	Annular or spiral thread, 4d to 60d, plain					1.60			1.60	1.76	
	1200	Galvanized					1.80			1.80	1.98	
	1400	Drywall nails, plain					.79			.79	.87	
	1600	Galvanized					1.54			1.54	1.69	
	1800	Finish nails, 4d to 10d, plain					.94			.94	1.03	
	2000	Galvanized					1.32			1.32	1.45	
	2100	Aluminum					4.10			4.10	4.51	
	2300	Flooring nails, hardened steel, 2d to 10d, plain					1.40			1.40	1.54	
	2400	Galvanized					2.25			2.25	2.48	
	2500	Gypsum lath nails, 1-1/8", 13 ga. flathead, blued					1.54			1.54	1.69	
	2600	Masonry nails, hardened steel, 3/4" to 3" long, plain					1.48			1.48	1.63	
	2700	Galvanized					1.87			1.87	2.06	
	2900	Roofing nails, threaded, galvanized					1.35			1.35	1.49	
	3100	Aluminum					4.80			4.80	5.30	
	3300	Compressed lead head, threaded, galvanized					1.50			1.50	1.65	
	3600	Siding nails, plain shank, galvanized					1.45			1.45	1.60	
	3800	Aluminum					4.11			4.11	4.52	
	5000	Add to prices above for cement coating					.10			.10	.11	
	5200	Zinc or tin plating					.13			.13	.14	
	5500	Vinyl coated sinkers, 8d to 16d				▼	.57			.57	.63	
650	0010	**NAILS** mat. only, for pneumatic tools, framing, per carton of 5000, 2"				Ea.	37			37	41	**650**
	0100	2-3/8"					42.50			42.50	46.50	
	0200	Per carton of 4000, 3"					37.50			37.50	41.50	
	0300	3-1/4"					40			40	44	
	0400	Per carton of 5000, 2-3/8", galv.					57.50			57.50	63.50	
	0500	Per carton of 4000, 3", galv.					65			65	71.50	
	0600	3-1/4", galv.					80.50			80.50	88.50	
	0700	Roofing, per carton of 7200, 1"					35			35	38.50	
	0800	1-1/4"					32.50			32.50	36	
	0900	1-1/2"					37.50			37.50	41.50	
	1000	1-3/4"				▼	45.50			45.50	50	
700	0010	**SHEET METAL SCREWS** Steel, standard, #8 x 3/4", plain				C	2.81			2.81	3.09	**700**
	0100	Galvanized					3.44			3.44	3.78	
	0300	#10 x 1", plain					3.76			3.76	4.14	
	0400	Galvanized					4.35			4.35	4.79	
	0600	With washers, #14 x 1", plain					10.20			10.20	11.20	
	0700	Galvanized					10.20			10.20	11.20	
	0900	#14 x 2", plain					10.20			10.20	11.20	
	1000	Galvanized					18.15			18.15	19.95	
	1500	Self-drilling, with washers, (pinch point) #8 x 3/4", plain					6.10			6.10	6.70	
	1600	Galvanized					6.10			6.10	6.70	
	1800	#10 x 3/4", plain					6.10			6.10	6.70	
	1900	Galvanized					6.10			6.10	6.70	
	3000	Stainless steel w/aluminum or neoprene washers, #14 x 1", plain					18.35			18.35	20	
	3100	#14 x 2", plain				▼	25			25	27.50	
750	0010	**WOOD SCREWS** #8, 1" long, steel				C	3.36			3.36	3.70	**750**
	0100	Brass					11.20			11.20	12.35	
	0200	#8, 2" long, steel					3.76			3.76	4.14	
	0300	Brass				▼	11.70			11.70	12.90	

For expanded coverage of these items see _Means Interior Cost Data 2005_

06090	Wood & Plastic Fastenings	CREW	DAILY OUTPUT	LABOR-HOURS	UNIT	2005 BARE COSTS				TOTAL INCL O&P	
						MAT.	LABOR	EQUIP.	TOTAL		
750											**750**
0400	#10, 1" long, steel				C	4.30			4.30	4.73	
0500	Brass					23			23	25	
0600	#10, 2" long, steel					7.65			7.65	8.45	
0700	Brass					40.50			40.50	44.50	
0800	#10, 3" long, steel					11.95			11.95	13.15	
1000	#12, 2" long, steel					4.89			4.89	5.40	
1100	Brass					16.30			16.30	17.95	
1500	#12, 3" long, steel					16.20			16.20	17.85	
2000	#12, 4" long, steel					29			29	32	
800											**800**
0010	**TIMBER CONNECTORS** Add up cost of each part for total										
0020	cost of connection										
0100	Connector plates, steel, with bolts, straight	2 Carp	75	.213	Ea.	18.75	7.30		26.05	32	
0110	Tee		50	.320		27.50	10.95		38.45	47.50	
0120	T-Strap, 14 gauge 12" x 8" x 2"		50	.320		27.50	10.95		38.45	47.50	
0150	Anchor plate, 7 gauge, 9" x 7"		75	.213		18.75	7.30		26.05	32	
0200	Bolts, machine, sq. hd. with nut & washer, 1/2" diameter, 4" long	1 Carp	140	.057		.36	1.96		2.32	3.45	
0300	7-1/2" long		130	.062		.63	2.11		2.74	3.97	
0500	3/4" diameter, 7-1/2" long		130	.062		1.66	2.11		3.77	5.10	
0610	Machine bolts, w/ nut, washer, 3/4" dia, 15" L, HD's & beam hangers		95	.084		3.05	2.88		5.93	7.85	
0800	Drilling bolt holes in timber, 1/2" diameter		450	.018	Inch		.61		.61	.95	
0900	1" diameter		350	.023	"		.78		.78	1.22	
1100	Framing anchors, 2 or 3 dimensional, 10 gauge, no nails incl.		175	.046	Ea.	.46	1.57		2.03	2.95	
1150	Framing anchors, 18 gauge, 4 1/2" x 2 3/4"		175	.046		.46	1.57		2.03	2.95	
1160	Framing anchors, 18 gauge, 4 1/2" x 3"		175	.046		.46	1.57		2.03	2.95	
1170	Clip anchors plates, 18 gauge, 12" x 1 1/8"		175	.046		.46	1.57		2.03	2.95	
1250	Holdowns, 3 gauge base, 10 gauge body		8	1		15.60	34.50		50.10	70.50	
1260	Holdowns, 7 gauge 11 1/16" x 3 1/4"		8	1		15.60	34.50		50.10	70.50	
1270	Holdowns, 7 gauge 14 3/8" x 3 1/8"		8	1		15.60	34.50		50.10	70.50	
1275	Holdowns, 12 gauge 8" x 2 1/2"		8	1		15.60	34.50		50.10	70.50	
1300	Joist and beam hangers, 18 ga. galv., for 2" x 4" joist		175	.046		.58	1.57		2.15	3.08	
1400	2" x 6" to 2" x 10" joist		165	.048		.67	1.66		2.33	3.33	
1600	16 ga. galv., 3" x 6" to 3" x 10" joist		160	.050		2.64	1.71		4.35	5.55	
1700	3" x 10" to 3" x 14" joist		160	.050		3.06	1.71		4.77	6.05	
1800	4" x 6" to 4" x 10" joist		155	.052		2.32	1.77		4.09	5.30	
1900	4" x 10" to 4" x 14" joist		155	.052		3.14	1.77		4.91	6.20	
2000	Two-2" x 6" to two-2" x 10" joists		150	.053		2.53	1.83		4.36	5.65	
2100	Two-2" x 10" to two-2" x 14" joists		150	.053		2.53	1.83		4.36	5.65	
2300	3/16" thick, 6" x 8" joist		145	.055		5.50	1.89		7.39	9	
2400	6" x 10" joist		140	.057		6.50	1.96		8.46	10.20	
2500	6" x 12" joist		135	.059		7.80	2.03		9.83	11.75	
2700	1/4" thick, 6" x 14" joist		130	.062		9.70	2.11		11.81	14	
2900	Plywood clips, extruded aluminum H clip, for 3/4" panels					.15			.15	.17	
3000	Galvanized 18 ga. back-up clip					.14			.14	.15	
3200	Post framing, 16 ga. galv. for 4" x 4" base, 2 piece	1 Carp	130	.062		5.45	2.11		7.56	9.30	
3300	Cap		130	.062		2.67	2.11		4.78	6.20	
3500	Rafter anchors, 18 ga. galv., 1-1/2" wide, 5-1/4" long		145	.055		.46	1.89		2.35	3.45	
3600	10-3/4" long		145	.055		.91	1.89		2.80	3.94	
3800	Shear plates, 2-5/8" diameter		120	.067		1.63	2.28		3.91	5.35	
3900	4" diameter		115	.070		3.75	2.38		6.13	7.85	
4000	Sill anchors, embedded in concrete or block, 18-5/8" long		115	.070		1.10	2.38		3.48	4.92	
4100	Spike grids, 4" x 4", flat or curved		120	.067		.41	2.28		2.69	4.01	
4400	Split rings, 2-1/2" diameter		120	.067		1.32	2.28		3.60	5	
4500	4" diameter		110	.073		2.04	2.49		4.53	6.10	
4550	Tie plate, 20 gauge, 7" x 3 1/8"		110	.073		2.04	2.49		4.53	6.10	
4560	Tie plate, 20 gauge, 5" x 4 1/8"		110	.073		2.04	2.49		4.53	6.10	

6
WOOD & PLASTICS

Important: See the Reference Section for critical supporting data - Reference Nos., Crews, & City Cost Indexes

06090	Wood & Plastic Fastenings	CREW	DAILY OUTPUT	LABOR-HOURS	UNIT	2005 BARE COSTS				TOTAL INCL O&P	
						MAT.	LABOR	EQUIP.	TOTAL		
800											**800**
4575	Twist straps, 18 gauge, 12" x 1 1/4"	1 Carp	110	.073	Ea.	2.04	2.49		4.53	6.10	
4580	Twist straps, 18 gauge, 16" x 1 1/4"		110	.073		2.04	2.49		4.53	6.10	
4600	Strap ties, 20 ga., 2 -1/16" wide, 12 13/16" long		180	.044		1.19	1.52		2.71	3.68	
4700	Strap ties, 16 ga., 1-3/8" wide, 12" long		180	.044		1.19	1.52		2.71	3.68	
4800	24" long		160	.050		1.64	1.71		3.35	4.47	
5000	Toothed rings, 2-5/8" or 4" diameter		90	.089		1.13	3.04		4.17	6	
5200	Truss plates, nailed, 20 gauge, up to 32' span		17	.471	Truss	8.15	16.10		24.25	34	
5400	Washers, 2" x 2" x 1/8"				Ea.	.26			.26	.29	
5500	3" x 3" x 3/16"				"	.67			.67	.74	
6101	Beam hangers, polymer painted										
6102	Bolted, 3 ga., (W x H x L)										
6104	3-1/4" x 9" x 12" top flange	1 Carp	1	8	C	6,375	274		6,649	7,450	
6106	5-1/4" x 9" x 12" top flange		1	8		6,650	274		6,924	7,725	
6108	5-1/4" x 11" x 11-3/4" top flange		1	8		14,900	274		15,174	16,800	
6110	6-7/8" x 9" x 12" top flange		1	8		6,850	274		7,124	7,975	
6112	6-7/8" x 11" x 13-1/2" top flange		1	8		15,700	274		15,974	17,700	
6114	8-7/8" x 11" x 15-1/2" top flange		1	8		16,800	274		17,074	18,900	
6116	Nailed, 3 ga., (W x H x L)										
6118	3-1/4" x 10-1/2" x 10" top flange	1 Carp	1.80	4.444	C	4,400	152		4,552	5,075	
6120	3-1/4" x 10-1/2" x 12" top flange		1.80	4.444		5,100	152		5,252	5,850	
6122	5-1/4" x 9-1/2" x 10" top flange		1.80	4.444		4,750	152		4,902	5,450	
6124	5-1/4" x 9-1/2" x 12" top flange		1.80	4.444		5,400	152		5,552	6,150	
6126	5-1/2" x 9-1/2" x 12" top flange		1.80	4.444		4,750	152		4,902	5,450	
6128	6-7/8" x 8-1/2" x 12" top flange		1.80	4.444		4,925	152		5,077	5,625	
6130	7-1/2" x 8-1/2" x 12" top flange		1.80	4.444		5,075	152		5,227	5,800	
6132	8-7/8" x 7-1/2" x 14" top flange		1.80	4.444		5,625	152		5,777	6,400	
6201	Beam and purlin hangers, galvanized, 12 ga.										
6202	Purlin or joist size, 3" x 8"	1 Carp	1.70	4.706	C	805	161		966	1,125	
6204	3" x 10"		1.70	4.706		905	161		1,066	1,250	
6206	3" x 12"		1.65	4.848		1,075	166		1,241	1,425	
6208	3" x 14"		1.65	4.848		1,250	166		1,416	1,625	
6210	3" x 16"		1.65	4.848		1,425	166		1,591	1,825	
6212	4" x 8"		1.65	4.848		805	166		971	1,150	
6214	4" x 10"		1.65	4.848		940	166		1,106	1,275	
6216	4" x 12"		1.60	5		1,000	171		1,171	1,375	
6218	4" x 14"		1.60	5		1,100	171		1,271	1,475	
6220	4" x 16"		1.60	5		1,250	171		1,421	1,650	
6224	6" x 10"		1.55	5.161		1,375	177		1,552	1,775	
6226	6" x 12"		1.55	5.161		1,475	177		1,652	1,900	
6228	6" x 14"		1.50	5.333		1,600	183		1,783	2,025	
6230	6" x 16"		1.50	5.333		1,825	183		2,008	2,275	
6300	Column bases										
6302	4 x 4, 16 ga.	1 Carp	1.80	4.444	C	1,125	152		1,277	1,450	
6306	7 ga.		1.80	4.444		2,075	152		2,227	2,525	
6314	6 x 6, 16 ga.		1.75	4.571		1,425	157		1,582	1,825	
6318	7 ga.		1.75	4.571		2,925	157		3,082	3,475	
6326	8 x 8, 7 ga.		1.65	4.848		4,875	166		5,041	5,625	
6330	8 x 10, 7 ga.		1.65	4.848		5,200	166		5,366	5,975	
6590	Joist hangers, heavy duty 12 ga., galvanized										
6592	2" x 4"	1 Carp	1.75	4.571	C	915	157		1,072	1,250	
6594	2" x 6"		1.65	4.848		980	166		1,146	1,325	
6595	2" x 6", 16 gauge		1.65	4.848		980	166		1,146	1,325	
6596	2" x 8"		1.65	4.848		1,050	166		1,216	1,400	
6597	2" x 8", 16 gauge		1.65	4.848		1,050	166		1,216	1,400	
6598	2" x 10"		1.65	4.848		1,125	166		1,291	1,475	
6600	2" x 12"		1.65	4.848		1,250	166		1,416	1,625	

WOOD & PLASTICS **6**

06090		Wood & Plastic Fastenings	CREW	DAILY OUTPUT	LABOR-HOURS	UNIT	2005 BARE COSTS				TOTAL INCL O&P	
							MAT.	LABOR	EQUIP.	TOTAL		
800	6622	(2) 2" x 6"	1 Carp	1.60	5	C	1,225	171		1,396	1,625	**800**
	6624	(2) 2" x 8"		1.60	5		1,325	171		1,496	1,725	
	6626	(2) 2" x 10"		1.55	5.161		1,475	177		1,652	1,900	
	6628	(2) 2" x 12"	↓	1.55	5.161	↓	1,800	177		1,977	2,250	
	6890	Purlin hangers, painted										
	6892	12 ga., 2" x 6"	1 Carp	1.80	4.444	C	1,125	152		1,277	1,450	
	6894	2" x 8"		1.80	4.444		1,175	152		1,327	1,525	
	6896	2" x 10"		1.80	4.444		1,200	152		1,352	1,550	
	6898	2" x 12"		1.75	4.571		1,300	157		1,457	1,700	
	6934	(2) 2" x 6"		1.70	4.706		1,125	161		1,286	1,500	
	6936	(2) 2" x 8"		1.70	4.706		1,250	161		1,411	1,625	
	6938	(2) 2" x 10"		1.70	4.706		1,375	161		1,536	1,750	
	6940	(2) 2" x 12"	↓	1.65	4.848	↓	1,475	166		1,641	1,875	
825	0010	**ROUGH HARDWARE** Average % of carpentry material, minimum					.50%					**825**
	0200	Maximum					1.50%					
	0210	In seismic or hurricane areas, up to					10%					
850	0010	**BRACING**										**850**
	0300	Let-in, "T" shaped, 22 ga. galv. steel, studs at 16" O.C.	1 Carp	5.80	1.379	C.L.F.	45	47		92	123	
	0400	Studs at 24" O.C.		6	1.333		45	45.50		90.50	121	
	0500	16 ga. galv. steel straps, studs at 16" O.C.		6	1.333		67	45.50		112.50	145	
	0600	Studs at 24" O.C.	↓	6.20	1.290	↓	67	44		111	143	

06100 | Rough Carpentry

06110		Wood Framing	CREW	DAILY OUTPUT	LABOR-HOURS	UNIT	2005 BARE COSTS				TOTAL INCL O&P	
							MAT.	LABOR	EQUIP.	TOTAL		
100	0010	**BLOCKING**										**100**
	2600	Miscellaneous, to wood construction										
	2620	2" x 4"	1 Carp	.17	47.059	M.B.F.	510	1,600		2,110	3,050	
	2625	Pneumatic nailed		.21	38.095		510	1,300		1,810	2,575	
	2660	2" x 8"		.27	29.630		620	1,025		1,645	2,250	
	2665	Pneumatic nailed	↓	.33	24.242	↓	620	830		1,450	1,975	
	2720	To steel construction										
	2740	2" x 4"	1 Carp	.14	57.143	M.B.F.	510	1,950		2,460	3,600	
	2780	2" x 8"	"	.21	38.095	"	620	1,300		1,920	2,700	
150	0010	**BRACING** Let-in, with 1" x 6" boards, studs @ 16" O.C.	1 Carp	1.50	5.333	C.L.F.	57.50	183		240.50	350	**150**
	0200	Studs @ 24" O.C.	"	2.30	3.478	"	57.50	119		176.50	250	
200	0010	**BRIDGING** Wood, for joists 16" O.C., 1" x 3"	1 Carp	1.30	6.154	C.Pr.	42.50	211		253.50	375	**200**
	0015	Pneumatic nailed		1.70	4.706		42.50	161		203.50	298	
	0100	2" x 3" bridging		1.30	6.154		39.50	211		250.50	375	
	0105	Pneumatic nailed		1.70	4.706		39.50	161		200.50	295	
	0300	Steel, galvanized, 18 ga., for 2" x 10" joists at 12" O.C.		1.30	6.154		98	211		309	440	
	0400	24" O.C.		1.40	5.714		118	196		314	435	
	0600	For 2" x 14" joists at 16" O.C.		1.30	6.154		116	211		327	455	
	0700	24" O.C.		1.40	5.714		190	196		386	515	
	0900	Compression type, 16" O.C., 2" x 8" joists		2	4		115	137		252	340	
	1000	2" x 12" joists	↓	2	4	↓	115	137		252	340	

Important: See the Reference Section for critical supporting data - Reference Nos., Crews, & City Cost Indexes

06110 | Wood Framing

		CREW	DAILY OUTPUT	LABOR-HOURS	UNIT	2005 BARE COSTS				TOTAL INCL O&P	
						MAT.	LABOR	EQUIP.	TOTAL		
505	0010	**FRAMING, BEAMS & GIRDERS**									505
	3500	Single, 2" x 6" R06100-010	2 Carp	.70	22.857	M.B.F.	535	785		1,320	1,825
	3505	Pneumatic nailed R06110-030		.81	19.704		535	675		1,210	1,650
	3520	2" x 8"		.86	18.605		620	635		1,255	1,675
	3525	Pneumatic nailed		1	16.048		620	550		1,170	1,525
	3540	2" x 10"		1	16		705	550		1,255	1,625
	3545	Pneumatic nailed		1.16	13.793		705	470		1,175	1,500
	3560	2" x 12"		1.10	14.545		800	500		1,300	1,650
	3565	Pneumatic nailed		1.28	12.539		800	430		1,230	1,550
	3580	2" x 14"		1.17	13.675		925	470		1,395	1,750
	3585	Pneumatic nailed		1.36	11.791		925	405		1,330	1,650
	3600	3" x 8"		1.10	14.545		1,300	500		1,800	2,200
	3620	3" x 10"		1.25	12.800		1,300	440		1,740	2,100
	3640	3" x 12"		1.35	11.852		1,300	405		1,705	2,075
	3660	3" x 14"	↓	1.40	11.429		1,350	390		1,740	2,075
	3680	4" x 8"	F-3	2.66	15.038		1,000	520	227	1,747	2,150
	3700	4" x 10"		3.16	12.658		1,000	440	191	1,631	2,000
	3720	4" x 12"		3.60	11.111		1,100	385	167	1,652	1,975
	3740	4" x 14"	↓	3.96	10.101		1,050	350	152	1,552	1,875
	4000	Double, 2" x 6"	2 Carp	1.25	12.800		535	440		975	1,275
	4005	Pneumatic nailed		1.45	11.034		535	380		915	1,175
	4020	2" x 8"		1.60	10		620	345		965	1,225
	4025	Pneumatic nailed		1.86	8.621		620	295		915	1,150
	4040	2" x 10"		1.92	8.333		705	285		990	1,225
	4045	Pneumatic nailed		2.23	7.185		705	246		951	1,150
	4060	2" x 12"		2.20	7.273		800	249		1,049	1,275
	4065	Pneumatic nailed		2.55	6.275		800	215		1,015	1,225
	4080	2" x 14"		2.45	6.531		925	224		1,149	1,375
	4085	Pneumatic nailed		2.84	5.634		925	193		1,118	1,325
	5000	Triple, 2" x 6"		1.65	9.697		535	330		865	1,100
	5005	Pneumatic nailed		1.91	8.377		535	287		822	1,025
	5020	2" x 8"		2.10	7.619		620	261		881	1,075
	5025	Pneumatic nailed		2.44	6.568		620	225		845	1,025
	5040	2" x 10"		2.50	6.400		705	219		924	1,125
	5045	Pneumatic nailed		2.90	5.517		705	189		894	1,075
	5060	2" x 12"		2.85	5.614		800	192		992	1,175
	5065	Wood framing		3.31	4.840		800	166		966	1,150
	5080	2" x 14"		3.15	5.079		925	174		1,099	1,300
	5085	Pneumatic nailed	↓	3.35	4.770	↓	925	163		1,088	1,275
510	0010	**FRAMING, CEILINGS**									510
	6400	Suspended, 2" x 3"	2 Carp	.50	32	M.B.F.	530	1,100		1,630	2,275
	6450	2" x 4"		.59	27.119		510	930		1,440	2,000
	6500	2" x 6"		.80	20		535	685		1,220	1,675
	6550	2" x 8"	↓	.86	18.605	↓	620	635		1,255	1,675
515	0010	**FRAMING, COLUMNS**									515
	0400	4" x 4"	2 Carp	.52	30.769	M.B.F.	940	1,050		1,990	2,675
	0420	4" x 6"		.55	29.091		1,000	995		1,995	2,650
	0440	4" x 8"		.59	27.119		1,000	930		1,930	2,550
	0460	6" x 6"		.65	24.615		1,475	845		2,320	2,950
	0480	6" x 8"		.70	22.857		1,450	785		2,235	2,825
	0500	6" x 10"	↓	.75	21.333	↓	1,900	730		2,630	3,225
520	0010	**FRAMING, HEAVY** Mill timber, beams, single 6" x 10"	2 Carp	1.10	14.545	M.B.F.	2,125	500		2,625	3,125	520
	0100	Single 8" x 16"	↓	1.20	13.333	↓	2,675	455		3,130	3,650	

For expanded coverage of these items see *Means Interior Cost Data 2005*

6 WOOD & PLASTICS

06110 | Wood Framing

		CREW	DAILY OUTPUT	LABOR-HOURS	UNIT	2005 BARE COSTS				TOTAL INCL O&P	
						MAT.	LABOR	EQUIP.	TOTAL		
520	0200	Built from 2" lumber, multiple 2" x 14"	2 Carp	.90	17.778	M.B.F.	925	610		1,535	1,975	**520**
	0210	Built from 3" lumber, multiple 3" x 6"		.70	22.857		1,275	785		2,060	2,625	
	0220	Multiple 3" x 8"		.80	20		1,300	685		1,985	2,500	
	0230	Multiple 3" x 10"		.90	17.778		1,300	610		1,910	2,375	
	0240	Multiple 3" x 12"		1	16		1,300	550		1,850	2,300	
	0250	Built from 4" lumber, multiple 4" x 6"		.80	20		1,000	685		1,685	2,175	
	0260	Multiple 4" x 8"		.90	17.778		1,000	610		1,610	2,050	
	0270	Multiple 4" x 10"		1	16		1,000	550		1,550	1,975	
	0280	Multiple 4" x 12"		1.10	14.545		1,100	500		1,600	1,975	
	0290	Columns, structural grade, 1500f, 4" x 4"		.60	26.667		1,725	915		2,640	3,325	
	0300	6" x 6"		.65	24.615		2,125	845		2,970	3,650	
	0400	8" x 8"		.70	22.857		2,300	785		3,085	3,775	
	0500	10" x 10"		.75	21.333		2,350	730		3,080	3,750	
	0600	12" x 12"		.80	20		2,375	685		3,060	3,675	
	0800	Floor planks, 2" thick, T & G, 2" x 6"		1.05	15.238		1,850	520		2,370	2,875	
	0900	2" x 10"		1.10	14.545		1,875	500		2,375	2,825	
	1100	3" thick, 3" x 6"		1.05	15.238		1,325	520		1,845	2,300	
	1200	3" x 10"		1.10	14.545		1,325	500		1,825	2,250	
	1400	Girders, structural grade, 12" x 12"		.80	20		1,925	685		2,610	3,175	
	1500	10" x 16"		1	16		1,875	550		2,425	2,900	
	2050	Roof planks, see division 06150-600										
	2300	Roof purlins, 4" thick, structural grade	2 Carp	1.05	15.238	M.B.F.	1,300	520		1,820	2,275	
	2500	Roof trusses, add timber connectors, division 06090-800	"	.45	35.556	"	1,275	1,225		2,500	3,300	
530	0010	**FRAMING, JOISTS** R06100-010										**530**
	2650	Joists, 2" x 4"	2 Carp	.83	19.277	M.B.F.	510	660		1,170	1,575	
	2655	Pneumatic nailed R06110-030		.96	16.667		510	570		1,080	1,450	
	2680	2" x 6"		1.25	12.800		535	440		975	1,275	
	2685	Pneumatic nailed		1.44	11.111		535	380		915	1,175	
	2700	2" x 8"		1.46	10.959		620	375		995	1,275	
	2705	Pneumatic nailed		1.68	9.524		620	325		945	1,200	
	2720	2" x 10"		1.49	10.738		705	370		1,075	1,350	
	2725	Pneumatic nailed		1.71	9.357		705	320		1,025	1,275	
	2740	2" x 12"		1.75	9.143		800	315		1,115	1,375	
	2745	Pneumatic nailed		2.01	7.960		800	273		1,073	1,300	
	2760	2" x 14"		1.79	8.939		925	305		1,230	1,500	
	2765	Pneumatic nailed		2.06	7.767		925	266		1,191	1,450	
	2780	3" x 6"		1.39	11.511		1,275	395		1,670	2,025	
	2790	3" x 8"		1.90	8.421		1,300	288		1,588	1,875	
	2800	3" x 10"		1.95	8.205		1,300	281		1,581	1,875	
	2820	3" x 12"		1.80	8.889		1,300	305		1,605	1,925	
	2840	4" x 6"		1.60	10		1,000	345		1,345	1,625	
	2850	4" x 8"		4.15	3.855		1,000	132		1,132	1,300	
	2860	4" x 10"		2	8		1,000	274		1,274	1,550	
	2880	4" x 12"		1.80	8.889		1,100	305		1,405	1,675	
	3000	Composite wood joist 9-1/2" deep		.90	17.778	M.L.F.	1,475	610		2,085	2,575	
	3010	11-1/2" deep		.88	18.182		1,550	625		2,175	2,700	
	3020	14" deep		.82	19.512		1,775	670		2,445	3,025	
	3030	16" deep		.78	20.513		2,550	705		3,255	3,900	
	4000	Open web joist 12" deep		.88	18.182		1,675	625		2,300	2,825	
	4010	14" deep		.82	19.512		1,950	670		2,620	3,200	
	4020	16" deep		.78	20.513		2,025	705		2,730	3,325	
	4030	18" deep		.74	21.622		2,050	740		2,790	3,400	
	6000	Composite rim joist, 1-1/4" x 9-1/2"		90	.178		1,825	6.10		1,831.10	2,000	
	6010	1-1/4" x 11-1/2"		.88	18.182		2,100	625		2,725	3,300	
	6020	1-1/4" x 14-1/2"		.82	19.512		2,575	670		3,245	3,875	

Important: See the Reference Section for critical supporting data - Reference Nos., Crews, & City Cost Indexes

06110	Wood Framing		CREW	DAILY OUTPUT	LABOR-HOURS	UNIT	MAT.	LABOR	EQUIP.	TOTAL	TOTAL INCL O&P	
530	6030	1-1/4" x 16-1/2" R06100-010	2 Carp	.78	20.513	M.L.F.	2,925	705		3,630	4,325	530
545	0010	**FRAMING, MISCELLANEOUS**										545
	8500	Firestops, 2" x 4"	2 Carp	.51	31.373	M.B.F.	510	1,075		1,585	2,225	
	8505	Pneumatic nailed		.62	25.806		510	885		1,395	1,925	
	8520	2" x 6"		.60	26.667		535	915		1,450	2,025	
	8525	Pneumatic nailed		.73	21.858		535	750		1,285	1,775	
	8540	2" x 8"		.60	26.667		620	915		1,535	2,100	
	8560	2" x 12"		.70	22.857		800	785		1,585	2,100	
	8600	Nailers, treated, wood construction, 2" x 4"		.53	30.189		730	1,025		1,755	2,400	
	8605	Pneumatic nailed		.64	25.157		730	860		1,590	2,150	
	8620	2" x 6"		.75	21.333		770	730		1,500	2,000	
	8625	Pneumatic nailed		.90	17.778		770	610		1,380	1,800	
	8640	2" x 8"		.93	17.204		750	590		1,340	1,750	
	8645	Pneumatic nailed		1.12	14.337		750	490		1,240	1,600	
	8660	Steel construction, 2" x 4"		.50	32		730	1,100		1,830	2,500	
	8680	2" x 6"		.70	22.857		770	785		1,555	2,075	
	8700	2" x 8"		.87	18.391		750	630		1,380	1,800	
	8760	Rough bucks, treated, for doors or windows, 2" x 6"		.40	40		770	1,375		2,145	2,975	
	8765	Pneumatic nailed		.48	33.333		770	1,150		1,920	2,625	
	8780	2" x 8"		.51	31.373		750	1,075		1,825	2,500	
	8785	Pneumatic nailed		.61	26.144		750	895		1,645	2,225	
	8800	Stair stringers, 2" x 10"		.22	72.727		705	2,500		3,205	4,650	
	8820	2" x 12"		.26	61.538		800	2,100		2,900	4,150	
	8840	3" x 10"		.31	51.613		1,300	1,775		3,075	4,175	
	8860	3" x 12"		.38	42.105		1,300	1,450		2,750	3,700	
	8870	Composite LSL, 1-1/4" x 11-1/2"		130	.123	L.F.	2.11	4.22		6.33	8.85	
	8880	1-1/4" x 14-1/2"		130	.123	"	2.57	4.22		6.79	9.40	
550	0010	**PARTITIONS** Wood stud with single bottom plate and										550
	0020	double top plate, no waste, std. & better lumber										
	0180	2" x 4" studs, 8' high, studs 12" O.C.	2 Carp	80	.200	L.F.	3.73	6.85		10.58	14.75	
	0185	12" O.C., pneumatic nailed		96	.167		3.73	5.70		9.43	13	
	0200	16" O.C.		100	.160		3.05	5.50		8.55	11.90	
	0205	16" O.C., pneumatic nailed		120	.133		3.05	4.57		7.62	10.45	
	0300	24" O.C.		125	.128		2.37	4.38		6.75	9.45	
	0305	24" O.C., pneumatic nailed		150	.107		2.37	3.65		6.02	8.30	
	0380	10' high, studs 12" O.C.		80	.200		4.40	6.85		11.25	15.50	
	0385	12" O.C., pneumatic nailed		96	.167		4.40	5.70		10.10	13.75	
	0400	16" O.C.		100	.160		3.56	5.50		9.06	12.45	
	0405	16" O.C., pneumatic nailed		120	.133		3.56	4.57		8.13	11	
	0500	24" O.C.		125	.128		2.71	4.38		7.09	9.85	
	0505	24" O.C., pneumatic nailed		150	.107		2.71	3.65		6.36	8.70	
	0580	12' high, studs 12" O.C.		65	.246		5.10	8.45		13.55	18.75	
	0585	12" O.C., pneumatic nailed		78	.205		5.10	7.05		12.15	16.55	
	0600	16" O.C.		80	.200		4.06	6.85		10.91	15.10	
	0605	16" O.C., pneumatic nailed		96	.167		4.06	5.70		9.76	13.35	
	0700	24" O.C.		100	.160		3.05	5.50		8.55	11.90	
	0705	24" O.C., pneumatic nailed		120	.133		3.05	4.57		7.62	10.45	
	0780	2" x 6" studs, 8' high, studs 12" O.C.		70	.229		5.90	7.85		13.75	18.70	
	0785	12" O.C., pneumatic nailed		84	.190		5.90	6.50		12.40	16.65	
	0800	16" O.C.		90	.178		4.83	6.10		10.93	14.80	
	0805	16" O.C., pneumatic nailed		108	.148		4.83	5.05		9.88	13.20	
	0900	24" O.C.		115	.139		3.76	4.77		8.53	11.55	
	0905	24" O.C., pneumatic nailed		138	.116		3.76	3.97		7.73	10.35	
	0980	10' high, studs 12" O.C.		70	.229		7	7.85		14.85	19.90	
	0985	12" O.C., pneumatic nailed		84	.190		7	6.50		13.50	17.85	

WOOD & PLASTICS 6

06110	Wood Framing	CREW	DAILY OUTPUT	LABOR-HOURS	UNIT	2005 BARE COSTS				TOTAL INCL O&P	
						MAT.	LABOR	EQUIP.	TOTAL		
550											**550**
1000	16" O.C.	2 Carp	90	.178	L.F.	5.65	6.10		11.75	15.70	
1005	16" O.C., pneumatic nailed		108	.148		5.65	5.05		10.70	14.10	
1100	24" O.C.		115	.139		4.30	4.77		9.07	12.15	
1105	24" O.C., pneumatic nailed		138	.116		4.30	3.97		8.27	10.95	
1180	12' high, studs 12" O.C.		55	.291		8.05	9.95		18	24.50	
1185	12" O.C., pneumatic nailed		66	.242		8.05	8.30		16.35	22	
1200	16" O.C.		70	.229		6.45	7.85		14.30	19.30	
1205	16" O.C., pneumatic nailed		84	.190		6.45	6.50		12.95	17.25	
1300	24" O.C.		90	.178		4.83	6.10		10.93	14.80	
1305	24" O.C., pneumatic nailed		108	.148		4.83	5.05		9.88	13.20	
1400	For horizontal blocking, 2" x 4", add		600	.027		.34	.91		1.25	1.79	
1500	2" x 6", add		600	.027		.54	.91		1.45	2.01	
1600	For openings, add	▼	250	.064	▼		2.19		2.19	3.41	
1700	Headers for above openings, material only, add				M.B.F.	620			620	680	
552	0010	**FRAMING, PORCH OR DECK**									**552**
0100	Treated lumber, posts or columns, 4" x 4"	2 Carp	390	.041	L.F.	1.38	1.41		2.79	3.71	
0110	4" x 6"		275	.058		3.03	1.99		5.02	6.45	
0120	4" x 8"		220	.073		6	2.49		8.49	10.50	
0130	Girder, single, 4" x 4"		675	.024		1.38	.81		2.19	2.78	
0140	4" x 6"		600	.027		3.03	.91		3.94	4.75	
0150	4" x 8"		525	.030		6	1.04		7.04	8.25	
0160	Double, 2" x 4"		625	.026		1	.88		1.88	2.47	
0170	2" x 6"		600	.027		1.59	.91		2.50	3.16	
0180	2" x 8"		575	.028		2.06	.95		3.01	3.75	
0190	2" x 10"		550	.029		3.04	1		4.04	4.90	
0200	2" x 12"		525	.030		3.90	1.04		4.94	5.90	
0210	Triple, 2" x 4"		575	.028		1.50	.95		2.45	3.14	
0220	2" x 6"		550	.029		2.38	1		3.38	4.17	
0230	2" x 8"		525	.030		3.09	1.04		4.13	5.05	
0240	2" x 10"		500	.032		4.56	1.10		5.66	6.70	
0250	2" x 12"		475	.034		5.85	1.15		7	8.25	
0260	Ledger, bolted 4' O.C., 2" x 4"		400	.040		.58	1.37		1.95	2.77	
0270	2" x 6"		395	.041		.87	1.39		2.26	3.11	
0280	2" x 8"		390	.041		1.10	1.41		2.51	3.40	
0300	2" x 12"		380	.042		2.01	1.44		3.45	4.46	
0310	Joists, 2" x 4"		1,250	.013		.50	.44		.94	1.23	
0320	2" x 6"		1,250	.013		.79	.44		1.23	1.55	
0330	2" x 8"		1,100	.015		1.03	.50		1.53	1.91	
0340	2" x 10"		900	.018		1.52	.61		2.13	2.63	
0350	2" x 12"	▼	875	.018		1.83	.63		2.46	2.99	
0360	Railings and trim , 1" x 4"	1 Carp	300	.027		.60	.91		1.51	2.08	
0370	2" x 2"		300	.027		.30	.91		1.21	1.75	
0380	2" x 4"		300	.027		.49	.91		1.40	1.96	
0390	2" x 6"		300	.027	▼	.77	.91		1.68	2.27	
0400	Decking, 1" x 4"		275	.029	S.F.	1.89	1		2.89	3.63	
0410	2" x 4"		300	.027		1.67	.91		2.58	3.26	
0420	2" x 6"		320	.025		1.68	.86		2.54	3.18	
0430	5/4" x 6"	▼	320	.025	▼	2.71	.86		3.57	4.31	
0440	Balusters, square, 2" x 2"	2 Carp	220	.073	L.F.	.31	2.49		2.80	4.22	
0450	Turned, 2" x 2"		140	.114		.41	3.91		4.32	6.55	
0460	Stair stringer, 2" x 10"		130	.123		1.52	4.22		5.74	8.25	
0470	2" x 12"		130	.123		1.83	4.22		6.05	8.55	
0480	Stair treads, 1" x 4"		140	.114		.64	3.91		4.55	6.80	
0490	2" x 4"		140	.114		.50	3.91		4.41	6.65	
0500	2" x 6"		160	.100		.75	3.43		4.18	6.20	
0510	5/4" x 6"	▼	160	.100	▼	1.25	3.43		4.68	6.75	

6 WOOD & PLASTICS

06110 | Wood Framing

		CREW	DAILY OUTPUT	LABOR-HOURS	UNIT	2005 BARE COSTS				TOTAL INCL O&P
						MAT.	LABOR	EQUIP.	TOTAL	
552	0520	Turned handrail post, 4" x 4"	2 Carp	64	.250	Ea.	13.05	8.55		21.60	27.50
	0530	Lattice panel, 4' x 8'		1,600	.010	S.F.	.66	.34		1	1.26
	0540	Cedar, posts or columns, 4" x 4"		390	.041	L.F.	1.34	1.41		2.75	3.67
	0550	4" x 6"		275	.058		2.44	1.99		4.43	5.80
	0560	4" x 8"		220	.073		5.85	2.49		8.34	10.30
	0800	Decking, 1" x 4"		550	.029		1.13	1		2.13	2.80
	0810	2" x 4"		600	.027		2.33	.91		3.24	3.99
	0820	2" x 6"		640	.025		4.25	.86		5.11	6
	0830	5/4" x 6"		640	.025		3.16	.86		4.02	4.80
	0840	Railings and trim, 1" x 4"		600	.027		1.13	.91		2.04	2.67
	0860	2" x 4"		600	.027		2.33	.91		3.24	3.99
	0870	2" x 6"		600	.027		4.25	.91		5.16	6.10
	0920	Stair treads, 1" x 4"		140	.114		1.13	3.91		5.04	7.35
	0930	2" x 4"		140	.114		2.33	3.91		6.24	8.65
	0940	2" x 6"		160	.100		4.25	3.43		7.68	10
	0950	5/4" x 6"		160	.100		3.16	3.43		6.59	8.80
	0980	Redwood, posts or columns, 4" x 4"		390	.041		3.73	1.41		5.14	6.30
	0990	4" x 6"		275	.058		6.75	1.99		8.74	10.50
	1000	4" x 8"	▼	220	.073	▼	10.05	2.49		12.54	15
	1240	Redwood decking, 1" x 4"	1 Carp	275	.029	S.F.	5.05	1		6.05	7.10
	1260	2" x 6"		340	.024		12.55	.81		13.36	15.05
	1270	5/4" x 6"	▼	320	.025	▼	8.10	.86		8.96	10.25
	1280	Railings and trim, 1" x 4"	2 Carp	600	.027	L.F.	1.48	.91		2.39	3.05
	1310	2" x 6"		600	.027		5.80	.91		6.71	7.75
	1420	Alternative decking, wood / plastic composite, 5/4" x 6"		640	.025		2.67	.86		3.53	4.27
	1430	Vinyl, 1-1/2" x 5-1/2"		640	.025		3.16	.86		4.02	4.81
	1440	1" x 4" square edge fir		550	.029		1.10	1		2.10	2.76
	1450	1" x 4" tongue and groove fir		450	.036		1.10	1.22		2.32	3.11
	1460	1" x 4" mahogany	▼	550	.029	▼	1.40	1		2.40	3.09
	1470	Accessories, joist hangers, 2" x 4"	1 Carp	160	.050	Ea.	.58	1.71		2.29	3.31
	1480	2" x 6" through 2" x 12"	"	150	.053		.67	1.83		2.50	3.59
	1530	Post footing, incl excav, backfill, tube form & concrete, 4' deep, 8" dia	F-7	12	2.667		10.10	81.50		91.60	138
	1540	10" diameter		11	2.909		14	88.50		102.50	153
	1550	12" diameter	▼	10	3.200	▼	18.40	97.50		115.90	172
555	0010	**FRAMING, ROOFS**									
	5250	Composite rafter, 9-1/2" deep	2 Carp	575	.028	L.F.	1.47	.95		2.42	3.09
	5260	11-1/2" deep		575	.028	"	1.56	.95		2.51	3.19
	6070	Fascia boards, 2" x 8"		.30	53.333	M.B.F.	620	1,825		2,445	3,525
	6080	2" x 10"		.30	53.333		705	1,825		2,530	3,625
	7000	Rafters, to 4 in 12 pitch, 2" x 6"		1	16		535	550		1,085	1,450
	7060	2" x 8"		1.26	12.698		620	435		1,055	1,350
	7300	Hip and valley rafters, 2" x 6"		.76	21.053		535	720		1,255	1,725
	7360	2" x 8"		.96	16.667		620	570		1,190	1,575
	7540	Hip and valley jacks, 2" x 6"		.60	26.667		535	915		1,450	2,025
	7600	2" x 8"	▼	.65	24.615	▼	620	845		1,465	2,000
	7780	For slopes steeper than 4 in 12, add						30%			
	7790	For dormers or complex roofs, add						50%			
	7800	Rafter tie, 1" x 4", #3	2 Carp	.27	59.259	M.B.F.	1,225	2,025		3,250	4,475
	7820	Ridge board, #2 or better, 1" x 6" **CN**		.30	53.333		1,825	1,825		3,650	4,850
	7840	1" x 8"		.37	43.243		1,825	1,475		3,300	4,325
	7860	1" x 10"		.42	38.095		1,825	1,300		3,125	4,025
	7880	2" x 6"		.50	32		535	1,100		1,635	2,300
	7900	2" x 8"		.60	26.667		620	915		1,535	2,100
	7920	2" x 10"		.66	24.242		705	830		1,535	2,075
	7940	Roof cants, split, 4" x 4"		.86	18.605		940	635		1,575	2,025
	7960	6" x 6"	▼	1.80	8.889		1,475	305		1,780	2,100

WOOD & PLASTICS 6

06110 | Wood Framing

			CREW	DAILY OUTPUT	LABOR-HOURS	UNIT	2005 BARE COSTS				TOTAL INCL O&P	
							MAT.	LABOR	EQUIP.	TOTAL		
555	7980	Roof curbs, untreated, 2" x 6"	2 Carp	.52	30.769	M.B.F.	535	1,050		1,585	2,250	555
	8000	2" x 12"	↓	.80	20	↓	800	685		1,485	1,950	
560	0010	**FRAMING, SILLS**										560
	4482	Ledgers, nailed, 2" x 4"	2 Carp	.50	32	M.B.F.	510	1,100		1,610	2,250	
	4484	2" x 6"		.60	26.667		535	915		1,450	2,025	
	4486	Bolted, not including bolts, 3" x 8"		.65	24.615		1,300	845		2,145	2,750	
	4488	3" x 12"		.70	22.857		1,300	785		2,085	2,675	
	4490	Mud sills, redwood, construction grade, 2" x 4"		.59	27.119		3,300	930		4,230	5,075	
	4492	2" x 6"		.78	20.513		3,300	705		4,005	4,725	
	4500	Sills, 2" x 4"		.40	40		510	1,375		1,885	2,675	
	4520	2" x 6"		.55	29.091		535	995		1,530	2,150	
	4540	2" x 8"		.67	23.881		620	820		1,440	1,950	
	4600	Treated, 2" x 4"		.36	44.444		730	1,525		2,255	3,175	
	4620	2" x 6"		.50	32		770	1,100		1,870	2,550	
	4640	2" x 8"		.60	26.667		750	915		1,665	2,250	
	4700	4" x 4"		.60	26.667		1,025	915		1,940	2,550	
	4720	4" x 6"		.70	22.857		1,500	785		2,285	2,875	
	4740	4" x 8"		.80	20		2,225	685		2,910	3,525	
	4760	4" x 10"	↓	.87	18.391	↓	2,225	630		2,855	3,425	
565	0010	**FRAMING, SLEEPERS**										565
	0300	On concrete, treated, 1" x 2"	2 Carp	.39	41.026	M.B.F.	600	1,400		2,000	2,850	
	0320	1" x 3"		.50	32		755	1,100		1,855	2,525	
	0340	2" x 4"		.99	16.162		730	555		1,285	1,675	
	0360	2" x 6"	↓	1.30	12.308	↓	770	420		1,190	1,500	
570	0010	**FRAMING, SOFFITS & CANOPIES**										570
	1300	Canopy or soffit framing, 1" x 4"	2 Carp	.30	53.333	M.B.F.	1,825	1,825		3,650	4,850	
	1340	1" x 8"		.50	32		1,825	1,100		2,925	3,725	
	1360	2" x 4"		.41	39.024		510	1,325		1,835	2,625	
	1400	2" x 8"		.67	23.881		620	820		1,440	1,950	
	1420	3" x 4"		.50	32		1,100	1,100		2,200	2,900	
	1460	3" x 8"	↓	.60	26.667	↓	1,300	915		2,215	2,850	
575	0010	**FRAMING, TREATED LUMBER**										575
	0100	2" x 4"				M.B.F.	730			730	805	
	0110	2" x 6"					770			770	850	
	0120	2" x 8"					750			750	825	
	0130	2" x 10"					890			890	980	
	0140	2" x 12"					955			955	1,050	
	0200	4" x 4"					1,025			1,025	1,125	
	0210	4" x 6"					1,500			1,500	1,650	
	0220	4" x 8"				↓	2,225			2,225	2,450	
590	0010	**FRAMING, WALLS**										590
	5860	Headers over openings, 2" x 6" R06100-010	2 Carp	.36	44.444	M.B.F.	535	1,525		2,060	2,975	
	5865	2" x 6", pneumatic nailed R06110-030		.43	37.209		535	1,275		1,810	2,575	
	5880	2" x 8"		.45	35.556		620	1,225		1,845	2,575	
	5885	2" x 8", pneumatic nailed		.54	29.630		620	1,025		1,645	2,250	
	5900	2" x 10"		.53	30.189		705	1,025		1,730	2,375	
	5905	2" x 10", pneumatic nailed		.67	23.881		705	820		1,525	2,050	
	5920	2" x 12"		.60	26.667		800	915		1,715	2,300	
	5925	2" x 12", pneumatic nailed		.72	22.222		800	760		1,560	2,050	
	5940	4" x 12"		.76	21.053		1,100	720		1,820	2,325	
	5945	4" x 12", pneumatic nailed		.92	17.391		1,100	595		1,695	2,125	
	5960	6" x 12"	↓	.84	19.048	↓	1,525	650		2,175	2,700	

6 WOOD & PLASTICS

Important: See the Reference Section for critical supporting data - Reference Nos., Crews, & City Cost Indexes

06110 | Wood Framing

			CREW	DAILY OUTPUT	LABOR-HOURS	UNIT	2005 BARE COSTS				TOTAL INCL O&P	
							MAT.	LABOR	EQUIP.	TOTAL		
590	5965	6" x 12", pneumatic nailed	2 Carp	1.01	15.873	M.B.F.	1,525	545		2,070	2,525	590
	6000	Plates, untreated, 2" x 3" R06100-010		.43	37.209		530	1,275		1,805	2,550	
	6005	2" x 3", pneumatic nailed		.52	30.769		530	1,050		1,580	2,225	
	6020	2" x 4" CN R06110-030		.53	30.189		510	1,025		1,535	2,150	
	6025	2" x 4", pneumatic nailed		.67	23.881		510	820		1,330	1,825	
	6040	2" x 6"		.75	21.333		535	730		1,265	1,750	
	6045	2" x 6", pneumatic nailed		.90	17.778		535	610		1,145	1,550	
	6120	Studs, 8' high wall, 2" x 3"		.60	26.667		530	915		1,445	2,000	
	6125	2" x 3", pneumatic nailed		.72	22.222		530	760		1,290	1,750	
	6140	2" x 4"		.92	17.391		510	595		1,105	1,500	
	6145	2" x 4", pneumatic nailed		1.10	14.493		510	495		1,005	1,325	
	6160	2" x 6"		1	16		535	550		1,085	1,450	
	6165	2" x 6", pneumatic nailed		1.20	13.333		535	455		990	1,300	
	6180	3" x 4"		.80	20		1,100	685		1,785	2,275	
	6185	3" x 4", pneumatic nailed		.96	16.667		1,100	570		1,670	2,100	
	8200	For 12' high walls, deduct						5%				
	8220	For stub wall, 6' high, add						20%				
	8240	3' high, add						40%				
	8250	For second story & above, add						5%				
	8300	For dormer & gable, add						15%				
600	0010	**FURRING** Wood strips, 1" x 2", on walls, on wood	1 Carp	550	.015	L.F.	.19	.50		.69	.99	600
	0015	On wood, pneumatic nailed		710	.011		.19	.39		.58	.81	
	0300	On masonry		495	.016		.19	.55		.74	1.07	
	0400	On concrete		260	.031		.19	1.05		1.24	1.85	
	0600	1" x 3", on walls, on wood		550	.015		.28	.50		.78	1.09	
	0605	On wood, pneumatic nailed		710	.011		.28	.39		.67	.91	
	0700	On masonry		495	.016		.28	.55		.83	1.17	
	0800	On concrete		260	.031		.28	1.05		1.33	1.95	
	0850	On ceilings, on wood		350	.023		.28	.78		1.06	1.53	
	0855	On wood, pneumatic nailed		450	.018		.28	.61		.89	1.26	
	0900	On masonry		320	.025		.28	.86		1.14	1.64	
	0950	On concrete		210	.038		.28	1.30		1.58	2.34	
700	0010	**GROUNDS** For casework, 1" x 2" wood strips, on wood	1 Carp	330	.024	L.F.	.19	.83		1.02	1.50	700
	0100	On masonry		285	.028		.19	.96		1.15	1.71	
	0200	On concrete		250	.032		.19	1.10		1.29	1.92	
	0400	For plaster, 3/4" deep, on wood		450	.018		.19	.61		.80	1.16	
	0500	On masonry		225	.036		.19	1.22		1.41	2.11	
	0600	On concrete		175	.046		.19	1.57		1.76	2.65	
	0700	On metal lath		200	.040		.19	1.37		1.56	2.34	

06120 | Structural Panels

			CREW	DAILY OUTPUT	LABOR-HOURS	UNIT	MAT.	LABOR	EQUIP.	TOTAL	TOTAL INCL O&P	
900	0010	**STRUCTURAL INSULATED PANELS**										900
	0100	Structural insul. panels, 7/16" OSB both faces, EPS insul, 3-5/8" T	F-3	2,075	.019	S.F.	3.72	.67	.29	4.68	5.45	
	0110	5-5/8" thick		1,725	.023		4.06	.80	.35	5.21	6.10	
	0120	7-3/8" thick		1,425	.028		4.34	.97	.42	5.73	6.75	
	0130	9-3/8" thick		1,125	.036		4.68	1.23	.54	6.45	7.65	
	0140	7/16" OSB one face, EPS insul, 3-5/8" thick		2,175	.018		1.87	.64	.28	2.79	3.34	
	0150	5-5/8" thick		1,825	.022		2.12	.76	.33	3.21	3.86	
	0160	7-3/8" thick		1,525	.026		2.35	.91	.40	3.66	4.42	
	0170	9-3/8" thick		1,225	.033		2.56	1.13	.49	4.18	5.10	
	0180	11-3/8" thick		925	.043		2.75	1.50	.65	4.90	6.05	
	0190	7/16" OSB - 1/2" GWB faces , EPS insul, 3-5/8" T		2,075	.019		3.06	.67	.29	4.02	4.72	
	0200	5-5/8" thick		1,725	.023		3.50	.80	.35	4.65	5.45	
	0210	7-3/8" thick		1,425	.028		4.12	.97	.42	5.51	6.50	
	0220	9-3/8" thick		1,125	.036		4.78	1.23	.54	6.55	7.75	

WOOD & PLASTICS 6

6 WOOD & PLASTICS

06120 | Structural Panels

			CREW	DAILY OUTPUT	LABOR-HOURS	UNIT	2005 BARE COSTS MAT.	LABOR	EQUIP.	TOTAL	TOTAL INCL O&P	
900	0230	11-3/8" thick	F-3	825	.048	S.F.	5.10	1.68	.73	7.51	9	900
	0240	7/16" OSB - 1/2" MRGWB faces , EPS insul, 3-5/8" T		2,075	.019		3.06	.67	.29	4.02	4.72	
	0250	5-5/8" thick	▼	1,725	.023		3.50	.80	.35	4.65	5.45	
	0300	For 1/2" GWB added to OSB skin, add					.60			.60	.66	
	0310	For 1/2" MRGWB added to OSB skin, add					.75			.75	.83	
	0320	For one T1-11 skin, add to OSB-OSB					.75			.75	.83	
	0330	For one 19/32" CDX skin, add to OSB-OSB				▼	.72			.72	.79	

06150 | Wood Decking

			CREW	DAILY OUTPUT	LABOR-HOURS	UNIT	MAT.	LABOR	EQUIP.	TOTAL	TOTAL INCL O&P	
600	0010	**ROOF DECKS**										600
	0020	For laminated decks, see division 06170-550										
	0200	For cementitious decks, see division 03510-200										
	0400	Cedar planks, 3" thick	2 Carp	320	.050	S.F.	6.40	1.71		8.11	9.70	
	0500	4" thick		250	.064		8.65	2.19		10.84	12.90	
	0700	Douglas fir, 3" thick		320	.050		2.40	1.71		4.11	5.30	
	0800	4" thick		250	.064		3.21	2.19		5.40	6.95	
	1000	Hemlock, 3" thick		320	.050		2.40	1.71		4.11	5.30	
	1100	4" thick		250	.064		3.20	2.19		5.39	6.95	
	1300	Western white spruce, 3" thick		320	.050		2.31	1.71		4.02	5.20	
	1400	4" thick	▼	250	.064	▼	3.08	2.19		5.27	6.80	

06160 | Sheathing

			CREW	DAILY OUTPUT	LABOR-HOURS	UNIT	MAT.	LABOR	EQUIP.	TOTAL	TOTAL INCL O&P	
800	0010	**SHEATHING** Plywood on roof, CDX										800
	0030	5/16" thick R06160-020	2 Carp	1,600	.010	S.F.	.63	.34		.97	1.22	
	0035	Pneumatic nailed		1,952	.008		.63	.28		.91	1.13	
	0050	3/8" thick R06110-030		1,525	.010		.65	.36		1.01	1.28	
	0055	Pneumatic nailed		1,860	.009		.65	.29		.94	1.18	
	0100	1/2" thick **CN**		1,400	.011		.62	.39		1.01	1.29	
	0105	Pneumatic nailed		1,708	.009		.62	.32		.94	1.18	
	0200	5/8" thick		1,300	.012		.78	.42		1.20	1.52	
	0205	Pneumatic nailed		1,586	.010		.78	.35		1.13	1.40	
	0300	3/4" thick		1,200	.013		.94	.46		1.40	1.74	
	0305	Pneumatic nailed		1,464	.011		.94	.37		1.31	1.61	
	0500	Plywood on walls with exterior CDX, 3/8" thick		1,200	.013		.65	.46		1.11	1.43	
	0505	Pneumatic nailed		1,488	.011		.65	.37		1.02	1.29	
	0600	1/2" thick		1,125	.014		.62	.49		1.11	1.44	
	0605	Pneumatic nailed		1,395	.011		.62	.39		1.01	1.29	
	0700	5/8" thick		1,050	.015		.78	.52		1.30	1.67	
	0705	Pneumatic nailed		1,302	.012		.78	.42		1.20	1.52	
	0800	3/4" thick		975	.016		.94	.56		1.50	1.91	
	0805	Pneumatic nailed	▼	1,209	.013	▼	.94	.45		1.39	1.74	
	1000	For shear wall construction, add						20%				
	1200	For structural 1 exterior plywood, add				S.F.	10%					
	1400	With boards, on roof 1" x 6" boards, laid horizontal	2 Carp	725	.022		1.25	.76		2.01	2.56	
	1500	Laid diagonal		650	.025		1.25	.84		2.09	2.69	
	1700	1" x 8" boards, laid horizontal		875	.018		1.11	.63		1.74	2.21	
	1800	Laid diagonal	▼	725	.022		1.11	.76		1.87	2.41	
	2000	For steep roofs, add						40%				
	2200	For dormers, hips and valleys, add					5%	50%				
	2400	Boards on walls, 1" x 6" boards, laid regular	2 Carp	650	.025		1.25	.84		2.09	2.69	
	2500	Laid diagonal		585	.027		1.25	.94		2.19	2.84	
	2700	1" x 8" boards, laid regular		765	.021		1.11	.72		1.83	2.35	
	2800	Laid diagonal		650	.025		1.11	.84		1.95	2.54	
	2850	Gypsum, weatherproof, 1/2" thick	▼	1,125	.014	▼	.29	.49		.78	1.08	

Important: See the Reference Section for critical supporting data - Reference Nos., Crews, & City Cost Indexes

		06160	Sheathing		CREW	DAILY OUTPUT	LABOR-HOURS	UNIT	2005 BARE COSTS MAT.	LABOR	EQUIP.	TOTAL	TOTAL INCL O&P	
800	2900		Sealed, 4/10" thick	R06160 -020	2 Carp	1,100	.015	S.F.	.45	.50		.95	1.28	800
	3000		Wood fiber, regular, no vapor barrier, 1/2" thick			1,200	.013		.55	.46		1.01	1.32	
	3100		5/8" thick	R06110 -030		1,200	.013		.72	.46		1.18	1.50	
	3300		No vapor barrier, in colors, 1/2" thick			1,200	.013		.78	.46		1.24	1.57	
	3400		5/8" thick			1,200	.013		.96	.46		1.42	1.77	
	3600		With vapor barrier one side, white, 1/2" thick			1,200	.013		.55	.46		1.01	1.32	
	3700		Vapor barrier 2 sides, 1/2" thick			1,200	.013		.77	.46		1.23	1.56	
	3800		Asphalt impregnated, 25/32" thick			1,200	.013		.28	.46		.74	1.02	
	3850		Intermediate, 1/2" thick			1,200	.013		.18	.46		.64	.91	
850	0010	SUBFLOOR	Plywood, CDX, 1/2" thick	R06160 -020	2 Carp	1,500	.011	SF Flr.	.62	.37		.99	1.25	850
	0015		Pneumatic nailed			1,860	.009		.62	.29		.91	1.14	
	0100		5/8" thick			1,350	.012		.78	.41		1.19	1.49	
	0105		Pneumatic nailed			1,674	.010		.78	.33		1.11	1.37	
	0200		3/4" thick			1,250	.013		.94	.44		1.38	1.71	
	0205		Pneumatic nailed			1,550	.010		.94	.35		1.29	1.58	
	0300		1-1/8" thick, 2-4-1 including underlayment			1,050	.015		1.19	.52		1.71	2.12	
	0450		1" x 8" S4S, laid regular			1,000	.016		1.11	.55		1.66	2.08	
	0460		Laid diagonal			850	.019		1.11	.64		1.75	2.23	
	0500		1" x 10" S4S, laid regular			1,100	.015		1.36	.50		1.86	2.28	
	0600		Laid diagonal			900	.018		1.36	.61		1.97	2.45	
900	0011	UNDERLAYMENT	Plywood, underlayment grade, 3/8" thick	R06160 -020	2 Carp	1,500	.011	SF Flr.	.87	.37		1.24	1.53	900
	0016		Pneumatic nailed			1,860	.009		.87	.29		1.16	1.42	
	0100		1/2" thick			1,450	.011		1.06	.38		1.44	1.76	
	0105		Pneumatic nailed			1,798	.009		1.06	.30		1.36	1.64	
	0200		5/8" thick			1,400	.011		1.22	.39		1.61	1.95	
	0205		Pneumatic nailed			1,736	.009		1.22	.32		1.54	1.83	
	0300		3/4" thick			1,300	.012		1.40	.42		1.82	2.20	
	0305		Pneumatic nailed			1,612	.010		1.40	.34		1.74	2.07	
	0500		Particle board, 3/8" thick			1,500	.011		.37	.37		.74	.98	
	0505		Pneumatic nailed			1,860	.009		.37	.29		.66	.87	
	0600		1/2" thick			1,450	.011		.39	.38		.77	1.02	
	0605		Pneumatic nailed			1,798	.009		.39	.30		.69	.90	
	0800		5/8" thick			1,400	.011		.49	.39		.88	1.15	
	0805		Pneumatic nailed			1,736	.009		.49	.32		.81	1.03	
	0900		3/4" thick			1,300	.012		.55	.42		.97	1.27	
	0905		Pneumatic nailed			1,612	.010		.55	.34		.89	1.14	
	1100		Hardboard, underlayment grade, 4' x 4', .215" thick			1,500	.011		.39	.37		.76	1	

06170 | Prefabricated Structural Wood

				CREW	DAILY OUTPUT	LABOR-HOURS	UNIT	MAT.	LABOR	EQUIP.	TOTAL	TOTAL INCL O&P	
550	0010	LAMINATED ROOF DECK	Pine or hemlock, 3" thick	2 Carp	425	.038	S.F.	3.02	1.29		4.31	5.35	550
	0100		4" thick		325	.049		4.02	1.69		5.71	7.05	
	0300		Cedar, 3" thick		425	.038		3.69	1.29		4.98	6.05	
	0400		4" thick		325	.049		4.70	1.69		6.39	7.80	
	0600		Fir, 3" thick		425	.038		3.08	1.29		4.37	5.40	
	0700		4" thick		325	.049		3.85	1.69		5.54	6.85	
600	0010	STRUCTURAL JOISTS	Fabricated "I" joists with wood flanges,										600
	0100		Plywood webs, incl. bridging & blocking, panels 24" O.C.										
	1200		15' to 24' span, 50 psf live load	F-5	2,400	.013	SF Flr.	1.67	.46		2.13	2.56	
	1300		55 psf live load		2,250	.014		1.78	.49		2.27	2.72	
	1400		24' to 30' span, 45 psf live load		2,600	.012		2.03	.43		2.46	2.91	
	1500		55 psf live load		2,400	.013		2.90	.46		3.36	3.91	
	1600		Tubular steel open webs, 45 psf, 24" O.C., 40' span	F-3	6,250	.006		1.80	.22	.10	2.12	2.43	
	1700		55' span		7,750	.005		1.75	.18	.08	2.01	2.30	

WOOD & PLASTICS 6

		06170	Prefabricated Structural Wood	CREW	DAILY OUTPUT	LABOR-HOURS	UNIT	MAT.	LABOR	EQUIP.	TOTAL	TOTAL INCL O&P	
600	1800		70' span	F-3	9,250	.004	SF Flr.	2.27	.15	.07	2.49	2.80	600
	1900		85 psf live load, 26' span	↓	2,300	.017	↓	2.11	.60	.26	2.97	3.54	
980	0010	**ROOF TRUSSES**											980
	0020	For timber connectors, see div. 06090-800	R06170 -100										
	0100	Fink (W) or King post type, 2'-0" O.C.											
	0200	Metal plate connected, 4 in 12 slope											
	0210		24' to 29' span	F-3	3,000	.013	SF Flr.	1.60	.46	.20	2.26	2.69	
	0300		30' to 43' span		3,000	.013		1.77	.46	.20	2.43	2.88	
	0400		44' to 60' span	↓	3,000	.013	↓	1.96	.46	.20	2.62	3.09	
	0700		Glued and nailed, add					50%					

		06180	Glued-Laminated Construction										
400	0010	**LAMINATED FRAMING** Not including decking											400
	0020	30 lb., short term live load, 15 lb. dead load											
	0200	Straight roof beams, 20' clear span, beams 8' O.C.		F-3	2,560	.016	SF Flr.	1.63	.54	.24	2.41	2.89	
	0300		Beams 16' O.C.		3,200	.013		1.17	.43	.19	1.79	2.17	
	0500		40' clear span, beams 8' O.C.		3,200	.013		3.12	.43	.19	3.74	4.31	
	0600		Beams 16' O.C.		3,840	.010		2.54	.36	.16	3.06	3.52	
	0800		60' clear span, beams 8' O.C.	F-4	2,880	.017		5.35	.56	.31	6.22	7.10	
	0900		Beams 16' O.C.	"	3,840	.013		3.99	.42	.23	4.64	5.30	
	1100		Tudor arches, 30' to 40' clear span, frames 8' O.C.	F-3	1,680	.024		7	.82	.36	8.18	9.35	
	1200		Frames 16' O.C.	"	2,240	.018		5.45	.62	.27	6.34	7.25	
	1400		50' to 60' clear span, frames 8' O.C.	F-4	2,200	.022		7.55	.74	.41	8.70	9.90	
	1500		Frames 16' O.C.		2,640	.018		6.40	.62	.34	7.36	8.35	
	1700		Radial arches, 60' clear span, frames 8' O.C.		1,920	.025		7.05	.85	.47	8.37	9.55	
	1800		Frames 16' O.C.		2,880	.017		5.40	.56	.31	6.27	7.15	
	2000		100' clear span, frames 8' O.C.		1,600	.030		7.30	1.01	.56	8.87	10.20	
	2100		Frames 16' O.C.		2,400	.020		6.40	.68	.37	7.45	8.50	
	2300		120' clear span, frames 8' O.C.		1,440	.033		9.70	1.13	.62	11.45	13.05	
	2400		Frames 16' O.C.	↓	1,920	.025		8.85	.85	.47	10.17	11.55	
	2600		Bowstring trusses, 20' O.C., 40' clear span	F-3	2,400	.017		4.37	.58	.25	5.20	6	
	2700		60' clear span	F-4	3,600	.013		3.92	.45	.25	4.62	5.30	
	2800		100' clear span		4,000	.012		5.55	.41	.22	6.18	7	
	2900		120' clear span	↓	3,600	.013		5.95	.45	.25	6.65	7.50	
	3000	For less than 1000 B.F., add						20%					
	3050	For over 5000 B.F., deduct						10%					
	3100	For premium appearance, add to S.F. prices						5%					
	3300	For industrial type, deduct						15%					
	3500	For stain and varnish, add						5%					
	3900	For 3/4" laminations, add to straight						25%					
	4100	Add to curved					↓	15%					
	4300	Alternate pricing method: (use nominal footage of											
	4310	components). Straight beams, camber less than 6"		F-3	3.50	11.429	M.B.F.	2,425	395	172	2,992	3,450	
	4400	Columns, including hardware			2	20		2,600	690	300	3,590	4,250	
	4600	Curved members, radius over 32'			2.50	16		2,650	555	241	3,446	4,050	
	4700	Radius 10' to 32'		↓	3	13.333		2,625	460	201	3,286	3,825	
	4900	For complicated shapes, add maximum						100%					
	5100	For pressure treating, add to straight						35%					
	5200	Add to curved					↓	45%					
	6000	Laminated veneer members, southern pine or western species											
	6050	1-3/4" wide x 5-1/2" deep		2 Carp	480	.033	L.F.	2.77	1.14		3.91	4.83	
	6100	9-1/2" deep			480	.033		3.45	1.14		4.59	5.60	
	6150	14" deep			450	.036		5.15	1.22		6.37	7.55	
	6200	18" deep		↓	450	.036	↓	6.95	1.22		8.17	9.55	
	6300	Parallel strand members, southern pine or western species											
	6350	1-3/4" wide x 9-1/4" deep		2 Carp	480	.033	L.F.	3.27	1.14		4.41	5.40	

Important: See the Reference Section for critical supporting data - Reference Nos., Crews, & City Cost Indexes

06180	Glued-Laminated Construction	CREW	DAILY OUTPUT	LABOR-HOURS	UNIT	2005 BARE COSTS				TOTAL INCL O&P
						MAT.	LABOR	EQUIP.	TOTAL	
400 6400	11-1/4" deep	2 Carp	450	.036	L.F.	4.02	1.22		5.24	6.30	**400**
6450	14" deep		400	.040		4.79	1.37		6.16	7.40
6500	3-1/2" wide x 9-1/4" deep		480	.033		7.95	1.14		9.09	10.55
6550	11-1/4" deep		450	.036		9.85	1.22		11.07	12.70
6600	14" deep		400	.040		11.65	1.37		13.02	15
6650	7" wide x 9-1/4" deep		450	.036		16.55	1.22		17.77	20
6700	11-1/4" deep		420	.038		20.50	1.30		21.80	25
6750	14" deep		400	.040		25	1.37		26.37	29.50

06200 | Finish Carpentry

06220	Millwork	CREW	DAILY OUTPUT	LABOR-HOURS	UNIT	2005 BARE COSTS				TOTAL INCL O&P
						MAT.	LABOR	EQUIP.	TOTAL	
200 0010	**MOLDINGS, BASE**										**200**
0500	Base, stock pine, 9/16" x 3-1/2"	1 Carp	240	.033	L.F.	1.53	1.14		2.67	3.46
0550	9/16" x 4-1/2"		200	.040		1.63	1.37		3	3.92
0561	Base shoe, oak, 3/4" x 1"		240	.033		1.04	1.14		2.18	2.92
400 0010	**MOLDINGS, CASINGS**										**400**
0090	Apron, stock pine, 5/8" x 2"	1 Carp	250	.032	L.F.	.99	1.10		2.09	2.80
0110	5/8" x 3-1/2"		220	.036		1.19	1.25		2.44	3.25
0300	Band, stock pine, 11/16" x 1-1/8"		270	.030		.54	1.01		1.55	2.17
0350	11/16" x 1-3/4"		250	.032		.86	1.10		1.96	2.66
0700	Casing, stock pine, 11/16" x 2-1/2"		240	.033		.85	1.14		1.99	2.72
0750	11/16" x 3-1/2"		215	.037		1.21	1.27		2.48	3.32
450 0010	**MOLDINGS, CEILINGS**										**450**
0600	Bed, stock pine, 9/16" x 1-3/4"	1 Carp	270	.030	L.F.	.65	1.01		1.66	2.30
0650	9/16" x 2"		240	.033		.92	1.14		2.06	2.79
1200	Cornice molding, stock pine, 9/16" x 1-3/4"		330	.024		.66	.83		1.49	2.02
1300	9/16" x 2-1/4"		300	.027		.83	.91		1.74	2.33
2400	Cove scotia, stock pine, 9/16" x 1-3/4"		270	.030		.60	1.01		1.61	2.24
2500	11/16" x 2-3/4"		255	.031		1.25	1.07		2.32	3.05
2600	Crown, stock pine, 9/16" x 3-5/8"		250	.032		1.67	1.10		2.77	3.55
2700	11/16" x 4-5/8"		220	.036		2.58	1.25		3.83	4.78
500 0010	**MOLDINGS, EXTERIOR**										**500**
1500	Cornice, boards, pine, 1" x 2"	1 Carp	330	.024	L.F.	.33	.83		1.16	1.65
1700	1" x 6"		250	.032		1.19	1.10		2.29	3.02
2000	1" x 12"		180	.044		1.77	1.52		3.29	4.32
2200	Three piece, built-up, pine, minimum		80	.100		2.20	3.43		5.63	7.75
2300	Maximum		65	.123		4.85	4.22		9.07	11.90
3000	Corner board, sterling pine, 1" x 4"		200	.040		.68	1.37		2.05	2.88
3100	1" x 6"		200	.040		1	1.37		2.37	3.23
3350	Fascia, sterling pine, 1" x 6"		250	.032		1	1.10		2.10	2.81
3370	1" x 8"		225	.036		1.56	1.22		2.78	3.62
3400	Trim, exterior, sterling pine, back band		250	.032		.68	1.10		1.78	2.46
3500	Casing		250	.032		1.80	1.10		2.90	3.69
3600	Crown		250	.032		1.73	1.10		2.83	3.61
3700	Porch rail with balusters		22	.364		14.15	12.45		26.60	35
3800	Screen		395	.020		1.09	.69		1.78	2.28
4100	Verge board, sterling pine, 1" x 4"		200	.040		.66	1.37		2.03	2.86

WOOD & PLASTICS 6

06220 | Millwork

		CREW	DAILY OUTPUT	LABOR-HOURS	UNIT	2005 BARE COSTS MAT.	LABOR	EQUIP.	TOTAL	TOTAL INCL O&P	
500	4200	1" x 6"	1 Carp	200	.040	L.F.	.99	1.37		2.36	3.22	**500**
	4300	2" x 6"		165	.048		1.61	1.66		3.27	4.36	
	4400	2" x 8"	↓	165	.048	↓	2.13	1.66		3.79	4.93	
	4700	For redwood trim, add					200%					
700	0010	**MOLDINGS, TRIM**										**700**
	0200	Astragal, stock pine, 11/16" x 1-3/4"	1 Carp	255	.031	L.F.	.96	1.07		2.03	2.73	
	0250	1-5/16" x 2-3/16"		240	.033		2.57	1.14		3.71	4.61	
	0800	Chair rail, stock pine, 5/8" x 2-1/2"		270	.030		1.12	1.01		2.13	2.81	
	0900	5/8" x 3-1/2"		240	.033		1.62	1.14		2.76	3.56	
	1000	Closet pole, stock pine, 1-1/8" diameter		200	.040		.79	1.37		2.16	3	
	1100	Fir, 1-5/8" diameter		200	.040		1.27	1.37		2.64	3.53	
	3300	Half round, stock pine, 1/4" x 1/2"		270	.030		.16	1.01		1.17	1.76	
	3350	1/2" x 1"	↓	255	.031	↓	.39	1.07		1.46	2.10	
	3400	Handrail, fir, single piece, stock, hardware not included										
	3450	1-1/2" x 1-3/4"	1 Carp	80	.100	L.F.	1.22	3.43		4.65	6.70	
	3470	Pine, 1-1/2" x 1-3/4"		80	.100		1.22	3.43		4.65	6.70	
	3500	1-1/2" x 2-1/2"		76	.105		1.27	3.61		4.88	7	
	3600	Lattice, stock pine, 1/4" x 1-1/8"		270	.030		.23	1.01		1.24	1.83	
	3700	1/4" x 1-3/4"		250	.032		.35	1.10		1.45	2.10	
	3800	Miscellaneous, custom, pine, 1" x 1"		270	.030		.32	1.01		1.33	1.93	
	3900	1" x 3"		240	.033		.65	1.14		1.79	2.50	
	4100	Birch or oak, nominal 1" x 1"		240	.033		.50	1.14		1.64	2.33	
	4200	Nominal 1" x 3"		215	.037		1.71	1.27		2.98	3.87	
	4400	Walnut, nominal 1" x 1"		215	.037		.82	1.27		2.09	2.89	
	4500	Nominal 1" x 3"		200	.040		2.46	1.37		3.83	4.84	
	4700	Teak, nominal 1" x 1"		215	.037		1.16	1.27		2.43	3.27	
	4800	Nominal 1" x 3"		200	.040		3.32	1.37		4.69	5.80	
	4900	Quarter round, stock pine, 1/4" x 1/4"		275	.029		.14	1		1.14	1.70	
	4950	3/4" x 3/4"		255	.031	↓	.36	1.07		1.43	2.07	
	5600	Wainscot moldings, 1-1/8" x 9/16", 2' high, minimum		76	.105	S.F.	8.95	3.61		12.56	15.45	
	5700	Maximum	↓	65	.123	"	18.30	4.22		22.52	26.50	
800	0010	**MOLDINGS, WINDOW AND DOOR**										**800**
	2800	Door moldings, stock, decorative, 1-1/8" wide, plain	1 Carp	17	.471	Set	29	16.10		45.10	57	
	2900	Detailed		17	.471	"	81	16.10		97.10	114	
	2960	Clear pine door jamb, no stops, 11/16" x 4-9/16"		240	.033	L.F.	2.30	1.14		3.44	4.31	
	3150	Door trim set, 1 head and 2 sides, pine, 2-1/2 wide		5.90	1.356	Opng.	14.45	46.50		60.95	88.50	
	3170	3-1/2" wide		5.30	1.509	"	20.50	51.50		72	103	
	3250	Glass beads, stock pine, 3/8" x 1/2"		275	.029	L.F.	.36	1		1.36	1.95	
	3270	3/8" x 7/8"		270	.030		.40	1.01		1.41	2.02	
	4850	Parting bead, stock pine, 3/8" x 3/4"		275	.029		.26	1		1.26	1.84	
	4870	1/2" x 3/4"		255	.031		.31	1.07		1.38	2.01	
	5000	Stool caps, stock pine, 11/16" x 3-1/2"		200	.040		1.53	1.37		2.90	3.81	
	5100	1-1/16" x 3-1/4"		150	.053	↓	2.67	1.83		4.50	5.80	
	5300	Threshold, oak, 3' long, inside, 5/8" x 3-5/8"		32	.250	Ea.	7.75	8.55		16.30	22	
	5400	Outside, 1-1/2" x 7-5/8"	↓	16	.500	"	35	17.15		52.15	65	
	5900	Window trim sets, including casings, header, stops,										
	5910	stool and apron, 2-1/2" wide, minimum	1 Carp	13	.615	Opng.	19.90	21		40.90	55	
	5950	Average		10	.800		23.50	27.50		51	68	
	6000	Maximum	↓	6	1.333	↓	58	45.50		103.50	135	
900	0010	**SOFFITS**										**900**
	0200	Soffits, pine, 1" x 4"	2 Carp	420	.038	L.F.	.41	1.30		1.71	2.48	
	0210	1" x 6"		420	.038		.58	1.30		1.88	2.66	
	0220	1" x 8"		420	.038		.70	1.30		2	2.80	
	0230	1" x 10"		400	.040		1.08	1.37		2.45	3.32	
	0240	1" x 12"	↓	400	.040		1.56	1.37		2.93	3.85	

6

WOOD & PLASTICS

Important: See the Reference Section for critical supporting data - Reference Nos., Crews, & City Cost Indexes

06220	Millwork		CREW	DAILY OUTPUT	LABOR-HOURS	UNIT	2005 BARE COSTS MAT.	LABOR	EQUIP.	TOTAL	TOTAL INCL O&P	
900	0250	STK cedar, 1" x 4"	2 Carp	420	.038	L.F.	.52	1.30		1.82	2.60	900
	0260	1" x 6"		420	.038		.83	1.30		2.13	2.94	
	0270	1" x 8"		420	.038		1.09	1.30		2.39	3.23	
	0280	1" x 10"		400	.040		1.31	1.37		2.68	3.57	
	0290	1" x 12"		400	.040	↓	1.61	1.37		2.98	3.90	
	1000	Exterior AC plywood, 1/4" thick		420	.038	S.F.	.78	1.30		2.08	2.89	
	1050	3/8" thick		420	.038		.87	1.30		2.17	2.99	
	1100	1/2" thick		420	.038		1.06	1.30		2.36	3.20	
	1150	Polyvinyl chloride, white, solid	1 Carp	230	.035		.66	1.19		1.85	2.59	
	1160	Perforated	"	230	.035	↓	.66	1.19		1.85	2.59	

06250	Prefinished Paneling											
200	0010	**PANELING, HARDBOARD**										200
	0050	Not incl. furring or trim, hardboard, tempered, 1/8" thick	2 Carp	500	.032	S.F.	.30	1.10		1.40	2.04	
	0100	1/4" thick		500	.032		.38	1.10		1.48	2.13	
	0300	Tempered pegboard, 1/8" thick		500	.032		.37	1.10		1.47	2.12	
	0400	1/4" thick		500	.032		.42	1.10		1.52	2.17	
	0600	Untempered hardboard, natural finish, 1/8" thick		500	.032		.33	1.10		1.43	2.07	
	0700	1/4" thick		500	.032		.32	1.10		1.42	2.06	
	0900	Untempered pegboard, 1/8" thick		500	.032		.33	1.10		1.43	2.07	
	1000	1/4" thick		500	.032		.37	1.10		1.47	2.12	
	1200	Plastic faced hardboard, 1/8" thick		500	.032		.55	1.10		1.65	2.32	
	1300	1/4" thick		500	.032		.73	1.10		1.83	2.51	
	1500	Plastic faced pegboard, 1/8" thick		500	.032		.52	1.10		1.62	2.28	
	1600	1/4" thick		500	.032		.64	1.10		1.74	2.41	
	1800	Wood grained, plain or grooved, 1/4" thick, minimum		500	.032		.48	1.10		1.58	2.24	
	1900	Maximum		425	.038	↓	1.02	1.29		2.31	3.13	
	2100	Moldings for hardboard, wood or aluminum, minimum		500	.032	L.F.	.33	1.10		1.43	2.07	
	2200	Maximum	↓	425	.038	"	.92	1.29		2.21	3.02	

500	0010	**PANELING, PLYWOOD** R06160 -020										500
	2400	Plywood, prefinished, 1/4" thick, 4' x 8' sheets										
	2410	with vertical grooves. Birch faced, minimum	2 Carp	500	.032	S.F.	.79	1.10		1.89	2.58	
	2420	Average		420	.038		1.21	1.30		2.51	3.36	
	2430	Maximum		350	.046		1.76	1.57		3.33	4.38	
	2600	Mahogany, African		400	.040		2.25	1.37		3.62	4.61	
	2700	Philippine (Lauan)		500	.032		.97	1.10		2.07	2.78	
	2900	Oak or Cherry, minimum		500	.032		1.89	1.10		2.99	3.79	
	3000	Maximum		400	.040		2.90	1.37		4.27	5.30	
	3200	Rosewood		320	.050		4.11	1.71		5.82	7.20	
	3400	Teak		400	.040		2.90	1.37		4.27	5.30	
	3600	Chestnut		375	.043		4.28	1.46		5.74	7	
	3800	Pecan		400	.040		1.85	1.37		3.22	4.17	
	3900	Walnut, minimum		500	.032		2.47	1.10		3.57	4.43	
	3950	Maximum		400	.040		4.68	1.37		6.05	7.30	
	4000	Plywood, prefinished, 3/4" thick, stock grades, minimum		320	.050		1.12	1.71		2.83	3.90	
	4100	Maximum		224	.071		4.83	2.45		7.28	9.10	
	4300	Architectural grade, minimum		224	.071		3.57	2.45		6.02	7.75	
	4400	Maximum		160	.100		5.45	3.43		8.88	11.35	
	4600	Plywood, "A" face, birch, V.C., 1/2" thick, natural		450	.036		1.69	1.22		2.91	3.76	
	4700	Select		450	.036		1.85	1.22		3.07	3.94	
	4900	Veneer core, 3/4" thick, natural		320	.050		1.78	1.71		3.49	4.63	
	5000	Select		320	.050		2.01	1.71		3.72	4.88	
	5200	Lumber core, 3/4" thick, natural		320	.050		2.68	1.71		4.39	5.60	
	5500	Plywood, knotty pine, 1/4" thick, A2 grade		450	.036		1.46	1.22		2.68	3.51	
	5600	A3 grade	↓	450	.036	↓	1.85	1.22		3.07	3.94	

WOOD & PLASTICS 6

06250 | Prefinished Paneling

		CREW	DAILY OUTPUT	LABOR-HOURS	UNIT	2005 BARE COSTS				TOTAL INCL O&P	
						MAT.	LABOR	EQUIP.	TOTAL		
500	5800	3/4" thick, veneer core, A2 grade	2 Carp	320	.050	S.F.	1.90	1.71		3.61	4.76	500
	5900	A3 grade		320	.050		2.13	1.71		3.84	5	
	6100	Aromatic cedar, 1/4" thick, plywood		400	.040		1.87	1.37		3.24	4.19	
	6200	1/4" thick, particle board		400	.040		.91	1.37		2.28	3.13	

(R06160-020)

06260 | Board Paneling

		CREW	DAILY OUTPUT	LABOR-HOURS	UNIT	MAT.	LABOR	EQUIP.	TOTAL	INCL O&P	
400	0010	**PANELING, BOARDS**									400
	6400	Wood board paneling, 3/4" thick, knotty pine	2 Carp	300	.053	S.F.	1.36	1.83		3.19	4.35
	6500	Rough sawn cedar		300	.053		1.74	1.83		3.57	4.76
	6700	Redwood, clear, 1" x 4" boards		300	.053		4.06	1.83		5.89	7.30
	6900	Aromatic cedar, closet lining, boards		275	.058		3.13	1.99		5.12	6.55

06270 | Closet/Utility Wood Shelving

		CREW	DAILY OUTPUT	LABOR-HOURS	UNIT	MAT.	LABOR	EQUIP.	TOTAL	INCL O&P	
200	0010	**SHELVING** Pine, clear grade, no edge band, 1" x 8"	1 Carp	115	.070	L.F.	1.76	2.38		4.14	5.65	200
	0100	1" x 10"		110	.073		2.24	2.49		4.73	6.35
	0200	1" x 12"		105	.076		3.01	2.61		5.62	7.35
	0600	Plywood, 3/4" thick with lumber edge, 12" wide		75	.107		.32	3.65		3.97	6.05
	0700	24" wide		70	.114		3.12	3.91		7.03	9.55
	0900	Bookcase, clear grade pine, shelves 12" O.C., 8" deep, /SF shelf		70	.114	S.F.	5.70	3.91		9.61	12.40
	1000	12" deep shelves		65	.123	"	9.80	4.22		14.02	17.30
	1200	Adjustable closet rod and shelf, 12" wide, 3' long		20	.400	Ea.	8.60	13.70		22.30	31
	1300	8' long		15	.533	"	21	18.25		39.25	51.50
	1500	Prefinished shelves with supports, stock, 8" wide		75	.107	L.F.	3.72	3.65		7.37	9.80
	1600	10" wide		70	.114	"	4.13	3.91		8.04	10.65

06400 | Architectural Woodwork

06410 | Custom Cabinets

		CREW	DAILY OUTPUT	LABOR-HOURS	UNIT	2005 BARE COSTS				TOTAL INCL O&P	
						MAT.	LABOR	EQUIP.	TOTAL		
100	0010	**CABINETS** Corner china cabinets, stock pine,										100
	0020	80" high, unfinished, minimum	2 Carp	6.60	2.424	Ea.	455	83		538	630
	0100	Maximum	"	4.40	3.636	"	1,000	125		1,125	1,300
	0300	Built-in drawer units, pine, 18" deep, 32" high, unfinished									
	0400	Minimum	2 Carp	53	.302	L.F.	120	10.35		130.35	148
	0500	Maximum	"	40	.400	"	146	13.70		159.70	183
	0700	Kitchen base cabinets, hardwood, not incl. counter tops,									
	0710	24" deep, 35" high, prefinished									
	0800	One top drawer, one door below, 12" wide	2 Carp	24.80	.645	Ea.	132	22		154	181
	0840	18" wide		23.30	.687		197	23.50		220.50	253
	0880	24" wide		22.30	.717		236	24.50		260.50	298
	1000	Four drawers, 12" wide		24.80	.645		315	22		337	380
	1040	18" wide		23.30	.687		269	23.50		292.50	335
	1060	24" wide		22.30	.717		292	24.50		316.50	360
	1200	Two top drawers, two doors below, 27" wide		22	.727		261	25		286	325
	1260	36" wide		20.30	.788		300	27		327	370
	1300	48" wide		18.90	.847		345	29		374	425
	1500	Range or sink base, two doors below, 30" wide		21.40	.748		231	25.50		256.50	294
	1540	36" wide		20.30	.788		259	27		286	325
	1580	48" wide		18.90	.847		291	29		320	365

6 WOOD & PLASTICS

06410 | Custom Cabinets

			CREW	DAILY OUTPUT	LABOR-HOURS	UNIT	2005 BARE COSTS				TOTAL INCL O&P	
							MAT.	LABOR	EQUIP.	TOTAL		
100	1800	For sink front units, deduct				Ea.	50			50	55	100
	2000	Corner base cabinets, 36" wide, standard	2 Carp	18	.889		380	30.50		410.50	470	
	2100	Lazy Susan with revolving door	"	16.50	.970		370	33		403	455	
	4000	Kitchen wall cabinets, hardwood, 12" deep with two doors										
	4050	12" high, 30" wide	2 Carp	24.80	.645	Ea.	145	22		167	195	
	4100	36" wide		24	.667		169	23		192	222	
	4400	15" high, 30" wide		24	.667		152	23		175	204	
	4440	36" wide		22.70	.705		172	24		196	227	
	4700	24" high, 30" wide		23.30	.687		189	23.50		212.50	245	
	4720	36" wide		22.70	.705		209	24		233	268	
	5000	30" high, one door, 12" wide		22	.727		128	25		153	179	
	5040	18" wide		20.90	.766		158	26		184	215	
	5060	24" wide		20.30	.788		178	27		205	238	
	5300	Two doors, 27" wide		19.80	.808		221	27.50		248.50	286	
	5340	36" wide		18.80	.851		244	29		273	315	
	5380	48" wide		18.40	.870		299	30		329	375	
	6000	Corner wall, 30" high, 24" wide		18	.889		146	30.50		176.50	208	
	6050	30" wide		17.20	.930		173	32		205	241	
	6100	36" wide		16.50	.970		188	33		221	258	
	6500	Revolving Lazy Susan		15.20	1.053		286	36		322	370	
	7000	Broom cabinet, 84" high, 24" deep, 18" wide		10	1.600		390	55		445	515	
	7500	Oven cabinets, 84" high, 24" deep, 27" wide		8	2		565	68.50		633.50	730	
	7750	Valance board trim		396	.040	L.F.	8.10	1.38		9.48	11.05	
	9000	For deluxe models of all cabinets, add					40%					
	9500	For custom built in place, add					25%	10%				
	9550	Rule of thumb, kitchen cabinets not including										
	9560	appliances & counter top, minimum	2 Carp	30	.533	L.F.	92	18.25		110.25	130	
	9600	Maximum	"	25	.640	"	240	22		262	298	
	9610	For metal cabinets, see division 12310-750										
210	0010	**CASEWORK, FRAMES**										**210**
	0050	Base cabinets, counter storage, 36" high, one bay										
	0100	18" wide	1 Carp	2.70	2.963	Ea.	103	101		204	271	
	0400	Two bay, 36" wide		2.20	3.636		157	125		282	365	
	1100	Three bay, 54" wide		1.50	5.333		187	183		370	490	
	2800	Book cases, one bay, 7' high, 18" wide		2.40	3.333		121	114		235	310	
	3500	Two bay, 36" wide		1.60	5		176	171		347	460	
	4100	Three bay, 54" wide		1.20	6.667		291	228		519	675	
	5100	Coat racks, one bay, 7' high, 24" wide		4.50	1.778		121	61		182	228	
	5300	Two bay, 48" wide		2.75	2.909		168	99.50		267.50	340	
	5800	Three bay, 72" wide		2.10	3.810		248	130		378	475	
	6100	Wall mounted cabinet, one bay, 24" high, 18" wide		3.60	2.222		66.50	76		142.50	193	
	6800	Two bay, 36" wide		2.20	3.636		97	125		222	300	
	7400	Three bay, 54" wide		1.70	4.706		121	161		282	385	
	8400	30" high, one bay, 18" wide		3.60	2.222		72.50	76		148.50	199	
	9000	Two bay, 36" wide		2.15	3.721		96	127		223	305	
	9400	Three bay, 54" wide		1.60	5		120	171		291	400	
	9800	Wardrobe, 7' high, single, 24" wide		2.70	2.963		133	101		234	305	
	9880	Partition & adjustable shelves, 48" wide		1.70	4.706		170	161		331	440	
	9950	Partition, adjustable shelves & drawers, 48" wide		1.40	5.714		254	196		450	585	
220	0010	**CABINET DOORS**										**220**
	2000	Glass panel, hardwood frame										
	2200	12" wide, 18" high	1 Carp	34	.235	Ea.	22	8.05		30.05	37	
	2600	30" high		32	.250		25.50	8.55		34.05	41.50	
	4450	18" wide, 18" high		32	.250		24	8.55		32.55	39.50	
	4550	30" high		29	.276		26.50	9.45		35.95	43.50	

WOOD & PLASTICS 6

6 WOOD & PLASTICS

06410 | Custom Cabinets

			CREW	DAILY OUTPUT	LABOR-HOURS	UNIT	MAT.	LABOR	EQUIP.	TOTAL	TOTAL INCL O&P	
220	5000	Hardwood, raised panel										**220**
	5100	12" wide, 18" high	1 Carp	16	.500	Ea.	33.50	17.15		50.65	63.50	
	5200	30" high		15	.533		40	18.25		58.25	72.50	
	5500	18" wide, 18" high		15	.533		37	18.25		55.25	69	
	5600	30" high	▼	14	.571	▼	45	19.55		64.55	80	
	6000	Plastic laminate on particle board										
	6100	12" wide, 18" high	1 Carp	25	.320	Ea.	15.20	10.95		26.15	34	
	6140	30" high		23	.348		24.50	11.90		36.40	45.50	
	6500	18" wide, 18" high		24	.333		22	11.40		33.40	42.50	
	6600	30" high	▼	22	.364		37.50	12.45		49.95	60.50	
	7000	Plywood, with edge band										
	7010	12" wide, 18" high	1 Carp	27	.296	Ea.	20.50	10.15		30.65	38.50	
	7120	30" high		25	.320		35	10.95		45.95	55.50	
	7650	18" wide, 18" high		26	.308		30	10.55		40.55	49.50	
	7750	30" high	▼	24	.333	▼	51.50	11.40		62.90	74.50	
230	0010	**CABINET HARDWARE**										**230**
	1000	Catches, minimum	1 Carp	235	.034	Ea.	.82	1.17		1.99	2.72	
	1040	Maximum	"	80	.100	"	5.05	3.43		8.48	10.95	
	2000	Door/drawer pulls, handles										
	2200	Handles and pulls, projecting, metal, minimum	1 Carp	160	.050	Ea.	3.79	1.71		5.50	6.85	
	2240	Maximum		68	.118		6.95	4.03		10.98	13.95	
	2300	Wood, minimum		160	.050		3.79	1.71		5.50	6.85	
	2340	Maximum		68	.118		6.95	4.03		10.98	13.95	
	2600	Flush, metal, minimum		160	.050		3.79	1.71		5.50	6.85	
	2640	Maximum		68	.118	▼	6.95	4.03		10.98	13.95	
	3000	Drawer tracks/glides, minimum		48	.167	Pr.	6.45	5.70		12.15	15.95	
	3040	Maximum		24	.333		18.70	11.40		30.10	38.50	
	4000	Cabinet hinges, minimum		160	.050		2.18	1.71		3.89	5.05	
	4040	Maximum	▼	68	.118	▼	7.65	4.03		11.68	14.70	
240	0010	**DRAWERS**										**240**
	0100	Solid hardwood front										
	1000	4" high, 12" wide	1 Carp	17	.471	Ea.	18.10	16.10		34.20	45	
	1200	18" wide	"	16	.500	"	25	17.15		42.15	54	
	2800	Plastic laminate on particle board front										
	3000	4" high, 12" wide	1 Carp	17	.471	Ea.	23.50	16.10		39.60	51	
	3200	18" wide	"	16	.500	"	27.50	17.15		44.65	56.50	
	5400	Plywood, flush panel front										
	6000	4" high, 12" wide	1 Carp	17	.471	Ea.	24.50	16.10		40.60	52	
	6200	18" wide	"	16	.500	"	29.50	17.15		46.65	59	
400	0010	**VANITIES**										**400**
	8000	Vanity bases, 2 doors, 30" high, 21" deep, 24" wide	2 Carp	20	.800	Ea.	177	27.50		204.50	237	
	8050	30" wide		16	1		203	34.50		237.50	277	
	8100	36" wide		13.33	1.200		271	41		312	360	
	8150	48" wide	▼	11.43	1.400		325	48		373	430	
	9000	For deluxe models of all vanities, add to above					40%					
	9500	For custom built in place, add to above				▼	25%	10%				

06415 | Countertops

			CREW	DAILY OUTPUT	LABOR-HOURS	UNIT	MAT.	LABOR	EQUIP.	TOTAL	TOTAL INCL O&P	
100	0010	**COUNTER TOP** Stock, plastic lam., 24" wide w/backsplash, min.	1 Carp	30	.267	L.F.	8.75	9.15		17.90	24	**100**
	0100	Maximum		25	.320		15.85	10.95		26.80	34.50	
	0300	Custom plastic, 7/8" thick, aluminum molding, no splash		30	.267		17.50	9.15		26.65	33.50	
	0400	Cove splash		30	.267		23	9.15		32.15	40	
	0600	1-1/4" thick, no splash		28	.286		20	9.80		29.80	37.50	
	0700	Square splash	▼	28	.286	▼	25	9.80		34.80	43	

Important: See the Reference Section for critical supporting data - Reference Nos., Crews, & City Cost Indexes

			DAILY	LABOR-		\multicolumn{4}{c}{2005 BARE COSTS}	TOTAL	
---	---	CREW	OUTPUT	HOURS	UNIT	MAT.	LABOR	EQUIP.	TOTAL	INCL O&P	
06415	**Countertops**										
100	0900	Square edge, plastic face, 7/8" thick, no splash	1 Carp	30	.267	L.F.	22	9.15		31.15	38.50	100
	1000	With splash	↓	30	.267		28	9.15		37.15	45.50	
	1200	For stainless channel edge, 7/8" thick, add					2.30			2.30	2.53	
	1300	1-1/4" thick, add					2.70			2.70	2.97	
	1500	For solid color suede finish, add				↓	2.25			2.25	2.48	
	1700	For end splash, add				Ea.	15			15	16.50	
	1900	For cut outs, standard, add, minimum	1 Carp	32	.250		3	8.55		11.55	16.65	
	2000	Maximum		8	1	↓	5	34.50		39.50	59	
	2100	Postformed, including backsplash and front edge		30	.267	L.F.	9	9.15		18.15	24	
	2110	Mitred, add		12	.667	Ea.		23		23	35.50	
	2200	Built-in place, 25" wide, plastic laminate		25	.320	L.F.	12	10.95		22.95	30.50	
	2300	Ceramic tile mosaic	↓	25	.320		26	10.95		36.95	45.50	
	2500	Marble, stock, with splash, 1/2" thick, minimum	1 Bric	17	.471		32	16.60		48.60	60.50	
	2700	3/4" thick, maximum	"	13	.615		80	21.50		101.50	121	
	2900	Maple, solid, laminated, 1-1/2" thick, no splash	1 Carp	28	.286		52.50	9.80		62.30	73	
	3000	With square splash		28	.286	↓	62	9.80		71.80	84	
	3200	Stainless steel		24	.333	S.F.	120	11.40		131.40	150	
	3400	Recessed cutting block with trim, 16" x 20" x 1"	↓	8	1	Ea.	61	34.50		95.50	121	
06430	**Stairs & Railings**										
500	0010	**RAILING** Custom design, architectural grade, hardwood, minimum	1 Carp	38	.211	L.F.	5.60	7.20		12.80	17.40	500
	0100	Maximum		30	.267		46.50	9.15		55.65	66	
	0300	Stock interior railing with spindles 6" O.C., 4' long		40	.200		42	6.85		48.85	56.50	
	0400	8' long	↓	48	.167		21	5.70		26.70	32	
620	0010	**STAIRS, PREFABRICATED**										620
	0100	Box stairs, prefabricated, 3'-0" wide										
	0110	Oak treads, up to 14 risers	2 Carp	39	.410	Riser	71.50	14.05		85.55	101	
	0600	With pine treads for carpet, up to 14 risers	"	39	.410	"	46	14.05		60.05	72.50	
	1100	For 4' wide stairs, add				Flight	25%					
	1550	Stairs, prefabricated stair handrail with balusters	1 Carp	30	.267	L.F.	60.50	9.15		69.65	81	
	1700	Basement stairs, prefabricated, pine treads										
	1710	Pine risers, 3' wide, up to 14 risers	2 Carp	52	.308	Riser	46	10.55		56.55	67	
	4000	Residential, wood, oak treads, prefabricated		1.50	10.667	Flight	930	365		1,295	1,600	
	4200	Built in place	↓	.44	36.364	"	1,800	1,250		3,050	3,950	
	4400	Spiral, oak, 4'-6" diameter, unfinished, prefabricated,										
	4500	incl. railing, 9' high	2 Carp	1.50	10.667	Flight	4,400	365		4,765	5,425	
630	0010	**STAIR PARTS** Balusters, turned, 30" high, pine, minimum	1 Carp	28	.286	Ea.	3.70	9.80		13.50	19.30	630
	0100	Maximum		26	.308		19	10.55		29.55	37.50	
	0300	30" high birch balusters, minimum		28	.286		6.30	9.80		16.10	22	
	0400	Maximum		26	.308		25.50	10.55		36.05	44.50	
	0600	42" high, pine balusters, minimum		27	.296		4.90	10.15		15.05	21	
	0700	Maximum		25	.320		27.50	10.95		38.45	47	
	0900	42" high birch balusters, minimum		27	.296		10.10	10.15		20.25	27	
	1000	Maximum	↓	25	.320		36.50	10.95		47.45	57	
	1050	Baluster, stock pine, 1-1/4" x 1-1/4"		240	.033	L.F.	4.65	1.14		5.79	6.90	
	1100	1-3/4" x 1-3/4"		220	.036	"	8.25	1.25		9.50	11.05	
	1200	Newels, 3-1/4" wide, starting, minimum		7	1.143	Ea.	38.50	39		77.50	103	
	1300	Maximum		6	1.333		300	45.50		345.50	400	
	1500	Landing, minimum		5	1.600		98	55		153	194	
	1600	Maximum		4	2	↓	330	68.50		398.50	470	
	1800	Railings, oak, built-up, minimum		60	.133	L.F.	30	4.57		34.57	40	
	1900	Maximum		55	.145		43.50	4.98		48.48	56	
	2100	Add for sub rail		110	.073		5	2.49		7.49	9.40	
	2300	Risers, beech, 3/4" x 7-1/2" high		64	.125		5.80	4.28		10.08	13	
	2400	Fir, 3/4" x 7-1/2" high		64	.125		1.60	4.28		5.88	8.40	
	2600	Oak, 3/4" x 7-1/2" high	↓	64	.125	↓	6.10	4.28		10.38	13.35	

For expanded coverage of these items see *Means Interior Cost Data 2005*

Wood & Plastics 6

			CREW	DAILY OUTPUT	LABOR-HOURS	UNIT	2005 BARE COSTS				TOTAL INCL O&P	
		06430 Stairs & Railings					MAT.	LABOR	EQUIP.	TOTAL		
630	2800	Pine, 3/4" x 7-1/2" high	1 Carp	66	.121	L.F.	3.05	4.15		7.20	9.80	**630**
	2850	Skirt board, pine, 1" x 10"		55	.145		2.90	4.98		7.88	10.95	
	2900	1" x 12"		52	.154		3.45	5.25		8.70	12	
	3000	Treads, oak, 1-1/4" x 10" wide, 3' long		18	.444	Ea.	80.50	15.20		95.70	113	
	3100	4' long, oak		17	.471		108	16.10		124.10	143	
	3300	1-1/4" x 11-1/2" wide, 3' long, oak		18	.444		29.50	15.20		44.70	55.50	
	3400	6' long, oak		14	.571		176	19.55		195.55	224	
	3600	Beech treads, add					40%					
	3800	For mitered return nosings, add				L.F.	3.90			3.90	4.29	
		06440 Wood Ornaments										
150	0010	**BEAMS, DECORATIVE** Rough sawn cedar, non-load bearing, 4" x 4"	2 Carp	180	.089	L.F.	1.35	3.04		4.39	6.25	**150**
	0100	4" x 6"		170	.094		2.60	3.22		5.82	7.85	
	0200	4" x 8"		160	.100		3.34	3.43		6.77	9	
	0300	4" x 10"		150	.107		4.64	3.65		8.29	10.80	
	0400	4" x 12"		140	.114		5.60	3.91		9.51	12.30	
	0500	8" x 8"		130	.123		7.85	4.22		12.07	15.20	
	1100	Beam connector plates see div. 06090-800										
350	0010	**GRILLES** and panels, hardwood, sanded										**350**
	0020	2' x 4' to 4' x 8', custom designs, unfinished, minimum	1 Carp	38	.211	S.F.	12.10	7.20		19.30	24.50	
	0050	Average		30	.267		26.50	9.15		35.65	43.50	
	0100	Maximum		19	.421		40.50	14.40		54.90	67	
	0300	As above, but prefinished, minimum		38	.211		12.10	7.20		19.30	24.50	
	0400	Maximum		19	.421		45.50	14.40		59.90	72.50	
400	0010	**LOUVERS** Redwood, 2'-0" diameter, full circle	1 Carp	16	.500	Ea.	136	17.15		153.15	177	**400**
	0100	Half circle		16	.500		130	17.15		147.15	170	
	0200	Octagonal		16	.500		104	17.15		121.15	141	
	0300	Triangular, 5/12 pitch, 5'-0" at base		16	.500		220	17.15		237.15	269	
500	0010	**FIREPLACE MANTELS** 6" molding, 6' x 3'-6" opening, minimum	1 Carp	5	1.600	Opng.	139	55		194	239	**500**
	0100	Maximum		5	1.600		173	55		228	276	
	0300	Prefabricated pine, colonial type, stock, deluxe		2	4		2,175	137		2,312	2,625	
	0400	Economy		3	2.667		375	91.50		466.50	550	
550	0010	**FIREPLACE MANTEL BEAMS** Rough texture wood, 4" x 8"	1 Carp	36	.222	L.F.	5.15	7.60		12.75	17.50	**550**
	0100	4" x 10"		35	.229	"	6.20	7.85		14.05	19	
	0300	Laminated hardwood, 2-1/4" x 10-1/2" wide, 6' long		5	1.600	Ea.	98.50	55		153.50	194	
	0400	8' long		5	1.600	"	137	55		192	236	
	0600	Brackets for above, rough sawn		12	.667	Pr.	9.05	23		32.05	45.50	
	0700	Laminated		12	.667	"	13.70	23		36.70	50.50	
700	0010	**COLUMNS**										**700**
	0050	Aluminum, round colonial, 6" diameter	2 Carp	80	.200	V.L.F.	18.10	6.85		24.95	30.50	
	0100	8" diameter		62.25	.257		25	8.80		33.80	41	
	0200	10" diameter		55	.291		30	9.95		39.95	48.50	
	0250	Fir, stock units, hollow round, 6" diameter		80	.200		19.65	6.85		26.50	32	
	0300	8" diameter		80	.200		20	6.85		26.85	32.50	
	0350	10" diameter		70	.229		25	7.85		32.85	39.50	
	0400	Solid turned, to 8' high, 3-1/2" diameter		80	.200		7.45	6.85		14.30	18.80	
	0500	4-1/2" diameter		75	.213		10.60	7.30		17.90	23	
	0600	5-1/2" diameter		70	.229		14.85	7.85		22.70	28.50	
	0800	Square columns, built-up, 5" x 5"		65	.246		13.80	8.45		22.25	28.50	
	0900	Solid, 3-1/2" x 3-1/2"		130	.123		6.35	4.22		10.57	13.55	
	1600	Hemlock, tapered, T & G, 12" diam, 10' high		100	.160		32	5.50		37.50	43.50	
	1700	16' high		65	.246		56	8.45		64.45	75	
	1900	10' high, 14" diameter		100	.160		81.50	5.50		87	98	
	2000	18' high		65	.246		77.50	8.45		85.95	98	

6

WOOD & PLASTICS

Important: See the Reference Section for critical supporting data - Reference Nos., Crews, & City Cost Indexes

				CREW	DAILY OUTPUT	LABOR-HOURS	UNIT	2005 BARE COSTS				TOTAL INCL O&P	
	06440	**Wood Ornaments**						MAT.	LABOR	EQUIP.	TOTAL		
700	2200	18″ diameter, 12′ high		2 Carp	65	.246	V.L.F.	109	8.45		117.45	133	700
	2300	20′ high			50	.320		105	10.95		115.95	133	
	2500	20″ diameter, 14′ high			40	.400		129	13.70		142.70	164	
	2600	20′ high			35	.457		134	15.65		149.65	172	
	2800	For flat pilasters, deduct						33%					
	3000	For splitting into halves, add					Ea.	95			95	105	
	4000	Rough sawn cedar posts, 4″ x 4″		2 Carp	250	.064	V.L.F.	2.58	2.19		4.77	6.25	
	4100	4″ x 6″			235	.068		3.84	2.33		6.17	7.85	
	4200	6″ x 6″			220	.073		5.80	2.49		8.29	10.25	
	4300	8″ x 8″			200	.080		5.95	2.74		8.69	10.80	
	06445	**Simulated Wood Ornaments**											
100	0010	**MILLWORK, HIGH DENSITY POLYMER**											100
	0100	Base, 9/16″ x 3-3/16″		1 Carp	230	.035	L.F.	1.34	1.19		2.53	3.33	
	0200	Casing, fluted, 5/8″ x 3-1/4″			215	.037		3.73	1.27		5	6.10	
	0300	Chair rail, 9/16″ x 2-1/4″			260	.031		1.94	1.05		2.99	3.77	
	0600	Cove, 13/16″ x 3-3/4″			260	.031		3.90	1.05		4.95	5.95	
	0700	Crown, 3/4″ x 3-13/16″			260	.031		5.50	1.05		6.55	7.70	
	0800	Half round, 15/16″ x 2″			240	.033		17.40	1.14		18.54	21	
	06470	**Screen, Blinds & Shutters**											
100	0010	**SHUTTERS, EXTERIOR** Aluminum, louvered, 1′-4″ wide, 3′-0″ long		1 Carp	10	.800	Pr.	42	27.50		69.50	88.50	100
	0400	6′-8″ long			9	.889		84	30.50		114.50	140	
	1000	Pine, louvered, primed, each 1′-2″ wide, 3′-3″ long			10	.800		79.50	27.50		107	130	
	1100	4′-7″ long			10	.800		107	27.50		134.50	161	
	1500	Each 1′-6″ wide, 3′-3″ long			10	.800		84.50	27.50		112	136	
	1600	4′-7″ long			10	.800		119	27.50		146.50	173	
	1620	Hemlock, louvered, 1′-2″ wide, 5′-7″ long			10	.800		130	27.50		157.50	186	
	1630	Each 1′-4″ wide, 2′-2″ long			10	.800		81	27.50		108.50	132	
	1670	4′-3″ long			10	.800		97	27.50		124.50	150	
	1690	5′-11″ long			10	.800		137	27.50		164.50	194	
	1700	Door blinds, 6′-9″ long, each 1′-3″ wide			9	.889		138	30.50		168.50	199	
	1710	1′-6″ wide			9	.889		148	30.50		178.50	211	
	1720	Hemlock, solid raised panel, each 1′-4″ wide, 3′-3″ long			10	.800		131	27.50		158.50	187	
	1740	4′-3″ long			10	.800		166	27.50		193.50	226	
	1770	5′-11″ long			10	.800		222	27.50		249.50	287	
	1800	Door blinds, 6′-9″ long, each 1′-3″ wide			9	.889		250	30.50		280.50	325	
	1900	1′-6″ wide			9	.889		272	30.50		302.50	350	
	2500	Polystyrene, solid raised panel, each 1′-4″ wide, 3′-3″ long			10	.800		55	27.50		82.50	103	
	2700	4′-7″ long			10	.800		66.50	27.50		94	116	
	4500	Polystyrene, louvered, each 1′-2″ wide, 3′-3″ long			10	.800		40	27.50		67.50	86.50	
	4750	5′-3″ long			10	.800		53	27.50		80.50	101	
	6000	Vinyl, louvered, each 1′-2″ x 4′-7″ long			10	.800		52	27.50		79.50	99.50	
	6200	Each 1′-4″ x 6′-8″ long			9	.889		79	30.50		109.50	135	
200	0010	**SHUTTERS, INTERIOR** Wood, louvered,											200
	0200	Two panel, 27″ wide, 36″ high		1 Carp	5	1.600	Set	105	55		160	202	
	0300	33″ wide, 36″ high			5	1.600		125	55		180	224	
	0500	47″ wide, 36″ high			5	1.600		182	55		237	286	
	1000	Four panel, 27″ wide, 36″ high			5	1.600		241	55		296	350	
	1100	33″ wide, 36″ high			5	1.600		241	55		296	350	
	1300	47″ wide, 36″ high			5	1.600		320	55		375	440	

WOOD & PLASTICS 6

					DAILY	LABOR-		UNIT	2005 BARE COSTS				TOTAL
		06510	**Struct Plastic Shapes & Plates**	CREW	OUTPUT	HOURS		UNIT	MAT.	LABOR	EQUIP.	TOTAL	INCL O&P
400	0010		**CASTINGS, FIBERGLASS**											**400**
	0100		Angle, 1" x 1" x 1/8" thick	2 Sswk	240	.067		L.F.	1.05	2.54		3.59	5.70
	0120		3" x 3" x 1/4" thick		200	.080			4.37	3.05		7.42	10.25
	0140		4" x 4" x 1/4" thick		200	.080			5.60	3.05		8.65	11.65
	0160		4" x 4" x 3/8" thick		200	.080			8.75	3.05		11.80	15.10
	0180		6" x 6" x 1/2" thick		160	.100	↓		17.15	3.82		20.97	25.50
	1000		Flat sheet, 1/8" thick		140	.114		S.F.	4.46	4.36		8.82	12.70
	1020		1/4" thick		120	.133			8.55	5.10		13.65	18.55
	1040		3/8" thick		100	.160			14.85	6.10		20.95	27.50
	1060		1/2" thick		80	.200	↓		22	7.65		29.65	37.50
	2000		Handrail, 42" high, 2" diam. rails pickets 5' O.C.		32	.500		L.F.	43.50	19.10		62.60	82
	3000		Round bar, 1/4" diam.		240	.067			.48	2.54		3.02	5.10
	3020		1/2" diam.		200	.080			1.15	3.05		4.20	6.70
	3040		3/4" diam.		200	.080			1.67	3.05		4.72	7.30
	3060		1" diam.		160	.100			2.97	3.82		6.79	10.10
	3080		1-1/4" diam.		160	.100			3.52	3.82		7.34	10.70
	3100		1-1/2" diam.		140	.114			4.31	4.36		8.67	12.55
	3500		Round tube, 1" diam. x 1/8" thick		240	.067			1.90	2.54		4.44	6.65
	3520		2" diam. x 1/4" thick		200	.080			4.67	3.05		7.72	10.60
	3540		3" diam. x 1/4" thick		160	.100			9.05	3.82		12.87	16.85
	4000		Square bar, 1/2" square		240	.067			3.68	2.54		6.22	8.60
	4020		1" square		200	.080			3.90	3.05		6.95	9.75
	4040		1-1/2" square		160	.100			7.20	3.82		11.02	14.80
	4500		Square tube, 1" x 1" x 1/8" thick		240	.067			1.69	2.54		4.23	6.40
	4520		2" x 2" x 1/8" thick		200	.080			3.19	3.05		6.24	8.95
	4540		3" x 3" x 1/4" thick		160	.100			8.70	3.82		12.52	16.40
	5000		Threaded rod, 3/8" diam.		320	.050			2.94	1.91		4.85	6.65
	5020		1/2" diam.		320	.050			3.46	1.91		5.37	7.20
	5040		5/8" diam.		280	.057			3.83	2.18		6.01	8.10
	5060		3/4" diam.		280	.057			4.36	2.18		6.54	8.70
	6000		Wide flange beam, 4" x 4" x 1/4" thick		120	.133			7.85	5.10		12.95	17.75
	6020		6" x 6" x 1/4" thick		100	.160			14.25	6.10		20.35	26.50
	6040		8" x 8" x 3/8" thick	↓	80	.200	↓		27	7.65		34.65	43.50

		06520	**Plastic Struct Assemblies**										
100	0010		**STAIR TREAD, FIBERGLASS** Isophthalic Resin, 10-1/2" deep											**100**
	0100		24" wide	2 Sswk	52	.308		Ea.	34	11.75		45.75	58.50
	0140		30" wide		52	.308			40.50	11.75		52.25	65.50
	0180		36" wide		52	.308			47.50	11.75		59.25	73
	0220		42" wide	↓	52	.308	↓		56	11.75		67.75	82.50
150	0010		**GRATING, FIBERGLASS**											**150**
	0100		Molded, green (for mod. corrosive environment)										
	0140		1" x 4" mesh, 1" thick	2 Sswk	400	.040		S.F.	10.50	1.53		12.03	14.30
	0180		1-1/2" square mesh, 1" thick		400	.040			15.60	1.53		17.13	19.90
	0220		1-1/4" thick		400	.040			12.45	1.53		13.98	16.45
	0260		1-1/2" thick		400	.040			22.50	1.53		24.03	27.50
	0300		2" square mesh, 2" thick	↓	320	.050	↓		21.50	1.91		23.41	27
	1000		Orange (for highly corrosive environment)										
	1040		1" x 4" mesh, 1" thick	2 Sswk	400	.040		S.F.	13.20	1.53		14.73	17.25
	1080		1-1/2" square mesh, 1" thick		400	.040			17.45	1.53		18.98	22
	1120		1-1/4" thick		400	.040			18.15	1.53		19.68	22.50
	1160		1-1/2" thick		400	.040			19.60	1.53		21.13	24
	1200		2" square mesh, 2" thick	↓	320	.050	↓		23.50	1.91		25.41	29
	3000		Pultruded, green (for mod. corrosive environment)										

Important: See the Reference Section for critical supporting data - Reference Nos., Crews, & City Cost Indexes

06520	Plastic Struct Assemblies	CREW	DAILY OUTPUT	LABOR-HOURS	UNIT	2005 BARE COSTS				TOTAL INCL O&P	
						MAT.	LABOR	EQUIP.	TOTAL		
150	3040	1" O.C. bar spacing, 1" thick	2 Sswk	400	.040	S.F.	15.60	1.53		17.13	19.90	**150**
	3080	1-1/2" thick		320	.050		17.35	1.91		19.26	22.50	
	3120	1-1/2" O.C. bar spacing, 1" thick		400	.040		12.70	1.53		14.23	16.70	
	3160	1-1/2" thick		400	.040		14.20	1.53		15.73	18.35	
	4000	Grating support legs, fixed height, no base				Ea.	41			41	45	
	4040	With base					36.50			36.50	40	
	4080	Adjustable to 60"					55.50			55.50	61	
175	0010	**FLOOR GRATING, FIBERGLASS**										**175**
	0100	Reinforced polyester, fire retardant, 1" x 4" grid, 1" thick	E-4	510	.063	S.F.	13.35	2.43	.16	15.94	19.20	
	0200	1-1/2" x 6" mesh, 1-1/2" thick		500	.064		15.65	2.47	.16	18.28	22	
	0300	With grit surface, 1-1/2" x 6" grid, 1-1/2" thick		500	.064		15.95	2.47	.16	18.58	22	

06620	Non-Structural Plastics	CREW	DAILY OUTPUT	LABOR-HOURS	UNIT	2005 BARE COSTS				TOTAL INCL O&P	
						MAT.	LABOR	EQUIP.	TOTAL		
600	0010	**NETTING, FLEXIBLE PLASTIC** 1/8" square mesh	4 Clab	4,000	.008	S.F.	.03	.21		.24	.36	**600**
	0100	1/4" square mesh		4,000	.008		.04	.21		.25	.37	
	0120	1/2" square mesh		4,000	.008		.04	.21		.25	.37	
	0140	5/8" x 3/4" mesh		4,000	.008		.04	.21		.25	.37	
	0160	1-1/4" x 1-1/2" mesh		4,000	.008		.05	.21		.26	.39	
	0200	4" square mesh		4,000	.008		.06	.21		.27	.40	
	1000	Poly clips				Ea.	.01			.01	.01	
810	0010	**SOLID SURFACE COUNTERTOPS**, Acrylic polymer										**810**
	0020	Pricing for orders of 100 L.F. or greater										
	0100	25" wide, solid colors	2 Carp	28	.571	L.F.	43.50	19.55		63.05	78.50	
	0200	Patterned colors		28	.571		55.50	19.55		75.05	91.50	
	0300	Premium patterned colors		28	.571		69.50	19.55		89.05	107	
	0400	With silicone attached 4" backsplash, solid colors		27	.593		48	20.50		68.50	84.50	
	0500	Patterned colors		27	.593		61	20.50		81.50	98.50	
	0600	Premium patterned colors		27	.593		75.50	20.50		96	115	
	0700	With hard seam attached 4" backsplash, solid colors		23	.696		48	24		72	90	
	0800	Patterned colors		23	.696		61	24		85	104	
	0900	Premium patterned colors		23	.696		75.50	24		99.50	121	
	1000	Pricing for order of 51 - 99 L.F.										
	1100	25" wide, solid colors	2 Carp	24	.667	L.F.	50.50	23		73.50	91	
	1200	Patterned colors		24	.667		64	23		87	106	
	1300	Premium patterned colors		24	.667		79.50	23		102.50	123	
	1400	With silicone attached 4" backsplash, solid colors		23	.696		55	24		79	97.50	
	1500	Patterned colors		23	.696		70	24		94	114	
	1600	Premium patterned colors		23	.696		87	24		111	133	
	1700	With hard seam attached 4" backsplash, solid colors		20	.800		55	27.50		82.50	103	
	1800	Patterned colors		20	.800		70	27.50		97.50	120	
	1900	Premium patterned colors		20	.800		87	27.50		114.50	139	
	2000	Pricing for order of 1 - 50 L.F.										
	2100	25" wide, solid colors	2 Carp	20	.800	L.F.	59	27.50		86.50	108	
	2200	Patterned colors		20	.800		75	27.50		102.50	125	
	2300	Premium patterned colors		20	.800		93.50	27.50		121	146	
	2400	With silicone attached 4" backsplash, solid colors		19	.842		65	29		94	117	

WOOD & PLASTICS 6

For expanded coverage of these items see *Means Interior Cost Data 2005*

			DAILY OUTPUT	LABOR-HOURS	UNIT	2005 BARE COSTS				TOTAL INCL O&P	
06620		**Non-Structural Plastics**	CREW			MAT.	LABOR	EQUIP.	TOTAL		
810	2500	Patterned colors	2 Carp	19	.842	L.F.	82	29		111	136	810
	2600	Premium patterned colors		19	.842		102	29		131	157	
	2700	With hard seam attached 4" backsplash, solid colors		4	4		65	137		202	285	
	2800	Patterned colors		15	1.067		82	36.50		118.50	148	
	2900	Premium patterned colors		15	1.067		102	36.50		138.50	169	
	3000	Sinks, pricing for order of 100 or greater units										
	3100	Single bowl, hard seamed, solid colors, 13" x 17"	1 Carp	3	2.667	Ea.	295	91.50		386.50	465	
	3200	10" x 15"		7	1.143		137	39		176	211	
	3300	Cutouts for sinks		8	1			34.50		34.50	53.50	
	3400	Sinks, pricing for order of 51 - 99 units										
	3500	Single bowl, hard seamed, solid colors, 13" x 17"	1 Carp	2.55	3.137	Ea.	340	107		447	540	
	3600	10" x 15"		6	1.333		157	45.50		202.50	244	
	3700	Cutouts for sinks		7	1.143			39		39	61	
	3800	Sinks, pricing for order of 1 - 50 units										
	3900	Single bowl, hard seamed, solid colors, 13" x 17"	1 Carp	2	4	Ea.	400	137		537	655	
	4000	10" x 15"		4.55	1.758		184	60		244	297	
	4100	Cutouts for sinks		5.25	1.524			52		52	81.50	
	4200	Cooktop cutouts, pricing for 100 or greater units		4	2		21.50	68.50		90	131	
	4300	51 - 99 units		3.40	2.353		25	80.50		105.50	154	
	4400	1 - 50 units		3	2.667		29	91.50		120.50	174	
850	0010	**VANITY TOPS**										850
	0015	Solid surface, center bowl, 17" x 19"	1 Carp	12	.667	Ea.	168	23		191	221	
	0020	19" x 25"		12	.667		204	23		227	260	
	0030	19" x 31"		12	.667		247	23		270	310	
	0040	19" x 37"		12	.667		288	23		311	350	
	0050	22" x 25"		10	.800		229	27.50		256.50	295	
	0060	22" x 31"		10	.800		267	27.50		294.50	335	
	0070	22" x 37"		10	.800		310	27.50		337.50	385	
	0080	22" x 43"		10	.800		355	27.50		382.50	435	
	0090	22" x 49"		10	.800		395	27.50		422.50	475	
	0110	22" x 55"		8	1		445	34.50		479.50	545	
	0120	22" x 61"		8	1		510	34.50		544.50	615	
	0220	Double bowl, 22" x 61"		8	1		575	34.50		609.50	685	
	0230	Double bowl, 22" x 73"		8	1		625	34.50		659.50	745	
	0240	For aggregate colors, add					35%					
	0250	For faucets and fittings see 15410-300										

For information about Means Estimating Seminars, see yellow pages 12 and 13 in back of book

Important: See the Reference Section for critical supporting data - Reference Nos., Crews, & City Cost Indexes

Division 7
Thermal & Moisture Protection

Estimating Tips

07100 Dampproofing & Waterproofing

- Be sure of the job specifications before pricing this subdivision. The difference in cost between waterproofing and dampproofing can be great. Waterproofing will hold back standing water. Dampproofing prevents the transmission of water vapor. Also included in this section are vapor retarding membranes.

07200 Thermal Protection

- Insulation and fireproofing products are measured by area, thickness, volume or R value. Specifications may only give what the specific R value should be in a certain situation. The estimator may need to choose the type of insulation to meet that R value.

07300 Shingles, Roof Tiles & Roof Coverings
07400 Roofing & Siding Panels

- Many roofing and siding products are bought and sold by the square. One square is equal to an area that measures 100 square feet.

This simple change in unit of measure could create a large error if the estimator is not observant. Accessories necessary for a complete installation must be figured into any calculations for both material and labor.

07500 Membrane Roofing
07600 Flashing & Sheet Metal
07700 Roof Specialties & Accessories

- The items in these subdivisions compose a roofing system. No one component completes the installation and all must be estimated. Built-up or single ply membrane roofing systems are made up of many products and installation trades. Wood blocking at roof perimeters or penetrations, parapet coverings, reglets, roof drains, gutters, downspouts, sheet metal flashing, skylights, smoke vents or roof hatches all need to be considered along with the roofing material. Several different installation trades will need to work together on the roofing system. Inherent difficulties in the scheduling and coordination of various trades must be accounted for when estimating labor costs.

07900 Joint Sealers

- To complete the weather-tight shell the sealants and caulkings must be estimated. Where different materials meet—at expansion joints, at flashing penetrations, and at hundreds of other locations throughout a construction project—they provide another line of defense against water penetration. Often, an entire system is based on the proper location and placement of caulking or sealants. The detail drawings that are included as part of a set of architectural plans, show typical locations for these materials. When caulking or sealants are shown at typical locations, this means the estimator must include them for all the locations where this detail is applicable. Be careful to keep different types of sealants separate, and remember to consider backer rods and primers if necessary.

Reference Numbers

Reference numbers are shown in bold squares at the beginning of some major classifications. These numbers refer to related items in the Reference Section. The reference information may be an estimating procedure, an alternate pricing method or technical information.

Note: Not all subdivisions listed here necessarily appear in this publication.

07060 | Selective Demolition

		CREW	DAILY OUTPUT	LABOR-HOURS	UNIT	2005 BARE COSTS				TOTAL INCL O&P
						MAT.	LABOR	EQUIP.	TOTAL	
110	0010 **SELECTIVE DEMOLITION, ROOFING AND SIDING** R02220-510									110
1000	Deck, roof, concrete plank	B-13	1,680	.033	S.F.		.96	.37	1.33	1.88
1100	Gypsum plank		3,900	.014			.41	.16	.57	.81
1150	Metal decking		3,500	.016			.46	.18	.64	.90
1200	Wood, boards, tongue and groove, 2" x 6"	2 Clab	960	.017			.45		.45	.69
1220	2" x 10"		1,040	.015			.41		.41	.64
1280	Standard planks, 1" x 6"		1,080	.015			.40		.40	.62
1320	1" x 8"		1,160	.014			.37		.37	.57
1340	1" x 12"		1,200	.013			.36		.36	.55
1350	Plywood, to 1" thick		2,000	.008			.21		.21	.33
2000	Gutters, aluminum or wood, edge hung	1 Clab	240	.033	L.F.		.89		.89	1.38
2100	Built-in		100	.080	"		2.14		2.14	3.32
2500	Roof accessories, plumbing vent flashing		14	.571	Ea.		15.25		15.25	23.50
2600	Adjustable metal chimney flashing		9	.889	"		23.50		23.50	37
2650	Coping, sheet metal, up to 12" wide		240	.033	L.F.		.89		.89	1.38
2660	Concrete, up to 12" wide	2 Clab	160	.100	"		2.67		2.67	4.16
3000	Roofing, built-up, 5 ply roof, no gravel	B-2	1,600	.025	S.F.		.68		.68	1.05
3001	Including gravel		890	.045			1.22		1.22	1.90
3100	Gravel removal, minimum		5,000	.008			.22		.22	.34
3120	Maximum		2,000	.020			.54		.54	.84
3400	Roof insulation board, up to 2" thick		3,900	.010			.28		.28	.43
3450	Roll roofing, cold adhesive	1 Clab	12	.667	Sq.		17.80		17.80	27.50
4000	Shingles, asphalt strip, 1 layer	B-2	3,500	.011	S.F.		.31		.31	.48
4100	Slate		2,500	.016			.43		.43	.67
4300	Wood		2,200	.018			.49		.49	.77
4500	Skylight to 10 S.F.	1 Clab	8	1	Ea.		26.50		26.50	41.50
5000	Siding, metal, horizontal		444	.018	S.F.		.48		.48	.75
5020	Vertical		400	.020			.53		.53	.83
5200	Wood, boards, vertical		400	.020			.53		.53	.83
5220	Clapboards, horizontal		380	.021			.56		.56	.87
5240	Shingles		350	.023			.61		.61	.95
5260	Textured plywood		725	.011			.29		.29	.46

07100 | Dampproofing and Waterproofing

07110 | Dampproofing

		CREW	DAILY OUTPUT	LABOR-HOURS	UNIT	2005 BARE COSTS				TOTAL INCL O&P
						MAT.	LABOR	EQUIP.	TOTAL	
100	0010 **BITUMINOUS ASPHALT COATING** For foundation									100
0030	Brushed on, below grade, 1 coat	1 Rofc	665	.012	S.F.	.07	.35		.42	.68
0100	2 coat		500	.016		.10	.47		.57	.91
0300	Sprayed on, below grade, 1 coat, 25.6 S.F./gal.		830	.010		.07	.28		.35	.56
0400	2 coat, 20.5 S.F./gal.		500	.016		.14	.47		.61	.96
0500	Asphalt coating, with fibers				Gal.	3.74			3.74	4.11
0600	Troweled on, asphalt with fibers, 1/16" thick	1 Rofc	500	.016	S.F.	.16	.47		.63	.98
0700	1/8" thick		400	.020		.29	.59		.88	1.32
1000	1/2" thick		350	.023		.94	.67		1.61	2.18
200	0010 **CEMENT PARGING** 2 coats, 1/2" thick, regular P.C. R07110-010	D-1	250	.064	S.F.	.17	1.99		2.16	3.22	200
0100	Waterproofed Portland cement	"	250	.064	"	.18	1.99		2.17	3.23

Important: See the Reference Section for critical supporting data - Reference Nos., Crews, & City Cost Indexes

7 THERMAL & MOISTURE PROTECTION

07130 | Sheet Waterproofing

		CREW	DAILY OUTPUT	LABOR-HOURS	UNIT	2005 BARE COSTS MAT.	LABOR	EQUIP.	TOTAL	TOTAL INCL O&P	
200	0010	**ELASTOMERIC WATERPROOFING**									**200**
	0050	Acrylic rubber, fluid applied, 20 mils thick	3 Rofc	1,000	.024	S.F.	1.59	.71		2.30	2.95	
	0060	50 mil, reinforced, stucco texture	"	600	.040		2.55	1.18		3.73	4.82	
	0090	EPDM, plain, 45 mils thick	2 Rofc	580	.028		.85	.81		1.66	2.33	
	0100	60 mils thick		570	.028		.92	.83		1.75	2.42	
	0300	Nylon reinforced sheets, 45 mils thick		580	.028		1.32	.81		2.13	2.84	
	0400	60 mils thick	▼	570	.028	▼	1.68	.83		2.51	3.26	
	0600	Vulcanizing splicing tape for above, 2" wide				C.L.F.	33.50			33.50	36.50	
	0700	4" wide				"	67			67	73.50	
	0900	Adhesive, bonding, 60 SF per gal				Gal.	13.60			13.60	14.95	
	1000	Splicing, 75 SF per gal				"	21			21	23	
	1200	Neoprene sheets, plain, 45 mils thick	2 Rofc	580	.028	S.F.	.93	.81		1.74	2.41	
	1300	60 mils thick		570	.028		1.48	.83		2.31	3.04	
	1500	Nylon reinforced, 45 mils thick		580	.028		1.20	.81		2.01	2.71	
	1600	60 mils thick		570	.028		1.36	.83		2.19	2.91	
	1800	120 mils thick	▼	500	.032	▼	2.25	.94		3.19	4.09	
	1900	Adhesive, splicing, 150 S.F. per gal. per coat				Gal.	18.15			18.15	19.95	
	2100	Fiberglass reinforced, fluid applied, 1/8" thick	2 Rofc	500	.032	S.F.	1.50	.94		2.44	3.26	
	2200	Polyethylene and rubberized asphalt sheets, 1/8" thick		550	.029		.54	.86		1.40	2.05	
	2210	Asphaltic hardboard protection board, 1/8" thick		500	.032		.30	.94		1.24	1.94	
	2220	Asphaltic hardboard protection board, 1/4" thick		450	.036		.52	1.05		1.57	2.36	
	2400	Polyvinyl chloride sheets, plain, 10 mils thick		580	.028		.14	.81		.95	1.54	
	2500	20 mils thick		570	.028		.22	.83		1.05	1.65	
	2700	30 mils thick	▼	560	.029	▼	.30	.84		1.14	1.76	
	3000	Adhesives, trowel grade, 40-100 SF per gal				Gal.	20			20	22	
	3100	Brush grade, 100-250 SF per gal.				"	20			20	22	
	3300	Bitumen modified polyurethane, fluid applied, 55 mils thick	2 Rofc	665	.024	S.F.	.65	.71		1.36	1.93	
	3600	Vinyl plastic, sprayed on, 25 to 40 mils thick	"	475	.034	"	.98	.99		1.97	2.77	
500	0010	**MEMBRANE WATERPROOFING** On slabs, 1 ply, felt	G-1	3,000	.019	S.F.	.19	.51	.10	.80	1.19	**500**
	0015	Membrane waterproofing on walls, 1 ply, felt, mopped		3,000	.019		.19	.51	.10	.80	1.19	
	0300	2 ply, felt		2,500	.022		.38	.62	.12	1.12	1.60	
	0400	Glass fiber fabric		1,650	.034		.44	.93	.18	1.55	2.28	
	0600	3 ply, felt		2,100	.027		.57	.73	.14	1.44	2.04	
	0700	Glass fiber fabric	▼	1,550	.036		.59	.99	.19	1.77	2.55	
	0900	For installation on walls, add						15%				
	1050	3/8" thick, add	2 Rofc	3,500	.005		.16	.13		.29	.41	
	1060	1/2" thick, add		3,500	.005	▼	.18	.13		.31	.43	
	1070	Fiberglass fabric, black, 20/10 mesh		116	.138	Sq.	9.95	4.06		14.01	17.85	
	1080	White, 20/10 mesh		116	.138	"	10.25	4.06		14.31	18.20	
	1100	1/16" urethane, troweled		200	.080	S.F.	.66	2.36		3.02	4.75	
	1200	Roller applied	▼	120	.133	"	.60	3.93		4.53	7.35	

07160 | Cement. & Reactive Waterproofing

		CREW	DAILY OUTPUT	LABOR-HOURS	UNIT	MAT.	LABOR	EQUIP.	TOTAL	TOTAL INCL O&P	
150	0010	**CEMENTITIOUS WATERPROOFING** One coat cement base										**150**
	0020	1/8" application, sprayed on	G-2	1,000	.024	S.F.	1.50	.68	.11	2.29	2.81	
	0030	2 coat, cementitious/metallic slurry, troweled, 1/4" thick	1 Cefi	2.48	3.226	C.S.F.	35	106		141	195	
	0040	3 coat, 3/8" thick		1.84	4.348		52.50	143		195.50	268	
	0050	4 coat, 1/2" thick	▼	1.20	6.667	▼	70	219		289	395	

07170 | Bentonite Waterproofing

		CREW	DAILY OUTPUT	LABOR-HOURS	UNIT	MAT.	LABOR	EQUIP.	TOTAL	TOTAL INCL O&P	
700	0010	**BENTONITE**, Panels, 4' x 4', 3/16" thick	1 Rofc	625	.013	S.F.	.74	.38		1.12	1.45	**700**
	0100	Rolls, 3/8" thick, with geotextile fabric both sides	"	550	.015	"	.94	.43		1.37	1.76	
	0300	Granular bentonite, 50 lb. bags (.625 C.F.)				Bag	18.65			18.65	20.50	
	0400	3/8" thick, troweled on	1 Rofc	475	.017	S.F.	.93	.50		1.43	1.88	

07170 | Bentonite Waterproofing

			CREW	DAILY OUTPUT	LABOR-HOURS	UNIT	2005 BARE COSTS				TOTAL INCL O&P	
							MAT.	LABOR	EQUIP.	TOTAL		
700	0500	Drain board, expanded polystyrene, binder encapsulated, 1-1/2" thick	1 Rohe	1,600	.005	S.F.	1.04	.11		1.15	1.32	700
	0510	2" thick		1,600	.005		1.38	.11		1.49	1.70	
	0520	3" thick		1,600	.005		2.07	.11		2.18	2.46	
	0530	4" thick		1,600	.005		2.75	.11		2.86	3.21	
	0600	With filter fabric, 1-1/2" thick		1,600	.005		1.25	.11		1.36	1.56	
	0625	2" thick		1,600	.005		1.65	.11		1.76	2	
	0650	3" thick		1,600	.005		2.34	.11		2.45	2.75	
	0675	4" thick		1,600	.005		3.03	.11		3.14	3.51	
	0700	Vapor retarder, see polyethelene, 07260-100										

07190 | Water Repellents

			CREW	DAILY OUTPUT	LABOR-HOURS	UNIT	MAT.	LABOR	EQUIP.	TOTAL	TOTAL INCL O&P	
700	0010	RUBBER COATING Water base liquid, roller applied	2 Rofc	7,000	.002	S.F.	.55	.07		.62	.73	700
	0200	Silicone or stearate, sprayed on CMU, 1 coat	1 Rofc	4,000	.002		.31	.06		.37	.44	
	0300	2 coats	"	3,000	.003		.63	.08		.71	.82	

07210 | Building Insulation

			CREW	DAILY OUTPUT	LABOR-HOURS	UNIT	2005 BARE COSTS				TOTAL INCL O&P	
							MAT.	LABOR	EQUIP.	TOTAL		
150	0010	BLOWN-IN INSULATION Ceilings, with open access										150
	0020	Cellulose, 3-1/2" thick, R13	G-4	5,000	.005	S.F.	.13	.13	.04	.30	.39	
	0030	5-3/16" thick, R19		3,800	.006		.21	.17	.06	.44	.56	
	0050	6-1/2" thick, R22		3,000	.008		.26	.22	.07	.55	.71	
	0100	8-11/16" thick, R30		2,600	.009		.34	.25	.08	.67	.85	
	0120	10-7/8" thick, R38		1,800	.013		.43	.36	.12	.91	1.17	
	1000	Fiberglass, 5" thick, R11		3,800	.006		.16	.17	.06	.39	.51	
	1050	6" thick, R13		3,000	.008		.22	.22	.07	.51	.66	
	1100	8-1/2" thick, R19		2,200	.011		.31	.30	.10	.71	.91	
	1200	10" thick, R22		1,800	.013		.36	.36	.12	.84	1.10	
	1300	12" thick, R26		1,500	.016		.43	.44	.14	1.01	1.31	
	2000	Mineral wool, 4" thick, R12		3,500	.007		.18	.19	.06	.43	.56	
	2050	6" thick, R17		2,500	.010		.20	.26	.09	.55	.73	
	2100	9" thick, R23		1,750	.014		.29	.38	.12	.79	1.04	
	2500	Wall installation, incl. drilling & patching from outside, two 1"										
	2510	diam. holes @ 16" O.C., top & mid-point of wall, add to above										
	2700	For masonry	G-4	415	.058	S.F.	.06	1.58	.52	2.16	3.10	
	2800	For wood siding		840	.029		.06	.78	.26	1.10	1.57	
	2900	For stucco/plaster		665	.036		.06	.99	.33	1.38	1.97	
400	0010	INSULATION FASTENERS										400
	0050	Clips, galv, 1" or 2" insulation				C	17.50			17.50	19.25	
	0100	6" insulation					30			30	33	
	0300	Screws with plates, galv, for 1" or 2" insulation					15.75			15.75	17.35	
	0500	For 6" insulation					28.50			28.50	31.50	
500	0010	POURED INSULATION Cellulose fiber, R3.8 per inch	1 Carp	200	.040	C.F.	.48	1.37		1.85	2.66	500
	0040	Ceramic type (perlite), R3.2 per inch		200	.040		1.65	1.37		3.02	3.95	
	0080	Fiberglass wool, R4 per inch		200	.040		.39	1.37		1.76	2.56	
	0100	Mineral wool, R3 per inch		200	.040		.33	1.37		1.70	2.49	

Important: See the Reference Section for critical supporting data - Reference Nos., Crews, & City Cost Indexes

07210 | Building Insulation

		CREW	DAILY OUTPUT	LABOR-HOURS	UNIT	2005 BARE COSTS				TOTAL INCL O&P	
						MAT.	LABOR	EQUIP.	TOTAL		
500	0300	Polystyrene, R4 per inch	1 Carp	200	.040	C.F.	2.23	1.37		3.60	4.58	**500**
	0400	Vermiculite or perlite, R2.7 per inch		200	.040		1.65	1.37		3.02	3.95	
	0700	Wood fiber, R3.85 per inch	↓	200	.040	↓	.61	1.37		1.98	2.80	
550	0010	**MASONRY INSULATION** Vermiculite or perlite, poured										**550**
	0100	In cores of concrete block, 4" thick wall, .115 CF/SF	D-1	4,800	.003	S.F.	.19	.10		.29	.37	
	0200	6" thick wall, .175 CF/SF		3,000	.005		.29	.17		.46	.57	
	0300	8" thick wall, .258 CF/SF		2,400	.007		.43	.21		.64	.79	
	0400	10" thick wall, .340 CF/SF		1,850	.009		.56	.27		.83	1.03	
	0500	12" thick wall, .422 CF/SF	↓	1,200	.013	↓	.70	.41		1.11	1.40	
	0550	For sand fill, deduct from above					70%					
	0600	Poured cavity wall, vermiculite or perlite, water repellant	D-1	250	.064	C.F.	1.65	1.99		3.64	4.85	
	0700	Foamed in place, urethane in 2-5/8" cavity	G-2	1,035	.023	S.F.	.40	.66	.10	1.16	1.56	
	0800	For each 1" added thickness, add	"	2,372	.010	"	.12	.29	.05	.46	.62	
600	0010	**PERIMETER INSULATION**										**600**
	0600	Polystyrene, expanded, 1" thick, R4	1 Carp	680	.012	S.F.	.19	.40		.59	.84	
	0700	2" thick, R8	"	675	.012	"	.41	.41		.82	1.08	
700	0010	**REFLECTIVE INSULATION**, aluminum foil on reinforced scrim	1 Carp	19	.421	C.S.F.	13.90	14.40		28.30	38	**700**
	0100	Reinforced with woven polyolefin		19	.421		16.90	14.40		31.30	41	
	0500	With single bubble air space, R8.8		15	.533	↓	27	18.25		45.25	58	
	0600	With double bubble air space, R9.8	↓	15	.533		29	18.25		47.25	60.50	
800	0010	**SPRAYED** Fibrous/cementitious, finished wall, 1" thick, R3.7	G-2	2,050	.012	S.F.	.22	.33	.05	.60	.81	**800**
	0100	Attic, 5.2" thick, R19	"	1,550	.015	"	.36	.44	.07	.87	1.15	
	0300	Foam type, incl. preparation										
	0600	3 #/CF, 1" thick, R3.8	G-2	770	.031	S.F.	.49	.88	.14	1.51	2.05	
	0700	2" thick, R7.5	"	475	.051	"	.98	1.43	.23	2.64	3.53	
900	0010	**WALL INSULATION, RIGID**										**900**
	0040	Fiberglass, 1.5#/CF, unfaced, 1" thick, R4.1	1 Carp	1,000	.008	S.F.	.31	.27		.58	.77	
	0060	1-1/2" thick, R6.2		1,000	.008		.41	.27		.68	.88	
	0080	2" thick, R8.3		1,000	.008		.47	.27		.74	.95	
	0120	3" thick, R12.4		800	.010		.57	.34		.91	1.16	
	0370	3#/CF, unfaced, 1" thick, R4.3		1,000	.008		.36	.27		.63	.83	
	0390	1-1/2" thick, R6.5		1,000	.008		.69	.27		.96	1.19	
	0400	2" thick, R8.7		890	.009		.83	.31		1.14	1.39	
	0420	2-1/2" thick, R10.9		800	.010		1.02	.34		1.36	1.65	
	0440	3" thick, R13		800	.010		1.21	.34		1.55	1.86	
	0520	Foil faced, 1" thick, R4.3		1,000	.008		.80	.27		1.07	1.31	
	0540	1-1/2" thick, R6.5		1,000	.008		1.08	.27		1.35	1.62	
	0560	2" thick, R8.7		890	.009		1.35	.31		1.66	1.97	
	0580	2-1/2" thick, R10.9		800	.010		1.60	.34		1.94	2.29	
	0600	3" thick, R13		800	.010		1.74	.34		2.08	2.44	
	0670	6#/CF, unfaced, 1" thick, R4.3		1,000	.008		.77	.27		1.04	1.28	
	0690	1-1/2" thick, R6.5		890	.009		1.19	.31		1.50	1.79	
	0700	2" thick, R8.7		800	.010		1.68	.34		2.02	2.38	
	0721	2-1/2" thick, R10.9		800	.010		1.84	.34		2.18	2.55	
	0741	3" thick, R13		730	.011		2.20	.38		2.58	3	
	0821	Foil faced, 1" thick, R4.3		1,000	.008		1.09	.27		1.36	1.63	
	0840	1-1/2" thick, R6.5		890	.009		1.57	.31		1.88	2.21	
	0850	2" thick, R8.7		800	.010		2.05	.34		2.39	2.79	
	0880	2-1/2" thick, R10.9		800	.010		2.46	.34		2.80	3.24	
	0900	3" thick, R13		730	.011		2.94	.38		3.32	3.81	
	1500	Foamglass, 1-1/2" thick, R4.5		800	.010		1.21	.34		1.55	1.86	
	1550	3" thick, R9	↓	730	.011	↓	2.92	.38		3.30	3.79	
	1600	Isocyanurate, 4' x 8' sheet, foil faced, both sides										

THERMAL & MOISTURE PROTECTION 7

07210 | Building Insulation

			CREW	DAILY OUTPUT	LABOR-HOURS	UNIT	2005 BARE COSTS MAT.	LABOR	EQUIP.	TOTAL	TOTAL INCL O&P	
900	1610	1/2" thick, R3.9	1 Carp	800	.010	S.F.	.29	.34		.63	.85	900
	1620	5/8" thick, R4.5		800	.010		.30	.34		.64	.86	
	1630	3/4" thick, R5.4		800	.010		.24	.34		.58	.79	
	1640	1" thick, R7.2		800	.010		.33	.34		.67	.89	
	1650	1-1/2" thick, R10.8		730	.011		.35	.38		.73	.97	
	1660	2" thick, R14.4 **CN**		730	.011		.46	.38		.84	1.09	
	1670	3" thick, R21.6		730	.011		1.11	.38		1.49	1.80	
	1680	4" thick, R28.8		730	.011		1.37	.38		1.75	2.09	
	1700	Perlite, 1" thick, R2.77		800	.010		.26	.34		.60	.82	
	1750	2" thick, R5.55		730	.011		.50	.38		.88	1.13	
	1900	Extruded polystyrene, 25 PSI compressive strength, 1" thick, R5		800	.010		.38	.34		.72	.95	
	1940	2" thick R10		730	.011		.75	.38		1.13	1.41	
	1960	3" thick, R15		730	.011		1.04	.38		1.42	1.72	
	2100	Expanded polystyrene, 1" thick, R3.85		800	.010		.17	.34		.51	.72	
	2120	2" thick, R7.69		730	.011		.42	.38		.80	1.04	
	2140	3" thick, R11.49		730	.011		.52	.38		.90	1.15	
950	0010	**WALL OR CEILING INSUL., NON-RIGID**										950
	0040	Fiberglass, kraft faced, batts or blankets										
	0060	3-1/2" thick, R11, 11" wide	1 Carp	1,150	.007	S.F.	.27	.24		.51	.67	
	0080	15" wide **CN**		1,600	.005		.27	.17		.44	.57	
	0100	23" wide		1,600	.005		.27	.17		.44	.57	
	0140	6" thick, R19, 11" wide		1,000	.008		.35	.27		.62	.82	
	0160	15" wide		1,350	.006		.35	.20		.55	.71	
	0180	23" wide		1,600	.005		.35	.17		.52	.66	
	0200	9" thick, R30, 15" wide		1,150	.007		.60	.24		.84	1.03	
	0220	23" wide		1,350	.006		.60	.20		.80	.98	
	0240	12" thick, R38, 15" wide		1,000	.008		.76	.27		1.03	1.27	
	0260	23" wide		1,350	.006		.76	.20		.96	1.16	
	0400	Fiberglass, foil faced, batts or blankets										
	0420	3-1/2" thick, R11, 15" wide	1 Carp	1,600	.005	S.F.	.40	.17		.57	.71	
	0440	23" wide		1,600	.005		.40	.17		.57	.71	
	0460	6" thick, R19, 15" wide		1,350	.006		.45	.20		.65	.82	
	0480	23" wide		1,600	.005		.45	.17		.62	.77	
	0500	9" thick, R30, 15" wide		1,150	.007		.71	.24		.95	1.15	
	0550	23" wide		1,350	.006		.71	.20		.91	1.10	
	0800	Fiberglass, unfaced, batts or blankets										
	0820	3-1/2" thick, R11, 15" wide	1 Carp	1,350	.006	S.F.	.24	.20		.44	.58	
	0830	23" wide		1,600	.005		.24	.17		.41	.53	
	0860	6" thick, R19, 15" wide		1,150	.007		.46	.24		.70	.88	
	0880	23" wide		1,350	.006		.46	.20		.66	.83	
	0900	9" thick, R30, 15" wide		1,000	.008		.60	.27		.87	1.09	
	0920	23" wide		1,150	.007		.60	.24		.84	1.03	
	0940	12" thick, R38, 15" wide		1,000	.008		.76	.27		1.03	1.27	
	0960	23" wide		1,150	.007		.76	.24		1	1.21	
	1300	Mineral fiber batts, kraft faced										
	1320	3-1/2" thick, R12	1 Carp	1,600	.005	S.F.	.28	.17		.45	.58	
	1340	6" thick, R19		1,600	.005		.38	.17		.55	.69	
	1380	10" thick, R30		1,350	.006		.57	.20		.77	.95	
	1850	Friction fit wire insulation supports, 16" O.C.		960	.008	Ea.	.05	.29		.34	.50	
	1900	For foil backing, add				S.F.	.04			.04	.04	

07220 | Roof and Deck Insulation

			CREW	DAILY OUTPUT	LABOR-HOURS	UNIT	MAT.	LABOR	EQUIP.	TOTAL	TOTAL INCL O&P	
700	0010	**ROOF DECK INSULATION**										700
	0020	Fiberboard low density, 1/2" thick R1.39	1 Rofc	1,000	.008	S.F.	.19	.24		.43	.61	
	0030	1" thick R2.78		800	.010		.34	.29		.63	.87	
	0080	1 1/2" thick R4.17		800	.010		.50	.29		.79	1.05	

Important: See the Reference Section for critical supporting data - Reference Nos., Crews, & City Cost Indexes

7 THERMAL & MOISTURE PROTECTION

07220	Roof and Deck Insulation	CREW	DAILY OUTPUT	LABOR-HOURS	UNIT	2005 BARE COSTS				TOTAL INCL O&P	
						MAT.	LABOR	EQUIP.	TOTAL		
700											**700**
0100	2" thick R5.56	1 Rofc	800	.010	S.F.	.68	.29		.97	1.25	
0110	Fiberboard high density, 1/2" thick R1.3		1,000	.008		.20	.24		.44	.62	
0120	1" thick R2.5		800	.010		.36	.29		.65	.90	
0130	1-1/2" thick R3.8		800	.010		.59	.29		.88	1.15	
0200	Fiberglass, 3/4" thick R2.78		1,000	.008		.46	.24		.70	.91	
0400	15/16" thick R3.70		1,000	.008		.61	.24		.85	1.07	
0460	1-1/16" thick R4.17		1,000	.008		.76	.24		1	1.24	
0600	1-5/16" thick R5.26		1,000	.008		1.05	.24		1.29	1.56	
0650	2-1/16" thick R8.33		800	.010		1.12	.29		1.41	1.73	
0700	2-7/16" thick R10		800	.010		1.28	.29		1.57	1.91	
1500	Foamglass, 1-1/2" thick R4.5		800	.010		1.26	.29		1.55	1.89	
1530	3" thick R9		700	.011	▼	2.57	.34		2.91	3.40	
1600	Tapered for drainage		600	.013	B.F.	1.10	.39		1.49	1.88	
1650	Perlite, 1/2" thick R1.32		1,050	.008	S.F.	.27	.22		.49	.68	
1655	3/4" thick R2.08		800	.010		.29	.29		.58	.82	
1660	1" thick R2.78		800	.010		.31	.29		.60	.84	
1670	1-1/2" thick R4.17		800	.010		.39	.29		.68	.93	
1680	2" thick R5.56		700	.011		.62	.34		.96	1.25	
1685	2-1/2" thick R6.67		700	.011	▼	.77	.34		1.11	1.42	
1690	Tapered for drainage		800	.010	B.F.	.59	.29		.88	1.15	
1700	Polyisocyanurate, 2#/CF density, 3/4" thick, R5.1		1,500	.005	S.F.	.30	.16		.46	.60	
1705	1" thick R7.14		1,400	.006		.33	.17		.50	.65	
1715	1-1/2" thick R10.87		1,250	.006		.35	.19		.54	.71	
1725	2" thick R14.29		1,100	.007		.45	.21		.66	.87	
1735	2-1/2" thick R16.67		1,050	.008		.50	.22		.72	.93	
1745	3" thick R21.74		1,000	.008		.73	.24		.97	1.20	
1755	3-1/2" thick R25		1,000	.008	▼	.74	.24		.98	1.21	
1765	Tapered for drainage	▼	1,400	.006	B.F.	.38	.17		.55	.71	
1900	Extruded Polystyrene										
1910	15 PSI compressive strength, 1" thick, R5	1 Rofc	1,500	.005	S.F.	.23	.16		.39	.52	
1920	2" thick, R10		1,250	.006		.35	.19		.54	.71	
1930	3" thick R15		1,000	.008		.68	.24		.92	1.15	
1932	4" thick R20		1,000	.008	▼	1.07	.24		1.31	1.58	
1934	Tapered for drainage		1,500	.005	B.F.	.35	.16		.51	.66	
1940	25 PSI compressive strength, 1" thick R5		1,500	.005	S.F.	.50	.16		.66	.82	
1942	2" thick R10		1,250	.006		.97	.19		1.16	1.39	
1944	3" thick R15		1,000	.008		1.47	.24		1.71	2.02	
1946	4" thick R20		1,000	.008	▼	1.96	.24		2.20	2.56	
1948	Tapered for drainage		1,500	.005	B.F.	.40	.16		.56	.71	
1950	40 psi compressive strength, 1" thick R5		1,500	.005	S.F.	.35	.16		.51	.66	
1952	2" thick R10		1,250	.006		.69	.19		.88	1.08	
1954	3" thick R15		1,000	.008		1.01	.24		1.25	1.51	
1956	4" thick R20		1,000	.008	▼	1.35	.24		1.59	1.89	
1958	Tapered for drainage		1,400	.006	B.F.	.50	.17		.67	.84	
1960	60 PSI compressive strength, 1" thick R5		1,450	.006	S.F.	.42	.16		.58	.74	
1962	2" thick R10		1,200	.007		.75	.20		.95	1.16	
1964	3" thick R15		975	.008		1.12	.24		1.36	1.64	
1966	4" thick R20		950	.008	▼	1.56	.25		1.81	2.14	
1968	Tapered for drainage		1,400	.006	B.F.	.61	.17		.78	.96	
2010	Expanded polystyrene, 1#/CF density, 3/4" thick R2.89		1,500	.005	S.F.	.19	.16		.35	.48	
2020	1" thick R3.85		1,500	.005		.19	.16		.35	.48	
2100	2" thick R7.69		1,250	.006		.41	.19		.60	.77	
2110	3" thick R11.49		1,250	.006		.70	.19		.89	1.09	
2120	4" thick R15.38		1,200	.007		.57	.20		.77	.96	
2130	5" thick R19.23		1,150	.007		.71	.21		.92	1.13	
2140	6" thick R23.26	▼	1,150	.007	▼	.83	.21		1.04	1.26	

THERMAL & MOISTURE PROTECTION 7

219

7

THERMAL & MOISTURE PROTECTION

07220 | Roof and Deck Insulation

		CREW	DAILY OUTPUT	LABOR-HOURS	UNIT	2005 BARE COSTS				TOTAL INCL O&P	
						MAT.	LABOR	EQUIP.	TOTAL		
700	2150	Tapered for drainage	1 Rofc	1,500	.005	B.F.	.34	.16		.50	.64	700
	2400	Composites with 2" EPS									
	2410	1" fiberboard	1 Rofc	950	.008	S.F.	.78	.25		1.03	1.28
	2420	7/16" oriented strand board		800	.010		.93	.29		1.22	1.52
	2430	1/2" plywood		800	.010		1	.29		1.29	1.60
	2440	1" perlite		800	.010		.82	.29		1.11	1.40
	2450	Composites with 1 1/2" polyisocyanurate									
	2460	1" fiberboard	1 Rofc	800	.010	S.F.	.84	.29		1.13	1.42
	2470	1" perlite		850	.009		.88	.28		1.16	1.44
	2480	7/16" oriented strand board		800	.010		1.01	.29		1.30	1.61

07240 | Ext. Insulation Finish Systems (EIFS)

		CREW	DAILY OUTPUT	LABOR-HOURS	UNIT	MAT.	LABOR	EQUIP.	TOTAL	TOTAL INCL O&P	
100	0010	**EXTERIOR INSULATION FINISH SYSTEM**										100
	0095	Field applied, 1" EPS insulation	J-1	295	.136	S.F.	1.97	4.02	.35	6.34	8.65
	0100	With 1/2" cement board sheathing		220	.182		2.72	5.40	.47	8.59	11.70
	0105	2" EPS insulation		295	.136		2.22	4.02	.35	6.59	8.90
	0110	With 1/2" cement board sheathing		220	.182		2.97	5.40	.47	8.84	12
	0115	3" EPS insulation		295	.136		2.32	4.02	.35	6.69	9.05
	0120	With 1/2" cement board sheathing		220	.182		3.07	5.40	.47	8.94	12.10
	0125	4" EPS insulation		295	.136		2.64	4.02	.35	7.01	9.40
	0130	With 1/2" cement board sheathing		220	.182		4.15	5.40	.47	10.02	13.25
	0140	Premium finish add		1,265	.032		.28	.94	.08	1.30	1.83
	0150	Heavy duty reinforcement add		914	.044		1.68	1.30	.11	3.09	3.94
	0160	2.5#/S.Y. metal lath substrate add	1 Lath	75	.107	S.Y.	2.16	3.35		5.51	7.35
	0170	3.4#/S.Y. metal lath substrate add	"	75	.107	"	2.25	3.35		5.60	7.45
	0180	Color or texture change,	J-1	1,265	.032	S.F.	.73	.94	.08	1.75	2.32
	0190	With substrate leveling base coat	1 Plas	530	.015		.73	.47		1.20	1.52
	0210	With substrate sealing base coat	1 Pord	1,224	.007		.07	.20		.27	.38
	0220	Prefab. panels, with hat channels 2 1/2" x 3/4", 2" EPS insul.	L-5	1,800	.031		7.70	1.19	.34	9.23	10.90
	0240	3" EPS insulation		1,800	.031		8.55	1.19	.34	10.08	11.85
	0250	4" EPS insulation		1,800	.031		9.55	1.19	.34	11.08	12.95
	0260	1-1/2" profile lightweight metal backing, 1" EPS insulation		1,800	.031		11.75	1.19	.34	13.28	15.40
	0270	2" EPS insulation		1,800	.031		13.10	1.19	.34	14.63	16.90
	0280	3" EPS insulation		1,800	.031		14.65	1.19	.34	16.18	18.55
	0290	4" EPS insulation		1,800	.031		16.35	1.19	.34	17.88	20.50
	0310	6"-16 Ga. steel stud back-up, 1" EPS insulation		1,800	.031		13.55	1.19	.34	15.08	17.35
	0320	2" EPS insulation		1,800	.031		15.15	1.19	.34	16.68	19.10
	0330	3" EPS insulation		1,800	.031		16.85	1.19	.34	18.38	21
	0340	4" EPS insulation		1,800	.031		18.85	1.19	.34	20.38	23
	0350	Premium finish add	J-1	1,265	.032		.27	.94	.08	1.29	1.82
	0360	Heavy duty reinforcement add	"	914	.044		1.65	1.30	.11	3.06	3.91
	0370	V groove shape in panel face				L.F.	.52			.52	.57
	0380	U groove shape in panel face				"	.69			.69	.76
	0440	For higher than one story, add						25%			

07260 | Vapor Retarders

		CREW	DAILY OUTPUT	LABOR-HOURS	UNIT	MAT.	LABOR	EQUIP.	TOTAL	TOTAL INCL O&P	
100	0010	**BUILDING PAPER** Aluminum and kraft laminated, foil 1 side	1 Carp	37	.216	Sq.	3.64	7.40		11.04	15.55	100
	0100	Foil 2 sides		37	.216		6.40	7.40		13.80	18.55
	0300	Asphalt, two ply, 30#, for subfloors		19	.421		13.95	14.40		28.35	38
	0400	Asphalt felt sheathing paper, 15#		37	.216		3.41	7.40		10.81	15.30
	0450	Housewrap, exterior, spun bonded polypropylene									
	0470	Small roll	1 Carp	3,800	.002	S.F.	.16	.07		.23	.29
	0480	Large roll	"	4,000	.002	"	.10	.07		.17	.22
	0500	Material only, 3' x 111.1' roll				Ea.	55			55	60.50
	0520	9' x 111.1' roll				"	100			100	110
	0600	Polyethylene vapor barrier, standard, .002" thick	1 Carp	37	.216	Sq.	.96	7.40		8.36	12.60

Important: See the Reference Section for critical supporting data - Reference Nos., Crews, & City Cost Indexes

07200 | Thermal Protection

07260 | Vapor Retarders

			CREW	DAILY OUTPUT	LABOR-HOURS	UNIT	2005 BARE COSTS MAT.	LABOR	EQUIP.	TOTAL	TOTAL INCL O&P	
100	0700	.004" thick	1 Carp	37	.216	Sq.	1.88	7.40		9.28	13.60	100
	0900	.006" thick		37	.216		2.75	7.40		10.15	14.60	
	1200	.010" thick		37	.216		5.10	7.40		12.50	17.15	
	1300	Clear reinforced, fire retardant, .008" thick		37	.216		8.55	7.40		15.95	21	
	1350	Cross laminated type, .003" thick		37	.216		6.65	7.40		14.05	18.85	
	1400	.004" thick		37	.216		7.40	7.40		14.80	19.70	
	1500	Red rosin paper, 5 sq rolls, 4 lb per square		37	.216		1.58	7.40		8.98	13.30	
	1600	5 lbs. per square		37	.216		2.04	7.40		9.44	13.80	
	1800	Reinf. waterproof, .002" polyethylene backing, 1 side		37	.216		5	7.40		12.40	17.05	
	1900	2 sides		37	.216		6.65	7.40		14.05	18.85	
	2100	Roof deck vapor barrier, class 1 metal decks	1 Rofc	37	.216		11.60	6.35		17.95	23.50	
	2200	For all other decks	"	37	.216		8.20	6.35		14.55	19.90	
	2400	Waterproofed kraft with sisal or fiberglass fibers, minimum	1 Carp	37	.216		5.40	7.40		12.80	17.50	
	2500	Maximum	"	37	.216		13.50	7.40		20.90	26.50	

07300 | Shingles, Roof Tiles and Roof Coverings

07310 | Shingles

			CREW	DAILY OUTPUT	LABOR-HOURS	UNIT	2005 BARE COSTS MAT.	LABOR	EQUIP.	TOTAL	TOTAL INCL O&P	
050	0010	**ALUMINUM** Shingles, mill finish, .019" thick	1 Carp	5	1.600	Sq.	145	55		200	246	050
	0100	.020" thick	"	5	1.600		159	55		214	261	
	0300	For colors, add					15.15			15.15	16.65	
	0600	Ridge cap, .024" thick	1 Carp	170	.047	L.F.	1.72	1.61		3.33	4.40	
	0700	End wall flashing, .024" thick		170	.047		1.38	1.61		2.99	4.03	
	0900	Valley section, .024" thick		170	.047		2.30	1.61		3.91	5.05	
	1000	Starter strip, .024" thick		400	.020		1.22	.69		1.91	2.41	
	1200	Side wall flashing, .024" thick		170	.047		1.37	1.61		2.98	4.02	
	1500	Gable flashing, .024" thick		400	.020		1.40	.69		2.09	2.61	
100	0010	**ASPHALT SHINGLES**										100
	0100	Standard strip shingles										
	0150	Inorganic, class A, 210-235 lb/sq **CN**	1 Rofc	5.50	1.455	Sq.	31	43		74	107	
	0155	Pneumatic nailed		7	1.143		31	33.50		64.50	91.50	
	0200	Organic, class C, 235-240 lb/sq		5	1.600		40.50	47		87.50	125	
	0205	Pneumatic nailed		6.25	1.280		40.50	37.50		78	109	
	0250	Standard, laminated multi-layered shingles										
	0300	Class A, 240-260 lb/sq	1 Rofc	4.50	1.778	Sq.	43	52.50		95.50	137	
	0305	Pneumatic nailed		5.63	1.422		43	42		85	119	
	0350	Class C, 260-300 lb/square, 4 bundles/square		4	2		49	59		108	154	
	0355	Pneumatic nailed		5	1.600		49	47		96	135	
	0400	Premium, laminated multi-layered shingles										
	0450	Class A, 260-300 lb, 4 bundles/sq	1 Rofc	3.50	2.286	Sq.	54.50	67.50		122	175	
	0455	Pneumatic nailed		4.37	1.831		54.50	54		108.50	152	
	0500	Class C, 300-385 lb/square, 5 bundles/square		3	2.667		65.50	78.50		144	206	
	0505	Pneumatic nailed		3.75	2.133		65.50	63		128.50	179	
	0800	#15 felt underlayment		64	.125		3.41	3.68		7.09	10.05	
	0825	#30 felt underlayment		58	.138		6.95	4.06		11.01	14.55	
	0850	Self adhering polyethylene and rubberized asphalt underlayment		22	.364		42	10.70		52.70	64.50	
	0900	Ridge shingles		330	.024	L.F.	.79	.71		1.50	2.09	
	0905	Pneumatic nailed		412.50	.019	"	.79	.57		1.36	1.84	
	1000	For steep roofs (7 to 12 pitch or greater), add						50%				

7 THERMAL & MOISTURE PROTECTION

07310	Shingles	CREW	DAILY OUTPUT	LABOR-HOURS	UNIT	2005 BARE COSTS				TOTAL INCL O&P	
						MAT.	LABOR	EQUIP.	TOTAL		
500	0010	**FIBER CEMENT** shingles, 16" x 9.35", 500 lb per square	1 Rofc	2.20	3.636	Sq.	244	107		351	450	**500**
	0110	Starters, 16" x 9.35"		3	2.667	C.L.F.	83	78.50		161.50	225	
	0120	Hip & ridge, 4.75" x 14"		1	8	"	600	236		836	1,050	
	0200	Shakes, 16" x 9.35", 550 lb per square		2.20	3.636	Sq.	221	107		328	425	
	0300	Hip & ridge, 4.75 x 14"		1	8	C.L.F.	600	236		836	1,050	
	0400	Hexagonal, 16" x 16"		3	2.667	Sq.	165	78.50		243.50	315	
	0500	Square, 16" x 16"	▼	3	2.667	"	148	78.50		226.50	297	
800	0010	**SLATE**, Buckingham, Virginia, black [R07310-020]	1 Rots	1.75	4.571	Sq.	550	136		686	835	**800**
	0100	3/16" - 1/4" thick		1.75	4.571		550	136		686	835	
	0200	1/4" thick		1.75	4.571		580	136		716	870	
	0900	Pennsylvania black, Bangor, #1 clear		1.75	4.571		450	136		586	725	
	1200	Vermont, unfading, green, mottled green		1.75	4.571		345	136		481	610	
	1300	Semi-weathering green & gray		1.75	4.571		330	136		466	595	
	1400	Purple		1.75	4.571		370	136		506	635	
	1500	Black or gray		1.75	4.571		340	136		476	605	
	1600	Red		1.75	4.571		1,100	136		1,236	1,450	
	1700	Variegated purple	▼	1.75	4.571	▼	385	136		521	655	
900	0010	**STEEL** Shingles, galvanized, 26 gauge	1 Rots	2.20	3.636	Sq.	152	108		260	350	**900**
	0200	24 gauge	"	2.20	3.636		160	108		268	360	
	0300	For colored galvanized shingles, add					41.50			41.50	46	
	0500	For 1" factory applied polystyrene insulation, add				▼	30.50			30.50	33.50	
980	0010	**WOOD** 16" No. 1 red cedar shingles, 5" exposure, on roof	1 Carp	2.50	3.200	Sq.	163	110		273	350	**980**
	0015	Pneumatic nailed		3.25	2.462		163	84.50		247.50	310	
	0200	7-1/2" exposure, on walls		2.05	3.902		108	134		242	325	
	0205	Pneumatic nailed		2.67	2.996		108	103		211	279	
	0300	18" No. 1 red cedar perfections, 5-1/2" exposure, on roof		2.75	2.909		160	99.50		259.50	330	
	0305	Pneumatic nailed		3.57	2.241		160	77		237	296	
	0500	7-1/2" exposure, on walls		2.25	3.556		118	122		240	320	
	0505	Pneumatic nailed		2.92	2.740		118	94		212	275	
	0600	Resquared, and rebutted, 5-1/2" exposure, on roof		3	2.667		199	91.50		290.50	360	
	0605	Pneumatic nailed		3.90	2.051		199	70.50		269.50	330	
	0900	7-1/2" exposure, on walls		2.45	3.265		146	112		258	335	
	0905	Pneumatic nailed	▼	3.18	2.516		146	86		232	295	
	1000	Add to above for fire retardant shingles, 16" long					30			30	33	
	1050	18" long				▼	28.50			28.50	31.50	
	1060	Preformed ridge shingles	1 Carp	400	.020	L.F.	1.65	.69		2.34	2.89	
	1100	Hand-split red cedar shakes, 1/2" thick x 24" long, 10" exp. on roof		2.50	3.200	Sq.	138	110		248	325	
	1105	Pneumatic nailed		3.25	2.462		138	84.50		222.50	283	
	1110	3/4" thick x 24" long, 10" exp. on roof		2.25	3.556		138	122		260	340	
	1115	Pneumatic nailed		2.92	2.740		138	94		232	298	
	1200	1/2" thick, 18" long, 8-1/2" exp. on roof		2	4		97.50	137		234.50	320	
	1205	Pneumatic nailed		2.60	3.077		97.50	105		202.50	271	
	1210	3/4" thick x 18" long, 8 1/2" exp. on roof		1.80	4.444		97.50	152		249.50	345	
	1215	Pneumatic nailed		2.34	3.419		97.50	117		214.50	289	
	1255	10" exp. on walls		2	4		110	137		247	335	
	1260	10" exposure on walls, pneumatic nailed	▼	2.60	3.077		110	105		215	285	
	1700	Add to above for fire retardant shakes, 24" long					30			30	33	
	1800	18" long				▼	30			30	33	
	1810	Ridge shakes	1 Carp	350	.023	L.F.	2.35	.78		3.13	3.81	
	2000	White cedar shingles, 16" long, extras, 5" exposure, on roof		2.40	3.333	Sq.	122	114		236	315	
	2005	Pneumatic nailed		3.12	2.564		122	88		210	272	
	2050	5" exposure on walls		2	4		122	137		259	350	
	2055	Pneumatic nailed		2.60	3.077		122	105		227	299	
	2100	7-1/2" exposure, on walls		2	4		87.50	137		224.50	310	
	2105	Pneumatic nailed	▼	2.60	3.077	▼	87.50	105		192.50	260	

Important: See the Reference Section for critical supporting data - Reference Nos., Crews, & City Cost Indexes

07310	Shingles	CREW	DAILY OUTPUT	LABOR-HOURS	UNIT	2005 BARE COSTS				TOTAL INCL O&P	
						MAT.	LABOR	EQUIP.	TOTAL		
980	2150	"B" grade, 5" exposure on walls	1 Carp	2	4	Sq.	110	137		247	335	980
	2155	Pneumatic nailed		2.60	3.077		110	105		215	285	
	2300	For 15# organic felt underlayment on roof, 1 layer, add		64	.125		3.41	4.28		7.69	10.40	
	2400	2 layers, add		32	.250		6.80	8.55		15.35	21	
	2600	For steep roofs (7/12 pitch or greater), add to above						50%				
	2700	Panelized systems, No.1 cedar shingles on 5/16" CDX plywood										
	2800	On walls, 8' strips, 7" or 14" exposure	2 Carp	700	.023	S.F.	3.20	.78		3.98	4.74	
	3500	On roofs, 8' strips, 7" or 14" exposure	1 Carp	3	2.667	Sq.	320	91.50		411.50	490	
	3505	Pneumatic nailed	"	4	2	"	320	68.50		388.50	455	

07320	Roof Tiles	CREW	DAILY OUTPUT	LABOR-HOURS	UNIT	2005 BARE COSTS				TOTAL INCL O&P	
						MAT.	LABOR	EQUIP.	TOTAL		
100	0010	ALUMINUM Tiles with accessories, .032" thick, mission tile	1 Carp	2.50	3.200	Sq.	550	110		660	775	100
	0200	Spanish tiles	"	3	2.667	"	380	91.50		471.50	555	
200	0010	CLAY TILE ASTM C1167, GR 1, severe weathering, acces. incl.										200
	0200	Lanai tile or Classic tile, 158 pc per sq	1 Rots	1.65	4.848	Sq.	395	144		539	680	
	0300	Americana, 158 pc per sq, most colors		1.65	4.848		510	144		654	805	
	0350	Green, gray or brown		1.65	4.848		490	144		634	785	
	0400	Blue		1.65	4.848		490	144		634	785	
	0600	Spanish tile, 171 pc per sq, red		1.80	4.444		260	132		392	510	
	0800	Blend		1.80	4.444		420	132		552	685	
	0900	Glazed white		1.80	4.444		500	132		632	775	
	1100	Mission tile, 192 pc per sq, machine scored finish, red		1.15	6.957		635	207		842	1,050	
	1700	French tile, 133 pc per sq, smooth finish, red		1.35	5.926		575	176		751	930	
	1750	Blue or green		1.35	5.926		685	176		861	1,050	
	1800	Norman black 317 pc per sq		1	8		805	238		1,043	1,300	
	2200	Williamsburg tile, 158 pc per sq, aged cedar		1.35	5.926		490	176		666	840	
	2250	Gray or green		1.35	5.926		490	176		666	840	
	2510	One piece mission tile, natural red, 75 pc per square		1.65	4.848		133	144		277	390	
	2530	Mission Tile, 134 pc per square		1.15	6.957		178	207		385	545	
300	0010	CONCRETE TILE Including installation of accessories										300
	0020	Corrugated, 13" x 16-1/2", 90 per sq, 950 lb per sq										
	0050	Earthtone colors, nailed to wood deck	1 Rots	1.35	5.926	Sq.	91	176		267	400	
	0150	Blues		1.35	5.926		104	176		280	415	
	0200	Greens		1.35	5.926		104	176		280	415	
	0250	Premium colors		1.35	5.926		153	176		329	470	
	0500	Shakes, 13" x 16-1/2", 90 per sq, 950 lb per sq										
	0600	All colors, nailed to wood deck	1 Rots	1.50	5.333	Sq.	185	158		343	475	
	1500	Accessory pieces, ridge & hip, 10" x 16-1/2", 8 lbs. each				Ea.	2.25			2.25	2.48	
	1700	Rake, 6-1/2" x 16-3/4", 9 lbs. each					2.25			2.25	2.48	
	1800	Mansard hip, 10" x 16-1/2", 9.2 lbs. each					2.25			2.25	2.48	
	1900	Hip starter, 10" x 16-1/2", 10.5 lbs. each					9.50			9.50	10.45	
	2000	3 or 4 way apex, 10" each side, 11.5 lbs. each					10.25			10.25	11.30	

07400 | Roofing and Siding Panels

07410	Metal Roof and Wall Panels	CREW	DAILY OUTPUT	LABOR-HOURS	UNIT	2005 BARE COSTS				TOTAL INCL O&P	
						MAT.	LABOR	EQUIP.	TOTAL		
100	0010	ALUMINUM ROOFING Corrugated or ribbed, .0155" thick, natural	G-3	1,200	.027	S.F.	.62	.89		1.51	2.06	100
	0300	Painted		1,200	.027		.89	.89		1.78	2.36	

THERMAL & MOISTURE PROTECTION 7

THERMAL & MOISTURE PROTECTION · 7

			CREW	DAILY OUTPUT	LABOR-HOURS	UNIT	MAT.	LABOR	EQUIP.	TOTAL	TOTAL INCL O&P	
07410		**Metal Roof and Wall Panels**						2005 BARE COSTS				
100	0400	Corrugated, .018" thick, on steel frame, natural finish	G-3	1,200	.027	S.F.	.80	.89		1.69	2.26	100
	0600	Painted		1,200	.027		.99	.89		1.88	2.47	
	0700	Corrugated, on steel frame, natural, .024" thick		1,200	.027		1.16	.89		2.05	2.66	
	0800	Painted, .024" thick		1,200	.027		1.40	.89		2.29	2.92	
	0900	.032" thick, natural		1,200	.027		1.38	.89		2.27	2.90	
	1200	painted		1,200	.027		1.91	.89		2.80	3.48	
	1300	V-Beam, on steel frame construction, .032" thick, natural		1,200	.027		1.55	.89		2.44	3.09	
	1500	Painted		1,200	.027		1.93	.89		2.82	3.50	
	1600	.040" thick, natural		1,200	.027		1.89	.89		2.78	3.46	
	1800	Painted		1,200	.027		2.30	.89		3.19	3.91	
	1900	.050" thick, natural		1,200	.027		2.29	.89		3.18	3.90	
	2100	Painted		1,200	.027		2.77	.89		3.66	4.43	
	2200	For roofing on wood frame, deduct		4,600	.007		.05	.23		.28	.42	
	2400	Ridge cap, .032" thick, natural		800	.040	L.F.	1.82	1.34		3.16	4.07	
500	0010	**MANSARD** Colored aluminum, with battens, .032" thick										500
	0600	Stock units, straight surfaces	1 Shee	115	.070	S.F.	2.13	2.80		4.93	6.65	
	0700	Concave or convex surfaces		75	.107	"	2.35	4.29		6.64	9.20	
	0800	For framing, to 5' high, add		115	.070	L.F.	2.35	2.80		5.15	6.90	
	0900	Soffits, to 1' wide		125	.064	S.F.	1.16	2.57		3.73	5.25	
690	0010	**METAL FACING PANELS**										690
	0400	Textured aluminum, 4' x 8' x 5/16" plywood backing, single face	2 Shee	375	.043	S.F.	2.24	1.72		3.96	5.10	
	0600	Double face		375	.043		3.45	1.72		5.17	6.45	
	0700	4' x 10' x 5/16" plywood backing, single face		375	.043		2.44	1.72		4.16	5.30	
	0900	Double face		375	.043		3.58	1.72		5.30	6.60	
	1000	4' x 12' x 5/16" plywood backing, single face		375	.043		3.14	1.72		4.86	6.10	
	1300	Smooth al, 1/4" panel, fluoropolymer finish, double face		375	.043		5.60	1.72		7.32	8.85	
	1350	Clear anodized finish, double face		375	.043		6.25	1.72		7.97	9.55	
	1400	Double face textured aluminum, structural panel, 1" EPS insulation		375	.043		3.78	1.72		5.50	6.80	
	1500	Accessories, outside corner	1 Shee	175	.046	L.F.	1.73	1.84		3.57	4.73	
	1600	Inside corner		175	.046		1.89	1.84		3.73	4.91	
	1800	Batten mounting clip		200	.040		.35	1.61		1.96	2.86	
	1900	Low profile batten		480	.017		.72	.67		1.39	1.82	
	2100	High profile batten		480	.017		1.21	.67		1.88	2.36	
	2200	Water table		200	.040		1.75	1.61		3.36	4.40	
	2400	Horizontal joint connector		200	.040		1.63	1.61		3.24	4.26	
	2500	Corner cap		200	.040		1.51	1.61		3.12	4.13	
	2700	H - moulding		480	.017		1.04	.67		1.71	2.17	
700	0010	**STEEL ROOFING** on steel frame, corrugated or ribbed, 30 ga galv	G-3	1,100	.029	S.F.	.77	.97		1.74	2.35	700
	0100	28 ga		1,050	.030		.81	1.02		1.83	2.47	
	0300	26 ga		1,000	.032		.97	1.07		2.04	2.72	
	0400	24 ga		950	.034		1.05	1.13		2.18	2.90	
	0600	Colored, 28 ga		1,050	.030		1.08	1.02		2.10	2.77	
	0700	26 ga		1,000	.032		1.15	1.07		2.22	2.92	
	0710	Flat profile, 1-3/4" standing seams, 10" wide, standard finish. 26 ga		1,000	.032		2.37	1.07		3.44	4.26	
	0715	24 ga		950	.034		2.75	1.13		3.88	4.77	
	0720	22 ga		900	.036		3.39	1.19		4.58	5.55	
	0725	Zinc aluminum alloy finish, 26 ga		1,000	.032		1.85	1.07		2.92	3.69	
	0730	24 ga		950	.034		2.21	1.13		3.34	4.17	
	0735	22 ga		900	.036		2.54	1.19		3.73	4.63	
	0740	12" wide, standard finish, 26 ga		1,000	.032		2.36	1.07		3.43	4.25	
	0745	24 ga		950	.034		2.74	1.13		3.87	4.75	
	0750	Zinc aluminum alloy finish, 26 ga		1,000	.032		1.75	1.07		2.82	3.58	
	0755	24 ga		950	.034		2.09	1.13		3.22	4.04	
	0840	Flat profile, 1" x 3/8" batten, 12" wide, standard finish, 26 ga		1,000	.032		1.84	1.07		2.91	3.67	
	0845	24 ga		950	.034		2.16	1.13		3.29	4.12	

Important: See the Reference Section for critical supporting data - Reference Nos., Crews, & City Cost Indexes

07410 | Metal Roof and Wall Panels

			CREW	DAILY OUTPUT	LABOR-HOURS	UNIT	MAT.	2005 BARE COSTS LABOR	EQUIP.	TOTAL	TOTAL INCL O&P	
700	0850	22 ga	G-3	900	.036	S.F.	2.60	1.19		3.79	4.70	700
	0855	Zinc aluminum alloy finish, 26 ga		1,000	.032		1.56	1.07		2.63	3.37	
	0860	24 ga		950	.034		1.73	1.13		2.86	3.64	
	0865	22 ga		900	.036		2.01	1.19		3.20	4.05	
	0870	16-1/2" wide, standard finish, 24 ga		950	.034		2.19	1.13		3.32	4.15	
	0875	22 ga		900	.036		2.46	1.19		3.65	4.55	
	0880	Zinc aluminum alloy finish, 24 ga		950	.034		1.65	1.13		2.78	3.56	
	0885	22 ga		900	.036		1.83	1.19		3.02	3.85	
	0890	Flat profile, 2" x 2" batten, 12" wide, standard finish, 26 ga		1,000	.032		2.11	1.07		3.18	3.97	
	0895	24 ga		950	.034		2.52	1.13		3.65	4.51	
	0900	22 ga		900	.036		3.09	1.19		4.28	5.25	
	0905	Zinc aluminum alloy finish, 26 ga		1,000	.032		1.74	1.07		2.81	3.56	
	0910	24 ga		950	.034		1.88	1.13		3.01	3.81	
	0915	22 ga		900	.036		2.32	1.19		3.51	4.39	
	0920	16-1/2" wide, standard finish, 24 ga		950	.034		2.34	1.13		3.47	4.31	
	0925	22 ga		900	.036		2.72	1.19		3.91	4.83	
	0930	Zinc aluminum alloy finish, 24 ga		950	.034		1.85	1.13		2.98	3.78	
	0935	22 ga		900	.036		2.11	1.19		3.30	4.16	
	1200	Ridge, galvanized, 10" wide		800	.040	L.F.	1.73	1.34		3.07	3.97	
	1210	20" wide		750	.043	"	3.04	1.43		4.47	5.55	

07420 | Plastic Roof and Wall Panels

			CREW	DAILY OUTPUT	LABOR-HOURS	UNIT	MAT.	LABOR	EQUIP.	TOTAL	TOTAL INCL O&P	
770	0010	**FIBERGLASS** Corrugated panels, roofing, 8 oz per SF	G-3	1,000	.032	S.F.	2.24	1.07		3.31	4.11	770
	0100	12 oz per SF		1,000	.032		3.23	1.07		4.30	5.20	
	0300	Corrugated siding, 6 oz per SF		880	.036		1.94	1.22		3.16	4.01	
	0400	8 oz per SF		880	.036		2.24	1.22		3.46	4.34	
	0500	Fire retardant		880	.036		3.09	1.22		4.31	5.30	
	0600	12 oz. siding, textured		880	.036		3.14	1.22		4.36	5.35	
	0700	Fire retardant		880	.036		4.14	1.22		5.36	6.45	
	0900	Flat panels, 6 oz per SF, clear or colors		880	.036		1.73	1.22		2.95	3.78	
	1100	Fire retardant, class A		880	.036		3	1.22		4.22	5.20	
	1300	8 oz per SF, clear or colors		880	.036		2.24	1.22		3.46	4.34	
	1700	Sandwich panels, fiberglass, 1-9/16" thick, panels to 20 SF		180	.178		21.50	5.95		27.45	33	
	1900	As above, but 2-3/4" thick, panels to 100 SF		265	.121		16.10	4.04		20.14	24	

07440 | Faced Panels

			CREW	DAILY OUTPUT	LABOR-HOURS	UNIT	MAT.	LABOR	EQUIP.	TOTAL	TOTAL INCL O&P	
200	0010	**EXPOSED AGGREGATE PANELS**										200
	1400	Fiberglass polymer back-up,										
	1500	Small size aggregate	F-3	445	.090	S.F.	4.62	3.11	1.35	9.08	11.40	
	1600	Medium size aggregate		445	.090		5.20	3.11	1.35	9.66	12	
	1700	Large size aggregate		445	.090		5.35	3.11	1.35	9.81	12.20	

07460 | Siding

			CREW	DAILY OUTPUT	LABOR-HOURS	UNIT	MAT.	LABOR	EQUIP.	TOTAL	TOTAL INCL O&P	
100	0010	**ALUMINUM SIDING** .019" thick, on steel construction, natural	G-3	775	.041	S.F.	.66	1.38		2.04	2.86	100
	0100	Painted		775	.041		.86	1.38		2.24	3.08	
	0400	Farm type, .021" thick on steel frame, natural		775	.041		.77	1.38		2.15	2.98	
	0600	Painted		775	.041		.94	1.38		2.32	3.16	
	0700	Industrial type, corrugated, on steel, .024" thick, mill		775	.041		.97	1.38		2.35	3.20	
	0900	Painted		775	.041		1.17	1.38		2.55	3.42	
	1000	.032" thick, mill		775	.041		1.42	1.38		2.80	3.69	
	1200	Painted		775	.041		1.80	1.38		3.18	4.11	
	1300	V-Beam, on steel frame, .032" thick, mill		775	.041		1.66	1.38		3.04	3.96	
	1500	Painted		775	.041		1.97	1.38		3.35	4.30	
	1600	.040" thick, mill		775	.041		2.01	1.38		3.39	4.34	
	1800	Painted		775	.041		2.37	1.38		3.75	4.74	

07460 | Siding

			CREW	DAILY OUTPUT	LABOR-HOURS	UNIT	2005 BARE COSTS				TOTAL INCL O&P	
							MAT.	LABOR	EQUIP.	TOTAL		
100	1900	.050" thick, mill	G-3	775	.041	S.F.	2.38	1.38		3.76	4.75	**100**
	2100	Painted		775	.041		2.85	1.38		4.23	5.25	
	2200	Ribbed, 3" profile, on steel frame, .032" thick, natural		775	.041		1.38	1.38		2.76	3.65	
	2400	Painted		775	.041		1.82	1.38		3.20	4.13	
	2500	.040" thick, natural		775	.041		1.70	1.38		3.08	4	
	2700	Painted		775	.041		1.99	1.38		3.37	4.32	
	2750	.050" thick, natural		775	.041		1.96	1.38		3.34	4.29	
	2760	Painted		775	.041		2.23	1.38		3.61	4.58	
	3300	For siding on wood frame, deduct from above		2,800	.011		.06	.38		.44	.66	
	3400	Screw fasteners, aluminum, self tapping, neoprene washer, 1"				M	125			125	138	
	3600	Stitch screws, self tapping, with neoprene washer, 5/8"				"	94			94	104	
	3630	Flashing, sidewall, .032" thick	G-3	800	.040	L.F.	1.79	1.34		3.13	4.04	
	3650	End wall, .040" thick		800	.040		2.09	1.34		3.43	4.37	
	3670	Closure strips, corrugated, .032" thick		800	.040		.53	1.34		1.87	2.65	
	3680	Ribbed, 4" or 8", .032" thick		800	.040		.53	1.34		1.87	2.65	
	3690	V-beam, .040" thick		800	.040		.72	1.34		2.06	2.86	
	3800	Horizontal, colored clapboard, 8" wide, plain	2 Carp	515	.031	S.F.	1.23	1.06		2.29	3.01	
	3900	Insulated		515	.031		1.52	1.06		2.58	3.33	
	4000	Vertical board & batten, colored, non-insulated		515	.031		1.08	1.06		2.14	2.85	
	4200	For simulated wood design, add					.08			.08	.09	
	4300	Corners for above, outside	2 Carp	515	.031	V.L.F.	1.70	1.06		2.76	3.53	
	4500	Inside corners	"	515	.031	"	.96	1.06		2.02	2.72	
	4600	Sandwich panels, 1" insulation, single story	G-3	395	.081	S.F.	4.61	2.71		7.32	9.25	
	4900	Multi-story	"	345	.093		6.45	3.10		9.55	11.90	
	5100	For baked enamel finish 1 side, add					.24			.24	.26	
	5200	See also Metal Facing Panels, division 07410-690										
300	0010	**FASCIA** Aluminum, reverse board and batten,										**300**
	0100	.032" thick, colored, no furring included	1 Shee	145	.055	S.F.	2.14	2.22		4.36	5.75	
	0300	Steel, galv and enameled, stock, no furring, long panels		145	.055		2.23	2.22		4.45	5.85	
	0600	Short panels		115	.070		3.38	2.80		6.18	8	
500	0010	**FIBER CEMENT SIDING**										**500**
	0020	Lap siding, 5/16" thick, 6" wide, smooth texture	2 Carp	415	.039	S.F.	1.01	1.32		2.33	3.17	
	0025	Woodgrain texture		415	.039		1.01	1.32		2.33	3.17	
	0030	7-1/2" wide, smooth texture		425	.038		.88	1.29		2.17	2.98	
	0035	Woodgrain texture		425	.038		.88	1.29		2.17	2.98	
	0040	8" wide, smooth texture		425	.038		.87	1.29		2.16	2.97	
	0045	Roughsawn texture		425	.038		.87	1.29		2.16	2.97	
	0050	9-1/2" wide, smooth texture		440	.036		.84	1.25		2.09	2.87	
	0055	Woodgrain texture		440	.036		.84	1.25		2.09	2.87	
	0060	12" wide, smooth texture		455	.035		.81	1.20		2.01	2.78	
	0065	Woodgrain texture		455	.035		.81	1.20		2.01	2.78	
	0070	Panel siding, 5/16" thick, smooth texture		750	.021		.73	.73		1.46	1.94	
	0075	Stucco texture		750	.021		.73	.73		1.46	1.94	
	0080	Grooved woodgrain texture		750	.021		.73	.73		1.46	1.94	
	0085	V - grooved woodgrain texture		750	.021		.73	.73		1.46	1.94	
	0090	Wood starter strip		400	.040	L.F.	.19	1.37		1.56	2.34	
600	0010	**VINYL SIDING** Solid PVC panels, 8" to 10" wide, plain	1 Carp	255	.031	S.F.	.69	1.07		1.76	2.43	**600**
	0100	with 3/8" insulation		255	.031		.76	1.07		1.83	2.51	
	0200	Soffit and fascia		205	.039		1.45	1.34		2.79	3.68	
	0300	Window and door trim moldings		185	.043	L.F.	.36	1.48		1.84	2.71	
	0500	Corner posts, outside corner		205	.039		1.15	1.34		2.49	3.35	
	0600	Inside corner		205	.039		.61	1.34		1.95	2.75	
750	0010	**SOFFIT** Aluminum, residential, stock units, .020" thick	1 Carp	210	.038	S.F.	1.01	1.30		2.31	3.14	**750**
	0100	Baked enamel on steel, 16 or 18 gauge		105	.076		3.74	2.61		6.35	8.15	
	0300	Polyvinyl chloride, white, solid		230	.035		.66	1.19		1.85	2.59	
	0400	Perforated		230	.035		.66	1.19		1.85	2.59	

7 THERMAL & MOISTURE PROTECTION

Important: See the Reference Section for critical supporting data - Reference Nos., Crews, & City Cost Indexes

07460	Siding		CREW	DAILY OUTPUT	LABOR-HOURS	UNIT	2005 BARE COSTS				TOTAL INCL O&P	
							MAT.	LABOR	EQUIP.	TOTAL		
750	0500	For colors, add				S.F.	.07			.07	.08	**750**
800	0010	**STEEL SIDING**, Beveled, vinyl coated, 8" wide, including fasteners	1 Carp	265	.030	S.F.	1.15	1.03		2.18	2.88	**800**
	0050	10" wide	"	275	.029		1.23	1		2.23	2.90	
	0080	Galv, corrugated or ribbed, on steel frame, 30 gauge	G-3	800	.040		.77	1.34		2.11	2.92	
	0100	28 gauge		795	.040		.81	1.35		2.16	2.97	
	0300	26 gauge		790	.041		1.13	1.36		2.49	3.33	
	0400	24 gauge		785	.041		1.14	1.36		2.50	3.36	
	0600	22 gauge		770	.042		1.31	1.39		2.70	3.59	
	0700	Colored, corrugated/ribbed, on steel frame, 10 yr fnsh, 28 ga.		800	.040		1.20	1.34		2.54	3.39	
	0900	26 gauge		795	.040		1.25	1.35		2.60	3.46	
	1000	24 gauge		790	.041		1.46	1.36		2.82	3.70	
	1020	20 gauge		785	.041		1.85	1.36		3.21	4.15	
	1200	Factory sandwich panel, 26 ga., 1" insulation, galvanized		380	.084		3.07	2.82		5.89	7.75	
	1300	Colored 1 side		380	.084		3.75	2.82		6.57	8.50	
	1500	Galvanized 2 sides		380	.084		4.66	2.82		7.48	9.50	
	1600	Colored 2 sides		380	.084		4.83	2.82		7.65	9.65	
	1800	Acrylic paint face, regular paint liner	↓	380	.084		3.64	2.82		6.46	8.35	
	1900	For 2" thick polystyrene, add					.56			.56	.62	
	2000	22 ga, galv, 2" insulation, baked enamel exterior	G-3	360	.089		7.10	2.97		10.07	12.45	
	2100	P.V.F. exterior finish	"	360	.089	↓	7.50	2.97		10.47	12.85	
900	0010	**WOOD SIDING, BOARDS**										**900**
	3200	Wood, cedar bevel, A grade, 1/2" x 6"	1 Carp	250	.032	S.F.	3.28	1.10		4.38	5.30	
	3300	1/2" x 8"		275	.029		2.92	1		3.92	4.76	
	3500	3/4" x 10", clear grade		300	.027		3.54	.91		4.45	5.30	
	3600	"B" grade		300	.027		2.80	.91		3.71	4.50	
	3800	Cedar, rough sawn, 1" x 4", A grade, natural		240	.033		2.67	1.14		3.81	4.72	
	3900	Stained		240	.033		3.01	1.14		4.15	5.10	
	4100	1" x 12", board & batten, #3 & Btr., natural		260	.031		2.02	1.05		3.07	3.86	
	4200	Stained		260	.031		2.35	1.05		3.40	4.23	
	4400	1" x 8" channel siding, #3 & Btr., natural		250	.032		1.96	1.10		3.06	3.87	
	4500	Stained		250	.032		2.23	1.10		3.33	4.16	
	4700	Redwood, clear, beveled, vertical grain, 1/2" x 4"		200	.040		3.12	1.37		4.49	5.55	
	4750	1/2" x 6"		225	.036		2.62	1.22		3.84	4.78	
	4800	1/2" x 8"		250	.032		2.12	1.10		3.22	4.04	
	5000	3/4" x 10"		300	.027		3.46	.91		4.37	5.25	
	5200	Channel siding, 1" x 10", B grade	↓	285	.028		2.23	.96		3.19	3.95	
	5250	Redwood, T&G boards, B grade, 1" x 4"	2 Carp	300	.053		2.66	1.83		4.49	5.80	
	5270	1" x 8"	"	375	.043		2.29	1.46		3.75	4.80	
	5400	White pine, rough sawn, 1" x 8", natural	1 Carp	275	.029		.67	1		1.67	2.29	
	5500	Stained	"	275	.029	↓	.99	1		1.99	2.64	
950	0010	**WOOD PRODUCT SIDING**										**950**
	0030	Lap siding, hardboard, 7/16" x 8", primed										
	0051	Wood grain texture finish	L-2	552	.029	S.F.	1.04	.87		1.91	2.49	
	0100	Panels, 7/16" thick, smooth, textured or grooved, primed	2 Carp	700	.023		.73	.78		1.51	2.02	
	0200	Stained		700	.023		.88	.78		1.66	2.19	
	0700	Particle board, overlaid, 3/8" thick		750	.021		.63	.73		1.36	1.83	
	0900	Plywood, medium density overlaid, 3/8" thick		750	.021		.84	.73		1.57	2.06	
	1000	1/2" thick		700	.023		.95	.78		1.73	2.27	
	1100	3/4" thick		650	.025		1.28	.84		2.12	2.72	
	1600	Texture 1-11, cedar, 5/8" thick, natural		675	.024		2.38	.81		3.19	3.88	
	1700	Factory stained		675	.024		2.21	.81		3.02	3.69	
	1900	Texture 1-11, fir, 5/8" thick, natural		675	.024		1.01	.81		1.82	2.37	
	2000	Factory stained		675	.024		1.08	.81		1.89	2.45	
	2050	Texture 1-11, S.Y.P., 5/8" thick, natural	↓	675	.024	↓	.81	.81		1.62	2.15	

THERMAL & MOISTURE PROTECTION 7

07460	Siding	CREW	DAILY OUTPUT	LABOR-HOURS	UNIT	2005 BARE COSTS				TOTAL INCL O&P	
						MAT.	LABOR	EQUIP.	TOTAL		
950 2100	Factory stained	2 Carp	675	.024	S.F.	.88	.81		1.69	2.23	**950**
2200	Rough sawn cedar, 3/8" thick, natural		675	.024		1.08	.81		1.89	2.45	
2300	Factory stained		675	.024		1.20	.81		2.01	2.58	
2500	Rough sawn fir, 3/8" thick, natural		675	.024		.58	.81		1.39	1.90	
2600	Factory stained		675	.024		.65	.81		1.46	1.98	
2800	Redwood, textured siding, 5/8" thick		675	.024		1.80	.81		2.61	3.24	
3000	Polyvinyl chloride coated, 3/8" thick	▼	750	.021	▼	.88	.73		1.61	2.11	

07510	Built-Up Bituminous Roofing	CREW	DAILY OUTPUT	LABOR-HOURS	UNIT	2005 BARE COSTS				TOTAL INCL O&P	
						MAT.	LABOR	EQUIP.	TOTAL		
050 0010	**ASPHALT** Coated felt, #30, 2 sq per roll, not mopped	1 Rofc	58	.138	Sq.	6.95	4.06		11.01	14.55	**050**
0200	#15, 4 sq per roll, plain or perforated, not mopped		58	.138		3.41	4.06		7.47	10.65	
0300	Roll roofing, smooth, #65		15	.533		6.35	15.70		22.05	34	
0500	#90		15	.533		17.70	15.70		33.40	46.50	
0520	Mineralized		15	.533		15.35	15.70		31.05	44	
0540	D.C. (Double coverage), 19" selvage edge	▼	10	.800	▼	29.50	23.50		53	72.50	
0580	Adhesive (lap cement)				Gal.	3.68			3.68	4.05	
0600	Steep, flat or dead level asphalt, 10 ton lots, bulk				Ton	260			260	286	
0800	Packaged				"	400			400	440	
300 0010	**BUILT-UP ROOFING** R07510 -020										**300**
0120	Asphalt flood coat with gravel/slag surfacing, not including										
0140	Insulation, flashing or wood nailers										
0200	Asphalt base sheet, 3 plies #15 asphalt felt, mopped	G-1	22	2.545	Sq.	48.50	70	13.55	132.05	187	
0350	On nailable decks		21	2.667		53.50	73.50	14.20	141.20	199	
0500	4 plies #15 asphalt felt, mopped		20	2.800		69	77	14.90	160.90	223	
0550	On nailable decks		19	2.947		62.50	81	15.70	159.20	224	
0700	Coated glass base sheet, 2 plies glass (type IV), mopped		22	2.545		50	70	13.55	133.55	189	
0850	3 plies glass, mopped		20	2.800		59.50	77	14.90	151.40	213	
0950	On nailable decks		19	2.947		56	81	15.70	152.70	217	
1100	4 plies glass fiber felt (type IV), mopped		20	2.800		72.50	77	14.90	164.40	227	
1150	On nailable decks		19	2.947		66	81	15.70	162.70	228	
1200	Coated & saturated base sheet, 3 plies #15 asph. felt, mopped		20	2.800		55	77	14.90	146.90	208	
1250	On nailable decks		19	2.947		51.50	81	15.70	148.20	212	
1300	4 plies #15 asphalt felt, mopped	▼	22	2.545	▼	64	70	13.55	147.55	204	
2000	Asphalt flood coat, smooth surface										
2200	Asphalt base sheet & 3 plies #15 asphalt felt, mopped	G-1	24	2.333	Sq.	52.50	64	12.40	128.90	180	
2400	On nailable decks		23	2.435		49	67	12.95	128.95	182	
2600	4 plies #15 asphalt felt, mopped		24	2.333		61.50	64	12.40	137.90	190	
2700	On nailable decks	▼	23	2.435	▼	58	67	12.95	137.95	192	
2900	Coated glass fiber base sheet, mopped, and 2 plies of										
2910	glass fiber felt (type IV)	G-1	25	2.240	Sq.	45.50	61.50	11.90	118.90	168	
3100	On nailable decks		24	2.333		43	64	12.40	119.40	170	
3200	3 plies, mopped		23	2.435		55.50	67	12.95	135.45	189	
3300	On nailable decks		22	2.545		52	70	13.55	135.55	191	
3800	4 plies glass fiber felt (type IV), mopped		23	2.435		65	67	12.95	144.95	200	
3900	On nailable decks		22	2.545		61.50	70	13.55	145.05	202	
4000	Coated & saturated base sheet, 3 plies #15 asph. felt, mopped	▼	24	2.333	▼	51	64	12.40	127.40	179	

7

THERMAL & MOISTURE PROTECTION

07510 | Built-Up Bituminous Roofing

		CREW	DAILY OUTPUT	LABOR-HOURS	UNIT	2005 BARE COSTS				TOTAL INCL O&P
						MAT.	LABOR	EQUIP.	TOTAL	
300 4200	On nailable decks	G-1	23	2.435	Sq.	47.50	67	12.95	127.45	180 **300**
4300	4 plies #15 organic felt, mopped [R07510-020]	↓	22	2.545	↓	60	70	13.55	143.55	199
4500	Coal tar pitch with gravel/slag surfacing									
4600	4 plies #15 tarred felt, mopped	G-1	21	2.667	Sq.	107	73.50	14.20	194.70	258
4800	3 plies glass fiber felt (type IV), mopped	"	19	2.947	"	87.50	81	15.70	184.20	251
5000	Coated glass fiber base sheet, and 2 plies of									
5010	glass fiber felt, (type IV), mopped	G-1	19	2.947	Sq.	87	81	15.70	183.70	251
5300	On nailable decks		18	3.111		77	85.50	16.55	179.05	249
5600	4 plies glass fiber felt (type IV), mopped		21	2.667		121	73.50	14.20	208.70	274
5800	On nailable decks		20	2.800	↓	111	77	14.90	202.90	269
400 0010	**CANTS** 4" x 4", treated timber, cut diagonally	1 Rofc	325	.025	L.F.	1.35	.73		2.08	2.73 **400**
0100	Foamglass		325	.025		2.15	.73		2.88	3.61
0300	Mineral or fiber, trapezoidal, 1"x 4" x 48"		325	.025		.17	.73		.90	1.43
0400	1-1/2" x 5-5/8" x 48"		325	.025	↓	.29	.73		1.02	1.56
700 0010	**FELT** Glass fibered, #15, no mopping	1 Rofc	58	.138	Sq.	4.09	4.06		8.15	11.40 **700**
0300	Base sheet, #45, channel vented		58	.138		16.50	4.06		20.56	25
0400	#50, coated		58	.138		8.95	4.06		13.01	16.70
0500	Cap, mineral surfaced		58	.138		16.50	4.06		20.56	25
0600	Flashing membrane, #65		16	.500		25	14.75		39.75	53
0800	Coal tar fibered, #15, no mopping		58	.138		7.50	4.06		11.56	15.15
0900	Asphalt felt, #15, 4 sq per roll, no mopping **CN**		58	.138		3.41	4.06		7.47	10.65
1100	#30, 2 sq per roll		58	.138		6.95	4.06		11.01	14.55
1200	Double coated, #33		58	.138		5.90	4.06		9.96	13.40
1400	#40, base sheet		58	.138		8.10	4.06		12.16	15.85
1450	Coated and saturated		58	.138		6.45	4.06		10.51	14
1500	Tarred felt, organic, #15, 4 sq rolls		58	.138		9.10	4.06		13.16	16.95
1550	#30, 2 sq roll		58	.138		15.60	4.06		19.66	24
1700	Add for mopping above felts, per ply, asphalt, 24 lb per sq	G-1	192	.292		4.80	8	1.55	14.35	20.50
1800	Coal tar mopping, 30 lb per sq		186	.301		8.75	8.30	1.60	18.65	25.50
1900	Flood coat, with asphalt, 60 lb per sq		60	.933		12	25.50	4.97	42.47	62
2000	With coal tar, 75 lb per sq	↓	56	1	↓	22	27.50	5.30	54.80	77

07520 | Cold Applied Bituminous Roofing

		CREW	DAILY OUTPUT	LABOR-HOURS	UNIT	MAT.	LABOR	EQUIP.	TOTAL	TOTAL INCL O&P
200 0010	**COLD APPLIED** 3-ply system (components listed below)	G-5	50	.800	Sq.		21.50	2.94	24.44	39.50 **200**
0100	Spunbond poly. fabric, 1.35 oz/SY, 36"W, 10.8 Sq/roll				Ea.	123			123	135
0200	49" wide, 14.6 Sq./roll					169			169	186
0300	2.10 oz./S.Y., 36" wide, 10.8 Sq./roll					185			185	203
0400	49" wide, 14.6 Sq./roll				↓	250			250	275
0500	Base & finish coat, 3 gal./Sq., 5 gal./can				Gal.	3.25			3.25	3.58
0600	Coating, ceramic granules, 1/2 Sq./bag				Ea.	11.50			11.50	12.65
0700	Aluminum, 2 gal./Sq.				Gal.	9.25			9.25	10.20
0800	Emulsion, fibered or non-fibered, 4 gal./Sq.				"	4.25			4.25	4.68

07530 | Elastomeric Membrane Roofing

		CREW	DAILY OUTPUT	LABOR-HOURS	UNIT	MAT.	LABOR	EQUIP.	TOTAL	TOTAL INCL O&P
350 0010	**ELASTOMERIC ROOFING**									**350**
0100	For Elastomeric waterproofing, see division 07130-200									
0110	Acrylic rubber, fluid applied, 20 mils thick	G-5	2,000	.020	S.F.	1.80	.53	.07	2.40	2.97
0120	50 mils, reinforced		1,200	.033		2.80	.89	.12	3.81	4.73
0130	For walking surface, add	↓	900	.044		.85	1.19	.16	2.20	3.14
0300	Hypalon neoprene, fluid applied, 20 mil thick, not-reinforced	G-1	1,135	.049		2.05	1.36	.26	3.67	4.86
0600	Non-woven polyester, reinforced		960	.058		2.07	1.60	.31	3.98	5.35
0700	5 coat neoprene deck, 60 mil thick, under 10,000 SF		325	.172		4.56	4.74	.92	10.22	14.05
0900	Over 10,000 SF		625	.090		4.25	2.46	.48	7.19	9.40
1300	Vinyl plastic traffic deck, sprayed, 2 to 4 mils thick		625	.090	↓	1.33	2.46	.48	4.27	6.20

THERMAL & MOISTURE PROTECTION 7

07530	Elastomeric Membrane Roofing	CREW	DAILY OUTPUT	LABOR-HOURS	UNIT	2005 BARE COSTS				TOTAL INCL O&P	
						MAT.	LABOR	EQUIP.	TOTAL		
350	1500 Vinyl and neoprene membrane traffic deck	G-1	1,550	.036	S.F.	1.41	.99	.19	2.59	3.45	350
800	0010 **SINGLE-PLY MEMBRANE**										800
	0800 Chlorosulfonated polyethylene-hypalon (CSPE), 45 mils,										
	0900 0.29 P.S.F., fully adhered	G-5	26	1.538	Sq.	123	41	5.65	169.65	212	
	1100 Loose-laid & ballasted with stone (10 P.S.F.)		51	.784		130	21	2.88	153.88	182	
	1200 Mechanically attached		35	1.143		125	30.50	4.19	159.69	195	
	1300 Plates with adhesive attachment		35	1.143		123	30.50	4.19	157.69	193	
	3500 Ethylene propylene diene monomer (EPDM), 45 mils, 0.28 P.S.F.										
	3600 Loose-laid & ballasted with stone (10 P.S.F.)	G-5	51	.784	Sq.	55.50	21	2.88	79.38	99.50	
	3700 Mechanically attached		35	1.143		47.50	30.50	4.19	82.19	109	
	3800 Fully adhered with adhesive		26	1.538		67.50	41	5.65	114.15	151	
	4500 60 mils, 0.40 P.S.F.										
	4600 Loose-laid & ballasted with stone (10 P.S.F.)	G-5	51	.784	Sq.	67.50	21	2.88	91.38	113	
	4700 Mechanically attached		35	1.143		58.50	30.50	4.19	93.19	121	
	4800 Fully adhered with adhesive		26	1.538		79	41	5.65	125.65	163	
	4810 45 mil, .28 PSF, membrane only **CN**					29.50			29.50	32.50	
	4820 60 mil, .40 PSF, membrane only					39.50			39.50	43.50	
	4850 Seam tape for membrane, 4" x 100' roll				Ea.	67.50			67.50	74	
	4900 Batten strips, 10' sections					2.47			2.47	2.72	
	4910 Cover tape for batten strips, 6" x 100' roll					127			127	140	
	4930 Plate anchors				M	124			124	137	
	4970 Adhesive for fully adhered systems, 60 S.F./gal.				Gal.	12.45			12.45	13.70	
	7500 Polyisobutylene (PIB), 100 mils, 0.57 P.S.F.										
	7600 Loose-laid & ballasted with stone/gravel (10 P.S.F.)	G-5	51	.784	Sq.	124	21	2.88	147.88	176	
	7700 Partially adhered with adhesive		35	1.143		156	30.50	4.19	190.69	228	
	7800 Hot asphalt attachment		35	1.143		148	30.50	4.19	182.69	220	
	7900 Fully adhered with contact cement		26	1.538		161	41	5.65	207.65	253	

07540	PVC Single Ply Membrane	CREW	DAILY OUTPUT	LABOR-HOURS	UNIT	MAT.	LABOR	EQUIP.	TOTAL	TOTAL INCL O&P	
800	0010 **POLYVINYL CHLORIDE ROOFING**										800
	8200 Heat welded seams										
	8700 Reinforced, 48 mils, 0.33 P.S.F.										
	8750 Loose-laid & ballasted with stone/gravel (12 P.S.F.)	G-5	51	.784	Sq.	94.50	21	2.88	118.38	143	
	8800 Mechanically attached		35	1.143		86	30.50	4.19	120.69	151	
	8850 Fully adhered with adhesive		26	1.538		115	41	5.65	161.65	203	
	8860 Reinforced, 60 mils, .40 P.S.F.										
	8870 Loose-laid & ballasted with stone/gravel (12 P.S.F.)	G-5	51	.784	Sq.	87.50	21	2.88	111.38	135	
	8880 Mechanically attached		35	1.143		79	30.50	4.19	113.69	143	
	8890 Fully adhered with adhesive		26	1.538		108	41	5.65	154.65	195	

07550	Modified Bit. Membrane Roofing	CREW	DAILY OUTPUT	LABOR-HOURS	UNIT	MAT.	LABOR	EQUIP.	TOTAL	TOTAL INCL O&P	
500	0010 **MODIFIED BITUMEN ROOFING** R07550 -030										500
	0020 Base sheet, #15 glass fiber felt, nailed to deck	1 Rofc	58	.138	Sq.	4.85	4.06		8.91	12.25	
	0030 Spot mopped to deck	G-1	295	.190		6.50	5.20	1.01	12.71	17.15	
	0040 Fully mopped to deck	"	192	.292		8.90	8	1.55	18.45	25	
	0050 #15 organic felt, nailed to deck	1 Rofc	58	.138		4.17	4.06		8.23	11.50	
	0060 Spot mopped to deck	G-1	295	.190		5.80	5.20	1.01	12.01	16.40	
	0070 Fully mopped to deck	"	192	.292		8.20	8	1.55	17.75	24.50	
	0080 SBS modified, granule surf cap sheet, polyester rein., mopped										
	0600 150 mils	G-1	2,000	.028	S.F.	.45	.77	.15	1.37	1.97	
	1100 160 mils		2,000	.028		.66	.77	.15	1.58	2.20	
	1500 Glass fiber reinforced, mopped, 160 mils		2,000	.028		.42	.77	.15	1.34	1.93	
	1600 Smooth surface cap sheet, mopped, 145 mils		2,100	.027		.42	.73	.14	1.29	1.87	
	1700 Smooth surface flashing, 145 mils		1,260	.044		.42	1.22	.24	1.88	2.80	
	1800 150 mils		1,260	.044		.41	1.22	.24	1.87	2.79	

Important: See the Reference Section for critical supporting data - Reference Nos., Crews, & City Cost Indexes

7

THERMAL & MOISTURE PROTECTION

07550 | Modified Bit. Membrane Roofing

			CREW	DAILY OUTPUT	LABOR-HOURS	UNIT	2005 BARE COSTS MAT.	LABOR	EQUIP.	TOTAL	TOTAL INCL O&P	
500	1900	Granular surface flashing, 150 mils R07550-030	G-1	1,260	.044	S.F.	.45	1.22	.24	1.91	2.84	**500**
	2000	160 mils	↓	1,260	.044		.66	1.22	.24	2.12	3.07	
	2100	APP mod., smooth surf. cap sheet, poly. reinf., torched, 160 mils	G-5	2,100	.019		.41	.51	.07	.99	1.40	
	2150	170 mils		2,100	.019		.46	.51	.07	1.04	1.46	
	2200	Granule surface cap sheet, poly. reinf., torched, 180 mils		2,000	.020		.50	.53	.07	1.10	1.54	
	2250	Smooth surface flashing, torched, 160 mils		1,260	.032		.41	.85	.12	1.38	2.03	
	2300	170 mils		1,260	.032		.46	.85	.12	1.43	2.09	
	2350	Granule surface flashing, torched, 180 mils	↓	1,260	.032		.50	.85	.12	1.47	2.13	
	2400	Fibrated aluminum coating	1 Rofc	3,800	.002	↓	.10	.06		.16	.22	

07570 | Foamed Coated Roofing

			CREW	DAILY OUTPUT	LABOR-HOURS	UNIT	MAT.	LABOR	EQUIP.	TOTAL	TOTAL INCL O&P	
350	0010	**FOAMED COATED ROOFING**										**350**
	1600	Polyurethane spray-on with 20 mil silicone rubber coating applied										
	1700	1" thick, R7, minimum	G-2A	875	.027	S.F.	1.34	.71	.12	2.17	2.79	
	1800	Maximim		805	.030		1.80	.77	.13	2.70	3.41	
	1900	2" thick, R14, minimum		685	.035		1.90	.91	.16	2.97	3.76	
	2000	Maximum		575	.042		2.41	1.08	.19	3.68	4.65	
	2100	3" thick, R21, minimum		500	.048		2.50	1.24	.22	3.96	5.05	
	2200	Maximum	↓	440	.055	↓	3.20	1.41	.24	4.85	6.15	

07580 | Roll Roofing

			CREW	DAILY OUTPUT	LABOR-HOURS	UNIT	MAT.	LABOR	EQUIP.	TOTAL	TOTAL INCL O&P	
200	0010	**ROLL ROOFING**										**200**
	0100	Asphalt, mineral surface										
	0200	1 ply #15 organic felt, 1 ply mineral surfaced										
	0300	Selvage roofing, lap 19", nailed & mopped	G-1	27	2.074	Sq.	39	57	11.05	107.05	152	
	0400	3 plies glass fiber felt (type IV), 1 ply mineral surfaced										
	0500	Selvage roofing, lapped 19", mopped	G-1	25	2.240	Sq.	61	61.50	11.90	134.40	185	
	0600	Coated glass fiber base sheet, 2 plies of glass fiber										
	0700	Felt (type IV), 1 ply mineral surfaced selvage										
	0800	Roofing, lapped 19", mopped	G-1	25	2.240	Sq.	66	61.50	11.90	139.40	191	
	0900	On nailable decks	"	24	2.333	"	61	64	12.40	137.40	190	
	1000	3 plies glass fiber felt (type III), 1 ply mineral surfaced										
	1100	Selvage roofing, lapped 19", mopped	G-1	25	2.240	Sq.	61	61.50	11.90	134.40	185	

07590 | Roof Maintenance and Repairs

			CREW	DAILY OUTPUT	LABOR-HOURS	UNIT	MAT.	LABOR	EQUIP.	TOTAL	TOTAL INCL O&P	
300	0010	**ROOF COATINGS** Asphalt **CN**				Gal.	2.95			2.95	3.25	**300**
	0200	Asphalt base, fibered aluminum coating					9.60			9.60	10.55	
	0300	Asphalt primer, 5 gallon				↓	3.92			3.92	4.31	
	0600	Coal tar pitch, 200 lb. barrels				Ton	580			580	640	
	0700	Tar roof cement, 5 gal. lots				Gal.	6.10			6.10	6.70	
	0800	Glass fibered roof & patching cement, 5 gallon				"	4.43			4.43	4.87	
	0900	Reinforcing glass membrane, 450 S.F./roll				Ea.	45.50			45.50	50	
	1000	Neoprene roof coating, 5 gal, 2 gal/sq				Gal.	22			22	24	
	1100	Roof patch & flashing cement, 5 gallon					18.05			18.05	19.85	
	1200	Roof resturant, glass fibered, 3 gal/sq					7.30			7.30	8.05	
	1300	Mineral rubber, 3 gal/sq				↓	4.71			4.71	5.20	

7

THERMAL & MOISTURE PROTECTION

			DAILY	LABOR-		2005 BARE COSTS				TOTAL	
07610	**Sheet Metal Roofing**	CREW	OUTPUT	HOURS	UNIT	MAT.	LABOR	EQUIP.	TOTAL	INCL O&P	
300 0010	**COPPER ROOFING** Batten seam, over 10 sq, 16 oz, 130 lb/sq	1 Shee	1.10	7.273	Sq.	405	292		697	895	**300**
0200	18 oz, 145 lb per sq		1	8		450	320		770	990	
0300	20 oz, 160 lb per sq		1	8		500	320		820	1,050	
0400	Standing seam, over 10 squares, 16 oz, 125 lb per sq		1.30	6.154		390	247		637	805	
0600	18 oz, 140 lb per sq		1.20	6.667		435	268		703	890	
0700	20 oz, 150 lb per sq		1.10	7.273		465	292		757	965	
0900	Flat seam, over 10 squares, 16 oz, 115 lb per sq		1.20	6.667		355	268		623	805	
1000	20 oz, 145 lb per sq		1.10	7.273		450	292		742	945	
1200	For abnormal conditions or small areas, add					25%	100%				
1300	For lead-coated copper, add					25%					
500 0010	**LEAD ROOFING** 5 lb. per SF, batten seam	1 Shee	1.20	6.667	Sq.	385	268		653	830	**500**
0100	Flat seam	"	1.30	6.154	"	385	247		632	800	
700 0010	**STAINLESS STEEL ROOFING** Type 304, batten seam, 28 gauge	1 Shee	1.20	6.667	Sq.	305	268		573	750	**700**
0100	26 gauge	"	1.15	6.957		380	280		660	850	
0200	For standing seam construction, deduct					2%					
0500	For flat seam construction, deduct					3%					
0800	For lead or terne coated stainless, 28 gauge, add					70			70	77	
0900	For 26 gauge, add					92.50			92.50	102	
900 0010	**ZINC** Copper alloy roofing, batten seam, .020" thick	1 Shee	1.20	6.667	Sq.	520	268		788	980	**900**
0100	.027" thick		1.15	6.957		630	280		910	1,125	
0300	.032" thick		1.10	7.273		710	292		1,002	1,225	
0400	.040" thick		1.05	7.619		835	305		1,140	1,400	
0600	For standing seam construction, deduct					2%					
0700	For flat seam construction, deduct					3%					

07650	**Flexible Flashing**										
600 0010	**FLASHING** Aluminum, mill finish, .013" thick	1 Rofc	145	.055	S.F.	.35	1.62		1.97	3.16	**600**
0030	.016" thick		145	.055		.51	1.62		2.13	3.33	
0060	.019" thick		145	.055		.65	1.62		2.27	3.49	
0100	.032" thick		145	.055		1.06	1.62		2.68	3.94	
0200	.040" thick		145	.055		1.45	1.62		3.07	4.37	
0300	.050" thick		145	.055		1.84	1.62		3.46	4.79	
0325	Mill finish 5" x 7" step flashing, .016" thick		1,920	.004	Ea.	.10	.12		.22	.32	
0350	Mill finish 12" x 12" step flashing, .016" thick		1,600	.005	"	.40	.15		.55	.69	
0400	Painted finish, add				S.F.	.24			.24	.26	
0500	Fabric-backed 2 sides, .004" thick	1 Rofc	330	.024		.93	.71		1.64	2.24	
0700	.005" thick		330	.024		1.09	.71		1.80	2.42	
0750	Mastic-backed, self adhesive		460	.017		2.26	.51		2.77	3.36	
0800	Mastic-coated 2 sides, .004" thick		330	.024		.93	.71		1.64	2.24	
1000	.005" thick		330	.024		1.09	.71		1.80	2.42	
1100	.016" thick		330	.024		1.20	.71		1.91	2.54	
1300	Asphalt flashing cement, 5 gallon				Gal.	3.98			3.98	4.38	
1600	Copper, 16 oz, sheets, under 1000 lbs.	1 Rofc	115	.070	S.F.	2.62	2.05		4.67	6.35	
1700	Over 4000 lbs.		155	.052		2.91	1.52		4.43	5.80	
1900	20 oz sheets, under 1000 lbs.		110	.073		3.90	2.14		6.04	7.95	
2000	Over 4000 lbs.		145	.055		3.62	1.62		5.24	6.75	
2200	24 oz sheets, under 1000 lbs.		105	.076		4.69	2.24		6.93	8.95	
2300	Over 4000 lbs.		135	.059		4.34	1.75		6.09	7.75	
2500	32 oz sheets, under 1000 lbs.		100	.080		6.20	2.36		8.56	10.85	
2600	Over 4000 lbs.		130	.062		5.80	1.81		7.61	9.50	
2700	W shape for valleys, 16 oz, 24" wide		100	.080	L.F.	5.90	2.36		8.26	10.50	
2800	Copper, paperbacked 1 side, 2 oz		330	.024	S.F.	.86	.71		1.57	2.17	
2900	3 oz		330	.024		1.12	.71		1.83	2.45	
3100	Paperbacked 2 sides, 2 oz		330	.024		.86	.71		1.57	2.17	
3150	3 oz		330	.024		1.11	.71		1.82	2.44	
3200	5 oz		330	.024		1.66	.71		2.37	3.05	

Important: See the Reference Section for critical supporting data - Reference Nos., Crews, & City Cost Indexes

07650 | Flexible Flashing

		CREW	DAILY OUTPUT	LABOR-HOURS	UNIT	MAT.	LABOR	EQUIP.	TOTAL	TOTAL INCL O&P	
3250	7 oz	1 Rofc	330	.024	S.F.	2.72	.71		3.43	4.21	600
3400	Mastic-backed 2 sides, copper, 2 oz		330	.024		1.02	.71		1.73	2.34	
3500	3 oz		330	.024		1.26	.71		1.97	2.61	
3700	5 oz		330	.024		1.85	.71		2.56	3.26	
3800	Fabric-backed 2 sides, copper, 2 oz		330	.024		1.09	.71		1.80	2.42	
4000	3 oz		330	.024		1.40	.71		2.11	2.76	
4100	5 oz		330	.024		1.90	.71		2.61	3.31	
4300	Copper-clad stainless steel, .015" thick, under 500 lbs.		115	.070		3.16	2.05		5.21	6.95	
4400	Over 2000 lbs.		155	.052		3.05	1.52		4.57	5.95	
4600	.018" thick, under 500 lbs.		100	.080		4.18	2.36		6.54	8.60	
4700	Over 2000 lbs.		145	.055	▼	3.06	1.62		4.68	6.15	
4900	Fabric, asphalt-saturated cotton, specification grade		35	.229	S.Y.	1.93	6.75		8.68	13.55	
5000	Utility grade		35	.229		1.22	6.75		7.97	12.80	
5200	Open-mesh fabric, saturated, 40 oz per S.Y.		35	.229		1.35	6.75		8.10	12.95	
5300	Close-mesh fabric, saturated, 17 oz per S.Y.		35	.229		1.42	6.75		8.17	13	
5500	Fiberglass, resin-coated		35	.229		1.16	6.75		7.91	12.75	
5600	Asphalt-coated, 40 oz per S.Y.		35	.229	▼	7.90	6.75		14.65	20	
5800	Lead, 2.5 lb. per SF, up to 12" wide		135	.059	S.F.	2.94	1.75		4.69	6.20	
5900	Over 12" wide		135	.059		3.25	1.75		5	6.55	
6100	Lead-coated copper, fabric-backed, 2 oz		330	.024		1.41	.71		2.12	2.77	
6200	5 oz		330	.024		1.62	.71		2.33	3	
6400	Mastic-backed 2 sides, 2 oz		330	.024		1.10	.71		1.81	2.43	
6500	5 oz		330	.024		1.37	.71		2.08	2.73	
6700	Paperbacked 1 side, 2 oz		330	.024		.95	.71		1.66	2.27	
6800	3 oz		330	.024		1.12	.71		1.83	2.45	
7000	Paperbacked 2 sides, 2 oz		330	.024		.98	.71		1.69	2.30	
7100	5 oz		330	.024		1.60	.71		2.31	2.98	
7300	Polyvinyl chloride, black, .010" thick		285	.028		.16	.83		.99	1.59	
7400	.020" thick		285	.028		.23	.83		1.06	1.66	
7600	.030" thick		285	.028		.30	.83		1.13	1.74	
7700	.056" thick		285	.028		.71	.83		1.54	2.19	
7900	Black or white for exposed roofs, .060" thick	▼	285	.028	▼	1.55	.83		2.38	3.12	
8060	PVC tape, 5" x 45 mils, for joint covers, 100 L.F./roll				Ea.	79.50			79.50	87	
8100	Rubber, butyl, 1/32" thick	1 Rofc	285	.028	S.F.	.70	.83		1.53	2.18	
8200	1/16" thick		285	.028		1.05	.83		1.88	2.57	
8300	Neoprene, cured, 1/16" thick		285	.028		1.48	.83		2.31	3.04	
8400	1/8" thick		285	.028		2.99	.83		3.82	4.70	
8500	Shower pan, bituminous membrane, 7 oz		155	.052		1.08	1.52		2.60	3.78	
8550	3 ply copper and fabric, 3 oz		155	.052		1.63	1.52		3.15	4.38	
8600	7 oz		155	.052		3.37	1.52		4.89	6.30	
8650	Copper, 16 oz		100	.080		3.11	2.36		5.47	7.45	
8700	Lead on copper and fabric, 5 oz		155	.052		1.62	1.52		3.14	4.37	
8800	7 oz		155	.052		2.93	1.52		4.45	5.80	
8850	Polyvinyl chloride, .030" thick		160	.050		.30	1.47		1.77	2.84	
8900	Stainless steel sheets, 32 ga, .010" thick		155	.052		2.16	1.52		3.68	4.97	
9000	28 ga, .015" thick		155	.052		2.68	1.52		4.20	5.55	
9100	26 ga, .018" thick		155	.052		3.25	1.52		4.77	6.15	
9200	24 ga, .025" thick	▼	155	.052	▼	4.22	1.52		5.74	7.25	
9290	For mechanically keyed flashing, add					40%					
9300	Stainless steel, paperbacked 2 sides, .005" thick	1 Rofc	330	.024	S.F.	1.90	.71		2.61	3.31	
9400	Terne coated stainless steel, .015" thick, 28 ga		155	.052		4.05	1.52		5.57	7.05	
9500	.018" thick, 26 ga		155	.052		4.57	1.52		6.09	7.65	
9600	Zinc and copper alloy (brass), .020" thick		155	.052		3.26	1.52		4.78	6.20	
9700	.027" thick		155	.052		4.37	1.52		5.89	7.40	
9800	.032" thick		155	.052		5.10	1.52		6.62	8.20	
9900	.040" thick	▼	155	.052	▼	6.20	1.52		7.72	9.45	

2005 BARE COSTS

THERMAL & MOISTURE PROTECTION 7

07710	Manufactured Roof Specialties	CREW	DAILY OUTPUT	LABOR-HOURS	UNIT	2005 BARE COSTS				TOTAL INCL O&P	
						MAT.	LABOR	EQUIP.	TOTAL		
400	**0010**	**DOWNSPOUTS** Aluminum 2" x 3", .020" thick, embossed	1 Shee	190	.042	L.F.	.68	1.69		2.37	3.35	**400**
0100	Enameled		190	.042		.97	1.69		2.66	3.67
0300	Enameled, .024" thick, 2" x 3"		180	.044		1.12	1.79		2.91	3.98
0400	3" x 4"		140	.057		1.73	2.30		4.03	5.45
0600	Round, corrugated aluminum, 3" diameter, .020" thick		190	.042		1.04	1.69		2.73	3.74
0700	4" diameter, .025" thick		140	.057	↓	1.49	2.30		3.79	5.15
0900	Wire strainer, round, 2" diameter		155	.052	Ea.	1.75	2.07		3.82	5.10
1000	4" diameter		155	.052		1.82	2.07		3.89	5.20
1200	Rectangular, perforated, 2" x 3"		145	.055		2.15	2.22		4.37	5.80
1300	3" x 4"		145	.055	↓	3.10	2.22		5.32	6.80
1500	Copper, round, 16 oz., stock, 2" diameter		190	.042	L.F.	4.55	1.69		6.24	7.60
1600	3" diameter		190	.042		4.50	1.69		6.19	7.55
1800	4" diameter		145	.055		4.91	2.22		7.13	8.80
1900	5" diameter		130	.062		6.75	2.47		9.22	11.20
2100	Rectangular, corrugated copper, stock, 2" x 3"		190	.042		3.91	1.69		5.60	6.90
2200	3" x 4"		145	.055		4.96	2.22		7.18	8.85
2400	Rectangular, plain copper, stock, 2" x 3"		190	.042		4.78	1.69		6.47	7.85
2500	3" x 4"		145	.055	↓	6.30	2.22		8.52	10.30
2700	Wire strainers, rectangular, 2" x 3"		145	.055	Ea.	2.80	2.22		5.02	6.50
2800	3" x 4"		145	.055		4.44	2.22		6.66	8.30
3000	Round, 2" diameter		145	.055		2.63	2.22		4.85	6.30
3100	3" diameter		145	.055		3.67	2.22		5.89	7.45
3300	4" diameter		145	.055		5.70	2.22		7.92	9.65
3400	5" diameter		115	.070	↓	8.20	2.80		11	13.35
3600	Lead-coated copper, round, stock, 2" diameter		190	.042	L.F.	5.50	1.69		7.19	8.65
3700	3" diameter		190	.042		6.35	1.69		8.04	9.55
3900	4" diameter		145	.055		8.55	2.22		10.77	12.80
4200	6" diameter, corrugated		105	.076		14.85	3.06		17.91	21
4300	Rectangular, corrugated, stock, 2" x 3"		190	.042		7.35	1.69		9.04	10.65
4500	Plain, stock, 2" x 3"		190	.042		7.35	1.69		9.04	10.70
4600	3" x 4"		145	.055		7.95	2.22		10.17	12.15
4800	Steel, galvanized, round, corrugated, 2" or 3" diam, 28 ga		190	.042		.75	1.69		2.44	3.43
4900	4" diameter, 28 gauge		145	.055		1.02	2.22		3.24	4.53
5100	5" diameter, 28 gauge		130	.062		1.54	2.47		4.01	5.50
5200	26 gauge		130	.062		1.36	2.47		3.83	5.30
5400	6" diameter, 28 gauge		105	.076		2.15	3.06		5.21	7.10
5500	26 gauge		105	.076		1.42	3.06		4.48	6.25
5700	Rectangular, corrugated, 28 gauge, 2" x 3"		190	.042		.54	1.69		2.23	3.19
5800	3" x 4"		145	.055		1.50	2.22		3.72	5.05
6000	Rectangular, plain, 28 gauge, galvanized, 2" x 3"		190	.042		.81	1.69		2.50	3.49
6100	3" x 4"		145	.055		1.23	2.22		3.45	4.76
6300	Epoxy painted, 24 gauge, corrugated, 2" x 3"		190	.042		1.05	1.69		2.74	3.76
6400	3" x 4"		145	.055	↓	1.91	2.22		4.13	5.50
6600	Wire strainers, rectangular, 2" x 3"		145	.055	Ea.	1.62	2.22		3.84	5.20
6700	3" x 4"		145	.055		2.61	2.22		4.83	6.30
6900	Round strainers, 2" or 3" diameter		145	.055		1.20	2.22		3.42	4.73
7000	4" diameter		145	.055		1.42	2.22		3.64	4.97
7200	5" diameter		145	.055		2.20	2.22		4.42	5.85
7300	6" diameter		115	.070	↓	2.63	2.80		5.43	7.20
7500	Steel pipe, black, extra heavy, 4" diameter		20	.400	L.F.	11.90	16.10		28	37.50
7600	6" diameter		18	.444		25.50	17.85		43.35	56
7800	Stainless steel tubing, schedule 5, 2" x 3" or 3" diameter		190	.042		18.50	1.69		20.19	23
7900	3" x 4" or 4" diameter		145	.055		23.50	2.22		25.72	29.50
8100	4" x 5" or 5" diameter		135	.059		48.50	2.38		50.88	57
8200	Vinyl, rectangular, 2" x 3"		210	.038	↓	.72	1.53		2.25	3.15
8300	Round, 2-1/2"	↓	220	.036	↓	.72	1.46		2.18	3.04

THERMAL & MOISTURE PROTECTION

7

Important: See the Reference Section for critical supporting data - Reference Nos., Crews, & City Cost Indexes

07710 | Manufactured Roof Specialties

		CREW	DAILY OUTPUT	LABOR-HOURS	UNIT	2005 BARE COSTS MAT.	LABOR	EQUIP.	TOTAL	TOTAL INCL O&P	
450	0010	**DRIP EDGE**, aluminum, .016" thick, 5" wide, mill finish	1 Carp	400	.020	L.F.	.25	.69		.94	1.35	**450**
	0100	White finish		400	.020		.22	.69		.91	1.31	
	0200	8" wide, mill finish		400	.020		.30	.69		.99	1.40	
	0300	Ice belt, 28" wide, mill finish		100	.080		3.42	2.74		6.16	8.05	
	0310	Vented, mill finish		400	.020		1.41	.69		2.10	2.62	
	0320	Painted finish		400	.020		1.54	.69		2.23	2.76	
	0400	Galvanized, 5" wide		400	.020		.22	.69		.91	1.31	
	0500	8" wide, mill finish		400	.020		.33	.69		1.02	1.43	
	0510	Rake edge, aluminum, 1-1/2" x 1-1/2"		400	.020		.13	.69		.82	1.21	
	0520	3-1/2" x 1-1/2"		400	.020		.19	.69		.88	1.28	
500	0010	**ELBOWS** Aluminum, 2" x 3", embossed	1 Shee	100	.080	Ea.	.91	3.22		4.13	5.95	**500**
	0100	Enameled		100	.080		1.76	3.22		4.98	6.90	
	0200	3" x 4", .025" thick, embossed		100	.080		3.20	3.22		6.42	8.45	
	0300	Enameled		100	.080		3.20	3.22		6.42	8.45	
	0400	Round corrugated, 3", embossed, .020" thick		100	.080		1.98	3.22		5.20	7.15	
	0500	4", .025" thick		100	.080		2.99	3.22		6.21	8.25	
	0600	Copper, 16 oz. round, 2" diameter		100	.080		11.15	3.22		14.37	17.25	
	0700	3" diameter		100	.080		5.95	3.22		9.17	11.45	
	0800	4" diameter		100	.080		9.05	3.22		12.27	14.90	
	1000	2" x 3" corrugated		100	.080		5.15	3.22		8.37	10.65	
	1100	3" x 4" corrugated		100	.080		7.65	3.22		10.87	13.35	
	1300	Vinyl, 2-1/2" diameter, 45° or 75°		100	.080		2	3.22		5.22	7.15	
	1400	Tee Y junction		75	.107		8.50	4.29		12.79	15.95	
550	0010	**GRAVEL STOP** Aluminum, .050" thick, 4" face height, mill finish	1 Shee	145	.055	L.F.	3.82	2.22		6.04	7.60	**550**
	0080	Duranodic finish		145	.055		3.69	2.22		5.91	7.45	
	0100	Painted		145	.055		4.26	2.22		6.48	8.10	
	0300	6" face height		135	.059		4.10	2.38		6.48	8.20	
	0350	Duranodic finish		135	.059		4.35	2.38		6.73	8.45	
	0400	Painted		135	.059		5.05	2.38		7.43	9.20	
	0600	8" face height		125	.064		5.15	2.57		7.72	9.60	
	0650	Duranodic finish		125	.064		4.98	2.57		7.55	9.45	
	0700	Painted		125	.064		5.05	2.57		7.62	9.55	
	0900	12" face height, .080 thick, 2 piece		100	.080		6.85	3.22		10.07	12.45	
	0950	Duranodic finish		100	.080		6.30	3.22		9.52	11.85	
	1000	Painted		100	.080		7.40	3.22		10.62	13.10	
	1350	Galv steel, 24 ga., 4" leg, plain, with continuous cleat, 4" face		145	.055		1.50	2.22		3.72	5.05	
	1360	6" face height		145	.055		2.27	2.22		4.49	5.90	
	1500	Polyvinyl chloride, 6" face height		135	.059		3.28	2.38		5.66	7.30	
	1600	9" face height		125	.064		3.87	2.57		6.44	8.20	
	1800	Stainless steel, 24 ga., 6" face height		135	.059		7.15	2.38		9.53	11.50	
	1900	12" face height		100	.080		15	3.22		18.22	21.50	
	2100	20 ga., 6" face height		135	.059		8.10	2.38		10.48	12.55	
	2200	12" face height		100	.080		17.10	3.22		20.32	24	
650	0010	**GUTTERS** Aluminum, stock units, 5" box, .027" thick, plain	1 Shee	120	.067	L.F.	1.24	2.68		3.92	5.50	**650**
	0100	Enameled		120	.067		1.09	2.68		3.77	5.30	
	0300	5" box type, .032" thick, plain		120	.067		1.22	2.68		3.90	5.45	
	0400	Enameled		120	.067		1.21	2.68		3.89	5.45	
	0600	5" x 6" combination fascia & gutter, .032" thick, enameled		60	.133		3.52	5.35		8.87	12.10	
	0700	Copper, half round, 16 oz, stock units, 4" wide		120	.067		4.10	2.68		6.78	8.65	
	0900	5" wide		120	.067		5.35	2.68		8.03	10	
	1000	6" wide		115	.070		6.25	2.80		9.05	11.20	
	1200	K type, 16 oz, stock, 4" wide		120	.067		5.05	2.68		7.73	9.65	
	1300	5" wide		120	.067		5.50	2.68		8.18	10.15	
	1500	Lead coated copper, half round, stock, 4" wide		120	.067		7.50	2.68		10.18	12.35	
	1600	6" wide		115	.070		10.40	2.80		13.20	15.75	

THERMAL & MOISTURE PROTECTION 7

		07710	**Manufactured Roof Specialties**	CREW	DAILY OUTPUT	LABOR-HOURS	UNIT	MAT.	LABOR	EQUIP.	TOTAL	TOTAL INCL O&P	
												2005 BARE COSTS	
650	1800		K type, stock, 4" wide	1 Shee	120	.067	L.F.	8.10	2.68		10.78	13	**650**
	1900		5" wide		120	.067		8.25	2.68		10.93	13.20	
	2100		Stainless steel, half round or box, stock, 4" wide		120	.067		4.65	2.68		7.33	9.20	
	2200		5" wide		120	.067		5	2.68		7.68	9.60	
	2400		Steel, galv, half round or box, 28 ga, 5" wide, plain		120	.067		1	2.68		3.68	5.20	
	2500		Enameled		120	.067		1.12	2.68		3.80	5.35	
	2700		26 ga, stock, 5" wide		120	.067		.95	2.68		3.63	5.15	
	2800		6" wide		120	.067		1.60	2.68		4.28	5.90	
	3000		Vinyl, O.G., 4" wide	1 Carp	110	.073		.85	2.49		3.34	4.82	
	3100		5" wide		110	.073		1	2.49		3.49	4.98	
	3200		4" half round, stock units		110	.073		.68	2.49		3.17	4.63	
	3250		Joint connectors				Ea.	1.36			1.36	1.50	
	3300		Wood, clear treated cedar, fir or hemlock, 3" x 4"	1 Carp	100	.080	L.F.	6.30	2.74		9.04	11.15	
	3400		4" x 5"	"	100	.080	"	7.30	2.74		10.04	12.25	
700	0010		**GUTTER GUARD** 6" wide strip, aluminum mesh	1 Carp	500	.016	L.F.	.37	.55		.92	1.26	**700**
	0100		Vinyl mesh		500	.016	"	.40	.55		.95	1.29	
750	0010		**REGLET** Aluminum, .025" thick, in concrete parapet		225	.036	L.F.	.94	1.22		2.16	2.93	**750**
	0100		Copper, 10 oz.		225	.036		1.62	1.22		2.84	3.68	
	0300		16 oz.		225	.036		2.16	1.22		3.38	4.28	
	0400		Galvanized steel, 24 gauge		225	.036		.69	1.22		1.91	2.66	
	0600		Stainless steel, .020" thick		225	.036		1.60	1.22		2.82	3.66	
	0700		Zinc and copper alloy, 20 oz.		225	.036		1.79	1.22		3.01	3.87	
	0900		Counter flashing for above, 12" wide, .032" aluminum	1 Shee	150	.053		1.21	2.14		3.35	4.63	
	1000		Copper, 10 oz.		150	.053		3.39	2.14		5.53	7.05	
	1200		16 oz.		150	.053		3.77	2.14		5.91	7.45	
	1300		Galvanized steel, .020" thick		150	.053		.63	2.14		2.77	3.99	
	1500		Stainless steel, .020" thick		150	.053		2.82	2.14		4.96	6.40	
	1600		Zinc and copper alloy, 20 oz.		150	.053		3.17	2.14		5.31	6.80	
800	0010		**EXPANSION JOINT**										**800**
	0300		Butyl or neoprene center with foam insulation, metal flanges										
	0400		Aluminum, .032" thick for openings to 2-1/2"	1 Rofc	165	.048	L.F.	7.55	1.43		8.98	10.75	
	0600		For joint openings to 3-1/2"		165	.048		8.80	1.43		10.23	12.15	
	0610		For joint openings to 5"		165	.048		10.60	1.43		12.03	14.10	
	0620		For joint openings to 8"		165	.048		16.75	1.43		18.18	21	
	0700		Copper, 16 oz. for openings to 2-1/2"		165	.048		10.60	1.43		12.03	14.10	
	0900		For joint openings to 3-1/2"		165	.048		12.20	1.43		13.63	15.85	
	0910		For joint openings to 5"		165	.048		14.30	1.43		15.73	18.15	
	0920		For joint openings to 8"		165	.048		21.50	1.43		22.93	26	
	1000		Galvanized steel, 26 ga. for openings to 2-1/2"		165	.048		6.40	1.43		7.83	9.45	
	1200		For joint openings to 3-1/2"		165	.048		7.55	1.43		8.98	10.75	
	1210		For joint openings to 5"		165	.048		9.45	1.43		10.88	12.85	
	1220		For joint openings to 8"		165	.048		16.05	1.43		17.48	20	
	1300		Lead-coated copper, 16 oz. for openings to 2-1/2"		165	.048		18.85	1.43		20.28	23.50	
	1500		For joint openings to 3-1/2"		165	.048		22.50	1.43		23.93	27	
	1600		Stainless steel, .018", for openings to 2-1/2"		165	.048		9.45	1.43		10.88	12.85	
	1800		For joint openings to 3-1/2"		165	.048		10.75	1.43		12.18	14.30	
	1810		For joint openings to 5"		165	.048		13.30	1.43		14.73	17.10	
	1820		For joint openings to 8"		165	.048		20.50	1.43		21.93	25	
	1900		Neoprene, double-seal type with thick center, 4-1/2" wide		125	.064		9.10	1.88		10.98	13.20	
	1950		Polyethylene bellows, with galv steel flat flanges		100	.080		3.52	2.36		5.88	7.90	
	1960		With galvanized angle flanges		100	.080		3.88	2.36		6.24	8.30	
	2000		Roof joint with extruded aluminum cover, 2"	1 Shee	115	.070		24.50	2.80		27.30	31	
	2100		Roof joint, plastic curbs, foam center, standard	1 Rofc	100	.080		8.85	2.36		11.21	13.75	
	2200		Large		100	.080		11.85	2.36		14.21	17	
	2300		Transitions, regular, minimum		10	.800	Ea.	76.50	23.50		100	124	
	2350		Maximum		4	2		97.50	59		156.50	207	

Important: See the Reference Section for critical supporting data - Reference Nos., Crews, & City Cost Indexes

7 THERMAL & MOISTURE PROTECTION

07710 | Manufactured Roof Specialties

		CREW	DAILY OUTPUT	LABOR-HOURS	UNIT	MAT.	LABOR	EQUIP.	TOTAL	TOTAL INCL O&P	
800	2400	Large, minimum	1 Rofc	9	.889	Ea.	112	26		138	168	800
	2450	Maximum	↓	3	2.667	↓	117	78.50		195.50	263	
	2500	Roof to wall joint with extruded aluminum cover	1 Shee	115	.070	L.F.	20.50	2.80		23.30	27	
	2600	See also divisions 03150-250 & 05810-350										
	2700	Wall joint, closed cell foam on PVC cover, 9" wide	1 Rofc	125	.064	L.F.	3.16	1.88		5.04	6.70	
	2800	12" wide	"	115	.070	"	3.57	2.05		5.62	7.40	

07720 | Roof Accessories

		CREW	DAILY OUTPUT	LABOR-HOURS	UNIT	MAT.	LABOR	EQUIP.	TOTAL	TOTAL INCL O&P	
480	0010	**PITCH POCKETS**										480
	0100	Adjustable, 4" to 7", welded corners, 4" deep	1 Rofc	48	.167	Ea.	10.50	4.91		15.41	19.90	
	0200	Side extenders, 6"	"	240	.033	"	1.70	.98		2.68	3.54	
500	0010	**ROOF VENTS** Mushroom for built-up roofs, aluminum	1 Rofc	30	.267	Ea.	24	7.85		31.85	40	500
	0100	PVC, 6" high	"	30	.267	"	27.50	7.85		35.35	44	
550	0010	**RIDGE VENT**										550
	0100	Aluminum strips, mill finish	1 Rofc	160	.050	L.F.	1.20	1.47		2.67	3.83	
	0150	Painted finish		160	.050	"	2.12	1.47		3.59	4.85	
	0200	Connectors		48	.167	Ea.	1.91	4.91		6.82	10.45	
	0300	End caps		48	.167	"	.93	4.91		5.84	9.35	
	0400	Galvanized strips		160	.050	L.F.	2.07	1.47		3.54	4.79	
	0430	Molded polyethylene, shingles not included		160	.050	"	2.55	1.47		4.02	5.30	
	0440	End plugs		48	.167	Ea.	.93	4.91		5.84	9.35	
	0450	Flexible roll, shingles not included	↓	160	.050	L.F.	1.99	1.47		3.46	4.70	
560	0010	**SNOW GUARDS**										560
	0100	Slate & asphalt shingle roofs	1 Rofc	160	.050	Ea.	8	1.47		9.47	11.30	
	0200	Standing seam metal roofs		48	.167		12.25	4.91		17.16	22	
	0300	Surface mount for metal roofs	↓	48	.167	↓	6.75	4.91		11.66	15.80	
700	0010	**ROOF HATCHES** With curb, 1" fiberglass insulation, 2'-6" x 3'-0"										700
	0500	Aluminum curb and cover	G-3	10	3.200	Ea.	510	107		617	725	
	0520	Galvanized steel curb and aluminum cover CN		10	3.200		350	107		457	550	
	0540	Galvanized steel curb and cover		10	3.200		415	107		522	620	
	0600	2'-6" x 4'-6", aluminum curb and cover		9	3.556		605	119		724	850	
	0800	Galvanized steel curb and aluminum cover		9	3.556		500	119		619	735	
	0900	Galvanized steel curb and cover		9	3.556		635	119		754	880	
	1100	4' x 4' aluminum curb and cover		8	4		620	134		754	885	
	1120	Galvanized steel curb and aluminum cover		8	4		500	134		634	755	
	1140	Galvanized steel curb and cover		8	4		580	134		714	845	
	1200	2'-6" x 8'-0", aluminum curb and cover		6.60	4.848		1,250	162		1,412	1,650	
	1400	Galvanized steel curb and aluminum cover		6.60	4.848		965	162		1,127	1,300	
	1500	Galvanized steel curb and cover	↓	6.60	4.848		875	162		1,037	1,200	
	1800	For plexiglass panels, 2'-6" x 3'-0", add to above				↓	345			345	380	
800	0010	**WALKWAY** For built-up roofs, asphalt impregnated, 3' x 6' x 1/2" thk	1 Rofc	400	.020	S.F.	1.13	.59		1.72	2.25	800
	0100	3' x 3' x 3/4" thick	"	400	.020		2.17	.59		2.76	3.39	
	0300	Concrete patio blocks, 2" thick, natural	1 Clab	115	.070		1.39	1.86		3.25	4.42	
	0400	Colors	"	115	.070	↓	1.75	1.86		3.61	4.82	
850	0010	**SMOKE HATCHES** Unlabeled, not including hand winch operator										850
	0200	For 3'-0" long, add to roof hatches from division 07720-700				Ea.	25%	5%				
	0250	For 4'-0" long, add to roof hatches from division 07720-700					20%	5%				
	0300	For 8'-0" long, add to roof hatches from division 07720-700				↓	10%	5%				
860	0010	**SMOKE VENT**, insulated, 4' x 4'										860
	0100	Aluminum cover and frame	G-3	13	2.462	Ea.	1,050	82.50		1,132.50	1,275	
	0200	Galvanized steel cover and frame		13	2.462		950	82.50		1,032.50	1,175	
	0300	4' x 8' aluminum cover and frame		8	4		1,425	134		1,559	1,775	
	0400	Galvanized steel cover and frame	↓	8	4	↓	1,250	134		1,384	1,575	

07720 | Roof Accessories

			CREW	DAILY OUTPUT	LABOR-HOURS	UNIT	2005 BARE COSTS				TOTAL INCL O&P	
							MAT.	LABOR	EQUIP.	TOTAL		
870	0010	**VENTS, ONE-WAY** For insul. decks, 1 per M.S.F., plastic, min.	1 Rofc	40	.200	Ea.	12.50	5.90		18.40	24	870
	0100	Maximum		20	.400		28.50	11.80		40.30	51.50	
	0300	Aluminum	↓	30	.267		12.50	7.85		20.35	27	
	0800	Polystyrene baffles, 12" wide for 16" O.C. rafter spacing	1 Carp	90	.089		.45	3.04		3.49	5.25	
	0900	For 24" O.C. rafter spacing	"	110	.073	↓	1.05	2.49		3.54	5.05	

07812 | Cementitious Fireproofing

			CREW	DAILY OUTPUT	LABOR-HOURS	UNIT	2005 BARE COSTS				TOTAL INCL O&P	
							MAT.	LABOR	EQUIP.	TOTAL		
600	0010	**SPRAYED** Mineral fiber or cementitious for fireproofing,										600
	0050	not incl tamping or canvas protection										
	0100	1" thick, on flat plate steel	G-2	3,000	.008	S.F.	.41	.23	.04	.68	.85	
	0200	Flat decking		2,400	.010		.41	.28	.04	.73	.95	
	0400	Beams		1,500	.016		.41	.45	.07	.93	1.24	
	0500	Corrugated or fluted decks		1,250	.019		.62	.54	.09	1.25	1.61	
	0700	Columns, 1-1/8" thick		1,100	.022		.47	.62	.10	1.19	1.57	
	0800	2-3/16" thick	↓	700	.034	↓	.88	.97	.15	2	2.63	
	0850	For tamping, add						10%				
	0900	For canvas protection, add	G-2	5,000	.005	S.F.	.06	.14	.02	.22	.30	
	1000	Acoustical sprayed, 1" thick, finished, straight work, minimum		520	.046		.44	1.31	.21	1.96	2.72	
	1100	Maximum		200	.120		.47	3.41	.54	4.42	6.30	
	1300	Difficult access, minimum		225	.107		.47	3.03	.48	3.98	5.70	
	1400	Maximum	↓	130	.185	↓	.51	5.25	.83	6.59	9.50	
	1500	Intumescent epoxy fireproofing on wire mesh, 3/16" thick										
	1550	1 hour rating, exterior use	G-2	136	.176	S.F.	5.40	5	.79	11.19	14.50	
	1600	Magnesium oxychloride, 35# to 40# density, 1/4" thick		3,000	.008		1.14	.23	.04	1.41	1.64	
	1650	1/2" thick		2,000	.012		2.28	.34	.05	2.67	3.09	
	1700	60# to 70# density, 1/4" thick		3,000	.008		1.50	.23	.04	1.77	2.04	
	1750	1/2" thick		2,000	.012		3.03	.34	.05	3.42	3.91	
	2000	Vermiculite cement, troweled or sprayed, 1/4" thick		3,000	.008		1.03	.23	.04	1.30	1.52	
	2050	1/2" thick	↓	2,000	.012	↓	2.04	.34	.05	2.43	2.82	

07840 | Firestopping

			CREW	DAILY OUTPUT	LABOR-HOURS	UNIT	2005 BARE COSTS				TOTAL INCL O&P	
							MAT.	LABOR	EQUIP.	TOTAL		
100	0010	**FIRESTOPPING**										100
	0100	Metallic piping, non insulated	R07800 -030									
	0110	Through walls, 2" diameter	1 Carp	16	.500	Ea.	9.95	17.15		27.10	37.50	
	0120	4" diameter		14	.571		15.20	19.55		34.75	47.50	
	0130	6" diameter		12	.667		20.50	23		43.50	58	
	0140	12" diameter		10	.800		36.50	27.50		64	82.50	
	0150	Through floors, 2" diameter		32	.250		6.05	8.55		14.60	20	
	0160	4" diameter		28	.286		8.70	9.80		18.50	25	
	0170	6" diameter		24	.333		11.40	11.40		22.80	30.50	
	0180	12" diameter	↓	20	.400	↓	19.25	13.70		32.95	42.50	
	0190	Metallic piping, insulated										
	0200	Through walls, 2" diameter	1 Carp	16	.500	Ea.	14.15	17.15		31.30	42	
	0210	4" diameter		14	.571		19.40	19.55		38.95	52	
	0220	6" diameter		12	.667		24.50	23		47.50	62.50	
	0230	12" diameter		10	.800		40.50	27.50		68	87	
	0240	Through floors, 2" diameter		32	.250		10.20	8.55		18.75	24.50	

Important: See the Reference Section for critical supporting data - Reference Nos., Crews, & City Cost Indexes

		07840	Firestopping	CREW	DAILY OUTPUT	LABOR-HOURS	UNIT	2005 BARE COSTS				TOTAL INCL O&P	
								MAT.	LABOR	EQUIP.	TOTAL		
100	0250		4″ diameter	1 Carp	28	.286	Ea.	12.85	9.80		22.65	29.50	100
	0260		6″ diameter		24	.333		15.60	11.40		27	35	
	0270		12″ diameter		20	.400		19.25	13.70		32.95	42.50	
	0280	Non metallic piping, non insulated											
	0290		Through walls, 2″ diameter	1 Carp	12	.667	Ea.	41	23		64	81	
	0300		4″ diameter		10	.800		51.50	27.50		79	99.50	
	0310		6″ diameter		8	1		72	34.50		106.50	133	
	0330		Through floors, 2″ diameter		16	.500		32	17.15		49.15	62	
	0340		4″ diameter		6	1.333		40	45.50		85.50	115	
	0350		6″ diameter		6	1.333		48	45.50		93.50	124	
	0370	Ductwork, insulated & non insulated, round											
	0380		Through walls, 6″ diameter	1 Carp	12	.667	Ea.	21	23		44	58.50	
	0390		12″ diameter		10	.800		41.50	27.50		69	88.50	
	0400		18″ diameter		8	1		67.50	34.50		102	128	
	0410		Through floors, 6″ diameter		16	.500		11.45	17.15		28.60	39	
	0420		12″ diameter		14	.571		21	19.55		40.55	53.50	
	0430		18″ diameter		12	.667		36.50	23		59.50	75.50	
	0440	Ductwork, insulated & non insulated, rectangular											
	0450		With stiffener/closure angle, through walls, 6″ x 12″	1 Carp	8	1	Ea.	17.35	34.50		51.85	72.50	
	0460		12″ x 24″		6	1.333		23	45.50		68.50	96.50	
	0470		24″ x 48″		4	2		65.50	68.50		134	179	
	0480		With stiffener/closure angle, through floors, 6″ x 12″		10	.800		9.35	27.50		36.85	53	
	0490		12″ x 24″		8	1		16.85	34.50		51.35	72	
	0500		24″ x 48″		6	1.333		33	45.50		78.50	108	
	0510	Multi trade openings											
	0520		Through walls, 6″ x 12″	1 Carp	2	4	Ea.	36.50	137		173.50	253	
	0530		12″ x 24″	″	1	8		147	274		421	585	
	0540		24″ x 48″	2 Carp	1	16		585	550		1,135	1,500	
	0550		48″ x 96″	″	.75	21.333		2,350	730		3,080	3,750	
	0560		Through floors, 6″ x 12″	1 Carp	2	4		36.50	137		173.50	253	
	0570		12″ x 24″	″	1	8		147	274		421	585	
	0580		24″ x 48″	2 Carp	.75	21.333		585	730		1,315	1,800	
	0590		48″ x 96″	″	.50	32		2,350	1,100		3,450	4,300	
	0600	Structural penetrations, through walls											
	0610		Steel beams, W8 x 10	1 Carp	8	1	Ea.	23	34.50		57.50	78.50	
	0620		W12 x 14		6	1.333		36.50	45.50		82	111	
	0630		W21 x 44		5	1.600		73	55		128	166	
	0640		W36 x 135		3	2.667		177	91.50		268.50	335	
	0650		Bar joists, 18″ deep		6	1.333		33.50	45.50		79	108	
	0660		24″ deep		6	1.333		41.50	45.50		87	117	
	0670		36″ deep		5	1.600		62.50	55		117.50	154	
	0680		48″ deep		4	2		73	68.50		141.50	187	
	0690	Construction joints, floor slab at exterior wall											
	0700		Precast, brick, block or drywall exterior											
	0710		2″ wide joint	1 Carp	125	.064	L.F.	5.20	2.19		7.39	9.10	
	0720		4″ wide joint	″	75	.107	″	10.40	3.65		14.05	17.15	
	0730		Metal panel, glass or curtain wall exterior											
	0740		2″ wide joint	1 Carp	40	.200	L.F.	12.30	6.85		19.15	24	
	0750		4″ wide joint	″	25	.320	″	16.80	10.95		27.75	35.50	
	0760		Floor slab to drywall partition											
	0770		Flat joint	1 Carp	100	.080	L.F.	5.10	2.74		7.84	9.85	
	0780		Fluted joint		50	.160		10.40	5.50		15.90	20	
	0790		Etched fluted joint		75	.107		6.75	3.65		10.40	13.15	
	0800		Floor slab to concrete/masonry partition											
	0810		Flat joint	1 Carp	75	.107	L.F.	11.45	3.65		15.10	18.30	
	0820		Fluted joint	″	50	.160	″	13.55	5.50		19.05	23.50	

R07800
-030

THERMAL & MOISTURE PROTECTION 7

07840	Firestopping		CREW	DAILY OUTPUT	LABOR-HOURS	UNIT	2005 BARE COSTS				TOTAL INCL O&P
							MAT.	LABOR	EQUIP.	TOTAL	
100	0830	Concrete/CMU wall joints	R07800 -030									100
	0840	1" wide		1 Carp	100	.080	L.F.	6.25	2.74		8.99	11.10
	0850	2" wide			75	.107		11.45	3.65		15.10	18.30
	0860	4" wide			50	.160		22	5.50		27.50	32.50
	0870	Concrete/CMU floor joints										
	0880	1" wide		1 Carp	200	.040	L.F.	3.12	1.37		4.49	5.55
	0890	2" wide			150	.053		5.70	1.83		7.53	9.15
	0900	4" wide			100	.080		10.90	2.74		13.64	16.25

07920	Joint Sealants	CREW	DAILY OUTPUT	LABOR-HOURS	UNIT	2005 BARE COSTS				TOTAL INCL O&P
						MAT.	LABOR	EQUIP.	TOTAL	
800	0010 **CAULKING AND SEALANTS**										800
	0020 Acoustical sealant, elastomeric, cartridges				Ea.	2.21			2.21	2.43
	0030 Backer rod, polyethylene, 1/4" diameter	1 Bric	4.60	1.739	C.L.F.	2.12	61.50		63.62	96
	0050 1/2" diameter		4.60	1.739		3.39	61.50		64.89	97
	0070 3/4" diameter		4.60	1.739		5.80	61.50		67.30	100
	0090 1" diameter		4.60	1.739		9.35	61.50		70.85	104
	0100 Acrylic latex caulk, white									
	0200 11 fl. oz cartridge				Ea.	1.88			1.88	2.07
	0500 1/4" x 1/2"	1 Bric	248	.032	L.F.	.15	1.14		1.29	1.90
	0600 1/2" x 1/2"		250	.032		.31	1.13		1.44	2.06
	0800 3/4" x 3/4"		230	.035		.69	1.23		1.92	2.63
	0900 3/4" x 1"		200	.040		.92	1.41		2.33	3.16
	1000 1" x 1"		180	.044		1.15	1.57		2.72	3.66
	1400 Butyl based, bulk				Gal.	22			22	24.50
	1500 Cartridges				"	27			27	29.50
	1700 Bulk, in place 1/4" x 1/2", 154 L.F./gal.	1 Bric	230	.035	L.F.	.14	1.23		1.37	2.03
	1800 1/2" x 1/2", 77 L.F./gal.	"	180	.044	"	.29	1.57		1.86	2.71
	2000 Latex acrylic based, bulk				Gal.	23			23	25.50
	2100 Cartridges					26			26	28.50
	2300 Polysulfide compounds, 1 component, bulk					44			44	48
	2600 1 or 2 component, in place, 1/4" x 1/4", 308 L.F./gal.	1 Bric	145	.055	L.F.	.14	1.94		2.08	3.12
	2700 1/2" x 1/4", 154 L.F./gal.		135	.059		.28	2.09		2.37	3.49
	2900 3/4" x 3/8", 68 L.F./gal.		130	.062		.64	2.17		2.81	4.01
	3000 1" x 1/2", 38 L.F./gal.		130	.062		1.15	2.17		3.32	4.57
	3200 Polyurethane, 1 or 2 component				Gal.	47.50			47.50	52.50
	3500 Bulk, in place, 1/4" x 1/4"	1 Bric	150	.053	L.F.	.15	1.88		2.03	3.03
	3600 1/2" x 1/4"		145	.055		.31	1.94		2.25	3.30
	3800 3/4" x 3/8", 68 L.F./gal.		130	.062		.70	2.17		2.87	4.07
	3900 1" x 1/2"		110	.073		1.24	2.56		3.80	5.25
	4100 Silicone rubber, bulk				Gal.	34.50			34.50	38
	4200 Cartridges				"	40.50			40.50	44.50
	4400 Neoprene gaskets, closed cell, adhesive, 1/8" x 3/8"	1 Bric	240	.033	L.F.	.20	1.17		1.37	2.01
	4500 1/4" x 3/4"		215	.037		.48	1.31		1.79	2.53
	4700 1/2" x 1"		200	.040		1.40	1.41		2.81	3.69
	4800 3/4" x 1-1/2"		165	.048		2.91	1.71		4.62	5.80
	5500 Resin epoxy coating, 2 component, heavy duty				Gal.	26			26	28.50
	5800 Tapes, sealant, P.V.C. foam adhesive, 1/16" x 1/4"				C.L.F.	4.60			4.60	5.05
	5900 1/16" x 1/2"					6.80			6.80	7.50

Important: See the Reference Section for critical supporting data - Reference Nos., Crews, & City Cost Indexes

07920	Joint Sealants		CREW	DAILY OUTPUT	LABOR-HOURS	UNIT	2005 BARE COSTS				TOTAL INCL O&P	
							MAT.	LABOR	EQUIP.	TOTAL		
800	5950	1/16" x 1"				C.L.F.	11.30			11.30	12.45	800
	6000	1/8" x 1/2"				↓	7.65			7.65	8.40	
	6200	Urethane foam, 2 component, handy pack, 1 C.F.				Ea.	27.50			27.50	30.50	
	6300	50.0 C.F. pack				C.F.	14.05			14.05	15.45	

For information about Means Estimating Seminars, see yellow pages 12 and 13 in back of book

THERMAL & MOISTURE PROTECTION 7

	CREW	DAILY OUTPUT	LABOR-HOURS	UNIT	2005 BARE COSTS				TOTAL INCL O&P
					MAT.	LABOR	EQUIP.	TOTAL	

Division 8
Doors & Windows

Estimating Tips

08100 Metal Doors & Frames

- Most metal doors and frames look alike, but there may be significant differences among them. When estimating these items be sure to choose the line item that most closely compares to the specification or door schedule requirements regarding:
  - type of metal
  - metal gauge
  - door core material
  - fire rating
  - finish

08200 Wood & Plastic Doors

- Wood and plastic doors vary considerably in price. The primary determinant is the veneer material. Lauan, birch and oak are the most common veneers. Other variables include the following:
  - hollow or solid core
  - fire rating
  - flush or raised panel
  - finish
- If the specifications require compliance with AWI (Architectural Woodwork Institute) standards or acoustical standards, the cost of the door may increase substantially. All wood doors are priced pre-mortised for hinges and predrilled for cylindrical locksets.

- Frequently doors, frames, and windows are unique in old buildings. Specified replacement units could be stock, custom (similar to the original) or exact reproduction. The estimator should work closely with a window consultant to determine any extra costs that may be associated with the unusual installation requirements.

08300 Specialty Doors

- There are many varieties of special doors, and they are usually priced per each. Add frames, hardware or operators required for a complete installation.

08510 Steel Windows

- Most metal windows are delivered preglazed. However, some metal windows are priced without glass. Refer to 08800 Glazing for glass pricing. The grade C indicates commercial grade windows, usually ASTM C-35.

08550 Wood Windows

- All wood windows are priced preglazed. The two glazing options priced are single pane float glass and insulating glass 1/2" thick. Add the cost of screens and grills if required.

08700 Hardware

- Hardware costs add considerably to the cost of a door. The most efficient method to determine the hardware requirements for a project is to review the door schedule. This schedule, in conjunction with the specifications, is all you should need to take off the door hardware.

- Door hinges are priced by the pair, with most doors requiring 1-1/2 pairs per door. The hinge prices do not include installation labor because it is included in door installation. Hinges are classified according to the frequency of use.

08800 Glazing

- Different openings require different types of glass. The three most common types are:
  - float
  - tempered
  - insulating
- Most exterior windows are glazed with insulating glass. Entrance doors and window walls, where the glass is less than 18" from the floor, are generally glazed with tempered glass. Interior windows and some residential windows are glazed with float glass.
- Energy efficient coatings are also available

08900 Glazed Curtain Wall

- Glazed curtain walls consist of the metal tube framing and the glazing material. The cost data in this subdivision is presented for the metal tube framing alone or the composite wall. If your estimate requires a detailed takeoff of the framing, be sure to add the glazing cost.

Reference Numbers

Reference numbers are shown in bold squares at the beginning of some major classifications. These numbers refer to related items in the Reference Section. The reference information may be an estimating procedure, an alternate pricing method or technical information.

Note: Not all subdivisions listed here necessarily appear in this publication.

			DAILY	LABOR-		2005 BARE COSTS				TOTAL	
08060 | Selective Demolition		CREW	OUTPUT	HOURS	UNIT	MAT.	LABOR	EQUIP.	TOTAL	INCL O&P	
110	0010	**SELECTIVE DEMOLITION, DOORS** R02220 -510										110
	0200	Doors, exterior, 1-3/4" thick, single, 3' x 7' high	1 Clab	16	.500	Ea.		13.35		13.35	21	
	0220	Double, 6' x 7' high		12	.667			17.80		17.80	27.50	
	0500	Interior, 1-3/8" thick, single, 3' x 7' high		20	.400			10.70		10.70	16.60	
	0520	Double, 6' x 7' high		16	.500			13.35		13.35	21	
	0700	Bi-folding, 3' x 6'-8" high		20	.400			10.70		10.70	16.60	
	0720	6' x 6'-8" high		18	.444			11.85		11.85	18.45	
	0900	Bi-passing, 3' x 6'-8" high		16	.500			13.35		13.35	21	
	0940	6' x 6'-8" high		14	.571			15.25		15.25	23.50	
	1500	Remove and reset, minimum	1 Carp	8	1			34.50		34.50	53.50	
	1520	Maximum		6	1.333			45.50		45.50	71	
	2000	Frames, including trim, metal		8	1			34.50		34.50	53.50	
	2200	Wood	2 Carp	32	.500			17.15		17.15	26.50	
	3000	Special doors, counter doors		6	2.667			91.50		91.50	142	
	3100	Double acting		10	1.600			55		55	85.50	
	3200	Floor door (trap type), or access type		8	2			68.50		68.50	107	
	3300	Glass, sliding, including frames		12	1.333			45.50		45.50	71	
	3400	Overhead, commercial, 12' x 12' high		4	4			137		137	213	
	3440	up to 20' x 16' high		3	5.333			183		183	285	
	3445	up to 35' x 30' high		1	16			550		550	855	
	3500	Residential, 9' x 7' high		8	2			68.50		68.50	107	
	3540	16' x 7' high		7	2.286			78.50		78.50	122	
	3600	Remove and reset, minimum		4	4			137		137	213	
	3620	Maximum		2.50	6.400			219		219	340	
	3700	Roll-up grille		5	3.200			110		110	171	
	3800	Revolving door		2	8			274		274	425	
	3900	Storefront swing door		3	5.333			183		183	285	
	5020	Remove Bay/Bow window		6	2.667			91.50		91.50	142	
	5032	Remove Skylight, plstc domes,flush/curb mtd	G-3	395	.081	S.F.		2.71		2.71	4.19	
	6600	Demo flexible transparent strip entrance	3 Shee	115	.209	SF Surf		8.40		8.40	12.90	
	7100	Remove per opening	2 Skwk	.50	32	Opng.		1,125		1,125	1,725	
	7110	Remove Automatic operators, industrial, sliding doors, to 12' wide	"	.40	40	"		1,400		1,400	2,175	
	7550	Hangar door demo	2 Sswk	220	.073	S.F.		2.77		2.77	4.96	
	7570	Demo Shock door	"	1.90	8.421	Opng.		320		320	575	
120	0010	**SELECTIVE DEMOLITION, WINDOWS** R02220 -510										120
	0200	Aluminum, including trim, to 12 S.F.	1 Clab	16	.500	Ea.		13.35		13.35	21	
	0240	To 25 S.F.		11	.727			19.40		19.40	30	
	0280	To 50 S.F.		5	1.600			42.50		42.50	66.50	
	0320	Storm windows, to 12 S.F.		27	.296			7.90		7.90	12.30	
	0360	To 25 S.F.		21	.381			10.15		10.15	15.85	
	0400	To 50 S.F.		16	.500			13.35		13.35	21	
	0600	Glass, minimum		200	.040	S.F.		1.07		1.07	1.66	
	0620	Maximum		150	.053	"		1.42		1.42	2.22	
	1000	Steel, including trim, to 12 S.F.		13	.615	Ea.		16.45		16.45	25.50	
	1020	To 25 S.F.		9	.889			23.50		23.50	37	
	1040	To 50 S.F.		4	2			53.50		53.50	83	
	2000	Wood, including trim, to 12 S.F.		22	.364			9.70		9.70	15.10	
	2020	To 25 S.F.		18	.444			11.85		11.85	18.45	
	2060	To 50 S.F.		13	.615			16.45		16.45	25.50	
	2065	To 180 S.F.		8	1			26.50		26.50	41.50	
	5020	Remove and reset window, minimum	1 Carp	6	1.333			45.50		45.50	71	
	5040	Average		4	2			68.50		68.50	107	
	5080	Maximum		2	4			137		137	213	

08110	Steel Doors and Frames	CREW	DAILY OUTPUT	LABOR-HOURS	UNIT	2005 BARE COSTS MAT.	LABOR	EQUIP.	TOTAL	TOTAL INCL O&P
200 0010	**COMMERCIAL STEEL DOORS** R08110 -020									**200**
0015	Flush, full panel, hollow core									
0020	1-3/8″ thick, 20 ga., 2'-0″x 6'-8″	2 Carp	20	.800	Ea.	168	27.50		195.50	228
0040	2'-8″ x 6'-8″		18	.889		173	30.50		203.50	238
0060	3'-0″ x 6'-8″		17	.941		176	32		208	243
0100	3'-0″ x 7'-0″		17	.941		182	32		214	250
0120	For vision lite, add					53			53	58
0140	For narrow lite, add					52.50			52.50	58
0320	Half glass, 20 ga., 2'-0″ x 6'-8″	2 Carp	20	.800		236	27.50		263.50	305
0340	2'-8″ x 6'-8″		18	.889		241	30.50		271.50	315
0360	3'-0″ x 6'-8″		17	.941		244	32		276	320
0400	3'-0″ x 7'-0″		17	.941		250	32		282	325
0410	1-3/8″ thick, 18 ga., 2'-0″x 6'-8″		20	.800		195	27.50		222.50	257
0420	3'-0″ x 6'-8″		17	.941		197	32		229	267
0425	3'-0″ x 7'-0″		17	.941		205	32		237	275
0450	For vision lite, add					53			53	58
0452	For narrow lite, add					52.50			52.50	58
0460	Half glass, 18 ga., 2'-0″ x 6'-8″	2 Carp	20	.800		263	27.50		290.50	335
0465	2'-8″ x 6'-8″		18	.889		271	30.50		301.50	345
0470	3'-0″ x 6'-8″		17	.941		265	32		297	340
0475	3'-0″ x 7'-0″		17	.941		273	32		305	350
0500	Hollow core, 1-3/4″ thick, full panel, 20 ga., 2'-8″ x 6'-8″		18	.889		188	30.50		218.50	255
0520	3'-0″ x 6'-8″		17	.941		176	32		208	243
0640	3'-0″ x 7'-0″		17	.941		197	32		229	266
0680	4'-0″ x 7'-0″		15	1.067		275	36.50		311.50	355
0700	4'-0″ x 8'-0″		13	1.231		325	42		367	425
1000	18 ga., 2'-8″ x 6'-8″		17	.941		191	32		223	260
1020	3'-0″ x 6'-8″		16	1		186	34.50		220.50	258
1120	3'-0″ x 7'-0″ CN		17	.941		219	32		251	290
1180	4'-0″ x 7'-0″		14	1.143		275	39		314	360
1200	4'-0″ x 8'-0″		17	.941		325	32		357	410
1212	For vision lite, add					53			53	58
1214	For narrow lite, add					52.50			52.50	58
1230	Half glass, 20 ga., 2'-8″ x 6'-8″	2 Carp	20	.800		244	27.50		271.50	310
1240	3'-0″ x 6'-8″		18	.889		244	30.50		274.50	315
1260	3'-0″ x 7'-0″		18	.889		252	30.50		282.50	325
1280	4'-0″ x 7'-0″		16	1		310	34.50		344.50	395
1300	4'-0″ x 8'-0″		13	1.231		395	42		437	500
1320	18 ga., 2'-8″ x 6'-8″		18	.889		268	30.50		298.50	345
1340	3'-0″ x 6'-8″		17	.941		265	32		297	340
1360	3'-0″ x 7'-0″		17	.941		273	32		305	350
1380	4'-0″ x 7'-0″		15	1.067		345	36.50		381.50	430
1400	4'-0″ x 8'-0″		14	1.143		395	39		434	495
1720	Insulated, 1-3/4″ thick, full panel, 18 ga., 3'-0″ x 6'-8″		15	1.067		197	36.50		233.50	274
1740	2'-8″ x 7'-0″		16	1		209	34.50		243.50	284
1760	3'-0″ x 7'-0″		15	1.067		211	36.50		247.50	289
1800	4'-0″ x 8'-0″		13	1.231		325	42		367	425
1805	For vision lite, add					53			53	58
1810	For narrow lite, add					52.50			52.50	58
1820	Half glass, 18 ga., 3'-0″ x 6'-8″	2 Carp	16	1		265	34.50		299.50	345
1840	2'-8″ x 7'-0″		17	.941		277	32		309	355
1860	3'-0″ x 7'-0″		16	1		305	34.50		339.50	390
1900	4'-0″ x 8'-0″		14	1.143		395	39		434	495
2000	For bottom louver, add					96.50			96.50	106
2020	For baked enamel finish, add					30%	15%			
2040	For galvanizing, add					15%				

DOORS & WINDOWS 8

08110 | Steel Doors and Frames

			CREW	DAILY OUTPUT	LABOR-HOURS	UNIT	2005 BARE COSTS MAT.	LABOR	EQUIP.	TOTAL	TOTAL INCL O&P	
250	0010	**DOOR FRAMES**										**250**
	0020	Steel channels with anchors and bar stops										
	0100	6" channel @ 8.2#/L.F., 3' x 7' door, weighs 150#	E-4	13	2.462	Ea.	158	95	6.25	259.25	350	
	0200	8" channel @ 11.5#/L.F., 6' x 8' door, weighs 275#		9	3.556		289	137	9.05	435.05	575	
	0300	8' x 12' door, weighs 400#		6.50	4.923		420	190	12.50	622.50	815	
	0400	10" channel @ 15.3#/L.F., 10' x 10' door, weighs 500# **CN**		6	5.333		525	206	13.55	744.55	965	
	0500	12' x 12' door, weighs 600#		5.50	5.818		630	225	14.80	869.80	1,100	
	0600	12" channel @ 20.7#/L.F., 12' x 12' door, weighs 825#		4.50	7.111		865	275	18.05	1,158.05	1,475	
	0700	12' x 16' door, weighs 1000#		4	8		1,050	310	20.50	1,380.50	1,725	
	0800	For frames without bar stops, light sections, deduct					15%					
	0900	Heavy sections, deduct					10%					
300	0010	**FIRE DOOR** R08110-020										**300**
	0015	Steel, flush, "B" label, 90 minute										
	0020	Full panel, 20 ga., 2'-0" x 6'-8"	2 Carp	20	.800	Ea.	203	27.50		230.50	266	
	0040	2'-8" x 6'-8"		18	.889		210	30.50		240.50	279	
	0060	3'-0" x 6'-8"		17	.941		210	32		242	281	
	0080	3'-0" x 7'-0"		17	.941		218	32		250	290	
	0140	18 ga., 3'-0" x 6'-8"		16	1		232	34.50		266.50	310	
	0160	2'-8" x 7'-0"		17	.941		245	32		277	320	
	0180	3'-0" x 7'-0"		16	1		239	34.50		273.50	315	
	0200	4'-0" x 7'-0"		15	1.067		310	36.50		346.50	395	
	0220	For "A" label, 3 hour, 18 ga., use same price as "B" label										
	0240	For vision lite, add				Ea.	49.50			49.50	54.50	
	0520	Flush, "B" label 90 min., composite, 20 ga., 2'-0" x 6'-8"	2 Carp	18	.889		267	30.50		297.50	340	
	0540	2'-8" x 6'-8"		17	.941		272	32		304	350	
	0560	3'-0" x 6'-8"		16	1		275	34.50		309.50	355	
	0580	3'-0" x 7'-0"		16	1		283	34.50		317.50	365	
	0640	Flush, "A" label 3 hour, composite, 18 ga., 3'-0" x 6'-8"		15	1.067		232	36.50		268.50	310	
	0660	2'-8" x 7'-0"		16	1		243	34.50		277.50	320	
	0680	3'-0" x 7'-0"		15	1.067		239	36.50		275.50	320	
	0700	4'-0" x 7'-0"		14	1.143		310	39		349	400	
600	0010	**RESIDENTIAL STEEL DOOR**										**600**
	0020	Prehung, insulated, exterior										
	0030	Embossed, full panel, 2'-8" x 6'-8"	2 Carp	17	.941	Ea.	193	32		225	263	
	0040	3'-0" x 6'-8"		15	1.067		189	36.50		225.50	265	
	0060	3'-0" x 7'-0"		15	1.067		249	36.50		285.50	330	
	0070	5'-4" x 6'-8", double		8	2		375	68.50		443.50	520	
	0220	Half glass, 2'-8" x 6'-8"		17	.941		230	32		262	305	
	0240	3'-0" x 6'-8"		16	1		234	34.50		268.50	310	
	0260	3'-0" x 7'-0"		16	1		286	34.50		320.50	370	
	0270	5'-4" x 6'-8", double		8	2		485	68.50		553.50	640	
	0720	Raised plastic face, full panel, 2'-8" x 6'-8"		16	1		249	34.50		283.50	330	
	0740	3'-0" x 6'-8"		15	1.067		251	36.50		287.50	335	
	0760	3'-0" x 7'-0"		15	1.067		253	36.50		289.50	335	
	0780	5'-4" x 6'-8", double		8	2		470	68.50		538.50	620	
	0820	Half glass, 2'-8" x 6'-8"		17	.941		278	32		310	355	
	0840	3'-0" x 6'-8"		16	1		282	34.50		316.50	365	
	0860	3'-0" x 7'-0"		16	1		310	34.50		344.50	395	
	0880	5'-4" x 6'-8", double		8	2		595	68.50		663.50	760	
	1320	Flush face, full panel, 2'-6" x 6'-8"		16	1		192	34.50		226.50	265	
	1340	3'-0" x 6'-8"		15	1.067		195	36.50		231.50	272	
	1360	3'-0" x 7'-0"		15	1.067		247	36.50		283.50	330	
	1380	5'-4" x 6'-8", double		8	2		405	68.50		473.50	550	
	1420	Half glass, 2'-8" x 6'-8"		17	.941		241	32		273	315	
	1440	3'-0" x 6'-8"		16	1		245	34.50		279.50	325	

8

DOORS & WINDOWS

08110	Steel Doors and Frames	CREW	DAILY OUTPUT	LABOR-HOURS	UNIT	2005 BARE COSTS				TOTAL INCL O&P	
						MAT.	LABOR	EQUIP.	TOTAL		
600	1460	3'-0" x 7'-0"	2 Carp	16	1	Ea.	284	34.50		318.50	365	**600**
	1480	5'-4" x 6'-8", double	↓	8	2		475	68.50		543.50	625	
	1500	Sidelight, full lite, 1'-0" x 6'-8"					199			199	219	
	1510	1'-0" x 6'-8"					240			240	264	
	1520	1'-0" x 6'-8"					184			184	202	
	1530	1'-0" x 6'-8"					223			223	246	
	2300	Interior, residential, closet, bi-fold, 6'-8" x 2'-0" wide	2 Carp	16	1		136	34.50		170.50	203	
	2330	3'-0" wide		16	1		152	34.50		186.50	222	
	2360	4'-0" wide		15	1.067		231	36.50		267.50	310	
	2400	5'-0" wide		14	1.143		268	39		307	355	
	2420	6'-0" wide	↓	13	1.231	↓	300	42		342	395	
820	0010	**STEEL FRAMES, KNOCK DOWN**										**820**
	0020	16 ga., up to 5-3/4" jamb depth										
	0025	6'-8" high, 3'-0" wide, single	2 Carp	16	1	Ea.	81.50	34.50		116	143	
	0028	6'-8" high, 3'-6" wide, single		16	1		72.50	34.50		107	133	
	0030	6'-8" high, 4'-0" wide, single		16	1		72.50	34.50		107	133	
	0040	6'-0" wide, double		14	1.143		95.50	39		134.50	166	
	0045	8'-0" wide, double		14	1.143		99.50	39		138.50	170	
	0100	7'-0" high, 3'-0" wide, single		16	1		83.50	34.50		118	145	
	0110	7'-0" high, 3'-6" wide, single		16	1		74.50	34.50		109	136	
	0112	7'-0" high, 4'-0" wide, single		16	1		127	34.50		161.50	193	
	0140	6'-0" wide, double		14	1.143		100	39		139	171	
	0145	8'-0" wide, double		14	1.143		93.50	39		132.50	164	
	1000	16 ga., up to 4-7/8" deep, 7'-0" H, 3'-0" W, single **CN**		16	1		81.50	34.50		116	143	
	1140	6'-0" wide, double		14	1.143		100	39		139	171	
	2800	14 ga., up to 3-7/8" deep, 7'-0" high, 3'-0" wide, single		16	1		83.50	34.50		118	146	
	2840	6'-0" wide, double		14	1.143		103	39		142	174	
	3000	14 ga., up to 5-3/4" deep, 6'-8" high, 3'-0" wide, single		16	1		91.50	34.50		126	155	
	3002	6'-8" high, 3'-6" wide, single		16	1		88	34.50		122.50	151	
	3005	6'-8" high, 4'-0" wide, single		16	1		88	34.50		122.50	151	
	3600	up to 5-3/4" jamb depth, 7'-0" high, 4'-0" wide, single		15	1.067		69	36.50		105.50	133	
	3620	6'-0" wide, double		12	1.333		108	45.50		153.50	190	
	3640	8'-0" wide, double		12	1.333		91	45.50		136.50	171	
	3700	8'-0" high, 4'-0" wide, single		15	1.067		91	36.50		127.50	157	
	3740	8'-0" wide, double		12	1.333		113	45.50		158.50	195	
	4000	6-3/4" deep, 7'-0" high, 4'-0" wide, single		15	1.067		97	36.50		133.50	164	
	4020	6'-0" wide, double		12	1.333		111	45.50		156.50	194	
	4040	8'-0", wide double		12	1.333		129	45.50		174.50	213	
	4100	8'-0" high, 4'-0" wide, single		15	1.067		108	36.50		144.50	176	
	4140	8'-0" wide, double		12	1.333		126	45.50		171.50	210	
	4400	8-3/4" deep, 7'-0" high, 4'-0" wide, single		15	1.067		101	36.50		137.50	168	
	4440	8'-0" wide, double		12	1.333		134	45.50		179.50	218	
	4500	8'-0" high, 4'-0" wide, single		15	1.067		119	36.50		155.50	187	
	4540	8'-0" wide, double	↓	12	1.333		139	45.50		184.50	224	
	4900	For welded frames, add					29.50			29.50	32.50	
	5400	14 ga., "B" label, up to 5-3/4" deep, 7'-0" high, 4'-0" wide, single	2 Carp	15	1.067		108	36.50		144.50	176	
	5440	8'-0" wide, double		12	1.333		126	45.50		171.50	210	
	5800	6-3/4" deep, 7'-0" high, 4'-0" wide, single		15	1.067		93.50	36.50		130	160	
	5840	8'-0" wide, double		12	1.333		140	45.50		185.50	224	
	6200	8-3/4" deep, 7'-0" high, 4'-0" wide, single		15	1.067		116	36.50		152.50	185	
	6240	8'-0" wide, double	↓	12	1.333	↓	149	45.50		194.50	235	
	6300	For "A" label use same price as "B" label										
	6400	For baked enamel finish, add					30%	15%				
	6500	For galvanizing, add					15%					
	6600	For Hospital Stop, add				Ea.	176			176	193	

DOORS & WINDOWS 8

08100 | Metal Doors and Frames

08110 | Steel Doors and Frames

			CREW	DAILY OUTPUT	LABOR-HOURS	UNIT	2005 BARE COSTS MAT.	LABOR	EQUIP.	TOTAL	TOTAL INCL O&P	
820	7900	Transom lite frames, fixed, add	2 Carp	155	.103	S.F.	28.50	3.54		32.04	36.50	820
	8000	Movable, add	"	130	.123	"	34.50	4.22		38.72	44	

08160 | Sliding Metal Doors and Grilles

			CREW	DAILY OUTPUT	LABOR-HOURS	UNIT	2005 BARE COSTS MAT.	LABOR	EQUIP.	TOTAL	TOTAL INCL O&P	
300	0010	**STEEL, SLIDING**										300
	0020	Up to 50' x 18', electric, standard duty, minimum	L-5	360	.156	S.F.	20	5.95	1.71	27.66	35	
	0100	Maximum		340	.165		34	6.30	1.81	42.11	50.50	
	0500	Heavy duty, minimum		297	.189		27.50	7.20	2.08	36.78	45	
	0600	Maximum	▼	277	.202	▼	72.50	7.70	2.23	82.43	96	

08180 | Metal Screen and Storm Doors

			CREW	DAILY OUTPUT	LABOR-HOURS	UNIT	2005 BARE COSTS MAT.	LABOR	EQUIP.	TOTAL	TOTAL INCL O&P	
100	0010	**STORM DOORS & FRAMES** Aluminum, residential,										100
	0020	combination storm and screen										
	0400	Clear anodic coating, 6'-8" x 2'-6" wide	2 Carp	15	1.067	Ea.	153	36.50		189.50	225	
	0420	2'-8" wide		14	1.143		176	39		215	255	
	0440	3'-0" wide	▼	14	1.143	▼	176	39		215	255	
	0500	For 7' door height, add					5%					
	1000	Mill finish, 6'-8" x 2'-6" wide	2 Carp	15	1.067	Ea.	203	36.50		239.50	281	
	1020	2'-8" wide		14	1.143		203	39		242	285	
	1040	3'-0" wide	▼	14	1.143		221	39		260	305	
	1100	For 7'-0" door, add					5%					
	1500	White painted, 6'-8" x 2'-6" wide	2 Carp	15	1.067	Ea.	202	36.50		238.50	280	
	1520	2'-8" wide		14	1.143		206	39		245	288	
	1540	3'-0" wide	▼	14	1.143	▼	215	39		254	298	
	1600	For 7'-0" door, add					5%					
	2000	Wood door & screen, see division 08210-930										

08200 | Wood and Plastic Doors

08210 | Wood Doors

			CREW	DAILY OUTPUT	LABOR-HOURS	UNIT	2005 BARE COSTS MAT.	LABOR	EQUIP.	TOTAL	TOTAL INCL O&P	
450	0010	**KALAMEIN**										450
	0020	Interior, flush type, 3' x 7'	2 Carp	4.30	3.721	Opng.	165	127		292	380	
720	0010	**PRE-HUNG DOORS**										720
	0300	Exterior, wood, comb. storm & screen, 6'-9" x 2'-6" wide	2 Carp	15	1.067	Ea.	266	36.50		302.50	350	
	0320	2'-8" wide		15	1.067		266	36.50		302.50	350	
	0340	3'-0" wide	▼	15	1.067	▼	273	36.50		309.50	355	
	0360	For 7'-0" high door, add					23			23	25.50	
	0370	For aluminum storm doors, see division 08180-100										
	1600	Entrance door, flush, birch, solid core										
	1620	4-5/8" solid jamb, 1-3/4" x 6'-8" x 2'-8" wide	2 Carp	16	1	Ea.	272	34.50		306.50	355	
	1640	3'-0" wide	"	16	1		279	34.50		313.50	360	
	1680	For 7'-0" high door, add				▼	16.50			16.50	18.15	
	2000	Entrance door, colonial, 6 panel pine										
	2020	4-5/8" solid jamb, 1-3/4" x 6'-8" x 2'-8" wide	2 Carp	16	1	Ea.	445	34.50		479.50	545	
	2040	3'-0" wide	"	16	1		470	34.50		504.50	575	
	2060	For 7'-0" high door, add					35			35	38	
	2200	For 5-5/8" solid jamb, add				▼	22.50			22.50	24.50	
	4000	Interior, passage door, 4-5/8" solid jamb										

8
DOORS & WINDOWS

Important: See the Reference Section for critical supporting data - Reference Nos., Crews, & City Cost Indexes

08210	Wood Doors	CREW	DAILY OUTPUT	LABOR-HOURS	UNIT	2005 BARE COSTS MAT.	LABOR	EQUIP.	TOTAL	TOTAL INCL O&P	
720	4400	Lauan, flush, solid core, 1-3/8" x 6'-8" x 2'-6" wide	2 Carp	20	.800	Ea.	167	27.50		194.50	227	**720**
	4420	2'-8" wide		20	.800		147	27.50		174.50	204	
	4440	3'-0" wide		19	.842		151	29		180	212	
	4600	Hollow core, 1-3/8" x 6'-8" x 2'-6" wide		20	.800		109	27.50		136.50	163	
	4620	2'-8" wide		20	.800		110	27.50		137.50	165	
	4640	3'-0" wide		19	.842		112	29		141	168	
	4700	For 7'-0" high door, add					21.50			21.50	24	
	5000	Birch, flush, solid core, 1-3/8" x 6'-8" x 2'-6" wide	2 Carp	20	.800		156	27.50		183.50	215	
	5020	2'-8" wide		20	.800		176	27.50		203.50	237	
	5040	3'-0" wide		19	.842		188	29		217	252	
	5200	Hollow core, 1-3/8" x 6'-8" x 2'-6" wide		20	.800		129	27.50		156.50	185	
	5220	2'-8" wide		20	.800		135	27.50		162.50	192	
	5240	3'-0" wide		19	.842		135	29		164	194	
	5280	For 7'-0" high door, add					15.45			15.45	17	
	5500	Hardboard paneled, 1-3/8" x 6'-8" x 2'-6" wide	2 Carp	20	.800		130	27.50		157.50	186	
	5520	2'-8" wide		20	.800		136	27.50		163.50	193	
	5540	3'-0" wide		19	.842		135	29		164	193	
	6000	Pine paneled, 1-3/8" x 6'-8" x 2'-6" wide	2 Carp	20	.800		223	27.50		250.50	289	
	6020	2'-8" wide		20	.800		241	27.50		268.50	310	
	6040	3'-0" wide		19	.842		250	29		279	320	
	6500	For 5-5/8" solid jamb, add					10.90			10.90	12	
	6520	For split jamb, deduct					13.30			13.30	14.65	
850	0010	**TIN CLAD**										**850**
	0020	3 ply, 6' x 7', double sliding, doors only	2 Carp	1	16	Opng.	1,300	550		1,850	2,275	
	1000	For electric operator, add	1 Elec	2	4	"	2,300	163		2,463	2,775	
900	0010	**WOOD DOOR, ARCHITECTURAL**										**900**
	0015	Flush, int., 1-3/8", 7 ply, hollow core,										
	0020	Lauan face, 2'-0" x 6'-8"	2 Carp	17	.941	Ea.	27.50	32		59.50	80	
	0040	2'-6" x 6'-8"		17	.941		31.50	32		63.50	84.50	
	0080	3'-0" x 6'-8"		17	.941		38	32		70	92	
	0100	4'-0" x 6'-8"		16	1		67	34.50		101.50	127	
	0120	Birch face, 2'-0" x 6'-8"		17	.941		44	32		76	98.50	
	0140	2'-6" x 6'-8"		17	.941		41.50	32		73.50	96	
	0180	3'-0" x 6'-8"		17	.941		47	32		79	102	
	0200	4'-0" x 6'-8"		16	1		91.50	34.50		126	154	
	0220	Oak face, 2'-0" x 6'-8"		17	.941		70	32		102	127	
	0240	2'-6" x 6'-8"		17	.941		74.50	32		106.50	132	
	0280	3'-0" x 6'-8"		17	.941		79.50	32		111.50	138	
	0300	4'-0" x 6'-8"		16	1		101	34.50		135.50	165	
	0320	Walnut face, 2'-0" x 6'-8"		17	.941		141	32		173	205	
	0340	2'-6" x 6'-8"		17	.941		144	32		176	208	
	0380	3'-0" x 6'-8"		17	.941		149	32		181	214	
	0400	4'-0" x 6'-8"		16	1		169	34.50		203.50	240	
	0430	For 7'-0" high, add					13.90			13.90	15.30	
	0440	For 8'-0" high, add					19.45			19.45	21.50	
	0480	For prefinishing, clear, add					30.50			30.50	33.50	
	0500	For prefinishing, stain, add					41.50			41.50	46	
	1320	M.D. overlay on hardboard, 2'-0" x 6'-8"	2 Carp	17	.941		86	32		118	145	
	1340	2'-6" x 6'-8"		17	.941		86	32		118	145	
	1380	3'-0" x 6'-8"		17	.941		102	32		134	162	
	1400	4'-0" x 6'-8"		16	1		140	34.50		174.50	208	
	1420	For 7'-0" high, add					7.65			7.65	8.40	
	1440	For 8'-0" high, add					20.50			20.50	22.50	
	1720	H.P. plastic laminate, 2'-0" x 6'-8"	2 Carp	16	1		205	34.50		239.50	280	
	1740	2'-6" x 6'-8"		16	1		205	34.50		239.50	280	

DOORS & WINDOWS 8

For expanded coverage of these items see *Means Interior Cost Data 2005*

08210		Wood Doors	CREW	DAILY OUTPUT	LABOR-HOURS	UNIT	2005 BARE COSTS				TOTAL INCL O&P	
							MAT.	LABOR	EQUIP.	TOTAL		
900	1780	3'-0" x 6'-8"	2 Carp	15	1.067	Ea.	235	36.50		271.50	315	900
	1800	4'-0" x 6'-8"	↓	14	1.143		325	39		364	420	
	1820	For 7'-0" high, add					8.05			8.05	8.85	
	1840	For 8'-0" high, add					20.50			20.50	22.50	
	2020	5 ply particle core, lauan face, 2'-6" x 6'-8"	2 Carp	15	1.067		70	36.50		106.50	134	
	2040	3'-0" x 6'-8"		14	1.143		73	39		112	142	
	2080	3'-0" x 7'-0"		13	1.231		81	42		123	155	
	2100	4'-0" x 7'-0"		12	1.333		94.50	45.50		140	175	
	2120	Birch face, 2'-6" x 6'-8"		15	1.067		80	36.50		116.50	145	
	2140	3'-0" x 6'-8"		14	1.143		87.50	39		126.50	158	
	2180	3'-0" x 7'-0"		13	1.231		89.50	42		131.50	164	
	2200	4'-0" x 7'-0"		12	1.333		109	45.50		154.50	191	
	2220	Oak face, 2'-6" x 6'-8"		15	1.067		88	36.50		124.50	154	
	2240	3'-0" x 6'-8"		14	1.143		97	39		136	168	
	2280	3'-0" x 7'-0"		13	1.231		99.50	42		141.50	176	
	2300	4'-0" x 7'-0"		12	1.333		122	45.50		167.50	205	
	2320	Walnut face, 2'-0" x 6'-8"		15	1.067		98	36.50		134.50	165	
	2340	2'-6" x 6'-8"		14	1.143		112	39		151	184	
	2380	3'-0" x 6'-8"		13	1.231		126	42		168	204	
	2400	4'-0" x 6'-8"	↓	12	1.333		164	45.50		209.50	252	
	2440	For 8'-0" high, add					24			24	26.50	
	2460	For 8'-0" high walnut, add					13.15			13.15	14.45	
	2480	For solid wood core, add					29.50			29.50	32.50	
	2720	For prefinishing, clear, add					19.15			19.15	21	
	2740	For prefinishing, stain, add					42.50			42.50	47	
	3320	M.D. overlay on hardboard, 2'-6" x 6'-8"	2 Carp	14	1.143		86	39		125	156	
	3340	3'-0" x 6'-8"		13	1.231		90	42		132	165	
	3380	3'-0" x 7'-0"		12	1.333		92	45.50		137.50	172	
	3400	4'-0" x 7'-0"	↓	10	1.600		112	55		167	209	
	3440	For 8'-0" height, add					25			25	27.50	
	3460	For solid wood core, add					32.50			32.50	36	
	3720	H.P. plastic laminate, 2'-6" x 6'-8"	2 Carp	13	1.231		126	42		168	205	
	3740	3'-0" x 6'-8"		12	1.333		143	45.50		188.50	228	
	3780	3'-0" x 7'-0"		11	1.455		148	50		198	241	
	3800	4'-0" x 7'-0"	↓	8	2		180	68.50		248.50	305	
	3840	For 8'-0" height, add					25			25	27.50	
	3860	For solid wood core, add					30.50			30.50	33.50	
	4000	Exterior, flush, solid wood stave core, birch, 1-3/4" x 7'-0" x 2'-6"	2 Carp	15	1.067		152	36.50		188.50	224	
	4020	2'-8" wide		15	1.067		159	36.50		195.50	231	
	4040	3'-0" wide		14	1.143		169	39		208	246	
	4100	Oak faced 1-3/4" x 7'-0" x 2'-6" wide		15	1.067		167	36.50		203.50	241	
	4120	2'-8" wide		15	1.067		179	36.50		215.50	253	
	4140	3'-0" wide		14	1.143		190	39		229	270	
	4200	Walnut faced, 1-3/4" x 7'-0" x 2'-6" wide		15	1.067		245	36.50		281.50	325	
	4220	2'-8" wide		15	1.067		256	36.50		292.50	340	
	4240	3'-0" wide	↓	14	1.143		267	39		306	355	
	4300	For 6'-8" high door, deduct from 7'-0" door				↓	13.65			13.65	15	
910	0010	**WOOD DOORS, DECORATOR**										910
	3000	Solid wood, 1-3/4" thick stile and rail										
	3020	Mahogany, 3'-0" x 7'-0", minimum	2 Carp	14	1.143	Ea.	925	39		964	1,075	
	3030	Maximum		10	1.600		1,275	55		1,330	1,475	
	3040	3'-6" x 8'-0", minimum		10	1.600		850	55		905	1,025	
	3050	Maximum		8	2		1,550	68.50		1,618.50	1,825	
	3100	Pine, 3'-0" x 7'-0", minimum		14	1.143		405	39		444	505	
	3110	Maximum	↓	10	1.600	↓	665	55		720	815	

Important: See the Reference Section for critical supporting data - Reference Nos., Crews, & City Cost Indexes

8 DOORS & WINDOWS

08210	Wood Doors	CREW	DAILY OUTPUT	LABOR-HOURS	UNIT	2005 BARE COSTS				TOTAL INCL O&P	
						MAT.	LABOR	EQUIP.	TOTAL		
910 3120	3'-6" x 8'-0", minimum	2 Carp	10	1.600	Ea.	625	55		680	775	**910**
3130	Maximum		8	2		1,050	68.50		1,118.50	1,250	
3200	Red oak, 3'-0" x 7'-0", minimum		14	1.143		1,375	39		1,414	1,575	
3210	Maximum		10	1.600		1,650	55		1,705	1,900	
3220	3'-6" x 8'-0", minimum		10	1.600		1,525	55		1,580	1,750	
3230	Maximum		8	2		2,725	68.50		2,793.50	3,100	
4000	Hand carved door, mahogany										
4020	3'-0" x 7'-0", minimum	2 Carp	14	1.143	Ea.	1,375	39		1,414	1,550	
4030	Maximum		11	1.455		2,700	50		2,750	3,050	
4040	3'-6" x 8'-0", minimum		10	1.600		1,725	55		1,780	1,975	
4050	Maximum		8	2		2,650	68.50		2,718.50	3,025	
4200	Red oak, 3'-0" x 7'-0", minimum		14	1.143		4,200	39		4,239	4,675	
4210	Maximum		11	1.455		11,500	50		11,550	12,800	
4220	3'-6" x 8'-0", minimum		10	1.600		4,725	55		4,780	5,275	
4280	For 6'-8" high door, deduct from 7'-0" door					30			30	33	
4400	For custom finish, add					320			320	350	
4600	Side light, mahogany, 7'-0" x 1'-6" wide, minimum	2 Carp	18	.889		770	30.50		800.50	900	
4610	Maximum		14	1.143		2,275	39		2,314	2,550	
4620	8'-0" x 1'-6" wide, minimum		14	1.143		1,450	39		1,489	1,650	
4630	Maximum		10	1.600		1,650	55		1,705	1,875	
4640	Side light, oak, 7'-0" x 1'-6" wide, minimum		18	.889		890	30.50		920.50	1,025	
4650	Maximum		14	1.143		1,575	39		1,614	1,800	
4660	8'-0" x 1-6" wide, minimum		14	1.143		840	39		879	985	
4670	Maximum		10	1.600		1,575	55		1,630	1,825	
6520	Interior cafe doors, 2'-6" opening, stock, panel pine		16	1		168	34.50		202.50	239	
6540	3'-0" opening		16	1		175	34.50		209.50	247	
6550	Louvered pine										
6560	2'-6" opening	2 Carp	16	1	Ea.	147	34.50		181.50	215	
8000	3'-0" opening		16	1		157	34.50		191.50	227	
8010	2'-6" opening, hardwood		16	1		260	34.50		294.50	340	
8020	3'-0" opening		16	1		286	34.50		320.50	370	
8800	Pre-hung doors, see division 08210-720										
920 0010	**WOOD DOORS, PANELED**										**920**
0020	Interior, six panel, hollow core, 1-3/8" thick										
0040	Molded hardboard, 2'-0" x 6'-8"	2 Carp	17	.941	Ea.	45.50	32		77.50	100	
0060	2'-6" x 6'-8"		17	.941		49	32		81	104	
0070	2'-8" x 6'-8"		17	.941		51.50	32		83.50	107	
0080	3'-0" x 6'-8"		17	.941		54.50	32		86.50	110	
0140	Embossed print, molded hardboard, 2'-0" x 6'-8"		17	.941		49	32		81	104	
0160	2'-6" x 6'-8"		17	.941		49	32		81	104	
0180	3'-0" x 6'-8"		17	.941		54.50	32		86.50	110	
0540	Six panel, solid, 1-3/8" thick, pine, 2'-0" x 6'-8"		15	1.067		119	36.50		155.50	188	
0560	2'-6" x 6'-8"		14	1.143		134	39		173	208	
0580	3'-0" x 6'-8"		13	1.231		154	42		196	235	
1020	Two panel, bored rail, solid, 1-3/8" thick, pine, 1'-6" x 6'-8"		16	1		217	34.50		251.50	293	
1040	2'-0" x 6'-8"		15	1.067		286	36.50		322.50	370	
1060	2'-6" x 6'-8"		14	1.143		325	39		364	420	
1340	Two panel, solid, 1-3/8" thick, fir, 2'-0" x 6'-8"		15	1.067		119	36.50		155.50	188	
1360	2'-6" x 6'-8"		14	1.143		134	39		173	208	
1380	3'-0" x 6'-8"		13	1.231		325	42		367	425	
1740	Five panel, solid, 1-3/8" thick, fir, 2'-0" x 6'-8"		15	1.067		213	36.50		249.50	292	
1760	2'-6" x 6'-8"		14	1.143		345	39		384	435	
1780	3'-0" x 6'-8"		13	1.231		345	42		387	440	
930 0010	**WOOD DOORS, RESIDENTIAL**										**930**
0200	Exterior, combination storm & screen, pine										

DOORS & WINDOWS **8**

930

08210	Wood Doors	CREW	DAILY OUTPUT	LABOR-HOURS	UNIT	MAT.	LABOR	EQUIP.	TOTAL	TOTAL INCL O&P
0260	2'-8" wide	2 Carp	10	1.600	Ea.	256	55		311	365
0280	3'-0" wide		9	1.778		261	61		322	380
0300	7'-1" x 3'-0" wide		9	1.778		284	61		345	405
0400	Full lite, 6'-9" x 2'-6" wide		11	1.455		266	50		316	370
0420	2'-8" wide		10	1.600		266	55		321	380
0440	3'-0" wide		9	1.778		273	61		334	395
0500	7'-1" x 3'-0" wide		9	1.778		295	61		356	420
0700	Dutch door, pine, 1-3/4" x 6'-8" x 2'-8" wide, minimum		12	1.333		595	45.50		640.50	725
0720	Maximum		10	1.600		630	55		685	775
0800	3'-0" wide, minimum		12	1.333		620	45.50		665.50	750
0820	Maximum		10	1.600		670	55		725	825
1000	Entrance door, colonial, 1-3/4" x 6'-8" x 2'-8" wide		16	1		330	34.50		364.50	420
1020	6 panel pine, 3'-0" wide		15	1.067		355	36.50		391.50	445
1100	8 panel pine, 2'-8" wide		16	1		585	34.50		619.50	700
1120	3'-0" wide	▼	15	1.067		500	36.50		536.50	605
1200	For tempered safety glass lites, (min of 2)add					62			62	68
1300	Flush, birch, solid core, 1-3/4" x 6'-8" x 2'-8" wide	2 Carp	16	1		91	34.50		125.50	154
1320	3'-0" wide		15	1.067		94	36.50		130.50	161
1350	7'-0" x 2'-8" wide		16	1		98	34.50		132.50	162
1360	3'-0" wide	▼	15	1.067		106	36.50		142.50	174
1380	For tempered safety glass lites, add				▼	92			92	101
2700	Interior, closet, bi-fold, w/hardware, no frame or trim incl.									
2720	Flush, birch, 6'-6" or 6'-8" x 2'-6" wide	2 Carp	13	1.231	Ea.	45.50	42		87.50	116
2740	3'-0" wide		13	1.231		49.50	42		91.50	120
2760	4'-0" wide		12	1.333		90	45.50		135.50	170
2780	5'-0" wide		11	1.455		91	50		141	178
2800	6'-0" wide		10	1.600		98	55		153	194
3000	Raised panel pine, 6'-6" or 6'-8" x 2'-6" wide		13	1.231		149	42		191	230
3020	3'-0" wide		13	1.231		171	42		213	254
3040	4'-0" wide		12	1.333		218	45.50		263.50	310
3060	5'-0" wide		11	1.455		252	50		302	355
3080	6'-0" wide		10	1.600		283	55		338	395
3200	Louvered, pine 6'-6" or 6'-8" x 2'-6" wide		13	1.231		93.50	42		135.50	169
3220	3'-0" wide		13	1.231		147	42		189	228
3240	4'-0" wide		12	1.333		164	45.50		209.50	251
3260	5'-0" wide		11	1.455		185	50		235	281
3280	6'-0" wide	▼	10	1.600	▼	204	55		259	310
4400	Bi-passing closet, incl. hardware and frame, no trim incl.									
4420	Flush, lauan, 6'-8" x 4'-0" wide	2 Carp	12	1.333	Opng.	156	45.50		201.50	243
4440	5'-0" wide		11	1.455		171	50		221	266
4460	6'-0" wide		10	1.600		183	55		238	287
4600	Flush, birch, 6'-8" x 4'-0" wide		12	1.333		192	45.50		237.50	282
4620	5'-0" wide		11	1.455		193	50		243	290
4640	6'-0" wide		10	1.600		230	55		285	340
4800	Louvered, pine, 6'-8" x 4'-0" wide		12	1.333		380	45.50		425.50	485
4820	5'-0" wide		11	1.455		360	50		410	475
4840	6'-0" wide		10	1.600		465	55		520	595
5000	Paneled, pine, 6'-8" x 4'-0" wide		12	1.333		360	45.50		405.50	470
5020	5'-0" wide		11	1.455		375	50		425	490
5040	6'-0" wide	▼	10	1.600	▼	445	55		500	575
6100	Folding accordion, closet, including track and frame									
6120	Vinyl, 2 layer, stock (see also division 10651-100)	2 Carp	400	.040	S.F.	2.93	1.37		4.30	5.35
6140	Woven mahogany and vinyl, stock		400	.040		1.87	1.37		3.24	4.19
6160	Wood slats with vinyl overlay, stock		400	.040		9.45	1.37		10.82	12.55
6180	Economy vinyl, stock		400	.040		1.65	1.37		3.02	3.95
6200	Rigid PVC	▼	400	.040	▼	4.40	1.37		5.77	6.95

930

8 DOORS & WINDOWS

Important: See the Reference Section for critical supporting data - Reference Nos., Crews, & City Cost Indexes

08210	Wood Doors	CREW	DAILY OUTPUT	LABOR-HOURS	UNIT	2005 BARE COSTS				TOTAL INCL O&P	
						MAT.	LABOR	EQUIP.	TOTAL		
930 6220	For custom partition, add					25%	10%				**930**
7310	Passage doors, flush, no frame included										
7320	Hardboard, hollow core, 1-3/8" x 6'-8" x 1'-6" wide	2 Carp	18	.889	Ea.	38	30.50		68.50	89.50	
7330	2'-0" wide		18	.889		38	30.50		68.50	89	
7340	2'-6" wide		18	.889		42	30.50		72.50	93.50	
7350	2'-8" wide		18	.889		44	30.50		74.50	96	
7360	3'-0" wide		17	.941		46.50	32		78.50	101	
7420	Lauan, hollow core, 1-3/8" x 6'-8" x 1'-6" wide		18	.889		26.50	30.50		57	77	
7440	2'-0" wide		18	.889		25.50	30.50		56	75.50	
7450	2'-4" wide		18	.889		28.50	30.50		59	79	
7460	2'-6" wide		18	.889		28.50	30.50		59	79	
7480	2'-8" wide		18	.889		30	30.50		60.50	80.50	
7500	3'-0" wide		17	.941		31.50	32		63.50	85	
7700	Birch, hollow core, 1-3/8" x 6'-8" x 1'-6" wide		18	.889		34	30.50		64.50	84.50	
7720	2'-0" wide		18	.889		37.50	30.50		68	88.50	
7740	2'-6" wide		18	.889		41.50	30.50		72	93.50	
7760	2'-8" wide		18	.889		43.50	30.50		74	95	
7780	**CN** 3'-0" wide		17	.941		47	32		79	102	
8000	Pine louvered, 1-3/8" x 6'-8" x 1'-6" wide		19	.842		89.50	29		118.50	144	
8020	2'-0" wide		18	.889		113	30.50		143.50	172	
8040	2'-6" wide		18	.889		123	30.50		153.50	183	
8060	2'-8" wide		18	.889		130	30.50		160.50	191	
8080	3'-0" wide		17	.941		139	32		171	203	
8300	Pine paneled, 1-3/8" x 6'-8" x 1'-6" wide		19	.842		101	29		130	156	
8320	2'-0" wide		18	.889		117	30.50		147.50	176	
8330	2'-4" wide		18	.889		128	30.50		158.50	189	
8340	2'-6" wide		18	.889		131	30.50		161.50	192	
8360	2'-8" wide		18	.889		142	30.50		172.50	204	
8380	3'-0" wide		17	.941		148	32		180	213	
8550	For over 20 doors, deduct					15%					
950 0010	**WOOD FIRE DOORS**										**950**
0020	Particle core, 7 face plys, "B" label,										
0040	1 hour, birch face, 1-3/4" x 2'-6" x 6'-8"	2 Carp	14	1.143	Ea.	281	39		320	370	
0080	3'-0" x 6'-8"		13	1.231		291	42		333	385	
0090	3'-0" x 7'-0"		12	1.333		305	45.50		350.50	405	
0100	4'-0" x 7'-0"		12	1.333		405	45.50		450.50	515	
0140	Oak face, 2'-6" x 6'-8"		14	1.143		282	39		321	370	
0180	3'-0" x 6'-8"		13	1.231		292	42		334	385	
0190	3'-0" x 7'-0"		12	1.333		305	45.50		350.50	405	
0200	4'-0" x 7'-0"		12	1.333		400	45.50		445.50	510	
0240	Walnut face, 2'-6" x 6'-8"		14	1.143		370	39		409	465	
0280	3'-0" x 6'-8"		13	1.231		380	42		422	480	
0290	3'-0" x 7'-0"		12	1.333		395	45.50		440.50	505	
0300	4'-0" x 7'-0"		12	1.333		535	45.50		580.50	655	
0440	M.D. overlay on hardboard, 2'-6" x 6'-8"		15	1.067		247	36.50		283.50	330	
0480	3'-0" x 6'-8"		14	1.143		257	39		296	345	
0490	3'-0" x 7'-0"		13	1.231		271	42		313	365	
0500	4'-0" x 7'-0"		12	1.333		330	45.50		375.50	435	
0540	H.P. plastic laminate, 2'-6" x 6'-8"		13	1.231		330	42		372	430	
0580	3'-0" x 6'-8"		12	1.333		345	45.50		390.50	450	
0590	3'-0" x 7'-0"		11	1.455		350	50		400	465	
0600	4'-0" x 7'-0"		10	1.600		445	55		500	575	
0740	90 minutes, birch face, 1-3/4" x 2'-6" x 6'-8"		14	1.143		243	39		282	330	
0780	3'-0" x 6'-8"		13	1.231		252	42		294	345	
0790	3'-0" x 7'-0"		12	1.333		300	45.50		345.50	400	
0800	4'-0" x 7'-0"		12	1.333		360	45.50		405.50	470	

DOORS & WINDOWS 8

For expanded coverage of these items see *Means Interior Cost Data 2005*

8

DOORS & WINDOWS

08210	Wood Doors	CREW	DAILY OUTPUT	LABOR-HOURS	UNIT	2005 BARE COSTS				TOTAL INCL O&P
						MAT.	LABOR	EQUIP.	TOTAL	
950	0840	Oak face, 2'-6" x 6'-8"	2 Carp	14	1.143	Ea.	253	39		292	340	950
0880	3'-0" x 6'-8"		13	1.231		263	42		305	355
0890	3'-0" x 7'-0"		12	1.333		276	45.50		321.50	375
0900	4'-0" x 7'-0"		12	1.333		380	45.50		425.50	490
0940	Walnut face, 2'-6" x 6'-8"		14	1.143		350	39		389	445
0980	3'-0" x 6'-8"		13	1.231		355	42		397	455
0990	3'-0" x 7'-0"		12	1.333		375	45.50		420.50	480
1000	4'-0" x 7'-0"		12	1.333		535	45.50		580.50	660
1140	M.D. overlay on hardboard, 2'-6" x 6'-8"		15	1.067		276	36.50		312.50	360
1180	3'-0" x 6'-8"		14	1.143		283	39		322	370
1190	3'-0" x 7'-0"		13	1.231		294	42		336	390
1200	4'-0" x 7'-0"		12	1.333		400	45.50		445.50	510
1240	For 8'-0" height, add					50.50			50.50	55.50
1260	For 8'-0" height walnut, add					67			67	73.50
1340	H.P. plastic laminate, 2'-6" x 6'-8"	2 Carp	13	1.231		345	42		387	445
1380	3'-0" x 6'-8"		12	1.333		360	45.50		405.50	465
1390	3'-0" x 7'-0"		11	1.455		365	50		415	480
1400	4'-0" x 7'-0"		10	1.600		475	55		530	605
2200	Custom architectural "B" label, flush, 1-3/4" thick, birch,									
2210	Solid core									
2220	2'-6" x 7'-0"	2 Carp	15	1.067	Ea.	248	36.50		284.50	330
2260	3'-0" x 7'-0"		14	1.143		257	39		296	345
2300	4'-0" x 7'-0"		13	1.231		360	42		402	460
2420	4'-0" x 8'-0"		11	1.455		435	50		485	560
2460	For 6'-8" high door, deduct from 7'-0" door					10.15			10.15	11.15
2480	For oak veneer, add					50%				
2500	For walnut veneer, add					75%				
960	0010	**WOOD FRAMES**										960
0400	Exterior frame, incl. ext. trim, pine, 5/4 x 4-9/16" deep	2 Carp	375	.043	L.F.	4.85	1.46		6.31	7.65
0420	5-3/16" deep		375	.043		7.60	1.46		9.06	10.65
0440	6-9/16" deep		375	.043		7.50	1.46		8.96	10.55
0600	Oak, 5/4 x 4-9/16" deep		350	.046		8.45	1.57		10.02	11.70
0620	5-3/16" deep		350	.046		9.50	1.57		11.07	12.90
0640	6-9/16" deep		350	.046		10.55	1.57		12.12	14.05
0800	Walnut, 5/4 x 4-9/16" deep		350	.046		9.90	1.57		11.47	13.35
0820	5-3/16" deep		350	.046		14.40	1.57		15.97	18.25
0840	6-9/16" deep		350	.046		16.95	1.57		18.52	21
1000	Sills, 8/4 x 8" deep, oak, no horns		100	.160		11.95	5.50		17.45	21.50
1020	2" horns		100	.160		13.30	5.50		18.80	23
1040	3" horns		100	.160		15.35	5.50		20.85	25.50
1100	8/4 x 10" deep, oak, no horns		90	.178		16.10	6.10		22.20	27
1120	2" horns		90	.178		17.95	6.10		24.05	29.50
1140	3" horns		90	.178		19.55	6.10		25.65	31
2000	Exterior, colonial, frame & trim, 3' opng., in-swing, minimum		22	.727	Ea.	279	25		304	345
2010	Average		21	.762		415	26		441	495
2020	Maximum		20	.800		940	27.50		967.50	1,075
2100	5'-4" opening, in-swing, minimum		17	.941		315	32		347	395
2120	Maximum		15	1.067		940	36.50		976.50	1,075
2140	Out-swing, minimum		17	.941		325	32		357	405
2160	Maximum		15	1.067		975	36.50		1,011.50	1,125
2400	6'-0" opening, in-swing, minimum		16	1		300	34.50		334.50	390
2420	Maximum		10	1.600		975	55		1,030	1,150
2460	Out-swing, minimum		16	1		325	34.50		359.50	410
2480	Maximum		10	1.600		1,200	55		1,255	1,400
2600	For two sidelights, add, minimum		30	.533	Opng.	310	18.25		328.25	375

Important: See the Reference Section for critical supporting data - Reference Nos., Crews, & City Cost Indexes

08200 | Wood and Plastic Doors

08210 | Wood Doors

		CREW	DAILY OUTPUT	LABOR-HOURS	UNIT	2005 BARE COSTS				TOTAL INCL O&P	
						MAT.	LABOR	EQUIP.	TOTAL		
960											960
2620	Maximum	2 Carp	20	.800	Opng.	995	27.50		1,022.50	1,150	
2700	Custom birch frame, 3'-0" opening		16	1	Ea.	179	34.50		213.50	251	
2750	6'-0" opening		16	1		270	34.50		304.50	350	
2900	Exterior, modern, plain trim, 3' opng., in-swing, minimum		26	.615		30	21		51	66.50	
2920	Average		24	.667		36	23		59	75	
2940	Maximum		22	.727	↓	44	25		69	87	
3000	Interior frame, pine, 11/16" x 3-5/8" deep		375	.043	L.F.	4.02	1.46		5.48	6.70	
3020	4-9/16" deep		375	.043		5.50	1.46		6.96	8.35	
3200	Oak, 11/16" x 3-5/8" deep		350	.046		3.59	1.57		5.16	6.40	
3220	4-9/16" deep		350	.046		3.87	1.57		5.44	6.70	
3240	5-3/16" deep		350	.046		3.98	1.57		5.55	6.80	
3400	Walnut, 11/16" x 3-5/8" deep		350	.046		6.05	1.57		7.62	9.10	
3420	4-9/16" deep		350	.046		6.40	1.57		7.97	9.45	
3440	5-3/16" deep		350	.046	↓	6.60	1.57		8.17	9.75	
3600	Pocket door frame		16	1	Ea.	53.50	34.50		88	113	
3800	Threshold, oak, 5/8" x 3-5/8" deep		200	.080	L.F.	2.33	2.74		5.07	6.85	
3820	4-5/8" deep		190	.084		2.96	2.88		5.84	7.75	
3840	5-5/8" deep	↓	180	.089	↓	5	3.04		8.04	10.25	
4000	For casing see division 06220-400 & 06220-800										

08220 | Plastic Doors

		CREW	DAILY OUTPUT	LABOR-HOURS	UNIT	2005 BARE COSTS				TOTAL INCL O&P	
						MAT.	LABOR	EQUIP.	TOTAL		
100	0010	**FIBERGLASS DOORS**									100
0020	Exterior, fiberglass, door, 2'-8"wide x 6'-8" high	2 Carp	15	1.067	Ea.	289	36.50		325.50	375	
0040	3'-0"wide x 6'-8" high		15	1.067		289	36.50		325.50	375	
0060	3'-0"wide x 7'-0" high		15	1.067		355	36.50		391.50	445	
0080	3'-0"wide x 6'-8" high, with two lites		15	1.067		350	36.50		386.50	440	
0100	3'-0"wide x 7'-0" high, with two lites		15	1.067		415	36.50		451.50	510	
0110	Half glass, 3'-0"wide x 6'-8" high		15	1.067		390	36.50		426.50	485	
0120	3'-0"wide x 6'-8" high, low e		15	1.067		415	36.50		451.50	510	
0130	3'-0"wide x 7'-0" high		15	1.067		455	36.50		491.50	555	
0140	3'-0"wide x 7'-0" high, low e	↓	15	1.067		480	36.50		516.50	585	
0150	Side Lights, 1'-0"wide x 6'-8" high,					265			265	291	
0160	1'-0"wide x 6'-8" high, low e					277			277	305	
0180	Side Lights, 1'-0"wide x 6'-8" high, full glass					305			305	335	
0190	1'-0"wide x 6'-8" high, low e				↓	325			325	360	

08260 | Sliding Wood and Plastic Doors

		CREW	DAILY OUTPUT	LABOR-HOURS	UNIT	2005 BARE COSTS				TOTAL INCL O&P	
						MAT.	LABOR	EQUIP.	TOTAL		
700	0010	**GLASS, SLIDING**, VINYL									700
0012	Vinyl clad, 1" insul. glass, 6'-0" x 6'-10" high	2 Carp	4	4	Opng.	1,150	137		1,287	1,500	
0030	6'-0" x 8'-0" high		4	4	Ea.	1,750	137		1,887	2,150	
0100	8'-0" x 6'-10" high		4	4	Opng.	1,800	137		1,937	2,200	
0500	3 leaf, 9'-0" x 6'-10" high		3	5.333		1,625	183		1,808	2,075	
0600	12'-0" x 6'-10" high	↓	3	5.333	↓	2,025	183		2,208	2,500	

08300 | Specialty Doors

08310 | Access Doors and Panels

		CREW	DAILY OUTPUT	LABOR-HOURS	UNIT	2005 BARE COSTS				TOTAL INCL O&P	
						MAT.	LABOR	EQUIP.	TOTAL		
100	0010	**ACCESS DOORS**									100
1000	Fire rated door with lock										

08310 | Access Doors and Panels

		CREW	DAILY OUTPUT	LABOR-HOURS	UNIT	2005 BARE COSTS				TOTAL INCL O&P
						MAT.	LABOR	EQUIP.	TOTAL	
100	1100	Metal, 12" x 12"	1 Carp	10	.800	Ea.	135	27.50		162.50	191	100
	1150	18" x 18"		9	.889		176	30.50		206.50	241
	1200	24" x 24"		9	.889		212	30.50		242.50	281
	1250	24" x 36"		8	1		285	34.50		319.50	370
	1300	24" x 48"		8	1		350	34.50		384.50	440
	1350	36" x 36"		7.50	1.067		425	36.50		461.50	520
	1400	48" x 48"		7.50	1.067		545	36.50		581.50	650
	1600	Stainless steel, 12" x 12"		10	.800		241	27.50		268.50	310
	1650	18" x 18"		9	.889		350	30.50		380.50	435
	1700	24" x 24"		9	.889		430	30.50		460.50	520
	1750	24" x 36"	▼	8	1	▼	545	34.50		579.50	655
	2000	Flush door for finishing									
	2100	Metal 8" x 8"	1 Carp	10	.800	Ea.	39	27.50		66.50	85
	2150	12" x 12"	"	10	.800	"	43.50	27.50		71	90
	3000	Recessed door for acoustic tile									
	3100	Metal, 12" x 12"	1 Carp	4.50	1.778	Ea.	59.50	61		120.50	160
	3150	12" x 24"		4.50	1.778		77.50	61		138.50	181
	3200	24" x 24"		4	2		104	68.50		172.50	221
	3250	24" x 36"	▼	4	2	▼	133	68.50		201.50	254
	4000	Recessed door for drywall									
	4100	Metal 12" x 12"	1 Carp	6	1.333	Ea.	66	45.50		111.50	144
	4150	12" x 24"		5.50	1.455		98	50		148	186
	4200	24" x 36"	▼	5	1.600	▼	156	55		211	258
	6000	Standard door									
	6100	Metal, 8" x 8"	1 Carp	10	.800	Ea.	34	27.50		61.50	80
	6150	12" x 12"		10	.800		39	27.50		66.50	85
	6200	18" x 18"		9	.889		53.50	30.50		84	107
	6250	24" x 24"		9	.889		69.50	30.50		100	124
	6300	24" x 36"		8	1		103	34.50		137.50	167
	6350	36" x 36"		8	1		125	34.50		159.50	192
	6500	Stainless steel, 8" x 8"		10	.800		68.50	27.50		96	118
	6550	12" x 12"		10	.800		91	27.50		118.50	143
	6600	18" x 18"		9	.889		169	30.50		199.50	234
	6650	24" x 24"	▼	9	.889		221	30.50		251.50	291
150	0010	**BULKHEAD CELLAR DOORS**										150
	0020	Steel, not incl. sides, 44" x 62"	1 Carp	5.50	1.455	Ea.	204	50		254	305
	0100	52" x 73"		5.10	1.569		227	53.50		280.50	335
	0500	With sides and foundation plates, 57" x 45" x 24"		4.70	1.702		266	58.50		324.50	385
	0600	42" x 49" x 51"	▼	4.30	1.860	▼	320	63.50		383.50	455
300	0010	**FLOOR, COMMERCIAL**										300
	0020	Aluminum tile, steel frame, one leaf, 2' x 2' opng.	2 Sswk	3.50	4.571	Opng.	365	174		539	710
	0050	3'-6" x 3'-6" opening		3.50	4.571		660	174		834	1,025
	0500	Double leaf, 4' x 4' opening		3	5.333		965	203		1,168	1,425
	0550	5' x 5' opening	▼	3	5.333	▼	1,425	203		1,628	1,950
350	0010	**FLOOR, INDUSTRIAL**										350
	0020	Steel 300 psf L.L., single leaf, 2' x 2', 175#	2 Sswk	6	2.667	Opng.	530	102		632	765
	0050	3' x 3' opening, 300#		5.50	2.909		730	111		841	1,000
	0300	Double leaf, 4' x 4' opening, 455#		5	3.200		1,100	122		1,222	1,425
	0350	5' x 5' opening, 645#		4.50	3.556		1,450	136		1,586	1,825
	1000	Aluminum, 300 psf L.L., single leaf, 2' x 2', 60#		6	2.667		520	102		622	755
	1050	3' x 3' opening, 100#		5.50	2.909		815	111		926	1,100
	1500	Double leaf, 4' x 4' opening, 160#		5	3.200		1,275	122		1,397	1,625
	1550	5' x 5' opening, 235#		4.50	3.556		1,700	136		1,836	2,125
	2000	Aluminum, 150 psf L.L., single leaf, 2' x 2', 60#	▼	6	2.667	▼	480	102		582	705

8 DOORS & WINDOWS

Important: See the Reference Section for critical supporting data - Reference Nos., Crews, & City Cost Indexes

08310 | Access Doors and Panels

		CREW	DAILY OUTPUT	LABOR-HOURS	UNIT	2005 BARE COSTS				TOTAL INCL O&P	
						MAT.	LABOR	EQUIP.	TOTAL		
350	2050	3' x 3' opening, 95#	2 Sswk	5.50	2.909	Opng.	710	111		821	980	**350**
	2500	Double leaf, 4' x 4' opening, 150#		5	3.200		1,125	122		1,247	1,475	
	2550	5' x 5' opening, 230#	▼	4.50	3.556	▼	1,525	136		1,661	1,925	

		CREW	DAILY OUTPUT	LABOR-HOURS	UNIT	MAT.	LABOR	EQUIP.	TOTAL	TOTAL INCL O&P	
400	0010	**KENNEL DOORS**										**400**
	0020	2 way, swinging type, 13" x 19" opening	2 Carp	11	1.455	Opng.	195	50		245	293	
	0100	17" x 29" opening	"	11	1.455	"	210	50		260	310	

08320 | Detention Doors and Frames

900	0010	**SECURITY GATES** See division 10610-100										**900**
950	0010	**VAULT FRONT** See division 11021-800										**950**

08330 | Coiling Doors and Grilles

			CREW	DAILY OUTPUT	LABOR-HOURS	UNIT	MAT.	LABOR	EQUIP.	TOTAL	TOTAL INCL O&P	
130	0010	**COUNTER DOORS**										**130**
	0020	Manual, incl. frm and hdwe, galv. stl., 4' roll-up, 6' long	2 Carp	2	8	Opng.	865	274		1,139	1,375	
	0300	Galvanized steel, UL label		1.80	8.889		995	305		1,300	1,575	
	0600	Stainless steel, 4' high roll-up, 6' long		2	8		1,250	274		1,524	1,800	
	0700	10' long		1.80	8.889		2,000	305		2,305	2,675	
	2000	Aluminum, 4' high, 4' long		2.20	7.273		855	249		1,104	1,325	
	2020	6' long		2	8		945	274		1,219	1,475	
	2040	8' long		1.90	8.421		1,100	288		1,388	1,650	
	2060	10' long		1.80	8.889		1,300	305		1,605	1,900	
	2080	14' long		1.40	11.429		1,900	390		2,290	2,675	
	2100	6' high, 4' long		2	8		945	274		1,219	1,475	
	2120	6' long		1.60	10		1,100	345		1,445	1,750	
	2140	10' long	▼	1.40	11.429	▼	1,525	390		1,915	2,275	
640	0010	**COILING GRILLE**										**640**
	0015	Aluminum, manual, incl. frame, mill finish										
	0020	Top coiling, 4' high, 4' long	2 Sswk	3.20	5	Opng.	1,125	191		1,316	1,575	
	0030	6' long		3.20	5		1,175	191		1,366	1,625	
	0040	8' long		2.40	6.667		1,325	254		1,579	1,900	
	0050	12' long		2.40	6.667		1,550	254		1,804	2,150	
	0060	16' long		1.60	10		1,925	380		2,305	2,800	
	0070	6' high, 4' long		3.20	5		1,200	191		1,391	1,650	
	0080	6' long		3.20	5		1,250	191		1,441	1,725	
	0090	8' long		2.40	6.667		1,400	254		1,654	1,975	
	0100	12' long		1.60	10		1,625	380		2,005	2,475	
	0110	16' long		1.20	13.333		2,100	510		2,610	3,200	
	0200	Side coiling, 8' high, 12' long		.60	26.667		1,725	1,025		2,750	3,725	
	0220	18' long		.50	32		2,625	1,225		3,850	5,075	
	0240	24' long		.40	40		2,675	1,525		4,200	5,675	
	0260	12' high, 12' long		.50	32		2,500	1,225		3,725	4,925	
	0280	18' long		.40	40		3,675	1,525		5,200	6,775	
	0300	24' long	▼	.28	57.143	▼	4,925	2,175		7,100	9,325	
700	0010	**ROLLING GRILLE SUPPORTS** Overhead framed	E-4	36	.889	L.F.	18.40	34.50	2.26	55.16	84	**700**
720	0010	**ROLLING SERVICE DOORS** Steel, manual, 20 ga., incl. hardware										**720**
	0050	8' x 8' high	2 Sswk	1.60	10	Ea.	770	380		1,150	1,525	
	0100	10' x 10' high		1.40	11.429		1,025	435		1,460	1,900	
	0200	20' x 10' high		1	16		2,100	610		2,710	3,400	
	0300	12' x 12' high		1.20	13.333		1,325	510		1,835	2,350	
	0400	20' x 12' high	▼	.90	17.778	▼	2,175	680		2,855	3,600	

DOORS & WINDOWS 8

For expanded coverage of these items see *Means Interior Cost Data 2005*

08330 | Coiling Doors and Grilles

		CREW	DAILY OUTPUT	LABOR-HOURS	UNIT	2005 BARE COSTS MAT.	LABOR	EQUIP.	TOTAL	TOTAL INCL O&P	
720	0500	14' x 14' high	2 Sswk	.80	20	Ea.	1,750	765		2,515	3,300	720
	0600	20' x 16' high		.60	26.667		2,475	1,025		3,500	4,550	
	0700	10' x 20' high		.50	32		1,500	1,225		2,725	3,825	
	1000	12' x 12', crank operated, crank on door side		.80	20		1,175	765		1,940	2,650	
	1100	Crank thru wall		.70	22.857		1,300	870		2,170	3,000	
	1300	For vision panel, add					214			214	235	
	1400	For 22 ga., deduct				S.F.	.68			.68	.75	
	1600	3' x 7' pass door within rolling steel door, new construction				Ea.	1,150			1,150	1,275	
	1700	Existing construction	2 Sswk	2	8		1,150	305		1,455	1,825	
	2000	Class A fire doors, manual, 20 ga., 8' x 8' high		1.40	11.429		1,000	435		1,435	1,875	
	2100	10' x 10' high		1.10	14.545		1,325	555		1,880	2,475	
	2200	20' x 10' high		.80	20		2,675	765		3,440	4,325	
	2300	12' x 12' high		1	16		1,775	610		2,385	3,050	
	2400	20' x 12' high		.80	20		3,075	765		3,840	4,750	
	2500	14' x 14' high		.60	26.667		2,300	1,025		3,325	4,375	
	2600	20' x 16' high		.50	32		3,850	1,225		5,075	6,400	
	2700	10' x 20' high		.40	40		2,425	1,525		3,950	5,400	
	3000	For 18 ga. doors, add				S.F.	.76			.76	.84	
	3300	For enamel finish, add				"	.90			.90	.99	
	3600	For safety edge bottom bar, pneumatic, add				L.F.	11.85			11.85	13.05	
	3700	Electric, add					22.50			22.50	24.50	
	4000	For weatherstripping, extruded rubber, jambs, add					7.70			7.70	8.45	
	4100	Hood, add					4.88			4.88	5.35	
	4200	Sill, add					2.79			2.79	3.07	
	4500	Motor operators, to 14' x 14' opening	2 Sswk	5	3.200	Ea.	730	122		852	1,025	
	4600	Over 14' x 14', jack shaft type	"	5	3.200		720	122		842	1,000	
	4700	For fire door, additional fusible link, add					13.95			13.95	15.35	

08340 | Special Function Doors

		CREW	DAILY OUTPUT	LABOR-HOURS	UNIT	2005 BARE COSTS MAT.	LABOR	EQUIP.	TOTAL	TOTAL INCL O&P	
100	0010	**COLD STORAGE**										100
	0020	Single, 20 ga. galvanized steel										
	0300	Horizontal sliding, 5' x 7', manual operation, 3.5" thick	2 Carp	2	8	Ea.	2,400	274		2,674	3,050	
	0400	4" thick		2	8		2,900	274		3,174	3,600	
	0500	6" thick		2	8		2,700	274		2,974	3,400	
	0800	5' x 7', power operation, 2" thick		1.90	8.421		4,500	288		4,788	5,400	
	0900	4" thick		1.90	8.421		4,575	288		4,863	5,475	
	1000	6" thick		1.90	8.421		5,200	288		5,488	6,175	
	1300	9' x 10', manual operation, 2" insulation		1.70	9.412		3,625	320		3,945	4,500	
	1400	4" insulation		1.70	9.412		3,750	320		4,070	4,625	
	1500	6" insulation		1.70	9.412		4,525	320		4,845	5,475	
	1800	Power operation, 2" insulation		1.60	10		6,250	345		6,595	7,400	
	1900	4" insulation		1.60	10		6,400	345		6,745	7,575	
	2000	6" insulation		1.70	9.412		7,250	320		7,570	8,475	
	2300	For stainless steel face, add					20%					
	3000	Hinged, lightweight, 3' x 7'-0", galvanized 1 face, 2" thick	2 Carp	2	8	Ea.	1,125	274		1,399	1,675	
	3050	4" thick		1.90	8.421		1,325	288		1,613	1,900	
	3300	Aluminum doors, 3' x 7'-0", 4" thick		1.90	8.421		1,075	288		1,363	1,625	
	3350	6" thick		1.40	11.429		1,900	390		2,290	2,700	
	3600	Stainless steel, 3' x 7'-0", 4" thick		1.90	8.421		1,375	288		1,663	1,975	
	3650	6" thick		1.40	11.429		2,300	390		2,690	3,150	
	3900	Painted, 3' x 7'-0", 4" thick		1.90	8.421		985	288		1,273	1,525	
	3950	6" thick		1.40	11.429		1,875	390		2,265	2,675	
	5000	Bi-parting, electric operated										
	5010	6' x 8' opening, galv. faces, 4" thick for cooler	2 Carp	.80	20	Opng.	5,875	685		6,560	7,550	
	5050	For freezer, 4" thick		.80	20		6,475	685		7,160	8,200	

Important: See the Reference Section for critical supporting data - Reference Nos., Crews, & City Cost Indexes

8 DOORS & WINDOWS

			CREW	DAILY OUTPUT	LABOR-HOURS	UNIT	2005 BARE COSTS				TOTAL INCL O&P	
							MAT.	LABOR	EQUIP.	TOTAL		
08340		**Special Function Doors**										
100	5300	For door buck framing and door protection, add	2 Carp	2.50	6.400	Opng.	480	219		699	870	100
	6000	Galvanized batten door, galvanized hinges, 4' x 7'		2	8		1,450	274		1,724	2,025	
	6050	6' x 8'		1.80	8.889		1,975	305		2,280	2,650	
	6500	Fire door, 3 hr., 6' x 8', single slide		.80	20		6,800	685		7,485	8,550	
	6550	Double, bi-parting		.70	22.857		9,825	785		10,610	12,000	
200	0010	**HANGAR DOOR**										200
	0020	Bi-fold, ovhd., 20 PSF wind load, incl. elec. oper.										
	0100	Without sheeting, 12' high x 40'	2 Sswk	240	.067	S.F.	12.90	2.54		15.44	18.75	
	0200	16' high x 60'		230	.070		16.45	2.65		19.10	23	
	0300	20' high x 80'		220	.073		16.70	2.77		19.47	23.50	
610	0010	**ACOUSTICAL DOORS**										610
	0020	Including framed seals, 3' x 7', wood, 27 STC rating	2 Carp	1.50	10.667	Ea.	370	365		735	975	
	0100	Steel, 40 STC rating		1.50	10.667		1,375	365		1,740	2,100	
	0200	45 STC rating		1.50	10.667		1,825	365		2,190	2,575	
	0300	48 STC rating		1.50	10.667		2,300	365		2,665	3,100	
	0400	52 STC rating		1.50	10.667		2,750	365		3,115	3,600	
700	0010	**DOUBLE ACTING, SWING**										700
	1000	.063" aluminum, 7'-0" high, 4'-0" wide	2 Carp	4.20	3.810	Pr.	1,750	130		1,880	2,125	
	1050	6'-0" wide	"	4	4	"	2,300	137		2,437	2,750	
	2000	Solid core wood, 3/4" thick, metal frame, stainless steel										
	2010	base plate, 7' high opening, 4' wide	2 Carp	4	4	Pr.	2,075	137		2,212	2,525	
	2050	7' wide	"	3.80	4.211	"	2,300	144		2,444	2,750	
710	0010	**GLASS DOOR, SWING**										710
	0020	Including hardware, 1/2" thick, tempered, 3' x 7' opening	2 Glaz	2	8	Opng.	1,700	264		1,964	2,275	
	0100	6' x 7' opening	"	1.40	11.429	"	3,300	375		3,675	4,200	
720	0010	**SHOCK ABSORBING DOORS**										720
	0020	Rigid, no frame, 1-1/2" thick, 5' x 7'	2 Sswk	1.90	8.421	Opng.	1,225	320		1,545	1,925	
	0100	8' x 8'		1.80	8.889		1,725	340		2,065	2,500	
	0500	Flexible, no frame, insulated, .16" thick, economy, 5' x 7'		2	8		1,500	305		1,805	2,200	
	0600	Deluxe		1.90	8.421		2,275	320		2,595	3,075	
	1000	8' x 8' opening, economy		2	8		2,325	305		2,630	3,125	
	1100	Deluxe		1.90	8.421		3,000	320		3,320	3,875	
740	0010	**TUBULAR STEEL SWING DOORS**										740
	0020	Tubular steel, 7' high, single, 3'-4" opening	2 Sswk	2.50	6.400	Ea.	550	244		794	1,050	
	0100	Double, 6'-0" opening	"	2	8	Pr.	695	305		1,000	1,300	
08360		**Overhead Doors**										
550	0010	**OVERHEAD, COMMERCIAL** Frames not included										550
	1000	Stock, sectional, heavy duty, wood, 1-3/4" thick, 8' x 8' high	2 Carp	2	8	Ea.	575	274		849	1,050	
	1100	10' x 10' high		1.80	8.889		865	305		1,170	1,425	
	1200	12' x 12' high		1.50	10.667		1,250	365		1,615	1,975	
	1300	Chain hoist, 14' x 14' high		1.30	12.308		1,825	420		2,245	2,650	
	1400	12' x 16' high		1	16		1,600	550		2,150	2,600	
	1500	20' x 8' high		1.30	12.270		1,500	420		1,920	2,275	
	1600	20' x 16' high		.65	24.615		3,750	845		4,595	5,450	
	1800	Center mullion openings, 8' high		4	4		595	137		732	870	
	1900	20' high		2	8		1,100	274		1,374	1,650	
	2100	For medium duty custom door, deduct					5%	5%				
	2150	For medium duty stock doors, deduct					10%	5%				
	2300	Fiberglass and aluminum, heavy duty, sectional, 12' x 12' high	2 Carp	1.50	10.667	Ea.	1,775	365		2,140	2,525	
	2450	Chain hoist, 20' x 20' high		.50	32		4,375	1,100		5,475	6,500	

DOORS & WINDOWS 8

		08360	Overhead Doors	CREW	DAILY OUTPUT	LABOR-HOURS	UNIT	2005 BARE COSTS MAT.	LABOR	EQUIP.	TOTAL	TOTAL INCL O&P	
550	2600		Steel, 24 ga. sectional, manual, 8' x 8' high	2 Carp	2	8	Ea.	505	274		779	980	**550**
	2650		10' x 10' high		1.80	8.889		695	305		1,000	1,250	
	2700		12' x 12' high		1.50	10.667		950	365		1,315	1,625	
	2800		Chain hoist, 20' x 14' high	↓	.70	22.857	↓	2,825	785		3,610	4,325	
	2850		For 1-1/4" rigid insulation and 26 ga. galv.										
	2860		back panel, add				S.F.	2.16			2.16	2.38	
	2900		For electric trolley operator, 1/3 H.P., to 12' x 12', add	1 Carp	2	4	Ea.	660	137		797	940	
	2950		Over 12' x 12', 1/2 H.P., add	"	1	8	"	720	274		994	1,225	
600	0010		**RESIDENTIAL GARAGE DOORS** Including hardware, no frame										**600**
	0050		Hinged, wood, custom, double door, 9' x 7'	2 Carp	4	4	Ea.	365	137		502	620	
	0070		16' x 7'		3	5.333		625	183		808	970	
	0200		Overhead, sectional, incl. hardware, fiberglass, 9' x 7', standard		5.28	3.030		565	104		669	780	
	0220		Deluxe		5.28	3.030		720	104		824	950	
	0300		16' x 7', standard		6	2.667		965	91.50		1,056.50	1,200	
	0320		Deluxe		6	2.667		1,200	91.50		1,291.50	1,475	
	0500		Hardboard, 9' x 7', standard		8	2		370	68.50		438.50	515	
	0520		Deluxe		8	2		480	68.50		548.50	630	
	0600		16' x 7', standard		6	2.667		700	91.50		791.50	910	
	0620		Deluxe		6	2.667		815	91.50		906.50	1,050	
	0700		Metal, 9' x 7', standard		5.28	3.030		450	104		554	655	
	0720		Deluxe		8	2		585	68.50		653.50	745	
	0800		16' x 7', standard		3	5.333		570	183		753	915	
	0820		Deluxe		6	2.667		855	91.50		946.50	1,075	
	0900		Wood, 9' x 7', standard		8	2		440	68.50		508.50	590	
	0920		Deluxe		8	2		1,275	68.50		1,343.50	1,500	
	1000		16' x 7', standard		6	2.667		890	91.50		981.50	1,125	
	1020		Deluxe		6	2.667		1,850	91.50		1,941.50	2,200	
	1800		Door hardware, sectional	1 Carp	4	2		195	68.50		263.50	320	
	1810		Door tracks only		4	2		95	68.50		163.50	211	
	1820		One side only	↓	7	1.143		63	39		102	131	
	3000		Swing-up, including hardware, fiberglass, 9' x 7', standard	2 Carp	8	2		610	68.50		678.50	775	
	3020		Deluxe		8	2		670	68.50		738.50	845	
	3100		16' x 7', standard		6	2.667		770	91.50		861.50	985	
	3120		Deluxe		6	2.667		830	91.50		921.50	1,050	
	3200		Hardboard, 9' x 7', standard		8	2		293	68.50		361.50	425	
	3220		Deluxe		8	2		390	68.50		458.50	530	
	3300		16' x 7', standard		6	2.667		410	91.50		501.50	590	
	3320		Deluxe		6	2.667		610	91.50		701.50	810	
	3400		Metal, 9' x 7', standard		8	2		320	68.50		388.50	460	
	3420		Deluxe		8	2		570	68.50		638.50	735	
	3500		16' x 7', standard		6	2.667		505	91.50		596.50	695	
	3520		Deluxe		6	2.667		810	91.50		901.50	1,025	
	3600		Wood, 9' x 7', standard		8	2		350	68.50		418.50	490	
	3620		Deluxe		8	2		620	68.50		688.50	790	
	3700		16' x 7', standard		6	2.667		610	91.50		701.50	810	
	3720		Deluxe	↓	6	2.667		860	91.50		951.50	1,075	
	3900		Door hardware only, swing up	1 Carp	4	2		102	68.50		170.50	219	
	3920		One side only		7	1.143		57	39		96	124	
	4000		For electric operator, economy, add		8	1		254	34.50		288.50	335	
	4100		Deluxe, including remote control	↓	8	1	↓	370	34.50		404.50	465	
	4500		For transmitter/receiver control , add to operator				Total	80.50			80.50	88.50	
	4600		Transmitters, additional				"	28			28	31	
800	0010		**TELESCOPING STEEL DOORS** Baked enamel finish										**800**
	1000		Overhead, .03" thick, electric operated, 10' x 10'	E-3	.80	30	Ea.	9,525	1,175	101	10,801	12,700	
	2000		20' x 10'		.60	40		12,700	1,550	135	14,385	16,800	
	3000		20' x 16'	↓	.40	60	↓	15,200	2,325	203	17,728	21,100	

Important: See the Reference Section for critical supporting data - Reference Nos., Crews, & City Cost Indexes

08300 | Specialty Doors

08370 | Vertical Lift Doors

		CREW	DAILY OUTPUT	LABOR-HOURS	UNIT	2005 BARE COSTS				TOTAL INCL O&P	
						MAT.	LABOR	EQUIP.	TOTAL		
950	0010 **VERTICAL LIFT DOORS**										950
	0020 Motorized, 14 ga. stl, incl., frm and ctrl pnl										
	0050 16' x 16' high	L-10	.50	48	Ea.	16,400	1,825	1,200	19,425	22,400	
	0100 10' x 20' high		1.30	18.462		22,700	705	465	23,870	26,700	
	0120 15' x 20' high		1.30	18.462		27,800	705	465	28,970	32,300	
	0140 20' x 20' high		1	24		32,600	915	605	34,120	38,000	
	0160 25' x 20' high		1	24		36,100	915	605	37,620	41,900	
	0170 32' x 24' high		.75	32		33,700	1,225	805	35,730	40,100	
	0180 20' x 25' high		1	24		37,300	915	605	38,820	43,300	
	0200 25' x 25' high		.70	34.286		42,000	1,300	860	44,160	49,400	
	0220 25' x 30' high		.70	34.286		49,100	1,300	860	51,260	57,000	
	0240 30' x 30' high		.70	34.286		56,500	1,300	860	58,660	65,000	
	0260 35' x 30' high		.70	34.286		62,000	1,300	860	64,160	71,000	

08380 | Traffic Doors

		CREW	DAILY OUTPUT	LABOR-HOURS	UNIT	MAT.	LABOR	EQUIP.	TOTAL	TOTAL INCL O&P	
480	0010 **FLEXIBLE TRANSPARENT STRIP ENTRANCE**										480
	0100 12" strip width, 2/3 overlap	3 Shee	135	.178	SF Surf	4.97	7.15		12.12	16.45	
	0200 Full overlap		115	.209		5.20	8.40		13.60	18.65	
	0220 8" strip width, 1/2 overlap		140	.171		4.82	6.90		11.72	15.90	
	0240 Full overlap		120	.200		6.35	8.05		14.40	19.35	
	0300 Add for suspension system, header mount				L.F.	7.35			7.35	8.10	
	0400 Wall mount				"	7.40			7.40	8.15	

08400 | Entrances and Storefronts

08410 | Metal-Framed Storefronts

		CREW	DAILY OUTPUT	LABOR-HOURS	UNIT	2005 BARE COSTS				TOTAL INCL O&P	
						MAT.	LABOR	EQUIP.	TOTAL		
110	0010 **ALUMINUM-FRAMED STOREFRONTS**										110
	0020 Entrance, 3' x 7' opening, clear anodized finish	2 Sswk	7	2.286	Opng.	243	87		330	425	
	0040 Bronze finish		7	2.286		300	87		387	485	
	0060 Black finish		7	2.286		340	87		427	530	
	0500 6' x 7' opening, clear finish		6	2.667		298	102		400	510	
	0520 Bronze finish		6	2.667		340	102		442	550	
	0540 Black finish		6	2.667		395	102		497	615	
	1000 With 3' high transoms, 3' x 10' opening, clear finish		6.50	2.462		360	94		454	565	
	1050 Bronze finish		6.50	2.462		395	94		489	605	
	1100 Black finish		6.50	2.462		465	94		559	685	
	1500 With 3' high transoms, 6' x 10' opening, clear finish		5.50	2.909		435	111		546	680	
	1550 Bronze finish		5.50	2.909		465	111		576	715	
	1600 Black finish		5.50	2.909		555	111		666	810	
120	0010 **ALUMINUM DOORS** Commercial entrance, no glazing										120
	0020 Incl. hinges, push/pull, deadlock, cyl., threshold										
	0800 Narrow stile, no glazing, standard hardware, pair of 2'-6" x 7'-0"	2 Carp	1.70	9.412	Pr.	730	320		1,050	1,300	
	1000 3'-0" x 7'-0", single		3	5.333	Ea.	480	183		663	815	
	1200 Pair of 3'-0" x 7'-0"		1.70	9.412	Pr.	960	320		1,280	1,550	
	1500 3'-6" x 7'-0", single		3	5.333	Ea.	515	183		698	855	
	2000 Medium stile, pair of 2'-6" x 7'-0"		1.70	9.412	Pr.	900	320		1,220	1,500	
	2100 3'-0" x 7'-0", single		3	5.333	Ea.	630	183		813	980	

8

DOORS & WINDOWS

For expanded coverage of these items see *Means Interior Cost Data 2005*

		08410	Metal-Framed Storefronts	CREW	DAILY OUTPUT	LABOR-HOURS	UNIT	2005 BARE COSTS				TOTAL INCL O&P	
								MAT.	LABOR	EQUIP.	TOTAL		
120	2200		Pair of 3'-0" x 7'-0"	2 Carp	1.70	9.412	Pr.	1,250	320		1,570	1,850	120
	2300		3'-6" x 7'-0", single	↓	5.33	3.002	Ea.	745	103		848	980	
	5000		Flush panel doors, pair of 2'-6" x 7'-0"	2 Sswk	2	8	Pr.	905	305		1,210	1,550	
	5050		3'-0" x 7'-0", single		2.50	6.400	Ea.	450	244		694	930	
	5100		Pair of 3'-0" x 7'-0"		2	8	Pr.	905	305		1,210	1,550	
	5150		3'-6" x 7'-0", single	↓	2.50	6.400	Ea.	535	244		779	1,025	
130	0010		**ALUMINUM DOORS & FRAMES** Entrance, narrow stile, including										130
	0015		Standard hardware, clear finish, not incl. glass, 2'-6" x 7'-0" opng.	2 Sswk	2	8	Ea.	695	305		1,000	1,300	
	0020		3'-0" x 7'-0" opening		2	8		535	305		840	1,125	
	0030		3'-6" x 7'-0" opening		2	8		515	305		820	1,125	
	0100		3'-0" x 10'-0" opening, 3' high transom		1.80	8.889		750	340		1,090	1,425	
	0200		3'-6" x 10'-0" opening, 3' high transom		1.80	8.889		740	340		1,080	1,425	
	0280		5'-0" x 7'-0" opening		2	8	↓	870	305		1,175	1,500	
	0300		6'-0" x 7'-0" opening		1.30	12.308	Pr.	820	470		1,290	1,750	
	0400		6'-0" x 10'-0" opening, 3' high transom		1.10	14.545		685	555		1,240	1,750	
	0420		7'-0" x 7'-0" opening		1	16	↓	855	610		1,465	2,050	
	0500		Wide stile, 2'-6" x 7'-0" opening		2	8	Ea.	700	305		1,005	1,325	
	0520		3'-0" x 7'-0" opening		2	8		715	305		1,020	1,325	
	0540		3'-6" x 7'-0" opening		2	8		760	305		1,065	1,375	
	0560		5'-0" x 7'-0" opening		2	8	↓	1,250	305		1,555	1,925	
	0580		6'-0" x 7'-0" opening		1.30	12.308	Pr.	1,175	470		1,645	2,150	
	0600		7'-0" x 7'-0" opening	↓	1	16	"	1,300	610		1,910	2,525	
	1100		For full vision doors, with 1/2" glass, add				Leaf	55%					
	1200		For non-standard size, add					67%					
	1300		Light bronze finish, add					36%					
	1400		Dark bronze finish, add					18%					
	1500		For black finish, add					36%					
	1600		Concealed panic device, add				↓	985			985	1,075	
	1700		Electric striker release, add				Opng.	254			254	279	
	1800		Floor check, add				Leaf	755			755	830	
	1900		Concealed closer, add				"	500			500	550	
	2000		Flush 3' x 7' Insulated, 12"x 12" lite, clear finish	2 Sswk	2	8	Ea.	920	305		1,225	1,550	
140	0010		**STOREFRONT SYSTEMS** Aluminum frame, clear 3/8" plate glass,										140
	0020		incl. 3' x 7' door with hardware (400 sq. ft. max. wall)										
	0500		Wall height to 12' high, commercial grade	2 Glaz	150	.107	S.F.	13.25	3.52		16.77	19.85	
	0600		Institutional grade		130	.123		16.65	4.06		20.71	24.50	
	0700		Monumental grade		115	.139		25	4.59		29.59	34.50	
	1000		6' x 7' door with hardware, commercial grade		135	.119		13.55	3.91		17.46	21	
	1100		Institutional grade		115	.139		18.50	4.59		23.09	27.50	
	1200		Monumental grade	↓	100	.160		34.50	5.30		39.80	45.50	
	1500		For bronze anodized finish, add					15%					
	1600		For black anodized finish, add					30%					
	1700		For stainless steel framing, add to monumental				↓	75%					
300	0010		**STAINLESS STEEL AND GLASS** Entrance unit, narrow stiles										300
	0020		3' x 7' opening, including hardware, minimum	2 Sswk	1.60	10	Opng.	4,600	380		4,980	5,725	
	0050		Average		1.40	11.429		4,975	435		5,410	6,250	
	0100		Maximum	↓	1.20	13.333		5,325	510		5,835	6,775	
	1000		For solid bronze entrance units, statuary finish, add					60%					
	1100		Without statuary finish, add				↓	45%					
	2000		Balanced doors, 3' x 7', economy	2 Sswk	.90	17.778	Ea.	6,225	680		6,905	8,075	
	2100		Premium	"	.70	22.857	"	10,700	870		11,570	13,400	

		08460	Automatic Entrance Doors										
600	0010		**SLIDING ENTRANCE** 12' x 7'-6" opng., 5' x 7' door, 2 way traf.,										600
	0020		mat activated, panic pushout, incl. operator & hardware,										

Important: See the Reference Section for critical supporting data - Reference Nos., Crews, & City Cost Indexes

08460 | Automatic Entrance Doors

		CREW	DAILY OUTPUT	LABOR-HOURS	UNIT	2005 BARE COSTS MAT.	LABOR	EQUIP.	TOTAL	TOTAL INCL O&P	
600	0030	not including glass or glazing	2 Glaz	.70	22.857	Opng.	5,900	755		6,655	7,650	600

08470 | Revolving Entrance Doors

			CREW	DAILY OUTPUT	LABOR-HOURS	UNIT	MAT.	LABOR	EQUIP.	TOTAL	TOTAL INCL O&P	
600	0010	REVOLVING DOORS Aluminum, 6'-6" to 7'-0" diameter,										600
	0020	6'-10" to 7' high, stock units, minimum	4 Sswk	.75	42.667	Opng.	15,500	1,625		17,125	20,000	
	0050	Average		.60	53.333		19,100	2,025		21,125	24,800	
	0100	Maximum		.45	71.111		25,200	2,725		27,925	32,600	
	1000	Stainless steel		.30	105		28,800	4,050		32,850	38,900	
	1100	Solid bronze		.15	213		33,600	8,150		41,750	51,500	
	1500	For automatic controls, add	2 Elec	2	8		11,700	325		12,025	13,300	

08480 | Balanced Entrance Doors

			CREW	DAILY OUTPUT	LABOR-HOURS	UNIT	MAT.	LABOR	EQUIP.	TOTAL	TOTAL INCL O&P	
150	0010	BALANCED DOORS										150
	0020	Hardware & frame, alum. & glass, 3' x 7', econ.	2 Sswk	.90	17.778	Ea.	4,975	680		5,655	6,700	
	0150	Premium	"	.70	22.857	"	6,175	870		7,045	8,350	

08490 | Sliding Storefronts

			CREW	DAILY OUTPUT	LABOR-HOURS	UNIT	MAT.	LABOR	EQUIP.	TOTAL	TOTAL INCL O&P	
100	0010	SLIDING PANELS										100
	0020	Mall fronts, aluminum & glass, 15' x 9' high	2 Glaz	1.30	12.308	Opng.	2,325	405		2,730	3,175	
	0100	24' x 9' high		.70	22.857		3,350	755		4,105	4,850	
	0200	48' x 9' high, with fixed panels		.90	17.778		6,250	585		6,835	7,750	
	0500	For bronze finish, add					17%					

08500 | Windows

08510 | Steel Windows

			CREW	DAILY OUTPUT	LABOR-HOURS	UNIT	MAT.	LABOR	EQUIP.	TOTAL	TOTAL INCL O&P	
700	0010	SCREENS										700
	0020	For metal sash, aluminum or bronze mesh, flat screen	2 Sswk	1,200	.013	S.F.	3.47	.51		3.98	4.73	
	0500	Wicket screen, inside window		1,000	.016		5.30	.61		5.91	6.95	
	0800	Security screen, aluminum frame with stainless steel cloth		1,200	.013		18.70	.51		19.21	21.50	
	0900	Steel grate, painted, on steel frame		1,600	.010		10.05	.38		10.43	11.75	
	1000	For solar louvers, add		160	.100		19.35	3.82		23.17	28.50	
	4000	See also division 05580-900										
750	0010	STEEL SASH Custom units, glazing and trim not included	R08510 -100									750
	0100	Casement, 100% vented	2 Sswk	200	.080	S.F.	39.50	3.05		42.55	49	
	0200	50% vented		200	.080		36	3.05		39.05	45	
	0300	Fixed		200	.080		25	3.05		28.05	33	
	1000	Projected, commercial, 40% vented		200	.080		41	3.05		44.05	50.50	
	1100	Intermediate, 50% vented		200	.080		44	3.05		47.05	54	
	1500	Industrial, horizontally pivoted		200	.080		42.50	3.05		45.55	52	
	1600	Fixed		200	.080		24	3.05		27.05	32	
	2000	Industrial security sash, 50% vented		200	.080		45.50	3.05		48.55	55.50	
	2100	Fixed		200	.080		37	3.05		40.05	46.50	
	2500	Picture window		200	.080		23.50	3.05		26.55	31	
	3000	Double hung		200	.080		44.50	3.05		47.55	54.50	

DOORS & WINDOWS 8

08510 | Steel Windows

			CREW	DAILY OUTPUT	LABOR-HOURS	UNIT	MAT.	LABOR	EQUIP.	TOTAL	TOTAL INCL O&P	
750	5000	Mullions for above, open interior face	R08510 -100	2 Sswk	240	.067	L.F.	7.95	2.54		10.49	13.30	750
	5100	With interior cover		↓	240	.067	"	13.20	2.54		15.74	19.05	
770	0010	**STEEL WINDOWS** Stock, including frame, trim and insul. glass											770
	0020	See also division 13128-700											
	1000	Custom units, double hung, 2'-8" x 4'-6" opening	R08510 -100	2 Sswk	12	1.333	Ea.	570	51		621	720	
	1100	2'-4" x 3'-9" opening			12	1.333		470	51		521	610	
	1500	Commercial projected, 3'-9" x 5'-5" opening			10	1.600		1,000	61		1,061	1,200	
	1600	6'-9" x 4'-1" opening			7	2.286		1,325	87		1,412	1,600	
	2000	Intermediate projected, 2'-9" x 4'-1" opening			12	1.333		560	51		611	705	
	2100	4'-1" x 5'-5" opening		↓	10	1.600	↓	1,125	61		1,186	1,350	

08520 | Aluminum Windows

			CREW	DAILY OUTPUT	LABOR-HOURS	UNIT	MAT.	LABOR	EQUIP.	TOTAL	TOTAL INCL O&P	
100	0010	**ALUMINUM SASH**											100
	0020	Stock, grade C, glaze & trim not incl., casement	2 Sswk	200	.080	S.F.	27.50	3.05		30.55	35.50	
	0050	Double hung		200	.080		28	3.05		31.05	36	
	0100	Fixed casement		200	.080		11.05	3.05		14.10	17.60	
	0150	Picture window		200	.080		11.90	3.05		14.95	18.55	
	0200	Projected window		200	.080		25.50	3.05		28.55	33.50	
	0250	Single hung		200	.080		13.30	3.05		16.35	20	
	0300	Sliding		200	.080	↓	17.20	3.05		20.25	24.50	
	1000	Mullions for above, tubular		240	.067	L.F.	4.35	2.54		6.89	9.35	
	2000	Custom aluminum sash, grade HC, glazing not included, minimum		200	.080	S.F.	32	3.05		35.05	40.50	
	2100	Maximum	↓	85	.188	"	41.50	7.20		48.70	59	
120	0010	**ALUMINUM WINDOWS** Incl. frame and glazing, Commercial grade											120
	0020	See also division 05120-440											
	1000	Stock units, casement, 3'-1" x 3'-2" opening	2 Sswk	10	1.600	Ea.	299	61		360	440	
	1050	Add for storms					60			60	66	
	1600	Projected, with screen, 3'-1" x 3'-2" opening	2 Sswk	10	1.600		213	61		274	345	
	1700	Add for storms					56.50			56.50	62	
	2000	4'-5" x 5'-3" opening	2 Sswk	8	2		300	76.50		376.50	465	
	2100	Add for storms					78.50			78.50	86.50	
	2500	Enamel finish windows, 3'-1" x 3'-2"	2 Sswk	10	1.600		191	61		252	320	
	2600	4'-5" x 5'-3"		8	2		287	76.50		363.50	450	
	3000	Single hung, 2' x 3' opening, enameled, standard glazed		10	1.600		143	61		204	267	
	3100	Insulating glass		10	1.600		174	61		235	300	
	3300	2'-8" x 6'-8" opening, standard glazed		8	2		305	76.50		381.50	470	
	3400	Insulating glass		8	2		390	76.50		466.50	565	
	3700	3'-4" x 5'-0" opening, standard glazed		9	1.778		196	68		264	335	
	3800	Insulating glass		9	1.778		276	68		344	425	
	3890	Awning type, 3' x 3' opening standard glass		14	1.143		350	43.50		393.50	465	
	3900	Insulating glass		14	1.143		375	43.50		418.50	495	
	3910	3' x 4' opening, standard glass		10	1.600		410	61		471	560	
	3920	Insulating glass		10	1.600		475	61		536	630	
	3930	3' x 5'-4" opening, standard glass		10	1.600		495	61		556	655	
	3940	Insulating glass		10	1.600		585	61		646	755	
	3950	4' x 5'-4" opening, standard glass		9	1.778		545	68		613	720	
	3960	Insulating glass		9	1.778		650	68		718	840	
	4000	Sliding aluminum, 3' x 2' opening, standard glazed		10	1.600		161	61		222	286	
	4100	Insulating glass		10	1.600		179	61		240	305	
	4300	5' x 3' opening, standard glazed		9	1.778		205	68		273	345	
	4400	Insulating glass		9	1.778		286	68		354	435	
	4600	8' x 4' opening, standard glazed		6	2.667		295	102		397	505	
	4700	Insulating glass		6	2.667		475	102		577	700	
	5000	9' x 5' opening, standard glazed		4	4		445	153		598	765	
	5100	Insulating glass	↓	4	4	↓	715	153		868	1,050	

Important: See the Reference Section for critical supporting data - Reference Nos., Crews, & City Cost Indexes

08520 | Aluminum Windows

			DAILY OUTPUT	LABOR-HOURS	UNIT	2005 BARE COSTS				TOTAL INCL O&P	
			CREW			MAT.	LABOR	EQUIP.	TOTAL		
120	5500	Sliding, with thermal barrier and screen, 6' x 4', 2 track	2 Sswk	8	2	Ea.	605	76.50		681.50	805	**120**
	5700	4 track	↓	8	2		740	76.50		816.50	950	
	6000	For above units with bronze finish, add					12%					
	6200	For installation in concrete openings, add				↓	5%					
500	0010	**JALOUSIES**										**500**
	0020	Aluminum incl. glazing & screens, stock, 1'-7" x 3'-2"	2 Sswk	10	1.600	Ea.	131	61		192	253	
	0100	2'-3" x 4'-0"		10	1.600		187	61		248	315	
	0200	3'-1" x 2'-0"		10	1.600		147	61		208	271	
	0300	3'-1" x 5'-3"		10	1.600		267	61		328	400	
	1000	Mullions for above, 2'-0" long		80	.200		10.05	7.65		17.70	24.50	
	1100	5'-3" long	↓	80	.200	↓	17.20	7.65		24.85	32.50	

08550 | Wood Windows

			CREW	DAILY OUTPUT	LABOR-HOURS	UNIT	MAT.	LABOR	EQUIP.	TOTAL	TOTAL INCL O&P	
100	0010	**AWNING WINDOW** Including frame, screens and grills										**100**
	0100	Average quality, builders model, 34" x 22", double insulated glass	1 Carp	10	.800	Ea.	209	27.50		236.50	273	
	0200	Low E glass		10	.800		221	27.50		248.50	286	
	0300	40" x 28", double insulated glass		9	.889		264	30.50		294.50	340	
	0400	Low E Glass		9	.889		280	30.50		310.50	360	
	0500	48" x 36", double insulated glass		8	1		385	34.50		419.50	480	
	0600	Low E glass		8	1		405	34.50		439.50	500	
	1000	34" x 22"		10	.800		223	27.50		250.50	288	
	1100	40" x 22"		10	.800		243	27.50		270.50	310	
	1200	36" x 28"		9	.889		259	30.50		289.50	335	
	1300	36" x 36"		9	.889		288	30.50		318.50	365	
	1400	48" x 28"		8	1		310	34.50		344.50	395	
	1500	60" x 36"		8	1		450	34.50		484.50	550	
	2000	Metal clad, deluxe, double insulated glass, 34" x 22"		10	.800		208	27.50		235.50	272	
	2100	40" x 22"		10	.800		244	27.50		271.50	310	
	2200	36" x 25"		9	.889		226	30.50		256.50	297	
	2300	40" x 30"		9	.889		283	30.50		313.50	360	
	2400	48" x 28"		8	1		289	34.50		323.50	375	
	2500	60" x 36"	↓	8	1	↓	310	34.50		344.50	395	
150	0010	**BOW-BAY WINDOW** Including frame, screens and grills,										**150**
	0020	end panels operable										
	1000	Bow type, casement, wood, bldrs mdl, 8' x 5' dbl insltd glass, 4 panel	2 Carp	10	1.600	Ea.	1,475	55		1,530	1,700	
	1050	Low E glass		10	1.600		1,200	55		1,255	1,400	
	1100	10'-0" x 5'-0", double insulated glass, 6 panels		6	2.667		1,250	91.50		1,341.50	1,525	
	1200	Low E glass, 6 panels		6	2.667		1,325	91.50		1,416.50	1,600	
	1300	Vinyl clad, bldrs model, double insulated glass, 6'-0" x 4'-0", 3 panel		10	1.600		1,000	55		1,055	1,175	
	1340	9'-0" x 4'-0", 4 panel		8	2		1,350	68.50		1,418.50	1,575	
	1380	10'-0" x 6'-0", 5 panels		7	2.286		2,150	78.50		2,228.50	2,500	
	1420	12'-0" x 6'-0", 6 panels		6	2.667		2,725	91.50		2,816.50	3,150	
	1600	Metal clad, casement, bldrs mdl, 6'-0" x 4'-0", dbl insltd gls, 3 panels		10	1.600		855	55		910	1,025	
	1640	9'-0" x 4'-0", 4 panels		8	2		1,200	68.50		1,268.50	1,425	
	1680	10'-0" x 5'-0", 5 panels		7	2.286		1,650	78.50		1,728.50	1,950	
	1720	12'-0" x 6'-0", 6 panels		6	2.667		2,300	91.50		2,391.50	2,675	
	2000	Bay window, casement, builders model, 8' x 5' dbl insul glass, 4 panels		10	1.600		1,425	55		1,480	1,625	
	2050	Low E glass,		10	1.600		1,725	55		1,780	1,950	
	2100	12'-0" x 6'-0", double insulated glass, 6 panels		6	2.667		1,775	91.50		1,866.50	2,100	
	2200	Low E glass		6	2.667		1,825	91.50		1,916.50	2,175	
	2280	6'-0" x 4'-0"		11	1.455		975	50		1,025	1,150	
	2300	Vinyl clad, premium, double insulated glass, 8'-0" x 5'-0"		10	1.600		1,150	55		1,205	1,325	
	2340	10'-0" x 5'-0"		8	2		1,625	68.50		1,693.50	1,875	
	2380	10'-0" x 6'-0"	↓	7	2.286		1,700	78.50		1,778.50	2,000	

For expanded coverage of these items see *Means Interior Cost Data 2005*

DOORS & WINDOWS 8

08550 | Wood Windows

Line		Description	CREW	DAILY OUTPUT	LABOR-HOURS	UNIT	MAT.	LABOR	EQUIP.	TOTAL	TOTAL INCL O&P
150	2420	12'-0" x 6'-0"	2 Carp	6	2.667	Ea.	2,025	91.50		2,116.50	2,375
	2600	Metal clad, deluxe, dbl insul. glass, 8'-0" x 5'-0" high, 4 panels		10	1.600		1,250	55		1,305	1,450
	2640	10'-0" x 5'-0" high, 5 panels		8	2		1,350	68.50		1,418.50	1,575
	2680	10'-0" x 6'-0" high, 5 panels		7	2.286		1,600	78.50		1,678.50	1,875
	2720	12'-0" x 6'-0" high, 6 panels		6	2.667		2,200	91.50		2,291.50	2,575
	3000	Double hung, bldrs. model, bay, 8' x 4' high, dbl insulated glass		10	1.600		995	55		1,050	1,175
	3050	Low E glass		10	1.600		1,075	55		1,130	1,250
	3100	9'-0" x 5'-0" high, doublel insulated glass		6	2.667		1,075	91.50		1,166.50	1,325
	3200	Low E glass		6	2.667		1,125	91.50		1,216.50	1,400
	3300	Vinyl clad, premium, double insulated glass, 7'-0" x 4'-6"		10	1.600		1,025	55		1,080	1,200
	3340	8'-0" x 4'-6"		8	2		1,050	68.50		1,118.50	1,250
	3380	8'-0" x 5'-0"		7	2.286		1,100	78.50		1,178.50	1,325
	3420	9'-0" x 5'-0"		6	2.667		1,125	91.50		1,216.50	1,400
	3600	Metal clad, deluxe, dbl insul. glass, 7'-0" x 4'-0" high		10	1.600		955	55		1,010	1,125
	3640	8'-0" x 4'-0" high		8	2		990	68.50		1,058.50	1,175
	3680	8'-0" x 5'-0" high		7	2.286		1,025	78.50		1,103.50	1,250
	3720	9'-0" x 5'-0" high		6	2.667		1,075	91.50		1,166.50	1,350
200	0010	**CASEMENT WINDOW** Including frame, screen, and grills									
	0100	Avg. quality, bldrs. model, 2'-0" x 3'-0" H, dbl. insulated glass	1 Carp	10	.800	Ea.	176	27.50		203.50	236
	0150	Low E glass CN		10	.800		282	27.50		309.50	355
	0200	2'-0" x 4'-6" high, double insulated glass		9	.889		228	30.50		258.50	299
	0250	Low E glass		9	.889		330	30.50		360.50	410
	0300	2'-3" x 6'-0" high, double insulated glass		8	1		395	34.50		429.50	490
	0350	Low E glass		8	1		325	34.50		359.50	415
	0522	Vinyl clad, premium, double insulated glass, 2'-0" x 3'-0"		10	.800		237	27.50		264.50	305
	0524	2'-0" x 4'-0"		9	.889		277	30.50		307.50	355
	0525	2'-0" x 5'-0"		8	1		315	34.50		349.50	405
	0528	2'-0" x 6'-0"		8	1		360	34.50		394.50	450
	8100	Metal clad, deluxe, dbl. insul. glass, 2'-0" x 3'-0" high		10	.800		175	27.50		202.50	236
	8120	2'-0" x 4'-0" high		9	.889		211	30.50		241.50	280
	8140	2'-0" x 5'-0" high		8	1		240	34.50		274.50	315
	8160	2'-0" x 6'-0" high		8	1		275	34.50		309.50	360
	8200	For multiple leaf units, deduct for stationary sash									
	8220	2' high				Ea.	18.20			18.20	20
	8240	4'-6" high					21			21	23
	8260	6' high					28			28	31
	8300	For installation, add per leaf						15%			
250	0010	**DOUBLE HUNG** Including frame, screens, and grills R08550 -010									
	0100	Avg. quality, bldrs. model, 2'-0" x 3'-0" high, dbl insul. glass	1 Carp	10	.800	Ea.	183	27.50		210.50	245
	0150	Low E glass		10	.800		180	27.50		207.50	241
	0200	3'-0" x 4'-0" high, double insulated glass		9	.889		244	30.50		274.50	315
	0250	Low E glass		9	.889		234	30.50		264.50	305
	0300	4'-0" x 4'-6" high, double insulated glass		8	1		277	34.50		311.50	360
	0350	Low E glass		8	1		300	34.50		334.50	385
	1000	Vinyl clad, premium, double insulated glass, 2'-6" x 3'-0"		10	.800		195	27.50		222.50	258
	1100	3'-0" x 3'-6"		10	.800		230	27.50		257.50	296
	1200	3'-0" x 4'-0"		9	.889		325	30.50		355.50	405
	1300	3'-0" x 4'-6"		9	.889		266	30.50		296.50	340
	1400	3'-0" x 5'-0"		8	1		278	34.50		312.50	360
	1500	3'-6" x 6'-0"		8	1		325	34.50		359.50	410
	2000	Metal clad, deluxe, dbl. insul. glass, 2'-6" x 3'-0" high		10	.800		186	27.50		213.50	247
	2100	3'-0" x 3'-6" high		10	.800		220	27.50		247.50	285
	2200	3'-0" x 4'-0" high		9	.889		234	30.50		264.50	305
	2300	3'-0" x 4'-6" high		9	.889		253	30.50		283.50	325
	2400	3'-0" x 5'-0" high		8	1		270	34.50		304.50	350

Important: See the Reference Section for critical supporting data - Reference Nos., Crews, & City Cost Indexes

08550	Wood Windows		CREW	DAILY OUTPUT	LABOR-HOURS	UNIT	2005 BARE COSTS				TOTAL INCL O&P	
							MAT.	LABOR	EQUIP.	TOTAL		
250	2500	3'-6" x 6'-0" high R08550-010	1 Carp	8	1	Ea.	330	34.50		364.50	415	250
650	0010	**PALLADIAN WINDOWS**										650
	0020	Vinyl clad, double insulated glass, including frame and grills										
	0040	3'-2" x 2'-0" high	2 Carp	11	1.455	Ea.	1,175	50		1,225	1,350	
	0060	3'-2" x 4'-10"		11	1.455		1,325	50		1,375	1,525	
	0080	3'-2" x 6'-4"		10	1.600		1,575	55		1,630	1,800	
	0100	4'-0" x 4'-0"		10	1.600		1,275	55		1,330	1,500	
	0120	4'-0" x 5'-4"	3 Carp	10	2.400		1,500	82		1,582	1,775	
	0140	4'-0" x 6'-0"		9	2.667		1,550	91.50		1,641.50	1,850	
	0160	4'-0" x 7'-4"		9	2.667		1,700	91.50		1,791.50	2,025	
	0180	5'-5" x 4'-10"		9	2.667		1,700	91.50		1,791.50	2,025	
	0200	5'-5" x 6'-10"		9	2.667		1,925	91.50		2,016.50	2,275	
	0220	5'-5" x 7'-9"		9	2.667		2,025	91.50		2,116.50	2,375	
	0240	6'-0" x 7'-11"		8	3		2,800	103		2,903	3,225	
	0260	8'-0" x 6'-0"		8	3		2,325	103		2,428	2,700	
670	0010	**PICTURE WINDOW** Including frame and grills										670
	0100	Average quality, bldrs. model, 3'-6" x 4'-0" high, dbl insulated glass	2 Carp	12	1.333	Ea.	268	45.50		313.50	365	
	0150	Low E glass		12	1.333		292	45.50		337.50	390	
	0200	4'-0" x 4'-6" high, double insulated glass		11	1.455		268	50		318	375	
	0250	Low E glass		11	1.455		294	50		344	405	
	0300	5'-0" x 4'-0" high, double insulated glass		11	1.455		330	50		380	445	
	0350	Low E glass		11	1.455		370	50		420	490	
	0400	6'-0" x 4'-6" high, double insulated glass		10	1.600		420	55		475	550	
	0450	Low E glass		10	1.600		475	55		530	605	
	1000	Vinyl clad, premium, dbl. insul. glass, 4'-0" x 4'-0"		12	1.333		430	45.50		475.50	545	
	1100	4'-0" x 6'-0"		11	1.455		635	50		685	780	
	1200	5'-0" x 6'-0"		10	1.600		915	55		970	1,075	
	1300	6'-0" x 6'-0"		10	1.600		930	55		985	1,100	
	2000	Metal clad, deluxe, dbl. insul. glass, 4'-0" x 4'-0" high		12	1.333		275	45.50		320.50	375	
	2100	4'-0" x 6'-0" high		11	1.455		405	50		455	525	
	2200	5'-0" x 6'-0" high		10	1.600		445	55		500	575	
	2300	6'-0" x 6'-0" high		10	1.600		515	55		570	650	
750	0010	**SLIDING WINDOW** Including frame, screen, and grills										750
	0100	Average quality, bldrs. model, 3'-0" x 3'-0" high, double insulated	1 Carp	10	.800	Ea.	144	27.50		171.50	201	
	0120	Low E glass		10	.800		182	27.50		209.50	243	
	0200	4'-0" x 3'-6" high, double insulated		9	.889		171	30.50		201.50	236	
	0220	Low E glass		9	.889		214	30.50		244.50	284	
	0300	6'-0" x 5'-0" high, double insulated		8	1		315	34.50		349.50	400	
	0320	Low E glass		8	1		375	34.50		409.50	470	
	1000	Vinyl clad, premium, dbl. insulated glass, 3'-0" x 3'-0"		10	.800		515	27.50		542.50	610	
	1050	4'-0" x 3'-6"		9	.889		640	30.50		670.50	755	
	1100	5'-0" x 4'-0"		9	.889		770	30.50		800.50	900	
	1150	6'-0" x 5'-0"		8	1		945	34.50		979.50	1,075	
	2000	Metal clad, deluxe, double insulated glass, 3'-0" x 3'-0" high		10	.800		290	27.50		317.50	365	
	2050	4'-0" x 3'-6" high		9	.889		355	30.50		385.50	440	
	2100	5'-0" x 4'-0" high		9	.889		430	30.50		460.50	520	
	2150	6'-0" x 5'-0" high		8	1		650	34.50		684.50	765	
770	0010	**WEATHERSTRIPPING** See division 08720										770
800	0010	**WINDOW GRILLE OR MUNTIN** Snap-in type										800
	0020	Standard pattern interior grills										
	2000	Wood, awning window, glass size 28" x 16" high	1 Carp	30	.267	Ea.	19.10	9.15		28.25	35.50	
	2060	44" x 24" high		32	.250		28	8.55		36.55	44	

DOORS & WINDOWS 8

			CREW	DAILY OUTPUT	LABOR-HOURS	UNIT	2005 BARE COSTS				TOTAL INCL O&P	
08550		**Wood Windows**					MAT.	LABOR	EQUIP.	TOTAL		
800	2100	Casement, glass size, 20" x 36" high	1 Carp	30	.267	Ea.	23.50	9.15		32.65	40.50	**800**
	2180	20" x 56" high		32	.250	↓	34	8.55		42.55	50.50	
	2200	Double hung, glass size, 16" x 24" high		24	.333	Set	41.50	11.40		52.90	64	
	2280	32" x 32" high		34	.235	"	116	8.05		124.05	141	
	2500	Picture, glass size, 48" x 48" high		30	.267	Ea.	133	9.15		142.15	160	
	2580	60" x 68" high		28	.286	"	97.50	9.80		107.30	122	
	2600	Sliding, glass size, 14" x 36" high		24	.333	Set	23	11.40		34.40	43.50	
	2680	36" x 36" high	↓	22	.364	"	35	12.45		47.45	57.50	
820	0010	**WOOD SASH** Including glazing but not including trim										**820**
	0050	Custom, 5'-0" x 4'-0", 1" dbl. glazed, 3/16" thick lites	2 Carp	3.20	5	Ea.	147	171		318	430	
	0100	1/4" thick lites		5	3.200		151	110		261	335	
	0200	1" thick, triple glazed		5	3.200		345	110		455	550	
	0300	7'-0" x 4'-6" high, 1" double glazed, 3/16" thick lites		4.30	3.721		350	127		477	585	
	0400	1/4" thick lites		4.30	3.721		395	127		522	635	
	0500	1" thick, triple glazed		4.30	3.721		455	127		582	700	
	0600	8'-6" x 5'-0" high, 1" double glazed, 3/16" thick lites		3.50	4.571		475	157		632	770	
	0700	1/4" thick lites		3.50	4.571		520	157		677	815	
	0800	1" thick, triple glazed	↓	3.50	4.571	↓	525	157		682	820	
	0900	Window frames only, based on perimeter length				L.F.	2.90			2.90	3.19	
	1200	Window sill, stock, per lineal foot					6.65			6.65	7.30	
	1250	Casing, stock				↓	2.31			2.31	2.54	
	3000	Replacement sash, double hung, double glazing, to 12 S.F.	1 Carp	64	.125	S.F.	16.05	4.28		20.33	24.50	
	3100	12 S.F. to 20 S.F.		94	.085		16	2.92		18.92	22	
	3200	20 S.F. and over	↓	106	.075		13.95	2.58		16.53	19.40	
	3800	Triple glazing for above, add				↓	2.13			2.13	2.34	
	7000	Sash, single lite, 2'-0" x 2'-0" high	1 Carp	20	.400	Ea.	41	13.70		54.70	66.50	
	7050	2'-6" x 2'-0" high		19	.421		44	14.40		58.40	71	
	7100	2'-6" x 2'-6" high		18	.444		46.50	15.20		61.70	75	
	7150	3'-0" x 2'-0" high	↓	17	.471	↓	58.50	16.10		74.60	89.50	
840	0010	**WOOD SCREENS**										**840**
	0020	Over 3 S.F., 3/4" frames	2 Carp	375	.043	S.F.	3.34	1.46		4.80	5.95	
	0100	1-1/8" frames	"	375	.043	"	5.95	1.46		7.41	8.85	

08560		**Plastic Windows**										
100	0010	**VINYL SINGLE HUNG WINDOWS**										**100**
	0100	Grids, low E, J fin, ext. jambs, 21" x 53"	2 Carp	18	.889	Ea.	130	30.50		160.50	191	
	0110	21" x 57"		17	.941		133	32		165	197	
	0120	21" x 65"		16	1		139	34.50		173.50	207	
	0130	25" x 41"		20	.800		123	27.50		150.50	178	
	0140	25" x 49"		18	.889		136	30.50		166.50	197	
	0150	25" x 57"		17	.941		139	32		171	203	
	0160	25" x 65"		16	1		145	34.50		179.50	213	
	0170	29" x 41"		18	.889		131	30.50		161.50	192	
	0180	29" x 53"		18	.889		140	30.50		170.50	202	
	0190	29" x 57"		17	.941		143	32		175	208	
	0200	29" x 65"		16	1		149	34.50		183.50	218	
	0210	33" x 41"		20	.800		135	27.50		162.50	192	
	0220	33" x 53"		18	.889		146	30.50		176.50	208	
	0230	33" x 57"		17	.941		149	32		181	214	
	0240	33" x 65"		16	1		155	34.50		189.50	224	
	0250	37" x 41"		20	.800		143	27.50		170.50	200	
	0260	37" x 53"		18	.889		153	30.50		183.50	216	
	0270	37" x 57"		17	.941		156	32		188	222	
	0280	37" x 65"	↓	16	1	↓	163	34.50		197.50	233	

Important: See the Reference Section for critical supporting data - Reference Nos., Crews, & City Cost Indexes

8 DOORS & WINDOWS

08560 | Plastic Windows

		CREW	DAILY OUTPUT	LABOR-HOURS	UNIT	MAT.	LABOR	EQUIP.	TOTAL	TOTAL INCL O&P	
200	**0010**	**VINYL DOUBLE HUNG WINDOWS**										**200**
	0100	Grids, low E, J fin, ext. jambs, 21" x 53"	2 Carp	18	.889	Ea.	149	30.50		179.50	212	
	0102	21" x 37"		18	.889		133	30.50		163.50	194	
	0104	21" x 41"		18	.889		136	30.50		166.50	198	
	0106	21" x 49"		18	.889		143	30.50		173.50	205	
	0110	21" x 57"		17	.941		152	32		184	217	
	0120	21" x 65"		16	1		158	34.50		192.50	228	
	0128	25" x 37"		20	.800		140	27.50		167.50	197	
	0130	25" x 41"		20	.800		143	27.50		170.50	201	
	0140	25" x 49"		18	.889		148	30.50		178.50	210	
	0145	25" x 53"		18	.889		154	30.50		184.50	217	
	0150	25" x 57"		17	.941		154	32		186	220	
	0160	25" x 65"		16	1		164	34.50		198.50	234	
	0162	25" x 69"		16	1		170	34.50		204.50	241	
	0164	25" x 77"		16	1		180	34.50		214.50	252	
	0168	29" x 37"		18	.889		145	30.50		175.50	207	
	0170	29" x 41"		18	.889		148	30.50		178.50	210	
	0172	29" x 49"		18	.889		155	30.50		185.50	218	
	0180	29" x 53"		18	.889		158	30.50		188.50	222	
	0190	29" x 57"		17	.941		162	32		194	228	
	0200	29" x 65"		16	1		168	34.50		202.50	239	
	0202	29" x 69"		16	1		174	34.50		208.50	246	
	0205	29" x 77"		16	1		184	34.50		218.50	257	
	0208	33" x 37"		20	.800		149	27.50		176.50	206	
	0210	33" x 41"		20	.800		152	27.50		179.50	210	
	0215	33" x 49"		20	.800		160	27.50		187.50	219	
	0220	33" x 53"		18	.889		163	30.50		193.50	228	
	0230	33" x 57"		17	.941		167	32		199	234	
	0240	33" x 65"		16	1		171	34.50		205.50	242	
	0242	33" x 69"		16	1		181	34.50		215.50	254	
	0246	33" x 77"		16	1		191	34.50		225.50	264	
	0250	37" x 41"		20	.800		156	27.50		183.50	214	
	0255	37" x 49"		20	.800		163	27.50		190.50	223	
	0260	37" x 53"		18	.889		170	30.50		200.50	235	
	0270	37" x 57"		17	.941		174	32		206	241	
	0280	37" x 65"		16	1		178	34.50		212.50	250	
	0282	37" x 69"		16	1		233	34.50		267.50	310	
	0286	37" x 77"	▼	16	1		244	34.50		278.50	320	
	0300	Solid vinyl, average quality, double insulated glass, 2'-0" x 3'-0"	1 Carp	10	.800		127	27.50		154.50	182	
	0310	3'-0" x 4'-0"		9	.889		158	30.50		188.50	222	
	0320	4'-0" x 4'-6"		8	1		190	34.50		224.50	263	
	0330	Premium, double insulated glass, 2'-6" x 3'-0"		10	.800		153	27.50		180.50	211	
	0340	3'-0" x 3'-6"		9	.889		177	30.50		207.50	243	
	0350	3'-0" x 4'-0"		9	.889		188	30.50		218.50	255	
	0360	3'-0" x 4'-6"		9	.889		193	30.50		223.50	260	
	0370	3'-0" x 5'-0"		8	1		198	34.50		232.50	272	
	0380	3'-6" x 6'-0"	▼	8	1	▼	218	34.50		252.50	293	
300	**0010**	**VINYL CASEMENT WINDOWS**										**300**
	0100	Grids, low E, J fin, ext. jambs, 1 lt, 21" x 41"	2 Carp	20	.800	Ea.	193	27.50		220.50	255	
	0110	21" x 47"		20	.800		210	27.50		237.50	274	
	0120	21" x 53"		20	.800		227	27.50		254.50	293	
	0128	24" x 35"		19	.842		185	29		214	249	
	0130	24" x 41"		19	.842		201	29		230	266	
	0140	24" x 47"		19	.842		218	29		247	285	
	0150	24" x 53"	▼	19	.842	▼	235	29		264	305	

8

DOORS & WINDOWS

08560 | Plastic Windows

		CREW	DAILY OUTPUT	LABOR-HOURS	UNIT	MAT.	LABOR	EQUIP.	TOTAL	TOTAL INCL O&P	
300	0158	28" x 35"	2 Carp	19	.842	Ea.	197	29		226	262	300
	0160	28" x 41"		19	.842		213	29		242	279	
	0170	28" x 47"		19	.842		230	29		259	298	
	0180	28" x 53"		19	.842		254	29		283	325	
	0184	28" x 59"		19	.842		258	29		287	330	
	0188	Two lites, 33" x 35"		18	.889		315	30.50		345.50	400	
	0190	33" x 41"		18	.889		340	30.50		370.50	425	
	0200	33" x 47"		18	.889		365	30.50		395.50	450	
	0210	33" x 53"		18	.889		390	30.50		420.50	480	
	0212	33" x 59"		18	.889		415	30.50		445.50	505	
	0215	33" x 72"		18	.889		430	30.50		460.50	520	
	0220	41" x 41"		18	.889		370	30.50		400.50	460	
	0230	41" x 47"		18	.889		395	30.50		425.50	485	
	0240	41" x 53"		17	.941		420	32		452	515	
	0242	41" x 59"		17	.941		445	32		477	540	
	0246	41" x 72"		17	.941		465	32		497	560	
	0250	47" x 41"		17	.941		375	32		407	460	
	0260	47" x 47"		17	.941		400	32		432	490	
	0270	47" x 53"		17	.941		425	32		457	515	
	0272	47" x 59"		17	.941		465	32		497	560	
	0280	56" x 41"		15	1.067		400	36.50		436.50	495	
	0290	56" x 47"		15	1.067		425	36.50		461.50	520	
	0300	56" x 53"		15	1.067		460	36.50		496.50	565	
	0302	56" x 59"		15	1.067		480	36.50		516.50	585	
	0310	56" x 72"		15	1.067		525	36.50		561.50	630	
	0340	Solid vinyl, premium, double insulated glass, 2'-0" x 3'-0" high	1 Carp	10	.800		170	27.50		197.50	230	
	0360	2'-0" x 4'-0" high		9	.889		209	30.50		239.50	278	
	0380	2'-0" x 5'-0" high		8	1		248	34.50		282.50	325	
400	0010	**VINYL PICTURE WINDOWS**										400
	0100	Grids, low E, J fin, ext. jambs, 33" x 47"	2 Carp	12	1.333	Ea.	194	45.50		239.50	284	
	0110	35" x 71"		12	1.333		205	45.50		250.50	296	
	0120	41" x 47"		12	1.333		225	45.50		270.50	320	
	0130	41" x 71"		12	1.333		244	45.50		289.50	340	
	0140	47" x 47"		12	1.333		254	45.50		299.50	350	
	0150	47" x 71"		11	1.455		267	50		317	370	
	0160	53" x 47"		11	1.455		250	50		300	355	
	0170	53" x 71"		11	1.455		261	50		311	365	
	0180	59" x 47"		11	1.455		283	50		333	390	
	0190	59" x 71"		11	1.455		305	50		355	415	
	0200	71" x 47"		10	1.600		315	55		370	430	
	0210	71" x 71"		10	1.600		330	55		385	450	

08580 | Special Function Windows

		CREW	DAILY OUTPUT	LABOR-HOURS	UNIT	MAT.	LABOR	EQUIP.	TOTAL	TOTAL INCL O&P	
900	0010	**STORM WINDOWS** Aluminum, residential										900
	0300	Basement, mill finish, incl. fiberglass screen										
	0320	1'-10" x 1'-0" high	2 Carp	30	.533	Ea.	28	18.25		46.25	59	
	0340	2'-9" x 1'-6" high	"	30	.533	"	30.50	18.25		48.75	62	
	1600	Double-hung, combination, storm & screen										
	2000	Average quality, clear anodic coating, 2'-0" x 3'-5" high	2 Carp	30	.533	Ea.	73.50	18.25		91.75	109	
	2020	2'-6" x 5'-0" high		28	.571		89	19.55		108.55	129	
	2040	4'-0" x 6'-0" high		25	.640		109	22		131	154	
	2400	White painted, 2'-0" x 3'-5" high		30	.533		72	18.25		90.25	108	
	2420	2'-6" x 5'-0" high		28	.571		79.50	19.55		99.05	118	
	2440	4'-0" x 6'-0" high		25	.640		87	22		109	130	
	2600	Mill finish, 2'-0" x 3'-5" high		30	.533		65.50	18.25		83.75	101	

8 DOORS & WINDOWS

Important: See the Reference Section for critical supporting data - Reference Nos., Crews, & City Cost Indexes

08500 | Windows

08580		Special Function Windows	CREW	DAILY OUTPUT	LABOR-HOURS	UNIT	| 2005 BARE COSTS MAT.	LABOR	EQUIP.	TOTAL	TOTAL INCL O&P	
900	2620	2'-6" x 5'-0" high	2 Carp	28	.571	Ea.	73.50	19.55		93.05	111	900
	2640	4'-0" x 6-8" high	↓	25	.640	↓	82	22		104	125	

08600 | Skylights

08620		Unit Skylights	CREW	DAILY OUTPUT	LABOR-HOURS	UNIT	2005 BARE COSTS MAT.	LABOR	EQUIP.	TOTAL	TOTAL INCL O&P	
400	0010	**PREFABRICATED** Glass block with metal frame, minimum	G-3	265	.121	S.F.	43.50	4.04		47.54	54	400
	0100	Maximum	"	160	.200	"	86.50	6.70		93.20	106	
800	0010	**SKYLIGHT** Plastic domes, flush or curb mounted, ten or										800
	0100	more units, curb not included										
	0300	Nominal size under 10 S.F., double	G-3	130	.246	S.F.	19.05	8.25		27.30	34	
	0400	Single		160	.200		13.80	6.70		20.50	25.50	
	0600	10 S.F. to 20 S.F., double		315	.102		16.65	3.40		20.05	23.50	
	0700	Single		395	.081		8.40	2.71		11.11	13.45	
	0900	20 S.F. to 30 S.F., double		395	.081		15.10	2.71		17.81	21	
	1000	Single		465	.069		10.60	2.30		12.90	15.20	
	1200	30 S.F. to 65 S.F., double		465	.069		11.35	2.30		13.65	16	
	1300	Single	↓	610	.052		14.80	1.75		16.55	18.95	
	1500	For insulated 4" curbs, double, add					25%					
	1600	Single, add				↓	30%					
	1800	For integral insulated 9" curbs, double, add					30%					
	1900	Single, add					40%					
	2120	Ventilating insulated plexiglass dome with										
	2130	curb mounting, 36" x 36"	G-3	12	2.667	Ea.	380	89		469	560	
	2150	52" x 52"		12	2.667		570	89		659	765	
	2160	28" x 52"		10	3.200		445	107		552	655	
	2170	36" x 52"	↓	10	3.200		480	107		587	695	
	2180	For electric opening system, add				↓	285			285	315	
	2200	Field fabricated, factory type, aluminum and wire glass	G-3	120	.267	S.F.	14.70	8.90		23.60	30	
	2300	Insulated safety glass with aluminum frame		160	.200		85.50	6.70		92.20	105	
	2400	Sandwich panels, fiberglass, for walls, 1-9/16" thick, to 250 SF		200	.160		15.55	5.35		20.90	25.50	
	2500	250 SF and up		265	.121		13.95	4.04		17.99	21.50	
	2700	As above, but for roofs, 2-3/4" thick, to 250 SF		295	.108		22.50	3.63		26.13	30	
	2800	250 SF and up	↓	330	.097	↓	18.35	3.24		21.59	25	

08700 | Hardware

08710		Door Hardware	CREW	DAILY OUTPUT	LABOR-HOURS	UNIT	2005 BARE COSTS MAT.	LABOR	EQUIP.	TOTAL	TOTAL INCL O&P	
100	0010	**AUTOMATIC OPENERS COMMERCIAL**										100
	0020	Pneumatic, incl opener, motion sens, control box, tubing, compressor										
	0050	For single swing door, per opening	2 Skwk	.80	20	Ea.	3,550	695		4,245	4,975	
	0100	Pair, per opening		.50	32	Opng.	5,900	1,125		7,025	8,225	
	1000	For single sliding door, per opening		.60	26.667		3,925	930		4,855	5,775	
	1300	Bi-parting pair	↓	.50	32		5,925	1,125		7,050	8,250	

08710 | Door Hardware

		CREW	DAILY OUTPUT	LABOR-HOURS	UNIT	MAT.	LABOR	EQUIP.	TOTAL	TOTAL INCL O&P	
100	1420	Electronic door opener incl motion sens, 12V control box, motor										**100**
	1450	For single swing door, per opening	2 Skwk	.80	20	Opng.	3,250	695		3,945	4,650	
	1500	Pair, per opening		.50	32		5,325	1,125		6,450	7,575	
	1600	For single sliding door, per opening		.60	26.667		3,525	930		4,455	5,325	
	1700	Bi-parting pair		.50	32		5,400	1,125		6,525	7,675	
	1750	Handicap actuator buttons, 2, including 12V DC wiring, add	1 Carp	1.50	5.333	Pr.	355	183		538	680	
120	0010	**AUTOMATIC OPENERS INDUSTRIAL,** sliding doors, to 6' wide	2 Skwk	.60	26.667	Opng.	4,625	930		5,555	6,550	**120**
	0200	To 12' wide	"	.40	40	"	5,650	1,400		7,050	8,400	
	0400	Over 12' wide, add per L.F. of excess				L.F.	620			620	680	
	1000	Swing doors, to 5' wide	2 Skwk	.80	20	Ea.	2,700	695		3,395	4,050	
	1860	Add for controls, wall pushbutton, 3 button		4	4		165	139		304	400	
	1870	Ceiling pull cord		4.30	3.721		142	130		272	360	
150	0010	**AVERAGE** Percentage for hardware, total job cost, minimum									.75%	**150**
	0050	Maximum									3.50%	
	0500	Total hardware for building, average distribution					85%	15%				
	1000	Door hardware, apartment, interior				Door	120			120	132	
	1500	Hospital bedroom, minimum					269			269	295	
	2000	Maximum					590			590	645	
	2100	Pocket door				Ea.	120			120	132	
	2250	School, single exterior, incl. lever, not incl. panic device				Door	395			395	435	
	2500	Single interior, regular use, no lever included					265			265	291	
	2550	Including handicap lever					360			360	395	
	2600	Heavy use, incl. lever and closer					465			465	510	
	2850	Stairway, single interior					660			660	725	
	3100	Double exterior, with panic device				Pr.	930			930	1,025	
	3600	Toilet, public, single interior				Door	146			146	160	
200	0010	**BOLTS, FLUSH**										**200**
	0020	Standard, concealed	1 Carp	7	1.143	Ea.	19.05	39		58.05	82	
	0800	Automatic fire exit	"	5	1.600		258	55		313	370	
	1600	For electric release, add	1 Elec	3	2.667		104	109		213	276	
	3000	Barrel, brass, 2" long	1 Carp	40	.200		4.09	6.85		10.94	15.15	
	3020	4" long		40	.200		2.79	6.85		9.64	13.70	
	3060	6" long		40	.200		8	6.85		14.85	19.45	
220	0010	**DOOR PROTECTORS**										**220**
	0020	1-3/4" x 3/4" U channel	2 Carp	80	.200	L.F.	17.15	6.85		24	29.50	
	0021	1-3/4" x 1-1/4" U channel		80	.200	"	8.05	6.85		14.90	19.50	
	1000	Tear drop, spring-stl, 8" high x 19" long		15	1.067	Ea.	74.50	36.50		111	139	
	1010	Tear drop, spring-stl, 8" high x 32" long		15	1.067		92	36.50		128.50	158	
	1100	8" high x19" long		15	1.067		190	36.50		226.50	266	
	1200	8" high x 32" long		15	1.067		236	36.50		272.50	315	
300	0010	**DOOR CLOSER** Rack and pinion	1 Carp	6.50	1.231	Ea.	127	42		169	205	**300**
	0020	Adjustable backcheck, 3 way mount, all sizes, regular arm		6	1.333		131	45.50		176.50	215	
	0040	Hold open arm		6	1.333		148	45.50		193.50	234	
	0100	Fusible link		6.50	1.231		115	42		157	193	
	0200	Non sized, regular arm		6	1.333		129	45.50		174.50	213	
	0240	Hold open arm		6	1.333		160	45.50		205.50	247	
	0400	4 way mount, non sized, regular arm		6	1.333		177	45.50		222.50	265	
	0440	Hold open arm		6	1.333		190	45.50		235.50	280	
	2000	Backcheck and adjustable power, hinge face mount										
	2010	All sizes, regular arm	1 Carp	6.50	1.231	Ea.	163	42		205	245	
	2040	Hold open arm		6.50	1.231		175	42		217	258	
	2400	Top jamb mount, all sizes, regular arm		6	1.333		163	45.50		208.50	250	
	2440	Hold open arm		6	1.333		175	45.50		220.50	263	
	2800	Top face mount, all sizes, regular arm		6.50	1.231		163	42		205	245	

Important: See the Reference Section for critical supporting data - Reference Nos., Crews, & City Cost Indexes

			DAILY	LABOR-		2005 BARE COSTS				TOTAL	
08710		**Door Hardware**	CREW	OUTPUT	HOURS	UNIT	MAT.	LABOR	EQUIP.	TOTAL	INCL O&P	
300	2840	Hold open arm	1 Carp	6.50	1.231	Ea.	175	42		217	258	**300**
	4000	Backcheck, overhead concealed, all sizes, regular arm		5.50	1.455		172	50		222	267	
	4040	Concealed arm		5	1.600		183	55		238	288	
	4400	Compact overhead, concealed, all sizes, regular arm		5.50	1.455		315	50		365	425	
	4440	Concealed arm		5	1.600		325	55		380	445	
	4800	Concealed in door, all sizes, regular arm		5.50	1.455		116	50		166	205	
	4840	Concealed arm		5	1.600		125	55		180	223	
	4900	Floor concealed, all sizes, single acting		2.20	3.636		147	125		272	355	
	4940	Double acting	▼	2.20	3.636		190	125		315	405	
	5000	For cast aluminum cylinder, deduct					15.60			15.60	17.15	
	5040	For delayed action, add					27.50			27.50	30	
	5080	For fusible link arm, add					11.30			11.30	12.40	
	5120	For shock absorbing arm, add					34			34	37	
	5160	For spring power adjustment, add					26			26	28.50	
	6000	Closer-holder, hinge face mount, all sizes, exposed arm	1 Carp	6.50	1.231		120	42		162	198	
	7000	Electronic closer-holder, hinge facemount, concealed arm		5	1.600		182	55		237	286	
	7400	With built-in detector	▼	5	1.600	▼	550	55		605	690	
320	0010	**DEADLOCKS** Mortise, heavy duty, outside key	1 Carp	9	.889	Ea.	124	30.50		154.50	184	**320**
	0020	Double cylinder		9	.889		137	30.50		167.50	198	
	0100	Medium duty, outside key		10	.800		96	27.50		123.50	148	
	0110	Double cylinder		10	.800		120	27.50		147.50	175	
	1000	Tubular, standard duty, outside key		10	.800		51.50	27.50		79	99.50	
	1010	Double cylinder		10	.800		66.50	27.50		94	116	
	1200	Night latch, outside key	▼	10	.800	▼	65	27.50		92.50	114	
340	0010	**DOORSTOPS** Holder and bumper, floor or wall	1 Carp	32	.250	Ea.	29.50	8.55		38.05	46	**340**
	1300	Wall bumper, 4" diameter, with rubber pad, aluminum		32	.250		8.90	8.55		17.45	23	
	1600	Door bumper, floor type, aluminum		32	.250		4.58	8.55		13.13	18.40	
	1900	Plunger type, door mounted		32	.250	▼	24.50	8.55		33.05	40.50	
400	0010	**ENTRANCE LOCKS** Cylinder, grip handle, deadlocking latch	1 Carp	9	.889	Ea.	113	30.50		143.50	173	**400**
	0020	Deadbolt		8	1		138	34.50		172.50	206	
	0100	Push and pull plate, dead bolt	▼	8	1		131	34.50		165.50	198	
	0900	For handicapped lever, add					143			143	158	
450	0010	**FLOOR CHECKS** For over 3' wide doors, single acting	1 Carp	2.50	3.200	Ea.	435	110		545	650	**450**
	0500	Double acting	"	2.50	3.200	"	555	110		665	780	
500	0010	**HASP** Steel, 3" assembly	1 Carp	26	.308	Ea.	2.48	10.55		13.03	19.15	**500**
	0020	4-1/2"		13	.615		3.18	21		24.18	36.50	
	0040	6"	▼	12.50	.640	▼	4.81	22		26.81	39.50	
520	0010	**HINGES** Full mortise, avg. freq., steel base, 4-1/2" x 4-1/2", USP				Pr.	19.80			19.80	22	**520**
	0100	5" x 5", USP					32.50			32.50	36	
	0200	6" x 6", USP					69.50			69.50	76.50	
	0400	Brass base, 4-1/2" x 4-1/2", US10					40.50			40.50	45	
	0500	5" x 5", US10					58.50			58.50	64.50	
	0600	6" x 6", US10					99.50			99.50	110	
	0800	Stainless steel base, 4-1/2" x 4-1/2", US32				▼	62.50			62.50	69	
	0900	For non removable pin, add				Ea.	2.27			2.27	2.50	
	0910	For floating pin, driven tips, add					2.60			2.60	2.86	
	0930	For hospital type tip on pin, add					11.25			11.25	12.35	
	0940	For steeple type tip on pin, add				▼	9.85			9.85	10.80	
	0950	Full mortise, high frequency, steel base, 3-1/2" x 3-1/2", US26D				Pr.	18.45			18.45	20.50	
	1000	4-1/2" x 4-1/2", USP					47.50			47.50	52	
	1100	5" x 5", USP					45.50			45.50	50	
	1200	6" x 6", USP					109			109	120	
	1400	Brass base, 3-1/2" x 3-1/2", US4	▼			▼	37.50			37.50	41.50	

Note (row 0400): R08700 -100

DOORS & WINDOWS 8

08710	Door Hardware	CREW	DAILY OUTPUT	LABOR-HOURS	UNIT	MAT.	LABOR	EQUIP.	TOTAL	TOTAL INCL O&P	
520	1430	4-1/2" x 4-1/2", US10				Pr.	65			65	71.50	**520**
	1500	5" x 5", US10					96			96	106	
	1600	6" x 6", US10					139			139	153	
	1800	Stainless steel base, 4-1/2" x 4-1/2", US32					104			104	114	
	1810	Stainless steel base, 5" x 4-1/2", US32					147			147	162	
	1930	For hospital type tip on pin, add				Ea.	6.70			6.70	7.35	
	1950	Full mortise, low frequency, steel base, 3-1/2" x 3-1/2", US26D				Pr.	9.65			9.65	10.65	
	2000	4-1/2" x 4-1/2", USP					9			9	9.90	
	2100	5" x 5", USP					24			24	26.50	
	2200	6" x 6", USP					48.50			48.50	53.50	
	2300	4-1/2" x 4-1/2", US3					14.25			14.25	15.70	
	2310	5" x 5", US3					35			35	38.50	
	2400	Brass bass, 4-1/2" x 4-1/2", US10					34			34	37.50	
	2500	5" x 5", US10					52			52	57	
	2800	Stainless steel base, 4-1/2" x 4-1/2", US32					58.50			58.50	64.50	
550	0010	**KICK PLATE** 6" high, for 3' door, stainless steel	1 Carp	15	.533	Ea.	26	18.25		44.25	57.50	**550**
	0500	Bronze		15	.533		33	18.25		51.25	64.50	
	2000	Aluminum, .050, with 3 beveled edges, 10" x 28"		15	.533		18.80	18.25		37.05	49	
	2010	10" x 30"		15	.533		20.50	18.25		38.75	51	
	2020	10" x 34"		15	.533		21.50	18.25		39.75	52	
	2040	10" x 38"		15	.533		24	18.25		42.25	55	
650	0010	**LOCKSET** Standard duty, cylindrical, with sectional trim										**650**
	0020	Non-keyed, passage	1 Carp	12	.667	Ea.	40.50	23		63.50	80.50	
	0100	Privacy		12	.667		50.50	23		73.50	91	
	0400	Keyed, single cylinder function CN		10	.800		70.50	27.50		98	120	
	0420	Hotel		8	1		100	34.50		134.50	164	
	0500	Lever handled, keyed, single cylinder function		10	.800		124	27.50		151.50	180	
	1000	Heavy duty with sectional trim, non-keyed, passages		12	.667		116	23		139	164	
	1100	Privacy		12	.667		147	23		170	198	
	1400	Keyed, single cylinder function		10	.800		174	27.50		201.50	235	
	1420	Hotel		8	1		259	34.50		293.50	340	
	1600	Communicating		10	.800		206	27.50		233.50	269	
	1690	For re-core cylinder, add					29.50			29.50	32	
	1700	Residential, interior door, minimum	1 Carp	16	.500		14.05	17.15		31.20	42	
	1720	Maximum		8	1		37.50	34.50		72	94.50	
	1800	Exterior, minimum		14	.571		31.50	19.55		51.05	65	
	1810	Average		8	1		65	34.50		99.50	125	
	1820	Maximum		8	1		132	34.50		166.50	199	
700	0010	**MORTISE LOCKSET** Comm., wrought knobs & full escutcheon trim										**700**
	0020	Non-keyed, passage, minimum	1 Carp	9	.889	Ea.	151	30.50		181.50	214	
	0030	Maximum		8	1		243	34.50		277.50	320	
	0040	Privacy, minimum		9	.889		160	30.50		190.50	224	
	0050	Maximum		8	1		263	34.50		297.50	345	
	0100	Keyed, office/entrance/apartment, minimum		8	1		184	34.50		218.50	256	
	0110	Maximum		7	1.143		315	39		354	405	
	0120	Single cylinder, typical, minimum		8	1		158	34.50		192.50	228	
	0130	Maximum		7	1.143		292	39		331	380	
	0200	Hotel, minimum		7	1.143		190	39		229	270	
	0210	Maximum		6	1.333		310	45.50		355.50	410	
	0300	Communication, double cylinder, minimum		8	1		190	34.50		224.50	263	
	0310	Maximum		7	1.143		246	39		285	330	
	1000	Wrought knobs and sectional trim, non-keyed, passage, minimum		10	.800		100	27.50		127.50	153	
	1010	Maximum		9	.889		197	30.50		227.50	265	
	1040	Privacy, minimum		10	.800		117	27.50		144.50	172	

R08700 -100

8 DOORS & WINDOWS

Important: See the Reference Section for critical supporting data - Reference Nos., Crews, & City Cost Indexes

08710 | Door Hardware

		CREW	DAILY OUTPUT	LABOR-HOURS	UNIT	2005 BARE COSTS MAT.	LABOR	EQUIP.	TOTAL	TOTAL INCL O&P	
700	1050	Maximum	1 Carp	9	.889	Ea.	210	30.50		240.50	280	700
	1100	Keyed, entrance, office/apartment, minimum		9	.889		173	30.50		203.50	239	
	1110	Maximum		8	1		251	34.50		285.50	330	
	1120	Single cylinder, typical, minimum		9	.889		167	30.50		197.50	231	
	1130	Maximum	▼	8	1	▼	243	34.50		277.50	320	
	2000	Cast knobs and full escutcheon trim										
	2010	Non-keyed, passage, minimum	1 Carp	9	.889	Ea.	212	30.50		242.50	281	
	2020	Maximum		8	1		345	34.50		379.50	435	
	2040	Privacy, minimum		9	.889		256	30.50		286.50	330	
	2050	Maximum		8	1		370	34.50		404.50	465	
	2120	Keyed, single cylinder, typical, minimum		8	1		256	34.50		290.50	335	
	2130	Maximum		7	1.143		405	39		444	505	
	2200	Hotel, minimum		7	1.143		285	39		324	375	
	2210	Maximum		6	1.333		510	45.50		555.50	630	
	3000	Cast knob and sectional trim, non-keyed, passage, minimum		10	.800		165	27.50		192.50	225	
	3010	Maximum		10	.800		330	27.50		357.50	405	
	3040	Privacy, minimum		10	.800		188	27.50		215.50	250	
	3050	Maximum		10	.800		330	27.50		357.50	405	
	3100	Keyed, office/entrance/apartment, minimum		9	.889		213	30.50		243.50	282	
	3110	Maximum		9	.889		335	30.50		365.50	420	
	3120	Single cylinder, typical, minimum		9	.889		213	30.50		243.50	282	
	3130	Maximum	▼	9	.889		415	30.50		445.50	505	
	3190	For re-core cylinder, add					29			29	31.50	
	3800	Cipher lockset	1 Carp	13	.615	▼	670	21		691	770	
	3900	Keyless, pushbutton type										
	4000	Residential/light commercial, deadbolt, standard	1 Carp	9	.889	Ea.	96.50	30.50		127	154	
	4010	Heavy duty		9	.889		114	30.50		144.50	174	
	4020	Industrial, heavy duty, with deadbolt		9	.889		223	30.50		253.50	293	
	4030	Key override		9	.889		247	30.50		277.50	320	
	4040	Lever activated handle		9	.889		271	30.50		301.50	345	
	4050	Key override		9	.889		300	30.50		330.50	380	
	4060	Double sided pushbutton type		8	1		495	34.50		529.50	600	
	4070	Key override	▼	8	1	▼	535	34.50		569.50	640	
750	0010	**PANIC DEVICE** For rim locks, single door, exit only	1 Carp	6	1.333	Ea.	350	45.50		395.50	455	750
	0020	Outside key and pull		5	1.600		400	55		455	525	
	0200	Bar and vertical rod, exit only		5	1.600		510	55		565	645	
	0210	Outside key and pull		4	2		610	68.50		678.50	775	
	0400	Bar and concealed rod		4	2		510	68.50		578.50	665	
	0600	Touch bar, exit only		6	1.333		405	45.50		450.50	515	
	0610	Outside key and pull		5	1.600		485	55		540	620	
	0700	Touch bar and vertical rod, exit only		5	1.600		550	55		605	695	
	0710	Outside key and pull		4	2		645	68.50		713.50	815	
	1000	Mortise, bar, exit only		4	2		455	68.50		523.50	605	
	1600	Touch bar, exit only		4	2		520	68.50		588.50	675	
	2000	Narrow stile, rim mounted, bar, exit only		6	1.333		545	45.50		590.50	670	
	2010	Outside key and pull		5	1.600		590	55		645	735	
	2200	Bar and vertical rod, exit only		5	1.600		560	55		615	705	
	2210	Outside key and pull		4	2		560	68.50		628.50	725	
	2400	Bar and concealed rod, exit only		3	2.667		655	91.50		746.50	860	
	3000	Mortise, bar, exit only		4	2		480	68.50		548.50	630	
	3600	Touch bar, exit only	▼	4	2	▼	695	68.50		763.50	870	
770	0010	**DOOR AND WINDOW ACCESSORIES**										770
	0050	Door closing coordinator, 36" (for paired openings up to 56")	1 Carp	8	1	Ea.	83.50	34.50		118	146	
	0060	48" (for paired openings up to 84")		8	1		89.50	34.50		124	152	
	0070	56" (for paired openings up to 96")	▼	8	1	▼	98.50	34.50		133	162	

Row 4030 marked with accessibility (wheelchair) symbol.

DOORS & WINDOWS 8

DOORS & WINDOWS

8

08710 | Door Hardware

		CREW	DAILY OUTPUT	LABOR-HOURS	UNIT	2005 BARE COSTS				TOTAL INCL O&P
						MAT.	LABOR	EQUIP.	TOTAL	
780	**0010 PUSH-PULL PLATE**									**780**
0100	Push plate, .050 thick, 4" x 16", aluminum	1 Carp	12	.667	Ea.	6.50	23		29.50	42.50
0500	Bronze		12	.667		16.45	23		39.45	53.50
1500	Pull handle and push bar, aluminum		11	.727		115	25		140	165
2000	Bronze		10	.800		149	27.50		176.50	206
3000	Push plate both sides, aluminum		14	.571		14	19.55		33.55	46
3500	Bronze		13	.615		35	21		56	71.50
4000	Door pull, designer style, cast aluminum, minimum		12	.667		61	23		84	103
5000	Maximum		8	1		310	34.50		344.50	395
6000	Cast bronze, minimum		12	.667		71.50	23		94.50	114
7000	Maximum		8	1		335	34.50		369.50	425
8000	Walnut, minimum		12	.667		55	23		78	96
9000	Maximum	▼	8	1	▼	305	34.50		339.50	390
800	**0010 SPECIAL HINGES**									**800**
0015	Paumelle, high frequency									
0020	Steel base, 6" x 4-1/2", US10				Pr.	119			119	131
0100	Bronze base, 5" x 4-1/2", US10					156			156	171
0200	Paumelle, average frequency, steel base, 4-1/2" x 3-1/2", US10				▼	80.50			80.50	89
0400	Olive knuckle, low frequency, brass base, 6" x 4-1/2", US10					142			142	156
1000	Electric hinge with concealed conductor, average frequency									
1010	Steel base, 4-1/2" x 4-1/2", US26D				Pr.	264			264	290
1100	Bronze base, 4-1/2" x 4-1/2", US26D				"	277			277	305
1200	Electric hinge with concealed conductor, high frequency									
1210	Steel base, 4-1/2" x 4-1/2", US26D				Pr.	196			196	216
1600	Double weight, 800 lb., steel base, removable pin, 5" x 6", USP					117			117	129
1700	Steel base-welded pin, 5" x 6", USP					134			134	148
1800	Triple weight, 2000 lb., steel base, welded pin, 5" x 6", USP					135			135	148
2000	Pivot reinf., high frequency, steel base, 7-3/4" door plate, USP					161			161	177
2200	Bronze base, 7-3/4" door plate, US10				▼	189			189	208
3000	Swing clear, full mortise, full or half surface, high frequency,									
3010	Steel base, 5" high, USP				Pr.	134			134	147
3200	Swing clear, full mortise, average frequency									
3210	Steel base, 4-1/2" high, USP				Pr.	106			106	117
4000	Wide throw, average frequency, steel base, 4-1/2" x 6", USP					81			81	89
4200	High frequency, steel base, 4-1/2" x 6", USP				▼	124			124	136
4600	Spring hinge, single acting, 6" flange, steel				Ea.	45.50			45.50	50.50
4700	Brass					80.50			80.50	88.50
4900	Double acting, 6" flange, steel					82.50			82.50	91
4950	Brass				▼	135			135	149
9000	Continuous hinge, steel, full mortise, heavy duty	2 Carp	64	.250	L.F.	10.80	8.55		19.35	25.50

08720 | Weatherstripping & Seals

		CREW	DAILY OUTPUT	LABOR-HOURS	UNIT	MAT.	LABOR	EQUIP.	TOTAL	TOTAL INCL O&P
100	**0010 ASTRAGALS** One piece overlapping									**100**
0400	Cadmium plated steel, flat, 3/16" x 2"	1 Carp	90	.089	L.F.	2.89	3.04		5.93	7.90
0600	Prime coated steel, flat, 1/8" x 3"		90	.089		4.01	3.04		7.05	9.15
0800	Stainless steel, flat, 3/32" x 1-5/8"		90	.089		15.05	3.04		18.09	21.50
1000	Aluminum, flat, 1/8" x 2"		90	.089		2.90	3.04		5.94	7.95
1200	Nail on, "T" extrusion		120	.067		.64	2.28		2.92	4.26
1300	Vinyl bulb insert		105	.076		1.04	2.61		3.65	5.20
1600	Screw on, "T" extrusion		90	.089		4.22	3.04		7.26	9.40
1700	Vinyl insert		75	.107		2.85	3.65		6.50	8.85
2000	"L" extrusion, neoprene bulbs		75	.107		1.64	3.65		5.29	7.50
2100	Neoprene sponge insert		75	.107		5	3.65		8.65	11.20
2200	Magnetic	▼	75	.107	▼	8.25	3.65		11.90	14.75

Important: See the Reference Section for critical supporting data - Reference Nos., Crews, & City Cost Indexes

08720	Weatherstripping & Seals	CREW	DAILY OUTPUT	LABOR-HOURS	UNIT	2005 BARE COSTS				TOTAL INCL O&P
						MAT.	LABOR	EQUIP.	TOTAL	
100 2400	Spring hinged security seal, with cam	1 Carp	75	.107	L.F.	5.30	3.65		8.95	11.55 **100**
2600	Spring loaded locking bolt, vinyl insert		45	.178		7.20	6.10		13.30	17.45
2800	Neoprene sponge strip, "Z" shaped, aluminum		60	.133		3.43	4.57		8	10.85
2900	Solid neoprene strip, nail on aluminum strip		90	.089		2.88	3.04		5.92	7.90
3000	One piece stile protection									
3020	Neoprene fabric loop, nail on aluminum strips	1 Carp	60	.133	L.F.	.51	4.57		5.08	7.65
3110	Flush mounted aluminum extrusion, 1/2" x 1-1/4"		60	.133		2.73	4.57		7.30	10.10
3140	3/4" x 1-3/8"		60	.133		3.47	4.57		8.04	10.90
3160	1-1/8" x 1-3/4"		60	.133		5.55	4.57		10.12	13.20
3300	Mortise, 9/16" x 3/4"		60	.133		2.94	4.57		7.51	10.35
3320	13/16" x 1-3/8"		60	.133		3.18	4.57		7.75	10.60
3600	Spring bronze strip, nail on type		105	.076		2.50	2.61		5.11	6.80
3620	Screw on, with retainer		75	.107		2.05	3.65		5.70	7.95
3800	Flexible stainless steel housing, pile insert, 1/2" door		105	.076		5.65	2.61		8.26	10.30
3820	3/4" door		105	.076		6.40	2.61		9.01	11.05
4000	Extruded aluminum retainer, flush mount, pile insert		105	.076		1.98	2.61		4.59	6.25
4080	Mortise, felt insert		90	.089		3.53	3.04		6.57	8.60
4160	Mortise with spring, pile insert		90	.089		2.76	3.04		5.80	7.80
4400	Rigid vinyl retainer, mortise, pile insert		105	.076		1.97	2.61		4.58	6.25
4600	Wool pile filler strip, aluminum backing		105	.076		1.98	2.61		4.59	6.25
5000	Two piece overlapping astragal, extruded aluminum retainer									
5010	Pile insert	1 Carp	60	.133	L.F.	2.57	4.57		7.14	9.95
5020	Vinyl bulb insert		60	.133		1.66	4.57		6.23	8.95
5040	Vinyl flap insert		60	.133		5.15	4.57		9.72	12.80
5060	Solid neoprene flap insert		60	.133		5.10	4.57		9.67	12.70
5080	Hypalon rubber flap insert		60	.133		5.20	4.57		9.77	12.85
5090	Snap on cover, pile insert		60	.133		5.95	4.57		10.52	13.65
5400	Magnetic aluminum, surface mounted		60	.133		20	4.57		24.57	29
5500	Interlocking aluminum, 5/8" x 1" neoprene bulb insert		45	.178		3.20	6.10		9.30	13
5600	Adjustable aluminum, 9/16" x 21/32", pile insert		45	.178		15.20	6.10		21.30	26
5790	For vinyl bulb, deduct					.39			.39	.43
5800	Magnetic, adjustable, 9/16" x 21/32"	1 Carp	45	.178		19.50	6.10		25.60	31
6000	Two piece stile protection									
6010	Cloth backed rubber loop, 1" gap, nail on aluminum strips	1 Carp	45	.178	L.F.	3.28	6.10		9.38	13.10
6040	Screw on aluminum strips		45	.178		5.10	6.10		11.20	15.10
6100	1-1/2" gap, screw on aluminum extrusion		45	.178		4.59	6.10		10.69	14.55
6240	Vinyl fabric loop, slotted aluminum extrusion, 1" gap		45	.178		1.62	6.10		7.72	11.30
6300	1-1/4" gap		45	.178		4.83	6.10		10.93	14.80
300 0010	**WEATHERSTRIPPING** Window, double hung, 3' x 5', zinc	1 Carp	7.20	1.111	Opng.	11.35	38		49.35	72 **300**
0100	Bronze		7.20	1.111		22.50	38		60.50	84
0500	As above but heavy duty, zinc		4.60	1.739		14.50	59.50		74	109
0600	Bronze		4.60	1.739		25	59.50		84.50	121
1000	Doors, wood frame, interlocking, for 3' x 7' door, zinc		3	2.667		13	91.50		104.50	156
1100	Bronze		3	2.667		20.50	91.50		112	165
1300	6' x 7' opening, zinc		2	4		14.20	137		151.20	229
1400	Bronze		2	4		27	137		164	243
1700	Wood frame, spring type, bronze									
1800	3' x 7' door	1 Carp	7.60	1.053	Opng.	16.10	36		52.10	73.50
1900	6' x 7' door	"	7	1.143	"	17.60	39		56.60	80.50
2200	Metal frame, spring type, bronze									
2300	3' x 7' door	1 Carp	3	2.667	Opng.	27.50	91.50		119	172
2400	6' x 7' door	"	2.50	3.200	"	38.50	110		148.50	214
2500	For stainless steel, spring type, add					133%				
2700	Metal frame, extruded sections, 3' x 7' door, aluminum	1 Carp	2	4	Opng.	37.50	137		174.50	255
2800	Bronze		2	4		95.50	137		232.50	320
3100	6' x 7' door, aluminum		1.20	6.667		48	228		276	410

DOORS & WINDOWS **8**

08720 | Weatherstripping & Seals

			CREW	DAILY OUTPUT	LABOR-HOURS	UNIT	2005 BARE COSTS				TOTAL INCL O&P	
							MAT.	LABOR	EQUIP.	TOTAL		
300	3200	Bronze	1 Carp	1.20	6.667	Opng.	112	228		340	480	300
	3500	Threshold weatherstripping										
	3650	Door sweep, flush mounted, aluminum	1 Carp	25	.320	Ea.	11.15	10.95		22.10	29.50	
	3700	Vinyl		25	.320		13.20	10.95		24.15	31.50	
	5000	Garage door bottom weatherstrip, 12' aluminum, clear		14	.571		17.85	19.55		37.40	50	
	5010	Bronze		14	.571		68	19.55		87.55	106	
	5050	Bottom protection, 12' aluminum, clear		14	.571		21	19.55		40.55	53.50	
	5100	Bronze		14	.571		84.50	19.55		104.05	124	
800	0010	**THRESHOLD** 3' long door saddles, aluminum	1 Carp	48	.167	L.F.	3.43	5.70		9.13	12.65	800
	0100	Aluminum, 8" wide, 1/2" thick		12	.667	Ea.	29	23		52	67.50	
	0500	Bronze		60	.133	L.F.	28.50	4.57		33.07	38.50	
	0600	Bronze, panic threshold, 5" wide, 1/2" thick		12	.667	Ea.	57	23		80	98	
	0700	Rubber, 1/2" thick, 5-1/2" wide		20	.400		31.50	13.70		45.20	56	
	0800	2-3/4" wide		20	.400		14.05	13.70		27.75	37	

08770 | Door/Window Accessories

			CREW	DAILY OUTPUT	LABOR-HOURS	UNIT	2005 BARE COSTS				TOTAL INCL O&P	
							MAT.	LABOR	EQUIP.	TOTAL		
100	0010	**AREA WINDOW WELL**, Galvanized steel, 20 ga., 3'-2" wide, 1' deep	1 Sswk	29	.276	Ea.	12.95	10.50		23.45	33	100
	0100	2' deep		23	.348		23	13.25		36.25	49	
	0300	16 ga., galv., 3'-2" wide, 1' deep		29	.276		18.25	10.50		28.75	39	
	0400	3' deep		23	.348		37	13.25		50.25	64	
	0600	Welded grating for above, 15 lbs., painted		45	.178		33	6.80		39.80	48	
	0700	Galvanized		45	.178		78.50	6.80		85.30	98	
	0900	Translucent plastic cap for above		60	.133		13.40	5.10		18.50	24	
200	0010	**DOOR PROTECTION** Acrylic and neoprene cover for door										200
	0020	frame, high impact type	1 Carp	100	.080	L.F.	7.50	2.74		10.24	12.50	
550	0010	**DETECTION SYSTEMS** See division 13851										550

08810 | Glass

			CREW	DAILY OUTPUT	LABOR-HOURS	UNIT	2005 BARE COSTS				TOTAL INCL O&P	
							MAT.	LABOR	EQUIP.	TOTAL		
100	0010	**ACOUSTICAL GLASS UNITS** 1 lite at 3/8", 1 lite at 3/16", for 1" thick	2 Glaz	100	.160	S.F.	21.50	5.30		26.80	31.50	100
	0100	For 4" thick	"	80	.200	"	37	6.60		43.60	50.50	
160	0010	**BEVELED GLASS** With design patterns, 1/4" thick, 1/2" bevel, minimum	2 Glaz	150	.107	S.F.	37	3.52		40.52	46	160
	0050	Average		125	.128		86.50	4.22		90.72	101	
	0100	Maximum		100	.160		153	5.30		158.30	176	
205	0010	**CURTAIN WALL** See division 08900										205
250	0010	**FACETED** Color tinted glass, 3/4" thick, minimum	2 Glaz	95	.168	S.F.	31	5.55		36.55	42.50	250
	0100	Maximum	"	75	.213	"	54.50	7.05		61.55	70.50	
260	0010	**FLOAT GLASS** 3/16" thick, clear, plain	2 Glaz	130	.123	S.F.	4.19	4.06		8.25	10.75	260
	0200	Tempered, clear		130	.123		4.96	4.06		9.02	11.60	
	0300	Tinted		130	.123		6.25	4.06		10.31	13.05	
	0600	1/4" thick, clear, plain **CN**		120	.133		4.98	4.40		9.38	12.15	
	0700	Tinted		120	.133		4.95	4.40		9.35	12.10	
	0800	Tempered, clear		120	.133		6.15	4.40		10.55	13.45	
	0900	Tinted		120	.133		8.40	4.40		12.80	15.90	
	1600	3/8" thick, clear, plain		75	.213		7.95	7.05		15	19.35	

Note for Float Glass row 0010: R08810-010

8

DOORS & WINDOWS

Important: See the Reference Section for critical supporting data - Reference Nos., Crews, & City Cost Indexes

			CREW	DAILY OUTPUT	LABOR-HOURS	UNIT	2005 BARE COSTS				TOTAL INCL O&P	
		08810 Glass					MAT.	LABOR	EQUIP.	TOTAL		
260	1700	Tinted	2 Glaz	75	.213	S.F.	9.90	7.05		16.95	21.50	**260**
	1800	Tempered, clear [R08810-010]		75	.213		12.40	7.05		19.45	24.50	
	1900	Tinted		75	.213		15.15	7.05		22.20	27.50	
	2200	1/2" thick, clear, plain		55	.291		15.85	9.60		25.45	32	
	2300	Tinted		55	.291		17.35	9.60		26.95	33.50	
	2400	Tempered, clear		55	.291		18.65	9.60		28.25	35	
	2500	Tinted		55	.291		23	9.60		32.60	39.50	
	2800	5/8" thick, clear, plain		45	.356		17.35	11.75		29.10	37	
	2900	Tempered, clear		45	.356		19.80	11.75		31.55	40	
	3200	3/4" thick, clear, plain		35	.457		22.50	15.10		37.60	47.50	
	3300	Tempered, clear		35	.457		26	15.10		41.10	51.50	
	3600	1" thick, clear, plain		30	.533		37	17.60		54.60	67.50	
	8900	For low emissivity coating for 3/16" and 1/4" only, add to above					15%					
270	0010	**FULL VISION** Window system with 3/4" glass mullions, 10' high	H-2	130	.185	S.F.	49.50	5.70		55.20	63	**270**
	0100	10' to 20' high, minimum		110	.218		52.50	6.75		59.25	68	
	0150	Average		100	.240		56.50	7.40		63.90	73.50	
	0200	Maximum		80	.300		63.50	9.25		72.75	84	
280	0010	**GLASS BLOCK** See division 04810-325										**280**
300	0010	**GLAZING VARIABLES** [R08810-010]										**300**
	0500	For high rise glazing, exterior, add per S.F. per story				S.F.					.09	
	0600	For glass replacement, add				"		100%				
	0700	For gasket settings, add				L.F.	3.71			3.71	4.08	
	0900	For sloped glazing, add				S.F.		25%				
	2000	Fabrication, polished edges, 1/4" thick				Inch	.32			.32	.35	
	2100	1/2" thick					.81			.81	.89	
	2500	Mitered edges, 1/4" thick					.81			.81	.89	
	2600	1/2" thick					1.31			1.31	1.44	
460	0010	**INSULATING GLASS** 2 lites 1/8" float, 1/2" thk, under 15 S.F.										**460**
	0020	Clear	2 Glaz	95	.168	S.F.	7.15	5.55		12.70	16.25	
	0100	Tinted		95	.168		10.50	5.55		16.05	19.95	
	0200	2 lites 3/16" float, for 5/8" thk unit, 15 to 30 S.F., clear [R08810-010]		90	.178		8.65	5.85		14.50	18.40	
	0300	Tinted		90	.178		8.90	5.85		14.75	18.65	
	0400	1" thk, dbl. glazed, 1/4" float, 30-70 S.F., clear **CN**		75	.213		12.05	7.05		19.10	24	
	0500	Tinted		75	.213		14.85	7.05		21.90	27	
	0600	1" thick double glazed, 1/4" float, 1/4" wire		75	.213		17.95	7.05		25	30.50	
	0700	1/4" float, 1/4" tempered		75	.213		17.85	7.05		24.90	30.50	
	0800	1/4" wire, 1/4" tempered		75	.213		24.50	7.05		31.55	37.50	
	0900	Both lites, 1/4" wire		75	.213		23.50	7.05		30.55	36.50	
	2000	Both lites, light & heat reflective		85	.188		19.80	6.20		26	31.50	
	2500	Heat reflective, film inside, 1" thick unit, clear		85	.188		17.35	6.20		23.55	28.50	
	2600	Tinted		85	.188		18.70	6.20		24.90	30	
	3000	Film on weatherside, clear, 1/2" thick unit		95	.168		12.40	5.55		17.95	22	
	3100	5/8" thick unit		90	.178		15.10	5.85		20.95	25.50	
	3200	1" thick unit		85	.188		17.05	6.20		23.25	28	
500	0010	**LAMINATED GLASS** Clear float, .03" vinyl, 1/4" thick	2 Glaz	90	.178	S.F.	8.50	5.85		14.35	18.20	**500**
	0100	3/8" thick		78	.205		13.55	6.75		20.30	25	
	0200	.06" vinyl, 1/2" thick		65	.246		15.85	8.10		23.95	29.50	
	1000	5/8" thick		90	.178		18.55	5.85		24.40	29.50	
	2000	Bullet-resisting, 1-3/16" thick, to 15 S.F.		16	1		51	33		84	106	
	2100	Over 15 S.F.		16	1		46	33		79	101	
	2500	2-1/4" thick, to 15 S.F.		12	1.333		61	44		105	134	
	2600	Over 15 S.F.		12	1.333		54	44		98	126	

DOORS & WINDOWS 8

8 DOORS & WINDOWS

		08810 | Glass	CREW	DAILY OUTPUT	LABOR-HOURS	UNIT	MAT.	LABOR	EQUIP.	TOTAL	TOTAL INCL O&P	
550	0010	**LEAD GLASS** For X-ray, see division 13091-600										550
600	0010	**OBSCURE GLASS** 1/8" thick, minimum	2 Glaz	140	.114	S.F.	7	3.77		10.77	13.40	600
	0100	Maximum		125	.128		8.40	4.22		12.62	15.60	
	0300	7/32" thick, minimum		120	.133		7.75	4.40		12.15	15.20	
	0400	Maximum	▼	105	.152	▼	9.90	5.05		14.95	18.50	
650	0010	**PATTERNED GLASS** Colored, 1/8" thick, minimum	2 Glaz	140	.114	S.F.	7.70	3.77		11.47	14.20	650
	0100	Maximum		125	.128		9.60	4.22		13.82	16.95	
	0300	7/32" thick, minimum		120	.133		9.05	4.40		13.45	16.65	
	0400	Maximum	▼	105	.152	▼	10.30	5.05		15.35	18.95	
675	0010	**REFLECTIVE GLASS** 1/4" float with fused metallic oxide, tinted	2 Glaz	115	.139	S.F.	10.30	4.59		14.89	18.30	675
	0500	1/4" float glass with reflective applied coating		115	.139		8.60	4.59		13.19	16.40	
	2000	Solar film on glass, not including glass, minimum		180	.089		4.15	2.93		7.08	9	
	2010	Window protection film, controls blast damage		80	.200		4.15	6.60		10.75	14.55	
	2050	Solar film on glass, not including glass, maximum	▼	225	.071	▼	9.60	2.35		11.95	14.10	
740	0010	**SANDBLASTED GLASS** Float glass, 1/8" thick	2 Glaz	160	.100	S.F.	6.85	3.30		10.15	12.55	740
	0100	3/16" thick		130	.123		7.60	4.06		11.66	14.50	
	0500	Plate glass, 1/4" thick		120	.133		7.90	4.40		12.30	15.35	
	0600	3/8" thick	▼	75	.213	▼	8.60	7.05		15.65	20	
760	0010	**SHEET GLASS** Gray, 1/8" thick	2 Glaz	160	.100	S.F.	3.58	3.30		6.88	8.95	760
	0200	1/4" thick		130	.123	"	4.69	4.06		8.75	11.30	
780	0010	**SPANDREL GLASS** 1/4" thick, standard colors, up to 1,000 S.F.		110	.145	S.F.	11.90	4.80		16.70	20.50	780
	0200	1,000 to 2,000 S.F.	▼	120	.133	"	11.15	4.40		15.55	18.90	
	0300	For custom colors, add				Total	10%					
	0500	For 3/8" thick, add				S.F.	7.10			7.10	7.85	
	1000	For double coated, 1/4" thick, add					2.65			2.65	2.92	
	1200	For insulation on panels, add					4.39			4.39	4.83	
	2000	Panels, insulated, with aluminum backed fiberglass, 1" thick	2 Glaz	120	.133		10.60	4.40		15	18.30	
	2100	2" thick	"	120	.133		12.30	4.40		16.70	20	
	2500	With galvanized steel backing, add				▼	3.66			3.66	4.03	
850	0010	**WINDOW GLASS** Clear float, stops, putty bed, 1/8" thick	2 Glaz	480	.033	S.F.	3.44	1.10		4.54	5.45	850
	0500	3/16" thick, clear		480	.033		4.23	1.10		5.33	6.30	
	0600	Tinted		480	.033		4.77	1.10		5.87	6.90	
	0700	Tempered	▼	480	.033	▼	5.75	1.10		6.85	8	
860	0010	**WINDOW WALL** See division 08910-900										860
900	0010	**WIRE GLASS** 1/4" thick, rough obscure (chicken wire)	2 Glaz	135	.119	S.F.	11.15	3.91		15.06	18.15	900
	1000	Polished wire, 1/4" thick, diamond, clear		135	.119		14.05	3.91		17.96	21.50	
	1500	Pinstripe, obscure	▼	135	.119	▼	14.10	3.91		18.01	21.50	

		08830 | Mirrors	CREW	DAILY OUTPUT	LABOR-HOURS	UNIT	MAT.	LABOR	EQUIP.	TOTAL	TOTAL INCL O&P	
100	0010	**MIRRORS** No frames, wall type, 1/4" plate glass, polished edge										100
	0100	Up to 5 S.F.	2 Glaz	125	.128	S.F.	6.70	4.22		10.92	13.75	
	0200	Over 5 S.F.		160	.100		6.45	3.30		9.75	12.10	
	0500	Door type, 1/4" plate glass, up to 12 S.F.		160	.100		6.95	3.30		10.25	12.65	
	1000	Float glass, up to 10 S.F., 1/8" thick		160	.100		3.91	3.30		7.21	9.30	
	1100	3/16" thick		150	.107		4.54	3.52		8.06	10.30	
	1500	12" x 12" wall tiles, square edge, clear		195	.082		1.59	2.71		4.30	5.85	
	1600	Veined		195	.082		4.26	2.71		6.97	8.80	
	2000	1/4" thick, stock sizes, one way transparent		125	.128		14.90	4.22		19.12	23	
	2010	Bathroom, unframed, laminated	▼	160	.100	▼	11.10	3.30		14.40	17.25	

Important: See the Reference Section for critical supporting data - Reference Nos., Crews, & City Cost Indexes

08800 | Glazing

08840 | Plastic Glazing

		CREW	DAILY OUTPUT	LABOR-HOURS	UNIT	2005 BARE COSTS				TOTAL INCL O&P	
						MAT.	LABOR	EQUIP.	TOTAL		
600 0010	**PLEXIGLASS ACRYLIC** Clear, masked, 1/8″ thick, cut sheets	2 Glaz	170	.094	S.F.	3.45	3.11		6.56	8.50	**600**
0200	Full sheets		195	.082		1.80	2.71		4.51	6.05	
0500	1/4″ thick, cut sheets		165	.097		6.10	3.20		9.30	11.55	
0600	Full sheets		185	.086		3.31	2.85		6.16	7.95	
0900	3/8″ thick, cut sheets		155	.103		11.15	3.41		14.56	17.40	
1000	Full sheets		180	.089		6	2.93		8.93	11.05	
1300	1/2″ thick, cut sheets		135	.119		12.85	3.91		16.76	20	
1400	Full sheets		150	.107		12.45	3.52		15.97	19	
1700	3/4″ thick, cut sheets		115	.139		45.50	4.59		50.09	57	
1800	Full sheets		130	.123		26.50	4.06		30.56	35	
2100	1″ thick, cut sheets		105	.152		51.50	5.05		56.55	64	
2200	Full sheets		125	.128		32	4.22		36.22	41.50	
3000	Colored, 1/8″ thick, cut sheets		170	.094		10.60	3.11		13.71	16.35	
3200	Full sheets		195	.082		6.85	2.71		9.56	11.65	
3500	1/4″ thick, cut sheets		165	.097		11.85	3.20		15.05	17.90	
3600	Full sheets		185	.086		8.10	2.85		10.95	13.20	
4000	Mirrors, untinted, cut sheets, 1/8″ thick		185	.086		4.98	2.85		7.83	9.80	
4200	1/4″ thick		180	.089		8	2.93		10.93	13.25	
650 0010	**POLYCARBONATE** Clear, masked, cut sheets, 1/8″ thick	2 Glaz	170	.094	S.F.	6	3.11		9.11	11.30	**650**
0500	3/16″ thick		165	.097		7.25	3.20		10.45	12.80	
1000	1/4″ thick		155	.103		8	3.41		11.41	13.95	
1500	3/8″ thick		150	.107		14.75	3.52		18.27	21.50	
900 0010	**VINYL GLASS** Steel mesh reinforced, stock sizes, .090″ thick	2 Glaz	170	.094	S.F.	8.10	3.11		11.21	13.60	**900**
0500	.120″ thick		170	.094		10.25	3.11		13.36	16	
1000	.250″ thick		155	.103		14.30	3.41		17.71	21	
1500	For non-standard sizes, add					15%					

08900 | Glazed Curtain Wall

08910 | Glazed Aluminum Curtain Wall

		CREW	DAILY OUTPUT	LABOR-HOURS	UNIT	2005 BARE COSTS				TOTAL INCL O&P	
						MAT.	LABOR	EQUIP.	TOTAL		
200 0010	**CURTAIN WALLS** Aluminum, stock, including glazing, minimum	H-1	205	.156	S.F.	22.50	5.55		28.05	33.50	**200**
0050	Average, single glazed		195	.164		29	5.85		34.85	41.50	
0150	Average, double glazed		180	.178		41	6.35		47.35	55.50	
0200	Maximum		160	.200		110	7.10		117.10	132	
700 0010	**TUBE FRAMING** For window walls and store fronts, aluminum, stock										**700**
0050	Plain tube frame, mill finish, 1-3/4″ x 1-3/4″	2 Glaz	103	.155	L.F.	6.55	5.15		11.70	14.95	
0150	1-3/4″ x 4″		98	.163		8.55	5.40		13.95	17.55	
0200	1-3/4″ x 4-1/2″ **CN**		95	.168		9.75	5.55		15.30	19.15	
0250	2″ x 6″		89	.180		14.35	5.95		20.30	24.50	
0350	4″ x 4″		87	.184		14.05	6.05		20.10	24.50	
0400	4-1/2″ x 4-1/2″		85	.188		16	6.20		22.20	27	
0450	Glass bead		240	.067		1.81	2.20		4.01	5.30	
1000	Flush tube frame, mill finish, 1/4″ glass, 1-3/4″ x 4″, open header		80	.200		8.40	6.60		15	19.25	
1050	Open sill		82	.195		7.30	6.45		13.75	17.80	
1100	Closed back header		83	.193		11.75	6.35		18.10	22.50	
1150	Closed back sill		85	.188		11.15	6.20		17.35	21.50	
1200	Vertical mullion, one piece		75	.213		12.40	7.05		19.45	24.50	
1250	Two piece		73	.219		13.30	7.25		20.55	25.50	
1300	90° or 180° vertical corner post		75	.213		21	7.05		28.05	33.50	
1400	1-3/4″ x 4-1/2″, open header		80	.200		10.25	6.60		16.85	21.50	

08910 | Glazed Aluminum Curtain Wall

			CREW	DAILY OUTPUT	LABOR-HOURS	UNIT	2005 BARE COSTS MAT.	LABOR	EQUIP.	TOTAL	TOTAL INCL O&P	
700	1450	Open sill	2 Glaz	82	.195	L.F.	8.45	6.45		14.90	19.05	**700**
	1500	Closed back header		83	.193		12.55	6.35		18.90	23.50	
	1550	Closed back sill		85	.188		12.05	6.20		18.25	22.50	
	1600	Vertical mullion, one piece		75	.213		13.45	7.05		20.50	25.50	
	1650	Two piece		73	.219		14.20	7.25		21.45	26.50	
	1700	90° or 180° vertical corner post		75	.213		14.55	7.05		21.60	26.50	
	2000	Flush tube frame, mill fin. for ins. glass, 2" x 4-1/2", open header		75	.213		11.45	7.05		18.50	23.50	
	2050	Open sill		77	.208		10.05	6.85		16.90	21.50	
	2100	Closed back header		78	.205		12.50	6.75		19.25	24	
	2150	Closed back sill		80	.200		12.35	6.60		18.95	23.50	
	2200	Vertical mullion, one piece		70	.229		13.85	7.55		21.40	26.50	
	2250	Two piece		68	.235		14.80	7.75		22.55	28	
	2300	90° or 180° vertical corner post		70	.229		14	7.55		21.55	27	
	5000	Flush tube frame, mill fin., thermal brk., 2-1/4"x 4-1/2", open header		74	.216		12.65	7.15		19.80	24.50	
	5050	Open sill		75	.213		11.05	7.05		18.10	23	
	5100	Vertical mullion, one piece		69	.232		15.25	7.65		22.90	28.50	
	5150	Two piece		67	.239		16.30	7.90		24.20	30	
	5200	90° or 180° vertical corner post		69	.232		14.65	7.65		22.30	27.50	
	6980	Door stop (snap in)	▼	380	.042	▼	2.50	1.39		3.89	4.85	
	7000	For joints, 90°, clip type, add				Ea.	20			20	22	
	7050	Screw spline joint, add					15.35			15.35	16.85	
	7100	For joint other than 90°, add				▼	32			32	35.50	
	8000	For bronze anodized aluminum, add					15%					
	8020	For black finish, add					27%					
	8050	For stainless steel materials, add					350%					
	8100	For monumental grade, add					50%					
	8150	For steel stiffener, add	2 Glaz	200	.080	L.F.	7.85	2.64		10.49	12.60	
	8200	For 2 to 5 stories, add per story				Story		5%				
900	0010	**WINDOW WALLS** Aluminum, stock, including glazing, minimum	H-2	160	.150	S.F.	27.50	4.64		32.14	37.50	**900**
	0050	Average		140	.171	"	33.50	5.30		38.80	45	
	0100	Maximum	▼	110	.218	S.F.	105	6.75		111.75	126	
	0500	For translucent sandwich wall systems, see div. 07420-770										
	0850	Cost of the above walls depends on material,										
	0860	finish, repetition, and size of units.										
	0870	The larger the opening, the lower the S.F. cost										
	1200	Double glazed acoustical window wall for airports,										
	1220	including 1" thick glass with 2" x 4-1/2" tube frame	H-2	40	.600	S.F.	62.50	18.55		81.05	97	

08950 | Translucent Wall & Roof Assemblies

			CREW	DAILY OUTPUT	LABOR-HOURS	UNIT	MAT.	LABOR	EQUIP.	TOTAL	TOTAL INCL O&P	
100	0010	**SKYROOFS** Translucent panels, 2-3/4" thick, under 5000 S.F.	G-3	395	.081	SF Hor.	20.50	2.71		23.21	26.50	**100**
	0100	Over 5000 S.F.		465	.069		18.10	2.30		20.40	23.50	
	0300	Continuous vaulted, semi-circular, to 8' wide, double glazed		145	.221		45.50	7.40		52.90	61.50	
	0400	Single glazed		160	.200		30.50	6.70		37.20	44.50	
	0600	To 20' wide, single glazed		175	.183		34.50	6.10		40.60	47.50	
	0700	Over 20' wide, single glazed		200	.160		39.50	5.35		44.85	51.50	
	0900	Motorized opening type, single glazed, 1/3 opening		145	.221		42	7.40		49.40	58	
	1000	Full opening	▼	130	.246	▼	48	8.25		56.25	66	
	1200	Pyramid type units, self-supporting, to 30' clear opening,										
	1300	square or circular, single glazed, minimum	G-3	200	.160	SF Hor.	22	5.35		27.35	33	
	1310	Average		165	.194		31.50	6.50		38	44.50	
	1400	Maximum		130	.246		45	8.25		53.25	62.50	
	1500	Grid type, 4' to 10' modules, single glass glazed, minimum		200	.160		29	5.35		34.35	40.50	
	1550	Maximum		128	.250		47.50	8.35		55.85	65	
	1600	Preformed acrylic, minimum		300	.107		34.50	3.57		38.07	43.50	
	1650	Maximum	▼	175	.183	▼	48.50	6.10		54.60	62.50	

Important: See the Reference Section for critical supporting data - Reference Nos., Crews, & City Cost Indexes

08950	Translucent Wall & Roof Assemblies	CREW	DAILY OUTPUT	LABOR-HOURS	UNIT	2005 BARE COSTS				TOTAL INCL O&P	
						MAT.	LABOR	EQUIP.	TOTAL		
100	1800	Skyroofs,dome type, self-supporting, to 100' clear opening, circular										100
	1900	Rise to span ratio = 0.20									
	1920	Minimum	G-3	197	.162	SF Hor.	14.45	5.45		19.90	24.50
	1950	Maximum		113	.283		48	9.45		57.45	67.50
	2100	Rise to span ratio = 0.33, minimum		169	.189		29	6.35		35.35	41.50
	2150	Maximum		101	.317		57	10.60		67.60	79
	2200	Rise to span ratio = 0.50, minimum		148	.216		43.50	7.25		50.75	59
	2250	Maximum		87	.368		63.50	12.30		75.80	88.50
	2400	Ridge units, continuous, to 8' wide, double		130	.246		111	8.25		119.25	135
	2500	Single		200	.160		76	5.35		81.35	92
	2700	Ridge and furrow units, over 4' O.C., double, minimum		200	.160		24	5.35		29.35	35
	2750	Maximum		120	.267		44.50	8.90		53.40	62.50
	2800	Single, minimum		214	.150		20.50	5		25.50	30.50
	2850	Maximum		153	.209	▼	43	7		50	58.50
	3000	Rolling roof, translucent panels, flat roof, mininum		253	.126	S.F.	18.40	4.23		22.63	26.50
	3030	Maximum		160	.200		37	6.70		43.70	51
	3100	Lean-to skyroof, long span, double, minimum		197	.162		24.50	5.45		29.95	35.50
	3150	Maximum		101	.317		49	10.60		59.60	70.50
	3300	Single, minimum		321	.100		18.45	3.33		21.78	25.50
	3350	Maximum	▼	160	.200	▼	30.50	6.70		37.20	44.50

For information about Means Estimating Seminars, see yellow pages 12 and 13 in back of book

DOORS & WINDOWS **8**

	CREW	DAILY OUTPUT	LABOR-HOURS	UNIT	2005 BARE COSTS				TOTAL INCL O&P
					MAT.	LABOR	EQUIP.	TOTAL	

Division 9
Finishes

Estimating Tips

General

- Room Finish Schedule: A complete set of plans should contain a room finish schedule. If one is not available, it would be well worth the time and effort to put one together. A room finish schedule should contain the room number, room name (for clarity), floor materials, base materials, wainscot materials, wainscot height, wall materials (for each wall), ceiling materials, ceiling height and special instructions.
- Surplus Finishes: Review the specifications to determine if there is any requirement to provide certain amounts of extra materials for the owner's maintenance department. In some cases the owner may require a substantial amount of materials, especially when it is a special order item or long lead time item.

09200 Plaster & Gypsum Board

- Lath is estimated by the square yard for both gypsum and metal lath, plus usually 5% allowance for waste. Furring, channels and accessories are measured by the linear foot. An extra foot should be allowed for each accessory miter or stop.
- Plaster is also estimated by the square yard. Deductions for openings vary by preference, from zero deduction to 50% of all openings over 2 feet in width. Some estimators deduct a percentage of the total yardage for openings. The estimator should allow one extra square foot for each linear foot of horizontal interior or exterior angle located below the ceiling level. Also, double the areas of small radius work.
- Each room should be measured, perimeter times maximum wall height. Floors and ceiling areas are equal to length times width.
- Drywall accessories, studs, track, and acoustical caulking are all measured by the linear foot. Drywall taping is figured by the square foot. Gypsum wallboard is estimated by the square foot. No material deductions should be made for door or window openings under 32 S.F. Coreboard can be

obtained in a 1″ thickness for solid wall and shaft work. Additions should be made to price out the inside or outside corners.
- Different types of partition construction should be listed separately on the quantity sheets. There may be walls with studs of various widths, double studded, and similar or dissimilar surface materials. Shaft work is usually different construction from surrounding partitions requiring separate quantities and pricing of the work.

09300 Tile
09400 Terrazzo

- Tile and terrazzo areas are taken off on a square foot basis. Trim and base materials are measured by the linear foot. Accent tiles are listed per each. Two basic methods of installation are used. Mud set is approximately 30% more expensive than the thin set. In terrazzo work, be sure to include the linear footage of embedded decorative strips, grounds, machine rubbing and power cleanup.

09600 Flooring

- Wood flooring is available in strip, parquet, or block configuration. The latter two types are set in adhesives with quantities estimated by the square foot. The laying pattern will influence labor costs and material waste. In addition to the material and labor for laying wood floors, the estimator must make allowances for sanding and finishing these areas unless the flooring is prefinished.
- Most of the various types of flooring are all measured on a square foot basis. Base is measured by the linear foot. If adhesive materials are to be quantified, they are estimated at a specified coverage rate by the gallon depending upon the specified type and the manufacturer's recommendations.
- Sheet flooring is measured by the square yard. Roll widths vary, so consideration should be given to use the most economical width, as waste must be figured into the total quantity. Consider also the installation methods available, direct glue down or stretched.

09700 Wall Finishes

- Wall coverings are estimated by the square foot. The area to be covered is measured, length by height of wall above baseboards, to calculate the square footage of each wall. This figure is divided by the number of square feet in the single roll which is being used. Deduct, in full, the areas of openings such as doors and windows. Where a pattern match is required allow 25%-30% waste. One gallon of paste should be sufficient to hang 12 single rolls of light to medium weight paper.

09800 Acoustical Treatment

- Acoustical systems fall into several categories. The takeoff of these materials should be by the square foot of area with a 5% allowance for waste. Do not forget about scaffolding, if applicable, when estimating these systems.

09900 Paints & Coatings

- A major portion of the work in painting involves surface preparation. Be sure to include cleaning, sanding, filling and masking costs in the estimate.
- Painting is one area where bids vary to a greater extent than almost any other section of a project. This arises from the many methods of measuring surfaces to be painted. The estimator should check the plans and specifications carefully to be sure of the required number of coats.
- Protection of adjacent surfaces is not included in painting costs. When considering the method of paint application, an important factor is the amount of protection and masking required. These must be estimated separately and may be the determining factor in choosing the method of application.

Reference Numbers

Reference numbers are shown in bold squares at the beginning of some major classifications. These numbers refer to related items in the Reference Section. The reference information may be an estimating procedure, an alternate pricing method or technical information.

Note: Not all subdivisions listed here necessarily appear in this publication.

		09060	Selective Demolition		CREW	DAILY OUTPUT	LABOR-HOURS	UNIT	2005 BARE COSTS				TOTAL INCL O&P	
									MAT.	LABOR	EQUIP.	TOTAL		
110	0010		**SELECTIVE DEMOLITION, CEILINGS**	R02220 -510										110
	0200		Ceiling, drywall, furred and nailed or screwed		2 Clab	800	.020	S.F.		.53		.53	.83	
	0220		On metal frame			760	.021			.56		.56	.87	
	0240		On suspension system, including system			720	.022			.59		.59	.92	
	1000		Plaster, lime and horse hair, on wood lath, incl. lath			700	.023			.61		.61	.95	
	1020		On metal lath			570	.028			.75		.75	1.17	
	1100		Gypsum, on gypsum lath			720	.022			.59		.59	.92	
	1120		On metal lath			500	.032			.85		.85	1.33	
	1200		Suspended ceiling, mineral fiber, 2'x2' or 2'x4'			1,500	.011			.28		.28	.44	
	1250		On suspension system, incl. system			1,200	.013			.36		.36	.55	
	1500		Tile, wood fiber, 12" x 12", glued			900	.018			.47		.47	.74	
	1540		Stapled			1,500	.011			.28		.28	.44	
	1580		On suspension system, incl. system			760	.021			.56		.56	.87	
	2000		Wood, tongue and groove, 1" x 4"			1,000	.016			.43		.43	.66	
	2040		1" x 8"			1,100	.015			.39		.39	.60	
	2400		Plywood or wood fiberboard, 4' x 8' sheets			1,200	.013			.36		.36	.55	
120	0010		**SELECTIVE DEMOLITION, FLOORING**	R02220 -510										120
	0200		Brick with mortar		2 Clab	475	.034	S.F.		.90		.90	1.40	
	0400		Carpet, bonded, including surface scraping			2,000	.008			.21		.21	.33	
	0440		Scrim applied			8,000	.002			.05		.05	.08	
	0480		Tackless			9,000	.002			.05		.05	.07	
	0600		Composition, acrylic or epoxy			400	.040			1.07		1.07	1.66	
	0700		Concrete, scarify skin		A-1A	225	.036			1.24	1.50	2.74	3.58	
	0800		Resilient, sheet goods		2 Clab	1,400	.011			.31		.31	.47	
	0820		For gym floors			900	.018			.47		.47	.74	
	0900		Vinyl composition tile, 12" x 12"			1,000	.016			.43		.43	.66	
	2000		Tile, ceramic, thin set			675	.024			.63		.63	.98	
	2020		Mud set			625	.026			.68		.68	1.06	
	2200		Marble, slate, thin set			675	.024			.63		.63	.98	
	2220		Mud set			625	.026			.68		.68	1.06	
	2600		Terrazzo, thin set			450	.036			.95		.95	1.48	
	2620		Mud set			425	.038			1.01		1.01	1.56	
	2640		Cast in place			300	.053			1.42		1.42	2.22	
	3000		Wood, block, on end		1 Carp	400	.020			.69		.69	1.07	
	3200		Parquet			450	.018			.61		.61	.95	
	3400		Strip flooring, interior, 2-1/4" x 25/32" thick			325	.025			.84		.84	1.31	
	3500		Exterior, porch flooring, 1" x 4"			220	.036			1.25		1.25	1.94	
	3800		Subfloor, tongue and groove, 1" x 6"			325	.025			.84		.84	1.31	
	3820		1" x 8"			430	.019			.64		.64	.99	
	3840		1" x 10"			520	.015			.53		.53	.82	
	4000		Plywood, nailed			600	.013			.46		.46	.71	
	4100		Glued and nailed			400	.020			.69		.69	1.07	
	8000		Remove flooring, bead blast, minimum		A-1A	1,000	.008			.28	.34	.62	.80	
	8100		Maximum			400	.020			.70	.85	1.55	2.01	
	8150		Mastic only			1,500	.005			.19	.23	.42	.54	
130	0010		**SELECTIVE DEMOLITION, WALLS AND PARTITIONS**	R02220 -510										130
	0100		Brick, 4" to 12" thick		B-9C	220	.182	C.F.		4.93	.64	5.57	8.35	
	0200		Concrete block, 4" thick			1,000	.040	S.F.		1.08	.14	1.22	1.85	
	0280		8" thick			810	.049			1.34	.17	1.51	2.27	
	0300		Exterior stucco 1" thick over mesh		B-9	3,200	.013			.34	.04	.38	.58	
	1000		Drywall, nailed		1 Clab	1,000	.008			.21		.21	.33	
	1020		Glued and nailed			900	.009			.24		.24	.37	
	1500		Fiberboard, nailed			900	.009			.24		.24	.37	
	1520		Glued and nailed			800	.010			.27		.27	.42	
	1568		Plenum barrier, sheet lead			300	.027			.71		.71	1.11	

09050 | Basic Material Finishes and Methods

09060 | Selective Demolition

			CREW	DAILY OUTPUT	LABOR-HOURS	UNIT	2005 BARE COSTS MAT.	LABOR	EQUIP.	TOTAL	TOTAL INCL O&P	
130	2000	Movable walls, metal, 5′ high	1 Clab	300	.027	S.F.		.71		.71	1.11	130
	2020	8′ high	↓	400	.020			.53		.53	.83	
	2200	Metal or wood studs, finish 2 sides, fiberboard	B-1	520	.046			1.26		1.26	1.97	
	2250	Lath and plaster		260	.092			2.53		2.53	3.93	
	2300	Plasterboard (drywall)		520	.046			1.26		1.26	1.97	
	2350	Plywood	↓	450	.053			1.46		1.46	2.27	
	3000	Plaster, lime and horsehair, on wood lath	1 Clab	400	.020			.53		.53	.83	
	3020	On metal lath		335	.024			.64		.64	.99	
	3400	Gypsum or perlite, on gypsum lath		410	.020			.52		.52	.81	
	3420	On metal lath	↓	300	.027			.71		.71	1.11	
	3600	Plywood, one side	B-1	1,500	.016			.44		.44	.68	
	3750	Terra cotta block and plaster, to 6″ thick	″	175	.137	↓		3.75		3.75	5.85	
	3800	Toilet partitions, slate or marble	1 Clab	5	1.600	Ea.		42.50		42.50	66.50	
	3820	Hollow metal	″	8	1	″		26.50		26.50	41.50	
	5000	Wallcovering, vinyl	1 Pape	700	.011	S.F.		.35		.35	.52	
	5040	Designer	″	480	.017	″		.51		.51	.76	

09100 | Metal Support Assemblies

09110 | Non-Load Bearing Wall Framing

			CREW	DAILY OUTPUT	LABOR-HOURS	UNIT	2005 BARE COSTS MAT.	LABOR	EQUIP.	TOTAL	TOTAL INCL O&P	
100	0010	**METAL STUDS AND TRACK**										100
	1600	Non-load bearing, galv, 8′ high, 25 ga. 1-5/8″ wide, 16″ O.C.	1 Carp	619	.013	S.F.	.33	.44		.77	1.05	
	1610	24″ O.C.		950	.008		.25	.29		.54	.72	
	1620	2-1/2″ wide, 16″ O.C.		613	.013		.37	.45		.82	1.10	
	1630	24″ O.C.		938	.009		.28	.29		.57	.76	
	1640	3-5/8″ wide, 16″ O.C.		600	.013		.43	.46		.89	1.18	
	1650	24″ O.C.		925	.009		.32	.30		.62	.82	
	1660	4″ wide, 16″ O.C.		594	.013		.55	.46		1.01	1.33	
	1670	24″ O.C.		925	.009		.42	.30		.72	.92	
	1680	6″ wide, 16″ O.C.		588	.014		.65	.47		1.12	1.45	
	1690	24″ O.C.		906	.009		.49	.30		.79	1.01	
	1700	20 ga. studs, 1-5/8″ wide, 16″ O.C.		494	.016		.62	.55		1.17	1.54	
	1710	24″ O.C.		763	.010		.46	.36		.82	1.07	
	1720	2-1/2″ wide, 16″ O.C.		488	.016		.70	.56		1.26	1.64	
	1730	24″ O.C.		750	.011		.53	.37		.90	1.15	
	1740	3-5/8″ wide, 16″ O.C.		481	.017		.81	.57		1.38	1.78	
	1750	24″ O.C.		738	.011		.61	.37		.98	1.25	
	1760	4″ wide, 16″ O.C.		475	.017		.91	.58		1.49	1.91	
	1770	24″ O.C.		738	.011		.69	.37		1.06	1.34	
	1780	6″ wide, 16″ O.C.		469	.017		1.15	.58		1.73	2.17	
	1790	24″ O.C.		725	.011		.86	.38		1.24	1.54	
	2000	Non-load bearing, galv, 10′ high, 25 ga. 1-5/8″ wide, 16″ O.C.		495	.016		.31	.55		.86	1.20	
	2100	24″ O.C.		760	.011		.23	.36		.59	.81	
	2200	2-1/2″ wide, 16″ O.C.		490	.016		.35	.56		.91	1.25	
	2250	24″ O.C.		750	.011		.26	.37		.63	.85	
	2300	3-5/8″ wide, 16″ O.C.		480	.017		.41	.57		.98	1.34	
	2350	24″ O.C.		740	.011		.30	.37		.67	.91	
	2400	4″ wide, 16″ O.C.		475	.017		.52	.58		1.10	1.48	
	2450	24″ O.C.		740	.011		.39	.37		.76	1.01	
	2500	6″ wide, 16″ O.C.	↓	470	.017	↓	.62	.58		1.20	1.59	

R02220 -510

FINISHES 9

09110	Non-Load Bearing Wall Framing	CREW	DAILY OUTPUT	LABOR-HOURS	UNIT	2005 BARE COSTS				TOTAL INCL O&P	
						MAT.	LABOR	EQUIP.	TOTAL		
100 2550	24" O.C.	1 Carp	725	.011	S.F.	.46	.38		.84	1.09	100
2600	20 ga. studs, 1-5/8" wide, 16" O.C.		395	.020		.58	.69		1.27	1.72	
2650	24" O.C.		610	.013		.43	.45		.88	1.17	
2700	2-1/2" wide, 16" O.C.		390	.021		.66	.70		1.36	1.82	
2750	24" O.C.		600	.013		.49	.46		.95	1.25	
2800	3-5/8" wide, 16" O.C. **CN**		385	.021		.77	.71		1.48	1.96	
2850	24" O.C.		590	.014		.57	.46		1.03	1.34	
2900	4" wide, 16" O.C.		380	.021		.87	.72		1.59	2.07	
2950	24" O.C.		590	.014		.64	.46		1.10	1.42	
3000	6" wide, 16" O.C.		375	.021		1.09	.73		1.82	2.34	
3050	24" O.C.		580	.014		.80	.47		1.27	1.62	
3060	Non-load bearing, galv, 12' high, 25 ga. 1-5/8" wide, 16" O.C.		413	.019		.29	.66		.95	1.35	
3070	24" O.C.		633	.013		.22	.43		.65	.91	
3080	2-1/2" wide, 16" O.C.		408	.020		.33	.67		1	1.42	
3090	24" O.C.		625	.013		.24	.44		.68	.95	
3100	3-5/8" wide, 16" O.C.		400	.020		.39	.69		1.08	1.50	
3110	24" O.C.		617	.013		.28	.44		.72	1	
3120	4" wide, 16" O.C.		396	.020		.50	.69		1.19	1.63	
3130	24" O.C.		617	.013		.37	.44		.81	1.09	
3140	6" wide, 16" O.C.		392	.020		.59	.70		1.29	1.74	
3150	24" O.C.		604	.013		.43	.45		.88	1.19	
3160	20 ga. studs, 1-5/8" wide, 16" O.C.		329	.024		.56	.83		1.39	1.91	
3170	24" O.C.		508	.016		.41	.54		.95	1.29	
3180	2-1/2" wide, 16" O.C.		325	.025		.63	.84		1.47	2.01	
3190	24" O.C.		500	.016		.46	.55		1.01	1.36	
3200	3-5/8" wide, 16" O.C.		321	.025		.74	.85		1.59	2.14	
3210	24" O.C.		492	.016		.54	.56		1.10	1.46	
3220	4" wide, 16" O.C.		317	.025		.83	.86		1.69	2.26	
3230	24" O.C.		492	.016		.60	.56		1.16	1.53	
3240	6" wide, 16" O.C.		313	.026		1.04	.88		1.92	2.50	
3250	24" O.C.	▼	483	.017	▼	.76	.57		1.33	1.71	
5000	Load bearing studs, see division 05410-400										

09120	Ceiling Suspension										
100 0010	**CEILING SUSPENSION SYSTEMS** For gypsum board or plaster										100
8000	Suspended ceilings, including carriers										
8200	1-1/2" carriers, 24" O.C. with:										
8300	7/8" channels, 16" O.C.	1 Lath	165	.048	S.F.	.43	1.52		1.95	2.73	
8320	24" O.C.		200	.040		.36	1.26		1.62	2.26	
8400	1-5/8" channels, 16" O.C.		155	.052		.52	1.62		2.14	2.98	
8420	24" O.C.	▼	190	.042	▼	.42	1.32		1.74	2.42	
8600	2" carriers, 24" O.C. with:										
8700	7/8" channels, 16" O.C.	1 Lath	155	.052	S.F.	.41	1.62		2.03	2.86	
8720	24" O.C.		190	.042		.34	1.32		1.66	2.34	
8800	1-5/8" channels, 16" O.C.		145	.055		.50	1.74		2.24	3.12	
8820	24" O.C.	▼	180	.044		.40	1.40		1.80	2.51	

09130	Acoustical Suspension										
100 0010	**CEILING SUSPENSION SYSTEMS** For boards and tile										100
0050	Class A suspension system, 15/16" T bar, 2' x 4' grid **CN**	1 Carp	800	.010	S.F.	.49	.34		.83	1.07	
0300	2' x 2' grid	"	650	.012		.61	.42		1.03	1.33	
0350	For 9/16" grid, add					.14			.14	.15	
0360	For fire rated grid, add					.08			.08	.09	
0370	For colored grid, add					.17			.17	.19	
0400	Concealed Z bar suspension system, 12" module	1 Carp	520	.015		.44	.53		.97	1.30	
0600	1-1/2" carrier channels, 4' O.C., add	"	470	.017	▼	.13	.58		.71	1.05	

Important: See the Reference Section for critical supporting data - Reference Nos., Crews, & City Cost Indexes

9 FINISHES

09100 | Metal Support Assemblies

09130 | Acoustical Suspension

			CREW	DAILY OUTPUT	LABOR-HOURS	UNIT	2005 BARE COSTS				TOTAL INCL O&P	
							MAT.	LABOR	EQUIP.	TOTAL		
100	0700	Carrier channels for ceilings with										**100**
	0900	recessed lighting fixtures, add	1 Carp	460	.017	S.F.	.23	.60		.83	1.18	
	1040	Hanging wire, 12 ga., 4' long		65	.123	C.S.F.	5.25	4.22		9.47	12.35	
	1080	8' long		65	.123	"	10.55	4.22		14.77	18.15	
	1200	Seismic bracing, 2-1/2" compression post with four bracing wires	▼	50	.160	Ea.	2.37	5.50		7.87	11.15	

09200 | Plaster & Gypsum Board

09205 | Furring & Lathing

			CREW	DAILY OUTPUT	LABOR-HOURS	UNIT	2005 BARE COSTS				TOTAL INCL O&P	
							MAT.	LABOR	EQUIP.	TOTAL		
530	0010	**FURRING** Beams & columns, 7/8" galvanized channels,										**530**
	0030	12" O.C.	1 Lath	155	.052	S.F.	.24	1.62		1.86	2.68	
	0050	16" O.C.		170	.047		.20	1.48		1.68	2.41	
	0070	24" O.C.		185	.043		.13	1.36		1.49	2.17	
	0100	Ceilings, on steel, 7/8" channels, galvanized, 12" O.C.		210	.038		.22	1.20		1.42	2.02	
	0300	16" O.C.		290	.028		.20	.87		1.07	1.51	
	0400	24" O.C.		420	.019		.13	.60		.73	1.04	
	0600	1-5/8" channels, galvanized, 12" O.C.		190	.042		.32	1.32		1.64	2.31	
	0700	16" O.C.		260	.031		.29	.97		1.26	1.75	
	0900	24" O.C.		390	.021		.19	.65		.84	1.17	
	1000	Walls, 7/8" channels, galvanized, 12" O.C.		235	.034		.22	1.07		1.29	1.83	
	1200	16" O.C.		265	.030		.20	.95		1.15	1.63	
	1300	24" O.C.		350	.023		.13	.72		.85	1.22	
	1500	1-5/8" channels, galvanized, 12" O.C.		210	.038		.32	1.20		1.52	2.13	
	1600	16" O.C.		240	.033		.29	1.05		1.34	1.87	
	1800	24" O.C.	▼	305	.026	▼	.19	.82		1.01	1.43	
540	0010	**GYPSUM LATH** Plain or perforated, nailed, 3/8" thick	1 Lath	85	.094	S.Y.	4.05	2.96		7.01	8.85	**540**
	0100	1/2" thick, nailed		80	.100		4.05	3.15		7.20	9.10	
	0300	Clipped to steel studs, 3/8" thick		75	.107		4.05	3.35		7.40	9.45	
	0400	1/2" thick		70	.114		4.05	3.59		7.64	9.80	
	0600	Firestop gypsum base, to steel studs, 3/8" thick		70	.114		3.51	3.59		7.10	9.20	
	0700	1/2" thick		65	.123		4.50	3.87		8.37	10.70	
	0900	Foil back, to steel studs, 3/8" thick		75	.107		3.96	3.35		7.31	9.35	
	1000	1/2" thick		70	.114		4.14	3.59		7.73	9.90	
	1500	For ceiling installations, add		216	.037			1.16		1.16	1.73	
	1600	For columns and beams, add	▼	170	.047	▼		1.48		1.48	2.19	
560	0010	**METAL LATH** Diamond, expanded, 2.5 lb. per S.Y., painted				S.Y.	2.85			2.85	3.14	**560**
	0100	Galvanized, 2.5 lb. per S.Y.					2.76			2.76	3.04	
	0300	3.4 lb. per S.Y., painted					3.40			3.40	3.74	
	0400	Galvanized					3.02			3.02	3.32	
	0600	For 15# asphalt sheathing paper, add					.31			.31	.34	
	0900	Flat rib, 1/8" high, 2.75 lb., painted					3.04			3.04	3.34	
	1000	Foil backed					3.65			3.65	4.02	
	1200	3.4 lb. per S.Y., painted					3.64			3.64	4	
	1300	Galvanized					4.13			4.13	4.54	
	1500	For 15# asphalt sheating paper, add					.31			.31	.34	
	1800	High rib, 3/8" high, 3.4 lb. per S.Y., painted					4.40			4.40	4.84	
	1900	Galvanized				▼	3.67			3.67	4.04	
	2400	High rib, 3/4" high, painted, .60 lb. per S.F.				S.F.	.42			.42	.46	
	2500	.75 lb. per S.F.				"	.90			.90	.99	

09205	Furring & Lathing	CREW	DAILY OUTPUT	LABOR-HOURS	UNIT	2005 BARE COSTS				TOTAL INCL O&P	
						MAT.	LABOR	EQUIP.	TOTAL		
560	2800	Stucco mesh, painted, 3.6 lb.				S.Y.	3.31			3.31	3.64	560
	3000	K-lath, perforated, absorbent paper, regular					3.21			3.21	3.53	
	3100	Heavy duty					3.79			3.79	4.17	
	3300	Waterproof, heavy duty, grade B backing					3.71			3.71	4.08	
	3400	Fire resistant backing					4.10			4.10	4.51	
	3600	2.5 lb. diamond painted, on wood framing, on walls	1 Lath	85	.094		2.85	2.96		5.81	7.55	
	3700	On ceilings		75	.107		2.85	3.35		6.20	8.10	
	3900	3.4 lb. diamond painted, on wood framing, on walls		80	.100		3.64	3.15		6.79	8.65	
	4000	On ceilings		70	.114		3.64	3.59		7.23	9.35	
	4200	3.4 lb. diamond painted, wired to steel framing		75	.107		3.64	3.35		6.99	8.95	
	4300	On ceilings		60	.133		3.64	4.19		7.83	10.20	
	4500	Columns and beams, wired to steel		40	.200		3.64	6.30		9.94	13.30	
	4600	Cornices, wired to steel		35	.229		3.64	7.20		10.84	14.65	
	4800	Screwed to steel studs, 2.5 lb.		80	.100		2.85	3.15		6	7.80	
	4900	3.4 lb.		75	.107		3.40	3.35		6.75	8.70	
	5100	Rib lath, painted, wired to steel, on walls, 2.5 lb.		75	.107		3.04	3.35		6.39	8.30	
	5200	3.4 lb.		70	.114		4.40	3.59		7.99	10.20	
	5400	4.0 lb.		65	.123		4.52	3.87		8.39	10.70	
	5500	For self-furring lath, add					.08			.08	.09	
	5700	Suspended ceiling system, incl. 3.4 lb. diamond lath, painted	1 Lath	15	.533		11.90	16.75		28.65	38	
	5800	Galvanized	"	15	.533		12.25	16.75		29	38.50	
	6000	Hollow metal stud partitions, 3.4 lb. painted lath both sides										
	6010	Non-load bearing, 25 ga., w/rib lath 2-1/2" studs, 12" O.C.	1 Lath	20.30	.394	S.Y.	12.75	12.40		25.15	32.50	
	6300	16" O.C.		21.10	.379		11.90	11.90		23.80	30.50	
	6350	24" O.C.		22.70	.352		11.10	11.10		22.20	28.50	
	6400	3-5/8" studs, 16" O.C.		19.50	.410		12.45	12.90		25.35	33	
	6600	24" O.C.		20.40	.392		11.50	12.35		23.85	31	
	6700	4" studs, 16" O.C.		20.40	.392		13.50	12.35		25.85	33	
	6900	24" O.C.		21.60	.370		12.30	11.65		23.95	31	
	7000	6" studs, 16" O.C.		19.50	.410		14.40	12.90		27.30	35	
	7100	24" O.C.		21.10	.379		12.90	11.90		24.80	32	
	7200	L.B. partitions, 16 ga., w/rib lath, 2-1/2" studs, 16" O.C.		20	.400		13.20	12.60		25.80	33	
	7300	3-5/8" studs, 16 ga.		19.70	.406		15.15	12.75		27.90	35.50	
	7500	4" studs, 16 ga.		19.50	.410		15.75	12.90		28.65	36.50	
	7600	6" studs, 16 ga.		18.70	.428		18.80	13.45		32.25	40.50	
600	0010	**SECURITY MESH**, expanded metal, flat, screwed to framing	2 Carp	1,500	.011	S.F.	1.25	.37		1.62	1.95	600
	0100	On walls, 3/4", 1.76 lb/S.F.										
	0110	1-1/2", 1.14 lb/S.F.		1,600	.010		.87	.34		1.21	1.49	
	0200	On ceilings, 3/4", 1.76 lb/S.F.		1,350	.012		1.25	.41		1.66	2.01	
	0210	1-1/2", 1.14 lb/S.F.		1,450	.011		.87	.38		1.25	1.55	
700	0010	**ACCESSORIES, PLASTER** Casing bead, expanded flange, galvanized	1 Lath	2.70	2.963	C.L.F.	38	93		131	180	700
	0900	Channels, cold rolled, 16 ga., 3/4" deep, galvanized					22			22	24.50	
	1200	1-1/2" deep, 16 ga., galvanized					32			32	35	
	1620	Corner bead, expanded bullnose, 3/4" radius, #10, galvanized	1 Lath	2.60	3.077		28	97		125	174	
	1650	#1, galvanized		2.55	3.137		47.50	98.50		146	198	
	1670	Expanded wing, 2-3/4" wide, galv. #1		2.65	3.019		30	95		125	174	
	1700	Inside corner, (corner rite) 3" x 3", painted		2.60	3.077		25.50	97		122.50	171	
	1750	Strip-ex, 4" wide, painted		2.55	3.137		22	98.50		120.50	171	
	1800	Expansion joint, 3/4" grounds, limited expansion, galv., 1 piece		2.70	2.963		92.50	93		185.50	239	
	2100	Extreme expansion, galvanized, 2 piece		2.60	3.077		162	97		259	320	

09210	Gypsum Plaster										
100	0010	**GYPSUM PLASTER** 80# bag, less than 1 ton				Bag	14.05			14.05	15.45	100
	0100	Over 1 ton				"	12.85			12.85	14.10	

09210 | Gypsum Plaster

	Line		Crew	Daily Output	Labor-Hours	Unit	Mat.	Labor	Equip.	Total	Total Incl O&P	
100	0300	2 coats, no lath included, on walls	J-1	105	.381	S.Y.	3.13	11.30	.98	15.41	21.50	100
	0400	On ceilings	"	92	.435		3.13	12.90	1.12	17.15	24.50	
	0600	2 coats on and incl. 3/8" gypsum lath on steel, on walls	J-2	97	.495		7.20	14.80	1.06	23.06	31.50	
	0700	On ceilings	"	83	.578		7.20	17.30	1.24	25.74	35.50	
	0900	3 coats, no lath included, on walls	J-1	87	.460		4.47	13.65	1.19	19.31	26.50	
	1000	On ceilings	"	78	.513		4.47	15.20	1.32	20.99	29.50	
	1200	3 coats on and including painted metal lath, on wood studs	J-2	86	.558		8.60	16.70	1.20	26.50	36.50	
	1300	On ceilings	"	76.50	.627		8.60	18.80	1.35	28.75	39.50	
	1600	For irregular or curved surfaces, add						30%				
	1800	For columns & beams, add						50%				
200	0010	GAUGING PLASTER 100 lb. bags, less than 1 ton				Bag	22.50			22.50	25	200
	0100	Over 1 ton	"				18.50			18.50	20.50	
300	0010	KEENES CEMENT In 100 lb. bags, less than 1 ton				Bag	23.50			23.50	25.50	300
	0100	Over 1 ton	"				22.50			22.50	25	
	0300	Finish only, add to plaster prices, standard	J-1	215	.186	S.Y.	2.19	5.50	.48	8.17	11.35	
	0400	High quality	"	144	.278	"	2.21	8.25	.72	11.18	15.75	
500	0010	PERLITE OR VERMICULITE PLASTER 100 lb. bags Under 200 bags				Bag	13.75			13.75	15.10	500
	0100	Over 200 bags	"				12.65			12.65	13.95	
	0300	2 coats, no lath included, on walls	J-1	92	.435	S.Y.	3.24	12.90	1.12	17.26	24.50	
	0400	On ceilings	"	79	.506		3.24	15	1.31	19.55	28	
	0600	2 coats, on and incl. 3/8" gypsum lath, on metal studs	J-2	84	.571		6.90	17.10	1.23	25.23	35	
	0700	On ceilings	"	70	.686		6.90	20.50	1.47	28.87	40	
	0900	3 coats, no lath included, on walls	J-1	74	.541		5.65	16.05	1.39	23.09	32.50	
	1000	On ceilings	"	63	.635		5.65	18.85	1.64	26.14	36.50	
	1200	3 coats, on and incl. painted metal lath, on metal studs	J-2	72	.667		10.05	19.95	1.43	31.43	43	
	1300	On ceilings		61	.787		10.05	23.50	1.69	35.24	48.50	
	1500	3 coats, on and incl. suspended metal lath ceiling		37	1.297		17.55	39	2.79	59.34	81.50	
	1700	For irregular or curved surfaces, add to above						30%				
	1800	For columns and beams, add to above						50%				
	1900	For soffits, add to ceiling prices						40%				
650	0010	PLASTER PARTITION WALL										650
	0400	Stud walls, 3.4 lb. metal lath, 3 coat gypsum plaster, 2 sides										
	0600	2" x 4" wood studs, 16" O.C.	J-2	315	.152	S.F.	3.13	4.56	.33	8.02	10.70	
	0700	2-1/2" metal studs, 25 ga., 12" O.C.		325	.148		2.90	4.42	.32	7.64	10.25	
	0800	3-5/8" metal studs, 25 ga., 16" O.C.		320	.150		2.92	4.49	.32	7.73	10.35	
	0900	Gypsum lath, 2 coat vermiculite plaster, 2 sides										
	1000	2" x 4" wood studs, 16" O.C.	J-2	355	.135	S.F.	3.49	4.05	.29	7.83	10.30	
	1200	2-1/2" metal studs, 25 ga., 12" O.C.		365	.132		3.09	3.94	.28	7.31	9.65	
	1300	3-5/8" metal studs, 25 ga., 16" O.C.		360	.133		3.18	3.99	.29	7.46	9.85	
900	0010	THIN COAT Plaster, 1 coat veneer, not incl. lath	J-1	3,600	.011	S.F.	.07	.33	.03	.43	.61	900
	1000	In 50 lb. bags				Bag	9.65			9.65	10.65	

09220 | Portland Cement Plaster

	Line		Crew	Daily Output	Labor-Hours	Unit	Mat.	Labor	Equip.	Total	Total Incl O&P	
200	0010	STUCCO, 3 coats 1" thick, float finish, with mesh, on wood frame	J-2	63	.762	S.Y.	5.35	23	1.64	29.99	42	200
	0100	On masonry construction, no mesh incl.	J-1	67	.597		2.05	17.70	1.54	21.29	31	
	0300	For trowel finish, add	1 Plas	170	.047			1.48		1.48	2.24	
	0400	For 3/4" thick, on masonry, deduct	J-1	880	.045		.56	1.35	.12	2.03	2.80	
	0600	For coloring and special finish, add, minimum		685	.058		.36	1.73	.15	2.24	3.20	
	0700	Maximum		200	.200		1.26	5.95	.52	7.73	10.95	
	0900	For soffits, add	J-2	155	.310		1.94	9.25	.67	11.86	16.90	
	1000	Exterior stucco, with bonding agent, 3 coats, on walls, no mesh incl.	J-1	200	.200		3.25	5.95	.52	9.72	13.15	
	1200	Ceilings		180	.222		3.25	6.60	.57	10.42	14.25	
	1300	Beams		80	.500		3.25	14.85	1.29	19.39	27.50	
	1500	Columns		100	.400		3.25	11.85	1.03	16.13	23	
	1600	Mesh, painted, nailed to wood, 1.8 lb.	1 Lath	60	.133		4.17	4.19		8.36	10.80	

FINISHES 9

09220	Portland Cement Plaster	CREW	DAILY OUTPUT	LABOR-HOURS	UNIT	2005 BARE COSTS				TOTAL INCL O&P	
						MAT.	LABOR	EQUIP.	TOTAL		
200 1800	3.6 lb.	1 Lath	55	.145	S.Y.	3.31	4.57		7.88	10.45	**200**
1900	Wired to steel, painted, 1.8 lb.		53	.151		4.17	4.75		8.92	11.65	
2100	3.6 lb.	↓	50	.160	↓	3.31	5.05		8.36	11.10	

09250	Gypsum Board										
200 0010	**CEMENTITIOUS BACKERBOARD**										**200**
0070	Cementitious backerboard, on floor, 3' x 4' x 1/2" sheets	2 Carp	525	.030	S.F.	1.04	1.04		2.08	2.78	
0080	3' x 5' x 1/2" sheets		525	.030		1.04	1.04		2.08	2.78	
0090	3' x 6' x 1/2" sheets		525	.030		.76	1.04		1.80	2.46	
0100	3' x 4' x 5/8" sheets		525	.030		1.08	1.04		2.12	2.82	
0110	3' x 5' x 5/8" sheets		525	.030		1.06	1.04		2.10	2.80	
0120	3' x 6' x 5/8" sheets		525	.030		1.07	1.04		2.11	2.80	
0150	On wall, 3' x 4' x 1/2" sheets		350	.046		1.04	1.57		2.61	3.59	
0160	3' x 5' x 1/2" sheets		350	.046		1.04	1.57		2.61	3.59	
0170	3' x 6' x 1/2" sheets		350	.046		.76	1.57		2.33	3.27	
0180	3' x 4' x 5/8" sheets		350	.046		1.08	1.57		2.65	3.63	
0190	3' x 5' x 5/8" sheets		350	.046		1.06	1.57		2.63	3.61	
0200	3' x 6' x 5/8" sheets		350	.046		1.07	1.57		2.64	3.61	
0250	On counter, 3' x 4' x 1/2" sheets		180	.089		1.04	3.04		4.08	5.90	
0260	3' x 5' x 1/2" sheets		180	.089		1.04	3.04		4.08	5.90	
0270	3' x 6' x 1/2" sheets		180	.089		.76	3.04		3.80	5.55	
0300	3' x 4' x 5/8" sheets		180	.089		1.08	3.04		4.12	5.95	
0310	3' x 5' x 5/8" sheets		180	.089		1.06	3.04		4.10	5.90	
0320	3' x 6' x 5/8" sheets	↓	180	.089	↓	1.07	3.04		4.11	5.90	
300 0010	**BLUEBOARD** For use with thin coat										**300**
0100	plaster application (see division 09210-900)										
1000	3/8" thick, on walls or ceilings, standard, no finish included	2 Carp	1,900	.008	S.F.	.18	.29		.47	.65	
1100	With thin coat plaster finish		875	.018		.25	.63		.88	1.26	
1400	On beams, columns, or soffits, standard, no finish included		675	.024		.21	.81		1.02	1.49	
1450	With thin coat plaster finish		475	.034		.28	1.15		1.43	2.11	
3000	1/2" thick, on walls or ceilings, standard, no finish included		1,900	.008		.21	.29		.50	.68	
3100	With thin coat plaster finish		875	.018		.28	.63		.91	1.29	
3300	Fire resistant, no finish included		1,900	.008		.21	.29		.50	.68	
3400	With thin coat plaster finish		875	.018		.28	.63		.91	1.29	
3450	On beams, columns, or soffits, standard, no finish included		675	.024		.24	.81		1.05	1.53	
3500	With thin coat plaster finish		475	.034		.31	1.15		1.46	2.14	
3700	Fire resistant, no finish included		675	.024		.24	.81		1.05	1.53	
3800	With thin coat plaster finish		475	.034		.31	1.15		1.46	2.14	
5000	5/8" thick, on walls or ceilings, fire resistant, no finish included		1,900	.008		.24	.29		.53	.71	
5100	With thin coat plaster finish		875	.018		.31	.63		.94	1.32	
5500	On beams, columns, or soffits, no finish included		675	.024		.28	.81		1.09	1.56	
5600	With thin coat plaster finish		475	.034		.35	1.15		1.50	2.18	
6000	For high ceilings, over 8' high, add		3,060	.005			.18		.18	.28	
6500	For over 3 stories high, add per story	↓	6,100	.003	↓		.09		.09	.14	
500 0010	**CEILINGS** Gypsum drywall, fire rated, finished										**500**
0100	Screwed to grid, channel or joists, 1/2" thick	2 Carp	765	.021	S.F.	.22	.72		.94	1.36	
0200	5/8" thick		765	.021		.25	.72		.97	1.40	
0300	Over 8' high, 1/2" thick		615	.026		.22	.89		1.11	1.63	
0400	5/8" thick	↓	615	.026	↓	.25	.89		1.14	1.67	
0600	Grid suspension system, direct hung										
0700	1-1/2" C.R.C., with 7/8" hi hat furring channel, 16" O.C.	2 Carp	600	.027	S.F.	.87	.91		1.78	2.38	
0800	24" O.C.		900	.018		.79	.61		1.40	1.82	
0900	3-5/8" C.R.C., with 7/8" hi hat furring channel, 16" O.C.		600	.027		.92	.91		1.83	2.43	
1000	24" O.C.	↓	900	.018	↓	.80	.61		1.41	1.83	

Important: See the Reference Section for critical supporting data - Reference Nos., Crews, & City Cost Indexes

9 FINISHES

09250 | Gypsum Board

		CREW	DAILY OUTPUT	LABOR-HOURS	UNIT	2005 BARE COSTS				TOTAL INCL O&P
						MAT.	LABOR	EQUIP.	TOTAL	
700	0010	**DRYWALL** Gypsum plasterboard, nailed or screwed									700
	0100	to studs unless otherwise noted R09250-100								
	0150	3/8" thick, on walls, standard, no finish included	2 Carp	2,000	.008	S.F.	.20	.27		.47	.65
	0200	On ceilings, standard, no finish included		1,800	.009		.20	.30		.50	.69
	0250	On beams, columns, or soffits, no finish included		675	.024		.20	.81		1.01	1.48
	0300	1/2" thick, on walls, standard, no finish included CN		2,000	.008		.22	.27		.49	.67
	0350	Taped and finished (level 4 finish)		965	.017		.25	.57		.82	1.16
	0390	With compound skim coat (level 5 finish)		775	.021		.28	.71		.99	1.41
	0400	Fire resistant, no finish included		2,000	.008		.22	.27		.49	.67
	0450	Taped and finished (level 4 finish)		965	.017		.25	.57		.82	1.16
	0490	With compound skim coat (level 5 finish)		775	.021		.28	.71		.99	1.41
	0500	Water resistant, no finish included		2,000	.008		.22	.27		.49	.67
	0550	Taped and finished (level 4 finish)		965	.017		.25	.57		.82	1.16
	0590	With compound skim coat (level 5 finish)		775	.021		.28	.71		.99	1.41
	0600	Prefinished, vinyl, clipped to studs		900	.018		.56	.61		1.17	1.57
	1000	On ceilings, standard, no finish included		1,800	.009		.22	.30		.52	.71
	1050	Taped and finished (level 4 finish)		765	.021		.25	.72		.97	1.40
	1090	With compound skim coat (level 5 finish)		610	.026		.28	.90		1.18	1.71
	1100	Fire resistant, no finish included		1,800	.009		.22	.30		.52	.71
	1150	Taped and finished (level 4 finish)		765	.021		.25	.72		.97	1.40
	1195	With compound skim coat (level 5 finish)		610	.026		.28	.90		1.18	1.71
	1200	Water resistant, no finish included		1,800	.009		.22	.30		.52	.71
	1250	Taped and finished (level 4 finish)		765	.021		.25	.72		.97	1.40
	1290	With compound skim coat (level 5 finish)		610	.026		.28	.90		1.18	1.71
	1500	On beams, columns, or soffits, standard, no finish included		675	.024		.25	.81		1.06	1.54
	1550	Taped and finished (level 4 finish)		475	.034		.25	1.15		1.40	2.08
	1590	With compound skim coat (level 5 finish)		540	.030		.28	1.01		1.29	1.89
	1600	Fire resistant, no finish included		675	.024		.25	.81		1.06	1.54
	1650	Taped and finished (level 4 finish)		475	.034		.25	1.15		1.40	2.08
	1690	With compound skim coat (level 5 finish)		540	.030		.28	1.01		1.29	1.89
	1700	Water resistant, no finish included		675	.024		.25	.81		1.06	1.54
	1750	Taped and finished (level 4 finish)		475	.034		.25	1.15		1.40	2.08
	1790	With compound skim coat (level 5 finish)		540	.030		.28	1.01		1.29	1.89
	2000	5/8" thick, on walls, standard, no finish included		2,000	.008		.24	.27		.51	.69
	2050	Taped and finished (level 4 finish)		965	.017		.27	.57		.84	1.18
	2090	With compound skim coat (level 5 finish)		775	.021		.30	.71		1.01	1.43
	2100	Fire resistant, no finish included		2,000	.008		.25	.27		.52	.71
	2150	Taped and finished (level 4 finish)		965	.017		.28	.57		.85	1.19
	2195	With compound skim coat (level 5 finish)		775	.021		.31	.71		1.02	1.45
	2200	Water resistant, no finish included		2,000	.008		.26	.27		.53	.72
	2250	Taped and finished (level 4 finish)		965	.017		.29	.57		.86	1.20
	2290	With compound skim coat (level 5 finish)		775	.021		.32	.71		1.03	1.46
	2300	Prefinished, vinyl, clipped to studs		900	.018		.65	.61		1.26	1.67
	3000	On ceilings, standard, no finish included		1,800	.009		.24	.30		.54	.73
	3050	Taped and finished (level 4 finish)		765	.021		.27	.72		.99	1.42
	3090	With compound skim coat (level 5 finish)		615	.026		.30	.89		1.19	1.72
	3100	Fire resistant, no finish included		1,800	.009		.25	.30		.55	.75
	3150	Taped and finished (level 4 finish)		765	.021		.28	.72		1	1.43
	3190	With compound skim coat (level 5 finish)		615	.026		.31	.89		1.20	1.74
	3200	Water resistant, no finish included		1,800	.009		.26	.30		.56	.76
	3250	Taped and finished (level 4 finish)		765	.021		.29	.72		1.01	1.44
	3290	With compound skim coat (level 5 finish)		615	.026		.32	.89		1.21	1.75
	3500	On beams, columns, or soffits, no finish included		675	.024		.28	.81		1.09	1.56
	3550	Taped and finished (level 4 finish)		475	.034		.31	1.15		1.46	2.14
	3590	With compound skim coat (level 5 finish)		380	.042		.35	1.44		1.79	2.64
	3600	Fire resistant, no finish included		675	.024		.29	.81		1.10	1.58

FINISHES 9

		09250 | Gypsum Board	CREW	DAILY OUTPUT	LABOR-HOURS	UNIT	2005 BARE COSTS				TOTAL INCL O&P	
							MAT.	LABOR	EQUIP.	TOTAL		
700	3650	Taped and finished (level 4 finish) R09250-100	2 Carp	475	.034	S.F.	.32	1.15		1.47	2.15	**700**
	3690	With compound skim coat (level 5 finish)		380	.042		.31	1.44		1.75	2.60	
	3700	Water resistant, no finish included		675	.024		.30	.81		1.11	1.59	
	3750	Taped and finished (level 4 finish)		475	.034		.33	1.15		1.48	2.17	
	3790	With compound skim coat (level 5 finish)		380	.042		.32	1.44		1.76	2.61	
	4000	Fireproofing, beams or columns, 2 layers, 1/2" thick, incl finish		330	.048		.47	1.66		2.13	3.11	
	4050	5/8" thick		300	.053		.56	1.83		2.39	3.47	
	4100	3 layers, 1/2" thick		225	.071		.69	2.44		3.13	4.55	
	4150	5/8" thick		210	.076		.84	2.61		3.45	4.99	
	5050	For 1" thick coreboard on columns		480	.033		.45	1.14		1.59	2.28	
	5100	For foil-backed board, add					.10			.10	.11	
	5200	For work over 8' high, add	2 Carp	3,060	.005			.18		.18	.28	
	5270	For textured spray, add	2 Lath	1,600	.010		.04	.31		.35	.51	
	5300	For over 3 stories high, add per story	2 Carp	6,100	.003			.09		.09	.14	
	5350	For finishing inner corners, add		950	.017	L.F.	.07	.58		.65	.97	
	5355	For finishing outer corners, add		1,250	.013		.17	.44		.61	.86	
	5500	For acoustical sealant, add per bead	1 Carp	500	.016		.03	.55		.58	.88	
	5550	Sealant, 1 quart tube				Ea.	4.89			4.89	5.40	
	5600	Sound deadening board, 1/4" gypsum	2 Carp	1,800	.009	S.F.	.22	.30		.52	.71	
	5650	1/2" wood fiber	"	1,800	.009	"	.38	.30		.68	.89	
800	0010	**HIGH ABUSE GYPSUM BOARD**, fiber reinforced, nailed or										**800**
	0100	screwed to studs unless otherwise noted										
	0110	1/2" thick, on walls, no finish included	2 Carp	1,800	.009	S.F.	.42	.30		.72	.93	
	0120	Taped and finished (level 4 finish)		870	.018		.45	.63		1.08	1.48	
	0130	With compound skim coat (level 5 finish)		700	.023		.48	.78		1.26	1.75	
	0150	On ceilings, no finish included		1,620	.010		.42	.34		.76	.99	
	0160	Taped and finished (level 4 finish)		690	.023		.45	.79		1.24	1.74	
	0170	With compound skim coat (level 5 finish)		550	.029		.48	1		1.48	2.08	
	0210	5/8" thick, on walls, no finish included		1,800	.009		.57	.30		.87	1.10	
	0220	Taped and finished (level 4 finish)		870	.018		.60	.63		1.23	1.64	
	0230	With compound skim coat (level 5 finish)		700	.023		.63	.78		1.41	1.92	
	0250	On ceilings, no finish included		1,620	.010		.57	.34		.91	1.16	
	0260	Taped and finished (level 4 finish)		690	.023		.60	.79		1.39	1.90	
	0270	With compound skim coat (level 5 finish)		550	.029		.63	1		1.63	2.25	
	0310	5/8" thick, on walls, very high impact, no finish included		1,800	.009		.68	.30		.98	1.22	
	0320	Taped and finished (level 4 finish)		870	.018		.71	.63		1.34	1.76	
	0330	With compound skim coat (level 5 finish)		700	.023		.74	.78		1.52	2.04	
	0350	On ceilings, no finish included		1,620	.010		.68	.34		1.02	1.28	
	0360	Taped and finished (level 4 finish)		690	.023		.71	.79		1.50	2.02	
	0370	With compound skim coat (level 5 finish)		550	.029		.74	1		1.74	2.37	
	0400	High abuse, gypsum core, paper face										
	0410	1/2" thick, on walls, no finish included	2 Carp	1,800	.009	S.F.	.52	.30		.82	1.04	
	0420	Taped and finished (level 4 finish)		870	.018		.55	.63		1.18	1.59	
	0430	With compound skim coat (level 5 finish)		700	.023		.58	.78		1.36	1.86	
	0450	On ceilings, no finish included		1,620	.010		.52	.34		.86	1.10	
	0460	Taped and finished (level 4 finish)		690	.023		.55	.79		1.34	1.85	
	0470	With compound skim coat (level 5 finish)		550	.029		.58	1		1.58	2.19	
	0510	5/8" thick, on walls, no finish included		1,800	.009		.58	.30		.88	1.11	
	0520	Taped and finished (level 4 finish)		870	.018		.61	.63		1.24	1.65	
	0530	With compound skim coat (level 5 finish)		700	.023		.64	.78		1.42	1.93	
	0550	On ceilings, no finish included		1,620	.010		.58	.34		.92	1.17	
	0560	Taped and finished (level 4 finish)		690	.023		.61	.79		1.40	1.91	
	0570	With compound skim coat (level 5 finish)		550	.029		.64	1		1.64	2.26	
	1000	For high ceilings, over 8' high, add		2,750	.006			.20		.20	.31	
	1010	For over 3 stories high, add per story		5,500	.003			.10		.10	.16	

Important: See the Reference Section for critical supporting data - Reference Nos., Crews, & City Cost Indexes

9 FINISHES

09260	Gypsum Board Systems	CREW	DAILY OUTPUT	LABOR-HOURS	UNIT	MAT.	LABOR	EQUIP.	TOTAL	TOTAL INCL O&P	
100	0010	**PARTITION WALL** Stud wall, 8' to 12' high										100
	0050	1/2", interior, gypsum drywall, standard, taped both sides										
	0500	Installed on and incl., 2" x 4" wood studs, 16" O.C.	2 Carp	310	.052	S.F.	.88	1.77		2.65	3.72	
	1000	Metal studs, NLB, 25 ga., 16" O.C., 3-5/8" wide		350	.046		.91	1.57		2.48	3.44	
	1200	6" wide		330	.048		1.12	1.66		2.78	3.82	
	1400	Water resistant, on 2" x 4" wood studs, 16" O.C.		310	.052		.88	1.77		2.65	3.72	
	1600	Metal studs, NLB, 25 ga., 16" O.C., 3-5/8" wide		350	.046		.91	1.57		2.48	3.44	
	1800	6" wide		330	.048		1.12	1.66		2.78	3.82	
	2000	Fire res., 2 layers, 1-1/2 hr., on 2" x 4" wood studs, 16" O.C.		210	.076		1.32	2.61		3.93	5.50	
	2200	Metal studs, NLB, 25 ga., 16" O.C., 3-5/8" wide		250	.064		1.35	2.19		3.54	4.89	
	2400	6" wide		230	.070		1.56	2.38		3.94	5.45	
	2600	Fire & water res., 2 layers, 1-1/2 hr., 2"x4" studs, 16" O.C.		210	.076		1.32	2.61		3.93	5.50	
	2800	Metal studs, NLB, 25 ga., 16" O.C., 3-5/8" wide		250	.064		1.35	2.19		3.54	4.89	
	3000	6" wide	▼	230	.070	▼	1.56	2.38		3.94	5.45	
	3200	5/8", interior, gypsum drywall, standard, taped both sides										
	3400	Installed on and including 2" x 4" wood studs, 16" O.C.	2 Carp	300	.053	S.F.	.92	1.83		2.75	3.86	
	3600	24" O.C.		330	.048		.84	1.66		2.50	3.51	
	3800	Metal studs, NLB, 25 ga., 16" O.C., 3-5/8" wide		340	.047		.95	1.61		2.56	3.55	
	4000	6" wide		320	.050		1.16	1.71		2.87	3.95	
	4200	24" O.C., 3-5/8" wide		360	.044		.84	1.52		2.36	3.30	
	4400	6" wide		340	.047		1	1.61		2.61	3.61	
	4800	Water resistant, on 2" x 4" wood studs, 16" O.C.		300	.053		.96	1.83		2.79	3.91	
	5000	24" O.C.		330	.048		.88	1.66		2.54	3.56	
	5200	Metal studs, NLB, 25 ga. 16" O.C., 3-5/8" wide		340	.047		.99	1.61		2.60	3.60	
	5400	6" wide		320	.050		1.20	1.71		2.91	3.99	
	5600	24" O.C., 3-5/8" wide		360	.044		.88	1.52		2.40	3.34	
	5800	6" wide		340	.047		1.04	1.61		2.65	3.65	
	6000	Fire res., 2 layers, 2 hr., on 2" x 4" wood studs, 16" O.C.		205	.078		1.36	2.67		4.03	5.65	
	6200	24" O.C.		235	.068		1.36	2.33		3.69	5.10	
	6400	Metal studs, NLB, 25 ga., 16" O.C., 3-5/8" wide		245	.065		1.56	2.24		3.80	5.20	
	6600	6" wide		225	.071		1.68	2.44		4.12	5.65	
	6800	24" O.C., 3-5/8" wide		265	.060		1.36	2.07		3.43	4.72	
	7000	6" wide		245	.065		1.52	2.24		3.76	5.15	
	7200	Fire & water res., 2 layers, 2 hr., 2" x 4" studs, 16" O.C.		205	.078		1.44	2.67		4.11	5.75	
	7400	24" O.C.		235	.068		1.36	2.33		3.69	5.10	
	7600	Metal studs, NLB, 25 ga., 16" O.C., 3-5/8" wide		245	.065		1.47	2.24		3.71	5.10	
	7800	6" wide		225	.071		1.68	2.44		4.12	5.65	
	8000	24" O.C., 3-5/8" wide		265	.060		1.36	2.07		3.43	4.72	
	8200	6" wide	▼	245	.065	▼	1.52	2.24		3.76	5.15	
	8600	1/2" blueboard, mesh tape both sides										
	8620	Installed on and including 2" x 4" wood studs, 16" O.C.	2 Carp	300	.053	S.F.	.92	1.83		2.75	3.86	
	8640	Metal studs, NLB, 25 ga., 16" O.C., 3-5/8" wide		340	.047		.95	1.61		2.56	3.55	
	8660	6" wide	▼	320	.050	▼	1.16	1.71		2.87	3.95	
	9000	Exterior, 1/2" gypsum sheathing, 1/2" gypsum finished, interior,										
	9100	including foil faced insulation, metal studs, 20 ga.										
	9200	16" O.C., 3-5/8" wide	2 Carp	290	.055	S.F.	1.71	1.89		3.60	4.82	
	9400	6" wide	"	270	.059	"	2.03	2.03		4.06	5.40	
800	0010	**SHAFT WALL** Cavity type on 25 ga. J track & C-H studs, 24" O.C.										800
	0030	1" thick coreboard wall liner on shaft side										
	0040	2-hour assembly with double layer										
	0060	5/8" fire rated gypsum board on room side	2 Carp	220	.073	S.F.	.93	2.49		3.42	4.90	
	0100	3-hour assembly with triple layer										
	0300	5/8" fire rated gypsum board on room side	2 Carp	180	.089	S.F.	1.17	3.04		4.21	6.05	
	0400	4-hour assembly, 1" coreboard, 5/8" fire rated gypsum board										
	0600	and 3/4" galv. metal furring channels, 24" O.C., with										

FINISHES **9**

09260 | Gypsum Board Systems

			CREW	DAILY OUTPUT	LABOR-HOURS	UNIT	2005 BARE COSTS MAT.	LABOR	EQUIP.	TOTAL	TOTAL INCL O&P	
800	0700	Double layer 5/8" fire rated gypsum board on room side	2 Carp	110	.145	S.F.	1.07	4.98		6.05	8.95	800
	0900	For taping & finishing, add per side	1 Carp	1,050	.008	"	.03	.26		.29	.44	
	1000	For insulation, see div. 07210-000										

09270 | Drywall Accessories

			CREW	DAILY OUTPUT	LABOR-HOURS	UNIT	2005 BARE COSTS MAT.	LABOR	EQUIP.	TOTAL	TOTAL INCL O&P	
100	0010	**ACCESSORIES, DRYWALL** Casing bead, galvanized steel	1 Carp	2.90	2.759	C.L.F.	17.45	94.50		111.95	166	100
	0100	Vinyl		3	2.667		17.40	91.50		108.90	161	
	0300	Corner bead, galvanized steel, 1" x 1"		4	2		11.55	68.50		80.05	120	
	0400	Corner bead, galvanized steel, 1-1/4" x 1-1/4"		3.50	2.286		20	78.50		98.50	144	
	0600	Vinyl corner bead		4	2		20.50	68.50		89	130	
	0900	Furring channel, galv. steel, 7/8" deep, standard		2.60	3.077		23	105		128	190	
	1000	Resilient		2.55	3.137		22.50	107		129.50	192	
	1100	J trim, galvanized steel, 1/2" wide		3	2.667		16.85	91.50		108.35	161	
	1120	5/8" wide		2.95	2.712		16.15	93		109.15	163	
	1140	L trim, galvanized		3	2.667		14.85	91.50		106.35	158	
	1150	U trim, galvanized		2.95	2.712		16.30	93		109.30	163	
	1160	Screws #6 x 1" A				M	7.15			7.15	7.85	
	1170	#6 x 1-5/8" A				"	9.35			9.35	10.30	
	1200	For stud partitions, see Divisions 05410-400 and 09110-100										
	1500	Z stud, galvanized steel, 1-1/2" wide	1 Carp	2.60	3.077	C.L.F.	34.50	105		139.50	202	
	1600	2" wide	"	2.55	3.137	"	44	107		151	216	

09280 | Gypsum Wallboard Repairs

			CREW	DAILY OUTPUT	LABOR-HOURS	UNIT	2005 BARE COSTS MAT.	LABOR	EQUIP.	TOTAL	TOTAL INCL O&P	
100	0010	**GYPSUM WALLBOARD REPAIRS**										100
	0100	Fill and sand, pin / nail holes	1 Carp	960	.008	Ea.		.29		.29	.44	
	0110	Screw head pops		480	.017			.57		.57	.89	
	0120	Dents, up to 2" square		48	.167		.01	5.70		5.71	8.90	
	0130	2" to 4" square		24	.333		.02	11.40		11.42	17.80	
	0140	Cut square, patch, sand and finish, holes, up to 2" square		12	.667		.02	23		23.02	35.50	
	0150	2" to 4" square		11	.727		.06	25		25.06	39	
	0160	4" to 8" square		10	.800		.16	27.50		27.66	42.50	
	0170	8" to 12" square		8	1		.31	34.50		34.81	54	

09300 | Tile

09310 | Ceramic Tile

			CREW	DAILY OUTPUT	LABOR-HOURS	UNIT	2005 BARE COSTS MAT.	LABOR	EQUIP.	TOTAL	TOTAL INCL O&P	
100	0010	**CERAMIC TILE**										100
	0050	Base, using 1' x 4" high pc. with 1" x 1" tiles, mud set	D-7	82	.195	L.F.	4.07	5.65		9.72	12.80	
	0100	Thin set	"	128	.125		3.87	3.63		7.50	9.60	
	0300	For 6" high base, 1" x 1" tile face, add					.64			.64	.70	
	0400	For 2" x 2" tile face, add to above					.34			.34	.37	
	0600	Cove base, 4-1/4" x 4-1/4" high, mud set	D-7	91	.176		3.12	5.10		8.22	10.95	
	0700	Thin set		128	.125		3.14	3.63		6.77	8.80	
	0900	6" x 4-1/4" high, mud set		100	.160		2.85	4.64		7.49	9.95	
	1000	Thin set		137	.117		2.98	3.39		6.37	8.25	
	1200	Sanitary cove base, 6" x 4-1/4" high, mud set		93	.172		3.60	4.99		8.59	11.30	
	1300	Thin set		124	.129		3.67	3.75		7.42	9.55	
	1500	6" x 6" high, mud set		84	.190		3.38	5.55		8.93	11.80	

9 FINISHES

Important: See the Reference Section for critical supporting data - Reference Nos., Crews, & City Cost Indexes

			DAILY OUTPUT	LABOR-HOURS	UNIT	2005 BARE COSTS				TOTAL INCL O&P	
		**09310	Ceramic Tile**				MAT.	LABOR	EQUIP.	TOTAL		
			CREW									
100	1600	Thin set	D-7	117	.137	L.F.	3.42	3.97		7.39	9.60	**100**
	1800	Bathroom accessories, average		82	.195	Ea.	9.70	5.65		15.35	18.95	
	1900	Bathtub, 5', rec. 4-1/4" x 4-1/4" tile wainscot, adhesive set 6' high		2.90	5.517		140	160		300	390	
	2100	7' high wainscot		2.50	6.400		160	186		346	450	
	2200	8' high wainscot		2.20	7.273		170	211		381	495	
	2400	Bullnose trim, 4-1/4" x 4-1/4", mud set		82	.195	L.F.	2.69	5.65		8.34	11.25	
	2500	Thin set		128	.125		2.63	3.63		6.26	8.25	
	2700	6" x 4-1/4" bullnose trim, mud set		84	.190		2.05	5.55		7.60	10.35	
	2800	Thin set		124	.129		2.05	3.75		5.80	7.75	
	3000	Floors, natural clay, random or uniform, thin set, color group 1		183	.087	S.F.	3.69	2.54		6.23	7.80	
	3100	Color group 2		183	.087		3.98	2.54		6.52	8.10	
	3255	Glazed, thin set, 6" x 6", color group 1		200	.080		3.14	2.32		5.46	6.85	
	3260	8" x 8" tile		250	.064		3.14	1.86		5	6.20	
	3270	12" x 12" tile		325	.049		3.94	1.43		5.37	6.45	
	3280	16" x 16" tile		550	.029		4.95	.84		5.79	6.70	
	3285	Border, 6" x 12" tile		275	.058		10	1.69		11.69	13.50	
	3290	3" x 12" tile		200	.080		28.50	2.32		30.82	34.50	
	3300	Porcelain type, 1 color, color group 2, 1" x 1"		183	.087		4.27	2.54		6.81	8.45	
	3310	2" x 2" or 2" x 1", thin set		190	.084		4.07	2.44		6.51	8.05	
	3350	For random blend, 2 colors, add					.79			.79	.87	
	3360	4 colors, add					1.12			1.12	1.23	
	3370	For color group 3, add					.46			.46	.51	
	3380	For abrasive non-slip tile, add					.45			.45	.50	
	4300	Specialty tile, 4-1/4" x 4-1/4" x 1/2", decorator finish	D-7	183	.087		8.95	2.54		11.49	13.60	
	4500	Add for epoxy grout, 1/16" joint, 1" x 1" tile		800	.020		.55	.58		1.13	1.46	
	4600	2" x 2" tile		820	.020		.50	.57		1.07	1.38	
	4800	Pregrouted sheets, walls, 4-1/4" x 4-1/4", 6" x 4-1/4"										
	4810	and 8-1/2" x 4-1/4", 4 S.F. sheets, silicone grout	D-7	240	.067	S.F.	4.23	1.94		6.17	7.50	
	5100	Floors, unglazed, 2 S.F. sheets,										
	5110	urethane adhesive	D-7	180	.089	S.F.	4.21	2.58		6.79	8.40	
	5400	Walls, interior, thin set, 4-1/4" x 4-1/4" tile CN		190	.084		2.09	2.44		4.53	5.90	
	5500	6" x 4-1/4" tile		190	.084		2.35	2.44		4.79	6.20	
	5700	8-1/2" x 4-1/4" tile		190	.084		3.32	2.44		5.76	7.25	
	5800	6" x 6" tile		200	.080		2.67	2.32		4.99	6.35	
	5810	8" x 8" tile		225	.071		3.19	2.06		5.25	6.55	
	5820	12" x 12" tile		300	.053		3.04	1.55		4.59	5.60	
	5830	16" x 16" tile		500	.032		3.29	.93		4.22	4.98	
	6000	Decorated wall tile, 4-1/4" x 4-1/4", minimum		270	.059		3.50	1.72		5.22	6.40	
	6100	Maximum		180	.089		39	2.58		41.58	46.50	
	6300	Exterior walls, frostproof, mud set, 4-1/4" x 4-1/4"		102	.157		4.02	4.55		8.57	11.10	
	6400	1-3/8" x 1-3/8"		93	.172		3.86	4.99		8.85	11.60	
	6600	Crystalline glazed, 4-1/4" x 4-1/4", mud set, plain		100	.160		3.28	4.64		7.92	10.40	
	6700	4-1/4" x 4-1/4", scored tile		100	.160		4.13	4.64		8.77	11.35	
	6900	6" x 6" plain		93	.172		3.31	4.99		8.30	11	
	7000	For epoxy grout, 1/16" joints, 4-1/4" tile, add		800	.020		.33	.58		.91	1.21	
	7200	For tile set in dry mortar, add		1,735	.009			.27		.27	.39	
	7300	For tile set in portland cement mortar, add		290	.055			1.60		1.60	2.35	
	9000	Regrout tile 4-1/2 x 4-1/2, or larger, wall	1 Tilf	100	.080		.13	2.62		2.75	3.98	
	9220	Floor	"	125	.064		.14	2.09		2.23	3.23	
300	0010	**CERAMIC TILE PANELS** Insulated, over 1000 S.F., 1-1/2" thick	D-7	220	.073	S.F.	8.85	2.11		10.96	12.80	**300**
	0100	2-1/2" thick	"	220	.073	"	9.45	2.11		11.56	13.50	

		**09330	Quarry Tile**									
100	0010	**QUARRY TILE** Base, cove or sanitary, 2" or 5" high, mud set										**100**
	0100	1/2" thick	D-7	110	.145	L.F.	3.88	4.22		8.10	10.45	

9

FINISHES

09330 | Quarry Tile

			CREW	DAILY OUTPUT	LABOR-HOURS	UNIT	MAT.	LABOR	EQUIP.	TOTAL	TOTAL INCL O&P	
100	0300	Bullnose trim, red, mud set, 6" x 6" x 1/2" thick	D-7	120	.133	L.F.	3.81	3.87		7.68	9.90	100
	0400	4" x 4" x 1/2" thick		110	.145		4.30	4.22		8.52	10.95	
	0600	4" x 8" x 1/2" thick, using 8" as edge		130	.123		3.78	3.57		7.35	9.40	
	0700	Floors, mud set, 1,000 S.F. lots, red, 4" x 4" x 1/2" thick		120	.133	S.F.	3.89	3.87		7.76	10	
	0900	6" x 6" x 1/2" thick		140	.114		2.99	3.32		6.31	8.15	
	1000	4" x 8" x 1/2" thick		130	.123		3.89	3.57		7.46	9.55	
	1300	For waxed coating, add					.62			.62	.68	
	1500	For colors other than green, add					.37			.37	.41	
	1600	For abrasive surface, add					.44			.44	.48	
	1800	Brown tile, imported, 6" x 6" x 3/4"	D-7	120	.133		4.62	3.87		8.49	10.80	
	1900	8" x 8" x 1"		110	.145		5.25	4.22		9.47	11.95	
	2100	For thin set mortar application, deduct		700	.023			.66		.66	.98	
	2200	For epoxy grout & mortar, 6" x 6" x 1/2", add		350	.046		1.63	1.33		2.96	3.74	
	2700	Stair tread, 6" x 6" x 3/4", plain		50	.320		4.24	9.30		13.54	18.30	
	2800	Abrasive		47	.340		4.77	9.90		14.67	19.75	
	3000	Wainscot, 6" x 6" x 1/2", thin set, red		105	.152		3.56	4.42		7.98	10.40	
	3100	Colors other than green		105	.152		3.97	4.42		8.39	10.85	
	3300	Window sill, 6" wide, 3/4" thick		90	.178	L.F.	4.47	5.15		9.62	12.50	
	3400	Corners		80	.200	Ea.	5	5.80		10.80	14.05	

09350 | Glass Mosaics

			CREW	DAILY OUTPUT	LABOR-HOURS	UNIT	MAT.	LABOR	EQUIP.	TOTAL	TOTAL INCL O&P	
100	0010	**GLASS MOSAICS** 3/4" tile on 12" sheets, standard grout										100
	0300	Color group 1 & 2	D-7	73	.219	S.F.	15.25	6.35		21.60	26	
	0350	Color group 3		73	.219		16.05	6.35		22.40	27	
	0400	Color group 4		73	.219		22	6.35		28.35	33.50	
	0450	Color group 5		73	.219		24.50	6.35		30.85	36.50	
	0500	Color group 6		73	.219		30	6.35		36.35	42.50	
	0600	Color group 7		73	.219		34.50	6.35		40.85	47.50	
	0700	Color group 8, golds, silvers & specialties		64	.250		50	7.25		57.25	65.50	

09370 | Metal Tile

			CREW	DAILY OUTPUT	LABOR-HOURS	UNIT	MAT.	LABOR	EQUIP.	TOTAL	TOTAL INCL O&P	
100	0010	**METAL TILE** 4' x 4' sheet, 24 ga., tile pattern, nailed										100
	0200	Stainless steel	2 Carp	512	.031	S.F.	22	1.07		23.07	26	
	0400	Aluminized steel	"	512	.031	"	11.90	1.07		12.97	14.75	

09400 | Terrazzo

09410 | Portland Cement Terrazzo

			CREW	DAILY OUTPUT	LABOR-HOURS	UNIT	MAT.	LABOR	EQUIP.	TOTAL	TOTAL INCL O&P	
100	0010	**PORTLAND CEMENT TERRAZZO**, cast-in-place R09400-100										100
	0020	Cove base, 6" high, 16ga. zinc	1 Mstz	20	.400	L.F.	2.92	13.05		15.97	22.50	
	0100	Curb, 6" high and 6" wide		6	1.333		4.66	43.50		48.16	69	
	0300	Divider strip for floors, 14 ga., 1-1/4" deep, zinc		375	.021		1.06	.70		1.76	2.19	
	0400	Brass		375	.021		1.89	.70		2.59	3.10	
	0600	Heavy top strip 1/4" thick, 1-1/4" deep, zinc		300	.027		1.49	.87		2.36	2.92	
	0900	Galv. bottoms, brass		300	.027		2.56	.87		3.43	4.10	
	1200	For thin set floors, 16 ga., 1/2" x 1/2", zinc		350	.023		.64	.75		1.39	1.80	
	1300	Brass		350	.023		1.28	.75		2.03	2.51	
	1500	Floor, bonded to concrete, 1-3/4" thick, gray cement	J-3	130	.123	S.F.	2.49	3.62	1.61	7.72	9.80	

Important: See the Reference Section for critical supporting data - Reference Nos., Crews, & City Cost Indexes

9 FINISHES

09410 | Portland Cement Terrazzo

		CREW	DAILY OUTPUT	LABOR-HOURS	UNIT	MAT.	LABOR	EQUIP.	TOTAL	TOTAL INCL O&P	
100	1600	White cement, mud set	J-3	130	.123	S.F.	2.83	3.62	1.61	8.06	10.20	100
	1800	Not bonded, 3" total thickness, gray cement		115	.139		3.11	4.09	1.82	9.02	11.40	
	1900	White cement, mud set	↓	115	.139	↓	3.39	4.09	1.82	9.30	11.75	
	2100	For Venetian terrazzo, 1" topping, add					50%	50%				
	2200	For heavy duty abrasive terrazzo, add					50%	50%				
	2700	Monolithic terrazzo, 1/2" thick										
	2710	10' panels	J-3	125	.128	S.F.	2.26	3.76	1.68	7.70	9.90	
	3000	Stairs, cast in place, pan filled treads		30	.533	L.F.	2.27	15.65	7	24.92	33	
	3100	Treads and risers	↓	14	1.143	"	4.71	33.50	14.95	53.16	71	
	3300	For stair landings, add to floor prices						50%				
	3400	Stair stringers and fascia	J-3	30	.533	S.F.	3.78	15.65	7	26.43	35	
	3600	For abrasive metal nosings on stairs, add		150	.107	L.F.	6.50	3.13	1.40	11.03	13.30	
	3700	For abrasive surface finish, add		600	.027	S.F.	.90	.78	.35	2.03	2.52	
	3900	For raised abrasive strips, add		150	.107	L.F.	.86	3.13	1.40	5.39	7.10	
	4000	Wainscot, bonded, 1-1/2" thick		30	.533	S.F.	2.89	15.65	7	25.54	34	
	4200	Epoxy terrazzo, 1/4" thick	↓	40	.400	"	4.23	11.75	5.25	21.23	27.50	
	4300	Stone chips, onyx gemstone, per 50 lb. bag				Bag	12			12	13.20	

09420 | Precast Terrazzo

		CREW	DAILY OUTPUT	LABOR-HOURS	UNIT	MAT.	LABOR	EQUIP.	TOTAL	TOTAL INCL O&P	
900	0010	TERRAZZO, PRECAST Base, 6" high, straight	1 Mstz	35	.229	L.F.	9.25	7.45		16.70	21	900
	0100	Cove		30	.267		10.60	8.70		19.30	24.50	
	0300	8" high base, straight		30	.267		9.45	8.70		18.15	23	
	0400	Cove	↓	25	.320		13.90	10.45		24.35	30.50	
	0600	For white cement, add					.38			.38	.42	
	0700	For 16 ga. zinc toe strip, add					1.39			1.39	1.53	
	0900	Curbs, 4" x 4" high	1 Mstz	19	.421		26.50	13.75		40.25	49.50	
	1000	8" x 8" high	"	15	.533	↓	31	17.40		48.40	59.50	
	1200	Floor tiles, non-slip, 1" thick, 12" x 12"	D-1	29	.552	S.F.	15.85	17.15		33	43.50	
	1300	1-1/4" thick, 12" x 12"		29	.552		17.60	17.15		34.75	45.50	
	1500	16" x 16"		23	.696		19.10	21.50		40.60	54	
	1600	1-1/2" thick, 16" x 16"	↓	21	.762		17.50	23.50		41	55.50	
	1800	For Venetian terrazzo, add					5.25			5.25	5.75	
	1900	For white cement, add				↓	.49			.49	.54	
	2400	Stair treads, 1-1/2" thick, non-slip, three line pattern	2 Mstz	70	.229	L.F.	35.50	7.45		42.95	50	
	2500	Nosing and two lines		70	.229		35.50	7.45		42.95	50	
	2700	2" thick treads, straight		60	.267		37.50	8.70		46.20	54	
	2800	Curved		50	.320		49	10.45		59.45	69	
	3000	Stair risers, 1" thick, to 6" high, straight sections		60	.267		8.50	8.70		17.20	22	
	3100	Cove		50	.320		12.65	10.45		23.10	29.50	
	3300	Curved, 1" thick, to 6" high, vertical		48	.333		17.25	10.90		28.15	35	
	3400	Cove		38	.421		33	13.75		46.75	56.50	
	3600	Stair tread and riser, single piece, straight, minimum		60	.267		45	8.70		53.70	62.50	
	3700	Maximum		40	.400		58	13.05		71.05	83	
	3900	Curved tread and riser, minimum		40	.400		63	13.05		76.05	88	
	4000	Maximum		32	.500		79.50	16.35		95.85	111	
	4200	Stair stringers, notched, 1" thick		25	.640		25.50	21		46.50	59	
	4300	2" thick		22	.727	↓	30.50	24		54.50	68.50	
	4500	Stair landings, structural, non-slip, 1-1/2" thick		85	.188	S.F.	28	6.15		34.15	40	
	4600	3" thick	↓	75	.213		39.50	6.95		46.45	54	
	4800	Wainscot, 12" x 12" x 1" tiles	1 Mstz	12	.667		5.65	22		27.65	38.50	
	4900	16" x 16" x 1-1/2" tiles	"	8	1	↓	12.30	32.50		44.80	61.50	

09430 | Conductive Terrazzo

		CREW	DAILY OUTPUT	LABOR-HOURS	UNIT	MAT.	LABOR	EQUIP.	TOTAL	TOTAL INCL O&P	
100	0010	CONDUCTIVE TERRAZZO										100
	2400	Bonded conductive floor for hospitals	J-3	90	.178	S.F.	3.39	5.20	2.33	10.92	14	

FINISHES 9

09400 | Terrazzo

09430 | Conductive Terrazzo

			CREW	DAILY OUTPUT	LABOR-HOURS	UNIT	2005 BARE COSTS MAT.	LABOR	EQUIP.	TOTAL	TOTAL INCL O&P	
100	2500	Epoxy terrazzo, 1/4" thick, minimum	J-3	100	.160	S.F.	3.84	4.70	2.10	10.64	13.45	100
	2550	Average		75	.213		4.24	6.25	2.79	13.28	16.95	
	2600	Maximum	↓	60	.267	↓	4.64	7.85	3.49	15.98	20.50	

09450 | Cast-in-Place Terrazzo

			CREW	DAILY OUTPUT	LABOR-HOURS	UNIT	MAT.	LABOR	EQUIP.	TOTAL	TOTAL INCL O&P	
200	0010	**TILE OR TERRAZZO BASE** Scratch coat only	1 Mstz	150	.053	S.F.	.37	1.74		2.11	2.97	200
	0500	Scratch and brown coat only	"	75	.107	"	.67	3.48		4.15	5.85	

09500 | Ceilings

09510 | Acoustical Ceilings

			CREW	DAILY OUTPUT	LABOR-HOURS	UNIT	2005 BARE COSTS MAT.	LABOR	EQUIP.	TOTAL	TOTAL INCL O&P	
700	0010	**SUSPENDED ACOUSTIC CEILING TILES**, Not including										700
	0100	suspension system										
	0300	Fiberglass boards, film faced, 2' x 2' or 2' x 4', 5/8" thick	1 Carp	625	.013	S.F.	.51	.44		.95	1.24	
	0400	3/4" thick		600	.013		1.14	.46		1.60	1.96	
	0500	3" thick, thermal, R11		450	.018		1.26	.61		1.87	2.34	
	0600	Glass cloth faced fiberglass, 3/4" thick		500	.016		1.62	.55		2.17	2.63	
	0700	1" thick		485	.016		1.79	.56		2.35	2.85	
	0820	1-1/2" thick, nubby face		475	.017		2.23	.58		2.81	3.35	
	1110	Mineral fiber tile, lay-in, 2' x 2' or 2' x 4', 5/8" thick, fine texture		625	.013		.43	.44		.87	1.15	
	1115	Rough textured		625	.013		1.06	.44		1.50	1.85	
	1125	3/4" thick, fine textured		600	.013		1.17	.46		1.63	2	
	1130	Rough textured		600	.013		1.47	.46		1.93	2.33	
	1135	Fissured		600	.013		1.74	.46		2.20	2.62	
	1150	Tegular, 5/8" thick, fine textured		470	.017		1.04	.58		1.62	2.05	
	1155	Rough textured		470	.017		1.35	.58		1.93	2.40	
	1165	3/4" thick, fine textured		450	.018		1.47	.61		2.08	2.57	
	1170	Rough textured		450	.018		1.66	.61		2.27	2.78	
	1175	Fissured	↓	450	.018	↓	2.59	.61		3.20	3.80	
	1180	For aluminum face, add					4.72			4.72	5.20	
	1185	For plastic film face, add					.75			.75	.83	
	1190	For fire rating, add					.35			.35	.39	
	1300	Mirror faced panels, 15/16" thick, 2' x 2'	1 Carp	500	.016		9.80	.55		10.35	11.65	
	1900	Eggcrate, acrylic, 1/2" x 1/2" x 1/2" cubes		500	.016		1.42	.55		1.97	2.41	
	2100	Polystyrene eggcrate, 3/8" x 3/8" x 1/2" cubes		510	.016		1.19	.54		1.73	2.15	
	2200	1/2" x 1/2" x 1/2" cubes		500	.016		1.60	.55		2.15	2.61	
	2400	Luminous panels, prismatic, acrylic		400	.020		1.73	.69		2.42	2.97	
	2500	Polystyrene		400	.020		.88	.69		1.57	2.04	
	2700	Flat white acrylic		400	.020		3.01	.69		3.70	4.38	
	2800	Polystyrene		400	.020		2.06	.69		2.75	3.34	
	3000	Drop pan, white, acrylic		400	.020		4.41	.69		5.10	5.90	
	3100	Polystyrene		400	.020		3.69	.69		4.38	5.15	
	3600	Perforated aluminum sheets, .024" thick, corrugated, painted		490	.016		1.76	.56		2.32	2.81	
	3700	Plain		500	.016		3.07	.55		3.62	4.23	
	3720	Mineral fiber, 24" x 24" or 48", reveal edge, painted, 5/8" thick		600	.013		.98	.46		1.44	1.79	
	3740	3/4" thick	↓	575	.014	↓	1.60	.48		2.08	2.50	
760	0010	**SUSPENDED CEILINGS, COMPLETE** Including standard										760
	0100	suspension system but not incl. 1-1/2" carrier channels										

Important: See the Reference Section for critical supporting data - Reference Nos., Crews, & City Cost Indexes

09510	Acoustical Ceilings	CREW	DAILY OUTPUT	LABOR-HOURS	UNIT	2005 BARE COSTS				TOTAL INCL O&P
						MAT.	LABOR	EQUIP.	TOTAL	
760 0600	Fiberglass ceiling board, 2' x 4' x 5/8", plain faced,	1 Carp	500	.016	S.F.	1	.55		1.55	1.95	**760**
0700	Offices, 2' x 4' x 3/4"		380	.021		1.63	.72		2.35	2.91
0800	Mineral fiber, on 15/16" T bar susp. 2' x 2' x 3/4" lay-in board		345	.023		1.78	.79		2.57	3.20
0810	2' x 4' x 5/8" tile		380	.021		.92	.72		1.64	2.13
0820	Tegular, 2' x 2' x 5/8" tile on 9/16" grid		250	.032		1.79	1.10		2.89	3.68
0830	2' x 4' x 3/4" tile		275	.029		2.10	1		3.10	3.86
0900	Luminous panels, prismatic, acrylic		255	.031		2.22	1.07		3.29	4.11
1200	Metal pan with acoustic pad, steel		75	.107		3.30	3.65		6.95	9.35
1300	Painted aluminum		75	.107		2.25	3.65		5.90	8.15
1500	Aluminum, degreased finish		75	.107		3.82	3.65		7.47	9.90
1600	Stainless steel		75	.107		7.15	3.65		10.80	13.60
1800	Tile, Z bar suspension, 5/8" mineral fiber tile		150	.053		1.55	1.83		3.38	4.56
1900	3/4" mineral fiber tile		150	.053		1.65	1.83		3.48	4.67
2400	For strip lighting, see division 16510-440									
2500	For rooms under 500 S.F., add				S.F.		25%			
900 0010	**CEILING TILE**, Stapled or cemented										**900**
0100	12" x 12" or 12" x 24", not including furring									
0600	Mineral fiber, vinyl coated, 5/8" thick **CN**	1 Carp	1,000	.008	S.F.	1.23	.27		1.50	1.78
0700	3/4" thick		1,000	.008		1.35	.27		1.62	1.92
0900	Fire rated, 3/4" thick, plain faced		1,000	.008		1.17	.27		1.44	1.72
1000	Plastic coated face		1,000	.008		1.28	.27		1.55	1.84
1200	Aluminum faced, 5/8" thick, plain		1,000	.008		1.12	.27		1.39	1.66
3700	Wall application of above, add		3,100	.003			.09		.09	.14
3900	For ceiling primer, add					.12			.12	.13
4000	For ceiling cement, add					.33			.33	.36

FINISHES 9

09600 | Flooring

09620	Specialty Flooring	CREW	DAILY OUTPUT	LABOR-HOURS	UNIT	2005 BARE COSTS				TOTAL INCL O&P
						MAT.	LABOR	EQUIP.	TOTAL	
100 0010	**ATHLETIC FLOORING**										**100**
3700	Polyethylene, in rolls, no base incl., landscape surfaces	1 Tilf	275	.029	S.F.	2.57	.95		3.52	4.23
3800	Nylon action surface, 1/8" thick		275	.029		2.76	.95		3.71	4.44
3900	1/4" thick		275	.029		3.98	.95		4.93	5.80
4000	3/8" thick		275	.029		5	.95		5.95	6.90
4100	Golf tee surface with foam back		235	.034		4.95	1.11		6.06	7.10
4200	Practice putting, knitted nylon surface		235	.034		4.20	1.11		5.31	6.25
4400	Polyurethane, thermoset, prefabricated in place, indoor									
4500	3/8" thick for basketball, gyms, etc.	1 Tilf	100	.080	S.F.	3.95	2.62		6.57	8.20
4600	1/2" thick for professional sports		95	.084		4.62	2.75		7.37	9.15
4700	Outdoor, 1/4" thick, smooth, for tennis		100	.080		3.54	2.62		6.16	7.75
4800	Rough, for track, 3/8" thick		95	.084		4.11	2.75		6.86	8.55
5000	Poured in place, indoor, with finish, 1/4" thick		80	.100		2.92	3.27		6.19	8
5050	3/8" thick		65	.123		3.54	4.02		7.56	9.80
5100	1/2" thick		50	.160		4.69	5.25		9.94	12.85
5500	Polyvinyl chloride, sheet goods for gyms, 1/4" thick		80	.100		4.04	3.27		7.31	9.25
5600	3/8" thick		60	.133		4.56	4.36		8.92	11.40

09631 | Brick Flooring

		CREW	DAILY OUTPUT	LABOR-HOURS	UNIT	2005 BARE COSTS				TOTAL INCL O&P
						MAT.	LABOR	EQUIP.	TOTAL	
100	0010	**BRICK FLOORING**								
	0020	Acid proof shales, red, 8" x 3-3/4" x 1-1/4" thick	D-7	.43	37.209	M	730	1,075		1,805	2,375
	0050	2-1/4" thick	D-1	.40	40		795	1,250		2,045	2,775
	0200	Acid proof clay brick, 8" x 3-3/4" x 2-1/4" thick	"	.40	40	↓	755	1,250		2,005	2,725
	0260	Cast ceramic, pressed, 4" x 8" x 1/2", unglazed	D-7	100	.160	S.F.	4.94	4.64		9.58	12.25
	0270	Glazed		100	.160		6.60	4.64		11.24	14.05
	0280	Hand molded flooring, 4" x 8" x 3/4", unglazed		95	.168		6.55	4.89		11.44	14.40
	0290	Glazed		95	.168		8.20	4.89		13.09	16.20
	0300	8" hexagonal, 3/4" thick, unglazed		85	.188		7.15	5.45		12.60	15.95
	0310	Glazed	↓	85	.188		12.95	5.45		18.40	22.50
	0400	Heavy duty industrial, cement mortar bed, 2" thick, not incl. brick	D-1	80	.200		.67	6.20		6.87	10.20
	0450	Acid proof joints, 1/4" wide	"	65	.246		1.13	7.65		8.78	12.90
	0500	Pavers, 8" x 4", 1" to 1-1/4" thick, red	D-7	95	.168		2.88	4.89		7.77	10.35
	0510	Ironspot	"	95	.168		4.07	4.89		8.96	11.70
	0540	1-3/8" to 1-3/4" thick, red	D-1	95	.168		2.78	5.25		8.03	11
	0560	Ironspot		95	.168		4.02	5.25		9.27	12.35
	0580	2-1/4" thick, red		90	.178		2.83	5.55		8.38	11.50
	0590	Ironspot	↓	90	.178	↓	4.38	5.55		9.93	13.20
	0800	For sidewalks and patios with pavers, see division 02780-200									
	0870	For epoxy joints, add	D-1	600	.027	S.F.	2.15	.83		2.98	3.63
	0880	For Furan underlayment, add	"	600	.027		1.78	.83		2.61	3.22
	0890	For waxed surface, steam cleaned, add	D-5	1,000	.008	↓	.15	.28		.43	.60

09635 | Marble Flooring

		CREW	DAILY OUTPUT	LABOR-HOURS	UNIT	MAT.	LABOR	EQUIP.	TOTAL	TOTAL INCL O&P
100	0010	**MARBLE** Thin gauge tile, 12" x 6", 3/8", White Carara	D-7	60	.267	S.F.	8.75	7.75		16.50	21
	0100	Travertine		60	.267		9.65	7.75		17.40	22
	0200	12" x 12" x 3/8", thin set, floors		60	.267		6.70	7.75		14.45	18.75
	0300	On walls	↓	52	.308	↓	8.90	8.95		17.85	23

09637 | Stone Flooring

		CREW	DAILY OUTPUT	LABOR-HOURS	UNIT	MAT.	LABOR	EQUIP.	TOTAL	TOTAL INCL O&P
100	0010	**SLATE TILE** Vermont, 6" x 6" x 1/4" thick, thin set	D-7	180	.089	S.F.	4.14	2.58		6.72	8.35
	0200	See also division 02780-600									
200	0010	**SLATE & STONE FLOORS** See division 02780-600									
300	0010	**CONCRETE FLOORS** And toppings, see division 03350-300									

09642 | Wood Athletic Flooring

		CREW	DAILY OUTPUT	LABOR-HOURS	UNIT	MAT.	LABOR	EQUIP.	TOTAL	TOTAL INCL O&P
100	0010	**WOOD ATHLETIC FLOORING**									
	0600	Gym floor, in mastic, over 2 ply felt, #2 & better									
	0700	25/32" thick maple	1 Carp	100	.080	S.F.	3.45	2.74		6.19	8.05
	0900	33/32" thick maple	↓	98	.082		3.98	2.80		6.78	8.75
	1000	For 1/2" corkboard underlayment, add		750	.011		.77	.37		1.14	1.42
	1300	For #1 grade maple, add				↓	.42			.42	.46
	1600	Maple flooring, over sleepers, #2 & better									
	1700	25/32" thick	1 Carp	85	.094	S.F.	3.73	3.22		6.95	9.10
	1900	33/32" thick	"	83	.096		4.34	3.30		7.64	9.90
	2000	For #1 grade, add					.44			.44	.48
	2200	For 3/4" subfloor, add	1 Carp	350	.023		.97	.78		1.75	2.29
	2300	With two 1/2" subfloors, 25/32" thick	"	69	.116	↓	4.67	3.97		8.64	11.35
	2500	Maple, incl. finish, #2 & btr., 25/32" thick, on rubber									
	2600	Sleepers, with two 1/2" subfloors	1 Carp	76	.105	S.F.	5	3.61		8.61	11.10
	2800	With steel spline, double connection to channels	"	73	.110		5.35	3.75		9.10	11.70
	2900	For 33/32" maple, add					.58			.58	.64
	3100	For #1 grade maple, add					.44			.44	.48
	3500	For termite proofing all of the above, add				↓	.22			.22	.24

Important: See the Reference Section for critical supporting data - Reference Nos., Crews, & City Cost Indexes

9 FINISHES

09600 | Flooring

09642 | Wood Athletic Flooring

		CREW	DAILY OUTPUT	LABOR-HOURS	UNIT	2005 BARE COSTS				TOTAL INCL O&P
						MAT.	LABOR	EQUIP.	TOTAL	
100	3700	Portable hardwood, prefinished panels	1 Carp	83	.096	S.F.	6.70	3.30		10	12.50	100
	3720	Insulated with polystyrene, 1" thick, add		165	.048		.58	1.66		2.24	3.23
	3750	Running tracks, Sitka spruce surface, 25/32" x 2-1/4"		62	.129		12.25	4.42		16.67	20.50
	3770	3/4" plywood surface, finished		100	.080		2.90	2.74		5.64	7.45
	3800	See also resilient gym floors, division 09620-100									

09643 | Wood Block Flooring

		CREW	DAILY OUTPUT	LABOR-HOURS	UNIT	MAT.	LABOR	EQUIP.	TOTAL	TOTAL INCL O&P
100	0010	**WOOD BLOCK FLOORING** End grain flooring, coated, 2" thick	1 Carp	295	.027	S.F.	2.98	.93		3.91	4.73	100
	0400	Natural finish, 1" thick, fir		125	.064		3.08	2.19		5.27	6.80
	0600	1-1/2" thick, pine		125	.064		3.03	2.19		5.22	6.75
	0700	2" thick, pine		125	.064		3.55	2.19		5.74	7.30

09644 | Wood Comp. Flooring

		CREW	DAILY OUTPUT	LABOR-HOURS	UNIT	MAT.	LABOR	EQUIP.	TOTAL	TOTAL INCL O&P
100	0010	**WOOD COMPOSITION** Gym floors										100
	0100	2-1/4" x 6-7/8" x 3/8", on 2" grout setting bed	D-7	150	.107	S.F.	5.05	3.10		8.15	10.15
	0200	Thin set, on concrete	"	250	.064		4.63	1.86		6.49	7.85
	0300	Sanding and finishing, add	1 Carp	200	.040		.68	1.37		2.05	2.88

09647 | Wood Parquet Flooring

		CREW	DAILY OUTPUT	LABOR-HOURS	UNIT	MAT.	LABOR	EQUIP.	TOTAL	TOTAL INCL O&P
100	0010	**WOOD PARQUET** flooring										100
	5200	Parquetry, standard, 5/16" thick, not incl. finish, oak, minimum	1 Carp	160	.050	S.F.	3.21	1.71		4.92	6.20
	5300	Maximum		100	.080		5.30	2.74		8.04	10.10
	5500	Teak, minimum		160	.050		4.40	1.71		6.11	7.50
	5600	Maximum		100	.080		7.70	2.74		10.44	12.70
	5650	13/16" thick, select grade oak, minimum		160	.050		8.50	1.71		10.21	12
	5700	Maximum		100	.080		12.90	2.74		15.64	18.40
	5800	Custom parquetry, including finish, minimum		100	.080		14.20	2.74		16.94	19.85
	5900	Maximum		50	.160		18.80	5.50		24.30	29
	6700	Parquetry, prefinished white oak, 5/16" thick, minimum		160	.050		3.50	1.71		5.21	6.50
	6800	Maximum		100	.080		7.15	2.74		9.89	12.15
	7000	Walnut or teak, parquetry, minimum		160	.050		4.77	1.71		6.48	7.90
	7100	Maximum		100	.080		8.30	2.74		11.04	13.40
	7200	Acrylic wood parquet blocks, 12" x 12" x 5/16",									
	7210	irradiated, set in epoxy	1 Carp	160	.050	S.F.	6.95	1.71		8.66	10.30

09648 | Wood Strip Flooring

		CREW	DAILY OUTPUT	LABOR-HOURS	UNIT	MAT.	LABOR	EQUIP.	TOTAL	TOTAL INCL O&P
100	0010	**WOOD** Fir, vertical grain, 1" x 4", not incl. finish, B & better	1 Carp	255	.031	S.F.	2.44	1.07		3.51	4.35	100
	0100	C grade & better		255	.031		2.29	1.07		3.36	4.19
	4000	Maple, strip, 25/32" x 2-1/4", not incl. finish, select		170	.047		4.65	1.61		6.26	7.60
	4100	#2 & better		170	.047		2.95	1.61		4.56	5.75
	4300	33/32" x 3-1/4", not incl. finish, #1 grade		170	.047		3.62	1.61		5.23	6.50
	4400	#2 & better		170	.047		3.22	1.61		4.83	6.05
	4600	Oak, white or red, 25/32" x 2-1/4", not incl. finish									
	4700	#1 common **CN**	1 Carp	170	.047	S.F.	3	1.61		4.61	5.80
	4900	Select quartered, 2-1/4" wide		170	.047		2.56	1.61		4.17	5.35
	5000	Clear		170	.047		3.72	1.61		5.33	6.60
	6100	Prefinished, white oak, prime grade, 2-1/4" wide		170	.047		6.20	1.61		7.81	9.30
	6200	3-1/4" wide		185	.043		7.95	1.48		9.43	11.05
	6400	Ranch plank		145	.055		7.70	1.89		9.59	11.40
	6500	Hardwood blocks, 9" x 9", 25/32" thick		160	.050		5.15	1.71		6.86	8.30
	7400	Yellow pine, 3/4" x 3-1/8", T & G, C & better, not incl. finish		200	.040		2.28	1.37		3.65	4.64
	7500	Refinish wood floor, sand, 2 cts poly, wax, soft wood, min.	1 Clab	400	.020		.69	.53		1.22	1.59
	7600	Hard wood, max		130	.062		1.04	1.64		2.68	3.70
	7800	Sanding and finishing, 2 coats polyurethane		295	.027		.69	.72		1.41	1.89

FINISHES 9

09648 | Wood Strip Flooring

		CREW	DAILY OUTPUT	LABOR-HOURS	UNIT	MAT.	LABOR	EQUIP.	TOTAL	TOTAL INCL O&P	
100	7900	Subfloor and underlayment, see division 06160										100
	8015	Transition molding, 2 1/4" wide, 5' long	1 Carp	19.20	.417	Ea.	12.90	14.25		27.15	36	
	8300	Floating floor, wood composition strip, complete.	1 Clab	133	.060	S.F.	3.38	1.61		4.99	6.20	
	8310	Floating floor components, T & G wood composite strips					3.05			3.05	3.35	
	8320	Film					.14			.14	.15	
	8330	Foam					.17			.17	.18	
	8340	Adhesive					.07			.07	.08	
	8350	Installation kit				↓	.16			.16	.18	
	8360	Trim, 2" wide x 3' long				L.F.	2.05			2.05	2.26	
	8370	Reducer moulding				"	4.15			4.15	4.57	

09651 | Resilient Base & Access.

		CREW	DAILY OUTPUT	LABOR-HOURS	UNIT	MAT.	LABOR	EQUIP.	TOTAL	TOTAL INCL O&P	
100	0010	**STAIR TREADS AND RISERS** See index for materials other										100
	0100	than rubber and vinyl										
	0300	Rubber, molded tread, 12" wide, 5/16" thick, black	1 Tilf	115	.070	L.F.	9.25	2.27		11.52	13.55	
	0400	Colors		115	.070		8.65	2.27		10.92	12.85	
	0600	1/4" thick, black		115	.070		8.35	2.27		10.62	12.55	
	0700	Colors		115	.070		8.65	2.27		10.92	12.85	
	0900	Grip strip safety tread, colors, 5/16" thick		115	.070		11.50	2.27		13.77	16	
	1000	3/16" thick		120	.067	↓	8.10	2.18		10.28	12.10	
	1200	Landings, smooth sheet rubber, 1/8" thick		120	.067	S.F.	3.90	2.18		6.08	7.50	
	1300	3/16" thick		120	.067	"	5	2.18		7.18	8.70	
	1500	Nosings, 3" wide, 3/16" thick, black		140	.057	L.F.	2.80	1.87		4.67	5.85	
	1600	Colors		140	.057		2.73	1.87		4.60	5.75	
	1800	Risers, 7" high, 1/8" thick, flat		250	.032		3.30	1.05		4.35	5.15	
	1900	Coved		250	.032		2.97	1.05		4.02	4.81	
	2100	Vinyl, molded tread, 12" wide, colors, 1/8" thick		115	.070		3.29	2.27		5.56	6.95	
	2200	1/4" thick		115	.070	↓	5.30	2.27		7.57	9.15	
	2300	Landing material, 1/8" thick		200	.040	S.F.	3.62	1.31		4.93	5.90	
	2400	Riser, 7" high, 1/8" thick, coved		175	.046	L.F.	2.12	1.49		3.61	4.53	
	2500	Tread and riser combined, 1/8" thick	↓	80	.100	"	6.10	3.27		9.37	11.55	
200	0010	**RESILIENT BASE**										200
	0800	Base, cove, rubber or vinyl, .080" thick										
	1100	Standard colors, 2-1/2" high	1 Tilf	315	.025	L.F.	.46	.83		1.29	1.73	
	1150	4" high		315	.025		.51	.83		1.34	1.78	
	1200	6" high		315	.025		.84	.83		1.67	2.14	
	1450	1/8" thick, standard colors, 2-1/2" high		315	.025		.54	.83		1.37	1.81	
	1500	4" high **CN**		315	.025	↓	.70	.83		1.53	1.99	
	1600	Corners, 2-1/2" high		315	.025	Ea.	1.17	.83		2	2.51	
	1630	4" high		315	.025		1.22	.83		2.05	2.56	
	1660	6" high	↓	315	.025	↓	1.59	.83		2.42	2.97	

09653 | Resilient Sheet Flooring

		CREW	DAILY OUTPUT	LABOR-HOURS	UNIT	MAT.	LABOR	EQUIP.	TOTAL	TOTAL INCL O&P	
100	0010	**RESILIENT SHEET FLOORING**										100
	5900	Rubber, sheet goods, 36" wide, 1/8" thick	1 Tilf	120	.067	S.F.	3.54	2.18		5.72	7.10	
	5950	3/16" thick		100	.080		5.05	2.62		7.67	9.40	
	6000	1/4" thick		90	.089		5.80	2.91		8.71	10.65	
	8000	Vinyl sheet goods, backed, .065" thick, minimum		250	.032		2.07	1.05		3.12	3.82	
	8050	Maximum		200	.040		2.64	1.31		3.95	4.82	
	8100	.080" thick, minimum		230	.035		2.23	1.14		3.37	4.12	
	8150	Maximum		200	.040		3.18	1.31		4.49	5.40	
	8200	.125" thick, minimum		230	.035		2.51	1.14		3.65	4.43	
	8250	Maximum	↓	200	.040	↓	3.98	1.31		5.29	6.30	
	8700	Adhesive cement, 1 gallon does 200 to 300 S.F.				Gal.	16.35			16.35	17.95	
	8800	Asphalt primer, 1 gallon per 300 S.F.					9.65			9.65	10.60	

Important: See the Reference Section for critical supporting data - Reference Nos., Crews, & City Cost Indexes

9

FINISHES

			DAILY	LABOR-		2005 BARE COSTS				TOTAL	
	09653	**Resilient Sheet Flooring**	CREW	OUTPUT	HOURS	UNIT	MAT.	LABOR	EQUIP.	TOTAL	INCL O&P
100	8900	Emulsion, 1 gallon per 140 S.F.				Gal.	12.25			12.25	13.50	100
	8950	Latex underlayment, liquid, fortified				↓	33			33	36.50

	09658	**Resilient Tile Flooring**									
100	0010	**RESILIENT TILE FLOORING**										100
	2200	Cork tile, standard finish, 1/8" thick	1 Tilf	315	.025	S.F.	3.75	.83		4.58	5.35
	2250	3/16" thick		315	.025		3.74	.83		4.57	5.35
	2300	5/16" thick		315	.025		5	.83		5.83	6.70
	2350	1/2" thick		315	.025		5.75	.83		6.58	7.55
	2500	Urethane finish, 1/8" thick		315	.025		4.72	.83		5.55	6.40
	2550	3/16" thick		315	.025		5.10	.83		5.93	6.85
	2600	5/16" thick		315	.025		6.35	.83		7.18	8.20
	2650	1/2" thick		315	.025		8.95	.83		9.78	11.05
	6050	Tile, marbleized colors, 12" x 12", 1/8" thick		400	.020		4.05	.65		4.70	5.40
	6100	3/16" thick		400	.020		5.45	.65		6.10	6.95
	6300	Special tile, plain colors, 1/8" thick		400	.020		4.39	.65		5.04	5.80
	6350	3/16" thick		400	.020		5.90	.65		6.55	7.45
	6410	Raised, radial or square, minimum		400	.020		6.90	.65		7.55	8.50
	6430	Maximum		400	.020		6.85	.65		7.50	8.50
	6450	For golf course, skating rink, etc., 1/4" thick		275	.029		6.75	.95		7.70	8.85
	6700	Synthetic turf, 3/8" thick	↓	90	.089	↓	3.62	2.91		6.53	8.25
	6750	Interlocking 2' x 2' squares, 1/2" thick, not									
	6810	cemented, for playgrounds, minimum	1 Tilf	210	.038	S.F.	2.99	1.25		4.24	5.10
	6850	Maximum		190	.042		7.70	1.38		9.08	10.45
	7000	Vinyl composition tile, 12" x 12", 1/16" thick		500	.016		.79	.52		1.31	1.64
	7050	Embossed		500	.016		1	.52		1.52	1.87
	7100	Marbleized		500	.016		1	.52		1.52	1.87
	7150	Solid		500	.016		1.11	.52		1.63	1.99
	7200	3/32" thick, embossed		500	.016		1.01	.52		1.53	1.88
	7250	Marbleized		500	.016		1.12	.52		1.64	2
	7300	Solid		500	.016		1.64	.52		2.16	2.57
	7350	1/8" thick, marbleized CN		500	.016		1.07	.52		1.59	1.95
	7400	Solid		500	.016		2.01	.52		2.53	2.98
	7450	Conductive		500	.016		3.76	.52		4.28	4.91
	7500	Vinyl tile, 12" x 12", .050" thick, minimum		500	.016		1.76	.52		2.28	2.71
	7550	Maximum		500	.016		3.45	.52		3.97	4.57
	7600	1/8" thick, minimum		500	.016		2.23	.52		2.75	3.22
	7650	Solid colors		500	.016		4.80	.52		5.32	6.05
	7700	Marbleized or Travertine pattern		500	.016		3.56	.52		4.08	4.69
	7750	Florentine pattern		500	.016		4.10	.52		4.62	5.30
	7800	Maximum	↓	500	.016	↓	8.40	.52		8.92	10

	09662	**Static Control Flooring**									
100	0010	**CONDUCTIVE RESILIENT FLOORING**										100
	1700	Conductive flooring, rubber tile, 1/8" thick	1 Tilf	315	.025	S.F.	2.88	.83		3.71	4.39
	1800	Homogeneous vinyl tile, 1/8" thick	"	315	.025	"	3.92	.83		4.75	5.55

	09670	**Fluid Applied Flooring**									
100	0010	**RESILIENT TILE UNDERLAYMENT**										100
	3600	Latex underlayment, 1/8" thk., cementitious for resilient flooring	1 Tilf	160	.050	S.F.	1.55	1.64		3.19	4.10

	09673	**Composition Flooring**									
100	0010	**COMPOSITION FLOORING**, cementitous acrylic, 1/4" thick	C-6	520	.092	S.F.	1.27	2.59	.09	3.95	5.50	100
	0100	3/8" thick	"	450	.107	↓	1.63	2.99	.11	4.73	6.50

FINISHES 9

09673 | Composition Flooring

		CREW	DAILY OUTPUT	LABOR-HOURS	UNIT	MAT.	LABOR	EQUIP.	TOTAL	TOTAL INCL O&P	
100	0200	Methyl methachrylate, 1/4" thick	C-8A	3,000	.016	S.F.	5	.47		5.47	6.20	100
	0210	1/8" thick	"	3,000	.016		2.75	.47		3.22	3.74	
	0300	Cupric oxychloride, on bond coat, minimum	C-6	480	.100		2.77	2.81	.10	5.68	7.50	
	0400	Maximum		420	.114		4.63	3.21	.11	7.95	10.15	
	0600	Epoxy, with colored quartz chips, broadcast, minimum		675	.071		2.19	2	.07	4.26	5.55	
	0700	Maximum		490	.098		2.66	2.75	.10	5.51	7.25	
	0900	Trowelled, minimum		560	.086		2.83	2.41	.09	5.33	6.90	
	1000	Maximum		480	.100		4.12	2.81	.10	7.03	8.95	
	1200	Heavy duty epoxy topping, 1/4" thick,										
	1300	500 to 1,000 S.F.	C-6	420	.114	S.F.	5.05	3.21	.11	8.37	10.60	
	1500	1,000 to 2,000 S.F.		450	.107		3.99	2.99	.11	7.09	9.10	
	1600	Over 10,000 S.F.		480	.100		3.69	2.81	.10	6.60	8.50	
	1800	Epoxy terrazzo, 1/4" thick, chemical resistant, minimum	J-3	200	.080		5.30	2.35	1.05	8.70	10.45	
	1900	Maximum		150	.107		8.20	3.13	1.40	12.73	15.15	
	2100	Conductive, minimum		210	.076		6.60	2.24	1	9.84	11.65	
	2200	Maximum		160	.100		8.95	2.94	1.31	13.20	15.60	
	2400	Mastic, hot laid, 2 coat, 1-1/2" thick, standard, minimum	C-6	690	.070		3.21	1.95	.07	5.23	6.60	
	2500	Maximum		520	.092		4.12	2.59	.09	6.80	8.60	
	2700	Acidproof, minimum		605	.079		4.12	2.23	.08	6.43	8.05	
	2800	Maximum		350	.137		5.70	3.85	.14	9.69	12.35	
	3000	Neoprene, trowelled on, 1/4" thick, minimum		545	.088		3.16	2.47	.09	5.72	7.40	
	3100	Maximum		430	.112		4.35	3.13	.11	7.59	9.75	
	3150	Polyacrylate terrazzo, 1/4" thick, minimum		735	.065		2.98	1.83	.07	4.88	6.15	
	3170	Maximum		480	.100		3.85	2.81	.10	6.76	8.65	
	3200	3/8" thick, minimum		620	.077		3.92	2.17	.08	6.17	7.75	
	3220	Maximum		480	.100		5.75	2.81	.10	8.66	10.80	
	3300	Conductive terrazzo, 1/4" thick, minimum		450	.107		6.95	2.99	.11	10.05	12.35	
	3330	Maximum		305	.157		8.90	4.42	.16	13.48	16.75	
	3350	3/8" thick, minimum		365	.132		9.30	3.69	.13	13.12	16	
	3370	Maximum		255	.188		11.45	5.30	.19	16.94	21	
	3450	Granite, conductive, 1/4" thick, minimum		695	.069		7.25	1.94	.07	9.26	11.05	
	3470	Maximum		420	.114		9.35	3.21	.11	12.67	15.35	
	3500	3/8" thick, minimum		695	.069		10.60	1.94	.07	12.61	14.70	
	3520	Maximum		380	.126		12.85	3.54	.13	16.52	19.70	
	3600	Polyester, with colored quartz chips, 1/16" thick, minimum		1,065	.045		2.89	1.26	.05	4.20	5.15	
	3700	Maximum		560	.086		4	2.41	.09	6.50	8.20	
	3900	1/8" thick, minimum		810	.059		3.17	1.66	.06	4.89	6.10	
	4000	Maximum		675	.071		4.47	2	.07	6.54	8.05	
	4200	Polyester, heavy duty, compared to epoxy, add		2,590	.019		1.21	.52	.02	1.75	2.15	
	4300	Polyurethane, with suspended vinyl chips, minimum		1,065	.045		6.15	1.26	.05	7.46	8.80	
	4500	Maximum		860	.056		8.85	1.57	.06	10.48	12.20	

09680 | Carpet

		CREW	DAILY OUTPUT	LABOR-HOURS	UNIT	MAT.	LABOR	EQUIP.	TOTAL	TOTAL INCL O&P	
600	0010	**CARPET PAD**, commercial grade										600
	9000	Sponge rubber pad, minimum	1 Tilf	150	.053	S.Y.	3.18	1.74		4.92	6.05	
	9100	Maximum		150	.053		8.25	1.74		9.99	11.65	
	9200	Felt pad, minimum		150	.053		3.57	1.74		5.31	6.50	
	9300	Maximum		150	.053		6.70	1.74		8.44	9.95	
	9400	Bonded urethane pad, minimum		150	.053		3.95	1.74		5.69	6.90	
	9500	Maximum		150	.053		6.75	1.74		8.49	10	
	9600	Prime urethane pad, minimum		150	.053		2.27	1.74		4.01	5.05	
	9700	Maximum		150	.053		4.19	1.74		5.93	7.15	
800	0010	**CARPET** Commercial grades, direct cement										800
	0700	Nylon, level loop, 26 oz., light to medium traffic CN	1 Tilf	75	.107	S.Y.	16.45	3.49		19.94	23	

Important: See the Reference Section for critical supporting data - Reference Nos., Crews, & City Cost Indexes

9 FINISHES

09680 | Carpet

		CREW	DAILY OUTPUT	LABOR-HOURS	UNIT	MAT.	LABOR	EQUIP.	TOTAL	TOTAL INCL O&P	
800							**2005 BARE COSTS**				**800**
0720	28 oz., light to medium traffic	1 Tilf	75	.107	S.Y.	15.45	3.49		18.94	22	
0900	32 oz., medium traffic		75	.107		23.50	3.49		26.99	30.50	
1100	40 oz., medium to heavy traffic		75	.107		34.50	3.49		37.99	43	
2920	Nylon plush, 30 oz., medium traffic		57	.140		17.20	4.59		21.79	25.50	
3000	36 oz., medium traffic		75	.107		21.50	3.49		24.99	29	
3100	42 oz., medium to heavy traffic		70	.114		21.50	3.74		25.24	29	
3200	46 oz., medium to heavy traffic		70	.114		30	3.74		33.74	38.50	
3300	54 oz., heavy traffic		70	.114		33.50	3.74		37.24	42.50	
3340	60 oz., heavy traffic		70	.114		47.50	3.74		51.24	58	
3665	Olefin, 24 oz., light to medium traffic		75	.107		10.05	3.49		13.54	16.20	
3670	26 oz., medium traffic		75	.107		10.30	3.49		13.79	16.50	
3680	28 oz., medium to heavy traffic		75	.107		12.15	3.49		15.64	18.55	
3700	32 oz., medium to heavy traffic		75	.107		14.95	3.49		18.44	21.50	
3730	42 oz., heavy traffic		70	.114		22	3.74		25.74	29.50	
4110	Wool, level loop, 40 oz., medium traffic		75	.107		74.50	3.49		77.99	87	
4500	50 oz., medium to heavy traffic		75	.107		76.50	3.49		79.99	89	
4700	Patterned, 32 oz., medium to heavy traffic		70	.114		75.50	3.74		79.24	88.50	
4900	48 oz., heavy traffic		70	.114		76.50	3.74		80.24	90	
5000	For less than full roll, add					25%					
5100	For small rooms, less than 12' wide, add						25%				
5200	For large open areas (no cuts), deduct						25%				
5600	For bound carpet baseboard, add	1 Tilf	300	.027	L.F.	1.20	.87		2.07	2.60	
5610	For stairs, not incl. price of carpet, add	"	30	.267	Riser		8.70		8.70	12.80	
5620	For borders and patterns, add to labor						18%				
8950	For tackless, stretched installation, add padding to above										
9850	For "branded" fiber, add				S.Y.	25%					
900	**CARPET TILE**										**900**
0010											
0100	Tufted nylon, 18" x 18", hard back, 20 oz.	1 Tilf	150	.053	S.Y.	20.50	1.74		22.24	25	
0110	26 oz.		150	.053		35.50	1.74		37.24	41.50	
0200	Cushion back, 20 oz.		150	.053		26	1.74		27.74	31	
0210	26 oz.		150	.053		40.50	1.74		42.24	47	
910	**CARPET MAINTENANCE** Steam clean, per cleaning, minimum	1 Clab	3,000	.003	S.F.	.07	.07		.14	.19	**910**
0010											
0500	Maximum	"	2,000	.004	"	.11	.11		.22	.29	

09720 | Wall Coverings

		CREW	DAILY OUTPUT	LABOR-HOURS	UNIT	MAT.	LABOR	EQUIP.	TOTAL	TOTAL INCL O&P	
100							**2005 BARE COSTS**				**100**
0010	**WALL COVERING** Including sizing, add 10%-30% waste at takeoff R09700-700										
0050	Aluminum foil	1 Pape	275	.029	S.F.	.87	.88		1.75	2.29	
0100	Copper sheets, .025" thick, vinyl backing		240	.033		4.66	1.01		5.67	6.65	
0300	Phenolic backing		240	.033		6.05	1.01		7.06	8.15	
0600	Cork tiles, light or dark, 12" x 12" x 3/16"		240	.033		2.55	1.01		3.56	4.33	
0700	5/16" thick		235	.034		3.10	1.03		4.13	4.96	
0900	1/4" basketweave		240	.033		4.76	1.01		5.77	6.75	
1000	1/2" natural, non-directional pattern		240	.033		5.35	1.01		6.36	7.40	
1100	3/4" natural, non-directional pattern		240	.033		9	1.01		10.01	11.40	
1200	Granular surface, 12" x 36", 1/2" thick		385	.021		1.03	.63		1.66	2.08	
1300	1" thick		370	.022		1.33	.66		1.99	2.44	
1500	Polyurethane coated, 12" x 12" x 3/16" thick		240	.033		3.21	1.01		4.22	5.05	

For expanded coverage of these items see *Means Interior Cost Data 2005*

FINISHES 9

09720 | Wall Coverings

			CREW	DAILY OUTPUT	LABOR-HOURS	UNIT	2005 BARE COSTS				TOTAL INCL O&P	
							MAT.	LABOR	EQUIP.	TOTAL		
100	1600	5/16" thick	1 Pape	235	.034	S.F.	4.57	1.03		5.60	6.60	100
	1800	Cork wallpaper, paperbacked, natural R09700-700		480	.017		1.83	.51		2.34	2.77	
	1900	Colors		480	.017		2.26	.51		2.77	3.25	
	2100	Flexible wood veneer, 1/32" thick, plain woods		100	.080		1.92	2.42		4.34	5.75	
	2200	Exotic woods	↓	95	.084	↓	2.92	2.55		5.47	7.05	
	2400	Gypsum-based, fabric-backed, fire										
	2500	resistant for masonry walls, minimum, 21 oz./S.Y.	1 Pape	800	.010	S.F.	.68	.30		.98	1.21	
	2700	Maximum, (small quantities)	"	640	.013		1.11	.38		1.49	1.79	
	2750	Acrylic, modified, semi-rigid PVC, .028" thick	2 Carp	330	.048		.92	1.66		2.58	3.60	
	2800	.040" thick	"	320	.050		1.21	1.71		2.92	4	
	3000	Vinyl wall covering, fabric-backed, lightweight, (12-15 oz./S.Y.)	1 Pape	640	.013		.58	.38		.96	1.21	
	3300	Medium weight, type 2, (20-24 oz./S.Y.)		480	.017		.72	.51		1.23	1.55	
	3400	Heavy weight, type 3, (28 oz./S.Y.)	↓	435	.018	↓	1.16	.56		1.72	2.12	
	3600	Adhesive, 5 gal. lots, (18SY/Gal.)				Gal.	8.95			8.95	9.80	
	3700	Wallpaper, average workmanship, solid pattern, low cost paper	1 Pape	640	.013	S.F.	.29	.38		.67	.89	
	3900	basic patterns (matching required), avg. cost paper		535	.015		.65	.45		1.10	1.40	
	4000	Paper at $85per double roll, quality workmanship	↓	435	.018	↓	1.53	.56		2.09	2.52	
	4100	Linen wall covering, paper backed										
	4150	Flame treatment, minimum				S.F.	.69			.69	.76	
	4180	Maximum					1.26			1.26	1.39	
	4200	Grass cloths with lining paper, minimum	1 Pape	400	.020		.63	.61		1.24	1.60	
	4300	Maximum	"	350	.023	↓	2.03	.69		2.72	3.27	

09770 | Special Wall Surfaces

			CREW	DAILY OUTPUT	LABOR-HOURS	UNIT	MAT.	LABOR	EQUIP.	TOTAL	TOTAL INCL O&P	
400	0010	**FIBERGLASS REINFORCED PLASTIC** panels, .090" thick, on walls										400
	0020	Adhesive mounted, embossed surface	2 Carp	640	.025	S.F.	1.38	.86		2.24	2.85	
	0030	Smooth surface		640	.025		1.62	.86		2.48	3.11	
	0040	Fire rated, embossed surface		640	.025		1.80	.86		2.66	3.31	
	0050	Nylon rivet mounted, on drywall, embossed surface		480	.033		1.28	1.14		2.42	3.19	
	0060	Smooth surface		480	.033		1.51	1.14		2.65	3.44	
	0070	Fire rated, embossed surface		480	.033		1.68	1.14		2.82	3.63	
	0080	On masonry, embossed surface		320	.050		1.28	1.71		2.99	4.08	
	0090	Smooth surface		320	.050		1.62	1.71		3.33	4.45	
	0100	Fire rated, embossed surface		320	.050		1.71	1.71		3.42	4.55	
	0110	Nylon rivet and adhesive mounted, on drywall, embossed surface		240	.067		1.51	2.28		3.79	5.20	
	0120	Smooth surface		240	.067		1.71	2.28		3.99	5.45	
	0130	Fire rated, embossed surface		240	.067		1.90	2.28		4.18	5.65	
	0140	On masonry, embossed surface		190	.084		1.51	2.88		4.39	6.15	
	0150	Smooth surface		190	.084		1.71	2.88		4.59	6.35	
	0160	Fire rated, embossed surface	↓	190	.084	↓	1.90	2.88		4.78	6.60	
	0170	For moldings add	1 Carp	250	.032	L.F.	.25	1.10		1.35	1.99	
	0180	On ceilings, for lay in grid system, embossed surface		400	.020	S.F.	1.38	.69		2.07	2.59	
	0190	Smooth surface		400	.020		1.62	.69		2.31	2.85	
	0200	Fire rated, embossed surface	↓	400	.020	↓	1.80	.69		2.49	3.05	
700	0010	**PANEL SYSTEM**										700
	0100	Raised panel, eng. wood core w/ wood veneer, std., paint grade	2 Carp	300	.053	S.F.	9.10	1.83		10.93	12.85	
	0110	Oak veneer		300	.053		15.50	1.83		17.33	19.90	
	0120	Maple veneer		300	.053		19.80	1.83		21.63	25	
	0130	Cherry veneer		300	.053		25	1.83		26.83	30.50	
	0300	Class I fire rated, paint grade		300	.053		14.55	1.83		16.38	18.85	
	0310	Oak veneer		300	.053		25	1.83		26.83	30.50	
	0320	Maple veneer		300	.053		31.50	1.83		33.33	38	
	0330	Cherry veneer		300	.053		45	1.83		46.83	52.50	
	0510	Beadboard, 5/8" MDF, standard, primed		300	.053		7.15	1.83		8.98	10.75	
	0520	Oak veneer, unfinished		300	.053		12.90	1.83		14.73	17	
	0530	Maple veneer, unfinished	↓	300	.053		15.25	1.83		17.08	19.60	

9 FINISHES

Important: See the Reference Section for critical supporting data - Reference Nos., Crews, & City Cost Indexes

09700 | Wall Finishes

09770 | Special Wall Surfaces

			CREW	DAILY OUTPUT	LABOR-HOURS	UNIT	2005 BARE COSTS				TOTAL INCL O&P	
							MAT.	LABOR	EQUIP.	TOTAL		
700	0610	Rustic paneling, 5/8" MDF, standard, maple veneer, unfinished	2 Carp	300	.053	S.F.	21.50	1.83		23.33	26.50	700
	5000	For prefinished paneling, see division 06250-500 & 06250-200										
750	0010	**SLATWALL PANELS AND ACCESSORIES**										750
	0100	Slatwall panel, 4' x 8' x 3/4" T, MDF, paint grade	1 Carp	500	.016	S.F.	1.34	.55		1.89	2.32	
	0110	Melamine finish		500	.016		2.04	.55		2.59	3.09	
	0120	High pressure plastic laminate finish	↓	500	.016		3.25	.55		3.80	4.43	
	0130	Aluminum channel inserts, add				↓	2.44			2.44	2.68	
	0200	Accessories, corner forms, 8' L				L.F.	3.47			3.47	3.82	
	0210	T-connector, 8' L					5.05			5.05	5.55	
	0220	J-mold, 8' L					1.29			1.29	1.42	
	0230	Edge cap, 8' L					.84			.84	.92	
	0240	Finish end cap, 8' L				↓	3.08			3.08	3.39	
	0300	Display hook, 4" L				Ea.	.95			.95	1.05	
	0310	6" L					1.01			1.01	1.11	
	0320	8" L					1.12			1.12	1.23	
	0330	10" L					1.23			1.23	1.35	
	0340	12" L					1.34			1.34	1.47	
	0350	Acrylic, 4" L					.78			.78	.86	
	0360	6" L					.90			.90	.99	
	0370	8" L					.95			.95	1.05	
	0380	10" L					1.06			1.06	1.17	
	0400	Waterfall hanger, metal, 12" - 16"					6.05			6.05	6.65	
	0410	Acrylic					8.60			8.60	9.50	
	0500	Shelf bracket, metal, 8"					4.70			4.70	5.15	
	0510	10"					5.05			5.05	5.55	
	0520	12"					5.40			5.40	5.90	
	0530	14"					6.05			6.05	6.65	
	0540	16"					6.70			6.70	7.40	
	0550	Acrylic, 8"					2.80			2.80	3.08	
	0560	10"					3.14			3.14	3.45	
	0570	12"					3.47			3.47	3.82	
	0580	14"					3.92			3.92	4.31	
	0600	Shelf, acrylic, 12" x 16" x 1/4"					23.50			23.50	26	
	0610	12" x 24" x 1/4"				↓	39			39	43	

09800 | Acoustical Treatment

09820 | Acoustical Insul/Sealants

			CREW	DAILY OUTPUT	LABOR-HOURS	UNIT	2005 BARE COSTS				TOTAL INCL O&P	
							MAT.	LABOR	EQUIP.	TOTAL		
500	0010	**SOUND ATTENUATION** Blanket, 1" thick	1 Carp	925	.009	S.F.	.24	.30		.54	.72	500
	0500	1-1/2" thick		920	.009		.24	.30		.54	.72	
	1000	2" thick		915	.009		.29	.30		.59	.79	
	1500	3" thick		910	.009		.40	.30		.70	.91	
	3000	Thermal or acoustical batt above ceiling, 2" thick		900	.009		.42	.30		.72	.93	
	3100	3" thick		900	.009		.63	.30		.93	1.16	
	3200	4" thick	↓	900	.009		.77	.30		1.07	1.32	
	3400	Urethane plastic foam, open cell, on wall, 2" thick	2 Carp	2,050	.008		2.56	.27		2.83	3.24	
	3500	3" thick		1,550	.010		3.40	.35		3.75	4.29	
	3600	4" thick		1,050	.015		4.77	.52		5.29	6.05	
	3700	On ceiling, 2" thick		1,700	.009		2.55	.32		2.87	3.31	
	3800	3" thick	↓	1,300	.012	↓	3.40	.42		3.82	4.40	

For expanded coverage of these items see *Means Interior Cost Data 2005*

09820 | Acoustical Insul/Sealants

			CREW	DAILY OUTPUT	LABOR-HOURS	UNIT	2005 BARE COSTS MAT.	LABOR	EQUIP.	TOTAL	TOTAL INCL O&P	
500	3900	4" thick	2 Carp	900	.018	S.F.	4.77	.61		5.38	6.20	500
	4000	Nylon matting 0.4" thick, with carbon black spinerette										
	4010	plus polyester fabric, on floor	J-4	4,000	.004	S.F.	1.90	.12		2.02	2.26	
	4200	Fiberglass reinf. backer board underlayment, 7/16" thick, on floor	"	800	.020	"	1.59	.58		2.17	2.60	

09830 | Acoustical Barriers

			CREW	DAILY OUTPUT	LABOR-HOURS	UNIT	2005 BARE COSTS MAT.	LABOR	EQUIP.	TOTAL	TOTAL INCL O&P	
100	0011	BARRIERS Plenum										100
	0600	Aluminum foil, fiberglass reinf., parallel with joists	1 Carp	275	.029	S.F.	.72	1		1.72	2.34	
	0700	Perpendicular to joists		180	.044		.73	1.52		2.25	3.17	
	0900	Aluminum mesh, kraft paperbacked		275	.029		.69	1		1.69	2.31	
	0970	Fiberglass batts, kraft faced, 3-1/2" thick		1,400	.006		.25	.20		.45	.58	
	0980	6" thick		1,300	.006		.42	.21		.63	.79	
	1000	Sheet lead, 1 lb., 1/64" thick, perpendicular to joists		150	.053		2.43	1.83		4.26	5.50	
	1100	Vinyl foam reinforced, 1/8" thick, 1.0 lb. per S.F.		150	.053		3.18	1.83		5.01	6.35	

09840 | Acoustical Wall Treatment

			CREW	DAILY OUTPUT	LABOR-HOURS	UNIT	2005 BARE COSTS MAT.	LABOR	EQUIP.	TOTAL	TOTAL INCL O&P	
100	0010	SOUND ABSORBING PANELS Perforated steel facing, painted with										100
	0100	fiberglass or mineral filler, no backs, 2-1/4" thick, modular										
	0200	space units, ceiling or wall hung, white or colored	1 Carp	100	.080	S.F.	8.90	2.74		11.64	14.05	
	0300	Fiberboard sound deadening panels, 1/2" thick	"	600	.013	"	.29	.46		.75	1.03	
	0500	Fiberglass panels, 4' x 8' x 1" thick, with										
	0600	glass cloth face for walls, cemented	1 Carp	155	.052	S.F.	6.20	1.77		7.97	9.55	
	0700	1-1/2" thick, dacron covered, inner aluminum frame,										
	0710	wall mounted	1 Carp	300	.027	S.F.	7.35	.91		8.26	9.50	
	0900	Mineral fiberboard panels, fabric covered, 30"x 108",										
	1000	3/4" thick, concealed spline, wall mounted	1 Carp	150	.053	S.F.	5.30	1.83		7.13	8.70	

09900 | Paints & Coatings

09910 | Paints

				CREW	DAILY OUTPUT	LABOR-HOURS	UNIT	2005 BARE COSTS MAT.	LABOR	EQUIP.	TOTAL	TOTAL INCL O&P	
100	0010	CABINETS AND CASEWORK	R09910 -220										100
	1000	Primer coat, oil base, brushwork		1 Pord	650	.012	S.F.	.05	.38		.43	.62	
	2000	Paint, oil base, brushwork, 1 coat			650	.012		.06	.38		.44	.64	
	3000	Stain, brushwork, wipe off			650	.012		.05	.38		.43	.62	
	4000	Shellac, 1 coat, brushwork			650	.012		.06	.38		.44	.63	
	4500	Varnish, 3 coats, brushwork, sand after 1st coat			325	.025		.17	.75		.92	1.32	
	5000	For latex paint, deduct						10%					
300	0010	DOORS AND WINDOWS, EXTERIOR	R09910 -220										300
	0100	Door frames & trim, only											
	0110	Brushwork, primer		1 Pord	512	.016	L.F.	.05	.48		.53	.78	
	0120	Finish coat, exterior latex			512	.016		.06	.48		.54	.78	
	0130	Primer & 1 coat, exterior latex			300	.027		.11	.82		.93	1.35	
	0140	Primer & 2 coats, exterior latex			265	.030		.17	.92		1.09	1.57	
	0150	Doors, flush, both sides, incl. frame & trim											
	0160	Roll & brush, primer		1 Pord	10	.800	Ea.	3.89	24.50		28.39	41.50	
	0170	Finish coat, exterior latex			10	.800		4.42	24.50		28.92	42	
	0180	Primer & 1 coat, exterior latex			7	1.143		8.30	35		43.30	61.50	
	0190	Primer & 2 coats, exterior latex			5	1.600		12.75	49		61.75	87.50	
	0200	Brushwork, stain, sealer & 2 coats polyurethane			4	2		15.80	61		76.80	109	

Important: See the Reference Section for critical supporting data - Reference Nos., Crews, & City Cost Indexes

9 FINISHES

09910 | Paints

		CREW	DAILY OUTPUT	LABOR-HOURS	UNIT	2005 BARE COSTS MAT.	LABOR	EQUIP.	TOTAL	TOTAL INCL O&P	
300											**300**
0210	Doors, French, both sides, 10-15 lite, incl. frame & trim R09910-220										
0220	Brushwork, primer	1 Pord	6	1.333	Ea.	1.95	41		42.95	63.50	
0230	Finish coat, exterior latex		6	1.333		2.21	41		43.21	64	
0240	Primer & 1 coat, exterior latex		3	2.667		4.15	81.50		85.65	128	
0250	Primer & 2 coats, exterior latex		2	4		6.25	122		128.25	191	
0260	Brushwork, stain, sealer & 2 coats polyurethane		2.50	3.200		5.65	98		103.65	153	
0270	Doors, louvered, both sides, incl. frame & trim										
0280	Brushwork, primer	1 Pord	7	1.143	Ea.	3.89	35		38.89	57	
0290	Finish coat, exterior latex		7	1.143		4.42	35		39.42	57.50	
0300	Primer & 1 coat, exterior latex		4	2		8.30	61		69.30	101	
0310	Primer & 2 coats, exterior latex		3	2.667		12.50	81.50		94	137	
0320	Brushwork, stain, sealer & 2 coats polyurethane		4.50	1.778		15.80	54.50		70.30	99.50	
0330	Doors, panel, both sides, incl. frame & trim										
0340	Roll & brush, primer	1 Pord	6	1.333	Ea.	3.89	41		44.89	66	
0350	Finish coat, exterior latex		6	1.333		4.42	41		45.42	66.50	
0360	Primer & 1 coat, exterior latex		3	2.667		8.30	81.50		89.80	132	
0370	Primer & 2 coats, exterior latex		2.50	3.200		12.50	98		110.50	161	
0380	Brushwork, stain, sealer & 2 coats polyurethane		3	2.667		15.80	81.50		97.30	140	
0400	Windows, per ext. side, based on 15 SF										
0410	1 to 6 lite										
0420	Brushwork, primer	1 Pord	13	.615	Ea.	.77	18.85		19.62	29.50	
0430	Finish coat, exterior latex		13	.615		.87	18.85		19.72	29.50	
0440	Primer & 1 coat, exterior latex		8	1		1.64	30.50		32.14	48	
0450	Primer & 2 coats, exterior latex		6	1.333		2.46	41		43.46	64	
0460	Stain, sealer & 1 coat varnish		7	1.143		2.24	35		37.24	55	
0470	7 to 10 lite										
0480	Brushwork, primer	1 Pord	11	.727	Ea.	.77	22.50		23.27	34.50	
0490	Finish coat, exterior latex		11	.727		.87	22.50		23.37	34.50	
0500	Primer & 1 coat, exterior latex		7	1.143		1.64	35		36.64	54.50	
0510	Primer & 2 coats, exterior latex		5	1.600		2.46	49		51.46	76	
0520	Stain, sealer & 1 coat varnish		6	1.333		2.24	41		43.24	64	
0530	12 lite										
0540	Brushwork, primer	1 Pord	10	.800	Ea.	.77	24.50		25.27	38	
0550	Finish coat, exterior latex		10	.800		.87	24.50		25.37	38	
0560	Primer & 1 coat, exterior latex		6	1.333		1.64	41		42.64	63.50	
0570	Primer & 2 coats, exterior latex		5	1.600		2.46	49		51.46	76	
0580	Stain, sealer & 1 coat varnish		6	1.333		2.23	41		43.23	64	
0590	For oil base paint, add					10%					
310	0010	**DOORS & WINDOWS, INTERIOR LATEX** R09910-220									**310**
0100	Doors flush, both sides, incl. frame & trim										
0110	Roll & brush, primer	1 Pord	10	.800	Ea.	3.38	24.50		27.88	40.50	
0120	Finish coat, latex		10	.800		3.65	24.50		28.15	41	
0130	Primer & 1 coat latex		7	1.143		7.05	35		42.05	60.50	
0140	Primer & 2 coats latex		5	1.600		10.50	49		59.50	85	
0160	Spray, both sides, primer		20	.400		3.56	12.25		15.81	22.50	
0170	Finish coat, latex		20	.400		3.83	12.25		16.08	22.50	
0180	Primer & 1 coat latex		11	.727		7.45	22.50		29.95	41.50	
0190	Primer & 2 coats latex		8	1		11.10	30.50		41.60	58	
0200	Doors, French, both sides, 10-15 lite, incl. frame & trim										
0210	Roll & brush, primer	1 Pord	6	1.333	Ea.	1.69	41		42.69	63.50	
0220	Finish coat, latex		6	1.333		1.83	41		42.83	63.50	
0230	Primer & 1 coat latex		3	2.667		3.52	81.50		85.02	127	
0240	Primer & 2 coats latex		2	4		5.25	122		127.25	190	
0260	Doors, louvered, both sides, incl. frame & trim										
0270	Roll & brush, primer	1 Pord	7	1.143	Ea.	3.38	35		38.38	56	
0280	Finish coat, latex		7	1.143		3.65	35		38.65	56.50	

FINISHES 9

		09910	Paints	CREW	DAILY OUTPUT	LABOR-HOURS	UNIT	2005 BARE COSTS				TOTAL INCL O&P	
								MAT.	LABOR	EQUIP.	TOTAL		
310	0290		Primer & 1 coat, latex R09910-220	1 Pord	4	2	Ea.	6.85	61		67.85	99.50	**310**
	0300		Primer & 2 coats, latex		3	2.667		10.70	81.50		92.20	135	
	0320		Spray, both sides, primer		20	.400		3.56	12.25		15.81	22.50	
	0330		Finish coat, latex		20	.400		3.83	12.25		16.08	22.50	
	0340		Primer & 1 coat, latex		11	.727		7.45	22.50		29.95	41.50	
	0350		Primer & 2 coats, latex		8	1		11.30	30.50		41.80	58.50	
	0360		Doors, panel, both sides, incl. frame & trim										
	0370		Roll & brush, primer	1 Pord	6	1.333	Ea.	3.56	41		44.56	65.50	
	0380		Finish coat, latex		6	1.333		3.65	41		44.65	65.50	
	0390		Primer & 1 coat, latex		3	2.667		7.05	81.50		88.55	131	
	0400		Primer & 2 coats, latex		2.50	3.200		10.70	98		108.70	159	
	0420		Spray, both sides, primer		10	.800		3.56	24.50		28.06	41	
	0430		Finish coat, latex		10	.800		3.83	24.50		28.33	41	
	0440		Primer & 1 coat, latex		5	1.600		7.45	49		56.45	81.50	
	0450		Primer & 2 coats, latex		4	2		11.30	61		72.30	104	
	0460		Windows, per interior side, based on 15 SF										
	0470		1 to 6 lite										
	0480		Brushwork, primer	1 Pord	13	.615	Ea.	.67	18.85		19.52	29	
	0490		Finish coat, enamel		13	.615		.72	18.85		19.57	29.50	
	0500		Primer & 1 coat enamel		8	1		1.39	30.50		31.89	47.50	
	0510		Primer & 2 coats enamel		6	1.333		2.11	41		43.11	64	
	0530		7 to 10 lite										
	0540		Brushwork, primer	1 Pord	11	.727	Ea.	.67	22.50		23.17	34	
	0550		Finish coat, enamel		11	.727		.72	22.50		23.22	34.50	
	0560		Primer & 1 coat enamel		7	1.143		1.39	35		36.39	54	
	0570		Primer & 2 coats enamel		5	1.600		2.11	49		51.11	76	
	0590		12 lite										
	0600		Brushwork, primer	1 Pord	10	.800	Ea.	.67	24.50		25.17	37.50	
	0610		Finish coat, enamel		10	.800		.72	24.50		25.22	38	
	0620		Primer & 1 coat enamel		6	1.333		1.39	41		42.39	63	
	0630		Primer & 2 coats enamel		5	1.600		2.11	49		51.11	76	
	0650		For oil base paint, add					10%					
320	0010		**DOORS AND WINDOWS, INTERIOR ALKYD (OIL BASE)** R09910-220										**320**
	0500		Flush door & frame, 3' x 7', oil, primer, brushwork	1 Pord	10	.800	Ea.	2.12	24.50		26.62	39.50	
	1000		Paint, 1 coat		10	.800		2.04	24.50		26.54	39	
	1400		Stain, brushwork, wipe off		18	.444		1.02	13.60		14.62	21.50	
	1600		Shellac, 1 coat, brushwork		25	.320		1.18	9.80		10.98	16	
	1800		Varnish, 3 coats, brushwork, sand after 1st coat		9	.889		3.61	27		30.61	45	
	2000		Panel door & frame, 3' x 7', oil, primer, brushwork		6	1.333		1.82	41		42.82	63.50	
	2200		Paint, 1 coat		6	1.333		2.04	41		43.04	63.50	
	2600		Stain, brushwork, panel door, 3' x 7', not incl. frame		16	.500		1.02	15.30		16.32	24	
	2800		Shellac, 1 coat, brushwork		22	.364		1.18	11.15		12.33	18.05	
	3000		Varnish, 3 coats, brushwork, sand after 1st coat		7.50	1.067		3.61	32.50		36.11	53	
	4400		Windows, including frame and trim, per side										
	4600		Colonial type, 6/6 lites, 2' x 3', oil, primer, brushwork	1 Pord	14	.571	Ea.	.29	17.50		17.79	27	
	5800		Paint, 1 coat		14	.571		.32	17.50		17.82	27	
	6200		3' x 5' opening, 6/6 lites, primer coat, brushwork		12	.667		.72	20.50		21.22	31.50	
	6400		Paint, 1 coat		12	.667		.80	20.50		21.30	31.50	
	6800		4' x 8' opening, 6/6 lites, primer coat, brushwork		8	1		1.54	30.50		32.04	47.50	
	7000		Paint, 1 coat		8	1		1.72	30.50		32.22	48	
	8000		Single lite type, 2' x 3', oil base, primer coat, brushwork		33	.242		.29	7.40		7.69	11.45	
	8200		Paint, 1 coat		33	.242		.32	7.40		7.72	11.50	
	8600		3' x 5' opening, primer coat, brushwork		20	.400		.72	12.25		12.97	19.20	
	8800		Paint, 1 coat		20	.400		.80	12.25		13.05	19.30	
	9200		4' x 8' opening, primer coat, brushwork		14	.571		1.54	17.50		19.04	28	
	9400		Paint, 1 coat		14	.571		1.72	17.50		19.22	28.50	

Important: See the Reference Section for critical supporting data - Reference Nos., Crews, & City Cost Indexes

			CREW	DAILY OUTPUT	LABOR-HOURS	UNIT	2005 BARE COSTS				TOTAL INCL O&P
		09910	Paints				MAT.	LABOR	EQUIP.	TOTAL	
400	0010	**FENCES**									**400**
	0100	Chain link or wire metal, one side, water base [R09910-220]									
	0110	Roll & brush, first coat	1 Pord	960	.008	S.F.	.06	.25		.31	.44
	0120	Second coat		1,280	.006		.05	.19		.24	.35
	0130	Spray, first coat		2,275	.004		.06	.11		.17	.22
	0140	Second coat		2,600	.003		.06	.09		.15	.20
	0150	Picket, water base									
	0160	Roll & brush, first coat	1 Pord	865	.009	S.F.	.06	.28		.34	.50
	0170	Second coat		1,050	.008		.06	.23		.29	.42
	0180	Spray, first coat		2,275	.004		.06	.11		.17	.23
	0190	Second coat		2,600	.003		.06	.09		.15	.21
	0200	Stockade, water base									
	0210	Roll & brush, first coat	1 Pord	1,040	.008	S.F.	.06	.24		.30	.42
	0220	Second coat		1,200	.007		.06	.20		.26	.38
	0230	Spray, first coat		2,275	.004		.06	.11		.17	.23
	0240	Second coat		2,600	.003		.06	.09		.15	.21
500	0010	**FLOORS, INTERIOR**									**500**
	0100	Concrete									
	0110	Brushwork, latex, block filler									
	0120	1st coat	1 Pord	975	.008	S.F.	.11	.25		.36	.50
	0130	2nd coat		1,150	.007		.07	.21		.28	.40
	0140	3rd coat		1,300	.006		.06	.19		.25	.35
	0150	Roll, latex, block filler									
	0160	1st coat	1 Pord	2,600	.003	S.F.	.15	.09		.24	.30
	0170	2nd coat		3,250	.002		.09	.08		.17	.21
	0180	3rd coat		3,900	.002		.07	.06		.13	.16
	0190	Spray, latex, block filler									
	0200	1st coat	1 Pord	2,600	.003	S.F.	.13	.09		.22	.28
	0210	2nd coat		3,250	.002		.07	.08		.15	.19
	0220	3rd coat		3,900	.002		.06	.06		.12	.15
620	0010	**MISCELLANEOUS, EXTERIOR**									**620**
	0015	For painting metals, see Div. 05950-650 [R09910-220]									
	0100	Railing, ext., decorative wood, incl. cap & baluster									
	0110	newels & spindles @ 12" O.C.									
	0120	Brushwork, stain, sand, seal & varnish									
	0130	First coat	1 Pord	90	.089	L.F.	.45	2.72		3.17	4.58
	0140	Second coat	"	120	.067	"	.45	2.04		2.49	3.56
	0150	Rough sawn wood, 42" high, 2"x2" verticals, 6" O.C.									
	0160	Brushwork, stain, each coat	1 Pord	90	.089	L.F.	.15	2.72		2.87	4.25
	0170	Wrought iron, 1" rail, 1/2" sq. verticals									
	0180	Brushwork, zinc chromate, 60" high, bars 6" O.C.									
	0190	Primer	1 Pord	130	.062	L.F.	.51	1.88		2.39	3.39
	0200	Finish coat		130	.062		.16	1.88		2.04	3
	0210	Additional coat		190	.042		.19	1.29		1.48	2.14
	0220	Shutters or blinds, single panel, 2'x4', paint all sides									
	0230	Brushwork, primer	1 Pord	20	.400	Ea.	.56	12.25		12.81	19
	0240	Finish coat, exterior latex		20	.400		.47	12.25		12.72	18.90
	0250	Primer & 1 coat, exterior latex		13	.615		.90	18.85		19.75	29.50
	0260	Spray, primer		35	.229		.82	7		7.82	11.40
	0270	Finish coat, exterior latex		35	.229		.99	7		7.99	11.60
	0280	Primer & 1 coat, exterior latex		20	.400		.88	12.25		13.13	19.35
	0290	For louvered shutters, add				S.F.	10%				
	0300	Stair stringers, exterior, metal									
	0310	Roll & brush, zinc chromate, to 14", each coat	1 Pord	320	.025	L.F.	.05	.77		.82	1.21
	0320	Rough sawn wood, 4" x 12"									
	0330	Roll & brush, exterior latex, each coat	1 Pord	215	.037	L.F.	.07	1.14		1.21	1.79

FINISHES **9**

09910 | Paints

		CREW	DAILY OUTPUT	LABOR-HOURS	UNIT	2005 BARE COSTS				TOTAL INCL O&P
						MAT.	LABOR	EQUIP.	TOTAL	
620 0340	Trellis/lattice, 2"x2" @ 3" O.C. with 2"x8" supports R09910-220									**620**
0350	Spray, latex, per side, each coat	1 Pord	475	.017	S.F.	.07	.52		.59	.85
0450	Decking, Ext., sealer, alkyd, brushwork, sealer coat		1,140	.007		.05	.21		.26	.38
0460	1st coat		1,140	.007		.06	.21		.27	.38
0470	2nd coat		1,300	.006		.04	.19		.23	.32
0500	Paint, alkyd, brushwork, primer coat		1,140	.007		.07	.21		.28	.40
0510	1st coat		1,140	.007		.07	.21		.28	.40
0520	2nd coat		1,300	.006		.05	.19		.24	.34
0600	Sand paint, alkyd, brushwork, 1 coat		150	.053		.09	1.63		1.72	2.55
630 0010	**MISCELLANEOUS, INTERIOR**									**630**
2400	Floors, conc./wood, oil base, primer/sealer coat, brushwork R09910-220	2 Pord	1,950	.008	S.F.	.06	.25		.31	.44
2450	Roller		5,200	.003		.06	.09		.15	.21
2600	Spray		6,000	.003		.06	.08		.14	.19
2650	Paint 1 coat, brushwork		1,950	.008		.05	.25		.30	.44
2800	Roller		5,200	.003		.06	.09		.15	.20
2850	Spray		6,000	.003		.06	.08		.14	.18
3000	Stain, wood floor, brushwork, 1 coat		4,550	.004		.05	.11		.16	.21
3200	Roller		5,200	.003		.05	.09		.14	.20
3250	Spray		6,000	.003		.05	.08		.13	.18
3400	Varnish, wood floor, brushwork		4,550	.004		.06	.11		.17	.22
3450	Roller		5,200	.003		.06	.09		.15	.21
3600	Spray		6,000	.003		.06	.08		.14	.19
3650	For dust proofing or anti skid, see division 03350-300									
3800	Grilles, per side, oil base, primer coat, brushwork	1 Pord	520	.015	S.F.	.10	.47		.57	.82
3850	Spray		1,140	.007		.10	.21		.31	.43
3920	Paint 2 coats, brushwork		325	.025		.21	.75		.96	1.36
3940	Spray		650	.012		.24	.38		.62	.83
5000	Pipe, to 4" diameter, primer or sealer coat, oil base, brushwork	2 Pord	1,250	.013	L.F.	.06	.39		.45	.66
5100	Spray		2,165	.007		.06	.23		.29	.40
5350	Paint 2 coats, brushwork		775	.021		.11	.63		.74	1.07
5400	Spray		1,240	.013		.13	.39		.52	.73
6300	To 16" diameter, primer or sealer coat, brushwork		310	.052		.24	1.58		1.82	2.64
6350	Spray		540	.030		.27	.91		1.18	1.66
6500	Paint 2 coats, brushwork		195	.082		.45	2.51		2.96	4.27
6550	Spray		310	.052		.50	1.58		2.08	2.92
7000	Trim, wood, incl. puttying, under 6" wide									
7200	Primer coat, oil base, brushwork	1 Pord	650	.012	L.F.	.02	.38		.40	.60
7250	Paint, 1 coat, brushwork		650	.012		.03	.38		.41	.60
7450	3 coats		325	.025		.08	.75		.83	1.22
7500	Over 6" wide, primer coat, brushwork		650	.012		.05	.38		.43	.62
7550	Paint, 1 coat, brushwork		650	.012		.05	.38		.43	.63
7650	3 coats		325	.025		.15	.75		.90	1.30
8000	Cornice, simple design, primer coat, oil base, brushwork		650	.012	S.F.	.05	.38		.43	.62
8250	Paint, 1 coat		650	.012		.05	.38		.43	.63
8350	Ornate design, primer coat		350	.023		.05	.70		.75	1.10
8400	Paint, 1 coat		350	.023		.05	.70		.75	1.11
8600	Balustrades, primer coat, oil base, brushwork		520	.015		.05	.47		.52	.76
8650	Paint, 1 coat		520	.015		.05	.47		.52	.77
8900	Trusses and wood frames, primer coat, oil base, brushwork		800	.010		.05	.31		.36	.51
8950	Spray		1,200	.007		.05	.20		.25	.37
9220	Paint 2 coats, brushwork		500	.016		.10	.49		.59	.85
9240	Spray		600	.013		.12	.41		.53	.74
9260	Stain, brushwork, wipe off		600	.013		.05	.41		.46	.66
9280	Varnish, 3 coats, brushwork		275	.029		.17	.89		1.06	1.53
9350	For latex paint, deduct					10%				

9

FINISHES

Important: See the Reference Section for critical supporting data - Reference Nos., Crews, & City Cost Indexes

			CREW	DAILY OUTPUT	LABOR-HOURS	UNIT	2005 BARE COSTS				TOTAL INCL O&P	
							MAT.	LABOR	EQUIP.	TOTAL		
700	0010	**SIDING EXTERIOR**, Alkyd (oil base)										**700**
	0450	Steel siding, oil base, paint 1 coat, brushwork [R09910-220]	2 Pord	2,015	.008	S.F.	.06	.24		.30	.43	
	0500	Spray		4,550	.004		.08	.11		.19	.25	
	0800	Paint 2 coats, brushwork		1,300	.012		.11	.38		.49	.69	
	1000	Spray		4,550	.004		.14	.11		.25	.31	
	1200	Stucco, rough, oil base, paint 2 coats, brushwork		1,300	.012		.11	.38		.49	.69	
	1400	Roller		1,625	.010		.12	.30		.42	.58	
	1600	Spray		2,925	.005		.12	.17		.29	.39	
	1800	Texture 1-11 or clapboard, oil base, primer coat, brushwork		1,300	.012		.09	.38		.47	.67	
	2000	Spray		4,550	.004		.09	.11		.20	.26	
	2400	Paint 2 coats, brushwork		810	.020		.16	.60		.76	1.09	
	2600	Spray		2,600	.006		.18	.19		.37	.48	
	3400	Stain 2 coats, brushwork		950	.017		.10	.52		.62	.88	
	4000	Spray		3,050	.005		.11	.16		.27	.36	
	4200	Wood shingles, oil base primer coat, brushwork		1,300	.012		.09	.38		.47	.66	
	4400	Spray		3,900	.004		.08	.13		.21	.28	
	5000	Paint 2 coats, brushwork		810	.020		.13	.60		.73	1.06	
	5200	Spray		2,275	.007		.13	.22		.35	.46	
	6500	Stain 2 coats, brushwork		950	.017		.10	.52		.62	.88	
	7000	Spray		2,660	.006		.13	.18		.31	.43	
	8000	For latex paint, deduct					10%					
	8100	For work over 12' H, from pipe scaffolding, add						15%				
	8200	For work over 12' H, from extension ladder, add						25%				
	8300	For work over 12' H, from swing staging, add						35%				
710	0010	**SIDING, MISC.** [R09910-220]										**710**
	0100	Aluminum siding										
	0110	Brushwork, primer	2 Pord	2,275	.007	S.F.	.05	.22		.27	.38	
	0120	Finish coat, exterior latex		2,275	.007		.04	.22		.26	.36	
	0130	Primer & 1 coat exterior latex		1,300	.012		.10	.38		.48	.68	
	0140	Primer & 2 coats exterior latex		975	.016		.14	.50		.64	.90	
	0150	Mineral Fiber shingles										
	0160	Brushwork, primer	2 Pord	1,495	.011	S.F.	.09	.33		.42	.59	
	0170	Finish coat, industrial enamel		1,495	.011		.10	.33		.43	.60	
	0180	Primer & 1 coat enamel		810	.020		.19	.60		.79	1.12	
	0190	Primer & 2 coats enamel		540	.030		.30	.91		1.21	1.68	
	0200	Roll, primer		1,625	.010		.10	.30		.40	.56	
	0210	Finish coat, industrial enamel		1,625	.010		.11	.30		.41	.57	
	0220	Primer & 1 coat enamel		975	.016		.21	.50		.71	.98	
	0230	Primer & 2 coats enamel		650	.025		.32	.75		1.07	1.48	
	0240	Spray, primer		3,900	.004		.08	.13		.21	.28	
	0250	Finish coat, industrial enamel		3,900	.004		.09	.13		.22	.29	
	0260	Primer & 1 coat enamel		2,275	.007		.17	.22		.39	.51	
	0270	Primer & 2 coats enamel		1,625	.010		.26	.30		.56	.74	
	0280	Waterproof sealer, first coat		4,485	.004		.07	.11		.18	.23	
	0290	Second coat		5,235	.003		.06	.09		.15	.21	
	0300	Rough wood incl. shingles, shakes or rough sawn siding										
	0310	Brushwork, primer	2 Pord	1,280	.013	S.F.	.11	.38		.49	.71	
	0320	Finish coat, exterior latex		1,280	.013		.07	.38		.45	.66	
	0330	Primer & 1 coat exterior latex		960	.017		.18	.51		.69	.97	
	0340	Primer & 2 coats exterior latex		700	.023		.25	.70		.95	1.33	
	0350	Roll, primer		2,925	.005		.15	.17		.32	.42	
	0360	Finish coat, exterior latex		2,925	.005		.08	.17		.25	.34	
	0370	Primer & 1 coat exterior latex		1,790	.009		.24	.27		.51	.67	
	0380	Primer & 2 coats exterior latex		1,300	.012		.32	.38		.70	.92	
	0390	Spray, primer		3,900	.004		.13	.13		.26	.33	
	0400	Finish coat, exterior latex		3,900	.004		.06	.13		.19	.26	

FINISHES 9

	09910	Paints	CREW	DAILY OUTPUT	LABOR-HOURS	UNIT	2005 BARE COSTS				TOTAL INCL O&P	
							MAT.	LABOR	EQUIP.	TOTAL		
710	0410	Primer & 1 coat exterior latex	2 Pord	2,600	.006	S.F.	.19	.19		.38	.49	**710**
	0420	Primer & 2 coats exterior latex		2,080	.008		.26	.24		.50	.63	
	0430	Waterproof sealer, first coat		4,485	.004		.12	.11		.23	.29	
	0440	Second coat		4,485	.004		.07	.11		.18	.23	
	0450	Smooth wood incl. butt, T&G, beveled, drop or B&B siding										
	0460	Brushwork, primer	2 Pord	2,325	.007	S.F.	.08	.21		.29	.41	
	0470	Finish coat, exterior latex		1,280	.013		.07	.38		.45	.66	
	0480	Primer & 1 coat exterior latex		800	.020		.15	.61		.76	1.09	
	0490	Primer & 2 coats exterior latex		630	.025		.22	.78		1	1.41	
	0500	Roll, primer		2,275	.007		.09	.22		.31	.42	
	0510	Finish coat, exterior latex		2,275	.007		.08	.22		.30	.40	
	0520	Primer & 1 coat exterior latex		1,300	.012		.17	.38		.55	.75	
	0530	Primer & 2 coats exterior latex		975	.016		.24	.50		.74	1.02	
	0540	Spray, primer		4,550	.004		.07	.11		.18	.24	
	0550	Finish coat, exterior latex		4,550	.004		.06	.11		.17	.23	
	0560	Primer & 1 coat exterior latex		2,600	.006		.14	.19		.33	.43	
	0570	Primer & 2 coats exterior latex		1,950	.008		.20	.25		.45	.60	
	0580	Waterproof sealer, first coat		5,230	.003		.07	.09		.16	.21	
	0590	Second coat		5,980	.003		.07	.08		.15	.19	
	0600	For oil base paint, add					10%					
800	0010	**TRIM, EXTERIOR**										**800**
	0100	Door frames & trim (see Doors, interior or exterior)										
	0110	Fascia, latex paint, one coat coverage										
	0120	1" x 4", brushwork	1 Pord	640	.013	L.F.	.02	.38		.40	.60	
	0130	Roll		1,280	.006		.02	.19		.21	.31	
	0140	Spray		2,080	.004		.01	.12		.13	.20	
	0150	1" x 6" to 1" x 10", brushwork		640	.013		.06	.38		.44	.65	
	0160	Roll		1,230	.007		.07	.20		.27	.37	
	0170	Spray		2,100	.004		.05	.12		.17	.23	
	0180	1" x 12", brushwork		640	.013		.06	.38		.44	.65	
	0190	Roll		1,050	.008		.07	.23		.30	.42	
	0200	Spray		2,200	.004		.05	.11		.16	.22	
	0210	Gutters & downspouts, metal, zinc chromate paint										
	0220	Brushwork, gutters, 5", first coat	1 Pord	640	.013	L.F.	.06	.38		.44	.64	
	0230	Second coat		960	.008		.05	.25		.30	.44	
	0240	Third coat		1,280	.006		.04	.19		.23	.34	
	0250	Downspouts, 4", first coat		640	.013		.06	.38		.44	.64	
	0260	Second coat		960	.008		.05	.25		.30	.44	
	0270	Third coat		1,280	.006		.04	.19		.23	.34	
	0280	Gutters & downspouts, wood										
	0290	Brushwork, gutters, 5", primer	1 Pord	640	.013	L.F.	.05	.38		.43	.64	
	0300	Finish coat, exterior latex		640	.013		.05	.38		.43	.64	
	0310	Primer & 1 coat exterior latex		400	.020		.11	.61		.72	1.04	
	0320	Primer & 2 coats exterior latex		325	.025		.17	.75		.92	1.31	
	0330	Downspouts, 4", primer		640	.013		.05	.38		.43	.64	
	0340	Finish coat, exterior latex		640	.013		.05	.38		.43	.64	
	0350	Primer & 1 coat exterior latex		400	.020		.11	.61		.72	1.04	
	0360	Primer & 2 coats exterior latex		325	.025		.08	.75		.83	1.22	
	0370	Molding, exterior, up to 14" wide										
	0380	Brushwork, primer	1 Pord	640	.013	L.F.	.06	.38		.44	.65	
	0390	Finish coat, exterior latex		640	.013		.06	.38		.44	.65	
	0400	Primer & 1 coat exterior latex		400	.020		.13	.61		.74	1.06	
	0410	Primer & 2 coats exterior latex		315	.025		.13	.78		.91	1.31	
	0420	Stain & fill		1,050	.008		.06	.23		.29	.41	
	0430	Shellac		1,850	.004		.07	.13		.20	.27	
	0440	Varnish		1,275	.006		.07	.19		.26	.37	

Reference: R09910-220

Important: See the Reference Section for critical supporting data - Reference Nos., Crews, & City Cost Indexes

09910	Paints	CREW	DAILY OUTPUT	LABOR-HOURS	UNIT	2005 BARE COSTS				TOTAL INCL O&P
						MAT.	LABOR	EQUIP.	TOTAL	
910	**0350**	**WALLS, MASONRY (CMU), EXTERIOR**									**910**
0360	Concrete masonry units (CMU), smooth surface									
0370	Brushwork, latex, first coat	1 Pord	640	.013	S.F.	.04	.38		.42	.62
0380	Second coat		960	.008		.03	.25		.28	.41
0390	Waterproof sealer, first coat		736	.011		.07	.33		.40	.58
0400	Second coat		1,104	.007		.05	.22		.27	.38
0410	Roll, latex, paint, first coat		1,465	.005		.04	.17		.21	.30
0420	Second coat		1,790	.004		.03	.14		.17	.24
0430	Waterproof sealer, first coat		1,680	.005		.07	.15		.22	.30
0440	Second coat		2,060	.004		.05	.12		.17	.23
0450	Spray, latex, paint, first coat		1,950	.004		.03	.13		.16	.23
0460	Second coat		2,600	.003		.03	.09		.12	.17
0470	Waterproof sealer, first coat		2,245	.004		.09	.11		.20	.26
0480	Second coat	▼	2,990	.003	▼	.03	.08		.11	.16
0490	Concrete masonry unit (CMU), porous									
0500	Brushwork, latex, first coat	1 Pord	640	.013	S.F.	.07	.38		.45	.66
0510	Second coat		960	.008		.04	.25		.29	.42
0520	Waterproof sealer, first coat		736	.011		.09	.33		.42	.60
0530	Second coat		1,104	.007		.04	.22		.26	.38
0540	Roll latex, first coat		1,465	.005		.05	.17		.22	.31
0550	Second coat		1,790	.004		.03	.14		.17	.25
0560	Waterproof sealer, first coat		1,680	.005		.09	.15		.24	.32
0570	Second coat		2,060	.004		.05	.12		.17	.24
0580	Spray latex, first coat		1,950	.004		.04	.13		.17	.23
0590	Second coat		2,600	.003		.03	.09		.12	.17
0600	Waterproof sealer, first coat		2,245	.004		.07	.11		.18	.24
0610	Second coat	▼	2,990	.003	▼	.04	.08		.12	.16
920	**0010**	**WALLS AND CEILINGS, Interior**									**920**
0100	Concrete, dry wall or plaster, oil base, primer or sealer coat R09910 -220									
0200	Smooth finish, brushwork	1 Pord	1,150	.007	S.F.	.05	.21		.26	.37
0240	Roller		2,040	.004		.05	.12		.17	.23
0300	Sand finish, brushwork		975	.008		.05	.25		.30	.43
0340	Roller		1,150	.007		.05	.21		.26	.38
0380	Spray		2,275	.004		.04	.11		.15	.20
0800	Paint 2 coats, smooth finish, brushwork		680	.012		.09	.36		.45	.64
0840	Roller		800	.010		.10	.31		.41	.57
0880	Spray		1,625	.005		.08	.15		.23	.32
0900	Sand finish, brushwork		605	.013		.09	.40		.49	.71
0940	Roller		1,020	.008		.10	.24		.34	.47
0980	Spray		1,700	.005		.08	.14		.22	.31
1200	Paint 3 coats, smooth finish, brushwork		510	.016		.14	.48		.62	.87
1240	Roller		650	.012		.14	.38		.52	.73
1280	Spray		1,625	.005		.12	.15		.27	.37
1600	Glaze coating, 5 coats, spray, clear		900	.009		.60	.27		.87	1.07
1640	Multicolor	▼	900	.009		.85	.27		1.12	1.35
1700	For latex paint, deduct					10%				
1800	For ceiling installations, add				▼		25%			
2000	Masonry or concrete block, oil base, primer or sealer coat									
2100	Smooth finish, brushwork	1 Pord	1,224	.007	S.F.	.05	.20		.25	.35
2180	Spray		2,400	.003		.07	.10		.17	.23
2200	Sand finish, brushwork		1,089	.007		.07	.22		.29	.42
2280	Spray		2,400	.003		.07	.10		.17	.23
2800	Paint 2 coats, smooth finish, brushwork		756	.011		.15	.32		.47	.65
2880	Spray		1,360	.006		.14	.18		.32	.42
2900	Sand finish, brushwork	▼	672	.012	▼	.15	.36		.51	.71

FINISHES 9

For expanded coverage of these items see *Means Interior Cost Data 2005*

09910 | Paints

		CREW	DAILY OUTPUT	LABOR-HOURS	UNIT	2005 BARE COSTS				TOTAL INCL O&P	
						MAT.	LABOR	EQUIP.	TOTAL		
920 2980	Spray	1 Pord	1,360	.006	S.F.	.14	.18		.32	.42	**920**
3600	Glaze coating, 5 coats, spray, clear	R09910 -220	900	.009		.60	.27		.87	1.07	
3620	Multicolor		900	.009		.85	.27		1.12	1.35	
4000	Block filler, 1 coat, brushwork		425	.019		.12	.58		.70	1.01	
4100	Silicone, water repellent, 2 coats, spray	▼	2,000	.004		.25	.12		.37	.46	
4120	For latex paint, deduct					10%					
8200	For work 8 - 15' H, add						10%				
8300	For work over 15' H, add	▼			▼		20%				
940 0010	**DRY FALL PAINTING**									.	**940**
0100	Walls										
0200	Wallboard and smooth plaster, one coat, brush	1 Pord	910	.009	S.F.	.04	.27		.31	.45	
0210	Roll		1,560	.005		.04	.16		.20	.29	
0220	Spray		2,600	.003		.04	.09		.13	.19	
0230	Two coats, brush		520	.015		.09	.47		.56	.81	
0240	Roll		877	.009		.09	.28		.37	.52	
0250	Spray		1,560	.005		.09	.16		.25	.34	
0260	Concrete or textured plaster, one coat, brush		747	.011		.04	.33		.37	.54	
0270	Roll		1,300	.006		.04	.19		.23	.33	
0280	Spray		1,560	.005		.04	.16		.20	.29	
0290	Two coats, brush		422	.019		.09	.58		.67	.97	
0300	Roll		747	.011		.09	.33		.42	.59	
0310	Spray		1,300	.006		.09	.19		.28	.38	
0320	Concrete block, one coat, brush		747	.011		.04	.33		.37	.54	
0330	Roll		1,300	.006		.04	.19		.23	.33	
0340	Spray		1,560	.005		.04	.16		.20	.29	
0350	Two coats, brush		422	.019		.09	.58		.67	.97	
0360	Roll		747	.011		.09	.33		.42	.59	
0370	Spray		1,300	.006		.09	.19		.28	.38	
0380	Wood, one coat, brush		747	.011		.04	.33		.37	.54	
0390	Roll		1,300	.006		.04	.19		.23	.33	
0400	Spray		877	.009		.04	.28		.32	.47	
0410	Two coats, brush		487	.016		.09	.50		.59	.86	
0420	Roll		747	.011		.09	.33		.42	.59	
0430	Spray	▼	650	.012	▼	.09	.38		.47	.67	
0440	Ceilings										
0450	Wallboard and smooth plaster, one coat, brush	1 Pord	600	.013	S.F.	.04	.41		.45	.66	
0460	Roll		1,040	.008		.04	.24		.28	.40	
0470	Spray		1,560	.005		.04	.16		.20	.29	
0480	Two coats, brush		341	.023		.09	.72		.81	1.18	
0490	Roll		650	.012		.09	.38		.47	.67	
0500	Spray		1,300	.006		.09	.19		.28	.38	
0510	Concrete or textured plaster, one coat, brush		487	.016		.04	.50		.54	.81	
0520	Roll		877	.009		.04	.28		.32	.47	
0530	Spray		1,560	.005		.04	.16		.20	.29	
0540	Two coats, brush		276	.029		.09	.89		.98	1.43	
0550	Roll		520	.015		.09	.47		.56	.81	
0560	Spray		1,300	.006		.09	.19		.28	.38	
0570	Structural steel, bar joists or metal deck, one coat, spray		1,560	.005		.04	.16		.20	.29	
0580	Two coats, spray	▼	1,040	.008	▼	.09	.24		.33	.45	

09930 | Stains/Transp. Finishes

		CREW	DAILY OUTPUT	LABOR-HOURS	UNIT	MAT.	LABOR	EQUIP.	TOTAL	TOTAL INCL O&P	
100 0010	**VARNISH** 1 coat + sealer, on wood trim, no sanding included	1 Pord	400	.020	S.F.	.07	.61		.68	1	**100**
0100	Hardwood floors, 2 coats, no sanding included, roller	"	1,890	.004	"	.13	.13		.26	.33	

9
FINISHES

Important: See the Reference Section for critical supporting data - Reference Nos., Crews, & City Cost Indexes

09963 | Glazed Coatings

		CREW	DAILY OUTPUT	LABOR-HOURS	UNIT	2005 BARE COSTS MAT.	LABOR	EQUIP.	TOTAL	TOTAL INCL O&P	
200	0010	**WALL COATINGS**										200
	0100	Acrylic glazed coatings, minimum	1 Pord	525	.015	S.F.	.26	.47		.73	.99	
	0200	Maximum		305	.026		.54	.80		1.34	1.80	
	0300	Epoxy coatings, minimum		525	.015		.33	.47		.80	1.06	
	0400	Maximum		170	.047		1.02	1.44		2.46	3.28	
	0600	Exposed aggregate, troweled on, 1/16" to 1/4", minimum		235	.034		.50	1.04		1.54	2.12	
	0700	Maximum (epoxy or polyacrylate)		130	.062		1.09	1.88		2.97	4.03	
	0900	1/2" to 5/8" aggregate, minimum		130	.062		1.01	1.88		2.89	3.94	
	1000	Maximum		80	.100		1.72	3.06		4.78	6.50	
	1500	Exposed aggregate, sprayed on, 1/8" aggregate, minimum		295	.027		.47	.83		1.30	1.77	
	1600	Maximum		145	.055		.87	1.69		2.56	3.50	
	1800	High build epoxy, 50 mil, minimum		390	.021		.56	.63		1.19	1.56	
	1900	Maximum		95	.084		.96	2.58		3.54	4.93	
	2100	Laminated epoxy with fiberglass, minimum		295	.027		.61	.83		1.44	1.92	
	2200	Maximum		145	.055		1.08	1.69		2.77	3.73	
	2400	Sprayed perlite or vermiculite, 1/16" thick, minimum		2,935	.003		.22	.08		.30	.37	
	2500	Maximum		640	.013		.62	.38		1	1.26	
	2700	Vinyl plastic wall coating, minimum		735	.011		.28	.33		.61	.81	
	2800	Maximum		240	.033		.68	1.02		1.70	2.28	
	3000	Urethane on smooth surface, 2 coats, minimum		1,135	.007		.21	.22		.43	.55	
	3100	Maximum		665	.012		.46	.37		.83	1.06	
	3600	Ceramic-like glazed coating, cementitious, minimum		440	.018		.40	.56		.96	1.28	
	3700	Maximum		345	.023		.67	.71		1.38	1.81	
	3900	Resin base, minimum		640	.013		.28	.38		.66	.89	
	4000	Maximum	▼	330	.024	▼	.45	.74		1.19	1.62	

09990 | Paint Restoration

		CREW	DAILY OUTPUT	LABOR-HOURS	UNIT	2005 BARE COSTS MAT.	LABOR	EQUIP.	TOTAL	TOTAL INCL O&P	
800	0010	**SANDING** and puttying interior trim, compared to										800
	0100	Painting 1 coat, on quality work				L.F.		100%				
	0300	Medium work						50%				
	0400	Industrial grade				▼		25%				
	0500	Surface protection, placement and removal										
	0510	Basic drop cloths	1 Pord	6,400	.001	S.F.		.04		.04	.06	
	0520	Masking with paper		800	.010		.03	.31		.34	.49	
	0530	Volume cover up (using plastic sheathing, or building paper)	▼	16,000	.001	▼		.02		.02	.02	
900	0010	**SURFACE PREPARATION, EXTERIOR**										900
	0015	Doors, per side, not incl. frames or trim										
	0020	Scrape & sand										
	0030	Wood, flush	1 Pord	616	.013	S.F.		.40		.40	.60	
	0040	Wood, detail		496	.016			.49		.49	.74	
	0050	Wood, louvered		280	.029			.87		.87	1.31	
	0060	Wood, overhead	▼	616	.013	▼		.40		.40	.60	
	0070	Wire brush										
	0080	Metal, flush	1 Pord	640	.013	S.F.		.38		.38	.58	
	0090	Metal, detail		520	.015			.47		.47	.71	
	0100	Metal, louvered		360	.022			.68		.68	1.02	
	0110	Metal or fibr., overhead		640	.013			.38		.38	.58	
	0120	Metal, roll up		560	.014			.44		.44	.66	
	0130	Metal, bulkhead	▼	640	.013	▼		.38		.38	.58	
	0140	Power wash, based on 2500 lb. operating pressure										
	0150	Metal, flush	B-9	2,240	.018	S.F.		.48	.06	.54	.82	
	0160	Metal, detail		2,120	.019			.51	.07	.58	.87	
	0170	Metal, louvered		2,000	.020			.54	.07	.61	.92	
	0180	Metal or fibr., overhead		2,400	.017			.45	.06	.51	.76	
	0190	Metal, roll up	▼	2,400	.017	▼		.45	.06	.51	.76	

FINISHES **9**

			CREW	DAILY OUTPUT	LABOR-HOURS	UNIT	2005 BARE COSTS				TOTAL INCL O&P	
09990		**Paint Restoration**					MAT.	LABOR	EQUIP.	TOTAL		
900	0200	Metal, bulkhead	B-9	2,200	.018	S.F.		.49	.06	.55	.84	**900**
	0400	Windows, per side, not incl. trim										
	0410	Scrape & sand										
	0420	Wood, 1-2 lite	1 Pord	320	.025	S.F.		.77		.77	1.15	
	0430	Wood, 3-6 lite		280	.029			.87		.87	1.31	
	0440	Wood, 7-10 lite		240	.033			1.02		1.02	1.53	
	0450	Wood, 12 lite		200	.040			1.22		1.22	1.84	
	0460	Wood, Bay / Bow		320	.025			.77		.77	1.15	
	0470	Wire brush										
	0480	Metal, 1-2 lite	1 Pord	480	.017	S.F.		.51		.51	.77	
	0490	Metal, 3-6 lite		400	.020			.61		.61	.92	
	0500	Metal, Bay / Bow		480	.017			.51		.51	.77	
	0510	Power wash, based on 2500 lb. operating pressure										
	0520	1-2 lite	B-9	4,400	.009	S.F.		.25	.03	.28	.42	
	0530	3-6 lite		4,320	.009			.25	.03	.28	.43	
	0540	7-10 lite		4,240	.009			.26	.03	.29	.44	
	0550	12 lite		4,160	.010			.26	.03	.29	.45	
	0560	Bay / Bow		4,400	.009			.25	.03	.28	.42	
	0600	Siding, scrape and sand, light=10-30%, med.=30-70%										
	0610	Heavy=70-100%, % of surface to sand										
	0650	Texture 1-11, light	1 Pord	480	.017	S.F.		.51		.51	.77	
	0660	Med.		440	.018			.56		.56	.84	
	0670	Heavy		360	.022			.68		.68	1.02	
	0680	Wood shingles, shakes, light		440	.018			.56		.56	.84	
	0690	Med.		360	.022			.68		.68	1.02	
	0700	Heavy		280	.029			.87		.87	1.31	
	0710	Clapboard, light		520	.015			.47		.47	.71	
	0720	Med.		480	.017			.51		.51	.77	
	0730	Heavy		400	.020			.61		.61	.92	
	0740	Wire brush										
	0750	Aluminum, light	1 Pord	600	.013	S.F.		.41		.41	.61	
	0760	Med.		520	.015			.47		.47	.71	
	0770	Heavy		440	.018			.56		.56	.84	
	0780	Pressure wash, based on 2500 lb.. operating pressure										
	0790	Stucco	B-9	3,080	.013	S.F.		.35	.05	.40	.60	
	0800	Aluminum or vinyl		3,200	.013			.34	.04	.38	.58	
	0810	Siding, masonry, brick & block		2,400	.017			.45	.06	.51	.76	
	1300	Miscellaneous, wire brush										
	1310	Metal, pedestrian gate	1 Pord	100	.080	S.F.		2.45		2.45	3.68	
	8000	For Chemical Washing, see Division 04930										
	8010	For Steam Cleaning, see Division 04930										
	8020	For Sand Blasting, see Division 05910-500 and 03350-350										
910	0010	**SURFACE PREPARATION, INTERIOR**										**910**
	0020	Doors										
	0030	Scrape & sand										
	0040	Wood, flush	1 Pord	616	.013	S.F.		.40		.40	.60	
	0050	Wood, detail		496	.016			.49		.49	.74	
	0060	Wood, louvered		280	.029			.87		.87	1.31	
	0070	Wire brush										
	0080	Metal, flush	1 Pord	640	.013	S.F.		.38		.38	.58	
	0090	Metal, detail		520	.015			.47		.47	.71	
	0100	Metal, louvered		360	.022			.68		.68	1.02	
	0110	Hand wash										
	0120	Wood, flush	1 Pord	2,160	.004	S.F.		.11		.11	.17	
	0130	Wood, detailed		2,000	.004			.12		.12	.18	
	0140	Wood, louvered		1,360	.006			.18		.18	.27	

Important: See the Reference Section for critical supporting data - Reference Nos., Crews, & City Cost Indexes

9 FINISHES

09990	Paint Restoration	CREW	DAILY OUTPUT	LABOR-HOURS	UNIT	2005 BARE COSTS				TOTAL INCL O&P	
						MAT.	LABOR	EQUIP.	TOTAL		
910	0150	Metal, flush	1 Pord	2,160	.004	S.F.		.11		.11	.17	910
	0160	Metal, detail		2,000	.004			.12		.12	.18	
	0170	Metal, louvered	↓	1,360	.006	↓		.18		.18	.27	
	0400	Windows, per side, not incl. trim										
	0410	Scrape & sand										
	0420	Wood, 1-2 lite	1 Pord	360	.022	S.F.		.68		.68	1.02	
	0430	Wood, 3-6 lite		320	.025			.77		.77	1.15	
	0440	Wood, 7-10 lite		280	.029			.87		.87	1.31	
	0450	Wood, 12 lite		240	.033			1.02		1.02	1.53	
	0460	Wood, Bay / Bow	↓	360	.022	↓		.68		.68	1.02	
	0470	Wire brush										
	0480	Metal, 1-2 lite	1 Pord	520	.015	S.F.		.47		.47	.71	
	0490	Metal, 3-6 lite		440	.018			.56		.56	.84	
	0500	Metal, Bay / Bow	↓	520	.015	↓		.47		.47	.71	
	0600	Walls, sanding, light=10-30%										
	0610	Med.=30-70%, heavy=70-100%, % of surface to sand										
	0650	Walls, sand										
	0660	Drywall, gypsum, plaster, light	1 Pord	3,077	.003	S.F.		.08		.08	.12	
	0670	Drywall, gypsum, plaster, med.		2,160	.004			.11		.11	.17	
	0680	Drywall, gypsum, plaster, heavy		923	.009			.27		.27	.40	
	0690	Wood, T&G, light		2,400	.003			.10		.10	.15	
	0700	Wood, T&G, med.		1,600	.005			.15		.15	.23	
	0710	Wood, T&G, heavy	↓	800	.010	↓		.31		.31	.46	
	0720	Walls, wash										
	0730	Drywall, gypsum, plaster	1 Pord	3,200	.002	S.F.		.08		.08	.12	
	0740	Wood, T&G		3,200	.002			.08		.08	.12	
	0750	Masonry, brick & block, smooth		2,800	.003			.09		.09	.13	
	0760	Masonry, brick & block, coarse	↓	2,000	.004	↓		.12		.12	.18	
	8000	For Chemical Washing, see Division 04930										
	8010	For Steam Cleaning, see Division 04930										
	8020	For Sand Blasting, see Division 03350-350 and 05910-500										

FINISHES 9

For information about Means Estimating Seminars, see yellow pages 12 and 13 in back of book

Division Notes

		CREW	DAILY OUTPUT	LABOR-HOURS	UNIT	2005 BARE COSTS				TOTAL INCL O&P
						MAT.	LABOR	EQUIP.	TOTAL	

Division 10
Specialties

Estimating Tips

General

- The items in this division are usually priced per square foot or each.
- Many items in Division 10 require some type of support system or special anchors that are not usually furnished with the item. The required anchors must be added to the estimate in the appropriate division.
- Some items in Division 10, such as lockers, may require assembly before installation. Verify the amount of assembly required. Assembly can often exceed installation time.

10150 Compartments & Cubicles

- Support angles and blocking are not included in the installation of toilet compartments, shower/dressing compartments or cubicles. Appropriate line items from Divisions 5 or 6 may need to be added to support the installations.
- Toilet partitions are priced by the stall. A stall consists of a side wall, pilaster and door with hardware. Toilet tissue holders and grab bars are extra.

10600 Partitions

- The required acoustical rating of a folding partition can have a significant impact on costs. Verify the sound transmission coefficient rating of the panel priced to the specification requirements.

10800 Toilet/Bath/Laundry Accessories

- Grab bar installation does not include supplemental blocking or backing to support the required load. When grab bars are installed at an existing facility provisions must be made to attach the grab bars to solid structure.

Reference Numbers

Reference numbers are shown in bold squares at the beginning of some major classifications. These numbers refer to related items in the Reference Section. The reference information may be an estimating procedure, an alternate pricing method or technical information.

Note: Not all subdivisions listed here necessarily appear in this publication.

10 SPECIALTIES

			CREW	DAILY OUTPUT	LABOR-HOURS	UNIT	2005 BARE COSTS MAT.	LABOR	EQUIP.	TOTAL	TOTAL INCL O&P	
10110	**Chalkboards**											
240	0010	**CHALKBOARDS** Porcelain enamel steel										240
	0100	Freestanding, reversible										
	0120	Economy, wood frame, 4' x 6'										
	0140	Chalkboard both sides				Ea.	545			545	600	
	0160	Chalkboard one side, cork other side				"	455			455	500	
	0200	Standard, lightweight satin finished aluminum, 4' x 6'										
	0220	Chalkboard both sides				Ea.	565			565	620	
	0240	Chalkboard one side, cork other side				"	490			490	540	
	0300	Deluxe, heavy duty extruded aluminum, 4' x 6'										
	0320	Chalkboard both sides				Ea.	1,175			1,175	1,300	
	0340	Chalkboard one side, cork other side				"	1,075			1,075	1,175	
	0400	Sliding chalkboards										
	0450	Vertical, one sliding board with back panel, wall mounted										
	0500	8' x 4'	2 Carp	8	2	Ea.	1,525	68.50		1,593.50	1,775	
	0520	8' x 8'		7.50	2.133		2,125	73		2,198	2,475	
	0540	8' x 12'	↓	7	2.286	↓	2,650	78.50		2,728.50	3,025	
	0600	Two sliding boards, with back panel										
	0620	8' x 4'	2 Carp	8	2	Ea.	2,325	68.50		2,393.50	2,650	
	0640	8' x 8'		7.50	2.133		3,200	73		3,273	3,650	
	0660	8' x 12'	↓	7	2.286	↓	3,975	78.50		4,053.50	4,500	
	0700	Horizontal, two track										
	0800	4' x 8', 2 sliding panels	2 Carp	8	2	Ea.	1,575	68.50		1,643.50	1,825	
	0820	4' x 12', 2 sliding panels		7.50	2.133		1,975	73		2,048	2,300	
	0840	4' x 16', 4 sliding panels	↓	7	2.286	↓	2,675	78.50		2,753.50	3,075	
	0900	Four track, four sliding panels										
	0920	4' x 8'	2 Carp	8	2	Ea.	2,525	68.50		2,593.50	2,875	
	0940	4' x 12'		7.50	2.133		3,200	73		3,273	3,625	
	0960	4' x 16'	↓	7	2.286	↓	4,050	78.50		4,128.50	4,575	
	1200	Vertical, motor operated										
	1400	One sliding panel with back panel										
	1450	10' x 4'	2 Carp	4	4	Ea.	5,775	137		5,912	6,575	
	1500	10' x 10'		3.75	4.267		6,700	146		6,846	7,600	
	1550	10' x 16'	↓	3.50	4.571	↓	7,650	157		7,807	8,650	
	1700	Two sliding panels with back panel										
	1750	10' x 4'	2 Carp	4	4	Ea.	9,250	137		9,387	10,400	
	1800	10' x 10'		3.75	4.267		10,500	146		10,646	11,700	
	1850	10' x 16'	↓	3.50	4.571	↓	11,800	157		11,957	13,200	
	2000	Three sliding panels with back panel										
	2100	10' x 4'	2 Carp	4	4	Ea.	12,100	137		12,237	13,500	
	2150	10' x 10'		3.75	4.267		14,000	146		14,146	15,600	
	2200	10' x 16'	↓	3.50	4.571	↓	15,800	157		15,957	17,600	
	2400	For projection screen, glass beaded, add				S.F.	10.80			10.80	11.90	
	2500	For remote control, 1 panel control, add				Ea.	455			455	500	
	2600	2 panel control, add				"	785			785	865	
	2800	For units without back panels, deduct				S.F.	4.99			4.99	5.50	
	2850	For liquid chalk porcelain panels, add				"	4.21			4.21	4.63	
	3000	Swing leaf, any comb. of chalkboard & cork, aluminum frame										
	3100	Floor style, 6 panels										
	3150	30" x 40" panels				Ea.	880			880	965	
	3200	48" x 40" panels				"	2,275			2,275	2,525	
	3300	Wall mounted, 6 panels										
	3400	30" x 40" panels	2 Carp	16	1	Ea.	810	34.50		844.50	950	
	3450	48" x 40" panels	"	16	1	"	1,700	34.50		1,734.50	1,900	
	3600	Extra panels for swing leaf units										
	3700	30" x 40" panels				Ea.	263			263	290	
	3750	48" x 40" panels				"	315			315	345	

Important: See the Reference Section for critical supporting data - Reference Nos., Crews, & City Cost Indexes

10110 | Chalkboards

			CREW	DAILY OUTPUT	LABOR-HOURS	UNIT	MAT.	LABOR	EQUIP.	TOTAL	TOTAL INCL O&P	
240	3900	Wall hung										**240**
	4000	Aluminum frame and chalktrough										
	4200	3' x 4'	2 Carp	16	1	Ea.	150	34.50		184.50	219	
	4300	3' x 5'		15	1.067		218	36.50		254.50	297	
	4500	4' x 8'		14	1.143		335	39		374	425	
	4600	4' x 12'		13	1.231		480	42		522	595	
	4700	Wood frame and chalktrough										
	4800	3' x 4'	2 Carp	16	1	Ea.	155	34.50		189.50	225	
	5000	3' x 5'		15	1.067		217	36.50		253.50	295	
	5100	4' x 5'		14	1.143		219	39		258	300	
	5300	4' x 8'		13	1.231		310	42		352	405	
	5400	Liquid chalk, white porcelain enamel, wall hung										
	5420	Deluxe units, aluminum trim and chalktrough										
	5450	4' x 4'	2 Carp	16	1	Ea.	191	34.50		225.50	264	
	5500	4' x 8'		14	1.143		335	39		374	425	
	5550	4' x 12'		12	1.333		455	45.50		500.50	570	
	5700	Wood trim and chalktrough										
	5900	4' x 4'	2 Carp	16	1	Ea.	440	34.50		474.50	540	
	6000	4' x 6'		15	1.067		530	36.50		566.50	635	
	6200	4' x 8'		14	1.143		615	39		654	735	
	6300	Liquid chalk, felt tip markers					1.50			1.50	1.65	
	6500	Erasers					1.53			1.53	1.68	
	6600	Board cleaner, 8 oz. bottle					4.13			4.13	4.54	

10120 | Tack & Visual Aid Boards

			CREW	DAILY OUTPUT	LABOR-HOURS	UNIT	MAT.	LABOR	EQUIP.	TOTAL	TOTAL INCL O&P	
350	0010	**CONTROL BOARDS** Magnetic, porcelain finish, 18" x 24", framed	2 Carp	8	2	Ea.	189	68.50		257.50	315	**350**
	0100	24" x 36"		7.50	2.133		275	73		348	420	
	0200	36" x 48"		7	2.286		365	78.50		443.50	525	
	0300	48" x 72"		6	2.667		810	91.50		901.50	1,025	
	0400	48" x 96"		5	3.200		1,000	110		1,110	1,275	
940	0010	**BULLETIN BOARD** Cork sheets, unbacked, no frame, 1/4" thick	2 Carp	290	.055	S.F.	3.40	1.89		5.29	6.70	**940**
	0100	1/2" thick		290	.055		5.80	1.89		7.69	9.30	
	0300	Fabric-face, no frame, on 7/32" cork underlay		290	.055		5.10	1.89		6.99	8.55	
	0400	On 1/4" cork on 1/4" hardboard		290	.055		6.15	1.89		8.04	9.75	
	0600	With edges wrapped		290	.055		6.60	1.89		8.49	10.20	
	0700	On 7/16" fire retardant core		290	.055		4.45	1.89		6.34	7.85	
	0900	With edges wrapped		290	.055		6.60	1.89		8.49	10.20	
	1000	Designer fabric only, cut to size					1.90			1.90	2.09	
	1200	1/4" vinyl cork, on 1/4" hardboard, no frame	2 Carp	290	.055		6.10	1.89		7.99	9.65	
	1300	On 1/4" coreboard		290	.055		6.20	1.89		8.09	9.75	
	2000	For map and display rail, economy, add		385	.042	L.F.	2.08	1.42		3.50	4.51	
	2100	Deluxe, add		350	.046	"	3.04	1.57		4.61	5.80	
	2120	Prefabricated, 1/4" cork, 3' x 5' with aluminum frame		16	1	Ea.	114	34.50		148.50	179	
	2140	Wood frame		16	1		147	34.50		181.50	216	
	2160	4' x 4' with aluminum frame		16	1		94.50	34.50		129	158	
	2180	Wood frame		16	1		130	34.50		164.50	197	
	2200	4' x 8' with aluminum frame		14	1.143		182	39		221	261	
	2210	With wood frame		14	1.143		121	39		160	194	
	2220	4' x 12' with aluminum frame		12	1.333		262	45.50		307.50	360	
	2230	Bulletin board case, single glass door, with lock										
	2240	36" x 24", economy	2 Carp	12	1.333	Ea.	198	45.50		243.50	288	
	2250	Deluxe		12	1.333		450	45.50		495.50	565	
	2260	42" x 30", economy		12	1.333		365	45.50		410.50	470	
	2270	Deluxe		12	1.333		465	45.50		510.50	580	

SPECIALTIES 10

10120 | Tack & Visual Aid Boards

		CREW	DAILY OUTPUT	LABOR-HOURS	UNIT	2005 BARE COSTS				TOTAL INCL O&P
						MAT.	LABOR	EQUIP.	TOTAL	
940	2300	Glass enclosed cabinets, alum., cork panel, hinged doors									940
	2400	3' x 3', 1 door	2 Carp	12	1.333	Ea.	510	45.50		555.50	630
	2500	4' x 4', 2 door		11	1.455		850	50		900	1,025
	2600	4' x 7', 3 door		10	1.600		1,400	55		1,455	1,625
	2800	4' x 10', 4 door		8	2		2,050	68.50		2,118.50	2,350
	2900	For lights, add per door opening	1 Elec	13	.615		154	25		179	208
	3100	Horizontal sliding units, 4 doors, 4' x 8' 8' x 4'	2 Carp	9	1.778		1,575	61		1,636	1,825
	3200	4' x 12'		7	2.286		1,975	78.50		2,053.50	2,300
	3400	8 doors, 4' x 16'		5	3.200		2,675	110		2,785	3,125
	3500	4' x 24'		4	4		3,600	137		3,737	4,200

10160 | Metal Toilet Compartments

		CREW	DAILY OUTPUT	LABOR-HOURS	UNIT	2005 BARE COSTS				TOTAL INCL O&P
						MAT.	LABOR	EQUIP.	TOTAL	
100	0010	**TOILET PARTITIONS, METAL**									100
	0110	Cubicles, ceiling hung									
	0200	Painted metal	2 Carp	4	4	Ea.	455	137		592	715
	0500	Stainless steel	"	4	4		1,175	137		1,312	1,525
	0600	For handicap units, incl. 52" grab bars, add					320			320	355
	0900	Floor and ceiling anchored									
	1000	Painted metal	2 Carp	5	3.200	Ea.	415	110		525	630
	1300	Stainless steel	"	5	3.200		1,350	110		1,460	1,675
	1400	For handicap units, incl. 52" grab bars, add					287			287	315
	1610	Floor mounted									
	1700	Painted metal	2 Carp	7	2.286	Ea.	465	78.50		543.50	635
	2000	Stainless steel	"	7	2.286		1,450	78.50		1,528.50	1,700
	2100	For handicap units, incl. 52" grab bars, add					287			287	315
	2200	For juvenile units, deduct					39.50			39.50	43.50
	2450	Floor mounted, headrail braced									
	2500	Painted metal	2 Carp	6	2.667	Ea.	415	91.50		506.50	600
	2800	Stainless steel	"	6	2.667		1,350	91.50		1,441.50	1,625
	2900	For handicap units, incl. 52" grab bars, add					287			287	315
	3000	Wall hung partitions, painted metal	2 Carp	7	2.286		525	78.50		603.50	700
	3300	Stainless steel	"	7	2.286		1,325	78.50		1,403.50	1,575
	3400	For handicap units, incl. 52" grab bars, add					287			287	315
	4000	Screens, entrance, floor mounted, 58" high, 48" wide									
	4200	Painted metal	2 Carp	15	1.067	Ea.	188	36.50		224.50	264
	4500	Stainless steel	"	15	1.067	"	715	36.50		751.50	845
	4650	Urinal screen, 18" wide									
	4700	Painted metal	2 Carp	8	2	Ea.	156	68.50		224.50	279
	5000	Stainless steel	"	8	2	"	515	68.50		583.50	670
	5100	Floor mounted, head rail braced									
	5300	Painted metal	2 Carp	8	2	Ea.	172	68.50		240.50	296
	5600	Stainless steel	"	8	2	"	600	68.50		668.50	765
	5750	Pilaster, flush									
	5800	Painted metal	2 Carp	10	1.600	Ea.	252	55		307	365
	6100	Stainless steel		10	1.600		450	55		505	580
	6300	Urinal screen, post braced, painted metal		10	1.600		260	55		315	370
	6600	Stainless steel		10	1.600		450	55		505	580
	6700	Wall hung, bracket supported									

Important: See the Reference Section for critical supporting data - Reference Nos., Crews, & City Cost Indexes

10 SPECIALTIES

10150 | Compartments & Cubicles

10160 | Metal Toilet Compartments

			CREW	DAILY OUTPUT	LABOR-HOURS	UNIT	2005 BARE COSTS MAT.	LABOR	EQUIP.	TOTAL	TOTAL INCL O&P	
100	6800	Painted metal	2 Carp	10	1.600	Ea.	262	55		317	375	100
	7100	Stainless steel		10	1.600		410	55		465	535	
	7400	Flange supported, painted metal		10	1.600		193	55		248	298	
	7700	Stainless steel		10	1.600		475	55		530	610	
	7800	Wedge type, painted metal		10	1.600		228	55		283	335	
	8100	Stainless steel		10	1.600		490	55		545	625	

10165 | Plastic Laminate Toilet Compartment

			CREW	DAILY OUTPUT	LABOR-HOURS	UNIT	MAT.	LABOR	EQUIP.	TOTAL	TOTAL INCL O&P	
100	0010	**TOILET PARTITIONS, PLASTIC LAMINATE**										100
	0110	Cubicles, ceiling hung										
	0300	Plastic laminate on particle board	2 Carp	4	4	Ea.	590	137		727	865	
	0600	For handicap units, incl. 52" grab bars, add				"	320			320	355	
	0900	Floor and ceiling anchored										
	1100	Plastic laminate on particle board	2 Carp	5	3.200	Ea.	665	110		775	900	
	1400	For handicap units, incl. 52" grab bars, add				"	287			287	315	
	1610	Floor mounted										
	1800	Plastic laminate on particle board	2 Carp	7	2.286	Ea.	585	78.50		663.50	760	
	2450	Floor mounted, headrail braced										
	2600	Plastic laminate on particle board CN	2 Carp	6	2.667	Ea.	770	91.50		861.50	990	
	3400	For handicap units, incl. 52" grab bars, add					287			287	315	
	4300	Entrance screen, floor mtd., plas. lam., 58" high, 48" wide	2 Carp	15	1.067		435	36.50		471.50	530	
	4800	Urinal screen, 18" wide, ceiling braced, plastic laminate		8	2		293	68.50		361.50	425	
	5400	Floor mounted, headrail braced		8	2		355	68.50		423.50	495	
	5900	Pilaster, flush, plastic laminate		10	1.600		360	55		415	480	
	6400	Post braced, plastic laminate		10	1.600		360	55		415	480	
	6700	Wall hung, bracket supported										
	6900	Plastic laminate on particle board	2 Carp	10	1.600	Ea.	167	55		222	269	
	7450	Flange supported										
	7500	Plastic laminate on particle board	2 Carp	10	1.600	Ea.	395	55		450	520	

10170 | Plastic Toilet Compartments

			CREW	DAILY OUTPUT	LABOR-HOURS	UNIT	MAT.	LABOR	EQUIP.	TOTAL	TOTAL INCL O&P	
100	0010	**TOILET PARTITIONS, PLASTIC**										100
	0110	Cubicles, ceiling hung										
	0250	Phenolic	2 Carp	4	4	Ea.	875	137		1,012	1,175	
	0600	For handicap units, incl. 52" grab bars, add				"	320			320	355	
	0900	Floor and ceiling anchored										
	1050	Phenolic	2 Carp	5	3.200	Ea.	920	110		1,030	1,200	
	1400	For handicap units, incl. 52" grab bars, add				"	287			287	315	
	1610	Floor mounted										
	1750	Phenolic	2 Carp	7	2.286	Ea.	990	78.50		1,068.50	1,225	
	2100	For handicap units, incl. 52" grab bars, add					287			287	315	
	2200	For juvenile units, deduct					39.50			39.50	43.50	
	2450	Floor mounted, headrail braced										
	2550	Phenolic	2 Carp	6	2.667	Ea.	890	91.50		981.50	1,125	

10180 | Stone Toilet Compartments

			CREW	DAILY OUTPUT	LABOR-HOURS	UNIT	MAT.	LABOR	EQUIP.	TOTAL	TOTAL INCL O&P	
100	0010	**TOILET PARTITIONS, STONE**										100
	0100	Cubicles, ceiling hung, marble	2 Marb	2	8	Ea.	1,475	266		1,741	2,025	
	0600	For handicap units, incl. 52" grab bars, add					320			320	355	
	0800	Floor & ceiling anchored, marble	2 Marb	2.50	6.400		1,600	212		1,812	2,100	
	1400	For handicap units, incl. 52" grab bars, add					287			287	315	
	1600	Floor mounted, marble	2 Marb	3	5.333		945	177		1,122	1,325	
	2400	Floor mounted, headrail braced, marble	"	3	5.333		905	177		1,082	1,275	
	2900	For handicap units, incl. 52" grab bars, add					287			287	315	

SPECIALTIES 10

			DAILY	LABOR-		2005 BARE COSTS				TOTAL	
	10180	**Stone Toilet Compartments**	CREW	OUTPUT	HOURS	UNIT	MAT.	LABOR	EQUIP.	TOTAL	INCL O&P
100	4100	Entrance screen, floor mounted marble, 58" high, 48" wide	2 Marb	9	1.778	Ea.	595	59		654	745	**100**
	4600	Urinal screen, 18" wide, ceiling braced, marble	D-1	6	2.667	↓	595	83		678	780
	5100	Floor mounted, head rail braced									
	5200	Marble	D-1	6	2.667	Ea.	520	83		603	700
	5700	Pilaster, flush, marble		9	1.778		660	55.50		715.50	815
	6200	Post braced, marble	↓	9	1.778	↓	655	55.50		710.50	805

	10185	**Shower/Dressing Compartments**									
100	0010	**PARTITIONS, SHOWER** Floor mounted, no plumbing										**100**
	0100	Cabinet, incl. base, no door, painted steel, 1" thick walls	2 Shee	5	3.200	Ea.	730	129		859	1,000
	0300	With door, fiberglass		4.50	3.556		590	143		733	870
	0600	Galvanized and painted steel, 1" thick walls		5	3.200		770	129		899	1,050
	0800	Stall, 1" thick wall, no base, enameled steel		5	3.200		830	129		959	1,125
	1200	Stainless steel	↓	5	3.200		1,325	129		1,454	1,675
	1400	For double entry type, no doors, deduct					10%				
	1500	Circular fiberglass, cabinet 36" diameter,	2 Shee	4	4		605	161		766	910
	1700	One piece, 36" diameter, less door		4	4		510	161		671	805
	1800	With door		3.50	4.571		840	184		1,024	1,200
	2000	Curved shell shower, no door needed	↓	3	5.333		730	214		944	1,125
	2300	For fiberglass seat, add to both above					104			104	114
	2400	Glass stalls, with doors, no receptors, chrome on brass	2 Shee	3	5.333		1,200	214		1,414	1,650
	2700	Anodized aluminum	"	4	4		830	161		991	1,150
	2900	Marble shower stall, stock design, with shower door	2 Marb	1.20	13.333		1,850	445		2,295	2,700
	3000	With curtain		1.30	12.308		1,625	410		2,035	2,425
	3200	Receptors, precast terrazzo, 32" x 32"		14	1.143		225	38		263	305
	3300	48" x 34"		9.50	1.684		385	56		441	510
	3500	Plastic, simulated terrazzo receptor, 32" x 32"		14	1.143		95	38		133	163
	3600	32" x 48"		12	1.333		140	44.50		184.50	222
	3800	Precast concrete, colors, 32" x 32"		14	1.143		188	38		226	265
	3900	48" x 48"	↓	8	2		200	66.50		266.50	320
	4100	Shower doors, economy plastic, 24" wide	1 Shee	9	.889		104	35.50		139.50	169
	4200	Tempered glass door, economy		8	1		182	40		222	262
	4400	Folding, tempered glass, aluminum frame		6	1.333		315	53.50		368.50	430
	4500	Sliding, tempered glass, 48" opening		6	1.333		207	53.50		260.50	310
	4700	Deluxe, tempered glass, chrome on brass frame, minimum		8	1		257	40		297	345
	4800	Maximum		1	8		700	320		1,020	1,275
	4850	On anodized aluminum frame, minimum		2	4		122	161		283	380
	4900	Maximum	↓	1	8	↓	410	320		730	950
	5100	Shower enclosure, tempered glass, anodized alum. frame									
	5120	2 panel & door, corner unit, 32" x 32"	1 Shee	2	4	Ea.	410	161		571	700
	5140	Neo-angle corner unit, 16" x 24" x 16"	"	2	4		755	161		916	1,075
	5200	Shower surround, 3 wall, polypropylene, 32" x 32"	1 Carp	4	2		276	68.50		344.50	410
	5220	PVC, 32" x 32"		4	2		259	68.50		327.50	390
	5240	Fiberglass		4	2		300	68.50		368.50	435
	5250	2 wall, polypropylene, 32" x 32"		4	2		215	68.50		283.50	345
	5270	PVC		4	2		267	68.50		335.50	400
	5290	Fiberglass	↓	4	2		297	68.50		365.50	430
	5300	Tub doors, tempered glass & frame, minimum	1 Shee	8	1		175	40		215	255
	5400	Maximum		6	1.333		405	53.50		458.50	530
	5600	Chrome plated, brass frame, minimum		8	1		229	40		269	315
	5700	Maximum		6	1.333		455	53.50		508.50	585
	5900	Tub/shower enclosure, temp. glass, alum. frame, minimum		2	4		310	161		471	585
	6200	Maximum		1.50	5.333		605	214		819	995
	6500	On chrome-plated brass frame, minimum		2	4		430	161		591	720
	6600	Maximum	↓	1.50	5.333		885	214		1,099	1,300
	6800	Tub surround, 3 wall, polypropylene	1 Carp	4	2	↓	174	68.50		242.50	298

Important: See the Reference Section for critical supporting data - Reference Nos., Crews, & City Cost Indexes

10 SPECIALTIES

10150 | Compartments & Cubicles

10185 | Shower/Dressing Compartments

			CREW	DAILY OUTPUT	LABOR-HOURS	UNIT	MAT.	LABOR	EQUIP.	TOTAL	TOTAL INCL O&P	
							2005 BARE COSTS					
100	6900	PVC	1 Carp	4	2	Ea.	264	68.50		332.50	395	100
	7000	Fiberglass, minimum		4	2		294	68.50		362.50	430	
	7100	Maximum	↓	3	2.667	↓	505	91.50		596.50	695	

10190 | Cubicles

			CREW	DAILY OUTPUT	LABOR-HOURS	UNIT	MAT.	LABOR	EQUIP.	TOTAL	TOTAL INCL O&P	
200	0010	PARTITIONS, HOSPITAL Curtain track, box channel, ceiling mounted	1 Carp	135	.059	L.F.	4.50	2.03		6.53	8.10	200
	0100	Suspended	"	100	.080	"	6.15	2.74		8.89	11	
	0300	Curtains, nylon mesh tops, fire resistant, 11 oz. per lineal yard										
	0310	Polyester oxford cloth, 9' ceiling height	1 Carp	425	.019	L.F.	9.95	.64		10.59	11.95	
	0500	8' ceiling height		425	.019		8.40	.64		9.04	10.25	
	0700	Designer oxford cloth	↓	425	.019	↓	21.50	.64		22.14	24.50	
	0800	I.V. track systems										
	0820	I.V. track, oval	1 Carp	135	.059	L.F.	4.37	2.03		6.40	7.95	
	0830	I.V. trolley		32	.250	Ea.	30	8.55		38.55	46.50	
	0840	I.V. pendent, (tree, 5 hook)	↓	32	.250	"	107	8.55		115.55	131	

10200 | Louvers & Vents

10210 | Wall Louvers

			CREW	DAILY OUTPUT	LABOR-HOURS	UNIT	MAT.	LABOR	EQUIP.	TOTAL	TOTAL INCL O&P	
							2005 BARE COSTS					
800	0010	LOUVERS Aluminum with screen, residential, 8" x 8"	1 Carp	38	.211	Ea.	9.35	7.20		16.55	21.50	800
	0100	12" x 12"		38	.211		10.20	7.20		17.40	22.50	
	0200	12" x 18"		35	.229		13.30	7.85		21.15	27	
	0250	14" x 24"		30	.267		17.50	9.15		26.65	33.50	
	0300	18" x 24"		27	.296		21	10.15		31.15	39.50	
	0500	24" x 30"		24	.333		25.50	11.40		36.90	46	
	0700	Triangle, adjustable, small		20	.400		25.50	13.70		39.20	49.50	
	0800	Large	↓	15	.533	↓	42	18.25		60.25	74.50	
	1200	Extruded aluminum, see division 15850-600										
	2100	Midget, aluminum, 3/4" deep, 1" diameter	1 Carp	85	.094	Ea.	.87	3.22		4.09	5.95	
	2150	3" diameter		60	.133		1.81	4.57		6.38	9.10	
	2200	4" diameter		50	.160		2.81	5.50		8.31	11.65	
	2250	6" diameter	↓	30	.267	↓	3.35	9.15		12.50	17.95	
	2300	Ridge vent strip, mill finish	1 Shee	155	.052	L.F.	2.53	2.07		4.60	5.95	
	2330	Soffit vent, continuous, 3" wide, aluminum, mill finish	1 Carp	200	.040		.49	1.37		1.86	2.67	
	2340	Baked enamel finish		200	.040	↓	.44	1.37		1.81	2.61	
	2400	Under eaves vent, aluminum, mill finish, 16" x 4"		48	.167	Ea.	1.78	5.70		7.48	10.85	
	2500	16" x 8"		48	.167		1.97	5.70		7.67	11.05	
	7000	Vinyl gable vent, 8" x 8"		38	.211		10.50	7.20		17.70	23	
	7020	12" x 12"		38	.211		21.50	7.20		28.70	35	
	7080	12" x 18"		35	.229		26	7.85		33.85	40.50	
	7200	18" x 24"	↓	30	.267	↓	29.50	9.15		38.65	47	

		Wall & Corner Guards	CREW	DAILY OUTPUT	LABOR-HOURS	UNIT	2005 BARE COSTS				TOTAL INCL O&P	
	10265						MAT.	LABOR	EQUIP.	TOTAL		
200	0010	**CORNER GUARDS** Steel angle w/anchors, 1" x 1" x 1/4", 1.5#/L.F.	2 Carp	160	.100	L.F.	5.50	3.43		8.93	11.40	200
	0100	2" x 2" x 1/4" angles, 3.2#/L.F.		150	.107		10.70	3.65		14.35	17.50	
	0200	3" x 3" x 5/16" angles, 6.1#/L.F.		140	.114		14.30	3.91		18.21	22	
	0300	4" x 4" x 5/16" angles, 8.2#/L.F.		120	.133		18.05	4.57		22.62	27	
	0350	For angles drilled and anchored to masonry, add					15%	120%				
	0370	Drilled and anchored to concrete, add					20%	170%				
	0400	For galvanized angles, add					35%					
	0450	For stainless steel angles, add				L.F.	100%					
	0500	Steel door track/wheel guards, 4' - 0" high	E-4	22	1.455	Ea.	122	56	3.69	181.69	239	
	0800	Pipe bumper for truck doors, 8' long, 6" diameter, filled		20	1.600		297	62	4.06	363.06	440	
	0900	8" diameter		20	1.600		455	62	4.06	521.06	615	
250	0010	**CORNER PROTECTION**										250
	0100	Stainless steel, 16 ga., adhesive mount, 3-1/2" leg	1 Sswk	80	.100	L.F.	18.10	3.82		21.92	27	
	0200	12 ga. stainless, adhesive mount	"	80	.100		31	3.82		34.82	41	
	0300	For screw mount, add						10%				
	0500	Vinyl acrylic, adhesive mount, 3" leg	1 Carp	128	.063		6.10	2.14		8.24	10.05	
	0550	1-1/2" leg		160	.050		3.50	1.71		5.21	6.50	
	0600	Screw mounted, 3" leg		80	.100		6.80	3.43		10.23	12.80	
	0650	1-1/2" leg		100	.080		4.08	2.74		6.82	8.75	
	0700	Clear plastic, screw mounted, 2-1/2"		60	.133		2.83	4.57		7.40	10.20	
	1000	Vinyl cover, alum. retainer, surface mount, 3" x 3"		48	.167		6.25	5.70		11.95	15.80	
	1050	2" x 2"		48	.167		4.96	5.70		10.66	14.35	
	1100	Flush mounted, 3" x 3"		32	.250		13.75	8.55		22.30	28.50	
	1150	2" x 2"		32	.250		10.15	8.55		18.70	24.50	
500	0010	**WALLGUARD**										500
	0400	Rub rail, vinyl, adhesive mounted	1 Carp	185	.043	L.F.	4	1.48		5.48	6.70	
	0500	Neoprene, aluminum backing, 1-1/2" x 2"		110	.073		6.85	2.49		9.34	11.40	
	1000	Trolley rail, PVC, clipped to wall, 5" high		185	.043		6.20	1.48		7.68	9.15	
	1050	8" high		180	.044		7.90	1.52		9.42	11.05	
	1200	Bed bumper, vinyl acrylic, alum. retainer, 21" long		10	.800	Ea.	37.50	27.50		65	84	
	1300	53" long with aligner		9	.889	"	111	30.50		141.50	170	
	1400	Bumper, vinyl cover, alum. retain., cush. mnt., 1-1/2" x 2-3/4"		80	.100	L.F.	11.35	3.43		14.78	17.80	
	1500	2" x 4-1/4"		80	.100		15.15	3.43		18.58	22	
	1600	Surface mounted, 1-3/4" x 3-5/8"		80	.100		10.25	3.43		13.68	16.60	
	2000	Crash rail, vinyl cover, alum. retainer, 1" x 4"		110	.073		11.25	2.49		13.74	16.30	
	2100	1" x 8"		90	.089		14.05	3.04		17.09	20	
	2150	Vinyl inserts, aluminum plate, 1" x 2-1/2"		110	.073		11.30	2.49		13.79	16.35	
	2200	1" x 5"		90	.089		15.55	3.04		18.59	22	
	3000	Handrail/bumper, vinyl cover, alum. retainer										
	3010	Bracket mounted, flat rail, 5-1/2"	1 Carp	80	.100	L.F.	13.60	3.43		17.03	20.50	
	3100	6-1/2"		80	.100		21.50	3.43		24.93	29	
	3200	Bronze bracket, 1-3/4" diam. rail		80	.100		17.45	3.43		20.88	24.50	

		Access Flooring	CREW	DAILY OUTPUT	LABOR-HOURS	UNIT	2005 BARE COSTS				TOTAL INCL O&P	
	10275						MAT.	LABOR	EQUIP.	TOTAL		
150	0010	**PEDESTAL ACCESS FLOORS** Computer room application, metal										150
	0020	Particle board or steel panels, no covering, under 6,000 S.F.	2 Carp	400	.040	S.F.	13.80	1.37		15.17	17.30	

Important: See the Reference Section for critical supporting data - Reference Nos., Crews, & City Cost Indexes

10270 | Access Flooring

10275 | Access Flooring

		CREW	DAILY OUTPUT	LABOR-HOURS	UNIT	2005 BARE COSTS				TOTAL INCL O&P
						MAT.	LABOR	EQUIP.	TOTAL	
150	0300	Metal covered, over 6,000 S.F.	2 Carp	450	.036	S.F.	8	1.22		9.22	10.70	**150**
	0400	Aluminum, 24" panels	↓	500	.032		29.50	1.10		30.60	34
	0600	For carpet covering, add					5.70			5.70	6.25
	0700	For vinyl floor covering, add					5.80			5.80	6.35
	0900	For high pressure laminate covering, add					3.85			3.85	4.24
	0910	For snap on stringer system, add	2 Carp	1,000	.016	↓	1.32	.55		1.87	2.30
	0950	Office applications, to 8" high, steel panels,									
	0960	no covering, over 6,000 S.F.	2 Carp	500	.032	S.F.	9.20	1.10		10.30	11.85
	1000	Machine cutouts after initial installation	1 Carp	10	.800	Ea.	4.21	27.50		31.71	47
	1050	Pedestals, 6" to 12"	2 Carp	85	.188		8.10	6.45		14.55	18.95
	1100	Air conditioning grilles, 4" x 12"	1 Carp	17	.471		58	16.10		74.10	89
	1150	4" x 18"	"	14	.571	↓	79.50	19.55		99.05	118
	1200	Approach ramps, minimum	2 Carp	85	.188	S.F.	22	6.45		28.45	34
	1300	Maximum	"	60	.267	"	29.50	9.15		38.65	46.50
	1500	Handrail, 2 rail, aluminum	1 Carp	15	.533	L.F.	85	18.25		103.25	122

10300 | Fireplaces & Stoves

10305 | Manufactured Fireplaces

		CREW	DAILY OUTPUT	LABOR-HOURS	UNIT	2005 BARE COSTS				TOTAL INCL O&P
						MAT.	LABOR	EQUIP.	TOTAL	
100	0010	**FIREPLACE, PREFABRICATED** Free standing or wall hung										**100**
	0100	with hood & screen, minimum	1 Carp	1.30	6.154	Ea.	1,150	211		1,361	1,575
	0150	Average		1	8		1,375	274		1,649	1,925
	0200	Maximum		.90	8.889	↓	3,350	305		3,655	4,175
	0500	Chimney dbl. wall, all stainless, over 8'-6", 7" diam., add		33	.242	V.L.F.	52	8.30		60.30	70
	0600	10" diameter, add		32	.250		55	8.55		63.55	74
	0700	12" diameter, add		31	.258		72	8.85		80.85	93
	0800	14" diameter, add		30	.267	↓	91	9.15		100.15	114
	1000	Simulated brick chimney top, 4' high, 16" x 16"		10	.800	Ea.	196	27.50		223.50	259
	1100	24" x 24"		7	1.143	"	365	39		404	460
	1500	Simulated logs, gas fired, 40,000 BTU, 2' long, minimum		7	1.143	Set	480	39		519	585
	1600	Maximum		6	1.333		670	45.50		715.50	805
	1700	Electric, 1,500 BTU, 1'-6" long, minimum		7	1.143		136	39		175	210
	1800	11,500 BTU, maximum		6	1.333	↓	293	45.50		338.50	390
	2000	Fireplace, built-in, 36" hearth, radiant		1.30	6.154	Ea.	585	211		796	975
	2100	Recirculating, small fan		1	8		840	274		1,114	1,350
	2150	Large fan		.90	8.889		1,550	305		1,855	2,200
	2200	42" hearth, radiant		1.20	6.667		750	228		978	1,175
	2300	Recirculating, small fan		.90	8.889		980	305		1,285	1,550
	2350	Large fan		.80	10		1,875	345		2,220	2,575
	2400	48" hearth, radiant		1.10	7.273		1,400	249		1,649	1,925
	2500	Recirculating, small fan		.80	10		1,725	345		2,070	2,425
	2550	Large fan		.70	11.429		2,675	390		3,065	3,550
	3000	See through, including doors		.80	10		2,200	345		2,545	2,950
	3200	Corner (2 wall)	↓	1	8	↓	1,100	274		1,374	1,625

10310 | Fireplace Specialties & Accessories

		CREW	DAILY OUTPUT	LABOR-HOURS	UNIT	2005 BARE COSTS				TOTAL INCL O&P
						MAT.	LABOR	EQUIP.	TOTAL	
100	0010	**FIREPLACE ACCESSORIES** Chimney screens, galv., 13" x 13" flue	1 Bric	8	1	Ea.	33.50	35.50		69	90.50	**100**
	0050	Galv., 24" x 24" flue	↓	5	1.600	↓	100	56.50		156.50	196

10300 | Fireplaces & Stoves

10310 | Fireplace Specialties & Accessories

			CREW	DAILY OUTPUT	LABOR-HOURS	UNIT	MAT.	LABOR	EQUIP.	TOTAL	TOTAL INCL O&P	
100	0200	Stainless steel, 13" x 13" flue	1 Bric	8	1	Ea.	264	35.50		299.50	345	**100**
	0250	20" x 20" flue		5	1.600		360	56.50		416.50	480	
	0400	Cleanout doors and frames, cast iron, 8" x 8"		12	.667		29	23.50		52.50	68	
	0450	12" x 12"		10	.800		33.50	28		61.50	79.50	
	0500	18" x 24"		8	1		105	35.50		140.50	169	
	0550	Cast iron frame, steel door, 24" x 30"		5	1.600		227	56.50		283.50	335	
	0800	Damper, rotary control, steel, 30" opening		6	1.333		64	47		111	142	
	0850	Cast iron, 30" opening		6	1.333		71	47		118	150	
	0880	36" opening		6	1.333		77	47		124	157	
	0900	48" opening		6	1.333		110	47		157	193	
	0920	60" opening		6	1.333		247	47		294	345	
	0950	72" opening		5	1.600		297	56.50		353.50	410	
	1000	84" opening, special order		5	1.600		635	56.50		691.50	785	
	1050	96" opening, special order		4	2		645	70.50		715.50	815	
	1200	Steel plate, poker control, 60" opening		8	1		225	35.50		260.50	300	
	1250	84" opening, special opening		5	1.600		410	56.50		466.50	535	
	1400	"Universal" type, chain operated, 32" x 20" opening		8	1		156	35.50		191.50	225	
	1450	48" x 24" opening		5	1.600		261	56.50		317.50	375	
	1600	Dutch Oven door and frame, cast iron, 12" x 15" opening		13	.615		92	21.50		113.50	134	
	1650	Copper plated, 12" x 15" opening		13	.615		177	21.50		198.50	227	
	1800	Fireplace forms, no accessories, 32" opening		3	2.667		525	94		619	725	
	1900	36" opening		2.50	3.200		635	113		748	870	
	2000	40" opening		2	4		765	141		906	1,050	
	2100	78" opening		1.50	5.333		1,100	188		1,288	1,475	
	2400	Squirrel and bird screens, galvanized, 8" x 8" flue		16	.500		36	17.65		53.65	67	
	2450	13" x 13" flue	▼	12	.667	▼	41.50	23.50		65	81.50	

10320 | Stoves

			CREW	DAILY OUTPUT	LABOR-HOURS	UNIT	MAT.	LABOR	EQUIP.	TOTAL	TOTAL INCL O&P	
100	0010	**WOODBURNING STOVES** Cast iron, minimum	2 Carp	1.30	12.308	Ea.	705	420		1,125	1,425	**100**
	0020	Average		1	16		1,300	550		1,850	2,275	
	0030	Maximum	▼	.80	20		2,175	685		2,860	3,450	
	0050	For gas log lighter, add				▼	39			39	43	

10340 | Manufactured Exterior Specialties

10342 | Cupolas

			CREW	DAILY OUTPUT	LABOR-HOURS	UNIT	MAT.	LABOR	EQUIP.	TOTAL	TOTAL INCL O&P	
100	0010	**CUPOLA** Stock units, pine, painted, 18" sq., 28" high, alum. roof	1 Carp	4.10	1.951	Ea.	141	67		208	259	**100**
	0100	Copper roof		3.80	2.105		150	72		222	277	
	0300	23" square, 33" high, aluminum roof		3.70	2.162		236	74		310	375	
	0400	Copper roof		3.30	2.424		252	83		335	405	
	0600	30" square, 37" high, aluminum roof		3.70	2.162		360	74		434	510	
	0700	Copper roof		3.30	2.424		380	83		463	550	
	0900	Hexagonal, 31" wide, 46" high, copper roof		4	2		540	68.50		608.50	700	
	1000	36" wide, 50" high, copper roof	▼	3.50	2.286		605	78.50		683.50	785	
	1200	For deluxe stock units, add to above					25%					
	1400	For custom built units, add to above				▼	50%	50%				

Important: See the Reference Section for critical supporting data - Reference Nos., Crews, & City Cost Indexes

10355	Flagpoles	CREW	DAILY OUTPUT	LABOR-HOURS	UNIT	2005 BARE COSTS				TOTAL INCL O&P
						MAT.	LABOR	EQUIP.	TOTAL	
400	0010 **FLAGPOLE**, Ground set										**400**
	0050 Not including base or foundation									
	0100 Aluminum, tapered, ground set 20' high	K-1	2	8	Ea.	820	244	88	1,152	1,375
	0200 25' high		1.70	9.412		965	287	104	1,356	1,600
	0300 30' high		1.50	10.667		975	325	117	1,417	1,700
	0400 35' high		1.40	11.429		1,400	350	126	1,876	2,200
	0500 40' high		1.20	13.333		2,000	405	147	2,552	3,025
	0600 50' high		1	16		2,525	490	176	3,191	3,725
	0700 60' high		.90	17.778		4,350	545	196	5,091	5,825
	0800 70' high		.80	20		7,550	610	220	8,380	9,500
	1100 Counterbalanced, internal halyard, 20' high		1.80	8.889		2,050	271	98	2,419	2,775
	1200 30' high		1.50	10.667		2,275	325	117	2,717	3,125
	1300 40' high		1.30	12.308		3,800	375	136	4,311	4,900
	1400 50' high		1	16		7,000	490	176	7,666	8,650
	2820 Aluminum, electronically operated, 30' high		1.40	11.429		3,975	350	126	4,451	5,050
	2840 35' high		1.30	12.308		4,075	375	136	4,586	5,225
	2860 39' high		1.10	14.545		4,450	445	160	5,055	5,750
	2880 45' high		1	16		5,675	490	176	6,341	7,200
	2900 50' high		.90	17.778		6,050	545	196	6,791	7,700
	3000 Fiberglass, tapered, ground set, 23' high		2	8		1,050	244	88	1,382	1,650
	3100 29"-7" high		1.50	10.667		1,275	325	117	1,717	2,025
	3200 36'-1" high		1.40	11.429		1,650	350	126	2,126	2,500
	3300 39'-5" high		1.20	13.333		2,150	405	147	2,702	3,175
	3400 49'-2" high		1	16		4,200	490	176	4,866	5,575
	3500 59' high	▼	.90	17.778	▼	6,275	545	196	7,016	7,950
	4300 Concrete, direct imbedded installation									
	4400 Internal halyard, 20' high	K-1	2.50	6.400	Ea.	1,350	195	70.50	1,615.50	1,850
	4500 25' high		2.50	6.400		1,500	195	70.50	1,765.50	2,025
	4600 30' high		2.30	6.957		1,775	212	76.50	2,063.50	2,375
	4700 40' high		2.10	7.619		2,625	233	84	2,942	3,350
	4800 50' high		1.90	8.421		3,900	257	92.50	4,249.50	4,775
	5000 60' high		1.80	8.889		5,875	271	98	6,244	7,000
	5100 70' high		1.60	10		7,175	305	110	7,590	8,500
	5200 80' high		1.40	11.429		9,475	350	126	9,951	11,100
	5300 90' high		1.20	13.333		13,000	405	147	13,552	15,100
	5500 100' high	▼	1	16	▼	13,400	490	176	14,066	15,700
	6400 Wood poles, tapered, clear vertical grain fir with tilting									
	6410 base, not incl. foundation, 4" butt, 25' high	K-1	1.90	8.421	Ea.	625	257	92.50	974.50	1,175
	6800 6" butt, 30' high	"	1.30	12.308	"	775	375	136	1,286	1,575
	7300 Foundations for flagpoles, including									
	7400 excavation and concrete, to 35' high poles	C-1	10	3.200	Ea.	500	104		604	710
	7600 40' to 50' high		3.50	9.143		910	296		1,206	1,450
	7700 Over 60' high	▼	2	16	▼	1,025	520		1,545	1,925
900	0010 **FLAGPOLE**, Structure mounted										**900**
	0100 Fiberglass, vertical wall set, 19'-8" long	K-1	1.50	10.667	Ea.	1,225	325	117	1,667	1,975
	0200 23' long		1.40	11.429		1,300	350	126	1,776	2,100
	0300 26'-3" long		1.30	12.308		1,700	375	136	2,211	2,575
	0800 19'-8" long outrigger		1.30	12.308		1,250	375	136	1,761	2,100
	1300 Aluminum, vertical wall set, tapered, with base, 20' high		1.20	13.333		925	405	147	1,477	1,825
	1400 29'-6" high		1	16		2,175	490	176	2,841	3,350
	2400 Outrigger poles with base, 12' long		1.30	12.308		860	375	136	1,371	1,675
	2500 14' long	▼	1	16	▼	1,375	490	176	2,041	2,475

SPECIALTIES 10

333

10410		Directories	CREW	DAILY OUTPUT	LABOR-HOURS	UNIT	2005 BARE COSTS				TOTAL INCL O&P
							MAT.	LABOR	EQUIP.	TOTAL	
100	0010	**DIRECTORY BOARDS**										**100**
	0050	Plastic, glass covered, 30" x 20"	2 Carp	3	5.333	Ea.	262	183		445	575
	0100	36" x 48"		2	8		770	274		1,044	1,275
	0300	Grooved cork, 30" x 20"		3	5.333		355	183		538	675
	0400	36" x 48"		2	8		310	274		584	765
	0600	Black felt, 30" x 20"		3	5.333		254	183		437	565
	0700	36" x 48"		2	8		375	274		649	835
	0900	Outdoor, weatherproof, black plastic, 36" x 24"		2	8		685	274		959	1,175
	1000	36" x 36"		1.50	10.667		790	365		1,155	1,450
	1800	Indoor, economy, open face, 18" x 24"		7	2.286		94.50	78.50		173	226
	1900	24" x 36"		7	2.286		137	78.50		215.50	272
	2000	36" x 24"		6	2.667		156	91.50		247.50	315
	2100	36" x 48"		6	2.667		234	91.50		325.50	400
	2400	Building directory, alum., black felt panels, 1 door, 24" x 18"		4	4		281	137		418	525
	2500	36" x 24"		3.50	4.571		320	157		477	600
	2600	48" x 32"		3	5.333		480	183		663	810
	2700	36" x 48", 2 door		2.50	6.400		590	219		809	990
	2800	36" x 60"		2	8		845	274		1,119	1,350
	2900	48" x 60"	▼	1	16		925	550		1,475	1,875
	3100	For bronze enamel finish, add					15%				
	3200	For bronze anodized finish, add					25%				
	3400	For illuminated directory, single door unit, add				▼	154			154	170
	3500	For 6" header panel, 6 letters per foot, add				L.F.	27			27	30

10430		**Exterior Signage**									
200	0010	**SIGNS** Letters, 2" high, 3/8" deep, cast bronze	1 Carp	24	.333	Ea.	22.50	11.40		33.90	43	**200**
	0140	1/2" deep, cast aluminum		18	.444		22.50	15.20		37.70	48.50
	0160	Cast bronze		32	.250		24	8.55		32.55	40
	0300	6" high, 5/8" deep, cast aluminum		24	.333		26	11.40		37.40	46.50
	0400	Cast bronze		24	.333		39	11.40		50.40	60.50
	0600	8" high, 3/4" deep, cast aluminum		14	.571		29.50	19.55		49.05	63
	0700	Cast bronze		20	.400		66	13.70		79.70	94
	0900	10" high, 1" deep, cast aluminum		18	.444		31	15.20		46.20	57.50
	1000	Bronze		18	.444		79.50	15.20		94.70	111
	1200	12" high, 1-1/4" deep, cast aluminum		12	.667		39.50	23		62.50	79
	1500	Cast bronze		18	.444		96	15.20		111.20	130
	1600	14" high, 2-5/16" deep, cast aluminum		12	.667		63	23		86	105
	1800	Fabricated stainless steel, 6" high, 2" deep		20	.400		113	13.70		126.70	146
	1900	12" high, 3" deep		18	.444		134	15.20		149.20	171
	2100	18" high, 3" deep		12	.667		196	23		219	251
	2200	24" high, 4" deep		10	.800		273	27.50		300.50	345
	2700	Acrylic, on high density foam,		20	.400		20	13.70		33.70	44
	2800	12" high, 2" deep	▼	18	.444		44	15.20		59.20	72
	3900	Plaques, custom, 20" x 30", for up to 450 letters, cast aluminum	2 Carp	4	4		815	137		952	1,100
	4000	Cast bronze		4	4		1,075	137		1,212	1,400
	4200	30" x 36", up to 900 letters cast aluminum		3	5.333		1,725	183		1,908	2,175
	4300	Cast bronze		3	5.333		2,225	183		2,408	2,725
	4500	36" x 48", for up to 1300 letters, cast bronze		2	8		3,200	274		3,474	3,950
	4800	Signs, reflective alum. street signs, dbl. face, 2-way, w/bracket		30	.533		67.50	18.25		85.75	103
	4900	4-way	▼	30	.533		105	18.25		123.25	145
	5100	Exit signs, 24 ga. alum., 14" x 12" surface mounted	1 Carp	30	.267		32	9.15		41.15	50
	5200	10" x 7"		20	.400		18.60	13.70		32.30	42
	5400	Bracket mounted, double face, 12" x 10"	▼	30	.267		31.50	9.15		40.65	49
	5500	Sticky back, stock decals, 14" x 10"	1 Clab	50	.160		8	4.27		12.27	15.45
	6000	Interior elec., wall mount, fiberglass panels, 2 lamps, 6"	1 Elec	8	1	▼	99	41		140	170

10 SPECIALTIES

10400 | Identification Devices

<table>
<tr><th colspan="2">10430 | Exterior Signage</th><th>CREW</th><th>DAILY OUTPUT</th><th>LABOR-HOURS</th><th>UNIT</th><th colspan="4">2005 BARE COSTS</th><th>TOTAL INCL O&P</th></tr>
<tr><th></th><th></th><th></th><th></th><th></th><th></th><th>MAT.</th><th>LABOR</th><th>EQUIP.</th><th>TOTAL</th><th></th></tr>
<tr><td>200 6100</td><td>8"</td><td>1 Elec</td><td>8</td><td>1</td><td>Ea.</td><td>44</td><td>41</td><td></td><td>85</td><td>109 200</td></tr>
<tr><td>6400</td><td>Replacement sign faces, 6" or 8"</td><td>1 Clab</td><td>50</td><td>.160</td><td>↓</td><td>24</td><td>4.27</td><td></td><td>28.27</td><td>33</td></tr>
</table>

10450 | Pedestrian Control Devices

<table>
<tr><th colspan="2">10455 | Turnstiles</th><th>CREW</th><th>DAILY OUTPUT</th><th>LABOR-HOURS</th><th>UNIT</th><th colspan="4">2005 BARE COSTS</th><th>TOTAL INCL O&P</th></tr>
<tr><th></th><th></th><th></th><th></th><th></th><th></th><th>MAT.</th><th>LABOR</th><th>EQUIP.</th><th>TOTAL</th><th></th></tr>
<tr><td>900 0010</td><td>**TURNSTILES** One way, 4 arm, 46" diameter, economy, manual</td><td>2 Carp</td><td>5</td><td>3.200</td><td>Ea.</td><td>269</td><td>110</td><td></td><td>379</td><td>465 900</td></tr>
<tr><td>0100</td><td>Electric</td><td></td><td>1.20</td><td>13.333</td><td></td><td>900</td><td>455</td><td></td><td>1,355</td><td>1,700</td></tr>
<tr><td>0300</td><td>High security, galv., 5'-5" diameter, 7' high, manual</td><td></td><td>1</td><td>16</td><td></td><td>2,325</td><td>550</td><td></td><td>2,875</td><td>3,425</td></tr>
<tr><td>0350</td><td>Electric</td><td></td><td>.60</td><td>26.667</td><td></td><td>3,025</td><td>915</td><td></td><td>3,940</td><td>4,750</td></tr>
<tr><td>0420</td><td>Three arm, 24" opening, light duty, manual</td><td></td><td>2</td><td>8</td><td></td><td>865</td><td>274</td><td></td><td>1,139</td><td>1,375</td></tr>
<tr><td>0450</td><td>Heavy duty</td><td></td><td>1.50</td><td>10.667</td><td></td><td>2,350</td><td>365</td><td></td><td>2,715</td><td>3,175</td></tr>
<tr><td>0460</td><td>Manual, with registering & controls, light duty</td><td></td><td>2</td><td>8</td><td></td><td>1,875</td><td>274</td><td></td><td>2,149</td><td>2,475</td></tr>
<tr><td>0470</td><td>Heavy duty</td><td></td><td>1.50</td><td>10.667</td><td></td><td>2,275</td><td>365</td><td></td><td>2,640</td><td>3,100</td></tr>
<tr><td>0480</td><td>Electric, heavy duty</td><td>↓</td><td>1.10</td><td>14.545</td><td></td><td>2,675</td><td>500</td><td></td><td>3,175</td><td>3,725</td></tr>
<tr><td>0500</td><td>For coin or token operating, add</td><td></td><td></td><td></td><td>↓</td><td>355</td><td></td><td></td><td>355</td><td>395</td></tr>
<tr><td>1200</td><td>One way gate with horizontal bars, 5'-5" diameter</td><td></td><td></td><td></td><td></td><td></td><td></td><td></td><td></td><td></td></tr>
<tr><td>1300</td><td>7' high, recreation or transit type</td><td>2 Carp</td><td>.80</td><td>20</td><td>Ea.</td><td>3,025</td><td>685</td><td></td><td>3,710</td><td>4,400</td></tr>
<tr><td>1500</td><td>For electronic counter, add</td><td></td><td></td><td></td><td>"</td><td>174</td><td></td><td></td><td>174</td><td>191</td></tr>
</table>

10500 | Lockers

<table>
<tr><th colspan="2">10505 | Metal Lockers</th><th>CREW</th><th>DAILY OUTPUT</th><th>LABOR-HOURS</th><th>UNIT</th><th colspan="4">2005 BARE COSTS</th><th>TOTAL INCL O&P</th></tr>
<tr><th></th><th></th><th></th><th></th><th></th><th></th><th>MAT.</th><th>LABOR</th><th>EQUIP.</th><th>TOTAL</th><th></th></tr>
<tr><td>500 0011</td><td>**LOCKERS** Steel, baked enamel</td><td></td><td></td><td></td><td></td><td></td><td></td><td></td><td></td><td>500</td></tr>
<tr><td>0110</td><td>Single tier box locker, 12" x 15" x 72"</td><td>1 Shee</td><td>8</td><td>1</td><td>Ea.</td><td>157</td><td>40</td><td></td><td>197</td><td>235</td></tr>
<tr><td>0120</td><td>18" x 15" x 72"</td><td></td><td>8</td><td>1</td><td></td><td>170</td><td>40</td><td></td><td>210</td><td>249</td></tr>
<tr><td>0130</td><td>12" x 18" x 72"</td><td></td><td>8</td><td>1</td><td></td><td>162</td><td>40</td><td></td><td>202</td><td>240</td></tr>
<tr><td>0140</td><td>18" x 18" x 72"</td><td></td><td>8</td><td>1</td><td></td><td>183</td><td>40</td><td></td><td>223</td><td>264</td></tr>
<tr><td>0410</td><td>Double tier, 12" x 15" x 36"</td><td></td><td>21</td><td>.381</td><td></td><td>167</td><td>15.30</td><td></td><td>182.30</td><td>208</td></tr>
<tr><td>0420</td><td>18" x 15" x 36"</td><td></td><td>21</td><td>.381</td><td></td><td>225</td><td>15.30</td><td></td><td>240.30</td><td>272</td></tr>
<tr><td>0430</td><td>12" x 18" x 36"</td><td></td><td>21</td><td>.381</td><td></td><td>175</td><td>15.30</td><td></td><td>190.30</td><td>216</td></tr>
<tr><td>0440</td><td>18" x 18" x 36"</td><td></td><td>21</td><td>.381</td><td></td><td>188</td><td>15.30</td><td></td><td>203.30</td><td>231</td></tr>
<tr><td>0500</td><td>Two person, 18" x 15" x 72"</td><td></td><td>8</td><td>1</td><td></td><td>251</td><td>40</td><td></td><td>291</td><td>340</td></tr>
<tr><td>0510</td><td>18" x 18" x 72"</td><td></td><td>8</td><td>1</td><td></td><td>257</td><td>40</td><td></td><td>297</td><td>345</td></tr>
<tr><td>0520</td><td>Duplex, 15" x 15" x 72"</td><td></td><td>8</td><td>1</td><td></td><td>246</td><td>40</td><td></td><td>286</td><td>330</td></tr>
<tr><td>0530</td><td>15" x 21" x 72"</td><td></td><td>8</td><td>1</td><td>↓</td><td>288</td><td>40</td><td></td><td>328</td><td>375</td></tr>
<tr><td>0600</td><td>5 tier box lockers, minimum</td><td></td><td>30</td><td>.267</td><td>Opng.</td><td>35.50</td><td>10.70</td><td></td><td>46.20</td><td>55.50</td></tr>
<tr><td>0700</td><td>Maximum</td><td></td><td>24</td><td>.333</td><td></td><td>44</td><td>13.40</td><td></td><td>57.40</td><td>69</td></tr>
<tr><td>0900</td><td>6 tier box lockers, minimum</td><td></td><td>36</td><td>.222</td><td></td><td>34.50</td><td>8.95</td><td></td><td>43.45</td><td>52</td></tr>
<tr><td>1000</td><td>Maximum</td><td></td><td>30</td><td>.267</td><td>↓</td><td>41.50</td><td>10.70</td><td></td><td>52.20</td><td>62</td></tr>
<tr><td>1100</td><td>Wire meshed wardrobe, floor. mtd., open front varsity type</td><td></td><td>7.50</td><td>1.067</td><td>Ea.</td><td>158</td><td>43</td><td></td><td>201</td><td>240</td></tr>
<tr><td>2100</td><td>Locker bench, laminated maple, top only</td><td></td><td>100</td><td>.080</td><td>L.F.</td><td>12.55</td><td>3.22</td><td></td><td>15.77</td><td>18.80</td></tr>
<tr><td>2200</td><td>Pedestals, steel pipe</td><td>↓</td><td>25</td><td>.320</td><td>Ea.</td><td>33.50</td><td>12.85</td><td></td><td>46.35</td><td>57</td></tr>
</table>

SPECIALTIES 10

10500 | Lockers

10505 | Metal Lockers

		CREW	DAILY OUTPUT	LABOR-HOURS	UNIT	2005 BARE COSTS				TOTAL INCL O&P
						MAT.	LABOR	EQUIP.	TOTAL	
500										**500**
2400	16-person locker unit with clothing rack									
2500	72 wide x 15″ deep x 72″ high	1 Shee	8	1	Ea.	450	40		490	555
2550	18″ deep	″	8	1	″	380	40		420	480
3000	Wall mounted lockers, 4 person, with coat bar									
3100	48″ wide x 18″ deep x 12″ high	1 Shee	8	1	Ea.	200	40		240	282
3250	Rack w/ 24 wire mesh baskets		1.50	5.333	Set	310	214		524	670
3260	30 baskets		1.25	6.400		244	257		501	665
3270	36 baskets		.95	8.421		305	340		645	860
3280	42 baskets		.80	10		535	400		935	1,200
3300	For built-in lock with 2 keys, add				Ea.	6.65			6.65	7.30
3600	For hanger rods, add				″	1.70			1.70	1.87

10520 | Fire Protection Specialties

10525 | Fire Prot. Specialties

		CREW	DAILY OUTPUT	LABOR-HOURS	UNIT	2005 BARE COSTS				TOTAL INCL O&P
						MAT.	LABOR	EQUIP.	TOTAL	
200	0010	**FIRE EQUIPMENT CABINETS** Not equipped, 20 ga. steel box,									**200**
	0040	recessed, D.S. glass in door, box size given									
	1000	Portable extinguisher, single, 8″ x 12″ x 27″, alum. door & frame	Q-12	8	2	Ea.	84.50	73.50		158	204
	1100	Steel door and frame	″	8	2	″	62.50	73.50		136	180
	3000	Hose rack assy., 1-1/2″ valve & 100′ hose, 24″ x 40″ x 5-1/2″									
	3100	Aluminum door and frame	Q-12	6	2.667	Ea.	206	98		304	375
	3200	Steel door and frame		6	2.667		134	98		232	296
	3300	Stainless steel door and frame		6	2.667		268	98		366	445
	4000	Hose rack assy., 2-1/2″ x 1-1/2″ valve, 100′ hose, 24″ x 40″ x 8″									
	4100	Aluminum door and frame	Q-12	6	2.667	Ea.	208	98		306	375
	4200	Steel door and frame		6	2.667		141	98		239	305
	4300	Stainless steel door and frame		6	2.667		272	98		370	445
	5000	Hose rack assy., 2-1/2″ x 1-1/2″ valve, 100′ hose									
	5100	Aluminum door and frame	Q-12	5	3.200	Ea.	265	117		382	470
	5200	Steel door and frame		5	3.200		154	117		271	345
	5300	Stainless steel door and frame		5	3.200		291	117		408	495
	8000	Valve cabinet for 2-1/2″ FD angle valve, 18″ x 18″ x 8″									
	8100	Aluminum door and frame	Q-12	12	1.333	Ea.	92	49		141	175
	8200	Steel door and frame		12	1.333		69.50	49		118.50	151
	8300	Stainless steel door and frame		12	1.333		121	49		170	207
300	0010	**FIRE EXTINGUISHERS**									**300**
	0120	CO2, portable with swivel horn, 5 lb.				Ea.	130			130	143
	0140	With hose and ″H″ horn, 10 lb.					160			160	176
	0160	15 lb.					190			190	209
	0180	20 lb.					220			220	242
	0360	Wheeled type, cart mounted, 50 lb.					1,200			1,200	1,325
	0400	100 lb.					1,525			1,525	1,675
	1000	Dry chemical, pressurized									
	1040	Standard type, portable, painted, 2-1/2 lb.				Ea.	27.50			27.50	30.50
	1060	5 lb.					60			60	66
	1080	10 lb.					80			80	88
	1100	20 lb.					110			110	121
	1120	30 lb.					157			157	173
	1300	Standard type, wheeled, 150 lb.					1,600			1,600	1,750

Important: See the Reference Section for critical supporting data - Reference Nos., Crews, & City Cost Indexes

10520 | Fire Protection Specialties

10525 | Fire Prot. Specialties

		CREW	DAILY OUTPUT	LABOR-HOURS	UNIT	MAT.	LABOR	EQUIP.	TOTAL	TOTAL INCL O&P	
300	**2000**	ABC all purpose type, portable, 2-1/2 lb.				Ea.	30			30	33	**300**
	2060	5 lb.					45			45	49.50	
	2080	9-1/2 lb.					68			68	75	
	2100	20 lb.					98			98	108	
	2300	Wheeled, 45 lb.					650			650	715	
	2360	150 lb.					1,600			1,600	1,750	
	3000	Dry chemical, outside cartridge to -65°F, painted, 9 lb.					200			200	220	
	3060	26 lb.					250			250	275	
	5000	Pressurized water, 2-1/2 gallon, stainless steel					85			85	93.50	
	5060	With anti-freeze					135			135	149	
	9400	Installation of extinguishers, 12 or more, on wood	1 Carp	30	.267			9.15		9.15	14.25	
	9420	On masonry or concrete	"	15	.533			18.25		18.25	28.50	

10530 | Protective Covers

10535 | Awnings & Canopies

		CREW	DAILY OUTPUT	LABOR-HOURS	UNIT	MAT.	LABOR	EQUIP.	TOTAL	TOTAL INCL O&P	
050	**0010**	**AWNINGS, FABRIC**										**050**
	0020	Including acrylic canvas and frame, standard design										
	0100	Door and window, slope, 3' high, 4' wide	1 Carp	4.50	1.778	Ea.	560	61		621	710	
	0110	6' wide		3.50	2.286		720	78.50		798.50	910	
	0120	8' wide		3	2.667		880	91.50		971.50	1,100	
	0200	Quarter round convex, 4' wide		3	2.667		870	91.50		961.50	1,100	
	0210	6' wide		2.25	3.556		1,125	122		1,247	1,450	
	0220	8' wide		1.80	4.444		1,375	152		1,527	1,750	
	0300	Dome, 4' wide		7.50	1.067		335	36.50		371.50	425	
	0310	6' wide		3.50	2.286		755	78.50		833.50	950	
	0320	8' wide		2	4		1,350	137		1,487	1,700	
	0350	Elongated dome, 4' wide		1.33	6.015		1,250	206		1,456	1,700	
	0360	6' wide		1.11	7.207		1,500	247		1,747	2,025	
	0370	8' wide		1	8		1,750	274		2,024	2,350	
	1000	Entry or walkway, peak, 12' long, 4' wide	2 Carp	.90	17.778		3,975	610		4,585	5,325	
	1010	6' wide		.60	26.667		6,125	915		7,040	8,150	
	1020	8' wide		.40	40		8,450	1,375		9,825	11,400	
	1100	Radius with dome end, 4' wide		1.10	14.545		3,025	500		3,525	4,100	
	1110	6' wide		.70	22.857		4,850	785		5,635	6,550	
	1120	8' wide		.50	32		6,875	1,100		7,975	9,275	
	2000	Retractable lateral arm awning, manual										
	2010	To 12' wide, 8' - 6" projection	2 Carp	1.70	9.412	Ea.	900	320		1,220	1,475	
	2020	To 14' wide, 8' - 6" projection		1.10	14.545		1,050	500		1,550	1,925	
	2030	To 19' wide, 8' - 6" projection		.85	18.824		1,425	645		2,070	2,575	
	2040	To 24' wide, 8' - 6" projection		.67	23.881		1,800	820		2,620	3,250	
	2050	Motor for above, add	1 Carp	2.67	3		770	103		873	1,000	
	3000	Patio/deck canopy with frame										
	3010	12' wide, 12' projection	2 Carp	2	8	Ea.	1,275	274		1,549	1,825	
	3020	16' wide, 14' projection	"	1.20	13.333		1,975	455		2,430	2,875	
	9000	For fire retardant canvas, add					7%					
	9010	For lettering or graphics, add					35%					
	9020	For painted or coated acrylic canvas, deduct					8%					
	9030	For translucent or opaque vinyl canvas, add					10%					
	9040	For 6 or more units, deduct					20%	15%				

10535 | Awnings & Canopies

		CREW	DAILY OUTPUT	LABOR-HOURS	UNIT	MAT.	LABOR	EQUIP.	TOTAL	TOTAL INCL O&P	
200	**0010**	**CANOPIES** Wall hung, .032", aluminum, prefinished, 8' x 10'	K-2	1.30	18.462	Ea.	1,725	645	136	2,506	3,175	**200**
	0300	8' x 20'		1.10	21.818		3,425	765	160	4,350	5,275	
	0500	10' x 10'		1.30	18.462		1,925	645	136	2,706	3,400	
	0700	10' x 20'		1.10	21.818		3,575	765	160	4,500	5,425	
	1000	12' x 20'		1	24		4,400	840	176	5,416	6,475	
	1360	12' x 30'		.80	30		6,450	1,050	220	7,720	9,150	
	1700	12' x 40'		.60	40		6,875	1,400	294	8,569	10,300	
	1900	For free standing units, add					20%	10%				
	2300	Aluminum entrance canopies, flat soffit, .032"										
	2500	3'-6" x 4'-0", clear anodized	2 Carp	4	4	Ea.	720	137		857	1,000	
	2700	Bronze anodized		4	4		1,275	137		1,412	1,625	
	3000	Polyurethane painted		4	4		1,025	137		1,162	1,350	
	3300	4'-6" x 10'-0", clear anodized		2	8		2,000	274		2,274	2,625	
	3500	Bronze anodized		2	8		2,550	274		2,824	3,225	
	3700	Polyurethane painted		2	8		2,125	274		2,399	2,775	
	4000	Wall downspout, 10 L.F., clear anodized	1 Carp	7	1.143		124	39		163	197	
	4300	Bronze anodized		7	1.143		216	39		255	298	
	4500	Polyurethane painted		7	1.143		185	39		224	265	
	7000	Carport, baked vinyl finish, .032", 20' x 10', no foundations, min.	K-2	4	6	Car	3,100	210	44	3,354	3,800	
	7250	Maximum		2	12	"	6,200	420	88	6,708	7,650	
	7500	Walkway cover, to 12' wide, stl., vinyl finish, .032",no fndtns., min.		250	.096	S.F.	18.10	3.36	.70	22.16	26.50	
	7750	Maximum		200	.120	"	19.70	4.20	.88	24.78	29.50	

10550 | Postal Specialties

10555 | Mail Delivery Systems

		CREW	DAILY OUTPUT	LABOR-HOURS	UNIT	MAT.	LABOR	EQUIP.	TOTAL	TOTAL INCL O&P	
600	**0010**	**MAIL BOXES** Horiz., key lock, 5"H x 6"W x 15"D, alum., rear load	1 Carp	34	.235	Ea.	31.50	8.05		39.55	47	**600**
	0100	Front loading		34	.235		35.50	8.05		43.55	51.50	
	0200	Double, 5"H x 12"W x 15"D, rear loading		26	.308		60	10.55		70.55	82.50	
	0300	Front loading		26	.308		63	10.55		73.55	86	
	0500	Quadruple, 10"H x 12"W x 15"D, rear loading		20	.400		112	13.70		125.70	145	
	0600	Front loading		20	.400		116	13.70		129.70	150	
	0800	Vertical, front load, 15"H x 5"W x 6"D, alum., per compartment		34	.235		26	8.05		34.05	41.50	
	0900	Bronze, duranodic finish		34	.235		44	8.05		52.05	61	
	1000	Steel, enameled		34	.235		32	8.05		40.05	47.50	
	1700	Alphabetical directories, 120 names		10	.800		114	27.50		141.50	169	
	1800	Letter collection box		6	1.333		590	45.50		635.50	720	
	1900	Letter slot, residential		20	.400		60	13.70		73.70	87.50	
	2000	Post office type		8	1		220	34.50		254.50	296	
	2200	Post office counter window, with grille		2	4		515	137		652	780	
	2250	Key keeper, single key, aluminum		26	.308		37	10.55		47.55	57.50	
	2300	Steel, enameled		26	.308		100	10.55		110.55	126	
700	**0010**	**MAIL CHUTES** Aluminum & glass, 14-1/4" wide, 4-5/8" deep	2 Shee	4	4	Floor	650	161		811	960	**700**
	0100	8-5/8" deep		3.80	4.211		790	169		959	1,125	
	0300	8-3/4" x 3-1/2", aluminum		5	3.200		600	129		729	860	
	0400	Bronze or stainless		4.50	3.556		930	143		1,073	1,250	
	0600	Lobby collection boxes, aluminum		5	3.200	Ea.	1,800	129		1,929	2,175	
	0700	Bronze or stainless		4.50	3.556	"	2,300	143		2,443	2,750	

10 SPECIALTIES

		10605	Wire Mesh Partitions	CREW	DAILY OUTPUT	LABOR-HOURS	UNIT	2005 BARE COSTS				TOTAL INCL O&P	
								MAT.	LABOR	EQUIP.	TOTAL		
100	0010		**PARTITIONS, WOVEN WIRE** For tool or stockroom enclosures										100
	0100		Channel frame, 1-1/2" diamond mesh, 10 ga. wire, painted										
	0300		Wall panels, 4'-0" wide, 7' high	2 Carp	25	.640	Ea.	109	22		131	154	
	0400		8' high		23	.696		117	24		141	166	
	0600		10' high		18	.889		139	30.50		169.50	201	
	0700		For 5' wide panels, add					5%					
	0900		Ceiling panels, 10' long, 2' wide	2 Carp	25	.640		93.50	22		115.50	137	
	1000		4' wide		15	1.067		114	36.50		150.50	183	
	1200		Panel with service window & shelf, 5' wide, 7' high		20	.800		299	27.50		326.50	375	
	1300		8' high		15	1.067		310	36.50		346.50	395	
	1500		Sliding doors, full height, 3' wide, 7' high		6	2.667		315	91.50		406.50	485	
	1600		10' high		5	3.200		330	110		440	535	
	1800		6' wide sliding door, 7' full height		5	3.200		400	110		510	610	
	1900		10' high		4	4		515	137		652	785	
	2100		Swinging doors, 3' wide, 7' high, no transom		6	2.667		212	91.50		303.50	375	
	2200		7' high, 3' transom		5	3.200		261	110		371	460	

10610 | Folding Gates

				CREW	DAILY OUTPUT	LABOR-HOURS	UNIT	MAT.	LABOR	EQUIP.	TOTAL	TOTAL INCL O&P	
100	0010		**SECURITY GATES** For roll up type, see division 08330-130										100
	0300		Scissors type folding gate, ptd. steel, single, 6-1/2" high, 5-1/2"wide	2 Sswk	4	4	Opng.	136	153		289	420	
	0350		6-1/2' wide		4	4		144	153		297	430	
	0400		7-1/2' wide		4	4		175	153		328	465	
	0600		Double gate, 8' high, 8' wide		2.50	6.400		222	244		466	680	
	0650		10' wide		2.50	6.400		260	244		504	720	
	0700		12' wide		2	8		360	305		665	940	
	0750		14' wide		2	8		415	305		720	1,000	
	0900		Door gate, folding steel, 4' wide, 61" high		4	4		69	153		222	350	
	1000		71" high		4	4		71.50	153		224.50	350	
	1200		81" high		4	4		77.50	153		230.50	360	
	1300		Window gates, 2' to 4' wide, 31" high		4	4		37	153		190	315	
	1500		55" high		3.75	4.267		53	163		216	350	
	1600		79" high		3.50	4.571		66	174		240	385	

10615 | Demountable Partitions

				CREW	DAILY OUTPUT	LABOR-HOURS	UNIT	MAT.	LABOR	EQUIP.	TOTAL	TOTAL INCL O&P	
100	0010		**PARTITIONS, MOVABLE OFFICE** Demountable, add for doors										100
	0100		Do not deduct door openings from total L.F.										
	0900		Demountable gypsum system on 2" to 2-1/2"										
	1000		steel studs, 9' high, 3" to 3-3/4" thick										
	1200		Vinyl clad gypsum **CN**	2 Carp	48	.333	L.F.	43.50	11.40		54.90	66	
	1300		Fabric clad gypsum		44	.364		108	12.45		120.45	138	
	1500		Steel clad gypsum		40	.400		117	13.70		130.70	151	
	1600		1.75 system, aluminum framing, vinyl clad hardboard,										
	1800		paper honeycomb core panel, 1-3/4" to 2-1/2" thick										
	1900		9' high	2 Carp	48	.333	L.F.	73	11.40		84.40	98.50	
	2100		7' high		60	.267		65.50	9.15		74.65	86.50	
	2200		5' high		80	.200		55.50	6.85		62.35	71.50	
	2250		Unitized gypsum system										
	2300		Unitized panel, 9' high, 2" to 2-1/2" thick										
	2350		Vinyl clad gypsum	2 Carp	48	.333	L.F.	94	11.40		105.40	121	
	2400		Fabric clad gypsum	"	44	.364	"	155	12.45		167.45	189	
	2500		Unitized mineral fiber system										
	2510		Unitized panel, 9' high, 2-1/4" thick, aluminum frame										
	2550		Vinyl clad mineral fiber	2 Carp	48	.333	L.F.	93.50	11.40		104.90	121	
	2600		Fabric clad mineral fiber	"	44	.364	"	139	12.45		151.45	172	
	2800		Movable steel walls, modular system										
	2900		Unitized panels, 9' high, 48" wide										

10 SPECIALTIES

10615 | Demountable Partitions

		CREW	DAILY OUTPUT	LABOR-HOURS	UNIT	2005 BARE COSTS				TOTAL INCL O&P
						MAT.	LABOR	EQUIP.	TOTAL	
100	3100	Baked enamel, pre-finished	2 Carp	60	.267	L.F.	106	9.15		115.15	131	100
	3200	Fabric clad steel		56	.286	"	153	9.80		162.80	183
	5310	Trackless wall, cork finish, semi-acoustic, 1-5/8" thick, minimum		325	.049	S.F.	31	1.69		32.69	36.50
	5320	Maximum		190	.084		29	2.88		31.88	36.50
	5330	Acoustic, 2" thick, minimum		305	.052		25	1.80		26.80	30.50
	5340	Maximum		225	.071		43	2.44		45.44	51.50
	5500	For acoustical partitions, add, minimum					1.79			1.79	1.97
	5550	Maximum					8.35			8.35	9.15
	5700	For doors, see Div. 08100 & 08200									
	5800	For door hardware, see div. 08700									
	6100	In-plant modular office system, w/prehung hollow core door									
	6200	3" thick polystyrene core panels									
	6250	12' x 12', 2 wall	2 Clab	3.80	4.211	Ea.	2,900	112		3,012	3,375
	6300	4 wall		1.90	8.421		4,050	225		4,275	4,800
	6350	16' x 16', 2 wall		3.60	4.444		4,525	119		4,644	5,175
	6400	4 wall		1.80	8.889		5,825	237		6,062	6,775

10630 | Port. Partitions/Screens/Panels

		CREW	DAILY OUTPUT	LABOR-HOURS	UNIT	MAT.	LABOR	EQUIP.	TOTAL	TOTAL INCL O&P
100	0010	**PARTITIONS, PORTABLE** Divider panels, free standing, fiber core										100
	0020	Fabric face straight									
	0100	3'-0" long, 4'-0" high	2 Carp	100	.160	L.F.	106	5.50		111.50	126
	0200	5'-0" high		90	.178		107	6.10		113.10	128
	0500	6'-0" high		75	.213		122	7.30		129.30	145
	0900	5'-0" long, 4'-0" high		175	.091		80.50	3.13		83.63	93.50
	1000	5'-0" high		150	.107		98.50	3.65		102.15	115
	1500	6"-0" high		125	.128		94.50	4.38		98.88	111
	1600	6'-0" long, 5'-0" high		162	.099		79	3.38		82.38	92.50
	3100	Curved, 3'-0" long, 5'-0" high		90	.178		83	6.10		89.10	101
	3150	6'-0" high		75	.213		95.50	7.30		102.80	116
	3200	Economical panels, fabric face, 4'-0" long, 5'-0" high		132	.121		36.50	4.15		40.65	46.50
	3250	6'-0" high		112	.143		39	4.89		43.89	50
	3300	5'-0" long, 5'-0" high		150	.107		30.50	3.65		34.15	39
	3350	6'-0" high		125	.128		35.50	4.38		39.88	46
	3380	3'-0" curved, 5'-0" high		90	.178		83	6.10		89.10	101
	3390	6'-0" high		75	.213		95.50	7.30		102.80	116
	3450	Acoustical panels, 60 to 90 NRC, 3'-0" long, 5'-0" high		90	.178		95.50	6.10		101.60	115
	3550	6'-0" high		75	.213		91.50	7.30		98.80	112
	3600	5'-0" long, 5'-0" high		150	.107		66.50	3.65		70.15	78.50
	3650	6'-0" high		125	.128		73.50	4.38		77.88	88
	3700	6'-0" long, 5'-0" high		162	.099		64.50	3.38		67.88	76
	3750	6'-0" high		138	.116		64.50	3.97		68.47	77
	3800	Economy acoustical panels, 40 NRC, 4'-0" long, 5'-0" high		132	.121		36.50	4.15		40.65	46.50
	3850	6'-0" high		112	.143		42	4.89		46.89	53.50
	3900	5'-0" long, 6'-0" high		125	.128		35.50	4.38		39.88	46
	3950	6'-0" long, 5'-0" high		162	.099		29.50	3.38		32.88	38
	4000	Metal chalkboard, 6'-6" high, chalkboard, 1 side		125	.128		86	4.38		90.38	102
	4100	Metal chalkboard, 2 sides		120	.133		98.50	4.57		103.07	115
	4300	Tackboard, both sides		123	.130		78	4.46		82.46	92.50

10651 | Accordion Folding Partitions

		CREW	DAILY OUTPUT	LABOR-HOURS	UNIT	MAT.	LABOR	EQUIP.	TOTAL	TOTAL INCL O&P
100	0010	**PARTITIONS, FOLDING ACCORDION**										100
	0100	Vinyl covered, over 150 S.F., frame not included									
	0300	Residential, 1.25 lb. per S.F., 8' maximum height	2 Carp	300	.053	S.F.	16.55	1.83		18.38	21
	0400	Commercial, 1.75 lb. per S.F., 8' maximum height		225	.071		16.55	2.44		18.99	22
	0600	2 lb. per S.F., 17' maximum height		150	.107		17.10	3.65		20.75	24.50
	0700	Industrial, 4 lb. per S.F., 20' maximum height		75	.213		29.50	7.30		36.80	44

Important: See the Reference Section for critical supporting data - Reference Nos., Crews, & City Cost Indexes

10600 | Partitions

10651 | Accordion Folding Partitions

		CREW	DAILY OUTPUT	LABOR-HOURS	UNIT	2005 BARE COSTS				TOTAL INCL O&P	
						MAT.	LABOR	EQUIP.	TOTAL		
100	0900	Acoustical, 3 lb. per S.F., 17' maximum height	2 Carp	100	.160	S.F.	23	5.50		28.50	34	100
	1200	5 lb. per S.F., 20' maximum height		95	.168		32	5.75		37.75	44	
	1300	5.5 lb. per S.F., 17' maximum height		90	.178		35.50	6.10		41.60	48.50	
	1400	Fire rated, 4.5 psf, 20' maximum height		160	.100		35.50	3.43		38.93	44.50	
	1500	Vinyl clad wood or steel, electric operation, 5.0 psf		160	.100		41.50	3.43		44.93	51	
	1900	Wood, non-acoustic, birch or mahogany, to 10' high		300	.053		21	1.83		22.83	26.50	

10653 | Folding Panel Partitions

		CREW	DAILY OUTPUT	LABOR-HOURS	UNIT	2005 BARE COSTS				TOTAL INCL O&P	
						MAT.	LABOR	EQUIP.	TOTAL		
200	0010	**PARTITIONS, FOLDING LEAF** Acoustic, wood										200
	0100	Vinyl faced, to 18' high, 6 psf, minimum	2 Carp	60	.267	S.F.	42	9.15		51.15	60.50	
	0150	Average		45	.356		50	12.20		62.20	74	
	0200	Maximum		30	.533		64.50	18.25		82.75	99.50	
	0400	Formica or hardwood finish, minimum		60	.267		43	9.15		52.15	62	
	0500	Maximum		30	.533		46	18.25		64.25	79	
	0600	Wood, low acoustical type, 4.5 psf, to 14' high		50	.320		31.50	10.95		42.45	51.50	
	1100	Steel, acoustical, 9 to 12 lb. per S.F., vinyl faced, minimum		60	.267		44.50	9.15		53.65	63.50	
	1200	Maximum		30	.533		54.50	18.25		72.75	88.50	
	1700	Aluminum framed, acoustical, to 12' high, 5.5 psf, minimum		60	.267		30	9.15		39.15	47.50	
	1800	Maximum		30	.533		36.50	18.25		54.75	68.50	
	2000	6.5 lb. per S.F., minimum		60	.267		31.50	9.15		40.65	49.50	
	2100	Maximum		30	.533		39	18.25		57.25	71.50	

10658 | Acoustic Air Wall

		CREW	DAILY OUTPUT	LABOR-HOURS	UNIT	2005 BARE COSTS				TOTAL INCL O&P	
						MAT.	LABOR	EQUIP.	TOTAL		
100	0010	**PARTITIONS, OPERABLE** Acoustic air wall, 1-5/8" thick, minimum	2 Carp	375	.043	S.F.	25.50	1.46		26.96	30.50	100
	0100	Maximum		365	.044		44	1.50		45.50	51	
	0300	2-1/4" thick, minimum		360	.044		29	1.52		30.52	34	
	0400	Maximum		330	.048		50.50	1.66		52.16	58	
	0600	For track type, add to above				L.F.	93.50			93.50	103	
	0700	Overhead track type, acoustical, 3" thick, 11 psf, minimum	2 Carp	350	.046	S.F.	64.50	1.57		66.07	73.50	
	0800	Maximum	"	300	.053	"	78	1.83		79.83	88.50	

10670 | Storage Shelving

10674 | Storage Shelving

		CREW	DAILY OUTPUT	LABOR-HOURS	UNIT	2005 BARE COSTS				TOTAL INCL O&P	
						MAT.	LABOR	EQUIP.	TOTAL		
500	0010	**SHELVING** Metal, industrial, cross-braced, 3' wide, 12" deep	1 Sswk	175	.046	SF Shlf	6.50	1.74		8.24	10.25	500
	0100	24" deep		330	.024		4.95	.92		5.87	7.10	
	0300	4' wide, 12" deep		185	.043		7.40	1.65		9.05	11.10	
	0400	24" deep		380	.021		7.15	.80		7.95	9.30	
	1200	Enclosed sides, cross-braced back, 3' wide, 12" deep		175	.046		16.15	1.74		17.89	21	
	1300	24" deep		290	.028		10.25	1.05		11.30	13.15	
	1500	Fully enclosed, sides and back, 3' wide, 12" deep		150	.053		14.15	2.03		16.18	19.20	
	1600	24" deep		255	.031		7.25	1.20		8.45	10.15	
	1800	4' wide, 12" deep		150	.053		6.35	2.03		8.38	10.60	
	1900	24" deep		290	.028		10.15	1.05		11.20	13.05	
	2200	Wide span, 1600 lb. capacity per shelf, 6' wide, 24" deep		380	.021		8.10	.80		8.90	10.35	
	2400	36" deep		440	.018		7.10	.69		7.79	9.05	
	2600	8' wide, 24" deep		440	.018		7.30	.69		7.99	9.25	
	2800	36" deep		520	.015		6.05	.59		6.64	7.70	

10670 | Storage Shelving

		10674	Storage Shelving	CREW	DAILY OUTPUT	LABOR-HOURS	UNIT	2005 BARE COSTS				TOTAL INCL O&P	
								MAT.	LABOR	EQUIP.	TOTAL		
500	4000		Pallet racks, steel frame 5,000 lb. capacity, 8' long, 36" deep	2 Sswk	450	.036	SF Shlf	10.05	1.36		11.41	13.50	500
	4200		42" deep		500	.032		8.80	1.22		10.02	11.90	
	4400		48" deep	↓	520	.031	↓	8.55	1.17		9.72	11.55	
600	0010		**PARTS BINS** Metal, gray baked enamel finish										600
	0100		6'-3" high, 3' wide										
	0300		12 bins, 18" wide x 12" high, 12" deep	2 Clab	10	1.600	Ea.	240	42.50		282.50	330	
	0400		24" deep		10	1.600		305	42.50		347.50	400	
	0600		72 bins, 6" wide x 6" high, 12" deep		8	2		425	53.50		478.50	555	
	0700		18" deep	↓	8	2	↓	920	53.50		973.50	1,075	
	1000		7'-3" high, 3' wide										
	1200		14 bins, 18" wide x 12" high, 12" deep	2 Clab	10	1.600	Ea.	296	42.50		338.50	390	
	1300		24" deep		10	1.600		395	42.50		437.50	500	
	1500		84 bins, 6" wide x 6" high, 12" deep		8	2		675	53.50		728.50	830	
	1600		24" deep	↓	8	2	↓	785	53.50		838.50	950	

10750 | Telephone Specialties

		10755	Telephone Enclosures	CREW	DAILY OUTPUT	LABOR-HOURS	UNIT	2005 BARE COSTS				TOTAL INCL O&P	
								MAT.	LABOR	EQUIP.	TOTAL		
400	0010		**TELEPHONE ENCLOSURE**										400
	0300		Shelf type, wall hung, minimum	2 Carp	5	3.200	Ea.	970	110		1,080	1,250	
	0400		Maximum		5	3.200		2,500	110		2,610	2,925	
	0600		Booth type, painted steel, indoor or outdoor, minimum		1.50	10.667		3,050	365		3,415	3,925	
	0700		Maximum (stainless steel)		1.50	10.667		10,200	365		10,565	11,800	
	1300		Outdoor, acoustical, on post		3	5.333		1,350	183		1,533	1,750	
	1400		Phone carousel, pedestal mounted with dividers		.60	26.667		5,100	915		6,015	7,050	
	1900		Outdoor, drive-up type, wall mounted		4	4		820	137		957	1,125	
	2000		Post mounted, stainless steel posts	↓	3	5.333	↓	1,250	183		1,433	1,675	
	2200		Directory shelf, wall mounted, stainless steel										
	2300		3 binders	2 Carp	8	2	Ea.	980	68.50		1,048.50	1,175	
	2500		4 binders		7	2.286		1,100	78.50		1,178.50	1,350	
	2600		5 binders		6	2.667		1,425	91.50		1,516.50	1,725	
	2800		Table type, stainless steel, 4 binders		8	2		1,150	68.50		1,218.50	1,375	
	2900		7 binders	↓	7	2.286	↓	1,425	78.50		1,503.50	1,700	

10800 | Toilet/Bath/Laundry Accessories

		10810	Toilet Accessories	CREW	DAILY OUTPUT	LABOR-HOURS	UNIT	2005 BARE COSTS				TOTAL INCL O&P	
								MAT.	LABOR	EQUIP.	TOTAL		
100	0010		**COMMERCIAL TOILET ACCESSORIES**										100
	0200		Curtain rod, stainless steel, 5' long, 1" diameter	1 Carp	13	.615	Ea.	30.50	21		51.50	66.50	
	0300		1-1/4" diameter		13	.615		29	21		50	64.50	
	0400		Diaper changing station, horizontal, wall mounted, plastic	↓	10	.800	↓	174	27.50		201.50	235	

10 SPECIALTIES

Important: See the Reference Section for critical supporting data - Reference Nos., Crews, & City Cost Indexes

			DAILY	LABOR-			2005 BARE COSTS				TOTAL	
	10810	**Toilet Accessories**	**CREW**	**OUTPUT**	**HOURS**	**UNIT**	**MAT.**	**LABOR**	**EQUIP.**	**TOTAL**	**INCL O&P**	
100	0500	Dispenser units, combined soap & towel dispensers,										**100**
	0510	mirror and shelf, flush mounted	1 Carp	10	.800	Ea.	330	27.50		357.50	410	
	0600	Towel dispenser and waste receptacle,										
	0610	18 gallon capacity	1 Carp	10	.800	Ea.	256	27.50		283.50	325	
	0800	Grab bar, straight, 1-1/4" diameter, stainless steel, 18" long		24	.333		18.75	11.40		30.15	38.50	
	0900	24" long		23	.348		21	11.90		32.90	41.50	
	1000	30" long		22	.364		21	12.45		33.45	43	
	1100	36" long		20	.400		19.75	13.70		33.45	43	
	1105	42" long		20	.400		20.50	13.70		34.20	44	
	1200	1-1/2" diameter, 24" long		23	.348		50.50	11.90		62.40	74	
	1300	36" long		20	.400		57	13.70		70.70	84	
	1310	42" long		18	.444		48	15.20		63.20	76.50	
	1500	Tub bar, 1-1/4" diameter, 24" x 36"		14	.571		87.50	19.55		107.05	127	
	1600	Plus vertical arm		12	.667		86.50	23		109.50	131	
	1900	End tub bar, 1" diameter, 90° angle, 16" x 32"		12	.667		118	23		141	166	
	2010	Tub/shower/toilet, 2-wall, 36" x 24"		12	.667		76	23		99	119	
	2300	Hand dryer, surface mounted, electric, 115 volt, 20 amp		4	2		485	68.50		553.50	635	
	2400	230 volt, 10 amp		4	2		485	68.50		553.50	635	
	2600	Hat and coat strip, stainless steel, 4 hook, 36" long		24	.333		48	11.40		59.40	71	
	2700	6 hook, 60" long		20	.400		84	13.70		97.70	114	
	3000	Mirror, with stainless steel 3/4" square frame, 18" x 24"		20	.400		65.50	13.70		79.20	93.50	
	3100	36" x 24"		15	.533		104	18.25		122.25	143	
	3200	48" x 24"		10	.800		159	27.50		186.50	217	
	3300	72" x 24"		6	1.333		196	45.50		241.50	287	
	3500	With 5" stainless steel shelf, 18" x 24"		20	.400		114	13.70		127.70	147	
	3600	36" x 24"		15	.533		161	18.25		179.25	206	
	3700	48" x 24"		10	.800		171	27.50		198.50	231	
	3800	72" x 24"		6	1.333		286	45.50		331.50	385	
	4100	Mop holder strip, stainless steel, 5 holders, 48" long		20	.400		80.50	13.70		94.20	110	
	4200	Napkin/tampon dispenser, recessed		15	.533		310	18.25		328.25	375	
	4300	Robe hook, single, regular		36	.222		4.95	7.60		12.55	17.30	
	4400	Heavy duty, concealed mounting		36	.222		11.10	7.60		18.70	24	
	4600	Soap dispenser, chrome, surface mounted, liquid		20	.400		44	13.70		57.70	69.50	
	4700	Powder		20	.400		42	13.70		55.70	67.50	
	5000	Recessed stainless steel, liquid		10	.800		143	27.50		170.50	200	
	5100	Powder		10	.800		174	27.50		201.50	234	
	5300	Soap tank, stainless steel, 1 gallon		10	.800		159	27.50		186.50	218	
	5400	5 gallon		5	1.600		219	55		274	325	
	5600	Shelf, stainless steel, 5" wide, 18 ga., 24" long		24	.333		41.50	11.40		52.90	64	
	5700	48" long		16	.500		76.50	17.15		93.65	111	
	5800	8" wide shelf, 18 ga., 24" long		22	.364		49	12.45		61.45	73.50	
	5900	48" long		14	.571		108	19.55		127.55	150	
	6000	Toilet seat cover dispenser, stainless steel, recessed		20	.400		102	13.70		115.70	134	
	6050	Surface mounted		15	.533		26	18.25		44.25	57	
	6100	Toilet tissue dispenser, surface mounted, SS, single roll		30	.267		10.80	9.15		19.95	26	
	6200	Double roll		24	.333		15.25	11.40		26.65	34.50	
	6400	Towel bar, stainless steel, 18" long		23	.348		35	11.90		46.90	57	
	6500	30" long		21	.381		56.50	13.05		69.55	83	
	6700	Towel dispenser, stainless steel, surface mounted		16	.500		41	17.15		58.15	72	
	6800	Flush mounted, recessed		10	.800		172	27.50		199.50	232	
	7000	Towel holder, hotel type, 2 guest size		20	.400		15.40	13.70		29.10	38.50	
	7200	Towel shelf, stainless steel, 24" long, 8" wide		20	.400		48.50	13.70		62.20	75	
	7400	Tumbler holder, tumbler only		30	.267		25	9.15		34.15	42	
	7500	Soap, tumbler & toothbrush		30	.267		22.50	9.15		31.65	39.50	
	7700	Wall urn ash receiver, surface mount, 11" long		12	.667		113	23		136	160	
	7800	7-1/2", long		18	.444		83	15.20		98.20	115	

SPECIALTIES **10**

10810 | Toilet Accessories

		CREW	DAILY OUTPUT	LABOR-HOURS	UNIT	MAT.	LABOR	EQUIP.	TOTAL	TOTAL INCL O&P	
						2005 BARE COSTS					
100	8000 Waste receptacles, stainless steel, with top, 13 gallon	1 Carp	10	.800	Ea.	187	27.50		214.50	248	100
	8100 36 gallon	↓	8	1	↓	345	34.50		379.50	435	

10820 | Bath Accessories

		CREW	DAILY OUTPUT	LABOR-HOURS	UNIT	MAT.	LABOR	EQUIP.	TOTAL	TOTAL INCL O&P	
400	0010 **MEDICINE CABINETS** With mirror, st. st. frame, 16" x 22", unlighted	1 Carp	14	.571	Ea.	69.50	19.55		89.05	107	400
	0100 Wood frame		14	.571		96.50	19.55		116.05	137	
	0300 Sliding mirror doors, 20" x 16" x 4-3/4", unlighted		7	1.143		86.50	39		125.50	156	
	0400 24" x 19" x 8-1/2", lighted		5	1.600		136	55		191	236	
	0600 Triple door, 30" x 32", unlighted, plywood body		7	1.143		214	39		253	297	
	0700 Steel body		7	1.143		282	39		321	370	
	0900 Oak door, wood body, beveled mirror, single door		7	1.143		125	39		164	199	
	1000 Double door		6	1.333		320	45.50		365.50	420	
	1200 Hotel cabinets, stainless, with lower shelf, unlighted		10	.800		173	27.50		200.50	233	
	1300 Lighted	↓	5	1.600	↓	256	55		311	370	

10885 | Scales

		CREW	DAILY OUTPUT	LABOR-HOURS	UNIT	MAT.	LABOR	EQUIP.	TOTAL	TOTAL INCL O&P	
						2005 BARE COSTS					
100	0010 **SCALES** Built-in floor scale, not incl. foundations										100
	0100 Dial type, 5 ton capacity, 8' x 6' platform	3 Carp	.50	48	Ea.	6,225	1,650		7,875	9,400	
	0300 9' x 7' platform		.40	60		8,075	2,050		10,125	12,100	
	0400 10 ton capacity, steel platform, 8' x 6' platform		.40	60		9,775	2,050		11,825	14,000	
	0600 9' x 7' platform	↓	.35	68.571	↓	9,175	2,350		11,525	13,800	
	0700 Truck scales, incl. steel weigh bridge,										
	0800 not including foundation, pits										
	1550 Digital, electronic, 100 ton capacity, steel deck 12' x 10' platform	3 Carp	.20	120	Ea.	10,900	4,100		15,000	18,300	
	1600 40' x 10' platform		.14	171		22,100	5,875		27,975	33,500	
	1640 60' x 10' platform		.13	184		27,300	6,325		33,625	40,000	
	1680 70' x 10' platform	↓	.12	200		30,500	6,850		37,350	44,300	
	2000 For standard automatic printing device, add					2,200			2,200	2,425	
	2100 For remote reading electronic system, add					2,000			2,000	2,200	
	2300 Concrete foundation pits for above, 8' x 6', 5 C.Y. required	C-1	.50	64		830	2,075		2,905	4,150	
	2400 14' x 6' platform, 10 C.Y. required		.35	91.429		1,225	2,950		4,175	5,950	
	2600 50' x 10' platform, 30 C.Y. required		.25	128		1,650	4,150		5,800	8,250	
	2700 70' x 10' platform, 40 C.Y. required	↓	.15	213		3,575	6,900		10,475	14,800	
	2750 Crane scales, dial, 1 ton capacity					885			885	975	
	2780 5 ton capacity					1,050			1,050	1,150	
	2800 Digital, 1 ton capacity					1,950			1,950	2,125	
	2850 10 ton capacity				↓	3,725			3,725	4,100	
	2900 Low profile electronic warehouse scale,										
	3000 not incl. printer, 4' x 4' platform, 10,000 lb. capacity	2 Carp	.30	53.333	Ea.	2,325	1,825		4,150	5,400	
	3300 5' x 7' platform, 10,000 lb. capacity		.25	64		3,575	2,200		5,775	7,375	
	3400 20,000 lb. capacity	↓	.20	80		4,625	2,750		7,375	9,375	
	3500 For printers, incl. time, date & numbering, add					1,150			1,150	1,275	
	3800 Portable, beam type, capacity 1000#, platform 18" x 24"					670			670	735	
	3900 Dial type, capacity 2000#, platform 24" x 24"					1,150			1,150	1,275	
	4000 Digital type, capacity 1000#, platform 24" x 30"					1,850			1,850	2,025	
	4100 Portable contractor truck scales, 50 ton cap., 40' x 10' platform					28,700			28,700	31,600	
	4200 60' x 10' platform				↓	25,700			25,700	28,300	

Important: See the Reference Section for critical supporting data - Reference Nos., Crews, & City Cost Indexes

10 SPECIALTIES

10905	Coat Racks/Wardrobes	CREW	DAILY OUTPUT	LABOR-HOURS	UNIT	2005 BARE COSTS				TOTAL INCL O&P
						MAT.	LABOR	EQUIP.	TOTAL	
500	**0010 COAT RACKS & WARDROBES**									**500**
0020	Floor model hat & coat racks, 6 hangers									
0050	Standing, beech wood, 21" x 21" x 72", chrome				Ea.	223			223	245
0100	18 gauge tubular steel, 21" x 21" x 69", wood walnut				"	255			255	280
0500	16 gauge steel frame, 22 gauge steel shelves									
0650	Single pedestal, 30" x 18" x 63"				Ea.	162			162	178
0800	Single face rack, 29" x 18-1/2" x 62"					164			164	181
0900	51" x 18-1/2" x 70"					289			289	320
0910	Double face rack, 39" x 26" x 70"					191			191	210
0920	63" x 26" x 70"					340			340	375
0940	For 2" ball casters, add				Set	77			77	84.50
1400	Utility hook strips, 3/8" x 2-1/2" x 18", 6 hooks	1 Carp	48	.167	Ea.	50	5.70		55.70	63.50
1500	34" long, 12 hooks	"	48	.167	"	48.50	5.70		54.20	62
1650	Wall mounted racks, 16 gauge steel frame, 22 gauge steel shelves									
1850	12" x 15" x 26", 6 hangers	1 Carp	32	.250	Ea.	120	8.55		128.55	144
2000	12" x 15" x 50", 12 hangers	"	32	.250	"	128	8.55		136.55	154
2150	Wardrobe cabinet, steel, baked enamel finish									
2300	36" x 21" x 78", incl. top shelf & hanger rod				Ea.	248			248	273

For information about Means Estimating Seminars, see yellow pages 12 and 13 in back of book

SPECIALTIES 10

Division Notes

	CREW	DAILY OUTPUT	LABOR-HOURS	UNIT	2005 BARE COSTS				TOTAL INCL O&P
					MAT.	LABOR	EQUIP.	TOTAL	

Division 11
Equipment

Estimating Tips
General
- The items in this division are usually priced per square foot or each. Many of these items are purchased by the owner for installation by the contractor. Check the specifications for responsibilities, and include time for receiving, storage, installation and mechanical and electrical hook-ups in the appropriate divisions.

- Many items in Division 11 require some type of support system that is not usually furnished with the item. Examples of these systems include blocking for the attachment of casework and support angles for ceiling hung projection screens. The required blocking or supports must be added to the estimate in the appropriate division.

- Some items in Division 11 may require assembly or electrical hook-ups. Verify the amount of assembly required or the need for a hard electrical connection and add the appropriate costs.

Reference Numbers
Reference numbers are shown in bold squares at the beginning of some major classifications. These numbers refer to related items in the Reference Section. The reference information may be an estimating procedure, an alternate pricing method or technical information.

Note: Not all subdivisions listed here necessarily appear in this publication.

11012 | Contractor Equipment

			CREW	DAILY OUTPUT	LABOR-HOURS	UNIT	2005 BARE COSTS				TOTAL INCL O&P	
							MAT.	LABOR	EQUIP.	TOTAL		
100	0010	CONTRACTOR EQUIPMENT See Div. 01590	R01590 -100									100

11013 | Floor/Wall Cleaning Equipment

			CREW	DAILY OUTPUT	LABOR-HOURS	UNIT	MAT.	LABOR	EQUIP.	TOTAL	TOTAL INCL O&P	
800	0010	VACUUM CLEANING										800
	0020	Central, 3 inlet, residential	1 Skwk	.90	8.889	Total	610	310		920	1,150	
	0200	Commercial		.70	11.429		1,150	400		1,550	1,900	
	0400	5 inlet system, residential		.50	16		925	560		1,485	1,900	
	0600	7 inlet system, commercial		.40	20		1,025	695		1,720	2,225	
	0800	9 inlet system, residential		.30	26.667		1,300	930		2,230	2,875	
	4010	Rule of thumb: First 1200 S.F., installed									1,125	
	4020	For each additional S.F., add				S.F.					.18	

11 EQUIPMENT

11021 | Safes and Vault Doors

			CREW	DAILY OUTPUT	LABOR-HOURS	UNIT	2005 BARE COSTS MAT.	LABOR	EQUIP.	TOTAL	TOTAL INCL O&P	
600	0010	SAFE										600
	0015	Office, 4 hr. rating, 30" x 18" x 18" inside				Ea.	3,125			3,125	3,425	
	0100	60" x 36" x 18"					6,775			6,775	7,450	
	0200	1 hr. rating, 30" x 18" x 18"					1,650			1,650	1,800	
	0250	40" x 18" x 18"					3,275			3,275	3,600	
	0300	60" x 36" x 18", double door					4,225			4,225	4,650	
	0400	Data, 4 hr. rating, 23-1/2" x 19-1/2" x 17" inside					3,025			3,025	3,325	
	0450	52" x 19" x 17"					5,875			5,875	6,450	
	0500	63" x 34" x 16", double door					9,100			9,100	10,000	
	0550	52" x 34" x 16", inside					8,550			8,550	9,400	
	0600	1 hr. rating, 27" x 19" x 16"					4,050			4,050	4,450	
	0700	63" x 34" x 16"					7,525			7,525	8,275	
	0750	Diskette, 1 hr., 14" x 12" x 11", inside					2,775			2,775	3,075	
	0800	Money, "B" label, 9" x 14" x 14"					395			395	435	
	0900	Tool resistive, 24" x 24" x 20"					2,675			2,675	2,925	
	1050	Tool and torch resistive, 24" x 24" x 20"					6,550			6,550	7,200	
	1150	Jewelers, 23" x 20" x 18"					8,350			8,350	9,200	
	1200	63" x 25" x 18"					14,600			14,600	16,000	
	1300	For handling into building, add, minimum	A-2	8.50	2.824			75.50	15.15	90.65	134	
	1400	Maximum	"	.78	30.769			820	165	985	1,450	
800	0010	VAULT DOOR										800
	0020	Door and frame, 32" x 78", clear opening										
	0100	1 hour test, 32" door, weighs 750 lbs.	2 Sswk	1.50	10.667	Opng.	3,375	405		3,780	4,425	
	0200	2 hour test, 32" door, weighs 950 lbs.		1.30	12.308		3,550	470		4,020	4,775	
	0250	40" door, weighs 1130 lbs.		1	16		4,075	610		4,685	5,575	
	0300	4 hour test, 32" door, weighs 1025 lbs.		1.20	13.333		3,900	510		4,410	5,175	
	0350	40" door, weighs 1140 lbs.		.90	17.778		4,525	680		5,205	6,200	
	0500	For stainless steel front, including frame, add to above					1,750			1,750	1,925	
	0550	Back, add					1,750			1,750	1,925	
	0600	For time lock, two movement, add	1 Elec	2	4	Ea.	1,350	163		1,513	1,725	
	0650	Three movement, add	"	2	4		1,825	163		1,988	2,275	
	0800	Day gate, painted, steel, 32" wide	2 Sswk	1.50	10.667		1,575	405		1,980	2,450	
	0850	40" wide		1.40	11.429		1,750	435		2,185	2,700	
	0900	Aluminum, 32" wide		1.50	10.667		2,375	405		2,780	3,350	

Important: See the Reference Section for critical supporting data - Reference Nos., Crews, & City Cost Indexes

11020 | Security & Vault Equipment

11021 | Safes and Vault Doors

			CREW	DAILY OUTPUT	LABOR-HOURS	UNIT	2005 BARE COSTS MAT.	LABOR	EQUIP.	TOTAL	TOTAL INCL O&P	
800	0950	40" wide	2 Sswk	1.40	11.429	Ea.	2,650	435		3,085	3,700	800
	2050	Security vault door, class I, 3' wide, 3 1/2" thick	E-24	.19	166	Opng.	12,700	6,225	3,225	22,150	28,200	
	2100	Class II, 3' wide, 7" thick		.19	166		14,800	6,225	3,225	24,250	30,400	
	2150	Class III, 9R, 3' wide, 10" thick, minimum		.13	250	↓	18,400	9,325	4,825	32,550	41,600	
	2160	Class V, type 1, 40" door		2.48	12.903	Ea.	5,300	480	249	6,029	6,925	
	2170	Class V, type 2, 40" door		2.48	12.903		5,250	480	249	5,979	6,900	
	2180	Day gate for class V vault	2 Sswk	2	8	↓	1,100	305		1,405	1,750	

11030 | Teller & Service Equipment

11038 | Bank Equipment

			CREW	DAILY OUTPUT	LABOR-HOURS	UNIT	2005 BARE COSTS MAT.	LABOR	EQUIP.	TOTAL	TOTAL INCL O&P	
150	0010	**BANK EQUIPMENT**										150
	0020	Alarm system, police	2 Elec	1.60	10	Ea.	4,025	410		4,435	5,050	
	0100	With vault alarm	"	.40	40		16,000	1,625		17,625	20,000	
	0400	Bullet resistant teller window, 44" x 60"	1 Glaz	.60	13.333		2,700	440		3,140	3,625	
	0500	48" x 60"	"	.60	13.333	↓	3,400	440		3,840	4,400	
	3000	Counters for banks, frontal only	2 Carp	1	16	Station	1,500	550		2,050	2,500	
	3100	Complete with steel undercounter	"	.50	32	"	2,950	1,100		4,050	4,925	
	4600	Door and frame, bullet-resistant, with vision panel, minimum	2 Sswk	1.10	14.545	Ea.	3,325	555		3,880	4,675	
	4700	Maximum		1.10	14.545		4,525	555		5,080	5,975	
	4800	Drive-up window, drawer & mike, not incl. glass, minimum		1	16		3,800	610		4,410	5,300	
	4900	Maximum		.50	32		7,675	1,225		8,900	10,600	
	5000	Night depository, with chest, minimum		1	16		5,725	610		6,335	7,400	
	5100	Maximum		.50	32		8,300	1,225		9,525	11,300	
	5200	Package receiver, painted		3.20	5		1,125	191		1,316	1,575	
	5300	Stainless steel	↓	3.20	5	↓	1,900	191		2,091	2,450	
	5400	Partitions, bullet-resistant, 1-3/16" glass, 8' high	2 Carp	10	1.600	L.F.	163	55		218	265	
	5450	Acrylic	"	10	1.600	"	310	55		365	425	
	5500	Pneumatic tube systems, 2 lane drive-up, complete	L-3	.25	64	Total	21,300	2,400		23,700	27,100	
	5550	With T.V. viewer	"	.20	80	"	41,200	3,000		44,200	49,900	
	5570	Safety deposit boxes, minimum	1 Sswk	44	.182	Opng.	47	6.95		53.95	64.50	
	5580	Maximum, 10" x 15" opening		19	.421		100	16.05		116.05	139	
	5590	Teller locker, average	↓	15	.533	↓	1,300	20.50		1,320.50	1,450	
	5600	Pass thru, bullet-res. window, painted steel, 24" x 36"	2 Sswk	1.60	10	Ea.	1,775	380		2,155	2,625	
	5700	48" x 48"		1.20	13.333		2,200	510		2,710	3,300	
	5800	72" x 40"	↓	.80	20	↓	2,725	765		3,490	4,375	
	5900	For stainless steel frames, add					20%					
	6100	Surveillance system, video camera, complete	2 Elec	1	16	Ea.	12,400	650		13,050	14,700	
	6110	For each additional camera, add				"	815			815	895	
	6120	CCTV system, see Div. 16850-600										
	6200	Twenty-four hour teller, single unit,										
	6300	automated deposit, cash and memo	L-3	.25	64	Ea.	37,500	2,400		39,900	45,000	
	7000	Vault front, see Div. 11021-800										

EQUIPMENT 11

		11041	Ecclesiastical Equipment	CREW	DAILY OUTPUT	LABOR-HOURS	UNIT	2005 BARE COSTS				TOTAL INCL O&P	
								MAT.	LABOR	EQUIP.	TOTAL		
250	0010		CHURCH EQUIPMENT										250
	0020		Altar, wood, custom design, plain	1 Carp	1.40	5.714	Ea.	1,825	196		2,021	2,300	
	0050		Deluxe	"	.20	40		8,800	1,375		10,175	11,800	
	0070		Granite or marble, average	2 Marb	.50	32		7,050	1,050		8,100	9,400	
	0090		Deluxe	"	.20	80		23,100	2,650		25,750	29,600	
	0100		Arks, prefabricated, plain	2 Carp	.80	20		6,950	685		7,635	8,700	
	0130		Deluxe, maximum	"	.20	80		87,000	2,750		89,750	100,000	
	0150		Baptistry, fiberglass, 3'-6" deep, x 13'-7" long,										
	0160		steps at both ends, incl. plumbing, minimum	L-8	1	20	Ea.	2,775	710		3,485	4,150	
	0200		Maximum	"	.70	28.571		5,400	1,025		6,425	7,525	
	0250		Add for filter, heater and lights					1,175			1,175	1,300	
	0300		Carillon, 4 octave (48 bells), with keyboard				System	630,000			630,000	693,000	
	0320		2 octave (24 bells)					262,500			262,500	289,000	
	0340		3 to 4 bell peal, minimum					58,000			58,000	63,500	
	0360		Maximum					420,000			420,000	462,000	
	0380		Cast bronze bell, average				Ea.	73,500			73,500	81,000	
	0400		Electronic, digital, minimum					13,100			13,100	14,400	
	0410		With keyboard, maximum					63,000			63,000	69,500	
	0500		Reconciliation room, wood, prefabricated, single, plain	1 Carp	.60	13.333		2,325	455		2,780	3,250	
	0550		Deluxe		.40	20		6,375	685		7,060	8,075	
	0650		Double, plain		.40	20		4,625	685		5,310	6,175	
	0700		Deluxe		.20	40		14,500	1,375		15,875	18,000	
	1000		Lecterns, wood, plain		5	1.600		655	55		710	805	
	1100		Deluxe		2	4		6,525	137		6,662	7,425	
	1500		Pews, bench type, hardwood, minimum		20	.400	L.F.	64	13.70		77.70	92	
	1550		Maximum		15	.533		127	18.25		145.25	169	
	1570		For kneeler, add					15.25			15.25	16.80	
	2000		Pulpits, hardwood, prefabricated, plain	1 Carp	2	4	Ea.	1,150	137		1,287	1,500	
	2100		Deluxe		1.60	5	"	7,925	171		8,096	9,000	
	2500		Railing, hardwood, average		25	.320	L.F.	144	10.95		154.95	175	
	2700		Safes, see division 11021-600										
	3000		Seating, individual, oak, contour, laminated	1 Carp	21	.381	Person	127	13.05		140.05	161	
	3100		Cushion seat		21	.381		115	13.05		128.05	148	
	3200		Fully upholstered		21	.381		103	13.05		116.05	134	
	3300		Combination, self-rising		21	.381		320	13.05		333.05	370	
	3500		For cherry, add					30%					
	4000		Steeples, translucent fiberglass, 30" square, 15' high	F-3	2	20	Ea.	3,600	690	300	4,590	5,350	
	4150		25' high		1.80	22.222		4,200	770	335	5,305	6,200	
	4350		Opaque fiberglass, 24" square, 14' high		2	20		3,000	690	300	3,990	4,700	
	4500		28' high		1.80	22.222		3,450	770	335	4,555	5,375	
	4600		Aluminum, baked finish, 14' high, 16" square					1,700			1,700	1,875	
	4620		20' high, 3'-6" base					4,725			4,725	5,200	
	4640		35' high, 8' base					17,900			17,900	19,700	
	4660		60' high, 14' base					37,600			37,600	41,300	
	4680		152' high, custom					396,500			396,500	436,000	
	4700		Porcelain enamel steeples, custom, 40' high	F-3	.50	80		11,400	2,775	1,200	15,375	18,200	
	4800		60' high	"	.30	133		19,800	4,600	2,000	26,400	31,100	
	5000		Wall cross, aluminum, extruded, 2" x 2" section	1 Carp	34	.235	L.F.	35	8.05		43.05	51	
	5150		4" x 4" section		29	.276		50	9.45		59.45	69.50	
	5300		Bronze, extruded, 1" x 2" section		31	.258		69	8.85		77.85	89.50	
	5350		2-1/2" x 2-1/2" section		34	.235		104	8.05		112.05	128	
	5450		Solid bar stock, 1/2" x 3" section		29	.276		137	9.45		146.45	166	
	5600		Fiberglass, stock		34	.235		28.50	8.05		36.55	43.50	
	5700		Stainless steel, 4" deep, channel section		29	.276		110	9.45		119.45	137	
	5800		4" deep box section		29	.276		151	9.45		160.45	181	

11 EQUIPMENT

Important: See the Reference Section for critical supporting data - Reference Nos., Crews, & City Cost Indexes

11050 | Library Equipment

11051	Library Equipment	CREW	DAILY OUTPUT	LABOR-HOURS	UNIT	2005 BARE COSTS				TOTAL INCL O&P
						MAT.	LABOR	EQUIP.	TOTAL	
400	0010 **LIBRARY EQUIPMENT**									400
0020	Bookshelf, mtl, 90" high, 10" shelf, dbl face	1 Carp	11.50	.696	L.F.	145	24		169	197
0300	Single face	"	12	.667	"	122	23		145	171
0600	For 8" shelving, subtract from above					10%				
0700	For 12" shelving, add to above					10%				
0800	For 42" high with countertop, subtract from above					20%				
1100	Magazine shelving, 82" high, 12" deep, single face	1 Carp	11.50	.696	L.F.	169	24		193	223
1400	Double face		11.50	.696	"	192	24		216	248
2500	Carrels, hardwood, 36" x 24", minimum		5	1.600	Ea.	635	55		690	785
2650	Maximum		4	2		820	68.50		888.50	1,000
2700	Card catalog file, 60 trays, complete					5,050			5,050	5,575
2720	Alternate method: each tray					84.50			84.50	93
3100	Chairs, wood					139			139	153
3500	Charging desk, built-in, with counter, plastic laminated top	1 Carp	7	1.143	L.F.	440	39		479	540
3700	Reading table, laminated top, 60" x 36"				Ea.	460			460	510
3800	Mobile compacted shelving, hand crank, 9'-0" high									
3820	Double face, including track, 3' section				Ea.	975			975	1,075
3840	For electrical operation, add					25%				

11060 | Theater & Stage Equipment

11063	Stage Equipment	CREW	DAILY OUTPUT	LABOR-HOURS	UNIT	2005 BARE COSTS				TOTAL INCL O&P
						MAT.	LABOR	EQUIP.	TOTAL	
600	0010 **STAGE EQUIPMENT**									600
0050	Control boards with dimmers and breakers, minimum	1 Elec	1	8	Ea.	11,400	325		11,725	13,100
0100	Average		.50	16		31,800	650		32,450	36,000
0150	Maximum		.20	40		104,000	1,625		105,625	116,500
0500	Curtain track, straight, light duty	2 Carp	20	.800	L.F.	20	27.50		47.50	64.50
0600	Heavy duty		18	.889		43	30.50		73.50	95
0700	Curved sections		12	1.333		133	45.50		178.50	217
1000	Curtains, velour, medium weight		600	.027	S.F.	6.50	.91		7.41	8.55
1150	Silica based yarn, fireproof		50	.320		12.50	10.95		23.45	31
1500	Flooring, portable oak parquet, 3' x 3' sections					11			11	12.10
1600	Cart to carry 225 S.F. of flooring				Ea.	330			330	360
2000	Lights, border, quartz, reflector, vented,									
2100	colored or white	1 Elec	20	.400	L.F.	142	16.30		158.30	181
2500	Spotlight, follow spot, with transformer, 2,100 watt	"	4	2	Ea.	1,125	81.50		1,206.50	1,375
2600	For no transformer, deduct					750			750	825
3000	Stationary spot, fresnel quartz, 6" lens	1 Elec	4	2		89.50	81.50		171	220
3100	8" lens		4	2		179	81.50		260.50	320
3500	Ellipsoidal quartz, 1,000W, 6" lens		4	2		271	81.50		352.50	420
3600	12" lens		4	2		450	81.50		531.50	615
4000	Strobe light, 1 to 15 flashes per second, quartz		3	2.667		620	109		729	840
4500	Color wheel, portable, five hole, motorized		4	2		143	81.50		224.50	279
5000	Stages, portable with steps, folding legs, stock, 8" high				SF Stg.	21.50			21.50	23.50
5100	16" high					21			21	23
5200	32" high					29.50			29.50	32.50
5300	40" high					49			49	53.50
6000	Telescoping platforms, extruded alum., straight, minimum	4 Carp	157	.204		23.50	7		30.50	36.50
6100	Maximum		77	.416		31.50	14.25		45.75	57
6500	Pie-shaped, minimum		150	.213		50.50	7.30		57.80	67

11060 | Theater & Stage Equipment

11063 | Stage Equipment

		CREW	DAILY OUTPUT	LABOR-HOURS	UNIT	2005 BARE COSTS MAT.	LABOR	EQUIP.	TOTAL	TOTAL INCL O&P	
600	6600	Maximum	4 Carp	70	.457	SF Stg.	56.50	15.65		72.15	86.50	600
	6800	For 3/4" plywood covered deck, deduct					3.20			3.20	3.52	
	7000	Band risers, steel frame, plywood deck, minimum	4 Carp	275	.116		25	3.99		28.99	33.50	
	7100	Maximum	"	138	.232		52.50	7.95		60.45	70.50	
	7500	Chairs for above, self-storing, minimum	2 Carp	43	.372	Ea.	82.50	12.75		95.25	111	
	7600	Maximum	"	40	.400	"	138	13.70		151.70	173	
	8000	Rule of thumb: total stage equipment, minimum	4 Carp	100	.320	SF Stg.	80	10.95		90.95	105	
	8100	Maximum	"	25	1.280	"	450	44		494	565	

11100 | Mercantile Equipment

11102 | Barber Shop Equipment

		CREW	DAILY OUTPUT	LABOR-HOURS	UNIT	2005 BARE COSTS MAT.	LABOR	EQUIP.	TOTAL	TOTAL INCL O&P	
150	0010	**BARBER EQUIPMENT**										150
	0020	Chair, hydraulic, movable, minimum	1 Carp	24	.333	Ea.	420	11.40		431.40	480	
	0050	Maximum	"	16	.500		2,750	17.15		2,767.15	3,050	
	0200	Wall hung styling station with mirrors, minimum	L-2	8	2		405	60		465	540	
	0300	Maximum	"	4	4		1,825	120		1,945	2,200	
	0500	Sink, hair washing basin, rough plumbing not incl.	1 Plum	8	1		279	41		320	365	
	1000	Sterilizer, liquid solution for tools					128			128	140	
	1100	Total equipment, rule of thumb, per chair, minimum	L-8	1	20		1,450	710		2,160	2,675	
	1150	Maximum	"	1	20		4,075	710		4,785	5,600	

11103 | Cash Register/Checking

		CREW	DAILY OUTPUT	LABOR-HOURS	UNIT	2005 BARE COSTS MAT.	LABOR	EQUIP.	TOTAL	TOTAL INCL O&P	
200	0010	**CHECKOUT COUNTER**										200
	0020	Supermarket conveyor, single belt	2 Clab	10	1.600	Ea.	2,200	42.50		2,242.50	2,475	
	0100	Double belt, power take-away		9	1.778		3,675	47.50		3,722.50	4,125	
	0400	Double belt, power take-away, incl. side scanning		7	2.286		4,575	61		4,636	5,125	
	0800	Warehouse or bulk type		6	2.667		5,350	71		5,421	5,975	
	1000	Scanning system, 2 lanes, w/registers, scan gun & memory				System	13,400			13,400	14,700	
	1100	10 lanes, single processor, full scan, with scales				"	127,000			127,000	140,000	
	2000	Register, restaurant, minimum				Ea.	540			540	595	
	2100	Maximum					2,375			2,375	2,625	
	2150	Store, minimum					540			540	595	
	2200	Maximum					2,375			2,375	2,625	

11104 | Display Cases & Systems

		CREW	DAILY OUTPUT	LABOR-HOURS	UNIT	2005 BARE COSTS MAT.	LABOR	EQUIP.	TOTAL	TOTAL INCL O&P	
700	0010	**REFRIGERATED FOOD CASES**										700
	0030	Dairy, multi-deck, 12' long	Q-5	3	5.333	Ea.	8,200	197		8,397	9,300	
	0100	For rear sliding doors, add					1,125			1,125	1,250	
	0200	Delicatessen case, service deli, 12' long, single deck	Q-5	3.90	4.103		5,500	151		5,651	6,275	
	0300	Multi-deck, 18 S.F. shelf display		3	5.333		6,750	197		6,947	7,725	
	0400	Freezer, self-contained, chest-type, 30 C.F.		3.90	4.103		4,025	151		4,176	4,650	
	0500	Glass door, upright, 78 C.F.		3.30	4.848		7,650	179		7,829	8,700	
	0600	Frozen food, chest type, 12' long		3.30	4.848		5,525	179		5,704	6,375	
	0700	Glass door, reach-in, 5 door		3	5.333		10,600	197		10,797	12,000	
	0800	Island case, 12' long, single deck		3.30	4.848		6,275	179		6,454	7,175	
	0900	Multi-deck		3	5.333		13,200	197		13,397	14,900	
	1000	Meat case, 12' long, single deck		3.30	4.848		4,550	179		4,729	5,275	

Important: See the Reference Section for critical supporting data - Reference Nos., Crews, & City Cost Indexes

11100 | Mercantile Equipment

		11104	Display Cases & Systems	CREW	DAILY OUTPUT	LABOR-HOURS	UNIT	2005 BARE COSTS MAT.	LABOR	EQUIP.	TOTAL	TOTAL INCL O&P	
700	1050		Multi-deck	Q-5	3.10	5.161	Ea.	7,825	190		8,015	8,875	**700**
	1100		Produce, 12' long, single deck	↓	3.30	4.848	↓	6,025	179		6,204	6,900	
	1200		Multi-deck	↓	3.10	5.161	↓	6,700	190		6,890	7,650	

11110 | Commercial Laundry & Dry Cleaning Equipment

		11119	Laundry Cleaning	CREW	DAILY OUTPUT	LABOR-HOURS	UNIT	2005 BARE COSTS MAT.	LABOR	EQUIP.	TOTAL	TOTAL INCL O&P	
450	0010	**LAUNDRY EQUIPMENT** Not incl. rough-in											**450**
	0500	Dryers, gas fired residential, 16 lb. capacity, average		1 Plum	3	2.667	Ea.	565	109		674	785	
	1000	Commercial, 30 lb. capacity, coin operated, single			3	2.667		2,575	109		2,684	3,000	
	1100	Double stacked			2	4		5,525	163		5,688	6,325	
	1500	Industrial, 30 lb. capacity			2	4		2,200	163		2,363	2,650	
	1600	50 lb. capacity		↓	1.70	4.706		2,875	192		3,067	3,450	
	2000	Dry cleaners, electric, 20 lb. capacity		L-1	.20	80		30,200	3,275		33,475	38,100	
	2050	25 lb. capacity			.17	94.118		39,300	3,850		43,150	49,100	
	2100	30 lb. capacity			.15	106		41,400	4,350		45,750	52,000	
	2150	60 lb. capacity		↓	.09	177		65,000	7,250		72,250	82,500	
	3500	Folders, blankets & sheets, minimum		1 Elec	.17	47.059		27,600	1,925		29,525	33,200	
	3700	King size with automatic stacker			.10	80		47,700	3,250		50,950	57,500	
	3800	For conveyor delivery, add			.45	17.778		5,675	725		6,400	7,325	
	4500	Ironers, institutional, 110", single roll		↓	.20	40		26,200	1,625		27,825	31,200	
	4700	Lint collector, ductwork not included, 8,000 to 10,000 C.F.M.		Q-10	.30	80		7,500	3,000		10,500	12,900	
	4800	Pressers, low capacity air operated		L-6	1.75	6.857		8,075	280		8,355	9,300	
	4820	Hand operated			1.75	6.857		6,950	280		7,230	8,075	
	4830	Extractor, low capacity			1.75	6.857		5,925	280		6,205	6,950	
	4840	Ironer 48", 240V			3.50	3.429		92,500	140		92,640	101,500	
	4860	Coin dry cleaner 20 lb.			1.75	6.857		23,100	280		23,380	25,800	
	4900	Spreader feeders, 240V, 2 station			.70	17.143		45,100	700		45,800	50,500	
	4920	4 station		↓	.35	34.286		55,500	1,400		56,900	63,000	
	5000	Washers, residential, 4 cycle, average		1 Plum	3	2.667		645	109		754	875	
	5300	Commercial, coin operated, average		"	3	2.667		1,050	109		1,159	1,325	
	6000	Combination washer/extractor, 20 lb. capacity		L-6	1.50	8		3,725	325		4,050	4,600	
	6100	30 lb. capacity			.80	15		7,200	610		7,810	8,850	
	6200	50 lb. capacity			.68	17.647		8,950	720		9,670	10,900	
	6300	75 lb. capacity			.30	40		18,500	1,625		20,125	22,800	
	6350	125 lb. capacity			.16	75		22,400	3,050		25,450	29,300	
	6380	Washer extractor/dryer, 110 lb., 240V			1	12		7,200	490		7,690	8,625	
	6400	Washer extractor, 135 lb, 240V			1	12		21,500	490		21,990	24,400	
	6450	Pass through			1	12		58,000	490		58,490	64,000	
	6500	200 lb.			1	12		55,500	490		55,990	62,000	
	6550	Pass through			1	12		60,000	490		60,490	67,000	
	6600	Hand operated presser			.70	17.143		5,200	700		5,900	6,775	
	6620	Mushroom press 115V		↓	.70	17.143	↓	6,175	700		6,875	7,825	

EQUIPMENT 11

			DAILY	LABOR-		2005 BARE COSTS				TOTAL	
11136		**Projection Screens**	CREW	OUTPUT	HOURS	UNIT	MAT.	LABOR	EQUIP.	TOTAL	INCL O&P
500	0010	**PROJECTION SCREENS** Wall or ceiling hung, matte white									500
	0100	Manually operated, economy	2 Carp	500	.032	S.F.	4.99	1.10		6.09	7.20
	0300	Intermediate		450	.036		5.85	1.22		7.07	8.30
	0400	Deluxe		400	.040		8.05	1.37		9.42	11.05
	0600	Electric operated, matte white, 25 S.F., economy		5	3.200	Ea.	730	110		840	975
	0700	Deluxe		4	4		1,625	137		1,762	2,025
	0900	50 S.F., economy		3	5.333		400	183		583	725
	1000	Deluxe		2	8		1,700	274		1,974	2,300
	1200	Heavy duty, electric operated, 200 S.F.		1.50	10.667		3,325	365		3,690	4,250
	1300	400 S.F.		1	16		4,100	550		4,650	5,375
	1500	Rigid acrylic in wall, for rear projection, 1/4" thick	2 Glaz	30	.533	S.F.	40.50	17.60		58.10	71
	1600	1/2" thick (maximum size 10' x 20')	"	25	.640	"	71	21		92	111
600	0010	**MOVIE EQUIPMENT**									600
	0020	Changeover, minimum				Ea.	400			400	435
	0100	Maximum					770			770	850
	0400	Film transport, incl. platters and autowind, minimum					4,300			4,300	4,725
	0500	Maximum					12,200			12,200	13,400
	0800	Lamphouses, incl. rectifiers, xenon, 1,000 watt	1 Elec	2	4		5,675	163		5,838	6,500
	0900	1,600 watt		2	4		6,050	163		6,213	6,900
	1000	2,000 watt		1.50	5.333		6,500	217		6,717	7,475
	1100	4,000 watt		1.50	5.333		8,050	217		8,267	9,175
	1400	Lenses, anamorphic, minimum					1,100			1,100	1,200
	1500	Maximum					2,500			2,500	2,725
	1800	Flat 35 mm, minimum					945			945	1,050
	1900	Maximum					1,475			1,475	1,625
	2200	Pedestals, for projectors					1,275			1,275	1,400
	2300	Console type					9,150			9,150	10,100
	2600	Projector mechanisms, incl. soundhead, 35 mm, minimum					9,500			9,500	10,400
	2700	Maximum					13,000			13,000	14,300
	3000	Projection screens, rigid, in wall, acrylic, 1/4" thick	2 Glaz	195	.082	S.F.	35.50	2.71		38.21	43.50
	3100	1/2" thick	"	130	.123	"	41.50	4.06		45.56	51.50
	3300	Electric operated, heavy duty, 400 S.F.	2 Carp	1	16	Ea.	2,525	550		3,075	3,625
	3320	Theatre projection screens, matte white, including frames	"	200	.080	S.F.	5.65	2.74		8.39	10.45
	3400	Also see division 11136-500									
	3700	Sound systems, incl. amplifier, mono, minimum	1 Elec	.90	8.889	Ea.	2,825	360		3,185	3,650
	3800	Dolby/Super Sound, maximum		.40	20		15,400	815		16,215	18,200
	4100	Dual system, 2 channel, front surround, minimum		.70	11.429		3,950	465		4,415	5,050
	4200	Dolby/Super Sound, 4 channel, maximum		.40	20		14,100	815		14,915	16,700
	4500	Sound heads, 35 mm					4,525			4,525	4,975
	4900	Splicer, wet type, minimum					635			635	695
	5000	Tape type, maximum					1,125			1,125	1,250
	5300	Speakers, recessed behind screen, minimum	1 Elec	2	4		905	163		1,068	1,250
	5400	Maximum	"	1	8		2,650	325		2,975	3,375
	5700	Seating, painted steel, upholstered, minimum	2 Carp	35	.457		113	15.65		128.65	149
	5800	Maximum	"	28	.571		360	19.55		379.55	430
	6100	Rewind tables, minimum					2,250			2,250	2,475
	6200	Maximum					4,025			4,025	4,425
	7000	For automation, varying sophistication, minimum	1 Elec	1	8	System	2,025	325		2,350	2,725
	7100	Maximum	2 Elec	.30	53.333	"	4,750	2,175		6,925	8,450

11140 | Vehicle Service Equipment

		11141 Service Station Equipment	CREW	DAILY OUTPUT	LABOR-HOURS	UNIT	2005 BARE COSTS MAT.	LABOR	EQUIP.	TOTAL	TOTAL INCL O&P	
100	0010	**COMPRESSED AIR EQUIPMENT**										**100**
	0030	Compressors, electric, 1-1/2 H.P., standard controls	L-4	1.50	16	Ea.	296	510		806	1,125	
	0550	Dual controls		1.50	16		475	510		985	1,325	
	0600	5 H.P., 115/230 volt, standard controls		1	24		1,700	760		2,460	3,050	
	0650	Dual controls	▼	1	24	▼	1,800	760		2,560	3,175	
200	0010	**FUEL DISPENSING EQUIPMENT**										**200**
	1100	Product dispenser with vapor recovery for 6 nozzles, installed, not										
	1110	including piping to storage tanks				Ea.	18,000			18,000	19,800	
300	0010	**LUBRICATION EQUIPMENT**										**300**
	3000	Lube equipment, 3 reel type, with pumps, not including piping	L-4	.50	48	Set	6,825	1,525		8,350	9,875	
400	0010	**SPRAY PAINTING EQUIPMENT**										**400**
	4000	Spray painting booth, 26′ long, complete	L-4	.40	60	Ea.	12,700	1,900		14,600	16,900	

11150 | Parking Control Equipment

		11156 Parking Equipment	CREW	DAILY OUTPUT	LABOR-HOURS	UNIT	2005 BARE COSTS MAT.	LABOR	EQUIP.	TOTAL	TOTAL INCL O&P	
600	0010	**PARKING EQUIPMENT**										**600**
	5000	Barrier gate with programmable controller	2 Elec	3	5.333	Ea.	3,050	217		3,267	3,675	
	5020	Industrial	"	3	5.333		4,150	217		4,367	4,900	
	5100	Card reader	1 Elec	2	4		1,675	163		1,838	2,075	
	5120	Proximity with customer display	2 Elec	1	16		5,050	650		5,700	6,525	
	5200	Cashier booth, average	B-22	1	30		8,800	940	238	9,978	11,400	
	5300	Collector station, pay on foot	2 Elec	.20	80		100,500	3,250		103,750	115,500	
	5320	Credit card only		.50	32		18,500	1,300		19,800	22,400	
	5500	Exit verifier	▼	1	16		16,000	650		16,650	18,600	
	5600	Fee computer	1 Elec	1.50	5.333		12,200	217		12,417	13,700	
	5700	Full sign, 4″ letters	"	2	4		1,100	163		1,263	1,475	
	5800	Inductive loop	2 Elec	4	4		155	163		318	415	
	5900	Ticket spitter with time/date stamp, standard		2	8		5,800	325		6,125	6,850	
	5920	Mag stripe encoding	▼	2	8		17,000	325		17,325	19,200	
	5950	Vehicle detector, microprocessor based	1 Elec	3	2.667		355	109		464	550	
	6000	Parking control software, minimum		.50	16		20,100	650		20,750	23,100	
	6020	Maximum	▼	.20	40	▼	83,500	1,625		85,125	94,500	

11160 | Loading Dock Equipment

		11161 Loading Dock Equipment	CREW	DAILY OUTPUT	LABOR-HOURS	UNIT	2005 BARE COSTS MAT.	LABOR	EQUIP.	TOTAL	TOTAL INCL O&P	
200	0010	**DOCK BUMPERS** Bolts not included										**200**
	0020	2″ x 6″ to 4″ x 8″, average	1 Carp	.30	26.667	M.B.F.	1,075	915		1,990	2,625	
400	0010	**LOADING DOCK**										**400**
	0020	Bumpers, rubber blocks 4-1/2″ thk, 10″ H, 14″ long	1 Carp	26	.308	Ea.	44.50	10.55		55.05	65.50	

EQUIPMENT 11

355

11161	Loading Dock Equipment	CREW	DAILY OUTPUT	LABOR-HOURS	UNIT	2005 BARE COSTS				TOTAL INCL O&P	
						MAT.	LABOR	EQUIP.	TOTAL		
400	0200	24" long	1 Carp	22	.364	Ea.	78.50	12.45		90.95	106	**400**
	0300	36" long		17	.471		83.50	16.10		99.60	117	
	0500	12" high, 14" long		25	.320		83	10.95		93.95	109	
	0550	24" long		20	.400		92	13.70		105.70	123	
	0600	36" long		15	.533		102	18.25		120.25	141	
	0800	Rubber blocks 6" thick, 10" high, 14" long		22	.364		74	12.45		86.45	101	
	0850	24" long		18	.444		101	15.20		116.20	135	
	0900	36" long		13	.615		120	21		141	164	
	0910	20" high, 11" long		13	.615		121	21		142	167	
	0920	Extruded rubber bumpers, T section, 22" x 22" x 3" thick		41	.195		46.50	6.70		53.20	62	
	0940	Molded rubber bumpers, 24" x 12" x 3" thick	▼	20	.400		42	13.70		55.70	67.50	
	1000	Welded installation of above bumpers	E-14	8	1		3.10	40	10.15	53.25	86.50	
	1100	For drilled anchors, add per anchor	1 Carp	36	.222	▼	5.55	7.60		13.15	17.95	
	1300	Steel bumpers, see Div. 10265-200										
	1350	Wood bumpers, see Div. 11161-200										
	2200	Dock boards, heavy duty, 60" x 60", aluminum, 5,000 lb. cap.				Ea.	1,200			1,200	1,325	
	2700	9,000 lb. capacity					1,125			1,125	1,250	
	3200	15,000 lb. capacity				▼	1,250			1,250	1,375	
	3600	Door seal for door perimeter, 12" x 12", vinyl covered	1 Carp	26	.308	L.F.	22.50	10.55		33.05	41	
	3900	Folding gates, see Div. 10610-100										
	4200	Platform lifter, 6' x 6', portable, 3,000 lb. capacity				Ea.	7,575			7,575	8,325	
	4250	4,000 lb. capacity					9,325			9,325	10,300	
	4400	Fixed, 6' x 8', 5,000 lb. capacity	E-16	.70	22.857		8,175	895	116	9,186	10,700	
	4500	Levelers, hinged for trucks, 10 ton capacity, 6' x 8'		1.08	14.815		4,125	580	75.50	4,780.50	5,650	
	4650	7' x 8'		1.08	14.815		3,975	580	75.50	4,630.50	5,500	
	4670	Air bag power operated, 10 ton cap., 6'x8'		1.08	14.815		4,775	580	75.50	5,430.50	6,400	
	4680	7' x 8'		1.08	14.815		4,800	580	75.50	5,455.50	6,425	
	4700	Hydraulic, 10 ton capacity, 6' x 8' **CN**		1.08	14.815		7,175	580	75.50	7,830.50	9,025	
	4800	7' x 8'	▼	1.08	14.815		7,725	580	75.50	8,380.50	9,625	
	5000	Lights for loading docks, single arm, 24" long	1 Elec	3.80	2.105		122	86		208	262	
	5700	Double arm, 60" long	"	3.80	2.105		145	86		231	287	
	5800	Loading dock safety restraints, manual style	E-16	1.08	14.815		2,625	580	75.50	3,280.50	4,000	
	5900	Automatic style	"	1.08	14.815		4,425	580	75.50	5,080.50	5,975	
	6200	Shelters, fabric, for truck or train, scissor arms, minimum	1 Carp	1	8		1,225	274		1,499	1,775	
	6300	Maximum	"	.50	16	▼	1,800	550		2,350	2,850	

11179	Waste Handling Equipment	CREW	DAILY OUTPUT	LABOR-HOURS	UNIT	2005 BARE COSTS				TOTAL INCL O&P	
						MAT.	LABOR	EQUIP.	TOTAL		
150	0010	**WASTE HANDLING**										**150**
	0020	Compactors, 115 volt, 250#/hr., chute fed	L-4	1	24	Ea.	9,475	760		10,235	11,600	
	0100	Hand fed		2.40	10		6,825	320		7,145	8,000	
	0300	Multi-bag, 230 volt, 600#/hr, chute fed		1	24		8,550	760		9,310	10,600	
	0400	Hand fed		1	24		7,325	760		8,085	9,250	
	0500	Containerized, hand fed, 2 to 6 C.Y. containers, 250#/hr.		1	24		8,975	760		9,735	11,100	
	0550	For chute fed, add per floor		1	24		1,025	760		1,785	2,300	
	1000	Heavy duty industrial compactor, 0.5 C.Y. capacity		1	24		6,075	760		6,835	7,850	
	1050	1.0 C.Y. capacity		1	24		9,350	760		10,110	11,500	
	1100	3 C.Y. capacity	▼	.50	48	▼	12,700	1,525		14,225	16,400	

Important: See the Reference Section for critical supporting data - Reference Nos., Crews, & City Cost Indexes

11170 | Solid Waste Handling Equipment

11179 | Waste Handling Equipment

		CREW	DAILY OUTPUT	LABOR-HOURS	UNIT	2005 BARE COSTS MAT.	LABOR	EQUIP.	TOTAL	TOTAL INCL O&P	
150	**1150**	5.0 C.Y. capacity	L-4	.50	48	Ea.	15,800	1,525		17,325	19,700	**150**
	1200	Combination shredder/compactor (5,000 lbs./hr.)	↓	.50	48		30,900	1,525		32,425	36,400	
	1400	For handling hazardous waste materials, 55 gallon drum packer, std.					14,300			14,300	15,800	
	1410	55 gallon drum packer w/HEPA filter					17,900			17,900	19,700	
	1420	55 gallon drum packer w/charcoal & HEPA filter					23,900			23,900	26,300	
	1430	All of the above made explosion proof, add					10,900			10,900	12,000	
	1500	Crematory, not including building, 1 place	Q-3	.20	160		51,000	6,225		57,225	65,500	
	1750	2 place		.10	320		72,500	12,500		85,000	98,500	
	4400	Incinerator, gas, not incl. chimney, elec. or pipe, 50#/hr., minimum		.80	40		19,900	1,550		21,450	24,200	
	4420	Maximum		.70	45.714		25,900	1,775		27,675	31,100	
	4440	200 lb. per hr., minimum (batch type)		.60	53.333		25,900	2,075		27,975	31,500	
	4460	Maximum (with feeder)		.50	64		50,500	2,500		53,000	59,500	
	4480	400 lb. per hr., minimum (batch type)		.30	106		30,600	4,150		34,750	40,000	
	4500	Maximum (with feeder)		.25	128		58,000	4,975		62,975	71,000	
	4520	800 lb. per hr., with feeder, minimum		.20	160		75,500	6,225		81,725	92,500	
	4540	Maximum		.17	188		103,000	7,325		110,325	124,500	
	4560	1,200 lb. per hr., with feeder, minimum		.15	213		109,000	8,300		117,300	132,500	
	4580	Maximum		.11	290		131,000	11,300		142,300	161,500	
	4600	2,000 lb. per hr., with feeder, minimum		.10	320		190,500	12,500		203,000	228,000	
	4620	Maximum		.05	640		320,500	24,900		345,400	390,000	
	4700	For heat recovery system, add, minimum		.25	128		62,000	4,975		66,975	75,500	
	4710	Add, maximum		.11	290		198,000	11,300		209,300	234,500	
	4720	For automatic ash conveyer, add		.50	64	↓	26,000	2,500		28,500	32,400	
	4750	Large municipal incinerators, incl. stack, minimum		.25	128	Ton/day	15,800	4,975		20,775	24,900	
	4850	Maximum	↓	.10	320	"	42,000	12,500		54,500	65,000	
	5500	Shredder, municipal use, 35 ton per hour				Ea.	235,500			235,500	259,000	
	5600	60 ton per hour					501,500			501,500	551,500	
	5750	Shredder & baler, 50 ton per day					470,000			470,000	517,000	
	5800	Shredder, industrial, minimum					18,500			18,500	20,400	
	5850	Maximum					99,500			99,500	109,000	
	5900	Baler, industrial, minimum					7,425			7,425	8,150	
	5950	Maximum				↓	433,500			433,500	476,500	
	6000	Transfer station compactor, with power unit										
	6050	and pedestal, not including pit, 50 ton per hour				Ea.	148,500			148,500	163,000	

11190 | Detention Equipment

11191 | Detention Equipment

		CREW	DAILY OUTPUT	LABOR-HOURS	UNIT	2005 BARE COSTS MAT.	LABOR	EQUIP.	TOTAL	TOTAL INCL O&P	
150	**0010**	**DETENTION EQUIPMENT**										**150**
	0500	Bar front, rolling, 7/8" bars, 4" O.C., 7' high, 5' wide, with hardware	E-4	2	16	Ea.	4,600	620	40.50	5,260.50	6,225	
	1000	Doors & frames, 3' x 7', complete, with hardware, single plate	↓	4	8	↓	3,450	310	20.50	3,780.50	4,375	
	1650	Double plate		4	8		4,275	310	20.50	4,605.50	5,275	
	2000	Cells, prefab., 5' to 6' wide, 7' to 8' high, 7' to 8' deep,										
	2010	bar front, cot, not incl. plumbing	E-4	1.50	21.333	Ea.	7,600	825	54	8,479	9,900	
	2500	Cot, bolted, single, painted steel	↓	20	1.600	↓	287	62	4.06	353.06	430	
	2700	Stainless steel		20	1.600		795	62	4.06	861.06	985	
	3000	Toilet apparatus including wash basin, average	L-8	1.50	13.333		2,850	475		3,325	3,850	
	4000	Visitor cubicle, vision panel, no intercom	E-4	2	16	↓	2,500	620	40.50	3,160.50	3,900	

EQUIPMENT 11

11300 | Fluid Waste Treatment & Disposal Equipment

11310 | Sewage & Sludge Pumps

			CREW	DAILY OUTPUT	LABOR-HOURS	UNIT	2005 BARE COSTS MAT.	LABOR	EQUIP.	TOTAL	TOTAL INCL O&P	
700	0010	**SEWAGE PUMPING STATIONS** Prefabricated steel, concrete										700
	0020	or fiberglass, 200 GPM	C-17D	.17	494	Total	34,100	17,400	2,925	54,425	68,000	
	0200	1,000 GPM		.07	1,200		48,100	42,300	7,100	97,500	126,500	
	0500	Add for generator unit, 200 GPM, steel		.34	247		24,600	8,725	1,475	34,800	42,200	
	0600	Concrete		.51	164		15,600	5,800	975	22,375	27,300	
	1000	Add for generator unit, 1,000 GPM, steel		.30	280		26,200	9,875	1,650	37,725	45,900	
	1200	Concrete	↓	.38	221		22,100	7,800	1,300	31,200	37,900	
	1500	For drilled water well, if required, add	B-23	.50	80	↓	6,375	2,175	6,700	15,250	17,800	

11390 | Pkg Sewage Treat Plants

			CREW	DAILY OUTPUT	LABOR-HOURS	UNIT	MAT.	LABOR	EQUIP.	TOTAL	TOTAL INCL O&P	
200	0010	**SEWAGE TREATMENT PLANTS,** not incl fencing or external piping										200
	0020	Steel packaged, blown air aeration plants										
	0100	1,000 GPD				Gal.				16.85	19.40	
	0200	5,000 GPD								11.25	12.90	
	0300	15,000 GPD								6.20	7.10	
	0400	30,000 GPD								5.85	6.70	
	0500	50,000 GPD								4.50	5.05	
	0600	100,000 GPD								3.95	4.50	
	0700	200,000 GPD								2.81	3.23	
	0800	500,000 GPD				↓				2.75	3.15	
	1000	Concrete, extended aeration, primary and secondary treatment										
	1010	10,000 GPD				Gal.				12.35	14.65	
	1100	30,000 GPD								6.20	7.15	
	1200	50,000 GPD								5.05	5.85	
	1400	100,000 GPD								3.94	4.55	
	1500	500,000 GPD				↓				2,081	3.26	
	1700	Municipal wastewater treatment facility										
	1720	1.0 MGD				Gal.				4.83	5.55	
	1740	1.5 MGD								4.78	5.50	
	1760	2.0 MGD								4.10	4.73	
	1780	3.0 MGD								3.20	3.71	
	1800	5.0 MGD				↓				2.92	3.37	
	2000	Holding tank system, not incl. excavation or backfill										
	2010	Recirculating chemical water closet	2 Plum	4	4	Ea.	750	163		913	1,075	
	2100	For voltage converter, add	"	16	1		199	41		240	281	
	2200	For high level alarm, add	1 Plum	7.80	1.026	↓	114	42		156	188	
900	0010	**WASTEWATER TREATMENT SYSTEM** Fiberglass, 1,000 gallon	B-21	1.29	21.705	Ea.	2,925	670	123	3,718	4,400	900
	0100	1,500 gallon	"	1.03	27.184	"	6,400	840	154	7,394	8,525	

11400 | Food Service Equipment

11405 | Food Storage Equipment

			CREW	DAILY OUTPUT	LABOR-HOURS	UNIT	2005 BARE COSTS MAT.	LABOR	EQUIP.	TOTAL	TOTAL INCL O&P	
110	0010	**FOOD STORAGE EQUIPMENT**										110
	2350	Cooler, reach-in, beverage, 6' long	Q-1	6	2.667	Ea.	3,550	98		3,648	4,075	
	4300	Freezers, reach-in, 44 C.F.		4	4		8,125	147		8,272	9,175	
	4500	68 C.F.	↓	3	5.333		9,150	196		9,346	10,400	
	4600	Freezer, pre-fab, 8' x 8' w/refrigeration	2 Carp	.45	35.556		7,575	1,225		8,800	10,200	
	4620	8' x 12'		.35	45.714		9,250	1,575		10,825	12,700	
	4640	8' x 16'		.25	64		11,600	2,200		13,800	16,200	
	4660	8' x 20'	↓	.17	94.118	↓	14,200	3,225		17,425	20,600	

Important: See the Reference Section for critical supporting data - Reference Nos., Crews, & City Cost Indexes

11405	Food Storage Equipment	CREW	DAILY OUTPUT	LABOR-HOURS	UNIT	2005 BARE COSTS				TOTAL INCL O&P	
						MAT.	LABOR	EQUIP.	TOTAL		
110 4680	Reach-in, 1 compartment	Q-1	4	4	Ea.	1,975	147		2,122	2,400	**110**
4700	2 compartment		3	5.333		4,075	196		4,271	4,775	
8300	Refrigerators, reach-in type, 44 C.F.		5	3.200		4,950	118		5,068	5,625	
8310	With glass doors, 68 C.F.	▼	4	4		7,150	147		7,297	8,075	
8320	Refrigerator, reach-in, 1 compartment	R-18	7.80	3.333		1,600	105		1,705	1,900	
8330	2 compartment		6.20	4.194		1,900	132		2,032	2,300	
8340	3 compartment	▼	5.60	4.643		2,800	146		2,946	3,325	
8350	Pre-fab, with refrigeration, 8' x 8'	2 Carp	.45	35.556		4,975	1,225		6,200	7,375	
8360	8' x 12'		.35	45.714		6,400	1,575		7,975	9,500	
8370	8' x 16'		.25	64		8,150	2,200		10,350	12,400	
8380	8' x 20'	▼	.17	94.118		10,000	3,225		13,225	16,000	
8390	Pass-thru/roll-in, 1 compartment	R-18	7.80	3.333		3,225	105		3,330	3,700	
8400	2 compartment		6.24	4.167		4,600	131		4,731	5,250	
8410	3 compartment		5.60	4.643		5,825	146		5,971	6,625	
8420	Walk-in, alum, door & floor only, no refrig, 6' x 6' x 7'-6"	2 Carp	1.40	11.429		6,325	390		6,715	7,550	
8430	10' x 6' x 7'-6"		.55	29.091		9,050	995		10,045	11,500	
8440	12' x 14' x 7'-6"		.25	64		12,600	2,200		14,800	17,200	
8450	12' x 20' x 7'-6"	▼	.17	94.118		15,400	3,225		18,625	22,000	
8460	Refrigerated cabinets, mobile					2,675			2,675	2,925	
8470	Refrigerator/freezer, reach-in, 1 compartment	R-18	5.60	4.643		4,100	146		4,246	4,725	
8480	2 compartment	"	4.80	5.417		5,925	170		6,095	6,750	
8600	Stainless steel shelving, louvered 4-tier, 20" x 3'	1 Clab	6	1.333		915	35.50		950.50	1,050	
8605	20" x 4'		6	1.333		1,300	35.50		1,335.50	1,475	
8610	20" x 6'		6	1.333		1,900	35.50		1,935.50	2,125	
8615	24" x 3'		6	1.333		1,200	35.50		1,235.50	1,350	
8620	24" x 4'		6	1.333		1,400	35.50		1,435.50	1,575	
8625	24" x 6'		6	1.333		2,025	35.50		2,060.50	2,275	
8630	Flat 4-tier, 20" x 3'		6	1.333		1,050	35.50		1,085.50	1,225	
8635	20" x 4'		6	1.333		1,275	35.50		1,310.50	1,450	
8640	20" x 5'		6	1.333		1,450	35.50		1,485.50	1,650	
8645	24" x 3'		6	1.333		1,125	35.50		1,160.50	1,275	
8650	24" x 4'		6	1.333		1,300	35.50		1,335.50	1,475	
8655	24" x 6'		6	1.333		2,750	35.50		2,785.50	3,075	
8700	Galvanized shelving, louvered 4-tier, 20" x 3'		6	1.333		465	35.50		500.50	565	
8705	20" x 4'		6	1.333		520	35.50		555.50	625	
8710	20" x 6'		6	1.333		540	35.50		575.50	650	
8715	24" x 3'		6	1.333		425	35.50		460.50	525	
8720	24" x 4'		6	1.333		555	35.50		590.50	665	
8725	24" x 6'		6	1.333		795	35.50		830.50	930	
8730	Flat 4-tier, 20" x 3'		6	1.333		330	35.50		365.50	420	
8735	20" x 4'		6	1.333		370	35.50		405.50	465	
8740	20" x 6'		6	1.333		670	35.50		705.50	790	
8745	24" x 3'		6	1.333		325	35.50		360.50	410	
8750	24" x 4'		6	1.333		495	35.50		530.50	600	
8755	24" x 6'		6	1.333		710	35.50		745.50	835	
8760	Stainless steel dunnage rack, 24" x 3'		8	1		640	26.50		666.50	745	
8765	24" x 4'		8	1		670	26.50		696.50	780	
8770	Galvanized dunnage rack, 24" x 3'		8	1		106	26.50		132.50	158	
8775	24" x 4'	▼	8	1	▼	136	26.50		162.50	192	
800 0010	**WINE CELLAR**, refrigerated, Redwood interior, carpeted, walk-in type										**800**
0020	6'-8" high, including racks										
0200	80"W x 48"D for 900 bottles	2 Carp	1.50	10.667	Ea.	2,925	365		3,290	3,775	
0250	80" W x 72" D for 1300 bottles		1.33	12.030		3,850	410		4,260	4,900	
0300	80" W x 94" D for 1900 bottles		1.17	13.675		5,000	470		5,470	6,225	
0400	80" W x 124" D for 2500 bottles	▼	1	16	▼	6,050	550		6,600	7,500	

EQUIPMENT 11

11405 | Food Storage Equipment

			CREW	DAILY OUTPUT	LABOR-HOURS	UNIT	2005 BARE COSTS				TOTAL INCL O&P	
							MAT.	LABOR	EQUIP.	TOTAL		
800	0600	Portable cabinets, red oak, reach-in temp.& humidity controlled										800
	0650	26-5/8"W x 26-1/2"D x 68"H for 235 bottles				Ea.	2,625			2,625	2,900	
	0660	32"W x 21-1/2"D x 73-1/2"H for 144 bottles					3,075			3,075	3,375	
	0670	32"W x 29-1/2"D x 73-1/2"H for 288 bottles					3,075			3,075	3,375	
	0680	39-1/2"W x 29-1/2"D x 86-1/2"H for 440 bottles					3,500			3,500	3,850	
	0690	52-1/2"W x 29-1/2"D x 73-1/2"H for 468 bottles					4,025			4,025	4,425	
	0700	52-1/2"W x 29-1/2"D x 86-1/2"H for 572 bottles					4,225			4,225	4,650	
	0730	Portable, red oak, can be built-in with glass door										
	0750	23-7/8"W x 24"D x 34-1/2"H for 50 bottles				Ea.	960			960	1,050	

11410 | Food Preparation Equipment

			CREW	DAILY OUTPUT	LABOR-HOURS	UNIT	MAT.	LABOR	EQUIP.	TOTAL	TOTAL INCL O&P	
110	0010	**FOOD PREPARATION EQUIPMENT**										110
	1700	Choppers, 5 pounds	R-18	7	3.714	Ea.	2,275	117		2,392	2,700	
	1720	16 pounds		5	5.200		2,575	163		2,738	3,075	
	1740	35 to 40 pounds		4	6.500		3,950	204		4,154	4,650	
	1840	Coffee brewer, 5 burners	1 Plum	3	2.667		1,075	109		1,184	1,350	
	1850	Coffee urn, twin 6 gallon urns		2	4		2,450	163		2,613	2,925	
	1860	Single, 3 gallon		3	2.667		2,700	109		2,809	3,150	
	3000	Fast food equipment, total package, minimum	6 Skwk	.08	600		133,000	20,900		153,900	178,500	
	3100	Maximum	"	.07	685		181,000	23,900		204,900	236,000	
	3800	Food mixers, 20 quarts	L-7	7	4		2,750	132		2,882	3,200	
	3850	40 quarts		5.40	5.185		6,275	171		6,446	7,175	
	3900	60 quarts		5	5.600		10,100	185		10,285	11,400	
	4040	80 quarts		3.90	7.179		10,600	237		10,837	12,100	
	4080	130 quarts		2.20	12.727		16,200	420		16,620	18,600	
	4100	Floor type, 20 quarts		15	1.867		2,800	61.50		2,861.50	3,175	
	4120	60 quarts		14	2		7,050	66		7,116	7,850	
	4140	80 quarts		12	2.333		9,300	77		9,377	10,300	
	4160	140 quarts		8.60	3.256		21,500	108		21,608	23,800	
	6700	Peelers, small	R-18	8	3.250		1,075	102		1,177	1,350	
	6720	Large	"	6	4.333		3,450	136		3,586	4,000	
	6800	Pulper/extractor, close coupled, 5 HP	1 Plum	1.90	4.211		3,050	172		3,222	3,625	
	8580	Slicer with table	R-18	9	2.889		3,975	91		4,066	4,525	

11415 | Food Delivery Carts and Conveyors

			CREW	DAILY OUTPUT	LABOR-HOURS	UNIT	MAT.	LABOR	EQUIP.	TOTAL	TOTAL INCL O&P	
110	0010	**FOOD DELIVERY CARTS**										110
	1650	Cabinet, heated, 1 compartment, reach-in	R-18	5.60	4.643	Ea.	2,200	146		2,346	2,650	
	1655	Pass-thru roll-in		5.60	4.643		3,775	146		3,921	4,375	
	1660	2 compartment, reach-in		4.80	5.417		5,575	170		5,745	6,375	
	1670	Mobile					2,625			2,625	2,900	
	6850	Mobile rack w/pan slide					930			930	1,025	
	9180	Tray and silver dispenser, mobile	1 Clab	16	.500		725	13.35		738.35	815	

11420 | Food Cooking Equipment

			CREW	DAILY OUTPUT	LABOR-HOURS	UNIT	MAT.	LABOR	EQUIP.	TOTAL	TOTAL INCL O&P	
110	0010	**COOKING EQUIPMENT**										110
	0020	Bake oven, gas, one section	Q-1	8	2	Ea.	4,225	73.50		4,298.50	4,750	
	0300	Two sections		7	2.286		8,625	84		8,709	9,625	
	0600	Three sections		6	2.667		12,200	98		12,298	13,600	
	0900	Electric convection, single deck	L-7	4	7		4,525	231		4,756	5,325	
	1300	Broiler, without oven, standard	Q-1	8	2		2,700	73.50		2,773.50	3,075	
	1550	Infra-red	L-7	4	7		6,175	231		6,406	7,150	
	4750	Fryer, with twin baskets, modular model	Q-1	7	2.286		1,475	84		1,559	1,725	
	5000	Floor model, on 6" legs	"	5	3.200		1,475	118		1,593	1,800	
	5100	Extra single basket, large					88.50			88.50	97.50	

Important: See the Reference Section for critical supporting data - Reference Nos., Crews, & City Cost Indexes

11420 | Food Cooking Equipment

		CREW	DAILY OUTPUT	LABOR-HOURS	UNIT	2005 BARE COSTS				TOTAL INCL O&P
						MAT.	LABOR	EQUIP.	TOTAL	
110	5300	Griddle, SS, 24" plate, w/4" legs, elec, 208V, 3 phase, 3' long	Q-1	7	2.286	Ea.	875	84		959	1,075	110
	5550	4' long	"	6	2.667		1,150	98		1,248	1,400
	6200	Iced tea brewer	1 Plum	3.44	2.326		595	95		690	795
	6350	Kettle, w/steam jacket, tilting, w/positive lock, SS, 20 gallons	L-7	7	4		5,350	132		5,482	6,100
	6600	60 gallons	"	6	4.667		6,825	154		6,979	7,775
	6900	Range, restaurant type, 6 burners and 1 standard oven, 36" wide	Q-1	7	2.286		1,775	84		1,859	2,100
	6950	Convection		7	2.286		2,450	84		2,534	2,800
	7150	2 standard ovens, 24" griddle, 60" wide		6	2.667		4,375	98		4,473	4,950
	7200	1 standard, 1 convection oven		6	2.667		5,300	98		5,398	5,975
	7450	Heavy duty, single 34" standard oven, open top		5	3.200		3,875	118		3,993	4,450
	7500	Convection oven		5	3.200		5,275	118		5,393	5,975
	7700	Griddle top		6	2.667		4,575	98		4,673	5,200
	7750	Convection oven	▼	6	2.667		6,125	98		6,223	6,875
	8850	Steamer, electric 27 KW	L-7	7	4		8,550	132		8,682	9,600
	9100	Electric, 10 KW or gas 100,000 BTU	"	5	5.600		4,125	185		4,310	4,800
	9150	Toaster, conveyor type, 16-22 slices per minute					995			995	1,100
	9160	Pop-up, 2 slot				▼	545			545	600
	9200	For deluxe models of above equipment, add					75%				
	9400	Rule of thumb: Equipment cost based									
	9410	on kitchen work area									
	9420	Office buildings, minimum	L-7	77	.364	S.F.	58	12		70	82.50
	9450	Maximum		58	.483		98	15.95		113.95	133
	9550	Public eating facilities, minimum		77	.364		76	12		88	102
	9600	Maximum		46	.609		124	20		144	167
	9750	Hospitals, minimum		58	.483		78	15.95		93.95	111
	9800	Maximum	▼	39	.718	▼	131	23.50		154.50	181

11425 | Hood and Ventilation Equipment

		CREW	DAILY OUTPUT	LABOR-HOURS	UNIT	MAT.	LABOR	EQUIP.	TOTAL	TOTAL INCL O&P
110	0010	**HOOD & VENTILATION EQUIPMENT**									110
	2970	Exhaust hood, sst, gutter on all sides, 4' x 4' x 2'	1 Carp	1.80	4.444	Ea.	3,050	152		3,202	3,600
	2980	4' x 4' x 7'	"	1.60	5		4,750	171		4,921	5,500
	7950	Hood fire protection system, minimum	Q-1	3	5.333		3,225	196		3,421	3,850
	8050	Maximum	"	1	16	▼	24,100	590		24,690	27,500

11430 | Food Dispensing Equipment

		CREW	DAILY OUTPUT	LABOR-HOURS	UNIT	MAT.	LABOR	EQUIP.	TOTAL	TOTAL INCL O&P
110	0010	**FOOD DISPENSING EQUIPMENT**									110
	1050	Butter pat dispenser	1 Clab	13	.615	Ea.	730	16.45		746.45	830
	1100	Bread dispenser, counter top		13	.615		695	16.45		711.45	790
	1900	Cup and glass dispenser, drop in		4	2		995	53.50		1,048.50	1,175
	1920	Disposable cup, drop in		16	.500		300	13.35		313.35	350
	2650	Dish dispenser, drop in, 12"		11	.727		1,125	19.40		1,144.40	1,275
	2660	Mobile	▼	10	.800		1,900	21.50		1,921.50	2,125
	3300	Food warmer, counter, 1.2 KW					605			605	665
	3550	1.6 KW					1,450			1,450	1,600
	3600	Well, hot food, built-in, rectangular, 12" x 20"	R-30	10	2.600		340	84		424	500
	3610	Circular, 7 qt		10	2.600		310	84		394	470
	3620	Refrigerated, 2 compartments		10	2.600		1,775	84		1,859	2,075
	3630	3 compartments		9	2.889		1,750	93		1,843	2,075
	3640	4 compartments		8	3.250		2,425	105		2,530	2,825
	4720	Frost cold plate	▼	9	2.889		12,000	93		12,093	13,200
	5700	Hot chocolate dispenser	1 Plum	4	2		815	81.50		896.50	1,025
	6250	Jet spray dispenser	R-18	4.50	5.778		2,475	182		2,657	3,000
	6300	Juice dispenser, concentrate	"	4.50	5.778		920	182		1,102	1,275
	6690	Milk dispenser, bulk, 2 flavor	R-30	8	3.250		1,125	105		1,230	1,375
	6695	3 flavor	"	8	3.250	▼	1,525	105		1,630	1,825

EQUIPMENT 11

11400 | Food Service Equipment

11430 | Food Dispensing Equipment

		CREW	DAILY OUTPUT	LABOR-HOURS	UNIT	2005 BARE COSTS				TOTAL INCL O&P	
						MAT.	LABOR	EQUIP.	TOTAL		
110	8800	Serving counter, straight	1 Carp	40	.200	L.F.	585	6.85		591.85	655	110
	8820	Curved section	"	30	.267	"	730	9.15		739.15	820	
	8825	Solid surface, see section 06620-810										
	8830	Soft serve ice cream machine, medium	R-18	11	2.364	Ea.	7,900	74.50		7,974.50	8,825	
	8840	Large	"	9	2.889	"	11,700	91		11,791	13,000	

11435 | Ice Machines

		CREW	DAILY OUTPUT	LABOR-HOURS	UNIT	2005 BARE COSTS				TOTAL INCL O&P	
110	0010	**ICE MACHINES**										110
	5800	Ice cube maker, 50 pounds per day	Q-1	6	2.667	Ea.	1,325	98		1,423	1,600	
	5900	250 pounds per day		1.20	13.333		1,750	490		2,240	2,650	
	6050	500 pounds per day		4	4		2,875	147		3,022	3,400	
	6060	With bin		1.20	13.333		2,050	490		2,540	2,975	
	6090	1000 pounds per day, with bin		1	16		6,225	590		6,815	7,725	
	6100	Ice flakers, 300 pounds per day		1.60	10		2,875	370		3,245	3,700	
	6120	600 pounds per day		.95	16.842		3,400	620		4,020	4,650	
	6130	1000 pounds per day		.75	21.333		4,700	785		5,485	6,325	
	6140	2000 pounds per day		.65	24.615		14,100	905		15,005	16,900	
	6160	Ice storage bin, 500 pound capacity	Q-5	1	16		1,675	590		2,265	2,725	
	6180	1000 pound	"	.56	28.571		2,950	1,050		4,000	4,800	

11440 | Cleaning and Disposal Equipment

		CREW	DAILY OUTPUT	LABOR-HOURS	UNIT	2005 BARE COSTS				TOTAL INCL O&P	
110	0010	**CLEANING & DISPOSAL EQUIPMENT**										110
	2700	Dishwasher, commercial, rack type										
	2720	10 to 12 racks per hour	Q-1	3.20	5	Ea.	3,650	184		3,834	4,300	
	2750	Semi-automatic 38 to 50 racks per hour	"	1.30	12.308		6,550	455		7,005	7,875	
	2800	Automatic, 190 to 230 racks per hour	L-6	.35	34.286		8,200	1,400		9,600	11,100	
	2820	235 to 275 racks per hour		.25	48		19,000	1,950		20,950	23,800	
	2840	8,750 to 12,500 dishes per hour		.10	120		41,200	4,900		46,100	52,500	
	2950	Dishwasher hood, canopy type	L-3A	10	1.200	L.F.	430	45		475	545	
	2960	Pant leg type	"	2.50	4.800	Ea.	6,125	180		6,305	7,025	
	5200	Garbage disposal 1.5 HP, 100 GPH	L-1	4.80	3.333		1,325	136		1,461	1,675	
	5210	3 HP, 120 GPH		4.60	3.478		2,150	142		2,292	2,550	
	5220	5 HP, 250 GPH		4.50	3.556		3,075	145		3,220	3,600	
	6750	Pot sink, 3 compartment	1 Plum	7.25	1.103	L.F.	580	45		625	705	
	6760	Pot washer, small		1.60	5	Ea.	16,000	204		16,204	17,900	
	6770	Large		1.20	6.667		35,700	272		35,972	39,700	
	9170	Trash compactor, small, up to 125 lb. compacted weight	L-4	4	6		16,400	191		16,591	18,400	
	9175	Large, up to 175 lb. compacted weight	"	3	8		20,100	254		20,354	22,500	

11450 | Residential Equipment

11454 | Residential Appliances

		CREW	DAILY OUTPUT	LABOR-HOURS	UNIT	2005 BARE COSTS				TOTAL INCL O&P	
						MAT.	LABOR	EQUIP.	TOTAL		
500	0010	**RESIDENTIAL APPLIANCES**										500
	0020	Cooking range, 30" free standing, 1 oven, minimum	2 Clab	10	1.600	Ea.	241	42.50		283.50	330	
	0050	Maximum		4	4		1,475	107		1,582	1,800	
	0150	2 oven, minimum		10	1.600		1,475	42.50		1,517.50	1,700	
	0200	Maximum		10	1.600		1,550	42.50		1,592.50	1,775	
	0350	Built-in, 30" wide, 1 oven, minimum	1 Elec	6	1.333		445	54.50		499.50	570	

Important: See the Reference Section for critical supporting data - Reference Nos., Crews, & City Cost Indexes

11454	Residential Appliances	CREW	DAILY OUTPUT	LABOR-HOURS	UNIT	2005 BARE COSTS				TOTAL INCL O&P
						MAT.	LABOR	EQUIP.	TOTAL	
500 0400	Maximum	2 Carp	2	8	Ea.	1,325	274		1,599	1,875
0500	2 oven, conventional, minimum		4	4		930	137		1,067	1,250
0550	1 conventional, 1 microwave, maximum	▼	2	8		1,475	274		1,749	2,050
0700	Free-standing, 1 oven, 21" wide range, minimum	2 Clab	10	1.600		247	42.50		289.50	340
0750	21" wide, maximum	"	4	4		273	107		380	465
0900	Counter top cook tops, 4 burner, standard, minimum	1 Elec	6	1.333		186	54.50		240.50	285
0950	Maximum		3	2.667		520	109		629	730
1050	As above, but with grille and griddle attachment, minimum		6	1.333		465	54.50		519.50	590
1100	Maximum		3	2.667		640	109		749	865
1200	Induction cooktop, 30" wide		3	2.667		520	109		629	730
1250	Microwave oven, minimum		4	2		82.50	81.50		164	212
1300	Maximum	▼	2	4		390	163		553	675
1750	Compactor, residential size, 4 to 1 compaction, minimum	1 Carp	5	1.600		415	55		470	540
1800	Maximum	"	3	2.667		470	91.50		561.50	655
2000	Deep freeze, 15 to 23 C.F., minimum	2 Clab	10	1.600		390	42.50		432.50	495
2050	Maximum		5	3.200		515	85.50		600.50	700
2200	30 C.F., minimum		8	2		750	53.50		803.50	910
2250	Maximum	▼	3	5.333		850	142		992	1,150
2450	Dehumidifier, portable, automatic, 15 pint					149			149	164
2550	40 pint					166			166	183
2750	Dishwasher, built-in, 2 cycles, minimum	L-1	4	4		248	163		411	515
2800	Maximum		2	8		289	325		614	810
2950	4 or more cycles, minimum		4	4		262	163		425	530
2960	Average		4	4		350	163		513	630
3000	Maximum	▼	2	8		540	325		865	1,075
3200	Dryer, automatic, minimum	L-2	3	5.333		282	159		441	560
3250	Maximum	"	2	8		790	239		1,029	1,250
3300	Garbage disposal, sink type, minimum	L-1	10	1.600		42.50	65.50		108	144
3350	Maximum	"	10	1.600		145	65.50		210.50	258
3550	Heater, electric, built-in, 1250 watt, ceiling type, minimum	1 Elec	4	2		72	81.50		153.50	201
3600	Maximum		3	2.667		118	109		227	292
3700	Wall type, minimum		4	2		102	81.50		183.50	234
3750	Maximum		3	2.667		136	109		245	310
3900	1500 watt wall type, with blower		4	2		127	81.50		208.50	260
3950	3000 watt	▼	3	2.667		258	109		367	445
4150	Hood for range, 2 speed, vented, 30" wide, minimum	L-3	5	3.200		37.50	120		157.50	224
4200	Maximum		3	5.333		595	199		794	960
4300	42" wide, minimum		5	3.200		225	120		345	430
4330	Custom		5	3.200		625	120		745	870
4350	Maximum	▼	3	5.333		760	199		959	1,150
4500	For ventless hood, 2 speed, add					15			15	16.50
4650	For vented 1 speed, deduct from maximum					39			39	43
4850	Humidifier, portable, 8 gallons per day					149			149	164
5000	15 gallons per day					179			179	197
5200	Icemaker, automatic, 20 lb. per day	1 Plum	7	1.143		360	46.50		406.50	465
5350	51 lb. per day	"	2	4		1,000	163		1,163	1,350
5500	Refrigerator, no frost, 10 C.F. to 12 C.F. minimum	2 Clab	10	1.600		440	42.50		482.50	550
5600	Maximum		6	2.667		680	71		751	860
5750	14 C.F. to 16 C.F., minimum		9	1.778		445	47.50		492.50	560
5800	Maximum		5	3.200		480	85.50		565.50	665
5950	18 C.F. to 20 C.F., minimum		8	2		510	53.50		563.50	645
6000	Maximum		4	4		840	107		947	1,100
6150	21 C.F. to 29 C.F., minimum		7	2.286		645	61		706	805
6200	Maximum	▼	3	5.333		2,075	142		2,217	2,500
6400	Sump pump cellar drainer, pedestal, 1/3 H.P., molded PVC base	1 Plum	3	2.667		87	109		196	260
6450	Solid brass	"	2	4	▼	179	163		342	445

EQUIPMENT 11

11454	Residential Appliances	CREW	DAILY OUTPUT	LABOR-HOURS	UNIT	2005 BARE COSTS				TOTAL INCL O&P	
						MAT.	LABOR	EQUIP.	TOTAL		
500	6460	Sump pump, see also division 15440-940									**500**
	6650	Washing machine, automatic, minimum	1 Plum	3	2.667	Ea.	279	109		388	470
	6700	Maximum	"	1	8		1,025	325		1,350	1,625
	6900	Water heater, electric, glass lined, 30 gallon, minimum	L-1	5	3.200		252	131		383	470
	6950	Maximum		3	5.333		350	218		568	710
	7100	80 gallon, minimum		2	8		485	325		.810	1,025
	7150	Maximum		1	16		670	655		1,325	1,725
	7180	Water heater, gas, glass lined, 30 gallon, minimum	2 Plum	5	3.200		345	131		476	575
	7220	Maximum		3	5.333		480	218		698	855
	7260	50 gallon, minimum		2.50	6.400		480	261		741	925
	7300	Maximum		1.50	10.667		665	435		1,100	1,400
	7310	Water heater, see also division 15480-200									
	7350	Water softener, automatic, to 30 grains per gallon	2 Plum	5	3.200	Ea.	430	131		561	670
	7400	To 100 grains per gallon	"	4	4		600	163		763	905
	7450	Vent kits for dryers	1 Carp	10	.800		12.65	27.50		40.15	56.50
550	0010	**DISAPPEARING STAIRWAY** No trim included									**550**
	0100	Custom grade, pine, 8'-6" ceiling, minimum	1 Carp	4	2	Ea.	104	68.50		172.50	222
	0150	Average		3.50	2.286		105	78.50		183.50	238
	0200	Maximum		3	2.667		174	91.50		265.50	335
	0500	Heavy duty, pivoted, from 7'-7" to 12'-10" floor to floor		3	2.667		340	91.50		431.50	515
	0600	16'-0" ceiling		2	4		1,150	137		1,287	1,475
	0800	Economy folding, pine, 8'-6" ceiling		4	2		95.50	68.50		164	212
	0900	9'-6" ceiling		4	2		103	68.50		171.50	221
	1000	Fire escape, galvanized steel, 8'-0" to 10'-4" ceiling	2 Carp	1	16		1,200	550		1,750	2,175
	1010	10'-6" to 13'-6" ceiling		1	16		1,525	550		2,075	2,525
	1100	Automatic electric, aluminum, floor to floor height, 8' to 9'		1	16		6,250	550		6,800	7,725
	1400	11' to 12'		.90	17.778		6,725	610		7,335	8,350
	1700	14' to 15'		.70	22.857		7,150	785		7,935	9,100

11460 | Unit Kitchens

			CREW	DAILY OUTPUT	LABOR-HOURS	UNIT	MAT.	LABOR	EQUIP.	TOTAL	INCL O&P
100	0010	**UNIT KITCHENS**									**100**
	1500	Combination range, refrigerator and sink, 30" wide, minimum	L-1	2	8	Ea.	730	325		1,055	1,300
	1550	Maximum		1	16		1,450	655		2,105	2,575
	1570	60" wide, average		1.40	11.429		2,400	465		2,865	3,350
	1590	72" wide, average		1.20	13.333		2,725	545		3,270	3,825
	1600	Office model, 48" wide **CN**		2	8		2,050	325		2,375	2,775
	1620	Refrigerator and sink only		2.40	6.667		2,050	272		2,322	2,675
	1640	Combination range, refrigerator, sink, microwave									
	1660	oven and ice maker	L-1	.80	20	Ea.	4,025	815		4,840	5,650

11471	Darkroom Processing	CREW	DAILY OUTPUT	LABOR-HOURS	UNIT	2005 BARE COSTS				TOTAL INCL O&P	
						MAT.	LABOR	EQUIP.	TOTAL		
700	0010	**DARKROOM EQUIPMENT**									**700**
	0020	Developing sink, 5" deep, 24" x 48"	Q-1	2	8	Ea.	4,225	294		4,519	5,100
	0050	48" x 52"		1.70	9.412		4,300	345		4,645	5,250
	0200	10" deep, 24" x 48"		1.70	9.412		5,275	345		5,620	6,325

11471 | Darkroom Processing

		CREW	DAILY OUTPUT	LABOR-HOURS	UNIT	2005 BARE COSTS				TOTAL INCL O&P	
						MAT.	LABOR	EQUIP.	TOTAL		
700	0250	24" x 108"	Q-1	1.50	10.667	Ea.	1,900	390		2,290	2,700	700
	0500	Dryers, dehumidified filtered air, 36" x 25" x 68" high	L-7	6	4.667		4,125	154		4,279	4,800	
	0550	48" x 25" x 68" high		5	5.600		7,600	185		7,785	8,650	
	2000	Processors, automatic, color print, minimum		4	7		10,700	231		10,931	12,200	
	2050	Maximum		.60	46.667		11,300	1,550		12,850	14,800	
	2300	Black and white print, minimum		2	14		8,100	460		8,560	9,650	
	2350	Maximum		.80	35		53,000	1,150		54,150	60,000	
	2600	Manual processor, 16" x 20" maximum print size		2	14		8,150	460		8,610	9,700	
	2650	20" x 24" maximum print size		1	28		7,675	925		8,600	9,875	
	3000	Viewing lites, 20" x 24"		6	4.667		335	154		489	605	
	3100	20" x 24" with color correction	▼	6	4.667		470	154		624	760	
	3500	Washers, round, minimum sheet 11" x 14"	Q-1	2	8		2,600	294		2,894	3,325	
	3550	Maximum sheet 20" x 24"		1	16		3,050	590		3,640	4,250	
	3800	Square, minimum sheet 20" x 24"		1	16		2,750	590		3,340	3,900	
	3900	Maximum sheet 50" x 56"	▼	.80	20	▼	4,275	735		5,010	5,800	
	4500	Combination tank sink, tray sink, washers, with										
	4510	dry side tables, average	Q-1	.45	35.556	Ea.	8,100	1,300		9,400	10,900	

11472 | Revolving Darkroom Doors

		CREW	DAILY OUTPUT	LABOR-HOURS	UNIT	2005 BARE COSTS				TOTAL INCL O&P	
						MAT.	LABOR	EQUIP.	TOTAL		
370	0010	**DARKROOM DOORS**										370
	0015	Revolving, standard, 2 way, 36" diameter	2 Carp	3.10	5.161	Opng.	1,725	177		1,902	2,175	
	0020	41" diameter		3.10	5.161		1,850	177		2,027	2,325	
	0050	3 way, 51" diameter		1.40	11.429		2,325	390		2,715	3,150	
	1000	4 way, 49" diameter		1.40	11.429		2,750	390		3,140	3,625	
	2000	Hinged safety, 2 way, 41" diameter		2.30	6.957		2,225	238		2,463	2,825	
	2500	3 way, 51" diameter		1.40	11.429		2,800	390		3,190	3,700	
	3000	Pop out safety, 2 way, 41" diameter		3.10	5.161		2,725	177		2,902	3,275	
	4000	3 way, 51" diameter		1.40	11.429		2,725	390		3,115	3,600	
	5000	Wheelchair-type, pop out, 51" diameter		1.40	11.429		2,750	390		3,140	3,625	
	5020	72" diameter	▼	.90	17.778	▼	5,975	610		6,585	7,525	
	9300	For complete dark rooms, see division 13035-300										

EQUIPMENT 11

11483 | Bowling Alleys

		CREW	DAILY OUTPUT	LABOR-HOURS	UNIT	2005 BARE COSTS				TOTAL INCL O&P	
						MAT.	LABOR	EQUIP.	TOTAL		
150	0010	**BOWLING ALLEYS** Including alley, pinsetter, scorer,										150
	0020	counters and misc. supplies, minimum	4 Carp	.20	160	Lane	35,100	5,475		40,575	47,100	
	0150	Average		.19	168		40,000	5,775		45,775	53,000	
	0300	Maximum	▼	.18	177		46,000	6,100		52,100	60,000	
	0400	Combo table ball rack, add					585			585	640	
	0600	For automatic scorer, add, minimum					6,350			6,350	7,000	
	0700	Maximum				▼	7,200			7,200	7,925	

11484 | Exercise Equipment

		CREW	DAILY OUTPUT	LABOR-HOURS	UNIT	2005 BARE COSTS				TOTAL INCL O&P	
						MAT.	LABOR	EQUIP.	TOTAL		
400	0010	**HEALTH CLUB EQUIPMENT**										400
	0020	Abdominal rack, 2 board capacity				Ea.	415			415	460	
	0050	Abdominal board, upholstered					465			465	510	
	0200	Bicycle trainer, minimum				▼	725			725	795	

11484	Exercise Equipment	CREW	DAILY OUTPUT	LABOR-HOURS	UNIT	2005 BARE COSTS				TOTAL INCL O&P
						MAT.	LABOR	EQUIP.	TOTAL	
400 0300	Deluxe, electric				Ea.	3,700			3,700	4,050	**400**
0400	Bar bell set, chrome plated steel, 25 lbs.					217			217	239
0420	100 lbs.					330			330	360
0450	200 lbs.					635			635	700
0500	Weight plates, cast iron, per lb.				Lb.	4.45			4.45	4.90
0520	Storage rack, 10 station				Ea.	700			700	770
0600	Circuit training apparatus, 12 machines minimum	2 Clab	1.25	12.800	Set	24,500	340		24,840	27,500
0700	Average		1	16		30,100	425		30,525	33,800
0800	Maximum		.75	21.333		35,700	570		36,270	40,100
0820	Dumbbell set, cast iron, with rack and 5 pair					685			685	755
0900	Squat racks	2 Clab	5	3.200	Ea.	680	85.50		765.50	885
1200	Multi-station gym machine, 5 station					7,075			7,075	7,775
1250	9 station					9,925			9,925	10,900
1280	Rowing machine, hydraulic					1,300			1,300	1,450
1300	Treadmill, manual					1,275			1,275	1,400
1320	Motorized					3,675			3,675	4,025
1340	Electronic					4,850			4,850	5,325
1360	Cardio-testing					6,900			6,900	7,600
1400	Treatment/massage tables, minimum					460			460	510
1420	Deluxe, with accessories					610			610	670
1600	For saunas, see Div. 13035-800									
1640	For steam baths, see Div. 13035-940									
1680	For whirlpool baths, see Div. 15418-100									
1720	For additional exercise equipment, see Div. 11486-700									

11486	Gymnasium Equipment									
700 0010	**SCHOOL EQUIPMENT**									**700**
0200	For exterior equipment, see Div. 02880									
0201	For chairs & desks, see Div. 12560-200									
0300	For chalkboards & bulletin boards, see Div. 10120-940 & 10110-240									
0400	For lockers, see Div. 10505-500									
1000	Basketball backstops, wall mtd., 6' extended, fixed, minimum	L-2	1	16	Ea.	1,025	480		1,505	1,875
1100	Maximum		1	16		1,500	480		1,980	2,400
1200	Swing up, minimum		1	16		1,200	480		1,680	2,075
1250	Maximum		1	16		2,150	480		2,630	3,125
1300	Portable, manual, heavy duty, spring operated		1.90	8.421		10,100	252		10,352	11,500
1400	Ceiling suspended, stationary, minimum		.78	20.513		1,800	615		2,415	2,925
1450	Fold up, with accessories, maximum		.40	40		5,425	1,200		6,625	7,825
1600	For electrically operated, add	1 Elec	1	8		1,775	325		2,100	2,425
2000	Benches, folding, in wall, 14' table, 2 benches	L-4	2	12	Set	570	380		950	1,225
3000	Bleachers, telescoping, manual to 15 tier, minimum	F-5	65	.492	Seat	67.50	17.10		84.60	101
3100	Maximum		60	.533		101	18.55		119.55	141
3300	16 to 20 tier, minimum		60	.533		162	18.55		180.55	208
3400	Maximum		55	.582		203	20		223	255
3600	21 to 30 tier, minimum		50	.640		169	22		191	221
3700	Maximum		40	.800		203	28		231	267
3900	For integral power operation, add, minimum	2 Elec	300	.053		34	2.17		36.17	40
4000	Maximum	"	250	.064		54	2.61		56.61	63.50
4100	Boxing ring, elevated, 22' x 22'	L-4	.10	240	Ea.	8,300	7,625		15,925	21,000
4110	For cellular plastic foam padding, add		.10	240		2,850	7,625		10,475	15,100
4120	Floor level, including posts and ropes only, 20' x 20'		.80	30		1,500	955		2,455	3,125
4130	Canvas, 30' x 30'		5	4.800		1,025	152		1,177	1,350
4150	Exercise equipment, bicycle trainer					385			385	425
4180	Chinning bar, adjustable, wall mounted	1 Carp	5	1.600		273	55		328	385
4200	Exercise ladder, 16' x 1'-7", suspended	L-2	3	5.333		1,200	159		1,359	1,575
4210	High bar, floor plate attached	1 Carp	4	2		1,025	68.50		1,093.50	1,225

Important: See the Reference Section for critical supporting data - Reference Nos., Crews, & City Cost Indexes

11 EQUIPMENT

11486 | Gymnasium Equipment

		CREW	DAILY OUTPUT	LABOR-HOURS	UNIT	2005 BARE COSTS				TOTAL INCL O&P
						MAT.	LABOR	EQUIP.	TOTAL	
700	4240	Parallel bars, adjustable	1 Carp	4	2	Ea.	2,250	68.50		2,318.50	2,575	700
	4270	Uneven parallel bars, adjustable	↓	4	2	↓	2,700	68.50		2,768.50	3,075
	4280	Wall mounted, adjustable	L-2	1.50	10.667	Set	775	320		1,095	1,350
	4300	Rope, ceiling mounted, 18' long	1 Carp	3.66	2.186	Ea.	202	75		277	340
	4330	Side horse, vaulting	↓	5	1.600		1,625	55		1,680	1,850
	4360	Treadmill, motorized, deluxe, training type	↓	5	1.600		5,175	55		5,230	5,775
	4390	Weight lifting multi-station, minimum	2 Clab	1	16		2,450	425		2,875	3,375
	4450	Maximum	"	.50	32	↓	13,500	855		14,355	16,200
	4480	For additional exercise equipment, see Div. 11484-400									
	4500	Gym divider curtain, mesh top, vinyl bottom, manual	L-4	500	.048	S.F.	7.35	1.52		8.87	10.45
	4700	Electric roll up	L-7	400	.070		10.75	2.31		13.06	15.40
	5500	Gym mats, 2" thick, naugahyde covered					3.36			3.36	3.70
	5600	Vinyl/nylon covered					5.05			5.05	5.55
	5800	Wall pads, 1-1/2" thick	2 Carp	640	.025		5.55	.86		6.41	7.45
	6000	Wrestling mats, 1" thick, heavy duty				↓	4.50			4.50	4.95
	7000	Scoreboards, baseball, minimum	R-3	1.30	15.385	Ea.	2,825	615	122	3,562	4,150
	7200	Maximum		.05	400		14,000	16,000	3,175	33,175	42,700
	7300	Football, minimum		.86	23.256		3,775	930	184	4,889	5,725
	7400	Maximum		.20	100		12,000	4,000	795	16,795	20,000
	7500	Basketball (one side), minimum		2.07	9.662		1,950	385	76.50	2,411.50	2,800
	7600	Maximum		.30	66.667		13,600	2,675	530	16,805	19,600
	7700	Hockey-basketball (four sides), minimum		.25	80		5,425	3,200	635	9,260	11,400
	7800	Maximum	↓	.15	133	↓	22,500	5,325	1,050	28,875	33,800

11488 | Shooting Ranges

		CREW	DAILY OUTPUT	LABOR-HOURS	UNIT	2005 BARE COSTS				TOTAL INCL O&P
						MAT.	LABOR	EQUIP.	TOTAL	
700	0010	**SHOOTING RANGE** Incl. bullet traps, target provisions, controls,										700
	0100	separators, ceiling system, etc. Not incl. structural shell									
	0200	Commercial	L-9	.64	56.250	Point	11,500	1,750		13,250	15,500
	0300	Law enforcement		.28	128		26,000	3,975		29,975	35,000
	0400	National Guard armories		.71	50.704		12,500	1,575		14,075	16,300
	0500	Reserve training centers		.71	50.704		10,500	1,575		12,075	14,100
	0600	Schools and colleges		.32	112		22,000	3,475		25,475	29,800
	0700	Major acadamies	↓	.19	189		40,000	5,850		45,850	53,500
	0800	For acoustical treatment, add					10%	10%			
	0900	For lighting, add					28%	25%			
	1000	For plumbing, add					5%	5%			
	1100	For ventilating system, add, minimum					40%	40%			
	1200	Add, average					25%	25%			
	1300	Add, maximum				↓	35%	35%			

11520 | Industrial Equipment

		CREW	DAILY OUTPUT	LABOR-HOURS	UNIT	2005 BARE COSTS				TOTAL INCL O&P
						MAT.	LABOR	EQUIP.	TOTAL	
250	0010	**DUST COLLECTION SYSTEMS** Commercial / industrial										250
	0120	Central vacuum units									
	0130	Includes stand, filters and motorized shaker									
	0200	500 CFM, 10" inlet, 2 HP	Q-20	2.40	8.333	Ea.	3,475	310		3,785	4,300
	0220	1000 CFM, 10" inlet, 3 HP		2.20	9.091		3,625	335		3,960	4,525
	0240	1500 CFM, 10" inlet, 5 HP	↓	2	10	↓	3,700	370		4,070	4,650

			CREW	DAILY OUTPUT	LABOR-HOURS	UNIT	2005 BARE COSTS				TOTAL INCL O&P	
11520		**Industrial Equipment**					MAT.	LABOR	EQUIP.	TOTAL		
250	0260	3000 CFM, 13" inlet, 10 HP	Q-20	1.50	13.333	Ea.	9,700	495		10,195	11,500	**250**
	0280	5000 CFM, 16" inlet, 2 @ 10 HP	↓	1	20	↓	10,200	740		10,940	12,400	
	1000	Vacuum tubing, galvanized										
	1100	2-1/8" OD, 16 ga.	Q-9	440	.036	L.F.	1.84	1.32		3.16	4.04	
	1110	2-1/2" OD, 16 ga.		420	.038		2.13	1.38		3.51	4.46	
	1120	3" OD, 16 ga.		400	.040		2.71	1.45		4.16	5.20	
	1130	3-1/2" OD, 16 ga.		380	.042		3.84	1.52		5.36	6.55	
	1140	4" OD, 16 ga.		360	.044		3.93	1.61		5.54	6.80	
	1150	5" OD, 14 ga.		320	.050		7.25	1.81		9.06	10.80	
	1160	6" OD, 14 ga.		280	.057		8.15	2.07		10.22	12.15	
	1170	8" OD, 14 ga.		200	.080		12.35	2.89		15.24	18	
	1180	10" OD, 12 ga.		160	.100		12.35	3.62		15.97	19.10	
	1190	12" OD, 12 ga.		120	.133		35.50	4.82		40.32	46.50	
	1200	14" OD, 12 ga.	↓	80	.200	↓	39	7.25		46.25	54	
	1940	Hose, flexible wire reinforced rubber										
	1956	3" dia.	Q-9	400	.040	L.F.	5.85	1.45		7.30	8.70	
	1960	4" dia.		360	.044		7.90	1.61		9.51	11.15	
	1970	5" dia.		320	.050		9.10	1.81		10.91	12.80	
	1980	6" dia.	↓	280	.057	↓	10.50	2.07		12.57	14.75	
	2000	90° Elbow, slip fit										
	2110	2-1/8" dia.	Q-9	70	.229	Ea.	7.70	8.25		15.95	21	
	2120	2-1/2" dia.		65	.246		10.75	8.90		19.65	25.50	
	2130	3" dia.		60	.267		14.55	9.65		24.20	31	
	2140	3-1/2" dia.		55	.291		18	10.55		28.55	36	
	2150	4" dia.		50	.320		22.50	11.60		34.10	42.50	
	2160	5" dia.		45	.356		42.50	12.85		55.35	66.50	
	2170	6" dia.		40	.400		58.50	14.45		72.95	87	
	2180	8" dia.	↓	30	.533	↓	111	19.30		130.30	152	
	2400	45° Elbow, slip fit										
	2410	2-1/8" dia.	Q-9	70	.229	Ea.	6.75	8.25		15	20	
	2420	2-1/2" dia.		65	.246		9.90	8.90		18.80	24.50	
	2430	3" dia.		60	.267		11.95	9.65		21.60	28	
	2440	3-1/2" dia.		55	.291		15.05	10.55		25.60	33	
	2450	4" dia.		50	.320		19.50	11.60		31.10	39.50	
	2460	5" dia.		45	.356		33	12.85		45.85	56	
	2470	6" dia.		40	.400		44.50	14.45		58.95	71.50	
	2480	8" dia.	↓	35	.457	↓	87.50	16.55		104.05	122	
	2800	90° TY, slip fit thru 6" dia										
	2810	2-1/8" dia.	Q-9	42	.381	Ea.	15.05	13.80		28.85	37.50	
	2820	2-1/2" dia.		39	.410		19.55	14.85		34.40	44.50	
	2830	3" dia.		36	.444		26	16.10		42.10	53	
	2840	3-1/2" dia.		33	.485		33	17.55		50.55	63.50	
	2850	4" dia.		30	.533		47.50	19.30		66.80	81.50	
	2860	5" dia.		27	.593		90	21.50		111.50	132	
	2870	6" dia.	↓	24	.667		123	24		147	172	
	2880	8" dia., butt end					260			260	286	
	2890	10" dia., butt end					470			470	515	
	2900	12" dia., butt end					470			470	515	
	2910	14" dia., butt end					855			855	940	
	2920	6" x 4" dia., butt end					74.50			74.50	82	
	2930	8" x 4" dia., butt end					141			141	156	
	2940	10" x 4" dia., butt end					185			185	204	
	2950	12" x 4" dia., butt end				↓	213			213	234	
	3100	90° Elbow, butt end segmented										
	3110	8" dia., butt end, segmented				Ea.	168			168	185	
	3120	10" dia., butt end, segmented				↓	276			276	305	

Important: See the Reference Section for critical supporting data - Reference Nos., Crews, & City Cost Indexes

11520	Industrial Equipment	CREW	DAILY OUTPUT	LABOR-HOURS	UNIT	2005 BARE COSTS				TOTAL INCL O&P	
						MAT.	LABOR	EQUIP.	TOTAL		
250	3130	12" dia., butt end, segmented				Ea.	350			350	385	**250**
	3140	14" dia., butt end, segmented				↓	440			440	480	
	3200	45° Elbow, butt end segmented										
	3210	8" dia., butt end, segmented				Ea.	143			143	157	
	3220	10" dia., butt end, segmented					199			199	219	
	3230	12" dia., butt end, segmented					204			204	225	
	3240	14" dia., butt end, segmented				↓	251			251	276	
	3400	All butt end fittings require one coupling per joint.										
	3410	Labor for fitting included with couplings.										
	3460	Compression coupling, galvanized, neoprene gasket										
	3470	2-1/8" dia.	Q-9	44	.364	Ea.	11.70	13.15		24.85	33	
	3480	2-1/2" dia.		44	.364		11.70	13.15		24.85	33	
	3490	3" dia.		38	.421		15.35	15.25		30.60	40.50	
	3500	3-1/2" dia.		35	.457		17.20	16.55		33.75	44.50	
	3510	4" dia.		33	.485		18.70	17.55		36.25	47.50	
	3520	5" dia.		29	.552		21.50	19.95		41.45	54	
	3530	6" dia.		26	.615		25	22.50		47.50	62	
	3540	8" dia.		22	.727		45.50	26.50		72	90.50	
	3550	10" dia.		20	.800		66.50	29		95.50	118	
	3560	12" dia.		18	.889		82.50	32		114.50	141	
	3570	14" dia.	↓	16	1	↓	117	36		153	185	
	3800	Air gate valves, galvanized										
	3810	2-1/8" dia.	Q-9	30	.533	Ea.	89	19.30		108.30	128	
	3820	2-1/2" dia.		28	.571		93.50	20.50		114	135	
	3830	3" dia.		26	.615		109	22.50		131.50	154	
	3840	4" dia.		23	.696		142	25		167	195	
	3850	6" dia.	↓	18	.889	↓	196	32		228	266	
300	0010	**EQUIPMENT INSTALLATION**										**300**
	0020	Industrial equipment, minimum	E-2	12	4.667	Ton		172	110	282	415	
	0200	Maximum	"	2	28	"		1,025	660	1,685	2,500	
850	0010	**VOCATIONAL SHOP EQUIPMENT**										**850**
	0020	Benches, work, wood, average	2 Carp	5	3.200	Ea.	390	110		500	600	
	0100	Metal, average		5	3.200		370	110		480	575	
	0400	Combination belt & disc sander, 6"		4	4		710	137		847	1,000	
	0700	Drill press, floor mounted, 12", 1/2 H.P.	↓	4	4		345	137		482	595	
	0800	Dust collector, not incl. ductwork, 6" diameter	1 Shee	1.10	7.273		2,725	292		3,017	3,450	
	1000	Grinders, double wheel, 1/2 H.P.	2 Carp	5	3.200		192	110		302	385	
	1300	Jointer, 4", 3/4 H.P.		4	4		985	137		1,122	1,300	
	1600	Kilns, 16 C.F., to 2000°		4	4		2,050	137		2,187	2,475	
	1900	Lathe, woodworking, 10", 1/2 H.P.		4	4		670	137		807	955	
	2200	Planer, 13" x 6"		4	4		1,225	137		1,362	1,575	
	2500	Potter's wheel, motorized		4	4		980	137		1,117	1,300	
	2800	Saws, band, 14", 3/4 H.P.		4	4		685	137		822	970	
	3100	Metal cutting band saw, 14"		4	4		1,025	137		1,162	1,350	
	3400	Radial arm saw, 10", 2 H.P.		4	4		655	137		792	940	
	3700	Scroll saw, 24"		4	4		1,400	137		1,537	1,750	
	4000	Table saw, 10", 3 H.P.		4	4		1,600	137		1,737	1,975	
	4300	Welder AC arc, 30 amp capacity	↓	4	4	↓	2,450	137		2,587	2,925	

EQUIPMENT **11**

11620 | Laboratory Equipment

		CREW	DAILY OUTPUT	LABOR-HOURS	UNIT	2005 BARE COSTS MAT.	LABOR	EQUIP.	TOTAL	TOTAL INCL O&P	
110	0010	**LABORATORY EQUIPMENT**										110
	0600	Fume hood, with countertop & base, not including HVAC										
	0610	Simple, minimum	2 Carp	5.40	2.963	L.F.	520	101		621	730	
	0620	Complex, including fixtures		2.40	6.667		1,225	228		1,453	1,700	
	0630	Special, maximum		1.70	9.412		1,275	320		1,595	1,900	
	0650	Ductwork, minimum	2 Shee	1	16	Hood	860	645		1,505	1,950	
	0660	Maximum	"	.50	32	"	4,350	1,275		5,625	6,775	
	0670	Service fixtures, average				Ea.	44			44	48.50	
	0680	For sink assembly with hot and cold water, add	1 Plum	1.40	5.714		590	233		823	1,000	
	0700	Glassware washer, undercounter, minimum	L-1	1.80	8.889		6,200	365		6,565	7,375	
	0710	Maximum	"	1	16		8,175	655		8,830	9,975	
	0750	Glove box, fiberglass, bacteriological					13,200			13,200	14,600	
	0760	Controlled atmosphere					8,750			8,750	9,625	
	0770	Radioisotope					11,300			11,300	12,500	
	0780	Carcinogenic					11,300			11,300	12,500	
	1000	Incubators, minimum					2,950			2,950	3,250	
	1010	Maximum					16,500			16,500	18,200	
	1200	Refrigerator, blood bank, 28.6 C.F. emergency signal					6,750			6,750	7,400	
	1210	Reach-in, 16.9 C.F.					6,900			6,900	7,575	
	1400	Safety equipment, eye wash, hand held					294			294	325	
	1450	Deluge shower					565			565	625	
	1600	Sink, one piece plastic, flask wash, hose, free standing	1 Plum	1.60	5		1,500	204		1,704	1,950	
	1610	Epoxy resin sink, 25" x 16" x 10"	"	2	4		166	163		329	430	
	1700	Thermometer, electric, portable					310			310	340	
	1800	Titration unit, four 2000 ml reservoirs					8,600			8,600	9,475	
	1850	Utensil washer-sanitizer	1 Plum	2	4		8,975	163		9,138	10,100	
	1950	Utility table, acid resistant top with drawers	2 Carp	30	.533	L.F.	123	18.25		141.25	164	
	8000	Alternate pricing method: as percent of lab furniture										
	8050	Installation, not incl. plumbing & duct work				% Furn.					22%	
	8100	Plumbing, final connections, simple system									10%	
	8110	Moderately complex system									15%	
	8120	Complex system									20%	
	8150	Electrical, simple system									10%	
	8160	Moderately complex system									20%	
	8170	Complex system									35%	

11700 | Medical Equipment

11710 | Medical Sterilizing Equipment

		CREW	DAILY OUTPUT	LABOR-HOURS	UNIT	2005 BARE COSTS MAT.	LABOR	EQUIP.	TOTAL	TOTAL INCL O&P	
110	0010	**MEDICAL STERILIZING EQUIPMENT**										110
	0700	Distiller, water, steam heated, 50 gal. capacity	1 Plum	1.40	5.714	Ea.	15,200	233		15,433	17,100	
	5600	Sterilizers, floor loading, 26" x 62" x 42", single door, steam					130,500			130,500	143,500	
	5650	Double door, steam					167,500			167,500	184,500	
	5800	General purpose, 20" x 20" x 38", single door					12,600			12,600	13,800	
	6000	Portable, counter top, steam, minimum					3,125			3,125	3,450	
	6020	Maximum					4,900			4,900	5,375	
	6050	Portable, counter top, gas, 17"x15"x32-1/2"					32,300			32,300	35,500	
	6150	Manual washer/sterilizer, 16"x16"x26"	1 Plum	2	4		44,300	163		44,463	48,900	
	6200	Steam generators, electric 10 KW to 180 KW, freestanding										

Important: See the Reference Section for critical supporting data - Reference Nos., Crews, & City Cost Indexes

11700 | Medical Equipment

11710 | Medical Sterilizing Equipment

			CREW	DAILY OUTPUT	LABOR-HOURS	UNIT	MAT.	LABOR	EQUIP.	TOTAL	TOTAL INCL O&P	
110	6250	Minimum	1 Elec	3	2.667	Ea.	6,575	109		6,684	7,400	110
	6300	Maximum	"	.70	11.429		23,700	465		24,165	26,800	
	8200	Bed pan washer-sanitizer	1 Plum	2	4		5,875	163		6,038	6,700	

11720 | Examination & Treatment Equip.

			CREW	DAILY OUTPUT	LABOR-HOURS	UNIT	MAT.	LABOR	EQUIP.	TOTAL	TOTAL INCL O&P	
110	0010	**EXAMINATION & TREATMENT EQUIPMENT**										110
	0300	Blood pressure unit, mercurial, wall				Ea.	125			125	137	
	0400	Diagnostic set, wall					535			535	590	
	4400	Scale, physician's, with height rod					262			262	288	
	6500	Surgery table, minor minimum	1 Sswk	.70	11.429		11,400	435		11,835	13,300	
	6520	Maximum	"	.50	16		21,400	610		22,010	24,700	
	6700	Surgical lights, doctor's office, single arm	2 Elec	2	8		1,125	325		1,450	1,700	
	6750	Dual arm	"	1	16		4,550	650		5,200	5,975	

11730 | Patient Care Equipment

			CREW	DAILY OUTPUT	LABOR-HOURS	UNIT	MAT.	LABOR	EQUIP.	TOTAL	TOTAL INCL O&P	
110	0010	**PATIENT CARE EQUIPMENT**										110
	0750	Exam room furnishings, average per room				Ea.	5,300			5,300	5,825	
	1800	Heat therapy unit, humidified, 26" x 78" x 28"				"	2,800			2,800	3,075	
	2100	Hubbard tank with accessories, stainless steel,										
	2110	125 GPM at 45 psi water pressure				Ea.	21,600			21,600	23,800	
	2150	For electric overhead hoist, add					2,350			2,350	2,600	
	2900	K-Module for heat therapy, 20 oz. capacity, 75° to 110°F					335			335	365	
	3600	Paraffin bath, 126°F, auto controlled					1,250			1,250	1,375	
	3900	Parallel bars for walking training, 12'-0"					1,025			1,025	1,150	
	4600	Station, dietary, medium, with ice					13,200			13,200	14,500	
	4700	Medicine					5,950			5,950	6,550	
	7000	Tables, physical therapy, walk off, electric	2 Carp	3	5.333		2,625	183		2,808	3,175	
	7150	Standard, vinyl top with base cabinets, minimum		3	5.333		1,075	183		1,258	1,450	
	7200	Maximum		2	8		3,175	274		3,449	3,925	
	8400	Whirlpool bath, mobile, sst, 18" x 24" x 60"					3,775			3,775	4,150	
	8450	Fixed, incl. mixing valves	1 Plum	2	4		7,575	163		7,738	8,575	

11740 | Dental Equipment

			CREW	DAILY OUTPUT	LABOR-HOURS	UNIT	MAT.	LABOR	EQUIP.	TOTAL	TOTAL INCL O&P	
200	0010	**DENTAL EQUIPMENT**										200
	0020	Central suction system, minimum	1 Plum	1.20	6.667	Ea.	2,900	272		3,172	3,575	
	0100	Maximum	"	.90	8.889		22,300	365		22,665	25,000	
	0300	Air compressor, minimum	1 Skwk	.80	10		3,175	350		3,525	4,025	
	0400	Maximum		.50	16		22,000	560		22,560	25,100	
	0600	Chair, electric or hydraulic, minimum		.50	16		3,625	560		4,185	4,850	
	0700	Maximum		.25	32		7,675	1,125		8,800	10,200	
	0800	Doctor's/assistant's stool, minimum					410			410	450	
	0850	Maximum					805			805	885	
	1000	Drill console with accessories, minimum	1 Skwk	1.60	5		3,275	174		3,449	3,875	
	1100	Maximum		1.60	5		9,325	174		9,499	10,600	
	2000	Light, ceiling mounted, minimum		8	1		1,875	35		1,910	2,125	
	2100	Maximum		8	1		3,800	35		3,835	4,225	
	2200	Unit light, minimum	2 Skwk	5.33	3.002		1,400	105		1,505	1,725	
	2210	Maximum		5.33	3.002		4,600	105		4,705	5,250	
	2220	Track light, minimum		3.20	5		2,500	174		2,674	3,025	
	2230	Maximum		3.20	5		5,475	174		5,649	6,300	
	2300	Sterilizers, steam portable, minimum					3,175			3,175	3,500	
	2350	Maximum					11,000			11,000	12,100	
	2600	Steam, institutional					13,700			13,700	15,000	
	2650	Dry heat, electric, portable, 3 trays					1,075			1,075	1,175	
	2700	Ultra-sonic cleaner, portable, minimum					860			860	945	

EQUIPMENT **11**

		11740	Dental Equipment	CREW	DAILY OUTPUT	LABOR-HOURS	UNIT	2005 BARE COSTS MAT.	LABOR	EQUIP.	TOTAL	TOTAL INCL O&P	
200	2750		Maximum (institutional)				Ea.	1,925			1,925	2,125	200
	3000		X-ray unit, wall, minimum	1 Skwk	4	2		4,350	69.50		4,419.50	4,900	
	3010		Maximum		4	2		8,225	69.50		8,294.50	9,150	
	3100		Panoramic unit	↓	.60	13.333		19,700	465		20,165	22,400	
	3105		Deluxe, minimum	2 Skwk	1.60	10		28,500	350		28,850	31,900	
	3110		Maximum	"	1.60	10		66,000	350		66,350	73,000	
	3500		Developers, X-ray, average	1 Plum	5.33	1.501		3,850	61.50		3,911.50	4,325	
	3600		Maximum	"	5.33	1.501	↓	7,125	61.50		7,186.50	7,950	

		11760	Operating Room Equipment										
110	0010	**OPERATING ROOM EQUIPMENT**											110
	5000		Scrub, surgical, stainless steel, single station, minimum	1 Plum	3	2.667	Ea.	2,125	109		2,234	2,525	
	5100		Maximum					7,075			7,075	7,775	
	6550		Major surgery table, minimum	1 Sswk	.50	16		25,800	610		26,410	29,500	
	6570		Maximum	"	.50	16		85,500	610		86,110	95,000	
	6800		Surgical lights, major operating room, dual head, minimum	2 Elec	1	16		14,400	650		15,050	16,800	
	6850		Maximum	"	1	16	↓	23,900	650		24,550	27,300	

		11770	Radiology Equipment										
110	0010	**RADIOLOGY EQUIPMENT**											110
	8700		X-ray, mobile, minimum				Ea.	11,000			11,000	12,100	
	8750		Maximum					62,000			62,000	68,000	
	8900		Stationary, minimum					34,400			34,400	37,800	
	8950		Maximum					179,000			179,000	197,000	
	9150		Developing processors, minimum					9,575			9,575	10,500	
	9200		Maximum				↓	25,100			25,100	27,600	

		11781	Mortuary Equipment										
110	0010	**MORTUARY EQUIPMENT**											110
	0015		Autopsy table, standard	1 Plum	1	8	Ea.	7,275	325		7,600	8,525	
	0020		Deluxe	"	.60	13.333		13,400	545		13,945	15,600	
	3200		Mortuary refrigerator, end operated, 2 capacity					11,500			11,500	12,700	
	3300		6 capacity				↓	21,400			21,400	23,500	

For information about Means Estimating Seminars, see yellow pages 12 and 13 in back of book

EQUIPMENT

Important: See the Reference Section for critical supporting data - Reference Nos., Crews, & City Cost Indexes

Division 12
Furnishings

Estimating Tips

General
- The items in this division are usually priced per square foot or each. Most of these items are purchased by the owner and placed by the supplier. Do not assume the items in Division 12 will be purchased and installed by the supplier. Check the specifications for responsibilities and include receiving, storage, installation and mechanical and electrical hook-ups in the appropriate divisions.

- Some items in this division require some type of support system that is not usually furnished with the item. Examples of these systems include blocking for the attachment of casework and heavy drapery rods. The required blocking must be added to the estimate in the appropriate division.

Reference Numbers
Reference numbers are shown in bold squares at the beginning of some major classifications. These numbers refer to related items in the Reference Section. The reference information may be an estimating procedure, an alternate pricing method or technical information.

Note: Not all subdivisions listed here necessarily appear in this publication.

		12310 │ Metal Casework	CREW	DAILY OUTPUT	LABOR-HOURS	UNIT	MAT.	LABOR	EQUIP.	TOTAL	TOTAL INCL O&P	
								2005 BARE COSTS				
100	0010	**KEY CABINETS**										**100**
	0020	Wall mounted, 60 key capacity	1 Carp	20	.400	Ea.	73.50	13.70		87.20	103	
	0200	Drawer type, 600 key capacity	1 Clab	15	.533		760	14.25		774.25	860	
	0300	2,400 key capacity		20	.400		3,600	10.70		3,610.70	4,000	
	0400	Tray type, 20 key capacity		50	.160		67.50	4.27		71.77	81	
	0500	50 key capacity		40	.200		127	5.35		132.35	148	
200	0010	**DISPLAY CASES** Free standing, all glass										**200**
	0020	Aluminum frame, 42" high x 36" x 12" deep	2 Carp	8	2	Ea.	460	68.50		528.50	610	
	0100	70" high x 48" x 18" deep	"	6	2.667		1,325	91.50		1,416.50	1,600	
	0500	For wood bases, add					9%					
	0600	For hardwood frames, deduct					8%					
	0700	For bronze, baked enamel finish, add					10%					
	2000	Wall mounted, glass front, aluminum frame										
	2010	Non-illuminated, one section 3' x 4' x 1'-4"	2 Carp	5	3.200	Ea.	1,150	110		1,260	1,425	
	2100	5' x 4' x 1'-4"		5	3.200		1,250	110		1,360	1,550	
	2200	6' x 4' x 1'-4"		4	4		1,425	137		1,562	1,800	
	2500	Two sections, 8' x 4' x 1'-4"		2	8		1,725	274		1,999	2,325	
	2600	10' x 4' x 1'-4"		2	8		1,900	274		2,174	2,500	
	3000	Three sections, 16' x 4' x 1'-4"		1.50	10.667	Section	3,275	365		3,640	4,175	
	3500	For fluorescent lights, add					260			260	286	
	4000	Table exhibit cases, 2' wide, 3' high, 4' long, flat top	2 Carp	5	3.200	Ea.	740	110		850	985	
	4100	3' wide, 3' high, 4' long, sloping top	"	3	5.333	"	1,125	183		1,308	1,525	
560	0010	**IRONING CENTER**										**560**
	0020	Including cabinet, board & light, minimum	1 Carp	2	4	Ea.	310	137		447	560	
	0100	Maximum, see also Div. 11119-450	"	1.50	5.333	"	505	183		688	840	
750	0010	**CASEWORK**										**750**
	0500	Hospital, base cabinets, laminated plastic	2 Carp	10	1.600	L.F.	211	55		266	320	
	1000	Stainless steel	"	10	1.600		289	55		344	405	
	1200	For all drawers, add					10.75			10.75	11.80	
	1300	Cabinet base trim, 4" high, enameled steel	2 Carp	200	.080		33.50	2.74		36.24	41.50	
	1400	Stainless steel		200	.080		52.50	2.74		55.24	62.50	
	1450	Counter top, laminated plastic, no backsplash		40	.400		36.50	13.70		50.20	61.50	
	1650	With backsplash		40	.400		45	13.70		58.70	71	
	1800	For sink cutout, add		12.20	1.311	Ea.		45		45	70	
	1900	Stainless steel counter top		40	.400	L.F.	101	13.70		114.70	133	
	2000	For drop-in stainless 43" x 21" sink, add				Ea.	565			565	620	
	2050	For solid surface countertops see Div. 06620-810										
	2100	Nurses station, door type, laminated plastic	2 Carp	10	1.600	L.F.	244	55		299	355	
	2200	Enameled steel		10	1.600		222	55		277	330	
	2300	Stainless steel		10	1.600		310	55		365	425	
	2400	For drawer type, add					99			99	109	
	2500	Wall cabinets, laminated plastic	2 Carp	15	1.067		158	36.50		194.50	230	
	2600	Enameled steel		15	1.067		185	36.50		221.50	260	
	2700	Stainless steel		15	1.067		300	36.50		336.50	385	
	2800	For glass doors, add					25			25	27.50	
	3500	Kitchen, base cabinets, metal, minimum	2 Carp	30	.533		54	18.25		72.25	88	
	3600	Maximum		25	.640		137	22		159	185	
	3700	Wall cabinets, metal, minimum		30	.533		54	18.25		72.25	88	
	3800	Maximum		25	.640		124	22		146	171	
	5000	School, 24" deep, metal, 84" high units		15	1.067		270	36.50		306.50	355	
	5150	Counter height units		20	.800		211	27.50		238.50	275	
	5450	Wood, custom fabricated, 32" high counter		20	.800		175	27.50		202.50	236	
	5600	Add for counter top		56	.286		19.55	9.80		29.35	37	
	5800	84" high wall units		15	1.067		183	36.50		219.50	258	
	6000	Laminated plastic finish is same price as wood										

12 FURNISHINGS

Important: See the Reference Section for critical supporting data - Reference Nos., Crews, & City Cost Indexes

12300 | Manufactured Casework

12350 | Specialty Casework

		CREW	DAILY OUTPUT	LABOR-HOURS	UNIT	2005 BARE COSTS				TOTAL INCL O&P
						MAT.	LABOR	EQUIP.	TOTAL	
500	**0010**	**LABORATORY CASEWORK**								**500**
	0020	Cabinets, base, door units, metal	2 Carp	18	.889	L.F.	154	30.50		184.50	218
	0300	Drawer units		18	.889		335	30.50		365.50	415
	0700	Tall storage cabinets, open, 7' high		20	.800		335	27.50		362.50	415
	0900	With glazed doors		20	.800		410	27.50		437.50	500
	1300	Wall cabinets, metal, 12-1/2" deep, open		20	.800		104	27.50		131.50	157
	1500	With doors		20	.800	▼	216	27.50		243.50	281
	1550	Counter tops, not incl. base cabinets, acidproof, minimum		82	.195	S.F.	25.50	6.70		32.20	38.50
	1600	Maximum		70	.229		35	7.85		42.85	50.50
	1650	Stainless steel	▼	82	.195	▼	86	6.70		92.70	105
	6300	Rule of thumb: lab furniture including installation & connection									
	6320	High school				S.F.					30.50
	6340	College									45
	6360	Clinical, health care									38.50
	6380	Industrial				▼					62

12400 | Furnishings & Accessories

12460 | Furnishing Accessories

		CREW	DAILY OUTPUT	LABOR-HOURS	UNIT	2005 BARE COSTS				TOTAL INCL O&P
						MAT.	LABOR	EQUIP.	TOTAL	
900	**0010**	**ASH/TRASH RECEIVERS**								**900**
	1000	Ash urn, cylindrical metal									
	1020	8" diameter, 20" high	1 Clab	60	.133	Ea.	116	3.56		119.56	134
	1060	10" diameter, 26" high	"	60	.133	"	163	3.56		166.56	185
	2000	Combination ash/trash urn, metal									
	2020	8" diameter, 20" high	1 Clab	60	.133	Ea.	108	3.56		111.56	125
	2050	10" diameter, 26" high	"	60	.133	"	182	3.56		185.56	206
	4000	Trash receptacle, metal									
	4020	8" diameter, 15" high	1 Clab	60	.133	Ea.	62	3.56		65.56	73.50
	4040	10" diameter, 18" high		60	.133		105	3.56		108.56	121
	5040	16" x 8" x 14" high	▼	60	.133	▼	32	3.56		35.56	40.50
	5500	Trash receptacle, plastic, with lid									
	5520	35 gallon	1 Clab	60	.133	Ea.	128	3.56		131.56	147
	5540	45 gallon		60	.133		156	3.56		159.56	178
	5550	Plastic recycling barrel, w/lid & wheels, 32 gal		60	.133		105	3.56		108.56	122
	5560	65 gal		60	.133		125	3.56		128.56	144
	5570	95 gal	▼	60	.133	▼	310	3.56		313.56	345

12483 | Floor Mats & Frames

		CREW	DAILY OUTPUT	LABOR-HOURS	UNIT	2005 BARE COSTS				TOTAL INCL O&P
						MAT.	LABOR	EQUIP.	TOTAL	
200	**0010**	**FLOOR MATS**								**200**
	0020	Recessed, in-laid black rubber, 3/8" thick, solid	1 Clab	155	.052	S.F.	20.50	1.38		21.88	25
	0050	Perforated		155	.052		17.50	1.38		18.88	21.50
	0100	1/2" thick, solid		155	.052		19.55	1.38		20.93	23.50
	0150	Perforated		155	.052		22.50	1.38		23.88	27
	0200	In colors, 3/8" thick, solid		155	.052		20	1.38		21.38	24
	0250	Perforated		155	.052		20	1.38		21.38	24
	0300	1/2" thick, solid		155	.052		29.50	1.38		30.88	34.50
	0350	Perforated		155	.052		29.50	1.38		30.88	34.50
	0500	Link mats, including nosings, aluminum, 3/8" thick **CN**	▼	155	.052	▼	18.50	1.38		19.88	22.50

12483 | Floor Mats & Frames

		CREW	DAILY OUTPUT	LABOR-HOURS	UNIT	2005 BARE COSTS				TOTAL INCL O&P
						MAT.	LABOR	EQUIP.	TOTAL	
200	0550	Black rubber with galvanized tie rods	1 Clab	155	.052	S.F.	14.40	1.38		15.78	18
	0600	Steel, galvanized, 3/8" thick		155	.052		8	1.38		9.38	10.95
	0650	Vinyl, in colors	↓	155	.052	↓	18.55	1.38		19.93	22.50
	0750	Add for nosings, rubber				L.F.	3.48			3.48	3.83
	0850	Recess frames for above mats, aluminum	1 Carp	100	.080		3.48	2.74		6.22	8.10
	0870	Bronze	"	100	.080	↓	5.05	2.74		7.79	9.80
	0900	Skate lock tile, 24" x 24" x 1/2" thick, rubber, black	1 Clab	125	.064	S.F.	14.55	1.71		16.26	18.65
	0950	Color		125	.064	"	18.55	1.71		20.26	23
	1000	12" x 24" border, black		75	.107	L.F.	16.55	2.85		19.40	22.50
	1100	Color		75	.107	"	22	2.85		24.85	29
	1150	12" x 12" outside corner, black		100	.080	S.F.	16.75	2.14		18.89	22
	1200	Color		100	.080		20.50	2.14		22.64	26
	1500	Duckboard, aluminum slats		155	.052		15.15	1.38		16.53	18.85
	1700	Hardwood strips on rubber base, to 54" wide		155	.052		12.10	1.38		13.48	15.45
	1800	Assembled with brass rods and vinyl spacers, to 48" wide		155	.052		18.40	1.38		19.78	22
	1850	Tire fabric, 3/4" thick		155	.052		10.55	1.38		11.93	13.75
	1900	Vinyl, 36" wide, in colors, hollow top & bottoms		155	.052		8.65	1.38		10.03	11.65
	1950	Solid top & bottom members	↓	155	.052	↓	8.65	1.38		10.03	11.65

12492 | Blinds and Shades

		CREW	DAILY OUTPUT	LABOR-HOURS	UNIT	2005 BARE COSTS				TOTAL INCL O&P
						MAT.	LABOR	EQUIP.	TOTAL	
100	0010	**BLINDS, INTERIOR**									100
	0020	Horizontal, 1" aluminum slats, solid color, stock	1 Carp	590	.014	S.F.	2.89	.46		3.35	3.90
	0090	Custom, minimum		590	.014		2.28	.46		2.74	3.23
	0100	Maximum		440	.018		6.50	.62		7.12	8.10
	0250	2" aluminum slats, custom, minimum		590	.014		3.80	.46		4.26	4.90
	0350	Maximum		440	.018		13	.62		13.62	15.25
	0450	Stock, minimum		590	.014		4.21	.46		4.67	5.35
	0500	Maximum		440	.018		6.85	.62		7.47	8.45
	0600	2" steel slats, stock, minimum		590	.014		1.60	.46		2.06	2.48
	0630	Maximum		440	.018		4.46	.62		5.08	5.90
	0750	Custom, minimum		590	.014		1.44	.46		1.90	2.30
	0850	Maximum		400	.020		7.15	.69		7.84	8.90
	1500	Vertical, 3" to 5" PVC or cloth strips, minimum **CN**		460	.017		7.85	.60		8.45	9.55
	1600	Maximum		400	.020		9.50	.69		10.19	11.50
	1800	4" aluminum slats, minimum		460	.017		3.12	.60		3.72	4.36
	1900	Maximum		400	.020		7.50	.69		8.19	9.30
	1950	Mylar mirror-finish strips, to 8" wide, minimum		460	.017		12.50	.60		13.10	14.70
	1970	Maximum		400	.020	↓	18.80	.69		19.49	21.50
	3000	Wood folding panels with movable louvers, 7" x 20" each		17	.471	Pr.	42	16.10		58.10	71.50
	3300	8" x 28" each		17	.471		61	16.10		77.10	92
	3450	9" x 36" each		17	.471		72.50	16.10		88.60	105
	3600	10" x 40" each		17	.471		82	16.10		98.10	115
	4000	Fixed louver type, stock units, 8" x 20" each		17	.471		63	16.10		79.10	94.50
	4150	10" x 28" each		17	.471		86.50	16.10		102.60	120
	4300	12" x 36" each		17	.471		110	16.10		126.10	146
	4450	18" x 40" each		17	.471		131	16.10		147.10	169
	5000	Insert panel type, stock, 7" x 20" each		17	.471		15.60	16.10		31.70	42
	5150	8" x 28" each		17	.471		28.50	16.10		44.60	56.50
	5300	9" x 36" each		17	.471		36.50	16.10		52.60	65
	5450	10" x 40" each		17	.471		39	16.10		55.10	68
	5600	Raised panel type, stock, 10" x 24" each		17	.471		109	16.10		125.10	145
	5650	12" x 26" each		17	.471		126	16.10		142.10	164
	5700	14" x 30" each		17	.471		143	16.10		159.10	182
	5750	16" x 36" each	↓	17	.471		160	16.10		176.10	201
	6000	For custom built pine, add					22%				
	6500	For custom built hardwood blinds, add				↓	42%				

12 FURNISHINGS

Important: See the Reference Section for critical supporting data - Reference Nos., Crews, & City Cost Indexes

12400 | Furnishings & Accessories

12492 | Blinds and Shades

		CREW	DAILY OUTPUT	LABOR-HOURS	UNIT	2005 BARE COSTS MAT.	LABOR	EQUIP.	TOTAL	TOTAL INCL O&P	
600	**0010**	**SHADES**										**600**
	0020	Basswood, roll-up, stain finish, 3/8" slats	1 Carp	300	.027	S.F.	9.85	.91		10.76	12.20	
	0200	7/8" slats		300	.027		9.30	.91		10.21	11.60	
	0300	Vertical side slide, stain finish, 3/8" slats		300	.027		14.85	.91		15.76	17.75	
	0400	7/8" slats	↓	300	.027		14.85	.91		15.76	17.75	
	0500	For fire retardant finishes, add					16%					
	0600	For "B" rated finishes, add					20%					
	0900	Mylar, single layer, non-heat reflective	1 Carp	685	.012		5.55	.40		5.95	6.70	
	1000	Double layered, heat reflective		685	.012		6.65	.40		7.05	7.90	
	1100	Triple layered, heat reflective	↓	685	.012	↓	9.15	.40		9.55	10.65	
	1200	For metal roller instead of wood, add per				Shade	3.35			3.35	3.69	
	1300	Vinyl coated cotton, standard	1 Carp	685	.012	S.F.	1.93	.40		2.33	2.74	
	1400	Lightproof decorator shades		685	.012		1.74	.40		2.14	2.53	
	1500	Vinyl, lightweight, 4 gauge		685	.012		.41	.40		.81	1.07	
	1600	Heavyweight, 6 gauge		685	.012		1.30	.40		1.70	2.05	
	1700	Vinyl laminated fiberglass, 6 ga., translucent		685	.012		1.92	.40		2.32	2.73	
	1800	Lightproof		685	.012		2.99	.40		3.39	3.91	
	3000	Woven aluminum, 3/8" thick, lightproof and fireproof	↓	350	.023	↓	4.28	.78		5.06	5.95	

12493 | Curtains and Drapes

		CREW	DAILY OUTPUT	LABOR-HOURS	UNIT	2005 BARE COSTS MAT.	LABOR	EQUIP.	TOTAL	TOTAL INCL O&P	
200	**0010**	**DRAPERY HARDWARE**										**200**
	0030	Standard traverse, per foot, minimum	1 Carp	59	.136	L.F.	1.93	4.64		6.57	9.35	
	0100	Maximum		51	.157	"	10.40	5.35		15.75	19.80	
	4000	Traverse rods, adjustable, 28" to 48"		22	.364	Ea.	17.05	12.45		29.50	38	
	4020	48" to 84"		20	.400		24.50	13.70		38.20	48.50	
	4040	66" to 120"		18	.444		28.50	15.20		43.70	55	
	4060	84" to 156"		16	.500		31	17.15		48.15	60.50	
	4080	100" to 180"		14	.571		35.50	19.55		55.05	69.50	
	4090	156" to 228"		13	.615		46	21		67	83.50	
	4100	228" to 312"		13	.615		55.50	21		76.50	94	
	4200	Double rods, adjustable, 30" to 48"		9	.889		37.50	30.50		68	88.50	
	4220	48" to 86"		9	.889		48.50	30.50		79	101	
	4240	86" to 150"		8	1		62	34.50		96.50	122	
	4260	100" to 180"		7	1.143		72.50	39		111.50	141	
	4300	Tray and curtain rod, adjustable, 30" to 48"		9	.889		28.50	30.50		59	78.50	
	4320	48" to 86"		9	.889		40	30.50		70.50	91.50	
	4340	86" to 150"		8	1		48	34.50		82.50	107	
	4360	100" to 180"	↓	7	1.143		51	39		90	118	
	4600	Valance, pinch pleated fabric, 12" deep, up to 54" long, minimum					34			34	37.50	
	4610	Maximum					85.50			85.50	94	
	4620	Up to 77" long, minimum					52.50			52.50	58	
	4630	Maximum					138			138	152	
	5000	Stationary rods, first 2 feet				↓	8.65			8.65	9.50	
	5020	Each additional foot, add				L.F.	3.40			3.40	3.74	

12500 | Furniture

12510 | Office Furniture

		CREW	DAILY OUTPUT	LABOR-HOURS	UNIT	2005 BARE COSTS MAT.	LABOR	EQUIP.	TOTAL	TOTAL INCL O&P	
580	**0010**	**MULTI-MEDIA BOARDS** See Div. 10110-240										**580**

12510 | Office Furniture

		CREW	DAILY OUTPUT	LABOR-HOURS	UNIT	2005 BARE COSTS MAT.	LABOR	EQUIP.	TOTAL	TOTAL INCL O&P	
600	**0010**	**OFFICE CASE GOODS**										**600**
	0020	Desks, 29" high, double pedestal, 30" x 60", metal, minimum				Ea.	325			325	355	
	0030	Maximum					1,050			1,050	1,150	
	0160	Wood, minimum					350			350	385	
	0180	Maximum					2,050			2,050	2,250	
	0600	Desks, single pedestal, 30" x 60", metal, minimum					261			261	287	
	0620	Maximum					855			855	940	
	0640	Wood, minimum					375			375	415	
	0650	Maximum					640			640	705	
	0670	Executive return, 24" x 42", with box, file, wood, minimum					249			249	273	
	0680	Maximum					640			640	705	
	0720	Desks, secretarial, 30" x 60", metal, minimum					215			215	237	
	0730	Maximum					485			485	530	
	0740	Return, 20" x 42", minimum					275			275	305	
	0750	Maximum					380			380	420	
	0800	Wood, minimum					360			360	395	
	0810	Maximum					2,075			2,075	2,275	
	0820	Return, 20" x 42", minimum					340			340	375	
	0830	Maximum					1,100			1,100	1,200	
	0900	Desktop organizer, 72" x 14" x 36" high, wood, minimum					214			214	236	
	0920	Maximum					275			275	305	
	0940	59" x 12" x 23" high, steel, minimum					187			187	205	
	0960	Maximum					245			245	269	
	2000	Chairs, office type, executive, minimum					276			276	305	
	2150	Maximum					1,750			1,750	1,925	
	2200	Management, minimum					244			244	269	
	2250	Maximum					1,600			1,600	1,775	
	2280	Task, minimum					159			159	175	
	2290	Maximum					685			685	750	
	2300	Arm kit, minimum					50.50			50.50	55.50	
	2320	Maximum				▼	97			97	107	
	6000	Table, conference										
	6050	Boat, 96" x 42", minimum				Ea.	610			610	670	
	6150	Maximum					3,800			3,800	4,200	
	6720	Rectangle, 96" x 42", minimum					710			710	780	
	6740	Maximum				▼	3,150			3,150	3,450	
675	**0010**	**POSTS**										**675**
	0020	Portable for pedestrian traffic control, standard, minimum				Ea.	94			94	103	
	0100	Maximum					145			145	160	
	0300	Deluxe posts, minimum					173			173	190	
	0400	Maximum				▼	285			285	315	
	0600	Ropes for above posts, plastic covered, 1-1/2" diameter				L.F.	9.35			9.35	10.30	
	0700	Chain core				"	9.60			9.60	10.55	

12520 | Seating

		CREW	DAILY OUTPUT	LABOR-HOURS	UNIT	MAT.	LABOR	EQUIP.	TOTAL	TOTAL INCL O&P	
550	**0010**	**SEATING**										**550**
	0100	Bleachers, See Div. 02870 & 11486										
	0101	Benches, See Div. 02870, 10505 & 11486										
	0500	Classroom, movable chair & desk type, minimum				Set					69.50	
	0600	Maximum				"					128	
	1000	Lecture hall, pedestal type, minimum	2 Carp	22	.727	Ea.	127	25		152	179	
	1200	Maximum		14.50	1.103		440	38		478	545	
	2000	Auditorium chair, all veneer construction		22	.727		140	25		165	193	
	2200	Veneer back, padded seat		22	.727		176	25		201	233	
	2350	Fully upholstered, spring seat	▼	22	.727	▼	178	25		203	234	

Important: See the Reference Section for critical supporting data - Reference Nos., Crews, & City Cost Indexes

12 FURNISHINGS

			DAILY	LABOR-		2005 BARE COSTS				TOTAL	
	12520	**Seating**	CREW	OUTPUT	HOURS	UNIT	MAT.	LABOR	EQUIP.	TOTAL	INCL O&P	
550	2450	For tablet arms, add				Ea.	46			46	51	**550**
	2500	For fire retardancy, CATB-133, add				↓	19			19	21	

	12540	**Hospitality Furniture**										
100	0010	**TABLES, FOLDING** Laminated plastic tops										**100**
	1000	Tubular steel legs with glides										
	1020	18" x 60", minimum				Ea.	164			164	180	
	1040	Maximum					750			750	825	
	1840	36" x 96", minimum					205			205	225	
	1860	Maximum					1,450			1,450	1,600	
	2000	Round, wood stained, plywood top, 60" diameter, minimum					298			298	330	
	2020	Maximum				↓	1,800			1,800	1,975	
500	0010	**FURNITURE, HOTEL**										**500**
	0020	Standard quality set, minimum				Room	1,525			1,525	1,675	
	0200	Maximum				"	7,900			7,900	8,675	
700	0010	**FURNITURE, RESTAURANT**										**700**
	0020	Bars, built-in, front bar	1 Carp	5	1.600	L.F.	214	55		269	320	
	0200	Back bar	"	5	1.600	"	155	55		210	257	
	0300	Booth seating see Div. 12640-150										
	2000	Chair, bentwood side chair, metal, minimum				Ea.	48.50			48.50	53.50	
	2020	Maximum					78			78	86	
	2600	Upholstered seat & back, arms, minimum					156			156	172	
	2620	Maximum				↓	335			335	365	

	12560	**Institutional Furniture**										
100	0010	**BANK FURNITURE** See Div. 12510-600										**100**
150	0010	**CHURCH FURNITURE** See Div. 11041-250										**150**
200	0010	**FURNITURE, SCHOOL**										**200**
	1000	Chair, molded plastic,										
	1100	Integral tablet arm, minimum				Ea.	91			91	100	
	1150	Maximum					230			230	253	
	2000	Desk, single pedestal, top book compartment, minimum					53.50			53.50	59	
	2020	Maximum					84			84	92.50	
	2200	Flip top, minimum					86			86	94.50	
	2220	Maximum				↓	102			102	112	
300	0010	**FURNITURE, DORMITORY**										**300**
	1000	Chest, four drawer, minimum				Ea.	380			380	420	
	1020	Maximum				"	520			520	570	
	1050	Built-in, minimum	2 Carp	13	1.231	L.F.	120	42		162	198	
	1150	Maximum		10	1.600		222	55		277	330	
	1200	Desk top, built-in, laminated plastic, 24" deep, minimum		50	.320		25	10.95		35.95	44.50	
	1300	Maximum		40	.400		75	13.70		88.70	104	
	1450	30" deep, minimum		50	.320		32	10.95		42.95	52	
	1550	Maximum		40	.400		140	13.70		153.70	176	
	1750	Dressing unit, built-in, minimum		12	1.333		180	45.50		225.50	269	
	1850	Maximum	↓	8	2	↓	540	68.50		608.50	700	
	8000	Rule of thumb: total cost for furniture, minimum				Student					2,025	
	8050	Maximum				"					3,850	
400	0010	**FURNITURE, HOSPITAL**										**400**
	0020	Beds, manual, minimum				Ea.	905			905	1,000	
	0100	Maximum				↓	2,500			2,500	2,750	
	0300	Manual and electric beds, minimum					1,150			1,150	1,250	

FURNISHINGS 12

12560	Institutional Furniture	CREW	DAILY OUTPUT	LABOR-HOURS	UNIT	2005 BARE COSTS				TOTAL INCL O&P	
						MAT.	LABOR	EQUIP.	TOTAL		
400											**400**
0400	Maximum				Ea.	2,525			2,525	2,775	
0600	All electric hospital beds, minimum					1,350			1,350	1,475	
0700	Maximum					3,650			3,650	4,025	
0900	Manual, nursing home beds, minimum					720			720	790	
1000	Maximum					1,500			1,500	1,650	
1020	Overbed table, laminated top, minimum					305			305	335	
1040	Maximum				▼	765			765	840	
1100	Patient wall systems, not incl. plumbing, minimum				Room	915			915	1,000	
1200	Maximum				"	1,700			1,700	1,850	
2000	Geriatric chairs, minimum				Ea.	405			405	445	
2020	Maximum				"	955			955	1,050	
700	0010	**FURNITURE, LIBRARY**									**700**
0100	Attendant desk, 36" x 62" x 29" high	1 Carp	16	.500	Ea.	2,275	17.15		2,292.15	2,525	
0200	Book display, "A" frame display, both sides, 42" x 42" x 60" high		16	.500		1,975	17.15		1,992.15	2,200	
0220	Table with bulletin board, 42" x 24" x 49" high		16	.500		1,400	17.15		1,417.15	1,575	
0800	Card catalogue, 30 tray unit		16	.500		2,450	17.15		2,467.15	2,725	
0840	60 tray unit	▼	16	.500		4,400	17.15		4,417.15	4,875	
0880	72 tray unit	2 Carp	16	1		5,325	34.50		5,359.50	5,900	
1000	Carrels, single face, initial unit	1 Carp	16	.500		660	17.15		677.15	750	
1500	Double face, initial unit	2 Carp	16	1		990	34.50		1,024.50	1,150	
4000	Dictionary stand, stationary	1 Carp	16	.500		690	17.15		707.15	780	
4020	Revolving		16	.500		273	17.15		290.15	325	
4200	Exhibit case, table style, 60" x 28" x 36"		11	.727		3,875	25		3,900	4,325	
7000	Tables, card catalog reference, 24" x 60" x 42"	▼	16	.500	▼	875	17.15		892.15	990	

12640	Booths & Tables	CREW	DAILY OUTPUT	LABOR-HOURS	UNIT	2005 BARE COSTS				TOTAL INCL O&P	
						MAT.	LABOR	EQUIP.	TOTAL		
150	0010	**BOOTHS**									**150**
1000	Banquet, upholstered seat and back, custom										
1500	Straight, minimum	2 Carp	40	.400	L.F.	131	13.70		144.70	166	
1520	Maximum		36	.444		255	15.20		270.20	305	
1600	"L" or "U" shape, minimum		35	.457		134	15.65		149.65	172	
1620	Maximum	▼	30	.533	▼	238	18.25		256.25	290	
1800	Upholstered outside finished backs for										
1810	single booths and custom banquets										
1820	Minimum	2 Carp	44	.364	L.F.	15.15	12.45		27.60	36	
1840	Maximum	"	40	.400	"	45.50	13.70		59.20	71.50	
3000	Fixed seating, one piece plastic chair and										
3010	plastic laminate table top										
3100	Two seat, 24" x 24" table, minimum	F-7	30	1.067	Ea.	530	32.50		562.50	630	
3120	Maximum		26	1.231		755	37.50		792.50	890	
3200	Four seat, 24" x 48" table, minimum		28	1.143		525	35		560	630	
3220	Maximum	▼	24	1.333	▼	895	40.50		935.50	1,050	
5000	Mount in floor, wood fiber core with										
5010	plastic laminate face, single booth										
5050	24" wide	F-7	30	1.067	Ea.	203	32.50		235.50	274	
5100	48" wide	"	28	1.143	"	258	35		293	335	

Important: See the Reference Section for critical supporting data - Reference Nos., Crews, & City Cost Indexes

12 FURNISHINGS

12830		Interior Planters	CREW	DAILY OUTPUT	LABOR-HOURS	UNIT	2005 BARE COSTS				TOTAL INCL O&P	
							MAT.	LABOR	EQUIP.	TOTAL		
600	0010	**PLANTERS**										600
	1000	Fiberglass, hanging, 12" diameter, 7" high				Ea.	39			39	42.50	
	1500	Rectangular, 48" long, 16" high x 15" wide					228			228	251	
	1650	60" long, 30" high, 28" wide					580			580	635	
	2000	Round, 12" diameter, 13" high					76.50			76.50	84	
	2050	25" high					84.50			84.50	93	
	5000	Square, 10" side, 20" high					103			103	113	
	5100	14" side, 15" high					107			107	118	
	6000	Metal bowl, 32" diameter, 8" high, minimum					305			305	340	
	6050	Maximum					395			395	435	
	8750	Wood, fiberglass liner, square										
	8780	14" square, 15" high, minimum				Ea.	197			197	217	
	8800	Maximum					243			243	267	
	9400	Plastic cylinder, molded, 10" diameter, 10" high					8			8	8.80	
	9500	11" diameter, 11" high					25.50			25.50	28	

For information about Means Estimating Seminars, see yellow pages 12 and 13 in back of book

FURNISHINGS 12

For expanded coverage of these items see *Means Interior Cost Data 2005*

Division Notes

	CREW	DAILY OUTPUT	LABOR-HOURS	UNIT	2005 BARE COSTS				TOTAL INCL O&P
					MAT.	LABOR	EQUIP.	TOTAL	

Division 13
Special Construction

Estimating Tips

General

- The items and systems in this division are usually estimated, purchased, supplied and installed as a unit by one or more subcontractors. The estimator must ensure that all parties are operating from the same set of specifications and assumptions and that all necessary items are estimated and will be provided. Many times the complex items and systems are covered but the more common ones such as excavation or a crane are overlooked for the very reason that everyone assumes nobody could miss them. The estimator should be the central focus and be able to ensure that all systems are complete.

- Another area where problems can develop in this division is at the interface between systems. The estimator must ensure, for instance, that anchor bolts, nuts and washers are estimated and included for the air-supported structures and pre-engineered buildings to be bolted to their foundations.

Utility supply is a common area where essential items or pieces of equipment can be missed or overlooked due to the fact that each subcontractor may feel it is the others' responsibility. The estimator should also be aware of certain items which may be supplied as part of a package but installed by others, and ensure that the installing contractor's estimate includes the cost of installation. Conversely, the estimator must also ensure that items are not costed by two different subcontractors, resulting in an inflated overall estimate.

13120 Pre-Engineered Structures

- The foundations and floor slab, as well as rough mechanical and electrical, should be estimated, as this work is required for the assembly and erection of the structure. Generally, as noted in the book, the pre-engineered building comes as a shell and additional features, such as windows and doors, must be included by the estimator. Here again, the estimator must have a clear understanding of the scope of each portion of the work and all the necessary interfaces.

13200 Storage Tanks

- The prices in this subdivision for above and below ground storage tanks do not include foundations or hold-down slabs. The estimator should refer to Divisions 2 and 3 for foundation system pricing. In addition to the foundations, required tank accessories such as tank gauges, leak detection devices, and additional manholes and piping must be added to the tank prices.

Reference Numbers

Reference numbers are shown in bold squares at the beginning of some major classifications. These numbers refer to related items in the Reference Section. The reference information may be an estimating procedure, an alternate pricing method or technical information.

Note: Not all subdivisions listed here necessarily appear in this publication.

13005	Selective demolition	CREW	DAILY OUTPUT	LABOR-HOURS	UNIT	2005 BARE COSTS				TOTAL INCL O&P
						MAT.	LABOR	EQUIP.	TOTAL	
011	0010	**SELECTIVE DEMOLITION, AIR SUPPORTED STRUCTURES**									011
	0020	Tank covers, scrim, dbl layer, vinyl poly w/ hdw, blower & controls									
	0050	Round and rectangular R02220-510	B-2	9,000	.004	S.F.		.12		.12	.19
	0100	Warehouse structures									
	0120	Whse, poly/vinyl fabric, 28 oz., w/tension cables & inflation sys	4 Clab	9,000	.004	SF Flr.		.10		.10	.15
	0150	Warehouse, reinforced vinyl, 12 oz., 3000 S.F.	"	5,000	.006			.17		.17	.27
	0200	12,000 to 24,000 S.F.	8 Clab	20,000	.003			.09		.09	.13
	0250	Warehouse, tedlar vinyl fabric, 28 oz. w/liner, to 3000 S.F.	4 Clab	5,000	.006			.17		.17	.27
	0300	12,000 to 24,000 S.F.	8 Clab	20,000	.003	▼		.09		.09	.13
	0350	Greenhouse/shelter, woven polyethylene with liner									
	0400	3000 S.F.	4 Clab	5,000	.006	SF Flr.		.17		.17	.27
	0450	12,000 to 24,000 S.F.	8 Clab	20,000	.003			.09		.09	.13
	0500	Tennis/gymnasium, poly/vinyl fabric, 28 oz., incl thermal liner	4 Clab	9,000	.004			.10		.10	.15
	0600	Stadium/Convention Ctr, teflon coated fibergalss, incl thermal liner	9 Clab	40,000	.002	▼		.05		.05	.07
	0700	Doors, air lock, 15' long, 10' x 10'	2 Carp	1.50	10.667	Ea.		365		365	570
	0720	15' x 15'		.80	20			685		685	1,075
	0750	Revolving personnel door, 6' dia. x 6'-6" high	▼	1.50	10.667	▼		365		365	570
035	0010	**SELECTIVE DEMOLITION, SPECIAL PURPOSE ROOMS** R02220-510									035
	0100	Audiometric rooms, under 500 S.F. surface	4 Carp	200	.160	SF Surf		5.50		5.50	8.55
	0110	Over 500 S.F. surface	"	240	.133	"		4.57		4.57	7.10
	0200	Clean rooms, 12' x 12' soft wall, class 100	1 Carp	.30	26.667	Ea.		915		915	1,425
	0210	Class 1000		.30	26.667			915		915	1,425
	0220	Class 10,000		.35	22.857			785		785	1,225
	0230	Class 100,000	▼	.35	22.857	▼		785		785	1,225
	0300	Darkrooms, shell complete, 8' high	2 Carp	220	.073	SF Flr.		2.49		2.49	3.88
	0310	12' high		110	.145	"		4.98		4.98	7.75
	0350	Darkrooms doors, mini-cylindrical, revolving		4	4	Ea.		137		137	213
	0400	Music room, practice modular	▼	140	.114	SF Surf		3.91		3.91	6.10
	0500	Refrigeration structures and finishes									
	0510	Wall finish, 2 coat portland cement plaster, 1/2" thick	1 Clab	200	.040	S.F.		1.07		1.07	1.66
	0520	Fiberglass panels, 1/8" thick		400	.020			.53		.53	.83
	0530	Ceiling finish, polystyrene plastic, 1" to 2" thick		500	.016			.43		.43	.66
	0540	4" thick		450	.018			.47		.47	.74
	0550	Refrigerator, prefab aluminum walk-in, 7'-6" high, 6' x 6' OD	2 Carp	100	.160	SF Flr.		5.50		5.50	8.55
	0560	10' x 10' OD		160	.100			3.43		3.43	5.35
	0570	Over 150 S.F.		200	.080			2.74		2.74	4.27
	0610	To 40 S.F.		140	.114			3.91		3.91	6.10
	0620	To 60 S.F.		175	.091			3.13		3.13	4.88
	0630	To 100 S.F.		220	.073			2.49		2.49	3.88
	0640	To 130 S.F.	▼	250	.064	▼		2.19		2.19	3.41
128	0010	**PRE-ENGD STEEL BLDGS DEMOLITION**									128
	0500	Pre-engd steel bldgs, rigid frame, clear span & multi post, excl salvage									
	0550	3500 to 7500 S.F	L-11	1,350	.024	SF Flr.		.73	1.05	1.78	2.33
	0600	7501 to 12,500 S.F		2,000	.016			.49	.71	1.20	1.57
	0650	12,500 S.F or greater	▼	2,200	.015	▼		.45	.64	1.09	1.43
	0700	Pre-engd steel building components									
	0710	Entrance canopy, including frame 4' x 4'	E-24	8	4	Ea.		149	77	226	340
	0720	4' x 8'	"	7	4.571			170	88	258	390
	0730	H.M doors, self framing, single leaf	2 Skwk	8	2			69.50		69.50	108
	0740	double leaf		5	3.200	▼		112		112	173
	0760	Gutter, eave type		600	.027	L.F.		.93		.93	1.45
	0770	Sash, single slide, double slide or fixed		24	.667	Ea.		23		23	36
	0780	Skylight, fiberglass, to 30 S.F.		16	1			35		35	54
	0785	Roof vents, circular, 12" to 24" diameter	▼	12	1.333	▼		46.50		46.50	72.50

T3 SPECIAL CONSTRUCTION

			DAILY	LABOR-			2005 BARE COSTS				TOTAL	
		13005	**Selective demolition**		CREW	OUTPUT	HOURS	UNIT	MAT.	LABOR	EQUIP.	TOTAL	INCL O&P	
128	0790	continuous, 10' long	2 Skwk	8	2	Ea.		69.50		69.50	108	**128**
	0900	Shelters, aluminum frame										
	0910	Aluminum frame, acrylic glazing, 3' x 9' x 8' high	2 Skwk	2	8	Ea.		279		279	435	
	0920	9' x 12' x 8' high	"	1.50	10.667	"		370		370	580	
201	0010	**STORAGE TANK DEMOLITION**										**201**
	0500	Steel tank, single wall, aboveground, not incl fdn, pumps or piping										
	0510	Single wall, 275 gallon R02220 -510	Q-1	3	5.333	Ea.		196		196	295	
	0520	550 thru 2000 gallon	B-34P	2	12			410	196	606	835	
	0530	5000 thru 10000 gallon	B-34Q	2	12			415	515	930	1,200	
	0540	15000 thru 30000 gallon	B-34N	2	4			110	223	333	415	
	0600	Steel tank, double wall, above ground not incl fdn, pumps & piping										
	0620	500 thru 2000 gallon	B-34P	2	12	Ea.		410	196	606	835	

			DAILY	LABOR-			2005 BARE COSTS				TOTAL	
		13011	**Air Supported Structures**		CREW	OUTPUT	HOURS	UNIT	MAT.	LABOR	EQUIP.	TOTAL	INCL O&P	
100	0010	**AIR SUPPORTED TANK COVERS**, vinyl polyester										**100**
	0100	scrim, double layer, with hardware, blower, standby & controls										
	0200	Round, 75' diameter	B-2	4,500	.009	S.F.	6.50	.24		6.74	7.55	
	0300	100' diameter		5,000	.008		6	.22		6.22	6.95	
	0400	150' diameter		5,000	.008		4.51	.22		4.73	5.30	
	0500	Rectangular, 20' x 20'		4,500	.009		14.65	.24		14.89	16.55	
	0600	30' x 40'		4,500	.009		14.65	.24		14.89	16.55	
	0700	50' x 60'		4,500	.009		14.65	.24		14.89	16.55	
	0800	For single wall construction, deduct, minimum					.42			.42	.46	
	0900	Maximum					1.39			1.39	1.53	
	1000	For maximum resistance to atmosphere or cold, add					.58			.58	.64	
	1100	For average shipping charges, add				Total	1,150			1,150	1,275	
200	0010	**AIR SUPPORTED STRUCTURES** R13011 -110	B-11B	1,000	.016	SF Flr.	.77	.48	.18	1.43	1.78	**200**
	0020	Site preparation, incl. anchor placement and utilities										
	0030	For concrete curb, see division 03310-240										
	0050	Warehouse, polyester/vinyl fabric, 28 oz., over 10 yr. life, welded										
	0060	Seams, tension cables, primary & auxiliary inflation system,										
	0070	airlock, personnel doors and liner										
	0100	5000 S.F.	4 Clab	5,000	.006	SF Flr.	16.25	.17		16.42	18.15	
	0250	12,000 S.F.	"	6,000	.005		11.75	.14		11.89	13.15	
	0400	24,000 S.F.	8 Clab	12,000	.005		8.60	.14		8.74	9.70	
	0500	50,000 S.F.	"	12,500	.005		7.45	.14		7.59	8.40	
	0700	12 oz. reinforced vinyl fabric, 5 yr. life, sewn seams,										
	0710	accordian door, including liner										
	0750	3000 S.F.	4 Clab	3,000	.011	SF Flr.	8.50	.28		8.78	9.80	
	0800	12,000 S.F.	"	6,000	.005		6.30	.14		6.44	7.10	
	0850	24,000 S.F.	8 Clab	12,000	.005		6.25	.14		6.39	7.05	
	0950	Deduct for single layer					.68			.68	.75	
	1000	Add for welded seams					.98			.98	1.08	
	1050	Add for double layer, welded seams included					2.02			2.02	2.22	
	1250	Tedlar/vinyl fabric, 28 oz., with liner, over 10 yr. life,										
	1260	incl. overhead and personnel doors										
	1300	3000 S.F.	4 Clab	3,000	.011	SF Flr.	15.80	.28		16.08	17.80	
	1450	12,000 S.F.	"	6,000	.005		11.10	.14		11.24	12.40	

SPECIAL CONSTRUCTION 13

		13011	Air Supported Structures	CREW	DAILY OUTPUT	LABOR-HOURS	UNIT	2005 BARE COSTS				TOTAL INCL O&P	
								MAT.	LABOR	EQUIP.	TOTAL		
200	1550		24,000 S.F.	8 Clab	12,000	.005	SF Flr.	8.60	.14		8.74	9.70	200
	1700		Deduct for single layer				↓	1.40			1.40	1.54	
	2250		Greenhouse/shelter, woven polyethylene with liner, 2 yr. life,										
	2260		sewn seams, including doors										
	2300		3000 S.F.	4 Clab	3,000	.011	SF Flr.	6.40	.28		6.68	7.50	
	2350		12,000 S.F.	"	6,000	.005		6.50	.14		6.64	7.35	
	2450		24,000 S.F.	8 Clab	12,000	.005	SF Flr.	5.45	.14		5.59	6.20	
	2550		Deduct for single layer				↓	.61			.61	.67	
	2600		Tennis/gymnasium, polyester/vinyl fabric, 28 oz., over 10 yr. life,										
	2610		including thermal liner, heat and lights										
	2650		7200 S.F.	4 Clab	6,000	.005	SF Flr.	15.20	.14		15.34	16.90	
	2750		13,000 S.F.	"	6,500	.005		11.40	.13		11.53	12.75	
	2850		Over 24,000 S.F.	8 Clab	12,000	.005		10.30	.14		10.44	11.55	
	2860		For low temperature conditions, add				↓	.69			.69	.76	
	2870		For average shipping charges, add				Total	3,025			3,025	3,325	
	2900		Thermal liner, translucent reinforced vinyl				SF Flr.	.95			.95	1.05	
	2950		Metalized mylar fabric and mesh, double liner				"	1.51			1.51	1.66	
	3050		Stadium/convention center, teflon coated fiberglass, heavy weight,										
	3060		over 20 yr. life, incl. thermal liner and heating system										
	3100		Minimum	9 Clab	26,000	.003	SF Flr.	39	.07		39.07	42.50	
	3110		Maximum	"	19,000	.004	"	45.50	.10		45.60	50	
	3400		Doors, air lock, 15' long, 10' x 10'	2 Carp	.80	20	Ea.	14,200	685		14,885	16,800	
	3600		15' x 15'	"	.50	32		20,600	1,100		21,700	24,300	
	3700		For each added 5' length, add					2,950			2,950	3,250	
	3900		Revolving personnel door, 6' diameter, 6'-6" high	2 Carp	.80	20	↓	10,400	685		11,085	12,600	
	4200		Double wall, self supporting, shell only, minimum				SF Flr.					21	
	4300		Maximum				"					39	

(R13011 -110 reference box beside line 1550)

		13035	Special Purpose Rooms	CREW	DAILY OUTPUT	LABOR-HOURS	UNIT	2005 BARE COSTS				TOTAL INCL O&P	
								MAT.	LABOR	EQUIP.	TOTAL		
150	0010		**ANECHOIC CHAMBERS** Standard units, 7' ceiling heights										150
	0100		Area for pricing is net inside dimensions										
	0300		200 cycles per second cutoff, 25 S.F. floor area				SF Flr.					1,500	
	0400		50 S.F.									1,100	
	0600		75 S.F.									1,050	
	0700		100 S.F.									1,125	
	0900		For 150 cycles per second cutoff, add				↓	30%	30%				
	1000		For 100 cycles per second cutoff, add					45%	45%				
180	0010		**AUDIOMETRIC ROOMS** Under 500 S.F. surface	4 Carp	98	.327	SF Surf	44	11.20		55.20	66	180
	0100		Over 500 S.F. surface	"	120	.267	"	42	9.15		51.15	60.50	
200	0010		**CLEAN ROOMS**										200
	1100		Clean room, soft wall, 12' x 12', Class 100	1 Carp	.18	44.444	Ea.	12,600	1,525		14,125	16,300	
	1110		Class 1,000		.18	44.444		9,775	1,525		11,300	13,200	
	1120		Class 10,000		.21	38.095		8,325	1,300		9,625	11,200	
	1130		Class 100,000	↓	.21	38.095	↓	7,725	1,300		9,025	10,500	
	2800		Ceiling grid support, slotted channel struts 4'-0" O.C., ea. way				S.F.					6.50	
	3000		Ceiling panel, vinyl coated foil on mineral substrate										
	3020		Sealed, non-perforated				S.F.					1.40	

13 SPECIAL CONSTRUCTION

13035	Special Purpose Rooms	CREW	DAILY OUTPUT	LABOR-HOURS	UNIT	2005 BARE COSTS MAT.	LABOR	EQUIP.	TOTAL	TOTAL INCL O&P	
200											**200**
4000	Ceiling panel seal, silicone sealant, 150 L.F./gal.	1 Carp	150	.053	L.F.	.22	1.83		2.05	3.09	
4100	Two sided adhesive tape	"	240	.033	"	.10	1.14		1.24	1.89	
4200	Clips, one per panel				Ea.	.88			.88	.97	
6000	HEPA filter,2'x4',99.97% eff.,3" dp beveled frame (silicone seal)					269			269	296	
6040	6" deep skirted frame (channel seal)					310			310	340	
6100	99.99% efficient, 3" deep beveled frame (silicone seal)					290			290	320	
6140	6" deep skirted frame (channel seal)					330			330	365	
6200	99.999% efficient, 3" deep beveled frame (silicone seal)					330			330	365	
6240	6" deep skirted frame (channel seal)				▼	375			375	410	
7000	Wall panel systems, including channel strut framing										
7020	Polyester coated aluminum, particle board				S.F.					20	
7100	Porcelain coated aluminum, particle board									35	
7400	Wall panel support, slotted channel struts, to 12' high				▼					18	
300	0010	**DARKROOMS** Shell, complete except for door, 64 S.F., 8' high	2 Carp	128	.125	SF Flr.	71.50	4.28		75.78	85	**300**
0100	12' high		64	.250		84	8.55		92.55	106	
0500	120 S.F. floor, 8' high		120	.133		52.50	4.57		57.07	65	
0600	12' high		60	.267		63.50	9.15		72.65	84	
0800	240 S.F. floor, 8' high		120	.133		36.50	4.57		41.07	47	
0900	12' high		60	.267	▼	50.50	9.15		59.65	70	
1200	Mini-cylindrical, revolving, unlined, 4' diameter		3.50	4.571	Ea.	3,500	157		3,657	4,100	
1400	5'-6" diameter	▼	2.50	6.400		5,500	219		5,719	6,400	
1600	Add for lead lining, inner cylinder, 1/32" thick					1,400			1,400	1,550	
1700	1/16" thick					1,650			1,650	1,800	
1800	Add for lead lining, inner and outer cylinder, 1/32" thick					2,600			2,600	2,875	
1900	1/16" thick				▼	2,725			2,725	3,000	
2000	For darkroom door, see division 11472-370										
500	0010	**MUSIC** Practice room, modular, perforated steel, under 500 S.F.	2 Carp	70	.229	SF Surf	27	7.85		34.85	41.50	**500**
0100	Over 500 S.F.	"	80	.200	"	23	6.85		29.85	35.50	
700	0010	**REFRIGERATION** Curbs, 12" high, 4" thick, concrete	2 Carp	58	.276	L.F.	3.20	9.45		12.65	18.20	**700**
1000	Doors, see division 08341-200										
2400	Finishes, 2 coat portland cement plaster, 1/2" thick	1 Plas	48	.167	S.F.	.88	5.25		6.13	8.90	
2500	For galvanized reinforcing mesh, add	1 Lath	335	.024		.59	.75		1.34	1.76	
2700	3/16" thick latex cement	1 Plas	88	.091		1.51	2.85		4.36	6	
2900	For glass cloth reinforced ceilings, add	"	450	.018		.36	.56		.92	1.25	
3100	Fiberglass panels, 1/8" thick	1 Carp	149.45	.054		1.99	1.83		3.82	5.05	
3200	Polystyrene, plastic finish ceiling, 1" thick		274	.029		1.84	1		2.84	3.58	
3400	2" thick		274	.029		2.09	1		3.09	3.86	
3500	4" thick	▼	219	.037		2.33	1.25		3.58	4.51	
3800	Floors, concrete, 4" thick	1 Cefi	93	.086		.87	2.83		3.70	5.10	
3900	6" thick	"	85	.094	▼	1.31	3.09		4.40	6	
4000	Insulation, 1" to 6" thick, cork				B.F.	.85			.85	.94	
4100	Urethane					.84			.84	.92	
4300	Polystyrene, regular					.57			.57	.63	
4400	Bead board				▼	.43			.43	.47	
4600	Installation of above, add per layer	2 Carp	657.60	.024	S.F.	.27	.83		1.10	1.60	
4700	Wall and ceiling juncture		298.90	.054	L.F.	1.36	1.83		3.19	4.36	
4900	Partitions, galvanized sandwich panels, 4" thick, stock		219.20	.073	S.F.	5.75	2.50		8.25	10.25	
5000	Aluminum or fiberglass	▼	219.20	.073	"	6.30	2.50		8.80	10.80	
5200	Prefab walk-in, 7'-6" high, aluminum, incl. door & floors,										
5210	not incl. partitions or refrigeration, 6' x 6' O.D. nominal	2 Carp	54.80	.292	SF Flr.	103	10		113	129	
5500	10' x 10' O.D. nominal		82.20	.195		82.50	6.65		89.15	101	
5700	12' x 14' O.D. nominal		109.60	.146		74	5		79	89.50	
5800	12' x 20' O.D. nominal	▼	109.60	.146	▼	64.50	5		69.50	79	
6100	For 8'-6" high, add					5%					

SPECIAL CONSTRUCTION 13

		13035	**Special Purpose Rooms**	CREW	DAILY OUTPUT	LABOR-HOURS	UNIT	2005 BARE COSTS				TOTAL INCL O&P	
								MAT.	LABOR	EQUIP.	TOTAL		
700	6300		Rule of thumb for complete units, w/o doors & refrigeration, cooler	2 Carp	146	.110	SF Flr.	93.50	3.75		97.25	109	**700**
	6400		Freezer		109.60	.146	▼	110	5		115	129	
	6600		Shelving, plated or galvanized, steel wire type		360	.044	SF Hor.	8.15	1.52		9.67	11.35	
	6700		Slat shelf type		375	.043		10.05	1.46		11.51	13.40	
	6900		For stainless steel shelving, add				▼	300%					
	7000		Vapor barrier, on wood walls	2 Carp	1,644	.010	S.F.	.12	.33		.45	.65	
	7200		On masonry walls	"	1,315	.012	"	.31	.42		.73	.99	
	7500		For air curtain doors, see division 15840-200										
800	0010		**SAUNA** Prefabricated, incl. heater & controls, 7' high, 6' x 4', C/C	L-7	2.20	12.727	Ea.	3,525	420		3,945	4,525	**800**
	0050		6' x 4', C/P		2	14		3,275	460		3,735	4,325	
	0400		6' x 5', C/C		2	14		3,950	460		4,410	5,075	
	0450		6' x 5', C/P		2	14		3,700	460		4,160	4,800	
	0600		6' x 6', C/C		1.80	15.556		4,200	515		4,715	5,425	
	0650		6' x 6', C/P		1.80	15.556		3,925	515		4,440	5,125	
	0800		6' x 9', C/C		1.60	17.500		5,275	580		5,855	6,700	
	0850		6' x 9', C/P		1.60	17.500		5,000	580		5,580	6,400	
	1000		8' x 12', C/C		1.10	25.455		8,175	840		9,015	10,300	
	1050		8' x 12', C/P		1.10	25.455		7,500	840		8,340	9,550	
	1200		8' x 8', C/C		1.40	20		6,225	660		6,885	7,875	
	1250		8' x 8', C/P		1.40	20		5,850	660		6,510	7,475	
	1400		8' x 10', C/C		1.20	23.333		6,925	770		7,695	8,800	
	1450		8' x 10', C/P		1.20	23.333		6,425	770		7,195	8,275	
	1600		10' x 12', C/C		1	28		8,650	925		9,575	11,000	
	1650		10' x 12', C/P	▼	1	28		7,825	925		8,750	10,100	
	1700		Door only, cedar, 2'x6', with tempered insulated glass window	2 Carp	3.40	4.706		465	161		626	760	
	1800		Prehung, incl. jambs, pulls & hardware	"	12	1.333		475	45.50		520.50	595	
	2500		Heaters only (incl. above), wall mounted, to 200 C.F.					460			460	505	
	2750		To 300 C.F.					555			555	615	
	3000		Floor standing, to 720 C.F., 10,000 watts, w/controls	1 Elec	3	2.667		1,450	109		1,559	1,725	
	3250		To 1,000 C.F., 16,000 watts	"	3	2.667	▼	1,500	109		1,609	1,800	
900	0010		**SPORT COURT** Floors, No. 2 & better maple, 25/32" thick				SF Flr.					6.65	**900**
	0100		Walls, laminated plastic bonded to galv. steel studs				SF Wall					8.45	
	0150		Laminated fiberglass surfacing, minimum				▼					2.10	
	0180		Maximum									2.37	
	0300		Squash, regulation court in existing building, minimum				Court					16,000	
	0400		Maximum				"					29,900	
	0450		Rule of thumb for components:										
	0470		Walls	3 Carp	.15	160	Court	11,100	5,475		16,575	20,700	
	0500		Floor	"	.25	96		4,975	3,300		8,275	10,600	
	0550		Lighting	2 Elec	.60	26.667		1,575	1,075		2,650	3,350	
	0600		Handball, racquetball court in existing building, minimum	C-1	.20	160		1,925	5,175		7,100	10,200	
	0800		Maximum	"	.10	320		25,800	10,400		36,200	44,500	
	0900		Rule of thumb for components: walls	3 Carp	.12	200		10,200	6,850		17,050	21,900	
	1000		Floor		.25	96		5,300	3,300		8,600	11,000	
	1100		Ceiling	▼	.33	72.727		2,200	2,500		4,700	6,300	
	1200		Lighting	2 Elec	.60	26.667	▼	2,725	1,075		3,800	4,625	
940	0010		**STEAM BATH** Heater, timer & head, single, to 140 C.F.	1 Plum	1.20	6.667	Ea.	975	272		1,247	1,475	**940**
	0500		To 300 C.F.	"	1.10	7.273		1,100	297		1,397	1,650	
	1000		Commercial size, with blow-down assembly, to 800 C.F.	1 Plum	.90	8.889		3,800	365		4,165	4,725	
	1500		To 2500 C.F.	"	.80	10		5,975	410		6,385	7,200	
	2000		Multiple, motels, apts., 2 baths, w/ blow-down assm., 500 C.F.	Q-1	1.30	12.308		3,750	455		4,205	4,800	
	2500		4 baths	"	.70	22.857		4,075	840		4,915	5,775	
	2700		Conversion unit for residential tub, including door				▼	3,075			3,075	3,375	

13080 | Sound, Vibration & Seismic Control

	13081	**Sound Control**	CREW	DAILY OUTPUT	LABOR-HOURS	UNIT	2005 BARE COSTS				TOTAL INCL O&P	
							MAT.	LABOR	EQUIP.	TOTAL		
100	0010	**ACOUSTICAL** Enclosure, 4" thick wall and ceiling panels										100
	0020	8# per S.F., up to 12' span	3 Carp	72	.333	SF Surf	27.50	11.40		38.90	48.50	
	0300	Better quality panels, 10.5# per S.F.		64	.375		31	12.85		43.85	54.50	
	0400	Reverb-chamber, 4" thick, parallel walls		60	.400		39	13.70		52.70	64.50	
	0600	Skewed wall, parallel roof, 4" thick panels		55	.436		44.50	14.95		59.45	72.50	
	0700	Skewed walls, skewed roof, 4" layers, 4" air space		48	.500		50	17.15		67.15	81.50	
	0900	Sound-absorbing panels, pntd mtl, 2'-6" x 8', under 1,000 S.F.		215	.112		10.35	3.82		14.17	17.30	
	1100	Over 1000 S.F.		240	.100		9.95	3.43		13.38	16.30	
	1200	Fabric faced	▼	240	.100		8.05	3.43		11.48	14.25	
	1500	Flexible transparent curtain, clear	3 Shee	215	.112		6.25	4.49		10.74	13.75	
	1600	50% foam		215	.112		8.70	4.49		13.19	16.45	
	1700	75% foam		215	.112		8.70	4.49		13.19	16.45	
	1800	100% foam	▼	215	.112	▼	8.70	4.49		13.19	16.45	
	3100	Audio masking system, including speakers, amplification										
	3110	and signal generator										
	3200	Ceiling mounted, 5,000 S.F.	2 Elec	2,400	.007	S.F.	1.16	.27		1.43	1.68	
	3300	10,000 S.F.		2,800	.006		1.01	.23		1.24	1.46	
	3400	Plenum mounted, 5,000 S.F.		3,800	.004		.90	.17		1.07	1.25	
	3500	10,000 S.F.	▼	4,400	.004	▼	.61	.15		.76	.89	

13090 | Radiation Protection

	13091	**X-Ray/Radio Freq Protection**	CREW	DAILY OUTPUT	LABOR-HOURS	UNIT	2005 BARE COSTS				TOTAL INCL O&P	
							MAT.	LABOR	EQUIP.	TOTAL		
600	0010	**SHIELDING LEAD**										600
	0100	Laminated lead in wood doors, 1/16" thick, no hardware				S.F.	27			27	29.50	
	0200	Lead lined door frame, not incl. hardware,										
	0210	1/16" thick lead, butt prepared for hardware	1 Lath	2.40	3.333	Ea.	345	105		450	530	
	0300	Lead sheets, 1/16" thick	2 Lath	135	.119	S.F.	5.20	3.73		8.93	11.20	
	0400	1/8" thick		120	.133		10.60	4.19		14.79	17.85	
	0500	Lead shielding, 1/4" thick		135	.119		18.55	3.73		22.28	26	
	0550	1/2" thick	▼	120	.133	▼	36.50	4.19		40.69	46.50	
	0600	Lead glass, 1/4" thick, 2.0 mm LE, 12" x 16"	2 Glaz	13	1.231	Ea.	262	40.50		302.50	350	
	0700	24" x 36"		8	2		1,025	66		1,091	1,225	
	0800	36" x 60"		2	8		3,000	264		3,264	3,700	
	0850	Window frame with 1/16" lead and voice passage, 36" x 60"		2	8		1,050	264		1,314	1,550	
	0870	24" x 36" frame		8	2	▼	600	66		666	760	
	0900	Lead gypsum board, 5/8" thick with 1/16" lead	▼	160	.100	S.F.	5.10	3.30		8.40	10.60	

SPECIAL CONSTRUCTION 13

		13091	X-Ray/Radio Freq Protection	CREW	DAILY OUTPUT	LABOR-HOURS	UNIT	2005 BARE COSTS				TOTAL INCL O&P	
								MAT.	LABOR	EQUIP.	TOTAL		
600	0910		1/8" lead **CN**	2 Glaz	140	.114	S.F.	10.20	3.77		13.97	16.95	600
	0930		1/32" lead	2 Lath	200	.080	↓	3.40	2.52		5.92	7.45	
	0950		Lead headed nails (average 1 lb. per sheet)				Lb.	6			6	6.60	
	1000		Butt joints in 1/8" lead or thicker, 2" batten strip x 7' long	2 Lath	240	.067	Ea.	14.55	2.10		16.65	19.10	
	1200		X-ray protection, average radiography or fluoroscopy										
	1210		room, up to 300 S.F. floor, 1/16" lead, minimum	2 Lath	.25	64	Total	5,725	2,025		7,750	9,275	
	1500		Maximum, 7'-0" walls	"	.15	106	"	7,150	3,350		10,500	12,900	
	1600		Deep therapy X-ray room, 250 KV capacity,										
	1800		up to 300 S.F. floor, 1/4" lead, minimum	2 Lath	.08	200	Total	17,500	6,300		23,800	28,600	
	1900		Maximum, 7'-0" walls	"	.06	266	"	22,700	8,375		31,075	37,400	
	2000		X-ray viewing panels, clear lead plastic										
	2010		7 mm thick, 0.3 mm LE, 2.3 lbs/S.F.	H-3	139	.115	S.F.	142	3.37		145.37	161	
	2020		12 mm thick, 0.5 mm LE, 3.9 lbs/S.F.		82	.195		192	5.70		197.70	220	
	2030		18 mm thick, 0.8mm LE, 5.9 lbs/S.F.		54	.296		208	8.70		216.70	242	
	2040		22 mm thick, 1.0 mm LE, 7.2 lbs/S.F.		44	.364		213	10.65		223.65	250	
	2050		35 mm thick, 1.5 mm LE, 11.5 lbs/S.F.		28	.571		238	16.75		254.75	288	
	2060		46 mm thick, 2.0 mm LE, 15.0 lbs/S.F.	↓	21	.762	↓	315	22.50		337.50	380	
	2090		For panels 12 S.F. to 48 S.F., add crating charge				Ea.					50	
	4000		X-ray barriers, modular, panels mounted within framework for										
	4002		attaching to floor, wall or ceiling, upper portion is clear lead										
	4005		plastic window panels 48"H, lower portion is opaque leaded										
	4008		steel panels 36"H, structural supports not incl.										
	4010		1-section barrier, 36"W x 84"H overall										
	4020		0.5 mm LE panels	H-3	6.40	2.500	Ea.	3,200	73		3,273	3,625	
	4030		0.8 mm LE panels		6.40	2.500		3,400	73		3,473	3,825	
	4040		1.0 mm LE panels		5.33	3.002		3,525	88		3,613	4,000	
	4050		1.5 mm LE panels	↓	5.33	3.002	↓	3,750	88		3,838	4,250	
	4060		2-section barrier, 72"W x 84"H overall										
	4070		0.5 mm LE panels	H-3	4	4	Ea.	6,625	117		6,742	7,450	
	4080		0.8 mm LE panels		4	4		7,000	117		7,117	7,875	
	4090		1.0 mm LE panels		3.56	4.494		7,150	132		7,282	8,050	
	5000		1.5 mm LE panels	↓	3.20	5	↓	7,725	146		7,871	8,725	
	5010		3-section barrier, 108"W x 84"H overall										
	5020		0.5 mm LE panels	H-3	3.20	5	Ea.	9,475	146		9,621	10,600	
	5030		0.8 mm LE panels		3.20	5		9,925	146		10,071	11,100	
	5040		1.0 mm LE panels		2.67	5.993		10,200	175		10,375	11,500	
	5050		1.5 mm LE panels	↓	2.46	6.504	↓	11,000	190		11,190	12,400	
	7000		X-ray barriers, mobile, mounted within framework w/casters on										
	7005		bottom, clear lead plastic window panels on upper portion,										
	7010		opaque on lower, 30"W x 75"H overall, incl. framework										
	7020		24"H upper w/0.5 mm LE, 48"H lower w/0.8 mm LE	1 Carp	16	.500	Ea.	2,175	17.15		2,192.15	2,425	
	7030		48"W x 75"H overall, incl. framework										
	7040		36"H upper w/0.5 mm LE, 36"H lower w/0.8 mm LE	1 Carp	16	.500	Ea.	3,850	17.15		3,867.15	4,250	
	7050		36"H upper w/1.0 mm LE, 36"H lower w/1.5 mm LE	"	16	.500	"	4,650	17.15		4,667.15	5,125	
	7060		72"W x 75"H overall, incl. framework										
	7070		36"H upper w/0.5 mm LE, 36"H lower w/0.8 mm LE	1 Carp	16	.500	Ea.	4,550	17.15		4,567.15	5,025	
	7080		36"H upper w/1.0 mm LE, 36"H lower w/1.5 mm LE	"	16	.500	"	5,750	17.15		5,767.15	6,350	
700	0010		**SHIELDING, RADIO FREQUENCY**										700
	0020		Prefabricated or screen-type copper or steel, minimum	2 Carp	180	.089	SF Surf	24	3.04		27.04	31	
	0100		Average		155	.103		26	3.54		29.54	34	
	0150		Maximum	↓	145	.110	↓	31	3.78		34.78	40	

T3 SPECIAL CONSTRUCTION

Important: See the Reference Section for critical supporting data - Reference Nos., Crews, & City Cost Indexes

		CREW	DAILY OUTPUT	LABOR HOURS	UNIT	2005 BARE COSTS MAT.	LABOR	EQUIP.	TOTAL	TOTAL INCL O&P	
13128	**Pre-Engineered Structures**										
060 0010	**PORTABLE BOOTHS** Prefab aluminum with doors, windows, ext. roof										060
0100	lights wiring & insulation, 15 S.F. building, O.D., painted, minimum				S.F.	268			268	295	
0300	30 S.F. building, minimum					184			184	202	
0400	50 S.F. building, minimum					137			137	150	
0600	80 S.F. building, minimum					110			110	121	
0700	100 S.F. building, minimum				↓	102			102	112	
0900	Acoustical booth, 27 Db @ 1,000 Hz, 15 S.F. floor				Ea.	3,025			3,025	3,325	
1000	7' x 7'-6", including light & ventilation					6,200			6,200	6,825	
1200	Ticket booth, galv. steel, not incl. foundations., 4' x 4'					5,125			5,125	5,650	
1300	4' x 6'				↓	6,000			6,000	6,600	
070 0010	**CONTROL TOWERS**, 12' x 10', incl. instruments, min.				Ea.					510,000	070
0100	Maximum									1,000,000	
0500	With standard 40' tower, average				↓					710,500	
1000	Temporary portable control towers, 8' x 12',										
1010	complete with one position communications, minimum				Ea.					275,000	
2000	For fixed facilities, depending on height, minimum									50,000	
2010	Maximum				↓					100,000	
160 0010	**TENSION STRUCTURES** Rigid steel/alum. frame, vyl. coated polyester										160
0100	fabric shell, 60' clear span, not incl. foundations or floors										
0200	6,000 S.F.	B-41	1,000	.044	SF Flr.	10.90	1.22	.22	12.34	14.15	
0300	12,000 S.F.	↓	1,100	.040		9.55	1.11	.20	10.86	12.45	
0400	80' clear span, 20,800 S.F.	↓	1,220	.036		9.75	1	.18	10.93	12.50	
0410	100' clear span, 10,000 S.F.	L-5	2,175	.026		10.85	.98	.28	12.11	13.95	
0430	26,000 S.F.		2,300	.024		9.60	.93	.27	10.80	12.50	
0450	36,000 S.F.		2,500	.022		9.40	.85	.25	10.50	12.10	
0460	120' clear span, 24,000 S.F.		3,000	.019		11.35	.71	.21	12.27	14	
0470	150' clear span, 30,000 S.F.	↓	6,000	.009		11.65	.36	.10	12.11	13.60	
0480	200' clear span, 40,000 S.F.	E-6	8,000	.016	↓	14.15	.60	.19	14.94	16.85	
0500	For roll-up door, 12' x 14', add	L-2	1	16	Ea.	4,275	480		4,755	5,450	
0600	For personnel doors, add, minimum				SF Flr.	5%					
0700	Add, maximum					15%					
0800	For site work, simple foundation, etc., add, minimum								1.25	1.95	
0900	Add, maximum				↓				2.75	3.05	
200 0010	**COMFORT STATIONS** Prefab., stock, w/doors, windows & fixt.										200
0100	Not incl. interior finish or electrical										
0300	Mobile, on steel frame, minimum				S.F.	36.50			36.50	40	
0350	Maximum					57.50			57.50	63.50	
0400	Permanent, including concrete slab, minimum	B-12J	50	.320		142	10	16.75	168.75	190	
0500	Maximum	"	43	.372	↓	206	11.65	19.50	237.15	266	
0600	Alternate pricing method, mobile, minimum				Fixture	1,775			1,775	1,950	
0650	Maximum					2,650			2,650	2,925	
0700	Permanent, minimum	B-12J	.70	22.857		10,200	715	1,200	12,115	13,600	
0750	Maximum	"	.50	32	↓	17,000	1,000	1,675	19,675	22,100	
300 0010	**DOMES**										300
0020	Revolv alum, elec drv for astronomy observation, shell only, stk units										
0600	10'-0" diameter, 800#, dome	2 Carp	.25	64	Ea.	10,200	2,200		12,400	14,600	
0700	Base		.67	23.881		3,525	820		4,345	5,150	
0900	18'-0" diameter, 2,500#, dome		.17	94.118		29,700	3,225		32,925	37,700	
1000	Base		.33	48.485		9,850	1,650		11,500	13,400	
1200	24'-0" diameter, 4,500#, dome		.08	200		55,000	6,850		61,850	71,000	
1300	Base	↓	.25	64	↓	17,700	2,200		19,900	22,900	
1500	Bulk storage, shell only, dual radius hemispher. arch, steel										
1600	framing, corrugated steel covering, 150' diameter	E-2	550	.102	SF Flr.	28	3.76	2.41	34.17	39.50	
1700	400' diameter	"	720	.078		22.50	2.87	1.84	27.21	32	
1800	Wood framing, wood decking, to 400' diameter	F-4	400	.120	↓	20.50	4.06	2.24	26.80	31	

SPECIAL CONSTRUCTION 13

		13128 | Pre-Engineered Structures	CREW	DAILY OUTPUT	LABOR-HOURS	UNIT	2005 BARE COSTS MAT.	LABOR	EQUIP.	TOTAL	TOTAL INCL O&P	
300	1900	Radial framed wood (2" x 6"), 1/2" thick										**300**
	2000	plywood, asphalt shingles, 50' diameter	F-3	2,000	.020	SF Flr.	38.50	.69	.30	39.49	44	
	2100	60' diameter		1,900	.021		29.50	.73	.32	30.55	34	
	2200	72' diameter		1,800	.022		26.50	.77	.33	27.60	30.50	
	2300	116' diameter		1,730	.023		23	.80	.35	24.15	27	
	2400	150' diameter	↓	1,500	.027	↓	20	.92	.40	21.32	24	
340	0010	**GEODESIC DOME** Shell only, interlocking plywood panels R13128-310										**340**
	0400	30' diameter	F-5	1.60	20	Ea.	19,300	695		19,995	22,300	
	0500	34' diameter		1.14	28.070		22,800	975		23,775	26,500	
	0600	39' diameter	↓	1	32		25,600	1,100		26,700	29,900	
	0700	45' diameter	F-3	1.13	35.556		31,500	1,225	535	33,260	37,200	
	0750	55' diameter		1	40		45,200	1,375	605	47,180	52,500	
	0800	60' diameter		1	40		58,500	1,375	605	60,480	67,500	
	0850	65' diameter	↓	.80	50	↓	71,500	1,725	755	73,980	82,500	
	1100	Aluminum panel, with 6" insulation										
	1200	100' diameter				SF Flr.					34	
	1300	500' diameter				"					28.50	
	1600	Aluminum framed, plexiglass closure panels										
	1700	40' diameter				SF Flr.					93.50	
	1800	200' diameter				"					77	
	2100	Aluminum framed, aluminum closure panels										
	2200	40' diameter				SF Flr.					88	
	2300	100' diameter									38.50	
	2400	200' diameter									33	
	2500	For VRP faced bonded fiberglass insulation, add				↓					10	
	2700	Aluminum framed, fiberglass sandwich panel closure										
	2800	6' diameter	2 Carp	150	.107	SF Flr.	33	3.65		36.65	42	
	2900	28' diameter	"	350	.046	"	30	1.57		31.57	35.50	
360	0010	**GARAGE COSTS** R13128-910										**360**
	0020	Public parking, average				Car					13,200	
	0100	See also division 17100-410										
	0300	Residential, prefab shell, stock, wood, single car, minimum	2 Carp	1	16	Total	3,175	550		3,725	4,350	
	0350	Maximum		.67	23.881		7,250	820		8,070	9,250	
	0400	Two car, minimum		.67	23.881		4,800	820		5,620	6,550	
	0450	Maximum	↓	.50	32	↓	9,575	1,100		10,675	12,200	
380	0010	**GARDEN HOUSE** Prefab wood, no floors or foundations										**380**
	0100	32 to 200 S.F., minimum	2 Carp	200	.080	SF Flr.	19.05	2.74		21.79	25.50	
	0300	Maximum	"	48	.333	"	34.50	11.40		45.90	56	
500	0010	**GRANDSTANDS** Permanent, municipal, including foundation										**500**
	0050	Steel understructure w/aluminum closed deck, minimum				Seat					130	
	0100	Maximum									225	
	0300	Steel, minimum									33	
	0400	Maximum									88	
	0600	Aluminum, extruded, stock design, minimum									70	
	0700	Maximum									115	
	0900	Composite, steel, wood and plastic, stock design, minimum									82.50	
	1000	Maximum				↓					154	
510	0010	**BLEACHERS**										**510**
	0020	Bleachers, outdoor, portable, 3 to 5 tiers, to 300' long, min	2 Sswk	120	.133	Seat	33.50	5.10		38.60	46	
	0100	Maximum, less than 15' long, prefabricated		80	.200		44	7.65		51.65	62	
	0200	6 to 20 tiers, minimum, up to 300' long		120	.133		40.50	5.10		45.60	53.50	
	0300	Max., under 15', (highly prefabricated, on wheels)	↓	80	.200	↓	58.50	7.65		66.15	78	
	0500	Permanent grandstands, wood seat, steel frame, 24" row										

13 SPECIAL CONSTRUCTION

Important: See the Reference Section for critical supporting data - Reference Nos., Crews, & City Cost Indexes

13128	Pre-Engineered Structures	CREW	DAILY OUTPUT	LABOR-HOURS	UNIT	2005 BARE COSTS				TOTAL INCL O&P	
						MAT.	LABOR	EQUIP.	TOTAL		
510	0600	3 to 15 tiers, minimum	2 Sswk	60	.267	Seat	100	10.15		110.15	128	**510**
	0700	Maximum		48	.333		110	12.70		122.70	144	
	0900	16 to 30 tiers, minimum		60	.267		116	10.15		126.15	145	
	0950	Average		55	.291		145	11.10		156.10	179	
	1000	Maximum		48	.333		173	12.70		185.70	214	
	1200	Seat backs only, 30" row, fiberglass		160	.100		20.50	3.82		24.32	29.50	
	1300	Steel and wood	▼	160	.100	▼	25.50	3.82		29.32	35	
	1400	NOTE: average seating is 1.5' in width										
540	0010	**GREENHOUSE** Shell only, stock units, not incl. 2' stub walls,										**540**
	0020	foundation, floors, heat or compartments										
	0300	Residential type, free standing, 8'-6" long x 7'-6" wide	2 Carp	59	.271	SF Flr.	39.50	9.30		48.80	58	
	0400	10'-6" wide		85	.188		30.50	6.45		36.95	43.50	
	0600	13'-6" wide		108	.148		27	5.05		32.05	38	
	0700	17'-0" wide		160	.100		30.50	3.43		33.93	39	
	0900	Lean-to type, 3'-10" wide		34	.471		35	16.10		51.10	63.50	
	1000	6'-10" wide	▼	58	.276	▼	27	9.45		36.45	44.50	
	1500	Commercial, custom, truss frame, incl. equip., plumbing, elec.,										
	1510	benches and controls, under 2,000 S.F., minimum				SF Flr.					30	
	1550	Maximum									43	
	1700	Over 5,000 S.F., minimum									26	
	1750	Maximum				▼					30	
	2000	Institutional, custom, rigid frame, including compartments and										
	2010	multi-controls, under 500 S.F., minimum				SF Flr.					77	
	2050	Maximum									112	
	2150	Over 2,000 S.F., minimum									44	
	2200	Maximum									65	
	2400	Concealed rigid frame, under 500 S.F., minimum									90	
	2450	Maximum									110	
	2550	Over 2,000 S.F., minimum									68	
	2600	Maximum									79	
	2800	Lean-to type, under 500 S.F., minimum									80	
	2850	Maximum									120	
	3000	Over 2,000 S.F., minimum				▼					44	
	3050	Maximum									73	
	3600	For 1/4" clear plate glass, add				SF Surf	1.58			1.58	1.74	
	3700	For 1/4" tempered glass, add				"	3.57			3.57	3.93	
	3900	For cooling, add, minimum				SF Flr.	2.39			2.39	2.63	
	4000	Maximum					5.90			5.90	6.50	
	4200	For heaters, 13.6 MBH, add					4.54			4.54	4.99	
	4300	60 MBH, add				▼	1.70			1.70	1.87	
	4500	For benches, 2' x 3'-6", add				SF Hor.	20.50			20.50	22.50	
	4600	3' x 10', add				S.F.	11			11	12.10	
	4800	For controls, add, minimum				Total	2,050			2,050	2,250	
	4900	Maximum				"	12,100			12,100	13,300	
	5100	For humidification equipment, add				M.C.F.	5.20			5.20	5.75	
	5200	For vinyl shading, add				S.F.	1.10			1.10	1.21	
	6000	Geodesic hemisphere, 1/8" plexiglass glazing										
	6050	8' diameter	2 Carp	2	8	Ea.	2,300	274		2,574	2,950	
	6150	24' diameter		.35	45.714		11,800	1,575		13,375	15,500	
	6250	48' diameter	▼	.20	80	▼	32,000	2,750		34,750	39,500	
580	0010	**HANGARS** Prefabricated steel T hangars, Galv. steel roof &										**580**
	0100	walls, incl. electric bi-folding doors, 4 or more units,										
	0110	not including floors or foundations, minimum	E-2	1,275	.044	SF Flr.	10.25	1.62	1.04	12.91	15.25	
	0130	Maximum		1,063	.053		10.65	1.95	1.25	13.85	16.40	
	0900	With bottom rolling doors, minimum		1,386	.040	▼	9.80	1.49	.96	12.25	14.35	
	1000	Maximum	▼	966	.058		10.65	2.14	1.37	14.16	16.90	

SPECIAL CONSTRUCTION 13

		CREW	DAILY OUTPUT	LABOR-HOURS	UNIT	2005 BARE COSTS				TOTAL INCL O&P
13128	**Pre-Engineered Structures**					MAT.	LABOR	EQUIP.	TOTAL	
580 1200	Alternate pricing method:									**580**
1300	Galv. roof and walls, electric bi-folding doors, minimum	E-2	1.06	52.830	Plane	12,700	1,950	1,250	15,900	18,600
1500	Maximum		.91	61.538		14,100	2,275	1,450	17,825	21,000
1600	With bottom rolling doors, minimum		1.25	44.800		10,100	1,650	1,050	12,800	15,100
1800	Maximum	↓	.97	57.732	↓	12,000	2,125	1,375	15,500	18,500
2000	Circular type, prefab., steel frame, plastic skin, electric									
2010	door, including foundations, 80' diameter,									
2020	for up to 5 light planes, minimum	E-2	.50	112	Total	63,000	4,150	2,650	69,800	79,500
2200	Maximum	"	.25	224	"	70,500	8,275	5,300	84,075	97,500
600 0010	**KIOSKS** Round, 5' diameter, 8' high, 1/4" fiberglass wall				Ea.	5,600			5,600	6,175 **600**
0100	1" insulated double wall, fiberglass					6,375			6,375	7,025
0500	Rectangular, 5' x 9', 7'-6" high, 1/4" fiberglass wall					8,150			8,150	8,975
0600	1" insulated double wall, fiberglass				↓	9,700			9,700	10,700
700 0010	**PRE-ENGINEERED STEEL BUILDINGS**									**700**
0100	Clear span rigid frame, 26 ga. colored roofing and siding	R13128 -210								
0150	20 wide, 10' eave height	E-2	425	.132	SF Flr.	7.60	4.87	3.12	15.59	20
0160	14' eave height		350	.160		8.05	5.90	3.79	17.74	23
0170	16' eave height		320	.175		8.45	6.45	4.14	19.04	25
0180	20' eave height		275	.204		9.35	7.55	4.82	21.72	28.50
0190	24' eave height		240	.233		10.65	8.60	5.50	24.75	32.50
0200	30' to 40' wide, 10' eave height		535	.105		6.30	3.87	2.48	12.65	16.30
0300	14' eave height		450	.124		6.70	4.60	2.94	14.24	18.50
0400	16' eave height		415	.135		7.05	4.99	3.19	15.23	19.80
0500	20' eave height		360	.156		7.70	5.75	3.68	17.13	22.50
0600	24' eave height		320	.175		8.65	6.45	4.14	19.24	25
0700	50' to 100' wide, 10' eave height		865	.065		5.45	2.39	1.53	9.37	11.80
0800	14' eave height		770	.073		5.80	2.69	1.72	10.21	12.85
0900	16' eave height		730	.077		6.15	2.84	1.82	10.81	13.60
1000	20' eave height		660	.085		6.60	3.14	2.01	11.75	14.85
1100	24' eave height	↓	605	.093	↓	7.25	3.42	2.19	12.86	16.20
1200	Clear span tapered beam frame, 26 ga. colored roofing/siding									
1300	30' wide, 10' eave height	E-2	535	.105	SF Flr.	7	3.87	2.48	13.35	17.05
1400	14' eave height		450	.124		7.70	4.60	2.94	15.24	19.60
1500	16' eave height		415	.135		8.25	4.99	3.19	16.43	21
1600	20' eave height		360	.156		9.10	5.75	3.68	18.53	24
1700	40' wide, 10' eave height		600	.093		6.40	3.45	2.21	12.06	15.40
1800	14' eave height		510	.110		7	4.06	2.60	13.66	17.50
1900	16' eave height		475	.118		7.30	4.36	2.79	14.45	18.55
2000	20' eave height		415	.135		8	4.99	3.19	16.18	21
2100	50' to 80' wide, 10' eave height		770	.073		6.10	2.69	1.72	10.51	13.20
2200	14' eave height		675	.083		6.50	3.07	1.96	11.53	14.55
2300	16' eave height		635	.088		6.75	3.26	2.09	12.10	15.35
2400	20' eave height	↓	565	.099	↓	7.55	3.66	2.35	13.56	17.20
2500	Single post 2-span frame, 26 ga. colored roofing and siding									
2600	80' wide, 14' eave height	E-2	740	.076	SF Flr.	5.20	2.80	1.79	9.79	12.50
2700	16' eave height		695	.081		5.55	2.98	1.91	10.44	13.30
2800	20' eave height		625	.090		6.05	3.31	2.12	11.48	14.70
2900	24' eave height		570	.098		6.55	3.63	2.32	12.50	16
3000	100' wide, 14' eave height		835	.067		5.05	2.48	1.59	9.12	11.55
3100	16' eave height		795	.070		5	2.60	1.67	9.27	11.80
3200	20' eave height		730	.077		5.85	2.84	1.82	10.51	13.25
3300	24' eave height		670	.084		6.30	3.09	1.98	11.37	14.45
3400	120' wide, 14' eave height		870	.064		5	2.38	1.52	8.90	11.25
3500	16' eave height		830	.067		5.25	2.49	1.60	9.34	11.80
3600	20' eave height	↓	765	.073	↓	5.65	2.71	1.73	10.09	12.80

Important: See the Reference Section for critical supporting data - Reference Nos., Crews, & City Cost Indexes

13 SPECIAL CONSTRUCTION

		CREW	DAILY OUTPUT	LABOR-HOURS	UNIT	2005 BARE COSTS				TOTAL INCL O&P	
13128	**Pre-Engineered Structures**					MAT.	LABOR	EQUIP.	TOTAL		
700 3700	24' eave height R13128 -210	E-2	705	.079	SF Flr.	6.15	2.94	1.88	10.97	13.90	**700**
3800	Double post 3-span frame, 26 ga. colored roofing and siding										
3900	150' wide, 14' eave height	E-2	925	.061	SF Flr.	4.65	2.24	1.43	8.32	10.50	
4000	16' eave height		890	.063		4.80	2.33	1.49	8.62	10.95	
4100	20' eave height		820	.068		5.10	2.52	1.62	9.24	11.70	
4200	24' eave height		765	.073		6	2.71	1.73	10.44	13.15	
4300	Triple post 4-span frame, 26 ga. colored roofing and siding										
4400	160' wide, 14' eave height	E-2	970	.058	SF Flr.	4.52	2.13	1.37	8.02	10.15	
4500	16' eave height		930	.060		4.73	2.23	1.42	8.38	10.60	
4600	20' eave height		870	.064		4.99	2.38	1.52	8.89	11.25	
4700	24' eave height		815	.069		5.90	2.54	1.63	10.07	12.65	
4800	200' wide, 14' eave height		1,030	.054		4.55	2.01	1.29	7.85	9.85	
4900	16' eave height		995	.056		4.69	2.08	1.33	8.10	10.20	
5000	20' eave height		935	.060		5.10	2.21	1.42	8.73	10.95	
5100	24' eave height		885	.063		5.95	2.34	1.50	9.79	12.15	
5200	Accessory items: add to the basic building cost above										
5250	Eave overhang, 2' wide, 26 ga., with soffit	E-2	360	.156	L.F.	22.50	5.75	3.68	31.93	38.50	
5300	4' wide, without soffit		300	.187		21.50	6.90	4.42	32.82	40.50	
5350	With soffit		250	.224		30	8.30	5.30	43.60	53	
5400	6' wide, without soffit		250	.224		29.50	8.30	5.30	43.10	52	
5450	With soffit		200	.280		38	10.35	6.60	54.95	67	
5500	Entrance canopy, incl. frame, 4' x 4'		25	2.240	Ea.	288	83	53	424	515	
5550	4' x 8'		19	2.947	"	395	109	69.50	573.50	700	
5600	End wall roof overhang, 4' wide, without soffit		850	.066	L.F.	14.55	2.43	1.56	18.54	22	
5650	With soffit		500	.112	"	21	4.14	2.65	27.79	33.50	
5700	Doors, H.M. self-framing, incl. butts, lockset and trim										
5750	Single leaf, 3070 (3' x 7'), economy	2 Sswk	5	3.200	Opng.	280	122		402	530	
5800	Deluxe		4	4		294	153		447	600	
5825	Glazed		4	4		335	153		488	645	
5850	3670 (3'-6" x 7')		4	4		345	153		498	655	
5900	4070 (4' x 7')		3	5.333		375	203		578	780	
5950	Double leaf, 6070 (6' x 7')		2	8		495	305		800	1,100	
6000	Glazed		2	8		605	305		910	1,225	
6050	Framing only, for openings, 3' x 7'	E-2	25	2.240		122	83	53	258	335	
6100	10' x 10'		21	2.667		400	98.50	63	561.50	680	
6150	For windows below, 2020 (2' x 2')		25	2.240		129	83	53	265	345	
6200	4030 (4' x 3')		22	2.545		158	94	60	312	405	
6250	Flashings, 26 ga., corner or eave, painted	2 Sswk	240	.067	L.F.	3.53	2.54		6.07	8.45	
6300	Galvanized		240	.067		2.90	2.54		5.44	7.75	
6350	Rake flashing, painted		240	.067		3.78	2.54		6.32	8.70	
6400	Galvanized		240	.067		3.18	2.54		5.72	8.05	
6450	Ridge flashing, 18" wide, painted		240	.067		4.67	2.54		7.21	9.70	
6500	Galvanized		240	.067		4.09	2.54		6.63	9.05	
6550	Gutter, eave type, 26 ga., painted		320	.050		4.85	1.91		6.76	8.75	
6600	Galvanized		320	.050		2.97	1.91		4.88	6.70	
6650	Valley type, between buildings, painted		120	.133		7.70	5.10		12.80	17.60	
6700	Galvanized		120	.133		8.25	5.10		13.35	18.15	
6750	Insulation, rated .6 lb density, vinyl faced										
6800	1-1/2" thick, R5	2 Carp	2,300	.007	S.F.	.24	.24		.48	.63	
6850	3" thick, R10		2,300	.007		.30	.24		.54	.70	
6900	4" thick, R13		2,300	.007		.41	.24		.65	.82	
6950	Foil/scrim/kraft (FSK) faced, 1-1/2" thick, R5		2,300	.007		.26	.24		.50	.66	
7000	2" thick, R6		2,300	.007		.32	.24		.56	.72	
7050	3" thick, R10		2,300	.007		.38	.24		.62	.79	
7100	4" thick, R13		2,300	.007		.47	.24		.71	.89	
7150	Met polyester/scrim/kraft (PSK) facing,1-1/2" thk,R5		2,300	.007		.42	.24		.66	.83	

SPECIAL CONSTRUCTION 13

13128	Pre-Engineered Structures	CREW	DAILY OUTPUT	LABOR-HOURS	UNIT	2005 BARE COSTS				TOTAL INCL O&P
						MAT.	LABOR	EQUIP.	TOTAL	
700 7200	2" thick, R6	2 Carp	2,300	.007	S.F.	.48	.24		.72	.90	**700**
7250	3" thick, R11	R13128 -210	2,300	.007		.49	.24		.73	.91
7300	4" thick, R13		2,300	.007		.58	.24		.82	1.01
7350	Vinyl/scrim/foil (VSF), 1-1/2" thick, R5		2,300	.007		.37	.24		.61	.78
7400	2" thick, R6		2,300	.007		.46	.24		.70	.88
7450	3" thick, R10		2,300	.007		.55	.24		.79	.98
7500	4" thick, R13	↓	2,300	.007	↓	.61	.24		.85	1.04
7650	Sash, single slide, glazed, with screens, 2020 (2'x 2')	E-1	22	1.091	Opng.	76	40.50	3.69	120.19	157
7700	3030 (3' x 3')		14	1.714		171	63.50	5.80	240.30	305
7750	4030 (4' x 3')		13	1.846		228	68.50	6.25	302.75	375
7800	6040 (6' x 4')		12	2		455	74	6.75	535.75	635
7850	Double slide sash, 3030 (3' x 3')		14	1.714		143	63.50	5.80	212.30	272
7900	6040 (6' x 4')		12	2		380	74	6.75	460.75	555
7950	Fixed glass, no screens, 3030 (3' x 3')		14	1.714		185	63.50	5.80	254.30	320
8000	6040 (6' x 4')		12	2		495	74	6.75	575.75	680
8050	Prefinished storm sash, 3030 (3' x 3')	↓	70	.343	↓	61.50	12.75	1.16	75.41	90.50
8100	Siding and roofing, see division 07300 & 07400									
8200	Skylight, fiberglass panels, to 30 S.F.	E-1	10	2.400	Ea.	97	89	8.10	194.10	268
8250	Larger sizes, add for excess over 30 S.F.	"	300	.080	S.F.	3.23	2.97	.27	6.47	8.90
8300	Roof vents, circular with damper, birdscreen									
8350	and operator hardware, painted									
8400	26 ga., 12" diameter	1 Sswk	4	2	Ea.	72	76.50		148.50	216
8450	20" diameter		3	2.667		148	102		250	345
8500	24 ga., 24" diameter		2	4		293	153		446	595
8550	Galvanized	↓	2	4		245	153		398	545
8600	Continuous, 26 ga., 10' long, 9" wide	2 Sswk	4	4		20.50	153		173.50	296
8650	12" wide	"	4	4	↓	25	153		178	300
800 0010	**SHELTERS** Aluminum frame, acrylic glazing, 3' x 9' x 8' high	2 Sswk	1.14	14.035	Ea.	2,550	535		3,085	3,750	**800**
0100	9' x 12' x 8' high	"	.73	21.918	"	4,400	835		5,235	6,350
840 0010	**SILOS** Concrete stave industrial, not incl. foundations, conical or										**840**
0100	sloping bottoms, 12' diameter, 35' high	D-8	.11	363	Ea.	18,100	11,600		29,700	37,600
0200	16' diameter, 45' high		.08	500		24,200	16,000		40,200	51,000
0400	25' diameter, 75' high	↓	.05	800		59,500	25,500		85,000	104,500
0500	Steel, factory fab., 30,000 gallon cap., painted, minimum	L-5	1	56		17,200	2,125	615	19,940	23,400
0700	Maximum		.50	112		27,400	4,275	1,225	32,900	39,000
0800	Epoxy lined, minimum		1	56		28,200	2,125	615	30,940	35,400
1000	Maximum	↓	.50	112	↓	35,700	4,275	1,225	41,200	48,100
880 0010	**SWIMMING POOL ENCLOSURE** Translucent, free standing,										**880**
0020	not including foundations, heat or light									
0200	Economy, minimum	2 Carp	200	.080	SF Hor.	11.90	2.74		14.64	17.35
0300	Maximum		100	.160		32.50	5.50		38	44
0400	Deluxe, minimum		100	.160		34.50	5.50		40	46.50
0600	Maximum	↓	70	.229		175	7.85		182.85	204
0700	For motorized roof, 40% opening, solid roof, add					6.45			6.45	7.10
0800	Skylight type roof, add					6.80			6.80	7.45
0900	Air-inflated, including blowers and heaters, minimum									3.50
1000	Maximum				↓					6.70

Important: See the Reference Section for critical supporting data - Reference Nos., Crews, & City Cost Indexes

13151	Swimming Pools	CREW	DAILY OUTPUT	LABOR-HOURS	UNIT	MAT.	LABOR	EQUIP.	TOTAL	TOTAL INCL O&P
200 0010	**SWIMMING POOLS** Residential in-ground, vinyl lined, concrete sides									**200**
0020	Sides including equipment, sand bottom	B-52	300	.187	SF Surf	10.60	5.85	1.31	17.76	22
0100	Metal or polystyrene sides	B-14	410	.117		8.85	3.29	.53	12.67	15.45
0200	Add for vermiculite bottom	R13128 -520				.68			.68	.75
0500	Gunite bottom and sides, white plaster finish									
0600	12' x 30' pool	B-52	145	.386	SF Surf	17.55	12.10	2.72	32.37	41
0720	16' x 32' pool		155	.361		15.85	11.35	2.54	29.74	38
0750	20' x 40' pool		250	.224		14.15	7.05	1.58	22.78	28
0810	Concrete bottom and sides, tile finish									
0820	12' x 30' pool	B-52	80	.700	SF Surf	17.75	22	4.93	44.68	59
0830	16' x 32' pool		95	.589		14.65	18.50	4.15	37.30	49
0840	20' x 40' pool		130	.431		11.65	13.50	3.03	28.18	37
1100	Motel, gunite with plaster finish, incl. medium									
1150	capacity filtration & chlorination	B-52	115	.487	SF Surf	21.50	15.30	3.43	40.23	51.50
1200	Municipal, gunite with plaster finish, incl. high									
1250	capacity filtration & chlorination	B-52	100	.560	SF Surf	28	17.60	3.94	49.54	62.50
1350	Add for formed gutters				L.F.	48.50			48.50	53.50
1360	Add for stainless steel gutters				"	144			144	158
1600	For water heating system, see division 15510-880									
1700	Filtration and deck equipment only, as % of total				Total				20%	20%
1800	Deck equipment, rule of thumb, 20' x 40' pool				SF Pool					1.30
1900	5000 S.F. pool				"					1.90
3000	Painting pools, preparation + 3 coats, 20' x 40' pool, epoxy	2 Pord	.33	48.485	Total	605	1,475		2,080	2,900
3100	Rubber base paint, 18 gallons	"	.33	48.485		460	1,475		1,935	2,725
3500	42' x 82' pool, 75 gallons, epoxy paint	3 Pord	.14	171		2,550	5,250		7,800	10,700
3600	Rubber base paint	"	.14	171		1,975	5,250		7,225	10,100
700 0010	**SWIMMING POOL EQUIPMENT** Diving stand, stainless steel, 3 meter	2 Carp	.40	40	Ea.	4,925	1,375		6,300	7,550 **700**
0300	1 meter		2.70	5.926		3,650	203		3,853	4,325
0600	Diving boards, 16' long, aluminum		2.70	5.926		2,525	203		2,728	3,100
0700	Fiberglass		2.70	5.926		2,025	203		2,228	2,550
0900	Filter system, sand or diatomite type, incl. pump, 6,000 gal./hr.	2 Plum	1.80	8.889	Total	1,125	365		1,490	1,775
1020	Add for chlorination system, 800 S.F. pool		3	5.333	Ea.	203	218		421	550
1040	5,000 S.F. pool		3	5.333	"	1,150	218		1,368	1,600
1100	Gutter system, stainless steel, with grating, stock,									
1110	contains supply and drainage system	E-1	20	1.200	L.F.	170	44.50	4.06	218.56	267
1120	Integral gutter and 5' high wall system, stainless steel	"	10	2.400	"	254	89	8.10	351.10	440
1200	Ladders, heavy duty, stainless steel, 2 tread	2 Carp	7	2.286	Ea.	490	78.50		568.50	660
1500	4 tread		6	2.667		610	91.50		701.50	815
1800	Lifeguard chair, stainless steel, fixed		2.70	5.926		1,300	203		1,503	1,750
1900	Portable					1,025			1,025	1,150
2100	Lights, underwater, 12 volt, with transformer, 300 watt	1 Elec	.40	20		145	815		960	1,375
2200	110 volt, 500 watt, standard		.40	20		130	815		945	1,375
2400	Low water cutoff type		.40	20		135	815		950	1,375
2800	Heaters, see division 15510-880									
3000	Pool covers, reinforced vinyl	3 Clab	1,800	.013	S.F.	.30	.36		.66	.88
3050	Automatic, electric									9.65
3100	Vinyl water tube	3 Clab	3,200	.007		.23	.20		.43	.56
3200	Maximum	"	3,000	.008		.47	.21		.68	.85
3250	Sealed air bubble polyethylene solar blanket					.20			.20	.22
3300	Slides, tubular, fiberglass, aluminum handrails & ladder, 5'-0", straight	2 Carp	1.60	10	Ea.	2,200	345		2,545	2,950
3320	8'-0", curved		3	5.333		6,075	183		6,258	6,975
3400	10'-0", curved		1	16		15,100	550		15,650	17,500
3420	12'-0", straight with platform		1.20	13.333		2,025	455		2,480	2,925
4500	Hydraulic lift, movable pool bottom, single ram									
4520	Under 1,000 S.F. area	L-9	.03	1,200	Ea.	80,000	37,100		117,100	147,500
4600	Four ram lift, over 1,000 S.F.	"	.02	1,800		96,500	55,500		152,000	195,500

SPECIAL CONSTRUCTION 13

13150 | Swimming Pools

		13151	Swimming Pools	CREW	DAILY OUTPUT	LABOR-HOURS	UNIT	2005 BARE COSTS				TOTAL INCL O&P	
								MAT.	LABOR	EQUIP.	TOTAL		
700	5000		Removable access ramp, stainless steel	2 Clab	2	8	Ea.	6,375	214		6,589	7,350	700
	5500		Removable stairs, stainless steel, collapsible	"	2	8	↓	2,600	214		2,814	3,175	

13170 | Tubs & Pools

		13171	Therapeutic Pools	CREW	DAILY OUTPUT	LABOR-HOURS	UNIT	2005 BARE COSTS				TOTAL INCL O&P	
								MAT.	LABOR	EQUIP.	TOTAL		
800	0010		**THERAPEUTIC POOLS** See division 15418										800

13175 | Ice Rinks

		13176	Ice Rinks	CREW	DAILY OUTPUT	LABOR-HOURS	UNIT	2005 BARE COSTS				TOTAL INCL O&P	
								MAT.	LABOR	EQUIP.	TOTAL		
500	0010		**ICE SKATING** Equipment incl. refrigeration, plumbing & cooling										500
	0020		coils & concrete slab, 85' x 200' rink										
	0300		55° system, 5 mos., 100 ton				Total					545,500	
	0700		90° system, 12 mos., 135 ton				"					605,500	
	1000		Dasher boards, 1/2" H.D. polyethylene faced steel frame, 3' acrylic										
	1020		screen at sides, 5' acrylic ends, 85' x 200'	F-5	.06	533	Ea.	115,000	18,500		133,500	155,500	
	1100		Fiberglass & aluminum construction, same sides and ends	"	.06	533		160,000	18,500		178,500	205,000	
	1200		Subsoil heating system (recycled from compressor), 85' x 200'	Q-7	.27	118		25,000	4,625		29,625	34,500	
	1300		Subsoil insulation, 2 lb. polystyrene with vapor barrier, 85' x 200'	2 Carp	.14	114	↓	30,000	3,925		33,925	39,100	

13200 | Storage Tanks

		13201	Storage Tanks	CREW	DAILY OUTPUT	LABOR-HOURS	UNIT	2005 BARE COSTS				TOTAL INCL O&P	
								MAT.	LABOR	EQUIP.	TOTAL		
200	0010		**ELEVATED STORAGE TANKS**, not incl pipe, pumps or foundation										200
	3000		Elevated water tanks, 100' to bottom capacity line, incl painting										
	3010		50,000 gallons				Ea.					193,764	
	3300		100,000 gallons									267,300	
	3400		250,000 gallons									368,851	
	3600		500,000 gallons									595,298	
	3700		750,000 gallons									820,577	
	3900		1,000,000 gallons				↓					824,180	
300	0010		**GROUND TANKS** Not incl. pipe or pumps, prestress conc., 250,000 gal.				Ea.					284,200	300
	0100		500,000 gallons									385,700	
	0300		1,000,000 gallons									548,100	
	0400		2,000,000 gallons				↓					828,240	

SPECIAL CONSTRUCTION T3

Important: See the Reference Section for critical supporting data - Reference Nos., Crews, & City Cost Indexes

		13201	Storage Tanks	CREW	DAILY OUTPUT	LABOR-HOURS	UNIT	2005 BARE COSTS				TOTAL INCL O&P	
								MAT.	LABOR	EQUIP.	TOTAL		
300	0600		4,000,000 gallons				Ea.					1,309,350	300
	0700		6,000,000 gallons									1,786,400	
	0750		8,000,000 gallons									2,273,600	
	0800		10,000,000 gallons									2,740,500	
	0900		Steel, ground level, ht/dia less than 1, not incl. fdn, 100,000 gallons									141,500	
	1000		250,000 gallons									159,910	
	1200		500,000 gallons									243,955	
	1250		750,000 gallons									290,645	
	1300		1,000,000 gallons									444,360	
	1500		2,000,000 gallons									717,858	
	1600		4,000,000 gallons									1,114,723	
	1800		6,000,000 gallons									1,546,606	
	1850		8,000,000 gallons									2,025,179	
	1900		10,000,000 gallons									2,784,500	
	2100		Steel standpipes, hgt/dia more than 1, 100' to overflow, no fdn				Ea.						
	2200		500,000 gallons									320,993	
	2400		750,000 gallons									391,029	
	2500		1,000,000 gallons									476,238	
	2700		1,500,000 gallons									665,333	
	2800		2,000,000 gallons									817,075	
	3000		Steel, storage, above ground, including cradles, coating,										
	3020		fittings, not including fdn, pumps or piping										
	3040		Single wall, interior, 275 gallon	Q-5	5	3.200	Ea.	305	118		423	515	
	3060		550 gallon	"	2.70	5.926		1,400	218		1,618	1,850	
	3080		1,000 gallon	Q-7	5	6.400		2,200	250		2,450	2,800	
	3100		1,500 gallon		4.75	6.737		3,325	263		3,588	4,050	
	3120		2,000 gallon		4.60	6.957		3,800	272		4,072	4,600	
	3140		5,000 gallon		3.20	10		5,175	390		5,565	6,275	
	3150		10,000 gallon		2	16		15,500	625		16,125	17,900	
	3160		15,000 gallon		1.70	18.824		16,400	735		17,135	19,100	
	3170		20,000 gallon		1.45	22.069		20,400	860		21,260	23,800	
	3180		25,000 gallon capacity		1.30	24.615		23,500	960		24,460	27,300	
	3190		30,000 gallon		1.10	29.091		28,100	1,125		29,225	32,600	
	3320		Double wall, 500 gallon capacity	Q-5	2.40	6.667		2,425	246		2,671	3,050	
	3330		2000 gallon capacity	Q-7	4.15	7.711		5,575	300		5,875	6,575	
	3340		4000 gallon capacity		3.60	8.889		9,925	345		10,270	11,400	
	3350		6000 gallon capacity		2.40	13.333		11,700	520		12,220	13,700	
	3360		8000 gallon capacity		2	16		15,000	625		15,625	17,400	
	3370		10000 gallon capacity		1.80	17.778		16,600	695		17,295	19,300	
	3380		15000 gallon capacity		1.50	21.333		25,200	835		26,035	29,000	
	3390		20000 gallon capacity		1.30	24.615		28,600	960		29,560	33,000	
	3400		25000 gallon capacity		1.15	27.826		34,800	1,075		35,875	39,900	
	3410		30000 gallon capacity		1	32		38,200	1,250		39,450	43,900	
	4000		Fixed roof oil storage tanks, steel, (1 BBL=42 GAL w/ fdn 3'D x 1'W)				Ea.						
	4200		5,000 barrels									165,600	
	4300		25,000 barrels									232,300	
	4500		55,000 barrels									393,300	
	4600		100,000 barrels									632,500	
	4800		150,000 barrels									891,250	
	4900		225,000 barrels									1,343,200	
	5100		Floating roof gasoline tanks, steel, 5,000 barrels									140,300	
	5200		25,000 barrels									313,950	
	5400		55,000 barrels									474,950	
	5500		100,000 barrels									718,750	
	5700		150,000 barrels									972,900	
	5800		225,000 barrels									1,348,950	

SPECIAL CONSTRUCTION 13

		CREW	DAILY OUTPUT	LABOR-HOURS	UNIT	2005 BARE COSTS				TOTAL INCL O&P	
13201	**Storage Tanks**					MAT.	LABOR	EQUIP.	TOTAL		
300											**300**
6000	Wood tanks, ground level, 2" cypress, 3,000 gallons	C-1	.19	168	Ea.	6,800	5,450		12,250	16,000	
6100	2-1/2" cypress, 10,000 gallons		.12	266		18,000	8,625		26,625	33,200	
6300	3" redwood or 3" fir, 20,000 gallons		.10	320		27,500	10,400		37,900	46,400	
6400	30,000 gallons		.08	400		34,000	12,900		46,900	57,500	
6600	45,000 gallons		.07	457		51,500	14,800		66,300	79,500	
6700	Larger sizes, minimum				Gal.					.66	
6900	Maximum				"					.81	
7000	Vinyl coated fabric pillow tanks, freestanding, 5,000 gallons	4 Clab	4	8	Ea.	3,625	214		3,839	4,325	
7100	Supporting embankment not included, 25,000 gallons	6 Clab	2	24		6,900	640		7,540	8,575	
7200	50,000 gallons	8 Clab	1.50	42.667		13,600	1,150		14,750	16,700	
7300	100,000 gallons	9 Clab	.90	80		28,200	2,125		30,325	34,300	
7400	150,000 gallons		.50	144		34,500	3,850		38,350	44,000	
7500	200,000 gallons		.40	180		40,400	4,800		45,200	52,000	
7600	250,000 gallons		.30	240		45,500	6,400		51,900	60,000	
800											**800**
0010	**UNDERGROUND STORAGE TANKS**										
0210	Fiberglass, underground, single wall, U.L. listed, not including										
0220	manway or hold-down strap										
0240	2,000 gallon capacity	Q-7	4.57	7.002	Ea.	2,525	273		2,798	3,175	
0250	4,000 gallon capacity		3.55	9.014		3,650	350		4,000	4,550	
0260	6,000 gallon capacity		2.67	11.985		4,075	470		4,545	5,175	
0280	10,000 gallon capacity		2	16		5,550	625		6,175	7,050	
0284	15,000 gallon capacity		1.68	19.048		8,075	745		8,820	10,000	
0290	20,000 gallon capacity		1.45	22.069		11,200	860		12,060	13,600	
0500	For manway, fittings and hold-downs, add					20%	15%				
1020	Fiberglass, underground, double wall, U.L. listed										
1030	includes manways, not incl. hold-down straps										
1040	600 gallon capacity	Q-5	2.42	6.612	Ea.	3,325	244		3,569	4,025	
1050	1,000 gallon capacity	"	2.25	7.111		4,650	262		4,912	5,525	
1060	2,500 gallon capacity	Q-7	4.16	7.692		6,800	300		7,100	7,950	
1070	3,000 gallon capacity		3.90	8.205		7,325	320		7,645	8,550	
1080	4,000 gallon capacity		3.64	8.791		8,700	345		9,045	10,100	
1090	6,000 gallon capacity		2.42	13.223		9,725	515		10,240	11,500	
1100	8,000 gallon capacity CN		2.08	15.385		10,800	600		11,400	12,800	
1110	10,000 gallon capacity		1.82	17.582		12,000	685		12,685	14,200	
1120	12,000 gallon capacity		1.70	18.824		14,800	735		15,535	17,400	
2210	Fiberglass, underground, single wall, U.L. listed, including										
2220	hold-down straps, no manways										
2240	2,000 gallon capacity	Q-7	3.55	9.014	Ea.	2,750	350		3,100	3,550	
2250	4,000 gallon capacity		2.90	11.034		3,900	430		4,330	4,925	
2260	6,000 gallon capacity		2	16		4,550	625		5,175	5,950	
2280	10,000 gallon capacity		1.60	20		6,025	780		6,805	7,800	
2284	15,000 gallon capacity		1.39	23.022		8,550	900		9,450	10,800	
2290	20,000 gallon capacity		1.14	28.070		11,900	1,100		13,000	14,800	
3020	Fiberglass, underground, double wall, U.L. listed										
3030	includes manways and hold-down straps										
3040	600 gallon capacity	Q-5	1.86	8.602	Ea.	3,550	315		3,865	4,400	
3050	1,000 gallon capacity	"	1.70	9.412		4,875	345		5,220	5,900	
3060	2,500 gallon capacity	Q-7	3.29	9.726		7,050	380		7,430	8,325	
3070	3,000 gallon capacity		3.13	10.224		7,575	400		7,975	8,925	
3080	4,000 gallon capacity		2.93	10.921		8,950	425		9,375	10,500	
3090	6,000 gallon capacity		1.86	17.204		10,200	670		10,870	12,200	
3100	8,000 gallon capacity		1.65	19.394		11,300	755		12,055	13,500	
3110	10,000 gallon capacity		1.48	21.622		12,500	845		13,345	15,100	
3120	12,000 gallon capacity		1.40	22.857		15,300	890		16,190	18,200	
5000	Steel underground, sti-P3, set in place, not incl. hold-down bars.										
5500	Excavation, pad, pumps and piping not included										

T3 SPECIAL CONSTRUCTION

Important: See the Reference Section for critical supporting data - Reference Nos., Crews, & City Cost Indexes

13200 | Storage Tanks

13201	Storage Tanks	CREW	DAILY OUTPUT	LABOR-HOURS	UNIT	2005 BARE COSTS MAT.	LABOR	EQUIP.	TOTAL	TOTAL INCL O&P
5510	Single wall, 500 gallon capacity, 7 gauge shell	Q-5	2.70	5.926	Ea.	1,000	218		1,218	1,425
5520	1,000 gallon capacity, 7 gauge shell	"	2.50	6.400		1,650	236		1,886	2,150
5530	2,000 gallon capacity, 1/4" thick shell	Q-7	4.60	6.957		3,025	272		3,297	3,725
5535	2,500 gallon capacity, 7 gauge shell	Q-5	3	5.333		3,675	197		3,872	4,350
5540	5,000 gallon capacity, 1/4" thick shell	Q-7	3.20	10		6,850	390		7,240	8,100
5580	15,000 gallon capacity, 5/16" thick shell		1.70	18.824		13,200	735		13,935	15,700
5600	20,000 gallon capacity, 5/16" thick shell		1.50	21.333		17,800	835		18,635	20,800
5610	25,000 gallon capacity, 3/8" thick shell		1.30	24.615		21,500	960		22,460	25,200
5620	30,000 gallon capacity, 3/8" thick shell		1.10	29.091		28,600	1,125		29,725	33,200
5630	40,000 gallon capacity, 3/8" thick shell		.90	35.556		37,400	1,400		38,800	43,200
5640	50,000 gallon capacity, 3/8" thick shell	▼	.80	40	▼	46,800	1,550		48,350	54,000

13280 | Hazardous Material Remediation

13281	Hazardous Material Remediation	CREW	DAILY OUTPUT	LABOR-HOURS	UNIT	2005 BARE COSTS MAT.	LABOR	EQUIP.	TOTAL	TOTAL INCL O&P
0010	**BULK ASBESTOS REMOVAL**									
0020	Includes disposable tools and 2 suits and 1 respirator filter/day/worker									
0100	Beams, W 10 x 19	A-9	235	.272	L.F.	.77	10.10		10.87	16.80
0110	W 12 x 22		210	.305		.86	11.30		12.16	18.80
0120	W 14 x 26		180	.356		1.01	13.20		14.21	22
0130	W 16 x 31		160	.400		1.14	14.85		15.99	25
0140	W 18 x 40		140	.457		1.30	17		18.30	28.50
0150	W 24 x 55		110	.582		1.65	21.50		23.15	36
0160	W 30 x 108		85	.753		2.14	28		30.14	46.50
0170	W 36 x 150		72	.889	▼	2.52	33		35.52	55
0200	Boiler insulation	▼	480	.133	S.F.	.44	4.95		5.39	8.30
0210	With metal lath add				%				50%	
0300	Boiler breeching or flue insulation	A-9	520	.123	S.F.	.35	4.57		4.92	7.60
0310	For active boiler, add				%				100%	
0400	Duct or AHU insulation	A-10B	440	.073	S.F.	.21	2.71		2.92	4.50
0500	Duct vibration isolation joints, up to 24 Sq. In. duct	A-9	56	1.143	Ea.	3.24	42.50		45.74	70.50
0520	25 Sq. In. to 48 Sq. In. duct		48	1.333		3.78	49.50		53.28	82
0530	49 Sq. In. to 76 Sq. In. duct		40	1.600	▼	4.54	59.50		64.04	99
0600	Pipe insulation, air cell type, up to 4" diameter pipe		900	.071	L.F.	.20	2.64		2.84	4.39
0610	4" to 8" diameter pipe		800	.080		.23	2.97		3.20	4.94
0620	10" to 12" diameter pipe		700	.091		.26	3.40		3.66	5.65
0630	14" to 16" diameter pipe		550	.116	▼	.33	4.32		4.65	7.15
0650	Over 16" diameter pipe		650	.098	S.F.	.28	3.66		3.94	6.05
0700	With glove bag up to 3" diameter pipe		300	.213	L.F.	3.15	7.95		11.10	15.95
1000	Pipe fitting insulation up to 4" diameter pipe		320	.200	Ea.	.57	7.45		8.02	12.35
1100	6" to 8" diameter pipe		304	.211		.60	7.80		8.40	13
1110	10" to 12" diameter pipe		192	.333		.95	12.40		13.35	20.50
1120	14" to 16" diameter pipe		128	.500	▼	1.42	18.60		20.02	31
1130	Over 16" diameter pipe		176	.364	S.F.	1.03	13.50		14.53	22.50
1200	With glove bag, up to 8" diameter pipe		100	.640	L.F.	6.55	24		30.55	44.50
2000	Scrape foam fireproofing from flat surface		2,400	.027	S.F.	.08	.99		1.07	1.64
2100	Irregular surfaces		1,200	.053		.15	1.98		2.13	3.30
3000	Remove cementitious material from flat surface		1,800	.036		.10	1.32		1.42	2.20
3100	Irregular surface	▼	1,400	.046	▼	.13	1.70		1.83	2.82

SPECIAL CONSTRUCTION 13

13281	Hazardous Material Remediation	CREW	DAILY OUTPUT	LABOR-HOURS	UNIT	2005 BARE COSTS				TOTAL INCL O&P	
						MAT.	LABOR	EQUIP.	TOTAL		
120 4000	Scrape acoustical coating/fireproofing, from ceiling	A-9	3,200	.020	S.F.	.06	.74		.80	1.23	**120**
5000	Remove VAT from floor by hand	↓	2,400	.027		.08	.99		1.07	1.64	
5100	By machine	A-11	4,800	.013	↓	.04	.50	.01	.55	.83	
5150	For 2 layers, add				%				50%		
6000	Remove contaminated soil from crawl space by hand	A-9	400	.160	C.F.	.45	5.95		6.40	9.90	
6100	With large production vacuum loader	A-12	700	.091	"	.26	3.40	.69	4.35	6.40	
7000	Radiator backing, not including radiator removal	A-9	1,200	.053	S.F.	.15	1.98		2.13	3.30	
8000	Cement-asbestos transite board	2 Asbe	1,000	.016		.13	.59		.72	1.08	
8100	Transite shingle siding	A-10B	750	.043		.20	1.59		1.79	2.73	
8200	Shingle roofing	"	2,000	.016		.07	.60		.67	1.02	
8250	Built-up, no gravel, non-friable	B-2	1,400	.029		.07	.77		.84	1.29	
8300	Asbestos millboard	2 Asbe	1,000	.016	↓	.08	.59		.67	1.03	
9000	For type C (supplied air) respirator equipment, add				%					10%	
125 0010	**ASBESTOS ABATEMENT WORK AREA** Containment and preparation.										**125**
0100	Pre-cleaning, HEPA vacuum and wet wipe, flat surfaces	A-10	12,000	.005	S.F.	.02	.20		.22	.33	
0200	Protect carpeted area, 2 layers 6 mil poly on 3/4" plywood	"	1,000	.064		1.50	2.38		3.88	5.40	
0300	Separation barrier, 2" x 4" @ 16", 1/2" plywood ea. side, 8' high	2 Carp	400	.040		1.25	1.37		2.62	3.51	
0310	12' high		320	.050		1.40	1.71		3.11	4.21	
0320	16' high		200	.080		1.50	2.74		4.24	5.90	
0400	Personnel decontam. chamber, 2" x 4" @ 16", 3/4" ply ea. side		280	.057		2.50	1.96		4.46	5.80	
0450	Waste decontam. chamber, 2" x 4" studs @ 16", 3/4" ply ea. side	↓	360	.044	↓	3	1.52		4.52	5.65	
0500	Cover surfaces with polyethelene sheeting										
0501	Including glue and tape										
0550	Floors, each layer, 6 mil	A-10	8,000	.008	S.F.	2.85	.30		3.15	3.61	
0551	4 mil		9,000	.007		1.94	.26		2.20	2.55	
0560	Walls, each layer, 6 mil		6,000	.011		2.84	.40		3.24	3.75	
0561	4 mil	↓	7,000	.009	↓	1.95	.34		2.29	2.69	
0570	For heights above 12', add						20%				
0575	For heights above 20', add						30%				
0580	For fire retardant poly, add					100%					
0590	For large open areas, deduct					10%	20%				
0600	Seal floor penetrations with foam firestop to 36 Sq. In.	2 Carp	200	.080	Ea.	6.25	2.74		8.99	11.15	
0610	36 Sq. In. to 72 Sq. In.		125	.128		12.50	4.38		16.88	20.50	
0615	72 Sq. In. to 144 Sq. In.		80	.200		25	6.85		31.85	38	
0620	Wall penetrations, to 36 square inches		180	.089		6.25	3.04		9.29	11.65	
0630	36 Sq. In. to 72 Sq. In.		100	.160		12.50	5.50		18	22.50	
0640	72 Sq. In. to 144 Sq. In.	↓	60	.267	↓	25	9.15		34.15	42	
0800	Caulk seams with latex	1 Carp	230	.035	L.F.	.15	1.19		1.34	2.03	
0900	Set up neg. air machine, 1-2k C.F.M. /25 M.C.F. volume	1 Asbe	4.30	1.860	Ea.		69		69	109	
130 0010	**DEMOLITION IN ASBESTOS CONTAMINATED AREA**										**130**
0200	Ceiling, including suspension system, plaster and lath	A-9	2,100	.030	S.F.	.09	1.13		1.22	1.89	
0210	Finished plaster, leaving wire lath		585	.109		.31	4.07		4.38	6.75	
0220	Suspended acoustical tile		3,500	.018		.05	.68		.73	1.13	
0230	Concealed tile grid system		3,000	.021		.06	.79		.85	1.32	
0240	Metal pan grid system		1,500	.043		.12	1.59		1.71	2.63	
0250	Gypsum board		2,500	.026	↓	.07	.95		1.02	1.58	
0260	Lighting fixtures up to 2' x 4'	↓	72	.889	Ea.	2.52	33		35.52	55	
0400	Partitions, non load bearing										
0410	Plaster, lath, and studs	A-9	690	.093	S.F.	.81	3.45		4.26	6.35	
0450	Gypsum board and studs	"	1,390	.046	"	.13	1.71		1.84	2.84	
9000	For type C (supplied air) respirator equipment, add				%					10%	
135 0010	**ASBESTOS ABATEMENT EQUIPMENT** and supplies, buy	R13281 -120									**135**
0200	Air filtration device, 2000 C.F.M.				Ea.	2,500			2,500	2,750	
0250	Large volume air sampling pump, minimum					365			365	400	
0260	Maximum	↓			↓	680			680	745	

			CREW	DAILY OUTPUT	LABOR-HOURS	UNIT	2005 BARE COSTS				TOTAL INCL O&P	
	13281	**Hazardous Material Remediation**					MAT.	LABOR	EQUIP.	TOTAL		
135	0300	Airless sprayer unit, 2 gun R13281 -120				Ea.	2,000			2,000	2,200	**135**
	0350	Light stand, 500 watt				▼	250			250	275	
	0400	Personal respirators										
	0410	Negative pressure, 1/2 face, dual operation, min.				Ea.	22.50			22.50	25	
	0420	Maximum					24			24	26.50	
	0450	P.A.P.R., full face, minimum					400			400	440	
	0460	Maximum					700			700	770	
	0470	Supplied air, full face, incl. air line, minimum					450			450	495	
	0480	Maximum					600			600	660	
	0500	Personnel sampling pump, minimum					450			450	495	
	0510	Maximum					750			750	825	
	1500	Power panel, 20 unit, incl. G.F.I.					1,800			1,800	1,975	
	1600	Shower unit, including pump and filters					1,125			1,125	1,250	
	1700	Supplied air system (type C)					10,000			10,000	11,000	
	1750	Vacuum cleaner, HEPA, 16 gal., stainless steel, wet/dry					1,000			1,000	1,100	
	1760	55 gallon					2,200			2,200	2,425	
	1800	Vacuum loader, 9-18 ton/hr					90,000			90,000	99,000	
	1900	Water atomizer unit, including 55 gal. drum					230			230	253	
	2000	Worker protection, whole body, foot, head cover & gloves, plastic					5.50			5.50	6.05	
	2500	Respirator, single use					10.50			10.50	11.55	
	2550	Cartridge for respirator					11.70			11.70	12.85	
	2570	Glove bag, 7 mil, 50" x 64"					8.50			8.50	9.35	
	2580	10 mil, 44" x 60"					8.40			8.40	9.25	
	3000	HEPA vacuum for work area, minimun					1,050			1,050	1,150	
	3050	Maximum					2,775			2,775	3,050	
	6000	Disposable polyethelene bags, 6 mil, 3 C.F.					1.15			1.15	1.27	
	6300	Disposable fiber drums, 3 C.F.					6.50			6.50	7.15	
	6400	Pressure sensitive caution lables, 3" x 5"					.88			.88	.97	
	6450	11" x 17"					6.50			6.50	7.15	
	6500	Negative air machine, 1800 C.F.M.				▼	775			775	855	
140	0010	**DECONTAMINATION CONTAINMENT AREA DEMOLITION** and clean-up										**140**
	0100	Spray exposed substrate with surfactant (bridging)										
	0200	Flat surfaces	A-9	6,000	.011	S.F.	.36	.40		.76	1.03	
	0250	Irregular surfaces		4,000	.016	"	.31	.59		.90	1.28	
	0300	Pipes, beams, and columns		2,000	.032	L.F.	.56	1.19		1.75	2.50	
	1000	Spray encapsulate polyethelene sheeting		8,000	.008	S.F.	.31	.30		.61	.81	
	1100	Roll down polyethelene sheeting		8,000	.008	"		.30		.30	.47	
	1500	Bag polyethelene sheeting		400	.160	Ea.	.77	5.95		6.72	10.25	
	2000	Fine clean exposed substrate, with nylon brush		2,400	.027	S.F.		.99		.99	1.56	
	2500	Wet wipe substrate		4,800	.013			.50		.50	.78	
	2600	Vacuum surfaces, fine brush	▼	6,400	.010	▼		.37		.37	.59	
	3000	Structural demolition										
	3100	Wood stud walls	A-9	2,800	.023	S.F.		.85		.85	1.34	
	3500	Window manifolds, not incl. window replacement		4,200	.015			.57		.57	.89	
	3600	Plywood carpet protection	▼	2,000	.032	▼		1.19		1.19	1.88	
	4000	Remove custom decontamination facility	A-10A	8	3	Ea.	15	112		127	193	
	4100	Remove portable decontamination facility	3 Asbe	12	2	"	12.50	74		86.50	131	
	5000	HEPA vacuum, shampoo carpeting	A-9	4,800	.013	S.F.	.05	.50		.55	.84	
	9000	Final cleaning of protected surfaces	A-10A	8,000	.003	"		.11		.11	.18	
145	0010	**OSHA TESTING**										**145**
	0100	Certified technician, minimum				Day					300	
	0110	Maximum				"					500	
	0200	Personal sampling, PCM analysis, NIOSH 7400, minimum	1 Asbe	8	1	Ea.	2.75	37		39.75	61.50	
	0210	Maximum	"	4	2	"	3	74		77	120	
	0300	Industrial hygenist, minimum				Day					400	

	13281	Hazardous Material Remediation	CREW	DAILY OUTPUT	LABOR-HOURS	UNIT	2005 BARE COSTS				TOTAL INCL O&P	
							MAT.	LABOR	EQUIP.	TOTAL		
145	0310	Maximum				Day					550	**145**
	1000	Cleaned area samples	1 Asbe	8	1	Ea.	2.63	37		39.63	61.50	
	1100	PCM air sample analysis, NIOSH 7400, minimum		8	1		30	37		67	91.50	
	1110	Maximum		4	2		3.09	74		77.09	120	
	1200	TEM air sample analysis, NIOSH 7402, minimum								100	125	
	1210	Maximum								400	500	
150	0010	**ENCAPSULATION WITH SEALANTS**										**150**
	0100	Ceilings and walls, minimum	A-9	21,000	.003	S.F.	.27	.11		.38	.48	
	0110	Maximum		10,600	.006		.42	.22		.64	.81	
	0200	Columns and beams, minimum		13,300	.005		.27	.18		.45	.58	
	0210	Maximum		5,325	.012		.47	.45		.92	1.23	
	0300	Pipes to 12" diameter including minor repairs, minimum		800	.080	L.F.	.37	2.97		3.34	5.10	
	0310	Maximum		400	.160	"	1.04	5.95		6.99	10.55	
155	0010	**WASTE PACKAGING, HANDLING, & DISPOSAL**										**155**
	0100	Collect and bag bulk material, 3 C.F. bags, by hand	A-9	400	.160	Ea.	1.15	5.95		7.10	10.65	
	0200	Large production vacuum loader	A-12	880	.073		.80	2.70	.55	4.05	5.75	
	1000	Double bag and decontaminate	A-9	960	.067		2.30	2.48		4.78	6.45	
	2000	Containerize bagged material in drums, per 3 C.F. drum	"	800	.080		6.50	2.97		9.47	11.85	
	3000	Cart bags 50' to dumpster	2 Asbe	400	.040			1.48		1.48	2.34	
	5000	Disposal charges, not including haul, minimum				C.Y.					50	
	5020	Maximum				"					175	
	5100	Remove refrigerant from system	1 Plum	40	.200	Lb.		8.15		8.15	12.30	
	9000	For type C (supplied air) respirator equipment, add				%					10%	
440	0010	**REMOVAL** Existing lead paint, by chemicals, per application										**440**
	0020	See also, Div. 13280, Haz. Mat'l. Abatement [R13281-460]										
	0050	Baseboard, to 6" wide	1 Pord	64	.125	L.F.	1.53	3.83		5.36	7.45	
	0070	To 12" wide		32	.250	"	3.01	7.65		10.66	14.80	
	0200	Balustrades, one side		28	.286	S.F.	3.41	8.75		12.16	16.90	
	1400	Cabinets, simple design		32	.250		2.99	7.65		10.64	14.80	
	1420	Ornate design		25	.320		3.84	9.80		13.64	18.90	
	1600	Cornice, simple design		60	.133		1.61	4.08		5.69	7.90	
	1620	Ornate design		20	.400		4.74	12.25		16.99	23.50	
	2800	Doors, one side, flush		84	.095		1.16	2.91		4.07	5.65	
	2820	Two panel		80	.100		1.20	3.06		4.26	5.90	
	2840	Four panel		45	.178		2.12	5.45		7.57	10.55	
	2880	For trim, one side, add		64	.125	L.F.	1.53	3.83		5.36	7.45	
	3000	Fence, picket, one side		30	.267	S.F.	3.21	8.15		11.36	15.80	
	3200	Grilles, one side, simple design		30	.267		3.21	8.15		11.36	15.80	
	3220	Ornate design		25	.320		3.84	9.80		13.64	18.90	
	4400	Pipes, to 4" diameter		90	.089	L.F.	1.10	2.72		3.82	5.30	
	4420	To 8" diameter		50	.160		1.91	4.90		6.81	9.45	
	4440	To 12" diameter		36	.222		2.68	6.80		9.48	13.15	
	4460	To 16" diameter		20	.400		4.77	12.25		17.02	23.50	
	4500	For hangers, add		40	.200	Ea.	2.39	6.10		8.49	11.85	
	4800	Siding		90	.089	S.F.	1.10	2.72		3.82	5.30	
	5000	Trusses, open		55	.145	SF Face	1.75	4.45		6.20	8.65	
	6200	Windows, one side only, double hung, 1/1 light, 24" x 48" high		4	2	Ea.	24	61		85	119	
	6220	30" x 60" high		3	2.667		32	81.50		113.50	159	
	6240	36" x 72" high		2.50	3.200		38.50	98		136.50	190	
	6280	40" x 80" high		2	4		48	122		170	237	
	6400	Colonial window, 6/6 light, 24" x 48" high		2	4		48	122		170	237	
	6420	30" x 60" high		1.50	5.333		64	163		227	315	
	6440	36" x 72" high		1	8		96	245		341	475	
	6480	40" x 80" high		1	8		96	245		341	475	
	6600	8/8 light, 24" x 48" high		2	4		48	122		170	237	

SPECIAL CONSTRUCTION — **13**

	13281	Hazardous Material Remediation	CREW	DAILY OUTPUT	LABOR-HOURS	UNIT	MAT.	LABOR	EQUIP.	TOTAL	TOTAL INCL O&P	
440	6620	40" x 80" high	1 Pord	1	8	Ea.	96	245		341	475	**440**
	6800	12/12 light, 24" x 48" high R13281-460		1	8		96	245		341	475	
	6820	40" x 80" high	↓	.75	10.667	↓	128	325		453	630	
	6840	Window frame & trim items, included in pricing above ↓										
460	0010	**LEAD PAINT ENCAPSULATION**, water based polymer coating,14 mil DFT										**460**
	0020	Interior, brushwork, trim, under 6"	1 Pord	240	.033	L.F.	2.26	1.02		3.28	4.02	
	0030	6" to 12" wide		180	.044		3	1.36		4.36	5.35	
	0040	Balustrades		300	.027		1.81	.82		2.63	3.22	
	0050	Pipe to 4" diameter		500	.016		1.09	.49		1.58	1.94	
	0060	To 8" diameter		375	.021		1.44	.65		2.09	2.56	
	0070	To 12" diameter		250	.032		2.16	.98		3.14	3.85	
	0080	To 16" diameter		170	.047	↓	3.18	1.44		4.62	5.65	
	0090	Cabinets, ornate design		200	.040	S.F.	2.72	1.22		3.94	4.83	
	0100	Simple design	↓	250	.032	"	2.16	.98		3.14	3.85	
	0110	Doors, 3'x 7', both sides, incl. frame & trim										
	0120	Flush	1 Pord	6	1.333	Ea.	27.50	41		68.50	92	
	0130	French, 10-15 lite		3	2.667		5.55	81.50		87.05	129	
	0140	Panel		4	2		33.50	61		94.50	129	
	0150	Louvered	↓	2.75	2.909	↓	30.50	89		119.50	168	
	0160	Windows, per interior side, per 15 S.F.										
	0170	1 to 6 lite	1 Pord	14	.571	Ea.	19.15	17.50		36.65	47.50	
	0180	7 to 10 lite		7.50	1.067		21	32.50		53.50	72	
	0190	12 lite		5.75	1.391		28.50	42.50		71	95	
	0200	Radiators		8	1	↓	67.50	30.50		98	120	
	0210	Grilles, vents		275	.029	S.F.	1.97	.89		2.86	3.51	
	0220	Walls, roller, drywall or plaster		1,000	.008		.54	.24		.78	.96	
	0230	With spunbonded reinforcing fabric		720	.011		.62	.34		.96	1.19	
	0240	Wood		800	.010		.68	.31		.99	1.21	
	0250	Ceilings, roller, drywall or plaster		900	.009		.62	.27		.89	1.09	
	0260	Wood		700	.011	↓	.77	.35		1.12	1.38	
	0270	Exterior, brushwork, gutters and downspouts		300	.027	L.F.	1.81	.82		2.63	3.22	
	0280	Columns		400	.020	S.F.	1.34	.61		1.95	2.39	
	0290	Spray, siding	↓	600	.013	"	.90	.41		1.31	1.60	
	0300	Miscellaneous										
	0310	Electrical conduit, brushwork, to 2" diameter	1 Pord	500	.016	L.F.	1.09	.49		1.58	1.94	
	0320	Brick, block or concrete, spray		500	.016	S.F.	1.09	.49		1.58	1.94	
	0330	Steel, flat surfaces and tanks to 12"		500	.016		1.09	.49		1.58	1.94	
	0340	Beams, brushwork		400	.020		1.34	.61		1.95	2.39	
	0350	Trusses	↓	400	.020	↓	1.34	.61		1.95	2.39	
600	0010	**MOLD ABATEMENT WORK AREA Con**tainment and preparation										**600**
	0100	Pre-cleaning, HEPA vacuum and wet wipe, flat surfaces	A-10	12,000	.005	S.F.	.02	.20		.22	.33	
	0300	Separation barrier, 2" x 4" @ 16", 1/2" plywood ea. side, 8' high	2 Carp	400	.040		1.25	1.37		2.62	3.51	
	0310	12' high		320	.050		1.40	1.71		3.11	4.21	
	0320	16' high		200	.080		1.50	2.74		4.24	5.90	
	0400	Personnel decontam. chamber, 2" x 4" @ 16", 3/4" ply ea. side		280	.057		2.50	1.96		4.46	5.80	
	0450	Waste decontam. chamber, 2" x 4" studs @ 16", 3/4" ply each side	↓	360	.044	↓	3	1.52		4.52	5.65	
	0500	Cover surfaces with polyethelene sheeting										
	0501	Including glue and tape										
	0550	Floors, each layer, 6 mil	A-10	8,000	.008	S.F.	2.85	.30		3.15	3.61	
	0551	4 mil		9,000	.007		1.94	.26		2.20	2.55	
	0560	Walls, each layer, 6 mil		6,000	.011		2.84	.40		3.24	3.75	
	0561	4 mil	↓	7,000	.009	↓	1.95	.34		2.29	2.69	
	0570	For heights above 12', add						20%				
	0575	For heights above 20', add						30%				
	0580	For fire retardant poly, add					100%					

13281 | Hazardous Material Remediation

		CREW	DAILY OUTPUT	LABOR-HOURS	UNIT	2005 BARE COSTS				TOTAL INCL O&P
						MAT.	LABOR	EQUIP.	TOTAL	
600	0590	For large open areas, deduct					10%	20%			**600**
	0600	Seal floor penetrations with foam firestop to 36 sq in	2 Carp	200	.080	Ea.	6.25	2.74		8.99	11.15
	0610	36 sq in to 72 sq in		125	.128		12.50	4.38		16.88	20.50
	0615	72 sq in to 144 sq in		80	.200		25	6.85		31.85	38
	0620	Wall penetrations, to 36 square inches		180	.089		6.25	3.04		9.29	11.65
	0630	36 Sq. in. to 72 sq. in.		100	.160		12.50	5.50		18	22.50
	0640	72 Sq. in. to 144 sq. in.	▼	60	.267	▼	25	9.15		34.15	42
	0800	Caulk seams with latex	1 Carp	230	.035	L.F.	.15	1.19		1.34	2.03
	0900	Set up neg. air machine, 1-2k C.F.M. /25 M.C.F. volume	1 Asbe	4.30	1.860	Ea.		69		69	109
610	0010	**DEMOLITION IN MOLD CONTAMINATED AREA**									**610**
	0200	Ceiling, including suspension system, plaster and lath	A-9	2,100	.030	S.F.	.09	1.13		1.22	1.89
	0210	Finished plaster, leaving wire lath		585	.109		.31	4.07		4.38	6.75
	0220	Suspended acoustical tile		3,500	.018		.05	.68		.73	1.13
	0230	Concealed tile grid system		3,000	.021		.06	.79		.85	1.32
	0240	Metal pan grid system		1,500	.043		.12	1.59		1.71	2.63
	0250	Gypsum board		2,500	.026		.07	.95		1.02	1.58
	0255	Plywood		2,500	.026	▼	.07	.95		1.02	1.58
	0260	Lighting fixtures up to 2' x 4'	▼	72	.889	Ea.	2.52	33		35.52	55
	0400	Partitions, non load bearing									
	0410	Plaster, lath, and studs	A-9	690	.093	S.F.	.81	3.45		4.26	6.35
	0450	Gypsum board and studs		1,390	.046		.13	1.71		1.84	2.84
	0465	Carpet & pad		1,390	.046	▼	.13	1.71		1.84	2.84
	0600	Pipe insulation, air cell type, up to 4" diameter pipe		900	.071	L.F.	.20	2.64		2.84	4.39
	0610	4" to 8" diameter pipe		800	.080		.23	2.97		3.20	4.94
	0620	10" to 12" diameter pipe		700	.091		.26	3.40		3.66	5.65
	0630	14" to 16" diameter pipe		550	.116	▼	.33	4.32		4.65	7.15
	0650	Over 16" diameter pipe	▼	650	.098	S.F.	.28	3.66		3.94	6.05
	9000	For type C (supplied air) respirator equipment, add				%					10%
750	0010	**MOLD ABATEMENT METHODS**									**750**
	0020	Initial inspection, average 3 bedroom home				Total					250
	0030	Initial inspection, average 5 bedroom home									350
	0040	Air sample test, each									250
	0050	Swab sample test, each									150
	0060	Tape sample test, each									150
	0070	After remediation air test, each									250
	0080	Mold abatement plan, average 3 bedroom home									1,250
	0090	Mold abatement plan, average 5 bedroom home									1,500
	0100	Packup & removal of contents, average 3 bedroom home, excl storage									10,000
	0110	Packup & removal of contents, average 5 bedroom home, excl storage				▼					25,000
	0120	For demolition in mold contaminated areas, see Div. 13281-130									
	0130	For personal protection equipment, see Div. 13281-135									

13600 | Solar and Wind Energy Equipment

13630 | Solar Collector Components

			CREW	DAILY OUTPUT	LABOR-HOURS	UNIT	2005 BARE COSTS				TOTAL INCL O&P
							MAT.	LABOR	EQUIP.	TOTAL	
200	0010	**SOLAR ENERGY**	R13600 -610									**200**
	0020	System/Package prices, not including connecting									

Important: See the Reference Section for critical supporting data - Reference Nos., Crews, & City Cost Indexes

13630 | Solar Collector Components

			CREW	DAILY OUTPUT	LABOR-HOURS	UNIT	2005 BARE COSTS				TOTAL INCL O&P	
							MAT.	LABOR	EQUIP.	TOTAL		
200	0030	pipe, insulation, or special heating/plumbing fixtures	R13600 -610									200
	0500	Hot water, standard package, low temperature										
	0540	1 collector, circulator, fittings, 65 gal. tank	Q-1	.50	32	Ea.	1,450	1,175		2,625	3,350	
	0580	2 collectors, circulator, fittings, 120 gal. tank		.40	40		2,050	1,475		3,525	4,450	
	0620	3 collectors, circulator, fittings, 120 gal. tank		.34	47.059		2,475	1,725		4,200	5,325	
	0700	Medium temperature package										
	0720	1 collector, circulator, fittings, 80 gal. tank	Q-1	.50	32	Ea.	1,550	1,175		2,725	3,475	
	0740	2 collectors, circulator, fittings, 120 gal. tank		.40	40		2,200	1,475		3,675	4,625	
	0780	3 collectors, circulator, fittings, 120 gal. tank		.30	53.333		4,000	1,950		5,950	7,350	
	0980	For each additional 120 gal. tank, add					700			700	770	
	2300	Circulators, air										
	2310	Blowers										
	2400	Reversible fan, 20" diameter, 2 speed	Q-9	18	.889	Ea.	108	32		140	169	
	2520	Space & DHW system, less duct work	"	.50	32		1,475	1,150		2,625	3,400	
	2870	1/12 HP, 30 GPM	Q-1	10	1.600		330	59		389	455	
	3000	Collector panels, air with aluminum absorber plate										
	3010	Wall or roof mount										
	3040	Flat black, plastic glazing										
	3080	4' x 8'	Q-9	6	2.667	Ea.	635	96.50		731.50	845	
	3200	Flush roof mount, 10' to 16' x 22" wide	"	96	.167	L.F.	52	6.05		58.05	66.50	
	3300	Collector panels, liquid with copper absorber plate										
	3330	Alum. frame, 4' x 8', 5/32" single glazing	Q-1	9.50	1.684	Ea.	545	62		607	695	
	3390	Alum. frame, 4' x 10', 5/32" single glazing		6	2.667		650	98		748	855	
	3450	Flat black, alum. frame, 3' x 8'		9	1.778		430	65.50		495.50	575	
	3500	4' x 8.5'		5.50	2.909		490	107		597	700	
	3520	4' x 10.5'		10	1.600		595	59		654	745	
	3540	4' x 12.5'		5	3.200		735	118		853	985	
	3600	Liquid, full wetted, plastic, alum. frame, 3' x 10'		5	3.200		195	118		313	390	
	3650	Collector panel mounting, flat roof or ground rack		7	2.286		45	84		129	176	
	3670	Roof clamps		70	.229	Set	1.61	8.40		10.01	14.40	
	3700	Roof strap, teflon	1 Plum	205	.039	L.F.	9.50	1.59		11.09	12.85	
	3900	Differential controller with two sensors										
	3930	Thermostat, hard wired	1 Plum	8	1	Ea.	87.50	41		128.50	158	
	4100	Five station with digital read-out	"	3	2.667	"	218	109		327	405	
	4300	Heat exchanger										
	4580	Fluid to fluid package includes two circulating pumps										
	4590	expansion tank, check valve, relief valve										
	4600	controller, high temperature cutoff and sensors	Q-1	2.50	6.400	Ea.	695	235		930	1,125	
	4650	Heat transfer fluid										
	4700	Propylene glycol, inhibited anti-freeze	1 Plum	28	.286	Gal.	8.80	11.65		20.45	27.50	
	8250	Water storage tank with heat exchanger and electric element										
	8300	80 gal. with 2" x 2 lb. density insulation	1 Plum	1.60	5	Ea.	835	204		1,039	1,225	
	8380	120 gal. with 2" x 2 lb. density insulation		1.40	5.714		940	233		1,173	1,375	
	8400	120 gal. with 2" x 2 lb. density insul., 40 S.F. heat coil		1.40	5.714		1,075	233		1,308	1,550	

13710 | Security Access

			CREW	DAILY OUTPUT	LABOR-HOURS	UNIT	2005 BARE COSTS				TOTAL INCL O&P	
							MAT.	LABOR	EQUIP.	TOTAL		
300	0010	ACCESS CONTROL										300
	0020	Card type, 1 time zone, minimum				Ea.	320			320	350	

SPECIAL CONSTRUCTION 13

13710 | Security Access

	Line	Description	CREW	DAILY OUTPUT	LABOR-HOURS	UNIT	MAT.	LABOR	EQUIP.	TOTAL	TOTAL INCL O&P	
300	0040	Maximum				Ea.	1,075			1,075	1,175	300
	0060	3 time zones, minimum					785			785	865	
	0080	Maximum				↓	1,800			1,800	2,000	
	0100	System with printer, and control console, 3 zones				Total	8,800			8,800	9,700	
	0120	6 zones				"	11,600			11,600	12,800	
	0140	For each door, minimum, add				Ea.	1,300			1,300	1,425	
	0160	Maximum, add				"	1,925			1,925	2,125	

13720 | Detection & Alarm

	Line	Description	CREW	DAILY OUTPUT	LABOR-HOURS	UNIT	MAT.	LABOR	EQUIP.	TOTAL	TOTAL INCL O&P	
065	0010	**DETECTION SYSTEMS**, not including wires & conduits										065
	0100	Burglar alarm, battery operated, mechanical trigger	1 Elec	4	2	Ea.	254	81.50		335.50	400	
	0200	Electrical trigger		4	2		305	81.50		386.50	455	
	0400	For outside key control, add		8	1		72	41		113	140	
	0600	For remote signaling circuitry, add		8	1		114	41		155	187	
	0800	Card reader, flush type, standard		2.70	2.963		850	121		971	1,125	
	1000	Multi-code		2.70	2.963		1,100	121		1,221	1,375	
	1200	Door switches, hinge switch		5.30	1.509		53.50	61.50		115	151	
	1400	Magnetic switch		5.30	1.509		63	61.50		124.50	161	
	1600	Exit control locks, horn alarm		4	2		315	81.50		396.50	470	
	1800	Flashing light alarm		4	2		355	81.50		436.50	515	
	2000	Indicating panels, 1 channel		2.70	2.963		335	121		456	550	
	2200	10 channel	2 Elec	3.20	5		1,150	204		1,354	1,550	
	2400	20 channel		2	8		2,250	325		2,575	2,950	
	2600	40 channel		1.14	14.035		4,075	570		4,645	5,350	
	2800	Ultrasonic motion detector, 12 volt	1 Elec	2.30	3.478		210	142		352	440	
	3000	Infrared photoelectric detector	"	2.30	3.478	↓	173	142		315	400	
	3594	Fire, alarm control panel										
	3600	4 zone	2 Elec	2	8	Ea.	940	325		1,265	1,500	
	3800	8 zone		1	16		1,400	650		2,050	2,525	
	4000	12 zone		.67	23.988		1,825	980		2,805	3,450	
	4020	Alarm device	1 Elec	8	1		122	41		163	195	
	4050	Actuating device		8	1		292	41		333	380	
	4200	Battery and rack		4	2		690	81.50		771.50	880	
	4400	Automatic charger		8	1		445	41		486	550	
	4600	Signal bell		8	1		49.50	41		90.50	115	
	4800	Trouble buzzer or manual station		8	1		37	41		78	101	
	5000	Detector, rate of rise		8	1		33.50	41		74.50	97.50	
	5200	Smoke detector, ceiling type		6.20	1.290		75	52.50		127.50	161	
	5400	Duct type		3.20	2.500		250	102		352	425	
	5600	Strobe and horn		5.30	1.509		95	61.50		156.50	197	
	5800	Fire alarm horn		6.70	1.194		36.50	48.50		85	113	
	6000	Door holder, electro-magnetic		4	2		77.50	81.50		159	207	
	6200	Combination holder and closer		3.20	2.500		430	102		532	625	
	6400	Code transmitter		4	2		690	81.50		771.50	880	
	6600	Drill switch		8	1		86.50	41		127.50	156	
	6800	Master box		2.70	2.963		3,100	121		3,221	3,575	
	7000	Break glass station		8	1		50	41		91	116	
	7800	Remote annunciator, 8 zone lamp		1.80	4.444		175	181		356	465	
	8000	12 zone lamp	2 Elec	2.60	6.154		300	251		551	705	
	8200	16 zone lamp	"	2.20	7.273	↓	300	296		596	770	

T3 SPECIAL CONSTRUCTION

Important: See the Reference Section for critical supporting data - Reference Nos., Crews, & City Cost Indexes

13834	Electric/Electronic Control	CREW	DAILY OUTPUT	LABOR-HOURS	UNIT	2005 BARE COSTS				TOTAL INCL O&P	
						MAT.	LABOR	EQUIP.	TOTAL		
200	0010 **CONTROL SYSTEMS, ELECTRONIC**										**200**
	0020 For electronic costs, add to division 13836-200				Ea.					15%	

13836 | Pneumatic Controls

		CREW	DAILY OUTPUT	LABOR-HOURS	UNIT	MAT.	LABOR	EQUIP.	TOTAL	TOTAL INCL O&P	
200	0010 **CONTROL SYSTEMS, PNEUMATIC** (Sub's quote incl. mat. & labor)										**200**
	0011 and nominal 50 Ft. of tubing. Add control panelboard if req'd.										
	0100 Heating and Ventilating, split system										
	0200 Mixed air control, economizer cycle, panel readout, tubing										
	0220 Up to 10 tons	Q-19	.68	35.294	Ea.	2,975	1,350		4,325	5,300	
	0240 For 10 to 20 tons		.63	37.915		3,175	1,450		4,625	5,675	
	0260 For over 20 tons	▼	.58	41.096	▼	3,450	1,575		5,025	6,150	
	0300 Heating coil, hot water, 3 way valve,										
	0320 Freezestat, limit control on discharge, readout	Q-5	.69	23.088	Ea.	2,200	850		3,050	3,700	
	0500 Cooling coil, chilled water, room										
	0520 Thermostat, 3 way valve	Q-5	2	8	Ea.	985	295		1,280	1,525	
	0600 Cooling tower, fan cycle, damper control,										
	0620 Control system including water readout in/out at panel	Q-19	.67	35.821	Ea.	3,900	1,375		5,275	6,350	
	1000 Unit ventilator, day/night operation,										
	1100 freezestat, ASHRAE, cycle 2	Q-19	.91	26.374	Ea.	2,175	1,000		3,175	3,875	
	2000 Compensated hot water from boiler, valve control,										
	2100 readout and reset at panel, up to 60 GPM	Q-19	.55	43.956	Ea.	4,050	1,675		5,725	6,950	
	2120 For 120 GPM		.51	47.059		4,325	1,800		6,125	7,475	
	2140 For 240 GPM		.49	49.180		4,525	1,875		6,400	7,800	
	3000 Boiler room combustion air, damper to 5 SF, controls		1.36	17.582		1,950	670		2,620	3,150	
	3500 Fan coil, heating and cooling valves, 4 pipe control system		3	8		880	305		1,185	1,425	
	3600 Heat exchanger system controls	▼	.86	27.907		1,900	1,075		2,975	3,700	
	4000 Pneumatic thermostat, including controlling room radiator valve	Q-5	2.43	6.593		590	243		833	1,000	
	4060 Pump control system	Q-19	3	8	▼	905	305		1,210	1,450	
	4500 Air supply for pneumatic control system										
	4600 Tank mounted duplex compressor, starter, alternator,										
	4620 piping, dryer, PRV station and filter										
	4630 1/2 HP	Q-19	.68	35.139	Ea.	7,275	1,350		8,625	10,000	
	4660 1-1/2 HP		.57	41.739		8,875	1,600		10,475	12,100	
	4690 5 HP	▼	.42	57.143	▼	21,100	2,175		23,275	26,500	

13851	Detection & Alarm	CREW	DAILY OUTPUT	LABOR-HOURS	UNIT	2005 BARE COSTS				TOTAL INCL O&P	
						MAT.	LABOR	EQUIP.	TOTAL		
050	0010 **CLOCKS**										**050**
	0080 12" diameter, single face	1 Elec	8	1	Ea.	75.50	41		116.50	144	
	0100 Double face	"	6.20	1.290	"	144	52.50		196.50	237	
055	0010 **CLOCK SYSTEMS**, not including wires & conduits										**055**
	0100 Time system components, master controller	1 Elec	.33	24.242	Ea.	1,625	990		2,615	3,275	
	0200 Program bell		8	1		53.50	41		94.50	120	
	0400 Combination clock & speaker		3.20	2.500		189	102		291	360	
	0600 Frequency generator		2	4		6,350	163		6,513	7,225	
	0800 Job time automatic stamp recorder, minimum		4	2		445	81.50		526.50	610	
	1000 Maximum	▼	4	2		680	81.50		761.50	870	
	1200 Time stamp for correspondence, hand operated				▼	340			340	375	

SPECIAL CONSTRUCTION **13**

13850 | Detection & Alarm

		13851	Detection & Alarm	CREW	DAILY OUTPUT	LABOR-HOURS	UNIT	2005 BARE COSTS MAT.	LABOR	EQUIP.	TOTAL	TOTAL INCL O&P	
055	1400		Fully automatic				Ea.	495			495	545	055
	1600		Master time clock system, clocks & bells, 20 room	4 Elec	.20	160		4,075	6,525		10,600	14,200	
	1800		50 room	"	.08	400		9,400	16,300		25,700	34,600	
	2000		Time clock, 100 cards in & out, 1 color	1 Elec	3.20	2.500		950	102		1,052	1,200	
	2200		2 colors		3.20	2.500		1,025	102		1,127	1,275	
	2400		With 3 circuit program device, minimum		2	4		375	163		538	660	
	2600		Maximum		2	4		545	163		708	845	
	2800		Metal rack for 25 cards		7	1.143		60	46.50		106.50	136	
	3000		Watchman's tour station		8	1		63	41		104	130	
	3200		Annunciator with zone indication		1	8		230	325		555	740	
	3400		Time clock with tape	▼	1	8	▼	610	325		935	1,150	
350	0010	**TANK LEAK DETECTION SYSTEMS** Liquid and vapor											350
	0100		For hydrocarbons and hazardous liquids/vapors										
	0120		Controller, data acquisition, incl. printer, modem, RS232 port										
	0140		24 channel, for use with all probes				Ea.	2,100			2,100	2,325	
	0160		9 channel, for external monitoring				"	1,625			1,625	1,800	
	0200		Probes										
	0210		Well monitoring										
	0220		Liquid phase detection				Ea.	735			735	805	
	0230		Hydrocarbon vapor, fixed position					675			675	745	
	0240		Hydrocarbon vapor, float mounted					600			600	660	
	0250		Both liquid and vapor hydrocarbon				▼	910			910	1,000	
	0300		Secondary containment, liquid phase										
	0310		Pipe trench/manway sump				Ea.	325			325	360	
	0320		Double wall pipe and manual sump					320			320	350	
	0330		Double wall fiberglass annular space					295			295	325	
	0340		Double wall steel tank annular space				▼	295			295	325	
	0500		Accessories										
	0510		Modem, non-dedicated phone line				Ea.	264			264	290	
	0600		Monitoring, internal										
	0610		Automatic tank gauge, incl. overfill				Ea.	1,050			1,050	1,150	
	0620		Product line				"	1,125			1,125	1,250	
	0700		Monitoring, special										
	0710		Cathodic protection				Ea.	630			630	690	
	0720		Annular space chemical monitor				"	855			855	940	

13900 | Fire Suppression

		13910	Basic Fire Protection Matl/Methd	CREW	DAILY OUTPUT	LABOR-HOURS	UNIT	2005 BARE COSTS MAT.	LABOR	EQUIP.	TOTAL	TOTAL INCL O&P	
400	0010	**FIRE HOSE AND EQUIPMENT**											400
	0200		Adapters, rough brass, straight hose threads										
	0220		One piece, female to male, rocker lugs										
	0240		1" x 1"				Ea.	26.50			26.50	29	
	2200		Hose, less couplings										
	2260		Synthetic jacket, lined, 300 lb. test, 1-1/2" diameter	Q-12	2,600	.006	L.F.	1.62	.23		1.85	2.12	
	2280		2-1/2" diameter		2,200	.007		2.70	.27		2.97	3.37	
	2360		High strength, 500 lb. test, 1-1/2" diameter		2,600	.006		1.68	.23		1.91	2.19	
	2380		2-1/2" diameter	▼	2,200	.007	▼	2.91	.27		3.18	3.60	
	2600		Hose rack, swinging, for 1-1/2" diameter hose,										

Important: See the Reference Section for critical supporting data - Reference Nos., Crews, & City Cost Indexes

		13910 | Basic Fire Protection Matl/Methd	CREW	DAILY OUTPUT	LABOR-HOURS	UNIT	MAT.	LABOR	EQUIP.	TOTAL	TOTAL INCL O&P	
400	2620	Enameled steel, 50' & 75' lengths of hose	Q-12	20	.800	Ea.	33	29.50		62.50	81	**400**
	2640	100' and 125' lengths of hose	"	20	.800	"	33	29.50		62.50	81	
	3750	Hydrants, wall, w/caps, single, flush, polished brass										
	3800	2-1/2" x 2-1/2"	Q-12	5	3.200	Ea.	120	117		237	310	
	3840	2-1/2" x 3"	"	5	3.200	"	161	117		278	355	
	3950	Double, flush, polished brass										
	4000	2-1/2" x 2-1/2" x 4"	Q-12	5	3.200	Ea.	320	117		437	530	
	4040	2-1/2" x 2-1/2" x 6"	"	4.60	3.478	↓	465	128		593	705	
	4200	For polished chrome, add					10%					
	4350	Double, projecting, polished brass										
	4400	2-1/2" x 2-1/2" x 4"	Q-12	5	3.200	Ea.	148	117		265	340	
	4450	2-1/2" x 2-1/2" x 6"	"	4.60	3.478	"	286	128		414	510	
	4460	Valve control, dbl. flush/projecting hydrant, cap &										
	4470	chain, ext. rod & cplg., escutcheon, polished brass	Q-12	8	2	Ea.	174	73.50		247.50	300	
	5600	Nozzles, brass										
	5620	Adjustable fog, 3/4" booster line				Ea.	70.50			70.50	77.50	
	5630	1" booster line					87			87	95.50	
	5640	1-1/2" leader line					91			91	100	
	5660	2-1/2" direct connection				↓	178			178	196	
	5780	For chrome plated, add					8%					
	5850	Electrical fire, adjustable fog, no shock										
	5900	1-1/2"				Ea.	268			268	295	
	5920	2-1/2"				↓	365			365	400	
	5980	For polished chrome, add					6%					
	6200	Heavy duty, comb. adj. fog and str. stream, with handle										
	6210	1" booster line				Ea.	256			256	282	
	7140	Standpipe connections, wall, w/plugs & chains										
	7160	Single, flush, brass, 2-1/2" x 2-1/2"	Q-12	5	3.200	Ea.	88.50	117		205.50	275	
	7180	2-1/2" x 3"	"	5	3.200	"	92	117		209	278	
	7240	For polished chrome, add					15%					
	7280	Double, flush, polished brass										
	7300	2-1/2" x 2-1/2" x 4"	Q-12	5	3.200	Ea.	297	117		414	500	
	7330	2-1/2" x 2-1/2" x 6"	"	4.60	3.478	"	405	128		533	640	
	7400	For polished chrome, add					15%					
	7440	For sill cock combination, add				Ea.	50			50	55.50	
	7900	Three way, flush, polished brass										
	7920	2-1/2" (3) x 4"	Q-12	4.80	3.333	Ea.	930	122		1,052	1,200	
	7930	2-1/2" (3) x 6"	"	4.80	3.333	↓	940	122		1,062	1,200	
	8000	For polished chrome, add				↓	9%					
	8020	Three way, projecting, polished brass										
	8040	2-1/2"(3) x 4"	Q-12	4.80	3.333	Ea.	620	122		742	865	
800	0010	**FIRE VALVES**										**800**
	0080	Wheel handle, 300 lb., 1-1/2"	1 Spri	12	.667	Ea.	30	27		57	74	
	0090	2-1/2"	"	7	1.143	"	51.50	46.50		98	127	
	0100	For polished brass, add					35%					
	0110	For polished chrome, add					50%					

13920 | Fire Pumps

			CREW	DAILY OUTPUT	LABOR-HOURS	UNIT	MAT.	LABOR	EQUIP.	TOTAL	TOTAL INCL O&P	
400	0010	**FIRE PUMPS** Including controller, fittings and relief valve										**400**
	0030	Diesel										
	0050	500 GPM, 50 psi, 27 HP, 4" pump	Q-13	.64	50	Ea.	48,700	1,950		50,650	56,500	
	0200	750 GPM, 50 psi, 44 HP, 5" pump		.60	53.333		52,000	2,075		54,075	60,000	
	0400	1000 GPM, 100 psi, 89 HP, 4" pump		.56	57.143		55,500	2,225		57,725	65,000	
	0700	2,000 GPM, 100 psi, 167 HP, 6" pump	↓	.34	94.118	↓	59,000	3,650		62,650	70,500	

SPECIAL CONSTRUCTION **13**

13920 | Fire Pumps

		CREW	DAILY OUTPUT	LABOR-HOURS	UNIT	2005 BARE COSTS				TOTAL INCL O&P	
						MAT.	LABOR	EQUIP.	TOTAL		
400	0950	3500 GPM, 100 psi, 300 HP, 10" pump	Q-13	.24	133	Ea.	104,000	5,175		109,175	122,500	400
	3000	Electric										
	3100	250 GPM, 55 psi, 15 HP, 3,550 RPM, 2" pump	Q-13	.70	45.714	Ea.	11,400	1,775		13,175	15,300	
	3200	500 GPM, 50 psi, 27 HP, 1770 RPM, 4" pump		.68	47.059		14,200	1,825		16,025	18,400	
	3350	750 GPM, 50 psi, 44 HP, 1,770 RPM, 5" pump		.64	50		20,700	1,950		22,650	25,700	
	3400	750 GPM, 100 psi, 66 HP, 3550 RPM, 4" pump	↓	.58	55.172		18,300	2,150		20,450	23,500	
	5000	For jockey pump 1", 3 HP, with control, add	Q-12	2	8	↓	2,550	293		2,843	3,250	

13930 | Wet-Pipe Fire Supp. Sprinklers

		CREW	DAILY OUTPUT	LABOR-HOURS	UNIT	MAT.	LABOR	EQUIP.	TOTAL	TOTAL INCL O&P	
400	0010	**SPRINKLER SYSTEM COMPONENTS**										400
	0600	Accelerator	1 Spri	8	1	Ea.	380	41		421	480	
	2600	Sprinkler heads, not including supply piping										
	2640	Dry, pendent, 1/2" orifice, 3/4" or 1" NPT										
	2700	15-1/4" to 18" length	1 Spri	14	.571	Ea.	53.50	23.50		77	94	
	2710	18-1/4" to 21" length		13	.615		56	25		81	100	
	2720	21-1/4" to 24" length		13	.615		60	25		85	104	
	2730	24-1/4" to 27" length		13	.615		62	25		87	106	
	3600	Foam-water, pendent or upright, 1/2" NPT	↓	12	.667	↓	56	27		83	103	
	3700	Standard spray, pendent or upright, brass, 135° to 286°F										
	3730	1/2" NPT, 7/16" orifice	1 Spri	16	.500	Ea.	6.75	20.50		27.25	38.50	
	3740	1/2" NPT, 1/2" orifice **CN**	"	16	.500		5.10	20.50		25.60	36.50	
	3860	For wax and lead coating, add					5.70			5.70	6.30	
	3880	For wax coating, add					2.81			2.81	3.09	
	3900	For lead coating, add				↓	3.38			3.38	3.72	
	3920	For 360°F, same cost										
	3930	For 400°F, add				Ea.	5.70			5.70	6.30	
	3940	For 500°F, add				"	5.70			5.70	6.30	
	4500	Sidewall, horizontal, brass, 135° to 286°F										
	4520	1/2" NPT, 1/2" orifice	1 Spri	16	.500	Ea.	6.75	20.50		27.25	38.50	
	4540	For 360°F, same cost										
	4800	Recessed pendent, brass, 135° to 286°F										
	4820	1/2" NPT, 3/8" orifice	1 Spri	10	.800	Ea.	16.10	32.50		48.60	67	
	4830	1/2" NPT, 7/16" orifice		10	.800		11.45	32.50		43.95	61.50	
	4840	1/2" NPT, 1/2" orifice	↓	10	.800	↓	13.50	32.50		46	64	
	6200	Valves										
	6500	Check, swing, C.I. body, brass fittings, auto. ball drip										
	6520	4" size	Q-12	3	5.333	Ea.	163	196		359	475	
	6800	Check, wafer, butterfly type, C.I. body, bronze fittings										
	6820	4" size	Q-12	4	4	Ea.	172	147		319	410	
	7000	Deluge, assembly, incl. trim, pressure										
	7020	operated relief, emergency release, gauges										
	7040	2" size	Q-12	2	8	Ea.	1,025	293		1,318	1,575	
	7060	3" size	"	1.50	10.667	"	1,225	390		1,615	1,950	

13960 | CO2 Fire Extinguishing

		CREW	DAILY OUTPUT	LABOR-HOURS	UNIT	MAT.	LABOR	EQUIP.	TOTAL	TOTAL INCL O&P	
200	0010	**AUTOMATIC FIRE SUPPRESSION SYSTEMS**										200
	0040	For detectors and control stations, see division 13850										
	1000	Dispersion nozzle, CO2, 3" x 5"	1 Plum	18	.444	Ea.	50	18.15		68.15	82.50	
	1100	FM200, 1-1/2"	"	14	.571		50	23.50		73.50	90	
	2000	Extinguisher, CO2 system, high pressure, 75 lb. cylinder	Q-1	6	2.667	↓	950	98		1,048	1,200	
	2400	FM200 system, filled, with mounting bracket										
	2540	196 lb. container	Q-1	4	4	Ea.	5,800	147		5,947	6,600	
	3000	Electro/mechanical release	L-1	4	4		125	163		288	380	
	3400	Manual pull station	1 Plum	6	1.333	↓	45	54.50		99.50	132	
	6000	Average FM200 system, minimum				C.F.	1.25			1.25	1.38	

Important: See the Reference Section for critical supporting data - Reference Nos., Crews, & City Cost Indexes

13960	CO2 Fire Extinguishing	CREW	DAILY OUTPUT	LABOR-HOURS	UNIT	2005 BARE COSTS				TOTAL INCL O&P	
						MAT.	LABOR	EQUIP.	TOTAL		
200	6020	Maximum				C.F.	2.50			2.50	2.75	200

For information about Means Estimating Seminars, see yellow pages 12 and 13 in back of book

Division Notes

		CREW	DAILY OUTPUT	LABOR-HOURS	UNIT	2005 BARE COSTS				TOTAL INCL O&P
						MAT.	LABOR	EQUIP.	TOTAL	

Division 14
Conveying Systems

Estimating Tips

General

- Many products in Division 14 will require some type of support or blocking for installation not included with the item itself. Examples are supports for conveyors or tube systems, attachment points for lifts, and footings for hoists or cranes. Add these supports in the appropriate division.

14100 Dumbwaiters
14200 Elevators

- Dumbwaiters and elevators are estimated and purchased in a method similar to buying a car. The manufacturer has a base unit with standard features. Added to this base unit price will be whatever options the owner or specifications require. Increased load capacity, additional vertical travel, additional stops, higher speed, and cab finish options are items to be considered. When developing an estimate for dumbwaiters and elevators, remember that some items needed by the installers may have to be included as part of the general contract.

Examples are:
- shaftway
- rail support brackets
- machine room
- electrical supply
- sill angles
- electrical connections
- pits
- roof penthouses
- pit ladders

Check the job specifications and drawings before pricing.

- Installation of elevators and handicapped lifts in historic structures can require significant additional costs. The associated structural requirements may involve cutting into and repairing finishes, mouldings, flooring, etc. The estimator must account for these special conditions.

14300 Escalators & Moving Walks

- Escalators and moving walks are specialty items installed by specialty contractors. There are numerous options associated with these items. For specific options contact a manufacturer or contractor. In a method similar to estimating dumbwaiters and elevators, you should verify the extent of general contract work and add items as necessary.

14400 Lifts
14500 Material Handling
14600 Hoists & Cranes

- Products such as correspondence lifts, conveyors, chutes, pneumatic tube systems, material handling cranes and hoists as well as other items specified in this subdivision may require trained installers. The general contractor might not have any choice as to who will perform the installation or when it will be performed. Long lead times are often required for these products, making early decisions in scheduling necessary.

Reference Numbers

Reference numbers are shown in bold squares at the beginning of some major classifications. These numbers refer to related items in the Reference Section. The reference information may be an estimating procedure, an alternate pricing method or technical information.

Note: Not all subdivisions listed here necessarily appear in this publication.

14100 | Dumbwaiters

14110 | Manual Dumbwaiters

			CREW	DAILY OUTPUT	LABOR-HOURS	UNIT	2005 BARE COSTS				TOTAL INCL O&P	
							MAT.	LABOR	EQUIP.	TOTAL		
400	0010	**DUMBWAITERS** 2 stop, hand, minimum	2 Elev	.75	21.333	Ea.	2,400	960		3,360	4,100	400
	0100	Maximum		.50	32	"	5,450	1,450		6,900	8,150	
	0300	For each additional stop, add	↓	.75	21.333	Stop	870	960		1,830	2,400	

14120 | Electric Dumbwaiters

			CREW	DAILY OUTPUT	LABOR-HOURS	UNIT	2005 BARE COSTS				TOTAL INCL O&P	
							MAT.	LABOR	EQUIP.	TOTAL		
400	0010	**DUMBWAITERS** 2 stop, electric, minimum	2 Elev	.13	123	Ea.	6,050	5,550		11,600	15,000	400
	0100	Maximum		.11	145	"	18,200	6,550		24,750	29,800	
	0600	For each additional stop, add	↓	.54	29.630	Stop	2,675	1,325		4,000	4,925	
	0750	45 lb capacity	2 Elec	.20	80	Ea.	7,025	3,250		10,275	12,600	

14200 | Elevators

14210 | Electric Traction Elevators

			CREW	DAILY OUTPUT	LABOR-HOURS	UNIT	2005 BARE COSTS				TOTAL INCL O&P	
							MAT.	LABOR	EQUIP.	TOTAL		
100	0010	**ELEVATOR SYSTEMS**										100
	7000	Residential, cab type, 1 floor, 2 stop, minimum	2 Elev	.20	80	Ea.	8,500	3,600		12,100	14,800	
	7100	Maximum		.10	160		14,300	7,200		21,500	26,600	
	7200	2 floor, 3 stop, minimum		.12	133		12,600	6,000		18,600	22,900	
	7300	Maximum	↓	.06	266	↓	20,600	12,000		32,600	40,600	
200	0010	**ELEVATORS**										200
	0020	For multi-story buildings, housing project, minimum	R14200 -100			% total					2.50%	
	0100	Maximum									4.50%	
	0300	Office building, minimum	R14200 -200								2.50%	
	0400	Maximum				↓					10%	
	0410	Holeless hydraulic passenger, 2000 lb capacity, 2 story	R14200 -300	2 Elev	.10	160	Ea.	29,800	7,200		37,000	43,600	
	0420	2500 lb capacity		.10	160		32,800	7,200		40,000	46,900	
	0425	Electric freight, base unit, 4000 lb, 200 fpm, 4 stop, std. fin.	R14200 -400	↓	.05	320		71,000	14,400		85,400	100,000	
	0450	For 5000 lb capacity, add					5,100			5,100	5,625	
	0500	For 6000 lb capacity, add					9,000			9,000	9,900	
	0525	For 7000 lb capacity, add					12,000			12,000	13,300	
	0550	For 8000 lb capacity, add					16,500			16,500	18,100	
	0575	For 10000 lb capacity, add					19,700			19,700	21,700	
	0600	For 12000 lb capacity, add					24,100			24,100	26,500	
	0625	For 16000 lb capacity, add					28,900			28,900	31,800	
	0650	For 20000 lb capacity, add					31,900			31,900	35,100	
	0675	For increased speed, 250 fpm, add					9,575			9,575	10,500	
	0700	300 fpm, geared electric, add					12,000			12,000	13,100	
	0725	350 fpm, geared electric, add					14,100			14,100	15,500	
	0750	400 fpm, geared electric, add					15,900			15,900	17,500	
	0775	500 fpm, gearless electric, add					20,400			20,400	22,400	
	0800	600 fpm, gearless electric, add					22,500			22,500	24,800	
	0825	700 fpm, gearless electric, add					26,200			26,200	28,800	
	0850	800 fpm, gearless electric, add					29,100			29,100	32,100	
	0875	For class "B" loading, add					1,725			1,725	1,900	
	0900	For class "C-1" loading, add					4,275			4,275	4,700	
	0925	For class "C-2" loading, add					5,100			5,100	5,600	
	0950	For class "C-3" loading, add				↓	7,000			7,000	7,700	
	0975	For travel over 40 V.L.F., add	2 Elev	7.25	2.207	V.L.F.	102	99.50		201.50	261	
	1000	For number of stops over 4, add	↓	.27	59.259	Stop	1,800	2,675		4,475	6,000	

Important: See the Reference Section for critical supporting data - Reference Nos., Crews, & City Cost Indexes

14 CONVEYING SYSTEMS

14200 | Elevators

14210 | Electric Traction Elevators

			CREW	DAILY OUTPUT	LABOR-HOURS	UNIT	MAT.	LABOR	EQUIP.	TOTAL	TOTAL INCL O&P	
200	1625	Electric pass., base unit, 2000 lb, 200 fpm, 4 stop, std. fin.	2 Elev	.05	320	Ea.	66,500	14,400		80,900	95,000	**200**
	1650	For 2500 lb capacity, add					2,725			2,725	3,000	
	1675	For 3000 lb capacity, add					4,150			4,150	4,575	
	1700	For 3500 lb capacity, add					5,775			5,775	6,350	
	1725	For 4000 lb capacity, add					6,175			6,175	6,800	
	1750	For 4500 lb capacity, add					8,125			8,125	8,925	
	1775	For 5000 lb capacity, add					10,300			10,300	11,400	
	1800	For increased speed, 250 fpm, geared electric, add					2,275			2,275	2,500	
	1825	300 fpm, geared electric, add					4,700			4,700	5,175	
	1850	350 fpm, geared electric, add					5,575			5,575	6,125	
	1875	400 fpm, geared electric, add					7,925			7,925	8,700	
	1900	500 fpm, gearless electric, add					37,200			37,200	40,900	
	1925	600 fpm, gearless electric, add					39,300			39,300	43,200	
	1950	700 fpm, gearless electric, add					43,000			43,000	47,300	
	1975	800 fpm, gearless electric, add					47,400			47,400	52,000	
	2000	For travel over 40 V.L.F., add	2 Elev	7.25	2.207	V.L.F.	104	99.50		203.50	263	
	2025	For number of stops over 4, add		.27	59.259	Stop	2,325	2,675		5,000	6,550	
	2400	Electric hospital, base unit, 4000 lb, 200 fpm, 4 stop, std fin.		.05	320	Ea.	73,000	14,400		87,400	101,500	
	2425	For 4500 lb capacity, add					5,025			5,025	5,525	
	2450	For 5000 lb capacity, add					6,575			6,575	7,225	
	2475	For increased speed, 250 fpm, geared electric, add					2,450			2,450	2,700	
	2500	300 fpm, geared electric, add					4,675			4,675	5,125	
	2525	350 fpm, geared electric, add					5,700			5,700	6,250	
	2550	400 fpm, geared electric, add					7,925			7,925	8,725	
	2575	500 fpm, gearless electric, add					35,600			35,600	39,100	
	2600	600 fpm, gearless electric, add					39,300			39,300	43,200	
	2625	700 fpm, gearless electric, add					42,900			42,900	47,200	
	2650	800 fpm, gearless electric, add					47,400			47,400	52,000	
	2675	For travel over 40 V.L.F., add	2 Elev	7.25	2.207	V.L.F.	104	99.50		203.50	263	
	2700	For number of stops over 4, add	"	.27	59.259	Stop	2,875	2,675		5,550	7,175	

R14200-100, R14200-200, R14200-300, R14200-400

14240 | Hydraulic Elevators

			CREW	DAILY OUTPUT	LABOR-HOURS	UNIT	MAT.	LABOR	EQUIP.	TOTAL	TOTAL INCL O&P	
200	0010	**HYDRAULIC ELEVATORS**										**200**
	1025	Hydraulic freight, base unit, 2000 lb, 50 fpm, 2 stop, std. fin.	2 Elev	.10	160	Ea.	35,100	7,200		42,300	49,400	
	1050	For 2500 lb capacity, add					2,550			2,550	2,800	
	1075	For 3000 lb capacity, add					4,050			4,050	4,450	
	1100	For 3500 lb capacity, add					6,125			6,125	6,750	
	1125	For 4000 lb capacity, add					6,550			6,550	7,200	
	1150	For 4500 lb capacity, add					7,500			7,500	8,250	
	1175	For 5000 lb capacity, add					10,200			10,200	11,200	
	1200	For 6000 lb capacity, add					10,800			10,800	11,900	
	1225	For 7000 lb capacity, add					16,500			16,500	18,100	
	1250	For 8000 lb capacity, add					18,500			18,500	20,300	
	1275	For 10000 lb capacity, add					19,600			19,600	21,600	
	1300	For 12000 lb capacity, add					23,700			23,700	26,100	
	1325	For 16000 lb capacity, add					30,900			30,900	34,000	
	1350	For 20000 lb capacity, add					34,300			34,300	37,700	
	1375	For increased speed, 100 fpm, add					735			735	805	
	1400	125 fpm, add					1,400			1,400	1,550	
	1425	150 fpm, add					2,675			2,675	2,950	
	1450	175 fpm, add					4,025			4,025	4,450	
	1475	For class "B" loading, add					1,675			1,675	1,850	
	1500	For class "C-1" loading, add					4,225			4,225	4,650	
	1525	For class "C-2" loading, add					5,075			5,075	5,575	
	1550	For class "C-3" loading, add					6,950			6,950	7,650	
	1575	For travel over 20 V.L.F., add	2 Elev	7.25	2.207	V.L.F.	340	99.50		439.50	525	

2005 BARE COSTS

417

		14240	**Hydraulic Elevators**	CREW	DAILY OUTPUT	LABOR-HOURS	UNIT	\multicolumn{4}{c}{2005 BARE COSTS}	TOTAL INCL O&P	
								MAT.	LABOR	EQUIP.	TOTAL		
200	1600		For number of stops over 2, add	2 Elev	.27	59.259	Stop	505	2,675		3,180	4,550	200
	2050		Hyd. pass., base unit, 1500 lb, 100 fpm, 2 stop, std. fin. **CN**	↓	.10	160	Ea.	29,700	7,200		36,900	43,500	
	2075		For 2000 lb capacity, add					1,000			1,000	1,100	
	2100		For 2500 lb capacity, add					1,975			1,975	2,175	
	2125		For 3000 lb capacity, add					3,550			3,550	3,900	
	2150		For 3500 lb capacity, add					5,275			5,275	5,800	
	2175		For 4000 lb capacity, add					6,325			6,325	6,975	
	2200		For 4500 lb capacity, add					7,625			7,625	8,400	
	2225		For 5000 lb capacity, add					11,500			11,500	12,700	
	2250		For increased speed, 125 fpm, add					1,125			1,125	1,250	
	2275		150 fpm, add					2,250			2,250	2,475	
	2300		175 fpm, add					3,650			3,650	4,000	
	2325		200 fpm, add				↓	5,300			5,300	5,850	
	2350		For travel over 12 V.L.F., add	2 Elev	7.25	2.207	V.L.F.	345	99.50		444.50	530	
	2375		For number of stops over 2, add	↓	.27	59.259	Stop	500	2,675		3,175	4,550	
	2725		Hydraulic hospital, base unit, 4000 lb, 100 fpm, 2 stop, std. fin.	↓	.10	160	Ea.	41,200	7,200		48,400	56,000	
	2750		For 4000 lb capacity, add					4,975			4,975	5,475	
	2775		For 4500 lb capacity, add					5,850			5,850	6,425	
	2800		For 5000 lb capacity, add					8,500			8,500	9,375	
	2825		For increased speed, 125 fpm, add					1,375			1,375	1,525	
	2850		150 fpm, add					2,325			2,325	2,575	
	2875		175 fpm, add					3,750			3,750	4,125	
	2900		200 fpm, add				↓	5,475			5,475	6,025	
	2925		For travel over 12 V.L.F., add	2 Elev	7.25	2.207	V.L.F.	227	99.50		326.50	400	
	2950		For number of stops over 2, add	"	.27	59.259	Stop	2,875	2,675		5,550	7,150	

		14270	**Custom Elevator Cabs**										
200	0010		**CAB FINISHES**										200
	3325		Passenger elevator cab finishes (based on 3500 lb cab size)										
	3350		Acrylic panel ceiling				Ea.	430			430	470	
	3375		Aluminum eggcrate ceiling					495			495	545	
	3400		Stainless steel doors					1,975			1,975	2,175	
	3425		Carpet flooring					360			360	395	
	3450		Epoxy flooring					295			295	325	
	3475		Quarry tile flooring					455			455	500	
	3500		Slate flooring					635			635	700	
	3525		Textured rubber flooring					116			116	127	
	3550		Stainless steel walls					2,575			2,575	2,850	
	3575		Stainless steel returns at door				↓	545			545	595	
	4450		Hospital elevator cab finishes (based on 3500# cab size)										
	4475		Aluminum eggcrate ceiling				Ea.	495			495	545	
	4500		Stainless steel doors					1,575			1,575	1,750	
	4525		Epoxy flooring					295			295	325	
	4550		Quarry tile flooring					450			450	500	
	4575		Textured rubber flooring					116			116	127	
	4600		Stainless steel walls					2,575			2,575	2,850	
	4625		Stainless steel returns at door				↓	545			545	595	

		14280	**Elevator Equipment and Controls**										
200	0010		**ELEVATOR CONTROLS AND DOORS**										200
	2975		Passenger elevator options										
	3000		2 car group automatic controls	2 Elev	.66	24.242	Ea.	2,375	1,100		3,475	4,250	
	3025		3 car group automatic controls		.44	36.364		3,550	1,650		5,200	6,375	
	3050		4 car group automatic controls		.33	48.485		5,900	2,175		8,075	9,750	
	3075		5 car group automatic controls	↓	.26	61.538	↓	8,025	2,775		10,800	13,000	

Important: See the Reference Section for critical supporting data - Reference Nos., Crews, & City Cost Indexes

		CREW	DAILY OUTPUT	LABOR-HOURS	UNIT	2005 BARE COSTS				TOTAL INCL O&P
14280	**Elevator Equipment and Controls**					MAT.	LABOR	EQUIP.	TOTAL	
200 3100	6 car group automatic controls	2 Elev	.22	72.727	Ea.	12,100	3,275		15,375	18,300 **200**
3125	Intercom service		3	5.333		315	240		555	705
3150	Duplex car selective collective		.66	24.242		2,600	1,100		3,700	4,525
3175	Center opening 1 speed doors		2	8		1,375	360		1,735	2,050
3200	Center opening 2 speed doors		2	8		1,725	360		2,085	2,450
3225	Rear opening doors (opposite front)		2	8		3,775	360		4,135	4,700
3250	Side opening 2 speed doors		2	8		3,400	360		3,760	4,275
3275	Automatic emergency power switching		.66	24.242		850	1,100		1,950	2,575
3300	Manual emergency power switching	▼	8	2		325	90		415	495
3625	Hall finishes, stainless steel doors					900			900	990
3650	Stainless steel frames					900			900	990
3675	12 month maintenance contract									2,640
3700	Signal devices, hall lanterns	2 Elev	8	2		335	90		425	500
3725	Position indicators, up to 3		9.40	1.702		232	76.50		308.50	370
3750	Position indicators, per each over 3	▼	32	.500		64	22.50		86.50	105
3775	High speed heavy duty door opener					1,625			1,625	1,800
3800	Variable voltage, O.H. gearless machine, min.	2 Elev	.16	100		23,900	4,500		28,400	33,100
3815	Maximum		.07	228		53,000	10,300		63,300	73,500
3825	Basement installed geared machine	▼	.33	48.485	▼	11,800	2,175		13,975	16,300
3850	Freight elevator options									
3875	Doors, bi-parting	2 Elev	.66	24.242	Ea.	4,075	1,100		5,175	6,125
3900	Power operated door and gate	"	.66	24.242		16,400	1,100		17,500	19,800
3925	Finishes, steel plate floor					715			715	785
3950	14 ga. 1/4" x 4' steel plate walls					1,750			1,750	1,900
3975	12 month maintenance contract									1,992
4000	Signal devices, hall lanterns	2 Elev	8	2		335	90		425	500
4025	Position indicators, up to 3		9.40	1.702		232	76.50		308.50	370
4050	Position indicators, per each over 3		32	.500		64	22.50		86.50	105
4075	Variable voltage basement installed geared machine	▼	.66	24.242	▼	13,200	1,100		14,300	16,200
4100	Hospital elevator options									
4125	2 car group automatic controls	2 Elev	.66	24.242	Ea.	2,375	1,100		3,475	4,250
4150	3 car group automatic controls		.44	36.364		3,550	1,650		5,200	6,375
4175	4 car group automatic controls		.33	48.485		5,900	2,175		8,075	9,775
4200	5 car group automatic controls		.26	61.538		8,050	2,775		10,825	13,000
4225	6 car group automatic controls		.22	72.727		12,200	3,275		15,475	18,300
4250	Intercom service		3	5.333		315	240		555	705
4275	Duplex car selective collective		.66	24.242		2,600	1,100		3,700	4,525
4300	Center opening 1 speed doors		2	8		1,375	360		1,735	2,050
4325	Center opening 2 speed doors		2	8		1,800	360		2,160	2,525
4350	Rear opening doors (opposite front)		2	8		3,775	360		4,135	4,700
4375	Side opening 2 speed doors		2	8		5,750	360		6,110	6,875
4400	Automatic emergency power switching		.66	24.242		850	1,100		1,950	2,575
4425	Manual emergency power switching	▼	8	2		325	90		415	495
4675	Hall finishes, stainless steel doors					900			900	990
4700	Stainless steel frames					900			900	990
4725	12 month maintenance contract									3,985
4750	Signal devices, hall lanterns	2 Elev	8	2		335	90		425	500
4775	Position indicators, up to 3		9.40	1.702		232	76.50		308.50	370
4800	Position indicators, per each over 3	▼	32	.500		64	22.50		86.50	105
4825	High speed heavy duty door opener					1,625			1,625	1,800
4850	Variable voltage, O.H. gearless machine, min.	2 Elev	.16	100		23,900	4,500		28,400	33,100
4865	Maximum		.07	228		53,000	10,300		63,300	73,500
4875	Basement installed geared machine	▼	.33	48.485	▼	11,800	2,175		13,975	16,300
5000	Drilling for piston, casing included, 18" diameter	B-48	80	.700	V.L.F.	32	21	42.50	95.50	114

CONVEYING SYSTEMS 14

14320 | Escalators

			CREW	DAILY OUTPUT	LABOR-HOURS	UNIT	2005 BARE COSTS				TOTAL INCL O&P
							MAT.	LABOR	EQUIP.	TOTAL	
300	0010	**ESCALATORS** R14320-100	M-1	.07	457	Ea.	62,000	19,600	900	82,500	99,000	300
	1000	Glass, 32" wide x 10' floor to floor height		.07	457		62,000	19,600	900	82,500	99,000	
	1010	48" wide x 10' floor to floor height		.07	457		68,500	19,600	900	89,000	105,500	
	1020	32" wide x 15' floor to floor height **CN**		.06	533		65,000	22,800	1,050	88,850	107,000	
	1030	48" wide x 15' floor to floor height		.06	533		69,000	22,800	1,050	92,850	111,500	
	1040	32" wide x 20' floor to floor height		.05	653		67,000	28,000	1,275	96,275	117,500	
	1050	48" wide x 20' floor to floor height		.05	653		72,000	28,000	1,275	101,275	122,500	
	1060	32" wide x 25' floor to floor height		.04	800		72,500	34,200	1,575	108,275	132,500	
	1070	48" wide x 25' floor to floor height		.04	800		77,500	34,200	1,575	113,275	138,000	
	1080	Enameled steel, 32" wide x 10' floor to floor height		.07	457		63,500	19,600	900	84,000	100,500	
	1090	48" wide x 10' floor to floor height		.07	457		68,500	19,600	900	89,000	105,500	
	1110	32" wide x 15' floor to floor height		.06	533		65,000	22,800	1,050	88,850	107,000	
	1120	48" wide x 15' floor to floor height		.06	533		69,000	22,800	1,050	92,850	111,500	
	1130	32" wide x 20' floor to floor height		.05	653		67,000	28,000	1,275	96,275	117,500	
	1140	48" wide x 20' floor to floor height		.05	653		72,000	28,000	1,275	101,275	122,500	
	1150	32" wide x 25' floor to floor height		.04	800		72,500	34,200	1,575	108,275	132,500	
	1160	48" wide x 25' floor to floor height		.04	800		77,500	34,200	1,575	113,275	138,000	
	1170	Stainless steel, 32" wide x 10' floor to floor height		.07	457		70,000	19,600	900	90,500	107,500	
	1180	48" wide x 10' floor to floor height		.07	457		73,000	19,600	900	93,500	111,000	
	1500	32" wide x 15' floor to floor height		.06	533		71,000	22,800	1,050	94,850	113,500	
	1700	48" wide x 15' floor to floor height		.06	533		74,500	22,800	1,050	98,350	117,500	
	2300	32" wide x 25' floor to floor height		.04	800		80,500	34,200	1,575	116,275	141,500	
	2500	48" wide x 25' floor to floor height		.04	800		85,500	34,200	1,575	121,275	147,000	

14360 | Moving Walks

			CREW	DAILY OUTPUT	LABOR-HOURS	UNIT	MAT.	LABOR	EQUIP.	TOTAL	TOTAL INCL O&P
500	0010	**MOVING RAMPS AND WALKS** Walk, 27" tread width, minimum R14360-200	M-1	6.50	4.923	L.F.	715	211	9.70	935.70	1,125	500
	0100	300' to 500', maximum		4.43	7.223		995	310	14.25	1,319.25	1,575	
	0300	48" tread width walk, minimum		4.43	7.223		1,625	310	14.25	1,949.25	2,250	
	0400	100' to 350', maximum		3.82	8.377		1,900	360	16.50	2,276.50	2,625	
	0600	Ramp, 12° incline, 36" tread width, minimum		5.27	6.072		1,325	260	11.95	1,596.95	1,850	
	0700	70' to 90' maximum		3.82	8.377		1,900	360	16.50	2,276.50	2,625	
	0900	48" tread width, minimum		3.57	8.964		1,925	385	17.65	2,327.65	2,725	
	1000	40' to 70', maximum		2.91	10.997		2,450	470	21.50	2,941.50	3,400	

14420 | Wheelchair Lifts

			CREW	DAILY OUTPUT	LABOR-HOURS	UNIT	2005 BARE COSTS				TOTAL INCL O&P
							MAT.	LABOR	EQUIP.	TOTAL	
100	0010	**CHAIR / WHEELCHAIR LIFT**										100
	7700	Stair climber (chair lift), single seat, minimum	2 Elev	1	16	Ea.	4,100	720		4,820	5,600	
	7800	Maximum		.20	80		5,650	3,600		9,250	11,600	
	8000	Wheelchair lift, minimum		1	16		5,625	720		6,345	7,275	
	8500	Maximum		.50	32		13,300	1,450		14,750	16,900	
	8700	Stair lift, minimum		1	16		11,100	720		11,820	13,400	
	8900	Maximum		.20	80		17,600	3,600		21,200	24,800	

14450 | Vehicle Lifts

			CREW	DAILY OUTPUT	LABOR-HOURS	UNIT	MAT.	LABOR	EQUIP.	TOTAL	TOTAL INCL O&P
700	0010	**HYDRAULIC LIFT AUTO/TRUCK**										700
	0200	Double post, 9000 lb frame	L-4	1.15	20.870	Ea.	5,150	665		5,815	6,700	

Important: See the Reference Section for critical supporting data - Reference Nos., Crews, & City Cost Indexes

14 CONVEYING SYSTEMS

14400 | Lifts

14450 | Vehicle Lifts

			CREW	DAILY OUTPUT	LABOR-HOURS	UNIT	2005 BARE COSTS				TOTAL INCL O&P	
							MAT.	LABOR	EQUIP.	TOTAL		
700	0400	26000 lb frame	L-4	.22	109	Ea.	22,900	3,475		26,375	30,600	700
	2200	Hoists, single post, 8,000# capacity, swivel arms		.40	60		4,300	1,900		6,200	7,700	
	2400	Two posts, adjustable frames, 11,000# capacity		.25	96		5,525	3,050		8,575	10,900	
	2500	24,000# capacity		.15	160		7,375	5,075		12,450	16,000	
	2700	7,500# capacity, frame supports		.50	48		6,150	1,525		7,675	9,150	
	2800	Four post, roll on ramp		.50	48		5,525	1,525		7,050	8,475	
	2810	Hydraulic lifts, above ground, 2 post, clear floor, 6000 lb cap		2.67	8.989		5,775	285		6,060	6,800	
	2815	9000 lb capacity		2.29	10.480		13,700	335		14,035	15,600	
	2820	15,000 lb capacity		2	12		15,500	380		15,880	17,600	
	2825	30,000 lb capacity		1.60	15		34,300	475		34,775	38,400	
	2830	4 post, ramp style, 25,000 lb capacity		2	12		13,400	380		13,780	15,300	
	2835	35,000 lb capacity		1	24		62,500	760		63,260	70,000	
	2840	50,000 lb capacity		1	24		70,000	760		70,760	78,000	
	2845	75,000 lb capacity	↓	1	24		81,000	760		81,760	90,500	
	2850	For drive thru tracks, add, minimum					860			860	945	
	2855	Maximum					1,475			1,475	1,625	
	2860	Ramp extensions, 3'(set of 2)					705			705	780	
	2865	Rolling jack platform					2,450			2,450	2,700	
	2870	Elec/hyd jacking beam					6,575			6,575	7,250	
	2880	Scissor lift, portable, 6000 lb capacity				↓	6,450			6,450	7,100	

14460 | Correspondence & Parcel Lifts

			CREW	DAILY OUTPUT	LABOR-HOURS	UNIT	MAT.	LABOR	EQUIP.	TOTAL	TOTAL INCL O&P	
200	0010	**CORRESPONDENCE LIFT**, 1 floor 2 stop, 25 lb capacity, electric	2 Elev	.20	80	Ea.	5,025	3,600		8,625	10,900	200
	0100	Hand, 5 lb capacity	"	.20	80	"	1,900	3,600		5,500	7,500	
600	0010	**PARCEL LIFT** 20" x 20", 100 lb capacity, electric, per floor	2 Mill	.25	64	Ea.	7,850	2,275		10,125	12,000	600

14500 | Material Transport

14510 | Material Handling

			CREW	DAILY OUTPUT	LABOR-HOURS	UNIT	2005 BARE COSTS				TOTAL INCL O&P	
							MAT.	LABOR	EQUIP.	TOTAL		
900	0010	**MOTORIZED CAR** Distribution systems, single track,										900
	0100	20 lb. per car capacity, material handling										
	0200	Minimum, 50' to 100'	4 Mill	.19	168	Station	25,100	5,975		31,075	36,400	
	0300	Maximum, 100' to 200'	"	.15	213	"	33,800	7,575		41,375	48,400	
	0400	Larger system, incl. hospital transport, track type,										
	0500	fully automated material handling system										
	0600	Minimum, complete system	4 Mill	.05	640	Ea.	111,000	22,700		133,700	155,500	
	0700	Maximum, with several stations	E-6	.05	2,560	"	648,000	96,000	30,400	774,400	914,000	

14550 | Conveyors

			CREW	DAILY OUTPUT	LABOR-HOURS	UNIT	MAT.	LABOR	EQUIP.	TOTAL	TOTAL INCL O&P	
350	0010	**MATERIAL HANDLING** Conveyors, gravity type 2" rollers, 3" O.C.										350
	0050	10' sections with 2 supports, 600 lb. capacity, 18" wide				Ea.	375			375	410	
	0100	24" wide					430			430	475	
	0150	1400 lb. capacity, 18" wide					345			345	380	
	0200	24" wide				↓	380			380	420	
	0350	Horizontal belt, center drive and takeup, 60 fpm										
	0400	16" belt, 26.5' length	2 Mill	.50	32	Ea.	2,625	1,125		3,750	4,575	
	0450	24" belt, 41.5' length		.40	40	↓	4,075	1,425		5,500	6,600	
	0500	61.5' length	↓	.30	53.333		5,450	1,900		7,350	8,775	
	0600	Inclined belt, 10' rise with horizontal loader and										

		14550	Conveyors	CREW	DAILY OUTPUT	LABOR-HOURS	UNIT	MAT.	LABOR	EQUIP.	TOTAL	TOTAL INCL O&P	
								\multicolumn{4}{c}{2005 BARE COSTS}		
350	0620		End idler assembly, 27.5' length, 18" belt	2 Mill	.30	53.333	Ea.	7,675	1,900		9,575	11,200	350
	0700		24" belt	"	.15	106	"	8,600	3,775		12,375	15,100	
	3600		Monorail, overhead, manual, channel type										
	3700		125 lb. per L.F.	1 Mill	26	.308	L.F.	8.50	10.90		19.40	25.50	
	3900		500 lb. per L.F.	"	21	.381	"	7.70	13.50		21.20	28.50	
	4000		Trolleys for above, 2 wheel, 125 lb. capacity				Ea.	53.50			53.50	59	
	4200		4 wheel, 250 lb. capacity					163			163	179	
	4300		8 wheel, 1,000 lb. capacity				↓	315			315	345	
900	0010		**VERTICAL CONVEYOR** Automatic selective										900
	0100		to 10 floors, base price	2 Mill	.13	120	Floor	22,900	4,275		27,175	31,500	
	0200		Add for electrical hook-up and testing	2 Elec	.50	32	"		1,300		1,300	1,950	

		14560	Chutes										
250	0010		**CHUTES** Linen or refuse, incl. sprinklers, 12' floor height										250
	0050		Aluminized steel, 16 ga., 18" diameter	2 Shee	3.50	4.571	Floor	860	184		1,044	1,225	
	0100		24" diameter		3.20	5		935	201		1,136	1,325	
	0200		30" diameter		3	5.333		1,025	214		1,239	1,450	
	0300		36" diameter		2.80	5.714		1,175	230		1,405	1,625	
	0400		Galvanized steel, 16 ga., 18" diameter		3.50	4.571		765	184		949	1,125	
	0500		24" diameter		3.20	5		860	201		1,061	1,250	
	0600		30" diameter		3	5.333		965	214		1,179	1,400	
	0700		36" diameter		2.80	5.714		1,150	230		1,380	1,600	
	0800		Stainless steel, 18" diameter		3.50	4.571		1,775	184		1,959	2,225	
	0900		24" diameter		3.20	5		1,875	201		2,076	2,375	
	1000		30" diameter		3	5.333		2,225	214		2,439	2,775	
	1005		36" diameter		2.80	5.714	↓	2,475	230		2,705	3,075	
	1200		Linen chute bottom collector, aluminized steel		4	4	Ea.	1,075	161		1,236	1,425	
	1300		Stainless steel		4	4		1,375	161		1,536	1,750	
	1500		Refuse, bottom hopper, aluminized steel, 18" diameter		3	5.333		695	214		909	1,100	
	1600		24" diameter		3	5.333		765	214		979	1,175	
	1800		36" diameter		3	5.333	↓	1,300	214		1,514	1,775	
	2900		Package chutes, spiral type, minimum		4.50	3.556	Floor	1,850	143		1,993	2,250	
	3000		Maximum	↓	1.50	10.667	"	4,825	430		5,255	5,975	

		14580	Pneumatic Tube Systems										
800	0010		**PNEUMATIC TUBE SYSTEM** Single tube, 2 stations, blower										800
	0020		100' long, stock										
	0100		3" diameter	2 Stpi	.12	133	Total	5,025	5,450		10,475	13,700	
	0300		4" diameter	"	.09	177	"	5,675	7,275		12,950	17,200	
	0400		Twin tube, two stations or more, conventional system										
	0600		2-1/2" round	2 Stpi	62.50	.256	L.F.	10.15	10.50		20.65	27	
	0700		3" round		46	.348		11.65	14.25		25.90	34.50	
	0900		4" round		49.60	.323		12.90	13.20		26.10	34	
	1000		4" x 7" oval		37.60	.426	↓	18.80	17.45		36.25	46.50	
	1050		Add for blower		2	8	System	3,975	330		4,305	4,875	
	1110		Plus for each round station, add		7.50	2.133	Ea.	450	87.50		537.50	625	
	1150		Plus for each oval station, add		7.50	2.133	"	450	87.50		537.50	625	
	1200		Alternate pricing method: base cost, minimum		.75	21.333	Total	4,875	875		5,750	6,675	
	1300		Maximum		.25	64	"	9,675	2,625		12,300	14,600	
	1500		Plus total system length, add, minimum		93.40	.171	L.F.	6.25	7		13.25	17.45	
	1600		Maximum		37.60	.426	"	18.50	17.45		35.95	46.50	
	1800		Completely automatic system, 4" round, 15 to 50 stations		.29	55.172	Station	15,300	2,250		17,550	20,300	
	2200		51 to 144 stations		.32	50		11,900	2,050		13,950	16,200	
	2400		6" round or 4" x 7" oval, 15 to 50 stations		.24	66.667		19,200	2,725		21,925	25,200	
	2800		51 to 144 stations	↓	.23	69.565	↓	16,100	2,850		18,950	22,000	

14 CONVEYING SYSTEMS

14600 | Hoists and Cranes

14610 | Fixed Hoists

			CREW	DAILY OUTPUT	LABOR-HOURS	UNIT	2005 BARE COSTS				TOTAL INCL O&P	
							MAT.	LABOR	EQUIP.	TOTAL		
500	0010	**MATERIAL HANDLING**										500
	1500	Cranes, portable hydraulic, floor type, 2,000 lb capacity				Ea.	1,800			1,800	1,975	
	1600	4,000 lb capacity					3,100			3,100	3,425	
	1800	Movable gantry type, 12′ to 15′ range, 2,000 lb capacity					2,775			2,775	3,050	
	1900	6,000 lb capacity					4,025			4,025	4,425	
	2100	Hoists, electric overhead, chain, hook hung, 15′ lift, 1 ton cap.					1,825			1,825	2,000	
	2200	3 ton capacity					2,275			2,275	2,500	
	2500	5 ton capacity					5,000			5,000	5,500	
	2600	For hand-pushed trolley, add					15%					
	2700	For geared trolley, add					30%					
	2800	For motor trolley, add					75%					
	3000	For lifts over 15′, 1 ton, add				L.F.	17.10			17.10	18.80	
	3100	5 ton, add				"	40			40	44	
	3300	Lifts, scissor type, portable, electric, 36″ high, 2,000 lb				Ea.	3,275			3,275	3,600	
	3400	48″ high, 4,000 lb				"	3,750			3,750	4,125	

14630 | Bridge Cranes

			CREW	DAILY OUTPUT	LABOR-HOURS	UNIT	MAT.	LABOR	EQUIP.	TOTAL	TOTAL INCL O&P	
300	0010	**CRANE RAIL** Box beam bridge, no equipment included	E-4	3,400	.009	Lb.	.91	.36	.02	1.29	1.68	300
	0200	Running track only, 104 lb per yard	"	5,600	.006	"	.48	.22	.01	.71	.94	
700	0010	**OVERHEAD BRIDGE CRANES**										700
	0100	1 girder, 20′ span, 3 ton	M-3	1	34	Ea.	19,100	1,275	146	20,521	23,100	
	0125	5 ton		1	34		20,900	1,275	146	22,321	25,100	
	0150	7.5 ton		1	34		24,900	1,275	146	26,321	29,500	
	0175	10 ton		.80	42.500		32,900	1,600	182	34,682	38,800	
	0200	15 ton		.80	42.500		42,300	1,600	182	44,082	49,100	
	0225	30′ span, 3 ton		1	34		19,900	1,275	146	21,321	24,000	
	0250	5 ton		1	34		21,900	1,275	146	23,321	26,100	
	0275	7.5 ton		1	34		26,200	1,275	146	27,621	30,900	
	0300	10 ton		.80	42.500		34,100	1,600	182	35,882	40,100	
	0325	15 ton		.80	42.500		44,200	1,600	182	45,982	51,000	
	0350	2 girder, 40′ span, 3 ton	M-4	.50	72		32,900	2,675	435	36,010	40,600	
	0375	5 ton		.50	72		34,400	2,675	435	37,510	42,300	
	0400	7.5 ton		.50	72		37,700	2,675	435	40,810	46,000	
	0425	10 ton		.40	90		44,100	3,350	545	47,995	54,000	
	0450	15 ton		.40	90		60,000	3,350	545	63,895	71,500	
	0475	25 ton		.30	120		71,000	4,450	725	76,175	85,500	
	0500	50′ span, 3 ton		.50	72		37,400	2,675	435	40,510	45,700	
	0525	5 ton		.50	72		38,900	2,675	435	42,010	47,300	
	0550	7.5 ton		.50	72		41,600	2,675	435	44,710	50,500	
	0575	10 ton		.40	90		48,000	3,350	545	51,895	58,500	
	0600	15 ton		.40	90		63,000	3,350	545	66,895	74,500	
	0625	25 ton		.30	120		74,500	4,450	725	79,675	89,000	

For information about Means Estimating Seminars, see yellow pages 12 and 13 in back of book

CONVEYING SYSTEMS **14**

Division Notes

	CREW	DAILY OUTPUT	LABOR-HOURS	UNIT	2005 BARE COSTS				TOTAL INCL O&P
					MAT.	LABOR	EQUIP.	TOTAL	

Division 15
Mechanical

Estimating Tips

15100 Building Services Piping
This subdivision is primarily basic pipe and related materials. The pipe may be used by any of the mechanical disciplines, i.e., plumbing, fire protection, heating, and air conditioning.

- The piping section lists the add to labor for elevated pipe installation. These adds apply to all elevated pipe, fittings, valves, insulation, etc., that are placed above 10′ high. CAUTION: the correct percentage may vary for the same pipe. For example, the percentage add for the basic pipe installation should be based on the maximum height that the craftsman must install for that particular section. If the pipe is to be located 14′ above the floor but it is suspended on threaded rod from beams, the bottom flange of which is 18′ high (4′ rods), then the height is actually 18′ and the add is 20%. The pipe coverer, however, does not have to go above the 14′ and so his add should be 10%.

- Most pipe is priced first as straight pipe with a joint (coupling, weld, etc.) every 10′ and a hanger usually every 10′. There are exceptions with hanger spacing such as: for cast iron pipe (5′) and plastic pipe (3 per 10′). Following each type of pipe there are several lines listing sizes and the amount to be subtracted to delete couplings and hangers. This is for pipe that is to be buried or supported together on trapeze hangers. The reason that the couplings are deleted is that these runs are usually long and frequently longer lengths of pipe are used. By deleting the couplings the estimator is expected to look up and add back the correct reduced number of couplings.

- When preparing an estimate it may be necessary to approximate the fittings. Fittings usually run between 25% and 50% of the cost of the pipe. The lower percentage is for simpler runs, and the higher number is for complex areas like mechanical rooms.

- For historic restoration projects, the systems must be as invisible as possible, and pathways must be sought for pipes, conduits, and ductwork. While installations in accessible spaces (such as basements and attics) are relatively straightforward to estimate, labor costs may be more difficult to determine when delivery systems must be concealed.

15400 Plumbing Fixtures & Equipment

- Plumbing fixture costs usually require two lines, the fixture itself and its "rough-in, supply and waste".

- In the Assemblies Section (Plumbing D2010) for the desired fixture, the System Components Group in the center of the page shows the fixture itself on the first line while the rest of the list (fittings, pipe, tubing, etc.) will total up to what we refer to in the Unit Price section as "Rough-in, supply, waste and vent". Note that for most fixtures we allow a nominal 5′ of tubing to reach from the fixture to a main or riser.

- Remember that gas and oil fired units need venting.

15500 Heat Generation Equipment

- When estimating the cost of an HVAC system, check to see who is responsible for providing and installing the temperature control system. It is possible to overlook controls, assuming that they would be included in the electrical estimate.

- When looking up a boiler be careful on specified capacity. Some manufacturers rate their products on output while others use input.

- Include HVAC insulation for pipe, boiler and duct (wrap and liner).

- Be careful when looking up mechanical items to get the correct pressure rating and connection type (thread, weld, flange).

15700 Heating/Ventilation/ Air Conditioning Equipment

- Combination heating and cooling units are sized by the air conditioning requirements. (See Reference No. R15710-020 for preliminary sizing guide.)

- A ton of air conditioning is nominally 400 CFM.

- Rectangular duct is taken off by the linear foot for each size, but its cost is usually estimated by the pound. Remember that SMACNA standards now base duct on internal pressure.

- Prefabricated duct is estimated and purchased like pipe: straight sections and fittings.

- Note that cranes or other lifting equipment are not included on any lines in Division 15. For example, if a crane is required to lift a heavy piece of pipe into place high above a gym floor, or to put a rooftop unit on the roof of a four-story building, etc., it must be added. Due to the potential for extreme variation—from nothing additional required, to a major crane or helicopter—we feel that including a nominal amount for "lifting contingency" would be useless and detract from the accuracy of the estimate. When using equipment rental from Means do not forget to include the cost of the operator(s).

Reference Numbers
Reference numbers are shown in bold squares at the beginning of some major classifications. These numbers refer to related items in the Reference Section. The reference information may be an estimating procedure, an alternate pricing method or technical information.

Note: Not all subdivisions listed here necessarily appear in this publication.

Note: **i2 Trade Service,** *in part, has been used as a reference source for some of the material prices used in Division 15.*

15051	Mechanical General	CREW	DAILY OUTPUT	LABOR-HOURS	UNIT	2005 BARE COSTS				TOTAL INCL O&P	
						MAT.	LABOR	EQUIP.	TOTAL		
100	0010 **BOILERS, GENERAL** Prices do not include flue piping, elec. wiring,										100
	0020 gas or oil piping, boiler base, pad, or tankless unless noted										
	0100 Boiler H.P.: 10 KW = 34 lbs/steam/hr = 33,475 BTU/hr.	R15500 -050									
	0120										
	0150 To convert SFR to BTU rating: Hot water, 150 x SFR;										
	0160 Forced hot water, 180 x SFR; steam, 240 x SFR										
700	0010 **PIPING** See also divisions 02500 & 02600 for site work										700
	1000 Add to labor for elevated installation										
	1080 10' to 15' high						10%				
	1100 15' to 20' high						20%				
	1120 20' to 25' high						25%				
	1140 25' to 30' high						35%				
	1160 30' to 35' high						40%				
	1180 35' to 40' high						50%				
	1200 Over 40' high						55%				

15055	Selective Mech Demolition										
300	0010 **HVAC DEMOLITION**										300
	0100 Air conditioner, split unit, 3 ton	Q-5	2	8	Ea.		295		295	445	
	0150 Package unit, 3 ton	Q-6	3	8	"		305		305	460	
	0298 Boilers										
	0300 Electric, up thru 148 kW	Q-19	2	12	Ea.		460		460	685	
	0310 150 thru 518 kW	"	1	24			915		915	1,375	
	0320 550 thru 2000 kW	Q-21	.40	80			3,100		3,100	4,650	
	0330 2070 kW and up	"	.30	106			4,150		4,150	6,200	
	0340 Gas and or oil, up thru 150 MBH	Q-7	2.20	14.545			570		570	855	
	0350 160 thru 2000 MBH		.80	40			1,550		1,550	2,350	
	0360 2100 thru 4500 MBH		.50	64			2,500		2,500	3,750	
	0370 4600 thru 7000 MBH		.30	106			4,175		4,175	6,250	
	0380 7100 thru 12,000 MBH		.16	200			7,800		7,800	11,700	
	0390 12,200 thru 25,000 MBH		.12	266			10,400		10,400	15,600	
	1000 Ductwork, 4" high, 8" wide	1 Clab	200	.040	L.F.		1.07		1.07	1.66	
	1100 6" high, 8" wide		165	.048			1.29		1.29	2.01	
	1200 10" high, 12" wide		125	.064			1.71		1.71	2.66	
	1300 12"-14" high, 16"-18" wide		85	.094			2.51		2.51	3.91	
	1400 18" high, 24" wide		67	.119			3.19		3.19	4.96	
	1500 30" high, 36" wide		56	.143			3.81		3.81	5.95	
	1540 72" wide		50	.160			4.27		4.27	6.65	
	3000 Mechanical equipment, light items. Unit is weight, not cooling.	Q-5	.90	17.778	Ton		655		655	985	
	3600 Heavy items	"	1.10	14.545	"		535		535	805	
	3700 Deduct for salvage (when applicable), minimum				Job					55	
	3710 Maximum				"					420	
600	0010 **PLUMBING DEMOLITION**										600
	1020 Fixtures, including 10' piping										
	1100 Bath tubs, cast iron	1 Plum	4	2	Ea.		81.50		81.50	123	
	1120 Fiberglass		6	1.333			54.50		54.50	82	
	1140 Steel		5	1.600			65.50		65.50	98	
	1200 Lavatory, wall hung		10	.800			32.50		32.50	49	
	1220 Counter top		8	1			41		41	61.50	
	1300 Sink, steel or cast iron, single		8	1			41		41	61.50	
	1320 Double		7	1.143			46.50		46.50	70	
	1400 Water closet, floor mounted		8	1			41		41	61.50	
	1420 Wall mounted		7	1.143			46.50		46.50	70	
	1500 Urinal, floor mounted		4	2			81.50		81.50	123	

Important: See the Reference Section for critical supporting data - Reference Nos., Crews, & City Cost Indexes

15055 | Selective Mech Demolition

			DAILY OUTPUT	LABOR-HOURS	UNIT	2005 BARE COSTS				TOTAL INCL O&P	
						MAT.	LABOR	EQUIP.	TOTAL		
600	1520	Wall mounted	1 Plum	7	1.143	Ea.		46.50		46.50	70	**600**
	1600	Water fountains, free standing		8	1			41		41	61.50	
	1620	Recessed		6	1.333	↓		54.50		54.50	82	
	2000	Piping, metal, up thru 1-1/2" diameter		200	.040	L.F.		1.63		1.63	2.46	
	2050	2" thru 3-1/2" diameter	▼	150	.053			2.18		2.18	3.27	
	2100	4" thru 6" diameter	2 Plum	100	.160			6.55		6.55	9.80	
	2150	8" thru 14" diameter	"	60	.267			10.90		10.90	16.35	
	2153	16" thru 20" diameter	Q-18	70	.343			13.10	1.15	14.25	21	
	2155	24" thru 26" diameter		55	.436			16.70	1.47	18.17	26.50	
	2156	30" thru 36" diameter	▼	40	.600	▼		23	2.02	25.02	36.50	
	2212	Deduct for salvage, aluminum scrap				Ton					525	
	2214	Brass scrap									420	
	2216	Copper scrap									1,025	
	2218	Lead scrap									235	
	2220	Steel scrap				▼					65	
	2250	Water heater, 40 gal.	1 Plum	6	1.333	Ea.		54.50		54.50	82	

15080 | Mechanical Insulation

			DAILY OUTPUT	LABOR-HOURS	UNIT	2005 BARE COSTS				TOTAL INCL O&P	
						MAT.	LABOR	EQUIP.	TOTAL		
200	0010	**DUCT INSULATION**										**200**
	0100	Rule of thumb, as a percentage of total mechanical costs				Job				10%		
	0110	Insulation req'd is based on the surface size/area to be covered										
	3000	Ductwork										
	3020	Blanket type, fiberglass, flexible										
	3030	Fire resistant liner, black coating one side										
	3050	1/2" thick, 2 lb. density	Q-14	380	.042	S.F.	.38	1.41		1.79	2.64	
	3060	1" thick, 1-1/2 lb. density		350	.046		.50	1.53		2.03	2.96	
	3070	1-1/2" thick, 1-1/2 lb. density		320	.050		.88	1.67		2.55	3.61	
	3080	2" thick, 1-1/2 lb. density	▼	300	.053	▼	1.04	1.78		2.82	3.95	
	3140	FRK vapor barrier wrap, .75 lb. density										
	3160	1" thick	Q-14	350	.046	S.F.	.25	1.53		1.78	2.69	
	3170	1-1/2" thick		320	.050		.27	1.67		1.94	2.94	
	3180	2" thick		300	.053		.38	1.78		2.16	3.23	
	3190	3" thick		260	.062		.52	2.06		2.58	3.81	
	3200	4" thick	▼	242	.066	▼	.74	2.21		2.95	4.29	
	3210	Vinyl jacket, same as FRK										
	3280	Unfaced, 1 lb. density										
	3310	1" thick	Q-14	360	.044	S.F.	.33	1.48		1.81	2.70	
	3320	1-1/2" thick		330	.048		.45	1.62		2.07	3.05	
	3330	2" thick	▼	310	.052	▼	.50	1.72		2.22	3.27	
	3490	Board type, fiberglass liner, 3 lb. density										
	3500	Fire resistant, black pigmented, 1 side										
	3520	1" thick	Q-14	150	.107	S.F.	1.44	3.56		5	7.20	
	3540	1-1/2" thick		130	.123		1.76	4.11		5.87	8.45	
	3560	2" thick	▼	120	.133	▼	2.12	4.45		6.57	9.40	
	3600	FRK vapor barrier										
	3620	1" thick	Q-14	150	.107	S.F.	1.57	3.56		5.13	7.35	
	3630	1-1/2" thick		130	.123		2.03	4.11		6.14	8.75	
	3640	2" thick	▼	120	.133		2.46	4.45		6.91	9.75	
	3680	No finish										
	3700	1" thick	Q-14	170	.094	S.F.	.84	3.14		3.98	5.90	
	3710	1-1/2" thick		140	.114		1.25	3.82		5.07	7.40	
	3720	2" thick		130	.123		1.57	4.11		5.68	8.25	
	3800	1/2" cement over 1" wire mesh, incl. corner bead		100	.160		1.50	5.35		6.85	10.10	
	3820	Glass cloth, pasted on		170	.094		2.43	3.14		5.57	7.65	
	3900	Weatherproof, non-metallic, 2 lb. per S.F.	▼	100	.160	▼	3.47	5.35		8.82	12.25	
	9600	Minimum labor/equipment charge	1 Stpi	4	2	Job		82		82	123	

MECHANICAL 15

			DAILY	LABOR-		2005 BARE COSTS				TOTAL	
15080		**Mechanical Insulation**	CREW	OUTPUT	HOURS	UNIT	MAT.	LABOR	EQUIP.	TOTAL	INCL O&P
400	0010	**EQUIPMENT INSULATION**										**400**
	0100	Rule of thumb, as a percentage of total mechanical costs				Job				10%	
	0110	Insulation req'd is based on the surface size/area to be covered									
	1000	Boiler, 1-1/2" calcium silicate, 1/2" cement finish	Q-14	50	.320	S.F.	3.30	10.70		14	20.50
	1020	2" fiberglass	"	80	.200	"	2.03	6.70		8.73	12.80
	2000	Breeching, 2" calcium silicate with 1/2" cement finish, no lath									
	2020	Rectangular	Q-14	42	.381	S.F.	3.87	12.70		16.57	24.50
	2040	Round	"	38.70	.413	"	4.33	13.80		18.13	27
600	0010	**PIPING INSULATION**										**600**
	0100	Rule of thumb, as a percentage of total mechanical costs				Job				10%	
	0110	Insulation req'd is based on the surface size/area to be covered									
	4000	Pipe covering (price copper tube one size less than IPS)									
	6600	Fiberglass, with all service jacket									
	6840	1" wall, 1/2" iron pipe size	Q-14	240	.067	L.F.	.71	2.23		2.94	4.29
	6870	1" iron pipe size		220	.073		.83	2.43		3.26	4.74
	6900	2" iron pipe size		200	.080		1.12	2.67		3.79	5.45
	6920	3" iron pipe size		180	.089		1.39	2.97		4.36	6.20
	6940	4" iron pipe size		150	.107		1.80	3.56		5.36	7.60
	7320	2" wall, 1/2" iron pipe size		220	.073		2.29	2.43		4.72	6.35
	7440	6" iron pipe size		100	.160		4.54	5.35		9.89	13.45
	7460	8" iron pipe size		80	.200		5.60	6.70		12.30	16.70
	7480	10" iron pipe size		70	.229		6.70	7.65		14.35	19.45
	7490	12" iron pipe size	▼	65	.246	▼	7.40	8.20		15.60	21
	7600	For fiberglass with standard canvas jacket, deduct					5%				
	7660	For fittings, add 3 L.F. for each fitting									
	7680	plus 4 L.F. for each flange of the fitting									
	7700	For equipment or congested areas, add						20%			
	7720	Finishes, for .010" aluminum jacket, add	Q-14	200	.080	S.F.	.53	2.67		3.20	4.80
	7740	For .016" aluminum jacket, add		200	.080		.68	2.67		3.35	4.97
	7760	For .010" stainless steel, add	▼	160	.100	▼	1.99	3.34		5.33	7.45
	7780	For single layer of felt, add					10%	10%			
	7879	Rubber tubing, flexible closed cell foam									
	7880	3/8" wall, 1/4" iron pipe size	1 Asbe	120	.067	L.F.	.28	2.47		2.75	4.21
	7910	1/2" iron pipe size		115	.070		.35	2.58		2.93	4.46
	7920	3/4" iron pipe size		115	.070		.40	2.58		2.98	4.51
	7930	1" iron pipe size		110	.073		.49	2.70		3.19	4.80
	7950	1-1/2" iron pipe size		110	.073		.72	2.70		3.42	5.05
	8100	1/2" wall, 1/4" iron pipe size		90	.089		.37	3.30		3.67	5.60
	8130	1/2" iron pipe size		89	.090		.57	3.33		3.90	5.90
	8140	3/4" iron pipe size		89	.090		.63	3.33		3.96	5.95
	8150	1" iron pipe size		88	.091		.68	3.37		4.05	6.05
	8170	1-1/2" iron pipe size		87	.092		1.03	3.41		4.44	6.55
	8180	2" iron pipe size		86	.093		1.20	3.45		4.65	6.75
	8200	3" iron pipe size		85	.094		1.72	3.49		5.21	7.40
	8300	3/4" wall, 1/4" iron pipe size		90	.089		.67	3.30		3.97	5.95
	8330	1/2" iron pipe size		89	.090		.81	3.33		4.14	6.15
	8340	3/4" iron pipe size		89	.090		.97	3.33		4.30	6.30
	8350	1" iron pipe size		88	.091		1.18	3.37		4.55	6.60
	8370	1-1/2" iron pipe size		87	.092		2.07	3.41		5.48	7.70
	8380	2" iron pipe size		86	.093		2.43	3.45		5.88	8.10
	8400	3" iron pipe size		85	.094		3.32	3.49		6.81	9.15
	8444	1" wall, 1/2" iron pipe size		86	.093		1.53	3.45		4.98	7.15
	8445	3/4" iron pipe size		84	.095		1.90	3.53		5.43	7.70
	8446	1" iron pipe size		84	.095		2.30	3.53		5.83	8.15
	8447	1-1/4" iron pipe size		82	.098		2.68	3.62		6.30	8.65
	8448	1-1/2" iron pipe size	▼	82	.098	▼	3.02	3.62		6.64	9

15 MECHANICAL

Important: See the Reference Section for critical supporting data - Reference Nos., Crews, & City Cost Indexes

15050 | Basic Materials & Methods

15080 | Mechanical Insulation

			CREW	DAILY OUTPUT	LABOR-HOURS	UNIT	2005 BARE COSTS				TOTAL INCL O&P	
							MAT.	LABOR	EQUIP.	TOTAL		
600	8449	2" iron pipe size	1 Asbe	80	.100	L.F.	4.70	3.71		8.41	11	600
	8450	2-1/2" iron pipe size	↓	80	.100	↓	6.15	3.71		9.86	12.60	
	8456	Rubber insulation tape, 1/8" x 2" x 30'				Ea.	10.80			10.80	11.90	

15100 | Building Services Piping

15106 | Glass Pipe & Fittings

			CREW	DAILY OUTPUT	LABOR-HOURS	UNIT	2005 BARE COSTS				TOTAL INCL O&P	
							MAT.	LABOR	EQUIP.	TOTAL		
120	0010	**PIPE, GLASS** Borosilicate, couplings & hangers 10' O.C.										120
	0020	Drainage										
	1100	1-1/2" diameter	Q-1	52	.308	L.F.	9.15	11.30		20.45	27	
	1120	2" diameter		44	.364		12.20	13.35		25.55	33.50	
	1140	3" diameter		39	.410		16.30	15.10		31.40	40.50	
	1160	4" diameter		30	.533		29.50	19.60		49.10	62	
	1180	6" diameter	↓	26	.615	↓	53.50	22.50		76	93	

15107 | Metal Pipe & Fittings

			CREW	DAILY OUTPUT	LABOR-HOURS	UNIT	2005 BARE COSTS				TOTAL INCL O&P	
							MAT.	LABOR	EQUIP.	TOTAL		
220	0010	**PIPE, BRASS** Plain end,										220
	0900	Field threaded, coupling & clevis hanger 10' O.C.										
	0920	Regular weight										
	0980	1/8" diameter	1 Plum	62	.129	L.F.	2.23	5.25		7.48	10.35	
	1120	1/2" diameter		48	.167		4.20	6.80		11	14.85	
	1140	3/4" diameter		46	.174		5.75	7.10		12.85	17	
	1160	1" diameter	↓	43	.186		8.30	7.60		15.90	20.50	
	1180	1-1/4" diameter	Q-1	72	.222		12.55	8.15		20.70	26	
	1200	1-1/2" diameter		65	.246		14.90	9.05		23.95	30	
	1220	2" diameter	↓	53	.302	↓	20.50	11.10		31.60	39	
320	0010	**PIPE, CAST IRON** Soil, on hangers 5' O.C.										320
	0020	Single hub, service wt., lead & oakum joints 10' O.C.										
	2120	2" diameter	Q-1	63	.254	L.F.	4.63	9.35		13.98	19.15	
	2140	3" diameter		60	.267		6.35	9.80		16.15	21.50	
	2160	4" diameter **CN**	↓	55	.291		8.10	10.70		18.80	25	
	2180	5" diameter	Q-2	76	.316		8.20	12.05		20.25	27	
	2200	6" diameter	"	73	.329		13.75	12.55		26.30	34	
	2220	8" diameter	Q-3	59	.542		21.50	21		42.50	55	
	2240	10" diameter		54	.593		39	23		62	77.50	
	2260	12" diameter	↓	48	.667		55.50	26		81.50	100	
	2320	For service weight, double hub, add					10%					
	2340	For extra heavy, single hub, add					48%	4%				
	2360	For extra heavy, double hub, add				↓	71%	4%				
	2400	Lead for caulking, (1#/dia in)	Q-1	160	.100	Lb.	.96	3.68		4.64	6.60	
	2420	Oakum for caulking, (1/8#/dia in)	"	40	.400	"	2.36	14.70		17.06	24.50	
	4000	No hub, couplings 10' O.C.										
	4100	1-1/2" diameter	Q-1	71	.225	L.F.	5.30	8.30		13.60	18.30	
	4120	2" diameter		67	.239		5.45	8.80		14.25	19.20	
	4160	4" diameter	↓	58	.276	↓	9.10	10.15		19.25	25.50	
420	0010	**PIPE, COPPER** Solder joints										420
	1000	Type K tubing, couplings & clevis hangers 10' O.C.										

MECHANICAL 15

			15107		**Metal Pipe & Fittings**	CREW	DAILY OUTPUT	LABOR-HOURS	UNIT	2005 BARE COSTS				TOTAL INCL O&P		
										MAT.	LABOR	EQUIP.	TOTAL			
420	1100		1/4" diameter			1 Plum	84	.095	L.F.	1.37	3.89		5.26	7.35		**420**
	1200		1" diameter				66	.121		3.97	4.95		8.92	11.80		
	1260		2" diameter			▼	40	.200	▼	9.95	8.15		18.10	23.50		
	2000		Type L tubing, couplings & hangers 10' O.C.													
	2100		1/4" diameter			1 Plum	88	.091	L.F.	.94	3.71		4.65	6.65		
	2120		3/8" diameter				84	.095		1.28	3.89		5.17	7.25		
	2140		1/2" diameter	CN			81	.099		1.50	4.03		5.53	7.70		
	2160		5/8" diameter				79	.101		2.23	4.14		6.37	8.65		
	2180		3/4" diameter				76	.105		2.20	4.30		6.50	8.85		
	2200		1" diameter				68	.118		3.10	4.81		7.91	10.60		
	2220		1-1/4" diameter				58	.138		4.25	5.65		9.90	13.15		
	2240		1-1/2" diameter				52	.154		5.40	6.30		11.70	15.40		
	2260		2" diameter			▼	42	.190		8.40	7.80		16.20	21		
	2280		2-1/2" diameter			Q-1	62	.258		12.75	9.50		22.25	28.50		
	2300		3" diameter				56	.286		17.25	10.50		27.75	35		
	2320		3-1/2" diameter				43	.372		23.50	13.70		37.20	46.50		
	2340		4" diameter				39	.410		29.50	15.10		44.60	55		
	2360		5" diameter			▼	34	.471		82.50	17.30		99.80	117		
	2380		6" diameter			Q-2	40	.600		103	23		126	148		
	2400		8" diameter			"	36	.667		180	25.50		205.50	236		
	2410		For other than full hard temper, add						▼	21%						
	2590		For silver solder, add								15%					
	4000		Type DWV tubing, couplings & hangers 10' O.C.													
	4100		1-1/4" diameter			1 Plum	60	.133	L.F.	3.67	5.45		9.12	12.25		
	4120		1-1/2" diameter				54	.148		4.56	6.05		10.61	14.10		
	4140		2" diameter			▼	44	.182		6.10	7.45		13.55	17.85		
	4160		3" diameter			Q-1	58	.276		10.90	10.15		21.05	27.50		
	4180		4" diameter				40	.400		19.15	14.70		33.85	43		
	4200		5" diameter			▼	36	.444		57	16.35		73.35	87		
	4220		6" diameter			Q-2	42	.571	▼	81.50	22		103.50	123		
500	0010		**PIPE, CORROSION RESISTANT** No couplings or hangers													**500**
	0020		Iron alloy, drain, mechanical joint													
	1000		1-1/2" diameter			Q-1	70	.229	L.F.	22.50	8.40		30.90	37.50		
	1100		2" diameter				66	.242		26	8.90		34.90	42		
	1120		3" diameter				60	.267		36	9.80		45.80	55		
	1140		4" diameter			▼	52	.308	▼	46.50	11.30		57.80	68		
	2980		Plastic, epoxy, fiberglass filament wound													
	3000		2" diameter			Q-1	62	.258	L.F.	5.20	9.50		14.70	19.95		
	3100		3" diameter				51	.314		7.10	11.55		18.65	25		
	3120		4" diameter				45	.356		9.40	13.10		22.50	30		
	3140		6" diameter			▼	32	.500	▼	14.40	18.40		32.80	43.50		
	3980		Polyester, fiberglass filament wound													
	4000		2" diameter			Q-1	62	.258	L.F.	9.85	9.50		19.35	25		
	4100		3" diameter				51	.314		13.20	11.55		24.75	32		
	4120		4" diameter				45	.356		14.90	13.10		28	36		
	4140		6" diameter			▼	32	.500		23.50	18.40		41.90	53.50		
	4980		Polypropylene, acid resistant, fire retardant, schedule 40													
	5000		1-1/2" diameter			Q-1	68	.235	L.F.	1.13	8.65		9.78	14.25		
	5100		2" diameter				62	.258		1.53	9.50		11.03	15.95		
	5120		3" diameter				51	.314		4.28	11.55		15.83	22		
	5140		4" diameter			▼	45	.356	▼	4.55	13.10		17.65	24.50		
	5980		Proxylene, fire retardant, Schedule 40													
	6000		1-1/2" diameter			Q-1	68	.235	L.F.	3.87	8.65		12.52	17.25		
	6100		2" diameter				62	.258		5.30	9.50		14.80	20		
	6120		3" diameter				51	.314		9.60	11.55		21.15	28		
	6140		4" diameter			▼	45	.356		13.65	13.10		26.75	34.50		

Important: See the Reference Section for critical supporting data - Reference Nos., Crews, & City Cost Indexes

15 MECHANICAL

15107	Metal Pipe & Fittings	CREW	DAILY OUTPUT	LABOR-HOURS	UNIT	2005 BARE COSTS				TOTAL INCL O&P
						MAT.	LABOR	EQUIP.	TOTAL	
620	**0010**	**PIPE, STEEL**										**620**
	0012	The steel pipe in this section does not include fittings such as ells,tees									
	0014	For fittings either add a % (usually 25 to 35%) or see									
	0015	the Mechanical or Plumbing Cost Data									
	0020	All pipe sizes are to Spec. A-53 unless noted otherwise	R15100 -050								
	0050	Schedule 40, threaded, with couplings, and clevis type									
	0060	hangers sized for covering, 10' O.C.									
	0540	Black, 1/4" diameter	1 Plum	66	.121	L.F.	2.09	4.95		7.04	9.75
	0550	3/8" diameter		65	.123		2.02	5.05		7.07	9.80
	0560	1/2" diameter		63	.127		1.28	5.20		6.48	9.20
	0570	3/4" diameter		61	.131		1.54	5.35		6.89	9.75
	0580	1" diameter		53	.151		2.16	6.15		8.31	11.65
	0590	1-1/4" diameter	Q-1	89	.180		2.75	6.60		9.35	12.95
	0600	1-1/2" diameter		80	.200		3.16	7.35		10.51	14.55
	0610	2" diameter CN		64	.250		4.13	9.20		13.33	18.35
	0620	2-1/2" diameter		50	.320		7	11.75		18.75	25.50
	0630	3" diameter		43	.372		9.05	13.70		22.75	30.50
	0640	3-1/2" diameter		40	.400		11.95	14.70		26.65	35
	0650	4" diameter		36	.444		14.25	16.35		30.60	40
	0660	5" diameter		26	.615		26	22.50		48.50	62.50
	0670	6" diameter	Q-2	31	.774		32.50	29.50		62	80.50
	1290	Galvanized, 1/4" diameter	1 Plum	66	.121		2.36	4.95		7.31	10.05
	1300	3/8" diameter		65	.123		2.69	5.05		7.74	10.50
	1310	1/2" diameter		63	.127		1.52	5.20		6.72	9.45
	1320	3/4" diameter		61	.131		1.76	5.35		7.11	10
	1330	1" diameter		53	.151		2.46	6.15		8.61	11.95
	1340	1-1/4" diameter	Q-1	89	.180		3.20	6.60		9.80	13.45
	1350	1-1/2" diameter		80	.200		3.71	7.35		11.06	15.15
	1360	2" diameter		64	.250		4.85	9.20		14.05	19.15
	1370	2-1/2" diameter		50	.320		7.90	11.75		19.65	26.50
	1380	3" diameter		43	.372		10.20	13.70		23.90	32
	1390	3-1/2" diameter		40	.400		16.50	14.70		31.20	40
	1400	4" diameter		36	.444		16.65	16.35		33	43
	1410	5" diameter		26	.615		27.50	22.50		50	64.50
	1420	6" diameter	Q-2	31	.774		38	29.50		67.50	86
	1430	8" diameter		27	.889		57	34		91	114
	1440	10" diameter		23	1.043		84	40		124	153
	1450	12" diameter		18	1.333		115	51		166	204
	2000	Welded, sch. 40, on yoke & roll hangers, sized for covering,									
	2010	10' O.C. (no hangers incl. for 14" diam. and up)									
	2040	Black, 1" diameter	Q-15	93	.172	L.F.	2.48	6.35	.87	9.70	13.20
	2070	2" diameter		61	.262		4.07	9.65	1.32	15.04	20.50
	2090	3" diameter		43	.372		6.45	13.70	1.88	22.03	29.50
	2110	4" diameter		37	.432		12	15.90	2.18	30.08	39.50
	2120	5" diameter		32	.500		21	18.40	2.52	41.92	53.50
	2130	6" diameter	Q-16	36	.667		26.50	25.50	2.24	54.24	70
	2140	8" diameter		29	.828		39	31.50	2.78	73.28	93.50
	2150	10" diameter		24	1		64	38	3.36	105.36	132
	2160	12" diameter		19	1.263		83.50	48	4.24	135.74	169
690	**0010**	**PIPE, GROOVED-JOINT STEEL FITTINGS & VALVES**										**690**
	0012	Fittings are ductile iron. Steel fittings noted.									
	0020	Pipe includes coupling & clevis type hanger 10' O.C.									
	1000	Schedule 40, black									
	1040	3/4" diameter	1 Plum	71	.113	L.F.	1.98	4.60		6.58	9.10
	1050	1" diameter		63	.127		1.85	5.20		7.05	9.85

MECHANICAL 15

15107	Metal Pipe & Fittings	CREW	DAILY OUTPUT	LABOR-HOURS	UNIT	2005 BARE COSTS				TOTAL INCL O&P	
						MAT.	LABOR	EQUIP.	TOTAL		
690											**690**
1060	1-1/4" diameter	1 Plum	58	.138	L.F.	2.37	5.65		8.02	11.05	
1070	1-1/2" diameter		51	.157		2.21	6.40		8.61	12.10	
1080	2" diameter	▼	40	.200		3.30	8.15		11.45	15.95	
1090	2-1/2" diameter	Q-1	57	.281		4.80	10.30		15.10	21	
1100	3" diameter		50	.320		6	11.75		17.75	24.50	
1110	4" diameter		45	.356		10.95	13.10		24.05	31.50	
1120	5" diameter	▼	37	.432		19.95	15.90		35.85	46	
1130	6" diameter	Q-2	42	.571	▼	25.50	22		47.50	61	
1800	Galvanized										
1840	3/4" diameter	1 Plum	71	.113	L.F.	2.19	4.60		6.79	9.30	
1850	1" diameter		63	.127		2.13	5.20		7.33	10.15	
1860	1-1/4" diameter		58	.138		2.76	5.65		8.41	11.50	
1870	1-1/2" diameter		51	.157		3.18	6.40		9.58	13.15	
1880	2" diameter	▼	40	.200		3.91	8.15		12.06	16.60	
1890	2-1/2" diameter	Q-1	57	.281		5.75	10.30		16.05	22	
1900	3" diameter		50	.320		7.25	11.75		19	25.50	
1910	4" diameter		45	.356		10	13.10		23.10	30.50	
1920	5" diameter	▼	37	.432		26	15.90		41.90	52.50	
1930	6" diameter	Q-2	42	.571	▼	33.50	22		55.50	70	
3990	Fittings: cplg. & labor required at joints not incl. in fitting										
3994	price. Add 1 per joint for installed price.										
4000	Elbow, 90° or 45°, painted										
4030	3/4" diameter	1 Plum	50	.160	Ea.	19.10	6.55		25.65	31	
4040	1" diameter		50	.160		10.15	6.55		16.70	21	
4050	1-1/4" diameter		40	.200		10.15	8.15		18.30	23.50	
4060	1-1/2" diameter		33	.242		10.15	9.90		20.05	26	
4070	2" diameter	▼	25	.320		10.15	13.05		23.20	31	
4080	2-1/2" diameter	Q-1	40	.400		10.15	14.70		24.85	33	
4090	3" diameter		33	.485		18	17.85		35.85	47	
4100	4" diameter		25	.640		19.60	23.50		43.10	57	
4110	5" diameter	▼	20	.800		47	29.50		76.50	96	
4120	6" diameter	Q-2	25	.960		55.50	36.50		92	116	
4250	For galvanized elbows, add				▼	26%					
4690	Tee, painted										
4700	3/4" diameter	1 Plum	38	.211	Ea.	20.50	8.60		29.10	35.50	
4740	1" diameter		33	.242		15.60	9.90		25.50	32	
4750	1-1/4" diameter		27	.296		15.60	12.10		27.70	35.50	
4760	1-1/2" diameter		22	.364		15.60	14.85		30.45	39.50	
4770	2" diameter	▼	17	.471		15.60	19.20		34.80	46	
4780	2-1/2" diameter	Q-1	27	.593		15.60	22		37.60	50	
4790	3" diameter		22	.727		22	27		49	64	
4800	4" diameter		17	.941		33	34.50		67.50	88.50	
4810	5" diameter	▼	13	1.231		76	45.50		121.50	152	
4820	6" diameter	Q-2	17	1.412		90	54		144	180	
4900	For galvanized tees, add				▼	24%					
4906	Couplings, rigid style, painted										
4908	1" diameter	1 Plum	100	.080	Ea.	7.45	3.27		10.72	13.10	
4909	1-1/4" diameter		100	.080		7.45	3.27		10.72	13.10	
4910	1-1/2" diameter		67	.119		7.45	4.88		12.33	15.55	
4912	2" diameter	▼	50	.160		7.65	6.55		14.20	18.25	
4914	2-1/2" diameter	Q-1	80	.200		8.80	7.35		16.15	21	
4916	3" diameter		67	.239		10.30	8.80		19.10	24.50	
4918	4" diameter		50	.320		14.60	11.75		26.35	34	
4920	5" diameter	▼	40	.400		18.90	14.70		33.60	43	
4922	6" diameter	Q-2	50	.480	▼	25.50	18.30		43.80	55.50	
4940	Flexible, standard, painted										

15 MECHANICAL

Important: See the Reference Section for critical supporting data - Reference Nos., Crews, & City Cost Indexes

		CREW	DAILY OUTPUT	LABOR-HOURS	UNIT	2005 BARE COSTS				TOTAL INCL O&P	
	15107 | **Metal Pipe & Fittings**					MAT.	LABOR	EQUIP.	TOTAL		
690	4950	3/4" diameter	1 Plum	100	.080	Ea.	5.40	3.27		8.67	10.80	**690**
	4960	1" diameter		100	.080		5.40	3.27		8.67	10.80	
	4970	1-1/4" diameter		80	.100		7.15	4.09		11.24	14	
	4980	1-1/2" diameter		67	.119		7.80	4.88		12.68	15.95	
	4990	2" diameter	▼	50	.160		8.25	6.55		14.80	18.90	
	5000	2-1/2" diameter	Q-1	80	.200		9.85	7.35		17.20	22	
	5010	3" diameter		67	.239		10.90	8.80		19.70	25	
	5020	3-1/2" diameter		57	.281		15.85	10.30		26.15	33	
	5030	4" diameter		50	.320		12.95	11.75		24.70	32	
	5040	5" diameter	▼	40	.400		24.50	14.70		39.20	48.50	
	5050	6" diameter	Q-2	50	.480	▼	28.50	18.30		46.80	59	
920	0010	**PIPE, STAINLESS STEEL**										**920**
	3500	Threaded, couplings and hangers 10' O.C.										
	3520	Schedule 40, type 304										
	3540	1/4" diameter	1 Plum	54	.148	L.F.	5.25	6.05		11.30	14.90	
	3550	3/8" diameter		53	.151		6.10	6.15		12.25	15.95	
	3560	1/2" diameter		52	.154		7.20	6.30		13.50	17.35	
	3580	1" diameter	▼	45	.178		12.05	7.25		19.30	24	
	3610	2" diameter	Q-1	57	.281		24.50	10.30		34.80	42.50	
	3640	4" diameter	Q-2	51	.471		76.50	17.95		94.45	111	
	3740	For small quantities, add				▼	10%					
	4250	Schedule 40, type 316										
	4290	1/4" diameter	1 Plum	54	.148	L.F.	6.15	6.05		12.20	15.85	
	4300	3/8" diameter		53	.151		7.35	6.15		13.50	17.35	
	4310	1/2" diameter		52	.154		9.50	6.30		15.80	19.95	
	4320	3/4" diameter		51	.157		11.80	6.40		18.20	22.50	
	4330	1" diameter	▼	45	.178		15.45	7.25		22.70	28	
	4360	2" diameter	Q-1	57	.281		31.50	10.30		41.80	50	
	4390	4" diameter	Q-2	51	.471		100	17.95		117.95	137	
	4490	For small quantities, add				▼	10%					
	15108 | **Plastic Pipe & Fittings**										
520	0010	**PIPE, PLASTIC**										**520**
	0020	Fiberglass reinforced, couplings 10' O.C., hangers 3 per 10'										
	0200	High strength										
	0240	2" diameter	Q-1	58	.276	L.F.	9.20	10.15		19.35	25.50	
	0260	3" diameter		51	.314		13.10	11.55		24.65	32	
	0280	4" diameter		47	.340		15.85	12.50		28.35	36.50	
	0300	6" diameter	▼	38	.421		24.50	15.50		40	50.50	
	0320	8" diameter	Q-2	48	.500	▼	39	19.05		58.05	71	
	1800	PVC, couplings 10' O.C., hangers 3 per 10'										
	1820	Schedule 40										
	1860	1/2" diameter	1 Plum	54	.148	L.F.	.98	6.05		7.03	10.20	
	1870	3/4" diameter		51	.157		1.06	6.40		7.46	10.80	
	1880	1" diameter		46	.174		1.17	7.10		8.27	12	
	1890	1-1/4" diameter		42	.190		1.34	7.80		9.14	13.20	
	1900	1-1/2" diameter	▼	36	.222		1.44	9.10		10.54	15.25	
	1910	2" diameter **CN**	Q-1	59	.271		1.63	9.95		11.58	16.80	
	1920	2-1/2" diameter		56	.286		2.20	10.50		12.70	18.20	
	1930	3" diameter		53	.302		2.87	11.10		13.97	19.85	
	1940	4" diameter		48	.333		3.67	12.25		15.92	22.50	
	1950	5" diameter		43	.372		4.86	13.70		18.56	26	
	1960	6" diameter	▼	39	.410	▼	6.50	15.10		21.60	29.50	
	4100	DWV type, schedule 40, couplings 10' O.C., hangers 3 per 10'										

MECHANICAL 15

15108	Plastic Pipe & Fittings	CREW	DAILY OUTPUT	LABOR-HOURS	UNIT	2005 BARE COSTS				TOTAL INCL O&P	
						MAT.	LABOR	EQUIP.	TOTAL		
520	4120	ABS										**520**
4140	1-1/4" diameter	1 Plum	42	.190	L.F.	1.08	7.80		8.88	12.90
4150	1-1/2" diameter	"	36	.222		1.09	9.10		10.19	14.85
4160	2" diameter	Q-1	59	.271		1.19	9.95		11.14	16.30
4400	PVC									
4410	1-1/4" diameter	1 Plum	42	.190	L.F.	1.19	7.80		8.99	13
4420	1-1/2" diameter	"	36	.222		1.20	9.10		10.30	14.95
4460	2" diameter	Q-1	59	.271		1.35	9.95		11.30	16.50
4470	3" diameter		53	.302		2.40	11.10		13.50	19.35
4480	4" diameter		48	.333		3.13	12.25		15.38	22
4490	6" diameter		39	.410		5.95	15.10		21.05	29
5360	CPVC, couplings 10' O.C., hangers 3 per 10'									
5380	Schedule 40									
5460	1/2" diameter	1 Plum	54	.148	L.F.	2.31	6.05		8.36	11.65
5470	3/4" diameter		51	.157		3.05	6.40		9.45	13
5480	1" diameter		46	.174		3.74	7.10		10.84	14.80
5490	1-1/4" diameter		42	.190		4.35	7.80		12.15	16.50
5500	1-1/2" diameter		36	.222		4.86	9.10		13.96	19
5510	2" diameter	Q-1	59	.271		6	9.95		15.95	21.50
5520	2-1/2" diameter		56	.286		9.55	10.50		20.05	26.50
5530	3" diameter		53	.302		11.75	11.10		22.85	29.50

15110	Valves	CREW	DAILY OUTPUT	LABOR-HOURS	UNIT	MAT.	LABOR	EQUIP.	TOTAL	TOTAL INCL O&P	
100	0010	**VALVES, BRASS**										**100**
0500	Gas cocks, threaded									
0530	1/2"	1 Plum	24	.333	Ea.	5.75	13.60		19.35	27
0540	3/4"		22	.364		10.80	14.85		25.65	34.50
0550	1"		19	.421		11.35	17.20		28.55	38.50
0560	1-1/4"		15	.533		28	22		50	64
160	0010	**VALVES, BRONZE**										**160**
1020	Angle, 150 lb., rising stem, threaded									
1030	1/8"	1 Plum	24	.333	Ea.	49.50	13.60		63.10	75
1040	1/4"		24	.333		49.50	13.60		63.10	75
1050	3/8"		24	.333		49.50	13.60		63.10	75
1060	1/2"		22	.364		49.50	14.85		64.35	77
1070	3/4"		20	.400		67	16.35		83.35	98
1080	1"		19	.421		95.50	17.20		112.70	131
1100	1-1/2"		13	.615		161	25		186	215
1110	2"		11	.727		260	29.50		289.50	330
1380	Ball, 150 psi, threaded									
1400	1/4"	1 Plum	24	.333	Ea.	7	13.60		20.60	28
1430	3/8"		24	.333		7	13.60		20.60	28
1450	1/2"		22	.364		7	14.85		21.85	30
1460	3/4"		20	.400		11.55	16.35		27.90	37
1470	1"		19	.421		14.55	17.20		31.75	42
1480	1-1/4"		15	.533		17.15	22		39.15	52
1490	1-1/2"		13	.615		26.50	25		51.50	67
1500	2"		11	.727		33	29.50		62.50	81
1750	Check, swing, class 150, regrinding disc, threaded									
1800	1/8"	1 Plum	24	.333	Ea.	25	13.60		38.60	48
1830	1/4"		24	.333		25	13.60		38.60	48
1840	3/8"		24	.333		26	13.60		39.60	49
1850	1/2"		24	.333		26	13.60		39.60	49
1860	3/4"		20	.400		35	16.35		51.35	63
1870	1"		19	.421		52	17.20		69.20	83

			CREW	DAILY OUTPUT	LABOR-HOURS	UNIT	2005 BARE COSTS				TOTAL INCL O&P
	15110	**Valves**					MAT.	LABOR	EQUIP.	TOTAL	
160	1880	1-1/4"	1 Plum	15	.533	Ea.	73	22		95	113
	1890	1-1/2"		13	.615		85.50	25		110.50	132
	1900	2"	↓	11	.727		125	29.50		154.50	183
	1910	2-1/2"	Q-1	15	1.067	↓	265	39		304	350
	2000	For 200 lb, add					5%	10%			
	2040	For 300 lb, add					15%	15%			
	2850	Gate, N.R.S., soldered, 125 psi									
	2900	3/8"	1 Plum	24	.333	Ea.	18.30	13.60		31.90	40.50
	2920	1/2"		24	.333		16	13.60		29.60	38
	2940	3/4"		20	.400		25	16.35		41.35	52
	2950	1"		19	.421		34.50	17.20		51.70	64
	2960	1-1/4"		15	.533		46.50	22		68.50	84
	2970	1-1/2"		13	.615		52.50	25		77.50	95.50
	2980	2"	↓	11	.727		83.50	29.50		113	137
	2990	2-1/2"	Q-1	15	1.067		195	39		234	274
	3000	3"	"	13	1.231	↓	256	45.50		301.50	350
	3850	Rising stem, soldered, 300 psi									
	3950	1"	1 Plum	19	.421	Ea.	59.50	17.20		76.70	91
	3980	2"	"	11	.727		161	29.50		190.50	222
	4000	3"	Q-1	13	1.231	↓	455	45.50		500.50	570
	4250	Threaded, class 150									
	4310	1/4"	1 Plum	24	.333	Ea.	27	13.60		40.60	50.50
	4320	3/8"		24	.333		27	13.60		40.60	50.50
	4330	1/2"		24	.333		25.50	13.60		39.10	48.50
	4340	3/4"		20	.400		30	16.35		46.35	57.50
	4350	1"		19	.421		40.50	17.20		57.70	70.50
	4360	1-1/4"		15	.533		53.50	22		75.50	91.50
	4370	1-1/2"		13	.615		67.50	25		92.50	113
	4380	2"	↓	11	.727		91.50	29.50		121	146
	4390	2-1/2"	Q-1	15	1.067		214	39		253	294
	4400	3"	"	13	1.231		299	45.50		344.50	400
	4500	For 300 psi, threaded, add					100%	15%			
	4540	For chain operated type, add				↓	15%				
	4850	Globe, class 150, rising stem, threaded									
	4920	1/4"	1 Plum	24	.333	Ea.	35.50	13.60		49.10	60
	4940	3/8" size		24	.333		39	13.60		52.60	63.50
	4950	1/2"		24	.333		39	13.60		52.60	63.50
	4960	3/4"		20	.400		53.50	16.35		69.85	83
	4970	1"		19	.421		83.50	17.20		100.70	118
	4980	1-1/4"		15	.533		131	22		153	177
	4990	1-1/2"		13	.615		160	25		185	214
	5000	2"	↓	11	.727		239	29.50		268.50	310
	5010	2-1/2"	Q-1	15	1.067		480	39		519	590
	5020	3"	"	13	1.231	↓	605	45.50		650.50	735
	5120	For 300 lb threaded, add					50%	15%			
	5600	Relief, pressure & temperature, self-closing, ASME, threaded									
	5640	3/4"	1 Plum	28	.286	Ea.	81.50	11.65		93.15	107
	5650	1"		24	.333		118	13.60		131.60	151
	5660	1-1/4"		20	.400		237	16.35		253.35	286
	5670	1-1/2"		18	.444		455	18.15		473.15	530
	5680	2"	↓	16	.500	↓	495	20.50		515.50	575
	5950	Pressure, poppet type, threaded									
	6000	1/2"	1 Plum	30	.267	Ea.	13.55	10.90		24.45	31.50
	6040	3/4"	"	28	.286	"	45	11.65		56.65	67
	6400	Pressure, water, ASME, threaded									
	6440	3/4"	1 Plum	28	.286	Ea.	47	11.65		58.65	69.50

MECHANICAL 15

15110 | Valves

			CREW	DAILY OUTPUT	LABOR-HOURS	UNIT	2005 BARE COSTS				TOTAL INCL O&P	
							MAT.	LABOR	EQUIP.	TOTAL		
160	6450	1"	1 Plum	24	.333	Ea.	93.50	13.60		107.10	124	160
	6460	1-1/4"		20	.400		153	16.35		169.35	193	
	6470	1-1/2"		18	.444		211	18.15		229.15	260	
	6480	2"		16	.500		305	20.50		325.50	365	
	6490	2-1/2"		15	.533		1,550	22		1,572	1,725	
	6900	Reducing, water pressure										
	6920	300 psi to 25-75 psi, threaded or sweat										
	6940	1/2"	1 Plum	24	.333	Ea.	147	13.60		160.60	182	
	6950	3/4"		20	.400		147	16.35		163.35	186	
	6960	1"		19	.421		227	17.20		244.20	276	
	6970	1-1/4"		15	.533		405	22		427	480	
	6980	1-1/2"		13	.615		615	25		640	715	
	8350	Tempering, water, sweat connections										
	8400	1/2"	1 Plum	24	.333	Ea.	52.50	13.60		66.10	78	
	8440	3/4"	"	20	.400	"	64	16.35		80.35	95	
	8650	Threaded connections										
	8700	1/2"	1 Plum	24	.333	Ea.	79	13.60		92.60	107	
	8740	3/4"		20	.400		244	16.35		260.35	294	
	8750	1"		19	.421		269	17.20		286.20	320	
	8760	1-1/4"		15	.533		430	22		452	505	
	8770	1-1/2"		13	.615		465	25		490	555	
	8780	2"		11	.727		700	29.50		729.50	815	
200	0010	**VALVES, IRON BODY**										200
	1020	Butterfly, wafer type, gear actuator, 200 lb.										
	1030	2"	1 Plum	14	.571	Ea.	113	23.50		136.50	159	
	1040	2-1/2"	Q-1	9	1.778		116	65.50		181.50	227	
	1050	3"		8	2		121	73.50		194.50	244	
	1060	4"		5	3.200		151	118		269	345	
	1070	5"	Q-2	5	4.800		182	183		365	475	
	1080	6"	"	5	4.800		205	183		388	500	
	1650	Gate, 125 lb., N.R.S.										
	2150	Flanged										
	2200	2"	1 Plum	5	1.600	Ea.	320	65.50		385.50	450	
	2240	2-1/2"	Q-1	5	3.200		330	118		448	535	
	2260	3"		4.50	3.556		365	131		496	600	
	2280	4"		3	5.333		525	196		721	875	
	2300	6"	Q-2	3	8		895	305		1,200	1,450	
	3550	OS&Y, 125 lb., flanged										
	3600	2"	1 Plum	5	1.600	Ea.	226	65.50		291.50	345	
	3660	3"	Q-1	4.50	3.556		263	131		394	485	
	3680	4"	"	3	5.333		375	196		571	705	
	3700	6"	Q-2	3	8		615	305		920	1,150	
	3900	For 175 lb, flanged, add					200%	10%				
	5450	Swing check, 125 lb., threaded										
	5470	1"	1 Plum	13	.615	Ea.	310	25		335	380	
	5500	2"	"	11	.727		400	29.50		429.50	485	
	5540	2-1/2"	Q-1	15	1.067		455	39		494	560	
	5550	3"		13	1.231		510	45.50		555.50	630	
	5560	4"		10	1.600		815	59		874	985	
	5950	Flanged										
	6000	2"	1 Plum	5	1.600	Ea.	156	65.50		221.50	269	
	6040	2-1/2"	Q-1	5	3.200		189	118		307	385	
	6050	3"		4.50	3.556		202	131		333	420	
	6060	4"		3	5.333		320	196		516	645	
	6070	6"	Q-2	3	8		545	305		850	1,050	

15 MECHANICAL

Important: See the Reference Section for critical supporting data - Reference Nos., Crews, & City Cost Indexes

15110	Valves	CREW	DAILY OUTPUT	LABOR-HOURS	UNIT	2005 BARE COSTS				TOTAL INCL O&P
						MAT.	LABOR	EQUIP.	TOTAL	
500 0010	**VALVES, PLASTIC**									**500**
1100	Angle, PVC, threaded									
1110	1/4"	1 Plum	26	.308	Ea.	49.50	12.55		62.05	73.50
1120	1/2"		26	.308		49.50	12.55		62.05	73.50
1130	3/4"		25	.320		58.50	13.05		71.55	84
1140	1"	↓	23	.348	↓	70	14.20		84.20	98.50
1150	Ball, PVC, socket or threaded, single union									
1200	1/4"	1 Plum	26	.308	Ea.	24	12.55		36.55	45.50
1220	3/8"		26	.308		24	12.55		36.55	45.50
1230	1/2"		26	.308		24	12.55		36.55	45.50
1240	3/4"		25	.320		28.50	13.05		41.55	51
1250	1"		23	.348		34.50	14.20		48.70	59.50
1260	1-1/4"		21	.381		57	15.55		72.55	86
1270	1-1/2"		20	.400		57	16.35		73.35	87
1280	2"	↓	17	.471		82	19.20		101.20	119
1360	For PVC, flanged, add				↓	100%	15%			
1650	CPVC, socket or threaded, single union									
1700	1/2"	1 Plum	26	.308	Ea.	35	12.55		47.55	57.50
1720	3/4"		25	.320		44	13.05		57.05	68
1730	1"		23	.348		52.50	14.20		66.70	79.50
1750	1-1/4"		21	.381		88	15.55		103.55	121
1760	1-1/2"	↓	20	.400		88	16.35		104.35	121
1840	For CPVC, flanged, add					65%	15%			
1880	For true union, socket or threaded, add				↓	50%	5%			
2050	Polypropylene, threaded									
2100	1/4"	1 Plum	26	.308	Ea.	36	12.55		48.55	58.50
2120	3/8"		26	.308		36	12.55		48.55	58.50
2130	1/2"		26	.308		36	12.55		48.55	58.50
2140	3/4"		25	.320		45.50	13.05		58.55	69.50
2150	1"		23	.348		53.50	14.20		67.70	80.50
2160	1-1/4"		21	.381		77.50	15.55		93.05	109
2170	1-1/2"		20	.400		89.50	16.35		105.85	123
2180	2"	↓	17	.471	↓	122	19.20		141.20	164
4850	Foot valve, PVC, socket or threaded									
4900	1/2"	1 Plum	34	.235	Ea.	49.50	9.60		59.10	69
4930	3/4"		32	.250		56	10.20		66.20	77
4940	1"		28	.286		73	11.65		84.65	97.50
4950	1-1/4"		27	.296		140	12.10		152.10	172
4960	1-1/2"	↓	26	.308	↓	140	12.55		152.55	173
6350	Y sediment strainer, PVC, socket or threaded									
6400	1/2"	1 Plum	26	.308	Ea.	36	12.55		48.55	58.50
6440	3/4"		24	.333		39	13.60		52.60	63.50
6450	1"		23	.348		47	14.20		61.20	73.50
6460	1-1/4"		21	.381		77.50	15.55		93.05	109
6470	1-1/2"	↓	20	.400	↓	77.50	16.35		93.85	110
700 0010	**VALVES, STEEL**									**700**
0800	Cast									
1350	Check valve, swing type, 150 lb., flanged									
1370	1"	1 Plum	10	.800	Ea.	365	32.50		397.50	455
1400	2"	"	8	1		515	41		556	625
1440	2-1/2"	Q-1	5	3.200		505	118		623	730
1450	3"		4.50	3.556		595	131		726	850
1460	4"	↓	3	5.333		875	196		1,071	1,250
1540	For 300 lb., flanged, add					50%	15%			
1548	For 600 lb., flanged, add				↓	110%	20%			

MECHANICAL 15

15110 | Valves

		CREW	DAILY OUTPUT	LABOR-HOURS	UNIT	2005 BARE COSTS				TOTAL INCL O&P	
						MAT.	LABOR	EQUIP.	TOTAL		
700	1950	Gate valve, 150 lb., flanged									700
	2000	2"	1 Plum	8	1	Ea.	540	41		581	655
	2040	2-1/2"	Q-1	5	3.200		765	118		883	1,025
	2050	3"		4.50	3.556		765	131		896	1,025
	2060	4"		3	5.333		930	196		1,126	1,325
	2070	6"	Q-2	3	8		1,475	305		1,780	2,075
	3650	Globe valve, 150 lb., flanged									
	3700	2"	1 Plum	8	1	Ea.	710	41		751	840
	3740	2-1/2"	Q-1	5	3.200		910	118		1,028	1,175
	3750	3"		4.50	3.556		910	131		1,041	1,200
	3760	4"		3	5.333		1,325	196		1,521	1,750
	3770	6"	Q-2	3	8		2,100	305		2,405	2,750
	5150	Forged									
	5650	Check valve, class 800, horizontal, socket									
	5698	Threaded									
	5700	1/4"	1 Plum	24	.333	Ea.	61	13.60		74.60	88
	5720	3/8"		24	.333		61	13.60		74.60	88
	5730	1/2"		24	.333		61	13.60		74.60	88
	5740	3/4"		20	.400		67	16.35		83.35	98
	5750	1"		19	.421		78	17.20		95.20	112
	5760	1-1/4"		15	.533		153	22		175	202

15120 | Piping Specialties

		CREW	DAILY OUTPUT	LABOR-HOURS	UNIT	MAT.	LABOR	EQUIP.	TOTAL	TOTAL INCL O&P	
120	0010	**AIR CONTROL**									120
	0030	Air separator, with strainer									
	0040	2" diameter	Q-5	6	2.667	Ea.	555	98.50		653.50	760
	0080	2-1/2" diameter		5	3.200		630	118		748	870
	0100	3" diameter		4	4		960	147		1,107	1,275
	0120	4" diameter		3	5.333		1,400	197		1,597	1,825
	0130	5" diameter	Q-6	3.60	6.667		1,775	255		2,030	2,325
	0140	6" diameter	"	3.40	7.059		2,125	270		2,395	2,725
160	0010	**AUTOMATIC AIR VENT**									160
	0020	Cast iron body, stainless steel internals, float type									
	0060	1/2" NPT inlet, 300 psi	1 Stpi	12	.667	Ea.	73	27.50		100.50	121
	0220	3/4" NPT inlet, 250 psi	"	10	.800		234	33		267	305
	0340	1-1/2" NPT inlet, 250 psi	Q-5	12	1.333		725	49		774	875
300	0010	**EXPANSION JOINTS**									300
	0100	Bellows type, neoprene cover, flanged spool									
	0140	6" face to face, 1-1/4" diameter	1 Stpi	11	.727	Ea.	210	30		240	276
	0160	1-1/2" diameter	"	10.60	.755		210	31		241	278
	0180	2" diameter	Q-5	13.30	1.203		214	44.50		258.50	300
	0190	2-1/2" diameter		12.40	1.290		222	47.50		269.50	315
	0200	3" diameter		11.40	1.404		252	51.50		303.50	355
	0480	10" face to face, 2" diameter		13	1.231		315	45.50		360.50	415
	0500	2-1/2" diameter		12	1.333		330	49		379	440
	0520	3" diameter		11	1.455		335	53.50		388.50	450
	0540	4" diameter		8	2		375	73.50		448.50	525
	0560	5" diameter		7	2.286		455	84		539	625
	0580	6" diameter		6	2.667		465	98.50		563.50	665
320	0010	**EXPANSION TANKS**									320
	1505	Fiberglass and steel single / double wall storage, see Div 13201									
	1510	Tank leak detection systems, see Div 13851-350									
	2000	Steel, liquid expansion, ASME, painted, 15 gallon capacity	Q-5	17	.941	Ea.	365	34.50		399.50	450

Important: See the Reference Section for critical supporting data - Reference Nos., Crews, & City Cost Indexes

15 MECHANICAL

15120	Piping Specialties	CREW	DAILY OUTPUT	LABOR-HOURS	UNIT	2005 BARE COSTS				TOTAL INCL O&P	
						MAT.	LABOR	EQUIP.	TOTAL		
320	2020	24 gallon capacity	Q-5	14	1.143	Ea.	370	42		412	470	**320**
	2040	30 gallon capacity		12	1.333		405	49		454	520	
	2060	40 gallon capacity		10	1.600		460	59		519	595	
	2080	60 gallon capacity		8	2		555	73.50		628.50	720	
	2100	80 gallon capacity		7	2.286		570	84		654	750	
	2120	100 gallon capacity		6	2.667		735	98.50		833.50	960	
	3000	Steel ASME expansion, rubber diaphragm, 19 gal. cap. accept.		12	1.333		1,425	49		1,474	1,650	
	3020	31 gallon capacity		8	2		1,600	73.50		1,673.50	1,850	
	3040	61 gallon capacity		6	2.667		2,250	98.50		2,348.50	2,625	
	3080	119 gallon capacity		4	4		2,425	147		2,572	2,875	
	3100	158 gallon capacity		3.80	4.211		3,350	155		3,505	3,925	
	3140	317 gallon capacity		2.80	5.714		5,075	211		5,286	5,900	
	3180	528 gallon capacity	▼	2.40	6.667	▼	8,200	246		8,446	9,400	
	5950											
350	0010	**FLEXIBLE CONNECTORS**, Corrugated, 7/8" O.D., 1/2" I.D.										**350**
	0050	Gas, seamless brass, steel fittings										
	0200	12" long	1 Plum	36	.222	Ea.	11.95	9.10		21.05	27	
	0220	18" long		36	.222		14.85	9.10		23.95	30	
	0240	24" long		34	.235		17.55	9.60		27.15	34	
	0280	36" long		32	.250		21	10.20		31.20	38.50	
	0340	60" long	▼	30	.267	▼	31.50	10.90		42.40	51.50	
	2000	Water, copper tubing, dielectric separators										
	2100	12" long	1 Plum	36	.222	Ea.	9.40	9.10		18.50	24	
	2260	24" long	"	34	.235	"	13.95	9.60		23.55	30	
370	0010	**FLEXIBLE METAL HOSE** Connectors, standard lengths										**370**
	0100	Bronze braided, bronze ends										
	0120	3/8" diameter x 12"	1 Stpi	26	.308	Ea.	15.40	12.60		28	36	
	0160	3/4" diameter x 12"		20	.400		21.50	16.40		37.90	48	
	0180	1" diameter x 18"		19	.421		27.50	17.25		44.75	56.50	
	0200	1-1/2" diameter x 18"		13	.615		44	25		69	86.50	
	0220	2" diameter x 18"	▼	11	.727	▼	53	30		83	103	
520	0010	**HYDRONIC HEATING CONTROL VALVES**										**520**
	0100	Radiator supply, 1/2" diameter	1 Stpi	24	.333	Ea.	40	13.65		53.65	64.50	
	0120	3/4" diameter		20	.400		40.50	16.40		56.90	69.50	
	0140	1" diameter		19	.421		53	17.25		70.25	84	
	0160	1-1/4" diameter	▼	15	.533	▼	74	22		96	115	
	0500	For low pressure steam, add					25%					
670	0010	**PRESSURE REGULATOR**										**670**
	3000	Steam, high capacity, bronze body, stainless steel trim										
	3020	Threaded, 1/2" diameter	1 Stpi	24	.333	Ea.	940	13.65		953.65	1,050	
	3030	3/4" diameter		24	.333		975	13.65		988.65	1,100	
	3040	1" diameter		19	.421		1,100	17.25		1,117.25	1,225	
	3060	1-1/4" diameter		15	.533		1,150	22		1,172	1,300	
	3080	1-1/2" diameter	▼	13	.615		1,325	25		1,350	1,500	
	3100	2" diameter		11	.727		1,625	30		1,655	1,825	
	3120	2-1/2" diameter	Q-5	12	1.333		2,025	49		2,074	2,300	
	3140	3" diameter	"	11	1.455	▼	2,325	53.50		2,378.50	2,625	
	3500	Flanged connection, iron body, 125 lb. W.S.P.										
	3520	3" diameter	Q-5	11	1.455	Ea.	2,550	53.50		2,603.50	2,875	
	3540	4" diameter	"	5	3.200	"	3,200	118		3,318	3,700	
760	0010	**STEAM TRAP**										**760**
	0030	Cast iron body, threaded										
	0040	Inverted bucket										
	0050	1/2" pipe size	1 Stpi	12	.667	Ea.	107	27.50		134.50	158	

MECHANICAL 15

		15120	Piping Specialties	CREW	DAILY OUTPUT	LABOR-HOURS	UNIT	MAT.	LABOR	EQUIP.	TOTAL	TOTAL INCL O&P	
760	0070		3/4" pipe size	1 Stpi	10	.800	Ea.	185	33		218	253	760
	0100		1" pipe size		9	.889		286	36.50		322.50	370	
	0120		1-1/4" pipe size	▼	8	1	▼	430	41		471	535	
	1000		Float & thermostatic, 15 psi										
	1010		3/4" pipe size	1 Stpi	16	.500	Ea.	87	20.50		107.50	127	
	1020		1" pipe size		15	.533		104	22		126	147	
	1040		1-1/2" pipe size		9	.889		185	36.50		221.50	259	
	1060		2" pipe size	▼	6	1.333	▼	340	54.50		394.50	455	
820	0010		**STRAINERS, Y TYPE** Bronze body										820
	0050		Screwed, 150 lb., 1/4 pipe size	1 Stpi	24	.333	Ea.	12.05	13.65		25.70	34	
	0070		3/8" pipe size		24	.333		15.90	13.65		29.55	38	
	0100		1/2" pipe size		20	.400		15.90	16.40		32.30	42	
	0140		1" pipe size		17	.471		18.70	19.25		37.95	49.50	
	0160		1-1/2" pipe size		14	.571		40.50	23.50		64	79.50	
	0180		2" pipe size		13	.615		53.50	25		78.50	97	
	0182		3" pipe size	▼	12	.667		340	27.50		367.50	415	
	0200		300 lb., 2-1/2" pipe size	Q-5	17	.941		286	34.50		320.50	365	
	0220		3" pipe size		16	1		565	37		602	680	
	0240		4" pipe size	▼	15	1.067	▼	1,300	39.50		1,339.50	1,475	
	0500		For 300 lb rating 1/4" thru 2", add					15%					
	1000		Flanged, 150 lb., 1-1/2" pipe size	1 Stpi	11	.727	Ea.	296	30		326	370	
	1020		2" pipe size	"	8	1		345	41		386	440	
	1030		2-1/2" pipe size	Q-5	5	3.200		530	118		648	760	
	1040		3" pipe size		4.50	3.556		655	131		786	915	
	1060		4" pipe size	▼	3	5.333		990	197		1,187	1,400	
	1100		6" pipe size	Q-6	3	8		2,650	305		2,955	3,350	
	1106		8" pipe size	"	2.60	9.231	▼	2,750	355		3,105	3,550	
	1500		For 300 lb rating, add					40%					
840	0010		**STRAINERS, Y TYPE** Iron body										840
	0050		Screwed, 250 lb., 1/4" pipe size	1 Stpi	20	.400	Ea.	6.65	16.40		23.05	32	
	0070		3/8" pipe size		20	.400		6.65	16.40		23.05	32	
	0100		1/2" pipe size		20	.400		6.65	16.40		23.05	32	
	0140		1" pipe size		16	.500		10.75	20.50		31.25	43	
	0160		1-1/2" pipe size		12	.667		17.75	27.50		45.25	60.50	
	0180		2" pipe size	▼	8	1		27.50	41		68.50	91.50	
	0220		3" pipe size	Q-5	11	1.455		151	53.50		204.50	247	
	0240		4" pipe size	"	5	3.200	▼	255	118		373	455	
	0500		For galvanized body, add					50%					
	1000		Flanged, 125 lb., 1-1/2" pipe size	1 Stpi	11	.727	Ea.	86	30		116	140	
	1020		2" pipe size	"	8	1		64	41		105	132	
	1040		3" pipe size	Q-5	4.50	3.556		84.50	131		215.50	290	
	1060		4" pipe size	"	3	5.333		155	197		352	465	
	1080		5" pipe size	Q-6	3.40	7.059		242	270		512	670	
	1100		6" pipe size	"	3	8	▼	295	305		600	785	
	1500		For 250 lb rating, add					20%					
	2000		For galvanized body, add					50%					
	2500		For steel body, add					40%					
920	0010		**VENTURI FLOW** Measuring device										920
	0050		1/2" diameter	1 Stpi	24	.333	Ea.	175	13.65		188.65	214	
	0120		1" diameter		19	.421		169	17.25		186.25	212	
	0140		1-1/4" diameter		15	.533		203	22		225	256	
	0160		1-1/2" diameter		13	.615		206	25		231	265	
	0180		2" diameter	▼	11	.727		222	30		252	289	
	0220		3" diameter	Q-5	14	1.143		380	42		422	485	
	0240		4" diameter	"	11	1.455	▼	595	53.50		648.50	735	

Important: See the Reference Section for critical supporting data - Reference Nos., Crews, & City Cost Indexes

15120 | Piping Specialties

		CREW	DAILY OUTPUT	LABOR-HOURS	UNIT	2005 BARE COSTS MAT.	LABOR	EQUIP.	TOTAL	TOTAL INCL O&P	
920	0280	6" diameter	Q-6	3.50	6.857	Ea.	780	262		1,042	1,250	**920**
	0500	For meter, add				↓	1,175			1,175	1,300	
940	0010	**WATER SUPPLY METERS**										**940**
	2000	Domestic/commercial, bronze										
	2020	Threaded										
	2060	5/8" diameter, to 20 GPM	1 Plum	16	.500	Ea.	40	20.50		60.50	74.50	
	2080	3/4" diameter, to 30 GPM		14	.571		67.50	23.50		91	110	
	2100	1" diameter, to 50 GPM	↓	12	.667	↓	94	27		121	144	
	2300	Threaded/flanged										
	2340	1-1/2" diameter, to 100 GPM	1 Plum	8	1	Ea.	330	41		371	425	
	2360	2" diameter, to 160 GPM	"	6	1.333	"	415	54.50		469.50	535	
	2600	Flanged, compound										
	2640	3" diameter, 320 GPM	Q-1	3	5.333	Ea.	1,950	196		2,146	2,425	
	2660	4" diameter, to 500 GPM		1.50	10.667		3,025	390		3,415	3,925	
	2680	6" diameter, to 1,000 GPM		1	16		4,350	590		4,940	5,650	
	2700	8" diameter, to 1,800 GPM	↓	.80	20	↓	8,600	735		9,335	10,600	

15140 | Domestic Water Piping

		CREW	DAILY OUTPUT	LABOR-HOURS	UNIT	2005 BARE COSTS MAT.	LABOR	EQUIP.	TOTAL	TOTAL INCL O&P	
100	0010	**BACKFLOW PREVENTER** Includes valves										**100**
	0020	and four test cocks, corrosion resistant, automatic operation										
	4000	Reduced pressure principle										
	4100	Threaded, bronze, valves are ball										
	4120	3/4" pipe size	1 Plum	16	.500	Ea.	188	20.50		208.50	238	
	4140	1" pipe size		14	.571		203	23.50		226.50	258	
	4150	1-1/4" pipe size		12	.667		345	27		372	420	
	4160	1-1/2" pipe size		10	.800		380	32.50		412.50	470	
	4180	2" pipe size	↓	7	1.143	↓	425	46.50		471.50	540	
	5000	Flanged, valves are OS&Y										
	5060	2-1/2" pipe size	Q-1	5	3.200	Ea.	2,175	118		2,293	2,575	
	5080	3" pipe size		4.50	3.556		2,300	131		2,431	2,725	
	5100	4" pipe size	↓	3	5.333		2,900	196		3,096	3,475	
	5120	6" pipe size	Q-2	3	8	↓	4,175	305		4,480	5,050	
	5600	Flanged, iron, valves are OS&Y										
	5660	2-1/2" pipe size	Q-1	5	3.200	Ea.	1,750	118		1,868	2,100	
	5680	3" pipe size		4.50	3.556		1,850	131		1,981	2,225	
	5700	4" pipe size	↓	3	5.333		2,375	196		2,571	2,925	
	5720	6" pipe size	Q-2	3	8		3,350	305		3,655	4,125	
	5740	8" pipe size		2	12		5,875	460		6,335	7,175	
	5760	10" pipe size	↓	1	24	↓	7,875	915		8,790	10,000	
600	0010	**VACUUM BREAKERS** Hot or cold water										**600**
	1030	Anti-siphon, brass										
	1040	1/4" size	1 Plum	24	.333	Ea.	21	13.60		34.60	43.50	
	1050	3/8" size		24	.333		21	13.60		34.60	43.50	
	1060	1/2" size		24	.333		16.35	13.60		29.95	38.50	
	1080	3/4" size		20	.400		28	16.35		44.35	55	
	1100	1" size		19	.421		43.50	17.20		60.70	74	
	1120	1-1/4" size		15	.533		76.50	22		98.50	117	
	1140	1-1/2" size		13	.615		89.50	25		114.50	137	
	1160	2" size	↓	11	.727	↓	140	29.50		169.50	199	
	1300	For polished chrome, (1/4" thru 1"), add					50%					
	1900	Vacuum relief, water service, bronze										
	2000	1/2" size	1 Plum	30	.267	Ea.	20.50	10.90		31.40	39	
800	0010	**WATER HAMMER ARRESTORS / SHOCK ABSORBERS**										**800**
	0490	Copper										

15140	Domestic Water Piping	CREW	DAILY OUTPUT	LABOR-HOURS	UNIT	2005 BARE COSTS				TOTAL INCL O&P	
						MAT.	LABOR	EQUIP.	TOTAL		
800	0500	3/4" male I.P.S. For 1 to 11 fixtures	1 Plum	12	.667	Ea.	14.50	27		41.50	57	**800**
	0600	1" male I.P.S., For 12 to 32 fixtures		8	1		37	41		78	103	
	0700	1-1/4" male I.P.S. For 33 to 60 fixtures		8	1		43	41		84	109	
	0800	1-1/2" male I.P.S. For 61 to 113 fixtures		8	1		58.50	41		99.50	126	
	0900	2" male I.P.S. For 114 to 154 fixtures		8	1		94	41		135	165	
	1000	2-1/2" male I.P.S. For 155 to 330 fixtures	↓	4	2	↓	266	81.50		347.50	415	

15150	Sanitary Waste and Vent Piping										
100	0010	**BACKWATER VALVES**, C.I. body										**100**
	6980	Bronze gate and automatic flapper valves										
	7000	3" and 4" pipe size	Q-1	13	1.231	Ea.	800	45.50		845.50	950	
	7100	5" and 6" pipe size	"	13	1.231	"	1,225	45.50		1,270.50	1,425	
	7240	Bronze flapper valve, bolted cover										
	7260	2" pipe size	Q-1	16	1	Ea.	235	37		272	315	
	7300	4" pipe size	"	13	1.231	↓	450	45.50		495.50	565	
	7340	6" pipe size	Q-2	17	1.412		650	54		704	795	
200	0010	**CLEANOUTS**										**200**
	0060	Floor type										
	0080	Round or square, scoriated nickel bronze top										
	0100	2" pipe size	1 Plum	10	.800	Ea.	77	32.50		109.50	134	
	0120	3" pipe size		8	1		116	41		157	189	
	0140	4" pipe size	↓	6	1.333	↓	116	54.50		170.50	209	
	0980	Round top, recessed for terrazzo										
	1000	2" pipe size	1 Plum	9	.889	Ea.	77	36.50		113.50	139	
	1080	3" pipe size		6	1.333		116	54.50		170.50	209	
	1100	4" pipe size	↓	4	2		116	81.50		197.50	250	
	1120	5" pipe size	Q-1	6	2.667	↓	147	98		245	310	
250	0010	**CLEANOUT TEE**										**250**
	0100	Cast iron, B&S, with countersunk plug										
	0200	2" pipe size	1 Plum	4	2	Ea.	105	81.50		186.50	239	
	0220	3" pipe size		3.60	2.222		115	91		206	262	
	0240	4" pipe size	↓	3.30	2.424		143	99		242	305	
	0280	6" pipe size	Q-1	5	3.200	↓	385	118		503	600	
	0500	For round smooth access cover, same price				↓						
	4000	Plastic, tees and adapters. Add plugs										
	4010	ABS, DWV										
	4020	Cleanout tee, 1-1/2" pipe size	1 Plum	15	.533	Ea.	3.41	22		25.41	37	
	4030	2" pipe size	Q-1	27	.593		4.50	22		26.50	38	
	4040	3" pipe size		21	.762		9	28		37	52	
	4050	4" pipe size	↓	16	1		16.70	37		53.70	74	
	4100	Cleanout plug, 1-1/2" pipe size	1 Plum	32	.250		.95	10.20		11.15	16.40	
	4110	2" pipe size	Q-1	56	.286		1.33	10.50		11.83	17.25	
	4120	3" pipe size		36	.444		3.77	16.35		20.12	28.50	
	4130	4" pipe size	↓	30	.533		2.48	19.60		22.08	32	
	4180	Cleanout adapter fitting, 1-1/2" pipe size	1 Plum	32	.250		.95	10.20		11.15	16.40	
	4190	2" pipe size	Q-1	56	.286		1.33	10.50		11.83	17.25	
	4200	3" pipe size		36	.444		3.77	16.35		20.12	28.50	
	4210	4" pipe size	↓	30	.533	↓	6.05	19.60		25.65	36	
	5000	PVC, DWV										
	5010	Cleanout tee, 1-1/2" pipe size	1 Plum	15	.533	Ea.	3.42	22		25.42	37	
	5020	2" pipe size	Q-1	27	.593		4.60	22		26.60	38	
	5030	3" pipe size		21	.762		9.20	28		37.20	52	
	5040	4" pipe size	↓	16	1		17.10	37		54.10	74.50	
	5090	Cleanout plug, 1-1/2" pipe size	1 Plum	32	.250		.77	10.20		10.97	16.20	
	5100	2" pipe size	Q-1	56	.286	↓	.95	10.50		11.45	16.85	

Important: See the Reference Section for critical supporting data - Reference Nos., Crews, & City Cost Indexes

15150 | Sanitary Waste and Vent Piping

		CREW	DAILY OUTPUT	LABOR-HOURS	UNIT	2005 BARE COSTS MAT.	LABOR	EQUIP.	TOTAL	TOTAL INCL O&P	
250	5110	3" pipe size	Q-1	36	.444	Ea.	1.98	16.35		18.33	26.50	**250**
	5120	4" pipe size		30	.533		2.94	19.60		22.54	32.50	
	5130	6" pipe size		24	.667		9	24.50		33.50	47	
	5170	Cleanout adapter fitting, 1-1/2" pipe size	1 Plum	32	.250		1.11	10.20		11.31	16.55	
	5180	2" pipe size	Q-1	56	.286		1.57	10.50		12.07	17.55	
	5190	3" pipe size		36	.444		4.47	16.35		20.82	29.50	
	5200	4" pipe size		30	.533		7.30	19.60		26.90	37.50	
	5210	6" pipe size		24	.667		16.45	24.50		40.95	55	
300	0010	**FLOOR AND AREA DRAINS**										**300**
	0400	Deck, auto park, C.I., 13" top										
	0440	3", 4", 5", and 6" pipe size	Q-1	8	2	Ea.	565	73.50		638.50	730	
	0480	For galvanized body, add				"	268			268	295	
	2000	Floor, medium duty, C.I., deep flange, 7" dia top										
	2040	2" and 3" pipe size	Q-1	12	1.333	Ea.	79	49		128	161	
	2080	For galvanized body, add					33.50			33.50	37	
	2120	For polished bronze top, add					37			37	41	
	2400	Heavy duty, with sediment bucket, C.I., 12" dia. loose grate										
	2420	2", 3", 4", 5", and 6" pipe size	Q-1	9	1.778	Ea.	272	65.50		337.50	400	
	2460	For polished bronze top, add				"	112			112	123	
	2500	Heavy duty, cleanout & trap w/bucket, C.I., 15" top										
	2540	2", 3", and 4" pipe size	Q-1	6	2.667	Ea.	2,525	98		2,623	2,925	
	2560	For galvanized body, add					580			580	640	
	2580	For polished bronze top, add					204			204	225	
400	0010	**INTERCEPTORS**										**400**
	0150	Grease, cast iron, 4 GPM, 8 lb. fat capacity	1 Plum	4	2	Ea.	615	81.50		696.50	805	
	0200	7 GPM, 14 lb. fat capacity		4	2		650	81.50		731.50	840	
	1000	10 GPM, 20 lb. fat capacity		4	2		760	81.50		841.50	965	
	1040	15 GPM, 30 lb. fat capacity		4	2		1,125	81.50		1,206.50	1,375	
	1060	20 GPM, 40 lb. fat capacity		3	2.667		1,375	109		1,484	1,700	
	1120	Fabricated steel, 50 GPM, 100 lb. fat capacity	Q-1	2	8		2,550	294		2,844	3,250	
	1160	100 GPM, 200 lb. fat capacity	"	2	8		5,750	294		6,044	6,775	
	1560	For chemical add-port, add					98			98	108	
	1580	For seepage pan, add					7%					
	3000	Hair, cast iron, 1-1/4" and 1-1/2" pipe connection	1 Plum	8	1	Ea.	175	41		216	255	
	3100	For chrome-plated cast iron, add					108			108	119	
	4000	Oil, fabricated steel, 10 GPM, 2" pipe size	1 Plum	4	2		1,050	81.50		1,131.50	1,275	
	4100	15 GPM, 2" or 3" pipe size		4	2		1,425	81.50		1,506.50	1,700	
	4120	20 GPM, 2" or 3" pipe size		3	2.667		1,725	109		1,834	2,075	
	4220	100 GPM, 3" pipe size	Q-1	2	8		5,750	294		6,044	6,775	
	6000	Solids, precious metals recovery, C.I., 1-1/4" to 2" pipe	1 Plum	4	2		264	81.50		345.50	415	
	6100	Dental Lab., large, C.I., 1-1/2" to 2" pipe	"	3	2.667		925	109		1,034	1,200	
600	0010	**SHOWER DRAINS**										**600**
	2780	Shower, with strainer, uniform diam. trap, bronze top										
	2800	2" and 3" pipe size	Q-1	8	2	Ea.	163	73.50		236.50	290	
	2820	4" pipe size	"	7	2.286		184	84		268	330	
	2840	For galvanized body, add					65			65	71.50	
800	0010	**TRAPS**										**800**
	0030	Cast iron, service weight										
	0050	Running P trap, without vent										
	1100	2"	Q-1	16	1	Ea.	25	37		62	82.50	
	1140	3"		14	1.143		40.50	42		82.50	108	
	1150	4"		13	1.231		70	45.50		115.50	145	
	1160	6"	Q-2	17	1.412		325	54		379	435	
	1180	Running trap, single hub, with vent										

15150 | Sanitary Waste and Vent Piping

		CREW	DAILY OUTPUT	LABOR-HOURS	UNIT	MAT.	LABOR	EQUIP.	TOTAL	TOTAL INCL O&P	
800	2080	3" pipe size, 3" vent	Q-1	14	1.143	Ea.	59	42		101	128	**800**
	2120	4" pipe size, 4" vent	"	13	1.231		80	45.50		125.50	156	
	2300	For double hub, vent, add					10%	20%				
	3000	P trap, B&S, 2" pipe size	Q-1	16	1		17.60	37		54.60	75	
	3040	3" pipe size	"	14	1.143		26.50	42		68.50	92	
	3350	Deep seal trap, B&S										
	3400	1-1/4" pipe size	Q-1	14	1.143	Ea.	29.50	42		71.50	95.50	
	3410	1-1/2" pipe size		14	1.143		29.50	42		71.50	95.50	
	3420	2" pipe size		14	1.143		20	42		62	85	
	3440	3" pipe size		12	1.333		32.50	49		81.50	109	
	4700	Copper, drainage, drum trap										
	4800	3" x 5" solid, 1-1/2" pipe size	1 Plum	16	.500	Ea.	38	20.50		58.50	72.50	
	4840	3" x 6" swivel, 1-1/2" pipe size	"	16	.500	"	60.50	20.50		81	97	
	5100	P trap, standard pattern										
	5200	1-1/4" pipe size	1 Plum	18	.444	Ea.	28	18.15		46.15	58.50	
	5240	1-1/2" pipe size		17	.471		27	19.20		46.20	59	
	5260	2" pipe size		15	.533		42	22		64	79	
	5280	3" pipe size		11	.727		101	29.50		130.50	156	
	5340	With cleanout and slip joint										
	5360	1-1/4" pipe size	1 Plum	18	.444	Ea.	27	18.15		45.15	57	
	5400	1-1/2" pipe size	"	17	.471	"	41.50	19.20		60.70	74.50	
900	0010	**VENT FLASHING, CAPS**										**900**
	0120	Vent caps										
	0140	Cast iron										
	0180	2-1/2" - 3-5/8" pipe	1 Plum	21	.381	Ea.	35	15.55		50.55	61.50	
	0190	4" - 4-1/8" pipe	"	19	.421	"	41.50	17.20		58.70	72	
	0900	Vent flashing										
	1000	Aluminum with lead ring										
	1020	1-1/4" pipe	1 Plum	20	.400	Ea.	6.65	16.35		23	32	
	1030	1-1/2" pipe		20	.400		7.15	16.35		23.50	32.50	
	1040	2" pipe		18	.444		7.15	18.15		25.30	35.50	
	1050	3" pipe		17	.471		7.90	19.20		27.10	37.50	
	1060	4" pipe		16	.500		9.55	20.50		30.05	41	
	1350	Copper with neoprene ring										
	1400	1-1/4" pipe	1 Plum	20	.400	Ea.	14.05	16.35		30.40	40	
	1430	1-1/2" pipe		20	.400		14.05	16.35		30.40	40	
	1440	2" pipe		18	.444		14.85	18.15		33	44	
	1450	3" pipe		17	.471		17.40	19.20		36.60	48	
	1460	4" pipe		16	.500		19.20	20.50		39.70	51.50	

15160 | Storm Drainage Piping

		CREW	DAILY OUTPUT	LABOR-HOURS	UNIT	MAT.	LABOR	EQUIP.	TOTAL	TOTAL INCL O&P	
500	0010	**STORM AREA DRAINS**										**500**
	0140	Cornice, C.I., 45° or 90° outlet										
	0180	1-1/2" & 2" pipe size	Q-1	14	1.143	Ea.	95	42		137	168	
	0200	3" and 4" pipe size	"	12	1.333		130	49		179	216	
	0260	For galvanized body, add					26			26	28.50	
	0280	For polished bronze dome, add					22.50			22.50	24.50	
	3860	Roof, flat metal deck, C.I. body, 12" C.I. dome										
	3890	3" pipe size	Q-1	14	1.143	Ea.	155	42		197	233	
	3920	6" pipe size	"	10	1.600	"	268	59		327	385	
	4620	Main, all aluminum, 12" low profile dome										
	4640	2", 3" and 4" pipe size	Q-1	14	1.143	Ea.	172	42		214	252	
	4980	Scupper floor, oblique strainer, C.I.										
	5000	6" x 7" top, 2", 3" and 4" pipe size	Q-1	16	1	Ea.	111	37		148	178	
	5100	8" x 12" top, 5" and 6" pipe size	"	14	1.143		216	42		258	300	

15100 | Building Services Piping

15160 | Storm Drainage Piping

		CREW	DAILY OUTPUT	LABOR-HOURS	UNIT	2005 BARE COSTS MAT.	LABOR	EQUIP.	TOTAL	TOTAL INCL O&P	
500	5160	For galvanized body, add				Ea.	40%					**500**
	5200	For polished bronze strainer, add				↓	85%					
	5980	Trench, floor, hvy duty, modular, C.I., 12" x 12" top										
	6000	2", 3", 4", 5", & 6" pipe size	Q-1	8	2	Ea.	350	73.50		423.50	495	
	6100	For polished bronze top, add	"			"	172			172	189	

15180 | Heating and Cooling Piping

		CREW	DAILY OUTPUT	LABOR-HOURS	UNIT	2005 BARE COSTS MAT.	LABOR	EQUIP.	TOTAL	TOTAL INCL O&P	
200	0010	**PUMPS, CIRCULATING** Heated or chilled water application										**200**
	0600	Bronze, sweat connections, 1/40 HP, in line										
	0640	3/4" size	Q-1	16	1	Ea.	120	37		157	188	
	1000	Flange connection, 3/4" to 1-1/2" size										
	1040	1/12 HP	Q-1	6	2.667	Ea.	325	98		423	505	
	1060	1/8 HP		6	2.667		560	98		658	760	
	1100	1/3 HP		6	2.667		630	98		728	835	
	1140	2" size, 1/6 HP		5	3.200		805	118		923	1,050	
	1180	2-1/2" size, 1/4 HP	↓	5	3.200	↓	1,050	118		1,168	1,325	
	2000	Cast iron, flange connection										
	2040	3/4" to 1-1/2" size, in line, 1/12 HP	Q-1	6	2.667	Ea.	213	98		311	380	
	2100	1/3 HP		6	2.667		395	98		493	580	
	2140	2" size, 1/6 HP		5	3.200		435	118		553	655	
	2180	2-1/2" size, 1/4 HP		5	3.200		570	118		688	800	
	2220	3" size, 1/4 HP	↓	4	4	↓	580	147		727	855	
	2600	For non-ferrous impeller, add					3%					
300	0010	**PUMPS, CONDENSATE RETURN SYSTEM**										**300**
	0200	Simplex, 3/4 H.P. mtr, float switch, controls, 10 Gal. C.I. rcvr, 6-15GPM	Q-1	1	16	Ea.	1,750	590		2,340	2,800	
	1000	Duplex, 2 pumps, 3/4 H.P. motors, float switch,										
	1060	alternator asssembly, 15 Gal. C.I. receiver	Q-1	.50	32	Ea.	4,800	1,175		5,975	7,050	
800	0010	**STEAM CONDENSATE METER**										**800**
	0100	500 lb. per hour	1 Stpi	14	.571	Ea.	2,300	23.50		2,323.50	2,550	
	0140	1500 lb. per hour	"	7	1.143	"	2,475	47		2,522	2,800	

15200 | Process Piping

15230 | Industrial Process Piping

		CREW	DAILY OUTPUT	LABOR-HOURS	UNIT	2005 BARE COSTS MAT.	LABOR	EQUIP.	TOTAL	TOTAL INCL O&P	
500	0010	**PUMPS, GENERAL UTILITY** With motor										**500**
	0200	Multi-stage, horizontal split, for boiler feed applications										
	0300	Two stage, 3" discharge x 4" suction, 75 HP	Q-7	.30	106	Ea.	15,700	4,175		19,875	23,500	
	0340	Four stage, 3" discharge x 4" suction, 150 HP	"	.18	177	"	25,800	6,950		32,750	38,700	
	2000	Single stage										
	2060	End suction, 1"D. x 2"S., 3 HP	Q-1	.50	32	Ea.	3,825	1,175		5,000	5,975	
	2100	1-1/2"D. x 3"S., 10 HP	"	.40	40		4,125	1,475		5,600	6,725	
	2140	2"D. x 3"S., 15 HP	Q-2	.60	40		4,475	1,525		6,000	7,225	
	3000	Double suction, 2"D. x 2-1/2"S., 10 HP	Q-1	.30	53.333		5,300	1,950		7,250	8,775	
	3060	3"D. x 4"S., 15 HP	Q-2	.46	52.174		5,475	2,000		7,475	9,000	
	3180	6"D. x 8"S., 60 HP	Q-3	.30	106		8,900	4,150		13,050	16,100	
	3190	75 HP, to 2500 GPM		.28	114		11,100	4,450		15,550	18,900	
	3220	100 HP, to 3000 GPM		.26	123		14,100	4,800		18,900	22,700	
	3240	150 HP, to 4000 GPM	↓	.24	133	↓	18,900	5,200		24,100	28,600	

		15410	Plumbing Fixtures	CREW	DAILY OUTPUT	LABOR-HOURS	UNIT	2005 BARE COSTS				TOTAL INCL O&P	
								MAT.	LABOR	EQUIP.	TOTAL		
040	0010	**FIXTURES** Includes trim fittings unless otherwise noted	R15100 -420										040
	0080	For rough-in, supply, waste, and vent, see add for each type											
	0120	For electric water coolers, see division 15413											
	0160	For color, unless otherwise noted, add					Ea.	20%					
200	0010	**CARRIERS/SUPPORTS** For plumbing fixtures											200
	0500	Drinking fountain, wall mounted											
	0600	Plate type with studs, top back plate	1 Plum	7	1.143	Ea.	35	46.50		81.50	109	
	0700	Top front and back plate		7	1.143		43.50	46.50		90	118	
	0800	Top & bottom, front & back plates, w/bearing jacks		7	1.143		63	46.50		109.50	140	
	3000	Lavatory, concealed arm										
	3050	Floor mounted, single										
	3100	High back fixture	1 Plum	6	1.333	Ea.	258	54.50		312.50	365	
	3200	Flat slab fixture		6	1.333		300	54.50		354.50	410	
	3220	Paraplegic		6	1.333		181	54.50		235.50	281	
	3250	Floor mounted, back to back										
	3300	High back fixtures	1 Plum	5	1.600	Ea.	370	65.50		435.50	505	
	3400	Flat slab fixtures		5	1.600		455	65.50		520.50	600	
	3430	Paraplegic		5	1.600		241	65.50		306.50	365	
	3500	Wall mounted, in stud or masonry										
	3600	High back fixture	1 Plum	6	1.333	Ea.	153	54.50		207.50	250	
	3700	Flat slab fixture	"	6	1.333	"	196	54.50		250.50	298	
	4600	Sink, floor mounted										
	4650	Exposed arm system										
	4700	Single heavy fixture	1 Plum	5	1.600	Ea.	330	65.50		395.50	460	
	4750	Single heavy sink with slab		5	1.600		266	65.50		331.50	390	
	4800	Back to back, standard fixtures		5	1.600		241	65.50		306.50	365	
	4850	Back to back, heavy fixtures		5	1.600		291	65.50		356.50	420	
	4900	Back to back, heavy sink with slab		5	1.600		291	65.50		356.50	420	
	4950	Exposed offset arm system										
	5000	Single heavy deep fixture	1 Plum	5	1.600	Ea.	410	65.50		475.50	550	
	5100	Plate type system										
	5200	With bearing jacks, single fixture	1 Plum	5	1.600	Ea.	360	65.50		425.50	495	
	5300	With exposed arms, single heavy fixture		5	1.600		460	65.50		525.50	605	
	5400	Wall mounted, exposed arms, single heavy fixture		5	1.600		167	65.50		232.50	282	
	6000	Urinal, floor mounted, 2" or 3" coupling, blowout type		6	1.333		275	54.50		329.50	380	
	6100	With fixture or hanger bolts, blowout or washout		6	1.333		196	54.50		250.50	298	
	6200	With bearing plate		6	1.333		196	54.50		250.50	298	
	6300	Wall mounted, plate type system		6	1.333		128	54.50		182.50	223	
	6980	Water closet, siphon jet										
	7000	Horizontal, adjustable, caulk										
	7040	Single, 4" pipe size	1 Plum	5.33	1.501	Ea.	272	61.50		333.50	390	
	7050	4" pipe size, paraplegic		5.33	1.501		272	61.50		333.50	390	
	7060	5" pipe size		5.33	1.501		360	61.50		421.50	485	
	7100	Double, 4" pipe size		5	1.600		505	65.50		570.50	655	
	7110	4" pipe size, paraplegic		5	1.600		505	65.50		570.50	655	
	7120	5" pipe size		5	1.600		630	65.50		695.50	795	
	7160	Horizontal, adjustable, extended, caulk										
	7180	Single, 4" pipe size	1 Plum	5.33	1.501	Ea.	340	61.50		401.50	465	
	7200	5" pipe size		5.33	1.501		485	61.50		546.50	625	
	7240	Double, 4" pipe size		5	1.600		565	65.50		630.50	720	
	7260	5" pipe size		5	1.600		805	65.50		870.50	990	
	7400	Vertical, adjustable, caulk or thread										
	7440	Single, 4" pipe size	1 Plum	5.33	1.501	Ea.	272	61.50		333.50	390	
	7460	5" pipe size		5.33	1.501		360	61.50		421.50	485	
	7480	6" pipe size		5	1.600		420	65.50		485.50	565	
	7520	Double, 4" pipe size		5	1.600		505	65.50		570.50	655	

Important: See the Reference Section for critical supporting data - Reference Nos., Crews, & City Cost Indexes

15 MECHANICAL

15410	Plumbing Fixtures	CREW	DAILY OUTPUT	LABOR-HOURS	UNIT	2005 BARE COSTS				TOTAL INCL O&P
						MAT.	LABOR	EQUIP.	TOTAL	
200 7540	5" pipe size	1 Plum	5	1.600	Ea.	630	65.50		695.50	795 **200**
7560	6" pipe size	↓	4	2	↓	695	81.50		776.50	890
7600	Vertical, adjustable, extended, caulk									
7620	Single, 4" pipe size	1 Plum	5.33	1.501	Ea.	272	61.50		333.50	390
7640	5" pipe size		5.33	1.501		360	61.50		421.50	485
7680	6" pipe size		5	1.600		450	65.50		515.50	600
7720	Double, 4" pipe size		5	1.600		505	65.50		570.50	655
7740	5" pipe size		5	1.600		630	65.50		695.50	795
7760	6" pipe size	↓	4	2	↓	675	81.50		756.50	870
7780	Water closet, blow out									
7800	Vertical offset, caulk or thread									
7820	Single, 4" pipe size	1 Plum	5.33	1.501	Ea.	370	61.50		431.50	500
7840	Double, 4" pipe size	"	5	1.600	"	635	65.50		700.50	800
7880	Vertical offset, extended, caulk									
7900	Single, 4" pipe size	1 Plum	5.33	1.501	Ea.	465	61.50		526.50	605
7920	Double, 4" pipe size	"	5	1.600	"	730	65.50		795.50	905
7960	Vertical, for floor mounted back-outlet									
7980	Single, 4" thread, 2" vent	1 Plum	5.33	1.501	Ea.	249	61.50		310.50	365
8000	Double, 4" thread, 2" vent	"	6	1.333	"	730	54.50		784.50	885
8040	Vertical, for floor mounted back-outlet, extended									
8060	Single, 4" caulk, 2" vent	1 Plum	6	1.333	Ea.	249	54.50		303.50	355
8080	Double, 4" caulk, 2" vent	"	6	1.333	"	730	54.50		784.50	885
8200	Water closet, residential									
8220	Vertical centerline, floor mount									
8240	Single, 3" caulk, 2" or 3" vent	1 Plum	6	1.333	Ea.	297	54.50		351.50	405
8260	4" caulk, 2" or 4" vent		6	1.333		385	54.50		439.50	500
8280	3" copper sweat, 3" vent		6	1.333		267	54.50		321.50	375
8300	4" copper sweat, 4" vent	↓	6	1.333	↓	325	54.50		379.50	435
8400	Vertical offset, floor mount									
8420	Single, 3" or 4" caulk, vent	1 Plum	4	2	Ea.	370	81.50		451.50	535
8440	3" or 4" copper sweat, vent		5	1.600		370	65.50		435.50	510
8460	Double, 3" or 4" caulk, vent		4	2		635	81.50		716.50	825
8480	3" or 4" copper sweat, vent	↓	5	1.600	↓	635	65.50		700.50	800
9000	Water cooler (electric), floor mounted									
9100	Plate type with bearing plate, single	1 Plum	6	1.333	Ea.	143	54.50		197.50	239
300 0010	**FAUCETS/FITTINGS**									**300**
0150	Bath, faucets, diverter spout combination, sweat	1 Plum	8	1	Ea.	69.50	41		110.50	138
0200	For integral stops, IPS unions, add				↓	73			73	80.50
0420	Bath, press-bal mix valve w/diverter, spout, shower hd, arm/flange	1 Plum	8	1	↓	114	41		155	187
0810	Bidet									
0812	Fitting, over the rim, swivel spray/pop-up drain	1 Plum	8	1	Ea.	138	41		179	214
0840	Flush valves, with vacuum breaker									
0850	Water closet									
0860	Exposed, rear spud	1 Plum	8	1	Ea.	112	41		153	185
0920	Urinal									
0930	Exposed, stall	1 Plum	8	1	Ea.	103	41		144	175
0940	Wall, (washout)		8	1		103	41		144	175
0950	Pedestal, top spud		8	1		105	41		146	177
0960	Concealed, stall		8	1		118	41		159	192
0970	Wall (washout)	↓	8	1	↓	118	41		159	192
0971	Automatic flush sensor and operator for									
0972	urinals or water closets	1 Plum	5.33	1.501	Ea.	340	61.50		401.50	465
1000	Kitchen sink faucets, top mount, cast spout		10	.800		51.50	32.50		84	106
1100	For spray, add		24	.333		11.05	13.60		24.65	32.50
2000	Laundry faucets, shelf type, IPS or copper unions	↓	12	.667	↓	42.50	27		69.50	87.50

MECHANICAL 15

		15410 Plumbing Fixtures	CREW	DAILY OUTPUT	LABOR-HOURS	UNIT	2005 BARE COSTS				TOTAL INCL O&P	
							MAT.	LABOR	EQUIP.	TOTAL		
300	2100	Lavatory faucet, centerset, without drain	1 Plum	10	.800	Ea.	37.50	32.50		70	90	300
	2810	Automatic sensor and operator, with faucet head		6.15	1.301		300	53		353	410	
	3000	Service sink faucet, cast spout, pail hook, hose end		14	.571		72	23.50		95.50	114	
	4000	Shower by-pass valve with union		18	.444		50.50	18.15		68.65	83	
	4200	Shower thermostatic mixing valve, concealed		8	1		242	41		283	330	
	5000	Sillcock, compact, brass, IPS or copper to hose		24	.333		4.74	13.60		18.34	25.50	

		15411 Commercial/Indust Fixtures										
400	0010	**HOT WATER DISPENSERS**										400
	0160	Commercial, 100 cup, 11.3 amp	1 Plum	14	.571	Ea.	291	23.50		314.50	355	
	3180	Household, 60 cup	"	14	.571	"	156	23.50		179.50	207	
500	0010	**HYDRANTS**										500
	0050	Wall type, moderate climate, bronze, encased										
	0200	3/4" IPS connection	1 Plum	16	.500	Ea.	300	20.50		320.50	360	
	0300	1" IPS connection	"	14	.571		299	23.50		322.50	365	
	0500	Anti-siphon type					259			259	285	
	1000	Non-freeze, bronze, exposed										
	1100	3/4" IPS connection, 4" to 9" thick wall	1 Plum	14	.571	Ea.	203	23.50		226.50	258	
	1120	10" to 14" thick wall		12	.667		220	27		247	283	
	1140	15" to 19" thick wall		12	.667		245	27		272	310	
	1160	20" to 24" thick wall		10	.800		265	32.50		297.50	340	
	1200	For 1" IPS connection, add					15%	10%				
	1240	For 3/4" adapter type vacuum breaker, add				Ea.	25			25	27.50	
	1280	For anti-siphon type, add				"	54			54	59.50	
	2000	Non-freeze bronze, encased, anti-siphon type										
	2100	3/4" IPS connection, 5" to 9" thick wall	1 Plum	14	.571	Ea.	271	23.50		294.50	335	
	2140	15" to 19" thick wall	"	12	.667	"	395	27		422	470	
	3000	Ground box type, bronze frame, 3/4" IPS connection										
	3080	Non-freeze, all bronze, polished face, set flush										
	3100	2 feet depth of bury	1 Plum	8	1	Ea.	375	41		416	470	
	3180	6 feet depth of bury		7	1.143		490	46.50		536.50	605	
	3220	8 feet depth of bury		5	1.600		545	65.50		610.50	700	
	3400	For 1" IPS connection, add					15%	10%				
	3550	For 2" connection, add					445%	24%				
	3600	For tapped drain port in box, add					34.50			34.50	37.50	
	5000	Moderate climate, all bronze, polished face										
	5020	and scoriated cover, set flush										
	5100	3/4" IPS connection	1 Plum	16	.500	Ea.	265	20.50		285.50	325	
	5120	1" IPS connection	"	14	.571		265	23.50		288.50	325	
	5200	For tapped drain port in box, add					34.50			34.50	37.50	
700	0010	**URINALS**										700
	3000	Wall hung, vitreous china, with hanger & self-closing valve										
	3100	Siphon jet type	Q-1	3	5.333	Ea.	298	196		494	625	
	3120	Blowout type		3	5.333		340	196		536	670	
	3300	Rough-in, supply, waste & vent		2.83	5.654		110	208		318	435	
	5000	Stall type, vitreous china, includes valve		2.50	6.400		510	235		745	915	
	5100	3" seam cover, add		12	1.333		146	49		195	235	
	5200	6" seam cover, add		12	1.333		161	49		210	251	
	6980	Rough-in, supply, waste and vent		1.99	8.040		152	296		448	615	
840	0010	**WASH FOUNTAINS**										840
	1900	Group, foot control										
	2000	Precast terrazzo, circular, 36" diam., 5 or 6 persons	Q-2	3	8	Ea.	2,775	305		3,080	3,500	
	2100	54" diameter for 8 or 10 persons		2.50	9.600		3,350	365		3,715	4,250	

Important: See the Reference Section for critical supporting data - Reference Nos., Crews, & City Cost Indexes

15 MECHANICAL

15411 | Commercial/Indust Fixtures

840			CREW	DAILY OUTPUT	LABOR-HOURS	UNIT	2005 BARE COSTS				TOTAL INCL O&P	840
							MAT.	LABOR	EQUIP.	TOTAL		
	2400	Semi-circular, 36" diam. for 3 persons	Q-2	3	8	Ea.	2,475	305		2,780	3,175	
	2500	54" diam. for 4 or 5 persons		2.50	9.600		3,050	365		3,415	3,925	
	2700	Quarter circle (corner), 54" for 3 persons		3.50	6.857		3,050	261		3,311	3,775	
	3000	Stainless steel, circular, 36" diameter		3.50	6.857		3,175	261		3,436	3,900	
	3100	54" diameter		2.80	8.571		4,175	325		4,500	5,100	
	3400	Semi-circular, 36" diameter		3.50	6.857		2,750	261		3,011	3,425	
	3500	54" diameter	▼	2.80	8.571	▼	3,600	325		3,925	4,475	
	5610	Group, infrared control, barrier free ♿										
	5614	Precast terrazzo										
	5620	Semi-circular 36" diam. for 3 persons	Q-2	3	8	Ea.	3,650	305		3,955	4,450	
	5630	46" diam. for 4 persons		2.80	8.571		4,075	325		4,400	5,000	
	5640	Circular, 54" diam. for 8 persons, button control	▼	2.50	9.600		5,975	365		6,340	7,125	
	5700	Rough-in, supply, waste and vent for above wash fountains	Q-1	1.82	8.791		148	325		473	650	
	6200	Duo for small washrooms, stainless steel		2	8		1,875	294		2,169	2,500	
	6500	Rough-in, supply, waste & vent for duo fountains	▼	2.02	7.921	▼	83.50	291		374.50	530	

15412 | Drinking Fountains

200			CREW	DAILY OUTPUT	LABOR-HOURS	UNIT	MAT.	LABOR	EQUIP.	TOTAL	TOTAL INCL O&P	200
	0010	**DRINKING FOUNTAIN** For connection to cold water supply [R15100-420]										
	1000	Wall mounted, non-recessed										
	1400	Bronze, with no back	1 Plum	4	2	Ea.	1,000	81.50		1,081.50	1,225	
	1800	Cast aluminum, enameled, for correctional institutions		4	2		625	81.50		706.50	810	
	2000	Fiberglass, 12" back, single bubbler unit		4	2		555	81.50		636.50	735	
	2040	Dual bubbler		3.20	2.500		810	102		912	1,050	
	2400	Precast stone, no back		4	2		505	81.50		586.50	680	
	2700	Stainless steel, single bubbler, no back		4	2		825	81.50		906.50	1,025	
	2740	With back		4	2		980	81.50		1,061.50	1,200	
	2780	Dual handle & wheelchair projection type ♿		4	2		505	81.50		586.50	680	
	2820	Dual level for handicapped type ♿	▼	3.20	2.500	▼	1,025	102		1,127	1,275	
	3300	Vitreous china										
	3340	7" back	1 Plum	4	2	Ea.	420	81.50		501.50	585	
	3940	For vandal-resistant bottom plate, add					52.50			52.50	57.50	
	3960	For freeze-proof valve system, add	1 Plum	2	4		410	163		573	695	
	3980	For rough-in, supply and waste, add	"	2.21	3.620	▼	65.50	148		213.50	294	
	4000	Wall mounted, semi-recessed										
	4200	Poly-marble, single bubbler	1 Plum	4	2	Ea.	680	81.50		761.50	870	
	4600	Stainless steel, satin finish, single bubbler		4	2		785	81.50		866.50	985	
	4900	Vitreous china, single bubbler		4	2		565	81.50		646.50	745	
	5980	For rough-in, supply and waste, add	▼	1.83	4.372	▼	65.50	179		244.50	340	
	6000	Wall mounted, fully recessed										
	6400	Poly-marble, single bubbler	1 Plum	4	2	Ea.	765	81.50		846.50	965	
	6800	Stainless steel, single bubbler		4	2		700	81.50		781.50	890	
	7560	For freeze-proof valve system, add		2	4		500	163		663	800	
	7580	For rough-in, supply and waste, add	▼	1.83	4.372		65.50	179		244.50	340	
	7600	Floor mounted, pedestal type										
	7700	Aluminum, architectural style, C.I. base	1 Plum	2	4	Ea.	460	163		623	750	
	7780	Wheelchair handicap unit		2	4		1,050	163		1,213	1,425	
	8400	Stainless steel, architectural style		2	4		1,200	163		1,363	1,550	
	8600	Enameled iron, heavy duty service, 2 bubblers		2	4		1,100	163		1,263	1,450	
	8660	4 bubblers		2	4		1,175	163		1,338	1,550	
	8880	For freeze-proof valve system, add		2	4		335	163		498	615	
	8900	For rough-in, supply and waste, add	▼	1.83	4.372	▼	65.50	179		244.50	340	
	9100	Deck mounted										
	9500	Stainless steel, circular receptor	1 Plum	4	2	Ea.	280	81.50		361.50	435	
	9760	White enameled steel, 14" x 9" receptor		4	2		252	81.50		333.50	400	
	9860	White enameled cast iron, 24" x 16" receptor	▼	3	2.667		242	109		351	430	

MECHANICAL 15

15412 | Drinking Fountains

			CREW	DAILY OUTPUT	LABOR-HOURS	UNIT	2005 BARE COSTS				TOTAL INCL O&P	
							MAT.	LABOR	EQUIP.	TOTAL		
200	9980	For rough-in, supply and waste, add R15100-420	1 Plum	1.83	4.372	Ea.	65.50	179		244.50	340	200

15413 | Electric Water Coolers

			CREW	DAILY OUTPUT	LABOR-HOURS	UNIT	MAT.	LABOR	EQUIP.	TOTAL	TOTAL INCL O&P	
900	0010	**WATER COOLER** R15100-430										900
	0030	See line 15413-900-9800 for rough-in, waste & vent										
	0040	for all water coolers										
	0100	Wall mounted, non-recessed										
	0140	4 GPH	Q-1	4	4	Ea.	500	147		647	770	
	0160	8 GPH, barrier free, sensor operated		4	4		630	147		777	915	
	0180	8.2 GPH		4	4		560	147		707	835	
	0600	For hot and cold water, add					144			144	158	
	0640	For stainless steel cabinet, add					73.50			73.50	81	
	1000	Dual height, 8.2 GPH	Q-1	3.80	4.211		805	155		960	1,125	
	1040	14.3 GPH	"	3.80	4.211		855	155		1,010	1,175	
	1240	For stainless steel cabinet, add					112			112	123	
	2600	Wheelchair type, 8 GPH	Q-1	4	4		1,425	147		1,572	1,800	
	3300	Semi-recessed, 8.1 GPH		4	4		840	147		987	1,150	
	3320	12 GPH		4	4		885	147		1,032	1,200	
	4600	Floor mounted, flush-to-wall										
	4640	4 GPH	1 Plum	3	2.667	Ea.	520	109		629	740	
	4680	8.2 GPH		3	2.667		575	109		684	795	
	4720	14.3 GPH		3	2.667		520	109		629	740	
	4960	14 GPH hot and cold water		3	2.667		620	109		729	850	
	4980	For stainless steel cabinet, add					112			112	123	
	5000	Dual height, 8.2 GPH	1 Plum	2	4		870	163		1,033	1,200	
	5040	14.3 GPH	"	2	4		900	163		1,063	1,250	
	5120	For stainless steel cabinet, add					147			147	162	
	9800	For supply, waste & vent, all coolers	1 Plum	2.21	3.620		65.50	148		213.50	294	

15414 | Emergency Fixtures

			CREW	DAILY OUTPUT	LABOR-HOURS	UNIT	MAT.	LABOR	EQUIP.	TOTAL	TOTAL INCL O&P	
200	0010	**INDUSTRIAL SAFETY FIXTURES** Rough-in not included										200
	1000	Eye wash fountain										
	1400	Plastic bowl, pedestal mounted	Q-1	4	4	Ea.	186	147		333	425	
	1600	Unmounted		4	4		132	147		279	365	
	1800	Wall mounted		4	4		182	147		329	420	
	2000	Stainless steel, pedestal mounted		4	4		278	147		425	525	
	2200	Unmounted		4	4		182	147		329	420	
	2400	Wall mounted		4	4		174	147		321	410	
	3000	Eye wash, portable, self-contained					375			375	410	
	4000	Eye and face wash, combination fountain										
	4200	Stainless steel, pedestal mounted	Q-1	4	4	Ea.	390	147		537	650	
	4400	Unmounted		4	4		300	147		447	550	
	4600	Wall mounted		4	4		305	147		452	555	
	5000	Shower, single head, drench, ball valve, pull, freestanding		4	4		255	147		402	500	
	5200	Horizontal or vertical supply		4	4		237	147		384	480	
	6000	Multi-nozzle, eye/face wash combination		4	4		480	147		627	750	
	6400	Multi-nozzle, 12 spray, shower only		4	4		1,175	147		1,322	1,500	
	6600	For freeze-proof, add		6	2.667		271	98		369	445	

15418 | Resi/Comm/Industrial Fixtures

			CREW	DAILY OUTPUT	LABOR-HOURS	UNIT	MAT.	LABOR	EQUIP.	TOTAL	TOTAL INCL O&P	
100	0010	**BATHS**										100
	0100	Tubs, recessed porcelain enamel on cast iron, with trim										
	0140	42" x 37"	Q-1	5	3.200	Ea.	855	118		973	1,125	
	0180	48" x 42"		4	4		1,500	147		1,647	1,875	

Important: See the Reference Section for critical supporting data - Reference Nos., Crews, & City Cost Indexes

			CREW	DAILY OUTPUT	LABOR-HOURS	UNIT	2005 BARE COSTS				TOTAL INCL O&P	
	15418	**Resi/Comm/Industrial Fixtures**					MAT.	LABOR	EQUIP.	TOTAL		
100	0220	72" x 36"	Q-1	3	5.333	Ea.	1,525	196		1,721	2,000	100
	2000	Enameled formed steel, 4'-6" long		5.80	2.759		330	101		431	520	
	2200	5' long		5.50	2.909		290	107		397	480	
	4000	Soaking, acrylic with pop-up drain, 60" x 32" x 21" deep		5.50	2.909		805	107		912	1,050	
	4100	60" x 48" x 18-1/2" deep	▼	5	3.200	▼	630	118		748	870	
	6000	Whirlpool, bath with vented overflow, molded fiberglass										
	6100	66" x 48" x 24"	Q-1	1	16	Ea.	2,350	590		2,940	3,475	
	9600	Rough-in, supply, waste and vent, for all above tubs, add	"	2.07	7.729	"	142	284		426	580	
400	0010	**LAUNDRY SINKS** With trim										400
	0020	Porcelain enamel on cast iron, black iron frame										
	0050	24" x 21", single compartment	Q-1	6	2.667	Ea.	345	98		443	525	
	0100	26" x 21", single compartment	"	6	2.667	"	269	98		367	445	
	2000	Molded stone, on wall hanger or legs										
	2020	22" x 23", single compartment	Q-1	6	2.667	Ea.	124	98		222	283	
	2100	45" x 21", double compartment	"	5	3.200	"	257	118		375	460	
	3000	Plastic, on wall hanger or legs										
	3020	18" x 23", single compartment	Q-1	6.50	2.462	Ea.	92	90.50		182.50	237	
	3300	40" x 24", double compartment		5.50	2.909		216	107		323	400	
	5000	Stainless steel, counter top, 22" x 17" single compartment		6	2.667		345	98		443	525	
	5100	22" x 22", single compartment		6	2.667		440	98		538	630	
	5200	33" x 22", double compartment		5	3.200		435	118		553	650	
	9600	Rough-in, supply, waste and vent, for all laundry sinks	▼	2.14	7.477	▼	104	275		379	530	
450	0010	**LAVATORIES** With trim, white unless noted otherwise	R15100 -420									450
	0500	Vanity top, porcelain enamel on cast iron										
	0600	20" x 18"	Q-1	6.40	2.500	Ea.	206	92		298	365	
	0640	33" x 19" oval		6.40	2.500		415	92		507	595	
	0720	19" round	▼	6.40	2.500	▼	221	92		313	380	
	0860	For color, add					25%					
	1000	Cultured marble, 19" x 17", single bowl	Q-1	6.40	2.500	Ea.	153	92		245	305	
	1040	25" x 19", single bowl	"	6.40	2.500	"	153	92		245	305	
	1580	For color, same price										
	1900	Stainless steel, self-rimming, 25" x 22", single bowl, ledge	Q-1	6.40	2.500	Ea.	235	92		327	395	
	1960	17" x 22", single bowl		6.40	2.500		229	92		321	390	
	2600	Steel, enameled, 20" x 17", single bowl		5.80	2.759		129	101		230	295	
	2660	19" round		5.80	2.759		127	101		228	293	
	2900	Vitreous china, 20" x 16", single bowl		5.40	2.963		222	109		331	410	
	2960	20" x 17", single bowl		5.40	2.963		146	109		255	325	
	3020	19" round, single bowl		5.40	2.963		144	109		253	320	
	3200	22" x 13", single bowl	▼	5.40	2.963	▼	231	109		340	420	
	3560	For color, add					50%					
	3580	Rough-in, supply, waste and vent for all above lavatories	Q-1	2.30	6.957	Ea.	90.50	256		346.50	485	
	4000	Wall hung										
	4040	Porcelain enamel on cast iron, 16" x 14", single bowl	Q-1	8	2	Ea.	320	73.50		393.50	460	
	4180	20" x 18", single bowl		8	2		237	73.50		310.50	370	
	4240	22" x 19", single bowl	▼	8	2		405	73.50		478.50	555	
	4580	For color, add					30%					
	6000	Vitreous china, 18" x 15", single bowl with backsplash	Q-1	7	2.286	Ea.	198	84		282	345	
	6180	19" x 19", corner style	"	8	2	"	420	73.50		493.50	575	
	6500	For color, add					30%					
	6960	Rough-in, supply, waste and vent for above lavatories	Q-1	1.66	9.639	Ea.	285	355		640	850	
500	0010	**SHOWERS**										500
	1500	Stall, with drain only. Add for valve and door/curtain										
	3000	Fiberglass, one piece, with 3 walls, 32" x 32" square	Q-1	2.40	6.667	Ea.	415	245		660	830	
	3100	36" x 36" square		2.40	6.667		460	245		705	875	
	3250	64" x 65-3/4" x 81-1/2" fold. seat, whlchr.		1.80	8.889		2,900	325		3,225	3,675	
	4000	Polypropylene, stall only, w/ molded-stone floor, 30" x 30"	▼	2	8	▼	330	294		624	800	

MECHANICAL 15

15418 | Resi/Comm/Industrial Fixtures

			CREW	DAILY OUTPUT	LABOR-HOURS	UNIT	MAT.	LABOR	EQUIP.	TOTAL	TOTAL INCL O&P	
500	4200	Rough-in, supply, waste and vent for above showers	Q-1	2.05	7.805	Ea.	97	287		384	535	500
	4520	Showers, fiberglass receptor only	1 Plum	8	1		161	41		202	239	
	5000	Built-in, head, arm, 4 GPM valve		4	2		91	81.50		172.50	223	
	5200	Head, arm, by-pass, integral stops, handles	↓	3.60	2.222	↓	224	91		315	385	
	6000	Group, w/pressure balancing valve, rough-in and rigging not included										
	6800	Column, 6 heads, no receptors, less partitions	Q-1	3	5.333	Ea.	2,025	196		2,221	2,525	
	6900	With stainless steel partitions		1	16		4,325	590		4,915	5,625	
	7600	5 heads, no receptors, less partitions		3	5.333		1,550	196		1,746	2,000	
	7620	4 heads (1 handicap) no receptors, less partitions		3	5.333		2,175	196		2,371	2,675	
	7700	With stainless steel partitions		1	16		1,550	590		2,140	2,575	
	8000	Wall, 2 heads, no receptors, less partitions		4	4		895	147		1,042	1,200	
	8100	With stainless steel partitions	↓	2	8	↓	2,175	294		2,469	2,825	
600	0010	**SINKS** With faucets and drain										600
	2000	Kitchen, counter top style, P.E. on C.I., 24" x 21" single bowl	Q-1	5.60	2.857	Ea.	206	105		311	385	
	2100	31" x 22" single bowl		5.60	2.857		252	105		357	435	
	2200	32" x 21" double bowl		4.80	3.333		288	123		411	500	
	2300	42" x 21" double bowl		4.80	3.333		480	123		603	715	
	3000	Stainless steel, self rimming, 19" x 18" single bowl		5.60	2.857		335	105		440	525	
	3100	25" x 22" single bowl		5.60	2.857		370	105		475	565	
	4000	Steel, enameled, with ledge, 24" x 21" single bowl		5.60	2.857		110	105		215	279	
	4100	32" x 21" double bowl	↓	4.80	3.333		147	123		270	345	
	4960	For color sinks except stainless steel, add					10%					
	4980	For rough-in, supply, waste and vent, counter top sinks	Q-1	2.14	7.477	↓	104	275		379	530	
	5000	Kitchen, raised deck, P.E. on C.I.										
	5100	32" x 21", dual level, double bowl	Q-1	2.60	6.154	Ea.	259	226		485	625	
	5700	For color, add					20%					
	5790	For rough-in, supply, waste & vent, sinks	Q-1	1.85	8.649		104	320		424	595	
	6650	Service, floor, corner, P.E. on C.I., 28" x 28"		4.40	3.636		530	134		664	785	
	6790	For rough-in, supply, waste & vent, floor service sinks		1.64	9.756		263	360		623	830	
	7000	Service, wall, P.E. on C.I., roll rim, 22" x 18"		4	4		445	147		592	710	
	7100	24" x 20"		4	4		460	147		607	730	
	8600	Vitreous china, 22" x 20"	↓	4	4		390	147		537	650	
	8960	For stainless steel rim guard, front or side, add					35			35	38.50	
	8980	For rough-in, supply, waste & vent, wall service sinks	Q-1	1.30	12.308	↓	410	455		865	1,125	
900	0010	**WATER CLOSETS**										900
	0030	For automatic flush, see 15410-300-0972										
	0150	Tank type, vitreous china, incl. seat, supply pipe w/stop										
	0200	Wall hung, one piece	Q-1	5.30	3.019	Ea.	390	111		501	595	
	0400	Two piece, close coupled		5.30	3.019		480	111		591	690	
	0960	For rough-in, supply, waste, vent and carrier		2.73	5.861		335	216		551	690	
	1000	Floor mounted, one piece		5.30	3.019		495	111		606	710	
	1100	Two piece, close coupled **CN**	↓	5.30	3.019	↓	163	111		274	345	
	1960	For color, add					30%					
	1980	For rough-in, supply, waste and vent	Q-1	3.05	5.246	Ea.	152	193		345	460	
	3000	Bowl only, with flush valve, seat										
	3100	Wall hung	Q-1	5.80	2.759	Ea.	345	101		446	535	
	3200	For rough-in, supply, waste and vent, single WC		2.56	6.250		350	230		580	730	
	3300	Floor mounted		5.80	2.759		310	101		411	495	
	3350	With wall outlet		5.80	2.759		485	101		586	690	
	3400	For rough-in, supply, waste and vent, single WC	↓	2.84	5.634	↓	168	207		375	495	

15440 | Plumbing Pumps

			CREW	DAILY OUTPUT	LABOR-HOURS	UNIT	MAT.	LABOR	EQUIP.	TOTAL	TOTAL INCL O&P	
240	0010	**PUMPS, PRESSURE BOOSTER SYSTEM**										240
	0200	Pump system, with diaphragm tank, control, press. switch										
	0300	1 HP pump	Q-1	1.30	12.308	Ea.	4,000	455		4,455	5,075	
	0400	1-1/2 HP pump	↓	1.25	12.800	↓	4,050	470		4,520	5,150	

Important: See the Reference Section for critical supporting data - Reference Nos., Crews, & City Cost Indexes

15440		Plumbing Pumps	CREW	DAILY OUTPUT	LABOR-HOURS	UNIT	2005 BARE COSTS				TOTAL INCL O&P	
							MAT.	LABOR	EQUIP.	TOTAL		
240	0420	2 HP pump	Q-1	1.20	13.333	Ea.	4,150	490		4,640	5,300	**240**
	0440	3 HP pump	↓	1.10	14.545		4,200	535		4,735	5,425	
	0460	5 HP pump	Q-2	1.50	16		4,650	610		5,260	6,050	
	0480	7-1/2 HP pump		1.42	16.901		5,175	645		5,820	6,675	
	0500	10 HP pump	↓	1.34	17.910	↓	5,425	685		6,110	7,000	
	1000	Pump/ energy storage system, diaphragm tank, 3 HP pump										
	1100	motor, PRV, switch, gauge, control center, flow switch										
	1200	125 lb. working pressure	Q-2	.70	34.286	Ea.	11,700	1,300		13,000	14,800	
	1300	250 lb. working pressure	"	.64	37.500	"	12,800	1,425		14,225	16,200	
400	0010	**PUMPS, GRINDER SYSTEM** Complete, incl. check valve, tank, std.										**400**
	0020	controls incl. alarm/disconnect panel w/wire. Excavation not included										
	0260	Simplex, 9 GPM at 60 PSIG, 70 gal. tank				Ea.	2,425			2,425	2,675	
	0300	For manway, 26" I.D., 18" high, add					310			310	340	
	0340	26" I.D., 36" high, add					425			425	470	
	0380	43" I.D., 4' high, add				↓	475			475	525	
800	0010	**PUMPS, SEWAGE EJECTOR** With operating and level controls										**800**
	0100	Simplex system incl. tank, cover, pump 15' head										
	0500	37 gal PE tank, 12 GPM, 1/2 HP, 2" discharge	Q-1	3.20	5	Ea.	375	184		559	690	
	0510	3" discharge		3.10	5.161		405	190		595	730	
	0530	87 GPM, .7 HP, 2" discharge		3.20	5		570	184		754	905	
	0540	3" discharge		3.10	5.161		620	190		810	965	
	0600	45 gal. coated stl tank, 12 GPM, 1/2 HP, 2" discharge		3	5.333		670	196		866	1,025	
	0610	3" discharge		2.90	5.517		695	203		898	1,075	
	0630	87 GPM, .7 HP, 2" discharge		3	5.333		855	196		1,051	1,225	
	0640	3" discharge		2.90	5.517		905	203		1,108	1,300	
	0660	134 GPM, 1 HP, 2" discharge		2.80	5.714		925	210		1,135	1,350	
	0680	3" discharge		2.70	5.926		975	218		1,193	1,400	
	0700	70 gal. PE tank, 12 GPM, 1/2 HP, 2" discharge		2.60	6.154		735	226		961	1,150	
	0710	3" discharge		2.40	6.667		785	245		1,030	1,225	
	0730	87 GPM, 0.7 HP, 2" discharge		2.50	6.400		940	235		1,175	1,375	
	0740	3" discharge		2.30	6.957		1,000	256		1,256	1,475	
	0760	134 GPM, 1 HP, 2" discharge		2.20	7.273		1,025	267		1,292	1,525	
	0770	3" discharge	↓	2	8	↓	1,100	294		1,394	1,650	
900	0010	**PUMPS, PEDESTAL SUMP** With float control										**900**
	0400	Molded PVC base, 21 GPM at 15' head, 1/3 HP	1 Plum	5	1.600	Ea.	87	65.50		152.50	194	
	0800	Iron base, 21 GPM at 15' head, 1/3 HP		5	1.600		110	65.50		175.50	219	
	1200	Solid brass, 21 GPM at 15' head, 1/3 HP	↓	5	1.600	↓	179	65.50		244.50	295	
940	0010	**PUMPS, SUBMERSIBLE** Sump										**940**
	7000	Sump pump, automatic										
	7100	Plastic, 1-1/4" discharge, 1/4 HP	1 Plum	6	1.333	Ea.	105	54.50		159.50	198	
	7140	1/3 HP		5	1.600		127	65.50		192.50	238	
	7160	1/2 HP		5	1.600		153	65.50		218.50	266	
	7180	1-1/2" discharge, 1/2 HP		4	2		171	81.50		252.50	310	
	7500	Cast iron, 1-1/4" discharge, 1/4 HP		6	1.333		122	54.50		176.50	216	
	7540	1/3 HP		6	1.333		144	54.50		198.50	240	
	7560	1/2 HP	↓	5	1.600	↓	170	65.50		235.50	285	

15460		Domestic Water Cond Equipment										
900	0010	**WATER SOFTENER**										**900**
	5800	Softener systems, automatic, intermediate sizes										
	5820	available, may be used in multiples.										
	6000	Hardness capacity between regenerations and flow										
	6100	150,000 grains, 37 GPM cont., 51 GPM peak	Q-1	1.20	13.333	Ea.	3,775	490		4,265	4,875	
	6200	300,000 grains, 81 GPM cont., 113 GPM peak	↓	1	16	↓	5,600	590		6,190	7,050	

MECHANICAL 15

15460	Domestic Water Cond Equipment	CREW	DAILY OUTPUT	LABOR-HOURS	UNIT	2005 BARE COSTS				TOTAL INCL O&P
						MAT.	LABOR	EQUIP.	TOTAL	
900 6300	750,000 grains, 160 GPM cont., 230 GPM peak	Q-1	.80	20	Ea.	7,300	735		8,035	9,150 **900**
6400	900,000 grains, 185 GPM cont., 270 GPM peak	↓	.70	22.857	↓	11,800	840		12,640	14,200

15480 | Domestic Water Heaters

		CREW	DAILY OUTPUT	LABOR-HOURS	UNIT	MAT.	LABOR	EQUIP.	TOTAL	TOTAL INCL O&P
200 0010	**WATER HEATERS**									**200**
0020	For solar, see division 13600									
1000	Residential, electric, glass lined tank, 5 yr, 10 gal., single element	1 Plum	2.30	3.478	Ea.	210	142		352	445
1040	20 gallon, single element		2.20	3.636		244	149		393	490
1060	30 gallon, double element		2.20	3.636		280	149		429	535
1080	40 gallon, double element		2	4		300	163		463	575
1100	52 gallon, double element		2	4		355	163		518	635
1180	120 gallon, double element	↓	1.40	5.714	↓	770	233		1,003	1,200
2000	Gas fired, foam lined tank, 10 yr, vent not incl.,									
2040	30 gallon	1 Plum	2	4	Ea.	385	163		548	670
2100	75 gallon		1.50	5.333		745	218		963	1,150
2120	100 gallon		1.30	6.154		1,150	251		1,401	1,625
3000	Oil fired, glass lined tank, 5 yr, vent not included, 30 gallon		2	4		745	163		908	1,075
3040	50 gallon		1.80	4.444		1,375	182		1,557	1,775
3060	70 gallon		1.50	5.333		1,525	218		1,743	2,000
3080	85 gallon	↓	1.40	5.714	↓	4,300	233		4,533	5,075
4000	Commercial, 100° rise. NOTE: for each size tank, a range of									
4010	heaters between the ones shown are available									
4020	Electric									
4100	5 gal., 3 kW, 12 GPH, 208V	1 Plum	2	4	Ea.	1,625	163		1,788	2,050
4120	10 gal., 6 kW, 25 GPH, 208V		2	4		1,800	163		1,963	2,225
4140	50 gal., 9 kW, 37 GPH, 208V		1.80	4.444		2,475	182		2,657	3,000
4160	50 gal., 36 kW, 148 GPH, 208V	↓	1.80	4.444		3,775	182		3,957	4,425
4300	200 gal., 15 kW, 61 GPH, 480V	Q-1	1.70	9.412		11,900	345		12,245	13,600
4320	200 gal., 120 kW , 490 GPH, 480V		1.70	9.412		16,300	345		16,645	18,400
4460	400 gal., 30 kW, 123 GPH, 480V	↓	1	16		16,400	590		16,990	18,900
5400	Modulating step control, 2-5 steps	1 Elec	5.30	1.509		1,525	61.50		1,586.50	1,775
5440	6-10 steps		3.20	2.500		1,900	102		2,002	2,250
5460	11-15 steps		2.70	2.963		2,350	121		2,471	2,750
5480	16-20 steps	↓	1.60	5	↓	2,775	204		2,979	3,350
6000	Gas fired, flush jacket, std. controls, vent not incl.									
6040	75 MBH input, 73 GPH	1 Plum	1.40	5.714	Ea.	1,475	233		1,708	1,975
6060	98 MBH input, 95 GPH		1.40	5.714		2,600	233		2,833	3,225
6080	120 MBH input, 110 GPH **CN**		1.20	6.667		2,825	272		3,097	3,500
6180	200 MBH input, 192 GPH		.60	13.333		4,300	545		4,845	5,575
6200	250 MBH input, 245 GPH		.50	16		4,675	655		5,330	6,125
6900	For low water cutoff, add		8	1		220	41		261	305
6960	For bronze body hot water circulator, add	↓	4	2	↓	1,150	81.50		1,231.50	1,400
8000	Oil fired, flush jacket, std. controls, vent not incl.									
8060	103 MBH gross output, 116 GPH	1 Plum	1.10	7.273	Ea.	980	297		1,277	1,525
8080	122 MBH gross output, 141 GPH		1	8		1,800	325		2,125	2,500
8100	137 MBH gross output, 161 GPH	↓	.80	10		1,875	410		2,285	2,675
8160	225 MBH gross output, 256 GPH	Q-1	.80	20		2,575	735		3,310	3,925
8180	262 MBH gross output, 315 GPH		.70	22.857		2,725	840		3,565	4,275
8280	735 MBH gross output, 880 GPH		.40	40		9,775	1,475		11,250	13,000
8300	840 MBH gross output, 1000 GPH	↓	.40	40		10,200	1,475		11,675	13,500
8900	For low water cutoff, add	1 Plum	8	1		220	41		261	305
8960	For bronze body hot water circulator, add	"	4	2	↓	805	81.50		886.50	1,000
900 0010	**HEAT TRANSFER PACKAGES** Complete, controls,									**900**
0020	expansion tank, converter, air separator									
1000	Hot water, 180°F enter, 200°F leaving, 15# steam									
1010	One pump system, 28 GPM	Q-6	.75	32	Ea.	11,600	1,225		12,825	14,700

Important: See the Reference Section for critical supporting data - Reference Nos., Crews, & City Cost Indexes

15400 | Plumbing Fixtures & Equipment

15480 | Domestic Water Heaters

			CREW	DAILY OUTPUT	LABOR-HOURS	UNIT	MAT.	LABOR	EQUIP.	TOTAL	TOTAL INCL O&P	
900	1020	35 GPM	Q-6	.70	34.286	Ea.	12,800	1,300		14,100	16,100	900
	1040	55 GPM		.65	36.923		14,600	1,400		16,000	18,200	
	1060	130 GPM		.55	43.636		18,100	1,675		19,775	22,400	
	1080	255 GPM		.40	60		24,900	2,300		27,200	30,900	
	1100	550 GPM	▼	.30	80	▼	31,700	3,050		34,750	39,400	

15490 | Pool and Fountain Equipment

			CREW	DAILY OUTPUT	LABOR-HOURS	UNIT	MAT.	LABOR	EQUIP.	TOTAL	TOTAL INCL O&P	
100	0010	**FOUNTAINS/AERATORS**										100
	0100	Pump w/controls										
	0200	Single phase, 100' cord, 1/2 H.P. pump	2 Skwk	4.40	3.636	Ea.	2,675	127		2,802	3,150	
	0300	3/4 H.P. pump		4.30	3.721		3,075	130		3,205	3,575	
	0400	1 H.P. pump		4.20	3.810		3,100	133		3,233	3,600	
	0500	1-1/2 H.P. pump		4.10	3.902		3,200	136		3,336	3,725	
	0600	2 H.P. pump		4	4		3,250	139		3,389	3,800	
	0700	Three phase, 200' cord, 5 H.P. pump		3.90	4.103		7,850	143		7,993	8,875	
	0800	7-1/2 H.P. pump		3.80	4.211		8,875	147		9,022	9,975	
	0900	10 H.P. pump		3.70	4.324		9,900	151		10,051	11,100	
	1000	15 H.P. pump		3.60	4.444		11,700	155		11,855	13,100	
	1100	Nozzles, minimum		8	2		197	69.50		266.50	325	
	1200	Maximum		8	2		259	69.50		328.50	395	
	1300	Lights w/mounting kits, 200 watt		18	.889		325	31		356	410	
	1400	300 watt		18	.889		360	31		391	445	
	1500	500 watt		18	.889		395	31		426	485	
	1600	Color blender	▼	12	1.333	▼	305	46.50		351.50	410	

15500 | Heat Generation Equipment

15510 | Heating Boilers and Accessories

			CREW	DAILY OUTPUT	LABOR-HOURS	UNIT	MAT.	LABOR	EQUIP.	TOTAL	TOTAL INCL O&P	
300	0010	**BOILERS, ELECTRIC, ASME** Standard controls and trim										300
	1000	Steam, 6 KW, 20.5 MBH	Q-19	1.20	20	Ea.	2,725	765		3,490	4,150	
	1160	60 KW, 205 MBH		1	24		4,600	915		5,515	6,425	
	1220	112 KW, 382 MBH		.75	32		6,300	1,225		7,525	8,750	
	1280	222 KW, 758 MBH	▼	.55	43.636		10,400	1,675		12,075	14,000	
	1380	518 KW, 1768 MBH	Q-21	.36	88.889		17,600	3,450		21,050	24,500	
	1480	814 KW, 2778 MBH		.25	128		24,600	4,975		29,575	34,600	
	1600	2,340 KW, 7984 MBH	▼	.16	200		50,500	7,775		58,275	67,000	
	2000	Hot water, 7.5 KW, 25.6 MBH	Q-19	1.30	18.462		2,900	705		3,605	4,225	
	2100	90 KW, 307 MBH		1.10	21.818		4,225	830		5,055	5,900	
	2220	296 KW, 1010 MBH	▼	.55	43.636		9,200	1,675		10,875	12,600	
	2500	1,036 KW, 3536 MBH	Q-21	.34	94.118		21,400	3,650		25,050	29,000	
	2680	2,400 KW, 8191 MBH		.25	128		44,500	4,975		49,475	56,500	
	2820	3,600 KW, 12,283 MBH	▼	.16	200	▼	63,000	7,775		70,775	80,500	
400	0010	**BOILERS, GAS FIRED** Natural or propane, standard controls										400
	1000	Cast iron, with insulated jacket										
	2000	Steam, gross output, 81 MBH	Q-7	1.40	22.857	Ea.	1,450	890		2,340	2,950	
	2080	203 MBH		.90	35.556		2,375	1,400		3,775	4,675	
	2180	400 MBH		.56	56.838		3,600	2,225		5,825	7,275	
	2240	765 MBH	▼	.43	74.419	▼	7,800	2,900		10,700	13,000	

MECHANICAL 15

15510	Heating Boilers and Accessories		CREW	DAILY OUTPUT	LABOR-HOURS	UNIT	2005 BARE COSTS				TOTAL INCL O&P	
							MAT.	LABOR	EQUIP.	TOTAL		
400	2320	1,875 MBH	Q-7	.30	106	Ea.	11,800	4,175		15,975	19,300	**400**
	2440	4,720 MBH		.15	207		22,700	8,100		30,800	37,100	
	2480	6,100 MBH		.13	246		56,000	9,600		65,600	76,000	
	2540	6,970 MBH		.10	320		62,500	12,500		75,000	88,000	
	3000	Hot water, gross output, 80 MBH		1.46	21.918		1,550	855		2,405	2,975	
	3140	320 MBH	CN	.80	40		2,800	1,550		4,350	5,425	
	3260	1,088 MBH		.40	80		6,825	3,125		9,950	12,200	
	3380	3,264 MBH		.18	179		17,700	7,025		24,725	30,000	
	3480	6,100 MBH		.13	250		56,000	9,750		65,750	76,500	
	3540	6,970 MBH		.09	359		63,000	14,000		77,000	90,500	
	4000	Steel, insulating jacket										
	4500	Steam, not including burner, gross output										
	5500	1,440 MBH	Q-6	.35	68.571	Ea.	19,100	2,625		21,725	25,000	
	5580	3,065 MBH	"	.18	133		27,200	5,100		32,300	37,600	
	5640	7,200 MBH	Q-7	.18	177		49,900	6,950		56,850	65,500	
	5720	17,990 MBH	"	.10	320		93,000	12,500		105,500	121,500	
	6000	Hot water, including burner & one zone valve, gross output										
	6020	72 MBH	Q-6	2	12	Ea.	2,025	460		2,485	2,925	
	6120	227 MBH		1.30	18.462		4,725	705		5,430	6,225	
	6180	480 MBH		.70	34.286		7,675	1,300		8,975	10,400	
	6240	960 MBH		.45	53.333		13,200	2,050		15,250	17,600	
	6340	3,000 MBH		.15	160		36,700	6,125		42,825	49,600	
	7000	For tankless water heater on smaller gas units, add					10%					
460	0010	**BOILERS, GAS/OIL** Combination with burners and controls										**460**
	1000	Cast iron with insulated jacket										
	2000	Steam, gross output, 720 MBH	Q-7	.43	74.074	Ea.	6,950	2,900		9,850	12,000	
	2080	1,600 MBH		.30	107		11,400	4,175		15,575	18,900	
	2140	2,700 MBH		.19	165		16,300	6,475		22,775	27,600	
	2280	5,520 MBH		.14	235		61,500	9,175		70,675	82,000	
	2340	6,390 MBH		.11	296		66,000	11,600		77,600	90,000	
	2380	6,970 MBH		.09	372		70,000	14,500		84,500	99,000	
	2900	Hot water, gross output										
	2910	200 MBH	Q-6	.61	39.024	Ea.	5,700	1,500		7,200	8,525	
	2920	300 MBH		.49	49.080		5,700	1,875		7,575	9,100	
	2930	400 MBH		.41	57.971		6,675	2,225		8,900	10,700	
	2940	500 MBH		.36	67.039		7,200	2,550		9,750	11,800	
	3000	584 MBH	Q-7	.44	72.072		8,800	2,825		11,625	13,900	
	3060	1,460 MBH		.28	113		14,300	4,425		18,725	22,300	
	3160	4,088 MBH		.16	195		33,400	7,625		41,025	48,100	
	3300	13,500 MBH, 403.3 BHP		.04	727		103,000	28,400		131,400	156,000	
	4000	Steel, insulated jacket, skid base, tubeless										
	4500	Steam, 150 psi gross output, 335 MBH, 10 BHP	Q-6	.54	44.037	Ea.	12,000	1,675		13,675	15,700	
	4560	670 MBH, 20 BHP		.36	65.934		14,800	2,525		17,325	20,100	
	4640	1,339 MBH, 40 BHP		.29	81.911		23,500	3,125		26,625	30,500	
	4720	2,511 MBH, 75 BHP		.17	141		32,000	5,400		37,400	43,300	
	5000	Hot water, gross output, 525 MBH		.45	52.980		11,400	2,025		13,425	15,700	
	5080	1,050 MBH		.35	67.989		18,400	2,600		21,000	24,200	
	5140	2,310 MBH		.20	120		34,300	4,575		38,875	44,600	
	5180	3,150 MBH		.16	146		45,700	5,600		51,300	58,500	
500	0010	**BOILERS, OIL FIRED** Standard controls, flame retention burner										**500**
	1000	Cast iron, with insulated flush jacket										
	2000	Steam, gross output, 109 MBH	Q-7	1.20	26.667	Ea.	1,925	1,050		2,975	3,700	
	2060	207 MBH		.90	35.556		2,775	1,400		4,175	5,125	
	2180	1,084 MBH		.38	85.106		7,025	3,325		10,350	12,700	
	2280	3,000 MBH		.19	170		15,800	6,650		22,450	27,400	

Important: See the Reference Section for critical supporting data - Reference Nos., Crews, & City Cost Indexes

15510	Heating Boilers and Accessories	CREW	DAILY OUTPUT	LABOR-HOURS	UNIT	2005 BARE COSTS				TOTAL INCL O&P	
						MAT.	LABOR	EQUIP.	TOTAL		
500	2380	5,520 MBH	Q-7	.14	235	Ea.	52,000	9,175		61,175	71,000	**500**
	2460	6,970 MBH	↓	.09	363	↓	66,500	14,200		80,700	94,500	
	3000	Hot water, same price as steam										
	4000	For tankless coil in smaller sizes, add					15%					
	5000	Steel, insulated jacket, burner										
	6000	Steam, full water leg construction, gross output										
	6020	144 MBH	Q-6	1.33	18.045	Ea.	4,050	690		4,740	5,475	
	6120	468 MBH		.70	34.483		7,450	1,325		8,775	10,200	
	6180	1,008 MBH		.39	61.381		11,600	2,350		13,950	16,200	
	6280	2,400 MBH	↓	.19	125	↓	22,100	4,775		26,875	31,500	
	6400	Larger sizes are same as steel, gas fired										
	7000	Hot water, gross output, 103 MBH	Q-6	1.60	15	Ea.	1,250	575		1,825	2,250	
	7120	420 MBH		.70	34.483		4,950	1,325		6,275	7,400	
	7200	840 MBH		.43	55.172		10,000	2,100		12,100	14,200	
	7260	1,680 MBH		.29	83.624		13,800	3,200		17,000	20,000	
	7320	3,150 MBH	↓	.13	184	↓	19,900	7,050		26,950	32,500	
	7340	For tankless coil in steam or hot water, add					7%					
880	0010	**SWIMMING POOL HEATERS** Not including wiring, external										**880**
	0020	piping, base or pad,										
	0060	Gas fired, input, 115 MBH	Q-6	3	8	Ea.	1,850	305		2,155	2,500	
	0100	135 MBH		2	12		2,100	460		2,560	3,000	
	0160	155 MBH		1.50	16		2,200	610		2,810	3,350	
	0200	190 MBH		1	24		2,875	915		3,790	4,525	
	0280	500 MBH		.40	60		6,950	2,300		9,250	11,100	
	0400	1,800 MBH		.14	171		19,400	6,550		25,950	31,200	
	0440	3,750 MBH	↓	.09	266		39,100	10,200		49,300	58,500	
	2000	Electric, 12 KW, 4,800 gallon pool	Q-19	3	8		1,725	305		2,030	2,350	
	2020	15 KW, 7,200 gallon pool		2.80	8.571		1,725	325		2,050	2,400	
	2040	24 KW, 9,600 gallon pool		2.40	10		2,325	380		2,705	3,150	
	2100	55 KW, 24,000 gallon pool	↓	1.20	20	↓	3,325	765		4,090	4,800	
15530		**Furnaces**										
400	0010	**FURNACES** Hot air heating, blowers, standard controls										**400**
	0020	not including gas, oil or flue piping										
	1000	Electric, UL listed										
	1020	10.2 MBH	Q-20	5	4	Ea.	315	148		463	570	
	1040	17.1 MBH		4.80	4.167		330	155		485	595	
	1060	27.3 MBH		4.60	4.348		395	161		556	680	
	1100	34.1 MBH	↓	4.40	4.545	↓	405	169		574	705	
	3000	Gas, AGA certified, upflow, direct drive models										
	3020	45 MBH input	Q-9	4	4	Ea.	435	145		580	705	
	3040	60 MBH input		3.80	4.211		580	152		732	875	
	3060	75 MBH input		3.60	4.444		615	161		776	925	
	3100	100 MBH input	↓	3.20	5	↓	655	181		836	1,000	
	6000	Oil, UL listed, atomizing gun type burner										
	6020	56 MBH output	Q-9	3.60	4.444	Ea.	745	161		906	1,075	
	6030	84 MBH output		3.50	4.571		775	165		940	1,100	
	6040	95 MBH output		3.40	4.706		790	170		960	1,125	
	6060	134 MBH output		3.20	5		1,100	181		1,281	1,475	
	6080	151 MBH output	↓	3	5.333	↓	1,225	193		1,418	1,625	
15540		**Fuel-Fired Heaters**										
300	0010	**DUCT FURNACES** Includes burner, controls, stainless steel										**300**
	0020	heat exchanger. Gas fired, electric ignition										

MECHANICAL 15

15540 | Fuel-Fired Heaters

		CREW	DAILY OUTPUT	LABOR-HOURS	UNIT	2005 BARE COSTS				TOTAL INCL O&P
						MAT.	LABOR	EQUIP.	TOTAL	
300	0030	Indoor installation									**300**
	0100	120 MBH output	Q-5	4	4	Ea.	1,425	147		1,572	1,800
	0130	200 MBH output		2.70	5.926		1,825	218		2,043	2,350
	0140	240 MBH output		2.30	6.957		1,975	256		2,231	2,550
	0180	320 MBH output	↓	1.60	10		2,375	370		2,745	3,175
	0300	For powered venter and adapter, add				↓	325			325	360
	0500	For required flue pipe, see 15550									
	1000	Outdoor installation, with vent cap									
	1020	75 MBH output	Q-5	4	4	Ea.	1,850	147		1,997	2,250
	1060	120 MBH output		4	4		2,375	147		2,522	2,825
	1100	187 MBH output		3	5.333		3,375	197		3,572	4,025
	1140	300 MBH output		1.80	8.889		4,400	330		4,730	5,350
	1180	450 MBH output	↓	1.40	11.429		4,900	420		5,320	6,000
	1400	For powered venter, add				↓	10%				
900	0010	**SPACE HEATERS** Cabinet, grilles, fan, controls, burner,									**900**
	0020	thermostat, no piping. For flue see 15550									
	1000	Gas fired, floor mounted									
	1100	60 MBH output	Q-5	10	1.600	Ea.	545	59		604	690
	1140	100 MBH output		8	2		595	73.50		668.50	765
	1180	180 MBH output		6	2.667		815	98.50		913.50	1,050
	2000	Suspension mounted, propeller fan, 20 MBH output		8.50	1.882		450	69.50		519.50	600
	2040	60 MBH output		7	2.286		555	84		639	735
	2060	80 MBH output		6	2.667		635	98.50		733.50	850
	2100	130 MBH output		5	3.200		825	118		943	1,075
	2240	320 MBH output	↓	2	8		1,700	295		1,995	2,325
	2500	For powered venter and adapter, add					282			282	310
	5000	Wall furnace, 17.5 MBH output	Q-5	6	2.667		570	98.50		668.50	775
	5020	24 MBH output		5	3.200		595	118		713	830
	5040	35 MBH output		4	4		805	147		952	1,100
	6000	Oil fired, suspension mounted, 94 MBH output		4	4		2,175	147		2,322	2,625
	6040	140 MBH output		3	5.333		2,325	197		2,522	2,850
	6060	184 MBH output	↓	3	5.333	↓	2,500	197		2,697	3,050

15550 | Breechings, Chimneys & Stacks

		CREW	DAILY OUTPUT	LABOR-HOURS	UNIT	MAT.	LABOR	EQUIP.	TOTAL	TOTAL INCL O&P
440	0010	**VENT CHIMNEY** Prefab metal, U.L. listed									**440**
	0020	Gas, double wall, galvanized steel									
	0080	3" diameter	Q-9	72	.222	V.L.F.	3.77	8.05		11.82	16.50
	0100	4" diameter		68	.235		4.71	8.50		13.21	18.30
	0120	5" diameter		64	.250		5.50	9.05		14.55	19.95
	0140	6" diameter		60	.267		6.40	9.65		16.05	22
	0160	7" diameter		56	.286		9.40	10.35		19.75	26.50
	0180	8" diameter		52	.308		10.45	11.15		21.60	28.50
	0200	10" diameter		48	.333		22	12.05		34.05	43
	0220	12" diameter		44	.364		29.50	13.15		42.65	52.50
	0260	16" diameter	↓	40	.400		67	14.45		81.45	96.50
	0300	20" diameter	Q-10	36	.667		102	25		127	151
	0480	38" diameter	"	24	1		325	37.50		362.50	420
	7800	All fuel, double wall, stainless steel, 6" diameter	Q-9	60	.267		32.50	9.65		42.15	50.50
	7802	7" diameter		56	.286		42	10.35		52.35	62
	7804	8" diameter		52	.308		49	11.15		60.15	71
	7806	10" diameter		48	.333		71.50	12.05		83.55	97
	7808	12" diameter		44	.364		96	13.15		109.15	125
	7810	14" diameter	↓	42	.381	↓	126	13.80		139.80	159
	9000	High temp. (2000° F), steel jacket, acid resist. refractory lining									
	9010	11 ga. galvanized jacket, U.L. listed									
	9020	Straight section, 48" long, 10" diameter	Q-10	13.30	1.805	Ea.	355	67.50		422.50	495

Important: See the Reference Section for critical supporting data - Reference Nos., Crews, & City Cost Indexes

15 MECHANICAL

15550	Breechings, Chimneys & Stacks	CREW	DAILY OUTPUT	LABOR-HOURS	UNIT	2005 BARE COSTS				TOTAL INCL O&P
						MAT.	LABOR	EQUIP.	TOTAL	
440										**440**
9040	18" diameter	Q-10	7.40	3.243	Ea.	545	122		667	785
9070	36" diameter	↓	2.70	8.889		1,150	335		1,485	1,800
9090	48" diameter	Q-11	2.70	11.852		1,700	455		2,155	2,575
9120	Tee section, 10" diameter	Q-10	4.40	5.455		755	205		960	1,150
9140	18" diameter		2.40	10		1,075	375		1,450	1,750
9170	36" diameter	↓	.80	30		3,525	1,125		4,650	5,600
9190	48" diameter	Q-11	.90	35.556		6,000	1,350		7,350	8,700
9220	Cleanout pier section, 10" diameter	Q-10	3.50	6.857		580	257		837	1,025
9240	18" diameter	↓	1.90	12.632		780	475		1,255	1,600
9270	36" diameter		.75	32		1,425	1,200		2,625	3,425
9290	48" diameter	Q-11	.75	42.667		2,175	1,625		3,800	4,925
9320	For drain, add					53%				
9330	Elbow, 30° and 45°, 10" diameter	Q-10	6.60	3.636		495	136		631	750
9350	18" diameter	↓	3.70	6.486		795	243		1,038	1,250
9380	36" diameter	↓	1.30	18.462		1,975	695		2,670	3,250
9400	48" diameter	Q-11	1.35	23.704		3,450	910		4,360	5,200
9430	For 60° and 90° elbow, add					112%				
9440	End cap, 10" diameter	Q-10	26	.923		845	34.50		879.50	985
9460	18" diameter		15	1.600		1,200	60		1,260	1,425
9490	36" diameter	↓	5	4.800		2,100	180		2,280	2,575
9510	48" diameter	Q-11	5.30	6.038		3,075	231		3,306	3,725
9540	Increaser (1 diameter), 10" diameter	Q-10	6.60	3.636		420	136		556	670
9560	18" diameter	↓	3.70	6.486		655	243		898	1,100
9590	36" diameter	↓	1.30	18.462		1,275	695		1,970	2,475
9592	42" diameter	Q-11	1.60	20		1,400	765		2,165	2,725
9594	48" diameter	↓	1.35	23.704		1,700	910		2,610	3,275
9596	54" diameter	↓	1.10	29.091	↓	2,050	1,125		3,175	3,975
9600	For expansion joints, add to straight section					7%				
9610	For 1/4" hot rolled steel jacket, add					157%				
9620	For 2950°F very high temperature, add					89%				
9630	26 ga. aluminized jacket, straight section, 48" long									
9640	10" diameter	Q-10	15.30	1.569	V.L.F.	210	59		269	320
9660	18" diameter		8.50	2.824		325	106		431	520
9690	36" diameter		3.10	7.742	↓	745	290		1,035	1,275
9810	Draw band, galv. stl., 11 gauge, 10" diameter		32	.750	Ea.	57	28		85	106
9820	12" diameter		30	.800		61	30		91	113
9830	18" diameter		26	.923		76	34.50		110.50	138
9860	36" diameter	↓	18	1.333		129	50		179	219
9880	48" diameter	Q-11	20	1.600	↓	222	61.50		283.50	340
600	**0010**	**INDUCED DRAFT FANS**								**600**
	1000	Breeching installation								
	1800	Hot gas, 600°F, variable pitch pulley and motor								
	1860	8" diam. inlet, 1/4 H.P., 1phase, 1120 CFM	Q-9	4	4	Ea.	1,600	145		1,745	1,975
	1900	12" diam. inlet, 3/4 H.P., 3 phase, 2960 CFM	↓	3	5.333		2,175	193		2,368	2,700
	1940	18" diam. inlet, 3 H.P., 3 phase, 9120 CFM		2	8		3,375	289		3,664	4,175
	1980	24" diam. inlet, 7-1/2 H.P., 3 phase, 17,760 CFM	↓	.80	20	↓	5,300	725		6,025	6,975
	2300	For multi-blade damper at fan inlet, add					20%				

		15620	Packaged Water Chillers	CREW	DAILY OUTPUT	LABOR-HOURS	UNIT	2005 BARE COSTS				TOTAL INCL O&P	
								MAT.	LABOR	EQUIP.	TOTAL		
100	0010		**ABSORPTION WATER CHILLERS**										**100**
	0020		Steam or hot water, water cooled										
	0050		100 ton	Q-7	.13	240	Ea.	119,000	9,375		128,375	145,000	
	0400		420 ton	"	.10	323	"	204,000	12,600		216,600	243,500	
600	0010		**CENTRIFUGAL/SCREW/RECIP. WATER CHILLERS,** W/ standard controls										**600**
	0020		Centrifugal liquid chiller, water cooled										
	0030		not including water tower										
	0100		Open drive, 2000 ton	Q-7	.07	477	Ea.	537,000	18,600		555,600	618,500	
	0110		Screw, liquid chiller, air cooled, insulated evaporator										
	0120		130 ton	Q-7	.14	228	Ea.	61,000	8,925		69,925	80,500	
	0124		160 ton		.13	246		75,000	9,600		84,600	97,000	
	0128		180 ton		.13	250		84,500	9,750		94,250	107,000	
	0132		210 ton		.12	258		93,000	10,100		103,100	117,000	
	0136		270 ton		.12	266		106,000	10,400		116,400	132,000	
	0140		320 ton		.12	275		133,000	10,800		143,800	162,500	
	0490		Reciprocating, packaged w/integral air cooled condenser, 15 ton cool		.37	86.486		13,300	3,375		16,675	19,700	
	0500		20 ton cooling		.34	94.955		17,600	3,700		21,300	25,000	
	0520		40 ton cooling		.30	108		23,900	4,225		28,125	32,700	
	0600		100 ton cooling	▼	.25	129	▼	51,500	5,025		56,525	64,000	
	0680		Water cooled, single compressor, semi-hermetic, tower not incl.										
	0700		2 to 5 ton cooling	Q-5	.57	28.021	Ea.	6,375	1,025		7,400	8,575	
	0740		8 ton cooling	"	.31	52.117		10,100	1,925		12,025	14,000	
	0760		10 ton cooling	Q-6	.36	67.039		10,300	2,550		12,850	15,200	
	0800		20 ton cooling	Q-7	.38	83.989		11,300	3,275		14,575	17,400	
	0820		30 ton cooling	"	.33	96.096	▼	12,900	3,750		16,650	19,800	
	0980		Water cooled, multiple compress., semi-hermetic, tower not incl.										
	1000		15 ton cooling	Q-6	.36	65.934	Ea.	11,000	2,525		13,525	15,900	
	1020		20 ton cooling	Q-7	.41	78.049		12,000	3,050		15,050	17,800	
	1060		30 ton cooling		.31	101		14,200	3,975		18,175	21,600	
	1100		50 ton cooling		.28	113		25,400	4,450		29,850	34,700	
	1160		100 ton cooling		.18	179		45,300	7,025		52,325	60,500	
	1180		120 ton cooling		.16	196		51,500	7,650		59,150	68,000	
	1200		140 ton cooling	▼	.16	202	▼	60,500	7,900		68,400	78,500	
	1450		Water cooled, dual compressors, direct drive, tower not incl.										
	1500		80 ton cooling	Q-7	.14	222	Ea.	47,800	8,675		56,475	65,500	
	1520		100 ton cooling		.14	228		55,500	8,925		64,425	74,500	
	1540		120 ton cooling		.14	231		61,500	9,050		70,550	81,000	
	1580		150 ton cooling		.13	240		74,500	9,400		83,900	96,000	
	1620		200 ton cooling		.13	250		91,000	9,750		100,750	114,500	
	1660		250 ton cooling	▼	.12	260	▼	105,000	10,200		115,200	131,000	

15640 | Packaged Cooling Towers

				CREW	DAILY OUTPUT	LABOR-HOURS	UNIT	MAT.	LABOR	EQUIP.	TOTAL	TOTAL INCL O&P	
400	0010		**COOLING TOWERS** Packaged units										**400**
	0070		Galvanized steel										
	0080		Induced draft, crossflow										
	0100		Vertical, belt drive, 61 tons	Q-6	90	.267	TonAC	82.50	10.20		92.70	106	
	0150		100 ton		100	.240		65.50	9.15		74.65	86	
	0200		115 ton		109	.220		64	8.40		72.40	83	
	0250		131 ton		120	.200		56.50	7.65		64.15	73.50	
	0260		162 ton	▼	132	.182	▼	47.50	6.95		54.45	63	
	1500		Induced air, double flow										
	1900		Vertical, gear drive, 167 ton	Q-6	126	.190	TonAC	88	7.30		95.30	108	
	2000		297 ton		129	.186		63	7.10		70.10	80	
	2100		582 ton		132	.182		52	6.95		58.95	68	
	2150		849 ton		142	.169		51.50	6.45		57.95	66	
	2200		1016 ton	▼	150	.160	▼	49	6.10		55.10	63	

15 MECHANICAL

Important: See the Reference Section for critical supporting data - Reference Nos., Crews, & City Cost Indexes

15600 | Refrigeration Equipment

15640 | Packaged Cooling Towers

		CREW	DAILY OUTPUT	LABOR-HOURS	UNIT	2005 BARE COSTS				TOTAL INCL O&P
						MAT.	LABOR	EQUIP.	TOTAL	
400 3000	For higher capacities, use multiples									**400**
3500	For pumps and piping, add	Q-6	38	.632	TonAC	41.50	24		65.50	82
4000	For absorption systems, add				"	75%	75%			
4500	For rigging, see division 01590-600									
5000	Fiberglass									
5010	Draw thru									
5100	60 ton	Q-6	1.50	16	Ea.	3,175	610		3,785	4,425
5120	125 ton		.99	24.242		6,500	925		7,425	8,550
5140	300 ton		.43	55.814		15,200	2,125		17,325	20,000
5160	600 ton		.22	109		27,400	4,175		31,575	36,400
5180	1000 ton	▼	.15	160	▼	47,000	6,125		53,125	60,500
6000	Stainless steel									
6010	Induced draft, crossflow, horizontal, belt drive									
6100	57 ton	Q-6	1.50	16	Ea.	8,350	610		8,960	10,100
6120	91 ton		.99	24.242		12,000	925		12,925	14,600
6140	111 ton		.43	55.814		14,200	2,125		16,325	18,800
6160	126 ton	▼	.22	109	▼	15,300	4,175		19,475	23,200

15660 | Liquid Coolers/Evap Condensers

		CREW	DAILY OUTPUT	LABOR-HOURS	UNIT	2005 BARE COSTS				TOTAL INCL O&P
						MAT.	LABOR	EQUIP.	TOTAL	
100 0010	**CONDENSERS** Ratings are for 30°F TD, R-22									**100**
0080	Air cooled, belt drive, propeller fan									
0240	50 ton	Q-6	.69	34.985	Ea.	10,500	1,325		11,825	13,600
0280	59 ton		.58	41.308		11,700	1,575		13,275	15,300
0320	73 ton		.47	51.173		13,800	1,950		15,750	18,200
0360	86 ton		.40	60.302		16,400	2,300		18,700	21,500
0380	88 ton	▼	.39	61.697	▼	17,500	2,350		19,850	22,900
1550	Air cooled, direct drive, propeller fan									
1590	1 ton	Q-5	3.80	4.211	Ea.	540	155		695	830
1600	1-1/2 ton		3.60	4.444		640	164		804	950
1620	2 ton		3.20	5		710	184		894	1,050
1640	5 ton		2	8		1,450	295		1,745	2,025
1660	10 ton		1.40	11.429		2,225	420		2,645	3,075
1690	16 ton		1.10	14.545		3,125	535		3,660	4,250
1720	26 ton	▼	.84	19.002		4,050	700		4,750	5,500
1760	41 ton	Q-6	.77	31.008		7,425	1,175		8,600	9,950
1800	63 ton	"	.54	44.037	▼	11,700	1,675		13,375	15,400

15700 | Heating/Ventilating/Air Conditioning Equipment

15710 | Heat Exchangers

		CREW	DAILY OUTPUT	LABOR-HOURS	UNIT	2005 BARE COSTS				TOTAL INCL O&P
						MAT.	LABOR	EQUIP.	TOTAL	
900 0010	**HEAT EXCHANGERS**									**900**
0016	Shell & tube type, 2 or 4 pass, 3/4" O.D. copper tubes,									
0020	C.I. heads, C.I. tube sheet, steel shell									
0100	Hot water 40°F to 180°F, by steam at 10 PSI									
0120	8 GPM	Q-5	6	2.667	Ea.	960	98.50		1,058.50	1,200
0140	10 GPM		5	3.200		1,450	118		1,568	1,775
0160	40 GPM		4	4		2,250	147		2,397	2,700
0180	64 GPM	▼	2	8	▼	3,425	295		3,720	4,225

For expanded coverage of these items see *Means Mechanical or Plumbing Cost Data 2005*

		15710	Heat Exchangers	CREW	DAILY OUTPUT	LABOR-HOURS	UNIT	MAT.	2005 BARE COSTS LABOR	EQUIP.	TOTAL	TOTAL INCL O&P	
900	0200		96 GPM	Q-5	1	16	Ea.	4,600	590		5,190	5,925	900
	0220		120 GPM	Q-6	1.50	16	↓	6,050	610		6,660	7,575	
	1000		Hot water 40°F to 140°F, by water at 200°F										
	1020		7 GPM	Q-5	6	2.667	Ea.	1,175	98.50		1,273.50	1,450	
	1040		16 GPM		5	3.200		1,675	118		1,793	2,025	
	1060		34 GPM		4	4		2,525	147		2,672	3,000	
	1100		74 GPM	↓	1.50	10.667	↓	4,600	395		4,995	5,650	
	3000	Plate type,											
	3100		400 GPM	Q-6	.80	30	Ea.	18,100	1,150		19,250	21,600	
	3120		800 GPM	"	.50	48		31,300	1,825		33,125	37,300	
	3140		1200 GPM	Q-7	.34	94.118		38,000	3,675		41,675	47,300	
	3160		1800 GPM	"	.24	133	↓	56,500	5,200		61,700	70,500	

		15720	Air Handling Units										
500	0010	**MAKE-UP AIR UNIT**											500
	0020	Indoor suspension, natural/LP gas, direct fired,											
	0030	standard control. For flue see division 15550-440											
	0040	70°F temperature rise, MBH is input											
	0100	2000 CFM, 168 MBH		Q-6	3	8	Ea.	6,425	305		6,730	7,525	
	0160	6000 CFM, 502 MBH			1.50	16		8,100	610		8,710	9,825	
	0220	12,000 CFM, 1005 MBH		↓	1	24		10,100	915		11,015	12,500	
	0300	24,000 CFM, 2007 MBH		Q-7	1	32		13,100	1,250		14,350	16,300	
	0400	50,000 CFM, 4180 MBH		"	.80	40		18,000	1,550		19,550	22,200	
	0600	For discharge louver assembly, add						10%					
	0700	For filters, add						20%					
	0800	For air shut-off damper section, add						10%					
	0900	For vertical unit, add					↓	10%					

		15730	Unitary Air Conditioning Equip										
200	0010	**COMPUTER ROOM UNITS**											200
	1000	Air cooled, includes remote condenser but not											
	1020	interconnecting tubing or refrigerant											
	1080	3 ton		Q-5	.50	32	Ea.	8,725	1,175		9,900	11,400	
	1120	5 ton			.45	35.556		11,700	1,300		13,000	14,900	
	1160	6 ton			.30	53.333		21,500	1,975		23,475	26,700	
	1200	8 ton			.27	59.259		24,400	2,175		26,575	30,100	
	1240	10 ton			.25	64		25,400	2,350		27,750	31,500	
	1280	15 ton			.22	72.727		28,300	2,675		30,975	35,100	
	1290	18 ton		↓	.20	80		32,300	2,950		35,250	39,900	
	1320	20 ton		Q-6	.29	82.759		33,800	3,175		36,975	42,000	
	1360	23 ton		"	.28	85.714	↓	35,500	3,275		38,775	43,900	
500	0010	**PACKAGED TERMINAL AIR CONDITIONER** Cabinet, wall sleeve,											500
	0100	louver, electric heat, thermostat, manual changeover, 208 V											
	0200	6,000 BTUH cooling, 8800 BTU heat		Q-5	6	2.667	Ea.	955	98.50		1,053.50	1,200	
	0220	9,000 BTUH cooling, 13,900 BTU heat			5	3.200		1,000	118		1,118	1,275	
	0240	12,000 BTUH cooling, 13,900 BTU heat			4	4		1,100	147		1,247	1,450	
	0260	15,000 BTUH cooling, 13,900 BTU heat		↓	3	5.333		1,300	197		1,497	1,750	
	0500	For hot water coil, increase heat by 10%, add						5%	10%				
	1000	For steam, increase heat output by 30%, add					↓	8%	10%				
600	0010	**ROOF TOP AIR CONDITIONERS** Standard controls, curb, economizer											600
	1000	Single zone, electric cool, gas heat											
	1090	2 ton cooling, 55 MBH heating		Q-5	.93	17.204	Ea.	2,550	635		3,185	3,750	
	1100	3 ton cooling, 60 MBH heating			.70	22.857		2,550	840		3,390	4,075	
	1120	4 ton cooling, 95 MBH heating			.61	26.403		3,700	975		4,675	5,525	
	1140	5 ton cooling, 112 MBH heating		↓	.56	28.520	↓	4,050	1,050		5,100	6,025	

R15700-020

15 MECHANICAL

		15730	Unitary Air Conditioning Equip	CREW	DAILY OUTPUT	LABOR-HOURS	UNIT	MAT.	LABOR	EQUIP.	TOTAL	TOTAL INCL O&P	
600	1145		6 ton cooling, 140 MBH heating	Q-5	.52	30.769	Ea.	4,750	1,125		5,875	6,925	600
	1150		7.5 ton cooling, 170 MBH heating	↓	.50	32.258		5,900	1,200		7,100	8,275	
	1160		10 ton cooling, 200 MBH heating	Q-6	.67	35.982		7,600	1,375		8,975	10,400	
	1170		12.5 ton cooling, 230 MBH heating		.63	37.975		8,650	1,450		10,100	11,700	
	1190		18 ton cooling, 330 MBH heating	↓	.52	45.889		14,100	1,750		15,850	18,200	
	1200		20 ton cooling, 360 MBH heating	Q-7	.67	47.976		15,800	1,875		17,675	20,200	
	1210		25 ton cooling, 450 MBH heating	↓	.56	57.554		19,500	2,250		21,750	24,800	
	1220		30 ton cooling, 540 MBH heating	↓	.47	68.376	↓	24,600	2,675		27,275	31,000	
	2000	Multizone, electric cool, gas heat, economizer											
	2100		15 ton cooling, 360 MBH heating	Q-7	.61	52.545	Ea.	73,500	2,050		75,550	84,000	
	2120		20 ton cooling, 360 MBH heating CN		.53	60.038		84,500	2,350		86,850	96,500	
	2200		40 ton cooling, 540 MBH heating		.28	113		114,000	4,450		118,450	132,000	
	2210		50 ton cooling, 540 MBH heating		.22	142		121,500	5,550		127,050	142,000	
	2220		70 ton cooling, 1500 MBH heating		.16	198		131,000	7,750		138,750	156,000	
	2240		80 ton cooling, 1500 MBH heating		.14	228		150,000	8,925		158,925	178,500	
	2260		90 ton cooling, 1500 MBH heating		.13	256		159,000	10,000		169,000	189,500	
	2280		105 ton cooling, 1500 MBH heating	↓	.11	290		173,500	11,400		184,900	208,000	
	2400	For hot water heat coil, deduct						5%					
	2500	For steam heat coil, deduct						2%					
	2600	For electric heat, deduct					↓	3%	5%				
840	0010	**SELF-CONTAINED SINGLE PACKAGE**											840
	0100	Air cooled, for free blow or duct, not incl. remote condenser											
	0200		3 ton cooling	Q-5	1	16	Ea.	2,575	590		3,165	3,725	
	0220		5 ton cooling	Q-6	1.20	20		3,050	765		3,815	4,525	
	0240		10 ton cooling	Q-7	1	32		5,250	1,250		6,500	7,650	
	0260		20 ton cooling		.90	35.556		12,300	1,400		13,700	15,600	
	0280		30 ton cooling	↓	.80	40		12,800	1,550		14,350	16,500	
	0340		60 ton cooling	Q-8	.40	80	↓	25,500	3,125	202	28,827	32,900	
	0490	For duct mounting no price change											
	0500	For steam heating coils, add					Ea.	10%	10%				
	1000	Water cooled for free blow or duct, not including tower											
	1010	Constant volume											
	1100		3 ton cooling	Q-6	1	24	Ea.	2,450	915		3,365	4,075	
	1120		5 ton cooling	"	1	24		3,175	915		4,090	4,875	
	1140		10 ton cooling	Q-7	.90	35.556		6,275	1,400		7,675	8,975	
	1160		20 ton cooling		.80	40		21,100	1,550		22,650	25,600	
	1180		30 ton cooling	↓	.70	45.714	↓	28,000	1,775		29,775	33,400	

		15740	Heat Pumps										
100	0010	**AIR-SOURCE HEAT PUMPS** (Not including interconnecting tubing)											100
	1000	Air to air, split system, not including curbs, pads, or ductwork											
	1020		2 ton cooling, 8.5 MBH heat @ 0°F	Q-5	1.20	13.333	Ea.	1,425	490		1,915	2,300	
	1060		5 ton cooling, 27 MBH heat @ 0°F	↓	.50	32		2,450	1,175		3,625	4,475	
	1080		7.5 ton cooling, 33 MBH heat @ 0°F		.30	53.333		5,450	1,975		7,425	8,925	
	1100		10 ton cooling, 50 MBH heat @ 0°F	Q-6	.38	63.158		7,300	2,425		9,725	11,700	
	1120		15 ton cooling, 64 MBH heat @ 0°F		.26	92.308		10,400	3,525		13,925	16,700	
	1130		20 ton cooling, 85 MBH heat @ 0°F		.20	120		13,600	4,575		18,175	21,900	
	1140		25 ton cooling, 119 MBH heat @ 0°F	↓	.20	120	↓	16,400	4,575		20,975	24,900	
	1300	Supplementary electric heat coil, included											
	1500	Single package, not including curbs, pads, or plenums											
	1520		2 ton cooling, 6.5 MBH heat @ 0°F	Q-5	1.50	10.667	Ea.	2,200	395		2,595	3,000	
	1580		4 ton cooling, 13 MBH heat @ 0°F		.96	16.667		3,075	615		3,690	4,325	
	1640		7.5 ton cooling, 35 MBH heat @ 0°F	↓	.40	40	↓	5,075	1,475		6,550	7,825	
800	0010	**WATER-SOURCE HEAT PUMPS** (Not including interconnecting tubing)											800
	2000	Water source to air, single package											

			DAILY OUTPUT	LABOR-HOURS	UNIT	2005 BARE COSTS				TOTAL INCL O&P	
15740		**Heat Pumps**	CREW			MAT.	LABOR	EQUIP.	TOTAL		
800	2100	1 ton cooling, 13 MBH heat @ 75°F	Q-5	2	8	Ea.	990	295		1,285	1,550	**800**
	2140	2 ton cooling, 19 MBH heat @ 75°F		1.70	9.412		1,125	345		1,470	1,750	
	2220	5 ton cooling, 29 MBH heat @ 75°F		.90	17.778		1,725	655		2,380	2,875	
	3960	For supplementary heat coil, add					10%					
	4000	For increase in capacity thru use										
	4020	of solar collector, size boiler at 60%										
15750		**Humidity Control Equipment**										
500	0010	**HUMIDIFIERS**										**500**
	0520	Steam, room or duct, filter, regulators, auto. controls, 220 V										
	0540	11 lb. per hour	Q-5	6	2.667	Ea.	2,175	98.50		2,273.50	2,550	
	0560	22 lb. per hour		5	3.200		2,400	118		2,518	2,800	
	0580	33 lb. per hour		4	4		2,450	147		2,597	2,925	
	0600	50 lb. per hour		4	4		3,025	147		3,172	3,550	
	0620	100 lb. per hour		3	5.333		3,600	197		3,797	4,275	
15760		**Terminal Heating & Cooling Units**										
100	0010	**COILS, FLANGED**										**100**
	0500	Chilled water cooling, 6 rows, 24" x 48"	Q-5	3.20	5	Ea.	2,750	184		2,934	3,300	
	1000	Direct expansion cooling, 6 rows, 24" x 48"		2.80	5.714		3,000	211		3,211	3,625	
	1500	Hot water heating, 1 row, 24" x 48"		4	4		1,100	147		1,247	1,425	
	2000	Steam heating, 1 row, 24" x 48"		3.06	5.229		1,550	193		1,743	2,025	
200	0010	**DUCT HEATERS** Electric, 480 V, 3 Ph										**200**
	0020	Finned tubular insert, 500°F										
	0100	8" wide x 6" high, 4.0 kW	Q-20	16	1.250	Ea.	540	46.50		586.50	665	
	0120	12" high, 8.0 kW		15	1.333		895	49.50		944.50	1,050	
	0140	18" high, 12.0 kW		14	1.429		1,250	53		1,303	1,450	
	0160	24" high, 16.0 kW		13	1.538		1,625	57		1,682	1,850	
	0180	30" high, 20.0 kW		12	1.667		1,975	62		2,037	2,275	
	0300	12" wide x 6" high, 6.7 kW		15	1.333		575	49.50		624.50	705	
	0360	24" high, 26.7 kW		12	1.667		1,675	62		1,737	1,925	
	0700	24" wide x 6" high, 17.8 kW		13	1.538		670	57		727	820	
	0760	24" high, 71.1 kW		10	2		2,075	74		2,149	2,425	
	8000	To obtain BTU multiply kW by 3413										
250	0010	**ELECTRIC HEATING**, not incl. conduit or feed wiring										**250**
	1100	Rule of thumb: Baseboard units, including control	1 Elec	4.40	1.818	kW	77	74		151	195	
	1300	Baseboard heaters, 2' long, 375 watt		8	1	Ea.	29	41		70	92.50	
	1400	3' long, 500 watt		8	1		34.50	41		75.50	98.50	
	1600	4' long, 750 watt		6.70	1.194		41	48.50		89.50	118	
	1800	5' long, 935 watt		5.70	1.404		48.50	57		105.50	139	
	2000	6' long, 1125 watt		5	1.600		54	65		119	157	
	2400	8' long, 1500 watt		4	2		68	81.50		149.50	196	
	2950	Wall heaters with fan, 120 to 277 volt										
	3600	Thermostats, integral	1 Elec	16	.500	Ea.	18	20.50		38.50	50.50	
	3800	Line voltage, 1 pole	"	8	1	"	22.50	41		63.50	85	
300	0010	**FAN COIL AIR CONDITIONING** Cabinet mounted, filters, controls										**300**
	0100	Chilled water, 1/2 ton cooling	Q-5	8	2	Ea.	520	73.50		593.50	685	
	0120	1 ton cooling		6	2.667		625	98.50		723.50	840	
	0140	1.5 ton cooling		5.50	2.909		715	107		822	950	
	0150	2 ton cooling		5.25	3.048		915	112		1,027	1,175	
	0180	3 ton cooling		4	4		1,500	147		1,647	1,850	
	0190	7.5 ton cooling		2.70	5.926		1,575	218		1,793	2,075	
	0262	For hot water coil, add					40%	10%				

Important: See the Reference Section for critical supporting data - Reference Nos., Crews, & City Cost Indexes

15760	Terminal Heating & Cooling Units	CREW	DAILY OUTPUT	LABOR-HOURS	UNIT	2005 BARE COSTS				TOTAL INCL O&P	
						MAT.	LABOR	EQUIP.	TOTAL		
300	0940	Direct expansion, for use w/air cooled condensing, 1.5 ton cooling	Q-5	5	3.200	Ea.	410	118		528	625	**300**
	1000	5 ton cooling	"	3	5.333		760	197		957	1,125	
	1040	10 ton cooling	Q-6	2.60	9.231		2,125	355		2,480	2,875	
	1060	20 ton cooling	"	.70	34.286		3,750	1,300		5,050	6,100	
500	0010	**HEATING & VENTILATING UNITS** Classroom										**500**
	0020	Includes filter, heating/cooling coils, standard controls										
	0080	750 CFM, 2 tons cooling	Q-6	2	12	Ea.	3,150	460		3,610	4,150	
	0120	1250 CFM, 3 tons cooling		1.40	17.143		3,825	655		4,480	5,200	
	0140	1500 CFM, 4 tons cooling	▼	.80	30		4,100	1,150		5,250	6,225	
	0500	For electric heat, add					35%					
	1000	For no cooling, deduct	▼			▼	25%	10%				
600	0010	**HYDRONIC HEATING** Terminal units, not incl. main supply pipe										**600**
	1000	Radiation										
	1100	Panel, baseboard, C.I., including supports, no covers	Q-5	46	.348	L.F.	24.50	12.80		37.30	46.50	
	1150	Fin tube, wall hung, 14″ slope top cover, with damper										
	1200	1-1/4″ copper tube, 4-1/4″ alum. fin	Q-5	38	.421	L.F.	28	15.50		43.50	54	
	1250	1-1/4″ steel tube, 4-1/4″ steel fin	"	36	.444	"	25.50	16.40		41.90	52.50	
	1500	Note: fin tube may also require corners, caps, etc.										
	1990	Convector unit, floor recessed, flush, with trim										
	2000	for under large glass wall areas, no damper	Q-5	20	.800	L.F.	25.50	29.50		55	72.50	
	2100	For unit with damper	"			"	36			36	40	
	3000	Radiators, cast iron										
	3100	Free standing or wall hung, 6 tube, 25″ high	Q-5	96	.167	Section	24	6.15		30.15	36	
	3200	4 tube, 19″ high	"	96	.167	"	16.65	6.15		22.80	27.50	
	3250	Adj. brackets, 2 per wall radiator up to 30 sections	1 Stpi	32	.250	Ea.	17.10	10.25		27.35	34	
	9500	To convert SFR to BTU rating: Hot water, 150 x SFR										
	9510	Forced hot water, 180 x SFR; steam, 240 x SFR										
700	0010	**INFRARED UNIT**										**700**
	0020	Gas fired, unvented, electric ignition, 100% shutoff.										
	0030	Piping and wiring not included										
	0060	Input, 15 MBH	Q-5	7	2.286	Ea.	360	84		444	525	
	0120	45 MBH		5	3.200		430	118		548	650	
	0160	60 MBH		4	4		465	147		612	730	
	0240	120 MBH	▼	2	8	▼	895	295		1,190	1,425	
800	0010	**UNIT HEATERS**										**800**
	3950	Unit heaters, propeller, 115 V 2 psi steam, 60°F entering air										
	4000	Horizontal, 12 MBH	Q-5	12	1.333	Ea.	257	49		306	355	
	4060	43.9 MBH		8	2		415	73.50		488.50	565	
	4140	96.8 MBH		6	2.667		600	98.50		698.50	810	
	4180	157.6 MBH		4	4		770	147		917	1,075	
	4240	286.9 MBH		2	8		1,300	295		1,595	1,875	
	4250	326.0 MBH		1.90	8.421		1,500	310		1,810	2,125	
	4260	364 MBH		1.80	8.889		1,850	330		2,180	2,550	
	4270	404 MBH	▼	1.60	10		2,025	370		2,395	2,800	
	4300	For vertical diffuser, add					143			143	157	
	4310	Vertical flow, 40 MBH	Q-5	11	1.455		410	53.50		463.50	530	
	4314	58.5 MBH		8	2		465	73.50		538.50	620	
	4326	131.0 MBH		4	4		650	147		797	935	
	4346	297.0 MBH	▼	1.80	8.889		1,025	330		1,355	1,625	
	4354	420 MBH,(460 V)	Q-6	1.80	13.333		1,550	510		2,060	2,475	
	4358	500 MBH, (460 V)		1.71	14.035		2,200	535		2,735	3,225	
	4362	570 MBH, (460 V)		1.40	17.143		3,025	655		3,680	4,300	
	4366	620 MBH, (460 V)		1.30	18.462		3,475	705		4,180	4,875	
	4370	960 MBH, (460 V)	▼	1.10	21.818	▼	6,100	835		6,935	7,950	

MECHANICAL T5

15810	Ducts		CREW	DAILY OUTPUT	LABOR-HOURS	UNIT	2005 BARE COSTS				TOTAL INCL O&P	
							MAT.	LABOR	EQUIP.	TOTAL		
100	0010	**METAL DUCTWORK**	R15810 -050									**100**
	0020	Fabricated rectangular, includes fittings, joints, supports,										
	0030	allowance for flexible connections, no insulation										
	0031	NOTE: Fabrication and installation are combined										
	0040	as LABOR cost. Approx. 25% fittings assumed.										
	0100	Aluminum, alloy 3003-H14, under 100 lb.		Q-10	75	.320	Lb.	2.50	12		14.50	21
	0110	100 to 500 lb.			80	.300		2	11.25		13.25	19.50
	0120	500 to 1,000 lb.			95	.253		1.70	9.50		11.20	16.45
	0140	1,000 to 2,000 lb.			120	.200		1.60	7.50		9.10	13.30
	0150	2,000 to 5,000 lb.	**CN**		130	.185		1.53	6.95		8.48	12.35
	0160	Over 5,000 lb.			145	.166		1.50	6.20		7.70	11.20
	0500	Galvanized steel, under 200 lb.			235	.102		1.04	3.83		4.87	7.05
	0520	200 to 500 lb.			245	.098		.84	3.68		4.52	6.55
	0540	500 to 1,000 lb.			255	.094		.74	3.53		4.27	6.25
	0560	1,000 to 2,000 lb.			265	.091		.59	3.40		3.99	5.90
	0570	2,000 to 5,000 lb.			275	.087		.56	3.27		3.83	5.65
	0580	Over 5,000 lb.	**CN**		285	.084		.54	3.16		3.70	5.45
	1000	Stainless steel, type 304, under 100 lb.			165	.145		2.47	5.45		7.92	11.10
	1020	100 to 500 lb.			175	.137		1.97	5.15		7.12	10.05
	1030	500 to 1,000 lb.			190	.126		1.67	4.74		6.41	9.15
	1040	1,000 to 2,000 lb.			200	.120		1.57	4.50		6.07	8.70
	1050	2,000 to 5,000 lb.			225	.107		1.50	4		5.50	7.80
	1060	Over 5,000 lb.			235	.102		1.47	3.83		5.30	7.50
	1100	For medium pressure ductwork, add							15%			
	1200	For high pressure ductwork, add							40%			
300	0010	**FIBEROUS GLASS DUCTWORK**	R15810 -050									**300**
	3490	Rigid fiberglass duct board, foil reinf. kraft facing										
	3500	Rectangular, 1" thick, alum. faced, (FRK), std. weight		Q-10	350	.069	SF Surf	.66	2.57		3.23	4.69
500	0010	**FLEXIBLE DUCTS**	R15810 -050									**500**
	1300	Flexible, coated fiberglass fabric on corr. resist. metal helix										
	1400	pressure to 12" (WG) UL-181										
	1500	Non-insulated, 3" diameter		Q-9	400	.040	L.F.	1.03	1.45		2.48	3.36
	1540	5" diameter			320	.050		1.29	1.81		3.10	4.20
	1560	6" diameter			280	.057		1.52	2.07		3.59	4.85
	1580	7" diameter			240	.067		1.81	2.41		4.22	5.70
	1600	8" diameter			200	.080		2.09	2.89		4.98	6.75
	1640	10" diameter			160	.100		2.63	3.62		6.25	8.45
	1660	12" diameter			120	.133		3.18	4.82		8	10.90
	1900	Insulated, 1" thick, PE jacket, 3" diameter			380	.042		1.14	1.52		2.66	3.59
	1910	4" diameter			340	.047		1.23	1.70		2.93	3.97
	1920	5" diameter			300	.053		1.35	1.93		3.28	4.46
	1940	6" diameter			260	.062		1.56	2.23		3.79	5.15
	1960	7" diameter			220	.073		1.76	2.63		4.39	6
	1980	8" diameter			180	.089		1.96	3.22		5.18	7.10
	2020	10" diameter			140	.114		2.40	4.14		6.54	9
	2040	12" diameter			100	.160		2.95	5.80		8.75	12.15

15820	Duct Accessories		CREW	DAILY OUTPUT	LABOR-HOURS	UNIT	MAT.	LABOR	EQUIP.	TOTAL	INCL O&P	
300	0010	**DUCT ACCESSORIES**										**300**
	0050	Air extractors, 12" x 4"		1 Shee	24	.333	Ea.	15.55	13.40		28.95	37.50
	0100	8" x 6"			22	.364		15.55	14.60		30.15	39.50
	0200	20" x 8"			16	.500		35	20		55	70
	0280	24" x 12"			10	.800		48	32		80	102
	1000	Duct access door, insulated, 6" x 6"			14	.571		11.65	23		34.65	48.50

Important: See the Reference Section for critical supporting data - Reference Nos., Crews, & City Cost Indexes

T5 MECHANICAL

15820 | Duct Accessories

		CREW	DAILY OUTPUT	LABOR-HOURS	UNIT	2005 BARE COSTS MAT.	LABOR	EQUIP.	TOTAL	TOTAL INCL O&P	
300	1020	10" x 10"	1 Shee	11	.727	Ea.	16.90	29		45.90	63.50	**300**
	1040	12" x 12"		10	.800		18.20	32		50.20	69.50	
	1050	12" x 18"		9	.889		22.50	35.50		58	79.50	
	1070	18" x 18"		8	1		32.50	40		72.50	98	
	1074	24" x 18"		8	1		32	40		72	97	
	2000	Fabrics for flexible connections, with metal edge		100	.080	L.F.	1.62	3.22		4.84	6.75	
	2100	Without metal edge		160	.050	"	1.07	2.01		3.08	4.27	
	3000	Fire damper, curtain type, 1-1/2 hr rated, vertical, 6" x 6"		24	.333	Ea.	13.40	13.40		26.80	35	
	3020	8" x 6"		22	.364		13.40	14.60		28	37.50	
	3240	16" x 14"		18	.444		32.50	17.85		50.35	63.50	
	3400	24" x 20"		8	1		41.50	40		81.50	108	
	5990	Multi-blade dampers, opposed blade, 8" x 6"		24	.333		19.95	13.40		33.35	42.50	
	5994	8" x 8"		22	.364		21	14.60		35.60	45.50	
	5996	10" x 10"		21	.381		23.50	15.30		38.80	49	
	6000	12" x 12"		21	.381		27.50	15.30		42.80	53.50	
	6020	12" x 18"		18	.444		36.50	17.85		54.35	68	
	6030	14" x 10"		20	.400		26.50	16.10		42.60	54	
	6031	14" x 14"		17	.471		32.50	18.90		51.40	64.50	
	6033	16" x 12"		17	.471		32.50	18.90		51.40	64.50	
	6035	16" x 16"		16	.500		40	20		60	75	
	6037	18" x 16"		15	.533		45	21.50		66.50	82.50	
	6038	18" x 18"		15	.533		48.50	21.50		70	86	
	6070	20" x 16"		14	.571		48.50	23		71.50	88.50	
	6072	20" x 20"		13	.615		57.50	24.50		82	101	
	6074	22" x 18"		14	.571		57.50	23		80.50	98.50	
	6076	24" x 16"		11	.727		56.50	29		85.50	107	
	6078	24" x 20"		8	1		67.50	40		107.50	136	
	6080	24" x 24"		8	1		79	40		119	149	
	6110	26" x 26"		6	1.333		91.50	53.50		145	184	
	6133	30" x 30"	Q-9	6.60	2.424		122	87.50		209.50	270	
	6135	32" x 32"		6.40	2.500		140	90.50		230.50	293	
	6180	48" x 36"		5.60	2.857		234	103		337	415	
	7000	Splitter damper assembly, self-locking, 1' rod	1 Shee	24	.333		17.40	13.40		30.80	39.50	
	7020	3' rod		22	.364		22.50	14.60		37.10	47.50	
	7040	4' rod		20	.400		25	16.10		41.10	52	
	7060	6' rod		18	.444		30.50	17.85		48.35	61	
	8000	Multi-blade dampers, parallel blade										
	8100	8" x 8"	1 Shee	24	.333	Ea.	53.50	13.40		66.90	79.50	
	8140	16" x 10"		20	.400		71	16.10		87.10	103	
	8200	24" x 16"		11	.727		92	29		121	146	
	8260	30" x 18"		7	1.143		127	46		173	211	
	9000	Silencers, noise control for air flow, duct				MCFM	48			48	53	

15830 | Fans

		CREW	DAILY OUTPUT	LABOR-HOURS	UNIT	2005 BARE COSTS MAT.	LABOR	EQUIP.	TOTAL	TOTAL INCL O&P	
100	0010	**FANS**										**100**
	0020	Air conditioning and process air handling										
	0030	Axial flow, compact, low sound, 2.5" S.P.										
	0050	3,800 CFM, 5 HP	Q-20	3.40	5.882	Ea.	3,775	218		3,993	4,475	
	0080	6,400 CFM, 5 HP		2.80	7.143		4,200	265		4,465	5,025	
	0100	10,500 CFM, 7-1/2 HP		2.40	8.333		5,250	310		5,560	6,250	
	0120	15,600 CFM, 10 HP		1.60	12.500		6,600	465		7,065	7,975	
	0200	In-line centrifugal, supply/exhaust booster										
	0220	aluminum wheel/hub, disconnect switch, 1/4" S.P.										
	0240	500 CFM, 10" diameter connection	Q-20	3	6.667	Ea.	900	247		1,147	1,375	
	0260	1,380 CFM, 12" diameter connection		2	10		955	370		1,325	1,625	
	0280	1,520 CFM, 16" diameter connection		2	10		1,050	370		1,420	1,725	

The R15700-040 reference box appears next to line 0010.

For expanded coverage of these items see *Means Mechanical or Plumbing Cost Data 2005*

15830 | Fans

			CREW	DAILY OUTPUT	LABOR-HOURS	UNIT	MAT.	LABOR	EQUIP.	TOTAL	TOTAL INCL O&P
100	**0300**	2,560 CFM, 18" diameter connection	R15700-040	Q-20	1	20	Ea.	1,125	740		1,865	2,375	**100**
	0320	3,480 CFM, 20" diameter connection			.80	25		1,350	925		2,275	2,900
	0326	5,080 CFM, 20" diameter connection			.75	26.667		1,475	990		2,465	3,100
	1500	Vaneaxial, low pressure, 2000 CFM, 1/2 HP			3.60	5.556		1,625	206		1,831	2,100
	1520	4,000 CFM, 1 HP			3.20	6.250		1,650	232		1,882	2,175
	1540	8,000 CFM, 2 HP			2.80	7.143		1,950	265		2,215	2,550
	2500	Ceiling fan, right angle, extra quiet, 0.10" S.P.										
	2520	95 CFM		Q-20	20	1	Ea.	143	37		180	214
	2540	210 CFM			19	1.053		169	39		208	246
	2560	385 CFM			18	1.111		214	41		255	299
	2580	885 CFM			16	1.250		420	46.50		466.50	535
	2600	1,650 CFM			13	1.538		585	57		642	725
	2620	2,960 CFM			11	1.818		780	67.50		847.50	960
	2640	For wall or roof cap, add		1 Shee	16	.500		143	20		163	188
	2660	For straight thru fan, add						10%				
	2680	For speed control switch, add		1 Elec	16	.500		78	20.50		98.50	116
	3000	Paddle blade air circulator, 3 speed switch										
	3020	42", 5,000 CFM high, 3000 CFM low		1 Elec	2.40	3.333	Ea.	73.50	136		209.50	283
	3040	52", 6,500 CFM high, 4000 CFM low		"	2.20	3.636	"	74.50	148		222.50	305
	3100	For antique white motor, same cost										
	3200	For brass plated motor, same cost										
	3300	For light adaptor kit, add					Ea.	27.50			27.50	30
	3500	Centrifugal, airfoil, motor and drive, complete										
	3520	1000 CFM, 1/2 HP		Q-20	2.50	8	Ea.	905	297		1,202	1,450
	3540	2,000 CFM, 1 HP			2	10		970	370		1,340	1,650
	3560	4,000 CFM, 3 HP			1.80	11.111		1,250	410		1,660	2,000
	3580	8,000 CFM, 7-1/2 HP			1.40	14.286		1,825	530		2,355	2,825
	3600	12,000 CFM, 10 HP			1	20		2,700	740		3,440	4,100
	4500	Corrosive fume resistant, plastic										
	4600	roof ventilators, centrifugal, V belt drive, motor										
	4620	1/4" S.P., 250 CFM, 1/4 HP		Q-20	6	3.333	Ea.	2,475	124		2,599	2,925
	4640	895 CFM, 1/3 HP			5	4		2,700	148		2,848	3,200
	4660	1630 CFM, 1/2 HP			4	5		3,200	185		3,385	3,800
	4680	2240 CFM, 1 HP			3	6.667		3,325	247		3,572	4,025
	5000	Utility set, centrifugal, V belt drive, motor										
	5020	1/4" S.P., 1200 CFM, 1/4 HP		Q-20	6	3.333	Ea.	2,550	124		2,674	3,000
	5040	1520 CFM, 1/3 HP			5	4		2,550	148		2,698	3,025
	5060	1850 CFM, 1/2 HP			4	5		2,550	185		2,735	3,100
	5080	2180 CFM, 3/4 HP			3	6.667		2,600	247		2,847	3,225
	5100	1/2" S.P., 3600 CFM, 1 HP			2	10		3,750	370		4,120	4,700
	5120	4250 CFM, 1-1/2 HP			1.60	12.500		3,800	465		4,265	4,875
	5140	4800 CFM, 2 HP			1.40	14.286		3,850	530		4,380	5,025
	6000	Propeller exhaust, wall shutter, 1/4" S.P.										
	6020	Direct drive, two speed										
	6100	375 CFM, 1/10 HP		Q-20	10	2	Ea.	256	74		330	395
	6120	730 CFM, 1/7 HP			9	2.222		285	82.50		367.50	440
	6140	1000 CFM, 1/8 HP			8	2.500		395	92.50		487.50	575
	6200	4720 CFM, 1 HP			5	4		635	148		783	925
	6300	V-belt drive, 3 phase										
	6320	6175 CFM, 3/4 HP		Q-20	5	4	Ea.	535	148		683	810
	6340	7500 CFM, 3/4 HP			5	4		560	148		708	845
	6360	10,100 CFM, 1 HP			4.50	4.444		685	165		850	1,000
	6380	14,300 CFM, 1-1/2 HP			4	5		805	185		990	1,175
	6650	Residential, bath exhaust, grille, back draft damper										
	6660	50 CFM		Q-20	24	.833	Ea.	30	31		61	80
	6670	110 CFM			22	.909		50	33.50		83.50	107

Important: See the Reference Section for critical supporting data - Reference Nos., Crews, & City Cost Indexes

15830	Fans		CREW	DAILY OUTPUT	LABOR-HOURS	UNIT	2005 BARE COSTS				TOTAL INCL O&P	
							MAT.	LABOR	EQUIP.	TOTAL		
100 6680	Light combination, squirrel cage, 100 watt, 70 CFM	R15700 -040	Q-20	24	.833	Ea.	61.50	31		92.50	115	**100**
6700	Light/heater combination, ceiling mounted											
6710	70 CFM, 1450 watt		Q-20	24	.833	Ea.	74.50	31		105.50	129	
6800	Heater combination, recessed, 70 CFM			24	.833		35.50	31		66.50	86	
6820	With 2 infrared bulbs			23	.870		53.50	32.50		86	109	
6900	Kitchen exhaust, grille, complete, 160 CFM			22	.909		62.50	33.50		96	120	
6910	180 CFM			20	1		73	37		110	137	
6920	270 CFM			18	1.111		95.50	41		136.50	168	
6930	350 CFM		▼	16	1.250	▼	105	46.50		151.50	186	
6940	Residential roof jacks and wall caps											
6944	Wall cap with back draft damper											
6946	3" & 4" dia. round duct		1 Shee	11	.727	Ea.	12.80	29		41.80	59	
6948	6" dia. round duct		"	11	.727	"	31	29		60	79	
6958	Roof jack with bird screen and back draft damper											
6960	3" & 4" dia. round duct		1 Shee	11	.727	Ea.	12.30	29		41.30	58.50	
6962	3-1/4" x 10" rectangular duct		"	10	.800	"	22.50	32		54.50	74.50	
6980	Transition											
6982	3-1/4" x 10" to 6" dia. round		1 Shee	20	.400	Ea.	13.85	16.10		29.95	40	
7000	Roof exhauster, centrifugal, aluminum housing, 12" galvanized											
7020	curb, bird screen, back draft damper, 1/4" S.P.											
7100	Direct drive, 320 CFM, 11" sq. damper		Q-20	7	2.857	Ea.	320	106		426	510	
7120	600 CFM, 11" sq. damper			6	3.333		325	124		449	545	
7140	815 CFM, 13" sq. damper			5	4		325	148		473	580	
7160	1450 CFM, 13" sq. damper			4.20	4.762		415	177		592	725	
7180	2050 CFM, 16" sq. damper			4	5		415	185		600	740	
7200	V-belt drive, 1650 CFM, 12" sq. damper			6	3.333		725	124		849	985	
7220	2750 CFM, 21" sq. damper			5	4		840	148		988	1,150	
7230	3500 CFM, 21" sq. damper			4.50	4.444		925	165		1,090	1,275	
7240	4910 CFM, 23" sq. damper			4	5		1,150	185		1,335	1,550	
7260	8525 CFM, 28" sq. damper			3	6.667		1,425	247		1,672	1,950	
7280	13,760 CFM, 35" sq. damper			2	10		1,975	370		2,345	2,750	
7300	20,558 CFM, 43" sq. damper		▼	1	20	▼	4,200	740		4,940	5,750	
7320	For 2 speed winding, add						15%					
7340	For explosionproof motor, add					Ea.	330			330	360	
7360	For belt driven, top discharge, add						15%					
7500	Utility set, steel construction, pedestal, 1/4" S.P.											
7520	Direct drive, 150 CFM, 1/8 HP		Q-20	6.40	3.125	Ea.	635	116		751	870	
7540	485 CFM, 1/6 HP			5.80	3.448		795	128		923	1,075	
7560	1950 CFM, 1/2 HP			4.80	4.167		935	155		1,090	1,250	
7580	2410 CFM, 3/4 HP			4.40	4.545		1,725	169		1,894	2,150	
7600	3328 CFM, 1-1/2 HP		▼	3	6.667	▼	1,925	247		2,172	2,475	
7680	V-belt drive, drive cover, 3 phase											
7700	800 CFM, 1/4 HP		Q-20	6	3.333	Ea.	460	124		584	700	
7720	1,300 CFM, 1/3 HP			5	4		485	148		633	760	
7740	2,000 CFM, 1 HP			4.60	4.348		575	161		736	875	
7760	2,900 CFM, 3/4 HP			4.20	4.762		775	177		952	1,125	
8500	Wall exhausters, centrifugal, auto damper, 1/8" S.P.											
8520	Direct drive, 610 CFM, 1/20 HP		Q-20	14	1.429	Ea.	207	53		260	310	
8540	796 CFM, 1/12 HP			13	1.538		214	57		271	325	
8560	822 CFM, 1/6 HP			12	1.667		340	62		402	465	
8580	1,320 CFM, 1/4 HP		▼	12	1.667	▼	340	62		402	470	
9500	V-belt drive, 3 phase											
9520	2,800 CFM, 1/4 HP		Q-20	9	2.222	Ea.	940	82.50		1,022.50	1,150	
9540	3,740 CFM, 1/2 HP		"	8	2.500	"	975	92.50		1,067.50	1,225	

MECHANICAL 15

15840 | Air Terminal Units

		CREW	DAILY OUTPUT	LABOR-HOURS	UNIT	MAT.	LABOR	EQUIP.	TOTAL	TOTAL INCL O&P	
200	0010 **AIR CURTAINS** Incl. motor starters, transformers,										**200**
	0050 door switches & temperature controls										
	0100 Shipping and receiving doors, unheated, minimal wind stoppage										
	0150 8' high, multiples of 3' wide	2 Shee	6	2.667	L.F.	320	107		427	515	
	0160 5' wide		10	1.600		310	64.50		374.50	445	
	0210 10' high, multiples of 4' wide		8	2		305	80.50		385.50	460	
	0250 12' high, 3'-6" wide		7	2.286		335	92		427	505	
	0260 12' wide		6	2.667		305	107		412	500	
	0350 16' high, 3'-6" wide		7	2.286		320	92		412	490	
	0360 12' wide	▼	6	2.667	▼	355	107		462	555	
	0500 Maximum wind stoppage										
	0550 10' high, multiples of 4' wide	2 Shee	8	2	L.F.	246	80.50		326.50	395	
	0650 14' high, multiples of 4' wide		8	2		246	80.50		326.50	395	
	0750 20' high, multiples of 8' wide	▼	6	2.667		855	107		962	1,100	
	1100 Heated, maximum wind stoppage, steam heat										
	1150 10' high, multiples of 4' wide	2 Shee	6	2.667	L.F.	425	107		532	630	
	1250 14' high, multiples of 4' wide		6	2.667		460	107		567	670	
	1350 20' high, multiples of 8' wide	▼	4	4	▼	1,300	161		1,461	1,675	
	1500 Customer entrance doors, unheated, minimal wind stoppage										
	1550 10' high, multiples of 3' wide	2 Shee	6	2.667	L.F.	282	107		389	475	
	1560 5' wide		10	1.600		305	64.50		369.50	435	
	1650 Maximum wind stoppage, 12' high, multiples of 4' wide	▼	8	2	▼	227	80.50		307.50	375	
	1700 Heated, minimal wind stoppage, electric heat										
	1750 8' high, multiples of 3' wide	2 Shee	6	2.667	L.F.	560	107		667	785	
	1760 Multiples of 5' wide		10	1.600		400	64.50		464.50	540	
	1850 10' high, multiples of 3' wide		6	2.667		480	107		587	695	
	1860 Multiples of 5' wide	▼	10	1.600	▼	525	64.50		589.50	675	
	1950 Maximum wind stoppage, steam heat										
	1960 12' high, multiples of 4' wide	2 Shee	8	2	L.F.	510	80.50		590.50	685	
	2000 Walk-in coolers and freezers, ambient air, minimal wind stoppage										
	2050 8' high, multiples of 3' wide	2 Shee	6	2.667	L.F.	214	107		321	400	
	2060 Multiples of 5' wide		10	1.600		231	64.50		295.50	355	
	2250 Maximum wind stoppage, 12' high, multiples of 3' wide		6	2.667		246	107		353	435	
	2450 Conveyor openings or service windows, unheated, 5' high		5	3.200		268	129		397	490	
	2460 Heated, electric, 5' high, 2'-6" wide	▼	5	3.200	▼	194	129		323	410	

15850 | Air Outlets & Inlets

		CREW	DAILY OUTPUT	LABOR-HOURS	UNIT	MAT.	LABOR	EQUIP.	TOTAL	TOTAL INCL O&P	
300	0010 **DIFFUSERS** Aluminum, opposed blade damper unless noted										**300**
	0100 Ceiling, linear, also for sidewall										
	0500 Perforated, 24" x 24" lay-in panel size, 6" x 6"	1 Shee	16	.500	Ea.	75.50	20		95.50	114	
	0520 8" x 8"		15	.533		77.50	21.50		99	118	
	0530 9" x 9"		14	.571		79	23		102	123	
	0540 10" x 10"		14	.571		79.50	23		102.50	123	
	0560 12" x 12"		12	.667		81.50	27		108.50	131	
	0590 16" x 16"		11	.727		96	29		125	150	
	0600 18" x 18"		10	.800		101	32		133	161	
	0610 20" x 20"		10	.800		116	32		148	177	
	0620 24" x 24"		9	.889		133	35.50		168.50	202	
	1000 Rectangular, 1 to 4 way blow, 6" x 6"		16	.500		40.50	20		60.50	76	
	1010 8" x 8"		15	.533		48.50	21.50		70	86	
	1014 9" x 9"		15	.533		52	21.50		73.50	90	
	1016 10" x 10"		15	.533		61	21.50		82.50	100	
	1020 12" x 6"		15	.533		67	21.50		88.50	107	
	1040 12" x 9"		14	.571		73	23		96	116	
	1060 12" x 12"	▼	12	.667	▼	71.50	27		98.50	120	

Important: See the Reference Section for critical supporting data - Reference Nos., Crews, & City Cost Indexes

15850 | Air Outlets & Inlets

		CREW	DAILY OUTPUT	LABOR-HOURS	UNIT	MAT.	LABOR	EQUIP.	TOTAL	TOTAL INCL O&P
300						2005 BARE COSTS					**300**
1070	14" x 6"	1 Shee	13	.615	Ea.	72	24.50		96.50	118
1074	14" x 14"		12	.667		93.50	27		120.50	144
1150	18" x 18"		9	.889		137	35.50		172.50	206
1160	21" x 21"		8	1		166	40		206	244
1170	24" x 12"		10	.800		124	32		156	186
1500	Round, butterfly damper, 6" diameter		18	.444		16.85	17.85		34.70	46
1520	8" diameter		16	.500		18.20	20		38.20	51
1540	10" diameter		14	.571		22	23		45	60
1560	12" diameter		12	.667		29.50	27		56.50	73.50
1580	14" diameter		10	.800		38	32		70	91.50
2000	T bar mounting, 24" x 24" lay-in frame, 6" x 6"		16	.500		82	20		102	121
2020	9" x 9"		14	.571		90.50	23		113.50	135
2040	12" x 12"		12	.667		117	27		144	170
2060	15" x 15"		11	.727		150	29		179	210
2080	18" x 18"		10	.800		157	32		189	222
6000	For steel diffusers instead of aluminum, deduct					10%				
500	0010 **GRILLES**										**500**
0020	Aluminum									
1000	Air return, 6" x 6"	1 Shee	26	.308	Ea.	13	12.35		25.35	33.50
1020	10" x 6"		24	.333		15.60	13.40		29	37.50
1080	16" x 8"		22	.364		23.50	14.60		38.10	48
1100	12" x 12"		22	.364		23.50	14.60		38.10	48
1120	24" x 12"		18	.444		41.50	17.85		59.35	73.50
1220	24" x 18"		16	.500		49.50	20		69.50	85.50
1280	36" x 24"		14	.571		99	23		122	145
3000	Filter grille with filter, 12" x 12"		24	.333		39	13.40		52.40	63.50
3020	18" x 12"		20	.400		57	16.10		73.10	87.50
3040	24" x 18"		18	.444		83	17.85		100.85	119
3060	24" x 24"		16	.500		109	20		129	151
6000	For steel grilles instead of aluminum in above, deduct					10%				
600	0010 **LOUVERS**										**600**
0100	Aluminum, extruded, with screen, mill finish									
1000	Brick vent, (see also division 04090-860)									
1100	Standard, 4" deep, 8" wide, 5" high	1 Shee	24	.333	Ea.	24	13.40		37.40	47
1200	Modular, 4" deep, 7-3/4" wide, 5" high		24	.333		25	13.40		38.40	48
1300	Speed brick, 4" deep, 11-5/8" wide, 3-7/8" high		24	.333		25	13.40		38.40	48
1400	Fuel oil brick, 4" deep, 8" wide, 5" high		24	.333		44	13.40		57.40	69
2000	Cooling tower and mechanical equip., screens, light weight		40	.200	S.F.	10.45	8.05		18.50	24
2020	Standard weight		35	.229		29	9.20		38.20	46
2500	Dual combination, automatic, intake or exhaust		20	.400		40	16.10		56.10	68.50
2520	Manual operation		20	.400		29.50	16.10		45.60	57
2540	Electric or pneumatic operation		20	.400		29.50	16.10		45.60	57
2560	Motor, for electric or pneumatic		14	.571	Ea.	350	23		373	420
3000	Fixed blade, continuous line									
3100	Mullion type, stormproof	1 Shee	28	.286	S.F.	29.50	11.50		41	50
3200	Stormproof		28	.286		29.50	11.50		41	50
3300	Vertical line		28	.286		39	11.50		50.50	60.50
3500	For damper to use with above, add					50%	30%			
3520	Motor, for damper, electric or pneumatic	1 Shee	14	.571	Ea.	350	23		373	420
4000	Operating, 45°, manual, electric or pneumatic		24	.333	S.F.	29.50	13.40		42.90	53
4100	Motor, for electric or pneumatic		14	.571	Ea.	350	23		373	420
4200	Penthouse, roof		56	.143	S.F.	17.15	5.75		22.90	27.50
4300	Walls		40	.200		41	8.05		49.05	58
5000	Thinline, under 4" thick, fixed blade		40	.200		16.60	8.05		24.65	30.50
5010	Finishes, applied by mfr. at additional cost, available in colors									
5020	Prime coat only, add				S.F.	2.42			2.42	2.66

	15850	**Air Outlets & Inlets**	CREW	DAILY OUTPUT	LABOR-HOURS	UNIT	2005 BARE COSTS				TOTAL INCL O&P	
							MAT.	LABOR	EQUIP.	TOTAL		
600	5040	Baked enamel finish coating, add				S.F.	4.44			4.44	4.88	**600**
	5060	Anodized finish, add					4.82			4.82	5.30	
	5080	Duranodic finish, add					8.75			8.75	9.60	
	5100	Fluoropolymer finish coating, add					13.65			13.65	15.05	
	9980	For small orders (under 10 pieces), add				↓	25%					
700	0010	**REGISTERS**										**700**
	0980	Air supply										
	1000	Ceiling/wall, O.B. damper, anodized aluminum										
	1010	One or two way deflection, adj. curved face bars										
	1020	8″ x 4″	1 Shee	26	.308	Ea.	20.50	12.35		32.85	42	
	1120	12″ x 12″		18	.444		34	17.85		51.85	64.50	
	1240	20″ x 6″		18	.444		34	17.85		51.85	64.50	
	1340	24″ x 8″		13	.615		45.50	24.50		70	88	
	1350	24″ x 18″	↓	12	.667		85	27		112	135	
	2700	Above registers in steel instead of aluminum, deduct				↓	10%					
	4000	Floor, toe operated damper, enameled steel										
	4020	4″ x 8″	1 Shee	32	.250	Ea.	17.75	10.05		27.80	35	
	4100	8″ x 10″		22	.364		21.50	14.60		36.10	46.50	
	4140	10″ x 10″		20	.400		25.50	16.10		41.60	52.50	
	4220	14″ x 14″		16	.500		90.50	20		110.50	131	
	4240	14″ x 20″	↓	15	.533	↓	121	21.50		142.50	166	
	4980	Air return										
	5000	Ceiling or wall, fixed 45° face blades										
	5010	Adjustable O.B. damper, anodized aluminum										
	5020	4″ x 8″	1 Shee	26	.308	Ea.	23.50	12.35		35.85	44.50	
	5060	6″ x 10″		19	.421		26	16.95		42.95	54.50	
	5280	24″ x 24″		11	.727		112	29		141	168	
	5300	24″ x 36″	↓	8	1	↓	174	40		214	254	
	6000	For steel construction instead of aluminum, deduct					10%					
800	0010	**VENTILATORS** Base & damper										**800**
	1280	Rotary ventilators, wind driven, galvanized										
	1300	4″ neck diameter	Q-9	20	.800	Ea.	52	29		81	102	
	1340	6″ neck diameter		16	1		58	36		94	120	
	1400	12″ neck diameter		10	1.600		84.50	58		142.50	182	
	1500	24″ neck diameter, 3,100 CFM		8	2		245	72.50		317.50	380	
	1540	36″ neck diameter, 5,500 CFM	↓	6	2.667	↓	720	96.50		816.50	940	
	2000	Stationary, gravity, syphon, galvanized										
	2160	6″ neck diameter, 66 CFM	Q-9	16	1	Ea.	45	36		81	105	
	2240	12″ neck diameter, 160 CFM		10	1.600		94.50	58		152.50	193	
	2340	24″ neck diameter, 900 CFM		8	2		257	72.50		329.50	395	
	2380	36″ neck diameter, 2,000 CFM		6	2.667		805	96.50		901.50	1,025	
	4200	Stationary mushroom, aluminum, 16″ orifice diameter		10	1.600		320	58		378	440	
	4220	26″ orifice diameter		6.15	2.602		470	94		564	665	
	4230	30″ orifice diameter		5.71	2.802		690	101		791	915	
	4240	38″ orifice diameter		5	3.200		990	116		1,106	1,250	
	4250	42″ orifice diameter		4.70	3.404		1,300	123		1,423	1,650	
	4260	50″ orifice diameter	↓	4.44	3.604	↓	1,550	130		1,680	1,925	
	5000	Relief vent										
	5500	Rectangular, aluminum, galvanized curb										
	5510	intake/exhaust, 0.05″ SP										
	5600	500 CFM, 12″ x 16″	Q-9	8	2	Ea.	390	72.50		462.50	540	
	5640	1000 CFM, 12″ x 24″		6.60	2.424		465	87.50		552.50	645	
	5680	3000 CFM, 20″ x 42″	↓	4	4		805	145		950	1,100	

T5 MECHANICAL

Important: See the Reference Section for critical supporting data - Reference Nos., Crews, & City Cost Indexes

15800 | Air Distribution

15860 | Air Cleaning Devices

			CREW	DAILY OUTPUT	LABOR-HOURS	UNIT	2005 BARE COSTS MAT.	LABOR	EQUIP.	TOTAL	TOTAL INCL O&P	
100	0010	**AIR FILTERS**										100
	0050	Activated charcoal type, full flow				MCFM	600			600	660	
	0060	Activated charcoal type, full flow, impregnated media 12" deep					175			175	193	
	0070	Activated charcoal type, HEPA filter & frame for field erection					175			175	193	
	0080	Activated charcoal type, HEPA filter-diffuser, ceiling install.				▼	250			250	275	
	2000	Electronic air cleaner, duct mounted										
	2150	400 - 1000 CFM	1 Shee	2.30	3.478	Ea.	695	140		835	980	
	2200	1000 - 1400 CFM		2.20	3.636		725	146		871	1,025	
	2250	1400 - 2000 CFM	▼	2.10	3.810	▼	800	153		953	1,125	
	2950	Mechanical media filtration units										
	3000	High efficiency type, with frame, non-supported				MCFM	45			45	49.50	
	3100	Supported type					55			55	60.50	
	4000	Medium efficiency, extended surface					5			5	5.50	
	4500	Permanent washable					20			20	22	
	5000	Renewable disposable roll				▼	120			120	132	
	5500	Throwaway glass or paper media type				Ea.	4.60			4.60	5.05	

15950 | Testing/Adjusting/Balancing

15955 | HVAC Test/Adjust/Balance

			CREW	DAILY OUTPUT	LABOR-HOURS	UNIT	2005 BARE COSTS MAT.	LABOR	EQUIP.	TOTAL	TOTAL INCL O&P	
100	0010	**BALANCING, AIR** (Subcontractor's quote incl. material & labor)										100
	0900	Heating and ventilating equipment										
	1000	Centrifugal fans, utility sets				Ea.					296.88	
	1100	Heating and ventilating unit									445.32	
	1200	In-line fan									445.32	
	1300	Propeller and wall fan									84.12	
	1400	Roof exhaust fan									197.92	
	2000	Air conditioning equipment, central station									643.24	
	2100	Built-up low pressure unit									593.76	
	2200	Built-up high pressure unit									692.72	
	2500	Multi-zone A.C. and heating unit									445.32	
	2600	For each zone over one, add									98.96	
	2700	Package A.C. unit									247.40	
	2800	Rooftop heating and cooling unit									346.36	
	3000	Supply, return, exhaust, registers & diffusers, avg. height ceiling									59.38	
	3100	High ceiling									89.06	
	3200	Floor height				▼					49.48	
900	0010	**BALANCING, WATER** (Subcontractor's quote incl. material & labor)										900
	0050	Air cooled condenser				Ea.					172.34	
	0100	Cabinet unit heater									59.09	
	0200	Chiller									418.54	
	0300	Convector									49.24	
	0500	Cooling tower									320.06	
	0600	Fan coil unit, unit ventilator									88.63	
	0700	Fin tube and radiant panels									98.48	
	0800	Main and duct re-heat coils									91.09	
	0810	Heat exchanger									91.09	
	1000	Pumps									216.66	
	1100	Unit heater				▼					68.94	

MECHANICAL 15

For information about Means Estimating Seminars, see yellow pages 12 and 13 in back of book

For expanded coverage of these items see *Means Mechanical or Plumbing Cost Data 2005*

		CREW	DAILY OUTPUT	LABOR-HOURS	UNIT	2005 BARE COSTS				TOTAL INCL O&P
						MAT.	LABOR	EQUIP.	TOTAL	

Division 16
Electrical

Estimating Tips

16060 Grounding & Bonding
- When taking off grounding system, identify separately the type and size of wire and list each unique type of ground connection.

16100 Wiring Methods
- Conduit should be taken off in three main categories: power distribution, branch power, and branch lighting, so the estimator can concentrate on systems and components, therefore making it easier to ensure all items have been accounted for.
- For cost modifications for elevated conduit installation, add the percentages to labor according to the height of installation and only the quantities exceeding the different height levels, not to the total conduit quantities.
- Remember that aluminum wiring of equal ampacity is larger in diameter than copper and may require larger conduit.
- If more than three wires at a time are being pulled, deduct percentages from the labor hours of that grouping of wires.

- The estimator should take the weights of materials into consideration when completing a takeoff. Topics to consider include: How will the materials be supported? What methods of support are available? How high will the support structure have to reach? Will the final support structure be able to withstand the total burden? Is the support material included or separate from the fixture, equipment and material specified?

16200 Electrical Power
- Do not overlook the costs for equipment used in the installation. If scaffolding or highlifts are available in the field, contractors may use them in lieu of the proposed ladders and rolling staging.

16400 Low-Voltage Distribution
- Supports and concrete pads may be shown on drawings for the larger equipment, or the support system may be just a piece of plywood for the back of a panelboard. In either case, it must be included in the costs.

16500 Lighting
- Fixtures should be taken off room by room, using the fixture schedule, specifications, and the ceiling plan. For large concentrations of lighting fixtures in the same area deduct the percentages from labor hours.

16700 Communications
16800 Sound & Video
- When estimating material costs for special systems, it is always prudent to obtain manufacturers' quotations for equipment prices and special installation requirements which will affect the total costs.

Reference Numbers
Reference numbers are shown in bold squares at the beginning of some major classifications. These numbers refer to related items in the Reference Section. The reference information may be an estimating procedure, an alternate pricing method or technical information.

Note: Not all subdivisions listed here necessarily appear in this publication.

Note: **i2 Trade Service,** *in part, has been used as a reference source for some of the material prices used in Division 16.*

16055	Selective Demolition	CREW	DAILY OUTPUT	LABOR-HOURS	UNIT	2005 BARE COSTS				TOTAL INCL O&P
						MAT.	LABOR	EQUIP.	TOTAL	
300	0010 **ELECTRICAL DEMOLITION**										300
	0020 Conduit to 15' high, including fittings & hangers									
	0100 Rigid galvanized steel, 1/2" to 1" diameter	1 Elec	242	.033	L.F.		1.35		1.35	2.01
	0120 1-1/4" to 2"	"	200	.040			1.63		1.63	2.43
	0140 2-1/2" to 3-1/2"	2 Elec	302	.053			2.16		2.16	3.21
	0160 4" to 6"	"	160	.100			4.08		4.08	6.05
	0200 Electric metallic tubing (EMT), 1/2" to 1"	1 Elec	394	.020			.83		.83	1.23
	0220 1-1/4" to 1-1/2"	"	326	.025			1		1	1.49
	0260 3-1/2" to 4"	2 Elec	310	.052			2.10		2.10	3.13
	0270 Armored cable, (BX) avg. 50' runs									
	0280 #14, 2 wire	1 Elec	690	.012	L.F.		.47		.47	.70
	0290 #14, 3 wire		571	.014			.57		.57	.85
	0300 #12, 2 wire		605	.013			.54		.54	.80
	0310 #12, 3 wire		514	.016			.63		.63	.94
	0320 #10, 2 wire		514	.016			.63		.63	.94
	0330 #10, 3 wire		425	.019			.77		.77	1.14
	0340 #8, 3 wire		342	.023			.95		.95	1.42
	0350 Non metallic sheathed cable (Romex)									
	0360 #14, 2 wire	1 Elec	720	.011	L.F.		.45		.45	.67
	0370 #14, 3 wire		657	.012			.50		.50	.74
	0380 #12, 2 wire		629	.013			.52		.52	.77
	0390 #10, 3 wire		450	.018			.72		.72	1.08
	0400 Wiremold raceway, including fittings & hangers									
	0420 No. 3000	1 Elec	250	.032	L.F.		1.30		1.30	1.94
	0440 No. 4000		217	.037			1.50		1.50	2.24
	0460 No. 6000		166	.048			1.96		1.96	2.92
	0465 Telephone/power pole		12	.667	Ea.		27		27	40.50
	0470 Non-metallic, straight section		480	.017	L.F.		.68		.68	1.01
	0500 Channels, steel, including fittings & hangers									
	0520 3/4" x 1-1/2"	1 Elec	308	.026	L.F.		1.06		1.06	1.58
	0540 1-1/2" x 1-1/2"		269	.030			1.21		1.21	1.80
	0560 1-1/2" x 1-7/8"		229	.035			1.42		1.42	2.12
	0600 Copper bus duct, indoor, 3 phase									
	0610 Including hangers & supports									
	0620 225 amp	2 Elec	135	.119	L.F.		4.83		4.83	7.20
	0640 400 amp		106	.151			6.15		6.15	9.15
	0660 600 amp		86	.186			7.60		7.60	11.30
	0680 1000 amp		60	.267			10.85		10.85	16.15
	0700 1600 amp		40	.400			16.30		16.30	24.50
	0720 3000 amp		10	1.600			65		65	97
	1300 Transformer, dry type, 1 ph, incl. removal of									
	1320 supports, wire & conduit terminations									
	1340 1 kVA	1 Elec	7.70	1.039	Ea.		42.50		42.50	63
	1420 75 kVA	2 Elec	2.50	6.400	"		261		261	390
	1440 3 Phase to 600V, primary									
	1460 3 kVA	1 Elec	3.85	2.078	Ea.		84.50		84.50	126
	1520 75 kVA	2 Elec	2.70	5.926			241		241	360
	1550 300 kVA	R-3	1.80	11.111			445	88	533	760
	1570 750 kVA	"	1.10	18.182			725	144	869	1,225
	1800 Wire, THW-THWN-THHN, removed from									
	1810 in place conduit, to 15' high									
	1830 #14	1 Elec	65	.123	C.L.F.		5		5	7.45
	1840 #12		55	.145			5.95		5.95	8.80
	1850 #10		45.50	.176			7.15		7.15	10.65
	1860 #8		40.40	.198			8.05		8.05	12
	1870 #6		32.60	.245			10		10	14.90

16 ELECTRICAL

Important: See the Reference Section for critical supporting data - Reference Nos., Crews, & City Cost Indexes

	16055	Selective Demolition	CREW	DAILY OUTPUT	LABOR-HOURS	UNIT	2005 BARE COSTS				TOTAL INCL O&P	
							MAT.	LABOR	EQUIP.	TOTAL		
300	1880	#4	2 Elec	53	.302	C.L.F.		12.30		12.30	18.30	300
	1890	#3		50	.320			13.05		13.05	19.40	
	1900	#2		44.60	.359			14.60		14.60	22	
	1910	1/0		33.20	.482			19.65		19.65	29	
	1920	2/0		29.20	.548			22.50		22.50	33	
	1930	3/0		25	.640			26		26	39	
	1940	4/0		22	.727			29.50		29.50	44	
	1950	250 kcmil		20	.800			32.50		32.50	48.50	
	1960	300 kcmil		19	.842			34.50		34.50	51	
	1970	350 kcmil		18	.889			36		36	54	
	1980	400 kcmil		17	.941			38.50		38.50	57	
	1990	500 kcmil	↓	16.20	.988	↓		40.50		40.50	60	
	2000	Interior fluorescent fixtures, incl. supports										
	2010	& whips, to 15' high										
	2100	Recessed drop-in 2' x 2', 2 lamp	2 Elec	35	.457	Ea.		18.65		18.65	27.50	
	2120	2' x 4', 2 lamp		33	.485			19.75		19.75	29.50	
	2140	2' x 4', 4 lamp		30	.533			21.50		21.50	32.50	
	2160	4' x 4', 4 lamp	↓	20	.800	↓		32.50		32.50	48.50	
	2180	Surface mount, acrylic lens & hinged frame										
	2200	1' x 4', 2 lamp	2 Elec	44	.364	Ea.		14.80		14.80	22	
	2220	2' x 2', 2 lamp		44	.364			14.80		14.80	22	
	2260	2' x 4', 4 lamp		33	.485			19.75		19.75	29.50	
	2280	4' x 4', 4 lamp	↓	23	.696	↓		28.50		28.50	42	
	2300	Strip fixtures, surface mount										
	2320	4' long, 1 lamp	2 Elec	53	.302	Ea.		12.30		12.30	18.30	
	2340	4' long, 2 lamp		50	.320			13.05		13.05	19.40	
	2360	8' long, 1 lamp		42	.381			15.50		15.50	23	
	2380	8' long, 2 lamp	↓	40	.400	↓		16.30		16.30	24.50	
	2400	Pendant mount, industrial, incl. removal										
	2410	of chain or rod hangers, to 15' high										
	2420	4' long, 2 lamp	2 Elec	35	.457	Ea.		18.65		18.65	27.50	
	2440	8' long, 2 lamp	"	27	.593	"		24		24	36	

	16060	Grounding & Bonding										
800	0010	**GROUNDING**										800
	0030	Rod, copper clad, 8' long, 1/2" diameter	1 Elec	5.50	1.455	Ea.	13.50	59.50		73	103	
	0050	3/4" diameter		5.30	1.509		28	61.50		89.50	123	
	0080	10' long, 1/2" diameter		4.80	1.667		17.55	68		85.55	120	
	0100	3/4" diameter		4.40	1.818		31	74		105	145	
	0130	15' long, 3/4" diameter		4	2	↓	82	81.50		163.50	212	
	0390	Bare copper wire, #8 stranded		11	.727	C.L.F.	11.95	29.50		41.45	57	
	0400	#6	↓	10	.800		20.50	32.50		53	71	
	0600	#2	2 Elec	10	1.600		47.50	65		112.50	150	
	0800	3/0		6.60	2.424		135	99		234	296	
	1000	4/0	↓	5.70	2.807		169	114		283	355	
	1200	250 kcmil	3 Elec	7.20	3.333	↓	191	136		327	410	
	1800	Water pipe ground clamps, heavy duty										
	2000	Bronze, 1/2" to 1" diameter	1 Elec	8	1	Ea.	14.40	41		55.40	76.50	
	2100	1-1/4" to 2" diameter		8	1		18.95	41		59.95	81.50	
	2200	2-1/2" to 3" diameter		6	1.333		41	54.50		95.50	126	
	2800	Brazed connections, #6 wire		12	.667		11.60	27		38.60	53.50	
	3000	#2 wire		10	.800		15.55	32.50		48.05	65.50	
	3100	3/0 wire		8	1		23.50	41		64.50	86	
	3200	4/0 wire		7	1.143		26.50	46.50		73	99	
	3400	250 kcmil wire		5	1.600		31	65		96	131	
	3600	500 kcmil wire	↓	4	2	↓	38.50	81.50		120	163	

ELECTRICAL 16

		16120	**Conductors & Cables**	CREW	DAILY OUTPUT	LABOR-HOURS	UNIT	2005 BARE COSTS				TOTAL INCL O&P
								MAT.	LABOR	EQUIP.	TOTAL	
120	0010	**ARMORED CABLE**										120
	0050		600 volt, copper (BX), #14, 2 conductor, solid	1 Elec	2.40	3.333	C.L.F.	58.50	136		194.50	267
	0100		3 conductor, solid		2.20	3.636		93	148		241	325
	0150		#12, 2 conductor, solid		2.30	3.478		59.50	142		201.50	277
	0200		3 conductor, solid		2	4		95	163		258	345
	0250		#10, 2 conductor, solid		2	4		108	163		271	360
	0300		3 conductor, solid		1.60	5		149	204		353	470
	0350		#8, 3 conductor, solid		1.30	6.154		268	251		519	670
	0400		3 conductor with PVC jacket, in cable tray, #6		3.10	2.581		320	105		425	510
	0450		#4	2 Elec	5.40	2.963		415	121		536	635
	0500		#2		4.60	3.478		550	142		692	815
	0550		#1		4	4		745	163		908	1,075
	0600		1/0		3.60	4.444		875	181		1,056	1,225
	0650		2/0		3.40	4.706		1,075	192		1,267	1,450
	0700		3/0		3.20	5		1,250	204		1,454	1,675
	0750		4/0		3	5.333		1,425	217		1,642	1,900
	0800		250 kcmil	3 Elec	3.60	6.667		1,600	272		1,872	2,150
	0850		350 kcmil		3.30	7.273		2,150	296		2,446	2,800
	0900		500 kcmil		3	8		2,775	325		3,100	3,525
	1050		5 kV, copper, 3 conductor with PVC jacket,									
	1060		non-shielded, in cable tray, #4	2 Elec	380	.042	L.F.	6.35	1.72		8.07	9.50
	1100		#2		360	.044		8.20	1.81		10.01	11.75
	1200		#1		300	.053		10.45	2.17		12.62	14.75
	1400		1/0		290	.055		12.10	2.25		14.35	16.65
	1600		2/0		260	.062		13.95	2.51		16.46	19.10
	2000		4/0		240	.067		18.65	2.72		21.37	24.50
	2100		250 kcmil	3 Elec	330	.073		25.50	2.96		28.46	32.50
	2150		350 kcmil		315	.076		31.50	3.10		34.60	39
	2200		500 kcmil		270	.089		38.50	3.62		42.12	48
	2400		15 kV, copper, 3 conductor with PVC jacket galv steel armored									
	2500		grounded neutral, in cable tray, #2	2 Elec	300	.053	L.F.	13.30	2.17		15.47	17.85
	2600		#1		280	.057		14.15	2.33		16.48	19
	2800		1/0		260	.062		16.35	2.51		18.86	21.50
	2900		2/0		220	.073		21.50	2.96		24.46	28
	3000		4/0		190	.084		24.50	3.43		27.93	32
	3100		250 kcmil	3 Elec	270	.089		27	3.62		30.62	35.50
	3150		350 kcmil		240	.100		32	4.08		36.08	41
	3200		500 kcmil		210	.114		43	4.66		47.66	54.50
	3400		15 kV, copper, 3 conductor with PVC jacket,									
	3450		ungrounded neutral, in cable tray, #2	2 Elec	260	.062	L.F.	14.30	2.51		16.81	19.50
	3500		#1		230	.070		15.85	2.84		18.69	21.50
	3600		1/0		200	.080		18.15	3.26		21.41	25
	3700		2/0		190	.084		22.50	3.43		25.93	29.50
	3800		4/0		160	.100		27	4.08		31.08	35.50
	4000		250 kcmil	3 Elec	210	.114		31.50	4.66		36.16	41.50
	4050		350 kcmil		195	.123		41.50	5		46.50	53
	4100		500 kcmil		180	.133		50.50	5.45		55.95	63.50
	9010		600 volt, copper (MC) steel clad, #14, 2 wire	1 Elec	2.40	3.333	C.L.F.	61.50	136		197.50	270
	9020		3 wire		2.20	3.636		95.50	148		243.50	325
	9040		#12, 2 wire		2.30	3.478		62.50	142		204.50	280
	9050		3 wire		2	4		96.50	163		259.50	350
	9070		#10, 2 wire		2	4		111	163		274	365
	9080		3 wire		1.60	5		171	204		375	495
	9100		#8, 2 wire, stranded		1.80	4.444		210	181		391	500
	9110		3 wire, stranded		1.30	6.154		325	251		576	730
	9200		600 volt, copper (MC) alum. clad, #14, 2 wire		2.65	3.019		61.50	123		184.50	251

Important: See the Reference Section for critical supporting data - Reference Nos., Crews, & City Cost Indexes

16120 | Conductors & Cables

			CREW	DAILY OUTPUT	LABOR-HOURS	UNIT	2005 BARE COSTS MAT.	LABOR	EQUIP.	TOTAL	TOTAL INCL O&P	
120	9210	3 wire	1 Elec	2.45	3.265	C.L.F.	95.50	133		228.50	305	**120**
	9220	4 wire		2.20	3.636		128	148		276	360	
	9230	#12, 2 wire		2.55	3.137		62.50	128		190.50	259	
	9240	3 wire		2.20	3.636		96.50	148		244.50	325	
	9250	4 wire		2	4		131	163		294	385	
	9260	#10, 2 wire		2.20	3.636		113	148		261	345	
	9270	3 wire		1.80	4.444		176	181		357	465	
	9280	4 wire		1.55	5.161		277	210		487	620	
230	0010	**CABLE TERMINATIONS**										**230**
	0015	Wire connectors, screw type, #22 to #14	1 Elec	260	.031	Ea.	.07	1.25		1.32	1.95	
	0020	#18 to #12		240	.033		.08	1.36		1.44	2.11	
	0025	#18 to #10		240	.033		.12	1.36		1.48	2.15	
	0030	Screw-on connectors, insulated, #18 to #12		240	.033		.09	1.36		1.45	2.12	
	0035	#16 to #10		230	.035		.11	1.42		1.53	2.23	
	0040	#14 to #8		210	.038		.24	1.55		1.79	2.57	
	0045	#12 to #6		180	.044		.40	1.81		2.21	3.14	
	0050	Terminal lugs, solderless, #16 to #10		50	.160		.46	6.50		6.96	10.20	
	0100	#8 to #4		30	.267		.64	10.85		11.49	16.85	
	0150	#2 to #1		22	.364		.86	14.80		15.66	23	
	0200	1/0 to 2/0		16	.500		2.22	20.50		22.72	33	
	0250	3/0		12	.667		2.95	27		29.95	44	
	0300	4/0		11	.727		2.95	29.50		32.45	47.50	
	0350	250 kcmil		9	.889		2.95	36		38.95	57.50	
	0400	350 kcmil		7	1.143		3.80	46.50		50.30	73.50	
	0450	500 kcmil		6	1.333		7.40	54.50		61.90	89	
	1600	Crimp 1 hole lugs, copper or aluminum, 600 volt										
	1620	#14	1 Elec	60	.133	Ea.	.32	5.45		5.77	8.45	
	1630	#12		50	.160		.45	6.50		6.95	10.20	
	1640	#10		45	.178		.45	7.25		7.70	11.30	
	1780	#8		36	.222		1.72	9.05		10.77	15.40	
	1800	#6		30	.267		2.78	10.85		13.63	19.20	
	2000	#4		27	.296		3	12.05		15.05	21.50	
	2200	#2		24	.333		4.07	13.60		17.67	24.50	
	2400	#1		20	.400		4.90	16.30		21.20	30	
	2600	2/0		15	.533		6.60	21.50		28.10	40	
	2800	3/0		12	.667		8.20	27		35.20	49.50	
	3000	4/0		11	.727		8.90	29.50		38.40	54	
	3200	250 kcmil		9	.889		10.60	36		46.60	65.50	
	3400	300 kcmil		8	1		11.90	41		52.90	73.50	
	3500	350 kcmil		7	1.143		12.30	46.50		58.80	83	
	3600	400 kcmil		6.50	1.231		15.30	50		65.30	91.50	
	3800	500 kcmil		6	1.333		18.60	54.50		73.10	102	
280	0010	**CONTROL CABLE**										**280**
	0020	600 volt, copper, #14 THWN wire with PVC jacket, 2 wires	1 Elec	9	.889	C.L.F.	14.95	36		50.95	70.50	
	0100	4 wires		7	1.143		24.50	46.50		71	96.50	
	0200	6 wires		6	1.333		44	54.50		98.50	130	
	0300	8 wires		5.30	1.509		54	61.50		115.50	151	
	0400	10 wires		4.80	1.667		63.50	68		131.50	171	
	0500	12 wires		4.30	1.860		72.50	76		148.50	193	
	0600	14 wires		3.80	2.105		82.50	86		168.50	219	
	0700	16 wires		3.50	2.286		97	93		190	246	
	0800	18 wires		3.30	2.424		101	99		200	258	
	0900	20 wires		3	2.667		124	109		233	298	
	1000	22 wires		2.80	2.857		127	116		243	315	
500	0010	**MINERAL INSULATED CABLE** 600 volt										**500**
	0100	1 conductor, #12	1 Elec	1.60	5	C.L.F.	225	204		429	555	

ELECTRICAL 16

16120 | Conductors & Cables

			CREW	DAILY OUTPUT	LABOR-HOURS	UNIT	2005 BARE COSTS				TOTAL INCL O&P	
							MAT.	LABOR	EQUIP.	TOTAL		
500	0200	#10	1 Elec	1.60	5	C.L.F.	293	204		497	625	**500**
	0400	#8		1.50	5.333		345	217		562	705	
	0500	#6		1.40	5.714		385	233		618	770	
	0600	#4	2 Elec	2.40	6.667		535	272		807	990	
	0800	#2		2.20	7.273		715	296		1,011	1,225	
	0900	#1		2.10	7.619		815	310		1,125	1,350	
	1000	1/0		2	8		940	325		1,265	1,500	
	1100	2/0		1.90	8.421		1,125	345		1,470	1,725	
	1200	3/0		1.80	8.889		1,325	360		1,685	2,000	
	1400	4/0		1.60	10		1,525	410		1,935	2,275	
	1410	250 kcmil	3 Elec	2.40	10		1,725	410		2,135	2,500	
	1420	350 kcmil		1.95	12.308		2,100	500		2,600	3,075	
	1430	500 kcmil		1.95	12.308		2,625	500		3,125	3,650	
550	0010	**NON-METALLIC SHEATHED CABLE** 600 volt										**550**
	0100	Copper with ground wire, (Romex)										
	0150	#14, 2 conductor	1 Elec	2.70	2.963	C.L.F.	17.20	121		138.20	199	
	0200	3 conductor		2.40	3.333		29	136		165	234	
	0250	#12, 2 conductor		2.50	3.200		25.50	130		155.50	222	
	0300	3 conductor		2.20	3.636		42	148		190	268	
	0350	#10, 2 conductor		2.20	3.636		51.50	148		199.50	278	
	0400	3 conductor		1.80	4.444		66	181		247	345	
	0450	#8, 3 conductor		1.50	5.333		124	217		341	460	
	0500	#6, 3 conductor		1.40	5.714		192	233		425	555	
	0550	SE type SER aluminum cable, 3 RHW and										
	0600	1 bare neutral, 3 #8 & 1 #8	1 Elec	1.60	5	C.L.F.	112	204		316	430	
	0650	3 #6 & 1 #6	"	1.40	5.714		127	233		360	485	
	0700	3 #4 & 1 #6	2 Elec	2.40	6.667		142	272		414	560	
	0750	3 #2 & 1 #4		2.20	7.273		209	296		505	670	
	0800	3 #1/0 & 1 #2		2	8		315	325		640	835	
	0850	3 #2/0 & 1 #1		1.80	8.889		370	360		730	950	
	0900	3 #4/0 & 1 #2/0		1.60	10		530	410		940	1,200	
700	0010	**SHIELDED CABLE** Splicing & terminations not included										**700**
	0040	Copper, XLP shielding, 5 kV, #6	2 Elec	4.40	3.636	C.L.F.	109	148		257	340	
	0050	#4		4.40	3.636		142	148		290	375	
	0100	#2		4	4		168	163		331	425	
	0200	#1		4	4		192	163		355	455	
	0400	1/0		3.80	4.211		212	172		384	490	
	0600	2/0		3.60	4.444		267	181		448	565	
	0800	4/0		3.20	5		345	204		549	685	
	1000	250 kcmil	3 Elec	4.50	5.333		430	217		647	795	
	1200	350 kcmil		3.90	6.154		555	251		806	985	
	1400	500 kcmil		3.60	6.667		675	272		947	1,150	
	1600	15 kV, ungrounded neutral, #1	2 Elec	4	4		234	163		397	500	
	1800	1/0		3.80	4.211		280	172		452	560	
	2000	2/0		3.60	4.444		320	181		501	620	
	2200	4/0		3.20	5		425	204		629	770	
	2400	250 kcmil	3 Elec	4.50	5.333		470	217		687	840	
	2600	350 kcmil		3.90	6.154		590	251		841	1,025	
	2800	500 kcmil		3.60	6.667		730	272		1,002	1,200	
900	0010	**WIRE** R16120-910										**900**
	0020	600 volt type THW, copper, solid, #14	1 Elec	13	.615	C.L.F.	3.80	25		28.80	41.50	
	0030	#12		11	.727		5.60	29.50		35.10	50	
	0040	#10		10	.800		8.80	32.50		41.30	58	
	0050	Stranded, #14 R16132-210		13	.615		4.70	25		29.70	42.50	
	0100	#12		11	.727		6.90	29.50		36.40	51.50	

Important: See the Reference Section for critical supporting data - Reference Nos., Crews, & City Cost Indexes

16 ELECTRICAL

16120 | Conductors & Cables

			CREW	DAILY OUTPUT	LABOR-HOURS	UNIT	2005 BARE COSTS MAT.	LABOR	EQUIP.	TOTAL	TOTAL INCL O&P	
900	0120	#10	1 Elec	10	.800	C.L.F.	10.45	32.50		42.95	60	900
	0140	#8		8	1		16.10	41		57.10	78	
	0160	#6	▼	6.50	1.231		26.50	50		76.50	104	
	0180	#4	2 Elec	10.60	1.509		40.50	61.50		102	136	
	0200	#3		10	1.600		50	65		115	152	
	0220	#2		9	1.778		63	72.50		135.50	178	
	0240	#1		8	2		80	81.50		161.50	209	
	0260	1/0		6.60	2.424		94	99		193	250	
	0280	2/0		5.80	2.759		118	112		230	296	
	0300	3/0		5	3.200		146	130		276	355	
	0350	4/0	▼	4.40	3.636		181	148		329	420	
	0400	250 kcmil	3 Elec	6	4		216	163		379	480	
	0420	300 kcmil		5.70	4.211		256	172		428	535	
	0450	350 kcmil		5.40	4.444		297	181		478	595	
	0480	400 kcmil		5.10	4.706		340	192		532	660	
	0490	500 kcmil	▼	4.80	5		420	204		624	765	
	0530	Aluminum, stranded, #8	1 Elec	9	.889		15.10	36		51.10	70.50	
	0540	#6	"	8	1		20.50	41		61.50	83.50	
	0560	#4	2 Elec	13	1.231		25.50	50		75.50	103	
	0580	#2		10.60	1.509		35	61.50		96.50	130	
	0600	#1		9	1.778		51	72.50		123.50	164	
	0620	1/0		8	2		61	81.50		142.50	189	
	0640	2/0		7.20	2.222		72	90.50		162.50	214	
	0680	3/0		6.60	2.424		89.50	99		188.50	246	
	0700	4/0	▼	6.20	2.581		99.50	105		204.50	267	
	0720	250 kcmil	3 Elec	8.70	2.759		122	112		234	300	
	0740	300 kcmil		8.10	2.963		168	121		289	365	
	0760	350 kcmil		7.50	3.200		171	130		301	380	
	0780	400 kcmil		6.90	3.478		200	142		342	430	
	0800	500 kcmil		6	4		220	163		383	485	
	0850	600 kcmil		5.70	4.211		279	172		451	560	
	0880	700 kcmil		5.10	4.706		320	192		512	640	
	0900	750 kcmil	▼	4.80	5		325	204		529	665	
	0920	Type THWN-THHN, copper, solid, #14	1 Elec	13	.615		3.80	25		28.80	41.50	
	0940	#12		11	.727		5.60	29.50		35.10	50	
	0960	#10		10	.800		8.80	32.50		41.30	58	
	1000	Stranded, #14		13	.615		4.70	25		29.70	42.50	
	1200	#12		11	.727		6.90	29.50		36.40	51.50	
	1250	#10		10	.800		10.45	32.50		42.95	60	
	1300	#8		8	1		16.10	41		57.10	78	
	1350	#6	▼	6.50	1.231		26.50	50		76.50	104	
	1400	#4	2 Elec	10.60	1.509		40.50	61.50		102	136	

R16120 -910 (at rows 0120/0140)
R16132 -210 (at rows 0160/0180)
CN (at row 1250)

16131 | Cable Trays

			CREW	DAILY OUTPUT	LABOR-HOURS	UNIT	MAT.	LABOR	EQUIP.	TOTAL	TOTAL INCL O&P	
105	0010	**CABLE TRAY LADDER TYPE** w/ ftngs & supports, 4" dp., to 15' elev.										105
	0160	Galvanized steel tray										
	0170	4" rung spacing, 6" wide	2 Elec	98	.163	L.F.	10.60	6.65		17.25	21.50	
	0200	12" wide		86	.186		12.75	7.60		20.35	25.50	
	0400	18" wide		82	.195		14.80	7.95		22.75	28	
	0600	24" wide		78	.205		17.05	8.35		25.40	31	
	3200	Aluminum tray, 4" deep, 6" rung spacing, 6" wide		134	.119		12.80	4.87		17.67	21.50	
	3220	12" wide		124	.129		14.30	5.25		19.55	23.50	
	3230	18" wide		114	.140		15.95	5.70		21.65	26	
	3240	24" wide	▼	106	.151	▼	18.55	6.15		24.70	29.50	
	9980	Allow. for tray ftngs., 5% min.-20% max.										

ELECTRICAL 16

16132 | Conduit & Tubing

		CREW	DAILY OUTPUT	LABOR-HOURS	UNIT	2005 BARE COSTS				TOTAL INCL O&P
						MAT.	LABOR	EQUIP.	TOTAL	
205										**205**
0010	**CONDUIT** To 15' high, includes 2 terminations, 2 elbows and									
0020	11 beam clamps per 100 L.F.									
0300	Aluminum, 1/2" diameter	1 Elec	100	.080	L.F.	1.83	3.26		5.09	6.85
0500	3/4" diameter		90	.089		2.45	3.62		6.07	8.10
0700	1" diameter		80	.100		3.25	4.08		7.33	9.60
1000	1-1/4" diameter		70	.114		4.35	4.66		9.01	11.75
1030	1-1/2" diameter		65	.123		5.20	5		10.20	13.20
1050	2" diameter		60	.133		7.10	5.45		12.55	15.90
1070	2-1/2" diameter		50	.160		11.20	6.50		17.70	22
1100	3" diameter	2 Elec	90	.178		15.20	7.25		22.45	27.50
1130	3-1/2" diameter		80	.200		19.65	8.15		27.80	33.50
1140	4" diameter		70	.229		23.50	9.30		32.80	40
1750	Rigid galvanized steel, 1/2" diameter	1 Elec	90	.089		2.60	3.62		6.22	8.25
1770	3/4" diameter		80	.100		2.96	4.08		7.04	9.30
1800	1" diameter		65	.123		4.40	5		9.40	12.30
1830	1-1/4" diameter		60	.133		6	5.45		11.45	14.70
1850	1-1/2" diameter		55	.145		6.90	5.95		12.85	16.40
1870	2" diameter		45	.178		8.85	7.25		16.10	20.50
1900	2-1/2" diameter		35	.229		16.50	9.30		25.80	32
1930	3" diameter	2 Elec	50	.320		20	13.05		33.05	41.50
1950	3-1/2" diameter		44	.364		24.50	14.80		39.30	49
1970	4" diameter		40	.400		28	16.30		44.30	55.50
2500	Steel, intermediate conduit (IMC), 1/2" diameter	1 Elec	100	.080		2.06	3.26		5.32	7.10
2530	3/4" diameter		90	.089		2.42	3.62		6.04	8.05
2550	1" diameter		70	.114		3.51	4.66		8.17	10.80
2570	1-1/4" diameter		65	.123		4.60	5		9.60	12.50
2600	1-1/2" diameter		60	.133		5.35	5.45		10.80	14
2630	2" diameter		50	.160		6.85	6.50		13.35	17.20
2650	2-1/2" diameter		40	.200		14	8.15		22.15	27.50
2670	3" diameter	2 Elec	60	.267		18.30	10.85		29.15	36
2700	3-1/2" diameter		54	.296		22.50	12.05		34.55	42.50
2730	4" diameter		50	.320		25.50	13.05		38.55	47.50
5000	Electric metallic tubing (EMT), 1/2" diameter	1 Elec	170	.047		.63	1.92		2.55	3.54
5020	3/4" diameter **CN**		130	.062		1.06	2.51		3.57	4.89
5040	1" diameter		115	.070		1.86	2.84		4.70	6.25
5060	1-1/4" diameter		100	.080		2.89	3.26		6.15	8
5080	1-1/2" diameter		90	.089		3.66	3.62		7.28	9.40
5100	2" diameter		80	.100		4.63	4.08		8.71	11.15
5120	2-1/2" diameter		60	.133		8.65	5.45		14.10	17.60
5140	3" diameter	2 Elec	100	.160		11.65	6.50		18.15	22.50
5160	3-1/2" diameter		90	.178		15.10	7.25		22.35	27.50
5180	4" diameter		80	.200		16.05	8.15		24.20	30
9900	Add to labor for higher elevated installation									
9910	15' to 20' high, add						10%			
9920	20' to 25' high, add						20%			
9930	25' to 30' high, add						25%			
9940	30' to 35' high, add						30%			
9950	35' to 40' high, add						35%			
9960	Over 40' high, add						40%			
9980	Allow. for cond. ftngs., 5% min.-20% max.									
230	0010	**CONDUIT IN CONCRETE SLAB** Including terminations,								**230**
	0020	fittings and supports								
	3230	PVC, schedule 40, 1/2" diameter	1 Elec	270	.030	L.F.	.49	1.21		1.70	2.34
	3250	3/4" diameter		230	.035		.58	1.42		2	2.75
	3270	1" diameter		200	.040		.77	1.63		2.40	3.28
	3300	1-1/4" diameter		170	.047		1.09	1.92		3.01	4.05

R16132-210

16 ELECTRICAL

16132	Conduit & Tubing	CREW	DAILY OUTPUT	LABOR-HOURS	UNIT	2005 BARE COSTS				TOTAL INCL O&P	
						MAT.	LABOR	EQUIP.	TOTAL		
230 3330	1-1/2" diameter	1 Elec	140	.057	L.F.	1.34	2.33		3.67	4.95	230
3350	2" diameter		120	.067		1.70	2.72		4.42	5.90	
4350	Rigid galvanized steel, 1/2" diameter		200	.040		2.36	1.63		3.99	5	
4400	3/4" diameter		170	.047		2.72	1.92		4.64	5.85	
4450	1" diameter		130	.062		4.15	2.51		6.66	8.30	
4500	1-1/4" diameter		110	.073		5.55	2.96		8.51	10.50	
4600	1-1/2" diameter		100	.080		6.45	3.26		9.71	11.95	
4800	2" diameter	▼	90	.089	▼	8.25	3.62		11.87	14.45	
240 0010	CONDUIT IN TRENCH Includes terminations and fittings										240
0020	Does not include excavation or backfill, see div. 02315										
0200	Rigid galvanized steel, 2" diameter	1 Elec	150	.053	L.F.	8	2.17		10.17	12.05	
0400	2-1/2" diameter	"	100	.080		15.45	3.26		18.71	22	
0600	3" diameter	2 Elec	160	.100		18.90	4.08		22.98	27	
0800	3-1/2" diameter		140	.114		23.50	4.66		28.16	33	
1000	4" diameter		100	.160		26.50	6.50		33	38.50	
1200	5" diameter		80	.200		55.50	8.15		63.65	73.50	
1400	6" diameter	▼	60	.267	▼	81.50	10.85		92.35	106	

16133	Multi-outlet Assemblies										
540 0010	TRENCH DUCT Steel with cover										540
0020	Standard adjustable, depths to 4"										
0100	Straight, single compartment, 9" wide	2 Elec	40	.400	L.F.	80	16.30		96.30	113	
0200	12" wide		32	.500		91	20.50		111.50	131	
0400	18" wide		26	.615		121	25		146	171	
0600	24" wide		22	.727		155	29.50		184.50	215	
0800	30" wide		20	.800		187	32.50		219.50	255	
1000	36" wide		16	1	▼	225	41		266	310	
1200	Horizontal elbow, 9" wide		5.40	2.963	Ea.	296	121		417	505	
1400	12" wide		4.60	3.478		340	142		482	585	
1600	18" wide		4	4		440	163		603	725	
1800	24" wide		3.20	5		615	204		819	980	
2000	30" wide		2.60	6.154		820	251		1,071	1,275	
2200	36" wide		2.40	6.667		1,075	272		1,347	1,575	
2400	Vertical elbow, 9" wide		5.40	2.963		103	121		224	293	
2600	12" wide		4.60	3.478		112	142		254	335	
2800	18" wide		4	4		129	163		292	385	
3000	24" wide		3.20	5		160	204		364	480	
3200	30" wide		2.60	6.154		177	251		428	570	
3400	36" wide		2.40	6.667		194	272		466	620	
3600	Cross, 9" wide		4	4		485	163		648	780	
3800	12" wide		3.20	5		515	204		719	870	
4000	18" wide		2.60	6.154		615	251		866	1,050	
4200	24" wide		2.20	7.273		780	296		1,076	1,300	
4400	30" wide		2	8		1,000	325		1,325	1,600	
4600	36" wide		1.80	8.889		1,275	360		1,635	1,950	
4800	End closure, 9" wide		14.40	1.111		30.50	45.50		76	101	
5000	12" wide		12	1.333		34.50	54.50		89	119	
5200	18" wide		10	1.600		53	65		118	156	
5400	24" wide		8	2		70	81.50		151.50	198	
5600	30" wide		6.60	2.424		87.50	99		186.50	244	
5800	36" wide		5.80	2.759		104	112		216	281	
6000	Tees, 9" wide		4	4		295	163		458	570	
6200	12" wide		3.60	4.444		340	181		521	645	
6400	18" wide		3.20	5		440	204		644	785	
6600	24" wide	▼	3	5.333	▼	630	217		847	1,025	

ELECTRICAL 16

For expanded coverage of these items see *Means Electrical Cost Data 2005*

			CREW	DAILY OUTPUT	LABOR-HOURS	UNIT	2005 BARE COSTS				TOTAL INCL O&P	
	16133	**Multi-outlet Assemblies**					MAT.	LABOR	EQUIP.	TOTAL		
540	6800	30" wide	2 Elec	2.60	6.154	Ea.	820	251		1,071	1,275	**540**
	7000	36" wide		2	8		1,075	325		1,400	1,650	
	7200	Riser, and cabinet connector, 9" wide		5.40	2.963		129	121		250	320	
	7400	12" wide		4.60	3.478		150	142		292	375	
	7600	18" wide		4	4		184	163		347	445	
	7800	24" wide		3.20	5		223	204		427	550	
	8000	30" wide		2.60	6.154		257	251		508	660	
	8200	36" wide		2	8		299	325		624	815	
	8400	Insert assembly, cell to conduit adapter, 1-1/4"	1 Elec	16	.500		51	20.50		71.50	86.50	
560	0010	**UNDERFLOOR DUCT**										**560**
	0100	Duct, 1-3/8" x 3-1/8" blank, standard	2 Elec	160	.100	L.F.	9.75	4.08		13.83	16.80	
	0200	1-3/8" x 7-1/4" blank, super duct		120	.133		19.50	5.45		24.95	29.50	
	0400	7/8" or 1-3/8" insert type, 24" O.C., 1-3/8" x 3-1/8", std.		140	.114		13.10	4.66		17.76	21.50	
	0600	1-3/8" x 7-1/4", super duct		100	.160		23	6.50		29.50	34.50	
	0800	Junction box, single duct, 1 level, 3-1/8"	1 Elec	4	2	Ea.	294	81.50		375.50	445	
	1000	Junction box, single duct, 1 level, 7-1/4"		2.70	2.963		340	121		461	555	
	1200	1 level, 2 duct, 3-1/8"		3.20	2.500		390	102		492	580	
	1400	Junction box, 1 level, 2 duct, 7-1/4"		2.30	3.478		1,000	142		1,142	1,300	
	1600	Triple duct, 3-1/8"		2.30	3.478		665	142		807	940	
	1800	Insert to conduit adapter, 3/4" & 1"		32	.250		24	10.20		34.20	41.50	
	2000	Support, single cell		27	.296		36	12.05		48.05	57.50	
	2200	Super duct		16	.500		36	20.50		56.50	70	
	2400	Double cell		16	.500		36	20.50		56.50	70	
	2600	Triple cell		11	.727		36	29.50		65.50	83.50	
	2800	Vertical elbow, standard duct		10	.800		64.50	32.50		97	120	
	3000	Super duct		8	1		64.50	41		105.50	132	
	3200	Cabinet connector, standard duct		32	.250		49	10.20		59.20	68.50	
	3400	Super duct		27	.296		49	12.05		61.05	71.50	
	3600	Conduit adapter, 1" to 1-1/4"		32	.250		49	10.20		59.20	68.50	
	3800	2" to 1-1/4"		27	.296		58.50	12.05		70.55	82	
	4000	Outlet, low tension (tele, computer, etc.)		8	1		68.50	41		109.50	136	
	4200	High tension, receptacle (120 volt)		8	1		68.50	41		109.50	136	
800	0010	**SURFACE RACEWAY**										**800**
	0090	Metal, straight section										
	0100	No. 500	1 Elec	100	.080	L.F.	.81	3.26		4.07	5.75	
	0110	No. 700		100	.080		.91	3.26		4.17	5.85	
	0400	No. 1500, small pancake		90	.089		1.65	3.62		5.27	7.20	
	0600	No. 2000, base & cover, blank		90	.089		1.61	3.62		5.23	7.15	
	0800	No. 3000, base & cover, blank		75	.107		3.25	4.35		7.60	10.05	
	1000	No. 4000, base & cover, blank		65	.123		5.25	5		10.25	13.25	
	1200	No. 6000, base & cover, blank		50	.160		8.80	6.50		15.30	19.40	
	2400	Fittings, elbows, No. 500		40	.200	Ea.	1.47	8.15		9.62	13.75	
	2800	Elbow cover, No. 2000		40	.200		2.80	8.15		10.95	15.25	
	2880	Tee, No. 500		42	.190		2.82	7.75		10.57	14.65	
	2900	No. 2000		27	.296		8.80	12.05		20.85	27.50	
	3000	Switch box, No. 500		16	.500		9.55	20.50		30.05	41	
	3400	Telephone outlet, No. 1500		16	.500		10.60	20.50		31.10	42	
	3600	Junction box, No. 1500		16	.500		7.40	20.50		27.90	38.50	
	3800	Plugmold wired sections, No. 2000										
	4000	1 circuit, 6 outlets, 3 ft. long	1 Elec	8	1	Ea.	27	41		68	90	
	4100	2 circuits, 8 outlets, 6 ft. long	"	5.30	1.509	"	45	61.50		106.50	141	

Important: See the Reference Section for critical supporting data - Reference Nos., Crews, & City Cost Indexes

16 ELECTRICAL

16134 | Wireway & Aux Gutters

		CREW	DAILY OUTPUT	LABOR-HOURS	UNIT	2005 BARE COSTS				TOTAL INCL O&P
						MAT.	LABOR	EQUIP.	TOTAL	
150	0010 **WIREWAY** to 15' high									150
	0100 Screw cover, NEMA 1 w/ fittings and supports, 2-1/2" x 2-1/2"	1 Elec	45	.178	L.F.	10.95	7.25		18.20	23
	0200 4" x 4"	"	40	.200		11.60	8.15		19.75	25
	0400 6" x 6"	2 Elec	60	.267		19.70	10.85		30.55	37.50
	0600 8" x 8"	"	40	.400		33	16.30		49.30	61
	4475 Screw cover, NEMA 3R w/ fittings and supports, 4" x 4"	1 Elec	36	.222		24.50	9.05		33.55	40.50
	4480 6" x 6"	2 Elec	55	.291		33	11.85		44.85	53.50
	4485 8" x 8"		36	.444		53	18.10		71.10	85
	4490 12" x 12"	▼	18	.889	▼	73.50	36		109.50	135
	9980 Allow. for wireway ftngs., 5% min.-20% max.									

16136 | Boxes

		CREW	DAILY OUTPUT	LABOR-HOURS	UNIT	2005 BARE COSTS				TOTAL INCL O&P
						MAT.	LABOR	EQUIP.	TOTAL	
600	0010 **OUTLET BOXES**									600
	0020 Pressed steel, octagon, 4"	1 Elec	20	.400	Ea.	1.82	16.30		18.12	26.50
	0060 Covers, blank		64	.125		.76	5.10		5.86	8.45
	0100 Extension rings		40	.200		3.01	8.15		11.16	15.45
	0150 Square, 4"		20	.400		2.14	16.30		18.44	27
	0200 Extension rings		40	.200		3.05	8.15		11.20	15.50
	0250 Covers, blank		64	.125		.86	5.10		5.96	8.55
	0300 Plaster rings		64	.125		1.68	5.10		6.78	9.45
	0650 Switchbox		27	.296		2.94	12.05		14.99	21
	1100 Concrete, floor, 1 gang		5.30	1.509		67.50	61.50		129	166
	2000 Poke-thru fitting, fire rated, for 3-3/4" floor		6.80	1.176		92.50	48		140.50	174
	2040 For 7" floor		6.80	1.176		92.50	48		140.50	174
	2100 Pedestal, 15 amp, duplex receptacle & blank plate		5.25	1.524		96	62		158	198
	2120 Duplex receptacle and telephone plate		5.25	1.524		96.50	62		158.50	199
	2140 Pedestal, 20 amp, duplex recept. & phone plate		5	1.600		97	65		162	203
	2200 Abandonment plate	▼	32	.250	▼	28	10.20		38.20	46
700	0010 **PULL BOXES & CABINETS**									700
	0100 Sheet metal, pull box, NEMA 1, type SC, 6" W x 6" H x 4" D	1 Elec	8	1	Ea.	9.75	41		50.75	71.50
	0200 8" W x 8" H x 4" D		8	1		13.35	41		54.35	75
	0300 10" W x 12" H x 6" D		5.30	1.509		23.50	61.50		85	118
	0400 16" W x 20" H x 8" D		4	2		88.50	81.50		170	219
	0500 20" W x 24" H x 8" D		3.20	2.500		104	102		206	266
	0600 24" W x 36" H x 8" D		2.70	2.963		147	121		268	340
	0650 Hinged cabinets, NEMA 1, 6" W x 6" H x 4" D		8	1		9.90	41		50.90	71.50
	0800 12" W x 16" H x 6" D		4.70	1.702		32.50	69.50		102	139
	1000 20" W x 20" H x 6" D		3.60	2.222		65.50	90.50		156	207
	1200 20" W x 20" H x 8" D		3.20	2.500		131	102		233	296
	1400 24" W x 36" H x 8" D		2.70	2.963		210	121		331	410
	1600 24" W x 42" H x 8" D	▼	2	4	▼	320	163		483	595
	2100 Pull box, NEMA 3R, type SC, raintight & weatherproof									
	2150 6" L x 6" W x 6" D	1 Elec	10	.800	Ea.	16.70	32.50		49.20	67
	2200 8" L x 6" W x 6" D		8	1		21	41		62	83.50
	2250 10" L x 6" W x 6" D		7	1.143		27.50	46.50		74	100
	2300 12" L x 12" W x 6" D		5	1.600		39	65		104	140
	2350 16" L x 16" W x 6" D		4.50	1.778		78	72.50		150.50	194
	2400 20" L x 20" W x 6" D		4	2		107	81.50		188.50	239
	2450 24" L x 18" W x 8" D		3	2.667		117	109		226	291
	2500 24" L x 24" W x 10" D		2.50	3.200		157	130		287	365
	2550 30" L x 24" W x 12" D		2	4		286	163		449	560
	2600 36" L x 36" W x 12" D	▼	1.50	5.333	▼	375	217		592	740
	2800 Cast iron, pull boxes for surface mounting									
	3000 NEMA 4, watertight & dust tight									
	3050 6" L x 6" W x 6" D	1 Elec	4	2	Ea.	158	81.50		239.50	295
	3100 8" L x 6" W x 6" D	▼	3.20	2.500	▼	214	102		316	390

ELECTRICAL 16

16136 | Boxes

		CREW	DAILY OUTPUT	LABOR-HOURS	UNIT	2005 BARE COSTS				TOTAL INCL O&P
						MAT.	LABOR	EQUIP.	TOTAL	
700	3150	10" L x 6" W x 6" D	1 Elec	2.50	3.200	Ea.	266	130		396	485	700
	3200	12" L x 12" W x 6" D		2.30	3.478		450	142		592	705
	3250	16" L x 16" W x 6" D		1.30	6.154		920	251		1,171	1,400
	3300	20" L x 20" W x 6" D		.80	10		1,725	410		2,135	2,475
	3350	24" L x 18" W x 8" D		.70	11.429		1,800	465		2,265	2,675
	3400	24" L x 24" W x 10" D		.50	16		2,750	650		3,400	4,000
	3450	30" L x 24" W x 12" D		.40	20		4,625	815		5,440	6,325
	3500	36" L x 36" W x 12" D		.20	40		5,725	1,625		7,350	8,725
	6000	J.I.C. wiring boxes, NEMA 12, dust tight & drip tight									
	6050	6" L x 8" W x 4" D	1 Elec	10	.800	Ea.	39	32.50		71.50	91.50
	6100	8" L x 10" W x 4" D		8	1		48.50	41		89.50	114
	6150	12" L x 14" W x 6" D		5.30	1.509		79	61.50		140.50	179
	6200	14" L x 16" W x 6" D		4.70	1.702		94.50	69.50		164	207
	6250	16" L x 20" W x 6" D		4.40	1.818		201	74		275	330
	6300	24" L x 30" W x 6" D		3.20	2.500		296	102		398	475
	6350	24" L x 30" W x 8" D		2.90	2.759		315	112		427	515
	6400	24" L x 36" W x 8" D		2.70	2.963		350	121		471	565
	6450	24" L x 42" W x 8" D		2.30	3.478		385	142		527	630
	6500	24" L x 48" W x 8" D		2	4		415	163		578	705
	7000	Cabinets, current transformer									
	7050	Single door, 24" H x 24" W x 10" D	1 Elec	1.60	5	Ea.	125	204		329	445
	7100	30" H x 24" W x 10" D		1.30	6.154		138	251		389	525
	7150	36" H x 24" W x 10" D		1.10	7.273		145	296		441	600
	7200	30" H x 30" W x 10" D		1	8		150	325		475	650
	7250	36" H x 30" W x 10" D		.90	8.889		207	360		567	770
	7300	36" H x 36" W x 10" D		.80	10		220	410		630	845
	7500	Double door, 48" H x 36" W x 10" D		.60	13.333		415	545		960	1,275
	7550	24" H x 24" W x 12" D		1	8		207	325		532	715

16139 | Residential Wiring

		CREW	DAILY OUTPUT	LABOR-HOURS	UNIT	2005 BARE COSTS				TOTAL INCL O&P
						MAT.	LABOR	EQUIP.	TOTAL	
700	0010	**RESIDENTIAL WIRING**									700
	0020	20' avg. runs and #14/2 wiring incl. unless otherwise noted									
	1000	Service & panel, includes 24' SE-AL cable, service eye, meter,									
	1010	Socket, panel board, main bkr., ground rod, 15 or 20 amp									
	1020	1-pole circuit breakers, and misc. hardware									
	1100	100 amp, with 10 branch breakers	1 Elec	1.19	6.723	Ea.	445	274		719	895
	1110	With PVC conduit and wire		.92	8.696		485	355		840	1,050
	1120	With RGS conduit and wire		.73	10.959		675	445		1,120	1,400
	1150	150 amp, with 14 branch breakers		1.03	7.767		690	315		1,005	1,225
	1170	With PVC conduit and wire		.82	9.756		775	400		1,175	1,450
	1180	With RGS conduit and wire		.67	11.940		1,125	485		1,610	1,975
	1200	200 amp, with 18 branch breakers	2 Elec	1.80	8.889		910	360		1,270	1,550
	1220	With PVC conduit and wire		1.46	10.959		990	445		1,435	1,775
	1230	With RGS conduit and wire		1.24	12.903		1,450	525		1,975	2,375
	1800	Lightning surge suppressor for above services, add	1 Elec	32	.250		41.50	10.20		51.70	60.50
	2000	Switch devices									
	2100	Single pole, 15 amp, Ivory, with a 1-gang box, cover plate,									
	2110	Type NM (Romex) cable	1 Elec	17.10	.468	Ea.	8.10	19.05		27.15	37.50
	2120	Type MC (BX) cable		14.30	.559		20.50	23		43.50	56.50
	2130	EMT & wire		5.71	1.401		29	57		86	117
	2150	3-way, #14/3, type NM cable		14.55	.550		12	22.50		34.50	46.50
	2170	Type MC cable		12.31	.650		29	26.50		55.50	71.50
	2180	EMT & wire		5	1.600		31	65		96	132
	2200	4-way, #14/3, type NM cable		14.55	.550		25.50	22.50		48	61.50
	2220	Type MC cable		12.31	.650		42.50	26.50		69	86.50
	2230	EMT & wire		5	1.600		45	65		110	146

Important: See the Reference Section for critical supporting data - Reference Nos., Crews, & City Cost Indexes

16139	Residential Wiring	CREW	DAILY OUTPUT	LABOR-HOURS	UNIT	2005 BARE COSTS				TOTAL INCL O&P
						MAT.	LABOR	EQUIP.	TOTAL	
700 2250	S.P., 20 amp, #12/2, type NM cable	1 Elec	13.33	.600	Ea.	14.05	24.50		38.55	52
2270	Type MC cable		11.43	.700		25	28.50		53.50	70
2280	EMT & wire		4.85	1.649		35	67		102	139
2290	S.P. rotary dimmer, 600W, no wiring		17	.471		16.20	19.20		35.40	46.50
2300	S.P. rotary dimmer, 600W, type NM cable		14.55	.550		19.65	22.50		42.15	55
2320	Type MC cable		12.31	.650		32	26.50		58.50	75
2330	EMT & wire		5	1.600		41	65		106	143
2350	3-way rotary dimmer, type NM cable		13.33	.600		17.35	24.50		41.85	55.50
2370	Type MC cable		11.43	.700		30	28.50		58.50	75.50
2380	EMT & wire		4.85	1.649		39	67		106	143
2400	Interval timer wall switch, 20 amp, 1-30 min., #12/2									
2410	Type NM cable	1 Elec	14.55	.550	Ea.	31	22.50		53.50	67.50
2420	Type MC cable		12.31	.650		39	26.50		65.50	82
2430	EMT & wire		5	1.600		52	65		117	154
2500	Decorator style									
2510	S.P., 15 amp, type NM cable	1 Elec	17.10	.468	Ea.	11.65	19.05		30.70	41.50
2520	Type MC cable		14.30	.559		24	23		47	60.50
2530	EMT & wire		5.71	1.401		32.50	57		89.50	121
2550	3-way, #14/3, type NM cable		14.55	.550		15.55	22.50		38.05	50.50
2570	Type MC cable		12.31	.650		32.50	26.50		59	75.50
2580	EMT & wire		5	1.600		34.50	65		99.50	135
2600	4-way, #14/3, type NM cable		14.55	.550		29	22.50		51.50	65.50
2620	Type MC cable		12.31	.650		46	26.50		72.50	90.50
2630	EMT & wire		5	1.600		48.50	65		113.50	150
2650	S.P., 20 amp, #12/2, type NM cable		13.33	.600		17.60	24.50		42.10	56
2670	Type MC cable		11.43	.700		29	28.50		57.50	74
2680	EMT & wire		4.85	1.649		38.50	67		105.50	143
2700	S.P., slide dimmer, type NM cable		17.10	.468		27	19.05		46.05	58
2720	Type MC cable		14.30	.559		39.50	23		62.50	77.50
2730	EMT & wire		5.71	1.401		48.50	57		105.50	138
2750	S.P., touch dimmer, type NM cable		17.10	.468		23	19.05		42.05	54
2770	Type MC cable		14.30	.559		35.50	23		58.50	73
2780	EMT & wire		5.71	1.401		44.50	57		101.50	134
2800	3-way touch dimmer, type NM cable		13.33	.600		41	24.50		65.50	81.50
2820	Type MC cable		11.43	.700		53.50	28.50		82	102
2830	EMT & wire		4.85	1.649		62.50	67		129.50	169
3000	Combination devices									
3100	S.P. switch/15 amp recpt., Ivory, 1-gang box, plate									
3110	Type NM cable	1 Elec	11.43	.700	Ea.	16.35	28.50		44.85	60.50
3120	Type MC cable		10	.800		29	32.50		61.50	80.50
3130	EMT & wire		4.40	1.818		38	74		112	152
3150	S.P. switch/pilot light, type NM cable		11.43	.700		17	28.50		45.50	61
3170	Type MC cable		10	.800		29.50	32.50		62	81
3180	EMT & wire		4.43	1.806		38.50	73.50		112	153
3190	2-S.P. switches, 2-#14/2, no wiring		14	.571		6.75	23.50		30.25	42
3200	2-S.P. switches, 2-#14/2, type NM cables		10	.800		19.05	32.50		51.55	69.50
3220	Type MC cable		8.89	.900		40	36.50		76.50	98.50
3230	EMT & wire		4.10	1.951		39.50	79.50		119	162
3250	3-way switch/15 amp recpt., #14/3, type NM cable		10	.800		23.50	32.50		56	74
3270	Type MC cable		8.89	.900		40.50	36.50		77	99
3280	EMT & wire		4.10	1.951		42.50	79.50		122	165
3300	2-3 way switches, 2-#14/3, type NM cables		8.89	.900		31.50	36.50		68	89
3320	Type MC cable		8	1		61	41		102	128
3330	EMT & wire		4	2		47	81.50		128.50	173
3350	S.P. switch/20 amp recpt., #12/2, type NM cable		10	.800		27.50	32.50		60	78.50
3370	Type MC cable		8.89	.900		35.50	36.50		72	93.50

ELECTRICAL **16**

16139	Residential Wiring	CREW	DAILY OUTPUT	LABOR-HOURS	UNIT	2005 BARE COSTS				TOTAL INCL O&P
						MAT.	LABOR	EQUIP.	TOTAL	
3380	EMT & wire	1 Elec	4.10	1.951	Ea.	48.50	79.50		128	172
3400	Decorator style									
3410	S.P. switch/15 amp recpt., type NM cable	1 Elec	11.43	.700	Ea.	19.90	28.50		48.40	64.50
3420	Type MC cable		10	.800		32.50	32.50		65	84
3430	EMT & wire		4.40	1.818		41.50	74		115.50	156
3450	S.P. switch/pilot light, type NM cable		11.43	.700		20.50	28.50		49	65
3470	Type MC cable		10	.800		33	32.50		65.50	85
3480	EMT & wire		4.40	1.818		42	74		116	157
3500	2-S.P. switches, 2-#14/2, type NM cables		10	.800		22.50	32.50		55	73.50
3520	Type MC cable		8.89	.900		43.50	36.50		80	103
3530	EMT & wire		4.10	1.951		43	79.50		122.50	166
3550	3-way/15 amp recpt., #14/3, type NM cable		10	.800		27	32.50		59.50	78
3570	Type MC cable		8.89	.900		44	36.50		80.50	103
3580	EMT & wire		4.10	1.951		46	79.50		125.50	169
3650	2-3 way switches, 2-#14/3, type NM cables		8.89	.900		35	36.50		71.50	93
3670	Type MC cable		8	1		64.50	41		105.50	132
3680	EMT & wire		4	2		50.50	81.50		132	177
3700	S.P. switch/20 amp recpt., #12/2, type NM cable		10	.800		31	32.50		63.50	82.50
3720	Type MC cable		8.89	.900		39	36.50		75.50	97
3730	EMT & wire	▼	4.10	1.951	▼	52	79.50		131.50	175
4000	Receptacle devices									
4010	Duplex outlet, 15 amp recpt., Ivory, 1-gang box, plate									
4015	Type NM cable	1 Elec	14.55	.550	Ea.	6.45	22.50		28.95	40.50
4020	Type MC cable		12.31	.650		19	26.50		45.50	60.50
4030	EMT & wire		5.33	1.501		27	61		88	121
4050	With #12/2, type NM cable		12.31	.650		8.05	26.50		34.55	48.50
4070	Type MC cable		10.67	.750		19.20	30.50		49.70	66.50
4080	EMT & wire		4.71	1.699		29	69		98	135
4100	20 amp recpt., #12/2, type NM cable		12.31	.650		15.45	26.50		41.95	56.50
4120	Type MC cable		10.67	.750		26.50	30.50		57	75
4130	EMT & wire	▼	4.71	1.699	▼	36.50	69		105.50	143
4140	For GFI see line 4300 below									
4150	Decorator style, 15 amp recpt., type NM cable	1 Elec	14.55	.550	Ea.	10	22.50		32.50	44.50
4170	Type MC cable		12.31	.650		22.50	26.50		49	64.50
4180	EMT & wire		5.33	1.501		31	61		92	125
4200	With #12/2, type NM cable		12.31	.650		11.60	26.50		38.10	52.50
4220	Type MC cable		10.67	.750		23	30.50		53.50	70.50
4230	EMT & wire		4.71	1.699		32.50	69		101.50	139
4250	20 amp recpt. #12/2, type NM cable		12.31	.650		19	26.50		45.50	60.50
4270	Type MC cable		10.67	.750		30	30.50		60.50	78.50
4280	EMT & wire		4.71	1.699		40	69		109	147
4300	GFI, 15 amp recpt., type NM cable		12.31	.650		35	26.50		61.50	78
4320	Type MC cable		10.67	.750		47.50	30.50		78	98
4330	EMT & wire		4.71	1.699		56	69		125	165
4350	GFI with #12/2, type NM cable		10.67	.750		36.50	30.50		67	85.50
4370	Type MC cable		9.20	.870		47.50	35.50		83	105
4380	EMT & wire		4.21	1.900		57.50	77.50		135	179
4400	20 amp recpt., #12/2 type NM cable		10.67	.750		38	30.50		68.50	87.50
4420	Type MC cable		9.20	.870		49.50	35.50		85	107
4430	EMT & wire		4.21	1.900		59.50	77.50		137	180
4500	Weather-proof cover for above receptacles, add	▼	32	.250	▼	4.40	10.20		14.60	20
4550	Air conditioner outlet, 20 amp-240 volt recpt.									
4560	30' of #12/2, 2 pole circuit breaker									
4570	Type NM cable	1 Elec	10	.800	Ea.	44.50	32.50		77	97
4580	Type MC cable		9	.889		59	36		95	119
4590	EMT & wire	▼	4	2	▼	64.50	81.50		146	192

700

16 ELECTRICAL

Important: See the Reference Section for critical supporting data - Reference Nos., Crews, & City Cost Indexes

16139 | Residential Wiring

		CREW	DAILY OUTPUT	LABOR-HOURS	UNIT	2005 BARE COSTS				TOTAL INCL O&P
						MAT.	LABOR	EQUIP.	TOTAL	
4600	Decorator style, type NM cable	1 Elec	10	.800	Ea.	48	32.50		80.50	102
4620	Type MC cable		9	.889		63	36		99	123
4630	EMT & wire	▼	4	2	▼	68.50	81.50		150	196
4650	Dryer outlet, 30 amp-240 volt recpt., 20' of #10/3									
4660	2 pole circuit breaker									
4670	Type NM cable	1 Elec	6.41	1.248	Ea.	54	51		105	135
4680	Type MC cable		5.71	1.401		65.50	57		122.50	157
4690	EMT & wire	▼	3.48	2.299		69	93.50		162.50	215
4700	Range outlet, 50 amp-240 volt recpt., 30' of #8/3									
4710	Type NM cable	1 Elec	4.21	1.900	Ea.	80.50	77.50		158	204
4720	Type MC cable		4	2		128	81.50		209.50	262
4730	EMT & wire		2.96	2.703		86	110		196	259
4750	Central vacuum outlet, Type NM cable		6.40	1.250		48	51		99	129
4770	Type MC cable		5.71	1.401		70	57		127	162
4780	EMT & wire	▼	3.48	2.299	▼	75	93.50		168.50	222
4800	30 amp-110 volt locking recpt., #10/2 circ. bkr.									
4810	Type NM cable	1 Elec	6.20	1.290	Ea.	58.50	52.50		111	143
4820	Type MC cable		5.40	1.481		82	60.50		142.50	181
4830	EMT & wire	▼	3.20	2.500	▼	82.50	102		184.50	243
4900	Low voltage outlets									
4910	Telephone recpt., 20' of 4/C phone wire	1 Elec	26	.308	Ea.	7.40	12.55		19.95	27
4920	TV recpt., 20' of RG59U coax wire, F type connector	"	16	.500	"	12.65	20.50		33.15	44.50
4950	Door bell chime, transformer, 2 buttons, 60' of bellwire									
4970	Economy model	1 Elec	11.50	.696	Ea.	54.50	28.50		83	102
4980	Custom model		11.50	.696		85.50	28.50		114	136
4990	Luxury model, 3 buttons	▼	9.50	.842	▼	231	34.50		265.50	305
6000	Lighting outlets									
6050	Wire only (for fixture), type NM cable	1 Elec	32	.250	Ea.	4.42	10.20		14.62	20
6070	Type MC cable		24	.333		14.10	13.60		27.70	35.50
6080	EMT & wire		10	.800		21	32.50		53.50	72
6100	Box (4"), and wire (for fixture), type NM cable		25	.320		9.40	13.05		22.45	30
6120	Type MC cable		20	.400		19.10	16.30		35.40	45.50
6130	EMT & wire	▼	11	.727	▼	26	29.50		55.50	73
6200	Fixtures (use with lines 6050 or 6100 above)									
6210	Canopy style, economy grade	1 Elec	40	.200	Ea.	26	8.15		34.15	40.50
6220	Custom grade		40	.200		48	8.15		56.15	65
6250	Dining room chandelier, economy grade		19	.421		78	17.15		95.15	112
6260	Custom grade		19	.421		230	17.15		247.15	279
6270	Luxury grade		15	.533		510	21.50		531.50	595
6310	Kitchen fixture (fluorescent), economy grade		30	.267		52	10.85		62.85	73
6320	Custom grade		25	.320		165	13.05		178.05	201
6350	Outdoor, wall mounted, economy grade		30	.267		27.50	10.85		38.35	46.50
6360	Custom grade		30	.267		104	10.85		114.85	130
6370	Luxury grade		25	.320		235	13.05		248.05	278
6410	Outdoor PAR floodlights, 1 lamp, 150 watt		20	.400		21.50	16.30		37.80	48
6420	2 lamp, 150 watt each		20	.400		36	16.30		52.30	64
6430	For infrared security sensor, add		32	.250		90	10.20		100.20	114
6450	Outdoor, quartz-halogen, 300 watt flood		20	.400		39	16.30		55.30	67.50
6600	Recessed downlight, round, pre-wired, 50 or 75 watt trim		30	.267		36	10.85		46.85	55.50
6610	With shower light trim		30	.267		45	10.85		55.85	65.50
6620	With wall washer trim		28	.286		54	11.65		65.65	77
6630	With eye-ball trim	▼	28	.286		54	11.65		65.65	77
6640	For direct contact with insulation, add					1.65			1.65	1.82
6700	Porcelain lamp holder	1 Elec	40	.200		3.60	8.15		11.75	16.10
6710	With pull switch		40	.200		3.86	8.15		12.01	16.40
6750	Fluorescent strip, 1-20 watt tube, wrap around diffuser, 24"	▼	24	.333	▼	53	13.60		66.60	78.50

700 (left margin) **700** (right margin)

ELECTRICAL 16 (right side tab)

16139 | Residential Wiring

		CREW	DAILY OUTPUT	LABOR-HOURS	UNIT	2005 BARE COSTS				TOTAL INCL O&P
						MAT.	LABOR	EQUIP.	TOTAL	
6760	1-40 watt tube, 48"	1 Elec	24	.333	Ea.	65	13.60		78.60	91.50
6770	2-40 watt tubes, 48"		20	.400		79	16.30		95.30	112
6780	With residential ballast		20	.400		89.50	16.30		105.80	123
6800	Bathroom heat lamp, 1-250 watt		28	.286		35	11.65		46.65	56
6810	2-250 watt lamps		28	.286		56	11.65		67.65	79
6820	For timer switch, see line 2400									
6900	Outdoor post lamp, incl. post, fixture, 35' of #14/2									
6910	Type NMC cable	1 Elec	3.50	2.286	Ea.	184	93		277	340
6920	Photo-eye, add		27	.296		30	12.05		42.05	51
6950	Clock dial time switch, 24 hr., w/enclosure, type NM cable		11.43	.700		53.50	28.50		82	102
6970	Type MC cable		11	.727		66	29.50		95.50	117
6980	EMT & wire		4.85	1.649		74.50	67		141.50	182
7000	Alarm systems									
7050	Smoke detectors, box, #14/3, type NM cable	1 Elec	14.55	.550	Ea.	29.50	22.50		52	66
7070	Type MC cable		12.31	.650		43.50	26.50		70	87.50
7080	EMT & wire		5	1.600		45.50	65		110.50	147
7090	For relay output to security system, add					12.10			12.10	13.30
8000	Residential equipment									
8050	Disposal hook-up, incl. switch, outlet box, 3' of flex									
8060	20 amp-1 pole circ. bkr., and 25' of #12/2									
8070	Type NM cable	1 Elec	10	.800	Ea.	22.50	32.50		55	73
8080	Type MC cable		8	1		35.50	41		76.50	99.50
8090	EMT & wire		5	1.600		46.50	65		111.50	149
8100	Trash compactor or dishwasher hook-up, incl. outlet box,									
8110	3' of flex, 15 amp-1 pole circ. bkr., and 25' of #14/2									
8120	Type NM cable	1 Elec	10	.800	Ea.	15.90	32.50		48.40	66
8130	Type MC cable		8	1		31	41		72	94.50
8140	EMT & wire		5	1.600		40.50	65		105.50	142
8150	Hot water sink dispensor hook-up, use line 8100									
8200	Vent/exhaust fan hook-up, type NM cable	1 Elec	32	.250	Ea.	4.42	10.20		14.62	20
8220	Type MC cable		24	.333		14.10	13.60		27.70	35.50
8230	EMT & wire		10	.800		21	32.50		53.50	72
8250	Bathroom vent fan, 50 CFM (use with above hook-up)									
8260	Economy model	1 Elec	15	.533	Ea.	22	21.50		43.50	56.50
8270	Low noise model		15	.533		31	21.50		52.50	66.50
8280	Custom model		12	.667		114	27		141	166
8300	Bathroom or kitchen vent fan, 110 CFM									
8310	Economy model	1 Elec	15	.533	Ea.	58	21.50		79.50	96.50
8320	Low noise model	"	15	.533	"	77	21.50		98.50	117
8350	Paddle fan, variable speed (w/o lights)									
8360	Economy model (AC motor)	1 Elec	10	.800	Ea.	103	32.50		135.50	162
8370	Custom model (AC motor)		10	.800		178	32.50		210.50	245
8380	Luxury model (DC motor)		8	1		350	41		391	445
8390	Remote speed switch for above, add		12	.667		25	27		52	68
8500	Whole house exhaust fan, ceiling mount, 36", variable speed									
8510	Remote switch, incl. shutters, 20 amp-1 pole circ. bkr.									
8520	30' of #12/2, type NM cable	1 Elec	4	2	Ea.	615	81.50		696.50	795
8530	Type MC cable		3.50	2.286		630	93		723	830
8540	EMT & wire		3	2.667		645	109		754	865
8600	Whirlpool tub hook-up, incl. timer switch, outlet box									
8610	3' of flex, 20 amp-1 pole GFI circ. bkr.									
8620	30' of #12/2, type NM cable	1 Elec	5	1.600	Ea.	85	65		150	191
8630	Type MC cable		4.20	1.905		93.50	77.50		171	219
8640	EMT & wire		3.40	2.353		103	96		199	257
8650	Hot water heater hook-up, incl. 1-2 pole circ. bkr., box;									
8660	3' of flex, 20' of #10/2, type NM cable	1 Elec	5	1.600	Ea.	25	65		90	125

Important: See the Reference Section for critical supporting data - Reference Nos., Crews, & City Cost Indexes

16139 | Residential Wiring

			CREW	DAILY OUTPUT	LABOR-HOURS	UNIT	2005 BARE COSTS MAT.	LABOR	EQUIP.	TOTAL	TOTAL INCL O&P	
700	8670	Type MC cable	1 Elec	4.20	1.905	Ea.	40.50	77.50		118	161	700
	8680	EMT & wire	↓	3.40	2.353	↓	40.50	96		136.50	188	
	9000	Heating/air conditioning										
	9050	Furnace/boiler hook-up, incl. firestat, local on-off switch										
	9060	Emergency switch, and 40' of type NM cable	1 Elec	4	2	Ea.	45	81.50		126.50	171	
	9070	Type MC cable		3.50	2.286		66	93		159	212	
	9080	EMT & wire	↓	1.50	5.333	↓	82	217		299	415	
	9100	Air conditioner hook-up, incl. local 60 amp disc. switch										
	9110	3' sealtite, 40 amp, 2 pole circuit breaker										
	9130	40' of #8/2, type NM cable	1 Elec	3.50	2.286	Ea.	158	93		251	315	
	9140	Type MC cable		3	2.667		233	109		342	420	
	9150	EMT & wire	↓	1.30	6.154	↓	183	251		434	575	
	9200	Heat pump hook-up, 1-40 & 1-100 amp 2 pole circ. bkr.										
	9210	Local disconnect switch, 3' sealtite										
	9220	40' of #8/2 & 30' of #3/2										
	9230	Type NM cable	1 Elec	1.30	6.154	Ea.	385	251		636	795	
	9240	Type MC cable		1.08	7.407		560	300		860	1,075	
	9250	EMT & wire	↓	.94	8.511	↓	425	345		770	985	
	9500	Thermostat hook-up, using low voltage wire										
	9520	Heating only	1 Elec	24	.333	Ea.	5.25	13.60		18.85	26	
	9530	Heating/cooling	"	20	.400	"	6.35	16.30		22.65	31.50	

16140 | Wiring Devices

			CREW	DAILY OUTPUT	LABOR-HOURS	UNIT	MAT.	LABOR	EQUIP.	TOTAL	TOTAL INCL O&P	
500	0010	**LOW VOLTAGE SWITCHING**										500
	3600	Relays, 120 V or 277 V standard	1 Elec	12	.667	Ea.	26	27		53	69	
	3800	Flush switch, standard		40	.200		9.05	8.15		17.20	22	
	4000	Interchangeable		40	.200		11.80	8.15		19.95	25	
	4100	Surface switch, standard		40	.200		6.60	8.15		14.75	19.45	
	4200	Transformer 115 V to 25 V		12	.667		93	27		120	143	
	4400	Master control, 12 circuit, manual		4	2		94	81.50		175.50	224	
	4500	25 circuit, motorized		4	2		102	81.50		183.50	233	
	4600	Rectifier, silicon		12	.667		30.50	27		57.50	74	
	4800	Switchplates, 1 gang, 1, 2 or 3 switch, plastic		80	.100		3	4.08		7.08	9.35	
	5000	Stainless steel		80	.100		8.10	4.08		12.18	14.95	
	5400	2 gang, 3 switch, stainless steel		53	.151		15.65	6.15		21.80	26.50	
	5500	4 switch, plastic		53	.151		6.70	6.15		12.85	16.50	
	5800	3 gang, 9 switch, stainless steel	↓	32	.250	↓	50	10.20		60.20	70	
910	0010	**WIRING DEVICES**										910
	0200	Toggle switch, quiet type, single pole, 15 amp **CN**	1 Elec	40	.200	Ea.	4.84	8.15		12.99	17.45	
	0600	3 way, 15 amp		23	.348		7	14.15		21.15	28.50	
	0900	4 way, 15 amp		15	.533		21	21.50		42.50	55.50	
	1650	Dimmer switch, 120 volt, incandescent, 600 watt, 1 pole		16	.500		10.80	20.50		31.30	42.50	
	2460	Receptacle, duplex, 120 volt, grounded, 15 amp		40	.200		1.14	8.15		9.29	13.40	
	2470	20 amp		27	.296		8.55	12.05		20.60	27.50	
	2490	Dryer, 30 amp		15	.533		10.35	21.50		31.85	44	
	2500	Range, 50 amp		11	.727		11.40	29.50		40.90	56.50	
	2600	Wall plates, stainless steel, 1 gang		80	.100		1.80	4.08		5.88	8.05	
	2800	2 gang		53	.151		4.10	6.15		10.25	13.65	
	3200	Lampholder, keyless		26	.308		9.70	12.55		22.25	29.50	
	3400	Pullchain with receptacle	↓	22	.364	↓	9.20	14.80		24	32	

16150 | Wiring Connections

			CREW	DAILY OUTPUT	LABOR-HOURS	UNIT	MAT.	LABOR	EQUIP.	TOTAL	TOTAL INCL O&P	
275	0010	**MOTOR CONNECTIONS**										275
	0020	Flexible conduit and fittings, 115 volt, 1 phase, up to 1 HP motor	1 Elec	8	1	Ea.	4.96	41		45.96	66	

ELECTRICAL 16

16100 | Wiring Methods

16150 | Wiring Connections

		CREW	DAILY OUTPUT	LABOR-HOURS	UNIT	2005 BARE COSTS				TOTAL INCL O&P	
						MAT.	LABOR	EQUIP.	TOTAL		
275	0120	230 volt, 10 HP motor, 3 phase	1 Elec	4.20	1.905	Ea.	9.05	77.50		86.55	126	275
	0200	25 HP motor		2.70	2.963		19.50	121		140.50	202	
	0400	50 HP motor		2.20	3.636		54.50	148		202.50	281	
	0600	100 HP motor		1.50	5.333		139	217		356	480	

16200 | Electrical Power

16220 | Motors & Generators

		CREW	DAILY OUTPUT	LABOR-HOURS	UNIT	2005 BARE COSTS				TOTAL INCL O&P	
						MAT.	LABOR	EQUIP.	TOTAL		
610	0010	**MOTORS** 230/460 volts, 60 HZ										610
	0050	Dripproof, premium efficiency, 1.15 service factor										
	0060	1800 RPM, 1/4 HP	1 Elec	5.33	1.501	Ea.	128	61		189	232	
	0070	1/3 HP		5.33	1.501		140	61		201	245	
	0080	1/2 HP		5.33	1.501		165	61		226	273	
	0090	3/4 HP		5.33	1.501		184	61		245	293	
	0100	1 HP		4.50	1.778		194	72.50		266.50	320	
	0250	5 HP		4.50	1.778		295	72.50		367.50	435	
	0350	10 HP		4	2		530	81.50		611.50	705	
	0450	20 HP　　　　　　**CN**	2 Elec	5.20	3.077		840	125		965	1,100	

16230 | Generator Assemblies

		CREW	DAILY OUTPUT	LABOR-HOURS	UNIT	2005 BARE COSTS				TOTAL INCL O&P	
						MAT.	LABOR	EQUIP.	TOTAL		
450	0010	**GENERATOR SET**										450
	0020	Gas or gasoline operated, includes battery,										
	0050	charger, muffler & transfer switch										
	0200	3 phase 4 wire, 277/480 volt, 7.5 kW	R-3	.83	24.096	Ea.	6,000	965	191	7,156	8,225	
	0300	11.5 kW		.71	28.169		8,500	1,125	223	9,848	11,300	
	0400	20 kW		.63	31.746		10,000	1,275	252	11,527	13,200	
	0500	35 kW		.55	36.364		12,000	1,450	288	13,738	15,700	
	0600	80 kW		.40	50		20,500	2,000	395	22,895	26,000	
	0700	100 kW		.33	60.606		22,500	2,425	480	25,405	29,000	
	0800	125 kW		.28	71.429		46,000	2,850	565	49,415	55,500	
	0900	185 kW		.25	80		61,000	3,200	635	64,835	72,500	
	2000	Diesel engine, including battery, charger,										
	2010	muffler, automatic transfer switch & day tank, 30 kW	R-3	.55	36.364	Ea.	16,000	1,450	288	17,738	20,100	
	2100	50 kW		.42	47.619		19,700	1,900	380	21,980	25,000	
	2200	75 kW		.35	57.143		25,700	2,275	455	28,430	32,200	
	2300	100 kW		.31	64.516		28,500	2,575	510	31,585	35,800	
	2400	125 kW		.29	68.966		30,000	2,750	545	33,295	37,700	
	2500	150 kW		.26	76.923		34,400	3,075	610	38,085	43,100	
	2600	175 kW		.25	80		37,500	3,200	635	41,335	46,800	
	2700	200 kW		.24	83.333		38,700	3,325	660	42,685	48,300	
	2800	250 kW		.23	86.957		45,600	3,475	690	49,765	56,000	
	2900	300 kW		.22	90.909		49,300	3,625	720	53,645	60,000	
	3000	350 kW		.20	100		55,500	4,000	795	60,295	68,500	
	3100	400 kW		.19	105		69,000	4,200	835	74,035	82,500	
	3200	500 kW		.18	111		86,500	4,450	880	91,830	102,500	

16270 | Transformers

		CREW	DAILY OUTPUT	LABOR-HOURS	UNIT	2005 BARE COSTS				TOTAL INCL O&P	
						MAT.	LABOR	EQUIP.	TOTAL		
200	0010	**DRY TYPE TRANSFORMER**										200
	0050	Single phase, 240/480 volt primary, 120/240 volt secondary										

16 ELECTRICAL

Important: See the Reference Section for critical supporting data - Reference Nos., Crews, & City Cost Indexes

16270	Transformers	CREW	DAILY OUTPUT	LABOR-HOURS	UNIT	2005 BARE COSTS				TOTAL INCL O&P
						MAT.	LABOR	EQUIP.	TOTAL	
200 0100	1 kVA	1 Elec	2	4	Ea.	175	163		338	435	**200**
0300	2 kVA		1.60	5		263	204		467	595
0500	3 kVA		1.40	5.714		325	233		558	705
0700	5 kVA		1.20	6.667		450	272		722	900
0900	7.5 kVA	2 Elec	2.20	7.273		625	296		921	1,125
1100	10 kVA		1.60	10		800	410		1,210	1,475
1300	15 kVA		1.20	13.333		1,100	545		1,645	2,000
1500	25 kVA		1	16		1,400	650		2,050	2,525
1700	37.5 kVA		.80	20		1,825	815		2,640	3,225
1900	50 kVA		.70	22.857		2,175	930		3,105	3,750
2100	75 kVA		.65	24.615		2,875	1,000		3,875	4,650
2190	480V primary 120/240V secondary, nonvent., 15 kVA		1.20	13.333		1,100	545		1,645	2,000
2200	25 kVA		.90	17.778		1,625	725		2,350	2,850
2210	37 kVA		.75	21.333		1,925	870		2,795	3,425
2220	50 kVA		.65	24.615		2,300	1,000		3,300	4,025
2300	3 phase, 480 volt primary 120/208 volt secondary									
2310	Ventilated, 3 kVA	1 Elec	1	8	Ea.	580	325		905	1,125
2700	6 kVA		.80	10		795	410		1,205	1,475
2900	9 kVA		.70	11.429		905	465		1,370	1,700
3100	15 kVA	2 Elec	1.10	14.545		1,200	595		1,795	2,200
3300	30 kVA		.90	17.778		1,400	725		2,125	2,625
3500	45 kVA		.80	20		1,700	815		2,515	3,075
3700	75 kVA		.70	22.857		2,550	930		3,480	4,175
3900	112.5 kVA	R-3	.90	22.222		3,375	890	176	4,441	5,250
4100	150 kVA		.85	23.529		4,400	940	187	5,527	6,450
4300	225 kVA		.65	30.769		5,975	1,225	244	7,444	8,675
4500	300 kVA		.55	36.364		7,575	1,450	288	9,313	10,800
4700	500 kVA		.45	44.444		12,500	1,775	350	14,625	16,800
4800	750 kVA		.35	57.143		22,000	2,275	455	24,730	28,100
600 0010	**OIL FILLED TRANSFORMER** Pad mounted, primary delta or Y,										**600**
0050	5 kV or 15 kV, with taps, 277/480 V secondary, 3 phase									
0100	150 kVA	R-3	.65	30.769	Ea.	7,025	1,225	244	8,494	9,825
0200	300 kVA		.45	44.444		9,300	1,775	350	11,425	13,200
0300	500 kVA		.40	50		14,000	2,000	395	16,395	18,800
0400	750 kVA		.38	52.632		16,700	2,100	415	19,215	22,000
0500	1000 kVA		.26	76.923		19,800	3,075	610	23,485	27,100
0600	1500 kVA		.23	86.957		23,500	3,475	690	27,665	31,800
0700	2000 kVA		.20	100		29,700	4,000	795	34,495	39,500
0800	3750 kVA		.16	125		55,500	5,000	990	61,490	70,000

16280	Power Filters & Conditioners									
300 0010	**CAPACITORS** Indoor										**300**
0020	240 volts, single & 3 phase, 0.5 kVAR	1 Elec	2.70	2.963	Ea.	295	121		416	505
0100	1.0 kVAR		2.70	2.963		355	121		476	570
0150	2.5 kVAR		2	4		425	163		588	710
0200	5.0 kVAR		1.80	4.444		495	181		676	810
0250	7.5 kVAR		1.60	5		560	204		764	920
0300	10 kVAR		1.50	5.333		675	217		892	1,075
0350	15 kVAR		1.30	6.154		930	251		1,181	1,400
0400	20 kVAR		1.10	7.273		1,150	296		1,446	1,725
0450	25 kVAR		1	8		1,375	325		1,700	1,975
1000	480 volts, single & 3 phase, 1 kVAR		2.70	2.963		268	121		389	475
1050	2 kVAR		2.70	2.963		310	121		431	520
1100	5 kVAR		2	4		390	163		553	675
1150	7.5 kVAR		2	4		425	163		588	715

CN

ELECTRICAL 16

16200 | Electrical Power

16280 | Power Filters & Conditioners

			CREW	DAILY OUTPUT	LABOR-HOURS	UNIT	2005 BARE COSTS				TOTAL INCL O&P	
							MAT.	LABOR	EQUIP.	TOTAL		
300	1200	10 kVAR	1 Elec	2	4	Ea.	470	163		633	765	300
	1250	15 kVAR		2	4		585	163		748	885	
	1300	20 kVAR		1.60	5		650	204		854	1,025	
	1350	30 kVAR		1.50	5.333		810	217		1,027	1,225	
	1400	40 kVAR		1.20	6.667		1,025	272		1,297	1,525	
	1450	50 kVAR		1.10	7.273		1,175	296		1,471	1,750	

16290 | Power Measurement & Control

			CREW	DAILY OUTPUT	LABOR-HOURS	UNIT	MAT.	LABOR	EQUIP.	TOTAL	TOTAL INCL O&P	
800	0010	**SWITCHBOARD INSTRUMENTS** 3 phase, 4 wire										800
	0100	AC indicating, ammeter & switch	1 Elec	8	1	Ea.	1,425	41		1,466	1,625	
	0200	Voltmeter & switch		8	1		1,425	41		1,466	1,625	
	0300	Wattmeter		8	1		2,825	41		2,866	3,150	
	0400	AC recording, ammeter		4	2		5,050	81.50		5,131.50	5,675	
	0500	Voltmeter		4	2		5,050	81.50		5,131.50	5,675	
	0600	Ground fault protection, zero sequence		2.70	2.963		4,450	121		4,571	5,075	
	0700	Ground return path		2.70	2.963		4,450	121		4,571	5,075	
	0800	3 current transformers, 5 to 800 amp		2	4		2,075	163		2,238	2,525	
	0900	1000 to 1500 amp		1.30	6.154		2,975	251		3,226	3,650	
	1200	2000 to 4000 amp		1	8		3,525	325		3,850	4,350	
	1300	Fused potential transformer, maximum 600 volt		8	1		780	41		821	920	

16400 | Low-Voltage Distribution

16410 | Encl Switches & Circuit Breakers

			CREW	DAILY OUTPUT	LABOR-HOURS	UNIT	2005 BARE COSTS				TOTAL INCL O&P	
							MAT.	LABOR	EQUIP.	TOTAL		
200	0010	**CIRCUIT BREAKERS** (in enclosure)										200
	0100	Enclosed (NEMA 1), 600 volt, 3 pole, 30 amp	1 Elec	3.20	2.500	Ea.	430	102		532	625	
	0200	60 amp		2.80	2.857		430	116		546	650	
	0400	100 amp		2.30	3.478		495	142		637	755	
	0600	225 amp		1.50	5.333		1,150	217		1,367	1,575	
	0700	400 amp	2 Elec	1.60	10		1,950	410		2,360	2,750	
	0800	600 amp		1.20	13.333		2,825	545		3,370	3,900	
	1000	800 amp		.94	17.021		3,675	695		4,370	5,075	
800	0010	**SAFETY SWITCHES**										800
	0100	General duty 240 volt, 3 pole NEMA 1, fusible, 30 amp	1 Elec	3.20	2.500	Ea.	78	102		180	238	
	0200	60 amp		2.30	3.478		132	142		274	355	
	0300	100 amp		1.90	4.211		227	172		399	505	
	0400	200 amp		1.30	6.154		490	251		741	910	
	0500	400 amp	2 Elec	1.80	8.889		1,225	360		1,585	1,900	
	0600	600 amp	"	1.20	13.333		2,300	545		2,845	3,325	
	2900	Heavy duty, 240 volt, 3 pole NEMA 1 fusible										
	2910	30 amp	1 Elec	3.20	2.500	Ea.	126	102		228	290	
	3000	60 amp		2.30	3.478		213	142		355	445	
	3300	100 amp		1.90	4.211		335	172		507	625	
	3500	200 amp		1.30	6.154		580	251		831	1,000	
	3700	400 amp	2 Elec	1.80	8.889		1,475	360		1,835	2,175	
	3900	600 amp	"	1.20	13.333		2,550	545		3,095	3,600	

16420 | Enclosed Controllers

			CREW	DAILY OUTPUT	LABOR-HOURS	UNIT	MAT.	LABOR	EQUIP.	TOTAL	TOTAL INCL O&P	
220	0010	**CONTROL STATIONS**										220
	0050	NEMA 1, heavy duty, stop/start	1 Elec	8	1	Ea.	121	41		162	195	

Important: See the Reference Section for critical supporting data - Reference Nos., Crews, & City Cost Indexes

16 ELECTRICAL

		16420	**Enclosed Controllers**	CREW	DAILY OUTPUT	LABOR-HOURS	UNIT	MAT.	LABOR	2005 BARE COSTS EQUIP.	TOTAL	TOTAL INCL O&P	
220	0100		Stop/start, pilot light	1 Elec	6.20	1.290	Ea.	165	52.50		217.50	261	**220**
	0200		Hand/off/automatic		6.20	1.290		90	52.50		142.50	177	
	0400		Stop/start/reverse	▼	5.30	1.509	▼	164	61.50		225.50	272	

		16440	**Swbds, Panels & Control Centers**										
660	0010		**MOTOR STARTERS & CONTROLS** R16220-610										**660**
	0050		Magnetic, FVNR, with enclosure and heaters, 480 volt										
	0100		5 HP, size 0	1 Elec	2.30	3.478	Ea.	209	142		351	440	
	0200		10 HP, size 1	"	1.60	5		235	204		439	565	
	0300		25 HP, size 2	2 Elec	2.20	7.273		440	296		736	925	
	0400		50 HP, size 3		1.80	8.889		720	360		1,080	1,325	
	0500		100 HP, size 4		1.20	13.333		1,600	545		2,145	2,550	
	0600		200 HP, size 5	▼	.90	17.778		3,725	725		4,450	5,175	
	0700		Combination, with motor circuit protectors, 5 HP, size 0	1 Elec	1.80	4.444		680	181		861	1,025	
	0800		10 HP, size 1	"	1.30	6.154		705	251		956	1,150	
	0900		25 HP, size 2	2 Elec	2	8		990	325		1,315	1,550	
	1000		50 HP, size 3		1.32	12.121		1,425	495		1,920	2,300	
	1200		100 HP, size 4	▼	.80	20		3,100	815		3,915	4,650	
	1400		Combination, with fused switch, 5 HP, size 0	1 Elec	1.80	4.444		520	181		701	840	
	1600		10 HP, size 1	"	1.30	6.154		555	251		806	985	
	1800		25 HP, size 2	2 Elec	2	8		900	325		1,225	1,475	
	2000		50 HP, size 3		1.32	12.121		1,525	495		2,020	2,400	
	2200		100 HP, size 4		.80	20		2,650	815		3,465	4,150	

700	0010		**PANELBOARD & LOAD CENTER CIRCUIT BREAKERS**										**700**
	0050		Bolt-on, 10,000 amp IC, 120 volt, 1 pole										
	0100		15 to 50 amp	1 Elec	10	.800	Ea.	11	32.50		43.50	60.50	
	0200		60 amp		8	1		11	41		52	72.50	
	0300		70 amp	▼	8	1	▼	21	41		62	83.50	
	0350		240 volt, 2 pole										
	0400		15 to 50 amp	1 Elec	8	1	Ea.	24	41		65	87	
	0500		60 amp		7.50	1.067		24	43.50		67.50	91	
	0600		80 to 100 amp		5	1.600		62	65		127	166	
	0700		3 pole, 15 to 60 amp		6.20	1.290		76.50	52.50		129	163	
	0800		70 amp		5	1.600		96.50	65		161.50	203	
	0900		80 to 100 amp		3.60	2.222		109	90.50		199.50	255	
	1000		22,000 amp I.C., 240 volt, 2 pole, 70 - 225 amp		2.70	2.963		470	121		591	695	
	1100		3 pole, 70 - 225 amp		2.30	3.478		525	142		667	785	
	1200		14,000 amp I.C., 277 volts, 1 pole, 15 - 30 amp		8	1		29	41		70	92.50	
	1300		22,000 amp I.C., 480 volts, 2 pole, 70 - 225 amp		2.70	2.963		470	121		591	695	
	1400		3 pole, 70 - 225 amp	▼	2.30	3.478	▼	580	142		722	850	

720	0010		**PANELBOARDS** (Commercial use)										**720**
	0050		NQOD, w/20 amp 1 pole bolt-on circuit breakers										
	0100		3 wire, 120/240 volts, 100 amp main lugs										
	0150		10 circuits	1 Elec	1	8	Ea.	410	325		735	935	
	0200		14 circuits		.88	9.091		480	370		850	1,075	
	0250		18 circuits		.75	10.667		525	435		960	1,225	
	0300		20 circuits	▼	.65	12.308		595	500		1,095	1,400	
	0350		225 amp main lugs, 24 circuits	2 Elec	1.20	13.333		675	545		1,220	1,550	
	0400		30 circuits		.90	17.778		785	725		1,510	1,925	
	0450		36 circuits		.80	20		895	815		1,710	2,200	
	0500		38 circuits		.72	22.222		960	905		1,865	2,400	
	0550		42 circuits	▼	.66	24.242		1,000	990		1,990	2,575	
	0600		4 wire, 120/208 volts, 100 amp main lugs, 12 circuits	1 Elec	1	8		465	325		790	995	
	0650		16 circuits	▼	.75	10.667	▼	535	435		970	1,225	

ELECTRICAL 16

16440	Swbds, Panels & Control Centers		CREW	DAILY OUTPUT	LABOR-HOURS	UNIT	2005 BARE COSTS				TOTAL INCL O&P	
							MAT.	LABOR	EQUIP.	TOTAL		
720	0700	20 circuits		1 Elec	.65	12.308	Ea.	625	500		1,125	1,425	720
	0750	24 circuits **CN**			.60	13.333		675	545		1,220	1,550	
	0800	30 circuits		↓	.53	15.094		780	615		1,395	1,775	
	0850	225 amp main lugs, 32 circuits		2 Elec	.90	17.778		880	725		1,605	2,050	
	0900	34 circuits			.84	19.048		900	775		1,675	2,150	
	0950	36 circuits			.80	20		920	815		1,735	2,250	
	1000	42 circuits		↓	.68	23.529	↓	1,025	960		1,985	2,550	
	1200	NEHB, w/20 amp, 1 pole bolt-on circuit breakers											
	1250	4 wire, 277/480 volts, 100 amp main lugs, 12 circuits		1 Elec	.88	9.091	Ea.	890	370		1,260	1,525	
	1300	20 circuits		"	.60	13.333		1,300	545		1,845	2,250	
	1350	225 amp main lugs, 24 circuits		2 Elec	.90	17.778		1,525	725		2,250	2,750	
	1400	30 circuits		↓	.80	20	↓	1,825	815		2,640	3,225	
	1450	36 circuits		↓	.72	22.222	↓	2,125	905		3,030	3,675	
	1600	NQOD panel, w/20 amp, 1 pole, circuit breakers											
	1650	3 wire, 120/240 volt with main circuit breaker											
	1700	100 amp main, 12 circuits		1 Elec	.80	10	Ea.	570	410		980	1,225	
	1750	20 circuits		"	.60	13.333		730	545		1,275	1,600	
	1800	225 amp main, 30 circuits		2 Elec	.68	23.529		1,400	960		2,360	2,975	
	1850	42 circuits			.52	30.769		1,625	1,250		2,875	3,650	
	1900	400 amp main, 30 circuits			.54	29.630		1,925	1,200		3,125	3,925	
	1950	42 circuits		↓	.50	32	↓	2,150	1,300		3,450	4,325	
	2000	4 wire, 120/208 volts with main circuit breaker											
	2050	100 amp main, 24 circuits		1 Elec	.47	17.021	Ea.	850	695		1,545	1,950	
	2100	30 circuits		"	.40	20		965	815		1,780	2,275	
	2200	225 amp main, 32 circuits		2 Elec	.72	22.222		1,625	905		2,530	3,150	
	2250	42 circuits			.56	28.571		1,775	1,175		2,950	3,675	
	2300	400 amp main, 42 circuits			.48	33.333		2,400	1,350		3,750	4,675	
	2350	600 amp main, 42 circuits		↓	.40	40	↓	3,575	1,625		5,200	6,350	
	2400	NEHB, with 20 amp, 1 pole circuit breaker											
	2450	4 wire, 277/480 volts with main circuit breaker											
	2500	100 amp main, 24 circuits		1 Elec	.42	19.048	Ea.	1,750	775		2,525	3,075	
	2550	30 circuits		"	.38	21.053		2,050	860		2,910	3,525	
	2600	225 amp main, 30 circuits		2 Elec	.72	22.222		2,575	905		3,480	4,200	
	2650	42 circuits		"	.56	28.571		3,200	1,175		4,375	5,225	
800	0010	**SWITCHBOARDS** distribution section											800
	0100	Aluminum bus bars, not including breakers											
	0200	120/208 or 277/480 volt, 4 wire, 600 amp		2 Elec	1	16	Ea.	1,450	650		2,100	2,550	
	0300	800 amp			.88	18.182		1,825	740		2,565	3,125	
	0400	1000 amp			.80	20		1,825	815		2,640	3,250	
	0500	1200 amp			.72	22.222		2,625	905		3,530	4,225	
	0600	1600 amp			.66	24.242		3,075	990		4,065	4,850	
	0700	2000 amp			.62	25.806		3,600	1,050		4,650	5,525	
	0800	2500 amp			.60	26.667		4,050	1,075		5,125	6,100	
	0900	3000 amp			.56	28.571		4,925	1,175		6,100	7,125	
	0950	4000 amp		↓	.52	30.769	↓	7,175	1,250		8,425	9,775	
820	0010	**SWITCHBOARDS** feeder section group mounted devices											820
	0030	Circuit breakers											
	0160	FA frame, 15 to 60 amp, 240 volt, 1 pole		1 Elec	8	1	Ea.	79	41		120	148	
	0280	FA frame, 70 to 100 amp, 240 volt, 1 pole			7	1.143		104	46.50		150.50	184	
	0420	KA frame, 70 to 225 amp			3.20	2.500		875	102		977	1,100	
	0430	LA frame, 125 to 400 amp			2.30	3.478		1,975	142		2,117	2,375	
	0460	MA frame, 450 to 600 amp			1.60	5		3,250	204		3,454	3,875	
	0470	700 to 800 amp			1.30	6.154		4,225	251		4,476	5,025	
	0480	MAL frame, 1000 amp			1	8		4,375	325		4,700	5,300	
	0490	PA frame, 1200 amp		↓	.80	10	↓	8,900	410		9,310	10,400	

Important: See the Reference Section for critical supporting data - Reference Nos., Crews, & City Cost Indexes

16440	Swbds, Panels & Control Centers	CREW	DAILY OUTPUT	LABOR-HOURS	UNIT	2005 BARE COSTS				TOTAL INCL O&P	
						MAT.	LABOR	EQUIP.	TOTAL		
820 0500	Branch circuit, fusible switch, 600 volt, double 30/30 amp	1 Elec	4	2	Ea.	850	81.50		931.50	1,050	**820**
0550	60/60 amp		3.20	2.500		865	102		967	1,100	
0600	100/100 amp		2.70	2.963		1,200	121		1,321	1,500	
0650	Single, 30 amp		5.30	1.509		535	61.50		596.50	680	
0700	60 amp		4.70	1.702		535	69.50		604.50	695	
0750	100 amp		4	2		810	81.50		891.50	1,025	
0800	200 amp		2.70	2.963		1,200	121		1,321	1,475	
0850	400 amp		2.30	3.478		2,250	142		2,392	2,675	
0900	600 amp		1.80	4.444		2,675	181		2,856	3,225	
0950	800 amp		1.30	6.154		4,375	251		4,626	5,175	
1000	1200 amp	▼	.80	10	▼	5,225	410		5,635	6,325	
840 0010	**SWITCHBOARDS** Incoming main service section										**840**
0100	Aluminum bus bars, not including CT's or PT's										
0200	No main disconnect, includes CT compartment										
0300	120/208 volt, 4 wire, 600 amp	2 Elec	1	16	Ea.	2,925	650		3,575	4,175	
0400	800 amp		.88	18.182		2,925	740		3,665	4,300	
0500	1000 amp		.80	20		3,500	815		4,315	5,075	
0600	1200 amp		.72	22.222		3,500	905		4,405	5,200	
0700	1600 amp		.66	24.242		3,500	990		4,490	5,325	
0800	2000 amp		.62	25.806		3,775	1,050		4,825	5,725	
1000	3000 amp	▼	.56	28.571	▼	4,975	1,175		6,150	7,200	
2000	Fused switch & CT compartment										
2100	120/208 volt, 4 wire, 400 amp	2 Elec	1.12	14.286	Ea.	3,775	580		4,355	5,025	
2200	600 amp		.94	17.021		4,475	695		5,170	5,925	
2300	800 amp		.84	19.048		7,875	775		8,650	9,825	
2400	1200 amp	▼	.68	23.529	▼	10,300	960		11,260	12,700	
2900	Pressure switch & CT compartment										
3000	120/208 volt, 4 wire, 800 amp	2 Elec	.80	20	Ea.	7,075	815		7,890	9,025	
3100	1200 amp		.66	24.242		13,700	990		14,690	16,600	
3200	1600 amp		.62	25.806		14,600	1,050		15,650	17,600	
3300	2000 amp	▼	.56	28.571	▼	15,500	1,175		16,675	18,700	
4400	Circuit breaker, molded case & CT compartment										
4600	3 pole, 4 wire, 600 amp	2 Elec	.94	17.021	Ea.	6,100	695		6,795	7,750	
4800	800 amp		.84	19.048		7,300	775		8,075	9,175	
5000	1200 amp	▼	.68	23.529	▼	9,950	960		10,910	12,400	
5100	Copper bus bars, not incl. CT's or PT's, add, minimum					15%					

16450	Enclosed Bus Assemblies										
320 0010	**COPPER BUS DUCT** 10 ft long										**320**
0050	3 pole 4 wire, plug-in/indoor, straight section, 225 amp	2 Elec	40	.400	L.F.	141	16.30		157.30	180	
1000	400 amp		32	.500		141	20.50		161.50	186	
1500	600 amp		26	.615		141	25		166	193	
2400	800 amp		20	.800		167	32.50		199.50	233	
2450	1000 amp		18	.889		185	36		221	257	
2500	1350 amp		16	1		264	41		305	350	
2510	1600 amp		12	1.333		299	54.50		353.50	410	
2520	2000 amp		10	1.600		380	65		445	510	
2550	Feeder, 600 amp		28	.571		123	23.50		146.50	171	
2600	800 amp		22	.727		150	29.50		179.50	209	
2700	1000 amp		20	.800		167	32.50		199.50	233	
2800	1350 amp		18	.889		246	36		282	325	
2900	1600 amp		14	1.143		282	46.50		328.50	380	
3000	2000 amp	▼	12	1.333	▼	360	54.50		414.50	475	
3100	Elbows, 225 amp	▼	4	4	Ea.	835	163		998	1,175	

ELECTRICAL 16

16450 | Enclosed Bus Assemblies

		CREW	DAILY OUTPUT	LABOR-HOURS	UNIT	MAT.	LABOR	EQUIP.	TOTAL	TOTAL INCL O&P	
320							2005 BARE COSTS				**320**
3200	400 amp	2 Elec	3.60	4.444	Ea.	835	181		1,016	1,200	
3300	600 amp		3.20	5		835	204		1,039	1,225	
3400	800 amp		2.80	5.714		905	233		1,138	1,350	
3500	1000 amp		2.60	6.154		1,000	251		1,251	1,475	
3600	1350 amp		2.40	6.667		1,175	272		1,447	1,700	
3700	1600 amp		2.20	7.273		1,300	296		1,596	1,875	
3800	2000 amp		1.80	8.889		1,600	360		1,960	2,300	
4000	End box, 225 amp		34	.471		113	19.20		132.20	153	
4100	400 amp		32	.500		113	20.50		133.50	155	
4200	600 amp		28	.571		113	23.50		136.50	159	
4300	800 amp		26	.615		113	25		138	162	
4400	1000 amp		24	.667		113	27		140	165	
4500	1350 amp		22	.727		113	29.50		142.50	168	
4600	1600 amp		20	.800		113	32.50		145.50	173	
4700	2000 amp		18	.889		138	36		174	206	
4800	Cable tap box end, 225 amp		3.20	5		840	204		1,044	1,225	
5000	400 amp		2.60	6.154		840	251		1,091	1,300	
5100	600 amp		2.20	7.273		840	296		1,136	1,375	
5200	800 amp		2	8		940	325		1,265	1,500	
5300	1000 amp		1.60	10		1,025	410		1,435	1,750	
5400	1350 amp		1.40	11.429		1,250	465		1,715	2,075	
5500	1600 amp		1.20	13.333		1,400	545		1,945	2,325	
5600	2000 amp		1	16		1,600	650		2,250	2,725	
5700	Switchboard stub, 225 amp		5.40	2.963		845	121		966	1,100	
5800	400 amp		4.60	3.478		845	142		987	1,150	
5900	600 amp		4	4		845	163		1,008	1,175	
6000	800 amp		3.20	5		1,025	204		1,229	1,425	
6100	1000 amp		3	5.333		1,175	217		1,392	1,625	
6200	1350 amp		2.60	6.154		1,525	251		1,776	2,050	
6300	1600 amp		2.40	6.667		1,725	272		1,997	2,300	
6400	2000 amp		2	8		2,075	325		2,400	2,775	
6490	Tee fittings, 225 amp		2.40	6.667		1,150	272		1,422	1,650	
6500	400 amp		2	8		1,150	325		1,475	1,725	
6600	600 amp		1.80	8.889		1,150	360		1,510	1,800	
6700	800 amp		1.60	10		1,325	410		1,735	2,050	
6800	1350 amp		1.20	13.333		1,900	545		2,445	2,900	
7000	1600 amp		1	16		2,175	650		2,825	3,350	
7100	2000 amp		.80	20		2,575	815		3,390	4,050	
7200	Plug-in fusible switches w/3 fuses, 600 volt, 3 pole, 30 amp	1 Elec	4	2		500	81.50		581.50	670	
7300	60 amp		3.60	2.222		565	90.50		655.50	755	
7400	100 amp		2.70	2.963		860	121		981	1,125	
7500	200 amp	2 Elec	3.20	5		1,525	204		1,729	2,000	
7600	400 amp		1.40	11.429		4,500	465		4,965	5,650	
7700	600 amp		.90	17.778		5,100	725		5,825	6,700	
7800	800 amp		.66	24.242		7,200	990		8,190	9,400	
7900	1200 amp		.50	32		13,500	1,300		14,800	16,900	
8000	Plug-in circuit breakers, molded case, 15 to 50 amp	1 Elec	4.40	1.818		475	74		549	635	
8100	70 to 100 amp	"	3.10	2.581		525	105		630	735	
8200	150 to 225 amp	2 Elec	3.40	4.706		1,425	192		1,617	1,850	
8300	250 to 400 amp		1.40	11.429		2,500	465		2,965	3,450	
8400	500 to 600 amp		1	16		3,400	650		4,050	4,700	
8500	700 to 800 amp		.64	25		4,175	1,025		5,200	6,125	
8600	900 to 1000 amp		.56	28.571		5,975	1,175		7,150	8,300	
8700	1200 amp		.44	36.364		7,175	1,475		8,650	10,100	

16 ELECTRICAL

Important: See the Reference Section for critical supporting data - Reference Nos., Crews, & City Cost Indexes

16510	Interior Luminaires	CREW	DAILY OUTPUT	LABOR-HOURS	UNIT	2005 BARE COSTS				TOTAL INCL O&P
						MAT.	LABOR	EQUIP.	TOTAL	
440 0010	**INTERIOR LIGHTING FIXTURES** Including lamps, mounting									**440**
0030	hardware and connections									
0100	Fluorescent, C.W. lamps, troffer, recess mounted in grid, RS									
0200	Acrylic lens, 1'W x 4'L, two 40 watt	1 Elec	5.70	1.404	Ea.	48	57		105	138
0210	1'W x 4'L, three 40 watt		5.40	1.481		56	60.50		116.50	152
0300	2'W x 2'L, two U40 watt		5.70	1.404		51.50	57		108.50	142
0400	2'W x 4'L, two 40 watt		5.30	1.509		51.50	61.50		113	148
0500	2'W x 4'L, three 40 watt		5	1.600		55	65		120	158
0600	2'W x 4'L, four 40 watt **CN**		4.70	1.702		58.50	69.50		128	167
0700	4'W x 4'L, four 40 watt	2 Elec	6.40	2.500		310	102		412	490
0800	4'W x 4'L, six 40 watt		6.20	2.581		320	105		425	505
0900	4'W x 4'L, eight 40 watt		5.80	2.759		325	112		437	525
0910	Acrylic lens, 1'W x 4'L, two 32 watt	1 Elec	5.70	1.404		56	57		113	147
0930	2'W x 2'L, two U32 watt		5.70	1.404		80	57		137	173
0940	2'W x 4'L, two 32 watt		5.30	1.509		72	61.50		133.50	171
0950	2'W x 4'L, three 32 watt		5	1.600		76.50	65		141.50	181
0960	2'W x 4'L, four 32 watt		4.70	1.702		78.50	69.50		148	190
1000	Surface mounted, RS									
1030	Acrylic lens with hinged & latched door frame									
1100	1'W x 4'L, two 40 watt	1 Elec	7	1.143	Ea.	73.50	46.50		120	151
1110	1'W x 4'L, three 40 watt		6.70	1.194		75.50	48.50		124	156
1200	2'W x 2'L, two U40 watt		7	1.143		79	46.50		125.50	156
1300	2'W x 4'L, two 40 watt		6.20	1.290		90.50	52.50		143	178
1400	2'W x 4'L, three 40 watt		5.70	1.404		91.50	57		148.50	185
1500	2'W x 4'L, four 40 watt		5.30	1.509		93.50	61.50		155	195
1600	4'W x 4'L, four 40 watt	2 Elec	7.20	2.222		425	90.50		515.50	605
1700	4'W x 4'L, six 40 watt		6.60	2.424		465	99		564	655
1800	4'W x 4'L, eight 40 watt		6.20	2.581		480	105		585	685
1900	2'W x 8'L, four 40 watt		6.40	2.500		163	102		265	330
2000	2'W x 8'L, eight 40 watt		6.20	2.581		187	105		292	365
2100	Strip fixture									
2130	Surface mounted									
2200	4' long, one 40 watt RS	1 Elec	8.50	.941	Ea.	28	38.50		66.50	88
2300	4' long, two 40 watt RS		8	1		30.50	41		71.50	94
2400	4' long, one 40 watt, SL		8	1		41	41		82	106
2500	4' long, two 40 watt, SL		7	1.143		56	46.50		102.50	132
2600	8' long, one 75 watt, SL	2 Elec	13.40	1.194		42.50	48.50		91	119
2700	8' long, two 75 watt, SL	"	12.40	1.290		51	52.50		103.50	135
2800	4' long, two 60 watt, HO	1 Elec	6.70	1.194		82.50	48.50		131	163
2900	8' long, two 110 watt, HO	2 Elec	10.60	1.509		86.50	61.50		148	187
3000	Pendent mounted, industrial, white porcelain enamel									
3100	4' long, two 40 watt, RS	1 Elec	5.70	1.404	Ea.	46.50	57		103.50	136
3200	4' long, two 60 watt, HO	"	5	1.600		74	65		139	178
3300	8' long, two 75 watt, SL	2 Elec	8.80	1.818		87.50	74		161.50	207
3400	8' long, two 110 watt, HO	"	8	2		112	81.50		193.50	245
3470	Troffer, air handling, 2'W x 4'L with four 40 watt, RS	1 Elec	4	2		81	81.50		162.50	210
3480	2'W x 2'L with two U40 watt RS		5.50	1.455		70.50	59.50		130	166
3490	Air connector insulated, 5" diameter		20	.400		55	16.30		71.30	85
3500	6" diameter		20	.400		56	16.30		72.30	86.50
4450	Incandescent, high hat can, round alzak reflector, prewired									
4470	100 watt	1 Elec	8	1	Ea.	58	41		99	125
4480	150 watt		8	1		85	41		126	154
4500	300 watt		6.70	1.194		195	48.50		243.50	288
4600	Square glass lens with metal trim, prewired									
4630	100 watt	1 Elec	6.70	1.194	Ea.	45	48.50		93.50	122
4700	200 watt		6.70	1.194		78.50	48.50		127	159

ELECTRICAL 16

16510 | Interior Luminaires

			CREW	DAILY OUTPUT	LABOR-HOURS	UNIT	MAT.	LABOR	EQUIP.	TOTAL	TOTAL INCL O&P
440	4800	300 watt	1 Elec	5.70	1.404	Ea.	122	57		179	219	**440**
	4900	Ceiling/wall, surface mounted, metal cylinder, 75 watt		10	.800		46	32.50		78.50	99
	4920	150 watt	↓	10	.800	↓	66	32.50		98.50	121
	5200	Ceiling, surface mounted, opal glass drum									
	5300	8", one 60 watt lamp	1 Elec	10	.800	Ea.	36	32.50		68.50	88
	5400	10", two 60 watt lamps		8	1		40.50	41		81.50	105
	5500	12", four 60 watt lamps		6.70	1.194		55	48.50		103.50	133
	6010	Vapor tight, incandescent, ceiling mounted, 200 watt		6.20	1.290		51	52.50		103.50	135
	6100	Fluorescent, surface mounted, 2 lamps, 4'L, RS, 40 watt	↓	3.20	2.500	↓	94.50	102		196.50	256
	6300	Explosionproof									
	6510	Incandescent, ceiling mounted, 200 watt	1 Elec	4	2	Ea.	670	81.50		751.50	855
	6600	Fluorescent, RS, 4' long, ceiling mounted, two 40 watt		2.70	2.963		1,675	121		1,796	2,025
	6850	Vandalproof, surface mounted, fluorescent, two 40 watt		3.20	2.500		201	102		303	375
	6860	Incandescent, one 150 watt	↓	8	1	↓	53	41		94	119
	7500	Ballast replacement, by weight of ballast, to 15' high									
	7520	Indoor fluorescent, less than 2 lbs.	1 Elec	10	.800	Ea.		32.50		32.50	48.50
	7540	Two 40W, watt reducer, 2 to 5 lbs.		9.40	.851		23	34.50		57.50	77
	7560	Two F96 slimline, over 5 lbs.		8	1		39	41		80	103
	7580	Vaportite ballast, less than 2 lbs.		9.40	.851			34.50		34.50	51.50
	7600	2 lbs. to 5 lbs.		8.90	.899			36.50		36.50	54.50
	7620	Over 5 lbs.		7.60	1.053			43		43	64
	7630	Electronic ballast for two tubes		8	1		34.50	41		75.50	98.50
	7640	Dimmable ballast one lamp		8	1		49.50	41		90.50	115
	7650	Dimmable ballast two-lamp	↓	7.60	1.053		80	43		123	152

16520 | Exterior Luminaires

			CREW	DAILY OUTPUT	LABOR-HOURS	UNIT	MAT.	LABOR	EQUIP.	TOTAL	TOTAL INCL O&P
300	0010	**EXTERIOR FIXTURES** With lamps										**300**
	0200	Wall mounted, incandescent, 100 watt	1 Elec	8	1	Ea.	28	41		69	91.50
	0400	Quartz, 500 watt		5.30	1.509		54	61.50		115.50	151
	1100	Wall pack, low pressure sodium, 35 watt		4	2		214	81.50		295.50	355
	1150	55 watt	↓	4	2	↓	255	81.50		336.50	400
	1200	Floodlights with ballast and lamp,									
	1400	pole mounted, pole not included									
	1950	Metal halide, 175 watt	1 Elec	2.70	2.963	Ea.	300	121		421	510
	2000	400 watt	2 Elec	4.40	3.636		350	148		498	605
	2200	1000 watt	"	4	4		505	163		668	800
	2340	High pressure sodium, 70 watt	1 Elec	2.70	2.963		207	121		328	405
	2400	400 watt	2 Elec	4.40	3.636		340	148		488	590
	2600	1000 watt	"	4	4		500	163		663	795
	2650	Roadway area luminaire, low pressure sodium, 135 watt	1 Elec	2	4		535	163		698	835
	2700	180 watt	"	2	4		565	163		728	865
	2750	Metal halide, 400 watt	2 Elec	4.40	3.636		445	148		593	705
	2760	1000 watt		4	4		500	163		663	795
	2780	High pressure sodium, 400 watt		4.40	3.636		460	148		608	725
	2790	1000 watt	↓	4	4	↓	525	163		688	825
	2800	Light poles, anchor base									
	2820	not including concrete bases									
	2840	Aluminum pole, 8' high	1 Elec	4	2	Ea.	520	81.50		601.50	690
	3000	20' high	R-3	2.90	6.897		675	276	54.50	1,005.50	1,225
	3200	30' high		2.60	7.692		1,325	310	61	1,696	1,975
	3400	35' high		2.30	8.696		1,425	350	69	1,844	2,175
	3600	40' high	↓	2	10		1,625	400	79.50	2,104.50	2,475
	3800	Bracket arms, 1 arm	1 Elec	8	1		88.50	41		129.50	158
	4000	2 arms		8	1		178	41		219	257
	4200	3 arms		5.30	1.509		267	61.50		328.50	385
	4400	4 arms		5.30	1.509		355	61.50		416.50	480

Important: See the Reference Section for critical supporting data - Reference Nos., Crews, & City Cost Indexes

16 ELECTRICAL

16520 | Exterior Luminaires

		CREW	DAILY OUTPUT	LABOR-HOURS	UNIT	2005 BARE COSTS				TOTAL INCL O&P	
						MAT.	LABOR	EQUIP.	TOTAL		
300	4500	Steel pole, galvanized, 8' high	1 Elec	3.80	2.105	Ea.	460	86		546	640	**300**
	4600	20' high	R-3	2.60	7.692		820	310	61	1,191	1,425
	4800	30' high		2.30	8.696		965	350	69	1,384	1,650
	5000	35' high		2.20	9.091		1,050	365	72	1,487	1,775
	5200	40' high	↓	1.70	11.765		1,300	470	93.50	1,863.50	2,225
	5400	Bracket arms, 1 arm	1 Elec	8	1		134	41		175	209
	5600	2 arms		8	1		208	41		249	290
	5800	3 arms		5.30	1.509		226	61.50		287.50	340
	6000	4 arms	↓	5.30	1.509	↓	315	61.50		376.50	435

16530 | Emergency Lighting

		CREW	DAILY OUTPUT	LABOR-HOURS	UNIT	MAT.	LABOR	EQUIP.	TOTAL	TOTAL INCL O&P	
320	0010	**EXIT AND EMERGENCY LIGHTING**										**320**
	0080	Exit light ceiling or wall mount, incandescent, single face	1 Elec	8	1	Ea.	40.50	41		81.50	106
	0100	Double face		6.70	1.194		46	48.50		94.50	123
	0200	L.E.D. standard, single face		8	1		64.50	41		105.50	132
	0220	Double face		6.70	1.194		67.50	48.50		116	147
	0240	L.E.D. w/battery unit, single face		4.40	1.818		109	74		183	230
	0260	Double face	↓	4	2	↓	111	81.50		192.50	243
	0300	Emergency light units, battery operated									
	0350	Twin sealed beam light, 25 watt, 6 volt each									
	0500	Lead battery operated	1 Elec	4	2	Ea.	114	81.50		195.50	247
	0700	Nickel cadmium battery operated		4	2		525	81.50		606.50	695
	0900	Self-contained fluorescent lamp pack	↓	10	.800	↓	127	32.50		159.50	189

16585 | Lamps

		CREW	DAILY OUTPUT	LABOR-HOURS	UNIT	MAT.	LABOR	EQUIP.	TOTAL	TOTAL INCL O&P	
600	0010	**LAMPS**										**600**
	0080	Fluorescent, rapid start, cool white, 2' long, 20 watt	1 Elec	1	8	C	310	325		635	825
	0100	4' long, 40 watt		.90	8.889		281	360		641	850
	0200	Slimline, 4' long, 40 watt		.90	8.889		670	360		1,030	1,275
	0210	4' long, 30 watt energy saver		.90	8.889		670	360		1,030	1,275
	0400	High output, 4' long, 60 watt		.90	8.889		890	360		1,250	1,525
	0410	8' long, 95 watt energy saver		.80	10		815	410		1,225	1,500
	0500	8' long, 110 watt		.80	10		815	410		1,225	1,500
	0512	2' long, T5, 14 watt energy saver		1	8		1,050	325		1,375	1,625
	0514	3' long, T5, 21 watt energy saver		.90	8.889		1,050	360		1,410	1,700
	0516	4' long, T5, 28 watt energy saver		.90	8.889		900	360		1,260	1,525
	0560	Twin tube compact lamp		.90	8.889		510	360		870	1,100
	0570	Double twin tube compact lamp		.80	10		1,250	410		1,660	1,950
	0600	Mercury vapor, mogul base, deluxe white, 100 watt		.30	26.667		2,575	1,075		3,650	4,475
	0700	250 watt		.30	26.667		3,425	1,075		4,500	5,375
	0800	400 watt		.30	26.667		2,750	1,075		3,825	4,650
	0900	1000 watt		.20	40		6,425	1,625		8,050	9,475
	1000	Metal halide, mogul base, 175 watt		.30	26.667		3,225	1,075		4,300	5,175
	1200	400 watt		.30	26.667		3,475	1,075		4,550	5,425
	1300	1000 watt		.20	40		9,100	1,625		10,725	12,400
	1350	High pressure sodium, 70 watt		.30	26.667		3,750	1,075		4,825	5,750
	1380	250 watt		.30	26.667		4,275	1,075		5,350	6,350
	1400	400 watt		.30	26.667		4,400	1,075		5,475	6,475
	1450	1000 watt		.20	40		13,200	1,625		14,825	16,900
	3000	Guards, fluorescent lamp, 4' long		1	8		750	325		1,075	1,300
	3200	8' long	↓	.90	8.889	↓	1,500	360		1,860	2,200

ELECTRICAL 16

16710 | Communication Circuits

		CREW	DAILY OUTPUT	LABOR-HOURS	UNIT	2005 BARE COSTS MAT.	LABOR	EQUIP.	TOTAL	TOTAL INCL O&P	
400	**0010**	**FIBER OPTICS**									**400**
	0020	Fiber optics cable only. Added costs depend on the type of fiber									
	0030	special connectors, optical modems, and networking parts.									
	0040	Specialized tools & techniques cause installation costs to vary.									
	0070	Cable, minimum, bulk simplex	1 Elec	8	1	C.L.F.	24	41		65	87	
	0080	Cable, maximum, bulk plenum quad	"	2.29	3.493	"	117	142		259	340	
	0150	Fiber optic jumper				Ea.	55.50			55.50	61	
	0200	Fiber optic pigtail					30			30	33	
	0300	Fiber optic connector	1 Elec	24	.333		16.90	13.60		30.50	38.50	
	0350	Fiber optic finger splice		32	.250		32	10.20		42.20	50	
	0400	Transceiver (low cost bi-directional)		8	1		425	41		466	530	
	0450	Multi-channel rack enclosure (10 modules)		2	4		475	163		638	770	
	0500	Fiber optic patch panel (12 ports)	▼	6	1.333	▼	178	54.50		232.50	277	

16720 | Tel and Intercomm Equipment

		CREW	DAILY OUTPUT	LABOR-HOURS	UNIT	2005 BARE COSTS MAT.	LABOR	EQUIP.	TOTAL	TOTAL INCL O&P	
600	**0010**	**NURSE CALL SYSTEMS**									**600**
	0100	Single bedside call station	1 Elec	8	1	Ea.	190	41		231	270	
	0200	Ceiling speaker station		8	1		55.50	41		96.50	122	
	0400	Emergency call station		8	1		94	41		135	165	
	0600	Pillow speaker		8	1		174	41		215	253	
	0800	Double bedside call station		4	2		180	81.50		261.50	320	
	1000	Duty station		4	2		143	81.50		224.50	278	
	1200	Standard call button		8	1		76	41		117	144	
	1400	Lights, corridor, dome or zone indicator	▼	8	1	▼	42.50	41		83.50	107	
	1600	Master control station for 20 stations	2 Elec	.65	24.615	Total	3,375	1,000		4,375	5,225	

16820 | Sound Reinforcement

		CREW	DAILY OUTPUT	LABOR-HOURS	UNIT	2005 BARE COSTS MAT.	LABOR	EQUIP.	TOTAL	TOTAL INCL O&P	
300	**0010**	**DOORBELL SYSTEM** Incl. transformer, button & signal									**300**
	0100	6" bell	1 Elec	4	2	Ea.	86	81.50		167.50	216	
	0200	Buzzer	"	4	2	"	68.50	81.50		150	196	
800	**0010**	**PUBLIC ADDRESS SYSTEM**									**800**
	0100	Conventional, office	1 Elec	5.33	1.501	Speaker	103	61		164	205	
	0200	Industrial	"	2.70	2.963	"	199	121		320	400	
	0400	Explosionproof system is 3 times cost of central control										
	0600	Installation costs run about 120% of material cost										
840	**0010**	**SOUND SYSTEM** not including rough-in wires, cables & conduits									**840**
	0100	Components, outlet, projector	1 Elec	8	1	Ea.	47.50	41		88.50	113	
	0200	Microphone		4	2		53	81.50		134.50	180	
	0400	Speakers, ceiling or wall		8	1		90	41		131	160	
	0600	Trumpets		4	2		167	81.50		248.50	305	
	0800	Privacy switch		8	1		67	41		108	134	
	1000	Monitor panel		4	2		298	81.50		379.50	450	
	1200	Antenna, AM/FM		4	2		166	81.50		247.50	305	
	1400	Volume control		8	1		67	41		108	134	
	1600	Amplifier, 250 watts	▼	1	8	▼	1,325	325		1,650	1,925	

16 ELECTRICAL

Important: See the Reference Section for critical supporting data - Reference Nos., Crews, & City Cost Indexes

16820	Sound Reinforcement	CREW	DAILY OUTPUT	LABOR-HOURS	UNIT	2005 BARE COSTS				TOTAL INCL O&P	
						MAT.	LABOR	EQUIP.	TOTAL		
840	1800	Cabinets	1 Elec	1	8	Ea.	650	325		975	1,200	**840**
	2000	Intercom, 25 station capacity, master station	2 Elec	2	8		1,550	325		1,875	2,200	
	2200	Remote station	1 Elec	8	1		125	41		166	199	
	2400	Intercom outlets		8	1		73.50	41		114.50	142	
	2600	Handset		4	2		243	81.50		324.50	390	
	2800	Emergency call system, 12 zones, annunciator		1.30	6.154		735	251		986	1,175	
	3000	Bell		5.30	1.509		75.50	61.50		137	175	
	3200	Light or relay		8	1		38	41		79	102	
	3400	Transformer		4	2		166	81.50		247.50	305	
	3600	House telephone, talking station		1.60	5		355	204		559	700	
	3800	Press to talk, release to listen	↓	5.30	1.509		83	61.50		144.50	183	
	4000	System-on button					50			50	55	
	4200	Door release	1 Elec	4	2		89	81.50		170.50	219	
	4400	Combination speaker and microphone		8	1		152	41		193	228	
	4600	Termination box		3.20	2.500		47.50	102		149.50	205	
	4800	Amplifier or power supply		5.30	1.509	↓	550	61.50		611.50	695	
	5000	Vestibule door unit		16	.500	Name	101	20.50		121.50	142	
	5200	Strip cabinet		27	.296	Ea.	190	12.05		202.05	227	
	5400	Directory	↓	16	.500	"	89.50	20.50		110	129	

16850	Television Equipment	CREW	DAILY OUTPUT	LABOR-HOURS	UNIT	MAT.	LABOR	EQUIP.	TOTAL	INCL O&P	
600	0010	**T.V. SYSTEMS** not including rough-in wires, cables & conduits										**600**
	0100	Master TV antenna system										
	0200	VHF reception & distribution, 12 outlets	1 Elec	6	1.333	Outlet	164	54.50		218.50	262	
	0400	30 outlets		10	.800		108	32.50		140.50	168	
	0600	100 outlets		13	.615		111	25		136	160	
	0800	VHF & UHF reception & distribution, 12 outlets		6	1.333		163	54.50		217.50	261	
	1000	30 outlets		10	.800		108	32.50		140.50	168	
	1200	100 outlets		13	.615		110	25		135	159	
	1400	School and deluxe systems, 12 outlets		2.40	3.333		215	136		351	440	
	1600	30 outlets		4	2		189	81.50		270.50	330	
	1800	80 outlets	↓	5.30	1.509	↓	181	61.50		242.50	291	
	2000	Closed circuit, surveillance, one station (camera & monitor)	2 Elec	2.60	6.154	Total	1,050	251		1,301	1,525	
	2200	For additional camera stations, add	1 Elec	2.70	2.963	Ea.	585	121		706	820	
	2400	Industrial quality, one station (camera & monitor)	2 Elec	2.60	6.154	Total	2,150	251		2,401	2,750	
	2600	For additional camera stations, add	1 Elec	2.70	2.963	Ea.	1,325	121		1,446	1,625	
	2610	For low light, add		2.70	2.963		1,075	121		1,196	1,350	
	2620	For very low light, add		2.70	2.963		7,850	121		7,971	8,800	
	2800	For weatherproof camera station, add		1.30	6.154		815	251		1,066	1,275	
	3000	For pan and tilt, add		1.30	6.154		2,100	251		2,351	2,700	
	3200	For zoom lens - remote control, add, minimum		2	4		1,950	163		2,113	2,400	
	3400	Maximum		2	4		7,100	163		7,263	8,075	
	3410	For automatic iris for low light, add	↓	2	4	↓	1,700	163		1,863	2,125	
	3600	Educational T.V. studio, basic 3 camera system, black & white,										
	3800	electrical & electronic equip. only, minimum	4 Elec	.80	40	Total	10,100	1,625		11,725	13,500	
	4000	Maximum (full console)		.28	114		43,000	4,650		47,650	54,000	
	4100	As above, but color system, minimum		.28	114		57,000	4,650		61,650	69,500	
	4120	Maximum	↓	.12	266	↓	247,500	10,900		258,400	288,500	
	4200	For film chain, black & white, add	1 Elec	1	8	Ea.	11,600	325		11,925	13,200	
	4250	Color, add		.25	32		14,100	1,300		15,400	17,500	
	4400	For video tape recorders, add, minimum	↓	1	8		2,425	325		2,750	3,150	
	4600	Maximum	4 Elec	.40	80	↓	20,200	3,250		23,450	27,100	

ELECTRICAL 16

For information about Means Estimating Seminars, see yellow pages 12 and 13 in back of book

Division Notes

	CREW	DAILY OUTPUT	LABOR-HOURS	UNIT	2005 BARE COSTS				TOTAL INCL O&P
					MAT.	LABOR	EQUIP.	TOTAL	

Division 17
Square Foot & Cubic Foot Costs

Estimating Tips

- The cost figures in Division 17 were derived from approximately 11,200 projects contained in the Means database of completed construction projects, and include the contractor's overhead and profit, but do not generally include architectural fees or land costs. The figures have been adjusted to January of the current year. New projects are added to our files each year, and outdated projects are discarded. For this reason, certain costs may not show a uniform annual progression. In no case are all subdivisions of a project listed.
- These projects were located throughout the U.S. and reflect a tremendous variation in square foot (S.F.) and cubic foot (C.F.) costs. This is due to differences, not only in labor and material costs, but also in individual owners' requirements. For instance, a bank in a large city would have different features than one in a rural area. This is true of all the different types of buildings analyzed. Therefore, caution should be exercised when using Division 17 costs. For example, for court houses, costs in the database are local court house costs and will not apply to the larger, more elaborate federal court houses. As a general rule, the projects in the 1/4 column do not include any site work or equipment, while the projects in the 3/4 column may include both equipment and site work. The median figures do not generally include site work.

- None of the figures "go with" any others. All individual cost items were computed and tabulated separately. Thus the sum of the median figures for Plumbing, HVAC and Electrical will not normally total up to the total Mechanical and Electrical costs arrived at by separate analysis and tabulation of the projects.
- Each building was analyzed as to total and component costs and percentages. The figures were arranged in ascending order with the results tabulated as shown. The 1/4 column shows that 25% of the projects had lower costs, 75% higher. The 3/4 column shows that 75% of the projects had lower costs, 25% had higher. The median column shows that 50% of the projects had lower costs, 50% had higher.
- There are two times when square foot costs are useful. The first is in the conceptual stage when no details are available. Then square foot costs make a useful starting point. The second is after the bids are in and the costs can be worked back into their appropriate units for information purposes. As soon as details become available in the project design, the square foot approach should be discontinued and the project priced as to its particular components. When more precision is required or for estimating the replacement cost of specific buildings, the current edition of *Means Square Foot Costs* should be used.

- In using the figures in Division 17, it is recommended that the median column be used for preliminary figures if no additional information is available. The median figures, when multiplied by the total city construction cost index figures (see City Cost Indexes) and then multiplied by the project size modifier in Reference Number R17100-100, should present a fairly accurate base figure, which would then have to be adjusted in view of the estimator's experience, local economic conditions, code requirements and the owner's particular requirements. There is no need to factor the percentage figures, as these should remain constant from city to city. All tabulations mentioning air conditioning had at least partial air conditioning.
- The editors of this book would greatly appreciate receiving cost figures on one or more of your recent projects which would then be included in the averages for next year. All cost figures received will be kept confidential except that they will be averaged with other similar projects to arrive at S.F. and C.F. cost figures for next year's book. See the last page of the book for details and the discount available for submitting one or more of your projects.

17100 | S.F. & C.F. Costs

		UNIT	UNIT COSTS 1/4	UNIT COSTS MEDIAN	UNIT COSTS 3/4	% OF TOTAL 1/4	% OF TOTAL MEDIAN	% OF TOTAL 3/4	
010	0010	**APARTMENTS** Low Rise (1 to 3 story) `R17100 -100`	S.F.	53.50	67.50	90				010
	0020	Total project cost	C.F.	4.80	6.35	7.85				
	0100	Site work	S.F.	4.58	6.25	10.60	6.05%	10.55%	14.05%	
	0500	Masonry		1.05	2.45	4.24	1.54%	3.67%	6.35%	
	1500	Finishes		5.70	7.80	9.65	9.05%	10.75%	12.85%	
	1800	Equipment		1.75	2.65	3.95	2.73%	4.03%	5.95%	
	2720	Plumbing		4.17	5.35	6.80	6.65%	8.95%	10.05%	
	2770	Heating, ventilating, air conditioning		2.66	3.28	4.82	4.20%	5.60%	7.60%	
	2900	Electrical		3.10	4.06	5.55	5.20%	6.65%	8.40%	
	3100	Total: Mechanical & Electrical		10.75	13.70	17.05	15.90%	18.05%	23%	
	9000	Per apartment unit, total cost	Apt.	49,900	75,500	112,500				
	9500	Total: Mechanical & Electrical	"	9,425	14,900	19,400				
020	0010	**APARTMENTS** Mid Rise (4 to 7 story) `R17100 -100`	S.F.	71	85.50	105				020
	0020	Total project costs	C.F.	5.55	7.65	10.45				
	0100	Site work	S.F.	2.84	5.65	11.10	5.25%	6.70%	9.20%	
	0500	Masonry		4.72	6.65	9.25	5.15%	7.40%	10.50%	
	1500	Finishes		8.95	12.45	14.50	10.55%	13.45%	17.70%	
	1800	Equipment		2.33	3.33	4.38	2.55%	3.48%	4.31%	
	2500	Conveying equipment		1.61	1.98	2.36	2.05%	2.29%	2.69%	
	2720	Plumbing		4.17	6.65	7.40	5.70%	7.20%	8.95%	
	2900	Electrical		4.90	6.70	7.80	6.65%	7.20%	8.95%	
	3100	Total: Mechanical & Electrical		14.15	18.80	23.50	18.85%	21%	25%	
	9000	Per apartment unit, total cost	Apt.	80,500	95,000	157,000				
	9500	Total: Mechanical & Electrical	"	15,200	17,600	22,200				
030	0010	**APARTMENTS** High Rise (8 to 24 story) `R17100 -100`	S.F.	80.50	97	119				030
	0020	Total project costs	C.F.	6.85	9.55	11.65				
	0100	Site work	S.F.	2.47	4.73	6.85	2.58%	4.84%	6.15%	
	0500	Masonry		4.66	8.50	10.40	4.74%	9.65%	11.05%	
	1500	Finishes		8.95	11.20	13.20	9.75%	11.80%	13.70%	
	1800	Equipment		2.59	3.19	4.22	2.78%	3.49%	4.35%	
	2500	Conveying equipment		1.83	2.78	3.97	2.23%	2.78%	3.37%	
	2720	Plumbing		5.95	7	9.80	6.80%	7.20%	10.45%	
	2900	Electrical		5.55	7	9.45	6.45%	7.65%	8.80%	
	3100	Total: Mechanical & Electrical		16.35	21	25	17.95%	22.50%	24.50%	
	9000	Per apartment unit, total cost	Apt.	83,500	92,000	128,000				
	9500	Total: Mechanical & Electrical	"	18,100	20,700	21,900				
040	0010	**AUDITORIUMS** `R17100 -100`	S.F.	83	113	159				040
	0020	Total project costs	C.F.	5.25	7.30	11				
	2720	Plumbing	S.F.	5.30	7.35	9.25	5.85%	7.20%	8.70%	
	2900	Electrical		6.75	9.35	12.50	6.80%	8.95%	11.30%	
	3100	Total: Mechanical & Electrical		13	17.30	30	24.50%	30.50%	31.50%	
050	0010	**AUTOMOTIVE SALES** `R17100 -100`	S.F.	60.50	84	103				050
	0020	Total project costs	C.F.	4.08	4.90	6.35				
	2720	Plumbing	S.F.	2.82	4.90	5.35	2.89%	6.05%	6.50%	
	2770	Heating, ventilating, air conditioning		4.36	6.65	7.20	4.61%	10%	10.35%	
	2900	Electrical		4.99	7.70	8.60	7.40%	9.95%	12.40%	
	3100	Total: Mechanical & Electrical		13.90	19.55	25	19.15%	20.50%	26%	
060	0010	**BANKS** `R17100 -100`	S.F.	121	151	189				060
	0020	Total project costs	C.F.	8.70	11.80	15.55				
	0100	Site work	S.F.	13.50	21	31	7.75%	12.95%	17.15%	
	0500	Masonry		6.30	11.85	22	3.36%	6.95%	10.80%	
	1500	Finishes		10.90	14.70	18.75	5.85%	8.55%	11.25%	
	1800	Equipment		4.85	10.05	21.50	1.34%	5.95%	10.65%	
	2720	Plumbing		3.81	5.45	7.95	2.82%	3.90%	4.93%	
	2770	Heating, ventilating, air conditioning		7.25	9.65	12.90	4.86%	7.15%	8.50%	
	2900	Electrical		11.50	15.35	19.95	8.25%	10.20%	12.20%	
	3100	Total: Mechanical & Electrical		27	36.50	44	17.75%	19.45%	23%	

17 SQUARE FOOT

See R17100-100 and City Cost Indexes in the Reference Section

17100 | S.F. & C.F. Costs			UNIT	UNIT COSTS			% OF TOTAL		
				1/4	MEDIAN	3/4	1/4	MEDIAN	3/4
060	3500	See also division 11020 & 11030								060
130	0010	**CHURCHES** R17100-100	S.F.	81.50	103	134				130
	0020	Total project costs	C.F.	5.05	6.40	8.45				
	1800	Equipment	S.F.	1.05	2.34	5.15	1.06%	2.33%	4.50%	
	2720	Plumbing		3.18	4.45	6.55	3.51%	4.96%	6.25%	
	2770	Heating, ventilating, air conditioning		7.45	9.70	14.25	7.50%	10%	12%	
	2900	Electrical		6.90	9.40	12.60	7.30%	8.75%	10.90%	
	3100	Total: Mechanical & Electrical	▼	18.75	27	38	18.25%	21.50%	24%	
	3500	See also division 11040								
150	0010	**CLUBS, COUNTRY** R17100-100	S.F.	87.50	106	133				150
	0020	Total project costs	C.F.	7.05	8.60	11.90				
	2720	Plumbing	S.F.	5.65	7.85	17.90	5.60%	7.90%	10%	
	2900	Electrical		6.90	9.85	13	7%	8.95%	11%	
	3100	Total: Mechanical & Electrical	▼	21	36.50	46	19%	26.50%	29.50%	
170	0010	**CLUBS, SOCIAL** Fraternal R17100-100	S.F.	70	100	131				170
	0020	Total project costs	C.F.	4.38	6.65	7.90				
	2720	Plumbing	S.F.	4.40	5.50	6.65	5.60%	6.90%	8.55%	
	2770	Heating, ventilating, air conditioning		6.35	7.70	9.85	8.20%	9.25%	14.40%	
	2900	Electrical		5.55	8.65	10.45	6.50%	9.50%	10.55%	
	3100	Total: Mechanical & Electrical	▼	15.55	27	29.50	21%	23%	37%	
180	0010	**CLUBS, Y.M.C.A.** R17100-100	S.F.	88	119	145				180
	0020	Total project costs	C.F.	4.05	6.80	10.10				
	2720	Plumbing	S.F.	5.55	11.05	12.40	5.65%	7.60%	10.85%	
	2900	Electrical		6.65	9.10	12.95	6.25%	8.65%	10.20%	
	3100	Total: Mechanical & Electrical	▼	19.70	25	33.50	18.40%	22.50%	29.50%	
190	0010	**COLLEGES** Classrooms & Administration R17100-100	S.F.	99	131	175				190
	0020	Total project costs	C.F.	7.15	10.10	15.90				
	0500	Masonry	S.F.	6.50	12.65	14.45	5.65%	8.25%	10.50%	
	2720	Plumbing		4.81	9.65	17.30	5.10%	6.60%	8.95%	
	2900	Electrical		8.05	12.40	14.90	7.70%	9.85%	12%	
	3100	Total: Mechanical & Electrical	▼	19.80	35.50	49	24%	28%	31.50%	
210	0010	**COLLEGES** Science, Engineering, Laboratories R17100-100	S.F.	165	193	239				210
	0020	Total project costs	C.F.	9.45	13.80	15.70				
	1800	Equipment	S.F.	9.20	21	23	2%	6.45%	12.65%	
	2900	Electrical		13.60	19.30	29.50	7.10%	9.40%	12.10%	
	3100	Total: Mechanical & Electrical	▼	50.50	60	93	28.50%	31.50%	41%	
	3500	See also division 11600								
230	0010	**COLLEGES** Student Unions R17100-100	S.F.	105	148	173				230
	0020	Total project costs	C.F.	5.90	7.70	9.95				
	3100	Total: Mechanical & Electrical	S.F.	27.50	42.50	50.50	23.50%	26%	29%	
250	0010	**COMMUNITY CENTERS** R17100-100	"	82.50	107	144				250
	0020	Total project costs	C.F.	5.70	8.10	10.50				
	1800	Equipment	S.F.	2.18	3.72	6.05	1.87%	3.12%	6%	
	2720	Plumbing		4.22	7.20	10.50	4.94%	7%	9.10%	
	2770	Heating, ventilating, air conditioning		6.95	10	13.75	6.95%	10.65%	13.05%	
	2900	Electrical		6.95	9.40	14.30	7.35%	9.10%	10.85%	
	3100	Total: Mechanical & Electrical	▼	26	31	44.50	22.50%	26.50%	32.50%	
280	0010	**COURT HOUSES** R17100-100	S.F.	124	144	166				280
	0020	Total project costs	C.F.	9.60	11.50	14.50				
	2720	Plumbing	S.F.	5.95	8.35	12	5.95%	7.45%	8.20%	
	2900	Electrical		11.70	14	17.50	8.55%	9.95%	11.50%	
	3100	Total: Mechanical & Electrical	▼	30	34	46.50	22.50%	29.50%	30.50%	

SQUARE FOOT 17

17100 | S.F. & C.F. Costs

				UNIT COSTS			% OF TOTAL		
			UNIT	1/4	MEDIAN	3/4	1/4	MEDIAN	3/4
300	0010	**DEPARTMENT STORES** R17100-100	S.F.	46.50	62.50	79			
	0020	Total project costs	C.F.	2.50	3.37	4.74			
	2720	Plumbing	S.F.	1.51	1.83	2.77	1.82%	4.21%	5.90%
	2770	Heating, ventilating, air conditioning		4.22	6.50	9.80	8.20%	9.10%	14.80%
	2900	Electrical		5.35	7.30	8.60	9.05%	12.15%	14.95%
	3100	Total: Mechanical & Electrical		9.35	12	21	13.20%	21.50%	50%
310	0010	**DORMITORIES** Low Rise (1 to 3 story) R17100-100	S.F.	76	111	135			
	0020	Total project costs	C.F.	5.25	8.05	12.05			
	2720	Plumbing	S.F.	5.30	7.10	8.95	8.05%	9%	9.65%
	2770	Heating, ventilating, air conditioning		5.60	6.70	8.95	4.61%	8.05%	10%
	2900	Electrical		5.65	8.55	10.75	6.55%	8.90%	9.55%
	3100	Total: Mechanical & Electrical		29.50	31	32.50	22%	26%	29%
	9000	Per bed, total cost	Bed	37,300	41,400	89,000			
320	0010	**DORMITORIES** Mid Rise (4 to 8 story) R17100-100	S.F.	108	141	171			
	0020	Total project costs	C.F.	11.90	13.10	15.65			
	2900	Electrical	S.F.	8	12.55	14.20	8.20%	9.45%	10.35%
	3100	Total: Mechanical & Electrical	"	24	32	42	19.50%	30.50%	34.50%
	9000	Per bed, total cost	Bed	15,400	35,100	72,500			
340	0010	**FACTORIES** R17100-100	S.F.	41	61	94			
	0020	Total project costs	C.F.	2.62	3.91	6.50			
	0100	Site work	S.F.	4.67	8.50	13.90	6.95%	11.45%	18.25%
	2720	Plumbing		2.29	4.10	6.80	3.73%	6.05%	8.10%
	2770	Heating, ventilating, air conditioning		4.29	6.15	8.30	5.25%	8.45%	11.35%
	2900	Electrical		5.15	8.15	12.25	8.30%	10.55%	14.20%
	3100	Total: Mechanical & Electrical		12.10	19.40	29.50	22%	28.50%	34.50%
360	0010	**FIRE STATIONS** R17100-100	S.F.	78	107	144			
	0020	Total project costs	C.F.	4.74	6.50	8.65			
	0500	Masonry	S.F.	11.90	21.50	28.50	8.60%	11.55%	16.15%
	1140	Roofing		2.64	7.15	8.15	1.90%	4.94%	5.05%
	1580	Painting		2.17	3.07	3.14	1.37%	1.57%	2.07%
	1800	Equipment		1.71	2.43	5.85	.81%	2.50%	4.07%
	2720	Plumbing		5.15	7.55	11.15	5.85%	7.35%	9.50%
	2770	Heating, ventilating, air conditioning		4.42	7.05	11	4.86%	7.25%	9.25%
	2900	Electrical		5.75	9.70	13.05	6.80%	8.75%	10.65%
	3100	Total: Mechanical & Electrical		28	33.50	39	18.75%	23%	27%
370	0010	**FRATERNITY HOUSES** and Sorority Houses R17100-100	S.F.	81	104	143			
	0020	Total project costs	C.F.	7.75	8.40	10.80			
	2720	Plumbing	S.F.	6.15	7	12.85	6.80%	8%	10.85%
	2900	Electrical		5.35	11.55	14.15	6.60%	9.90%	10.65%
	3100	Total: Mechanical & Electrical		14.60	20.50	25		15.10%	15.90%
380	0010	**FUNERAL HOMES** R17100-100	S.F.	85.50	117	212			
	0020	Total project costs	C.F.	8.75	9.70	18.65			
	2900	Electrical	S.F.	3.78	6.95	8.15	3.58%	4.44%	5.95%
	3100	Total: Mechanical & Electrical	"	13.50	20	26.50	12.90%	12.90%	12.90%
390	0010	**GARAGES, COMMERCIAL** (Service) R17100-100	S.F.	48	76	103			
	0020	Total project costs	C.F.	3.26	4.71	6.85			
	1800	Equipment	S.F.	2.73	6.15	9.55	2.69%	4.62%	6.80%
	2720	Plumbing		3.36	5.15	9.45	5.45%	7.85%	10.65%
	2730	Heating & ventilating		4.41	6	8.30	5.25%	6.85%	8.20%
	2900	Electrical		4.67	7.15	10.20	7.15%	9.25%	10.85%
	3100	Total: Mechanical & Electrical		10.45	19.40	29	13.60%	17.40%	27%
400	0010	**GARAGES, MUNICIPAL** (Repair) R17100-100	S.F.	71	95	134			
	0020	Total project costs	C.F.	4.44	5.65	9.20			

17 SQUARE FOOT

17100 | S.F. & C.F. Costs		UNIT	UNIT COSTS			% OF TOTAL		
			1/4	MEDIAN	3/4	1/4	MEDIAN	3/4
400	0500	Masonry	S.F.	6.70	13.05	20.50	5.60%	9.15%	12.50%	**400**
	2720	Plumbing		3.19	6.15	11.55	3.59%	6.70%	7.95%	
	2730	Heating & ventilating		5.45	7.90	15.25	6.15%	7.45%	13.50%	
	2900	Electrical		5.25	8.25	11.90	6.65%	8.15%	11.15%	
	3100	Total: Mechanical & Electrical	↓	16.90	29.50	49	21.50%	25.50%	28.50%	
410	0010	**GARAGES, PARKING** [R17100 -100]	S.F.	27.50	40	69.50				**410**
	0020	Total project costs	C.F.	2.60	3.53	5.15				
	2720	Plumbing	S.F.	.78	1.22	1.88	1.72%	2.70%	3.85%	
	2900	Electrical		1.52	1.84	2.87	4.33%	5.20%	6.25%	
	3100	Total: Mechanical & Electrical	↓	2.42	4.33	5.70	6.90%	8.80%	11.05%	
	3200									
	9000	Per car, total cost	Car	11,700	14,700	18,700				
430	0010	**GYMNASIUMS** [R17100 -100]	S.F.	77	102	125				**430**
	0020	Total project costs	C.F.	3.84	5.20	6.40				
	1800	Equipment	S.F.	1.62	3.31	6.60	2.03%	3.35%	6.70%	
	2720	Plumbing		4.87	6	7.45	4.95%	6.75%	7.75%	
	2770	Heating, ventilating, air conditioning		5.25	8	16.05	5.80%	9.80%	11.10%	
	2900	Electrical		5.30	7.80	9.90	6.60%	8.50%	10.35%	
	3100	Total: Mechanical & Electrical	↓	21	29.50	33.50	19.60%	26%	29.50%	
	3500	See also division 11480								
460	0010	**HOSPITALS** [R17100 -100]	S.F.	152	181	275				**460**
	0020	Total project costs	C.F.	11.20	13.95	19.95				
	1800	Equipment	S.F.	3.75	7.20	12.45	1.10%	2.68%	5%	
	2720	Plumbing		12.75	17.80	23	7.60%	9.10%	10.85%	
	2770	Heating, ventilating, air conditioning		18.70	24	32.50	7.80%	12.95%	16.65%	
	2900	Electrical		16.15	21	32.50	9.85%	11.75%	14%	
	3100	Total: Mechanical & Electrical	↓	45.50	60	98.50	26.50%	33%	36.50%	
	9000	Per bed or person, total cost	Bed	76,500	171,500	223,500				
	9900	See also division 11700								
480	0010	**HOUSING** For the Elderly [R17100 -100]	S.F.	72.50	92	113				**480**
	0020	Total project costs	C.F.	5.20	7.20	9.20				
	0100	Site work	S.F.	5.40	8.20	11.55	5.05%	7.90%	12.10%	
	0500	Masonry		2.22	8.30	12.10	1.30%	6.05%	11%	
	1800	Equipment		1.76	2.42	3.85	1.88%	3.23%	4.43%	
	2510	Conveying systems		1.77	2.38	3.22	1.78%	2.20%	2.81%	
	2720	Plumbing		5.40	6.90	9	8.15%	9.55%	10.50%	
	2730	Heating, ventilating, air conditioning		2.77	3.92	5.85	3.30%	5.60%	7.25%	
	2900	Electrical		5.40	7.35	9.45	7.30%	8.50%	10.25%	
	3100	Total: Mechanical & Electrical	↓	18.65	22	29.50	18.10%	22.50%	29%	
	9000	Per rental unit, total cost	Unit	67,500	79,000	88,000				
	9500	Total: Mechanical & Electrical	"	14,900	17,300	20,200				
500	0010	**HOUSING** Public (Low Rise) [R17100 -100]	S.F.	61	85	111				**500**
	0020	Total project costs	C.F.	5.45	6.80	8.45				
	0100	Site work	S.F.	7.80	11.20	18.15	8.35%	11.75%	16.50%	
	1800	Equipment		1.66	2.71	4.13	2.26%	3.03%	4.24%	
	2720	Plumbing		4.41	5.85	7.35	7.15%	9.05%	11.60%	
	2730	Heating, ventilating, air conditioning		2.22	4.30	4.71	4.26%	6.05%	6.45%	
	2900	Electrical		3.70	5.50	7.65	5.10%	6.55%	8.25%	
	3100	Total: Mechanical & Electrical	↓	17.55	22.50	25.50	14.50%	17.55%	26.50%	
	9000	Per apartment, total cost	Apt.	67,000	76,500	96,000				
	9500	Total: Mechanical & Electrical	"	14,300	17,600	19,600				
510	0010	**ICE SKATING RINKS** [R17100 -100]	S.F.	54.50	122	134				**510**
	0020	Total project costs	C.F.	3.84	3.93	4.53				
	2720	Plumbing	S.F.	1.95	3.66	3.74	3.12%	3.23%	5.65%	
	2900	Electrical	↓	5.60	8.60	9.10	6.30%	10.15%	15.05%	

SQUARE FOOT 17

17100 | S.F. & C.F. Costs

			UNIT	UNIT COSTS			% OF TOTAL		
				1/4	MEDIAN	3/4	1/4	MEDIAN	3/4
510	3100	Total: Mechanical & Electrical	S.F.	9.35	13.20	16.50	18.95%	18.95%	18.95%	**510**
520	0010	**JAILS** R17100 -100	S.F.	159	205	265				**520**
	0020	Total project costs	C.F.	15	20	24.50				
	1800	Equipment	S.F.	6.40	18.35	31	2.77%	4.07%	11.90%	
	2720	Plumbing		16.20	20.50	27	7%	8.90%	13.35%	
	2770	Heating, ventilating, air conditioning		14.35	19.15	37	7.50%	9.45%	17.75%	
	2900	Electrical		17.50	22	27.50	8.20%	11.45%	14.70%	
	3100	Total: Mechanical & Electrical		43.50	79	94	27.50%	30%	34%	
530	0010	**LIBRARIES** R17100 -100	S.F.	99	126	165				**530**
	0020	Total project costs	C.F.	6.90	8.35	11				
	0500	Masonry	S.F.	6.85	13.90	23.50	5.80%	8.25%	12.60%	
	1800	Equipment		1.37	3.69	5.75	.41%	1.50%	4.16%	
	2720	Plumbing		3.71	5.40	7.30	3.43%	4.60%	5.70%	
	2770	Heating, ventilating, air conditioning		8.20	13.90	18.10	7.80%	10.95%	12.80%	
	2900	Electrical		10.10	13	16.65	8.35%	10.40%	11.95%	
	3100	Total: Mechanical & Electrical		30.50	38	47.50	19.65%	23%	26.50%	
540	0010	**LIVING, ASSISTED** R17100 -100	S.F.	94	111	129				**540**
	0020	Total project costs	C.F.	7.85	9.15	10.40				
	0500	Masonry	S.F.	2.73	3.26	3.83	2.37%	3.16%	3.86%	
	1800	Equipment		2.11	2.46	3.15	2.12%	2.45%	2.87%	
	2720	Plumbing		7.80	10.45	10.80	6.05%	8.15%	10.60%	
	2770	Heating, ventilating, air conditioning		9.25	9.65	10.60	7.95%	9.35%	9.70%	
	2900	Electrical		9.15	10.25	11.65	9%	10.30%	10.70%	
	3100	Total: Mechanical & Electrical		26.50	30	34.50	26%	29%	31.50%	
550	0010	**MEDICAL CLINICS** R17100 -100	S.F.	93	117	149				**550**
	0020	Total project costs	C.F.	6.95	9.40	12.10				
	1800	Equipment	S.F.	2.56	5.35	8.35	1.10%	3.13%	6.35%	
	2720	Plumbing		6.30	8.85	12	6.15%	8.40%	10.10%	
	2770	Heating, ventilating, air conditioning		7.50	9.85	14.45	6.65%	8.85%	11.35%	
	2900	Electrical		8.15	11.50	15.05	8.10%	10%	12.20%	
	3100	Total: Mechanical & Electrical		26	36	48.50	22.50%	27.50%	33.50%	
	3500	See also division 11700								
570	0010	**MEDICAL OFFICES** R17100 -100	S.F.	89	110	135				**570**
	0020	Total project costs	C.F.	6.65	9	12.15				
	1800	Equipment	S.F.	3.09	5.80	8.25	1.33%	5.15%	7.05%	
	2720	Plumbing		4.92	7.60	10.20	5.60%	6.80%	8.50%	
	2770	Heating, ventilating, air conditioning		5.95	8.75	11.35	6.15%	8.05%	9.70%	
	2900	Electrical		7.10	10.35	14.50	7.65%	9.80%	11.70%	
	3100	Total: Mechanical & Electrical		19.20	27.50	41	19.35%	23.50%	30.50%	
590	0010	**MOTELS** R17100 -100	S.F.	56	82.50	107				**590**
	0020	Total project costs	C.F.	5	6.70	10.95				
	2720	Plumbing	S.F.	5.70	7.25	8.65	9.45%	10.60%	12.55%	
	2770	Heating, ventilating, air conditioning		3.47	5.15	9.30	5.60%	5.60%	10%	
	2900	Electrical		5.35	6.80	8.65	7.45%	9.20%	10.80%	
	3100	Total: Mechanical & Electrical		18.05	22.50	38.50	18.50%	21%	25.50%	
	5000									
	9000	Per rental unit, total cost	Unit	28,600	54,500	58,500				
	9500	Total: Mechanical & Electrical	"	5,575	8,425	9,800				
600	0010	**NURSING HOMES** R17100 -100	S.F.	89	114	139				**600**
	0020	Total project costs	C.F.	6.90	8.65	11.80				
	1800	Equipment	S.F.	2.77	3.78	6.15	2.02%	3.62%	4.99%	
	2720	Plumbing		7.85	11.45	13.80	8.75%	10.10%	12.70%	
	2770	Heating, ventilating, air conditioning		7.95	12.05	16	9.70%	11.45%	11.80%	
	2900	Electrical		8.75	11.10	14.85	9.50%	10.60%	12.50%	

See R17100-100 and City Cost Indexes in the Reference Section

17 SQUARE FOOT

17100 | S.F. & C.F. Costs

			UNIT	UNIT COSTS			% OF TOTAL		
				1/4	MEDIAN	3/4	1/4	MEDIAN	3/4
600	3100	Total: Mechanical & Electrical	S.F.	21	29	49	26%	29.50%	30.50%	**600**
	9000	Per bed or person, total cost	Bed	39,600	50,000	63,500				
610	0010	**OFFICES** Low Rise (1 to 4 story) R17100-100	S.F.	73.50	95	124				**610**
	0020	Total project costs	C.F.	5.30	7.30	9.60				
	0100	Site work	S.F.	5.90	10	14.65	6.20%	9.70%	13.60%	
	0500	Masonry	"	2.47	5.65	10.30	2.55%	5.30%	8%	
	1800	Equipment	S.F.	.89	1.57	4.43	.91%	1.50%	3.93%	
	2720	Plumbing		2.65	4.10	5.95	3.66%	4.50%	6.10%	
	2770	Heating, ventilating, air conditioning		5.85	8.30	11.95	7.20%	10.40%	11.70%	
	2900	Electrical		6.05	8.60	12.15	7.45%	9.65%	11.40%	
	3100	Total: Mechanical & Electrical	↓	15.80	22	33	18.25%	22.50%	27%	
620	0010	**OFFICES** Mid Rise (5 to 10 story) R17100-100	S.F.	78.50	95	129				**620**
	0020	Total project costs	C.F.	5.55	7.10	10.05				
	2720	Plumbing	S.F.	2.37	3.83	5.30	2.83%	3.74%	4.50%	
	2770	Heating, ventilating, air conditioning		5.95	8.55	13.60	7.65%	9.40%	11%	
	2900	Electrical		5.85	7.45	9.90	6.35%	7.80%	10%	
	3100	Total: Mechanical & Electrical	↓	14.90	18.80	37.50	19.15%	21%	26%	
630	0010	**OFFICES** High Rise (11 to 20 story) R17100-100	S.F.	96.50	122	150				**630**
	0020	Total project costs	C.F.	6.75	8.45	12.10				
	2900	Electrical	S.F.	5.85	7.15	10.60	5.80%	7.85%	10.50%	
	3100	Total: Mechanical & Electrical	"	18.85	25.50	42.50	16.90%	23.50%	34%	
640	0010	**POLICE STATIONS** R17100-100	S.F.	116	152	192				**640**
	0020	Total project costs	C.F.	9.20	11.30	15.45				
	0500	Masonry	S.F.	10.90	19.20	25.50	6.70%	10.55%	11.35%	
	1800	Equipment		1.84	8.10	12.80	1.43%	4.07%	6.70%	
	2720	Plumbing		6.70	12.90	16.05	5.65%	6.90%	10.75%	
	2770	Heating, ventilating, air conditioning		10.10	13.45	18.25	5.85%	10.55%	11.70%	
	2900	Electrical		12.60	18.90	24	9.80%	11.85%	14.80%	
	3100	Total: Mechanical & Electrical	↓	41.50	50	67.50	28.50%	32%	32.50%	
650	0010	**POST OFFICES** R17100-100	S.F.	87.50	113	140				**650**
	0020	Total project costs	C.F.	5.50	6.95	8.40				
	2720	Plumbing	S.F.	4.11	5.10	6.45	4.24%	5.30%	5.60%	
	2770	Heating, ventilating, air conditioning		6.45	7.95	8.85	6.65%	7.15%	9.35%	
	2900	Electrical		7.55	10.60	12.60	7.25%	9%	11%	
	3100	Total: Mechanical & Electrical	↓	22	28.50	32.50	16.25%	18.80%	22%	
660	0010	**POWER PLANTS** R17100-100	S.F.	635	835	1,550				**660**
	0020	Total project costs	C.F.	17.45	38	81.50				
	2900	Electrical	S.F.	45	95	142	9.30%	12.75%	21.50%	
	8100	Total: Mechanical & Electrical	"	112	365	815	32.50%	32.50%	52.50%	
670	0010	**RELIGIOUS EDUCATION** R17100-100	S.F.	73.50	94	118				**670**
	0020	Total project costs	C.F.	4.09	5.85	7.55				
	2720	Plumbing	S.F.	3.08	4.36	6.15	4.40%	5.30%	7.10%	
	2770	Heating, ventilating, air conditioning		7.80	8.80	12.45	10.05%	11.45%	12.35%	
	2900	Electrical		5.85	7.70	10.80	7.60%	9.10%	10.35%	
	3100	Total: Mechanical & Electrical	↓	20	31	33.50	22%	23%	26%	
690	0010	**RESEARCH** Laboratories and facilities R17100-100	S.F.	110	158	230				**690**
	0020	Total project costs	C.F.	8.60	16.60	19.35				
	1800	Equipment	S.F.	4.89	9.65	23.50	.90%	4.58%	8.80%	
	2720	Plumbing		11.05	14.15	22.50	6.15%	8.30%	10.80%	
	2770	Heating, ventilating, air conditioning		9.90	33.50	39.50	7.25%	16.50%	17.50%	
	2900	Electrical		12.70	21.50	36	9.45%	11.15%	15.40%	
	3100	Total: Mechanical & Electrical	↓	36	71	106	29.50%	37%	45.50%	
700	0010	**RESTAURANTS** R17100-100	S.F.	106	137	179				**700**
	0020	Total project costs	C.F.	9	11.80	15.45				
	1800	Equipment	S.F.	6.90	16.95	25.50	6.10%	13%	15.65%	
	2720	Plumbing	↓	8.55	10.25	14.60	6.10%	8.15%	9%	

SQUARE FOOT 17

				UNIT COSTS			% OF TOTAL			
17100	**S.F. & C.F. Costs**		UNIT	1/4	MEDIAN	3/4	1/4	MEDIAN	3/4	
700	2770	Heating, ventilating, air conditioning	S.F.	11.15	14.85	19.35	9.20%	12%	12.40%	**700**
	2900	Electrical		11.30	13.95	18.10	8.35%	10.55%	11.55%	
	3100	Total: Mechanical & Electrical	↓	35	37	47.50	19.25%	24%	29.50%	
	9000	Per seat unit, total cost	Seat	3,925	5,225	6,175				
	9500	Total: Mechanical & Electrical	"	980	1,300	1,550				
720	0010	**RETAIL STORES** R17100-100	S.F.	50	67	88.50				**720**
	0020	Total project costs	C.F.	3.38	4.82	6.70				
	2720	Plumbing	S.F.	1.81	3.02	5.15	3.26%	4.60%	6.80%	
	2770	Heating, ventilating, air conditioning		3.91	5.35	8.05	6.75%	8.75%	10.15%	
	2900	Electrical		4.48	6.25	8.90	7.30%	9.90%	11.65%	
	3100	Total: Mechanical & Electrical	↓	11.95	15.35	20.50	17.10%	21.50%	23.50%	
740	0010	**SCHOOLS** Elementary R17100-100	S.F.	80.50	99.50	120				**740**
	0020	Total project costs	C.F.	5.35	6.80	8.80				
	0500	Masonry	S.F.	7.30	11.65	17.15	5.45%	10.50%	14.05%	
	1800	Equipment		2.46	4.19	7.45	1.83%	3.33%	5.05%	
	2720	Plumbing		4.67	6.60	8.85	5.70%	7.15%	9.35%	
	2730	Heating, ventilating, air conditioning		7	11.20	15.60	8.15%	10.80%	15.20%	
	2900	Electrical		7.55	9.90	12.50	8.45%	10.05%	11.80%	
	3100	Total: Mechanical & Electrical	↓	27	33	39.50	24%	27.50%	30.50%	
	9000	Per pupil, total cost	Ea.	7,850	14,000	41,400				
	9500	Total: Mechanical & Electrical	"	2,650	3,350	12,000				
760	0010	**SCHOOLS** Junior High & Middle R17100-100	S.F.	81.50	102	120				**760**
	0020	Total project costs	C.F.	5.35	6.90	7.75				
	0500	Masonry	S.F.	9	12.30	15.35	8.85%	12.10%	14.30%	
	1800	Equipment		2.68	4.32	6.70	1.89%	3.47%	5.20%	
	2720	Plumbing		5.35	6.05	7.80	5.50%	6.90%	8.15%	
	2770	Heating, ventilating, air conditioning		6.15	11.80	16.50	8.75%	12.75%	17.45%	
	2900	Electrical		7.75	9.85	12.20	7.90%	9.30%	10.60%	
	3100	Total: Mechanical & Electrical	↓	23.50	33	38.50	22.50%	25.50%	29.50%	
	9000	Per pupil, total cost	Ea.	10,700	14,400	18,800				
780	0010	**SCHOOLS** Senior High R17100-100	S.F.	85.50	108	136				**780**
	0020	Total project costs	C.F.	5.55	7.75	12.85				
	1800	Equipment	S.F.	2.30	5.40	8	1.88%	3.29%	4.80%	
	2720	Plumbing		4.47	5.95	13.45	5.70%	7%	8.35%	
	2770	Heating, ventilating, air conditioning		10	11.45	22	8.95%	11.60%	15%	
	2900	Electrical		8.55	11.15	17.90	8.35%	10.10%	11.95%	
	3100	Total: Mechanical & Electrical	↓	30	34	57	23%	26.50%	28.50%	
	9000	Per pupil, total cost	Ea.	8,225	14,000	21,000				
800	0010	**SCHOOLS** Vocational R17100-100	S.F.	70.50	100	125				**800**
	0020	Total project costs	C.F.	4.39	6.30	8.70				
	0500	Masonry	S.F.	4.15	10.25	15.65	3.53%	4.61%	10.95%	
	1800	Equipment	"	2.12	3	7.60	1.24%	3.13%	4.68%	
	2720	Plumbing	S.F.	4.60	6.90	10.35	5.40%	6.90%	8.55%	
	2770	Heating, ventilating, air conditioning		6.30	11.75	19.70	8.60%	11.90%	14.65%	
	2900	Electrical		7.20	10.05	14.20	8.45%	10.95%	13.20%	
	3100	Total: Mechanical & Electrical	↓	25.50	28.50	48.50	23.50%	29.50%	31%	
	9000	Per pupil, total cost	Ea.	9,850	26,300	39,300				
830	0010	**SPORTS ARENAS** R17100-100	S.F.	60	82.50	127				**830**
	0020	Total project costs	C.F.	3.35	6	7.75				
	2720	Plumbing	S.F.	3.58	5.45	11.50	4.35%	6.35%	9.40%	
	2770	Heating, ventilating, air conditioning		7.70	9.15	12.65	8.80%	10.20%	13.55%	
	2900	Electrical		6.50	8.75	11.30	8.60%	9.90%	12.25%	
	3100	Total: Mechanical & Electrical	↓	16	28.50	37	21.50%	25%	27.50%	
850	0010	**SUPERMARKETS** R17100-100	S.F.	56	66	80				**850**
	0020	Total project costs	C.F.	3.18	3.84	5.80				

17 SQUARE FOOT

		17100 | S.F. & C.F. Costs		UNIT	UNIT COSTS			% OF TOTAL		
					1/4	MEDIAN	3/4	1/4	MEDIAN	3/4
850	2720	Plumbing		S.F.	3.19	4.10	4.67	5.40%	6%	7.45%	850
	2770	Heating, ventilating, air conditioning			4.69	6.25	7.60	8.60%	8.65%	9.60%	
	2900	Electrical			7.15	8.20	9.70	10.40%	12.45%	13.60%	
	3100	Total: Mechanical & Electrical			18.35	19.95	28	20.50%	26.50%	31%	
860	0010	**SWIMMING POOLS**	R17100 -100	S.F.	92.50	155	305				860
	0020	Total project costs		C.F.	7.40	9.25	10.05				
	2720	Plumbing		S.F.	8.55	9.75	13.60	4.80%	9.70%	20.50%	
	2900	Electrical			6.95	11.25	16.40	6.50%	7.25%	7.60%	
	3100	Total: Mechanical & Electrical			16.95	43	58.50	11.15%	14.10%	23.50%	
870	0010	**TELEPHONE EXCHANGES**	R17100 -100	S.F.	123	180	228				870
	0020	Total project costs		C.F.	7.65	12.25	16.85				
	2720	Plumbing		S.F.	5.20	8.20	11.70	4.52%	5.80%	6.90%	
	2770	Heating, ventilating, air conditioning			12	24	30	11.80%	16.05%	18.40%	
	2900	Electrical			12.50	19.80	35	10.90%	14%	17.85%	
	3100	Total: Mechanical & Electrical			37	70	99	29.50%	33.50%	44.50%	
910	0010	**THEATERS**	R17100 -100	S.F.	77	99	146				910
	0020	Total project costs		C.F.	3.56	5.25	7.75				
	2720	Plumbing		S.F.	2.57	2.79	11.40	2.92%	4.70%	6.80%	
	2770	Heating, ventilating, air conditioning			7.50	9.10	11.25	8%	12.25%	13.40%	
	2900	Electrical			6.75	9.10	18.55	8.05%	9.95%	12.25%	
	3100	Total: Mechanical & Electrical			17.35	26	53.50	23%	26.50%	27.50%	
940	0010	**TOWN HALLS** City Halls & Municipal Buildings	R17100 -100	S.F.	90.50	116	153				940
	0020	Total project costs		C.F.	7.85	9.70	13.25				
	2720	Plumbing		S.F.	3.60	6.75	12.40	4.31%	5.95%	7.95%	
	2770	Heating, ventilating, air conditioning			6.50	12.90	18.85	7.05%	9.05%	13.45%	
	2900	Electrical			8.20	11.95	15.95	8.05%	9.50%	12.05%	
	3100	Total: Mechanical & Electrical			27.50	32.50	42.50	22%	26.50%	31%	
970	0010	**WAREHOUSES** And Storage Buildings	R17100 -100	S.F.	34	49.50	68.50				970
	0020	Total project costs		C.F.	1.85	2.63	4.36				
	0100	Site work		S.F.	3.32	6.60	9.95	6.05%	12.95%	19.85%	
	0500	Masonry			2.01	4.57	9.85	3.73%	7.40%	12.30%	
	1800	Equipment			.52	1.11	6.25	.91%	1.82%	5.55%	
	2720	Plumbing			1.07	1.92	3.67	2.90%	4.80%	6.55%	
	2730	Heating, ventilating, air conditioning			1.29	3.45	4.63	2.41%	5%	8.90%	
	2900	Electrical			2.03	2.69	5.95	5.15%	7.20%	10.10%	
	3100	Total: Mechanical & Electrical			5.30	8.15	17.80	12.75%	18.90%	26%	
990	0010	**WAREHOUSE & OFFICES** Combination	R17100 -100	S.F.	39.50	52.50	73				990
	0020	Total project costs		C.F.	2.02	2.94	4.34				
	1800	Equipment		S.F.	.69	1.33	2.05	.52%	1.21%	2.40%	
	2720	Plumbing			1.53	2.71	4.07	3.74%	4.76%	6.30%	
	2770	Heating, ventilating, air conditioning			2.41	3.77	5.30	5%	5.65%	10.05%	
	2900	Electrical			2.65	3.93	6.20	5.85%	8%	10%	
	3100	Total: Mechanical & Electrical			7.30	10.45	17.25	14.40%	19.95%	24.50%	

For information about Means Estimating Seminars, see yellow pages 12 and 13 in back of book

SQUARE FOOT 17

Reference Section

All the reference information is in one section making it easy to find what you need to know . . . and easy to use the book on a daily basis. This section is visually identified by a vertical gray bar on the edge of pages.

In the reference number information that follows, you'll see the background that relates to the "reference numbers" that appeared in the Unit Price Sections. You'll find reference tables, explanations and estimating information that support how we arrived at the unit price data. Also included are alternate pricing methods, technical data and estimating procedures along with information on design and economy in construction.

Also in this Reference Section, we've included Change Orders, information on pricing changes to contract documents; Crew Listings, a full listing of all the crews, equipment and their costs; Historical Cost Indexes for cost comparisons over time; City Cost Indexes and Location Factors for adjusting costs to the region you are in; and an explanation of all Abbreviations used in the book.

Table of Contents

R01100-005 Tips for Accurate Estimating

1. Use pre-printed or columnar forms for orderly sequence of dimensions and locations and for recording telephone quotations.

2. Use only the front side of each paper or form except for certain pre-printed summary forms.

3. Be consistent in listing dimensions: For example, length x width x height. This helps in rechecking to ensure that, the total length of partitions is appropriate for the building area.

4. Use printed (rather than measured) dimensions where given.

5. Add up multiple printed dimensions for a single entry where possible.

6. Measure all other dimensions carefully.

7. Use each set of dimensions to calculate multiple related quantities.

8. Convert foot and inch measurements to decimal feet when listing. Memorize decimal equivalents to .01 parts of a foot (1/8″ equals approximately .01′).

9. Do not "round off" quantities until the final summary.

10. Mark drawings with different colors as items are taken off.

11. Keep similar items together, different items separate.

12. Identify location and drawing numbers to aid in future checking for completeness.

13. Measure or list everything on the drawings or mentioned in the specifications.

14. It may be necessary to list items not called for to make the job complete.

15. Be alert for: Notes on plans such as N.T.S. (not to scale); changes in scale throughout the drawings; reduced size drawings; discrepancies between the specifications and the drawings.

16. Develop a consistent pattern of performing an estimate. For example:
    a. Start the quantity takeoff at the lower floor and move to the next higher floor.
    b. Proceed from the main section of the building to the wings.
    c. Proceed from south to north or vice versa, clockwise or counterclockwise.
    d. Take off floor plan quantities first, elevations next, then detail drawings.

17. List all gross dimensions that can be either used again for different quantities, or used as a rough check of other quantities for verification (exterior perimeter, gross floor area, individual floor areas, etc.).

18. Utilize design symmetry or repetition (repetitive floors, repetitive wings, symmetrical design around a center line, similar room layouts, etc.). Note: Extreme caution is needed here so as not to omit or duplicate an area.

19. Do not convert units until the final total is obtained. For instance, when estimating concrete work, keep all units to the nearest cubic foot, then summarize and convert to cubic yards.

20. When figuring alternatives, it is best to total all items involved in the basic system, then total all items involved in the alternates. Therefore you work with positive numbers in all cases. When adds and deducts are used, it is often confusing whether to add or subtract a portion of an item; especially on a complicated or involved alternate.

R01100-020 Construction Time Requirements

Table at left is average construction time in months for different types of building projects. Table at right is the construction time in months for different size projects. Design time runs 25% to 40% of construction time.

Type Building	Construction Time	Project Value	Construction Time
Industrial Buildings	12 Months	Under $1,400,000	10 Months
Commercial Buildings	15 Months	Up to $3,800,000	15 Months
Research & Development	18 Months	Up to $19,000,000	21 Months
Institutional Buildings	20 Months	over $19,000,000	28 Months

R01100-040 Builder's Risk Insurance

Builder's Risk Insurance is insurance on a building during construction. Premiums are paid by the owner or the contractor. Blasting, collapse and underground insurance would raise total insurance costs above those listed. Floater policy for materials delivered to the job runs $.75 to $1.25 per $100 value. Contractor equipment insurance runs $.50 to $1.50 per $100 value. Insurance for miscellaneous tools to $1,500 value runs from $3.00 to $7.50 per $100 value.

Tabulated below are New England Builder's Risk insurance rates in dollars per $100 value for $1,000 deductible. For $25,000 deductible, rates can be reduced 13% to 34%. On contracts over $1,000,000, rates may be lower than those tabulated. Policies are written annually for the total completed value in place. For "all risk" insurance (excluding flood, earthquake and certain other perils) add $.025 to total rates below.

Coverage	Frame Construction (Class 1) Range		Average	Brick Construction (Class 4) Range		Average	Fire Resistive (Class 6) Range		Average
Fire Insurance	$.350 to	$ .850	$.600	$.158 to	$.189	$.174	$.052 to	$.080	$.070
Extended Coverage	.115 to	.200	.158	.080 to	.105	.101	.081 to	.105	.100
Vandalism	.012 to	.016	.014	.008 to	.011	.011	.008 to	.011	.010
Total Annual Rate	$.477 to	$1.066	$.772	$.246 to	$.305	$.286	$.141 to	$.196	$.180

R01100-050　General Contractor's Overhead

There are two distinct types of overhead on a construction project: Project Overhead and Main Office Overhead. Project Overhead includes those costs at a construction site not directly associated with the installation of construction materials. Examples of Project Overhead costs include the following:

1. Superintendent
2. Construction office and storage trailers
3. Temporary sanitary facilities
4. Temporary utilities
5. Security fencing
6. Photographs
7. Clean up
8. Performance and payment bonds

The above Project Overhead items are also referred to as General Requirements and therefore are estimated in Division 1. Division 1 is the first division listed in the CSI MasterFormat but it is usually the last division estimated. The sum of the costs in Divisions 1 through 16 is referred to as the sum of the direct costs.

All construction projects also include indirect costs. The primary components of indirect costs are the contractor's Main Office Overhead and profit. The amount of the Main Office Overhead expense varies depending on the the following:

1. Owner's compensation
2. Project managers and estimator's wages
3. Clerical support wages
4. Office rent and utilities
5. Corporate legal and accounting costs
6. Advertising
7. Automobile expenses
8. Association dues
9. Travel and entertainment expenses

These costs are usually calculated as a percentage of annual sales volume. This percentage can range from 35% for a small contractor doing less than $500,000 to 5% for a large contractor with sales in excess of $100 million.

R01100-060 Workers' Compensation Insurance Rates by Trade

The table below tabulates the national averages for Workers' Compensation insurance rates by trade and type of building. The average "Insurance Rate" is multiplied by the "% of Building Cost" for each trade. This produces the "Workers' Compensation Cost" by % of total labor cost, to be added for each trade by building type to determine the weighted average Workers' Compensation rate for the building types analyzed.

Trade	Insurance Rate (% Labor Cost) Range		Average	% of Building Cost Office Bldgs.	Schools & Apts.	Mfg.	Workers' Compensation Office Bldgs.	Schools & Apts.	Mfg.
Excavation, Grading, etc.	4.2 % to	19.5%	10.4%	4.8%	4.9%	4.5%	.50%	.51%	.47%
Piles & Foundations	7.7 to	76.7	22.5	7.1	5.2	8.7	1.60	1.17	1.96
Concrete	5.2 to	35.8	16.1	5.0	14.8	3.7	.81	2.38	.60
Masonry	5.1 to	28.3	15.1	6.9	7.5	1.9	1.04	1.13	.29
Structural Steel	7.8 to	103.8	38.6	10.7	3.9	17.6	4.13	1.51	6.79
Miscellaneous & Ornamental Metals	4.9 to	24.8	12.6	2.8	4.0	3.6	.35	.50	.45
Carpentry & Millwork	7.0 to	53.2	18.4	3.7	4.0	0.5	.68	.74	.09
Metal or Composition Siding	5.3 to	32.1	17.0	2.3	0.3	4.3	.39	.05	.73
Roofing	7.8 to	78.9	33.1	2.3	2.6	3.1	.76	.86	1.03
Doors & Hardware	4.5 to	24.9	11.3	0.9	1.4	0.4	.10	.16	.05
Sash & Glazing	4.6 to	33.7	13.9	3.5	4.0	1.0	.49	.56	.14
Lath & Plaster	4.3 to	40.2	14.9	3.3	6.9	0.8	.49	1.03	.12
Tile, Marble & Floors	2.4 to	31.8	9.7	2.6	3.0	0.5	.25	.29	.05
Acoustical Ceilings	2.4 to	29.7	10.9	2.4	0.2	0.3	.26	.02	.03
Painting	4.5 to	33.7	13.1	1.5	1.6	1.6	.20	.21	.21
Interior Partitions	7.0 to	53.2	18.4	3.9	4.3	4.4	.72	.79	.81
Miscellaneous Items	2.6 to	137.6	16.9	5.2	3.7	9.7	.88	.63	1.64
Elevators	2.8 to	42.8	7.6	2.1	1.1	2.2	.16	.08	.17
Sprinklers	3.2 to	22.1	8.8	0.5	—	2.0	.04	—	.18
Plumbing	2.7 to	12.4	8.0	4.9	7.2	5.2	.39	.58	.42
Heat., Vent., Air Conditioning	4.0 to	27.2	11.5	13.5	11.0	12.9	1.55	1.27	1.48
Electrical	2.6 to	12.7	6.5	10.1	8.4	11.1	.66	.55	.72
Total	2.4 % to	137.6%	—	100.0%	100.0%	100.0%	16.45%	15.02%	18.43%

Overall Weighted Average 16.63%

Workers' Compensation Insurance Rates by States

The table below lists the weighted average Workers' Compensation base rate for each state with a factor comparing this with the national average of 16.2%.

State	Weighted Average	Factor	State	Weighted Average	Factor	State	Weighted Average	Factor
Alabama	25.2%	156	Kentucky	17.4%	107	North Dakota	11.4%	70
Alaska	19.7	122	Louisiana	26.2	162	Ohio	12.8	79
Arizona	7.4	46	Maine	19.3	119	Oklahoma	13.7	85
Arkansas	13.7	85	Maryland	17.7	109	Oregon	12.6	78
California	17.4	107	Massachusetts	12.4	77	Pennsylvania	13.4	83
Colorado	12.3	76	Michigan	17.6	109	Rhode Island	19.6	121
Connecticut	23.1	143	Minnesota	25.1	155	South Carolina	12.9	80
Delaware	14.2	88	Mississippi	16.1	99	South Dakota	14.2	88
District of Columbia	17.3	107	Missouri	17.6	109	Tennessee	17.9	110
Florida	25.1	155	Montana	18.9	117	Texas	13.5	83
Georgia	21.3	131	Nebraska	18.8	116	Utah	11.4	70
Hawaii	15.7	97	Nevada	13.9	86	Vermont	18.6	115
Idaho	10.4	64	New Hampshire	21.3	131	Virginia	10.8	67
Illinois	17.2	106	New Jersey	10.5	65	Washington	10.4	64
Indiana	5.4	33	New Mexico	16.3	101	West Virginia	12.4	77
Iowa	11.8	73	New York	13.4	83	Wisconsin	15.2	94
Kansas	8.5	52	North Carolina	13.2	81	Wyoming	7.9	49

Weighted Average for U.S. is 15.5% of payroll = 100%

Rates in the following table are the base or manual costs per $100 of payroll for Workers' Compensation in each state. Rates are usually applied to straight time wages only and not to premium time wages and bonuses.

The weighted average skilled worker rate for 35 trades is 16.2%. For bidding purposes, apply the full value of Workers' Compensation directly to total labor costs, or if labor is 38%, materials 42% and overhead and profit 20% of total cost, carry 38/80 x 16.2% =7.7% of cost (before overhead and profit) into overhead. Rates vary not only from state to state but also with the experience rating of the contractor.

Rates are the most current available at the time of publication.

GENERAL REQUIREMENTS

REFERENCE NOS.

RO1100-060 Workers' Compensation Insurance Rates by Trade and State (cont.)

State	Carpentry — 3 stories or less	Carpentry — interior cab. work	Carpentry — general	Concrete Work — NOC	Concrete Work — flat (flr., sdwk.)	Electrical Wiring — inside	Excavation — earth NOC	Excavation — rock	Glaziers	Insulation Work	Lathing	Masonry	Painting & Decorating	Pile Driving	Plastering	Plumbing	Roofing	Sheet Metal Work (HVAC)	Steel Erection — door & sash	Steel Erection — inter., ornam.	Steel Erection — structure	Steel Erection — NOC	Tile Work — (interior ceramic)	Waterproofing	Wrecking
	5651	5437	5403	5213	5221	5190	6217	6217	5462	5479	5443	5022	5474	6003	5480	5183	5551	5538	5102	5102	5040	5057	5348	9014	5701
AL	26.93	15.28	34.89	12.39	9.90	8.95	13.69	13.69	33.66	17.12	12.78	27.57	33.72	34.63	40.23	11.91	60.85	27.15	23.11	23.11	51.16	39.09	13.16	6.65	51.16
AK	17.02	14.74	13.84	12.07	11.18	12.70	19.54	19.54	19.86	25.41	9.78	24.13	16.28	59.80	19.42	11.79	38.40	12.13	12.24	12.24	42.33	18.60	8.48	10.79	42.33
AZ	7.71	5.37	12.63	6.33	3.44	4.18	4.83	4.83	7.59	9.87	4.70	6.40	5.23	9.47	15.76	3.77	11.96	4.71	7.92	7.92	16.58	7.75	2.37	2.55	50.31
AR	13.04	11.01	16.99	13.77	6.64	6.11	9.42	9.42	15.49	33.95	8.65	11.34	10.50	12.82	12.93	5.69	21.95	10.28	7.70	7.70	37.63	28.19	7.10	3.75	37.63
CA	28.28	9.07	28.28	13.18	13.18	9.93	7.88	7.88	16.37	23.34	10.90	14.55	19.45	21.18	18.12	11.59	39.79	15.33	13.55	13.55	23.93	21.91	7.74	19.45	21.91
CO	17.36	9.17	12.24	14.36	7.69	5.32	10.44	10.44	9.28	13.68	5.57	13.41	10.77	17.70	10.81	8.51	24.47	12.11	7.39	7.39	31.30	15.53	7.99	5.78	31.30
CT	20.96	18.42	31.08	29.77	13.28	8.63	11.18	11.18	15.10	24.22	17.24	25.81	19.56	27.54	27.61	10.37	51.40	14.91	18.42	18.42	70.81	35.84	12.65	6.69	50.86
DE	15.39	15.39	12.92	13.03	7.13	6.79	10.03	10.03	11.96	12.92	13.83	12.33	16.85	21.38	13.83	7.94	27.84	10.69	13.02	13.02	29.92	13.02	10.51	12.33	29.92
DC	13.05	10.16	15.52	17.41	20.87	6.52	10.32	10.32	27.59	11.07	8.44	22.34	10.52	18.70	13.70	12.43	21.23	9.29	17.97	17.97	54.79	20.38	31.80	4.05	54.79
FL	30.55	21.73	31.10	30.26	15.05	10.74	13.79	13.79	23.24	22.05	13.08	23.21	21.38	58.93	36.28	10.71	46.17	18.10	14.78	14.78	58.02	39.86	11.26	9.25	58.02
GA	32.05	17.69	24.16	15.16	10.44	8.97	16.86	16.86	15.69	22.02	17.16	20.30	17.13	31.16	19.94	10.77	46.74	17.52	13.97	13.97	44.46	48.01	9.64	8.55	44.46
HI	17.38	11.62	28.20	13.74	12.27	7.10	7.73	7.73	20.55	21.19	10.94	17.87	11.33	20.15	15.61	6.06	33.24	8.02	11.11	11.11	31.71	21.70	9.78	11.54	31.71
ID	11.48	6.50	13.42	8.40	5.87	5.01	6.24	6.24	9.48	7.60	5.27	9.01	7.40	13.14	9.74	4.67	29.72	9.37	9.80	9.80	30.78	13.76	6.47	4.09	30.78
IL	16.38	13.48	17.63	29.67	11.05	8.51	10.49	10.49	15.50	11.58	9.74	16.53	11.34	28.71	11.92	9.79	28.32	14.36	14.87	14.87	52.98	26.69	14.11	4.36	52.98
IN	5.31	4.50	7.02	5.18	3.17	2.64	4.17	4.17	4.59	4.15	2.43	5.07	4.48	7.71	4.30	2.70	10.93	4.03	4.94	4.94	16.12	9.00	3.08	2.55	16.12
IA	11.67	5.80	10.47	12.55	6.82	4.18	5.98	5.98	13.66	9.22	5.45	8.79	7.73	10.70	9.26	5.92	18.71	7.17	9.92	9.92	51.91	36.28	5.29	4.20	30.67
KS	10.58	8.42	10.47	7.12	6.36	3.72	4.73	4.73	6.49	7.80	4.37	7.45	6.06	9.36	8.73	5.40	20.70	6.76	7.83	7.83	23.76	12.28	4.75	3.48	23.76
KY	15.79	12.91	19.21	20.50	6.16	7.64	17.08	17.08	21.66	24.14	11.20	10.81	11.40	18.49	14.90	6.85	48.37	13.64	11.61	11.61	40.21	30.64	12.56	4.72	40.21
LA	23.14	24.93	53.17	26.43	15.61	9.88	17.49	17.49	20.14	21.43	24.51	28.33	29.58	31.25	22.37	8.64	77.12	21.20	18.98	18.98	51.76	24.21	13.77	13.78	66.41
ME	13.43	10.42	43.59	24.29	11.12	5.03	12.19	12.19	13.35	15.39	14.03	17.26	16.07	31.44	17.95	7.32	32.07	8.82	15.08	15.08	37.84	61.81	11.66	6.06	37.84
MD	13.71	5.95	10.55	18.82	8.70	5.15	12.91	12.91	25.52	20.07	11.37	13.31	9.37	26.22	14.21	8.58	48.72	12.74	13.66	13.66	56.25	37.81	8.03	7.14	26.80
MA	9.93	6.07	16.09	17.95	8.10	3.69	6.49	6.49	7.55	13.48	5.76	13.04	7.54	14.31	5.34	4.55	38.25	7.47	12.35	12.35	35.13	27.39	9.24	3.32	28.82
MI	21.68	13.59	19.80	22.12	9.35	5.71	11.69	11.69	13.87	12.38	13.59	19.33	14.93	39.06	15.59	8.24	41.33	10.41	11.36	11.36	39.06	29.15	11.31	6.40	39.06
MN	21.84	21.96	43.39	16.70	14.64	6.94	16.68	16.68	17.42	11.95	21.02	20.28	17.69	26.83	21.02	10.70	78.89	12.72	14.64	14.64	103.82	38.11	14.59	6.96	132.92
MS	16.92	15.26	19.71	11.11	8.98	6.80	12.22	12.22	12.02	14.11	8.02	13.18	12.86	19.38	22.03	8.15	37.48	21.88	11.67	11.67	33.42	32.27	9.89	6.93	33.42
MO	21.84	12.24	16.70	15.66	11.37	7.09	10.05	10.05	11.75	18.49	12.47	16.14	13.64	21.93	16.25	9.37	33.06	14.20	12.51	12.51	59.20	40.85	9.59	7.02	59.20
MT	22.03	10.79	17.11	12.89	9.98	5.54	17.20	17.20	11.51	16.19	19.02	14.58	12.03	76.71	13.82	8.93	57.20	9.44	9.73	9.73	41.23	17.93	6.54	5.33	41.23
NE	23.15	10.97	20.95	28.60	9.65	8.32	12.20	12.20	15.50	24.47	11.30	18.50	12.32	23.35	14.75	11.17	36.72	13.05	13.15	13.15	52.27	38.35	9.97	6.15	50.40
NV	19.78	7.76	13.18	10.57	11.57	6.18	11.21	11.21	12.58	13.22	7.97	11.38	10.62	14.55	12.30	8.93	21.55	18.17	12.87	12.87	31.33	33.90	6.64	6.15	44.29
NH	21.98	14.43	24.90	35.84	11.10	6.02	17.16	17.16	14.70	33.70	8.60	24.98	15.72	17.80	18.65	10.22	60.89	14.43	14.01	14.01	54.76	25.50	17.84	6.36	54.76
NJ	11.76	8.69	11.76	9.82	7.77	4.22	7.74	7.74	6.91	11.95	10.39	12.07	9.66	13.10	10.39	6.14	29.59	6.69	9.94	9.94	19.67	10.61	4.42	4.65	23.42
NM	28.63	8.03	18.67	17.07	9.38	6.85	9.33	9.33	14.38	12.83	7.20	13.69	12.57	18.59	11.22	8.64	32.00	11.70	24.78	24.78	43.20	25.65	7.98	6.81	43.20
NY	17.66	6.98	15.91	17.16	13.57	6.32	10.02	10.02	12.02	9.09	5.11	19.20	13.75	16.81	8.43	8.53	33.44	17.47	9.63	9.63	18.80	19.14	9.43	6.09	11.29
NC	16.34	11.24	15.16	11.77	7.00	8.63	8.65	8.65	10.30	11.88	8.28	10.32	9.75	13.68	15.22	7.78	26.83	11.26	7.71	7.71	52.96	19.19	5.79	4.27	52.96
ND	9.60	9.60	9.60	5.66	5.66	4.00	5.39	5.39	9.60	9.60	9.04	8.10	6.27	20.12	9.04	5.35	20.74	5.35	20.12	20.12	20.12	20.12	9.60	20.24	14.48
OH	6.32	11.48	9.82	11.66	10.37	5.54	8.42	8.42	9.22	14.76	29.68	12.20	15.20	25.34	4.41	7.14	22.43	8.85	10.53	10.53	25.27	18.05	9.73	5.02	25.27
OK	13.44	9.11	12.70	12.65	6.63	5.82	10.25	10.25	16.10	19.50	8.46	10.98	9.09	20.29	12.21	7.24	22.32	9.49	13.78	13.78	39.78	24.01	6.95	5.88	39.78
OR	19.10	9.61	15.51	13.55	8.43	4.65	9.70	9.70	15.02	8.08	7.48	15.36	12.79	15.81	10.75	5.89	23.26	10.49	9.53	9.53	32.00	13.92	11.23	4.97	32.00
PA	13.85	13.85	12.80	14.90	9.67	6.74	8.44	8.44	9.60	12.80	11.67	12.18	14.33	17.72	11.67	7.51	27.83	8.19	15.17	15.17	26.43	15.17	8.22	12.18	26.43
RI	19.53	11.65	18.07	18.23	16.24	4.43	10.38	10.38	12.85	22.78	11.97	25.11	24.13	37.66	17.25	8.52	33.92	10.29	14.07	14.07	59.49	37.50	14.36	7.70	78.79
SC	19.10	13.03	19.24	12.88	6.19	7.02	8.32	8.32	12.22	10.69	7.27	9.55	11.19	16.64	18.89	6.88	28.98	11.42	9.49	9.49	21.80	23.66	6.23	4.65	21.80
SD	17.83	7.27	14.55	18.00	6.34	5.44	13.24	13.24	10.58	13.24	8.39	8.01	12.63	24.87	11.78	8.53	21.26	9.64	11.53	11.53	53.25	20.15	6.68	4.20	53.25
TN	31.79	14.64	24.26	19.40	9.95	8.72	14.06	14.06	12.09	14.84	13.05	17.02	14.59	17.43	17.47	11.33	38.38	14.51	13.38	13.38	45.26	25.19	9.73	6.57	45.26
TX	16.89	11.33	13.25	12.10	8.85	7.36	10.49	10.49	11.61	15.68	9.53	13.68	9.89	13.95	20.99	8.06	22.80	14.58	11.23	11.23	33.39	14.78	7.29	8.09	17.86
UT	9.45	6.23	10.52	7.68	7.99	7.05	6.38	6.38	9.55	10.95	12.32	13.75	15.29	17.24	10.60	6.41	26.27	6.30	8.99	8.99	23.51	23.51	6.06	5.83	26.53
VT	19.90	11.34	19.46	34.13	12.01	6.10	10.61	10.61	21.25	25.54	10.85	19.53	10.41	23.91	18.23	10.10	28.73	12.62	13.94	13.94	47.05	37.23	8.52	9.59	47.05
VA	11.97	7.81	9.72	10.47	5.32	4.43	6.53	6.53	7.60	7.96	13.79	8.17	9.95	13.55	7.16	5.52	21.62	8.30	11.34	11.34	39.01	16.86	6.32	3.09	39.01
WA	9.13	9.13	9.13	8.17	8.17	3.12	8.54	8.54	13.74	8.92	9.13	13.27	9.39	20.04	11.57	5.04	19.54	4.60	12.96	12.96	9.39	9.59	10.76	10.23	9.59
WV	12.30	12.30	12.30	27.37	27.37	5.86	8.90	8.90	6.99	13.15	13.15	11.94	12.35	8.90	12.35	5.48	14.63	6.99	10.89	10.89	10.89	10.89	13.15	13.15	48.62
WI	12.40	10.86	19.86	13.12	11.66	5.32	6.74	6.74	16.53	14.04	10.60	18.99	14.82	17.24	13.95	6.31	43.57	8.01	14.30	14.30	34.84	21.34	14.42	5.80	34.84
WY	7.77	7.77	7.77	7.77	7.77	7.77	7.77	7.77	7.77	7.77	7.77	7.77	7.77	7.77	7.77	7.77	7.77	7.77	7.77	7.77	7.77	7.77	7.77	7.77	7.77
AVG.	16.96	11.33	18.42	16.07	9.94	6.46	10.43	10.43	13.91	15.53	10.87	15.06	13.12	22.45	14.88	7.96	33.14	11.46	12.57	12.57	38.60	24.71	9.66	7.12	39.48

R01100-060 Workers' Compensation (cont.) (Canada in Canadian dollars)

Province		Alberta	British Columbia	Manitoba	Ontario	New Brunswick	Newfndld. & Labrador	Northwest Territories	Nova Scotia	Prince Edward Island	Quebec	Saskat-chewan	Yukon
Carpentry—3 stories or less	Rate	10.80	6.66	4.57	4.83	4.63	10.05	4.25	7.67	6.15	14.69	7.01	4.36
	Code	42143	721028	40102	723	422	4226	4-41	4226	401	80110	B1317	202
Carpentry—interior cab. work	Rate	2.59	5.94	4.57	4.83	4.21	6.35	4.25	5.74	3.41	14.69	3.65	4.36
	Code	42133	721021	40102	723	427	4270	4-41	4274	402	80110	B11-27	202
CARPENTRY—general	Rate	10.80	6.66	4.57	4.83	4.63	6.35	4.25	7.67	6.15	14.69	7.01	4.36
	Code	42143	721028	40102	723	422	4299	4-41	4226	401	80110	B1317	202
CONCRETE WORK—NOC	Rate	7.26	8.48	7.60	16.47	4.63	10.05	4.25	4.83	6.15	15.84	7.01	4.36
	Code	42104	721010	40110	748	422	4224	4-41	4224	401	80100	B13-14	203
CONCRETE WORK—flat (flr. sidewalk)	Rate	7.26	8.48	7.60	16.47	4.63	10.05	4.25	4.83	6.15	15.84	7.01	4.36
	Code	42104	721010	40110	748	422	4224	4-41	4224	401	80100	B13-14	203
ELECTRICAL Wiring—inside	Rate	2.86	2.67	2.52	3.03	2.43	3.19	3.00	2.23	3.41	7.29	3.65	4.36
	Code	42124	721019	40203	704	426	4261	4-46	4261	402	80170	B11-05	206
EXCAVATION—earth NOC	Rate	3.59	4.29	3.84	4.20	3.69	5.55	3.50	4.11	3.58	8.80	4.02	4.36
	Code	40604	721031	40706	711	421	4214	4-43	4214	404	80030	R11-06	207
EXCAVATION—rock	Rate	3.59	4.29	3.84	4.20	3.69	5.55	3.50	4.11	3.58	8.80	4.02	4.36
	Code	40604	721031	40706	711	421	4214	4-43	4214	404	80030	R11-06	207
GLAZIERS	Rate	4.84	3.00	4.57	8.12	6.12	5.97	4.25	7.67	3.41	14.59	7.01	4.36
	Code	42121	715020	40109	751	423	4233	4-41	4233	402	80150	B13-04	212
INSULATION WORK	Rate	3.67	6.94	4.57	8.12	6.12	5.97	4.25	7.67	6.15	14.69	6.20	4.36
	Code	42184	721029	40102	751	423	4234	4-41	4234	401	80110	B12-07	202
LATHING	Rate	8.30	11.20	4.57	4.83	4.21	6.35	4.25	5.74	3.41	14.69	7.01	4.36
	Code	42135	721033	40102	723	427	4279	4-41	4271	402	80110	B13-16	202
MASONRY	Rate	7.26	11.20	4.57	12.21	6.12	5.97	4.25	7.67	6.15	15.84	7.01	4.36
	Code	42102	721037	40102	741	423	4231	4-41	4231	401	80100	B13-18	202
PAINTING & DECORATING	Rate	5.64	6.94	3.45	6.83	4.21	6.35	4.25	5.74	3.41	14.69	6.20	4.36
	Code	42111	721041	40105	719	427	4275	4-41	4275	402	80110	B12-01	202
PILE DRIVING	Rate	7.26	13.07	3.84	5.84	4.63	10.05	3.50	4.83	6.15	8.80	7.01	4.36
	Code	42159	722004	40706	732	422	4221	4-43	4221	401	80030	B13-10	202
PLASTERING	Rate	8.30	11.20	5.13	6.83	4.21	6.35	4.25	5.74	3.41	14.69	6.20	4.36
	Code	42135	721042	40108	719	427	4271	4-41	4271	402	80110	B12-21	202
PLUMBING	Rate	2.86	5.48	2.88	3.83	3.74	3.89	3.00	2.23	3.41	7.75	3.65	4.36
	Code	42122	721043	40204	707	424	4241	4-46	4241	402	80160	B11-01	214
ROOFING	Rate	11.09	10.45	6.52	12.34	8.00	10.05	4.25	9.49	6.15	23.29	7.01	4.36
	Code	42118	721036	40403	728	430	4236	4-41	4236	401	80130	B13-20	202
SHEET METAL WORK (HVAC)	Rate	2.86	5.48	6.52	3.83	3.74	3.89	3.50	3.23	3.41	7.75	3.65	6.60
	Code	42117	721043	40402	707	424	4244	4-46	4244	402	80160	B11-07	208
STEEL ERECTION—door & sash	Rate	3.67	13.07	9.89	16.47	4.63	10.05	4.25	7.67	6.15	29.23	7.01	4.36
	Code	42106	722005	40502	748	422	4227	4-41	4227	401	80080	B13-22	202
STEEL ERECTION—inter., ornam.	Rate	3.67	13.07	9.89	16.47	4.63	10.05	4.25	7.67	6.15	29.23	7.01	4.36
	Code	42106	722005	40502	748	422	4227	4-41	4227	401	80080	B13-22	202
STEEL ERECTION—structure	Rate	3.67	13.07	9.89	16.47	4.63	10.05	4.25	7.67	6.15	29.23	7.01	4.36
	Code	42106	722005	40502	748	422	4227	4-41	4227	401	80080	B13-22	202
STEEL ERECTION—NOC	Rate	3.67	13.07	9.89	16.47	4.63	10.05	4.25	7.67	6.15	29.23	7.01	4.36
	Code	42106	722005	40502	748	422	4227	4-41	4227	401	80080	B13-22	202
TILE WORK—inter. (ceramic)	Rate	4.86	4.70	2.17	6.83	4.21	6.35	4.25	5.74	3.41	14.69	7.01	4.36
	Code	42113	721054	40103	719	427	4276	4-41	4276	402	80110	B13-01	202
WATERPROOFING	Rate	5.64	6.94	4.57	4.83	6.12	6.35	4.25	7.67	3.41	23.29	6.20	4.36
	Code	42139	721016	40102	723	423	4299	4-41	4239	402	80130	B12-17	202
WRECKING	Rate	3.59	6.53	6.46	16.47	3.69	5.55	3.50	4.11	6.15	14.69	7.01	4.36
	Code	40604	721005	40106	748	421	4211	4-43	4211	401	80110	B13-09	202

GENERAL REQUIREMENTS

REFERENCE NOS.

R01100-070 Contractor's Overhead & Profit

Below are the **average** installing contractor's percentage mark-ups applied to base labor rates to arrive at typical billing rates.

Column A: Labor rates are based on union wages averaged for 30 major U.S. cities. Base rates including fringe benefits are listed hourly and daily. These figures are the sum of the wage rate and employer-paid fringe benefits such as vacation pay, employer-paid health and welfare costs, pension costs, plus appropriate training and industry advancement funds costs.

Column B: Workers' Compensation rates are the national average of state rates established for each trade.

Column C: Column C lists average fixed overhead figures for all trades. Included are Federal and State Unemployment costs set at 6.2%; Social Security Taxes (FICA) set at 7.65%; Builder's Risk Insurance costs set at 0.44%; and Public Liability costs set at 2.02%. All the percentages except those for Social Security Taxes vary from state to state as well as from company to company.

Columns D and E: Percentages in Columns D and E are based on the presumption that the installing contractor has annual billing of $4,000,000 and up. Overhead percentages may increase with smaller annual billing. The overhead percentages for any given contractor may vary greatly and depend on a number of factors, such as the contractor's annual volume, engineering and logistical support costs, and staff requirements. The figures for overhead and profit will also vary depending on the type of job, the job location, and the prevailing economic conditions. All factors should be examined very carefully for each job.

Column F: Column F lists the total of Columns B, C, D, and E.

Column G: Column G is Column A (hourly base labor rate) multiplied by the percentage in Column F (O&P percentage).

Column H: Column H is the total of Column A (hourly base labor rate) plus Column G (Total O&P).

Column I: Column I is Column H multiplied by eight hours.

		A		B	C	D	E	F	G	H	I
		Base Rate Incl. Fringes		Workers' Comp. Ins.	Average Fixed Overhead	Overhead	Profit	Total Overhead & Profit		Rate with O & P	
Abbr.	Trade	Hourly	Daily					%	Amount	Hourly	Daily
Skwk	Skilled Workers Average (35 trades)	$34.85	$278.80	16.2%	16.3%	13.0%	10.0%	55.5%	$19.35	$54.20	$433.60
	Helpers Average (5 trades)	25.55	204.40	18.2		11.0		55.5	14.20	39.75	318.00
	Foreman Average, Inside ($0.50 over trade)	35.35	282.80	16.2		13.0		55.5	19.60	54.95	439.60
	Foreman Average, Outside ($2.00 over trade)	36.85	294.80	16.2		13.0		55.5	20.45	57.30	458.40
Clab	Common Building Laborers	26.70	213.60	18.4		11.0		55.7	14.85	41.55	332.40
Asbe	Asbestos/Insulation Workers/Pipe Coverers	37.10	296.80	15.5		16.0		57.8	21.45	58.55	468.40
Boil	Boilermakers	42.90	343.20	13.7		16.0		56.0	24.00	66.90	535.20
Bric	Bricklayers	35.25	282.00	15.1		11.0		52.4	18.45	53.70	429.60
Brhe	Bricklayer Helpers	26.90	215.20	15.1		11.0		52.4	14.10	41.00	328.00
Carp	Carpenters	34.25	274.00	18.4		11.0		55.7	19.10	53.35	426.80
Cefi	Cement Finishers	32.85	262.80	9.9		11.0		47.2	15.50	48.35	386.80
Elec	Electricians	40.75	326.00	6.5		16.0		48.8	19.90	60.65	485.20
Elev	Elevator Constructors	45.05	360.40	7.6		16.0		49.9	22.50	67.55	540.40
Eqhv	Equipment Operators, Crane or Shovel	35.90	287.20	10.4		14.0		50.7	18.20	54.10	432.80
Eqmd	Equipment Operators, Medium Equipment	34.65	277.20	10.4		14.0		50.7	17.55	52.20	417.60
Eqlt	Equipment Operators, Light Equipment	33.05	264.40	10.4		14.0		50.7	16.75	49.80	398.40
Eqol	Equipment Operators, Oilers	30.10	240.80	10.4		14.0		50.7	15.25	45.35	362.80
Eqmm	Equipment Operators, Master Mechanics	36.20	289.60	10.4		14.0		50.7	18.35	54.55	436.40
Glaz	Glaziers	33.00	264.00	13.9		11.0		51.2	16.90	49.90	399.20
Lath	Lathers	31.45	251.60	10.9		11.0		48.2	15.15	46.60	372.80
Marb	Marble Setters	33.20	265.60	15.1		11.0		52.4	17.40	50.60	404.80
Mill	Millwrights	35.50	284.00	10.4		11.0		47.7	16.95	52.45	419.60
Mstz	Mosaic and Terrazzo Workers	32.65	261.20	9.7		11.0		47.0	15.35	48.00	384.00
Pord	Painters, Ordinary	30.60	244.80	13.1		11.0		50.4	15.40	46.00	368.00
Psst	Painters, Structural Steel	31.00	248.00	46.8		11.0		84.1	26.05	57.05	456.40
Pape	Paper Hangers	30.30	242.40	13.1		11.0		50.4	15.25	45.55	364.40
Pile	Pile Drivers	33.30	266.40	22.5		16.0		64.8	21.60	54.90	439.20
Plas	Plasterers	31.35	250.80	14.9		11.0		52.2	16.35	47.70	381.60
Plah	Plasterer Helpers	27.10	216.80	14.9		11.0		52.2	14.15	41.25	330.00
Plum	Plumbers	40.85	326.80	8.0		16.0		50.3	20.55	61.40	491.20
Rodm	Rodmen (Reinforcing)	37.95	303.60	24.7		14.0		65.0	24.65	62.60	500.80
Rofc	Roofers, Composition	29.45	235.60	33.1		11.0		70.4	20.75	50.20	401.60
Rots	Roofers, Tile and Slate	29.70	237.60	33.1		11.0		70.4	20.90	50.60	404.80
Rohe	Roofer Helpers (Composition)	21.60	172.80	33.1		11.0		70.4	15.20	36.80	294.40
Shee	Sheet Metal Workers	40.20	321.60	11.5		16.0		53.8	21.65	61.85	494.80
Spri	Sprinkler Installers	40.75	326.00	8.8		16.0		51.1	20.80	61.55	492.40
Stpi	Steamfitters or Pipefitters	40.95	327.60	8.0		16.0		50.3	20.60	61.55	492.40
Ston	Stone Masons	35.30	282.40	15.1		11.0		52.4	18.50	53.80	430.40
Sswk	Structural Steel Workers	38.15	305.20	38.6		14.0		78.9	30.10	68.25	546.00
Tilf	Tile Layers	32.70	261.60	9.7		11.0		47.0	15.35	48.05	384.40
Tilh	Tile Layer Helpers	25.35	202.80	9.7		11.0		47.0	11.90	37.25	298.00
Trlt	Truck Drivers, Light	26.80	214.40	15.2		11.0		52.5	14.05	40.85	326.80
Trhv	Truck Drivers, Heavy	27.55	220.40	15.2		11.0		52.5	14.45	42.00	336.00
Sswl	Welders, Structural Steel	38.15	305.20	38.6		14.0		78.9	30.10	68.25	546.00
Wrck	*Wrecking	26.70	213.60	39.5	▼	11.0	▼	76.8	20.50	47.20	377.60

*Not included in Averages.

GENERAL REQUIREMENTS 1

REFERENCE NOS.

R01100-080 Performance Bond

This table shows the cost of a Performance Bond for a construction job scheduled to be completed in 12 months. Add 1% of the premium cost per month for jobs requiring more than 12 months to complete. The rates are "standard" rates offered to contractors that the bonding company considers financially sound and capable of doing the work. Preferred rates are offered by some bonding companies based upon financial strength of the contractor. Actual rates vary from contractor to contractor and from bonding company to bonding company. Contractors should prequalify through a bonding agency before submitting a bid on a contract that requires a bond.

Contract Amount	Building Construction Class B Projects			Highways & Bridges					
				Class A New Construction			Class A-1 Highway Resurfacing		
First $ 100,000 bid	$25.00 per M			$15.00 per M			$9.40 per M		
Next 400,000 bid	$ 2,500	plus	$15.00 per M	$ 1,500	plus	$10.00 per M	$ 940	plus	$7.20 per M
Next 2,000,000 bid	8,500	plus	10.00 per M	5,500	plus	7.00 per M	3,820	plus	5.00 per M
Next 2,500,000 bid	28,500	plus	7.50 per M	19,500	plus	5.50 per M	15,820	plus	4.50 per M
Next 2,500,000 bid	47,250	plus	7.00 per M	33,250	plus	5.00 per M	28,320	plus	4.50 per M
Over 7,500,000 bid	64,750	plus	6.00 per M	45,750	plus	4.50 per M	39,570	plus	4.00 per M

R01100-090 Sales Tax by State

State sales tax on materials is tabulated below (5 states have no sales tax). Many states allow local jurisdictions, such as a county or city, to levy additional sales tax.

Some projects may be sales tax exempt, particularly those constructed with public funds.

State	Tax (%)	State	Tax (%)	State	Tax (%)	State	Tax (%)
Alabama	4	Illinois	6.25	Montana	0	Rhode Island	7
Alaska	0	Indiana	6	Nebraska	5.5	South Carolina	5
Arizona	5.6	Iowa	5	Nevada	6.5	South Dakota	4
Arkansas	5.125	Kansas	5.3	New Hampshire	0	Tennessee	7
California	7.25	Kentucky	6	New Jersey	6	Texas	6.25
Colorado	2.9	Louisiana	4	New Mexico	5	Utah	4.75
Connecticut	6	Maine	5	New York	4.25	Vermont	6
Delaware	0	Maryland	5	North Carolina	4.5	Virginia	5
District of Columbia	5.75	Massachusetts	5	North Dakota	5	Washington	6.5
Florida	6	Michigan	6	Ohio	6	West Virginia	6
Georgia	4	Minnesota	6.5	Oklahoma	4.5	Wisconsin	5
Hawaii	4	Mississippi	7	Oregon	0	Wyoming	4
Idaho	6	Missouri	4.225	Pennsylvania	6	Average	4.86 %

Sales Tax by Province (Canada)

GST - a value-added tax, which the government imposes on most goods and services provided in or imported into Canada. PST - a retail sales tax, which five of the provinces impose on the price of most goods and some services. QST - a value-added tax, similar to the federal GST, which Quebec imposes. HST - Three provinces have combined their retail sales tax with the federal GST into one harmonized tax.

Province	PST (%)	QST (%)	GST(%)	HST(%)
Alberta	0	0	7	0
British Columbia	7.5	0	7	0
Manitoba	7	0	7	0
New Brunswick	0	0	0	15
Newfoundland	0	0	0	15
Northwest Territories	0	0	7	0
Nova Scotia	0	0	0	15
Ontario	8	0	7	0
Prince Edward Island	10	0	7	0
Quebec	0	7.5	7	0
Saskatchewan	6	0	7	0
Yukon	0	0	7	0

R01100-100 Unemployment Taxes and Social Security Taxes

Mass. State Unemployment tax ranges from 1.12% to 10.96% plus an experience rating assessment the following year, on the first $14,000 of wages. Federal Unemployment tax is 6.2% of the first $7,000 of wages. This is reduced by a credit for payment to the state. The minimum Federal Unemployment tax is 0.8% after all credits.

Combined rates in Mass. thus vary from 1.92% to 11.76% of the first $14,000 of wages. Combined average U.S. rate is about 6.2% of the first $7,000. Contractors with permanent workers will pay less since the average annual wages for skilled workers is $34.85 x 2,000 hours or about $69,700 per year. The average combined rate for U.S. would thus be 6.2% x $7,000 ÷ $69,700 = 0.6% of total wages for permanent employees.

Rates vary not only from state to state but also with the experience rating of the contractor.

Social Security (FICA) for 2005 is estimated at time of publication to be 7.65% of wages up to $87,900.

R01100-110   Overtime

One way to improve the completion date of a project or eliminate negative float from a schedule is to compress activity duration times. This can be achieved by increasing the crew size or working overtime with the proposed crew.

To determine the costs of working overtime to compress activity duration times, consider the following examples. Below is an overtime efficiency and cost chart based on a five, six, or seven day week with an eight through twelve hour day. Payroll percentage increases for time and one half and double time are shown for the various working days.

| Days per Week | Hours per Day | Production Efficiency | | | | | Payroll Cost Factors | |
		1 Week	2 Weeks	3 Weeks	4 Weeks	Average 4 Weeks	@ 1-1/2 Times	@ 2 Times
5	8	100%	100%	100%	100%	100 %	100 %	100 %
	9	100	100	95	90	96.25	105.6	111.1
	10	100	95	90	85	91.25	110.0	120.0
	11	95	90	75	65	81.25	113.6	127.3
	12	90	85	70	60	76.25	116.7	133.3
6	8	100	100	95	90	96.25	108.3	116.7
	9	100	95	90	85	92.50	113.0	125.9
	10	95	90	85	80	87.50	116.7	133.3
	11	95	85	70	65	78.75	119.7	139.4
	12	90	80	65	60	73.75	122.2	144.4
7	8	100	95	85	75	88.75	114.3	128.6
	9	95	90	80	70	83.75	118.3	136.5
	10	90	85	75	65	78.75	121.4	142.9
	11	85	80	65	60	72.50	124.0	148.1
	12	85	75	60	55	68.75	126.2	152.4

R01107-010 Architectural Fees

Tabulated below are typical percentage fees by project size, for good professional architectural service. Fees may vary from those listed depending upon degree of design difficulty and economic conditions in any particular area.

Rates can be interpolated horizontally and vertically. Various portions of the same project requiring different rates should be adjusted proportionally. For alterations, add 50% to the fee for the first $500,000 of project cost and add 25% to the fee for project cost over $500,000.

Architectural fees tabulated below include Structural, Mechanical and Electrical Engineering Fees. They do not include the fees for special

consultants such as kitchen planning, security, acoustical, interior design, etc.

Civil Engineering fees are included in the Architectural fee for project sites requiring minimal design such as city sites. However, separate Civil Engineering fees must be added when utility connections require design, drainage calculations are needed, stepped foundations are required, or provisions are required to protect adjacent wetlands.

Building Types	Total Project Size in Thousands of Dollars						
	100	250	500	1,000	5,000	10,000	50,000
Factories, garages, warehouses, repetitive housing	9.0%	8.0%	7.0%	6.2%	5.3%	4.9%	4.5%
Apartments, banks, schools, libraries, offices, municipal buildings	12.2	12.3	9.2	8.0	7.0	6.6	6.2
Churches, hospitals, homes, laboratories, museums, research	15.0	13.6	12.7	11.9	9.5	8.8	8.0
Memorials, monumental work, decorative furnishings	—	16.0	14.5	13.1	10.0	9.0	8.3

R01107-030 Engineering Fees

Typical **Structural Engineering Fees** based on type of construction and total project size. These fees are included in Architectural Fees.

Type of Construction	Total Project Size (in thousands of dollars)			
	$500	$500-$1,000	$1,000-$5,000	Over $5000
Industrial buildings, factories & warehouses	Technical payroll times 2.0 to 2.5	1.60%	1.25%	1.00%
Hotels, apartments, offices, dormitories, hospitals, public buildings, food stores		2.00%	1.70%	1.20%
Museums, banks, churches and cathedrals		2.00%	1.75%	1.25%
Thin shells, prestressed concrete, earthquake resistive		2.00%	1.75%	1.50%
Parking ramps, auditoriums, stadiums, convention halls, hangars & boiler houses		2.50%	2.00%	1.75%
Special buildings, major alterations, underpinning & future expansion		Add to above 0.5%	Add to above 0.5%	Add to above 0.5%

For complex reinforced concrete or unusually complicated structures, add 20% to 50%.

Typical **Mechanical and Electrical Engineering Fees** are based on the size of the subcontract. The fee structure for Mechanical Engineering is shown below; Electrical Engineering fees range from 2 to 5%. These fees are included in Architectural Fees.

Type of Construction	Subcontract Size							
	$25,000	$50,000	$100,000	$225,000	$350,000	$500,000	$750,000	$1,000,000
Simple structures	6.4%	5.7%	4.8%	4.5%	4.4%	4.3%	4.2%	4.1%
Intermediate structures	8.0	7.3	6.5	5.6	5.1	5.0	4.9	4.8
Complex structures	12.0	9.0	9.0	8.0	7.5	7.5	7.0	7.0

For renovations, add 15% to 25% to applicable fee.

R01540-100 Steel Tubular Scaffolding

On new construction, tubular scaffolding is efficient up to 60' high or five stories. Above this it is usually better to use a hung scaffolding if construction permits. Swing scaffolding operations may interfere with tenants. In this case, the tubular is more practical at all heights.

In repairing or cleaning the front of an existing building the cost of tubular scaffolding per S.F. of building front increases as the height increases above the first tier. The first tier cost is relatively high due to leveling and alignment.

The minimum efficient crew for erection is three workers. For heights over 50', a crew of four is more efficient. Use two or more on top and two at the bottom for handing up or hoisting. Four workers can erect and dismantle about nine frames per hour up to five stories. From five to eight stories they will average six frames per hour. With 7' horizontal spacing this will run about 400 S.F. and 265 S.F. of wall surface, respectively. Time for placing planks must be added to the above. On heights above 50', five planks can be placed per labor-hour.

The table below shows the number of pieces required to erect tubular steel scaffolding for 1000 S.F. of building frontage. This area is made up of a scaffolding system that is 12 frames (11 bays) long by 2 frames high.

For jobs under twenty-five frames, add 50% to rental cost. Rental rates will be lower for jobs over three months duration. Large quantities for long periods can reduce rental rates by 20%.

Description of Component	CSI Line Item	Number of Pieces for 1000 S.F. of Building Front	Unit
5' Wide Standard Frame, 6'-4" High	01540-750-2200	24	Ea.
Leveling Jack & Plate	01540-750-2650	24	
Cross Brace	01540-750-2500	44	
Side Arm Bracket, 21"	01540-750-2700	12	
Guardrail Post	01540-750-2550	12	
Guardrail, 7' section	01540-750-2600	22	
Stairway Section	01540-750-2900	2	
Stairway Starter Bar	01540-750-2910	1	
Stairway Inside Handrail	01540-750-2920	2	
Stairway Outside Handrail	01540-750-2930	2	
Walk-Thru Frame Guardrail	01540-750-2940	2	

Scaffolding is often used as falsework over 15' high during construction of cast-in-place concrete beams and slabs. Two foot wide scaffolding is generally used for heavy beam construction. The span between frames depends upon the load to be carried with a maximum span of 5'.

Heavy duty shoring frames with a capacity of 10,000#/leg can be spaced up to 10' O.C. depending upon form support design and loading.

Scaffolding used as horizontal shoring requires less than half the material required with conventional shoring.

On new construction, erection is done by carpenters.

Rolling towers supporting horizontal shores can reduce labor and speed the job. For maintenance work, catwalks with spans up to 70' can be supported by the rolling towers.

R01540-200 Pump Staging

Pump staging is generally not available for rent. The table below shows the number of pieces required to erect pump staging for 2400 S.F. of building frontage. This area is made up of a pump jack system that is 3 poles (2 bays) wide by 2 poles high.

Item	CSI Line Item	Number of Pieces for 2400 S.F. of Building Front	Unit
Aluminum pole section, 24' long	01540-550-0200	6	Ea.
Aluminum splice joint, 6' long	01540-550-0600	3	
Aluminum foldable brace	01540-550-0900	3	
Aluminum pump jack	01540-550-0700	3	
Aluminum support for workbench/back safety rail	01540-550-1000	3	
Aluminum scaffold plank/workbench, 14" wide x 24' long	01540-550-1100	4	
Safety net, 22' long	01540-550-1250	2	
Aluminum plank end safety rail	01540-550-1200	2	

The cost in place for this 2400 S.F. will depend on how many uses are realized during the life of the equipment. Several options are given in Division 01540-550.

R01590-100   Contractor Equipment

Rental Rates shown in Division 01590 pertain to late model high quality machines in excellent working condition, rented from equipment dealers. Rental rates from contractors may be substantially lower than the rental rates from equipment dealers depending upon economic conditions; for older, less productive machines, reduce rates by a maximum of 15%. Any overtime must be added to the base rates. For shift work, rates are lower. Usual rule of thumb is 150% of one shift rate for two shifts; 200% for three shifts.

For periods of less than one week, operated equipment is usually more economical to rent than renting bare equipment and hiring an operator.

Costs to move equipment to a job site (mobilization) or from a job site (demobilization) are not included in rental rates in Division 01590, nor in any Equipment costs on any Unit Price line items or crew listings. These costs can be found in section 02305-250. If a piece of equipment is already at a job site, it is not appropriate to utilize mob/demob costs in an estimate again.

Rental rates vary throughout the country with larger cities generally having lower rates. Lease plans for new equipment are available for periods in excess of six months with a percentage of payments applying toward purchase.

Monthly rental rates vary from 2% to 5% of the cost of the equipment depending on the anticipated life of the equipment and its wearing parts. Weekly rates are about 1/3 the monthly rates and daily rental rates about 1/3 the weekly rate.

The hourly operating costs for each piece of equipment include costs to the user such as fuel, oil, lubrication, normal expendables for the equipment, and a percentage of mechanic's wages chargeable to maintenance. The hourly operating costs listed do not include the operator's wages.

The daily cost for equipment used in the standard crews is figured by dividing the weekly rate by five, then adding eight times the hourly operating cost to give the total daily equipment cost, not including the operator. This figure is in the right hand column of Division 01590 under Crew Equipment Cost/Day.

Pile Driving rates shown for pile hammer and extractor do not include leads, crane, boiler or compressor. Vibratory pile driving requires an added field specialist during set-up and pile driving operation for the electric model. The hydraulic model requires a field specialist for set-up only. Up to 125 reuses of sheet piling are possible using vibratory drivers. For normal conditions, crane capacity for hammer type and size are as follows.

Crane Capacity	Hammer Type and Size		
	Air or Steam	Diesel	Vibratory
25 ton	to 8,750 ft.-lb.		70 H.P.
40 ton	15,000 ft.-lb.	to 32,000 ft.-lb.	170 H.P.
60 ton	25,000 ft.-lb.		300 H.P.
100 ton		112,000 ft.-lb.	

Cranes should be specified for the job by size, building and site characteristics, availability, performance characteristics, and duration of time required.

Backhoes & Shovels rent for about the same as equivalent size cranes but maintenance and operating expense is higher. Crane operators rate must be adjusted for high boom heights. Average adjustments: for 150' boom add 2% per hour; over 185', add 4% per hour; over 210', add 6% per hour; over 250', add 8% per hour and over 295', add 12% per hour.

Tower Cranes of the climbing or static type have jibs from 50' to 200' and capacities at maximum reach range from 4,000 to 14,000 pounds. Lifting capacities increase up to maximum load as the hook radius decreases.

Typical rental rates, based on purchase price are about 2% to 3% per month.

Erection and dismantling runs between 500 and 2000 labor hours. Climbing operation takes 10 labor hours per 20' climb. Crane dead time is about 5 hours per 40' climb. If crane is bolted to side of the building add cost of ties and extra mast sections. Climbing cranes have from 80' to 180' of mast while static cranes have 80' to 800' of mast.

Truck Cranes can be converted to tower cranes by using tower attachments. Mast heights over 400' have been used. See Division 01590-600 for rental rates of high boom cranes.

A single 100' high material **Hoist and Tower** can be erected and dismantled in about 400 labor hours; a double 100' high hoist and tower in about 600 labor hours. Erection times for additional heights are 3 and 4 labor hours per vertical foot respectively up to 150', and 4 to 5 labor hours per vertical foot over 150' high. A 40' high portable Buck hoist takes about 160 labor hours to erect and dismantle. Additional heights take 2 labor hours per vertical foot to 80' and 3 labor hours per vertical foot for the next 100'. Most material hoists do not meet local code requirements for carrying personnel.

A 150' high **Personnel Hoist** requires about 500 to 800 labor hours to erect and dismantle. Budget erection time at 5 labor hours per vertical foot for all trades. Local code requirements or labor scarcity requiring overtime can add up to 50% to any of the above erection costs.

Earthmoving Equipment: The selection of earthmoving equipment depends upon the type and quantity of material, moisture content, haul distance, haul road, time available, and equipment available. Short haul cut and fill operations may require dozers only, while another operation may require excavators, a fleet of trucks, and spreading and compaction equipment. Stockpiled material and granular material are easily excavated with front end loaders. Scrapers are most economically used with hauls between 300' and 1-1/2 miles if adequate haul roads can be maintained. Shovels are often used for blasted rock and any material where a vertical face of 8' or more can be excavated. Special conditions may dictate the use of draglines, clamshells, or backhoes. Spreading and compaction equipment must be matched to the soil characteristics, the compaction required and the rate the fill is being supplied.

R01590-150   Heavy Lifting

Hydraulic Climbing Jacks

The use of hydraulic heavy lift systems is an alternative to conventional type crane equipment. The lifting, lowering, pushing, or pulling mechanism is a hydraulic climbing jack moving on a square steel jackrod from 1-5/8" to 4" square, or a steel cable. The jackrod or cable can be vertical or horizontal, stationary or movable, depending on the individual application. When the jackrod is stationary, the climbing jack will climb the rod and push or pull the load along with itself. When the climbing jack is stationary, the jackrod is movable with the load attached to the end and the climbing jack will lift or lower the jackrod with the attached load. The heavy lift system is normally operated by a single control lever located at the hydraulic pump.

The system is flexible in that one or more climbing jacks can be applied wherever a load support point is required, and the rate of lift synchronized.

Economic benefits have been demonstrated on projects such as: erection of ground assembled roofs and floors, complete bridge spans, girders and trusses, towers, chimney liners and steel vessels, storage tanks, and heavy machinery. Other uses are raising and lowering offshore work platforms, caissons, tunnel sections and pipelines.

R02065-300   Bituminous Paving

City	Sidewalk Mix Bituminous Asphalt per Ton*	Sidewalks (2") 9.2 S.Y./ton				Pavement (3") 6.13 S.Y./ton			
		Cost per S.Y.			Per Ton	Cost per S.Y.			Per Ton
		Material*	Installation	Total	Total	Material*	Installation	Total	Total
Atlanta	$30.00	$3.26	$1.94	$5.20	$47.84	$4.89	$ .98	$5.87	$35.98
Baltimore	34.73	3.78	1.89	5.67	52.16	5.67	.96	6.63	40.64
Boston	39.50	4.29	2.01	6.30	57.96	6.44	1.02	7.46	45.73
Buffalo	34.60	3.76	1.89	5.65	51.98	5.64	.96	6.60	40.46
Chicago	31.75	3.45	2.08	5.53	50.88	5.18	1.06	6.24	38.25
Cincinnati	41.00	4.46	2.20	6.66	61.27	6.69	1.12	7.81	47.88
Cleveland	29.50	3.21	2.16	5.37	49.40	4.81	1.10	5.91	36.23
Columbus	30.50	3.32	2.07	5.39	49.59	4.98	1.05	6.03	36.96
Dallas	29.25	3.18	1.75	4.93	45.36	4.77	.89	5.66	34.70
Denver	30.00	3.26	2.16	5.42	49.86	4.89	1.09	5.98	36.66
Detroit	32.00	3.48	1.97	5.45	50.14	5.22	1.00	6.22	38.13
Houston	34.50	3.75	1.80	5.55	51.06	5.63	.91	6.54	40.09
Indianapolis	29.75	3.23	2.03	5.26	48.39	4.85	1.03	5.88	36.04
Kansas City	29.50	3.21	1.98	5.19	47.75	4.81	1.00	5.81	35.62
Los Angeles	36.00	3.91	2.18	6.09	56.03	5.87	1.11	6.98	42.79
Memphis	38.75	4.21	1.86	6.07	55.84	6.32	.94	7.26	44.50
Milwaukee	32.50	3.53	1.89	5.42	49.86	5.30	.96	6.26	38.37
Minneapolis	31.38	3.41	2.21	5.62	51.70	5.12	1.12	6.24	38.25
Nashville	27.75	3.02	2.02	5.04	46.37	4.53	1.02	5.55	34.02
New Orleans	33.75	3.67	1.78	5.45	50.14	5.51	.90	6.41	39.29
New York City	49.50	5.38	2.33	7.71	70.93	8.08	1.18	9.26	56.76
Philadelphia	32.50	3.53	1.92	5.45	50.14	5.30	.97	6.27	38.44
Phoenix	30.00	3.26	2.13	5.39	49.59	4.89	1.08	5.97	36.60
Pittsburgh	35.00	3.80	2.21	6.01	55.29	5.71	1.12	6.83	41.87
St. Louis	32.00	3.48	1.96	5.44	50.05	5.22	.99	6.21	38.07
San Antonio	33.25	3.61	1.80	5.41	49.77	5.42	.91	6.33	38.80
San Diego	37.00	4.02	2.04	6.06	55.75	6.04	1.04	7.08	43.40
San Francisco	40.25	4.38	2.25	6.63	61.00	6.57	1.14	7.71	47.26
Seattle	39.50	4.29	2.30	6.59	60.63	6.44	1.16	7.60	46.59
Washington, D.C.	35.41	3.85	1.82	5.67	52.16	5.78	.92	6.70	41.07
Average	$34.00	$3.70	$2.02	$5.72	$52.62	$5.55	$1.02	$6.57	$40.27

Assumed density is 145 lb. per C.F.

*Includes delivery within 20 miles for quantities over 300 tons only.

Table below shows quantities and bare costs for 1000 S.Y. of Bituminous Paving.

Item	Sidewalks, 2" Thick (02775-275-0020)		Roads and Parking Areas, 3" Thick (02740-310-0460)	
	Quantities	Cost	Quantities	Cost
Bituminous asphalt	108.7 tons @ $34.00 per ton	$3,696.00	163.1 tons@ $34.00 per ton	$5,546.00
Installation using	Crew B-37 @ $1,455.80 /720 SY/day x 1000	2,021.94	Crew B-25B @ $5,023.80 /4900SY/ day x 1000	1,025.30
Total per 1000 S.Y.		$5,717.94		$6,571.30
Total per S.Y.		$ 5.72		$ 6.57
Total per Ton		$ 52.62		$ 40.27

SITE CONSTRUCTION 2

REFERENCE NOS. 2

R02115-200   Underground Storage Tank Removal

Underground Storage Tank Removal can be divided into two categories: Non-Leaking and Leaking. Prior to removing an underground storage tank, tests should be made, with the proper authorities present, to determine whether a tank has been leaking or the surrounding soil has been contaminated.

To safely remove Liquid Underground Storage Tanks:
1. Excavate to the top of the tank.
2. Disconnect all piping.
3. Open all tank vents and access ports.
4. Remove all liquids and/or sludge.
5. Purge the tank with an inert gas.
6. Provide access to the inside of the tank and clean out the interior using proper personal protective equipment (PPE).
7. Excavate soil surrounding the tank using proper PPE for on-site personnel.
8. Pull and properly dispose of the tank.
9. Clean up the site of all contaminated material.
10. Install new tanks or close the excavation.

R02220-510   Demolition Defined

Whole Building Demolition (Division 02220-110) - Demolition of the whole building with no concern for any particular building element, component, or material type being demolished. This type of demolition is accomplished with large pieces of construction equipment that break up the structure, load it into trucks and haul it to a disposal site, but disposal or dump fees are not included (Divisions 02220-320 and 02220-330). Demolition of below-grade foundation elements, such as footings, foundation walls, grade beams, slabs on grade, etc. (Division 02220-130), is not included. Certain mechanical equipment containing flammable liquids or ozone-depleting refrigerants, electric lighting elements, communication equipment components, and other building elements may contain hazardous waste, and must be removed, either selectively or carefully, as hazardous waste before the building can be demolished (Division 13281).

Foundation Demolition (Division 02220-130) - Demolition of below-grade foundation footings, foundation walls, grade beams, and slabs on grade. This type of demolition is accomplished by hand or pneumatic hand tools, and does not include saw cutting (Division 02220-360), or handling, loading, hauling, or disposal of the debris (Divisions 02220-320, 02220-330 and 02220-350).

Gutting (Division 02220-340) - Removal of building interior finishes and electrical/mechanical systems down to the load-bearing and sub-floor elements of the rough building frame, with no concern for any particular building element, component, or material type being demolished. This type of demolition is accomplished by hand or pneumatic hand tools, and includes loading into trucks, but not hauling, disposal or dump fees (Divisions 02220-320 and 02220-330); scaffolding (01540-750); or shoring (03150-600). Certain mechanical equipment containing flammable liquids or ozone-depleting refrigerants, electric lighting elements, communication equipment components, and other building elements may contain hazardous waste, and must be removed, either selectively or carefully, as hazardous waste, before the building is gutted (Division 13281).

Selective Demolition - Demolition of a selected building element, component, or finish, with some concern for surrounding or adjacent elements, components, or finishes (see the first Subdivision (s) at the beginning of Divisions 3 through 16). This type of demolition is accomplished by hand or pneumatic hand tools, and does not include handling, loading, storing, hauling, or disposal of the debris (Divisions 02220-320, 02220-330 and 02220-350); scaffolding (Division 01540-750); or shoring (Division 03150-600). "Gutting" methods may be used in order to save time, but damage that was caused to surrounding or adjacent elements, components, or finishes may have to be repaired at a later time.

Careful Removal - Removal of a piece of service equipment, building element or component, or material type, with great concern for both the removed item and surrounding or adjacent elements, components or finishes. The purpose of careful removal may be to protect the removed item for later re-use, preserve a higher salvage value of the removed item, or replace an item while taking care to protect surrounding or adjacent elements, components, connections, or finishes from cosmetic and/or structural damage. An approximation of the time required to perform this type of removal is 1/3 to 1/2 the time it would take to install a new item of like kind (see Reference Numbers R15050-720 and R16055-300). This type of removal is accomplished by hand or pneumatic hand tools, and does not include loading, hauling, or storing the removed item (Divisions 01840-100 and 02220-350); scaffolding (Division 01540-750); shoring (Division 03150-600); or lifting equipment (Division 01590-600).

Cutout Demolition (Division 02220-310) - Demolition of a small quantity of floor, wall, roof, or other assembly, with concern for the appearance and structural integrity of the surrounding materials. This type of demolition is accomplished by hand or pneumatic hand tools, and does not include saw cutting (Division 02220-360); handling, loading, hauling, or disposal of debris (Divisions 02220-320, 02220-330 and 02220-350); scaffolding (Division 01540-750); or shoring (Division 03150-600).

Rubbish Handling (Division 02220-350) - Work activities that involve handling, loading or hauling of debris. Generally, the cost of rubbish handling must be added to the cost of all types of demolition, with the exception of whole building demolition.

Minor Site Demolition (Division 02220-240) - Demolition of site elements outside the footprint of a building. This type of demolition is accomplished by hand or pneumatic hand tools, or with larger pieces of construction equipment, and may include loading a removed item onto a truck (check the Crew for equipment used). It does not include saw cutting (Division 02220-360), hauling or disposal of debris (Divisions 02220-320 and 02220-330), and, sometimes, handling or loading (Division 02220-350).

R02240-900 Wellpoints

A single stage wellpoint system is usually limited to dewatering an average 15' depth below normal ground water level. Multi-stage systems are employed for greater depth with the pumping equipment installed only at the lowest header level. Ejectors, with unlimited lift capacity, can be economical when two or more stages of wellpoints can be replaced or when horizontal clearance is restricted, such as in deep trenches or tunneling projects, and where low water flows are expected. Wellpoints are usually spaced on 2-1/2' to 10' centers along a header pipe. Wellpoint spacing, header size, and pump size are all determined by the expected flow, as dictated by soil conditions.

In almost all soils encountered in wellpoint dewatering, the wellpoints may be jetted into place. Cemented soils and stiff clays may require sand wicks about 12" in diameter around each wellpoint to increase efficiency and eliminate weeping into the excavation. These sand wicks require 1/2 to 3 C.Y. of washed filter sand and are installed by using a 12" diameter steel casing and hole puncher jetted into the ground 2' deeper than the wellpoint. Rock may require predrilled holes.

Labor required for the complete installation and removal of a single stage wellpoint system is in the range of 3/4 to 2 labor-hours per linear foot of header, depending upon jetting conditions, wellpoint spacing, etc.

Continuous pumping is necessary except in some free draining soil where temporary flooding is permissible (as in trenches which are backfilled after each day's work). Good practice requires provision of a stand-by pump during the continuous pumping operation.

Systems for continuous trenching below the water table should be installed three to four times the length of expected daily progress to insure uninterrupted digging, and header pipe size should not be changed during the job.

For pervious free draining soils, deep wells in place of wellpoints may be economical because of lower installation and maintenance costs. Daily production ranges between two to three wells per day, for 25' to 40' depths, to one well per day for depths over 50'.

Detailed analysis and estimating for any dewatering problem is available at no cost from wellpoint manufacturers. Major firms will quote "sufficient equipment" quotes or their affiliates offer lump sum proposals to cover complete dewatering responsibility.

Description for 200' System with 8" Header	Quantities	1st Month Unit	1st Month Total	Thereafter Unit	Thereafter Total
Equipment & Material					
Wellpoints 25' long, 2" diameter @ 5' O.C.	40 Each	$ 31.00	$ 1,240.00	$ 23.25	$ 930.00
Header pipe, 8" diameter	200 L.F.	5.90	1,180.00	2.95	590.00
Discharge pipe, 8" diameter	100 L.F.	3.96	396.00	2.57	257.00
8" valves	3 Each	151.00	453.00	98.15	294.45
Combination Jetting & Wellpoint pump (standby)	1 Each	2,450.00	2,450.00	1,715.00	1,715.00
Wellpoint pump, 8" diameter	1 Each	2,400.00	2,400.00	1,680.00	1,680.00
Transportation to and from site	1 Day	697.00	697.00	—	—
Fuel 30 days x 60 gal./day	1800 Gallons	2.00	3,600.00	2.00	3,600.00
Lubricants for 30 days x 16 lbs./day	480 Lbs.	1.50	720.00	1.50	720.00
Sand for points	40 C.Y.	21.35	854.00	—	—
Equipment & Materials Subtotal			$13,990.00		$ 9,786.45
Labor					
Technician to supervise installation	1 Week	$ 36.20	$ 1,448.00	—	—
Labor for installation and removal of system	300 Labor-hours	26.70	8,010.00	—	—
4 Operators straight time 40 hrs./wk. for 4.33 wks.	693 Hrs.	33.05	22,903.65	$ 33.05	$22,903.65
4 Operators overtime 2 hrs./wk. for 4.33 wks.	35 Hrs.	66.10	2,313.50	66.10	2,313.50
Bare Labor Subtotal			$34,675.15		$25,217.15
Total Bare Cost			$48,665.15		$35,003.60
Monthly Bare Cost/L.F. Header			$ 243.33		$ 175.02

R02250-400 Wood Sheet Piling

Wood sheet piling may be used for depths to 20' where there is no ground water. If moderate ground water is encountered Tongue & Groove sheeting will help to keep it out. When considerable ground water is present, steel sheeting must be used.

For estimating purposes on trench excavation, sizes are as follows:

Depth	Sheeting	Wales	Braces	B.F. per S.F.
To 8'	3 x 12's	6 x 8's, 2 line	6 x 8's, @ 10'	4.0 @ 8'
8' x 12'	3 x 12's	10 x 10's, 2 line	10 x 10's, @ 9'	5.0 average
12' to 20'	3 x 12's	12 x 12's, 3 line	12 x 12's, @ 8'	7.0 average

Sheeting to be toed in at least 2' depending upon soil conditions. A five person crew with an air compressor and sheeting driver can drive and brace 440 SF/day at 8' deep, 360 SF/day at 12' deep, and 320 SF/day at 16' deep. For normal soils, piling can be pulled in 1/3 the time to install. Pulling difficulty increases with the time in the ground. Production can be increased by high pressure jetting. Figures below assume 50% of lumber is salvaged and includes pulling costs. Some jurisdictions require an equipment operator in addition to Crew B-31.

Sheeting Pulled

Daily Cost Crew B-31	L.H./ Day	Hourly Cost	Daily Cost	8' Depth, 440 S.F./Day		16' Depth, 320 S.F./Day	
				To Drive (1 Day)	To Pull (1/3 Day)	To Drive (1 Day)	To Pull (1/3 Day)
1 Foreman	8	$28.70	$ 229.60	$ 229.60	$ 76.46	$ 229.60	$ 76.46
3 Laborers	24	26.70	640.80	640.80	213.39	640.80	213.39
1 Carpenter	8	34.25	274.00	274.00	91.24	274.00	91.24
1 Air Compressor			112.40	112.40	37.43	112.40	37.43
1 Sheeting Driver			7.20	7.20	2.40	7.20	2.40
2 -50 Ft. Air Hoses, 1-1/2" Diam.			25.50	25.50	8.49	25.50	8.49
Lumber (50% salvage)				1.76 MBF 616.00		2.24 MBF 784.00	
Bare Total			$1,289.50	$1,905.50	$429.41	$2,073.50	$429.41
Bare Total/S.F.				$ 4.33	$ .98	$ 4.71	$ 1.34
Bare Total (Drive and Pull)/S.F.				$ 5.31		$ 6.05	

Sheeting Left in Place

Daily Cost	8' Depth 440 S.F./Day		10' Depth 400 S.F./Day		12' Depth 360 S.F./Day		16' Depth 320 S.F./Day		18' Depth 305 S.F./Day		20' Depth 280 S.F./Day	
Crew B-31		$1,273.40		$1,273.40		$1,273.40		$1,273.40		$1,273.40		$1,273.40
Lumber	1.76 MBF	1,232.00	1.8 MBF	1,260.00	1.8 MBF	1,260.00	2.24 MBF	1,568.00	2.1 MBF	1,470.00	1.9 MBF	1,330.00
Bare Total in Place		$2,505.40		$2,533.40		$2,533.40		$2,841.40		$2,743.40		$2,603.40
Bare Total/S.F.		$ 5.69		$ 6.33		$ 7.04		$ 8.88		$ 8.99		$ 9.30
Bare Total/M.B.F.		$1,423.52		$1,407.44		$1,407.44		$1,268.48		$1,306.38		$1,370.21

R02250-450 Steel Sheet Piling

Limiting weights are 22 to 38#/S.F. of wall surface with 27#/S.F. average for usual types and sizes. (Weights of piles themselves are from 30.7#/L.F. to 57#/L.F. but they are 15″ to 21″ wide.) Lightweight sections 12″ to 28″ wide from 3 ga. to 12 ga. thick are also available for shallow excavations. Piles may be driven two at a time with an impact or vibratory hammer (use vibratory to pull) hung from a crane without leads. A reasonable estimate of the life of steel sheet piling is 10 uses with up to 125 uses possible if a vibratory hammer is used. Used piling costs from 50% to 80% of new piling depending on location and market conditions. Sheet piling and H piles

can be rented for about 30% of the delivered mill price for the first month and 5% per month thereafter. Allow 1 labor-hour per pile for cleaning and trimming after driving. These costs increase with depth and hydrostatic head. Vibratory drivers are faster in wet granular soils and are excellent for pile extraction. Pulling difficulty increases with the time in the ground and may cost more than driving. It is often economical to abandon the sheet piling, especially if it can be used as the outer wall form. Allow about 1/3 additional length or more, for toeing into ground. Add bracing, waler and strut costs. Waler costs can equal the cost per ton of sheeting.

Cost of Sheet Piling & Production Rate by Ton & S.F.						
Depth of Excavation	**15′ Depth**		**20′ Depth**		**25′ Depth**	
Description of Pile	**22 psf = 90.9 S.F./Ton**		**27 psf = 74 S.F./Ton**		**38 psf = 52.6 S.F./Ton**	
Type of operation:	Drive &	Drive &	Drive &	Drive &	Drive &	Drive &
Left in Place or Removed	Left	Extract*	Left	Extract*	Left	Extract*
Labor & Equip. to Drive 1 ton	$ 412.06	$742.40	$ 343.97	$680.06	$ 234.44	$424.23
Piling (75% Salvage)	880.00	220.00	880.00	220.00	880.00	220.00
Bare Cost/Ton in Place	$1,292.06	$962.40	$1,223.97	$900.06	$1,114.44	$644.23
Production Rate, Tons/Day	10.81	6.00	12.95	6.55	19.00	10.50
Bare Cost, S.F. in Place (incl. 33% toe in)	$ 14.21	$ 10.59	$ 16.54	$ 12.16	$ 21.19	$ 12.25
Production Rate, S.F./Day (incl. 33% toe in)	983	545	959	485	1000	553

Crew B-40	Bare Costs		Inc. Subs O&P			
	Hr.	Daily	Hr.	Daily		
1 Pile Driver Foreman (Out)	$35.30	$ 282.40	$58.15	$ 465.20	**Bare Installation Cost per Ton**	
4 Pile Drivers	33.30	1,065.60	54.90	1,756.80	for 15′ Deep Excavation plus 33% toe-in	
2 Equip. Oper. (crane)	35.90	574.40	54.10	865.60	$4,454 per day/10.81 tons = $412.03 per ton	
1 Equip. Oper. Oiler	30.10	240.80	45.35	362.80	**Bare Installation Cost per S.F.**	
1 Crane, 40 Ton		888.20		977.00	$\frac{2000\#}{22\#/S.F.}$ = 90.9 S.F. x 10.81 tons = 983 S.F. per day	
Vibratory Hammer & Gen.		1,403.00		1,543.30		
64 L.H., Daily Totals		$4,454.40		$5,970.70	$4,454 / 983 S.F. per day = $4.53 per S.F.	

*For driving & extracting, two mobilizations & demobilizations will be necessary

2

SITE CONSTRUCTION

R02250-900　Vibroflotation and Vibro Replacement Soil Compaction

Vibroflotation is a proprietary system of compacting sandy soils in place to increase relative density to about 70%. Typical bearing capacities attained will be 6000 psf for saturated sand and 12,000 psf for dry sand. Usual range is 4000 to 8000 psf capacity. Costs in the front of the book are for a vertical foot of compacted cylinder 6′ to 10′ in diameter.

Vibro replacement is a proprietary system of improving cohesive soils in place to increase bearing capacity. Most silts and clays above or below the water table can be strengthened by installation of stone columns.

The process consists of radial displacement of the soil by vibration. The created hole is then backfilled in stages with coarse granular fill which is thoroughly compacted and displaced into the surrounding soil in the form of a column.

The total project cost would depend on the number and depth of the compacted cylinders. The installing company guarantees relative soil density of the sand cylinders after compaction and the bearing capacity of the soil after the replacement process. Detailed estimating information is available from the installer at no cost.

R02315-300　Compacting Backfill

Compaction of fill in embankments, around structures, in trenches, and under slabs is important to control settlement. Factors affecting compaction are:

1. Soil gradation
2. Moisture content
3. Equipment used
4. Depth of fill per lift
5. Density required

The costs for testing and soil analyses are listed in Division 01450-500. Also, see Division 02315 for further backfill, borrow, and compaction costs.

Example:

Compact granular fill around a building foundation using a 21″ wide x 24″ vibratory plate in 8″ lifts. Operator moves at 50 FPM working a 50 minute hour to develop 95% Modified Proctor Density with 4 passes.

Production Rate:

$$\frac{1.75' \text{ plate width x 50 F.P.M. x 50 min./hr. x .67' lift}}{27 \text{ C.F. per C.Y.}} = 108.5 \text{ C.Y./hr.}$$

Production Rate for 4 Passes:

$$\frac{108.5 \text{ C.Y.}}{4 \text{ passes}} = 27.125 \text{ C.Y./hr. x 8 hrs.} = 217 \text{ C.Y./day}$$

	Compacting 217 C.Y. with 21″ Wide Vibratory Plate	L.H./ Day	Hourly Cost	Daily Cost	C.Y. Cost
1	Laborer	8	$26.70	$213.60	$ .98
1	Vibratory Plate Compactor			36.80	.17
Bare Total for 217 C.Y./day				$250.40	$1.15

REFERENCE NOS.

R02315-400 Excavating

The selection of equipment used for structural excavation and bulk excavation or for grading is determined by the following factors.
1. Quantity of material.
2. Type of material.
3. Depth or height of cut.
4. Length of haul.
5. Condition of haul road.
6. Accessibility of site.
7. Moisture content and dewatering requirements.
8. Availability of excavating and hauling equipment.

Some additional costs must be allowed for hand trimming the sides and bottom of concrete pours and other excavation below the general excavation.

Number of B.C.Y. per truck = 1.5 C.Y. bucket x 8 passes = 12 loose C.Y.

$$= 12 \times \frac{100}{118} = 10.2 \text{ B.C.Y. per truck}$$

Truck Haul Cycle:

Load truck 8 passes	=	4 minutes
Haul distance 1 mile	=	9 minutes
Dump time	=	2 minutes
Return 1 mile	=	7 minutes
Spot under machine	=	1 minute
		23 minute cycle

When planning excavation and fill, the following should also be considered.
1. Swell factor.
2. Compaction factor.
3. Moisture content.
4. Density requirements.

A typical example for scheduling and estimating the cost of excavation of a 15' deep basement on a dry site when the material must be hauled off the site, is outlined below.

Assumptions:
1. Swell factor, 18%.
2. No mobilization or demobilization.
3. Allowance included for idle time and moving on job.
4. No dewatering, sheeting, or bracing.
5. No truck spotter or hand trimming.

Fleet Haul Production per day in B.C.Y.

$$4 \text{ trucks} \times \frac{50 \text{ min. hour}}{23 \text{ min. haul cycle}} \times 8 \text{ hrs.} \times 10.2 \text{ B.C.Y.}$$

$$= 4 \times 2.2 \times 8 \times 10.2 = 718 \text{ B.C.Y./day}$$

Excavating Cost with a 1-1/2 C.Y. Hydraulic Excavator 15' Deep, 2 Mile Round Trip Haul		L.H./Day	Hourly Cost	Daily Cost	Subtotal	Unit Price
1	Equipment Operator	8	$35.90	$ 287.20		
1	Oiler	8	26.70	213.60		
4	Truck Drivers	32	27.55	881.60	$1,382.40	$1.92
1	Hydraulic Excavator			720.40		
4	Dump Trucks			1,906.40	2,626.80	$3.65
Bare Total for 720 B.C.Y.				$4,009.20	$4,009.20	$5.57

Description		1-1/2 C.Y. Hyd. Backhoe 15' Deep		1-1/2 C.Y. Power Shovel 7' Bank		1-1/2 C.Y. Dragline 7' Deep		2-1/2 C.Y. Trackloader Stockpile
Operator (and Oiler, if required)		$ 500.80		$ 500.80		$ 500.80		$ 500.80
Truck Drivers	3 Ea.	661.20	4 Ea.	881.60	3 Ea.	661.20	4 Ea.	881.60
Equipment Rental		720.40		956.00		1,236.60		551.80
20 C.Y. Trailer Dump Trucks	3 Ea.	1,429.80	4 Ea.	1,906.40	3 Ea.	1,429.80	4 Ea.	1,906.40
Bare Total Cost per Day		$3,312.20		$4,244.80		$3,828.40		$3,840.60
Daily Production, C.Y. Bank Measure		720.00		960.00		640.00		1,000.00
Bare Cost per C.Y.		$ 4.60		$ 4.42		$ 5.98		$ 3.84

Add the mobilization and demobilization costs to the total excavation costs. When equipment is rented for more than three days, there is often no mobilization charge by the equipment dealer. On larger jobs outside of urban areas, scrapers can move earth economically provided a dump site or fill area and adequate haul roads are available. Excavation within sheeting bracing or cofferdam bracing is usually done with a clamshell and production is low, since the clamshell may have to be guided by hand between the bracing. When excavating or filling an area enclosed with a wellpoint system, add 10% to 15% to the cost to allow for restricted access. When estimating earth excavation quantities for structures, allow work space outside the building footprint for construction of the foundation, and a slope of 1:1 unless sheeting is used.

R02315-450   Excavating Equipment

The table below lists THEORETICAL hourly production in C.Y./hr. bank measure for some typical excavation equipment. Figures assume 50 minute hours, 83% job efficiency, 100% operator efficiency, 90° swing and properly sized hauling units, which must be modified for adverse digging and loading conditions. Actual production costs in the front of the book average about 50% of the theoretical values listed here.

Equipment	Soil Type	B.C.Y. Weight	% Swell	1 C.Y.	1-1/2 C.Y.	2 C.Y.	2-1/2 C.Y.	3 C.Y.	3-1/2 C.Y.	4 C.Y.
Hydraulic Excavator "Backhoe" 15' Deep Cut	Moist loam, sandy clay	3400 lb.	40%	85	125	175	220	275	330	380
	Sand and gravel	3100	18	80	120	160	205	260	310	365
	Common earth	2800	30	70	105	150	190	240	280	330
	Clay, hard, dense	3000	33	65	100	130	170	210	255	300
Power Shovel Optimum Cut (Ft.)	Moist loam, sandy clay	3400	40	170 (6.0)	245 (7.0)	295 (7.8)	335 (8.4)	385 (8.8)	435 (9.1)	475 (9.4)
	Sand and gravel	3100	18	165 (6.0)	225 (7.0)	275 (7.8)	325 (8.4)	375 (8.8)	420 (9.1)	460 (9.4)
	Common earth	2800	30	145 (7.8)	200 (9.2)	250 (10.2)	295 (11.2)	335 (12.1)	375 (13.0)	425 (13.8)
	Clay, hard, dense	3000	33	120 (9.0)	175 (10.7)	220 (12.2)	255 (13.3)	300 (14.2)	335 (15.1)	375 (16.0)
Drag Line Optimum Cut (Ft.)	Moist loam, sandy clay	3400	40	130 (6.6)	180 (7.4)	220 (8.0)	250 (8.5)	290 (9.0)	325 (9.5)	385 (10.0)
	Sand and gravel	3100	18	130 (6.6)	175 (7.4)	210 (8.0)	245 (8.5)	280 (9.0)	315 (9.5)	375 (10.0)
	Common earth	2800	30	110 (8.0)	160 (9.0)	190 (9.9)	220 (10.5)	250 (11.0)	280 (11.5)	310 (12.0)
	Clay, hard, dense	3000	33	90 (9.3)	130 (10.7)	160 (11.8)	190 (12.3)	225 (12.8)	250 (13.3)	280 (12.0)

Equipment	Soil Type	B.C.Y. Weight	% Swell	Wheel Loaders				Track Loaders		
				3 C.Y.	4 C.Y.	6 C.Y.	8 C.Y.	2-1/4 C.Y.	3 C.Y.	4 C.Y.
Loading Tractors	Moist loam, sandy clay	3400	40	260	340	510	690	135	180	250
	Sand and gravel	3100	18	245	320	480	650	130	170	235
	Common earth	2800	30	230	300	460	620	120	155	220
	Clay, hard, dense	3000	33	200	270	415	560	110	145	200
	Rock, well-blasted	4000	50	180	245	380	520	100	130	180

R02455-900 Wood Bearing Piles

Untreated Southern Yellow Pine is most generally used for pile foundations cut off below the low water line. These are driven with the bark on. All piles cut off above the low water line should be treated with a preservative or encased in concrete for the section above water.

Item	Unit Cost	Units	Quantity	Total Cost	Cost/Pile	Cost/L.F.
50′ Treated Piles, 13″ Butt, 7″ Tip	$ 8.75	L.F.	200	$ 87,500.00	$437.50	$ 8.75
Installation, Crew B-19	3,690.40	Day	13	47,975.20	239.88	4.80
Mobilization & Demobilization, Crew B-19	3,690.40	Day	3	11,071.20	55.36	1.11
Transportation of Equip. One Way	744.20	Day	1	744.20	3.72	.07
Bare Totals				$147,290.60	$736.46	$14.73

The above figures are based on driving 800 L.F. daily which can be considered average. Time is included for moving rig, cutoff & ordinary delays. A general observation is that the cost of a pile in place complete is about two times the cost of pile only. See also equipment rental division 01590 and R01590-100 for equipment capacities.

R02465-600 Caissons

The three principal types of caissons are:

(1) Belled Caissons, which except for shallow depths and poor soil conditions, are generally recommended. They provide more bearing than shaft area. Because of its conical shape, no horizontal reinforcement of the bell is required.

(2) Straight Shaft Caissons are used where relatively light loads are to be supported by caissons that rest on high value bearing strata. While the shaft is larger in diameter than for belled types this is more than offset by the saving in time and labor.

(3) Keyed Caissons are used when extremely heavy loads are to be carried. A keyed or socketed caisson transfers its load into rock by a combination of end-bearing and shear reinforcing of the shaft. The most economical shaft often consists of a steel casing, a steel wide flange core and concrete. Allowable compressive stresses of .225 f′c for concrete, 16,000 psi for the wide flange core, and 9,000 psi for the steel casing are commonly used. The usual range of shaft diameter is 18″ to 84″. The number of sizes specified for any one project should be limited due to the problems of casing and auger storage. When hand work is to be performed, shaft diameters should not be less than 32″. When inspection of borings is required a minimum shaft diameter of 30″ is recommended. Concrete caissons are intended to be poured against earth excavation so permanent forms which add to cost should not be used if the excavation is clean and the earth sufficiently impervious to prevent excessive loss of concrete.

Soil Conditions for Belling		
Good	Requires Handwork	Not Recommended
Clay	Hard Shale	Silt
Sandy Clay	Limestone	Sand
Silty Clay	Sandstone	Gravel
Clayey Silt	Weathered Mica	Igneous Rock
Hard-pan		
Soft Shale		
Decomposed Rock		

R02465-800 Pressure Injected Footings

Pressure Injected Footings are end bearing foundation units consisting of expanded bulbous bases formed by high energy blows, and concrete shafts either cased or uncased, to transfer the load.

The bulb must be formed in granular, rock or hardpan soils. Bearing is achieved at a predetermined depth so that the length of the shaft is usually less than for friction piles. High load capacity reduces the number of units and caps required.

Mobilization and demobilization costs below assume maximum distance of 50 miles.

100 Uncased Pressure Injected Footings 25′ Long	Quantity Unit	Unit Cost	Total	Cost/Pile	Cost/L.F.
Crew B-44, 8 piles/day	13 Days	$3,080.45	$40,045.85	$400.46	$16.02
Mobilization & Demobilization	2 Days	3,080.45	6,160.90	61.61	2.46
Transportation one way	1.5 Days	744.20	1,116.30	11.16	.45
Total Installation			$47,323.05	$473.23	$18.93
Shafts: 100 ea. 17″ diam., 25′ deep	155 C.Y.	$ 117.97	$18,285.35	$182.85	$ 7.31
Pressure Bulbs: 100 ea. @ .75 C.Y.	75 C.Y.	117.97	8,847.75	88.48	3.54
Waste: 15% of above	35 C.Y.	117.97	4,128.95	41.29	1.65
Total Bare Material Cost			$31,262.05	$312.62	$12.50
Total Bare Cost in Place			$78,585.10	$785.85	$31.43

R02510-800 Piping Designations

There are several systems currently in use to describe pipe and fittings. The following paragraphs will help to identify and clarify classifications of piping systems used for water distribution.

Piping may be classified by schedule. Piping schedules include 5S, 10S, 10, 20, 30, Standard, 40, 60, Extra Strong, 80, 100, 120, 140, 160 and Double Extra Strong. These schedules are dependent upon the pipe wall thickness. The wall thickness of a particular schedule may vary with pipe size.

Ductile iron pipe for water distribution is classified by Pressure Classes such as Class 150, 200, 250, 300 and 350. These classes are actually the rated water working pressure of the pipe in pounds per square inch (psi). The pipe in these pressure classes is designed to withstand the rated water working pressure plus a surge allowance of 100 psi.

The American Water Works Association (AWWA) provides standards for various types of **plastic pipe.** C-900 is the specification for polyvinyl chloride (PVC) piping used for water distribution in sizes ranging from 4″ through 12″. C-901 is the specification for polyethylene (PE) pressure pipe, tubing and fittings used for water distribution in sizes ranging from 1/2″ through 3″. C-905 is the specification for PVC piping sizes 14″ and greater.

PVC pressure-rated pipe is identified using the standard dimensional ratio (SDR) method. This method is defined by the American Society for Testing and Materials (ASTM) Standard D 2241. This pipe is available in SDR numbers 64, 41, 32.5, 26, 21, 17, and 13.5. Pipe with an SDR of 64 will have the thinnest wall while pipe with an SDR of 13.5 will have the thickest wall. When the pressure rating (PR) of a pipe is given in psi, it is based on a line supplying water at 73 degrees F.

The National Sanitation Foundation (NSF) seal of approval is applied to products that can be used with potable water. These products have been tested to ANSI/NSF Standard 14.

Valves and strainers are classified by American National Standards Institute (ANSI) Classes. These Classes are 125, 150, 200, 250, 300, 400, 600, 900, 1500 and 2500. Within each class there is an operating pressure range dependent upon temperature. Design parameters should be compared to the appropriate material dependent, pressure-temperature rating chart for accurate valve selection.

R02510-810 Concrete Pipe

Prices given are for inside 20 mile delivery zone. Add $2.15 per ton of pipe for each additional 10 miles. Minimum truckload is 10 tons. The non-reinforced pipe listed in the front of the book is designation ASTM

C14-59 extra strength. The reinforced pipe listed is ASTM C76-65T class 3, no gaskets. The installation cost given includes shaping bottom of the trench, placing the pipe, and backfilling and tamping to the top of the pipe only.

R02920-500 Seeding

The type of grass is determined by light, shade and moisture content of soil plus intended use. Fertilizer should be disked 4″ before seeding. For steep slopes disk five tons of mulch and lay two tons of hay or straw on surface per acre after seeding. Surface mulch can be staked, lightly disked or tar emulsion sprayed. Material for mulch can be wood chips, peat moss, partially

rotted hay or straw, wood fibers and sprayed emulsions. Hemp seed blankets with fertilizer are also available. For spring seeding, watering is necessary. Late fall seeding may have to be reseeded in the spring. Hydraulic seeding, power mulching, and aerial seeding can be used on large areas.

R02920-500 Cost of Trees: Based on Pin Oak (Quercus palustris)

Tree Diameter	Normal Height	Catalog List Price of Tree	Guying Material	Bare Equipment Charge	Bare Installation Labor	Bare Total
2 to 3 inch	14 feet	$ 146	$15.35	$ 54.24	$ 91.20	$ 306.79
3 to 4 inch	16 feet	323	18.35	90.40	152.00	583.75
4 to 5 inch	18 feet	373	75.50	108.48	182.40	739.38
6 to 7 inch	22 feet	813	80.00	135.60	228.00	1,256.60
8 to 9 inch	26 feet	1,023	90.00	180.80	304.00	1,597.80

Installation Time & Cost for Planting Trees, Bare Costs													
Ball Size Diam. X Depth	Soil in Ball	Weight of Ball	Hole Diam. Req'd	Hole Excavation	Amount of Soil Displ.	Topsoil Handled	Time Required in Labor-Hours					Cost	
							Dig & Lace	Handle Ball	Dig Hole	Plant & Prune	Water & Guy	Total L.H.	Crew	Bare Total per Tree
Inches	C.F.	Lbs.	Feet	C.F.	C.F.	C.F.								
12 x 12	0.70	56.00	2.00	4.00	3.00	11.00	.25	.17	.33	.25	.07	1.10	1 Clab	$ 29.37
18 x 16	2.00	160.00	2.50	8.00	6.00	21.00	.50	.33	.47	.35	.08	1.70	2 Clab	45.39
24 x 18	4.00	320.00	3.00	13.00	9.00	38.00	1.00	.67	1.08	.82	.20	3.80	3 Clab	101.46
30 x 21	7.50	600.00	4.00	27.00	19.50	76.00	.82	.71	.79	1.22	.26	3.80		143.68
36 x 24	12.50	980.00	4.50	38.00	25.50	114.00	1.08	.95	1.11	1.32	.30	4.76		179.98
42 x 27	19.00	1,520.00	5.50	64.00	45.00	185.00	1.90	1.27	1.87	1.43	.34	6.80		257.11
48 x 30	28.00	2,040.00	6.00	85.00	57.00	254.00	2.41	1.60	2.06	1.55	.39	8.00		302.48
54 x 33	38.50	3,060.00	7.00	127.00	88.50	370.00	2.86	1.90	2.39	1.76	.45	9.40	B-6	355.41
60 x 36	52.00	4,160.00	7.50	159.00	107.00	474.00	3.26	2.17	2.73	2.00	.51	10.70		404.57
66 x 39	68.00	5,440.00	8.00	196.00	128.00	596.00	3.61	2.41	3.07	2.26	.58	11.90		449.94
72 x 42	87.00	7,160.00	9.00	267.00	180.00	785.00	3.90	2.60	3.71	2.78	.70	13.70		518.00

SITE CONSTRUCTION

2

REFERENCE NOS.

R03110-010   Wall Form Materials

Aluminum Forms

Approximate weight is 3 lbs. per S.F.C.A. Standard widths are available from 4" to 36" with 36" most common. Standard lengths of 2', 4', 6' to 8' are available. Forms are lightweight and fewer ties are needed with the wider widths. The form face is either smooth or textured.

Cost of bare forms per S.F.C.A. with different facing surface is listed below. Forms may also be rented.

Purchase Cost Per S.F.								
Finish	**3' x 8'**	**2' x 8'**	**12" x 8'**	**6" x 8'**	**3' x 4'**	**2' x 4'**	**12" x 4'**	**6" x 4'**
Smooth Aluminum (.096" Face Sheet)	$11.57	$14.28	$19.28	$32.13	$11.21	$15.57	$20.55	$31.92
Textured Brick Aluminum	$13.19	$19.50	$28.56	$35.91	$15.28	$20.11	$25.74	$41.23

Metal Framed Plywood Forms

Manufacturers claim over 75 reuses of plywood and over 300 reuses of steel frames. Many specials such as corners, fillers, pilasters, etc. are available. Monthly rental is generally about 15% of purchase price for first month and 9% per month thereafter with 90% of rental applied to purchase for the first month and decreasing percentages thereafter. Aluminum framed forms cost 25% to 30% more than steel framed.

After the first month, extra days may be prorated from the monthly charge. Rental rates do not include ties, accessories, cleaning, loss of hardware or freight in and out. Approximate weight is 5 lbs. per S.F. for steel; 3 lbs. per S.F. for aluminum.

Forms can be rented with option to buy.

Plywood Forms, Job Fabricated

There are two types of plywood used for concrete forms.

1. Exterior plyform which is completely waterproof. This is face oiled to facilitate stripping. Ten reuses can be expected with this type with 25 reuses possible.
2. An overlaid type consists of a resin fiber fused to exterior plyform. No oiling is required except to facilitate cleaning. This is available in both high density (HDO) and medium density overlaid (MDO). Using HDO, 50 reuses can be expected with 200 possible.

Plyform is available in 5/8" and 3/4" thickness. High density overlaid is available in 3/8", 1/2", 5/8" and 3/4" thickness.

5/8" thick is sufficient for most building forms, while 3/4" is best on heavy construction.

For prices on plywood and framing lumber see R06100-010 and R06160-020.

Plywood Forms, Modular, Prefabricated

There are many plywood forming systems without frames. Most of these are manufactured from 1-1/8" (HDO) plywood and have some hardware attached. These are used principally for foundation walls 8' or less high. With care and maintenance, 100 reuses can be attained with decreasing quality of surface finish.

Steel Forms

Approximate weight is 6-1/2 lbs. per S.F.C.A. including accessories. Standard widths are available from 2" to 24", with 24" most common. Standard lengths are from 2' to 8', with 4' the most common. Forms are easily ganged into modular units.

Forms are usually leased for 15% of the purchase price per month prorated daily over 30 days.

Rental may be applied to sale price and usually rental forms are bought. With careful handling and cleaning 200 to 400 reuses are possible.

Straight wall gang forms up to 12' x 20' or 8' x 30' can be fabricated. These crane handled forms usually lease for approx. 9% per month.

Individual job analysis is available from the manufacturer at no charge.

R03110-020　Floor Pans and Domes

For 8' to 15' Ceiling Heights Using Crew C-2 at Bare Cost per Labor-Hour		20" Pans (03110-420)		19" Domes (03110-420)	
		1 Use	4 Use	1 Use	4 Use
Purchase 20" pans or rent 19" domes		$2,596.00	$ 649.00	$229.00	$183.20
Adjustable shores and accessories		75.00	60.00	75.00	60.00
110 S.F. 3/4" plyform		159.50	51.84	159.50	51.84
210 B.F. supporting joists, girts and braces		106.68	34.67	106.68	34.67
Labor handle, place, strip, oil and clean pans and domes	1.5 L.H. 50.00		1.0 L.H. 33.33	1.8 L.H. 59.99	1.2 L.H. 40.00
Make up, erect, remove wood decking	10 L.H. 333.30		8.5 L.H. 283.31	10.0 L.H. 333.30	8.5 L.H. 283.31
Bare Total per 100 S.F. of floor area		$3,320.48	$1,112.15	$963.47	$653.02

The figures above are for closed deck forming. For open deck forming deduct $36 per 100 S.F. for four uses.

For slab height from 15' to 20' add $56 per 100 S.F. and for 20' to 35' add $75 per 100 S.F., both for four uses.

R03110-030　Slipforms

The slipform method of forming may be used for forming circular silo and multi-celled storage bin type structures over 30' high, and building core shear walls over eight stories high. The shear walls, usually enclose elevator shafts, stairwells, mechanical spaces, and toilet rooms. Reuse of the form on duplicate structures will reduce the height necessary and spread the cost of building the form. Slipform systems can be used to cast chimneys, towers, piers, dams, underground shafts or other structures capable of being extruded.

Slipforms are usually 4' high and are raised semi-continuously by jacks climbing on rods which are embedded in the concrete. The jacks are powered by a hydraulic, pneumatic, or electric source and are available in 3, 6, and 22 ton capacities. Interior work decks and exterior scaffolds must be provided for placing inserts, embedded items, reinforcing steel, and concrete. Scaffolds below the form for finishers may be required. The interior work decks are often used as roof slab forms on silos and bin work. Form raising rates will range from 6" to 20" per hour for silos; 6" to 30" per hour for buildings; and 6" to 48" per hour for shaft work.

Reinforcing bars and stressing strands are usually hoisted by crane or gin pole and the concrete material can be hoisted by crane, winch-powered skip, or pumps. The slipform system is operated on a continuous 24 hour day when a monolithic structure is desired. For least cost, the system is operated only during normal working hours.

Placing concrete will range from 0.5 to 1.5 labor-hours per C.Y. Bucks, blockouts, keyways, weldplates, etc. are extra.

R03110-040 Forms for Reinforced Concrete

Design Economy

Avoid many sizes in proportioning beams and columns.

From story to story avoid changing column dimensions. Gain strength by adding steel or using a richer mix. If a change in size of column is necessary, vary one dimension only to minimize form alterations. Keep beams and columns the same width.

From floor to floor in a multi-story building vary beam depth, not width, as that will leave slab panel form unchanged. It is cheaper to vary the strength of a beam from floor to floor by means of steel area than by 2" changes in either width or depth.

Cost Factors

Material includes the cost of lumber, cost of rent for metal pans or forms if used, nails, form ties, form oil, bolts and accessories.

Labor includes the cost of carpenters to make up, erect, remove and repair, plus common labor to clean and move. Having carpenters remove forms minimizes repairs.

Improper alignment and condition of forms will increase finishing cost. When forms are heavily oiled, concrete surfaces must be neutralized before finishing. Special curing compounds will cause spillages to spall off in first frost. Gang forming methods will reduce costs on large projects.

Materials Used

Boards are seldom used unless their architectural finish is required. Generally, steel, fiberglass and plywood are used for contact surfaces. Labor on plywood is 10% less than with boards. The plywood is backed up with

2 x 4's at 12" to 32" O.C. Walers are generally 2 - 2 x 4's. Column forms are held together with steel yokes or bands. Shoring is with adjustable shoring or scaffolding for high ceilings.

Reuse

Floor and column forms can be reused four or possibly five times without excessive repair. Remember to allow for 10% waste on each reuse.

When modular sized wall forms are made, up to twenty uses can be expected with exterior plyform.

When forms are reused, the cost to erect, strip, clean and move will not be affected. 10% replacement of lumber should be included and about one hour of carpenter time for repairs on each reuse per 100 S.F.

The reuse cost for certain accessory items normally rented on a monthly basis will be lower than the cost for the first use.

After fifth use, new material required plus time needed for repair prevent form cost from dropping further and it may go up. Much depends on care in stripping, the number of special bays, changes in beam or column sizes and other factors.

Costs for multiple use of formwork may be developed as follows:

2 Uses	3 Uses	4 Uses
$\dfrac{(\text{1st Use + Reuse})}{2} = \text{avg. cost/2 uses}$	$\dfrac{(\text{1st Use + 2 Reuse})}{3} = \text{avg. cost/3 uses}$	$\dfrac{(\text{1st use + 3 Reuse})}{4} = \text{avg. cost/4 uses}$

R03110-050 Forms In Place

This section assumes that all cuts are made with power saws, that adjustable shores are employed and that maximum use is made of commercial form ties and accessories. Bare costs are used in the table below.

BEAM AND GIRDER, INTERIOR, 12" Wide (Line 03110-405-2000)			First Use			Reuse		
Item	Cost	Unit	Quantity	Material	Installation	Quantity	Material	Installation
5/8" exterior plyform	$1,364.00	M.S.F.	115 S.F.	$156.86		11.5 S.F.	$15.69	
Lumber	508.00	M.B.F.	200 B.F.	101.60		20.0 B.F.	10.16	
Accessories, incl. adjustable shores			Allow	25.00		Allow	25.00	
Make up, crew C-2	33.33	L.H.	6.4 L.H.		$213.31	1.0 L.H.		$ 33.33
Erect and strip			8.3 L.H.		276.64	8.3 L.H.		276.64
Clean and move			1.3 L.H.		43.33	1.3 L.H.		43.33
Bare Total per 100 S.F.C.A.			16.0 L.H.	$283.46	$533.28	10.6 L.H.	$50.85	$353.30

For structural steel frame with beams encased, subtract 1.2 labor-hours, and 25% of the lumber per 100 S.F.C.A.

BOX CULVERT, 5' to 8' Square or Rectangular (Line 03110-415-0010)			First Use			Reuse		
Item	Cost	Unit	Quantity	Material	Installation	Quantity	Material	Installation
3/4" exterior plyform	$1,450.00	M.S.F.	110 S.F.	$159.50		11.0 S.F.	$15.95	
Lumber	508.00	M.B.F.	170 B.F.	86.36		17.0 B.F.	8.64	
Accessories			Allow	20.00		Allow	20.00	
Build in place, crew C-1	32.36	L.H.	14.5 L.H.		$469.22	1.0 L.H.		$ 32.36
Strip and salvage			4.3 L.H.		139.15	4.3 L.H.		139.15
Bare Total per 100 S.F.C.A.			18.8 L.H.	$265.86	$608.37	18.8 L.H.	$44.59	$171.51

R03110-050 Forms in Place (cont.)

COLUMNS, 24" x 24" (Line 03110-410-6500)

Item	Cost	Unit	First Use Quantity	First Use Material	First Use Installation	Reuse Quantity	Reuse Material	Reuse Installation
5/8" exterior plyform	$1,364.00	M.S.F.	120 S.F.	$163.68		12.0 S.F.	$16.37	
Lumber	508.00	M.B.F.	125 B.F.	63.50		12.5 B.F.	6.35	
Clamps, chamfer strips and accessories			Allow	25.00		Allow	25.00	
Make up, crew C-1	32.36	L.H.	5.8 L.H.		$187.69	1.0 L.H.		$ 32.36
Erect and strip			9.8 L.H.		317.13	9.8 L.H.		317.13
Clean and move			1.2 L.H.		38.83	1.2 L.H.		38.83
Bare Total per 100 S.F.C.A.			16.8 L.H.	$252.18	$543.65	12.0 L.H.	$47.72	$388.32

FLAT SLAB WITH DROP PANELS (Line 03110-420-2000)

Item	Cost	Unit	First Use Quantity	First Use Material	First Use Installation	Reuse Quantity	Reuse Material	Reuse Installation
5/8" exterior plyform	$1,364.00	M.S.F.	115 S.F.	$156.86		11.5 S.F.	$15.69	
Lumber	508.00	M.B.F.	210 B.F.	106.68		21.0 B.F.	10.67	
Accessories, incl. adjustable shores			Allow	25.00		Allow	25.00	
Make up, crew C-2	33.33	L.H.	3.5 L.H.		$116.66	1.0 L.H.		$ 33.33
Erect and strip			6.0 L.H.		199.98	6.0 L.H.		199.98
Clean and move			1.2 L.H.		40.00	1.2 L.H.		40.00
Bare Total per 100 S.F.C.A.			10.7 L.H.	$288.54	$356.64	8.2 L.H.	$51.36	$273.31

Drop panels included but column caps figure with columns.

FOOTINGS, SPREAD (Line 03110-430-5000)

Item	Cost	Unit	First Use Quantity	First Use Materials	First Use Installation	Reuse Quantity	Reuse Material	Reuse Installation
Lumber	$508.00	M.B.F.	260 B.F.	$132.08		26 B.F.	$13.21	
Accessories			Allow	7.50		Allow	7.50	
Make up, crew C-1	32.36	L.H.	4.7 L.H.		$152.09	1.0 L.H.		$ 32.36
Erect and strip			4.2 L.H.		135.91	4.2 L.H.		135.91
Clean and move			1.6 L.H.		51.78	1.6 L.H.		51.78
Bare Total per 100 S.F.C.A.			10.5 L.H.	$139.58	$339.78	6.8 L.H.	$20.71	$220.05

FOUNDATION WALL, 8' High (Line 03110-455-2000)

Item	Cost	Unit	First Use Quantity	First Use Material	First Use Installation	Reuse Quantity	Reuse Material	Reuse Installation
5/8" exterior plyform	$1,364.00	M.S.F.	110 S.F.	$150.04		11.0 S.F.	$15.00	
Lumber	508.00	M.B.F.	140 B.F.	71.12		14.0 B.F.	7.11	
Accessories			Allow	25.00		Allow	25.00	
Make up, crew C-2	33.33	L.H.	5.0 L.H.		$166.65	1.0 L.H.		$ 33.33
Erect and strip			6.5 L.H.		216.65	6.5 L.H.		216.65
Clean and move			1.5 L.H.		50.00	1.5 L.H.		50.00
Bare Total per 100 S.F.C.A.			13.0 L.H.	$246.16	$433.30	9.0 L.H.	$47.11	$299.98

PILE CAPS, Square or Rectangular (Line 03110-430-3000)

Item	Cost	Unit	First Use Quantity	First Use Material	First Use Installation	Reuse Quantity	Reuse Material	Reuse Installation
5/8" exterior plyform	$1,364.00	M.S.F.	110 S.F.	$150.04		11.0 S.F.	$15.00	
Lumber	508.00	M.B.F.	160 B.F.	81.28		16.0 B.F.	8.13	
Accessories			Allow	7.50		Allow	7.50	
Make up, crew C-1	32.36	L.H.	4.5 L.H.		$145.62	1.0 L.H.		$ 32.36
Erect and strip			5.0 L.H.		161.80	5.0 L.H.		161.80
Clean and move			1.5 L.H.		48.54	1.5 L.H.		48.54
Bare Total per 100 S.F.C.A.			11.0 L.H.	$238.82	$355.96	7.5 L.H.	$30.63	$242.70

STAIRS, Average Run (Inclined Length x Width) (Line 03110-450-0010)

Item	Cost	Unit	First Use Quantity	First Use Materials	First Use Installation	Reuse Quantity	Reuse Material	Reuse Installation
5/8" exterior plyform	$1,364.00	M.S.F.	110 S.F.	$150.04		11.0 S.F.	$15.00	
Lumber	508.00	M.B.F.	425 B.F.	215.90		42.5 B.F.	21.59	
Accessories			Allow	20.00		Allow	20.00	
Build in place, crew C-2	33.33	L.H.	25.0 L.H.		$833.25	1.0 L.H.		$ 33.33
Strip and salvage			4.0 L.H.		133.32	4.0 L.H.		133.32
Bare Total per 100 S.F.C.A.			29.0 L.H.	$385.94	$966.57	29.0 L.H.	$56.59	$166.65

3 CONCRETE

REFERENCE NOS.

R03110-060 Formwork Labor Hours

Item	Unit	Hours Required			Total Hours	Multiple Use		
		Fabricate	Erect & Strip	Clean & Move	1 Use	2 Use	3 Use	4 Use
Beam and Girder, interior beams, 12" wide	100 S.F.	6.4	8.3	1.3	16.0	13.3	12.4	12.0
Hung from steel beams		5.8	7.7	1.3	14.8	12.4	11.6	11.2
Beam sides only, 36" high		5.8	7.2	1.3	14.3	11.9	11.1	10.7
Beam bottoms only, 24" wide		6.6	13.0	1.3	20.9	18.1	17.2	16.7
Box out for openings		9.9	10.0	1.1	21.0	16.6	15.1	14.3
Buttress forms, to 8' high		6.0	6.5	1.2	13.7	11.2	10.4	10.0
Centering, steel, 3/4" rib lath			1.0		1.0			
3/8" rib lath or slab form	↓		0.9		0.9			
Chamfer strip or keyway	100 L.F.		1.5		1.5	1.5	1.5	1.5
Columns, fiber tube 8" diameter			20.6		20.6			
12"			21.3		21.3			
16"			22.9		22.9			
20"			23.7		23.7			
24"			24.6		24.6			
30"	↓		25.6		25.6			
Round Steel, 12" diameter			22.0		22.0	22.0	22.0	22.0
16"			25.6		25.6	25.6	25.6	25.6
20"			30.5		30.5	30.5	30.5	30.5
24"	↓		37.7		37.7	37.7	37.7	37.7
Plywood 8" x 8"	100 S.F.	7.0	11.0	1.2	19.2	16.2	15.2	14.7
12" x 12"		6.0	10.5	1.2	17.7	15.2	14.4	14.0
16" x 16"		5.9	10.0	1.2	17.1	14.7	13.8	13.4
24" x 24"		5.8	9.8	1.2	16.8	14.4	13.6	13.2
Steel framed plywood 8" x 8"			10.0	1.0	11.0	11.0	11.0	11.0
12" x 12"			9.3	1.0	10.3	10.3	10.3	10.3
16" x 16"			8.5	1.0	9.5	9.5	9.5	9.5
24" x 24"			7.8	1.0	8.8	8.8	8.8	8.8
Drop head forms, plywood		9.0	12.5	1.5	23.0	19.0	17.7	17.0
Coping forms		8.5	15.0	1.5	25.0	21.3	20.0	19.4
Culvert, box			14.5	4.3	18.8	18.8	18.8	18.8
Curb forms, 6" to 12" high, on grade		5.0	8.5	1.2	14.7	12.7	12.1	11.7
On elevated slabs	↓	6.0	10.8	1.2	18.0	15.5	14.7	14.3
Edge forms to 6" high, on grade	100 L.F.	2.0	3.5	0.6	6.1	5.6	5.4	5.3
7" to 12" high	100 S.F.	2.5	5.0	1.0	8.5	7.8	7.5	7.4
Equipment foundations		10.0	18.0	2.0	30.0	25.5	24.0	23.3
Flat slabs, including drops		3.5	6.0	1.2	10.7	9.5	9.0	8.8
Hung from steel		3.0	5.5	1.2	9.7	8.7	8.4	8.2
Closed deck for domes		3.0	5.8	1.2	10.0	9.0	8.7	8.5
Open deck for pans		2.2	5.3	1.0	8.5	7.9	7.7	7.6
Footings, continuous, 12" high		3.5	3.5	1.5	8.5	7.3	6.8	6.6
Spread, 12" high		4.7	4.2	1.6	10.5	8.7	8.0	7.7
Pile caps, square or rectangular		4.5	5.0	1.5	11.0	9.3	8.7	8.4
Grade beams, 24" deep		2.5	5.3	1.2	9.0	8.3	8.0	7.9
Lintel or Sill forms		8.0	17.0	2.0	27.0	23.5	22.3	21.8
Spandrel beams, 12" wide		9.0	11.2	1.3	21.5	17.5	16.2	15.5
Stairs			25.0	4.0	29.0	29.0	29.0	29.0
Trench forms in floor		4.5	14.0	1.5	20.0	18.3	17.7	17.4
Walls, Plywood, at grade, to 8' high		5.0	6.5	1.5	13.0	11.0	9.7	9.5
8' to 16'		7.5	8.0	1.5	17.0	13.8	12.7	12.1
16' to 20'		9.0	10.0	1.5	20.5	16.5	15.2	14.5
Foundation walls, to 8' high		4.5	6.5	1.0	12.0	10.3	9.7	9.4
8' to 16' high		5.5	7.5	1.0	14.0	11.8	11.0	10.6
Retaining wall to 12' high, battered		6.0	8.5	1.5	16.0	13.5	12.7	12.3
Radial walls to 12' high, smooth		8.0	9.5	2.0	19.5	16.0	14.8	14.3
2' chords		7.0	8.0	1.5	16.5	13.5	12.5	12.0
Prefabricated modular, to 8' high		—	4.3	1.0	5.3	5.3	5.3	5.3
Steel, to 8' high		—	6.8	1.2	8.0	8.0	8.0	8.0
8' to 16' high		—	9.1	1.5	10.6	10.3	10.2	10.2
Steel framed plywood to 8' high		—	6.8	1.2	8.0	7.5	7.3	7.2
8' to 16' high	↓	—	9.3	1.2	10.5	9.5	9.2	9.0

R03210-010 Reinforcing Steel Weights and Measures

Bar Designation No.**	Nominal Weight Lb./Ft.	U.S. Customary Units Nominal Dimensions*			SI Units Nominal Dimensions*			
		Diameter in.	Cross Sectional Area, in.²	Perimeter in.	Nominal Weight kg/m	Diameter mm	Cross Sectional Area, cm²	Perimeter mm
3	.376	.375	.11	1.178	.560	9.52	.71	29.9
4	.668	.500	.20	1.571	.994	12.70	1.29	39.9
5	1.043	.625	.31	1.963	1.552	15.88	2.00	49.9
6	1.502	.750	.44	2.356	2.235	19.05	2.84	59.8
7	2.044	.875	.60	2.749	3.042	22.22	3.87	69.8
8	2.670	1.000	.79	3.142	3.973	25.40	5.10	79.8
9	3.400	1.128	1.00	3.544	5.059	28.65	6.45	90.0
10	4.303	1.270	1.27	3.990	6.403	32.26	8.19	101.4
11	5.313	1.410	1.56	4.430	7.906	35.81	10.06	112.5
14	7.650	1.693	2.25	5.320	11.384	43.00	14.52	135.1
18	13.600	2.257	4.00	7.090	20.238	57.33	25.81	180.1

* The nominal dimensions of a deformed bar are equivalent to those of a plain round bar having the same weight per foot as the deformed bar.
** Bar numbers are based on the number of eighths of an inch included in the nominal diameter of the bars.

R03210-020 Metric Rebar Specification - ASTM A615-81

Grade 300 (300 MPa* = 43,560 psi; +8.7% vs. Grade 40)				
Grade 400 (400 MPa* = 58,000 psi; −3.4% vs. Grade 60)				
Bar No.	Diameter mm	Area mm²	Equivalent in.²	Comparison with U.S. Customary Bars
10M	11.3	100	.16	Between #3 & #4
15M	16.0	200	.31	#5 (.31 in.²)
20M	19.5	300	.47	#6 (.44 in.²)
25M	25.2	500	.78	#8 (.79 in.²)
30M	29.9	700	1.09	#9 (1.00 in.²)
35M	35.7	1000	1.55	#11 (1.56 in.²)
45M	43.7	1500	2.33	#14 (2.25 in.²)
55M	56.4	2500	3.88	#18 (4.00 in.²)

* MPa = megapascals

R03210-040 Weight of Steel Reinforcing Per Square Foot of Wall (PSF)

Reinforced Weights: The table below suggests the weights per square foot for reinforcing steel in walls. Weights are approximate and will be the same for all grades of steel bars. For bars in two directions, add weights for each size and spacing.

C/C Spacing in Inches	Bar Size								
	#3 Wt. (PSF)	#4 Wt. (PSF)	#5 Wt. (PSF)	#6 Wt. (PSF)	#7 Wt. (PSF)	#8 Wt. (PSF)	#9 Wt. (PSF)	#10 Wt. (PSF)	#11 Wt. (PSF)
2"	2.26	4.01	6.26	9.01	12.27				
3"	1.50	2.67	4.17	6.01	8.18	10.68	13.60	17.21	21.25
4"	1.13	2.01	3.13	4.51	6.13	8.10	10.20	12.91	15.94
5"	.90	1.60	2.50	3.60	4.91	6.41	8.16	10.33	12.75
6"	.752	1.34	2.09	3.00	4.09	5.34	6.80	8.61	10.63
8"	.564	1.00	1.57	2.25	3.07	4.01	5.10	6.46	7.97
10"	.451	.802	1.25	1.80	2.45	3.20	4.08	5.16	6.38
12"	.376	.668	1.04	1.50	2.04	2.67	3.40	4.30	5.31
18"	.251	.445	.695	1.00	1.32	1.78	2.27	2.86	3.54
24"	.188	.334	.522	.751	1.02	1.34	1.70	2.15	2.66
30"	.150	.267	.417	.600	.817	1.07	1.36	1.72	2.13
36"	.125	.223	.348	.501	.681	.890	1.13	1.43	1.77
42"	.107	.191	.298	.429	.584	.753	.97	1.17	1.52
48"	.094	.167	.261	.376	.511	.668	.85	1.08	1.33

R03210-050 Minimum Wall Reinforcement Weight (PSF)

This table lists the approximate minimum wall reinforcement weights per S.F. according to the specification of .12% of gross area for vertical bars and .20% of gross area for horizontal bars.

Location	Wall Thickness	Bar Size	Horizontal Steel Spacing C/C	Sq. In. Req'd per S.F.	Total Wt. per S.F.	Bar Size	Vertical Steel Spacing C/C	Sq. In. Req'd per S.F.	Total Wt. per S.F.	Horizontal & Vertical Steel Total Weight per S.F.
Both Faces	10"	#4	18"	.24	.89#	#3	18"	.14	.50#	1.39#
	12"	#4	16"	.29	1.00	#3	16"	.17	.60	1.60
	14"	#4	14"	.34	1.14	#3	13"	.20	.69	1.84
	16"	#4	12"	.38	1.34	#3	11"	.23	.82	2.16
	18"	#5	17"	.43	1.47	#4	18"	.26	.89	2.36
One Face	6"	#3	9"	.15	.50	#3	18"	.09	.25	.75
	8"	#4	12"	.19	.67	#3	11"	.12	.41	1.08
	10"	#5	15"	.24	.83	#4	16"	.14	.50	1.34

R03210-070   Bend, Place and Tie Reinforcing

Placing and tying by rodmen for footings and slabs runs from nine hrs. per ton for heavy bars to fifteen hrs. per ton for light bars. For beams, columns, and walls, production runs from eight hrs. per ton for heavy bars to twenty hrs. per ton for light bars. Overall average for typical reinforced concrete buildings is about fourteen hrs. per ton. These production figures include placing of accessories and usual inserts. Equipment handling is necessary for the larger size bars so that installation costs for the very heavy bars will not decrease proportionately.

The use of commercially available reinforcing accessories improves placing accuracy, and although this carries an associated material cost (included in the bare cost of an installed reinforcing line item), it is offset by reduced labor time. Allow 15% of the cost of delivered bent rods, or $120.00 for each ton of bars.

Installation costs for splicing reinforcing bars includes allowance for equipment to hold the bars in place while splicing as well as necessary scaffolding for iron workers.

CONCRETE 3

REFERENCE NOS.

R03210-080   Reinforcing Key Prices

Costs given below are for over 50 tons delivered and bent from fabricators in metropolitan area. Over 500 tons, subtract $16 per ton. Figures are for domestic A615, grade 60. Economic conditions can alter the prices substantially.

City	Bare Costs Per Ton					Costs Incl. Subs O & P		
	Material		Installation			Material	Installation	Total per Ton
	50 Tons Fabricated	Allow for Accessories	Unload, Sort and Pile	Place & Tie	in Place	Incl. 10% Mark-Up	Include Mark-up	
Atlanta	$830	$125	$21	$ 588	$1,563	$1,050	$1,001	$2,051
Baltimore	905	136	25	707	1,773	1,145	1,204	2,348
Boston	905	136	33	933	2,006	1,145	1,588	2,733
Buffalo	805	121	28	807	1,761	1,018	1,374	2,393
Chicago	775	116	34	977	1,903	980	1,663	2,644
Cincinnati	830	125	25	713	1,693	1,050	1,214	2,264
Cleveland	750	113	29	831	1,723	949	1,415	2,364
Columbus	705	106	25	711	1,547	892	1,211	2,103
Dallas	855	128	16	466	1,466	1,082	794	1,875
Denver	805	121	20	559	1,505	1,018	953	1,971
Detroit	855	128	31	885	1,899	1,082	1,506	2,588
Houston	750	113	18	520	1,401	949	885	1,834
Indianapolis	750	113	25	717	1,605	949	1,221	2,170
Kansas City	980	147	28	798	1,953	1,240	1,359	2,599
Los Angeles	805	121	31	883	1,840	1,018	1,504	2,522
Memphis	705	106	20	555	1,386	892	945	1,837
Milwaukee	725	109	28	803	1,665	917	1,367	2,285
Minneapolis	815	122	32	911	1,880	1,031	1,551	2,582
Nashville	705	106	19	530	1,360	892	903	1,795
New Orleans	805	121	16	450	1,392	1,018	767	1,785
New York	805	121	51	1,445	2,421	1,018	2,460	3,478
Philadelphia	805	121	37	1,040	2,002	1,018	1,771	2,789
Phoenix	775	116	22	635	1,549	980	1,081	2,062
Pittsburgh	775	116	29	813	1,733	980	1,384	2,364
St. Louis	665	100	29	813	1,606	841	1,384	2,226
San Antonio	805	121	15	430	1,371	1,018	732	1,751
San Diego	725	109	31	883	1,748	917	1,504	2,421
San Francisco	855	128	31	883	1,897	1,082	1,504	2,585
Seattle	775	116	29	820	1,741	980	1,397	2,377
Washington, D.C.	955	143	23	660	1,781	1,208	1,123	2,331
Average per ton	$800	$120	$27	$ 759	$1,706	$1,012	$1,292	$2,304

Note: For field fabrication, add $85 to $110 per ton to shop fabrication costs. Place and tie costs do not include crane charge.

Bare Material Costs		Total Installed Costs		
Item	Cost per Ton	Description	Bare Costs	Cost Incl. Subs O&P
Base price at mill	$588	Unload, sort & pile, C-5 crew @ 0.56 Hours	$ 27	$ 40
Size and length extras	59	Place & tie, 4 Rodmen @ 20 Hours	759	1,252
Freight from mill	27			
Warehouse handling & storage	22	Total Installation Cost	$ 786	$1,292
Shearing and shop bending	44			
Drafting, detailing and listing	35	Bare Material Cost	$ 920	
Trucking to job site	25	Material Cost + 10%		$1,012
Reinforcing Delivered	$800	Total Installed Cost	$1,706	$2,304
Accessories Delivered	120			
Total Material Delivered	$920			

Engineering or design fees are not included above.

Material and erection costs can be considerably higher than those above for small jobs consisting primarily of smaller bars.

R03220-030   Common Stock Styles of Welded Wire Fabric

This table provides some of the basic specifications, sizes, and weights of welded wire fabric used for reinforcing concrete.

New Designation		Old Designation		Steel Area per Foot				Approximate Weight per 100 S.F.	
Spacing — Cross Sectional Area (in.) — (Sq. in. 100)		Spacing — Wire Gauge (in.) — (AS & W)		Longitudinal		Transverse			
				in.	cm	in.	cm	lbs	kg
Rolls	6 x 6 — W1.4 x W1.4	6 x 6 — 10 x 10		.028	.071	.028	.071	21	9.53
	6 x 6 — W2.0 x W2.0	6 x 6 — 8 x 8	1	.040	.102	.040	.102	29	13.15
	6 x 6 — W2.9 x W2.9	6 x 6 — 6 x 6		.058	.147	.058	.147	42	19.05
	6 x 6 — W4.0 x W4.0	6 x 6 — 4 x 4		.080	.203	.080	.203	58	26.91
	4 x 4 — W1.4 x W1.4	4 x 4 — 10 x 10		.042	.107	.042	.107	31	14.06
	4 x 4 — W2.0 x W2.0	4 x 4 — 8 x 8	1	.060	.152	.060	.152	43	19.50
	4 x 4 — W2.9 x W2.9	4 x 4 — 6 x 6		.087	.227	.087	.227	62	28.12
	4 x 4 — W4.0 x W4.0	4 x 4 — 4 x 4		.120	.305	.120	.305	85	38.56
Sheets	6 x 6 — W2.9 x W2.9	6 x 6 — 6 x 6		.058	.147	.058	.147	42	19.05
	6 x 6 — W4.0 x W4.0	6 x 6 — 4 x 4		.080	.203	.080	.203	58	26.31
	6 x 6 — W5.5 x W5.5	6 x 6 — 2 x 2	2	.110	.279	.110	.279	80	36.29
	4 x 4 — W1.4 x W1.4	4 x 4 — 4 x 4		.120	.305	.120	.305	85	38.56

NOTES: 1. Exact W—number size for 8 gauge is W2.1
2. Exact W—number size for 2 gauge is W5.4

CONCRETE 3

R03310-010 Proportionate Quantities

The tables below show both quantities per S.F. of floor areas as well as form and reinforcing quantities per C.Y. Unusual structural requirements would increase the ratios below. High strength reinforcing would reduce the steel weights. Figures are for 3000 psi concrete and 60,000 psi reinforcing unless specified otherwise.

Type of Construction	Live Load	Span	Per S.F. of Floor Area				Per C.Y. of Concrete		
			Concrete	Forms	Reinf.	Pans	Forms	Reinf.	Pans
Flat Plate	50 psf	15 Ft.	.46 C.F.	1.06 S.F.	1.71 lb.		62 S.F.	101 lb.	
		20	.63	1.02	2.40		44	104	
		25	.79	1.02	3.03		35	104	
	100	15	.46	1.04	2.14		61	126	
		20	.71	1.02	2.72		39	104	
		25	.83	1.01	3.47		33	113	
Flat Plate (waffle construction) 20" domes	50	20	.43	1.00	2.10	.84 S.F.	63	135	53 S.F.
		25	.52	1.00	2.90	.89	52	150	46
		30	.64	1.00	3.70	.87	42	155	37
	100	20	.51	1.00	2.30	.84	53	125	45
		25	.64	1.00	3.20	.83	42	135	35
		30	.76	1.00	4.40	.81	36	160	29
Waffle Construction 30" domes	50	25	.69	1.06	1.83	.68	42	72	40
		30	.74	1.06	2.39	.69	39	87	39
		35	.86	1.05	2.71	.69	33	85	39
		40	.78	1.00	4.80	.68	35	165	40
Flat Slab (two way with drop panels)	50	20	.62	1.03	2.34		45	102	
		25	.77	1.03	2.99		36	105	
		30	.95	1.03	4.09		29	116	
	100	20	.64	1.03	2.83		43	119	
		25	.79	1.03	3.88		35	133	
		30	.96	1.03	4.66		29	131	
	200	20	.73	1.03	3.03		38	112	
		25	.86	1.03	4.23		32	133	
		30	1.06	1.03	5.30		26	135	
One Way Joists 20" Pans	50	15	.36	1.04	1.40	.93	78	105	70
		20	.42	1.05	1.80	.94	67	120	60
		25	.47	1.05	2.60	.94	60	150	54
	100	15	.38	1.07	1.90	.93	77	140	66
		20	.44	1.08	2.40	.94	67	150	58
		25	.52	1.07	3.50	.94	55	185	49
One Way Joists 8" x 16" filler blocks	50	15	.34	1.06	1.80	.81 Ea.	84	145	64 Ea.
		20	.40	1.08	2.20	.82	73	145	55
		25	.46	1.07	3.20	.83	63	190	49
	100	15	.39	1.07	1.90	.81	74	130	56
		20	.46	1.09	2.80	.82	64	160	48
		25	.53	1.10	3.60	.83	56	190	42
One Way Beam & Slab	50	15	.42	1.30	1.73		84	111	
		20	.51	1.28	2.61		68	138	
		25	.64	1.25	2.78		53	117	
	100	15	.42	1.30	1.90		84	122	
		20	.54	1.35	2.69		68	154	
		25	.69	1.37	3.93		54	145	
	200	15	.44	1.31	2.24		80	137	
		20	.58	1.40	3.30		65	163	
		25	.69	1.42	4.89		53	183	
Two Way Beam & Slab	100	15	.47	1.20	2.26		69	130	
		20	.63	1.29	3.06		55	131	
		25	.83	1.33	3.79		43	123	
	200	15	.49	1.25	2.70		41	149	
		20	.66	1.32	4.04		54	165	
		25	.88	1.32	6.08		41	187	

R03310-010 Proportionate Quantities (cont.)

Item	Size		Forms	Reinforcing	Minimum	Maximum
	4000 psi Concrete and 60,000 psi Reinforcing—Form and Reinforcing Quantities per C.Y.					
Columns (square tied)	10" x 10"		130 S.F.C.A.	#5 to #11	220 lbs.	875 lbs.
	12" x 12"		108	#6 to #14	200	955
	14" x 14"		92	#7 to #14	190	900
	16" x 16"		81	#6 to #14	187	1082
	18" x 18"		72	#6 to #14	170	906
	20" x 20"		65	#7 to #18	150	1080
	22" x 22"		59	#8 to #18	153	902
	24" x 24"		54	#8 to #18	164	884
	26" x 26"		50	#9 to #18	169	994
	28" x 28"		46	#9 to #18	147	864
	30" x 30"		43	#10 to #18	146	983
	32" x 32"		40	#10 to #18	175	866
	34" x 34"		38	#10 to #18	157	772
	36" x 36"		36	#10 to #18	175	852
	38" x 38"		34	#10 to #18	158	765
	40" x 40"		32	#10 to #18	143	692

Item	Size	Form	Spiral	Reinforcing	Minimum	Maximum
Columns (spirally reinforced)	12" diameter	34.5 L.F.	190 lbs.	#4 to #11	165 lbs.	1505 lb.
		34.5	190	#14 & #18	—	1100
	14"	25	170	#4 to #11	150	970
		25	170	#14 & #18	800	1000
	16"	19	160	#4 to #11	160	950
		19	160	#14 & #18	605	1080
	18"	15	150	#4 to #11	160	915
		15	150	#14 & #18	480	1075
	20"	12	130	#4 to #11	155	865
		12	130	#14 & #18	385	1020
	22"	10	125	#4 to #11	165	775
		10	125	#14 & #18	320	995
	24"	9	120	#4 to #11	195	800
		9	120	#14 & #18	290	1150
	26"	7.3	100	#4 to #11	200	729
		7.3	100	#14 & #18	235	1035
	28"	6.3	95	#4 to #11	175	700
		6.3	95	#14 & #18	200	1075
	30"	5.5	90	#4 to #11	180	670
		5.5	90	#14 & #18	175	1015
	32"	4.8	85	#4 to #11	185	615
		4.8	85	#14 & #18	155	955
	34"	4.3	80	#4 to #11	180	600
		4.3	80	#14 & #18	170	855
	36"	3.8	75	#4 to #11	165	570
		3.8	75	#14 & #18	155	865
	40"	3.0	70	#4 to #11	165	500
		3.0	70	#14 & #18	145	765

CONCRETE 3

REFERENCE NOS.

R03310-010 Proportionate Quantities (cont.)

3000 psi Concrete and 60,000 psi Reinforcing—Form and Reinforcing Quantities per C.Y.

Item	Type	Loading	Height	C.Y./L.F.	Forms/C.Y.	Reinf./C.Y.
Retaining Walls	Cantilever	Level Backfill	4 Ft.	0.2 C.Y.	49 S.F.	35 lbs.
			8	0.5	42	45
			12	0.8	35	70
			16	1.1	32	85
			20	1.6	28	105
		Highway Surcharge	4	0.3	41	35
			8	0.5	36	55
			12	0.8	33	90
			16	1.2	30	120
			20	1.7	27	155
		Railroad Surcharge	4	0.4	28	45
			8	0.8	25	65
			12	1.3	22	90
			16	1.9	20	100
			20	2.6	18	120
	Gravity, with Vertical Face	Level Backfill	4	0.4	37	None
			7	0.6	27	
			10	1.2	20	
		Sloping Backfill	4	0.3	31	
			7	0.8	21	
			10	1.6	15	↓

		Live Load in Kips per Linear Foot							
	Span	Under 1 Kip		2 to 3 Kips		4 to 5 Kips		6 to 7 Kips	
		Forms	Reinf.	Forms	Reinf.	Forms	Reinf.	Forms	Reinf.
Beams	10 Ft.	—	—	90 S.F.	170 #	85 S.F.	175 #	75 S.F.	185 #
	16	130 S.F.	165 #	85	180	75	180	65	225
	20	110	170	75	185	62	200	51	200
	26	90	170	65	215	62	215	—	—
	30	85	175	60	200	—	—	—	—

Item	Size	Type	Forms per C.Y.	Reinforcing per C.Y.
Spread Footings	Under 1 C.Y.	1,000 psf soil	24 S.F.	44 lbs.
		5,000	24	42
		10,000	24	52
	1 C.Y. to 5 C.Y.	1,000	14	49
		5,000	14	50
		10,000	14	50
	Over 5 C.Y.	1,000	9	54
		5,000	9	52
		10,000	9	56
Pile Caps (30 Ton Concrete Piles)	Under 5 C.Y.	shallow caps	20	65
		medium	20	50
		deep	20	40
	5 C.Y. to 10 C.Y.	shallow	14	55
		medium	15	45
		deep	15	40
	10 C.Y. to 20 C.Y.	shallow	11	60
		medium	11	45
		deep	12	35
	Over 20 C.Y.	shallow	9	60
		medium	9	45
		deep	10	40

CONCRETE 3

REFERENCE NOS.

R03310-010 Proportionate Quantities (cont.)

			3000 psi Concrete and 60,000 psi Reinforcing — Form and Reinforcing Quantities per C.Y.			
Item	Size	Pile Spacing	50 T Pile	100 T Pile	50 T Pile	100 T Pile
Pile Caps (Steel H Piles)	Under 5 C.Y.	24" O.C.	24 S.F.	24 S.F.	75 lbs.	90 lbs.
		30"	25	25	80	100
		36"	24	24	80	110
	5 C.Y. to 10 C.Y.	24"	15	15	80	110
		30"	15	15	85	110
		36"	15	15	75	90
	Over 10 C.Y.	24"	13	13	85	90
		30"	11	11	85	95
		36"	10	10	85	90

		8" Thick		10" Thick		12" Thick		15" Thick	
	Height	Forms	Reinf.	Forms	Reinf.	Forms	Reinf.	Forms	Reinf.
Basement Walls	7 Ft.	81 S.F.	44 lbs.	65 S.F.	45 lbs.	54 S.F.	44 lbs.	41 S.F.	43 lbs.
	8		44		45		44		43
	9		46		45		44		43
	10		57		45		44		43
	12		83		50		52		43
	14		116		65		64		51
	16				86		90		65
	18						106		70

R03310-020 Materials for One C.Y. of Concrete

This is an approximate method of figuring quantities of cement, sand and coarse aggregate for a field mix with waste allowance included.

With crushed gravel as coarse aggregate, to determine barrels of cement required, divide 10 by total mix; that is, for 1:2:4 mix, 10 divided by 7 = 1-3/7 barrels.

If the coarse aggregate is crushed stone, use 10-1/2 instead of 10 as given for gravel.

To determine tons of sand required, multiply barrels of cement by parts of sand and then by 0.2; that is, for the 1:2:4 mix, as above, 1-3/7 x 2 x .2 = .57 tons.

Tons of crushed gravel are in the same ratio to tons of sand as parts in the mix, or 4/2 x .57 = 1.14 tons.

1 bag cement = 94#	1 C.Y. sand or crushed gravel = 2700#	1 C.Y. crushed stone = 2575#
4 bags = 1 barrel	1 ton sand or crushed gravel = 20 C.F.	1 ton crushed stone = 21 C.F.

Average carload of cement is 692 bags; of sand or gravel is 56 tons.

Do not stack stored cement over 10 bags high.

R03310-030 Metric Equivalents of Cement Content for Concrete Mixes

94 Pound Bags per Cubic Yard	Kilograms per Cubic Meter	94 Pound Bags per Cubic Yard	Kilograms per Cubic Meter
1.0	55.77	7.0	390.4
1.5	83.65	7.5	418.3
2.0	111.5	8.0	446.2
2.5	139.4	8.5	474.0
3.0	167.3	9.0	501.9
3.5	195.2	9.5	529.8
4.0	223.1	10.0	557.7
4.5	251.0	10.5	585.6
5.0	278.8	11.0	613.5
5.5	306.7	11.5	641.3
6.0	334.6	12.0	669.2
6.5	362.5	12.5	697.1

a. If you know the cement content in pounds per cubic yard, multiply by .5933 to obtain kilograms per cubic meter.

b. If you know the cement content in 94 pound bags per cubic yard, multiply by 55.77 to obtain kilograms per cubic meter.

R03310-040 Metric Equivalents of Common Concrete Strengths
(to convert other psi values to megapascals, multiply by 0.006895)

U.S. Values psi	SI Value Megapascals	Non-SI Metric Value kgf/cm²*
2000	14	140
2500	17	175
3000	21	210
3500	24	245
4000	28	280
4500	31	315
5000	34	350
6000	41	420
7000	48	490
8000	55	560
9000	62	630
10,000	69	705

* kilograms force per square centimeter

R03310-050 Quantities of Cement, Sand and Stone for One C.Y. of Concrete per Various Mixes

This table can be used to determine the quantities of the ingredients for smaller quantities of site mixed concrete.

Concrete (C.Y.)	Mix = 1:1:1-3/4			Mix = 1:2:2.25			Mix = 1:2.25:3			Mix = 1:3:4		
	Cement (sacks)	Sand (C.Y.)	Stone (C.Y.)	Cement (sacks)	Sand (C.Y.)	Stone (C.Y.)	Cement (sacks)	Sand (C.Y.)	Stone (C.Y.)	Cement (sacks)	Sand (C.Y.)	Stone (C.Y.)
1	10	.37	.63	7.75	.56	.65	6.25	.52	.70	5	.56	.74
2	20	.74	1.26	15.50	1.12	1.30	12.50	1.04	1.40	10	1.12	1.48
3	30	1.11	1.89	23.25	1.68	1.95	18.75	1.56	2.10	15	1.68	2.22
4	40	1.48	2.52	31.00	2.24	2.60	25.00	2.08	2.80	20	2.24	2.96
5	50	1.85	3.15	38.75	2.80	3.25	31.25	2.60	3.50	25	2.80	3.70
6	60	2.22	3.78	46.50	3.36	3.90	37.50	3.12	4.20	30	3.36	4.44
7	70	2.59	4.41	54.25	3.92	4.55	43.75	3.64	4.90	35	3.92	5.18
8	80	2.96	5.04	62.00	4.48	5.20	50.00	4.16	5.60	40	4.48	5.92
9	90	3.33	5.67	69.75	5.04	5.85	56.25	4.68	6.30	45	5.04	6.66
10	100	3.70	6.30	77.50	5.60	6.50	62.50	5.20	7.00	50	5.60	7.40
11	110	4.07	6.93	85.25	6.16	7.15	68.75	5.72	7.70	55	6.16	8.14
12	120	4.44	7.56	93.00	6.72	7.80	75.00	6.24	8.40	60	6.72	8.88
13	130	4.82	8.20	100.76	7.28	8.46	81.26	6.76	9.10	65	7.28	9.62
14	140	5.18	8.82	108.50	7.84	9.10	87.50	7.28	9.80	70	7.84	10.36
15	150	5.56	9.46	116.26	8.40	9.76	93.76	7.80	10.50	75	8.40	11.10
16	160	5.92	10.08	124.00	8.96	10.40	100.00	8.32	11.20	80	8.96	11.84
17	170	6.30	10.72	131.76	9.52	11.06	106.26	8.84	11.90	85	9.52	12.58
18	180	6.66	11.34	139.50	10.08	11.70	112.50	9.36	12.60	90	10.08	13.32
19	190	7.04	11.98	147.26	10.64	12.36	118.76	9.84	13.30	95	10.64	14.06
20	200	7.40	12.60	155.00	11.20	13.00	125.00	10.40	14.00	100	11.20	14.80
21	210	7.77	13.23	162.75	11.76	13.65	131.25	10.92	14.70	105	11.76	15.54
22	220	8.14	13.86	170.05	12.32	14.30	137.50	11.44	15.40	110	12.32	16.28
23	230	8.51	14.49	178.25	12.88	14.95	143.75	11.96	16.10	115	12.88	17.02
24	240	8.88	15.12	186.00	13.44	15.60	150.00	12.48	16.80	120	13.44	17.76
25	250	9.25	15.75	193.75	14.00	16.25	156.25	13.00	17.50	125	14.00	18.50
26	260	9.64	16.40	201.52	14.56	16.92	162.52	13.52	18.20	130	14.56	19.24
27	270	10.00	17.00	209.26	15.12	17.56	168.76	14.04	18.90	135	15.02	20.00
28	280	10.36	17.64	217.00	15.68	18.20	175.00	14.56	19.60	140	15.68	20.72
29	290	10.74	18.28	224.76	16.24	18.86	181.26	15.08	20.30	145	16.24	21.46

R03310-060 Concrete Material Net Prices

Costs below are C.Y. of concrete delivered; per ton of bulk cement; per bag cement delivered T.L.L.; per ton for stone and sand aggregates loaded at plant (no trucking included) and per 4 C.F. bag for perlite or vermiculite aggregate delivered T.L.L.

City	Ready Mix Concrete		Cement T.L. Lots		Aggregates per Ton			Vermiculite or Perlite 4 C.F. Bag
	Regular Weight per C.Y.		Bulk	Bags	Crushed Stone			
	3000 psi	5000 psi	per Ton	per Bag	1-1/2"	3/4"	Sand	
Atlanta	$ 81.50	$ 90.00	$ 88.95	$ 7.60	$18.20	$18.75	$17.75	$6.15
Baltimore	80.25	88.00	81.95	7.00	13.10	13.50	15.75	5.65
Boston	86.00	96.00	105.35	9.00	14.55	15.00	13.25	7.30
Buffalo	96.25	104.00	87.20	7.45	13.90	14.35	16.75	6.05
Chicago	91.75	100.00	83.70	7.15	18.65	19.25	16.25	5.80
Cincinnati	67.75	77.00	101.25	8.65	9.70	10.00	9.75	7.00
Cleveland	78.60	88.00	97.75	8.35	16.50	17.00	16.25	6.75
Columbus	73.25	83.00	107.70	9.20	13.10	13.50	11.25	7.45
Dallas	78.00	88.00	108.25	9.25	23.30	24.00	23.25	7.50
Denver	74.50	85.00	114.70	9.80	14.05	14.50	15.25	7.95
Detroit	82.50	91.00	88.35	7.55	12.75	13.15	11.75	6.10
Houston	71.00	79.00	88.95	7.60	20.60	21.25	19.75	6.15
Indianapolis	80.25	90.00	107.10	9.15	11.95	12.30	11.25	7.40
Kansas City	76.75	85.00	93.05	7.95	11.95	12.30	11.75	6.45
Los Angeles	71.00	81.00	110.00	9.40	12.95	13.35	15.25	7.60
Memphis	72.25	81.00	96.55	8.25	17.45	18.00	17.25	6.70
Milwaukee	83.75	94.00	105.35	9.00	13.10	13.50	16.25	7.30
Minneapolis	88.25	98.00	107.10	9.15	14.55	15.00	12.50	7.40
Nashville	75.50	84.00	89.55	7.65	12.35	12.75	14.75	6.20
New Orleans	73.25	83.00	98.90	8.45	19.40	20.00	16.50	6.85
New York City	98.50	107.00	90.10	7.70	23.30	24.00	24.25	6.25
Philadelphia	78.00	86.00	88.95	7.60	16.50	17.00	16.75	6.15
Phoenix	80.25	89.00	92.45	7.90	12.15	12.50	13.75	6.40
Pittsburgh	80.25	89.00	98.30	8.40	16.00	16.50	15.75	6.80
St. Louis	68.75	78.00	94.80	8.10	12.35	12.75	13.25	6.55
San Antonio	64.25	72.00	86.00	7.35	12.60	13.00	13.25	5.95
San Diego	90.50	100.00	103.00	8.80	15.75	16.25	19.25	7.15
San Francisco	111.25	122.00	117.65	10.05	21.35	22.00	23.75	8.15
Seattle	82.50	89.00	69.05	5.90	18.45	19.00	19.75	4.80
Washington, D.C.	94.00	103.00	93.65	8.00	15.05	15.50	17.75	6.50
Average	$ 81.00	$ 90.00	$ 96.50	$ 8.25	$15.50	$16.00	$16.00	$6.68

R03310-080 Field Mix Concrete

Presently most building jobs are built with ready mix concrete except for isolated locations and some larger jobs requiring over 10,000 C.Y. where land is readily available for setting up a temporary batch plant.

The most economical mix is a controlled mix using local aggregate proportioned by trial to give the required strength with the least cost of material.

Costs tabulated below are based on a job that requires 10,000 C.Y. of concrete to be placed within an eight month period.

28 Day Strength	3000 psi		2250 psi	
Mix Proportions	1:2.5:3		1:2.75:4	
Cement, trucked in bulk	6-1/4 bags @ 94 lb./bag	$24.97	5 bags @ 94 lb./bag	$19.98
Sand	.68 tons	9.07	.7 tons	9.34
Stone, aggregate, 3/4" to 1-1/2"	.9 tons	13.39	1.02 tons	15.18
Set up plant, foundations, piping, etc.		1.07		1.07
Plant, trucks and equipment rental		10.23		10.23
Labor to mix and transport to forms		10.85		10.85
Supervision		5.74		5.74
Bare Cost per C.Y. FOB Forms		$75.32		$72.39

See R03310-060 for material prices in the major cities.

CONCRETE 3

REFERENCE NOS.

R03310-090 Placing Ready Mixed Concrete

For ground pours allow for 5% waste when figuring quantities.

Prices in the front of the book assume normal deliveries. If deliveries are made before 8 A.M. or after 5 P.M. or on Saturday afternoons add $28 per C.Y. Large volume discounts are not included in prices in front of book.

For the lower floors without truck access, concrete may be wheeled in rubber tired buggies, conveyer handled, crane handled or pumped. Pumping is economical if there is top steel. Conveyers are more efficient for thick slabs. Concrete pump with an operator can be rented from $500 per day for up to 25 C.Y. to $1700 per day for a 400 C.Y. pour. Figures include travel time if done at straight time. Pumping lightweight concrete costs an extra $2.75 per C.Y.

At higher floors the rubber tired buggies may be hoisted by a hoisting tower then wheeled to location. Placement by a conveyer is limited to three floors and is best for high volume pours. Pumped concrete is best when building has no crane access. Concrete may be pumped directly as high as thirty-six stories using special pumping techniques. Normal maximum height is about fifteen stories.

Best pumping aggregate is screened and graded bank gravel rather than crushed stone.

Pumping downward is more difficult than pumping upwards. Horizontal distance from pump to pour may increase preparation time prior to pour. Placing by cranes, either mobile, climbing or tower types continues as the most efficient method for high rise concrete buildings.

	Cost per C.Y. for Wheeled Concrete, Dumped Only (Add to appropriate placing cost)							
	10 C.F. Walking Cart				18 C.F. Riding Cart			
Item	Hourly Cost	Wheeled up to 50 ft.	Wheeled up to 150 ft.	Wheeled up to 250 ft.	Hourly Cost	Wheeled up to 50 ft.	Wheeled up to 150 ft.	Wheeled up to 250 ft.
Laborer	$26.70	$6.68	$ 8.92	$11.88	$26.70	$2.67	$3.47	$4.54
.125 Labor foreman	3.59	.90	1.20	1.60	3.59	.36	.47	.61
Concrete cart	6.23	1.56	2.08	2.77	.95	.09	.12	.16
Total Cost/C.Y.		$9.14	$12.20	$16.25		$3.12	$4.06	$5.31
Hourly production		4 C.Y.				10 C.Y.		

R03310-100　Average C.Y. of Concrete

Rubbing and floor finish not included — 4 uses of forms assumed, 5 story building.

Item	Description	Strength in psi	Cost for 1 C.Y. of Concrete				Total
			Ready Mix	Place	Forms	Reinforcing	
Beams	10' span	4000	$84.00	$64.75	$349.83	$139.30	$637.88
5 kip/L.F.	25'		84.00	43.16	319.41	171.14	617.71
Beam & Slab, 1 way	15'	4000	84.00	19.92	455.28	76.37	635.57
125 psf Sup. L.	25'		84.00	17.27	292.68	96.40	490.35
Beam & Slab, 2 way	15'	4000	84.00	19.92	382.95	81.38	568.25
125 psf Sup. L.	25'		84.00	17.27	238.65	77.00	416.92
Columns,	16" x 16"	4000	84.00	43.16	426.06	441.26	994.48
square tied	24" x 24"		84.00	28.15	284.04	364.70	760.89
Columns, Tied	16" diameter	4000	84.00	43.16	208.62	421.78	757.56
reinforced	24" diameter		84.00	28.15	134.64	417.60	664.39
Flat Plate,	15' span	4000	84.00	23.54	228.75	78.88	415.17
125 psf Sup. L.	25'		84.00	19.92	123.75	70.74	298.41
Flat Slab with drops,	20'	4000	84.00	23.54	174.15	74.49	356.18
125 psf Sup. L.	30'		84.00	19.92	117.45	82.01	303.38
Grade Wall,	8" thick	3000	81.00	15.50	348.30	27.10	471.90
8' high	15" thick		81.00	13.28	176.30	26.49	297.07
Fiberglass Pan Joists, 24"	15' span	4000	84.00	19.92	514.98	28.95	647.85
125 psf Sup. L.	25' span		84.00	17.27	427.92	49.90	579.09
Pile Caps	under 5 C.Y.	3000	81.00	15.50	70.80	30.47	197.77
"	over 10 C.Y.		81.00	6.49	35.40	28.13	151.02
Slab	4" thick	3500	82.00	12.68	18.40	26.73	139.81
on Grade	6" thick		82.00	8.45	12.00	17.82	120.27
Spread	under 1 C.Y.	3000	81.00	25.36	72.00	32.48	210.84
Footings	over 5 C.Y.		81.00	12.68	27.00	38.12	158.80
Strip	9" x 18" plain	3000	81.00	12.68	90.72	—	184.40
Footings	12" x 36" reinforced		81.00	9.00	45.36	28.24	163.60
Waffle, 24" x 24"	20' span	4000	84.00	19.92	276.90	96.30	477.12
125 psf Sup. L.	30' span		84.00	17.27	220.10	91.80	413.17

*Placement by direct chute assumed. All others, placement by pump assumed.

When form and reinforcing quantity ratios are available compute the C.Y. costs directly from the unit prices in the front of the book. See also R03310-010 for relative quantities per C.Y. for additional items. Higher strength concrete would change the ratios below. The tables below show typical examples of how the total bare costs per C.Y. have been derived. Ready mix concrete prices are U.S. average and equipment handling is assumed.

Beams 5 kip/L.F.	10' Span				25' Span			
	Material		Installation		Material		Installation	
4000 psi concrete	1 C.Y.	$ 84.00	1 C.Y.	$ 64.75	1 C.Y.	$ 84.00	1 C.Y.	$ 43.16
Formwork, 4 uses	69 S.F. @　1.09	75.21	69 S.F. @　3.98	274.62	63 S.F. @　1.09	68.67	63 S.F.@　3.98	250.74
Reinforcing steel	175 # @　.42	72.80	175 # @　.38	66.50	215 # @　.42	89.44	215 # @　.38	81.70
Total per C.Y.		$232.01		$405.87		$242.11		$375.60

Beam & Slab, One Way	125 psf, 15' Span				125 psf, 25' Span			
	Material		Installation		Material		Installation	
4000 psi concrete	1 C.Y.	$ 84.00	1 C.Y.	$ 19.92	1 C.Y.	$ 84.00	1 C.Y.	$ 17.27
Formwork, 4 uses	84 S.F. @　1.14	95.76	84 S.F. @　4.28	359.52	54 S.F. @　1.14	61.56	54 S.F. @　4.28	231.12
Reinforcing steel	122 # @　.42	5.75	122 # @　.21	25.62	154 # @　.42	64.06	154 # @　.21	32.34
Total per C.Y.		$230.51		$405.06		$209.62		$280.73

Beam & Slab, Two Way	125 psf, 15' Span				125 psf, 25' Span			
	Material		Installation		Material		Installation	
4000 psi concrete	1 C.Y.	$ 84.00	1 C.Y.	$ 19.92	1 C.Y.	$ 84.00	1 C.Y.	$ 17.27
Formwork, 4 uses	69 S.F. @　1.05	72.45	69 S.F. @　4.50	31.50	43 S.F. @　1.05	45.15	43 S.F. @　4.50	193.50
Reinforcing steel	130 # @　.42	54.08	130 # @　.21	27.30	123 # @　.42	51.17	123 # @　.21	25.83
Total per C.Y.		$210.53		$357.72		$180.32		$236.60

R03310-100   Average C.Y. of Concrete (cont.)

Columns, Square Tied	16" Square				24" Square			
	Material		Installation		Material		Installation	
4000 psi concrete	1 C.Y.	$ 84.00	1 C.Y.	$ 43.16	1 C.Y.	$ 84.00	1 C.Y.	$ 28.15
Formwork, 4 uses	81 S.F. @ .99	80.19	81 S.F. @ 4.27	345.87	54 S.F. @ .99	53.46	54 S.F. @ 4.27	230.58
Reinforcing steel, avg.	634 # @ .42	263.74	634 # @ .28	177.52	524 # @ .42	217.98	524 # @ .28	146.72
Total per C.Y.		$427.93		$566.55		$355.44		$405.45

Note: Reinforcing of 16" and 24" square columns can vary from 144 lb. to 972 lb. per C.Y.

Columns, Round Tied Reinforced	16" Diameter				24" Diameter			
	Material		Installation		Material		Installation	
4000 psi concrete	1 C.Y.	$ 84.00	1 C.Y.	$ 43.16	1 C.Y.	$ 84.00	1 C.Y.	$ 28.15
Formwork, fiber forms	19 L.F. @ 3.58	68.02	19 L.F. @ 7.40	140.60	9 L.F. @ 6.99	62.91	9 L.F. @ 7.97	71.73
Reinforcing steel, avg.	606 # @ .42	252.10	606 # @ .28	169.68	600 # @ .42	249.60	600 # @ .28	168.00
Ties	70 # @ .83	58.10	70 # @ .40	28.00	18 # @ .83	14.94	18 # @ .40	7.20
Total per C.Y.		$462.22		$381.44		$411.45		$275.08

Note: Reinforcing of 16" and 24" diameter columns vary from 160 lb. to 1150 lb. per C.Y.

Flat Plate	125 psf, 15' Span				125 psf, 25' Span			
	Material		Installation		Material		Installation	
4000 psi concrete	1 C.Y.	$ 84.00	1 C.Y.	$ 23.54	1 C.Y.	$ 84.00	1 C.Y.	$ 19.92
Formwork, 4 uses	61 S.F. @ .90	54.90	61 S.F. @ 2.85	173.85	33 S.F. @ .90	29.70	33 S.F. @ 2.85	94.05
Reinforcing steel	126 # @ .42	52.42	126 # @ .21	26.46	113 # @ .42	47.01	113 # @ .21	23.73
Total per C.Y.		$191.32		$223.85		$160.71		$137.70

Flat Slab with Drops	125 psf, 20' Span				125 psf, 30' Span			
	Material		Installation		Material		Installation	
4000 psi concrete	1 C.Y.	$ 84.00	1 C.Y.	$ 23.54	1 C.Y.	$ 84.00	1 C.Y.	$ 19.92
Formwork, 4 uses	43 S.F. @ 1.11	47.73	43 S.F. @ 2.94	126.42	29 S.F. @ 1.11	32.19	29 S.F. @ 2.94	85.26
Reinforcing steel	119 # @ .42	49.50	119 # @ .21	24.99	131 # @ .42	54.50	131 # @ .21	27.51
Total per C.Y.		$181.23		$174.95		$170.69		$132.69

Grade Wall, 8' High	8" Thick				15" Thick			
	Material		Installation		Material		Installation	
3000 psi concrete	1 C.Y.	$ 81.00	1 C.Y.	$ 15.50	1 C.Y.	$ 81.00	1 C.Y.	$ 13.28
Formwork, 4 uses	81 S.F. @ .97	78.57	81 S.F. @ 3.33	269.73	41 S.F. @ .97	39.77	41 S.F. @ 3.33	136.53
Reinforcing steel	44 # @ .42	18.30	44 # @ .20	8.80	43 # @ .42	17.89	43 # @ .20	8.60
Total per C.Y.		$177.87		$294.03		$138.66		$158.41

Fiberglass Pan Joists, 24" One Way	125 psf, 15' Span				125 psf, 25' Span			
	Material		Installation		Material		Installation	
4000 psi concrete	1 C.Y.	$ 84.00	1 C.Y.	$ 19.92	1 C.Y.	$ 84.00	1 C.Y.	$ 17.27
Formwork, 4 uses	49 S.F. @ .93	45.57	49 S.F. @ 3.33	163.17	40 S.F. @ .93	37.20	40 S.F. @ 3.33	133.20
Reinforcing steel	47 # @ .42	19.55	47 # @ .20	9.40	81 # @ .42	33.70	81 # @ .20	16.20
19" fiberglass pans	44 S.F. @ 6.49	285.56	44 S.F. @ .47	20.68	37 S.F. @ 6.49	240.13	37 # @ .47	17.39
Total per C.Y.		$434.68		$213.17		$395.03		$184.06

Pile Caps	Under 5 C.Y.				Over 10 C.Y.			
	Material		Installation		Material		Installation	
3000 psi concrete	1.05 C.Y.	$ 81.00	1 C.Y.	$15.50	1.05 C.Y.	$ 81.00	1 C.Y.	$ 6.49
Formwork, 4 uses	20 S.F. @ .83	16.60	20 S.F. @ 2.71	54.20	10 S.F. @ .83	8.30	10 S.F. @ 2.71	27.10
Reinforcing steel	52 # @ .42	21.63	52 # @ .17	8.84	48 # @ .42	19.97	48 # @ .17	8.16
Total per C.Y.		$119.23		$78.54		$109.27		$41.75

Slab on Grade	4" Thick				6" Thick			
	Material		Installation		Material		Installation	
3500 psi concrete	1.05 C.Y.	$82.00	1 C.Y.	$12.68	1.05 C.Y.	$82.00	1 C.Y.	$ 8.45
Formwork, 4 uses	9.2 L.F. @ .27	2.48	9.2 L.F. @ 1.73	15.92	6 L.F. @ .27	1.62	6 L.F. @ 1.73	10.38
W.W. Fabric	81 S.F. @ .19	15.39	81 S.F. @ .14	11.34	54 S.F. @ .19	10.26	54 S.F. @ .14	7.56
Total per C.Y.		$99.87		$39.94		$93.88		$26.39

R03310-100 Average C.Y. of Concrete (cont.)

Spread Footings	Under 1 C.Y.						Over 5 C.Y.					
	Material			Installation			Material			Installation		
3000 psi concrete	1.05 C.Y.		$ 81.00	1.05 C.Y.		$25.36	1.05 C.Y.		$ 81.00	1 C.Y.		$12.68
Formwork, 4 uses	24 S.F. @	.50	12.00	24 S.F. @	2.50	60.00	9 S.F. @	.50	4.50	9 S.F. @	2.50	22.50
Reinforcing steel	46 # @	.42	19.14	46 # @	.29	13.34	54 # @	.42	22.46	54 # @	.29	15.66
Total per C.Y.			$112.14			$98.70			$107.96			$50.84

Strip Footings	9' x 18" Plain						12" x 36" Reinforced					
	Material			Installation			Material			Installation		
3000 psi concrete	1.05 C.Y.		$81.00	1.05 C.Y.		$12.68	1.05 C.Y.		$ 81.00	1 C.Y.		$ 9.00
Formwork, 4 uses	36 S.F. @	.38	13.68	36 S.F. @	2.14	77.04	18 S.F. @	.38	6.84	18 S.F. @	2.14	38.52
Reinforcing steel			—			—	40 # @	.42	16.64	40 # @	.29	11.60
Total per C.Y.			$94.68			$89.72			$104.48			$59.12

Waffle, 24" x 24"	125 psf, 20' Span						125 psf, 30' Span					
	Material			Installation			Material			Installation		
4000 psi concrete	1 C.Y.		$ 84.00	1 C.Y.		$ 19.92	1 C.Y.		$ 84.00	1 C.Y.		$ 17.27
Formwork, 4 uses	39 S.F. @	.93	36.27	39 S.F. @	3.40	132.60	31 S.F. @	.93	28.83	31 S.F. @	3.40	105.40
Reinforcing steel & mesh	107 # @	.61	65.27	107 # @	.29	31.03	102 # @	.61	62.22	102 # @	.29	29.58
19" x 19" Fiberglass Domes	39 S.F. @	2.29	89.31	39 S.F. @	.48	18.72	31 S.F. @	2.29	70.99	31 S.F. @	.48	14.88
Total per C.Y.			$274.85			$202.27			$246.04			$167.13

CONCRETE **3**

REFERENCE NOS.

R03310-120   Lift Slabs

The cost advantage of the lift slab method is due to placing all concrete, reinforcing steel, inserts and electrical conduit at ground level and in reduction of formwork. Minimum economical project size is about 30,000 S.F. Slabs may be tilted for parking garage ramps.

It is now used in all types of buildings and has gone up to 22 stories high in apartment buildings. Current trend is to use post-tensioned flat plate slabs with spans from 22' to 35'. Cylindrical void forms are used when deep slabs are required. One pound of prestressing steel is about equal to seven pounds of conventional reinforcing.

To be considered cured for stressing and lifting, a slab must have attained 75% of design strength. Seven days are usually sufficient with four to five days

possible if high early strength cement is used. Slabs can be stacked using two coats of a non-bonding agent to insure that slabs do not stick to each other. Lifting is done by companies specializing in this work. Lift rate is 5' to 15' per hour with an average of 10' per hour. Total areas up to 33,000 S.F. have been lifted at one time. 24 to 36 jacking columns are common. Most economical bay sizes are 24' to 28' with four to fourteen stories most efficient. Continuous design reduces reinforcing steel cost. Use of post-tensioned slabs allows larger bay sizes.

R03410-030   Prestressed Precast Concrete Structural Units

See also R03410-090 for post-tensioned prestressed concrete.

Type	Location	Depth	Span in Ft.		Live Load Lb. per S.F.
Double Tee	Floor	28" to 34"	60 to 80		50 to 80
	Roof	12" to 24"	30 to 50		40
	Wall	Width 8'	Up to 55' high		Wind
Multiple Tee	Roof	8" to 12"	15 to 40		40
	Floor	8" to 12"	15 to 30		100
Plank	Roof		Roof	Floor	
		4"	13	12	40 for Roof
		6"	22	18	
or	or	8"	26	25	
		10"	33	29	100 for
	Floor	12"	42	32	Floor
Single Tee	Roof	28"	40		
		32"	80		
		36"	100		40
		48"	120		
AASHO Girder	Bridges	Type 4	100		
		5	110		Highway
		6	125		
Box Beam	Bridges	15"	40		
		27"	to		Highway
		33"	100		

The majority of precast projects today utilize double tees rather than single tees because of speed and ease of installation. As a result casting beds at manufacturing plants are normally formed for double tees. Single tee projects will therefore require an initial set up charge to be spread over the individual single tee costs.

For floors, a 2" to 3" topping is field cast over the shapes. For roofs, insulating concrete or rigid insulation is placed over the shapes.

Member lengths up to 40' are standard haul, 40' to 60' require special permits and lengths over 60' must be escorted. Over width and/or over length can add up to 100% on hauling costs.

Large heavy members may require two cranes for lifting which would increase erection costs by about 45%. An eight man crew can install 12 to 20 double tees, or 45 to 70 quad tees or planks per day.

Grouting of connections must also be included.

Several system buildings utilizing precast members are available. Heights can go up to 22 stories for apartment buildings. Optimum design ratio is 3 S.F. of surface to 1 S.F. of floor area.

R03410-090 Prestressed Concrete, Post-tensioned

In post-tensioned concrete the steel tendons are tensioned after the concrete has reached about 3/4 of its ultimate strength. The cableways are grouted after tensioning to provide bond between the steel and concrete. If bond is to be prevented, the tendons are coated with a corrosion-preventative grease and wrapped with waterproof paper or plastic. Bonded tendons are usually used when ultimate strength (beams & girders) is a controlling factor.

High strength concrete is used to fully utilize the steel, thereby reducing the size and weight of the member. A plasticizing agent may be added to reduce water content. Maximum size aggregate ranges from 1/2" to 1-1/2" depending on the spacing of the tendons.

The types of steel commonly used are bars and strands. Job conditions determine which is best suited. Bars are best for vertical prestresses since they are easy to support. The trend is for steel manufacturers to supply a finished package, cut to length, which reduces field preparation to a minimum.

Bars vary from 3/4" to 1-3/8" diameter. Table below gives time in labor-hours per tendon and the labor cost per pound for placing, tensioning and grouting (if required) a 75' beam. Tendons used in buildings are not usually grouted; tendons for bridges usually are grouted. For strands the table indicates the labor-hours and labor cost per pound for typical prestressed units 100' long. Simple span beams usually require one end stressing regardless of lengths. Continuous beams are usually stressed from two ends. Long slabs are poured from the center outward and stressed in 75' increments after the initial 150' center pour. Prices below do not include subcontractor's overhead and profit.

Length	100' Beam		75' Beam		100' Slab	
Type Steel	Strand		Bars		Strand	
Diameter	0.5"		3/4"	1-3/8"	0.5"	0.6"
Number	4	12	1	1	1	1
Force in Kips	100	300	42	143	25	35
Preparation & Placing Cables	3.6	7.4	0.9	2.9	0.9	1.1
Stressing Cables	2.0	2.4	0.8	1.6	0.5	0.5
Grouting, if required	2.5	3.0	0.6	1.3		
Total Labor Hours	8.1	12.8	2.3	5.8	1.4	1.6
Prestressing Steel Weights (Lbs.)	215	640	115	380	53	74
Labor cost per lb. Bonded	$1.31	$.70	$.70	$.53		
Non-bonded					$1.02	$.83

Table title: Labor Hours per Tendon and Labor Costs per Pound of Prestress Steel

Flat Slab construction — 4000 psi concrete with span to depth ratio between 36 and 44. Two way post-tensioned steel averages 1.0 lb. per S.F. for 24' to 28' bays (usually strand) and additional reinforcing steel averages .5 lb. per S.F.

Pan and Joist construction — 4000 psi concrete with span to depth ratio 28 to 30. Post-tensioned steel averages .8 lb. per S.F. and reinforcing steel about 1.0 lb. per S.F. Placing and stressing averages 40 hours per ton of total material.

Beam construction — 4000 to 5000 psi concrete. Steel weights vary greatly.

Labor cost per pound goes down as the size and length of the tendon increases. The primary economic consideration is the cost per kip for the member.

Post-tensioning becomes feasible for beams and girders over 30' long; for continuous two-way slabs over 20' clear; also in transferring upper building loads over longer spans at lower levels. Post-tension suppliers will provide engineering services at no cost to the user. Substantial economies are possible by using post-tensioned Lift Slabs. See R03310-120 for Lift Slabs.

R03450-010 Precast Concrete Wall Panels

Panels are either solid or insulated with plain, colored or textured finishes. Transportation is an important cost factor. Prices below are based on delivery within 50 miles of a plant including fabricators' overhead and profit. Engineering data is available from fabricators to assist with construction details. Usual minimum job size for economical use of panels is about 5000 S.F. Small jobs can double the prices below. For large, highly repetitive jobs, deduct up to 15% from the prices below.

Panel Cost and Maximum Size   Base price for panels based on 50 S.F. or larger panels is as follows:

Thickness	Cost per S.F.	Maximum Size	Thickness	Cost per S.F.	Maximum Size
3″	$ 9.95	50 S.F.	6″	$14.65	300 S.F.
4″	10.45	150 S.F.	7″	16.70	300 S.F.
5″	12.55	200 S.F.	8″	18.30	300 S.F.

2″ thick panels cost about the same as 3″ thick panels and maximum panel size is less. For building panels faced with granite, marble or stone, add the material prices from Division 04850 to the plain panel price above. There is a growing trend toward aggregate facings and broken rib finish rather than plain gray concrete panels.

Composite Panels   Add to above panel prices for core insulation, mesh, shear ties and loose hardware per S.F.

Type	1″ Thick	1-1/2″ Thick	2″ Thick
Fiberglass	$.40	$.71	$.90
Polystyrene (E.P.S. Board)	.38	.56	.74

Erection   Table shows cost ranges for erection including Subcontractor's O & P using a six person crew, crane, operator & oiler.

Panel Size L x H	Area per Panel	Low Rise Hyd. 55 Ton Crane Daily Production Range Pieces	Area	Erection Cost Per Piece	Per S.F.	High Rise Daily Production Range Pieces	Area	55 Ton Crane Piece Cost	S.F. Cost	90 Ton Crane Piece Cost	S.F. Cost	150 Ton Crane Piece Cost	S.F. Cost
4′ x 4′	16 S.F.	10	160 S.F.	$ 529	$33.08	9	144 S.F.	$588	$36.76	$ 641	$40.04	$ 655	$40.95
		20	320	265	16.54	18	288	294	18.38	320	20.02	328	20.48
4′ x 8′	32	10	320	529	16.54	9	288	588	18.38	641	20.02	655	20.48
		19	608	279	8.71	17	544	311	9.73	339	10.60	347	10.84
8′ x 8′	64	9	576	588	9.19	8	512	662	10.34	721	11.26	737	11.52
		18	1152	294	4.59	16	1024	331	5.17	360	5.63	369	5.76
10′ x 10′	100	8	800	662	6.62	7	700	756	7.56	824	8.24	842	8.42
		15	1500	353	3.53	14	1400	378	3.78	412	4.12	421	4.21
15′ x 10′	150	7	1050	756	5.04	6	900	—	—	961	6.41	983	6.55
		12	1800	441	2.94	11	1650	—	—	524	3.49	536	3.57
20′ x 10′	200	6	1200	882	4.41	5	1000	—	—	1153	5.77	1179	5.90
		8	1600	662	3.31	7	1400	—	—	824	4.12	842	4.21
30′ x 10′	300	5	1500	1059	3.53	4	1200	—	—	1441	4.80	1474	4.91
		7	2100	756	2.52	7	2100	—	—	824	2.75	842	2.81

Total Cost in Place for Low Rise Construction Including Subcontractor's O & P

Description	Gray 4′ x 8′ x 4″	20′ x 10′ x 6″	White Face 4′ x 8′ x 4″	20′ x 10′ x 6″	Exposed Aggregate 4′ x 8′ x 4″	20′ x 10′ x 6″
Panel, steel form, broomed finish	$10.45	$14.65	$12.33	$16.53	$11.33	$15.53
Caulking, grout, etc.	.93	.37	.93	.37	.93	.37
Erect, plumb, align (from above)	16.54	4.41	16.54	4.41	16.54	4.41
Total in place per S.F.	$27.92	$19.43	$29.80	$21.31	$28.80	$20.31

No allowance has been made for supporting steel framework. On one story buildings, panels may rest on grade beams and require only wind bracing and fasteners. On multi-story buildings panels can span from column to column and floor to floor. Plastic designed steel framed structures may have large deflections which slow down erection and raise costs.

Large panels are more economical than small panels on a S.F. basis. When figuring areas include all protrusions, returns, etc. Overhangs can triple erection costs. Panels over 45′ have been produced. Larger flat units should be prestressed. Vacuum lifting of smooth finish panels eliminates inserts and can speed erection.

R03470-020 Tilt Up Concrete Panels

The advantage of tilt up construction is in the low cost of forms and the placing of concrete and reinforcing. Panels up to 75' high and 5-1/2" thick have been tilted using strongbacks. Tilt up has been used for one to five story buildings and is well suited for warehouses, stores, offices, schools and residences.

The panels are cast in forms on the floor slab. Most jobs use 5-1/2" thick solid reinforced concrete panels. Sandwich panels with a layer of insulating materials are also used. Where dampness is a factor, lightweight aggregate is used. Optimum panel size is 300 to 500 S.F.

Slabs are usually poured with 3000 psi concrete which permits tilting seven days after pouring. Slabs may be stacked on top of each other and are separated from each other by either two coats of bond breaker or a film of polyethylene. Use of high early strength cement allows tilting two days

after a pour. Tilting up is done with a roller outrigger crane with a capacity of at least 1-1/2 times the weight of the panel at the required reach. Exterior precast columns can be set at the same time as the panels; interior precast columns can be set first and the panels clipped directly to them. The use of cast-in-place concrete columns is diminishing due to shrinkage problems. Structural steel columns are sometimes used if crane rails are planned. Panels can be clipped to the columns or lowered between the flanges. Steel channels with anchors may be used as edge forms for the slab. When the panels are lifted the channels form an integral steel column to take structural loads. Roof loads can be carried directly by the panels for wall heights to 14'. For soft ground requiring mats for cranes, add 100% to costs below.

Below are typical costs per S.F., for panels of 300 to 500 S.F. 20' high, not including contractor's overhead and profit.

Item	5-1/2" Thick			7-1/2" Thick		
	Material	Installation	Total	Material	Installation	Total
Prepare pouring surface	$ .03	$ .07	$ .10	$ .03	$ .07	$ .10
Erect, strip side forms	.08	.48	.56	.14	.62	.76
Place concrete, 3000 psi	1.56	.20	1.76	2.13	.23	2.36
Steel trowel finish & curing	.05	.53	.58	.05	.53	.58
Reinforcing, inserts & misc. items	1.33	.66	1.99	1.55	.66	2.21
Panel erection and aligning	—	2.29	2.29	—	2.29	2.29
Total Panel Cost per S.F. of Wall	$ 3.05	$ 4.23	$ 7.28	$ 3.90	$ 4.40	$ 8.30
Site precast concrete columns, add	1.30	1.58	2.88	1.90	2.41	4.31
Total Panel & Column Cost per S.F. of Wall	$ 4.35	$ 5.81	$ 10.16	$ 5.80	$ 6.81	$ 12.61
Panels only, per C.Y.	$179.67	$249.19	$428.86	$168.48	$190.08	$358.56

Requirements of local building codes may be a limiting factor and should be checked. Building floor slabs should be poured first and should be a minimum of 5" thick with 100% compaction of soil or 6" thick with less than 100% compaction.

Setting times as fast as nine minutes per panel have been observed, but a safer expectation would be four panels per hour with a crane and a four man setting crew. If crane erects from inside building, some provision must be made to get crane out after walls are erected. Good yarding procedure is important to minimize delays. Equalizing three point lifting beams and self-releasing pick-up hooks speed erection. If panels must be carried to their final location, setting time per panel will be increased and erection costs

may fall in the erection cost range of architectural precast wall panels. Placing panels into slots formed in continuous footers will speed erection.

Reinforcing should be with #5 bars with vertical bars on the bottom. If surface is to be sandblasted, stainless steel chairs should be used to prevent rust staining.

Use of a broom finish is popular since the unavoidable surface blemishes are concealed.

Precast columns run from three to five times the C.Y. price of the panels only.

R03520-010   Lightweight Concrete

Vermiculite or Perlite come in bags of 4 C.F. under various trade names. Weight is about 8 lbs. per C.F. For insulating roof fill use 1:6 mix. For structural deck use 1:4 mix over gypsum boards, steeltex, steel centering, etc. supported by closely spaced joists or bulb trees. For structural slabs use 1:3:2 vermiculite sand concrete over steeltex, metal lath, steel centering, etc. on joists spaced 2'-0" O.C. for maximum L.L. of 80 P.S.F. Use same mix for slab base fill over steel flooring or regular reinforced concrete slab when tile, terrazzo or other finish is to be laid over.

For slabs on grade use 1:3:2 mix when tile, etc. finish is to be laid over. If radiant heating units are installed use a 1:6 mix for a base. After coils are in place, cover with a regular granolithic finish (mix 1:3:2) to a minimum depth of 1-1/2" over top of units.

Reinforce all slabs with 6 x 6 or 10 x 10 welded wire mesh.

Vermiculite concrete can be purchased ready mixed but the following breakdown is included for field mix. Prices given below are for a one story building and assume 50 C.Y. or more. For less than 50 C.Y. add 10%. For over one story add 5% per C.Y. Screed finish cost is included below.

See Section 03310-220 for prices of ready mix lightweight concrete.

Quantities per C.Y., Field Mix	1:6 Mix Insulating Roof Fill		1:3:2 Mix Lightweight Structural Concrete	
Portland cement	5.0 bags	$ 41.25	6.2 bags	$ 51.15
Vermiculite or Perlite	7.5 bags	50.10		
treated type			4.7 bags	32.90
Sand			12.5 C.F.	10.00
Plant and Water		7.51		7.51
Labor, machine mix, hoist and place, Crew C-8	1.12 L.H.	47.56	.91 L.H.	38.64
Total bare cost in place, per C.Y.		$146.42		$140.20

R04060-100 Cement Mortar (material only)

Type N - 1:1:6 mix by volume. Use everywhere above grade except as noted below.

- 1:3 mix using conventional masonry cement which saves handling two separate bagged materials.

Type M - 1:1/4:3 mix by volume, or 1 part cement, 1/4 (10% by wt.) lime, 3 parts sand. Use for heavy loads and where earthquakes or hurricanes may occur. Also for reinforced brick, sewers, manholes and everywhere below grade.

Cost and Mix Proportions of Various Types of Mortar

Components	Type Mortar and Mix Proportions by Volume										
	M		S		N		O		K	PM	PL
	1:1:6	1:1/4:3	1/2:1:4	1:1/2:4	1:3	1:1:6	1:3	1:2:9	1:3:12	1:1:6	1:1/2:4
Portland cement	$ 8.25	$ 8.25	$ 4.13	$ 8.25	—	$ 8.25	—	$ 8.25	$ 8.25	$ 8.25	$ 8.25
Masonry cement	6.15	—	6.15	—	$ 6.15	—	$ 6.15	—	—	6.15	—
Lime	—	1.45	—	2.89	—	5.78	—	11.56	17.34	—	2.89
Masonry sand*	4.28	2.14	2.85	2.85	2.14	4.28	2.14	6.42	8.56	4.28	2.85
Mixing machine incl. fuel**	3.70	1.85	2.46	2.46	1.85	3.70	1.85	5.54	7.39	3.70	2.46
Total for Materials	$22.38	$13.69	$15.59	$16.45	$10.14	$22.01	$10.14	$31.77	$41.54	$22.38	$16.45
Total C.F.	6	3	4	4	3	6	3	9	12	6	4
Approximate Cost per C.F.	$ 3.73	$ 4.56	$ 3.90	$ 4.11	$ 3.38	$ 3.67	$ 3.38	$ 3.53	$ 3.46	$ 3.73	$ 4.11

*Includes 10 mile haul
**Based on a daily rental, 10 C.F., 25 H.P. mixer, mix 200 C.F./Day

Mix Proportions by Volume, Compressive Strength and Cost of Mortar

Where Used	Mortar Type	Allowable Proportions by Volume				Compressive Strength @ 28 days	Cost per Cubic Foot
		Portland Cement	Masonry Cement	Hydrated Lime	Masonry Sand		
Plain Masonry		1	1	—	6		$3.73
	M	1	—	1/4	3	2500 psi	4.56
		1/2	1	—	4		3.90
	S	1	—	1/4 to 1/2	4	1800 psi	4.11
		—	1	—	3		3.38
	N	1	—	1/2 to 1-1/4	6	750 psi	3.67
		—	1	—	3		3.38
	O	1	—	1-1/4 to 2-1/2	9	350 psi	3.53
	K	1	—	2-1/2 to 4	12	75 psi	3.46
Reinforced Masonry	PM	1	1	—	6	2500 psi	3.73
	PL	1	—	1/4 to 1/2	4	2500 psi	4.11

Note: The total aggregate should be between 2.25 to 3 times the sum of the cement and lime used.

The labor cost to mix the mortar is included in the labor cost on brickwork.

Machine mixing is usually specified on jobs of any size. There is a large price saving over hand mixing and mortar is more uniform.

There are two types of mortar color used. Prices in Section 04060-540 are for the inert additive type with about 100 lbs. per M brick as the typical quantity required. These colors are also available in smaller batch size bags (1 lb. to 15 lb.) which can be placed directly into the mixer without measuring. The other type is premixed and replaces the masonry cement. Dark green color has the highest cost.

R04060-200   Miscellaneous Mortar (material only)

Quantities		Glass Block Mortar		Gypsum Cement Mortar	
White Portland cement, 94 Lb bag	7 bags	$136.15			
Gypsum cement, 80 Lb bag				11.25 bags	$258.75
Lime, 50 Lb bag	280 lbs.	32.37			
Sand*	1 C.Y.	19.25		1 C.Y.	19.25
Mixing machine and fuel**		16.65			16.65
Total per C.Y.		$204.42			$294.65
Approximate Total per C.F.		$ 7.57			$ 10.91

* Includes 10 mile haul

** Based on a daily rental, 10 C.F., 25 HP mixer, mix 200 C.F./Day = 7.4 C.Y./Day

R04080-500   Masonry Reinforcing

Horizontal joint reinforcing helps prevent wall cracks where wall movement may occur and in many locations is required by code. Horizontal joint reinforcing is generally not considered to be structural reinforcing and an unreinforced wall may still contain joint reinforcing.

Reinforcing strips come in 10′ and 12′ lengths and in truss and ladder shapes, with and without drips. Field labor runs between 2.7 to 5.3 hours per 1000 L.F. for wall thicknesses up to 12″.

The wire meets ASTM A82 for cold drawn steel wire and the typical size is 9 ga. sides and ties with 3/16″ diameter also available. Typical finish is mill galvanized with zinc coating at .10 oz. per S.F. Class I (.40 oz. per S.F.) and Class III (.80 oz per S.F.) are also available, as is hot dipped galvanizing at 1.50 oz. per S.F.

R04210-055   Industrial Chimneys

Foundation requirements in C.Y. of concrete for various sized chimneys.

Size Chimney	2 Ton Soil	3 Ton Soil	Size Chimney	2 Ton Soil	3 Ton Soil	Size Chimney	2 Ton Soil	3 Ton Soil
75′ x 3′-0″	13 C.Y.	11 C.Y.	160′ x 6′-6″	86 C.Y.	76 C.Y.	300′ x 10′-0″	325 C.Y.	245 C.Y.
85′ x 5′-6″	19	16	175′ x 7′-0″	108	95	350′ x 12′-0″	422	320
100′ x 5′-0″	24	20	200′ x 6′-0″	125	105	400′ x 14′-0″	520	400
125′ x 5′-6″	43	36	250′ x 8′-0″	230	175	500′ x 18′-0″	725	575

R04210-100   Economy in Bricklaying

Have adequate supervision. Be sure bricklayers are always supplied with materials so there is no waiting. Place best bricklayers at corners and openings.

Use only screened sand for mortar. Otherwise, labor time will be wasted picking out pebbles. Use seamless metal tubs for mortar as they do not leak or catch the trowel. Locate stack and mortar for easy wheeling.

Have brick delivered for stacking. This makes for faster handling, reduces chipping and breakage, and requires less storage space. Many dealers will deliver select common in 2′ x 3′ x 4′ pallets or face brick packaged. This affords quick handling with a crane or forklift and easy tonging in units of ten, which reduces waste.

Use wider bricks for one wythe wall construction. Keep scaffolding away from wall to allow mortar to fall clear and not stain wall.

On large jobs develop specialized crews for each type of masonry unit.

Consider designing for prefabricated panel construction on high rise projects.

Avoid excessive corners or openings. Each opening adds about 50% to labor cost for area of opening.

Bolting stone panels and using window frames as stops reduces labor costs and speeds up erection.

R04210-120 Common and Face Brick Prices

Prices are based on truckload lot purchases for Common Building Brick and Facing Brick. Prices are per M, (thousand), brick.

City	Material			Installation				Total			
	Brick per M Delivered		Mortar 3/8″ Joint	Common in 8″ Wall		Face Brick, 4″ Veneer		Common in 8″ Wall		Face Brick, 4″ Veneer	
	Common	Face		Bare Costs	Incl. O & P	Bare Costs	Incl. O & P	Bare Costs	Incl. O & P	Bare Costs	Incl. O & P
Atlanta	$315	$438	$46.00 for 8″ Wall and $38.00 for 4″ Wall	$ 472	$ 719	$ 566	$ 862	$ 842	$1,089	$1,055	$1,351
Baltimore	230	325		572	871	686	1,045	855	1,154	1,059	1,418
Boston	395	550		1,028	1,566	1,233	1,879	1,481	2,019	1,838	2,484
Buffalo	410	575		846	1,289	1,015	1,547	1,314	1,757	1,645	2,177
Chicago	275	388		908	1,384	1,090	1,661	1,237	1,713	1,528	2,099
Cincinnati	260	367		682	1,039	818	1,246	996	1,353	1,234	1,662
Cleveland	270	375		772	1,176	926	1,412	1,096	1,500	1,350	1,836
Columbus	250	350		660	1,005	792	1,206	964	1,309	1,191	1,605
Dallas	305	425		460	701	552	841	820	1,061	1,028	1,317
Denver	275	388		555	846	666	1,015	884	1,175	1,104	1,453
Detroit	330	465		865	1,319	1,038	1,582	1,251	1,705	1,555	2,099
Houston	270	375		473	721	568	866	797	1,045	992	1,290
Indianapolis	305	425		681	1,038	818	1,246	1,041	1,398	1,294	1,722
Kansas City	245	344		774	1,180	929	1,416	1,072	1,478	1,321	1,808
Los Angeles	255	355		805	1,227	966	1,473	1,114	1,536	1,370	1,877
Memphis	225	313		581	886	698	1,063	859	1,164	1,058	1,423
Milwaukee	290	405		795	1,212	954	1,454	1,140	1,557	1,409	1,909
Minneapolis	320	450		867	1,322	1,041	1,586	1,243	1,698	1,543	2,088
Nashville	195	275		488	743	585	892	735	990	906	1,213
New Orleans	395	555		440	671	528	805	893	1,124	1,138	1,415
New York City	285	400		1,126	1,716	1,351	2,059	1,466	2,056	1,801	2,509
Philadelphia	440	615		894	1,363	1,073	1,635	1,393	1,862	1,744	2,306
Phoenix	350	490		508	774	609	929	915	1,181	1,152	1,472
Pittsburgh	205	290		689	1,050	826	1,260	946	1,307	1,163	1,597
St. Louis	250	350		813	1,238	975	1,486	1,117	1,542	1,374	1,885
San Antonio	215	300		476	726	571	871	743	993	918	1,218
San Diego	270	380		753	1,148	904	1,377	1,077	1,472	1,333	1,806
San Francisco	470	655		909	1,385	1,091	1,662	1,439	1,915	1,804	2,375
Seattle	430	600		792	1,207	950	1,448	1,281	1,696	1,606	2,104
Washington, D.C.	220	305		591	901	709	1,081	864	1,174	1,061	1,433
Average	$300	$420	▼	$ 710	$1,080	$ 850	$1,295	$1,065	$1,435	$1,320	$1,765

Common building brick manufactured according to ASTM C62 and facing brick manufactured according to ASTM C216 are the two standard bricks available for general building use.

Building brick is made in three grades; SW, where high resistance to damage caused by cyclic freezing is required; MW, where moderate resistance to cyclic freezing is needed; and NW, where little resistance to cyclic freezing is needed. Facing brick is made in only the two grades SW and MW. Additionally, facing brick is available in three types; FBS, for general use; FBX, for general use where a higher degree of precision and lower permissible variation in size than FBS is needed; and FBA, for general use to produce characteristic architectural effects resulting from non-uniformity in size and texture of the units.

In figuring above installation costs, a D-8 Crew (with a daily output of 1.5 M) was used for the 4″ veneer. A D-8 Crew (with a daily output of 1.8 M) was used for the 8″ solid wall.

In figuring the total cost including overhead and profit, an allowance of 10% was added to the sum of the cost of the brick and mortar. Also, 3% breakage was included for both the bare costs and the costs with overhead and profit. If bricks are delivered palletized with 280 to 300 per pallet, or packaged, allow only 1-1/2% for breakage. Then add $10 per M to the cost of brick and deduct two hours helper time. The net result is a savings of $30 to $40 per M in place. Packaged or palletized delivery is practical when

a job is big enough to have a crane or other equipment available to handle a package of brick. This is so on all industrial work but not always true on small commercial buildings.

The prices above are for bricks used in commercial, apartment house or industial construction. If it is possible to obtain the price of the actual brick to be used, it should be done and substituted in the table. The use of buff and gray face is increasing, and there is a continuing trend to the Norman, Roman, Jumbo and SCR brick.

See R04210-500 for brick quantities per S.F. and mortar quantities per M brick. (Average prices for the various sizes are listed in Division 4.)

Common red clay brick for backup is not used that often. Concrete block is the most usual backup material with occasional use of sand lime or cement brick. Sand lime cost about $15 per M less than red clay and cement brick are about $5 per M less than red clay. These figures may be substituted in the common brick breakdown for the cost of these items in place, as labor is about the same. Building brick is commonly used in solid walls for strength and as a fire stop.

Brick panels built on the ground and then crane erected to the upper floors have proven to be economical. This allows the work to be done under cover and without scaffolding.

R04210-180 Brick in Place

Table below is for common bond with 3/8" concave joints and includes 3% waste for brick and 25% waste for mortar. Crew costs are bare costs.

Item	8" Common Brick Wall 8" x 2-2/3" x 4"		Select Common Face 8" x 2-2/3" x 4"		Red Face Brick 8" x 2-2/3" x 4"	
1030 brick delivered		$ 306.94		$ 365.65		$ 384.19
Type N mortar	12.5 C.F.	45.88	10.3 C.F.	37.80	10.3 C.F.	37.80
Installation using indicated crew	Crew D-8 @ 0.556 Days	709.68	Crew D-8 @ 0.667 Days	851.36	Crew D-8 @ 0.667 Days	851.36
Total per M in place		$1,062.50		$1,254.81		$1,273.35
Total per S.F. of wall	13.5 bricks/S.F.	$ 14.34	6.75 bricks/S.F.	$ 8.47	6.75 bricks/S.F.	$ 8.60

R04210-500 Brick, Block & Mortar Quantities

Type Brick	Nominal Size (incl. mortar) L H W	Modular Coursing	Number of Brick per S.F.	C.F. of Mortar per M Bricks, Waste Included 3/8" Joint	1/2" Joint	Bond Type	Description	Factor
Standard	8 x 2-2/3 x 4	3C=8"	6.75	10.3	12.9	Common	full header every fifth course	+20%
Economy	8 x 4 x 4	1C=4"	4.50	11.4	14.6		full header every sixth course	+16.7%
Engineer	8 x 3-1/5 x 4	5C=16"	5.63	10.6	13.6	English	full header every second course	+50%
Fire	9 x 2-1/2 x 4-1/2	2C=5"	6.40	550 # Fireclay	—	Flemish	alternate headers every course	+33.3%
Jumbo	12 x 4 x 6 or 8	1C=4"	3.00	23.8	30.8		every sixth course	+5.6%
Norman	12 x 2-2/3 x 4	3C=8"	4.50	14.0	17.9	Header = W x H exposed		+100%
Norwegian	12 x 3-1/5 x 4	5C=16"	3.75	14.6	18.6	Rowlock = H x W exposed		+100%
Roman	12 x 2 x 4	2C=4"	6.00	13.4	17.0	Rowlock stretcher = L x W exposed		+33.3%
SCR	12 x 2-2/3 x 6	3C=8"	4.50	21.8	28.0	Soldier = H x L exposed		—
Utility	12 x 4 x 4	1C=4"	3.00	15.4	19.6	Sailor = W x L exposed		-33.3%

Running Bond — Number of Brick per S.F. of Wall - Single Wythe with 3/8" Joints. For Other Bonds Standard Size Add to S.F. Quantities in Table to Left.

Concrete Blocks Nominal Size	Approximate Weight per S.F. Standard	Lightweight	Blocks per 100 S.F.	Mortar per M block, waste included Partitions	Back up
2" x 8" x 16"	20 PSF	15 PSF	113	27 C.F.	36 C.F.
4"	30	20		41	51
6"	42	30		56	66
8"	55	38		72	82
10"	70	47		87	97
12"	85	55		102	112

R04220-200 Concrete Block

Concrete masonry units, 8″ high x 16″ long, sand aggregate, 3/8 joints for partitions,
tooled joints two sides, 113 blocks per 100 S.F., bare costs.

City	Material Per Block, Delivered 4″ Thick	8″ Thick	113 Block, Delivered 4″ Thick	8″ Thick	Mortar 3/8″ Joint	Bare Installation 4″ Thick	8″ Thick	Bare Total Per 100 S.F. 4″ Thick	8″ Thick
Atlanta	$ .60	$1.00	$ 67.80	$113.00	$16.88 for 4″ wall	$197.77	$226.63	$282.45	$369.36
Baltimore	.54	.89	61.02	100.57	$29.73 for 8″ wall	239.69	274.67	317.59	404.97
Boston	.68	1.13	76.84	127.69		432.06	495.10	525.78	652.52
Buffalo	.68	1.12	76.84	126.56		354.78	406.54	448.50	562.83
Chicago	.59	.98	66.67	110.74		380.93	436.51	464.48	576.98
Cincinnati	.66	1.10	74.58	124.30		285.84	327.56	377.30	481.59
Cleveland	.65	1.07	73.45	120.91		323.74	370.98	414.07	521.62
Columbus	.57	.94	64.41	106.22		276.67	317.05	357.96	453.00
Dallas	.71	1.17	80.23	132.21		192.96	221.12	290.07	383.06
Denver	1.10	1.83	124.30	206.79		232.85	266.83	374.03	503.35
Detroit	.83	1.37	93.79	154.81		362.88	415.84	473.55	600.38
Houston	.66	1.10	74.58	124.30		198.52	227.48	289.98	381.51
Indianapolis	.77	1.27	87.01	143.51		285.77	327.47	389.66	500.71
Kansas City	.74	1.22	83.62	137.86		324.82	372.22	425.32	539.81
Los Angeles	.57	.95	64.41	107.35		337.74	387.02	419.03	524.10
Memphis	.65	1.08	73.45	122.04		243.81	279.39	334.14	431.16
Milwaukee	.65	1.07	73.45	120.91		333.47	382.13	423.80	532.77
Minneapolis	.65	1.08	73.45	122.04		363.76	416.84	454.09	568.61
Nashville	.67	1.11	75.71	125.43		204.50	234.34	297.09	389.50
New Orleans	.74	1.23	83.62	138.99		184.63	211.57	285.13	380.29
New York City	.65	1.08	73.45	122.04		472.28	541.20	562.61	692.97
Philadelphia	.55	.91	62.15	102.83		374.96	429.68	453.99	562.24
Phoenix	.49	.81	55.37	91.53		212.96	244.04	285.21	365.30
Pittsburgh	.60	1.00	67.80	113.00		288.85	330.99	373.53	473.72
St. Louis	.80	1.33	90.40	150.29		340.81	390.55	448.09	570.57
San Antonio	.72	1.20	81.36	135.60		199.67	228.81	297.91	394.14
San Diego	.53	.88	59.89	99.44		315.82	361.90	392.59	491.07
San Francisco	1.15	1.90	129.95	214.70		381.15	436.77	527.98	681.20
Seattle	1.02	1.69	115.26	190.97		332.15	380.61	464.29	601.31
Washington D.C.	.75	1.24	84.75	140.12		247.97	284.15	349.60	454.00
Average	$ .70	$1.16	$ 79.10	$131.08		$297.46	$340.87	$393.44	$501.68

Cost for 100 S.F. of 8″ x 16″ Concrete Block Partitions to Four Stories High, Tooled Joints Two Sides

8″ x 16″ Sand Aggregate	4″ Thick Block		8″ Thick Block		12″ Thick Block	
113 block delivered		$ 79.10		$131.08		$184.19
Mortar Type N, 1:3	4.6 C.F.	16.88	8.1 C.F.	29.73	11.5 C.F.	42.21
Installation Crew	D-8 @ 0.233 days	297.46	D-8 @ .267 days	340.87	D-9 @ .294 days	438.59
Total bare cost per 100 S.F.		$393.44		$501.68		$664.99
Add for filling cores solid	6.7 C.F.	$138.18	25.8 C.F.	$248.75	42.2 C.F.	$313.18

Cost for 100 S.F. of 8″ x 16″ Concrete Block Backup, Tooled Joints One Side

8″ x 16″ Sand Aggregate	4″ Thick Block		8″ Thick Block		12″ Thick Block	
113 block delivered		$ 79.10		$131.08		$184.19
Mortar Type N, 1:3	5.8 C.F.	21.29	9.3 C.F.	34.13	12.7 C.F.	46.61
Installation crew	D-8 @ .217 days	277.03	D-8 @ .250 days	319.16	D-9 @ .323 days	481.85
Total bare cost per 100 S.F.		$377.42		$484.37		$712.65

Special block: corner, jamb and head block are same price as ordinary block of same size. Tabulated on the next page are national average prices per block. Labor on specials is about the same as equal sized regular block. Bond beam and 16″ high lintel blocks cost 30% more than regular units of equal size. Lintel blocks are 8″ long and 8″ or 16″ high. Costs in individual cities may be factored from the table above.

Use of motorized mortar spreader box will speed construction of continuous walls. Hollow non-load bearing units are made according to ASTM C129 and hollow load bearing units according to ASTM C90.

R04930-100   Cleaning Face Brick

On smooth brick a person can clean 70 S.F. an hour; on rough brick 50 S.F. per hour. Use one gallon muriatic acid to 20 gallons of water for 1000 S.F. Do not use acid solution until wall is at least seven days old, but a mild soap solution may be used after two days. Commercial cleaners cost from $9 to $12 per gallon.

Time has been allowed for clean-up in brick prices.

5 METALS

R05080-310   Coating Structural Steel

On field-welded jobs, the shop-applied primer coat is necessarily omitted. All painting must be done in the field and usually consists of red oxide rust inhibitive paint or an aluminum paint. The table below shows paint coverage and daily production for field painting.

See Division 05950-650 for hot-dipped galvanizing and for field-applied cold galvanizing and other paints and protective coatings.

See Division 05910-500 for steel surface preparation treatments such as wire brushing, pressure washing and sand blasting.

Type Construction	Surface Area per Ton	Coat	One Gallon Covers		In 8 Hrs. Person Covers		Average per Ton Spray	
			Brush	Spray	Brush	Spray	Gallons	Labor-hours
Light Structural	300 S.F. to 500 S.F.	1st	500 S.F.	455 S.F.	640 S.F.	2000 S.F.	0.9 gals.	1.6 L.H.
		2nd	450	410	800	2400	1.0	1.3
		3rd	450	410	960	3200	1.0	1.0
Medium	150 S.F. to 300 S.F.	All	400	365	1600	3200	0.6	0.6
Heavy Structural	50 S.F. to 150 S.F.	1st	400	365	1920	4000	0.2	0.2
		2nd	400	365	2000	4000	0.2	0.2
		3rd	400	365	2000	4000	0.2	0.2
Weighted Average	225 S.F.	All	400	365	1350	3000	0.6	0.6

R05090-510   High Strength Bolts

Common bolts (A307) are usually used in secondary connections (see Division 05090-150).

High strength bolts (A325 and A490) are usually specified for primary connections such as column splices, beam and girder connections to columns, column bracing, connections for supports of operating equipment or of other live loads which produce impact or reversal of stress, and in structures carrying cranes of over 5-ton capacity.

Allow 20 field bolts per ton of steel for a 6 story office building, apartment house or light industrial building. For 6 to 12 stories allow 25 bolts per ton, and above 12 stories, 30 bolts per ton. On power stations, 20 to 25 bolts per ton are needed.

R05090-520   Welded Structural Steel

Usual weight reductions with welded design run 10% to 20% compared with bolted or riveted connections. This amounts to about the same total cost compared with bolted structures since field welding is more expensive than bolts. For normal spans of 18' to 24' figure 6 to 7 connections per ton.

Trusses — For welded trusses add 4% to weight of main members for connections. Up to 15% less steel can be expected in a welded truss compared to one that is shop bolted. Cost of erection is the same whether shop bolted or welded.

General — Typical electrodes for structural steel welding are E6010, E6011, E60T and E70T. Typical buildings vary between 2# to 8# of weld rod per ton of steel. Buildings utilizing continuous design require about three times as much welding as conventional welded structures. In estimating field erection by welding, it is best to use the average linear feet of weld per ton to arrive at the welding cost per ton. The type, size and position of the weld will have a direct bearing on the cost per linear foot. A typical field welder will deposit 1.8# to 2# of weld rod per hour manually. Using semiautomatic methods can increase production by as much as 50% to 75%.

REFERENCE NOS.

R05120-210 Structural Steel

1 to 2 Story Building	Bare Costs for 100 Ton Job with High Strength Bolts					Including Subs O&P		
City	Fabricated and Delivered	Unloading and Sorting	Erection Equipment	Field Erection Labor	Total per Ton in Place	Material	Installation	Total
Atlanta	$1,400	$29	$111	$197	$1,737	$1,540	$ 526	$2,066
Baltimore	1,675	35	137	237	2,084	1,843	637	2,480
Boston	1,775	47	174	312	2,308	1,953	834	2,787
Buffalo	2,050	40	129	270	2,489	2,255	696	2,951
Chicago	1,800	49	167	327	2,343	1,980	856	2,836
Cincinnati	1,500	37	122	248	1,907	1,650	644	2,294
Cleveland	1,500	42	118	278	1,938	1,650	702	2,352
Columbus	1,600	36	113	238	1,987	1,760	614	2,374
Dallas	1,625	23	116	156	1,920	1,788	448	2,236
Denver	1,675	28	121	187	2,011	1,843	518	2,361
Detroit	1,950	48	143	319	2,460	2,145	814	2,959
Houston	1,700	26	124	174	2,024	1,870	494	2,364
Indianapolis	1,650	36	108	240	2,034	1,815	613	2,428
Kansas City	1,925	40	156	267	2,388	2,118	721	2,839
Los Angeles	2,125	44	136	296	2,601	2,338	758	3,096
Memphis	1,725	28	131	186	2,070	1,898	527	2,425
Milwaukee	1,750	40	126	269	2,185	1,925	691	2,616
Minneapolis	1,625	46	193	305	2,169	1,788	840	2,628
Nashville	1,800	27	127	178	2,132	1,980	506	2,486
New Orleans	1,725	23	104	151	2,003	1,898	426	2,324
New York City	2,150	72	192	484	2,898	2,365	1,206	3,571
Philadelphia	1,950	54	173	359	2,536	2,145	929	3,074
Phoenix	1,525	32	98	213	1,868	1,678	546	2,224
Pittsburgh	1,650	41	170	272	2,133	1,815	747	2,562
St. Louis	1,800	41	172	272	2,285	1,980	749	2,729
San Antonio	1,625	22	94	144	1,885	1,788	400	2,188
San Diego	1,900	44	136	296	2,376	2,090	758	2,848
San Francisco	2,050	44	147	296	2,537	2,255	770	3,025
Seattle	1,625	41	124	275	2,065	1,788	702	2,490
Washington, DC	1,650	33	149	221	2,053	1,815	618	2,433
U.S. Average	$1,750	$38	$137	$253	$2,178	$1,925	$ 650	$2,575

Adjustments to the Base Price Above	Bare Costs	Incl. Subs O&P
For 3 to 6 story building, add per ton	$ 36	$ 75
For 7 to 15 stories, add per ton	93	150
For over 15 stories, add per ton	177	250
For field welded connections, add per ton	188	275
For multi-story masonry wall bearing construction, add to erection costs	30%	30%

General Average per Ton in Place for A36 Steel (High Strength Steel Base Price May Be Substituted)					
Item	Description	Bare Costs			Incl. Subs O&P
		Itemized	Summary		
Material	Base price	$ 650			
	Extras and delivery to shop	200			
	Drafting	110	$1,750		$1,925
	Shop fabrication & warehouse rehandling	695			
	Shop coat paint	43			
	Trucking to job site	52			
Installation	Unload and shake out, 1.0 hours @ $38.15	38			
	Erect and plumb, 5.0 hours @ $38.15	189	$ 428		$ 650
	Field bolt, 1.7 hours @ $38.15	64			
	Crane & minor erection equipment	137			
	Total per ton in place	$2,178	$2,178		$2,575

R05120-220 Steel Estimating Quantities

One estimate on erection is that a crane can handle 35 to 60 pieces per day. Say the average is 45. With usual sizes of beams, girders, and columns, this would amount to about 20 tons per day. The type of connection greatly affects the speed of erection. Moment connections for continuous design slow down production and increase erection costs.

Short open web bar joists can be set at the rate of 75 to 80 per day, with 50 per day being the average for setting long span joists.

After main members are calculated, add the following for usual allowances: base plates 2% to 3%; column splices 4% to 5%; and miscellaneous details 4% to 5%, for a total of 10% to 13% in addition to main members.

The ratio of column to beam tonnage varies depending on type of steels used, typical spans, story heights and live loads.

It is more economical to keep the column size constant and to vary the strength of the column by using high strength steels. This also saves floor space. Buildings have recently gone as high as ten stories with 8″ high strength columns. For light columns under W8X31 lb. sections, concrete filled steel columns are economical.

High strength steels may be used in columns and beams to save floor space and to meet head room requirements. High strength steels in some sizes sometimes require long lead times.

Round, square and rectangular columns, both plain and concrete filled, are readily available and save floor area, but are higher in cost per pound than rolled columns. For high unbraced columns, tube columns may be less expensive.

Below are average minimum figures for the weights of the structural steel frame for different types of buildings using A36 steel, rolled shapes and simple joints. For economy in domes, rise to span ratio = .13. Open web joist framing systems will reduce weights by 10% to 40%. Composite design can reduce steel weight by up to 25% but additional concrete floor slab thickness may be required. Continuous design can reduce the weights up to 20%. There are many building codes with different live load requirements and different structural requirements, such as hurricane and earthquake loadings which can alter the figures.

*See R13128-310 for Domes and Thin Shell Structures.

Structural Steel Weights per S.F. of Floor Area									
Type of Building	No. of Stories	Avg. Spans	L.L. #/S.F.	Lbs. Per S.F.	Type of Building	No. of Stories	Avg. Spans	L.L. #/S.F.	Lbs. Per S.F.
Steel Frame Mfg.	1	20′x20′	40	8	Apartments	2-8	20′x20′	40	8
		30′x30′		13		9-25			14
		40′x40′		18	Office	to 10	Various	80	10
Parking garage	4	Various	80	8.5		20			18
Domes (Schwedler)*	1	200′	30	10		30			26
		300′		15		over 50			35

RSMeans
Consulting Services

*Leaders in Cost Management
for the Construction Industry*

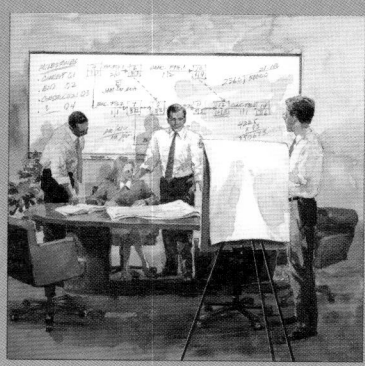

RSMeans leads the construction industry with standards and benchmarks for construction and facility cost management. RSMeans also offers executives and managers, in both public and private sectors, consulting services for strategic, technical and research cost solutions . . .

Strategic Audits & Best Practices Research — RSMeans Engineers and Senior Consultants approach each client engagement with sector-specific knowledge of processes, technology and construction that impact capital planning, facilities management and ownership of complex organizations. RSMeans Engineers include licensed engineers in areas of civil, structural, electrical, mechanical and architectural.

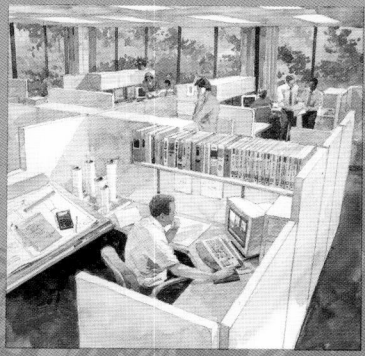

Predictive Cost Models — for sectors (e.g., government, healthcare, retail and banking) with large building portfolios, as well as similar structures in multiple locations, parametric cost models that streamline cost processes and predict construction costs for different locations in the U.S.

Job Order Contracting — a delivery and contracting method for construction remodeling and renovation that reduces construction cycle times and administrative overhead costs. RSMeans provides consulting in managing JOC contracts, training in JOC cost estimating, a JOC database and custom software particularly applicable to municipal, educational, government and military facilities.

Software Applications & Custom Unit Price Databases — web designed cost calculators (examples include permit costs, facility maintenance, life cycle costs, and international indices) apply RSMeans vast construction and facilities databases. RSMeans is the developer of custom cost estimating databases and national indices for federal government clients as well as licenses cost data from the RSMeans 27 annual cost guides.

Benchmark Studies & Surveys — RSMeans studies apply Reed Construction Data-RSMeans proprietary databases to investigate various areas of client inquiry. Examples include local market condition studies, time & motion studies, validation cost estimates, and life cycle cost studies. Specifically for building product manufacturers, analytic consulting includes retrospective and forecast metrics for analysis of construction volume and web-based anaylsis tools for market sizing.

**CORPORATE • HEALTHCARE • EDUCATION • GOVERNMENT • MUNICIPAL
RETAIL • ASSOCIATIONS • BUILDING PRODUCT MANUFACTURERS**

For more information contact RSMeans at 781-422-5101
or visit our website at www.rsmeans.com

 Reed Construction Data.

R05120-225 Common Structural Steel Specifications

ASTM A992 (formerly A36, then A572 Grade 50) is the all-purpose carbon grade steel widely used in building and bridge construction.

The other high-strength steels listed below may each have certain advantages over ASTM A992 stuctural carbon steel, depending on the application. They have proven to be economical choices where, due to lighter members, the reduction of dead load and the associated savings in shipping cost can be significant.

ASTM A588 atmospheric weathering, high-strength low-alloy steels can be used in the bare (uncoated) condition, where exposure to normal atmosphere causes a tightly adherant oxide to form on the surface protecting the steel from further oxidation. ASTM A242 corrosion-resistant, high-strength low-alloy steels have enhanced atmospheric corrosion resistance of at least two times that of carbon structural steels with copper, or four times that of carbon structural steels without copper. The reduction or elimination of maintenance resulting from the use of these steels often offsets their higher initial cost.

Steel Type	ASTM Designation	Minimum Yield Stress in KSI	Shapes Available
Carbon	A36	36	All structural shape groups, and plates & bars up thru 8″ thick
	A529	42	Structural shape group 1, and plates & bars up thru 1/2″ thick
High-Strength Low-Alloy Manganese-Vanadium	A441	40	Plates & bars over 4″ up thru 8″ thick
		42	Structural shape groups 4 & 5, and plates & bars over 1-1/2″ up thru 4″ thick
		46	Structural shape group 3, and plates & bars over 3/4″ up thru 1-1/2″ thick
		50	Structural shape groups 1 & 2, and plates & bars up thru 3/4″ thick
High-Strength Low-Alloy Columbium-Vanadium	A572	42	All structural shape groups, and plates & bars up thru 6″ thick
		50	All structural shape groups, and plates & bars up thru 4″ thick
		60	Structural shape groups 1 & 2, and plates & bars up thru 1-1/4″ thick
		65	Structural shape group 1, and plates & bars up thru 1-1/4″ thick
High-Strength Low-Alloy Columbium-Vanadium	A992	50	All structural shape groups
Corrosion-Resistant High-Strength Low-Alloy	A242	42	Structural shape groups 4 & 5, and plates & bars over 1-1/2″ up thru 4″ thick
		46	Structural shape group 3, and plates & bars over 3/4″ up thru 1-1/2″ thick
		50	Structural shape groups 1 & 2, and plates & bars up thru 3/4″ thick
Weathering High-Strength Low-Alloy	A588	42	Plates & bars over 5″ up thru 8″ thick
		46	Plates & bars over 4″ up thru 5″ thick
		50	All structural shape groups, and plates & bars up thru 4″ thick
Quenched and Tempered Low-Alloy	A852	70	Plates & bars up thru 4″ thick
Quenched and Tempered Alloy	A514	90	Plates & bars over 2-1/2″ up thru 6″ thick
		100	Plates & bars up thru 2-1/2″ thick

R05120-230 High Strength Steels

The mill price of high strength steels may be higher than A992 carbon steel but their proper use can achieve overall savings thru total reduced weights. For columns with L/r over 100, A992 steel is best; under 100, high strength steels are economical. For heavy columns, high strength steels are economical when cover plates are eliminated. There is no economy using high strength steels for clip angles or supports or for beams where deflection governs. Thinner members are more economical than thick.

See section 05120-680 for size and specification extras. See also R05120-210 for typical in place costs of A992 steel. The per ton erection and fabricating costs of the high strength steels will be higher than for A992 since the same number of pieces, but less weight, will be installed.

R05120-235   Common Steel Sections

The upper portion of this table shows the name, shape, common designation and basic characteristics of commonly used steel sections. The lower portion explains how to read the designations used for the above illustrated common sections.

Shape & Designation	Name & Characteristics	Shape & Designation	Name & Characteristics
W	W Shape — Parallel flange surfaces	MC	Miscellaneous Channel — Infrequently rolled by some producers
S	American Standard Beam (I Beam) — Sloped inner flange	L	Angle — Equal or unequal legs, constant thickness
M	Miscellaneous Beams — Cannot be classified as W, HP or S; infrequently rolled by some producers	T	Structural Tee — Cut from W, M or S on center of web
C	American Standard Channel — Sloped inner flange	HP	Bearing Pile — Parallel flanges and equal flange and web thickness

Common drawing designations follow:

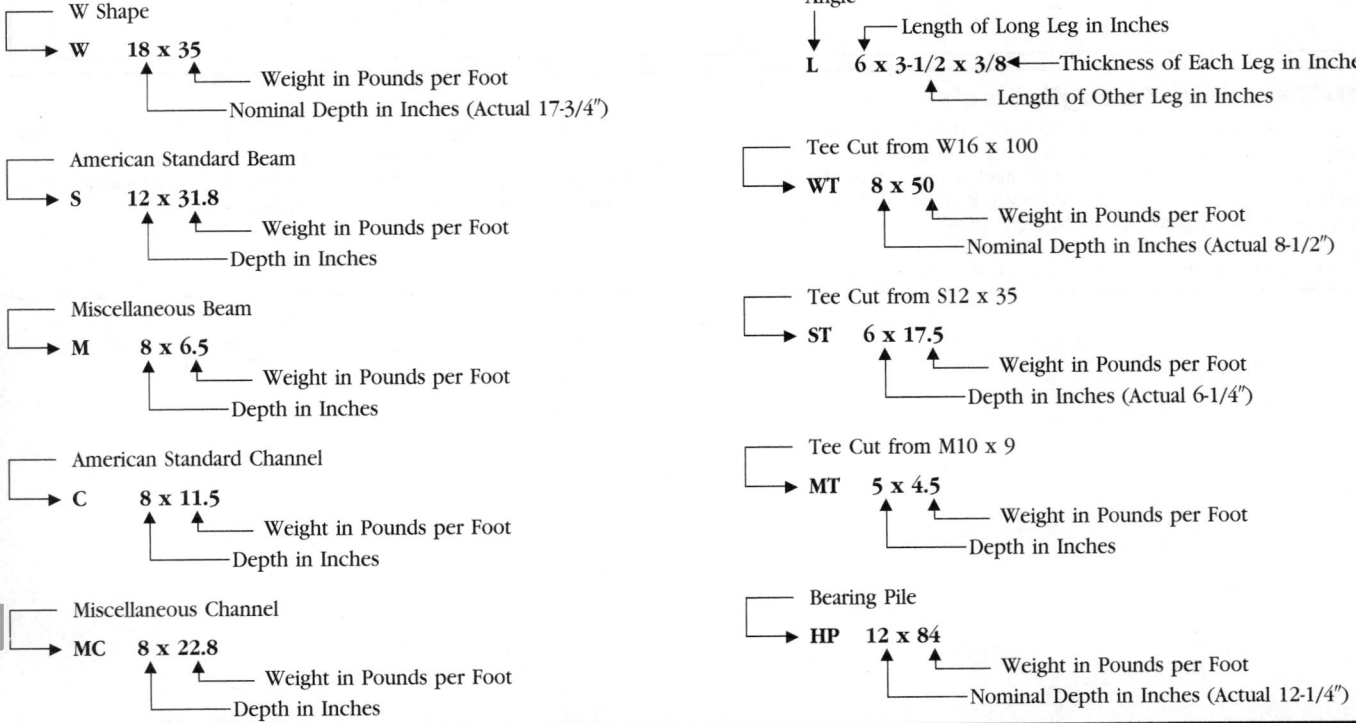

W Shape
W   18 x 35
— Weight in Pounds per Foot
— Nominal Depth in Inches (Actual 17-3/4")

American Standard Beam
S   12 x 31.8
— Weight in Pounds per Foot
— Depth in Inches

Miscellaneous Beam
M   8 x 6.5
— Weight in Pounds per Foot
— Depth in Inches

American Standard Channel
C   8 x 11.5
— Weight in Pounds per Foot
— Depth in Inches

Miscellaneous Channel
MC   8 x 22.8
— Weight in Pounds per Foot
— Depth in Inches

Angle
L   6 x 3-1/2 x 3/8
— Length of Long Leg in Inches
— Thickness of Each Leg in Inches
— Length of Other Leg in Inches

Tee Cut from W16 x 100
WT   8 x 50
— Weight in Pounds per Foot
— Nominal Depth in Inches (Actual 8-1/2")

Tee Cut from S12 x 35
ST   6 x 17.5
— Weight in Pounds per Foot
— Depth in Inches (Actual 6-1/4")

Tee Cut from M10 x 9
MT   5 x 4.5
— Weight in Pounds per Foot
— Depth in Inches

Bearing Pile
HP   12 x 84
— Weight in Pounds per Foot
— Nominal Depth in Inches (Actual 12-1/4")

R05120-245 Installation Time for Structural Steel Building Components

The following tables show the expected average installation times for various structural steel shapes. Table A presents installation times for columns, Table B for beams, Table C for light framing and bolts, and Table D for structural steel for various project types.

Table A

Description	Labor-Hours	Unit
Columns		
Steel, Concrete Filled		
3-1/2" Diameter	.933	Ea.
6-5/8" Diameter	1.120	Ea.
Steel Pipe		
3" Diameter	.933	Ea.
8" Diameter	1.120	Ea.
12" Diameter	1.244	Ea.
Structural Tubing		
4" x 4"	.966	Ea.
8" x 8"	1.120	Ea.
12" x 8"	1.167	Ea.
W Shape 2 Tier		
W8 x 31	.052	L.F.
W8 x 67	.057	L.F.
W10 x 45	.054	L.F.
W10 x 112	.058	L.F.
W12 x 50	.054	L.F.
W12 x 190	.061	L.F.
W14 x 74	.057	L.F.
W14 x 176	.061	L.F.

Table B

Description	Labor-Hours	Unit	Labor-Hours	Unit
Beams, W Shape				
W6 x 9	.949	Ea.	.093	L.F.
W10 x 22	1.037	Ea.	.085	L.F.
W12 x 26	1.037	Ea.	.064	L.F.
W14 x 34	1.333	Ea.	.069	L.F.
W16 x 31	1.333	Ea.	.062	L.F.
W18 x 50	2.162	Ea.	.088	L.F.
W21 x 62	2.222	Ea.	.077	L.F.
W24 x 76	2.353	Ea.	.072	L.F.
W27 x 94	2.581	Ea.	.067	L.F.
W30 x 108	2.857	Ea.	.067	L.F.
W33 x 130	3.200	Ea.	.071	L.F.
W36 x 300	3.810	Ea.	.077	L.F.

Table C

Description	Labor-Hours	Unit
Light Framing		
Angles 4" and Larger	.055	lbs.
Less than 4"	.091	lbs.
Channels 8" and Larger	.048	lbs.
Less than 8"	.072	lbs.
Cross Bracing Angles	.055	lbs.
Rods	.034	lbs.
Hanging Lintels	.069	lbs.
High Strength Bolts in Place		
3/4" Bolts	.070	Ea.
7/8" Bolts	.076	Ea.

Table D

Description	Labor-Hours	Unit	Labor-Hours	Unit
Apartments, Nursing Homes, etc.				
1-2 Stories	4.211	Piece	7.767	Ton
3-6 Stories	4.444	Piece	7.921	Ton
7-15 Stories	4.923	Piece	9.014	Ton
Over 15 Stories	5.333	Piece	9.209	Ton
Offices, Hospitals, etc.				
1-2 Stories	4.211	Piece	7.767	Ton
3-6 Stories	4.741	Piece	8.889	Ton
7-15 Stories	4.923	Piece	9.014	Ton
Over 15 Stories	5.120	Piece	9.209	Ton
Industrial Buildings				
1 Story	3.478	Piece	6.202	Ton

R05120-250   Subpurlins

Bulb tee subpurlins are structural members designed to support and reinforce a variety of roof deck systems such as precast cement fiber roof deck tiles, monolithic roof deck systems, and gypsum or lightweight concrete over formboard. Other uses include interstitial service ceiling systems, wall panel systems, and joist anchoring in bond beams. See Division 05120-140 for pricing on a square foot basis at 32-5/8" O.C. Maximum span is based on a 3-span condition with a total allowable vertical load of 40 psf.

R05120-810   Dimensions and Weights of Sheet Steel

Gauge No.	Approximate Thickness				Weight		
	Inches (in fractions)	Inches (in decimal parts)		Millimeters	per S.F. in Ounces	per S.F. in Lbs.	per Square Meter in Kg.
	Wrought Iron	Wrought Iron	Steel	Steel			
0000000	1/2"	.5	.4782	12.146	320	20.000	97.650
000000	15/32"	.46875	.4484	11.389	300	18.750	91.550
00000	7/16"	.4375	.4185	10.630	280	17.500	85.440
0000	13/32"	.40625	.3886	9.870	260	16.250	79.330
000	3/8"	.375	.3587	9.111	240	15.000	73.240
00	11/32"	.34375	.3288	8.352	220	13.750	67.130
0	5/16"	.3125	.2989	7.592	200	12.500	61.030
1	9/32"	.28125	.2690	6.833	180	11.250	54.930
2	17/64"	.265625	.2541	6.454	170	10.625	51.880
3	1/4"	.25	.2391	6.073	160	10.000	48.820
4	15/64"	.234375	.2242	5.695	150	9.375	45.770
5	7/32"	.21875	.2092	5.314	140	8.750	42.720
6	13/64"	.203125	.1943	4.935	130	8.125	39.670
7	3/16"	.1875	.1793	4.554	120	7.500	36.320
8	11/64"	.171875	.1644	4.176	110	6.875	33.570
9	5/32"	.15625	.1495	3.797	100	6.250	30.520
10	9/64"	.140625	.1345	3.416	90	5.625	27.460
11	1/8"	.125	.1196	3.038	80	5.000	24.410
12	7/64"	.109375	.1046	2.657	70	4.375	21.360
13	3/32"	.09375	.0897	2.278	60	3.750	18.310
14	5/64"	.078125	.0747	1.897	50	3.125	15.260
15	9/128"	.0713125	.0673	1.709	45	2.813	13.730
16	1/16"	.0625	.0598	1.519	40	2.500	12.210
17	9/160"	.05625	.0538	1.367	36	2.250	10.990
18	1/20"	.05	.0478	1.214	32	2.000	9.765
19	7/160"	.04375	.0418	1.062	28	1.750	8.544
20	3/80"	.0375	.0359	.912	24	1.500	7.324
21	11/320"	.034375	.0329	.836	22	1.375	6.713
22	1/32"	.03125	.0299	.759	20	1.250	6.103
23	9/320"	.028125	.0269	.683	18	1.125	5.490
24	1/40"	.025	.0239	.607	16	1.000	4.882
25	7/320"	.021875	.0209	.531	14	.875	4.272
26	3/160"	.01875	.0179	.455	12	.750	3.662
27	11/640"	.0171875	.0164	.417	11	.688	3.357
28	1/64"	.015625	.0149	.378	10	.625	3.052

R05310-100 Decking Descriptions

General - All Deck Products

Steel deck is made by cold forming structural grade sheet steel into a repeating pattern of parallel ribs. The strength and stiffness of the panels are the result of the ribs and the material properties of the steel. Deck lengths can be varied to suit job conditions, but because of shipping considerations, are usually less than 40 feet. Standard deck width varies with the product used but full sheets are usually 12″, 18″, 24″, 30″, or 36″. Deck is typically furnished in a standard width with the ends cut square. Any cutting for width, such as at openings or for angular fit, is done at the job site.

Deck is typically attached to the building frame with arc puddle welds, self-drilling screws, or powder or pneumatically driven pins. Sheet to sheet fastening is done with screws, button punching (crimping), or welds.

Composite Floor Deck

After installation and adequate fastening, floor deck serves several purposes. It (a) acts as a working platform, (b) stabilizes the frame, (c) serves as a concrete form for the slab, and (d) reinforces the slab to carry the design loads applied during the life of the building. Composite decks are distinguished by the presence of shear connector devices as part of the deck. These devices are designed to mechanically lock the concrete and deck together so that the concrete and the deck work together to carry subsequent floor loads. These shear connector devices can be rolled-in embossments, lugs, holes, or wires welded to the panels. The deck profile can also be used to interlock concrete and steel.

Composite deck finishes are either galvanized (zinc coated) or phosphatized/painted. Galvanized deck has a zinc coating on both the top and bottom surfaces. The phosphatized/painted deck has a bare (phosphatized) top surface that will come into contact with the concrete. This bare top surface can be expected to develop rust before the concrete is placed. The bottom side of the deck has a primer coat of paint.

Composite floor deck is normally installed so the panel ends do not overlap on the supporting beams. Shear lugs or panel profile shape often prevent a tight metal to metal fit if the panel ends overlap; the air gap caused by overlapping will prevent proper fusion with the structural steel supports when the panel end laps are shear stud welded.

Adequate end bearing of the deck must be obtained as shown on the drawings. If bearing is actually less in the field than shown on the drawings, further investigation is required.

Roof Deck

Roof deck is not designed to act compositely with other materials. Roof deck acts alone in transferring horizontal and vertical loads into the building frame. Roof deck rib openings are usually narrower than floor deck rib openings. This provides adequate support of rigid thermal insulation board.

Roof deck is typically installed to endlap approximately 2″ over supports. However, it can be butted (or lapped more than 2″) to solve field fit problems. Since designers frequently use the installed deck system as part of the horizontal bracing system (the deck as a diaphragm), any fastening substitution or change should be approved by the designer. Continuous perimeter support of the deck is necessary to limit edge deflection in the finished roof and may be required for diaphragm shear transfer.

Standard roof deck finishes are galvanized or primer painted. The standard factory applied paint for roof deck is a primer paint and is not intended to weather for extended periods of time. Field painting or touching up of abrasions and deterioration of the primer coat or other protective finishes is the responsibility of the contractor.

Cellular Deck

Cellular deck is made by attaching a bottom steel sheet to a roof deck or composite floor deck panel. Cellular deck can be used in the same manner as floor deck. Electrical, telephone, and data wires are easily run through the chase created between the deck panel and the bottom sheet.

When used as part of the electrical distribution system, the cellular deck must be installed so that the ribs line up and create a smooth cell transition at abutting ends. The joint that occurs at butting cell ends must be taped or otherwise sealed to prevent wet concrete from seeping into the cell. Cell interiors must be free of welding burrs, or other sharp intrusions, to prevent damage to wires.

When used as a roof deck, the bottom flat plate is usually left exposed to view. Care must be maintained during erection to keep good alignment and prevent damage.

Cellular deck is sometimes used with the flat plate on the top side to provide a flat working surface. Installation of the deck for this purpose requires special methods for attachment to the frame because the flat plate, now on the top, can prevent direct access to the deck material that is bearing on the structural steel. It may be advisable to treat the flat top surface to prevent slipping.

Cellular deck is always furnished galvanized or painted over galvanized.

Form Deck

Form deck can be any floor or roof deck product used as a concrete form. Connections to the frame are by the same methods used to anchor floor and roof deck. Welding washers are recommended when welding deck that is less than 20 gauge thickness.

Form deck is furnished galvanized, prime painted, or uncoated. Galvanized deck must be used for those roof deck systems where form deck is used to carry a lightweight insulating concrete fill.

R05650-100 Single Track R.R. Siding

The costs for a single track RR siding in section 05655-700 include the components shown in the table below.

Description of Component	CSI Line Item	Qty. per L.F. of Track	Unit
Ballast, 1-1/2" crushed stone	02060-150-0320	.667	C.Y.
6" x 8" x 8'-6" Treated timber ties, 22" O.C.	05655-700-1600	.545	Ea.
Tie plates, 2 per tie	05655-750-0300	1.091	Ea.
Track rail	05655-750-1000	2.000	L.F.
Spikes, 6", 4 per tie	05655-750-0200	2.182	Ea.
Splice bars w/ bolts, lock washers & nuts, @ 33' O.C.	05655-750-0100	.061	Pair
Crew B-14 @ 57 L.F./Day		.018	Day

R05650-200 Single Track, Steel Ties, Concrete Bed

The costs for a R.R. siding with steel ties and a concrete bed in section 05655-700 include the components shown in the table below.

Description of Component	CSI Line Item	Qty. per L.F. of Track	Unit
Concrete bed, 9' wide, 10" thick	03310-240-3950	.278	C.Y.
Ties, W6x16 x 6'-6" long, @ 30" O.C.	05120-640-0120	.400	Ea.
Tie plates, 4 per tie	05655-750-0300	1.600	Ea.
Track rail	05655-750-1000	2.000	L.F.
Tie plate bolts, 1", 8 per tie	05655-750-0020	3.200	Ea.
Splice bars w/bolts, lock washers & nuts, @ 33' O.C.	05655-750-0100	.061	Pair
Crew B-14 @ 22 L.F./Day		.045	Day

R06100-010 Thirty City Lumber Prices

Prices for boards are for #2 or better or sterling, whichever is in best supply. Dimension lumber is "Standard or Better" either Southern Yellow Pine (S.Y.P.), Spruce-Pine-Fir (S.P.F.), Hem-Fir (H.F.) or Douglas Fir (D.F.). The species of lumber used in a geographic area is listed by city. Plyform is 3/4" BB oil sealed fir or S.Y.P. whichever prevails locally, 3/4" CDX is S.Y.P. or Fir.

These are prices at the time of publication and should be checked against the current market price. Relative differences between cities will stay approximately constant.

City	Species	Contractor Purchases per M.B.F. S4S Dimensions 2"x4"	2"x6"	2"x8"	2"x10"	2"x12"	4"x4"	Boards 1"x6"	1"x12"	Contractor Purchases per M.S.F. 3/4" Ext. Plyform	3/4" Thick CDX T&G
Atlanta	S.Y.P.	$428	$453	$523	$595	$675	$ 793	$ 970	$1,316	$1,223	$1,067
Baltimore	S.P.F.	536	566	654	743	843	991	1212	1645	1529	1334
Boston	S.P.F.	517	547	631	718	814	957	1171	1588	1476	1288
Buffalo	S.P.F.	519	549	634	721	818	961	1175	1594	1482	1293
Chicago	H.F.	519	549	634	721	818	961	1175	1594	1482	1293
Cincinnati	S.P.F.	536	566	654	743	843	991	1212	1645	1529	1334
Cleveland	S.P.F.	520	550	635	722	819	963	1178	1598	1485	1295
Columbus	S.P.F.	530	561	647	736	835	982	1201	1629	1514	1321
Dallas	S.Y.P.	431	456	527	599	680	798	977	1325	1232	1074
Denver	H.F.	517	547	631	718	814	957	1171	1588	1476	1288
Detroit	H.F.	531	562	649	738	837	983	1203	1632	1517	1323
Houston	S.Y.P.	427	452	521	593	673	790	967	1312	1219	1064
Indianapolis	S.P.F.	524	554	640	728	826	970	1187	1610	1497	1306
Kansas City	D.F.	517	547	631	718	814	957	1171	1588	1476	1288
Los Angeles	D.F.	527	557	644	732	831	976	1194	1620	1505	1313
Memphis	S.Y.P.	508	537	620	705	800	940	1150	1560	1450	1265
Milwaukee	H.F.	519	549	634	721	818	961	1175	1594	1482	1293
Minneapolis	H.F.	517	547	631	718	814	957	1171	1588	1476	1288
Nashville	S.Y.P.	508	537	620	705	800	940	1150	1560	1450	1265
New Orleans	S.Y.P.	440	465	537	611	693	814	996	1352	1256	1096
New York City	S.P.F.	532	563	650	739	839	985	1205	1635	1520	1326
Philadelphia	S.P.F.	532	563	650	739	839	985	1205	1635	1520	1326
Phoenix	D.F.	536	566	654	743	843	991	1212	1645	1529	1334
Pittsburgh	S.P.F.	530	561	647	736	835	982	1201	1629	1514	1321
St. Louis	H.F.	524	554	640	728	826	970	1187	1610	1497	1306
San Antonio	S.Y.P.	436	461	532	605	686	806	987	1338	1244	1085
San Diego	D.F.	533	563	650	739	839	986	1206	1636	1520	1326
San Francisco	D.F.	508	537	620	705	800	940	1150	1560	1450	1265
Seattle	D.F.	503	532	614	698	792	931	1139	1544	1436	1252
Washington, DC	S.P.F.	541	572	660	751	852	1001	1224	1661	1544	1347
Average		$508	$537	$620	$705	$800	$ 940	$1,150	$1,560	$1,450	$1,265

To convert square feet of surface to board feet, 4% waste included.

S4S Size	Multiply S.F. by	T & G Size	Multiply S.F. by	Flooring Size	Multiply S.F. by
				25/32" x 2-1/4"	1.37
1 x 4	1.18	1 x 4	1.27	25/32" x 3-1/4"	1.29
1 x 6	1.13	1 x 6	1.18	15/32" x 1-1/2"	1.54
1 x 8	1.11	1 x 8	1.14	1" x 3"	1.28
1 x 10	1.09	2 x 6	2.36	1" x 4"	1.24

WOOD & PLASTICS 6

REFERENCE NOS.

R06110-030 Lumber Product Material Prices

The price of forest products fluctuates widely from location to location and from season to season depending upon economic conditions. The table below indicates National Average material prices in effect Jan. 1 of this book year. The table shows relative differences between various sizes, grades and species. These percentage differentials remain fairly constant even though lumber prices in general may change significantly during the year.

Availability of certain items depends upon geographic location and must be checked prior to firm price bidding.

		National Average Contractor Price, Quantity Purchase					Heavy Timbers, Fir	
		Dimension Lumber, S4S, #2 & Better, KD						
	Species	2"x4"	2"x6"	2"x8"	2"x10"	2"x12"		
Framing	Douglas Fir	$ 608	$ 540	$ 567	$ 527	$ 594	3" x 4" thru 3" x 12"	$872
Lumber	Spruce	543	473	527	533	578	4" x 4" thru 4" x 12"	872
per MBF	Southern Yellow Pine	540	358	459	493	709	6" x 6" thru 6" x 12"	1,060
	Hem-Fir	486	473	493	520	473	8" x 8" thru 8" x 12"	1,222
							10" x 10" and 10" x 12"	1,285

		S4S "D" Quality or Clear, KD					S4S # 2 & Better or Sterling, KD				
	Species	1"x4"	1"x6"	1"x8"	1"x10"	1"x12"	Species	1"x4"	1"x6"	1"x8"	1"x10"	1"x12"
	Sugar Pine	$1,485	$1,667	$1,688	$2,329	$3,071	Sugar Pine	$648	$770	$ 790	$ 797	$1,073
Boards per MBF	Idaho Pine	979	986	1,033	1,013	1,087	Idaho Pine	628	986	1,033	1,013	1,019
*See also	Engleman Spruce	905	1,168	1,127	1,202	1,715	Engleman Spruce	702	736	682	675	992
Cedar Siding	So. Yellow Pine	662	878	965	898	1,121	So. Yellow Pine	513	608	689	554	1,121
	Ponderosa Pine	1,033	1,175	1,046	1,249	2,025	Ponderosa Pine	689	648	655	662	945

R06160-020   Plywood

There are two types of plywood used in construction: interior, which is moisture resistant but not waterproofed, and exterior, which is waterproofed.

The grade of the exterior surface of the plywood sheets is designated by the first letter: A, for smooth surface with patches allowed; B, for solid surface with patches and plugs allowed; C, which may be surface plugged or may have knot holes up to 1″ wide; and D, which is used only for interior type plywood and may have knot holes up to 2-1/2″ wide. "Structural Grade" is specifically designed for engineered applications such as box beams. All CC & DD grades have roof and floor spans marked on them.

Underlayment grade plywood runs from 1/4″ to 1-1/4″ thick. Thicknesses 5/8″ and over have optional tongue and groove joints which eliminates the need for blocking the edges. Underlayment 19/32″ and over may be referred to as Sturd-i-Floor.

The price of plywood can fluctuate widely due to geographic and economic conditions. When one or two local prices are known, the relative prices for other types and sizes may be found by direct factoring of the prices in the table below.

Typical uses for various plywood grades are as follows:

AA-AD Interior — cupboards, shelving, paneling, furniture

BB Plyform — concrete form plywood

CDX — wall and roof sheathing

Structural — box beams, girders, stressed skin panels

AA-AC Exterior — fences, signs, siding, soffits, etc.

Underlayment — base for resilient floor coverings

Overlaid HDO — high density for concrete forms & highway signs

Overlaid MDO — medium density for painting, siding, soffits & signs

303 Siding — exterior siding, textured, striated, embossed, etc.

Grade	National Average Price in Lots of 10 MSF, per MSF-January					
	Type	4′x8′	Type	4′x8′	4′x10′	
Sanded Grade	1/4″ Interior AD	$ 779	1/4″ Exterior AC	$ 780	$ 779	
	3/8″	869	3/8″	870	869	
	1/2″	1,059	1/2″	1,060	1,059	
	5/8″	1,215	5/8″	1,215	1,215	
	3/4″	1,399	3/4″	1,399	1,339	
	1″	2,014	1″	2,014	2,014	
	1-1/4″	2,196	Exterior AA, add	225	230	
	Interior AA, add	225	Exterior AB, add	180	195	
Unsanded Grade 4′ x 8′ Sheets	5/16″ CDX	$ 630	CD Structural 1	Underlayment		
	3/8″	652		3/8″, 4′x8′ sheets	$ 869	
	1/2″	619	3/8″ 4′x8′ sheets 340	1/2″	1,059	
	5/8″	780	1/2″ 592	5/8″T&G	800	
	3/4″	940	3/4″ 840	3/4″T&G	1,050	
	3/4″ T&G	1,050		1-1/8″ 2-4-1T&G	1,185	
Form Plywood	5/8″ Exterior, oiled BB, plyform	$1,364	5/8″ HDO (overlay 2 sides)		$2,340	
	3/4″ Exterior, oiled BB, plyform	1,450	3/4″ HDO (overlay 2 sides)		2,410	
Overlaid 4′x8′ Sheets	Overlay 2 Sides MDO		Overlay 1 Side MDO			
	3/8″ thick	$1,345	3/8″ thick		$1,250	
	1/2″	1,576	1/2″		1,316	
	5/8″	1,776	5/8″		1,360	
	3/4″	2,034	3/4″		1,400	
303 Siding	Fir, rough sawn, natural finish, 3/8″ thick	$ 730	Texture 1-11	5/8″ thick, Fir	$1,320	
	Redwood	2,514		Redwood	2,650	
	Cedar	2,380		Cedar	2,150	
	Southern Yellow Pine	585		Southern Yellow	1,070	
Waferboard/O.S.B.	1/4″ sheathing	$ 335	19/32″ T&G		$ 780	
	7/16″ sheathing	624	23/32″ T&G		937	

For 2 MSF to 10 MSF, add 10%. For less than 2 MSF, add 15%.

R06170-100 Wood Roof Trusses

Loading figures represent live load. An additional load of 10 psf on the
top chord and 10 psf on the bottom chord is included in the truss
design. Spacing is 24″ O.C.

Span in Feet	Cost per Truss for Different Live Loads and Roof Pitches					
	Flat	4 in 12 Pitch		5 in 12 Pitch		8 in 12 Pitch
	40 psf	30 psf	40 psf	30 psf	40 psf	30 psf
20	$ 64	$ 54	$ 51	$ 59	$ 54	$ 60
22	70	57	55	62	57	65
24	77	62	47	64	55	70
26	80	72	52	73	60	72
28	88	73	60	75	65	78
30	111	81	67	92	72	82
32	119	99	72	93	130	115
34	126	99	106	113	113	124
36	133	105	112	115	115	130
38	141	120	120	121	122	140
40	148	124	122	121	122	146

R07110-010   1/2″ Pargetting (rough dampproofing plaster)

1:2-1/2 Mix, 4.5 C.F. Covers 100 S.F., 2 Coats, Waste Included	Regular Portland Cement		Waterproofed Portland Cement	
1.7 Lbs. integral waterproofing admixture			$.85 per lb.	$ 1.45
1.7 Bags Portland cement	$8.25 per bag	$ 14.03	8.25 per bag	14.03
4.25 C.F. sand	15.80 per C.Y.	2.49	15.80 per C.Y.	2.49
0.40 Days of labor to mix and apply (crew D-1)	497.20 per day	198.88	497.20 per day	198.88
Total Bare Cost per 100 S.F.		$215.40		$216.85

R07310-020   Roof Slate

16″, 18″ and 20″ are standard lengths, and slate usually comes in
random widths. For standard 3/16″ thickness use 1-1/2″ copper
nails. Allow for 3% breakage.

Quantities per Square	Unfading Vermont Colored	Weathering Sea Green	Buckingham, Virginia Black
Slate delivered (incl. punching)	$346.00	$330.00	$550.00
# 30 Felt, 2.5 lbs. copper nails	19.45	19.45	19.45
Slate roofer 4.6 hrs. @ $29.70 per hr.	136.62	136.62	136.62
Total Bare Cost per Square	$502.07	$486.07	$706.07

R07510-020 Built-Up Roofing

Asphalt is available in kegs of 100 lbs. each; coal tar pitch in 560 lb. kegs. Prepared roofing felts are available in a wide range of sizes, weights & characteristics. However, the most commonly used are #15 (432 S.F. per roll, 13 lbs. per square) and #30 (216 S.F. per roll, 27 lbs. per square).

Inter-ply bitumen varies from 24 lbs. per sq. (asphalt) to 30 lbs. per sq. (coal tar) per ply, ± 25%. Flood coat bitumen also varies from 60 lbs. per sq. (asphalt) to 75 lbs. per sq. (coal tar), ± 25%. Expendable equipment (mops, brooms, screeds, etc.) runs about 16% of the bitumen cost. For new, inexperienced crews this factor may be much higher.

Rigid insulation board is typically applied in two layers. The first is mechanically attached to nailable decks or spot or solid mopped to non-nailable decks; the second layer is then spot or solid mopped to the first layer. Membrane application follows the insulation, except in protected membrane roofs, where the membrane goes down first and the insulation on top, followed with ballast (stone or concrete pavers). Insulation and related labor costs are NOT included in Div. 07510-300; they can be found in Div. 07220-700.

Prices shown are bare costs only.

4-Ply Felt on Insulated Deck	Glass Fiber Felt w/ Hot Asphalt		Organic Felt w/ Hot Coal Tar	
4 ply #15 felt (1/4 roll per ply)	$4.09 per ply	$ 16.36	$9.12 per ply	$ 36.48
156 Lbs. hot asphalt/195 lbs. hot coal tar	$400.00 per ton	31.20	$582.00 per ton	56.75
500 Lbs. roofing stone (3/8″ to 1/2″)	$15.10 per ton	3.78	$15.10 per ton	3.78
Installation, crew G-1, @ $1,837.50 per day	0.050 days	91.88	0.048 days	88.20
Total Bare Cost per Square in-place		$143.22		$185.21

R07550-030 Modified Bitumen Roofing

The cost of modified bitumen roofing is highly dependent on the type of installation that is planned. Installation is based on the type of modifier used in the bitumen. The two most popular modifiers are atactic polypropylene (APP) and styrene butadiene styrene (SBS). The modifiers are added to heated bitumen during the manufacturing process to change its characteristics. A polyethylene, polyester or fiberglass reinforcing sheet is then sandwiched between layers of this bitumen. When completed, the result is a pre-assembled, built-up roof that has increased elasticity and weatherablility. Some manufacturers include a surfacing material such as ceramic or mineral granules, metal particles or sand.

The preferred method of adhering SBS-modified bitumen roofing to the substrate is with hot-mopped asphalt (much the same as built-up roofing). This installation method requires a tar kettle/pot to heat the asphalt, as well as the labor, tools and equipment necessary to distribute and spread the hot asphalt.

The alternative method for applying APP and SBS modified bitumen is as follows. A skilled installer uses a torch to melt a small pool of bitumen off the membrane. This pool must form across the entire roll for proper adhesion. The installer must unroll the roofing at a pace slow enough to melt the bitumen, but fast enough to prevent damage to the rest of the membrane.

Modified bitumen roofing provides the advantages of both built-up and single-ply roofing. Labor costs are reduced over those of built-up roofing because only a single ply is necessary. The elasticity of single-ply roofing is attained with the reinforcing sheet and polymer modifiers. Modifieds have some self-healing characteristics and because of their multi-layer construction, they offer the reliability and safety of built-up roofing.

R07800-030 Firestopping

Firestopping is the sealing of structural, mechanical, electrical and other penetrations through fire-rated assemblies. The basic components of firestop systems are safing insulation and firestop sealant on both sides of wall penetrations and the top side of floor penetrations.

Pipe penetrations are assumed to be through concrete, grout, or joint compound and can be sleeved or unsleeved. Costs for the penetrations and sleeves are not included. An annular space of 1″ is assumed. Escutcheons are not included.

Metallic pipe is assumed to be copper, aluminum, cast iron or similar metallic material. Insulated metallic pipe is assumed to be covered with a thermal insulating jacket of varying thickness and materials.

Non-metallic pipe is assumed to be PVC, CPVC, FR Polypropylene or similar plastic piping material. Intumescent firestop sealant or wrap strips are included. Collars on both sides of wall penetrations and a sheet metal plate on the underside of floor penetrations are included.

Ductwork is assumed to be sheet metal, stainless steel or similar metallic material. Duct penetrations are assumed to be through concrete, grout or joint compound. Costs for penetrations and sleeves are not included. An annular space of 1/2″ is assumed.

Multi-trade openings include costs for sheet metal forms, firestop mortar, wrap strips, collars and sealants as necessary.

Structural penetrations joints are assumed to be 1/2″ or less. CMU walls are assumed to be within 1-1/2″ of metal deck. Drywall walls are assumed to be tight to the underside of metal decking.

Metal panel, glass or curtain wall systems include a spandrel area of 5′ filled with mineral wool foil-faced insulation. Fasteners and stiffeners are included.

R08110-020 Steel Door Selection Guide

Standard steel doors are classified into four levels, as recommended by the Steel Door Institute in the chart below. Each of the four levels offers a range of construction models and designs, to meet architectural requirements for preference and appearance, including full flush, seamless, and stile & rail. Recommended minimum gauge requirements are also included.

For complete standard steel door construction specifications and available sizes, refer to the Steel Door Institute Technical Data Series, ANSI A250.8-98 (SDI-100), and ANSI A250.4-94 Test Procedure and Acceptance Criteria for Physical Endurance of Steel Door and Hardware Reinforcements.

Level		Model	Construction	For Full Flush or Seamless		
				Min. Gauge	Thickness (in)	Thickness (mm)
I	Standard Duty	1	Full Flush	20	0.032	0.8
		2	Seamless			
II	Heavy Duty	1	Full Flush	18	0.042	1.0
		2	Seamless			
III	Extra Heavy Duty	1	Full Flush	16	0.053	1.3
		2	Seamless			
		3	*Stile & Rail			
IV	Maximum Duty	1	Full Flush	14	0.067	1.6
		2	Seamless			

*Stiles & rails are 16 gauge; flush panels, when specified, are 18 gauge.

R08510-100 Steel Sash

Ironworker crew will erect 25 S.F. or 1.3 sash unit per hour, whichever is less.

Mechanic will point 30 L.F. per hour.

Painter will paint 90 S.F. per coat per hour.

Glazier production depends on light size.

Allow 1 lb. special steel sash putty per 16″ x 20″ light.

R08550-010 Double Hung Windows - Tilt Wash

Ponderosa pine and vinyl clad sash, exterior primed with double insulated, low E glass.

Description	2'-0″ x 3'-0″				3'-0″ x 4'-0″			
	Wood		Vinyl Clad		Wood		Vinyl Clad	
Window, 2 lights w/ screens & grilles		$239.50		$278.90		$313.50		$372.00
Interior trim set		$ 19.90		$ 19.90		$ 23.50		$ 23.50
Carpenter @ $34.25 per hr.	1.7 hr.	58.23	1.7 hr.	58.23	2 hr.	68.50	2 hr.	68.50
Complete bare cost in place		$317.63		$357.03		$405.50		$464.00

R08700-100 Hinges

All closer equipped doors should have ball bearing hinges. Lead lined or extremely heavy doors require special strength hinges. Usually 1-1/2 pair of hinges are used per door up to 7'-6" high openings. Table below shows typical hinge requirements.

Use Frequency	Type Hinge Required	Type of Opening	Type of Structure
High	Heavy weight ball bearing	Entrances	Banks, Office buildings, Schools, Stores & Theaters
		Toilet Rooms	Office buildings and Schools
Average	Standard weight ball bearing	Entrances	Dwellings
		Corridors	Office buildings and Schools
		Toilet Rooms	Stores
Low	Plain bearing	Interior	Dwellings

Door Thickness	Weight of Doors in Pounds per Square Foot				
	White Pine	Oak	Hollow Core	Solid Core	Hollow Metal
1-3/8"	3 psf	6 psf	1-1/2 psf	3-1/2 — 4 psf	6-1/2 psf
1-3/4"	3-1/2	7	2-1/2	4-1/2 — 5-1/4	6-1/2
2-1/4"	4-1/2	9	—	5-1/2 — 6-3/4	6-1/2

R08810-010 Glazing Labor

Glass sizes are estimated by the "united inch" (height + width). Table below shows the number of lights glazed in an eight hour period by the crew size indicated, for glass up to 1/4" thick. Square or nearly square lights are more economical on a S.F. basis. Long slender lights will have a high S.F. installation cost. For insulated glass reduce production by 33%. For 1/2" float glass reduce production by 50%. Production time for glazing with two glaziers per day averages: 1/4" float glass 120 S.F.; 1/2" float glass 55 S.F.; 1/2" insulated glass 95 S.F.; 3/4" insulated glass 75 S.F.

Glazing Method	United Inches per Light							
	40"	60"	80"	100"	135"	165"	200"	240"
Number of Men in Crew	1	1	1	1	2	3	3	4
Industrial sash, putty	60	45	24	15	18	—	—	—
With stops, putty bed	50	36	21	12	16	8	4	3
Wood stops, rubber	40	27	15	9	11	6	3	2
Metal stops, rubber	30	24	14	9	9	6	3	2
Structural glass	10	7	4	3	—	—	—	—
Corrugated glass	12	9	7	4	4	4	3	—
Storefronts	16	15	13	11	7	6	4	4
Skylights, putty glass	60	36	21	12	16	—	—	—
Thiokol set	15	15	11	9	9	6	3	2
Vinyl set, snap on	18	18	13	12	12	7	5	4
Maximum area per light	2.8 S.F.	6.3 S.F.	11.1 S.F.	17.4 S.F.	31.6 S.F.	47 S.F.	69 S.F.	100 S.F.
Daily Bare Crew Cost	$264.00	$264.00	$264.00	$264.00	$528.00	$792.00	$792.00	$1,056.00

R09250-050 Lath, Plaster and Gypsum Board

Gypsum board lath is available in 3/8″ thick x 16″ wide x 4′ long sheets as a base material for multi-layer plaster applications. It is also available as a base for either multi-layer or veneer plaster applications in 1/2″ and 5/8″ thick–4′ wide x 8′, 10′ or 12′ long sheets. Fasteners are screws or blued ring shank nails for wood framing and screws for metal framing.

Metal lath is available in diamond mesh pattern with flat or self-furring profiles. Paper backing is available for applications where excessive plaster waste needs to be avoided. A slotted mesh ribbed lath should be used in areas where the span between structural supports is greater than normal. Most metal lath comes in 27″ x 96″ sheets. Diamond mesh weighs 1.75, 2.5 or 3.4 pounds per square yard, slotted mesh lath weighs 2.75 or 3.4 pounds per square yard. Metal lath can be nailed, screwed or tied in place.

Many **accessories** are available. Corner beads, flat reinforcing strips, casing beads, control and expansion joints, furring brackets and channels are some examples. Note that accessories are not included in plaster or stucco line items.

Plaster is defined as a material or combination of materials that when mixed with a suitable amount of water, forms a plastic mass or paste. When applied to a surface, the paste adheres to it and subsequently hardens, preserving in a rigid state the form or texture imposed during the period of elasticity.

Gypsum plaster is made from ground calcined gypsum. It is mixed with aggregates and water for use as a base coat plaster.

Vermiculite plaster is a fire-retardant plaster covering used on steel beams, concrete slabs and other heavy construction materials. Vermiculite is a group name for certain clay minerals, hydrous silicates or aluminum, magnesium and iron that have been expanded by heat.

Perlite plaster is a plaster using perlite as an aggregate instead of sand. Perlite is a volcanic glass that has been expanded by heat.

Gauging plaster is a mix of gypsum plaster and lime putty that when applied produces a quick drying finish coat.

Veneer plaster is a one or two component gypsum plaster used as a thin finish coat over special gypsum board.

Keenes cement is a white cementitious material manufactured from gypsum that has been burned at a high temperature and ground to a fine powder. Alum is added to accelerate the set. The resulting plaster is hard and strong and accepts and maintains a high polish, hence it is used as a finishing plaster.

Stucco is a Portland cement based plaster used primarily as an exterior finish.

Plaster is used on both interior and exterior surfaces. Generally it is applied in multiple-coat systems. A three-coat system uses the terms scratch, brown and finish to identify each coat. A two-coat system uses base and finish to describe each coat. Each type of plaster and application system has attributes that are chosen by the designer to best fit the intended use.

Gypsum Plaster Quantities for 100 S.Y.	2 Coat, 5/8″ Thick		3 Coat, 3/4″ Thick		
	Base	Finish	Scratch	Brown	Finish
	1:3 Mix	2:1 Mix	1:2 Mix	1:3 Mix	2:1 Mix
Gypsum plaster	1,300 lb.		1,350 lb.	650 lb.	
Sand	1.75 C.Y.		1.85 C.Y.	1.35 C.Y.	
Finish hydrated lime		340 lb.			340 lb.
Gauging plaster		170 lb.			170 lb.

Vermiculite or Perlite Plaster Quantities for 100 S.Y.	2 Coat, 5/8″ Thick		3 Coat, 3/4″ Thick		
	Base	Finish	Scratch	Brown	Finish
Gypsum plaster	1,250 lb.		1,450 lb.	800 lb.	
Vermiculite or perlite	7.8 bags		8.0 bags	3.3 bags	
Finish hydrated lime		340 lb.			340 lb.
Gauging plaster		170 lb.			170 lb.

Stucco–Three-Coat System Quantities for 100 S.Y.	On Wood Frame	On Masonry
Portland cement	29 bags	21 bags
Sand	2.6 C.Y.	2.0 C.Y.
Hydrated lime	180 lb.	120 lb.

R09250-100 Levels of Gypsum Drywall Finish

In the past, contract documents often used phrases such as "industry standard" and "workmanlike finish" to specify the expected quality of gypsum board wall and ceiling installations. The vagueness of these descriptions led to unacceptable work and disputes.

In order to resolve this problem, four major trade associations concerned with the manufacture, erection, finish and decoration of gypsum board wall and ceiling systems have developed an industry-wide *Recommended Levels of Gypsum Board Finish*.

The finish of gypsum board walls and ceilings for specific final decoration is dependent on a number of factors. A primary consideration is the location of the surface and the degree of decorative treatment desired. Painted and unpainted surfaces in warehouses and other areas where appearance is normally not critical may simply require the taping of wallboard joints and 'spotting' of fastener heads. Blemish-free, smooth, monolithic surfaces often intended for painted and decorated walls and ceilings in habitated structures, ranging from single-family dwellings through monumental buildings, require additional finishing prior to the application of the final decoration.

Other factors to be considered in determining the level of finish of the gypsum board surface are (1) the type of angle of surface illumination (both natural and artificial lighting), and (2) the paint and method of application or the type and finish of wallcovering specified as the final decoration. Critical lighting conditions, gloss paints, and thin wallcoverings require a higher level of gypsum board finish than do heavily textured surfaces which are subsequently painted or surfaces which are to be decorated with heavy grade wallcoverings.

The following descriptions were developed jointly by the Association of the Wall and Ceiling Industries-International (AWCI), Ceiling & Interior Systems Construction Association (CISCA), Gypsum Association (GA), and Painting and Decorating Contractors of America (PDCA) as a guide.

Level 0: No taping, finishing, or accessories required. This level of finish may be useful in temporary construction or whenever the final decoration has not been determined.

Level 1: All joints and interior angles shall have tape set in joint compound. Surface shall be free of excess joint compound. Tool marks and ridges are acceptable. Frequently specified in plenum areas above ceilings, in attics, in areas where the assembly would generally be concealed or in building service corridors, and other areas not normally open to public view.

Level 2: All joints and interior angles shall have tape embedded in joint compound and wiped with a joint knife leaving a thin coating of joint compound over all joints and interior angles. Fastener heads and accessories shall be covered with a coat of joint compound. Surface shall be free of excess joint compound. Tool marks and ridges are acceptable. Joint compound applied over the body of the tape at the time of tape embedment shall be considered a separate coat of joint compound and shall satisfy the conditions of this level. Specified where water-resistant gypsum backing board is used as a substrate for tile; may be specified in garages, warehouse storage, or other similar areas where surface appearance is not of primary concern.

Level 3: All joints and interior angles shall have tape embedded in joint compound and one additional coat of joint compound applied over all joints and interior angles. Fastener heads and accessories shall be covered with two separate coats of joint compound. All joint compound shall be smooth and free of tool marks and ridges. Typically specified in appearance areas which are to receive heavy- or medium-texture (spray or hand applied) finishes before final painting, or where heavy-grade wallcoverings are to be applied as the final decoration. This level of finish is not recommended where smooth painted surfaces or light to medium wallcoverings are specified.

Level 4: All joints and interior angles shall have tape embedded in joint compound and two separate coats of joint compound applied over all flat joints and one separate coat of joint compound applied over interior angles. Fastener heads and accessories shall be covered with three separate coats of joint compound. All joint compound shall be smooth and free of tool marks and ridges. This level should be specified where flat paints, light textures, or wallcoverings are to be applied. In critical lighting areas, flat paints applied over light textures tend to reduce joint photographing. Gloss, semi-gloss, and enamel paints are not recommended over this level of finish. The weight, texture, and sheen level of wallcoverings applied over this level of finish should be carefully evaluated. Joints and fasteners must be adequately concealed if the wallcovering material is lightweight, contains limited pattern, has a gloss finish, or any combination of these finishes is present. Unbacked vinyl wallcoverings are not recommended over this level of finish.

Level 5: All joints and interior angles shall have tape embedded in joint compound and two separate coats of joint compound applied over all flat joints and one separate coat of joint compound applied over interior angles. Fastener heads and accessories shall be covered with three separate coats of joint compound. A thin skim coat of joint compound or a material manufactured especially for this purpose, shall be applied to the entire surface. The surface shall be smooth and free of tool marks and ridges. This level of finish is highly recommended where gloss, semi-gloss, enamel, or nontextured flat paints are specified or where severe lighting conditions occur. This highest quality finish is the most effective method to provide a uniform surface and minimize the possibility of joint photographing and of fasteners showing through the final decoration.

R09400-100 Terrazzo Floor

5000 S.F. 5/8" Terrazzo Topping Quantities for 100 S.F.	Bonded to Concrete 1-1/8" Bed, 1:4 Mix			Not Bonded 2-1/8" Bed and 1/4" Sand		
	Quantities	Bare Cost	Incl. O & P	Quantities	Bare Cost	Incl. O & P
Portland cement @ $8.25 per 94 lb. bag	6 bags	$ 49.50	$ 54.45	8 bags	$ 66.00	$ 72.60
Sand @ $21.35 per C.Y.	10 C.F.	7.91	8.70	20 C.F.	15.81	17.39
Div. strips, 4' sq. @ $.85 per L.F.	55 L.F.	46.75	51.43	55 L.F.	46.75	51.43
Terrazzo fill @ $12.00 per 50 lb. bag	12 bags	144.00	158.40	12 bags	144.00	158.40
15 lb. tarred felt @ $3.66 per C.S.F.	—	—	—	1 C.S.F.	3.66	4.03
Mesh 2 x 2 #14 @ $28.00 per C.S.F.	—	—	—	1 C.S.F.	28.00	30.80
Crew J-3 @ $679.60 & $921.35 per day	0.77 days	523.29	709.44	0.87 days	591.25	801.57
Total per 100 S.F. in place	Bonded	$771.45	$982.42	Not Bonded	$895.47	$1,136.22

2' x 2' panels require 1 L.F. divider strip per S.F.; 6' x 6' panels need .33 L.F. per S.F.

R09700-700 Wall Covering

Quantities for 100 S.F.		Medium Price Paper			Expensive Paper		
		Quantities	Bare Cost	Incl. O & P	Quantities	Bare Cost	Incl. O & P
Paper @ $30.00 and $57.00 per double roll		1.6 dbl. rolls	$ 48.00	$ 52.80	1.6 dbl. rolls	$ 91.20	$100.32
Wall sizing @ $16.50 per gallon		0.25 gallon	4.13	4.54	0.25 gallon	4.13	4.54
Vinyl wall paste @ $ 8.60 per gallon		0.6 gallon	5.16	5.68	0.6 gallon	5.16	5.68
Apply sizing @ $30.30 and $45.55 per hour		0.3 hour	9.09	13.67	0.3 hour	9.09	13.67
Apply paper @ $30.30 and $45.55 per hour		1.2 hours	36.36	54.66	1.5 hours	45.45	68.33
Total cost in place per 100 S.F.			$102.74	$131.35		$155.03	$192.54
Total cost in place per double roll			$ 64.21	$ 82.09		$ 96.89	$120.34

Most wallpapers now come in double rolls only.
To remove old paper, allow 1.3 hours per 100 S.F.

R09910-220 Painting

Item	Coat	One Gallon Covers			In 8 Hours a Laborer Covers			Labor-Hours per 100 S.F.		
		Brush	Roller	Spray	Brush	Roller	Spray	Brush	Roller	Spray
Paint wood siding	prime	250 S.F.	225 S.F.	290 S.F.	1150 S.F.	1300 S.F.	2275 S.F.	.695	.615	.351
	others	270	250	290	1300	1625	2600	.615	.492	.307
Paint exterior trim	prime	400	—	—	650	—	—	1.230	—	—
	1st	475	—	—	800	—	—	1.000	—	—
	2nd	520	—	—	975	—	—	.820	—	—
Paint shingle siding	prime	270	255	300	650	975	1950	1.230	.820	.410
	others	360	340	380	800	1150	2275	1.000	.695	.351
Stain shingle siding	1st	180	170	200	750	1125	2250	1.068	.711	.355
	2nd	270	250	290	900	1325	2600	.888	.603	.307
Paint brick masonry	prime	180	135	160	750	800	1800	1.066	1.000	.444
	1st	270	225	290	815	975	2275	.981	.820	.351
	2nd	340	305	360	815	1150	2925	.981	.695	.273
Paint interior plaster or drywall	prime	400	380	495	1150	2000	3250	.695	.400	.246
	others	450	425	495	1300	2300	4000	.615	.347	.200
Paint interior doors and windows	prime	400	—	—	650	—	—	1.230	—	—
	1st	425	—	—	800	—	—	1.000	—	—
	2nd	450	—	—	975	—	—	.820	—	—

R13011-110 Air Supported Structures

Air supported structures are made from fabrics that can be classified into two groups: temporary and permanent. Temporary fabrics include nylon, woven polyethylene, vinyl film, and vinyl coated dacron. These have lifespans that range from five to fifteen plus years. The cost per square foot includes a fabric shell, tension cables, primary and back-up inflation systems and doors. The lower cost structures are used for construction shelters, bulk storage and pond covers. The more expensive are used for recreational structures and warehouses.

Permanent fabrics are teflon coated fiberglass. The life of this structure is twenty plus years. The high cost limits its application to architectural designed structures which call for a clear span covered area, such as stadiums and convention centers. Both temporary and permanent structures are available in translucent fabrics which eliminates the need for daytime lighting.

Areas to be covered vary from 10,000 S.F. to any area up to 1000 foot wide by any length. Height restrictions range from a maximum of 1/2 of width to a minimum of 1/6 of the width. Erection of even the largest of the temporary structures requires no more than a week.

Centrifugal fans provide the inflation necessary to support the structure during application of live loads. Airlocks are usually used at large entrances to prevent loss of static pressure. Some manufacturers employ propeller fans which generate sufficient airflow (30,000 CFM) to eliminate the need for airlocks. These fans may also be automatically controlled to resist high wind conditions, regulate humidity (air changes), and provide cooling and heat.

Insulation can be provided with the addition of a second or even third interior liner, creating a dead air space with an "R" value of four to nine. Some structures allow for the liner to be collapsed into the outer shell to enable the internal heat to melt accumulated snow. For cooling or air conditioning, the exterior face of the liner can be aluminized to reflect the sun's heat.

R13128-210 Pre-engineered Steel Buildings

These buildings are manufactured by many companies and normally erected by franchised dealers throughout the U.S. The four basic types are: Rigid Frames, Truss type, Post and Beam and the Sloped Beam type. Most popular roof slope is low pitch of 1″ in 12″. The minimum economical area of these buildings is about 3000 S.F. of floor area. Bay sizes are usually 20′ to 24′ but can go as high as 30′ with heavier girts and purlins. Eave heights are usually 12′ to 24′ with 18′ to 20′ most typical.

Material prices shown in Division 13128-700 are bare costs for the building shell only and do not include floors, foundations, anchor bolts, interior finishes or utilities. Costs assume at least three bays of 24′ each, a 1″ in 12″ roof slope, and they are based on 30 psf roof load and 20 psf wind load (wind load is a function of wind speed, building height, and terrain characteristics; this should be determined by a Registered Structural Engineer) and no unusual requirements. Costs include the structural frame, 26 ga. non-insulated colored corrugated or ribbed roofing and siding panels, fasteners, closures, trim and flashing but no allowance for insulation, doors, windows, skylights, gutters or downspouts. Very large projects would generally cost less for materials than the prices shown. For roof panel substitutions, see Section 07410-700. For wall panel substitutions, see Section 07460-800.

Conditions at the site, weather, shape and size of the building, and labor availability will affect the erection cost of the building.

R13128-310 Dome Structures

Steel — The four types are Lamella, Schwedler, Arch and Geodesic. For maximum economy, rise should be about 15 to 20% of diameter. Most common diameters are in the 200′ to 300′ range. Lamella domes weigh about 5 P.S.F. of floor area less than Schwedler domes. Schwedler dome weight in lbs. per S.F. approaches .046 times the diameter. Domes below 125′ diameter weigh .07 times diameter and the cost per ton of steel is higher. See R05120-220 for estimating weight.

Wood — Small domes are of sawn lumber, larger ones are laminated. In larger sizes, triaxial and triangular cost about the same; radial domes cost more. Radial domes are economical in the 60′ to 70′ diameter range. Most economical range of all types is 80′ to 200′ diameters. Diameters can run over 400′. All costs are quoted above the foundation. Prices include 2″ decking and a tension tie ring in place.

Plywood — Stock prefab geodesic domes are available with diameters from 24′ to 60′.

Fiberglass — Aluminum framed translucent sandwich panels with spans from 5′ to 45′ are commercially available.

Aluminum — Stressed skin aluminum panels form geodesic domes with spans ranging from 82′ to 232′. An aluminum space truss, triangulated or nontriangulated, with aluminum or clear acrylic closure panels can be used for clear spans of 40′ to 415′.

R13128-520 Swimming Pools

Pool prices given per square foot of surface area include pool structure, filter and chlorination equipment, pumps, related piping, ladders/steps, maintenance kit, skimmer and vacuum system. Decks and electrical service to equipment are not included.

Residential in-ground pool construction can be divided into two categories: vinyl lined and gunite. Vinyl lined pool walls are constructed of different materials including wood, concrete, plastic or metal. The bottom is often graded with sand over which the vinyl liner is installed. Vermiculite or soil cement bottoms may be substituted for an added cost.

Gunite pool construction is used both in residential and municipal installations. These structures are steel reinforced for strength and finished with a white cement limestone plaster.

Municipal pools will have a higher cost because plumbing codes require more expensive materials, chlorination equipment and higher filtration rates.

Municipal pools greater than 1,800 S.F. require gutter systems to control waves. This gutter may be formed into the concrete wall. Often a vinyl/stainless steel gutter or gutter/wall system is specified, which will raise the pool cost.

Competition pools usually require tile bottoms and sides with contrasting lane striping, which will also raise the pool cost.

SPECIAL CONSTRUCTION 13

REFERENCE NOS.

R13128-910 Public Garage Costs

Cars	Per Car	Levels	Per Car	Levels	Per Car	Levels	Per S.F.
200 to 400	$12,500	3	$13,100	6	$12,800	Surface	$34.50
400 to 600	12,800	4	13,100	7	13,300	2	35.25
600 to 800	13,400	5	12,800	8 to 12	13,900	3 or 4	43.15
800 to 1200	13,900					5 or 6	42.55
Average	$13,200	Average per Car (310 to 320 S.F. per car)			$13,200	7 or 8	44.75

The above costs refer to open wall construction. For enclosed structures, costs must be added for sprinkler and ventilation systems.

R13281-120 Asbestos Removal Process

Asbestos removal is accomplished by a specialty contractor who understands the federal and state regulations regarding the handling and disposal of the material. The process of asbestos removal is divided into many individual steps. An accurate estimate can be calculated only after all the steps have been priced.

The steps are generally as follows:

1. Obtain an asbestos abatement plan from an industrial hygienist.
2. Monitor the air quality in and around the removal area and along the path of travel between the removal area and transport area. This establishes the background contamination.
3. Construct a two part decontamination chamber at entrance to removal area.
4. Install a HEPA filter to create a negative pressure in the removal area.
5. Install wall, floor and ceiling protection as required by the plan, usually 2 layers of fireproof 6 mil polyethylene.
6. Industrial hygienist visually inspects work area to verify compliance with plan.
7. Provide temporary supports for conduit and piping affected by the removal process.
8. Proceed with asbestos removal and bagging process. Monitor air quality as described in Step #2. Discontinue operations when contaminate levels exceed applicable standards.
9. Document the legal disposal of materials in accordance with EPA standards.
10. Thoroughly clean removal area including all ledges, crevices and surfaces.
11. Post abatement inspection by industrial hygienist to verify plan compliance.
12. Provide a certificate from a licensed industrial hygienist attesting that contaminate levels are within acceptable standards before returning area to regular use.

R13281-460 Lead Paint Remediation Methods

Lead paint remediation can be accomplished by the following methods.
1. Abrasive blast
2. Chemical stripping
3. Power tool cleaning with vacuum collection system
4. Encapsulation
5. Remove and replace
6. Enclosure

Each of these methods has strengths and weakness depending on the specific circumstances of the project. The following is an overview of each method.

1. **Abrasive blasting** is usually accomplished with sand or recyclable metallic blast. Before work can begin, the area must be contained to ensure the blast material with lead does not escape to the atmosphere. The use of vacuum blast greatly reduces the containment requirements. Lead abatement equipment that may be associated with this work includes a negative air machine. In addition, it is necessary to have an industrial hygienist monitor the project on a continual basis. When the work is complete, the spent blast sand with lead must be disposed of as a hazardous material. If metallic shot was used, the lead is separated from the shot and disposed of as hazardous material. Worker protection includes disposable clothing and respiratory protection.

2. **Chemical stripping** requires strong chemicals be applied to the surface to remove the lead paint. Before the work can begin, the area under/adjacent to the work area must be covered to catch the chemical and removed lead. After the chemical is applied to the painted surface it is usually covered with paper. The chemical is left in place for the specified period, then the paper with lead paint is pulled or scraped off. The process may require several chemical applications. The paper with chemicals and lead paint adhered to it, plus the containment and loose scrapings collected by a HEPA (High Efficiency Particulate Air Filter) vac, must be disposed of as a hazardous material. The chemical stripping process usually requires a neutralizing agent and several wash downs after the paint is removed. Worker protection includes a neoprene or other compatible protective clothing and respiratory protection with face shield. An industrial hygienist is required intermittently during the process.

3. **Power tool cleaning** is accomplished using shrouded needle blasting guns. The shrouding with different end configurations is held up against the surface to be cleaned. The area is blasted with hardened needles and the shroud captures the lead with a HEPA vac and deposits it in a holding tank. An industrial hygienist monitors the project, protective clothing and a respirator is required until air samples prove otherwise. When the work is complete the lead must be disposed of as a hazardous material.

4. **Encapsulation** is a method that leaves the well bonded lead paint in place after the peeling paint has been removed. Before the work can begin, the area under/adjacent to the work must be covered to catch the scrapings. The scraped surface is then washed with a detergent and rinsed. The prepared surface is covered with approximately 10 mils of paint. A reinforcing fabric can also be embedded in the paint covering. The scraped paint and containment must be disposed of as a hazardous material. Workers must wear protective clothing and respirators.

5. **Remove and replace** is an effective way to remove lead paint from windows, gypsum walls and concrete masonry surfaces. The painted materials are removed and new materials are installed. Workers should wear a respirator and tyvek suit. The demolished materials must be disposed of as hazardous waste if it fails the TCLP (Toxicity Characteristic Leachate Process) test.

6. **Enclosure** is the process that permanently seals lead painted materials in place. This process has many applications such as covering lead painted drywall with new drywall, covering exterior construction with tyvek paper then residing, or covering lead painted structural members with aluminum or plastic. The seams on all enclosing materials must be securely sealed. An industrial hygienist monitors the project, and protective clothing and a respirator is required until air samples prove otherwise.

All the processes require clearance monitoring and wipe testing as required by the hygienist.

SPECIAL CONSTRUCTION — REFERENCE NOS. — 13

R13600-610   Solar Heating (Space and Hot Water)

Collectors should face as close to due South as possible, however, variations of up to 20 degrees on either side of true South are acceptable. Local climate and collector type may influence the choice between east or west deviations. Obviously they should be located so they are not shaded from the sun's rays. Incline collectors at a slope of latitude minus 5 degrees for domestic hot water and latitude plus 15 degrees for space heating.

Flat plate collectors consist of a number of components as follows: Insulation to reduce heat loss through the bottom and sides of the collector. The enclosure which contains all the components in this assembly is usually weatherproof and prevents dust, wind and water from coming in contact with the absorber plate. The cover plate usually consists of one or more layers of a variety of glass or plastic and reduces the reradiation by creating an air space which traps the heat between the cover and the absorber plates.

The absorber plate must have a good thermal bond with the fluid passages. The absorber plate is usually metallic and treated with a surface coating which improves absorptivity. Black or dark paints or selective coatings are used for this purpose, and the design of this passage and plate combination helps determine a solar system's effectiveness.

Heat transfer fluid passage tubes are attached above and below or integral with an absorber plate for the purpose of transferring thermal energy from the absorber plate to a heat transfer medium. The heat exchanger is a device for transferring thermal energy from one fluid to another.

Piping and storage tanks should be well insulated to minimize heat losses.

Size domestic water heating storage tanks to hold 20 gallons of water per user, minimum, plus 10 gallons per dishwasher or washing machine. For domestic water heating an optimum collector size is approximately 3/4 square foot of area per gallon of water storage. For space heating of residences and small commercial applications the collector is commonly sized between 30% and 50% of the internal floor area. For space heating of large commercial applications, collector areas less than 30% of the internal floor area can still provide significant heat reductions.

A supplementary heat source is recommended for Northern states for December through February.

The solar energy transmission per square foot of collector surface varies greatly with the material used. Initial cost, heat transmittance and useful life are obviously interrelated.

SPECIAL CONSTRUCTION | **13**

REFERENCE NOS.

R14200-100   Freight Elevators

Capacities run from 2,000 lbs. to over 100,000 lbs. with 3,000 lbs. to 10,000 lbs. most common. Travel speeds are generally lower and control less intricate than on passenger elevators. Unit prices in division 14210-200 are for hydraulic and geared elevators.

R14200-200   Elevator Selective Costs   See R14200-400 for cost development.

	Passenger		Freight		Hospital	
A. Base Unit	**Hydraulic**	**Electric**	**Hydraulic**	**Electric**	**Hydraulic**	**Electric**
Capacity	1,500 lb.	2,000 lb.	2,000 lb.	4,000 lb.	4,000 lb.	4,000 lb.
Speed	100 F.P.M.	200 F.P.M.	100 F.P.M.	200 F.P.M.	100 F.P.M.	200 F.P.M.
#Stops/Travel Ft.	2/12	4/40	2/20	4/40	2/20	4/40
Push Button Oper.	Yes	Yes	Yes	Yes	Yes	Yes
Telephone Box & Wire	"	"	"	"	"	"
Emergency Lighting	"	"	No	No	"	"
Cab	Plastic Lam. Walls	Plastic Lam. Walls	Painted Steel	Painted Steel	Plastic Lam. Walls	Plastic Lam. Walls
Cove Lighting	Yes	Yes	No	No	Yes	Yes
Floor	V.C.T.	V.C.T.	Wood w/Safety Treads	Wood w/Safety Treads	V.C.T.	V.C.T.
Doors, & Speedside Slide	Yes	Yes	Yes	Yes	Yes	Yes
Gates, Manual	No	No	No	No	No	No
Signals, Lighted Buttons	Car and Hall	Car and Hall	Car and Hall	Car and Hall	Car and Hall	Car and Hall
O.H. Geared Machine	N.A.	Yes	N.A.	Yes	N.A.	Yes
Variable Voltage Contr.	"	"	N.A.	"	"	"
Emergency Alarm	Yes	"	Yes	"	Yes	"
Class "A" Loading	N.A.	N.A.	"	"	N.A.	N.A.

R14200-300   Passenger Elevators

Electric elevators are used generally but hydraulic elevators can be used for lifts up to 70′ and where large capacities are required. Hydraulic speeds are limited to 200 F.P.M. but cars are self leveling at the stops. On low rises, hydraulic installation runs about 15% less than standard electric types but on higher rises this installation cost advantage is reduced. Maintenance of hydraulic elevators is about the same as electric type but underground portion is not included in the maintenance contract.

In electric elevators there are several control systems available, the choice of which will be based upon elevator use, size, speed and cost criteria. The two types of drives are geared for low speeds and gearless for 450 F.P.M. and over.

The tables on the preceding pages illustrate typical installed costs of the various types of elevators available.

R14200-400   Elevator Cost Development

Requirement: One passenger elevator, five story hydraulic, 2,500 lb. capacity, 12′ floor to floor, speed 150 F.P.M., emergency power switching and maintenance contract.

Description	Adjustment		Unit Cost	Total Cost
A. Base Elevator, Hyd Pass, Base Unit, 1500 lb, 100 fpm, 2 Stops, Std. Finish				$43,500
B. Capacity Adjustment (2,500 lb.)	1	Ea.	$2,175	2,175
C. Excess Travel Over Base (4 x 12′) = 48′ — (12′ for Base Unit) =	36	V.L.F.	530	19,080
D. Stops Over Base 5 — (2 for Base Unit) =	3	Stops	4,550	13,650
E. Speed Adjustment (150 F.P.M.)	1	Ea.	2,475	2,475
F. Options:				
1. Intercom Service	1	Ea.	705	705
2. Emergency Power Switching, Automatic	1	Ea.	2,575	2,575
3. Stainless Steel Entrance Doors	5	Ea.	2,175	10,875
4. Maintenance Contract (12 Months)	1	Ea.	2,650	2,650
5. Position Indicators	2	Ea.	370	740
Total Cost				$98,425

R14320-100   Escalators

Moving stairs can be used for buildings where 600 or more people are to be carried to the second floor or beyond. Freight cannot be carried on escalators and at least one elevator must be available for this function.

Carrying capacity is 5,000 to 8,000 people per hour. Power requirement is 2 to 3 KW per hour and incline angle is 30°.

R14360-200   Moving Ramps and Walks

These are a specialized form of conveyor 3' to 6' wide with capacities of 3,600 to 18,000 persons per hour. Maximum speed is 140 F.P.M. and normal incline is 0° to 15°.

Local codes will determine the maximum angle. Outdoor units would require additional weather protection.

R15050-720   Demolition (Selective vs. Removal for Replacement)

Demolition can be divided into two basic categories.

One type of demolition involves the removal of material with no concern for its replacement. The labor-hours to estimate this work are found in Div. 15055 under "Selective Demolition". It is selective in that individual items or all the material installed as a system or trade grouping such as plumbing or heating systems are removed. This may be accomplished by the easiest way possible, such as sawing, torch cutting, or sledge hammer as well as simple unbolting.

The second type of demolition is the removal of some item for repair or replacement. This removal may involve careful draining, opening of unions, disconnecting and tagging of electrical connections, capping of pipes/ducts to prevent entry of debris or leakage of the material contained as well as transport of the item away from its in-place location to a truck/dumpster. An approximation of the time required to accomplish this type of demolition is to use half of the time indicated as necessary to install a new unit. For example; installation of a new pump might be listed as requiring 6 labor-hours so if we had to estimate the removal of the old pump we would allow an additional 3 hours for a total of 9 hours. That is, the complete replacement of a defective pump with a new pump would be estimated to take 9 labor-hours.

MECHANICAL 15

REFERENCE NOS.

R15100-050   Pipe Material Considerations

1. Malleable fittings should be used for gas service.
2. Malleable fittings are used where there are stresses/strains due to expansion and vibration.
3. Cast fittings may be broken as an aid to disassembling of heating lines frozen by long use, temperature and minerals.
4. Cast iron pipe is extensively used for underground and submerged service.
5. Type M (light wall) copper tubing is available in hard temper only and is used for nonpressure and less severe applications than K and L.
6. Type L (medium wall) copper tubing, available hard or soft for interior service.
7. Type K (heavy wall) copper tubing, available in hard or soft temper for use where conditions are severe. For underground and interior service.
8. Hard drawn tubing requires fewer hangers or supports but should not be bent. Silver brazed fittings are recommended, however soft solder is normally used.
9. Type DMV (very light wall) copper tubing designed for drainage, waste and vent plus other non-critical pressure services.

Domestic/Imported Pipe and Fittings Cost

The prices shown in this publication for steel/cast iron pipe and steel, cast iron, malleable iron fittings are based on domestic production sold at the normal trade discounts. The above listed items of foreign manufacture may be available at prices of 1/3 to 1/2 those shown. Some imported items after minor machining or finishing operations are being sold as domestic to further complicate the system.

Caution: Most pipe prices in this book also include a coupling and pipe hangers which for the larger sizes can add significantly to the per foot cost and should be taken into account when comparing "book cost" with quoted supplier's cost.

R15100-420   Plumbing Fixture Installation Time

Item	Rough-In	Set	Total Hours	Item	Rough-In	Set	Total Hours
Bathtub	5	5	10	Shower head only	2	1	3
Bathtub and shower, cast iron	6	6	12	Shower drain	3	1	4
Fire hose reel and cabinet	4	2	6	Shower stall, slate		15	15
Floor drain to 4 inch diameter	3	1	4	Slop sink	5	3	8
Grease trap, single, cast iron	5	3	8	Test 6 fixtures			14
Kitchen gas range		4	4	Urinal, wall	6	2	8
Kitchen sink, single	4	4	8	Urinal, pedestal or floor	6	4	10
Kitchen sink, double	6	6	12	Water closet and tank	4	3	7
Laundry tubs	4	2	6	Water closet and tank, wall hung	5	3	8
Lavatory wall hung	5	3	8	Water heater, 45 gals. gas, automatic	5	2	7
Lavatory pedestal	5	3	8	Water heaters, 65 gals. gas, automatic	5	2	7
Shower and stall	6	4	10	Water heaters, electric, plumbing only	4	2	6

Fixture prices in front of book are based on the cost per fixture set in place. The rough-in cost, which must be added for each fixture, includes carrier, if required, some supply, waste and vent pipe connecting fittings and stops. The lengths of rough-in pipe are nominal runs which would connect to the larger runs and stacks. The supply runs and DWV runs and stacks must be accounted for in separate entries. In the eastern half of the United States it is common for the plumber to carry these to a point 5′ outside the building.

R15100-430   Water Cooler Application

Type of Service	Requirement
Office, School or Hospital	12 persons per gallon per hour
Office, Lobby or Department Store	4 or 5 gallons per hour per fountain
Light manufacturing	7 persons per gallon per hour
Heavy manufacturing	5 persons per gallon per hour
Hot heavy manufacturing	4 persons per gallon per hour
Hotel	.08 gallons per hour per room
Theatre	1 gallon per hour per 100 seats

Mechanical | **R155** | **Heating**

R15500-035   Heating (42° degrees latitude)

$$\text{Approximate S.F. radiation} = \frac{\text{S.F. sash}}{2} + \frac{\text{S.F. wall} + \text{roof} - \text{sash}}{20} + \frac{\text{C.F. building}}{200}$$

15
MECHANICAL

REFERENCE NOS.

R15500-050   Factor for Determining Heat Loss for Various Types of Buildings

General: While the most accurate estimates of heating requirements would naturally be based on detailed information about the building being considered, it is possible to arrive at a reasonable approximation using the following procedure:

1. Calculate the cubic volume of the room or building.
2. Select the appropriate factor from Table 1 below. Note that the factors apply only to inside temperatures listed in the first column and to 0°F outside temperature.
3. If the building has bad north and west exposures, multiply the heat loss factor by 1.1.
4. If the outside design temperature is other than 0°F, multiply the factor from Table 1 by the factor from Table 2.
5. Multiply the cubic volume by the factor selected from Table 1. This will give the estimated BTUH heat loss which must be made up to maintain inside temperature.

Table 1 — Building Type	Conditions	Qualifications	Loss Factor*
Factories & Industrial Plants General Office Areas at 70°F	One Story	Skylight in Roof	6.2
		No Skylight in Roof	5.7
	Multiple Story	Two Story	4.6
		Three Story	4.3
		Four Story	4.1
		Five Story	3.9
		Six Story	3.6
	All Walls Exposed	Flat Roof	6.9
		Heated Space Above	5.2
	One Long Warm Common Wall	Flat Roof	6.3
		Heated Space Above	4.7
	Warm Common Walls on Both Long Sides	Flat Roof	5.8
		Heated Space Above	4.1
Warehouses at 60°F	All Walls Exposed	Skylights in Roof	5.5
		No Skylight in Roof	5.1
		Heated Space Above	4.0
	One Long Warm Common Wall	Skylight in Roof	5.0
		No Skylight in Roof	4.9
		Heated Space Above	3.4
	Warm Common Walls on Both Long Sides	Skylight in Roof	4.7
		No Skylight in Roof	4.4
		Heated Space Above	3.0

*Note: This table tends to be conservative particularly for new buildings designed for minimum energy consumption.

Table 2 — Outside Design Temperature Correction Factor (for Degrees Fahrenheit)									
Outside Design Temperature	50	40	30	20	10	0	-10	-20	-30
Correction Factor	.29	.43	.57	.72	.86	1.00	1.14	1.28	1.43

R15700-020 Air Conditioning Requirements

BTU's per hour per S.F. of floor area and S.F. per ton of air conditioning.

Type of Building	BTU per S.F.	S.F. per Ton	Type of Building	BTU per S.F.	S.F. per Ton	Type of Building	BTU per S.F.	S.F. per Ton
Apartments, Individual	26	450	Dormitory, Rooms	40	300	Libraries	50	240
Corridors	22	550	Corridors	30	400	Low Rise Office, Exterior	38	320
Auditoriums & Theaters	40	300/18*	Dress Shops	43	280	Interior	33	360
Banks	50	240	Drug Stores	80	150	Medical Centers	28	425
Barber Shops	48	250	Factories	40	300	Motels	28	425
Bars & Taverns	133	90	High Rise Office—Ext. Rms.	46	263	Office (small suite)	43	280
Beauty Parlors	66	180	Interior Rooms	37	325	Post Office, Individual Office	42	285
Bowling Alleys	68	175	Hospitals, Core	43	280	Central Area	46	260
Churches	36	330/20*	Perimeter	46	260	Residences	20	600
Cocktail Lounges	68	175	Hotel, Guest Rooms	44	275	Restaurants	60	200
Computer Rooms	141	85	Corridors	30	400	Schools & Colleges	46	260
Dental Offices	52	230	Public Spaces	55	220	Shoe Stores	55	220
Dept. Stores, Basement	34	350	Industrial Plants, Offices	38	320	Shop'g. Ctrs., Supermarkets	34	350
Main Floor	40	300	General Offices	34	350	Retail Stores	48	250
Upper Floor	30	400	Plant Areas	40	300	Specialty	60	200

*Persons per ton
12,000 BTU = 1 ton of air conditioning

R15700-040 Recommended Ventilation Air Changes

Table below lists range of time in minutes per change for various types of facilities.

Assembly Halls	2-10	Dance Halls	2-10	Laundries	1-3
Auditoriums	2-10	Dining Rooms	3-10	Markets	2-10
Bakeries	2-3	Dry Cleaners	1-5	Offices	2-10
Banks	3-10	Factories	2-5	Pool Rooms	2-5
Bars	2-5	Garages	2-10	Recreation Rooms	2-10
Beauty Parlors	2-5	Generator Rooms	2-5	Sales Rooms	2-10
Boiler Rooms	1-5	Gymnasiums	2-10	Theaters	2-8
Bowling Alleys	2-10	Kitchens-Hospitals	2-5	Toilets	2-5
Churches	5-10	Kitchens-Restaurant	1-3	Transformer Rooms	1-5

CFM air required for changes = Volume of room in cubic feet ÷ Minutes per change.

R15810-050   Ductwork

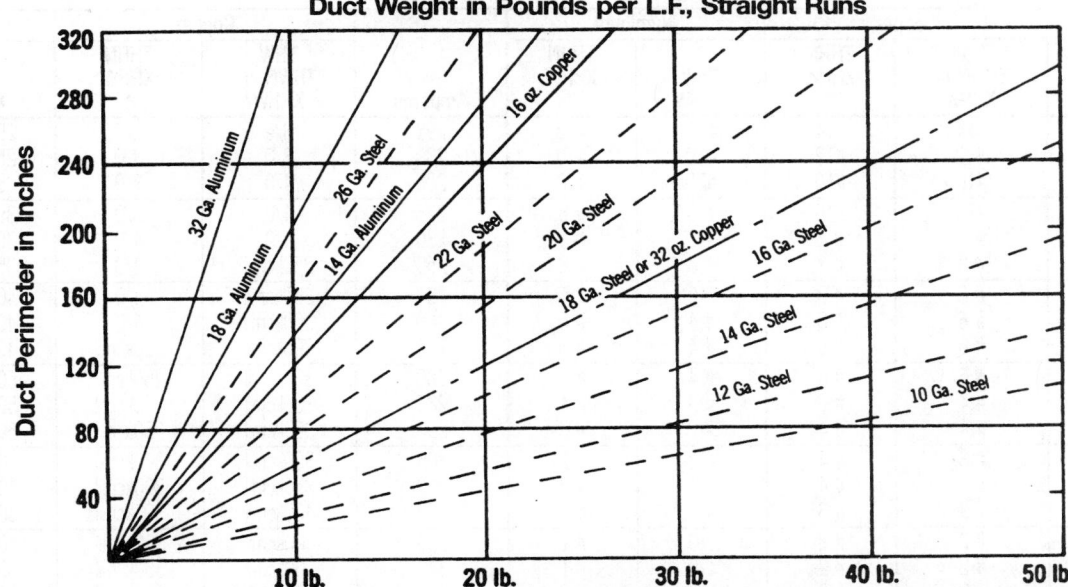

Duct Weight in Pounds per L.F., Straight Runs

Add to the above for fittings; 90° elbow is 3 L.F.; 45° elbow is 2.5 L.F.; offset is 4 L.F.; transition offset is 6 L.F.; square-to-round transition is 4 L.F.; 90° reducing elbow is 5 L.F. For bracing and waste, add 20% to aluminum and copper, 15% to steel.

R16120-910 Minimum Copper and Aluminum Wire Size Allowed for Various Types of Insulation

	Minimum Wire Sizes								
	Copper		Aluminum			Copper		Aluminum	
Amperes	THW THWN or XHHW	THHN XHHW *	THW XHHW	THHN XHHW *	Amperes	THW THWN or XHHW	THHN XHHW *	THW XHHW	THHN XHHW *
15A	#14	#14	#12	#12	195	3/0	2/0	250kcmil	4/0
20	#12	#12	#10	#10	200	3/0	3/0	250kcmil	4/0
25	#10	#10	#10	#10	205	4/0	3/0	250kcmil	4/0
30	#10	#10	#8	#8	225	4/0	3/0	300kcmil	250kcmil
40	#8	#8	#8	#8	230	4/0	4/0	300kcmil	250kcmil
45	#8	#8	#6	#8	250	250kcmil	4/0	350kcmil	300kcmil
50	#8	#8	#6	#6	255	250kcmil	4/0	400kcmil	300kcmil
55	#6	#8	#4	#6	260	300kcmil	4/0	400kcmil	350kcmil
60	#6	#6	#4	#6	270	300kcmil	250kcmil	400kcmil	350kcmil
65	#6	#6	#4	#4	280	300kcmil	250kcmil	500kcmil	350kcmil
75	#4	#6	#3	#4	285	300kcmil	250kcmil	500kcmil	400kcmil
85	#4	#4	#2	#3	290	350kcmil	250kcmil	500kcmil	400kcmil
90	#3	#4	#2	#2	305	350kcmil	300kcmil	500kcmil	400kcmil
95	#3	#4	#1	#2	310	350kcmil	300kcmil	500kcmil	500kcmil
100	#3	#3	#1	#2	320	400kcmil	300kcmil	600kcmil	500kcmil
110	#2	#3	1/0	#1	335	400kcmil	350kcmil	600kcmil	500kcmil
115	#2	#2	1/0	#1	340	500kcmil	350kcmil	600kcmil	500kcmil
120	#1	#2	1/0	1/0	350	500kcmil	350kcmil	700kcmil	500kcmil
130	#1	#2	2/0	1/0	375	500kcmil	400kcmil	700kcmil	600kcmil
135	1/0	#1	2/0	1/0	380	500kcmil	400kcmil	750kcmil	600kcmil
150	1/0	#1	3/0	2/0	385	600kcmil	500kcmil	750kcmil	600kcmil
155	2/0	1/0	3/0	3/0	420	600kcmil	500kcmil		700kcmil
170	2/0	1/0	4/0	3/0	430		500kcmil		750kcmil
175	2/0	2/0	4/0	3/0	435		600kcmil		750kcmil
180	3/0	2/0	4/0	4/0	475		600kcmil		

*Dry Locations Only

Notes:

1. Size #14 to 4/0 is in AWG units (American Wire Gauge).
2. Size 250 to 750 is in kcmil units (Thousand Circular Mils).
3. Use next higher ampere value if exact value is not listed in table.
4. For loads that operate continuously increase ampere value by 25% to obtain proper wire size.
5. Refer to Table R16120-905 for the maximum circuit length for the various size wires.
6. Table R16120-910 has been written for estimating purpose only, based on ambient temperature of 30 °C (86°F); for ambient temperature other than 30° C (86° F), ampacity correction factors will be applied.

R16132-210 Conductors in Conduit

Table below lists maximum number of conductors for various sized conduit using THW, TW or THWN insulations.

Copper Wire Size	1/2"			3/4"			1"			1-1/4"			1-1/2"			2"			2-1/2"			3"		3-1/2"		4"	
	TW	THW	THWN	TW	THW	THWN	TW	THW	THWN	TW	THW	THWN	TW	THW	THWN	TW	THW	THWN	TW	THW	THWN	THW	THWN	THW	THWN	THW	THWN
#14	9	6	13	15	10	24	25	16	39	44	29	69	60	40	94	99	65	154	142	93		143		192			
#12	7	4	10	12	8	18	19	13	29	35	24	51	47	32	70	78	53	114	111	76	164	117		157			
#10	5	4	6	9	6	11	15	11	18	26	19	32	36	26	44	60	43	73	85	61	104	95	160	127		163	
#8	2	1	3	4	3	5	7	5	9	12	10	16	17	13	22	28	22	36	40	32	51	49	79	66	106	85	136
#6		1	1		2	4		4	6		7	11		10	15		16	26		23	37	36	57	48	76	62	98
#4		1	1		1	2		3	4		5	7		7	9		12	16		17	22	27	35	36	47	47	60
#3		1	1		1	1		2	3		4	6		6	8		10	13		15	19	23	29	31	39	40	51
#2		1	1		1	1		2	3		4	5		5	7		9	11		13	16	20	25	27	33	34	43
#1					1	1		1	1		3	3		4	5		6	8		9	12	14	18	19	25	25	32
1/0					1	1		1	1		2	3		3	4		5	7		8	10	12	15	16	21	21	27
2/0					1	1		1	1		1	2		3	3		5	6		7	8	10	13	14	17	18	22
3/0					1	1		1	1		1	1		2	3		4	5		6	7	9	11	12	14	15	18
4/0						1		1	1		1	1		1	2		3	4		5	6	7	9	10	12	13	15
250 kcmil								1	1		1	1		1	1		2	3		4	4	6	7	8	10	10	12
300								1	1		1	1		1	1		2	3		3	4	5	6	7	8	9	11
350									1		1	1		1	1		1	2		3	3	4	5	6	7	8	9
400											1	1		1	1		1	1		2	3	4	5	5	6	7	8
500											1	1		1	1		1	1		1	2	3	4	4	5	6	7
600												1		1	1		1	1		1	1	3	3	4	4	5	5
700														1	1		1	1		1	1	2	3	3	4	4	5
750														1	1		1	1		1	1	2	2	3	3	4	4

R16220-610 Motors

230/460 Volt A.C., three phase 60 cycle ball bearing squirrel cage
induction motors, NEMA design B Standard Line, including
installation and Subs O & P. (No conduit, wire, or terminations)

Horsepower	Synchronous Speed in RPM	Drip Proof, Premium Efficiency Grade 1.15 Service Factor			Totally Enclosed, Premium Efficiency Grade 1.0 or 1.15 Service Factor		
		Motor Only	With Manual Starter	With Magnetic Starter	Motor Only	With Manual Starter	With Magnetic Starter
1	1,200	$ 356	$ 662	$ 687	$ 407	$ 712	$ 737
	1,800	321	627	652	364	672	692
2	1,200	458	767	900	473	473	915
	1,800	359	667	800	413	722	855
	3,600	362	667	805	404	712	845
3	1,200	578	887	1,020	593	902	1,035
	1,800	360	667	800	448	757	890
	3,600	398	707	840	453	762	895
5	1,200	673	1,115	1,240	823	1,265	1,390
	1,800	433	875	1,000	498	940	1,065
	3,600	433	875	1,000	538	980	1,105
7.5	1,800	586	1,030	1,510	676	1,120	1,605
10		706	1,150	1,635	796	1,240	1,710
15		892	—	1,815	1,057	—	1,990
20		1,112	—	2,450	1,287	—	2,625
25		1,269	—	2,610	1,544	—	2,885
30		1,452	—	2,790	1,777	—	3,115
40		2,118	—	4,675	2,268	—	4,825
50		2,430	—	5,000	2,805	—	5,375
60		3,120	—	8,300	4,095	—	9,275
75		3,555	—	8,725	5,180	—	10,350
100		4,690	—	9,875	6,215	—	11,400
125		5,520	—	18,625	8,945	—	22,125
150		6,960	—	20,125	10,160	—	23,325
200		8,920	—	22,100	12,170	—	25,300

R17100-100 Square Foot Project Size Modifier

One factor that affects the S.F. cost of a particular building is the size. In general, for buildings built to the same specifications in the same locality, the larger building will have the lower S.F. cost. This is due mainly to the decreasing contribution of the exterior walls plus the economy of scale usually achievable in larger buildings. The Area Conversion Scale shown below will give a factor to convert costs for the typical size building to an adjusted cost for the particular project.

The Square Foot Base Size lists the median costs, most typical project size in our accumulated data and the range in size of the projects.

The Size Factor for your project is determined by dividing your project area in S.F. by the typical project size for the particular Building Type. With this factor, enter the Area Conversion Scale at the appropriate Size Factor and determine the appropriate cost multiplier for your building size.

Example: Determine the cost per S.F. for a 100,000 S.F. Mid-rise apartment building.

$$\frac{\text{Proposed building area} = 100,000 \text{ S.F.}}{\text{Typical size from below} = 50,000 \text{ S.F.}} = 2.00$$

Enter Area Conversion scale at 2.0, intersect curve, read horizontally the appropriate cost multiplier of .94. Size adjusted cost becomes .94 x $85.50 = $80.35 based on national average costs.

Note: For Size Factors less than .50, the Cost Multiplier is 1.1
For Size Factors greater than 3.5, the Cost Multiplier is .90

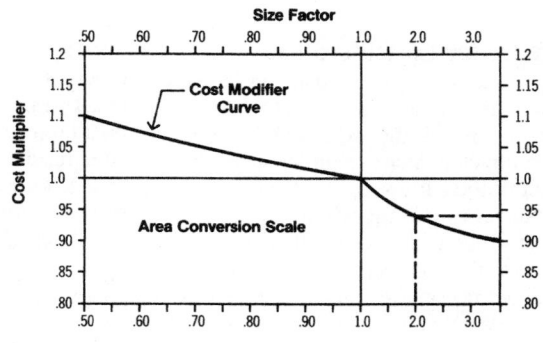

Square Foot Base Size							
Building Type	**Median Cost per S.F.**	**Typical Size Gross S.F.**	**Typical Range Gross S.F.**	**Building Type**	**Median Cost per S.F.**	**Typical Size Gross S.F.**	**Typical Range Gross S.F.**
Apartments, Low Rise	$ 67.50	21,000	9,700 - 37,200	Jails	$205.00	40,000	5,500 - 145,000
Apartments, Mid Rise	85.50	50,000	32,000 - 100,000	Libraries	126.00	12,000	7,000 - 31,000
Apartments, High Rise	97.00	145,000	95,000 - 600,000	Living, Assisted	111.00	32,300	23,500 - 50,300
Auditoriums	113.00	25,000	7,600 - 39,000	Medical Clinics	117.00	7,200	4,200 - 15,700
Auto Sales	84.00	20,000	10,800 - 28,600	Medical Offices	110.00	6,000	4,000 - 15,000
Banks	151.00	4,200	2,500 - 7,500	Motels	82.50	40,000	15,800 - 120,000
Churches	103.00	17,000	2,000 - 42,000	Nursing Homes	114.00	23,000	15,000 - 37,000
Clubs, Country	106.00	6,500	4,500 - 15,000	Offices, Low Rise	95.00	20,000	5,000 - 80,000
Clubs, Social	100.00	10,000	6,000 - 13,500	Offices, Mid Rise	95.00	120,000	20,000 - 300,000
Clubs, YMCA	119.00	28,300	12,800 - 39,400	Offices, High Rise	122.00	260,000	120,000 - 800,000
Colleges (Class)	131.00	50,000	15,000 - 150,000	Police Stations	152.00	10,500	4,000 - 19,000
Colleges (Science Lab)	193.00	45,600	16,600 - 80,000	Post Offices	113.00	12,400	6,800 - 30,000
College (Student Union)	148.00	33,400	16,000 - 85,000	Power Plants	835.00	7,500	1,000 - 20,000
Community Center	107.00	9,400	5,300 - 16,700	Religious Education	94.00	9,000	6,000 - 12,000
Court Houses	144.00	32,400	17,800 - 106,000	Research	158.00	19,000	6,300 - 45,000
Dept. Stores	62.50	90,000	44,000 - 122,000	Restaurants	137.00	4,400	2,800 - 6,000
Dormitories, Low Rise	111.00	25,000	10,000 - 95,000	Retail Stores	67.00	7,200	4,000 - 17,600
Dormitories, Mid Rise	141.00	85,000	20,000 - 200,000	Schools, Elementary	99.50	41,000	24,500 - 55,000
Factories	61.00	26,400	12,900 - 50,000	Schools, Jr. High	102.00	92,000	52,000 - 119,000
Fire Stations	107.00	5,800	4,000 - 8,700	Schools, Sr. High	108.00	101,000	50,500 - 175,000
Fraternity Houses	104.00	12,500	8,200 - 14,800	Schools, Vocational	100.00	37,000	20,500 - 82,000
Funeral Homes	117.00	10,000	4,000 - 20,000	Sports Arenas	82.50	15,000	5,000 - 40,000
Garages, Commercial	76.00	9,300	5,000 - 13,600	Supermarkets	66.00	44,000	12,000 - 60,000
Garages, Municipal	95.00	8,300	4,500 - 12,600	Swimming Pools	155.00	20,000	10,000 - 32,000
Garages, Parking	40.00	163,000	76,400 - 225,300	Telephone Exchange	180.00	4,500	1,200 - 10,600
Gymnasiums	102.00	19,200	11,600 - 41,000	Theaters	99.00	10,500	8,800 - 17,500
Hospitals	181.00	55,000	27,200 - 125,000	Town Halls	116.00	10,800	4,800 - 23,400
House (Elderly)	92.00	37,000	21,000 - 66,000	Warehouses	49.50	25,000	8,000 - 72,000
Housing (Public)	85.00	36,000	14,400 - 74,400	Warehouse & Office	52.50	25,000	8,000 - 72,000
Ice Rinks	122.00	29,000	27,200 - 33,600				

SQUARE FOOT 17

REFERENCE NOS.

Change Orders

Change Order Considerations

A Change Order is a written document, usually prepared by the design professional, and signed by the owner, the architect/engineer and the contractor. A change order states the agreement of the parties to: an addition, deletion, or revision in the work; an adjustment in the contract sum, if any; or an adjustment in the contract time, if any. Change orders, or "extras" in the construction process occur after execution of the construction contract and impact architects/engineers, contractors and owners.

Change orders that are properly recognized and managed can ensure orderly, professional and profitable progress for all who are involved in the project. There are many causes for change orders and change order requests. In all cases, change orders or change order requests should be addressed promptly and in a precise and prescribed manner. The following paragraphs include information regarding change order pricing and procedures.

The Causes of Change Orders

Reasons for issuing change orders include:

- Unforeseen field conditions that require a change in the work
- Correction of design discrepancies, errors or omissions in the contract documents
- Owner-requested changes, either by design criteria, scope of work, or project objectives
- Completion date changes for reasons unrelated to the construction process
- Changes in building code interpretations, or other public authority requirements that require a change in the work
- Changes in availability of existing or new materials and products

Procedures

Properly written contract documents must include the correct change order procedures for all parties—owners, design professionals and contractors—to follow in order to avoid costly delays and litigation.

Being "in the right" is not always a sufficient or acceptable defense. The contract provisions requiring notification and documentation must be adhered to within a defined or reasonable time frame.

The appropriate method of handling change orders is by a written proposal and acceptance by all parties involved. Prior to starting work on a project, all parties should identify their authorized agents who may sign and accept change orders, as well as any limits placed on their authority.

Time may be a critical factor when the need for a change arises. For such cases, the contractor might be directed to proceed on a "time and materials" basis, rather than wait for all paperwork to be processed—a delay that could impede progress. In this situation, the contractor must still follow the prescribed change order procedures, including but not limited to, notification and documentation.

All forms used for change orders should be dated and signed by the proper authority. Lack of documentation can be very costly, especially if legal judgments are to be made and if certain field personnel are no longer available. For time and material change orders, the contractor should keep accurate daily records of all labor and material allocated to the change. Forms that can be used to document change order work are available in *Means Forms for Building Construction Professionals.*

Owners or awarding authorities who do considerable and continual building construction (such as the federal government) realize the inevitability of change orders for numerous reasons, both predictable and unpredictable. As a result, the federal government, the American Institute of Architects (AIA), the Engineers Joint Contract Documents Committee (EJCDC) and other contractor, legal and technical organizations have developed standards and procedures to be followed by all parties to achieve contract continuance and timely completion, while being financially fair to all concerned.

In addition to the change order standards put forth by industry associations, there are also many books available on the subject.

Pricing Change Orders

When pricing change orders, regardless of their cause, the most significant factor is *when* the change occurs. The need for a change may be perceived in the field or requested by the architect/engineer *before* any of the actual installation has begun, or may evolve or appear *during* construction when the item of work in question is partially installed. In the latter cases, the original sequence of construction is disrupted, along with all contiguous and supporting systems. Change orders cause the greatest impact when they occur *after* the installation has been completed and must be uncovered, or even replaced. Post-completion changes may be caused by necessary design changes, product failure, or changes in the owner's requirements that are not discovered until the building or the systems begin to function.

Specified procedures of notification and record keeping must be adhered to and enforced regardless of the stage of construction: *before, during,* or *after* installation. Some bidding documents anticipate change orders by requiring that unit prices including overhead and profit percentages—for additional as well as deductible changes—be listed. Generally these unit prices do not fully take into account the ripple effect, or impact on other trades, and should be used for general guidance only.

When pricing change orders, it is important to classify the time frame in which the change occurs. There are two basic time frames for change orders: *pre-installation change orders,* which occur before the start of construction, and *post-installation change orders,* which involve reworking after the original installation. Change orders that occur between these stages may be priced according to the extent of work completed using a combination of techniques developed for pricing *pre-* and *post-installation* changes.

The following factors are the basis for a check list to use when preparing a change order estimate.

Factors To Consider When Pricing Change Orders

As an estimator begins to prepare a change order, the following questions should be reviewed to determine their impact on the final price.

General

- Is the change order work *pre-installation* or *post-installation*?

  Change order work costs vary according to how much of the installation has been completed. Once workers have the project scoped in their mind, even though they have not started, it can be difficult to refocus.

Consequently they may spend more than the normal amount of time understanding the change. Also, modifications to work in place such as trimming or refitting usually take more time than was initially estimated. The greater the amount of work in place, the more reluctant workers are to change it. Psychologically they may resent the change and as a result the rework takes longer than normal. Post-installation change order estimates must include demolition of existing work as required to accomplish the change. If the work is performed at a later time, additional obstacles such as building finishes may be present which must be

protected. Regardless of whether the change occurs pre-installation or post-installation, attempt to isolate the identifiable factors and price them separately. For example, add shipping costs that may be required pre-installation or any demolition required post-installation. Then analyze the potential impact on productivity of psychological and/or learning curve factors and adjust the output rates accordingly. One approach is to break down the typical workday into segments and quantify the impact on each segment. The following chart may be useful as a guide:

	Activities (Productivity) Expressed as Percentages of a Workday		
Task	Means Mechanical Cost Data (for New Construction)	Pre-Installation Change Orders	Post-Installation Change Orders
1. Study plans	3%	6%	6%
2. Material procurement	3%	3%	3%
3. Receiving and storing	3%	3%	3%
4. Mobilization	5%	5%	5%
5. Site movement	5%	5%	8%
6. Layout and marking	8%	10%	12%
7. Actual installation	64%	59%	54%
8. Clean-up	3%	3%	3%
9. Breaks—non-productive	6%	6%	6%
Total	100%	100%	100%

Change Order Installation Efficiency

The labor-hours expressed (for new construction) are based on average installation time, using an efficiency level of approximately 60-65%. For change order situations, adjustments to this efficiency level should reflect the daily labor-hour allocation for that particular occurrence.

If any of the specific percentages expressed in the above chart do not apply to a particular project situation, then those percentage points should be reallocated to the appropriate task(s). Example: Using data for new construction, assume there is no new material being utilized. The percentages for Tasks 2 and 3 would therefore be reallocated to other tasks. If the time required for Tasks 2 and 3 can now be applied to installation, we can add the time allocated for *Material Procurement* and *Receiving and Storing* to the *Actual Installation* time for new construction, thereby increasing the Actual Installation percentage.

This chart shows that, due to reduced productivity, labor costs will be higher than those for new construction by 5% to 15% for pre-installation change orders and by 15% to 25% for post-installation change orders. Each job and change order is unique and must be examined individually. Many factors, covered elsewhere in this section, can each have a significant impact on productivity and change order costs. All such factors should be considered in every case.

- Will the change substantially delay the original completion date?

  A significant change in the project may cause the original completion date to be extended. The extended schedule may subject the contractor to new wage rates dictated by relevant labor contracts. Project supervision and other project overhead must also be extended beyond the original completion date. The schedule extension may also put installation into a new weather season. For example, underground piping scheduled for October installation was delayed until January. As a result, frost penetrated the trench area, thereby changing the degree of difficulty of the task. Changes and delays may have a ripple effect throughout the project. This effect must be analyzed and negotiated with the owner.

- What is the net effect of a deduct change order?

  In most cases, change orders resulting in a deduction or credit reflect only bare costs. The contractor may retain the overhead and profit based on the original bid.

Materials

- Will you have to pay more or less for the new material, required by the change order, than you paid for the original purchase?

  The same material prices or discounts will usually apply to materials purchased for change orders as new construction. In some instances, however, the contractor may forfeit the advantages of competitive pricing for change orders. Consider the following example:

  A contractor purchased over $20,000 worth of fan coil units for an installation, and obtained the maximum discount. Some time later it was determined the project required an additional matching unit. The contractor has to purchase this unit from the original supplier to ensure a match. The supplier at this time may not discount the unit because of the small quantity, and the fact that he is no longer in a competitive situation. The impact of quantity on purchase can add between 0% and 25% to material prices and/or subcontractor quotes.

- If materials have been ordered or delivered to the job site, will they be subject to a cancellation charge or restocking fee?

  Check with the supplier to determine if ordered materials are subject to a cancellation charge. Delivered materials not used as result of a change order may be subject to a restocking fee if returned to the supplier. Common restocking charges run between 20% and 40%. Also, delivery charges to return the goods to the supplier must be added.

Labor

- How efficient is the existing crew at the actual installation?

  Is the same crew that performed the initial work going to do the change order? Possibly the change consists of the installation of a unit identical to one already installed; therefore the change should take less time. Be sure to consider this potential productivity increase and modify the productivity rates accordingly.

- If the crew size is increased, what impact will that have on supervision requirements?

  Under most bargaining agreements or management practices, there is a point at which a working foreman is replaced by a nonworking foreman. This replacement increases project overhead by adding a nonproductive worker. If additional workers are added to accelerate the project or to perform changes while maintaining the schedule, be sure to add additional supervision time if warranted. Calculate the hours involved and the additional cost directly if possible.

- What are the other impacts of increased crew size?

  The larger the crew, the greater the potential for productivity to decrease. Some of the factors that cause this productivity loss are: overcrowding (producing restrictive conditions in the working space), and possibly a shortage of any special tools and equipment required. Such factors affect not only the crew working on the elements directly involved in the change order, but other crews whose movement may also be hampered.

As the crew increases, check its basic composition for changes by the addition or deletion of apprentices or nonworking foreman and quantify the potential effects of equipment shortages or other logistical factors.

- As new crews, unfamiliar with the project, are brought onto the site, how long will it take them to become oriented to the project requirements?

  The orientation time for a new crew to become 100% effective varies with the site and type of project. Orientation is easiest at a new construction site, and most difficult at existing, very restrictive renovation sites. The type of work also affects orientation time. When all elements of the work are exposed, such as concrete or masonry work, orientation is decreased. When the work is concealed or less visible, such as existing electrical systems, orientation takes longer. Usually orientation can be accomplished in one day or less. Costs for added orientation should be itemized and added to the total estimated cost.

- How much actual production can be gained by working overtime?

  Short term overtime can be used effectively to accomplish more work in a day. However, as overtime is scheduled to run beyond several weeks, studies have shown marked decreases in output. The following chart shows the effect of long term overtime on worker efficiency. If the anticipated change requires extended overtime to keep the job on schedule, these factors can be used as a guide to predict the impact on time and cost. Add project overhead, particularly supervision, that may also be incurred.

Days per Week	Hours per Day	Production Efficiency					Payroll Cost Factors	
		1 Week	2 Weeks	3 Weeks	4 Weeks	Average 4 Weeks	@ 1-1/2 Times	@ 2 Times
5	8	100%	100%	100%	100%	100%	100%	100%
	9	100	100	95	90	96.25	105.6	111.1
	10	100	95	90	85	91.25	110.0	120.0
	11	95	90	75	65	81.25	113.6	127.3
	12	90	85	70	60	76.25	116.7	133.3
6	8	100	100	95	90	96.25	108.3	116.7
	9	100	95	90	85	92.50	113.0	125.9
	10	95	90	85	80	87.50	116.7	133.3
	11	95	85	70	65	78.75	119.7	139.4
	12	90	80	65	60	73.75	122.2	144.4
7	8	100	95	85	75	88.75	114.3	128.6
	9	95	90	80	70	83.75	118.3	136.5
	10	90	85	75	65	78.75	121.4	142.9
	11	85	80	65	60	72.50	124.0	148.1
	12	85	75	60	55	68.75	126.2	152.4

Effects of Overtime

Caution: Under many labor agreements, Sundays and holidays are paid at a higher premium than the normal overtime rate.

The use of long-term overtime is counterproductive on almost any construction job; that is, the longer the period of overtime, the lower the actual production rate. Numerous studies have been conducted, and while they have resulted in slightly different numbers, all reach the same conclusion. The figure above tabulates the effects of overtime work on efficiency.

As illustrated, there can be a difference between the *actual* payroll cost per hour and the *effective* cost per hour for overtime work. This is due to the reduced production efficiency with the increase in weekly hours beyond 40. This difference between actual and effective cost results from overtime work over a prolonged period. Short-term overtime work does not result in as great a reduction in efficiency, and in such cases, effective cost may not vary significantly from the actual payroll cost. As the total hours per week are increased on a regular basis, more time is lost because of fatigue, lowered morale, and an increased accident rate.

As an example, assume a project where workers are working 6 days a week, 10 hours per day. From the figure above (based on productivity studies), the average effective productive hours over a four-week period are:

$$0.875 \times 60 = 52.5$$

Depending upon the locale and day of week, overtime hours may be paid at time and a half or double time. For time and a half, the overall (average) *actual* payroll cost (including regular and overtime hours) is determined as follows:

$$\frac{40 \text{ reg. hrs.} + (20 \text{ overtime hrs.} \times 1.5)}{60 \text{ hrs.}} = 1.167$$

Based on 60 hours, the payroll cost per hour will be 116.7% of the normal rate at 40 hours per week. However, because the effective production (efficiency) for 60 hours is reduced to the equivalent of 52.5 hours, the effective cost of overtime is calculated as follows:

For time and a half:

$$\frac{40 \text{ reg. hrs.} + (20 \text{ overtime hrs.} \times 1.5)}{52.5 \text{ hrs.}} = 1.33$$

Installed cost will be 133% of the normal rate (for labor).

Thus, when figuring overtime, the actual cost per unit of work will be higher than the apparent overtime payroll dollar increase, due to the reduced productivity of the longer workweek. These efficiency calculations are true only for those cost factors determined by hours worked. Costs that are applied weekly or monthly, such as equipment rentals, will not be similarly affected.

Equipment

- What equipment is required to complete the change order?

Change orders may require extending the rental period of equipment already on the job site, or the addition of special equipment brought in to accomplish the change work. In either case, the additional rental charges and operator labor charges must be added.

Summary

The preceding considerations and others you deem appropriate should be analyzed and applied to a change order estimate. The impact of each should be quantified and listed on the estimate to form an audit trail.

Change orders that are properly identified, documented, and managed help to ensure the orderly, professional and profitable progress of the work. They also minimize potential claims or disputes at the end of the project.

REFERENCE

Crew No.	Bare Costs Hr.	Daily	Incl. Subs O & P Hr.	Daily	Cost Per Labor-Hour Bare Costs	Incl. O&P
Crew A-1	Hr.	Daily	Hr.	Daily	Bare Costs	Incl. O&P
1 Building Laborer	$26.70	$213.60	$41.55	$332.40	$26.70	$41.55
1 Concrete saw, gas manual		49.80		54.80	6.23	6.85
8 L.H., Daily Totals		$263.40		$387.20	$32.93	$48.40
Crew A-1A	Hr.	Daily	Hr.	Daily	Bare Costs	Incl. O&P
1 Skilled Worker	$34.85	$278.80	$54.20	$433.60	$34.85	$54.20
1 Shot Blaster, 20"		338.35		372.20	42.29	46.52
8 L.H., Daily Totals		$617.15		$805.80	$77.14	$100.72
Crew A-1B	Hr.	Daily	Hr.	Daily	Bare Costs	Incl. O&P
1 Building Laborer	$26.70	$213.60	$41.55	$332.40	$26.70	$41.55
1 Concr. saw, gas, self-prop.		114.40		125.85	14.30	15.73
8 L.H., Daily Totals		$328.00		$458.25	$41.00	$57.28
Crew A-1C	Hr.	Daily	Hr.	Daily	Bare Costs	Incl. O&P
1 Building Laborer	$26.70	$213.60	$41.55	$332.40	$26.70	$41.55
1 Brush saw		19.20		21.10	2.40	2.64
8 L.H., Daily Totals		$232.80		$353.50	$29.10	$44.19
Crew A-1D	Hr.	Daily	Hr.	Daily	Bare Costs	Incl. O&P
1 Building Laborer	$26.70	$213.60	$41.55	$332.40	$26.70	$41.55
1 Vibrating plate, gas, 18"		25.20		27.70	3.15	3.47
8 L.H., Daily Totals		$238.80		$360.10	$29.85	$45.02
Crew A-1E	Hr.	Daily	Hr.	Daily	Bare Costs	Incl. O&P
1 Building Laborer	$26.70	$213.60	$41.55	$332.40	$26.70	$41.55
1 Vibrating plate, gas, 21"		36.80		40.50	4.60	5.06
8 L.H., Daily Totals		$250.40		$372.90	$31.30	$46.61
Crew A-1F	Hr.	Daily	Hr.	Daily	Bare Costs	Incl. O&P
1 Building Laborer	$26.70	$213.60	$41.55	$332.40	$26.70	$41.55
1 Rammer/tamper, gas, 8"		34.80		38.30	4.35	4.79
8 L.H., Daily Totals		$248.40		$370.70	$31.05	$46.34
Crew A-1G	Hr.	Daily	Hr.	Daily	Bare Costs	Incl. O&P
1 Building Laborer	$26.70	$213.60	$41.55	$332.40	$26.70	$41.55
1 Rammer/tamper, gas, 8"		34.80		38.30	4.35	4.79
8 L.H., Daily Totals		$248.40		$370.70	$31.05	$46.34
Crew A-1H	Hr.	Daily	Hr.	Daily	Bare Costs	Incl. O&P
1 Building Laborer	$26.70	$213.60	$41.55	$332.40	$26.70	$41.55
1 Pressure washer		55.00		60.50	6.88	7.56
8 L.H., Daily Totals		$268.60		$392.90	$33.58	$49.11
Crew A-1J	Hr.	Daily	Hr.	Daily	Bare Costs	Incl. O&P
1 Building Laborer	$26.70	$213.60	$41.55	$332.40	$26.70	$41.55
1 Rototiller		84.50		92.95	10.56	11.61
8 L.H., Daily Totals		$298.10		$425.35	$37.26	$53.16
Crew A-1K	Hr.	Daily	Hr.	Daily	Bare Costs	Incl. O&P
1 Building Laborer	$26.70	$213.60	$41.55	$332.40	$26.70	$41.55
1 Lawn aerator		23.95		26.35	2.99	3.29
8 L.H., Daily Totals		$237.55		$358.75	$29.69	$44.84
Crew A-1L	Hr.	Daily	Hr.	Daily	Bare Costs	Incl. O&P
1 Building Laborer	$26.70	$213.60	$41.55	$332.40	$26.70	$41.55
1 Power blower/vacuum		23.95		26.35	2.99	3.29
8 L.H., Daily Totals		$237.55		$358.75	$29.69	$44.84

Crew No.	Bare Costs Hr.	Daily	Incl. Subs O & P Hr.	Daily	Cost Per Labor-Hour Bare Costs	Incl. O&P
Crew A-1M	Hr.	Daily	Hr.	Daily	Bare Costs	Incl. O&P
1 Building Laborer	$26.70	$213.60	$41.55	$332.40	$26.70	$41.55
1 Snow blower		84.50		92.95	10.56	11.61
8 L.H., Daily Totals		$298.10		$425.35	$37.26	$53.16
Crew A-2	Hr.	Daily	Hr.	Daily	Bare Costs	Incl. O&P
2 Laborers	$26.70	$427.20	$41.55	$664.80	$26.73	$41.32
1 Truck Driver (light)	26.80	214.40	40.85	326.80		
1 Light Truck, 1.5 Ton		128.80		141.70	5.37	5.90
24 L.H., Daily Totals		$770.40		$1133.30	$32.10	$47.22
Crew A-2A	Hr.	Daily	Hr.	Daily	Bare Costs	Incl. O&P
2 Laborers	$26.70	$427.20	$41.55	$664.80	$26.73	$41.32
1 Truck Driver (light)	26.80	214.40	40.85	326.80		
1 Light Truck, 1.5 Ton		128.80		141.70		
1 Concrete Saw		114.40		125.85	10.13	11.15
24 L.H., Daily Totals		$884.80		$1259.15	$36.86	$52.47
Crew A-3	Hr.	Daily	Hr.	Daily	Bare Costs	Incl. O&P
1 Truck Driver (heavy)	$27.55	$220.40	$42.00	$336.00	$27.55	$42.00
1 Dump Truck, 12 Ton		326.60		359.25	40.83	44.91
8 L.H., Daily Totals		$547.00		$695.25	$68.38	$86.91
Crew A-3A	Hr.	Daily	Hr.	Daily	Bare Costs	Incl. O&P
1 Truck Driver (light)	$26.80	$214.40	$40.85	$326.80	$26.80	$40.85
1 Pickup Truck (4x4)		85.20		93.70	10.65	11.72
8 L.H., Daily Totals		$299.60		$420.50	$37.45	$52.57
Crew A-3B	Hr.	Daily	Hr.	Daily	Bare Costs	Incl. O&P
1 Equip. Oper. (medium)	$34.65	$277.20	$52.20	$417.60	$31.10	$47.10
1 Truck Driver (heavy)	27.55	220.40	42.00	336.00		
1 Dump Truck, 16 Ton		476.60		524.25		
1 F.E. Loader, 3 C.Y.		305.00		335.50	48.85	53.74
16 L.H., Daily Totals		$1279.20		$1613.35	$79.95	$100.84
Crew A-3C	Hr.	Daily	Hr.	Daily	Bare Costs	Incl. O&P
1 Equip. Oper. (light)	$33.05	$264.40	$49.80	$398.40	$33.05	$49.80
1 Wheeled Skid Steer Loader		197.80		217.60	24.73	27.20
8 L.H., Daily Totals		$462.20		$616.00	$57.78	$77.00
Crew A-3D	Hr.	Daily	Hr.	Daily	Bare Costs	Incl. O&P
1 Truck Driver, Light	$26.80	$214.40	$40.85	$326.80	$26.80	$40.85
1 Pickup Truck (4x4)		85.20		93.70		
1 Flatbed Trailer, 25 Ton		88.00		96.80	21.65	23.82
8 L.H., Daily Totals		$387.60		$517.30	$48.45	$64.67
Crew A-3E	Hr.	Daily	Hr.	Daily	Bare Costs	Incl. O&P
1 Equip. Oper. (crane)	$35.90	$287.20	$54.10	$432.80	$31.73	$48.05
1 Truck Driver (heavy)	27.55	220.40	42.00	336.00		
1 Pickup Truck (4x4)		85.20		93.70	5.33	5.86
16 L.H., Daily Totals		$592.80		$862.50	$37.06	$53.91
Crew A-3F	Hr.	Daily	Hr.	Daily	Bare Costs	Incl. O&P
1 Equip. Oper. (crane)	$35.90	$287.20	$54.10	$432.80	$31.73	$48.05
1 Truck Driver (heavy)	27.55	220.40	42.00	336.00		
1 Pickup Truck (4x4)		85.20		93.70		
1 Tractor, 6x2, 40 Ton Cap.		301.60		331.75		
1 Lowbed Trailer, 75 Ton		170.20		187.20	34.81	38.29
16 L.H., Daily Totals		$1064.60		$1381.45	$66.54	$86.34

Crew No.	Bare Costs		Incl. Subs O & P		Cost Per Labor-Hour	
Crew A-3G	Hr.	Daily	Hr.	Daily	Bare Costs	Incl. O&P
1 Equip. Oper. (crane)	$35.90	$287.20	$54.10	$432.80	$31.73	$48.05
1 Truck Driver (heavy)	27.55	220.40	42.00	336.00		
1 Pickup Truck (4x4)		85.20		93.70		
1 Tractor, 6x4, 45 Ton Cap.		353.60		388.95		
1 Lowbed Trailer, 75 Ton		170.20		187.20	38.06	41.87
16 L.H., Daily Totals		$1116.60		$1438.65	$69.79	$89.92

Crew A-4	Hr.	Daily	Hr.	Daily	Bare Costs	Incl. O&P
2 Carpenters	$34.25	$548.00	$53.35	$853.60	$33.03	$50.90
1 Painter, Ordinary	30.60	244.80	46.00	368.00		
24 L.H., Daily Totals		$792.80		$1221.60	$33.03	$50.90

Crew A-5	Hr.	Daily	Hr.	Daily	Bare Costs	Incl. O&P
2 Laborers	$26.70	$427.20	$41.55	$664.80	$26.71	$41.47
.25 Truck Driver (light)	26.80	53.60	40.85	81.70		
.25 Light Truck, 1.5 Ton		32.20		35.40	1.79	1.97
18 L.H., Daily Totals		$513.00		$781.90	$28.50	$43.44

Crew A-6	Hr.	Daily	Hr.	Daily	Bare Costs	Incl. O&P
1 Instrument Man	$34.85	$278.80	$54.20	$433.60	$33.93	$52.05
1 Rodman/Chainman	33.00	264.00	49.90	399.20		
1 Laser Transit/Level		60.75		66.85	3.79	4.17
16 L.H., Daily Totals		$603.55		$899.65	$37.72	$56.22

Crew A-7	Hr.	Daily	Hr.	Daily	Bare Costs	Incl. O&P
1 Chief Of Party	$42.90	$343.20	$66.90	$535.20	$36.92	$57.00
1 Instrument Man	34.85	278.80	54.20	433.60		
1 Rodman/Chainman	33.00	264.00	49.90	399.20		
1 Laser Transit/Level		60.75		66.85	2.53	2.78
24 L.H., Daily Totals		$946.75		$1434.85	$39.45	$59.78

Crew A-8	Hr.	Daily	Hr.	Daily	Bare Costs	Incl. O&P
1 Chief Of Party	$42.90	$343.20	$66.90	$535.20	$35.94	$55.23
1 Instrument Man	34.85	278.80	54.20	433.60		
2 Rodmen/Chainmen	33.00	528.00	49.90	798.40		
1 Laser Transit/Level		60.75		66.85	1.90	2.09
32 L.H., Daily Totals		$1210.75		$1834.05	$37.84	$57.32

Crew A-9	Hr.	Daily	Hr.	Daily	Bare Costs	Incl. O&P
1 Asbestos Foreman	$37.60	$300.80	$59.35	$474.80	$37.16	$58.65
7 Asbestos Workers	37.10	2077.60	58.55	3278.80		
64 L.H., Daily Totals		$2378.40		$3753.60	$37.16	$58.65

Crew A-10	Hr.	Daily	Hr.	Daily	Bare Costs	Incl. O&P
1 Asbestos Foreman	$37.60	$300.80	$59.35	$474.80	$37.16	$58.65
7 Asbestos Workers	37.10	2077.60	58.55	3278.80		
64 L.H., Daily Totals		$2378.40		$3753.60	$37.16	$58.65

Crew A-10A	Hr.	Daily	Hr.	Daily	Bare Costs	Incl. O&P
1 Asbestos Foreman	$37.60	$300.80	$59.35	$474.80	$37.27	$58.82
2 Asbestos Workers	37.10	593.60	58.55	936.80		
24 L.H., Daily Totals		$894.40		$1411.60	$37.27	$58.82

Crew A-10B	Hr.	Daily	Hr.	Daily	Bare Costs	Incl. O&P
1 Asbestos Foreman	$37.60	$300.80	$59.35	$474.80	$37.23	$58.75
3 Asbestos Workers	37.10	890.40	58.55	1405.20		
32 L.H., Daily Totals		$1191.20		$1880.00	$37.23	$58.75

Crew No.	Bare Costs		Incl. Subs O & P		Cost Per Labor-Hour	
Crew A-10C	Hr.	Daily	Hr.	Daily	Bare Costs	Incl. O&P
3 Asbestos Workers	$37.10	$890.40	$58.55	$1405.20	$37.10	$58.55
1 Flatbed Truck		128.80		141.70	5.37	5.90
24 L.H., Daily Totals		$1019.20		$1546.90	$42.47	$64.45

Crew A-10D	Hr.	Daily	Hr.	Daily	Bare Costs	Incl. O&P
2 Asbestos Workers	$37.10	$593.60	$58.55	$936.80	$35.05	$54.14
1 Equip. Oper. (crane)	35.90	287.20	54.10	432.80		
1 Equip. Oper. Oiler	30.10	240.80	45.35	362.80		
1 Hydraulic Crane, 33 Ton		641.60		705.75	20.05	22.06
32 L.H., Daily Totals		$1763.20		$2438.15	$55.10	$76.20

Crew A-11	Hr.	Daily	Hr.	Daily	Bare Costs	Incl. O&P
1 Asbestos Foreman	$37.60	$300.80	$59.35	$474.80	$37.16	$58.65
7 Asbestos Workers	37.10	2077.60	58.55	3278.80		
2 Chipping Hammers		32.00		35.20	.50	.55
64 L.H., Daily Totals		$2410.40		$3788.80	$37.66	$59.20

Crew A-12	Hr.	Daily	Hr.	Daily	Bare Costs	Incl. O&P
1 Asbestos Foreman	$37.60	$300.80	$59.35	$474.80	$37.16	$58.65
7 Asbestos Workers	37.10	2077.60	58.55	3278.80		
1 Large Prod. Vac. Loader		480.00		528.00	7.50	8.25
64 L.H., Daily Totals		$2858.40		$4281.60	$44.66	$66.90

Crew A-13	Hr.	Daily	Hr.	Daily	Bare Costs	Incl. O&P
1 Equip. Oper. (light)	$33.05	$264.40	$49.80	$398.40	$33.05	$49.80
1 Large Prod. Vac. Loader		480.00		528.00	60.00	66.00
8 L.H., Daily Totals		$744.40		$926.40	$93.05	$115.80

Crew B-1	Hr.	Daily	Hr.	Daily	Bare Costs	Incl. O&P
1 Labor Foreman (outside)	$28.70	$229.60	$44.70	$357.60	$27.37	$42.60
2 Laborers	26.70	427.20	41.55	664.80		
24 L.H., Daily Totals		$656.80		$1022.40	$27.37	$42.60

Crew B-1A	Hr.	Daily	Hr.	Daily	Bare Costs	Incl. O&P
1 Laborer Foreman	$28.70	$229.60	$44.70	$357.60	$27.37	$42.60
2 Laborers	26.70	427.20	41.55	664.80		
2 Cutting Torches		36.00		39.60		
2 Gases		129.60		142.55	6.90	7.59
24 L.H., Daily Totals		$822.40		$1204.55	$34.27	$50.19

Crew B-1B	Hr.	Daily	Hr.	Daily	Bare Costs	Incl. O&P
1 Laborer Foreman	$28.70	$229.60	$44.70	$357.60	$29.50	$45.48
2 Laborers	26.70	427.20	41.55	664.80		
1 Equip. Oper. (crane)	35.90	287.20	54.10	432.80		
2 Cutting Torches		36.00		39.60		
2 Gases		129.60		142.55		
1 Hyd. Crane, 12 Ton		602.60		662.85	24.01	26.41
32 L.H., Daily Totals		$1712.20		$2300.20	$53.51	$71.89

Crew B-2	Hr.	Daily	Hr.	Daily	Bare Costs	Incl. O&P
1 Labor Foreman (outside)	$28.70	$229.60	$44.70	$357.60	$27.10	$42.18
4 Laborers	26.70	854.40	41.55	1329.60		
40 L.H., Daily Totals		$1084.00		$1687.20	$27.10	$42.18

Crew B-3

Crew No.	Bare Costs Hr.	Daily	Incl. Subs O & P Hr.	Daily	Cost Per Labor-Hour Bare Costs	Incl. O&P
1 Labor Foreman (outside)	$28.70	$229.60	$44.70	$357.60	$28.64	$44.00
2 Laborers	26.70	427.20	41.55	664.80		
1 Equip. Oper. (med.)	34.65	277.20	52.20	417.60		
2 Truck Drivers (heavy)	27.55	440.80	42.00	672.00		
1 F.E. Loader, T.M., 2.5 C.Y.		788.00		866.80		
2 Dump Trucks, 16 Ton		953.20		1048.50	36.28	39.90
48 L.H., Daily Totals		$3116.00		$4027.30	$64.92	$83.90

Crew B-3A

Crew No.	Bare Costs Hr.	Daily	Incl. Subs O & P Hr.	Daily	Cost Per Labor-Hour Bare Costs	Incl. O&P
4 Laborers	$26.70	$854.40	$41.55	$1329.60	$28.29	$43.68
1 Equip. Oper. (med.)	34.65	277.20	52.20	417.60		
1 Hyd. Excavator, 1.5 C.Y.		720.40		792.45	18.01	19.81
40 L.H., Daily Totals		$1852.00		$2539.65	$46.30	$63.49

Crew B-3B

Crew No.	Bare Costs Hr.	Daily	Incl. Subs O & P Hr.	Daily	Cost Per Labor-Hour Bare Costs	Incl. O&P
2 Laborers	$26.70	$427.20	$41.55	$664.80	$28.90	$44.33
1 Equip. Oper. (med.)	34.65	277.20	52.20	417.60		
1 Truck Driver (heavy)	27.55	220.40	42.00	336.00		
1 Backhoe Loader, 80 H.P.		256.40		282.05		
1 Dump Truck, 16 Ton		476.60		524.25	22.91	25.20
32 L.H., Daily Totals		$1657.80		$2224.70	$51.81	$69.53

Crew B-3C

Crew No.	Bare Costs Hr.	Daily	Incl. Subs O & P Hr.	Daily	Cost Per Labor-Hour Bare Costs	Incl. O&P
3 Laborers	$26.70	$640.80	$41.55	$997.20	$28.69	$44.21
1 Equip. Oper. (med.)	34.65	277.20	52.20	417.60		
1 F.E. Crawler Ldr, 4 C.Y.		1094.00		1203.40	34.19	37.61
32 L.H., Daily Totals		$2012.00		$2618.20	$62.88	$81.82

Crew B-4

Crew No.	Bare Costs Hr.	Daily	Incl. Subs O & P Hr.	Daily	Cost Per Labor-Hour Bare Costs	Incl. O&P
1 Labor Foreman (outside)	$28.70	$229.60	$44.70	$357.60	$27.18	$42.15
4 Laborers	26.70	854.40	41.55	1329.60		
1 Truck Driver (heavy)	27.55	220.40	42.00	336.00		
1 Tractor, 4 x 2, 195 H.P.		208.20		229.00		
1 Platform Trailer		119.40		131.35	6.83	7.51
48 L.H., Daily Totals		$1632.00		$2383.55	$34.01	$49.66

Crew B-5

Crew No.	Bare Costs Hr.	Daily	Incl. Subs O & P Hr.	Daily	Cost Per Labor-Hour Bare Costs	Incl. O&P
1 Labor Foreman (outside)	$28.70	$229.60	$44.70	$357.60	$29.26	$45.04
4 Laborers	26.70	854.40	41.55	1329.60		
2 Equip. Oper. (med.)	34.65	554.40	52.20	835.20		
1 Air Compr., 250 C.F.M.		112.40		123.65		
2 Air Tools & Accessories		19.60		21.55		
2-50 Ft. Air Hoses, 1.5" Dia.		9.40		10.35		
1 F.E. Loader, T.M., 2.5 C.Y.		788.00		866.80	16.60	18.26
56 L.H., Daily Totals		$2567.80		$3544.75	$45.86	$63.30

Crew B-5A

Crew No.	Bare Costs Hr.	Daily	Incl. Subs O & P Hr.	Daily	Cost Per Labor-Hour Bare Costs	Incl. O&P
1 Foreman	$28.70	$229.60	$44.70	$357.60	$28.86	$44.35
6 Laborers	26.70	1281.60	41.55	1994.40		
2 Equip. Oper. (med.)	34.65	554.40	52.20	835.20		
1 Equip. Oper. (light)	33.05	264.40	49.80	398.40		
2 Truck Drivers (heavy)	27.55	440.80	42.00	672.00		
1 Air Compr. 365 C.F.M.		146.80		161.50		
2 Pavement Breakers		19.60		21.55		
8 Air Hoses w/Coup.,1"		29.20		32.10		
2 Dump Trucks, 12 Ton		653.20		718.50	8.84	9.73
96 L.H., Daily Totals		$3619.60		$5191.25	$37.70	$54.08

Crew B-5B

Crew No.	Bare Costs Hr.	Daily	Incl. Subs O & P Hr.	Daily	Cost Per Labor-Hour Bare Costs	Incl. O&P
1 Powderman	$34.85	$278.80	$54.20	$433.60	$31.13	$47.43
2 Equip. Oper. (med.)	34.65	554.40	52.20	835.20		
3 Truck Drivers (heavy)	27.55	661.20	42.00	1008.00		
1 F.E. Ldr. 2-1/2 CY		305.00		335.50		
3 Dump Trucks, 16 Ton		1429.80		1572.80		
1 Air Compr. 365 C.F.M.		146.80		161.50	39.20	43.12
48 L.H., Daily Totals		$3376.00		$4346.60	$70.33	$90.55

Crew B-5C

Crew No.	Bare Costs Hr.	Daily	Incl. Subs O & P Hr.	Daily	Cost Per Labor-Hour Bare Costs	Incl. O&P
3 Laborers	$26.70	$640.80	$41.55	$997.20	$29.48	$45.04
1 Equip. Oper. (medium)	34.65	277.20	52.20	417.60		
2 Truck Drivers (heavy)	27.55	440.80	42.00	672.00		
1 Equip. Oper. (crane)	35.90	287.20	54.10	432.80		
1 Equip. Oper. Oiler	30.10	240.80	45.35	362.80		
2 Dump Trucks, 16 Ton		953.20		1048.50		
1 F.E. Crawler Ldr, 4 C.Y.		1094.00		1203.40		
1 Hyd. Crane, 25 Ton		575.00		632.50	40.97	45.07
64 L.H., Daily Totals		$4509.00		$5766.80	$70.45	$90.11

Crew B-6

Crew No.	Bare Costs Hr.	Daily	Incl. Subs O & P Hr.	Daily	Cost Per Labor-Hour Bare Costs	Incl. O&P
2 Laborers	$26.70	$427.20	$41.55	$664.80	$28.82	$44.30
1 Equip. Oper. (light)	33.05	264.40	49.80	398.40		
1 Backhoe Loader, 48 H.P.		215.80		237.40	8.99	9.89
24 L.H., Daily Totals		$907.40		$1300.60	$37.81	$54.19

Crew B-6A

Crew No.	Bare Costs Hr.	Daily	Incl. Subs O & P Hr.	Daily	Cost Per Labor-Hour Bare Costs	Incl. O&P
.5 Labor Foreman (outside)	$28.70	$114.80	$44.70	$178.80	$30.28	$46.44
1 Laborer	26.70	213.60	41.55	332.40		
1 Equip. Oper. (med.)	34.65	277.20	52.20	417.60		
1 Vacuum Trk.,5000 Gal.		331.50		364.65	16.58	18.23
20 L.H., Daily Totals		$937.10		$1293.45	$46.86	$64.67

Crew B-6B

Crew No.	Bare Costs Hr.	Daily	Incl. Subs O & P Hr.	Daily	Cost Per Labor-Hour Bare Costs	Incl. O&P
2 Labor Foreman (out)	$28.70	$459.20	$44.70	$715.20	$27.37	$42.60
4 Laborers	26.70	854.40	41.55	1329.60		
1 Winch Truck		317.00		348.70		
1 Flatbed Truck		128.80		141.70		
1 Butt Fusion Machine		432.80		476.10	18.30	20.13
48 L.H., Daily Totals		$2192.20		$3011.30	$45.67	$62.73

Crew B-7

Crew No.	Bare Costs Hr.	Daily	Incl. Subs O & P Hr.	Daily	Cost Per Labor-Hour Bare Costs	Incl. O&P
1 Labor Foreman (outside)	$28.70	$229.60	$44.70	$357.60	$28.36	$43.85
4 Laborers	26.70	854.40	41.55	1329.60		
1 Equip. Oper. (med.)	34.65	277.20	52.20	417.60		
1 Chipping Machine		164.80		181.30		
1 F.E. Loader, T.M., 2.5 C.Y.		788.00		866.80		
2 Chain Saws, 36"		66.80		73.50	21.24	23.37
48 L.H., Daily Totals		$2380.80		$3226.40	$49.60	$67.22

Crew B-7A

Crew No.	Bare Costs Hr.	Daily	Incl. Subs O & P Hr.	Daily	Cost Per Labor-Hour Bare Costs	Incl. O&P
2 Laborers	$26.70	$427.20	$41.55	$664.80	$28.82	$44.30
1 Equip. Oper. (light)	33.05	264.40	49.80	398.40		
1 Rake w/Tractor		191.50		210.65		
2 Chain Saws, 18"		38.40		42.25	9.58	10.54
24 L.H., Daily Totals		$921.50		$1316.10	$38.40	$54.84

Crews

Crew No.	Bare Costs Hr.	Bare Costs Daily	Incl. Subs O & P Hr.	Incl. Subs O & P Daily	Cost Per Labor-Hour Bare Costs	Cost Per Labor-Hour Incl. O&P
Crew B-8						
1 Labor Foreman (outside)	$28.70	$229.60	$44.70	$357.60	$29.58	$45.19
2 Laborers	26.70	427.20	41.55	664.80		
2 Equip. Oper. (med.)	34.65	554.40	52.20	835.20		
1 Equip. Oper. Oiler	30.10	240.80	45.35	362.80		
2 Truck Drivers (heavy)	27.55	440.80	42.00	672.00		
1 Hyd. Crane, 25 Ton		616.80		678.50		
1 F.E. Loader, T.M., 2.5 C.Y.		788.00		866.80		
2 Dump Trucks, 16 Ton		953.20		1048.50	36.84	40.53
64 L.H., Daily Totals		$4250.80		$5486.20	$66.42	$85.72

Crew No.	Bare Costs Hr.	Bare Costs Daily	Incl. Subs O & P Hr.	Incl. Subs O & P Daily	Cost Per Labor-Hour Bare Costs	Cost Per Labor-Hour Incl. O&P
Crew B-9						
1 Labor Foreman (outside)	$28.70	$229.60	$44.70	$357.60	$27.10	$42.18
4 Laborers	26.70	854.40	41.55	1329.60		
1 Air Compr., 250 C.F.M.		112.40		123.65		
2 Air Tools & Accessories		19.60		21.55		
2-50 Ft. Air Hoses, 1.5" Dia.		9.40		10.35	3.54	3.89
40 L.H., Daily Totals		$1225.40		$1842.75	$30.64	$46.07

Crew No.	Bare Costs Hr.	Bare Costs Daily	Incl. Subs O & P Hr.	Incl. Subs O & P Daily	Cost Per Labor-Hour Bare Costs	Cost Per Labor-Hour Incl. O&P
Crew B-9A						
2 Laborers	$26.70	$427.20	$41.55	$664.80	$26.98	$41.70
1 Truck Driver (heavy)	27.55	220.40	42.00	336.00		
1 Water Tanker		116.80		128.50		
1 Tractor		208.20		229.00		
2-50 Ft. Disch. Hoses		5.00		5.50	13.75	15.13
24 L.H., Daily Totals		$977.60		$1363.80	$40.73	$56.83

Crew No.	Bare Costs Hr.	Bare Costs Daily	Incl. Subs O & P Hr.	Incl. Subs O & P Daily	Cost Per Labor-Hour Bare Costs	Cost Per Labor-Hour Incl. O&P
Crew B-9B						
2 Laborers	$26.70	$427.20	$41.55	$664.80	$26.98	$41.70
1 Truck Driver (heavy)	27.55	220.40	42.00	336.00		
2-50 Ft. Disch. Hoses		5.00		5.50		
1 Water Tanker		116.80		128.50		
1 Tractor		208.20		229.00		
1 Pressure Washer		46.20		50.80	15.68	17.24
24 L.H., Daily Totals		$1023.80		$1414.60	$42.66	$58.94

Crew No.	Bare Costs Hr.	Bare Costs Daily	Incl. Subs O & P Hr.	Incl. Subs O & P Daily	Cost Per Labor-Hour Bare Costs	Cost Per Labor-Hour Incl. O&P
Crew B-9C						
1 Labor Foreman (outside)	$28.70	$229.60	$44.70	$357.60	$27.10	$42.18
4 Laborers	26.70	854.40	41.55	1329.60		
1 Air Compr., 250 C.F.M.		112.40		123.65		
2-50 Ft. Air Hoses, 1.5" Dia.		9.40		10.35		
2 Breaker, Pavement, 60 lb.		19.60		21.55	3.54	3.89
40 L.H., Daily Totals		$1225.40		$1842.75	$30.64	$46.07

Crew No.	Bare Costs Hr.	Bare Costs Daily	Incl. Subs O & P Hr.	Incl. Subs O & P Daily	Cost Per Labor-Hour Bare Costs	Cost Per Labor-Hour Incl. O&P
Crew B-9D						
1 Labor Foreman (Outside)	$28.70	$229.60	$44.70	$357.60	$27.10	$42.18
4 Common Laborers	26.70	854.40	41.55	1329.60		
1 Air Compressor, 250 CFM		112.40		123.65		
2 Air hoses, 1.5" x 50'		9.40		10.35		
2 Air tamper		55.30		60.85	4.43	4.87
40 L.H., Daily Totals		$1261.10		$1882.05	$31.53	$47.05

Crew No.	Bare Costs Hr.	Bare Costs Daily	Incl. Subs O & P Hr.	Incl. Subs O & P Daily	Cost Per Labor-Hour Bare Costs	Cost Per Labor-Hour Incl. O&P
Crew B-10						
1 Equip. Oper. (med.)	$34.65	$277.20	$52.20	$417.60	$32.00	$48.65
.5 Laborer	26.70	106.80	41.55	166.20		
12 L.H., Daily Totals		$384.00		$583.80	$32.00	$48.65

Crew No.	Bare Costs Hr.	Bare Costs Daily	Incl. Subs O & P Hr.	Incl. Subs O & P Daily	Cost Per Labor-Hour Bare Costs	Cost Per Labor-Hour Incl. O&P
Crew B-10A						
1 Equip. Oper. (med.)	$34.65	$277.20	$52.20	$417.60	$32.00	$48.65
.5 Laborer	26.70	106.80	41.55	166.20		
1 Walk behind compactor, 7.5 HP		122.80		135.10	10.23	11.26
12 L.H., Daily Totals		$506.80		$718.90	$42.23	$59.91

Crew No.	Bare Costs Hr.	Bare Costs Daily	Incl. Subs O & P Hr.	Incl. Subs O & P Daily	Cost Per Labor-Hour Bare Costs	Cost Per Labor-Hour Incl. O&P
Crew B-10B						
1 Equip. Oper. (med.)	$34.65	$277.20	$52.20	$417.60	$32.00	$48.65
.5 Laborer	26.70	106.80	41.55	166.20		
1 Dozer, 200 H.P.		919.60		1011.55	76.63	84.30
12 L.H., Daily Totals		$1303.60		$1595.35	$108.63	$132.95

Crew No.	Bare Costs Hr.	Bare Costs Daily	Incl. Subs O & P Hr.	Incl. Subs O & P Daily	Cost Per Labor-Hour Bare Costs	Cost Per Labor-Hour Incl. O&P
Crew B-10C						
1 Equip. Oper. (med.)	$34.65	$277.20	$52.20	$417.60	$32.00	$48.65
.5 Laborer	26.70	106.80	41.55	166.20		
1 Dozer, 200 H.P.		919.60		1011.55		
1 Vibratory Roller, Towed		591.60		650.75	125.93	138.53
12 L.H., Daily Totals		$1895.20		$2246.10	$157.93	$187.18

Crew No.	Bare Costs Hr.	Bare Costs Daily	Incl. Subs O & P Hr.	Incl. Subs O & P Daily	Cost Per Labor-Hour Bare Costs	Cost Per Labor-Hour Incl. O&P
Crew B-10D						
1 Equip. Oper. (med.)	$34.65	$277.20	$52.20	$417.60	$32.00	$48.65
.5 Laborer	26.70	106.80	41.55	166.20		
1 Dozer, 200 H.P.		919.60		1011.55		
1 Sheepsft. Roller, Towed		610.20		671.20	127.48	140.23
12 L.H., Daily Totals		$1913.80		$2266.55	$159.48	$188.88

Crew No.	Bare Costs Hr.	Bare Costs Daily	Incl. Subs O & P Hr.	Incl. Subs O & P Daily	Cost Per Labor-Hour Bare Costs	Cost Per Labor-Hour Incl. O&P
Crew B-10E						
1 Equip. Oper. (med.)	$34.65	$277.20	$52.20	$417.60	$32.00	$48.65
.5 Laborer	26.70	106.80	41.55	166.20		
1 Tandem Roller, 5 Ton		107.40		118.15	8.95	9.85
12 L.H., Daily Totals		$491.40		$701.95	$40.95	$58.50

Crew No.	Bare Costs Hr.	Bare Costs Daily	Incl. Subs O & P Hr.	Incl. Subs O & P Daily	Cost Per Labor-Hour Bare Costs	Cost Per Labor-Hour Incl. O&P
Crew B-10F						
1 Equip. Oper. (med.)	$34.65	$277.20	$52.20	$417.60	$32.00	$48.65
.5 Laborer	26.70	106.80	41.55	166.20		
1 Tandem Roller, 10 Ton		179.20		197.10	14.93	16.43
12 L.H., Daily Totals		$563.20		$780.90	$46.93	$65.08

Crew No.	Bare Costs Hr.	Bare Costs Daily	Incl. Subs O & P Hr.	Incl. Subs O & P Daily	Cost Per Labor-Hour Bare Costs	Cost Per Labor-Hour Incl. O&P
Crew B-10G						
1 Equip. Oper. (med.)	$34.65	$277.20	$52.20	$417.60	$32.00	$48.65
.5 Laborer	26.70	106.80	41.55	166.20		
1 Sheepsft. Roll., 130 H.P.		794.80		874.30	66.23	72.86
12 L.H., Daily Totals		$1178.80		$1458.10	$98.23	$121.51

Crew No.	Bare Costs Hr.	Bare Costs Daily	Incl. Subs O & P Hr.	Incl. Subs O & P Daily	Cost Per Labor-Hour Bare Costs	Cost Per Labor-Hour Incl. O&P
Crew B-10H						
1 Equip. Oper. (med.)	$34.65	$277.20	$52.20	$417.60	$32.00	$48.65
.5 Laborer	26.70	106.80	41.55	166.20		
1 Diaphr. Water Pump, 2"		50.60		55.65		
1-20 Ft. Suction Hose, 2"		2.55		2.80		
2-50 Ft. Disch. Hoses, 2"		3.80		4.20	4.75	5.22
12 L.H., Daily Totals		$440.95		$646.45	$36.75	$53.87

Crew No.	Bare Costs Hr.	Bare Costs Daily	Incl. Subs O & P Hr.	Incl. Subs O & P Daily	Cost Per Labor-Hour Bare Costs	Cost Per Labor-Hour Incl. O&P
Crew B-10I						
1 Equip. Oper. (med.)	$34.65	$277.20	$52.20	$417.60	$32.00	$48.65
.5 Laborer	26.70	106.80	41.55	166.20		
1 Diaphr. Water Pump, 4"		74.40		81.85		
1-20 Ft. Suction Hose, 4"		5.05		5.55		
2-50 Ft. Disch. Hoses, 4"		7.10		7.80	7.21	7.93
12 L.H., Daily Totals		$470.55		$679.00	$39.21	$56.58

Crew No.	Bare Costs Hr.	Bare Costs Daily	Incl. Subs O & P Hr.	Incl. Subs O & P Daily	Cost Per Labor-Hour Bare Costs	Cost Per Labor-Hour Incl. O&P
Crew B-10J						
1 Equip. Oper. (med.)	$34.65	$277.20	$52.20	$417.60	$32.00	$48.65
.5 Laborer	26.70	106.80	41.55	166.20		
1 Centr. Water Pump, 3"		53.60		58.95		
1-20 Ft. Suction Hose, 3"		4.25		4.70		
2-50 Ft. Disch. Hoses, 3"		5.00		5.50	5.24	5.76
12 L.H., Daily Totals		$446.85		$652.95	$37.24	$54.41

Crew No.	Bare Costs Hr.	Daily	Incl. Subs O & P Hr.	Daily	Cost Per Labor-Hour Bare Costs	Incl. O&P
Crew B-10K	Hr.	Daily	Hr.	Daily	Bare Costs	Incl. O&P
1 Equip. Oper. (med.)	$34.65	$277.20	$52.20	$417.60	$32.00	$48.65
.5 Laborer	26.70	106.80	41.55	166.20		
1 Centr. Water Pump, 6"		217.00		238.70		
1-20 Ft. Suction Hose, 6"		12.80		14.10		
2-50 Ft. Disch. Hoses, 6"		17.80		19.60	20.63	22.70
12 L.H., Daily Totals		$631.60		$856.20	$52.63	$71.35
Crew B-10L	Hr.	Daily	Hr.	Daily	Bare Costs	Incl. O&P
1 Equip. Oper. (med.)	$34.65	$277.20	$52.20	$417.60	$32.00	$48.65
.5 Laborer	26.70	106.80	41.55	166.20		
1 Dozer, 80 H.P.		314.80		346.30	26.23	28.86
12 L.H., Daily Totals		$698.80		$930.10	$58.23	$77.51
Crew B-10M	Hr.	Daily	Hr.	Daily	Bare Costs	Incl. O&P
1 Equip. Oper. (med.)	$34.65	$277.20	$52.20	$417.60	$32.00	$48.65
.5 Laborer	26.70	106.80	41.55	166.20		
1 Dozer, 300 H.P.		1195.00		1314.50	99.58	109.54
12 L.H., Daily Totals		$1579.00		$1898.30	$131.58	$158.19
Crew B-10N	Hr.	Daily	Hr.	Daily	Bare Costs	Incl. O&P
1 Equip. Oper. (med.)	$34.65	$277.20	$52.20	$417.60	$32.00	$48.65
.5 Laborer	26.70	106.80	41.55	166.20		
1 F.E. Loader, T.M., 1.5 C.Y.		312.60		343.85	26.05	28.66
12 L.H., Daily Totals		$696.60		$927.65	$58.05	$77.31
Crew B-100	Hr.	Daily	Hr.	Daily	Bare Costs	Incl. O&P
1 Equip. Oper. (med.)	$34.65	$277.20	$52.20	$417.60	$32.00	$48.65
.5 Laborer	26.70	106.80	41.55	166.20		
1 F.E. Loader, T.M., 2.25 C.Y.		551.80		607.00	45.98	50.58
12 L.H., Daily Totals		$935.80		$1190.80	$77.98	$99.23
Crew B-10P	Hr.	Daily	Hr.	Daily	Bare Costs	Incl. O&P
1 Equip. Oper. (med.)	$34.65	$277.20	$52.20	$417.60	$32.00	$48.65
.5 Laborer	26.70	106.80	41.55	166.20		
1 F.E. Loader, T.M., 2.5 C.Y.		788.00		866.80	65.67	72.23
12 L.H., Daily Totals		$1172.00		$1450.60	$97.67	$120.88
Crew B-10Q	Hr.	Daily	Hr.	Daily	Bare Costs	Incl. O&P
1 Equip. Oper. (med.)	$34.65	$277.20	$52.20	$417.60	$32.00	$48.65
.5 Laborer	26.70	106.80	41.55	166.20		
1 F.E. Loader, T.M., 5 C.Y.		1094.00		1203.40	91.17	100.28
12 L.H., Daily Totals		$1478.00		$1787.20	$123.17	$148.93
Crew B-10R	Hr.	Daily	Hr.	Daily	Bare Costs	Incl. O&P
1 Equip. Oper. (med.)	$34.65	$277.20	$52.20	$417.60	$32.00	$48.65
.5 Laborer	26.70	106.80	41.55	166.20		
1 F.E. Loader, W.M., 1 C.Y.		193.80		213.20	16.15	17.77
12 L.H., Daily Totals		$577.80		$797.00	$48.15	$66.42
Crew B-10S	Hr.	Daily	Hr.	Daily	Bare Costs	Incl. O&P
1 Equip. Oper. (med.)	$34.65	$277.20	$52.20	$417.60	$32.00	$48.65
.5 Laborer	26.70	106.80	41.55	166.20		
1 F.E. Loader, W.M., 1.5 C.Y.		241.00		265.10	20.08	22.09
12 L.H., Daily Totals		$625.00		$848.90	$52.08	$70.74
Crew B-10T	Hr.	Daily	Hr.	Daily	Bare Costs	Incl. O&P
1 Equip. Oper. (med.)	$34.65	$277.20	$52.20	$417.60	$32.00	$48.65
.5 Laborer	26.70	106.80	41.55	166.20		
1 F.E. Loader, W.M., 2.5 C.Y.		305.00		335.50	25.42	27.96
12 L.H., Daily Totals		$689.00		$919.30	$57.42	$76.61

Crew No.	Bare Costs Hr.	Daily	Incl. Subs O & P Hr.	Daily	Cost Per Labor-Hour Bare Costs	Incl. O&P
Crew B-10U	Hr.	Daily	Hr.	Daily	Bare Costs	Incl. O&P
1 Equip. Oper. (med.)	$34.65	$277.20	$52.20	$417.60	$32.00	$48.65
.5 Laborer	26.70	106.80	41.55	166.20		
1 F.E. Loader, W.M., 5.5 C.Y.		700.40		770.45	58.37	64.20
12 L.H., Daily Totals		$1084.40		$1354.25	$90.37	$112.85
Crew B-10V	Hr.	Daily	Hr.	Daily	Bare Costs	Incl. O&P
1 Equip. Oper. (med.)	$34.65	$277.20	$52.20	$417.60	$32.00	$48.65
.5 Laborer	26.70	106.80	41.55	166.20		
1 Dozer, 700 H.P.		3139.00		3452.90	261.58	287.74
12 L.H., Daily Totals		$3523.00		$4036.70	$293.58	$336.39
Crew B-10W	Hr.	Daily	Hr.	Daily	Bare Costs	Incl. O&P
1 Equip. Oper. (med.)	$34.65	$277.20	$52.20	$417.60	$32.00	$48.65
.5 Laborer	26.70	106.80	41.55	166.20		
1 Dozer, 105 H.P.		453.80		499.20	37.82	41.60
12 L.H., Daily Totals		$837.80		$1083.00	$69.82	$90.25
Crew B-10X	Hr.	Daily	Hr.	Daily	Bare Costs	Incl. O&P
1 Equip. Oper. (med.)	$34.65	$277.20	$52.20	$417.60	$32.00	$48.65
.5 Laborer	26.70	106.80	41.55	166.20		
1 Dozer, 410 H.P.		1524.00		1676.40	127.00	139.70
12 L.H., Daily Totals		$1908.00		$2260.20	$159.00	$188.35
Crew B-10Y	Hr.	Daily	Hr.	Daily	Bare Costs	Incl. O&P
1 Equip. Oper. (med.)	$34.65	$277.20	$52.20	$417.60	$32.00	$48.65
.5 Laborer	26.70	106.80	41.55	166.20		
1 Vibratory Drum Roller		344.80		379.30	28.73	31.61
12 L.H., Daily Totals		$728.80		$963.10	$60.73	$80.26
Crew B-11A	Hr.	Daily	Hr.	Daily	Bare Costs	Incl. O&P
1 Equipment Oper. (med.)	$34.65	$277.20	$52.20	$417.60	$30.68	$46.88
1 Laborer	26.70	213.60	41.55	332.40		
1 Dozer, 200 H.P.		919.60		1011.55	57.48	63.22
16 L.H., Daily Totals		$1410.40		$1761.55	$88.16	$110.10
Crew B-11B	Hr.	Daily	Hr.	Daily	Bare Costs	Incl. O&P
1 Equipment Oper. (light)	$33.05	$264.40	$49.80	$398.40	$29.88	$45.68
1 Laborer	26.70	213.60	41.55	332.40		
1 Air Powered Tamper		27.65		30.40		
1 Air Compr. 365 C.F.M.		146.80		161.50		
2-50 Ft. Air Hoses, 1.5" Dia.		9.40		10.35	11.49	12.64
16 L.H., Daily Totals		$661.85		$933.05	$41.37	$58.32
Crew B-11C	Hr.	Daily	Hr.	Daily	Bare Costs	Incl. O&P
1 Equipment Oper. (med.)	$34.65	$277.20	$52.20	$417.60	$30.68	$46.88
1 Laborer	26.70	213.60	41.55	332.40		
1 Backhoe Loader, 48 H.P.		215.80		237.40	13.49	14.84
16 L.H., Daily Totals		$706.60		$987.40	$44.17	$61.72
Crew B-11J	Hr.	Daily	Hr.	Daily	Bare Costs	Incl. O&P
1 Equipment Oper. (med.)	$34.65	$277.20	$52.20	$417.60	$30.68	$46.88
1 Laborer	26.70	213.60	41.55	332.40		
1 Grader, 30,000 Lbs.		456.60		502.25		
1 Ripper, beam & 1 shank		71.40		78.55	33.00	36.30
16 L.H., Daily Totals		$1018.80		$1330.80	$63.68	$83.18

CREWS

Crew No.	Bare Costs		Incl. Subs O & P		Cost Per Labor-Hour	

Crew B-11K	Hr.	Daily	Hr.	Daily	Bare Costs	Incl. O&P
1 Equipment Oper. (med.)	$34.65	$277.20	$52.20	$417.60	$30.68	$46.88
1 Laborer	26.70	213.60	41.55	332.40		
1 Trencher, 8' D., 16" W.		1420.00		1562.00	88.75	97.63
16 L.H., Daily Totals		$1910.80		$2312.00	$119.43	$144.51

Crew B-11L	Hr.	Daily	Hr.	Daily	Bare Costs	Incl. O&P
1 Equipment Oper. (med.)	$34.65	$277.20	$52.20	$417.60	$30.68	$46.88
1 Laborer	26.70	213.60	41.55	332.40		
1 Grader, 30,000 Lbs.		456.60		502.25	28.54	31.39
16 L.H., Daily Totals		$947.40		$1252.25	$59.22	$78.27

Crew B-11M	Hr.	Daily	Hr.	Daily	Bare Costs	Incl. O&P
1 Equipment Oper. (med.)	$34.65	$277.20	$52.20	$417.60	$30.68	$46.88
1 Laborer	26.70	213.60	41.55	332.40		
1 Backhoe Loader, 80 H.P.		256.40		282.05	16.03	17.63
16 L.H., Daily Totals		$747.20		$1032.05	$46.71	$64.51

Crew B-11N	Hr.	Daily	Hr.	Daily	Bare Costs	Incl. O&P
1 Labor Foreman	$28.70	$229.60	$44.70	$357.60	$29.26	$44.57
2 Equipment Operators (med.)	34.65	554.40	52.20	835.20		
6 Truck Drivers (hvy.)	27.55	1322.40	42.00	2016.00		
1 F.E. Loader, 5.5 C.Y.		700.40		770.45		
1 Dozer, 400 H.P.		1524.00		1676.40		
6 Off Hwy. Tks. 50 Ton		7086.00		7794.60	129.31	142.24
72 L.H., Daily Totals		$11416.80		$13450.25	$158.57	$186.81

Crew B-11Q	Hr.	Daily	Hr.	Daily	Bare Costs	Incl. O&P
1 Equipment Operator (med.)	$34.65	$277.20	$52.20	$417.60	$32.00	$48.65
.5 Laborer	26.70	106.80	41.55	166.20		
1 Dozer, 140 H.P.		572.00		629.20	47.67	52.43
12 L.H., Daily Totals		$956.00		$1213.00	$79.67	$101.08

Crew B-11R	Hr.	Daily	Hr.	Daily	Bare Costs	Incl. O&P
1 Equipment Operator (med.)	$34.65	$277.20	$52.20	$417.60	$32.00	$48.65
.5 Laborer	26.70	106.80	41.55	166.20		
1 Dozer, 215 H.P.		919.60		1011.55	76.63	84.30
12 L.H., Daily Totals		$1303.60		$1595.35	$108.63	$132.95

Crew B-11S	Hr.	Daily	Hr.	Daily	Bare Costs	Incl. O&P
1 Equipment Operator (med.)	$34.65	$277.20	$52.20	$417.60	$32.00	$48.65
.5 Laborer	26.70	106.80	41.55	166.20		
1 Dozer, 300 H.P.		1195.00		1314.50		
1 Ripper, Beam & 1 Shank		71.40		78.55	105.53	116.09
12 L.H., Daily Totals		$1650.40		$1976.85	$137.53	$164.74

Crew B-11T	Hr.	Daily	Hr.	Daily	Bare Costs	Incl. O&P
1 Equipment Operator (med.)	$34.65	$277.20	$52.20	$417.60	$32.00	$48.65
.5 Laborer	26.70	106.80	41.55	166.20		
1 Dozer, 410 H.P.		1524.00		1676.40		
1 Ripper, Beam & 2 Shanks		80.20		88.20	133.68	147.05
12 L.H., Daily Totals		$1988.20		$2348.40	$165.68	$195.70

Crew B-11U	Hr.	Daily	Hr.	Daily	Bare Costs	Incl. O&P
1 Equipment Operator (med.)	$34.65	$277.20	$52.20	$417.60	$32.00	$48.65
.5 Laborer	26.70	106.80	41.55	166.20		
1 Dozer, 520 H.P.		2035.00		2238.50	169.58	186.54
12 L.H., Daily Totals		$2419.00		$2822.30	$201.58	$235.19

Crew B-11V	Hr.	Daily	Hr.	Daily	Bare Costs	Incl. O&P
3 Laborers	$26.70	$640.80	$41.55	$997.20	$26.70	$41.55
1 Walk behind compactor, 7.5 HP		122.80		135.10	5.12	5.63
24 L.H., Daily Totals		$763.60		$1132.30	$31.82	$47.18

Crew B-11W	Hr.	Daily	Hr.	Daily	Bare Costs	Incl. O&P
1 Equipment Operator (med.)	$34.65	$277.20	$52.20	$417.60	$28.07	$42.81
1 Common Laborer	26.70	213.60	41.55	332.40		
10 Truck Drivers, Heavy	27.55	2204.00	42.00	3360.00		
1 Dozer, 200 H.P.		919.60		1011.55		
1 Vib. roller, smth, towed, 23 Ton		591.60		650.75		
10 Dump Truck, 10 Ton		3266.00		3592.60	49.76	54.74
96 L.H., Daily Totals		$7472.00		$9364.90	$77.83	$97.55

Crew B-11Y	Hr.	Daily	Hr.	Daily	Bare Costs	Incl. O&P
1 Labor Foreman (Outside)	$28.70	$229.60	$44.70	$357.60	$29.57	$45.45
5 Common Laborers	26.70	1068.00	41.55	1662.00		
3 Equipment Operator (med.)	34.65	831.60	52.20	1252.80		
1 Dozer, 80 H.P.		314.80		346.30		
2 Walk behind compactor, 7.5 HP		245.60		270.15		
4 Vibratory plate, gas, 21"		147.20		161.90	9.83	10.81
72 L.H., Daily Totals		$2836.80		$4050.75	$39.40	$56.26

Crew B-12A	Hr.	Daily	Hr.	Daily	Bare Costs	Incl. O&P
1 Equip. Oper. (crane)	$35.90	$287.20	$54.10	$432.80	$31.30	$47.83
1 Laborer	26.70	213.60	41.55	332.40		
1 Hyd. Excavator, 1 C.Y.		557.80		613.60	34.86	38.35
16 L.H., Daily Totals		$1058.60		$1378.80	$66.16	$86.18

Crew B-12B	Hr.	Daily	Hr.	Daily	Bare Costs	Incl. O&P
1 Equip. Oper. (crane)	$35.90	$287.20	$54.10	$432.80	$31.30	$47.83
1 Laborer	26.70	213.60	41.55	332.40		
1 Hyd. Excavator, 1.5 C.Y.		720.40		792.45	45.03	49.53
16 L.H., Daily Totals		$1221.20		$1557.65	$76.33	$97.36

Crew B-12C	Hr.	Daily	Hr.	Daily	Bare Costs	Incl. O&P
1 Equip. Oper. (crane)	$35.90	$287.20	$54.10	$432.80	$31.30	$47.83
1 Laborer	26.70	213.60	41.55	332.40		
1 Hyd. Excavator, 2 C.Y.		909.80		1000.80	56.86	62.55
16 L.H., Daily Totals		$1410.60		$1766.00	$88.16	$110.38

Crew B-12D	Hr.	Daily	Hr.	Daily	Bare Costs	Incl. O&P
1 Equip. Oper. (crane)	$35.90	$287.20	$54.10	$432.80	$31.30	$47.83
1 Laborer	26.70	213.60	41.55	332.40		
1 Hyd. Excavator, 3.5 C.Y.		2020.00		2222.00	126.25	138.88
16 L.H., Daily Totals		$2520.80		$2987.20	$157.55	$186.71

Crew B-12E	Hr.	Daily	Hr.	Daily	Bare Costs	Incl. O&P
1 Equip. Oper. (crane)	$35.90	$287.20	$54.10	$432.80	$31.30	$47.83
1 Laborer	26.70	213.60	41.55	332.40		
1 Hyd. Excavator, .5 C.Y.		332.40		365.65	20.78	22.85
16 L.H., Daily Totals		$833.20		$1130.85	$52.08	$70.68

Crew B-12F	Hr.	Daily	Hr.	Daily	Bare Costs	Incl. O&P
1 Equip. Oper. (crane)	$35.90	$287.20	$54.10	$432.80	$31.30	$47.83
1 Laborer	26.70	213.60	41.55	332.40		
1 Hyd. Excavator, .75 C.Y.		474.00		521.40	29.63	32.59
16 L.H., Daily Totals		$974.80		$1286.60	$60.93	$80.42

Crew B-12G

Crew No.	Bare Costs Hr.	Bare Costs Daily	Incl. Subs O&P Hr.	Incl. Subs O&P Daily	Cost Per Labor-Hour Bare Costs	Cost Per Labor-Hour Incl. O&P
1 Equip. Oper. (crane)	$35.90	$287.20	$54.10	$432.80	$31.30	$47.83
1 Laborer	26.70	213.60	41.55	332.40		
1 Power Shovel, .5 C.Y.		508.10		558.90		
1 Clamshell Bucket, .5 C.Y.		33.80		37.20	33.87	37.25
16 L.H., Daily Totals		$1042.70		$1361.30	$65.17	$85.08

Crew B-12H

Crew No.	Bare Costs Hr.	Bare Costs Daily	Incl. Subs O&P Hr.	Incl. Subs O&P Daily	Cost Per Labor-Hour Bare Costs	Cost Per Labor-Hour Incl. O&P
1 Equip. Oper. (crane)	$35.90	$287.20	$54.10	$432.80	$31.30	$47.83
1 Laborer	26.70	213.60	41.55	332.40		
1 Power Shovel, 1 C.Y.		833.20		916.50		
1 Clamshell Bucket, 1 C.Y.		43.40		47.75	54.79	60.27
16 L.H., Daily Totals		$1377.40		$1729.45	$86.09	$108.10

Crew B-12I

Crew No.	Bare Costs Hr.	Bare Costs Daily	Incl. Subs O&P Hr.	Incl. Subs O&P Daily	Cost Per Labor-Hour Bare Costs	Cost Per Labor-Hour Incl. O&P
1 Equip. Oper. (crane)	$35.90	$287.20	$54.10	$432.80	$31.30	$47.83
1 Laborer	26.70	213.60	41.55	332.40		
1 Power Shovel, .75 C.Y.		630.45		693.50		
1 Dragline Bucket, .75 C.Y.		18.80		20.70	40.58	44.63
16 L.H., Daily Totals		$1150.05		$1479.40	$71.88	$92.46

Crew B-12J

Crew No.	Bare Costs Hr.	Bare Costs Daily	Incl. Subs O&P Hr.	Incl. Subs O&P Daily	Cost Per Labor-Hour Bare Costs	Cost Per Labor-Hour Incl. O&P
1 Equip. Oper. (crane)	$35.90	$287.20	$54.10	$432.80	$31.30	$47.83
1 Laborer	26.70	213.60	41.55	332.40		
1 Gradall, 3 Ton, .5 C.Y.		837.80		921.60	52.36	57.60
16 L.H., Daily Totals		$1338.60		$1686.80	$83.66	$105.43

Crew B-12K

Crew No.	Bare Costs Hr.	Bare Costs Daily	Incl. Subs O&P Hr.	Incl. Subs O&P Daily	Cost Per Labor-Hour Bare Costs	Cost Per Labor-Hour Incl. O&P
1 Equip. Oper. (crane)	$35.90	$287.20	$54.10	$432.80	$31.30	$47.83
1 Laborer	26.70	213.60	41.55	332.40		
1 Gradall, 3 Ton, 1 C.Y.		971.00		1068.10	60.69	66.76
16 L.H., Daily Totals		$1471.80		$1833.30	$91.99	$114.59

Crew B-12L

Crew No.	Bare Costs Hr.	Bare Costs Daily	Incl. Subs O&P Hr.	Incl. Subs O&P Daily	Cost Per Labor-Hour Bare Costs	Cost Per Labor-Hour Incl. O&P
1 Equip. Oper. (crane)	$35.90	$287.20	$54.10	$432.80	$31.30	$47.83
1 Laborer	26.70	213.60	41.55	332.40		
1 Power Shovel, .5 C.Y.		508.10		558.90		
1 F.E. Attachment, .5 C.Y.		47.40		52.15	34.72	38.19
16 L.H., Daily Totals		$1056.30		$1376.25	$66.02	$86.02

Crew B-12M

Crew No.	Bare Costs Hr.	Bare Costs Daily	Incl. Subs O&P Hr.	Incl. Subs O&P Daily	Cost Per Labor-Hour Bare Costs	Cost Per Labor-Hour Incl. O&P
1 Equip. Oper. (crane)	$35.90	$287.20	$54.10	$432.80	$31.30	$47.83
1 Laborer	26.70	213.60	41.55	332.40		
1 Power Shovel, .75 C.Y.		630.45		693.50		
1 F.E. Attachment, .75 C.Y.		51.80		57.00	42.64	46.90
16 L.H., Daily Totals		$1183.05		$1515.70	$73.94	$94.73

Crew B-12N

Crew No.	Bare Costs Hr.	Bare Costs Daily	Incl. Subs O&P Hr.	Incl. Subs O&P Daily	Cost Per Labor-Hour Bare Costs	Cost Per Labor-Hour Incl. O&P
1 Equip. Oper. (crane)	$35.90	$287.20	$54.10	$432.80	$31.30	$47.83
1 Laborer	26.70	213.60	41.55	332.40		
1 Power Shovel, 1 C.Y.		833.20		916.50		
1 F.E. Attachment, 1 C.Y.		58.60		64.45	55.74	61.31
16 L.H., Daily Totals		$1392.60		$1746.15	$87.04	$109.14

Crew B-12O

Crew No.	Bare Costs Hr.	Bare Costs Daily	Incl. Subs O&P Hr.	Incl. Subs O&P Daily	Cost Per Labor-Hour Bare Costs	Cost Per Labor-Hour Incl. O&P
1 Equip. Oper. (crane)	$35.90	$287.20	$54.10	$432.80	$31.30	$47.83
1 Laborer	26.70	213.60	41.55	332.40		
1 Power Shovel, 1.5 C.Y.		888.20		977.00		
1 F.E. Attachment, 1.5 C.Y.		67.80		74.60	59.75	65.73
16 L.H., Daily Totals		$1456.80		$1816.80	$91.05	$113.56

Crew B-12P

Crew No.	Bare Costs Hr.	Bare Costs Daily	Incl. Subs O&P Hr.	Incl. Subs O&P Daily	Cost Per Labor-Hour Bare Costs	Cost Per Labor-Hour Incl. O&P
1 Equip. Oper. (crane)	$35.90	$287.20	$54.10	$432.80	$31.30	$47.83
1 Laborer	26.70	213.60	41.55	332.40		
1 Crawler Crane, 40 Ton		888.20		977.00		
1 Dragline Bucket, 1.5 C.Y.		30.60		33.65	57.43	63.17
16 L.H., Daily Totals		$1419.60		$1775.85	$88.73	$111.00

Crew B-12Q

Crew No.	Bare Costs Hr.	Bare Costs Daily	Incl. Subs O&P Hr.	Incl. Subs O&P Daily	Cost Per Labor-Hour Bare Costs	Cost Per Labor-Hour Incl. O&P
1 Equip. Oper. (crane)	$35.90	$287.20	$54.10	$432.80	$31.30	$47.83
1 Laborer	26.70	213.60	41.55	332.40		
1 Hyd. Excavator, 5/8 C.Y.		432.40		475.65	27.03	29.73
16 L.H., Daily Totals		$933.20		$1240.85	$58.33	$77.56

Crew B-12R

Crew No.	Bare Costs Hr.	Bare Costs Daily	Incl. Subs O&P Hr.	Incl. Subs O&P Daily	Cost Per Labor-Hour Bare Costs	Cost Per Labor-Hour Incl. O&P
1 Equip. Oper. (crane)	$35.90	$287.20	$54.10	$432.80	$31.30	$47.83
1 Laborer	26.70	213.60	41.55	332.40		
1 Hyd. Excavator, 1.5 C.Y.		720.40		792.45	45.03	49.53
16 L.H., Daily Totals		$1221.20		$1557.65	$76.33	$97.36

Crew B-12S

Crew No.	Bare Costs Hr.	Bare Costs Daily	Incl. Subs O&P Hr.	Incl. Subs O&P Daily	Cost Per Labor-Hour Bare Costs	Cost Per Labor-Hour Incl. O&P
1 Equip. Oper. (crane)	$35.90	$287.20	$54.10	$432.80	$31.30	$47.83
1 Laborer	26.70	213.60	41.55	332.40		
1 Hyd. Excavator, 2.5 C.Y.		1215.00		1336.50	75.94	83.53
16 L.H., Daily Totals		$1715.80		$2101.70	$107.24	$131.36

Crew B-12T

Crew No.	Bare Costs Hr.	Bare Costs Daily	Incl. Subs O&P Hr.	Incl. Subs O&P Daily	Cost Per Labor-Hour Bare Costs	Cost Per Labor-Hour Incl. O&P
1 Equip. Oper. (crane)	$35.90	$287.20	$54.10	$432.80	$31.30	$47.83
1 Laborer	26.70	213.60	41.55	332.40		
1 Crawler Crane, 75 Ton		1170.00		1287.00		
1 F.E. Attachment, 3 C.Y.		90.20		99.20	78.76	86.64
16 L.H., Daily Totals		$1761.00		$2151.40	$110.06	$134.47

Crew B-12V

Crew No.	Bare Costs Hr.	Bare Costs Daily	Incl. Subs O&P Hr.	Incl. Subs O&P Daily	Cost Per Labor-Hour Bare Costs	Cost Per Labor-Hour Incl. O&P
1 Equip. Oper. (crane)	$35.90	$287.20	$54.10	$432.80	$31.30	$47.83
1 Laborer	26.70	213.60	41.55	332.40		
1 Crawler Crane, 75 Ton		1170.00		1287.00		
1 Dragline Bucket, 3 C.Y.		47.40		52.15	76.09	83.70
16 L.H., Daily Totals		$1718.20		$2104.35	$107.39	$131.53

Crew B-13

Crew No.	Bare Costs Hr.	Bare Costs Daily	Incl. Subs O&P Hr.	Incl. Subs O&P Daily	Cost Per Labor-Hour Bare Costs	Cost Per Labor-Hour Incl. O&P
1 Labor Foreman (outside)	$28.70	$229.60	$44.70	$357.60	$28.79	$44.34
4 Laborers	26.70	854.40	41.55	1329.60		
1 Equip. Oper. (crane)	35.90	287.20	54.10	432.80		
1 Equip. Oper. Oiler	30.10	240.80	45.35	362.80		
1 Hyd. Crane, 25 Ton		616.80		678.50	11.01	12.12
56 L.H., Daily Totals		$2228.80		$3161.30	$39.80	$56.46

Crew B-13A

Crew No.	Bare Costs Hr.	Bare Costs Daily	Incl. Subs O&P Hr.	Incl. Subs O&P Daily	Cost Per Labor-Hour Bare Costs	Cost Per Labor-Hour Incl. O&P
1 Foreman	$28.70	$229.60	$44.70	$357.60	$29.50	$45.17
2 Laborers	26.70	427.20	41.55	664.80		
2 Equipment Operators	34.65	554.40	52.20	835.20		
2 Truck Drivers (heavy)	27.55	440.80	42.00	672.00		
1 Crane, 75 Ton		1170.00		1287.00		
1 F.E. Lder, 3.75 C.Y.		1094.00		1203.40		
2 Dump Trucks, 12 Ton		653.20		718.50	52.09	57.30
56 L.H., Daily Totals		$4569.20		$5738.50	$81.59	$102.47

Crew No.	Bare Costs Hr.	Daily	Incl. Subs O & P Hr.	Daily	Cost Per Labor-Hour Bare Costs	Incl. O&P
Crew B-13B	Hr.	Daily	Hr.	Daily	Bare Costs	Incl. O&P
1 Labor Foreman (outside)	$28.70	$229.60	$44.70	$357.60	$28.79	$44.34
4 Laborers	26.70	854.40	41.55	1329.60		
1 Equip. Oper. (crane)	35.90	287.20	54.10	432.80		
1 Equip. Oper. Oiler	30.10	240.80	45.35	362.80		
1 Hyd. Crane, 55 Ton		895.40		984.95	15.99	17.59
56 L.H., Daily Totals		$2507.40		$3467.75	$44.78	$61.93
Crew B-13C	Hr.	Daily	Hr.	Daily	Bare Costs	Incl. O&P
1 Labor Foreman (outside)	$28.70	$229.60	$44.70	$357.60	$28.79	$44.34
4 Laborers	26.70	854.40	41.55	1329.60		
1 Equip. Oper. (crane)	35.90	287.20	54.10	432.80		
1 Equip. Oper. Oiler	30.10	240.80	45.35	362.80		
1 Crawler Crane, 100 Ton	.	1519.00		1670.90	27.13	29.84
56 L.H., Daily Totals		$3131.00		$4153.70	$55.92	$74.18
Crew B-14	Hr.	Daily	Hr.	Daily	Bare Costs	Incl. O&P
1 Labor Foreman (outside)	$28.70	$229.60	$44.70	$357.60	$28.09	$43.45
4 Laborers	26.70	854.40	41.55	1329.60		
1 Equip. Oper. (light)	33.05	264.40	49.80	398.40		
1 Backhoe Loader, 48 H.P.		215.80		237.40	4.50	4.95
48 L.H., Daily Totals		$1564.20		$2323.00	$32.59	$48.40
Crew B-15	Hr.	Daily	Hr.	Daily	Bare Costs	Incl. O&P
1 Equipment Oper. (med)	$34.65	$277.20	$52.20	$417.60	$29.46	$44.85
.5 Laborer	26.70	106.80	41.55	166.20		
2 Truck Drivers (heavy)	27.55	440.80	42.00	672.00		
2 Dump Trucks, 16 Ton		953.20		1048.50		
1 Dozer, 200 H.P.		919.60		1011.55	66.89	73.57
28 L.H., Daily Totals		$2697.60		$3315.85	$96.35	$118.42
Crew B-16	Hr.	Daily	Hr.	Daily	Bare Costs	Incl. O&P
1 Labor Foreman (outside)	$28.70	$229.60	$44.70	$357.60	$27.41	$42.45
2 Laborers	26.70	427.20	41.55	664.80		
1 Truck Driver (heavy)	27.55	220.40	42.00	336.00		
1 Dump Truck, 16 Ton		476.60		524.25	14.89	16.38
32 L.H., Daily Totals		$1353.80		$1882.65	$42.30	$58.83
Crew B-17	Hr.	Daily	Hr.	Daily	Bare Costs	Incl. O&P
2 Laborers	$26.70	$427.20	$41.55	$664.80	$28.50	$43.73
1 Equip. Oper. (light)	33.05	264.40	49.80	398.40		
1 Truck Driver (heavy)	27.55	220.40	42.00	336.00		
1 Backhoe Loader, 48 H.P.		215.80		237.40		
1 Dump Truck, 12 Ton		326.60		359.25	16.95	18.65
32 L.H., Daily Totals		$1454.40		$1995.85	$45.45	$62.38
Crew B-17A	Hr.	Daily	Hr.	Daily	Bare Costs	Incl. O&P
2 Laborer Foremen	$28.70	$459.20	$44.70	$715.20	$28.93	$45.02
6 Laborers	26.70	1281.60	41.55	1994.40		
1 Skilled Worker Foreman	36.85	294.80	57.30	458.40		
1 Skilled Worker	34.85	278.80	54.20	433.60		
80 L.H., Daily Totals		$2314.40		$3601.60	$28.93	$45.02
Crew B-18	Hr.	Daily	Hr.	Daily	Bare Costs	Incl. O&P
1 Labor Foreman (outside)	$28.70	$229.60	$44.70	$357.60	$27.37	$42.60
2 Laborers	26.70	427.20	41.55	664.80		
1 Vibrating Compactor		36.80		40.50	1.53	1.69
24 L.H., Daily Totals		$693.60		$1062.90	$28.90	$44.29

Crew No.	Bare Costs Hr.	Daily	Incl. Subs O & P Hr.	Daily	Cost Per Labor-Hour Bare Costs	Incl. O&P
Crew B-19	Hr.	Daily	Hr.	Daily	Bare Costs	Incl. O&P
1 Pile Driver Foreman	$35.30	$282.40	$58.15	$465.20	$33.80	$53.91
4 Pile Drivers	33.30	1065.60	54.90	1756.80		
2 Equip. Oper. (crane)	35.90	574.40	54.10	865.60		
1 Equip. Oper. Oiler	30.10	240.80	45.35	362.80		
1 Crane, 40 Ton & Access.		888.20		977.00		
60 L.F. Pile Leads		75.00		82.50		
1 Hammer, Diesel, 22k Ft-Lb		564.00		620.40	23.91	26.30
64 L.H., Daily Totals		$3690.40		$5130.30	$57.71	$80.21
Crew B-19A	Hr.	Daily	Hr.	Daily	Bare Costs	Incl. O&P
1 Pile Driver Foreman	$35.30	$282.40	$58.15	$465.20	$33.80	$53.91
4 Pile Drivers	33.30	1065.60	54.90	1756.80		
2 Equip. Oper. (crane)	35.90	574.40	54.10	865.60		
1 Equip. Oper. Oiler	30.10	240.80	45.35	362.80		
1 Crawler Crane, 75 Ton		1170.00		1287.00		
60 L.F. Leads, 25K Ft. Lbs.		120.00		132.00		
1 Hammer, Diesel, 41k Ft-Lb		662.60		728.85	30.51	33.56
64 L.H., Daily Totals		$4115.80		$5598.25	$64.31	$87.47
Crew B-20	Hr.	Daily	Hr.	Daily	Bare Costs	Incl. O&P
1 Labor Foreman (out)	$28.70	$229.60	$44.70	$357.60	$30.08	$46.82
1 Skilled Worker	34.85	278.80	54.20	433.60		
1 Laborer	26.70	213.60	41.55	332.40		
24 L.H., Daily Totals		$722.00		$1123.60	$30.08	$46.82
Crew B-20A	Hr.	Daily	Hr.	Daily	Bare Costs	Incl. O&P
1 Labor Foreman	$28.70	$229.60	$44.70	$357.60	$32.24	$49.20
1 Laborer	26.70	213.60	41.55	332.40		
1 Plumber	40.85	326.80	61.40	491.20		
1 Plumber Apprentice	32.70	261.60	49.15	393.20		
32 L.H., Daily Totals		$1031.60		$1574.40	$32.24	$49.20
Crew B-21	Hr.	Daily	Hr.	Daily	Bare Costs	Incl. O&P
1 Labor Foreman (out)	$28.70	$229.60	$44.70	$357.60	$30.91	$47.86
1 Skilled Worker	34.85	278.80	54.20	433.60		
1 Laborer	26.70	213.60	41.55	332.40		
.5 Equip. Oper. (crane)	35.90	143.60	54.10	216.40		
.5 S.P. Crane, 5 Ton		158.50		174.35	5.66	6.23
28 L.H., Daily Totals		$1024.10		$1514.35	$36.57	$54.09
Crew B-21A	Hr.	Daily	Hr.	Daily	Bare Costs	Incl. O&P
1 Labor Foreman	$28.70	$229.60	$44.70	$357.60	$32.97	$50.18
1 Laborer	26.70	213.60	41.55	332.40		
1 Plumber	40.85	326.80	61.40	491.20		
1 Plumber Apprentice	32.70	261.60	49.15	393.20		
1 Equip. Oper. (crane)	35.90	287.20	54.10	432.80		
1 S.P. Crane, 12 Ton		505.80		556.40	12.65	13.91
40 L.H., Daily Totals		$1824.60		$2563.60	$45.62	$64.09
Crew B-21B	Hr.	Daily	Hr.	Daily	Bare Costs	Incl. O&P
1 Laborer Foreman	$28.70	$229.60	$44.70	$357.60	$28.94	$44.69
3 Laborers	26.70	640.80	41.55	997.20		
1 Equip. Oper. (crane)	35.90	287.20	54.10	432.80		
1 Hyd. Crane, 12 Ton		602.60		662.85	15.07	16.57
40 L.H., Daily Totals		$1760.20		$2450.45	$44.01	$61.26

Crew B-21C

Crew No.	Bare Costs Hr.	Daily	Incl. Subs O & P Hr.	Daily	Cost Per Labor-Hour Bare Costs	Incl. O&P
1 Laborer Foreman	$28.70	$229.60	$44.70	$357.60	$28.79	$44.34
4 Laborers	26.70	854.40	41.55	1329.60		
1 Equip. Oper. (crane)	35.90	287.20	54.10	432.80		
1 Equip. Oper. Oiler	30.10	240.80	45.35	362.80		
2 Cutting Torches		36.00		39.60		
2 Gases		129.60		142.55		
1 Crane, 90 Ton		1325.00		1457.50	26.62	29.28
56 L.H., Daily Totals		$3102.60		$4122.45	$55.41	$73.62

Crew B-22

Crew No.	Bare Costs Hr.	Daily	Incl. Subs O & P Hr.	Daily	Cost Per Labor-Hour Bare Costs	Incl. O&P
1 Labor Foreman (out)	$28.70	$229.60	$44.70	$357.60	$31.25	$48.27
1 Skilled Worker	34.85	278.80	54.20	433.60		
1 Laborer	26.70	213.60	41.55	332.40		
.75 Equip. Oper. (crane)	35.90	215.40	54.10	324.60		
.75 S.P. Crane, 5 Ton		237.75		261.50	7.93	8.72
30 L.H., Daily Totals		$1175.15		$1709.70	$39.18	$56.99

Crew B-22A

Crew No.	Bare Costs Hr.	Daily	Incl. Subs O & P Hr.	Daily	Cost Per Labor-Hour Bare Costs	Incl. O&P
1 Labor Foreman (out)	$28.70	$229.60	$44.70	$357.60	$30.29	$46.86
1 Skilled Worker	34.85	278.80	54.20	433.60		
2 Laborers	26.70	427.20	41.55	664.80		
.75 Equipment Oper. (crane)	35.90	215.40	54.10	324.60		
.75 Crane, 5 Ton		237.75		261.50		
1 Generator, 5 KW		32.20		35.40		
1 Butt Fusion Machine		432.80		476.10	18.49	20.34
38 L.H., Daily Totals		$1853.75		$2553.60	$48.78	$67.20

Crew B-22B

Crew No.	Bare Costs Hr.	Daily	Incl. Subs O & P Hr.	Daily	Cost Per Labor-Hour Bare Costs	Incl. O&P
1 Skilled Worker	$34.85	$278.80	$54.20	$433.60	$30.78	$47.88
1 Laborer	26.70	213.60	41.55	332.40		
1 Electro Fusion Machine		171.80		189.00	10.74	11.81
16 L.H., Daily Totals		$664.20		$955.00	$41.52	$59.69

Crew B-23

Crew No.	Bare Costs Hr.	Daily	Incl. Subs O & P Hr.	Daily	Cost Per Labor-Hour Bare Costs	Incl. O&P
1 Labor Foreman (outside)	$28.70	$229.60	$44.70	$357.60	$27.10	$42.18
4 Laborers	26.70	854.40	41.55	1329.60		
1 Drill Rig, Wells		3179.00		3496.90		
1 Light Truck, 3 Ton		176.20		193.80	83.88	92.27
40 L.H., Daily Totals		$4439.20		$5377.90	$110.98	$134.45

Crew B-23A

Crew No.	Bare Costs Hr.	Daily	Incl. Subs O & P Hr.	Daily	Cost Per Labor-Hour Bare Costs	Incl. O&P
1 Labor Foreman (outside)	$28.70	$229.60	$44.70	$357.60	$30.02	$46.15
1 Laborer	26.70	213.60	41.55	332.40		
1 Equip. Operator (medium)	34.65	277.20	52.20	417.60		
1 Drill Rig, Wells		3179.00		3496.90		
1 Pickup Truck, 3/4 Ton		78.00		85.80	135.71	149.28
24 L.H., Daily Totals		$3977.40		$4690.30	$165.73	$195.43

Crew B-23B

Crew No.	Bare Costs Hr.	Daily	Incl. Subs O & P Hr.	Daily	Cost Per Labor-Hour Bare Costs	Incl. O&P
1 Labor Foreman (outside)	$28.70	$229.60	$44.70	$357.60	$30.02	$46.15
1 Laborer	26.70	213.60	41.55	332.40		
1 Equip. Operator (medium)	34.65	277.20	52.20	417.60		
1 Drill Rig, Wells		3179.00		3496.90		
1 Pickup Truck, 3/4 Ton		78.00		85.80		
1 Pump, Cntfgl, 6"		217.00		238.70	144.75	159.23
24 L.H., Daily Totals		$4194.40		$4929.00	$174.77	$205.38

Crew B-24

Crew No.	Bare Costs Hr.	Daily	Incl. Subs O & P Hr.	Daily	Cost Per Labor-Hour Bare Costs	Incl. O&P
1 Cement Finisher	$32.85	$262.80	$48.35	$386.80	$31.27	$47.63
1 Laborer	26.70	213.60	41.55	332.40		
1 Carpenter	34.25	274.00	53.35	426.80		
24 L.H., Daily Totals		$750.40		$1146.00	$31.27	$47.63

Crew B-25

Crew No.	Bare Costs Hr.	Daily	Incl. Subs O & P Hr.	Daily	Cost Per Labor-Hour Bare Costs	Incl. O&P
1 Labor Foreman	$28.70	$229.60	$44.70	$357.60	$29.05	$44.74
7 Laborers	26.70	1495.20	41.55	2326.80		
3 Equip. Oper. (med.)	34.65	831.60	52.20	1252.80		
1 Asphalt Paver, 130 H.P.		1590.00		1749.00		
1 Tandem Roller, 10 Ton		179.20		197.10		
1 Roller, Pneumatic Wheel		241.80		266.00	22.85	25.14
88 L.H., Daily Totals		$4567.40		$6149.30	$51.90	$69.88

Crew B-25B

Crew No.	Bare Costs Hr.	Daily	Incl. Subs O & P Hr.	Daily	Cost Per Labor-Hour Bare Costs	Incl. O&P
1 Labor Foreman	$28.70	$229.60	$44.70	$357.60	$29.52	$45.36
7 Laborers	26.70	1495.20	41.55	2326.80		
4 Equip. Oper. (medium)	34.65	1108.80	52.20	1670.40		
1 Asphalt Paver, 130 H.P.		1590.00		1749.00		
2 Rollers, Steel Wheel		358.40		394.25		
1 Roller, Pneumatic Wheel		241.80		266.00	22.81	25.10
96 L.H., Daily Totals		$5023.80		$6764.05	$52.33	$70.46

Crew B-25C

Crew No.	Bare Costs Hr.	Daily	Incl. Subs O & P Hr.	Daily	Cost Per Labor-Hour Bare Costs	Incl. O&P
1 Labor Foreman	$28.70	$229.60	$44.70	$357.60	$29.68	$45.63
3 Laborers	26.70	640.80	41.55	997.20		
2 Equip. Oper. (medium)	34.65	554.40	52.20	835.20		
1 Asphalt Paver, 130 H.P.		1590.00		1749.00		
1 Rollers, Steel Wheel		179.20		197.10	36.86	40.54
48 L.H., Daily Totals		$3194.00		$4136.10	$66.54	$86.17

Crew B-26

Crew No.	Bare Costs Hr.	Daily	Incl. Subs O & P Hr.	Daily	Cost Per Labor-Hour Bare Costs	Incl. O&P
1 Labor Foreman (outside)	$28.70	$229.60	$44.70	$357.60	$29.91	$46.27
6 Laborers	26.70	1281.60	41.55	1994.40		
2 Equip. Oper. (med.)	34.65	554.40	52.20	835.20		
1 Rodman (reinf.)	37.95	303.60	62.60	500.80		
1 Cement Finisher	32.85	262.80	48.35	386.80		
1 Grader, 30,000 Lbs.		456.60		502.25		
1 Paving Mach. & Equip.		1910.00		2101.00	26.89	29.58
88 L.H., Daily Totals		$4998.60		$6678.05	$56.80	$75.85

Crew B-27

Crew No.	Bare Costs Hr.	Daily	Incl. Subs O & P Hr.	Daily	Cost Per Labor-Hour Bare Costs	Incl. O&P
1 Labor Foreman (outside)	$28.70	$229.60	$44.70	$357.60	$27.20	$42.34
3 Laborers	26.70	640.80	41.55	997.20		
1 Berm Machine		216.20		237.80	6.76	7.43
32 L.H., Daily Totals		$1086.60		$1592.60	$33.96	$49.77

Crew B-28

Crew No.	Bare Costs Hr.	Daily	Incl. Subs O & P Hr.	Daily	Cost Per Labor-Hour Bare Costs	Incl. O&P
2 Carpenters	$34.25	$548.00	$53.35	$853.60	$31.73	$49.42
1 Laborer	26.70	213.60	41.55	332.40		
24 L.H., Daily Totals		$761.60		$1186.00	$31.73	$49.42

Crew B-29

Crew No.	Bare Costs Hr.	Daily	Incl. Subs O & P Hr.	Daily	Cost Per Labor-Hour Bare Costs	Incl. O&P
1 Labor Foreman (outside)	$28.70	$229.60	$44.70	$357.60	$28.79	$44.34
4 Laborers	26.70	854.40	41.55	1329.60		
1 Equip. Oper. (crane)	35.90	287.20	54.10	432.80		
1 Equip. Oper. Oiler	30.10	240.80	45.35	362.80		
1 Gradall, 3 Ton, 1/2 C.Y.		837.80		921.60	14.96	16.46
56 L.H., Daily Totals		$2449.80		$3404.40	$43.75	$60.80

Crews

Crew No.	Bare Costs		Incl. Subs O & P		Cost Per Labor-Hour	

Crew B-30

	Hr.	Daily	Hr.	Daily	Bare Costs	Incl. O&P
1 Equip. Oper. (med.)	$34.65	$277.20	$52.20	$417.60	$29.92	$45.40
2 Truck Drivers (heavy)	27.55	440.80	42.00	672.00		
1 Hyd. Excavator, 1.5 C.Y.		720.40		792.45		
2 Dump Trucks, 16 Ton		953.20		1048.50	69.73	76.71
24 L.H., Daily Totals		$2391.60		$2930.55	$99.65	$122.11

Crew B-31

	Hr.	Daily	Hr.	Daily	Bare Costs	Incl. O&P
1 Labor Foreman (outside)	$28.70	$229.60	$44.70	$357.60	$28.61	$44.54
3 Laborers	26.70	640.80	41.55	997.20		
1 Carpenter	34.25	274.00	53.35	426.80		
1 Air Compr., 250 C.F.M.		112.40		123.65		
1 Sheeting Driver		7.20		7.90		
2-50 Ft. Air Hoses, 1.5" Dia.		9.40		10.35	3.23	3.55
40 L.H., Daily Totals		$1273.40		$1923.50	$31.84	$48.09

Crew B-32

	Hr.	Daily	Hr.	Daily	Bare Costs	Incl. O&P
1 Laborer	$26.70	$213.60	$41.55	$332.40	$32.66	$49.54
3 Equip. Oper. (med.)	34.65	831.60	52.20	1252.80		
1 Grader, 30,000 Lbs.		456.60		502.25		
1 Tandem Roller, 10 Ton		179.20		197.10		
1 Dozer, 200 H.P.		919.60		1011.55	48.61	53.47
32 L.H., Daily Totals		$2600.60		$3296.10	$81.27	$103.01

Crew B-32A

	Hr.	Daily	Hr.	Daily	Bare Costs	Incl. O&P
1 Laborer	$26.70	$213.60	$41.55	$332.40	$32.00	$48.65
2 Equip. Oper. (medium)	34.65	554.40	52.20	835.20		
1 Grader, 30,000 Lbs.		456.60		502.25		
1 Roller, Vibratory, 29,000 Lbs.		433.20		476.50	37.08	40.78
24 L.H., Daily Totals		$1657.80		$2146.35	$69.08	$89.43

Crew B-32B

	Hr.	Daily	Hr.	Daily	Bare Costs	Incl. O&P
1 Laborer	$26.70	$213.60	$41.55	$332.40	$32.00	$48.65
2 Equip. Oper. (medium)	34.65	554.40	52.20	835.20		
1 Dozer, 200 H.P.		919.60		1011.55		
1 Roller, Vibratory, 29,000 Lbs.		433.20		476.50	56.37	62.00
24 L.H., Daily Totals		$2120.80		$2655.65	$88.37	$110.65

Crew B-32C

	Hr.	Daily	Hr.	Daily	Bare Costs	Incl. O&P
1 Labor Foreman	$28.70	$229.60	$44.70	$357.60	$31.01	$47.40
2 Laborers	26.70	427.20	41.55	664.80		
3 Equip. Oper. (medium)	34.65	831.60	52.20	1252.80		
1 Grader, 30,000 Lbs.		456.60		502.25		
1 Roller, Steel Wheel		179.20		197.10		
1 Dozer, 200 H.P.		919.60		1011.55	32.40	35.64
48 L.H., Daily Totals		$3043.80		$3986.10	$63.41	$83.04

Crew B-33A

	Hr.	Daily	Hr.	Daily	Bare Costs	Incl. O&P
1 Equip. Oper. (med.)	$34.65	$277.20	$52.20	$417.60	$32.38	$49.16
.5 Laborer	26.70	106.80	41.55	166.20		
.25 Equip. Oper. (med.)	34.65	69.30	52.20	104.40		
1 Scraper, Towed, 7 C.Y.		150.75		165.80		
1.25 Dozer, 300 H.P.		1493.75		1643.15	117.47	129.21
14 L.H., Daily Totals		$2097.80		$2497.15	$149.85	$178.37

Crew B-33B

	Hr.	Daily	Hr.	Daily	Bare Costs	Incl. O&P
1 Equip. Oper. (med.)	$34.65	$277.20	$52.20	$417.60	$32.38	$49.16
.5 Laborer	26.70	106.80	41.55	166.20		
.25 Equip. Oper. (med.)	34.65	69.30	52.20	104.40		
1 Scraper, Towed, 10 C.Y.		169.20		186.10		
1.25 Dozer, 300 H.P.		1493.75		1643.15	118.78	130.66
14 L.H., Daily Totals		$2116.25		$2517.45	$151.16	$179.82

Crew B-33C

	Hr.	Daily	Hr.	Daily	Bare Costs	Incl. O&P
1 Equip. Oper. (med.)	$34.65	$277.20	$52.20	$417.60	$32.38	$49.16
.5 Laborer	26.70	106.80	41.55	166.20		
.25 Equip. Oper. (med.)	34.65	69.30	52.20	104.40		
1 Scraper, Towed, 12 C.Y.		169.20		186.10		
1.25 Dozer, 300 H.P.		1493.75		1643.15	118.78	130.66
14 L.H., Daily Totals		$2116.25		$2517.45	$151.16	$179.82

Crew B-33D

	Hr.	Daily	Hr.	Daily	Bare Costs	Incl. O&P
1 Equip. Oper. (med.)	$34.65	$277.20	$52.20	$417.60	$32.38	$49.16
.5 Laborer	26.70	106.80	41.55	166.20		
.25 Equip. Oper. (med.)	34.65	69.30	52.20	104.40		
1 S.P. Scraper, 14 C.Y.		1552.00		1707.20		
.25 Dozer, 300 H.P.		298.75		328.65	132.20	145.42
14 L.H., Daily Totals		$2304.05		$2724.05	$164.58	$194.58

Crew B-33E

	Hr.	Daily	Hr.	Daily	Bare Costs	Incl. O&P
1 Equip. Oper. (med.)	$34.65	$277.20	$52.20	$417.60	$32.38	$49.16
.5 Laborer	26.70	106.80	41.55	166.20		
.25 Equip. Oper. (med.)	34.65	69.30	52.20	104.40		
1 S.P. Scraper, 24 C.Y.		2317.00		2548.70		
.25 Dozer, 300 H.P.		298.75		328.65	186.84	205.52
14 L.H., Daily Totals		$3069.05		$3565.55	$219.22	$254.68

Crew B-33F

	Hr.	Daily	Hr.	Daily	Bare Costs	Incl. O&P
1 Equip. Oper. (med.)	$34.65	$277.20	$52.20	$417.60	$32.38	$49.16
.5 Laborer	26.70	106.80	41.55	166.20		
.25 Equip. Oper. (med.)	34.65	69.30	52.20	104.40		
1 Elev. Scraper, 11 C.Y.		819.40		901.35		
.25 Dozer, 300 H.P.		298.75		328.65	79.87	87.85
14 L.H., Daily Totals		$1571.45		$1918.20	$112.25	$137.01

Crew B-33G

	Hr.	Daily	Hr.	Daily	Bare Costs	Incl. O&P
1 Equip. Oper. (med.)	$34.65	$277.20	$52.20	$417.60	$32.38	$49.16
.5 Laborer	26.70	106.80	41.55	166.20		
.25 Equip. Oper. (med.)	34.65	69.30	52.20	104.40		
1 Elev. Scraper, 20 C.Y.		1648.00		1812.80		
.25 Dozer, 300 H.P.		298.75		328.65	139.05	152.96
14 L.H., Daily Totals		$2400.05		$2829.65	$171.43	$202.12

Crew B-33H

	Hr.	Daily	Hr.	Daily	Bare Costs	Incl. O&P
.25 Laborer	$26.70	$53.40	$41.55	$83.10	$33.28	$50.36
1 Equipment Operator (med.)	34.65	277.20	52.20	417.60		
.2 Equipment Operator (med.)	34.65	55.44	52.20	83.52		
1 Scraper, 32-44 C.Y.		2728.00		3000.80		
.2 Dozer, 400 H.P.		304.80		335.30	261.45	287.59
11. L.H., Daily Totals		$3418.84		$3920.32	$294.73	$337.95

Crew B-33J

	Hr.	Daily	Hr.	Daily	Bare Costs	Incl. O&P
1 Equipment Operator (med.)	$34.65	$277.20	$52.20	$417.60	$34.65	$52.20
1 Scraper 17 C.Y.		1552.00		1707.20	194.00	213.40
8 L.H., Daily Totals		$1829.20		$2124.80	$228.65	$265.60

Crew B-34A

	Hr.	Daily	Hr.	Daily	Bare Costs	Incl. O&P
1 Truck Driver (heavy)	$27.55	$220.40	$42.00	$336.00	$27.55	$42.00
1 Dump Truck, 12 Ton		326.60		359.25	40.83	44.91
8 L.H., Daily Totals		$547.00		$695.25	$68.38	$86.91

Crew No.	Bare Costs		Incl. Subs O & P		Cost Per Labor-Hour	
Crew B-34B	Hr.	Daily	Hr.	Daily	Bare Costs	Incl. O&P
1 Truck Driver (heavy)	$27.55	$220.40	$42.00	$336.00	$27.55	$42.00
1 Dump Truck, 16 Ton		476.60		524.25	59.58	65.53
8 L.H., Daily Totals		$697.00		$860.25	$87.13	$107.53

Crew No.	Bare Costs		Incl. Subs O & P		Cost Per Labor-Hour	
Crew B-34C	Hr.	Daily	Hr.	Daily	Bare Costs	Incl. O&P
1 Truck Driver (heavy)	$27.55	$220.40	$42.00	$336.00	$27.55	$42.00
1 Truck Tractor, 40 Ton		301.60		331.75		
1 Dump Trailer, 16.5 C.Y.		103.00		113.30	50.58	55.63
8 L.H., Daily Totals		$625.00		$781.05	$78.13	$97.63

Crew No.	Bare Costs		Incl. Subs O & P		Cost Per Labor-Hour	
Crew B-34D	Hr.	Daily	Hr.	Daily	Bare Costs	Incl. O&P
1 Truck Driver (heavy)	$27.55	$220.40	$42.00	$336.00	$27.55	$42.00
1 Truck Tractor, 40 Ton		301.60		331.75		
1 Dump Trailer, 20 C.Y.		115.80		127.40	52.18	57.39
8 L.H., Daily Totals		$637.80		$795.15	$79.73	$99.39

Crew No.	Bare Costs		Incl. Subs O & P		Cost Per Labor-Hour	
Crew B-34E	Hr.	Daily	Hr.	Daily	Bare Costs	Incl. O&P
1 Truck Driver (heavy)	$27.55	$220.40	$42.00	$336.00	$27.55	$42.00
1 Truck, Off Hwy., 25 Ton		912.20		1003.40	114.03	125.43
8 L.H., Daily Totals		$1132.60		$1339.40	$141.58	$167.43

Crew No.	Bare Costs		Incl. Subs O & P		Cost Per Labor-Hour	
Crew B-34F	Hr.	Daily	Hr.	Daily	Bare Costs	Incl. O&P
1 Truck Driver (heavy)	$27.55	$220.40	$42.00	$336.00	$27.55	$42.00
1 Truck, Off Hwy., 22 C.Y.		933.40		1026.75	116.68	128.34
8 L.H., Daily Totals		$1153.80		$1362.75	$144.23	$170.34

Crew No.	Bare Costs		Incl. Subs O & P		Cost Per Labor-Hour	
Crew B-34G	Hr.	Daily	Hr.	Daily	Bare Costs	Incl. O&P
1 Truck Driver (heavy)	$27.55	$220.40	$42.00	$336.00	$27.55	$42.00
1 Truck, Off Hwy., 34 C.Y.		1181.00		1299.10	147.63	162.39
8 L.H., Daily Totals		$1401.40		$1635.10	$175.18	$204.39

Crew No.	Bare Costs		Incl. Subs O & P		Cost Per Labor-Hour	
Crew B-34H	Hr.	Daily	Hr.	Daily	Bare Costs	Incl. O&P
1 Truck Driver (heavy)	$27.55	$220.40	$42.00	$336.00	$27.55	$42.00
1 Truck, Off Hwy., 42 C.Y.		1276.00		1403.60	159.50	175.45
8 L.H., Daily Totals		$1496.40		$1739.60	$187.05	$217.45

Crew No.	Bare Costs		Incl. Subs O & P		Cost Per Labor-Hour	
Crew B-34J	Hr.	Daily	Hr.	Daily	Bare Costs	Incl. O&P
1 Truck Driver (heavy)	$27.55	$220.40	$42.00	$336.00	$27.55	$42.00
1 Truck, Off Hwy., 60 C.Y.		1649.00		1813.90	206.13	226.74
8 L.H., Daily Totals		$1869.40		$2149.90	$233.68	$268.74

Crew No.	Bare Costs		Incl. Subs O & P		Cost Per Labor-Hour	
Crew B-34K	Hr.	Daily	Hr.	Daily	Bare Costs	Incl. O&P
1 Truck Driver (heavy)	$27.55	$220.40	$42.00	$336.00	$27.55	$42.00
1 Truck Tractor, 240 H.P.		353.60		388.95		
1 Low Bed Trailer		170.20		187.20	65.48	72.02
8 L.H., Daily Totals		$744.20		$912.15	$93.03	$114.02

Crew No.	Bare Costs		Incl. Subs O & P		Cost Per Labor-Hour	
Crew B-34N	Hr.	Daily	Hr.	Daily	Bare Costs	Incl. O&P
1 Truck Driver (heavy)	$27.55	$220.40	$42.00	$336.00	$27.55	$42.00
1 Dump Truck, 12 Ton		326.60		359.25		
1 Flatbed Trailer, 40 Ton		119.40		131.35	55.75	61.33
8 L.H., Daily Totals		$666.40		$826.60	$83.30	$103.33

Crew No.	Bare Costs		Incl. Subs O & P		Cost Per Labor-Hour	
Crew B-34P	Hr.	Daily	Hr.	Daily	Bare Costs	Incl. O&P
1 Pipe Fitter	$40.95	$327.60	$61.55	$492.40	$34.13	$51.53
1 Truck Driver (light)	26.80	214.40	40.85	326.80		
1 Equip. Oper. (medium)	34.65	277.20	52.20	417.60		
1 Flatbed Truck, 3 Ton		176.20		193.80		
1 Backhoe Loader, 48 H.P.		215.80		237.40	16.33	17.97
24 L.H., Daily Totals		$1211.20		$1668.00	$50.46	$69.50

Crew No.	Bare Costs		Incl. Subs O & P		Cost Per Labor-Hour	
Crew B-34Q	Hr.	Daily	Hr.	Daily	Bare Costs	Incl. O&P
1 Pipe Fitter	$40.95	$327.60	$61.55	$492.40	$34.55	$52.17
1 Truck Driver (light)	26.80	214.40	40.85	326.80		
1 Eqip. Oper. (crane)	35.90	287.20	54.10	432.80		
1 Flatbed Trailer, 25 Ton		88.00		96.80		
1 Dump Truck, 12 Ton		326.60		359.25		
1 Hyd. Crane, 25 Ton		616.80		678.50	42.98	47.27
24 L.H., Daily Totals		$1860.60		$2386.55	$77.53	$99.44

Crew No.	Bare Costs		Incl. Subs O & P		Cost Per Labor-Hour	
Crew B-34R	Hr.	Daily	Hr.	Daily	Bare Costs	Incl. O&P
1 Pipe Fitter	$40.95	$327.60	$61.55	$492.40	$34.55	$52.17
1 Truck Driver (light)	26.80	214.40	40.85	326.80		
1 Eqip. Oper. (crane)	35.90	287.20	54.10	432.80		
1 Flatbed Trailer, 25 Ton		88.00		96.80		
1 Dump Truck, 12 Ton		326.60		359.25		
1 Hyd. Crane, 25 Ton		616.80		678.50		
1 Hyd. Excavator, 1 C.Y.		557.80		613.60	66.22	72.84
24 L.H., Daily Totals		$2418.40		$3000.15	$100.77	$125.01

Crew No.	Bare Costs		Incl. Subs O & P		Cost Per Labor-Hour	
Crew B-34S	Hr.	Daily	Hr.	Daily	Bare Costs	Incl. O&P
2 Pipe Fitters	$40.95	$655.20	$61.55	$984.80	$36.34	$54.80
1 Truck Driver (heavy)	27.55	220.40	42.00	336.00		
1 Eqip. Oper. (crane)	35.90	287.20	54.10	432.80		
1 Flatbed Trailer, 40 Ton		119.40		131.35		
1 Truck Tractor, 40 Ton		301.60		331.75		
1 Truck Crane, 80 Ton		995.00		1094.50		
1 Hyd. Excavator, 2 C.Y.		909.80		1000.80	72.68	79.95
32 L.H., Daily Totals		$3488.60		$4312.00	$109.02	$134.75

Crew No.	Bare Costs		Incl. Subs O & P		Cost Per Labor-Hour	
Crew B-34T	Hr.	Daily	Hr.	Daily	Bare Costs	Incl. O&P
2 Pipe Fitters	$40.95	$655.20	$61.55	$984.80	$36.34	$54.80
1 Truck Driver (heavy)	27.55	220.40	42.00	336.00		
1 Eqip. Oper. (crane)	35.90	287.20	54.10	432.80		
1 Flatbed Trailer, 40 Ton		119.40		131.35		
1 Truck Tractor, 40 Ton		301.60		331.75		
1 Truck Crane, 80 Ton		995.00		1094.50	44.25	48.68
32 L.H., Daily Totals		$2578.80		$3311.20	$80.59	$103.48

Crew No.	Bare Costs		Incl. Subs O & P		Cost Per Labor-Hour	
Crew B-35	Hr.	Daily	Hr.	Daily	Bare Costs	Incl. O&P
1 Laborer Foreman (out)	$28.70	$229.60	$44.70	$357.60	$32.85	$50.22
1 Skilled Worker	34.85	278.80	54.20	433.60		
1 Welder (plumber)	40.85	326.80	61.40	491.20		
1 Laborer	26.70	213.60	41.55	332.40		
1 Equip. Oper. (crane)	35.90	287.20	54.10	432.80		
1 Equip. Oper. Oiler	30.10	240.80	45.35	362.80		
1 Electric Welding Mach.		80.65		88.70		
1 Hyd. Excavator, .75 C.Y.		474.00		521.40	11.56	12.71
48 L.H., Daily Totals		$2131.45		$3020.50	$44.41	$62.93

Crew No.	Bare Costs		Incl. Subs O & P		Cost Per Labor-Hour	
Crew B-35A	Hr.	Daily	Hr.	Daily	Bare Costs	Incl. O&P
1 Laborer Foreman (out)	$28.70	$229.60	$44.70	$357.60	$31.97	$48.98
2 Laborers	26.70	427.20	41.55	664.80		
1 Skilled Worker	34.85	278.80	54.20	433.60		
1 Welder (plumber)	40.85	326.80	61.40	491.20		
1 Equip. Oper. (crane)	35.90	287.20	54.10	432.80		
1 Equip. Oper. Oiler	30.10	240.80	45.35	362.80		
1 Welder, 300 amp		81.20		89.30		
1 Crane, 75 Ton		1170.00		1287.00	22.34	24.58
56 L.H., Daily Totals		$3041.60		$4119.10	$54.31	$73.56

	Hr.	Daily	Hr.	Daily	Bare Costs	Incl. O&P
Crew B-36						
1 Labor Foreman (outside)	$28.70	$229.60	$44.70	$357.60	$30.28	$46.44
2 Laborers	26.70	427.20	41.55	664.80		
2 Equip. Oper. (med.)	34.65	554.40	52.20	835.20		
1 Dozer, 200 H.P.		919.60		1011.55		
1 Aggregate Spreader		35.00		38.50		
1 Tandem Roller, 10 Ton		179.20		197.10	28.35	31.18
40 L.H., Daily Totals		$2345.00		$3104.75	$58.63	$77.62

	Hr.	Daily	Hr.	Daily	Bare Costs	Incl. O&P
Crew B-36A						
1 Labor Foreman (outside)	$28.70	$229.60	$44.70	$357.60	$31.53	$48.09
2 Laborers	26.70	427.20	41.55	664.80		
4 Equip. Oper. (med.)	34.65	1108.80	52.20	1670.40		
1 Dozer, 200 H.P.		919.60		1011.55		
1 Aggregate Spreader		35.00		38.50		
1 Roller, Steel Wheel		179.20		197.10		
1 Roller, Pneumatic Wheel		241.80		266.00	24.56	27.02
56 L.H., Daily Totals		$3141.20		$4205.95	$56.09	$75.11

	Hr.	Daily	Hr.	Daily	Bare Costs	Incl. O&P
Crew B-36B						
1 Labor Foreman (outside)	$28.70	$229.60	$44.70	$357.60	$31.03	$47.33
2 Laborers	26.70	427.20	41.55	664.80		
4 Equip. Oper. (medium)	34.65	1108.80	52.20	1670.40		
1 Truck Driver, Heavy	27.55	220.40	42.00	336.00		
1 Grader, 30,000 Lbs.		456.60		502.25		
1 F.E. Loader, crl, 1.5 C.Y.		368.80		405.70		
1 Dozer, 300 H.P.		1195.00		1314.50		
1 Roller, Vibratory		433.20		476.50		
1 Truck, Tractor, 240 H.P.		353.60		388.95		
1 Water Tanker, 5000 Gal.		116.80		128.50	45.69	50.26
64 L.H., Daily Totals		$4910.00		$6245.20	$76.72	$97.59

	Hr.	Daily	Hr.	Daily	Bare Costs	Incl. O&P
Crew B-36C						
1 Labor Foreman (outside)	$28.70	$229.60	$44.70	$357.60	$32.04	$48.66
3 Equip. Oper. (medium)	34.65	831.60	52.20	1252.80		
1 Truck Driver, Heavy	27.55	220.40	42.00	336.00		
1 Grader, 30,000 Lbs.		456.60		502.25		
1 Dozer, 300 H.P.		1195.00		1314.50		
1 Roller, Vibratory		433.20		476.50		
1 Truck, Tractor, 240 H.P.		353.60		388.95		
1 Water Tanker, 5000 Gal.		116.80		128.50	63.88	70.27
40 L.H., Daily Totals		$3836.80		$4757.10	$95.92	$118.93

	Hr.	Daily	Hr.	Daily	Bare Costs	Incl. O&P
Crew B-37						
1 Labor Foreman (outside)	$28.70	$229.60	$44.70	$357.60	$28.09	$43.45
4 Laborers	26.70	854.40	41.55	1329.60		
1 Equip. Oper. (light)	33.05	264.40	49.80	398.40		
1 Tandem Roller, 5 Ton		107.40		118.15	2.24	2.46
48 L.H., Daily Totals		$1455.80		$2203.75	$30.33	$45.91

Crew No.	Bare Costs		Incl. Subs O & P		Cost Per Labor-Hour	
Crew B-38	Hr.	Daily	Hr.	Daily	Bare Costs	Incl. O&P
1 Labor Foreman (outside)	$28.70	$229.60	$44.70	$357.60	$29.96	$45.96
2 Laborers	26.70	427.20	41.55	664.80		
1 Equip. Oper. (light)	33.05	264.40	49.80	398.40		
1 Equip. Oper. (medium)	34.65	277.20	52.20	417.60		
1 Backhoe Loader, 48 H.P.		215.80		237.40		
1 Hyd.Hammer, (1200 lb)		112.00		123.20		
1 F.E. Loader, (170 H.P.)		445.60		490.15		
1 Pavt. Rem. Bucket		51.80		57.00	20.63	22.69
40 L.H., Daily Totals		$2023.60		$2746.15	$50.59	$68.65

	Hr.	Daily	Hr.	Daily	Bare Costs	Incl. O&P
Crew B-39						
1 Labor Foreman (outside)	$28.70	$229.60	$44.70	$357.60	$28.09	$43.45
4 Laborers	26.70	854.40	41.55	1329.60		
1 Equip. Oper. (light)	33.05	264.40	49.80	398.40		
1 Air Compr., 250 C.F.M.		112.40		123.65		
2 Air Tools & Accessories		19.60		21.55		
2-50 Ft. Air Hoses, 1.5" Dia.		9.40		10.35	2.95	3.24
48 L.H., Daily Totals		$1489.80		$2241.15	$31.04	$46.69

	Hr.	Daily	Hr.	Daily	Bare Costs	Incl. O&P
Crew B-40						
1 Pile Driver Foreman (out)	$35.30	$282.40	$58.15	$465.20	$33.80	$53.91
4 Pile Drivers	33.30	1065.60	54.90	1756.80		
2 Equip. Oper. (crane)	35.90	574.40	54.10	865.60		
1 Equip. Oper. Oiler	30.10	240.80	45.35	362.80		
1 Crane, 40 Ton		888.20		977.00		
1 Vibratory Hammer & Gen.		1403.00		1543.30	35.80	39.38
64 L.H., Daily Totals		$4454.40		$5970.70	$69.60	$93.29

	Hr.	Daily	Hr.	Daily	Bare Costs	Incl. O&P
Crew B-40B						
1 Laborer Foreman	$28.70	$229.60	$44.70	$357.60	$29.13	$44.80
3 Laborers	26.70	640.80	41.55	997.20		
1 Equip. Oper. (crane)	35.90	287.20	54.10	432.80		
1 Equip. Oper. Oiler	30.10	240.80	45.35	362.80		
1 Crane, 40 Ton		951.90		1047.10	19.83	21.81
48 L.H., Daily Totals		$2350.30		$3197.50	$48.96	$66.61

	Hr.	Daily	Hr.	Daily	Bare Costs	Incl. O&P
Crew B-41						
1 Labor Foreman (outside)	$28.70	$229.60	$44.70	$357.60	$27.64	$42.87
4 Laborers	26.70	854.40	41.55	1329.60		
.25 Equip. Oper. (crane)	35.90	71.80	54.10	108.20		
.25 Equip. Oper. Oiler	30.10	60.20	45.35	90.70		
.25 Crawler Crane, 40 Ton		222.05		244.25	5.05	5.55
44 L.H., Daily Totals		$1438.05		$2130.35	$32.69	$48.42

	Hr.	Daily	Hr.	Daily	Bare Costs	Incl. O&P
Crew B-42						
1 Labor Foreman (outside)	$28.70	$229.60	$44.70	$357.60	$29.96	$47.33
4 Laborers	26.70	854.40	41.55	1329.60		
1 Equip. Oper. (crane)	35.90	287.20	54.10	432.80		
1 Equip. Oper. Oiler	30.10	240.80	45.35	362.80		
1 Welder	38.15	305.20	68.25	546.00		
1 Hyd. Crane, 25 Ton		616.80		678.50		
1 Gas Welding Machine		81.20		89.30		
1 Horz. Boring Csg. Mch.		372.40		409.65	16.73	18.40
64 L.H., Daily Totals		$2987.60		$4206.25	$46.69	$65.73

Crew No.	Bare Costs		Incl. Subs O & P		Cost Per Labor-Hour	
Crew B-43	Hr.	Daily	Hr.	Daily	Bare Costs	Incl. O&P
1 Labor Foreman (outside)	$28.70	$229.60	$44.70	$357.60	$29.13	$44.80
3 Laborers	26.70	640.80	41.55	997.20		
1 Equip. Oper. (crane)	35.90	287.20	54.10	432.80		
1 Equip. Oper. Oiler	30.10	240.80	45.35	362.80		
1 Drill Rig & Augers		3179.00		3496.90	66.23	72.85
48 L.H., Daily Totals		$4577.40		$5647.30	$95.36	$117.65
Crew B-44	Hr.	Daily	Hr.	Daily	Bare Costs	Incl. O&P
1 Pile Driver Foreman	$35.30	$282.40	$58.15	$465.20	$33.38	$53.44
4 Pile Drivers	33.30	1065.60	54.90	1756.80		
2 Equip. Oper. (crane)	35.90	574.40	54.10	865.60		
1 Laborer	26.70	213.60	41.55	332.40		
1 Crane, 40 Ton, & Access.		888.20		977.00		
45 L.F. Leads, 15K Ft. Lbs.		56.25		61.90	14.79	16.27
64 L.H., Daily Totals		$3080.45		$4458.90	$48.17	$69.71
Crew B-45	Hr.	Daily	Hr.	Daily	Bare Costs	Incl. O&P
1 Equip. Oper. (med.)	$34.65	$277.20	$52.20	$417.60	$31.10	$47.10
1 Truck Driver (heavy)	27.55	220.40	42.00	336.00		
1 Dist. Tank Truck, 3K Gal.		233.80		257.20		
1 Tractor, 4 x 2, 250 H.P.		291.60		320.75	32.84	36.12
16 L.H., Daily Totals		$1023.00		$1331.55	$63.94	$83.22
Crew B-46	Hr.	Daily	Hr.	Daily	Bare Costs	Incl. O&P
1 Pile Driver Foreman	$35.30	$282.40	$58.15	$465.20	$30.33	$48.77
2 Pile Drivers	33.30	532.80	54.90	878.40		
3 Laborers	26.70	640.80	41.55	997.20		
1 Chain Saw, 36" Long		33.40		36.75	.70	.77
48 L.H., Daily Totals		$1489.40		$2377.55	$31.03	$49.54
Crew B-47	Hr.	Daily	Hr.	Daily	Bare Costs	Incl. O&P
1 Blast Foreman	$28.70	$229.60	$44.70	$357.60	$29.48	$45.35
1 Driller	26.70	213.60	41.55	332.40		
1 Equip. Oper. (light)	33.05	264.40	49.80	398.40		
1 Crawler Type Drill, 4"		664.40		730.85		
1 Air Compr., 600 C.F.M.		297.80		327.60		
2-50 Ft. Air Hoses, 3" Dia.		35.40		38.95	41.57	45.72
24 L.H., Daily Totals		$1705.20		$2185.80	$71.05	$91.07
Crew B-47A	Hr.	Daily	Hr.	Daily	Bare Costs	Incl. O&P
1 Drilling Foreman	$28.70	$229.60	$44.70	$357.60	$31.57	$48.05
1 Equip. Oper. (heavy)	35.90	287.20	54.10	432.80		
1 Oiler	30.10	240.80	45.35	362.80		
1 Quarry Drill		822.60		904.85	34.28	37.70
24 L.H., Daily Totals		$1580.20		$2058.05	$65.85	$85.75
Crew B-47C	Hr.	Daily	Hr.	Daily	Bare Costs	Incl. O&P
1 Laborer	$26.70	$213.60	$41.55	$332.40	$29.88	$45.68
1 Equip. Oper. (light)	33.05	264.40	49.80	398.40		
1 Air Compressor, 750 CFM		317.40		349.15		
2-50' Air Hose, 3"		35.40		38.95		
1 Air Track Drill, 4"		664.40		730.85	63.58	69.93
16 L.H., Daily Totals		$1495.20		$1849.75	$93.46	$115.61
Crew B-47E	Hr.	Daily	Hr.	Daily	Bare Costs	Incl. O&P
1 Laborer Foreman	$28.70	$229.60	$44.70	$357.60	$27.20	$42.34
3 Laborers	26.70	640.80	41.55	997.20		
1 Truck, Flatbed, 3 Ton		176.20		193.80	5.51	6.06
32 L.H., Daily Totals		$1046.60		$1548.60	$32.71	$48.40

Crew No.	Bare Costs		Incl. Subs O & P		Cost Per Labor-Hour	
Crew B-47G	Hr.	Daily	Hr.	Daily	Bare Costs	Incl. O&P
1 Laborer Foreman	$28.70	$229.60	$44.70	$357.60	$28.79	$44.40
2 Laborers	26.70	427.20	41.55	664.80		
1 Equip. Oper. (light)	33.05	264.40	49.80	398.40		
1 Air Track Drill, 4"		664.40		730.85		
1 Air Compr., 600 C.F.M.		297.80		327.60		
2-50 Ft. Air Hoses, 3" Dia.		35.40		38.95		
1 Grout Pump		101.40		111.55	34.34	37.78
32 L.H., Daily Totals		$2020.20		$2629.75	$63.13	$82.18
Crew B-48	Hr.	Daily	Hr.	Daily	Bare Costs	Incl. O&P
1 Labor Foreman (outside)	$28.70	$229.60	$44.70	$357.60	$29.69	$45.51
3 Laborers	26.70	640.80	41.55	997.20		
1 Equip. Oper. (crane)	35.90	287.20	54.10	432.80		
1 Equip. Oper. Oiler	30.10	240.80	45.35	362.80		
1 Equip. Oper. (light)	33.05	264.40	49.80	398.40		
1 Centr. Water Pump, 6"		217.00		238.70		
1-20 Ft. Suction Hose, 6"		12.80		14.10		
1-50 Ft. Disch. Hose, 6"		8.90		9.80		
1 Drill Rig & Augers		3179.00		3496.90	61.03	67.13
56 L.H., Daily Totals		$5080.50		$6308.30	$90.72	$112.64
Crew B-49	Hr.	Daily	Hr.	Daily	Bare Costs	Incl. O&P
1 Labor Foreman (outside)	$28.70	$229.60	$44.70	$357.60	$30.95	$47.99
3 Laborers	26.70	640.80	41.55	997.20		
2 Equip. Oper. (crane)	35.90	574.40	54.10	865.60		
2 Equip. Oper. Oilers	30.10	481.60	45.35	725.60		
1 Equip. Oper. (light)	33.05	264.40	49.80	398.40		
2 Pile Drivers	33.30	532.80	54.90	878.40		
1 Hyd. Crane, 25 Ton		616.80		678.50		
1 Centr. Water Pump, 6"		217.00		238.70		
1-20 Ft. Suction Hose, 6"		12.80		14.10		
1-50 Ft. Disch. Hose, 6"		8.90		9.80		
1 Drill Rig & Augers		3179.00		3496.90	45.85	50.43
88 L.H., Daily Totals		$6758.10		$8660.80	$76.80	$98.42
Crew B-50	Hr.	Daily	Hr.	Daily	Bare Costs	Incl. O&P
2 Pile Driver Foremen	$35.30	$564.80	$58.15	$930.40	$32.31	$51.71
6 Pile Drivers	33.30	1598.40	54.90	2635.20		
2 Equip. Oper. (crane)	35.90	574.40	54.10	865.60		
1 Equip. Oper. Oiler	30.10	240.80	45.35	362.80		
3 Laborers	26.70	640.80	41.55	997.20		
1 Crane, 40 Ton		888.20		977.00		
60 L.F. Leads, 15K Ft. Lbs.		75.00		82.50		
1 Hammer, 15K Ft. Lbs.		359.80		395.80		
1 Air Compr., 600 C.F.M.		297.80		327.60		
2-50 Ft. Air Hoses, 3" Dia.		35.40		38.95		
1 Chain Saw, 36" Long		33.40		36.75	15.11	16.62
112 L.H., Daily Totals		$5308.80		$7649.80	$47.42	$68.33
Crew B-51	Hr.	Daily	Hr.	Daily	Bare Costs	Incl. O&P
1 Labor Foreman (outside)	$28.70	$229.60	$44.70	$357.60	$27.05	$41.96
4 Laborers	26.70	854.40	41.55	1329.60		
1 Truck Driver (light)	26.80	214.40	40.85	326.80		
1 Light Truck, 1.5 Ton		128.80		141.70	2.68	2.95
48 L.H., Daily Totals		$1427.20		$2155.70	$29.73	$44.91

Crew No.	Bare Costs			Incl. Subs O & P		Cost Per Labor-Hour	

Crew B-52

	Hr.	Daily	Hr.	Daily	Bare Costs	Incl. O&P
1 Carpenter Foreman	$36.25	$290.00	$56.45	$451.60	$31.39	$48.55
1 Carpenter	34.25	274.00	53.35	426.80		
3 Laborers	26.70	640.80	41.55	997.20		
1 Cement Finisher	32.85	262.80	48.35	386.80		
.5 Rodman (reinf.)	37.95	151.80	62.60	250.40		
.5 Equip. Oper. (med.)	34.65	138.60	52.20	208.80		
.5 F.E. Ldr., T.M., 2.5 C.Y.		394.00		433.40	7.04	7.74
56 L.H., Daily Totals		$2152.00		$3155.00	$38.43	$56.29

Crew B-53

	Hr.	Daily	Hr.	Daily	Bare Costs	Incl. O&P
1 Equip. Oper. (light)	$33.05	$264.40	$49.80	$398.40	$33.05	$49.80
1 Trencher, Chain, 12 H.P.		48.00		52.80	6.00	6.60
8 L.H., Daily Totals		$312.40		$451.20	$39.05	$56.40

Crew B-54

	Hr.	Daily	Hr.	Daily	Bare Costs	Incl. O&P
1 Equip. Oper. (light)	$33.05	$264.40	$49.80	$398.40	$33.05	$49.80
1 Trencher, Chain, 40 H.P.		213.40		234.75	26.68	29.34
8 L.H., Daily Totals		$477.80		$633.15	$59.73	$79.14

Crew B-54A

	Hr.	Daily	Hr.	Daily	Bare Costs	Incl. O&P
.17 Labor Foreman (outside)	$28.70	$39.03	$44.70	$60.79	$33.79	$51.11
1 Equipment Operator (med.)	34.65	277.20	52.20	417.60		
1 Wheel Trencher, 67 H.P.		816.20		897.80	87.20	95.92
9.36 L.H., Daily Totals		$1132.43		$1376.19	$120.99	$147.03

Crew B-54B

	Hr.	Daily	Hr.	Daily	Bare Costs	Incl. O&P
.25 Labor Foreman (outside)	$28.70	$57.40	$44.70	$89.40	$33.46	$50.70
1 Equipment Operator (med.)	34.65	277.20	52.20	417.60		
1 Wheel Trencher, 150 H.P.		1460.00		1606.00	146.00	160.60
10 L.H., Daily Totals		$1794.60		$2113.00	$179.46	$211.30

Crew B-55

	Hr.	Daily	Hr.	Daily	Bare Costs	Incl. O&P
2 Laborers	$26.70	$427.20	$41.55	$664.80	$26.73	$41.32
1 Truck Driver (light)	26.80	214.40	40.85	326.80		
1 Auger, 4" to 36" Dia		604.60		665.05		
1 Flatbed 3 Ton Truck		176.20		193.80	32.53	35.79
24 L.H., Daily Totals		$1422.40		$1850.45	$59.26	$77.11

Crew B-56

	Hr.	Daily	Hr.	Daily	Bare Costs	Incl. O&P
1 Laborer	$26.70	$213.60	$41.55	$332.40	$29.88	$45.68
1 Equip. Oper. (light)	33.05	264.40	49.80	398.40		
1 Crawler Type Drill, 4"		664.40		730.85		
1 Air Compr., 600 C.F.M.		297.80		327.60		
1-50 Ft. Air Hose, 3" Dia.		17.70		19.45	61.24	67.37
16 L.H., Daily Totals		$1457.90		$1808.70	$91.12	$113.05

Crew B-57

	Hr.	Daily	Hr.	Daily	Bare Costs	Incl. O&P
1 Labor Foreman (outside)	$28.70	$229.60	$44.70	$357.60	$30.19	$46.18
2 Laborers	26.70	427.20	41.55	664.80		
1 Equip. Oper. (crane)	35.90	287.20	54.10	432.80		
1 Equip. Oper. (light)	33.05	264.40	49.80	398.40		
1 Equip. Oper. Oiler	30.10	240.80	45.35	362.80		
1 Power Shovel, 1 C.Y.		833.20		916.50		
1 Clamshell Bucket, 1 C.Y.		43.40		47.75		
1 Centr. Water Pump, 6"		217.00		238.70		
1-20 Ft. Suction Hose, 6"		12.80		14.10		
20-50 Ft. Disch. Hoses, 6"		178.00		195.80	26.76	29.43
48 L.H., Daily Totals		$2733.60		$3629.25	$56.95	$75.61

Crew B-58

	Hr.	Daily	Hr.	Daily	Bare Costs	Incl. O&P
2 Laborers	$26.70	$427.20	$41.55	$664.80	$28.82	$44.30
1 Equip. Oper. (light)	33.05	264.40	49.80	398.40		
1 Backhoe Loader, 48 H.P.		215.80		237.40		
1 Small Helicopter, w/pilot		2153.00		2368.30	98.70	108.57
24 L.H., Daily Totals		$3060.40		$3668.90	$127.52	$152.87

Crew B-59

	Hr.	Daily	Hr.	Daily	Bare Costs	Incl. O&P
1 Truck Driver (heavy)	$27.55	$220.40	$42.00	$336.00	$27.55	$42.00
1 Truck, 30 Ton		208.20		229.00		
1 Water tank, 6000 Gal.		116.80		128.50	40.63	44.69
8 L.H., Daily Totals		$545.40		$693.50	$68.18	$86.69

Crew B-59A

	Hr.	Daily	Hr.	Daily	Bare Costs	Incl. O&P
2 Laborers	$26.70	$427.20	$41.55	$664.80	$26.98	$41.70
1 Truck Driver (heavy)	27.55	220.40	42.00	336.00		
1 Water tank, 5K w/pum		116.80		128.50		
1 Truck, 30 Ton		208.20		229.00	13.54	14.90
24 L.H., Daily Totals		$972.60		$1358.30	$40.52	$56.60

Crew B-60

	Hr.	Daily	Hr.	Daily	Bare Costs	Incl. O&P
1 Labor Foreman (outside)	$28.70	$229.60	$44.70	$357.60	$30.60	$46.69
2 Laborers	26.70	427.20	41.55	664.80		
1 Equip. Oper. (crane)	35.90	287.20	54.10	432.80		
2 Equip. Oper. (light)	33.05	528.80	49.80	796.80		
1 Equip. Oper. Oiler	30.10	240.80	45.35	362.80		
1 Crawler Crane, 40 Ton		888.20		977.00		
45 L.F. Leads, 15K Ft. Lbs.		56.25		61.90		
1 Backhoe Loader, 48 H.P.		215.80		237.40	20.76	22.83
56 L.H., Daily Totals		$2873.85		$3891.10	$51.36	$69.52

Crew B-61

	Hr.	Daily	Hr.	Daily	Bare Costs	Incl. O&P
1 Labor Foreman (outside)	$28.70	$229.60	$44.70	$357.60	$28.37	$43.83
3 Laborers	26.70	640.80	41.55	997.20		
1 Equip. Oper. (light)	33.05	264.40	49.80	398.40		
1 Cement Mixer, 2 C.Y.		162.00		178.20		
1 Air Compr., 160 C.F.M.		87.40		96.15	6.24	6.86
40 L.H., Daily Totals		$1384.20		$2027.55	$34.61	$50.69

Crew B-62

	Hr.	Daily	Hr.	Daily	Bare Costs	Incl. O&P
2 Laborers	$26.70	$427.20	$41.55	$664.80	$28.82	$44.30
1 Equip. Oper. (light)	33.05	264.40	49.80	398.40		
1 Loader, Skid Steer		153.40		168.75	6.39	7.03
24 L.H., Daily Totals		$845.00		$1231.95	$35.21	$51.33

Crew B-63

	Hr.	Daily	Hr.	Daily	Bare Costs	Incl. O&P
4 Laborers	$26.70	$854.40	$41.55	$1329.60	$27.97	$43.20
1 Equip. Oper. (light)	33.05	264.40	49.80	398.40		
1 Loader, Skid Steer		153.40		168.75	3.84	4.22
40 L.H., Daily Totals		$1272.20		$1896.75	$31.81	$47.42

Crew B-64

	Hr.	Daily	Hr.	Daily	Bare Costs	Incl. O&P
1 Laborer	$26.70	$213.60	$41.55	$332.40	$26.75	$41.20
1 Truck Driver (light)	26.80	214.40	40.85	326.80		
1 Power Mulcher (small)		103.20		113.50		
1 Light Truck, 1.5 Ton		128.80		141.70	14.50	15.95
16 L.H., Daily Totals		$660.00		$914.40	$41.25	$57.15

Crew No.	Bare Costs Hr.	Daily	Incl. Subs O & P Hr.	Daily	Cost Per Labor-Hour Bare Costs	Incl. O&P
Crew B-65	Hr.	Daily	Hr.	Daily	Bare Costs	Incl. O&P
1 Laborer	$26.70	$213.60	$41.55	$332.40	$26.75	$41.20
1 Truck Driver (light)	26.80	214.40	40.85	326.80		
1 Power Mulcher (large)		184.00		202.40		
1 Light Truck, 1.5 Ton		128.80		141.70	19.55	21.51
16 L.H., Daily Totals		$740.80		$1003.30	$46.30	$62.71
Crew B-66	Hr.	Daily	Hr.	Daily	Bare Costs	Incl. O&P
1 Equip. Oper. (light)	$33.05	$264.40	$49.80	$398.40	$33.05	$49.80
1 Backhoe Ldr. w/Attchmt.		173.20		190.50	21.65	23.82
8 L.H., Daily Totals		$437.60		$588.90	$54.70	$73.62
Crew B-67	Hr.	Daily	Hr.	Daily	Bare Costs	Incl. O&P
1 Millwright	$35.50	$284.00	$52.45	$419.60	$34.28	$51.13
1 Equip. Oper. (light)	33.05	264.40	49.80	398.40		
1 Forklift		233.00		256.30	14.56	16.02
16 L.H., Daily Totals		$781.40		$1074.30	$48.84	$67.15
Crew B-68	Hr.	Daily	Hr.	Daily	Bare Costs	Incl. O&P
2 Millwrights	$35.50	$568.00	$52.45	$839.20	$34.68	$51.57
1 Equip. Oper. (light)	33.05	264.40	49.80	398.40		
1 Forklift		233.00		256.30	9.71	10.68
24 L.H., Daily Totals		$1065.40		$1493.90	$44.39	$62.25
Crew B-69	Hr.	Daily	Hr.	Daily	Bare Costs	Incl. O&P
1 Labor Foreman (outside)	$28.70	$229.60	$44.70	$357.60	$29.13	$44.80
3 Laborers	26.70	640.80	41.55	997.20		
1 Equip Oper. (crane)	35.90	287.20	54.10	432.80		
1 Equip Oper. Oiler	30.10	240.80	45.35	362.80		
1 Truck Crane, 80 Ton		995.00		1094.50	20.73	22.80
48 L.H., Daily Totals		$2393.40		$3244.90	$49.86	$67.60
Crew B-69A	Hr.	Daily	Hr.	Daily	Bare Costs	Incl. O&P
1 Labor Foreman	$28.70	$229.60	$44.70	$357.60	$29.38	$44.93
3 Laborers	26.70	640.80	41.55	997.20		
1 Equip. Oper. (medium)	34.65	277.20	52.20	417.60		
1 Concrete Finisher	32.85	262.80	48.35	386.80		
1 Curb Paver		582.80		641.10	12.14	13.36
48 L.H., Daily Totals		$1993.20		$2800.30	$41.52	$58.29
Crew B-69B	Hr.	Daily	Hr.	Daily	Bare Costs	Incl. O&P
1 Labor Foreman	$28.70	$229.60	$44.70	$357.60	$29.38	$44.93
3 Laborers	26.70	640.80	41.55	997.20		
1 Equip. Oper. (medium)	34.65	277.20	52.20	417.60		
1 Cement Finisher	32.85	262.80	48.35	386.80		
1 Curb/Gutter Paver		725.00		797.50	15.10	16.61
48 L.H., Daily Totals		$2135.40		$2956.70	$44.48	$61.54
Crew B-70	Hr.	Daily	Hr.	Daily	Bare Costs	Incl. O&P
1 Labor Foreman (outside)	$28.70	$229.60	$44.70	$357.60	$30.39	$46.56
3 Laborers	26.70	640.80	41.55	997.20		
3 Equip. Oper. (med.)	34.65	831.60	52.20	1252.80		
1 Motor Grader, 30,000 Lb.		456.60		502.25		
1 Grader Attach., Ripper		71.40		78.55		
1 Road Sweeper, S.P.		445.00		489.50		
1 F.E. Loader, 1-3/4 C.Y.		241.00		265.10	21.68	23.85
56 L.H., Daily Totals		$2916.00		$3943.00	$52.07	$70.41

Crew No.	Bare Costs Hr.	Daily	Incl. Subs O & P Hr.	Daily	Cost Per Labor-Hour Bare Costs	Incl. O&P
Crew B-71	Hr.	Daily	Hr.	Daily	Bare Costs	Incl. O&P
1 Labor Foreman (outside)	$28.70	$229.60	$44.70	$357.60	$30.39	$46.56
3 Laborers	26.70	640.80	41.55	997.20		
3 Equip. Oper. (med.)	34.65	831.60	52.20	1252.80		
1 Pvmt. Profiler, 750 H.P.		4266.00		4692.60		
1 Road Sweeper, S.P.		445.00		489.50		
1 F.E. Loader, 1-3/4 C.Y.		241.00		265.10	88.43	97.27
56 L.H., Daily Totals		$6654.00		$8054.80	$118.82	$143.83
Crew B-72	Hr.	Daily	Hr.	Daily	Bare Costs	Incl. O&P
1 Labor Foreman (outside)	$28.70	$229.60	$44.70	$357.60	$30.93	$47.27
3 Laborers	26.70	640.80	41.55	997.20		
4 Equip. Oper. (med.)	34.65	1108.80	52.20	1670.40		
1 Pvmt. Profiler, 750 H.P.		4266.00		4692.60		
1 Hammermill, 250 H.P.		1344.00		1478.40		
1 Windrow Loader		791.20		870.30		
1 Mix Paver 165 H.P.		1656.00		1821.60		
1 Roller, Pneu. Tire, 12 T.		241.80		266.00	129.67	142.64
64 L.H., Daily Totals		$10278.20		$12154.10	$160.60	$189.91
Crew B-73	Hr.	Daily	Hr.	Daily	Bare Costs	Incl. O&P
1 Labor Foreman (outside)	$28.70	$229.60	$44.70	$357.60	$31.92	$48.60
2 Laborers	26.70	427.20	41.55	664.80		
5 Equip. Oper. (med.)	34.65	1386.00	52.20	2088.00		
1 Road Mixer, 310 H.P.		1546.00		1700.60		
1 Roller, Tandem, 12 Ton		179.20		197.10		
1 Hammermill, 250 H.P.		1344.00		1478.40		
1 Motor Grader, 30,000 Lb.		456.60		502.25		
.5 F.E. Loader, 1-3/4 C.Y.		120.50		132.55		
.5 Truck, 30 Ton		104.10		114.50		
.5 Water Tank 5000 Gal.		58.40		64.25	59.51	65.46
64 L.H., Daily Totals		$5851.60		$7300.05	$91.43	$114.06
Crew B-74	Hr.	Daily	Hr.	Daily	Bare Costs	Incl. O&P
1 Labor Foreman (outside)	$28.70	$229.60	$44.70	$357.60	$31.14	$47.38
1 Laborer	26.70	213.60	41.55	332.40		
4 Equip. Oper. (med.)	34.65	1108.80	52.20	1670.40		
2 Truck Drivers (heavy)	27.55	440.80	42.00	672.00		
1 Motor Grader, 30,000 Lb.		456.60		502.25		
1 Grader Attach., Ripper		71.40		78.55		
2 Stabilizers, 310 H.P.		2090.00		2299.00		
1 Flatbed Truck, 3 Ton		176.20		193.80		
1 Chem. Spreader, Towed		54.60		60.05		
1 Vibr. Roller, 29,000 Lb.		433.20		476.50		
1 Water Tank 5000 Gal.		116.80		128.50		
1 Truck, 30 Ton		208.20		229.00	56.36	62.00
64 L.H., Daily Totals		$5599.80		$7000.05	$87.50	$109.38
Crew B-75	Hr.	Daily	Hr.	Daily	Bare Costs	Incl. O&P
1 Labor Foreman (outside)	$28.70	$229.60	$44.70	$357.60	$31.65	$48.15
1 Laborer	26.70	213.60	41.55	332.40		
4 Equip. Oper. (med.)	34.65	1108.80	52.20	1670.40		
1 Truck Driver (heavy)	27.55	220.40	42.00	336.00		
1 Motor Grader, 30,000 Lb.		456.60		502.25		
1 Grader Attach., Ripper		71.40		78.55		
2 Stabilizers, 310 H.P.		2090.00		2299.00		
1 Dist. Truck, 3000 Gal.		233.80		257.20		
1 Vibr. Roller, 29,000 Lb.		433.20		476.50	58.66	64.53
56 L.H., Daily Totals		$5057.40		$6309.90	$90.31	$112.68

Crew No.	Bare Costs		Incl. Subs O & P		Cost Per Labor-Hour	
					Bare Costs	Incl. O&P
Crew B-76	Hr.	Daily	Hr.	Daily	Bare Costs	Incl. O&P
1 Dock Builder Foreman	$35.30	$282.40	$58.15	$465.20	$33.74	$54.02
5 Dock Builders	33.30	1332.00	54.90	2196.00		
2 Equip. Oper. (crane)	35.90	574.40	54.10	865.60		
1 Equip. Oper. Oiler	30.10	240.80	45.35	362.80		
1 Crawler Crane, 50 Ton		1206.00		1326.60		
1 Barge, 400 Ton		273.20		300.50		
1 Hammer, 15K Ft. Lbs.		359.80		395.80		
60 L.F. Leads, 15K Ft. Lbs.		75.00		82.50		
1 Air Compr., 600 C.F.M.		297.80		327.60		
2-50 Ft. Air Hoses, 3″ Dia.		35.40		38.95	31.25	34.38
72 L.H., Daily Totals		$4676.80		$6361.55	$64.99	$88.40
Crew B-76A	Hr.	Daily	Hr.	Daily	Bare Costs	Incl. O&P
1 Laborer Foreman	$28.70	$229.60	$44.70	$357.60	$28.53	$43.99
5 Laborers	26.70	1068.00	41.55	1662.00		
1 Equip. Oper. (crane)	35.90	287.20	54.10	432.80		
1 Equip. Oper. Oiler	30.10	240.80	45.35	362.80		
1 Crawler Crane, 50 Ton		1206.00		1326.60		
1 Barge, 400 Ton		273.20		300.50	23.11	25.42
64 L.H., Daily Totals		$3304.80		$4442.30	$51.64	$69.41
Crew B-77	Hr.	Daily	Hr.	Daily	Bare Costs	Incl. O&P
1 Labor Foreman	$28.70	$229.60	$44.70	$357.60	$27.12	$42.04
3 Laborers	26.70	640.80	41.55	997.20		
1 Truck Driver (light)	26.80	214.40	40.85	326.80		
1 Crack Cleaner, 25 H.P.		44.40		48.85		
1 Crack Filler, Trailer Mtd.		157.00		172.70		
1 Flatbed Truck, 3 Ton		176.20		193.80	9.44	10.38
40 L.H., Daily Totals		$1462.40		$2096.95	$36.56	$52.42
Crew B-78	Hr.	Daily	Hr.	Daily	Bare Costs	Incl. O&P
1 Labor Foreman	$28.70	$229.60	$44.70	$357.60	$27.05	$41.96
4 Laborers	26.70	854.40	41.55	1329.60		
1 Truck Driver (light)	26.80	214.40	40.85	326.80		
1 Paint Striper, S.P.		134.20		147.60		
1 Flatbed Truck, 3 Ton		176.20		193.80		
1 Pickup Truck, 3/4 Ton		78.00		85.80	8.09	8.90
48 L.H., Daily Totals		$1686.80		$2441.20	$35.14	$50.86
Crew B-79	Hr.	Daily	Hr.	Daily	Bare Costs	Incl. O&P
1 Labor Foreman	$28.70	$229.60	$44.70	$357.60	$27.12	$42.04
3 Laborers	26.70	640.80	41.55	997.20		
1 Truck Driver (light)	26.80	214.40	40.85	326.80		
1 Thermo. Striper, T.M.		549.20		604.10		
1 Flatbed Truck, 3 Ton		176.20		193.80		
2 Pickup Truck, 3/4 Ton		156.00		171.60	22.04	24.24
40 L.H., Daily Totals		$1966.20		$2651.10	$49.16	$66.28
Crew B-80	Hr.	Daily	Hr.	Daily	Bare Costs	Incl. O&P
1 Labor Foreman	$28.70	$229.60	$44.70	$357.60	$28.81	$44.23
1 Laborer	26.70	213.60	41.55	332.40		
1 Truck Driver (light)	26.80	214.40	40.85	326.80		
1 Equip. Oper. (light)	33.05	264.40	49.80	398.40		
1 Flatbed Truck, 3 Ton		176.20		193.80		
1 Fence Post Auger, T.M.		357.40		393.15	16.68	18.34
32 L.H., Daily Totals		$1455.60		$2002.15	$45.49	$62.57
Crew B-80A	Hr.	Daily	Hr.	Daily	Bare Costs	Incl. O&P
3 Laborers	$26.70	$640.80	$41.55	$997.20	$26.70	$41.55
1 Flatbed Truck, 3 Ton		176.20		193.80	7.34	8.08
24 L.H., Daily Totals		$817.00		$1191.00	$34.04	$49.63

Crew No.	Bare Costs		Incl. Subs O & P		Cost Per Labor-Hour	
					Bare Costs	Incl. O&P
Crew B-80B	Hr.	Daily	Hr.	Daily	Bare Costs	Incl. O&P
3 Laborers	$26.70	$640.80	$41.55	$997.20	$28.29	$43.61
1 Equip. Oper. (light)	33.05	264.40	49.80	398.40		
1 Crane, Flatbed Mnt.		209.60		230.55	6.55	7.21
32 L.H., Daily Totals		$1114.80		$1626.15	$34.84	$50.82
Crew B-80C	Hr.	Daily	Hr.	Daily	Bare Costs	Incl. O&P
2 Laborers	$26.70	$427.20	$41.55	$664.80	$26.73	$41.32
1 Truck Driver (light)	26.80	214.40	40.85	326.80		
1 Light Truck, 1.5 Ton		128.80		141.70		
1 Manual fence post auger, gas		5.00		5.50	5.58	6.13
24 L.H., Daily Totals		$775.40		$1138.80	$32.31	$47.45
Crew B-81	Hr.	Daily	Hr.	Daily	Bare Costs	Incl. O&P
1 Laborer	$26.70	$213.60	$41.55	$332.40	$29.63	$45.25
1 Equip. Oper. (med.)	34.65	277.20	52.20	417.60		
1 Truck Driver (heavy)	27.55	220.40	42.00	336.00		
1 Hydromulcher, T.M.		196.00		215.60		
1 Tractor Truck, 4x2		208.20		229.00	16.84	18.53
24 L.H., Daily Totals		$1115.40		$1530.60	$46.47	$63.78
Crew B-82	Hr.	Daily	Hr.	Daily	Bare Costs	Incl. O&P
1 Laborer	$26.70	$213.60	$41.55	$332.40	$29.88	$45.68
1 Equip. Oper. (light)	33.05	264.40	49.80	398.40		
1 Horiz. Borer, 6 H.P.		62.60		68.85	3.91	4.30
16 L.H., Daily Totals		$540.60		$799.65	$33.79	$49.98
Crew B-83	Hr.	Daily	Hr.	Daily	Bare Costs	Incl. O&P
1 Tugboat Captain	$34.65	$277.20	$52.20	$417.60	$30.68	$46.88
1 Tugboat Hand	26.70	213.60	41.55	332.40		
1 Tugboat, 250 H.P.		440.60		484.65	27.54	30.29
16 L.H., Daily Totals		$931.40		$1234.65	$58.22	$77.17
Crew B-84	Hr.	Daily	Hr.	Daily	Bare Costs	Incl. O&P
1 Equip. Oper. (med.)	$34.65	$277.20	$52.20	$417.60	$34.65	$52.20
1 Rotary Mower/Tractor		228.80		251.70	28.60	31.46
8 L.H., Daily Totals		$506.00		$669.30	$63.25	$83.66
Crew B-85	Hr.	Daily	Hr.	Daily	Bare Costs	Incl. O&P
3 Laborers	$26.70	$640.80	$41.55	$997.20	$28.46	$43.77
1 Equip. Oper. (med.)	34.65	277.20	52.20	417.60		
1 Truck Driver (heavy)	27.55	220.40	42.00	336.00		
1 Aerial Lift Truck, 80′		496.80		546.50		
1 Brush Chipper, 130 H.P.		164.80		181.30		
1 Pruning Saw, Rotary		8.75		9.65	16.76	18.43
40 L.H., Daily Totals		$1808.75		$2488.25	$45.22	$62.20
Crew B-86	Hr.	Daily	Hr.	Daily	Bare Costs	Incl. O&P
1 Equip. Oper. (med.)	$34.65	$277.20	$52.20	$417.60	$34.65	$52.20
1 Stump Chipper, S.P.		70.70		77.75	8.84	9.72
8 L.H., Daily Totals		$347.90		$495.35	$43.49	$61.92
Crew B-86A	Hr.	Daily	Hr.	Daily	Bare Costs	Incl. O&P
1 Equip. Oper. (medium)	$34.65	$277.20	$52.20	$417.60	$34.65	$52.20
1 Grader, 30,000 Lbs.		456.60		502.25	57.08	62.78
8 L.H., Daily Totals		$733.80		$919.85	$91.73	$114.98
Crew B-86B	Hr.	Daily	Hr.	Daily	Bare Costs	Incl. O&P
1 Equip. Oper. (medium)	$34.65	$277.20	$52.20	$417.60	$34.65	$52.20
1 Dozer, 200 H.P.		919.60		1011.55	114.95	126.45
8 L.H., Daily Totals		$1196.80		$1429.15	$149.60	$178.65

Crew No.	Bare Costs		Incl. Subs O & P		Cost Per Labor-Hour	

Crew B-87

	Hr.	Daily	Hr.	Daily	Bare Costs	Incl. O&P
1 Laborer	$26.70	$213.60	$41.55	$332.40	$33.06	$50.07
4 Equip. Oper. (med.)	34.65	1108.80	52.20	1670.40		
2 Feller Bunchers, 50 H.P.		864.80		951.30		
1 Log Chipper, 22" Tree		1062.00		1168.20		
1 Dozer, 105 H.P.		453.80		499.20		
1 Chainsaw, Gas, 36" Long		33.40		36.75	60.35	66.39
40 L.H., Daily Totals		$3736.40		$4658.25	$93.41	$116.46

Crew B-88

	Hr.	Daily	Hr.	Daily	Bare Costs	Incl. O&P
1 Laborer	$26.70	$213.60	$41.55	$332.40	$33.51	$50.68
6 Equip. Oper. (med.)	34.65	1663.20	52.20	2505.60		
2 Feller Bunchers, 50 H.P.		864.80		951.30		
1 Log Chipper, 22" Tree		1062.00		1168.20		
2 Log Skidders, 50 H.P.		1490.80		1639.90		
1 Dozer, 105 H.P.		453.80		499.20		
1 Chainsaw, Gas, 36" Long		33.40		36.75	69.73	76.70
56 L.H., Daily Totals		$5781.60		$7133.35	$103.24	$127.38

Crew B-89

	Hr.	Daily	Hr.	Daily	Bare Costs	Incl. O&P
1 Equip. Oper. (light)	$33.05	$264.40	$49.80	$398.40	$29.93	$45.33
1 Truck Driver (light)	26.80	214.40	40.85	326.80		
1 Truck, Stake Body, 3 Ton		176.20		193.80		
1 Concrete Saw		114.40		125.85		
1 Water Tank, 65 Gal.		13.60		14.95	19.01	20.91
16 L.H., Daily Totals		$783.00		$1059.80	$48.94	$66.24

Crew B-89A

	Hr.	Daily	Hr.	Daily	Bare Costs	Incl. O&P
1 Skilled Worker	$34.85	$278.80	$54.20	$433.60	$30.78	$47.88
1 Laborer	26.70	213.60	41.55	332.40		
1 Core Drill (large)		103.05		113.35	6.44	7.08
16 L.H., Daily Totals		$595.45		$879.35	$37.22	$54.96

Crew B-89B

	Hr.	Daily	Hr.	Daily	Bare Costs	Incl. O&P
1 Equip. Oper. (light)	$33.05	$264.40	$49.80	$398.40	$29.93	$45.33
1 Truck Driver, Light	26.80	214.40	40.85	326.80		
1 Wall Saw, Hydraulic, 10 H.P.		73.30		80.65		
1 Generator, Diesel, 100 KW		176.60		194.25		
1 Water Tank, 65 Gal.		13.60		14.95		
1 Flatbed Truck, 3 Ton		176.20		193.80	27.48	30.23
16 L.H., Daily Totals		$918.50		$1208.85	$57.41	$75.56

Crew B-90

	Hr.	Daily	Hr.	Daily	Bare Costs	Incl. O&P
1 Labor Foreman (outside)	$28.70	$229.60	$44.70	$357.60	$28.75	$44.12
3 Laborers	26.70	640.80	41.55	997.20		
2 Equip. Oper. (light)	33.05	528.80	49.80	796.80		
2 Truck Drivers (heavy)	27.55	440.80	42.00	672.00		
1 Road Mixer, 310 H.P.		1546.00		1700.60		
1 Dist. Truck, 2000 Gal.		204.80		225.30	27.36	30.09
64 L.H., Daily Totals		$3590.80		$4749.50	$56.11	$74.21

Crew B-90A

	Hr.	Daily	Hr.	Daily	Bare Costs	Incl. O&P
1 Labor Foreman	$28.70	$229.60	$44.70	$357.60	$31.53	$48.09
2 Laborers	26.70	427.20	41.55	664.80		
4 Equip. Oper. (medium)	34.65	1108.80	52.20	1670.40		
2 Graders, 30,000 Lbs.		913.20		1004.50		
1 Roller, Steel Wheel		179.20		197.10		
1 Roller, Pneumatic Wheel		241.80		266.00	23.83	26.21
56 L.H., Daily Totals		$3099.80		$4160.40	$55.36	$74.30

Crew B-90B

	Hr.	Daily	Hr.	Daily	Bare Costs	Incl. O&P
1 Labor Foreman	$28.70	$229.60	$44.70	$357.60	$31.01	$47.40
2 Laborers	26.70	427.20	41.55	664.80		
3 Equip. Oper. (medium)	34.65	831.60	52.20	1252.80		
1 Roller, Steel Wheel		179.20		197.10		
1 Roller, Pneumatic Wheel		241.80		266.00		
1 Road Mixer, 310 H.P.		1546.00		1700.60	40.98	45.08
48 L.H., Daily Totals		$3455.40		$4438.90	$71.99	$92.48

Crew B-91

	Hr.	Daily	Hr.	Daily	Bare Costs	Incl. O&P
1 Labor Foreman (outside)	$28.70	$229.60	$44.70	$357.60	$31.03	$47.33
2 Laborers	26.70	427.20	41.55	664.80		
4 Equip. Oper. (med.)	34.65	1108.80	52.20	1670.40		
1 Truck Driver (heavy)	27.55	220.40	42.00	336.00		
1 Dist. Truck, 3000 Gal.		233.80		257.20		
1 Aggreg. Spreader, S.P.		771.20		848.30		
1 Roller, Pneu. Tire, 12 Ton		241.80		266.00		
1 Roller, Steel, 10 Ton		179.20		197.10	22.28	24.51
64 L.H., Daily Totals		$3412.00		$4597.40	$53.31	$71.84

Crew B-92

	Hr.	Daily	Hr.	Daily	Bare Costs	Incl. O&P
1 Labor Foreman (outside)	$28.70	$229.60	$44.70	$357.60	$27.20	$42.34
3 Laborers	26.70	640.80	41.55	997.20		
1 Crack Cleaner, 25 H.P.		44.40		48.85		
1 Air Compressor		55.40		60.95		
1 Tar Kettle, T.M.		40.95		45.05		
1 Flatbed Truck, 3 Ton		176.20		193.80	9.90	10.90
32 L.H., Daily Totals		$1187.35		$1703.45	$37.10	$53.24

Crew B-93

	Hr.	Daily	Hr.	Daily	Bare Costs	Incl. O&P
1 Equip. Oper. (med.)	$34.65	$277.20	$52.20	$417.60	$34.65	$52.20
1 Feller Buncher, 50 H.P.		432.40		475.65	54.05	59.46
8 L.H., Daily Totals		$709.60		$893.25	$88.70	$111.66

Crew B-94A

	Hr.	Daily	Hr.	Daily	Bare Costs	Incl. O&P
1 Laborer	$26.70	$213.60	$41.55	$332.40	$26.70	$41.55
1 Diaph. Water Pump, 2"		50.60		55.65		
1-20 Ft. Suction Hose, 2"		2.55		2.80		
2-50 Ft. Disch. Hoses, 2"		3.80		4.20	7.12	7.83
8 L.H., Daily Totals		$270.55		$395.05	$33.82	$49.38

Crew B-94B

	Hr.	Daily	Hr.	Daily	Bare Costs	Incl. O&P
1 Laborer	$26.70	$213.60	$41.55	$332.40	$26.70	$41.55
1 Diaph. Water Pump, 4"		74.40		81.85		
1-20 Ft. Suction Hose, 4"		5.05		5.55		
2-50 Ft. Disch. Hoses, 4"		7.10		7.80	10.82	11.90
8 L.H., Daily Totals		$300.15		$427.60	$37.52	$53.45

Crew B-94C

	Hr.	Daily	Hr.	Daily	Bare Costs	Incl. O&P
1 Laborer	$26.70	$213.60	$41.55	$332.40	$26.70	$41.55
1 Centr. Water Pump, 3"		53.60		58.95		
1-20 Ft. Suction Hose, 3"		4.25		4.70		
2-50 Ft. Disch. Hoses, 3"		5.00		5.50	7.86	8.64
8 L.H., Daily Totals		$276.45		$401.55	$34.56	$50.19

Crew B-94D

	Hr.	Daily	Hr.	Daily	Bare Costs	Incl. O&P
1 Laborer	$26.70	$213.60	$41.55	$332.40	$26.70	$41.55
1 Centr. Water Pump, 6"		217.00		238.70		
1-20 Ft. Suction Hose, 6"		12.80		14.10		
2-50 Ft. Disch. Hoses, 6"		17.80		19.60	30.95	34.05
8 L.H., Daily Totals		$461.20		$604.80	$57.65	$75.60

Crew No.	Bare Costs		Incl. Subs O & P		Cost Per Labor-Hour	
Crew B-95A	Hr.	Daily	Hr.	Daily	Bare Costs	Incl. O&P
1 Equip. Oper. (crane)	$35.90	$287.20	$54.10	$432.80	$31.30	$47.83
1 Laborer	26.70	213.60	41.55	332.40		
1 Hyd. Excavator, 5/8 C.Y.		432.40		475.65	27.03	29.73
16 L.H., Daily Totals		$933.20		$1240.85	$58.33	$77.56

Crew No.	Bare Costs		Incl. Subs O & P		Cost Per Labor-Hour	
Crew B-95B	Hr.	Daily	Hr.	Daily	Bare Costs	Incl. O&P
1 Equip. Oper. (crane)	$35.90	$287.20	$54.10	$432.80	$31.30	$47.83
1 Laborer	26.70	213.60	41.55	332.40		
1 Hyd. Excavator, 1.5 C.Y.		720.40		792.45	45.03	49.53
16 L.H., Daily Totals		$1221.20		$1557.65	$76.33	$97.36

Crew No.	Bare Costs		Incl. Subs O & P		Cost Per Labor-Hour	
Crew B-95C	Hr.	Daily	Hr.	Daily	Bare Costs	Incl. O&P
1 Equip. Oper. (crane)	$35.90	$287.20	$54.10	$432.80	$31.30	$47.83
1 Laborer	26.70	213.60	41.55	332.40		
1 Hyd. Excavator, 2.5 C.Y.		1215.00		1336.50	75.94	83.53
16 L.H., Daily Totals		$1715.80		$2101.70	$107.24	$131.36

Crew No.	Bare Costs		Incl. Subs O & P		Cost Per Labor-Hour	
Crew C-1	Hr.	Daily	Hr.	Daily	Bare Costs	Incl. O&P
3 Carpenters	$34.25	$822.00	$53.35	$1280.40	$32.36	$50.40
1 Laborer	26.70	213.60	41.55	332.40		
32 L.H., Daily Totals		$1035.60		$1612.80	$32.36	$50.40

Crew No.	Bare Costs		Incl. Subs O & P		Cost Per Labor-Hour	
Crew C-2	Hr.	Daily	Hr.	Daily	Bare Costs	Incl. O&P
1 Carpenter Foreman (out)	$36.25	$290.00	$56.45	$451.60	$33.33	$51.90
4 Carpenters	34.25	1096.00	53.35	1707.20		
1 Laborer	26.70	213.60	41.55	332.40		
48 L.H., Daily Totals		$1599.60		$2491.20	$33.33	$51.90

Crew No.	Bare Costs		Incl. Subs O & P		Cost Per Labor-Hour	
Crew C-2A	Hr.	Daily	Hr.	Daily	Bare Costs	Incl. O&P
1 Carpenter Foreman (out)	$36.25	$290.00	$56.45	$451.60	$33.09	$51.01
3 Carpenters	34.25	822.00	53.35	1280.40		
1 Cement Finisher	32.85	262.80	48.35	386.80		
1 Laborer	26.70	213.60	41.55	332.40		
48 L.H., Daily Totals		$1588.40		$2451.20	$33.09	$51.01

Crew No.	Bare Costs		Incl. Subs O & P		Cost Per Labor-Hour	
Crew C-3	Hr.	Daily	Hr.	Daily	Bare Costs	Incl. O&P
1 Rodman Foreman	$39.95	$319.60	$65.90	$527.20	$34.78	$56.15
4 Rodmen (reinf.)	37.95	1214.40	62.60	2003.20		
1 Equip. Oper. (light)	33.05	264.40	49.80	398.40		
2 Laborers	26.70	427.20	41.55	664.80		
3 Stressing Equipment		40.80		44.90		
.5 Grouting Equipment		76.40		84.05	1.83	2.01
64 L.H., Daily Totals		$2342.80		$3722.55	$36.61	$58.16

Crew No.	Bare Costs		Incl. Subs O & P		Cost Per Labor-Hour	
Crew C-4	Hr.	Daily	Hr.	Daily	Bare Costs	Incl. O&P
1 Rodman Foreman	$39.95	$319.60	$65.90	$527.20	$38.45	$63.43
3 Rodmen (reinf.)	37.95	910.80	62.60	1502.40		
3 Stressing Equipment		40.80		44.90	1.28	1.40
32 L.H., Daily Totals		$1271.20		$2074.50	$39.73	$64.83

Crew No.	Bare Costs		Incl. Subs O & P		Cost Per Labor-Hour	
Crew C-5	Hr.	Daily	Hr.	Daily	Bare Costs	Incl. O&P
1 Rodman Foreman	$39.95	$319.60	$65.90	$527.20	$36.82	$59.39
4 Rodmen (reinf.)	37.95	1214.40	62.60	2003.20		
1 Equip. Oper. (crane)	35.90	287.20	54.10	432.80		
1 Equip. Oper. Oiler	30.10	240.80	45.35	362.80		
1 Hyd. Crane, 25 Ton		616.80		678.50	11.01	12.12
56 L.H., Daily Totals		$2678.80		$4004.50	$47.83	$71.51

Crew No.	Bare Costs		Incl. Subs O & P		Cost Per Labor-Hour	
Crew C-6	Hr.	Daily	Hr.	Daily	Bare Costs	Incl. O&P
1 Labor Foreman (outside)	$28.70	$229.60	$44.70	$357.60	$28.06	$43.15
4 Laborers	26.70	854.40	41.55	1329.60		
1 Cement Finisher	32.85	262.80	48.35	386.80		
2 Gas Engine Vibrators		48.00		52.80	1.00	1.10
48 L.H., Daily Totals		$1394.80		$2126.80	$29.06	$44.25

Crew No.	Bare Costs		Incl. Subs O & P		Cost Per Labor-Hour	
Crew C-7	Hr.	Daily	Hr.	Daily	Bare Costs	Incl. O&P
1 Labor Foreman (outside)	$28.70	$229.60	$44.70	$357.60	$28.87	$44.22
5 Laborers	26.70	1068.00	41.55	1662.00		
1 Cement Finisher	32.85	262.80	48.35	386.80		
1 Equip. Oper. (med.)	34.65	277.20	52.20	417.60		
1 Equip. Oper. (oiler)	30.10	240.80	45.35	362.80		
2 Gas Engine Vibrators		48.00		52.80		
1 Concrete Bucket, 1 C.Y.		16.00		17.60		
1 Hyd. Crane, 55 Ton		895.40		984.95	13.33	14.66
72 L.H., Daily Totals		$3037.80		$4242.15	$42.20	$58.88

Crew No.	Bare Costs		Incl. Subs O & P		Cost Per Labor-Hour	
Crew C-7A	Hr.	Daily	Hr.	Daily	Bare Costs	Incl. O&P
1 Labor Foreman (outside)	$28.70	$229.60	$44.70	$357.60	$27.16	$42.06
5 Laborers	26.70	1068.00	41.55	1662.00		
2 Truck Drivers(Heavy)	27.55	440.80	42.00	672.00		
2 Conc. Transit Mixers		1480.80		1628.90	23.14	25.45
64 L.H., Daily Totals		$3219.20		$4320.50	$50.30	$67.51

Crew No.	Bare Costs		Incl. Subs O & P		Cost Per Labor-Hour	
Crew C-7B	Hr.	Daily	Hr.	Daily	Bare Costs	Incl. O&P
1 Labor Foreman (outside)	$28.70	$229.60	$44.70	$357.60	$28.53	$43.99
5 Laborers	26.70	1068.00	41.55	1662.00		
1 Equipment Operator (heavy)	35.90	287.20	54.10	432.80		
1 Equipment Oiler	30.10	240.80	45.35	362.80		
1 Conc. Bucket, 2 C.Y.		24.60		27.05		
1 Truck Crane, 165 Ton		1912.00		2103.20	30.26	33.29
64 L.H., Daily Totals		$3762.20		$4945.45	$58.79	$77.28

Crew No.	Bare Costs		Incl. Subs O & P		Cost Per Labor-Hour	
Crew C-7C	Hr.	Daily	Hr.	Daily	Bare Costs	Incl. O&P
1 Labor Foreman (outside)	$28.70	$229.60	$44.70	$357.60	$28.94	$44.61
5 Laborers	26.70	1068.00	41.55	1662.00		
2 Equipment Operator (medium)	34.65	554.40	52.20	835.20		
2 Wheel Loader, 4 C.Y.		891.20		980.30	13.93	15.32
64 L.H., Daily Totals		$2743.20		$3835.10	$42.87	$59.93

Crew No.	Bare Costs		Incl. Subs O & P		Cost Per Labor-Hour	
Crew C-7D	Hr.	Daily	Hr.	Daily	Bare Costs	Incl. O&P
1 Labor Foreman (outside)	$28.70	$229.60	$44.70	$357.60	$28.12	$43.52
5 Laborers	26.70	1068.00	41.55	1662.00		
1 Equipment Operator (med.)	34.65	277.20	52.20	417.60		
1 Concrete Conveyer		152.80		168.10	2.73	3.00
56 L.H., Daily Totals		$1727.60		$2605.30	$30.85	$46.52

Crew No.	Bare Costs		Incl. Subs O & P		Cost Per Labor-Hour	
Crew C-8	Hr.	Daily	Hr.	Daily	Bare Costs	Incl. O&P
1 Labor Foreman (outside)	$28.70	$229.60	$44.70	$357.60	$29.88	$45.36
3 Laborers	26.70	640.80	41.55	997.20		
2 Cement Finishers	32.85	525.60	48.35	773.60		
1 Equip. Oper. (med.)	34.65	277.20	52.20	417.60		
1 Concrete Pump (small)		704.20		774.60	12.58	13.83
56 L.H., Daily Totals		$2377.40		$3320.60	$42.46	$59.19

Crew No.	Bare Costs		Incl. Subs O & P		Cost Per Labor-Hour	
Crew C-8A	Hr.	Daily	Hr.	Daily	Bare Costs	Incl. O&P
1 Labor Foreman (outside)	$28.70	$229.60	$44.70	$357.60	$29.08	$44.23
3 Laborers	26.70	640.80	41.55	997.20		
2 Cement Finishers	32.85	525.60	48.35	773.60		
48 L.H., Daily Totals		$1396.00		$2128.40	$29.08	$44.23

CREWS

Crew No.	Bare Costs		Incl. Subs O & P		Cost Per Labor-Hour	

Crew C-8B

Crew C-8B	Hr.	Daily	Hr.	Daily	Bare Costs	Incl. O&P
1 Labor Foreman (outside)	$28.70	$229.60	$44.70	$357.60	$28.69	$44.31
3 Laborers	26.70	640.80	41.55	997.20		
1 Equipment Operator	34.65	277.20	52.20	417.60		
1 Vibrating Screed		43.05		47.35		
1 Vibratory Roller		433.20		476.50		
1 Dozer, 200 H.P.		919.60		1011.55	34.90	38.38
40 L.H., Daily Totals		$2543.45		$3307.80	$63.59	$82.69

Crew C-8C

Crew C-8C	Hr.	Daily	Hr.	Daily	Bare Costs	Incl. O&P
1 Labor Foreman (outside)	$28.70	$229.60	$44.70	$357.60	$29.38	$44.93
3 Laborers	26.70	640.80	41.55	997.20		
1 Cement Finisher	32.85	262.80	48.35	386.80		
1 Equipment Operator (med.)	34.65	277.20	52.20	417.60		
1 Shotcrete Rig, 12 CY/hr		223.80		246.20	4.66	5.13
48 L.H., Daily Totals		$1634.20		$2405.40	$34.04	$50.06

Crew C-8D

Crew C-8D	Hr.	Daily	Hr.	Daily	Bare Costs	Incl. O&P
1 Labor Foreman (outside)	$28.70	$229.60	$44.70	$357.60	$30.33	$46.01
1 Laborer	26.70	213.60	41.55	332.40		
1 Cement Finisher	32.85	262.80	48.35	386.80		
1 Equipment Operator (light)	33.05	264.40	49.80	398.40		
1 Compressor, 250 CFM		112.40		123.65		
2 Hoses, 1", 50'		7.30		8.05	3.74	4.11
32 L.H., Daily Totals		$1090.10		$1606.90	$34.07	$50.12

Crew C-8E

Crew C-8E	Hr.	Daily	Hr.	Daily	Bare Costs	Incl. O&P
1 Labor Foreman (outside)	$28.70	$229.60	$44.70	$357.60	$30.33	$46.01
1 Laborer	26.70	213.60	41.55	332.40		
1 Cement Finisher	32.85	262.80	48.35	386.80		
1 Equipment Operator (light)	33.05	264.40	49.80	398.40		
1 Compressor, 250 CFM		112.40		123.65		
2 Hoses, 1", 50'		7.30		8.05		
1 Concrete Pump (small)		704.20		774.60	25.75	28.32
32 L.H., Daily Totals		$1794.30		$2381.50	$56.08	$74.33

Crew C-10

Crew C-10	Hr.	Daily	Hr.	Daily	Bare Costs	Incl. O&P
1 Laborer	$26.70	$213.60	$41.55	$332.40	$30.80	$45.85
2 Cement Finishers	32.85	525.60	48.35	773.60		
24 L.H., Daily Totals		$739.20		$1106.00	$30.80	$45.85

Crew C-10B

Crew C-10B	Hr.	Daily	Hr.	Daily	Bare Costs	Incl. O&P
3 Laborers	$26.70	$640.80	$41.55	$997.20	$29.16	$44.13
2 Cement Finishers	32.85	525.60	48.35	773.60		
1 Concrete mixer, 10 CF		123.20		135.50		
2 Concrete finisher, 48" dia		49.20		54.10	4.31	4.74
40 L.H., Daily Totals		$1338.80		$1960.40	$33.47	$48.87

Crew C-11

Crew C-11	Hr.	Daily	Hr.	Daily	Bare Costs	Incl. O&P
1 Struc. Steel Foreman	$40.15	$321.20	$71.85	$574.80	$37.23	$64.53
6 Struc. Steel Workers	38.15	1831.20	68.25	3276.00		
1 Equip. Oper. (crane)	35.90	287.20	54.10	432.80		
1 Equip. Oper. Oiler	30.10	240.80	45.35	362.80		
1 Truck Crane, 150 Ton		1445.00		1589.50	20.07	22.08
72 L.H., Daily Totals		$4125.40		$6235.90	$57.30	$86.61

Crew C-12

Crew C-12	Hr.	Daily	Hr.	Daily	Bare Costs	Incl. O&P
1 Carpenter Foreman (out)	$36.25	$290.00	$56.45	$451.60	$33.60	$52.03
3 Carpenters	34.25	822.00	53.35	1280.40		
1 Laborer	26.70	213.60	41.55	332.40		
1 Equip. Oper. (crane)	35.90	287.20	54.10	432.80		
1 Hyd. Crane, 12 Ton		602.60		662.85	12.55	13.81
48 L.H., Daily Totals		$2215.40		$3160.05	$46.15	$65.84

Crew C-13

Crew C-13	Hr.	Daily	Hr.	Daily	Bare Costs	Incl. O&P
1 Struc. Steel Worker	$38.15	$305.20	$68.25	$546.00	$36.85	$63.28
1 Welder	38.15	305.20	68.25	546.00		
1 Carpenter	34.25	274.00	53.35	426.80		
1 Gas Welding Machine		81.20		89.30	3.38	3.72
24 L.H., Daily Totals		$965.60		$1608.10	$40.23	$67.00

Crew C-14

Crew C-14	Hr.	Daily	Hr.	Daily	Bare Costs	Incl. O&P
1 Carpenter Foreman (out)	$36.25	$290.00	$56.45	$451.60	$33.21	$51.96
5 Carpenters	34.25	1370.00	53.35	2134.00		
4 Laborers	26.70	854.40	41.55	1329.60		
4 Rodmen (reinf.)	37.95	1214.40	62.60	2003.20		
2 Cement Finishers	32.85	525.60	48.35	773.60		
1 Equip. Oper. (crane)	35.90	287.20	54.10	432.80		
1 Equip. Oper. Oiler	30.10	240.80	45.35	362.80		
1 Crane, 80 Ton, & Tools		995.00		1094.50	6.91	7.60
144 L.H., Daily Totals		$5777.40		$8582.10	$40.12	$59.56

Crew C-14A

Crew C-14A	Hr.	Daily	Hr.	Daily	Bare Costs	Incl. O&P
1 Carpenter Foreman (out)	$36.25	$290.00	$56.45	$451.60	$34.28	$53.75
16 Carpenters	34.25	4384.00	53.35	6828.80		
4 Rodmen (reinf.)	37.95	1214.40	62.60	2003.20		
2 Laborers	26.70	427.20	41.55	664.80		
1 Cement Finisher	32.85	262.80	48.35	386.80		
1 Equip. Oper. (med.)	34.65	277.20	52.20	417.60		
1 Gas Engine Vibrator		24.00		26.40		
1 Concrete Pump (small)		704.20		774.60	3.64	4.01
200 L.H., Daily Totals		$7583.80		$11553.80	$37.92	$57.76

Crew C-14B

Crew C-14B	Hr.	Daily	Hr.	Daily	Bare Costs	Incl. O&P
1 Carpenter Foreman (out)	$36.25	$290.00	$56.45	$451.60	$34.22	$53.53
16 Carpenters	34.25	4384.00	53.35	6828.80		
4 Rodmen (reinf.)	37.95	1214.40	62.60	2003.20		
2 Laborers	26.70	427.20	41.55	664.80		
2 Cement Finishers	32.85	525.60	48.35	773.60		
1 Equip. Oper. (med.)	34.65	277.20	52.20	417.60		
1 Gas Engine Vibrator		24.00		26.40		
1 Concrete Pump (small)		704.20		774.60	3.50	3.85
208 L.H., Daily Totals		$7846.60		$11940.60	$37.72	$57.38

Crew C-14C

Crew C-14C	Hr.	Daily	Hr.	Daily	Bare Costs	Incl. O&P
1 Carpenter Foreman (out)	$36.25	$290.00	$56.45	$451.60	$32.66	$51.14
6 Carpenters	34.25	1644.00	53.35	2560.80		
2 Rodmen (reinf.)	37.95	607.20	62.60	1001.60		
4 Laborers	26.70	854.40	41.55	1329.60		
1 Cement Finisher	32.85	262.80	48.35	386.80		
1 Gas Engine Vibrator		24.00		26.40	.21	.24
112 L.H., Daily Totals		$3682.40		$5756.80	$32.87	$51.38

Crew No.	Bare Costs		Incl. Subs O & P		Cost Per Labor-Hour	
Crew C-14D	**Hr.**	**Daily**	**Hr.**	**Daily**	**Bare Costs**	**Incl. O&P**
1 Carpenter Foreman (out)	$36.25	$290.00	$56.45	$451.60	$33.98	$53.01
18 Carpenters	34.25	4932.00	53.35	7682.40		
2 Rodmen (reinf.)	37.95	607.20	62.60	1001.60		
2 Laborers	26.70	427.20	41.55	664.80		
1 Cement Finisher	32.85	262.80	48.35	386.80		
1 Equip. Oper. (med.)	34.65	277.20	52.20	417.60		
1 Gas Engine Vibrator		24.00		26.40		
1 Concrete Pump (small)		704.20		774.60	3.64	4.01
200 L.H., Daily Totals		$7524.60		$11405.80	$37.62	$57.02
Crew C-14E	**Hr.**	**Daily**	**Hr.**	**Daily**	**Bare Costs**	**Incl. O&P**
1 Carpenter Foreman (out)	$36.25	$290.00	$56.45	$451.60	$33.59	$53.29
2 Carpenters	34.25	548.00	53.35	853.60		
4 Rodmen (reinf.)	37.95	1214.40	62.60	2003.20		
3 Laborers	26.70	640.80	41.55	997.20		
1 Cement Finisher	32.85	262.80	48.35	386.80		
1 Gas Engine Vibrator		24.00		26.40	.27	.30
88 L.H., Daily Totals		$2980.00		$4718.80	$33.86	$53.59
Crew C-14F	**Hr.**	**Daily**	**Hr.**	**Daily**	**Bare Costs**	**Incl. O&P**
1 Laborer Foreman (out)	$28.70	$229.60	$44.70	$357.60	$31.02	$46.20
2 Laborers	26.70	427.20	41.55	664.80		
6 Cement Finishers	32.85	1576.80	48.35	2320.80		
1 Gas Engine Vibrator		24.00		26.40	.33	.37
72 L.H., Daily Totals		$2257.60		$3369.60	$31.35	$46.57
Crew C-14G	**Hr.**	**Daily**	**Hr.**	**Daily**	**Bare Costs**	**Incl. O&P**
1 Laborer Foreman (out)	$28.70	$229.60	$44.70	$357.60	$30.50	$45.69
2 Laborers	26.70	427.20	41.55	664.80		
4 Cement Finishers	32.85	1051.20	48.35	1547.20		
1 Gas Engine Vibrator		24.00		26.40	.43	.47
56 L.H., Daily Totals		$1732.00		$2596.00	$30.93	$46.16
Crew C-14H	**Hr.**	**Daily**	**Hr.**	**Daily**	**Bare Costs**	**Incl. O&P**
1 Carpenter Foreman (out)	$36.25	$290.00	$56.45	$451.60	$33.71	$52.55
2 Carpenters	34.25	548.00	53.35	853.60		
1 Rodman (reinf.)	37.95	303.60	62.60	500.80		
1 Laborer	26.70	213.60	41.55	332.40		
1 Cement Finisher	32.85	262.80	48.35	386.80		
1 Gas Engine Vibrator		24.00		26.40	.50	.55
48 L.H., Daily Totals		$1642.00		$2551.60	$34.21	$53.10
Crew C-15	**Hr.**	**Daily**	**Hr.**	**Daily**	**Bare Costs**	**Incl. O&P**
1 Carpenter Foreman (out)	$36.25	$290.00	$56.45	$451.60	$32.06	$49.60
2 Carpenters	34.25	548.00	53.35	853.60		
3 Laborers	26.70	640.80	41.55	997.20		
2 Cement Finishers	32.85	525.60	48.35	773.60		
1 Rodman (reinf.)	37.95	303.60	62.60	500.80		
72 L.H., Daily Totals		$2308.00		$3576.80	$32.06	$49.60
Crew C-16	**Hr.**	**Daily**	**Hr.**	**Daily**	**Bare Costs**	**Incl. O&P**
1 Labor Foreman (outside)	$28.70	$229.60	$44.70	$357.60	$31.67	$49.19
3 Laborers	26.70	640.80	41.55	997.20		
2 Cement Finishers	32.85	525.60	48.35	773.60		
1 Equip. Oper. (med.)	34.65	277.20	52.20	417.60		
2 Rodmen (reinf.)	37.95	607.20	62.60	1001.60		
1 Concrete Pump (small)		704.20		774.60	9.78	10.76
72 L.H., Daily Totals		$2984.60		$4322.20	$41.45	$59.95

Crew No.	Bare Costs		Incl. Subs O & P		Cost Per Labor-Hour	
Crew C-17	**Hr.**	**Daily**	**Hr.**	**Daily**	**Bare Costs**	**Incl. O&P**
2 Skilled Worker Foremen	$36.85	$589.60	$57.30	$916.80	$35.25	$54.82
8 Skilled Workers	34.85	2230.40	54.20	3468.80		
80 L.H., Daily Totals		$2820.00		$4385.60	$35.25	$54.82
Crew C-17A	**Hr.**	**Daily**	**Hr.**	**Daily**	**Bare Costs**	**Incl. O&P**
2 Skilled Worker Foremen	$36.85	$589.60	$57.30	$916.80	$35.26	$54.81
8 Skilled Workers	34.85	2230.40	54.20	3468.80		
.125 Equip. Oper. (crane)	35.90	35.90	54.10	54.10		
.125 Crane, 80 Ton, & Tools		124.38		136.80	1.54	1.69
81 L.H., Daily Totals		$2980.28		$4576.50	$36.80	$56.50
Crew C-17B	**Hr.**	**Daily**	**Hr.**	**Daily**	**Bare Costs**	**Incl. O&P**
2 Skilled Worker Foremen	$36.85	$589.60	$57.30	$916.80	$35.27	$54.80
8 Skilled Workers	34.85	2230.40	54.20	3468.80		
.25 Equip. Oper. (crane)	35.90	71.80	54.10	108.20		
.25 Crane, 80 Ton, & Tools		248.75		273.65		
.25 Walk Behind Power Tools		6.15		6.75	3.11	3.42
82 L.H., Daily Totals		$3146.70		$4774.20	$38.38	$58.22
Crew C-17C	**Hr.**	**Daily**	**Hr.**	**Daily**	**Bare Costs**	**Incl. O&P**
2 Skilled Worker Foremen	$36.85	$589.60	$57.30	$916.80	$35.27	$54.79
8 Skilled Workers	34.85	2230.40	54.20	3468.80		
.375 Equip. Oper. (crane)	35.90	107.70	54.10	162.30		
.375 Crane, 80 Ton & Tools		373.13		410.45	4.50	4.95
83 L.H., Daily Totals		$3300.83		$4958.35	$39.77	$59.74
Crew C-17D	**Hr.**	**Daily**	**Hr.**	**Daily**	**Bare Costs**	**Incl. O&P**
2 Skilled Worker Foremen	$36.85	$589.60	$57.30	$916.80	$35.28	$54.79
8 Skilled Workers	34.85	2230.40	54.20	3468.80		
.5 Equip. Oper. (crane)	35.90	143.60	54.10	216.40		
.5 Crane, 80 Ton & Tools		497.50		547.25	5.92	6.51
84 L.H., Daily Totals		$3461.10		$5149.25	$41.20	$61.30
Crew C-17E	**Hr.**	**Daily**	**Hr.**	**Daily**	**Bare Costs**	**Incl. O&P**
2 Skilled Worker Foremen	$36.85	$589.60	$57.30	$916.80	$35.25	$54.82
8 Skilled Workers	34.85	2230.40	54.20	3468.80		
1 Hyd. Jack with Rods		79.00		86.90	.99	1.09
80 L.H., Daily Totals		$2899.00		$4472.50	$36.24	$55.91
Crew C-18	**Hr.**	**Daily**	**Hr.**	**Daily**	**Bare Costs**	**Incl. O&P**
.125 Labor Foreman (out)	$28.70	$28.70	$44.70	$44.70	$26.92	$41.90
1 Laborer	26.70	213.60	41.55	332.40		
1 Concrete Cart, 10 C.F.		49.80		54.80	5.53	6.09
9 L.H., Daily Totals		$292.10		$431.90	$32.45	$47.99
Crew C-19	**Hr.**	**Daily**	**Hr.**	**Daily**	**Bare Costs**	**Incl. O&P**
.125 Labor Foreman (out)	$28.70	$28.70	$44.70	$44.70	$26.92	$41.90
1 Laborer	26.70	213.60	41.55	332.40		
1 Concrete Cart, 18 C.F.		76.80		84.50	8.53	9.39
9 L.H., Daily Totals		$319.10		$461.60	$35.45	$51.29
Crew C-20	**Hr.**	**Daily**	**Hr.**	**Daily**	**Bare Costs**	**Incl. O&P**
1 Labor Foreman (outside)	$28.70	$229.60	$44.70	$357.60	$28.71	$44.08
5 Laborers	26.70	1068.00	41.55	1662.00		
1 Cement Finisher	32.85	262.80	48.35	386.80		
1 Equip. Oper. (med.)	34.65	277.20	52.20	417.60		
2 Gas Engine Vibrators		48.00		52.80		
1 Concrete Pump (small)		704.20		774.60	11.75	12.93
64 L.H., Daily Totals		$2589.80		$3651.40	$40.46	$57.01

Crew No.	Bare Costs		Incl. Subs O & P		Cost Per Labor-Hour	

Crew C-21

	Hr.	Daily	Hr.	Daily	Bare Costs	Incl. O&P
1 Labor Foreman (outside)	$28.70	$229.60	$44.70	$357.60	$28.71	$44.08
5 Laborers	26.70	1068.00	41.55	1662.00		
1 Cement Finisher	32.85	262.80	48.35	386.80		
1 Equip. Oper. (med.)	34.65	277.20	52.20	417.60		
2 Gas Engine Vibrators		48.00		52.80		
1 Concrete Conveyer		152.80		168.10	3.14	3.45
64 L.H., Daily Totals		$2038.40		$3044.90	$31.85	$47.53

Crew C-22

	Hr.	Daily	Hr.	Daily	Bare Costs	Incl. O&P
1 Rodman Foreman	$39.95	$319.60	$65.90	$527.20	$38.10	$62.62
4 Rodmen (reinf.)	37.95	1214.40	62.60	2003.20		
.125 Equip. Oper. (crane)	35.90	35.90	54.10	54.10		
.125 Equip. Oper. Oiler	30.10	30.10	45.35	45.35		
.125 Hyd. Crane, 25 Ton		77.10		84.80	1.84	2.02
42 L.H., Daily Totals		$1677.10		$2714.65	$39.94	$64.64

Crew C-23

	Hr.	Daily	Hr.	Daily	Bare Costs	Incl. O&P
2 Skilled Worker Foremen	$36.85	$589.60	$57.30	$916.80	$34.88	$53.93
6 Skilled Workers	34.85	1672.80	54.20	2601.60		
1 Equip. Oper. (crane)	35.90	287.20	54.10	432.80		
1 Equip. Oper. Oiler	30.10	240.80	45.35	362.80		
1 Crane, 90 Ton		1325.00		1457.50	16.56	18.22
80 L.H., Daily Totals		$4115.40		$5771.50	$51.44	$72.15

Crew C-23A

	Hr.	Daily	Hr.	Daily	Bare Costs	Incl. O&P
1 Labor Foreman (outside)	$28.70	$229.60	$44.70	$357.60	$29.62	$45.45
2 Laborers	26.70	427.20	41.55	664.80		
1 Equip. Oper. (crane)	35.90	287.20	54.10	432.80		
1 Equip. Oper. Oiler	30.10	240.80	45.35	362.80		
1 Crane, 100 ton capacity		1519.00		1670.90		
3 Conc. bucket, 8 C.Y.		465.60		512.15	49.62	54.58
40 L.H., Daily Totals		$3169.40		$4001.05	$79.24	$100.03

Crew C-24

	Hr.	Daily	Hr.	Daily	Bare Costs	Incl. O&P
2 Skilled Worker Foremen	$36.85	$589.60	$57.30	$916.80	$34.88	$53.93
6 Skilled Workers	34.85	1672.80	54.20	2601.60		
1 Equip. Oper. (crane)	35.90	287.20	54.10	432.80		
1 Equip. Oper. Oiler	30.10	240.80	45.35	362.80		
1 Truck Crane, 150 Ton		1445.00		1589.50	18.06	19.87
80 L.H., Daily Totals		$4235.40		$5903.50	$52.94	$73.80

Crew C-25

	Hr.	Daily	Hr.	Daily	Bare Costs	Incl. O&P
2 Rodmen (reinf.)	$37.95	$607.20	$62.60	$1001.60	$29.78	$49.70
2 Rodmen Helpers	21.60	345.60	36.80	588.80		
32 L.H., Daily Totals		$952.80		$1590.40	$29.78	$49.70

Crew C-27

	Hr.	Daily	Hr.	Daily	Bare Costs	Incl. O&P
2 Cement Finishers	$32.85	$525.60	$48.35	$773.60	$32.85	$48.00
1 Concrete Saw		114.40		125.85	7.15	7.87
16 L.H., Daily Totals		$640.00		$899.45	$40.00	$55.87

Crew C-28

	Hr.	Daily	Hr.	Daily	Bare Costs	Incl. O&P
1 Cement Finisher	$32.85	$262.80	$48.35	$386.80	$32.85	$48.00
1 Portable Air Compressor		17.30		19.05	2.16	2.37
8 L.H., Daily Totals		$280.10		$405.85	$35.01	$50.37

Crew D-1

	Hr.	Daily	Hr.	Daily	Bare Costs	Incl. O&P
1 Bricklayer	$35.25	$282.00	$53.70	$429.60	$31.08	$47.35
1 Bricklayer Helper	26.90	215.20	41.00	328.00		
16 L.H., Daily Totals		$497.20		$757.60	$31.08	$47.35

Crew D-2

	Hr.	Daily	Hr.	Daily	Bare Costs	Incl. O&P
3 Bricklayers	$35.25	$846.00	$53.70	$1288.80	$32.12	$49.05
2 Bricklayer Helpers	26.90	430.40	41.00	656.00		
.5 Carpenter	34.25	137.00	53.35	213.40		
44 L.H., Daily Totals		$1413.40		$2158.20	$32.12	$49.05

Crew D-3

	Hr.	Daily	Hr.	Daily	Bare Costs	Incl. O&P
3 Bricklayers	$35.25	$846.00	$53.70	$1288.80	$32.02	$48.85
2 Bricklayer Helpers	26.90	430.40	41.00	656.00		
.25 Carpenter	34.25	68.50	53.35	106.70		
42 L.H., Daily Totals		$1344.90		$2051.50	$32.02	$48.85

Crew D-4

	Hr.	Daily	Hr.	Daily	Bare Costs	Incl. O&P
1 Bricklayer	$35.25	$282.00	$53.70	$429.60	$30.53	$46.38
2 Bricklayer Helpers	26.90	430.40	41.00	656.00		
1 Equip. Oper. (light)	33.05	264.40	49.80	398.40		
1 Grout Pump, 50 C.F./hr		107.75		118.55		
1 Hoses & Hopper		15.20		16.70		
1 Accessories		11.90		13.10	4.22	4.64
32 L.H., Daily Totals		$1111.65		$1632.35	$34.75	$51.02

Crew D-5

	Hr.	Daily	Hr.	Daily	Bare Costs	Incl. O&P
1 Bricklayer	$35.25	$282.00	$53.70	$429.60	$35.25	$53.70
8 L.H., Daily Totals		$282.00		$429.60	$35.25	$53.70

Crew D-6

	Hr.	Daily	Hr.	Daily	Bare Costs	Incl. O&P
3 Bricklayers	$35.25	$846.00	$53.70	$1288.80	$31.20	$47.59
3 Bricklayer Helpers	26.90	645.60	41.00	984.00		
.25 Carpenter	34.25	68.50	53.35	106.70		
50 L.H., Daily Totals		$1560.10		$2379.50	$31.20	$47.59

Crew D-7

	Hr.	Daily	Hr.	Daily	Bare Costs	Incl. O&P
1 Tile Layer	$32.70	$261.60	$48.05	$384.40	$29.03	$42.65
1 Tile Layer Helper	25.35	202.80	37.25	298.00		
16 L.H., Daily Totals		$464.40		$682.40	$29.03	$42.65

Crew D-8

	Hr.	Daily	Hr.	Daily	Bare Costs	Incl. O&P
3 Bricklayers	$35.25	$846.00	$53.70	$1288.80	$31.91	$48.62
2 Bricklayer Helpers	26.90	430.40	41.00	656.00		
40 L.H., Daily Totals		$1276.40		$1944.80	$31.91	$48.62

Crew D-9

	Hr.	Daily	Hr.	Daily	Bare Costs	Incl. O&P
3 Bricklayers	$35.25	$846.00	$53.70	$1288.80	$31.08	$47.35
3 Bricklayer Helpers	26.90	645.60	41.00	984.00		
48 L.H., Daily Totals		$1491.60		$2272.80	$31.08	$47.35

Crew D-10

	Hr.	Daily	Hr.	Daily	Bare Costs	Incl. O&P
1 Bricklayer Foreman	$37.25	$298.00	$56.75	$454.00	$32.44	$49.31
1 Bricklayer	35.25	282.00	53.70	429.60		
2 Bricklayer Helpers	26.90	430.40	41.00	656.00		
1 Equip. Oper. (crane)	35.90	287.20	54.10	432.80		
1 Truck Crane, 12.5 Ton		505.80		556.40	12.65	13.91
40 L.H., Daily Totals		$1803.40		$2528.80	$45.09	$63.22

Crew D-11

	Hr.	Daily	Hr.	Daily	Bare Costs	Incl. O&P
1 Bricklayer Foreman	$37.25	$298.00	$56.75	$454.00	$33.13	$50.48
1 Bricklayer	35.25	282.00	53.70	429.60		
1 Bricklayer Helper	26.90	215.20	41.00	328.00		
24 L.H., Daily Totals		$795.20		$1211.60	$33.13	$50.48

Crew D-12

Crew No.	Bare Costs		Incl. Subs O & P		Cost Per Labor-Hour	
Crew D-12	**Hr.**	**Daily**	**Hr.**	**Daily**	**Bare Costs**	**Incl. O&P**
1 Bricklayer Foreman	$37.25	$298.00	$56.75	$454.00	$31.58	$48.11
1 Bricklayer	35.25	282.00	53.70	429.60		
2 Bricklayer Helpers	26.90	430.40	41.00	656.00		
32 L.H., Daily Totals		$1010.40		$1539.60	$31.58	$48.11

Crew D-13	**Hr.**	**Daily**	**Hr.**	**Daily**	**Bare Costs**	**Incl. O&P**
1 Bricklayer Foreman	$37.25	$298.00	$56.75	$454.00	$32.74	$49.98
1 Bricklayer	35.25	282.00	53.70	429.60		
2 Bricklayer Helpers	26.90	430.40	41.00	656.00		
1 Carpenter	34.25	274.00	53.35	426.80		
1 Equip. Oper. (crane)	35.90	287.20	54.10	432.80		
1 Truck Crane, 12.5 Ton		505.80		556.40	10.54	11.59
48 L.H., Daily Totals		$2077.40		$2955.60	$43.28	$61.57

Crew E-1	**Hr.**	**Daily**	**Hr.**	**Daily**	**Bare Costs**	**Incl. O&P**
1 Welder Foreman	$40.15	$321.20	$71.85	$574.80	$37.12	$63.30
1 Welder	38.15	305.20	68.25	546.00		
1 Equip. Oper. (light)	33.05	264.40	49.80	398.40		
1 Gas Welding Machine		81.20		89.30	3.38	3.72
24 L.H., Daily Totals		$972.00		$1608.50	$40.50	$67.02

Crew E-2	**Hr.**	**Daily**	**Hr.**	**Daily**	**Bare Costs**	**Incl. O&P**
1 Struc. Steel Foreman	$40.15	$321.20	$71.85	$574.80	$36.96	$63.47
4 Struc. Steel Workers	38.15	1220.80	68.25	2184.00		
1 Equip. Oper. (crane)	35.90	287.20	54.10	432.80		
1 Equip. Oper. Oiler	30.10	240.80	45.35	362.80		
1 Crane, 90 Ton		1325.00		1457.50	23.66	26.03
56 L.H., Daily Totals		$3395.00		$5011.90	$60.62	$89.50

Crew E-3	**Hr.**	**Daily**	**Hr.**	**Daily**	**Bare Costs**	**Incl. O&P**
1 Struc. Steel Foreman	$40.15	$321.20	$71.85	$574.80	$38.82	$69.45
1 Struc. Steel Worker	38.15	305.20	68.25	546.00		
1 Welder	38.15	305.20	68.25	546.00		
1 Gas Welding Machine		81.20		89.30	3.38	3.72
24 L.H., Daily Totals		$1012.80		$1756.10	$42.20	$73.17

Crew E-4	**Hr.**	**Daily**	**Hr.**	**Daily**	**Bare Costs**	**Incl. O&P**
1 Struc. Steel Foreman	$40.15	$321.20	$71.85	$574.80	$38.65	$69.15
3 Struc. Steel Workers	38.15	915.60	68.25	1638.00		
1 Gas Welding Machine		81.20		89.30	2.54	2.79
32 L.H., Daily Totals		$1318.00		$2302.10	$41.19	$71.94

Crew E-5	**Hr.**	**Daily**	**Hr.**	**Daily**	**Bare Costs**	**Incl. O&P**
2 Struc. Steel Foremen	$40.15	$642.40	$71.85	$1149.60	$37.52	$65.27
5 Struc. Steel Workers	38.15	1526.00	68.25	2730.00		
1 Equip. Oper. (crane)	35.90	287.20	54.10	432.80		
1 Welder	38.15	305.20	68.25	546.00		
1 Equip. Oper. Oiler	30.10	240.80	45.35	362.80		
1 Crane, 90 Ton		1325.00		1457.50		
1 Gas Welding Machine		81.20		89.30	17.58	19.34
80 L.H., Daily Totals		$4407.80		$6768.00	$55.10	$84.61

Crew No.	Bare Costs		Incl. Subs O & P		Cost Per Labor-Hour	
Crew E-6	**Hr.**	**Daily**	**Hr.**	**Daily**	**Bare Costs**	**Incl. O&P**
3 Struc. Steel Foremen	$40.15	$963.60	$71.85	$1724.40	$37.56	$65.46
9 Struc. Steel Workers	38.15	2746.80	68.25	4914.00		
1 Equip. Oper. (crane)	35.90	287.20	54.10	432.80		
1 Welder	38.15	305.20	68.25	546.00		
1 Equip. Oper. Oiler	30.10	240.80	45.35	362.80		
1 Equip. Oper. (light)	33.05	264.40	49.80	398.40		
1 Crane, 90 Ton		1325.00		1457.50		
1 Gas Welding Machine		81.20		89.30		
1 Air Compr., 160 C.F.M.		87.40		96.15		
2 Impact Wrenches		24.00		26.40	11.86	13.04
128 L.H., Daily Totals		$6325.60		$10047.75	$49.42	$78.50

Crew E-7	**Hr.**	**Daily**	**Hr.**	**Daily**	**Bare Costs**	**Incl. O&P**
1 Struc. Steel Foreman	$40.15	$321.20	$71.85	$574.80	$37.52	$65.27
4 Struc. Steel Workers	38.15	1220.80	68.25	2184.00		
1 Equip. Oper. (crane)	35.90	287.20	54.10	432.80		
1 Equip. Oper. Oiler	30.10	240.80	45.35	362.80		
1 Welder Foreman	40.15	321.20	71.85	574.80		
2 Welders	38.15	610.40	68.25	1092.00		
1 Crane, 90 Ton		1325.00		1457.50		
2 Gas Welding Machines		162.40		178.65	18.59	20.45
80 L.H., Daily Totals		$4489.00		$6857.35	$56.11	$85.72

Crew E-8	**Hr.**	**Daily**	**Hr.**	**Daily**	**Bare Costs**	**Incl. O&P**
1 Struc. Steel Foreman	$40.15	$321.20	$71.85	$574.80	$37.27	$64.53
4 Struc. Steel Workers	38.15	1220.80	68.25	2184.00		
1 Welder Foreman	40.15	321.20	71.85	574.80		
4 Welders	38.15	1220.80	68.25	2184.00		
1 Equip. Oper. (crane)	35.90	287.20	54.10	432.80		
1 Equip. Oper. Oiler	30.10	240.80	45.35	362.80		
1 Equip. Oper. (light)	33.05	264.40	49.80	398.40		
1 Crane, 90 Ton		1325.00		1457.50		
4 Gas Welding Machines		324.80		357.30	15.86	17.45
104 L.H., Daily Totals		$5526.20		$8526.40	$53.13	$81.98

Crew E-9	**Hr.**	**Daily**	**Hr.**	**Daily**	**Bare Costs**	**Incl. O&P**
2 Struc. Steel Foremen	$40.15	$642.40	$71.85	$1149.60	$37.56	$65.46
5 Struc. Steel Workers	38.15	1526.00	68.25	2730.00		
1 Welder Foreman	40.15	321.20	71.85	574.80		
5 Welders	38.15	1526.00	68.25	2730.00		
1 Equip. Oper. (crane)	35.90	287.20	54.10	432.80		
1 Equip. Oper. Oiler	30.10	240.80	45.35	362.80		
1 Equip. Oper. (light)	33.05	264.40	49.80	398.40		
1 Crane, 90 Ton		1325.00		1457.50		
5 Gas Welding Machines		406.00		446.60	13.52	14.88
128 L.H., Daily Totals		$6539.00		$10282.50	$51.08	$80.34

Crew E-10	**Hr.**	**Daily**	**Hr.**	**Daily**	**Bare Costs**	**Incl. O&P**
1 Welder Foreman	$40.15	$321.20	$71.85	$574.80	$39.15	$70.05
1 Welder	38.15	305.20	68.25	546.00		
1 Gas Welding Machines		81.20		89.30		
1 Truck, 3 Ton		176.20		193.80	16.09	17.70
16 L.H., Daily Totals		$883.80		$1403.90	$55.24	$87.75

Crew E-11	**Hr.**	**Daily**	**Hr.**	**Daily**	**Bare Costs**	**Incl. O&P**
2 Painters, Struc. Steel	$31.00	$496.00	$57.05	$912.80	$30.44	$51.36
1 Building Laborer	26.70	213.60	41.55	332.40		
1 Equip. Oper. (light)	33.05	264.40	49.80	398.40		
1 Air Compressor 250 C.F.M.		112.40		123.65		
1 Sand Blaster		15.20		16.70		
1 Sand Blasting Accessories		11.90		13.10	4.36	4.80
32 L.H., Daily Totals		$1113.50		$1797.05	$34.80	$56.16

Crew No.	Bare Costs Hr.	Daily	Incl. Subs O & P Hr.	Daily	Cost Per Labor-Hour Bare Costs	Incl. O&P
Crew E-12	Hr.	Daily	Hr.	Daily	Bare Costs	Incl. O&P
1 Welder Foreman	$40.15	$321.20	$71.85	$574.80	$36.60	$60.83
1 Equip. Oper. (light)	33.05	264.40	49.80	398.40		
1 Gas Welding Machine		81.20		89.30	5.08	5.58
16 L.H., Daily Totals		$666.80		$1062.50	$41.68	$66.41
Crew E-13	Hr.	Daily	Hr.	Daily	Bare Costs	Incl. O&P
1 Welder Foreman	$40.15	$321.20	$71.85	$574.80	$37.78	$64.50
.5 Equip. Oper. (light)	33.05	132.20	49.80	199.20		
1 Gas Welding Machine		81.20		89.30	6.77	7.44
12 L.H., Daily Totals		$534.60		$863.30	$44.55	$71.94
Crew E-14	Hr.	Daily	Hr.	Daily	Bare Costs	Incl. O&P
1 Welder Foreman	$40.15	$321.20	$71.85	$574.80	$40.15	$71.85
1 Gas Welding Machine		81.20		89.30	10.15	11.17
8 L.H., Daily Totals		$402.40		$664.10	$50.30	$83.02
Crew E-16	Hr.	Daily	Hr.	Daily	Bare Costs	Incl. O&P
1 Welder Foreman	$40.15	$321.20	$71.85	$574.80	$39.15	$70.05
1 Welder	38.15	305.20	68.25	546.00		
1 Gas Welding Machine		81.20		89.30	5.08	5.58
16 L.H., Daily Totals		$707.60		$1210.10	$44.23	$75.63
Crew E-17	Hr.	Daily	Hr.	Daily	Bare Costs	Incl. O&P
1 Structural Steel Foreman	$40.15	$321.20	$71.85	$574.80	$39.15	$70.05
1 Structural Steel Worker	38.15	305.20	68.25	546.00		
1 Power Tool		5.20		5.70	.33	.36
16 L.H., Daily Totals		$631.60		$1126.50	$39.48	$70.41
Crew E-18	Hr.	Daily	Hr.	Daily	Bare Costs	Incl. O&P
1 Structural Steel Foreman	$40.15	$321.20	$71.85	$574.80	$37.85	$65.76
3 Structural Steel Workers	38.15	915.60	68.25	1638.00		
1 Equipment Operator (med.)	34.65	277.20	52.20	417.60		
1 Crane, 20 Ton		797.10		876.80	19.93	21.92
40 L.H., Daily Totals		$2311.10		$3507.20	$57.78	$87.68
Crew E-19	Hr.	Daily	Hr.	Daily	Bare Costs	Incl. O&P
1 Structural Steel Worker	$38.15	$305.20	$68.25	$546.00	$37.12	$63.30
1 Structural Steel Foreman	40.15	321.20	71.85	574.80		
1 Equip. Oper. (light)	33.05	264.40	49.80	398.40		
1 Power Tool		5.20		5.70		
1 Crane, 20 Ton		797.10		876.80	33.43	36.77
24 L.H., Daily Totals		$1693.10		$2401.70	$70.55	$100.07
Crew E-20	Hr.	Daily	Hr.	Daily	Bare Costs	Incl. O&P
1 Structural Steel Foreman	$40.15	$321.20	$71.85	$574.80	$37.11	$64.07
5 Structural Steel Workers	38.15	1526.00	68.25	2730.00		
1 Equip. Oper. (crane)	35.90	287.20	54.10	432.80		
1 Oiler	30.10	240.80	45.35	362.80		
1 Power Tool		5.20		5.70		
1 Crane, 40 Ton		951.90		1047.10	14.95	16.45
64 L.H., Daily Totals		$3332.30		$5153.20	$52.06	$80.52
Crew E-22	Hr.	Daily	Hr.	Daily	Bare Costs	Incl. O&P
1 Skilled Worker Foreman	$36.85	$294.80	$57.30	$458.40	$35.52	$55.23
2 Skilled Workers	34.85	557.60	54.20	867.20		
24 L.H., Daily Totals		$852.40		$1325.60	$35.52	$55.23

Crew No.	Bare Costs Hr.	Daily	Incl. Subs O & P Hr.	Daily	Cost Per Labor-Hour Bare Costs	Incl. O&P
Crew E-24	Hr.	Daily	Hr.	Daily	Bare Costs	Incl. O&P
3 Structural Steel Workers	$38.15	$915.60	$68.25	$1638.00	$37.28	$64.24
1 Equipment Operator (medium)	34.65	277.20	52.20	417.60		
1-25 Ton Crane		616.80		678.50	19.28	21.20
32 L.H., Daily Totals		$1809.60		$2734.10	$56.56	$85.44
Crew E-25	Hr.	Daily	Hr.	Daily	Bare Costs	Incl. O&P
1 Welder Foreman	$40.15	$321.20	$71.85	$574.80	$40.15	$71.85
1 Cutting Torch		18.00		19.80		
1 Gases		64.80		71.30	10.35	11.39
8 L.H., Daily Totals		$404.00		$665.90	$50.50	$83.24
Crew F-3	Hr.	Daily	Hr.	Daily	Bare Costs	Incl. O&P
4 Carpenters	$34.25	$1096.00	$53.35	$1707.20	$34.58	$53.50
1 Equip. Oper. (crane)	35.90	287.20	54.10	432.80		
1 Hyd. Crane, 12 Ton		602.60		662.85	15.07	16.57
40 L.H., Daily Totals		$1985.80		$2802.85	$49.65	$70.07
Crew F-4	Hr.	Daily	Hr.	Daily	Bare Costs	Incl. O&P
4 Carpenters	$34.25	$1096.00	$53.35	$1707.20	$33.83	$52.14
1 Equip. Oper. (crane)	35.90	287.20	54.10	432.80		
1 Equip. Oper. Oiler	30.10	240.80	45.35	362.80		
1 Hyd. Crane, 55 Ton		895.40		984.95	18.65	20.52
48 L.H., Daily Totals		$2519.40		$3487.75	$52.48	$72.66
Crew F-5	Hr.	Daily	Hr.	Daily	Bare Costs	Incl. O&P
1 Carpenter Foreman	$36.25	$290.00	$56.45	$451.60	$34.75	$54.13
3 Carpenters	34.25	822.00	53.35	1280.40		
32 L.H., Daily Totals		$1112.00		$1732.00	$34.75	$54.13
Crew F-6	Hr.	Daily	Hr.	Daily	Bare Costs	Incl. O&P
2 Carpenters	$34.25	$548.00	$53.35	$853.60	$31.56	$48.78
2 Building Laborers	26.70	427.20	41.55	664.80		
1 Equip. Oper. (crane)	35.90	287.20	54.10	432.80		
1 Hyd. Crane, 12 Ton		602.60		662.85	15.07	16.57
40 L.H., Daily Totals		$1865.00		$2614.05	$46.63	$65.35
Crew F-7	Hr.	Daily	Hr.	Daily	Bare Costs	Incl. O&P
2 Carpenters	$34.25	$548.00	$53.35	$853.60	$30.48	$47.45
2 Building Laborers	26.70	427.20	41.55	664.80		
32 L.H., Daily Totals		$975.20		$1518.40	$30.48	$47.45
Crew G-1	Hr.	Daily	Hr.	Daily	Bare Costs	Incl. O&P
1 Roofer Foreman	$31.45	$251.60	$53.60	$428.80	$27.49	$46.86
4 Roofers, Composition	29.45	942.40	50.20	1606.40		
2 Roofer Helpers	21.60	345.60	36.80	588.80		
1 Application Equipment		146.60		161.25		
1 Tar Kettle/Pot		52.50		57.75		
1 Crew Truck		98.80		108.70	5.32	5.85
56 L.H., Daily Totals		$1837.50		$2951.70	$32.81	$52.71
Crew G-2	Hr.	Daily	Hr.	Daily	Bare Costs	Incl. O&P
1 Plasterer	$31.35	$250.80	$47.70	$381.60	$28.38	$43.50
1 Plasterer Helper	27.10	216.80	41.25	330.00		
1 Building Laborer	26.70	213.60	41.55	332.40		
1 Grouting Equipment		107.75		118.55	4.49	4.94
24 L.H., Daily Totals		$788.95		$1162.55	$32.87	$48.44

Crew No.	Bare Costs		Incl. Subs O & P		Cost Per Labor-Hour	

Crew G-3

	Bare Costs		Incl. Subs O & P		Cost Per Labor-Hour	
Crew G-3	Hr.	Daily	Hr.	Daily	Bare Costs	Incl. O&P
2 Sheet Metal Workers	$40.20	$643.20	$61.85	$989.60	$33.45	$51.70
2 Building Laborers	26.70	427.20	41.55	664.80		
32 L.H., Daily Totals		$1070.40		$1654.40	$33.45	$51.70

Crew G-4	Hr.	Daily	Hr.	Daily	Bare Costs	Incl. O&P
1 Labor Foreman (outside)	$28.70	$229.60	$44.70	$357.60	$27.37	$42.60
2 Building Laborers	26.70	427.20	41.55	664.80		
1 Light Truck, 1.5 Ton		128.80		141.70		
1 Air Compr., 160 C.F.M.		87.40		96.15	9.01	9.91
24 L.H., Daily Totals		$873.00		$1260.25	$36.38	$52.51

Crew G-5	Hr.	Daily	Hr.	Daily	Bare Costs	Incl. O&P
1 Roofer Foreman	$31.45	$251.60	$53.60	$428.80	$26.71	$45.52
2 Roofers, Composition	29.45	471.20	50.20	803.20		
2 Roofer Helpers	21.60	345.60	36.80	588.80		
1 Application Equipment		146.60		161.25	3.67	4.03
40 L.H., Daily Totals		$1215.00		$1982.05	$30.38	$49.55

Crew G-6A	Hr.	Daily	Hr.	Daily	Bare Costs	Incl. O&P
2 Roofers Composition	$29.45	$471.20	$50.20	$803.20	$29.45	$50.20
1 Small Compressor		12.15		13.35		
2 Pneumatic Nailers		39.00		42.90	3.19	3.51
16 L.H., Daily Totals		$522.35		$859.45	$32.64	$53.71

Crew G-7	Hr.	Daily	Hr.	Daily	Bare Costs	Incl. O&P
1 Carpenter	$34.25	$274.00	$53.35	$426.80	$34.25	$53.35
1 Small Compressor		12.15		13.35		
1 Pneumatic Nailer		19.50		21.45	3.94	4.34
8 L.H., Daily Totals		$305.65		$461.60	$38.19	$57.69

Crew H-1	Hr.	Daily	Hr.	Daily	Bare Costs	Incl. O&P
2 Glaziers	$33.00	$528.00	$49.90	$798.40	$35.58	$59.08
2 Struc. Steel Workers	38.15	610.40	68.25	1092.00		
32 L.H., Daily Totals		$1138.40		$1890.40	$35.58	$59.08

Crew H-2	Hr.	Daily	Hr.	Daily	Bare Costs	Incl. O&P
2 Glaziers	$33.00	$528.00	$49.90	$798.40	$30.90	$47.12
1 Building Laborer	26.70	213.60	41.55	332.40		
24 L.H., Daily Totals		$741.60		$1130.80	$30.90	$47.12

Crew H-3	Hr.	Daily	Hr.	Daily	Bare Costs	Incl. O&P
1 Glazier	$33.00	$264.00	$49.90	$399.20	$29.28	$44.83
1 Helper	25.55	204.40	39.75	318.00		
16 L.H., Daily Totals		$468.40		$717.20	$29.28	$44.83

Crew J-1	Hr.	Daily	Hr.	Daily	Bare Costs	Incl. O&P
3 Plasterers	$31.35	$752.40	$47.70	$1144.80	$29.65	$45.12
2 Plasterer Helpers	27.10	433.60	41.25	660.00		
1 Mixing Machine, 6 C.F.		103.00		113.30	2.58	2.83
40 L.H., Daily Totals		$1289.00		$1918.10	$32.23	$47.95

Crew J-2	Hr.	Daily	Hr.	Daily	Bare Costs	Incl. O&P
3 Plasterers	$31.35	$752.40	$47.70	$1144.80	$29.95	$45.37
2 Plasterer Helpers	27.10	433.60	41.25	660.00		
1 Lather	31.45	251.60	46.60	372.80		
1 Mixing Machine, 6 C.F.		103.00		113.30	2.15	2.36
48 L.H., Daily Totals		$1540.60		$2290.90	$32.10	$47.73

Crew J-3	Hr.	Daily	Hr.	Daily	Bare Costs	Incl. O&P
1 Terrazzo Worker	$32.65	$261.20	$48.00	$384.00	$29.38	$43.18
1 Terrazzo Helper	26.10	208.80	38.35	306.80		
1 Terrazzo Grinder, Electric		70.00		77.00		
1 Terrazzo Mixer		139.60		153.55	13.10	14.41
16 L.H., Daily Totals		$679.60		$921.35	$42.48	$57.59

Crew J-4	Hr.	Daily	Hr.	Daily	Bare Costs	Incl. O&P
1 Tile Layer	$32.70	$261.60	$48.05	$384.40	$29.03	$42.65
1 Tile Layer Helper	25.35	202.80	37.25	298.00		
16 L.H., Daily Totals		$464.40		$682.40	$29.03	$42.65

Crew K-1	Hr.	Daily	Hr.	Daily	Bare Costs	Incl. O&P
1 Carpenter	$34.25	$274.00	$53.35	$426.80	$30.53	$47.10
1 Truck Driver (light)	26.80	214.40	40.85	326.80		
1 Truck w/Power Equip.		176.20		193.80	11.01	12.11
16 L.H., Daily Totals		$664.60		$947.40	$41.54	$59.21

Crew K-2	Hr.	Daily	Hr.	Daily	Bare Costs	Incl. O&P
1 Struc. Steel Foreman	$40.15	$321.20	$71.85	$574.80	$35.03	$60.32
1 Struc. Steel Worker	38.15	305.20	68.25	546.00		
1 Truck Driver (light)	26.80	214.40	40.85	326.80		
1 Truck w/Power Equip.		176.20		193.80	7.34	8.08
24 L.H., Daily Totals		$1017.00		$1641.40	$42.37	$68.40

Crew L-1	Hr.	Daily	Hr.	Daily	Bare Costs	Incl. O&P
1 Electrician	$40.75	$326.00	$60.65	$485.20	$40.80	$61.03
1 Plumber	40.85	326.80	61.40	491.20		
16 L.H., Daily Totals		$652.80		$976.40	$40.80	$61.03

Crew L-2	Hr.	Daily	Hr.	Daily	Bare Costs	Incl. O&P
1 Carpenter	$34.25	$274.00	$53.35	$426.80	$29.90	$46.55
1 Carpenter Helper	25.55	204.40	39.75	318.00		
16 L.H., Daily Totals		$478.40		$744.80	$29.90	$46.55

Crew L-3	Hr.	Daily	Hr.	Daily	Bare Costs	Incl. O&P
1 Carpenter	$34.25	$274.00	$53.35	$426.80	$37.36	$57.30
.5 Electrician	40.75	163.00	60.65	242.60		
.5 Sheet Metal Worker	40.20	160.80	61.85	247.40		
16 L.H., Daily Totals		$597.80		$916.80	$37.36	$57.30

Crew L-3A	Hr.	Daily	Hr.	Daily	Bare Costs	Incl. O&P
1 Carpenter Foreman (outside)	$36.25	$290.00	$56.45	$451.60	$37.57	$58.25
.5 Sheet Metal Worker	40.20	160.80	61.85	247.40		
12 L.H., Daily Totals		$450.80		$699.00	$37.57	$58.25

Crew L-4	Hr.	Daily	Hr.	Daily	Bare Costs	Incl. O&P
2 Skilled Workers	$34.85	$557.60	$54.20	$867.20	$31.75	$49.38
1 Helper	25.55	204.40	39.75	318.00		
24 L.H., Daily Totals		$762.00		$1185.20	$31.75	$49.38

Crew L-5	Hr.	Daily	Hr.	Daily	Bare Costs	Incl. O&P
1 Struc. Steel Foreman	$40.15	$321.20	$71.85	$574.80	$38.11	$66.74
5 Struc. Steel Workers	38.15	1526.00	68.25	2730.00		
1 Equip. Oper. (crane)	35.90	287.20	54.10	432.80		
1 Hyd. Crane, 25 Ton		616.80		678.50	11.01	12.12
56 L.H., Daily Totals		$2751.20		$4416.10	$49.12	$78.86

Crew L-5A

	Bare Costs Hr.	Daily	Incl. Subs O & P Hr.	Daily	Bare Costs	Incl. O&P
1 Structural Steel Foreman	$40.15	$321.20	$71.85	$574.80	$38.09	$65.61
2 Structural Steel Workers	38.15	610.40	68.25	1092.00		
1 Equip. Oper. (crane)	35.90	287.20	54.10	432.80		
1 Crane, SP, 25 Ton		575.00		632.50	17.97	19.77
32 L.H., Daily Totals		$1793.80		$2732.10	$56.06	$85.38

Crew L-6

	Bare Costs Hr.	Daily	Incl. Subs O & P Hr.	Daily	Bare Costs	Incl. O&P
1 Plumber	$40.85	$326.80	$61.40	$491.20	$40.82	$61.15
.5 Electrician	40.75	163.00	60.65	242.60		
12 L.H., Daily Totals		$489.80		$733.80	$40.82	$61.15

Crew L-7

	Bare Costs Hr.	Daily	Incl. Subs O & P Hr.	Daily	Bare Costs	Incl. O&P
2 Carpenters	$34.25	$548.00	$53.35	$853.60	$33.02	$51.02
1 Building Laborer	26.70	213.60	41.55	332.40		
.5 Electrician	40.75	163.00	60.65	242.60		
28 L.H., Daily Totals		$924.60		$1428.60	$33.02	$51.02

Crew L-8

	Bare Costs Hr.	Daily	Incl. Subs O & P Hr.	Daily	Bare Costs	Incl. O&P
2 Carpenters	$34.25	$548.00	$53.35	$853.60	$35.57	$54.96
.5 Plumber	40.85	163.40	61.40	245.60		
20 L.H., Daily Totals		$711.40		$1099.20	$35.57	$54.96

Crew L-9

	Bare Costs Hr.	Daily	Incl. Subs O & P Hr.	Daily	Bare Costs	Incl. O&P
1 Labor Foreman (inside)	$27.20	$217.60	$42.35	$338.80	$30.92	$49.78
2 Building Laborers	26.70	427.20	41.55	664.80		
1 Struc. Steel Worker	38.15	305.20	68.25	546.00		
.5 Electrician	40.75	163.00	60.65	242.60		
36 L.H., Daily Totals		$1113.00		$1792.20	$30.92	$49.78

Crew L-10

	Bare Costs Hr.	Daily	Incl. Subs O & P Hr.	Daily	Bare Costs	Incl. O&P
1 Structural Steel Foreman	$40.15	$321.20	$71.85	$574.80	$38.07	$64.73
1 Structural Steel Worker	38.15	305.20	68.25	546.00		
1 Equip. Oper. (crane)	35.90	287.20	54.10	432.80		
1 Hyd. Crane, 12 Ton		602.60		662.85	25.11	27.62
24 L.H., Daily Totals		$1516.20		$2216.45	$63.18	$92.35

Crew L-11

	Bare Costs Hr.	Daily	Incl. Subs O & P Hr.	Daily	Bare Costs	Incl. O&P
2 Wreckers	$26.70	$427.20	$47.20	$755.20	$30.59	$49.58
1 Equip. Oper. (crane)	35.90	287.20	54.10	432.80		
1 Equip. Oper. (light)	33.05	264.40	49.80	398.40		
1 Hyd. Excavator, 2.5 C.Y.		1215.00		1336.50		
1 Skid steer loader		197.80		217.60	44.15	48.57
32 L.H., Daily Totals		$2391.60		$3140.50	$74.74	$98.15

Crew M-1

	Bare Costs Hr.	Daily	Incl. Subs O & P Hr.	Daily	Bare Costs	Incl. O&P
3 Elevator Constructors	$45.05	$1081.20	$67.55	$1621.20	$42.80	$64.18
1 Elevator Apprentice	36.05	288.40	54.05	432.40		
5 Hand Tools		63.00		69.30	1.97	2.17
32 L.H., Daily Totals		$1432.60		$2122.90	$44.77	$66.35

Crew M-3

	Bare Costs Hr.	Daily	Incl. Subs O & P Hr.	Daily	Bare Costs	Incl. O&P
1 Electrician Foreman (out)	$42.75	$342.00	$63.60	$508.80	$37.46	$56.42
1 Common Laborer	26.70	213.60	41.55	332.40		
.25 Equipment Operator, Medium	34.65	69.30	52.20	104.40		
1 Elevator Constructor	45.05	360.40	67.55	540.40		
1 Elevator Apprentice	36.05	288.40	54.05	432.40		
.25 Crane, SP, 4 x 4, 20 ton		145.35		159.90	4.28	4.70
34 L.H., Daily Totals		$1419.05		$2078.30	$41.74	$61.12

Crew M-4

	Bare Costs Hr.	Daily	Incl. Subs O & P Hr.	Daily	Bare Costs	Incl. O&P
1 Electrician Foreman (out)	$42.75	$342.00	$63.60	$508.80	$37.12	$55.91
1 Common Laborer	26.70	213.60	41.55	332.40		
.25 Equipment Operator, Crane	35.90	71.80	54.10	108.20		
.25 Equipment Operator, Oiler	30.10	60.20	45.35	90.70		
1 Elevator Constructor	45.05	360.40	67.55	540.40		
1 Elevator Apprentice	36.05	288.40	54.05	432.40		
.25 Crane, Hyd, SP, 4WD, 40 Ton		217.40		239.15	6.04	6.64
36 L.H., Daily Totals		$1553.80		$2252.05	$43.16	$62.55

Crew Q-1

	Bare Costs Hr.	Daily	Incl. Subs O & P Hr.	Daily	Bare Costs	Incl. O&P
1 Plumber	$40.85	$326.80	$61.40	$491.20	$36.78	$55.28
1 Plumber Apprentice	32.70	261.60	49.15	393.20		
16 L.H., Daily Totals		$588.40		$884.40	$36.78	$55.28

Crew Q-1C

	Bare Costs Hr.	Daily	Incl. Subs O & P Hr.	Daily	Bare Costs	Incl. O&P
1 Plumber	$40.85	$326.80	$61.40	$491.20	$36.07	$54.25
1 Plumber Apprentice	32.70	261.60	49.15	393.20		
1 Equip. Oper. (medium)	34.65	277.20	52.20	417.60		
1 Trencher, Chain		1420.00		1562.00	59.17	65.08
24 L.H., Daily Totals		$2285.60		$2864.00	$95.24	$119.33

Crew Q-2

	Bare Costs Hr.	Daily	Incl. Subs O & P Hr.	Daily	Bare Costs	Incl. O&P
2 Plumbers	$40.85	$653.60	$61.40	$982.40	$38.13	$57.32
1 Plumber Apprentice	32.70	261.60	49.15	393.20		
24 L.H., Daily Totals		$915.20		$1375.60	$38.13	$57.32

Crew Q-3

	Bare Costs Hr.	Daily	Incl. Subs O & P Hr.	Daily	Bare Costs	Incl. O&P
1 Plumber Foreman (inside)	$41.35	$330.80	$62.15	$497.20	$38.94	$58.53
2 Plumbers	40.85	653.60	61.40	982.40		
1 Plumber Apprentice	32.70	261.60	49.15	393.20		
32 L.H., Daily Totals		$1246.00		$1872.80	$38.94	$58.53

Crew Q-4

	Bare Costs Hr.	Daily	Incl. Subs O & P Hr.	Daily	Bare Costs	Incl. O&P
1 Plumber Foreman (inside)	$41.35	$330.80	$62.15	$497.20	$38.94	$58.53
1 Plumber	40.85	326.80	61.40	491.20		
1 Welder (plumber)	40.85	326.80	61.40	491.20		
1 Plumber Apprentice	32.70	261.60	49.15	393.20		
1 Electric Welding Mach.		80.65		88.70	2.52	2.77
32 L.H., Daily Totals		$1326.65		$1961.50	$41.46	$61.30

Crew Q-5

	Bare Costs Hr.	Daily	Incl. Subs O & P Hr.	Daily	Bare Costs	Incl. O&P
1 Steamfitter	$40.95	$327.60	$61.55	$492.40	$36.85	$55.38
1 Steamfitter Apprentice	32.75	262.00	49.20	393.60		
16 L.H., Daily Totals		$589.60		$886.00	$36.85	$55.38

Crew Q-6

	Bare Costs Hr.	Daily	Incl. Subs O & P Hr.	Daily	Bare Costs	Incl. O&P
2 Steamfitters	$40.95	$655.20	$61.55	$984.80	$38.22	$57.43
1 Steamfitter Apprentice	32.75	262.00	49.20	393.60		
24 L.H., Daily Totals		$917.20		$1378.40	$38.22	$57.43

Crew Q-7

	Bare Costs Hr.	Daily	Incl. Subs O & P Hr.	Daily	Bare Costs	Incl. O&P
1 Steamfitter Foreman (inside)	$41.45	$331.60	$62.30	$498.40	$39.03	$58.65
2 Steamfitters	40.95	655.20	61.55	984.80		
1 Steamfitter Apprentice	32.75	262.00	49.20	393.60		
32 L.H., Daily Totals		$1248.80		$1876.80	$39.03	$58.65

Crew Q-8	Hr.	Daily	Hr.	Daily	Bare Costs	Incl. O&P
1 Steamfitter Foreman (inside)	$41.45	$331.60	$62.30	$498.40	$39.03	$58.65
1 Steamfitter	40.95	327.60	61.55	492.40		
1 Welder (steamfitter)	40.95	327.60	61.55	492.40		
1 Steamfitter Apprentice	32.75	262.00	49.20	393.60		
1 Electric Welding Mach.		80.65		88.70	2.52	2.77
32 L.H., Daily Totals		$1329.45		$1965.50	$41.55	$61.42

Crew Q-9	Hr.	Daily	Hr.	Daily	Bare Costs	Incl. O&P
1 Sheet Metal Worker	$40.20	$321.60	$61.85	$494.80	$36.18	$55.65
1 Sheet Metal Apprentice	32.15	257.20	49.45	395.60		
16 L.H., Daily Totals		$578.80		$890.40	$36.18	$55.65

Crew Q-10	Hr.	Daily	Hr.	Daily	Bare Costs	Incl. O&P
2 Sheet Metal Workers	$40.20	$643.20	$61.85	$989.60	$37.52	$57.72
1 Sheet Metal Apprentice	32.15	257.20	49.45	395.60		
24 L.H., Daily Totals		$900.40		$1385.20	$37.52	$57.72

Crew Q-11	Hr.	Daily	Hr.	Daily	Bare Costs	Incl. O&P
1 Sheet Metal Foreman (inside)	$40.70	$325.60	$62.60	$500.80	$38.31	$58.94
2 Sheet Metal Workers	40.20	643.20	61.85	989.60		
1 Sheet Metal Apprentice	32.15	257.20	49.45	395.60		
32 L.H., Daily Totals		$1226.00		$1886.00	$38.31	$58.94

Crew Q-12	Hr.	Daily	Hr.	Daily	Bare Costs	Incl. O&P
1 Sprinkler Installer	$40.75	$326.00	$61.55	$492.40	$36.68	$55.40
1 Sprinkler Apprentice	32.60	260.80	49.25	394.00		
16 L.H., Daily Totals		$586.80		$886.40	$36.68	$55.40

Crew Q-13	Hr.	Daily	Hr.	Daily	Bare Costs	Incl. O&P
1 Sprinkler Foreman (inside)	$41.25	$330.00	$62.35	$498.80	$38.84	$58.68
2 Sprinkler Installers	40.75	652.00	61.55	984.80		
1 Sprinkler Apprentice	32.60	260.80	49.25	394.00		
32 L.H., Daily Totals		$1242.80		$1877.60	$38.84	$58.68

Crew Q-14	Hr.	Daily	Hr.	Daily	Bare Costs	Incl. O&P
1 Asbestos Worker	$37.10	$296.80	$58.55	$468.40	$33.40	$52.70
1 Asbestos Apprentice	29.70	237.60	46.85	374.80		
16 L.H., Daily Totals		$534.40		$843.20	$33.40	$52.70

Crew Q-15	Hr.	Daily	Hr.	Daily	Bare Costs	Incl. O&P
1 Plumber	$40.85	$326.80	$61.40	$491.20	$36.78	$55.28
1 Plumber Apprentice	32.70	261.60	49.15	393.20		
1 Electric Welding Mach.		80.65		88.70	5.04	5.54
16 L.H., Daily Totals		$669.05		$973.10	$41.82	$60.82

Crew Q-16	Hr.	Daily	Hr.	Daily	Bare Costs	Incl. O&P
2 Plumbers	$40.85	$653.60	$61.40	$982.40	$38.13	$57.32
1 Plumber Apprentice	32.70	261.60	49.15	393.20		
1 Electric Welding Mach.		80.65		88.70	3.36	3.70
24 L.H., Daily Totals		$995.85		$1464.30	$41.49	$61.02

Crew Q-17	Hr.	Daily	Hr.	Daily	Bare Costs	Incl. O&P
1 Steamfitter	$40.95	$327.60	$61.55	$492.40	$36.85	$55.38
1 Steamfitter Apprentice	32.75	262.00	49.20	393.60		
1 Electric Welding Mach.		80.65		88.70	5.04	5.54
16 L.H., Daily Totals		$670.25		$974.70	$41.89	$60.92

Crew Q-17A	Hr.	Daily	Hr.	Daily	Bare Costs	Incl. O&P
1 Steamfitter	$40.95	$327.60	$61.55	$492.40	$36.53	$54.95
1 Steamfitter Apprentice	32.75	262.00	49.20	393.60		
1 Equip. Oper. (crane)	35.90	287.20	54.10	432.80		
1 Truck Crane, 12 Ton		602.60		662.85		
1 Electric Welding Mach.		80.65		88.70	28.47	31.32
24 L.H., Daily Totals		$1560.05		$2070.35	$65.00	$86.27

Crew Q-18	Hr.	Daily	Hr.	Daily	Bare Costs	Incl. O&P
2 Steamfitters	$40.95	$655.20	$61.55	$984.80	$38.22	$57.43
1 Steamfitter Apprentice	32.75	262.00	49.20	393.60		
1 Electric Welding Mach.		80.65		88.70	3.36	3.70
24 L.H., Daily Totals		$997.85		$1467.10	$41.58	$61.13

Crew Q-19	Hr.	Daily	Hr.	Daily	Bare Costs	Incl. O&P
1 Steamfitter	$40.95	$327.60	$61.55	$492.40	$38.15	$57.13
1 Steamfitter Apprentice	32.75	262.00	49.20	393.60		
1 Electrician	40.75	326.00	60.65	485.20		
24 L.H., Daily Totals		$915.60		$1371.20	$38.15	$57.13

Crew Q-20	Hr.	Daily	Hr.	Daily	Bare Costs	Incl. O&P
1 Sheet Metal Worker	$40.20	$321.60	$61.85	$494.80	$37.09	$56.65
1 Sheet Metal Apprentice	32.15	257.20	49.45	395.60		
.5 Electrician	40.75	163.00	60.65	242.60		
20 L.H., Daily Totals		$741.80		$1133.00	$37.09	$56.65

Crew Q-21	Hr.	Daily	Hr.	Daily	Bare Costs	Incl. O&P
2 Steamfitters	$40.95	$655.20	$61.55	$984.80	$38.85	$58.24
1 Steamfitter Apprentice	32.75	262.00	49.20	393.60		
1 Electrician	40.75	326.00	60.65	485.20		
32 L.H., Daily Totals		$1243.20		$1863.60	$38.85	$58.24

Crew Q-22	Hr.	Daily	Hr.	Daily	Bare Costs	Incl. O&P
1 Plumber	$40.85	$326.80	$61.40	$491.20	$36.78	$55.28
1 Plumber Apprentice	32.70	261.60	49.15	393.20		
1 Truck Crane, 12 Ton		602.60		662.85	37.66	41.43
16 L.H., Daily Totals		$1191.00		$1547.25	$74.44	$96.71

Crew Q-22A	Hr.	Daily	Hr.	Daily	Bare Costs	Incl. O&P
1 Plumber	$40.85	$326.80	$61.40	$491.20	$34.04	$51.55
1 Plumber Apprentice	32.70	261.60	49.15	393.20		
1 Laborer	26.70	213.60	41.55	332.40		
1 Equip. Oper. (crane)	35.90	287.20	54.10	432.80		
1 Truck Crane, 12 Ton		602.60		662.85	18.83	20.71
32 L.H., Daily Totals		$1691.80		$2312.45	$52.87	$72.26

Crew Q-23	Hr.	Daily	Hr.	Daily	Bare Costs	Incl. O&P
1 Plumber Foreman	$42.85	$342.80	$64.40	$515.20	$39.45	$59.33
1 Plumber	40.85	326.80	61.40	491.20		
1 Equip. Oper. (medium)	34.65	277.20	52.20	417.60		
1 Power Tools		5.20		5.70		
1 Crane, 20 Ton		797.10		876.80	33.43	36.77
24 L.H., Daily Totals		$1749.10		$2306.50	$72.88	$96.10

Crew R-1	Hr.	Daily	Hr.	Daily	Bare Costs	Incl. O&P
1 Electrician Foreman	$41.25	$330.00	$61.40	$491.20	$35.77	$53.81
3 Electricians	40.75	978.00	60.65	1455.60		
2 Helpers	25.55	408.80	39.75	636.00		
48 L.H., Daily Totals		$1716.80		$2582.80	$35.77	$53.81

Crew No.	Bare Costs		Incl. Subs O & P		Cost Per Labor-Hour	

Crew R-1A	Hr.	Daily	Hr.	Daily	Bare Costs	Incl. O&P
1 Electrician	$40.75	$326.00	$60.65	$485.20	$33.15	$50.20
1 Helper	25.55	204.40	39.75	318.00		
16 L.H., Daily Totals		$530.40		$803.20	$33.15	$50.20

Crew R-2	Hr.	Daily	Hr.	Daily	Bare Costs	Incl. O&P
1 Electrician Foreman	$41.25	$330.00	$61.40	$491.20	$35.79	$53.85
3 Electricians	40.75	978.00	60.65	1455.60		
2 Helpers	25.55	408.80	39.75	636.00		
1 Equip. Oper. (crane)	35.90	287.20	54.10	432.80		
1 S.P. Crane, 5 Ton		317.00		348.70	5.66	6.23
56 L.H., Daily Totals		$2321.00		$3364.30	$41.45	$60.08

Crew R-3	Hr.	Daily	Hr.	Daily	Bare Costs	Incl. O&P
1 Electrician Foreman	$41.25	$330.00	$61.40	$491.20	$39.98	$59.64
1 Electrician	40.75	326.00	60.65	485.20		
.5 Equip. Oper. (crane)	35.90	143.60	54.10	216.40		
.5 S.P. Crane, 5 Ton		158.50		174.35	7.93	8.72
20 L.H., Daily Totals		$958.10		$1367.15	$47.91	$68.36

Crew R-4	Hr.	Daily	Hr.	Daily	Bare Costs	Incl. O&P
1 Struc. Steel Foreman	$40.15	$321.20	$71.85	$574.80	$39.07	$67.45
3 Struc. Steel Workers	38.15	915.60	68.25	1638.00		
1 Electrician	40.75	326.00	60.65	485.20		
1 Gas Welding Machine		81.20		89.30	2.03	2.23
40 L.H., Daily Totals		$1644.00		$2787.30	$41.10	$69.68

Crew R-5	Hr.	Daily	Hr.	Daily	Bare Costs	Incl. O&P
1 Electrician Foreman	$41.25	$330.00	$61.40	$491.20	$35.27	$53.12
4 Electrician Linemen	40.75	1304.00	60.65	1940.80		
2 Electrician Operators	40.75	652.00	60.65	970.40		
4 Electrician Groundmen	25.55	817.60	39.75	1272.00		
1 Crew Truck		98.80		108.70		
1 Tool Van		122.15		134.35		
1 Pickup Truck, 3/4 Ton		78.00		85.80		
.2 Crane, 55 Ton		179.08		197.00		
.2 Crane, 12 Ton		120.52		132.55		
.2 Auger, Truck Mtd.		635.80		699.40		
1 Tractor w/Winch		260.00		286.00	16.98	18.68
88 L.H., Daily Totals		$4597.95		$6318.20	$52.25	$71.80

Crew R-6	Hr.	Daily	Hr.	Daily	Bare Costs	Incl. O&P
1 Electrician Foreman	$41.25	$330.00	$61.40	$491.20	$35.27	$53.12
4 Electrician Linemen	40.75	1304.00	60.65	1940.80		
2 Electrician Operators	40.75	652.00	60.65	970.40		
4 Electrician Groundmen	25.55	817.60	39.75	1272.00		
1 Crew Truck		98.80		108.70		
1 Tool Van		122.15		134.35		
1 Pickup Truck, 3/4 Ton		78.00		85.80		
.2 Crane, 55 Ton		179.08		197.00		
.2 Crane, 12 Ton		120.52		132.55		
.2 Auger, Truck Mtd.		635.80		699.40		
1 Tractor w/Winch		260.00		286.00		
3 Cable Trailers		490.65		539.70		
.5 Tensioning Rig		160.00		176.00		
.5 Cable Pulling Rig		934.00		1027.40	34.99	38.49
88 L.H., Daily Totals		$6182.60		$8061.30	$70.26	$91.61

Crew R-7	Hr.	Daily	Hr.	Daily	Bare Costs	Incl. O&P
1 Electrician Foreman	$41.25	$330.00	$61.40	$491.20	$28.17	$43.36
5 Electrician Groundmen	25.55	1022.00	39.75	1590.00		
1 Crew Truck		98.80		108.70	2.06	2.26
48 L.H., Daily Totals		$1450.80		$2189.90	$30.23	$45.62

Crew R-8	Hr.	Daily	Hr.	Daily	Bare Costs	Incl. O&P
1 Electrician Foreman	$41.25	$330.00	$61.40	$491.20	$35.77	$53.81
3 Electrician Linemen	40.75	978.00	60.65	1455.60		
2 Electrician Groundmen	25.55	408.80	39.75	636.00		
1 Pickup Truck, 3/4 Ton		78.00		85.80		
1 Crew Truck		98.80		108.70	3.68	4.05
48 L.H., Daily Totals		$1893.60		$2777.30	$39.45	$57.86

Crew R-9	Hr.	Daily	Hr.	Daily	Bare Costs	Incl. O&P
1 Electrician Foreman	$41.25	$330.00	$61.40	$491.20	$33.21	$50.29
1 Electrician Lineman	40.75	326.00	60.65	485.20		
2 Electrician Operators	40.75	652.00	60.65	970.40		
4 Electrician Groundmen	25.55	817.60	39.75	1272.00		
1 Pickup Truck, 3/4 Ton		78.00		85.80		
1 Crew Truck		98.80		108.70	2.76	3.04
64 L.H., Daily Totals		$2302.40		$3413.30	$35.97	$53.33

Crew R-10	Hr.	Daily	Hr.	Daily	Bare Costs	Incl. O&P
1 Electrician Foreman	$41.25	$330.00	$61.40	$491.20	$38.30	$57.29
4 Electrician Linemen	40.75	1304.00	60.65	1940.80		
1 Electrician Groundman	25.55	204.40	39.75	318.00		
1 Crew Truck		98.80		108.70		
3 Tram Cars		345.75		380.35	9.26	10.19
48 L.H., Daily Totals		$2282.95		$3239.05	$47.56	$67.48

Crew R-11	Hr.	Daily	Hr.	Daily	Bare Costs	Incl. O&P
1 Electrician Foreman	$41.25	$330.00	$61.40	$491.20	$38.12	$57.09
4 Electricians	40.75	1304.00	60.65	1940.80		
1 Equip. Oper. (crane)	35.90	287.20	54.10	432.80		
1 Common Laborer	26.70	213.60	41.55	332.40		
1 Crew Truck		98.80		108.70		
1 Crane, 12 Ton		602.60		662.85	12.53	13.78
56 L.H., Daily Totals		$2836.20		$3968.75	$50.65	$70.87

Crew R-12	Hr.	Daily	Hr.	Daily	Bare Costs	Incl. O&P
1 Carpenter Foreman	$34.75	$278.00	$54.10	$432.80	$31.94	$50.38
4 Carpenters	34.25	1096.00	53.35	1707.20		
4 Common Laborers	26.70	854.40	41.55	1329.60		
1 Equip. Oper. (med.)	34.65	277.20	52.20	417.60		
1 Steel Worker	38.15	305.20	68.25	546.00		
1 Dozer, 200 H.P.		919.60		1011.55		
1 Pickup Truck, 3/4 Ton		78.00		85.80	11.34	12.47
88 L.H., Daily Totals		$3808.40		$5530.55	$43.28	$62.85

Crew R-13	Hr.	Daily	Hr.	Daily	Bare Costs	Incl. O&P
1 Electrician Foreman	$41.25	$330.00	$61.40	$491.20	$38.59	$57.57
3 Electricians	40.75	978.00	60.65	1455.60		
.25 Equip. Oper. (crane)	35.90	71.80	54.10	108.20		
1 Equipment Oiler	30.10	240.80	45.35	362.80		
.25-1 Hyd. Crane, 33 Ton		160.40		176.45	3.82	4.20
42 L.H., Daily Totals		$1781.00		$2594.25	$42.41	$61.77

Crew No.	Bare Costs		Incl. Sub O & P		Cost Per Labor-Hour	
Crew R-15	Hr.	Daily	Hr.	Daily	Bare Costs	Incl. O&P
1 Electrician Foreman	$41.25	$330.00	$61.40	$491.20	$39.55	$58.97
4 Electricians	40.75	1304.00	60.65	1940.80		
1 Equipment Operator	33.05	264.40	49.80	398.40		
1 Aerial Lift Truck		252.60		277.85	5.26	5.79
48 L.H., Daily Totals		$2151.00		$3108.25	$44.81	$64.76
Crew R-18	Hr.	Daily	Hr.	Daily	Bare Costs	Incl. O&P
.25 Electrician Foreman	$41.25	$82.50	$61.40	$122.80	$31.43	$47.85
1 Electrician	40.75	326.00	60.65	485.20		
2 Helpers	25.55	408.80	39.75	636.00		
26 L.H., Daily Totals		$817.30		$1244.00	$31.43	$47.85
Crew R-19	Hr.	Daily	Hr.	Daily	Bare Costs	Incl. O&P
.5 Electrician Foreman	$41.25	$165.00	$61.40	$245.60	$40.85	$60.80
2 Electricians	40.75	652.00	60.65	970.40		
20 L.H., Daily Totals		$817.00		$1216.00	$40.85	$60.80
Crew R-21	Hr.	Daily	Hr.	Daily	Bare Costs	Incl. O&P
1 Electrician Foreman	$41.25	$330.00	$61.40	$491.20	$40.72	$60.63
3 Electricians	40.75	978.00	60.65	1455.60		
.1 Equip. Oper. (med.)	34.65	27.72	52.20	41.76		
.1 Hyd. Crane 25 Ton		57.50		63.25	1.75	1.93
32. L.H., Daily Totals		$1393.22		$2051.81	$42.47	$62.56
Crew R-22	Hr.	Daily	Hr.	Daily	Bare Costs	Incl. O&P
.66 Electrician Foreman	$41.25	$217.80	$61.40	$324.19	$34.30	$51.79
2 Helpers	25.55	408.80	39.75	636.00		
2 Electricians	40.75	652.00	60.65	970.40		
37.28 L.H., Daily Totals		$1278.60		$1930.59	$34.30	$51.79
Crew R-30	Hr.	Daily	Hr.	Daily	Bare Costs	Incl. O&P
.25 Electrician	$42.75	$85.50	$63.60	$127.20	$32.26	$49.12
1 Electrician	40.75	326.00	60.65	485.20		
2 Laborers, (Semi-Skilled)	26.70	427.20	41.55	664.80		
26 L.H., Daily Totals		$838.70		$1277.20	$32.26	$49.12
Crew R-31	Hr.	Daily	Hr.	Daily	Bare Costs	Incl. O&P
1 Electrician	$40.75	$326.00	$60.65	$485.20	$40.75	$60.65
1 Core Drill, Elec, 2.5 HP		50.30		55.35	6.28	6.91
8 L.H., Daily Totals		$376.30		$540.55	$47.03	$67.56
Crew W-41E	Hr.	Daily	Hr.	Daily	Bare Costs	Incl. O&P
1 Laborers, (Semi-Skilled)	$26.70	$213.60	$41.55	$332.40	$35.59	$54.06
1 Plumber	40.85	326.80	61.40	491.20		
.5 Plumber	42.85	171.40	64.40	257.60		
20 L.H., Daily Totals		$711.80		$1081.20	$35.59	$54.06

Historical Cost Indexes

The table below lists both the Means Historical Cost Index based on Jan. 1, 1993 = 100 as well as the computed value of an index based on Jan. 1, 2005 costs. Since the Jan. 1, 2005 figure is estimated, space is left to write in the actual index figures as they become available through either the quarterly "Means Construction Cost Indexes" or as printed in the "Engineering News-Record." To compute the actual index based on Jan. 1, 2005 = 100, divide the Historical Cost Index for a particular year by the actual Jan. 1, 2005 Construction Cost Index. Space has been left to advance the index figures as the year progresses.

Year	Historical Cost Index Jan. 1, 1993 = 100		Current Index Based on Jan. 1, 2005 = 100		Year	Historical Cost Index Jan. 1, 1993 = 100	Current Index Based on Jan. 1, 2005 = 100		Year	Historical Cost Index Jan. 1, 1993 = 100	Current Index Based on Jan. 1, 2005 = 100	
	Est.	Actual	Est.	Actual		Actual	Est.	Actual		Actual	Est.	Actual
Oct 2005					July 1990	94.3	63.5		July 1972	34.8	23.4	
July 2005					1989	92.1	62.0		1971	32.1	21.6	
April 2005					1988	89.9	60.5		1970	28.7	19.3	
Jan 2005	148.5		100.0	100.0	1987	87.7	59.0		1969	26.9	18.1	
July 2004		143.7	96.8		1986	84.2	56.7		1968	24.9	16.8	
2003		132.0	88.9		1985	82.6	55.6		1967	23.5	15.8	
2002		128.7	86.7		1984	82.0	55.2		1966	22.7	15.3	
2001		125.1	84.2		1983	80.2	54.0		1965	21.7	14.6	
2000		120.9	81.4		1982	76.1	51.3		1964	21.2	14.3	
1999		117.6	79.2		1981	70.0	47.1		1963	20.7	13.9	
1998		115.1	77.5		1980	62.9	42.4		1962	20.2	13.6	
1997		112.8	76.0		1979	57.8	38.9		1961	19.8	13.3	
1996		110.2	74.2		1978	53.5	36.0		1960	19.7	13.3	
1995		107.6	72.5		1977	49.5	33.3		1959	19.3	13.0	
1994		104.4	70.3		1976	46.9	31.6		1958	18.8	12.7	
1993		101.7	68.5		1975	44.8	30.2		1957	18.4	12.4	
1992		99.4	67.0		1974	41.4	27.9		1956	17.6	11.9	
1991		96.8	65.2		1973	37.7	25.4		1955	16.6	11.2	

Adjustments to Costs

The Historical Cost Index can be used to convert National Average building costs at a particular time to the approximate building costs for some other time.

Example:

Estimate and compare construction costs for different years in the same city.

To estimate the National Average construction cost of a building in 1970, knowing that it cost $900,000 in 2005:

INDEX in 1970 = 28.7

INDEX in 2005 = 148.5

Note: The City Cost Indexes for Canada can be used to convert U.S. National averages to local costs in Canadian dollars.

Time Adjustment using the Historical Cost Indexes:

$$\frac{\text{Index for Year A}}{\text{Index for Year B}} \times \text{Cost in Year B} = \text{Cost in Year A}$$

$$\frac{\text{INDEX 1970}}{\text{INDEX 2005}} \times \text{Cost 2005} = \text{Cost 1970}$$

$$\frac{28.7}{148.5} \times \$900,000 = .193 \times \$900,000 = \$173,700$$

The construction cost of the building in 1970 is $173,700.

How to Use the City Cost Indexes

What you should know before you begin

Means City Cost Indexes (CCI) are an extremely useful tool to use when you want to compare costs from city to city and region to region.

This publication contains average construction cost indexes for 719 U.S. and Canadian cities covering over 930 three-digit zip code locations, as listed directly under each city.

Keep in mind that a City Cost Index number is a *percentage ratio* of a specific city's cost to the national average cost of the same item at a stated time period.

In other words, these index figures represent relative construction *factors* (or, if you prefer, multipliers) for Material and Installation costs, as well as the weighted average for Total In Place costs for each CSI MasterFormat division. Installation costs include both labor and equipment rental costs. When estimating equipment rental rates only, for a specific location, use 01590 EQUIPMENT RENTAL index.

The 30 City Average Index is the average of 30 major U.S. cities and serves as a National Average.

Index figures for both material and installation are based on the 30 major city average of 100 and represent the cost relationship as of July 1, 2003. The index for each division is computed from representative material and labor quantities for that division. The weighted average for each city is a weighted total of the components listed above it, but does not include relative productivity between trades or cities.

As changes occur in local material prices, labor rates and equipment rental rates, the impact of these changes should be accurately measured by the change in the City Cost Index for each particular city (as compared to the 30 City Average).

> *Therefore, if you know (or have estimated) building costs in one city today, you can easily convert those costs to expected building costs in another city.*
>
> *In addition, by using the Historical Cost Index, you can easily convert National Average building costs at a particular time to the approximate building costs for some other time. The City Cost Indexes can then be applied to calculate the costs for a particular city.*

Quick Calculations

Location Adjustment Using the City Cost Indexes:

$$\frac{\text{Index for City A}}{\text{Index for City B}} \times \text{Cost in City B} = \text{Cost in City A}$$

Time Adjustment for the National Average Using the Historical Cost Index:

$$\frac{\text{Index for Year A}}{\text{Index for Year B}} \times \text{Cost in Year B} = \text{Cost in Year A}$$

Adjustment from the National Average:

$$\frac{\text{Index for City A}}{100} \times \text{National Average Cost} = \text{Cost in City A}$$

Since each of the other RSMeans publications contains many different items, any *one* item multiplied by the particular city index may give incorrect results. However, the larger the number of items compiled, the closer the results should be to actual costs for that particular city.

The City Cost Indexes for Canadian cities are calculated using Canadian material and equipment prices and labor rates, in Canadian dollars. Therefore, indexes for Canadian cities can be used to convert U.S. National Average prices to local costs in Canadian dollars.

How to use this section

1. Compare costs from city to city.

In using the Means Indexes, remember that an index number is not a fixed number but a *ratio:* It's a percentage ratio of a building component's cost at any stated time to the National Average cost of that same component at the same time period. Put in the form of an equation:

$$\frac{\text{Specific City Cost}}{\text{National Average Cost}} \times 100 = \text{City Index Number}$$

Therefore, when making cost comparisons between cities, do not subtract one city's index number from the index number of another city and read the result as a percentage difference. Instead, divide one city's index number by that of the other city. The resulting number may then be used as a multiplier to calculate cost differences from city to city.

The formula used to find cost differences between cities for the purpose of comparison is as follows:

$$\frac{\text{City A Index}}{\text{City B Index}} \times \text{City B Cost (Known)} = \text{City A Cost (Unknown)}$$

In addition, you can use *Means CCI* to calculate and compare costs division by division between cities using the same basic formula. (Just be sure that you're comparing similar divisions.)

2. Compare a specific city's construction costs with the National Average.

When you're studying construction location feasibility, it's advisable to compare a prospective project's cost index with an index of the National Average cost.

For example, divide the weighted average index of construction costs of a specific city by that of the 30 City Average, which = 100.

$$\frac{\text{City Index}}{100} = \% \text{ of National Average}$$

As a result, you get a ratio that indicates the relative cost of construction in that city in comparison with the National Average.

3. Covert U.S. National Average to actual costs in Canadian City.

$$\frac{\text{Index for Canadian City}}{100} \times \text{National Average Cost} = \text{Cost in Canadian City in \$ CAN}$$

4. Adjust construction cost data based on a National Average.

When you use a source of construction cost data which is based on a National Average (such as *Means cost data publications*), it is necessary to adjust those costs to a specific location.

$$\frac{\text{City Index}}{100} \times \frac{\text{"Book" Cost Based on}}{\text{National Average Costs}} = \frac{\text{City Cost}}{\text{(Unknown)}}$$

5. When applying the City Cost Indexes to demolition projects, use the appropriate division installation index. For example, for removal of existing doors and windows, use the Division 8 index.

What you might like to know about how we developed the Indexes

To create a reliable index, RSMeans researched the building type most often constructed in the United States and Canada. Because it was concluded that no one type of building completely represented the building construction industry, nine different types of buildings were combined to create a composite model.

The exact material, labor and equipment quantities are based on detailed analysis of these nine building types, then each quantity is weighted in proportion to expected usage. These various material items, labor hours, and equipment rental rates are thus combined to form a composite building representing as closely as possible the actual usage of materials, labor and equipment used in the North American Building Construction Industry.

The following structures were chosen to make up that composite model:

1. Factory, 1 story
2. Office, 2–4 story
3. Store, Retail
4. Town Hall, 2–3 story
5. High School, 2–3 story
6. Hospital, 4–8 story
7. Garage, Parking
8. Apartment, 1–3 story
9. Hotel/Motel, 2–3 story

For the purposes of ensuring the timeliness of the data, the components of the index for the composite model have been streamlined. They currently consist of:

- specific quantities of 66 commonly used construction materials;
- specific labor-hours for 21 building construction trades; and
- specific days of equipment rental for 6 types of construction equipment (normally used to install the 66 material items by the 21 trades.)

A sophisticated computer program handles the updating of all costs for each city on a quarterly basis. Material and equipment price quotations are gathered quarterly from 719 cities in the United States and Canada. These prices and the latest negotiated labor wage rates for 21 different building trades are used to compile the quarterly update of the City Cost Index.

The 30 major U.S. cities used to calculate the National Average are:

Atlanta, GA	Memphis, TN
Baltimore, MD	Milwaukee, WI
Boston, MA	Minneapolis, MN
Buffalo, NY	Nashville, TN
Chicago, IL	New Orleans, LA
Cincinnati, OH	New York, NY
Cleveland, OH	Philadelphia, PA
Columbus, OH	Phoenix, AZ
Dallas, TX	Pittsburgh, PA
Denver, CO	St. Louis, MO
Detroit, MI	San Antonio, TX
Houston, TX	San Diego, CA
Indianapolis, IN	San Francisco, CA
Kansas City, MO	Seattle, WA
Los Angeles, CA	Washington, DC

F.Y.I.: The CSI MasterFormat Divisions

1. General Requirements
2. Site Construction
3. Concrete
4. Masonry
5. Metals
6. Wood & Plastics
7. Thermal & Moisture Protection
8. Doors & Windows
9. Finishes
10. Specialties
11. Equipment
12. Furnishings
13. Special Construction
14. Conveying Systems
15. Mechanical
16. Electrical

The information presented in the CCI is organized according to the Construction Specifications Institute (CSI) MasterFormat.

What the CCI does not indicate

The weighted average for each city is a total of the components listed above weighted to reflect typical usage, but it does *not* include the productivity variations between trades or cities.

In addition, the CCI does not take into consideration factors such as the following:

- managerial efficiency
- competitive conditions
- automation
- restrictive union practices
- unique local requirements
- regional variations due to specific building codes

DIVISION		UNITED STATES 30 CITY AVERAGE			ANNISTON 362			BIRMINGHAM 350 - 352			BUTLER 369			DECATUR 356			DOTHAN 363		
		MAT.	INST.	TOTAL	MAT.	INST.	TOTAL	MAT.	INST.	TOTAL	MAT.	INST.	TOTAL	MAT.	INST.	TOTAL	MAT.	INST.	TOTAL
01590	EQUIPMENT RENTAL	.0	100.0	100.0	.0	101.2	101.2	.0	101.3	101.3	.0	97.7	97.7	.0	101.2	101.2	.0	97.7	97.7
02	SITE CONSTRUCTION	100.0	100.0	100.0	90.9	91.8	91.6	84.6	93.1	90.9	103.7	85.6	90.3	83.0	91.5	89.3	101.2	85.6	89.6
03100	CONCRETE FORMS & ACCESSORIES	100.0	100.0	100.0	92.0	35.6	43.1	95.0	75.6	78.2	87.5	46.8	52.2	97.1	49.8	56.2	98.0	46.6	53.5
03200	CONCRETE REINFORCEMENT	100.0	100.0	100.0	92.1	86.5	89.3	92.1	88.2	90.1	97.3	57.5	77.2	92.1	55.1	73.4	97.3	57.5	77.2
03300	CAST-IN-PLACE CONCRETE	100.0	100.0	100.0	90.6	40.6	69.7	92.8	68.1	82.5	88.4	48.6	71.8	90.1	56.1	75.9	88.4	48.5	71.7
03	CONCRETE	100.0	100.0	100.0	94.1	48.9	71.5	90.9	76.4	83.7	95.2	51.2	73.2	89.8	54.7	72.3	94.6	51.0	72.9
04	MASONRY	100.0	100.0	100.0	85.0	35.2	53.7	85.9	76.0	79.7	89.1	36.6	56.2	85.9	51.9	64.6	90.0	36.6	56.5
05	METALS	100.0	100.0	100.0	94.2	90.7	93.1	94.6	95.5	94.9	93.1	78.7	88.4	96.0	79.9	90.7	93.1	78.5	88.3
06	WOOD & PLASTICS	100.0	100.0	100.0	91.3	35.0	61.9	96.7	76.0	85.9	86.0	48.3	66.3	96.7	49.0	71.8	98.1	48.3	72.1
07	THERMAL & MOISTURE PROTECTION	100.0	100.0	100.0	95.6	40.9	68.7	95.7	80.9	88.4	95.6	54.7	75.5	95.4	60.5	78.2	95.6	49.3	72.8
08	DOORS & WINDOWS	100.0	100.0	100.0	94.5	47.2	82.2	98.5	78.5	93.3	94.5	51.2	83.3	98.5	49.7	85.8	94.5	51.2	83.3
09200	PLASTER & GYPSUM BOARD	100.0	100.0	100.0	96.2	33.7	56.6	103.1	75.9	85.9	94.6	47.4	64.7	100.5	48.1	67.3	102.7	47.4	67.7
095,098	CEILINGS & ACOUSTICAL TREATMENT	100.0	100.0	100.0	95.3	33.7	58.4	102.7	75.9	86.6	95.3	47.4	66.6	100.7	48.1	69.2	95.3	47.4	66.6
09600	FLOORING	100.0	100.0	100.0	101.3	35.6	84.0	104.4	50.5	90.1	107.6	28.3	86.6	104.4	52.6	90.7	115.6	34.2	94.0
097,099	WALL FINISHES, PAINTS & COATINGS	100.0	100.0	100.0	95.3	33.0	57.5	95.3	70.8	80.4	95.3	56.0	71.4	95.3	52.3	69.2	95.3	56.0	71.4
09	FINISHES	100.0	100.0	100.0	97.3	34.7	64.8	100.8	70.1	84.8	100.6	44.0	71.1	99.8	50.2	74.0	103.9	45.0	73.3
10 - 14	TOTAL DIV. 10000 - 14000	100.0	100.0	100.0	100.0	49.0	89.2	100.0	86.6	97.1	100.0	52.1	89.8	100.0	54.2	90.3	100.0	52.1	89.8
15	MECHANICAL	100.0	100.0	100.0	100.9	38.5	73.6	100.3	69.6	86.9	97.5	41.2	72.8	100.2	50.0	78.2	97.5	41.2	72.8
16	ELECTRICAL	100.0	100.0	100.0	94.9	26.7	61.6	96.3	66.6	81.8	96.7	37.5	67.8	96.4	49.4	73.5	95.6	37.5	67.3
01 - 16	WEIGHTED AVERAGE	100.0	100.0	100.0	96.0	47.7	74.1	96.4	76.5	87.4	96.2	50.5	75.5	96.3	57.6	78.8	96.5	50.5	75.6

DIVISION		ALABAMA EVERGREEN 364			GADSDEN 359			HUNTSVILLE 357 - 358			JASPER 355			MOBILE 365 - 366			MONTGOMERY 360 - 361		
		MAT.	INST.	TOTAL	MAT.	INST.	TOTAL	MAT.	INST.	TOTAL	MAT.	INST.	TOTAL	MAT.	INST.	TOTAL	MAT.	INST.	TOTAL
01590	EQUIPMENT RENTAL	.0	97.7	97.7	.0	101.2	101.2	.0	101.2	101.2	.0	101.2	101.2	.0	101.2	101.2	.0	97.7	97.7
02	SITE CONSTRUCTION	104.2	85.8	90.6	89.6	92.1	91.5	82.7	92.9	90.2	89.2	91.9	91.2	93.7	86.1	88.0	94.1	86.6	88.6
03100	CONCRETE FORMS & ACCESSORIES	83.2	48.3	52.9	87.7	45.8	51.4	97.1	68.6	72.4	94.3	37.6	45.2	97.1	53.0	58.9	95.9	48.1	54.5
03200	CONCRETE REINFORCEMENT	97.3	57.5	77.2	97.4	87.0	92.2	92.1	77.4	84.7	92.1	86.7	89.4	95.0	53.9	74.2	95.0	86.7	90.8
03300	CAST-IN-PLACE CONCRETE	88.4	51.1	72.8	90.2	51.9	74.2	87.8	66.2	78.8	99.9	46.3	77.5	92.7	54.7	76.8	94.3	49.4	75.6
03	CONCRETE	95.2	52.6	74.0	94.0	57.4	75.7	88.7	70.7	79.7	97.1	51.8	74.5	91.5	55.4	73.5	92.2	57.5	74.9
04	MASONRY	89.2	41.1	59.0	84.4	43.5	58.7	85.8	65.2	72.8	81.2	38.0	54.1	86.3	52.3	65.0	86.9	36.3	55.2
05	METALS	93.1	78.5	88.3	94.3	92.8	93.8	96.0	91.0	94.4	94.2	91.5	93.4	94.8	79.6	89.8	94.6	92.1	93.8
06	WOOD & PLASTICS	81.0	48.3	63.9	86.7	44.1	64.4	96.7	68.4	81.9	93.6	35.8	63.4	96.7	52.7	73.7	95.3	48.3	70.7
07	THERMAL & MOISTURE PROTECTION	95.5	51.2	73.8	95.4	65.1	80.5	95.4	75.1	85.4	95.5	45.8	71.1	95.4	69.5	82.7	95.1	63.0	79.3
08	DOORS & WINDOWS	94.5	51.2	83.3	94.5	53.3	83.8	98.5	65.5	89.9	94.4	52.2	83.5	98.5	53.2	86.8	98.5	58.2	88.0
09200	PLASTER & GYPSUM BOARD	91.4	47.6	63.5	94.2	43.1	61.8	100.5	68.1	80.0	96.8	34.6	57.4	100.5	52.0	69.7	100.5	47.4	66.9
095,098	CEILINGS & ACOUSTICAL TREATMENT	95.3	47.4	66.6	95.3	43.1	64.0	102.7	68.1	81.9	94.5	34.6	58.6	102.7	52.0	72.3	102.7	47.4	69.6
09600	FLOORING	104.8	28.3	84.6	99.2	45.2	84.9	104.4	54.2	91.1	102.4	34.4	84.4	114.7	54.7	98.8	112.9	28.3	90.5
097,099	WALL FINISHES, PAINTS & COATINGS	95.3	56.0	71.4	95.3	63.7	76.1	95.3	64.2	76.4	95.3	40.5	62.0	100.1	56.9	73.9	95.3	56.0	71.4
09	FINISHES	99.3	45.1	71.1	96.3	46.1	70.2	100.3	65.0	81.9	97.4	36.3	65.6	104.8	53.1	77.9	104.0	43.9	72.8
10 - 14	TOTAL DIV. 10000 - 14000	100.0	53.9	90.2	100.0	77.1	95.1	100.0	84.5	96.7	100.0	49.9	89.3	100.0	69.0	93.4	100.0	77.1	95.1
15	MECHANICAL	97.5	43.6	73.9	102.9	44.2	77.2	100.3	69.8	86.9	102.9	60.3	84.3	99.8	70.8	87.2	99.8	41.7	74.4
16	ELECTRICAL	94.3	37.5	66.6	95.9	66.6	81.6	96.4	70.5	83.8	96.0	27.0	62.3	96.4	51.3	74.4	96.3	69.1	83.0
01 - 16	WEIGHTED AVERAGE	95.8	51.8	75.8	96.3	60.0	79.8	96.3	73.3	85.8	96.7	53.5	77.1	97.1	63.0	81.6	97.0	58.4	79.5

DIVISION		ALABAMA PHENIX CITY 368			SELMA 367			TUSCALOOSA 354			ALASKA ANCHORAGE 995 - 996			FAIRBANKS 997			JUNEAU 998		
		MAT.	INST.	TOTAL	MAT.	INST.	TOTAL	MAT.	INST.	TOTAL	MAT.	INST.	TOTAL	MAT.	INST.	TOTAL	MAT.	INST.	TOTAL
01590	EQUIPMENT RENTAL	.0	97.7	97.7	.0	97.7	97.7	.0	101.2	101.2	.0	118.4	118.4	.0	118.4	118.4	.0	118.4	118.4
02	SITE CONSTRUCTION	108.2	86.7	92.3	101.0	86.6	90.3	83.2	91.9	89.6	143.5	133.8	136.3	127.1	133.8	132.1	139.1	133.8	135.1
03100	CONCRETE FORMS & ACCESSORIES	91.8	42.4	49.0	88.9	47.4	53.0	97.0	39.7	47.3	132.4	113.9	116.4	134.1	119.7	121.6	133.8	113.9	116.6
03200	CONCRETE REINFORCEMENT	97.2	78.1	87.6	97.3	86.6	91.9	92.1	87.0	89.5	146.2	106.6	126.2	122.9	106.7	114.7	108.7	106.6	107.7
03300	CAST-IN-PLACE CONCRETE	88.4	49.9	72.3	88.4	49.3	72.0	91.4	47.1	72.9	195.7	114.6	161.8	163.1	115.1	143.0	196.5	114.6	162.3
03	CONCRETE	97.9	53.5	75.8	93.9	57.1	75.5	90.4	53.0	71.8	153.5	112.2	132.9	129.8	115.0	122.4	147.9	112.2	130.1
04	MASONRY	89.1	41.6	59.3	93.3	36.7	57.8	86.0	38.6	56.3	217.0	121.0	156.7	210.3	121.0	154.3	219.4	121.0	157.7
05	METALS	93.1	89.0	91.8	93.1	91.3	92.5	95.2	92.7	94.4	130.1	102.2	121.0	130.1	102.5	121.1	130.1	102.2	121.0
06	WOOD & PLASTICS	90.9	40.2	64.5	87.8	48.3	67.2	96.7	38.3	66.2	117.8	111.5	114.5	118.0	118.9	118.5	117.8	111.5	114.5
07	THERMAL & MOISTURE PROTECTION	96.0	63.7	80.1	95.4	54.7	75.4	95.3	61.8	78.8	198.5	115.8	157.8	194.6	118.6	157.2	195.2	115.8	156.2
08	DOORS & WINDOWS	94.5	51.7	83.4	94.5	58.2	85.1	98.5	58.0	88.0	126.2	108.8	121.7	123.3	112.9	120.6	123.3	108.8	119.6
09200	PLASTER & GYPSUM BOARD	97.4	39.2	60.5	95.8	47.4	65.2	100.5	37.2	60.4	132.6	111.6	119.3	132.6	119.3	124.1	132.6	111.6	119.3
095,098	CEILINGS & ACOUSTICAL TREATMENT	95.3	39.2	61.6	95.3	47.4	66.6	102.7	37.2	63.4	132.1	111.6	119.8	132.1	119.3	124.4	132.1	111.6	119.8
09600	FLOORING	110.6	28.3	88.8	108.5	28.3	87.3	104.4	41.5	87.7	164.9	127.0	154.9	164.9	127.0	154.9	164.9	127.0	154.9
097,099	WALL FINISHES, PAINTS & COATINGS	95.3	56.0	71.4	95.3	56.0	71.4	95.3	48.2	66.7	168.1	111.4	133.7	168.1	125.0	141.9	168.1	111.4	133.7
09	FINISHES	102.2	39.4	69.6	100.8	44.0	71.2	100.2	39.2	68.5	154.8	116.0	134.6	152.6	121.9	136.6	153.4	116.0	134.0
10 - 14	TOTAL DIV. 10000 - 14000	100.0	76.4	95.0	100.0	52.1	89.8	100.0	74.5	94.6	100.0	112.8	102.7	100.0	113.7	102.9	100.0	112.8	102.7
15	MECHANICAL	97.5	42.2	73.3	97.5	41.5	73.0	99.8	36.7	72.2	100.9	108.3	104.2	100.9	115.9	107.4	100.9	108.3	104.2
16	ELECTRICAL	96.2	62.8	79.9	95.4	37.5	67.2	96.4	66.6	81.9	149.1	112.3	131.2	150.9	112.3	132.1	150.9	112.3	132.1
01 - 16	WEIGHTED AVERAGE	96.8	56.4	78.5	96.1	53.0	76.5	96.3	56.4	78.2	134.5	113.8	125.1	130.7	116.9	124.4	133.6	113.8	124.6

DIVISION		ALASKA KETCHIKAN 999			ARIZONA CHAMBERS 865			ARIZONA FLAGSTAFF 860			ARIZONA GLOBE 855			ARIZONA KINGMAN 864			ARIZONA MESA/TEMPE 852		
		MAT.	INST.	TOTAL	MAT.	INST.	TOTAL	MAT.	INST.	TOTAL	MAT.	INST.	TOTAL	MAT.	INST.	TOTAL	MAT.	INST.	TOTAL
01590	EQUIPMENT RENTAL	.0	118.4	118.4	.0	94.9	94.9	.0	94.9	94.9	.0	98.2	98.2	.0	94.9	94.9	.0	98.2	98.2
02	SITE CONSTRUCTION	190.1	133.8	148.4	64.9	100.3	91.1	83.5	100.9	96.4	97.9	103.8	102.3	64.9	101.0	91.6	88.3	103.9	99.9
03100	CONCRETE FORMS & ACCESSORIES	125.5	113.8	115.4	96.6	55.9	61.3	102.8	61.5	67.1	94.0	55.9	61.0	94.2	56.3	61.4	97.3	63.1	67.7
03200	CONCRETE REINFORCEMENT	113.4	106.6	109.9	105.7	52.9	79.0	105.5	73.5	89.3	104.2	70.9	87.4	105.8	72.7	89.1	104.8	72.9	88.7
03300	CAST-IN-PLACE CONCRETE	317.0	114.5	232.4	91.7	64.2	80.2	91.8	78.3	86.2	97.6	64.6	83.8	91.4	64.4	80.1	98.3	71.4	87.1
03	CONCRETE	222.4	112.2	167.4	97.5	58.2	77.9	116.7	69.5	93.1	106.4	61.8	84.1	97.1	62.2	79.7	99.2	67.8	83.5
04	MASONRY	220.5	121.0	158.0	103.2	52.2	71.2	102.9	56.2	73.6	108.1	52.2	73.0	103.2	62.7	77.8	108.2	52.8	73.4
05	METALS	130.1	102.1	121.0	93.4	57.6	81.7	93.8	67.9	85.4	94.3	65.3	84.8	94.0	66.3	84.9	94.5	69.2	86.2
06	WOOD & PLASTICS	108.4	111.5	110.0	102.3	54.6	77.4	108.6	60.1	83.3	96.1	54.7	74.5	98.3	54.6	75.5	99.6	68.7	83.5
07	THERMAL & MOISTURE PROTECTION	199.0	115.8	158.1	104.9	62.9	84.2	107.3	66.9	87.4	106.4	59.7	83.4	104.8	62.8	84.2	105.4	63.8	85.0
08	DOORS & WINDOWS	123.7	108.8	119.8	102.1	52.7	89.3	102.2	64.2	92.3	98.8	57.4	88.1	102.2	57.6	90.7	98.9	65.2	90.2
09200	PLASTER & GYPSUM BOARD	127.4	111.6	117.4	87.8	53.3	65.9	90.8	59.0	70.6	89.6	53.3	66.6	83.6	53.3	64.4	91.6	67.7	76.4
095,098	CEILINGS & ACOUSTICAL TREATMENT	128.1	111.6	118.2	90.5	53.3	72.2	101.3	59.0	75.9	95.6	53.3	70.2	101.3	53.3	72.5	95.6	67.7	78.8
09600	FLOORING	164.9	127.0	154.9	92.7	51.6	81.8	95.2	51.8	83.8	95.8	51.6	84.1	91.7	70.1	86.0	97.3	64.6	88.7
097,099	WALL FINISHES, PAINTS & COATINGS	168.1	111.4	133.7	94.6	48.8	66.8	94.6	48.8	66.8	105.0	48.8	70.9	94.6	48.8	66.8	105.0	53.1	73.5
09	FINISHES	154.7	116.0	134.6	93.8	53.8	73.0	96.7	57.6	76.4	97.4	54.0	74.8	93.1	57.0	74.3	97.2	62.0	78.9
10 - 14	TOTAL DIV. 10000 - 14000	100.0	112.8	102.7	100.0	71.1	93.9	100.0	72.6	94.2	100.0	71.6	94.0	100.0	71.1	93.9	100.0	69.0	93.4
15	MECHANICAL	98.8	106.0	101.9	97.2	73.8	87.0	100.2	77.4	90.2	95.8	67.7	83.5	97.2	74.0	87.1	100.2	67.9	86.1
16	ELECTRICAL	150.9	112.3	132.1	97.9	46.2	72.7	96.9	48.0	73.1	93.5	46.2	70.4	97.9	46.2	72.7	90.6	60.3	75.8
01 - 16	WEIGHTED AVERAGE	143.2	113.4	129.6	96.8	62.5	81.2	100.5	67.6	85.5	98.2	62.9	82.2	96.8	65.6	82.0	97.8	67.7	84.1

DIVISION		ARIZONA PHOENIX 850,853			ARIZONA PRESCOTT 863			ARIZONA SHOW LOW 859			ARIZONA TUCSON 856 - 857			ARKANSAS BATESVILLE 725			ARKANSAS CAMDEN 717		
		MAT.	INST.	TOTAL	MAT.	INST.	TOTAL	MAT.	INST.	TOTAL	MAT.	INST.	TOTAL	MAT.	INST.	TOTAL	MAT.	INST.	TOTAL
01590	EQUIPMENT RENTAL	.0	98.8	98.8	.0	94.9	94.9	.0	98.2	98.2	.0	98.2	98.2	.0	86.0	86.0	.0	86.0	86.0
02	SITE CONSTRUCTION	88.8	104.7	100.6	71.2	100.3	92.7	100.1	103.8	102.8	85.1	104.4	99.4	73.1	83.7	81.0	74.0	83.3	80.9
03100	CONCRETE FORMS & ACCESSORIES	98.4	69.8	73.6	98.2	56.0	61.6	101.8	56.5	62.5	97.9	69.2	73.1	80.8	44.7	49.5	80.5	31.5	38.1
03200	CONCRETE REINFORCEMENT	103.2	74.2	88.5	105.5	72.5	88.8	104.8	72.3	88.4	86.6	73.5	80.0	95.8	48.9	72.1	97.1	26.9	61.6
03300	CAST-IN-PLACE CONCRETE	98.4	78.9	90.3	91.7	64.2	80.2	97.6	64.8	83.9	101.1	78.7	91.8	78.5	46.3	65.1	80.5	34.4	61.2
03	CONCRETE	98.8	73.6	86.2	102.6	62.0	82.3	108.4	62.4	85.4	97.4	73.1	85.3	79.8	46.9	63.4	81.4	32.8	57.1
04	MASONRY	96.0	68.1	78.5	103.3	58.8	75.4	108.1	54.1	74.2	97.8	56.2	71.7	101.6	42.4	64.5	110.9	36.7	64.3
05	METALS	96.0	70.9	87.8	93.8	66.7	85.0	94.1	66.5	85.1	95.3	68.7	86.6	93.8	60.1	82.8	93.8	50.9	79.8
06	WOOD & PLASTICS	100.6	70.5	84.9	103.8	54.6	78.1	104.2	54.7	78.4	99.8	70.5	84.5	83.0	45.9	63.6	82.7	32.5	56.5
07	THERMAL & MOISTURE PROTECTION	105.3	70.8	88.4	105.6	61.7	84.0	106.7	63.1	85.2	106.6	64.2	85.8	97.8	45.8	72.3	97.6	34.6	66.6
08	DOORS & WINDOWS	100.0	69.9	92.2	102.2	58.3	90.8	97.9	57.7	87.5	96.0	69.9	89.2	96.4	43.1	82.6	92.6	28.5	76.0
09200	PLASTER & GYPSUM BOARD	93.9	69.6	78.5	87.9	53.3	66.0	93.6	53.3	68.1	92.0	69.6	77.8	80.1	45.1	57.9	80.1	31.3	49.2
095,098	CEILINGS & ACOUSTICAL TREATMENT	102.7	69.6	82.8	99.6	53.3	71.8	95.6	53.3	70.2	95.6	69.6	80.0	85.8	45.1	61.4	85.8	31.3	53.1
09600	FLOORING	97.6	67.5	89.7	93.6	51.6	82.5	99.3	58.8	88.6	96.8	51.8	84.9	106.7	50.4	91.8	106.6	16.7	82.8
097,099	WALL FINISHES, PAINTS & COATINGS	105.0	60.4	77.9	94.6	48.8	66.8	105.0	48.8	70.9	102.6	48.8	70.0	96.4	44.6	65.0	96.4	34.8	59.1
09	FINISHES	99.5	68.0	83.1	94.3	53.8	73.2	99.2	55.2	76.3	96.9	63.8	79.6	91.4	46.0	67.8	91.4	29.7	59.3
10 - 14	TOTAL DIV. 10000 - 14000	100.0	74.4	94.6	100.0	71.1	93.9	100.0	71.6	94.0	100.0	74.4	94.6	100.0	47.7	88.8	100.0	43.7	88.0
15	MECHANICAL	100.2	77.5	90.2	100.2	73.6	88.6	95.8	74.0	86.3	100.1	79.1	90.4	95.7	42.3	72.3	95.7	32.6	68.1
16	ELECTRICAL	99.6	65.7	83.1	96.7	46.2	72.1	90.9	60.9	76.3	92.5	60.9	77.1	98.1	47.7	73.5	94.9	39.8	68.0
01 - 16	WEIGHTED AVERAGE	98.8	74.3	87.7	98.2	64.7	83.0	98.3	66.9	84.1	97.0	69.8	84.6	93.5	49.9	73.7	93.4	40.1	69.2

DIVISION		ARKANSAS FAYETTEVILLE 727			ARKANSAS FORT SMITH 729			ARKANSAS HARRISON 726			ARKANSAS HOT SPRINGS 719			ARKANSAS JONESBORO 724			ARKANSAS LITTLE ROCK 720 - 722		
		MAT.	INST.	TOTAL	MAT.	INST.	TOTAL	MAT.	INST.	TOTAL	MAT.	INST.	TOTAL	MAT.	INST.	TOTAL	MAT.	INST.	TOTAL
01590	EQUIPMENT RENTAL	.0	86.0	86.0	.0	86.0	86.0	.0	86.0	86.0	.0	86.0	86.0	.0	107.8	107.8	.0	86.0	86.0
02	SITE CONSTRUCTION	72.7	83.7	80.8	77.6	83.8	82.2	77.9	83.7	82.2	77.6	83.1	81.6	100.5	99.2	99.6	77.4	83.8	82.1
03100	CONCRETE FORMS & ACCESSORIES	76.0	40.7	45.5	97.7	42.0	49.4	85.8	44.7	50.2	77.6	29.5	36.0	84.5	47.8	52.7	91.8	59.2	63.5
03200	CONCRETE REINFORCEMENT	95.9	47.2	71.3	96.9	74.6	85.6	95.4	48.9	71.9	95.3	25.6	60.1	92.7	49.6	70.9	97.1	72.2	84.5
03300	CAST-IN-PLACE CONCRETE	78.6	45.8	64.9	89.8	66.8	80.2	87.2	46.3	70.1	82.4	33.4	62.0	85.6	57.1	73.7	89.8	67.0	80.3
03	CONCRETE	79.5	44.6	62.1	87.4	57.4	72.4	86.4	46.9	66.7	84.4	31.4	57.9	84.0	53.0	68.6	87.0	64.7	75.9
04	MASONRY	91.2	43.0	61.0	97.5	53.3	69.8	101.9	42.4	64.6	82.2	28.4	48.4	92.8	46.9	64.0	95.9	53.3	69.2
05	METALS	93.8	53.3	82.5	95.7	71.3	87.7	94.6	60.1	83.3	93.8	49.6	79.3	90.0	75.9	85.4	91.6	70.8	84.8
06	WOOD & PLASTICS	78.1	41.0	58.7	100.4	40.6	69.2	88.5	45.9	66.3	79.4	29.5	53.3	86.4	48.8	66.8	97.4	63.3	79.6
07	THERMAL & MOISTURE PROTECTION	98.8	47.7	73.7	99.4	48.4	74.3	98.3	45.8	72.5	97.9	35.5	67.2	109.1	51.6	80.8	97.9	50.7	74.7
08	DOORS & WINDOWS	96.4	41.2	82.1	97.2	45.8	83.9	97.2	43.1	83.2	92.5	28.6	76.0	98.8	47.9	85.6	97.2	58.4	87.1
09200	PLASTER & GYPSUM BOARD	76.9	40.1	53.6	85.4	39.6	56.4	84.1	45.1	59.4	78.1	28.2	46.5	92.3	47.7	64.0	85.4	62.9	71.1
095,098	CEILINGS & ACOUSTICAL TREATMENT	85.8	40.1	58.4	91.2	39.6	60.3	90.2	45.1	63.1	85.8	28.2	51.2	89.0	47.7	64.2	91.2	62.9	74.2
09600	FLOORING	103.1	50.4	89.1	115.8	70.0	103.7	110.0	50.4	94.2	105.0	50.4	90.6	77.3	45.6	68.9	117.1	70.0	104.6
097,099	WALL FINISHES, PAINTS & COATINGS	96.4	36.1	59.8	96.4	58.3	73.3	96.4	44.6	65.0	96.4	37.2	60.5	84.9	52.4	65.2	96.4	59.1	74.2
09	FINISHES	89.9	41.9	65.0	96.4	48.4	71.4	94.3	46.0	69.2	90.9	34.1	61.4	87.7	47.9	67.0	96.7	61.9	78.6
10 - 14	TOTAL DIV. 10000 - 14000	100.0	63.2	92.2	100.0	63.4	92.2	100.0	47.7	88.9	100.0	43.4	88.0	100.0	54.9	90.4	100.0	66.4	92.8
15	MECHANICAL	95.7	38.8	70.8	100.1	43.2	75.2	95.7	42.3	72.3	95.7	33.7	68.5	100.2	43.1	75.2	100.1	58.4	81.8
16	ELECTRICAL	92.5	44.0	68.8	95.6	64.4	80.4	96.8	47.7	72.8	96.9	38.9	68.6	102.7	47.8	75.9	95.3	70.0	83.0
01 - 16	WEIGHTED AVERAGE	92.1	48.1	72.1	96.0	56.8	78.2	94.8	49.9	74.4	92.6	39.6	68.5	95.5	55.0	77.1	95.2	64.3	81.2

COST INDEXES

ARKANSAS / CALIFORNIA

	DIVISION	PINE BLUFF 716			RUSSELLVILLE 728			TEXARKANA 718			WEST MEMPHIS 723			ALHAMBRA 917 - 918			ANAHEIM 928		
		MAT.	INST.	TOTAL	MAT.	INST.	TOTAL	MAT.	INST.	TOTAL	MAT.	INST.	TOTAL	MAT.	INST.	TOTAL	MAT.	INST.	TOTAL
01590	EQUIPMENT RENTAL	.0	86.0	86.0	.0	86.0	86.0	.0	86.8	86.8	.0	107.8	107.8	.0	97.0	97.0	.0	102.1	102.1
02	SITE CONSTRUCTION	79.6	83.8	82.7	74.6	83.7	81.3	95.3	84.6	87.4	108.1	99.2	101.5	100.1	107.3	105.4	103.2	109.0	107.5
03100	CONCRETE FORMS & ACCESSORIES	77.2	58.9	61.4	81.7	40.6	46.1	85.0	39.6	45.7	90.6	47.8	53.6	116.4	115.6	115.7	105.0	121.5	119.3
03200	CONCRETE REINFORCEMENT	97.1	72.1	84.5	96.5	50.9	73.4	96.6	48.1	72.1	92.7	49.6	70.9	108.1	112.7	110.4	95.6	112.5	104.1
03300	CAST-IN-PLACE CONCRETE	82.4	66.9	75.9	82.4	46.2	67.3	89.8	44.6	70.9	89.7	57.1	76.1	106.5	112.1	108.8	107.2	117.2	111.4
03	CONCRETE	85.0	64.5	74.8	82.7	45.4	64.1	84.3	43.9	64.1	90.3	53.0	71.7	112.1	112.9	112.5	107.8	117.4	112.6
04	MASONRY	118.2	53.3	77.5	96.9	42.4	62.7	97.5	32.8	56.9	79.5	46.9	59.0	123.7	110.7	115.6	89.0	111.4	103.1
05	METALS	94.3	70.6	86.6	93.8	60.7	83.0	86.9	59.4	77.9	89.3	75.9	84.9	91.3	96.7	93.1	110.9	99.8	107.2
06	WOOD & PLASTICS	78.9	63.3	70.7	84.4	40.6	61.5	88.2	42.3	64.3	92.6	48.8	69.8	100.8	115.6	108.6	94.9	121.2	108.7
07	THERMAL & MOISTURE PROTECTION	98.0	50.7	74.8	99.1	45.3	72.7	98.9	42.2	71.0	109.7	51.6	81.1	106.6	108.0	107.3	121.2	115.0	118.1
08	DOORS & WINDOWS	92.5	58.4	83.7	96.4	40.7	82.0	97.6	41.1	83.0	98.8	47.9	85.6	95.6	113.1	100.1	102.8	116.2	106.2
09200	PLASTER & GYPSUM BOARD	77.7	62.9	68.3	80.1	39.6	54.5	83.3	41.4	56.7	94.3	47.7	64.7	96.9	116.2	109.1	98.1	121.7	113.1
095,098	CEILINGS & ACOUSTICAL TREATMENT	86.8	62.9	72.4	85.8	39.6	58.1	93.9	41.4	62.4	88.0	47.7	63.8	96.6	116.2	108.3	119.6	121.7	120.9
09600	FLOORING	104.8	70.0	95.6	106.3	50.4	91.5	107.6	43.5	90.6	80.2	45.6	71.0	107.4	103.7	106.4	123.1	103.7	118.0
097,099	WALL FINISHES, PAINTS & COATINGS	96.4	59.9	74.2	96.4	44.6	65.0	96.4	36.8	60.2	84.9	52.4	65.2	112.2	105.0	107.8	111.4	110.7	111.0
09	FINISHES	91.1	61.9	75.9	91.5	42.8	66.2	95.0	40.1	66.4	89.2	47.9	67.7	104.5	112.4	108.6	114.3	117.4	115.9
10 - 14	TOTAL DIV. 10000 - 14000	100.0	66.4	92.8	100.0	47.1	88.7	100.0	39.3	87.1	100.0	54.9	90.4	100.0	103.8	100.8	100.0	114.8	103.2
15	MECHANICAL	100.1	45.6	76.3	95.7	37.0	70.1	100.1	38.1	73.0	95.9	43.1	72.8	95.9	106.1	100.4	100.2	108.8	103.8
16	ELECTRICAL	95.0	70.0	82.8	95.6	47.7	72.3	96.8	39.8	69.0	104.2	47.8	76.7	118.9	108.1	113.6	88.6	107.9	98.0
01 - 16	WEIGHTED AVERAGE	95.3	61.7	80.0	93.4	48.0	72.8	94.7	45.3	72.3	95.1	55.0	76.9	102.8	108.2	105.3	102.9	111.1	106.6

CALIFORNIA

	DIVISION	BAKERSFIELD 932 - 933			BERKELEY 947			EUREKA 955			FRESNO 936 - 938			INGLEWOOD 903 - 905			LONG BEACH 906 - 908		
		MAT.	INST.	TOTAL	MAT.	INST.	TOTAL	MAT.	INST.	TOTAL	MAT.	INST.	TOTAL	MAT.	INST.	TOTAL	MAT.	INST.	TOTAL
01590	EQUIPMENT RENTAL	.0	99.5	99.5	.0	103.3	103.3	.0	99.2	99.2	.0	99.5	99.5	.0	96.6	96.6	.0	96.6	96.6
02	SITE CONSTRUCTION	108.1	106.0	106.6	146.7	104.5	115.5	118.9	103.4	107.4	109.6	106.0	106.9	91.2	104.8	101.2	98.4	104.8	103.1
03100	CONCRETE FORMS & ACCESSORIES	97.5	121.1	117.9	121.4	130.9	129.6	115.8	123.1	122.1	101.2	127.2	123.7	108.8	115.5	114.6	102.4	115.5	113.8
03200	CONCRETE REINFORCEMENT	108.9	112.5	110.7	100.8	113.4	107.2	105.6	112.7	109.2	92.2	112.8	102.6	113.0	112.6	112.8	111.9	112.6	112.3
03300	CAST-IN-PLACE CONCRETE	102.3	116.1	108.1	143.4	114.5	131.3	111.0	108.6	110.0	112.0	113.5	112.6	80.2	112.9	93.8	91.3	112.9	100.3
03	CONCRETE	107.1	116.8	112.0	125.2	120.9	123.1	120.0	115.0	117.5	109.3	118.7	114.0	96.9	113.2	105.0	106.3	113.2	109.8
04	MASONRY	107.5	111.2	109.8	140.4	119.5	127.3	113.5	107.0	109.4	111.1	111.2	111.2	82.1	110.9	100.2	91.3	110.9	103.6
05	METALS	105.4	99.0	103.3	110.9	104.8	108.9	110.5	96.8	106.0	111.1	100.2	107.6	106.0	97.9	103.3	105.8	98.0	103.3
06	WOOD & PLASTICS	86.0	121.3	104.4	116.9	132.6	125.1	108.4	129.8	119.6	99.1	129.8	115.1	95.7	115.5	106.1	88.7	115.5	102.7
07	THERMAL & MOISTURE PROTECTION	105.5	108.9	107.2	112.8	122.5	117.6	112.1	100.4	106.3	101.3	110.5	105.8	111.2	109.7	110.5	111.5	109.7	110.6
08	DOORS & WINDOWS	101.6	113.3	104.6	106.3	124.5	111.0	102.8	113.0	105.4	101.9	118.6	106.2	90.8	113.0	96.5	90.7	113.0	96.5
09200	PLASTER & GYPSUM BOARD	98.1	121.7	113.1	101.9	132.9	121.5	103.3	130.4	120.5	96.1	130.4	117.9	100.3	116.2	110.4	97.1	116.2	109.2
095,098	CEILINGS & ACOUSTICAL TREATMENT	119.6	121.7	120.9	111.0	132.9	124.1	122.7	130.4	127.3	119.6	130.4	126.1	107.4	116.2	112.7	107.4	116.2	112.7
09600	FLOORING	117.3	93.4	111.0	121.7	122.8	122.0	122.2	108.5	118.6	133.8	138.6	135.1	116.1	103.7	112.8	112.2	106.1	110.6
097,099	WALL FINISHES, PAINTS & COATINGS	110.9	92.5	99.7	115.6	136.3	128.1	110.3	61.1	80.5	134.6	100.1	113.7	99.6	105.0	102.9	99.6	105.0	102.9
09	FINISHES	114.5	113.6	114.0	117.1	130.7	124.2	118.3	115.9	117.0	120.9	127.5	124.3	107.9	112.4	110.2	106.6	112.8	109.8
10 - 14	TOTAL DIV. 10000 - 14000	100.0	119.2	104.1	100.0	131.1	106.6	100.0	126.2	105.6	100.0	128.8	106.1	100.0	103.5	100.7	100.0	103.5	100.7
15	MECHANICAL	100.2	99.6	99.9	95.9	125.6	108.9	95.8	99.1	97.2	100.2	111.6	105.2	95.7	106.1	100.3	95.7	106.1	100.3
16	ELECTRICAL	89.5	93.6	91.5	103.3	134.4	118.5	97.7	82.5	90.3	87.5	97.1	92.2	99.5	108.1	103.7	99.2	108.1	103.6
01 - 16	WEIGHTED AVERAGE	102.6	106.2	104.2	110.1	122.5	115.7	106.3	103.7	105.1	104.3	111.9	107.8	98.5	108.2	102.9	100.0	108.3	103.8

CALIFORNIA

	DIVISION	LOS ANGELES 900 - 902			MARYSVILLE 959			MODESTO 953			MOJAVE 935			OAKLAND 946			OXNARD 930		
		MAT.	INST.	TOTAL	MAT.	INST.	TOTAL	MAT.	INST.	TOTAL	MAT.	INST.	TOTAL	MAT.	INST.	TOTAL	MAT.	INST.	TOTAL
01590	EQUIPMENT RENTAL	.0	98.5	98.5	.0	99.2	99.2	.0	99.2	99.2	.0	99.5	99.5	.0	103.3	103.3	.0	98.1	98.1
02	SITE CONSTRUCTION	98.1	107.8	105.3	114.4	105.1	107.5	107.9	105.3	106.0	101.8	106.0	104.9	155.8	104.6	117.9	109.7	103.8	105.3
03100	CONCRETE FORMS & ACCESSORIES	105.5	121.4	119.3	103.4	127.0	123.9	98.4	127.0	123.2	111.8	114.3	114.0	107.4	133.5	130.0	103.4	121.5	119.1
03200	CONCRETE REINFORCEMENT	113.0	112.7	112.9	105.6	112.6	109.1	109.5	112.6	111.1	110.9	112.4	111.7	103.1	113.4	108.3	108.9	112.2	110.6
03300	CAST-IN-PLACE CONCRETE	87.3	115.4	99.1	124.1	113.3	119.6	111.0	113.4	112.0	93.3	113.9	101.9	139.5	118.9	130.9	108.6	116.4	111.9
03	CONCRETE	103.1	116.7	109.9	121.4	118.5	120.0	111.4	118.6	115.0	101.4	113.0	107.2	126.4	123.6	125.0	110.5	117.0	113.8
04	MASONRY	97.7	115.3	108.8	114.4	108.0	110.5	111.6	108.2	109.5	107.8	106.8	107.2	148.4	122.5	132.1	112.3	105.9	108.2
05	METALS	112.8	98.4	108.1	110.0	99.6	106.6	106.9	99.9	104.6	107.1	99.0	104.4	105.1	104.9	105.0	105.2	99.0	103.1
06	WOOD & PLASTICS	91.8	120.9	107.0	94.5	129.8	112.9	89.6	129.8	110.6	100.6	114.2	107.7	101.9	134.8	119.1	94.0	121.3	108.3
07	THERMAL & MOISTURE PROTECTION	111.6	117.0	114.3	111.3	108.9	110.1	110.7	108.3	109.5	106.3	106.4	106.3	113.0	124.0	118.4	110.5	110.0	110.3
08	DOORS & WINDOWS	96.7	116.0	101.7	102.0	119.8	106.6	100.8	120.6	106.0	97.3	109.4	100.4	106.4	125.7	111.4	100.4	116.2	104.5
09200	PLASTER & GYPSUM BOARD	101.6	121.7	114.4	96.3	130.4	117.9	98.5	130.4	118.7	107.6	114.4	111.9	94.7	135.1	120.3	98.1	121.7	113.1
095,098	CEILINGS & ACOUSTICAL TREATMENT	119.8	121.7	121.0	121.0	130.4	126.6	115.6	130.4	124.5	121.0	114.4	117.0	114.8	135.1	127.0	119.6	121.7	120.9
09600	FLOORING	114.0	103.7	111.3	116.7	112.7	115.7	117.3	110.9	115.6	124.7	93.4	116.4	116.1	122.8	117.9	117.3	103.7	113.7
097,099	WALL FINISHES, PAINTS & COATINGS	99.6	110.7	106.4	110.3	112.3	111.5	110.3	102.0	105.3	110.3	91.0	98.6	115.6	136.3	128.1	110.3	102.7	105.7
09	FINISHES	110.8	117.1	114.1	114.7	124.1	119.6	113.5	122.6	118.2	118.1	108.3	113.0	115.9	132.6	124.6	114.3	116.5	115.5
10 - 14	TOTAL DIV. 10000 - 14000	100.0	114.1	103.0	100.0	128.9	106.2	100.0	128.8	106.1	100.0	116.8	103.6	100.0	132.3	106.9	100.0	115.1	103.2
15	MECHANICAL	100.1	108.4	103.7	95.8	105.7	100.1	100.2	111.6	105.2	95.8	97.4	96.5	100.2	127.0	111.9	100.2	108.4	103.8
16	ELECTRICAL	96.5	112.9	104.5	94.4	123.0	108.4	97.0	102.1	99.5	86.8	93.6	90.1	102.6	134.4	118.1	94.6	106.1	100.2
01 - 16	WEIGHTED AVERAGE	102.7	111.9	106.8	105.3	113.4	109.0	104.4	111.6	107.7	100.8	103.7	102.1	110.6	123.9	116.6	103.9	109.5	106.4

CALIFORNIA

	DIVISION	PALM SPRINGS 922			PALO ALTO 943			PASADENA 910 - 912			REDDING 960			RICHMOND 948			RIVERSIDE 925		
		MAT.	INST.	TOTAL	MAT.	INST.	TOTAL	MAT.	INST.	TOTAL	MAT.	INST.	TOTAL	MAT.	INST.	TOTAL	MAT.	INST.	TOTAL
01590	EQUIPMENT RENTAL	.0	100.6	100.6	.0	103.3	103.3	.0	97.0	97.0	.0	99.2	99.2	.0	103.3	103.3	.0	100.6	100.6
02	SITE CONSTRUCTION	93.8	106.2	103.0	141.3	104.5	114.1	96.7	107.4	104.7	114.4	105.2	107.6	154.8	104.6	117.7	100.8	106.7	105.1
03100	CONCRETE FORMS & ACCESSORIES	100.8	112.1	110.6	104.8	131.1	127.6	103.1	115.6	114.0	103.1	126.9	123.7	124.9	133.4	132.3	105.4	121.4	119.3
03200	CONCRETE REINFORCEMENT	111.0	112.4	111.7	100.8	113.4	107.2	109.0	112.7	110.8	105.6	112.6	109.1	100.8	113.4	107.2	107.9	112.5	110.2
03300	CAST-IN-PLACE CONCRETE	97.6	110.4	102.9	121.2	109.7	116.4	101.0	112.1	105.6	123.0	113.3	119.0	139.5	118.9	130.9	106.1	117.2	110.7
03	CONCRETE	102.8	110.9	106.8	112.3	119.4	115.8	107.0	112.9	109.9	120.2	118.4	119.3	127.2	123.6	125.4	109.3	117.4	113.3
04	MASONRY	84.6	107.1	98.7	116.3	119.5	118.3	105.8	110.7	108.9	114.4	108.2	110.5	140.2	122.0	128.7	85.6	111.0	101.5
05	METALS	111.8	99.1	107.6	103.2	105.0	103.8	91.4	96.8	93.1	110.3	99.4	106.7	103.2	104.8	103.7	111.1	99.6	107.4
06	WOOD & PLASTICS	89.5	111.3	100.8	99.0	132.7	116.6	86.7	115.6	101.8	94.2	129.8	112.8	121.1	134.8	128.3	94.9	121.2	108.7
07	THERMAL & MOISTURE PROTECTION	118.4	109.0	113.8	112.3	124.7	118.4	106.2	108.0	107.0	111.3	108.5	109.9	113.3	123.8	118.5	118.9	112.0	115.5
08	DOORS & WINDOWS	99.6	109.9	102.3	106.4	124.6	111.1	95.6	113.1	100.1	103.2	119.8	107.5	106.4	125.7	111.4	102.8	116.2	106.2
09200	PLASTER & GYPSUM BOARD	93.3	111.5	104.9	93.1	133.0	118.4	91.7	116.2	107.2	96.8	130.4	118.1	103.6	135.1	123.6	97.0	121.7	112.7
095,098	CEILINGS & ACOUSTICAL TREATMENT	111.1	111.5	111.4	112.8	133.0	124.9	96.6	116.2	108.3	126.7	130.4	128.9	112.8	135.1	126.2	115.1	121.7	119.1
09600	FLOORING	119.4	103.7	115.3	115.2	122.8	117.2	102.2	103.7	102.6	116.6	106.8	114.0	123.7	122.8	123.4	121.8	103.7	117.0
097,099	WALL FINISHES, PAINTS & COATINGS	107.8	105.0	106.1	115.6	134.5	127.0	112.2	105.0	107.8	110.3	112.3	111.5	115.6	136.3	128.1	107.8	110.7	109.6
09	FINISHES	109.6	109.6	109.6	114.2	130.5	122.7	102.0	112.4	107.4	116.1	123.1	119.8	119.0	132.6	126.1	112.1	117.4	114.8
10 - 14	TOTAL DIV. 10000 - 14000	100.0	103.7	100.8	100.0	131.1	106.6	100.0	103.8	100.8	100.0	128.9	106.1	100.0	132.3	106.9	100.0	114.8	103.1
15	MECHANICAL	95.8	103.5	99.2	95.9	138.2	114.4	95.9	106.1	100.4	100.2	105.7	102.6	95.9	125.1	108.6	100.1	108.4	103.7
16	ELECTRICAL	92.0	102.2	97.0	102.4	141.1	121.3	116.1	108.1	112.2	97.2	102.7	99.9	103.1	130.3	116.4	88.7	102.2	95.3
01 - 16	WEIGHTED AVERAGE	100.5	105.8	102.9	105.5	125.9	114.8	100.5	108.3	104.0	106.8	110.4	108.4	109.5	122.8	115.6	102.6	109.9	105.9

CALIFORNIA

	DIVISION	SACRAMENTO 942,956 - 958			SALINAS 939			SAN BERNARDINO 923 - 924			SAN DIEGO 919 - 921			SAN FRANCISCO 940 - 941			SAN JOSE 951		
		MAT.	INST.	TOTAL	MAT.	INST.	TOTAL	MAT.	INST.	TOTAL	MAT.	INST.	TOTAL	MAT.	INST.	TOTAL	MAT.	INST.	TOTAL
01590	EQUIPMENT RENTAL	.0	102.9	102.9	.0	99.5	99.5	.0	100.6	100.6	.0	97.1	97.1	.0	108.6	108.6	.0	100.0	100.0
02	SITE CONSTRUCTION	119.5	110.9	113.2	125.3	106.0	111.1	78.4	106.5	99.2	102.2	100.6	101.0	158.2	111.2	123.4	148.5	99.7	112.4
03100	CONCRETE FORMS & ACCESSORIES	105.9	127.6	124.7	107.5	127.8	125.1	109.8	113.6	113.1	106.2	111.0	110.3	107.8	134.8	131.2	105.7	133.8	130.0
03200	CONCRETE REINFORCEMENT	95.3	112.9	104.2	109.5	113.1	111.3	107.9	112.4	110.2	105.6	112.4	109.0	117.5	113.9	115.7	96.2	113.6	105.0
03300	CAST-IN-PLACE CONCRETE	120.2	114.3	117.8	107.3	113.8	110.0	73.2	112.6	89.7	110.8	104.9	108.3	139.4	120.7	131.6	128.5	118.6	124.4
03	CONCRETE	115.8	119.1	117.5	121.4	119.1	120.3	81.6	112.3	96.9	113.0	108.4	110.7	128.7	124.9	126.8	118.1	123.8	120.9
04	MASONRY	119.0	112.2	114.7	107.3	115.8	112.6	92.7	111.0	104.2	98.1	106.4	103.3	148.8	129.0	136.4	147.2	125.5	133.6
05	METALS	100.4	99.8	100.2	110.2	101.2	107.2	111.1	99.1	107.2	108.8	98.8	105.6	111.2	107.1	109.9	104.5	106.9	105.3
06	WOOD & PLASTICS	96.7	130.0	114.1	99.1	129.8	115.1	99.2	111.3	105.5	100.3	109.2	104.9	101.9	135.0	119.2	101.0	134.5	118.5
07	THERMAL & MOISTURE PROTECTION	123.8	112.8	118.4	107.9	120.8	114.2	117.2	110.7	114.0	113.8	104.1	109.0	113.0	130.8	121.7	107.2	129.5	118.2
08	DOORS & WINDOWS	117.7	120.8	118.5	101.8	123.0	107.3	99.7	109.9	102.3	103.9	107.7	104.9	110.8	128.8	115.5	92.8	126.7	101.6
09200	PLASTER & GYPSUM BOARD	92.1	130.4	116.4	99.6	130.4	119.1	99.3	111.5	107.1	97.6	109.2	104.9	97.6	135.1	121.4	97.9	135.1	121.5
095,098	CEILINGS & ACOUSTICAL TREATMENT	114.8	130.4	124.2	121.0	130.4	126.6	115.6	111.5	113.1	103.7	109.2	107.0	123.5	135.1	130.5	106.2	135.1	123.5
09600	FLOORING	119.0	115.6	118.1	118.3	122.8	119.5	124.0	103.7	118.6	110.7	103.5	108.8	116.1	128.7	119.5	114.3	128.7	118.1
097,099	WALL FINISHES, PAINTS & COATINGS	112.9	116.2	114.9	111.3	139.4	128.3	107.8	105.0	106.1	109.4	105.0	106.7	115.6	151.7	137.5	112.0	139.4	128.6
09	FINISHES	114.5	125.5	120.2	116.7	128.9	123.1	111.9	110.6	111.2	107.8	109.1	108.5	118.6	136.0	127.7	111.8	134.0	123.3
10 - 14	TOTAL DIV. 10000 - 14000	100.0	129.6	106.3	100.0	128.9	106.1	100.0	105.2	101.1	100.0	112.5	102.7	100.0	133.1	107.0	100.0	131.7	106.9
15	MECHANICAL	100.2	111.2	105.0	95.8	111.7	102.8	95.8	105.6	100.1	100.3	106.8	103.1	100.2	151.6	122.7	100.2	141.3	118.2
16	ELECTRICAL	97.9	102.8	100.3	87.4	118.0	102.3	92.0	99.8	95.8	96.0	98.7	97.4	102.6	152.1	126.8	102.5	142.9	122.2
01 - 16	WEIGHTED AVERAGE	106.9	113.1	109.7	104.5	116.1	109.8	98.3	106.7	102.2	103.9	105.0	104.4	112.6	133.8	122.2	107.3	128.4	116.9

CALIFORNIA

	DIVISION	SAN LUIS OBISPO 934			SAN MATEO 944			SAN RAFAEL 949			SANTA ANA 926 - 927			SANTA BARBARA 931			SANTA CRUZ 950		
		MAT.	INST.	TOTAL	MAT.	INST.	TOTAL	MAT.	INST.	TOTAL	MAT.	INST.	TOTAL	MAT.	INST.	TOTAL	MAT.	INST.	TOTAL
01590	EQUIPMENT RENTAL	.0	99.5	99.5	.0	103.3	103.3	.0	103.9	103.9	.0	100.6	100.6	.0	99.5	99.5	.0	100.0	100.0
02	SITE CONSTRUCTION	116.0	106.2	108.8	151.3	104.6	116.8	133.5	111.1	116.9	91.9	106.4	102.6	109.6	106.1	107.0	148.9	99.4	112.3
03100	CONCRETE FORMS & ACCESSORIES	114.0	113.7	113.8	112.5	133.5	130.7	119.1	133.1	131.2	110.6	112.2	112.0	104.1	121.4	119.1	105.7	127.9	124.9
03200	CONCRETE REINFORCEMENT	110.9	112.5	111.7	100.8	113.6	107.3	101.5	113.5	107.5	111.6	112.5	112.0	108.9	112.5	110.7	119.2	113.2	116.2
03300	CAST-IN-PLACE CONCRETE	115.3	111.7	113.8	135.0	119.0	128.3	157.2	117.5	140.6	93.6	110.4	100.6	108.2	116.3	111.6	128.2	116.0	123.1
03	CONCRETE	119.3	112.0	115.7	123.0	123.7	123.3	147.9	122.7	135.3	100.2	110.9	105.6	110.4	117.0	113.7	121.7	120.2	120.9
04	MASONRY	109.4	111.1	110.4	139.8	125.5	130.9	114.6	125.5	121.4	81.5	107.5	97.8	107.9	111.1	109.9	149.0	116.0	128.3
05	METALS	107.8	98.9	104.9	103.0	105.2	103.8	103.8	101.0	102.9	111.2	99.3	107.3	105.7	99.3	103.6	111.3	105.4	110.5
06	WOOD & PLASTICS	103.0	111.4	107.4	107.7	134.8	121.8	107.9	134.5	121.8	100.9	111.3	106.3	94.0	121.3	108.3	101.0	129.9	116.1
07	THERMAL & MOISTURE PROTECTION	107.7	107.7	107.7	112.9	125.1	118.9	128.2	123.4	125.8	119.0	108.9	114.0	106.8	108.9	107.8	107.2	120.9	113.9
08	DOORS & WINDOWS	99.5	107.9	101.7	106.3	125.7	111.3	116.6	125.6	118.9	98.9	109.9	101.7	101.6	116.2	105.4	95.0	123.1	102.2
09200	PLASTER & GYPSUM BOARD	108.8	111.5	110.5	97.5	135.1	121.4	100.9	135.1	122.6	99.8	111.5	107.2	98.1	121.7	113.1	101.9	130.4	120.0
095,098	CEILINGS & ACOUSTICAL TREATMENT	121.0	111.5	115.3	112.8	135.1	126.2	121.7	135.1	129.7	115.6	111.5	113.1	119.6	121.7	120.9	117.4	130.4	125.2
09600	FLOORING	126.1	100.4	119.3	118.2	122.8	119.4	128.9	128.7	128.8	124.7	103.7	119.2	117.3	100.4	112.9	114.3	122.8	116.5
097,099	WALL FINISHES, PAINTS & COATINGS	110.3	98.5	103.1	115.6	134.5	127.0	111.3	135.4	125.9	107.8	105.0	106.1	110.3	102.7	105.7	112.0	139.4	128.6
09	FINISHES	119.6	109.4	114.3	116.3	132.4	124.6	119.4	133.3	126.6	113.2	109.6	111.3	114.6	116.0	115.3	115.1	129.1	122.4
10 - 14	TOTAL DIV. 10000 - 14000	100.0	118.0	103.8	100.0	132.4	106.9	100.0	131.4	106.7	100.0	103.7	100.8	100.0	115.1	103.2	100.0	129.4	106.2
15	MECHANICAL	95.8	105.7	100.1	95.9	128.2	110.0	95.8	137.9	114.2	95.8	103.5	99.2	100.2	108.4	103.8	100.2	111.8	105.2
16	ELECTRICAL	86.8	97.2	91.9	102.4	138.4	120.0	96.0	104.2	100.0	92.0	107.6	99.6	85.8	107.2	96.3	101.4	118.0	109.5
01 - 16	WEIGHTED AVERAGE	103.8	106.3	105.0	108.4	125.1	116.0	110.5	122.3	115.9	100.4	106.7	103.2	102.7	110.3	106.1	109.6	116.1	112.6

COST INDEXES

CALIFORNIA / COLORADO

	DIVISION	SANTA ROSA 954			STOCKTON 952			SUSANVILLE 961			VALLEJO 945			VAN NUYS 913 - 916			ALAMOSA 811		
		MAT.	INST.	TOTAL	MAT.	INST.	TOTAL	MAT.	INST.	TOTAL	MAT.	INST.	TOTAL	MAT.	INST.	TOTAL	MAT.	INST.	TOTAL
01590	EQUIPMENT RENTAL	.0	99.8	99.8	.0	99.2	99.2	.0	99.2	99.2	.0	103.5	103.5	.0	97.0	97.0	.0	96.9	96.9
02	SITE CONSTRUCTION	107.4	105.4	105.9	107.6	105.3	105.9	122.4	105.2	109.6	116.2	111.1	112.4	114.9	107.4	109.4	132.3	94.8	104.6
03100	CONCRETE FORMS & ACCESSORIES	102.1	131.6	127.7	103.8	126.9	123.8	104.8	127.0	124.1	107.0	131.7	128.4	110.9	115.6	115.0	101.1	81.8	84.4
03200	CONCRETE REINFORCEMENT	106.6	113.7	110.2	109.5	112.6	111.1	105.6	112.6	109.1	102.7	113.5	108.2	109.0	112.7	110.8	114.5	84.1	99.1
03300	CAST-IN-PLACE CONCRETE	121.9	114.9	119.0	108.2	113.4	110.3	111.9	113.3	112.5	125.1	115.4	121.1	106.5	112.1	108.9	102.3	86.3	95.6
03	CONCRETE	121.1	121.4	121.2	110.4	118.5	114.5	122.0	118.5	120.2	119.4	121.4	120.4	122.3	112.9	117.6	116.2	83.8	100.0
04	MASONRY	112.6	125.0	120.4	111.9	108.2	109.5	113.4	108.1	110.1	84.4	121.6	107.7	123.7	110.7	115.6	140.3	82.1	103.8
05	METALS	111.5	102.7	108.6	107.1	99.8	104.7	109.3	99.6	106.1	103.8	101.4	103.0	90.5	96.8	92.5	96.0	85.1	92.5
06	WOOD & PLASTICS	90.8	134.2	113.5	95.9	129.8	113.6	96.4	129.8	113.8	95.1	134.5	115.6	95.4	115.6	105.9	100.6	84.2	92.0
07	THERMAL & MOISTURE PROTECTION	119.6	123.2	121.3	110.8	107.5	109.2	112.2	108.9	110.6	126.1	121.6	123.9	107.6	108.0	107.8	106.4	85.0	95.9
08	DOORS & WINDOWS	100.5	128.4	107.7	100.8	120.6	106.0	103.1	119.8	107.4	119.2	128.6	121.6	95.4	113.1	100.0	93.5	88.2	92.1
09200	PLASTER & GYPSUM BOARD	94.9	135.1	120.4	98.5	130.4	118.7	97.6	130.4	118.4	94.9	135.1	120.4	95.0	116.2	108.4	80.8	83.4	82.4
095,098	CEILINGS & ACOUSTICAL TREATMENT	115.6	135.1	127.3	119.6	130.4	126.1	121.0	130.4	126.6	123.7	135.1	130.5	93.9	116.2	107.3	96.9	83.4	88.8
09600	FLOORING	120.1	128.7	122.4	117.3	110.9	115.6	117.3	112.7	116.1	123.7	128.7	125.0	105.2	103.7	104.8	111.6	71.5	101.0
097,099	WALL FINISHES, PAINTS & COATINGS	107.8	138.0	126.1	110.3	101.4	104.9	110.3	136.0	125.9	112.1	136.0	126.6	112.2	105.0	107.8	117.1	31.2	65.0
09	FINISHES	111.9	132.9	122.8	114.4	122.6	118.7	115.8	126.8	121.5	116.0	132.8	124.8	103.7	112.4	108.3	102.6	74.5	88.0
10 - 14	TOTAL DIV. 10000 - 14000	100.0	129.1	106.2	100.0	128.9	106.1	100.0	128.8	106.1	100.0	129.9	106.4	100.0	103.8	100.8	100.0	91.7	98.2
15	MECHANICAL	95.8	137.1	113.8	100.2	101.3	100.7	95.8	106.7	100.5	100.2	118.4	108.2	95.9	106.1	100.4	95.8	71.2	85.0
16	ELECTRICAL	92.3	104.2	98.1	97.0	109.6	103.2	97.6	102.7	100.1	92.2	123.1	107.2	116.1	108.1	112.2	89.3	82.5	86.0
01 - 16	WEIGHTED AVERAGE	104.7	121.5	112.4	104.5	110.5	107.2	106.1	111.1	108.4	105.6	120.4	112.3	103.8	108.3	105.8	101.6	81.3	92.4

COLORADO

	DIVISION	BOULDER 803			COLORADO SPRINGS 808 - 809			DENVER 800 - 802			DURANGO 813			FORT COLLINS 805			FORT MORGAN 807		
		MAT.	INST.	TOTAL	MAT.	INST.	TOTAL	MAT.	INST.	TOTAL	MAT.	INST.	TOTAL	MAT.	INST.	TOTAL	MAT.	INST.	TOTAL
01590	EQUIPMENT RENTAL	.0	96.8	96.8	.0	95.2	95.2	.0	100.4	100.4	.0	96.9	96.9	.0	96.8	96.8	.0	96.8	96.8
02	SITE CONSTRUCTION	93.0	99.0	97.4	95.0	97.1	96.5	93.6	106.4	103.0	125.6	94.8	102.8	104.3	99.5	100.8	94.8	99.6	98.4
03100	CONCRETE FORMS & ACCESSORIES	104.0	81.8	84.8	92.6	85.9	86.8	101.9	87.1	89.1	107.9	81.7	85.2	101.3	80.4	83.2	104.3	81.8	84.8
03200	CONCRETE REINFORCEMENT	113.8	84.1	98.8	113.0	84.6	98.6	113.0	84.7	98.7	114.5	84.1	99.1	113.9	83.0	98.3	114.0	84.1	98.9
03300	CAST-IN-PLACE CONCRETE	94.2	85.2	90.5	98.0	88.1	93.9	91.7	88.5	90.3	117.6	86.3	104.5	106.4	83.5	96.9	92.5	85.1	89.4
03	CONCRETE	104.7	83.5	94.1	108.6	86.5	97.5	102.6	87.2	94.9	118.9	83.8	101.3	115.0	82.1	98.6	103.3	83.4	93.4
04	MASONRY	104.0	82.4	90.4	108.7	82.6	92.3	108.4	85.8	94.2	126.2	82.1	98.6	122.8	60.9	84.0	121.3	82.4	96.9
05	METALS	95.3	84.9	91.9	97.9	88.1	94.7	100.2	88.4	96.3	96.0	85.0	92.5	96.1	83.4	92.0	95.1	85.0	91.8
06	WOOD & PLASTICS	104.8	83.6	93.7	93.9	88.3	91.0	103.3	88.8	95.7	109.2	84.2	96.1	103.0	83.6	92.9	105.2	83.6	93.9
07	THERMAL & MOISTURE PROTECTION	103.2	82.9	93.2	103.8	85.8	95.0	103.2	84.3	93.9	106.3	85.0	95.8	103.8	72.5	88.4	103.1	82.1	92.8
08	DOORS & WINDOWS	95.9	87.9	93.9	98.5	90.6	96.5	99.9	91.5	97.7	100.8	88.2	97.6	95.9	87.9	93.8	95.9	87.9	93.8
09200	PLASTER & GYPSUM BOARD	99.6	83.4	89.3	87.6	88.1	87.9	97.6	88.8	92.0	90.8	83.4	86.1	96.0	83.4	88.0	100.8	83.4	89.8
095,098	CEILINGS & ACOUSTICAL TREATMENT	92.1	83.4	86.9	99.2	88.1	92.5	98.3	88.8	92.6	96.9	83.4	88.8	92.1	83.4	86.9	92.1	83.4	86.9
09600	FLOORING	112.9	97.0	108.7	107.8	89.1	102.9	108.8	101.0	106.7	116.7	71.5	104.8	108.6	71.5	98.8	113.5	71.5	102.4
097,099	WALL FINISHES, PAINTS & COATINGS	106.7	77.7	89.1	106.7	53.6	74.5	106.7	77.7	89.1	117.1	31.2	65.0	106.7	53.2	74.2	106.7	70.8	84.9
09	FINISHES	99.7	83.8	91.4	98.2	83.0	90.3	99.6	88.8	94.0	104.9	74.5	89.1	98.3	75.9	86.7	100.0	78.6	88.9
10 - 14	TOTAL DIV. 10000 - 14000	100.0	90.2	97.9	100.0	91.3	98.1	100.0	91.5	98.2	100.0	91.6	98.2	100.0	89.1	97.7	100.0	90.2	97.9
15	MECHANICAL	95.8	84.0	90.6	100.2	81.3	91.9	100.2	86.4	94.1	95.8	72.8	85.7	100.2	82.5	92.4	95.8	83.0	90.2
16	ELECTRICAL	92.3	93.0	92.6	95.3	90.8	93.1	97.1	95.4	96.3	88.8	65.4	77.4	92.3	93.0	92.6	92.5	93.0	92.7
01 - 16	WEIGHTED AVERAGE	97.6	86.7	92.7	100.1	86.6	94.0	100.3	90.3	95.8	102.0	79.2	91.6	101.0	82.7	92.7	98.4	85.9	92.7

COLORADO

	DIVISION	GLENWOOD SPRINGS 816			GOLDEN 804			GRAND JUNCTION 815			GREELEY 806			MONTROSE 814			PUEBLO 810		
		MAT.	INST.	TOTAL	MAT.	INST.	TOTAL	MAT.	INST.	TOTAL	MAT.	INST.	TOTAL	MAT.	INST.	TOTAL	MAT.	INST.	TOTAL
01590	EQUIPMENT RENTAL	.0	100.0	100.0	.0	96.8	96.8	.0	100.0	100.0	.0	96.8	96.8	.0	98.4	98.4	.0	96.9	96.9
02	SITE CONSTRUCTION	141.5	102.5	112.7	106.4	100.0	101.6	124.7	102.1	108.0	91.7	98.2	96.5	134.9	98.3	107.8	116.8	96.5	101.7
03100	CONCRETE FORMS & ACCESSORIES	97.5	81.7	83.9	95.8	81.7	83.6	106.7	81.0	84.4	98.7	49.5	56.1	97.1	81.4	83.5	103.3	86.1	88.4
03200	CONCRETE REINFORCEMENT	113.4	84.1	98.6	114.0	84.1	98.9	113.7	83.0	98.2	113.8	82.4	97.9	113.2	84.0	98.5	109.5	84.6	96.9
03300	CAST-IN-PLACE CONCRETE	102.2	85.6	95.3	92.6	85.2	89.5	113.2	84.6	101.2	89.0	60.8	77.2	102.2	85.4	95.2	101.5	89.3	96.4
03	CONCRETE	121.1	83.6	102.4	114.3	83.4	98.9	115.3	82.6	99.0	100.0	60.5	80.3	112.4	83.3	97.9	105.2	87.1	96.1
04	MASONRY	111.2	82.4	93.1	124.6	82.4	98.1	148.3	63.4	95.0	116.4	40.4	68.7	118.1	82.1	95.5	107.1	82.6	91.7
05	METALS	95.8	85.9	92.6	95.2	85.2	92.0	96.9	82.6	92.3	96.1	81.7	91.4	94.8	84.6	91.5	98.4	89.2	95.4
06	WOOD & PLASTICS	95.5	83.8	89.4	97.5	83.6	90.2	106.4	83.8	94.6	100.2	48.3	73.1	95.5	84.0	89.5	103.1	88.7	95.6
07	THERMAL & MOISTURE PROTECTION	106.2	81.3	93.9	104.4	80.4	92.6	104.9	68.6	87.0	102.9	60.7	82.2	106.4	85.0	95.9	104.2	85.5	95.0
08	DOORS & WINDOWS	100.0	88.0	96.9	95.9	87.9	93.8	100.8	88.0	97.5	95.9	68.7	88.8	101.0	88.1	97.7	95.2	90.8	94.1
09200	PLASTER & GYPSUM BOARD	97.9	83.4	88.7	93.9	83.4	87.3	106.7	83.4	91.9	94.7	47.1	64.5	77.1	83.4	81.1	85.2	88.1	87.0
095,098	CEILINGS & ACOUSTICAL TREATMENT	96.0	83.4	88.4	92.1	83.4	86.9	96.0	83.4	88.4	92.1	47.1	65.1	96.9	83.4	88.8	97.6	88.1	95.9
09600	FLOORING	110.6	65.9	98.8	106.0	71.5	96.9	116.1	71.5	104.4	107.2	71.5	97.8	114.0	59.2	99.5	112.5	101.0	109.5
097,099	WALL FINISHES, PAINTS & COATINGS	117.0	70.8	89.0	106.7	77.7	89.1	117.0	77.7	93.2	106.7	33.1	62.1	117.1	31.2	65.0	117.1	49.6	76.2
09	FINISHES	104.9	77.7	90.8	97.9	79.4	88.3	106.3	79.5	92.4	97.1	50.2	72.7	102.8	71.9	86.7	104.6	85.3	94.6
10 - 14	TOTAL DIV. 10000 - 14000	100.0	90.6	98.0	100.0	90.2	97.9	100.0	90.5	98.0	100.0	80.3	95.8	100.0	91.0	98.1	100.0	92.3	98.4
15	MECHANICAL	95.8	81.5	89.5	95.8	83.9	90.6	100.1	72.5	88.0	100.2	76.7	89.9	95.8	72.8	85.7	100.1	71.3	87.5
16	ELECTRICAL	86.5	83.9	85.3	92.5	93.0	92.7	88.5	65.4	77.3	92.3	93.0	92.6	88.5	65.4	77.3	89.3	82.6	86.0
01 - 16	WEIGHTED AVERAGE	101.4	84.6	93.8	99.9	86.2	93.7	103.8	77.6	91.9	98.5	71.1	86.0	100.6	79.0	90.8	100.0	83.9	92.6

COST INDEXES

City Cost Indexes

COLORADO / CONNECTICUT

DIVISION		SALIDA 812 MAT.	INST.	TOTAL	BRIDGEPORT 066 MAT.	INST.	TOTAL	BRISTOL 060 MAT.	INST.	TOTAL	HARTFORD 061 MAT.	INST.	TOTAL	MERIDEN 064 MAT.	INST.	TOTAL	NEW BRITAIN 060 MAT.	INST.	TOTAL
01590	EQUIPMENT RENTAL	.0	98.4	98.4	.0	101.6	101.6	.0	101.6	101.6	.0	101.6	101.6	.0	102.1	102.1	.0	101.6	101.6
02	SITE CONSTRUCTION	125.3	98.6	105.6	102.3	104.4	103.8	101.4	104.3	103.6	101.8	104.3	103.7	99.4	105.2	103.7	101.6	104.3	103.6
03100	CONCRETE FORMS & ACCESSORIES	106.5	81.5	84.8	101.0	110.7	109.4	101.0	110.3	109.1	99.9	110.3	108.9	100.7	110.5	109.2	101.2	110.4	109.1
03200	CONCRETE REINFORCEMENT	113.0	84.0	98.4	108.2	122.7	115.5	108.2	122.6	115.5	108.2	122.6	115.5	108.2	122.7	115.5	108.2	122.6	115.5
03300	CAST-IN-PLACE CONCRETE	117.2	85.5	103.9	105.2	117.7	110.4	98.6	117.6	106.5	99.7	117.6	107.2	94.8	117.7	104.3	100.1	117.6	107.4
03	CONCRETE	114.2	83.4	99.8	108.8	115.3	112.0	105.6	115.1	110.3	106.1	115.1	110.6	103.8	115.2	109.5	106.4	115.1	110.7
04	MASONRY	150.7	82.1	107.7	108.2	117.3	113.9	99.9	117.3	110.8	100.0	117.3	110.9	99.6	117.3	110.7	100.0	117.3	110.9
05	METALS	94.6	84.7	91.3	101.2	120.2	107.4	101.2	119.9	107.3	106.3	119.9	110.7	98.1	120.1	105.3	97.5	119.9	104.8
06	WOOD & PLASTICS	104.7	84.0	93.9	99.4	108.0	103.8	99.4	108.0	103.8	99.4	108.0	103.8	99.4	108.0	103.8	99.4	108.0	103.8
07	THERMAL & MOISTURE PROTECTION	105.0	85.0	95.2	101.6	117.5	109.4	101.8	113.6	107.6	100.3	113.6	106.9	101.8	113.6	107.6	101.8	115.2	108.4
08	DOORS & WINDOWS	93.6	88.1	92.2	107.1	121.0	110.7	107.1	113.0	108.6	107.1	113.0	108.6	109.7	121.0	112.6	107.1	113.0	108.6
09200	PLASTER & GYPSUM BOARD	80.4	83.4	82.3	104.4	107.2	106.2	104.4	107.2	106.2	104.4	107.2	106.2	106.3	107.2	106.9	104.4	107.2	106.2
095,098	CEILINGS & ACOUSTICAL TREATMENT	96.9	83.4	88.8	101.1	107.2	104.8	101.1	107.2	104.8	101.1	107.2	104.8	106.3	107.2	106.9	101.1	107.2	104.8
09600	FLOORING	119.6	59.2	103.6	99.8	115.1	103.9	99.8	115.1	103.9	99.8	115.1	103.9	99.8	115.1	103.9	99.8	115.1	103.9
097,099	WALL FINISHES, PAINTS & COATINGS	117.1	31.2	65.0	92.8	109.6	103.0	92.8	109.6	103.0	92.8	109.6	103.0	92.8	109.6	103.0	92.8	109.6	103.0
09	FINISHES	104.1	71.9	87.3	102.7	110.7	106.8	102.7	110.7	106.9	102.7	110.7	106.9	104.3	110.7	107.6	102.7	110.7	106.9
10 - 14	TOTAL DIV. 10000 - 14000	100.0	91.1	98.1	100.0	110.1	102.1	100.0	110.1	102.1	100.0	110.1	102.1	100.0	110.1	102.1	100.0	110.1	102.1
15	MECHANICAL	95.8	71.1	85.0	100.1	104.0	101.8	100.1	103.9	101.8	100.1	103.9	101.8	95.7	104.0	99.3	100.1	103.9	101.8
16	ELECTRICAL	88.7	82.5	85.7	100.3	105.8	103.0	100.3	108.4	104.2	99.9	103.1	101.5	100.2	108.4	104.2	100.3	108.4	104.3
01 - 16	WEIGHTED AVERAGE	101.6	81.1	92.3	102.7	110.8	106.4	101.9	110.6	105.9	102.7	109.9	106.0	100.6	111.1	105.4	101.4	110.7	105.6

CONNECTICUT

DIVISION		NEW HAVEN 065 MAT.	INST.	TOTAL	NEW LONDON 063 MAT.	INST.	TOTAL	NORWALK 068 MAT.	INST.	TOTAL	STAMFORD 069 MAT.	INST.	TOTAL	WATERBURY 067 MAT.	INST.	TOTAL	WILLIMANTIC 062 MAT.	INST.	TOTAL
01590	EQUIPMENT RENTAL	.0	102.1	102.1	.0	102.1	102.1	.0	101.6	101.6	.0	101.6	101.6	.0	101.6	101.6	.0	101.6	101.6
02	SITE CONSTRUCTION	101.5	105.2	104.2	93.8	105.2	102.2	102.0	104.4	103.8	102.7	104.4	103.9	101.8	104.4	103.7	102.1	104.3	103.8
03100	CONCRETE FORMS & ACCESSORIES	100.8	110.6	109.3	100.7	110.3	109.0	101.0	110.9	109.4	101.0	110.9	109.3	101.0	110.6	109.2	101.0	110.3	109.0
03200	CONCRETE REINFORCEMENT	108.2	122.7	115.5	84.9	122.6	103.9	108.2	122.8	115.6	108.2	122.8	115.6	108.2	122.7	115.5	108.2	122.6	115.5
03300	CAST-IN-PLACE CONCRETE	101.9	117.7	108.5	86.8	117.6	99.7	103.5	119.3	110.1	105.2	119.4	111.1	105.2	117.7	110.4	98.2	105.8	101.4
03	CONCRETE	121.6	115.2	118.4	93.2	115.1	104.1	107.9	115.9	111.9	108.8	116.0	112.4	108.8	115.2	112.0	105.5	111.0	108.2
04	MASONRY	100.2	117.3	110.9	98.5	117.3	110.3	100.3	119.0	112.1	100.4	119.0	112.1	100.4	117.3	111.0	99.9	117.3	110.8
05	METALS	97.7	120.2	105.1	97.4	119.9	104.8	101.2	120.4	107.5	101.2	120.7	107.5	101.2	120.1	107.4	100.9	119.8	107.1
06	WOOD & PLASTICS	99.4	108.0	103.8	99.4	108.0	103.8	99.4	108.0	103.8	99.4	108.0	103.8	99.4	108.0	103.8	99.4	108.0	103.8
07	THERMAL & MOISTURE PROTECTION	101.9	115.2	108.5	101.8	113.6	107.6	101.8	118.2	109.9	101.8	118.2	109.9	101.8	115.2	108.4	101.1	111.8	106.3
08	DOORS & WINDOWS	107.1	121.0	110.7	110.5	112.4	111.0	107.1	121.0	110.7	107.1	121.0	110.7	107.1	121.0	110.7	110.5	121.0	113.2
09200	PLASTER & GYPSUM BOARD	104.4	107.2	106.2	104.4	107.2	106.2	104.4	107.2	106.2	104.4	107.2	106.2	104.4	107.2	106.2	104.4	107.2	106.2
095,098	CEILINGS & ACOUSTICAL TREATMENT	101.1	107.2	104.8	99.1	107.2	104.0	101.1	107.2	104.8	101.1	107.2	104.8	101.1	107.2	104.8	99.1	107.2	104.0
09600	FLOORING	99.8	115.1	103.9	99.8	115.1	103.9	99.8	115.1	103.9	99.8	115.1	103.9	99.8	115.1	103.9	99.8	115.1	103.9
097,099	WALL FINISHES, PAINTS & COATINGS	92.8	109.6	103.0	92.8	109.6	103.0	92.8	109.6	103.0	92.8	109.6	103.0	92.8	109.6	103.0	92.8	109.6	103.0
09	FINISHES	102.8	110.7	106.9	101.8	110.7	106.4	102.7	110.7	106.9	102.8	110.7	106.9	102.6	110.7	106.8	102.3	110.7	106.7
10 - 14	TOTAL DIV. 10000 - 14000	100.0	110.1	102.1	100.0	110.1	102.1	100.0	110.1	102.1	100.0	110.2	102.2	100.0	110.1	102.1	100.0	110.1	102.1
15	MECHANICAL	100.1	104.0	101.8	95.7	103.9	99.3	100.1	104.0	101.8	100.1	104.0	101.8	100.1	104.0	101.8	100.1	103.9	101.8
16	ELECTRICAL	100.2	108.4	104.2	97.4	108.4	102.8	100.3	108.4	104.2	100.3	137.7	118.5	99.8	105.8	102.7	100.3	103.1	101.6
01 - 16	WEIGHTED AVERAGE	103.2	111.2	106.8	98.6	110.7	104.1	102.2	111.5	106.4	102.3	115.6	108.4	102.2	110.7	106.1	102.2	109.6	105.6

D.C. / DELAWARE / FLORIDA

DIVISION		WASHINGTON 200 - 205 MAT.	INST.	TOTAL	DOVER 199 MAT.	INST.	TOTAL	NEWARK 197 MAT.	INST.	TOTAL	WILMINGTON 198 MAT.	INST.	TOTAL	DAYTONA BEACH 321 MAT.	INST.	TOTAL	FORT LAUDERDALE 333 MAT.	INST.	TOTAL
01590	EQUIPMENT RENTAL	.0	102.3	102.3	.0	118.2	118.2	.0	118.2	118.2	.0	118.4	118.4	.0	97.7	97.7	.0	89.3	89.3
02	SITE CONSTRUCTION	100.3	89.5	92.3	101.1	111.1	108.5	101.1	111.1	108.5	87.5	111.4	105.1	115.6	86.7	94.2	101.3	73.8	81.0
03100	CONCRETE FORMS & ACCESSORIES	98.0	80.7	83.0	97.0	101.4	100.8	97.0	101.4	100.8	96.7	101.4	100.7	98.2	69.8	73.6	95.6	68.4	72.0
03200	CONCRETE REINFORCEMENT	105.6	89.5	97.5	101.3	99.0	100.2	102.2	99.0	100.6	102.2	99.0	100.6	95.0	85.5	90.2	95.0	72.9	83.8
03300	CAST-IN-PLACE CONCRETE	114.7	88.1	103.6	83.8	94.4	88.3	83.8	94.4	88.3	78.8	94.4	85.3	89.9	70.5	81.8	94.3	69.4	83.9
03	CONCRETE	111.2	86.2	98.7	100.0	99.5	99.8	100.2	99.5	99.9	97.8	99.5	98.6	90.4	74.2	82.3	92.3	70.9	81.6
04	MASONRY	92.2	82.0	85.8	105.0	88.8	94.8	103.0	88.8	94.0	108.5	88.8	96.1	86.7	64.7	72.9	86.9	67.6	74.8
05	METALS	107.4	107.8	107.5	105.0	115.1	108.3	105.0	115.1	108.3	101.8	115.1	106.1	96.4	95.9	96.2	96.3	89.9	94.2
06	WOOD & PLASTICS	95.0	80.9	87.7	95.3	103.2	99.4	95.3	103.2	99.4	95.3	103.2	99.4	97.8	72.5	84.6	93.3	65.8	79.0
07	THERMAL & MOISTURE PROTECTION	98.3	81.7	90.2	100.9	104.3	102.6	101.1	104.3	102.7	100.6	104.3	102.4	95.4	69.7	82.7	95.4	74.7	85.2
08	DOORS & WINDOWS	102.6	90.7	99.5	94.6	106.2	97.6	94.6	106.2	97.6	94.4	106.2	97.4	100.9	69.9	92.8	98.5	63.3	89.4
09200	PLASTER & GYPSUM BOARD	101.1	80.5	88.0	113.2	103.1	106.8	113.2	103.1	106.8	112.3	103.1	106.5	100.5	72.3	82.6	100.1	65.5	78.1
095,098	CEILINGS & ACOUSTICAL TREATMENT	99.3	80.5	88.0	108.1	103.1	105.1	108.1	103.1	105.1	103.6	103.1	103.3	102.7	72.3	84.5	102.7	65.5	80.4
09600	FLOORING	113.7	102.8	110.8	83.9	98.3	87.7	83.9	98.3	87.7	83.7	98.3	87.5	118.6	73.5	106.7	118.6	61.0	103.4
097,099	WALL FINISHES, PAINTS & COATINGS	113.0	87.8	97.7	90.7	97.0	94.5	90.7	97.0	94.5	90.7	97.0	94.5	111.2	72.1	87.5	107.4	49.9	72.5
09	FINISHES	101.2	85.3	92.9	103.4	100.2	101.7	103.4	100.2	101.8	102.2	100.2	101.2	108.4	70.8	88.9	106.4	64.5	84.6
10 - 14	TOTAL DIV. 10000 - 14000	100.0	97.5	99.5	100.0	100.0	100.0	100.0	100.0	100.0	100.0	100.0	100.0	100.0	82.5	96.3	100.0	89.4	97.8
15	MECHANICAL	100.0	90.9	96.0	100.3	116.6	107.4	100.3	116.6	107.4	100.3	116.6	107.4	99.8	72.6	87.9	99.8	69.7	86.7
16	ELECTRICAL	97.7	97.6	97.6	97.3	104.8	100.9	97.3	104.8	100.9	97.2	104.8	100.9	96.3	67.0	82.0	96.3	75.2	86.0
01 - 16	WEIGHTED AVERAGE	102.0	90.8	96.9	100.6	105.7	102.9	100.5	105.7	102.8	99.5	105.7	102.3	98.4	74.5	87.5	97.7	72.3	86.2

COST INDEXES

City Cost Indexes

FLORIDA

DIVISION		FORT MYERS 339,341			GAINESVILLE 326,344			JACKSONVILLE 320,322			LAKELAND 338			MELBOURNE 329			MIAMI 330 - 332,340		
		MAT.	INST.	TOTAL	MAT.	INST.	TOTAL	MAT.	INST.	TOTAL	MAT.	INST.	TOTAL	MAT.	INST.	TOTAL	MAT.	INST.	TOTAL
01590	EQUIPMENT RENTAL	.0	97.7	97.7	.0	97.7	97.7	.0	97.7	97.7	.0	97.7	97.7	.0	97.7	97.7	.0	89.3	89.3
02	SITE CONSTRUCTION	112.8	86.7	93.5	125.8	86.2	96.5	115.7	87.2	94.6	114.9	86.3	93.7	123.2	87.2	96.6	100.6	73.6	80.6
03100	CONCRETE FORMS & ACCESSORIES	90.7	74.2	76.4	91.9	52.9	58.1	98.0	53.3	59.3	86.2	74.8	76.3	92.5	73.7	76.2	95.8	68.4	72.0
03200	CONCRETE REINFORCEMENT	96.1	89.2	92.6	100.9	50.7	75.5	95.0	51.2	72.9	98.5	90.6	94.5	96.1	85.5	90.8	95.0	72.9	83.8
03300	CAST-IN-PLACE CONCRETE	98.4	60.6	82.6	103.3	49.6	80.9	98.0	58.7	77.4	100.6	61.6	84.3	104.5	75.8	92.5	91.9	69.6	82.5
03	CONCRETE	93.2	73.5	83.3	101.2	53.1	77.2	90.8	56.5	73.7	95.0	74.2	84.6	99.6	77.8	88.7	91.1	70.9	81.0
04	MASONRY	80.6	52.5	62.9	100.1	43.6	64.7	86.2	51.4	64.4	95.1	75.7	83.0	83.9	73.6	77.4	84.8	68.0	74.3
05	METALS	99.1	94.2	97.5	95.6	79.2	90.3	96.0	80.2	90.8	99.0	95.5	97.9	105.3	96.4	102.4	101.7	89.3	97.6
06	WOOD & PLASTICS	90.0	77.4	83.4	91.2	52.1	70.8	97.8	52.1	74.0	85.2	77.4	81.1	91.6	72.5	81.7	93.3	65.8	79.0
07	THERMAL & MOISTURE PROTECTION	95.1	56.3	76.0	95.7	55.1	75.7	95.7	58.0	77.2	95.1	62.3	79.0	95.6	76.3	86.1	99.3	71.4	85.6
08	DOORS & WINDOWS	99.5	73.4	92.8	98.9	49.1	86.0	100.9	49.7	87.6	99.5	71.1	92.1	100.0	75.4	93.7	98.5	63.9	89.5
09200	PLASTER & GYPSUM BOARD	95.8	77.4	84.1	96.0	51.3	67.7	100.5	51.3	69.3	93.6	77.4	83.3	96.2	72.3	81.1	100.1	65.5	78.1
095,098	CEILINGS & ACOUSTICAL TREATMENT	95.3	77.4	84.6	94.5	51.3	68.6	102.7	51.3	71.9	94.5	77.4	84.2	97.3	72.3	82.3	102.7	65.5	80.4
09600	FLOORING	114.4	41.8	95.2	115.0	35.9	94.1	118.6	50.1	100.5	111.4	53.0	96.0	115.3	73.5	104.2	126.0	61.7	109.0
097,099	WALL FINISHES, PAINTS & COATINGS	111.2	46.5	71.9	111.2	44.8	70.9	111.2	48.1	72.9	111.2	46.5	71.9	111.2	92.6	99.9	107.4	49.9	72.5
09	FINISHES	104.5	66.1	84.5	105.5	48.3	75.7	108.4	51.5	78.8	103.2	68.3	85.1	106.0	75.4	90.0	108.6	64.6	85.7
10 - 14	TOTAL DIV. 10000 - 14000	100.0	73.7	94.4	100.0	80.1	95.8	100.0	77.5	95.2	100.0	73.8	94.4	100.0	86.0	97.0	100.0	89.4	97.8
15	MECHANICAL	97.5	53.6	78.3	98.6	66.7	84.6	99.8	51.4	78.6	97.5	73.9	87.2	99.8	77.7	90.2	99.8	74.2	88.6
16	ELECTRICAL	98.4	40.4	70.1	96.6	41.9	69.9	96.1	67.2	82.0	96.6	46.9	72.3	96.2	73.9	85.3	96.7	73.3	85.3
01 - 16	WEIGHTED AVERAGE	97.9	64.3	82.6	99.7	58.7	81.1	98.3	61.0	81.4	98.5	72.3	86.6	100.5	79.0	90.7	98.7	72.9	87.0

FLORIDA

DIVISION		ORLANDO 327 - 328,347			PANAMA CITY 324			PENSACOLA 325			SARASOTA 342			ST. PETERSBURG 337			TALLAHASSEE 323		
		MAT.	INST.	TOTAL	MAT.	INST.	TOTAL	MAT.	INST.	TOTAL	MAT.	INST.	TOTAL	MAT.	INST.	TOTAL	MAT.	INST.	TOTAL
01590	EQUIPMENT RENTAL	.0	97.7	97.7	.0	97.7	97.7	.0	97.7	97.7	.0	97.7	97.7	.0	97.7	97.7	.0	97.7	97.7
02	SITE CONSTRUCTION	116.2	86.4	94.7	130.2	83.9	95.9	127.8	86.3	97.1	117.0	86.3	94.3	116.5	85.7	93.7	117.0	85.8	93.9
03100	CONCRETE FORMS & ACCESSORIES	98.0	71.3	74.9	97.0	27.7	37.0	87.4	51.1	55.9	98.2	74.3	77.5	95.5	47.2	53.7	97.9	39.0	46.9
03200	CONCRETE REINFORCEMENT	95.0	81.8	88.3	99.2	50.1	74.4	101.7	50.4	75.8	95.0	90.5	92.7	98.5	57.6	77.8	95.0	50.6	72.6
03300	CAST-IN-PLACE CONCRETE	97.7	70.4	86.3	95.4	34.6	70.0	95.4	54.9	78.5	102.9	61.4	85.6	101.7	55.3	82.3	94.1	48.4	75.0
03	CONCRETE	92.0	74.2	83.1	99.0	36.1	67.6	98.0	54.0	76.0	96.5	74.0	85.3	96.4	53.7	75.1	92.3	46.5	69.5
04	MASONRY	90.9	64.7	74.5	91.0	27.6	51.2	88.1	49.9	64.1	87.6	75.7	80.1	134.0	49.3	80.8	89.1	38.8	57.5
05	METALS	102.5	94.1	99.8	96.2	65.6	86.2	96.2	79.6	90.8	99.4	95.0	98.0	99.6	80.6	93.4	89.4	78.5	85.9
06	WOOD & PLASTICS	97.8	75.2	86.0	96.6	27.8	60.7	86.3	51.4	68.1	97.8	77.4	87.2	95.0	47.2	70.0	96.2	37.4	65.5
07	THERMAL & MOISTURE PROTECTION	95.7	69.9	83.1	96.0	31.4	64.2	95.7	52.8	74.6	95.4	62.3	79.1	95.3	49.6	72.8	95.7	46.5	71.5
08	DOORS & WINDOWS	100.9	69.6	92.8	98.4	26.7	79.9	98.4	50.2	85.9	100.9	70.5	93.0	99.5	46.5	85.8	99.5	41.0	84.3
09200	PLASTER & GYPSUM BOARD	103.1	75.0	85.3	98.4	26.4	52.7	94.0	50.6	66.5	100.5	77.4	85.8	97.6	46.3	65.1	100.5	36.2	59.7
095,098	CEILINGS & ACOUSTICAL TREATMENT	102.7	75.0	86.1	96.5	26.4	54.4	96.5	50.6	69.0	100.7	77.4	86.7	96.5	46.3	66.4	102.7	36.2	62.8
09600	FLOORING	118.6	73.5	106.7	118.0	19.2	91.9	112.2	53.2	96.6	118.6	52.8	101.2	116.9	53.0	100.0	118.6	38.9	97.5
097,099	WALL FINISHES, PAINTS & COATINGS	111.2	55.2	77.2	111.2	24.6	58.6	111.2	55.9	77.6	111.2	46.5	71.9	111.2	46.5	71.9	111.2	39.6	67.8
09	FINISHES	108.8	70.4	88.9	107.5	25.4	64.8	105.0	51.8	77.3	107.9	68.3	87.3	106.0	48.0	75.8	108.6	38.2	72.0
10 - 14	TOTAL DIV. 10000 - 14000	100.0	82.7	96.3	100.0	46.0	88.5	100.0	51.5	89.7	100.0	73.7	94.4	100.0	53.9	90.2	100.0	63.9	92.3
15	MECHANICAL	99.8	63.8	84.1	99.8	25.2	67.2	99.8	50.0	78.1	99.8	54.1	79.8	99.8	49.9	78.0	99.8	39.9	73.6
16	ELECTRICAL	96.7	44.6	71.3	95.3	33.4	65.1	98.4	58.8	.79.1	96.3	36.6	67.2	96.6	48.1	73.0	96.7	41.3	69.7
01 - 16	WEIGHTED AVERAGE	99.8	69.3	86.0	99.5	37.9	71.5	99.2	58.1	80.5	99.6	66.7	84.6	101.6	55.8	80.8	97.6	49.2	75.6

FLORIDA / GEORGIA

DIVISION		TAMPA 335 - 336,346			WEST PALM BEACH 334,349			ALBANY 317			ATHENS 306			ATLANTA 300 - 303,399			AUGUSTA 308 - 309		
		MAT.	INST.	TOTAL	MAT.	INST.	TOTAL	MAT.	INST.	TOTAL	MAT.	INST.	TOTAL	MAT.	INST.	TOTAL	MAT.	INST.	TOTAL
01590	EQUIPMENT RENTAL	.0	97.7	97.7	.0	89.3	89.3	.0	90.1	90.1	.0	92.1	92.1	.0	92.8	92.8	.0	92.1	92.1
02	SITE CONSTRUCTION	116.7	86.4	94.3	97.9	74.0	80.2	101.6	76.2	82.8	105.1	92.8	96.0	101.2	95.6	97.0	97.8	92.6	94.0
03100	CONCRETE FORMS & ACCESSORIES	99.4	74.9	78.2	99.7	67.9	72.2	97.1	45.9	52.8	91.9	42.0	48.7	96.0	78.8	81.1	93.1	51.9	57.4
03200	CONCRETE REINFORCEMENT	95.0	90.6	92.8	97.7	72.5	85.0	94.6	91.7	93.1	99.4	89.2	94.2	98.7	93.5	96.1	99.8	77.3	88.4
03300	CAST-IN-PLACE CONCRETE	99.5	61.6	83.7	89.7	65.3	79.5	96.0	46.2	75.2	99.6	49.3	78.6	99.6	74.8	89.2	94.1	48.9	75.2
03	CONCRETE	95.0	74.3	84.7	89.7	69.2	79.4	93.0	54.1	74.6	101.0	54.0	77.6	98.5	80.0	89.3	94.2	56.1	75.2
04	MASONRY	87.7	75.7	80.2	86.5	61.5	70.8	88.9	39.9	58.1	75.7	55.4	62.9	90.9	70.6	78.1	91.2	40.2	59.2
05	METALS	99.7	95.6	98.4	95.7	88.9	93.5	95.6	88.8	93.4	92.5	73.9	86.4	93.3	80.4	89.1	92.2	69.6	-84.8
06	WOOD & PLASTICS	99.3	77.4	87.9	98.2	65.8	81.3	96.7	42.5	68.4	92.1	36.6	63.1	96.4	80.9	88.3	93.4	53.3	72.5
07	THERMAL & MOISTURE PROTECTION	95.7	62.3	79.3	95.1	66.1	80.8	95.5	54.1	75.1	92.2	52.2	72.5	92.3	75.5	84.0	91.7	49.8	71.1
08	DOORS & WINDOWS	100.9	71.2	93.2	97.6	63.3	88.7	98.5	50.1	85.9	94.1	46.7	81.8	100.0	77.5	94.2	94.1	54.7	83.9
09200	PLASTER & GYPSUM BOARD	100.5	77.4	85.8	102.3	65.5	78.9	100.2	41.5	63.0	111.5	35.3	63.2	113.1	80.8	92.7	111.9	52.5	74.2
095,098	CEILINGS & ACOUSTICAL TREATMENT	102.7	77.4	87.5	95.3	65.5	77.4	100.2	41.5	65.6	107.9	35.3	64.4	107.9	80.8	91.7	108.9	52.5	75.1
09600	FLOORING	118.6	53.0	101.3	121.3	50.9	102.7	118.6	32.5	95.8	82.4	58.6	76.1	84.0	78.0	82.4	82.8	41.4	71.9
097,099	WALL FINISHES, PAINTS & COATINGS	111.2	46.5	71.9	107.4	46.1	70.2	107.4	41.4	67.3	90.3	34.5	56.4	90.3	84.5	86.8	90.3	39.3	59.4
09	FINISHES	108.5	68.3	87.6	105.6	62.0	82.9	106.2	41.5	72.6	95.8	42.9	68.3	96.2	79.3	87.4	95.7	48.2	71.0
10 - 14	TOTAL DIV. 10000 - 14000	100.0	85.3	96.9	100.0	89.3	97.7	100.0	77.5	95.2	100.0	53.6	90.1	100.0	84.6	96.7	100.0	67.6	93.1
15	MECHANICAL	99.8	73.9	88.5	97.5	74.5	87.4	99.8	49.7	77.9	95.7	69.9	84.4	100.0	80.4	91.5	100.0	66.3	85.3
16	ELECTRICAL	95.7	48.1	72.5	96.4	52.0	74.7	91.9	60.4	76.5	96.2	63.3	80.1	96.6	82.1	89.5	96.7	49.3	73.6
01 - 16	WEIGHTED AVERAGE	99.4	72.8	87.3	96.6	68.5	83.8	97.3	56.8	78.9	95.2	62.1	80.1	97.4	80.7	89.8	96.0	59.2	79.3

COST INDEXES

City Cost Indexes

GEORGIA

DIVISION		COLUMBUS 318 - 319			DALTON 307			GAINESVILLE 305			MACON 310 - 312			SAVANNAH 313 - 314			STATESBORO 304		
		MAT.	INST.	TOTAL	MAT.	INST.	TOTAL	MAT.	INST.	TOTAL	MAT.	INST.	TOTAL	MAT.	INST.	TOTAL	MAT.	INST.	TOTAL
01590	EQUIPMENT RENTAL	.0	90.1	90.1	.0	106.1	106.1	.0	92.1	92.1	.0	102.6	102.6	.0	91.2	91.2	.0	91.2	91.2
02	SITE CONSTRUCTION	101.5	76.6	83.1	106.0	95.8	98.5	104.3	92.1	95.3	102.3	93.8	96.0	102.1	77.1	83.6	106.5	74.6	82.9
03100	CONCRETE FORMS & ACCESSORIES	97.0	58.7	63.8	83.1	46.4	51.3	95.7	45.2	52.0	96.4	55.5	61.0	97.4	52.8	58.8	77.0	42.1	46.8
03200	CONCRETE REINFORCEMENT	95.0	92.7	93.9	98.9	55.3	76.9	99.2	68.5	83.7	96.3	91.9	94.1	96.0	77.5	86.6	98.4	33.3	65.5
03300	CAST-IN-PLACE CONCRETE	95.7	55.0	78.7	96.8	36.1	71.4	104.7	49.8	81.8	94.4	45.8	74.1	92.7	51.6	75.5	99.4	43.9	76.2
03	CONCRETE	92.9	65.1	79.0	100.1	45.9	73.0	102.6	51.8	77.2	92.5	60.4	76.4	91.7	58.4	75.1	100.1	43.0	71.6
04	MASONRY	88.9	57.2	69.0	78.8	30.0	48.2	86.0	32.6	52.5	102.8	38.4	62.4	91.9	53.9	68.0	78.3	40.6	54.6
05	METALS	95.2	92.3	94.3	94.0	67.5	85.4	91.8	63.5	82.5	90.9	90.8	90.9	91.2	84.3	89.0	97.3	69.0	88.1
06	WOOD & PLASTICS	96.7	58.6	76.8	75.6	50.2	62.4	96.2	42.4	68.1	104.4	57.5	79.9	110.2	50.2	78.9	68.1	44.1	55.6
07	THERMAL & MOISTURE PROTECTION	95.1	61.9	78.8	97.1	41.6	69.8	92.1	45.7	69.3	93.9	56.8	75.7	95.5	53.0	74.6	93.0	38.7	66.3
08	DOORS & WINDOWS	98.5	60.9	88.7	95.3	37.6	80.3	94.0	40.5	80.2	96.9	59.3	87.2	99.6	50.9	87.0	96.0	34.3	80.1
09200	PLASTER & GYPSUM BOARD	100.2	58.0	73.5	96.6	49.4	66.6	113.1	41.3	67.6	106.2	56.9	75.0	100.5	49.4	68.1	96.9	43.0	62.8
095,098	CEILINGS & ACOUSTICAL TREATMENT	101.8	58.0	75.6	112.9	49.4	74.8	107.9	41.3	68.0	97.7	56.9	73.2	102.7	49.4	70.7	114.0	43.0	71.4
09600	FLOORING	118.6	54.8	101.7	83.4	14.0	65.0	83.9	14.0	65.4	92.5	38.2	78.1	118.6	48.8	100.1	96.7	48.8	84.0
097,099	WALL FINISHES, PAINTS & COATINGS	107.4	49.1	72.0	81.5	41.4	57.1	90.3	36.1	57.4	109.1	48.4	72.3	108.1	49.1	72.3	88.9	42.9	61.0
09	FINISHES	106.1	56.2	80.2	101.8	40.5	69.9	96.3	37.5	65.7	94.1	51.1	71.7	106.7	51.4	77.9	105.0	44.1	73.3
10 - 14	TOTAL DIV. 10000 - 14000	100.0	79.6	95.7	100.0	26.7	84.4	100.0	54.4	90.3	100.0	77.6	95.2	100.0	70.6	93.8	100.0	50.8	89.5
15	MECHANICAL	99.8	50.1	78.1	95.7	34.1	68.8	95.7	70.0	84.4	99.8	41.9	74.5	99.8	47.8	77.1	96.2	37.3	70.4
16	ELECTRICAL	92.3	50.9	72.1	106.9	75.1	91.4	96.2	63.3	80.1	90.0	59.6	75.2	93.8	58.3	76.8	97.3	30.9	64.9
01 - 16	WEIGHTED AVERAGE	97.2	61.7	81.1	97.5	50.9	76.3	95.8	57.4	78.4	95.7	59.2	79.1	97.1	58.8	79.7	97.1	45.0	73.4

DIVISION		GEORGIA VALDOSTA 316			WAYCROSS 315			HAWAII HILO 967			HONOLULU 968			STATES & POSS., GUAM 969			IDAHO BOISE 836 - 837		
		MAT.	INST.	TOTAL	MAT.	INST.	TOTAL	MAT.	INST.	TOTAL	MAT.	INST.	TOTAL	MAT.	INST.	TOTAL	MAT.	INST.	TOTAL
01590	EQUIPMENT RENTAL	.0	90.1	90.1	.0	90.1	90.1	.0	99.3	99.3	.0	99.3	99.3	.0	175.9	175.9	.0	101.7	101.7
02	SITE CONSTRUCTION	111.8	76.4	85.6	108.4	74.7	83.4	134.0	106.5	113.7	139.1	106.5	115.0	174.7	105.6	123.6	79.9	102.7	96.8
03100	CONCRETE FORMS & ACCESSORIES	84.2	45.7	50.8	86.4	42.9	48.7	106.4	144.6	139.5	106.8	144.6	139.5	108.4	50.3	58.1	97.7	82.3	84.4
03200	CONCRETE REINFORCEMENT	97.0	44.0	70.2	97.0	66.8	81.7	109.9	126.2	118.1	108.9	126.2	117.6	1391.8	27.8	702.6	108.2	78.0	92.3
03300	CAST-IN-PLACE CONCRETE	94.0	52.2	76.5	106.2	44.4	80.4	207.1	129.5	174.7	199.7	129.5	170.3	179.5	110.4	150.6	95.9	91.0	93.9
03	CONCRETE	97.0	49.3	73.2	100.8	49.6	75.2	157.7	134.6	146.2	154.0	134.6	144.3	354.1	67.7	211.1	104.2	84.4	94.3
04	MASONRY	94.8	49.0	66.1	95.7	40.6	61.2	132.6	129.8	130.9	132.4	129.8	130.8	182.7	34.0	89.4	137.5	74.5	98.0
05	METALS	94.9	72.5	87.6	94.2	79.4	89.4	119.2	111.2	116.6	145.3	111.2	134.1	156.0	79.8	131.1	99.7	77.6	92.5
06	WOOD & PLASTICS	82.3	41.5	61.0	85.4	44.0	63.8	97.8	148.7	124.4	97.7	148.7	124.3	106.7	50.0	77.1	95.7	81.9	88.5
07	THERMAL & MOISTURE PROTECTION	95.2	57.4	76.7	95.0	40.9	68.4	110.8	130.0	120.2	113.0	130.0	121.4	123.3	60.1	92.3	96.9	80.0	88.6
08	DOORS & WINDOWS	93.8	38.7	79.5	93.8	44.0	80.9	100.6	140.1	110.8	105.8	140.1	114.7	100.4	43.1	85.6	94.7	77.8	90.3
09200	PLASTER & GYPSUM BOARD	91.2	40.4	59.0	94.0	43.0	61.7	98.6	149.8	131.1	116.7	149.8	137.7	2921.6	35.1	1091.9	83.2	81.2	81.9
095,098	CEILINGS & ACOUSTICAL TREATMENT	96.5	40.4	62.8	94.5	43.0	63.6	117.4	149.8	136.8	119.6	149.8	137.7	256.5	35.1	123.7	107.6	81.2	91.7
09600	FLOORING	110.4	39.0	91.6	112.0	32.5	91.0	145.3	134.9	142.5	168.4	134.9	159.5	118.2	34.2	96.0	99.8	56.3	88.3
097,099	WALL FINISHES, PAINTS & COATINGS	107.4	35.8	64.0	107.4	42.9	68.3	104.2	148.5	131.1	104.9	148.5	131.4	110.3	41.3	68.4	103.7	52.2	72.5
09	FINISHES	101.8	42.7	71.1	101.9	40.7	70.0	123.4	144.6	134.4	134.0	144.6	139.5	584.3	46.4	304.5	98.5	74.5	86.0
10 - 14	TOTAL DIV. 10000 - 14000	100.0	69.4	93.5	100.0	50.8	89.5	100.0	121.1	104.5	100.0	121.1	104.5	100.0	104.1	100.9	100.0	86.4	97.1
15	MECHANICAL	99.8	42.1	74.6	97.0	43.7	73.7	100.2	117.8	107.9	100.2	117.8	107.9	103.0	36.2	73.8	100.1	77.1	90.1
16	ELECTRICAL	90.6	30.2	61.1	94.4	55.0	75.2	105.2	124.1	114.4	108.0	124.1	115.9	117.9	32.7	76.3	93.5	78.3	86.1
01 - 16	WEIGHTED AVERAGE	97.0	49.1	75.2	97.2	51.6	76.5	115.3	125.7	120.0	121.0	125.7	123.1	193.9	54.3	130.5	100.2	80.5	91.2

DIVISION		IDAHO COEUR D'ALENE 838			IDAHO FALLS 834			LEWISTON 835			POCATELLO 832			TWIN FALLS 833			ILLINOIS BLOOMINGTON 617		
		MAT.	INST.	TOTAL	MAT.	INST.	TOTAL	MAT.	INST.	TOTAL	MAT.	INST.	TOTAL	MAT.	INST.	TOTAL	MAT.	INST.	TOTAL
01590	EQUIPMENT RENTAL	.0	94.6	94.6	.0	101.7	101.7	.0	94.6	94.6	.0	101.7	101.7	.0	101.7	101.7	.0	101.6	101.6
02	SITE CONSTRUCTION	78.6	95.7	91.2	77.7	101.1	95.0	84.1	96.9	93.6	81.1	102.6	97.0	87.8	100.6	97.2	93.0	95.0	94.5
03100	CONCRETE FORMS & ACCESSORIES	105.3	39.4	48.3	90.3	36.2	43.4	111.5	67.5	73.4	97.8	81.8	84.0	98.5	35.2	43.7	83.8	107.4	104.2
03200	CONCRETE REINFORCEMENT	115.6	93.1	104.2	110.1	54.9	82.2	115.6	97.7	106.5	108.5	77.5	92.8	110.4	49.1	79.4	100.2	99.9	100.0
03300	CAST-IN-PLACE CONCRETE	100.1	75.8	89.9	88.1	56.2	74.8	104.0	86.7	96.8	95.2	90.8	93.3	97.7	45.2	75.7	96.8	103.5	99.6
03	CONCRETE	108.2	63.0	85.6	93.7	47.7	70.7	111.9	80.0	96.0	100.9	84.0	92.4	108.4	42.3	75.4	96.3	105.0	100.6
04	MASONRY	137.9	53.9	85.2	131.2	32.0	69.0	138.3	86.6	105.8	132.9	66.5	91.2	136.3	39.0	75.2	108.2	108.5	108.4
05	METALS	93.1	83.6	90.0	107.8	65.5	94.0	92.6	88.0	91.1	107.8	76.4	97.6	107.8	62.2	92.9	95.6	105.9	99.5
06	WOOD & PLASTICS	97.7	34.3	64.6	88.1	34.0	59.9	103.9	60.5	81.2	95.7	81.9	88.5	96.4	35.1	64.4	82.7	104.2	93.9
07	THERMAL & MOISTURE PROTECTION	166.7	64.4	116.4	96.5	48.8	73.0	167.0	80.7	124.6	97.1	70.5	84.1	98.2	46.7	72.9	96.1	104.9	100.4
08	DOORS & WINDOWS	114.9	48.9	97.8	98.0	39.7	82.9	114.8	67.5	102.6	94.7	71.0	88.6	98.0	36.0	82.0	89.6	100.7	92.5
09200	PLASTER & GYPSUM BOARD	132.0	32.4	68.8	77.8	32.0	48.8	134.0	59.3	86.7	83.2	81.2	81.9	82.9	33.1	51.4	96.1	104.1	101.2
095,098	CEILINGS & ACOUSTICAL TREATMENT	128.4	32.4	70.8	97.8	32.0	58.3	128.4	59.3	87.0	107.6	81.2	91.7	100.4	33.1	60.1	90.0	104.1	98.5
09600	FLOORING	134.5	59.9	114.8	96.5	56.3	85.9	137.7	93.5	126.1	100.1	56.3	88.5	101.5	56.3	89.5	89.0	106.8	93.7
097,099	WALL FINISHES, PAINTS & COATINGS	129.6	74.3	96.0	103.7	42.9	66.8	129.6	74.3	96.0	103.7	54.3	73.7	103.7	32.0	60.2	91.2	102.9	96.6
09	FINISHES	155.6	45.0	98.1	93.9	39.6	65.7	157.0	71.7	112.6	98.6	74.7	86.2	97.7	38.4	66.9	90.6	106.8	99.0
10 - 14	TOTAL DIV. 10000 - 14000	100.0	59.6	91.4	100.0	49.6	89.3	100.0	88.3	97.5	100.0	86.4	97.1	100.0	48.6	89.1	100.0	94.7	98.9
15	MECHANICAL	100.0	47.3	76.9	101.1	49.9	78.7	101.3	85.9	94.6	100.1	77.1	90.0	100.1	36.4	72.2	95.7	100.0	97.6
16	ELECTRICAL	80.8	37.4	59.6	79.8	73.9	76.9	79.1	85.1	82.0	88.7	73.9	81.5	81.0	31.5	56.9	96.1	89.4	92.9
01 - 16	WEIGHTED AVERAGE	107.3	56.8	84.4	98.2	55.3	78.7	108.1	83.3	96.8	100.3	78.3	90.3	100.8	45.8	75.8	95.5	101.2	98.1

ILLINOIS

DIVISION		CARBONDALE 629			CENTRALIA 628			CHAMPAIGN 618 - 619			CHICAGO 606 - 608			DECATUR 625			EAST ST. LOUIS 620 - 622		
		MAT.	INST.	TOTAL	MAT.	INST.	TOTAL	MAT.	INST.	TOTAL	MAT.	INST.	TOTAL	MAT.	INST.	TOTAL	MAT.	INST.	TOTAL
01590	EQUIPMENT RENTAL	.0	109.0	109.0	.0	109.0	109.0	.0	102.4	102.4	.0	91.5	91.5	.0	102.4	102.4	.0	109.0	109.0
02	SITE CONSTRUCTION	101.0	95.9	97.2	101.3	96.4	97.7	102.2	96.0	97.6	96.9	90.9	92.5	87.2	96.0	93.7	103.5	96.3	98.2
03100	CONCRETE FORMS & ACCESSORIES	93.8	102.7	101.5	96.3	94.1	94.4	91.1	104.1	102.4	101.6	140.3	135.2	97.1	103.2	102.4	91.0	102.4	100.9
03200	CONCRETE REINFORCEMENT	99.0	89.3	94.1	99.0	102.5	100.8	100.2	96.4	98.3	100.7	139.3	120.2	97.0	99.8	98.4	98.9	102.5	100.7
03300	CAST-IN-PLACE CONCRETE	90.6	88.1	89.5	91.0	107.5	97.9	112.3	106.8	110.0	113.2	136.6	123.0	98.9	103.5	100.8	92.6	108.3	99.2
03	CONCRETE	87.4	96.2	91.8	87.9	101.6	94.7	108.6	104.0	106.3	111.4	137.9	124.6	97.7	103.1	100.4	88.6	105.5	97.1
04	MASONRY	73.8	98.6	89.3	73.8	106.0	94.0	131.4	92.7	107.1	93.4	134.3	119.1	68.6	91.8	83.2	73.9	107.9	95.2
05	METALS	94.8	112.0	100.4	94.8	119.2	102.8	95.6	105.0	98.7	96.2	122.1	104.7	97.9	106.4	100.7	95.6	118.6	103.1
06	WOOD & PLASTICS	97.7	102.8	100.4	100.5	89.4	94.7	90.5	102.1	96.6	103.3	139.5	122.2	98.8	102.1	100.5	94.9	99.9	97.5
07	THERMAL & MOISTURE PROTECTION	92.3	91.5	91.9	92.3	104.3	98.2	96.9	101.8	99.3	100.0	131.3	115.4	97.2	94.0	95.6	92.3	101.5	96.8
08	DOORS & WINDOWS	86.0	105.8	91.1	86.0	102.0	90.1	90.3	100.5	93.0	103.4	139.3	112.7	96.3	101.4	97.6	86.1	107.7	91.7
09200	PLASTER & GYPSUM BOARD	101.3	102.7	102.2	102.5	89.0	93.9	99.7	102.0	101.2	102.8	140.5	126.7	104.4	102.0	102.9	100.5	99.8	100.1
095,098	CEILINGS & ACOUSTICAL TREATMENT	90.0	102.7	97.6	90.0	89.0	89.4	90.0	102.0	97.2	106.3	140.5	126.8	97.1	102.0	100.1	90.0	99.8	95.9
09600	FLOORING	106.3	101.6	105.1	107.6	110.9	108.5	92.4	102.5	95.1	83.2	127.6	94.9	96.4	84.6	93.3	105.1	104.5	104.9
097,099	WALL FINISHES, PAINTS & COATINGS	99.3	81.9	88.8	99.3	90.4	93.9	91.2	98.7	95.7	80.5	129.5	110.2	91.2	97.6	95.1	99.3	90.4	93.9
09	FINISHES	95.0	100.8	98.0	95.5	95.4	95.5	92.7	103.4	98.3	93.1	137.4	116.1	95.8	99.4	97.7	94.5	100.2	97.5
10 - 14	TOTAL DIV. 10000 - 14000	100.0	93.7	98.5	100.0	92.7	98.5	100.0	94.0	98.7	100.0	126.9	105.7	100.0	93.5	98.6	100.0	94.8	98.9
15	MECHANICAL	95.6	94.2	95.0	95.6	88.3	92.4	95.7	100.0	97.6	100.0	123.2	110.1	100.0	101.4	100.6	100.0	95.8	98.2
16	ELECTRICAL	95.2	92.8	94.0	96.5	91.7	94.2	99.2	97.8	98.5	98.0	119.9	108.7	98.8	91.3	95.1	96.1	91.7	94.0
01 - 16	WEIGHTED AVERAGE	92.8	97.8	95.1	93.1	98.0	95.3	99.0	100.0	99.5	99.7	125.8	111.6	96.5	98.5	97.4	94.2	101.1	97.3

ILLINOIS

DIVISION		EFFINGHAM 624			GALESBURG 614			JOLIET 604			KANKAKEE 609			LA SALLE 613			NORTH SUBURBAN 600 - 603		
		MAT.	INST.	TOTAL	MAT.	INST.	TOTAL	MAT.	INST.	TOTAL	MAT.	INST.	TOTAL	MAT.	INST.	TOTAL	MAT.	INST.	TOTAL
01590	EQUIPMENT RENTAL	.0	102.4	102.4	.0	101.6	101.6	.0	89.3	89.3	.0	89.3	89.3	.0	101.6	101.6	.0	89.3	89.3
02	SITE CONSTRUCTION	94.2	95.8	95.4	95.3	94.9	95.0	96.8	89.9	91.7	90.6	88.7	89.2	94.8	95.4	95.2	96.0	89.5	91.2
03100	CONCRETE FORMS & ACCESSORIES	102.1	103.3	103.1	91.0	107.2	105.0	102.5	131.8	127.8	94.2	111.4	109.1	107.2	106.3	106.5	101.6	126.3	123.0
03200	CONCRETE REINFORCEMENT	100.2	99.5	99.9	99.6	99.9	99.7	100.7	128.0	114.5	101.6	116.4	109.1	99.7	108.7	104.3	100.7	138.1	119.6
03300	CAST-IN-PLACE CONCRETE	98.6	79.4	90.6	99.7	105.3	102.0	113.1	121.2	116.4	105.4	108.4	106.7	99.6	112.7	105.0	113.2	124.3	117.8
03	CONCRETE	98.7	94.8	96.8	99.1	105.5	102.3	111.4	126.6	118.9	104.8	110.9	107.8	100.2	109.4	104.8	111.3	127.0	119.2
04	MASONRY	77.3	89.1	84.7	108.4	108.5	108.5	94.9	121.5	111.6	92.9	125.0	112.9	108.4	105.5	106.6	93.4	121.4	111.0
05	METALS	95.6	105.3	98.7	95.6	107.2	99.4	94.1	114.3	100.7	94.1	106.6	98.2	95.7	113.8	101.6	95.3	117.6	102.5
06	WOOD & PLASTICS	101.3	102.1	101.8	90.4	104.5	97.7	104.9	132.2	119.2	96.2	111.6	104.3	106.5	104.7	105.6	103.3	125.3	114.8
07	THERMAL & MOISTURE PROTECTION	96.6	94.4	95.6	96.3	102.8	99.5	99.8	123.0	111.2	98.7	113.1	105.8	96.6	99.8	98.2	99.9	120.5	110.0
08	DOORS & WINDOWS	90.2	101.4	93.1	89.6	98.5	91.9	101.4	132.5	109.4	93.1	115.8	99.0	89.7	103.3	93.2	101.4	131.4	109.2
09200	PLASTER & GYPSUM BOARD	103.7	102.0	102.6	99.7	104.4	102.7	100.0	133.1	121.0	96.0	111.9	106.1	106.1	104.7	105.2	102.8	125.9	117.5
095,098	CEILINGS & ACOUSTICAL TREATMENT	90.0	102.0	97.2	90.0	104.4	98.6	106.3	133.1	122.3	106.3	111.9	109.6	90.0	104.7	98.8	106.3	125.9	118.0
09600	FLOORING	97.6	102.5	98.9	92.3	103.5	95.3	82.7	116.2	91.6	79.8	111.6	88.2	99.9	103.1	100.8	83.2	116.2	91.9
097,099	WALL FINISHES, PAINTS & COATINGS	91.2	87.6	89.0	91.2	91.2	91.2	78.3	115.5	100.9	78.3	82.8	81.1	91.2	100.2	96.6	80.5	113.8	100.7
09	FINISHES	94.4	102.1	98.4	92.2	105.2	99.0	92.5	127.9	110.9	90.7	107.4	99.4	95.5	104.5	100.2	93.1	123.2	108.8
10 - 14	TOTAL DIV. 10000 - 14000	100.0	93.9	98.7	100.0	92.7	98.4	100.0	123.2	104.9	100.0	106.7	101.4	100.0	90.9	98.1	100.0	112.5	102.7
15	MECHANICAL	95.7	99.9	97.5	95.7	102.3	98.6	100.1	117.7	107.8	95.7	105.2	99.9	95.7	95.7	95.7	100.0	114.3	106.2
16	ELECTRICAL	96.8	92.8	94.8	97.0	86.0	91.6	96.3	106.3	101.2	91.7	102.7	97.0	94.3	102.7	98.4	96.2	109.8	102.8
01 - 16	WEIGHTED AVERAGE	94.9	97.3	96.0	96.2	100.8	98.3	99.0	117.2	107.3	95.4	105.3	99.9	96.5	102.7	99.3	99.1	116.3	106.9

ILLINOIS

DIVISION		PEORIA 615 - 616			QUINCY 623			ROCKFORD 610 - 611			ROCK ISLAND 612			SOUTH SUBURBAN 605			SPRINGFIELD 626 - 627		
		MAT.	INST.	TOTAL	MAT.	INST.	TOTAL	MAT.	INST.	TOTAL	MAT.	INST.	TOTAL	MAT.	INST.	TOTAL	MAT.	INST.	TOTAL
01590	EQUIPMENT RENTAL	.0	101.6	101.6	.0	102.4	102.4	.0	101.6	101.6	.0	101.6	101.6	.0	89.3	89.3	.0	102.4	102.4
02	SITE CONSTRUCTION	95.9	95.0	95.3	93.0	95.6	94.9	95.0	95.6	95.4	93.6	94.5	94.2	96.0	89.5	91.2	93.3	96.0	95.3
03100	CONCRETE FORMS & ACCESSORIES	95.4	108.0	106.3	99.5	100.7	100.6	99.7	113.7	111.9	93.0	104.6	103.1	101.6	126.3	123.0	99.0	103.8	103.2
03200	CONCRETE REINFORCEMENT	97.0	99.8	98.5	99.7	100.1	99.9	97.0	130.2	113.8	99.6	99.7	99.7	100.7	138.1	119.6	97.0	99.8	98.4
03300	CAST-IN-PLACE CONCRETE	96.6	106.9	100.9	98.7	100.9	99.6	98.9	98.8	98.9	97.5	96.1	96.9	113.2	124.3	117.8	91.9	104.2	97.0
03	CONCRETE	96.6	106.4	101.5	98.3	101.1	99.5	97.9	110.8	104.9	97.3	101.1	99.0	111.3	127.0	119.2	94.5	103.6	99.0
04	MASONRY	112.3	108.7	110.1	97.0	91.9	93.8	83.1	110.8	100.5	108.2	98.5	102.1	93.4	121.4	111.0	69.8	92.8	84.3
05	METALS	97.9	107.4	101.0	95.6	106.8	99.3	97.9	123.4	106.2	95.7	106.6	99.3	95.3	117.6	102.5	100.5	106.6	102.5
06	WOOD & PLASTICS	98.8	104.3	101.7	98.7	102.1	100.5	98.8	112.0	105.7	92.4	105.3	99.2	103.3	125.3	114.8	101.5	102.1	101.8
07	THERMAL & MOISTURE PROTECTION	97.1	105.3	101.1	96.6	94.8	95.7	97.1	111.2	104.1	96.3	98.0	97.1	99.9	120.5	110.0	96.7	98.5	97.6
08	DOORS & WINDOWS	96.3	102.6	98.0	91.1	101.4	93.8	96.3	114.8	101.1	89.6	99.5	92.2	101.4	131.4	109.2	96.3	101.4	97.6
09200	PLASTER & GYPSUM BOARD	104.4	104.2	104.3	102.5	102.0	102.2	104.4	112.1	109.3	100.5	105.3	103.5	102.8	125.9	117.5	104.4	102.0	102.9
095,098	CEILINGS & ACOUSTICAL TREATMENT	97.1	104.2	101.4	90.0	105.3	99.2	90.0	112.1	106.1	90.0	105.3	99.2	106.3	125.9	118.0	97.1	102.0	100.1
09600	FLOORING	96.4	99.7	97.3	96.4	96.6	96.4	96.4	93.3	95.6	93.6	89.6	92.5	83.2	116.2	91.9	96.6	84.7	93.4
097,099	WALL FINISHES, PAINTS & COATINGS	91.2	96.3	94.3	91.2	82.3	85.8	91.2	108.8	101.9	91.2	91.2	91.2	80.5	113.8	100.7	91.2	90.2	90.6
09	FINISHES	95.9	105.4	100.8	93.8	98.4	96.2	95.9	109.3	102.9	92.7	100.9	97.0	93.1	123.2	108.8	95.9	98.8	97.4
10 - 14	TOTAL DIV. 10000 - 14000	100.0	95.3	99.0	100.0	94.4	98.8	100.0	103.9	100.8	100.0	90.4	98.0	100.0	112.5	102.7	100.0	93.9	98.7
15	MECHANICAL	100.0	102.1	101.0	95.7	95.7	95.7	100.0	105.7	102.5	95.7	98.1	96.7	100.0	114.3	106.2	100.0	103.7	101.7
16	ELECTRICAL	98.1	88.7	93.5	94.3	79.7	87.2	98.1	110.0	103.9	89.7	87.0	88.4	96.2	109.8	102.8	98.7	89.1	94.0
01 - 16	WEIGHTED AVERAGE	98.7	101.7	100.1	95.6	95.4	95.5	97.4	109.4	102.8	95.1	97.7	96.3	99.1	116.3	106.9	96.8	98.9	97.8

COST INDEXES

City Cost Indexes

INDIANA

DIVISION		ANDERSON 460 MAT.	INST.	TOTAL	BLOOMINGTON 474 MAT.	INST.	TOTAL	COLUMBUS 472 MAT.	INST.	TOTAL	EVANSVILLE 476-477 MAT.	INST.	TOTAL	FORT WAYNE 467-468 MAT.	INST.	TOTAL	GARY 463-464 MAT.	INST.	TOTAL
01590	EQUIPMENT RENTAL	.0	96.7	96.7	.0	85.9	85.9	.0	85.9	85.9	.0	121.3	121.3	.0	96.7	96.7	.0	96.7	96.7
02	SITE CONSTRUCTION	87.6	96.6	94.2	77.1	95.6	90.8	74.3	95.5	90.0	82.1	130.1	117.6	88.6	96.5	94.4	88.2	98.3	95.7
03100	CONCRETE FORMS & ACCESSORIES	96.1	75.6	78.3	99.9	76.1	79.3	93.6	73.2	75.9	92.6	75.8	78.0	94.7	75.0	77.6	96.2	100.6	100.0
03200	CONCRETE REINFORCEMENT	93.8	81.2	87.4	83.9	81.0	82.5	84.3	81.0	82.6	92.0	81.2	86.6	93.8	78.0	85.8	93.8	99.6	96.7
03300	CAST-IN-PLACE CONCRETE	102.4	80.8	93.4	101.6	77.3	91.5	101.1	76.4	90.8	97.0	90.4	94.2	108.8	76.2	95.2	107.0	101.5	104.7
03	CONCRETE	96.7	79.1	87.9	103.1	77.2	90.2	102.4	75.6	89.0	103.4	82.1	92.8	99.6	76.6	88.2	98.9	100.9	99.9
04	MASONRY	91.3	77.5	82.6	90.6	75.9	81.4	90.5	75.9	81.3	86.8	81.2	83.3	91.6	77.7	82.9	92.7	93.1	93.0
05	METALS	90.9	89.1	90.3	93.4	75.1	87.4	93.4	74.7	87.3	86.2	86.3	86.2	90.9	87.1	89.7	90.9	101.1	94.2
06	WOOD & PLASTICS	113.2	74.1	92.8	118.9	74.7	95.8	112.9	70.9	91.0	97.7	72.2	84.4	112.9	73.6	92.4	111.2	100.9	105.8
07	THERMAL & MOISTURE PROTECTION	101.0	76.1	88.7	91.5	80.6	86.1	91.0	80.2	85.7	94.8	84.1	89.5	100.7	82.4	91.7	100.0	98.0	99.0
08	DOORS & WINDOWS	99.3	78.1	93.8	103.2	78.4	96.8	98.6	76.3	92.8	95.6	75.8	90.5	99.3	74.7	92.9	99.3	102.5	100.1
09200	PLASTER & GYPSUM BOARD	96.7	74.0	82.3	98.2	74.7	83.3	94.1	70.9	79.4	91.8	70.6	78.4	95.9	73.5	81.7	91.4	101.5	97.8
095,098	CEILINGS & ACOUSTICAL TREATMENT	89.4	74.0	80.2	81.7	74.7	77.5	81.7	70.9	75.2	86.4	70.6	77.0	89.4	73.5	79.9	89.4	101.6	96.7
09600	FLOORING	87.3	88.6	87.6	103.9	88.6	99.9	98.1	88.6	95.6	97.0	83.6	93.5	87.3	79.9	85.3	87.3	99.4	90.5
097,099	WALL FINISHES, PAINTS & COATINGS	92.1	67.6	77.3	89.8	86.2	87.7	89.8	86.2	87.7	96.5	88.1	91.4	92.1	74.7	81.6	92.1	101.8	98.0
09	FINISHES	88.9	76.8	82.6	94.7	79.2	86.7	92.4	77.0	84.4	92.9	77.9	85.1	88.7	75.6	81.8	88.1	100.5	94.6
10-14	TOTAL DIV. 10000-14000	100.0	81.7	96.1	100.0	81.2	96.0	100.0	80.8	95.9	100.0	84.4	96.7	100.0	83.6	96.5	100.0	98.5	99.7
15	MECHANICAL	99.9	81.0	91.7	99.6	82.9	92.3	95.3	81.8	89.4	99.8	85.0	93.3	99.9	78.6	90.6	99.9	94.7	97.6
16	ELECTRICAL	85.2	92.8	88.9	102.4	80.5	91.7	101.6	83.9	93.0	95.5	83.7	89.7	86.3	80.8	83.6	93.4	102.6	97.9
01-16	WEIGHTED AVERAGE	94.6	83.3	89.5	98.2	80.7	90.3	96.3	80.2	90.0	95.2	86.8	91.4	95.1	80.6	88.5	95.8	98.7	97.1

INDIANA

DIVISION		INDIANAPOLIS 461-462 MAT.	INST.	TOTAL	KOKOMO 469 MAT.	INST.	TOTAL	LAFAYETTE 479 MAT.	INST.	TOTAL	LAWRENCEBURG 470 MAT.	INST.	TOTAL	MUNCIE 473 MAT.	INST.	TOTAL	NEW ALBANY 471 MAT.	INST.	TOTAL
01590	EQUIPMENT RENTAL	.0	91.5	91.5	.0	96.7	96.7	.0	85.9	85.9	.0	108.1	108.1	.0	96.2	96.2	.0	96.0	96.0
02	SITE CONSTRUCTION	87.5	100.4	97.0	84.4	94.5	93.3	75.0	95.5	90.1	72.9	115.4	104.3	77.2	95.8	90.9	69.6	99.6	91.8
03100	CONCRETE FORMS & ACCESSORIES	96.6	86.8	88.1	99.2	75.5	78.7	90.9	73.7	76.0	89.5	70.7	73.2	90.5	75.3	77.3	88.3	68.3	71.0
03200	CONCRETE REINFORCEMENT	93.2	85.4	89.2	84.9	81.3	83.1	83.9	79.1	81.5	83.2	84.4	83.8	92.9	81.1	86.9	84.5	80.5	82.5
03300	CAST-IN-PLACE CONCRETE	98.5	87.6	93.9	101.4	81.3	93.0	101.7	80.9	93.0	95.1	75.5	86.9	106.8	80.1	95.7	98.2	73.4	87.8
03	CONCRETE	98.8	86.4	92.6	93.6	79.2	86.4	102.5	77.0	89.8	95.5	75.7	85.6	101.8	78.7	90.3	101.1	72.9	87.0
04	MASONRY	98.2	86.6	90.9	91.0	80.1	84.2	96.7	77.7	84.8	75.7	75.7	75.7	92.8	77.5	83.2	82.9	67.4	73.2
05	METALS	92.8	80.1	88.7	88.0	88.8	88.2	92.3	73.9	86.3	88.5	89.5	88.8	95.1	88.8	93.0	90.7	82.6	88.1
06	WOOD & PLASTICS	110.1	86.9	97.0	116.5	74.1	94.4	109.3	71.7	89.7	96.4	68.1	81.7	111.0	73.9	91.6	98.5	67.2	82.1
07	THERMAL & MOISTURE PROTECTION	99.9	85.9	93.0	100.5	76.7	88.8	91.0	79.0	85.1	93.5	79.8	86.8	93.1	76.4	84.9	84.9	68.5	76.8
08	DOORS & WINDOWS	106.9	86.3	101.6	93.5	78.1	89.5	97.2	76.2	91.8	97.4	73.4	91.2	98.2	78.0	93.0	95.3	73.1	89.5
09200	PLASTER & GYPSUM BOARD	94.8	86.5	89.5	101.3	74.0	84.0	89.2	71.7	78.1	82.0	67.8	73.0	91.5	74.0	80.4	90.6	66.6	75.4
095,098	CEILINGS & ACOUSTICAL TREATMENT	93.9	86.5	89.4	88.5	74.0	79.8	73.6	71.7	72.4	91.8	67.8	77.4	82.5	74.0	77.4	86.4	66.6	74.6
09600	FLOORING	87.0	93.4	88.7	91.6	81.4	88.9	96.9	88.6	94.7	72.1	88.6	76.5	96.2	88.6	94.2	95.2	66.5	87.6
097,099	WALL FINISHES, PAINTS & COATINGS	92.1	89.3	90.4	92.1	81.6	85.7	89.8	84.2	86.4	91.7	79.6	84.3	89.8	67.6	76.4	96.5	71.9	81.6
09	FINISHES	90.0	88.4	89.1	90.5	76.9	83.4	89.3	77.2	83.0	85.3	74.6	79.8	91.4	76.7	83.7	92.3	68.0	79.7
10-14	TOTAL DIV. 10000-14000	100.0	85.1	96.8	100.0	81.6	96.1	100.0	81.0	96.0	100.0	57.2	90.9	100.0	81.2	96.0	100.0	56.9	90.8
15	MECHANICAL	99.9	88.6	95.0	95.5	81.9	89.6	95.3	78.2	87.8	96.0	78.3	88.3	99.6	81.0	91.4	95.5	74.3	86.2
16	ELECTRICAL	105.1	92.8	99.1	89.3	81.5	85.5	101.1	81.8	91.7	92.9	73.6	83.5	90.1	76.4	83.4	94.3	73.3	84.1
01-16	WEIGHTED AVERAGE	98.8	88.7	94.2	92.9	82.2	88.0	95.9	79.5	88.4	92.2	80.1	86.7	96.1	80.8	89.2	93.7	74.8	85.1

INDIANA / IOWA

DIVISION		SOUTH BEND 465-466 MAT.	INST.	TOTAL	TERRE HAUTE 478 MAT.	INST.	TOTAL	WASHINGTON 475 MAT.	INST.	TOTAL	BURLINGTON 526 MAT.	INST.	TOTAL	CARROLL 514 MAT.	INST.	TOTAL	CEDAR RAPIDS 522-524 MAT.	INST.	TOTAL
01590	EQUIPMENT RENTAL	.0	106.8	106.8	.0	121.3	121.3	.0	121.3	121.3	.0	99.5	99.5	.0	99.5	99.5	.0	95.5	95.5
02	SITE CONSTRUCTION	87.7	96.1	93.9	83.4	130.0	117.9	83.5	128.7	116.9	85.6	94.2	92.0	76.7	93.4	89.1	87.6	94.6	92.8
03100	CONCRETE FORMS & ACCESSORIES	100.3	77.6	80.6	94.1	77.7	79.9	94.8	75.0	77.7	97.2	68.6	72.5	82.5	51.3	55.5	103.6	78.2	81.6
03200	CONCRETE REINFORCEMENT	93.8	76.4	85.0	92.0	80.7	86.3	85.0	65.7	75.3	92.6	79.5	86.0	93.3	53.9	73.4	93.3	82.6	87.9
03300	CAST-IN-PLACE CONCRETE	99.6	82.4	92.4	94.0	90.4	92.5	102.3	91.3	97.7	107.5	53.4	84.9	107.5	60.4	87.8	107.8	80.1	96.2
03	CONCRETE	91.4	80.5	86.0	106.3	82.9	94.6	111.9	79.1	95.5	99.5	66.3	82.9	98.1	56.1	77.2	99.7	80.4	90.1
04	MASONRY	89.8	77.4	82.0	88.7	78.2	82.1	86.9	81.2	83.3	99.3	63.3	76.7	102.2	47.9	68.3	106.2	79.6	89.5
05	METALS	90.9	100.9	94.2	86.9	86.4	86.8	82.2	76.9	80.5	85.1	88.2	86.1	85.2	74.3	81.6	87.2	92.2	88.8
06	WOOD & PLASTICS	113.1	77.0	94.3	99.9	76.2	87.5	100.3	72.1	85.6	102.7	73.5	87.5	86.5	53.1	69.1	109.6	76.9	92.5
07	THERMAL & MOISTURE PROTECTION	99.7	82.3	91.1	94.9	81.5	88.3	95.0	82.4	88.8	98.7	65.9	82.6	99.1	55.3	77.6	100.0	80.3	90.3
08	DOORS & WINDOWS	92.7	77.0	88.6	96.2	79.0	91.7	93.1	73.0	87.9	93.0	69.6	86.9	98.3	49.4	85.7	99.3	80.2	94.4
09200	PLASTER & GYPSUM BOARD	96.3	77.0	84.0	91.8	74.7	81.0	90.1	70.5	77.7	100.6	72.9	83.0	93.7	51.9	67.2	104.4	76.6	86.8
095,098	CEILINGS & ACOUSTICAL TREATMENT	89.4	77.0	82.0	86.4	74.7	79.4	75.7	70.5	72.6	107.0	72.9	86.5	107.0	51.9	74.0	112.4	76.6	90.9
09600	FLOORING	87.3	77.4	84.7	97.0	87.3	94.5	98.2	87.3	95.1	106.5	94.3	103.2	99.3	43.4	84.7	123.0	58.2	105.9
097,099	WALL FINISHES, PAINTS & COATINGS	92.1	81.0	85.4	96.5	87.7	91.2	96.5	88.1	91.4	103.5	87.8	93.9	103.5	41.1	65.6	107.7	77.4	89.4
09	FINISHES	88.8	77.8	83.1	92.9	80.2	86.3	90.5	78.6	84.3	104.7	67.6	85.4	100.6	48.5	73.5	111.8	73.5	91.8
10-14	TOTAL DIV. 10000-14000	100.0	77.6	95.2	100.0	85.4	96.9	100.0	84.3	96.7	100.0	76.1	94.9	100.0	72.5	94.1	100.0	81.8	96.1
15	MECHANICAL	99.9	76.9	89.8	99.8	78.6	90.5	95.5	84.5	90.6	96.0	71.9	85.4	96.0	53.6	77.4	100.4	83.6	93.0
16	ELECTRICAL	99.8	83.4	91.8	94.1	80.6	87.5	94.6	83.6	89.3	99.1	76.4	88.0	99.8	42.0	71.6	97.0	82.1	89.7
01-16	WEIGHTED AVERAGE	95.0	82.6	89.4	96.0	85.3	91.1	94.1	85.2	90.1	96.0	73.7	85.9	95.8	57.0	78.2	98.8	82.7	91.5

COST INDEXES

City Cost Indexes

IOWA

DIVISION		COUNCIL BLUFFS 515			CRESTON 508			DAVENPORT 527 - 528			DECORAH 521			DES MOINES 500 - 503,509			DUBUQUE 520		
		MAT.	INST.	TOTAL	MAT.	INST.	TOTAL	MAT.	INST.	TOTAL	MAT.	INST.	TOTAL	MAT.	INST.	TOTAL	MAT.	INST.	TOTAL
01590	EQUIPMENT RENTAL	.0	95.2	95.2	.0	99.5	99.5	.0	99.5	99.5	.0	99.5	99.5	.0	101.5	101.5	.0	94.1	94.1
02	SITE CONSTRUCTION	92.8	91.2	91.6	76.7	94.4	89.8	85.8	98.4	95.1	84.3	93.9	91.4	79.3	99.9	94.6	85.7	91.5	90.0
03100	CONCRETE FORMS & ACCESSORIES	81.5	58.4	61.5	83.1	71.2	72.8	103.3	90.8	92.5	94.0	48.3	54.4	105.3	78.1	81.8	83.3	67.3	69.5
03200	CONCRETE REINFORCEMENT	95.2	78.5	86.8	92.7	78.2	85.4	93.3	95.7	94.5	92.6	54.1	73.2	93.3	79.3	86.2	92.0	82.3	87.1
03300	CAST-IN-PLACE CONCRETE	112.1	74.3	96.3	107.2	63.0	88.7	103.8	100.9	102.6	104.6	56.9	84.7	108.1	78.8	95.8	105.5	87.9	98.2
03	CONCRETE	101.5	68.8	85.2	97.9	70.5	84.2	97.8	95.7	96.7	97.4	53.7	75.6	98.8	79.4	89.1	96.1	78.2	87.1
04	MASONRY	105.8	73.9	85.8	102.2	52.0	70.7	103.8	87.6	93.7	120.6	44.6	72.9	100.0	80.3	87.6	107.2	72.0	85.1
05	METALS	92.4	89.5	91.5	85.1	85.7	85.3	87.2	101.6	91.9	85.3	73.9	81.6	87.3	92.3	89.0	85.8	91.4	87.6
06	WOOD & PLASTICS	85.2	54.5	69.2	87.1	77.0	81.8	109.6	89.1	98.9	99.3	49.3	73.2	111.1	77.0	93.3	87.2	64.7	75.5
07	THERMAL & MOISTURE PROTECTION	99.2	66.8	83.2	100.2	59.5	80.2	99.4	88.3	93.9	99.0	49.5	74.7	100.3	78.5	89.6	99.5	69.5	84.7
08	DOORS & WINDOWS	98.3	61.6	88.8	111.5	74.0	101.8	99.3	91.9	97.4	96.6	51.6	84.9	99.3	81.1	94.6	98.4	75.4	92.4
09200	PLASTER & GYPSUM BOARD	93.7	53.5	68.3	93.7	76.4	82.7	104.4	88.9	94.6	99.4	48.0	66.8	101.3	76.4	85.5	94.1	64.1	75.1
095,098	CEILINGS & ACOUSTICAL TREATMENT	107.0	53.5	74.9	107.0	76.4	88.7	112.4	88.9	98.3	107.0	48.0	71.6	110.6	76.4	90.1	107.0	64.1	81.3
09600	FLOORING	98.1	47.5	84.7	99.6	43.4	84.7	109.4	82.3	102.2	105.8	61.6	94.1	109.2	43.4	91.8	112.0	40.5	93.1
097,099	WALL FINISHES, PAINTS & COATINGS	99.2	72.1	82.8	103.5	50.7	71.5	103.5	94.9	98.3	103.5	43.5	67.1	103.5	78.5	88.3	106.5	77.4	88.9
09	FINISHES	100.8	56.4	77.7	100.6	64.7	81.9	107.4	89.4	98.0	104.3	50.3	76.2	105.6	71.0	87.6	105.5	62.2	83.0
10 - 14	TOTAL DIV. 10000 - 14000	100.0	74.3	94.5	100.0	77.2	95.2	100.0	86.3	97.1	100.0	54.8	90.4	100.0	82.5	96.3	100.0	78.7	95.5
15	MECHANICAL	100.4	76.9	90.1	96.0	55.9	78.4	100.4	94.9	98.0	96.0	46.5	74.3	100.4	80.7	91.7	100.4	78.9	91.0
16	ELECTRICAL	101.7	81.3	91.7	95.1	54.8	75.5	95.5	89.1	92.4	97.0	41.8	70.1	97.5	83.2	90.5	100.4	70.3	85.7
01 - 16	WEIGHTED AVERAGE	99.1	74.7	88.0	96.6	66.3	82.8	97.8	93.1	95.7	96.8	54.5	77.6	97.7	82.4	90.7	97.7	76.5	88.1

IOWA

DIVISION		FORT DODGE 505			MASON CITY 504			OTTUMWA 525			SHENANDOAH 516			SIBLEY 512			SIOUX CITY 510 - 511		
		MAT.	INST.	TOTAL	MAT.	INST.	TOTAL	MAT.	INST.	TOTAL	MAT.	INST.	TOTAL	MAT.	INST.	TOTAL	MAT.	INST.	TOTAL
01590	EQUIPMENT RENTAL	.0	99.5	99.5	.0	99.5	99.5	.0	94.1	94.1	.0	95.2	95.2	.0	99.5	99.5	.0	99.5	99.5
02	SITE CONSTRUCTION	81.8	93.1	90.1	81.8	93.9	90.8	86.0	89.4	88.5	91.0	89.1	89.6	93.3	93.4	93.3	94.7	95.4	95.2
03100	CONCRETE FORMS & ACCESSORIES	83.9	54.2	58.1	89.6	49.1	54.5	91.7	58.4	62.9	83.6	55.1	58.9	84.3	44.0	49.4	103.6	62.2	67.7
03200	CONCRETE REINFORCEMENT	92.7	53.5	72.9	92.6	69.4	80.9	92.6	78.5	85.5	95.2	76.9	85.9	95.2	66.6	80.7	93.3	67.3	80.2
03300	CAST-IN-PLACE CONCRETE	100.6	48.0	78.6	100.6	57.2	82.5	108.2	63.0	89.3	108.3	52.1	84.9	106.2	52.2	83.6	106.9	57.2	86.1
03	CONCRETE	93.5	53.1	73.3	93.8	57.0	75.4	99.0	64.9	81.9	99.2	59.3	79.3	98.2	52.5	75.4	99.3	62.5	80.9
04	MASONRY	100.1	28.1	54.9	114.2	50.5	74.2	103.8	52.3	72.1	105.5	36.3	62.1	126.0	40.8	72.6	99.1	59.5	74.3
05	METALS	85.2	73.0	81.2	85.2	81.7	84.1	85.1	86.6	85.6	91.7	82.7	88.8	85.3	79.2	83.3	87.2	83.7	86.0
06	WOOD & PLASTICS	87.8	60.4	73.5	93.8	49.3	70.6	96.5	59.6	77.2	87.2	60.5	73.3	88.2	43.8	65.0	109.6	60.7	84.1
07	THERMAL & MOISTURE PROTECTION	99.3	36.9	68.6	98.6	50.9	75.2	99.8	60.1	80.3	98.2	45.7	72.4	98.7	52.4	75.9	99.4	57.4	78.8
08	DOORS & WINDOWS	103.8	51.3	90.2	94.0	55.7	84.1	98.3	65.9	89.9	87.6	61.0	80.8	94.3	50.4	82.9	99.3	62.3	89.7
09200	PLASTER & GYPSUM BOARD	94.1	59.3	72.1	96.1	48.0	65.6	97.8	58.8	73.1	94.1	59.8	72.4	94.1	42.3	61.3	104.4	59.6	76.0
095,098	CEILINGS & ACOUSTICAL TREATMENT	107.0	59.3	78.4	107.0	48.0	71.6	107.0	58.8	78.1	107.0	59.8	78.7	107.0	42.3	68.2	112.4	59.6	80.7
09600	FLOORING	101.2	61.6	90.7	104.0	61.6	92.8	115.9	65.1	102.6	98.9	44.7	84.6	100.3	44.4	85.6	109.9	61.3	97.0
097,099	WALL FINISHES, PAINTS & COATINGS	103.5	18.7	52.0	103.5	39.8	64.8	106.5	78.5	89.5	99.2	28.1	56.1	103.5	41.1	65.6	104.7	63.0	79.4
09	FINISHES	102.0	52.6	76.3	103.1	49.9	75.5	107.2	61.3	83.3	101.1	50.9	75.0	102.9	43.3	71.9	108.6	61.7	84.2
10 - 14	TOTAL DIV. 10000 - 14000	100.0	71.2	93.9	100.0	55.8	90.6	100.0	73.5	94.4	100.0	71.5	93.9	100.0	67.6	93.1	100.0	77.7	95.2
15	MECHANICAL	96.0	68.0	83.7	96.0	69.4	84.3	96.0	72.5	85.7	96.0	35.4	69.5	96.0	42.3	72.5	100.4	82.0	92.3
16	ELECTRICAL	101.3	55.8	79.1	100.4	57.7	79.6	99.0	76.4	88.0	97.0	33.3	65.9	97.0	42.0	70.2	97.0	77.6	87.6
01 - 16	WEIGHTED AVERAGE	96.2	59.4	79.5	96.0	63.3	81.2	96.9	70.7	85.0	96.1	52.2	76.2	97.0	52.9	77.0	98.3	73.3	87.0

IOWA / KANSAS

DIVISION		SPENCER 513			WATERLOO 506 - 507			BELLEVILLE 669			COLBY 677			DODGE CITY 678			EMPORIA 668		
		MAT.	INST.	TOTAL	MAT.	INST.	TOTAL	MAT.	INST.	TOTAL	MAT.	INST.	TOTAL	MAT.	INST.	TOTAL	MAT.	INST.	TOTAL
01590	EQUIPMENT RENTAL	.0	99.5	99.5	.0	99.5	99.5	.0	103.5	103.5	.0	103.5	103.5	.0	103.5	103.5	.0	101.6	101.6
02	SITE CONSTRUCTION	93.4	93.3	93.3	86.3	94.5	92.4	110.0	92.7	97.2	107.4	92.7	96.5	109.9	92.7	97.1	101.0	90.1	92.9
03100	CONCRETE FORMS & ACCESSORIES	91.6	43.7	50.1	104.2	49.3	56.6	93.9	43.4	50.1	101.4	43.4	51.1	93.4	43.4	50.1	84.1	39.8	45.7
03200	CONCRETE REINFORCEMENT	95.2	43.5	69.1	93.3	82.1	87.6	105.1	58.5	81.5	107.7	58.5	82.8	105.1	58.5	81.5	103.6	51.5	77.2
03300	CAST-IN-PLACE CONCRETE	106.2	52.1	83.6	107.8	48.4	83.0	117.3	57.4	92.3	112.8	57.4	89.7	114.8	57.4	90.8	113.4	43.6	84.2
03	CONCRETE	98.7	47.9	73.4	99.7	56.6	78.2	116.4	52.6	84.6	113.8	52.6	83.3	114.8	52.6	83.8	108.8	45.0	76.9
04	MASONRY	126.0	40.8	72.6	100.7	59.5	74.8	94.5	42.2	61.7	95.8	42.2	62.2	104.2	42.2	65.3	100.2	41.7	63.5
05	METALS	85.3	68.6	79.8	87.2	90.2	88.2	94.0	78.2	88.8	94.3	78.2	89.1	95.3	78.2	89.7	93.8	74.6	87.5
06	WOOD & PLASTICS	96.4	43.8	68.9	110.3	49.3	78.5	92.4	43.4	66.8	100.1	43.4	70.5	91.7	43.4	66.5	82.9	39.7	60.3
07	THERMAL & MOISTURE PROTECTION	100.0	46.1	73.5	99.1	53.6	76.7	98.2	49.2	74.1	98.2	49.2	74.1	98.2	49.2	74.1	96.3	41.9	69.6
08	DOORS & WINDOWS	108.0	43.8	91.4	94.9	61.5	86.3	96.1	47.0	83.4	96.2	47.0	83.5	96.2	47.0	83.5	93.7	42.6	80.5
09200	PLASTER & GYPSUM BOARD	97.8	42.3	62.6	104.4	48.0	68.6	101.6	41.7	63.6	105.7	41.7	65.1	100.8	41.7	63.3	96.8	37.9	59.5
095,098	CEILINGS & ACOUSTICAL TREATMENT	107.0	42.3	68.2	112.4	48.0	73.7	88.2	41.7	60.3	88.2	41.7	60.3	88.2	41.7	60.3	88.2	37.9	58.0
09600	FLOORING	103.7	44.4	88.0	111.1	61.6	98.0	95.9	51.2	84.1	99.8	51.2	86.9	95.7	51.2	83.9	90.2	47.5	78.9
097,099	WALL FINISHES, PAINTS & COATINGS	103.5	41.1	65.6	104.7	36.8	63.5	91.2	49.9	66.2	91.2	49.9	66.2	91.2	49.9	66.2	91.2	49.9	66.2
09	FINISHES	104.4	43.3	72.6	108.0	49.5	77.6	94.1	45.1	68.6	95.8	45.1	69.4	93.9	45.1	68.5	91.0	41.6	65.3
10 - 14	TOTAL DIV. 10000 - 14000	100.0	71.3	93.9	100.0	72.1	94.1	100.0	50.2	89.4	100.0	50.2	89.4	100.0	50.2	89.4	100.0	45.9	88.5
15	MECHANICAL	96.0	42.2	72.5	100.4	46.9	77.0	95.7	47.9	74.8	95.7	47.9	74.8	100.0	47.9	77.2	95.7	66.6	82.9
16	ELECTRICAL	98.5	42.0	70.9	97.0	63.8	80.8	102.8	50.3	77.2	102.8	50.3	77.2	99.7	50.3	75.6	100.1	75.0	87.9
01 - 16	WEIGHTED AVERAGE	98.8	51.0	77.1	97.7	62.0	81.5	99.2	54.8	79.0	99.2	54.8	79.0	100.3	54.8	79.6	97.4	59.4	80.1

COST INDEXES

643

KANSAS

DIVISION	FORT SCOTT 667			HAYS 676			HUTCHINSON 675			INDEPENDENCE 673			KANSAS CITY 660 - 662			LIBERAL 679		
	MAT.	INST.	TOTAL	MAT.	INST.	TOTAL	MAT.	INST.	TOTAL	MAT.	INST.	TOTAL	MAT.	INST.	TOTAL	MAT.	INST.	TOTAL
01590 EQUIPMENT RENTAL	.0	102.7	102.7	.0	103.5	103.5	.0	103.5	103.5	.0	103.5	103.5	.0	100.1	100.1	.0	103.5	103.5
02 SITE CONSTRUCTION	97.7	91.7	93.3	112.9	92.7	97.9	90.3	93.0	92.3	110.7	93.0	97.6	91.0	90.9	90.9	113.2	93.0	98.2
03100 CONCRETE FORMS & ACCESSORIES	102.9	73.8	77.7	98.6	43.4	50.7	87.1	30.1	37.7	111.8	30.5	41.4	99.0	91.9	92.8	93.9	29.6	38.2
03200 CONCRETE REINFORCEMENT	103.0	56.5	79.5	105.1	58.5	81.5	105.1	48.7	76.6	104.5	49.3	76.6	99.8	82.4	91.0	106.6	48.7	77.3
03300 CAST-IN-PLACE CONCRETE	105.2	61.7	87.0	88.9	57.4	75.7	82.3	42.2	65.6	115.3	42.9	85.0	90.0	91.6	90.7	88.9	42.0	69.3
03 CONCRETE	104.3	67.2	85.8	105.3	52.6	79.0	88.2	39.7	64.0	116.3	40.2	78.3	96.2	90.5	93.3	107.5	39.4	73.5
04 MASONRY	101.8	60.0	75.6	103.9	42.2	65.2	95.5	40.5	61.0	93.0	51.4	66.9	103.5	87.5	93.5	102.3	40.5	63.5
05 METALS	93.7	78.7	88.8	94.0	78.2	88.8	93.8	74.3	87.4	93.7	75.0	87.6	100.1	94.5	98.3	94.2	73.8	87.5
06 WOOD & PLASTICS	102.5	81.0	91.3	97.1	43.4	69.1	86.0	28.2	55.8	110.7	28.2	67.6	98.5	92.3	95.2	92.4	28.2	58.9
07 THERMAL & MOISTURE PROTECTION	97.4	63.4	80.7	98.6	49.2	74.3	96.8	43.1	70.4	98.3	43.5	71.4	96.1	90.1	93.2	98.8	38.0	68.9
08 DOORS & WINDOWS	93.7	71.3	87.9	96.1	47.0	83.4	96.1	36.3	80.6	93.7	36.3	78.8	95.1	89.4	93.6	96.2	36.3	80.7
09200 PLASTER & GYPSUM BOARD	104.1	80.3	89.0	104.1	41.7	64.5	99.2	26.0	52.8	111.3	26.0	57.2	97.0	91.9	93.8	101.6	26.0	53.7
095,098 CEILINGS & ACOUSTICAL TREATMENT	88.2	80.3	83.4	88.2	41.7	60.3	88.2	26.0	50.9	88.2	26.0	50.9	89.1	91.9	90.5	88.2	26.0	50.9
09600 FLOORING	106.3	51.5	91.8	98.6	51.2	86.0	92.3	51.2	81.4	105.0	51.2	90.8	86.7	93.1	88.4	95.9	51.2	84.1
097,099 WALL FINISHES, PAINTS & COATINGS	93.6	49.9	67.1	91.2	49.9	66.2	91.2	49.9	66.2	91.2	49.9	66.2	99.3	89.5	93.4	91.2	49.9	66.2
09 FINISHES	97.0	68.5	82.2	95.6	45.1	69.3	91.5	35.0	62.1	98.2	35.3	65.5	92.2	91.2	91.7	94.6	35.0	63.6
10 - 14 TOTAL DIV. 10000 - 14000	100.0	62.7	92.1	100.0	50.2	89.4	100.0	46.6	88.6	100.0	47.1	88.7	100.0	84.4	96.7	100.0	46.4	88.6
15 MECHANICAL	95.7	46.1	74.0	95.7	47.9	74.8	95.7	39.1	70.9	95.7	37.6	70.3	99.9	88.2	94.8	95.7	33.0	68.3
16 ELECTRICAL	99.3	75.1	87.5	101.7	50.3	76.7	96.8	68.1	82.8	99.0	75.0	87.3	104.4	102.1	103.3	99.7	39.8	70.5
01 - 16 WEIGHTED AVERAGE	97.5	66.8	83.6	98.6	54.8	78.7	94.5	51.1	74.7	99.0	53.0	78.1	98.7	91.6	95.5	98.4	45.6	74.4

KANSAS / KENTUCKY

DIVISION	SALINA 674			TOPEKA 664 - 666			WICHITA 670 - 672			ASHLAND 411 - 412			BOWLING GREEN 421 - 422			CAMPTON 413 - 414		
	MAT.	INST.	TOTAL	MAT.	INST.	TOTAL	MAT.	INST.	TOTAL	MAT.	INST.	TOTAL	MAT.	INST.	TOTAL	MAT.	INST.	TOTAL
01590 EQUIPMENT RENTAL	.0	103.5	103.5	.0	101.6	101.6	.0	103.5	103.5	.0	98.7	98.7	.0	96.0	96.0	.0	103.5	103.5
02 SITE CONSTRUCTION	99.9	93.0	94.8	92.9	90.4	91.1	94.1	93.4	93.6	103.8	86.3	90.8	69.5	100.3	92.3	77.9	99.5	93.9
03100 CONCRETE FORMS & ACCESSORIES	89.2	46.0	51.8	98.5	45.2	52.3	95.3	58.1	63.1	85.3	105.1	102.5	83.8	82.9	83.0	86.9	34.0	41.1
03200 CONCRETE REINFORCEMENT	104.5	80.0	92.1	97.0	94.5	95.8	97.0	81.1	89.0	86.0	104.4	95.3	83.3	74.5	78.9	84.1	63.9	73.9
03300 CAST-IN-PLACE CONCRETE	99.6	49.1	78.5	91.0	53.6	75.4	86.8	57.6	74.6	89.3	101.9	94.6	88.6	82.0	85.8	98.9	40.3	74.4
03 CONCRETE	102.0	55.1	78.6	94.1	59.0	76.6	91.9	63.5	77.7	96.7	104.5	100.6	94.9	81.1	88.0	98.8	42.9	70.9
04 MASONRY	118.5	40.6	69.7	98.1	64.0	76.7	91.8	65.8	75.5	92.3	106.8	101.4	93.9	79.6	84.9	92.3	30.7	53.7
05 METALS	95.1	88.2	92.8	100.5	96.9	99.3	100.5	90.3	97.2	89.8	112.6	97.2	91.2	80.0	87.5	90.7	69.1	83.6
06 WOOD & PLASTICS	87.9	46.7	66.4	95.7	41.3	67.3	93.7	56.8	74.4	79.6	104.5	92.6	93.3	82.3	87.5	91.7	32.9	61.0
07 THERMAL & MOISTURE PROTECTION	97.5	54.7	76.5	97.4	71.3	84.6	97.1	65.8	81.7	87.3	100.9	94.0	84.7	80.2	82.5	95.2	41.7	68.9
08 DOORS & WINDOWS	96.1	52.5	84.8	96.3	60.2	87.0	96.3	63.5	87.8	93.5	100.0	95.2	95.3	76.0	90.3	96.8	49.9	84.7
09200 PLASTER & GYPSUM BOARD	100.0	45.1	65.2	102.3	39.5	62.5	102.3	55.4	72.6	63.9	104.8	89.8	87.4	82.2	84.1	87.4	30.3	51.2
095,098 CEILINGS & ACOUSTICAL TREATMENT	88.2	45.1	62.3	89.1	39.5	59.4	89.1	55.4	68.9	81.5	104.8	95.5	86.4	82.2	83.9	86.4	30.3	52.8
09600 FLOORING	93.7	37.4	78.8	97.2	47.5	84.1	96.4	79.3	91.9	77.2	106.4	84.9	92.9	63.0	85.0	94.2	43.8	80.9
097,099 WALL FINISHES, PAINTS & COATINGS	91.2	39.3	59.7	91.2	69.5	78.0	91.2	63.4	74.3	99.2	100.5	100.0	96.5	75.6	83.8	96.5	33.6	58.4
09 FINISHES	92.6	42.9	66.7	93.8	46.0	68.9	93.6	62.2	77.3	80.5	105.4	93.4	91.1	78.5	84.5	91.6	34.8	64.2
10 - 14 TOTAL DIV. 10000 - 14000	100.0	66.1	92.8	100.0	70.0	93.6	100.0	70.7	93.8	100.0	90.2	97.9	100.0	64.2	92.4	100.0	56.2	90.7
15 MECHANICAL	100.0	36.3	72.1	100.0	70.6	87.1	100.0	67.9	86.0	95.4	95.2	95.3	99.8	88.6	94.9	95.5	35.7	69.4
16 ELECTRICAL	99.4	72.6	86.3	103.3	75.1	89.5	101.5	72.6	87.4	91.0	84.9	88.0	94.5	84.2	89.5	92.0	27.6	60.6
01 - 16 WEIGHTED AVERAGE	99.0	57.2	80.1	98.5	69.3	85.2	97.7	71.1	85.6	92.6	98.6	95.3	94.4	83.6	89.5	94.2	45.1	71.9

KENTUCKY

DIVISION	CORBIN 407 - 409			COVINGTON 410			ELIZABETHTOWN 427			FRANKFORT 406			HAZARD 417 - 418			HENDERSON 424		
	MAT.	INST.	TOTAL	MAT.	INST.	TOTAL	MAT.	INST.	TOTAL	MAT.	INST.	TOTAL	MAT.	INST.	TOTAL	MAT.	INST.	TOTAL
01590 EQUIPMENT RENTAL	.0	103.5	103.5	.0	108.1	108.1	.0	96.0	96.0	.0	103.5	103.5	.0	103.5	103.5	.0	121.3	121.3
02 SITE CONSTRUCTION	75.4	99.9	93.6	74.3	116.7	105.7	64.5	100.1	90.8	76.6	101.6	95.1	76.0	101.2	94.7	73.1	129.8	115.0
03100 CONCRETE FORMS & ACCESSORIES	83.7	33.7	40.4	82.5	89.1	88.2	78.4	79.6	79.4	94.4	59.2	63.9	83.7	33.9	40.5	90.5	82.4	83.5
03200 CONCRETE REINFORCEMENT	83.7	63.8	73.7	82.8	99.7	91.4	83.7	94.9	89.4	84.9	75.6	80.2	84.5	63.9	74.1	83.4	95.5	89.5
03300 CAST-IN-PLACE CONCRETE	93.5	40.8	71.5	94.6	85.8	90.9	80.1	79.0	79.7	93.5	68.4	83.0	95.1	40.8	72.4	78.3	94.9	85.2
03 CONCRETE	93.9	42.9	68.5	97.1	90.3	93.7	87.2	82.4	84.8	94.9	66.0	80.5	95.5	43.1	69.3	92.2	89.2	90.7
04 MASONRY	92.1	30.9	53.7	107.1	96.0	100.1	77.3	77.0	77.1	91.1	72.1	79.2	89.1	30.9	52.6	98.1	92.1	94.3
05 METALS	90.6	69.6	83.7	88.5	98.1	91.6	90.6	88.6	89.9	98.1	79.9	92.2	90.7	69.8	83.9	82.1	91.5	85.1
06 WOOD & PLASTICS	88.2	32.9	59.3	89.0	86.6	87.7	87.7	79.2	83.3	99.9	57.6	77.9	88.2	32.9	59.3	95.6	79.1	87.0
07 THERMAL & MOISTURE PROTECTION	94.9	41.7	68.8	93.7	94.8	94.2	84.3	79.0	81.7	95.1	68.1	81.8	95.0	41.7	68.8	94.3	93.8	94.0
08 DOORS & WINDOWS	96.1	39.2	81.4	98.5	89.3	96.1	95.3	84.1	92.4	95.4	65.8	87.8	97.3	39.2	82.2	93.6	84.1	91.1
09200 PLASTER & GYPSUM BOARD	85.7	30.3	50.6	78.1	86.9	83.7	85.3	79.0	81.3	91.8	55.7	68.9	85.7	30.3	50.6	87.7	77.8	81.4
095,098 CEILINGS & ACOUSTICAL TREATMENT	86.4	30.3	52.8	89.1	86.9	87.8	86.4	79.0	82.0	86.4	55.7	68.0	86.4	30.3	52.8	75.7	77.8	77.0
09600 FLOORING	92.3	43.8	79.5	69.4	99.5	77.4	89.8	70.9	84.8	97.9	53.5	86.2	92.3	43.8	79.5	96.2	86.3	93.6
097,099 WALL FINISHES, PAINTS & COATINGS	96.5	33.6	58.4	91.7	90.0	90.7	96.5	81.3	87.3	96.5	50.7	68.8	96.5	33.6	58.4	96.5	99.6	98.4
09 FINISHES	90.6	34.8	61.6	83.4	90.8	87.2	89.6	77.8	83.5	93.2	56.4	74.1	90.7	34.8	61.6	89.0	84.6	86.7
10 - 14 TOTAL DIV. 10000 - 14000	100.0	56.2	90.7	100.0	98.7	99.7	100.0	84.4	96.7	100.0	68.4	93.3	100.0	56.2	90.7	100.0	80.1	95.8
15 MECHANICAL	95.5	35.2	69.1	96.1	92.2	94.4	95.7	80.2	88.9	99.9	62.3	83.5	95.5	35.3	69.2	95.7	80.6	89.1
16 ELECTRICAL	92.3	27.7	60.8	94.7	82.9	88.9	91.8	80.7	86.4	94.5	61.0	78.2	92.0	27.7	60.6	93.9	79.6	86.9
01 - 16 WEIGHTED AVERAGE	93.3	44.6	71.2	94.0	93.6	93.8	91.1	82.8	87.3	96.1	68.5	83.6	93.5	44.8	71.4	91.9	89.2	90.7

COST INDEXES

KENTUCKY

DIVISION		LEXINGTON 403 - 405			LOUISVILLE 400 - 402			OWENSBORO 423			PADUCAH 420			PIKEVILLE 415 - 416			SOMERSET 425 - 426		
		MAT.	INST.	TOTAL	MAT.	INST.	TOTAL	MAT.	INST.	TOTAL	MAT.	INST.	TOTAL	MAT.	INST.	TOTAL	MAT.	INST.	TOTAL
01590	EQUIPMENT RENTAL	.0	103.5	103.5	.0	96.0	96.0	.0	121.3	121.3	.0	121.3	121.3	.0	98.7	98.7	.0	103.5	103.5
02	SITE CONSTRUCTION	77.3	102.4	95.8	67.0	100.2	91.5	82.1	129.6	117.2	75.6	129.8	115.7	113.6	84.1	91.8	69.4	99.5	91.7
03100	CONCRETE FORMS & ACCESSORIES	97.6	70.3	74.0	92.5	82.6	83.9	88.7	64.7	67.9	86.2	81.0	81.7	95.2	59.3	64.1	84.3	34.2	40.9
03200	CONCRETE REINFORCEMENT	92.0	96.6	94.3	92.0	97.8	94.9	83.4	93.9	88.7	84.0	88.0	86.0	86.4	56.0	71.1	83.7	63.9	73.7
03300	CAST-IN-PLACE CONCRETE	95.7	80.1	89.2	92.7	80.5	87.6	91.4	76.7	85.3	83.4	89.9	86.1	98.2	72.1	87.3	78.3	72.4	75.8
03	CONCRETE	97.3	79.0	88.2	95.5	84.8	90.1	104.3	74.8	89.5	96.6	85.5	91.1	110.4	64.8	87.6	82.1	53.9	68.0
04	MASONRY	90.8	48.1	64.0	92.4	81.8	85.8	90.8	52.0	66.4	93.8	95.2	94.7	91.3	55.6	68.9	84.7	30.7	50.8
05	METALS	92.6	89.3	91.5	100.0	90.3	96.8	83.1	87.4	84.5	81.0	88.1	83.3	89.7	85.1	88.2	90.6	69.3	83.6
06	WOOD & PLASTICS	104.3	73.3	88.1	102.8	82.3	92.1	93.5	63.6	77.9	91.1	78.9	84.7	89.8	56.5	72.4	89.2	32.9	59.8
07	THERMAL & MOISTURE PROTECTION	95.3	78.7	87.2	85.0	82.0	83.5	94.8	67.5	81.4	94.4	83.8	89.2	88.2	55.2	72.0	94.3	41.7	68.4
08	DOORS & WINDOWS	96.2	80.1	92.0	96.2	87.1	93.8	93.6	75.3	88.8	92.7	79.8	89.4	94.3	59.9	85.4	96.1	49.9	84.2
09200	PLASTER & GYPSUM BOARD	94.8	71.8	80.2	95.9	82.2	87.2	86.5	61.8	70.9	85.7	77.5	80.5	67.5	55.4	59.9	87.0	30.3	51.0
095,098	CEILINGS & ACOUSTICAL TREATMENT	91.8	71.8	79.8	91.8	82.2	86.0	75.7	61.8	67.4	75.7	77.5	76.8	81.5	55.4	65.9	86.4	30.3	52.8
09600	FLOORING	97.9	43.8	83.6	96.4	74.8	90.7	95.3	63.0	86.8	94.0	63.0	85.8	82.3	52.1	74.3	92.6	43.8	79.7
097,099	WALL FINISHES, PAINTS & COATINGS	96.5	59.7	74.2	96.5	84.5	89.2	96.5	99.6	98.4	96.5	75.8	84.0	99.2	66.5	79.4	96.5	33.6	58.4
09	FINISHES	95.0	63.7	78.7	94.4	81.1	87.5	89.0	67.4	77.8	88.2	76.7	82.2	83.1	58.2	70.2	90.4	34.8	61.5
10 - 14	TOTAL DIV. 10000 - 14000	100.0	81.9	96.2	100.0	85.3	96.9	100.0	84.3	96.7	100.0	74.2	94.5	100.0	60.3	91.6	100.0	56.2	90.7
15	MECHANICAL	99.8	48.8	77.5	99.8	87.3	94.3	99.8	53.4	79.5	95.7	86.1	91.5	95.4	56.8	78.5	95.7	35.3	69.3
16	ELECTRICAL	94.9	50.8	73.4	94.9	84.2	89.7	94.0	82.3	88.3	96.2	83.8	90.1	93.8	49.3	72.1	92.4	27.6	60.8
01 - 16	WEIGHTED AVERAGE	95.9	66.9	82.7	96.3	86.4	91.8	94.2	74.3	85.2	92.2	88.8	90.7	95.1	62.1	80.1	91.5	46.5	71.0

LOUISIANA

DIVISION		ALEXANDRIA 713 - 714			BATON ROUGE 707 - 708			HAMMOND 704			LAFAYETTE 705			LAKE CHARLES 706			MONROE 712		
		MAT.	INST.	TOTAL	MAT.	INST.	TOTAL	MAT.	INST.	TOTAL	MAT.	INST.	TOTAL	MAT.	INST.	TOTAL	MAT.	INST.	TOTAL
01590	EQUIPMENT RENTAL	.0	86.8	86.8	.0	86.2	86.2	.0	87.1	87.1	.0	87.1	87.1	.0	86.2	86.2	.0	86.8	86.8
02	SITE CONSTRUCTION	102.8	85.7	90.1	109.3	85.3	91.5	109.5	86.8	92.7	110.3	86.6	92.8	110.9	85.0	91.8	102.8	85.2	89.8
03100	CONCRETE FORMS & ACCESSORIES	80.0	44.3	49.1	96.9	48.4	54.9	77.5	60.6	62.9	96.3	40.5	48.0	97.4	49.2	55.7	79.3	44.9	49.5
03200	CONCRETE REINFORCEMENT	98.4	63.8	80.9	93.6	51.2	72.2	92.3	64.9	78.5	93.6	51.2	72.2	93.6	51.2	72.2	97.4	64.0	80.5
03300	CAST-IN-PLACE CONCRETE	93.9	44.2	73.2	85.9	52.3	71.9	86.9	51.5	72.1	86.4	51.6	71.9	91.0	53.8	75.5	93.9	48.1	74.8
03	CONCRETE	89.0	49.1	69.1	93.5	51.1	72.3	92.5	58.9	75.7	93.7	47.4	70.6	96.0	52.0	74.0	88.8	50.6	69.7
04	MASONRY	115.3	54.0	76.9	96.6	55.6	70.9	96.1	62.6	75.1	96.1	54.5	70.0	94.4	56.6	70.7	110.2	58.1	77.5
05	METALS	85.7	74.4	82.0	95.7	65.5	85.8	90.8	73.3	85.1	90.1	65.8	82.2	90.1	65.7	82.1	85.7	73.0	81.5
06	WOOD & PLASTICS	82.5	43.9	62.3	103.8	48.3	74.8	82.5	62.8	72.2	103.8	38.4	69.7	102.0	49.0	74.3	81.7	44.3	62.2
07	THERMAL & MOISTURE PROTECTION	99.4	54.2	77.2	97.4	55.2	76.7	100.0	64.4	82.5	100.8	53.1	77.4	100.7	55.3	78.4	99.4	57.0	78.5
08	DOORS & WINDOWS	99.3	50.0	86.6	104.3	48.6	89.9	99.5	65.3	90.7	104.3	43.3	88.5	104.3	50.6	90.4	99.3	55.0	87.9
09200	PLASTER & GYPSUM BOARD	79.1	43.0	56.2	99.1	47.5	66.4	90.0	62.3	72.4	99.1	37.3	60.0	99.1	48.2	66.8	78.7	43.4	56.3
095,098	CEILINGS & ACOUSTICAL TREATMENT	90.3	43.0	61.9	96.7	47.5	67.2	97.4	62.3	76.4	95.7	37.3	60.7	96.7	48.2	67.6	90.3	43.4	62.2
09600	FLOORING	105.0	68.9	95.5	109.4	67.6	98.4	97.7	60.5	87.9	109.7	45.3	92.6	109.7	52.8	94.6	104.6	43.6	88.5
097,099	WALL FINISHES, PAINTS & COATINGS	96.4	42.1	63.5	98.7	48.3	68.1	98.7	63.9	77.6	98.7	48.3	68.1	98.7	44.5	65.8	96.4	50.3	68.4
09	FINISHES	93.2	48.1	69.7	102.1	51.8	75.9	97.5	60.9	78.5	101.9	41.1	70.3	102.2	49.0	74.5	93.0	44.3	67.7
10 - 14	TOTAL DIV. 10000 - 14000	100.0	71.6	94.0	100.0	75.3	94.7	100.0	68.3	93.3	100.0	73.6	94.4	100.0	75.5	94.8	100.0	65.4	92.6
15	MECHANICAL	100.1	38.5	73.1	100.0	43.7	75.4	95.7	60.6	80.3	100.0	47.9	77.2	100.0	59.3	82.2	100.1	52.0	79.1
16	ELECTRICAL	92.4	56.4	74.8	92.8	50.4	72.1	92.1	55.1	74.0	93.2	60.0	77.0	92.8	63.1	78.4	94.1	55.9	75.4
01 - 16	WEIGHTED AVERAGE	95.6	54.7	77.0	98.4	55.1	78.7	95.4	63.9	81.1	97.7	55.0	78.3	97.8	60.0	80.7	95.5	57.5	78.2

LOUISIANA / MAINE

DIVISION		NEW ORLEANS 700 - 701			SHREVEPORT 710 - 711			THIBODAUX 703			AUGUSTA 043			BANGOR 044			BATH 045		
		MAT.	INST.	TOTAL	MAT.	INST.	TOTAL	MAT.	INST.	TOTAL	MAT.	INST.	TOTAL	MAT.	INST.	TOTAL	MAT.	INST.	TOTAL
01590	EQUIPMENT RENTAL	.0	87.4	87.4	.0	86.8	86.8	.0	87.1	87.1	.0	101.6	101.6	.0	101.6	101.6	.0	101.6	101.6
02	SITE CONSTRUCTION	113.4	87.8	94.5	101.5	85.1	89.3	112.0	86.7	93.3	83.4	102.3	97.4	83.3	100.7	96.2	81.5	102.4	96.9
03100	CONCRETE FORMS & ACCESSORIES	97.2	69.5	73.2	98.6	47.6	54.4	90.8	55.1	59.9	99.6	64.9	69.6	93.9	80.3	82.1	89.2	65.1	68.3
03200	CONCRETE REINFORCEMENT	93.6	65.0	79.2	96.9	63.7	80.1	92.3	64.9	78.4	88.1	112.7	100.5	88.1	113.0	100.7	87.2	112.7	100.1
03300	CAST-IN-PLACE CONCRETE	90.1	60.0	77.5	92.6	51.8	75.5	93.3	55.2	77.4	81.5	63.2	73.9	81.5	65.1	74.7	81.5	63.2	73.9
03	CONCRETE	95.5	65.8	80.7	88.8	53.0	70.9	97.5	57.7	77.6	98.6	73.3	86.0	97.7	80.7	89.2	97.9	73.4	85.6
04	MASONRY	96.3	64.3	76.2	102.4	51.1	70.2	120.3	61.4	83.4	96.1	52.3	68.6	113.7	60.2	80.1	120.7	51.8	77.5
05	METALS	103.1	75.1	94.0	82.1	72.2	78.8	90.9	72.9	85.0	87.9	84.9	86.9	87.5	83.1	86.1	86.4	85.0	85.9
06	WOOD & PLASTICS	98.4	72.7	85.0	102.2	48.2	74.0	91.5	55.2	72.5	98.5	63.4	80.2	92.3	82.6	87.2	86.9	63.4	74.6
07	THERMAL & MOISTURE PROTECTION	101.6	65.2	83.7	98.3	55.6	77.3	101.8	61.2	81.8	102.1	52.5	77.7	102.0	59.5	81.1	101.9	56.9	79.8
08	DOORS & WINDOWS	104.9	69.4	95.7	97.2	52.5	85.6	105.3	61.2	93.9	103.9	62.1	93.1	103.8	74.3	96.2	103.8	63.9	93.5
09200	PLASTER & GYPSUM BOARD	97.9	72.5	81.8	85.4	47.4	61.3	95.2	54.5	69.4	102.8	61.4	76.6	98.9	81.2	87.6	95.6	61.4	73.9
095,098	CEILINGS & ACOUSTICAL TREATMENT	95.7	72.5	81.8	91.2	47.4	65.0	97.4	54.5	71.7	92.2	61.4	73.7	89.5	81.2	84.5	87.1	61.4	71.9
09600	FLOORING	110.1	51.8	94.7	115.6	64.6	102.1	105.9	57.6	93.1	99.8	43.9	85.0	97.3	54.7	86.1	95.2	43.9	81.7
097,099	WALL FINISHES, PAINTS & COATINGS	100.5	63.9	78.3	96.4	42.1	63.5	100.5	63.9	78.3	92.8	36.4	58.6	92.8	33.5	56.8	92.8	36.4	58.6
09	FINISHES	102.3	65.8	83.3	97.4	49.7	72.6	100.9	56.0	77.6	99.1	57.1	77.3	97.2	70.9	83.5	95.6	57.1	75.6
10 - 14	TOTAL DIV. 10000 - 14000	100.0	80.3	95.8	100.0	65.7	92.7	100.0	77.0	95.1	100.0	63.9	92.3	100.0	83.0	96.4	100.0	63.9	92.3
15	MECHANICAL	100.0	65.9	85.1	100.1	54.6	80.2	95.7	60.6	80.3	100.1	72.3	88.0	100.1	71.5	87.6	95.7	72.3	85.5
16	ELECTRICAL	94.0	70.5	82.6	92.9	67.0	80.3	90.8	66.4	78.9	100.2	80.9	90.8	98.5	80.9	89.9	96.7	80.9	89.0
01 - 16	WEIGHTED AVERAGE	100.2	69.8	86.4	94.7	59.7	78.8	98.1	64.4	82.8	97.7	72.3	86.2	98.0	76.8	88.3	96.7	72.4	85.7

COST INDEXES

645

MAINE

DIVISION		HOULTON 047			KITTERY 039			LEWISTON 042			MACHIAS 046			PORTLAND 040 - 041			ROCKLAND 048		
		MAT.	INST.	TOTAL	MAT.	INST.	TOTAL	MAT.	INST.	TOTAL	MAT.	INST.	TOTAL	MAT.	INST.	TOTAL	MAT.	INST.	TOTAL
01590	EQUIPMENT RENTAL	.0	101.6	101.6	.0	101.6	101.6	.0	101.6	101.6	.0	101.6	101.6	.0	101.6	101.6	.0	101.6	101.6
02	SITE CONSTRUCTION	82.9	102.4	97.3	77.9	102.4	96.0	82.0	100.7	95.8	82.4	102.3	97.1	80.8	100.7	95.5	79.1	102.3	96.3
03100	CONCRETE FORMS & ACCESSORIES	97.7	79.4	81.8	89.8	65.1	68.4	99.7	80.3	82.9	94.6	64.9	68.9	98.8	80.3	82.7	95.7	56.1	61.4
03200	CONCRETE REINFORCEMENT	88.1	112.7	100.5	86.6	112.7	99.8	108.2	113.0	110.7	88.1	112.7	100.5	108.2	113.0	110.7	88.1	112.8	100.6
03300	CAST-IN-PLACE CONCRETE	81.5	63.3	73.9	81.5	63.2	73.9	92.8	65.1	81.2	81.5	63.2	73.9	84.5	65.1	76.4	83.2	63.2	74.8
03	CONCRETE	98.7	79.8	89.3	93.2	73.4	83.3	102.8	80.7	91.7	98.3	73.3	85.8	98.8	80.7	89.7	95.3	69.4	82.4
04	MASONRY	96.5	51.8	68.5	110.8	51.8	73.8	97.9	60.2	74.2	96.5	51.8	68.5	95.4	60.2	73.3	90.2	53.2	67.0
05	METALS	86.6	85.1	86.1	86.3	85.0	85.9	90.7	83.1	88.2	86.6	84.8	86.0	92.0	83.1	89.1	86.5	85.0	86.0
06	WOOD & PLASTICS	96.4	82.6	89.2	87.4	63.4	74.8	98.5	82.6	90.2	93.3	63.4	77.7	98.5	82.6	90.2	94.2	51.4	71.9
07	THERMAL & MOISTURE PROTECTION	102.1	58.9	80.9	101.5	56.9	79.5	101.8	59.5	81.0	102.0	56.8	79.8	101.6	59.5	80.9	101.7	56.0	79.2
08	DOORS & WINDOWS	103.9	74.3	96.2	103.8	63.9	93.4	107.1	74.3	98.6	103.9	61.6	92.9	107.1	74.3	98.6	103.8	55.1	91.2
09200	PLASTER & GYPSUM BOARD	101.6	81.2	88.6	95.6	61.4	73.9	105.2	81.2	89.9	100.4	61.4	75.7	105.2	81.2	89.9	99.7	49.1	67.6
095,098	CEILINGS & ACOUSTICAL TREATMENT	90.2	81.2	84.8	87.5	61.4	71.9	101.1	81.2	89.1	90.2	61.4	72.9	101.1	81.2	89.1	87.5	49.1	64.5
09600	FLOORING	98.7	43.9	84.2	95.5	43.9	81.8	99.8	54.7	87.9	97.7	21.6	77.6	99.8	54.7	87.9	98.0	43.9	83.7
097,099	WALL FINISHES, PAINTS & COATINGS	92.8	36.4	58.6	92.8	36.4	58.6	92.8	33.5	56.8	92.8	36.4	58.6	92.8	33.5	56.8	92.8	36.4	58.6
09	FINISHES	98.2	68.4	82.7	95.3	57.1	75.5	101.4	70.9	85.5	97.7	53.2	74.6	101.4	70.9	85.6	96.7	50.1	72.5
10 - 14	TOTAL DIV. 10000 - 14000	100.0	66.4	92.9	100.0	63.9	92.3	100.0	83.0	96.4	100.0	63.9	92.3	100.0	83.0	96.4	100.0	62.4	92.0
15	MECHANICAL	95.7	72.4	85.5	95.7	72.3	85.5	100.1	71.5	87.6	95.7	72.3	85.5	100.1	71.5	87.6	95.7	72.3	85.5
16	ELECTRICAL	100.3	80.9	90.8	96.7	80.9	89.0	100.3	80.9	90.8	100.3	80.9	90.8	100.4	80.9	90.9	100.2	80.9	90.8
01 - 16	WEIGHTED AVERAGE	96.5	75.5	87.0	95.6	72.4	85.1	99.2	76.8	89.0	96.3	71.8	85.2	98.8	76.8	88.8	95.5	70.6	84.2

MAINE / MARYLAND

DIVISION		MAINE WATERVILLE 049			ANNAPOLIS 214			BALTIMORE 210 - 212			COLLEGE PARK 207 - 208			CUMBERLAND 215			EASTON 216		
		MAT.	INST.	TOTAL	MAT.	INST.	TOTAL	MAT.	INST.	TOTAL	MAT.	INST.	TOTAL	MAT.	INST.	TOTAL	MAT.	INST.	TOTAL
01590	EQUIPMENT RENTAL	.0	101.6	101.6	.0	99.2	99.2	.0	103.4	103.4	.0	102.2	102.2	.0	99.2	99.2	.0	99.2	99.2
02	SITE CONSTRUCTION	83.1	102.3	97.3	96.6	88.8	90.8	97.6	93.3	94.4	97.6	88.2	90.6	89.0	88.6	88.7	95.5	85.6	88.2
03100	CONCRETE FORMS & ACCESSORIES	88.6	41.4	47.7	101.6	59.0	64.7	101.7	74.8	78.4	82.8	69.0	70.9	91.4	79.3	80.9	88.9	36.0	43.1
03200	CONCRETE REINFORCEMENT	88.1	112.7	100.5	95.9	87.7	91.8	95.9	87.8	91.8	105.6	79.4	92.3	84.8	76.4	80.6	84.0	21.7	52.5
03300	CAST-IN-PLACE CONCRETE	81.5	63.2	73.9	98.3	77.7	89.7	98.3	78.7	90.1	115.8	72.2	97.6	87.8	83.1	85.9	97.5	44.6	75.4
03	CONCRETE	99.3	62.8	81.1	99.5	72.3	85.9	99.5	79.7	89.6	110.9	73.4	92.2	88.8	81.0	84.9	96.1	38.2	67.2
04	MASONRY	106.9	52.3	72.7	93.3	74.6	81.6	92.8	74.6	81.4	105.4	70.1	83.3	91.6	75.9	81.8	104.9	38.5	63.2
05	METALS	86.6	84.8	86.0	99.2	98.4	99.0	100.3	99.3	99.9	93.2	95.8	94.1	96.4	92.1	95.0	96.6	62.7	85.5
06	WOOD & PLASTICS	86.1	31.7	57.7	97.9	54.7	75.3	97.9	75.7	86.3	79.1	71.0	74.8	87.2	78.8	82.8	84.3	36.3	59.2
07	THERMAL & MOISTURE PROTECTION	102.1	52.7	77.8	94.2	79.2	86.9	94.2	81.7	88.1	97.9	75.5	86.9	93.8	79.7	86.9	93.9	42.7	68.7
08	DOORS & WINDOWS	103.9	44.9	88.6	91.9	72.4	86.8	93.8	83.3	91.1	97.1	75.3	91.5	93.0	84.5	90.8	91.1	33.2	76.1
09200	PLASTER & GYPSUM BOARD	95.9	28.9	53.4	108.7	53.8	73.9	108.7	75.3	87.5	91.4	70.2	78.0	104.7	78.6	88.2	103.0	34.9	59.9
095,098	CEILINGS & ACOUSTICAL TREATMENT	90.2	28.9	53.4	104.1	53.8	74.0	104.1	75.3	86.8	90.2	70.2	78.3	104.1	78.6	88.8	104.1	34.9	62.6
09600	FLOORING	95.0	43.9	81.5	92.8	82.2	90.0	92.8	82.2	90.0	104.4	67.4	94.6	88.3	78.8	85.8	87.4	16.7	68.7
097,099	WALL FINISHES, PAINTS & COATINGS	92.8	36.4	58.6	96.0	71.1	80.9	96.0	71.1	80.9	113.0	83.8	95.3	96.0	56.2	71.8	96.0	26.9	54.0
09	FINISHES	96.4	38.5	66.3	97.2	62.7	79.3	97.3	75.2	85.8	94.8	70.0	81.9	94.9	76.5	85.3	94.7	32.4	62.3
10 - 14	TOTAL DIV. 10000 - 14000	100.0	59.9	91.5	100.0	84.0	96.6	100.0	87.2	97.3	100.0	71.3	93.9	100.0	90.1	97.9	100.0	52.5	89.9
15	MECHANICAL	95.7	72.3	85.5	100.0	85.6	93.7	100.0	85.7	93.7	95.6	82.1	89.7	95.6	74.8	86.5	95.6	29.0	66.5
16	ELECTRICAL	100.3	80.9	90.8	105.0	84.5	95.0	107.1	90.7	99.1	96.8	92.6	94.7	103.8	82.1	93.2	103.3	59.3	81.8
01 - 16	WEIGHTED AVERAGE	96.8	67.3	83.4	98.8	79.9	90.2	99.4	84.8	92.7	98.0	80.6	90.1	95.5	80.8	88.8	96.9	45.4	73.5

MARYLAND / MASSACHUSETTS

DIVISION		ELKTON 219			HAGERSTOWN 217			SALISBURY 218			SILVER SPRING 209			WALDORF 206			BOSTON 020 - 022, 024		
		MAT.	INST.	TOTAL	MAT.	INST.	TOTAL	MAT.	INST.	TOTAL	MAT.	INST.	TOTAL	MAT.	INST.	TOTAL	MAT.	INST.	TOTAL
01590	EQUIPMENT RENTAL	.0	99.2	99.2	.0	99.2	99.2	.0	99.2	99.2	.0	93.6	93.6	.0	93.6	93.6	.0	108.4	108.4
02	SITE CONSTRUCTION	83.5	86.7	85.9	87.4	88.9	88.5	95.5	85.8	88.3	86.7	80.7	82.3	93.0	81.0	84.1	88.4	108.9	103.6
03100	CONCRETE FORMS & ACCESSORIES	96.1	66.6	70.5	90.3	76.1	78.0	105.4	34.0	43.6	92.0	63.0	66.9	99.6	55.8	61.7	105.1	135.6	131.5
03200	CONCRETE REINFORCEMENT	84.0	64.0	73.9	84.8	76.3	80.5	84.0	63.8	73.8	104.5	80.1	92.2	105.2	63.8	84.2	107.1	143.7	125.6
03300	CAST-IN-PLACE CONCRETE	78.9	52.1	67.7	83.6	61.7	74.4	97.5	43.9	75.1	118.6	69.0	97.9	132.8	62.3	103.4	105.2	143.5	121.2
03	CONCRETE	81.9	61.9	72.0	85.3	72.2	78.8	97.2	44.3	70.8	108.9	69.5	89.3	119.9	60.9	90.4	111.7	138.9	125.3
04	MASONRY	91.2	42.4	60.6	97.0	75.9	83.8	104.3	36.1	61.5	104.2	70.5	83.1	88.8	66.7	74.9	116.7	144.8	134.3
05	METALS	96.7	74.5	89.4	96.5	91.4	94.8	96.7	65.5	86.5	97.5	91.6	95.6	97.6	82.2	92.5	103.3	125.9	110.7
06	WOOD & PLASTICS	92.0	73.6	82.4	85.9	75.6	80.5	102.1	32.1	65.5	86.2	63.2	74.2	93.7	56.5	74.3	102.3	136.3	120.0
07	THERMAL & MOISTURE PROTECTION	93.4	51.8	73.0	93.4	71.6	82.7	94.3	60.3	77.6	100.1	80.0	90.2	100.8	75.7	88.5	102.4	141.9	121.8
08	DOORS & WINDOWS	91.1	60.6	83.2	91.1	77.1	87.5	91.4	27.8	74.9	90.7	71.3	85.7	91.5	55.6	82.2	103.1	135.4	111.4
09200	PLASTER & GYPSUM BOARD	106.3	73.3	85.4	103.5	75.3	85.6	110.7	30.6	59.9	98.7	63.2	76.2	101.5	56.4	72.9	107.8	136.5	126.0
095,098	CEILINGS & ACOUSTICAL TREATMENT	104.1	73.3	85.6	105.1	75.3	87.2	104.1	30.6	60.0	99.3	63.2	77.7	99.3	56.4	73.6	100.1	136.5	121.9
09600	FLOORING	90.2	63.9	83.2	87.9	78.8	85.5	94.4	63.9	86.4	111.6	67.4	99.9	116.4	63.9	102.5	100.4	153.6	114.5
097,099	WALL FINISHES, PAINTS & COATINGS	96.0	75.2	83.3	96.0	41.1	62.7	96.0	30.3	56.1	118.5	83.8	97.4	118.5	70.8	89.5	95.9	150.7	129.1
09	FINISHES	95.4	68.2	81.2	94.7	72.9	83.4	98.0	38.1	66.8	97.0	65.3	80.5	99.1	58.2	77.8	101.7	141.0	122.1
10 - 14	TOTAL DIV. 10000 - 14000	100.0	58.8	91.2	100.0	89.7	97.8	100.0	51.3	89.6	100.0	67.9	93.2	100.0	55.1	90.5	100.0	122.5	104.8
15	MECHANICAL	95.6	75.6	86.8	100.0	85.7	93.7	95.6	64.1	81.8	95.6	81.9	89.6	95.6	74.8	86.5	100.1	125.8	111.4
16	ELECTRICAL	105.0	65.8	85.9	103.6	82.1	93.1	102.1	62.4	82.7	92.7	92.6	92.6	90.6	92.6	91.5	99.3	123.3	111.0
01 - 16	WEIGHTED AVERAGE	94.6	67.2	82.2	96.0	80.8	89.1	97.4	54.8	78.0	97.3	78.2	88.6	98.0	73.1	86.7	102.9	130.5	115.4

		MASSACHUSETTS																
	DIVISION	BROCKTON			BUZZARDS BAY			FALL RIVER			FITCHBURG			FRAMINGHAM			GREENFIELD		
		023			025			027			014			017			013		
		MAT.	INST.	TOTAL	MAT.	INST.	TOTAL	MAT.	INST.	TOTAL	MAT.	INST.	TOTAL	MAT.	INST.	TOTAL	MAT.	INST.	TOTAL
01590	EQUIPMENT RENTAL	.0	103.8	103.8	.0	103.8	103.8	.0	105.0	105.0	.0	101.6	101.6	.0	102.9	102.9	.0	101.6	101.6
02	SITE CONSTRUCTION	86.1	104.8	100.0	77.1	104.8	97.6	85.2	105.0	99.8	78.6	104.7	98.0	75.5	104.4	96.9	82.1	103.7	98.0
03100	CONCRETE FORMS & ACCESSORIES	104.7	120.9	118.8	101.8	120.9	118.3	104.7	121.2	119.0	95.5	116.7	113.8	103.0	120.7	118.3	93.6	105.1	103.5
03200	CONCRETE REINFORCEMENT	108.2	142.8	125.7	86.8	125.6	106.4	108.2	125.6	117.0	86.8	139.2	113.3	86.7	143.3	115.3	90.9	121.1	106.2
03300	CAST-IN-PLACE CONCRETE	100.2	129.2	112.3	83.2	140.8	107.3	96.9	141.6	115.5	87.2	138.9	108.8	87.2	108.3	96.0	89.7	116.6	100.9
03	CONCRETE	109.0	127.2	118.1	92.1	127.9	110.0	107.4	128.4	117.9	90.2	127.8	108.9	93.3	120.0	106.6	94.1	111.5	102.8
04	MASONRY	112.6	127.3	121.9	103.0	140.1	126.2	112.4	140.0	129.7	99.2	138.0	123.5	105.3	138.2	125.9	103.7	112.4	109.2
05	METALS	100.0	121.6	107.1	95.3	115.2	101.8	100.0	115.7	105.2	95.2	117.5	102.5	95.2	121.9	103.9	96.9	105.8	99.8
06	WOOD & PLASTICS	101.3	120.4	111.3	98.1	120.4	109.8	101.3	120.8	111.4	94.0	114.8	104.9	100.6	120.1	110.8	91.9	104.9	98.7
07	THERMAL & MOISTURE PROTECTION	102.1	130.1	115.9	101.3	128.8	114.8	102.1	127.9	114.8	101.4	128.5	114.7	101.5	129.8	115.4	101.4	110.5	105.9
08	DOORS & WINDOWS	101.0	123.2	106.8	96.5	119.7	102.5	101.0	119.9	105.9	105.9	118.9	109.2	96.9	122.8	103.6	105.7	110.5	107.0
09200	PLASTER & GYPSUM BOARD	102.5	120.1	113.6	97.4	120.1	111.8	102.5	120.1	113.6	98.9	114.3	108.6	101.3	120.1	113.2	100.7	104.2	102.9
095,098	CEILINGS & ACOUSTICAL TREATMENT	102.1	120.1	112.9	88.5	120.1	107.4	102.1	120.1	112.9	87.5	114.3	103.6	87.5	120.1	107.0	99.1	104.2	102.1
09600	FLOORING	100.9	153.6	114.8	99.4	153.6	113.7	100.7	153.6	114.7	97.3	153.6	112.2	99.3	153.6	113.6	96.5	110.9	100.3
097,099	WALL FINISHES, PAINTS & COATINGS	95.3	119.2	109.8	95.3	119.2	109.8	95.3	119.2	109.8	92.8	119.2	108.8	94.1	119.2	109.3	92.8	99.0	96.6
09	FINISHES	101.5	127.0	114.8	96.5	127.0	112.4	101.5	127.2	114.9	95.8	123.7	110.3	96.6	126.7	112.2	98.9	105.7	102.4
10 - 14	TOTAL DIV. 10000 - 14000	100.0	118.7	104.0	100.0	118.7	104.0	100.0	119.6	104.2	100.0	103.0	100.6	100.0	117.9	103.8	100.0	103.7	100.8
15	MECHANICAL	100.1	105.5	102.5	95.7	105.3	99.9	100.1	105.4	102.4	94.3	108.8	101.7	96.3	122.7	107.8	96.3	97.8	96.9
16	ELECTRICAL	98.3	96.3	97.3	95.8	102.2	98.9	97.8	102.2	99.9	100.7	104.5	102.5	97.6	123.3	110.1	100.7	88.7	94.9
01 - 16	WEIGHTED AVERAGE	101.4	115.5	107.8	95.8	116.9	105.4	101.1	117.0	108.4	97.0	116.9	106.0	96.4	122.8	108.4	98.3	103.3	100.5

		MASSACHUSETTS																
	DIVISION	HYANNIS			LAWRENCE			LOWELL			NEW BEDFORD			PITTSFIELD			SPRINGFIELD		
		026			019			018			027			012			010 - 011		
		MAT.	INST.	TOTAL	MAT.	INST.	TOTAL	MAT.	INST.	TOTAL	MAT.	INST.	TOTAL	MAT.	INST.	TOTAL	MAT.	INST.	TOTAL
01590	EQUIPMENT RENTAL	.0	103.8	103.8	.0	103.8	103.8	.0	101.6	101.6	.0	105.0	105.0	.0	101.6	101.6	.0	101.6	101.6
02	SITE CONSTRUCTION	82.7	104.8	99.1	86.7	104.8	100.1	85.8	104.8	99.9	84.8	105.0	99.7	86.9	103.6	99.2	86.3	103.7	99.2
03100	CONCRETE FORMS & ACCESSORIES	95.4	120.9	117.5	104.5	121.4	119.1	101.2	121.6	118.9	104.7	121.2	119.0	101.2	101.2	101.2	101.4	109.8	108.7
03200	CONCRETE REINFORCEMENT	86.8	125.6	106.4	107.4	130.6	119.1	108.2	130.7	119.6	108.2	125.6	117.0	90.2	105.9	98.1	108.2	124.8	116.6
03300	CAST-IN-PLACE CONCRETE	91.4	140.8	112.0	101.0	139.5	117.1	91.8	139.6	111.8	94.0	141.6	113.9	100.2	116.0	106.8	95.5	117.3	104.6
03	CONCRETE	98.6	127.9	113.3	109.2	128.7	118.9	100.0	128.6	114.3	106.1	128.4	117.2	101.0	106.7	103.9	101.8	114.5	108.2
04	MASONRY	111.5	140.1	129.4	111.9	138.4	128.5	97.8	138.5	123.3	112.2	140.0	129.6	98.4	112.1	107.0	98.1	114.1	108.2
05	METALS	96.3	115.2	102.5	97.3	117.6	103.9	97.3	114.9	103.0	100.0	115.7	105.2	97.1	99.4	97.8	99.9	107.5	102.4
06	WOOD & PLASTICS	91.3	120.4	106.5	101.3	120.4	111.3	100.4	120.4	110.9	101.3	120.8	111.4	100.4	100.4	100.4	100.4	110.6	105.7
07	THERMAL & MOISTURE PROTECTION	101.6	128.8	115.0	102.1	134.4	118.0	101.8	134.4	117.8	102.1	127.9	114.8	101.9	109.1	105.4	101.8	111.0	106.3
08	DOORS & WINDOWS	97.2	119.7	103.0	101.0	119.8	105.9	107.1	119.8	110.4	101.0	119.9	105.9	107.1	103.9	106.2	107.1	114.5	109.0
09200	PLASTER & GYPSUM BOARD	94.2	120.1	110.6	105.2	120.1	114.6	105.2	120.1	114.6	102.5	120.1	113.6	105.2	99.4	101.5	105.2	110.0	108.2
095,098	CEILINGS & ACOUSTICAL TREATMENT	90.5	120.1	108.2	101.1	120.1	112.5	101.1	120.1	112.5	102.1	120.1	112.9	101.1	99.4	100.1	101.1	110.0	106.4
09600	FLOORING	96.9	153.6	111.9	99.8	153.6	114.0	99.8	153.6	114.0	100.7	153.6	114.7	100.0	126.7	107.1	99.7	126.7	106.9
097,099	WALL FINISHES, PAINTS & COATINGS	95.3	119.2	109.8	92.9	119.2	108.8	92.8	119.2	108.8	95.3	119.2	109.8	92.8	99.0	96.5	94.6	99.0	97.3
09	FINISHES	96.1	127.0	112.1	101.2	127.0	114.6	101.2	127.0	114.6	101.4	127.2	114.8	101.2	105.6	103.5	101.3	112.1	106.9
10 - 14	TOTAL DIV. 10000 - 14000	100.0	118.7	104.0	100.0	118.8	104.0	100.0	118.8	104.0	100.0	119.6	104.2	100.0	102.7	100.6	100.0	104.8	101.0
15	MECHANICAL	100.1	105.3	102.4	100.1	109.6	104.3	100.1	122.8	110.0	100.1	105.4	102.4	100.1	95.3	98.0	100.1	98.4	99.4
16	ELECTRICAL	96.5	102.2	99.3	99.8	123.3	111.3	100.3	123.3	111.5	99.3	102.2	100.7	100.3	88.7	94.7	100.3	88.0	94.3
01 - 16	WEIGHTED AVERAGE	98.3	116.9	106.7	101.1	121.0	110.2	100.0	123.5	110.7	101.1	117.0	108.4	100.2	101.1	100.6	100.7	105.1	102.7

		MASSACHUSETTS			MICHIGAN														
	DIVISION	WORCESTER			ANN ARBOR			BATTLE CREEK			BAY CITY			DEARBORN			DETROIT		
		015 - 016			481			490			487			481			482		
		MAT.	INST.	TOTAL	MAT.	INST.	TOTAL	MAT.	INST.	TOTAL	MAT.	INST.	TOTAL	MAT.	INST.	TOTAL	MAT.	INST.	TOTAL
01590	EQUIPMENT RENTAL	.0	101.6	101.6	.0	112.2	112.2	.0	105.1	105.1	.0	112.2	112.2	.0	112.2	112.2	.0	97.7	97.7
02	SITE CONSTRUCTION	85.8	104.8	99.8	78.1	95.2	90.7	82.5	88.1	86.6	70.4	93.9	87.8	77.9	95.5	90.9	91.5	97.2	95.8
03100	CONCRETE FORMS & ACCESSORIES	101.8	121.0	118.5	97.6	110.4	108.7	96.7	88.8	89.9	97.7	92.3	93.0	97.4	123.0	119.6	99.0	123.1	119.9
03200	CONCRETE REINFORCEMENT	108.2	142.5	125.6	97.1	124.3	110.9	92.0	88.4	90.2	97.1	123.4	110.4	97.1	125.1	111.3	96.5	125.1	110.9
03300	CAST-IN-PLACE CONCRETE	91.8	139.0	111.5	94.3	111.2	101.4	95.6	100.6	97.7	90.2	93.6	91.6	92.1	122.0	104.6	101.2	122.0	109.9
03	CONCRETE	100.1	130.4	115.2	93.0	113.9	103.4	97.2	92.0	94.6	91.0	96.1	95.3	91.9	123.3	107.6	96.2	121.9	109.1
04	MASONRY	97.8	138.0	123.0	98.5	106.2	103.3	97.7	84.3	89.3	98.1	86.1	90.6	98.3	122.1	113.2	97.3	122.1	112.9
05	METALS	100.0	119.1	106.2	100.6	126.0	108.9	91.5	84.9	89.3	101.3	122.1	108.1	100.7	127.9	109.6	105.2	105.7	105.4
06	WOOD & PLASTICS	100.9	120.4	111.1	101.4	109.6	105.7	99.9	88.2	93.8	101.4	92.4	96.7	101.4	123.0	112.7	102.3	123.0	113.1
07	THERMAL & MOISTURE PROTECTION	101.8	129.0	115.2	97.7	109.2	103.4	91.3	84.4	87.9	95.9	88.9	92.4	96.6	124.2	110.2	95.2	124.2	109.5
08	DOORS & WINDOWS	107.1	126.5	112.1	96.5	111.1	100.3	90.9	83.1	88.9	96.5	98.2	96.9	96.5	118.4	102.2	98.0	120.3	103.8
09200	PLASTER & GYPSUM BOARD	105.2	120.1	114.6	102.4	108.6	106.3	93.2	83.0	86.8	102.4	90.9	95.1	102.4	122.3	115.1	102.4	122.3	115.1
095,098	CEILINGS & ACOUSTICAL TREATMENT	101.1	120.1	112.5	95.6	108.6	103.4	91.8	83.0	86.5	96.6	90.9	93.2	95.6	122.3	111.6	96.6	122.3	112.0
09600	FLOORING	99.8	153.6	114.0	87.7	109.8	93.5	96.6	92.0	95.4	87.4	59.5	80.1	87.2	123.8	96.9	87.3	123.8	97.0
097,099	WALL FINISHES, PAINTS & COATINGS	92.8	119.2	108.8	86.5	103.3	96.7	96.5	86.6	90.5	86.5	81.1	83.2	86.5	114.0	103.1	88.3	114.0	103.9
09	FINISHES	101.2	127.0	114.6	93.9	109.5	102.0	94.3	89.1	91.6	93.8	84.4	88.9	93.8	122.5	108.7	94.8	122.5	109.2
10 - 14	TOTAL DIV. 10000 - 14000	100.0	108.3	101.8	100.0	111.3	102.4	100.0	103.8	100.8	100.0	96.5	99.2	100.0	114.9	103.2	100.0	114.9	103.2
15	MECHANICAL	100.1	108.8	103.9	99.9	103.3	101.4	99.8	90.9	95.9	99.9	91.4	96.2	99.9	116.9	107.3	99.9	119.7	108.6
16	ELECTRICAL	100.3	104.5	102.3	94.1	87.9	91.1	94.6	83.6	89.2	93.2	83.2	88.3	94.1	116.4	105.0	95.9	116.3	105.8
01 - 16	WEIGHTED AVERAGE	100.5	118.4	108.6	96.9	105.7	100.9	95.4	88.1	92.1	96.4	93.3	95.0	96.7	118.3	106.5	98.7	116.9	107.0

City Cost Indexes

MICHIGAN

DIVISION		FLINT 484-485 MAT.	INST.	TOTAL	GAYLORD 497 MAT.	INST.	TOTAL	GRAND RAPIDS 493,495 MAT.	INST.	TOTAL	IRON MOUNTAIN 498-499 MAT.	INST.	TOTAL	JACKSON 492 MAT.	INST.	TOTAL	KALAMAZOO 491 MAT.	INST.	TOTAL
01590	EQUIPMENT RENTAL	.0	112.2	112.2	.0	107.6	107.6	.0	105.1	105.1	.0	94.2	94.2	.0	107.6	107.6	.0	105.1	105.1
02	SITE CONSTRUCTION	69.0	94.2	87.6	78.9	85.3	83.6	82.2	87.8	86.3	85.0	95.5	92.7	98.1	87.4	90.1	82.7	88.1	86.7
03100	CONCRETE FORMS & ACCESSORIES	101.1	93.5	94.5	93.7	64.5	68.4	97.2	79.1	81.5	88.1	87.1	87.2	91.2	90.0	90.1	96.7	88.5	89.6
03200	CONCRETE REINFORCEMENT	97.1	124.0	110.7	85.7	118.7	102.4	92.0	86.7	89.3	85.6	101.9	93.8	83.2	123.7	103.7	92.0	88.3	90.1
03300	CAST-IN-PLACE CONCRETE	94.9	93.6	94.4	95.4	77.5	87.9	95.9	99.9	97.6	113.0	88.4	102.7	95.3	95.7	95.5	97.6	100.4	98.8
03	CONCRETE	93.5	100.2	96.9	93.3	80.4	86.9	97.3	87.1	92.2	103.0	90.4	96.7	88.0	99.1	93.6	100.5	91.8	96.2
04	MASONRY	98.5	92.6	94.8	109.6	76.9	89.1	94.9	56.0	70.5	94.8	88.2	90.7	88.8	92.4	91.1	97.9	84.3	89.4
05	METALS	100.7	123.3	108.1	92.9	111.6	99.1	92.9	80.6	88.9	92.5	92.8	92.6	93.1	120.6	102.1	91.5	84.5	89.2
06	WOOD & PLASTICS	104.9	94.1	99.3	92.8	61.5	76.4	98.2	79.7	88.5	88.7	86.2	87.4	91.6	87.6	89.5	99.9	88.2	93.8
07	THERMAL & MOISTURE PROTECTION	95.9	92.5	94.2	90.3	69.0	79.8	91.7	62.4	77.3	92.7	81.8	87.3	89.4	95.2	92.2	91.3	84.4	87.9
08	DOORS & WINDOWS	96.5	100.2	97.4	92.6	68.8	86.4	94.2	71.0	88.2	97.2	76.8	91.9	91.6	93.5	92.1	90.9	83.1	88.9
09200	PLASTER & GYPSUM BOARD	104.0	92.6	96.8	95.1	58.8	72.1	93.2	74.2	81.2	63.6	86.4	78.1	94.4	85.7	88.9	93.2	83.0	86.8
095,098	CEILINGS & ACOUSTICAL TREATMENT	95.6	92.6	93.8	89.1	58.8	70.9	91.8	74.2	81.2	89.8	86.4	87.8	90.9	85.7	87.8	91.8	83.0	86.5
09600	FLOORING	87.4	74.9	84.1	90.1	62.8	82.9	96.6	44.2	82.8	108.9	91.0	104.2	88.3	86.5	87.9	96.6	70.1	89.6
097,099	WALL FINISHES, PAINTS & COATINGS	86.5	88.1	87.5	91.7	55.9	70.0	96.5	42.7	63.8	116.4	46.6	74.1	91.7	94.9	93.6	96.5	86.6	90.5
09	FINISHES	93.6	89.0	91.2	94.4	62.0	77.5	94.3	68.5	80.9	96.2	83.5	89.6	95.1	89.2	92.0	94.3	84.7	89.3
10-14	TOTAL DIV. 10000-14000	100.0	96.0	99.2	100.0	89.8	97.8	100.0	101.4	100.3	100.0	91.6	98.2	100.0	97.4	99.5	100.0	103.7	100.8
15	MECHANICAL	99.9	93.6	97.1	95.7	82.4	89.9	99.8	57.0	81.1	95.6	88.5	92.5	95.7	93.2	94.6	99.8	86.3	93.9
16	ELECTRICAL	94.1	88.9	91.5	92.6	61.4	77.4	94.6	61.3	78.4	99.1	78.9	89.2	96.2	87.8	92.1	94.2	71.8	83.2
01-16	WEIGHTED AVERAGE	96.7	96.2	96.5	94.6	77.9	87.0	95.8	70.2	84.2	96.5	87.1	92.2	93.9	94.7	94.3	95.7	84.9	90.8

MICHIGAN / MINNESOTA

DIVISION		LANSING 488-489 MAT.	INST.	TOTAL	MUSKEGON 494 MAT.	INST.	TOTAL	ROYAL OAK 480,483 MAT.	INST.	TOTAL	SAGINAW 486 MAT.	INST.	TOTAL	TRAVERSE CITY 496 MAT.	INST.	TOTAL	BEMIDJI 566 MAT.	INST.	TOTAL
01590	EQUIPMENT RENTAL	.0	112.2	112.2	.0	105.1	105.1	.0	94.6	94.6	.0	112.2	112.2	.0	94.2	94.2	.0	98.2	98.2
02	SITE CONSTRUCTION	84.1	94.1	91.5	80.5	88.0	86.1	81.0	94.5	91.0	71.2	93.9	88.0	72.4	94.2	88.5	83.3	97.3	93.6
03100	CONCRETE FORMS & ACCESSORIES	101.1	91.5	92.8	97.4	85.8	87.3	93.5	111.7	109.2	97.6	92.2	92.9	88.1	62.1	65.6	85.6	89.8	89.2
03200	CONCRETE REINFORCEMENT	97.1	123.8	110.6	92.5	88.1	90.3	88.0	120.2	104.3	97.1	123.4	110.4	86.9	87.3	87.1	95.3	108.7	102.1
03300	CAST-IN-PLACE CONCRETE	94.3	94.7	94.5	95.3	98.3	96.5	82.5	115.2	96.2	93.1	93.5	93.3	88.1	67.8	79.7	102.1	94.7	99.0
03	CONCRETE	93.2	99.7	96.4	95.4	89.8	92.6	80.5	113.3	96.9	92.4	99.5	95.9	84.9	69.5	77.2	94.6	96.3	95.4
04	MASONRY	91.9	94.6	93.6	95.1	80.1	85.7	92.4	114.5	106.2	100.0	86.1	91.2	93.0	77.1	83.0	100.6	100.4	100.5
05	METALS	99.1	122.9	106.8	89.2	84.1	87.5	104.3	96.6	101.8	100.7	121.9	107.7	92.5	85.0	90.0	88.2	123.0	99.6
06	WOOD & PLASTICS	104.3	89.6	96.6	96.7	85.2	90.7	96.8	111.1	104.2	97.4	92.4	94.8	88.7	59.3	73.3	77.7	87.6	82.9
07	THERMAL & MOISTURE PROTECTION	96.7	92.5	94.6	90.4	78.7	84.7	94.4	114.2	104.1	96.5	88.9	92.8	91.6	69.2	80.6	100.0	89.5	94.9
08	DOORS & WINDOWS	96.5	97.7	96.8	90.1	82.0	87.8	96.6	108.7	99.7	96.0	98.2	96.6	97.2	60.7	87.7	99.2	109.7	101.9
09200	PLASTER & GYPSUM BOARD	105.1	87.9	94.2	84.1	79.9	81.4	99.9	110.0	106.3	102.4	90.9	95.1	63.6	58.8	60.5	106.9	87.6	94.7
095,098	CEILINGS & ACOUSTICAL TREATMENT	95.6	87.9	91.0	93.6	79.9	85.3	94.8	110.0	103.9	95.6	90.9	92.8	89.8	58.8	71.2	123.6	87.6	102.0
09600	FLOORING	95.8	83.0	92.4	95.8	81.5	92.0	84.7	109.8	91.3	87.7	59.5	80.2	108.8	62.8	96.7	102.7	126.4	108.9
097,099	WALL FINISHES, PAINTS & COATINGS	100.4	94.4	96.8	95.7	61.9	75.2	88.3	100.1	95.4	86.5	81.1	83.2	116.4	57.1	80.5	99.2	91.5	94.6
09	FINISHES	98.2	89.7	93.8	92.7	82.1	87.2	92.6	110.2	101.8	93.7	84.4	88.9	95.4	60.6	77.3	108.1	96.2	101.9
10-14	TOTAL DIV. 10000-14000	100.0	97.1	99.4	100.0	103.0	100.6	100.0	102.4	100.5	100.0	96.5	99.2	100.0	84.1	96.6	100.0	90.4	98.0
15	MECHANICAL	99.9	93.2	97.0	99.7	89.7	95.4	95.8	101.4	98.2	99.9	91.0	96.0	95.6	81.7	89.6	96.2	91.2	94.0
16	ELECTRICAL	92.3	88.2	90.3	94.7	67.3	81.4	95.9	108.4	102.0	93.5	83.2	88.5	94.7	60.6	78.1	104.3	92.4	98.5
01-16	WEIGHTED AVERAGE	96.7	96.1	96.4	94.3	83.6	89.5	94.9	106.2	100.0	96.5	93.2	95.0	93.4	73.9	84.6	97.3	97.7	97.5

MINNESOTA

DIVISION		BRAINERD 564 MAT.	INST.	TOTAL	DETROIT LAKES 565 MAT.	INST.	TOTAL	DULUTH 556-558 MAT.	INST.	TOTAL	MANKATO 560 MAT.	INST.	TOTAL	MINNEAPOLIS 553-555 MAT.	INST.	TOTAL	ROCHESTER 559 MAT.	INST.	TOTAL
01590	EQUIPMENT RENTAL	.0	101.5	101.5	.0	98.2	98.2	.0	101.8	101.8	.0	101.5	101.5	.0	106.1	106.1	.0	101.8	101.8
02	SITE CONSTRUCTION	82.6	103.1	97.8	81.7	97.6	93.5	82.7	102.5	97.4	79.5	102.7	96.7	82.6	108.8	102.0	81.8	101.7	96.5
03100	CONCRETE FORMS & ACCESSORIES	87.8	92.2	91.6	82.2	104.3	101.3	103.5	123.7	121.0	97.4	97.0	97.1	104.2	137.5	133.0	104.0	104.5	104.4
03200	CONCRETE REINFORCEMENT	94.2	108.7	101.5	95.3	108.7	102.0	95.3	109.0	102.2	94.1	129.3	111.9	95.4	130.0	112.9	95.3	129.2	112.4
03300	CAST-IN-PLACE CONCRETE	111.0	99.2	106.1	99.1	98.2	98.7	109.6	104.2	107.3	102.2	87.7	96.1	107.9	126.9	115.8	105.8	98.2	102.6
03	CONCRETE	98.6	98.9	98.8	92.2	103.9	98.0	100.2	114.7	107.4	94.3	101.0	97.6	101.0	132.8	116.9	98.4	108.1	103.2
04	MASONRY	124.6	107.1	113.6	122.9	106.9	112.8	107.9	113.8	111.6	111.4	110.7	111.0	107.9	136.5	125.9	107.2	109.5	108.6
05	METALS	89.6	122.9	100.4	88.2	122.6	99.5	93.9	125.5	104.2	89.4	134.0	104.0	96.7	139.3	110.8	93.7	137.1	107.9
06	WOOD & PLASTICS	95.2	87.3	91.1	73.9	104.4	89.8	114.7	127.0	121.2	106.1	95.5	100.6	115.2	136.5	126.3	115.2	103.7	109.2
07	THERMAL & MOISTURE PROTECTION	98.5	103.1	100.8	99.8	83.8	92.0	101.5	116.6	108.9	99.0	98.8	98.9	101.3	136.3	118.5	101.3	103.7	102.5
08	DOORS & WINDOWS	86.7	109.5	92.6	99.1	118.8	104.2	96.8	124.6	104.0	91.6	121.4	99.3	99.9	143.7	111.2	96.8	125.9	104.3
09200	PLASTER & GYPSUM BOARD	88.2	87.6	87.8	104.9	104.8	104.9	102.3	128.3	118.8	93.4	96.0	95.1	102.5	137.8	124.9	102.0	104.3	103.5
095,098	CEILINGS & ACOUSTICAL TREATMENT	72.1	87.6	81.4	123.6	104.8	112.3	93.4	128.3	114.3	72.1	96.0	86.4	94.2	137.8	120.4	92.5	104.3	99.6
09600	FLOORING	101.4	126.4	108.0	101.2	126.4	107.9	105.8	121.7	110.0	103.9	120.0	108.1	103.2	126.4	109.3	105.6	82.1	99.4
097,099	WALL FINISHES, PAINTS & COATINGS	95.0	91.5	92.9	99.2	91.5	94.6	93.3	103.0	99.2	103.4	103.5	103.5	99.6	127.1	116.3	95.7	103.5	100.4
09	FINISHES	92.0	97.6	94.9	107.3	107.6	107.5	100.6	122.4	111.9	93.9	101.8	98.0	100.4	134.8	118.3	100.3	100.6	100.5
10-14	TOTAL DIV. 10000-14000	100.0	92.3	98.4	100.0	95.0	98.9	100.0	99.3	99.8	100.0	92.0	98.3	100.0	109.4	102.0	100.0	97.4	99.4
15	MECHANICAL	95.3	94.7	95.0	96.2	94.4	95.4	100.2	94.8	97.8	95.3	91.4	93.6	100.1	122.2	109.8	100.1	99.2	99.7
16	ELECTRICAL	102.5	95.1	98.9	104.2	71.3	88.1	105.8	95.1	100.6	108.4	89.3	99.1	107.7	112.4	110.0	105.7	89.4	97.8
01-16	WEIGHTED AVERAGE	96.1	101.0	98.3	98.0	99.1	98.5	99.7	108.9	103.8	96.3	102.1	99.0	100.7	126.8	112.6	99.4	105.1	102.0

DIVISION		MINNESOTA														MISSISSIPPI		
		SAINT PAUL 550 - 551			ST. CLOUD 563			THIEF RIVER FALLS 567			WILLMAR 562			WINDOM 561			BILOXI 395		
		MAT.	INST.	TOTAL	MAT.	INST.	TOTAL	MAT.	INST.	TOTAL	MAT.	INST.	TOTAL	MAT.	INST.	TOTAL	MAT.	INST.	TOTAL
01590	EQUIPMENT RENTAL	.0	101.8	101.8	.0	101.5	101.5	.0	98.2	98.2	.0	101.5	101.5	.0	101.5	101.5	.0	98.3	98.3
02	SITE CONSTRUCTION	84.7	103.3	98.5	78.3	105.2	98.2	81.1	97.2	93.0	77.6	100.0	94.2	72.1	98.8	91.8	103.3	86.2	90.6
03100	CONCRETE FORMS & ACCESSORIES	94.5	130.9	126.0	84.5	126.8	121.1	84.7	89.3	88.7	84.2	60.5	63.7	89.5	56.5	60.9	95.5	42.5	49.6
03200	CONCRETE REINFORCEMENT	91.9	129.9	111.1	94.3	129.2	111.9	95.6	108.7	102.2	93.9	128.5	111.4	93.9	127.7	111.0	95.0	62.4	78.6
03300	CAST-IN-PLACE CONCRETE	108.9	126.0	116.0	97.8	124.0	108.7	98.0	89.4	94.4	99.4	84.0	93.0	86.0	60.0	75.1	104.5	46.3	80.2
03	CONCRETE	102.5	129.5	116.0	90.0	126.7	108.4	91.7	94.2	93.0	90.1	83.1	86.6	81.3	73.0	77.2	97.0	49.5	73.3
04	MASONRY	117.7	136.5	129.5	109.2	122.2	117.4	100.4	100.4	100.4	111.9	88.0	96.9	122.6	65.1	86.5	89.3	38.1	57.2
05	METALS	93.2	139.2	108.3	90.0	136.2	105.1	88.2	122.5	99.5	89.4	126.7	101.6	89.3	125.6	101.2	92.1	83.2	89.2
06	WOOD & PLASTICS	104.4	128.1	116.8	91.7	124.5	108.8	76.6	87.6	82.4	91.3	57.9	73.9	97.0	57.9	76.6	96.6	42.4	68.3
07	THERMAL & MOISTURE PROTECTION	101.2	133.9	117.3	98.6	117.3	107.8	99.9	79.0	89.6	98.4	61.8	80.4	98.5	52.7	76.0	95.2	47.3	71.6
08	DOORS & WINDOWS	94.1	139.1	105.8	91.6	137.2	103.4	99.2	109.7	101.9	89.2	82.9	87.5	93.5	82.9	90.8	98.5	48.4	85.5
09200	PLASTER & GYPSUM BOARD	96.3	129.4	117.3	87.0	125.8	111.6	106.5	87.6	94.5	87.0	57.3	68.2	89.0	57.3	68.9	103.7	41.3	64.2
095,098	CEILINGS & ACOUSTICAL TREATMENT	90.7	129.4	113.9	72.1	125.8	104.3	123.6	87.6	102.0	72.1	57.3	63.3	72.1	57.3	63.3	102.7	41.3	65.9
09600	FLOORING	98.6	126.4	106.0	97.4	126.4	105.1	102.3	126.4	108.7	99.2	113.5	103.0	102.1	113.5	105.1	118.6	41.4	98.2
097,099	WALL FINISHES, PAINTS & COATINGS	99.6	120.5	112.3	103.4	127.1	117.8	99.2	91.5	94.6	99.2	91.5	94.6	99.2	103.5	101.9	107.4	36.5	64.4
09	FINISHES	97.5	129.1	113.9	90.9	127.0	109.7	107.8	96.2	101.8	91.2	70.9	80.6	92.0	69.8	80.5	106.9	41.1	72.7
10 - 14	TOTAL DIV. 10000 - 14000	100.0	107.9	101.7	100.0	102.8	100.6	100.0	90.3	97.9	100.0	62.9	92.1	100.0	59.2	91.3	100.0	55.2	90.5
15	MECHANICAL	100.1	113.2	105.8	99.7	120.5	108.8	96.2	90.9	93.9	95.3	94.1	94.8	95.3	77.4	87.5	99.8	55.9	80.6
16	ELECTRICAL	106.5	106.1	106.3	102.5	106.1	104.3	100.5	71.3	86.2	102.5	78.6	90.8	108.4	89.3	99.0	96.7	61.1	79.3
01 - 16	WEIGHTED AVERAGE	99.8	122.0	109.8	95.6	120.5	106.9	96.4	94.0	95.3	94.5	87.7	91.4	95.1	81.4	88.9	97.9	56.6	79.1

DIVISION		MISSISSIPPI																
		CLARKSDALE 386			COLUMBUS 397			GREENVILLE 387			GREENWOOD 389			JACKSON 390 - 392			LAUREL 394		
		MAT.	INST.	TOTAL	MAT.	INST.	TOTAL	MAT.	INST.	TOTAL	MAT.	INST.	TOTAL	MAT.	INST.	TOTAL	MAT.	INST.	TOTAL
01590	EQUIPMENT RENTAL	.0	98.3	98.3	.0	98.3	98.3	.0	98.3	98.3	.0	98.3	98.3	.0	98.3	98.3	.0	98.3	98.3
02	SITE CONSTRUCTION	100.6	85.3	89.3	102.4	85.7	90.0	107.4	86.0	91.5	104.2	85.1	90.1	99.9	86.0	89.6	108.6	84.3	90.6
03100	CONCRETE FORMS & ACCESSORIES	84.4	18.9	27.7	82.0	30.5	37.4	80.3	35.1	41.1	95.4	20.8	30.8	95.3	40.0	47.4	82.4	24.0	31.8
03200	CONCRETE REINFORCEMENT	102.4	17.6	59.5	101.7	31.8	66.4	103.0	44.9	73.6	102.4	47.7	74.7	95.0	48.7	71.6	102.4	31.2	66.4
03300	CAST-IN-PLACE CONCRETE	101.9	27.5	70.8	106.6	38.1	78.0	104.9	40.5	78.0	109.6	27.4	75.3	102.5	43.0	77.7	104.2	30.7	73.5
03	CONCRETE	95.3	24.1	59.8	98.7	35.6	67.2	100.2	40.9	70.6	101.5	30.6	66.1	96.0	44.7	70.4	100.9	30.0	65.5
04	MASONRY	89.2	20.4	46.0	112.4	24.1	57.0	131.8	39.5	73.9	89.7	19.0	45.4	91.3	39.5	58.8	109.1	24.7	56.1
05	METALS	89.7	60.7	80.2	89.6	67.7	82.4	90.4	74.9	85.4	89.7	73.2	84.3	92.1	76.8	87.1	89.7	66.0	81.9
06	WOOD & PLASTICS	83.7	17.6	49.2	81.4	31.6	55.4	79.0	34.4	55.7	96.7	19.8	56.5	96.7	40.8	67.5	81.7	24.9	52.0
07	THERMAL & MOISTURE PROTECTION	94.9	28.4	62.2	95.0	28.3	62.2	95.3	42.4	69.3	95.4	23.8	60.2	95.0	43.6	69.8	95.2	27.0	61.7
08	DOORS & WINDOWS	97.8	21.9	78.2	97.8	29.7	80.2	97.8	39.2	82.7	97.8	31.5	80.7	98.9	43.8	84.6	94.3	25.9	76.6
09200	PLASTER & GYPSUM BOARD	94.0	15.8	44.4	93.6	30.2	53.4	91.2	33.1	54.4	102.0	18.1	48.8	103.7	39.8	63.2	91.2	23.4	48.2
095,098	CEILINGS & ACOUSTICAL TREATMENT	94.5	15.8	47.3	94.5	30.2	55.9	96.5	33.1	58.5	94.5	18.1	48.7	102.7	39.8	65.0	94.5	23.4	51.8
09600	FLOORING	111.5	24.1	88.4	110.2	20.8	86.5	109.1	35.4	89.6	118.6	21.7	93.0	118.6	40.9	98.1	108.1	19.6	84.7
097,099	WALL FINISHES, PAINTS & COATINGS	107.4	15.3	51.5	107.4	22.6	55.9	107.4	35.9	64.0	107.4	20.9	54.9	107.4	35.9	64.0	107.4	22.6	55.9
09	FINISHES	101.2	19.3	58.6	100.9	28.3	63.2	101.1	34.9	66.7	104.8	20.8	61.1	106.9	39.9	72.0	100.4	23.6	60.4
10 - 14	TOTAL DIV. 10000 - 14000	100.0	43.9	88.1	100.0	46.6	88.6	100.0	53.4	90.1	100.0	44.3	88.1	100.0	54.3	90.3	100.0	28.0	84.7
15	MECHANICAL	97.9	19.5	63.6	97.5	26.5	66.4	99.8	35.6	71.7	97.9	20.5	64.0	99.8	35.3	71.6	97.5	22.6	64.7
16	ELECTRICAL	96.1	25.2	61.5	94.3	27.6	61.8	96.1	38.1	67.8	96.1	25.7	61.7	97.5	38.1	68.5	95.7	25.7	61.6
01 - 16	WEIGHTED AVERAGE	96.0	31.7	66.8	97.2	37.8	70.2	99.4	45.9	75.1	97.3	34.3	68.7	97.9	47.5	75.0	97.3	34.4	68.7

DIVISION		MISSISSIPPI									MISSOURI								
		MCCOMB 396			MERIDIAN 393			TUPELO 388			BOWLING GREEN 633			CAPE GIRARDEAU 637			CHILLICOTHE 646		
		MAT.	INST.	TOTAL	MAT.	INST.	TOTAL	MAT.	INST.	TOTAL	MAT.	INST.	TOTAL	MAT.	INST.	TOTAL	MAT.	INST.	TOTAL
01590	EQUIPMENT RENTAL	.0	98.3	98.3	.0	98.3	98.3	.0	98.3	98.3	.0	108.3	108.3	.0	108.3	108.3	.0	103.3	103.3
02	SITE CONSTRUCTION	94.3	85.6	87.9	98.7	86.0	89.3	98.7	85.3	88.8	92.3	92.6	92.5	93.6	92.5	92.8	100.9	91.4	93.8
03100	CONCRETE FORMS & ACCESSORIES	82.0	58.8	62.0	78.8	34.1	40.1	81.0	24.4	32.0	92.5	76.1	78.3	84.4	74.9	76.2	88.4	81.3	82.2
03200	CONCRETE REINFORCEMENT	103.0	41.1	71.7	101.7	48.1	74.6	100.2	48.4	74.0	92.7	95.3	94.0	93.9	84.7	89.2	107.9	88.7	98.2
03300	CAST-IN-PLACE CONCRETE	92.5	41.5	71.2	99.0	43.0	75.6	101.9	39.6	75.9	85.5	84.9	85.3	84.6	84.5	84.6	93.4	72.9	84.9
03	CONCRETE	88.9	51.0	70.0	93.0	41.9	67.5	94.7	36.4	65.6	86.5	84.5	85.5	85.5	81.8	83.7	99.9	80.5	90.2
04	MASONRY	114.6	39.7	67.6	89.0	29.4	51.6	121.5	31.4	65.0	117.9	92.4	101.9	114.0	75.8	90.0	106.4	81.0	90.5
05	METALS	89.8	72.7	84.2	90.4	75.8	85.6	89.6	74.9	84.8	91.4	115.0	99.1	92.1	109.3	97.7	96.9	94.7	96.2
06	WOOD & PLASTICS	81.4	66.2	73.5	77.7	33.0	54.4	79.6	23.2	50.1	88.7	73.2	80.6	80.2	72.3	76.1	89.0	84.1	86.5
07	THERMAL & MOISTURE PROTECTION	94.5	44.1	69.7	94.7	40.2	67.9	94.8	35.3	65.6	95.6	93.2	94.5	95.5	79.8	87.7	98.5	79.0	88.9
08	DOORS & WINDOWS	97.9	54.7	86.7	97.8	37.9	82.3	97.8	29.9	80.2	93.9	78.1	89.9	93.9	75.0	89.1	90.6	76.8	87.1
09200	PLASTER & GYPSUM BOARD	93.6	65.8	76.0	91.2	31.7	53.5	91.2	21.6	47.1	100.1	72.3	82.5	96.1	71.4	80.4	97.7	83.4	88.6
095,098	CEILINGS & ACOUSTICAL TREATMENT	94.5	65.8	77.3	96.5	31.7	57.6	94.5	21.6	50.8	85.5	72.3	77.6	85.5	71.4	77.0	84.1	83.4	83.7
09600	FLOORING	110.2	24.1	87.4	108.0	28.7	87.0	109.5	26.5	87.6	96.5	78.2	91.6	92.4	78.2	88.7	89.1	56.1	80.4
097,099	WALL FINISHES, PAINTS & COATINGS	107.4	33.8	62.8	107.4	35.4	63.7	107.4	35.9	64.0	100.9	104.4	103.0	100.9	78.5	87.3	92.9	59.3	72.5
09	FINISHES	100.5	51.5	75.0	100.2	32.6	65.0	100.2	25.7	61.4	91.8	78.4	84.8	90.0	74.9	82.1	92.1	75.5	83.4
10 - 14	TOTAL DIV. 10000 - 14000	100.0	68.3	93.3	100.0	53.4	90.1	100.0	45.2	88.3	100.0	70.3	93.7	100.0	70.2	93.7	100.0	69.4	93.5
15	MECHANICAL	97.5	31.4	68.6	99.8	35.3	71.6	98.1	25.5	66.3	95.7	94.2	95.1	100.1	93.8	97.4	95.7	51.9	76.6
16	ELECTRICAL	93.0	62.8	78.3	95.7	60.2	78.4	95.8	25.2	61.3	102.4	83.8	93.3	102.4	108.8	105.5	99.5	43.8	72.3
01 - 16	WEIGHTED AVERAGE	95.8	53.3	76.5	96.0	47.8	74.1	97.4	38.4	70.6	95.6	89.4	92.8	96.1	89.3	93.0	97.0	70.7	85.1

MISSOURI

DIVISION		COLUMBIA 652			FLAT RIVER 636			HANNIBAL 634			HARRISONVILLE 647			JEFFERSON CITY 650 - 651			JOPLIN 648		
		MAT.	INST.	TOTAL	MAT.	INST.	TOTAL	MAT.	INST.	TOTAL	MAT.	INST.	TOTAL	MAT.	INST.	TOTAL	MAT.	INST.	TOTAL
01590	EQUIPMENT RENTAL	.0	109.0	109.0	.0	108.3	108.3	.0	108.3	108.3	.0	103.3	103.3	.0	109.0	109.0	.0	106.9	106.9
02	SITE CONSTRUCTION	101.5	94.5	96.3	95.0	92.6	93.2	89.6	91.8	91.2	92.7	93.9	93.6	103.5	94.0	96.4	100.7	98.6	99.1
03100	CONCRETE FORMS & ACCESSORIES	85.0	72.4	74.0	99.5	83.8	85.9	90.6	72.5	74.9	85.1	95.5	94.1	99.5	64.1	68.8	101.2	65.4	70.2
03200	CONCRETE REINFORCEMENT	100.7	112.5	106.7	93.9	107.7	100.9	92.3	85.8	89.0	107.4	97.7	102.5	99.4	114.5	107.0	111.3	73.6	92.2
03300	CAST-IN-PLACE CONCRETE	88.5	68.9	80.3	88.3	85.7	87.2	80.9	80.6	80.8	95.7	98.7	96.9	92.6	71.8	83.9	101.2	73.1	89.5
03	CONCRETE	85.9	80.5	83.2	89.2	90.5	89.8	83.1	79.5	81.3	96.4	97.6	97.0	89.3	78.2	83.8	99.9	70.8	85.4
04	MASONRY	131.1	83.4	101.2	115.6	77.6	91.7	108.9	79.3	90.3	100.5	92.6	95.5	93.8	83.4	87.3	98.6	60.4	74.6
05	METALS	92.7	121.1	102.0	91.3	119.4	100.5	91.4	108.9	97.1	97.4	105.5	100.1	92.5	122.0	102.2	99.8	89.8	96.5
06	WOOD & PLASTICS	88.7	68.3	78.1	96.3	82.7	89.2	86.5	72.3	79.1	84.6	95.1	90.1	104.3	57.1	79.7	100.7	63.6	81.3
07	THERMAL & MOISTURE PROTECTION	92.4	81.4	87.0	95.9	90.6	93.3	95.4	84.9	90.3	97.3	97.3	97.3	92.6	80.6	86.7	98.6	63.8	81.5
08	DOORS & WINDOWS	90.9	91.7	91.1	93.9	86.8	92.1	93.9	75.2	89.1	91.0	95.8	92.3	87.7	86.1	87.3	91.9	71.0	86.5
09200	PLASTER & GYPSUM BOARD	98.1	67.3	78.6	104.2	82.1	90.1	98.5	71.4	81.3	90.0	94.6	93.0	105.7	55.7	74.0	103.9	62.3	77.5
095,098	CEILINGS & ACOUSTICAL TREATMENT	90.0	67.3	76.4	85.5	82.1	83.4	85.5	71.4	77.0	84.1	94.6	90.4	90.0	55.7	69.5	87.6	62.3	72.4
09600	FLOORING	101.9	90.5	98.9	100.0	78.2	94.2	95.4	78.2	90.9	85.3	90.5	86.7	110.0	90.5	104.8	111.7	52.6	96.1
097,099	WALL FINISHES, PAINTS & COATINGS	99.3	74.8	84.5	100.9	78.5	87.3	100.9	84.5	91.0	97.2	93.7	95.1	99.3	74.8	84.5	92.3	46.4	64.4
09	FINISHES	93.1	73.3	82.8	93.5	81.4	87.2	91.1	74.1	82.2	89.6	94.2	92.0	96.8	66.7	81.1	99.0	60.6	79.0
10 - 14	TOTAL DIV. 10000 - 14000	100.0	91.0	98.1	100.0	72.3	94.1	100.0	67.7	93.1	100.0	91.0	98.1	100.0	89.5	97.8	100.0	79.8	95.7
15	MECHANICAL	100.0	92.2	96.6	95.7	95.1	95.4	95.7	90.7	93.5	95.6	97.7	96.5	100.0	95.6	98.1	100.1	57.9	81.6
16	ELECTRICAL	99.4	82.4	91.1	106.8	108.8	107.8	101.2	83.8	92.7	105.8	104.6	105.2	101.2	82.4	92.0	97.6	63.6	81.1
01 - 16	WEIGHTED AVERAGE	96.9	88.1	92.9	96.6	93.7	95.3	94.4	85.1	90.2	96.7	97.7	97.2	95.9	87.3	92.0	98.7	69.0	85.2

MISSOURI

DIVISION		KANSAS CITY 640 - 641			KIRKSVILLE 635			POPLAR BLUFF 639			ROLLA 654 - 655			SEDALIA 653			SIKESTON 638		
		MAT.	INST.	TOTAL	MAT.	INST.	TOTAL	MAT.	INST.	TOTAL	MAT.	INST.	TOTAL	MAT.	INST.	TOTAL	MAT.	INST.	TOTAL
01590	EQUIPMENT RENTAL	.0	104.8	104.8	.0	99.6	99.6	.0	102.0	102.0	.0	109.0	109.0	.0	100.1	100.1	.0	102.0	102.0
02	SITE CONSTRUCTION	94.3	98.0	97.0	88.5	87.9	88.1	75.1	92.4	87.9	101.0	93.1	95.1	92.4	90.2	90.8	78.4	92.5	88.8
03100	CONCRETE FORMS & ACCESSORIES	100.2	105.2	104.5	82.1	62.0	64.7	82.1	74.7	75.7	93.7	84.5	85.7	91.0	67.3	70.5	83.3	74.8	75.9
03200	CONCRETE REINFORCEMENT	105.8	113.4	109.6	93.0	85.5	89.2	95.9	88.6	92.2	101.3	61.1	81.0	99.8	87.9	93.8	95.3	78.7	86.9
03300	CAST-IN-PLACE CONCRETE	94.1	107.5	99.7	88.2	69.6	80.5	67.9	67.8	67.8	90.6	76.2	84.6	94.6	72.5	85.4	72.5	77.4	74.6
03	CONCRETE	95.9	107.9	101.9	98.4	70.3	84.4	77.0	76.0	76.5	87.8	78.6	83.2	99.7	74.2	87.0	80.3	77.5	78.9
04	MASONRY	103.0	105.7	104.7	121.4	74.3	91.8	113.3	65.0	83.0	107.5	51.9	72.6	113.7	59.3	79.6	113.3	70.9	86.7
05	METALS	107.5	115.0	109.9	91.1	96.9	93.0	91.6	99.8	94.3	92.5	92.5	92.5	91.0	98.3	93.4	91.9	95.1	93.0
06	WOOD & PLASTICS	100.1	104.4	102.4	74.3	60.8	67.2	73.4	75.9	74.7	97.5	94.4	95.9	90.8	61.4	75.5	74.7	75.9	75.3
07	THERMAL & MOISTURE PROTECTION	97.8	107.1	102.4	100.0	81.1	90.7	99.0	79.4	89.4	92.7	63.3	78.3	96.6	79.8	88.3	99.2	72.3	86.0
08	DOORS & WINDOWS	98.0	107.0	100.3	97.7	69.1	90.3	98.6	80.8	94.0	90.9	74.3	86.6	94.2	72.9	88.7	98.6	69.8	91.2
09200	PLASTER & GYPSUM BOARD	100.3	104.3	102.8	92.2	59.5	71.4	92.9	75.0	81.5	101.3	94.1	96.7	93.9	60.2	72.5	94.1	75.0	82.0
095,098	CEILINGS & ACOUSTICAL TREATMENT	93.9	104.3	100.1	82.9	59.5	68.8	85.5	75.0	79.2	90.0	94.1	92.5	88.2	60.2	71.4	85.5	75.0	79.2
09600	FLOORING	91.7	90.5	91.4	74.7	57.6	70.2	87.7	65.1	81.8	106.3	49.1	91.1	83.1	90.5	85.1	88.2	52.6	78.8
097,099	WALL FINISHES, PAINTS & COATINGS	97.2	118.8	110.3	96.5	46.4	66.1	95.9	90.5	92.6	99.3	83.4	89.7	99.3	93.7	95.9	95.9	66.7	78.2
09	FINISHES	95.3	103.6	99.6	87.7	58.1	72.3	89.4	74.5	81.6	95.0	79.1	86.7	90.6	73.0	81.5	89.9	69.5	79.3
10 - 14	TOTAL DIV. 10000 - 14000	100.0	93.8	98.7	100.0	62.5	92.0	100.0	65.4	92.6	100.0	67.5	93.1	100.0	84.1	96.6	100.0	66.4	92.9
15	MECHANICAL	100.0	107.1	103.1	95.8	86.5	91.7	95.8	85.7	91.4	95.6	76.1	87.1	95.5	91.6	93.8	95.8	86.7	91.8
16	ELECTRICAL	109.0	108.1	108.6	100.4	83.7	92.3	101.1	108.8	104.9	98.0	82.3	90.3	97.9	104.5	101.1	100.2	108.8	104.4
01 - 16	WEIGHTED AVERAGE	101.1	106.3	103.4	96.7	78.3	88.3	93.9	85.0	89.8	95.0	77.8	87.2	96.1	84.3	90.7	94.3	84.3	89.8

DIVISION		MISSOURI									MONTANA								
		SPRINGFIELD 656 - 658			ST. JOSEPH 644 - 645			ST. LOUIS 630 - 631			BILLINGS 590 - 591			BUTTE 597			GREAT FALLS 594		
		MAT.	INST.	TOTAL	MAT.	INST.	TOTAL	MAT.	INST.	TOTAL	MAT.	INST.	TOTAL	MAT.	INST.	TOTAL	MAT.	INST.	TOTAL
01590	EQUIPMENT RENTAL	.0	102.7	102.7	.0	103.3	103.3	.0	109.5	109.5	.0	98.5	98.5	.0	98.2	98.2	.0	98.2	98.2
02	SITE CONSTRUCTION	95.3	93.2	93.7	95.8	93.5	94.1	93.3	96.5	95.7	85.2	96.1	93.3	92.2	94.7	94.0	95.7	95.5	95.5
03100	CONCRETE FORMS & ACCESSORIES	100.2	65.6	70.2	100.1	86.0	87.9	99.6	105.7	104.9	98.1	65.3	69.7	85.4	62.9	65.9	99.4	66.5	70.9
03200	CONCRETE REINFORCEMENT	97.0	111.7	104.4	104.5	97.3	100.9	85.8	115.6	100.9	93.2	75.9	84.5	101.2	76.0	88.4	93.3	75.9	84.5
03300	CAST-IN-PLACE CONCRETE	101.5	63.6	85.6	94.0	100.6	96.8	84.6	108.2	94.5	119.0	68.6	97.9	122.5	68.7	100.0	129.4	58.6	99.9
03	CONCRETE	99.2	74.7	87.0	95.6	93.9	94.8	85.3	109.4	97.3	104.6	69.4	87.0	104.6	68.4	86.5	109.7	66.5	88.1
04	MASONRY	87.4	65.4	73.6	102.4	81.0	89.0	95.2	111.2	105.2	120.1	66.4	86.4	120.1	67.4	87.0	123.7	71.5	90.9
05	METALS	96.6	105.6	99.5	103.3	104.0	103.6	96.7	126.0	106.3	105.7	86.8	99.5	98.5	86.5	94.6	102.2	86.6	97.1
06	WOOD & PLASTICS	99.1	66.6	82.1	100.7	85.9	93.0	96.2	103.6	100.0	101.8	65.1	82.6	88.9	63.6	75.7	104.5	65.3	84.0
07	THERMAL & MOISTURE PROTECTION	96.5	70.1	83.5	98.4	90.3	94.4	95.4	106.2	100.7	99.9	68.5	84.5	99.2	67.4	83.6	100.1	68.5	84.5
08	DOORS & WINDOWS	96.3	78.2	91.6	96.8	92.9	95.8	92.9	112.3	97.9	99.1	64.1	90.0	95.8	63.2	87.4	99.3	64.2	90.2
09200	PLASTER & GYPSUM BOARD	102.5	65.5	79.0	105.3	85.2	92.6	105.2	103.5	104.1	104.4	64.5	79.1	94.8	63.0	74.6	104.4	64.7	79.2
095,098	CEILINGS & ACOUSTICAL TREATMENT	90.0	65.5	75.3	93.0	85.2	88.3	90.9	103.5	98.4	112.4	64.5	83.6	109.7	63.0	81.7	112.4	64.7	83.8
09600	FLOORING	104.9	52.6	91.1	94.0	84.2	91.4	99.7	98.5	99.4	107.8	51.2	92.9	99.1	40.8	83.7	107.8	55.3	94.0
097,099	WALL FINISHES, PAINTS & COATINGS	93.6	66.4	77.1	92.9	80.2	85.2	100.9	108.7	105.6	99.2	61.9	76.6	99.2	42.1	64.6	99.2	47.4	67.8
09	FINISHES	96.6	62.6	78.9	96.3	84.5	90.2	94.9	104.1	99.7	106.6	61.9	83.4	101.8	56.0	78.0	106.7	61.9	83.4
10 - 14	TOTAL DIV. 10000 - 14000	100.0	80.5	95.9	100.0	87.3	97.3	100.0	101.1	100.2	100.0	72.0	94.0	100.0	70.8	93.8	100.0	73.3	94.3
15	MECHANICAL	100.0	67.7	85.9	100.1	90.2	95.8	100.1	104.4	102.0	100.4	77.8	90.5	100.4	70.5	87.3	100.4	79.6	91.3
16	ELECTRICAL	102.6	68.7	86.1	105.9	83.4	94.9	105.3	114.1	109.6	95.0	74.7	85.1	101.6	69.6	86.0	95.0	75.1	85.3
01 - 16	WEIGHTED AVERAGE	98.2	74.5	87.4	100.1	89.6	95.3	96.7	108.6	102.1	102.0	74.4	89.5	101.0	71.1	87.4	102.6	74.9	90.0

MONTANA

DIVISION		HAVRE 595 MAT.	INST.	TOTAL	HELENA 596 MAT.	INST.	TOTAL	KALISPELL 599 MAT.	INST.	TOTAL	MILES CITY 593 MAT.	INST.	TOTAL	MISSOULA 598 MAT.	INST.	TOTAL	WOLF POINT 592 MAT.	INST.	TOTAL
01590	EQUIPMENT RENTAL	.0	98.2	98.2	.0	98.2	98.2	.0	98.2	98.2	.0	98.2	98.2	.0	98.2	98.2	.0	98.2	98.2
02	SITE CONSTRUCTION	99.4	95.2	96.3	97.5	95.3	95.9	82.7	94.6	91.5	88.9	95.2	93.6	75.6	94.6	89.6	106.0	95.2	98.0
03100	CONCRETE FORMS & ACCESSORIES	77.1	62.5	64.5	99.5	65.9	70.4	89.1	58.9	63.0	97.1	62.6	67.2	89.1	59.0	63.1	88.8	61.2	64.9
03200	CONCRETE REINFORCEMENT	102.0	66.1	83.9	96.6	65.9	81.0	104.0	81.6	92.7	101.5	66.0	83.6	102.9	78.4	90.5	103.0	66.1	84.4
03300	CAST-IN-PLACE CONCRETE	131.9	63.7	103.4	131.9	67.1	104.8	106.3	65.1	89.1	116.4	63.5	94.3	90.2	65.1	79.7	130.4	58.9	100.6
03	CONCRETE	112.4	64.6	88.5	111.4	67.2	89.3	94.7	66.4	80.6	101.8	64.5	83.2	83.6	65.8	74.7	116.2	62.3	89.3
04	MASONRY	121.0	69.6	88.8	119.2	65.7	85.6	119.1	54.9	78.9	126.0	67.7	89.4	145.0	60.5	91.9	127.0	69.6	91.0
05	METALS	95.1	81.3	90.5	101.3	81.2	94.8	94.9	87.7	92.6	94.4	81.3	90.1	95.3	86.6	92.5	94.4	81.3	90.1
06	WOOD & PLASTICS	79.9	61.7	70.4	104.6	65.3	84.1	93.4	59.8	75.9	101.1	61.7	80.5	93.4	59.8	75.9	92.3	59.8	75.4
07	THERMAL & MOISTURE PROTECTION	99.6	65.0	82.6	100.0	68.2	84.4	98.8	67.5	83.4	99.2	68.8	84.3	98.3	70.9	84.8	100.2	65.9	83.3
08	DOORS & WINDOWS	95.9	59.5	86.5	98.8	61.6	89.2	95.9	63.5	87.5	95.4	59.5	86.1	95.9	62.5	87.2	95.4	58.5	85.8
09200	PLASTER & GYPSUM BOARD	90.8	61.0	71.9	103.7	64.7	79.0	96.9	59.1	72.9	103.0	61.0	76.3	96.9	59.1	72.9	99.0	59.1	73.7
095,098	CEILINGS & ACOUSTICAL TREATMENT	109.7	61.0	80.5	109.7	64.7	82.7	109.7	59.1	79.3	107.0	61.0	79.4	109.7	59.1	79.3	107.0	59.1	78.2
09600	FLOORING	95.6	44.9	82.2	107.8	53.6	93.5	101.3	63.7	91.3	107.6	50.3	92.4	101.3	63.7	91.3	103.1	44.9	87.7
097,099	WALL FINISHES, PAINTS & COATINGS	99.2	46.8	67.4	99.2	52.6	71.0	99.2	47.4	67.8	99.2	44.2	65.9	99.2	47.4	67.8	99.2	44.2	65.9
09	FINISHES	100.6	57.2	78.0	106.0	62.0	83.1	102.3	58.2	79.4	104.7	57.9	80.4	102.0	58.2	79.2	103.7	55.8	78.8
10 - 14	TOTAL DIV. 10000 - 14000	100.0	58.3	91.1	100.0	59.2	91.3	100.0	55.2	90.5	100.0	58.2	91.1	100.0	55.2	90.5	100.0	58.1	91.1
15	MECHANICAL	96.0	78.4	88.3	100.4	74.7	89.1	96.0	70.2	84.7	96.0	78.3	88.3	100.4	70.2	87.2	96.0	78.4	88.3
16	ELECTRICAL	95.0	63.2	79.5	95.0	74.1	84.8	98.7	72.3	85.9	95.0	65.4	80.6	99.7	72.3	86.3	95.0	65.4	80.6
01 - 16	WEIGHTED AVERAGE	99.6	70.7	86.5	102.4	72.3	88.7	97.8	69.8	85.1	98.8	71.0	86.2	98.7	70.3	85.8	100.8	70.5	87.0

NEBRASKA

DIVISION		ALLIANCE 693 MAT.	INST.	TOTAL	COLUMBUS 686 MAT.	INST.	TOTAL	GRAND ISLAND 688 MAT.	INST.	TOTAL	HASTINGS 689 MAT.	INST.	TOTAL	LINCOLN 683 - 685 MAT.	INST.	TOTAL	MCCOOK 690 MAT.	INST.	TOTAL
01590	EQUIPMENT RENTAL	.0	96.6	96.6	.0	101.6	101.6	.0	101.6	101.6	.0	101.6	101.6	.0	101.6	101.6	.0	101.6	101.6
02	SITE CONSTRUCTION	94.6	95.5	95.3	93.5	89.2	90.3	98.1	91.0	92.9	96.9	89.8	91.7	89.5	91.1	90.6	97.7	89.2	91.4
03100	CONCRETE FORMS & ACCESSORIES	87.4	26.5	34.6	97.2	25.6	35.2	96.6	50.7	56.9	100.6	47.1	54.3	101.3	46.8	54.1	93.9	31.9	40.2
03200	CONCRETE REINFORCEMENT	115.8	46.3	80.7	106.2	51.6	78.6	105.6	74.5	89.9	105.6	51.6	78.3	97.0	75.4	86.1	107.7	74.2	90.7
03300	CAST-IN-PLACE CONCRETE	106.6	37.4	77.7	109.1	47.6	83.4	115.5	58.7	91.8	115.5	52.4	89.2	104.3	61.2	86.3	115.5	37.6	83.0
03	CONCRETE	118.7	35.1	76.9	104.1	40.0	72.1	108.5	59.3	83.9	108.8	51.2	80.0	100.6	58.5	79.6	108.6	43.5	76.1
04	MASONRY	107.8	29.9	58.9	113.2	30.9	61.5	105.5	49.9	70.6	114.7	40.5	68.1	95.8	66.7	77.5	103.1	29.8	57.1
05	METALS	99.8	57.0	85.8	91.6	72.7	85.4	92.8	83.3	89.7	94.3	72.9	87.3	97.9	84.8	93.6	94.5	80.4	89.9
06	WOOD & PLASTICS	84.8	25.4	53.8	96.2	22.7	57.8	95.5	45.5	69.4	99.5	45.5	71.3	100.1	39.8	68.6	93.3	31.5	61.0
07	THERMAL & MOISTURE PROTECTION	102.5	35.6	69.7	96.7	32.7	65.3	96.9	57.5	77.5	97.0	47.3	72.6	97.3	60.3	79.1	96.8	36.2	67.0
08	DOORS & WINDOWS	90.2	29.0	74.4	90.0	34.2	75.6	90.0	50.1	79.7	90.0	46.6	78.8	95.5	49.7	83.6	90.1	38.8	76.8
09200	PLASTER & GYPSUM BOARD	92.2	23.2	48.5	100.4	20.3	49.7	100.0	43.8	64.4	101.6	43.8	65.0	104.4	37.9	62.3	100.0	29.5	55.3
095,098	CEILINGS & ACOUSTICAL TREATMENT	90.5	23.2	50.2	88.2	20.3	47.5	88.2	43.8	61.6	88.2	43.8	61.6	97.1	37.9	61.6	88.2	29.5	53.0
09600	FLOORING	94.1	29.8	77.1	94.8	37.8	79.7	94.6	36.3	79.2	96.2	37.8	80.7	96.4	43.7	82.5	93.5	29.8	76.7
097,099	WALL FINISHES, PAINTS & COATINGS	144.6	22.6	70.4	91.2	51.8	67.3	91.2	42.8	61.9	91.2	51.8	67.3	91.2	43.2	62.1	91.2	24.6	50.8
09	FINISHES	93.7	26.1	58.6	93.0	29.0	59.7	93.1	45.7	68.4	93.8	45.3	68.5	96.3	43.9	69.0	92.7	29.8	60.0
10 - 14	TOTAL DIV. 10000 - 14000	100.0	49.8	89.3	100.0	63.5	92.2	100.0	73.9	94.4	100.0	71.0	93.8	100.0	73.2	94.3	100.0	48.8	89.1
15	MECHANICAL	95.9	27.2	65.9	95.7	67.0	83.1	100.0	77.5	90.2	95.7	73.1	85.8	100.0	77.6	90.2	95.7	67.3	83.2
16	ELECTRICAL	92.7	18.6	56.6	92.4	44.0	68.8	91.2	66.6	79.2	90.6	44.0	67.9	96.6	66.6	81.9	94.7	18.6	57.6
01 - 16	WEIGHTED AVERAGE	98.9	37.0	70.8	96.0	51.2	75.6	97.2	66.2	83.1	97.0	58.5	79.5	98.0	67.7	84.2	96.8	48.8	75.0

NEBRASKA / NEVADA

DIVISION		NORFOLK 687 MAT.	INST.	TOTAL	NORTH PLATTE 691 MAT.	INST.	TOTAL	OMAHA 680 - 681 MAT.	INST.	TOTAL	VALENTINE 692 MAT.	INST.	TOTAL	CARSON CITY 897 MAT.	INST.	TOTAL	ELKO 898 MAT.	INST.	TOTAL
01590	EQUIPMENT RENTAL	.0	90.8	90.8	.0	101.6	101.6	.0	90.8	90.8	.0	94.0	94.0	.0	101.7	101.7	.0	101.7	101.7
02	SITE CONSTRUCTION	77.7	88.5	85.7	98.8	89.8	92.2	78.9	89.7	86.9	82.5	92.3	89.8	63.0	104.1	93.4	57.6	104.0	92.0
03100	CONCRETE FORMS & ACCESSORIES	82.5	47.1	51.9	96.5	47.3	53.8	96.1	72.4	75.5	83.7	23.7	31.7	94.3	103.0	101.8	106.4	91.9	93.8
03200	CONCRETE REINFORCEMENT	106.4	59.3	82.6	107.1	74.2	90.5	101.6	76.0	88.7	107.7	41.6	74.3	113.7	118.6	116.2	115.6	118.4	117.0
03300	CAST-IN-PLACE CONCRETE	109.8	57.9	88.1	115.5	52.5	89.2	108.3	75.9	94.8	101.7	44.0	77.6	111.6	85.8	100.8	100.2	83.4	93.2
03	CONCRETE	102.5	53.9	78.3	108.7	55.6	82.2	102.0	74.4	88.2	105.6	35.4	70.6	109.3	99.8	104.5	101.6	94.0	97.8
04	MASONRY	119.9	63.5	84.5	91.1	40.5	59.4	100.9	74.5	84.9	102.5	31.0	57.6	135.0	80.6	100.9	135.6	81.4	101.6
05	METALS	94.9	67.3	85.8	93.0	83.2	89.8	97.3	76.1	90.3	105.9	57.6	90.1	93.4	104.5	97.0	92.6	103.2	96.1
06	WOOD & PLASTICS	79.4	44.8	61.4	95.4	45.5	69.3	94.7	73.3	83.5	80.1	22.0	49.8	91.2	106.4	99.1	105.5	92.8	98.9
07	THERMAL & MOISTURE PROTECTION	95.2	56.3	76.1	96.8	47.4	72.5	92.5	70.4	81.6	96.1	32.0	64.6	102.6	89.4	96.1	102.5	83.3	93.1
08	DOORS & WINDOWS	92.1	47.7	80.6	89.3	53.2	80.0	98.8	67.0	90.6	92.2	27.7	75.5	94.4	112.7	99.1	97.0	97.3	97.1
09200	PLASTER & GYPSUM BOARD	99.0	43.8	64.0	100.5	43.8	64.5	106.9	73.0	85.4	103.4	20.3	50.7	81.3	106.4	97.2	91.8	92.4	92.2
095,098	CEILINGS & ACOUSTICAL TREATMENT	102.5	43.8	67.3	90.0	43.8	62.3	107.9	73.0	87.0	106.7	20.3	54.9	100.4	106.4	104.0	100.4	92.4	95.6
09600	FLOORING	117.2	45.4	98.2	94.5	37.8	79.5	122.7	48.4	103.1	120.0	45.4	100.3	101.2	66.7	92.1	107.5	66.7	96.7
097,099	WALL FINISHES, PAINTS & COATINGS	148.5	51.8	89.8	91.2	51.8	67.3	148.5	73.6	103.0	148.5	34.8	79.5	103.7	81.6	90.3	103.7	81.6	90.3
09	FINISHES	110.5	46.3	77.1	93.6	45.3	68.4	114.6	67.9	90.3	113.4	28.2	69.1	96.1	94.0	95.0	99.2	86.2	92.5
10 - 14	TOTAL DIV. 10000 - 14000	100.0	69.2	93.5	100.0	55.2	90.5	100.0	75.9	94.9	100.0	40.3	87.3	100.0	114.9	103.2	100.0	91.5	98.2
15	MECHANICAL	95.4	73.1	85.6	100.0	73.4	88.4	99.8	78.4	90.4	95.3	24.3	64.2	100.1	92.9	96.9	98.0	84.5	92.1
16	ELECTRICAL	91.6	55.4	73.9	93.0	44.0	69.1	90.7	82.1	86.5	90.7	44.0	67.9	88.6	88.0	88.3	88.1	70.1	79.3
01 - 16	WEIGHTED AVERAGE	97.7	62.5	81.7	96.8	59.9	80.0	99.1	76.7	89.0	99.2	39.7	72.2	98.4	95.6	97.2	97.5	87.9	93.1

COST INDEXES

651

DIVISION		NEVADA									NEW HAMPSHIRE							
		ELY 893			LAS VEGAS 889 - 891			RENO 894 - 895			CHARLESTON 036			CLAREMONT 037			CONCORD 032 - 033		
		MAT.	INST.	TOTAL	MAT.	INST.	TOTAL	MAT.	INST.	TOTAL	MAT.	INST.	TOTAL	MAT.	INST.	TOTAL	MAT.	INST.	TOTAL
01590	EQUIPMENT RENTAL	.0	101.7	101.7	.0	101.7	101.7	.0	101.7	101.7	.0	101.6	101.6	.0	101.6	101.6	.0	101.6	101.6
02	SITE CONSTRUCTION	62.7	103.2	92.6	62.5	106.3	94.9	62.7	104.1	93.3	79.8	98.5	93.6	74.2	98.5	92.2	86.8	100.2	96.7
03100	CONCRETE FORMS & ACCESSORIES	97.7	74.0	77.2	93.2	107.9	105.9	94.3	102.9	101.8	87.2	38.7	45.2	94.2	38.7	46.1	88.8	63.9	67.2
03200	CONCRETE REINFORCEMENT	114.3	109.3	111.7	108.1	118.7	113.4	108.1	118.6	113.4	86.6	60.2	73.2	86.6	60.2	73.2	86.6	91.9	89.3
03300	CAST-IN-PLACE CONCRETE	107.6	80.3	96.2	105.3	108.6	106.7	113.7	85.8	102.1	97.6	54.8	79.7	89.6	54.8	75.1	94.8	97.4	95.9
03	CONCRETE	109.8	83.2	96.5	105.3	109.8	107.6	109.4	99.7	104.6	102.6	49.2	76.0	94.5	49.2	71.9	99.4	81.0	90.2
04	MASONRY	140.7	72.6	97.9	125.8	89.8	103.2	134.3	80.6	100.6	92.2	42.7	61.1	92.5	42.7	61.2	95.6	92.8	93.8
05	METALS	92.6	97.9	94.3	94.0	105.9	97.9	93.8	104.5	97.3	94.7	67.9	86.0	94.7	67.9	86.0	95.3	87.1	92.6
06	WOOD & PLASTICS	95.2	73.6	83.9	90.9	106.4	99.0	91.2	106.4	99.1	84.9	38.0	60.4	92.5	38.0	64.0	86.2	55.9	70.4
07	THERMAL & MOISTURE PROTECTION	103.3	77.6	90.7	105.1	98.7	102.0	102.6	89.4	96.1	100.9	46.6	74.2	100.8	46.6	74.2	101.6	89.0	95.4
08	DOORS & WINDOWS	96.9	84.5	93.7	94.7	112.7	99.4	94.7	111.1	99.0	103.7	41.6	87.6	105.0	41.6	88.6	105.0	65.4	94.7
09200	PLASTER & GYPSUM BOARD	87.4	72.6	78.0	81.9	106.4	97.5	83.2	106.4	97.9	94.0	35.4	56.8	97.6	35.4	58.2	94.0	53.8	68.5
095,098	CEILINGS & ACOUSTICAL TREATMENT	100.4	72.6	83.8	105.8	106.4	106.2	107.6	106.4	106.9	87.5	35.4	56.3	87.5	35.4	56.3	87.5	53.8	67.3
09600	FLOORING	104.9	74.7	96.9	101.2	79.5	95.4	101.2	66.7	92.1	93.8	41.9	80.1	96.7	41.9	82.2	94.4	103.7	96.8
097,099	WALL FINISHES, PAINTS & COATINGS	103.7	81.6	90.3	103.7	105.1	104.6	103.7	81.6	90.3	92.8	33.3	56.7	92.8	33.3	56.7	92.8	90.6	91.4
09	FINISHES	98.3	74.4	85.8	97.5	101.9	99.8	98.2	94.0	96.0	93.9	38.0	64.8	94.9	38.0	65.3	94.9	72.5	83.3
10 - 14	TOTAL DIV. 10000 - 14000	100.0	76.1	94.9	100.0	110.2	102.2	100.0	114.9	103.2	100.0	49.0	89.2	100.0	49.0	89.2	100.0	80.0	95.7
15	MECHANICAL	98.0	79.7	90.0	100.1	111.1	104.9	100.1	92.9	96.9	95.7	41.1	71.8	95.7	41.1	71.8	100.1	86.1	94.0
16	ELECTRICAL	88.2	70.0	79.4	90.5	105.5	97.8	88.6	88.0	88.3	98.0	36.6	68.0	98.0	36.6	68.0	100.2	74.5	87.7
01 - 16	WEIGHTED AVERAGE	98.7	81.2	90.7	98.1	105.5	101.5	98.7	95.5	97.3	97.0	49.3	75.4	96.3	49.3	75.0	98.6	82.9	91.4

DIVISION		NEW HAMPSHIRE															NEW JERSEY		
		KEENE 034			LITTLETON 035			MANCHESTER 031			NASHUA 030			PORTSMOUTH 038			ATLANTIC CITY 082,084		
		MAT.	INST.	TOTAL	MAT.	INST.	TOTAL	MAT.	INST.	TOTAL	MAT.	INST.	TOTAL	MAT.	INST.	TOTAL	MAT.	INST.	TOTAL
01590	EQUIPMENT RENTAL	.0	101.6	101.6	.0	101.6	101.6	.0	101.6	101.6	.0	101.6	101.6	.0	101.6	101.6	.0	99.4	99.4
02	SITE CONSTRUCTION	86.9	98.6	95.5	74.1	98.1	91.9	86.7	100.2	96.7	88.0	100.2	97.0	82.2	99.2	94.8	90.6	105.2	101.4
03100	CONCRETE FORMS & ACCESSORIES	92.6	39.8	46.9	103.9	45.1	53.0	100.9	63.9	68.9	101.4	63.9	68.9	88.8	59.4	63.3	111.3	123.5	121.8
03200	CONCRETE REINFORCEMENT	86.6	71.7	79.1	87.4	43.1	65.0	108.2	91.9	100.0	108.2	91.9	100.0	86.6	91.8	89.2	82.5	110.5	96.6
03300	CAST-IN-PLACE CONCRETE	98.1	55.2	80.1	88.0	49.1	71.7	101.0	97.4	99.5	92.8	97.4	94.7	88.0	90.9	89.2	83.2	128.8	102.3
03	CONCRETE	102.4	52.0	77.3	93.9	46.9	70.4	106.8	81.0	93.9	102.9	81.0	91.9	93.8	76.6	85.2	95.9	121.7	108.8
04	MASONRY	96.6	42.7	62.8	105.4	41.2	65.1	97.9	92.8	94.7	98.1	92.8	94.8	93.5	82.2	86.4	100.8	125.0	116.0
05	METALS	95.2	73.4	88.1	95.2	60.3	83.8	100.0	87.1	95.8	99.9	87.1	95.7	96.3	85.0	92.6	95.0	100.2	96.7
06	WOOD & PLASTICS	90.7	38.0	63.2	102.7	43.6	71.8	100.4	55.9	77.2	100.4	55.9	77.2	86.2	55.9	70.4	111.6	122.9	117.5
07	THERMAL & MOISTURE PROTECTION	101.5	48.5	75.5	100.9	51.2	76.5	101.6	89.0	95.4	102.1	89.0	95.6	101.5	92.1	96.9	101.7	122.6	112.0
08	DOORS & WINDOWS	102.1	51.0	88.9	106.1	41.2	89.3	107.1	65.4	96.3	107.1	65.4	96.3	108.0	59.4	95.4	104.6	115.2	107.4
09200	PLASTER & GYPSUM BOARD	96.8	35.4	57.9	110.9	41.1	66.7	105.2	53.8	72.6	105.2	53.8	72.6	94.0	53.8	68.5	107.7	122.6	117.1
095,098	CEILINGS & ACOUSTICAL TREATMENT	87.5	35.4	56.3	87.5	41.1	59.7	101.1	53.8	72.7	101.1	53.8	72.7	89.5	53.8	68.1	87.5	122.6	108.5
09600	FLOORING	96.0	68.7	88.8	106.6	41.9	89.5	100.0	103.7	101.0	99.8	103.7	100.8	94.4	103.7	96.8	104.0	132.6	111.6
097,099	WALL FINISHES, PAINTS & COATINGS	92.8	33.3	56.7	92.8	68.0	77.7	92.8	90.6	91.4	92.8	90.6	91.4	92.8	40.3	61.0	92.8	119.7	109.1
09	FINISHES	95.9	43.5	68.6	99.8	46.3	72.0	101.3	72.5	86.4	101.6	72.5	86.5	95.1	64.2	79.0	100.3	125.0	113.1
10 - 14	TOTAL DIV. 10000 - 14000	100.0	69.7	93.5	100.0	51.7	89.7	100.0	80.0	95.7	100.0	80.0	95.7	100.0	75.8	94.9	100.0	117.8	103.8
15	MECHANICAL	95.7	44.0	73.1	95.7	76.1	87.1	100.1	86.1	94.0	100.1	86.1	94.0	100.1	80.4	91.5	99.7	109.9	104.2
16	ELECTRICAL	98.0	36.6	68.0	99.0	74.5	87.0	100.4	74.5	87.8	100.2	74.5	87.7	98.5	74.5	86.8	96.1	117.6	106.6
01 - 16	WEIGHTED AVERAGE	97.6	52.5	77.1	97.7	62.0	81.5	101.2	82.8	92.9	100.9	82.8	92.7	98.0	78.4	89.1	98.6	115.7	106.4

DIVISION		NEW JERSEY																	
		CAMDEN 081			DOVER 078			ELIZABETH 072			HACKENSACK 076			JERSEY CITY 073			LONG BRANCH 077		
		MAT.	INST.	TOTAL	MAT.	INST.	TOTAL	MAT.	INST.	TOTAL	MAT.	INST.	TOTAL	MAT.	INST.	TOTAL	MAT.	INST.	TOTAL
01590	EQUIPMENT RENTAL	.0	99.4	99.4	.0	101.6	101.6	.0	101.6	101.6	.0	101.6	101.6	.0	99.4	99.4	.0	98.9	98.9
02	SITE CONSTRUCTION	91.3	105.0	101.4	101.6	105.3	104.4	105.3	105.3	105.3	101.9	104.8	104.0	91.3	104.9	101.3	95.9	105.3	102.9
03100	CONCRETE FORMS & ACCESSORIES	101.4	123.2	120.3	97.8	122.8	119.4	111.5	122.7	121.2	97.8	122.6	119.3	101.4	123.7	120.7	102.1	123.7	120.8
03200	CONCRETE REINFORCEMENT	108.2	104.9	106.6	83.4	118.6	101.2	83.4	118.6	101.2	83.4	118.6	101.2	108.2	118.6	113.5	83.4	118.5	101.1
03300	CAST-IN-PLACE CONCRETE	80.7	128.7	100.7	103.5	128.6	114.0	88.9	128.5	105.4	101.1	128.5	112.6	80.7	130.0	101.3	89.7	128.1	105.7
03	CONCRETE	97.1	120.6	108.8	103.7	123.1	113.4	99.4	123.1	111.2	101.8	123.0	112.4	97.1	123.8	110.5	101.2	123.1	112.2
04	MASONRY	90.3	125.0	112.1	95.3	123.0	112.6	112.4	123.0	119.0	99.5	123.0	114.2	90.3	125.5	112.4	104.5	123.9	116.7
05	METALS	99.8	97.9	99.2	95.0	108.7	99.4	96.0	108.6	100.1	95.0	108.5	99.4	99.9	105.6	101.7	95.1	105.4	98.4
06	WOOD & PLASTICS	100.4	122.9	112.1	99.5	122.9	111.7	115.3	122.9	119.3	99.5	122.9	111.7	100.4	122.9	112.1	100.4	122.9	112.1
07	THERMAL & MOISTURE PROTECTION	101.4	122.1	111.6	102.1	126.0	113.9	102.4	126.0	114.0	101.8	119.7	110.6	101.4	127.1	114.1	101.7	129.5	115.3
08	DOORS & WINDOWS	107.1	113.8	108.8	110.2	118.2	112.3	108.0	118.2	110.7	107.2	118.2	110.1	107.1	118.2	109.9	103.0	118.2	106.9
09200	PLASTER & GYPSUM BOARD	105.2	122.6	116.2	100.9	122.6	114.6	108.9	122.6	117.6	100.9	122.6	114.6	105.2	122.6	116.2	102.9	122.6	115.4
095,098	CEILINGS & ACOUSTICAL TREATMENT	101.1	122.6	114.0	87.5	122.6	108.5	89.5	122.6	109.3	87.5	122.6	108.5	101.1	122.6	114.0	87.5	122.6	108.5
09600	FLOORING	99.8	132.6	108.5	98.6	132.6	107.6	104.5	132.6	111.9	98.6	132.6	107.6	99.8	132.6	108.5	100.1	132.6	108.7
097,099	WALL FINISHES, PAINTS & COATINGS	92.8	119.7	109.1	92.7	119.7	109.1	92.7	119.7	109.1	92.7	119.7	109.1	92.8	119.7	109.1	92.8	119.7	109.1
09	FINISHES	101.8	125.0	113.9	97.8	124.1	111.5	101.6	124.1	113.3	97.7	124.1	111.4	101.8	124.8	113.8	98.7	124.7	112.2
10 - 14	TOTAL DIV. 10000 - 14000	100.0	117.8	103.8	100.0	125.0	105.1	100.0	124.0	105.1	100.0	124.1	105.1	100.0	125.1	105.3	100.0	117.2	103.7
15	MECHANICAL	100.1	109.6	104.3	99.8	125.5	111.0	100.1	121.4	109.4	99.8	125.5	111.0	100.1	126.8	111.8	99.8	125.7	111.1
16	ELECTRICAL	100.6	117.6	108.9	97.9	128.9	113.0	98.5	128.9	113.3	97.9	132.5	114.8	102.2	132.5	117.0	97.6	119.5	108.3
01 - 16	WEIGHTED AVERAGE	100.0	115.2	106.9	100.0	121.5	109.8	101.1	120.7	110.0	99.7	121.7	109.7	100.2	122.5	110.3	99.4	120.0	108.7

COST INDEXES

City Cost Indexes

NEW JERSEY

DIVISION		NEWARK 070 - 071			NEW BRUNSWICK 088 - 089			PATERSON 074 - 075			POINT PLEASANT 087			SUMMIT 079			TRENTON 085 - 086		
		MAT.	INST.	TOTAL	MAT.	INST.	TOTAL	MAT.	INST.	TOTAL	MAT.	INST.	TOTAL	MAT.	INST.	TOTAL	MAT.	INST.	TOTAL
01590	EQUIPMENT RENTAL	.0	101.6	101.6	.0	98.9	98.9	.0	101.6	101.6	.0	98.9	98.9	.0	101.6	101.6	.0	98.9	98.9
02	SITE CONSTRUCTION	109.8	105.3	106.5	103.6	105.2	104.8	103.7	104.8	104.5	104.9	105.4	105.2	102.9	105.3	104.7	93.0	105.4	102.1
03100	CONCRETE FORMS & ACCESSORIES	99.8	122.7	119.7	105.0	122.7	120.3	100.4	122.6	119.6	98.8	123.4	120.1	101.2	122.7	119.8	100.3	123.4	120.3
03200	CONCRETE REINFORCEMENT	108.2	118.6	113.5	83.4	118.5	101.1	108.2	118.6	113.5	83.4	117.9	100.8	83.4	118.6	101.2	108.2	114.7	111.5
03300	CAST-IN-PLACE CONCRETE	94.2	128.6	108.5	102.8	126.6	112.8	102.8	128.5	113.6	102.8	119.9	109.9	85.9	128.5	103.7	95.5	119.9	105.7
03	CONCRETE	103.4	123.1	113.2	113.4	122.2	117.8	107.6	123.0	115.3	113.0	120.1	116.5	95.9	123.1	109.5	104.1	119.4	111.8
04	MASONRY	100.1	123.0	114.5	98.3	121.3	112.8	95.8	123.0	112.9	90.4	116.0	106.5	98.1	123.0	113.7	89.9	116.0	106.3
05	METALS	99.8	108.6	102.7	95.0	105.4	98.4	94.7	108.5	99.2	95.0	104.0	98.0	95.0	108.6	99.4	94.6	101.2	96.8
06	WOOD & PLASTICS	102.2	122.9	113.0	104.8	122.8	114.2	102.2	122.9	113.0	97.8	122.8	110.8	103.4	122.9	113.6	100.4	122.8	112.1
07	THERMAL & MOISTURE PROTECTION	101.7	126.0	113.7	101.9	128.4	114.9	102.2	119.7	110.8	101.9	126.7	114.1	102.7	126.0	114.2	100.0	116.0	107.9
08	DOORS & WINDOWS	113.1	118.2	114.4	98.8	118.2	103.8	113.1	118.2	114.4	100.9	118.2	105.4	115.8	118.2	116.4	107.1	116.4	109.5
09200	PLASTER & GYPSUM BOARD	105.2	122.6	116.2	104.5	122.6	116.0	105.2	122.6	116.2	100.9	122.6	114.6	102.9	122.6	115.4	105.2	122.6	116.2
095,098	CEILINGS & ACOUSTICAL TREATMENT	101.1	122.6	114.0	87.5	122.6	108.5	101.1	122.6	114.0	87.5	122.6	108.5	87.5	122.6	108.5	101.1	122.6	114.0
09600	FLOORING	100.0	132.6	108.6	101.5	132.6	109.7	99.8	132.6	108.5	98.6	117.0	103.5	100.1	132.6	108.7	100.0	117.0	104.5
097,099	WALL FINISHES, PAINTS & COATINGS	92.7	119.7	109.1	92.8	119.7	109.1	92.7	119.7	109.1	92.8	119.7	109.1	92.7	119.7	109.1	92.8	119.7	109.1
09	FINISHES	102.4	124.1	113.7	100.0	124.1	112.5	102.0	124.1	113.5	98.6	121.8	110.7	98.8	124.1	112.0	101.9	121.8	112.3
10 - 14	TOTAL DIV. 10000 - 14000	100.0	124.1	105.1	100.0	123.8	105.1	100.0	124.0	105.1	100.0	105.1	101.1	100.0	124.0	105.1	100.0	117.7	103.8
15	MECHANICAL	100.1	122.5	109.9	99.7	124.2	110.5	100.1	125.4	111.2	99.7	120.1	108.7	99.8	121.4	109.2	100.1	126.1	111.5
16	ELECTRICAL	102.1	132.5	116.9	96.7	119.5	107.8	102.2	128.9	115.2	96.1	119.5	107.5	98.5	128.9	113.3	100.8	129.8	114.9
01 - 16	WEIGHTED AVERAGE	102.5	121.4	111.1	100.3	119.4	108.9	101.8	121.2	110.6	99.8	116.8	107.5	100.1	120.7	109.4	99.9	119.1	108.6

		NEW JERSEY VINELAND 080,083			NEW MEXICO ALBUQUERQUE 870 - 872			CARRIZOZO 883			CLOVIS 881			FARMINGTON 874			GALLUP 873		
DIVISION		MAT.	INST.	TOTAL	MAT.	INST.	TOTAL	MAT.	INST.	TOTAL	MAT.	INST.	TOTAL	MAT.	INST.	TOTAL	MAT.	INST.	TOTAL
01590	EQUIPMENT RENTAL	.0	99.4	99.4	.0	116.3	116.3	.0	116.3	116.3	.0	116.3	116.3	.0	116.3	116.3	.0	116.3	116.3
02	SITE CONSTRUCTION	95.1	105.0	102.4	80.2	111.5	103.3	100.7	111.5	108.7	88.2	111.4	105.4	85.7	111.5	104.8	94.2	111.5	107.0
03100	CONCRETE FORMS & ACCESSORIES	95.9	123.6	119.9	96.1	69.0	72.6	96.1	69.0	72.6	96.1	68.7	72.4	96.1	69.0	72.6	96.1	69.0	72.6
03200	CONCRETE REINFORCEMENT	82.5	110.5	96.6	108.8	67.2	87.8	116.6	67.2	91.6	117.9	59.3	88.3	118.7	67.2	92.7	113.7	67.2	90.2
03300	CAST-IN-PLACE CONCRETE	89.7	128.9	106.1	104.5	76.4	92.8	96.9	76.4	88.3	96.8	76.3	88.2	105.1	76.4	93.1	98.8	76.4	89.5
03	CONCRETE	100.7	121.8	111.2	105.2	72.2	88.7	119.7	72.2	96.0	108.4	70.6	89.5	109.0	72.2	90.6	114.9	72.2	93.6
04	MASONRY	87.7	125.0	111.1	122.3	66.3	87.2	115.3	66.3	84.6	115.3	66.3	84.6	127.3	66.3	89.1	115.4	66.3	84.6
05	METALS	94.9	100.3	96.7	100.2	88.0	96.2	97.3	88.0	94.2	97.0	84.0	92.7	97.9	88.0	94.7	97.3	88.0	94.3
06	WOOD & PLASTICS	94.8	122.9	109.4	95.7	70.1	82.4	95.7	70.1	82.4	95.7	70.1	82.4	95.7	70.1	82.4	95.7	70.1	82.4
07	THERMAL & MOISTURE PROTECTION	101.3	122.6	111.8	101.6	72.6	87.3	104.4	72.6	88.7	102.7	72.6	87.9	102.5	72.6	87.8	103.9	72.6	88.5
08	DOORS & WINDOWS	100.4	115.2	104.2	94.7	72.5	89.0	93.6	72.5	88.1	93.8	70.0	87.6	97.6	72.5	91.1	97.6	72.5	91.1
09200	PLASTER & GYPSUM BOARD	99.7	122.6	114.2	85.4	68.8	74.8	80.4	68.8	73.0	80.4	68.8	73.0	80.4	68.8	73.0	80.4	68.8	73.0
095,098	CEILINGS & ACOUSTICAL TREATMENT	87.5	122.6	108.5	105.8	68.8	83.6	96.9	68.8	80.0	96.9	68.8	80.0	96.9	68.8	80.0	96.9	68.8	80.0
09600	FLOORING	97.6	132.6	106.9	101.2	65.0	91.6	101.2	65.0	91.6	101.2	65.0	91.6	101.2	65.0	91.6	101.2	65.0	91.6
097,099	WALL FINISHES, PAINTS & COATINGS	92.8	119.7	109.1	103.7	60.7	77.6	103.7	60.7	77.6	103.7	60.7	77.6	103.7	60.7	77.6	103.7	60.7	77.6
09	FINISHES	97.5	125.0	111.8	98.5	67.2	82.2	97.3	67.2	81.7	96.2	67.2	81.1	95.9	67.2	81.0	96.8	67.2	81.4
10 - 14	TOTAL DIV. 10000 - 14000	100.0	117.8	103.8	100.0	76.4	95.0	100.0	76.4	95.0	100.0	76.4	95.0	100.0	76.4	95.0	100.0	76.4	95.0
15	MECHANICAL	99.7	119.6	108.5	100.1	72.9	88.2	97.6	72.9	86.8	97.6	72.5	86.6	100.1	72.9	88.2	97.4	72.9	86.7
16	ELECTRICAL	96.1	117.6	106.6	81.8	76.3	79.1	81.8	76.3	79.1	79.7	76.3	78.0	80.2	76.3	78.3	79.5	76.3	78.0
01 - 16	WEIGHTED AVERAGE	97.8	117.7	106.8	98.3	76.8	88.6	99.0	76.8	88.9	96.9	76.1	87.4	98.6	76.8	88.7	98.3	76.8	88.6

NEW MEXICO

DIVISION		LAS CRUCES 880			LAS VEGAS 877			ROSWELL 882			SANTA FE 875			SOCORRO 878			TRUTH/CONSEQUENCES 879		
		MAT.	INST.	TOTAL	MAT.	INST.	TOTAL	MAT.	INST.	TOTAL	MAT.	INST.	TOTAL	MAT.	INST.	TOTAL	MAT.	INST.	TOTAL
01590	EQUIPMENT RENTAL	.0	86.7	86.7	.0	116.3	116.3	.0	116.3	116.3	.0	116.3	116.3	.0	116.3	116.3	.0	86.7	86.7
02	SITE CONSTRUCTION	91.2	86.7	87.9	84.6	111.5	104.5	90.3	111.4	105.9	79.9	111.5	103.3	81.8	111.5	103.7	101.8	86.7	90.6
03100	CONCRETE FORMS & ACCESSORIES	93.5	67.2	70.7	96.1	69.0	72.6	96.1	68.9	72.5	96.1	69.0	72.6	96.1	69.0	72.6	93.5	67.4	70.9
03200	CONCRETE REINFORCEMENT	112.8	58.8	85.5	115.6	67.2	91.1	117.9	59.3	88.3	116.6	67.2	91.6	117.9	67.2	92.3	111.3	58.9	84.8
03300	CAST-IN-PLACE CONCRETE	91.4	66.7	81.1	102.2	76.4	91.4	96.8	76.4	88.3	98.8	76.4	89.4	100.1	76.4	90.2	109.8	66.8	91.8
03	CONCRETE	87.3	66.2	76.7	106.3	72.2	89.3	109.3	70.7	90.0	103.8	72.2	88.0	105.0	72.2	88.7	99.0	66.3	82.7
04	MASONRY	110.9	60.7	79.4	115.7	65.7	84.3	127.4	66.3	89.1	119.4	66.3	86.1	115.5	66.3	84.7	113.2	61.9	81.1
05	METALS	97.8	75.2	90.4	97.1	88.0	94.1	97.9	84.1	93.4	97.9	88.0	94.6	97.3	88.0	94.3	96.9	75.5	89.9
06	WOOD & PLASTICS	87.4	68.7	77.6	95.7	70.1	82.4	95.7	70.1	82.4	95.7	70.1	82.4	95.7	70.1	82.4	87.4	68.7	77.6
07	THERMAL & MOISTURE PROTECTION	88.7	65.5	77.3	102.0	72.4	87.4	102.0	72.6	87.9	101.9	72.6	87.5	101.9	72.6	87.5	89.4	65.5	77.8
08	DOORS & WINDOWS	87.1	69.2	82.5	93.8	72.5	88.3	93.6	70.0	87.5	93.7	72.5	88.2	93.6	72.5	88.1	87.0	69.2	82.4
09200	PLASTER & GYPSUM BOARD	82.7	68.8	73.9	80.4	68.8	73.0	80.4	68.8	73.0	80.4	68.8	73.0	80.4	68.8	73.0	82.7	68.8	73.9
095,098	CEILINGS & ACOUSTICAL TREATMENT	96.8	68.8	80.0	96.9	68.8	80.0	96.9	68.8	80.0	96.9	68.8	80.0	96.9	68.8	80.0	96.8	68.8	80.0
09600	FLOORING	133.4	65.2	115.4	101.2	65.0	91.6	101.2	65.0	91.6	101.2	65.0	91.6	101.2	65.0	91.6	133.4	67.4	116.0
097,099	WALL FINISHES, PAINTS & COATINGS	97.1	60.7	75.0	103.7	60.7	77.6	103.7	60.7	77.6	103.7	60.7	77.6	103.7	60.7	77.6	97.1	60.7	75.0
09	FINISHES	111.0	66.7	87.7	95.8	67.2	80.9	96.3	67.2	81.2	95.7	67.2	80.9	95.7	67.2	80.9	111.6	66.6	88.2
10 - 14	TOTAL DIV. 10000 - 14000	100.0	72.3	94.1	100.0	76.4	95.0	100.0	76.4	95.0	100.0	76.4	95.0	100.0	76.4	95.0	100.0	72.4	94.1
15	MECHANICAL	100.3	72.4	88.1	97.4	72.9	86.7	100.1	72.8	88.2	100.1	72.9	88.2	97.4	72.9	86.7	97.1	72.4	86.3
16	ELECTRICAL	82.5	60.4	71.7	81.8	76.3	79.1	80.9	76.3	78.7	81.8	76.3	79.1	80.0	76.3	78.2	85.0	76.3	80.8
01 - 16	WEIGHTED AVERAGE	95.8	69.1	83.7	96.8	76.8	87.7	98.5	76.1	88.4	97.3	76.8	88.0	96.4	76.8	87.5	97.1	71.6	85.5

City Cost Indexes

Table 1

DIVISION		NEW MEXICO			NEW YORK														
		TUCUMCARI			ALBANY			BINGHAMTON			BRONX			BROOKLYN			BUFFALO		
		884			120 - 122			137 - 139			104			112			140 - 142		
		MAT.	INST.	TOTAL	MAT.	INST.	TOTAL	MAT.	INST.	TOTAL	MAT.	INST.	TOTAL	MAT.	INST.	TOTAL	MAT.	INST.	TOTAL
01590	EQUIPMENT RENTAL	.0	116.3	116.3	.0	115.7	115.7	.0	115.8	115.8	.0	117.3	117.3	.0	117.2	117.2	.0	93.1	93.1
02	SITE CONSTRUCTION	87.9	111.4	105.3	71.8	106.5	97.5	93.8	89.8	90.9	124.3	127.8	126.9	119.2	130.3	127.4	96.1	93.4	94.1
03100	CONCRETE FORMS & ACCESSORIES	96.1	68.7	72.4	98.7	92.1	93.0	103.9	76.5	80.2	108.5	167.5	159.6	111.2	167.1	159.7	100.3	113.7	111.9
03200	CONCRETE REINFORCEMENT	115.6	59.3	87.1	103.1	89.8	96.4	102.0	87.8	94.8	105.1	187.0	146.5	103.3	184.5	144.3	104.0	105.9	105.0
03300	CAST-IN-PLACE CONCRETE	96.8	76.3	88.2	88.7	100.5	93.6	105.9	86.7	97.9	107.9	156.0	128.0	107.9	154.0	127.2	118.7	119.3	118.9
03	CONCRETE	107.8	70.6	89.2	101.3	95.5	98.4	101.1	84.2	92.6	108.6	165.3	136.9	113.4	164.0	138.7	108.2	113.3	110.7
04	MASONRY	129.4	66.3	89.8	89.3	97.0	94.1	106.0	79.0	89.1	94.6	155.8	133.0	115.7	155.8	140.8	104.9	119.8	114.2
05	METALS	97.0	84.0	92.7	96.9	104.9	99.5	93.4	114.1	100.2	95.8	139.4	110.0	102.4	138.5	114.2	100.4	94.8	98.5
06	WOOD & PLASTICS	95.7	70.1	82.4	98.3	91.3	94.6	107.3	75.0	90.4	106.2	171.8	140.4	110.1	171.4	142.1	104.3	112.9	108.8
07	THERMAL & MOISTURE PROTECTION	102.6	72.6	87.8	90.8	92.8	91.8	101.5	80.3	91.1	105.0	153.2	128.7	104.8	152.6	128.3	98.9	110.7	104.7
08	DOORS & WINDOWS	93.5	70.0	87.4	95.7	85.0	92.9	90.7	73.4	86.2	92.1	165.3	111.0	89.9	164.5	109.2	92.7	103.3	95.5
09200	PLASTER & GYPSUM BOARD	80.4	68.8	73.0	111.0	90.8	98.2	116.3	73.7	89.3	106.3	173.3	148.8	109.8	173.3	150.1	102.4	113.0	109.1
095,098	CEILINGS & ACOUSTICAL TREATMENT	96.9	68.8	80.0	99.8	90.8	94.4	99.8	73.7	84.2	82.6	173.3	137.0	80.3	173.3	136.1	94.6	113.0	105.6
09600	FLOORING	101.2	65.0	91.6	86.8	94.1	88.7	97.6	80.2	93.0	96.0	158.9	112.6	104.3	158.9	118.7	97.7	121.0	103.9
097,099	WALL FINISHES, PAINTS & COATINGS	103.7	60.7	77.6	84.9	78.0	80.7	90.1	79.1	83.4	97.2	146.7	127.2	112.6	146.7	133.3	94.8	114.8	106.9
09	FINISHES	96.2	67.2	81.1	97.5	91.0	94.1	98.8	77.0	87.5	99.9	164.8	133.6	107.8	164.4	137.2	95.7	115.9	106.2
10 - 14	TOTAL DIV. 10000 - 14000	100.0	76.4	95.0	100.0	87.3	97.3	100.0	89.3	97.7	100.0	125.5	105.4	100.0	124.4	105.2	100.0	108.0	101.7
15	MECHANICAL	97.6	72.5	86.6	100.4	88.9	95.3	100.6	82.4	92.6	100.3	158.3	125.7	99.8	158.2	125.4	99.8	97.4	98.8
16	ELECTRICAL	81.8	76.3	79.1	103.1	89.4	96.4	101.1	80.2	90.9	100.7	161.5	130.4	102.0	172.0	136.1	99.7	97.6	98.7
01 - 16	WEIGHTED AVERAGE	97.8	76.1	87.9	97.9	93.9	96.1	98.6	84.5	92.2	100.2	155.2	125.2	103.4	156.5	127.5	99.9	104.7	102.1

Table 2 — NEW YORK

DIVISION		ELMIRA			FAR ROCKAWAY			FLUSHING			GLENS FALLS			HICKSVILLE			JAMAICA		
		148 - 149			116			113			128			115,117,118			114		
		MAT.	INST.	TOTAL	MAT.	INST.	TOTAL	MAT.	INST.	TOTAL	MAT.	INST.	TOTAL	MAT.	INST.	TOTAL	MAT.	INST.	TOTAL
01590	EQUIPMENT RENTAL	.0	116.0	116.0	.0	117.2	117.2	.0	117.2	117.2	.0	115.7	115.7	.0	117.2	117.2	.0	117.2	117.2
02	SITE CONSTRUCTION	91.3	88.9	89.6	123.1	130.3	128.4	123.1	130.3	128.4	62.0	106.0	94.6	112.1	130.3	125.6	117.0	130.3	126.8
03100	CONCRETE FORMS & ACCESSORIES	82.1	76.0	76.8	94.7	167.1	157.4	99.3	167.1	158.0	83.8	87.1	86.7	90.5	149.8	141.9	99.3	167.1	158.0
03200	CONCRETE REINFORCEMENT	105.3	85.5	95.3	103.3	184.5	144.3	105.1	184.5	145.2	102.8	89.7	96.1	103.3	184.3	144.2	103.3	184.5	144.3
03300	CAST-IN-PLACE CONCRETE	111.0	81.3	98.6	116.7	154.0	132.3	116.7	154.0	132.3	80.5	96.2	87.1	99.1	151.6	121.0	107.9	154.0	127.2
03	CONCRETE	101.7	82.0	91.9	119.7	164.0	141.8	120.3	164.0	142.1	90.8	91.8	91.3	104.6	155.5	130.0	112.6	164.0	138.3
04	MASONRY	101.3	73.6	83.9	120.0	155.8	142.4	113.8	155.8	140.1	91.7	90.2	90.8	109.5	152.4	136.4	117.7	155.8	141.6
05	METALS	95.6	118.7	103.2	102.5	138.5	114.2	102.5	138.5	114.2	93.5	104.5	97.1	103.5	137.5	114.6	102.5	138.5	114.3
06	WOOD & PLASTICS	86.2	74.5	80.1	92.5	171.4	133.7	97.7	171.4	136.1	82.2	87.9	85.1	88.5	151.2	121.2	97.7	171.4	136.1
07	THERMAL & MOISTURE PROTECTION	99.5	76.7	88.3	104.5	152.6	128.1	104.6	152.6	128.2	89.6	89.6	89.6	104.0	145.1	124.2	104.4	152.6	128.1
08	DOORS & WINDOWS	93.6	72.7	88.2	88.3	164.5	108.1	88.3	164.5	108.1	89.3	83.2	87.7	88.3	153.6	105.2	88.3	164.5	108.1
09200	PLASTER & GYPSUM BOARD	96.5	73.6	82.0	103.0	173.3	147.6	105.4	173.3	148.5	100.0	87.3	91.9	102.1	152.7	134.2	105.4	173.3	148.5
095,098	CEILINGS & ACOUSTICAL TREATMENT	91.0	73.6	80.5	80.3	173.3	136.1	80.3	173.3	136.1	92.8	87.3	89.5	79.3	152.7	123.3	80.3	173.3	136.1
09600	FLOORING	87.8	77.5	85.0	98.7	158.9	114.6	100.4	158.9	115.9	78.6	92.3	82.2	97.4	90.1	95.5	100.4	158.9	115.9
097,099	WALL FINISHES, PAINTS & COATINGS	93.0	80.8	85.6	112.6	146.7	133.3	112.6	146.7	133.3	84.9	78.0	80.7	112.6	146.7	133.3	112.6	156.3	139.1
09	FINISHES	90.7	76.2	83.2	105.5	164.4	136.1	106.3	164.4	136.5	91.0	87.3	89.0	104.0	137.5	121.4	105.9	165.5	136.9
10 - 14	TOTAL DIV. 10000 - 14000	100.0	88.2	97.5	100.0	124.4	105.2	100.0	124.4	105.2	100.0	69.2	93.4	100.0	120.4	104.3	100.0	124.4	105.2
15	MECHANICAL	95.7	81.9	89.6	99.5	158.2	122.9	99.5	158.2	122.9	96.1	80.9	89.6	99.8	144.9	119.6	99.5	158.2	122.9
16	ELECTRICAL	98.4	88.7	93.7	108.7	161.5	134.5	108.7	161.5	134.5	96.4	89.4	93.0	101.6	152.1	126.3	100.7	161.5	130.3
01 - 16	WEIGHTED AVERAGE	96.5	84.8	91.2	103.7	155.0	127.0	103.6	155.0	127.0	92.8	89.9	91.5	101.2	145.0	121.1	101.7	155.2	126.0

Table 3 — NEW YORK

DIVISION		JAMESTOWN			KINGSTON			LONG ISLAND CITY			MONTICELLO			MOUNT VERNON			NEW ROCHELLE		
		147			124			111			127			105			108		
		MAT.	INST.	TOTAL	MAT.	INST.	TOTAL	MAT.	INST.	TOTAL	MAT.	INST.	TOTAL	MAT.	INST.	TOTAL	MAT.	INST.	TOTAL
01590	EQUIPMENT RENTAL	.0	89.5	89.5	.0	117.2	117.2	.0	117.2	117.2	.0	117.2	117.2	.0	117.3	117.3	.0	117.3	117.3
02	SITE CONSTRUCTION	92.7	89.3	90.2	118.8	127.8	125.5	120.7	130.3	127.8	114.1	127.8	124.2	133.2	125.2	127.3	131.8	125.2	127.0
03100	CONCRETE FORMS & ACCESSORIES	82.2	82.4	82.3	85.5	105.3	102.6	104.5	167.1	158.7	94.1	105.1	103.6	95.9	136.1	130.7	114.6	136.1	133.2
03200	CONCRETE REINFORCEMENT	105.5	93.2	99.3	103.1	130.9	117.2	103.3	184.5	144.3	102.4	130.9	116.8	103.9	183.5	144.1	104.0	183.5	144.2
03300	CAST-IN-PLACE CONCRETE	115.1	113.7	114.5	108.7	116.2	111.8	111.3	154.0	129.2	101.8	116.1	107.8	120.7	134.2	126.4	120.7	134.2	126.4
03	CONCRETE	104.7	95.0	99.9	113.4	113.6	113.5	116.0	164.0	140.0	108.0	113.5	110.8	119.1	143.1	131.1	118.8	143.1	131.0
04	MASONRY	109.3	87.7	95.8	107.2	118.1	114.0	112.6	155.8	139.7	99.9	118.1	111.3	100.8	134.1	121.7	100.8	134.1	121.7
05	METALS	93.0	85.4	90.5	102.4	113.1	105.9	102.4	138.5	114.2	102.4	113.0	105.9	95.5	134.3	108.2	95.8	134.3	108.4
06	WOOD & PLASTICS	84.9	74.8	79.6	83.5	100.9	92.6	103.5	171.4	138.9	92.8	100.9	97.0	93.0	135.7	115.3	113.9	135.7	125.3
07	THERMAL & MOISTURE PROTECTION	98.8	89.7	94.3	106.9	124.7	115.7	104.6	152.6	128.2	106.5	124.7	115.4	106.4	137.5	121.7	106.6	137.5	121.8
08	DOORS & WINDOWS	93.4	79.1	89.7	93.6	114.5	99.0	88.3	164.5	108.1	88.4	114.5	95.2	92.1	145.1	105.8	92.2	145.1	105.9
09200	PLASTER & GYPSUM BOARD	91.4	73.8	80.3	97.9	100.9	99.8	108.2	173.3	149.5	101.5	100.9	101.2	99.5	136.3	122.8	112.4	136.3	127.5
095,098	CEILINGS & ACOUSTICAL TREATMENT	86.5	73.8	78.9	74.9	100.9	90.5	80.3	173.3	136.1	74.9	100.9	90.5	79.9	136.3	113.7	79.9	136.3	113.7
09600	FLOORING	90.6	82.5	88.5	91.3	94.6	92.2	102.0	158.9	117.0	94.2	94.6	94.3	88.0	158.9	106.8	95.5	158.9	112.3
097,099	WALL FINISHES, PAINTS & COATINGS	94.8	80.4	86.1	110.2	129.0	121.6	112.6	146.7	133.3	110.2	100.2	104.1	94.8	146.7	126.3	94.8	146.7	126.3
09	FINISHES	89.9	81.0	85.3	100.8	104.1	102.5	107.0	164.4	136.9	101.9	100.8	101.3	96.5	141.6	120.0	100.4	141.6	121.8
10 - 14	TOTAL DIV. 10000 - 14000	100.0	93.3	98.6	100.0	116.1	103.4	100.0	124.4	105.2	100.0	116.1	103.4	100.0	117.6	103.7	100.0	117.6	103.7
15	MECHANICAL	95.5	83.5	90.2	96.0	106.9	100.7	99.8	158.2	125.4	96.0	106.4	100.5	96.1	127.0	109.6	96.1	127.0	109.6
16	ELECTRICAL	97.3	79.0	88.3	98.0	123.1	110.2	101.2	161.5	130.6	98.0	123.1	110.2	97.3	143.1	119.6	97.3	143.1	119.6
01 - 16	WEIGHTED AVERAGE	96.6	85.4	91.5	101.1	114.3	107.1	103.1	155.0	126.7	99.6	113.8	106.1	100.2	135.4	116.2	100.7	135.4	116.5

NEW YORK

DIVISION		NEW YORK 100 - 102			NIAGARA FALLS 143			PLATTSBURGH 129			POUGHKEEPSIE 125 - 126			QUEENS 110			RIVERHEAD 119		
		MAT.	INST.	TOTAL	MAT.	INST.	TOTAL	MAT.	INST.	TOTAL	MAT.	INST.	TOTAL	MAT.	INST.	TOTAL	MAT.	INST.	TOTAL
01590	EQUIPMENT RENTAL	.0	117.6	117.6	.0	89.5	89.5	.0	97.9	97.9	.0	117.2	117.2	.0	117.2	117.2	.0	117.2	117.2
02	SITE CONSTRUCTION	134.2	128.5	129.9	94.9	90.7	91.8	86.5	102.2	98.1	115.1	128.0	124.6	115.4	130.3	126.4	112.3	130.4	125.7
03100	CONCRETE FORMS & ACCESSORIES	113.8	180.5	171.6	82.1	112.5	108.4	88.8	79.5	80.7	85.5	121.2	116.4	90.8	167.1	156.9	94.6	149.9	142.5
03200	CONCRETE REINFORCEMENT	111.2	187.4	149.7	104.0	100.4	102.2	107.1	89.3	98.1	103.1	131.1	117.2	105.1	184.5	145.2	105.3	184.3	145.2
03300	CAST-IN-PLACE CONCRETE	121.6	160.8	137.9	119.1	118.0	118.7	98.4	92.2	95.8	105.3	118.0	110.6	102.5	154.0	124.0	97.2	151.7	120.0
03	CONCRETE	120.2	172.8	146.5	107.3	111.2	109.2	104.8	85.3	95.0	110.4	121.4	115.9	108.0	164.0	136.0	103.7	155.5	129.6
04	MASONRY	105.3	161.1	140.3	117.9	117.9	117.9	88.5	84.1	85.8	100.6	120.9	113.3	106.9	155.8	137.6	116.0	152.4	138.9
05	METALS	108.1	140.9	118.8	95.7	90.5	94.0	98.2	79.6	92.1	102.5	113.9	106.2	102.4	138.5	114.2	103.6	137.9	114.8
06	WOOD & PLASTICS	111.4	187.9	151.4	84.8	109.9	97.9	90.6	77.8	83.9	83.5	120.0	102.6	88.6	171.4	131.8	92.4	151.2	123.1
07	THERMAL & MOISTURE PROTECTION	105.5	160.3	132.4	98.9	103.5	101.2	98.7	84.9	91.9	106.8	127.9	117.2	104.1	152.6	127.9	104.0	145.1	124.2
08	DOORS & WINDOWS	97.7	176.6	118.1	93.4	99.5	95.0	98.7	77.7	93.2	93.6	124.9	101.7	88.3	164.5	108.1	88.4	153.6	105.2
09200	PLASTER & GYPSUM BOARD	115.8	189.9	162.8	91.4	110.0	103.2	121.2	76.3	92.7	97.9	120.6	112.3	102.1	173.3	147.3	103.0	152.7	134.5
095,098	CEILINGS & ACOUSTICAL TREATMENT	109.4	189.9	157.7	86.5	110.0	100.6	99.8	76.3	85.7	74.9	120.6	102.3	80.3	173.3	136.1	79.3	152.7	123.3
09600	FLOORING	97.4	171.3	117.0	90.4	110.7	95.8	101.0	94.1	99.1	91.3	123.5	99.8	97.4	158.9	113.7	98.7	90.1	96.5
097,099	WALL FINISHES, PAINTS & COATINGS	97.2	150.2	129.3	94.8	107.9	102.7	103.4	74.2	85.7	110.2	100.2	104.1	112.6	146.7	133.3	112.6	146.7	133.3
09	FINISHES	108.9	176.8	144.2	90.0	112.3	101.6	100.2	81.4	90.4	100.6	118.8	110.1	104.4	164.4	135.6	104.4	137.5	121.6
10 - 14	TOTAL DIV. 10000 - 14000	100.0	151.7	111.0	100.0	100.7	100.1	100.0	62.5	92.0	100.0	119.7	104.2	100.0	124.4	105.2	100.0	120.4	104.3
15	MECHANICAL	100.3	163.2	127.8	95.5	101.1	97.9	95.9	80.8	89.3	96.0	109.5	101.9	99.8	158.2	125.4	99.9	144.9	119.6
16	ELECTRICAL	107.8	178.5	142.3	96.0	94.0	95.0	90.7	78.9	84.9	98.0	123.1	110.2	101.6	161.5	130.8	102.2	152.1	126.6
01 - 16	WEIGHTED AVERAGE	106.6	163.5	132.4	97.6	102.8	100.0	97.0	82.8	90.6	100.3	119.0	108.8	101.4	155.0	125.8	101.6	145.1	121.4

NEW YORK

DIVISION		ROCHESTER 144 - 146			SCHENECTADY 123			STATEN ISLAND 103			SUFFERN 109			SYRACUSE 130 - 132			UTICA 133 - 135		
		MAT.	INST.	TOTAL	MAT.	INST.	TOTAL	MAT.	INST.	TOTAL	MAT.	INST.	TOTAL	MAT.	INST.	TOTAL	MAT.	INST.	TOTAL
01590	EQUIPMENT RENTAL	.0	116.5	116.5	.0	115.7	115.7	.0	117.3	117.3	.0	117.3	117.3	.0	115.7	115.7	.0	115.7	115.7
02	SITE CONSTRUCTION	73.9	107.2	98.6	71.5	106.3	97.3	138.3	127.8	130.5	128.2	125.3	126.1	92.6	106.6	103.0	69.6	105.6	96.2
03100	CONCRETE FORMS & ACCESSORIES	100.2	94.2	95.0	103.5	91.0	92.7	95.3	167.5	157.9	107.0	130.4	127.2	102.7	87.4	89.5	104.0	83.3	86.1
03200	CONCRETE REINFORCEMENT	105.0	86.5	95.7	101.8	89.8	95.7	105.1	187.0	146.5	104.0	131.6	118.0	103.1	88.4	95.7	103.1	86.9	94.9
03300	CAST-IN-PLACE CONCRETE	112.9	100.5	107.8	97.3	99.0	98.0	120.7	156.0	135.5	117.0	130.2	122.5	98.3	93.7	96.4	89.9	91.2	90.4
03	CONCRETE	113.6	96.0	104.8	105.5	94.5	100.0	121.0	165.3	143.1	115.4	129.7	122.5	104.4	90.8	97.6	102.3	87.8	95.1
04	MASONRY	103.0	98.0	99.8	90.8	94.1	92.9	108.2	155.8	138.1	100.7	128.6	118.2	96.9	92.4	94.1	89.4	89.1	89.2
05	METALS	96.8	107.3	100.2	97.0	104.9	99.6	93.7	139.4	108.6	93.7	114.2	100.4	96.8	104.2	99.3	94.9	103.2	97.6
06	WOOD & PLASTICS	101.1	93.8	97.3	103.9	91.3	97.3	91.5	171.8	133.4	105.4	133.6	120.1	103.9	86.8	95.0	103.9	82.8	92.9
07	THERMAL & MOISTURE PROTECTION	95.6	96.9	96.2	91.0	91.6	91.3	105.4	153.2	128.9	106.4	134.9	120.4	99.1	93.2	96.2	91.0	92.0	91.5
08	DOORS & WINDOWS	96.0	86.2	93.5	95.7	85.0	92.9	92.1	165.3	111.0	92.1	132.3	102.5	93.6	80.7	90.3	95.7	77.8	91.0
09200	PLASTER & GYPSUM BOARD	104.6	93.6	97.6	111.0	90.8	98.2	99.8	173.3	146.4	105.1	134.1	123.5	111.0	86.2	95.3	111.0	82.0	92.6
095,098	CEILINGS & ACOUSTICAL TREATMENT	99.1	93.6	95.8	99.8	90.8	94.4	82.6	173.3	137.0	79.9	134.1	112.4	99.8	86.2	91.6	99.8	82.0	89.1
09600	FLOORING	86.8	85.8	86.6	86.8	94.1	88.7	91.9	158.9	109.6	91.5	62.7	83.9	88.2	89.4	88.6	86.8	89.5	87.5
097,099	WALL FINISHES, PAINTS & COATINGS	92.5	98.4	96.1	84.9	78.0	80.7	97.2	146.7	127.2	94.8	103.4	100.0	90.3	86.8	88.2	84.9	84.6	84.8
09	FINISHES	96.8	93.1	94.9	97.3	90.3	93.6	98.8	164.8	133.1	98.0	114.4	106.5	98.6	87.6	92.9	97.3	84.3	90.5
10 - 14	TOTAL DIV. 10000 - 14000	100.0	91.5	98.2	100.0	86.2	97.1	100.0	125.5	105.4	100.0	115.1	103.2	100.0	97.7	99.5	100.0	84.1	96.6
15	MECHANICAL	99.8	88.7	95.0	100.4	87.4	94.7	100.3	158.3	125.7	96.1	116.9	105.2	100.4	84.3	93.3	100.4	80.8	91.8
16	ELECTRICAL	103.4	89.1	96.4	100.5	89.4	95.1	100.7	161.5	130.4	104.3	133.2	118.4	100.5	82.5	91.7	97.1	81.7	89.6
01 - 16	WEIGHTED AVERAGE	100.1	94.9	97.7	98.2	93.0	95.8	102.1	155.2	126.2	100.5	123.7	111.0	99.0	90.5	95.2	97.0	87.8	92.8

NEW YORK / NORTH CAROLINA

DIVISION		WATERTOWN 136			WHITE PLAINS 106			YONKERS 107			ASHEVILLE 287 - 288			CHARLOTTE 281 - 282			DURHAM 277		
		MAT.	INST.	TOTAL	MAT.	INST.	TOTAL	MAT.	INST.	TOTAL	MAT.	INST.	TOTAL	MAT.	INST.	TOTAL	MAT.	INST.	TOTAL
01590	EQUIPMENT RENTAL	.0	115.7	115.7	.0	117.3	117.3	.0	117.3	117.3	.0	93.3	93.3	.0	93.3	93.3	.0	99.7	99.7
02	SITE CONSTRUCTION	77.5	107.6	99.8	123.8	125.2	124.9	133.5	125.1	127.3	101.6	72.3	79.9	101.8	72.3	80.0	101.6	82.2	87.2
03100	CONCRETE FORMS & ACCESSORIES	85.8	90.6	90.0	113.4	136.1	133.0	113.6	136.3	133.3	94.3	43.0	49.8	101.7	43.0	50.9	96.2	43.3	50.4
03200	CONCRETE REINFORCEMENT	103.7	88.4	96.0	104.0	183.5	144.2	108.1	183.5	146.2	95.1	39.5	67.0	95.5	39.5	67.2	95.5	54.7	74.9
03300	CAST-IN-PLACE CONCRETE	104.7	85.7	96.8	107.1	134.2	118.4	119.9	134.3	125.9	96.3	47.3	75.8	98.7	47.4	77.3	96.5	47.5	76.0
03	CONCRETE	115.2	89.5	102.4	108.1	143.1	125.6	119.0	143.3	131.1	99.5	45.8	72.7	100.5	45.8	73.2	99.1	48.8	74.0
04	MASONRY	90.6	94.0	92.7	100.0	134.1	121.4	104.8	134.1	123.2	79.1	37.1	52.7	84.0	37.1	54.6	80.8	37.1	53.4
05	METALS	94.9	103.9	97.9	94.8	134.3	107.7	104.0	134.6	114.0	93.0	74.5	87.0	95.5	74.5	88.7	98.1	81.1	92.6
06	WOOD & PLASTICS	84.1	91.8	88.2	112.3	135.7	124.5	112.2	135.7	124.5	93.1	44.6	67.8	102.0	44.6	72.0	95.4	44.6	68.9
07	THERMAL & MOISTURE PROTECTION	91.2	93.6	92.4	106.2	137.5	121.6	106.6	137.5	121.8	96.8	42.2	70.0	96.9	43.2	70.5	97.5	42.8	70.6
08	DOORS & WINDOWS	95.7	82.0	92.1	92.2	145.1	105.9	95.8	147.6	109.2	92.2	41.3	79.0	96.3	41.3	82.1	96.3	47.1	83.6
09200	PLASTER & GYPSUM BOARD	102.6	91.4	95.5	108.4	136.3	126.1	115.3	136.3	128.6	100.7	42.9	64.0	107.2	42.9	66.4	107.2	42.9	66.4
095,098	CEILINGS & ACOUSTICAL TREATMENT	99.8	91.4	94.7	79.9	136.3	113.7	107.6	136.3	124.8	91.9	42.9	62.5	98.1	42.9	65.0	98.1	42.9	65.0
09600	FLOORING	79.2	89.5	81.9	93.6	158.9	110.9	93.2	158.9	110.6	101.3	41.4	85.5	105.1	41.4	88.3	105.3	41.4	88.4
097,099	WALL FINISHES, PAINTS & COATINGS	84.9	73.4	78.0	94.8	146.7	126.3	94.8	146.7	126.3	108.5	39.1	66.4	108.5	39.1	66.4	108.5	39.1	66.4
09	FINISHES	94.3	88.3	91.2	98.7	141.6	121.0	107.0	141.6	125.0	96.4	42.3	68.3	100.0	42.3	70.0	100.1	42.3	70.0
10 - 14	TOTAL DIV. 10000 - 14000	100.0	83.7	96.5	100.0	117.6	103.7	100.0	141.6	108.9	100.0	62.5	92.0	100.0	62.5	92.0	100.0	65.2	92.6
15	MECHANICAL	100.4	77.0	90.1	100.6	127.0	112.1	100.6	127.0	112.1	100.0	43.1	75.1	100.0	43.1	75.1	100.0	43.3	75.2
16	ELECTRICAL	100.5	79.0	90.0	97.3	143.1	119.6	104.3	143.1	123.2	97.9	40.4	69.9	99.7	36.3	68.8	97.4	38.6	68.8
01 - 16	WEIGHTED AVERAGE	98.7	88.4	94.0	99.9	135.4	116.0	105.1	136.2	119.2	96.3	48.3	74.5	98.2	47.8	75.3	97.9	50.3	76.3

City Cost Indexes

NORTH CAROLINA

DIVISION		ELIZABETH CITY 279			FAYETTEVILLE 283			GASTONIA 280			GREENSBORO 270,272-274			HICKORY 286			KINSTON 285		
		MAT.	INST.	TOTAL	MAT.	INST.	TOTAL	MAT.	INST.	TOTAL	MAT.	INST.	TOTAL	MAT.	INST.	TOTAL	MAT.	INST.	TOTAL
01590	EQUIPMENT RENTAL	.0	104.8	104.8	.0	99.7	99.7	.0	93.3	93.3	.0	99.7	99.7	.0	99.7	99.7	.0	99.7	99.7
02	SITE CONSTRUCTION	106.1	83.0	89.0	99.7	82.2	86.7	101.3	72.3	79.8	101.5	82.2	87.2	100.6	81.1	86.1	99.3	81.8	86.3
03100	CONCRETE FORMS & ACCESSORIES	80.6	25.8	33.1	92.0	43.3	49.8	101.9	43.0	50.8	96.3	43.3	50.4	89.3	24.9	33.5	85.2	25.1	33.1
03200	CONCRETE REINFORCEMENT	94.5	47.4	70.7	94.5	54.7	74.4	95.5	39.5	67.2	95.5	54.7	74.9	95.1	21.3	57.8	94.5	21.4	57.6
03300	CAST-IN-PLACE CONCRETE	96.6	34.0	70.5	93.2	47.5	74.1	94.1	47.3	74.6	95.7	47.5	75.6	96.2	36.3	71.2	92.9	37.3	69.7
03	CONCRETE	99.3	35.0	67.2	96.7	48.8	72.8	98.3	45.8	72.1	98.7	48.8	73.8	99.2	30.6	65.0	96.1	31.2	63.7
04	MASONRY	94.5	26.7	52.0	84.4	37.1	54.7	82.0	37.1	53.8	80.6	37.1	53.3	69.2	26.5	42.4	75.2	26.9	44.9
05	METALS	92.6	76.1	87.2	98.1	81.1	92.5	93.5	74.5	87.3	98.7	81.1	92.9	93.3	67.7	84.9	92.0	69.2	84.6
06	WOOD & PLASTICS	78.2	26.1	51.0	90.4	44.6	66.5	102.0	44.6	72.0	95.4	44.6	68.9	87.2	25.3	54.9	83.0	25.3	52.9
07	THERMAL & MOISTURE PROTECTION	96.8	26.7	62.4	97.2	42.8	70.4	97.2	43.2	70.7	97.5	42.8	70.6	97.2	29.4	63.9	97.0	30.0	64.1
08	DOORS & WINDOWS	93.1	28.9	76.5	92.3	47.1	80.6	96.3	41.3	82.1	96.3	47.1	83.6	92.2	24.9	74.8	92.3	25.3	75.0
09200	PLASTER & GYPSUM BOARD	97.2	23.0	50.2	101.0	42.9	64.1	107.2	42.9	66.4	107.2	42.9	66.4	100.7	23.0	51.5	99.1	23.0	50.9
095,098	CEILINGS & ACOUSTICAL TREATMENT	98.1	23.0	53.1	92.8	42.9	62.8	98.1	42.9	65.0	98.1	42.9	65.0	91.9	23.0	50.6	96.3	23.0	52.3
09600	FLOORING	94.8	16.2	74.0	101.4	41.4	85.6	105.3	28.4	85.0	105.3	41.4	88.4	101.2	14.8	78.3	97.9	22.1	77.9
097,099	WALL FINISHES, PAINTS & COATINGS	108.5	13.3	50.8	108.5	39.1	66.4	108.5	39.1	66.4	108.5	39.1	66.4	108.5	15.6	52.2	108.5	17.2	53.1
09	FINISHES	95.7	22.0	57.4	96.7	42.3	68.4	100.0	40.1	68.8	100.1	42.3	70.0	96.5	21.5	57.5	96.2	23.4	58.3
10 - 14	TOTAL DIV. 10000 - 14000	100.0	63.6	92.3	100.0	65.2	92.6	100.0	62.5	92.0	100.0	62.5	92.0	100.0	57.8	91.0	100.0	60.4	91.6
15	MECHANICAL	95.7	27.9	66.0	100.0	43.3	75.2	100.0	43.1	75.1	100.0	43.3	75.2	95.7	19.2	62.2	95.7	19.9	62.5
16	ELECTRICAL	98.4	29.8	64.9	94.8	38.6	67.4	97.4	36.3	67.6	98.0	38.6	69.1	95.7	29.8	63.5	95.5	29.8	63.4
01 - 16	WEIGHTED AVERAGE	96.1	38.6	70.0	96.7	50.3	75.6	97.2	47.5	74.6	98.0	50.3	76.3	94.5	34.9	67.5	94.2	35.8	67.6

NORTH CAROLINA / NORTH DAKOTA

DIVISION		MURPHY 289			RALEIGH 275-276			ROCKY MOUNT 278			WILMINGTON 284			WINSTON-SALEM 271			BISMARCK 585		
		MAT.	INST.	TOTAL	MAT.	INST.	TOTAL	MAT.	INST.	TOTAL	MAT.	INST.	TOTAL	MAT.	INST.	TOTAL	MAT.	INST.	TOTAL
01590	EQUIPMENT RENTAL	.0	93.3	93.3	.0	99.7	99.7	.0	99.7	99.7	.0	93.3	93.3	.0	99.7	99.7	.0	98.2	98.2
02	SITE CONSTRUCTION	102.9	71.1	79.4	102.6	82.2	87.5	104.3	81.1	87.2	102.5	72.3	80.2	101.9	82.2	87.3	88.4	96.0	94.0
03100	CONCRETE FORMS & ACCESSORIES	102.6	24.9	35.3	99.0	43.3	50.7	88.0	29.8	37.6	95.8	43.3	50.3	97.6	43.1	50.4	96.4	51.0	57.0
03200	CONCRETE REINFORCEMENT	94.6	21.2	57.5	95.5	54.7	74.9	94.5	21.3	57.5	95.8	54.7	75.0	95.5	39.5	67.2	101.2	75.5	88.2
03300	CAST-IN-PLACE CONCRETE	99.6	32.3	71.5	102.2	47.5	79.3	94.4	31.4	68.1	95.9	47.5	75.6	98.7	47.4	77.3	104.9	56.3	84.6
03	CONCRETE	102.5	29.3	66.0	102.0	48.8	75.4	100.3	31.1	65.8	99.4	48.8	74.2	100.2	45.8	73.1	99.1	58.5	78.9
04	MASONRY	71.8	24.1	41.9	85.5	37.1	55.1	75.4	28.5	46.0	69.7	37.1	49.2	80.8	37.1	53.4	104.0	66.9	80.7
05	METALS	91.1	67.3	83.3	95.9	81.1	91.1	92.0	67.5	84.0	92.5	81.1	88.8	95.9	74.6	88.9	92.0	81.2	88.5
06	WOOD & PLASTICS	102.8	25.3	62.3	98.6	44.6	70.4	85.9	30.9	57.2	94.9	44.6	68.6	95.4	44.6	68.9	88.9	46.2	66.6
07	THERMAL & MOISTURE PROTECTION	97.2	31.4	64.9	97.3	42.8	70.6	97.5	28.6	63.6	96.9	42.8	70.3	97.5	42.8	70.6	100.5	56.9	79.1
08	DOORS & WINDOWS	92.2	28.7	75.7	93.0	47.1	81.1	92.3	27.1	75.4	92.3	47.1	80.6	96.3	43.1	82.5	99.5	50.5	86.8
09200	PLASTER & GYPSUM BOARD	105.6	23.0	53.2	107.2	42.9	66.4	99.0	28.7	54.5	101.8	42.9	64.4	107.2	42.9	66.4	115.0	45.0	70.6
095,098	CEILINGS & ACOUSTICAL TREATMENT	91.9	23.0	50.6	98.1	42.9	65.0	92.8	28.7	54.4	92.8	42.9	62.8	98.1	42.9	65.0	134.3	45.0	80.7
09600	FLOORING	105.8	23.6	84.1	105.3	41.4	88.4	99.4	15.6	77.2	102.3	41.4	86.2	105.3	41.4	88.4	108.4	78.4	100.4
097,099	WALL FINISHES, PAINTS & COATINGS	108.5	17.8	53.5	108.5	39.1	66.4	108.5	29.7	60.7	108.5	39.1	66.4	108.5	39.1	66.4	99.2	38.9	62.6
09	FINISHES	98.6	23.3	59.4	100.1	42.3	70.1	96.2	27.5	60.4	97.0	42.3	68.6	100.1	42.3	70.0	113.9	53.8	82.6
10 - 14	TOTAL DIV. 10000 - 14000	100.0	57.9	91.0	100.0	65.2	92.6	100.0	62.0	91.9	100.0	65.2	92.6	100.0	62.5	92.0	100.0	74.2	94.5
15	MECHANICAL	95.7	19.3	62.2	100.0	43.3	75.2	95.7	19.3	62.2	100.0	43.3	75.2	100.0	43.1	75.1	100.6	64.2	84.7
16	ELECTRICAL	98.8	29.8	65.1	98.0	38.6	69.1	100.3	29.8	65.9	98.5	38.6	69.3	98.0	38.6	69.1	94.4	70.1	82.5
01 - 16	WEIGHTED AVERAGE	95.5	34.0	67.6	97.9	50.3	76.3	95.4	36.2	68.5	96.0	49.4	74.8	97.8	49.1	75.7	99.2	66.9	84.5

NORTH DAKOTA

DIVISION		DEVILS LAKE 583			DICKINSON 586			FARGO 580-581			GRAND FORKS 582			JAMESTOWN 584			MINOT 587		
		MAT.	INST.	TOTAL	MAT.	INST.	TOTAL	MAT.	INST.	TOTAL	MAT.	INST.	TOTAL	MAT.	INST.	TOTAL	MAT.	INST.	TOTAL
01590	EQUIPMENT RENTAL	.0	98.2	98.2	.0	98.2	98.2	.0	98.2	98.2	.0	98.2	98.2	.0	98.2	98.2	.0	98.2	98.2
02	SITE CONSTRUCTION	92.7	94.1	93.7	100.3	94.1	95.7	87.8	96.0	93.9	96.3	94.1	94.7	91.7	94.1	93.5	93.9	96.0	95.5
03100	CONCRETE FORMS & ACCESSORIES	100.7	44.9	52.4	88.5	44.7	50.5	97.4	51.4	57.5	93.1	44.8	51.3	90.3	44.5	50.6	88.1	57.8	61.9
03200	CONCRETE REINFORCEMENT	101.2	75.7	88.3	102.2	53.4	77.5	93.3	74.6	83.8	99.6	81.6	90.5	101.8	64.2	82.8	103.1	75.6	89.2
03300	CAST-IN-PLACE CONCRETE	118.7	54.4	91.8	107.4	54.3	85.2	109.9	58.6	88.5	107.4	54.4	85.2	117.2	54.3	90.9	107.4	54.9	85.5
03	CONCRETE	108.6	55.1	81.9	107.9	50.9	79.4	109.5	59.3	84.4	105.1	56.1	80.6	107.1	52.7	79.9	103.7	61.1	82.4
04	MASONRY	111.6	67.1	83.7	114.3	60.8	80.7	106.3	50.0	71.0	105.6	66.2	80.9	124.7	43.9	74.0	106.7	68.2	82.5
05	METALS	92.0	79.0	87.8	91.9	70.4	84.9	94.4	79.8	89.6	92.0	81.0	88.4	92.0	73.2	85.8	92.3	81.3	88.7
06	WOOD & PLASTICS	93.4	42.4	66.8	80.6	42.4	60.7	88.9	46.8	66.9	85.3	42.4	62.9	82.7	42.4	61.7	80.3	55.4	67.3
07	THERMAL & MOISTURE PROTECTION	101.0	57.3	79.5	101.6	54.4	78.4	101.2	54.2	78.1	101.2	57.1	79.5	100.7	51.2	76.4	100.9	58.1	79.9
08	DOORS & WINDOWS	99.5	45.7	85.6	99.4	40.5	84.2	99.4	50.9	86.9	99.5	47.1	85.9	99.5	43.3	84.9	99.6	55.6	88.2
09200	PLASTER & GYPSUM BOARD	120.2	41.2	70.1	111.4	41.2	66.9	115.0	45.7	71.1	113.0	41.2	67.5	112.6	41.2	67.3	111.4	54.5	75.4
095,098	CEILINGS & ACOUSTICAL TREATMENT	134.3	41.2	78.5	134.3	41.2	78.5	134.3	45.7	81.2	134.3	41.2	78.5	134.3	41.2	78.5	134.3	54.5	86.5
09600	FLOORING	112.7	46.9	95.3	104.1	46.9	88.9	108.2	46.9	91.9	106.6	46.9	90.8	105.2	46.9	89.8	103.9	81.0	97.8
097,099	WALL FINISHES, PAINTS & COATINGS	99.2	29.7	57.1	99.2	29.7	57.1	99.2	73.6	83.7	99.2	29.7	57.1	99.2	29.7	57.1	99.2	32.6	58.8
09	FINISHES	116.2	42.6	77.9	112.9	42.6	76.4	113.7	51.7	81.4	113.6	42.6	76.7	112.8	42.6	76.3	112.3	59.0	84.6
10 - 14	TOTAL DIV. 10000 - 14000	100.0	48.1	89.0	100.0	48.3	89.0	100.0	74.2	94.5	100.0	48.3	89.0	100.0	71.2	93.9	100.0	75.4	94.8
15	MECHANICAL	96.2	66.1	83.0	96.2	65.7	82.8	100.6	69.7	87.0	100.6	46.9	77.1	96.2	46.8	74.6	100.6	61.8	83.6
16	ELECTRICAL	93.7	47.7	71.3	101.5	74.2	88.2	94.2	69.8	82.3	96.8	71.4	84.4	93.7	47.7	71.3	99.7	71.4	85.9
01 - 16	WEIGHTED AVERAGE	100.0	61.0	82.3	100.8	62.3	83.3	100.8	65.9	85.0	100.4	60.7	82.4	100.0	54.2	79.2	100.5	68.2	85.8

		NORTH DAKOTA			OHIO															
	DIVISION	WILLISTON			AKRON			ATHENS			CANTON			CHILLICOTHE			CINCINNATI		
		588			442 - 443			457			446 - 447			456			451 - 452		
		MAT.	INST.	TOTAL	MAT.	INST.	TOTAL	MAT.	INST.	TOTAL	MAT.	INST.	TOTAL	MAT.	INST.	TOTAL	MAT.	INST.	TOTAL
01590	EQUIPMENT RENTAL	.0	98.2	98.2	.0	97.4	97.4	.0	90.5	90.5	.0	97.4	97.4	.0	102.8	102.8	.0	102.5	102.5
02	SITE CONSTRUCTION	94.4	94.1	94.2	100.8	106.8	105.2	87.7	94.3	92.6	100.9	106.9	105.4	77.1	108.8	100.6	74.9	108.7	99.9
03100	CONCRETE FORMS & ACCESSORIES	95.1	44.6	51.4	98.1	99.0	98.9	91.0	92.8	92.6	98.1	86.1	87.7	93.5	90.4	90.8	95.9	90.9	91.6
03200	CONCRETE REINFORCEMENT	104.1	53.4	78.5	94.1	98.0	96.1	96.5	88.7	92.5	94.1	84.1	89.0	93.1	86.4	89.7	98.7	86.7	92.6
03300	CAST-IN-PLACE CONCRETE	107.4	54.3	85.2	99.1	106.2	102.0	101.5	105.0	103.0	100.1	103.4	101.5	92.1	93.7	92.8	84.8	91.5	87.6
03	CONCRETE	105.1	50.8	78.0	99.4	100.2	99.8	103.3	95.3	99.3	99.8	91.0	95.4	98.1	90.8	94.4	92.9	90.4	91.6
04	MASONRY	101.4	60.8	75.9	90.0	102.4	97.8	77.2	101.7	92.6	90.7	93.3	92.4	85.3	98.1	93.3	84.9	95.9	91.8
05	METALS	92.1	70.3	85.0	88.2	82.6	86.3	92.5	76.4	87.2	88.2	76.0	84.2	84.9	87.9	85.9	86.8	89.6	87.7
06	WOOD & PLASTICS	86.9	42.4	63.7	92.2	97.8	95.2	82.4	90.8	86.8	92.6	84.1	88.2	94.0	87.5	90.6	96.4	89.5	92.8
07	THERMAL & MOISTURE PROTECTION	101.2	54.4	78.2	104.0	100.4	102.2	100.9	86.6	93.9	104.7	95.9	100.3	91.3	95.4	93.3	88.9	97.8	93.3
08	DOORS & WINDOWS	99.5	40.5	84.3	104.2	98.9	102.8	95.0	80.4	91.2	98.6	77.9	93.3	88.5	84.0	87.3	95.9	87.6	93.8
09200	PLASTER & GYPSUM BOARD	113.0	41.2	67.5	95.2	97.3	96.6	98.5	90.2	93.2	96.0	83.2	87.9	99.8	87.4	92.0	101.2	89.5	93.8
095,098	CEILINGS & ACOUSTICAL TREATMENT	134.3	41.2	78.5	91.9	97.3	95.2	100.3	90.2	94.2	91.9	83.2	86.7	94.6	87.4	90.3	95.5	89.5	91.9
09600	FLOORING	107.7	46.9	91.6	103.2	91.7	100.2	129.1	101.0	121.6	103.4	83.5	98.1	106.1	100.4	104.6	107.2	101.8	105.8
097,099	WALL FINISHES, PAINTS & COATINGS	99.2	29.7	57.1	107.1	122.7	116.5	108.4	55.1	76.1	107.1	85.8	94.2	104.8	99.6	101.6	104.8	96.1	99.5
09	FINISHES	113.8	42.6	76.8	99.5	100.2	99.8	101.8	91.1	96.2	99.7	84.6	91.8	97.5	93.2	95.2	98.1	93.4	95.6
10 - 14	TOTAL DIV. 10000 - 14000	100.0	48.3	89.0	100.0	92.6	98.4	100.0	75.8	94.8	100.0	80.1	95.8	100.0	84.6	96.7	100.0	94.7	98.9
15	MECHANICAL	96.2	62.7	81.5	99.8	101.4	100.5	95.4	72.2	85.2	99.8	85.4	93.5	96.0	92.6	94.5	99.8	91.6	96.2
16	ELECTRICAL	97.2	74.2	86.0	97.9	90.5	94.3	102.4	58.4	80.9	97.1	90.1	93.7	99.6	85.1	92.5	99.5	83.0	91.4
01 - 16	WEIGHTED AVERAGE	99.3	61.7	82.2	97.7	98.0	97.8	96.5	82.3	90.1	97.2	88.5	93.2	93.5	92.4	93.0	94.7	92.3	93.6

		OHIO																		
	DIVISION	CLEVELAND			COLUMBUS			DAYTON			HAMILTON			LIMA			LORAIN		
		441			430 - 432			453 - 454			450			458			440		
		MAT.	INST.	TOTAL	MAT.	INST.	TOTAL	MAT.	INST.	TOTAL	MAT.	INST.	TOTAL	MAT.	INST.	TOTAL	MAT.	INST.	TOTAL
01590	EQUIPMENT RENTAL	.0	97.7	97.7	.0	96.4	96.4	.0	96.8	96.8	.0	102.8	102.8	.0	93.6	93.6	.0	97.4	97.4
02	SITE CONSTRUCTION	100.8	106.8	105.3	86.5	101.8	97.8	73.7	108.2	99.2	74.0	108.7	99.7	82.7	94.4	91.4	100.2	106.0	104.5
03100	CONCRETE FORMS & ACCESSORIES	98.3	105.6	104.6	98.6	86.7	88.3	95.8	80.2	82.3	95.9	89.9	90.7	91.0	87.3	87.8	98.2	106.2	105.1
03200	CONCRETE REINFORCEMENT	94.6	98.3	96.5	76.1	86.5	81.4	98.7	82.1	90.3	98.7	86.7	92.6	96.5	82.4	89.4	94.1	98.1	96.1
03300	CAST-IN-PLACE CONCRETE	97.2	114.1	104.3	89.6	93.7	91.3	77.3	89.2	82.3	84.6	87.4	85.8	93.3	102.1	97.0	94.3	108.9	100.4
03	CONCRETE	98.6	106.1	102.3	92.2	88.7	90.5	89.3	83.4	86.4	92.8	88.5	90.7	96.6	91.1	93.9	97.1	104.4	100.8
04	MASONRY	94.2	110.6	104.5	95.4	93.8	94.4	84.3	85.3	84.9	84.8	93.7	90.4	106.9	90.8	96.8	87.0	100.5	95.5
05	METALS	89.6	85.8	88.3	96.5	82.0	91.8	86.1	77.3	83.3	86.2	89.4	87.3	92.5	81.4	88.9	88.7	83.5	87.0
06	WOOD & PLASTICS	91.5	102.4	97.2	104.3	84.8	94.1	97.5	78.2	87.4	96.4	89.5	92.8	82.3	85.1	83.8	92.2	108.3	100.6
07	THERMAL & MOISTURE PROTECTION	102.9	114.0	108.3	99.4	95.6	97.5	94.5	87.6	91.1	91.4	96.8	94.1	102.5	102.6	102.5	104.6	108.1	106.3
08	DOORS & WINDOWS	95.0	101.3	96.7	99.5	84.1	95.5	96.2	78.1	91.5	93.8	87.6	92.2	95.0	83.3	92.0	98.6	104.6	100.2
09200	PLASTER & GYPSUM BOARD	94.3	102.0	99.2	99.3	84.2	89.8	101.2	77.9	86.5	101.2	89.5	93.8	98.5	84.2	89.5	95.2	108.1	103.4
095,098	CEILINGS & ACOUSTICAL TREATMENT	90.1	102.0	97.2	96.2	84.2	89.0	96.5	77.9	85.3	95.5	89.5	91.9	99.3	84.2	90.3	91.9	108.1	101.6
09600	FLOORING	103.0	112.2	105.4	97.6	91.0	95.9	110.0	84.0	103.1	107.2	101.8	105.8	128.2	95.4	119.5	103.4	109.5	105.0
097,099	WALL FINISHES, PAINTS & COATINGS	107.1	122.4	116.4	99.3	99.6	99.5	104.8	92.2	97.1	104.8	89.9	95.7	108.4	78.5	90.3	107.1	122.4	116.4
09	FINISHES	98.9	108.3	103.8	98.9	88.4	93.5	99.1	81.6	90.0	98.0	92.1	94.9	101.0	87.3	93.9	99.5	109.2	104.6
10 - 14	TOTAL DIV. 10000 - 14000	100.0	108.1	101.7	100.0	94.5	98.8	100.0	91.6	98.2	100.0	93.8	98.7	100.0	89.1	97.7	100.0	105.5	101.2
15	MECHANICAL	99.8	108.1	103.5	99.9	92.6	96.7	100.7	84.4	93.6	100.4	90.4	96.0	95.4	87.3	91.9	99.8	89.4	95.3
16	ELECTRICAL	98.0	109.7	103.7	99.7	87.2	93.6	97.4	80.7	89.3	97.6	76.3	87.2	102.6	82.2	92.7	97.2	90.8	94.1
01 - 16	WEIGHTED AVERAGE	97.1	106.0	101.1	97.8	90.5	94.5	94.4	84.9	90.1	94.3	90.4	92.5	97.1	87.9	92.9	96.8	98.1	97.4

		OHIO																		
	DIVISION	MANSFIELD			MARION			SPRINGFIELD			STEUBENVILLE			TOLEDO			YOUNGSTOWN		
		448 - 449			433			455			439			434 - 436			444 - 445		
		MAT.	INST.	TOTAL	MAT.	INST.	TOTAL	MAT.	INST.	TOTAL	MAT.	INST.	TOTAL	MAT.	INST.	TOTAL	MAT.	INST.	TOTAL
01590	EQUIPMENT RENTAL	.0	97.4	97.4	.0	96.1	96.1	.0	96.8	96.8	.0	101.6	101.6	.0	99.4	99.4	.0	97.4	97.4
02	SITE CONSTRUCTION	96.0	106.7	103.9	82.1	102.1	96.9	74.2	107.3	98.7	123.6	112.0	115.0	85.8	102.9	98.4	100.7	107.2	105.5
03100	CONCRETE FORMS & ACCESSORIES	86.5	105.2	102.7	94.6	86.0	87.2	95.8	85.1	86.6	95.2	92.6	93.0	98.6	98.5	98.5	98.1	92.4	93.2
03200	CONCRETE REINFORCEMENT	85.6	84.3	84.9	70.2	86.4	78.4	98.7	82.1	90.3	68.3	94.5	81.5	76.1	95.6	85.9	94.1	98.0	96.0
03300	CAST-IN-PLACE CONCRETE	91.8	96.1	93.6	81.9	93.6	86.8	81.0	89.8	84.7	88.9	99.8	93.5	89.6	108.4	97.5	98.2	105.1	101.1
03	CONCRETE	91.5	97.0	94.3	84.6	88.2	86.4	91.1	85.8	88.4	88.4	94.6	91.5	92.2	100.9	96.5	98.9	97.0	97.9
04	MASONRY	89.2	97.8	94.6	97.9	95.2	96.2	84.6	86.4	85.7	84.3	96.0	91.6	105.3	104.7	104.9	90.0	98.1	95.1
05	METALS	88.9	77.0	85.0	95.4	79.4	90.2	86.1	77.2	83.2	91.5	81.4	88.2	96.3	89.5	94.1	88.2	83.0	86.5
06	WOOD & PLASTICS	79.8	108.3	94.7	100.2	84.8	92.1	98.7	84.4	91.2	96.0	90.5	93.1	104.3	97.2	100.6	92.2	90.2	91.2
07	THERMAL & MOISTURE PROTECTION	103.8	98.5	101.2	98.7	88.2	93.5	94.4	88.7	91.6	111.0	98.6	104.9	101.0	109.3	105.1	104.8	100.9	102.9
08	DOORS & WINDOWS	97.9	93.8	96.8	94.1	81.6	90.8	94.2	81.4	90.9	94.9	91.6	94.0	97.5	93.6	96.5	98.6	96.3	98.0
09200	PLASTER & GYPSUM BOARD	90.3	108.1	101.6	97.7	84.2	89.2	101.2	84.2	90.5	98.6	89.5	92.8	99.3	97.0	97.8	95.2	89.5	91.6
095,098	CEILINGS & ACOUSTICAL TREATMENT	94.6	108.1	102.7	96.2	84.2	89.0	96.5	84.2	89.1	93.1	89.5	91.0	96.2	97.0	96.7	91.9	89.5	90.4
09600	FLOORING	98.2	97.4	98.0	95.8	91.0	94.5	110.0	84.0	103.1	124.4	92.6	116.0	96.8	100.8	97.8	103.4	97.9	102.0
097,099	WALL FINISHES, PAINTS & COATINGS	107.1	73.3	86.6	99.3	62.8	77.2	104.8	92.2	97.1	116.9	101.0	107.3	99.4	110.4	106.1	107.1	103.6	105.0
09	FINISHES	97.7	101.4	99.6	97.9	84.4	90.9	99.1	85.5	92.1	115.0	93.1	103.6	98.6	100.1	99.4	99.6	94.0	96.7
10 - 14	TOTAL DIV. 10000 - 14000	100.0	88.2	97.5	100.0	76.1	94.9	100.0	92.9	98.5	100.0	96.9	99.3	100.0	93.4	98.6	100.0	91.9	98.3
15	MECHANICAL	95.5	89.4	92.8	95.5	90.0	93.1	100.7	85.0	93.8	95.9	86.9	91.9	99.9	99.9	99.9	99.8	90.7	95.8
16	ELECTRICAL	95.1	85.7	90.5	94.0	85.7	89.9	97.4	84.0	90.9	85.7	101.1	93.2	99.5	99.4	99.4	97.2	95.4	96.3
01 - 16	WEIGHTED AVERAGE	94.5	93.4	94.0	94.4	88.3	91.6	94.4	86.7	90.9	95.7	94.3	95.1	98.0	99.6	98.7	97.1	94.8	96.0

City Cost Indexes

		OHIO			OKLAHOMA													
	DIVISION	ZANESVILLE			ARDMORE			CLINTON			DURANT			ENID			GUYMON		
		437 - 438			734			736			747			737			739		
		MAT.	INST.	TOTAL	MAT.	INST.	TOTAL	MAT.	INST.	TOTAL	MAT.	INST.	TOTAL	MAT.	INST.	TOTAL	MAT.	INST.	TOTAL
01590	EQUIPMENT RENTAL	.0	96.1	96.1	.0	78.2	78.2	.0	77.1	77.1	.0	77.1	77.1	.0	77.1	77.1	.0	77.1	77.1
02	SITE CONSTRUCTION	84.6	102.5	97.9	105.6	90.8	94.6	107.0	89.1	93.8	99.8	88.9	91.7	108.8	89.1	94.3	112.4	88.0	94.3
03100	CONCRETE FORMS & ACCESSORIES	90.9	85.1	85.9	91.1	48.1	53.8	89.8	41.2	47.7	81.7	48.1	52.6	93.7	40.4	47.5	98.1	25.0	34.8
03200	CONCRETE REINFORCEMENT	69.7	93.8	81.9	96.8	82.7	89.6	97.4	82.7	89.9	97.7	69.3	83.4	96.7	82.7	89.6	97.4	35.0	65.9
03300	CAST-IN-PLACE CONCRETE	86.3	91.9	88.6	95.8	48.6	76.1	92.6	48.5	74.2	89.6	48.5	72.4	92.6	51.6	75.5	92.7	31.9	67.3
03	CONCRETE	88.1	88.6	88.4	90.7	54.8	72.8	90.0	51.7	70.9	85.7	52.3	69.0	90.6	52.4	71.5	93.1	29.7	61.5
04	MASONRY	94.2	81.6	86.3	94.5	60.2	73.0	122.0	60.2	83.2	93.2	60.9	72.9	103.5	60.2	76.3	97.6	23.9	51.4
05	METALS	97.0	81.8	92.0	88.2	67.3	81.4	88.2	67.3	81.4	88.2	61.5	79.5	89.2	67.3	82.0	88.6	40.2	72.8
06	WOOD & PLASTICS	96.2	84.8	90.2	94.4	47.9	70.1	93.3	38.6	64.8	84.4	47.9	65.3	97.4	37.5	66.1	101.5	24.5	61.3
07	THERMAL & MOISTURE PROTECTION	98.8	91.0	95.0	99.5	61.4	80.7	99.6	60.4	80.3	98.8	60.5	80.0	99.8	60.3	80.4	100.2	30.4	65.9
08	DOORS & WINDOWS	94.1	85.2	91.8	95.6	57.4	85.7	95.7	52.3	84.5	95.6	54.1	84.9	95.6	51.5	84.2	95.8	26.8	77.9
09200	PLASTER & GYPSUM BOARD	95.7	84.2	88.4	81.9	47.2	59.9	81.5	37.8	53.8	77.5	47.2	58.3	82.7	36.6	53.5	83.5	23.3	45.3
095,098	CEILINGS & ACOUSTICAL TREATMENT	96.2	84.2	89.0	80.4	47.2	60.5	80.4	37.8	54.8	80.4	47.2	60.5	81.4	36.6	54.6	83.1	23.3	47.2
09600	FLOORING	93.7	91.0	93.0	112.1	56.3	97.4	110.4	53.7	95.4	105.9	63.2	94.6	112.9	53.7	97.2	115.3	31.8	93.2
097,099	WALL FINISHES, PAINTS & COATINGS	99.3	67.2	79.9	96.4	57.1	72.6	96.4	57.1	72.6	96.4	57.1	72.6	96.4	57.1	72.6	96.4	20.1	50.1
09	FINISHES	97.1	84.1	90.3	93.7	49.4	70.7	93.3	43.5	67.4	91.0	50.9	70.1	94.6	42.8	67.6	96.2	25.6	59.5
10 - 14	TOTAL DIV. 10000 - 14000	100.0	82.1	96.2	100.0	65.3	92.6	100.0	64.1	92.4	100.0	65.5	92.7	100.0	63.9	92.3	100.0	59.5	91.4
15	MECHANICAL	95.5	81.9	89.5	95.7	63.7	81.7	95.7	63.7	81.7	95.7	63.9	81.8	100.1	63.7	84.2	95.7	28.2	66.2
16	ELECTRICAL	93.8	71.4	82.9	94.4	68.5	81.8	95.3	68.5	82.3	96.8	63.7	80.7	95.3	68.5	82.3	96.8	20.5	59.6
01 - 16	WEIGHTED AVERAGE	94.8	83.9	89.8	94.2	63.3	80.2	95.7	61.6	80.2	93.4	61.7	79.0	96.1	61.6	80.4	95.6	33.9	67.6

		OKLAHOMA																	
	DIVISION	LAWTON			MCALESTER			MIAMI			MUSKOGEE			OKLAHOMA CITY			PONCA CITY		
		735			745			743			744			730 - 731			746		
		MAT.	INST.	TOTAL	MAT.	INST.	TOTAL	MAT.	INST.	TOTAL	MAT.	INST.	TOTAL	MAT.	INST.	TOTAL	MAT.	INST.	TOTAL
01590	EQUIPMENT RENTAL	.0	78.2	78.2	.0	77.1	77.1	.0	86.8	86.8	.0	86.8	86.8	.0	78.5	78.5	.0	77.1	77.1
02	SITE CONSTRUCTION	104.2	90.8	94.3	92.0	88.5	89.4	94.2	86.1	88.2	94.8	85.2	87.7	105.5	91.4	95.0	100.7	89.1	92.1
03100	CONCRETE FORMS & ACCESSORIES	97.5	55.4	61.0	79.3	50.2	54.1	93.9	59.4	64.0	99.0	36.6	44.9	98.6	47.7	54.5	88.9	48.5	53.9
03200	CONCRETE REINFORCEMENT	96.9	82.7	89.7	97.4	52.0	74.5	95.9	82.7	89.2	96.7	39.3	67.7	96.9	82.7	89.7	96.7	82.6	89.6
03300	CAST-IN-PLACE CONCRETE	89.6	51.7	73.7	78.5	51.0	67.0	82.3	50.6	69.1	83.3	39.7	65.1	96.9	55.0	79.4	92.0	42.4	71.3
03	CONCRETE	87.3	59.1	73.2	77.4	50.6	64.0	81.6	61.5	71.6	83.1	39.3	61.3	90.9	56.8	73.8	87.7	52.8	70.3
04	MASONRY	97.5	60.2	74.1	112.0	60.6	79.8	96.0	61.6	74.4	113.4	43.2	69.4	99.3	62.4	76.1	87.6	60.9	70.8
05	METALS	92.6	67.4	84.4	88.2	51.1	76.1	88.1	82.5	86.3	89.1	59.1	79.3	94.5	67.3	85.6	88.1	67.0	81.2
06	WOOD & PLASTICS	100.4	57.6	78.1	81.7	52.0	66.2	97.3	62.4	79.1	102.4	37.1	68.3	102.2	46.2	73.0	92.5	48.6	69.6
07	THERMAL & MOISTURE PROTECTION	99.5	62.4	81.2	98.3	61.4	80.2	98.9	63.4	81.5	99.3	45.5	72.9	98.5	62.1	80.6	99.1	60.3	80.0
08	DOORS & WINDOWS	97.2	62.5	88.2	95.6	51.1	84.1	95.6	65.6	87.9	95.6	36.6	80.4	97.2	56.2	86.6	95.6	57.8	85.8
09200	PLASTER & GYPSUM BOARD	85.4	57.3	67.6	76.3	51.5	60.6	82.7	62.0	69.6	86.6	36.1	54.5	85.4	45.5	60.1	81.5	48.0	60.3
095,098	CEILINGS & ACOUSTICAL TREATMENT	91.2	57.3	70.9	80.4	51.5	63.1	80.4	62.0	69.4	91.2	36.1	58.1	91.2	45.5	63.8	80.4	48.0	61.0
09600	FLOORING	115.8	53.7	99.4	104.6	53.7	91.1	113.9	63.2	100.5	117.1	43.3	97.5	115.8	53.7	99.4	110.0	53.7	95.1
097,099	WALL FINISHES, PAINTS & COATINGS	96.4	57.1	72.6	96.4	50.0	68.2	96.4	68.6	79.6	96.4	36.3	59.9	96.4	57.1	72.6	96.4	57.1	72.6
09	FINISHES	97.8	54.6	75.3	89.9	50.6	69.5	93.5	61.0	76.6	97.7	37.7	66.5	97.9	48.4	72.1	92.8	49.5	70.3
10 - 14	TOTAL DIV. 10000 - 14000	100.0	66.5	92.9	100.0	66.1	92.8	100.0	68.5	93.3	100.0	63.3	92.2	100.0	65.9	92.7	100.0	65.6	92.7
15	MECHANICAL	100.1	63.7	84.2	95.7	39.8	71.2	95.7	62.4	81.1	100.1	30.3	69.5	100.1	64.9	84.7	95.7	61.9	80.9
16	ELECTRICAL	96.8	68.5	83.0	95.4	67.9	82.0	96.7	68.0	82.7	95.0	35.1	65.8	96.3	68.5	82.7	95.0	60.6	78.2
01 - 16	WEIGHTED AVERAGE	96.5	65.0	82.2	92.9	56.1	76.2	93.3	67.1	81.4	95.7	43.7	72.1	97.3	64.0	82.1	93.4	61.5	78.9

		OKLAHOMA												OREGON					
	DIVISION	POTEAU			SHAWNEE			TULSA			WOODWARD			BEND			EUGENE		
		749			748			740 - 741			738			977			974		
		MAT.	INST.	TOTAL	MAT.	INST.	TOTAL	MAT.	INST.	TOTAL	MAT.	INST.	TOTAL	MAT.	INST.	TOTAL	MAT.	INST.	TOTAL
01590	EQUIPMENT RENTAL	.0	86.0	86.0	.0	77.1	77.1	.0	86.8	86.8	.0	77.1	77.1	.0	99.5	99.5	.0	99.5	99.5
02	SITE CONSTRUCTION	75.4	84.6	82.2	103.8	88.9	92.7	101.2	86.2	90.1	107.6	89.1	94.0	121.6	104.8	109.2	108.8	104.8	105.8
03100	CONCRETE FORMS & ACCESSORIES	86.2	48.9	53.9	81.5	47.0	51.6	98.7	48.2	54.9	89.9	41.1	47.6	107.5	105.9	106.1	103.5	105.8	105.5
03200	CONCRETE REINFORCEMENT	97.8	82.7	90.2	96.7	57.2	76.7	96.9	82.6	89.7	96.7	82.7	89.6	101.7	99.5	100.6	106.0	99.5	102.7
03300	CAST-IN-PLACE CONCRETE	82.4	50.1	68.9	94.9	47.7	75.2	90.8	50.4	73.9	92.6	48.6	74.2	109.5	107.9	108.8	106.0	107.9	106.8
03	CONCRETE	83.2	56.6	69.9	89.3	49.2	69.3	87.9	56.4	72.2	90.3	51.7	71.0	118.7	105.0	111.9	108.8	105.0	106.9
04	MASONRY	96.9	61.0	74.3	113.3	59.3	79.4	96.8	62.4	75.2	91.4	60.2	71.8	114.2	101.4	106.2	112.0	101.4	105.3
05	METALS	88.2	82.3	86.2	88.1	55.4	77.4	92.0	81.9	88.7	88.3	67.3	81.4	93.8	97.2	94.9	94.4	97.0	95.3
06	WOOD & PLASTICS	89.5	48.7	68.2	84.2	47.6	65.1	101.1	47.6	73.1	93.5	38.5	64.8	98.5	105.7	102.3	94.1	105.7	100.1
07	THERMAL & MOISTURE PROTECTION	99.1	61.4	80.6	99.1	61.1	80.4	98.8	59.9	79.7	99.7	60.4	80.4	111.4	95.1	103.4	110.3	92.6	101.6
08	DOORS & WINDOWS	95.6	58.2	86.0	95.6	47.5	83.2	97.2	56.5	86.7	95.7	52.3	84.4	98.4	106.3	100.5	98.8	106.3	100.7
09200	PLASTER & GYPSUM BOARD	80.7	48.0	60.0	77.5	46.9	58.1	85.4	46.8	60.9	82.2	37.6	54.0	96.3	105.7	102.3	94.3	105.7	101.5
095,098	CEILINGS & ACOUSTICAL TREATMENT	80.4	48.0	61.0	80.4	46.9	60.3	91.2	46.8	64.6	83.1	37.6	55.8	101.3	105.7	103.9	105.3	105.7	105.5
09600	FLOORING	109.1	63.2	97.0	105.8	42.7	89.1	115.6	55.9	99.8	110.5	56.3	96.2	117.7	103.1	113.8	115.8	103.1	112.4
097,099	WALL FINISHES, PAINTS & COATINGS	96.4	57.1	72.6	96.4	45.1	65.3	96.4	53.6	70.4	96.4	57.1	72.6	115.8	74.6	90.8	115.8	74.6	90.8
09	FINISHES	91.1	51.5	70.5	91.1	45.2	67.3	97.3	49.3	72.3	94.2	43.9	68.0	111.1	102.1	106.4	110.2	102.1	106.0
10 - 14	TOTAL DIV. 10000 - 14000	100.0	66.0	92.8	100.0	64.9	92.5	100.0	66.8	92.9	100.0	64.1	92.4	100.0	108.2	101.7	100.0	108.2	101.7
15	MECHANICAL	95.7	61.6	80.8	95.7	62.9	81.4	100.1	62.6	83.7	95.7	63.7	81.7	95.8	103.9	99.4	100.2	103.9	101.8
16	ELECTRICAL	95.1	68.0	81.9	96.9	68.5	83.1	96.8	45.8	71.9	96.7	68.5	83.0	98.6	97.7	98.2	97.3	97.7	97.5
01 - 16	WEIGHTED AVERAGE	92.5	64.3	79.7	94.9	60.1	79.1	96.3	61.2	80.4	94.5	61.7	79.6	102.5	102.2	102.4	101.7	102.1	101.9

OREGON

DIVISION		KLAMATH FALLS 976			MEDFORD 975			PENDLETON 978			PORTLAND 970 - 972			SALEM 973			VALE 979		
		MAT.	INST.	TOTAL	MAT.	INST.	TOTAL	MAT.	INST.	TOTAL	MAT.	INST.	TOTAL	MAT.	INST.	TOTAL	MAT.	INST.	TOTAL
01590	EQUIPMENT RENTAL	.0	99.5	99.5	.0	99.5	99.5	.0	96.4	96.4	.0	99.5	99.5	.0	99.5	99.5	.0	96.4	96.4
02	SITE CONSTRUCTION	126.9	104.8	110.5	118.4	104.8	108.3	114.7	97.2	101.7	109.9	104.8	106.1	108.4	104.8	105.7	100.5	96.7	97.7
03100	CONCRETE FORMS & ACCESSORIES	99.4	105.5	104.7	98.3	105.6	104.6	101.0	105.7	105.0	104.9	106.3	106.1	104.8	106.0	105.9	108.6	104.6	105.1
03200	CONCRETE REINFORCEMENT	101.7	99.4	100.6	103.4	99.4	101.4	101.0	99.8	100.4	106.9	99.9	103.3	107.0	99.9	103.4	98.5	99.7	99.1
03300	CAST-IN-PLACE CONCRETE	109.5	103.2	106.9	109.5	107.8	108.7	110.2	104.6	107.9	113.8	108.0	111.4	109.9	107.9	109.1	86.8	103.3	93.7
03	CONCRETE	121.8	103.2	112.5	115.6	104.8	110.2	100.2	103.8	102.0	112.8	105.3	109.0	110.9	105.1	108.0	83.9	102.8	93.4
04	MASONRY	130.7	101.4	112.3	108.9	101.4	104.2	118.6	107.8	111.9	113.2	107.8	109.8	116.7	107.8	111.1	116.8	106.8	110.5
05	METALS	93.8	96.7	94.8	94.0	96.8	94.9	101.3	97.2	99.9	95.6	98.0	96.4	94.8	97.7	95.8	101.2	95.1	99.2
06	WOOD & PLASTICS	89.1	105.7	97.7	87.8	105.7	97.1	91.6	105.8	99.0	95.3	105.7	100.7	95.3	105.7	100.7	100.7	105.8	103.4
07	THERMAL & MOISTURE PROTECTION	111.7	89.6	100.8	111.1	89.6	100.5	100.2	90.6	95.4	110.1	102.6	106.4	110.1	98.3	104.3	99.4	93.3	96.4
08	DOORS & WINDOWS	98.4	106.3	100.5	101.5	106.3	102.7	94.6	106.4	97.6	96.3	106.3	98.9	98.1	106.3	100.2	94.5	99.5	95.8
09200	PLASTER & GYPSUM BOARD	93.3	105.7	101.1	92.5	105.7	100.9	76.0	105.7	94.8	92.3	105.7	100.8	92.3	105.7	100.8	81.6	105.7	96.9
095,098	CEILINGS & ACOUSTICAL TREATMENT	111.1	105.7	107.9	115.1	105.7	109.5	69.4	105.7	91.2	105.3	105.7	105.5	105.3	105.7	105.5	69.4	105.7	91.2
09600	FLOORING	114.0	103.1	111.1	113.3	103.1	110.6	80.4	103.1	86.4	115.8	103.1	112.4	115.8	103.1	112.4	82.6	103.1	88.0
097,099	WALL FINISHES, PAINTS & COATINGS	115.8	62.8	83.6	115.8	62.8	83.6	96.5	72.5	81.9	115.8	74.6	90.8	115.8	74.6	90.8	96.5	74.6	83.2
09	FINISHES	112.3	100.7	106.3	112.3	100.7	106.3	75.9	102.0	89.5	110.0	102.1	105.9	109.8	102.1	105.8	76.5	102.0	89.8
10 - 14	TOTAL DIV. 10000 - 14000	100.0	108.1	101.7	100.0	108.1	101.7	100.0	104.3	100.9	100.0	108.2	101.7	100.0	108.2	101.7	100.0	108.1	101.7
15	MECHANICAL	95.8	103.9	99.3	100.2	103.9	101.8	97.6	101.9	99.5	100.2	109.5	104.2	100.2	104.0	101.8	97.6	81.9	90.7
16	ELECTRICAL	97.4	87.7	92.7	100.6	87.7	94.3	88.2	93.5	90.8	97.8	105.8	101.7	97.4	97.7	97.6	88.2	76.7	82.6
01 - 16	WEIGHTED AVERAGE	103.7	100.1	102.1	103.4	100.3	102.0	96.6	100.7	98.5	102.2	105.4	103.7	102.2	103.0	102.5	94.4	93.7	94.1

PENNSYLVANIA

DIVISION		ALLENTOWN 181			ALTOONA 166			BEDFORD 155			BRADFORD 167			BUTLER 160			CHAMBERSBURG 172		
		MAT.	INST.	TOTAL	MAT.	INST.	TOTAL	MAT.	INST.	TOTAL	MAT.	INST.	TOTAL	MAT.	INST.	TOTAL	MAT.	INST.	TOTAL
01590	EQUIPMENT RENTAL	.0	115.7	115.7	.0	115.7	115.7	.0	113.9	113.9	.0	115.7	115.7	.0	115.7	115.7	.0	114.9	114.9
02	SITE CONSTRUCTION	91.4	106.4	102.5	95.3	106.6	103.7	99.0	105.0	103.5	90.6	106.6	102.4	86.4	107.6	102.1	85.9	104.0	99.3
03100	CONCRETE FORMS & ACCESSORIES	101.9	105.2	104.8	83.9	86.3	86.0	84.7	85.8	85.6	86.8	82.5	83.0	85.7	98.4	96.7	85.8	82.8	83.2
03200	CONCRETE REINFORCEMENT	103.1	102.0	102.5	100.0	97.1	98.5	98.8	84.8	91.7	102.0	94.3	98.1	100.6	106.0	103.4	100.4	91.6	96.0
03300	CAST-IN-PLACE CONCRETE	89.0	94.9	91.5	99.3	86.9	94.1	103.5	81.4	94.3	94.8	89.7	92.7	87.6	97.5	91.8	96.3	80.8	89.8
03	CONCRETE	98.8	102.1	100.5	94.0	90.0	92.0	97.3	85.5	91.4	100.6	88.8	94.7	85.6	100.8	93.2	106.5	85.2	95.9
04	MASONRY	94.0	90.7	91.9	96.6	80.7	86.6	96.8	83.8	88.7	92.9	80.2	84.9	98.8	95.6	96.8	95.0	81.2	86.3
05	METALS	97.3	120.2	104.8	91.0	117.0	99.5	94.3	109.3	99.2	95.7	115.4	102.1	90.8	123.7	101.5	95.7	111.7	100.9
06	WOOD & PLASTICS	103.2	108.5	105.9	80.4	88.2	84.4	81.8	87.0	84.5	86.9	80.0	83.3	82.1	98.0	90.4	91.0	84.3	87.5
07	THERMAL & MOISTURE PROTECTION	99.1	109.8	104.3	97.6	92.0	94.9	97.2	90.4	93.9	99.0	91.9	95.5	97.2	101.0	99.1	97.6	77.7	87.8
08	DOORS & WINDOWS	93.6	105.3	96.6	88.1	95.2	90.0	91.3	91.4	91.3	94.0	88.3	92.5	88.1	107.6	93.2	90.5	84.4	88.9
09200	PLASTER & GYPSUM BOARD	108.4	108.4	108.4	100.7	87.6	92.4	91.3	86.3	88.1	101.6	79.2	87.4	101.1	97.7	99.0	106.0	83.6	91.8
095,098	CEILINGS & ACOUSTICAL TREATMENT	90.0	108.4	101.1	94.5	87.6	90.3	89.8	86.3	87.7	92.8	79.2	84.6	95.5	97.7	96.8	92.8	83.6	87.3
09600	FLOORING	88.2	83.4	87.0	82.3	59.5	76.3	87.9	60.4	80.6	81.7	82.3	81.9	83.2	88.8	84.7	85.7	60.3	79.0
097,099	WALL FINISHES, PAINTS & COATINGS	90.3	93.3	92.1	85.4	111.7	101.4	96.3	89.4	92.1	90.3	88.6	89.2	85.4	102.7	95.9	90.3	76.4	81.8
09	FINISHES	95.8	100.0	98.0	94.2	83.3	88.5	92.4	80.7	86.3	93.6	81.9	87.5	94.3	96.5	95.5	94.7	77.8	85.9
10 - 14	TOTAL DIV. 10000 - 14000	100.0	96.8	99.3	100.0	99.7	99.9	100.0	100.4	100.1	100.0	101.1	100.2	100.0	104.9	101.0	100.0	97.3	99.4
15	MECHANICAL	100.4	99.8	100.1	99.8	84.7	93.2	95.7	85.7	91.3	96.0	88.9	92.9	95.5	94.2	94.9	96.1	84.9	91.2
16	ELECTRICAL	100.1	92.4	96.3	89.3	102.0	95.5	92.9	102.0	97.3	94.4	102.0	98.1	89.8	102.0	95.7	92.8	79.6	86.4
01 - 16	WEIGHTED AVERAGE	98.0	101.2	99.4	94.3	93.3	93.8	95.0	91.8	93.5	95.9	93.3	94.7	92.3	101.6	96.5	96.1	87.2	92.0

PENNSYLVANIA

DIVISION		DOYLESTOWN 189			DUBOIS 158			ERIE 164 - 165			GREENSBURG 156			HARRISBURG 170 - 171			HAZLETON 182		
		MAT.	INST.	TOTAL	MAT.	INST.	TOTAL	MAT.	INST.	TOTAL	MAT.	INST.	TOTAL	MAT.	INST.	TOTAL	MAT.	INST.	TOTAL
01590	EQUIPMENT RENTAL	.0	92.7	92.7	.0	113.9	113.9	.0	115.7	115.7	.0	113.9	113.9	.0	114.9	114.9	.0	115.7	115.7
02	SITE CONSTRUCTION	106.6	88.6	93.3	103.7	105.5	105.0	92.0	107.1	103.2	95.1	106.4	103.5	82.5	104.9	99.1	84.6	106.1	100.5
03100	CONCRETE FORMS & ACCESSORIES	82.6	125.8	120.0	83.9	87.5	87.0	101.2	96.9	97.4	92.3	98.4	97.6	94.4	85.2	86.4	80.1	84.3	83.7
03200	CONCRETE REINFORCEMENT	99.6	135.5	117.8	98.2	97.3	97.7	102.0	94.3	98.1	98.2	106.0	102.1	103.1	96.9	100.0	100.1	101.8	100.9
03300	CAST-IN-PLACE CONCRETE	84.1	97.0	89.5	99.8	93.5	97.1	97.6	84.9	92.3	96.0	97.2	96.5	98.3	89.2	94.5	84.1	84.8	84.4
03	CONCRETE	94.2	117.2	105.7	98.1	92.8	95.4	93.1	93.5	93.3	92.5	100.6	96.6	100.9	90.3	95.6	91.1	89.3	90.2
04	MASONRY	97.2	122.8	113.3	97.0	87.1	90.8	86.4	92.3	90.1	107.5	95.0	99.6	94.6	84.3	88.1	105.9	87.2	94.2
05	METALS	94.8	120.0	103.0	94.3	116.5	101.6	91.2	116.7	99.5	94.3	122.5	103.5	99.3	117.4	105.2	97.0	119.2	104.3
06	WOOD & PLASTICS	81.9	127.7	105.8	80.7	87.0	84.0	99.6	98.1	98.8	89.8	98.0	94.1	96.8	84.4	90.4	80.3	83.3	81.8
07	THERMAL & MOISTURE PROTECTION	96.5	124.3	110.2	97.6	96.4	97.0	97.8	96.5	97.2	97.1	100.5	98.8	102.8	103.3	103.0	98.3	96.9	97.6
08	DOORS & WINDOWS	95.3	135.0	105.6	91.3	95.1	92.3	88.3	94.3	89.8	91.2	107.6	95.5	93.6	93.1	93.5	94.3	94.0	94.2
09200	PLASTER & GYPSUM BOARD	99.1	128.2	117.5	90.5	86.3	87.8	108.4	97.8	101.7	93.8	97.7	96.3	108.4	83.7	92.8	99.5	82.6	88.8
095,098	CEILINGS & ACOUSTICAL TREATMENT	89.2	128.2	112.6	89.8	86.3	87.7	90.0	97.8	94.7	88.9	97.7	94.2	90.0	83.7	86.2	91.0	82.6	85.9
09600	FLOORING	72.5	108.2	82.0	87.5	105.1	92.2	90.2	88.6	89.8	91.7	88.8	91.0	88.5	78.4	85.8	78.9	82.1	79.7
097,099	WALL FINISHES, PAINTS & COATINGS	89.6	91.2	90.6	96.3	102.7	100.2	96.9	88.6	91.8	96.3	102.7	100.2	90.3	87.0	88.3	90.3	95.3	93.3
09	FINISHES	86.9	119.2	103.7	92.5	90.5	91.5	97.2	94.5	95.8	93.5	97.4	95.5	94.8	83.4	88.9	91.6	83.9	87.6
10 - 14	TOTAL DIV. 10000 - 14000	100.0	94.8	98.9	100.0	101.3	100.3	100.0	104.0	100.9	100.0	104.8	101.0	100.0	98.0	99.6	100.0	92.1	98.3
15	MECHANICAL	95.5	120.7	106.5	95.7	87.1	92.0	99.8	92.9	96.8	95.7	85.3	91.1	100.4	90.3	95.9	96.0	85.8	91.6
16	ELECTRICAL	93.4	114.5	103.7	93.4	102.0	97.6	90.7	86.4	88.6	93.4	102.0	97.6	99.1	83.7	91.6	95.2	89.3	92.3
01 - 16	WEIGHTED AVERAGE	94.8	116.5	104.6	95.2	95.7	95.5	94.3	96.2	95.1	95.1	99.6	97.1	98.1	92.3	95.5	95.3	92.4	94.0

COST INDEXES

PENNSYLVANIA

DIVISION		INDIANA 157			JOHNSTOWN 159			KITTANNING 162			LANCASTER 175 - 176			LEHIGH VALLEY 180			MONTROSE 188		
		MAT.	INST.	TOTAL	MAT.	INST.	TOTAL	MAT.	INST.	TOTAL	MAT.	INST.	TOTAL	MAT.	INST.	TOTAL	MAT.	INST.	TOTAL
01590	EQUIPMENT RENTAL	.0	113.9	113.9	.0	113.9	113.9	.0	115.7	115.7	.0	114.9	114.9	.0	115.7	115.7	.0	115.7	115.7
02	SITE CONSTRUCTION	93.5	106.2	102.9	99.5	105.9	104.2	89.0	107.6	102.8	77.9	104.8	97.8	88.7	106.4	101.8	87.4	105.5	100.8
03100	CONCRETE FORMS & ACCESSORIES	85.4	87.9	87.6	83.9	88.9	88.2	85.7	98.7	96.9	88.1	84.2	84.7	94.3	100.8	99.9	81.1	85.3	84.7
03200	CONCRETE REINFORCEMENT	97.4	105.9	101.7	98.8	105.7	102.3	100.6	106.2	103.4	100.1	96.8	98.4	100.1	105.7	103.0	104.9	101.9	103.4
03300	CAST-IN-PLACE CONCRETE	94.2	97.1	95.4	104.4	90.0	98.4	91.1	97.6	93.8	81.9	83.8	82.7	91.1	95.0	92.7	89.3	85.7	87.8
03	CONCRETE	90.3	95.9	93.1	97.9	93.8	95.8	88.3	101.0	94.6	93.8	88.0	90.9	97.8	100.9	99.3	96.3	90.0	93.1
04	MASONRY	93.6	95.0	94.5	94.4	87.3	89.9	101.8	97.2	98.9	100.0	81.7	88.5	94.0	90.6	91.9	94.0	90.0	91.1
05	METALS	94.4	122.2	103.5	94.3	120.3	102.8	90.8	124.2	101.7	95.7	116.7	102.6	97.0	122.0	105.2	95.8	116.8	102.7
06	WOOD & PLASTICS	82.7	84.3	83.5	80.7	88.1	84.6	82.1	98.0	90.4	93.9	84.4	88.9	95.0	102.4	98.9	81.1	83.8	82.5
07	THERMAL & MOISTURE PROTECTION	96.9	98.0	97.4	97.2	95.8	96.6	97.3	101.5	99.4	97.0	98.8	97.9	98.9	103.9	101.3	98.3	96.5	97.4
08	DOORS & WINDOWS	91.3	96.2	92.5	91.3	97.8	92.9	88.1	107.6	93.2	90.5	90.7	90.5	94.3	105.6	97.2	90.7	94.3	91.6
09200	PLASTER & GYPSUM BOARD	91.7	83.6	86.5	90.2	87.6	88.5	101.1	97.7	99.0	108.1	83.7	92.6	104.4	102.2	103.0	100.0	83.1	89.3
095,098	CEILINGS & ACOUSTICAL TREATMENT	89.8	83.6	86.1	88.9	87.6	88.1	95.5	97.7	96.8	92.8	83.7	87.4	91.0	102.2	97.7	92.8	83.1	86.9
09600	FLOORING	88.4	101.2	91.8	87.5	75.8	84.4	83.2	105.1	89.0	86.8	83.8	86.0	84.6	105.1	90.0	79.6	75.6	78.5
097,099	WALL FINISHES, PAINTS & COATINGS	96.3	102.7	100.2	96.3	111.7	105.6	85.4	102.7	95.9	90.3	75.6	81.3	90.3	87.8	88.8	90.3	95.3	93.3
09	FINISHES	92.3	90.8	91.5	91.9	88.1	89.9	94.4	99.5	97.1	94.9	81.8	88.1	94.2	99.6	97.0	92.5	84.2	88.2
10 - 14	TOTAL DIV. 10000 - 14000	100.0	103.1	100.7	100.0	101.9	100.4	100.0	104.9	101.0	100.0	97.4	99.4	100.0	96.0	99.2	100.0	88.3	97.5
15	MECHANICAL	95.7	85.2	91.1	95.7	87.9	92.3	95.5	99.0	97.0	96.1	89.5	93.2	96.0	99.7	97.6	96.0	87.1	92.2
16	ELECTRICAL	93.4	102.0	97.6	93.4	102.0	97.6	89.3	102.0	95.5	94.0	58.6	76.8	95.2	105.4	100.2	94.5	84.5	89.6
01 - 16	WEIGHTED AVERAGE	93.9	97.3	95.4	94.9	96.2	95.5	92.8	103.2	95.5	94.9	87.5	91.5	96.0	102.7	99.0	94.9	92.0	93.6

PENNSYLVANIA

DIVISION		NEW CASTLE 161			NORRISTOWN 194			OIL CITY 163			PHILADELPHIA 190 - 191			PITTSBURGH 150 - 152			POTTSVILLE 179		
		MAT.	INST.	TOTAL	MAT.	INST.	TOTAL	MAT.	INST.	TOTAL	MAT.	INST.	TOTAL	MAT.	INST.	TOTAL	MAT.	INST.	TOTAL
01590	EQUIPMENT RENTAL	.0	115.7	115.7	.0	95.3	95.3	.0	115.7	115.7	.0	94.7	94.7	.0	115.5	115.5	.0	114.9	114.9
02	SITE CONSTRUCTION	86.8	107.6	102.2	94.3	95.0	94.8	85.3	106.7	101.2	101.1	94.8	96.4	98.5	109.1	106.3	80.8	104.7	98.5
03100	CONCRETE FORMS & ACCESSORIES	85.7	98.2	96.5	81.4	122.5	117.0	85.7	84.3	84.4	99.9	134.0	129.5	101.5	98.8	99.1	78.4	82.0	81.5
03200	CONCRETE REINFORCEMENT	99.3	105.5	102.5	100.8	128.5	114.8	100.6	105.2	102.9	104.0	135.7	120.0	99.3	106.2	102.8	99.2	94.5	96.8
03300	CAST-IN-PLACE CONCRETE	88.5	98.5	92.6	79.2	121.9	97.0	85.9	94.1	89.3	97.1	128.4	110.2	99.8	97.3	98.7	87.2	85.3	86.4
03	CONCRETE	86.1	100.9	93.5	94.5	123.1	108.8	84.4	93.1	88.8	107.0	131.9	119.4	95.9	100.9	98.4	97.5	87.1	92.3
04	MASONRY	96.9	94.4	95.4	110.4	110.8	110.7	97.7	89.0	92.2	97.0	129.3	117.2	89.6	97.4	94.5	93.7	82.4	86.6
05	METALS	90.8	123.2	101.4	100.6	120.2	107.0	90.8	121.0	100.7	104.5	127.7	112.1	95.6	123.0	104.6	95.9	115.4	102.3
06	WOOD & PLASTICS	82.1	98.0	90.4	78.4	122.3	101.3	82.1	80.9	81.5	97.9	133.8	116.6	100.1	98.0	99.0	83.3	81.3	82.2
07	THERMAL & MOISTURE PROTECTION	97.2	98.7	98.0	100.1	122.3	111.0	97.1	94.7	95.9	101.2	132.8	116.7	97.5	101.1	99.2	97.1	97.1	97.1
08	DOORS & WINDOWS	88.1	107.5	93.1	88.6	128.9	99.0	88.1	88.3	88.2	96.7	139.5	107.8	93.8	107.6	97.4	90.5	90.7	90.6
09200	PLASTER & GYPSUM BOARD	101.1	97.7	99.0	96.6	122.8	113.2	101.1	80.1	87.8	105.1	134.6	123.8	99.3	97.7	98.3	102.8	80.5	88.7
095,098	CEILINGS & ACOUSTICAL TREATMENT	95.5	97.7	96.8	101.9	122.8	114.4	95.5	80.1	86.2	101.9	134.6	121.5	89.8	97.7	94.5	92.8	80.5	85.4
09600	FLOORING	83.2	75.0	81.0	77.6	125.5	90.3	83.2	93.4	85.9	85.7	134.9	98.7	96.3	106.2	98.9	82.6	79.8	81.8
097,099	WALL FINISHES, PAINTS & COATINGS	85.4	102.7	95.9	92.5	130.5	115.6	85.4	102.7	95.9	92.5	141.7	122.3	96.3	117.6	109.2	90.3	95.3	93.3
09	FINISHES	94.3	93.7	94.0	97.4	123.6	111.0	94.2	86.0	90.0	101.2	135.1	118.9	96.0	101.6	98.9	93.0	82.5	87.5
10 - 14	TOTAL DIV. 10000 - 14000	100.0	104.9	101.0	100.0	114.0	103.0	100.0	102.1	100.4	100.0	126.5	105.6	100.0	104.8	101.0	100.0	85.5	96.9
15	MECHANICAL	95.5	91.5	93.8	95.7	119.8	106.3	95.5	90.5	93.3	100.1	132.3	114.2	100.1	100.6	100.3	96.1	84.8	91.1
16	ELECTRICAL	89.8	99.6	94.6	93.1	109.7	101.2	91.4	98.7	95.0	97.2	134.2	115.2	95.4	102.0	98.6	92.4	81.9	87.3
01 - 16	WEIGHTED AVERAGE	92.3	100.2	95.9	96.5	116.5	105.6	92.3	95.8	93.9	100.8	128.9	113.6	96.7	103.8	99.9	94.6	89.3	92.2

PENNSYLVANIA

DIVISION		READING 195 - 196			SCRANTON 184 - 185			STATE COLLEGE 168			STROUDSBURG 183			SUNBURY 178			UNIONTOWN 154		
		MAT.	INST.	TOTAL	MAT.	INST.	TOTAL	MAT.	INST.	TOTAL	MAT.	INST.	TOTAL	MAT.	INST.	TOTAL	MAT.	INST.	TOTAL
01590	EQUIPMENT RENTAL	.0	118.2	118.2	.0	115.7	115.7	.0	114.9	114.9	.0	115.7	115.7	.0	115.7	115.7	.0	113.9	113.9
02	SITE CONSTRUCTION	98.7	110.7	107.6	91.9	106.3	102.6	82.8	105.4	99.5	86.4	105.3	100.4	90.4	105.9	101.9	93.9	106.4	103.2
03100	CONCRETE FORMS & ACCESSORIES	99.4	86.2	88.0	102.0	84.5	86.9	84.7	86.3	86.1	87.7	92.5	91.9	91.7	78.2	80.0	77.1	98.2	95.4
03200	CONCRETE REINFORCEMENT	102.2	103.1	102.7	103.1	101.9	102.5	101.3	97.1	99.2	103.4	101.7	102.6	102.0	91.8	96.9	98.2	105.7	102.0
03300	CAST-IN-PLACE CONCRETE	71.0	91.3	79.5	93.0	88.7	91.2	89.7	77.5	84.6	87.6	81.3	85.0	95.4	83.7	90.5	94.2	97.1	95.4
03	CONCRETE	94.3	92.7	93.5	100.7	90.8	95.8	100.7	86.9	93.8	95.1	91.7	93.4	101.4	84.3	92.9	89.9	100.5	95.2
04	MASONRY	98.2	87.2	91.3	94.3	92.8	93.4	99.5	81.2	88.0	91.5	90.6	91.0	92.9	82.0	86.1	108.6	90.2	97.0
05	METALS	100.9	120.7	107.4	99.4	119.5	106.0	95.6	117.1	102.6	97.1	116.5	103.4	95.6	113.9	101.6	94.1	121.9	103.2
06	WOOD & PLASTICS	98.1	84.4	90.9	103.2	81.9	92.1	89.8	88.2	88.9	88.2	94.0	91.2	92.1	76.6	84.0	73.4	98.0	86.2
07	THERMAL & MOISTURE PROTECTION	101.2	105.4	103.3	98.9	99.1	99.0	97.3	96.6	96.9	98.6	84.1	91.5	98.8	96.1	97.5	96.8	99.2	98.0
08	DOORS & WINDOWS	94.4	92.5	93.9	93.6	93.2	93.5	90.4	95.2	91.7	94.3	90.8	93.4	90.6	87.4	89.8	91.2	107.6	95.4
09200	PLASTER & GYPSUM BOARD	108.7	83.8	92.9	111.0	81.2	92.1	104.5	87.6	93.8	101.9	93.6	96.6	102.9	75.7	85.7	87.0	97.7	93.8
095,098	CEILINGS & ACOUSTICAL TREATMENT	90.2	83.8	86.4	99.8	81.2	88.6	90.1	87.6	88.6	89.2	93.6	91.8	90.1	75.7	81.5	88.9	97.7	94.2
09600	FLOORING	83.9	87.6	84.9	88.2	82.1	86.6	85.2	64.9	79.8	82.2	67.0	78.2	83.7	55.5	76.2	84.3	83.6	84.1
097,099	WALL FINISHES, PAINTS & COATINGS	90.7	98.8	95.6	90.3	95.3	93.3	90.3	111.7	103.3	90.3	85.4	87.3	90.3	95.3	93.3	96.3	102.7	100.2
09	FINISHES	98.3	86.8	92.3	98.5	84.4	91.2	93.5	84.5	88.8	92.6	87.4	89.9	93.7	75.4	84.2	90.2	95.5	92.9
10 - 14	TOTAL DIV. 10000 - 14000	100.0	92.1	98.3	100.0	93.0	98.5	100.0	97.0	99.4	100.0	84.5	96.7	100.0	84.7	96.8	100.0	104.8	101.0
15	MECHANICAL	100.3	107.3	103.4	100.4	87.4	94.7	96.1	88.1	92.6	96.0	87.4	92.3	96.0	85.4	91.4	95.7	89.5	93.0
16	ELECTRICAL	98.9	90.9	95.0	100.1	84.5	92.5	93.2	102.0	97.5	95.2	109.7	102.3	93.2	81.9	87.7	90.8	102.0	96.2
01 - 16	WEIGHTED AVERAGE	98.6	98.5	98.6	98.8	92.9	96.1	95.5	93.7	94.7	95.3	95.8	95.5	95.5	87.9	92.0	94.0	99.6	96.5

City Cost Indexes

PENNSYLVANIA

DIVISION		WASHINGTON 153			WELLSBORO 169			WESTCHESTER 193			WILKES-BARRE 186 - 187			WILLIAMSPORT 177			YORK 173 - 174		
		MAT.	INST.	TOTAL	MAT.	INST.	TOTAL	MAT.	INST.	TOTAL	MAT.	INST.	TOTAL	MAT.	INST.	TOTAL	MAT.	INST.	TOTAL
01590	EQUIPMENT RENTAL	.0	113.9	113.9	.0	115.7	115.7	.0	95.3	95.3	.0	115.7	115.7	.0	115.7	115.7	.0	114.9	114.9
02	SITE CONSTRUCTION	94.0	106.5	103.2	94.5	105.3	102.5	100.5	92.2	94.3	84.2	106.2	100.5	81.6	104.9	98.8	81.6	104.8	98.8
03100	CONCRETE FORMS & ACCESSORIES	85.6	98.5	96.8	86.1	75.9	77.3	89.1	124.9	120.1	90.7	84.6	85.4	87.7	69.7	72.1	82.4	79.7	80.1
03200	CONCRETE REINFORCEMENT	98.2	106.0	102.1	101.3	101.8	101.5	99.8	109.4	104.7	102.0	101.8	101.9	101.3	86.9	94.0	102.0	96.8	99.4
03300	CAST-IN-PLACE CONCRETE	94.2	97.3	95.5	94.0	81.9	89.0	87.6	119.8	101.1	84.1	85.3	84.6	80.7	79.2	80.1	87.6	83.9	86.1
03	CONCRETE	90.4	100.7	95.6	103.1	84.5	93.8	102.6	119.8	111.2	92.1	89.6	90.9	88.8	78.1	83.4	98.9	86.0	92.5
04	MASONRY	94.0	95.0	94.6	99.3	80.8	87.7	105.5	120.6	115.0	106.8	88.1	95.1	85.4	79.2	81.5	93.9	81.7	86.2
05	METALS	94.1	122.6	103.4	95.6	117.2	102.7	100.6	110.6	103.9	95.8	119.2	103.4	95.7	111.5	100.8	96.8	117.0	103.4
06	WOOD & PLASTICS	82.7	98.0	90.7	86.2	74.8	80.2	86.5	127.6	107.9	91.1	83.3	87.0	88.2	68.4	77.8	87.4	78.0	82.5
07	THERMAL & MOISTURE PROTECTION	96.9	100.5	98.7	99.2	92.0	95.7	100.7	123.4	111.8	98.3	97.3	97.8	98.1	92.1	95.1	97.2	98.7	97.9
08	DOORS & WINDOWS	91.2	107.6	95.4	94.0	89.4	92.8	88.6	113.1	95.0	90.6	91.5	90.9	90.6	72.4	85.9	90.5	89.6	90.3
09200	PLASTER & GYPSUM BOARD	91.4	97.7	95.4	100.9	73.8	83.7	99.8	128.2	117.8	103.2	82.6	90.1	102.8	67.2	80.3	104.4	77.1	87.1
095,098	CEILINGS & ACOUSTICAL TREATMENT	88.9	97.7	94.2	90.1	73.8	80.3	101.9	128.2	117.7	92.8	82.6	86.7	92.8	67.2	77.4	91.8	77.1	83.0
09600	FLOORING	88.4	98.3	91.0	81.5	66.3	77.5	81.2	108.2	88.3	83.3	82.1	83.0	82.2	64.9	77.6	84.2	83.8	84.1
097,099	WALL FINISHES, PAINTS & COATINGS	96.3	102.7	100.2	90.3	95.3	93.3	92.5	136.8	119.4	90.3	95.3	93.3	90.3	95.3	93.3	90.3	76.4	81.8
09	FINISHES	92.1	98.4	95.4	93.0	75.3	83.8	99.3	123.7	112.0	93.9	84.7	89.1	93.4	70.5	81.5	93.5	79.0	86.0
10 - 14	TOTAL DIV. 10000 - 14000	100.0	104.8	101.0	100.0	83.8	96.6	100.0	118.9	104.0	100.0	92.4	98.4	100.0	93.4	98.6	100.0	96.6	99.3
15	MECHANICAL	95.7	96.9	96.2	96.0	84.4	90.9	95.7	125.6	108.8	96.0	86.3	91.8	96.0	82.7	90.2	100.4	89.5	95.6
16	ELECTRICAL	92.8	102.0	97.3	94.4	79.9	87.3	93.0	101.8	97.3	95.2	80.4	88.0	93.6	62.1	78.2	94.0	71.7	83.1
01 - 16	WEIGHTED AVERAGE	93.8	102.1	97.6	96.5	87.4	92.4	97.6	115.6	105.7	95.3	91.4	93.5	93.4	81.8	88.2	96.2	88.6	92.7

DIVISION		PUERTO RICO — SAN JUAN 009			RHODE ISLAND — NEWPORT 028			PROVIDENCE 029			SOUTH CAROLINA — AIKEN 298			BEAUFORT 299			CHARLESTON 294		
		MAT.	INST.	TOTAL	MAT.	INST.	TOTAL	MAT.	INST.	TOTAL	MAT.	INST.	TOTAL	MAT.	INST.	TOTAL	MAT.	INST.	TOTAL
01590	EQUIPMENT RENTAL	.0	89.1	89.1	.0	103.4	103.4	.0	103.4	103.4	.0	99.3	99.3	.0	99.3	99.3	.0	99.3	99.3
02	SITE CONSTRUCTION	115.6	92.1	98.2	81.6	103.8	98.0	80.9	104.0	98.0	117.3	83.4	92.2	112.3	80.5	88.8	95.2	81.2	84.9
03100	CONCRETE FORMS & ACCESSORIES	94.1	20.1	30.0	104.7	115.4	114.0	103.5	115.4	113.8	99.2	69.5	73.4	97.4	29.7	38.8	96.3	38.4	46.1
03200	CONCRETE REINFORCEMENT	194.4	12.5	102.5	108.2	117.2	112.8	108.2	117.3	112.8	96.6	68.9	82.6	95.7	30.7	62.9	95.5	65.6	80.4
03300	CAST-IN-PLACE CONCRETE	105.2	31.4	74.4	81.5	113.2	94.7	86.1	113.2	97.4	72.6	71.5	72.1	72.6	41.2	59.5	85.1	49.3	70.2
03	CONCRETE	113.9	23.3	68.7	100.1	114.5	107.3	102.3	114.6	108.4	103.3	71.3	87.3	100.2	36.0	68.2	93.7	49.1	71.4
04	MASONRY	219.3	17.3	92.6	105.2	126.5	118.6	107.7	126.5	119.5	73.3	61.7	66.0	87.2	25.7	48.6	88.6	36.1	55.7
05	METALS	115.2	31.3	87.8	100.0	110.3	103.3	100.0	110.3	103.3	91.8	87.6	90.4	91.8	65.9	83.3	93.3	81.2	89.4
06	WOOD & PLASTICS	91.1	20.4	54.2	101.2	113.9	107.8	101.2	113.9	107.8	98.7	71.3	84.4	96.8	29.9	61.9	95.4	37.7	65.3
07	THERMAL & MOISTURE PROTECTION	167.7	23.8	97.0	102.0	114.4	108.1	100.6	114.4	107.4	98.6	69.0	84.0	98.2	34.0	66.6	97.1	43.8	70.9
08	DOORS & WINDOWS	153.8	17.1	118.5	101.0	115.1	104.7	101.0	115.1	104.7	92.2	68.0	85.9	92.2	30.1	76.1	96.3	42.6	82.4
09200	PLASTER & GYPSUM BOARD	252.1	18.0	103.7	102.0	113.4	109.2	102.0	113.4	109.2	104.0	70.3	82.6	108.0	27.8	57.2	107.2	35.7	61.9
095,098	CEILINGS & ACOUSTICAL TREATMENT	343.0	18.0	148.1	98.3	113.4	107.4	100.3	113.4	108.2	91.9	70.3	78.9	98.1	27.8	56.0	98.1	35.7	60.7
09600	FLOORING	200.8	17.1	152.3	100.7	133.9	109.5	100.7	133.9	109.5	104.2	17.6	81.3	105.9	32.1	86.4	105.3	41.0	88.3
097,099	WALL FINISHES, PAINTS & COATINGS	201.8	15.8	89.0	95.3	113.4	106.3	95.3	113.4	106.3	108.5	73.2	87.1	108.5	26.9	59.0	108.5	39.5	66.7
09	FINISHES	251.5	19.3	130.7	100.5	119.0	110.1	100.9	119.0	110.3	99.9	60.3	79.3	102.2	29.8	64.5	100.3	38.4	68.1
10 - 14	TOTAL DIV. 10000 - 14000	100.0	21.0	83.2	100.0	102.8	100.6	100.0	102.8	100.6	100.0	72.5	94.2	100.0	40.4	87.3	100.0	63.8	92.3
15	MECHANICAL	102.4	15.9	64.5	100.1	112.5	105.5	100.1	112.5	105.5	95.7	66.3	82.8	95.7	26.4	65.3	100.0	44.6	75.8
16	ELECTRICAL	133.6	15.7	76.1	99.3	101.4	100.3	99.0	101.4	100.1	95.7	71.6	84.0	98.9	26.5	63.6	97.5	33.2	66.1
01 - 16	WEIGHTED AVERAGE	136.6	26.3	86.5	99.9	112.4	105.6	100.2	112.4	105.8	95.9	70.4	84.3	96.7	37.4	69.8	96.8	48.9	75.1

SOUTH CAROLINA / SOUTH DAKOTA

DIVISION		COLUMBIA 290 - 292			FLORENCE 295			GREENVILLE 296			ROCK HILL 297			SPARTANBURG 293			ABERDEEN 574		
		MAT.	INST.	TOTAL	MAT.	INST.	TOTAL	MAT.	INST.	TOTAL	MAT.	INST.	TOTAL	MAT.	INST.	TOTAL	MAT.	INST.	TOTAL
01590	EQUIPMENT RENTAL	.0	99.3	99.3	.0	99.3	99.3	.0	99.3	99.3	.0	99.3	99.3	.0	99.3	99.3	.0	98.2	98.2
02	SITE CONSTRUCTION	94.5	81.2	84.7	105.8	81.2	87.6	100.8	80.8	86.0	98.1	80.5	85.1	100.7	80.8	86.0	84.2	93.6	91.1
03100	CONCRETE FORMS & ACCESSORIES	100.9	40.7	48.7	82.0	40.6	46.1	96.1	40.3	47.8	94.3	28.6	37.4	99.8	40.3	48.3	97.0	40.1	47.7
03200	CONCRETE REINFORCEMENT	95.5	65.6	80.4	95.2	65.6	80.2	95.1	40.0	67.2	95.9	19.0	57.1	95.1	40.0	67.2	99.8	49.3	74.3
03300	CAST-IN-PLACE CONCRETE	80.3	49.0	67.2	72.5	49.2	62.8	72.5	49.1	62.8	72.5	36.4	57.4	72.5	49.1	62.8	102.3	51.2	81.0
03	CONCRETE	91.7	50.0	70.9	94.3	50.0	72.2	93.0	45.2	69.1	91.0	31.7	61.4	93.2	45.2	69.2	98.4	47.1	72.8
04	MASONRY	87.3	37.1	55.8	73.3	36.1	50.0	71.2	36.1	49.2	94.8	23.7	50.2	73.3	36.1	50.0	107.7	58.0	76.5
05	METALS	93.3	81.1	89.3	92.3	81.0	88.6	92.3	71.6	85.5	91.8	61.2	81.8	92.3	71.6	85.5	98.3	70.8	89.3
06	WOOD & PLASTICS	101.0	40.9	69.6	79.5	40.9	59.4	95.2	40.9	66.9	93.3	28.9	59.7	99.5	40.9	68.9	101.8	38.2	68.6
07	THERMAL & MOISTURE PROTECTION	97.0	43.7	70.8	97.4	44.2	71.2	97.4	44.2	71.2	97.2	29.6	64.0	97.4	44.2	71.3	99.0	52.2	76.0
08	DOORS & WINDOWS	96.3	44.4	82.8	92.3	44.4	79.9	92.2	38.4	78.3	92.2	26.8	75.3	92.2	38.4	78.3	95.0	41.1	81.1
09200	PLASTER & GYPSUM BOARD	107.2	39.1	64.1	95.8	39.1	59.8	102.4	39.1	62.3	101.6	26.8	54.1	104.4	39.1	63.0	106.3	36.8	62.3
095,098	CEILINGS & ACOUSTICAL TREATMENT	98.1	39.1	62.7	92.8	39.1	60.6	91.9	39.1	60.2	91.9	26.8	52.8	91.9	39.1	60.2	112.0	36.8	66.9
09600	FLOORING	105.1	41.0	88.2	95.4	41.0	81.0	102.7	41.8	86.6	101.6	31.7	83.1	104.4	41.8	87.9	108.7	58.7	95.5
097,099	WALL FINISHES, PAINTS & COATINGS	108.5	39.5	66.7	108.5	39.5	66.7	108.5	39.5	66.7	108.5	26.9	59.0	108.5	39.5	66.7	99.2	35.6	60.6
09	FINISHES	100.3	40.4	69.1	95.3	40.4	66.7	97.9	40.5	68.1	97.3	29.1	61.8	98.7	40.5	68.4	106.8	42.6	73.4
10 - 14	TOTAL DIV. 10000 - 14000	100.0	64.2	92.4	100.0	64.2	92.4	100.0	64.2	92.4	100.0	61.3	91.8	100.0	64.2	92.4	100.0	57.5	91.0
15	MECHANICAL	100.0	37.6	72.7	100.0	37.6	72.7	100.0	37.5	72.7	95.7	18.8	62.0	100.0	37.5	72.7	100.2	40.1	73.9
16	ELECTRICAL	98.0	35.0	67.3	95.7	20.0	58.8	97.6	31.8	65.5	97.6	19.7	59.6	97.6	31.8	65.5	97.3	60.8	79.5
01 - 16	WEIGHTED AVERAGE	96.6	48.3	74.7	95.0	46.1	72.8	95.2	46.0	72.9	95.0	33.8	67.2	95.5	46.0	73.0	99.4	54.6	79.0

COST INDEXES

661

SOUTH DAKOTA

DIVISION		MITCHELL 573			MOBRIDGE 576			PIERRE 575			RAPID CITY 577			SIOUX FALLS 570 - 571			WATERTOWN 572		
		MAT.	INST.	TOTAL	MAT.	INST.	TOTAL	MAT.	INST.	TOTAL	MAT.	INST.	TOTAL	MAT.	INST.	TOTAL	MAT.	INST.	TOTAL
01590	EQUIPMENT RENTAL	.0	98.2	98.2	.0	98.2	98.2	.0	98.2	98.2	.0	98.2	98.2	.0	99.4	99.4	.0	98.2	98.2
02	SITE CONSTRUCTION	81.3	93.6	90.4	81.3	93.6	90.4	82.5	93.6	90.7	82.7	93.3	90.6	83.7	95.5	92.4	81.2	93.6	90.4
03100	CONCRETE FORMS & ACCESSORIES	96.0	41.4	48.7	85.5	40.0	46.1	95.3	41.6	48.8	104.4	37.0	46.0	95.4	42.6	49.6	81.5	40.0	45.5
03200	CONCRETE REINFORCEMENT	99.2	49.1	73.9	101.8	49.3	75.2	99.4	61.2	80.1	93.3	61.3	77.1	93.3	61.3	77.1	96.6	49.3	72.7
03300	CAST-IN-PLACE CONCRETE	99.4	49.5	78.6	99.4	51.1	79.3	99.4	46.9	77.5	98.7	44.3	76.0	102.8	49.8	80.7	99.4	51.2	79.3
03	CONCRETE	96.2	47.1	71.7	95.9	47.0	71.5	96.1	48.5	72.4	95.4	45.6	70.6	95.6	50.0	72.8	94.7	47.0	70.9
04	MASONRY	95.8	56.2	70.9	104.2	58.0	75.2	104.2	56.2	74.0	104.1	52.2	71.6	101.5	56.2	73.0	129.1	58.7	84.9
05	METALS	97.6	70.4	88.7	97.7	70.7	88.9	98.3	75.9	91.0	100.1	76.2	92.3	100.2	76.3	92.4	97.6	70.8	88.8
06	WOOD & PLASTICS	100.6	40.3	69.1	89.2	38.2	62.5	99.8	40.3	68.8	106.1	36.1	69.5	99.8	41.0	69.1	84.7	38.2	60.4
07	THERMAL & MOISTURE PROTECTION	98.8	49.7	74.6	98.8	52.2	75.9	99.2	49.9	75.0	99.6	48.1	74.3	99.1	49.5	74.7	98.5	52.4	75.8
08	DOORS & WINDOWS	93.4	41.8	80.1	96.5	40.2	81.9	98.3	45.7	84.7	99.3	43.4	84.8	99.3	46.0	85.5	93.4	40.2	79.7
09200	PLASTER & GYPSUM BOARD	104.8	39.0	63.1	99.9	36.8	59.9	104.4	39.0	63.0	106.3	34.6	60.9	106.3	39.7	64.1	96.8	36.8	58.7
095,098	CEILINGS & ACOUSTICAL TREATMENT	109.4	39.0	67.2	112.0	36.8	66.9	109.4	39.0	67.2	116.5	34.6	67.4	116.5	39.7	70.4	109.4	36.8	65.8
09600	FLOORING	108.2	58.7	95.1	102.5	58.7	90.9	107.8	41.8	90.4	107.8	73.6	98.8	107.8	72.1	98.4	100.6	58.7	89.5
097,099	WALL FINISHES, PAINTS & COATINGS	99.2	40.3	63.5	99.2	35.6	60.6	99.2	40.7	63.7	99.2	40.7	63.7	99.2	40.7	63.7	99.2	35.6	60.6
09	FINISHES	105.7	44.4	73.8	103.9	42.6	72.0	105.5	41.0	72.0	107.6	43.7	74.4	107.6	47.5	76.4	102.3	42.6	71.2
10 - 14	TOTAL DIV. 10000 - 14000	100.0	50.1	89.4	100.0	57.4	90.9	100.0	68.4	93.3	100.0	66.0	92.8	100.0	68.6	93.3	100.0	57.4	90.9
15	MECHANICAL	95.9	37.0	70.1	95.9	41.3	72.0	100.2	38.4	73.2	100.2	35.8	72.1	100.2	37.4	72.7	95.9	41.3	72.0
16	ELECTRICAL	95.9	45.5	71.3	97.3	40.5	69.6	93.7	51.9	73.3	94.2	51.9	73.6	93.6	73.7	83.9	95.1	40.5	68.5
01 - 16	WEIGHTED AVERAGE	96.9	51.6	76.3	97.5	51.9	76.8	98.7	53.7	78.2	99.3	52.5	78.0	99.1	57.8	80.3	97.9	52.0	77.0

TENNESSEE

DIVISION		CHATTANOOGA 373 - 374			COLUMBIA 384			COOKEVILLE 385			JACKSON 383			JOHNSON CITY 376			KNOXVILLE 377 - 379		
		MAT.	INST.	TOTAL	MAT.	INST.	TOTAL	MAT.	INST.	TOTAL	MAT.	INST.	TOTAL	MAT.	INST.	TOTAL	MAT.	INST.	TOTAL
01590	EQUIPMENT RENTAL	.0	104.5	104.5	.0	97.7	97.7	.0	97.7	97.7	.0	104.9	104.9	.0	97.4	97.4	.0	97.4	97.4
02	SITE CONSTRUCTION	103.2	96.5	98.2	92.0	84.9	86.8	98.7	85.4	88.8	100.6	95.7	97.0	113.1	85.2	92.5	89.9	85.3	86.5
03100	CONCRETE FORMS & ACCESSORIES	98.7	47.1	54.0	80.9	52.3	56.1	81.1	41.1	46.4	88.7	38.4	45.1	83.5	47.9	52.7	97.8	47.9	54.6
03200	CONCRETE REINFORCEMENT	87.9	44.6	66.0	89.6	46.7	67.9	89.6	48.9	69.1	89.6	44.3	66.7	88.5	44.8	66.4	87.9	44.8	66.1
03300	CAST-IN-PLACE CONCRETE	100.8	49.4	79.3	90.9	53.1	75.1	103.0	50.6	81.1	100.5	42.7	76.4	81.0	54.1	69.8	94.6	49.6	75.8
03	CONCRETE	92.6	49.3	71.0	92.7	53.3	73.0	102.2	47.9	75.1	93.9	43.1	68.5	97.6	51.3	74.5	89.9	49.8	69.9
04	MASONRY	96.9	44.4	64.0	111.1	52.8	74.5	106.1	40.7	65.0	111.4	34.5	63.2	108.8	44.2	68.3	75.1	44.2	55.7
05	METALS	95.4	77.7	89.6	91.0	78.7	87.0	91.1	79.4	87.3	92.8	75.9	87.3	92.7	77.3	87.7	96.0	77.6	90.0
06	WOOD & PLASTICS	102.3	48.1	74.0	71.2	55.6	63.1	71.5	40.7	55.4	87.2	38.2	61.6	76.1	49.2	62.0	91.5	49.2	69.4
07	THERMAL & MOISTURE PROTECTION	100.3	51.7	76.4	92.6	58.1	75.6	93.2	48.3	71.1	99.3	40.6	70.5	96.1	49.3	73.1	93.7	49.5	72.0
08	DOORS & WINDOWS	101.4	48.2	87.6	92.7	49.9	81.6	92.7	44.6	80.3	102.1	40.6	86.2	97.2	49.3	84.8	93.7	49.3	82.2
09200	PLASTER & GYPSUM BOARD	84.2	47.2	60.8	88.6	55.0	67.3	88.6	39.7	57.6	89.3	37.1	56.2	90.1	48.4	63.6	98.0	48.4	66.6
095,098	CEILINGS & ACOUSTICAL TREATMENT	98.1	47.2	67.6	86.4	55.0	67.6	86.4	39.7	58.4	95.1	37.1	60.3	93.4	48.4	66.4	96.1	48.4	67.5
09600	FLOORING	104.0	51.6	90.1	91.7	26.8	74.6	91.9	25.0	74.2	92.8	25.0	74.8	98.0	53.1	86.1	103.6	53.1	90.2
097,099	WALL FINISHES, PAINTS & COATINGS	109.4	47.4	71.8	95.8	44.1	64.4	95.8	48.2	66.9	96.5	32.1	57.5	107.2	55.7	75.9	107.2	55.7	75.9
09	FINISHES	98.6	47.9	72.3	95.0	47.5	70.3	95.5	38.2	65.7	95.6	35.2	64.2	99.2	49.6	73.4	94.6	49.6	71.2
10 - 14	TOTAL DIV. 10000 - 14000	100.0	55.4	90.5	100.0	54.7	90.4	100.0	61.1	91.7	100.0	52.3	89.8	100.0	63.7	92.3	100.0	63.7	92.3
15	MECHANICAL	99.8	46.5	76.5	97.3	54.1	78.4	97.3	45.7	74.7	100.0	57.6	81.4	99.8	61.3	83.0	99.8	61.3	83.0
16	ELECTRICAL	103.3	54.1	79.3	93.1	45.9	70.1	94.6	44.7	70.3	100.7	52.3	77.1	91.6	52.4	72.4	96.8	61.2	79.4
01 - 16	WEIGHTED AVERAGE	98.8	55.8	79.3	95.0	56.9	77.7	96.3	51.4	75.9	98.5	53.3	78.0	97.6	58.2	79.7	94.9	59.2	78.7

TENNESSEE / TEXAS

DIVISION		MCKENZIE 382			MEMPHIS 375,380 - 381			NASHVILLE 370 - 372			ABILENE 795 - 796			AMARILLO 790 - 791			AUSTIN 786 - 787		
		MAT.	INST.	TOTAL	MAT.	INST.	TOTAL	MAT.	INST.	TOTAL	MAT.	INST.	TOTAL	MAT.	INST.	TOTAL	MAT.	INST.	TOTAL
01590	EQUIPMENT RENTAL	.0	97.7	97.7	.0	102.2	102.2	.0	105.5	105.5	.0	86.8	86.8	.0	86.8	86.8	.0	85.9	85.9
02	SITE CONSTRUCTION	97.9	85.0	88.3	95.7	91.8	92.8	95.0	99.2	98.1	101.5	85.4	89.6	101.9	86.3	90.3	89.8	85.3	86.5
03100	CONCRETE FORMS & ACCESSORIES	90.5	36.7	43.9	99.3	67.1	71.4	97.8	65.2	69.5	95.4	43.9	50.8	97.6	53.2	59.1	94.1	56.6	61.6
03200	CONCRETE REINFORCEMENT	89.8	35.8	62.5	102.2	70.5	86.2	87.1	67.0	76.9	94.1	53.6	73.6	94.1	56.1	74.9	98.7	54.0	76.1
03300	CAST-IN-PLACE CONCRETE	100.8	44.3	77.2	88.9	71.2	81.5	94.4	69.7	84.1	94.2	43.6	73.0	99.0	49.9	78.5	90.0	50.8	73.6
03	CONCRETE	100.9	41.3	71.1	89.6	70.7	80.2	90.0	68.6	79.3	88.9	46.7	67.8	91.3	53.4	72.4	81.8	54.7	68.3
04	MASONRY	109.1	42.6	67.4	84.4	78.0	80.4	85.7	65.9	73.3	100.4	54.4	71.5	102.4	52.9	71.3	94.7	56.4	70.7
05	METALS	91.1	73.0	85.2	97.0	96.1	96.7	96.7	91.9	95.1	95.7	67.7	86.5	95.7	68.9	86.9	94.0	66.7	85.1
06	WOOD & PLASTICS	81.9	35.9	57.9	97.4	67.7	81.9	98.4	66.2	81.6	97.1	43.4	69.0	98.4	55.7	76.1	94.6	59.8	76.4
07	THERMAL & MOISTURE PROTECTION	93.2	44.5	69.3	96.6	74.9	85.9	97.0	66.7	82.1	99.4	51.0	75.6	102.0	50.1	76.5	96.1	55.4	76.1
08	DOORS & WINDOWS	92.7	36.6	78.2	100.9	69.8	92.9	98.5	67.5	90.5	92.9	46.4	80.9	92.9	51.4	82.2	94.5	59.2	85.4
09200	PLASTER & GYPSUM BOARD	93.5	34.7	56.2	95.3	67.3	77.5	92.3	65.7	75.4	85.4	42.5	58.2	85.4	55.1	66.2	90.7	59.3	70.8
095,098	CEILINGS & ACOUSTICAL TREATMENT	86.4	34.7	55.4	106.3	67.3	82.9	96.5	65.7	78.0	91.2	42.5	62.0	91.2	55.1	69.6	88.0	59.3	70.8
09600	FLOORING	95.5	51.3	83.8	98.3	47.1	84.7	106.8	77.5	99.1	115.8	69.1	103.4	115.6	66.7	102.7	101.4	51.2	88.1
097,099	WALL FINISHES, PAINTS & COATINGS	95.8	56.2	71.8	99.0	60.6	75.7	112.7	70.1	86.9	95.2	56.6	71.8	95.2	39.2	61.3	92.2	44.0	63.0
09	FINISHES	97.1	41.3	68.1	97.3	62.1	79.0	104.0	68.0	85.3	97.3	49.6	72.5	97.3	54.3	74.9	92.1	54.2	72.4
10 - 14	TOTAL DIV. 10000 - 14000	100.0	36.6	86.5	100.0	79.1	95.6	100.0	70.8	93.8	100.0	73.9	94.5	100.0	66.0	92.8	100.0	66.2	92.8
15	MECHANICAL	97.3	46.2	75.0	99.8	76.6	89.7	99.8	73.6	88.3	100.1	46.8	76.8	100.1	52.9	79.5	99.9	56.9	81.1
16	ELECTRICAL	94.4	46.5	71.1	95.8	78.5	87.4	102.9	68.2	86.0	96.9	50.0	74.0	97.4	61.5	79.9	99.0	69.4	84.6
01 - 16	WEIGHTED AVERAGE	96.5	49.7	75.3	96.6	77.2	87.8	98.0	74.0	87.1	96.8	54.5	77.6	97.3	59.1	79.9	94.9	61.8	79.8

City Cost Indexes

TEXAS

DIVISION		BEAUMONT 776-777			BROWNWOOD 768			BRYAN 778			CHILDRESS 792			CORPUS CHRISTI 783-784			DALLAS 752-753		
		MAT.	INST.	TOTAL	MAT.	INST.	TOTAL	MAT.	INST.	TOTAL	MAT.	INST.	TOTAL	MAT.	INST.	TOTAL	MAT.	INST.	TOTAL
01590	EQUIPMENT RENTAL	.0	88.7	88.7	.0	86.8	86.8	.0	88.7	88.7	.0	86.8	86.8	.0	94.7	94.7	.0	98.1	98.1
02	SITE CONSTRUCTION	96.5	87.4	89.7	108.4	84.1	90.4	86.2	88.6	88.0	114.0	84.8	92.4	124.9	80.6	92.1	128.3	86.3	97.2
03100	CONCRETE FORMS & ACCESSORIES	103.1	54.5	61.0	93.4	29.7	38.3	79.7	44.2	48.9	93.7	51.6	57.2	97.7	39.9	47.6	93.6	59.3	63.9
03200	CONCRETE REINFORCEMENT	92.5	45.7	68.9	94.5	27.5	60.6	95.0	54.0	74.3	94.3	42.9	68.3	97.8	52.0	74.7	96.5	58.4	77.2
03300	CAST-IN-PLACE CONCRETE	86.2	56.3	73.7	101.6	38.7	75.3	69.0	68.2	68.7	96.6	48.1	76.3	106.7	47.0	81.8	97.1	57.4	80.5
03	CONCRETE	85.6	54.3	70.0	95.6	33.7	64.7	72.1	55.4	63.7	96.3	49.5	72.9	91.6	46.6	69.1	88.2	60.0	74.2
04	MASONRY	104.9	59.1	76.1	128.6	51.2	80.0	147.0	64.5	95.3	104.9	51.1	71.2	86.0	49.4	63.1	102.7	61.3	76.8
05	METALS	97.1	65.7	86.8	93.8	52.4	80.3	97.4	70.2	88.5	93.6	60.9	82.9	93.5	79.9	89.1	93.0	84.3	90.2
06	WOOD & PLASTICS	110.9	55.1	81.7	95.6	28.5	60.6	77.1	36.7	56.0	96.5	55.7	75.2	107.2	39.3	71.8	96.8	60.2	77.7
07	THERMAL & MOISTURE PROTECTION	103.3	59.6	81.8	99.6	42.5	71.5	96.4	60.9	79.0	100.2	47.1	74.1	99.1	47.8	73.9	94.3	60.9	77.9
08	DOORS & WINDOWS	97.6	51.0	85.5	89.0	29.5	73.6	99.4	47.0	85.8	90.0	48.8	79.4	101.6	41.4	86.0	105.6	56.7	93.0
09200	PLASTER & GYPSUM BOARD	90.5	54.5	67.7	84.2	27.2	48.0	79.2	35.6	51.6	84.2	55.1	65.7	90.7	38.2	57.4	91.7	59.5	71.3
095,098	CEILINGS & ACOUSTICAL TREATMENT	93.0	54.5	69.9	85.8	27.2	50.6	86.5	35.6	56.0	85.8	55.1	67.4	88.0	38.2	58.1	94.6	59.5	73.6
09600	FLOORING	110.1	72.6	100.2	114.4	42.5	95.4	82.6	68.2	78.8	113.3	44.5	95.1	114.7	50.6	97.7	105.1	62.0	93.7
097,099	WALL FINISHES, PAINTS & COATINGS	91.4	52.9	68.0	95.2	31.5	56.6	87.2	64.0	73.1	95.2	41.7	62.7	105.7	43.1	67.7	104.7	56.1	75.2
09	FINISHES	91.8	57.5	74.0	95.9	31.9	62.6	79.7	49.0	63.7	96.1	49.7	72.0	98.0	41.4	68.6	99.9	59.3	78.8
10-14	TOTAL DIV. 10000-14000	100.0	78.2	95.4	100.0	47.5	88.8	100.0	67.5	93.1	100.0	65.2	92.6	100.0	74.7	94.6	100.0	78.8	95.5
15	MECHANICAL	100.0	64.4	84.4	95.7	30.8	67.3	95.7	73.0	85.7	95.7	51.7	76.5	99.9	47.8	77.1	99.8	67.0	85.4
16	ELECTRICAL	95.0	65.3	80.5	93.3	29.4	62.1	94.9	68.6	82.0	96.9	40.1	69.2	97.2	53.6	75.9	98.7	64.7	82.1
01-16	WEIGHTED AVERAGE	96.6	63.5	81.6	96.9	40.7	71.4	94.5	65.2	81.2	96.5	53.5	76.9	97.7	54.1	77.8	98.7	67.1	84.3

TEXAS

DIVISION		DEL RIO 788			DENTON 762			EASTLAND 764			EL PASO 798-799,885			FORT WORTH 760-761			GALVESTON 775		
		MAT.	INST.	TOTAL	MAT.	INST.	TOTAL	MAT.	INST.	TOTAL	MAT.	INST.	TOTAL	MAT.	INST.	TOTAL	MAT.	INST.	TOTAL
01590	EQUIPMENT RENTAL	.0	85.9	85.9	.0	93.9	93.9	.0	86.8	86.8	.0	86.8	86.8	.0	86.8	86.8	.0	99.2	99.2
02	SITE CONSTRUCTION	110.2	83.1	90.2	109.1	77.7	85.9	111.7	84.3	91.4	102.0	85.0	89.4	101.4	85.7	89.8	126.1	86.1	96.5
03100	CONCRETE FORMS & ACCESSORIES	91.7	26.1	34.9	101.2	39.3	47.6	94.6	35.0	43.0	94.8	48.5	54.7	96.3	58.6	63.6	89.0	66.5	69.5
03200	CONCRETE REINFORCEMENT	98.7	19.4	58.6	95.9	53.6	74.5	94.7	44.6	69.4	94.1	50.7	72.2	94.1	58.3	76.0	94.4	64.7	79.4
03300	CAST-IN-PLACE CONCRETE	114.9	33.4	80.9	78.5	51.1	67.1	107.4	39.5	79.0	96.5	38.2	72.1	92.6	52.4	75.8	91.7	69.4	82.4
03	CONCRETE	105.4	28.7	67.1	76.4	48.1	62.3	99.8	39.2	69.6	89.9	46.2	68.1	88.2	57.1	72.7	86.9	68.8	77.8
04	MASONRY	101.2	46.2	66.7	137.2	52.9	84.3	98.8	50.2	68.3	99.4	49.4	68.0	95.7	61.3	74.1	100.8	64.6	78.1
05	METALS	92.7	47.7	78.0	93.4	80.5	89.2	93.6	55.4	81.1	95.5	63.7	85.1	92.4	70.0	85.1	99.1	90.7	96.4
06	WOOD & PLASTICS	90.5	26.4	57.0	106.5	39.4	71.5	102.0	35.7	67.4	97.1	52.1	73.6	103.8	60.0	80.9	93.7	66.4	79.4
07	THERMAL & MOISTURE PROTECTION	96.4	35.2	66.3	98.1	49.2	74.1	100.1	42.7	71.9	98.9	52.1	75.9	99.8	55.5	78.0	94.8	66.2	80.7
08	DOORS & WINDOWS	95.3	22.7	76.5	104.9	44.2	89.2	78.0	33.0	66.3	92.9	47.5	81.2	87.5	56.6	79.5	104.9	66.0	94.8
09200	PLASTER & GYPSUM BOARD	86.5	25.1	47.6	88.6	38.3	56.7	84.2	34.6	52.7	85.4	51.4	63.9	85.4	59.5	69.0	86.2	66.0	73.4
095,098	CEILINGS & ACOUSTICAL TREATMENT	84.4	25.1	48.8	92.0	38.3	59.8	85.8	34.6	55.1	91.2	51.4	67.4	91.2	59.5	72.2	90.1	66.0	75.6
09600	FLOORING	98.0	20.4	77.5	106.5	49.6	91.4	148.8	30.6	117.5	115.8	66.4	102.7	150.1	49.6	123.5	97.0	68.2	89.4
097,099	WALL FINISHES, PAINTS & COATINGS	92.2	27.7	53.1	106.2	31.9	61.1	96.4	32.9	57.9	95.2	37.7	60.3	96.4	55.9	71.9	99.8	60.7	76.1
09	FINISHES	90.9	25.7	57.0	93.2	39.8	65.4	106.5	34.2	68.9	97.3	51.2	73.3	107.8	56.6	81.2	89.8	66.1	77.5
10-14	TOTAL DIV. 10000-14000	100.0	44.6	88.2	100.0	49.4	89.2	100.0	48.1	89.0	100.0	64.3	92.4	100.0	78.3	95.4	100.0	83.2	96.4
15	MECHANICAL	95.5	20.1	62.5	95.7	49.4	75.4	95.7	31.2	67.5	100.1	35.7	71.9	100.1	61.0	83.0	95.6	71.2	84.9
16	ELECTRICAL	98.8	20.9	60.8	96.2	47.6	72.5	93.2	36.6	65.6	96.2	56.9	77.0	96.1	62.4	79.6	96.0	57.8	77.3
01-16	WEIGHTED AVERAGE	97.1	34.2	68.6	96.9	53.2	77.0	95.9	43.3	72.0	96.7	52.4	76.6	96.4	63.3	81.4	97.0	70.7	85.0

TEXAS

DIVISION		GIDDINGS 789			GREENVILLE 754			HOUSTON 770-772			HUNTSVILLE 773			LAREDO 780			LONGVIEW 756		
		MAT.	INST.	TOTAL	MAT.	INST.	TOTAL	MAT.	INST.	TOTAL	MAT.	INST.	TOTAL	MAT.	INST.	TOTAL	MAT.	INST.	TOTAL
01590	EQUIPMENT RENTAL	.0	85.9	85.9	.0	95.3	95.3	.0	98.9	98.9	.0	88.7	88.7	.0	85.9	85.9	.0	88.8	88.8
02	SITE CONSTRUCTION	96.4	83.6	86.9	119.1	81.2	91.1	123.8	85.8	95.7	105.9	86.2	91.3	90.3	85.1	86.5	111.2	89.3	95.0
03100	CONCRETE FORMS & ACCESSORIES	89.9	35.0	42.4	84.4	29.6	36.9	91.4	66.7	70.0	88.0	36.4	43.3	91.7	40.2	47.1	79.9	30.5	37.2
03200	CONCRETE REINFORCEMENT	99.4	45.5	72.1	96.9	27.6	61.9	94.2	64.8	79.3	95.2	30.3	62.4	98.7	52.3	75.3	95.9	25.3	60.3
03300	CAST-IN-PLACE CONCRETE	97.3	39.4	73.1	88.4	39.0	67.8	88.8	69.5	80.8	94.9	47.0	74.9	82.1	63.3	74.3	102.8	36.1	75.0
03	CONCRETE	87.4	39.4	63.4	81.8	34.6	58.2	84.9	68.9	76.9	91.9	40.1	66.0	81.6	51.4	66.6	94.4	32.7	63.6
04	MASONRY	108.6	52.2	73.2	160.3	51.5	92.0	100.6	65.3	78.4	145.3	35.1	76.1	95.0	55.4	70.1	156.0	48.0	88.2
05	METALS	92.2	55.7	80.3	90.8	66.4	82.8	101.5	90.9	98.1	97.3	54.8	83.4	94.9	65.6	85.3	84.0	51.5	73.4
06	WOOD & PLASTICS	89.6	35.7	61.5	87.5	28.6	56.8	96.2	66.4	80.6	86.8	38.1	61.3	90.5	39.2	63.7	80.5	30.9	54.6
07	THERMAL & MOISTURE PROTECTION	96.8	43.3	70.5	93.9	43.8	69.3	94.7	67.6	81.3	97.8	40.2	69.5	95.1	52.6	74.2	95.0	38.0	67.0
08	DOORS & WINDOWS	94.5	33.0	78.6	104.4	29.6	85.1	107.1	66.0	96.5	99.3	32.9	82.1	95.1	42.0	81.4	94.1	28.5	77.1
09200	PLASTER & GYPSUM BOARD	85.0	34.6	53.1	87.5	27.2	49.3	88.6	66.0	74.3	84.1	37.1	54.3	87.4	38.2	56.2	83.1	29.7	49.3
095,098	CEILINGS & ACOUSTICAL TREATMENT	83.5	34.6	54.2	91.0	27.2	52.7	94.6	66.0	77.4	86.5	37.1	56.9	88.0	38.2	58.1	86.5	29.7	52.5
09600	FLOORING	97.8	30.6	80.0	99.8	49.6	86.5	98.2	68.2	90.3	86.1	30.6	71.4	97.9	50.6	85.4	106.4	43.7	89.8
097,099	WALL FINISHES, PAINTS & COATINGS	92.2	32.9	56.2	104.7	27.1	57.6	99.8	64.0	78.1	87.2	33.8	54.8	92.2	57.5	71.2	96.0	36.2	59.7
09	FINISHES	89.7	34.0	60.7	96.3	32.6	63.2	96.8	66.5	81.0	82.4	35.4	57.9	90.8	43.0	65.9	100.5	33.4	65.6
10-14	TOTAL DIV. 10000-14000	100.0	46.2	88.6	100.0	47.4	88.8	100.0	83.2	96.4	100.0	46.3	88.6	100.0	64.4	92.4	100.0	26.0	84.3
15	MECHANICAL	95.5	31.1	67.3	95.4	30.3	66.9	100.0	73.2	88.3	95.7	29.4	66.7	99.8	43.1	75.0	95.3	29.9	66.7
16	ELECTRICAL	95.9	36.6	67.0	95.3	29.4	63.2	97.7	68.6	83.5	94.9	35.9	66.1	98.9	63.9	81.8	94.9	57.2	76.5
01-16	WEIGHTED AVERAGE	94.5	43.4	71.3	98.3	41.9	72.7	99.1	72.8	87.2	97.6	41.6	72.1	94.9	55.0	76.8	97.5	43.9	73.1

COST INDEXES

663

City Cost Indexes

TEXAS

DIVISION		LUBBOCK 793 - 794			LUFKIN 759			MCALLEN 785			MCKINNEY 750			MIDLAND 797			ODESSA 797		
		MAT.	INST.	TOTAL	MAT.	INST.	TOTAL	MAT.	INST.	TOTAL	MAT.	INST.	TOTAL	MAT.	INST.	TOTAL	MAT.	INST.	TOTAL
01590	EQUIPMENT RENTAL	.0	96.9	96.9	.0	88.8	88.8	.0	94.7	94.7	.0	95.3	95.3	.0	96.9	96.9	.0	86.8	86.8
02	SITE CONSTRUCTION	131.4	83.5	96.0	104.8	89.3	93.3	129.3	80.6	93.2	114.2	82.0	90.4	135.3	83.5	97.0	101.7	85.8	89.9
03100	CONCRETE FORMS & ACCESSORIES	94.9	43.2	50.1	83.5	42.7	48.1	96.5	37.5	45.4	83.3	45.1	50.2	99.2	41.3	49.0	95.3	41.1	48.3
03200	CONCRETE REINFORCEMENT	95.4	53.5	74.2	97.5	33.7	65.3	98.1	51.9	74.8	96.9	53.6	75.0	96.5	53.4	74.7	94.1	53.4	73.6
03300	CAST-IN-PLACE CONCRETE	94.3	50.0	75.8	92.0	41.7	71.0	116.0	45.2	86.4	82.9	52.8	70.3	100.2	47.0	78.0	94.2	45.8	73.9
03	CONCRETE	88.0	49.5	68.8	87.8	41.8	64.9	97.1	44.9	71.1	77.8	51.1	64.5	92.2	47.6	69.9	88.9	46.1	67.5
04	MASONRY	99.8	51.2	69.3	119.7	41.0	70.4	100.5	49.4	68.5	172.4	56.3	99.6	118.0	51.6	76.3	100.4	51.5	69.7
05	METALS	99.2	82.5	93.8	90.9	59.2	80.6	92.5	79.4	88.2	90.7	79.3	87.0	97.8	82.0	92.7	95.1	66.9	85.8
06	WOOD & PLASTICS	97.4	43.5	69.2	89.0	44.3	65.7	105.7	36.6	69.6	86.0	45.2	64.7	102.6	41.4	70.6	97.1	41.2	67.9
07	THERMAL & MOISTURE PROTECTION	90.6	52.3	71.8	94.7	43.5	69.5	99.2	45.9	73.0	93.7	53.3	73.9	91.0	48.6	70.1	99.4	47.6	73.9
08	DOORS & WINDOWS	104.2	44.6	88.8	76.1	41.1	67.0	100.1	40.5	84.7	104.4	47.3	89.6	103.2	43.0	87.6	92.9	42.9	80.0
09200	PLASTER & GYPSUM BOARD	86.1	42.4	58.4	83.7	43.5	58.2	90.7	35.4	55.6	86.3	44.2	59.6	87.9	40.2	57.7	85.4	40.2	56.8
095,098	CEILINGS & ACOUSTICAL TREATMENT	93.9	42.4	63.0	76.7	43.5	56.8	88.0	35.4	56.4	91.0	44.2	62.9	89.3	40.2	59.9	91.2	40.2	60.6
09600	FLOORING	107.2	42.1	90.0	143.2	40.8	116.2	114.0	63.9	100.8	99.3	49.6	86.2	109.0	41.5	91.2	115.8	41.5	96.2
097,099	WALL FINISHES, PAINTS & COATINGS	106.8	35.2	63.4	96.0	40.8	62.5	105.7	31.1	60.5	104.7	50.8	72.0	106.8	34.8	63.1	95.2	35.2	58.8
09	FINISHES	99.3	41.3	69.1	108.9	42.1	74.2	98.1	41.2	68.5	95.7	46.2	70.0	99.3	40.0	68.5	97.3	39.9	67.4
10 - 14	TOTAL DIV. 10000 - 14000	100.0	73.6	94.4	100.0	65.4	92.6	100.0	64.4	92.4	100.0	51.5	89.7	100.0	63.9	92.3	100.0	63.5	92.2
15	MECHANICAL	99.7	50.1	78.0	95.3	34.8	68.9	99.5	33.7	68.5	95.4	51.2	76.1	95.3	39.1	70.7	100.1	39.1	73.4
16	ELECTRICAL	95.5	47.0	71.9	96.1	50.2	73.7	96.5	37.1	67.5	95.3	57.2	76.8	95.5	43.8	70.3	97.0	43.8	71.0
01 - 16	WEIGHTED AVERAGE	98.8	54.9	78.9	95.1	48.1	73.7	97.8	48.2	75.2	98.2	57.1	79.5	99.1	51.4	77.4	96.7	49.9	75.4

TEXAS

DIVISION		PALESTINE 758			SAN ANGELO 769			SAN ANTONIO 781 - 782			TEMPLE 765			TEXARKANA 755			TYLER 757		
		MAT.	INST.	TOTAL	MAT.	INST.	TOTAL	MAT.	INST.	TOTAL	MAT.	INST.	TOTAL	MAT.	INST.	TOTAL	MAT.	INST.	TOTAL
01590	EQUIPMENT RENTAL	.0	88.8	88.8	.0	86.8	86.8	.0	88.8	88.8	.0	86.8	86.8	.0	88.8	88.8	.0	88.8	88.8
02	SITE CONSTRUCTION	113.9	88.9	95.4	104.1	86.2	90.9	90.0	89.8	89.9	91.2	84.8	86.5	98.0	89.7	91.9	110.7	89.8	95.2
03100	CONCRETE FORMS & ACCESSORIES	74.0	35.0	40.2	93.8	39.0	46.3	91.8	57.9	62.4	97.8	37.3	45.4	91.4	36.9	44.2	85.4	35.7	42.4
03200	CONCRETE REINFORCEMENT	95.2	45.3	70.0	94.4	53.4	73.7	105.4	54.3	79.6	94.5	53.3	73.7	95.1	58.1	76.8	95.9	58.1	76.8
03300	CAST-IN-PLACE CONCRETE	84.1	39.1	65.3	95.8	44.1	74.2	80.6	70.6	76.4	78.5	50.2	66.7	84.7	45.5	68.4	101.0	44.4	77.4
03	CONCRETE	87.4	39.2	63.3	91.2	44.5	67.9	82.0	62.3	72.2	78.8	45.8	62.3	82.1	45.0	63.6	94.1	44.1	69.1
04	MASONRY	114.5	49.5	73.7	125.3	51.1	78.8	94.9	65.3	76.3	136.4	54.3	84.9	176.9	51.7	98.3	165.7	51.7	94.1
05	METALS	90.7	54.5	78.9	93.9	64.6	84.4	95.6	69.6	87.1	93.6	64.8	84.2	83.9	68.3	78.8	90.6	68.6	83.4
06	WOOD & PLASTICS	78.4	35.7	56.1	95.9	39.4	66.4	90.5	56.5	72.8	105.0	35.7	68.8	92.8	35.2	62.7	90.7	33.2	60.7
07	THERMAL & MOISTURE PROTECTION	95.3	44.4	70.3	99.3	45.7	73.0	95.1	66.0	80.8	98.8	48.6	74.1	94.5	47.2	71.2	95.1	51.4	73.7
08	DOORS & WINDOWS	76.0	33.7	65.1	89.0	41.9	76.8	97.0	56.7	86.6	75.8	38.6	66.2	94.0	41.7	80.5	76.0	40.6	66.8
09200	PLASTER & GYPSUM BOARD	78.1	34.6	50.5	84.2	38.4	55.1	87.4	55.9	67.5	85.4	34.6	53.2	87.6	34.1	53.7	83.7	32.0	51.0
095,098	CEILINGS & ACOUSTICAL TREATMENT	76.7	34.6	51.5	85.8	38.4	57.3	88.0	55.9	68.8	85.8	34.6	55.1	83.9	34.1	54.0	76.7	32.0	49.9
09600	FLOORING	132.4	30.6	105.5	114.5	37.1	94.1	97.9	57.8	87.3	151.9	50.9	125.2	116.5	67.9	103.7	145.4	45.2	118.9
097,099	WALL FINISHES, PAINTS & COATINGS	96.0	32.9	57.7	95.2	36.3	59.5	92.2	57.5	71.2	96.4	36.3	59.9	96.0	42.4	63.5	96.0	42.4	63.5
09	FINISHES	105.5	33.8	68.2	95.7	37.7	65.5	90.8	57.1	73.3	106.5	38.7	71.2	102.9	42.4	71.4	109.9	36.6	71.8
10 - 14	TOTAL DIV. 10000 - 14000	100.0	48.0	88.9	100.0	62.3	92.0	100.0	70.7	93.8	100.0	47.9	88.9	100.0	72.0	94.1	100.0	71.8	94.0
15	MECHANICAL	95.3	31.2	67.3	95.7	45.1	73.6	99.8	69.5	86.6	95.7	37.1	70.1	95.3	40.8	71.5	95.3	62.9	81.1
16	ELECTRICAL	92.5	41.5	67.6	96.9	34.9	66.7	99.0	63.9	81.9	94.3	63.2	79.1	96.0	57.8	77.4	94.9	55.7	75.8
01 - 16	WEIGHTED AVERAGE	94.1	44.3	71.5	96.6	49.0	75.0	95.2	66.8	82.3	94.8	51.4	75.1	97.2	53.0	77.2	98.1	56.4	79.2

TEXAS / UTAH

DIVISION		VICTORIA 779			WACO 766 - 767			WAXAHACHIE 751			WHARTON 774			WICHITA FALLS 763			LOGAN 843		
		MAT.	INST.	TOTAL	MAT.	INST.	TOTAL	MAT.	INST.	TOTAL	MAT.	INST.	TOTAL	MAT.	INST.	TOTAL	MAT.	INST.	TOTAL
01590	EQUIPMENT RENTAL	.0	97.7	97.7	.0	86.8	86.8	.0	95.3	95.3	.0	99.2	99.2	.0	86.8	86.8	.0	100.7	100.7
02	SITE CONSTRUCTION	132.1	81.8	94.9	100.4	85.7	89.5	117.3	81.9	91.1	140.4	83.7	98.4	101.1	85.4	89.5	92.0	100.8	98.5
03100	CONCRETE FORMS & ACCESSORIES	89.8	42.8	49.1	96.0	41.9	49.2	83.3	50.1	54.6	83.1	39.1	44.9	96.0	44.2	51.1	101.0	61.5	66.8
03200	CONCRETE REINFORCEMENT	90.7	33.8	61.9	94.1	53.8	73.8	96.9	53.6	75.0	94.3	65.9	80.0	94.1	53.8	73.8	108.5	80.7	94.5
03300	CAST-IN-PLACE CONCRETE	102.9	43.4	78.1	85.0	54.0	72.1	87.5	43.3	69.0	105.8	51.5	83.1	90.8	49.5	73.5	88.3	72.8	81.8
03	CONCRETE	93.4	43.6	68.5	84.5	49.4	67.0	80.9	50.1	65.5	96.8	50.4	73.6	87.3	48.8	68.1	108.4	69.3	88.9
04	MASONRY	118.4	41.1	69.9	97.2	59.9	73.8	160.9	54.1	93.9	101.8	36.0	60.5	97.6	58.0	72.7	121.3	62.3	84.3
05	METALS	97.4	77.5	90.9	95.6	67.5	86.4	90.8	79.2	87.0	99.1	84.1	94.2	95.5	68.3	86.6	95.9	77.4	92.3
06	WOOD & PLASTICS	97.5	44.5	69.8	102.8	38.3	69.1	86.0	53.3	68.9	87.2	40.3	62.7	102.8	43.3	71.8	88.1	59.7	73.3
07	THERMAL & MOISTURE PROTECTION	99.6	44.9	72.7	100.1	51.4	76.1	93.9	50.3	72.4	95.2	50.1	73.0	100.1	55.0	77.9	102.2	68.7	85.8
08	DOORS & WINDOWS	104.8	43.0	88.8	87.5	40.2	75.3	104.4	51.8	90.8	104.9	45.1	89.4	87.5	46.3	76.9	89.3	61.8	82.2
09200	PLASTER & GYPSUM BOARD	84.9	43.5	58.6	85.4	37.3	54.9	86.5	52.6	65.0	83.0	39.2	55.2	85.4	42.4	58.1	80.6	58.4	66.5
095,098	CEILINGS & ACOUSTICAL TREATMENT	91.0	43.5	62.5	91.2	37.3	58.8	91.9	52.6	68.3	90.1	39.2	59.6	91.2	42.4	61.9	97.8	58.4	74.1
09600	FLOORING	96.3	40.8	81.6	150.3	38.6	120.7	99.3	40.8	83.9	94.4	30.6	77.5	150.7	75.3	130.8	101.2	62.3	90.9
097,099	WALL FINISHES, PAINTS & COATINGS	100.0	40.8	64.1	96.4	36.3	59.9	104.7	40.8	65.9	99.8	33.8	59.8	98.2	57.5	73.5	103.7	49.6	70.9
09	FINISHES	88.2	42.3	64.3	107.8	39.4	72.2	96.2	47.8	71.0	89.3	36.9	62.0	108.0	50.9	78.3	97.0	59.6	77.5
10 - 14	TOTAL DIV. 10000 - 14000	100.0	51.1	89.6	100.0	75.5	94.8	100.0	52.8	90.0	100.0	49.4	89.2	100.0	64.4	92.4	100.0	75.8	94.9
15	MECHANICAL	95.7	34.8	69.0	100.1	56.6	81.1	95.4	50.0	75.5	95.6	28.9	66.4	100.1	51.0	78.6	100.1	71.3	87.5
16	ELECTRICAL	100.6	50.3	76.0	97.0	63.2	80.5	95.3	63.4	79.8	99.5	36.0	68.5	100.1	57.4	79.3	88.6	76.4	82.7
01 - 16	WEIGHTED AVERAGE	99.0	49.1	76.4	96.6	57.8	79.0	98.1	57.8	79.8	98.8	46.5	75.1	97.4	57.2	79.1	98.9	72.1	86.7

DIVISION		UTAH											VERMONT					
		OGDEN			PRICE			PROVO			SALT LAKE CITY			BELLOWS FALLS			BENNINGTON		
		842,844			845			846 - 847			840 - 841			051			052		
		MAT.	INST.	TOTAL	MAT.	INST.	TOTAL	MAT.	INST.	TOTAL	MAT.	INST.	TOTAL	MAT.	INST.	TOTAL	MAT.	INST.	TOTAL
01590	EQUIPMENT RENTAL	.0	100.7	100.7	.0	99.4	99.4	.0	99.4	99.4	.0	100.7	100.7	.0	101.6	101.6	.0	101.6	101.6
02	SITE CONSTRUCTION	80.1	100.8	95.4	89.6	97.7	95.6	88.4	98.8	96.1	79.8	100.7	95.3	76.7	98.3	92.7	76.2	98.3	92.6
03100	CONCRETE FORMS & ACCESSORIES	101.0	61.5	66.8	103.5	33.7	43.0	102.5	61.5	67.0	99.5	61.5	66.6	101.2	40.9	48.9	98.5	40.6	48.4
03200	CONCRETE REINFORCEMENT	113.4	80.7	96.9	116.4	73.3	94.6	117.4	80.7	98.9	110.7	80.7	95.6	86.6	81.5	84.0	86.6	81.5	84.0
03300	CAST-IN-PLACE CONCRETE	89.6	72.8	82.6	88.4	48.0	71.5	88.4	72.8	81.9	97.2	72.8	87.0	94.6	56.4	78.7	94.6	56.3	78.6
03	CONCRETE	99.2	69.3	84.3	109.9	47.1	78.5	108.4	69.4	88.9	116.9	69.4	93.2	99.0	54.6	76.8	98.9	54.4	76.7
04	MASONRY	114.6	62.3	81.8	127.3	42.9	74.4	127.3	62.3	86.5	132.9	62.3	88.6	96.4	39.3	60.5	105.9	39.3	64.1
05	METALS	99.9	77.4	92.6	97.1	72.8	89.2	97.8	77.5	91.1	103.1	77.5	94.7	95.0	73.5	88.0	94.9	73.2	87.8
06	WOOD & PLASTICS	88.1	59.7	73.3	91.1	31.4	59.9	89.7	59.7	74.0	88.3	59.7	73.4	100.4	38.7	68.2	97.2	38.7	66.7
07	THERMAL & MOISTURE PROTECTION	100.7	68.7	85.0	104.7	54.9	80.2	104.7	68.7	87.0	103.8	68.7	86.6	101.0	57.9	79.8	100.9	49.2	75.5
08	DOORS & WINDOWS	89.3	61.8	82.2	93.7	43.4	80.7	93.7	61.8	85.4	89.3	61.8	82.2	103.8	47.1	89.1	103.7	44.6	88.4
09200	PLASTER & GYPSUM BOARD	80.6	58.4	66.5	82.2	29.3	48.7	81.0	58.4	66.7	80.6	58.4	66.5	102.1	36.1	60.2	100.1	36.1	59.5
095,098	CEILINGS & ACOUSTICAL TREATMENT	97.8	58.4	74.1	97.8	29.3	56.7	97.8	58.4	74.1	97.8	58.4	74.1	87.5	36.1	56.7	87.5	36.1	56.7
09600	FLOORING	101.2	62.3	90.9	102.4	46.3	87.6	102.0	62.3	91.5	100.8	62.3	90.6	99.8	39.8	83.9	98.5	39.8	83.0
097,099	WALL FINISHES, PAINTS & COATINGS	103.7	49.6	70.9	103.7	49.6	70.9	103.7	63.3	79.2	103.7	63.3	79.2	92.8	35.3	57.9	92.8	35.3	57.9
09	FINISHES	95.9	59.6	77.0	98.0	36.2	65.9	97.6	61.1	78.6	96.2	61.1	77.9	96.6	39.1	66.7	95.9	39.1	66.4
10 - 14	TOTAL DIV. 10000 - 14000	100.0	75.8	94.9	100.0	57.3	90.9	100.0	75.8	94.9	100.0	75.8	94.9	100.0	50.7	89.5	100.0	50.7	89.5
15	MECHANICAL	100.1	71.3	87.5	97.8	41.9	73.3	100.1	71.3	87.5	100.2	71.3	87.6	95.7	37.5	70.2	95.7	41.3	71.9
16	ELECTRICAL	88.7	76.4	82.7	93.0	45.3	69.8	88.8	76.4	82.8	92.5	76.4	84.6	100.2	36.9	69.3	100.2	36.9	69.3
01 - 16	WEIGHTED AVERAGE	97.2	72.1	85.8	99.6	51.1	77.6	99.5	72.1	87.1	101.2	72.3	88.1	97.5	50.3	76.0	97.8	50.6	76.4

DIVISION		VERMONT																
		BRATTLEBORO			BURLINGTON			GUILDHALL			MONTPELIER			RUTLAND			ST. JOHNSBURY		
		053			054			059			056			057			058		
		MAT.	INST.	TOTAL	MAT.	INST.	TOTAL	MAT.	INST.	TOTAL	MAT.	INST.	TOTAL	MAT.	INST.	TOTAL	MAT.	INST.	TOTAL
01590	EQUIPMENT RENTAL	.0	101.6	101.6	.0	101.6	101.6	.0	101.6	101.6	.0	101.6	101.6	.0	101.6	101.6	.0	101.6	101.6
02	SITE CONSTRUCTION	77.5	98.3	92.9	79.6	99.1	94.0	75.8	97.9	92.2	78.2	99.1	93.6	79.6	99.1	94.0	75.8	97.9	92.2
03100	CONCRETE FORMS & ACCESSORIES	101.5	40.8	48.9	91.4	51.9	57.2	98.0	36.5	44.8	95.4	52.0	57.8	101.8	52.0	58.6	95.4	36.7	44.5
03200	CONCRETE REINFORCEMENT	85.7	81.5	83.6	108.2	56.0	81.8	87.4	35.7	61.3	86.1	56.0	70.9	108.2	56.0	81.8	85.7	54.9	70.2
03300	CAST-IN-PLACE CONCRETE	97.6	56.4	80.4	97.2	64.9	83.7	91.7	51.9	75.1	97.6	64.9	83.9	92.6	64.9	81.1	91.7	51.9	75.1
03	CONCRETE	101.5	54.5	78.0	104.3	57.7	81.0	96.7	42.6	69.7	101.1	57.8	79.5	102.8	57.8	80.3	96.2	45.9	71.1
04	MASONRY	105.4	39.3	63.9	98.7	62.0	75.7	105.7	41.1	65.2	89.7	62.0	72.3	86.5	62.0	71.1	132.8	41.1	75.3
05	METALS	94.9	73.4	87.9	101.7	67.6	90.5	95.0	56.2	82.3	94.9	67.7	86.0	99.9	67.7	89.4	95.0	56.3	82.3
06	WOOD & PLASTICS	100.7	38.7	68.3	88.6	50.1	68.5	96.3	35.1	64.3	92.6	50.1	70.4	100.9	50.1	74.4	92.6	35.1	62.6
07	THERMAL & MOISTURE PROTECTION	101.0	50.3	76.1	101.4	56.9	79.5	100.7	44.1	72.9	100.8	58.1	79.8	101.2	58.1	80.0	100.6	44.1	72.8
08	DOORS & WINDOWS	103.7	47.1	89.1	107.1	46.9	91.5	103.8	33.7	85.6	103.7	46.9	89.0	107.1	46.9	91.5	103.8	33.7	85.6
09200	PLASTER & GYPSUM BOARD	102.1	36.1	60.2	103.5	47.8	68.2	107.3	32.3	59.8	112.1	47.8	71.3	103.5	47.8	68.2	112.1	32.3	61.6
095,098	CEILINGS & ACOUSTICAL TREATMENT	87.5	36.1	56.7	94.9	47.8	66.6	87.5	32.3	54.4	87.5	47.8	63.7	94.9	47.8	66.6	87.5	32.3	54.4
09600	FLOORING	99.9	39.8	84.0	99.8	72.0	92.4	103.4	39.8	86.6	107.6	72.0	98.2	99.8	72.0	92.4	107.6	39.8	89.6
097,099	WALL FINISHES, PAINTS & COATINGS	92.8	35.3	57.9	92.8	41.1	61.4	92.8	27.4	53.1	92.8	41.1	61.4	92.8	41.1	61.4	92.8	28.7	53.9
09	FINISHES	96.7	39.1	66.8	98.8	53.8	75.4	98.4	35.3	65.6	100.5	53.8	76.2	98.7	53.8	75.4	100.3	35.5	66.6
10 - 14	TOTAL DIV. 10000 - 14000	100.0	50.7	89.5	100.0	84.8	96.8	100.0	49.0	89.2	100.0	84.8	96.8	100.0	84.8	96.8	100.0	49.0	89.2
15	MECHANICAL	95.7	41.3	71.9	100.1	68.5	86.3	95.7	64.7	82.2	95.7	68.5	83.8	100.1	68.5	86.3	95.7	64.7	82.2
16	ELECTRICAL	100.2	36.9	69.3	100.9	67.4	84.6	100.2	28.7	65.3	100.2	67.4	84.2	100.2	67.4	84.2	100.2	28.7	65.3
01 - 16	WEIGHTED AVERAGE	98.2	50.8	76.7	100.8	66.0	85.0	97.8	50.1	76.1	97.7	66.1	83.3	99.8	66.1	84.5	99.2	50.6	77.1

DIVISION		VERMONT			VIRGINIA													
		WHITE RIVER JCT.			ALEXANDRIA			ARLINGTON			BRISTOL			CHARLOTTESVILLE			CULPEPER		
		050			223			222			242			229			227		
		MAT.	INST.	TOTAL	MAT.	INST.	TOTAL	MAT.	INST.	TOTAL	MAT.	INST.	TOTAL	MAT.	INST.	TOTAL	MAT.	INST.	TOTAL
01590	EQUIPMENT RENTAL	.0	101.6	101.6	.0	101.2	101.2	.0	99.7	99.7	.0	99.7	99.7	.0	104.8	104.8	.0	99.7	99.7
02	SITE CONSTRUCTION	79.6	98.6	93.7	112.3	86.1	92.9	122.9	82.7	93.2	106.2	80.9	87.5	110.7	83.7	90.7	110.5	82.2	89.6
03100	CONCRETE FORMS & ACCESSORIES	95.5	42.0	49.1	91.7	75.0	77.2	90.5	71.9	74.4	85.3	33.9	40.7	83.4	41.4	47.0	80.5	66.5	68.3
03200	CONCRETE REINFORCEMENT	86.6	32.3	59.1	84.8	81.8	83.3	95.7	65.1	80.3	95.8	64.3	79.9	95.2	70.9	82.9	95.7	58.5	76.9
03300	CAST-IN-PLACE CONCRETE	97.6	53.2	79.0	99.6	82.7	92.6	96.9	74.3	87.5	96.5	42.5	74.0	100.4	54.1	81.1	99.3	43.4	75.9
03	CONCRETE	103.2	44.9	74.1	101.8	80.0	90.9	106.7	72.6	89.7	102.8	44.6	73.8	103.4	53.2	78.3	100.2	58.4	79.3
04	MASONRY	118.4	47.4	73.8	85.8	74.6	78.8	98.5	68.6	79.8	86.7	31.7	52.2	110.5	43.3	68.3	99.2	35.3	59.1
05	METALS	95.0	55.8	82.2	102.4	96.0	100.3	101.1	87.7	96.7	100.2	81.5	94.1	100.5	88.6	96.6	100.5	82.3	94.5
06	WOOD & PLASTICS	94.0	41.6	66.7	94.5	75.4	84.5	91.0	75.4	82.9	82.9	33.4	57.1	80.8	40.3	59.7	79.6	75.4	77.4
07	THERMAL & MOISTURE PROTECTION	101.1	41.4	71.8	96.0	78.9	87.6	98.1	72.7	85.6	97.5	41.1	69.8	97.0	49.8	73.8	97.2	48.4	73.2
08	DOORS & WINDOWS	103.8	39.3	87.1	96.3	77.5	91.4	94.2	72.6	88.6	97.6	42.0	83.2	95.6	46.2	82.8	95.9	68.1	88.7
09200	PLASTER & GYPSUM BOARD	98.9	39.1	61.0	107.2	74.6	86.5	103.4	74.6	85.1	98.7	31.4	56.0	97.5	37.6	59.6	98.2	74.6	83.2
095,098	CEILINGS & ACOUSTICAL TREATMENT	87.5	39.1	58.5	98.1	74.6	84.0	92.8	74.6	81.9	91.9	31.4	55.6	91.9	37.6	59.3	92.8	74.6	81.9
09600	FLOORING	97.3	39.8	82.1	105.3	91.2	101.6	103.5	70.5	94.8	99.1	35.3	82.2	97.5	41.6	82.7	97.5	69.8	90.2
097,099	WALL FINISHES, PAINTS & COATINGS	92.8	21.8	49.7	119.6	86.4	99.4	119.6	86.4	99.4	108.5	34.0	63.3	108.5	56.9	77.2	119.6	40.6	71.7
09	FINISHES	95.7	39.2	66.3	100.7	79.1	89.5	99.3	73.5	85.9	95.9	33.4	63.4	95.2	42.0	67.5	95.9	66.0	80.3
10 - 14	TOTAL DIV. 10000 - 14000	100.0	50.8	89.5	100.0	88.0	97.4	100.0	72.8	94.2	100.0	61.6	91.8	100.0	71.1	93.9	100.0	58.8	91.2
15	MECHANICAL	95.7	42.5	72.5	100.0	85.8	93.8	100.0	82.2	92.2	95.7	54.4	77.6	95.7	60.5	80.3	95.7	39.7	71.2
16	ELECTRICAL	100.2	32.9	67.3	98.5	94.4	96.5	96.0	92.1	94.1	97.9	30.6	65.1	97.8	69.6	84.1	100.1	49.3	75.3
01 - 16	WEIGHTED AVERAGE	99.0	47.9	75.8	99.6	84.6	92.8	100.2	79.4	90.7	97.8	48.7	75.5	98.9	60.5	81.5	98.3	56.6	79.4

VIRGINIA

DIVISION		FAIRFAX 220 - 221			FARMVILLE 239			FREDERICKSBURG 224 - 225			GRUNDY 246			HARRISONBURG 228			LYNCHBURG 245		
		MAT.	INST.	TOTAL	MAT.	INST.	TOTAL	MAT.	INST.	TOTAL	MAT.	INST.	TOTAL	MAT.	INST.	TOTAL	MAT.	INST.	TOTAL
01590	EQUIPMENT RENTAL	.0	99.7	99.7	.0	104.8	104.8	.0	99.7	99.7	.0	99.7	99.7	.0	99.7	99.7	.0	99.7	99.7
02	SITE CONSTRUCTION	121.7	82.8	92.9	106.9	83.2	89.4	110.0	82.3	89.5	104.0	80.5	86.6	118.5	82.3	91.7	104.7	81.0	87.2
03100	CONCRETE FORMS & ACCESSORIES	83.8	72.1	73.6	97.4	28.1	37.3	83.8	46.7	51.7	88.8	29.9	37.8	79.3	28.8	35.5	85.4	37.1	43.6
03200	CONCRETE REINFORCEMENT	95.7	84.2	89.9	95.8	47.3	71.3	96.5	80.2	88.2	94.5	47.7	70.8	95.7	58.4	76.8	95.2	64.4	79.6
03300	CAST-IN-PLACE CONCRETE	96.9	78.7	89.3	100.4	47.3	78.2	98.4	48.8	77.7	96.5	41.8	73.7	96.9	60.8	81.8	96.5	47.8	76.2
03	CONCRETE	106.3	77.8	92.1	102.5	40.0	71.3	99.9	55.5	77.7	101.5	39.2	70.4	103.9	47.6	75.8	101.4	47.8	74.6
04	MASONRY	98.5	68.2	79.5	94.5	33.3	56.1	97.9	36.3	59.3	88.9	34.3	54.7	97.3	33.8	57.5	103.2	34.5	60.1
05	METALS	100.6	96.8	99.3	100.4	65.6	89.1	100.5	91.2	97.5	100.2	68.9	90.0	100.5	82.0	94.4	100.4	81.8	94.3
06	WOOD & PLASTICS	82.9	75.4	79.0	96.6	26.9	60.2	82.9	47.4	64.3	86.4	28.8	56.4	78.3	23.7	49.8	82.9	36.4	58.6
07	THERMAL & MOISTURE PROTECTION	97.9	73.1	85.7	97.1	34.9	66.5	97.2	46.3	72.2	97.4	37.7	68.1	97.6	53.2	75.8	97.2	43.1	70.6
08	DOORS & WINDOWS	94.2	78.6	90.2	95.9	26.9	78.1	95.6	57.8	85.8	97.6	28.9	79.8	95.9	39.9	81.4	95.9	43.6	82.4
09200	PLASTER & GYPSUM BOARD	99.0	74.6	83.5	105.6	23.8	53.8	99.0	45.7	65.2	99.9	26.7	53.5	97.5	21.4	49.2	98.7	34.4	58.0
095,098	CEILINGS & ACOUSTICAL TREATMENT	92.8	74.6	81.9	91.9	23.8	51.1	92.8	45.7	64.5	91.9	26.7	52.8	91.9	21.4	49.6	91.9	34.4	57.4
09600	FLOORING	99.4	60.8	89.2	105.3	56.8	92.5	99.4	69.8	91.6	100.8	31.8	82.6	97.1	69.8	89.8	99.1	38.6	83.1
097,099	WALL FINISHES, PAINTS & COATINGS	119.6	87.2	99.9	108.5	35.9	64.5	119.6	40.6	71.7	108.5	34.0	63.3	119.6	44.2	73.9	108.5	34.0	63.3
09	FINISHES	97.4	71.8	84.1	98.5	34.0	64.9	96.5	49.5	72.1	96.4	30.2	61.9	96.0	36.4	65.0	95.7	36.3	64.8
10 - 14	TOTAL DIV. 10000 - 14000	100.0	72.6	94.2	100.0	47.4	88.8	100.0	62.2	92.0	100.0	46.7	88.7	100.0	59.5	91.4	100.0	62.9	92.1
15	MECHANICAL	95.7	82.2	89.7	95.7	27.5	65.8	95.7	77.4	87.7	95.7	35.8	69.5	95.7	40.4	71.5	95.7	59.9	80.0
16	ELECTRICAL	98.9	94.4	96.7	95.3	45.2	70.9	96.2	92.1	94.2	97.9	59.4	79.1	98.1	37.8	68.7	98.9	30.0	65.2
01 - 16	WEIGHTED AVERAGE	99.2	81.2	91.0	98.1	42.3	72.8	97.8	68.0	84.3	97.8	45.8	74.2	98.6	48.1	75.7	98.4	51.0	76.8

VIRGINIA

DIVISION		NEWPORT NEWS 236			NORFOLK 233 - 235			PETERSBURG 238			PORTSMOUTH 237			PULASKI 243			RICHMOND 230 - 232		
		MAT.	INST.	TOTAL	MAT.	INST.	TOTAL	MAT.	INST.	TOTAL	MAT.	INST.	TOTAL	MAT.	INST.	TOTAL	MAT.	INST.	TOTAL
01590	EQUIPMENT RENTAL	.0	104.9	104.9	.0	105.6	105.6	.0	104.8	104.8	.0	104.8	104.8	.0	99.7	99.7	.0	104.8	104.8
02	SITE CONSTRUCTION	105.5	84.5	90.0	104.8	85.6	90.6	109.0	84.9	91.2	104.1	84.1	89.3	103.5	80.4	86.4	106.1	84.9	90.4
03100	CONCRETE FORMS & ACCESSORIES	96.4	58.0	63.2	100.7	58.3	64.0	89.1	57.3	61.6	84.5	58.2	61.7	88.8	28.9	36.9	97.3	57.3	62.7
03200	CONCRETE REINFORCEMENT	95.5	72.0	83.6	95.5	72.0	83.6	95.2	71.2	83.1	95.2	72.0	83.5	94.5	47.7	70.8	95.5	71.2	83.2
03300	CAST-IN-PLACE CONCRETE	97.6	56.2	80.3	100.5	56.3	82.0	100.4	56.0	81.9	96.6	56.0	79.7	96.5	40.3	73.1	103.8	56.0	83.8
03	CONCRETE	99.6	61.5	80.6	101.3	61.7	81.5	103.4	61.0	82.2	98.3	61.5	79.9	101.5	38.2	69.9	102.6	61.0	81.8
04	MASONRY	89.6	55.3	68.1	96.2	55.3	70.6	102.2	50.8	70.0	94.6	55.3	69.9	84.2	31.8	51.3	88.5	50.8	64.9
05	METALS	102.4	89.6	98.2	101.5	89.8	97.7	100.5	89.6	96.9	101.4	89.1	97.4	100.3	68.8	90.0	104.4	89.6	99.5
06	WOOD & PLASTICS	95.4	61.5	77.7	100.7	61.5	80.2	86.9	59.5	72.6	82.2	61.5	71.4	86.4	28.8	56.4	96.5	59.5	77.2
07	THERMAL & MOISTURE PROTECTION	97.1	52.3	75.1	96.8	52.4	75.0	97.0	54.3	76.0	97.0	52.4	75.1	97.4	36.6	67.5	96.6	54.3	75.8
08	DOORS & WINDOWS	96.3	58.8	86.6	96.3	61.5	87.3	95.6	56.6	85.5	96.3	61.5	87.3	97.6	28.9	79.8	96.3	56.6	86.0
09200	PLASTER & GYPSUM BOARD	107.2	59.4	76.9	107.2	59.4	76.9	100.2	57.4	73.0	98.8	59.4	73.8	99.9	26.7	53.5	107.2	57.4	75.6
095,098	CEILINGS & ACOUSTICAL TREATMENT	98.1	59.4	74.9	98.1	59.4	74.9	92.8	57.4	71.5	98.1	59.4	74.9	91.9	26.7	52.8	98.1	57.4	73.7
09600	FLOORING	105.3	41.1	88.4	105.1	41.1	88.2	100.4	65.4	91.2	96.5	41.1	81.9	100.8	31.8	82.6	105.1	65.4	94.6
097,099	WALL FINISHES, PAINTS & COATINGS	108.5	43.5	69.1	108.5	64.1	81.6	108.5	56.9	77.2	108.5	64.1	81.6	108.5	34.0	63.3	108.5	56.9	77.2
09	FINISHES	100.1	53.5	75.9	100.1	55.8	77.1	96.6	58.9	77.0	96.3	55.8	75.2	96.4	29.5	61.6	100.0	58.9	78.6
10 - 14	TOTAL DIV. 10000 - 14000	100.0	70.0	93.6	100.0	70.0	93.6	100.0	74.9	94.7	100.0	70.0	93.6	100.0	45.7	88.4	100.0	74.9	94.7
15	MECHANICAL	100.0	58.5	81.9	100.0	62.3	83.5	95.7	62.3	81.0	100.0	62.3	83.5	95.7	28.1	66.1	100.0	62.3	83.5
16	ELECTRICAL	97.9	60.6	79.7	98.0	62.5	80.7	98.1	69.6	84.2	96.3	62.5	79.8	97.9	59.4	79.1	98.5	69.6	84.4
01 - 16	WEIGHTED AVERAGE	99.2	63.6	83.0	99.6	65.2	84.0	98.6	65.9	83.8	98.4	65.0	83.2	97.5	43.7	73.1	99.9	65.9	84.5

VIRGINIA / WASHINGTON

DIVISION		ROANOKE 240 - 241			STAUNTON 244			WINCHESTER 226			CLARKSTON 994			EVERETT 982			OLYMPIA 985		
		MAT.	INST.	TOTAL	MAT.	INST.	TOTAL	MAT.	INST.	TOTAL	MAT.	INST.	TOTAL	MAT.	INST.	TOTAL	MAT.	INST.	TOTAL
01590	EQUIPMENT RENTAL	.0	99.7	99.7	.0	104.8	104.8	.0	99.7	99.7	.0	90.0	90.0	.0	104.0	104.0	.0	104.0	104.0
02	SITE CONSTRUCTION	103.2	81.0	86.8	107.4	83.4	89.6	117.2	81.7	91.0	103.1	88.4	92.2	94.3	115.2	109.8	97.5	115.5	110.8
03100	CONCRETE FORMS & ACCESSORIES	96.2	36.9	44.8	88.6	36.3	43.3	82.0	38.9	44.7	112.8	77.7	82.4	106.4	101.3	102.0	97.0	102.4	101.6
03200	CONCRETE REINFORCEMENT	95.5	64.6	79.9	95.2	47.4	71.0	95.1	84.3	89.7	107.4	88.0	97.6	107.9	93.6	100.7	108.4	93.6	101.0
03300	CAST-IN-PLACE CONCRETE	109.6	47.7	83.8	100.4	48.1	78.6	96.9	59.8	81.4	117.1	83.7	103.2	97.8	103.4	100.2	102.7	109.6	105.6
03	CONCRETE	105.3	47.7	76.6	102.5	44.5	73.6	103.3	56.5	79.9	108.5	81.7	95.1	98.5	100.0	99.2	101.6	102.6	102.1
04	MASONRY	90.0	40.0	58.6	100.1	35.6	59.6	94.5	33.7	56.4	113.5	82.8	94.2	124.6	99.1	110.9	124.6	101.3	110.0
05	METALS	102.2	82.0	95.6	100.4	78.9	93.4	100.5	91.2	97.5	94.2	82.2	90.3	105.4	87.3	99.4	99.8	88.0	96.0
06	WOOD & PLASTICS	95.4	36.4	64.6	86.4	36.4	60.3	81.3	37.1	58.2	95.1	76.7	85.5	97.2	101.4	99.4	86.7	101.4	94.3
07	THERMAL & MOISTURE PROTECTION	97.0	45.9	71.9	97.0	41.5	69.7	97.8	54.5	76.5	171.2	77.6	125.2	107.2	98.5	102.9	107.0	98.0	102.6
08	DOORS & WINDOWS	96.3	43.6	82.6	95.9	40.0	81.4	97.7	53.3	86.2	111.5	81.0	103.6	98.6	98.2	98.5	98.5	99.1	98.7
09200	PLASTER & GYPSUM BOARD	107.2	34.4	61.1	99.9	33.6	57.9	99.0	35.2	58.6	119.1	75.8	91.7	104.4	101.3	102.5	98.6	101.3	100.3
095,098	CEILINGS & ACOUSTICAL TREATMENT	98.1	34.4	59.9	91.9	33.6	57.0	92.8	35.2	58.3	103.1	75.8	86.8	114.7	101.3	106.7	110.9	101.3	105.2
09600	FLOORING	105.3	36.9	87.3	100.1	37.2	83.5	98.7	69.8	91.1	110.5	61.5	97.6	120.1	101.6	115.2	112.6	101.6	109.7
097,099	WALL FINISHES, PAINTS & COATINGS	108.5	34.0	63.3	108.5	28.0	59.7	119.6	28.6	64.4	113.5	72.9	88.8	113.4	81.4	94.0	113.4	95.6	102.5
09	FINISHES	100.0	36.0	66.7	96.2	35.8	64.7	96.8	42.3	68.5	121.7	73.7	96.7	114.8	99.3	106.8	111.0	101.5	106.0
10 - 14	TOTAL DIV. 10000 - 14000	100.0	62.9	92.1	100.0	62.6	92.0	100.0	54.1	90.2	100.0	89.5	97.8	100.0	95.7	99.1	100.0	105.5	101.2
15	MECHANICAL	100.0	40.7	74.1	95.7	35.4	69.3	95.7	62.4	81.1	96.3	82.6	90.3	99.9	95.7	98.1	100.0	95.1	97.8
16	ELECTRICAL	97.9	35.0	67.2	96.8	38.3	68.3	96.6	34.3	66.2	97.9	86.7	92.5	104.4	95.3	100.0	104.2	97.1	100.7
01 - 16	WEIGHTED AVERAGE	99.8	48.4	76.5	98.2	46.5	74.8	98.5	55.4	78.9	104.7	82.3	94.6	104.0	98.3	101.4	102.8	99.7	101.4

COST INDEXES

City Cost Indexes

WASHINGTON

DIVISION		RICHLAND 993			SEATTLE 980 - 981,987			SPOKANE 990 - 992			TACOMA 983 - 984			VANCOUVER 986			WENATCHEE 988		
		MAT.	INST.	TOTAL	MAT.	INST.	TOTAL	MAT.	INST.	TOTAL	MAT.	INST.	TOTAL	MAT.	INST.	TOTAL	MAT.	INST.	TOTAL
01590	EQUIPMENT RENTAL	.0	90.0	90.0	.0	103.8	103.8	.0	90.0	90.0	.0	104.0	104.0	.0	97.2	97.2	.0	104.0	104.0
02	SITE CONSTRUCTION	104.4	89.5	93.3	98.4	113.3	109.4	105.2	89.5	93.6	97.1	115.5	110.7	109.6	100.9	103.1	109.0	114.0	112.7
03100	CONCRETE FORMS & ACCESSORIES	112.9	81.2	85.4	97.1	103.0	102.2	119.8	81.4	86.5	97.1	102.5	101.8	97.9	97.0	97.1	98.5	76.7	79.6
03200	CONCRETE REINFORCEMENT	103.1	88.0	95.5	106.7	93.9	100.2	103.8	88.1	95.8	106.7	93.7	100.1	107.5	93.5	100.4	107.5	92.9	100.1
03300	CAST-IN-PLACE CONCRETE	117.4	85.4	104.1	102.7	109.9	105.7	121.6	85.5	106.5	100.6	109.7	104.4	112.3	102.6	108.3	102.7	83.0	94.5
03	CONCRETE	108.0	83.8	95.9	101.3	103.2	102.2	110.7	83.9	97.3	100.3	102.7	101.5	110.5	98.0	104.2	108.7	81.9	95.3
04	MASONRY	113.3	84.1	95.0	125.6	103.9	112.0	114.9	84.8	96.0	125.4	101.3	110.3	125.9	98.3	108.6	128.5	84.0	100.6
05	METALS	94.3	82.5	90.4	107.0	90.6	101.6	96.9	82.8	92.3	107.0	88.2	100.8	104.2	89.6	99.5	104.8	84.9	98.3
06	WOOD & PLASTICS	95.3	80.5	87.6	87.7	101.4	94.8	104.6	80.5	92.0	86.7	101.4	94.3	81.0	96.7	89.2	87.8	74.1	80.7
07	THERMAL & MOISTURE PROTECTION	172.6	80.1	127.1	107.2	102.1	104.7	169.8	81.7	126.5	107.0	100.0	103.6	109.6	91.1	100.5	107.9	78.2	93.3
08	DOORS & WINDOWS	112.0	75.9	102.6	100.8	99.1	100.3	111.7	77.3	102.8	99.2	99.1	99.2	96.2	94.5	95.8	98.8	73.3	92.2
09200	PLASTER & GYPSUM BOARD	119.1	79.8	94.2	97.8	101.3	100.0	123.9	79.8	95.9	100.0	101.3	100.8	98.3	96.9	97.4	101.4	73.3	83.6
095,098	CEILINGS & ACOUSTICAL TREATMENT	107.1	79.8	90.7	118.0	101.3	108.0	107.1	79.8	90.7	118.2	101.3	108.1	113.0	96.9	103.4	110.9	73.3	88.4
09600	FLOORING	110.8	45.2	93.5	112.0	105.8	110.4	114.4	76.1	104.3	112.6	101.6	109.7	118.5	83.1	109.2	116.0	73.7	104.8
097,099	WALL FINISHES, PAINTS & COATINGS	113.5	74.1	89.6	113.4	95.6	102.6	113.5	76.8	91.3	113.4	95.6	102.6	121.5	70.5	90.6	113.4	74.1	89.6
09	FINISHES	122.8	73.1	96.9	112.5	102.4	107.2	124.6	79.7	101.2	113.0	101.5	107.0	110.9	91.4	100.8	113.3	75.0	93.4
10 - 14	TOTAL DIV. 10000 - 14000	100.0	88.9	97.6	100.0	105.6	101.2	100.0	88.9	97.6	100.0	105.5	101.2	100.0	85.6	96.9	100.0	94.2	98.8
15	MECHANICAL	100.5	94.5	97.9	99.9	114.1	106.1	100.5	84.6	93.5	100.0	95.1	97.9	100.1	100.2	100.2	95.6	80.5	89.0
16	ELECTRICAL	94.9	86.7	90.9	104.2	109.1	106.6	93.2	83.3	88.4	104.2	97.1	100.8	115.3	103.5	109.5	105.0	82.7	94.1
01 - 16	WEIGHTED AVERAGE	105.5	85.1	96.2	104.3	105.8	105.0	106.2	83.7	96.0	104.1	99.7	102.1	106.0	97.2	102.0	104.3	84.0	95.1

DIVISION		WASHINGTON YAKIMA 989			WEST VIRGINIA BECKLEY 258 - 259			BLUEFIELD 247 - 248			BUCKHANNON 262			CHARLESTON 250 - 253			CLARKSBURG 263 - 264		
		MAT.	INST.	TOTAL	MAT.	INST.	TOTAL	MAT.	INST.	TOTAL	MAT.	INST.	TOTAL	MAT.	INST.	TOTAL	MAT.	INST.	TOTAL
01590	EQUIPMENT RENTAL	.0	104.0	104.0	.0	99.7	99.7	.0	99.7	99.7	.0	99.7	99.7	.0	99.7	99.7	.0	99.7	99.7
02	SITE CONSTRUCTION	100.5	114.0	110.5	98.1	84.9	88.3	97.9	84.8	88.2	104.0	84.6	89.6	101.0	85.9	89.8	104.5	84.6	89.8
03100	CONCRETE FORMS & ACCESSORIES	97.5	96.2	96.4	84.1	92.2	91.1	85.7	91.1	90.4	84.9	87.9	87.5	103.9	93.8	95.1	82.0	87.9	87.1
03200	CONCRETE REINFORCEMENT	107.2	92.5	99.8	94.1	87.9	91.0	94.1	72.2	83.1	94.6	90.5	92.6	95.5	88.0	91.7	94.6	92.7	93.6
03300	CAST-IN-PLACE CONCRETE	107.5	83.5	97.5	94.4	105.5	99.0	94.4	97.2	95.6	94.1	99.7	96.5	95.5	108.2	100.8	103.1	99.7	101.7
03	CONCRETE	105.2	90.6	97.9	97.0	96.6	96.8	97.1	90.4	93.8	100.5	93.2	96.9	99.1	98.2	98.7	104.1	93.6	98.8
04	MASONRY	117.4	73.2	89.7	90.1	91.2	90.8	85.9	91.2	89.2	96.4	90.5	92.7	87.2	93.7	91.3	98.2	90.5	93.3
05	METALS	105.3	82.9	98.0	100.4	99.2	100.0	100.4	92.6	97.9	100.6	100.8	100.7	102.4	100.2	101.7	100.6	101.5	100.9
06	WOOD & PLASTICS	87.2	101.4	94.6	82.6	93.4	88.2	84.3	93.4	89.1	83.6	86.2	84.9	103.6	93.4	98.3	80.2	86.2	83.3
07	THERMAL & MOISTURE PROTECTION	107.2	78.2	93.0	97.1	89.7	93.4	97.2	84.6	91.0	97.5	87.8	92.7	97.0	90.7	93.9	97.4	87.8	92.7
08	DOORS & WINDOWS	98.8	85.4	95.3	96.9	86.0	94.1	97.8	82.5	93.8	97.8	82.7	93.9	97.5	86.0	94.5	97.8	85.4	94.6
09200	PLASTER & GYPSUM BOARD	99.8	101.3	100.8	96.0	93.0	94.1	96.4	93.0	94.3	96.9	85.6	89.7	106.8	93.0	98.1	95.3	85.6	89.1
095,098	CEILINGS & ACOUSTICAL TREATMENT	112.9	101.3	106.0	89.2	93.0	91.5	89.2	93.0	91.5	91.0	85.6	87.8	96.3	93.0	94.3	91.0	85.6	87.8
09600	FLOORING	114.0	67.9	101.8	95.9	104.2	98.1	96.5	104.2	98.5	96.1	105.0	98.5	105.1	104.2	104.9	94.8	105.0	97.5
097,099	WALL FINISHES, PAINTS & COATINGS	113.4	76.8	91.2	108.5	46.3	70.8	108.5	46.3	70.8	108.5	89.0	96.7	108.5	96.6	101.3	108.5	89.0	96.7
09	FINISHES	112.2	89.5	100.4	93.4	90.1	91.7	93.6	90.1	91.8	94.4	91.1	92.7	99.6	96.3	97.9	93.7	91.1	92.4
10 - 14	TOTAL DIV. 10000 - 14000	100.0	100.7	100.2	100.0	75.7	94.8	100.0	75.5	94.8	100.0	100.0	100.0	100.0	101.4	100.3	100.0	100.2	100.0
15	MECHANICAL	100.0	93.6	97.2	95.7	83.8	90.5	95.7	68.7	83.9	95.7	94.9	95.3	100.0	85.4	93.6	95.7	88.8	92.7
16	ELECTRICAL	106.9	86.7	97.0	94.5	89.4	92.0	96.9	57.4	77.6	98.2	98.7	98.4	97.9	89.4	93.8	98.2	98.7	98.4
01 - 16	WEIGHTED AVERAGE	104.2	90.0	97.8	96.4	89.5	93.3	96.6	80.2	89.1	97.9	93.2	95.8	99.1	92.0	95.9	98.3	92.2	95.5

WEST VIRGINIA

DIVISION		GASSAWAY 266			HUNTINGTON 255 - 257			LEWISBURG 249			MARTINSBURG 254			MORGANTOWN 265			PARKERSBURG 261		
		MAT.	INST.	TOTAL	MAT.	INST.	TOTAL	MAT.	INST.	TOTAL	MAT.	INST.	TOTAL	MAT.	INST.	TOTAL	MAT.	INST.	TOTAL
01590	EQUIPMENT RENTAL	.0	99.7	99.7	.0	99.7	99.7	.0	99.7	99.7	.0	99.7	99.7	.0	99.7	99.7	.0	99.7	99.7
02	SITE CONSTRUCTION	101.3	85.1	89.4	102.4	86.6	90.7	113.7	84.9	92.4	101.6	83.2	88.0	98.7	84.9	88.5	107.1	85.8	91.3
03100	CONCRETE FORMS & ACCESSORIES	84.2	93.3	92.1	97.3	95.2	95.5	82.5	91.8	90.5	84.1	74.0	75.4	82.5	87.9	87.2	87.0	88.2	88.1
03200	CONCRETE REINFORCEMENT	94.6	84.7	89.6	95.5	88.8	92.1	94.6	72.3	83.4	94.1	74.8	84.4	94.6	92.7	93.6	94.1	90.2	92.1
03300	CAST-IN-PLACE CONCRETE	98.5	107.6	102.3	103.0	108.8	105.4	94.5	105.3	99.0	98.9	94.9	97.2	94.1	99.7	96.5	96.2	95.3	95.8
03	CONCRETE	100.5	97.2	98.9	102.2	99.2	100.7	107.2	93.5	100.4	100.6	82.2	91.4	96.9	93.6	95.2	102.3	91.8	97.1
04	MASONRY	99.5	92.1	94.8	89.5	94.6	92.7	88.4	91.2	90.1	92.4	82.5	86.1	116.8	90.5	100.3	76.5	85.5	82.1
05	METALS	100.5	98.9	100.0	102.4	100.5	101.8	100.5	93.3	98.2	100.7	89.0	96.9	100.6	101.6	100.9	101.0	100.3	100.8
06	WOOD & PLASTICS	82.8	93.4	88.3	95.4	94.7	95.0	80.6	93.4	87.3	82.6	72.7	77.5	80.6	86.2	83.5	84.6	86.7	85.7
07	THERMAL & MOISTURE PROTECTION	97.1	89.9	93.6	97.4	91.2	94.4	98.2	89.7	94.0	97.4	72.8	85.3	97.2	87.8	92.6	97.4	86.0	91.8
08	DOORS & WINDOWS	95.7	85.3	93.0	96.3	86.9	93.8	97.8	82.5	93.8	99.2	66.4	90.7	99.2	85.4	95.7	96.9	86.0	92.7
09200	PLASTER & GYPSUM BOARD	96.5	93.0	94.3	105.6	94.4	98.5	95.3	93.0	93.8	96.5	71.8	80.8	95.3	85.6	89.1	99.3	86.2	91.0
095,098	CEILINGS & ACOUSTICAL TREATMENT	91.0	93.0	92.2	91.9	94.4	93.4	91.0	93.0	92.2	91.0	71.8	79.5	91.0	85.6	87.8	91.0	86.2	88.1
09600	FLOORING	95.9	104.2	98.1	105.1	109.0	106.1	94.9	104.2	97.4	95.9	70.1	89.0	94.9	105.0	97.6	99.8	95.1	98.5
097,099	WALL FINISHES, PAINTS & COATINGS	108.5	96.6	101.3	108.5	93.2	99.2	108.5	46.3	70.8	108.5	54.7	75.8	108.5	89.0	96.7	108.5	100.2	103.4
09	FINISHES	94.1	96.0	95.1	98.1	97.9	98.0	94.6	90.1	92.3	94.0	71.3	82.2	93.4	91.1	92.2	95.9	90.7	93.2
10 - 14	TOTAL DIV. 10000 - 14000	100.0	101.0	100.2	100.0	102.0	100.4	100.0	75.5	94.8	100.0	71.8	94.0	100.0	89.4	97.7	100.0	100.3	100.1
15	MECHANICAL	95.7	89.8	93.1	100.0	82.2	92.3	95.7	83.6	90.4	95.7	83.3	90.3	95.7	94.9	95.3	100.0	90.0	95.6
16	ELECTRICAL	98.2	89.4	93.9	97.9	81.9	90.1	94.5	57.4	76.4	99.8	78.6	89.5	98.4	98.7	98.5	98.3	93.5	96.0
01 - 16	WEIGHTED AVERAGE	97.7	92.3	95.3	99.3	90.9	95.5	98.1	83.9	91.6	98.0	80.0	89.8	98.4	93.2	96.0	98.3	90.6	94.8

COST INDEXES

WEST VIRGINIA / WISCONSIN

DIVISION		WEST VIRGINIA								WISCONSIN							
		PETERSBURG 268			ROMNEY 267			WHEELING 260			BELOIT 535			EAU CLAIRE 547			GREEN BAY 541-543
		MAT.	INST.	TOTAL	MAT.	INST.	TOTAL	MAT.	INST.	TOTAL	MAT.	INST.	TOTAL	MAT.	INST.	TOTAL	MAT.	INST.	TOTAL
01590	EQUIPMENT RENTAL	.0	99.7	99.7	.0	99.7	99.7	.0	99.7	99.7	.0	100.0	100.0	.0	100.5	100.5	.0	98.2	98.2
02	SITE CONSTRUCTION	98.1	85.3	88.7	100.9	85.3	89.4	107.6	85.6	91.3	88.1	104.9	100.5	84.7	102.4	97.8	87.2	98.3	95.4
03100	CONCRETE FORMS & ACCESSORIES	86.1	77.5	78.7	81.6	77.1	77.7	88.8	88.2	88.3	98.4	95.3	95.7	99.8	94.8	95.5	110.9	94.0	96.2
03200	CONCRETE REINFORCEMENT	94.1	83.8	88.9	94.6	72.7	83.6	93.5	92.7	93.1	89.5	110.9	100.3	91.8	100.2	96.1	90.1	89.5	89.8
03300	CAST-IN-PLACE CONCRETE	94.1	99.6	96.4	98.5	99.4	98.9	96.2	99.9	97.7	105.8	97.5	102.3	100.1	94.8	97.9	103.5	96.7	100.7
03	CONCRETE	97.0	87.3	92.2	100.3	85.0	92.7	102.3	93.8	98.1	98.2	99.2	98.7	96.4	96.2	96.3	97.6	94.4	96.0
04	MASONRY	92.4	90.5	91.2	88.7	76.9	81.3	99.6	90.5	93.9	102.0	99.1	100.2	92.3	96.2	94.7	123.1	95.2	105.6
05	METALS	100.7	98.0	99.8	100.7	92.9	98.1	101.2	101.9	101.4	92.6	105.1	96.7	91.3	102.5	94.9	93.8	97.5	95.0
06	WOOD & PLASTICS	84.6	72.7	78.4	79.5	72.7	76.0	86.4	86.2	86.3	109.4	93.8	101.2	112.8	93.9	102.9	118.8	93.9	105.8
07	THERMAL & MOISTURE PROTECTION	97.3	78.7	88.2	97.4	75.7	86.7	97.6	89.0	93.4	97.2	100.2	98.7	98.8	88.2	93.5	100.7	86.4	93.7
08	DOORS & WINDOWS	99.3	73.2	92.5	99.2	70.9	91.8	97.8	85.4	94.6	101.8	98.9	101.0	100.7	93.5	98.9	99.1	87.0	96.0
09200	PLASTER & GYPSUM BOARD	96.9	71.8	81.0	94.5	71.8	80.1	99.7	85.6	90.8	103.3	94.1	97.4	106.3	94.1	98.5	97.9	94.1	95.5
095,098	CEILINGS & ACOUSTICAL TREATMENT	91.0	71.8	79.5	91.0	71.8	79.5	91.0	85.6	87.8	90.2	94.1	92.5	97.2	94.1	95.3	91.2	94.1	92.9
09600	FLOORING	97.1	105.0	99.2	94.7	79.0	90.5	100.8	105.0	101.9	93.6	110.6	98.1	90.9	105.2	94.7	108.8	105.2	107.9
097,099	WALL FINISHES, PAINTS & COATINGS	108.5	83.5	93.4	108.5	83.5	93.4	108.5	89.0	96.7	93.3	91.7	92.4	89.5	84.8	86.7	99.2	75.8	85.0
09	FINISHES	94.3	82.6	88.2	93.4	77.3	85.0	96.3	91.1	93.6	95.6	97.7	96.7	97.5	93.8	95.5	100.9	94.5	97.6
10-14	TOTAL DIV. 10000-14000	100.0	87.6	97.4	100.0	98.4	99.7	100.0	103.6	100.8	100.0	92.6	98.4	100.0	92.8	98.5	100.0	92.4	98.4
15	MECHANICAL	95.7	98.6	96.9	95.7	98.2	96.8	100.0	94.9	97.8	99.8	90.4	95.7	100.2	86.7	94.3	100.6	83.2	93.0
16	ELECTRICAL	101.3	87.6	94.6	100.6	87.6	94.2	95.7	98.7	97.2	95.8	87.6	91.8	102.3	85.2	93.9	96.9	84.8	91.0
01-16	WEIGHTED AVERAGE	97.7	89.1	93.8	97.7	86.4	92.6	99.3	93.7	96.8	97.7	96.4	97.1	97.7	93.3	95.7	99.5	90.9	95.6

WISCONSIN

DIVISION		KENOSHA 531			LA CROSSE 546			LANCASTER 538			MADISON 537			MILWAUKEE 530,532			NEW RICHMOND 540
		MAT.	INST.	TOTAL	MAT.	INST.	TOTAL	MAT.	INST.	TOTAL	MAT.	INST.	TOTAL	MAT.	INST.	TOTAL	MAT.	INST.	TOTAL
01590	EQUIPMENT RENTAL	.0	97.7	97.7	.0	100.5	100.5	.0	100.0	100.0	.0	100.0	100.0	.0	86.1	86.1	.0	101.5	101.5
02	SITE CONSTRUCTION	92.8	101.6	99.3	78.9	102.4	96.3	87.1	104.8	100.2	87.5	104.6	100.1	88.8	93.3	92.1	80.2	103.5	97.4
03100	CONCRETE FORMS & ACCESSORIES	108.5	101.2	102.2	83.9	94.3	92.9	97.4	93.8	94.3	99.2	94.9	95.5	101.5	108.6	107.7	93.1	99.9	99.0
03200	CONCRETE REINFORCEMENT	89.3	97.9	93.7	91.5	88.2	89.9	90.6	88.3	89.4	89.5	88.5	89.0	89.5	98.3	93.9	89.3	100.2	94.8
03300	CAST-IN-PLACE CONCRETE	115.2	98.3	108.1	90.0	94.2	91.7	105.2	96.8	101.7	103.8	97.3	101.1	103.8	104.4	104.1	104.2	95.9	100.7
03	CONCRETE	103.3	99.7	101.5	87.9	93.5	90.7	97.9	94.0	95.9	97.3	94.7	96.0	97.7	104.5	101.1	94.4	98.8	96.6
04	MASONRY	98.3	106.4	103.4	91.5	96.2	94.4	102.0	99.1	100.2	102.4	99.1	100.4	101.9	115.1	110.2	116.9	96.4	104.1
05	METALS	93.2	100.8	95.7	91.2	96.8	93.0	90.7	94.0	91.8	94.3	95.2	94.6	96.0	92.0	94.7	91.5	101.9	94.9
06	WOOD & PLASTICS	114.7	99.9	107.0	95.1	93.9	94.5	108.5	93.8	100.8	109.5	93.8	101.3	112.1	108.0	110.0	100.9	101.1	101.0
07	THERMAL & MOISTURE PROTECTION	96.8	101.3	99.0	98.0	87.3	92.7	96.9	82.1	89.6	96.5	95.2	95.9	95.8	105.6	100.6	98.7	99.9	99.3
08	DOORS & WINDOWS	96.9	100.5	97.9	100.7	83.7	96.3	97.1	84.8	93.9	101.8	93.7	99.7	104.0	105.0	104.3	88.7	97.4	91.0
09200	PLASTER & GYPSUM BOARD	94.4	100.4	98.2	99.0	94.1	95.9	101.4	94.1	96.7	103.3	94.1	97.4	104.2	108.5	106.9	89.9	101.7	97.4
095,098	CEILINGS & ACOUSTICAL TREATMENT	84.3	100.4	93.9	95.4	94.1	94.6	83.1	94.1	89.7	90.2	94.1	92.5	93.8	108.5	102.6	91.2	94.1	92.9
09600	FLOORING	110.0	110.6	110.1	83.5	105.2	89.3	92.9	110.6	97.6	93.6	106.0	96.9	96.3	110.2	100.0	101.5	105.2	102.5
097,099	WALL FINISHES, PAINTS & COATINGS	103.6	91.9	96.5	89.5	69.8	77.6	93.3	87.8	90.0	93.3	91.7	92.4	95.7	102.6	99.9	103.4	84.8	92.1
09	FINISHES	98.5	102.2	100.4	93.2	94.1	93.7	93.4	95.9	94.7	95.6	97.0	96.3	97.6	108.8	103.4	92.1	98.5	95.4
10-14	TOTAL DIV. 10000-14000	100.0	91.7	98.2	100.0	92.8	98.5	100.0	63.8	92.3	100.0	92.6	98.4	100.0	94.4	98.8	100.0	90.6	98.0
15	MECHANICAL	100.1	94.2	97.5	100.2	86.2	94.1	95.5	86.2	91.4	99.8	89.7	95.4	99.8	95.7	98.0	95.3	86.4	91.4
16	ELECTRICAL	95.8	85.9	91.0	102.5	85.2	94.1	95.3	85.2	90.4	96.6	85.2	91.1	96.9	101.8	99.3	100.3	85.2	92.9
01-16	WEIGHTED AVERAGE	98.2	97.8	98.0	96.0	91.7	94.0	95.6	91.4	93.7	97.9	93.9	96.1	98.7	101.6	100.0	95.5	94.6	95.1

WISCONSIN

DIVISION		OSHKOSH 549			PORTAGE 539			RACINE 534			RHINELANDER 545			SUPERIOR 548			WAUSAU 544
		MAT.	INST.	TOTAL	MAT.	INST.	TOTAL	MAT.	INST.	TOTAL	MAT.	INST.	TOTAL	MAT.	INST.	TOTAL	MAT.	INST.	TOTAL
01590	EQUIPMENT RENTAL	.0	98.2	98.2	.0	100.0	100.0	.0	100.0	100.0	.0	98.2	98.2	.0	101.5	101.5	.0	98.2	98.2
02	SITE CONSTRUCTION	79.6	98.3	93.4	78.2	104.8	97.8	87.9	105.5	100.9	90.9	98.3	96.3	77.2	103.6	96.7	75.7	98.3	92.4
03100	CONCRETE FORMS & ACCESSORIES	91.2	93.8	93.4	89.2	94.2	93.5	98.7	101.3	100.9	88.3	93.7	93.0	91.2	91.7	91.6	90.6	93.7	93.3
03200	CONCRETE REINFORCEMENT	90.3	89.4	89.8	90.7	88.5	89.6	89.5	97.9	93.7	90.4	92.6	91.5	89.3	93.0	91.2	90.4	88.2	89.3
03300	CAST-IN-PLACE CONCRETE	96.0	96.7	96.3	90.3	98.7	93.8	103.8	98.0	101.4	108.9	96.7	103.8	98.2	96.5	97.5	89.5	96.6	92.5
03	CONCRETE	89.7	94.3	92.0	86.0	94.9	90.4	97.2	99.6	98.4	101.3	94.9	98.1	89.1	94.0	91.5	84.9	94.0	89.5
04	MASONRY	104.4	95.2	98.6	100.9	97.5	98.7	101.9	106.4	104.7	123.7	95.3	105.9	116.2	101.9	107.3	103.9	95.2	98.4
05	METALS	91.8	97.3	93.6	91.4	95.2	92.6	94.3	100.8	96.4	91.7	98.5	93.9	92.3	99.5	94.7	91.5	96.7	93.2
06	WOOD & PLASTICS	96.9	93.9	95.3	98.9	93.8	96.2	109.7	99.9	104.6	93.6	93.9	93.7	99.0	89.0	93.8	96.0	93.9	94.9
07	THERMAL & MOISTURE PROTECTION	99.5	88.1	93.9	96.1	94.5	95.3	97.2	100.8	99.0	100.4	83.3	92.0	98.3	100.7	99.5	99.2	82.1	90.8
08	DOORS & WINDOWS	94.3	87.0	92.4	97.3	93.7	96.4	101.8	100.5	101.5	94.3	87.9	92.7	88.7	89.0	88.7	99.2	85.3	95.6
09200	PLASTER & GYPSUM BOARD	87.4	94.1	91.6	96.8	94.1	95.1	103.3	100.4	101.4	86.2	94.1	91.2	89.7	89.3	89.5	86.6	94.1	91.3
095,098	CEILINGS & ACOUSTICAL TREATMENT	91.2	94.1	92.9	84.0	94.1	90.0	90.2	100.4	96.3	91.2	94.1	92.9	70.3	89.3	81.7	91.2	94.1	92.9
09600	FLOORING	99.2	105.2	100.8	89.1	110.6	94.8	93.6	110.6	98.1	98.3	105.2	100.1	102.7	114.1	105.7	99.0	105.2	100.7
097,099	WALL FINISHES, PAINTS & COATINGS	99.2	75.0	84.5	93.3	87.8	90.0	93.3	88.0	90.1	99.2	79.8	87.4	95.0	96.0	95.6	99.2	82.0	88.8
09	FINISHES	96.2	93.4	94.7	91.4	96.6	94.1	95.6	101.7	98.8	96.4	94.8	95.6	91.9	96.1	94.1	95.8	95.2	95.5
10-14	TOTAL DIV. 10000-14000	100.0	87.6	97.4	100.0	85.0	96.8	100.0	91.8	98.2	100.0	72.4	94.1	100.0	89.4	97.7	100.0	92.4	98.4
15	MECHANICAL	96.2	83.1	90.5	99.8	94.2	97.4	95.7	93.9	94.8	96.2	85.7	91.6	95.3	87.8	92.0	96.2	85.5	91.5
16	ELECTRICAL	100.9	84.1	92.7	98.7	85.2	92.1	95.7	93.9	94.8	100.3	82.8	91.8	104.7	94.3	99.6	101.9	92.9	97.1
01-16	WEIGHTED AVERAGE	95.5	90.5	93.3	94.2	93.3	93.8	97.8	99.2	98.4	98.0	90.7	94.7	95.4	94.8	95.1	94.9	90.9	93.1

WYOMING

DIVISION		CASPER 826			CHEYENNE 820			NEWCASTLE 827			RAWLINS 823			RIVERTON 825			ROCK SPRINGS 829 - 831		
		MAT.	INST.	TOTAL	MAT.	INST.	TOTAL	MAT.	INST.	TOTAL	MAT.	INST.	TOTAL	MAT.	INST.	TOTAL	MAT.	INST.	TOTAL
01590	EQUIPMENT RENTAL	.0	101.7	101.7	.0	101.7	101.7	.0	101.7	101.7	.0	101.7	101.7	.0	101.7	101.7	.0	101.7	101.7
02	SITE CONSTRUCTION	80.9	100.3	95.2	80.7	100.2	95.2	81.1	99.8	94.9	96.8	99.8	99.0	90.2	99.9	97.4	83.8	99.8	95.6
03100	CONCRETE FORMS & ACCESSORIES	97.8	43.6	50.8	98.5	43.4	50.8	92.5	30.5	38.8	97.5	30.5	39.5	91.0	35.5	43.0	99.3	30.6	39.8
03200	CONCRETE REINFORCEMENT	115.6	52.0	83.5	108.5	52.1	80.0	117.3	42.4	79.4	116.8	42.4	79.2	117.9	42.5	79.8	117.9	51.8	84.5
03300	CAST-IN-PLACE CONCRETE	99.4	58.0	82.1	99.4	57.9	82.1	99.9	43.6	76.3	100.0	43.6	76.4	99.9	44.7	76.9	99.9	43.6	76.4
03	CONCRETE	104.0	50.9	77.5	102.9	50.8	76.9	104.1	38.4	71.3	117.3	38.4	77.9	112.1	41.0	76.6	104.7	40.2	72.5
04	MASONRY	109.8	39.3	65.5	107.0	41.4	65.9	108.1	32.1	60.4	108.1	32.1	60.4	108.1	34.7	62.0	174.9	32.1	85.3
05	METALS	97.3	63.0	86.1	98.1	63.0	86.6	94.5	58.0	82.5	94.5	58.0	82.6	94.6	58.3	82.7	95.1	61.6	84.1
06	WOOD & PLASTICS	95.5	41.8	67.5	95.5	41.8	67.5	90.3	29.6	58.6	95.5	29.6	61.1	88.9	35.2	60.9	98.4	29.6	62.5
07	THERMAL & MOISTURE PROTECTION	102.4	54.3	78.8	102.4	54.9	79.1	102.9	41.3	72.6	105.2	41.3	73.8	104.3	44.3	74.8	103.0	41.3	72.7
08	DOORS & WINDOWS	93.5	44.4	80.8	94.3	44.4	81.4	100.3	35.5	83.5	99.9	35.5	83.3	100.1	38.6	84.2	100.6	37.7	84.4
09200	PLASTER & GYPSUM BOARD	80.4	40.0	54.8	80.4	40.0	54.8	78.0	27.5	46.0	80.4	27.5	46.8	77.6	33.2	49.4	86.4	27.5	49.1
095,098	CEILINGS & ACOUSTICAL TREATMENT	96.9	40.0	62.8	96.9	40.0	62.8	96.9	27.5	55.3	96.9	27.5	55.3	96.9	33.2	58.7	96.9	27.5	55.3
09600	FLOORING	101.2	43.1	85.8	101.2	69.2	92.7	98.3	42.2	83.5	101.1	42.2	85.6	97.6	42.2	83.0	103.5	42.2	87.3
097,099	WALL FINISHES, PAINTS & COATINGS	103.7	56.9	75.3	103.6	56.9	75.3	103.7	34.6	61.8	103.7	34.6	61.8	103.7	31.7	60.0	103.7	34.6	61.8
09	FINISHES	95.9	43.9	68.8	95.8	49.2	71.6	94.6	32.3	62.2	97.1	32.3	63.4	95.2	35.8	64.3	97.4	32.3	63.6
10 - 14	TOTAL DIV. 10000 - 14000	100.0	75.8	94.8	100.0	75.7	94.8	100.0	50.6	89.5	100.0	50.6	89.5	100.0	52.1	89.8	100.0	50.6	89.5
15	MECHANICAL	100.1	68.4	86.2	100.1	44.8	75.9	97.8	31.5	68.8	97.8	31.5	68.8	97.8	40.2	72.6	100.1	31.6	70.1
16	ELECTRICAL	88.2	62.8	75.8	88.2	52.6	70.8	89.0	45.3	67.7	89.0	45.3	67.7	89.0	45.3	67.7	87.6	45.3	67.0
01 - 16	WEIGHTED AVERAGE	97.6	60.1	80.5	97.5	54.7	78.0	97.2	44.1	73.1	99.4	44.1	74.3	98.4	47.3	75.2	101.4	44.8	75.7

WYOMING / CANADA

DIVISION		SHERIDAN 828			WHEATLAND 822			WORLAND 824			YELLOWSTONE NAT'L PA 821			BARRIE, ONTARIO			BATHURST, NEW BRUNSWICK		
		MAT.	INST.	TOTAL	MAT.	INST.	TOTAL	MAT.	INST.	TOTAL	MAT.	INST.	TOTAL	MAT.	INST.	TOTAL	MAT.	INST.	TOTAL
01590	EQUIPMENT RENTAL	.0	101.7	101.7	.0	101.7	101.7	.0	101.7	101.7	.0	101.7	101.7	.0	100.4	100.4	.0	98.6	98.6
02	SITE CONSTRUCTION	87.4	100.2	96.9	86.6	100.0	96.5	82.3	99.8	95.3	82.4	100.1	95.5	114.0	99.0	102.9	94.2	94.6	94.5
03100	CONCRETE FORMS & ACCESSORIES	91.5	43.2	49.6	94.8	38.2	45.8	94.9	30.6	39.2	94.9	35.4	43.4	122.7	82.8	88.2	98.0	58.9	64.1
03200	CONCRETE REINFORCEMENT	117.9	52.1	84.7	117.3	42.5	79.5	117.9	51.8	84.5	120.0	51.8	85.6	179.8	89.1	133.9	148.1	60.1	103.6
03300	CAST-IN-PLACE CONCRETE	99.9	57.8	82.3	108.9	45.1	82.3	99.9	43.6	76.4	99.9	56.9	81.9	174.7	81.7	135.8	144.0	57.1	107.7
03	CONCRETE	110.0	50.7	80.4	111.3	42.4	76.9	104.4	40.2	72.4	104.8	46.9	75.9	155.9	83.8	119.9	137.9	59.1	98.6
04	MASONRY	108.1	41.6	66.4	108.7	35.5	62.8	108.1	32.1	60.4	108.1	38.0	64.1	169.1	91.6	120.5	158.1	60.8	97.0
05	METALS	94.6	62.8	84.2	94.4	58.5	82.7	94.6	61.6	83.8	95.0	62.0	84.2	107.8	88.7	101.6	105.3	69.6	93.6
06	WOOD & PLASTICS	89.2	41.8	64.5	92.5	38.3	64.2	92.5	29.6	59.7	92.5	32.8	61.3	120.4	81.4	100.0	93.9	59.0	75.7
07	THERMAL & MOISTURE PROTECTION	103.9	54.9	79.8	103.3	45.7	75.0	103.0	41.3	72.6	102.4	47.5	75.4	110.1	85.1	97.8	104.2	60.0	82.5
08	DOORS & WINDOWS	100.5	45.5	86.3	98.9	40.3	83.7	100.5	37.7	84.2	93.7	39.4	79.7	91.0	81.9	88.7	85.9	52.7	77.3
09200	PLASTER & GYPSUM BOARD	77.6	40.0	53.7	78.4	36.4	51.8	78.4	27.5	46.1	78.8	30.7	48.3	158.7	81.0	109.4	169.7	57.9	98.8
095,098	CEILINGS & ACOUSTICAL TREATMENT	96.9	40.0	62.8	96.9	36.4	60.6	96.9	27.5	55.3	98.7	30.7	57.9	95.7	81.0	86.9	96.6	57.9	73.4
09600	FLOORING	97.7	47.5	84.5	99.7	42.2	84.5	99.7	42.2	84.5	99.7	52.1	87.1	132.7	91.9	121.9	114.0	43.9	95.4
097,099	WALL FINISHES, PAINTS & COATINGS	103.7	56.9	75.3	103.7	34.6	61.8	103.7	34.6	61.8	103.7	34.6	61.8	109.9	84.6	94.5	109.2	48.6	72.4
09	FINISHES	95.0	44.8	68.9	95.4	38.1	65.6	95.1	32.3	62.4	95.6	37.3	65.3	120.1	84.9	101.8	114.8	55.1	83.7
10 - 14	TOTAL DIV. 10000 - 14000	100.0	75.7	94.8	100.0	73.5	94.4	100.0	50.6	89.5	100.0	57.2	90.9	140.0	70.2	125.2	140.0	60.7	123.1
15	MECHANICAL	97.8	43.1	73.9	97.8	37.3	71.4	97.8	31.6	68.9	97.8	39.8	72.5	103.8	81.4	94.0	103.7	57.1	83.3
16	ELECTRICAL	89.0	45.3	67.7	89.0	45.3	67.7	89.0	45.3	67.7	88.2	45.3	67.3	124.6	91.4	108.4	126.0	62.5	95.0
01 - 16	WEIGHTED AVERAGE	98.1	52.8	77.5	98.1	48.0	75.4	97.3	44.8	73.5	96.7	49.1	75.1	120.0	86.7	104.9	115.2	62.8	91.4

CANADA

DIVISION		BRANDON, MANITOBA			BRANTFORD, ONTARIO			BRIDGEWATER, NS			CALGARY, ALBERTA			CAP-DE-LA-MADELEINE, PQ			CHARLESBOURG, QUEBEC		
		MAT.	INST.	TOTAL	MAT.	INST.	TOTAL	MAT.	INST.	TOTAL	MAT.	INST.	TOTAL	MAT.	INST.	TOTAL	MAT.	INST.	TOTAL
01590	EQUIPMENT RENTAL	.0	102.9	102.9	.0	100.4	100.4	.0	98.6	98.6	.0	107.0	107.0	.0	99.2	99.2	.0	99.2	99.2
02	SITE CONSTRUCTION	109.1	98.4	101.2	111.6	99.5	102.6	96.8	95.8	96.0	113.8	105.1	107.3	91.9	96.8	95.5	91.9	96.8	95.5
03100	CONCRETE FORMS & ACCESSORIES	121.8	67.2	74.5	122.6	89.8	94.2	90.8	66.2	69.5	122.1	76.4	82.5	129.9	81.4	87.9	129.9	81.4	87.9
03200	CONCRETE REINFORCEMENT	161.3	56.5	108.3	174.6	87.7	130.7	148.1	49.5	98.3	161.3	66.7	113.5	148.1	83.8	115.6	148.1	83.8	115.6
03300	CAST-IN-PLACE CONCRETE	157.5	71.2	121.4	170.4	101.1	141.5	175.2	66.4	129.8	191.6	87.1	147.9	137.4	89.5	117.4	137.4	89.5	117.4
03	CONCRETE	137.8	67.3	102.6	153.0	93.4	123.2	150.1	63.9	107.1	160.8	78.9	119.9	133.5	84.8	109.2	133.5	84.8	109.2
04	MASONRY	163.7	63.3	100.7	165.9	95.9	121.5	160.3	67.2	101.9	180.3	74.1	113.7	160.5	81.5	110.9	160.5	81.5	110.9
05	METALS	107.9	75.6	97.4	107.1	89.9	101.5	106.3	72.3	95.2	135.7	85.6	119.3	105.6	85.5	99.0	105.6	85.5	99.0
06	WOOD & PLASTICS	116.3	67.9	91.0	118.6	88.4	102.8	85.0	65.6	74.8	116.3	76.0	95.3	131.7	81.2	105.3	131.7	81.2	105.3
07	THERMAL & MOISTURE PROTECTION	104.7	68.9	87.1	109.9	90.1	100.2	104.9	67.1	86.3	125.9	79.9	103.3	104.9	85.7	95.5	104.9	85.7	95.5
08	DOORS & WINDOWS	91.8	61.3	83.9	90.9	87.5	90.0	82.5	61.1	77.0	91.8	73.2	87.0	91.8	76.7	87.9	91.8	76.7	87.9
09200	PLASTER & GYPSUM BOARD	134.3	66.6	91.4	158.7	88.2	114.0	162.6	64.7	100.5	151.0	74.8	102.7	196.1	80.7	122.9	196.1	80.7	122.9
095,098	CEILINGS & ACOUSTICAL TREATMENT	95.7	66.6	78.3	95.7	88.2	91.2	95.7	64.7	77.1	107.3	74.8	87.8	95.7	80.7	86.7	95.7	80.7	86.7
09600	FLOORING	132.6	65.3	114.8	132.6	91.9	121.9	110.0	62.6	97.4	132.6	86.2	120.4	132.6	92.6	122.1	132.6	92.6	122.1
097,099	WALL FINISHES, PAINTS & COATINGS	109.2	55.3	76.5	109.2	92.9	99.3	109.2	60.4	79.6	109.1	76.5	89.3	109.2	86.3	95.3	109.2	86.3	95.3
09	FINISHES	115.6	65.8	89.7	118.9	90.8	104.3	112.2	65.1	87.7	121.0	78.0	98.2	123.7	84.3	103.2	123.7	84.3	103.2
10 - 14	TOTAL DIV. 10000 - 14000	140.0	63.6	123.7	140.0	72.5	125.6	140.0	62.3	123.5	140.0	85.6	128.4	140.0	79.2	127.1	140.0	79.2	127.1
15	MECHANICAL	103.7	68.4	88.3	103.7	84.1	95.2	103.7	68.9	88.5	101.5	77.6	91.1	104.1	74.6	91.2	104.1	74.6	91.2
16	ELECTRICAL	125.3	70.3	98.5	124.6	90.9	108.2	128.0	65.0	97.3	117.3	86.3	102.2	120.3	73.3	97.4	120.3	73.3	97.4
01 - 16	WEIGHTED AVERAGE	117.1	70.7	96.0	119.2	90.3	106.1	116.5	69.2	95.0	124.6	82.0	105.2	116.0	81.4	100.3	116.0	81.4	100.3

COST INDEXES

CANADA

DIVISION		CHARLOTTETOWN, PEI			CHICOUTIMI, QUEBEC			CORNER BROOK, NFLD			CORNWALL, ONTARIO			DALHOUSIE, NB			DARTMOUTH, NOVA SCOTIA		
		MAT.	INST.	TOTAL	MAT.	INST.	TOTAL	MAT.	INST.	TOTAL	MAT.	INST.	TOTAL	MAT.	INST.	TOTAL	MAT.	INST.	TOTAL
01590	EQUIPMENT RENTAL	.0	98.6	98.6	.0	99.2	99.2	.0	98.6	98.6	.0	100.4	100.4	.0	98.6	98.6	.0	98.6	98.6
02	SITE CONSTRUCTION	104.9	94.0	96.8	91.3	96.8	95.3	112.5	94.5	99.2	109.5	99.0	101.7	94.2	94.6	94.5	105.1	95.8	98.2
03100	CONCRETE FORMS & ACCESSORIES	91.3	53.8	58.8	129.9	81.2	87.8	100.5	56.9	62.8	119.9	82.9	87.9	98.0	58.9	64.1	90.9	66.2	69.5
03200	CONCRETE REINFORCEMENT	148.1	49.6	98.3	111.8	83.7	97.6	148.1	51.5	99.3	174.6	87.4	130.6	148.1	60.1	103.6	148.1	49.5	98.3
03300	CAST-IN-PLACE CONCRETE	166.1	56.4	120.3	135.6	89.5	116.3	179.6	64.7	131.6	153.3	92.3	127.8	144.0	57.1	107.7	177.1	66.4	130.9
03	CONCRETE	154.4	54.7	104.6	126.7	84.7	105.7	164.6	59.5	112.1	144.6	87.2	116.0	137.9	59.1	98.6	151.0	63.9	107.6
04	MASONRY	159.8	58.7	96.4	159.5	81.5	110.5	160.7	60.0	97.5	164.6	88.0	116.6	158.0	60.8	97.0	172.2	67.2	106.4
05	METALS	106.4	65.7	93.1	105.2	85.4	98.7	108.5	70.3	96.0	107.1	88.4	101.0	105.3	69.6	93.6	108.5	72.3	96.6
06	WOOD & PLASTICS	85.3	53.3	68.6	131.7	81.2	105.3	95.0	56.0	74.6	117.2	82.2	99.0	93.9	59.0	75.7	85.0	65.6	74.8
07	THERMAL & MOISTURE PROTECTION	105.5	57.7	82.0	104.9	85.7	95.5	107.8	59.4	84.0	109.6	84.1	97.1	107.0	60.0	83.9	104.9	67.1	86.3
08	DOORS & WINDOWS	94.1	47.1	82.0	91.4	70.7	86.1	98.1	53.8	86.6	91.8	81.6	89.2	85.9	52.7	77.3	82.5	61.1	77.0
09200	PLASTER & GYPSUM BOARD	166.8	52.0	94.1	196.1	80.7	122.9	170.0	54.8	97.0	228.8	81.8	135.6	169.7	57.9	98.8	166.0	64.7	101.7
095,098	CEILINGS & ACOUSTICAL TREATMENT	109.9	52.0	75.2	95.7	80.7	86.7	97.4	54.8	71.9	98.3	81.8	88.4	96.6	57.9	73.4	108.1	64.7	82.1
09600	FLOORING	110.0	58.5	96.4	132.6	92.6	122.1	114.9	53.3	98.6	132.6	90.6	121.5	114.0	43.9	95.4	110.0	62.6	97.4
097,099	WALL FINISHES, PAINTS & COATINGS	109.2	40.5	67.5	109.2	86.3	95.3	109.2	58.2	78.2	109.2	86.5	95.4	109.2	48.6	72.4	109.2	60.4	79.6
09	FINISHES	117.0	53.5	84.0	123.7	84.3	103.2	116.1	56.3	85.0	129.7	85.0	106.4	114.8	55.1	83.7	115.7	65.1	89.4
10 - 14	TOTAL DIV. 10000 - 14000	140.0	60.1	123.0	140.0	79.2	127.1	140.0	60.8	123.2	140.0	69.8	125.1	140.0	60.7	123.1	140.0	62.3	123.5
15	MECHANICAL	103.7	52.3	81.2	103.7	74.6	91.0	103.7	58.7	84.0	104.1	81.9	94.4	103.7	57.1	83.3	103.7	68.9	88.5
16	ELECTRICAL	119.8	52.9	87.1	120.2	73.3	97.4	120.2	58.9	90.3	122.8	91.9	107.7	125.3	62.5	94.6	124.9	65.0	95.7
01 - 16	WEIGHTED AVERAGE	117.8	58.7	91.0	115.0	81.1	99.6	120.1	62.8	94.1	119.1	86.9	104.5	115.2	62.8	91.4	117.7	69.2	95.7

CANADA

DIVISION		EDMONTON, ALBERTA			FORT MCMURRAY, ALBERTA			FREDERICTON, NB			GATINEAU, QUEBEC			GRANBY, QUEBEC			HALIFAX, NOVA SCOTIA		
		MAT.	INST.	TOTAL	MAT.	INST.	TOTAL	MAT.	INST.	TOTAL	MAT.	INST.	TOTAL	MAT.	INST.	TOTAL	MAT.	INST.	TOTAL
01590	EQUIPMENT RENTAL	.0	107.0	107.0	.0	103.7	103.7	.0	98.6	98.6	.0	99.2	99.2	.0	99.2	99.2	.0	98.6	98.6
02	SITE CONSTRUCTION	117.5	105.1	108.3	108.8	101.5	103.4	93.7	94.6	94.3	91.7	96.8	95.5	92.2	96.8	95.6	97.2	95.8	96.1
03100	CONCRETE FORMS & ACCESSORIES	119.8	76.4	82.2	119.3	76.3	82.0	120.4	59.1	67.3	129.9	81.3	87.8	129.9	81.3	87.8	91.0	66.3	69.6
03200	CONCRETE REINFORCEMENT	161.3	66.7	113.5	161.3	66.7	113.5	148.1	60.2	103.7	156.8	83.7	119.9	156.8	83.7	119.9	148.1	49.7	98.3
03300	CAST-IN-PLACE CONCRETE	206.9	87.1	156.9	157.5	86.1	127.7	137.1	57.1	103.7	135.6	89.5	116.4	140.1	89.5	119.0	178.4	68.3	132.4
03	CONCRETE	168.0	78.9	123.5	144.4	78.5	111.5	136.3	59.3	97.8	134.0	84.8	109.4	136.2	84.8	110.5	151.7	64.6	108.2
04	MASONRY	181.6	74.1	114.1	160.7	74.0	106.3	157.5	62.3	97.8	160.4	81.5	110.9	160.7	81.5	111.0	172.3	69.6	107.9
05	METALS	136.5	85.6	119.9	107.9	85.4	100.6	105.2	70.0	93.7	105.6	85.4	99.0	105.6	85.4	99.0	120.8	72.5	105.0
06	WOOD & PLASTICS	112.9	76.0	93.6	112.9	75.9	93.6	119.4	59.0	87.9	131.7	81.2	105.3	131.7	81.2	105.3	85.0	65.6	74.8
07	THERMAL & MOISTURE PROTECTION	126.2	79.9	103.4	122.6	79.0	101.2	107.5	61.0	84.6	104.9	85.7	95.5	104.9	85.7	95.5	105.4	67.7	86.9
08	DOORS & WINDOWS	91.8	73.2	87.0	91.8	73.1	87.0	85.8	51.7	77.0	91.8	72.3	86.8	91.8	72.3	86.8	82.5	61.1	77.0
09200	PLASTER & GYPSUM BOARD	153.8	74.8	103.7	140.6	74.8	98.9	189.2	57.9	106.0	151.0	80.7	106.4	153.7	80.7	107.4	166.0	64.7	101.7
095,098	CEILINGS & ACOUSTICAL TREATMENT	125.1	74.8	94.9	95.7	74.8	83.1	104.6	57.9	76.6	95.7	80.7	86.7	95.7	80.7	86.7	108.1	64.7	82.1
09600	FLOORING	132.6	86.2	120.4	132.6	86.2	120.4	127.8	43.9	105.6	132.6	92.6	122.1	132.6	92.6	122.1	110.0	62.6	97.4
097,099	WALL FINISHES, PAINTS & COATINGS	109.2	76.5	89.4	109.2	76.5	89.4	109.2	61.9	80.5	109.2	86.3	95.3	109.2	86.3	95.3	109.2	60.4	79.6
09	FINISHES	125.9	78.0	101.0	116.8	77.9	96.6	123.8	56.5	88.8	117.2	84.3	100.1	117.6	84.3	100.3	115.7	65.1	89.4
10 - 14	TOTAL DIV. 10000 - 14000	140.0	85.6	128.4	140.0	85.3	128.4	140.0	60.7	123.1	140.0	79.2	127.1	140.0	79.2	127.1	140.0	62.3	123.5
15	MECHANICAL	101.7	77.6	91.2	103.7	77.6	92.3	103.7	64.5	86.6	104.1	74.6	91.2	103.7	74.6	91.0	101.3	69.0	87.1
16	ELECTRICAL	118.2	86.3	102.6	114.2	86.3	100.5	133.4	77.8	106.3	120.3	73.3	97.4	121.5	73.3	98.0	127.3	70.0	99.3
01 - 16	WEIGHTED AVERAGE	126.2	82.0	106.2	116.8	81.6	100.8	117.1	66.9	94.3	115.4	81.1	99.9	115.8	81.1	100.1	119.3	70.3	97.0

CANADA

DIVISION		HAMILTON, ONTARIO			HULL, QUEBEC			JOLIETTE, QUEBEC			KAMLOOPS, BC			KINGSTON, ONTARIO			KITCHENER, ONTARIO		
		MAT.	INST.	TOTAL	MAT.	INST.	TOTAL	MAT.	INST.	TOTAL	MAT.	INST.	TOTAL	MAT.	INST.	TOTAL	MAT.	INST.	TOTAL
01590	EQUIPMENT RENTAL	.0	103.1	103.1	.0	99.2	99.2	.0	99.2	99.2	.0	103.5	103.5	.0	103.1	103.1	.0	103.0	103.0
02	SITE CONSTRUCTION	110.0	103.5	105.2	91.7	96.8	95.5	92.4	96.8	95.6	114.7	100.7	104.4	109.5	103.3	104.9	99.2	102.9	101.9
03100	CONCRETE FORMS & ACCESSORIES	121.5	86.8	91.4	129.9	81.3	87.8	129.9	81.4	87.9	117.5	84.1	88.6	120.0	83.1	88.0	115.8	79.5	84.4
03200	CONCRETE REINFORCEMENT	174.6	89.1	131.4	156.8	83.7	119.9	148.1	83.8	115.6	116.4	81.6	98.8	174.6	87.4	130.6	110.3	89.0	99.5
03300	CAST-IN-PLACE CONCRETE	157.2	94.9	131.2	135.6	89.5	116.4	141.2	89.5	119.6	122.2	93.7	110.3	153.3	92.3	127.8	146.7	76.5	117.3
03	CONCRETE	146.6	90.1	118.4	134.0	84.8	109.4	135.3	84.8	110.1	140.5	87.1	113.8	144.6	87.3	116.0	124.4	80.5	102.5
04	MASONRY	167.2	91.2	119.5	160.4	81.5	110.9	160.8	81.5	111.0	167.6	88.7	118.1	171.3	88.1	119.1	162.8	87.1	115.3
05	METALS	139.4	89.2	123.0	105.6	85.4	99.0	105.6	85.5	99.0	107.8	85.7	100.6	108.7	88.5	102.1	116.4	88.7	107.3
06	WOOD & PLASTICS	117.7	85.6	100.9	131.7	81.2	105.3	131.7	81.2	105.3	98.4	82.6	90.1	117.2	82.3	99.0	111.2	77.9	93.8
07	THERMAL & MOISTURE PROTECTION	110.7	88.3	100.7	104.9	85.7	95.5	104.9	85.7	95.5	120.7	82.8	102.1	109.6	85.2	97.6	108.8	83.9	96.5
08	DOORS & WINDOWS	91.8	85.7	90.2	91.8	72.3	86.8	91.8	76.7	87.9	87.7	85.9	85.9	91.8	81.3	89.1	83.2	79.9	82.4
09200	PLASTER & GYPSUM BOARD	190.9	85.2	123.9	151.0	80.7	106.4	196.1	80.7	122.9	139.1	81.7	102.7	233.1	81.9	137.3	157.4	77.3	106.6
095,098	CEILINGS & ACOUSTICAL TREATMENT	112.6	85.2	96.2	95.7	80.7	86.7	95.7	80.7	86.7	95.7	81.7	87.3	114.4	81.9	94.9	107.3	77.3	89.3
09600	FLOORING	132.6	91.9	121.9	132.6	92.6	122.1	132.6	92.6	122.1	130.0	51.7	109.3	132.6	90.6	121.5	128.3	91.9	118.7
097,099	WALL FINISHES, PAINTS & COATINGS	109.2	99.2	103.2	109.2	86.3	95.3	109.2	86.3	95.3	109.2	79.0	90.9	109.2	94.4	100.2	109.2	90.9	98.1
09	FINISHES	127.7	89.2	107.7	117.2	84.3	100.1	123.7	84.3	103.2	116.9	77.9	96.6	134.3	85.9	109.1	119.7	83.0	100.6
10 - 14	TOTAL DIV. 10000 - 14000	140.0	91.5	129.1	140.0	79.2	127.1	140.0	79.2	127.1	140.0	89.0	129.1	140.0	69.8	125.1	140.0	89.3	129.2
15	MECHANICAL	101.7	85.5	94.7	103.7	74.6	91.0	103.7	74.6	91.0	103.7	77.7	92.4	104.1	82.0	94.4	101.3	81.9	92.8
16	ELECTRICAL	130.4	93.9	112.6	124.0	73.3	99.3	121.5	73.3	98.0	126.6	82.8	105.2	122.8	90.5	107.1	125.1	90.7	108.3
01 - 16	WEIGHTED AVERAGE	124.7	90.6	109.2	115.8	81.1	100.1	116.3	81.4	100.4	117.8	84.3	102.6	120.1	87.3	105.2	115.6	86.3	102.3

COST INDEXES

CANADA

DIVISION		LAVAL, QUEBEC			LETHBRIDGE, ALBERTA			LLOYDMINSTER, ALBERTA			LONDON, ONTARIO			MEDICINE HAT, ALBERTA			MONCTON, NEW BRUNSWICK		
		MAT.	INST.	TOTAL	MAT.	INST.	TOTAL	MAT.	INST.	TOTAL	MAT.	INST.	TOTAL	MAT.	INST.	TOTAL	MAT.	INST.	TOTAL
01590	EQUIPMENT RENTAL	.0	99.2	99.2	.0	103.7	103.7	.0	103.7	103.7	.0	103.2	103.2	.0	103.7	103.7	.0	98.6	98.6
02	SITE CONSTRUCTION	92.2	96.8	95.6	109.3	101.5	103.5	108.8	101.5	103.4	110.0	103.2	105.0	107.6	101.5	103.1	93.6	94.6	94.3
03100	CONCRETE FORMS & ACCESSORIES	130.1	81.3	87.8	121.7	76.2	82.3	119.3	76.3	82.0	127.2	82.4	88.4	121.6	76.2	82.3	98.0	58.9	64.1
03200	CONCRETE REINFORCEMENT	156.8	83.7	119.9	161.3	66.7	113.5	161.3	66.7	113.5	123.3	87.5	105.2	161.3	66.7	113.5	148.1	60.1	103.6
03300	CAST-IN-PLACE CONCRETE	140.1	89.5	119.0	171.6	86.1	135.9	157.5	86.1	127.7	157.2	92.5	130.2	157.5	86.1	127.6	138.7	57.1	104.6
03	CONCRETE	136.2	84.8	110.5	151.3	78.5	114.9	144.4	78.5	111.5	138.6	87.0	112.9	144.6	78.5	111.6	135.4	59.1	97.3
04	MASONRY	160.7	81.5	111.0	161.7	74.0	106.7	160.7	74.0	106.3	167.3	88.4	117.8	160.7	74.0	106.3	157.7	60.8	96.9
05	METALS	105.6	85.4	99.0	107.9	85.4	100.5	107.9	85.4	100.6	124.5	88.8	112.8	107.9	85.4	100.5	105.3	69.6	93.6
06	WOOD & PLASTICS	131.7	81.2	105.3	116.3	75.9	95.2	112.9	75.9	93.6	117.7	80.6	98.3	116.3	75.9	95.2	93.9	59.0	75.7
07	THERMAL & MOISTURE PROTECTION	105.4	85.7	95.7	122.7	79.0	101.2	121.2	79.0	100.5	114.1	86.5	100.5	125.6	79.0	102.7	107.0	60.0	83.9
08	DOORS & WINDOWS	91.8	72.3	86.8	91.8	73.1	87.0	91.8	73.1	87.0	92.7	82.0	90.0	91.8	73.1	87.0	85.9	52.7	77.3
09200	PLASTER & GYPSUM BOARD	156.3	80.7	108.4	142.6	74.8	99.6	140.6	74.8	98.9	191.4	80.2	120.9	142.6	74.8	99.6	169.7	57.9	98.8
095,098	CEILINGS & ACOUSTICAL TREATMENT	95.7	80.7	86.7	95.7	74.8	83.1	95.7	74.8	83.1	114.4	80.2	93.9	95.7	74.8	83.1	96.6	57.9	73.4
09600	FLOORING	132.6	92.6	122.1	132.6	86.2	120.4	132.6	86.2	120.4	134.7	91.9	123.4	132.6	86.2	120.4	114.0	43.9	95.4
097,099	WALL FINISHES, PAINTS & COATINGS	109.2	86.3	95.3	109.1	76.5	89.3	109.2	76.5	89.4	109.2	99.2	103.2	109.1	76.5	89.3	109.2	48.6	72.4
09	FINISHES	118.0	84.3	100.4	116.9	77.9	96.7	116.8	77.9	96.6	128.9	85.9	106.5	116.9	77.9	96.7	114.8	55.1	83.7
10 - 14	TOTAL DIV. 10000 - 14000	140.0	79.2	127.1	140.0	85.3	128.4	140.0	85.3	128.4	140.0	90.2	129.4	140.0	85.3	128.4	140.0	60.7	123.1
15	MECHANICAL	101.3	74.6	89.6	103.7	75.0	91.2	104.1	77.6	92.5	101.8	82.1	93.2	103.7	75.0	91.2	103.7	57.1	83.3
16	ELECTRICAL	121.9	73.3	98.2	114.2	86.3	100.6	114.2	86.3	100.6	130.7	90.9	111.3	114.2	86.3	100.6	130.4	62.5	97.3
01 - 16	WEIGHTED AVERAGE	115.4	81.1	99.8	117.7	81.0	101.1	116.9	81.6	100.8	121.8	88.0	106.5	117.0	81.0	100.6	115.5	62.8	91.6

CANADA

DIVISION		MONTREAL, QUEBEC			MOOSE JAW, SASKATCHEWAN			NEWCASTLE, NB			NEW GLASGOW, NOVA SCOTIA			NORTH BAY, ONTARIO			OSHAWA, ONTARIO		
		MAT.	INST.	TOTAL	MAT.	INST.	TOTAL	MAT.	INST.	TOTAL	MAT.	INST.	TOTAL	MAT.	INST.	TOTAL	MAT.	INST.	TOTAL
01590	EQUIPMENT RENTAL	.0	101.0	101.0	.0	98.6	98.6	.0	98.6	98.6	.0	98.6	98.6	.0	100.4	100.4	.0	103.0	103.0
02	SITE CONSTRUCTION	91.9	97.1	95.7	108.8	94.7	98.4	94.2	94.6	94.5	97.1	95.8	96.1	111.6	98.7	102.1	111.5	103.1	105.3
03100	CONCRETE FORMS & ACCESSORIES	130.4	81.6	88.1	102.3	56.0	62.2	98.0	58.9	64.1	90.8	66.2	69.5	122.6	80.7	86.3	122.7	83.2	88.5
03200	CONCRETE REINFORCEMENT	159.4	83.8	121.2	113.8	65.8	89.6	148.1	60.1	103.6	148.1	49.5	98.3	174.6	87.0	130.3	174.6	89.7	131.7
03300	CAST-IN-PLACE CONCRETE	137.5	91.1	118.1	152.8	65.7	116.4	144.0	57.1	107.7	177.1	66.4	130.9	170.4	80.2	132.7	169.7	81.8	133.0
03	CONCRETE	135.4	85.5	110.5	126.8	61.9	94.4	137.9	59.1	98.6	151.0	63.9	107.6	153.0	82.0	117.5	152.6	84.1	118.4
04	MASONRY	163.9	81.5	112.2	159.2	60.3	97.1	158.1	60.8	97.0	160.5	67.2	102.0	165.9	84.3	114.7	165.8	90.5	118.5
05	METALS	125.1	85.9	112.3	104.7	72.8	94.3	105.3	69.6	93.6	106.3	72.3	95.2	107.1	88.2	101.0	107.2	89.0	101.2
06	WOOD & PLASTICS	131.7	81.5	105.5	95.6	54.5	74.2	93.9	59.0	75.7	85.0	65.6	74.8	118.6	80.9	98.9	118.6	81.4	99.2
07	THERMAL & MOISTURE PROTECTION	106.2	86.4	96.5	104.0	61.7	83.2	107.0	60.0	83.9	104.9	67.1	86.3	109.9	81.9	96.1	110.1	84.4	97.5
08	DOORS & WINDOWS	91.8	72.4	86.8	86.7	53.7	78.2	85.9	52.7	77.3	82.5	61.1	77.0	90.9	79.3	87.9	90.9	83.1	88.9
09200	PLASTER & GYPSUM BOARD	159.4	80.7	109.5	135.1	53.3	83.3	169.7	57.9	98.8	162.6	64.7	100.5	158.7	80.4	109.1	159.9	81.0	109.9
095,098	CEILINGS & ACOUSTICAL TREATMENT	107.3	80.7	91.3	95.7	53.3	70.3	96.6	57.9	73.4	95.7	64.7	77.1	95.7	80.4	86.5	100.1	81.0	88.6
09600	FLOORING	132.6	92.6	122.1	120.0	58.9	103.9	114.0	43.9	95.4	110.0	62.6	97.4	132.6	90.6	121.5	132.6	94.3	122.5
097,099	WALL FINISHES, PAINTS & COATINGS	109.2	86.3	95.3	109.2	62.9	81.1	109.2	48.6	72.4	109.2	60.4	79.6	109.2	85.9	95.0	109.2	104.3	106.2
09	FINISHES	121.3	84.5	102.1	112.0	57.0	83.4	114.8	55.1	83.7	112.2	65.1	87.7	118.9	83.4	100.5	120.2	87.6	103.2
10 - 14	TOTAL DIV. 10000 - 14000	140.0	80.0	127.2	140.0	60.0	123.0	140.0	60.7	123.1	140.0	62.3	123.5	140.0	68.4	124.8	140.0	90.7	129.5
15	MECHANICAL	101.8	74.7	89.9	104.1	60.0	85.7	103.7	56.7	83.3	103.7	68.9	88.5	103.7	80.1	93.4	101.3	83.5	93.5
16	ELECTRICAL	120.9	73.3	97.7	125.9	63.4	95.4	125.3	62.5	94.6	120.9	65.0	93.6	124.6	91.7	108.6	126.2	91.4	109.2
01 - 16	WEIGHTED AVERAGE	118.8	81.4	101.8	114.2	64.8	91.8	115.2	62.8	91.4	115.8	69.2	94.6	119.2	85.0	103.6	118.9	84.4	105.1

CANADA

DIVISION		OTTAWA, ONTARIO			OWEN SOUND, ONTARIO			PETERBOROUGH, ONTARIO			PORTAGE LA PRAIRIE, MB			PRINCE ALBERT, SS			PRINCE GEORGE, BC		
		MAT.	INST.	TOTAL	MAT.	INST.	TOTAL	MAT.	INST.	TOTAL	MAT.	INST.	TOTAL	MAT.	INST.	TOTAL	MAT.	INST.	TOTAL
01590	EQUIPMENT RENTAL	.0	103.0	103.0	.0	100.4	100.4	.0	100.4	100.4	.0	102.9	102.9	.0	98.6	98.6	.0	103.5	103.5
02	SITE CONSTRUCTION	110.3	103.1	105.0	114.0	98.8	102.8	111.6	99.0	102.3	109.6	98.4	101.3	104.9	94.8	97.5	118.6	100.7	105.4
03100	CONCRETE FORMS & ACCESSORIES	120.2	83.2	88.2	122.7	79.2	85.0	122.6	81.6	87.1	121.8	67.2	74.5	102.3	55.9	62.1	108.2	84.1	87.3
03200	CONCRETE REINFORCEMENT	174.6	87.5	130.6	179.8	89.1	133.9	174.6	87.5	130.6	161.3	56.5	108.3	118.9	65.8	92.1	116.4	81.6	98.8
03300	CAST-IN-PLACE CONCRETE	159.6	92.4	131.5	174.7	76.3	133.6	170.4	81.7	133.4	157.5	71.2	121.4	145.1	65.6	111.9	153.3	93.7	128.4
03	CONCRETE	147.6	87.4	117.6	155.9	80.3	118.2	153.0	83.0	118.0	137.8	67.3	102.6	124.0	61.8	92.9	154.7	87.1	120.9
04	MASONRY	167.4	88.0	117.6	169.1	89.5	119.1	165.9	90.5	118.6	163.7	63.3	100.7	158.7	60.3	97.0	169.9	88.7	119.0
05	METALS	116.5	88.8	107.4	107.8	88.5	101.5	107.1	88.7	101.1	107.9	75.6	97.4	104.7	72.7	94.3	107.8	85.7	100.6
06	WOOD & PLASTICS	117.2	82.2	99.0	120.4	77.9	98.2	118.6	79.7	98.3	116.3	67.9	91.0	95.6	54.5	74.2	98.4	82.6	90.1
07	THERMAL & MOISTURE PROTECTION	110.9	85.1	98.2	110.1	82.7	96.6	109.9	86.4	98.3	104.7	68.9	87.1	103.9	60.7	82.7	114.5	82.8	99.0
08	DOORS & WINDOWS	91.8	82.8	89.5	91.0	78.6	87.8	90.9	80.9	88.3	91.8	61.3	83.9	85.3	53.7	77.1	87.7	80.9	85.9
09200	PLASTER & GYPSUM BOARD	231.0	81.8	136.4	158.7	77.3	107.1	158.7	79.2	108.3	134.3	66.6	91.4	135.1	53.3	83.3	136.5	81.7	101.8
095,098	CEILINGS & ACOUSTICAL TREATMENT	106.4	81.8	91.6	95.7	77.3	84.7	95.7	79.2	85.8	95.7	66.6	78.3	95.7	53.3	70.3	95.7	81.7	87.3
09600	FLOORING	132.6	90.6	121.5	132.7	91.9	121.9	132.6	90.6	121.5	132.6	65.3	114.8	120.0	58.9	103.9	127.0	51.7	107.1
097,099	WALL FINISHES, PAINTS & COATINGS	109.2	92.9	99.3	109.9	84.6	94.5	109.2	88.0	96.3	109.2	55.3	76.5	109.2	53.6	75.5	109.2	79.0	90.9
09	FINISHES	132.0	85.7	107.9	120.1	82.3	100.4	118.9	84.0	100.8	115.7	65.8	89.8	112.0	55.9	82.9	115.7	77.9	96.0
10 - 14	TOTAL DIV. 10000 - 14000	140.0	88.2	129.0	140.0	68.9	124.9	140.0	70.0	125.1	140.0	63.6	123.7	140.0	60.1	123.0	140.0	89.0	129.1
15	MECHANICAL	101.8	82.4	93.3	103.8	80.2	93.5	103.7	83.3	94.8	103.7	68.4	88.3	104.1	62.0	85.7	103.7	77.7	92.4
16	ELECTRICAL	121.9	91.9	107.3	126.7	90.5	109.1	124.6	91.4	108.4	125.3	70.3	98.5	125.9	63.4	95.4	124.2	82.8	104.0
01 - 16	WEIGHTED AVERAGE	120.7	88.1	105.9	120.3	85.0	104.2	119.2	86.7	104.4	117.1	70.7	96.0	113.6	64.7	91.4	119.1	84.3	103.3

COST INDEXES

City Cost Indexes

CANADA

DIVISION		QUEBEC, QUEBEC			RED DEER, ALBERTA			REGINA, SASKATCHEWAN			RIMOUSKI, QUEBEC			ROUYN-NORANDA, QUEBEC			SAINT HYACINTHE, QUEBEC		
		MAT.	INST.	TOTAL	MAT.	INST.	TOTAL	MAT.	INST.	TOTAL	MAT.	INST.	TOTAL	MAT.	INST.	TOTAL	MAT.	INST.	TOTAL
01590	EQUIPMENT RENTAL	.0	101.4	101.4	.0	103.7	103.7	.0	98.6	98.6	.0	99.2	99.2	.0	99.2	99.2	.0	99.2	99.2
02	SITE CONSTRUCTION	93.7	97.2	96.2	107.6	101.5	103.1	110.0	94.7	98.7	92.1	96.8	95.6	91.7	96.8	95.5	92.2	96.8	95.6
03100	CONCRETE FORMS & ACCESSORIES	130.2	81.8	88.3	135.2	76.2	84.1	102.4	56.0	62.2	129.9	81.2	87.8	129.9	81.3	87.8	129.9	81.3	87.8
03200	CONCRETE REINFORCEMENT	148.1	83.8	115.6	161.3	66.7	113.5	113.8	65.8	89.6	111.8	83.7	97.6	156.8	83.7	119.9	156.8	83.7	119.9
03300	CAST-IN-PLACE CONCRETE	151.5	91.6	126.5	157.5	86.1	127.6	162.8	65.7	122.2	142.6	89.5	120.4	135.6	89.5	116.4	140.1	89.5	119.0
03	CONCRETE	140.2	85.7	113.0	145.5	78.5	112.1	131.6	61.9	96.8	130.0	84.7	107.4	134.0	84.8	109.4	136.2	84.8	110.5
04	MASONRY	164.9	81.5	112.6	160.7	74.0	106.3	166.9	60.7	100.3	160.0	81.5	110.7	160.4	81.5	110.9	160.7	81.5	111.0
05	METALS	120.7	86.2	109.4	107.9	85.4	100.5	105.7	72.8	94.9	105.2	85.4	98.7	105.6	85.4	99.0	105.6	85.4	99.0
06	WOOD & PLASTICS	132.3	81.6	105.8	116.3	75.9	95.2	95.6	54.5	74.2	131.7	81.2	105.3	131.7	81.2	105.3	131.7	81.2	105.3
07	THERMAL & MOISTURE PROTECTION	105.7	86.6	96.3	135.1	79.0	107.5	106.5	61.8	84.5	104.9	85.7	95.5	104.9	85.7	95.5	105.4	85.7	95.7
08	DOORS & WINDOWS	91.8	79.6	88.7	91.8	73.1	87.0	86.7	53.7	78.2	91.4	70.7	86.1	91.8	72.3	86.8	91.8	72.3	86.8
09200	PLASTER & GYPSUM BOARD	198.1	80.7	123.7	142.6	74.8	99.6	166.3	53.3	94.7	195.3	80.7	122.6	150.3	80.7	106.2	155.6	80.7	108.1
095,098	CEILINGS & ACOUSTICAL TREATMENT	97.4	80.7	87.4	95.7	74.8	83.1	123.3	53.3	81.3	93.0	80.7	85.6	93.0	80.7	85.6	93.0	80.7	85.6
09600	FLOORING	132.6	92.6	122.1	134.7	86.2	121.9	120.0	58.9	103.9	132.6	92.6	122.1	132.6	92.6	122.1	132.6	92.6	122.1
097,099	WALL FINISHES, PAINTS & COATINGS	109.2	86.3	95.3	109.1	76.5	89.3	109.2	62.9	81.1	109.2	86.3	95.3	109.2	86.3	95.3	109.2	86.3	95.3
09	FINISHES	124.5	84.5	103.7	117.6	77.9	97.0	123.6	57.0	88.9	122.9	84.3	102.8	116.4	84.3	99.7	117.2	84.3	100.1
10 - 14	TOTAL DIV. 10000 - 14000	140.0	80.2	127.3	140.0	85.3	128.4	140.0	60.0	123.0	140.0	79.2	127.1	140.0	79.2	127.1	140.0	79.2	127.1
15	MECHANICAL	101.7	74.7	89.9	103.7	75.0	91.2	101.5	62.0	84.2	103.7	74.6	91.0	103.7	74.6	91.0	101.3	74.6	89.6
16	ELECTRICAL	125.7	73.3	100.2	114.2	86.3	100.6	126.3	63.4	95.6	121.5	73.3	98.0	121.5	73.3	98.0	121.9	73.3	98.2
01 - 16	WEIGHTED AVERAGE	119.6	81.8	102.5	117.4	81.0	100.9	116.0	64.9	92.8	115.5	81.1	99.9	115.4	81.1	99.9	115.3	81.1	99.8

CANADA

DIVISION		SAINT JOHN, N B			SARNIA, ONTARIO			SASKATOON, SASKATCHEWAN			SAULT STE MARIE, ON			SHERBROOKE, QUEBEC			SOREL, QUEBEC		
		MAT.	INST.	TOTAL	MAT.	INST.	TOTAL	MAT.	INST.	TOTAL	MAT.	INST.	TOTAL	MAT.	INST.	TOTAL	MAT.	INST.	TOTAL
01590	EQUIPMENT RENTAL	.0	98.6	98.6	.0	100.4	100.4	.0	98.6	98.6	.0	100.4	100.4	.0	99.2	99.2	.0	99.2	99.2
02	SITE CONSTRUCTION	94.3	94.6	94.5	110.0	99.1	101.9	105.2	94.8	97.5	99.7	98.7	99.0	92.2	96.8	95.6	92.4	96.8	95.6
03100	CONCRETE FORMS & ACCESSORIES	120.4	59.1	67.3	121.1	88.3	92.7	102.4	55.9	62.1	108.7	80.9	84.7	129.9	81.3	87.8	129.9	81.4	87.9
03200	CONCRETE REINFORCEMENT	148.1	60.2	103.7	123.3	88.9	105.9	118.9	65.8	92.1	112.2	87.0	99.4	156.8	83.7	119.9	148.1	83.8	115.6
03300	CAST-IN-PLACE CONCRETE	141.9	57.1	106.5	157.2	93.7	130.7	148.0	65.6	113.6	141.2	80.3	115.8	140.1	89.5	119.0	141.2	89.5	119.6
03	CONCRETE	138.6	59.3	99.0	138.2	90.3	114.3	125.3	61.8	93.6	121.6	82.1	101.9	136.2	84.8	110.5	135.3	84.8	110.1
04	MASONRY	176.1	62.3	104.7	176.7	93.0	124.2	167.1	60.7	100.3	162.3	84.3	113.4	160.7	81.5	111.0	160.8	81.5	111.0
05	METALS	105.2	70.0	93.7	107.1	89.0	101.2	105.7	72.7	94.9	106.6	88.5	100.7	105.6	85.4	99.0	105.6	85.5	99.0
06	WOOD & PLASTICS	119.4	59.0	87.9	117.7	87.5	101.9	94.2	54.5	73.5	103.7	80.9	91.8	131.7	81.2	105.3	131.7	81.2	105.3
07	THERMAL & MOISTURE PROTECTION	107.5	61.0	84.6	110.0	90.3	100.3	104.9	60.8	83.2	108.3	83.4	96.1	104.9	85.7	95.5	104.9	85.7	95.5
08	DOORS & WINDOWS	85.8	51.7	77.0	92.7	84.5	90.6	85.7	53.7	77.5	83.7	80.5	82.9	91.8	72.3	86.8	91.8	76.7	87.9
09200	PLASTER & GYPSUM BOARD	189.2	57.9	106.0	187.6	87.2	123.9	147.7	53.3	87.9	144.6	80.4	103.9	153.0	80.7	107.1	195.3	80.7	122.6
095,098	CEILINGS & ACOUSTICAL TREATMENT	104.6	57.9	76.6	100.1	87.2	92.4	123.3	53.3	81.3	95.7	80.4	86.5	93.0	80.7	85.6	93.0	80.7	85.6
09600	FLOORING	127.8	43.9	105.6	132.6	95.8	122.9	120.0	58.9	103.9	123.6	90.6	114.9	132.6	92.6	122.1	132.6	92.6	122.1
097,099	WALL FINISHES, PAINTS & COATINGS	109.2	61.9	80.5	109.2	99.5	103.3	109.2	53.6	75.5	109.2	92.2	98.9	109.2	86.3	95.3	109.2	86.3	95.3
09	FINISHES	123.8	56.5	88.8	124.3	91.3	107.1	120.7	55.9	87.0	113.5	84.1	98.2	116.8	84.3	99.9	122.9	84.3	102.8
10 - 14	TOTAL DIV. 10000 - 14000	140.0	60.7	123.1	140.0	71.5	125.4	140.0	60.1	123.0	140.0	88.9	129.1	140.0	79.2	127.1	140.0	79.2	127.1
15	MECHANICAL	103.7	64.5	86.6	103.7	88.2	96.9	101.4	62.0	84.2	103.7	77.5	92.3	104.1	74.6	91.2	103.7	74.6	91.0
16	ELECTRICAL	133.2	77.8	106.2	130.3	93.6	112.4	126.5	63.4	95.7	127.8	91.7	110.2	121.5	73.3	98.0	121.5	73.3	98.0
01 - 16	WEIGHTED AVERAGE	118.3	66.9	94.9	119.4	90.7	106.3	114.7	64.7	92.0	114.0	85.2	100.9	115.8	81.1	100.1	116.3	81.4	100.4

CANADA

DIVISION		ST CATHARINES, ONTARIO			ST JEROME, QUEBEC			ST JOHNS, NEWFOUNDLAND			SUDBURY, ONTARIO			SUMMERSIDE, PEI			SYDNEY, NOVA SCOTIA		
		MAT.	INST.	TOTAL	MAT.	INST.	TOTAL	MAT.	INST.	TOTAL	MAT.	INST.	TOTAL	MAT.	INST.	TOTAL	MAT.	INST.	TOTAL
01590	EQUIPMENT RENTAL	.0	100.4	100.4	.0	99.2	99.2	.0	98.6	98.6	.0	100.4	100.4	.0	98.6	98.6	.0	98.6	98.6
02	SITE CONSTRUCTION	99.6	98.8	99.0	91.7	96.8	95.5	112.5	94.5	99.2	99.7	98.7	99.0	105.0	94.0	96.9	92.0	95.8	94.8
03100	CONCRETE FORMS & ACCESSORIES	113.3	81.2	85.5	129.9	81.3	87.8	100.6	56.9	62.8	108.7	80.9	84.7	91.3	53.8	58.8	90.8	66.2	69.5
03200	CONCRETE REINFORCEMENT	111.2	89.0	100.0	156.8	83.7	119.9	148.1	51.5	99.3	112.2	87.0	99.4	148.1	49.6	98.3	148.1	49.5	98.3
03300	CAST-IN-PLACE CONCRETE	140.1	80.9	115.4	135.6	89.5	116.4	179.6	64.7	131.6	141.2	80.3	115.8	168.0	56.4	121.4	136.4	66.4	107.2
03	CONCRETE	121.2	82.8	102.1	134.0	84.8	109.4	164.6	59.5	112.1	121.6	82.1	101.9	155.3	54.7	105.1	131.7	63.9	97.9
04	MASONRY	162.3	84.5	113.5	160.4	81.5	110.9	160.7	60.0	97.5	162.3	84.3	113.4	159.8	58.7	96.3	157.4	67.2	100.8
05	METALS	106.6	88.8	100.8	105.6	85.4	99.0	108.6	70.3	96.0	106.6	88.5	100.7	106.4	65.7	93.1	106.3	72.3	95.2
06	WOOD & PLASTICS	108.5	82.0	94.7	131.7	81.2	105.3	95.0	56.0	74.6	103.7	80.9	91.8	85.3	53.3	68.6	85.0	65.6	74.8
07	THERMAL & MOISTURE PROTECTION	108.8	85.1	97.1	104.9	85.7	95.5	109.9	59.4	85.1	108.3	83.4	96.1	105.5	57.7	82.0	104.9	67.1	86.3
08	DOORS & WINDOWS	82.7	82.5	82.7	91.8	72.3	86.8	98.1	53.8	86.6	83.7	80.5	82.9	94.1	47.1	82.0	82.5	61.1	77.0
09200	PLASTER & GYPSUM BOARD	143.3	81.6	104.2	150.3	80.7	106.2	172.3	54.8	97.8	144.6	80.4	103.9	163.0	52.0	92.7	162.6	64.7	100.5
095,098	CEILINGS & ACOUSTICAL TREATMENT	100.1	81.6	89.0	93.0	80.7	85.6	106.4	54.8	75.4	95.7	80.4	86.5	95.7	52.0	69.5	95.7	64.7	77.1
09600	FLOORING	126.5	91.9	117.4	132.6	92.6	122.1	114.9	53.3	98.6	123.6	90.6	114.9	110.0	58.5	96.4	110.0	62.6	97.4
097,099	WALL FINISHES, PAINTS & COATINGS	109.2	92.9	99.3	109.2	86.3	95.3	109.2	58.2	78.2	109.2	92.2	98.9	109.2	40.5	67.5	109.2	60.4	79.6
09	FINISHES	115.3	84.7	99.4	116.4	84.3	99.7	118.7	56.3	86.2	113.5	84.1	98.2	112.9	53.5	82.0	112.2	65.1	87.7
10 - 14	TOTAL DIV. 10000 - 14000	140.0	68.0	124.7	140.0	79.2	127.1	140.0	60.8	123.2	140.0	88.9	129.1	140.0	60.1	123.0	140.0	62.3	123.5
15	MECHANICAL	101.3	80.3	92.1	103.7	74.6	91.0	101.7	58.7	82.9	103.8	77.5	92.3	103.7	52.3	81.2	103.7	68.9	88.5
16	ELECTRICAL	128.2	91.9	110.5	122.8	73.3	98.7	120.7	58.9	90.6	124.6	91.7	108.6	119.0	52.9	86.7	120.9	65.0	93.6
01 - 16	WEIGHTED AVERAGE	113.6	85.6	100.9	115.6	81.1	100.0	120.0	62.8	94.0	113.6	85.2	100.7	117.5	58.7	90.8	113.3	69.2	93.3

CANADA

DIVISION		THUNDER BAY, ONTARIO			TIMMINS, ONTARIO			TORONTO, ONTARIO			TROIS RIVIERES, QUEBEC			TRURO, NOVA SCOTIA			VANCOUVER, B C		
		MAT.	INST.	TOTAL	MAT.	INST.	TOTAL	MAT.	INST.	TOTAL	MAT.	INST.	TOTAL	MAT.	INST.	TOTAL	MAT.	INST.	TOTAL
01590	EQUIPMENT RENTAL	.0	100.4	100.4	.0	100.4	100.4	.0	103.2	103.2	.0	99.2	99.2	.0	98.6	98.6	.0	111.3	111.3
02	SITE CONSTRUCTION	105.3	98.9	100.6	111.6	98.7	102.1	111.2	103.9	105.8	92.4	96.8	95.6	97.1	95.8	96.1	115.5	107.5	109.6
03100	CONCRETE FORMS & ACCESSORIES	122.7	82.3	87.7	122.6	80.7	86.3	123.1	93.3	97.3	129.9	81.4	87.9	90.8	66.2	69.5	114.6	84.9	88.9
03200	CONCRETE REINFORCEMENT	99.4	88.4	93.8	174.6	87.0	130.3	170.9	90.0	130.0	148.1	83.8	115.6	148.1	49.5	98.3	161.3	81.7	121.1
03300	CAST-IN-PLACE CONCRETE	154.2	92.6	128.5	170.4	80.2	132.7	163.2	102.1	137.7	141.2	89.5	119.6	177.1	66.4	130.9	153.2	94.6	128.7
03	CONCRETE	130.0	87.1	108.6	153.0	82.0	117.5	149.0	95.7	122.4	135.3	84.8	110.1	151.0	63.9	107.6	146.0	88.0	117.0
04	MASONRY	163.0	85.4	114.3	165.9	84.3	114.7	186.3	99.1	131.6	160.8	81.5	111.0	160.5	67.2	102.0	175.3	88.8	121.0
05	METALS	106.4	87.9	100.4	107.1	88.2	101.0	130.5	91.1	117.6	105.6	85.5	99.0	106.3	72.3	95.2	141.7	90.9	125.1
06	WOOD & PLASTICS	118.6	82.4	99.7	118.6	80.9	98.9	118.6	91.7	104.6	131.7	81.2	105.3	85.0	65.6	74.8	114.3	82.7	97.8
07	THERMAL & MOISTURE PROTECTION	109.1	84.3	96.9	109.9	81.9	96.1	111.0	95.0	103.1	104.9	85.7	95.5	104.9	67.1	86.3	114.4	89.5	102.2
08	DOORS & WINDOWS	81.7	82.1	81.8	90.9	79.3	87.9	90.9	90.8	90.9	91.8	76.7	87.9	82.5	61.1	77.0	94.3	81.2	90.9
09200	PLASTER & GYPSUM BOARD	169.3	82.0	113.9	158.7	80.4	109.1	173.3	91.5	121.5	195.3	80.7	122.6	162.6	64.7	100.5	147.7	81.7	105.9
095,098	CEILINGS & ACOUSTICAL TREATMENT	95.7	82.0	87.5	95.7	80.4	86.5	120.6	91.5	103.2	93.0	80.7	85.6	95.7	64.7	77.1	109.0	81.7	92.7
09600	FLOORING	132.6	52.9	111.6	132.6	90.6	121.5	132.6	97.5	123.3	132.6	92.6	122.1	110.0	62.6	97.4	132.6	89.2	121.2
097,099	WALL FINISHES, PAINTS & COATINGS	109.2	94.9	100.5	109.2	85.9	95.0	112.2	104.3	107.4	109.2	86.3	95.3	109.2	60.4	79.6	109.1	98.7	102.8
09	FINISHES	120.1	79.1	98.8	118.9	83.4	100.5	127.5	95.4	110.8	122.9	84.3	102.8	112.2	65.1	87.7	121.6	86.7	103.4
10 - 14	TOTAL DIV. 10000 - 14000	140.0	69.0	124.9	140.0	68.4	124.8	140.0	93.7	130.2	140.0	79.2	127.1	140.0	62.3	123.5	140.0	89.3	129.2
15	MECHANICAL	101.3	81.0	92.4	103.7	80.1	93.4	101.7	91.4	97.2	103.7	74.6	91.0	103.7	68.9	88.5	101.7	84.6	94.2
16	ELECTRICAL	125.1	90.1	108.0	127.8	91.7	110.2	129.8	94.6	112.6	121.5	73.3	98.0	120.9	65.0	93.6	132.9	82.8	108.5
01 - 16	WEIGHTED AVERAGE	114.8	85.4	101.5	119.6	85.0	103.9	124.4	95.0	111.0	116.3	81.4	100.4	115.8	69.2	94.6	125.6	88.2	108.6

CANADA

DIVISION		VICTORIA, B C			WHITEHORSE, YUKON			WINDSOR, ONTARIO			WINNIPEG, MANITOBA			YARMOUTH, NOVA SCOTIA			YELLOWKNIFE, NWT		
		MAT.	INST.	TOTAL	MAT.	INST.	TOTAL	MAT.	INST.	TOTAL	MAT.	INST.	TOTAL	MAT.	INST.	TOTAL	MAT.	INST.	TOTAL
01590	EQUIPMENT RENTAL	.0	107.3	107.3	.0	99.2	99.2	.0	100.4	100.4	.0	105.3	105.3	.0	98.6	98.6	.0	98.7	98.7
02	SITE CONSTRUCTION	118.6	106.7	109.8	105.8	94.9	97.7	95.6	99.1	98.2	109.6	100.3	102.7	96.8	95.8	96.0	105.9	97.7	99.8
03100	CONCRETE FORMS & ACCESSORIES	108.2	84.1	87.3	102.4	55.1	61.4	122.7	82.9	88.2	122.3	67.3	74.7	90.8	66.2	69.5	102.4	75.6	79.2
03200	CONCRETE REINFORCEMENT	116.4	81.6	98.8	118.9	63.8	91.1	109.0	87.6	98.2	161.3	56.5	108.4	148.1	49.5	98.3	118.9	67.7	93.0
03300	CAST-IN-PLACE CONCRETE	153.3	93.7	128.4	152.8	64.9	116.0	143.5	93.7	122.7	157.5	72.5	122.0	175.2	66.4	129.8	152.8	81.9	123.2
03	CONCRETE	154.7	87.1	120.9	127.6	60.8	94.3	123.1	87.7	105.5	144.4	67.8	106.1	150.1	63.9	107.1	127.6	76.6	102.2
04	MASONRY	169.9	88.7	119.0	159.3	58.8	96.2	162.4	91.7	118.1	177.7	63.4	106.0	160.3	67.2	101.9	159.3	71.3	104.1
05	METALS	107.8	85.7	100.6	105.8	71.9	94.8	106.5	89.1	100.8	139.3	75.8	118.5	106.3	72.3	95.2	106.6	78.5	97.4
06	WOOD & PLASTICS	98.4	82.6	90.1	95.6	53.8	73.8	118.6	81.0	99.0	116.3	68.0	91.1	85.0	65.6	74.8	95.6	76.8	85.8
07	THERMAL & MOISTURE PROTECTION	114.5	82.8	99.0	103.8	59.8	82.2	108.8	88.1	98.6	106.1	70.2	88.5	104.9	67.1	86.3	105.3	76.1	90.9
08	DOORS & WINDOWS	87.7	80.9	85.9	85.3	52.5	76.8	81.5	81.8	81.5	91.8	61.3	83.9	82.5	61.1	77.0	85.3	66.6	80.5
09200	PLASTER & GYPSUM BOARD	136.5	81.7	101.8	121.8	52.5	77.8	164.0	80.5	111.1	153.3	66.6	98.4	162.6	64.7	100.5	124.5	76.2	93.9
095,098	CEILINGS & ACOUSTICAL TREATMENT	95.7	81.7	87.3	95.7	52.5	69.8	95.7	80.5	86.6	107.3	66.6	82.9	95.7	64.7	77.1	95.7	76.2	84.0
09600	FLOORING	127.0	51.7	107.1	120.0	57.1	103.4	132.6	92.6	122.1	132.6	65.3	114.8	110.0	62.6	97.4	120.0	87.6	111.5
097,099	WALL FINISHES, PAINTS & COATINGS	109.2	79.0	90.9	109.2	52.7	74.9	109.2	94.5	100.3	109.2	55.3	76.5	109.2	60.4	79.6	109.2	76.9	89.6
09	FINISHES	115.7	77.9	96.0	110.1	54.9	81.4	119.1	85.9	101.8	121.5	65.9	92.6	112.2	65.1	87.7	110.5	77.3	93.2
10 - 14	TOTAL DIV. 10000 - 14000	140.0	89.0	129.1	140.0	59.4	122.9	140.0	70.3	125.2	140.0	63.9	123.8	140.0	62.3	123.5	140.0	84.6	128.2
15	MECHANICAL	103.7	77.7	92.4	103.7	60.9	85.0	101.3	83.6	93.5	101.7	68.5	87.1	103.7	68.9	88.5	103.7	77.5	92.3
16	ELECTRICAL	124.8	82.8	104.3	125.9	62.8	95.2	133.5	91.9	113.2	130.9	70.3	101.3	120.9	65.0	93.6	127.2	87.6	107.9
01 - 16	WEIGHTED AVERAGE	119.1	84.8	103.6	114.0	63.8	91.2	114.7	88.0	102.6	124.3	71.0	100.1	115.6	69.2	94.5	114.3	79.8	98.6

COST INDEXES

Costs shown in *Means cost data publications* are based on National Averages for materials and installation. To adjust these costs to a specific location, simply multiply the base cost by the factor and divide by 100 for that city. The data is arranged alphabetically by state and postal zip code numbers. For a city not listed, use the factor for a nearby city with similar economic characteristics.

STATE/ZIP	CITY	MAT.	INST.	TOTAL
ALABAMA				
350-352	Birmingham	96.4	76.5	87.4
354	Tuscaloosa	96.3	56.4	78.2
355	Jasper	96.7	53.5	77.1
356	Decatur	96.3	57.6	78.8
357-358	Huntsville	96.3	73.3	85.8
359	Gadsden	96.3	60.0	79.8
360-361	Montgomery	97.0	58.4	79.5
362	Anniston	96.0	47.7	74.1
363	Dothan	96.5	50.5	75.6
364	Evergreen	95.8	51.8	75.8
365-366	Mobile	97.1	63.0	81.6
367	Selma	96.1	53.0	76.5
368	Phenix City	96.8	56.4	78.5
369	Butler	96.2	50.5	75.5
ALASKA				
995-996	Anchorage	134.5	113.8	125.1
997	Fairbanks	130.7	116.9	124.4
998	Juneau	133.6	113.8	124.6
999	Ketchikan	143.2	113.4	129.6
ARIZONA				
850,853	Phoenix	98.8	74.3	87.7
852	Mesa/Tempe	97.8	67.7	84.1
855	Globe	98.2	62.9	82.2
856-857	Tucson	97.0	69.8	84.6
859	Show Low	98.3	66.9	84.1
860	Flagstaff	100.5	67.6	85.5
863	Prescott	98.2	64.7	83.0
864	Kingman	96.8	65.6	82.6
865	Chambers	96.8	62.5	81.2
ARKANSAS				
716	Pine Bluff	95.3	61.7	80.0
717	Camden	93.4	40.1	69.2
718	Texarkana	94.7	45.3	72.3
719	Hot Springs	92.6	39.6	68.5
720-722	Little Rock	95.2	64.3	81.2
723	West Memphis	95.1	55.0	76.9
724	Jonesboro	95.5	55.0	77.1
725	Batesville	93.5	49.9	73.7
726	Harrison	94.8	49.9	74.4
727	Fayetteville	92.1	48.1	72.1
728	Russellville	93.4	48.0	72.8
729	Fort Smith	96.0	56.8	78.2
CALIFORNIA				
900-902	Los Angeles	102.7	111.9	106.8
903-905	Inglewood	98.5	108.2	102.9
906-908	Long Beach	100.0	108.3	103.8
910-912	Pasadena	100.5	108.3	104.0
913-916	Van Nuys	103.8	108.3	105.8
917-918	Alhambra	102.8	108.2	105.3
919-921	San Diego	103.9	105.0	104.4
922	Palm Springs	100.5	105.8	102.9
923-924	San Bernardino	98.3	106.7	102.2
925	Riverside	102.6	109.9	105.9
926-927	Santa Ana	100.4	106.7	103.2
928	Anaheim	102.9	111.1	106.6
930	Oxnard	103.9	109.5	106.4
931	Santa Barbara	102.7	110.3	106.1
932-933	Bakersfield	102.6	106.2	104.2
934	San Luis Obispo	103.8	106.3	105.0
935	Mojave	100.8	103.7	102.1
936-938	Fresno	104.3	111.9	107.8
939	Salinas	104.5	116.1	109.8
940-941	San Francisco	112.6	133.8	122.2
942,956-958	Sacramento	106.9	113.1	109.7
943	Palo Alto	105.5	125.9	114.8
944	San Mateo	108.4	125.1	116.0
945	Vallejo	105.6	120.4	112.3
946	Oakland	110.6	123.9	116.6
947	Berkeley	110.1	122.5	115.7
948	Richmond	109.5	122.8	115.6
949	San Rafael	110.5	122.3	115.9
950	Santa Cruz	109.6	116.1	112.6

STATE/ZIP	CITY	MAT.	INST.	TOTAL
CALIFORNIA (CONT'D)				
951	San Jose	107.3	128.4	116.9
952	Stockton	104.5	110.5	107.2
953	Modesto	104.4	111.6	107.7
954	Santa Rosa	104.7	121.5	112.4
955	Eureka	106.3	103.7	105.1
959	Marysville	105.3	113.4	109.0
960	Redding	106.8	110.4	108.4
961	Susanville	106.1	111.1	108.4
COLORADO				
800-802	Denver	100.3	90.3	95.8
803	Boulder	97.6	86.7	92.7
804	Golden	99.9	86.2	93.7
805	Fort Collins	101.0	82.7	92.7
806	Greeley	98.5	71.1	86.0
807	Fort Morgan	98.4	85.9	92.7
808-809	Colorado Springs	100.1	86.6	94.0
810	Pueblo	100.0	83.9	92.6
811	Alamosa	101.6	81.3	92.4
812	Salida	101.6	81.1	92.3
813	Durango	102.0	79.2	91.6
814	Montrose	100.6	79.0	90.8
815	Grand Junction	103.8	77.6	91.9
816	Glenwood Springs	101.4	84.6	93.8
CONNECTICUT				
060	New Britain	101.4	110.7	105.6
061	Hartford	102.7	109.9	106.0
062	Willimantic	102.2	109.6	105.6
063	New London	98.6	110.7	104.1
064	Meriden	100.6	111.1	105.4
065	New Haven	103.2	111.2	106.8
066	Bridgeport	102.7	110.8	106.4
067	Waterbury	102.2	110.7	106.1
068	Norwalk	102.2	111.5	106.4
069	Stamford	102.3	115.6	108.4
D.C.				
200-205	Washington	102.0	90.8	96.9
DELAWARE				
197	Newark	100.5	105.7	102.8
198	Wilmington	99.5	105.7	102.3
199	Dover	100.6	105.7	102.9
FLORIDA				
320,322	Jacksonville	98.3	61.0	81.4
321	Daytona Beach	98.4	74.5	87.5
323	Tallahassee	97.6	49.2	75.6
324	Panama City	99.5	37.9	71.5
325	Pensacola	99.2	58.1	80.5
326,344	Gainesville	99.7	58.7	81.1
327-328,347	Orlando	99.8	69.3	86.0
329	Melbourne	100.5	79.0	90.7
330-332,340	Miami	98.7	72.9	87.0
333	Fort Lauderdale	97.7	72.3	86.2
334,349	West Palm Beach	96.6	68.5	83.8
335-336,346	Tampa	99.4	72.8	87.3
337	St. Petersburg	101.6	55.8	80.8
338	Lakeland	98.5	72.3	86.6
339,341	Fort Myers	97.9	64.3	82.6
342	Sarasota	99.6	66.7	84.6
GEORGIA				
300-303,399	Atlanta	97.4	80.7	89.8
304	Statesboro	97.1	45.0	73.4
305	Gainesville	95.8	57.4	78.4
306	Athens	95.2	62.1	80.1
307	Dalton	97.5	50.9	76.3
308-309	Augusta	96.0	59.2	79.3
310-312	Macon	95.7	59.2	79.1
313-314	Savannah	97.1	58.8	79.7
315	Waycross	97.2	51.6	76.5
316	Valdosta	97.0	49.1	75.2
317	Albany	97.3	56.8	78.9
318-319	Columbus	97.2	61.7	81.1

Location Factors

STATE/ZIP	CITY	MAT.	INST.	TOTAL
HAWAII				
967	Hilo	115.3	125.7	120.0
968	Honolulu	121.0	125.7	123.1
STATES & POSS.				
969	Guam	193.9	54.3	130.5
IDAHO				
832	Pocatello	100.3	78.3	90.3
833	Twin Falls	100.8	45.8	75.8
834	Idaho Falls	98.2	55.3	78.7
835	Lewiston	108.1	83.3	96.8
836-837	Boise	100.2	80.5	91.2
838	Coeur d'Alene	107.3	56.8	84.4
ILLINOIS				
600-603	North Suburban	99.1	116.3	106.9
604	Joliet	99.0	117.2	107.3
605	South Suburban	99.1	116.3	106.9
606-608	Chicago	99.7	125.8	111.6
609	Kankakee	95.4	105.3	99.9
610-611	Rockford	97.4	109.4	102.8
612	Rock Island	95.1	97.7	96.3
613	La Salle	96.5	102.7	99.3
614	Galesburg	96.2	100.8	98.3
615-616	Peoria	98.7	101.7	100.1
617	Bloomington	95.5	101.2	98.1
618-619	Champaign	99.0	100.0	99.5
620-622	East St. Louis	94.2	101.1	97.3
623	Quincy	95.6	95.4	95.5
624	Effingham	94.9	97.3	96.0
625	Decatur	96.5	98.5	97.4
626-627	Springfield	96.8	98.9	97.8
628	Centralia	93.1	98.0	95.3
629	Carbondale	92.8	97.8	95.1
INDIANA				
460	Anderson	94.6	83.3	89.5
461-462	Indianapolis	98.8	88.7	94.2
463-464	Gary	95.8	98.7	97.1
465-466	South Bend	95.0	82.6	89.4
467-468	Fort Wayne	95.1	80.6	88.5
469	Kokomo	92.9	82.2	88.0
470	Lawrenceburg	92.2	80.1	86.7
471	New Albany	93.7	74.8	85.1
472	Columbus	96.3	80.2	89.0
473	Muncie	96.1	80.8	89.2
474	Bloomington	98.2	80.7	90.3
475	Washington	94.1	85.2	90.1
476-477	Evansville	95.2	86.8	91.4
478	Terre Haute	96.0	85.3	91.1
479	Lafayette	95.9	79.5	88.4
IOWA				
500-503,509	Des Moines	97.7	82.4	90.7
504	Mason City	96.0	63.3	81.2
505	Fort Dodge	96.2	59.4	79.5
506-507	Waterloo	97.7	62.0	81.5
508	Creston	96.6	66.3	82.8
510-511	Sioux City	98.3	73.3	87.0
512	Sibley	97.0	52.9	77.0
513	Spencer	98.8	51.0	77.1
514	Carroll	95.8	57.0	78.2
515	Council Bluffs	99.1	74.7	88.0
516	Shenandoah	96.1	52.2	76.2
520	Dubuque	97.7	76.5	88.1
521	Decorah	96.8	54.5	77.6
522-524	Cedar Rapids	98.8	82.7	91.5
525	Ottumwa	96.9	70.7	85.0
526	Burlington	96.0	73.7	85.9
527-528	Davenport	97.8	93.1	95.7
KANSAS				
660-662	Kansas City	98.7	91.6	95.5
664-666	Topeka	98.5	69.3	85.2
667	Fort Scott	97.5	66.8	83.6
668	Emporia	97.4	59.4	80.1
669	Belleville	99.2	54.8	79.0
670-672	Wichita	97.7	71.1	85.6
673	Independence	99.0	53.0	78.1
674	Salina	99.0	57.2	80.1
675	Hutchinson	94.5	51.1	74.7
676	Hays	98.6	54.8	78.7
677	Colby	99.2	54.8	79.0

STATE/ZIP	CITY	MAT.	INST.	TOTAL
KANSAS (CONT'D)				
678	Dodge City	100.3	54.8	79.6
679	Liberal	98.4	45.6	74.4
KENTUCKY				
400-402	Louisville	96.3	86.4	91.8
403-405	Lexington	95.9	66.9	82.7
406	Frankfort	96.1	68.5	83.6
407-409	Corbin	93.3	44.6	71.2
410	Covington	94.0	93.6	93.8
411-412	Ashland	92.6	98.6	95.3
413-414	Campton	94.2	45.1	71.9
415-416	Pikeville	95.1	62.1	80.1
417-418	Hazard	93.5	44.8	71.4
420	Paducah	92.2	88.8	90.7
421-422	Bowling Green	94.4	83.6	89.5
423	Owensboro	94.2	74.3	85.2
424	Henderson	91.9	89.2	90.7
425-426	Somerset	91.5	46.5	71.0
427	Elizabethtown	91.1	82.8	87.3
LOUISIANA				
700-701	New Orleans	100.2	69.8	86.4
703	Thibodaux	98.1	64.4	82.8
704	Hammond	95.4	63.9	81.1
705	Lafayette	97.7	55.0	78.3
706	Lake Charles	97.8	60.0	80.7
707-708	Baton Rouge	98.4	55.1	78.7
710-711	Shreveport	94.7	59.7	78.8
712	Monroe	95.5	57.5	78.2
713-714	Alexandria	95.6	54.7	77.0
MAINE				
039	Kittery	95.6	72.4	85.1
040-041	Portland	98.8	76.8	88.8
042	Lewiston	99.2	76.8	89.0
043	Augusta	97.7	72.3	86.2
044	Bangor	98.0	76.8	88.3
045	Bath	96.7	72.4	85.7
046	Machias	96.3	71.8	85.2
047	Houlton	96.5	75.5	87.0
048	Rockland	95.5	70.6	84.2
049	Waterville	96.8	67.3	83.4
MARYLAND				
206	Waldorf	98.0	73.1	86.7
207-208	College Park	98.0	80.6	90.1
209	Silver Spring	97.3	78.2	88.6
210-212	Baltimore	99.4	84.8	92.7
214	Annapolis	98.8	79.9	90.2
215	Cumberland	95.5	80.8	88.8
216	Easton	96.9	45.4	73.5
217	Hagerstown	96.0	80.8	89.1
218	Salisbury	97.4	54.8	78.0
219	Elkton	94.6	67.2	82.2
MASSACHUSETTS				
010-011	Springfield	100.7	105.1	102.7
012	Pittsfield	100.2	101.1	100.6
013	Greenfield	98.3	103.3	100.5
014	Fitchburg	97.0	116.9	106.0
015-016	Worcester	100.5	118.4	108.6
017	Framingham	96.4	122.8	108.4
018	Lowell	100.0	123.5	110.7
019	Lawrence	101.1	121.0	110.2
020-022, 024	Boston	102.9	130.5	115.4
023	Brockton	101.4	115.5	107.8
025	Buzzards Bay	95.8	116.9	105.4
026	Hyannis	98.3	116.9	106.7
027	New Bedford	101.1	117.0	108.4
MICHIGAN				
480,483	Royal Oak	94.9	106.2	100.0
481	Ann Arbor	96.9	105.7	100.9
482	Detroit	98.7	116.9	107.0
484-485	Flint	96.7	96.2	96.5
486	Saginaw	96.5	93.2	95.0
487	Bay City	96.4	93.3	95.0
488-489	Lansing	96.7	96.1	96.4
490	Battle Creek	95.4	88.1	92.1
491	Kalamazoo	95.7	84.9	90.8
492	Jackson	93.9	94.7	94.3
493,495	Grand Rapids	95.8	70.2	84.2
494	Muskegon	94.3	83.6	89.5

STATE/ZIP	CITY	MAT.	INST.	TOTAL
MICHIGAN (CONT'D)				
496	Traverse City	93.4	73.9	84.6
497	Gaylord	94.6	77.9	87.0
498-499	Iron Mountain	96.5	87.1	92.2
MINNESOTA				
550-551	Saint Paul	99.8	122.0	109.8
553-555	Minneapolis	100.7	126.8	112.6
556-558	Duluth	99.7	108.9	103.8
559	Rochester	99.4	105.1	102.0
560	Mankato	96.3	102.1	99.0
561	Windom	95.1	81.4	88.9
562	Willmar	94.5	87.7	91.4
563	St. Cloud	95.6	120.5	106.9
564	Brainerd	96.1	101.0	98.3
565	Detroit Lakes	98.0	99.1	98.5
566	Bemidji	97.3	97.7	97.5
567	Thief River Falls	96.4	94.0	95.3
MISSISSIPPI				
386	Clarksdale	96.0	31.7	66.8
387	Greenville	99.4	45.9	75.1
388	Tupelo	97.4	38.4	70.6
389	Greenwood	97.3	34.3	68.7
390-392	Jackson	97.9	47.5	75.0
393	Meridian	96.0	47.8	74.1
394	Laurel	97.3	34.4	68.7
395	Biloxi	97.9	56.6	79.1
396	McComb	95.8	53.3	76.5
397	Columbus	97.2	37.8	70.2
MISSOURI				
630-631	St. Louis	96.7	108.6	102.1
633	Bowling Green	95.6	89.4	92.8
634	Hannibal	94.4	85.1	90.2
635	Kirksville	96.7	78.3	88.3
636	Flat River	96.6	93.7	95.3
637	Cape Girardeau	96.1	89.3	93.0
638	Sikeston	94.3	84.3	89.8
639	Poplar Bluff	93.9	85.0	89.8
640-641	Kansas City	101.1	106.3	103.4
644-645	St. Joseph	100.1	89.6	95.3
646	Chillicothe	97.0	70.7	85.1
647	Harrisonville	96.7	97.7	97.2
648	Joplin	98.7	69.0	85.2
650-651	Jefferson City	95.9	87.3	92.0
652	Columbia	96.9	88.1	92.9
653	Sedalia	96.1	84.3	90.7
654-655	Rolla	95.0	77.8	87.2
656-658	Springfield	98.2	74.5	87.4
MONTANA				
590-591	Billings	102.0	74.4	89.5
592	Wolf Point	100.8	70.5	87.0
593	Miles City	98.8	71.0	86.2
594	Great Falls	102.6	74.9	90.0
595	Havre	99.6	70.7	86.5
596	Helena	102.4	72.3	88.7
597	Butte	101.0	71.1	87.4
598	Missoula	98.7	70.3	85.8
599	Kalispell	97.8	69.8	85.1
NEBRASKA				
680-681	Omaha	99.1	76.7	89.0
683-685	Lincoln	98.0	67.7	84.2
686	Columbus	96.0	51.2	75.6
687	Norfolk	97.7	62.5	81.7
688	Grand Island	97.2	66.2	83.1
689	Hastings	97.0	58.5	79.5
690	Mccook	96.8	48.8	75.0
691	North Platte	96.8	59.9	80.0
692	Valentine	99.2	39.7	72.2
693	Alliance	98.9	37.0	70.8
NEVADA				
889-891	Las Vegas	98.1	105.5	101.5
893	Ely	98.7	81.2	90.7
894-895	Reno	98.7	95.5	97.3
897	Carson City	98.4	95.6	97.2
898	Elko	97.5	87.9	93.1
NEW HAMPSHIRE				
030	Nashua	100.9	82.8	92.7
031	Manchester	101.2	82.8	92.9

STATE/ZIP	CITY	MAT.	INST.	TOTAL
NEW HAMPSHIRE (CONT'D)				
032-033	Concord	98.6	82.8	91.4
034	Keene	97.6	52.5	77.1
035	Littleton	97.7	62.0	81.5
036	Charleston	97.0	49.3	75.4
037	Claremont	96.3	49.3	75.0
038	Portsmouth	98.0	78.4	89.1
NEW JERSEY				
070-071	Newark	102.5	121.4	111.1
072	Elizabeth	101.1	120.7	110.0
073	Jersey City	100.2	122.5	110.3
074-075	Paterson	101.8	121.2	110.6
076	Hackensack	99.7	121.7	109.7
077	Long Branch	99.4	120.0	108.7
078	Dover	100.0	121.5	109.8
079	Summit	100.1	120.7	109.4
080,083	Vineland	97.8	117.7	106.8
081	Camden	100.0	115.2	106.9
082,084	Atlantic City	98.6	115.7	106.4
085-086	Trenton	99.9	119.1	108.6
087	Point Pleasant	99.8	116.8	107.5
088-089	New Brunswick	100.3	119.4	108.9
NEW MEXICO				
870-872	Albuquerque	98.3	76.8	88.6
873	Gallup	98.3	76.8	88.6
874	Farmington	98.6	76.8	88.7
875	Santa Fe	97.3	76.8	88.0
877	Las Vegas	96.8	76.8	87.7
878	Socorro	96.4	76.8	87.5
879	Truth/Consequences	97.1	71.6	85.5
880	Las Cruces	95.8	69.1	83.7
881	Clovis	96.9	76.1	87.4
882	Roswell	98.5	76.1	88.4
883	Carrizozo	99.0	76.8	88.9
884	Tucumcari	97.8	76.1	87.9
NEW YORK				
100-102	New York	106.6	163.5	132.4
103	Staten Island	102.1	155.2	126.2
104	Bronx	100.2	155.2	125.2
105	Mount Vernon	100.2	135.4	116.2
106	White Plains	99.9	135.4	116.0
107	Yonkers	105.1	136.2	119.2
108	New Rochelle	100.7	135.4	116.5
109	Suffern	100.5	123.7	111.0
110	Queens	101.4	155.0	125.8
111	Long Island City	103.1	155.0	126.7
112	Brooklyn	103.4	156.5	127.5
113	Flushing	103.6	155.0	127.0
114	Jamaica	101.7	155.2	126.0
115,117,118	Hicksville	101.2	145.0	121.1
116	Far Rockaway	103.7	155.0	127.0
119	Riverhead	101.6	145.1	121.4
120-122	Albany	97.9	93.9	96.1
123	Schenectady	98.2	93.0	95.8
124	Kingston	101.1	114.3	107.1
125-126	Poughkeepsie	100.3	119.0	108.8
127	Monticello	99.6	113.8	106.1
128	Glens Falls	92.8	89.9	91.5
129	Plattsburgh	97.0	82.8	90.6
130-132	Syracuse	99.0	90.5	95.2
133-135	Utica	97.0	87.8	92.8
136	Watertown	98.7	88.4	94.0
137-139	Binghamton	98.6	84.5	92.2
140-142	Buffalo	99.9	104.7	102.1
143	Niagara Falls	97.6	102.8	100.0
144-146	Rochester	100.1	94.9	97.7
147	Jamestown	96.6	85.4	91.5
148-149	Elmira	96.5	84.8	91.2
NORTH CAROLINA				
270,272-274	Greensboro	98.0	50.3	76.3
271	Winston-Salem	97.8	49.1	75.7
275-276	Raleigh	97.9	50.3	76.3
277	Durham	97.9	50.3	76.3
278	Rocky Mount	95.4	36.2	68.5
279	Elizabeth City	96.1	38.6	70.0
280	Gastonia	97.2	47.5	74.6
281-282	Charlotte	98.2	47.8	75.3
283	Fayetteville	96.7	50.3	75.6
284	Wilmington	96.0	49.4	74.8
285	Kinston	94.2	35.8	67.6

Location Factors

STATE/ZIP	CITY	MAT.	INST.	TOTAL
NORTH CAROLINA (CONT'D)				
286	Hickory	94.5	34.9	67.5
287-288	Asheville	96.3	48.3	74.5
289	Murphy	95.5	34.0	67.6
NORTH DAKOTA				
580-581	Fargo	100.8	65.9	85.0
582	Grand Forks	100.4	60.7	82.4
583	Devils Lake	100.0	61.0	82.3
584	Jamestown	100.0	54.2	79.2
585	Bismarck	99.2	66.9	84.5
586	Dickinson	100.8	62.3	83.3
587	Minot	100.5	68.2	85.8
588	Williston	99.3	61.7	82.2
OHIO				
430-432	Columbus	97.8	90.5	94.5
433	Marion	94.4	88.3	91.6
434-436	Toledo	98.0	99.6	98.7
437-438	Zanesville	94.8	83.9	89.8
439	Steubenville	95.7	94.3	95.1
440	Lorain	96.8	98.1	97.4
441	Cleveland	97.1	106.0	101.1
442-443	Akron	97.7	98.0	97.8
444-445	Youngstown	97.1	94.8	96.0
446-447	Canton	97.2	88.5	93.2
448-449	Mansfield	94.5	93.4	94.0
450	Hamilton	94.3	90.4	92.5
451-452	Cincinnati	94.7	92.3	93.6
453-454	Dayton	94.4	84.9	90.1
455	Springfield	94.4	86.7	90.9
456	Chillicothe	93.5	92.4	93.0
457	Athens	96.5	82.3	90.1
458	Lima	97.1	87.9	92.9
OKLAHOMA				
730-731	Oklahoma City	97.3	64.0	82.1
734	Ardmore	94.2	63.3	80.2
735	Lawton	96.5	65.0	82.2
736	Clinton	95.7	61.6	80.2
737	Enid	96.1	61.6	80.4
738	Woodward	94.5	61.7	79.6
739	Guymon	95.6	33.9	67.6
740-741	Tulsa	96.3	61.2	80.4
743	Miami	93.3	67.1	81.4
744	Muskogee	95.7	43.7	72.1
745	Mcalester	92.9	56.1	76.2
746	Ponca City	93.4	61.5	78.9
747	Durant	93.4	61.7	79.0
748	Shawnee	94.9	60.1	79.1
749	Poteau	92.5	64.3	79.7
OREGON				
970-972	Portland	102.2	105.4	103.7
973	Salem	102.2	103.0	102.5
974	Eugene	101.7	102.1	101.9
975	Medford	103.4	100.3	102.0
976	Klamath Falls	103.7	100.1	102.1
977	Bend	102.5	102.2	102.4
978	Pendleton	96.6	100.7	98.5
979	Vale	94.4	93.7	94.1
PENNSYLVANIA				
150-152	Pittsburgh	96.7	103.8	99.9
153	Washington	93.8	102.1	97.6
154	Uniontown	94.0	99.6	96.5
155	Bedford	95.0	91.8	93.5
156	Greensburg	95.1	99.6	97.1
157	Indiana	93.9	97.3	95.4
158	Dubois	95.2	95.7	95.5
159	Johnstown	94.9	96.2	95.5
160	Butler	92.3	101.6	96.5
161	New Castle	92.3	100.2	95.9
162	Kittanning	92.8	103.2	97.5
163	Oil City	92.3	95.8	93.9
164-165	Erie	94.3	96.2	95.1
166	Altoona	94.3	93.3	93.8
167	Bradford	95.9	93.3	94.7
168	State College	95.5	93.7	94.7
169	Wellsboro	96.5	87.4	92.4
170-171	Harrisburg	98.1	92.3	95.5
172	Chambersburg	96.1	87.2	92.0
173-174	York	96.2	88.6	92.7
175-176	Lancaster	94.9	87.5	91.5

STATE/ZIP	CITY	MAT.	INST.	TOTAL
PENNSYLVANIA (CONT'D)				
177	Williamsport	93.4	81.8	88.2
178	Sunbury	95.5	87.9	92.0
179	Pottsville	94.6	89.3	92.2
180	Lehigh Valley	96.0	102.7	99.0
181	Allentown	98.0	101.2	99.4
182	Hazleton	95.3	92.4	94.0
183	Stroudsburg	95.3	95.8	95.5
184-185	Scranton	98.8	92.9	96.1
186-187	Wilkes-Barre	95.3	91.4	93.5
188	Montrose	94.9	92.0	93.6
189	Doylestown	94.8	116.5	104.6
190-191	Philadelphia	100.8	128.9	113.6
193	Westchester	97.6	115.6	105.7
194	Norristown	96.5	116.5	105.6
195-196	Reading	98.6	98.5	98.6
PUERTO RICO				
009	San Juan	136.6	26.3	86.5
RHODE ISLAND				
028	Newport	99.9	112.4	105.6
029	Providence	100.2	112.4	105.8
SOUTH CAROLINA				
290-292	Columbia	96.6	48.3	74.7
293	Spartanburg	95.5	46.0	73.0
294	Charleston	96.8	48.9	75.1
295	Florence	95.0	46.1	72.8
296	Greenville	95.2	46.0	72.9
297	Rock Hill	95.0	33.8	67.2
298	Aiken	95.9	70.4	84.3
299	Beaufort	96.7	37.4	69.8
SOUTH DAKOTA				
570-571	Sioux Falls	99.1	57.8	80.3
572	Watertown	97.9	52.0	77.0
573	Mitchell	96.9	51.6	76.3
574	Aberdeen	99.4	54.6	79.0
575	Pierre	98.7	53.7	78.2
576	Mobridge	97.5	51.9	76.8
577	Rapid City	99.3	52.5	78.0
TENNESSEE				
370-372	Nashville	98.0	74.0	87.1
373-374	Chattanooga	98.8	55.8	79.3
375,380-381	Memphis	96.6	77.2	87.8
376	Johnson City	97.6	58.2	79.7
377-379	Knoxville	94.9	59.2	78.7
382	Mckenzie	96.5	49.7	75.3
383	Jackson	98.5	53.3	78.0
384	Columbia	95.0	56.9	77.7
385	Cookeville	96.3	51.4	75.9
TEXAS				
750	Mckinney	98.2	57.1	79.5
751	Waxahackie	98.1	57.8	79.8
752-753	Dallas	98.7	67.1	84.3
754	Greenville	98.3	41.9	72.7
755	Texarkana	97.2	53.0	77.2
756	Longview	97.5	43.9	73.1
757	Tyler	98.1	56.4	79.2
758	Palestine	94.1	44.3	71.5
759	Lufkin	95.1	48.1	73.7
760-761	Fort Worth	96.4	63.3	81.4
762	Denton	96.9	53.2	77.0
763	Wichita Falls	97.4	57.2	79.1
764	Eastland	95.9	43.3	72.0
765	Temple	94.8	51.4	75.1
766-767	Waco	96.6	57.8	79.0
768	Brownwood	96.9	40.7	71.4
769	San Angelo	96.6	49.0	75.0
770-772	Houston	99.1	72.8	87.2
773	Huntsville	97.6	41.6	72.1
774	Wharton	98.8	46.5	75.1
775	Galveston	97.0	70.7	85.0
776-777	Beaumont	96.6	63.5	81.6
778	Bryan	94.5	65.2	81.2
779	Victoria	99.0	49.1	76.4
780	Laredo	94.9	55.0	76.8
781-782	San Antonio	95.2	66.8	82.3
783-784	Corpus Christi	97.7	54.1	77.8
785	Mc Allen	97.8	48.2	75.2
786-787	Austin	94.9	61.8	79.8

Location Factors

STATE/ZIP	CITY	MAT.	INST.	TOTAL
TEXAS (CONT'D)				
788	Del Rio	97.1	34.2	68.6
789	Giddings	94.5	43.4	71.3
790-791	Amarillo	97.3	59.1	79.9
792	Childress	96.5	53.5	76.9
793-794	Lubbock	98.8	54.9	78.9
795-796	Abilene	96.8	54.5	77.6
797	Midland	99.1	51.4	77.4
798-799,885	El Paso	96.7	52.4	76.6
UTAH				
840-841	Salt Lake City	101.2	72.3	88.1
842,844	Ogden	97.2	72.1	85.8
843	Logan	98.9	72.1	86.7
845	Price	99.6	51.1	77.6
846-847	Provo	99.5	72.1	87.1
VERMONT				
050	White River Jct.	99.0	47.9	75.8
051	Bellows Falls	97.5	50.3	76.0
052	Bennington	97.8	50.6	76.4
053	Brattleboro	98.2	50.8	76.7
054	Burlington	100.8	66.0	85.0
056	Montpelier	97.7	66.1	83.3
057	Rutland	99.8	66.1	84.5
058	St. Johnsbury	99.2	50.6	77.1
059	Guildhall	97.8	50.1	76.1
VIRGINIA				
220-221	Fairfax	99.2	81.2	91.0
222	Arlington	100.2	79.4	90.7
223	Alexandria	99.6	84.6	92.8
224-225	Fredericksburg	97.8	68.0	84.3
226	Winchester	98.5	55.4	78.9
227	Culpeper	98.3	56.6	79.4
228	Harrisonburg	98.6	48.1	75.7
229	Charlottesville	98.9	60.5	81.5
230-232	Richmond	99.9	65.9	84.5
233-235	Norfolk	99.6	65.2	84.0
236	Newport News	99.2	63.6	83.0
237	Portsmouth	98.4	65.0	83.2
238	Petersburg	98.6	65.9	83.8
239	Farmville	98.1	42.3	72.8
240-241	Roanoke	99.8	48.4	76.5
242	Bristol	97.8	48.7	75.5
243	Pulaski	97.5	43.7	73.1
244	Staunton	98.2	46.5	74.8
245	Lynchburg	98.4	51.0	76.8
246	Grundy	97.8	45.8	74.2
WASHINGTON				
980-981,987	Seattle	104.3	105.8	105.0
982	Everett	104.0	98.3	101.4
983-984	Tacoma	104.1	99.7	102.1
985	Olympia	102.8	99.7	101.4
986	Vancouver	106.0	97.2	102.0
988	Wenatchee	104.3	84.0	95.1
989	Yakima	104.2	90.0	97.8
990-992	Spokane	106.2	83.7	96.0
993	Richland	105.5	85.1	96.2
994	Clarkston	104.7	82.3	94.6
WEST VIRGINIA				
247-248	Bluefield	96.6	80.2	89.1
249	Lewisburg	98.1	83.9	91.6
250-253	Charleston	99.1	92.0	95.9
254	Martinsburg	98.0	80.0	89.8
255-257	Huntington	99.3	90.9	95.5
258-259	Beckley	96.4	89.5	93.3
260	Wheeling	99.3	93.7	96.8
261	Parkersburg	98.3	90.6	94.8
262	Buckhannon	97.9	93.2	95.8
263-264	Clarksburg	98.3	92.2	95.5
265	Morgantown	98.4	93.2	96.0
266	Gassaway	97.7	92.3	95.3
267	Romney	97.7	86.4	92.6
268	Petersburg	97.7	89.1	93.8
WISCONSIN				
530,532	Milwaukee	98.7	101.6	100.0
531	Kenosha	98.2	97.8	98.0
534	Racine	97.8	99.2	98.4
535	Beloit	97.7	96.4	97.1
537	Madison	97.9	93.9	96.1

STATE/ZIP	CITY	MAT.	INST.	TOTAL
(CONT'D)				
538	Lancaster	95.6	91.4	93.7
539	Portage	94.2	93.3	93.8
540	New Richmond	95.5	94.6	95.1
541-543	Green Bay	99.5	90.9	95.6
544	Wausau	94.9	90.9	93.1
545	Rhinelander	98.0	90.7	94.7
546	La Crosse	96.0	91.7	94.0
547	Eau Claire	97.7	93.3	95.7
548	Superior	95.4	94.8	95.1
549	Oshkosh	95.5	90.5	93.3
WYOMING				
820	Cheyenne	97.5	54.7	78.0
821	Yellowstone Nat'l Park	96.7	49.1	75.1
822	Wheatland	98.1	48.0	75.4
823	Rawlins	99.4	44.1	74.3
824	Worland	97.3	44.8	73.5
825	Riverton	98.4	47.3	75.2
826	Casper	97.6	60.1	80.5
827	Newcastle	97.2	44.1	73.1
828	Sheridan	98.1	52.8	77.5
829-831	Rock Springs	101.4	44.8	75.7
CANADIAN FACTORS (reflect Canadian currency)				
ALBERTA				
	Calgary	124.6	82.0	105.2
	Edmonton	126.2	82.0	106.2
	Fort McMurray	116.8	81.6	100.8
	Lethbridge	117.7	81.0	101.1
	Lloydminster	116.9	81.6	100.8
	Medicine Hat	117.0	81.0	100.6
	Red Deer	117.4	81.0	100.9
BRITISH COLUMBIA				
	Kamloops	117.8	84.3	102.6
	Prince George	119.1	84.3	103.3
	Vancouver	125.6	88.2	108.6
	Victoria	119.1	84.8	103.6
MANITOBA				
	Brandon	117.1	70.7	96.0
	Portage la Prairie	117.1	70.7	96.0
	Winnipeg	124.3	71.0	100.1
NEW BRUNSWICK				
	Bathurst	115.2	62.8	91.4
	Dalhousie	115.2	62.8	91.4
	Fredericton	117.1	66.9	94.3
	Moncton	115.5	62.8	91.6
	Newcastle	115.2	62.8	91.4
	Saint John	118.3	66.9	94.9
NEWFOUNDLAND				
	Corner Brook	120.1	62.8	94.1
	St. John's	120.0	62.8	94.0
NORTHWEST TERRITORIES				
	Yellowknife	114.3	79.8	98.6
NOVA SCOTIA				
	Dartmouth	117.7	69.2	95.7
	Halifax	119.3	70.3	97.0
	New Glasgow	115.8	69.2	94.6
	Sydney	113.3	69.2	93.3
	Yarmouth	115.6	69.2	94.5
ONTARIO				
	Barrie	120.0	86.7	104.9
	Brantford	119.2	90.3	106.1
	Cornwall	119.1	86.9	104.5
	Hamilton	124.7	90.6	109.2
	Kingston	120.1	87.3	105.2
	Kitchener	115.6	86.3	102.3
	London	121.8	88.0	106.5
	North Bay	119.2	85.0	103.6
	Oshawa	118.9	88.4	105.1
	Ottawa	120.7	88.1	105.9
	Owen Sound	120.3	85.0	104.2
	Peterborough	119.2	86.7	104.4
	Sarnia	119.4	90.7	106.3
	St. Catharines	113.6	85.6	100.9
	Sudbury	113.6	85.2	100.7

STATE/ZIP	CITY	MAT.	INST.	TOTAL
ONTARIO (CONT'D)				
	Thunder Bay	114.8	85.4	101.5
	Toronto	124.4	95.0	111.0
	Windsor	114.7	88.0	102.6
PRINCE EDWARD ISLAND				
	Charlottetown	117.8	58.7	91.0
	Summerside	117.5	58.7	90.8
QUEBEC				
	Cap-de-la-Madeleine	116.0	81.4	100.3
	Charlesbourg	116.0	81.4	100.3
	Chicoutimi	115.0	81.1	99.6
	Gatineau	115.4	81.1	99.9
	Laval	115.4	81.1	99.8
	Montreal	118.8	81.4	101.8
	Quebec	119.6	81.8	102.5
	Sherbrooke	115.8	81.1	100.1
	Trois Rivieres	116.3	81.4	100.4
SASKATCHEWAN				
	Moose Jaw	114.2	64.8	91.8
	Prince Albert	113.6	64.7	91.4
	Regina	116.0	64.9	92.8
	Saskatoon	114.7	64.7	92.0
YUKON				
	Whitehorse	114.0	63.8	91.2

Abbreviations

A	Area Square Feet; Ampere	Cab.	Cabinet	Demob.	Demobilization
ABS	Acrylonitrile Butadiene Stryrene; Asbestos Bonded Steel	Cair.	Air Tool Laborer	d.f.u.	Drainage Fixture Units
A.C.	Alternating Current; Air-Conditioning; Asbestos Cement; Plywood Grade A & C	Calc	Calculated	D.H.	Double Hung
		Cap.	Capacity	DHW	Domestic Hot Water
		Carp.	Carpenter	Diag.	Diagonal
		C.B.	Circuit Breaker	Diam.	Diameter
A.C.I.	American Concrete Institute	C.C.A.	Chromate Copper Arsenate	Distrib.	Distribution
AD	Plywood, Grade A & D	C.C.F.	Hundred Cubic Feet	Dk.	Deck
Addit.	Additional	cd	Candela	D.L.	Dead Load; Diesel
Adj.	Adjustable	cd/sf	Candela per Square Foot	DLH	Deep Long Span Bar Joist
af	Audio-frequency	CD	Grade of Plywood Face & Back	Do.	Ditto
A.G.A.	American Gas Association	CDX	Plywood, Grade C & D, exterior glue	Dp.	Depth
Agg.	Aggregate			D.P.S.T.	Double Pole, Single Throw
A.H.	Ampere Hours	Cefi.	Cement Finisher	Dr.	Driver
A hr.	Ampere-hour	Cem.	Cement	Drink.	Drinking
A.H.U.	Air Handling Unit	CF	Hundred Feet	D.S.	Double Strength
A.I.A.	American Institute of Architects	C.F.	Cubic Feet	D.S.A.	Double Strength A Grade
AIC	Ampere Interrupting Capacity	CFM	Cubic Feet per Minute	D.S.B.	Double Strength B Grade
Allow.	Allowance	c.g.	Center of Gravity	Dty.	Duty
alt.	Altitude	CHW	Chilled Water; Commercial Hot Water	DWV	Drain Waste Vent
Alum.	Aluminum			DX	Deluxe White, Direct Expansion
a.m.	Ante Meridiem	C.I.	Cast Iron	dyn	Dyne
Amp.	Ampere	C.I.P.	Cast in Place	e	Eccentricity
Anod.	Anodized	Circ.	Circuit	E	Equipment Only; East
Approx.	Approximate	C.L.	Carload Lot	Ea.	Each
Apt.	Apartment	Clab.	Common Laborer	E.B.	Encased Burial
Asb.	Asbestos	Clam	Common maintenance laborer	Econ.	Economy
A.S.B.C.	American Standard Building Code	C.L.F.	Hundred Linear Feet	E.C.Y	Embankment Cubic Yards
Asbe.	Asbestos Worker	CLF	Current Limiting Fuse	EDP	Electronic Data Processing
A.S.H.R.A.E.	American Society of Heating, Refrig. & AC Engineers	CLP	Cross Linked Polyethylene	EIFS	Exterior Insulation Finish System
		cm	Centimeter	E.D.R.	Equiv. Direct Radiation
A.S.M.E.	American Society of Mechanical Engineers	CMP	Corr. Metal Pipe	Eq.	Equation
		C.M.U.	Concrete Masonry Unit	Elec.	Electrician; Electrical
A.S.T.M.	American Society for Testing and Materials	CN	Change Notice	Elev.	Elevator; Elevating
		Col.	Column	EMT	Electrical Metallic Conduit; Thin Wall Conduit
Attchmt.	Attachment	CO₂	Carbon Dioxide		
Avg.	Average	Comb.	Combination	Eng.	Engine, Engineered
A.W.G.	American Wire Gauge	Compr.	Compressor	EPDM	Ethylene Propylene Diene Monomer
AWWA	American Water Works Assoc.	Conc.	Concrete		
Bbl.	Barrel	Cont.	Continuous; Continued	EPS	Expanded Polystyrene
B. & B.	Grade B and Better; Balled & Burlapped	Corr.	Corrugated	Eqhv.	Equip. Oper., Heavy
		Cos	Cosine	Eqlt.	Equip. Oper., Light
B. & S.	Bell and Spigot	Cot	Cotangent	Eqmd.	Equip. Oper., Medium
B. & W.	Black and White	Cov.	Cover	Eqmm.	Equip. Oper., Master Mechanic
b.c.c.	Body-centered Cubic	C/P	Cedar on Paneling	Eqol.	Equip. Oper., Oilers
B.C.Y.	Bank Cubic Yards	CPA	Control Point Adjustment	Equip.	Equipment
BE	Bevel End	Cplg.	Coupling	ERW	Electric Resistance Welded
B.F	Board Feet	C.P.M.	Critical Path Method	E.S.	Energy Saver
Bg. cem.	Bag of Cement	CPVC	Chlorinated Polyvinyl Chloride	Est.	Estimated
BHP	Boiler Horsepower; Brake Horsepower	C.Pr.	Hundred Pair	esu	Electrostatic Units
		CRC	Cold Rolled Channel	E.W.	Each Way
B.I.	Black Iron	Creos.	Creosote	EWT	Entering Water Temperature
Bit.; Bitum.	Bituminous	Crpt.	Carpet & Linoleum Layer	Excav.	Excavation
Bk.	Backed	CRT	Cathode-ray Tube	Exp.	Expansion, Exposure
Bkrs.	Breakers	CS	Carbon Steel, Constant Shear Bar Joist	Ext.	Exterior
Bldg.	Building			Extru.	Extrusion
Blk.	Block	Csc	Cosecant	f.	Fiber stress
Bm.	Beam	C.S.F.	Hundred Square Feet	F	Fahrenheit; Female; Fill
Boil.	Boilermaker	CSI	Construction Specifications Institute	Fab.	Fabricated
B.P.M.	Blows per Minute			FBGS	Fiberglass
BR	Bedroom	C.T.	Current Transformer	F.C.	Footcandles
Brg.	Bearing	CTS	Copper Tube Size	f.c.c.	Face-centered Cubic
Brhe.	Bricklayer Helper	Cu	Copper, Cubic	f'c.	Compressive Stress in Concrete; Extreme Compressive Stress
Bric.	Bricklayer	Cu. Ft.	Cubic Foot		
Brk.	Brick	cw	Continuous Wave	F.E.	Front End
Brng.	Bearing	C.W.	Cool White; Cold Water	FEP	Fluorinated Ethylene Propylene (Teflon)
Brs.	Brass	Cwt.	100 Pounds		
Brz.	Bronze	C.W.X.	Cool White Deluxe	F.G.	Flat Grain
Bsn.	Basin	C.Y.	Cubic Yard (27 cubic feet)	F.H.A.	Federal Housing Administration
Btr.	Better	C.Y./Hr.	Cubic Yard per Hour	Fig.	Figure
BTU	British Thermal Unit	Cyl.	Cylinder	Fin.	Finished
BTUH	BTU per Hour	d	Penny (nail size)	Fixt.	Fixture
B.U.R.	Built-up Roofing	D	Deep; Depth; Discharge	Fl. Oz.	Fluid Ounces
BX	Interlocked Armored Cable	Dis.;Disch.	Discharge	Flr.	Floor
c	Conductivity, Copper Sweat	Db.	Decibel	F.M.	Frequency Modulation; Factory Mutual
C	Hundred; Centigrade	Dbl.	Double		
C/C	Center to Center, Cedar on Cedar	DC	Direct Current	Fmg.	Framing
		DDC	Direct Digital Control	Fndtn.	Foundation

Fori.	Foreman, Inside	I.W.	Indirect Waste	M.C.F.	Thousand Cubic Feet
Foro.	Foreman, Outside	J	Joule	M.C.F.M.	Thousand Cubic Feet per Minute
Fount.	Fountain	J.I.C.	Joint Industrial Council	M.C.M.	Thousand Circular Mils
FPM	Feet per Minute	K	Thousand; Thousand Pounds;	M.C.P.	Motor Circuit Protector
FPT	Female Pipe Thread		Heavy Wall Copper Tubing, Kelvin	MD	Medium Duty
Fr.	Frame	K.A.H.	Thousand Amp. Hours	M.D.O.	Medium Density Overlaid
F.R.	Fire Rating	KCMIL	Thousand Circular Mils	Med.	Medium
FRK	Foil Reinforced Kraft	KD	Knock Down	MF	Thousand Feet
FRP	Fiberglass Reinforced Plastic	K.D.A.T.	Kiln Dried After Treatment	M.F.B.M.	Thousand Feet Board Measure
FS	Forged Steel	kg	Kilogram	Mfg.	Manufacturing
FSC	Cast Body; Cast Switch Box	kG	Kilogauss	Mfrs.	Manufacturers
Ft.	Foot; Feet	kgf	Kilogram Force	mg	Milligram
Ftng.	Fitting	kHz	Kilohertz	MGD	Million Gallons per Day
Ftg.	Footing	Kip.	1000 Pounds	MGPH	Thousand Gallons per Hour
Ft. Lb.	Foot Pound	KJ	Kiljoule	MH, M.H.	Manhole; Metal Halide; Man-Hour
Furn.	Furniture	K.L.	Effective Length Factor	MHz	Megahertz
FVNR	Full Voltage Non-Reversing	K.L.F.	Kips per Linear Foot	Mi.	Mile
FXM	Female by Male	Km	Kilometer	MI	Malleable Iron; Mineral Insulated
Fy.	Minimum Yield Stress of Steel	K.S.F.	Kips per Square Foot	mm	Millimeter
g	Gram	K.S.I.	Kips per Square Inch	Mill.	Millwright
G	Gauss	kV	Kilovolt	Min., min.	Minimum, minute
Ga.	Gauge	kVA	Kilovolt Ampere	Misc.	Miscellaneous
Gal.	Gallon	K.V.A.R.	Kilovar (Reactance)	ml	Milliliter, Mainline
Gal./Min.	Gallon per Minute	KW	Kilowatt	M.L.F.	Thousand Linear Feet
Galv.	Galvanized	KWh	Kilowatt-hour	Mo.	Month
Gen.	General	L	Labor Only; Length; Long;	Mobil.	Mobilization
G.F.I.	Ground Fault Interrupter		Medium Wall Copper Tubing	Mog.	Mogul Base
Glaz.	Glazier	Lab.	Labor	MPH	Miles per Hour
GPD	Gallons per Day	lat	Latitude	MPT	Male Pipe Thread
GPH	Gallons per Hour	Lath.	Lather	MRT	Mile Round Trip
GPM	Gallons per Minute	Lav.	Lavatory	ms	Millisecond
GR	Grade	lb.; #	Pound	M.S.F.	Thousand Square Feet
Gran.	Granular	L.B.	Load Bearing; L Conduit Body	Mstz.	Mosaic & Terrazzo Worker
Grnd.	Ground	L. & E.	Labor & Equipment	M.S.Y.	Thousand Square Yards
H	High; High Strength Bar Joist;	lb./hr.	Pounds per Hour	Mtd.	Mounted
	Henry	lb./L.F.	Pounds per Linear Foot	Mthe.	Mosaic & Terrazzo Helper
H.C.	High Capacity	lbf/sq.in.	Pound-force per Square Inch	Mtng.	Mounting
H.D.	Heavy Duty; High Density	L.C.L.	Less than Carload Lot	Mult.	Multi; Multiply
H.D.O.	High Density Overlaid	L.C.Y.	Loose Cubic Yards	M.V.A.	Million Volt Amperes
Hdr.	Header	Ld.	Load	M.V.A.R.	Million Volt Amperes Reactance
Hdwe.	Hardware	LE	Lead Equivalent	MV	Megavolt
Help.	Helpers Average	LED	Light Emitting Diode	MW	Megawatt
HEPA	High Efficiency Particulate Air	L.F.	Linear Foot	MXM	Male by Male
	Filter	Lg.	Long; Length; Large	MYD	Thousand Yards
Hg	Mercury	L & H	Light and Heat	N	Natural; North
HIC	High Interrupting Capacity	LH	Long Span Bar Joist	nA	Nanoampere
HM	Hollow Metal	L.H.	Labor Hours	NA	Not Available; Not Applicable
H.O.	High Output	L.L.	Live Load	N.B.C.	National Building Code
Horiz.	Horizontal	L.L.D.	Lamp Lumen Depreciation	NC	Normally Closed
H.P.	Horsepower; High Pressure	L-O-L	Lateralolet	N.E.M.A.	National Electrical Manufacturers
H.P.F.	High Power Factor	lm	Lumen		Assoc.
Hr.	Hour	lm/sf	Lumen per Square Foot	NEHB	Bolted Circuit Breaker to 600V.
Hrs./Day	Hours per Day	lm/W	Lumen per Watt	N.L.B.	Non-Load-Bearing
HSC	High Short Circuit	L.O.A.	Length Over All	NM	Non-Metallic Cable
Ht.	Height	log	Logarithm	nm	Nanometer
Htg.	Heating	L.P.	Liquefied Petroleum; Low Pressure	No.	Number
Htrs.	Heaters	L.P.F.	Low Power Factor	NO	Normally Open
HVAC	Heating, Ventilation & Air-	LR	Long Radius	N.O.C.	Not Otherwise Classified
	Conditioning	L.S.	Lump Sum	Nose.	Nosing
Hvy.	Heavy	Lt.	Light	N.P.T.	National Pipe Thread
HW	Hot Water	Lt. Ga.	Light Gauge	NQOD	Combination Plug-on/Bolt on
Hyd.;Hydr.	Hydraulic	L.T.L.	Less than Truckload Lot		Circuit Breaker to 240V.
Hz.	Hertz (cycles)	Lt. Wt.	Lightweight	N.R.C.	Noise Reduction Coefficient
I.	Moment of Inertia	L.V.	Low Voltage	N.R.S.	Non Rising Stem
I.C.	Interrupting Capacity	M	Thousand; Material; Male;	ns	Nanosecond
ID	Inside Diameter		Light Wall Copper Tubing	nW	Nanowatt
I.D.	Inside Dimension; Identification	M²CA	Meters Squared Contact Area	OB	Opposing Blade
I.F.	Inside Frosted	m/hr; M.H.	Man-hour	OC	On Center
I.M.C.	Intermediate Metal Conduit	mA	Milliampere	OD	Outside Diameter
In.	Inch	Mach.	Machine	O.D.	Outside Dimension
Incan.	Incandescent	Mag. Str.	Magnetic Starter	ODS	Overhead Distribution System
Incl.	Included; Including	Maint.	Maintenance	O.G.	Ogee
Int.	Interior	Marb.	Marble Setter	O.H.	Overhead
Inst.	Installation	Mat; Mat'l.	Material	O & P	Overhead and Profit
Insul.	Insulation/Insulated	Max.	Maximum	Oper.	Operator
I.P.	Iron Pipe	MBF	Thousand Board Feet	Opng.	Opening
I.P.S.	Iron Pipe Size	MBH	Thousand BTU's per hr.	Orna.	Ornamental
I.P.T.	Iron Pipe Threaded	MC	Metal Clad Cable	OSB	Oriented Strand Board

O. S. & Y.	Outside Screw and Yoke	Rsr	Riser
Ovhd.	Overhead	RT	Round Trip
OWG	Oil, Water or Gas	S.	Suction; Single Entrance; South
Oz.	Ounce	SCFM	Standard Cubic Feet per Minute
P.	Pole; Applied Load; Projection	Scaf.	Scaffold
p.	Page	Sch.; Sched.	Schedule
Pape.	Paperhanger	S.C.R.	Modular Brick
P.A.P.R.	Powered Air Purifying Respirator	S.D.	Sound Deadening
PAR	Parabolic Reflector	S.D.R.	Standard Dimension Ratio
Pc., Pcs.	Piece, Pieces	S.E.	Surfaced Edge
P.C.	Portland Cement; Power Connector	Sel.	Select
P.C.F.	Pounds per Cubic Foot	S.E.R.;	Service Entrance Cable
P.C.M.	Phase Contract Microscopy	S.E.U.	Service Entrance Cable
P.E.	Professional Engineer;	S.F.	Square Foot
	Porcelain Enamel;	S.F.C.A.	Square Foot Contact Area
	Polyethylene; Plain End	S.F. Flr.	Square Foot of Floor
Perf.	Perforated	S.F.G.	Square Foot of Ground
Ph.	Phase	S.F. Hor.	Square Foot Horizontal
P.I.	Pressure Injected	S.F.R.	Square Feet of Radiation
Pile.	Pile Driver	S.F. Shlf.	Square Foot of Shelf
Pkg.	Package	S4S	Surface 4 Sides
Pl.	Plate	Shee.	Sheet Metal Worker
Plah.	Plasterer Helper	Sin.	Sine
Plas.	Plasterer	Skwk.	Skilled Worker
Pluh.	Plumbers Helper	SL	Saran Lined
Plum.	Plumber	S.L.	Slimline
Ply.	Plywood	Sldr.	Solder
p.m.	Post Meridiem	SLH	Super Long Span Bar Joist
Pntd.	Painted	S.N.	Solid Neutral
Pord.	Painter, Ordinary	S-O-L	Socketolet
pp	Pages	sp	Standpipe
PP; PPL	Polypropylene	S.P.	Static Pressure; Single Pole; Self-
P.P.M.	Parts per Million		Propelled
Pr.	Pair	Spri.	Sprinkler Installer
P.E.S.B.	Pre-engineered Steel Building	spwg	Static Pressure Water Gauge
Prefab.	Prefabricated	Sq.	Square; 100 Square Feet
Prefin.	Prefinished	S.P.D.T.	Single Pole, Double Throw
Prop.	Propelled	SPF	Spruce Pine Fir
PSF; psf	Pounds per Square Foot	S.P.S.T.	Single Pole, Single Throw
PSI; psi	Pounds per Square Inch	SPT	Standard Pipe Thread
PSIG	Pounds per Square Inch Gauge	Sq. Hd.	Square Head
PSP	Plastic Sewer Pipe	Sq. In.	Square Inch
Pspr.	Painter, Spray	S.S.	Single Strength; Stainless Steel
Psst.	Painter, Structural Steel	S.S.B.	Single Strength B Grade
P.T.	Potential Transformer	sst	Stainless Steel
P. & T.	Pressure & Temperature	Sswk.	Structural Steel Worker
Ptd.	Painted	Sswl.	Structural Steel Welder
Ptns.	Partitions	St.; Stl.	Steel
Pu	Ultimate Load	S.T.C.	Sound Transmission Coefficient
PVC	Polyvinyl Chloride	Std.	Standard
Pvmt.	Pavement	STK	Select Tight Knot
Pwr.	Power	STP	Standard Temperature & Pressure
Q	Quantity Heat Flow	Stpi.	Steamfitter, Pipefitter
Quan.; Qty.	Quantity	Str.	Strength; Starter; Straight
Q.C.	Quick Coupling	Strd.	Stranded
r	Radius of Gyration	Struct.	Structural
R	Resistance	Sty.	Story
R.C.P.	Reinforced Concrete Pipe	Subj.	Subject
Rect.	Rectangle	Subs.	Subcontractors
Reg.	Regular	Surf.	Surface
Reinf.	Reinforced	Sw.	Switch
Req'd.	Required	Swbd.	Switchboard
Res.	Resistant	S.Y.	Square Yard
Resi.	Residential	Syn.	Synthetic
Rgh.	Rough	S.Y.P.	Southern Yellow Pine
RGS	Rigid Galvanized Steel	Sys.	System
R.H.W.	Rubber, Heat & Water Resistant;	t.	Thickness
	Residential Hot Water	T	Temperature; Ton
rms	Root Mean Square	Tan	Tangent
Rnd.	Round	T.C.	Terra Cotta
Rodm.	Rodman	T & C	Threaded and Coupled
Rofc.	Roofer, Composition	T.D.	Temperature Difference
Rofp.	Roofer, Precast	T.E.M.	Transmission Electron Microscopy
Rohe.	Roofer Helpers (Composition)	TFE	Tetrafluoroethylene (Teflon)
Rots.	Roofer, Tile & Slate	T. & G.	Tongue & Groove;
R.O.W.	Right of Way		Tar & Gravel
RPM	Revolutions per Minute	Th.; Thk.	Thick
R.S.	Rapid Start	Thn.	Thin

Thrded	Threaded
Tilf.	Tile Layer, Floor
Tilh.	Tile Layer, Helper
THHN	Nylon Jacketed Wire
THW.	Insulated Strand Wire
THWN;	Nylon Jacketed Wire
T.L.	Truckload
T.M.	Track Mounted
Tot.	Total
T-O-L	Threadolet
T.S.	Trigger Start
Tr.	Trade
Transf.	Transformer
Trhv.	Truck Driver, Heavy
Trlr	Trailer
Trlt.	Truck Driver, Light
TV	Television
T.W.	Thermoplastic Water Resistant
	Wire
UCI	Uniform Construction Index
UF	Underground Feeder
UGND	Underground Feeder
U.H.F.	Ultra High Frequency
U.L.	Underwriters Laboratory
Unfin.	Unfinished
URD	Underground Residential
	Distribution
US	United States
USP	United States Primed
UTP	Unshielded Twisted Pair
V	Volt
V.A.	Volt Amperes
V.C.T.	Vinyl Composition Tile
VAV	Variable Air Volume
VC	Veneer Core
Vent.	Ventilation
Vert.	Vertical
V.F.	Vinyl Faced
V.G.	Vertical Grain
V.H.F.	Very High Frequency
VHO	Very High Output
Vib.	Vibrating
V.L.F.	Vertical Linear Foot
Vol.	Volume
VRP	Vinyl Reinforced Polyester
W	Wire; Watt; Wide; West
w/	With
W.C.	Water Column; Water Closet
W.F.	Wide Flange
W.G.	Water Gauge
Wldg.	Welding
W. Mile	Wire Mile
W-O-L	Weldolet
W.R.	Water Resistant
Wrck.	Wrecker
W.S.P.	Water, Steam, Petroleum
WT., Wt.	Weight
WWF	Welded Wire Fabric
XFER	Transfer
XFMR	Transformer
XHD	Extra Heavy Duty
XHHW; XLPE	Cross-Linked Polyethylene Wire
	Insulation
XLP	Cross-linked Polyethylene
Y	Wye
yd	Yard
yr	Year
Δ	Delta
%	Percent
~	Approximately
Ø	Phase
@	At
#	Pound; Number
<	Less Than
>	Greater Than

INDEX

685

loading 355
shelter 356
truck 356
Dog command 18
Dome drain 444
fiberglass 94
geodesic 392, 393
metal 94
roof 283
structures 585
Domes 391
Door accordion 252
acoustical 259, 341
air curtain 470
air lock 386
aluminum 261, 262
and grilles coiling 257
and panels access 255
and window matls 244
automatic entrance 262
balanced 263
bell residential 489
bell system 502
bi-fold 247, 252
bi-passing closet 252
birch 253
blind 209
bulkhead 256
bullet resistant 349
bumper 272, 273, 278
cabinet 205, 206
cafe 251
catch 206
chain hoist 259
cleanout 332
closer 272
closet 247, 252
closing coordinator 275
cold storage 258
combination storm 252
commercial 245, 254
composite 245
conveyor 470
cooler 470
counter 257
darkroom 365
demolition 244
double 246
duct access 466
dutch 252
dutch oven 332
entrance 252, 262
fire 246, 253, 258, 408
fire rated access 256
flexible 259
floor 256, 257
flush 249, 252
flush wood 249
folding 340
folding accordion 252
frame 246, 247, 254, 395
frame grout 124
frame interior 255
frame lead lined 389
garage 259, 260
glass 255, 259, 262
handle 206
hangar 259, 393
hardware 271, 272
industrial 257
jamb molding 202
kennel 257
kick plate 274
knob 274
labeled 246, 253
metal 245, 260, 395

metal access 256
mirror 280
molding 202
moulded 251
opener 244, 258, 260, 271, 272
operator 244, 272
overhead 259, 260
overhead commercial 259
panel 251
paneled 246
partition 340
passage 249, 253
plastic 248, 255
prefinished 249, 250
pre-hung 248, 255
prison 357
protection 278
pull 276
refrigerator 258
release 503
removal 244
residential 246, 251, 254
residential garage 260
revolving 263, 365, 386
revolving entrance 263
rolling 257
roof 237
sauna 388
seal 356
sectional 259
sectional overhead 260
security vault 349
shower 328
side light 255
sidelight 247
sill 202, 254, 278
sliding 248, 255
special 255, 259
stain 312
stainless steel (access) 256
steel 245, 246, 260, 580
stop 273, 282
storm 248
swing 259
switch burglar alarm 408
telescoping 260
threshold 255
traffic 261
varnish 312
vault 348
vertical lift 261
weatherstrip 277
weatherstrip garage 278
wire partition 339
wood 248-250, 254
Doors & windows interior paint . 311
and frames detention 257
Dormer gable 197
Dormitory S.F. & C.F. 508
Double acting door 259
brick 126
hung window 263, 266
tee beam 116
wall pipe 62, 63
weight hinge 276
Dovetail anchor 100
Dowel support 95
Dowels reinforcing 107
Downspout 234, 235, 338
aluminum 234
copper 234
elbow 235
lead coated copper 234
steel 234
strainer 234
Dozer 21

backfill 45, 46
excavating bulk 47
excavation 22, 47, 48
Draft inducer breeching 459
Dragline 47, 52
bucket 20
Drain 443-445
board 216
deck 443
dome 444
floor 443
main 444
pipe 41
scupper 444
sediment bucket 443
shower 443
trench 445
Drainage accessory 235
mat 376
pipe 61, 66, 67, 429, 430
site 65
trap 443, 444
Drains roof 444
Drapery hardware 377
Drawer 206
kitchen 204
track 206
type cabinet 374
wood 206
Drawings as-built 29
Dredge 52
hydraulic 52
Dredging 52
Dressing unit 379
Drill auger 35
bit 35
concrete 34
console dental 371
core 11, 19, 34
earth 34, 57
hammer 23
hole 44
rig 34
rig mobilization 57
rock 34, 46
rotary 23
shop 369
steel 23
track 22
wood 188
Drilled concrete piers and shafts . 58
pier 58
Drilling anchors 144
ceiling 144
concrete 144
horizontal 54
only rock 46
plaster 144
quarry 46
rock bolt 46
steel 144
Drinking bubbler 449
fountain 449, 450
fountain deck rough-in 450
fountain floor rough-in 449
fountain handicap 449
fountain support 446
fountain wall rough-in 449
Drip edge 235
edge aluminum 235
Dripproof motor 492
Driven pile 54
Driver post 26
sheeting 23
Drive-up window 349
Driveway 71

removal 37
Drop pan ceiling 300
panel formwork 94
Drum trap copper 444
Dry chemical extinguisher 337
fall painting 318
pipe sprinkler head 412
transformer 493
wall leaded 390
Drycleaner 353
Dryer clothes 363
commercial clothes 353
darkroom 365
hand 343
receptacle 489
residential 353
vent 364
Drywall 293
accessories 296
ceiling 292
column 293
cutout 38
demolition 286
finish 583
frame 247
gypsum 293, 294
high abuse 294
nail 187
painting 317
partition 295
prefinished 293
removal 286
screw 296
Duck tarpaulin 17
Duckboard 376
Duct access door 466
accessories 466
connection fabric flexible 467
electric 483, 497
electrical underfloor 484
fitting trench 483
fitting underfloor 484
flexible insulated 466
flexible noninsulated 466
furnace 457, 458
grille 331
heater electric 464
humidifier 464
HVAC 466
insulation 427
liner 427
mechanical 466
silencer 467
steel trench 483
trench 483
underfloor 484
underground 64
utility 64
Ductile iron fitting 59
iron fitting mech joint 59
iron grooved joint 432
iron pipe 59
Ducts flexible 466
Ductwork 466, 593
aluminum 466
demolition 426
fabric coated flexible 466
fabricated 466
fiberglass 466
galvanized 466
laboratory 370
metal 466
rectangular 466
rigid 466
Dumbbell 366
waterstop 103

Index

Index

Reed Construction Data, Inc.

Reed Construction Data, Inc., a leading worldwide provider of total construction information solutions, is comprised of three main product groups designed specifically to help construction professionals advance their businesses with timely, accurate and actionable project, product, and cost data. Reed Construction Data is a division of Reed Business Information, a member of the Reed Elsevier plc group of companies.

The *Project, Product, and Cost & Estimating* divisions offer a variety of innovative products and services designed for the full spectrum of design, construction, and manufacturing professionals. Through it's *International* companies, Reed Construction Data's reputation for quality construction market data is growing worldwide.

Cost Information
RSMeans, the undisputed market leader and authority on construction costs, publishes current cost and estimating information in annual cost books and on the CostWorks CD-ROM. RSMeans furnishes the construction industry with a rich library of complementary reference books and a series of professional seminars that are designed to sharpen professional skills and maximize the effective use of cost estimating and management tools. RSMeans also provides construction cost consulting for Owners, Manufacturers, Designers, and Contractors.

Project Data
Reed Construction Data provides complete, accurate and relevant project information through all stages of construction. Customers are supplied industry data through leads, project reports, contact lists, plans and specifications surveys, market penetration analyses and sales evaluation reports. Any of these products can pinpoint a county, look at a state, or cover the country. Data is delivered via paper, e-mail, CD-ROM or the Internet.

Building Product Information
The First Source suite of products is the only integrated building product information system offered to the commercial construction industry for comparing and specifying building products. These print and online resources include *First Source,* CSI's SPEC-DATA™, CSI's MANU-SPEC™, First Source CAD, and Manufacturer Catalogs. Written by industry professionals and organized using CSI's MasterFormat™, construction professionals use this information to make better design decisions.

FirstSourceONL.com combines Reed Construction Data's project, product and cost data with news and information from Reed Business Information's *Building Design & Construction* and *Consulting-Specifying Engineer,* this industry-focused site offers easy and unlimited access to vital information for all construction professionals.

International
BIMSA/Mexico provides construction project news, product information, cost-data, seminars and consulting services to construction professionals in Mexico. Its subsidiary, PRISMA, provides job costing software.

Byggfakta Scandinavia AB, founded in 1936, is the parent company for the leaders of customized construction market data for Denmark, Estonia, Finland, Norway and Sweden. Each company fully covers the local construction market and provides information across several platforms including subscription, ad-hoc basis, electronically and on paper.

Reed Construction Data Canada serves the Canadian construction market with reliable and comprehensive project and product information services that cover all facets of construction. Core services include: *BuildSource, BuildSpec, BuildSelect,* product selection and specification tools available in print and on the Internet; Building Reports, a national construction project lead service; CanaData, statistical and forecasting information; *Daily Commercial News,* a construction newspaper reporting on news and projects in Ontario; and *Journal of Commerce,* reporting news in British Columbia and Alberta.

Cordell Building Information Services, with its complete range of project and cost and estimating services, is Australia's specialist in the construction information industry. Cordell provides in-depth and historical information on all aspects of construction projects and estimation, including several customized reports, construction and sales leads, and detailed cost information among others.

For more information, please visit our Web site at www.reedconstructiondata.com.

Reed Construction Data, Inc., Corporate Office
30 Technology Parkway South
Norcross, GA 30092-2912
(800) 322-6996
(800) 895-8661 (fax)
info@reedbusiness.com
www.reedconstructiondata.com

Means Project Cost Report

By filling out and returning the Project Description, you can receive a discount of $20.00 off any one of the Means products advertised in the following pages. The cost information required includes all items marked (✔) except those where no costs occurred. The sum of all major items should equal the Total Project Cost.

$20.00 Discount per product for each report you submit.

DISCOUNT PRODUCTS AVAILABLE—FOR U.S. CUSTOMERS ONLY—STRICTLY CONFIDENTIAL

Project Description (No remodeling projects, please.)

✔ Type Building _____

✔ Location _____

  Capacity _____

✔ Frame _____

✔ Exterior _____

✔ Basement: full ☐ partial ☐ none ☐ crawl ☐

✔ Height in Stories _____

✔ Total Floor Area _____

  Ground Floor Area _____

✔ Volume in C.F. _____

  % Air Conditioned _____ Tons _____

  Comments _____

Owner _____

Architect _____

General Contractor _____

✔ Bid Date _____

  Typical Bay Size _____

✔ Labor Force: _____ % Union _____ % Non-Union

✔ Project Description (Circle one number in each line)
  1. Economy 2. Average 3. Custom 4. Luxury
  1. Square 2. Rectangular 3. Irregular 4. Very Irregular

	✔ **Total Project Cost**			$
A	✔ **General Conditions**			$
B	✔ **Site Work**			$
BS	**Site Clearing & Improvement**			
BE	*Excavation*	(	C.Y.)	
BF	*Caissons & Piling*	(	L.F.)	
BU	*Site Utilities*			
BP	*Roads & Walks Exterior Paving*	(	S.Y.)	
C	✔ **Concrete**			$
C	*Cast in Place*	(	C.Y.)	
CP	*Precast*	(	S.F.)	
D	✔ **Masonry**			$
DB	*Brick*	(	M)	
DC	*Block*	(	M)	
DT	*Tile*	(	S.F.)	
DS	*Stone*	(	S.F.)	
E	✔ **Metals**			$
ES	*Structural Steel*	(	Tons)	
EM	*Misc. & Ornamental Metals*			
F	✔ **Wood & Plastics**			$
FR	*Rough Carpentry*	(	MBF)	
FF	*Finish Carpentry*			
FM	*Architectural Millwork*			
G	✔ **Thermal & Moisture Protection**			$
GW	*Waterproofing-Dampproofing*	(	S.F.)	
GN	*Insulation*	(	S.F.)	
GR	*Roofing & Flashing*	(	S.F.)	
GM	*Metal Siding/Curtain Wall*	(	S.F.)	
H	✔ **Doors and Windows**			$
HD	*Doors*	(	Ea.)	
HW	*Windows*	(	S.F.)	
HH	*Finish Hardware*			
HG	*Glass & Glazing*	(	S.F.)	
HS	*Storefronts*	(	S.F.)	

J	✔ **Finishes**			$
JL	*Lath & Plaster*	(	S.Y.)	
JD	*Drywall*	(	S.F.)	
JM	*Tile & Marble*	(	S.F.)	
JT	*Terrazzo*	(	S.F.)	
JA	*Acoustical Treatment*	(	S.F.)	
JC	*Carpet*	(	S.Y.)	
JF	*Hard Surface Flooring*	(	S.F.)	
JP	*Painting & Wall Covering*	(	S.F.)	
K	✔ **Specialties**			$
KB	*Bathroom Partitions & Access.*	(	S.F.)	
KF	*Other Partitions*	(	S.F.)	
KL	*Lockers*	(	Ea.)	
L	✔ **Equipment**			$
LK	*Kitchen*			
LS	*School*			
LO	*Other*			
M	✔ **Furnishings**			$
MW	*Window Treatment*			
MS	*Seating*	(	Ea.)	
N	✔ **Special Construction**			$
NA	*Acoustical*	(	S.F.)	
NB	*Prefab. Bldgs.*	(	S.F.)	
NO	*Other*			
P	✔ **Conveying Systems**			$
PE	*Elevators*	(	Ea.)	
PS	*Escalators*	(	Ea.)	
PM	*Material Handling*			
Q	✔ **Mechanical**			$
QP	*Plumbing*	(No. of fixtures	)	
QS	*Fire Protection (Sprinklers)*			
QF	*Fire Protection (Hose Standpipes)*			
QB	*Heating, Ventilating & A.C.*			
QH	*Heating & Ventilating*	(BTU Output	)	
QA	*Air Conditioning*	(	Tons)	
R	✔ **Electrical**			$
RL	*Lighting*	(	S.F.)	
RP	*Power Service*			
RD	*Power Distribution*			
RA	*Alarms*			
RG	*Special Systems*			
S	✔ **Mech./Elec. Combined**			$

Product Name _____

Product Number _____

Your Name _____

Title _____

Company _____

☐ Company

☐ Home Street Address _____

City, State, Zip _____

☐ Please send _____ forms.

Please specify the Means product you wish to receive. Complete the address information as requested and return this form with your check (product cost less $20.00) to address below.

RSMeans Company, Inc.,
Square Foot Costs Department
P.O. Box 800
Kingston, MA 02364-9988

Reed Construction Data/RSMeans . . . a tradition of excellence in Construction Cost Information and Services since 1942.

For more information visit Means Web Site at www.rsmeans.com

Table of Contents

Book Selection Guide

The following table provides definitive information on the content of each cost data publication. The number of lines of data provided in each unit price or assemblies division, as well as the number of reference tables and crews is listed for each book. The presence of other elements such as an historical cost index, city cost indexes, square foot models or cross-referenced index is also indicated. You can use the table to help select the Means' book that has the quantity and type of information you most need in your work.

Unit Cost Divisions	Building Construction Costs	Mechanical	Electrical	Repair & Remodel.	Square Foot	Site Work Landsc.	Assemblies	Interior	Concrete Masonry	Open Shop	Heavy Construc.	Light Commercial	Facil. Construc.	Plumbing	Western Construction Costs	Residential
1	1104	822	907	1013		1045		847	1014	1102	1051	752	1528	874	1102	710
2	2505	1468	460	1450		7794		547	1387	2467	4978	692	4305	1584	2490	772
3	1401	113	100	726		1259		201	1782	1396	1404	229	1312	82	1399	248
4	835	18	0	647		657		574	1056	811	594	407	1058	0	821	334
5	1820	239	193	957		768		927	691	1788	1023	805	1802	316	1802	752
6	1489	82	78	1475		452		1406	318	1481	608	1626	1596	47	1826	1757
7	1264	158	74	1241		468		489	408	1264	349	956	1318	168	1264	752
8	1839	28	0	1907		313		1673	645	1821	51	1227	2034	0	1841	1193
9	1595	47	0	1424		233		1688	360	1547	159	1337	1809	47	1586	1216
10	861	47	25	498		193		700	170	863	0	394	906	233	861	217
11	1018	322	169	502		137		813	44	925	107	218	1173	291	925	104
12	307	0	0	47		210		1416	27	298	0	62	1435	0	298	64
13	1153	1005	375	494		388		884	75	1136	279	484	1783	928	1119	193
14	345	36	0	258		36		292	0	344	30	12	343	35	343	6
15	1987	13138	636	1779		1569		1174	59	1996	1766	1217	10865	9695	2017	826
16	1277	470	10063	1002		739		1108	55	1265	760	1081	9834	415	1226	552
17	427	354	427	0		0		0	0	427	0	0	427	356	427	0
Totals	21227	18347	13507	15420		16261		14739	8091	20961	13159	11499	43528	15071	21347	9696

Assembly Divisions	Building Construction Costs	Mechanical	Electrical	Repair & Remodel.	Square Foot	Site Work Landsc.	Assemblies	Interior	Concrete Masonry	Open Shop	Heavy Construc.	Light Commercial	Facil. Construc.	Plumbing	Western Construction Costs	Asm Div	Residential
A		19	0	192	150	540	612	0	550		542	149	24	0		1	374
B		0	0	809	2480	0	5590	333	1914		0	2024	144	0		2	217
C		0	0	635	862	0	1220	1568	145		0	767	238	0		3	588
D		1031	780	693	1823	0	2439	753	0		0	1310	1027	896		4	871
E		0	0	85	255	0	292	5	0		0	255	5	0		5	393
F		0	0	0	123	0	126	0	0		0	123	3	0		6	358
G		465	172	332	111	1856	584	0	482		432	110	113	560		7	299
																8	760
																9	80
																10	0
																11	0
																12	0
Totals		1515	952	2746	5804	2396	10863	2659	3091		974	4738	1554	1456			3940

Reference Section	Building Construction Costs	Mechanical	Electrical	Repair & Remodel.	Square Foot	Site Work Landsc.	Assemblies	Interior	Concrete Masonry	Open Shop	Heavy Construc.	Light Commercial	Facil. Construc.	Plumbing	Western Construction Costs	Residential
Tables	130	45	85	69	4	80	219	46	71	130	61	57	104	51	131	42
Models					102							43				32
Crews	410	410	410	391		410		410	410	393	410	393	391	410	410	393
City Cost Indexes	yes	yes	yes	yes	yes	yes	yes	yes	yes	yes	yes	yes	yes	yes	yes	yes
Historical Cost Indexes	yes	yes	yes	yes	yes	yes	yes	yes	yes	yes	yes	yes	yes	yes	yes	no

1

Annual Cost Guides

For more information
visit Means Web Site
at www.rsmeans.con

Means
Building Construction
Cost Data 2005

Available in Both Softbound and Looseleaf Editions

The "Bible" of the industry comes in the standard softcover edition or the looseleaf edition.

Many customers enjoy the convenience and flexibility of the looseleaf binder, which increases the usefulness of *Means Building Construction Cost Data 2005* by making it easy to add and remove pages. You can insert your own cost information pages, so everything is in one place. Copying pages for faxing is easier also. Whichever edition you prefer, softbound or the convenient looseleaf edition, you'll be eligible to receive *The Change Notice* FREE. Current subscribers receive *The Change Notice* via e-mail.

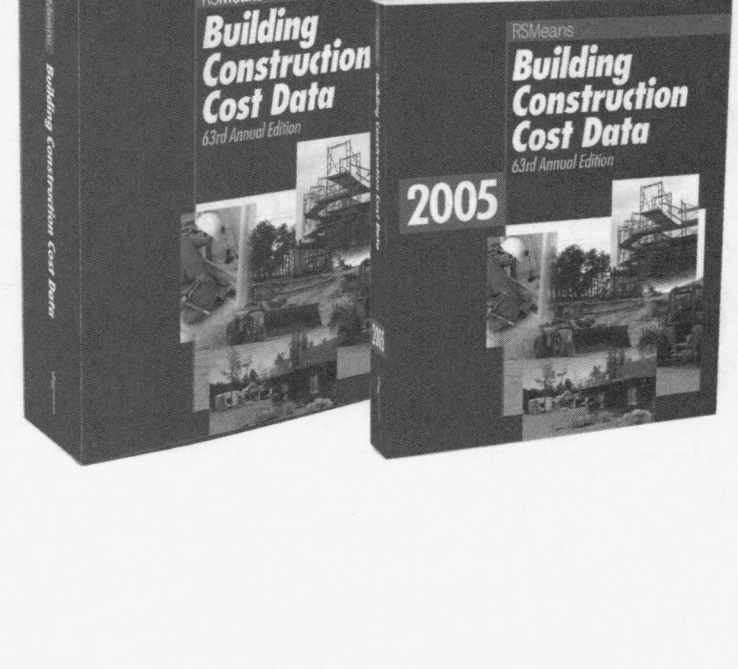

$115.95 per copy, Softbound
Catalog No. 60015

$144.95 per copy, Looseleaf
Catalog No. 61015

Means Building Construction Cost Data 2005

Offers you unchallenged unit price reliability in an easy-to-use arrangement. Whether used for complete, finished estimates or for periodic checks, it supplies more cost facts better and faster than any comparable source. Over 23,000 unit prices for 2005. The City Cost Indexes cover over 930 areas, for indexing to any project location in North America. Order and get *The Change Notice* FREE. You'll have year-long access to the Means Estimating **HOTLINE** FREE with your subscription. Expert assistance when using Means data is just a phone call away.

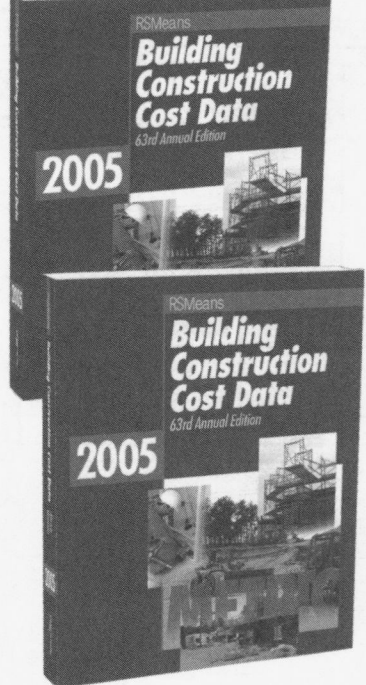

$115.95 per copy
Over 700 pages, illustrated, available Oct. 2004
Catalog No. 60015

Means Building Construction Cost Data 2005

Metric Version

The Federal Government has stated that all federal construction projects must now use metric documentation. The *Metric Version* of *Means Building Construction Cost Data 2005* is presented in metric measurements covering all construction areas. Don't miss out on these billion dollar opportunities. Make the switch to metric today.

$115.95 per copy
Over 700 pages, illus., available Dec. 2004
Catalog No. 63015

For more information
visit Means Web Site
at www.rsmeans.com

Annual Cost Guides

Means Mechanical Cost Data 2005

• HVAC • Controls

Total unit and systems price guidance for mechanical construction...materials, parts, fittings, and complete labor cost information. Includes prices for piping, heating, air conditioning, ventilation, and all related construction.

Plus new 2005 unit costs for:
• Over 2500 installed HVAC/controls assemblies
• "On Site" Location Factors for over 930 cities and towns in the U.S. and Canada
• Crews, labor and equipment

$115.95 per copy
Over 600 pages, illustrated, available Oct. 2004
Catalog No. 60025

Means Plumbing Cost Data 2005

Comprehensive unit prices and assemblies for plumbing, irrigation systems, commercial and residential fire protection, point-of-use water heaters, and the latest approved materials. This publication and its companion, *Means Mechanical Cost Data*, provide full-range cost estimating coverage for all the mechanical trades.

$115.95 per copy
Over 550 pages, illustrated, available Oct. 2004
Catalog No. 60215

Means Electrical Cost Data 2005

Pricing information for every part of electrical cost planning: More than 15,000 unit and systems costs with design tables; clear specifications and drawings; engineering guides and illustrated estimating procedures; complete labor-hour and materials costs for better scheduling and procurement; the latest electrical products and construction methods.
• A Variety of Special Electrical Systems including Cathodic Protection
• Costs for maintenance, demolition, HVAC/ mechanical, specialties, equipment, and more

$115.95 per copy
Over 450 pages, illustrated, available Oct. 2004
Catalog No. 60035

Means Electrical Change Order Cost Data 2005

You are provided with electrical unit prices exclusively for pricing change orders based on the recent, direct experience of contractors and suppliers. Analyze and check your own change order estimates against the experience others have had doing the same work. It also covers productivity analysis and change order cost justifications. With useful information for calculating the effects of change orders and dealing with their administration.

$115.95 per copy
Over 450 pages, available Oct. 2004
Catalog No. 60235

Means Facilities Maintenance & Repair Cost Data 2005

Published in a looseleaf format, *Means Facilities Maintenance & Repair Cost Data* gives you a complete system to manage and plan your facility repair and maintenance costs and budget efficiently. Guidelines for auditing a facility and developing an annual maintenance plan. Budgeting is included, along with reference tables on cost and management and information on frequency and productivity of maintenance operations.

The only nationally recognized source of maintenance and repair costs. Developed in cooperation with the Army Corps of Engineers.

$252.95 per copy
Over 600 pages, illustrated, available Dec. 2004
Catalog No. 60305

Means Square Foot Costs 2005

It's Accurate and Easy To Use!

• **Updated 2005 price information,** based on nationwide figures from suppliers, estimators, labor experts and contractors.
• "How-to-Use" Sections, with **clear examples** of commercial, residential, industrial, and institutional structures.
• Realistic graphics, offering true-to-life illustrations of building projects.
• Extensive information on using square foot cost data, including **sample estimates** and **alternate pricing methods.**

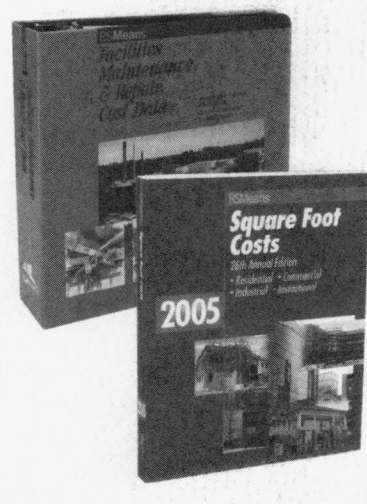

$126.95 per copy
Over 450 pages, illustrated, available Nov. 2004
Catalog No. 60055

For more information
visit Means Web Site
at www.rsmeans.com

Annual Cost Guides

Means Repair & Remodeling Cost Data 2005

Commercial/Residential

You can use this valuable tool to estimate commercial and residential renovation and remodeling.

Includes: New costs for hundreds of unique methods, materials and conditions that only come up in repair and remodeling. PLUS:

- Unit costs for over 16,000 construction components
- Installed costs for over 90 assemblies
- Costs for 300+ construction crews
- Over 930 "On Site" localization factors for the U.S. and Canada.

$99.95 per copy
Over 650 pages, illustrated, available Nov. 2004
Catalog No. 60045

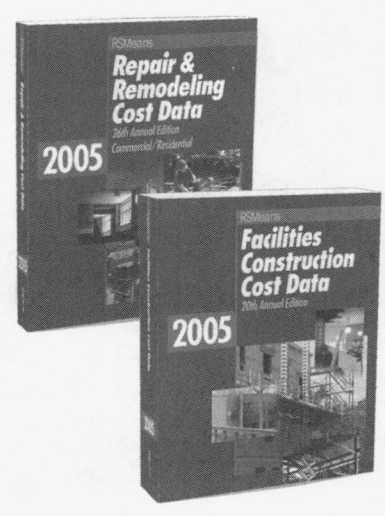

Means Facilities Construction Cost Data 2005

For the maintenance and construction of commercial, industrial, municipal, and institutional properties. Costs are shown for new and remodeling construction and are broken down into materials, labor, equipment, overhead, and profit. Special emphasis is given to sections on mechanical, electrical, furnishings, site work, building maintenance, finish work, and demolition. More than 45,000 unit costs, plus assemblies and reference sections are included.

$279.95 per copy
Over 1200 pages, illustrated, available Nov. 2004
Catalog No. 60205

Means Residential Cost Data 2005

Contains square foot costs for 30 basic home models with the look of today, plus hundreds of custom additions and modifications you can quote right off the page. With costs for the 100 residential systems you're most likely to use in the year ahead. Complete with blank estimating forms, sample estimates and step-by-step instructions.

$99.95 per copy
Over 600 pages, illustrated, available Oct. 2004
Catalog No. 60175

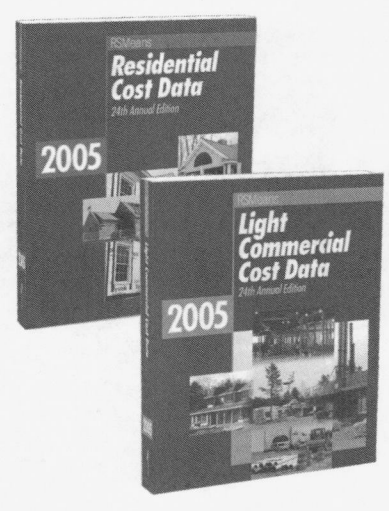

Means Light Commercial Cost Data 2005

Specifically addresses the light commercial market, which is an increasingly specialized niche in the industry. Aids you, the owner/designer/contractor, in preparing all types of estimates, from budgets to detailed bids. Includes new advances in methods and materials. Assemblies section allows you to evaluate alternatives in the early stages of design/planning.

Over 13,000 unit costs for 2005 ensure you have the prices you need...when you need them.

$99.95 per copy
Over 650 pages, illustrated, available Nov. 2004
Catalog No. 60185

Means Assemblies Cost Data 2005

Means Assemblies Cost Data 2005 takes the guesswork out of preliminary or conceptual estimates. Now you don't have to try to calculate the assembled cost by working up individual components costs. We've done all the work for you.

Presents detailed illustrations, descriptions, specifications and costs for every conceivable building assembly—240 types in all—arranged in the easy-to-use UNIFORMAT II system. Each illustrated "assembled" cost includes a complete grouping of materials and associated installation costs including the installing contractor's overhead and profit.

$189.95 per copy
Over 600 pages, illustrated, available Oct. 2004
Catalog No. 60065

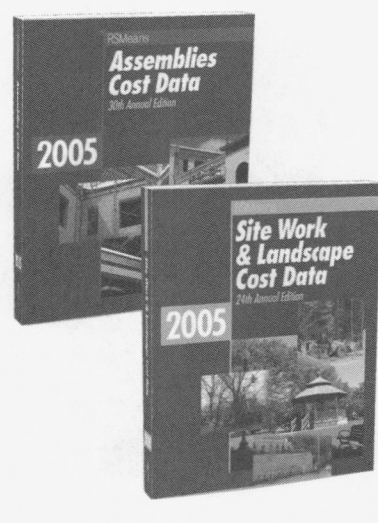

Means Site Work & Landscape Cost Data 2005

Means Site Work & Landscape Cost Data 2005 is organized to assist you in all your estimating needs. Hundreds of fact-filled pages help you make accurate cost estimates efficiently.

Updated for 2005!

- Demolition features—including ceilings, doors, electrical, flooring, HVAC, millwork, plumbing, roofing, walls and windows
- State-of-the-art segmental retaining walls
- Flywheel trenching costs and details
- Updated Wells section
- Landscape materials, flowers, shrubs and trees

$115.95 per copy
Over 600 pages, illustrated, available Nov. 2004
Catalog No. 60285

For more information
visit Means Web Site
at www.rsmeans.com

Annual Cost Guides

Means Open Shop Building Construction Cost Data 2005

The latest costs for accurate budgeting and estimating of new commercial and residential construction... renovation work... change orders... cost engineering. *Means Open Shop BCCD* will assist you to...
• Develop benchmark prices for change orders
• Plug gaps in preliminary estimates, budgets
• Estimate complex projects
• Substantiate invoices on contracts
• Price ADA-related renovations

$115.95 per copy
Over 700 pages, illustrated, available Dec. 2004
Catalog No. 60155

Means Heavy Construction Cost Data 2005

A comprehensive guide to heavy construction costs. Includes costs for highly specialized projects such as tunnels, dams, highways, airports, and waterways. Information on different labor rates, equipment, and material costs is included. Has unit price costs, systems costs, and numerous reference tables for costs and design. Valuable not only to contractors and civil engineers, but also to government agencies and city/town engineers.

$115.95 per copy
Over 450 pages, illustrated, available Nov. 2004
Catalog No. 60165

Means Building Construction Cost Data 2005
Western Edition

This regional edition provides more precise cost information for western North America. Labor rates are based on union rates from 13 western states and western Canada. Included are western practices and materials not found in our national edition: tilt-up concrete walls, glu-lam structural systems, specialized timber construction, seismic restraints, landscape and irrigation systems.

$115.95 per copy
Over 650 pages, illustrated, available Dec. 2004
Catalog No. 60225

Means Heavy Construction Cost Data 2005
Metric Version

Make sure you have the Means industry standard metric costs for the federal, state, municipal and private marketplace. With thousands of up-to-date metric unit prices in tables by CSI standard divisions. Supplies you with assemblies costs using the metric standard for reliable cost projections in the design stage of your project. Helps you determine sizes, material amounts, and has tips for handling metric estimates.

$115.95 per copy
Over 450 pages, illustrated, available Dec. 2004
Catalog No. 63165

Means Construction Cost Indexes 2005

Who knows what 2005 holds? What materials and labor costs will change unexpectedly? By how much?
• Breakdowns for 316 major cities.
• National averages for 30 key cities.
• Expanded five major city indexes.
• Historical construction cost indexes.

$252.95 per year
$63.00 individual quarters
Catalog No. 60145 A,B,C,D

Means Interior Cost Data 2005

Provides you with prices and guidance needed to make accurate interior work estimates. Contains costs on materials, equipment, hardware, custom installations, furnishings, labor costs . . . every cost factor for new and remodel commercial and industrial interior construction, including updated information on office furnishings, plus more than 50 reference tables. For contractors, facility managers, owners.

$115.95 per copy
Over 600 pages, illustrated, available Oct. 2004
Catalog No. 60095

Means Concrete & Masonry Cost Data 2005

Provides you with cost facts for virtually all concrete/masonry estimating needs, from complicated formwork to various sizes and face finishes of brick and block, all in great detail. The comprehensive unit cost section contains more than 8,500 selected entries. Also contains an assemblies cost section, and a detailed reference section which supplements the cost data.

$105.95 per copy
Over 450 pages, illustrated, available Dec. 2004
Catalog No. 60115

Means Labor Rates for the Construction Industry 2005

Complete information for estimating labor costs, making comparisons and negotiating wage rates by trade for over 300 cities (United States and Canada). With 46 construction trades listed by local union number in each city, and historical wage rates included for comparison. No similar book is available through the trade.

Each city chart lists the county and is alphabetically arranged with handy visual flip tabs for quick reference.

$253.95 per copy
Over 300 pages, available Dec. 2004
Catalog No. 60125

For more information
visit Means Web Site
at www.rsmeans.com

Reference Books

Building Security: Strategies & Costs
By David Owen

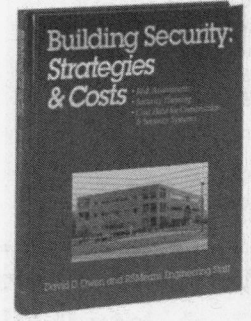

Unauthorized systems access, terrorism, hurricanes and tornadoes, sabotage, vandalism, fire, explosions, and other threats… All are considerations as building owners, facility managers, and design and construction professionals seek ways to meet security needs.

This comprehensive resource will help you evaluate your facility's security needs — and design and budget for the materials and devices needed to fulfill them. The text and cost data will help you to:

• Identify threats, probability of occurrence, and the potential losses— to determine and address your real vulnerabilities.
• Perform a detailed risk assessment of an existing facility, and prioritize and budget for security enhancement.
• Evaluate and price security systems and construction solutions, so you can make cost-effective choices.

Includes over 130 pages of Means Cost Data for installation of security systems and materials, plus a review of more than 50 security devices and construction solutions—how they work, and how they compare.

$89.95 per copy
Over 400 pages, Hardcover
Catalog No. 37339

Green Building: Project Planning & Cost Estimating
By RSMeans and Contributing Authors

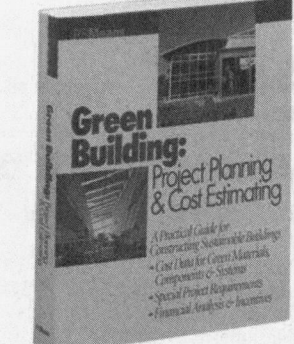

Written by a team of leading experts in sustainable design, this new book is a complete guide to planning and estimating green building projects, a growing trend in building design and construction - commercial, industrial, institutional and residential. It explains:

• All the different criteria for "green-ness"
• What criteria your building needs to meet to get a LEED, Energy Star, or other recognized rating for green buildings
• How the project team works differently on a green versus a traditional building project
• How to select and specify green products
• How to evaluate the cost and value of green products versus conventional ones — not only for their first (installation) cost, but their cost over time (in maintenance and operation).

Features an extensive Green Building Cost Data section, which details the available products, how they are specified, and how much they cost.

$89.95 per copy
350 pages, illustrated, Hardcover
Catalog No. 67338

Life Cycle Costing for Facilities
By Alphonse Dell'Isola and Dr. Steven Kirk

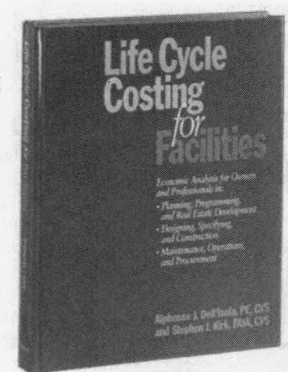

Guidance for achieving higher quality design and construction projects at lower costs!

Facility designers and owners are frustrated with cost-cutting efforts that yield the cheapest product, but sacrifice quality. Life Cycle Costing, properly done, enables them to achieve both—high quality, incorporating innovative design, and costs that meet their budgets.

The authors show how LCC can work for a broad variety of projects - from several types of buildings, to roads and bridges, to HVAC and electrical upgrades, to materials and equipment procurement. Case studies include:

• Health care and nursing facilities
• College campus and high schools
• Office buildings, courthouses, and banks
• Exterior walls, elevators, lighting, HVAC, and more

The book's extensive cost section provides maintenance and replacement costs for facility elements.

$99.95 per copy
450 pages, Hardcover
Catalog No. 67341

For more information
visit Means Web Site
at www.rsmeans.com

Reference Books

Value Engineering: Practical Applications

. . . For Design, Construction, Maintenance & Operations
By Alphonse Dell'Isola, PE

A tool for immediate application—for engineers, architects, facility managers, owners, and contractors. Includes: making the case for VE—the management briefing, integrating VE into planning and budgeting, conducting life cycle costing, integrating VE into the design process, using VE methodology in design review and consultant selection, case studies, VE workbook, and a life cycle costing program on disk.

$79.95 per copy
Over 450 pages, illustrated, Hardcover
Catalog No. 67319

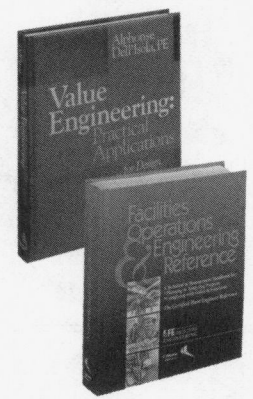

Facilities Operations & Engineering Reference

By the Association for Facilities Engineering and RSMeans

An all-in-one technical referance for planning and managing facility projects and solving day-to-day operations problems. Selected as the official Certified Plant Engineer reference, this handbook covers financial analysis, maintenance, HVAC and energy efficiency, and more.

$109.95 per copy
Over 700 pages, illustrated, Hardcover
Catalog No. 67318

The Building Professional's Guide to Contract Documents

3rd Edition
By Waller S. Poage, AIA, CSI, CVS

This comprehensive treatment of Contract Documents is an important reference for owners, design professionals, contractors, and students.
- Structure your Documents for Maximum Efficiency
- Effectively communicate construction requirements to all concerned
- Understand the Roles and Responsibilities of Construction Professionals
- Improve Methods of Project Delivery

$64.95 per copy, 400 pages
Diagrams and construction forms, Hardcover
Catalog No. 67261A

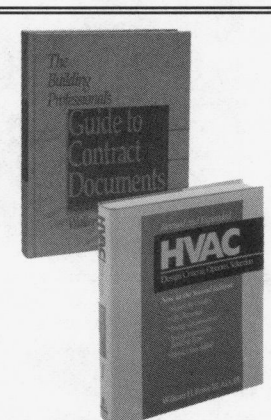

HVAC: Design Criteria, Options, Selection

Expanded 2nd Edition
By William H. Rowe III, AIA, PE

Includes Indoor Air Quality, CFC Removal, Energy Efficient Systems, and Special Systems by Building Type. Helps you solve a wide range of HVAC system design and selection problems effectively and economically. Gives you clear explanations of the latest ASHRAE standards.

$84.95 per copy
Over 600 pages, illustrated, Hardcover
Catalog No. 67306

Cost Planning & Estimating for Facilities Maintenance

In this unique book, a team of facilities management authorities shares their expertise at:
- Evaluating and budgeting maintenance operations
- Maintaining & repairing key building components
- Applying *Means Facilities Maintenance & Repair Cost Data* to your estimating

Covers special maintenance requirements of the 10 major building types.

$89.95 per copy
Over 475 pages, Hardcover
Catalog No. 67314

Facilities Maintenance Management

By Gregory H. Magee, PE

Now you can get successful management methods and techniques for all aspects of facilities maintenance. This comprehensive reference explains and demonstrates successful management techniques for all aspects of maintenance, repair, and improvements for buildings, machinery, equipment, and grounds. Plus, guidance for outsourcing and managing internal staffs.

$86.95 per copy
Over 280 pages with illustrations, Hardcover
Catalog No. 67249

Builder's Essentials: Advanced Framing Methods

By Scot Simpson

A highly illustrated, "framer-friendly" approach to advanced framing elements. Provides expert, but easy-to-interpret, instruction for laying out and framing complex walls, roofs, and stairs, and special requirements for earthquake and hurricane protection. Also helps bring framers up to date on the latest building code changes, and provides tips on the lead framer's role and responsibilities, how to prepare for a job, and how to get the crew started.

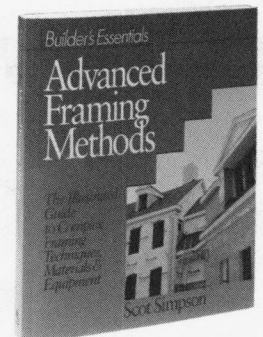

$24.95 per copy
250 pages, illustrated, Softcover
Catalog No. 67330

Reference Books

For more information
visit Means Web Site
at www.rsmeans.com

Interior Home Improvement Costs,

New 9th Edition

Estimates for the most popular remodeling and repair projects—from small, do-it-yourself jobs—to major renovations and new construction. Includes: Kitchens & Baths; New Living Space from Your Attic, Basement or Garage; New Floors, Paint & Wallpaper; Tearing Out or Building New Walls; Closets, Stairs & Fireplaces; New Energy-Saving Improvements, Home Theatres, and More!

$24.95 per copy
250 pages, illustrated, Softcover
Catalog No. 67308E

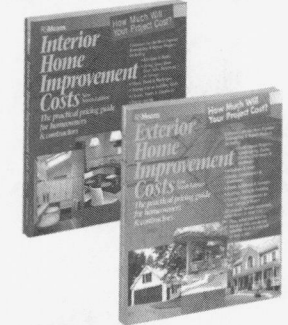

Exterior Home Improvement Costs

New 9th Edition

Estimates for the most popular remodeling and repair projects—from small, do-it-yourself jobs, to major renovations and new construction. Includes: Curb Appeal Projects—Landscaping, Patios, Porches, Driveways and Walkways; New Windows and Doors; Decks, Greenhouses, and Sunrooms; Room Additions and Garages; Roofing, Siding, and Painting; "Green" Improvements to Save Energy & Water

$24.95 per copy
Over 275 pages, illustrated, softcover
Catalog No. 67309E

Builder's Essentials: Plan Reading & Material Takeoff

By Wayne J. DelPico

For Residential and Light Commercial Construction

A valuable tool for understanding plans and specs, and accurately calculating material quantities. Step-by-step instructions and takeoff procedures based on a full set of working drawings.

$35.95 per copy
Over 420 pages, Softcover
Catalog No. 67307

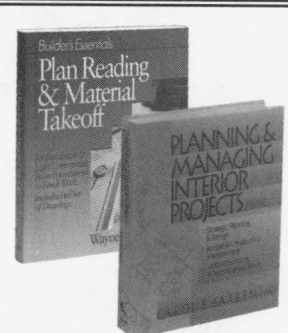

Planning & Managing Interior Projects

2nd Edition

By Carol E. Farren, CFM

Addresses changes in technology and business, guiding you through commercial design and construction from initial client meetings to post-project administration. Includes: evaluating space requirements, alternative work models, telecommunications and data management, and environmental issues.

$69.95 per copy
Over 400 pages, illustrated, Hardcover
Catalog No. 67245A

Builder's Essentials: Best Business Practices for Builders & Remodelers:

An Easy-to-Use Checklist System

By Thomas N. Frisby

A comprehensive guide covering all aspects of running a construction business, with more than 40 user–friendly checklists. This book provides expert guidance on: increasing your revenue and keeping more of your profit; planning for long-term growth; keeping good employees and managing subcontractors.

$29.95 per copy
Over 220 pages, Softcover
Catalog No. 67329

Builder's Essentials: Framing & Rough Carpentry 2nd Edition

By Scot Simpson

A complete training manual for apprentice and experienced carpenters. Develop and improve your skills with "framer-friendly," easy-to-follow instructions, and step-by-step illustrations. Learn proven techniques for framing walls, floors, roofs, stairs, doors, and windows. Updated guidance on standards, building codes, safety requirements, and more. Also available in Spanish!

$24.95 per copy
Over 150 pages, Softcover
Catalog No. 67298A   Spanish Catalog No. 67298AS

How to Estimate with Means Data & CostWorks

By RSMeans and Saleh Mubarak

New 2nd Edition!

Learn estimating techniques using Means cost data. Includes an instructional version of Means CostWorks CD–ROM with Sample Building Plans. The step-by-step guide takes you through all the major construction items. Over 300 sample estimating problems are included.

$59.95 per copy
Over 190 pages, Softcover
Catalog No. 67324A

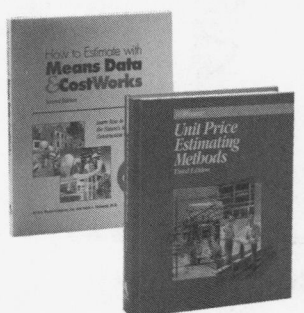

Unit Price Estimating Methods

New 3rd Edition

This new edition includes up-to-date cost data and estimating examples, updated to reflect changes to the CSI numbering system and new features of Means cost data. It describes the most productive, universally accepted ways to estimate, and uses checklists and forms to illustrate shortcuts and timesavers. A model estimate demonstrates procedures. A new chapter explores computer estimating alternatives.

$59.95 per copy
Over 350 pages, illustrated, Hardcover
Catalog No. 67303A

Means Landscape Estimating Methods

4th Edition

By Sylvia H. Fee

This revised edition offers expert guidance for preparing accurate estimates for new landscape construction and grounds maintenance. Includes a complete project estimate featuring the latest equipment and methods, and **two chapters on Life Cycle Costing, and Landscape Maintenance Estimating.**

$62.95 per copy
Over 300 pages, illustrated, Hardcover
Catalog No. 67295B

Means Environmental Remediation Estimating Methods, 2nd Edition

By Richard R. Rast

Guidelines for estimating 50 standard remediation technologies. Use it to prepare preliminary budgets, develop estimates, compare costs and solutions, estimate liability, review quotes, negotiate settlements.

A valuable support tool for *Means Environmental Remediation Unit Price* and *Assemblies* books.

$99.95 per copy
Over 750 pages, illustrated, Hardcover
Catalog No. 64777A

For more information
visit Means Web Site
at www.rsmeans.com

Reference Books

Means Illustrated Construction Dictionary, Condensed Edition, 2nd Edition
By RSMeans

Recognized in the industry as the best resource of its kind, this has been further enhanced with updates to existing terms and the addition of hundreds of new terms and illustrations . . . in keeping with the most recent developments in the industry.
The best portable dictionary for office or field use—an essential tool for contractors, architects, insurance and real estate personnel, homeowners, and anyone who needs quick, clear definitions for construction terms.

$59.95 per copy
Over 500 pages
Catalog No. 67282A

Means Repair and Remodeling Estimating
New 4th Edition
By Edward B. Wetherill & RSMeans

Focuses on the unique problems of estimating renovations of existing structures. It helps you determine the true costs of remodeling through careful evaluation of architectural details and a site visit.
New section on disaster restoration costs.

$69.95 per copy
Over 450 pages, illustrated, Hardcover
Catalog No. 67265B

Facilities Planning & Relocation
New, lower price and user-friendly format.
By David D. Owen

A complete system for planning space needs and managing relocations. Includes step-by-step manual, over 50 forms, and extensive reference section on materials and furnishings.

$89.95 per copy
Over 450 pages, Softcover
Catalog No. 67301

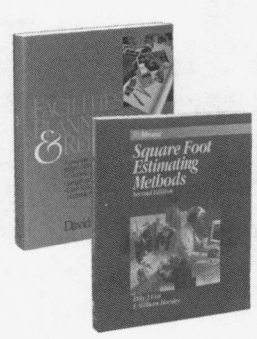

Means Square Foot & Assemblies Estimating Methods
3rd Edition!

Develop realistic Square Foot and Assemblies Costs for budgeting and construction funding. The new edition features updated guidance on square foot and assemblies estimating using UNIFORMAT II. An essential reference for anyone who performs conceptual estimates.

$69.95 per copy
Over 300 pages, illustrated, Hardcover
Catalog No. 67145B

Means Electrical Estimating Methods 3rd Edition

Expanded new edition includes sample estimates and cost information in keeping with the latest version of the CSI MasterFormat and UNIFORMAT II. Complete coverage of Fiber Optic and Uninterruptible Power Supply electrical systems, broken down by components and explained in detail. Includes a new chapter on computerized estimating methods. A practical companion to *Means Electrical Cost Data*.

$64.95 per copy
Over 325 pages, Hardcover
Catalog No. 67230A

Means Mechanical Estimating Methods 3rd Edition

This guide assists you in making a review of plans, specs, and bid packages with suggestions for takeoff procedures, listings, substitutions and pre-bid scheduling. Includes suggestions for budgeting labor and equipment usage. Compares materials and construction methods to allow you to select the best option.

$64.95 per copy
Over 350 pages, illustrated, Hardcover
Catalog No. 67294A

Means Spanish/English Construction Dictionary
By RSMeans, The International Conference of Building Officials (ICBO), and Rolf Jensen & Associates (RJA)

Designed to facilitate communication among Spanish- and English-speaking construction personnel—improving performance and job-site safety. Features the most common words and phrases used in the construction industry, with easy-to-follow pronunciations. Includes extensive building systems and tools illustrations.

$22.95 per copy
250 pages, illustrated, Softcover
Catalog No. 67327

Project Scheduling & Management for Construction
New 3rd Edition
By David R. Pierce, Jr.

A comprehensive yet easy-to-follow guide to construction project scheduling and control—from vital project management principles through the latest scheduling, tracking, and controlling techniques. The author is a leading authority on scheduling with years of field and teaching experience at leading academic institutions. Spend a few hours with this book and come away with a solid understanding of this essential management topic.

$64.95 per copy
Over 300 pages, illustrated, Hardcover
Catalog No. 67247B

Reference Books

For more information
visit Means Web Site
at www.rsmeans.com

Concrete Repair and Maintenance Illustrated
By Peter H. Emmons
$69.95 per copy
Catalog No. 67146

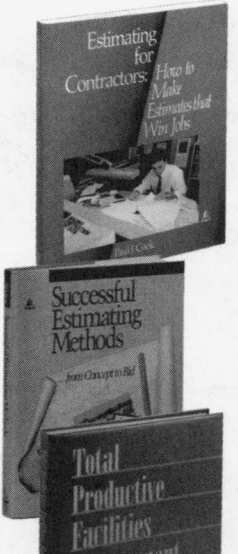

Estimating for Contractors
How to Make Estimates that Win Jobs
By Paul J. Cook
$35.95 per copy
Catalog No. 67160

Superintending for Contractors:
How to Bring Jobs in On-Time, On-Budget
By Paul J. Cook
$35.95 per copy
Catalog No. 67233

Successful Estimating Methods:
From Concept to Bid
By John D. Bledsoe, PhD, PE
$32.48 per copy
Catalog No. 67287

HVAC Systems Evaluation
By Harold R. Colen, PE
$84.95 per copy
Catalog No. 67281

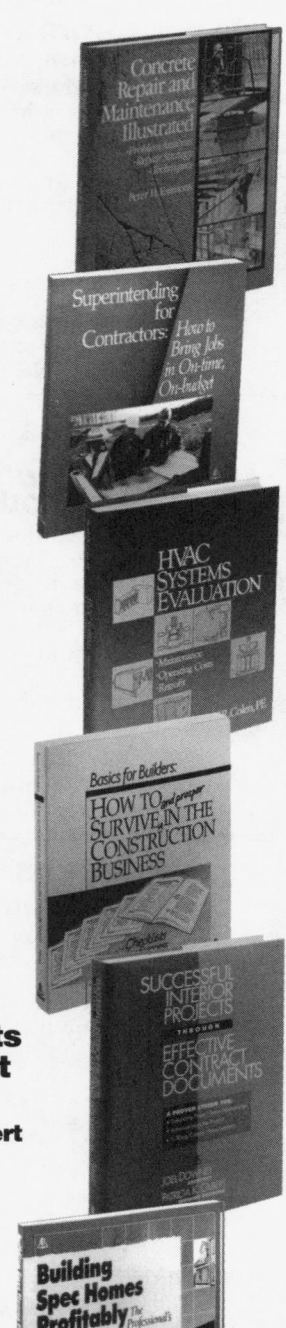

Total Productive Facilities Management
By Richard W. Sievert, Jr.
$29.98 per copy
Over 270 pages, illustrated, Hardcover
Catalog No. 67321

Basics for Builders: How to Survive and Prosper in Construction
By Thomas N. Frisby
$34.95 per copy
Catalog No. 67273

Means Productivity Standards for Construction
Expanded Edition (Formerly Man-Hour Standards)
$49.98 per copy
Over 800 pages, Hardcover
Catalog No. 67236A

Successful Interior Projects Through Effective Contract Documents
By Joel Downey & Patricia K. Gilbert
Now $24.98 per copy, limited quantity
Catalog No. 67313

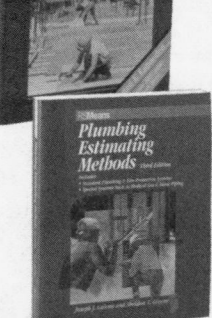

Means Plumbing Estimating Methods, 3rd Edition
By Joseph Galeno and Sheldon Greene
Now $29.98 per copy
Catalog No. 67283B

Building Spec Homes Profitably
By Kenneth V. Johnson
$29.95 per copy
Catalog No. 67312

For more information
visit Means Web Site
at www.rsmeans.com

Reference Books

Preventive Maintenance Guidelines for School Facilities

By John C. Maciha

A complete PM program for K-12 schools that ensures sustained security, safety, property integrity, user satisfaction, and reasonable ongoing expenditures.

Includes schedules for weekly, monthly, semiannual, and annual maintenance with hard copy and electronic forms.

$149.95 per copy
Over 225 pages, Hardcover
Catalog No. 67326

Preventive Maintenance for Higher Education Facilities

By Applied Management Engineering, Inc.

An easy-to-use system to help facilities professionals establish the value of PM, and to develop and budget for an appropriate PM program for their college or university. Features interactive campus building models typical of those found in different-sized higher education facilities, and PM checklists linked to each piece of equipment or system in hard copy and electronic format.

$149.95 per copy
150 pages, Hardcover
Catalog No. 67337

Historic Preservation: Project Planning & Estimating

By Swanke Hayden Connell Architects

Expert guidance on managing historic restoration, rehabilitation, and preservation building projects and determining and controlling their costs. Includes:

• How to determine whether a structure qualifies as historic
• Where to obtain funding and other assistance
• How to evaluate and repair more than 75 historic building materials

$99.95 per copy
Over 675 pages, Hardcover
Catalog No. 67323

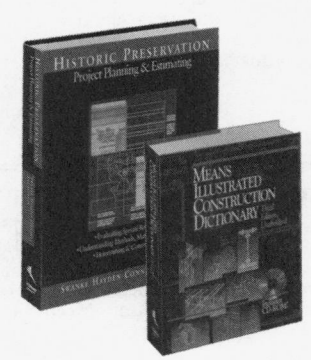

Means Illustrated Construction Dictionary, 3rd Edition

Long regarded as the Industry's finest, the Means *Illustrated Construction Dictionary* is now even better.
With the addition of over 1,000 new terms and hundreds of new illustrations, it is the clear choice for the most comprehensive and current information. **The companion CD-ROM that comes with this new edition adds many extra features: larger graphics, expanded definitions, and links to both CSI MasterFormat numbers and product information.**

$99.95 per copy
Over 790 pages, Illustrated, Hardcover
Catalog No. 67292A

Designing & Building with the IBC, 2nd Edition

By Rolf Jensen & Associates, Inc.

This updated comprehensive guide helps building professionals make the transition to the 2003 *International Building Code*®. Includes a side-by-side code comparison of the IBC 2003 to the IBC 2000 and the three primary model codes, a quick-find index, and professional code commentary. With illustrations, abbreviations key, and an extensive Resource section.

$99.95 per copy
Over 875 pages
Catalog No. 67328A

Residential & Light Commercial Construction Standards, 2nd Edition

By RSMeans and Contributing Authors

New, updated second edition of this unique collection of industry standards that define quality construction. For contractors, subcontractors, owners, developers, architects, engineers, attorneys, and insurance personnel, this book provides authoritative requirements and recommendations compiled from the nation's leading professional associations, industry publications, and building code organizations.

$59.95 per copy
600 pages, illustrated, Softcover
Catalog No. 67322A

Means Estimating Handbook, 2nd Edition

By RSMeans

Updated new Second Edition answers virtually any estimating technical question - all organized by CSI MasterFormat. This comprehensive reference covers the full spectrum of technical data required to estimate construction costs. The book includes information on sizing, productivity, equipment requirements, code-mandated specifications, design standards and engineering factors.

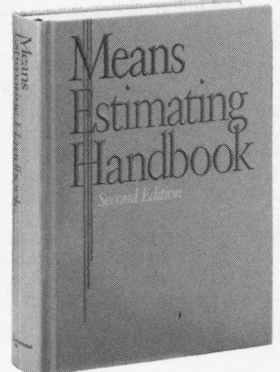

$99.95 per copy
Over 900 pages, Hardcover
Catalog No. 67276A

**For more information
visit Means Web Site
at www.rsmeans.com**

Seminars

Means CostWorks Training

This one-day seminar course has been designed with the intention of assisting both new and existing users to become more familiar with CostWorks program. The class is broken into two unique sections: (1) A one-half day presentation on the function of each icon; and each student will be shown how to use the software to develop a cost estimate. (2) Hands-on estimating exercises that will ensure that each student thoroughly understands how to use CostWorks. You must bring your own laptop computer to this course.

CostWorks Benefits/Features:
- Estimate in your own spreadsheet format
- Power of Means National Database
- Database automatically regionalized
- Save time with keyword searches
- Save time by establishing common estimate items in "Bookmark" files
- Customize your spreadsheet template
- Hot key to Product Manufacturers' listings and specs
- Merge capability for networking environments
- View crews and assembly components
- AutoSave capability
- Enhanced sorting capability

Unit Price Estimating

This interactive two-day seminar teaches attendees how to interpret project information and process it into final, detailed estimates with the greatest accuracy level.

The single most important credential an estimator can take to the job is the ability to visualize construction in the mind's eye, and thereby estimate accurately.

Some Of What You'll Learn:
- Interpreting the design in terms of cost
- The most detailed, time tested methodology for accurate "pricing"
- Key cost drivers—material, labor, equipment, staging and subcontracts
- Understanding direct and indirect costs for accurate job cost accounting and change order management

Who Should Attend: Corporate and government estimators and purchasers, architects, engineers…and others needing to produce accurate project estimates.

Square Foot and Assemblies Cost Estimating

This two-day course teaches attendees how to quickly deliver accurate square foot estimates using limited budget and design information.

Some Of What You'll Learn:
- How square foot costing gets the estimate done faster
- Taking advantage of a "systems" or "assemblies" format
- The Means "building assemblies/square foot cost approach"
- How to create a very reliable preliminary and systems estimate using bare-bones design information

Who Should Attend: Facilities managers, facilities engineers, estimators, planners, developers, construction finance professionals…and others needing to make quick, accurate construction cost estimates at commercial, government, educational and medical facilities.

Repair and Remodeling Estimating

This two-day seminar emphasizes all the underlying considerations unique to repair/remodeling estimating and presents the correct methods for generating accurate, reliable R&R project costs using the unit price and assemblies methods.

Some Of What You'll Learn:
- Estimating considerations—like labor-hours, building code compliance, working within existing structures, purchasing materials in smaller quantities, unforeseen deficiencies
- Identifies problems and provides solutions to estimating building alterations
- Rules for factoring in minimum labor costs, accurate productivity estimates and allowances for project contingencies
- R&R estimating examples are calculated using unite prices and assemblies data

Who Should Attend: Facilities managers, plant engineers, architects, contractors, estimators, builders…and others who are concerned with the proper preparation and/or evaluation of repair and remodeling estimates.

Mechanical and Electrical Estimating

This two-day course teaches attendees how to prepare more accurate and complete mechanical/electrical estimates, avoiding the pitfalls of omission and double-counting, while understanding the composition and rationale within the Means Mechanical/Electrical database.

Some Of What You'll Learn:
- The unique way mechanical and electrical systems are interrelated
- M&E estimates, conceptual, planning, budgeting and bidding stages
- Order of magnitude, square foot, assemblies and unit price estimating
- Comparative cost analysis of equipment and design alternatives

Who Should Attend: Architects, engineers, facilities managers, mechanical and electrical contractors…and others needing a highly reliable method for developing, understanding and evaluating mechanical and electrical contracts.

Plan Reading and Material Takeoff

This two-day program teaches attendees to read and understand construction documents and to use them in the preparation of material takeoffs.

Some of What You'll Learn:
- Skills necessary to read and understand typical contract documents—blueprints and specifications
- Details and symbols used by architects and engineers
- Construction specifications' importance in conjunction with blueprints
- Accurate takeoff of construction materials and industry-accepted takeoff methods

Who Should Attend: Facilities managers, construction supervisors, office managers…and other responsible for the execution and administration of a construction project including government, medical, commercial, educational or retail facilities.

Facilities Maintenance and Repair Estimating

This two-day course teaches attendees how to plan, budget, and estimate the cost of ongoing and preventive maintenance and repair for existing buildings and grounds.

Some Of What You'll Learn:
- The most financially favorable maintenance, repair and replacement scheduling and estimating
- Auditing and value engineering facilities
- Preventive planning and facilities upgrading
- Determining both in-house and contract-out service costs; annual, asset-protecting M&R plan

Who Should Attend: Facility managers, maintenance supervisors, buildings and grounds superintendents, plant managers, planners, estimators…and others involved in facilities planning and budgeting.

Scheduling and Project Management

This two-day course teaches attendees the most current and proven scheduling and management techniques needed to bring projects in on time and on budget.

Some Of What You'll Learn:
- Crucial phases of planning and scheduling
- How to establish project priorities, develop realistic schedules and management techniques
- Critical Path and Precedence Methods
- Special emphasis on cost control

Who Should Attend: Construction project managers, supervisors, engineers, estimators, contractors…and others who want to improve their project planning, scheduling and management skills.

Advanced Project Management

This two-day seminar will teach you how to effectively manage and control the entire design-build process and allow you to take home tangible skills that will be immediately applicable on existing projects.

Some Of What You'll Learn:
- Value engineering, bonding, fast-tracking and bid package creation
- How estimates and schedules can be integrated to provide advanced project management tools
- Cost engineering, quality control, productivity measurement and improvement
- Front loading a project and predicting its cash flow

Who Should Attend: Owners, project managers, architectural and engineering managers, construction managers, contractors…and anyone else who is responsible for the timely design and completion of construction projects.

Seminars

For more information
visit Means Web Site
at www.rsmeans.com

2005 Means Seminar Schedule

Note: Call for exact dates and details.

Location	Dates
Las Vegas, NV	March
Washington, DC	April
Phoenix, AZ	April
Denver, CO	May
San Francisco, CA	June
Philadelphia, PA	June
Washington, DC	September
Dallas, TX	September
Las Vegas, NV	October
Orlando, FL	November
Atlantic City, NJ	November
San Diego, CA	December

Registration Information

Register Early... Save up to $100! Register 30 days before the start date of a seminar and save $100 off your total fee. *Note: This discount can be applied only once per order. It cannot be applied to team discount registrations or any other special offer.*

How to Register Register by phone today! Means toll-free number for making reservations is: **1-800-334-3509.**

Individual Seminar Registration Fee $895. Individual CostWorks Training Registration Fee $349. To register by mail, complete the registration form and return with your full fee to: Seminar Division, Reed Construction Data, RSMeans Seminars, 63 Smiths Lane, Kingston, MA 02364.

Federal Government Pricing All Federal Government employees save 25% off regular seminar price. Other promotional discounts cannot be combined with Federal Government discount.

Team Discount Program Two to four seminar registrations: Call for pricing.

Multiple Course Discounts When signing up for two or more courses, call for pricing.

Refund Policy Cancellations will be accepted up to ten days prior to the seminar start. There are no refunds for cancellations received later than ten working days prior to the first day of the seminar. A $150 processing fee will be applied for all cancellations. Written notice of cancellation is required . Substitutions can be made at anytime before the session starts. **No-shows are subject to the full seminar fee.**

AACE Approved Courses The RSMeans Construction Estimating and Management Seminars described and offered to you here have each been approved for 14 hours (1.4 recertification credits) of credit by the AACE International Certification Board toward meeting the continuing education requirements for re-certification as a Certified Cost Engineer/Certified Cost Consultant.

AIA Continuing Education We are registered with the AIA Continuing Education System (AIA/CES) and are committed to developing quality learning activities in accordance with the CES criteria. RSMeans seminars meet the AIA/CES criteria for Quality Level 2. AIA members will receive (14) learning units (LUs) for each two-day RSMeans Course.

NASBA CPE Sponsor Credits We are part of the National Registry of CPE Sponsors. Attendees may be eligible for (16) CPE credits.

Daily Course Schedule The first day of each seminar session begins at 8:30 A.M. and ends at 4:30 P.M. The second day is 8:00 A.M.–4:00 P.M. Participants are urged to bring a hand-held calculator since many actual problems will be worked out in each session.

Continental Breakfast Your registration includes the cost of a continental breakfast, a morning coffee break, and an afternoon break. These informal segments will allow you to discuss topics of mutual interest with other members of the seminar. (You are free to make your own lunch and dinner arrangements.)

Hotel/Transportation Arrangements RSMeans has arranged to hold a block of rooms at each hotel hosting a seminar. To take advantage of special group rates when making your reservation, be sure to mention that you are attending the Means Seminar. You are, of course, free to stay at the lodging place of your choice. (**Hotel reservations and transportation arrangements should be made directly by seminar attendees.**)

Important Class sizes are limited, so please register as soon as possible.

Note: Pricing subject to change.

Registration Form Call 1-800-334-3509 to register or FAX 1-800-632-6732. Visit our Web site www.rsmeans.com

Please register the following people for the Means Construction Seminars as shown here. Full payment or deposit is enclosed, and we understand that we must make our own hotel reservations if overnight stays are necessary.

☐ Full payment of $ _____ enclosed.

☐ Bill me

Name of Registrant(s)
(To appear on certificate of completion)

_____

_____

_____

P.O. #: _____
GOVERNMENT AGENCIES MUST SUPPLY PURCHASE ORDER NUMBER

Firm Name _____

Address _____

City/State/Zip _____

Telephone No. fax No. _____

E-mail Address _____

Charge our registration(s) to: ☐ MasterCard ☐ VISA ☐ American Express ☐ Discover

Account No. _____ Exp. Date _____

Cardholder's Signature _____

Seminar Name	City	Dates

Please mail check to: Seminar Division, Reed Construction Data, RSMeans Seminars, 63 Smiths Lane, P.O.Box 800, Kingston, MA 02364 USA

MeansData™

CONSTRUCTION COSTS FOR SOFTWARE APPLICATIONS
Your construction estimating software is only as good as your cost data.

A proven construction cost database is a mandatory part of any estimating package. The following list of softwa providers can offer you MeansData™ as an added feature for their estimating systems. See the table below for w types of products and services they offer (match their numbers). Visit online at **www.rsmeans.com/demosource/** for more information and free demos. Or call their numbers listed below.

1. **3D International**
713-871-7000
venegas@3di.com

2. **4Clicks-Solutions, LLC**
719-574-7721
mbrown@4clicks-solutions.com

3. **Aepco, Inc.**
301-670-4642
blueworks@aepco.com

4. **Applied Flow Technology**
800-589-4943
info@aft.com

5. **ArenaSoft Estimating**
888-370-8806
info@arenasoft.com

6. **ARES Corporation**
925-299-6700
sales@arescorporation.com

7. **BSD - Building Systems Design, Inc.**
888-273-7638
bsd@bsdsoftlink.com

8. **CMS - Computerized Micro Solutions**
800-255-7407
cms@proest.com

9. **Corecon Technologies, Inc.**
714-895-7222
sales@corecon.com

10. **CorVet Systems**
301-622-9069
sales@corvetsys.com

11. **Discover Software**
727-559-0161
sales@discoversoftware.com

12. **Estimating Systems, Inc.**
800-967-8572
esipulsar@adelphia.net

13. **MAESTRO Estimator Schwaab Technology Solutions, Inc.**
281-578-3039
Stefan@schwaabtech.com

14. **Magellan K-12**
936-447-1744
sam.wilson@magellan-K12.com

15. **MC^2 - Management Computer**
800-225-5622
vkeys@mc2-ice.com

16. **Maximus Asset Solutions**
800-659-9001
assetsolutions@maximus.com

17. **Prime Time**
610-964-8200
dick@primetime.com

18. **Prism Computer Corp.**
800-774-7622
famis@prismcc.com

19. **Quest Solutions, Inc.**
800-452-2342
info@questsolutions.com

20. **RIB Software (Americas), Inc.**
800-945-7093
san@rib-software.com

21. **Shaw Beneco Enterprises, Inc.**
877-719-4748
inquire@beneco.com

22. **Timberline Software Corp.**
800-628-6583
product.info@timberline.com

23. **TMA SYSTEMS, Inc.**
800-862-1130
sales@tmasys.com

24. **US Cost, Inc.**
800-372-4003
sales@uscost.com

25. **Vanderweil Facility Advisors**
617-451-5100
info@VFA.com

26. **Vertigraph, Inc.**
800-989-4243
info-request@vertigraph.com

27. **WinEstimator, Inc.**
800-950-2374
sales@winest.com

TYPE	1	2	3	4	5	6	7	8	9	10	11	12	13	14	15	16	17	18	19	20	21	22	23	24	25	26	27
BID					●		●	●			●	●				●		●		●	●	●				●	●
Estimating		●			●	●	●	●	●	●	●	●	●		●		●		●	●	●	●		●		●	●
DOC/JOC/SABER		●			●		●	●			●	●				●					●	●					●
IDIQ		●									●	●				●					●						●
Asset Mgmt.											●			●		●	●								●	●	●
Facility Mgmt.	●		●								●		●	●	●	●	●	●		●			●		●		●
Project Mgmt.	●	●					●			●	●						●		●	●	●						
TAKE-OFF					●						●			●			●		●	●		●		●		●	●
EARTHWORK											●			●			●									●	●
Pipe Flow				●							●						●										
HVAC/Plumbing					●			●		●	●						●										
Roofing					●			●		●	●																
Design	●										●						●										●
Other Offers/Links:																											
Accounting/HR		●			●						●									●	●	●					
Scheduling		●									●		●							●				●		●	●
CAD											●			●								●		●			
PDA																						●		●			●
Lt. Versions		●						●									●					●				●	
Consulting	●	●			●			●		●					●	●				●		●					●
Training		●			●			●		●	●				●	●	●			●	●	●	●	●	●	●	●

Reseller applications now being accepted. Call Carol Polio Ext. 5107.

FOR MORE INFORMATION
CALL 1-800-448-8182, EXT. 5107 OR FAX 1-800-632-6732